C0-ARK-179

3 9087 0279 4260 8

UNIVERSITY OF ROCHESTER LIBRARIES

DATE DUE

GASTRULATION

From Cells to Embryo

Access to the On-line Supplemental Data and Gastrulation Movies

This print edition of *Gastrulation: From Cells to Embryo* is associated with a Web site (http://www.gastrulation.org) that will add to and extend the contents of the book. Specific supplementary data sets and movies referred to in the text can be found at this site. Access to the site is open, and no registration is required. For assistance with the Web site or additional queries, please use E-mail addresses provided at the site.

GASTRULATION

From Cells to Embryo

EDITED BY

CLAUDIO D. STERN

University College London

www.gastrulation.org

COLD SPRING HARBOR LABORATORY PRESS

Cold Spring Harbor, New York

GASTRULATION
From Cells to Embryos

All rights reserved.
©2004 Text and illustrations by individual chapter authors; compilation by Cold Spring Harbor
Laboratory Press, Cold Spring Harbor, New York
Printed in the United States of America

Publisher	John Inglis
Acquisition and Developmental Editor	David Crotty
Editorial Development Manager	Jan Argentine
Project Coordinator	Inez Sialiano
Production Editor	Pat Barker
Desktop Editor	Danny deBruin
Production Manager	Denise Weiss
Cover Designer	Mike Albano

Front cover: Collage of images representing gastrulation. Clockwise from upper left: Stages of gastrulation of the chick embryo (Chapter 15, Fig. 3); magnified view of midsagittal section of the amphibian embryo showing events that form the archenteron and elongate the dorsal body axis (Chapter 13, Fig. 2E); inductive interaction and modulation of signaling leading to patterning in the early-streak mouse embryo (Chapter 16, Fig. 8E); late gastrula of *Ischnochiton magdalenis* (Chapter 5, Fig. 13); regionalization of tissue fates in the late-streak mouse embryo (Chapter 16, Fig. 8D); Haeckel's gastraea (Chapter 51, Fig. 3C); midsagittal section of the amphibian embryo showing events that form the archenteron and elongate the dorsal body axis (Chapter 13, Fig. 2D); late gastrula stage of *Physa fontinalis* (Chapter 5, Fig. 15); sea urchin embryo expressing GFP coupled to a transmembrane sequence delineates cell boundaries at the beginning of gastrulation (Chapter 9, Fig. 5B); transitional phase between Haeckel's blastaea and gastraea (Chapter 51, Fig. 3B).

Back cover: Presumptive fates and movements of *Xenopus* gastrulation shown in midsagittal plane (Chapter 13, Fig. 1).

Library of Congress Cataloging-in-Publication Data

Gastrulation : from cells to embryo / edited by Claudio D. Stern.
    p. cm.
  Includes bibliographical references (p.).
  ISBN 0-87969-707-5 (hard cover : alk. paper)
  1. Gastrulation. I. Stern, C. D. (Claudio D.)
  QL955.G34 2004
  571.8'65--dc22

                                           2004009562

10 9 8 7 6 5 4 3 2 1

Authorization to photocopy items for internal or personal use, or the Internet or personal use of specific clients, is granted by Cold Spring Harbor Laboratory Press, provided that the appropriate fee is paid directly to the Copyright Clearance Center (CCC). Write or call CCC at 222 Rosewood Drive, Danvers, MA 01923 (508-750-8400) for information about fees and regulations. Prior to photocopying items for educational classroom use, contact CCC at the above address. Additional information on CCC can be obtained at CCC Online at http://www.copyright.com/

All Cold Spring Harbor Laboratory Press publications may be ordered directly from Cold Spring Harbor Laboratory Press, 500 Sunnyside Blvd., Woodbury, New York 11797-2924. Phone: 1-(800) 843-4388 in Continental U.S. and Canada. All other locations: (516) 422-4100. FAX: (516) 422-4097. E-mail: cshpress@cshl.edu. For a complete catalog of all Cold Spring Harbor Laboratory Press publications, visit our World Wide Web Site http://www.cshlpress.com/

QS
604
G 255
2004

*Our real teacher has been and still
is the embryo, who is,
incidentally, the only teacher
who is always right*

Viktor Hamburger (1900–2001)

CONTENTS

PREFACE

There is an intangible something about the process of gastrulation that invariably generates enormous awe and curiosity. It is for good reason that of all embryonic processes, it has been said that "*the most important event in your life is not birth, marriage, or death but gastrulation*" (Lewis Wolpert, quoted in numerous places, including a poster that was widely distributed some years ago). Arguably it is during gastrulation that epigenesis (as opposed to preformation) can be seen most dramatically: The embryo generates huge complexity from a very simple starting form. It is also gastrulation that allows the otherwise rather arbitrary words "morphogenesis"[1] and "pattern formation" to be defined. And it is also during gastrulation that some critical inductive processes can be demonstrated very clearly.

My interest in gastrulation is visceral. As I arrived to start my Ph.D., my supervisor asked me on what topic I would like to work—the choices: *Dictyostelium* cell migration, *Acetabularia* regeneration, axolotl limb regeneration, or "the chick." Within 24 hours, I had chosen to study the cell movements and cell interactions driving chick gastrulation, and I have stayed with it ever since (and after almost 30 years of working on it I can't claim to understand it well enough just yet). A similar bug must have bitten David Crotty from Cold Spring Harbor Laboratory Press, since he told me in 2000 that he had always dreamed of publishing a book on gastrulation. Despite having declined many offers to edit other books, I could not resist this one.

The next challenge was to do justice to such a complex topic. In a subject that is constantly growing by the addition of new information, a book is only valuable as a lasting scholarly source for earlier work. Thus, the first decision was to ask all contributing authors to cite primary literature in preference to reviews, and to try to give due credit to those that made the first observation, rather than just the latest. It is unfortunate how the culture in biomedical sciences has changed during the last few decades in this respect.

A second decision concerned the viewpoint from which to describe gastrulation. One might take a descriptive embryological view, comparing different organisms; or a cellular view, concentrating on the basic types of cell behaviors seen during gastrulation; or a molecular view, focusing on common molecular players known to play a role; or, indeed, an evolutionary or an historical view. Each of these views leads to a completely different image of gastrulation. It quickly became obvious that these are complementary and that we can only approach an understanding of this process by moving from one hill to the next, observing the same process from different angles.

The next problem to solve was whether to concentrate on the so-called "model organisms" (with the problem that our colleagues differ greatly in which organisms should belong to this aristocracy) or whether to include at least some groups which deserve more study or which add different insights. At a recent meeting, Lewis Wolpert pronounced the view that we should concentrate on fewer species so that we stand a chance of understanding anything. This generated some heated discussion, and I hope that by the end of it he had changed his mind. It became clear that the processes we are trying to understand are complex, and that restricting our study to a few species because they facilitate a particular, currently trendy, experimental approach will only generate a partial view of the processes. Even more serious is the problem that we would not have a hope of understanding any evolutionary issue if we were to concentrate exclusively on half a dozen, mostly peculiar, species. Although it is impossible to cover all species, or even all groups, the benefit from not concentrating on just a handful is overwhelming.

Then we had to look at cellular events, and to choose which molecules to discuss in more depth. Here the decision was much more difficult. As with model species, specific cellular events and molecules are subject to trends. Thus, there is a current obsession with "convergent extension," as if this

---

[1] A literal translation of "morphogenesis" yields something similar to "pattern formation." However, the former word is used to describe a process by which form is generated mainly through cell movements or tissue reorganizations, whereas the latter generates form by local changes in cell identity within a field of cells, without major movements. Gastrulation is the archetypal example of morphogenesis, whereas the development of periodic patterns (feathers or the ommatidia of the compound eye of the fly) illustrates pattern formation.

is the only movement of which cells are capable, and similar disproportionate emphasis on genes encoding secreted factors and, to a slightly lesser extent, a few transcription factors. In this case, I chose to focus mainly on the processes and molecules about which we know most. The reason is that these two sections of the book ("Cellular Events of Gastrulation" and "The Molecular Biology of Gastrulation") serve the main purpose of providing different standpoints from that of comparative embryology, and because had we not done this, the book would have to be many volumes long and have some extremely short chapters for those events and molecules that have been understudied.

A final decision concerned the presentation of the book. A multi-author book can take advantage of the extensive experience of the contributing authors in their respective fields, but the cost is a diversity of styles of writing and illustration. Keeping some of this diversity is just as interesting (from social and historical points of view—this is an historical cross-section of the study of gastrulation as it stands at present) as exemplifying the biological diversity being discussed. But a consequence of this is that the book can be too disjointed for a reader who is new to the subject, and too difficult even for the expert to navigate. To attempt to overcome this, I chose a single color scheme for germ layers and cell types which is used throughout the book, and even more dictatorially decided to edit all chapters to achieve some compromise between diversity and continuity. I am very grateful to all authors for allowing me to butcher their scholarly work. Additionally, supplemental data, including movies of gastrulation, can be found on the book's accompanying Web site, www.gastrulation.org.

Perhaps the most remarkable lesson to come out of this volume concerns the great diversity of strategies that are used by different species to do what appears to be the same thing. Thus, one species forms its mesoderm by ingression of individual cells, another species generates it by involution, yet another by a combination of the two. Genetic pathways are also conserved, but how they are deployed (if at all) to control particular events differs very greatly among different groups of animals. Yet there are also some general principles that apply to many, if not all, multicellular animals. To what extent the currently perceptible principles are truly general, and to what extent they reflect trends in experimental and intellectual approaches, will only emerge with considerably more hindsight, but there is no question that these are indeed exciting times to be studying gastrulation, and development in general.

This book really belongs to the authors of the individual chapters. It is they who provided their wisdom, their scholarship and their vision. All of them have been exceedingly supportive of the ideas behind the book, perhaps because they all share the same fascination with gastrulation as described earlier. I am enormously grateful for their patience, their collaboration, and their hard work and hope that the final product will fully justify the time they invested on their contributions. Finally, thanks are due to the staff of Cold Spring Harbor Laboratory Press, in particular, David Crotty for having had the initial idea for this book, John Inglis, Executive Director of the Press, Jan Argentine, Editorial Development Manager, and Inez Sialiano for spending so much time supporting all the editorial activities, and the production team, Denise Weiss, Pat Barker, and Danny deBruin, for doing such a fine job of keeping the book within its covers.

CLAUDIO STERN

GLOSSARY

This list of definitions is intended to guide the beginner through some of the complex terminology of embryology and to introduce some of the major concepts. To the expert, this glossary might also be a source of some amusement, perhaps irritation. To those of the latter group, I apologize. Although I have tried to be as precise and as inclusive as possible, the usage of these terms differs considerably in practice, particularly when used by people working on different animals. Terms in **boldface** within definitions refer to terms listed within this glossary.

Claudio Stern

Anlage (plural: **Anlagen**) (see **primordium**)

axial mesoderm (see also **chorda mesoderm, head process, Hensen's node, notochord**) The portion of the **mesoderm** underlying the midline of the embryo. In the strict sense, the axial mesoderm comprises only the **head process** under the brain and the **notochord** posterior to the ear vesicle. Some authors, however, also include the **paraxial mesoderm** as an axial component. In amphibian embryos (see Chapter 26), the term axial mesoderm is only rarely used. Instead, authors use the term dorsal mesoderm to describe the axial components; this is because after folding of the embryonic body, the axial mesoderm comes to lie most dorsally in the embryo. The axial mesoderm arises from **Hensen's node** in higher vertebrates and from the **dorsal lip of the blastopore** in amphibians.

axis Using Cartesian coordinates to map the embryo, three orthogonal axes can be defined: dorsoventral (also called axial-lateral in some animals; see Chapter 26), rostrocaudal (also called anterior–posterior in some animals; see Chapter 27), and left/right (see Chapter 28). One reason for the confusing terminology is that during development, embryos change their shape substantially. Thus, **blastoderm**-type embryos (e.g., avian and most mammalian embryos) generate the final body plan by folding so that the tissues that will ultimately occupy the most ventral positions start out very laterally, whereas tissues that start out along the midline become dorsal (e.g., neural tube, **notochord**). In humans, which walk erect, the anterior–posterior axis is the dorsoventral (rather than rostrocaudal) axis.

blastoderm Literally, a sheet or layer (Greek suffix *-derm* = skin) of embryonic cells. The term describes a specific, early stage of development of some organisms. In *Drosophila* and other

insects (see Chapter 7), the blastoderm stage is characterized by a single layer of nuclei arranged on the surface of the embryo, before the movements of **gastrulation** begin. The stage is subdivided into two: the *syncytial blastoderm stage,* before cell membranes form between the individual superficial nuclei, and the *cellular blastoderm stage,* after cells become defined by the appearance of membranes. In some vertebrates, the blastoderm is the group of developmental stages during which the embryo is flat, before body folding begins, and multicellular. In avian embryos, for example, the blastoderm first comprises a single layer of cells, the **epiblast**. Two further layers of cells are then formed: the **hypoblast** ventrally and the **mesoderm** between the other two. This tri-laminar embryo is still called the blastoderm, usually until the middle of the process of **neurulation**, when the embryo becomes more three-dimensional by the appearance of the **head fold**.

blastopore (see also **dorsal lip of the blastopore**) The site at which **endoderm** and/or **mesoderm** formation begins in amphibians and other (mostly spherical) embryos. In *Xenopus* the blastopore is first seen as a crescent-shaped lip on the dorsal side of the embryo, close to the equator. This lip gradually expands to form a large circular slit, which later gradually closes. For details of the movements in this region, see Chapters 13 and 19. In teleosts and amniotes there is no obvious circular blastopore. Teleosts have a marginal zone that includes the **shield**, whereas amniotes have a **primitive streak**.

cell determination (see **determination**)

cell differentiation (see **differentiation**)

cell fate (see **fate**)

cell lineage The collection of descendants of a single cell during normal development. The term is more usually reserved for the experimental situation of following, or mapping, these descendants and their ultimate **fates** by direct observation or by some marking method that allows the progeny of the parent cell to be identified. The expression "cell lineage analysis" does not formally differ from **fate mapping**, but in practice the first is used for mapping the progeny of *single cells*, whereas the latter implies that the progeny of a *group of cells* is studied.

celom (see **coelom**)

chemotaxis (see also **galvanotaxis, haptotaxis**) Guidance of cell movements by attraction or repulsion to a chemical cue, usually assumed to be soluble and at least to some extent diffusible in the extracellular space. For discussion of the possible roles of chemotaxis in gastrulation, see Chapter 22.

chimera (**chimaera**) (plural **chimerae** or **chimeras**) (see also **mosaic**) A composite embryo containing cells or tissues with two or more different genotypes (cf. **mosaic**). Chimeras can be produced experimentally to follow genetically marked cells within the embryo. The genetic marker may consist of cells of a different species. The word derives from Greek mythology; the chimera was a monster whose parents were Typhon and Echidna, and which consisted of three parts: a lion's head, a she-goat in the middle, and a snake for a tail.

chorda mesoderm (**chordamesoderm**) (see also **axial mesoderm, head process, notochord**) The rod of **axial mesoderm** that characterizes the embryos of **chordates**. It comprises the **head process** in the head and the **notochord** in the trunk behind the ear. This specialized **mesoderm** is thought to have appeared during evolution (in now-extinct groups of protochordates such as the *Solutes* and the *Mitrates*) to confer rigidity to the animal. The term is only rarely used; more frequently the terms **notochord** and **head process** are used instead.

chordate The group of animals (Phylum *Chordata*) whose embryos possess a **notochord**. The group includes all vertebrates, and other forms, such as the Tunicates, which do not possess an internal skeleton.

clonal analysis (see **cell lineage**)

clonal restriction (see also **cell lineage, clone, compartment**) In the general sense, one or more **clones** (def. 1) of cells that are *confined* or *limited* in some way. In practice, the term is usually used only to signify that the descendants of a particular group of progenitor cells are confined to well-defined *spatial* limits which they cannot cross. The expression is central to the definition of **cell lineage compartments**, which are thought to be important in defining the boundaries of **morphogenetic fields**.

clone (1 Embryol.) A group of cells derived from a single common progenitor cell. (2) Two or more individuals of a given species with identical genotypes. Thus, monozygotic twins are clonally related because the two members of the pair are derived from the same fertilized egg. (3 Molecular biol.) cDNA or genomic DNA inserted into a plasmid that can be propagated in a suitable bacterial, viral, or eukaryotic host cell.

coelom The cavity between **somatopleure** and **splanchnopleure**, the two layers of cells derived from the **lateral plate mesoderm**. It characterizes higher animals.

commitment (see also **determination**) Most authors use the terms commitment and **determination** interchangeably, to mean an *irreversible* decision to differentiate in a particular direction. Some authors, however, use commitment to identify a determinative decision that is not irreversible.

compartment (see also **cell lineage, clonal restriction**) A term introduced in the mid-1970s by García-Bellido, Lawrence, and their colleagues. The formal definition of a **cell lineage** compartment is: *The collection of all surviving descendants of a group of founder cells which are also confined to a spatial domain within well-defined boundaries.* Thus, a compartment is a polyclone, or a group of **clones** where cells never mix with cells of neighboring compartments. The boundaries between compartments may or may not correspond to the boundaries between visible structures, but are thought to be defined by limits of expression of particular genes, such as that of *engrailed* in *Drosophila*. The boundaries between adjacent compartments define the spatial limits of **morphogenetic fields**. During development, embryos become subdivided into an increasing number of compartments.

competence The ability of a cell or tissue to respond to a particular inducing signal (**induction**). Competence is believed to depend largely on the expression of appropriate receptors specific for the inducing molecule(s) by the responding cells.

convergence–extension (or **convergent extension**) A specific type of **morphogenetic** movement that causes simultaneous narrowing along the mediolateral **axis** (convergence) and elongation along the rostrocaudal **axis** (extension). In the best-studied cases, and therefore strictly in all cases of true convergence–extension, the process is driven by cell intercalation (see Chapters 13, 19, and 20). Note from the discussion of these chapters that not all extensions and not all convergences are convergent extension, since each of these movements can have different consequences depending on local constraints (e.g., extension can be driven by thinning of a sheet as well as by convergence).

cytoplasmic determinant A substance, organelle, or property that becomes allocated unequally between the two daughters of a cell division. This mechanism is thought to be able to generate two daughter cells with different **developmental potential** as a result of a single cell division. Although usually the term is used to refer to cytoplasmic substances or organelles, the determinant could be in the cell membrane or associated with the nucleus.

determinant (see **cytoplasmic determinant**)

determination (see also **commitment, developmental potential, fate**) An irreversible decision made by an undifferentiated cell which restricts its **differentiation** into a smaller subset of cell types. A cell becomes determined when its **developmental potential** no longer differs from its **fate**.

developmental potential (see also **commitment, determination, fate**) The range of all the *possible* phenotypes that can be produced by the descendants of a cell or group of cells. The concept differs from **fate**: The **fate** of a cell or group of cells is the range of phenotypes that *actually are* produced by those cells in normal development, whereas their developmental potential includes the phenotypes that the cells can give rise to in all possible circumstances. For example, consider a cell in the posterior **primitive streak**. The **fate** of this cell (in normal development) is to become part of the extraembryonic **mesoderm** (extraembryonic membranes, blood, etc.). However, if the cell is transplanted to the anterior part of the **primitive streak**, it will contribute to **axial mesoderm/endoderm** and not to extraembryonic **mesoderm**. The **fate** of this cell is to form extraembryonic **mesoderm**, but its **developmental potential** also includes other regions of the **mesoderm** and **endoderm**. The difference between these two concepts is central to the definition of **commitment** or **determination**: A cell becomes determined when its developmental potential no longer differs from its **fate**.

differentiation The expression or manifestation of the **fate** of a cell. A differentiated cell expresses characteristic proteins and has a clearly defined morphology that identify it as a member of a defined histological type. The process of differentiation is probably always irreversible.

dorsal lip of the blastopore (see also **blastopore, Hensen's node, shield**) In amphibians, the most animal/dorsal edge of the **blastopore**, where formation of the whole **blastopore** begins. It is molecularly and embryologically similar to the **shield** of teleosts and to **Hensen's node** in amniotes. The experiment of transplantation of the dorsal lip of the blastopore between two different species of amphibians by Spemann and Mangold in 1924 led to the concept of neural **induction**.

ectoderm (see also **endoderm, epiblast, germ layers, hypoblast, mesoderm**) (Greek: *ecto*=outer, *-derm*=skin). One of the three germ layers of the early embryo. In early development of higher vertebrates, it gives rise to cells of the other germ layers (embryonic **endoderm** and **mesoderm**). After gastrulation, the ectoderm can give rise to the skin (epidermis) and to the nervous system, depending on its position in the embryo and whether it has been subjected to the influence of neural **induction**.

embryonic shield (see **shield**)

endoderm (see also **ectoderm, germ layers, hypoblast, mesoderm**) (Greek: *endo*=inner, *-derm*=skin). One of the three "definitive" **germ layers** of the early embryo. In vertebrates and other chordates, it gives rise to the digestive system and some of its associated organs.

endothelium (plural **endothelia**) A histological term originally used to describe epithelia derived from the **endoderm** (see **germ layer**). Now more commonly used to refer to the innermost cell layer lining the blood vessels, which is of mesodermal origin (see also **mesothelium**).

Entwicklungsmechanik (see **experimental embryology**)

epiblast (see also **ectoderm, germ layers**) A term mainly used for amniote embryos (reptiles, birds, mammals). It is the amniote equivalent of the **ectoderm**, the layer of cells that after **gastrulation** gives rise to the skin and nervous system.

epiboly The morphogenetic movement of a layer of cells sliding or expanding over another layer of cells or a substrate.

epigenesis (Literally, "above the genes"). Historically, epigenesis was a doctrine (opposed to **preformation**) which maintained that the embryo develops de novo rather than unfolding from a preformed homunculus. More recently, epigenesis has been used to describe mechanisms that can control gene expression without themselves being controlled by genes. One example might be **cytoplasmic determinants**. *"Genetics proposes, epigenetics disposes."*

experimental embryology In the 1880s, Wilhelm Roux introduced the term Entwicklungsmechanik, now loosely translated as "experimental embryology" to describe the study of the rules governing developmental processes by carefully designed transplantation experiments.

fate The range of tissues or cell types *normally* generated by a given cell or tissue in the embryo. Compare with the concept of **developmental potential**, which is the range of tissues or cell types that a given cell *is capable* of generating.

fate map A spatial map of the **fates** of different regions of an embryo at a particular stage of development. Usually applied to *groups* of cells. The term **cell lineage** is more often used to refer to the collection of **fates** generated by *single* cells.

field (see **morphogenetic field**)

form (see also **morphogenesis, pattern formation**) The external and internal morphological characteristics of an embryo or **morphogenetic field**.

galvanotaxis Active, directional migration of cells in response to an electric field. It can be *positive* if cells migrate toward the cathode or *negative* if cells migrate toward the anode. The term excludes *electrophoresis*, which is the *passive* movement of cells in response to an electric field. The distinction is based on whether the cell's locomotory mechanism (cytoskeleton, attachment to substrate, etc.) is involved (galvanotaxis) or whether cells are moved in the field solely because of their net charge. Despite much hope for this mechanism in the 1970s and 1980s, there is little or no evidence that it is a major guidance cue for cell movements during **gastrulation** in the embryo.

gastrula (Greek: *gastr-*=stomach). The stage of development during which the **mesoderm** and **endoderm** begin to form and the embryo becomes three-layered. It is accompanied by massive cell movements that reorganize cells within the embryo and by the appearance of a characteristic structure around which the **mesendoderm** forms. In some organisms this structure is the **blastopore**. In amniotes the structure is the **primitive streak**. **Gastrulation** movements include **ingression, involution,** or **invagination** of cells derived from the **ectoderm** to give rise to the **mesoderm.**

gastrulation (see **gastrula**)

germ layers One of the three so-called "primary" layers of cells in the early embryo: **ectoderm, mesoderm,** and **endoderm.** In mammalian embryos, the term is usually used to describe only those tissues that contribute to the embryo proper and therefore excludes the trophectoderm, which only contributes to placental structures. Early embryologists had an exaggerated view of the importance of the traditional germ layers; this led to the introduction of terms like **endothelium** and **mesothelium.**

haptotaxis (adj.: **haptotactic**) Active, directional migration of cells up or down a gradient of adhesion laid down on their substrate. The term was introduced by Carter in the 1960s. Although this is an attractive mechanism for guiding cell migration, there is as yet little or no solid evidence that it plays a major role in guiding the normal movements of **gastrulation.**

head fold In amniote embryos, a ventrally directed movement of all three **germ layers** in the head region, which defines the entrance to the foregut and the outline of the head. Formation of the head fold also defines the *septum transversum,* the precursor of the diaphragm, which subdivides the trunk into thorax and abdomen and which also carries the **primordia** of the liver and heart. It should not be confused with **head process,** which is the cephalic part of the **notochord.**

head process (see also **chorda mesoderm, notochord**) The portion of the **notochord** in the head of amniote embryos, lying anterior to the otic (ear) vesicle. It is derived from cells in **Hensen's node** and extends anteriorly as a rod. In amniote embryos it underlies the developing brain between the middle of the hindbrain (boundary between rhombomeres 4 and 5) and the diencephalic part of the forebrain. It should not be confused with **head fold.**

Hensen's node A bulbous accumulation of cells at the anterior (rostral) tip of the **primitive streak,** appearing at the end of **gastrulation** in avian and mammalian embryos. It contains cells whose progeny contribute to the **head process, notochord,** embryonic **endoderm,** and the medial portion of the **somites.** Its superficial cells also contribute to the floor plate of the neural tube. Hensen's node comprises cells of all three embryonic **germ layers.** It is considered to be the amniote equivalent of the amphibian **dorsal lip of the blastopore** because if it is transplanted to another region of a host embryo, it will induce (**induction**) a supernumer-

ary embryonic **axis** or supernumerary neural tube. It is named after Viktor Hensen, who first described this structure in mammalian embryos (rabbit and guinea pig) in 1876.

homeogenetic induction (see **homoiogenetic induction**)

homoiogenetic induction Induction of a cell or tissue by a previously induced cell or tissue. It is usually used only to describe **induction** of the *same* cell type as the inducing tissue. A commonly cited example is the **induction** of neural plate from recently induced neural plate: If a piece of amphibian or avian **ectoderm** that has been exposed to the influence of inducing **mesoderm** is placed adjacent to a piece of **ectoderm** that has not, the induced property (ability to form a neural plate) is transferred to the host **ectoderm.**

hypoblast (see also **endoderm, germ layers**) (Greek: *hypo-*=under). Mostly used for amniote embryos (reptiles, birds, mammals), a transitory layer of cells that contributes only to extraembryonic structures. The term is not equivalent to **endoderm,** which is usually used for the embryonic layer that gives rise to the digestive system of the adult. In avian embryos, the hypoblast gives rise to the yolk sac stalk. The same term is used slightly differently in teleost embryos (see Chapter 12), where it refers more generically to the deeper layers of cells.

induction (embryonic) Defined by John Gurdon as " . . . *an interaction between one (inducing) tissue and another (responding) tissue, as a result of which the responding tissue undergoes a change in its direction of* **differentiation**" (Gurdon *Development* 99: 285 [1987]). The term "direction of **differentiation**" can perhaps best be replaced by "**fate.**" Embryonic inductions have been classified by this author as "permissive" or "instructive," depending on whether the tissue has only one possible **fate** open to it but requires an external signal to continue to differentiate in this direction (permissive), or whether the external signal(s) allows the responding tissue to choose between two or more alternative **fates** (instructive). One might argue that only instructive inductions are true inductions.

ingression (see also **invagination, involution**) The penetration of superficial cells into the interior of the embryo. Usually the term implies that the penetrating cells do so individually rather than as a sheet (as opposed to **invagination** or **involution**). Characterized by the presence of a transitory, *bottle cell* morphology, this process accompanies the early stages of **gastrulation.**

intermediate mesoderm The portion of the **mesoderm** of the trunk of vertebrate embryos lying between the **paraxial mesoderm** and the **lateral plate mesoderm.** The intermediate mesoderm contains cells that give rise to the two types of embryonic kidney, pronephros and mesonephros, and their ducts. In higher vertebrates, the pronephros and mesonephros, are both transitory, but their ducts (Wolffian and Müllerian) contribute to the adult urogenital system. There is no intermediate mesoderm in the head of higher vertebrates.

invagination (see also **ingression, involution**) The penetration of a layer of cells into the interior of the embryo. The term implies that part of the surface of a more or less spherical embryo is pushed into the interior of the sphere, as if poking a finger into a rubber ball. It should be distinguished from **ingression** and **involution.**

involution (see also **ingression, invagination**) The penetration of a layer of cells into deeper regions of the embryo by rolling against the inner surface of another cell layer (**epiboly, invagination**). It characterizes the process of **gastrulation** in amphibian and other embryos. Should not be confused with **ingression** or **invagination.**

lateral plate mesoderm The portion of the **mesoderm** of the trunk of vertebrate embryos lying lateral to the **intermediate mesoderm.** The lateral plate mesoderm starts to form during **gastrulation** and later subdivides into two plates: one dorsal, called the **somatopleure,** and one ventral, called the **splanchnopleure.** The gap between them is called the **coelom,** a cavity that characterizes higher animals. The lateral plate mesoderm gives rise to the circulatory system including the heart, and contributes to many other organs and tissues.

lineage (see **cell lineage**)

lineage compartments (see **compartments**)

mesenchyme (plural: **mesenchymata**) A rather vague term used by histologists to describe embryonic tissues consisting of stellate or fibroblastic cells loosely arranged in extracellular matrix. It does not have to be of mesodermal origin, nor is it necessarily a single tissue type.

mesoderm (see also **ectoderm, endoderm, epiblast, germ layers**) (Greek: *meso-*=middle; *-derm*=skin). One of the three **germ layers** of the early embryo. In vertebrates and other chordates, it gives rise to most of the musculoskeletal system, to the circulatory system, and to most of the internal organs of the adult except the gut and parts of its associated organs. It arises during the process of **gastrulation,** when it forms a layer situated between the other two **germ layers, ectoderm** and **endoderm.** It is usually subdivided into tissues with different developmental **fates** depending on their proximity to the embryonic **axis: axial mesoderm** (=**chorda mesoderm; head process** and **notochord**), **paraxial mesoderm, intermediate mesoderm,** and **lateral plate mesoderm.**

mesothelium (plural: **mesothelia**) A histological term used to describe epithelia derived from the **mesoderm** (see **germ layer**). The term is now becoming obsolete (see also **endothelium**).

morphogenesis (adj.: **morphogenetic**) **and pattern formation** (Greek: *morpho-*=form; *-genesis*=formation, origin). Literally, the word appears to be a Greek translation of **pat-** tern formation. However, in embryology the terms are usually used differently. Morphogenesis is the formation of embryonic pattern that includes considerable cell rearrangements (as occurs during **gastrulation**), whereas pattern formation is embryonic form arising from cells changing their **fates** or shapes in situ, without much cell rearrangement (an example is the formation of the pattern of bones of the vertebrate limb, which condense from the loose **mesoderm** cells within the limb bud).

morphogenetic field (see also **compartment**) A portion of an embryo contained within well-defined boundaries, which can develop independently, without instructive influences from the rest of the embryo. An important property of morphogenetic fields is that they are capable of **regulation;** that is, any portion of the field can regenerate the whole field. As development proceeds, fields subdivide, becoming smaller and more numerous. C.H. Waddington first defined the term in the 1920s.

mosaic (see also **chimera, regulation**) An embryo containing cells of two different phenotypes (cf. **chimera**). The term is used in two different contexts: (1) A *naturally occurring* condition in which gene expression varies among cells of the same histological cell type. An example is the mammalian female body, in which one of the two X chromosomes is inactivated and the genes on it are not expressed. If the inactivation affects a different X chromosome of the pair in different cells, some cells will express genes on one of the chromosomes of the pair, other cells will express genes situated on the other member of the pair. The effect of mosaicism can sometimes be visible externally, such as in individuals with different-colored eyes. (2) An embryo incapable of **regulation.** The term "mosaic" is used because the lack of ability to regulate implies that different cells are determined to different **fates.** During embryonic development, the ability to regulate is gradually lost, as the state of **determination** of cells becomes progressively more restricted.

mosaic embryos (see **mosaic,** def. 2)

neurulation The developmental process in vertebrate embryos that follows **gastrulation,** which results in the formation of the neural plate and terminates with its closure into the neural tube.

notochord (see also **axial mesoderm, chorda mesoderm, head process**) The portion of the **mesoderm** underlying the midline of the embryo. In the strict sense, the notochord comprises only the **axial mesoderm** posterior to the ear vesicle. Some authors, however, also include the **head process** as a component of the notochord.

organizer A group of embryonic cells which emits instructive signals that both **induce** (i.e., change the **fate** of the responding cells) and pattern their neighbors. During vertebrate gastrulation, the Spemann organizer (**shield, dorsal lip of the blastopore, Hensen's node**) is the clasical example; other organizers include the isthmus (a constriction between midbrain and hindbrain), which emits signals to other components of the caudal CNS on either side.

pattern formation (see **morphogenesis and pattern formation**)

paraxial mesoderm In vertebrate embryos, the **mesoderm** lying between the **notochord** (or **axial mesoderm**) and the **intermediate mesoderm**. The paraxial mesoderm of the trunk gives rise to the **somites**, which in turn contribute cells to the dermis, skeletal muscles, and axial skeleton of the adult. In the head, the paraxial mesoderm contains fewer cells and its fate is less well understood; among other tissues, it probably contributes to some facial muscles.

placode A localized thickening of an epithelial sheet, usually circular in shape. Applied especially to the **primordia** of cranial sensory organs (lens, ear, olfactory, epibranchial).

preformation (see also **epigenesis**) In the 17th century, those who subscribed to the doctrine of preformation (cf. **epigenesis**) believed that each sperm (or egg) contained within it a diminutive but complete individual, the *homunculus*, and that further development only consisted of growth. Preformationists were divided into two camps: those who believed that the homunculus was contained within the sperm (*spermists*), and those who believed that homunculi were present in the egg or oocyte (*ovists*).

primitive streak The amniotic equivalent of the **blastopore**: the region acting as a center for **ingression** of the **mesoderm** and definitive **endoderm**. Rather than a ring or indentation through which **ingression** occurs, the lateral sides of the blastoporal ring have been compressed to form a line (streak), with dorsal fates at one end (**Hensen's node**) and ventral (posterior streak, also called *nodus posterior*) at the other. Thus, **Hensen's node** is the equivalent of the **dorsal lip of the blastopore** in animals with a primitive streak (reptiles, birds, mammals).

primordium (plural: **primordia**; =**Anlage**; plural **Anlagen**) A structure or region that is the precursor of an adult structure or organ. The German word **Anlage** is often used as a synonym.

regulation (see also **mosaic**) The ability of a portion of a **morphogenetic field** to reconstitute the entire field. At early stages of development, when the entire embryo is a single field, **regulation** defines the ability of parts of some embryos to reconstitute the entire embryo. In 2-cell-stage amphibians, for example, separation of the two blastomeres can allow each blastomere to produce a whole embryo.

shield (or **embryonic shield**) In teleosts, the slightly thickened dorsal side of the marginal zone. The shield is molecularly and embryologically similar to the **dorsal lip of the blastopore** (amphibians) and to **Hensen's node** (amniotes).

somatopleure (see also **lateral plate mesoderm, splanchnopleure**) The dorsal sheet of **lateral plate mesoderm** that forms the outer wall of the **coelom**. The **mesoderm** of the limb bud in higher vertebrates (apart from the skeletal muscles and dermis) is derived from this layer.

somite A condensation of the **paraxial mesoderm** in vertebrate embryos. Somites form in rostral to caudal sequence along the **axis**, one on each side of the midline. In some fishes and in amphibians, the somites first form as a chevron shape, whereas in amniotes and other fishes, it is initially an epithelial sphere. Later, each somite subdivides into dermomyotome and sclerotome. The former gives rise to hypaxial muscle (ventral body wall and limbs) by delamination from its lateral/ventral edge and to dermis. The sclerotome contributes to the axial skeleton and to some blood vessels. Later still, the dermomyotome subdivides further into myotome (origin of the epaxial muscles, situated dorsally) and dermatome (future dermis).

splanchnopleure (see also **lateral plate mesoderm, somatopleure**) The ventral sheet of **lateral plate mesoderm** that forms the inner wall of the **coelom**.

subduction A relatively recently introduced term to describe the **morphogenetic** movements causing the "sinking in" of one layer with respect to another, as seen in some urodeles (see Chapters 3 and 19).

syncytium (plural: **syncytia**; adj.: **syncytial**) A cell containing many nuclei (a multinucleate cell).

teratology The branch of embryology concerned with the study of "monsters" (Greek: *teratos*), or abnormalities of development, although the term is used more commonly to describe experiments studying the effects of noxious substances on embryonic development. The immunologist Peter Medawar pronounced: *"The classification and investigation of abnormal embryos . . . are the subject matter of 'teratology,' a word that suggests pretensions to the stature of a science (a designation not really deserved) . . . , but teratology has not —as had at one time been hoped—thrown a flood of light upon developmental processes, and it has not helped us very notably in the interpretation of normal development. Teratology is more deeply in debt to embryology than the other way around."*

SUCKING IN THE GUT
A BRIEF HISTORY OF EARLY STUDIES ON GASTRULATION

S. Brauckmann[1] and S.F. Gilbert[2]

[1]*Konrad Lorenz Institut, A-3422 Altenberg, Austria;* [2]*Biology Department, Swarthmore College, Swarthmore, Pennsylvania 19081*

INTRODUCTION

The concept of gastrulation has a history far deeper than Haeckel's original use of the term (Haeckel 1874). Moreover, even after the concept was so named, it took nearly a century to reconcile conflicting notions of gastrulation into a coherent picture. These concepts included the formation of the three germ layers, the formation of the gut, the formation of the mesoderm, and the coordinated patterns of cell migration. It is from one of these concepts—the formation of the gut—that we get the name gastrulation. Indeed, the infolding of surface cells of the blastoderm to form the archenteron linked development with digestion during much of the 19th century. Haeckel proclaimed the universality of the animal kingdom through the digestive capabilities of all organisms, and Metchnikoff's discussions of mesoderm formation linked the phagocytic digestion of the mesodermal cells with the extracellular digestion of the larva.

In our historical study, we narrate the discoveries of and reflections on the early stages of embryogenesis, starting with the natural history of Mauro Rusconi and ending with Johannes Holtfreter. We introduce the dispute between Rusconi and Karl Ernst von Baer on the embryogenesis of the frog and how their successors, among them Krohn, Kowalevsky, Haeckel, and Metchnikoff, attempted to understand the phenomenon of gastrulation. Although the organisms studied and the methods used to study them have changed, these researchers were still dealing with the same crucial questions involving the mechanisms of gastru-

lation and their implications for animal classification. As Chapter 51 demonstrates, these are ongoing concerns despite the enormous amount of data we now possess.

The discovery of the germ layers in vertebrates (Pander 1817; Baer 1828; Remak 1855) and the homology of the corresponding primary anlagen in invertebrates (Siebold 1846; Kowalevsky 1866a,b) propelled *Entwickelungsgeschichte* to an important part of Darwinian morphology (Ospovat 1976; Gould 1977; Nyhardt 1995; Jahn 1998). In this way, the entire animal kingdom—both vertebrates and invertebrates—was united through their developmental processes. However, the manners by which these animals formed these germ layers appear to differ dramatically. The studies of gastrulation attempted to make sense of the different ways that different groups of animals produced these homologous germ layers.

GUT FORMATION IN FROGS AND SEA URCHINS

This historical sketch of gastrulation begins in natural history, specifically with that of the Italian anatomist and naturalist Mauro Rusconi (1776–1849). Rusconi observed the development of the common pond frog to determine whether this animal breathed with gills or lungs. His point of departure was a taxonomic question, as he wanted to find the right place for this animal in a natural scheme of classification. He attempted to answer this question by studying the frog's development and anatomy. In these studies (Rusconi 1823, 1854), he corrected Cuvier's observations concerning the length and form of the digestive canal, confirmed that the frog breathes atmospheric air, and proposed it as the link between reptiles and fishes. In 1826, he accurately described the early developmental stages of the green frog (*Rana*) and particularly the external appearance of an opening to form the presumptive anus, later called Rusconi's anus (Fig. 1A) (Rusconi 1826; Beetschen 2001). In a posthumously published paper, Rusconi studied the development of the salamander *Triton terrestre*, and he depicted its first cleavage stages, the first traces of the mesoderm, and a *tache circulaire*, the germinal ring. When this round spot had disappeared, he saw through his microscope that its pigment granules formed a circle about a rounded streak that opened to a largely curved line that we now call the blastopore.

In 1834, Karl Ernst von Baer (1792–1870) published an article on frog development in Müller's influential *Archiv für Anatomie, Physiologie, und wissenschaftliche Medicin* in which he first corrected the observations of French embryologists who claimed to have discovered an orifice in the germinal area and in the outer cells (Baer 1834; see Oppenheimer 1967; Raikov 1968; Churchill 1991). (Like many others, Baer had believed that the cells originated from the germinal area [*Keimbahn*] and that the outer portion of the egg constituted the cytoplasm from which the blastomeres derive.) Baer questioned Rusconi's interpretation that the anus appeared before the mouth opened. With his magnifying glass, Baer clearly noticed that the dorsal pole of the embryo pushes to "the remains of an uncovered spot" (the remnant of the blastocoel), but he denied Rusconi's assertion that this spot was the anus by arguing that such formations are teratological forms. Later, Alexander Kowalevsky supported Baer's hypothesis about the "teratological anus" when studying the development of sturgeons. He showed that in several embryonic anomalies, the blastopore plug protruded outward and ended up at the spinal neural tube canal as Baer had described it (Kowalevsky et al. 1870).

Baer and Rusconi also disagreed on the development of the alimentary canal (Baer 1834; Rusconi 1836, 1854), on the cause of nuclear division, and on the number of the branchial slits. Baer thought that the cell-forming area of the amphibian embryo expanded from the center of the egg to the periphery, and penetrated the outer *stratum granulosum* before it finally dissolved in the ovum between yolk plug and yolk where the blastoderm will emerge. According to Baer's observation, this form developed after the fourth cleavage, and not before it, as Rusconi had interpreted it (Baer 1834). Furthermore, Baer introduced the concept of the embryonic axes and dealt with their relationship to the yolk distribution and how these axes differ in the developmental stages of different species.

When studying unsegmented frog eggs, he distinguished two hemispheres that are visibly separated by the distribution of pigment. These two halves occupied a constant position with respect to gravity. The egg was more pigmented on the animal hemisphere, whereas the pigment was nearly absent in the area surrounding the pole of the vegetal hemisphere, evidently determined by the greater

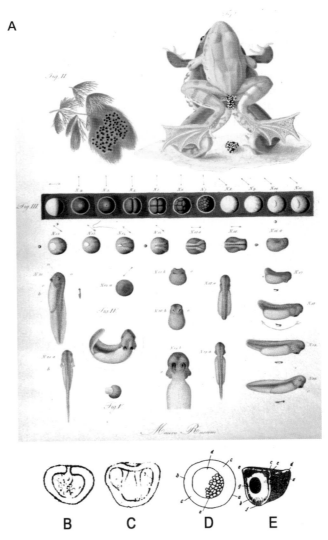

Figure 1. Early depictions of the gastrula. (*A*) Figure from Rusconi (1826; Table 1) showing the blastopore (Rusconi's anus) in the frog embryo. (*B, C*) Figures from Derbès (1847) showing the origins of the archenteron in the sea urchin *Echinus lividus*. (*D, E*) Krohn's (1849) figures of the sea urchin mesenchyme blastula and primitive gut. (*D*) Larva shortly after hatching. *e*, Accumulation of "nucleus-like formations" from which the fibrous tissue is later formed. (*E*) Cross-section of the same stage of development. *h*, Rudiment of the digestive tract in immature form; *k*, its rudiment in later form. (*D, E* Reprinted, with permission, from Blyakher 1982.)

abundance of yolk. He proposed, therefore, to think of the frog's egg as a sphere with two poles that are connected by a main axis, the early cleavage planes being related to its meridian. The first two cleavage planes pass through it in two meridians intersecting each other at a right angle, and the third is transverse horizontally to it (Baer 1834; Wilson 1928). This polarity was subsequently found to be expressed in one way or another in the eggs of all other amphibian species.

The Rusconi–Baer dispute on whether the blastopore is the anus or the mouth, and their disagreement about the axiality of the embryonic egg, influenced the research program of their younger colleagues, Alphonse Derbès (1802–1857) and the Russian-German anatomist and embryologist David August Krohn (1803–1891). Derbès, a French marine biologist from Marseille, studied the development of the sea urchin *Echinus esculentus*. Around 20 hours after fertilization, he saw that the spherical form of the egg had elongated and developed an orifice communicating with the rudiments of an alimentary canal, before it first moved "forward and then upwards such that the mouth looks toward the zenith" (Derbès 1847). What Derbès here described is the formation of the primitive gut, without realizing that it is formed by the multiplication of cells from the blastula surface (Fig. 1B,C). However, Krohn did identify the sources of the primitive gut, and he corrected Derbès's erroneous statement concerning the mouth (Krohn 1851a). In fact, Derbès reported a shape that resembled Haeckel's gastraea in echinoderms (Fig. 1D,E), but he did not place this observation into its larger phylogenetic context.

David Krohn investigated marine animals, about which he published more than ten monographs during his numerous travels around the Mediterranean Sea. He was the first biologist who described the primary mesenchyme and the invaginative gastrulation in echinoderms (Krohn 1849; Neigebaur 1860). His work on ascidians, which he, following Baer's suggestion (Baer 1847), inseminated artificially, represents the first systematic description of their embryogenesis. When examining the fertilized egg of *Echinus lividus*, Krohn observed an empty vesicle, which he, like Baer, considered the nucleus of the fertilized ovum, and dark bodies in the cavity (then considered as nuclei in the germinal vesicle; now considered as cells in the blastocoel). The number of these nuclei quickly increased until they filled half the cavity (Fig. 1D). When they were evenly distributed inside the cavity, they partially piled up to form the rudiments of the calciferous skeleton, forcing the nucleus to an angular shape. Then they transformed into the reticular tissue of the primary mesenchyme.

Krohn further detected a small hole-like depression in the center of the blunt (vegetal) pole, which widens and thickens, and he meticulously described how the outer layer gradually invaginates into the cavity of the body. This protrusion then progressively submerges deeper into the body cavity and elongates into a sac-like canal, stretching through the whole body cavity (i.e., forming the archenteron). In his own words: "The blind sac thus created ... is nothing less than the embryonic (primitive) gut. The site of the primary ingrowth forms an orifice which is, in effect, the anus" (Krohn 1849). Two years later he extended this idea to jellyfish, when he analyzed *Aurelia*. This coelenterate also forms its anus at the vegetal pole, leading to a baggy canal, the nutrition tube, which invaginates into the body and finally broadens like an ampulla (Krohn 1851b; see Smith 1891). The similarity to the developmental phase of the larva of echinoderms was obvious, and Kowalevsky later openly acknowledged the importance of Krohn's work on development to his own studies in comparative embryology (Krohn 1852).

The cellular perspective was brought into the study of embryology by the physician Robert Remak (1815–1865), who identified the neural folds that Christian Pander (1817) had noted in chick embryos, as ectoderm (Remak 1852; see Schmiedebach 1995). The term was introduced one year later by George James Allman researching the development of *Tubularia* zoophytes (Allman 1853). He described the ectoderm as the external layer of the polyps, and the endoderm covering the walls of an uninterrupted cavity, for the primordial germ layers; Huxley named the third layer the *mesoderm* (Huxley 1871). Remak further discovered that the frog's egg is a cell, and he reinterpreted Baer's layers when noticing that the larger cells of the vegetal hemisphere correspond to Baer's formative (vegetative, endodermal) layer, and, conversely, the smaller cells of the dorsal hemisphere to the nutritional (animal, ectodermal) layer. Based on his observations, he opted for mesodermal development from the endoderm whereby the mesoderm further divides into dorsal and ventral plates by forming a cavity, which Oscar Hertwig designated the *coelom* (Hertwig and Hertwig 1882; Hertwig 1883; see Gourko et al. 2000). Ironically, at least when recalling the dispute between Baer and Rusconi, Remak identified Rusconi's elliptic cavity as the archenteron that enlarges at the expense of Baer's cavity (blastocoel) (Remak 1855).

ALEXANDER O. KOWALEVSKY AND THE HOMOLOGY OF DEVELOPMENT

In 1865, Alexander Onufreevich Kowalevsky (1840–1901) observed the development of two embryonic layers in *Amphioxus* larvae, and in the next year he published his seminal *Entwickelungsgeschichte der einfachen Ascidien* (Kowalevsky 1866b; see Piliptchuk 1990; Mikhailov and Gilbert 2002). This article on the development of ascidians was considered a landmark, and the *Quarterly Journal of*

Microscopical Science, which as a rule published only original papers, issued a translation. Baer wrote in his letter to Max Schultze, then the editor of the *Archiv für mikroskopische Anatomie*, that Kowalevsky's interpretations were confirmed and partially refined by Kupffer, and he cited Darwin's references to Kowalevsky (Baer 1873; Darwin 1883; see Russell 1916; Blyakher 1982). Kowalevsky described here the developmental stages of *Phallusia mammillata* and *Ascidia intestinalis*, depicting the developmental phases with the camera lucida. He also continued Krohn's studies on the formation of the alimentary canal, describing the formation of the *Leibeshöhle* (for which Haeckel coined "gastrula" a few years later), as well as the alimentary canal and the notochord.

After the first cleavage, he saw that Baer's cavity had significantly decreased, and that the embryo elongated and now resembled that of *Amphioxus*, also the topic of his thesis (Kowalevsky 1873; see Conklin 1934). Figure 2 shows how the aperture *a* develops into the anus in the same way as Kowalevsky had already described in *Phoronix*, *Amphioxus, Echinus, Ophiura, Limnaeus*, the frog, and *Petromyzon* (Kowalevsky 1866b). First, a protrusion, which later becomes part of the alimentary canal, forms in the dorsal cell layer. Its opening narrows, and at the opposite end two rims arise which close up to a tube. Under the skin, Kowalevsky could observe that the tube was open at one side before it closed completely. In the two-layered blastula the internal cells proliferate faster and narrow Baer's cavity (*h*), and thus, elongate the egg (Fig. 2A). The internal large cells differentiate and distribute themselves irregularly in the embryonic body. During the elongation of Baer's cavity (blastocoel), the opening slowly moves to one side until it is plugged. As the cells of the internal layer proliferate at the same time and push against the cells of the alimentary cavity, the latter are spread in two layers at each side. On the dorsal end two rims unfold, merging at the ventral end and disappearing from the anterior side. Kowalevsky confusingly called the anterior side, where the cavity *h* is, the abdomen because in his view it represented the "front side" of the body during the larval stage. In the cross-section (Fig. 2B), the lips are completely closed and the furrow *f* is the cavity of the neural plate (medullary groove), which later becomes the embryonic nervous system. The cell movements press the cavity *h* to an elongated cleft while the wall of the anterior side bends up and folds around the (presumptive) nervous system. The cleft then breaks through the external epithelial layer and opens up to the mouth aperture, whereas the other opening becomes the anus.

Until the formation of the neural grooves and their closing to form the neural tube, which, he emphasized, characterized the development of all vertebrates, the archenteron, or alimentary canal, is the same in vertebrates and invertebrates. The main characteristics of vertebrates

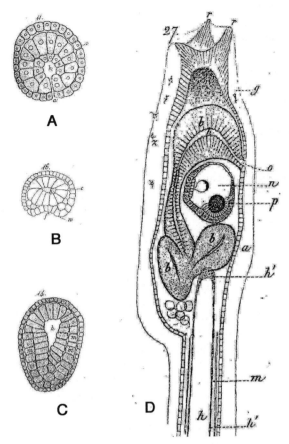

Figure 2. Kowalevsky's (1866b) depictions of the gastrula formation in *Amphioxus* and its homology with gastrula of ascidians. (*A*) Egg of *Ascidia intestinalis*, enlargement 303×. *c*, Remnant of Baer's cavity (blastocoel); *a*, orifice of Baer's cavity; *h*, invagination cavity that is constricted by increase of cells. (*B*) As in *A*, cross-section to show the height of the folds; *h*, archenteron cavity; *w*, neural folds (*Rückenwülste*); *c*, remnant of Baer's cavity (blastocoel); *f*, groove between the neural folds. (*C*) Egg of *Ascidia intestinalis*, enlargement 303×, horizontal cross-section; *h*, cavity that is encircled by multiplying cells; *m*, the cells that lie between the Chorda (*Achsenstrang*) and the epithelium and later develop to muscle fibers. (*D*) The anterior part of the *Ascidia intestinalis* larva. *h*, Chorda (*Achsenzylinder*), *h'*, its sheath with nuclei and protoplasma; *m*, muscles; *o*, mouth opening.

are the two dorsal lips (*Rückenwülste*) that develop into a solid streak (notochord). Even if the notochord of the *Ascidian* larva does not extend as far forward as that of the vertebrates, it develops in the same way as in *Amphioxus*. Thus, he equated the axis cylinder (Fig. 2C, D) of the tail of the ascidians with the notochord (chorda dorsalis) of *Amphioxus*, which is functionally and genetically homologous. If all animals that possess a segmentation cavity (blastocoel) invaginate, then the developmental modus of ascidians and vertebrates, as Kowalevsky argued, is homologous and they must share a common ancestor (Kowalevsky 1866b). Nevertheless, he openly confessed that his own observations perplexed him, since they forced him to accept "facts" that he thought of as paradoxical.

Carl Kupffer (1829–1902) did not believe Kowalevsky's discoveries. Hence, he started to investigate another tunicate, *Ascidia cannia*, which lives in the Baltic Sea near Kiel. In his letter to Max Schultze, Kupffer cautiously conceded that the gap between vertebrates and invertebrates might be bridged to formulate a phylogenetic theory of homology (Kupffer 1870). Kupffer also thought that the formation of the cavity of the neural tube developed by a protrusion of the blastoporal lips from the posterior end, and he thought that the notochord developed from two cell layers. However, he believed that the second invagination had its opening at the anterior pole, which is the mouth, and not the anus, as Kowalevsky had stated. Furthermore, he showed that the ventral (blastoporal) lips adjoin each other such that the cleft of the protrusive opening indicates the neural groove.

Kowalevsky tested Kupffer's corrections and acknowledged them, except for the statement that the intestinal sac invaginated at the posterior pole. He then reinvestigated the development of the ascidians and found that his former hypothesis on the opening, whether it is the mouth or the anus, was only true in rare cases. Even if Kowalevsky did not repeat his erroneous interpretation later, Haeckel, who based his gastraea theory on Kowalevsky's findings, continued to invoke it as a firmly established fact in his popularization of Kowalevsky's discoveries (Tauber and Chernyak 1991).

In the same year, Kowalevsky also studied the developmental stages of comb jellies, in particular *Eschscholtzia cordata (Callianira bialata)*. Here he observed a delaminative process, but interpreted it as invagination (Kowalevsky 1866a, Figs. 18 and 25). Furthermore, he observed cells at the bottom of the epithelial layer protruding into a secretory tissue and thought that these cells often anastomosed to form a syncytium. He clearly believed that this peculiar tissue secretion played a great part in *Ctenophora* development, and he extrapolated his discovery as significant for all embryonic *Bildung* in the *Coelenterata*, and even called this kind of syncytial tissue formation a universal law (Kowalevsky 1866b).

In 1870, Kowalevsky and Ovshennikov published their studies on sturgeons, which they thought linked bony fishes and frogs to amphibians (Kowalevsky et al. 1870). At the end of the first day, they observed how Rusconi's anus is formed: First an equatorial groove appeared at the marginal zone between the small and large cleavage cells. When the blastocoel disappeared, Rusconi's anus resembled a narrow cleft that was closed by a primitive streak. The shape of the egg changed, and the neural plates enclosed the primitive streak, the embryonic shield separated into two parts, the neural plates rose above and folded around the primitive streak, leaving the cleft-like aperture at the back side

(remains of Rusconi's anus). They concluded that the egg looks like an ascidian egg after segmentation, based on the dorsal epithelium that encircles the *Keimhöhle* (blastocoel).

To sum up, Kowalevsky studied early embryonic development, characterized an invaginated two-layered sac, and discovered homologies in the development of the lancelet (a chordate) and the tunicates (previously considered mollusks) (Tauber and Chernyak 1991). No doubt, Kowalevsky deserves the credit for having furnished Haeckel with the decisive ontogenetic data for the gastraea theory, namely the comparative analysis of the separation of the endoderm. After Haeckel had generalized this observation in his gastraea theory, *Amphioxus* became the model organism to study vertebrate embryology (Russell 1916). Haeckel and Darwin were enthusiastic about his discovery that ascidians pass through a larval stage in which they develop, like *Amphioxus*, a notochord. Both believed that it proved that vertebrates had arisen from a form resembling the larva of this animal (Junker 1998).

HAECKEL'S GASTRAEA AND THE PRIMACY OF PHYLOGENY

Ernst Haeckel (1834–1919) published his *Gastraea* theory in 1874, but in his monograph on calciferous sponges two years earlier, he already had formulated a preliminary version of it (Haeckel 1872; see Grell 1979; Richardson and Keuck 2002). Haeckel aimed at demonstrating the monophyletic origin of all Metazoa. He began by postulating that the two germ layers (the ectoderm and endoderm) are homologous, respectively, in all Metazoa and that an archigastrula (as an independent form of life) existed from which all Metazoa are derived (Haeckel 1874). Haeckel openly acknowledged that he could not have formulated his gastraea theory, or sketched it, without Kowalevsky's studies (in particular the discovery of the "homologous" ontogeny of *Amphioxus* and ascidians), which he considered the most important of all embryological investigations (Haeckel 1874, 1877).

Haeckel formulated his *Gastraea* theory in the following manner: After morula formation, the initially homogeneous cells differentiate into round cells that migrate into the inner cell cluster, and into cylindrical cells that remain on the surface of the morula, or blastula. Then the cylindrical cells develop a flagellum at the outer cell surface and change themselves into a flagellated organism, formerly called the *planula*. The term planula was introduced by John Graham Dalyell (1775–1851) for the ciliated larva of hydromedusa and later extended to encompass the similar developmental phases of other lower animals, e.g., worms and echinoderms (Dalyell 1847). To avoid a terminological confusion, Haeckel (1872) now redefined the planula stage as a globular ciliated body without a cavity or a mouth opening. Then he defined the gastrula as

> the developmental stage, which develops from the planula in sponges. ... I understand by gastrula a globular or spheroidal, egg-shaped or somewhat elongated round body, which contains an inner cavity with an external opening (primordial gastrula with orifice). The wall of this cavity consists of two different cell layers or blasts, an external, luminous layer and an internal, thick layer.

In Figure 3, Haeckel shows that the vesicle-like embryo encloses a single-axial cavity, the archenteron, or, in Haeckel's terminology, the protogaster, which opens at the vegetative pole to the circular ring of the blastopore, or protostoma. The wall of the archenteron cavity consists of two different adjacent cell layers, the primordial germ layer, the endoderm (red) and ectoderm (blue). During gastrulation the ectodermal cells, which are more numerous and smaller than the endodermal cells, proliferate faster than the endoderm cells and grow over the latter in an "overgrowth" process. This process meant, according to Haeckel, that the vegetative hemisphere invaginates into the animal hemisphere of the blastoderm and forms the protogaster (*Urdarm*, archenteron), which characterizes the formation of the gastrula (Haeckel 1877). Although the overgrowth was in effect an epiboly, he concluded that gastrulation always passes through an invagination that forms the embryonic axes. Haeckel himself was absolutely convinced, and never changed his mind, that all gastrulae develop by invagination, and that many—if not all—descriptions of delaminative gastrulation were due to observational mistakes (Haeckel 1875).

Next, Baer's cavity (the blastocoel) disappears gradually as the endodermal cells push against the invaginated ectoderm until large cells (macromeres) differentiate from these initial two germ layers to compose the mesoderm (Haeckel 1875). He adopted Baer's hypothesis on the formation of

the mesoderm, although Remak had corrected it over 20 years ago to the effect that the mesoderm arises from the endoderm. However, according to Haeckel, it explained in the best way the development of the secondary germ layers phylogenetically. An important pillar of the theory was the archigastrula, which he illustrated as the ontogenetic phase when the embryo has formed a hollow axial sphere, filled with liquid, composed of a layer of homogeneous epithelial cells enveloping Baer's cavity without the formative axes and an opening at the vegetal pole. The gastrula consists of the closely contiguous cell layers, the ectoderm and the endoderm, and it always develops by invagination. The earliest form of cleavage and the resulting archigastrula are conserved, according to Haeckel, in *Amphioxus* alone. When carefully reading Haeckel's description of the archigastrula, the reader will realize at once that he merely repeated Kowalevsky's observations. In addition, the majority of the illustrations which Haeckel used in his gastraea articles (Haeckel 1874, 1875, 1877) are copies of Kowalevsky's figures (Kowalevsky 1866a,b, 1871). Haeckel's innovation was to mark ectoderm and endoderm with red and blue color.

A phylogenetic consequence of the gastraea theory was the division of the animal kingdom into Protozoa and Metazoa. Moreover, Haeckel concluded that "all attempts to establish homologies between the different parts of the metazoan and protozoan organisms are futile" (Haeckel 1877). The reason there were no homologies between these two great divisions was that "the metazoans, which possess an ectoderm and endoderm, have ontogenetically a gastrula and are phylogenetically descended from a gastrula." When van Beneden proposed a division into three kingdoms to introduce a mesozoan group *planulades*, the primordial germ layers of which are developed by delamination and not by invagination (van Beneden 1876), Haeckel refuted it categorically since it contravened the invaginative dogma of the gastraea theory.

Haeckel proudly emphasized that Ray Lankester (whose refreshing irony he apparently did not grasp) had agreed on the homology of the germ layers in vertebrates and invertebrates. He also appreciated that they both agreed on the importance of homologies for phylogenetic systematics, even if their opinions differed widely from each other on the role of the secondary germ layers and the coelom (Haeckel 1874; Lankester 1875, 1876). In fact, Lankester was not really convinced of Haeckel's gastraea theory, especially since it did not explain the development of nutritional yolk. He conceded in his article that Haeckel had introduced some useful terms to designate phenomena that had been clearly recognized already some years before. However, he severely criticized Haeckel for maintaining a "discrete though disappointing silence" on the most pressing problems, namely the origin of the coelom and the blastopore (Lankester 1876). Another crucial issue Lankester pointed

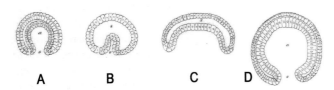

Figure 3. Haeckel's (1875) depiction of the gastrula. (*A*) Archigastrula of a calciferous sponge (*Asculmis armata*); *a*, archenteron cavity (*Protogaster*); *o*, blastoporus that Haeckel called *enteron* (*Protostoma*); the ectoderm is blue marked, and the endoderm red marked, the nutritional yolk is mostly hatched in red. (*B*) Archigastrula of a medusa (*Pelagia*). (Adapted from Kowalevsky 1873.) (*C*) Archiblastula invaginata of *Amphioxus*; *s*, Baer's cavity. (Adapted from Kowalevsky 1866b.) (*D*) Archigastrula of *Amphioxus*, completely developed. (Adapted from Kowalevsky 1866b; Haeckel 1875.)

to explicitly was the origin of the mesoderm in different phyla, whether from endoderm or ectoderm. In his interpretation, the gastrula consists of a simple two-cell-layered sac, which is bounded by the two primary germ layers, and opens to the exterior by the blastopore. It is here in the space between endoderm and ectoderm that the first cells are separated from the two primary germ layers to give rise to the mesoderm, and not from the invaginated ectoderm, as Haeckel stated.

A phalanx of well-known anatomists and embryologists, among them Carl Claus, Carl Semper, Alexander Agassiz, Wilhelm His, and Alexander Goette, attacked Haeckel for the embryological simplicity of the gastraea theory (Nyhart 1995; Russell 1916). Additionally, Kowalevsky, as already mentioned, did not argue for a complete homology of the two primary germ layers for all animal phyla after 1866, and he considered the intestinal glandular epithelium of insects and the endoderm of hydroids as developmentally specific differences. Moreover, he interpreted the secondary germ layers in a different way from Haeckel, who adopted van Beneden's view of the germ layers as "formative cells" (van Beneden 1870). Whether the attacks bothered Haeckel is not known, but they might have had an impact. Hence, from the 1880s he abandoned the exclusiveness of the gastraea theory and claimed that it was meant as a heuristic principle that might furnish a causal explanation, but that it was not a theory (Haeckel 1889; see McMurrich 1890).

METCHNIKOFF'S *PARENCHYMELLA* HYPOTHESES AND BÜTSCHLI'S *PLAKULA* HYPOTHESES

Eli Metchnikoff (1845–1916), who received the Nobel Prize for his work on the immunology of phagocytes in 1908, started his scientific career as an embryologist in St. Petersburg in the 1860s (see Moulin 1991; Tauber and Chernyak 1991). Early on, he investigated principally the ontogeny of echinoderms and ascidians with somewhat different results from those of Kowalevsky, with whom he shared a competitive friendship over years. His research questions centered around two crucial topics of 19th-century embryology; namely, how the mesoderm develops and where the blastopore emerges. In ascidians he demonstrated that during gastrulation one-half of the embryo invaginates, and confirmed its bilaterally symmetrical formation and the neural folds (Metchnikoff 1869). He partly corroborated Kowalevsky's results to the effect that as the ascidian embryo elongates, the dorsal end invaginates toward the ventral end, and that the aperture diminishes when moving from the dorsal to the ventral pole. Furthermore, he agreed with the statement of the homology of the secondary (gut)

cavity in echinoderms and ascidians when referring to Kowalevsky's studies of the *Coelenterata* (Metchnikoff 1869, 1870).

While he sharply criticized the gastraea theory as "old natural philosophy hidden inside a bombastic terminology" (Tauber and Chernyak 1991), Metchnikoff studied carefully the cleavage and larva formation of several hydromedusae to disprove Haeckel's dictum of invaginative gastrulation and the existence of an archigastrula as well (Metchnikoff 1871, 1881). In particular, he polemized against the strong belief in invagination as the one and only modus of gastrulation because this conviction meets difficulties in the embryogenesis of the lowest metazoa, and, especially, in the origin of the migratory cells (Metchnikoff 1882). Summarizing a long series of experimental observations, Metchnikoff was not able to corroborate the existence of an archigastrula in *Salmneta flavescens* or *Lunoctacantha*, nor an exclusively invaginative process of gastrulation, since he observed, for instance, in *Carmarina hastata* and *Polyxenia flavescens* a delaminative process.

In 1885 the *Zeitschrift für wissenschaftliche Zoologie* published Metchnikoff's comparative studies on gastrulation and mesoderm formation in the ctenophore *Callianira bialata*, which his friend Kowalevsky had investigated nearly 20 years earlier, and on which Haeckel had written a monograph. In his comparative embryological studies, Metchnikoff describes the formation of the mesoderm in *Callianira* (Metchnikoff 1885). When the number of the micromeres has increased to 30–50, the 4 ventral macromeres divide first and then the lateral macromeres, as Kowalevsky had already observed in his Russian thesis (Kowalevsky 1873). In this stage of 16 macromeres, the outer (ectodermal) spread ventrally and grow over the large cells. At the same time, the macromeres flatten, and when they have finished their longitudinal division, their nuclei migrate from the dorsal cell pole ventrally (Figs. 4A,B) (Metchnikoff 1885). The macromeres bud off micromeres (mesoderm) at their ventral surface. This is the first phase of the formation of the mesoderm. As the ectoderm arrives at the ventral side of the embryo during its epiboly, more ventral mesodermal micromeres form from the endodermal cells. The macromeres then invaginate inward, forming the endoderm, and "drag" the mesoderm into the inside of the embryo. During invagination the mesodermal cells migrate deeply into the gastrula cavity where they stay sac-like, until they move to the dorsal pole of the embryo, changing their shape to their former plate-like form (Metchnikoff 1885). Metchnikoff also uses these observations to sharply criticize and refute the coelom theory of the Hertwig brothers.

In these ctenophores, Metchnikoff discovered that spindle cells form a parenchymatous tissue when moving into the cleavage cavity. In 1871 Metchnikoff observed that this

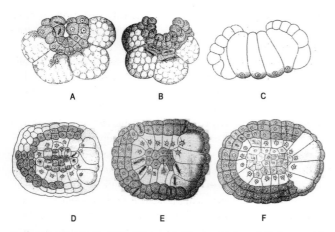

Figure 4. Metchnikoff (1885) on *Callianira bialata*. (*A*) Stage with nearly 48 ectodermal cells; (*B*) a similar stage with macromere division; (*C*) optical section. (*D–F*) Further stages of mesoderm formation (parenchymatous tissue). (Reprinted from Metchnikoff 1885.)

second pattern of embryonic layer formation, which formed a larva (without a blastocoel), resembled the larvae of sponges, the *Parenchymella* larvae, which had flagellated surface cells surrounding nonflagellated cells in the center (Metchnikoff 1882). He concluded that the future endoderm was formed by delamination of an inner mass, the parenchyma of which splits to form the archenteric cavity and blastopore by perforating at one pole. To strengthen the *Parenchymella* hypothesis, he appealed to *Trichoplax*, besides the studies on the development of *Coelenterata*. In a similar way, Goette argued that in *Volvox*, the cells, which migrate into the inside, were originally primitive germ cells and later changed to endodermal cells. Metchnikoff concluded that the gastrula of the *Ctenophora* results from invagination as a result of overgrowth, instead of stating that it also might develop through the epiboly of the surface cells. His conclusion is that the blastopore develops as the mouth cavity, and the mesoderm results from macromere budding, and not from invaginated ectoderm cells as Haeckel had claimed.

To complement our short narrative on phylogenetic theories of embryology, we need to mention Otto Bütschli (1848–1920), who was one of the first biologists to invoke surface tension as a causal factor in biological processes. In 1884, he suggested the *Plakula* theory, which he formulated when he heard that a zoologist in Innsbruck had discovered a little animal, *Trichoplax adhaerens*, on the side of an aquarium. It perfectly matched the imagined organism of Bütschli's phylogenetic theory; namely, a bilaterally symmetrical organism creeping about on the sea floor. When it ingests food, its ventral surface invaginates to increase the surface area while feeding, which resembles a reversible gastrulation process. In the 1960s, the protozoologist Karl Grell rediscovered the theory when a colleague sent him this

organism on algae from the Red Sea. *Trichoplax adhaerens* often rears up to form a temporary digestive cavity, and thus, resembles a gastraea. Grell described the creature as probably the most primitive multicellular animal known and considered it in a group separate from the eumetazoan animals (Bütschli 1880; see Grell 1968). This notion has been disputed by Willmer (1990), who sees *Trichoplax* as a planula-grade organism, containing parenchymal cells.

A BRIEF INTERLUDE ON OBSERVATION AND EXPERIMENT

The second half of the 19th century saw a controversy concerning the relative importance of observation and experimentation. For physiologists such as Carl Ludwig, Emil du Bois-Reymond, Ernst Brücke, and Hermann Helmholtz, it was unthinkable not to use experiments to answer their research questions. When Wilhelm Roux insisted on the experimental practice in biology, comparative embryologists and anatomists more or less ignored it. For many, it was merely a matter of which observations took priority. Thus, while Roux championed experimentation over observation (but did not discount the importance of comparative observations), Oscar Hertwig (Querner 1998) thought that experimentation should be used to support the detailed observations of normal development. To bolster his argument, he pointed toward the inorganic sciences, which modified their experimental object in the artificial environment of a laboratory to gain knowledge. Because the fundamental questions in embryology involved the detailed description of developing stages, it was not until the late 1890s that a new generation of embryologists shifted the focus of their field from searching for homologies between the phyla to the problems of how such transitions—the formation and placement of the blastopore, and emergence of the germ layers, and the mechanisms of folding—take place.

The technical improvements in staining methods, the microscope, and the microtome were extremely important in changing the focus of embryology. Before the 1860s one could only normally study the development of internal structures as far as the transparency of the eggs allowed, because the histological techniques, using a mixture of stearin and white wax, were not yet developed (Stricker 1864; see Beetschen 2001; Hopwood 2004). Thus, the embryologists were constrained to observe the external surface of living larvae, as Rusconi did. However, whether this technical limit really was the principal restriction in studying the role of embryonic layers in organogenesis might be questioned. Around 1875, aniline dyes were invented and their use improved. One could now use the hematoxylin/eosin method wherein the nucleus was stained deep blue with hematoxylin and the cytoplasm a pale pink with eosin, giving a better picture of cell structure. At the same time, Zeiss Jena invented better microscopes, e.g, the apochromatic objective, in 1886, that became the standard until the mid 1950s. It was now possible to cut thin slices of embryonic material with microtomes, which supplanted the usual free-hand method, to investigate the formation of the germ layers, especially of the mesoderm, for which cuts of many successive stages are necessary to show the mesoderm at each side of the primitive streak. A further improvement to investigate cell movements during gastrulation was invented by Friedrich Kopsch in the 1890s when he took time-lapse micrographs of frog cells which clearly demonstrated the nuclear movements and the movements of ectodermal cells (Kopsch 1895, 1900; Hamecher 1905; see Beetschen 2001).

ON THE FORMATION OF THE BLASTOPORE

In the second half of the 19th century, comparative embryologists, anatomists, and histologists still discussed at length the development of the mesoderm and the primitive streak in chick ontogeny. It was thought that either the mesoderm forms after ectoderm and endoderm has formed, or that the mesoderm develops from the ectoderm and then differentiates into the endoderm. Concerning the formation of the mesoderm in the chick, one opted for either one of two methods. In one view, ectodermal overgrowth into the presumptive primitive streak caused both layers to shift centrifugally and thereby form the mesoderm (Kölliker 1876). In the other view, supported by Goette, ectodermal ingrowth into the presumptive primitive streak occurred by the active pressing and migration of lateral cells to the middle (Goette 1875). The first solution was formulated by August Rauber (1847–1917), who discovered that the two-layered chick embryo represents a blastodisc. He was the first anatomist to emphasize the formation of the blastodisc in birds where the dorsal side closes in upon the ventral side (Mocek 1998; see Brauckmann 1999). In his investigation of chick gastrulation, he proposed a compromise for the formation of the primitive streak and the primitive groove.

Like Haeckel, Rauber explained the occurrence of the first embryonic anlage in the dorsal end of the germinal disc of birds and fishes phylogenetically, and stated that it rests on the relationship between medullary groove and blastopore. In chick gastrulation he observed that the mesoderm initiates the differentiation of endoderm and ectoderm and finally forms a complete vesicle, similar to that of the selachians, and different from the frog, where the blastopore functions as its primitive streak (Balfour 1881; Johnson 1884; Robinson and Assheton 1891). Rauber further showed that the endoderm consists of a one-layer membrane of flattened cells, and it differentiates earlier than the ectoderm, which folds over into the subjacent cell mass. He saw the primitive streak appearing as a small white mass of cells that proliferated ventrally from the ectoderm, transversing the area pellucida from a marginal position to the center. Indeed, the region of the epiblast giving rise to the primitive streak is still referred to as "Rauber's sickle," although "Koller's sickle" is still the more widely used term (see Callebaut and Van Neuten 1994). At the same time, the primitive groove develops, pervading the primitive streak as a thin furrow, and moves to the marginal lip until it reaches, together with the primitive streak, the area pellucida (Rauber 1874, 1876).

In an amphibian gastrula, the wide-open blastopore is bounded by a circular rim under which the prospective mesoderm is rolled. The corresponding structure in birds consists of the primitive streak, which Rauber and others considered the homolog of the merged lateral lips of a long, drawn-out blastopore (Fig. 5) (Rauber 1875). Only at the anterior end of the streak, where the merged lateral lips bend over into each other, is there some trace of a real blastopore in the form of a slight depression, called the primitive pit, which is overhung by a knob-like structure corresponding to the dorsal lip of the amphibian blastopore (Hensen's node). It proliferates anteriorly under the surface and later forms the notochordal plate (Rauber 1876; Patterson 1909a,b; Lillie 1919). In Rauber's interpretation, the primitive streak of the chick represents the homolog of the blastopore lip, and the primitive groove, or medullary tube, corresponds to the blastopore opening, as Kowalevsky had demonstrated in ascidian development (Kowalevsky 1866b; Rauber 1876; see Whitman 1883).

Rauber (1875) may also have been the first person to describe the transition of the mammalian blastocyst into the bilaminar disc and to note its primitive streak (see Balfour 1881). The thin layer of polar trophectoderm above

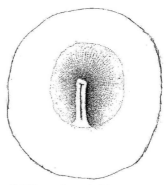

Figure 5. Rauber (1876) on the primitive groove and the primitive streak. 12-hour chicken embryo, the primitive streak shows the primitive groove, anterior and posterior end of the primitive streak are not cut by the primitive groove, star-like figure of ectodermal cells or area pellucida around the head part of the primitive streak. (Reprinted from Rauber 1876).

the inner cell mass is sometimes referred to as "Rauber's layer."

Nevertheless, and despite the meticulous observations of Rauber and other embryologists, nearly 20 years later the blastopore question was still not answered. For instance, when Hertwig stated that the medullary groove (neural plate) and blastopore develop independently of each other in amphibians, the anatomist Hermann Braus (1868–1924) refuted it with his gastrulation data on the newt *Triton alpestris* (Braus 1901). In this salamander, the medullary groove starts at the dorsal blastopore, forms a hardly visible hollow near the blastopore, shortens gradually into the deepening medullary groove, and disappears after two hours when the neural tube has closed.

Another more controversial approach to the enigma of gastrulation was the concrescence theory of the Swiss anatomist Wilhelm His (1831–1904) (His 1889; Eyclesheymer 1898; Summer 1900; see Mocek 1998; Hopwood 1999). The concrescence theory claimed that the medullary (neural) layer is pushed from the meridian to the ventral end of the egg by folding down on both sides. The two discs then fuse with each other at the bottom in the median plane in the cephalo-caudal direction. By this theory, His (1874) could also explain the migration of the blastopore at 170° in the same direction over the ventral side of the egg. However, His was mistaken in concluding that the longitudinal axes develop by closure of the blastopore lips, an error Rauber avoided; Sedgwick Minot agreed with Rauber (Rauber 1883a,b; Minot 1889). Nevertheless, the issue of concres-

cence was not solved until 1929, when Walter Vogt published his seminal work on amphibian gastrulation, with the vital staining method demonstrating that the cells from the surface migrate through the blastopore lip to become specific endodermal and mesodermal territories.

Nevertheless, the phylogenetic aspects were still debated by biologists, e.g., Berthold Hatschek (1854–1941; whose illustration of the developmental stages of *Amphioxus* became the standard of nearly every textbook on embryology) and Richard Semon (1859–1918), who were both former students of Haeckel and focused their research on the formation of the blastopore. When studying the development of the lungfish *Ceratodus* during his expedition to Australia in 1899, Semon described the developing blastopore as a small, nearly linear cleft on the base in the middle region between center and equator (Semon 1900). (These fish had been known only from Mesozoic fossils; so this "living fossil" with well-developed lungs which had survived in Australia was viewed as a significant transitionary form.) For Hatschek the blastopore lips of *Amphioxus* grow together in a line that forms the largest part of the later medullary groove (neural plate) in the lowest chordates (Fig. 6) (Hatschek 1882, 1888).

A third idea on blastopore formation was proposed by Edouard Van Beneden (1846-1910), who had studied the behavior of chromosomes at germ-cell formation. He found that during rabbit gastrulation the hypoblast cells divide faster than the epiblast cells at the region (the posterior marginal zone) which eventually forms the blastopore

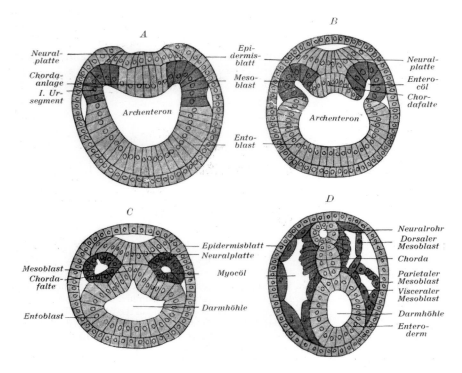

Figure 6. Hatschek's 1888 depiction of the *Amphioxus* neurulae, wherein the neural plate forms from the blastopore lips. Cross-sections through *Amphioxus* larvae; the left figures show the developing first primary segments, and the right figures the developing fifth segment. (Reprinted from Bonnet-Peter 1929; originally in Hatschek 1888.)

(van Beneden 1875). Keibel acknowledged the correctness of van Beneden's observations, but he criticized the interpretation because van Beneden, in effect, had merely rediscovered Rauber's epithelial tissue and renamed it the permanent epiblast (Keibel 1887). Keibel proposed that the yolk sac of mammals, which disappears during gastrulation, sufficiently explained the position of the blastopore, as well as the same homologous forms in mammals with the reversed infolding of the germ layers. To illustrate his argument, Keibel referred to Hatschek's illustration of the development of the medullary tube in *Amphioxus*. To solve these puzzling observations, Oscar Hertwig (1849–1922) suggested calling all areas at the beginning of invagination where the blastopore lips fold in, an aperture. Then the blastopore expands from the ventral-anterior end of the anlage to the anus, covering the presumptive dorsal area of the embryo (Hertwig 1892). (Haeckel [1896] would later expand his gastraea theory into a theory of larval development, as well, stating that deuterostomes and protostomes arose from a common "vermiform" type of larva. This idea would get molecular support from the observations of Arendt and colleagues [2001].)

In 1905 Hubrecht and Keibel summarized the data of the descriptive experiments with an approving reference to Haeckel's definition of the gastrula as a two-layer germ stage and a focus on cell migration (Hubrecht 1905; Keibel 1900, 1905). However, both refuted Haeckel's insistence on invagination and emphasized, as Ray Lankester had already done, that delaminative gastrulation begins at the blastopore. Both referred to the newest gastrulation data of H.V. Wilson (1863–1939) on *Chorophilus feriarum*. Wilson's microphotographic observations confirmed the involutive gastrulation pattern whereby the ectoderm rolls over the blastoporal lips, and the yolk cells disappear beneath these lips during the folding process (Wilson 1900, 1901; see Beetschen 2001).

The main and most important statement of Hubrecht and Keibel was, however, to differentiate between the gastrulation of vertebrates and invertebrates in a contradiction to Haeckel's gastraea theory. According to them, the question as to how a two-layered gastrula develops from a one-layered blastula, and which process is more primitive, invagination or delamination, has to be answered with two definitions, one for vertebrates and the other one for invertebrates, presupposing that gastrulation is an invaginative process. To formulate a definition that holds for all metazoa, they defined gastrulation as a process by which an alimentary endoderm differentiates from an ectoderm. However, according to Hubrecht (1905) and Keibel (1905), in mammals the endoderm always develops by delamination from a one-layered cell complex. When the blastopore has been formed, the gastrulation process is already finished, since the formation of the bilateral-symmetrical dorsal side and the chorda succeeds gastrulation (Hertwig

1888; Hubrecht 1905; Keibel 1905). Thus, their efforts did not really solve the blastoporal issue, but their idea to shift it to another developmental stage per definitionem deserves respect. Whether their programmatic summary had the impact both demanded might be questioned, however. Hence, five years later Goette argued that delamination and invagination, even if they seem to be different, are identical processes during gastrulation (Goette 1910; Bütschli 1915).

THE EXPERIMENTAL GASTRULATION OF THE SEA URCHIN OR THE CHALLENGE OF CELLULAR SHAPING

In his advocacy for His's concrescence theory, Roux looked to experimentation to decide the questions of gastrulation (Roux 1905; see Sander 1990, 1991; Mocek 1998). His point of departure was Pflüger's experiments on the influence of gravitation on early development (Pflüger 1883; Roux 1902). With the famous experiments in which he pricked the relatively colorless frog eggs with hot needles, he thought to demonstrate that axial organs encompass the developing gastrula in a ring shape, whereas the dorsal end materializes by a median connection of the right and left side above the intestinal cells of the yolk. Roux killed one of the blastomeres of the two-cell stage of the frog. The uninjured half developed in some cases into a well-formed half-larva. His results were later explained by Brachet (1921) and further developed by Driesch and Morgan on *Ctenophora beroë* (Driesch and Morgan 1896; Morgan 1897).

Since the dorsal side of the embryo corresponds to the dorsal vegetative side of the gastrula, Roux had a gastrula concept that differed significantly from the former ones. The community of experimental embryology argued a long time about it, particularly since the pricking experiments on the amphibian egg led in different hands to different conclusions. Roux and Morgan were convinced that the ventral lip remains stationary and the dorsal lip moves forward 120° over the yolk to fuse with it (Morgan 1894; Roux 1902). In 1890 Schulze characterized the debate as fruitless, since the inadequate and insufficient experimental methods of that time were unsuitable for solving the problem (Schultze 1894).

In 1891 a more elaborate experiment to solve the problem of gastrulation was started by Curt Herbst (1866–1946), who observed that the blastomeres of the cleaving sea urchin egg spontaneously separate from each other after a brief exposure to calcium-free seawater (Herbst 1893; see Oppenheimer 1970). This produced an elegant method for isolating undamaged blastomeres and for following their development independently. Herbst demonstrated that the skeleton-forming cells influence the formation of the ectoderm by "formative stimuli" (Herbst 1893). In another series of experiments he discovered the phenomenon of exogastrulation by treating the eggs during early segmentation with small amounts of lithium salts added to seawater (Herbst 1893, 1900). Three decades later, Holtfreter described exogastrulation as an autonomous segregation of the ectoderm from the endoderm and mesoderm, confirming that even under abnormal conditions the basic events of convergence and epiboly are effective. However, even in the exogastrulae, the morphogenetic movements such as epiboly, blastoporal groove formation, dorsal convergence, and ventral divergence are very similar to those in the normal gastrula (Holtfreter 1933). Herbst's chemical transformation experiments were later taken up by Moore and Burt with sugar and supplemented by osmotic pressure to study how the gastral plate was pushed outward (Moore 1930, 1941; Moore and Burt 1939).

In 1900 Theodor Boveri (1862–1915) studied the development of *Paracentrotus lividus* and rediscovered the pigment ring, already described by Rusconi (Rusconi 1823; Boveri 1901a; see Baltzer 1967). The pigment band that Selenka had described as a kind of polar stratification attributing it to the vegetal half (Selenka 1883) played a considerable role in the study of the organization of the sea-urchin egg. In applying Herbst's technique to test Driesch's hypothesis of the equipotential system, Boveri demonstrated that the pigment ring contains almost all the vegetative cells (see Driesch 1892, 1899, 1900). When he investigated it more closely, he recognized that the egg axis determined by the pigment ring coincided with the meridian axis of the gastrula. During gastrulation, Boveri observed that the curvature of the vegetal cell plate reversed from convex to concave and, at the same time, invaginated into the animal roof, which modifies the spherical gastrula of the sea urchin into a cup-like shape. Based on his experimental results, he formulated the hypothesis of an animal-vegetative gradient with a "priority region," which functions nearest the vegetal pole to determine the developmental stages of all other regions (Boveri 1901b). Another approach to indicate how the orientation of the axes triggers and determines the different developmental stages was conducted by John Wilfrid Jenkinson (1871–1915), who also showed that Driesch's concept of the equipotential system and the regulation potency of the embryonic complex was not applicable to the mesenchymal gastrulae (Jenkinson 1913).

Nearly 40 years later, Sven Hörstadius (1898–1996) reinvestigated Boveri's data with a more refined technique that used very fine glass needles (cutting the eggs with an accurate orientation) and the agar-staining method developed by Vogt (Detwiler 1917; Vogt 1929; Hörstadius 1937a,b). In a series of experiments on the development of sea urchins, he also hoped to solve the crucial issue of the temporal determination of the invagination process. The main objective was to improve on the data of Morgan and Driesch, who had estimated the invaginating material by counting nuclei, but could not give decisive results about the spatio-temporal determination of gastrulation (Driesch and Morgan 1896; von Ubisch 1925). Hörstadius now stained groups of blastomeres either by pushing them against agar, or by isolating and transplanting them back to the same position. With this technique he could trace the fate of the various cell groups through the blastula and gastrula into the later larva at their exact temporal determination.

Hörstadius's data showed that the fertilized egg of *Paracentrotus lividus* has an orange belt of about the same position as the pigment ring, but these two belts are not identical. At gastrulation, starting exactly 16 hours after fertilization, he demonstrated that the whole orange field was invaginated, and the process even affected the whole disc in thinning its wall, without any pressure from the surrounding ectoderm (Hörstadius 1935). Then he discovered that the edge of the invaginating endoderm coincides approximately with the boundary between the cell derivatives and the upper macromeres. The remaining cells derived from a plate of cells, which was composed of the upper halves of the former mesomeres surrounding the former animal pole, and which now formed the near-equatorial daughter cells of the mesomeres, and the lower and upper macromeres that occupy the outer surface of the gastrula and form the ectoderm. The invaginated endoderm, first a blind sac, later bends ventrally and, at its tip, breaks through into the stomodeum. Formerly it was thought that during invagination a disc which became the primordial archenteron layer was first lifted in, but Hörstadius found that invagination proceeds from the invaginating region to the periphery. Thus, he corrected the old invaginative explanation of gastrulation, namely, that the archenteron material was gradually pushed into the blastocoel by an enlarging ectoderm, thereby changing its shape into a tube, as Boveri had assumed and von Ubisch had repeated (von Ubisch 1925; Hörstadius 1937a,b, 1973; see Lewis 1947).

PHYSICAL MODELS SIMULATING GASTRULATION

The physical forces involved in the invagination of the cell layer to mold the gastrula were discussed repeatedly around 1900 (Rhumbler 1899, 1902; Gurwitsch 1909, 1914; Assheton 1916; see Thompson 1942), but the several explanations were somewhat conflicting. Bütschli suggested that the movements of chromosomes toward the poles might be caused by surface tensions highest in the equatorial region. He supported the hypothesis by his gelatin-bubble experiments which confirmed that aster and spindle chromosomes can be simulated in dropping a hot melted-gelatin solution in two adjoining air bubbles. When the air bubbles cool, they contract and exert streams and flows whenever the surface tension of a liquid drop is locally diminished (Bütschli 1890, 1892; Erlanger 1897; see Wilson 1928).

Ludwig Rhumbler (1864–1939) developed Bütschli's view further and suggested that infolding resulted from a change in shape of the vegetal cells caused directly by the ectoplasmic layer. His approach tried to reduce the phenomenon of gastrulation to an intracellular mass relocation, which he explained with the wedge-like form of the blastula cells (Rhumbler 1902, 1903, 1923). Even if this assumption was not quite correct, Rhumbler's influence on further attempts to explain gastrulation with physical models is significant, particularly on Holtfreter's research on cell aggregation. Another attempt to simulate gastrula-like development was done by the chemist Emil Hatschek (1868–1944), who formulated a colloidal model that suggested differential swelling of cells as the mechanism of gastrulation; i.e., the shaping into a cup-like gastrula. It used the fact that a colloidal cell plate has certain physical properties in common with colloids, such as gelatin, that swell in water (E. Hatschek 1918, 1922). He dropped gelatin containing potassium ferrocyanide into copper sulfate or a tannin solution and produced biconcave bodies simulating gastrula formation.

Like Hatschek, Josef Spek (1895–1964) also simulated gastrulation with a gelatin model that explained gastrulation on the basis of gradient actions on the endoderm (Spek 1918, 1931). He postulated a differential swelling between the inside and outside of the endodermal layer, due to a special arrangement of lypophilic and lypophobic colloids in the endodermal plate. Spek used a composite strip of two layers of gelatin with different densities; when immersed in water, this composite strip absorbed water differentially and curved to simulate the behavior of a portion of the endodermal plate of the invaginating gastrula. However, his approach could not stand the test of biological reality and was disproved by Moore's experiments on *Dendraster excentricus* (Moore 1941). Moore's result showed that the turgor of the blastocoel fluid had been increased by rearing the eggs in a seawater–sucrose medium, as mentioned above. When at the beginning of the gastrula stage the eggs were returned to seawater, their blastocoel exhibited a strong swelling, and the invagination was suppressed (Beetschen 2001). However, the composite strip model has been revitalized to provide a biomechanical explanation for the inward puckering of the vegetal plate during the initial stages of archenteron formation (Lane et al. 1993). The mechanical approach to gastrulation studies is only recently being revisited as a research program.

FOLLOWING CELL MIGRATION: VOGT, PASTEELS, AND HOLTFRETER

Meanwhile, Vogt's experiments on the toad *Bombinator* with the vital staining method revolutionized comparative embryology and formed the basis for interpreting experi-

mental embryology (Vogt 1923a,b, 1925, 1929; see Spemann 1942). He placed pieces of agar saturated with vital dyes upon the surface of the embryo. These dyes penetrated the cell membrane and colored the surface of the late blastula. Vogt traced the stain-labeled cells at later developmental phases, e.g., the formation of the hypoblast of the roof and the sides of the archenteron, and he constructed more or less idealized fate maps. Vogt's experiments firmly established that the three germ layers at the end of gastrulation are derived from distinct sectors of the blastula (see Fig. 7). Among others, Vogt confirmed that the marginal belt of the frog embryo, which is homologous to the blastopore rim of the sea urchin, is widest on the presumptive dorsal side of the germ and narrowest on the opposite side. He observed that in amphibians the rolling in of the vegetative part of the germ starts from the blastoporal groove and forms in the vegetative hemisphere near the widest portion of the marginal belt. Around this groove the surface materials bend under, with the wide part of the

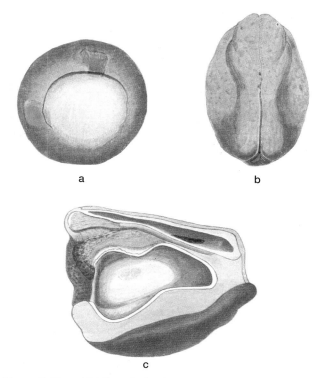

Figure 7. Vogt (1929) and the fate mapping of gastrulation in *Bombinator*. Staining of the dorsal and lateral blastoporal lip; (*a*) after staining; (*b*) closed neurula, the delicate segment of the red marker glimmers through from the inside; (*c*) the same stage opened up from the right side; the enteron (*Dachentoderm*) is stripped of mesoderm and Chorda; the dividing line is marked by dotted double lines; the Chorda is taken off from the mesoderm and slightly rolled up; the red mark is drawn from the dorsal lip (*Seitenlippe*) over the endoderm to the axial mesoderm; the blue mark dyes the posterior third of the Chorda and the ventral side of the medullary anlage. (Reprinted from Vogt 1929.)

marginal belt coming first. When more and more cell regions are "drawn" into the dorsal lip of the blastopore, the lip is forced to extend laterally in the shape of a horseshoe until it finally completely surrounds the vegetal yolk plug in the form of a circle. In this invaginative process the marginal belt proceeds much faster into the interior than does the rest of the surface. With considerable stretching, it applies itself closely to the overlying ectoderm so as to form an inner lining for the latter. Within an area roughly coextensive with the substratum, the ectoderm then begins to transform into the medullary plate (Vogt 1929).

Vogt stated that the attempts by others to find an unequivocal definition of the blastopore failed because of use of the same word for different phenomena. For example, Keibel and Hubrecht defined gastrulation as the formation of the endoderm, an idea that Brachet further developed. However, for Vogt, the mesoderm and the chorda arise phylogenetically from a specific edge of the blastopore and receive their support from the ectoderm, not from the primary endoderm. As opposed to Hubrecht and Keibel, he postulated that the mesoderm does not develop from the ectoderm or endoderm alone. Nevertheless, he was convinced that the gastraea theory can be maintained with regard to the germ layers of the vertebrates, and criticized Ruffini for having "thrown away the good with the bad" when completely refuting the gastraea theory (Ruffini 1925; Vogt 1929). He believed in the usefulness of the gastraea theory, which he modified in arguing that the hypothetical primitive form was a three-layered structure, namely a two-layered gastrula with an enclosing edge that phylogenetically acquired a specific importance. Thus, Vogt stated that the mesoderm arose dorsally through convergence of ectoderm and endoderm. His staining method was applied and advanced by Ludwig Gräper, who combined the fate mapping with time-lapse cinematography of chick embryos (Gräper 1929).

With Vogt's results, the presumptive areas of the main organs of *Amblystoma* seemed to be definitely established. However, Wintrebert, Nakamura, and Pasteels offered divergent views on *Discoglossus* and *Bombinator* (Wintrebert 1932; Nakamura 1938, 1942; Pasteels 1940, 1941). Due to Vogt's work, it was known that anurans and urodeles differ significantly in the process of their gastrulation. For example, in urodeles the lateral and dorsal parts of the head gut are located above the blastopore, and the lateral blastoporal lips follow the boundary between mesoderm and endoderm. However, in anurans the material of the lateral crests lies above the blastoporal furrow, which separates the lateral and basal parts of the presumptive gut. His fate map of *Bombinator* showed the blastopore joining the inferior limit of the mesoblast beyond 120° from the sagittal plane. Neither Wintrebert nor Pasteels could confirm or support the angle, and Nakamura also challenged Vogt's

data by establishing that the narrow blastoporal pit does not appear at the beginning of gastrulation. According to Vogt's data, the lateral parts of the embryo contain material for the hind part of the body. However, Nakamura confirmed that these lips hold the material for the middle trunk. Furthermore, he correctly situated the caudal somites above the limit of invagination because their territory penetrates into the middle layer in a specific way such that it is temporarily incorporated in the posterior part of the neural plate.

Pasteels stained 72 eggs of *Amblystoma* and *Discoglossus*, which were taken at the exact beginning of gastrulation, because he wanted to determine the exact form of the notochordal anlage and the distribution of the posterior somites. He determined precisely the presumptive notochordal and somitic areas and mapped the presumptive ectodermal territories (Pasteels 1941; Dettlaf 1946). Apparently, the article by Pasteels was referred to, but not really understood; e.g., for Nieuwkoop, Pasteels and Nakamura supplemented, but did not correct, Vogt's fate maps (Nieuwkoop et al. 1981; see Lovtrup 1975). Nieuwkoop himself reinterpreted Pasteels's centrifugation experiments later in his experiments which demonstrated the importance of the blastocoel in separating the ectoderm and endoderm, preventing their premature interactions (Nieuwkoop and Florschütz 1950; Nieuwkoop 1969).

Paul Weiss (1889–1989) gave a concise summary of the knowledge of gastrulation until the late 1930s in his seminal textbook *Principles of Development*. There he widened the focus from gastrulation movements to the developmental specificity of cells during and after gastrulation as, e.g., the invagination of the marginal zone around the rim of the blastopore, the expansion of the presumptive ectoderm, and the sinking of the vegetative material into the interior to form the endoderm (Weiss 1939). Each phase has its own motor and can proceed independently of the others, as Mangold had shown (Mangold 1923); if the same experiments are repeated with fragments taken from an embryo past the stage of gastrulation, the grafts have lost their plasticity to stretch, roll in, or sink in, as the embryonic one did (cf. Spemann). Two decades later, Weiss applied this knowledge on the determination of developmental stages to his experiments on cellular self-aggregation which, together with the work on cell adhesion of Townes and Holtfreter, paved the way for much of modern developmental biology (Townes and Holtfreter 1955).

The most thorough analysis of the specificity of the first stages of the gastrula was carried out by Johannes Holtfreter (1901–1992), who used Vogt's maps as a basis of reference (Holtfreter 1943, 1944). The difference between Vogt and Holtfreter was made evident in their approach toward the cell and its capacity to move. Vogt thought of the amphibian cells as passively carried along, comparable to the parts of an amoeba (Vogt 1923a,b). For Holtfreter, the cell was part of a dynamic system, wherein supracellular forces controlled and, to a certain extent, initiated, these cell movements. Holtfreter explored the functions of surfaces in regulating cell spreading, aggregation, tissue affinity, and cell polarity, and he became an important link in relating gastrulation to molecular processes (Hamburger 1988; Keller 1996).

Concerning gastrulation, Holtfreter extensively studied the cellular folding and shaping of the endoderm and mesoderm (Holtfreter 1938a,b, 1943) and confirmed, as many before him, that the three presumptive germ-layer regions differ significantly in their developmental capacity. His figures, which resemble Oscar Hertwig's drawings of 1883, show that the ectoderm consists of several layers: a darkly pigmented upper layer that adheres to its lateral cell walls, and one or two inner layers of spherical lighter cells that are interconnected by filaments. The peripheral meso-derm cells of the dorsal zone become flask-shaped as they approach the blastopore. The arrangement of the inner mesodermal cells resembles that of the inner ectoderm. The elongated cells of the mesoderm appear along the line of the ventral blastopore lip. By the way, Holtfreter also emphasized that the invagination of the ventral blastoporal lip resembles the process of a scattered immigration of individual mesoderm cells that characterizes the echino-derm gastrula—a reminiscence of the phylogenetic topics of the 19th century (Holtfreter 1943).

The new results Holtfreter offered here pertain to the surface coating (extracellular matrix) which firmly cements the peripheral cells. He finally established that gastrulation is a problem of mass movements, and not of growth, as pre-viously postulated by Rhumbler and Vogt (Holtfreter 1943). For Holtfreter the eccentric pattern of epiboly in a gastrula represents a position effect, in which the move-ment of the presumptive neural and epidermal material results from the geometrical position of the cells and their position inside the gastrula (Holtfreter 1939, 1943). To avoid misinterpretations, this statement is not a precursor of Wolpert's positional information, but a reformulation of the developmental field conception of Gurwitsch and Weiss (Beloussov et al. 1997; Brauckmann 2004).

When discussing at length the crucial issue of invagina-tion, Holtfreter referred to Rhumbler's histological data demonstrating that invagination is mainly due to an activi-ty of the endoderm itself, provided by a gradient of surface tension between the inner and outer milieu of the blastula. A decrease in surface tension at the inner side of the cells produced by chemical properties of the blastocoel interior transformed the vegetal surface cells into the shape of a wedge or club. This entails an inpocketing of the vegetal area. The results of his physical modeling of gastrulation, e.g., with calcium ions and oil drops, furnished his argu-ments that surface tension plays a predominant part in the ectodermal spreading process and in the morphogenetic movements of the cells (Schmitt 1941; Holtfreter 1943). Holtfreter's brilliant experiments refuted Goette, His, Spek, Vogt, and Dalcq-Pasteels and established instead that con-tracted endodermal cells of the blastopore can penetrate into an adequate cellular substratum as soon as they are bottle-shaped. Thus, invagination is mainly a problem of how the blastocoel is adsorbed, and of cell polarity, and of interstices between the inner cells. One of his most impor-tant conclusions from these experiments was that, if culti-vated in vitro, any part of the dorso-marginal zone, derived either from the notochordal or the somitic area, may yield notochord, somites, nervous tissue, and epidermis.

CONCLUSION

In the 19th century, gastrulation was considered to com-prise the formation of a two-layered embryo enclosing an archenteron. All early gastrulation research was an attempt to classify animals; the homologies of germ layers and the homologies of gastrulation movements preoccupied these researchers. The biology of that time looked first to the sheet folding, or to be precise, to the germ layers, their for-mation, and movements, and coined terms for the different embryonic stages. Even nowadays every embryologist will concede that cell migration and sheet folding are major forces shaping the embryo (Stern 1992; Narasimha and Leptin 2000). Despite the variety of disciplines involved in the research on gastrulation, e.g., anatomy, embryology, histology, phylogeny, developmental mechanics—to men-tion just a few, the 19th century shows that very often the same biologists and anatomists participated in all of these specialties. It also explains why the scientists in question interpreted their observations and experiments in such a controversial way, and why it was so difficult to formulate a terminology for processes that happen in a rotating organ-ism in which cell migration was only one of the phenome-na shaping this organism. Even if one named the phenom-ena or documented the processes one had observed, their acceptance and experimental confirmation often generated heated disputes and controversies on how to interpret what seemed to be self-evident. It was left to the molecular age to define the concepts concisely and to make them feasible to experimental data.

Vital staining helped to demonstrate that the two-lay-ered stage passed through by protochordates and a few anamniote vertebrates was not the gastrula as originally defined. Its gastrocoel was still partially lined by chordame-soderm, and, therefore, was not yet an archenteron (e.g., chick). The gastraea theory had a long life until it finally collapsed when Dalcq and Gérard published their classic revision of Brachet's textbook (Dalcq and Gérard 1935; see

Thieffry 2001). According to these authors, gastrulation could only be defined as an ensemble of processes that, when completed, would have formed an archenteric cavity and brought the materials which would form the principal organs to the places where they would develop.

Today, we are looking at gastrulation with probes that identify regions of gene expression. Again, we search for the homologies that can enable us to reaffirm the monophyletic origins of animals and the mechanisms of gastrulation across the animal phyla.

ACKNOWLEDGMENTS

The authors thank N. Hopwood, K. Kull, and D. Thieffry for their help in finding biographical information and copies of original articles. S.B. thanks in particular T. Söderqvist for the Rusconi figure. Both authors thank J. M. Opitz for his enlightening numerous discussions on gastrulation and its vexed history. S.B. researched this chapter as a visiting scientist at the University of Utah School of Medicine, with support from a Research Fellowship of the German Science Foundation (DFG, BR-1812/7-1). S.F.G. is supported by a National Science Foundation grant (IBN-0079341).

REFERENCES

Allman G.J. 1853. On the anatomy and physiology of *Cordylophora*: A contribution to our knowledge of the Tubularian zoophytes. *Philos. Trans. R. Soc. Lond.* **143:** 367–384.

Arendt D., Technau U., and Wittbrodt J. 2001. Evolution of the bilaterian larval foregut. *Nature* **409:** 81–85.

Assheton R. 1916. *The geometrical relations of the nuclei in an invaginating gastrula.* Cambridge University Press, Cambridge, United Kingdom.

von Baer K.E.v. 1828 (1837). *Entwickelungsgeschichte der Thiere. Beobachtung und Reflexion.* Bd. 1 und 2. Borntraeger, Königsberg, Germany.

———. 1834. Die Metamorphose des Eies der Batrachier vor der Erscheinung des Embryo und Folgerungen aus ihr für die Theorie der Erzeugung. *Müller's Archiv. Anat. Physiol.* **1:** 481–509.

———. 1847. Auszug aus einem Berichte des Akademikers v. Baer aus Triest vom 1. November 1845. *Bull. Phys. Math. Acad. Sci. St. Petersburg* **15:** 231–240.

———. 1873. Entwickelt sich die Larve der einfachen Ascidien in der ersten Zeit nach dem Typus der Wirbelthiere. *Mém. Acad. Imp. Sci. St. Petersburg* **8:** 1–35.

Balfour F.M. 1881. *A treatise of comparative embryology,* vol. 2. Macmillan, London, United Kingdom.

Baltzer F. 1967. *Theodor Boveri. Life and work of a great biologist: 1862–1915* (transl. by D. Rudnick). University of California Press, Berkeley and Los Angeles.

Beetschen J.-C. 2001. Amphibian gastrulation: History and evolution of a 125 year-old concept. *Int. J. Dev. Biol.* **45:** 771–795.

Beloussov L.V., Opitz J.M., and Gilbert S.F. 1997. Life of Alexander G. Gurwitsch and his relevant contribution to the theory of morphogenetic fields. *Int. J. Dev. Biol.* **41:** 771–777.

Blyakher L.Y. 1982. *History of embryology in Russia. From the middle of the eighteenth to the middle of the nineteenth century.* Smithsonian Institution Press, Washington, D.C.

Bonnet R. and Peter K. 1926. *Lehrbuch der Entwicklungsgeschichte.* Paul Parey Verlag, Berlin, Germany.

Boveri T. 1901a. Die Polarität von Oocyte: Ei und Larve des *Strongylocentrotus lividus. Zool. Jahrb. Abt. Anat. Ontol.* **14:** 630–653.

———. 1901b. Die Polarität des Seeigeleies. *Verh. Phys. Med. Ges. Würzb.* **34:** 145–176.

Brachet A. 1921. *Traité d'Embryologie des Vertébrés.* Masson, Paris, France.

Brauckmann S. 1999. A history of causal morphology: Reinhard Mocek's Die werdende Form (essay review). *Am. J. Med. Genet.* **87:** 281–285.

———. 2004. Paul A. Weiss, 1898-1989: The cell engineer. In *Creating a tradition of biomedical research* (ed. D.H. Stapleton). Rockefeller University Press, New York. (In press.) [Q in proofs: Update]

Braus H. 1901. Rückenrinne und Rückennaht der Tritongastrula. *Anat. Anz.* **20:** 239–240.

Bütschli O. 1880 (1889). Protozoa. In *Klassen und Ordnungen des Tierreichs* (ed. H.G. Bronn), Bd. 1. F.C. Winter, Leipzig, Germany.

———. 1890. Experimental imitation of protoplasmic movement. *Q. J. Microsc. Sci.* **31:** 99–103.

———. 1892. *Über mikroskopische Schäume und das Protoplasma.* Leipzig, Germany.

———. 1915. Bemerkungen zur mechanischen Erklärung der Gastrula-Invagination. *Sitzungsber. Akad. Wiss. Heidelberg* IV.

Callebaut M. amd Van Nueten E. 1994. Rauber's (Koller's) sickle: The early gastrulation organizer of the avian blastoderm. *Eur. J. Morphol.* **32:** 35–48.

Churchill F.B. 1991. Karl Ernst von Baer. The rise of classical descriptive embryology. *Dev. Biol.* **7:** 5–12.

Conklin E.G. 1934. Development of amphioxus. *J. Morphol.* **54:** 69–152.

Dalcq G. and Gerard P. 1935. La regulation dans le germ et son interpretation. *C.R. Soc. Biol.* **119:** 141–152.

Dalyell J.G. 1847. *Rare and remarkable animals of Scotland,* vol. 1., J. van Voorst, London, United Kingdom.

Darwin C. 1883 (1936). *Descent of man, and selection in relation to sex.* The Modern Library, New York.

Derbès A. 1847. Observations sur le mécanisme et les phénomènes qui accompagnent la formation de l'embryon chez l'Oursin comestible. *Ann. Sci. Nat.* (3 Sér.) **8:** 80–98.

Detlaf T. A. 1946. The topographic map of presumptive areas in Anura revised with reference to precision (in Russian). *C.R. Doklady Acad. Sci. USSR* **54:** 277–280.

Detwiler S.R. 1917. On the use of Nile Blue sulphate in embryonic tissue transplantation. *Anat. Rec.* **13:** 493–497.

Driesch H. 1892. Entwicklungsmechanische Studien I. Der Werth der beiden ersten Furchungszellen in der Echinodermenentwicklung. Experimentelle Erzeugung von Theil- und Doppelbildungen. *Z. Wiss. Zool.* **53:** 160–178.

———. 1899. Studien über das Regulationsvermögen der Organismen. 3. Notizen über die Auflösung und Wiederbildung des Skelettes von Echinidenlarven. *Wilhelm Roux' Arch. Entwicklungsmech. Org.* **9:** 103–139.

———. 1900. Die isolirten Blastomeren des Echinidenkeimes. Eine Nachprüfung und Erweiterung früherer Untersuchungen. *Arch. Entwicklungsmech. Org.* **10:** 361–410.

Driesch H. and Morgan T.H. 1896. Zur Analysis der ersten

Entwickelungsstadien des Ctenophoreneies. I. Von der Entwickelung einzelner Ctenophorenblastomeren. *Wilhelm Roux' Arch. Entwicklungsmech. Org.* **2:** 204–215.

Erlanger R. 1897 (1898). Zur Kenntnis der Zell- und Kernteilung. *Biol. Cbl.* **17:** 745–752 (**18:** 1–11).

Eyclesheymer A.E 1898. The location of the basis of the amphibian embryo. *J. Morphol.* **14:** 467–480.

Goette A. 1875. *Die Entwickelungsgeschichte der Unke (Bombinator igneus) als Grundlage einer vergleichenden Morphologie der Wirbelthiere.* Leopold Voss, Leipzig, Germany.

———. 1910. *Entwickelungsgeschichte der Tiere.* de Gruyter, Berlin and Leipzig, Germany.

Gould S.J. 1977. Ontogeny and phylogeny. The Belknap Press, Cambridge, Massachusetts.

Gourko H., Williamson D.I., and Tauber A.F., eds. 2000. *The evolutionary-biology papers of Elie Metchnikoff.* Boston Stud. Philos. Sci., vol. 212. Kluwer, Dordrecht, The Netherlands.

Gräper L. 1929. Die Primitventwicklung des Hühnchens nach stereokinematischen Untersuchungen, kontrolliert durch vitale Farbmarkierung und verglichen mit der Entwicklung anderer Wirbeltiere. *Wilhelm Roux' Arch. Entwicklungsmech. Org.* **116:** 382–429.

Grell K.G. 1968. Protozoologie, 2nd edition. Springer Verlag, Berlin, Germany.

———. 1979. Die Gastraea-Theorie. *Medizinhistorisches J.* **14:** 275–291.

Gurwitsch A.G. 1909. Prämissen und anstossgebende Faktoren der Furchung und Zelltheilung. *Arch. Zellforsch.* **2:** 495–548.

———. 1914. Der Vererbungsmechanismus der Form. *Wilhelm Roux' Arch. Entwicklungsmech. Org.* **39:** 514–577.

Haeckel E. 1872. Die Kalkschwämme. Eine Monographie. Bd. I. Biologie der Kalkschwämme (Calcispongae oder Grantien). Georg Reimer, Berlin, Germany.

———. 1874. Die Gastraea-Theorie: die phylogenetische Classification des Thierreichs und die Homologie der Keimblätter. *Jenaische Z. Naturwiss.* **8:** 1–55.

———. 1875. Die Gastrula und die Eifurchung der Thiere. *Jenaische Z. Naturwiss.* **9:** 402–510.

———. 1877. Nachtraege zur Gastraea-Theorie. In *Biologische Studien.* pp. 225–270. Hermann Dufft, Jena, Germany.

———. 1889. *Natürliche Schöpfungsgeschichte. Gemeinverständliche wissenschaftliche Vorträge über die Entwickelungs-Lehre im Allgemeinen und diejenige von Darwin: Goethe und Lamarck im Besonderen,* Georg Reimer, Berlin, Germany.

———. 1896. *Systematische Phylogenie. 2. Teil: Systematische Phylogenie der wirbellosen Thiere (Invertebrata),* pp. 259–347. G. Reimer, Berlin.

Hamburger V. 1988. *The heritage of experimental embryology.* Oxford University Press, New York.

Hamecher Jr., H. 1905. Ueber die Lage des kopfbildenden Teils und der Wachstumzone für Rumpf und Schwanz (F. Kopsch) zum Blastoporusrande bei *Rana fusca. Int. Monatsschr. Anat. Physiol.* **21:** 85–125.

Hatschek B. 1882. Studien über die Entwicklung des Amphioxus. *Arb. Zool. Inst. Wien* **4:** 1–88.

———. 1888 (1891). *Lehrbuch der Zoologie. Eine morphologische Übersicht des Thierreiches zur Einführung in das Studium dieser Wissenschaft,* G. Fischer, Jena, Germany.

Hatschek E. 1918. On forms assumed by a gelatinising liquid in various coagulating solutions. *Proc. R. Soc. Lond. A Math. Phys. Sci.* **94:** 303–316.

———. 1922. Structures in elastic gels caused by the formation of semi-permeable membranes. *Biochem. J.* **16:** 475–479.

Herbst C. 1893. Experimentelle Untersuchungen über den Einfluss der veränderten chemischen Zusammensetzung des umgebenden Mediums auf die Entwicklung der Thiere. II. Weiteres über die morphologische Wirkung der Lithiumsalze und ihre theoretische Bedeutung. *Mitt. Zool. Stn. Neapel* **11:** 136–220.

———. 1900. Über das Auseinandergehen von Furchungs- und Gewebszellen im kalkfreien Medium. *Arch. Entwicklungsmech. Org.* **9:** 424–463.

Hertwig O. 1883. *Die Entwicklung des mittleren Keimblattes der Wirbelthiere. Studien zur Blättertheorie,* Bd. V. G. Fischer, Jena, Germany.

———. 1888 (1910). *Lehrbuch der Entwicklungsgeschichte des Menschen und der Wirbelthiere,* 9th edition. G. Fischer, Jena, Germany.

———. 1892. Urmund und spina bifida. *Arch. Mikr. Anat.* **39:** 353–503.

Hertwig O. and Hertwig R. 1882. Die Coelomtheorie. Versuch einer Erklärung des mittleren Keimblattes. *Jenaische Z. Naturwiss.* **8:** 1–150.

His W. 1874. Unsere Körperform und das phzsiologische Problem ihrer Entstehung. Leipzig, Veit. [letter 7 p. 83].

His W. 1889. On the principles of animal morphology. *Proc. R. Soc. Edinb. Sect. B Biol. Sci.* **15:** 287–298.

Holtfreter J. 1933. Die totale Exogastrulation. Eine Selbstablösung des Ektoderms vom Entomesoderm. Entwicklung und funktionelles Verhalten nervenloser Organe. *Arch. Entwicklungsmech. Org.* **129:** 669–793.

———. 1938a. Differenzierungspotenzen isolierter Teile der Urodelengastrula. *Wilhelm Roux' Arch. Entwicklungsmech. Org.* **138:** 522–656 (English. transl., Hamburger V. 1996. *Dev. Dyn.* **205:** 223–244.)

———. 1938b. Differenzierungspotenzen isolierter Teile der Anurengastrula. *Roux's Arch. Entwicklungsmech. Org.* **138:** 657–738 (English transl., Hamburger V. 1996. Dev. Dyn. **205:** 217–222.)

———. 1939. Gewebeaffinität: ein Mittel der embryonalen Formbildung. *Arch. Exp. Zellforsch.* **23:** 169–209 (English transl., Willier B.H. and Oppenheimer J., eds. 1964. *Foundations of experimental embryology,* pp. 186–225. Prentice Hall, Englewood Cliffs, New Jersey).

———. 1943. A study of the mechanics of gastrulation. Part I. *J. Exp. Zool.* **94:** 261–318.

———. 1944. A study of the mechanics of gastrulation. Part II. *J. Exp. Zool.* **95:** 171–212.

Hopwood N. 1999. "Giving body" to embryos: Modeling, mechanism, and the microtome in late nineteenth-century anatomy. *ISIS* **90:** 462–496.

———. 2004. Embryology. In *The Cambridge history of science: Life and earth sciences since 1800* (ed. P.J. Bowler and J.V. Pickstone), vol. 6, Cambridge University Press, Cambridge, United Kingdom. (In press.)

Hörstadius S. 1935. Über die Determination im Verlaufe der Eiachse bei Seeigeln. *Pubbl. Stn. Zool. Napoli* **14:** 251–479.

———. 1937a. Über die zeitliche Determination im Keim von *Paracentrotus lividus. Wilhelm Roux' Arch. Entwicklungsmech. Org.* **135:** 1–40.

———. 1937b. Weitere Studien über die Determination im Verlaufe der Eiachse bei Seeigeln. *Wilhelm Roux' Arch. Entwicklungsmech. Org.* **135:** 40–69.

———. 1973. *Experimental embryology of echinoderms.* Clarendon Press, Oxford, United Kingdom.

Hubrecht A.A.W. 1905. Die Gastrulation der Wirbeltiere: *Anat. Anz.* **26:** 353–366 (English transl., The gastrulation of the vertebrates. *Q. J. Microsc. Sci.* **49:** 403–419 [1905]).

Huxley T.H. 1871. *A manual of the anatomy of vertebrated animals.* Churchill, London, United Kingdom.

Jahn I., ed. 1998. *Geschichte der Biologie. Theorien, Methoden, Institutionen, Kurzbiographien.* G. Fischer, Jena-Stuttgart, Germany.

Jenkinson J.W. 1913. *Vertebrate embryology: Comprising the early history of the embryo and its fetal membranes.* Clarendon Press, Oxford, United Kingdom.

Johnson A. 1884. On the fate of the blastopore and the presence of a primitive streak in the newt (*Triton cristatus*). *Q. J. Microsc. Sci.* **24:** 659–672.

Junker T. 1998. Charles Darwin und die Evolutionstheorie des 19. Jahrhunderts. In *Geschichte der Biologie. Theorien. Methoden, Institutionen, Kurzbiographien* (ed. I. Jahn), pp. 356–385. G. Fischer, Stuttgart-Leipzig, Germany.

Keibel F. 1887. Van Beneden's Blastoporen und die Rauber'sche Deckschicht. *Anat. Anz.* **2:** 769–773.

———. 1900. Die Gastrulation und die Keimblattbildung der Wirbeltiere. *Ergeb. Anat. Entwicklungsgesch.* **10:** 1002–1119.

———. 1905. Zur Gastrulationsfrage. *Anat. Anz.* **29:** 366–368.

Keller R. E. 1996. Holtfreter revisited: Unsolved problems in amphibian morphogenesis. *Dev. Dyn.* **205:** 257–264.

Kölliker R.A. v. 1876. *Entwickelungsgeschichte des Menschen und der höheren Thiere,* 2nd edition. Wilhelm Engelmann, Leipzig, Germany.

Kopsch F. 1895. Die Zellenbewegungen während des Gastrulationsprocesses an den Eiern vom Axolotl und vom braunen Grasfrosch. *Sitz. ber. Gesell. Naturforsch. Freunde zu Berlin.* **1895:** 21–30.

———. 1900. Ueber das Verhältnis der embryonalen Axen zu den drei ersten Furchungsebenen beim Frosch. *Intern. Monatsschr. Anat. Physiol.* **17:** 1–25.

Kowalevsky A. 1866a. Entwickelungsgeschichte der Rippenquallen. *Mém. Acad. Imp. Sci. St. Petersburg,* VII Série. Tome (no. 4) **X.**

———. 1866b. Entwickelungsgeschichte der einfachen Ascidien. *Mém. Acad. Imp. Sci. St. Petersburg,* VII Série. Tome (no. 15) **X** (*Q. J. Microsc. Sci.* **10:** 59–69 [1870]).

———. 1871. Weitere Studien über die Entwickelung der einfachen Ascidien. *Arch. Mikr. Anat.* **7:** 101–130.

———. 1873. *Über die Entwicklung der Coelenterata.* Mittheilungen der kaiserl. Gesellschaft der Liebhaber der Naturlehre, Anthropologie und Ethnographie, Moskau (in Russian).

Kowalevsky A., Ovshennikow P., and Wagner N. 1870. Die Entwickelungsgeschichte der Störe. *Bull. Acad. Imp. Sci. St. Petersburg* **14:** 318–324.

Krohn A. 1849. *Beitrag zur Entwickelungsgeschichte der Seeigellarven.* Heidelberg, Germany.

———. 1851a. Bemerkungen über einige Echinodermlarven. *Arch. Anat. Physiol.* **18:** 353–357.

———. 1851b. Beobachtungen aus der Entwickelungsgeschichte der Holothurien und Seeigel. *Arch. Anat. Physiol.* **18:** 344–352.

———. 1852. Über die Entwickelung der Ascidien. *Arch. Anat. Physiol.* **22:** 312–333 ([1853] On the development of the ascidians. In *Scientific memoirs: Natural history* [ed. A.H. Henfrey and T.H. Huxley], pp. 312–329, Taylor & Francis, London, United Kingdom).

Kupffer C.V. 1870. Die Stammverwandtschaft zwischen Ascidien und Wirbelthieren. *Arch. Mikr. Anat.* **6:** 115–172.

Lane M.C., Koehl M.A., Wilt F., and Keller R. 1993. A role for regulated secretion of apical extracellular matrix during epithelial invagination in the sea urchin. *Development* **117:** 1049–1060.

Lankester E.R. 1875. On the invaginate planula or diploblastic phase of *Paludina vivipara. Q. J. Microsc. Sci.* **15:** 159–166.

———. 1876. An account of Professor Haeckel's recent additions to the gastraea theory. *Q. J. Microsc. Sci.* **16:** 51–66.

Lewis W.H. 1947. Mechanics of invagination. *Anat. Rec.* **97:** 139–156.

Lillie F.R. 1919. *The development of the chick. An introduction to embryology.* Henry Holt, New York.

Løvtrup S. 1975. Fate maps and gastrulation in Amphibia—A critique of current views. *Can. J. Zool.* **53:** 473–479.

Mangold O. 1923. Transplantationsversuche zur Frage der Spezifität und der Bildung der Keimblätter. *Arch. Mikr. Anat. Entwicklungsmech. Org.* **100:** 198–301.

McMurrich J.P. 1890. *The gastraea theory and its successors.* Marine Biological Laboratory Biological Lectures, Woods Hole, p. 79-106.

Metchnikoff E. 1869. Entwickelungsgeschichtliche Beiträge. I-IX. *Bull. Acad. Imp. Sci. St. Petersburg,* Tome (no. 8) **XIII.**

———. 1870. Studien über die Entwickelung der Echinodermen und Nemertinen. *Mém. Acad. Imp. Sci. St. Petersburg,* VII Série: Tome (no. 8) **XIV.**

———. 1871. Ueber die Entwickelung einiger Coelenteraten. I.-II. *Bull. Acad. Imp. Sci. St. Petersburg* **15:** 95–100.

———. 1881 (1885). Comparative embryological studies (cited in Gourko et al. [2000], pp. 92–143). Kluwer, Dordrecht, The Netherlands.

———. 1882. Vergleichend-embryologische Studien 3. Über die Gastrula einiger Metazoen, *Z. Wiss. Zool.* **37:** 286–313.

———. 1885. Vergleichend-embryologische Studien. *Z. Wiss. Zool.* **42:** 649–673.

Mikhailov A.T. and Gilbert S.F. 2002. From development to evolution: The re-establishment of the "Alexander Kowalevsky Medal". *Int. J. Dev. Biol.* **46:** 693–698.

Minot C.S. 1889. The concrescence theory of the vertebrate embryo. *Am. Nat.* **24:** 501–516, 617–629, 702–719.

Mocek R. 1998. *Die werdende Form. Morphogenese: Evolution und Selbstorganisation in der Entwicklungsbiologie der zweiten Hälfte des 19. Jahrhunderts.* Acta Biohistorica, Bd. 3, Basilisken Presse, Marburg, Germany.

Moore A.R. 1930. On the invagination of the gastrula, *Protoplasma* **9:** 25–33.

———. 1941. On the mechanics of gastrulation in *Dendraster excentricus. J. Exp. Zool.* **87:** 101–111.

Moore A.R. and Burt A.S. 1939. On the locus and nature of the forces causing gastrulation in the embryos of *Dendraster exentricus. J. Exp. Zool.* **82:** 159–171.

Morgan T.H. 1894. The formation of the embryo of the frog. *Anat. Anz.* **9:** 697–705.

———. 1897. *The development of the frog's egg. An introduction to experimental embryology.* Macmillan, New York.

Moulin A.M. 1991. *Le dernier langage de la médecine. Histoire de l'immunologie de Pasteur au Sida,* Presses Universitaires de France, Paris.

Nakamura O. 1938. Tail formation in the Urodele. *Zool. Mag.* **50:** 442–446.

———. 1942. Die Entwicklung der hinteren Körperhälfte bei Urodelen. *Annot. Zool. Jpn.* **21:** 169–235.

Narasimha M. and Leptin M. 2000. Cell movements during gastrulation: Come in and be induced. *Trends Cell Biol.* **10:** 169–72.

Neigebaur J.D.F. 1860. August David Krohn. In *Geschichte der*

Kaiserlichen Leopoldino-Carolinischen deutschen Akademie der Naturforscher während des zweiten Jahrhunderts ihres Bestehens. Frommann, Jena, Germany.

Nieuwkoop P.D. 1969. The formation of the mesoderm in Urodelean Amphibians. I. Induction by the endoderm. *Wilhelm Roux' Arch. Entwicklungsmech. Org.* **162:** 341–373.

Nieuwkoop P.D. and Florschütz P.A. 1950. Quelques caractères spéciaux de la gastrulation et de la neurulation de l'oeuf de *Xenopus laevis*, Daud. et de quelques autres Anoures. 1 ère partie. Etude descriptive. *Arch. Biol.* **61:** 113–150.

Nieuwkoop P.D., Johnen A.G., and Albers B. 1981. *The epigenetic nature of early chordate development. Inductive interaction and competence.* Cambridge University Press, Cambridge, United Kingdom.

Nyhart L.K. 1995. *Biology takes form. Animal morphology and the German universities, 1800–1900.* Chicago University Press, Illinois.

Oppenheimer J.M. 1967. K. E. von Baer's beginning insights into causal-analytical relationships during development. In *Essays in the history of embryology and biology,* pp. 295–307. MIT Press, Cambridge, Massachusetts.

———. 1970. Some diverse background for Curt Herbst's idea about embryonic induction. *Bull. Hist. Med.* **44:** 241–250.

Ospovat D. 1976. The influence of Karl Ernst von Baer's embryology: 1828–1859: A reappraisal in the light of Richard Owen's and William B. Carpenter's paleontological application of von Baer's law. *J. Hist. Biol.* **9:** 1–28.

Pander C. 1817. *Beitraege zur Entwicklungsgeschichte des Hühnchens im Eye.* Würzburg, Germany.

Pasteels J. 1940. Un apercu comparatif de la gastrulation chez les Chordes. *Biol. Rev.* **15:** 59–106.

———. 1941. New observations concerning the maps of presumptive areas of the young amphibian gastrula (*Ambystoma* and *Discoglossus*). *J. Exp. Zool.* **89:** 255–281.

Patterson J.T. 1909a. Gastrulation in the pigeon's egg. A morphological and experimental study. *J. Morphol.* **20:** 65–124.

———. 1909b. An experimental study on the development of the vascular area of the chick blastoderm. *Biol. Bull.* **16:** 83–90.

Pflüger E. 1883. Ueber den Einfluss der Schwerkraft auf die Teilung der Zellen und auf die Entwicklung des Embryo. Zweite Abhandlung. *Arch. Gesamte Physiol.* **32:** 1–79.

Philiptchuk O.J. 1990. *Alexander Onufrievitch Kovalevsky.* Naukowa Dumka, Kiew, Ukraine.

Querner H. 1998. Der Übergang zur experimentellen Methode in der Biologie des 19. Jahrhunderts. In *Geschichte der Biologie. Theorien, Methoden, Institutionen, Kurzbiographien* (ed. I. Jahn), pp. 420–430. G. Fischer, Jena-Stuttgart, Germany.

Raikov B.E. 1968. *Karl Ernst von Baer, 1792–1876. Sein Leben und sein Werk.* Acta Historica Leopoldina, Nr. 5, Joh. A. Barth, Leipzig, Germany.

Rauber A. 1874 (1875). Ueber die embryonale Anlage des Hühnchens. II. Die Gastrula des Hühnerkeims. *Centralbl. Med. Wiss.* **12:** 786–788 (**13:** 49–52, 257–259).

———. 1875. Die erste Entwicluing d. Kaninchens. *Sitzungsber. D. naturfor. Gessell z. Leipzig;* quoted in Balfour (op. cit., 1881).

———. 1876. Primitivrinne und Urmund. Beitrag zur Entwickelungsgeschichte des Hühnchens. *Morphol. Jahrbuch* **2:** 550–576.

———. 1883a. Neue Grundlegungen zur Kenntnis der Zelle. *Morphol. Jahrbuch* **8:** 233–338.

———. 1883b. Noch ein Blastoporus. *Zool. Anz.* **6:** 143–147, 163–167.

Remak R. 1852. Über extracellulare Entstehung thierischer Zellen und über Vermehrung derselben durch Theilung. *Arch. Path. Anat. Physiol. Wiss. Med.* **19:** 47–92.

———. 1855. *Untersuchungen über die Entwicklung der Wirbelthiere.* Berlin, Germany.

Rhumbler L. 1899. Physikalische Analyse von Lebenserscheinungen der Zelle. III. Mechanik der Pigmentzusammenhäufungen in den Embryonalzellen der Amphibieneier. *Arch. Entwicklungsmech. Org.* **9:** 63–102.

———. 1902. Zur Mechanik des Gastrulationsvorganges, insbesondere der Invagination. Eine entwicklungsmechanische Studie. *Arch. Entwicklungsmech. Org.* **14:** 401–476.

———. 1903. Mechanische Erklärung der Ähnlichkeit zwischen magnetischen Kraftliniensystemen und Zellteilungsfiguren. *Arch. Entwicklungsmech. Org.* **16:** 476–535.

———. 1923. Methodik der Nachahmung von Lebensvorgängen durch physikalische Konstellationen. In *Handbuch der biologischen Arbeitsmethoden: Methodik der Entwicklungsmechanik* (ed. E. Abderhalden), pp. 219–440. Urban & Schwarzenberg, Berlin-Wien, Germany.

Richardson M.K. and Keuck G. 2002. Haeckel's ABC of evolution and development. *Biol. Rev. Camb. Philos Soc.* **77:** 495–528.

Robinson A. and Assheton R. 1891. The formation and fate of the primitive streak: With observations on the archenteron and germinal layers of *Rana temporaria. Q. J. Microsc. Sci.* **32:** 451–504.

Roux W. 1902. Bemerkungen über die Achsenbestimmung des Froschembryos und die Gastrulation des Froscheies. *Arch. Entwicklungsmech. Org.* **14:** 600–624.

———. 1905. Die Entwicklungsmechanik. Ein neuer Zweig der biologischen Wissenschaft. In *Verh. Ges. Deutscher Naturforscher und Ärzte: Erster Teil,* pp. 23–39. W. Engelmann, Leipzig, Germany.

Ruffini A. 1925. *Fisiogenia. La biodinamica dello sviluppo ed i fondamentali problemi morfologici dell'embriologia generale.* F. Vallardi, Milano, Italy.

Rusconi M. 1823. Observations on the natural history and structure of the aquatic salamander. Printed for A. Constable, Edinburgh, United Kingdom.

———. 1826. *Développement de la grenouille commune depuis le moment de sa naissance jusque à son état parfait.* P.E. Giusti, Milan, Italy.

———. 1836. Erwiderung auf einige kritische Bemerkungen des Herrn v. Baer über Rusconi's Entwickelungsgeschichte des Froscheies. In Briefen an Hrn. Prof. E. H. Weber. *Arch. Anat. Physiol. Wiss. Med.* **3:** 205–224.

———. 1854. *Histoire naturelle: développment et métamorphose de la salamandre terrestre,* par Mauro Rusconi. Ouvrage posthume inédit publié par Joseph Morganit, Chez Bizzoni Libraire, Pavia, Italy.

Russell E.S. 1916 (1982). *Form and function. A contribution to the history of animal morphology.* Murray, London, United Kingdom ([1982] new introduction by G.V. Lauder, Chicago University Press, Illinois).

Sander K. 1990. Von der Keimplasmatheorie zur synergetischen Musterbildung—Einhundert Jahre entwicklungsbiologischer Ideengeschichte. *Verh. Dtsch. Zool. Ges.* **83:** 133–177.

———. 1991. When seeing is believing: Wilhelm Roux's misconceived fate map. *Wilhelm Roux's Arch. Dev. Biol.* **200:** 177–179.

Schmidt F.O. 1941. Some protein patterns in cells. *Growth Third Growth Symp.* **5:** 1–20.

Schmiedebach H.-P. 1995. Robert Remak (1815-1865): *ein jüdischer Arzt im Spannungsfeld von Wissenschaft und Politik.* (Medizin in Geschichte und Kultur, Bd. 18). G. Fischer, Stuttgart, Germany.

Schultze O. 1894. Die künstliche Erzeugung von Doppelbildungen bei

Froschlarven mit Hilfe abnormer Gravitationswirkung. *Arch. Entwicklungsmech. Org.* **1:** 269–305.

Selenka E. 1883. *Die Keimblätter der Echinodermen,* I. 2. C.W. Kreidel, Wiesbaden, Germany.

Semon R. 1900. Die äussere Entwickelung des *Ceratodus* Forsteri. In *The Australian bush and on the coast of the Coral Sea: Being the experiences and observations of a naturalist in Australia: New Guinea and the Moluccas.* Macmillan, New York.

Siebold C.T.E. 1846. *Lehrbuch der vergleichenen Anatomie.* Bd. 1. Wirbellose Tiere, Veit, Berlin, Germany.

Smith F. 1891. The gastrulation of *Aurelia flavidula.* Contributions from the Zoological Laboratory, XXIX. *Harv. Univ. Mus. Compar. Zool. Bull.* **22:** 115–125.

Spek J. 1918. Differenzen im Quellungszustand der Plasmakolloide als eine Ursache der Gastrulainvagination, sowie der Einstülpungen und Faltungen von Zellplatten überhaupt. *Kolloidchemie. Beihefte* **9:** 259–399.

———. 1931. Die Zelle als morphologisches System. In *Lehrbuch der allgemeinen Physiologie* (ed. E. Gellhorn). Thieme, Leipzig, Germany, pp. 431–455.

Spemann H. 1942. Walther Vogt zum Gedächtnis. *Arch. Entwicklungsmech. Org.* **141:** 1–14.

Stern C.D. 1992. Vertebrate gastrulation. *Curr. Opin. Genet. Dev.* **4:** 556–561.

Stricker S. 1864. Mittheilungen über die selbstständigen Bewegungen embryonaler Zellen. *Sitz. Ber. Wien. Akad.* **49:** 471–476.

Summer F.B. 1900. Kupffer's vesicle and its relation to gastrulation and concrescence. *N.Y. Acad. Sci.* **2:** 47–83.

Tauber A.I. and Chernyak L. 1991. *Metchnikoff and the origins of immunology: From metaphor to theory.* Oxford University Press, Oxford, United Kingdom.

Thieffry D. 2001. Rationalizing early embryogenesis in the 1930s: Albert Dalcq on gradients and fields. *J. Hist. Biol.* **34:** 149–181.

Thompson D.W. 1942. *On growth and form.* The University Press, Cambridge, United Kingdom.

Townes P.L. and Holtfreter J. 1955. Directed movements and selective adhesion of embryonic amphibian cells. *J. Exp. Zool.* **128:** 53–120.

van Beneden E. 1870. *Recherches sur la composition de la significance de l'oeuf.* F. Hayez, Bruxelles, Belgium.

———. 1875. La maturation de l'oeuf: la fécondation et les pre-

mieères phases du développement embryonnaire des Mammifères, d'apreès des recherches faites chez le lapin. Communication préliminaire. *Bull. Acad. R. Belg.* **40:** 686–736.

———. 1876. Recherches sur les *Dicyemides:* survivants actuels d'un embranchment des Mesoziaires. *Bull. Acad. R. Belg.* **41:** 1160–1205; **42:** 35–97 (*Q. J. Microsc. Sci.* **17:** 132–145 [1877]).

Vogt W. 1923a. Weitere Versuche mit vitaler Farbmarkierung und farbiger Transplantation zur Analyse der Primitiventwicklung von Triton. *Anat. Anz. Erg.* **57:** 30–38.

———. 1923b. Morphologische und physiologische Fragen der Primitiventwicklung: Versuche zu ihrer Lösung mittels vitaler Farbmarkierung. *Sitz. Ber. Gesell. Morphol. Physiol. München* **35:** 22–32.

———. 1925. Gestaltungsanalyse am Amphibienkeim mit örtlicher Vitalfärbung. Vorwort über Wege und Ziele. I. Methodik und Wirkungsweise der örtlichen Vitalfärbung mit Agar als Farbträger. *Arch. Entwicklungsmech. Org.* **106:** 542–610.

———. 1929. Gestaltungsanalyse am Amphibienkeim mit örtlicher Vitalfärbung. II Teil. Gastrulation und Mesodermbildung bei Urodelen und Anuren. *Arch. Entwicklungsmech. Org.* **120:** 384–706.

Von Ubisch L. 1925. Entwicklungsphysiologische Studien am Seeigelkeim. I. Über die Beziehungen der ersten Furchungsebene zur Larvensymmetrie und die prospektive Bedeutung der Eibezirke. *Z. Wiss. Zool.* **124:** 361.

Weiss P.A. 1939. *Principles of development.* Henry Holt, New York.

Whitman C.O. 1883. A rare form of the blastoderm of the chick, and its bearing on the question of the formation of the vertebrate embryo. *Q. J. Microsc. Sci.* **23:** 376–398.

Willmer P. 1990. *Invertebrate relationships.* Cambridge University Press, New York.

Wilson E.B. 1928. *Genes, cells and organisms,* 3rd edition (ed. J.A. Moore). Macmillan, New York.

Wilson H.V. 1900. Formation of the blastopore in the frog egg. *Anat. Anz.* **18:** 209–239.

———. 1901. Closure of the blastopore in the normally placed frog egg. *Anat. Anz.* **20:** 123–128.

Wintrebert P. 1932. La ligne primitive des Amphibiens: Phase nouvelle du développement révélée par les marques colorées. *C.R. Acad. Sci.* **194:** 2164–2166.

PART I

THE EMBRYOLOGY OF GASTRULATION

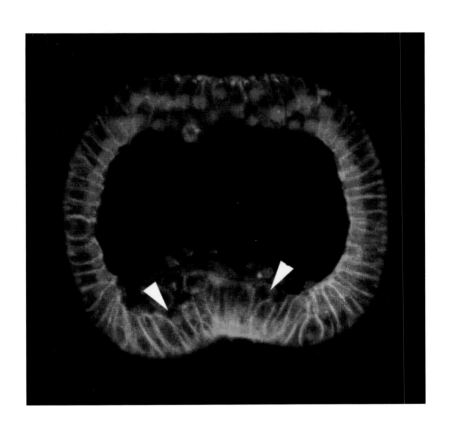

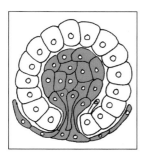

GASTRULATION IN SPONGES

S.P. Leys

Department of Biological Sciences, The University of Alberta, Edmonton, Alberta, Canada T6G 2E9

INTRODUCTION

The meaning of the term "gastrula" was established by Haeckel in 1872 after he had studied sponge development. He used the term to refer to a stage in embryogenesis that represented the ancestral diploblastic metazoan that had formed a primitive gut (the archenteron) and a mouth (the blastopore) by invagination (Haeckel 1872). He proposed that all metazoans passed through this stage, and that the embryonic germ layers were therefore homologous among all metazoans, a concept he termed the Gastraea Theory (Haeckel 1874). Ironically, because his theory hinged on finding a common feeding stage during the development of all metazoans, what he called the sponge gastrula was in fact the adult, feeding sponge. His contemporaries, Lankester (1877) and Metchnikoff (1874), carried out their own studies of sponge and cnidarian development and concluded that since ingression rather than invagination is the more common mode of formation of a bilayered embryo in basal metazoan groups, it was more likely that the ancestral metazoan lacked a mouth and gut and had a solid core of cells that could phagocytose material passed in from the outer layer. In other words, they argued that the formation of a diploblastic organism was not necessarily tied to the formation of the gut.

In recent texts (e.g., Brusca et al. 1997; Gilbert 1997), gastrulation is defined as the highly integrated movements of cells of the blastula that form a multilayered embryo in which the outer layer, the ectoderm, gives rise to the future skin and nervous tissues, and the inner layer, the endoderm, gives rise to many other components of the adult organism including the skeleton and organs, in particular, the gut.

Gastrulation therefore comprises two elements: the formation of a multilayered organism and the formation of a digestive system. Although in most organisms these events occur simultaneously, in sponges and in many cnidarians (Chapter 3) these two processes are separate. Furthermore, in many groups of these two phyla, we do not have evidence of how the second part occurs. Consequently, understanding what gastrulation is in basal metazoan phyla becomes a problem of the homology of cell or tissue layers between embryo and adult.

The problem of homology of the tissue layers between embryo or larva and adults in basal metazoans led Hyman (1940) to suggest that "the names ectoderm and entoderm [*sic*] be limited to developmental stages, and that a uniform terminology be adopted for the covering and digestive epithelia of adult invertebrates." The terms "epidermis" and "gastrodermis" were introduced to "avoid embryological implications, i.e., the gastrodermis need not necessarily be entodermal throughout." In this chapter, I give evidence that the formation of a multilayered organism by the coordinated rearrangement of cells in the blastula does occur in sponges. The various mechanisms of reorganizing the cells of the blastula are comparable to what is readily recognized as gastrulation in many cnidarians, and so the formation of the multilayered embryo during embryogenesis in sponges must be construed as gastrulation, and the cell layers must be therefore equated with ectoderm and endoderm that arise during development of other metazoans. Until we have more knowledge of the specific fate of embryonic cell layers during metamorphosis in sponges and in other basal metazoan phyla, we must view the process of embryogenesis as independent from metamorphosis, and only homologize cell layers of the fully differentiated sponge larva (or pre-adult) with those of other metazoan larvae.

GASTRULATION IN SPONGES— AN HISTORICAL PERSPECTIVE

Since Haeckel first studied embryogenesis in calcareous sponges (Haeckel 1870, 1872), opinion has been completely divided as to what constitutes gastrulation in the Porifera. Reviews by Tuzet (1963), Lévi (1963), Borojevic (1970), and most recently, Efremova (1997) provide different perspectives on the problem. There are three main schools of thought.

One school believes that sponges do not undergo gastrulation because sponge cells remain undifferentiated for the lifetime of the animal and do not form tissues at any time (Rasmont 1979; Ereskovsky and Korotkova 1997). Some proponents of this school suggest that sponges arose from a separate unicellular ancestor and as such are not metazoans. A second school believes that gastrulation occurs at metamorphosis when the outer layer of the larva migrates in to form what in this case is considered to be the gastrodermis of the adult (Lévi 1963; Tuzet 1963, 1973; Brien 1967, 1973; Fell 1974; Simpson 1984). Yet a third school believes that gastrulation occurs by ingression, epiboly, delamination, or invagination of cells of the blastula to form the germ layers of a bilayered embryo (Lévi 1956; Borojevic 1970; Efremova 1997; Boury-Esnault et al. 1999). Followers of this school either believe that at metamorphosis the germ layers invert such that the cell layers of adult sponges are reversed compared to those of other metazoans (Delage 1892; Delage and Herouard 1899; Brien 1967), or they regard the "inversion of layers" at metamorphosis as a reorganization of already differentiated cellular material and therefore as phylogenetically unimportant (Borojevic 1970; Amano and Hori 1996; Efremova 1997; Leys and Degnan 2002). In the subsequent paragraphs, I discuss the evidence for each of these schools of thought.

SPONGES ARE METAZOANS

The evidence, now overwhelming, that sponges are metazoans comes from molecular biology, but clues for metazoan monophyly were first sought in the mechanisms of cell–cell adhesion since this was an obvious unifying feature of multicellular animals. We now know that sponges possess type IV collagen (Exposito and Garrone 1990; Boute et al. 1996), a form unique to metazoans, where it is integral in forming the basement membranes that give epithelial tissues their cohesiveness. Sponges not only show species-specific and self/non-self recognition, but they also show alloimmune memory (Bigger et al. 1982) and possess high-

ly polymorphic Ig-like molecules such as receptor tyrosine kinase (for review, see Fernandez-Busquets and Burger 1999; Müller et al. 1999). An array of molecules that are involved in cell–cell adhesion (e.g., lectins, cadherins, fibronectin) and in numerous metabolic pathways (e.g., protein kinase C, HSP70) have been cloned and sequenced, and all results concur that the Metazoa are monophyletic (Müller 2003). Although the first sequences obtained from rRNA were too short to provide confidence in phylogenetic analyses of relationships among basal metazoan phyla (Rodrigo et al. 1994; Borchiellini et al. 2000), recently these sequences have been extended for a dozen species of sponges. Interestingly, although the new rRNA data support the monophyly of the Metazoa, they also suggest that the Porifera themselves are paraphyletic (Collins 1998; Schütze et al. 1999; Borchiellini et al. 2001; Medina et al. 2001), and that the Calcarea are most closely related to other metazoans, a conclusion also reached by the analysis of genes for PKC (Kruse et al. 1998).

Although texts frequently describe sponges as having a "cellular level of organization," molecular and morphological evidence indicates that these animals do have epithelial tissues like other metazoans. In adult sponges, two epithelial layers, the exopinacoderm and endopinacoderm, sandwich a collagenous mesohyl that contains a variety of amoeboid cells. Many sponge larvae are clearly bilayered, and a third layer containing both collagenous extracellular matrix and cells has been documented to lie between the two layers (see, e.g., Woollacott 1993; Leys and Degnan 2001). Tissues are considered to be layers of cells that have polarity, are tied together by adherens junctions, and are bordered on one side by a basement membrane. Cells in the outer layer of sponge larvae are clearly polarized and are joined by septate junctions (Woollacott 1990, 1993). Septate junctions also link pinacocytes and choanocytes that form the inner and outer epithelia of sponges (Feige 1969; Ledger 1975; Green and Bergquist 1982; Mackie and Singla 1983). Type IV collagen is present in at least one homoscleromorph sponge, where it is localized to basal laminae of epithelial cells (Boute et al. 1996). Although a well-defined basal lamina is not evident in the majority of sponges, short-chain collagens with some homologies to type IV collagen have been found in the freshwater sponge *Ephydatia mulleri* (Exposito et al. 1991) and could be in other groups.

SPONGE EMBRYOGENESIS: FORMATION OF THE DIPLOBLASTIC ORGANISM

The formation of a bilayered embryo by delamination, epiboly, ingression, egression, and even invagination, can be found in different groups of the Porifera (Fig. 1). In the

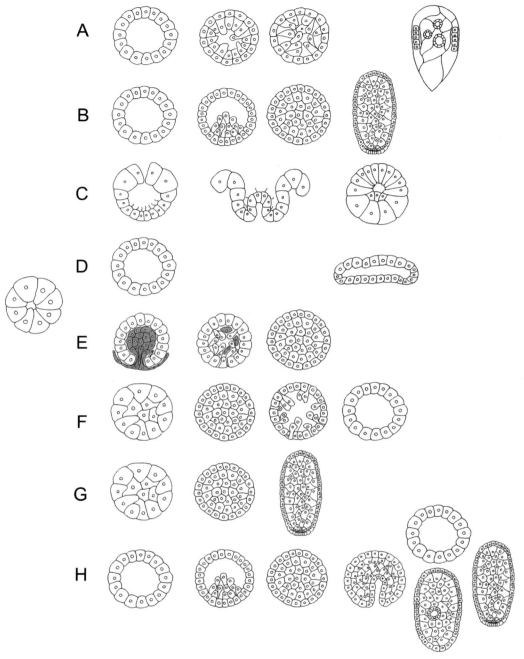

Figure 1. Types of gastrulation found in the Porifera. (*A*) *Oopsacas minuta*, Hexactinellida. (*B*) Calcinea, Calcarea. (*C*) Calcaronea Calcarea. (*D–F*) Demospongiae. (*D*) *Polymastia robusta*, Tetractinomorpha. (*E*) *Chondrosia reniformis*, Tetractinomorpha. (*F*) *Oscarella lobularis*, Homoscleromorpha. (*G*) *Reniera* sp., Ceractinomorpha. (*H*) *Halisarca dujardini*, Ceractinomorpha.

majority of species a solid bilayered larva is formed, but a few larval types have a small internal cavity (see Fig. 1), and at least two groups have choanocyte chambers in the free-swimming larva. Although embryogenesis cannot be generalized for any one group, for the purpose of this review, which seeks to compare mechanisms of gastrulation among metazoan phyla, the most common mode of embryogenesis for each group is summarized, and the reader is referred to additional works for further detail (Ijima 1901, 1904; Lévi 1956; Borojevic 1969; Tuzet, 1970; Maldonado and Bergquist 2002).

In the hexactinellid (glass) sponge *Oopsacas minuta*, cleavage occurs to form a hollow blastula (Boury-Esnault et al. 1999) (Figs. 1A, 2A–C). After the sixth division, cleavage becomes unequal and cleavage planes are tangential to the surface of the embryo. The macromeres formed in this way migrate into the blastocoel forming a bilayered embryo by delamination. Soon after, the macromeres give rise to multinucleate reticulate tissues, and the free-swimming larva is mostly syncytial (Fig. 2C). Observations by Ijima (1901, 1904) and Okada (1928) suggest that embryogenesis in other hexactinellids is similar.

Calcareous sponges have two very distinct larval types, the calciblastula (sometimes called a coeloblastula) of the calcinea and the amphiblastula of the calcaronea. In calcinean sponges, cleavage leads to a hollow ciliated blastula in which some cells lose their cilia and migrate inward,

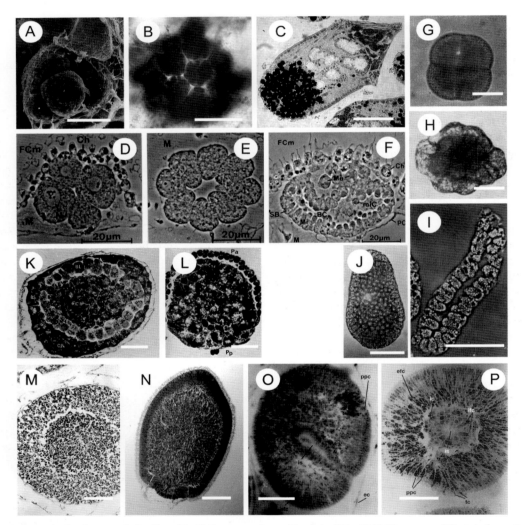

Figure 2. Stages in development in Porifera. (*A–C*) *Oopsacas minuta*, Hexactinellida. (*A,B*) Cleavage and blastula formation. (*C*) The syncytial trichimella larva. Bars, *A–C*: 50 μm. (*D–F*) *Scypha ciliata*, Calcaronea. (*D,E*) "Stomo" blastula formation; (*F*) Everting blastula. Bars *D–F*: 20 μm. (*G–J*) *Polymastia robusta*, Tetractinomorpha. (*G,H*) Blastula formation. (*I*) Epon section of a creeping coeloblastula larva. (*J*) Whole mount of a coeloblastula larva. Bars, *G–J*: 50 μm. (*K*) Yolk-filled coeloblastula of *Chondrosia reniformis*, Tetractinomorpha. (*L*) Free-swimming larva of *C. reniformis*. Bars, *K,L*: 20 μm. (*M, N*) *Reniera* sp., Ceractinomorpha. (*L*) Gastrula, (*M*) Parenchymella larva. Bars, *M,N*: 100 μm. (*O, P*) *Halisarca dujardini*, Ceractinomorpha. (*N*) Invagination of the stereogastrula, (*O*) Disphaerula larva. Bars, *O,P*: 40 μm. (*A*, Reprinted, with permission, from Boury-Esnault et al. 1999; *B,C*, courtesy of N. Boury-Esnault; *D–F*, reprinted, with permission, from Franzen, 1988 [© Springer-Verlag]; *G-J*, reprinted, with permission, from Borojevic 1966; *K, L*, reprinted, with permission, from Lévi and Lévi 1956; *O, P*, reprinted, with permission, from Ereskovsky and Gonobobleva 2000 [© Publications Scientifiques du Muséum national d'Histoire naturelle, Paris].)

eventually filling the blastocoel by unipolar and multipolar ingression (Fig. 1B) (Minchin 1896; Johnson 1979). Embryogenesis in calcaronean sponges is very unusual and is sometimes compared to the sequence of events seen during development in the Volvocales, but no phylogenetic significance is attributed to this. According to Tuzet (1970) and Franzen (1988), *Sycon* embryos at the 16-cell stage consist of 8 large macromeres that abut the basal portion of choanocytes, and 8 micromeres that have inwardly directed cilia. These embryos open via a pore at the apical side (termed by some a blastopore) to the choanocyte epithelium and then turn inside out so that the cilia now face externally (Figs. 1C, 2D–F). Continued cell division and differentiation lead to a larva that has large nonciliated globular cells on the posterior half, columnar ciliated cells on the anterior half, and a very small larval cavity. Understanding the significance of this unusual mode of embryogenesis is complicated by the fact that inward-facing cilia are rarely seen in another calcaronean sponge (Gallissian and Vacelet 1992) and that electron micrographs of these stages are lacking.

The class Demospongiae encompasses a highly diverse group of sponges. The tetractinomorph demosponges are a group that is largely oviparous. Embryogenesis is best-known from two species. In *Polymastia,* a ciliated, hollow blastula (coeloblastula or clavablastula) is formed and remains as a single layer through development to a creeping larva until settlement and metamorphosis (Borojevic 1967) (Figs. 1D, 2G–J). In *Chondrosia,* the coeloblastula is bilayered but the blastocoel is filled by the immigration of yolk-rich follicular cells from the parent (Lévi and Lévi 1976). During larval development, the follicular cells are gradually replaced by cells differentiating from the outer cell layer of the blastula (Figs. 1E, 2K–L). One tetractinomorph species is unusual in lacking a larval stage altogether. *Tetilla* embryos attach to the substrate immediately after fertilization and undergo cleavage to form a solid blastula which develops directly into a juvenile sponge (Watanabe 1978). Although the outer cells of the blastula form the pinacoderm, they remain indistinguishable from the rest of the cells in the embryo until the pinacoderm is formed. Embryogenesis in homoscleromorph sponges is known from species of *Oscarella.* Cleavage results in a solid blastula that becomes hollow by the egression (migrating outward) of cells (Fig. 1F). The larva is released as a hollow ciliated coeloblastula (cinctoblastula of Boury-Esnault and Rutzler 1997). Many of the frequently encountered intertidal demosponges are members of the subclass Ceractinomorpha. Embryogenesis in this subclass generally leads to a solid blastula (stereoblastula). Mixed delamination gives rise to a stereogastrula, and subsequently to a highly differentiated diploblastic parenchymella larva that has a mesohyl between the two layers (Figs. 1G, 2M,N)

(see, e.g., Fell and Jacob 1979; Leys and Degnan 2002; Maldonado and Bergquist 2002). Embryogenesis in halisarcid demosponges (also subclass Ceractinomorpha) is very intriguing. *Halisarca* forms from a bilayered embryo by multipolar ingression. When the blastocoel is filled, the outer layer of ciliated cells invaginates to form one and sometimes two ciliated internal cavities, giving the larva the name "disphaerula" (Figs. 1H, 2O,P) (Lévi 1956, 1963; Ereskovsky and Gonobobleva 2000). According to the latter authors, however, *Halisarca* can also form parenchymellae and coeloblastulae larvae.

METAMORPHOSIS: FORMATION OF THE CHOANOCYTE EPITHELIUM

Early authors claimed that upon settlement of the calcareous amphiblastula on the ciliated anterior pole, the ciliated cells invaginated, thereby forming the gut of the adult sponge (e.g., Metchnikoff 1874; Schulze 1878). Although no ultrastructural work confirms that this exact process occurs in the Calcarea, numerous studies have reported that the anterior and ciliated cells of sponge larvae move inward to become the choanocytes of the flagellated chambers (Delage 1892; Minchin 1896; Dubosq and Tuzet 1937; Lévi 1956; Brien 1967; Borojevic 1970; Amano and Hori 1993, 1996; for review, see Efremova 1997). Because the formation of the choanocytes is equated with the formation of the gastrodermis of other metazoans, many investigators have concluded that the inward migration of ciliated cells to become the choanocytes is gastrulation.

Recently, two studies using cell tracers have confirmed that the ciliated cells can migrate inward to form the choanocytes at metamorphosis in some demosponge parenchymellae and calcareous amphiblastulae (Amano and Hori 1996; Leys and Degnan 2002), but there are three fundamental problems with the argument that this constitutes gastrulation. First, the sponge larva is already a highly differentiated diploblastic organism that is even capable of nutrient uptake (Jaeckle 1995). Second, there is ample evidence that the choanocytes do not arise from the outer ciliated cells at metamorphosis in many groups of sponges. At the time of settlement, larvae of freshwater sponges (Ceractinomorpha) and hexactinellids already have flagellated chambers (Wielspütz and Saller 1990; Boury-Esnault et al. 1999). Freshwater sponges lose these chambers and form choanocytes de novo at settlement and metamorphosis; the fate of the flagellated chambers in hexactinellid larvae is unknown. Other larvae discharge their locomotory cilia at settlement (Bergquist and Green 1977), and yet others form choanocytes from amoeboid nonciliated cells (Misevic et al. 1990). Although experiments have demonstrated that the ciliated epithelium isolated from a sponge larva is only able to differentiate into choanocytes (Borojevic and Lévi 1965; Borojevic 1966), it has also been shown that choanocytes can differentiate from cells of the central mass (Borojevic 1966) and can form de novo from a purified fraction of archaeocytes from adult sponges (Buscema et al. 1980). Third, and most importantly, it is debatable whether the choanocytes can be considered homologous to the digestive epithelium of other metazoans. The outer surface of the sponge is known as the pinacoderm. The inner surface is formed by collar cells (choanocytes) that pump the water through the animal and by the cells lining the water canals, known as the endopinacoderm. All sponge epithelia, both external and internal, are phagocytic (Willenz and Van de Vyver 1982), and at the same time, all epithelia of the adult sponge are in contact with the environment and are sensory. Stimuli to any portion may cause the slow contraction of the whole sponge or cessation of pumping of water through the animal (for review, see Mackie 1979). Because sponge epithelia are multifunctional, it is difficult to homologize only the choanocyte chambers with the digestive epithelia of higher animals.

ARE THE GERM LAYERS OF SPONGES INVERTED?

From the above accounts it is clear that sponges undergo the same kinds of complex cell rearrangements of blastomeres during embryogenesis to form a multilayered larva as occurs in cnidarians. Many embryologists have considered these rearrangements to constitute gastrulation, and the layers thus formed to constitute the embryonic germ layers (Delage 1892; Delage and Herouard 1899; Tuzet 1970; Borojevic 1970; Efremova 1997). The majority of sponge larvae are solid and bilayered, like the planulae of many hydrozoans, and as in many planulae, a feeding system is not put in place until metamorphosis is complete. Thus, the formation of the primary germ layers is independent of formation of the gut. For those who share the view of Delage (1892), the immigration of the ciliated cells to form the choanocyte chambers is not gastrulation, but it does imply that the ectoderm is now in the place of the endoderm, and thus constitutes an inversion of the germ layers. If the germ layers do form during embryogenesis, what becomes of these layers at metamorphosis?

The cell migrations that occur at metamorphosis are massive. In the handful of calcareous sponges and demosponges where metamorphosis has been studied by electron microscopy, the epithelial cells of the sponge larva resorb or eject the locomotory cilia and migrate through a mass of cells to take up residence along the walls of choanocyte chambers and elsewhere in the sponge (Boury-Esnault

1976; Amano and Hori 1996; Leys and Degnan 2002). Because they do line the choanocyte chamber, it is easy to understand why it might be thought that the germ layers are inverted. Yet, because the rearrangement of cells that occurs at metamorphosis involves the dedifferentiation of already differentiated ciliated epithelial cells, and their transdifferentiation into the flagellated cells of the choanocyte epithelium and into other cells, the choanocyte epithelia can no longer be considered to be embryological layers (Amano and Hori 1996; Leys and Degnan 2002). The epithelia of the adult sponge are the epidermis and gastrodermis of the adult sponge, and as Hyman (1940) noted, the gastrodermis need not be of endodermal origin throughout.

How does the process of metamorphosis in sponges compare with that of cnidarians? Remarkably little is known of the fate of the germ layers at metamorphosis in cnidarians and other basal metazoan phyla. Cnidarian larvae are oblong, ciliated, bilayered planulae that may be solid or hollow. In many anthozoans the gut is already present in the planula, where it may be formed by invagination of an already bilayered embryo (e.g., Chia and Spaulding 1972). In contrast, most hydrozoan planulae have no digestive cavity or mouth until settlement and metamorphosis (Tardent 1978; Chapter 3). The homology of layers between embryo and adult is very problematic. If the gut is formed by invagination of the outer ciliated cells in an already diploblastic embryo, then the cells forming the ciliated lining of the gut were originally ectoderm as were the choanocytes in sponges. In many invertebrate larvae, however, portions of the digestive apparatus or epithelium form de novo (e.g., gastropods: Page and Pedersen 1998; echinoids: Chia and Burke 1978; Cnidaria: Martin et al. 1983; Stricker 1985), as it does in some sponges (Saller and Weissenfels 1985; Misevic et al. 1990). Hence, in many phyla, gastrulation is a two-step process as it is in sponges.

CONCLUSION: CURRENT PROBLEMS AND RECENT PROGRESS

Since the last major review of gastrulation in this phylum (Lévi 1963), four significant discoveries have changed our understanding of relationships within the Porifera. Developmental data must now be reexamined with these relationships in mind.

First, hexactinellid sponges have been placed in a subphylum of the Porifera due to the largely syncytial structure of their tissues (Reiswig and Mackie 1983). Yet Boury-Esnault and colleagues (1999) have shown that the embryos appear to be cellular at least until gastrulation, implying that hexactinellids may be more closely related to demosponges than previously thought. Future work will need to determine when multinucleate tissues first form during embryogenesis and whether cells of the embryos are nevertheless connected by cytoplasmic bridges and the unusual plugged junctions (Mackie 1981) that are particular to this group of sponges.

Second, if we are to believe rRNA data that calcareous sponges are more closely related to other metazoans than to other sponges, we must reconsider whether the developmental mechanisms found in the Calcarea are ancestral to the rest of the Metazoa. This requires a reexamination of embryogenesis in both calcinean and calcaronean sponges.

Third, doubts have been shed as to the monophyly of the Demospongiae (Boury-Esnault and Sole-Cava 2002). Not only do the Homoscleromorpha clearly possess a basement membrane in the adults, but similar structures have recently been found in the larvae (N. Boury-Esnault, pers. comm.). The very unusual development of the halisarcid ceractinomorph sponges needs thorough reexamination to understand whether invagination to form the disphaerula larva has any phylogenetic significance.

Last, use of high-resolution transmission electron microscopy (Amano and Hori 1993, 1996, 2001) and cell labeling techniques (Misevic et al. 1990; Leys and Degnan 2002) have shown that cell movements at metamorphosis in calcareous sponges and demosponges do not correspond to a simple inversion of layers. Such techniques need to be applied to the study of metamorphosis in other poriferan groups and the Cnidaria, and more importantly, to the study of embryogenesis in oviparous species of the Porifera. If these techniques can be coupled with in situ hybridization techniques as recently demonstrated with probes to an array of developmental regulatory genes (Degnan et al. 2002), we can expect exciting new information on poriferan development.

ACKNOWLEDGMENTS

I thank Radovan Borojevic, Claude Lévi, Alexander Ereskovsky, and Nicole Boury-Esnault for permission to reprint figures. I am especially indebted to Nicole for many stimulating discussions on this topic, for her comments on this manuscript, and for permission to publish Figures 2B and 2C.

REFERENCES

Amano S. and Hori I. 1993. Metamorphosis of calcareous sponges II. Cell rearrangement and differentiation in metamorphosis. *Invertebr. Reprod. Dev.* **24:** 13–26.

———. 1996. Transdifferentiation of larval flagellated cells to choanocytes in the metamorphosis of the demosponge *Haliclona permollis. Biol. Bull.* **190:** 161–172.

———. 2001. Metamorphosis of coeloblastula performed by the multipotential larval flagellated cells in the calcareous sponge *Leucosolenia laxa. Biol. Bull.* **200:** 20–32.

Bergquist P.R. and Green C.R. 1977. An ultrastructural study of settlement and metamorphosis in sponge larvae. *Cah. Biol. Mar.* **18:** 289–302.

Bigger C.H., Jokiel P.L., Hildemann W.H., and Johnston I.S. 1982. Characterization of alloimmune memory in a sponge. *J. Immunol.* **129:**1570–1572.

Borchiellini C., Chombard C., Lafay B., and Boury-Esnault N. 2000. Molecular systematics of sponges (Porifera). *Hydrobiologia* **420:** 15–27.

Borchiellini C., Manuel M., Alivon E., Boury-Esnault N., Vacelet J., and Le Parco Y. 2001. Sponge paraphyly and the origin of Metazoa. *Evol. Biol.* **14:** 171–179.

Borojevic R. 1966. Étude experimentale de la différenciation des cellules de l'éponge au cours de son développement. *Dev. Biol.* **14:** 130–153.

———. 1967. La ponte et le développement de *Polymastia robusta* (Démosponges). *Cah. Biol. Mar.* **8:** 1–6.

———. 1969. Étude du développement et de la différenciation cellulaire d'éponges calcaires calcinéennes (genres *Clathrina* et *Ascandra*). *Ann. Embryol. Morphog.* **2:** 15–36.

———. 1970. Différenciation cellulaire dans l'embyogénèse et la morphogénèse chez les spongiaires. *Proc. Zool. Soc. Lond.* **25:** 467–490.

Borojevic R. and Lévi C. 1965. Morphogénèse expérimentale d'une éponge à partir de cellules de la larve nageante dissociée. *Z. Zellforsch. Mikrosk. Anat.* **68:** 57–69.

Boury-Esnault N. 1976. Ultrastructure de la larve parenchymella d'*Hamigera hamigera* (Schmidt) (Démosponge, Poecilosclerida). Origine des cellules grises. *Cah. Biol. Mar.* **17:** 9–20.

Boury-Esnault N. and Rutzler K. 1997. Thesaurus of sponge morphology. *Smithson. Contrib. Zool.* **596:** 1–55.

Boury-Esnault N. and Sole-Cava A.M. 2002. The present state and future perspectives in molecular natural history of sponges. *Boll. Mus. Ist. Biol. Univ. Genova* **66–67:** 33–34.

Boury-Esnault N., Efremova S., Bézac C., and Vacelet J. 1999. Reproduction of a hexactinellid sponge: First description of gastrulation by cellular delamination in the Porifera. *Invertebr. Reprod. Dev.* **35:** 187–201.

Boute N., Exposito J.Y., Boury-Esnault N., Vacelet J., Nor N., Miyazaki K., Yoshizato K., and Garrone R. 1996. Type IV collagen in sponges, the missing link in basement membrane ubiquity. *Biol. Cell.* **88:** 37–44.

Brien P. 1967. Les éponges. Leur nature metazoaire, leur gastrulation, leur état colonial. *Ann. Soc. R. Zool. Belg.* **97:** 197–235.

———. 1973. Les Démosponges. In *Traité de Zoologie* (ed. P.-P. Grassé), pp. 133–461. Masson et Cie, Paris.

Brusca G.J., Brusca R.C., and Gilbert S.F. 1997. Characteristics of metazoan development. In *Embryology, constructing the organism* (ed. S.F. Gilbert and A.M. Raunio), pp. 3–19. Sinauer, Sunderland, Massachusetts.

Buscema M., De Sutter D., and Van de Vyver G. 1980. Ultrastructural study of differentiation processes during aggregation of purified sponge archaeocytes. *Wilhelm Roux's Arch.* **188:** 45–53.

Chia F.-S. and Burke R.D. 1978. Echinoderm metamorphosis: Fate of larval structures. In *Settlement and metamorphosis of marine invertebrate larvae* (ed. F.-S. Chia and M.E. Rice), pp. 219–234. Elsevier, New York.

Chia F.-S. and Spaulding J.G. 1972. Development and juvenile growth of the sea anemone *Tealia crassicornis*. *Biol. Bull.* **142**: 206–218.

Collins A.G. 1998. Evaluating multiple alternative hypotheses for the origin of Bilateria: An analysis of 18S rRNA molecular evidence. *Proc. Natl. Acad. Sci.* **95**: 15458–15463.

Degnan B.M., Larroux C., Leys S.P., Liubicich D., Hinman V.F., Gongora M., Elliott M., and Woerheide G. 2002. Expression of developmental genes in sponge larvae: Evidence for a shared regulatory architecture in all metazoans. *Boll. Mus. Ist. Biol. Univ. Genova* **66–67**: 54.

Delage Y. 1892. Embryogenie des éponges. Développement post-Larvaire. *Arch. Zool. Exp. Gén.* **10**: 345–498.

Delage Y. and Herouard E. 1899. *Traité de Zoologie concrete. Mésozoaires, Spongiaires*. Schleicher Frères, Éditeurs, Paris.

Dubosq O. and Tuzet O. 1937. L'ovogenèse, la fécondation et les premiers stades du développement des éponges calcaires. *Arch. Zool. Exp. Gén.* **79**: 157–316.

Efremova S.M. 1997. Once more on the position among Metazoa—Gastrulation and germinal layers of sponges. *Berliner Geowiss. Abh.* **20**: 7–15.

Ereskovsky A.V. and Gonobobleva E.L. 2000. New data on embryonic development of *Halisarca dujardini* Johnston, 1842 (Demospongiae: Halisarcida). *Zoosystema* **22**: 355–368.

Ereskovsky A.V. and Korotkova G.P. 1997. The reasons of sponge sexual morphogenesis peculiarities. *Berl. Geowiss. Abh.* **20**: 25–33.

Exposito J.Y. and Garrone R. 1990. Characterization of a fibrillar collagen gene in sponges reveals the early evolutionary appearance of two collagen gene families. *Proc. Natl. Acad. Sci.* **87**: 6669–6673.

Exposito J.Y., Le Guellec D., Lu Q., and Garrone R. 1991. Short chain collagens in sponges are encoded by a family of closely related genes. *J. Biol. Chem.* **266**: 21923–21928.

Feige W. 1969. Die Feinstruktur der Epithelien von Ephydatia fluviatilis. *Zool. Jb. Anat. Bd.* **86**: 177–237.

Fell P.E. 1974. Porifera. In *Reproduction of marine invertebrates* (ed. A.C. Giese and J.S. Pearse), pp. 51–132. Academic Press, New York.

Fell P.E. and Jacob W.F. 1979. Reproduction and development of *Halichondria* sp. in the Mystic estuary, Connecticut. *Biol. Bull.* **156**: 62–75.

Fernàndez-Busquets X. and Burger M.M. 1999. Cell adhesion and histocompatibility in sponges. *Microsc. Res. Tech.* **44**: 204–218.

Franzen W. 1988. Oogenesis and larval development of *Scypha ciliata* (Porifera, Calcarea). *Zoomorphology* **107**: 349–357.

Gallissian M.F. and Vacelet J. 1992. Ultrastructure of the oocyte and embryo of the calcified sponge, *Petrobiona massiliana* (Porifera, Calcarea). *Zoomorphology* **112**: 133–141.

Gilbert S.F. 1997. *Developmental biology*. Sinauer, Sunderland, Massachusetts.

Green C.R. and Bergquist P.R. 1982. Phylogenetic relationships within the invertebrata in relation to the structure of septate junctions and the development of 'occluding' junctional types. *J. Cell Sci.* **53**: 279–305.

Haeckel E. 1870. Ueber den Organismus der Schwäme und ihre Verwandtschaft mit den Corallen. *Jenaische Ztschr. Naturw.* **5**:207–235.

———. 1872. *Die Kalkschwäme. Eine Monographie*. Bd. **1–3**. Verlag von Georg Reimer, Berlin.

———. 1874. Die Gastraea-Theorie, die phylogenetische Classification des Thierreichs und die Homologie der Keimblätter. *Jenaische Ztschr. Naturw.* **8**: 1–55.

Hyman L.H. 1940. *The invertebrates*, pp. 248–283. McGraw-Hill, New York.

Ijima I. 1901. Studies on the Hexactinellida. Contribution I (Euplectellidae). *J. Coll. Sci. Imp. Univ. Tokyo* **15**: 1–299.

———. 1904. Studies on the Hexactinellida. Contribution IV. (Rossellidae). *J. Coll. Sci. imp. Univ. Tokyo* **28**: 13–307.

Jaeckle W.B. 1995. Transport and metabolism of alanine and palmitic acid by field-collected larvae of *Tedania ignis* (Porifera, Demospongiae): Estimated consequences of limited label translocation. *Biol. Bull.* **189**: 159–167.

Johnson M.F. 1979. Gametogenesis and embryonic development in the calcareous sponges *Clathrina coriacea* and *C. blanca* from Santa Catalina Island, California. *Bull. South. Calif. Acad. Sci.* **78**: 183–191.

Kruse M., Leys S.P., Müller I.M., and Müller W.E.G. 1998. Phylogenetic position of the Hexactienllida within the phylum Porifera based on the amino acid sequence of the protein kinase C from *Rhabdocalyptus dawsoni*. *J. Mol. Evol.* **46**: 721–728.

Lankaster E.R. 1877. Notes on the anatomy and classification of the animal kingdom comprising a revision of speculations relative to the origin and significance of the germ-layers. *Q. J. Microsc. Sci.* **17**: 399–441.

Ledger P.W. 1975. Septate junctions in the calcareous sponge *Sycon ciliatum*. *Tissue Cell* **7**: 13–81.

Lévi C. 1956. Étude des *Halisarca* de Roscoff. Émbryologie et systématique des démosponges. *Arch. Zool. Exp. Gén.* **93**: 1–184.

———. 1963. Gastrulation and larval phylogeny in sponges. In *The lower metazoa: Comaparative biology and phylogeny* (ed. E.C. Dougherty), pp. 375–382. University of California Press, Los Angeles.

Lévi C. and Lévi P. 1976. Embryogénèse de *Chondrosia reniformis* (Nardo), démosponge ovipare, et transmission des bactéries symbiotiques. *Ann. Sci. Natur. Zool.* **18**: 367–380.

Leys S.P. and Degnan B.M. 2001. Cytological basis of photoresponsive behaviour in a sponge larva. *Biol. Bull.* **201**: 323–338.

———. 2002. Embryogenesis and metamorphosis in a haplosclerid demosponge: Gastrulation and transdifferentiation of larval ciliated cells to choanocytes. *Invertebr. Biol.* **121**: 171–189.

Mackie G.O. 1979. Is there a conduction system in sponges? *Colloques Int. Cent. Natn. Rech. Sci.* **291**: 145–151.

———. 1981. Plugged syncytial interconnections in hexactinellid sponges. *J. Cell Biol.* **91**: 103a.

Mackie G.O. and Singla C.L. 1983. Studies on hexactinellid sponges. I. Histology of *Rhabdocalyptus dawsoni* (Lambe, 1873). *Philos. Trans. R. Soc. Lond. B Biol. Sci.* **301**: 365–400.

Maldonado M. and Bergquist P.R. 2002. Phylum Porifera. In *Atlas of marine invertebrate larvae* (ed. C.M. Young et al.), pp. 21–50. Academic Press, San Diego, California.

Martin V., Chia F.-S., and Koss R. 1983. A fine structural study of metamorphosis of the hydrozoan *Mitrocomella polydiademata*. *J. Morphol.* **176**: 261–287.

Medina M., Collins A.G., Silberman J., and Sogin M.L. 2001. Evaluating hypotheses of basal animal phylogeny using complete sequences of large and small subunit rRNA. *Proc. Natl. Acad. Sci.* **98**: 9707–9712.

Metchnikoff E. 1874. Zur Entwicklungsgeschichte der Kalkschwämme. *Z. Wiss. Zool.* **24**: 1–14.

Minchin E.A. 1896. Note on the larva and post larval development of *Leucosolenia variabilis*, H. sp., with remarks on the development of other Asconidae. *Proc. R. Soc. Lond. Ser. B* **40**: 42–52.

Misevic G.N., Schlup V., and Burger M.M. 1990. Larval metamorphosis of *Microciona prolifera*: Evidence against the reversal of layers. In *New perspectives in sponge biology* (ed. K. Rutzler), pp. 182–187.

Smithsonian Institution Press, Washington, D.C.

Müller W.E.G. 2003. The origin of metazoan complexity. Porifera as integrated animals. *Integrative and Comparative Biology.* **43:** 3–10.

Müller W.E.G., Koziol C., Müller I.M., and Wiens M. 1999. Towards an understanding of the molecular basis of immune responses in sponges: The marine demosponge *Geodia cydonium* as a model. *Microsc. Res. Tech.* **44:** 219–236.

Okada Y. 1928. On the development of a hexactinellid sponge, *Farrea sollasii. J. Fac. Sci. Imp. Univ. Tokyo* **4:** 1–29.

Page L.R. and Pedersen R.V.K. 1998. Transformation of phytoplanktivorous larvae into predatory carnivores during the development of *Polynices lewisii* (Mollusca, Caenogastropoda). *Invertebr. Biol.* **117:** 208–220.

Rasmont R. 1979. Les éponges: des metazoaires et des societes de cellules. *Colloques. Int. Cent. Natn. Rech. Sci.* **291:** 21–29.

Reiswig H.M. and Mackie G.O. 1983. Studies on hexactinellid sponges III. The taxonomic status of Hexactinellida within the Porifera. *Philos. Trans. R. Soc. Lond. B Biol. Sci.* **301:** 419–428.

Rodrigo A.G., Bergquist P.R., Bergquist P.L., and Reeves R.A. 1994. Are sponges animals? An investigation into the vagaries of phylogenetic inference. In *Sponges in time and space* (ed. R.W.M. van Soest et al.), pp. 47–54. A.A. Balkema, Rotterdam, The Netherlands.

Saller U. and Weissenfels N. 1985. The development of *Spongilla lacustris* from the oocyte to the free larva (Porifera, Spongillidae). *Zoomorphology* **105:** 367–374.

Schulze F.E. 1878. Undersuchungen über den Bau und die Entwicklung der Spongien. 5. Mitt. Die Metamorphose von *Sycandra raphanus. Z. Wiss. Zool.* **31:** 262–295.

Schütze J., Krasko A., Custodio M.R., Efremova S.M., Müller I.M., and Müller W.E.G. 1999. Evolutionary relationships of Metazoa within the eukaryotes based on molecular data from Porifera. *Proc. R. Soc. Lond. B* **266:** 63–73.

Simpson T.L. 1984. *The cell biology of sponges.* Springer-Verlag, New York.

Stricker S.A. 1985. An ultrastructural study of larval settlement in the sea anemone *Urticina crassicornis* (Cnidaria, Actinaria). *J. Morphol.* **186:** 237–253.

Tardent P. 1978. Coelenterata, Cnidaria. In *Einleitung zum Gesamtwerk Morphogenetische Arbeitsmethoden und Begriffssysteme* (ed. F. Seidel), pp. 69–415. Gustav Fischer, Stuttgart.

Tuzet O. 1963. The phylogeny of sponges according to embryological, histological and serological data, and their affinities with the Protozoa and the Cnidaria. In *The lower metazoa* (ed. E.C. Dougherty), pp. 129–148. University of California Press, Los Angeles.

———. 1970. La polarité de l'oeuf et la symmétrie de la larve des éponges calcaires. *Symp. Zool. Soc. Lond.* **25:** 437–448.

———. 1973. Éponges calcaires. In *Traité de zoologie* (ed. P.-P. Grassé), pp. 27–132. Masson et Cie, Paris.

Watanabe Y. 1978. The development of two species of *Tetilla* (Demosponge). *Nat. Sci. Rep., Ochanomizu Univ.* **29:** 71–106.

Wielspütz C. and Saller U. 1990. The metamorphosis of the parenchymula-larva of *Ephydatia fluviatilis* (Porifera, Spongillidae). *Zoomorphology* **109:** 173–177.

Willenz P. and Van de Vyver G. 1982. Endocytosis of latex beads by the exopinacoderm in the fresh water sponge *Ephydatia fluviatilis:* An *in vitro* and *in situ* study in SEM and TEM. *J. Ultrastr. Res.* **79:** 294–306.

Woollacott R.M. 1990. Structure and swimming behavior of the larva of *Halichondria melanadocia* (Porifera: Demospongiae). *J. Morphol.* **205:** 135–145.

———. 1993. Structure and swimming behavior of the larva of *Haliclona tubifera* (Porifera: Demospongiae). *J. Morphol.* **218:** 301–321.

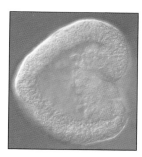

GASTRULATION IN THE CNIDARIA AND CTENOPHORA

C.A. Byrum and M.Q. Martindale

Kewalo Marine Lab, Pacific Biomedical Research Center, University of Hawai'i, Honolulu, Hawaii 96813

INTRODUCTION

Compared to many members of the animal kingdom, cnidarians and ctenophores have fairly simple body plans. Together, these two phyla are referred to as the Radiata, since they appear to be radially symmetrical along their major longitudinal body axis, the oral–aboral axis. Members of the Radiata are also referred to as diploblasts. T.H. Huxley (1849) demonstrated that cnidarian embryos only generate two germ layers, ectoderm and endoderm (called entoderm in cnidarians), thus lacking the mesodermal layer characteristic of the more commonly studied triploblasts. The apparent simplicity of diploblastic development naturally fed into Ernst Haeckel's views that "ontogeny is a brief and rapid recapitulation of phylogeny" (1866) and led Lankester (1873) to suggest that animals should be classified on the basis of the number of germ layers present as the Homoblastica (protozoans), the Diploblastica (cnidarians/ctenophores), and the Triploblastica (all other animals). One might thus infer that gastrulation in diploblasts is a relatively simple process resulting in only two, rather than three, germ layers, but this is not the case. Not only are there more modes of gastrulation in cnidarians than in any other animal group, but every pattern of gastrulation seen in triploblasts is also displayed in cnidarians. In contrast to the diverse forms of gastrulation seen in cnidarians, members of the phylum Ctenophora display only one form of gastrulation, epiboly. Indeed, ctenophore development bears little resemblance to that of cnidarians.

If the ancestral state is to be established, it is of utmost importance to determine the phylogenetic relationship of ctenophores and cnidarians to the rest of the triploblastic bilaterian metazoans. Traditional views based on symmetry properties, morphology, and development placed ctenophores as the sister group to triploblastic bilaterians (Schram 1991; Nielsen et al. 1996; Peterson and Eernisse 2001), however, molecular phylogenetic analyses (Fig. 1A) using a small number of ribosomal genes have placed the cnidarians in this position (Wainright et al. 1993; Kim et al. 1999; Medina et al. 2001; Podar et al. 2001). It is imperative to determine the sister group of the bilaterians both to identify the ancestral mechanism of gastrulation and to identify the molecular basis for the origin of the mesodermal germ layer. In this chapter, we summarize what is known about gastrulation movements in cnidarians and ctenophores, with special emphasis on the phylogenetic distribution of gastrulation patterns and directions of ongoing research.

PHYLUM CNIDARIA

General Characteristics

The cnidarians are a phylum of highly successful and diverse forms consisting of two major groups, the basal anthozoans (Anthozoa [sea anemones, corals, sea pens, sea pansies, and cerianthids]) and the more derived medusozoans (Scyphozoa [many of the larger jellyfish species], Cubozoa [box jellies/sea wasps], and Hydrozoa [*Hydra*, hydroids, fire corals, Portuguese man-of-war, etc.]) (Fig. 1B). These animals are classified as cnidarians based largely on similarities in their body plans and the possession of unique cell types called cnidocytes (Hyman 1940; Brusca

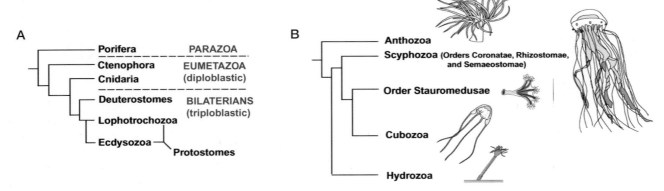

Figure 1. (*A*) The phylogenetic relationship of ctenophores and cnidarians to the bilaterians (Kim et al. 1999). (*B*) The phylogenetic relationships of the major cnidarian clades (Collins 2002).

and Brusca 1990; Ruppert and Barnes 1994). All cnidarians are diploblastic, lacking a well-defined mesodermal cell layer (Fig. 2A). The outer epidermis forms an epithelial sheet that is continuous with the gastrodermis, the cell layer lining the gastric cavity (or coelenteron). The coelenteron has a single opening that serves as both the mouth and anus, and is surrounded by rows of tentacles. In the gastrodermis, on tentacle surfaces, and occasionally in other areas, are specialized cells called cnidocytes. These cells are used to sting (nematocytes) or stick to (collocytes) prey and are characteristic of all members in this phylum. Between the epidermis and gastrodermis, cnidarians have a jelly-like layer called the mesoglea that may or may not contain mesenchyme-like cells (depending on the class). The gastrodermis in cnidarians is a bifunctional mesentodermal germ layer with each cell possessing both absorptive and epitheliomuscular functions. The apical surface of a gastrodermal epitheliomuscular cell is highly folded, increasing the absorptive area for greater uptake of organic materials. The basal region of the cell is modified for muscular activity. Cell processes extending from the cell's basal region are contractile and can generate planar forces along the epithelial sheet, either parallel to the oral–aboral axis (longitudinal muscles) or transverse to the body axis (circular muscles). The gastrodermis itself can be highly folded to form mesenteries that house epitheliomuscular cell processes and sites of gamete formation. Epitheliomuscular cells also appear in the tentacles and in the epidermis of more derived medusozoans.

Unlike most animals, cnidarians lack a centralized nervous system and polarized nerves. Instead, the nervous system is arranged in a diffuse nerve net. Cnidarians typically have two nerve nets, one lying basal to the ectoderm, and a second, less complex, nerve net underlying the gastrodermis. Nerve cell bodies lie within the epithelial layer of origin, and impulses propagate bidirectionally from synapses.

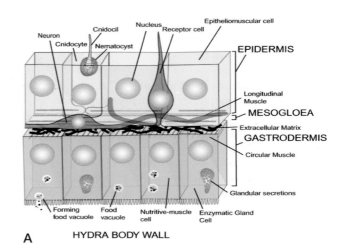

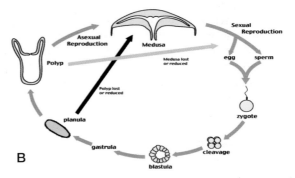

Figure 2. (*A*) Longitudinal section through the body wall of *Hydra* sp. The epidermis, mesogloea, and gastrodermis are found in all cnidarians. Reprinted, with permission of Brooks/Cole, a division of Thomson Learning: www.thomsonrights.com. (*B*) A generalized cnidarian life cycle. (*A*, Modified from Ruppert and Barnes 1994.)

Cnidarian Life History

Cnidarians have a complex life history (Fig. 2B). The sexes are generally separate (Hyman 1940), and embryogenesis consists of fertilization, cleavage, gastrulation, and morphogenesis to a benthic polyp (hydroid). In anthozoans, the polyp is the adult sexual form, and, in most cnidarians, the polyps are capable of extensive asexual reproduction, and some can make large clonal colonies (e.g., coral reefs). Some hydrozoan species have polymorphic colonies, that is, different morphological forms exist within the colony; each polyp is adapted to perform a specialized function (e.g., feeding, defense, reproduction). In most medusozoans, the benthic polyp is an immature asexual form that buds a free-swimming medusa ("jellyfish"). The medusa later reproduces sexually, releasing its eggs and/or sperm into the surrounding seawater. In more derived medusoids, the polyp stage is lost and the embryo develops directly to the medusoid stage (Fig. 2B). In other species, the medusa remains attached to the polyp as part of a multifunctional colony or is lost. Thus, both polyp and medusoid forms can be generated either sexually or asexually. It is not yet clear whether the molecular programs controlling these two very different patterning mechanisms are identical.

Cnidarian Development

Prior to fertilization, cnidarian eggs lack the primary animal–vegetal axis characteristic of triploblastic bilaterians at this developmental stage. In cnidarians, instead of being set up maternally, body axes are determined at the time of first cleavage (Freeman 1981). The site of sperm entry normally defines the site of the first unipolar cleavage and will become the oral end of the polyp, but this site is labile and can be changed experimentally (Freeman 1981). In many species, the oral pole is the site of gastrulation; however, in some it is difficult to identify "the" site of gastrulation. In general, cleavage in cnidarians is random, varying from individual to individual, although some embryos, usually direct-developing medusozoans, have relatively well-defined cleavage patterns through the first three divisions (Mergner 1957; Freeman 1983). Cleavage results in formation of a blastula (either a solid stereoblastula or a hollow coeloblastula) that later gastrulates to produce a ciliated, cigar-shaped stage called the planula. This swimming stage is often referred to as a planula "larva," but in most cases it does not feed, eventually settling and metamorphosing into a benthic polyp, with the posterior pole becoming the mouth/anus.

Cnidarian development produces several unique challenges to understanding gastrulation and germ layer formation, including the lack of a primary egg axis, the role of the cleavage program in establishing axial properties, the regulative nature of cnidarian development/regeneration, and the formation of a single bifunctional mesentoderm. Many forms of gastrulation are found in cnidarians. These are reviewed below, and the phylogenetic distribution of each is discussed in the following section.

How Cnidarians Gastrulate

Within the phylum Cnidaria, at least nine forms of gastrulation have been described. These include two forms of ingression (unipolar and multipolar ingression), three forms of delamination (blastula, morula, and syncytial delamination), invagination, epiboly, and at least two mixed forms of gastrulation (mixed delamination and forms combining ingression and invagination). Excellent reviews of these processes are found in Mergner (1971), Tardent (1978), Fioroni (1979), Siewing (1969), and Kume and Dan (1968). Much of what is known comes from Metschnikoff (1886).

Ingression

Ingression occurs in species with a coeloblastula and, occasionally, in groups forming a stereoblastula (Thomas et al. 1987). During ingression, the presumptive entodermal cells lose contact with neighboring blastodermal cells and individually migrate into the blastocoel. Cell migration, not tangential cell division, is responsible for formation of the presumptive entoderm. Ingression is the primary morphogenetic movement involved in both "unipolar ingression" (Fig. 3A) and "multipolar ingression" (Fig. 3B).

In embryos gastrulating by unipolar ingression, cells enter the blastocoel from the area that will become the oralmost region of the animal. In species with a coeloblastula, the blastula is originally spherical, but several hours prior to gastrulation the blastula becomes ciliated and elongates along the oral–aboral body axis. Oral cells become more columnar than aboral or aboral–lateral cells, and, at the onset of gastrulation, individual cells from oral areas begin to ingress and, later, those from oral–lateral areas (Byrum 2001a,b). In the ingressing cell, the nucleus shifts basally and the cell lengthens along its apical–basal cell axis. In some cases, these cells undergo drastic cell shape changes, constricting apically and expanding basally. The cell eventually loses contact with neighboring cells, sends out cell processes, and exhibits amoeboid movements. These cells first accumulate inside the blastocoel closest to the thickened end and continue to migrate in from the oral area until the blastocoel is completely occluded. The resulting post-gastrula stage has two cell layers: a central mass of loosely arranged cells that will later form entoderm, surrounded by an exterior cell layer that will become the ectoderm. The gastrovascular cavity forms when cells shift or

Unipolar Ingression

Multipolar Ingression

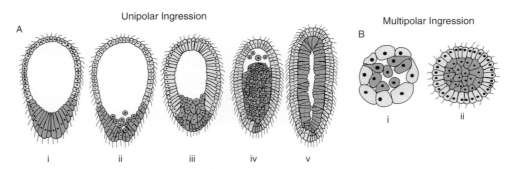

Figure 3. Modes of cnidarian gastrulation involving ingression of cells. (*A*) Unipolar ingression in the hydrozoan *Aequorea forskalea*. Cells migrate into the blastocoel from the oral pole of the embryo. (*i*) Preingression, (*ii*) early ingression, (*iii*) early mid-ingression, (*iv*) late ingression, and (v) planula. (*B*) Multipolar ingression in the hydrozoan *Aeginopsis mediterranea*. Cells migrate into the blastocoel from various positions in the blastoderm. (*i*) Blastula undergoing multipolar ingression. (*ii*) Post-gastrula. (*A*, Redrawn from Mergner 1971; *B*, modified from Metschnikoff 1886.)

disintegrate to form a central lumen. In some species, cells that do not contribute to the lumen separate from this structure and are utilized as food by the larva (Mergner 1971). Later (usually during metamorphosis) the archenteron fuses with the oral ectoderm to form the mouth.

In species gastrulating by multipolar ingression (Fig. 3B), cells enter the blastocoel in a similar manner, but simultaneously move into the blastocoel from all sides of the embryo. Again, the planula is a two-layered embryo consisting of a solid mass of presumptive entodermal cells surrounded by an exterior layer of ciliated presumptive ectoderm. Cells later reorganize to form the gastrovascular cavity. Multipolar ingression as the sole mode of gastrulation is far less common than unipolar ingression. It is most often seen accompanying blastula delamination in cases of mixed delamination (see below, Mixed Forms of Gastrulation).

Delamination

Cells can also enter the blastocoel by delamination. In these species, the blastomeres contributing to the entoderm undergo mitotic division with the spindles oriented perpendicular to the embryo surface. One descendant of the dividing cell remains in the outer layer and the other enters the blastocoel. In some cases (i.e., *Geryonia fungiformis*) the cells divide simultaneously and quickly adhere to each other to establish a gastrovascular cavity (Mergner 1971). In other species (i.e., *Liriope mucronata*) individual cells delaminate at different times, filling the blastocoel gradually (Metschnikoff 1886). These cells later assimilate to form the gastrovascular cavity. This latter form is frequently seen (e.g., in *L. mucronata*) when delamination occurs in addition to ingression (see below, Mixed Forms of Gastrulation).

Among the Cnidaria, delamination can occur at different stages of development. In "blastula delamination"

(referred to as primary delamination, Metschnikoff [1886]), the two layers form during the coeloblastula stage (Fig. 4A). Blastula delamination commonly occurs in association with ingression, but, in cnidarians, it is rarely the only form of gastrulation.

In species that gastrulate by "morula delamination" (referred to as secondary delamination by Metschnikoff [1886]), the embryo never forms a large blastocoel cavity. Instead, cells delaminate very early in development, producing a stereoblastula. Delamination is not simultaneous, but occurs at different times in individual cells. Instead of forming the presumptive entoderm during the blastula stage, these cells are produced much earlier, during cleavage, forming a solid, two-layered embryo (Fig. 4B). The presumptive entodermal and ectodermal cells are similar in appearance, and differences only become obvious after differentiation.

The plasticity of gastrulation movements in cnidarians is displayed in the hydrozoan *Gonothyraea loveni* (Wulfert 1902). In *Gonothyraea*, the embryos develop in a sessile gonophore (an asexually produced medusoid that remains attached to the polyp). Usually these embryos develop into a coeloblastula with a large blastocoel and subsequently gastrulate by multipolar ingression. Occasionally, however, when embryos are crowded in the gonophore, there is little space for the blastocoel, and the blastoderm delaminates during cleavage, forming a two-layered embryo by morula delamination.

A third form of delamination, "syncytial delamination," has been observed in several species of the hydrozoan genus *Eudendrium* (Fig. 4C) (Mergner 1957, 1971). These animals have extremely yolky eggs, forcing the nuclei to divide without cytokinesis. After fertilization, the zygote nucleus divides such that one nucleus remains near the embryo surface in the presumptive oral region (peripheral nucleus) and the other nucleus moves toward the interior (interior nucleus).

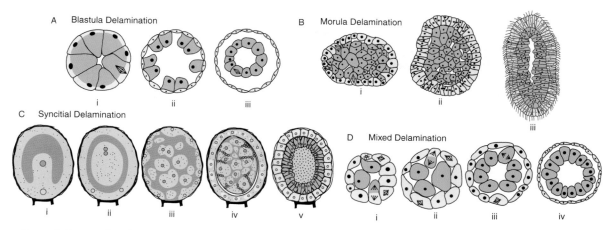

Figure 4. Forms of cnidarian gastrulation involving cell delamination. (*A*) Blastula delamination in the hydrozoan *Geryonia fungiformis*. Cells divide perpendicular to the outer surface to form a new entodermal layer. (*B*) Morula delamination in the hydrozoan *Clava squamata*. Similar to blastula delamination, but cell divisions begin earlier, preventing formation of a coeloblastula. Cells do not divide simultaneously to form two distinct layers, but division occurs at different times in different cells. (*i*) Late "morula" with two delaminating cells. (*ii*) Delamination is complete when the entoderm and ectoderm have differentiated and the mesolamella has been secreted. (*iii*) Hatching-stage planula. (*C*) Syncytial delamination in the hydrozoan *Eudendrium racemosum*. This process is similar to morula delamination, but embryos are syncytial during early cleavage and gastrulation. (*i*) First nuclear division in the zygote occurs perpendicular to embryo surface. One nucleus moves toward the interior of the embryo and the other remains at the exterior surface. (*ii–iii*) Interior nuclei divide to form other interior nuclei, and exterior nuclei divide to form other exterior nuclei. Interior nuclei will later form entoderm and vitellophages. Exterior nuclei will become ectoderm. (*iv*) Mesoglea begins to form and cytoplasm accumulates around peripheral nuclei (*yellow*). Cell boundaries begin to form around exterior cells. (*v*) Older gastrula in which cellularization is complete. Ectoderm and entoderm have formed and yolk is being absorbed in the center of the embryo. (*D*) Mixed delamination (both delamination and multipolar ingression take place) in the hydrozoan *Liriope mucronata*. (*i–ii*) Early stages of mixed delamination. Note that cells migrate into the blastocoel and radial mitotic cleavages are occurring. (*iii*) The presumptive entodermal cells interact to form an entodermal vesicle. (*iv*) Differentiation of the two germ layers has occurred and the mesolamella forms. Green areas represent entoderm or presumptive entoderm. Yellow areas are ectoderm or presumptive ectoderm. (*D-iv*, Redrawn from Balfour 1880; remaining illustrations redrawn from Mergner 1971.)

This is *the* delamination event in syncytial delamination. Descendants of the interior daughter nucleus become entoderm or vitellophages, and descendants of the peripheral nucleus become ectoderm. By the 32-cell stage, the peripheral nuclei are distributed throughout exterior portions of the embryo, and a ball of yolk surrounds each nucleus in the embryo (both interior and peripheral nuclei). By late cleavage, cytoplasm begins to accumulate around the peripheral nuclei, first in the oral areas, and, by the pre-gastrula stage, around the more aboral peripheral nuclei. The peripheral nuclei surrounded by plasma zones separate from the central region containing the interior nuclei and yolk (presumptive entoderm). Cell boundaries begin to appear between the peripheral cells, and the mesoglea forms, first in oral regions and later in more aboral regions, distinguishing the ectodermal layer from the presumptive entoderm. After the embryo becomes a swimming planula, cells in the interior undergo cellularization, and the gastric cavity forms from a cleft in the entodermal cell mass.

The anthomedusa *Turritopsis nutricula* shares some similarities with *Eudendrium* in its post-cleavage stages (Brooks and Rittenhouse 1907; Mergner 1971). In *Turritopsis*, cell boundaries are lost after an initial cleavage phase. Cellularization of the peripheral cells in this syncytium occurs just as it does in *Eudendrium* and, again, the entoderm is not cellularized until the embryo is a swimming planula. Despite similarities to later stages of syncytial delamination, the early stages of gastrulation in *Turritopsis* are not consistent with syncytial delamination. In actuality, *Turritopsis* gastrulates by morula delamination (this just occurs in a syncytium at later stages).

Invagination

Invagination is thought to be similar to ingression, but it involves groups of cells acting together, whereas ingression involves cells acting individually. During invagination (Fig. 5A), the cells forming entoderm undergo shape changes similar to ingressing cells, constricting apically and expanding basally, but the invaginating cells do not detach from their neighbors. Because these shape changes occur simultaneously, the blastoderm buckles into the blastocoel, form-

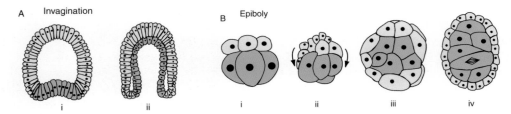

Figure 5. Two different modes of cnidarian gastrulation: invagination and epiboly. (*A*) Invagination in the anthozoan *Metridium marginatum*. (*i*) Early gastrulation, invagination starts with a flattening of the oral pole, (*ii*) late gastrula, invagination complete. (*B*) Epiboly in the hydrozoan *Aglaura hemistoma*. (*i*) Eight-cell embryo. (*ii*) Presumptive ectodermal cells begin to move over the presumptive entodermal cells. (*iii*) A later "morula" stage. (*iv*) The two-layered larva. Green areas represent entoderm or presumptive entoderm. Yellow areas are ectoderm or presumptive ectoderm. Invagination redrawn from Tardent (1978). Epiboly redrawn from Metschnikoff (1886).

ing the presumptive entoderm. Additional morphogenetic activities may also be important, but detailed cellular analysis has not been done. Invagination is the primary mode of gastrulation in many anthozoans and scyphozoans, but has not been observed in hydrozoans. The hydrozoan *Halocordyle disticha* was reported to gastrulate by invagination (Martin and Thomas 1984; Martin 1987), but investigators later found that the detected indentation was formed by unipolar ingression of a stereoblastula, not by invagination (Thomas et al. 1987).

In anthozoans that invaginate, the embryo is often very yolky. This yolk, as well as anucleate portions of cells that delaminate from the blastoderm (e.g., *Metridium marginatum, Pachycerianthus multiplicatus, Urticina crassicornis, Actinia equina, A. bermudensis, Bolocera turdiae,* and *Sagartia troglodytes*), is thought to nourish the embryo (Mergner 1971). In some invaginating embryos, nutritive materials in the blastocoel are transferred to the lumen of the archenteron. In *Pachycerianthus multiplicatus*, yolk flows through intercellular interstices in archenteron cells to accumulate inside the archenteron. Here these products are digested and absorbed by the embryo (Nyholm 1949; Siewing 1969).

The appearance of the archenteron differs in species depending on the yolk content of the embryo (Tardent 1978). If low levels of yolk are present, the archenteron usually forms a single extension from the blastoderm (i.e., *Metridium marginatum*). When yolk levels are high, the archenteron may appear bifurcated because the yolk impedes archenteron extension.

Epiboly

Epiboly (Fig. 5B) has been observed in the hydrozoans *Aglantha digitale* (Freeman 1983) and *Aglaura hemistoma* (Metschnikoff 1886). (Metschnikoff referred to this as secondary delamination, but based on similarities to gastrulation in *Aglantha*, it is more likely that *Aglaura* gastrulates by epiboly; see discussion in Freeman paper.) In *Aglantha*, blas-

tomeres of the 4-cell stage are equal in size. During formation of the 8-cell embryo, cleavage is unequal, resulting in the production of four large yolk-laden macromeres overlaid by four smaller micromeres. Cells derived from the macromeres later contribute to the entoderm, and the micromeres form ectoderm. At the onset of gastrulation, micromere descendants (presumptive ectodermal cells) flatten, and the micromere layer spreads over the large, yolky macromere descendants as a coherent cell layer. Simultaneously, macromere descendants are also thought to delaminate, each cell dividing to form an external ectodermal cell (free of endoplasmic granules) and an internal entodermal cell.

Mixed Forms of Gastrulation

In many species, gastrulation involves more than one morphogenetic activity. Hydrozoans often gastrulate by mixed delamination, a form of gastrulation in which both multipolar ingression and delamination occur (Fig. 4D). Mixed delamination can begin early in development: at the 16-cell stage (*Hydractinia echinata;* Bunting 1894), the 64-cell stage (*Cordylophora lacustris*; Morgenstern 1901), or the 128-cell stage (some species of *Hydra*; Brauer 1891). Mixed delamination is seen "associated with a distinct trend towards shortened development as a result of advancing entoderm formation to increasingly earlier cleavage stages" (Mergner 1971). In the hydrozoan *Liriope mucronata*, germ layer formation begins after the 16-cell stage (Metschnikoff 1886). Whole blastoderm cells migrate into the blastocoel during early development, forming a stereoblastula or a blastula with a very little blastocoel space. Later, the external blastoderm cells divide perpendicular to the cell surface, producing one daughter cell that will contribute to the entoderm and a second daughter cell that remains in the ectoderm. Eventually, the interior cells differentiate from those on the exterior surface and coalesce to form a gastrovascular cavity suspended in the blastocoel.

Often, species that gastrulate by mixed delamination are mistakenly reported to gastrulate by multipolar ingression. This has occurred several times in the hydrozoan literature. In fact, it may be that the mixed reports concerning modes of gastrulation in *Hydra* are due to this sort of misinterpretation (Brauer 1891; Mergner 1971; Martin 1997).

A second mixed form of gastrulation involves both ingression and invagination. In several scyphozoan species, ingression precedes invagination. During cleavage, blastoderm cells migrate into the blastocoel, where they are thought to act as nutritive sources (Siewing 1969; Mergner 1971). These cells degenerate before the onset of invagination; thus the gastrovascular cavity is only derived from the invaginating archenteron. The anthozoan *Halcampa duodecimcirrata* also utilizes both ingression and invagination. In this case, however, multipolar ingression begins after invagination is under way, and the ingressing cells contribute to the entoderm. Cells ingress from the external blastoderm, cross a yolky region, and form an epithelium that joins the archenteron (Nyholm 1949).

In other cases, gastrulation by invagination is followed by the ingression of cells from the archenteron tip. The ingressed cells later coalesce to produce an epithelial layer that integrates with the archenteron to form a gastrovascular cavity. In this case, the embryo is said to gastrulate by a combination of invagination and polar ingression. This gastrulation mode is seen in the scyphozoan *Linuiche unguicula* (Conklin 1909) and may also occur in the anthozoan *Nematostella vectensis* (personal observations).

PHYLOGENETIC DISTRIBUTION OF CNIDARIAN FORMS OF GASTRULATION

Different forms of gastrulation appear multiple times throughout the phylum Cnidaria. This is especially true of blastula, morula, and mixed delamination, as well as unipolar ingression. Other forms of gastrulation (e.g., epiboly and syncytial delamination) are much more restricted in distribution. Thus, it appears that mechanisms of gastrulation can be easily coopted and/or reacquired by species within an order. The presence of multiple forms of gastrulation in the phylum Cnidaria begs the question of how gastrulation evolved in this group. Using data gathered from the literature and recent molecular phylogenies of the Medusozoa (Collins 2002) and the Anthozoa (Daly et al. 2002), the phylogenetic distribution of different modes of gastrulation was analyzed (Fig. 6, Table 1; for a summary of the species and references used to generate these figures, see files Supplemental Text and Gastrulation Types, www.gastrulation.org).

The class Anthozoa is most likely to be the basal group within the Cnidaria (Fig. 1B), and forms of gastrulation are more homogeneous in the Anthozoa than in any other cnidarian group. Most anthozoans (12 of 17 described

species) gastrulate by invagination. In two other species, invagination is combined with ingression of cells, either from the tip of the archenteron (invagination + unipolar ingression) or from the external blastoderm (invagination + multipolar ingression). Because invagination is so common in the Anthozoa, both Haeckel (1874) and Mergner (1971) proposed that invagination was used by the earliest cnidarians. It is interesting, however, that two other forms of gastrulation (morula and blastula delamination) appear in groups thought to be basal anthozoans. Morula delamination occurs in the only alcyonacean studied, and blastula delamination is utilized by pennatulaceans (only two species have been studied). Investigators have pointed out that morula and blastula delaminations are rare in anthozoans, usually restricted to species with yolky eggs (Mergner 1971; Tardent 1978). Perhaps other alcyonaceans and pennatulaceans utilize invagination, and the species in Figure 6 evolved delamination secondarily. A more comprehensive examination of the basal taxa may help to resolve questions about the nature of early cnidarian gastrulation.

Most scyphozoans also gastrulate by invagination (10 of 14 described species). Although this form of gastrulation is not reported in the Stauromedusae, it is the primary mode of gastrulation in the remaining orders (the Coronatae, Rhizostomae, and Semaeostomae). Often, invagination is preceded by ingression, but in most cases, the ingressing cells do not form entoderm but are believed to be nutritive. The remaining scyphozoans gastrulate either by unipolar (all Stauromedusae sampled) or multipolar (members of the Coronatae and Semaeostomae) ingression. Why these species deviate is not known, but based on the egg diameters and gastrulation modes of nine scyphozoans, Berrill (1949) hypothesized that some modes of gastrulation could be predicted on the basis of egg size. He said that scyphozoans with small eggs should undergo unipolar ingression and those with larger eggs should invaginate. Although his hypothesis is interesting, most of Berrill's cases of unipolar ingression were subsequently found to utilize invagination during later developmental stages. In addition, the Stauromedusae form a distinct clade from the Scyphozoa (Collins 2002). A more plausible explanation for the absence of invagination in the Stauromedusae may be that these animals evolved from ancestors that did not utilize invagination.

Invagination has not been reported in the class Cubozoa (although the small sample size in our data set, *n* = 2 (see files Supplemental Text and Gastrulation Types, www.gastrulation.org), makes it impossible to say whether this is a true trend). In one case, *Carybdea rastonii*, entoderm forms by multipolar ingression and, in the other, *Tripedalia cystophora*, gastrulation occurs by blastula delamination.

The class Hydrozoa is the most derived and diverse cnidarian class. More forms of gastrulation occur in this group than in any other cnidarian class. Modes of gastrula-

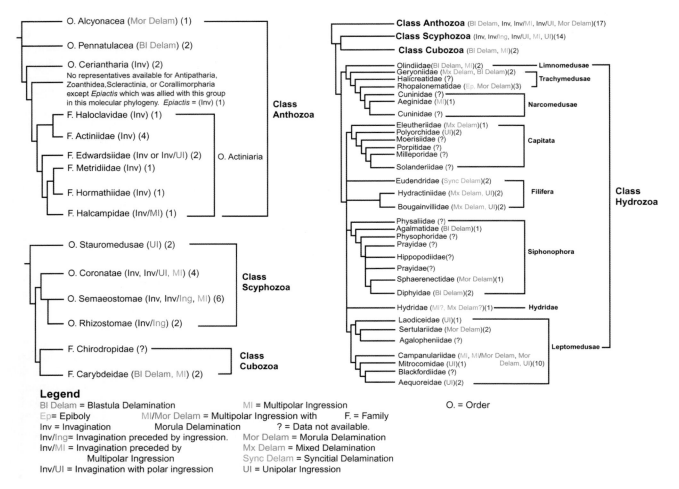

Legend
BI Delam = Blastula Delamination
Ep= Epiboly
Inv = Invagination
Inv/Ing= Invagination preceded by ingression.
Inv/MI = Invagination preceded by
        Multipolar Ingression
Inv/UI = Invagination with polar ingression

MI = Multipolar Ingression
MI/Mor Delam = Multipolar Ingression with
        Morula Delamination
Morula Delamination
Mor Delam = Morula Delamination
Mx Delam = Mixed Delamination
Sync Delam = Syncitial Delamination
UI = Unipolar Ingression

F. = Family
? = Data not available.

O. = Order

Figure 6. Modes of gastrulation plotted on current cnidarian phylogenies (Collins 2002; Daly et al. 2002) to search for evolutionary trends in the evolution of cnidarian development. Note that invagination is the primary form of gastrulation in anthozoans and scyphozoans, but this form of gastrulation is absent in the cubozoans and hydrozoans. Epiboly and syncytial delamination each appear one time in cnidarian families, but other forms (unipolar ingression, multipolar ingression, mixed delamination, blastula delamination, and morula delamination) appear many times throughout the phylogeny, and some of these forms may have been lost and reacquired more than once (based on the examination of species-level phylogenies).

tion include blastula delamination, morula delamination, syncytial delamination, mixed delamination, unipolar ingression, multipolar ingression, and epiboly. Invagination has not been described in hydrozoans.

The class Hydrozoa consists of two clades, the Trachylina and the Hydroidolina (Collins 2002). Trachylina is more basal and consists of the orders Limnomedusae, Narcomedusae, Trachymedusae, Actinulidae, and Laingiomedusae. Hydroidolina consists of five groups: Capitata, Filifera, Hydridae, Leptomedusae, and Siphonophorae.

Within the Trachylina, many species gastrulate by some form of delamination: blastula, morula, or mixed delamination. Multipolar ingression has also been observed (in a narcomedusa and a limnomedusa), but it is possible that these are actually cases of mixed delamination. Reports of

multipolar ingression are often later found to be cases of mixed delamination. The only instances of epiboly in cnidarians ($n = 2$) occur in the Trachymedusae (*Aglantha digitale* and *Aglaura hemistoma*). Gastrulation by syncytial delamination or unipolar ingression has not been observed.

In the Hydrolina (Capitata, Filifera, and Leptomedusae), germ layer formation often involves unipolar ingression. Like the trachylinans, many hydrolinans also use forms of delamination. This is especially true of the siphonophores, which only gastrulate by morula or blastula delamination. Reports of blastula delamination were not found in other hydrolinans, but mixed delamination (blastula delamination + multipolar ingression) is quite common in the Capitata, Filifera, and Hydridae. Morula delamination occurs in the Filifera, Leptomedusae, and Siphonophora,

Table 1. *Modes of gastrulation in cnidarians by order*

	Unipolar ingression (%)	Multipolar ingression (%)	Mixed delamination (%)	Blastula delamination (%)	Morula delamination (%)	Syncytial delamination (%)	Invagination (%)	Epiboly (%)
Class Hydrozoa								
Trachylina								
O. Actinulidae (2)					100			
O. Limnomedusae (2)		50		50				
O. Narcomedusae (1)		100						
O. Trachymedusae (5)			20	20	20			40
Hydroidolina								
O. Capitata (7)	43		57					
O. Filifera (12)	33		25		25	17		
O. Hydridae (1)		50	50					
O. Leptomedusae (22)	57	9	4.5		29.5			
O. Siphonophora (6)				50	50			
Class Scyphozoa								
(Paraphyletic?)								
O. Coronatae (2)		25					75 (1 of 3 w/polar ingression)	
O. Rhizostomae (2)							100 (all preceded by ingression)	
O. Semaeostomae (6)		17					83 (4 of 5 preceded by ingression)	
O. Stauromedusae (2)	100							
Class Cubozoa								
O. Cubmedusae (2)		50			50			
Class Anthozoa								
sCl. Ceriantipatharia								
O. Ceriantharia (2)							100	
sCl. Hexacorallia/Zoantharia								
O. Actiniaria (12)							100 (2 of 12 also use unipolar or multipolar ingression)	
sCl. Octocorallia/Alcyonaria								
O. Alcyonacea (1)					100			
O. Pennatulacea (2)				100				

Numbers in parentheses after the order indicate the number of species that have been studied within that order. In each order, the percent showing each mode of gastrulation is given. In the Siphonophora, for example, gastrulation has been reported in six species, 50% of these cases gastrulated by blastula delamination and 50% gastrulated by morula delamination. A detailed version of this table appears as the file Gastrulation Types; www.gastrulation.org.

and the only known cases of syncytial delamination occur in the Filifera (Family Eudendriidae). Careful examination of species- and family-level phylogenies suggests that forms of gastrulation can appear multiple times in the same clade (Byrum 2001b).

Life-history characteristics (e.g., brooding, direct development, egg size) undoubtedly influence the variable patterns seen throughout the Cnidaria. To test Berrill's hypothesis that egg size correlates to mode of gastrulation, we collected egg-size data from the literature and compared egg sizes to modes of gastrulation (Fig. 7). Raw data and a summary of average egg sizes for each class are presented in the Web site data (see file Egg Size, www.gastrulation.org). Egg sizes varied considerably within each mode of gastrulation. For example, unipolar ingression can occur in embryos with eggs as small as 30 μm in diameter or as large as 180

μm. Invagination occurs in eggs 60 μm to 750 μm in diameter, and blastula delamination occurs in eggs 70 μm to 600 μm in diameter. Because egg sizes for the Anthozoa tend to be much larger than those for others, comparisons were done within a single class. In the hydrozoa, unipolar ingression tends to occur in small to medium size eggs, but not usually in very large eggs. Our primary conclusion from this preliminary work is that many modes of gastrulation can occur over a large range of egg sizes. Certain types of gastrulation may tend toward larger or smaller egg sizes, but it will be necessary to collect more data to detect true trends.

In addition, life-history data were compiled for hydrozoan species to determine whether the mode of gastrulation correlates to certain life-history patterns (Table 2 and files Life History Data, Extra Brooding Data, and Brooding, www.gastrulation.org). The parameters included were

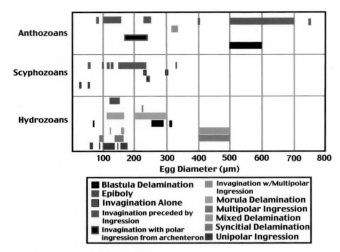

Figure 7. Egg sizes of cnidarian species using different forms of gastrulation. Detailed information and references can be found in supplemental information on the Web site (file Egg Size, www.gastrulation.org).

gonophore type (free medusae or sessile gonophore), parental care (free spawning or brooding), and type of life cycle (polyp to medusa, polyp to polyp, or medusa to medusa). Unipolar ingression is more common in species that produce both free medusae and polyps in their life cycle (these species also tend to free-spawn rather than brood). Morula delamination (and possibly mixed and blastula delamination) is more common in species that form sessile gonophores or lack a medusa stage (these species usually brood). Thus, within the Hydrozoa, life-history pattern appears to loosely correlate to some modes of gastrulation (unipolar ingression and morula delamination). It is not clear whether the life cycles lacking a polyp stage correlate to a particular mode of gastrulation.

In summary, the factors influencing evolution of cnidarian gastrulation may become more apparent with careful characterization of this process in weakly sampled taxa and generation of highly resolved cnidarian phylogenies. The prevalence of invagination in anthozoans suggests that this may be the ancestral mode of gastrulation for cnidarians, but additional taxon sampling is clearly required to unravel the influence of life-history strategies on modes of gastrulation.

MODERN INVESTIGATIONS OF CNIDARIAN GASTRULATION

Most studies concerning cnidarian gastrulation were carried out in the late 1800s and early 1900s. Metschnikoff's descriptions of gastrulation in hydrozoans and a few scyphozoans (1886) have been particularly influential. Other important investigations include those of Kowalevsky (1874), Claus (1883), Kowalevsky and Marion (1883), Müller-Calé (1913), Uchida (1926, 1927), and later, Nyholm (1943, 1949), Mergner (1971), Weiler-Stolt (1960), and Bodo and Bouillon (1968) (for a more comprehensive bibliography see file Supplemental Text, www.gastrulation.org).

Gastrulation in the Hydrozoan *Phialidium (Clytia) gregarium*

Experimental embryological techniques have recently been used to characterize cell activities during gastrulation in *P. gregarium* (Byrum 2001a,b), a pattern previously categorized as occurring by unipolar ingression. These experiments utilized fluorescent labeling techniques (Fig. 8), confocal, scanning electron, and transmission electron microscopy to show:

Table 2. *Comparison of life history aspects to mode of gastrulation in hydrozoans (shown as number of species)*

	Polyp and free medusa present in life cycle	Polyp present, but medusa stage reduced (sessile gonophores) or lost	Medusa present, but polyp stage lost
Unipolar ingression	17.5[a]	0	0
Multipolar ingression	2.5[a]	0.5[b]	1.0
Mixed delamination	3.0	6.0	1.0
Morula delamination	1.0	11.5[b]	3.0
Blastula delamination	1.0	3.0	1.0
Syncytial delamination	0	2.0	0
Epiboly	0	0	2.0
Total # of species	25.0	23.0	8.0

[a]The value 0.5 was used here because the individual used an odd intermediate form of unipolar ingression that spreads throughout the blastoderm to become multipolar ingression.

[b]The value 0.5 was used here because the species used either multipolar ingression or morula delamination (depending on the conditions).

Unipolar ingression is only found in species that have both a polyp and a free medusa in the life cycle. Morula delamination (and possibly mixed and blastula delamination) is more common when the medusa stage is reduced or lost.

1. Most cells contributing to the entoderm originate from the oral-most quarter of the embryo and enter the blastocoel by ingression. These ingressing cells change shape, elongating and occasionally constricting apically while expanding basally.

2. An external and internal fibrillar extracellular matrix (ECM) is present in gastrulating embryos. This fibrillar network may provide a sink for signaling molecules (Dinbergs et al. 1996; Schroeder et al. 2003) or act as a structural framework for cell adhesion.

3. Delamination may play a minor role in entoderm formation. During gastrulation, most cell divisions are oriented parallel to the embryo surface, but occasionally individual cells delaminate (as detected by perpendicular orientation of mitotic spindles).

Delaminating cells are not restricted to one area of the blastoderm but can occur in any area. Delaminating cells may contribute to the entoderm of *P. gregarium* or they may be reintegrated into the ectoderm at a later time. Further studies will be needed to test this. Low levels of delamination may also occur in other species thought to gastrulate solely by unipolar ingression.

4. Morphometric analyses indicate that the increase in length-to-width ratio of the embryo during gastrulation may involve convergent extension (see Chapter 19).

5. Cell-sorting studies showed that, during gastrulation, cells in oral areas of the embryo acquire higher adhesive affinities than do those from aboral regions. These changes in adhesive affinity may allow oral cells to occupy more central areas of the embryo relative to the aboral cells.

MOLECULAR INVESTIGATIONS OF GASTRULATION IN CNIDARIANS

Molecular investigations into gastrulation and germ layer formation in cnidarians are in their infancy. This is due in part to the fact that few "model" cnidarian species have been developed. The two primary groups being investigated are hydrozoans and anthozoans. The most commonly studied hydrozoans are the freshwater hydroid, *Hydra,* and the marine hydroid *Podocoryne carnea. Hydra* species are not amenable to studies of gastrulation because most species propagate largely by asexual budding and when they do produce embryos, they encyst in a hard egg shell for several weeks or months (Martin 1997; Martin et al. 1997). In contrast, a great deal of progress on the molecular basis of development has been made in *Podocoryne.* This species has a biphasic life cycle, and much research has focused on formation of the pelagic medusa. The two anthozoans examined include the coral *Acropora* and the starlet sea anemone *Nematostella vectensis. Acropora,* like many corals, undergoes massive spawning events on one or a few nights a year, thus making it difficult to study experimentally. *Nematostella,* on the other hand, is relatively easy to breed in the laboratory and is becoming a cnidarian model for molecular developmental studies.

Studies of gastrulation in *Nematostella* have focused on two areas, understanding the origins of the site of gastrulation, and understanding the genes involved in mesentoderm germ layer formation. As mentioned previously, cnidarian embryos have no meaningful polarity until first cleavage, yet only hours later, these embryos establish the cells that will move to the center of the embryo and give rise to the bifunctional mesentodermal germ layer (gastroder-

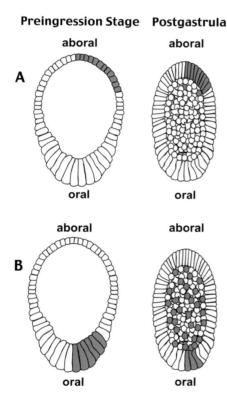

Figure 8. Cell-marking experiments demonstrating that the majority of the entoderm formed by unipolar ingression is derived from oral and oral–lateral regions of the hydrozoan *Phialidium gregarium.* During early cleavage, a single cell was labeled with Di-I, and distribution of this lipophilic dye (which is not transferred to unlabeled cells) was compared in the same individual prior to unipolar ingression (pre-ingression stage) and at the end of gastrulation (post-gastrula). (*A*) When the dye was in aboral areas of the preingression embryo, few marked cells contributed to the entodermal mass by the end of gastrulation. (*B*) When cells from the oral-most quarter of the embryo were marked, the resulting post-gastrula usually contained many marked entodermal cells (Byrum 2001a,b).

mis). This layer forms primarily from an invagination of the epithelial sheet at the site of the future mouth (Fig. 9); however, the results of molecular studies suggest that this may not be the only mechanism (see below).

One likely source of axial information involved in mesentoderm formation is the Wnt signaling pathway. Evidence for the involvement of the Wnt signaling pathway, including in situ hybridizations for several pathway members in *Hydra* (*Wnt*, *TCF/LEF*, and β-*catenin*), have shown that these genes are expressed in the hypostome/oral region of both budding and regenerating animals (Hobmayer et al. 2000). Unfortunately, these genes have not been examined during *Hydra* embryogenesis. Asexual budding and regeneration occur after both entodermal and ectodermal germ layers are already present, and so it is not appropriate to assess the role of these genes in the generation of germ layers, but their localized expression patterns are certainly suggestive of a role in axial organization.

Wikramanayake et al. (2003) have investigated the expression of β-catenin protein in the anthozoan *Nematostella* and have obtained evidence supporting a causal role of the canonical Wnt signaling pathway in the establishment of endomesoderm in developing embryos. Injection of GFP-labeled β-*catenin* RNA into uncleaved *Nematostella* embryos reveals that β-catenin protein is selectively degraded at the future aboral pole and accumulates in the nuclei of cells destined to enter the blastopore at gastrulation (Fig. 10). Furthermore, injection of constructs designed to interfere with nuclear β-catenin function

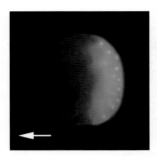

Figure 10. Fluorescent image of a *N. vectensis* embryo (5 hpf) in which synthetic GFP-labeled β-*catenin* mRNA was injected into the uncleaved embryo. Although the protein was initially expressed throughout the entire embryo, by this stage, protein expression is restricted to one hemisphere and protein is accumulating in nuclei of presumptive mesentodermal cells (A.H. Wikramanayake and M.Q. Martindale, unpubl.). The future swimming direction is indicated by the arrow, and the oral pole is oriented toward the right.

inhibits gastrulation. These data reveal that the mechanism for generating new germ layers was already present in cnidarians, without the production of a definitive mesodermal layer. Because cnidarian embryos have no overt animal–vegetal axis prior to first cleavage, the selective stabilization of β-catenin at the future site of gastrulation must be established by a polarizing event after fertilization. There is evidence that the site of polar body formation is the source of at least one diffusible molecule (Freeman and Miller 1982), so the Wnt pathway appears to be a good candidate for establishment and/or elaboration of the primary embryonic axis and the future site of gastrulation.

Other members of the Wnt signaling pathway are still being elucidated in cnidarians, but one downstream target of the Wnt pathway, *Brachyury*, has been studied in both *Hydra* (Technau and Bode 1999) and *Nematostella* (Scholz and Technau 2003). In these cnidarians, *Brachyury* is expressed around the future oral pole prior to gastrulation, and appears to be highly conserved as a marker for the blastopore in all metazoans, although its role is unclear. It is likely that Brachyury is involved in changes in cell behavior associated with gastrulation movements (Wilson et al. 1995; Wilson and Beddington 1997; Tada and Smith 2000; Gross and McClay 2001).

GENES INVOLVED IN MESENTODERM FORMATION IN OTHER METAZOANS

Readers of this volume will be impressed with the degree to which genes involved in mesentoderm formation appear to be conserved in protostomes and deuterostomes (bilaterians). Orthologs to many of these genes have been recovered in cnidarians and are currently being investigated to determine what role they may play in animals without true

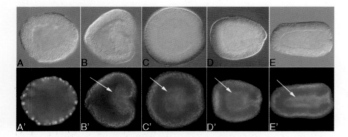

Figure 9. Gastrulation in *Nematostella vectensis*. DIC images along the top row (*A–E*) are oriented with the direction of swimming facing the left side and the presumptive oral pole toward the right. Corresponding Hoechst-stained fluorescent images (*A′–E′*) indicate positions of nuclei in both ectodermal and entodermal layers. (*A*) Blastula, 4.5 hpf (hours postfertilization at 23°C). Cells remain on exterior surrounding yolky fragments in the blastocoel. (*B*) By 10.5 hpf, embryo has invaginated at future oral pole and cells begin to move into the yolky interior from archenteron tip (presumably by polar ingression). (*C*) Embryo ~16 hpf slightly flattened under coverslip showing presumptive entodermal cells loosely aggregated in center of the embryo. A connection to the oral pole remains. (*D*) A 22-hpf embryo beginning to lengthen and decrease in diameter. Cells that migrated into the interior appear more cohesive, forming an early entodermal layer. (*E*) By 33 hpf, length-to-width ratio of the embryo has increased, aiding directed motility.

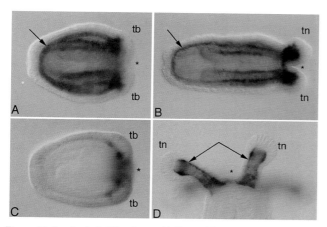

Figure 11. In situ hybridizations with "muscle" genes *Tropomyosin* (*A, B*) and *Muscle actin* (*C, D*) in young polyps of *N. vectensis*. The presumptive mouth is indicated with an *. The expression of both genes is restricted to the entodermal gastrodermis (*arrows*), particularly in the tentacles (tn) and tentacle buds (tb). If the staining reaction is allowed to proceed in *C* and *D*, the entire gastrodermis becomes heavily labeled, indicating an increased transcript level in the developing tentacles.

mesoderm (Groger et al. 1999; Spring et al. 2000, 2002; M.Q. Martindale et al., unpubl.). For example, in the anthozoan *Nematostella vectensis*, where myoepithelial cells reside primarily in the gastrodermis, genes involved in the formation of myofibrils (e.g., *Tropomyosin, Muscle actin*) are expressed in the gastrodermal layer (Fig. 11). A number of developmental regulatory genes are also expressed in the developing gastrodermis (Fig. 12). In *Nematostella*, genes such as *Snail, Dpp*, and an *HNF-forkhead* ortholog are first expressed at the blastula stage in presumptive mesentoderm before it enters the blastocoel, suggesting that these genes play a role in early invagination and/or cell fate specification. In *Nematostella* as well as another anthozoan (the

coral *Acropora*), *Dpp* is expressed asymmetrically around the blastopore lip (Hayward et al. 2002), indicating that molecular differences begin at this time in an axis orthogonal to the oral–aboral axis. Expression of *GATA* and *Mef2*, on the other hand, occurs in individual cells over relatively broad domains (Fig. 12C,D). *GATA*, but not *Mef2*, continues to be expressed in the cells that form mesentoderm, suggesting that these cells may represent a distinct population of mesentodermal precursors that ingress individually rather than invaginate as a sheet of cells. Others, such as *Twist, Otx*, and *Mox*, are expressed in discrete regions of mesentodermal cells after they enter the blastocoel and may be involved in regional differentiation of gastrodermal tissue (M.Q. Martindale et al., unpubl.).

Even before the exact functions of these genes are determined in *Nematostella*, these early molecular studies reveal two important issues. First, many of the genes normally associated with mesendoderm development in bilaterian embryos are expressed in the mesentodermal layer in this anthozoan cnidarian. It will be exciting to see whether the details of the formation and patterning of this layer in *Nematostella* yield any insight into the situation seen in triploblastic forms. For example, two simplistic scenarios could be presented for the origins of a new "middle" germ layer (Fig. 13). Signaling by overlying ectoderm might polarize mesentodermal cells such that transcription factors controlling "muscle" genes become localized to the basal surface where myoepithelial processes are found, whereas "gut" genes are found toward the lumen of the coelenteron (Fig. 13, top). Alternatively, distinct cell types expressing different networks of genes in the gastrodermis might sort out into different layers (Fig. 13, bottom).

Second, there is an underlying molecular complexity to the tissues in developing cnidarian embryos that is not apparent at the morphological level. For example, cells of

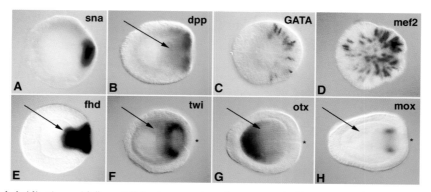

Figure 12. In situ hybridizations with "muscle" developmental regulatory genes in *N. vectensis* embryos (Martindale et al. 2004). The presumptive oral pole is located to the right in all images. Some genes such as *Snail, Dpp*, and *HNF-forkhead* (*A, B, E*) are expressed at the future oral pole before mesentodermal cells enter the blastocoel and continue to be expressed through the polyp stage. Whereas these genes appear to be expressed in involuting cells, *GATA* (*C*) and *Mef2* (*D*) are expressed in individual cells that appear to ingress into the blastocoel. *Twist, Otx*, and *Mox* expression begins later, after cells have already entered the blastocoel, and are expressed in discrete regional domains of the gastrodermis. Arrows indicate gastrodermal layer.

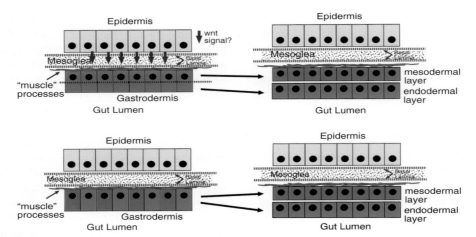

Figure 13. Hypothetical scenarios for the formation of a new mesodermal germ layer from a bifunctional mesentoderm. (*Top*) Factors responsible for formation of an absorptive lining of the gut and myoepithelial basal contractile machinery might become localized to opposite sides of the functional gastrodermis by an oriented cell division. Perhaps this signal is a Wnt-like diffusible substance secreted by the epidermis, either before or after the presumptive gastrodermis enters the blastocoel. Alternatively, the gastrodermis might be subdivided into distinct regions that are reassembled into discrete endodermal and mesodermal layers (*bottom*).

the aboral gastrodermis express *Otx* whereas those around the mouth do not, and a small number of cells around the mouth express *Twist* and *Mox* but not those aborally. It will be interesting to determine whether this molecular complexity corresponds to hitherto unrecognized physiological differences in cell-type-specific function of gastrodermal cells and whether this molecular complexity is the fodder for increasing cellular complexity in an evolutionary sense.

PHYLUM CTENOPHORA

Ctenophores are abundant gelatinous marine carnivores found throughout the world. The ctenophore body plan bears some resemblance to that of cnidarians. The longitudinal axis of both taxa is the oral–aboral axis, and the gut is functionally blind, with the mouth also serving as the anus (Fig. 14). Virtually all ctenophores are pelagic with a simple

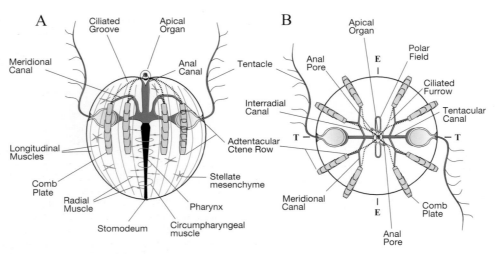

Figure 14. Ctenophore body plan in lateral (*A*) and aboral (*B*) views. The major longitudinal axis is the oral–aboral axis with the apical sensory organ located at the aboral pole. These animals swim through the water via coordinated metachronal beating of the comb rows and feed by catching prey in their sticky tentacles. Note that, unlike cnidarians, there are definitive contractile muscle cells dispersed throughout the body between the ectoderm and endoderm (*green*). (E) Esophageal plane or rotational symmetry; (T) tentacular plane of rotational symmetry.

life history; embryos develop directly into miniature adults. Pelagic adult ctenophores bear little resemblance to pelagic medusoid cnidarians, however, with different modes of locomotion, prey capture, and nervous system organization.

One of the most fundamental differences between adult ctenophores and cnidarians, however, is the presence of definitive muscle cells in ctenophores. These contractile cells are present in distinct locations of the body; parietal cells under the ectodermal body wall, longitudinal muscles in the ectodermally derived pharynx/esophagus, visceral muscles associated with the endodermal gut, circumpharyngeal and radial muscles that course through the mesoglea, as well as muscle cells in the tentacles. Muscle cell progenitors never form a definitive mesodermal layer, hence the diploblastic designation of both ctenophores and cnidarians (Hyman 1940). Nevertheless, both endodermal and mesodermal cell fates become segregated from ectoderm prior to the gastrulation process (Martindale and Henry 1999).

Ctenophore development is also significantly different from that of cnidarians (for review, see Martindale and Henry 1997). Unlike cnidarians that have highly variable cleavage programs and a regulative mode of development, ctenophores have a phylum-specific cleavage program (Fig. 15) that not only generates identifiable cells, but is also causally involved with the precocious specification of blastomere fate. There is a single mode of gastrulation in ctenophores: epiboly. Micromeres born at the vegetal pole move toward the future oral pole on the surface of larger macromeres (Figs. 15–17). The blastopore forms where these cells meet and corresponds to the position of the mouth. Intracellular fate mapping of identified blastomeres (Martindale and Henry 1999) has revealed that these micromeres generate exclusively ectodermal derivatives. Mesoderm arises from a second group of micromeres born at the oral (animal) pole of the macromeres (Fig. 15), and endoderm is generated by the remaining macromeres. Fate

mapping has also shown that differences in the embryonic origin of some muscle cells arise as early as the 4-cell stage; two blastomeres give rise to circumesophageal muscle cells and the other two do not (Martindale and Henry 1995, 1999). Thus, the origin of mesoderm is patterned early in ctenophore development and is endomesodermal, as it is in other basal metazoans (Henry et al. 2000). This developmental information is particularly important because many ctenophore muscle cells are intimately integrated into ectodermal derivatives, such as the tentacles, esophagus/pharynx, and outer epidermis. This may have led investigators to mistakenly assume that ctenophore muscles were ectomesodermal in origin.

Cell-labeling experiments (J.Q. Henry and M.Q. Martindale, unpubl.) show that the descendants of aboral micromeres generate clones that spread orally, perhaps by a convergent extension-like behavior (Figs. 15–17). M and E blastomeres give rise to two and three successive rounds (respectively) of oral micromeres (Fig. 15) that produce discrete regions of the juvenile body plan (Martindale and Henry 1999). Clones derived from aboral micromeres divide before and during their migration toward the oral pole. Labeled clones initially stay together during early gastrulation movements, but then separate to give rise to discrete structures such as ctene rows and tentacle bulbs (Fig. 17D). Therefore, epiboly movements toward the oral pole during gastrulation explain some, but not all, of the morphogenetic movements of these micromere descendants.

Essentially nothing is known mechanistically about the process of gastrulation in ctenophores. For example, it is not known whether the micromeres migrate on the surface of the macromeres or whether the macromeres play an active role in micromere movements. Cell divisions have been observed in migrating micromeres (Fig. 16), but it is not known whether they are causally involved in the gastrulation process. Cell-labeling experiments have shown

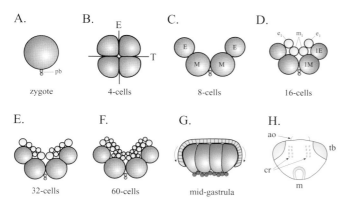

Figure 15. Schematic diagram of ctenophore development. All views seen from lateral view except *B*. (*A*) Unipolar first cleavage normally occurs under the polar bodies and is causally involved in establishing the oral pole. (*B*) Each of the first four blastomeres gives rise to one quadrant of the adult, with two blastomeres (shaded *red*) giving rise to different complements of mesodermal derivatives than the other two. (*C*) The E and M lineages are generated at the 8-cell stage, and each gives rise to distinct sets of muscle cells. (*D–F*) Small micromeres, generated at the aboral pole, give rise exclusively to ectodermal derivatives such as the comb plate cilia. (*G*) Ectodermal cells migrate over the larger macromeres by epiboly toward the future mouth. Mesodermal micromeres (*red*) are born at the site of the blastopore/mouth and migrate internally. (*H*) Within 24 hours, the juvenile adult is formed. Tentacle bulbs (tb) initially form superficially but later invaginate. The apical organ forms opposite the mouth. Ectodermal esophagus/pharynx connects the mouth to the endodermal gut.

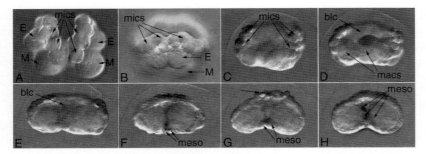

Figure 16. DIC micrographs of the gastrulation process in the lobate ctenophore *Mnemiopsis leidyi* (oral pole faces down in all images). (*A*) During early cleavage, ectodermal micromeres are born at the aboral pole of the E and M macromeres. (*B*) These micromeres form a ring circling the larger mesentodermal macromeres. (*C*) Ectodermal micromeres divide and form a cap of cells prior to gastrulation movements. (*D*) Micromeres form a single layer of cells creating a blastocoel (blc) as they begin to move over the macromeres. (*E*) Macromeres are completely covered by proliferating micromeres. (*F*) Micromeres at the aboral pole continue to divide while mesodermal micromeres are born at the oral pole. (*G*) Overlying micromeres continue to round up and divide (*arrow*) as the presumptive mesodermal cells begin to sink into the blastopore. (*H*) Mesodermal micromeres migrate up into the blastocoel and disperse.

extensive mixing between descendants of the distinct micromere lineages (Martindale and Henry 1999), so it is likely that these movements are a consequence of their precocious determination of cell fate.

SUMMARY/CONCLUSIONS

The phylogenetic position of ctenophores and cnidarians near the base of the metazoan tree, and the fact that neither group has a well-defined mesodermal germ layer, suggest that there are opportunities to understand the evolutionary origin of this important germ layer and to learn more about ancestral mechanisms of gastrulation and cell rearrangements. Many genes involved in mesodermal development are present in the gastrodermis of anthozoan embryos, suggesting that mesoderm originally evolved from endodermal precursors, a result supported by fate mapping in ctenophores (Martindale and Henry 1999) and acoel flatworms (Henry et al. 2000). An understanding of the details of the genetic regulatory circuits that control expression of mesendodermal genes in cnidarians and ctenophores will undoubtedly yield important insights into the origins of mesodermal structures in other metazoans.

Of critical importance is identification of the direction of evolutionary change. Is the extensive variation of developmental patterns seen in the Cnidaria a palimpsest of all the patterns seen in triploblastic forms, or do these developmental patterns represent independent variants from a relatively unconstrained "simple" program in "simple" animals? Is the precocious determination of cell fates and gastrulation by epiboly, as displayed in ctenophores, pleisiomorphic for triploblastic ancestors or the result of a recent evolutionary bottleneck secondarily reducing variation? Are the muscle cells of ctenophores homologous to those of triploblasts? If

ctenophores display ancestral developmental features (muscle cells, gastrulation by epiboly) and cnidarians are the sister group to bilaterians, did cnidarians lose the ability to form distinct mesodermal cell types? We clearly need to

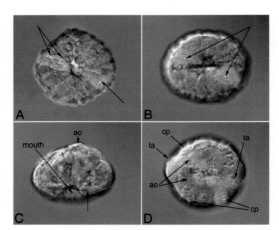

Figure 17. DIC and fluorescent micrographs of *M. leidyi* embryos at various stages of gastrulation in which two of the four ectodermal e_1 micromeres were labeled by the intracellular injection of Di-I (*red*) dissolved in soybean oil (Martindale and Henry 1999). The residual oil droplets are visible in panels *A* and *B* (*arrows*). *A*, *B*, and *D* are seen from the aboral pole, *C* from lateral view. (*A*) Cells are stacked on top of one another following the initial cleavage period. (*B*) As cells divide and become smaller, they form an epithelium over the larger macromeres. (*C*) Lateral view showing the expansion of the e_1 micromere clones over the larger macromeres from the apical organ (ao) at the aboral pole to a region close to, but not up to the presumptive mouth (*arrow*). Other micromere clones (e_2, e_3, m_2) give rise to the esophagus and edges of the mouth (Martindale and Henry 1999). (*D*) Labeled clones do not just give rise to spatially localized regions of the juvenile adult, but disperse to give rise to regions of the apical organ (ao), comb rows (cr), and components of the tentacle apparatus (ta).

know more cellular and molecular details of the early development of these groups and their evolutionary relationships in order to make reliable hypotheses about the origin (or loss?) of mesodermal tissues.

ACKNOWLEDGMENTS

This work was funded, in part, by National Science Foundation grant IBN-0110532 to Dr. Athula Wikramanayake, and grants from the National Aeronautics and Space Administration and NSF to M.Q.M. We also thank Jonathan Q. Henry, Kathleen Kendall, Carolina Livi, and the MBL Embryology course for contributing to figures for this paper.

REFERENCES

Balfour F.M. 1880. On the structure and homologies of the germinal layers of the embryo. *Q. J. Microsci. Sci.* **20:** 247–273.

Berrill N.J. 1949. Developmental analysis of scyphomedusae. *Biol. Rev.* **24:** 393–410.

Bodo F. and Bouillon J. 1968. Étude histologique du développement embryonnaire de quelques hydroméduses de Roscoff: *Phialidium hemisphaericum* (L.), *Obelia* sp. Péron et Lesueur, *Sarsia eximia* (Allman), *Podocoryne carnea* (Sars), *Gonionemus vertens* Agassiz. *Cah. Biol. Mar.* **9:** 69–104.

Brauer A. 1891. Über die Entwicklung von Hydra. *Z. Wiss. Zool.* **52:** 167–216.

Brooks W.K. and Rittenhouse S. 1907. On *Turritopsis nutricola* (McCrady). *Proc. Boston Soc. Nat. Hist.* **33:** 429–460.

Brusca R.C. and Brusca G.J. 1990. *Invertebrates.* Sinauer, Sunderland, Massachusetts.

Bunting M. 1894. The origin of the sex-cells in *Hydractinia* and *Podocoryne*, and the development of *Hydractinia*. *J. Morphol.* **9:** 203–236.

Byrum C.A. 2001a. "A cellular analysis of gastrulation by unipolar ingression in the hydrozoan *Phialidium* (*Clytia*) *gregarium*." Ph.D. thesis. University of Texas at Austin.

———. 2001b. An analysis of hydrozoan gastrulation by unipolar ingression. *Dev. Biol.* **240:** 627–640.

Claus C. 1883. Untersuchungen über die Organisation und Entwicklung der Medusen. Prag und Leipzig.

Collins A.G. 2002. Phylogeny of Medusozoa and the evolution of cnidarian life cycles. *J. Evol. Biol.* **15:** 418–432.

Conklin E.G. 1909. Two peculiar actinian larvae from Tortugas Florida. *Carnegie Inst. Washington* **103:** 171–186.

Daly M., Lipscomb D.L., and Allard M.W. 2002. A simple test: evaluating explanations for the relative simplicity of the Edwardsiidae (Cnidaria: Anthozoa). *Evolution* **56:** 502–510.

Dinbergs I.D., Brown L, and Edelman E.R. 1996. Cellular response to transforming growth factor beta 1 and basic fibroblast growth factor depends on release kinetics and extracellular matrix interactions. *J. Biol. Chem.* **271:** 29822–29829.

Fioroni V.P. 1979. Abänderungen des Gastrulationsverlaufs und ihre phylogenetische Bedeutung. In *Erlanger Symp. Ontogenie Evolutionsforsch: Ontogenie unid Phylogenie* (ed. R. Siewing), pp. 100–119. Parey, Hamburg, Germany.

Freeman G. 1981. The cleavage initiation site establishes the posterior pole of the hydrozoan embryo. *Wilhelm Roux's Arch.* **190:** 123–135.

———. 1983. Experimental studies on embryogenesis in hydrozoans (*Trachylina* and *Siphonophora*) with direct development. *Biol. Bull.* **165:** 591–618.

Freeman G. and Miller R.L. 1982. Hydrozoan eggs can only be fertilized at the site of polar body formation. *Dev. Biol.* **94:** 142–152.

Groger H., Callaerts P., Gehring W.J., and Schmid V. 1999. Gene duplication and recruitment of a specific tropomyosin into striated muscle cells in the jellyfish *Podocoryne carnea*. *J. Exp. Zool.* **285:** 378–386.

Gross J.M. and McClay D.R. 2001. The role of Brachyury (T) during gastrulation movements in the sea urchin *Lytechinus variegatus*. *Dev. Biol.* **239:** 132–147.

Haeckel E. 1866. *Generelle Morphologie der Organismen: Allgemeine Grundzüge der organischen formen-Wissenschaft, mechanisch begründet durch die von Charles Darwin reformite Descendenz-Theorie.* Georg Reimer, Berlin.

———. 1874. Die Gastraea-Theorie, die phylogenetische Klassification des Tierreiches und Homologie der Keimblätter. *Jena Z. Naturwiss* **8:** 1–55.

Hayward D.C., Samuel G., Pontynen P.C., Catmull J., Saint R., Miller D.J., and Ball E.E. 2002. Localized expression of a dpp/BMP2/4 ortholog in a coral embryo. *Proc. Natl. Acad. Sci.* **99:** 8106–8111.

Henry J.Q., Boyer B., and Martindale M.Q. 2000. The unique developmental program of the acoel flatworm, *Neochildia fusca*. *Dev. Biol.* **220:** 285–295.

Hobmayer B., Rentzsch F., Kuhn K., Happel C.M., von Laue C.C., Snyder P., Rothbacher U., and Holstein T.W. 2000. WNT signalling molecules act in axis formation in the diploblastic metazoan Hydra. *Nature* **407:** 186–189.

Huxley T. 1849. On the anatomy and affinities of the family of the Medusae. *Phil. Trans. R. Soc.* **139:** 413–434.

Hyman L.H. 1940. *The invertebrates: I. Protozoa through* Ctenophora. McGraw-Hill, New York.

Kim J., Kim W., and Cunningham C.W. 1999. A new perspective on lower metazoan relationships from 18S rDNA sequences. *Mol. Biol. Evol.* **16:** 423–427.

Kowalesvky A. 1874. Untersuchung über die Entwicklung der Coelenteraten. *Göttingen Nachrichten*, 154–158.

Kowalevsky A. and Marion A.F. 1883. Histoire embryogénique des Alcyonaires. *Ann. Mus. Hist. Nat. Marseille*: **1.**

Kume M. and Dan K. 1968. *Invertebrate embryology.* NOLIT Publishing, Belgrade, Yugoslavia.

Lankester E. 1873. On the primitive cell layers of the embryo as the basis of genealogical classification of animals, and on the origin of vascular and lymph systems. *Ann. Mag. Nat. Hist. Ser. 4* **11:** 321–328.

Martin V.J. 1987. A morphological examination of gastrulation in a marine athecate hydrozoan. *Biol. Bull.* **173:** 324–334.

———. 1997. Cnidarians, the jellyfish and hydras. In *Embryology: Constructing the organism* (ed. S.F. Gilbert and A.M. Raunio), pp. 57–86. Sinauer, Sunderland, Massachusetts.

Martin V. and Thomas M.B. 1984. An analysis of early development in *Pennaria tiarella*. *Am. Zool.* **23:** 797.

Martin V.J., Littlefield C.L., Archer W.E., and Bode H.R. 1997. Embryogenesis in hydra. *Biol. Bull.* **192:** 345–363.

Martindale M.Q. and Henry J.Q. 1995. Diagonal development: Establishment of the anal axis in the ctenophore *Mnemiopsis leidyi*. *Biol. Bull.* **189:** 190–192.

———. 1997. Ctenophorans, the comb jellies. In *Embryology:*

Constructing the organism (ed. S.F. Gilbert and A.M. Raunio), pp. 87–111. Sinauer, Sunderland, Massachusetts.

———. 1999. Intracellular fate mapping in a basal metazoan, the ctenophore *Mnemiopsis leidyi*, reveals the origins of mesoderm and the existence of indeterminate cell lineages. *Dev. Biol.* **214:** 243–257.

Martindale M.Q., Pang K., and Finnerty J.R. 2004. Investigating the origins of triploblasty: "Mesodermal" gene expression in a diploblastic animal, the sea anemone, *Nematostella vectensis* (Phylum, Cnidaria; Class Anthozoa). *Development* (in press).

Medina M., Collins A.G., Silberman J.D., and Sogin M.L. 2001. Evaluating hypotheses of basal animal phylogeny using complete sequences of large and small subunit rRNA. *Proc. Natl. Acad. Sci.* **98:** 9707–9712.

Mergner H. 1957. Die Ei- und Embryonalentwicklung von *Eudendrium racemosum* Cavolini. *Zool. Jahrb. Abt. Anat. Ontog. Tiere* **76:** 63–164.

———. 1971. Cnidaria. In *Experimental embryology of marine and fresh-water invertebrates* (ed. G. Reveberi), pp. 1–84. North-Holland Publisher, Amsterdam, The Netherlands.

Metschnikoff E. 1886. Embryologische Studien an Medusen: Ein Beitrag zur Genealogie der Primitive Organe. *Wien*, 45–71.

Morgenstern P. 1901. Untersuchungen über die Entwicklung von *Cordylophora lacustris*. *Z. Wiss. Zool.* **70**.

Müller-Calé K. 1913. Zur Entwicklungsgeschichte einiger Thecaphoren. *Zool. Jahrb. Abt. Anat. Ontog. Tiere* **37:** 83–112.

Nielsen C., Scharff N., and Eibye-Jacobsen D. 1996. Cladistic analyses of the animal kingdom. *Biol. J. Linn. Soc.* **57:** 385–410.

Nyholm K.G. 1943. Zur Entwicklung und Entwicklungsbiologie der Ceriantharien und Actinien. *Zool. Bijdr.* **22:** 87–248.

———. 1949. On the development and dispersal of *Athenaria actinia* with special reference to *Halcampa duodecimcirrata* M.Sars. *Zool. Bijdr.* **27:** 467–505.

Peterson K.J. and Eernisse D.J. 2001. Animal phyology and the ancestry of bilaterians: inferences from morphology and 18S rDNA gene sequences. *Evol. Dev.* **3:** 170–205.

Podar M., Haddock S.H., Sogin M.L., and Harbison G.R. 2001. A molecular phylogenetic framework for the phylum Ctenophora using 18S rRNA genes. *Mol. Phylogenet. Evol.* **21:** 218–230.

Ruppert E.E. and Barnes R. 1994. *Invertebrate zoology*. Saunders, New York.

Scholz C.B. and Technau U. 2003. The ancestral role of Brachyury: Expression of NemBra1 in the basal cnidarian *Nematostella vectensis* (Anthozoa). *Dev. Genes Evol.* **212:** 563–570.

Schram F.R. 1991. Cladistic analysis of metazoan phyla and the placement of fossil problematica. In *The early evolution of metazoa and the significance of problematic taxa* (ed. A. Simonetta and S. Conway Morris), pp. 35–46. Cambridge University Press, Cambridge, United Kingdom.

Schroeder J.A., Jackson L.F., Lee D.C., and Camenisch T.D. 2003. Form and function of developing heart valves: Coordination by extracellular matrix and growth factor signaling. *J. Mol. Med.* **81:** 392–403.

Siewing R. 1969. Keimblätterentwicklung. In *Lehrbuch der Vergleichenden Entwicklungsgeschichte der Tiere*, pp. 134–299. Verlag Paul Parey, Hamburg, Germany.

Spring J., Yanze N., Josch C., Middel A.M., Winninger B., and Schmid V. 2002. Conservation of Brachyury, Mef2, and Snail in the myogenic lineage of jellyfish: a connection to the mesoderm of bilateria. *Dev. Biol.* **244:** 372–384.

Spring J., Yanze N., Middel A.M., Stierwald M., Groger H., and Schmid V. 2000. The mesoderm specification factor twist in the life cycle of jellyfish. *Dev. Biol.* **228:** 363–375.

Tada M. and Smith J.C. 2000. *Xwnt11* is a target of *Xenopus* Brachyury: Regulation of gastrulation movements via Dishevelled, but not through the canonical Wnt pathway. *Development* **127:** 2227–2238.

Tardent P. 1978. Entwicklungsperioden. In *Morphogenese der Tiere: Einleitung zum Gesamtwerk Morphogenetische Arbeitsmethoden und Begriffssysteme (Coelenterata, Cnidaria.)* (ed. F. Seidel), pp. 199–222. Fischer Verlag, Jena, Germany.

Technau U. and Bode H.R. 1999. *HyBra1*, a *Brachyury* homologue, acts during head formation in *Hydra*. *Development* **126:** 999–1010.

Thomas M.B., Edwards N.C., and Norris T.A. 1987. Gastrulation in *Halocordyle disticha* (Hydrozoa, Athecata). *Int. J. Invertebr. Reprod. Dev.* **12:** 91–102.

Uchida T. 1926. The anatomy and development of rhizostome medusa, *Mastigias papua* L. Agassiz, with observations on the phylogeny of Rhizostomae. *J. Fac. Sci. Tokyo Univ. Sec. 4 Zool.* **1:** 45–95.

———. 1927. Studies on Japanese Hydromedusae. 1. Anthomedusae. *J. Fac. Sci. Tokyo Univ. Sect. 4 Zool.* **1:** 145–241.

Wainright P.O., Hinkle G., Sogin M.L., and Stickel S.K. 1993. Monophyletic origins of the metazoa: An evolutionary link with fungi. *Science* **260:** 340–342.

Weiler-Stolt B. 1960. Über die Beutung der Interstitiellen Zellen für die Entwicklung und Fortpflanzung Mariner Hydroiden. *Wilhelm Roux's Arch.* **152:** 398–454.

Wikramanayake A.H., Hong M., Lee P.N., Pang K., Byrum C., Bince J.M., Xu R., and Martindale M.Q. 2003. An ancient role for nuclear -catenin in the evolution of axial polarity and germ layer segregation. *Nature* **426:** 446–450.

Wilson V. and Beddington R. 1997. Expression of T protein in the primitive streak is necessary and sufficient for posterior mesoderm movement and somite differentiation. *Dev. Biol.* **192:** 45–58.

Wilson V., Manson L., Skarnes W.C., and Beddington R.S. 1995. The T gene is necessary for normal mesodermal morphogenetic cell movements during gastrulation. *Development* **121:** 877–886.

Wulfert J. 1902. Die Embryonalentwicklung von *Gonothyraea loveni* Allm. *Z. Wiss. Zool.* **71:** 296–327.

Gastrulation in Nematodes

J. Nance[1] and J.R. Priess[1,2]

[1]*Department of Basic Sciences, Fred Hutchinson Cancer Research Center, Seattle, Washington 98109;* [2]*Howard Hughes Medical Institute, Seattle, Washington 98109*

INTRODUCTION

Nematodes comprise a class of some of the most successful animals; they are found in abundance in a wide range of habitats, and there are estimates of as many as a million species. Different species can be one millimeter to several feet in length as adults, yet all share a remarkably simple and efficient body plan (Fig. 1). The body wall is an unsegmented monolayer of epidermal cells that secrete the cuticle covering the animal. Body movement is generated by four longitudinal rows of muscle cells that underlie the epidermal cells. The digestive tract consists of a mouth, a muscular pharynx (esophagus), an intestine, and a rectum that form a tube through the center of the body cavity. In coelomate animals, the embryonic body cavity is called a coelom and is separated from the outer body wall and inner digestive tract by a mesodermally derived epithelium. Nematodes are classified as pseudocoelomates because the embryonic body cavity (the pseudocoelom) contacts the digestive tract directly. Locomotion and behavior are controlled by a simple central nervous system, with ganglia in the head and tail that are connected by longitudinal nerve tracts.

Cell movements during gastrulation establish the basic nematode body plan. We address here two basic features of gastrulation in the model species *Caenorhabditis elegans*. First, how do adherent, apparently equivalent cells create the internal space that becomes the blastocoel? Second, how do surface cells move into the blastocoel? Finally, we compare gastrulation in *C. elegans* with that in two distantly related nematodes.

GASTRULATION IN *C. ELEGANS*

Overview of Embryogenesis

C. elegans embryos are enclosed by a vitelline membrane and eggshell. Thus, all morphogenetic events involve a redistribution of, rather than an increase in, cellular mass. Embryogenesis is remarkably rapid; a newly fertilized egg

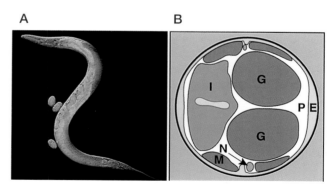

Figure 1. Nematode body plan. (*A*) Photomicrograph of adult hermaphrodite *Caenorhabditis elegans* next to three embryos. The head is upper left and the adult is ~1 mm in length. (*B*) Schematic cross section near the middle of an adult showing the major organs; dorsal is up. Two lobes of the gonad (G) are shown in purple, the epidermis (E) in yellow, the intestine (I) in green with a central microvilli-lined lumen in gray, each of the four rows of muscles (M) in red, and the longitudinal nerve tracts (N) in blue. The pseudocoelom (P) is shown in white and surrounds internal organs, and the cuticle (*thick black circle*) lines the exterior of the animal. The photomicrograph was provided by Greg Hermann; the schematic diagram was drawn from electron micrograph data available from Wormatlas (www.wormatlas.org).

hatches as a first-stage larva in only 12–15 hours, and the early cell cycles last only ~15–20 minutes. In contrast, embryonic cell cycles in some other nematodes can last hours or days. The cell division pattern and the pattern of cell differentiation are essentially invariant in *C. elegans* (Sulston et al. 1983; Schnabel et al. 1997). The first few divisions generate a set of cells called founder cells (Fig. 2A). The E founder cell produces the entire intestine (endoderm), P4 produces only germ cells, and D produces only muscles. The AB, MS, and C founder cells produce a mixture of ectodermal and mesodermal cell types.

Gastrulation involves two basic types of cell movements: ingression and epiboly. During early embryogenesis, the endodermal, mesodermal, and germ cell precursors ingress from the ventral surface into the blastocoel. The blastocoel becomes the body cavity, or pseudocoelom. Most of the epidermal (skin) cells are born on the dorsal surface of the embryo. Following the ingression of the last mesodermal cells, the epidermal cells spread over the surface of the embryo by epiboly and meet at the ventral midline, thus encasing the body in skin. The epidermal cells contract around the circumference of the body, changing the body from an ellipsoidal ball of cells into a long thin tube.

Formation of the Blastocoel

During the first few embryonic cleavages, the innermost surfaces of cells separate from each other to generate a small internal cavity or blastocoel (Fig. 3C) (Nance and Priess 2002). The blastocoel reaches its maximum size at the 26-cell stage of embryogenesis, and even then is less than half the volume of a typical ingressing cell. Thus, in order for cells to enter the interior of the embryo, it may be essential that neighboring cells redistribute their mass over the outer surface of an ingressing cell (see below). As the innermost, or basal, surfaces of cells separate to form the blastocoel, the lateral surfaces remain tightly adherent. Separation of the basal surfaces occurs as early as the 4-cell stage of embryogenesis when tiny discontinuous gaps appear between the basal surfaces of cells. Thus, understanding how an embryonic cell can differentiate between its various surfaces is central to understanding how the blastocoel forms.

Cell Polarity

Early *C. elegans* embryos lack the types of specialized cellular junctions or basement membranes that typically subdivide the surfaces of epithelial cells into apical, lateral, or basal domains. For example, HMR-1/E-cadherin is a component of adherens junctions in the epithelial cells of late embryos but is distributed uniformly across the inner surfaces of early embryonic cells (Krieg et al. 1978; Costa et al. 1998; Nance and Priess 2002). Cell isolation and cell recombination experiments have shown that embryos have a great

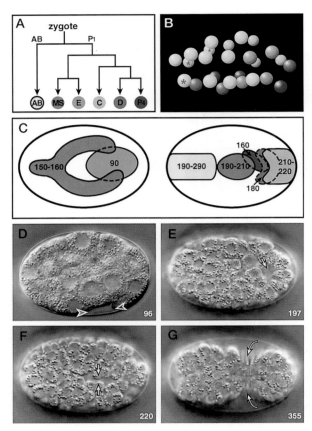

Figure 2. Cell lineages and gastrulation. (*A*) Lineage diagram of early cell divisions that produce the founder cells. One founder cell (AB) is born at the 2-cell stage; the remaining five founder cells descend from P1, the sister of AB. (*B*) Lateral view (dorsal is up) of a three-dimensional model of nuclei at the 26-cell stage. All descendants of a founder cell are colored as in *A*. Descendants of AB that produce ingressing cells are indicated with red asterisks, whereas AB descendants that remain on the surface are shown in unmarked yellow. (*C*) The ventral surface of an embryo shown at two successive times in development. Groups of cells that ingress are colored according to origin, and the times of ingression are listed. The left diagram indicates ingression of the E daughters (*green*) and a subset of MS descendants (*red*; described later in this text as "wishbone cells"). (*D–G*) Photomicrographs of representative stages of gastrulation; times at the lower right of each panel are in minutes after the 2-cell stage. (*D*) 28-cell embryo; lateral view as in *B*. An MS descendant (*red arrowhead*; MSap) and P4 (*purple arrowhead*) are shown spreading across the apical surfaces of the E daughters (*green asterisks*). (*E*) Ventral view showing MS "central" descendants (*red asterisks*) and a subset of C descendants (*orange asterisks*) prior to ingression. The arrow indicates the cleft created by ingression of the D descendants. (*F*) Ventral view showing part of the cleft created by ingression of the MS descendants; neighboring cells (*arrows*) will eventually move to cover the cleft. (*G*) Ventral view showing epiboly of the epidermal cells (*arrows*) that eventually cover remaining cells on the ventral surface. Anterior is left in all panels depicting embryos or nuclei. (Modified, with permission, from Nance and Priess 2002 [© Company of Biologists Ltd.].)

deal of plasticity in determining where the blastocoel forms, and hence, in determining which cell surfaces ultimately become apical, lateral, or basal (Nance and Priess

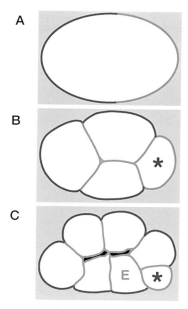

Figure 3. Asymmetry of PAR proteins and formation of the blastocoel. Panels illustrate lateral sections through a 1-cell embryo (*A*), a late 4-cell embryo (*B*), and an 8-cell embryo (*C*). Anterior is left, dorsal is up. The PAR proteins undergo dramatic redistributions during the first few cleavages as described in the text. The locations of the "anterior" PAR proteins are indicated in blue and the "posterior" PAR proteins in pink. Note that the distribution of the PAR proteins in the germ-cell precursors (*purple asterisks*) differs from that of the somatic precursors. The blastocoel is shown in black. The daughters of the E founder cell are the first cells to ingress.

2002). For example, if the AB founder cell is isolated from a 2-cell embryo and grown in culture, it generates a ball of cells surrounding a central blastocoel. Moreover, when two embryos at the 4-cell stage are joined head to head, juxtaposing their apical surfaces, an ectopic blastocoel forms at the point of contact. This result indicates that a presumptive apical surface can be respecified through cell contacts to become basal, and implies a similar respecification of the presumptive basal and lateral surfaces.

The distribution of the PAR proteins in *C. elegans* provides evidence for cell-surface differentiation during the first few cleavages. The PAR proteins were identified originally for their role in anterior/posterior polarity of the embryo (for review, see Bowerman 1998). Shortly after fertilization of the egg, anterior PAR proteins (PAR-3, PAR-6, and PKC-3) associate with the anterior cortex of the zygote while posterior PAR proteins (such as PAR-2) associate with the posterior cortex (Fig. 3A). The anterior PAR proteins, and homologous proteins in other animals, form a complex that is required for certain aspects of cell polarity in *C. elegans, Drosophila,* and vertebrates (for review, see Ohno 2001; Wodarz 2002).

During the 2-cell and early 4-cell stages, the anterior PAR proteins are associated uniformly with the entire cortex of each cell except for the germ-line precursor. These

proteins disappear from the cortex at regions of cell contact late in the 4-cell stage, but remain at the contact-free (apical) surface (Fig. 3B,C) (Etemad-Moghadam et al. 1995; Tabuse et al. 1998; Hung and Kemphues 1999; Nance and Priess 2002). In a reciprocal pattern, the posterior protein PAR-2 associates with the cortex only at sites of cell contact (the lateral and basal surfaces) (Boyd et al. 1996; Nance and Priess 2002). When embryos are recombined as described above, the PAR proteins redistribute to the appropriate new outer/inner surfaces (Nance and Priess 2002). Thus, cell contacts appear to cause the anterior/posterior PAR proteins to switch to an outer/inner, or apicobasal, polarity. This apicobasal PAR polarity persists through subsequent divisions as the blastocoel forms and the first embryonic cells ingress (see below).

Genetic analysis of PAR function in apicobasal polarity is complicated by their essential role in earlier anterior–posterior polarity. To circumvent this problem, the apical proteins PAR-3 and PAR-6 were tagged with a peptide from the PIE-1 protein (Nance et al. 2003). This peptide, called the ZF1 domain, promotes protein degradation after the 2-cell stage, and thus after the requirement for PAR-3 and PAR-6 in anterior–posterior polarity (Reese et al. 2000). *par* mutant embryos that express PAR[ZF1] hybrid proteins are called *par(ZF1)* embryos and have normal anterior/posterior polarity (Nance et al. 2003). After the 4-cell stage, however, the PAR[ZF1] proteins disappear and the basal PAR-2 protein inappropriately associates with the apical surface. Separations develop between lateral cell surfaces that normally occur only between basal surfaces, indicating that PAR-3 and PAR-6 are required for the normal, asymmetric pattern of cell adhesion.

Regulation of Ingression by Cell Fate

Cell ingressions into the blastocoel begin at the 26-cell stage of embryogenesis. At this stage, each cell has a rounded, apical surface containing cortical PAR-3 and other "anterior" PAR proteins. The basal surface of the cell abuts the blastocoel, and its lateral surfaces contact other cells. The first cells to ingress are the daughters of the E founder cell (asterisks, Fig. 4A–D; see movie 4-1 at www.gastrulation.org). Several studies have provided evidence that it is the fate of the E daughters that causes them to ingress, rather than their position in the embryo or their specific contacts with neighboring cells. Indeed, the E daughters undergo cell movements resembling normal ingression after surrounding cells are lysed or removed (Fig. 4I–P) (Laufer et al. 1980; Junkersdorf and Schierenberg 1992; Lee and Goldstein 2003). In contrast, all known mutations that prevent specification of the endodermal fate also prevent or delay ingression of the E descendants.

The endodermal fate of the E daughters is specified by a combination of maternally and embryonically expressed transcription factors and by a signaling pathway related to

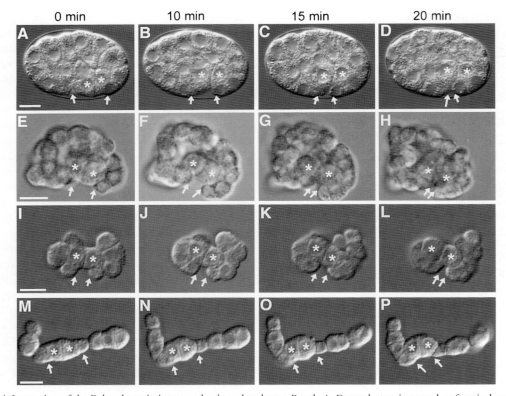

Figure 4. Ingression of the E daughters in intact and cultured embryos. Panels *A–D* are photomicrographs of optical sections of embryos taken through the plane of the blastocoel. Cells that appear to be in the center of the embryo in *A* (26-cell stage) are located at the periphery of the embryo on the bottom focal plane; these cells are visible through the shallow blastocoel. The E daughters (*asterisks*) are initially on the ventral surface (*bottom*), but ingress into the center of the embryo. MS descendants and P4 (*arrows*) spread over the E daughters during ingression. Panels within a row depict the same embryo at 5-minute intervals; 0 min indicates the start of gastrulation movements. Similar movements of the E daughters, the MS descendants, and P4 are observed when the eggshell and the vitelline membranes surrounding the embryo are removed (*E–H*), or when the E, MS, and P4 cells are produced from an isolated P1 cell (*I–L* and *M–P*). (Reprinted, with permission, from Lee and Goldstein 2003 [© Company of Biologists Ltd.].)

the Wnt pathway in vertebrates (for review, see Maduro and Rothman 2002). These events cause the E cell to express the GATA transcription factor END-1, a master regulator of endodermal differentiation. The first apparent defect in mutant embryos that fail to undergo endodermal specification is that the normal 49-minute cell cycle of the E daughters is shortened to 39 minutes (Zhu et al. 1997). These results suggest that factors like END-1 expand the cell cycle of the E daughters in normal development. Expansion of the cell cycle may be a general strategy to allow proteins that function in mitosis to be used instead for the morphogenetic movements of cells. In *Drosophila*, for example, the cell cycle regulator *tribbles* is critical for proper gastrulation (Chapter 7). All of the known *C. elegans* mutants with ingression defects, such as *emb-5, emb-23,* and *gad-1*, fail to fully expand the cell cycle of the E daughters (Schierenberg et al. 1980; Denich et al. 1984; Bucher and Seydoux 1994; Knight and Wood 1998). The specific functions of these genes are not known, although the EMB-5 protein is relat-

ed to yeast proteins implicated in transcriptional control (Winston 2001).

Cellular Mechanisms of Ingression

The ingressing E daughters lack obvious lamellipodia or filopodia (Lee and Goldstein 2003), although smaller protrusive structures may not be detectable by current techniques. Interestingly, proteins in the *C. elegans* Arp2/3 complex are required for ingression of the E daughters (Severson et al. 2002). Studies on other systems have shown that this complex functions in the nucleation and branching of microfilaments and that the complex localizes to the leading edge of crawling cells (for review, see Higgs and Pollard 2001). The most obvious change in ingressing cells is the flattening of their apical surfaces before and during ingression (Fig. 5A,B) (Nance and Priess 2002). Flattening appears to result from an actomyosin-based contraction rather than from tension on the ingressing cells from their

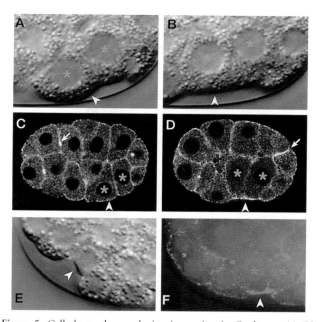

Figure 5. Cell shape change during ingression in *C. elegans*. (*A–D*) Endodermal cells (*green asterisks*) just prior to ingression (*A,C*) and during ingression (*B,D*). (*A,B*) Photomicrographs of live embryos. The apical surfaces (*arrowhead*) of endodermal cells change from rounded (*A*) to flattened (*B*) as cells begin to ingress. (*C,D*) Fixed embryos immunostained for non-muscle myosin/NMY-2. Prior to ingression of the endodermal cells (*C*), NMY-2 levels at the apical surfaces of endodermal cells (*arrowhead*) are comparable to levels at the apical surfaces of other cells. During endodermal cell ingression (*D*), NMY-2 levels at the apical surfaces of the endodermal cells (*arrowhead*) are enriched. NMY-2 also is enriched at the contractile ring in dividing cells (*arrow* in *D*) and in midbody remnants of cell divisions (*arrow* in *C*). (*E,F*) Ingression of MS descendants (*red asterisks*). (*E*) Ingressing MS descendant in a living embryo. The apical surface (*arrowhead*) is flattened. (*F*) Accumulation of NMY-2 (*green*) at the apical surface (*arrowhead*) of two ingressing MS descendants. Nuclei are shown in blue. Panels *A, B, E,* and *F* are shown at twice the magnification of panels *C* and *D*. (Reprinted, with permission, from Nance and Priess 2002 [© Company of Biologists Ltd.].)

that aspects of the cellular mechanisms of gastrulation may be conserved (Chapter 7).

How does NMY-2 become enriched apically, and is this apical enrichment essential for ingression in *C. elegans*? *par(ZF1)* embryos depleted of PAR-3 or PAR-6 (see above) have normal cytoplasmic levels of NMY-2, but ingressing cells do not show an apical enrichment of NMY-2 (Nance et al. 2003). Surprisingly, cells continue to ingress in *par(ZF1)* embryos, but the apical surfaces of the cells show less flattening than in wild-type embryos, and ingression occurs at a much slower speed. Thus, the PAR proteins appear to enhance a mechanism of cell ingression that is otherwise independent of PAR functions.

Ingression of Mesodermal and Germ Cells

Ingression of the various mesodermal precursors and the germ-cell precursors occurs two or more cell cycles after the ingression of the E daughters in an interval of about 200 minutes (Nance and Priess 2002). All of the ingressing cells enter the body cavity from the ventral surface of the embryo (summarized in Fig. 2C); these cells are either born on the ventral surface or move there prior to ingression (Fig. 2B). Cells that ingress immediately after the E daughters are relatively large, and the neighboring cells spread rapidly over their surfaces concomitant with ingression. Cells that ingress later are much smaller, and the neighboring cells do not immediately spread over their surfaces, thus creating a transient gap (the gastrulation cleft; see arrows in Fig. 2F). These ingressing cells show apical flattening and apical accumulation of non-muscle myosin, similar to that of the E daughters (Fig. 5E,F).

Cell-fate specification appears to play an important role in determining the pattern of the late ingressions. For example, multiple descendants of AB that have the same fate ingress at nearly the same time, although these cells can be widely separated across the ventral surface (Nance and Priess 2002). The ingression pattern of AB descendants is determined in part by the cell-fate regulator PHA-4 (Mango et al. 1994; Horner et al. 1998). Recent microarray studies have identified several presumptive targets of PHA-4; however, none of these has yet been implicated in ingression (Gaudet and Mango 2002).

The ingression of MS descendants also is controlled by cell fate; however, cell interactions can have a major role in determining the ingression pattern (Nance and Priess 2002). The MS descendants form a contiguous group of cells, yet subgroups of these cells ingress at two different times (indicated by red in Fig. 2C). An outer wishbone-shaped group of cells ("wishbone cells") ingresses first, while cells at the center of the wishbone ("central cells") are delayed by an additional cell cycle. The central cells overlie endodermal cells whereas wishbone cells do not, suggesting that central cell

neighbors. Laser-killing of MS, the precursor of most of the cells that flank and spread over the ingressing E daughters, does not prevent flattening (Nance and Priess 2002). Most significantly, microspheres applied to the surface of the E daughters move away from, rather than toward, the flanking cells (Lee and Goldstein 2003). Ingressing cells show elevated levels of non-muscle myosin (NMY-2) at their apical surfaces (Fig. 5C,D) (Nance and Priess 2002), and microfilament inhibitors as well as an inhibitor of myosin light-chain kinase prevent ingression (Lee and Goldstein 2003). Flattening of the apical surface displaces cytoplasm towards the basal surface. This may facilitate movement into the blastocoel, or it may simply allow the spreading of neighboring cells over the ingressing cell. Cells that invaginate during *Drosophila* gastrulation also show apical flattening and accumulation of non-muscle myosin, suggesting ingression might be blocked physically (Fig. 6A,B). However, the delay in central cell ingression occurs even after ingression of the endodermal cells is prevented (Fig. 6C). Cell fate alone is not sufficient to explain the different ingression times of the central and wishbone cells, because a mutation that transforms central cell fates into wishbone fates does not abolish the delay in central cell ingression (Fig. 6D). However, if this fate transformation occurs when endodermal ingression is prevented, the central and wishbone cells ingress simultaneously (Fig. 6E). Thus, differences in cell fate appear to subdivide a group of otherwise equivalent cells, preventing certain cells from trying to ingress when there is not sufficient space in the blastocoel.

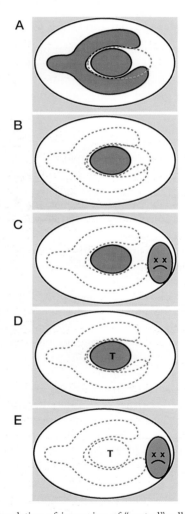

Figure 6. Regulation of ingression of "central" cells. Schematic diagrams are shown of the ventral surfaces of embryos. (*A*) After the E daughters ingress (*dotted green*), the "wishbone" and "central" groups of MS descendants are on the ventral surface. The wishbone group ingresses before the central group in normal development (*B*), after the E founder cell is killed (*C*), and after the central cell fates are transformed to wishbone fates (*D*). However, the wishbone and central groups ingress together when the central cell fates are transformed to wishbone fates and the E founder cell is killed.

Epiboly

Cell ingressions move endodermal, mesodermal, germ-cell precursors, and some neural precursors into the body cavity. After these movements, the ventral surface is occupied predominantly by neuroblasts, and the dorsal surface is occupied by epidermal cells (commonly called "hypodermal cells" in nematodes). The final phase of gastrulation involves the epidermal cells spreading across the ventral surface of the embryo, thus encasing the entire body in epidermis. In contrast to ingressing cells, the migrating epidermal cells are linked by adherens junctions and spread as a continuous sheet of cells ("epiboly"). This epiboly, termed epidermal enclosure or ventral enclosure, has been described in several recent reviews (Chin-Sang and Chisholm 2000; Simske and Hardin 2001). Epidermal cells are descended from the AB and C founder cells, and most are born on the dorsal surface (Fig. 2B). Prior to epiboly, the epidermal cells are organized in six longitudinal rows. The two dorsalmost rows of cells intercalate, resulting in five rows (Fig. 7) (Sulston et al. 1983; Williams-Masson et al. 1998). Intercalation is not essential for enclosure, since mutations in the transcription factor *die-1* disrupt dorsal intercalation but not enclosure (Heid et al. 2001). Enclosure requires 40 minutes and occurs in three distinct steps (Fig. 7) (Williams-Masson et al. 1997). Two left/right pairs of epidermal cells, called the "leading" cells, are essential for epiboly. These cells extend filopodia and initiate movement of the epidermal sheet by migrating ventrally over the surface of the embryo. Upon meeting at the ventral midline, the left/right pairs of leading cells link

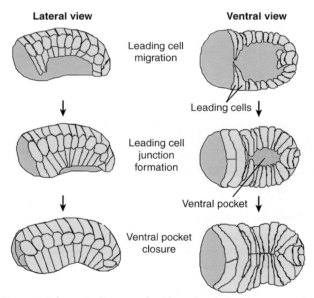

Figure 7. Schematic diagram of epidermal epiboly. Three successive stages of epiboly are shown from lateral and ventral views; anterior is left. (Modified, with permission, from Simske and Hardin 2001 [©Wiley].)

together through the formation of adherens junctions. Next, the more posterior left/right epidermal cells become wedge-shaped and move together to enclose the ventral surface, possibly through a purse-string-like contraction.

The Cytoskeleton and Cellular Junctions in Epiboly

Actin microfilaments are prominent in migrating leading cells and are required for proper enclosure. When embryos are exposed to the microfilament-destabilizing drug cytochalasin D, the left/right leading cells do not link together (Williams-Masson et al. 1997). In diverse cell types, microfilament dynamics are regulated by small GTPases such as Rac, Rho, and Cdc42 (for review, see Ridley et al. 1999; Bishop and Hall 2000). Depletion of the *C. elegans* homologs of Rac-1 (CED-10) or two Rac-1-interacting proteins (GEX-2 and GEX-3) blocks epidermal migration (Soto et al. 2002), suggesting that the Rac signaling pathway regulates actin dynamics during migration of the epidermal sheet. The microfilament cytoskeleton is linked to cell junctions through adherens junctions that contain E-cadherin (HMR-1), α-catenin (HMP-1), and β-catenin (HMP-2) (for review, see Nagafuchi 2001). Depletion of these proteins prevents the left/right pairs of leading cells from forming stable contacts (Fig. 8) (Costa et al. 1998; Raich et al. 1999). A similar phenotype is observed in embryos lacking LET-413, which is required for the proper assembly of adherens junctions (Legouis et al. 2000). APR-1 is a *C. elegans* homolog of the adenomatous polyposis coli (APC) protein, which can regulate β-catenin (Hoier et al. 2000). Adherens junctions form in *apr-1* mutants, but epidermal cells can fail to migrate. Together these observations indicate that migration of epidermal cells can occur in the absence of adherens junctions, but that cell adherence at the ventral midline requires the proper assembly and perhaps regulation of adherens junction

components. Mutations in the *evl-20* gene, encoding a small GTPase of the ARF (ADP-ribosylation factor) family, do not cause obvious defects in the microfilament cytoskeleton, but result in substantially reduced numbers of microtubules (Antoshechkin and Han 2002). These mutations either prevent enclosure or prevent stable junctions between left/right pairs of ventral cells, suggesting that the microtubule cytoskeleton may have a role in epiboly.

Interactions between Migrating Epidermal Cells and Neuroblasts

The ventral mesodermal precursors and neuroblasts do not appear to simply provide a passive surface for epidermal epiboly. If the gastrulation cleft persists inappropriately on the ventral surface, as occurs in *efn-1* and *vab-1* mutants, the left/right pairs of ventral epidermal cells often do not reach the ventral midline (Fig. 9). The *efn-1* gene encodes an ephrin, and *vab-1* encodes the only known Eph receptor

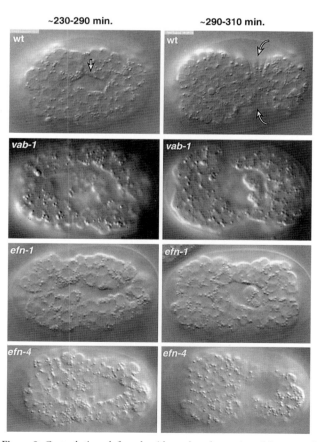

Figure 9. Gastrulation cleft and epidermal enclosure in wild-type and mutant embryos. Photomicrographs are shown of the ventral surfaces of live embryos at either 230–290 minutes (*left* column) or 290–300 minutes (*right* column). Times are after the 2-cell stage. The gastrulation cleft (*arrow*) closes by 300 minutes in wild-type embryos, but can remain open in *vab-1*, *efn-1*, or *efn-4* mutants. Photomicrographs courtesy of Andrew Chisholm.

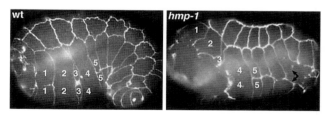

Figure 8. Cell positions after epidermal cell epiboly in *C. elegans* wild-type (wt) and *hmp-1* mutant embryos. Ventro-lateral view; anterior is left. Embryos are immunostained to visualize adherens junctions (*white lines* surrounding cells). In the wild-type embryo, left/right pairs of ventral epidermal cell pairs have met at the midline and formed junctions or fused. In the *hmp-1* mutant embryo, pairs 1–3 failed to meet. Similar defects are observed with posterior pairs of ventral cells (*double arrow*). (Modified, with permission, from Costa et al. 1998 [©The Rockefeller University Press].)

in *C. elegans* (George et al. 1998; Chin-Sang et al. 1999; Wang et al. 1999). Studies in other systems have shown that ephrins and Eph receptors mediate cell–cell interactions required for morphogenetic events such as cell migration, axon guidance, or the formation of regional boundaries (for review, see Kullander and Klein 2002). *efn-1* is expressed in clusters of neuroblasts adjacent to *vab-1*-expressing cells, some of which contact migrating epidermal leading cells (Chin-Sang et al. 1999; Wang et al. 1999). *efn-1* and *vab-1* appear to function in a partially redundant fashion with *efn-4*, which encodes a divergent ephrin, and *ptp-3*, which encodes a LAR-like tyrosine phosphatase. Double and triple combinations of mutations in these genes can cause more severe enclosure defects than mutations in any one gene (Fig. 9) (Chin-Sang et al. 2002; Harrington et al. 2002).

Mutations in the genes *mab-20* and *kal-1* cause enclosure defects similar, but not identical, to those caused by mutations in *vab-1* or *efn-1* (Roy et al. 2000; Chin-Sang et al. 2002; Rugarli et al. 2002). For example, *mab-20* mutant embryos can have defects in sealing of the gastrulation cleft and have abnormally positioned epidermal cells, some of which have abnormal, actin-rich protrusions (Chin-Sang et al. 2002; Rugarli et al. 2002). MAB-20 is a homolog of Semaphorin-2a (Roy et al. 2000), and semaphorins can function as cell guidance cues (for review, see Tamagnone and Comoglio 2000). *kal-1* is expressed in ventral neuroblasts and encodes a homolog of the Kallmann syndrome gene, which is required for specific neuronal migrations in humans (Rugarli et al. 2002). Although the *vab-1*, *efn-1*, *efn-4*, *ptp-3*, *mab-20*, and *kal-1* genes are required for epiboly, it is not yet clear whether they function primarily in preparing the ventral surface for epiboly, or whether they have additional roles during epiboly itself.

GASTRULATION IN OTHER NEMATODES

Early embryonic development of nematodes can differ substantially between taxa (see Chitwood and Chitwood 1974; Malakhov 1994). For example, after the first cleavage of the embryo, the posterior cell in *Acrobeloides nanus* retains the potential to generate an entire worm, whereas the corresponding cell in *C. elegans* produces only a highly restricted set of cell types (Wiegner and Schierenberg 1999). The endodermal cell fate may be the only cell fate that is specified during the first few cell divisions of *Enoplus brevis*, which belongs to a class of nematodes (Enoplia) that is considered ancestral (Malakhov 1994). In contrast to nematodes such as *C. elegans*, early *E. brevis* cells divide synchronously and symmetrically with variable positions. When random, individual cells in an 8-cell *E. brevis* embryo are injected with dye and the embryo is allowed to develop, only one invariant pattern of cell differentiation is observed: Either none of the endodermal cells is labeled, or only (and all of) the endodermal cells are labeled (Voronov and Panchin 1998). Thus, the early specification of the endodermal cells, the first cells to ingress, appears to be an ancient feature of nematode embryology.

Despite the variablity of early cell specification, gastrulation is remarkably similar within the nematodes and may represent a "phylotypic," or most similar, stage of embryogenesis (Schierenberg 2001). The first cells to ingress are always the endodermal cells, whether this is one cell (ex. *Plectus* [Lahl et al. 2003]), two cells (*C. elegans* and many

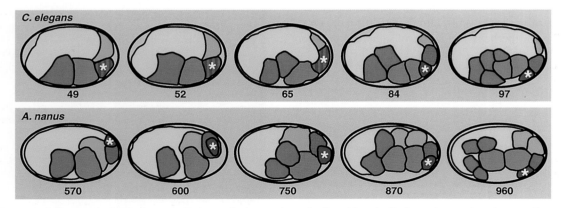

Figure 10. Cell repositioning in *A. nanus*. Each row depicts tracings of cell positions in *C. elegans* (*top row*) or *A. nanus* (*bottom row*) at comparable stages in early development (84-minute *C. elegans* stage advanced to show position of P4 prior to gastrulation). Anterior is left, dorsal is up. Cells are colored by origin according to Fig. 2A; the striped cell in *C. elegans* is the parent of P4 and D. Outlines of individual AB descendants (*yellow*) are not shown. In the final panel of each row, gastrulation has begun with the ingression of the endodermal cells (*green*). A P4 cell (*asterisk*) born on the dorsal surface in *A. nanus* is repositioned ventrally prior to gastrulation to adopt a configuration similar to that in *C. elegans*. Tracings of *A. nanus* cell positions courtesy of Frauke Skiba and Einhard Schierenberg.

other species), or four cells (*Rhabdias bufonis* [Spieler and Schierenberg 1995]). Mesodermal precursors ingress from the ventral surface after the endodermal cells and typically leave behind a central cleft that is sealed by the lateral movement of neighboring cells. Finally, the dorsal epidermis migrates ventrally to encase the embryo.

Ingression of the germ-line precursors may be associated with the endodermal cells in many nematode species. In both *C. elegans* and *A. nanus*, the germ-line precursor P4 is positioned next to the endodermal cells (Fig. 10). P4 and the endodermal cells are born adjacent to each other in *C. elegans*. However, in *A. nanus* these cells are often born distant to each other, and are brought together by a unique series of cell movements (asterisk, Fig. 10). Transcription is essential for ingression of the endodermal cells in *C. elegans*, yet the germ-line precursors are transcriptionally quiescent. Thus, ingression of the germ-line precursors may require a distinct mechanism, possibly exploiting an association with the endodermal cells.

FUTURE DIRECTIONS

Although studies of gastrulation are only beginning in *C. elegans*, this system offers several advantages that make it an attractive model for a detailed molecular analysis. These include the ability to culture and manipulate individual embryonic cells, the rapid development and optical clarity of the embryo, the availability of the complete cell lineage, and facile forward and reverse genetics. Determination of the complete genome sequence has spurred the development of new tools that allow genome-wide analyses. For example, DNA microarrays have been used to compare the expression of genes at different developmental stages and in various mutants (for examples, see Hill et al. 2000; Reinke et al. 2000). Such microarrays could be used to identify genes downstream of cell-fate regulators such as *end-1* that are important for the cell movements of gastrulation. Gene function can be assayed rapidly by RNA inhibition (RNAi) using double-stranded RNA (dsRNA). A library of bacterial strains expressing dsRNA for almost all genes is now available (Fraser et al. 2000). This powerful reagent could be used to survey the genome for genes involved in specific steps of gastrulation. These and other technological advances, in combination with comparative studies in other nematode species, are sure to enhance our understanding of nematode gastrulation.

ACKNOWLEDGMENTS

We thank Einhard Schierenberg for helpful comments on the manuscript. We also thank Andrew Chisholm, Bob Goldstein, Greg Hermann, Jen-Yi Lee, Einhard Schierenberg, Jeff Simske, and Frauke Skiba for supplying images or figures.

REFERENCES

Antoshechkin I. and Han M. 2002. The *C. elegans evl-20* gene is a homolog of the small GTPase ARL2 and regulates cytoskeleton dynamics during cytokinesis and morphogenesis. *Dev. Cell* **2:** 579–591.

Bishop A.L. and Hall A. 2000. Rho GTPases and their effector proteins. *Biochem. J.* **348:** 241–255.

Bowerman B. 1998. Maternal control of pattern formation in early *Caenorhabditis elegans* embryos. *Curr. Top. Dev. Biol.* **39:** 73–117.

Boyd L., Guo S., Levitan D., Stinchcomb D.T., and Kemphues K.J. 1996. PAR-2 is asymmetrically distributed and promotes association of P granules and PAR-1 with the cortex in *C. elegans* embryos. *Development* **122:** 3075–3084.

Bucher E.A. and Seydoux G. 1994. Gastrulation in the nematode *Caenorhabditis elegans*. *Sem. Dev. Biol.* **5:** 121–130.

Chin-Sang I.D. and Chisholm A.D. 2000. Form of the worm: Genetics of epidermal morphogenesis in *C. elegans*. *Trends Genet.* **16:** 544–551.

Chin-Sang I.D., Moseley S.L., Harrington R.J., George S.E., and Chisholm A.D. 2002. The divergent *C. elegans* ephrin *efn-4* functions in embryonic morphogenesis in a pathway independent of the VAB-1 Eph receptor. *Development* **129:** 5499–5510.

Chin-Sang I.D., George S.E., Ding M., Moseley S.L., Lynch A.S., Chisholm A.D., and Wolf F.W. 1999. The ephrin VAB-2/EFN-1 functions in neuronal signaling to regulate epidermal morphogenesis in *C. elegans*. *Cell* **99:** 781–790.

Chitwood B.G. and Chitwood M.B. 1974. *Introduction to nematology*, 2nd edition. University Park Press, Baltimore, Maryland.

Costa M., Raich W., Agbunag C., Leung B., Hardin J., and Priess J.R. 1998. A putative catenin-cadherin system mediates morphogenesis of the *Caenorhabditis elegans* embryo. *J. Cell Biol.* **141:** 297–308.

Denich K.T.R., Schierenberg E., Ishenghi E., and Cassada R. 1984. Cell-lineage and developmental defects of temperature-sensitive embryonic arrest mutants of the nematode *Caenorhabditis elegans*. *Roux's Arch. Dev. Biol.* **193:** 164–179.

Etemad-Moghadam B., Guo S., and Kemphues K.J. 1995. Asymmetrically distributed PAR-3 protein contributes to cell polarity and spindle alignment in early *C. elegans* embryos. *Cell* **83:** 743–752.

Fraser A.G., Kamath R.S., Zipperlen P., Martinez-Campos M., Sohrmann M., and Ahringer J. 2000. Functional genomic analysis of *C. elegans* chromosome I by systematic RNA interference. *Nature* **408:** 325–330.

Gaudet J. and Mango S.E. 2002. Regulation of organogenesis by the *Caenorhabditis elegans* FoxA protein PHA-4. *Science* **295:** 821–825.

George S.E., Simokat K., Hardin J., and Chisholm A.D. 1998. The VAB-1 Eph receptor tyrosine kinase functions in neural and epithelial morphogenesis in *C. elegans*. *Cell* **92:** 633–643.

Harrington R.J., Gutch M.J., Hengartner M.O., Tonks N.K., and Chisholm A.D. 2002. The *C. elegans* LAR-like receptor tyrosine phosphatase PTP-3 and the VAB-1 Eph receptor tyrosine kinase have partly redundant functions in morphogenesis. *Development* **129:** 2141–2153.

Heid P.J., Raich W.B., Smith R., Mohler W.A., Simokat K., Gendreau

S.B., Rothman J.H., and Hardin J. 2001. The zinc finger protein DIE-1 is required for late events during epithelial cell rearrangement in *C. elegans*. *Dev. Biol.* **236:** 165–180.

Higgs H.N. and Pollard T.D. 2001. Regulation of actin filament network formation through Arp2/3 complex: Activation by a diverse array of proteins. *Annu. Rev. Biochem.* **70:** 649–676.

Hill A.A., Hunter C.P., Tsung B.T., Tucker-Kellogg G., and Brown E.L. 2000. Genomic analysis of gene expression in *C. elegans*. *Science* **290:** 809–812.

Hoier E.F., Mohler W.A., Kim S.K., and Hajnal A. 2000. The *Caenorhabditis elegans* APC-related gene *apr-1* is required for epithelial cell migration and *Hox* gene expression. *Genes Dev.* **14:** 874–886.

Horner M.A., Quintin S., Domeier M.E., Kimble J., Labouesse M., and Mango S.E. 1998. *pha-4*, an HNF-3 homolog, specifies pharyngeal organ identity in *Caenorhabditis elegans*. *Genes Dev.* **12:** 1947–1952.

Hung T.J. and Kemphues K.J. 1999. PAR-6 is a conserved PDZ domain-containing protein that colocalizes with PAR-3 in *Caenorhabditis elegans* embryos. *Development* **126:** 127–135.

Junkersdorf B. and Schierenberg E. 1992. Embryogenesis in *C. elegans* after elimination of individual blastomeres or induced alteration of the cell division order. *Roux's Arch. Dev. Biol.* **202:** 17–22.

Knight J.K. and Wood W.B. 1998. Gastrulation initiation in *Caenorhabditis elegans* requires the function of *gad-1*, which encodes a protein with WD repeats. *Dev. Biol.* **198:** 253–265.

Krieg C., Cole T., Deppe U., Schierenberg E., Schmitt D., Yoder B., and von Ehrenstein G. 1978. The cellular anatomy of embryos of the nematode *Caenorhabditis elegans*. *Dev. Biol.* **65:** 193–215.

Kullander K. and Klein R. 2002. Mechanisms and functions of Eph and ephrin signalling. *Nat. Rev. Mol. Cell Biol.* **3:** 475–486.

Lahl V., Halama C., and Schierenberg E. 2003. Comparative and experimental embryogenesis of Plectidae (Nematoda). *Dev. Genes Evol.* **213:** 18–27.

Laufer J.S., Bazzicalupo P., and Wood W.B. 1980. Segregation of developmental potential in early embryos of *Caenorhabditis elegans*. *Cell* **19:** 569–577.

Lee J.-Y. and Goldstein B. 2003. Mechanisms of cell positioning during *C. elegans* gastrulation. *Development* **130:** 307–320.

Legouis R., Gansmuller A., Sookhareea S., Bosher J.M., Ballie D.L., and Labouesse M. 2000. LET-413 is a basolateral protein required for the assembly of adherens junctions in *Caenorhabditis elegans*. *Nat. Cell Biol.* **2:** 415–422.

Maduro M.F. and Rothman J.H. 2002. Making worm guts: the gene regulatory network of the *Caenorhabditis elegans* endoderm. *Dev. Biol.* **246:** 68–85.

Malakhov V.V. 1994. *Nematodes*. Smithsonian Institution Press, Washington, D.C.

Mango S.E., Lambie E.J., and Kimble J. 1994. The *pha-4* gene is required to generate the pharyngeal primordium of *Caenorhabditis elegans*. *Development* **120:** 3019–3031.

Nagafuchi A. 2001. Molecular architecture of adherens junctions. *Curr. Opin. Cell Biol.* **13:** 600–603.

Nance J. and Priess J.R. 2002. Cell polarity and gastrulation in *C. elegans*. *Development* **129:** 387–397.

Nance J., Munro E.M., and Priess J.R. 2003. *C. elegans* PAR-3 and PAR-6 are required for apicobasal asymmetries associated with cell adhesion and gastrulation. *Development* **130:** 5339–5350.

Ohno S. 2001. Intercellular junctions and cellular polarity: The PAR-aPKC complex, a conserved core cassette playing fundamental roles in cell polarity. *Curr. Opin. Cell Biol.* **13:** 641–648.

Raich W.B., Agbunag C., and Hardin J. 1999. Rapid epithelial-sheet sealing in the *Caenorhabditis elegans* embryo requires cadherin-dependent filopodial priming. *Curr. Biol.* **9:** 1139–1146.

Reese K.J., Dunn M.A., Waddle J.A., and Seydoux G. 2000. Asymmetric segregation of PIE-1 in *C. elegans* is mediated by two complementary mechanisms that act through separate PIE-1 protein domains. *Mol. Cell* **6:** 445–455.

Reinke V., Smith H.E., Nance J., Wang J., Doren C.V., Begley R., Jones S.J.M., Davis E.B., Scherer S., Ward S., and Kim S.K. 2000. A global profile of germline gene expression in *C. elegans*. *Mol. Cell* **6:** 605–616.

Ridley A.J., Allen W.E., Peppelenbosch M., and Jones G.E. 1999. Rho family proteins and cell migration. *Biochem. Soc. Symp.* **65:** 111–123.

Roy P.J., Zheng H., Warren C.E., and Culotti J.G. 2000. *mab-20* encodes Semaphorin-2a and is required to prevent ectopic cell contacts during epidermal morphogenesis in *Caenorhabditis elegans*. *Development* **127:** 755–767.

Rugarli E.I., Di Schiavi E., Hilliard M.A., Arbucci S., Ghezzi C., Facciolli A., Coppola G., Ballabio A., and Bazzicalupo P. 2002. The Kallmann syndrome gene homolog in *C. elegans* is involved in epidermal morphogenesis and neurite branching. *Development* **129:** 1283–1294.

Schierenberg E. 2001. Three sons of fortune: early embryogenesis, evolution and ecology of nematodes. *BioEssays* **23:** 841–847.

Schierenberg E., Miwa J., and von Ehrenstein G. 1980. Cell lineages and developmental defects of temperature-sensitive embryonic arrest mutants in *Caenorhabditis elegans*. *Dev. Biol.* **76:** 141–159.

Schnabel R., Hutter H., Moerman D., and Schnabel H. 1997. Assessing normal embryogenesis in *Caenorhabditis elegans* using a 4D microscope: Variability of development and regional specification. *Dev. Biol.* **184:** 234–265.

Severson A.F., Baillie D.L., and Bowerman B. 2002. A formin homology protein and a profilin are required for cytokinesis and Arp2/3-independent assembly of cortical microfilaments in the early *Caenorhabditis elegans* embryo. *Curr. Biol.* **12:** 2066–2075.

Simske J.S. and Hardin J. 2001. Getting into shape: epidermal morphogenesis in *Caenorhabditis elegans* embryos. *BioEssays* **23:** 12–23.

Soto M.C., Qadota H., Kasuya K., Inoue M., Tsuboi D., Mello C.C., and Kaibuchi K. 2002. The GEX-2 and GEX-3 proteins are required for tissue morphogenesis and cell migrations in *C. elegans*. *Genes Dev.* **16:** 620–632.

Spieler M. and Schierenberg E. 1995. On the development of the alternating free-living and parasitic generations of the nematode *Rhabdias bufonis*. *Invert. Reprod. Dev.* **28:** 193–203.

Sulston J.E., Schierenberg E., White J.G., and Thomson J.N. 1983. The embryonic cell lineage of the nematode *Caenorhabditis elegans*. *Dev. Biol.* **100:** 64–119.

Tabuse Y., Izumi Y., Piano F., Kemphues K.J., Miwa J., and Ohno S. 1998. Atypical protein kinase C cooperates with PAR-3 to establish embryonic polarity in *Caenorhabditis elegans*. *Development* **125:** 3607–3614.

Tamagnone L. and Comoglio P.M. 2000. Signalling by semaphorin receptors: Cell guidance and beyond. *Trends Cell Biol.* **10:** 377–383.

Voronov D.A. and Panchin Y.V. 1998. Cell lineage in marine nematode *Enoplus brevis*. *Development* **125:** 143–150.

Wang X., Roy P.J., Holland S.J., Zhang L.W., Culotti J.G., and Pawson T. 1999. Multiple ephrins control cell organization in *C. elegans* using kinase-dependent and -independent functions of the VAB-1 Eph receptor. *Mol. Cell* **4:** 903–913.

Wiegner O. and Schierenberg E. 1999. Regulative development in a nematode embryo: A hierarchy of cell fate transformations. *Dev. Biol.* **215:** 1–12.

Williams-Masson E.M., Malik A.N., and Hardin J. 1997. An actin-mediated two-step mechanism is required for ventral enclosure of the *C. elegans* hypodermis. *Development* **124:** 2889–2901.

Williams-Masson E.M., Heid P.J., Lavin C.A., and Hardin J. 1998. The cellular mechanism of epithelial rearrangement during morpho-genesis of the *Caenorhabditis elegans* dorsal hypodermis. *Dev. Biol.* **204:** 263–276.

Winston F. 2001. Control of eukaryotic transcription elongation. *Genome Biol.* **2:** Reviews 1006.

Wodarz A. 2002. Establishing cell polarity in development. *Nat. Cell Biol.* **4:** E39–E44.

Zhu J., Hill R.J., Heid P.J., Fukuyama M., Sugimoto A., Priess J.R., and Rothman J.H. 1997. *end-1* encodes an apparent GATA factor that specifies the endoderm precursor in *Caenorhabditis elegans* embryos. *Genes Dev.* **11:** 2883–2896.

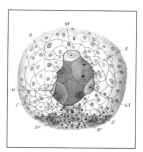

Gastrulation in the Molluscan Embryo

J.A.M. van den Biggelaar and W.J.A.G. Dictus

Department of Developmental Biology, Utrecht University, 3584 CH Utrecht, The Netherlands

INTRODUCTION

Gastrulation in mollusks has not been investigated systematically. Fortunately, a fairly complete idea of gastrulation can be derived from careful cell lineage studies in a number of molluscan species. In addition, some experimental investigations on cell competence and axis formation have contributed significantly to our knowledge about the process of gastrulation. The transition from a radial to a bilaterally symmetrical organization and the development of a dorsoventrally patterned embryo is fully realized during gastrulation. It now appears that the way in which mollusks gastrulate has much more in common with gastrulation in the better-known model systems than has been realized before.

THE PRE-GASTRULA STAGES

Spiral Cleavage

A defining feature of all molluscan taxa (cephalopods excepted) is the spiral character of their early cleavage divisions. From the 1- to the 4-cell stage, three different spiral cleavage patterns can be observed: equal cleavage, unequal cleavage, and equal cleavage associated with the formation of a polar lobe. Equal cleavage presumably represents the ancestral form (Heath 1899; Freeman and Lundelius 1992; van den Biggelaar and Haszprunar 1996). Therefore, this is used as the model pattern. Because the spiral character of the cleavages plays a predominant role in the specification of the developmental fate of the blastomeres, spiral cleavage is described first.

During the first cleavage, the spiral character is difficult to notice. During the second cleavage, however, it is more obvious as both cells divide into a minimally more animal and a slightly more vegetal cell. The pair of diametrically opposed upper blastomeres usually forms a small cross-furrow at the animal pole. The lower pair of diametrically opposed blastomeres forms the vegetal cross-furrow. The two blastomeres at the animal cross-furrow represent the presumptive lateral quadrants, whereas the blastomeres at the vegetal cross-furrow represent the two presumptive median quadrants. The spiral character and the regular alternation of its chirality become fully apparent from the third cleavage onward. The oblique position of the spindles then becomes much more prominent and, during the successive cleavages, each tier of blastomeres divides into an animal and a vegetal tier of cells. After the fifth cleavage, the alternation of dextral and sinistral cleavages is broken by the appearance of bilateral divisions.

In mollusks it is the general rule to denominate the most vegetal tier of cells as macromeres and their offspring as micromeres. The first generation of micromeres is indicated as 1a–1d and the first generation of macromeres as 1A–1D. During the following divisions the macromeres will form at least three generations of micromeres, successively denominated as 2a–2d, 3a–3d, and 4a–4d. The corresponding generations of macromeres are indicated as 2A–2D, 3A–3D, and 4A–4D. All the derivatives of a given quartet of micromeres are indicated with the character of that quartet. The blastomeres derived from 1a–1d are all indicated with the lettering 1a–1d. After each division a superscript is added, 1 for the animal and 2 for the vegetal tier of daughter cells. For instance, 1a–1d divide into $1a^1$-$1d^1$, and $1a^2$–$1d^2$. At the following division the former cells will divide into $1a^{11}$–$1d^{11}$, and $1a^{12}$–$1d^{12}$, and the latter into $1a^{21}$–$1d^{21}$ and $1a^{22}$–$1d^{22}$. Because of the continued spiral character of the successive divisions, the zygote is gradually subdivided into an increasing number of tiers of four cells along the animal–vegetal axis. Given the symmetrical distribution of the ooplasmic components along the animal–vegetal axis, each of the successive divisions is qualitative, and each tier of blastomeres has a different composition associated with predictable cell lineage contributions to the later body plan (van den Biggelaar and Guerrier 1983). Thus, the mosaic character of molluscan development can at least partly be explained as a result of spiral cleavage.

Blastula

After five synchronous cleavages, the embryo reaches the 32-cell blastula stage, e.g., in the embryo of *Patella* (Wilson 1904; van den Biggelaar 1977) or *Ischnochiton* (Heath 1899) and *Acanthochiton* (van den Biggelaar 1996). Up to that stage, the embryo maintains its quadri-radial organization along the animal–vegetal axis (Fig. 1a,b). Irrespective of differences in length of cleavage cycles in different species, this axial symmetry is broken during the interval between the formation of the third and fourth quartet of micromeres. Then, the four macromeres simultaneously intrude into the inside of the embryo and make contact with the opposite animal micromeres. Finally, a configuration is reached in which the contacts with the animal micromeres are limited to one of the two vegetal cross-furrow macromeres (Fig. 1c,d). Once this macromere has attained its position in the center of the embryo, it becomes induced to become the progenitor of the mesoderm, the 3D macromere. This crucial perturbation of radial symmetry associated with 3D specification is the first indication of dorsoventral symmetry. All other deviations of radial symmetry appear to be 3D-dependent. 3D functions as a dorsoventral organizer.

Fate Map of the Blastula

Fate maps of the molluscan blastula can be derived from cell lineage studies in a number of species. The derivatives of the first quartet (1a–1d) form the ectoderm of the head and part of the prototroch (Fig. 2). The prototroch is an equatorial band of ciliated cells that functions as the larval locomotory system. It separates the pretrochal head from the posttrochal trunk region. The second quartet cells also form a part of the prototroch, although the main part of their lineages is involved in the formation of the ectoderm of the trunk. The lineage of 2b also forms part of the anterior mesoderm, as well as the third quartet cells 3a and 3b, left and right from 2b (Fig. 3). Together with derivatives of

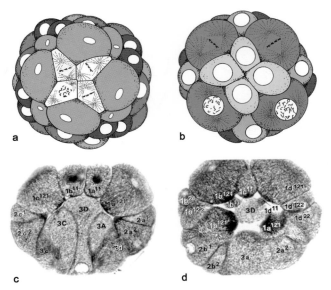

Figure 1. (*a*) 36-Cell embryo of *Acanthochiton crinitus*, animal view. (*b*) Same stage, vegetal view. (*c*) 36-Cell embryo, frontal section showing the mushroom-like 3D in touch with the overlying first quartet micromeres. (*d*) Same stage, section at right angles to the egg axis at the level of the tip of 3D. (*White*) Most animal micromeres $1m^{11}$; (*gray*) the following two tiers $1m^{121}$ and $1m^{122}$; (*brown*) the presumptive prototroch cells $1m^{21}$ and $1m^{22}$; (*red*) the animal tier of second quartet cells $2m^1$; (*blue*) the vegetal tier of second quartet cells $2m^2$; (*green*) the third quartet micromeres 3m; (*yellow*) the macromeres of the third generation 3M. (*a, b,* Reprinted, with permission, from van den Biggelaar et al. 2002; *c,d,* reprinted, with permission, from van den Biggelaar 1996 [©Elsevier].)

2a–2d, the micromeres 3c and 3d develop the ectoderm of the trunk (Dictus and Damen 1997). The endoderm is formed by macromeres 3A–3D. In addition to endoderm, 3D produces the mesentoblast 4d or M, the progenitor of the posterior mesodermal bands (Fig. 3).

SEGREGATION OF DEVELOPMENTAL POTENTIAL TO PROSPECTIVE REGIONS

Experimental Evidence for Segregation of Developmental Potential along the Animal–Vegetal Axis in Equal Cleaving Species

Blastomere deletion and transplantation experiments demonstrate that developmental potential differs in the animal–vegetal direction. Blastomeres at the same level along the animal–vegetal axis can be exchanged and are able to compensate for a deleted counterpart. Blastomere transplantations between different levels result in acquisition of a tier-specific developmental fate at new, ectopic positions (Arnolds et al. 1983; Render 1991, 1997).

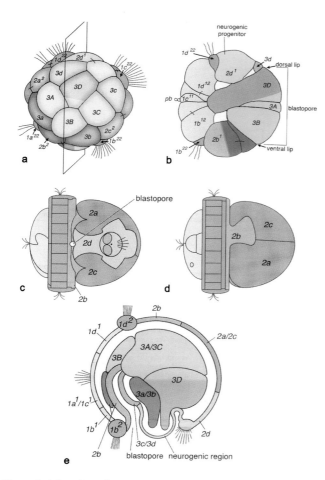

Figure 2. Migration of presumptive regions based on cell lineages in the embryo of *Patella vulgata*. (*a, b*) Prior to the initiation of gastrulation. (*a*) Vegetal view; (*b*) median section of same as in *a*; (*c*) 24–28-hour trochophore larva, ventral view; (*d*) dorsal view; (*e*) sagital section. (*a,b,e,* Reprinted, with permission, from van den Biggelaar et al. 2002 [©Elsevier]; *c,d,* reprinted, with permission, from Dictus and Damen 1997 [©Elsevier].)

The most direct evidence for the segregation of developmental potential has been shown in embryos of the polar lobe-forming species *Ilyanassa obsoleta*. This gastropod has orthologs for the dorsoventral patterning genes *Eve*, *dpp*, and *Tld*, called *IoEve, IoDpp,* and *IoTld,* respectively (Lambert and Nagy 2002). During the early cleavage cycles, mRNAs of these genes are localized to the centrosomes, which leads to an asymmetric distribution to the daughter cells. In 4-cell embryos, *IoDpp* mRNA is distributed diffusively in the cytoplasm of each blastomere. During cytokinesis at the transition to the 8-cell stage, it is localized to the centrosomes of macromeres 1A–1D (Fig. 4). Finally, mRNA of *IoDpp* is limited to 3a–3d and macromere 3D. At the 4-cell stage each blastomere has *IoEve* mRNA, whereas at the 8-cell stage it is limited to the first quartet 1a–1d (Fig. 5). At

the fourth division, it is segregated into the vegetal daughter cells, the trochoblast $1a^2$–$1d^2$. In the 24-cell embryo, only the four micromeres $1a^1$–$1d^1$ have obtained *IoTld* mRNA. These observations not only demonstrate the significance of the mitotic apparatus for the asymmetric allocation of pattern-controlling molecules, but also the significance of a normal invariant cleavage pattern. If this succession is disturbed, embryos develop abnormally. For instance, in *Patella* it has been shown that a normal succession of the first three cleavage cycles is required for normal development of the trochoblasts (Janssen-Dommerholt et al. 1983) and the expression of a trochoblast-specific tubulin gene (Damen et al. 1996). Once the third division has taken place, it is no longer possible to suppress ciliation and tubulin expression.

Factors associated with dorsoventral patterning appear to be concentrated at the vegetal pole. In species with polar lobe formation, these factors are initially concentrated in the polar lobe and finally segregated into the D quadrant. Experimental evidence for the dorsoventral patterning role of the polar lobe was first demonstrated by Wilson (1904) in the embryo of the scaphopod *Dentalium*. By removing the first polar lobe, Wilson obtained completely radialized embryos. Comparable results of lobe deletion experiments have been obtained in *Ilyanassa* (Clement 1952), *Crepidula* (Cather and Verdonk 1974; Cather et al. 1976), and again in *Dentalium* (van Dongen 1976a,b; van Dongen and Geilenkirchen 1974, 1975). Segregation of the polar lobe substances into more than one quadrant results in the development of twinned embryos (Guerrier et al. 1978; Render 1989). Similarly, equalization of the first cleavage in unequally cleaving species such as the marine bivalves, *Pholas* and *Spisula,* leads to the formation of double embryos (Guerrier 1970).

Cell Fate Specification by Cellular Interaction

In his discussion of the specific role of 2d in the development of *Ischnochiton*, Heath (1899) emphasized that 2d "arises from a cell which in origin, size and position is similar to the other second quartet cells, and that in all probability its relative excessive development has been acquired owing to the shifting of the mouth and formation of a ventral surface." In other words, differences in the developmental fate of equivalent cells depend on the establishment of the dorsoventral axis. A similar reasoning must lead to the conclusion that also in *Patella* the induction of one of the initially equivalent four macromeres, 3D, is the result of its positioning in the center of the embryo (van den Biggelaar and Guerrier 1979). The successive interactions between 3D and the animal micromeres appear to be essential for dorsoventral patterning of the embryo. After deletion of all

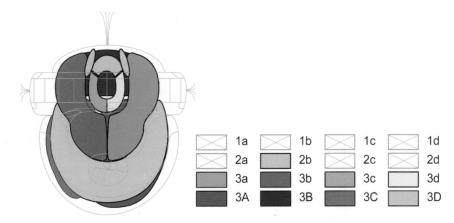

Figure 3. Internal position of the derivatives of the progenitor cells of the anterior mesoderm 2b, 3a, and 3b; of the posterior mesoderm 3D; of the endoderm 3A–3D. (Modified, with permission, from Dictus and Damen 1997 [©Elsevier].)

four first quartet cells in 8-cell embryos of *Patella* (van den Biggelaar and Guerrier 1979), *Acanthochiton* (van den Biggelaar 1996), and *Haminoea* (Boring 1989), the cell pattern remains radially symmetrical. In *Lymnaea* the presence of one or two first quartet micromeres appears to be sufficient to allow a more or less normal development (Morrill et al. 1973).

The inductive role of the animal micromeres in the specification of the organizer 3D has also been analyzed by chemical inhibition of cell contacts. If embryos of *Lymnaea* are treated with cytochalasin B during the period in which the centralized macromere is induced to become 3D, the normal dorsoventral cleavage pattern fails to appear (Martindale et al. 1985). Monensin interferes with the development of extracellular matrices. In monensin-treated embryos of *Patella*, the normal contact between the macromeres and the micromeres is inhibited, and radial-

ized embryos develop (Kühtreiber and van Dongen 1989; Damen and Dictus 1996). In equally cleaving species the organizer 3D can only start its dorsoventral patterning role after it has been specified itself.

In polar lobe-forming species the D quadrant is already specified at the 4-cell stage; however, Clement (1962) has shown that in the lobe-forming species *Ilyanassa obsoleta*, the D quadrant only becomes an effective organizing center after the formation of the third generation of macromeres 3A–3D. It also appeared that not all lobe-dependent structures are determined at the same time. It thus may be concluded that, irrespective of a precocious specification of the D quadrant, its organizing role only becomes apparent shortly after the formation of 3D. Similarly, in the equally cleaving embryo of *Lymnaea*, UV deletion of 3D at progressively later stages after its specification interferes less and less with normal development (Martindale 1986).

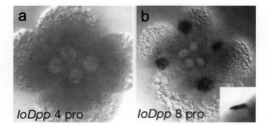

Figure 4. (*a*) 4-Cell stage of *Ilyanassa obsoleta* showing diffuse distribution of *IoDpp* mRNA; nuclei light blue; (*b*) 8-cell stage, macromeres in prophase prior to the formation of the second quartet of micromeres. *IoDpp* is located on a portion of the cortex that will be part of the second quartet. (*Inset*) Lateral view of *IoDpp* on the cortex of 1D. (Reprinted, with permission, from Lambert and Nagy 2002 [©Nature Publishing Group].)

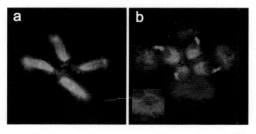

Figure 5. Localization of *IoEve* mRNA in *Ilyanassa obsoleta*. (*a*) During prometaphase of third division, *IoEve* mRNA (in *red*) is localized to granules on the cortex in the regions of each cell closest to the animal pole, which will become part of the animal daughter cells. (*b*) Early 8-cell embryo with *IoEve* mRNA only in the animal micromeres. (*Inset*) Section through the negative 1D cell. (Reprinted, with permission, from Lambert and Nagy 2002 [©Nature Publishing Group].)

Organizer Activation

The molecular basis of induction of 3D in species with equal or unequal cleavage is largely unknown. By simple reasoning one may expect the following. During the successive cleavages, maternal dorsoventral patterning mRNAs are segregated to all four quadrants. They become localized at the vegetal pole and become finally located in each of the four macromeres after the formation of the third quartet of micromeres. Then, only one of these macromeres, the presumptive 3D, becomes centralized, and as a result of its exclusive contacts with the animal micromeres, these maternal dorsoventral patterning mRNAs become activated in this cell only. This idea draws us to look closer at the so-called ectosomes (granules rich in RNA) that were first described in *Physa* (Wierzejski 1905). In the 8-cell embryo of *Physa,* the ectosomal material is limited to macromeres 1A–1D. During prophase, preceding the formation of the second quartet micromeres, the ectosomes appear to be located at one side of each of the four nuclei (Fig. 6a). This unilateral location is such that the ectosomes will become segregated into macromeres 2A–2D. This becomes clear during telophase when the ectosomes appear to be scattered within the asters of the macromeres, with the exclusion of the second quartet cells (Fig. 6b). At the beginning of the 24-cell stage, the ectosomes are located around the vegetal cross-furrow (Fig. 6c). Then they gradually move toward the central tips of the four macromeres (Fig. 6d–f). In *Lymnaea,* similar granules have been described by Raven (1946) and Minganti (1950). In *Lymnaea,* the ectosomes become dispersed only in the centralized macromere 3D, whereas in 3A, 3B, and 3C, they coalesce (van den Biggelaar 1976). The segregation pattern of the ectosomes strongly resembles the segregation pattern of mRNAs as described in *Ilyanassa* (Lambert and Nagy 2002). It is very likely that the ectosomes contain a maternal dorsoventral patterning controlling RNA which is only activated in 3D, as a result of which it obtains its organizing capacities. In species with unequal quadrants, these messengers are segregated into the polar lobe and finally into 3D. Activation of these pattern-controlling molecules, however, occurs exactly at the same developmental stage, i.e., between the formation of the third and fourth quartet micromeres.

The activation of the organizing capacities in one out of four equivalent macromeres following fifth cleavage probably reflects the ancestral mode of D quadrant specification (Freeman and Lundelius 1992; van den Biggelaar and Haszprunar 1996). It therefore may be expected that the evolution toward the specification of the organizer by segregation of dorsoventral patterning molecules to only one of the quadrants has occurred progressively. The early development of the caenogastropods *Littorina* (Moor 1973) and *Hydrobia* (J.A.M. van den Biggelaar, unpubl.) may rep-

resent an intermediate stage. In both species a polar lobe with specific cytoplasmic inclusions is formed during each of the first five cleavage cycles; i.e., preceding 3D formation. As soon as 3D has been formed, the polar lobe granules move toward the tip of 3D (Fig. 7), just like the ectosomes in *Physa* and *Lymnaea.*

It remains to be demonstrated whether after a precocious specification of the D quadrant as found in unequally cleaving species, the organizing role of 3D still requires an activation by the first quartet cells. In *Dentalium* the presence of the first quartet micromeres seems to be neces-

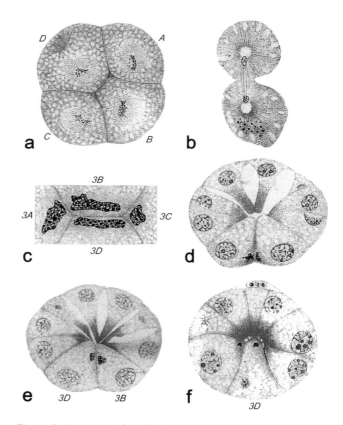

Figure 6. Ectosomes (*dark blue*) in embryos of *Physa fontinalis.* (*a*) The four macromeres of an 8-cell stage in prophase preparing the formation of the second quartet of micromeres. The ectosomes are located in the direct vicinity of the nuclei facing the side toward which the macromeres will be formed. Note that *Physa* is a sinistral species in which the second quartet will be formed clockwise. (*b*) One macromere after formation of the second quartet micromere with the ectosomes exclusively segregated in the astrosphere of the lower macromere. (*c*) Position of the ectosomes at the vegetal cross-furrow of the macromeres 3A–3D of a 24-cell embryo. (*d*) Meridianal section of a 24-cell embryo with the ectosomes at the vegetal pole. (*e*) 24-Cell stage before centralization of one of the macromeres. The ectosomes have migrated along the cell walls into the tips of the macromeres. (*f*) Late 24-cell embryo after centralization of 3D. Ectosomes in the tips of the macromeres. (Reprinted from Wierzejski 1905.)

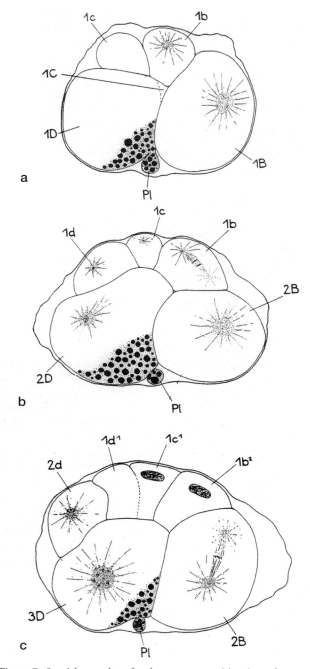

Figure 7. Special granules of unknown composition in embryos of *Littorina littoralis* concentrated in and near the polar lobes during the 3rd, 4th, and 5th cleavages. (*a*) Formation of third quartet of micromeres. (*b*) Formation of second quartet. (*c*) Formation of third quartet. (Reprinted, with permission, from Moor 1973 [©Urban & Fisher].)

does not interfere strongly with the organizing capacity of 3D. Unfortunately, the internal cell contacts of 3D with the animal micromeres in the embryo of *Ilyanassa* are unknown. In the related polar lobe, forming species *Nassarius reticulatus*, however, 3D is centralized and is the only macromere in contact with the overlying animal micromeres (Fig. 8) (J.A.M. van den Biggelaar, unpubl.). In the polar lobe-forming species *Bithynia*, 3D even makes no contact with the animal micromeres (J.A.M. van den Biggelaar, unpubl.), and after deletion of the first quartet, dorsoventrally organized embryos develop (van Dam and Verdonk 1982). These observations may indicate that during evolution the specification of the organizer became more and more independent of cellular interactions.

Direct evidence for a signal exchange between 3D and surrounding micromeres has been obtained in *Ilyanassa* (Lambert and Nagy 2001). The mitogen-associated protein kinase (MAPK) signaling pathway is activated in the cells that require a signal from 3D for normal differentiation. Prior to the formation of 3D, no activated MAPK could be detected. Within about 10–20 minutes after its formation, 3D shows a weak activation of MAPK (Fig. 9a,b). This period corresponds with the period in which the presumed interactions between 3D and the micromeres start. After division of 3D into 4d and 4D, active MAPK is limited to the mesentoblast 4d. Activation then extends further in animal and dorsal directions (Fig. 9c–h).

As the organizing role of 3D of the *Ilyanassa* embryo depends on the segregation of the contents of the polar lobe to the D quadrant (Clement 1952), Lambert and Nagy (2001) have analyzed whether the distribution of activated MAPK is lobe-dependent. After removal of the polar lobe, the normal dorsoventral pattern of MAPK activation was weakened, delayed, and radialized. In addition, deletion of

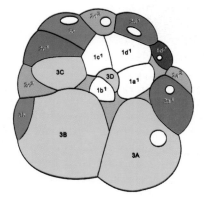

Figure 8. Section through a 24-cell embryo of *Nassarius reticulatus*, slightly oblique to the animal–vegetal egg axis. Note the tip of 3D surrounded by the micromeres 1a¹–1d¹. Same color code as in Fig. 1; (cleavage cavity) *gray*.

sary for the activation of the dorsoventral patterning role of 3D (E. Edsinger, pers. comm.). However, in *Ilyanassa* deletion of up to three first quartet micromeres (Clement 1967)

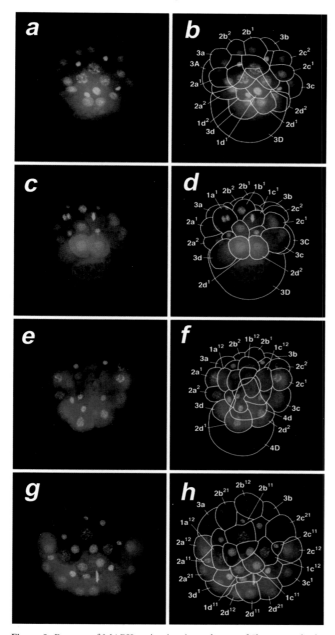

3D interfered with the proper temporal and spatial activation of MAPK signaling. Blocking MAPK activation with the MAPK inhibitor U0126 resulted in lobeless-like development.

MAPK signaling has also been studied in the chiton *Chaetopleura*, the closely related limpets *Patella* and *Tectura* (Lartillot et al. 2002a; Lambert and Nagy 2003), and the pulmonate snail *Lymnaea*, all equally cleaving species. In each of these species the MAPK pathway is only activated in 3D, not in the animal cells nor in any other micromere (Figs. 10 and 11). In *Patella*, however, besides a strong MAPK signal in 3D, a weak signal was found in 3A, 3B, and 3C. In conclusion, in equally cleaving species from three different molluscan taxa, MAPK is mainly activated in 3D. In species with equal cleavage, no MAPK signal from the inducing micromeres to the organizer, nor from the organizer to other blastomeres, has been observed. In embryos of *Patella* and *Tectura*, it appeared that inhibition of the MAPK pathway disrupts 3D differentiation (Lartillot 2001).

GASTRULATION

During gastrulation, dramatic reallocations of cells occur. The most reliable data on the rearrangement of cells during gastrulation can be derived from cell lineage studies with cell markers as known for *Patella* (Damen 1994; Dictus and Damen 1997) and *Ilyanassa* (Render 1991, 1997). The fate map of the various presumptive regions in the blastula of *Patella* is shown in Figure 2a,b; their final positions at the end of gastrulation are shown in Figure 2c,d and Figure 3. In agreement with the classical literature, it appears that the first quartet micromeres only develop ectoderm of the head region and part of the prototroch. The second quartet partly contributes to the prototroch, the posttrochal ectoderm and part of the anterior mesoderm. The lineage of 2b develops into a ring of ectodermal cells along the posterior edge of the prototroch, the anterior–dorsal part of the ectoderm involved in the development of the shell field, part of the

Figure 9. Pattern of MAPK activation in embryos of *Ilyanassa obsoleta* from the 24- to the 39-cell stage. Left column nuclei in green and activated MAPK in blue. The corresponding images with cell boundaries and cell denominations are depicted in the right column. The age of the embryos is given in minutes after the formation of 3D. (*a,b*) 30 minutes, the nucleus of 3D underlies $2d^1$. (*c,d*) 60 minutes. Strong MAPK activation in 3D, $2d^1$, and $2d^2$. (*e,f*) 100 minutes, shortly after the formation of 4d. MAPK activation found in $2a^1$, $2a^2$, $2c^1$, $2c^2$, $2d^1$, $2d^2$, 3c, and 3d. Weak activity has been found in $1a^{12}–1c^{12}$; 4d is also active, its outline is shown underneath the micromeres. 4D is negative. (*g,h*) 150 minutes after the formation of the third quartet, 39-cell stage. Activation is detected in $1c^{11}$, $1d^{11}$, $2a^{12}$, $2a^{21}$, $2c^{12}$, $2d^{11}$, $2d^{12}$, $2d^{21}$, $3c^1$, and $3d^1$. The mesentoblast 4d is in metaphase. (Reprinted, with permission, from Lambert and Nagy 2001.)

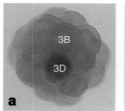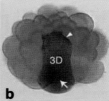

Figure 10. Localization of MAPK in a 32-cell *Patella vulgata* embryo. (*a*) Vegetal view; (*b*) side view. Labeling is strongest in the nucleus of 3D (*arrow*) and where 3D touches the animal micromeres (*arrowhead*) (Reprinted, with permission, from Lartillot et al. 2002b.)

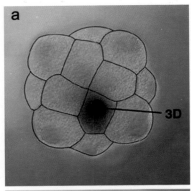

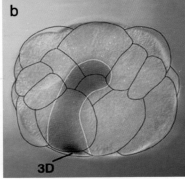

Figure 11. MAPK activation in a 36-cell embryo of the chiton *Chaetopleura apiculata*. Active MAPK is only found in 3D. (*a*) Vegetal view; (*b*) lateral view. (Reprinted, with permission, from Lambert and Nagy 2003 [© Elsevier].)

mesoderm at the ventral side of the head region. The mesodermal part is continuous with the upper roof of the stomodeum. The derivatives of the two lateral second quartet cells 2a and 2c form the lateral sides of the foot, left and right from the ventral midline (Fig. 2c). Dorsally their progenies meet each other and form the major part of the shell field (Fig. 2d). The progeny of 2d, the so-called first somatoblast, form the ventral midline of the foot, the telotroch and the posterior part of the shell field (Fig. 2c,e). It does not contribute to the formation of the stomodeum. Apparently due to dorsal extension of the derivatives of 2a and 2c, the descendants of 2d have been pushed in a more posterior–ventral position.

In a mirror-like fashion the two third quartet cells 3a and 3b at the ventral lip of the blastopore form a rod-like stripe of mesoderm along the mesoderm derived from 2b toward the foot (Fig. 2e and Fig. 3). Their dorsal counterparts 3c and 3d do not develop mesoderm. They contribute in a mirror way to the development of the anterior part of the ectoderm left and right from the mouth (Figs. 2c and 3). In connection with their original position at the dorsal border of the blastopore at the onset of gastrulation, 3c and 3d form the bottom of the stomodeum. Generally, the deriva-

tives of each individual blastomere form a continuous group of cells, irrespective of variations in developmental fate within their progenies. In the embryo of *Patella*, however, the derivatives of 3c and 3d appear to be split into a major anterior group near the stomodeum and a few cells near the posterior tip of the embryo.

The derivatives of 3A, 3B, and 3C only contribute to the formation of the endoderm. Macromere 3D develops into endoderm and two bands of cells diverging anteriorly. The major part of these two bands represents the two mesodermal bands (Figs. 3 and 14).

Comparison of the cell lineage studies of *Trochus* (Robert 1902), *Ischnochiton* (Heath 1899), *Crepidula* (Conklin 1897), and *Physa* (Wierzejski 1905) may reflect the intermediate steps from the beginning till the end of gastrulation. The margins of the early blastopore are formed by the boundary line that separates the presumptive ectodermal from the mesendodermal lineages. Initially the blastopore is exactly opposite of the animal pole. Gastrulation starts with the internalization of the macromeres 3A–3D. Soon afterward, the ectodermal derivatives of the second and third quartet begin to overgrow the endodermal cells. As a result of the epibolic movements of the second and third quartet micromeres, the blastopore is narrowing especially at the dorsal side because of the early internalization of 4d. The cells surrounding the blastopore in the 145-cell embryo of *Trochus* are shown in Figure 12a. During the following stages the inter-radial third quartet cells around the blastopore coalesce more and more and the radially located second quartet micromeres become excluded from the blastopore (Fig. 12b). In *Trochus,* the vegetal-most second quartet cells 2a^{22}–2d^{22} are shifted to the outer rim of the blastoporal region. In *Ischnochiton* the four groups of inter-radial third quartet cells also approach each other and partly overgrow the neighboring second quartet. The circular blastopore thus obtains the form of a cross, with the four arms of second quartet cells as deep grooves between the four groups of third quartet cells (Fig. 13). The bottoms of these grooves are occupied by the stomatoblasts of the second quartet. Especially in the dorsal quadrant, the groove may extend halfway up to the posterior–dorsal edge of the prototroch. In *Ischnochiton* the posterior groove can easily be misinterpreted as a posterior part of the blastopore (Fig. 13). In *Trochus* as well as in *Ischnochiton,* the derivatives of all four vegetal second quartet cells 2a^2–2d^2 contribute to the development of the stomodeum. In *Ilyanassa* the stomodeum is formed by descendants of 2a, 2b, and 2c; derivatives of 2d do not develop into stomodeal cells (Render 1997).

At the beginning of gastrulation the blastopore is situated just opposite the animal pole (Fig. 2a,b). During later stages it shifts in antero–ventral direction, finally to become located just under the mid-ventral part of the prototroch

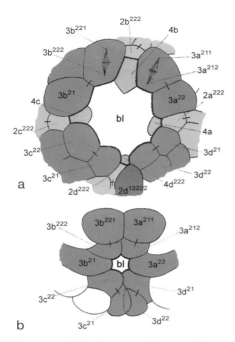

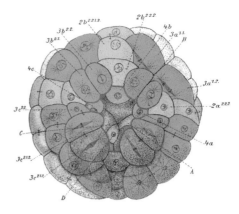

Figure 12. Blastopore of *Trochus magus*. (*a*) 145-cell stage. The presumptive stomatoblasts of the third quartet (*green*) are still separated from each other by the vegetal-most cells of the second quartet $2a^{22}$–$2d^{22}$ (*blue*). (*b*) Early trochophore stage; the blastopore is almost closed by the concrescence of the third quartet cells. (Reprinted, with permission, from Robert 1902 [©CNRS Editions].)

Figure 13. Late gastrula of *Ischnochiton magdalensis*, ~120 cells. The inter-radially located vegetal-most pairs of third quartet cells $3a^{21}$–$3a^{22}$ (*yellow*), $3b^{21}$–$3b^{22}$ (*yellow*), $3c^{21}$–$3c^{22}$ (*yellow*), $3d^{21}$–$3d^{22}$ start to overgrow the radial rows of second quartet cells $2a^{222}$, $2a^{2211}$–$2a^{2212}$; $2b^{222}$, $2b^{2211}$–$2b^{2212}$; $2c^{222}$, $2c^{2211}$–$2c^{2212}$; $2d^{222}$, $2d^{2211}$–$2d^{2212}$. In this way four radial grooves are formed and the blastopore (*red outline*) gets the form of a cross. The posterior groove extends far in posterior direction just above the two mesentoblasts (*pink*). (Reprinted, with permission, from Heath 1899 [©Urban & Fischer].) Color code: Cells derived from $2a^{22}$–$2d^{22}$ blue; 3rd quartet cells green; 4a–4c and the two mesentoblasts M^1 and M^2 orange, macromeres 4A–4D (marked as A–D) yellow. M^1 and M^2 in the lower half of the embryo, left and right from the blue row of cells representing $2d^{222}$, $2d^{2212}$, $2d^{2211}$.

(Fig. 2c,d). The function of this shift becomes evident if one realizes that the prototroch functions not only as a locomotory organ, but also as a collecting band in filter feeding (Nielsen 2001). The ventral shift of the blastopore is mainly achieved by two different but simultaneous cell movements: (1) a complex rearrangement of cells between the prototroch and the dorsal blastoporal lip and (2) the inward movement of presumptive anterior mesodermal cells between the ventral blastoporal lip and the ventral side of the prototroch (Dictus and Damen 1997).

In *Crepidula,* Conklin (1897) describes that in the posterior end of the embryo, divisions proceed so rapidly that in a very short time span there are many more cells at the posterior than at the anterior side. The center of proliferation lies just ventro–posterior to the domain of the presumptive shell gland, and thus almost immediately over the developing mesodermal bands (Fig. 14c). From this mitotic center more or less regular rows of cells extend anteriorly along the ventral midline. At the posterior side of these rows there appear to be three or four larger cells, which according to Conklin in *Crepidula,* correspond with the ciliated cells of the anal tuft or telotroch as observed in *Patella* (Smith 1935). The anal cells mark the place of contact between the anlage of the hindgut and the ectoderm, where

later the anus will break through. In embryos of the equally cleaving species *Patella* (Dictus and Damen 1997), *Trochus* (Robert 1902), and *Physa* (Wierzejski 1905), the expansion of the lineages of 2a and 2c toward the dorsal side of the embryo force the 2d lineage into a more ventral–lateral position along the midline of the embryo. In addition, part of the progeny of 2b displaces the 2d lineage from the dorsal–posterior edge of the prototroch in posterior direction. As a result, the blastopore is pushed anteriorly along the ventral midline toward a position just underneath the ventral side of the prototroch.

The posttrochal region is most affected by the gastrulation movements; therefore, the analysis of further development is limited to that region. Posteriorly from the stomodeum, at either side of the ventral midline, a pair of ectodermal swellings develops, initially separated by a ventral groove. These swellings form the anlage of the foot (Fig. 15). At the dorsal side a more or less circular field surrounded by a thickened edge represents the anlage of the shell field (Fig. 16). Internally the stomodeal cells are in contact with the derivatives of the endodermal cells. From the endoderm a cord of smaller cells extends posteriorly and makes contact with the posterior ectoderm where the anus will develop. In *Teredo* the cells surrounding the anal region develop a ring of ciliated cells, the telotroch (Baba 1938, 1940). In other species only a pair of anal cells provid-

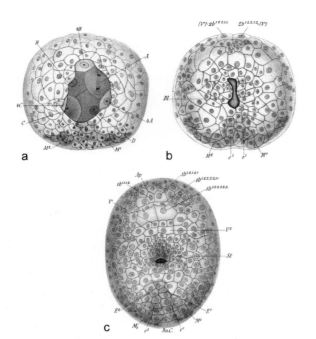

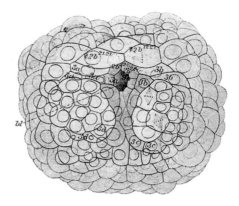

Figure 15. Late gastrula stage of *Physa fontinalis*. The blastopore (bl) is almost closed and shifted anteriorly where it is placed just ventrally from the prototroch cells 2b^{1221} and 2b^{2121} . The paired primordium of the foot is visible as two lateral swellings posterior to the mouth. (Reprinted from Wierzejski 1905.)

Figure 14. Three successive stages in the gastrulation of *Crepidula fornicata*. Macromeres 4A–4D indicated as A–D; fourth quartet cells 4a–4d indicated as 4A–4D. Endodermal cells derived from the mesentoblast 4d: blue; mesoblasts M^1 and M^2 and their mesodermal progeny: pink. (*a*) Midway gastrulation, note the internal position of the M^1 and M^2 and their small-celled mesodermal derivatives left and right from the small-celled endodermal cells derived from the enteroblasts E^1 and E^2 (*blue*) underneath the posterior region of small cells in the midline zone derived from descendants of 2d. (*b*) Late gastrula with slit-like blastopore. M^1 and M^2 have produced each a row of mesodermal cells. (*c*) Closure of the blastopore and beginning of the outgrowth of the stomodeum (shadowed) in anterior direction between the blastopore and the velum cells V^1 and V^2; M cells at the posterior end of the ventral midline. From the anal cells (An.C) many rows of cells radiate over the ventral surface toward the blastopore. (Ap) Apical plate. (Reprinted from Conklin 1897.)

lip is due to the specification of the D quadrant. It may be assumed that if D quadrant specification is suppressed, there will be no dorsal extension, resulting in development of a radially symmetric gastrula in which the blastopore maintains its position directly opposite the animal pole. This corresponds exactly with gastrulation in lobeless embryos of *Dentalium* (van Dongen 1976a,b). Lobeless embryos lack the organizer 3D, and all quadrants behave like the B quadrant. Similarly, in embryos of equally cleaving species, suppression of 3D induction results in the development of radialized embryos with the blastopore in its original position directly opposite the animal pole (Fig. 17). During normal development, the lineages of the B quadrant escape the influence of the organizer and repre-

ed with small cilia are found at this place; e.g., in *Physa* (Wierzejski 1905), *Lymnaea* (Raven 1946), and *Crepidula* (Conklin 1897). In *Patella* (Smith 1935; Dictus and Damen 1997), two cells with long and stiff cilia represent the telotroch and are found close to the edge of the shell field where the anus will break through. In the nudibranch *Aplysia* the anal cells are derivatives of 2d^{12}, although Carazzi (1905) erroneously derives them from 2d^{22}. For other species the origin of the anal cells is not always clear. So far, no direct proof can be found for a contribution of the blastopore to the formation of the anus (Fioroni 1966). It remains to be demonstrated whether the blastopore in the molluscan ancestor might have contributed to both the mouth and the anus, as discussed by Lartillot et al. (2002a).

As discussed above, the extension of the region between the dorsal part of the prototroch and the dorsal blastoporal

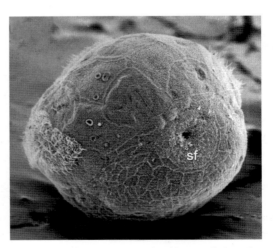

Figure 16. Embryo of *Lymnaea stagnalis* after completion of gastrulation, seen from the dorsal side. Head region upper half, trunk region lower half. (sf) Shell field.

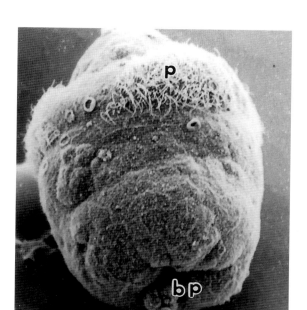

Figure 17. Radialized gastrula of *Lymnaea palustris*. Lateral view of 28-hour-old embryo resulting from cytochalasin treatment during the 24-cell stage, the interval between the formation of the third and fourth quartet of micromeres. (p) Prototroch cells, (bp) blastopore. (Courtesy of J.B. Morrill.)

sent a "default" pattern; i.e., embryos have a fourfold repetition of the cleavage pattern of the B quadrant (Martindale et al. 1985; Kühtreiber et al. 1988).

Expression of Pattern-controlling Genes during the Development of the Dorsoventral and Anterior–Posterior Axes

Just as in the chordate embryo, in mollusks gastrulation and dorsoventral axis formation are closely related. In both groups the dorsal blastoporal lip plays a crucial role in anterior–posterior patterning. In the molluscan embryo, the specification and activation of the organizer 3D results in the formation of the mesentoblast 4d (M). This M cell divides into left and right mesentoblasts, M_l and M_r, respectively. These are the progenitors of the left and right mesodermal bands from which the posterior mesoderm arises (Figs. 3 and 14). The anterior mesoderm is derived from 2b, 3a, and 3b. This anterior mesoderm develops independent of the organizer and extends far into the head region. In the early gastrula the two mesentoblasts are located at the dorsal blastoporal lip. Simultaneously with the gradual displacement of the blastopore, the cells at the ventral side are progressively replaced by derivatives of 2d, the first somatoblast. Inside the embryo, the two mesodermal teloblasts bud off a number of small cells in anterior direction. In this way, the two mesodermal bands are formed (Fig. 14a–c).

The elongation of the mesodermal bands follows the extension of the population of cells left and right along the ventral midline. Thus, the anterior parts of the mesodermal bands are in touch with earlier, and more posterior parts with later, formed offspring of the mesodermal teloblasts (Fig. 14a–c).

If one assumes that the organizer role of 3D passes to 4d, and from 4d to the two mesodermal teloblasts (M_l and M_r), then this situation resembles the development of the mesoderm in the chordate embryo. In chordate embryos, early organizer activity produces nervous tissue of anterior character. Later in development, part of this anterior neural tissue becomes posteriorized (Nieuwkoop 1991; Doniach 1993). This morphogenetic resemblance is strengthened by the apparent conservation of the molecular mechanisms involved in specification of the dorsoventral axis in mollusks and chordates. A large number of patterning genes have conserved functions throughout the animal kingdom. For example, in a number of organisms, the gene *Brachyury* is involved in axial patterning (Technau 2001; Chapter 41), as in the mouse (Beddington et al. 1992), in annelids (Arendt et al. 2001) and in *Patella* (Lartillot et al. 2002b). In *Patella* the homolog of the *Brachyury* gene, *PvuBra*, is first expressed in 3D, right after its centralization, thus in the organizer at the dorsal blastoporal lip, just as in the chordate embryo (Fig. 18). Shortly before the transition to the 64-cell stage, expression is propagated to derivatives of 2d, 3c, and 3d bordering the dorsal blastoporal lip. This activity is maintained during the greater part of gastrulation (Fig. 18b–d, f–h). After division of 3D, expression is stronger in the mesentoblast 4d than in 4D. During the following stages in both cells the expression is strongly reduced. A similar expression pattern of the *Brachyury* homolog has been observed in trochophores of the annelid *Platynereis* (Arendt et al. 2001). The morphogenetic function of *PvuBra* expression has been investigated by blocking the activation of MAPK (Lartillot et al. 2002b): 3D was normally centralized but it did not express *PvuBra*, and the cleavage pattern was radialized. Apparently, the MAPK signaling pathway and expression of *Brachyury* are essential for normal dorsoventral patterning.

During axis formation throughout the animal kingdom, from hydra (Martinez et al. 1997) to arthropods (Akiyama-Oda and Oda 2003) and vertebrates (Dirksen and Jamrich 1992; Sasaki and Hogan 1993), the specification of the anterior parts requires the function of the gene *forkhead*. Therefore, it is thought to play a conserved role. *Patella* appears to have a homolog for *forkhead*, *PvuFkh* (Lartillot et al. 2002b). This is first expressed immediately after the induction of 3D and appears to be limited to the presumptive endodermal macromeres 3A, 3B, and 3C, whereas no expression has been detected in 3D (Fig. 19a,c). The expression pattern of *PvuFkh* mirrors the expression of

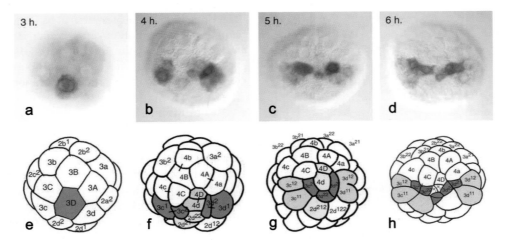

Figure 18. Expression of *PvuBra* during early development of *Patella vulgata*. (*a–d*) In situ hybridizations using antisense *PvuBra* antisense probe. (*e–h*) Schematic drawings indicating the position of the positive blastomeres. (*a, e*) Embryo between late 32- and 40-cell stage. 3D shows strong expression. (*b, f*) 64-cell stage, expression in 4d and 4D and in 2d², 3c, and 3d, dorsally from 4d. (*c, g*) Beginning of 88-cell stage, strong expression in 3c² and 3d², 2d²², weaker expression in the earlier positive cells 4d and 4D, and in the newly positive cells 3c¹¹, 3c¹², 3d¹¹, 3d¹², and 2d²¹². (*d, h*) Blastula after the first division of 4d into the two mesentoblast cells M₁ and M₂. Extension of the rim of positive cells around the dorsal side of the blastopore. (Modified, with permission, from Lartillot et al. 2002b [©Company of Biologists].)

PvuBra (Fig.18). Toward the sixth cleavage, *PvuFkh* expression strongly increases in the derivatives 4a–4c and 4A–4C; no expression has been detected in the posterior endodermal macromere 4D (Fig. 19b,d). This is striking, because in *Xenopus* there is also no expression of *forkhead* in the posterior endoderm. At the 64-cell stage the expression pattern of *PvuFkh* extends to the presumptive mesodermal cells 2a²²–2c²², 3a²¹, 3a²², and 3b²¹, 3b²² at the anterior side of the blastopore. After gastrulation the expression of *PvuFkh* is extended to all blastomeres around the stomodeum. Weak expression has been found in the internalized endodermal cells. Taken together, the expression of *PvuBra* at the posterior dorsal side of the blastopore and of *PvuFkf* at the anterior ventral side may be interpreted as an inherited character of a common ancestor of the radiata and the bilateria.

Finally, the role of the newly formed mesoderm in specification of surrounding structures should be discussed. In the chordate embryo the patterning role of the mesoderm for the development of the central nervous system is obvious. In the discussion above of the organizing role of the molluscan organizer 3D, it has also become apparent that, in the embryos of *Ilyanassa* (Clement 1962) and *Lymnaea* (Martindale 1986), the later the stage at which 3D is ablated, the less severe the resulting deficiencies. Direct evidence for a causal relation between mesoderm development and neurulation in mollusks does not exist.

It appears to be a general rule that in the molluscan embryo, gangliogenesis progresses in an anterior–posterior direction, to form a system of paired ganglia. The paired structure of the nervous system corresponds well with the presence of two mesodermal bands. The cerebral and the pedal ganglia develop first. Both are anterior parts of

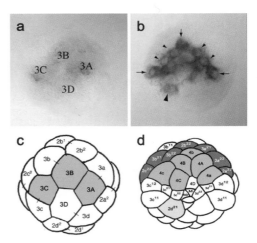

Figure 19. Expression of *PvuFkh* in the embryo of *Patella vulgata*. (*a*) Mid 32-cell stage with expression in the presumptive endodermal macromeres 3A, 3B, and 3C, but not in the presumptive mesendodermal 3D macromere. (*b*) Embryo just after the formation of the two M cells. The daughter cells of the macromeres 3A, 3B, and 3C: 4a, 4b, 4c, 4A, 4B, and 4C are still positive. Of the two cells derived from 3D, only the endodermal macromere 4D is positive, whereas the mesentoblast 4d is negative. The expression of *PvuFkH* is further extended to the cells 3a²¹, 3a²², 3b²¹, 3b²², 2a²², 2b²², and 2c²², forming a rim of blastomeres ventral to the positive macromere lineages. (*c* and *d*) Blastomere denominations of the embryos in *a* and *b*, respectively. (Reprinted, with permission, from Lartillot et al. 2002a [Springer-Verlag Gmb H].)

the central nervous system (Raven 1966). The cerebral ganglia develop in juxtaposition with the anterior mesoderm anterior from the mouth, whereas the pedal ganglia develop directly posterior from the mouth and right and left from the primordium of the foot (van Dongen 1976b; Wierzejski 1905). The formation of two ganglia in the ventral part of the head region is probably causally related with the mesodermal derivatives of 2b, 3a, and 3b. Micromeres 3c and 3d do not develop mesoderm, but only contribute to the development of the ectoderm in the trunk region. The mesodermal function of 2b, 3a, and 3b is probably the "default" character. The ectodermal fate of 3c and 3d is then acquired under the influence of 3D. Thus, in the absence of the organizer, 3c and 3d might also produce mesoderm, involute and induce the head ectoderm to develop two additional cerebral ganglia. That is exactly what happens in lobeless embryos of *Dentalium*. The head pattern is radialized with four cerebral ganglia (van Dongen 1976b). Lobeless embryos of *Bithynia* also produce mesoderm and cerebral ganglia (Cather et al. 1976).

CONCLUSION

Gastrulation in the protostomian molluscan embryo resembles gastrulation in the chordate embryo. This may be the result of common descent. In the common ancestor, gastrulation probably started at the vegetal pole with the blastopore directly opposite the animal pole. During gastrulation, an elongation of the embryonic axis may have taken place in the direction of the animal–vegetal axis. It is likely that the ancestor only had one opening, like the recent coelenterates. The evolutionary transition from such a radially symmetrical tube-like body plan with only one opening to organisms with a bilaterally symmetrical architecture with two openings probably did not arise suddenly. It may be assumed that this tube-like animal already did develop a bilateral symmetry, such as can be observed in the organization of the sea anemones. Such a relatively simple bilateral symmetry with a dorsal and a ventral side implies that the blastopore must have had a dorsal and a ventral lip. The later evolution to animals with a digestive tract with mouth and anus might have occurred in three different ways: (1) After extension of the body axis, the blastopore has maintained its posterior position and becomes the anus, while the mouth had to be developed later (Deuterostomia). (2) Extension is limited to the dorsal side and the mouth is shifted anteriorly toward the ventral side of the embryo. The anus is then formed later (Protostomia). (3) Extension of the primary body axis is limited to the dorsal side, associated with the transformation of the original circular blastopore into a longitudinal slit. The lateral lips of the blastopore converge and finally fuse. In this way, the original blastopore is transformed into a tube with an anterior

mouth opening and a posterior anal opening (amphistomy) (Arendt and Nübler-Jung 1997, 1999; Holland 2000; van den Biggelaar et al. 2002).

The above discussion of gastrulation in mollusks leads to the conclusion that gastrulation in protostomians and deuterostomians have essential features in common. In both groups the dorsal blastoporal lip has an organizer function, expresses homologous genes, and uses a conserved signaling pathway.

ACKNOWLEDGMENT

The encouraging and critical remarks made by Claus Nielsen, Lex Nederbragt, Eric Edsinger, and Nicolai Lartillot are gratefully acknowledged.

REFERENCES

Akiyama-Oda Y. and Oda H. 2003. Early patterning of the spider embryo: A cluster of mesenchymal cells at the cumulus produces Dpp signals received by germ disc epithelial cells. *Development* **130:** 1735–1747.

Arendt D. and Nübler-Jung K. 1997. Dorsal or ventral: Similarities in fate maps and gastrulation patterns in annelids, arthropods and chordates. *Mech. Dev.* **61:** 7–21.

———. 1999. Comparison of early nerve cord development in insects and vertebrates. *Development* **126:** 2309–2325.

Arendt D., Technau U., and Wittbrodt J. 2001. Evolution of the bilaterian larval foregut. *Nature* **409:** 81–85.

Arnolds W.J.A., van den Biggelaar J.A.M., and Verdonk N.H. 1983. Spatial aspects of cell interactions involved in the determination of dorsoventral polarity in the equally cleaving gastropods and regulative abilities of their embryos, as studied by micromere deletions in *Lymnaea* and *Patella*. *Roux's Arch. Dev. Biol.* **192:** 75–85.

Baba S. 1938. The later development of a solenogastre, *Epimenia verrucosa* (Nierstrasz). *J. Dep. Agric. Kyusyu Imper. Univ.* **6:** 21–40.

———. 1940. The early development of a solenogastre *Epimenia verrucosa* (Nierstrasz). *Annot. Zool. Jpn.* **19:** 107–113.

Beddington R.S., Rashbass P., and Wilson V. 1992. *Brachyury*—A gene affecting mouse gastrulation and early organogenesis. *Dev. Suppl.* **1992:** 157–165.

Boring L. 1989. Cell-cell interactions determine the dorsoventral axis in embryos of an equally cleaving opisthobranch mollusk. *Dev. Biol.* **136:** 239–253.

Carazzi D. 1905. L'embriologia dell'Aplysia e i problemi fondamentali dell'embriologia comparata. *Arch. Ital. Anat. Embrol.* **4:** 231–504 (Pls. 29–50).

Cather J.N. and Verdonk N.H. 1974. The development of *Bithynia tentaculata* (Prosobranchia, Gastropoda) after removal of the polar lobe. *J. Embryol. Exp. Morphol.* **31:** 415–422.

Cather J.N., Verdonk N.H., and Dohmen M.R. 1976. Role of the vegetal body in the regulation of development in *Bithynia tentaculata* (Prosobranchia, Gastropoda). *Am. Zool.* **16:** 455–468.

Clement A.C. 1952. Experimental studies on germinal localization in *Ilyanassa*. I. The role of the polar lobe in determination of the cleavage pattern and its influence in later development. *J. Exp. Zool.* **121:** 593–626.

———. 1962. Development of *Ilyanassa* following removal of the D macromere at successive cleavage stages. *J. Exp. Zool.* **149:** 193–216.

———. 1967. The embryonic value of the micromeres in *Ilyanassa obsoleta*, as determined by deletion experiments. I. The first quartet cells. *J. Exp. Zool.* **166:** 77–88.

Conklin E.G. 1897. The embryology of *Crepidula*, a contribution to the cell lineage and early development of some marine gastropods. *J. Morphol.* **13:** 1–226 (Pls. 1-8).

Damen P. 1994. "Cell lineage, and specification of developmental fate and dorsoventral organisation in the mollusc *Patella vulgata.*" Ph.D. thesis, University of Utrecht, The Netherlands.

Damen P. and Dictus W.J.A.G. 1996. Organiser role of the stem cell of the mesoderm in prototroch patterning in *Patella vulgata* (Mollusca, Gastropoda). *Mech. Dev.* **56:** 41–60.

Damen W., Klerkx A.H.E.M., and van Loon A.E. 1996. Micromere formation at third cleavage is decisive for trochoblast specification in the embryogenesis of *Patella vulgata. Dev. Biol.* **178:** 238–250.

Dictus W.J.A.G. and Damen P. 1997. Cell-lineage and clonal-contribution map of the trochophore larva of *Patella vulgata. Mech. Dev.* **62:** 213–226.

Dirksen M.L. and Jamrich M. 1992. A novel, activin-inducible, blastopore lip-specific gene of *Xenopus laevis* contains a fork head DNA-binding domain. *Genes Dev.* **6:** 599–608.

Doniach T. 1993. Planar and vertical induction of anteroposterior patterning during the development of the amphibian central nervous system. *J. Neurobiol.* **24:** 1256–1275.

Fioroni P. 1966. Zur Morphologie und Embryogenese der Darmtraktes und der transitorischen Organe bei Prosobranchiern (Mollusca, Gastropoda). *Rev. Suisse Zool.* **73:** 621–876.

Freeman G. and Lundelius J.W. 1992. Evolutionary implications of the mode of D-quadrant specification in coelomates with spiral cleavage. *J. Evol. Biol.* **5:** 205–247.

Guerrier P. 1970. Les charactères de la segmentation et de la détermination de la polarité dorsoventrale dans le développement de quelque Spiralia. III. *Pholas dactylus* et *Spisula subtruncata* (Mollusques, Lamellibranchs). *J. Embryol. Exp. Morphol.* **23:** 667–692.

Guerrier P., van den Biggelaar J.A.M., van Dongen C.A.M., and Verdonk N.H. 1978. Significance of the polar lobe for the determination of dorsoventral polarity in *Dentalium vulgare* (da Costa). *Dev. Biol.* **63:** 233–242.

Heath H. 1899. The development of *Ischnochiton. Zool. Jahrb. Anat.* **12:** 567–656 (Pls. 31–35).

Holland L.Z. 2000. Body-plan evolution in the Bilateria: early anteroposterior patterning and the deuterostome-protostome dichotomy. *Curr. Opin. Genet. Dev.* **10:** 434–442.

Janssen-Dommerholt C., van Wijk R., and Geilenkirchen W.L.M. 1983. Restriction of developmental potential and trochoblast ciliation in *Patella* embryos. *J. Embryol. Exp. Morphol.* **74:** 69–77.

Kühtreiber W.M. and van Dongen C.A.M. 1989. Microinjection of lectins, hyaluronidase, and hyaluronidase fragments interferes with cleavage delay and mesoderm induction in embryos of *Patella vulgata. Dev. Biol.* **132:** 436–441.

Kühtreiber W.M., van Till E.H., and van Dongen C.A.M. 1988. Monensin interferes with the determination of the mesodermal cell line in embryos of *Patella vulgata. Roux's Arch. Dev. Biol.* **197:** 10–18.

Lambert J.D. and Nagy L.M. 2001. MAPK signalling by the D quadrant embryonic organizer of the molluscs *Ilyanassa obsoleta. Development* **128:** 45–56.

———. 2002. Asymmetric inheritance of centrosomally localized mRNAs during embryonic cleavages. *Nature* **420:** 682–686.

———. 2003. The MAPK cascade in equally cleaving spiralian embryos. *Dev. Biol.* **263:** 231–241.

Lartillot N. 2001. "Une approche comparative de la gastrulation chez les Bilateria: étude des gènes *Brachyury* et *fork head* chez un mollusque, *Patella vulgata.*" Ph.D. thesis, Université de Paris-Sud, Orsay.

Lartillot N., Le Gouar M., and Adoutte A. 2002a. Expression patterns of fork head and goosecoid homologues in the mollusc *Patella vulgata* supports the ancestry of the anterior mesendoderm across Bilateria. *Dev. Genes Evol.* **212:** 551–561.

Lartillot N., Lespinet O., Vervoort M., and Adoutte A. 2002b. Expression pattern of *Brachyury* in the molluscs *Patella vulgata* suggests a conserved role in the establishment of the AP axis in Bilateria. *Development* **129:** 1411–1421.

Martindale M. 1986. The 'organizing' role of the D quadrant in an equal-cleaving spiralian, *Lymnaea stagnalis* as studied by UV laser deletion of macromeres at intervals between third and fourth quartet formation. *Int. J. Invertebr. Reprod. Dev.* **9:** 229–242.

Martindale M., Doe CQ., and Morrill J.B. 1985. The role of animal-vegetal interaction with respect to the determination of dorsoventral polarity in the equal-cleaving spiralian, *Lymnaea palustris. Roux's Arch. Dev. Biol.* **194:** 281–295.

Martinez D.E., Dirksen M.-L., Bode P.M., Jamrich M., Steele R.E., and Bode H.R. 1997. Budhead, a fork head/HNF-3 homolog, is expressed during axis formation and head specification in hydra. *Dev. Biol.* **192:** 523–536.

Minganti A. 1950. Acidi nucleici e fosfatasi nello sviluppo della *Limnaea. Riv. Biol.* **42:** 295–313.

Moor B. 1973. Zur frühen Furchung des Eies von *Littorina littorea* L. (Gastropoda Prosobranchia) Ein neues Beispiel von Pollappenbildung) *Zool. Jahrb. Anat.* **91:** 546–573.

Morrill J.B. 1997. Cellular patterns and morphogenesis in early development of freshwater pulmonate snails, *Lymnaea* and *Physa* (Gastropoda, Mollusca). In *Reproductive biology of invertebrates* (ed. K.G. Adiyodi and R.G. Adiyodi), vol. 7, pp. 67–107. Oxford & IBH Publishing Co., New Dehli.

Morrill J.B., Blair C.A., and Larsen W. 1973. Regulative development in the pulmonate gastropod, *Lymnaea palustris*, as determined by blastomere deletion experiments. *J. Exp. Zool.* **183:** 47–55.

Nielsen C. 2001. *Animal evolution: Interrelationships of the living phyla*, 2nd edition. Oxford University Press, New York.

Nieuwkoop P.D. 1991. IIA. Pattern formation in the developing central nervous system (CNS) of the amphibians and birds. *Proc. Kon. Ned. Akad. Wet.* **94:** 111–120.

Raven C.P. 1946 The development of the egg of *Limnaea stagnalis* L. from the first cleavage till the trochophore stage, with special reference to its "chemical embryology". *Arch. Néerl. Zool.* **7:** 353–434.

———. 1966. *Morphogenesis: The analysis of molluscan development*, 2nd edition Pergamon Press, Oxford, United Kingdom.

Render J. 1989. Development of *Ilyanassa obsoleta* after equal distribution of polar lobe material at first cleavage. *Dev. Biol.* **132:** 241–250.

———. 1991. Fate maps of the first quartet micromeres in the gastropod *Ilyanassa obsoleta. Development* **113:** 495–501.

———. 1997. Cell fate maps in the *Ilyanassa obsoleta* embryo beyond the third division. *Dev. Biol.* **189:** 301–310.

Robert A. 1902. Recherche sur le développement des troques. *Arch. Zool. Exp. Gén.* (3rd Ser.) **10:** 269–538.

Sasaki H. and Hogan B.L. 1993. Differential expression of multiple fork head related genes during gastrulation and axial pattern formation in the mouse embryo. *Development* **118:** 47–59.

Smith F.G.W. 1935. The development of *Patella vulgata. Philos. Trans. Roy. Soc. Lond. B* **225:** 95–125.

Technau U. 2001. Brachyury, the blastopore and the evolution of the mesoderm. *BioEssays* **23:** 788–794.

van Dam W.I. and Verdonk N.H. 1982. The morphogenetic significance of the first quartet micromeres for the development of the snail *Bithynia tentaculata. Roux's Arch. Dev. Biol.* **191:** 112–118.

van den Biggelaar J.A.M. 1976. The fate of maternal RNA containing ectosomes in relation to the appearance of dorsoventrality in the pond snail, *Lymnaea stagnalis. Proc. Kon. Ned. Akad. Wet. Ser. C Biol. Med. Sci.* **79:** 421–426.

———. 1977. Development of dorso-ventral polarity and mesentoblast determination in *Patella vulgata. J. Morphol.* **154:** 157–186.

———. 1996. Cleavage pattern and mesentoblast formation in *Acanthochiton crinitus. Dev. Biol.* **74:** 423–430.

van den Biggelaar J.A.M. and Guerrier P. 1979. Dorsoventral polarity and mesentoblast determination as concomitant results on cellular interactions in the mollusk *Patella vulgata. Dev. Biol.* **68:** 462–471.

———. 1983. Origin of spatial organization. In *The mollusca* (ed. N.H. Verdonk et al.), vol. 3, pp. 179–213. Academic Press, New York.

van den Biggelaar J.A.M. and Haszprunar G. 1996. Cleavage patterns and mesentoblast formation in the Gastropoda: An evolutionary perspective. *Evolution* **50:** 1520–1540.

van den Biggelaar J.A.M., Edsinger Gonzales E., and Schram F.R. 2002. The improbability of dorso-ventral axis inversion during animal evolution, as presumed by Geoffroy Saint Hilaire. *Contrib. Zool.* **71:** 29–36.

van Dongen C.A.M. 1976a. The development of *Dentalium* with special reference to the significance of the polar lobe. V and VI. Differentiation of the cell pattern in lobeless embryos of *Dentalium vulgare* (da Costa) during late larval development. *Proc. Kon. Ned. Akad. Wet. C Biol. Med. Sci.* **79:** 245–266.

———. 1976b. The development of *Dentalium* with special reference to the significance of the polar lobe. VII. Organogenesis and histogenesis in lobeless embryos of *Dentalium vulgare* (da Costa) as compared to normal development. *Proc. Kon. Ned. Akad. Wet. C Biol. Med. Sci.* **79:** 454–465.

van Dongen C.A.M. and Geilenkirchen W.L.M. 1974. The development of *Dentalium* with special reference to the significance of the polar lobe. *Proc. Kon. Ned. Akad. Wet. C Biol. Med. Sci.* **77:** 57–100.

———. 1975. The development of *Dentalium* with special reference to the significance of the polar lobe. IV. Division chronology and development of the cell pattern in *Dentalium dentale* after removal of the polar lobe at first cleavage. *Proc. Kon. Ned. Akad. Wet. C Biol. Med. Sci.* **78:** 358–375.

Wierzejski A. 1905. Embryologie von *Physa fontinalis* L. *Z. Wiss. Zool.* **83:** 502–706 (Pls. 18–26).

Wilson E.B. 1904. Experimental studies in germinal localization. *J. Exp. Zool.* **1:** 197–268.

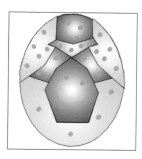

Gastrulation in Crustaceans
Germ Layers and Cell Lineages

M. Gerberding and N.H. Patel

Department of Integrative Biology, Department of Molecular Cell Biology, and Howard Hughes Medical Institute, University of California, Berkeley, California 94720

INTRODUCTION

Arthropod embryos progress through a well-conserved germ-band stage where they share many morphological characteristics. There is, however, a great deal of variation in the events leading up to the germ-band stage. Nowhere is this variation more evident than in the crustaceans, where the variety of patterns of cleavage and gastrulation are greater than in any other arthropod group. Indeed, the first major papers on crustacean embryology, published in 1877 and 1879, described the process of gastrulation in the crayfish and the water flea and revealed how different this process could be within this group of arthropods (Reichenbach 1877; Grobben 1879). Thanks to morphological studies in the following decades, descriptions of early development are available for most of the crustacean subgroups, and various modes of gastrulation are documented in a wide variety of crustacea (Shiino 1957; Weygoldt 1993). This variation begs the question as to how the different types of early development have evolved and what aspects are homologous versus convergent. For now, however, the relationship between crustacean subgroups and between crustaceans and other arthropod groups remains controversial, and there is no consensus as to the ancestral mode of development in this group (Scholtz 1997, 2000; Walossek 1999; Brusca and Brusca 2002). Therefore, we continue to organize the data according to the oldest and simplest of all groupings that splits the crustaceans into what is termed non-malacostracan and malacostracan crustaceans. This simple binary split, plus some subdivisions within the two

major crustacean groups, helps organize our discussion of crustacean gastrulation (see Table 1).

It is also important to note that although many of the past morphological studies of crustacean gastrulation were

Table 1. *Major grouping of crustaceans*

Non-malacostracans
    Branchiopoda
        Anostraca (brine and fairy shrimp): *Artemia salina* (brine shrimp, also known as sea monkeys)
        Cladocera (water fleas): *Leptodora kindti, Polyphemus pediculus, Daphnia pulex*
    Maxillopoda
        Copepoda: *Cyclops viridis, Lernaeocera*
        Ostracoda: *Cypris incongruens, Cyprideis litoralis*
        Cirripedia (barnacles): *Lepas, Balanus, Mitella, Teraclita, Ibla, Scalpellum*
Malacostracans
    Syncarida: *Anaspides tasmaniae*
    Decapoda: *Sicyonia ingentis* (prawn = shrimp), *Astacus* (crayfish)
    Leptostraca: *Nebalia bipes*
    Hoplocarida (mantis shrimps): *Squilla oratoria*
    Peracarida
        Mysidacea (opossum shrimps): *Hemimysis lamornae*
        Isopoda: *Jaera* (sea slaters), *Armadillidium* (pillbugs), *Asellus, Porcellio* (sowbugs)
        Amphipoda (beach hoppers): *Orchestia cavimana, Parhyale hawaiensis*

The phylogeny of crustaceans is not well resolved, but the species discussed in this chapter can be grouped as shown in this table. Names in italics are at the level of genera and species, and some common names are given in parentheses.

very carefully done, gastrulation is a dynamic process that involves the movement of cells and, as such, is difficult to fully comprehend by looking at discrete stages. Even the most detailed morphological papers struggle to give a satisfying account about the origin of the cells of the different germ layers and where they go. Recently, cell labeling experiments have been started in two amphipod crustaceans, and for the first time, cell lineage and cell fate in crustaceans can be followed from the first cell division until hatching in living crustacean embryos (Gerberding et al. 2002; Scholtz and Wolff 2002; Wolff and Scholtz 2002). In addition, molecular studies of crustacean gastrulation are just beginning, and these will help us to further understand the mechanisms and evolution of crustacean gastrulation.

BASIC CHARACTERS OF EARLY CRUSTACEAN DEVELOPMENT

Before reviewing details of gastrulation in specific species of crustaceans, it is useful first to describe some of the general properties of early crustacean development and gastrulation. This also allows us to introduce a number of descriptive terms that have been used by various authors in their studies of crustacean embryogenesis.

Total versus Superficial Cleavage

Depending on the species, early crustacean embryos undergo total cleavage, superficial cleavage, or a mix that starts out with total cleavage and switches to superficial cleavage and vice versa. Embryos with different modes of cleavage show different behaviors at gastrulation. In the case of total cleavage, cell divisions distribute all the material of the egg among the blastomeres, and then gastrulation takes place as a movement of some cells toward the center of the egg. In the case of superficial cleavage, cell division results in the formation of a superficial monolayer of cells overlying a yolk mass, and gastrulation creates a multilayered embryo, which still sits on the surface of the egg. For both total and superficial cleavage, gastrulation can occur through either invagination or ingression. During invagination, an inner layer of mesendoderm cells forms a pouch that is continuous with an outer layer of prospective ectoderm. When ingression occurs, single cells or groups of cells depart from the blastula layer to form the precursors of the mesoderm and endoderm. Embryos that display so-called mixed cleavages switch to superficial cleavage before gastrulation (e.g., *Cyprideis*), but in some cases the switch is made after gastrulation begins (e.g., *Daphnia*). Superficial, mixed, and total cleavage are found in closely related taxa, suggesting a relatively easy switch in evolution between the different modes of early development. Comparing gastrulation between these taxa can tell us something about the evolutionary history of this process. In some cases, certain arrangements of cells, lineage patterns, and movement are conserved between species, suggesting some ancestral characteristics, whereas other aspects show remarkable plasticity (see examples below for water fleas, isopods, and amphipods). Interestingly, neither egg size nor the amount of yolk correlates with the presence or absence of total cleavage. Yolky eggs as big as 1 mm in diameter show total cleavage (*Anaspides*), whereas eggs as small as 250 μm show superficial cleavage (*Leptodora*).

Fate Maps

Fate maps are commonly used to show the location of the material for the prospective germ layers before, during, and after gastrulation, and certainly most developmental biologists are familiar with the usefulness of fate maps in model systems such as *Drosophila* (Hartenstein et al. 1985). Fate maps of embryos prior to gastrulation are essential to understand the movement of cells during gastrulation. Their use in the study of crustacean gastrulation has been limited, however, because most crustacean fate maps depict the arrangement of cells at stages after gastrulation (Manton 1928; Shiino 1957; Anderson 1973). However, there are several crustacean groups whose embryos show total cleavage and a fixed cleavage pattern (see below). In these species, it has been possible to trace back the origin of the germ layers and to generate fate maps for stages before gastrulation. It is hoped that this level of analysis can also be extended to crustaceans displaying superficial cleavages and those without fixed lineage patterns.

Random versus Fixed Cleavage Patterns and Invariant Cell Lineages

In embryos where invariant early cell lineages are found, the pedigree of a given cell predicts its fate even before gastrulation starts. Invariant cleavage and invariant cell lineage are not necessarily the same and can be defined in more than one way, but for our purpose of looking at gastrulation and germ layer formation, the two are linked, because an invariant cell lineage requires that there be an invariant cleavage pattern (although the presence of an invariant cleavage pattern does not necessarily mean that there is an invariant cell lineage pattern). It does not appear that invariant lineages occur in the context of superficial cleavage.

There is a large body of morphological data for crustaceans that display total cleavage up through gastrulation. In most cases, several features such as differences in cell size, spindle direction, and rate of cell division are visible and can be used to identify individual cells and to follow their daughters. Many of these descriptions strongly suggest the presence of invariant cell lineages for the germ layers (for review, see Shiino 1957; Hertzler 2002). In the remainder of cases, either no such features are present and lineages cannot be detected (Müller-Cale 1913; Benesch 1969), or the features are present but they do not suggest an invariant lineage (Weygoldt 1960). More recently, several studies take advantage of modern techniques to label early blastomeres and follow the fate of clones beyond gastrulation for the 2- and 4-cell stage of decapod shrimps, and for the 8- and 16-cell stage of amphipods.

Other Problems of Terminology

In summarizing crustacean gastrulation, there are two complex characters dealt with in many studies that we only discuss briefly. The first character concerns the different kinds of mesoderm that have been described. Crustaceans have two kinds of somatic mesoderm, naupliar and post-naupliar. Several studies describe separate origins for the two kinds of mesoderm, most clearly for malacostracans where the post-naupliar mesoderm is generated by stem cells, called mesoteloblasts. However, many studies do not discriminate between the naupliar and post-naupliar mesoderm. In addition, there is also the distinction between somatic and visceral mesoderm. The somatic mesoderm is often distinguishable earlier than the visceral, since identifying the latter depends on the organogenesis of a gut. For this and other reasons, several studies concentrate on somatic mesoderm and do not address the origin of the visceral mesoderm. Because of the disparities in the data as to how populations of mesoderm cells are subdivided, we give priority to how mesoderm is made in general in different taxa and use a single header for all mesoderm.

The second complication concerns the endoderm. An obvious role of the endoderm is in the generation of the endodermal part of the gut and its connection to the ectodermal stomodeum and proctodeum. In yolky eggs, however, there is an additional need to enclose and digest the yolk. There are several ways that crustaceans form and internalize so-called endodermal yolk cells, but the homology of those cells and of the process of their internalization is unclear (Fioroni 1970). We consider the variation found in the generation of yolk cells to be of secondary relevance for this chapter and bypass it in our subsequent descriptions of crustacean gastrulation.

MORPHOLOGICAL ANALYSES OF CRUSTACEAN GASTRULATION

The Non-Malacostracan Crustaceans

Gastrulation by Invagination and Ingression: The Brine Shrimps, Water Fleas, Copepods, and Ostracods

Brine shrimps. The embryos of the arguably two most phylogenetically basal crustacean groups, remipeds and cephalocarids, have not been described. The remainder of

the crustaceans are classified as branchiopods, maxillopods, and malacostracans (Ax 1999; Walossek 1999; Richter 2002), and their embryos have been described previously (see Table 1). Among the branchiopods, the brine shrimp *Artemia* is considered the most primitive taxon. Cleavage in *Artemia* is total and perfectly equal, so there is no morphological feature that allows one unambiguously to identify cells and to track them over time. Gastrulation has two phases. The first is a posterior invagination of prospective mesoderm and putative germ cells (Fig. 1A), the second an anterior ingression of prospective endoderm (Fig. 1A′) (Benesch 1969). This gastrulation of mesoderm precursors and endoderm precursors with internalization at two locations and in two phases is also found in other non-malacostracan as well as malacostracan crustaceans, although in some cases the movements of prospective mesoderm and endoderm occur at the same time (see below). (Note: For the sake of simplicity, we from now on refer to mesoderm precursors or prospective mesoderm as simply mesoderm when talking about these cells prior to gastrulation, and likewise for endoderm and ectoderm.)

Water fleas. The comparison of the development of water fleas (cladocerans) is informative, and representatives of the raptorial and filter-feeding subgroups have been studied. One representative of the raptorial taxon Gymnomera is *Leptodora kindti*. The development of *Leptodora* is different from the development of the remainder of the water fleas. Early cleavage is superficial and gastrulation is a mix of epiboly and delamination (Samter 1900; Gerberding 1997). Initially, the blastopore is wide and oval. The ectoderm of the head is in front of it, the putative mesendoderm is at its anterior edge, the yolk cells fill its center up to its posterior edge, and the ectoderm of the trunk follows beyond its posterior edge. In an epiboly process, anterior and posterior ectoderm gradually slides over the smaller mesendoderm cells and the larger yolk cells, thereby closing the blastopore (Fig. 1B). In a parallel delamination process, the majority of mesendoderm cells can be seen dividing radially (Fig. 1B′) (Gerberding 1997). Such radial cell divisions are also found outside the branchipods in the isopod *Porcellio* and the shrimp *Sicyonia* (see below).

The development of another representative of the Gymnomera, *Polyphemus pediculus*, is the same as that of the two other taxa of the water fleas, the filter-feeding Anomopoda and Ctenopoda (Grobben 1879; Kühn 1913; Baldass 1941). They display total or mixed cleavage and a wide variation in the amount of yolk, but they all show the same invariant cleavage pattern. During initial cleavages, separate progenitors for ectoderm, mesoderm, endoderm, and the germ line are established (Fig. 1C). The fate of endoderm and germ-line cells separates from the ectoderm after the fourth division, the mesoderm fate from the ectoderm fate after the ninth division. Depending on the

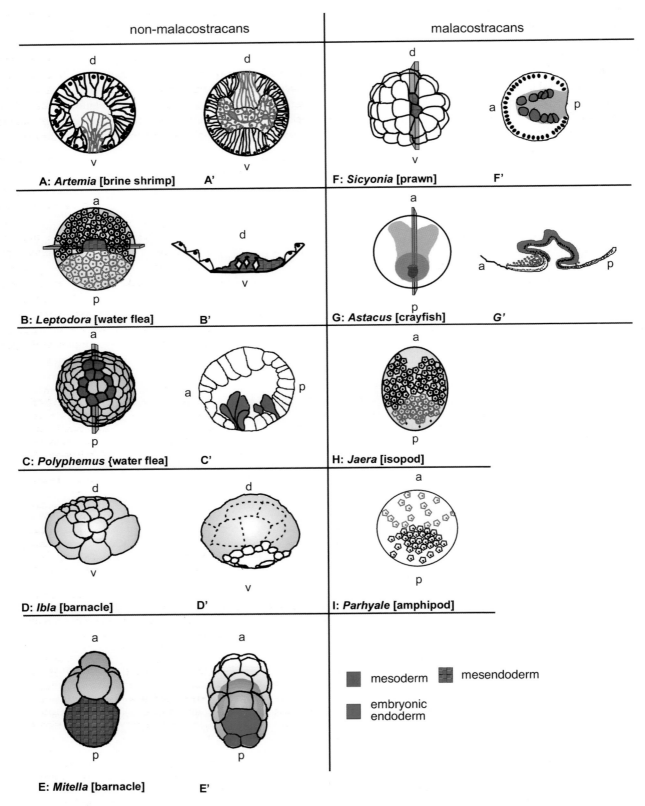

Figure 1. (*See facing page for legend.*)

amount of yolk in different species, the progenitors for the inner layers either invaginate by means of a blastopore and a groove, or they ingress as single cells from the blastoderm (Fig. 1C′) (Grobben 1879; Kühn 1913; Baldass 1941). In *Daphnia*, the endoderm is situated anterior to the mesoderm (Baldass 1941). The other descriptions for water fleas are less clear in this respect. However, the same arrangement as in *Daphnia* is found in *Artemia* (see above). In malacostracans, the reverse arrangement for these precursor cells is found (see below).

Copepods. The copepods and ostracods are maxillopods; it is unclear which of the two taxa is more phylogenetically primitive. The situation in the copepod *Cyclops viridis* is very similar to that in the water fleas with regard to cell lineage, but not to gastrulation (Fuchs 1914). There is a central pair of sister cells that give rise to both endoderm and germ line at the vegetal pole and a ring of mesoderm cells encircling this pair. At this time point, it is not possible to demarcate the prospective anterior–posterior axis. At gastrulation, the endoderm and germ line ingress first, and the adjacent mesoderm cells follow in an invagination process (Fuchs 1914), which is the reverse of the order seen in water fleas. In parasitic copepods, early embryonic development is quite different, possibly due in part to the unusual parasitic life history for these species. In *Lernaeocera*, there is a highly unequal cleavage setting up one large macromere with all the yolk and three micromeres without yolk that form a cap. Gastrulation is by epiboly as the micromeres grow and enclose the macromere. There are no data, however, to address whether any invariant cell lineage patterns exist prior to gastrulation in *Lernaeocera* (Köhler 1976).

Ostracods. In the ostracods *Cypris incongruens* and *Cyprideis litoralis*, no pattern of invariant cell lineage has been documented for the early embryo (Müller-Calé 1913; Weygoldt 1960). Cleavage is total and perfectly equal. Gastrulation happens comparatively late after ten rounds of cell division. The lack of distinctive features and the small size of the cells make it difficult to track any cell lineage and cell fate for individual blastoderm cells or for groups of cells. Gastrulation has two phases. First, mesoderm invaginates at the future posterior pole. Second, a ventral rim extends from the blastopore toward the future anterior end and the endoderm immigrates at the anterior tip of the rim (Weygoldt 1960). The situation is complicated by the pres-

Figure 1. Morphological analysis of gastrulation and the origin of germ layers in crustaceans. (*A–E*) Non-malacostracans. (*F–H*) Malacostracans. Orientation is coded as (d) dorsal, (v) ventral, (a) anterior, (p) posterior. (A, A′, B′, C′, F′, and G′) show sections of embryos. All other panels show whole mounts. Horizontal and vertical dark gray slices in the whole-mount panels indicate the planes of the corresponding sections. (*A*) The brine shrimp *Artemia*. Cleavage is total. (*A*) Transverse section of the 1st phase of gastrulation showing the ventral ingression of mesoderm at the prospective posterior end. (*A′*) Transverse section of the 2nd phase showing the dorsal ingression of endoderm at the prospective anterior end (see Benesch 1969). (*B*) The water flea *Leptodora*. Cleavage is superficial. (*B*) Whole mount of gastrulation. The blastopore is at the level of the division between prospective head and trunk, and gastrulation occurs through an epiboly of the ectoderm to cover the mesendoderm and yolk cells. (*B′*) Transverse section of the same stage revealing an additional delamination process of radial divisions of the mesendodermal material (see Gerberding 1997). (*C*) The water flea *Polyphemus*. Cleavage is total. (*C*) Whole mount of blastoderm at 118-cell stage. The endoderm progenitors are located in front of the mesoderm progenitors. (*C′*) Sagittal section of gastrulation at the 236-cell stage. The endoderm as well as the mesoderm ingresses (see Kühn 1913). (*D*) The barnacle *Ibla*. Cleavage is total and unequal. (*D*) Early blastula, 30–40 cells. Cell size is similar among micromeres and macromeres, respectively. (*D′*) Blastopore closure. Gastrulation is via epiboly as the micromeres overgrow the macromeres. There are, however, no data on cell fates in this embryo (see Anderson 1965). (*E*) The barnacle *Mitella*. Cleavage is total and unequal. (*E*) 8 cells. The mesendoderm arises from the biggest macromere. (*E′*) 32-cell stage. The blastopore is at the prospective posterior end. Gastrulation is an ingression of the endoderm progenitor, followed by two mesoderm progenitors (see Shiino 1957). (*F*) The shrimp *Sicyonia*. Cleavage is total. (*F*) 62-cell stage. The blastopore is at the prospective posterior end. Gastrulation is driven by cell divisions that are oriented toward the center of the egg. The first two cells that gastrulate are the progenitors for yolk endoderm and mesendoderm. Once inside, the mesendoderm progenitor gives rise to separate progenitors for endoderm, mesoderm, and germ line. (*F′*) Sagittal section of a later stage. The archenteron is shaded. After getting inside, the mesendoderm cell has divided and given rise to progenitors for endoderm and trunk mesoderm. The cells next to blastopore have followed them and moved internally as well to generate head mesoderm (see Hertzler and Clark 1992; Hertzler 2002). (*G*) The crayfish *Astacus*. Cleavage is superficial. (*G*) At gastrulation, invagination starts at a round, wide blastopore at the prospective posterior end. Mesoderm and endoderm descend from the same area of the blastoderm. At the stage shown, the blastopore is about halfway to its closure. (*G′*) Sagittal section of the same stage. Mesoderm and endoderm have separated at this time. The mesoderm ingresses as individual cells medially, mostly in front of the blastopore and less so behind it. The endoderm invaginates and forms a pouch (see Reichenbach 1877 and 1886). (*H*) The isopod *Jaera*. Cleavage is superficial. Ectoderm, mesoderm, endoderm, and yolk cells show early cytological differences. The blastopore is at the future posterior end, and gastrulation occurs via ingression. Prospective mesoderm is located anterior to the prospective endoderm (see McMurrich 1895). (*I*) The amphipod *Parhyale*. Cleavage is total. The material for the germ layers is generated by separate progenitors from the 8-cell stage. The blastopore is at the prospective anterior, with the mesoderm located anterior of the ectoderm and posterior of the endoderm (see Gerberding et al. 2002). See also Fig. 2.

ence of yolk and a blastocoel inside the yolk (Müller-Calé 1913; Weygoldt 1960).

Gastrulation by Epiboly of Micromeres: The Barnacles

In the cirripeds, the general condition is that the first two cleavages are equal and the third is an unequal cleavage that sets up four macromeres and four micromeres. The micromeres remain on the surface and overgrow the central macromeres. More specifically, there are two types of cleavages as shown in studies of *Lepas, Balanus, Mitella,* and *Teraclita* on one hand, and *Ibla* and *Scalpellum* on the other (Bigelow 1902; Delsman 1917; Anderson 1965, 1969; Kaufmann 1965). In the case of *Lepas, Balanus, Mitella,* and *Teraclita,* one macromere is considerably bigger than the other three; following the fourth division, this macromere divides to the single common progenitor for all endoderm and the primary mesoderm and to a sister cell that joins the rest of the blastomeres and their fate of mixed ectoderm/secondary mesoderm (Fig. 1D). The fifth division is again unequal and separates a larger endoderm progenitor and a smaller mesoderm progenitor. The endoderm progenitor is internalized first at the posterior pole by epiboly as the other cells overgrow it. The progenitor for primary mesoderm divides repeatedly at the surface before it ingresses (Fig. 1D′). Last, the secondary mesoderm cells ingress (Bigelow 1902; Delsman 1917; Shiino 1957; Anderson 1969). In contrast, in the case of *Ibla* and *Scalpellum,* there is no size difference between the macromeres. Gastrulation takes place by epiboly as the micromeres proliferate and overgrow the macromeres (Anderson 1965; Kaufmann 1965).

The Malacostracan Crustaceans

Gastrulation by Ingression into a Hollow Blastula: Anaspides

The Syncarida are a phylogenetically primitive malacostracan group with two taxa, the groundwater-dwelling Bathynellacea that have not been studied developmentally, and the Anaspidacea that are geographically restricted to the southern hemisphere. A representative of the Anaspidacea, *Anaspides tasmaniae,* has been studied and found to possess a unique pattern of early development: The egg is large, but cleavage is total and generates a hollow blastula; the rate of early cell divisions is slow, about one per day, and gastrulation is followed by an extended period of inactivity (Hickman 1936). Gastrulation has two phases. Starting at the 16-cell stage, mesoderm cells slip into the cavity of the blastula at the prospective posterior end. When this is done, endoderm invaginates also at the prospective posterior end, but more anterior to the mesoderm.

Gastrulation by Oriented Cell Division of Mesendoderm: Sicyonia ingentis

Two more groups of malacostracan crustaceans that show total cleavage are the Euphausiaceae such as krill and related species, as well as the Dendrobrachiatae, which are a subgroup of shrimps. For krill, there is only one study of embryonic development (Taube 1909). In dendrobrachiate shrimps, there are several studies of early development, and gastrulation has explicitly been addressed for the species *Sicyonia ingentis* using anti-tubulin antibodies, injection of tracers, and confocal microscopy (Zilch 1978; Hertzler and Clark 1992; Hertzler 2002). *Sicyonia,* therefore, is the reference for development in malacostracans after total cleavage. In *Sicyonia,* gastrulation takes place after five tangential cleavages have set up 32 blastomeres. Gastrulation starts as two cells invaginate at the posterior blastopore. The two cells arrest division, and later, both have mesendodermal fate: One gives rise to the yolk endoderm, the other to the endoderm proper and the main mesoderm (Fig. 1F). Gastrulation is then driven by additional mesoderm cells around the blastopore. They switch the position of their spindles during division from tangential to radial and divide toward the center of the egg and push the two mesendoderm cells inward. Gastrulation continues by the invagination of the mesoderm at the blastopore as well as by the continuation of radial cell divisions within the walls of the newly formed archenteron (Fig. 1F′) (Hertzler and Clark 1992; see caption for details). So far, this is the only documented example for oriented cell division initiating gastrulation in crustaceans (for examples of radial divisions that play minor roles in gastrulation in *Leptodora* and *Porcellio,* see above and below). The data on gastrulation in krill very much resemble those for *Sicyonia,* but recent phylogenetic analysis does not support a close relationship between the two groups (Taube 1909; Richter and Scholtz 2001).

Gastrulation by Ingression at a Posterior Blastopore: The Malacostracan Crustaceans with Superficial Cleavage

There are several common features in the gastrulation of most malacostracans that initially have superficial cleavages. Over time, data have accumulated for representatives of each clade of these malacostracans; *Nebalia bipes* (Leptostraca), *Squilla squilla* (Hoplocarida), *Astacus fluviatilis* (Decapoda), *Hemimysis lamornae* (Mysidacea, Peracarida), *Jaera, Asellus,* and *Porcellio* (all Isopoda, Peracarida) (Reichenbach 1877, 1886; McMurrich 1895; Manton 1928, 1934; Shiino 1942). These species share several common features including the arrangement of germ layer precursors prior to gastrulation, the location of the blastopore, as well as the movements of gastrulation.

Generally, the mesoderm is located in front of the endoderm, the blastopore is at the posterior end of the germ band, and gastrulation is by ingression (Manton 1934; Shiino 1957; Richter and Scholtz 2001). There are several differences between species including the generation of mesoderm and endoderm from one mixed or two separate cell populations as well as to the presence or absence of a blastopore or a blastopore lip. In the crayfish *Astacus*, gastrulation starts as an invagination at a round, wide blastopore at the prospective posterior end. Mesoderm and endoderm descend from the same posterior area of the blastoderm. Later, mesoderm and endoderm separate. The mesoderm cells ingress medially as individual cells, most of them anterior of the blastopore and few of them posterior. The endoderm invaginates and forms a pouch (Fig. 1G and G′) (Reichenbach 1877, 1886).

For isopods, several taxa have been looked at and the sampling reveals plasticity in the mechanism of gastrulation (McMurrich 1895). The differences are found in two aspects, first in the time that mesendoderm can be recognized, and second in the time of the separation between mesoderm and endoderm. At one end of this range, in *Jaera*, ectoderm, mesendoderm, and yolk cells show cytological differences as early as the 16-cell stage. Mesoderm and endoderm separate (into separate lineages) after the 64-cell stage (Fig. 1H). *Asellus* falls into the middle of the range as the mesendoderm becomes apparent after the 128-cell stage and the mesoderm and endoderm separate even later. *Porcellio* and *Armadillidium* are at the other end of the range as the mesendoderm is discernible at the 64-cell stage, but mesoderm and endoderm do not separate until after gastrulation (McMurrich 1895). In *Porcellio*, time-lapse microscopy using Nomarski optics has been done to analyze gastrulation in vivo by following proliferation and migration of unlabeled cells (Hejnol 2002). The experiments demonstrate that gastrulation is a mix of ingression, invagination, and delamination. It takes place after about 32 cells have aggregated to a germ disc. The main gastrulation movement is an ingression of cells sinking into the egg. In addition, some tangential and radial cell divisions in the outer, ectodermal layer also contribute to the population of internal cells (Hejnol 2002).

GASTRULATION BY INGRESSION AT AN ANTERIOR BLASTOPORE: IN VIVO ANALYSES OF MESODERM AND ENDODERM CLONES IN TWO AMPHIPODS

Amphipods are another group of malacostracan crustaceans that show total cleavage. This pattern of total cleavage appears to have evolved secondarily from the superficial

cleavage that is found in the Peracarida that are most closely related to amphipods, such as isopods and mysids (see above and Table 1). Studies on amphipod gastrulation date back quite far but give conflicting accounts as to what parts of the embryo are contributed to by the individual macromeres and micromeres (Langenbeck 1898; Weygoldt 1958; Rappaport 1960; Scholtz 1990). Recently, experiments have tracked the origin and movements of the germ layers by labeling macromeres and micromeres with tracers in two amphipods, *Parhyale hawaiensis* and *Orchestia cavimana* (Gerberding et al. 2002; Wolff and Scholtz 2002). This detailed knowledge of the cell lineage puts the understanding of amphipod development ahead of that of other crustaceans. An in-depth analysis of gastrulation can be carried out because cells can be followed through gastrulation so their fate is certain and not just inferred, and the analysis can be done by looking at clones in fixed specimens with whole mounts and sections as well as by recording the behavior of clones using time-lapse video microscopy (Gerberding et al. 2002; Scholtz and Wolff 2002; Wolff and Scholtz 2002; A.L. Price and N.H. Patel, unpubl.).

The germ layers in *Parhyale* and *Orchestia* segregate surprisingly early into separate lineages. After only three cleavages, each of the resulting eight cells normally only contributes progeny to a single germ layer. The analysis of *Parhyale* (Gerberding et al. 2002) indicates that first three cleavages set up eight blastomeres, four bigger macromeres and four smaller micromeres. Three macromeres generate ectoderm, and the remaining macromere generates visceral mesoderm. One micromere gives rise to the germ line, two micromeres form the somatic mesoderm, and the fourth generates the endoderm. The data on the lineages in *Orchestia* are similar to *Parhyale* for the ectoderm and germ line, but differ somewhat for the mesoderm and endoderm lineages. In the gastrulation of both *Parhyale* and *Orchestia*, the prospective material for the inner germ layers is located anterior of the ectoderm (as compared to posterior of the ectoderm in all other malacostracan crustaceans, see above). In both species, the inner germ layers comprise several clones, namely the somatic mesoderm of the head, the somatic mesoderm of the trunk, the visceral mesoderm, and the endoderm (Gerberding et al. 2002; Wolff and Scholtz 2002).

In *Parhyale* (Fig. 2), two phases of gastrulation are described (Gerberding et al. 2002). The first phase involves the formation of a round pit named a "rosette" that has deep (somewhat internal) cells and cells on the surface. The deep cells of the rosette are cells of the visceral mesoderm (Mv) and germ-line clones (g) as they begin to sink in; the outer cells are ectoderm cells. Eventually, the ectoderm cells completely cover the inner cells. In this first phase of gastrulation, the germ disc is formed at the anterior end of the egg. The rosette marks the prospective anterior end of the

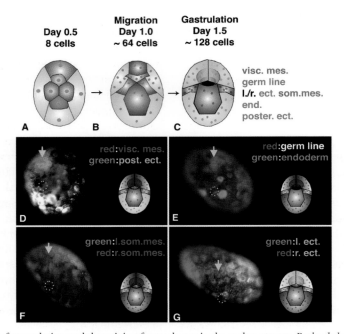

Figure 2. In vivo analysis of gastrulation and the origin of germ layers in the malacostracan *Parhyale hawaiensis*. Early dye labeling of cells allows cells and their progeny (clones) to be followed in vivo until hatching. (*A–C*) Schematic drawings showing the distribution of clones at the 8-cell stage at day 0.5, during migration at day 1, and at gastrulation at day 1.5 (total development time to hatching is 10 days). Dorsal views, anterior up, posterior down. (*A*) The third cleavage is unequal and generates four macromeres and four micromeres. Each of the macromeres and micromeres contributes to only one germ layer. Macromere clones give rise to visceral mesoderm (*red* macromere) and three parts of ectoderm, the anterior right (*right yellow* macromere), the anterior left (*left yellow* macromere), and the posterior (*bottom yellow* macromere). Micromere clones give rise to germ line (*gray* micromere), right and left somatic mesoderm (*right* and *left red* micromeres, respectively), and endoderm (*green* micromere). (*B*) From day 0.5 to day 1.5, there are few cell divisions and the cells migrate extensively. The area covered by ectoderm expands, while the area taken up by the visceral mesoderm shrinks. (*C*) Gastrulation has two phases. Before gastrulation, all clones for internal germ layers are located in separate areas anterior to the ectoderm when seen from a dorsal view. In the first phase of gastrulation at day 1.5, the visceral mesoderm precursors and the germ-line cells ingress medially, and the two mesoderm clones ingress laterally and migrate underneath the ectoderm. In a second phase at day 3, the endoderm ingresses and joins the visceral mesoderm to form a sheath (not shown). Although the posterior ectoderm may give the impression of being located far anterior, it is mostly ventral and will migrate all the way around ventrally to lie posterior to the rest of the ectoderm once the germ band begins to form. (*D–G*) Labeled embryos in vivo at gastrulation. Schematic drawings and embryos side by side. Different pairs of opposing macromeres or micromeres were separately labeled at the 8-cell stage with green and red fluorescent dyes and then viewed at gastrulation (day 1.5). To ease comparisons between embryos in vivo, two landmarks are given in each panel: gray circles indicate the location of the center of the endoderm clone, gray arrows point to the anterior margin of the blastopore (= edge of the rosette). Schematic cartoons use same colors as in *A–C*. (*D*) Two macromere clones: posterior ectoderm green and visceral mesoderm red. The posterior ectoderm has started to migrate toward the ventral side. The visceral mesoderm is the deeper part of a two-layered rosette of cells. (*E*) Two micromere clones: endoderm green and germ line red. The endoderm spreads out on the dorsal side. The germ-line is part of the outer cells of the early rosette. (*F*) Two micromere clones: right somatic mesoderm red and left somatic mesoderm green. (*G*) Two macromere clones: anterior right ectoderm red and anterior left ectoderm green. The visceral mesoderm and germ-line clones will come to lie under these macromere clones.

embryonic germ disc. The somatic mesoderm cells of the trunk (mr and ml) migrate underneath the ectoderm as single cells immediately next to the right and the left of the rosette (Gerberding et al. 2002; A.L. Price and N.H. Patel, unpubl.). In vivo imaging of mesoderm formation reveals an extremely dynamic process in which individual mesoderm precursors actively migrate to the edge of the ectoderm precursor cell sheet, and then slide under the ectoderm. The cells then split into two populations, one

giving rise to the head mesoderm and a second giving rise to the eight mesenteloblasts which are the stem cells that form all the somatic mesoderm posterior to the head (A.L. Price and N.H. Patel, unpubl.). What is thought to be the endoderm (en) remains on the dorsal surface during this phase of gastrulation. When the germ disc is completed and begins elongation on the ventral side (marking the very start of the germ-band stage), the endoderm joins the inner layers on the ventral side in a second phase of gastrulation.

In *Orchestia*, three phases of gastrulation are described (Scholtz and Wolff 2002; Wolff and Scholtz 2002). The first phase is an ingression of several daughters of macromeres becoming vitellophages. It has no described counterpart in *Parhyale*. The second gastrulation phase of *Orchestia* is the formation of a gastrulation center. This center comprises the germ-line clone "g" as it begins to sink in, and this phase is coincident with the formation of the germ disc. Thus, this second phase in *Orchestia* is comparable to the first phase described for *Parhyale* above, and the *Orchestia* gastrulation center is equivalent to the rosette of *Parhyale*. The third phase in *Orchestia* involves the propagation of a longitudinal groove throughout the germ disc that starts at the gastrulation center. The mesendoderm clone invaginates along the groove, and the somatic trunk mesoderm of mr and ml immigrates at both sides of the germ disc. Comparing the two data sets, there is no description of an invagination of mesendoderm in *Parhyale*, and there is no description of a late ingression of the endoderm in *Orchestia* (Gerberding et al. 2002; Scholtz and Wolff 2002). For now, it is an open question as to how many of these differences are real and the result of divergent evolution and how many of them are an artificial result of differences in the way cells are labeled and clones are analyzed. The absence of an invagination groove in *Parhyale* and its presence in *Orchestia* might reflect evolutionary differences in the relative numbers of cells in the embryos as gastrulation begins.

Amphipod embryonic development is clearly derived in many aspects, so it is unclear as to how many of the data from the two amphipods are relevant to crustaceans in general. Nevertheless, the techniques developed to study early pattern formation and morphogenesis in these amphipod species offer many exciting possibilities to analyze crustacean gastrulation in great detail. The similarities found between the gastrulation of *Parhyale* and *Orchestia* are striking, but the differences are equally interesting as they may have much to tell us about how development evolves between closely related taxa.

THE EVOLUTION OF GASTRULATION IN CRUSTACEANS

From our perspective, postulating or rejecting homologies of parts of an embryonic process such as gastrulation depends on how many details in the temporal and spatial order are taken into consideration. As more details of time and space are included, the two events look less similar—e.g., the ingression of mesendoderm in two species might take place in embryos that differ in the total number of cells or occur at different locations along the future anterior–posterior axis. To us, none of the characters found in malacostracans look strikingly similar to those found in non-malacostracans.

For the malacostracans alone, however, a case can be made that traits of gastrulation can be added to the number of morphological homologies that already support the malacostracans as a clade (Scholtz 2000; Richter and Scholtz 2001). Most of them cleave superficially and gastrulate at the posterior end of the germ band by ingression. If we assume this to be ancestral, total cleavage and gastrulation by either ingression or invagination have evolved secondarily within the malacostracans: in syncarids, in the dendrobrachiate shrimps, in krill, and in amphipods. The posterior location of the blastopore was maintained in all cases except for the amphipods, where it is shifted anteriorly.

For the non-malacostracans alone, many attempts to homologize different total cleavages and modes of gastrulation have been made at a time when they were classified as "entomostracans" (Bigelow 1902; Kühn 1913; Müller-Calé 1913; Fuchs 1914; Baldass 1941). The homologies are based on the observation that the progeny derived from one of the cells at the 4-cell stage are delayed in their divisions and that the cells of this clone have a mesendodermal fate that at times also includes the germ-line fate. However, any attempt to homologize these cells must explain why these cells are found at opposite poles in different taxa; they are found at the prospective posterior end in cirripeds and at the prospective anterior end in water fleas, copepods, and ostracods (Weygoldt 1960 and references therein).

The summary of gastrulation presented here is largely descriptive because this is, for the most part, the current state of our knowledge for crustaceans. This will certainly soon change. Given our understanding of the molecular and genetic underpinnings of gastrulation in many model systems, it is logical to begin applying molecular and manipulative methods to analyze crustacean gastrulation at a more mechanistic level, and such studies have been initiated in *Parhyale* (A.L. Price and N.H. Patel, unpubl.). No doubt many of the molecular mechanisms controlling cell morphogenesis during gastrulation will be similar between crustaceans and other animals, and even events that appear to be quite different between various crustacean species will probably also share certain similarities at the molecular level. Likewise, the apparent differences in the relative positions of different tissue precursors before gastrulation suggest that pre-gastrulation mechanisms for patterning may differ substantially between species, but no doubt will still utilize a number of similar genetic components. The challenge, however, will be explaining how such common mechanisms can be used to create such diversity in the events of gastrulation and early patterning between closely related species.

Finally, we must also ask what role gastrulation in crustaceans plays in body patterning and cell differentiation. Various data indicate that for at least some crustaceans, some level of cell differentiation already exists before gastrulation. For example, the germ line of *Parhyale* can be distinguished by the opacity of its cytoplasm, and other precursors by their rates of cell division before gastrulation. In other species of crustaceans, such differences are not noticeable. Ablation and other manipulative experiments will be required to define the extent of cell commitment to specific fates before, during, and after gastrulation. Even if commitment to germ-layer fate is set before gastrulation, it seems quite likely that the events of gastrulation could still be generating important subdivisions within these tissues. Whether other elements of body patterning and axial information are also established by gastrulation remains to be investigated as well.

All of these potential future studies should also help us to understand the evolutionary history of gastrulation within crustaceans, and how crustacean gastrulation relates to this process in other arthropods and other phyla. Beyond the descriptive level, much more is currently known about gastrulation in many other groups of animals, as evidenced by the other chapters in this book. On the other hand, the descriptive studies to date suggest a remarkable level of variation in the events of gastrulation and early pattern formation in the crustaceans, making them an excellent group for further exploration. We encourage developmental biologists to dig into the diversity of early crustacean development.

REFERENCES

Anderson D.T. 1965. Embryonic and larval development and segment formation in *Ibla quadrivalvis* (Cuv.) Cirripedia. *Austr. J. Zool.* **13:** 1–15.

———. 1969. On the embryology of the cirrepede crustaceans *Tetraclita rosea* (Krauss), *Tetraclita purpurascens* (Wood), *Chthamalus antennatus* (Darwin) and *Chamaesipho columna* (Spengler) and some considerations of crustacean phylogenetic relationships. *Philos. Trans. R. Soc. London B* **256:** 183–235.

———. 1973. *Embryology and phylogeny in annelids and arthropods.* Pergamon Press, New York.

Ax P. 1999. *Multicellular animals,* vol. 2. Springer, New York.

Baldass F. 1941. Die Entwicklung von *Daphnia pulex. Zool. Jahrb. Anat.* **67:** 1–60.

Benesch R. 1969. Zur Ontogenie und Morphologie von *Artemia salina* L. *Zool. Jahrb. Anat.* **86:** 307–458.

Bigelow M. A. 1902. The early development of *Lepas. Bull. Mus. Comp. Zool.* **40:** 61–144.

Brusca R.C. and Brusca G.J. 2002. *Invertebrates,* 2nd edition. Sinauer, Sunderland, Massachusetts.

Delsman H.C. 1917. Die Embryonalentwicklung von *Balanus balanoides* Linn. *Tijdschr. Nederl. Dierk. Ver.* **15:** 419–520.

Fioroni P. 1970. Am Dotteraufschluss beteiligte Organe und Zelltypen bei höheren Krebsen: Der Versuch zu einer einheitlichen Terminologie. *Zool. Jahrb. Anat.* **87:** 481–522.

Fuchs F. 1914. Die Keimblätterentwicklung von *Cyclops viridis* Jurine. *Zool. Jahrb. Anat.* **38:** 103–156.

Gerberding M. 1997. Germ band formation and early neurogenesis of

Leptodora kindti (Cladocera): First evidence for neuroblasts in the entomostracan crustaceans. *Inverterbr. Reprod. Dev.* **32:** 63–73.

Gerberding M., Browne W.E., and Patel N.H. 2002. Cell lineage analysis of the amphipod crustacean *Parhyale hawaiensis* reveals an early restriction of cell fates. *Development* **129:** 5789–5801.

Grobben C. 1879. Die Entwicklungsgeschichte der *Moina rectirostris. Arb. Zool. Inst. Wien* **2:** 203–268.

Hartenstein V., Technau G.M., and Campos-Ortega J.A. 1985. Fate-mapping in wild-type *Drosophila melanogaster.* III. A fate map of the blastoderm. *Roux's Arch. Dev. Biol.* **194:** 213–216.

Hejnol A. 2002. "Der postnaupliale Keimstreif von *Porcellio scaber* und *Orchestia cavimana* (Crustacea, Peracarida): Zelllinie, Genexpression und Beginn der Morphogenese." Ph.D. thesis, Humboldt University, Berlin.

Hertzler P.L. 2002. Development of the mesendoderm in the dendrobranchiate shrimp *Sicyonia ingentis. Arthropod Struct. Dev.* **31:** 33–49.

Hertzler P.L. and Clark W.H.J. 1992. Cleavage and gastrulation in the shrimp *Sicyonia ingentis*: invagination is accompanied by oriented cell division. *Development* **116:** 127–140.

Hickman V. 1936. The embryology of the syncarid crustacean *Anaspides tasmaniae. Pap. R. Soc. Tasmania* **1936:** 1–35.

Kaufmann R. 1965. Zur Embryonal- und Larvalentwicklung von *Scalpellum scalpellum* L. (Crust. Cirr.). *Z. Morphol. Oekol.* **55:** 161–232.

Köhler H.-J. 1976. Embryologische Untersuchungen an Copepoden: Die Entwicklung von *Lenaeocera branchialis* L. 1767 (Crustacea, Lernaeoidea, Lernaeidae). *Zool. Jahrb. Anat.* **95:** 448–504.

Kühn A. 1913. Die Sonderung der Keimesbezirke in der Entwicklung der Sommereier von *Polyphemus. Zool. Jahrb. Anat.* **35:** 243–340.

Langenbeck C. 1898. Formation of the germ layers in the amphipod *Microdeutopus gryllatalpa* Costa. *J. Morphol.* **14:** 301–336.

Manton S.M. 1928. On the embryology of a mysid crustacean *Hemimysis lamornae. Philos. Trans. R. Soc. London B* **216:** 363–463.

———. 1934. On the embryology of *Nebalia bipes. Philos. Trans. R. Soc. London B* **223:** 168–238.

McMurrich J.P. 1895. Embryology of the isopod Crustacea. *J. Morphol.* **11:** 63–154.

Müller-Calé C. 1913. Über die Entwicklung von *Cypris incongruens. Zool. Jahrb. Anat.* **36:** 113–170.

Rappaport R. 1960. The origin and formation of blastoderm cells of gammarid Crustacea. *J. Exp. Zool.* **144:** 43–59.

Reichenbach H. 1877. Die Embryoanlage und erste Entwicklung des Flusskrebses. *Z. Wiss. Zool.* **29:** 123–196.

———. 1886. Studien zur Entwicklungsgeschichte des Flusskrebses. *Abh. Senckenb. Natforsch. Ges.* **14:** 1–137.

Richter S. 2002. The Tetraconata concept: Hexapod-crustacean relationships and the phylogeny of Crustaciea. *Orig. Divers. Evol.* **2:** 217–237.

Richter S. and Scholtz G. 2001. Phylogenetic analysis of the Malacostraca. *J. Zool. Syst. Evol. Res.* **39:** 113–136.

Samter M. 1900. Studien zur Entwicklungsgeschichte der *Leptodora hyalina* Lillj. *Z. Wiss. Zool.* **68:** 196–260.

Scholtz G. 1990. The formation, differentiation and segmentation of the post-naupliar germ band of the amphipod *Gammarus pulex* L. Crustacea Malacostraca Peracarida. *Proc. R. Soc. Lond. B* **239:** 163–211.

———. 1997. Cleavage, germ band formation and head segmentation: The ground pattern of the Euarthropoda. In *Arthropod relationships* (ed. R. Fortey and R. Thomas), pp. 317–332. Chapman &

Hall, London, United Kingdom.

———. 2000. Evolution of the nauplius stage in malacostracan crustaceans. *J. Zool. Syst. Evol. Res.* **38:** 175–187.

Scholtz G. and Wolff C. 2002. Cleavage, gastrulation and germ disc formation of the amphipod *Orchestia cavimana* (Crustacea, Malacostraca, Peracarida). *Contrib. Zool.* **71:** 9–28.

Shiino S.M. 1942. Studies on the embryology of *Squilla oratoria* de Haan. *Mem. Coll. Sci. Kyoto Univ. B* **27:** 77–169.

———. 1957. Crustacea. In *Invertebrate embryology* (ed. M. Kume and K. Dan), pp. 332–438. Bai Fu Kan Press, Tokyo, Japan; English transl. (1988) Garland, New York.

Taube E. 1909. Beiträge zur Entwicklungsgeschichte der Euphausiden. I. Die Furchung der Eier bis zur Gastrulation. *Z. Wiss. Zool.* **92:** 427–464.

Walossek D. 1999. On the Cambrian diversity of Crustacea. In *Crustaceans and the biodiversity crisis. Crustacean issues* (ed. F. Schram and J. von Vaupel Klein), vol. 12, pp. 2–27. Balkema, Amsterdam, The Netherlands.

Weygoldt P. 1958. Die Embryonalentwicklung des Amphipoden *Gammarus pulex pulex* (L.). *Zool. Jahrb. Anat.* **77:** 51–110.

———. 1960. Embryologische Untersuchungen an Ostrakoden: Die Entwicklung von *Cyprideis litoralis* (G. S. Brady) (Ostracoda, Podocopa, Cytheridae). *Zool. Jahrb. Anat.* **78:** 369–426.

———. 1993. Le développement embryonnaire. In *Traité de Zoologie, Crustacés*, Tome VII, Fascicule I (ed. J. Forest), pp. 807–899. Masson, Paris, France.

Wolff C. and Scholtz G. 2002. Cell lineage, axis formation, and the origin of germ layers in the amphipod crustacean *Orchestia cavimana. Dev. Biol.* **250:** 44–58.

Zilch R. 1978. Embryologische Untersuchungen an der holoblastischen Ontogenese von *Penaeus trisulcatus* Leach (Crustacea, Decapoda). *Zoomorphology* **90:** 67–100.

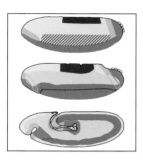

GASTRULATION IN *DROSOPHILA*

M. Leptin

Institut für Genetik, Universität zu Köln, D-50931 Köln, Germany

INTRODUCTION

Gastrulation comprises two major processes, the determination of the fates of the participating cells, and the morphogenetic processes which these cells undergo. Of all organisms, *Drosophila* is probably the one in which the mechanisms of early cell fate determination are understood most comprehensively. The maternal-effect genes that control patterning of the embryo along its anterior–posterior and dorsoventral axes have been identified, their mechanisms of action have been determined, and many of the zygotic genes that are regulated by their activities are known. The extremely precise early embryonic fate maps are based both on high-resolution expression patterns of these target genes and on classical embryological experiments such as single-cell labeling, transplantation, and UV or laser ablation (see, e.g., Lohs-Schardin et al. 1979; Hartenstein et al. 1985; Technau and Campos-Ortega 1985; Klapper et al. 1998). Gastrulation movements have also been described in great detail using time-lapse video microscopy and reconstructions from scanning EM and light microscopic images (Fullilove et al. 1978). Much less is known about the molecular mechanisms that drive the movements.

REGIONS OF THE EMBRYO THAT PARTICIPATE IN GASTRULATION

Fate Map of the Prospective Germ Layers

The fate map of a 3-hour-old *Drosophila* embryo is shown in Figure 1B. By convention, the *Drosophila* egg or embryo is shown with its anterior end to the left and its dorsal side up. The anterior–posterior and dorsoventral polarities are readily recognizable as morphological asymmetries which

are acquired by the egg during its development in the ovary of the mother. By 3 hours of development, at the cellular blastoderm stage, the embryo consists of about 6000 cells of identical appearance that form a tall columnar epithelium in which the apical surfaces of the cells face outward. The future germ cells are an exception, and these are set aside earlier and lie at the posterior pole of the embryo, on the surface of the blastoderm epithelium (Fig. 1A). By this stage, the embryo is already finely divided into territories with specified fates, as judged by gene expression patterns and cell marking and transplantation experiments.

The prospective mesoderm covers most of the ventral side of the embryo. It occupies about a quarter of the circumference and 70% of the length of the embryo. The endoderm arises from two completely separate regions. The primordium for the anterior part of the future midgut lies immediately anterior to the mesoderm primordium, also on the ventral side of the embryo. The region that will form the posterior part of the midgut corresponds to the posterior cap of the embryo. These two regions develop independently of each other under partly separate genetic control, and join each other by cell migration late during embryogenesis.

The ectodermal primordium can be subdivided into regions with different fates. A band of cells adjacent to the posterior endodermal primordium will give rise to the hindgut. The foregut will arise from the part of the anterior tip of the blastoderm that lies above the anterior endodermal primordium. The two large areas on each side of the embryo consist of the neuroectoderm, located ventrally, next to the mesectoderm, and the lateral and dorsal ectoderm. Neuroblasts will delaminate from the neuroectodermal epithelium toward the interior of the embryo, while the lateral and dorsal ectoderm will form most of the larval epi-

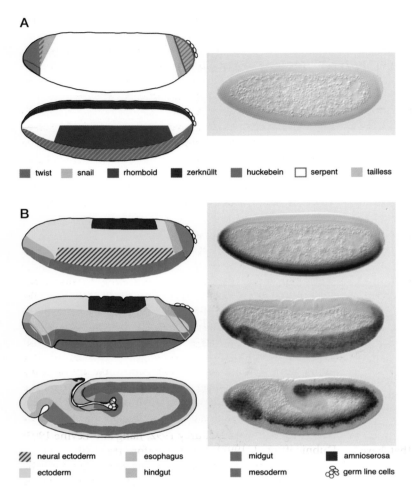

Figure 1. The blastoderm fate map, gene expression patterns, and gastrulation movements. The right column shows embryos that were fixed and stained with antibodies against Twist. The left column shows diagrams of gene expression patterns (*A*) or the fates of regions (*B*) at approximately the same stage as the embryos on the right. All embryos are shown with their anterior end to the left and the dorsal side up. An embryo is about 0.5 mm long. The ages of the embryos are ~2 hours, 3 hours, 3 hours and 15 minutes, 5 hours (from top to bottom). The youngest embryo on the right is only beginning to synthesize Twist protein on the ventral side. (*A*) (*Left*) Diagram of gene expression patterns at a stage shortly after the one shown on the right. The top diagram shows the genes that define the borders between germ layers along the anterior–posterior axis, the bottom diagram those that mark the dorsoventral borders. Graded distributions of gene products at the edges of the expression domains are not shown. (*B*) (*Left*) Fate map of the *Drosophila* embryo. For the first embryo, a surface view is shown, for the others, sections through the middle of the embryo. Thus, the ectoderm is not visible; instead the yolk is shown in light gray. (*Right*) The cells on the ventral side of the blastoderm stage embryo (*top*) express Twist, a nuclear protein. The Twist domain extends from the anterior to the posterior tip of the embryo and includes the prospective mesoderm. The ventral cells invaginate, and soon afterward, the posterior midgut invagination is formed at the posterior tip of the embryo, carrying the germ-line precursors (the pole cells) dorsally (*middle*). The dorsal epithelium shows a series of transverse folds. The germ-line precursors are carried into the body as the germ band extends (*bottom*). The extending germ band pushes the opening of the posterior midgut invagination toward the head, while the invagination deepens, forming a long sac. The mesoderm has spread out to form a single cell layer.

dermis as well as the tracheal system. The dorsal side of the embryo will develop into the amnioserosa.

Determination of Cell Fates in the Embryo

Gene Expression Patterns in the Prospective Germ Layers

The borders of the future germ layers can be visualized by the expression patterns of several genes at the cellular blastoderm stage. Along the dorsoventral axis, the most ventral cells express *snail* and *twist*, the mesodermal determinants (Thisse et al. 1987; Alberga et al. 1991). The sharp lateral border of the *snail* domain corresponds to the edge of the mesoderm primordium, whereas *twist* expression extends beyond this border, albeit at very low levels (Kosman et al. 1991; Leptin 1991). Immediately adjacent to the *snail*-expressing cells, a single line of cells expresses the gene *single-minded* (*sim*), which is specific for the mesectodermal cells that will later form the midline glia cells (Crews et al.

1988). Expression of *zen* in a band of cells on the dorsal side marks the future amnioserosa (Rushlow et al. 1987). Between the *sim* and *zen* domains lies the ectoderm. No single gene is known that is expressed specifically in the ectoderm; all genes that are expressed throughout the ectoderm are also expressed in the amnioserosa (for example, those coding for components of the epithelial junctional complexes, like *crumbs*, or neurogenic genes like *Delta*), and in the absence of *snail*, they are also expressed in the mesoderm (Leptin 1991). This suggests that "ectodermal" gene expression might represent a developmental ground state for the blastoderm.

The prospective mesoderm covers most, but not all, of the length of the embryo. *twist* expression extends beyond the mesoderm both at the anterior and posterior ends and covers the entire ventral side of the embryo. In contrast, the posterior border of *snail* expression coincides with the posterior border of the mesoderm primordium (Reuter and Leptin 1994); however, at the anterior end, *snail* expression, like *twist*, extends beyond the border of the prospective mesoderm all the way to the anterior tip of the embryo. The future endoderm is marked by *serpent*, which is expressed on the ventral side of the anterior tip (the anterior midgut primordium) and the posterior cap (the posterior midgut primordium) of the embryo (Rehorn et al. 1996).

The expression domains of two gap genes, *huckebein* (*hkb*) and *tailless* (*tll*) (Strecker et al. 1986; Pignoni et al. 1990; Weigel et al. 1990; Reuter and Leptin 1994), mark the borders of germ layers along the anterior–posterior axis. *hkb* is expressed at the posterior pole in the region that will form the posterior midgut, and that corresponds to the posterior *serpent* expression domain. On the ventral side, its anterior border abuts the posterior border of the *snail* expression domain. *tll* overlaps the *hkb* expression domain but extends further anteriorly, including the ectodermal territory that will invaginate with the posterior midgut and later form the hindgut. *hkb* and *tll* are also expressed at the anterior tip of the embryo. Here, *hkb* expression overlaps the *snail* expression domain, but its posterior border again marks the border of the endoderm and coincides with the anterior border of the prospective mesoderm. The anterior endoderm arises from the region in which *hkb*, *snail*, and *twist* are coexpressed, which corresponds precisely to the anterior *serpent* expression domain (Rehorn et al. 1996).

Maternal Control of Zygotic Gene Transcription

The expression domains described above are the result of a cascade of signaling and transcriptional events that have occurred during the first 3 hours of development. Even before it is laid, the egg already contains asymmetrically localized molecules that will ultimately determine gene expression patterns in the embryo (for review, see St Johnston and Nüsslein-Volhard 1992). Briefly, four sets of maternal-effect genes control the dorsoventral and anterior–posterior axes of the embryo. The anterior part of the egg is controlled by anteriorly localized *bicoid* RNA, which will generate a gradient of the transcription factor Bicoid with its highest point anteriorly. Together with the "posterior group" genes, which establish a gradient of the translational regulator Nanos, Bicoid sets up the expression domains of the segmentation genes, specifically, of the gap genes *hunchback*, *Krüppel*, *knirps*, and *giant* (Rivera-Pomar and Jäckle 1996). The third set of genes, the "terminal group," also contributes to anterior–posterior patterning. This system acts through the activation of the receptor tyrosine kinase Torso to regulate the transcription of terminal gap genes such as *hkb* and *tll*, expressed in the tips of the embryo (Fig. 1A). Torso is activated specifically at the termini of the egg due to the localized release of its ligand. The Torso signal counteracts the activities of the transcriptional repressors Capicua and Groucho via the activation of the MAPK cascade, causing the degradation of Capicua, there-

by releasing repression of *hkb* and *tll* expression (Paroush et al. 1997; Jimenez et al. 2000). The difference in the expression domains of the two genes is created at least in part by differences in their sensitivities to the strength of the Ras and MAPK signal (Greenwood and Struhl 1997). The transcriptional activator for *hkb* and *tll* is not known.

The fourth group of genes is responsible for the subdivision of the embryo along the dorsoventral axis. The transcription factor Dorsal, initially localized in the cytoplasm throughout the egg, becomes transported into the nuclei on the ventral side of the embryo under the influence of signaling from the transmembrane receptor Toll, but remains excluded from nuclei on the dorsal side, where Toll is not active (Roth et al. 1989; Rushlow et al. 1989). Thus, a dorsoventral gradient of active Dorsal is set up. Dorsal can act as a transcriptional activator with different thresholds of activation depending on the affinity of binding sites in the promoters of different genes expressed along the dorsoventral axis. Genes that have only low-affinity binding sites, like *snail* and *twist*, are expressed only in the regions with the highest levels of nuclear Dorsal; i.e., in the prospective mesoderm (Pan et al. 1991; Jiang et al. 1991). The expression domains of genes with high-affinity binding sites for Dorsal in their promoters extend further dorsally, into the ectoderm, as, for example, in the case of *rhomboid* (Ip et al. 1992a). Dorsal can also act as a repressor on promoters on which it interacts with the corepressors Cut, Groucho, or Dead ringer (Pan and Courey 1992; Jiang and Levine 1993; Dubnicoff et al. 1997). *zen* is therefore transcribed (by an as-yet-unidentified activator) only in regions in which the level of Dorsal is so low that it cannot repress *zen*; i.e., in the dorsal part of the embryo.

Interactions between Zygotic Transcription Factors

The effect of the localized activation of the maternally provided molecules is the localized transcription of genes from the zygotic genome. The first set of target genes, mentioned in the previous section, are initially transcribed in domains with fuzzy borders that still reflect the graded distributions of activity of the maternal components. These borders are then sharpened by interactions among the zygotic genes. In the case of the mesoderm, both *snail* and *twist* contain binding sites for Dorsal in their promoters and are directly activated by Dorsal in the most ventral nuclei (Jiang et al. 1991; Pan et al. 1991; Ip et al. 1992b). In a feedback loop, Twist then acts together with Dorsal to increase and maintain its own transcription and also to transcribe high levels of *snail* over the whole width of the prospective mesoderm (Ip et al. 1992b). For *snail*, this cooperative activation leads to a very sharp on–off border, whereas *twist* has a graded distribution that reaches beyond the border of the mesoderm, perhaps because of the ability of its promoter to respond to low levels of Dorsal. The posterior border of the *snail* expression domain is also determined by a zygotically expressed gene: In *hkb* mutant embryos, the expression domain of *snail* extends to the posterior pole (Reuter and Leptin 1994). In addition, *hkb* is necessary for the expression of the endoderm selector gene *serpent* (Rehorn et al. 1996).

Most of the genes that are expressed in the ventral ectoderm are excluded from the mesoderm via repression by Snail (Kosman et al. 1991; Leptin 1991). For some, it has been shown that repression is mediated by direct binding of Snail to their promoters. For example, *rhomboid,* which is expressed in a graded distribution from the edge of the mesoderm to approximately the middle of the ectodermal region, has high-affinity binding sites for Dorsal as well as binding sites for Snail (Ip et al. 1992a). The Dorsal-binding sites ensure its transcription throughout the ventral half of the embryo, whereas Snail blocks transcription in the mesoderm.

The strikingly sharp, single-cell-wide line of *sim* is entirely dependent on other zygotically expressed genes, although the ubiquitous, maternally provided transcription factor Suppressor of Hairless (Su[H]) participates in its transcription. The *sim* promoter has binding sites for Dorsal, Twist, Snail, and Su(H) (Kasai et al. 1992; Kasai et al. 1998; Morel et al. 2001). In the absence of *twist* and *snail*, no expression of *sim* is seen, and cell communication via Notch is needed to allow Su(H) to act on the *sim* promoter (Leptin 1991; Morel et al. 2001).

Finally, extensive interactions between the "dpp group" genes and their products pattern the ectoderm (for review, see Rusch and Levine 1996; Rushlow and Roth 1996). The dpp group includes *dpp*, *sog*, and *tolloid*, whose homologs *BMP4*, *chordin*, and *tolloid* also regulate dorsoventral patterning in vertebrates, as well as *zen*, *screw*, and *twisted gastrulation* (for review, see Lall and Patel 2001).

DESCRIPTION OF GASTRULATION MOVEMENTS

Most researchers have a clear feeling about which points in the development of an organism are before the gastrulation stage (for example, morula, blastula, or blastoderm stages) and by which stage gastrulation is over. However, the precise beginning and especially the end are not always agreed upon. In this chapter (as in most others in this book), gastrulation is used for the sum of those processes that translocate the future endodermal and mesodermal cells from the surface of the blastoderm-stage embryo into the inside, and the ectodermal movements that occur in parallel. This does not include the cell migration and spreading events that turn the groups of internalized cells into single cell layers, since the endoderm does not complete these until late in

embryogenesis. Thus, a "gastrula" state of organization, at which three germ layers can be clearly identified, is reached very late in *Drosophila* (stage 12/13). More commonly, the time at which the endoderm has been fully internalized is considered the end of gastrulation (stage 9/10) (Fig. 1B, bottom panel).

The cellular blastoderm is derived from the syncytial blastoderm, in which 6000 nuclei are lined up at the cortex of the embryo. Cells are generated from this syncytium by invaginations of the plasma membrane that grow between the nuclei and eventually enclose them to form cells. As soon as this process is complete on the ventral side of the embryo (the cells on the dorsal side lag behind by a few minutes), gastrulation begins. Four mechanistically and genetically separable sets of major, parallel morphogenetic movements can be distinguished (Poulson 1950; Sonnenblick 1950; Mahowald 1963; Turner and Mahowald 1977; Fullilove et al. 1978; Campos-Ortega and Hartenstein 1985; Leptin and Grunewald 1990; Sweeton et al. 1991; Wieschaus et al. 1991):

- On the ventral side, the mesoderm invaginates together with part of the anterior endoderm primordium. More anteriorly, the stomodeal invagination soon follows, internalizing the remainder of the endoderm.

- At the posterior end, the posterior endoderm invaginates, carrying the pole cells with it.

- A deep, transverse furrow behind the head region and several folds on the dorsal side are formed.

- Cell intercalations in the ectoderm result in the lengthening of the trunk region, which results in the displacement of the posterior end of the embryo toward the dorsal side.

These movements begin around 3 hours after fertilization and occur over a period of approximately 2 hours.

Mesoderm Invagination

Internalization of the mesoderm is the most extensively studied of the processes of gastrulation. It begins with the formation of a shallow indentation along the ventral side of the embryo, called the ventral furrow, which rapidly deepens until the epithelium sinks into the interior of the embryo (Fig. 2). Once inside, the cells of the epithelium disperse and migrate on the ectoderm, spreading away from the site of invagination, thus forming a single cell layer. This process has been analyzed and documented by scanning and transmission electron microscopy, time-lapse video microscopy, and light microscopy of fixed, stained, and sectioned embryos, giving rise to a detailed description of the cellular events that accompany, and probably cause, the invagination of the mesoderm.

The first microscopically discernible event, most clearly seen in scanning electron micrographs, is a smoothing of

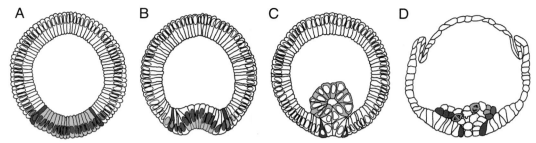

Figure 2. Drawings of cross sections of *Drosophila* embryos at four stages of mesoderm invagination. Dorsal is up. (*A*) Cellular blastoderm stage. An epithelium has formed in which all nuclei are located at the apical side of the cell, i.e., the periphery of the embryo. Twist, shown in red, is expressed in the nuclei of the prospective mesoderm at high levels, but also in a graded distribution reaching into the ventral ectoderm. The edge of the Snail expression domain, shown in blue, corresponds precisely to the lateral edge of the mesoderm. The mesectodermal cell at each side is marked in gray. (*B*) The central cells in which Fog is expressed (marked in *turquoise*) have begun to constrict at their apical ends, creating an indentation in the epithelium. (*C*) The ventral part of the epithelium has invaginated completely, but the epithelial structure is still intact. The central part expresses the FGF-receptor Heartless and the signal transduction molecule Dof (marked in *yellow*), which trigger the flattening of the mesoderm onto the ectoderm. (*D*) The epithelium has dissociated into mesenchymal cells that undergo mitosis (*light green*). Cells contacting the ectoderm activate PAPK (shown in *dark green*) under the control of the FGF receptor.

the ventral surface of the embryo, reflecting a flattening of each apical surface of each ventral cell, while the rest of the embryo retains its cobblestone appearance caused by the dome-shaped surfaces of the individual blastoderm cells (Turner and Mahowald 1977; Leptin and Grunewald 1990; Sweeton et al. 1991). Within minutes, cell flattening is followed by a reduction in the apical circumference of the ventral cells and the appearance of membrane blebs on the apical sides of the cells. In time-lapse video recordings where the outlines of cells have been fluorescently marked, it can be seen that not all cells constrict their surfaces simultaneously, but that a stochastically distributed, small number of cells begin restrictions and are gradually followed by others over a period of approximately 10 minutes (Kam et al. 1991; Oda and Tsukita 2001). This behavior is restricted to the middle; i.e., the most ventral part of the mesoderm primordium, covering perhaps 60% of the whole width of the primordium. The more lateral cells never constrict, indicating a subdivision of the mesoderm into two distinct populations. When the majority of cells have constricted, a very rapid invagination of the whole epithelium ensues.

Time-lapse recordings, as well as cross-sections through ventral furrows of fixed and stained embryos, show that apical constriction is accompanied by drastic cell shape changes and a movement of the nucleus away from its original position adjacent to the apical membrane to a new position within the basal region of the cell. Thus, ventral cells are converted from a columnar shape to a wedge shape (Leptin and Grunewald 1990; Sweeton et al. 1991). Throughout this period, the epithelial character of the region is maintained. Mathematical modeling shows that apical constriction resulting in this type of cell shape change is sufficient to cause an invagination (Odell et al. 1981), as is also intuitively obvious, since a series of wedge-shaped cells attached to each other cannot produce any other shape than a bent epithelium. Once a deep invagination has been formed, the cells shorten along their apical–basal axes (Sweeton et al. 1991). The furrow sinks or is pulled into the interior of the embryo, and the sides of the mesoderm primordium meet at the ventral midline.

In the next step of mesoderm morphogenesis, the basal sides of the cells that form the invaginated mesodermal tube (i.e., the sides that face the ectoderm) make contact with the basal sides of the ectodermal cells and the tube flattens against the ectoderm while, simultaneously, cell junctions between mesodermal cells are dissolved. All cells of the mesoderm then divide and migrate outward on the ectoderm, forming a single cell layer, the mesodermal germ layer.

Internalization of the Endoderm

The anterior portion of the ventral furrow does not give rise to mesoderm, but constitutes part of the primordium of the anterior midgut (Fig. 1B). The endodermal cells invaginate together with the mesoderm and then become mesenchymal and migratory.

A few minutes after the ventral furrow has begun to invaginate, a series of very similar cell shape changes occur at the posterior pole of the embryo (Sweeton et al. 1991). Cells flatten and then constrict their apical sides while their nuclei migrate away from the apical membrane. An indentation is formed, which contains the pole cells that lie on the surface of the posterior pole. At the same time, germ band extension (see below) is beginning to displace the posterior pole toward the dorsal side. As the indentation of the posterior pole epithelium deepens and moves first dorsally and then anteriorly, it carries the pole cells into the interior of the embryo, eventually forming a long sac-like epithelial structure.

Once the posterior midgut has invaginated, it disperses into single cells that migrate toward the anterior along the underlying mesoderm until the cells reach their counterparts migrating from the anterior end toward the center of the embryo (Reuter et al. 1993; Tepass and Hartenstein 1994). Unlike the creation of the mesodermal cell layer, which is complete within ~2 hours, the formation of the endodermal cell layer occurs over a longer period. The posterior half of the endoderm primordium reaches the cells of the anterior half in the middle of the embryo after ~10 hours of development. Together, these two populations become epithelial again late in development and enclose the central yolk mass to form the midgut.

Ectodermal Morphogenetic Movements

The Cephalic Furrow

Although the ectoderm is largely passive with respect to the formation of germ layers, it is engaged in various morphogenetic movements that occur in parallel with gastrulation. At the same time as the ventral surface of the embryo begins to flatten, a dorsoventral furrow becomes visible at the posterior end of the head region. It is first seen in lateral regions, but soon encircles the whole embryo. A band of three cells' width participates in the formation of this invagination which involves cell shape changes that differ completely from those seen in the mesoderm and endoderm (Vincent et al. 1997). First, a single row of cells move their nuclei toward their basal ends and shorten along their apical–basal axis, leading to a slight displacement inward from the surface of the embryo. These cells have been named "initiator cells." The apical surfaces of the neighboring rows of cells then turn toward and into the cleft created by the initiator cells. Their apical–basal axes also shorten, and the apical sides of the cells on each side of the initiator row come to face each other. A deep furrow with the initiator cells at the base is created.

The fold is prefigured by a difference in the appearance of the blastoderm epithelium. Cells in the region of the future furrow are taller, and the cytoplasmic connections to the underlying yolk mass are thinner, than in neighboring cells (Blankenship and Wieschaus 2001). The furrow disappears during later development, and its function is not known, but a role in separating the patterning regions of the trunk and the head has been put forward (Vincent et al. 1997).

Germ-band Extension and Dorsal Folds

After the mesoderm has invaginated, the ventral side of the embryo begins to extend backward, pushing the posterior

pole dorsally. The "germ band" (i.e., the main part of the trunk region of the embryo consisting of ectoderm and the underlying mesoderm) then extends around the posterior pole of the embryo, continuing to displace the posterior midgut invagination until the latter comes to lie directly behind the head. The cellular basis of this movement was completely unknown until a careful time-lapse study revealed that dorsoventrally directed cell intercalation within the ectoderm was associated with germ-band extension (Irvine and Wieschaus 1994). Irvine and Wieschaus observed the movement of cells in the lateral ectoderm relative to each other during germ-band extension and compared their behavior in various mutants. They found that cells which were initially adjacent to each other became separated from their anterior and posterior neighbors. Over a period of less than an hour, one or more cells inserted between them by intercalation from the dorsal side in a movement resembling convergent extension as it had been described in vertebrates (Gerhart and Keller 1986). This results in nearly doubling the length of the germ band. Since the egg is enveloped in a tough shell, the only way to accommodate this extra length is by doubling up the germ band around the posterior pole of the egg, pushing the true posterior end of the embryo, the posterior midgut invagination with the prospective germ cells, toward the dorsal side and then anteriorly.

While the germ band extends, the epithelium on the dorsal side also undergoes a number of changes. A series of transverse folds appear, which are probably not simply a passive reaction to germ-band extension movements, but rather seem to be genetically regulated, since they always appear in precisely the same order and position. The epithelium then becomes very thin and folds up between the advancing posterior edge of the germ band and the head fold. This tissue is the amnioserosa, which later, when germ-band extension is reversed, covers the dorsal side of the embryo but is ultimately internalized and undergoes histolysis after the germ band has grown over it and covered the whole circumference of the embryo.

THE GENETIC CONTROL OF GASTRULATION MOVEMENTS

Many of the genes that control the events of gastrulation are known. Two levels of control may be distinguished. Genes that determine the fate of a set of cells will necessarily also control their morphogenetic behavior, since this constitutes part of their fate. The genes *snail, twist, hkb,* and *tll* are examples. Obviously, the loss of cell's fate also entails the loss of some or all aspects of its behavior, so this may not tell us much about the mechanisms by which the cell executes its behavior. However, mutations in these genes have nevertheless been tremendously helpful in analyzing gastrulation events. A second set of genes encodes proteins that are directly involved in mediating cell behavior, such as adhesion molecules, actin, and small GTPases. These gene products are often ubiquitously distributed but activated or modified locally under the control of the fate-determining genes. The link between the two levels of control is still not very well understood.

All gastrulation movements are genetically separable; i.e., mutants exist that affect any one of the movements without affecting the others. This suggests that they are regulated independently. However, we show below that interactions between them exist.

Ventral Furrow Formation

Genetic Control

Two genes are essential in the zygote for the mesoderm to form: *snail* and *twist* (Simpson 1983; Leptin and Grunewald 1990; Sweeton et al. 1991). *snail* encodes a zinc-finger protein that acts as a repressor of ectodermal genes in the mesoderm primordium, whereas *twist* encodes a bHLH protein that acts as an activator of genes in the mesoderm (Kosman et al. 1991; Leptin 1991). A third gene, *folded gastrulation (fog)*, which codes for a secreted peptide (Costa et al. 1994), is necessary for the ventral furrow to form efficiently and rapidly, but even in its absence, the furrow eventually invaginates (Sweeton et al. 1991).

Mutations in both *snail* and *twist* affect the cell shape changes that lead to internalization of the mesoderm (Leptin and Grunewald 1990; Sweeton et al. 1991). In *snail* mutant embryos, no apical constriction is seen. However, all of the mutant mesodermal cells shorten along their basal–apical axes, leading to a thinning of the ventral epithelium and the appearance of irregular folds. In *twist* mutants, the ventral cells narrow at their apical ends, but not to the full extent of becoming wedge-shaped, as in wild-type embryos, and apical flattening does not occur. The nuclei do not move toward the basal end of the cells. These shape changes are accompanied by the formation of a narrow, fairly deep furrow. However, this furrow is not stable. The cells do not shorten along their basal–apical axes, and the more lateral cells do not make contact with their partners from the opposite side of the furrow. By the end of germ-band extension, all ventral cells are still at the surface of the embryo. The cells remain at the surface of the embryo throughout germ-band extension. Since Twist controls the expression of *snail*, the phenotype of *twist* mutant embryos may be partly explained by the loss of *snail*. Indeed, if the level of *snail* expression is raised, the defects in *twist* mutants can be partly suppressed (Ip et al. 1994). Conversely, the residual ability of morphogenetic activity in *twist* mutants is due to the early, Twist-independent *snail* expression. This follows from the finding that in embryos mutant for both *snail* and *twist*, no mesoderm-specific cell-shape changes

occur at all. Thus, these two genes together control all aspects of mesodermal cell-shape changes, but judged from their individual phenotypes, they have different, largely nonoverlapping functions, showing that the various processes during the cell-shape changes are controlled separately. For example, cell shortening does not depend on previous apical constriction or nuclear movement, as it can occur in *snail* mutants, in which the cells do not constrict and nuclei remain near the apical ends of the cells.

In *fog* mutant embryos, all aspects of the typical cell-shape changes of the future mesodermal cell can be observed. However, they do not happen at the correct time in all cells (Sweeton et al. 1991). In wild-type embryos, cell-shape changes begin in a small number of stochastically distributed cells within the mesoderm primordium (with cells near the ventral midline having a higher chance of being the first). Over a period of 10 minutes, increasing numbers of cells begin their constrictions until eventually the sum of all constrictions leads to the formation of the furrow. In *fog* mutants, the first cells appear to begin their shape changes at the correct time. However, the remaining cells seem to follow much more slowly than in the wild type, and the time it takes for a sufficient number of cells to become wedge-shaped is much longer (Sweeton et al. 1991; Oda and Tsukita 2001). The furrow therefore forms about 15 minutes later than in the wild-type embryo. This phenotype can be interpreted in different ways. It could indicate that the first constricting cells normally send out a signal that triggers surrounding cells also to constrict rapidly. This would be consistent with the biochemical nature of the Fog protein as a secreted peptide (Costa et al. 1994); i.e., a potential cell communication molecule. Alternatively, the same phenotype would be observed if the sequence of initiation of cell-shape changes were normal, but each cell underwent its sequence of changes much too slowly, so that they would not be detectable as soon as in the wild type (only fully constricted cell apices can be identified reliably; intermediate shapes are hard to distinguish from the normal variation). In either scenario, the role of *fog* is to increase the speed and efficiency of furrow formation, rather than to enable the cell-shape changes as such. The mechanism by which it does this may be to aid apical flattening, since ectopic expression of Fog can induce apical flattening in the ectoderm of the embryo (Morize et al. 1998).

Timing, Contribution of Other Regions, Autonomy

It has been proposed that large indentations in epithelia can be created by one central cell alone, or a single line of cells, initially being triggered to constrict. The stretching of neighboring cells caused by this first constriction would then act as a trigger for the neighbors' constrictions, which in turn would lead to stretching and constriction of the next set of neighbors (Odell et al. 1981). In this way, a wave of constrictions would be propagated through the field of cells, causing it to invaginate. Such a mechanism is unlikely to operate in the ventral furrow. First, the stochastic sequence in which cells in the prospective mesoderm begin their constrictions is inconsistent with the proposed model. Second, single cells or small groups of cells at any position in the epithelium have the ability to constrict and become internalized, irrespective of their neighbors' behavior (Leptin and Roth 1994). When nuclei from a wild-type embryo were transplanted into the future mesodermal region of *snail twist* mutant embryos, they became incorporated into the mutant epithelium and underwent their normal cell-shape changes, regardless of the position in the epithelium in which they found themselves. This indicates that every cell in the epithelium is able to follow the program of cell-shape changes independent of the activities of its neighbors.

Although the force that forms the furrow may be generated by apical constrictions in the mesoderm primordium, it is conceivable that activities of other parts of the embryo might help to internalize the furrow. For example, an expansion of the ectoderm might push the sides of the mesoderm toward the ventral midline, helping to deepen the furrow. The phenotypes of mutant embryos in which the cell fates of the ectoderm are transformed argue against this. The dominant maternal effect mutation *Toll[10B]* leads to a change in the activity of Dorsal in the progeny of mutant mothers such that Twist and Snail are expressed throughout the embryo (Roth et al. 1989). In many embryos, the levels of Twist and Snail are very high throughout, and all cells begin to flatten and attempt to constrict. Obviously, no furrow can be formed in this situation. However, in other embryos of this genotype, there is still a gradient of Twist, and constrictions begin at the point of highest Twist concentration, on the ventral side. This furrow succeeds in invaginating, even though the cells that would normally develop as ectoderm now express mesodermal markers and do not undergo the morphogenetic behavior typical of ectodermal cells (Leptin 1991). This supports the notion that the forces for the invagination of the furrow are generated in the furrow itself.

If the mesodermal cells follow only their own program to make the ventral furrow, what is it that triggers this program? The epithelium that forms the embryo before gastrulation begins is created by the subdivision of a syncytial layer of nuclei under the surface of the egg by plasma membranes growing between the nuclei and enclosing them to form cells. Completion of cellularization is immediately followed by the first signs of ventral furrow formation, suggesting a possible causal link. However, mutants in which cellularization fails to proceed normally nevertheless begin to form a furrow and attempt cell-shape changes (Straub et al. 1996). In cases of drastic failure in cellularization, the shape changes cannot be accomplished fully, since apical constriction and movement

of the nuclei away from the cortex destroy the partially formed cells. Yet, the mechanisms driving cell-shape changes must have been activated in the absence of complete cellularization, which shows that cellularization cannot be the signal that triggers furrow formation.

Together, these observations indicate that the information and forces required for furrow formation lie within the mesoderm primordium itself, and even more stringently, within the individual cells of the primordium. In other words, furrow formation is a process based entirely on cell-autonomous activities; it is the sum of the activities of individual cells that leads to the invagination. If there is no external signal that triggers furrow formation, then the signal must also be provided within each cell. The most likely timing mechanism is the accumulation of sufficient levels of one or more gene products under transcriptional regulation by Twist and Snail. Once these have reached a certain threshold, they may be able to modify components, leading to cell-shape change. This is consistent with the finding that reduction of Snail by half (in *snail/+* heterozygous embryos) leads to a significant delay in furrow formation, suggesting that in this situation, threshold levels of crucial components are reached later than in the wild type.

Permissive versus Instructive Mechanisms

In a search for further genes that direct ventral furrow formation, it was found that embryos homozygous for deletions in four different regions of the genome exhibited slight defects in furrow formation (Grosshans and Wieschaus 2000; Seher and Leptin 2000). None of these was as strong as the defects in *twist* or *snail* mutants, suggesting that no single gene is responsible for the effects mediated by Snail and Twist. Rather, they resembled the defects in *fog* mutants, in that furrow formation was delayed, but not completely abolished. For two of these regions, the genes whose loss caused the mutant phenotype were identified. *tribbles* codes for a protein with homology to the SNF1 family of kinases and is expressed transiently in the mesoderm primordium, in addition to many other places (Grosshans and Wieschaus 2000; Mata et al. 2000; Seher and Leptin 2000). *frühstart* codes for a small basic protein with a similar expression pattern (Grosshans et al. 2003). Closer analysis of mutant embryos revealed that instead of undergoing their normal cell-shape changes, the ventral cells were dividing. The proliferation of mesodermal cells, induced by the mesodermal expression of the cell cycle regulator String (Edgar and O'Farrell 1989), normally does not begin until the furrow has invaginated completely. The premature mitoses in *tribbles* mutants can be interpreted in two ways: Either correct cell-shape changes normally suppress mitosis, or conversely, mitosis needs to be suppressed for shape changes to be allowed to proceed normally. These possibilities can be distinguished by abolishing mitoses in the *tribbles* mutant embryos by eliminating the function of the cell cycle regulator String/cdc25 (Edgar and O'Farrell 1989). If the primary defect in *tribbles* is an inability to change shapes, then this should still be manifest in *tribbles string* double-mutant embryos—i. e., even in the absence of cell division. If, on the other hand, the premature mitoses are the cause of the failure to make a ventral furrow, then double-mutant embryos should be able to make a furrow. The latter was found to be the case (Seher and Leptin 2000). This shows that the function of *tribbles* is to block premature mitoses in the mesoderm.

Paradoxically, *string* is specifically expressed in the mesoderm under the control of Snail and Twist (Edgar et al. 1994), as is its inhibitor *tribbles*. Thus, *Drosophila* expresses an activator of mitosis in the mesoderm, but simultaneously expresses a regulator to interfere with mitosis and allow cells to change their shapes instead. To explain this apparently inefficient way of building an embryo, it may be worth considering mesoderm invagination in other insects. *Drosophila* is a long-germ insect in which the whole anterior–posterior axis is established before gastrulation begins. Gastrulation itself is an extremely rapid process, which internalizes the mesoderm simultaneously over its whole length within less than half an hour. Immediately after the mesoderm has been internalized, several rounds of cell division are needed to generate the number of cells necessary to produce all mesodermal derivatives. In contrast, short-germ insects internalize their mesoderm gradually, beginning in the anterior segments and proceeding posteriorly as new segments are added (see Chapter 8). The process is slower and less regular than in *Drosophila*, and mitoses can occur at the same time. Perhaps this is a sufficiently reliable mechanism in short-germ insects because there is sufficient time for all mesodermal cells to ingress or invaginate even if the process is slightly disrupted by concomitant mitoses, whereas in *Drosophila* the first mitosis has to be delayed to ensure speed and reliability of furrow formation. Interestingly, *tribbles* is not an essential gene under laboratory conditions, showing that a sufficient number of embryos survive even in the absence of an ideal gastrulation process.

Posterior and Anterior Midgut Formation

Only one gene is known to be essential in the zygote for invagination of the posterior midgut, *fog*. As described above, loss of *fog* leads to delays in the formation of the ventral furrow. However, *fog* was originally discovered because it has dramatic effects on germ-band extension (Zusman and Wieschaus 1985). The posterior end of the embryo does not move dorsally and then anteriorly, but remains at its original place. The extending germ band therefore buck-les and folds between the head and the posterior pole of the embryo rather than extending onto the dorsal side. It was found that the posterior end not only failed to move, but also failed to form an invagination (Sweeton et al. 1991). The failure to form an invagination is almost certainly the cause for the failure in germ-band extension, rather than the two defects being independent. First, mosaic analysis showed that the lethality associated with loss of *fog* is only expressed when the posterior end of the embryo is mutant (Zusman and Wieschaus 1985). Second, cloning of the gene allowed its expression pattern to be determined, and it turned out that *fog* is expressed transiently and very specifically only in the mesoderm primordium and the posterior midgut primordium (Costa et al. 1994). Thus, *fog* is not needed anywhere else in the embryo, and the effects on germ-band extension can only be caused by defects in the mesoderm or posterior midgut region. Further work showed that the germ-band extension defects in *fog* mutants can be completely suppressed by expression of *fog* only in the posterior midgut (T. Seher and M. Leptin, unpubl.). This shows that *fog* is essential for posterior midgut invagination, and the invagination is essential to permit germ-band extension to proceed.

fog expression in the posterior midgut primordium is controlled by the two gap genes that pattern the posterior end of the embryo, *hkb* and *tll*. Posterior midgut invagination is reduced in *hkb* and *tll* mutant embryos, and in the double mutants it is barely detectable. Simultaneously, *fog* expression disappears. This would be consistent with the two gap genes acting exclusively through *fog* to mediate PMG invagination, but it remains to be tested whether further, as-yet-unknown, factors under the control of *hkb* or *tll* also participate.

Although the anterior and the posterior midgut primordia are spatially separated, at least two of the regulators of their development, the zinc-finger transcription factor Hkb and the GATA factor Serpent, are nevertheless shared. Serpent defines both the anterior and the posterior midgut primordium, and in *serpent* mutants, the midgut primordia are transformed toward ectodermal cell fates (Reuter 1994; Rehorn et al. 1996). Hkb acts as the transcriptional regulator at the top of the genetic cascade of endoderm development and has two discernible functions. First, it is essential for the expression of *serpent* (Reuter 1994) both anteriorly and posteriorly, and of *fog* posteriorly (Costa et al. 1994), and therefore controls both early and late morphogenesis of the endoderm. Second, the ventral furrow and some mesoderm-specific gene expression patterns are extended anteriorly in *hkb* mutant embryos, and such embryos do not form an anterior midgut (Weigel et al. 1990; Reuter and Leptin 1994). However, in the anterior region, Hkb does not function by repressing *snail* expression, as it does in the posterior midgut. On the contrary, both Twist and Snail are needed together with Hkb for the proper morphogenesis of the anterior midgut. Hkb must therefore act by diverting Twist and Snail function from their mesodermal targets toward targets required for the endodermal developmental pathway.

The Cephalic Furrow

The band of cells that form the cephalic furrow overlap the most anterior stripe of expression of the segmentation gene *even-skipped*(*eve*) (Vincent et al. 1997). The position of the first *eve* stripe is determined by the gradient of Bicoid. In embryos with reduced or increased levels of Bicoid, the position of the first *eve* stripe is shifted anteriorly or posteriorly, and so is the head furrow. Furthermore, the head fold does not form in *eve* mutant embryos. Thus, Eve seems to be directly or indirectly responsible for the formation of the fold. However, it cannot be sufficient, since no folds are made in the other six *eve*-expressing stripes. In a search for other genes that are needed for the formation of the head fold, one further gene was found that had a similar early phenotype, *buttonhead* (*btd*) (Vincent et al. 1997). The zinc-finger transcription factor Btd is needed for aspects of patterning in the anterior part of the egg. Its expression domain overlaps that of the first *eve* stripe, but not that of the other six stripes. It is needed for the full expression of *eve* in the first stripe, but it is not clear whether it acts exclusively through *eve* to control the head fold, or whether it controls other, *eve*-independent processes that act in concert with *eve*.

A further gene that contributes to the formation of the cephalic furrow is the segmentation gene *paired* (Blankenship and Wieschaus 2001). Whereas the *eve* expression domain overlaps the posterior part of the furrow, the first stripe of the *paired* domain is precisely centered on the furrow and corresponds to the domain of cells whose connecting stalks to the underlying yolk are narrower than those in neighboring cells. This morphological difference depends on the function of Paired. In *paired* mutant embryos, all stalks are of the same diameter. Furthermore, the formation of the cephalic furrow is delayed, suggesting that the differences in blastoderm cell shapes aid later morphogenetic movements.

Germ-band Extension

Many parts of the embryo participate in germ-band extension. Whereas cell intercalation in the lateral ectoderm appears to provide the crucial part of the driving force, germ-band extension can only occur properly if two other parts of the embryo participate. The dorsal epithelium must fold and move out of the way of the posterior midgut primordium as the latter is displaced from its original posterior position anteriorly and dorsally to a new position just behind the head. We have seen that the formation of the posterior midgut invagination itself is necessary for the posterior end to be displaceable and for germ-band extension movements to occur in the proper geometry. The study of mutants has helped to distinguish between the roles of the various tissues.

The behavior of the ectoderm cells whose dorsoventral intercalation movements are thought to drive extension was compared in various mutants (Irvine and Wieschaus 1994). Surprisingly, no defects in intercalation were found in mutant embryos in which fate determination along the dorsoventral axis of the embryo was disrupted. In contrast, in mutants that affected anterior–posterior patterning, germ-band extension and cell intercalation were reduced to varying degrees. Mutants in which cell fates within a segment were changed (segment polarity mutants) extended their germ band normally. However, mutants in which segments were missing showed reduced extension, with the most extreme reduction caused by the most extensive loss of segmental subdivision (for example, as a result of combined loss of gap gene function, or combined maternal-effect mutations), whereas loss of fewer segments (in pair-rule mutants) had weaker effects.

Although the zygotic dorsoventral patterning genes do not seem to control cell intercalation, they do affect germ-band extension movements via their effect on the differentiation of the amnioserosa. The dorsal blastoderm epithelium normally makes three transverse folds at the beginning of germ-band extension and then undergoes a major morphogenetic change from a tall columnar to a very thin squamous epithelium that can be pushed back like a curtain behind the head fold. This differentiation of the amnioserosa depends on the dpp group genes, which include *dpp* and *zen*. In embryos that are mutant for these genes, the dorsal epithelium remains tall and columnar. When the posterior midgut begins to move dorsally and anteriorly, the defective amnioserosa cannot be displaced, and the posterior midgut stalls, or moves only part of its normal distance, and in the mutants with the strongest defects (e.g., *dpp*), it is pushed underneath the dorsal epithelium. This shows that the driving force for germ-band extension is still intact, and that the role of the amnioserosa for germ-band extension is a permissive one.

The same can probably be argued for the role of the invagination of the posterior midgut in germ-band extension. In embryos in which the invagination does not form, the germ band does not extend onto the dorsal side of the embryo. However, the germ band nevertheless lengthens and becomes twisted like a corkscrew or folds up like an accordion between the head furrow and the posterior end of the embryo, indicating that the driving force is still intact. Thus, the formation of the posterior midgut invagination is also a permissive function for germ-band extension.

Other sources of force contributing to the displacement of the posterior pole have been suggested. Based on a shearing movement between the blastoderm epithelium on the dorsal side of the embryo and the underlying yolk mass that can be deduced from electron microscopic images, it had been proposed that contractile forces in the cytoskeleton at the surface of the yolk might draw the posterior midgut invagination anteriorly (Rickoll and Counce 1980). Although a contribution of such forces cannot be excluded, they are unlikely to constitute the major mechanism driving germ-band extension.

It is unlikely that there is any specific morphogenetic purpose for germ-band extension. Instead, it has been suggested that germ-band extension is the result of a process that is required to generate sufficiently wide segmental units for patterning mechanisms to function within each segment primordium (Irvine and Wieschaus 1994). In the *Drosophila* embryo and in other long-germ insects, all segmental units are created simultaneously, in contrast to short-germ insects where most of the segments are added by growth. The limited number of cells along the length of the embryo at the blastoderm stage allows a band of only two to three cells' width to be allocated to each segment. However, proper segmental patterning requires at least four bands of cells with different gene expression patterns within each segment. It was proposed that the role of cell intercalation in a dorsoventral direction is to widen the segments along the anterior–posterior axis. This is supported by the result that mutants with fewer, and therefore wider, segments show significant reductions in cell intercalation and germ-band extension.

A detailed study of the relative positions of groups of cells during germ-band extension has revealed that only one round of cell intercalation occurs, and that it is associated with a well defined, stereotypic change in the junctions between neighboring cells (Fig. 3). Briefly, within the hexagonal array of cells that constitutes the cellular blastoderm, those cell boundaries that run along the dorsoventral axis (i.e., boundaries between anterior and posterior neighbors) are shortened to a point. This pulls two cells that were initially separate along the dorsoventral axis toward each other. These cells then establish contact, separating the two cells that initially touched each other in the anterior–posterior boundary (T. Lecuit, pers. comm.). It is thought that this movement is driven by myosin, which is enriched at anterior–posterior boundaries before germ-band extension begins.

SUBCELLULAR MACHINERY AND MATERNAL VERSUS ZYGOTIC CONTROL

For a full understanding of gastrulation, it is necessary not only to know the mechanisms by which cells are determined or what the genes are at the top of the regulatory cascades that control cell behavior, but also to know the molecules that are involved in the execution of morphogenetic cell behavior. In the case of the mesoderm and the posterior midgut, we would like to know how apical constriction, nuclear movement, and cell shortening are brought about, and in the case of germ-band extension, how cells intercalate and how the direction of intercalation is determined. So far, very little is known about these processes.

Since we know that Fog is a secreted peptide, we must assume that a signal transduction event is involved in the temporal control of furrow formation, and in enabling the cell-shape changes in the posterior midgut. Although the receptor for Fog is not known, there is good reason to believe that it is a serpentine receptor. This is based on the mutant phenotype of the maternal effect gene *concertina* (*cta*) (Parks and Wieschaus 1991). Embryos produced by homozygous mutant *cta* mothers show exactly the same defects as *fog* mutant embryos. This, together with the fact that *cta* encodes an α subunit of a heterotrimeric G-protein, suggests that *fog* and *cta* act in the same pathway, with Fog serving as the ligand for the unidentified serpentine receptor whose signal is transmitted through Cta. The downstream targets for *cta* are not known.

It is interesting, however, that Cta belongs to the G-α13 class of G-protein α subunits, and that these have been shown to interact with GTP exchange factors for rho family small GTPases in vertebrates (Hart et al. 1998; Kozasa et al. 1998). A member of this family in *Drosophila*, RhoGEF2, is the only other protein that has been found to play an important part in gastrulation (Barrett et al. 1997; Häcker and Perrimon 1998). Eggs are provided with a large maternal component of RhoGEF2. When this is removed (by making homozygous mutant germ lines in the mothers that are themselves heterozygous for the mutation), ventral furrow formation and posterior midgut invagination are com-

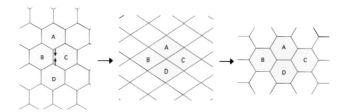

Figure 3. Cell intercalation during germ-band extension. Diagram of a group of cells in the ventral ectoderm at the cellular blastoderm stage. The connections between four cells are marked. The arrows indicate a postulated contractile force of the cortical actin cytoskeleton in two neighboring cells. This force is restricted to cell boundaries that run along the dorsoventral axis, and results in cells A and D approaching each other, so that the former contact between cells B and C is reduced to a point. This eventually allows cells A and D to establish a joint boundary.

Maternal Factors   Zygotic factors

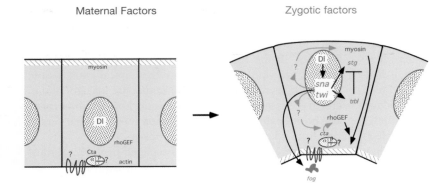

Figure 4. Zygotic factors acting on maternally supplied molecules to induce the invagination of the ventral furrow. The majority of the molecules present in the embryo at the blastoderm stage are supplied by the mother, or are derived from maternally provided RNA. With a few exceptions, they are present in all cells of the blastoderm embryo. Shown in blue are some of the proteins demonstrated or assumed to be involved in gastrulation. They can be activated locally under the control of genes (*red*) transcribed from the zygotic genome in specific regions and at specific times (see Fig. 1).

pletely blocked. No apical constrictions occur in either the mesoderm or the posterior midgut primordium. A reasonable hypothesis would therefore be that serpentine receptor signaling via Cta stimulates Rho, and Rho in turn modifies the actin cytoskeleton to cause apical constrictions. The problem with this model is that the defects caused by loss of RhoGEF2 are much more severe than those caused by loss of Cta. Thus, there must be mechanisms in addition to signaling via Cta that activate RhoGEF2, if Cta acts on RhoGEF2 at all (Fig. 3).

The fact that a RhoGEF is involved supports the obvious assumption that cytoskeletal rearrangements must be important for the cell-shape changes during ventral furrow and posterior midgut invagination. Another indication for this is that myosin relocates from the basal part of the cells, where it is needed during cellularization, to the apical sides of constricting ventral and posterior cells (Young et al. 1991; Leptin et al. 1992; Hunter and Wieschaus 2000). However, it has so far not been possible to test a role for myosin and the actin cytoskeleton by genetic means in vivo because both molecules are needed during the production of the egg as well as the stages of embryogenesis that immediately precede gastrulation. Therefore, eggs lacking myosin or actin cannot be obtained. Even studies using drugs are essentially impossible, because drugs that disrupt the actin cytoskeleton block cellularization.

The maternally supplied components discussed above (Cta, RhoGEF2, and Rho, cytoskeletal components, and myosin) are ubiquitously distributed throughout the egg, yet cell-shape changes occur only in precisely specified regions, indicating that there must be region-specific triggers for the activation of the maternal proteins. In the case of Cta, it is the regionally restricted expression of Fog that provides the necessary and sufficient trigger for activation (Morize et al. 1998), but the main trigger for RhoGEF2 is not known. Since no furrow formation occurs in the absence of Snail and Twist, it is likely that the regional activation of RhoGEF is caused by the product of one or more of their target genes. These have so far eluded their identi-

fication. The elucidation of the complete pathway from the morphogens that set up cell fates to the morphogenesis of the structures formed by the cells remains a challenge for those studying gastrulation.

ACKNOWLEDGMENTS

I thank Jörg Grosshans, Arno Müller, Rolf Reuter, Siegfried Roth, and Thomas Seher for discussions and helpful comments on the manuscript.

REFERENCES

Alberga A., Boulay J.-L., Kempe E., Dennefeld C., and Haenlin M. 1991. The snail gene required for mesoderm formation is expressed dynamically in derivatives of all three germ layers. *Development* **111:** 983–992.

Barrett K., Leptin M., and Settleman J. 1997. The Rho GTPase and a putative RhoGEF mediate a signaling pathway for the cell shape changes in *Drosophila* gastrulation. *Cell* **91:** 905–915.

Blankenship J.T. and Wieschaus E. 2001. Two new roles for the *Drosophila* AP patterning system in early morphogenesis. *Development* **128:** 5129–5138.

Campos-Ortega J.A. and Hartenstein V. 1985. *The embryonic development of* Drosophila melanogaster. Springer-Verlag, Berlin.

Costa M., Wilson E., and Wieschaus E. 1994. A putative cell signal encoded by the *folded gastrulation* gene coordinates cell shape changes during *Drosophila* gastrulation. *Cell* **76:** 1075–1089.

Crews S.T., Thomas J.B., and Goodman C.S. 1988. The *Drosophila* single-minded gene encodes a nuclear protein with sequence similarity to the *per* gene product. *Cell* **52:** 143–151.

Dubnicoff T., Valentine S.A., Chen G., Shi T., Lengyel J. A., Paroush Z., and Courey A.J. 1997. Conversion of dorsal from an activator to a repressor by the global corepressor groucho. *Genes Dev.* **11:** 2952–2957.

Edgar B.A. and O'Farrell P.H. 1989. Genetic control of cell division patterns in the *Drosophila* embryo. *Cell* **57:** 177–187.

Edgar B.A., Lehman D.A., and O'Farrell P.H. 1994. Transcriptional regulation of string (cdc25): A link between developmental programming and the cell cycle. *Development* **120:** 3131–3143.

Fullilove S.L., Jacobson A.G., and Turner F.R. 1978. Embryonic development: Descriptive. In *The genetics and biology of* Drosophila (ed.

M. Ashburner and T.R.F. Wright), vol. 2c, pp. 105-227. Academic Press, New York.

Gerhart J. and Keller R. 1986. Region-specific cell activities in amphibian gastrulation. *Annu. Rev. Cell Biol.* **2:** 201–229.

Greenwood S. and Struhl G. 1997. Different levels of Ras activity can specify distinct transcriptional and morphological consequences in early *Drosophila* embryos. *Development* **124:** 4879–4886.

Grosshans J. and Wieschaus E. 2000. A genetic link between morphogenesis and cell division during formation of the ventral furrow in *Drosophila. Cell* **101:** 523–531.

Grosshans J., Müller H.-A.J., and Wieschaus E. 2003. Control of cleavage cycles in *Drosophila* embryos by frühstart. *Dev. Cell* **5:** 285–294.

Häcker U. and Perrimon N. 1998. DRhoGEF2 encodes a member of the Dbl family of oncogenes and controls cell shape changes during gastrulation in *Drosophila. Genes Dev.* **12:** 274–84.

Hart M., Jiang X., Kozasa T., Roscoe W., Singer W., Gilman A., Sternweis P., and Bollag G. 1998. Direct stimulation of the guanine nucleotide exchange activity of p115 RhoGEF by Galpha13. *Science* **280:** 2112–2114.

Hartenstein V., Technau G.M., and Campos-Ortega J.A. 1985. Fate-mapping in wild-type *Drosophila melanogaster*. III. A fate map of the blastoderm. *Roux's Arch. Dev. Biol.* **194:** 213–216.

Hunter C. and Wieschaus E. 2000. Regulated expression of nullo is required for the formation of distinct apical and basal adherens junctions in the *Drosophila* blastoderm. *J. Cell Biol.* **150:** 391–401.

Ip Y.T., Maggert K., and Levine M. 1994. Uncoupling gastrulation and mesoderm differentiation in the *Drosophila* embryo. *EMBO J.* **13:** 5826–5834.

Ip Y.T., Park R.E., Kosman D., Bier E., and Levine M. 1992a. The dorsal gradient morphogen regulates stripes of rhomboid expression in the presumptive neuroectoderm of the *Drosophila* embryo. *Genes Dev.* **6:** 1728–1739.

Ip Y.T., Park R.E., Kosman D., Yazdanbakhsh K., and Levine M. 1992b. dorsal-twist interactions establish snail expression in the presumptive mesoderm of the *Drosophila* embryo. *Genes Dev.* **6:** 1518–1530.

Irvine K.D. and Wieschaus E. 1994. Cell intercalation during *Drosophila* germband extension and its regulation by pair-rule segmentation genes. *Development* **120:** 827–842.

Jiang J. and Levine M. 1993. Binding affinities and cooperative interactions with bHLH activators delimit threshold responses to the dorsal gradient morphogen. *Cell* **72:** 741–752.

Jiang J., Kosman D., Ip Y.T., and Levine M. 1991. The dorsal morphogen gradient regulates the mesoderm determinant twist in early *Drosophila* embryos. *Genes Dev.* **5:** 1881–1891.

Jimenez G., Guichet A., Ephrussi A., and Casanova J. 2000. Relief of gene repression by torso RTK signaling: Role of capicua in *Drosophila* terminal and dorsoventral patterning. *Genes Dev.* **14:** 224–231.

Kam Z., Minden J.S., Agard D.A., Sedat J.W., and Leptin M. 1991. *Drosophila* gastrulation: Analysis of cell behaviour in living embryos by three-dimensional fluorescence microscopy. *Development* **112:** 365–370.

Kasai Y., Stahl S., and Crews S. 1998. Specification of the *Drosophila* CNS midline cell lineage: Direct control of single-minded transcription by dorsal/ventral patterning genes. *Gene Expr.* **7:** 171–189.

Kasai Y., Nambu J.R., Lieberman P.M., and Crews S.T. 1992. Dorsal-ventral patterning in *Drosophila*: DNA binding of snail protein to the single-minded gene. *Proc. Natl. Acad. Sci.* **89:** 3414–3418.

Klapper R., Holz A., and Janning W. 1998. Fate map and cell lineage relationships of thoracic and abdominal mesodermal anlagen in *Drosophila melanogaster. Mech. Dev.* **71:** 77–87.

Kosman D., Ip Y.T., Levine M., and Arora K. 1991. Establishment of the mesoderm-neuroectoderm boundary in the *Drosophila* embryo. *Science* **254:** 118–122.

Kozasa T., Jiang X., Hart M., Sternweis P., Singer W., Gilman A., Bollag G., and Sternweis P. 1998. p115 RhoGEF, a GTPase activating protein for Galpha12 and Galpha13. *Science* **280:** 2109–2111.

Lall S. and Patel N. H. 2001. Conservation and divergence in molecular mechanisms of axis formation. *Annu. Rev. Genet.* **35:** 407–437.

Leptin M. 1991. *twist* and *snail* as positive and negative regulators during *Drosophila* mesoderm development. *Genes Dev.* **5:** 1568–1576.

Leptin M. and Grunewald B. 1990. Cell shape changes during gastrulation in *Drosophila. Development* **110:** 73–84.

Leptin M. and Roth S. 1994. Autonomy and non-autonomy in *Drosophila* mesoderm differentiation and morphogenesis. *Development* **120:** 853–859.

Leptin M., Casal J., Grunewald B., and Reuter R. 1992. Cellular and genetic events during early *Drosophila* mesoderm formation. *Development Suppl.* **1992:** 23–31.

Lohs-Schardin M., Cremer C., and Nüsslein-Volhard C. 1979. A fate map for the larval epidermis of *Drosophila melanogaster*: Localized cuticle defects following irradiation of the blastoderm with an ultraviolet laser microbeam. *Dev. Biol.* **73:** 239–255.

Mahowald A.P. 1963. Electron microscopy of the formation of the cellular blastoderm in *Drosophila melanogaster. Exp. Cell Res.* **32:** 457–468.

Mata J., Curado S., Ephrussi A., and Rørth P. 2000. Tribbles coordinates mitosis and morphogenesis in *Drosophila* by regulating string/CDC25 proteolysis. *Cell* **101:** 511–522.

Morel V., Lecourtois M., Massiani O., Maier D., Preiss A., and Schweisguth F. 2001. Transcriptional repression by suppressor of hairless involves the binding of a hairless-dCtBP complex in *Drosophila. Curr. Biol.* **11:** 789–792.

Morize P., Christiansen A.E., Costa M., Parks S., and Wieschaus E. 1998. Hyperactivation of the folded gastrulation pathway induces specific cell shape changes. *Development* **125:** 589–597.

Oda H. and Tsukita S. 2001. Real-time imaging of cell-cell adherens junctions reveals that *Drosophila* mesoderm invagination begins with two phases of apical constriction of cells. *J. Cell Sci.* **114:** 493–501.

Odell G. M., Oster G., Alberch P., and Burnside B. 1981. The mechanical basis of morphogenesis: Epithelial folding and invagination. *Dev. Biol.* **85:** 446–462.

Pan D. and Courey A. 1992. The same dorsal binding site mediates both activation and repression in a context-dependent manner. *EMBO J.* **11:** 1837–1842.

Pan D.J., Huang J.D., and Courey A.J. 1991. Functional analysis of the *Drosophila* twist promoter reveals a dorsal-binding ventral activator region. *Genes Dev.* **5:** 1892–1901.

Parks S. and Wieschaus E. 1991. The *Drosophila* gastrulation gene *concertina* encodes a Gα-like protein. *Cell* **64:** 447–458.

Paroush Z., Wainwright S.M., and Ish-Horowicz D. 1997. Torso signalling regulates terminal patterning in *Drosophila* by antagonising Groucho-mediated repression. *Development* **124:** 3827–3834.

Pignoni F., Baldarelli R M., Steingrimsson E., Diaz R.J., Patapoutian A., Merriam J.R., and Lengyel J.A. 1990. The *Drosophila* gene *tailless* is expressed at the embryonic termini and is a member of the steroid receptor superfamily. *Cell* **62:** 151–163.

Poulson D.F. 1950. Histogenesis, organogenesis and differentiation in the embryo of *Drosophila melanogaster* (Meigen). In *Biology of Drosophila* (ed. M. Demerec), pp. 168–274. Wiley, New York.

Rehorn K.P., Thelen H., Michelson A.M., and Reuter R. 1996. A molecular aspect of hematopoiesis and endoderm development common to vertebrates and *Drosophila*. *Development* **122:** 4023–4031.

Reuter R. 1994. The gene *serpent* has homeotic properties and specifies endoderm versus ectoderm within the *Drosophila* gut. *Development* **120:** 1123–1135.

Reuter R. and Leptin M. 1994. Interacting functions of *snail*, *twist* and *huckebein* during the early development of germ layers in *Drosophila*. *Development* **120:** 1137–1150.

Reuter R., Grunewald B., and Leptin M. 1993. A role for the mesoderm in endodermal migration and morphogenesis. *Development* **119:** 1135–1145.

Rickoll W.L. and Counce S.J. 1980. Morphogenesis in the embryo of *Drosophila melanogaster* - germ band extension. *Roux's Arch. Dev. Biol.* **188:** 163–177.

Rivera-Pomar R. and Jäckle H. 1996. From gradients to stripes in Drosophila embryogenesis: Filling in the gaps. *Trends Genet.* **12:** 478–483.

Roth S., Stein D., and Nüsslein-Volhard C. 1989. A gradient of nuclear localization of the dorsal protein determines dorsoventral pattern in the *Drosophila* embryo. *Cell* **59:** 1189–1202.

Rusch J. and Levine M. 1996. Threshold responses to the dorsal regulatory gradient and the subdivision of primary tissue territories in the *Drosophila* embryo. *Curr. Opin. Genet. Dev.* **6:** 416–423.

Rushlow C. and Roth S. 1996. The role of the dpp-group genes in dorsoventral patterning of the *Drosophila* embryo. *Adv. Dev. Biol.* **4:** 27–82.

Rushlow C., Frasch M., Doyle H., and Levine M. 1987. Maternal regulation of *zerknullt*: A homoeobox gene controlling differentiation of dorsal tissues in *Drosophila*. *Nature* **330:** 583–586.

Rushlow C.A., Han K., Manley J.L., and Levine M. 1989. The graded distribution of the dorsal morphogen is initiated by selective nuclear transport in *Drosophila*. *Cell* **59:** 1165–1177.

Seher T.C. and Leptin M. 2000. Tribbles, a cell-cycle brake that coordinates proliferation and morphogenesis during *Drosophila* gastrulation. *Curr. Biol.* **10:** 623–629.

Simpson P. 1983. Maternal-zygotic gene interactions during formation of the dorsoventral pattern in *Drosophila* embryos. *Genetics* **105:** 615–632.

Sonnenblick B.P. 1950. The early embryology of *Drosophila melanogaster*. In *Biology of* Drosophila (ed. M. Demerec), pp. 62–167. Wiley, New York.

St Johnston D. and Nüsslein-Volhard C. 1992. The origin of pattern and polarity in the *Drosophila* embryo. *Cell* **68:** 201–219.

Straub K.L., Stella M.C., and Leptin M. 1996. The gelsolin-related flightless I protein is required for actin distribution during cellularisation in *Drosophila*. *J. Cell Sci.* **109:** 263–270.

Strecker T.R., Kongsuwan K., Lengyel J.A., and Merriam J.R. 1986. The zygotic mutant *tailless* affects the anterior and posterior ectodermal regions of the *Drosophila* embryo. *Dev. Biol.* **113:** 64–76.

Sweeton D., Parks S., Costa M., and Wieschaus E. 1991. Gastrulation in *Drosophila*: The formation of the ventral furrow and posterior midgut invaginations. *Development* **112:** 775–789.

Technau G.M. and Campos-Ortega J.A. 1985. Fate-mapping in wild-type *Drosophila melanogaster*. II. Injections of horseradish peroxidase in cells of the early gastrula stage. *Roux's Arch. Dev. Biol.* **194:** 196–212.

Tepass U. and Hartenstein V. 1994. Epithelium formation in the Drosophila midgut depends on the interaction of endoderm and mesoderm. *Development* **120:** 579–590.

Thisse B., el Messal M., and Perrin-Schmitt F. 1987. The twist gene: Isolation of a *Drosophila* zygotic gene necessary for the establishment of dorsoventral pattern. *Nucleic Acids Res.* **15:** 3439–3453.

Turner F.R. and Mahowald A.P. 1977. Scanning electron microscopy of *Drosophila* embryogenesis. II. Gastrulation and segmentation. *Dev. Biol.* **49:** 403–416.

Vincent A., Blankenship J., and Wieschaus E. 1997. Integration of the head and trunk segmentation systems controls cephalic furrow formation in *Drosophila*. *Development* **124:** 3747–3754.

Weigel D., Jürgens G., Klingler M., and Jäckle H. 1990. Two gap genes mediate maternal terminal pattern information in *Drosophila*. *Science* **248:** 495–498.

Wieschaus E., Sweeton D., and Costa M. 1991. Convergence and extension during germ band elongation in *Drosophila* embryos. In *Gastrulation: Movements, patterns and molecules* (ed. R. Keller et al.), pp. 213–224. Plenum Press, New York.

Young P.E., Pesacreta T.C., and Kiehart D.P. 1991. Dynamic changes in the distribution of cytoplasmic myosin during *Drosophila* embryogenesis. *Development* **111:** 1–14.

Zusman S. and Wieschaus E. 1985. Requirements for autosomal gene activity during gastrulation in *Drosophila melanogaster*. *Dev. Biol.* **111:** 359–371.

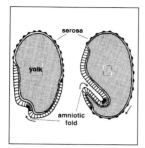

GASTRULATION IN OTHER INSECTS

S. Roth

Institut für Entwicklungsbiologie, Universität Köln, 50923 Köln, Germany

INTRODUCTION

The definition of gastrulation in insects was a topic of much controversy in the first half of the 20th century (for extensive coverage of this debate, see Johannsen and Butt 1941; Schwalm 1988). This was mainly due to the fact that the formation of the inner germ layer, the endoderm, a defining feature of gastrulation in most other animal groups, occurs in a very unusual way in insects. As a result, there is no early gastrula stage composed of three germ layers. This has been pointed out for the higher dipteran *Drosophila melanogaster* (see Chapter 7), but the feature is not a specialty of the higher dipterans, which in many other respects deviate from the archetypical mode of insect development. It is, in fact, one of the most conserved characters of insect embryogenesis, found in all pterygote insect orders analyzed so far. There have only been a few recent comparative studies on endoderm formation in insects that use molecular markers derived from *Drosophila* to reevaluate the data obtained by classical morphological observations. Because of this, in this chapter only the basic facts regarding this topic are summarized.

The main focus of this chapter is on those aspects of early morphogenetic movements in insects which are clearly different from *Drosophila*. These are particularly evident if we define gastrulation to include not only the movements resulting in the formation of the inner cell layers of mesoderm and endoderm, but also the movements of the embryonic and extraembryonic ectoderm that occur in parallel. The latter constitute one of the most obvious differences between the higher dipterans and all other major insect groups. Whereas higher dipterans form only one extraembryonic covering, the amnioserosa, which protects the embryo on the dorsal side prior to dorsal closure, other insects form two complete extraembryonic coverings, the amnion and the serosa (Fig. 1) (Schwalm 1988). Their topological relation to the embryo is similar to that found between amnion, chorion, and embryo proper in amniote vertebrates.

In most insect groups, the specification of the extraembryonic ectoderm is preceded by a differentiation within the uniform blastoderm into a region giving rise to the outer extraembryonic covering, the serosa, and one giving rise to both the embryo proper and the amnion (Fig. 2). This distinction is morphologically apparent prior to gastrulation. No corresponding pre-gastrulation stage exists in *Drosophila*. Following Anderson (1972), I refer to this stage as the "differentiated blastoderm." The region of the differentiated blastoderm giving rise to the embryo proper and the amnion has been referred to by a variety of names depending on author and organism. Among them are germ anlage, germ disc, blastodisc, germ band, embryonic rudiment, germinal rudiment, and embryonic primordium. In this chapter, the term germ anlage is used to avoid designations that refer only to the future embryo. Following the terminology common to the *Drosophila* literature, the term "germ band" is reserved for the metameric region of the embryo after it has become multilayered during gastrulation.

The following description of early morphogenetic movements in insects is focused on those species for which

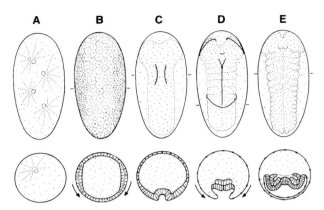

Figure 1. Schematic representation of the major events in early insect embryogenesis. The top row shows a ventral view of an embryo with the anterior end facing up; the bottom row shows cross sections at the levels indicated by bars in the top row. (*A*) Pre-blastoderm. The nuclei divide within the yolk mass before they migrate to the periphery. (*B*) Differentiated blastoderm. Initially, the blastoderm consists of an even layer of cells, but soon regional differences in cell density and shape indicate a differentiation into presumptive serosa and germ anlage. In many insects, cell condensations start in lateral regions and move toward the ventral side where they fuse to produce the germ anlage proper. (*C*) Mesoderm internalization. Frequently, a ventral furrow is first visible at a particular position along the anterior–posterior axis and subsequently expands along the entire ventral midline. (*D*) Early germ band. The amniotic folds form over the head lobes and over the abdomen. They approach each other and finally fuse. (*E*) Later germ band. The serosa and amnion have become independent epithelial layers. The serosa covers the entire yolk and embryo. The amnion covers the embryo at the ventral side. The fluid-filled space delimited by embryo and amnion is called the amniotic cavity. (Modified, with permission, from Sander 1976b.)

molecular data are available, in addition to morphological data. Currently, our knowledge is most detailed for the red flour beetle *Tribolium castaneum* and the grasshopper *Schistocerca gregaria* (and *americana*). Comparison with other insects is used to point out evolutionary trends and to evaluate whether the mechanisms found in *Tribolium* and *Schistocerca* are ancestral or derived.

FORMATION OF BLASTODERM AND GERM ANLAGE

In contrast to their arthropod sister group, the crustaceans, insects have adopted a terrestrial lifestyle. No doubt, many characteristic features of egg structure and early embryogenesis are evolutionary inventions linked to the transition from water to land. Insect eggs are large in comparison to the maternal body size, often more than a millimeter in one dimension (Fig. 2), and rich in yolk and other maternally supplied macromolecules to provide a nutritional supply for the development of complex larval forms. They possess

highly elaborated eggshells (chorion and vitelline membranes), which provide mechanical stability, protection from desiccation, chemical resistance, and a variety of other physiological functions with adaptive value to particular sites of egg deposition (e.g., respiration; Hinton 1981). As in *Drosophila*, many insect eggs show a clear axial organization with bilateral symmetry (Gutzeit and Sander 1985). The positions of specialized eggshell structures, like micropyles (pores for sperm entry), opercula (hatching devices), and respiratory appendages permit the identification of the anterior and posterior poles, and differences in the curvature of the elongated egg surface allow identification of the dorsal and ventral sides. Of course, the definition of the egg axes ultimately goes back to how the embryonic axes arise inside the egg. In the eggs depicted in Figure 2, anterior is to the left and ventral faces down. The germ anlagen possessing defined axes of polarity are shaded in gray. The left parts of the germ anlagen always harbor the presumptive head regions, and a narrow stripe of cells straddling the ventral midline (defined by the plane of bilateral symmetry of the egg) will give rise to the presumptive mesoderm (Fig. 3B). If the germ anlage is small relative to the total egg size (Kleinkeim, Krause 1939), its position along the anterior–posterior axis of the egg may vary (e.g., compare the three orthopteran eggs in Fig. 2), but it usually is located close to the posterior pole. Sometimes the anlage is confined to the posterior extremity of the egg (Fig. 2), as in the mayflies (Ephemeroptera; Bohle 1969), *Schistocerca*, or the walking sticks (Phasmatodea; Krause 1939). In these cases, it becomes difficult to align the primary embryonic axes with the egg axes, since the anterior–posterior axis of the embryo appears to run along the dorsoventral axis of the egg.

Despite variations in egg size and size of the germ anlagen, all groups of pterygote insects begin embryogenesis in the same manner (Schwalm 1988; secondary modifications due to viviparity and parasitism are mentioned later). The first nuclear divisions occur synchronously within the center of the egg without concomitant cytokinesis. This creates a syncytial stage in which a cloud of nuclei, each surrounded by an island of cytoplasm, populates the yolk mass. Each unit of nucleus plus surrounding cytoplasm is called an "energid." Two modes of blastoderm formation can be distinguished. In *Tribolium*, as in *Drosophila*, the nuclear cloud in the center of the egg first expands along the anterior–posterior axis, so that the number of energids subsequently migrating to the periphery is uniform over the entire egg surface (Stanley and Grundman 1970). Consequently, the resulting blastoderm layer has nearly uniform nuclear densities (Fig. 4A) (Handel et al. 2000). This mode of blastoderm formation appears to be ancestral for the winged insects, as it is also found in the primitive orders of mayflies (Ephemeroptera; Bohle 1969) and dragonflies (Odonata; Seidel 1935). In contrast, the huge yolk-rich eggs of many orders of Polyneoptera are not uniform-

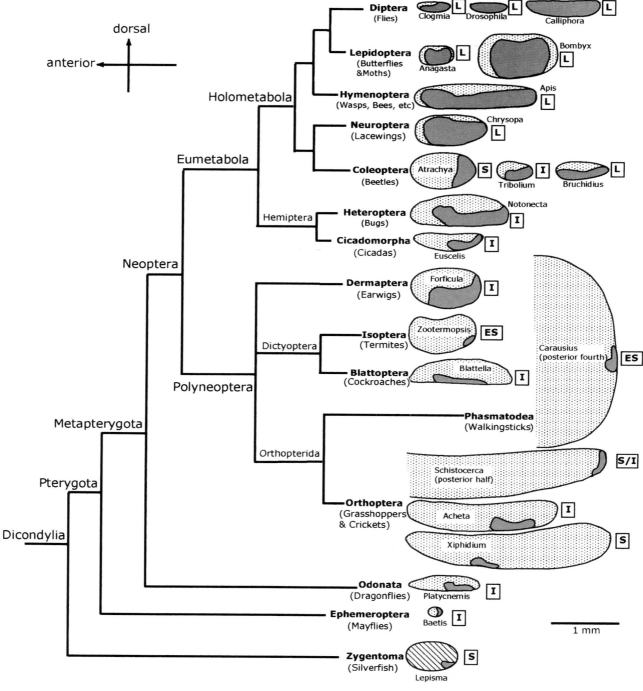

Figure 2. The phylogeny of insect eggs and germ types. The eggs are drawn to scale. The presumptive serosa is shown stippled and the germ anlage in gray. The hatched lines used for *Lepisma* indicate that the silverfish do not form a serosa, proper. Close to the embryos the presumed germ-type designation is shown in boxed capital letters. (*L*) Long-germ type. (*I*) Intermediate-germ type. (*S*) Short-germ type. (*ES*) Extreme short-germ type. See text for definition and critique of these terms. (After Krause 1939; Bohle 1969; Jura 1972; Anderson 1972; Sander 1976a; Schwalm 1988). The phylogeny is taken from Willmann (2003).

ly populated by nuclei before blastoderm formation starts (Fig. 2; Anderson 1972; Schwalm 1988). In *Schistocerca* and *Locusta* (Fig. 5A) (Roonwal 1936; Ho et al. 1997), the energids first reach the surface at the posterior pole. While formation and differentiation of the blastoderm ensue in this region, energids progressively invade the anterior parts

A

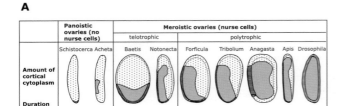

	Panoistic ovaries (no nurse cells)		Meroistic ovaries (nurse cells)							
			telotrophic		polytrophic					
	Schistocerca	Acheta	Baetis	Notonecta	Forficula	Tribolium	Anagasta	Apis	Drosophila
Amount of cortical cytoplasm										
Duration of embryo-genesis (days)	20	28	46	18	14	6	6	3	1

B

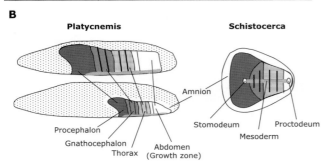

Platycnemis   Schistocerca

Amnion
Procephalon   Stomodeum   Proctodeum
Gnathocephalon   Mesoderm
Thorax   Abdomen (Growth zone)

Figure 3. Egg types and oogenesis. Fate maps of two germ anlagen. (*A*) The eggs are drawn such that they all have the same length. This depiction stresses differences in the relative sizes of germ anlage (*gray*) and serosa (*stippled*). The thickness of the cortical cytoplasm is indicated. Eggs rich in cortical cytoplasm are derived from oocytes that are connected to nurse cells via cytoplasmic bridges. In general, a correlation holds between content of cortical cytoplasm, size of the germ anlage, and speed of development. The slow development of *Baetis* is due to temperature (6°C) and an embryonic diapause. Only embryos with large germ anlagen possess a complete representation of all segment primordia at blastoderm stage (long-germ type). (Modified from Bier 1969; Büning 1994.) (*B*) Fate maps of germ anlagen from the dragonfly *Platycnemis* and the grasshopper *Schistocerca*. The fate map of *Platycnemis* is based on local UV irradiation (Seidel 1935). The upper embryo is at the uniform blastoderm stage, and the lower is shown after condensation of the germ anlage. The fate map for *Schistocerca* is based on the early expression of the segment polarity gene *wingless* (Dearden and Akam 2001).

of the egg surface. By the time a complete cell layer has formed at the anterior pole, the posteriorly localized embryo has already started to gastrulate. Since the germ anlage forms where the energids first populate the egg surface, the temporal course of blastoderm formation presages the subsequent differentiation events.

In light of the diffusion-based mechanisms that govern early pattern formation in *Drosophila* (Rivera-Pomar and Jäckle 1996), it is important to find out whether a syncytial blastoderm stage precedes cellularization in other insects. Unfortunately, this problem has not been fully addressed by functional studies. At the ultrastructural level, even embryos of other holometabolous insect orders, like those of lepidopterans, show remarkable differences in early blastoderm architecture compared to *Drosophila*. In *Bombyx*, all energids that reach the egg surface become immediately surrounded by plasma membranes in a process resembling pole cell formation in *Drosophila* (Takesue et al. 1980). In

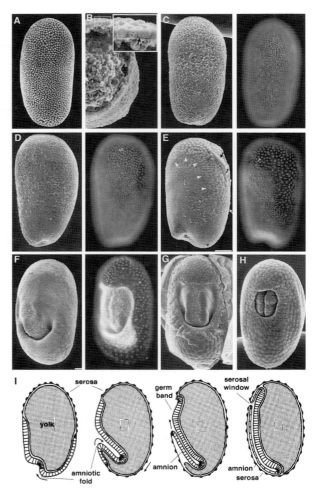

Figure 4. Formation of germ anlage and extraembryonic coverings in *Tribolium*. (*A–H*) Scanning electron microscopic (SEM) photographs and in *C–F* at the right side of the SEM photograph, a fluorescence photograph of the same embryo stained with DAPI to visualize the cell nuclei. (*A*) Embryo at the uniform blastoderm stage. (*B*) Anterior fracture showing the thickness of the uniform blastoderm cell layer before germ anlage formation. (*C–F*) The progressive condensation of the germ anlage is accompanied by flattening of the serosa cells. The condensation reaches its maximum when the posterior amniotic fold and the ventral furrow are well visible (*F*). (*G*) Anterior amniotic fold begins to form. (*H*) Serosal window shortly before closure. (*I*) Schematic drawings of longitudinal sections that explain the development of serosa and amnion. For easier understanding, the internalization of the mesoderm is omitted. (Modified, with permission, from Handel et al. 2000 [© Springer-Verlag].)

Schistocerca, isolated energids populate the egg surface. Only a subset of these energids becomes surrounded by cell membranes (Fig. 5A) (Ho et al. 1997). As a result, a discontinuous layer of isolated cells precedes the formation of a continuous blastoderm layer. Despite such extreme morphological differences compared to *Drosophila*, the existence of mechanisms that allow long-range diffusion during early patterning cannot be excluded. First, in *Schistocerca* a syncytial layer of yolk nuclei forms beneath

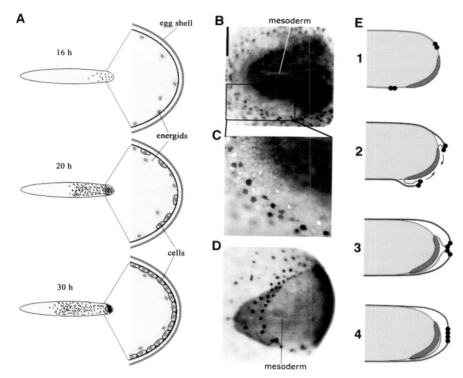

Figure 5. Formation of germ anlage and extraembryonic coverings in *Schistocerca*. (*A*) Diagram summarizing the early development of *Schistocerca gregaria*. At 16 hours, energids but no cells are found at the posterior pole of the egg. Four hours later, the first cells form at the posterior egg surface. The cell layer is, however, not contiguous. Energids are present in the gaps between adjacent cells. By 30 hours, the germ anlage has formed as a contiguous cell layer. Underneath the germ anlage a layer of syncytial energids is present. (*B*) *zerknüllt* expression in the "necklace cells" that surround the heart-shaped embryo at the time of mesoderm internalization. The necklace cells appear to mark the boundary between amnion and serosa. (*C*) Close-up of the area boxed in *B* which also shows the distribution of cell nuclei (fluorescently labeled appearing white and blue). (*D*) The formation of the posterior amniotic fold. The necklace cells move over the embryo, dragging both the serosa and the amnion over the posterior pole. (*E*) Schematic drawing of morphogenetic movements associated with the formation and separation of the amnion and serosa. The embryo (*blue*) lies at the posterior pole of the egg. The amnion is shown in pink, the necklace cells as black circles, and the serosa in green. The necklace cells form the leading edge of the amniotic folds. This results in a ring of necklace cells that contracts over the embryo to give a complete serosal membrane. The amnion seals and detaches from the serosa, wrapping the ventral surface of the embryo. (*A*, Modified, with permission, from Ho et al. 1997 [© Company of Biologists]; *B–E*, modified, with permission, from Dearden et al. 2000.)

the cellularized blastoderm layer (Roonwal 1936; Ho et al. 1997). Here, diffusible factors could spread, and subsequently, signaling events from this "yolk syncytial" layer to the overlaying blastoderm might take place. Second, on the basis of morphological data, Roonwal (1936) suggested for *Locusta*, a close relative of *Schistocerca*, that within a short time window before gastrulation the cells of the germ anlage form a secondary syncytium. Findings such as these indicate that investigations of ultrastructure and dye coupling should be extended beyond the early blastoderm stage to include also the germ anlage, which, after all, represents the most important stage for embryonic patterning.

Except for those species, like *Schistocerca*, in which the blastoderm emerges asynchronously, the formation of the germ anlage starts out from a uniform layer of blastoderm

cells. During the morphogenetic events that produce the germ anlage, the blastoderm remains a monolayer; however, cell density and cell shape start to change along both egg axes. Again, *Tribolium* may serve as an example, since similar events can be seen in most other insect orders, including the basal mayflies (Ephemeroptera; Bohle 1969) and dragonflies (Odonata; Seidel 1935; Ando 1962). In *Tribolium*, the late blastoderm cells (interphase of cycle 13) form a typical "cobblestone" pattern and are almost evenly spaced with a small decrease in cell density toward both egg poles (Fig. 4A,B) (Handel et al. 2000). Soon afterward, the prospective serosa can be distinguished from the germ anlage as a dorsally tilted anterior cap of large domed cells (Fig. 4C). These cells stop dividing, become polyploid, and assume a flattened, stretched-out appearance (Fig. 4D–F).

While the serosa expands dorsally and toward the posterior pole, the region of the germ anlage condenses toward the ventral side of the embryo. This condensation process, together with continued mitotic activity, increases the cell density within the germ anlage. In contrast to the flattened cells of the serosa, the cells of the germ rudiment form a high, columnar epithelium.

A notable variation to this theme has been described in many hemimetabolous insect orders (Schwalm 1988) and needs to be mentioned because it might constitute the ancestral condition. In these orders, cell condensations first occur in independent lateral positions at both sides of the blastoderm (Fig. 1). Subsequently, the two groups of condensing cells move toward the ventral midline of the egg, where they fuse with each other to form the germ anlage. Such a mechanism requires a large-scale patterning system operating in the blastoderm embryo, which provides the spatial information to generate a bilaterally symmetric pattern of cell condensation and directed migration.

The cellular mechanisms responsible for the condensation of the germ anlage have not been thoroughly analyzed. In *Tribolium*, condensation might be driven mainly through antagonistic cell-shape changes in which the flattening of one cell layer is accompanied by an increase in thickness of the other. However, this mechanism is insufficient to account for the ventral movement of the lateral condensation zones mentioned above.

The formation of the germ anlage by cell condensation poses problems for descriptive and experimental embryology. First, the drawings in Figure 2 have a certain degree of arbitrariness. The embryos in Figure 2 are depicted after the condensation of the germ anlage has started, but before amniotic folds and ventral furrow have formed. In the case of *Tribolium*, the size of the germ anlage during the earliest stages of condensation is larger (Fig. 4C). On the other hand, condensation continues during early gastrulation such that the embryo ultimately becomes confined to a narrow rectangle at the ventral side of the egg (Fig. 4F). Second, it is not easy to define the developmental stage for which a blastoderm fate map should be drawn corresponding to that of *Drosophila*. This is, however, crucial for comparative aspects of embryonic patterning. The early experimental work of Seidel on *Platycnemis* (Seidel 1935) shows that it is possible to establish a fate map of the blastoderm before condensation (Fig. 3B). Although it is unlikely that the detailed fate map he generated implies corresponding details of cell determination at the early blastoderm stage, his data show that condensation is not accompanied by considerable mixture of cells. Thus, even if the fates of cells were not determined, the cells might have been subjected to long-range morphogenetic information, which they faithfully carry to the germ anlage during the condensation process.

SIZE VARIATIONS AND FATE MAPS OF THE GERM ANLAGEN

Insect eggs vary widely with regard to the relative sizes of germ anlage and egg (Figs. 2 and 3). *Schistocerca* represents one extreme with a germ anlage spanning 400 μm in anterior–posterior dimension located at the posterior pole of a 6–8-mm-long egg. *Tribolium* represents an intermediate case where the germ anlage (300 μm in anterior–posterior dimension) occupies about half of the egg surface. *Drosophila* shows the other extreme in which the germ anlage (400 μm) fills out the entire egg surface. This comparison shows that the absolute sizes of the germ anlagen vary much less than their sizes in relation to the egg. Thus, one could imagine that the major difference between various insect embryos is the size of the extraembryonic tissue (serosa), whereas the embryonic anlagen are similar.

However, this is not the case. Considerable differences are found with regard to the fate maps of the anterior–posterior axis. Presumably, all germ anlagen harbor regions giving rise to the terminal anlagen, including the unsegmented procephalon and stomodeum at the anterior, and the proctodeum at the posterior pole (Fig. 3B) (Anderson 1972; Sander 1976a, and see below). Differences occur with regard to the intervening segmented regions. To classify the different modes of embryogenesis resulting from these variations, Krause (1939) distinguished short, semi-long, and long germ types of development (the term "intermediate" is now used instead of semi-long [Sander 1976a]). The crucial differences lie in the number of individual segment primordia, which can be distinguished before the germ anlage starts to elongate (Sander 1976a). In the long germ type (**L**), the germ anlage is proportionally subdivided to represent the entire segmented region of the later embryo; thus, prospective regions for each individual segment can be distinguished. In the extreme short germ type (**ES**), no individual segment primordia are believed to be present; all segments supposedly emerge from a narrow growth zone in an anterior-to-posterior sequence. The germ anlage of the less extreme short germ type (**S**) is assumed to possess only presumptive head segments. In the intermediate germ type (**I**), the head and thorax segment primordia are present, leaving only the abdominal segments to be formed from a growth zone. Thus, following this classification, the main difference between germ types is the amount of differential growth required to generate the segmented body parts, with extreme short germ embryos requiring most, and long germ embryos none.

In very few cases have fate maps been established experimentally. If the fate maps are based on the local irradiation of small parts of the germ anlage, the region with proportional representation of the segments gives rise to defects in single segments, the presumptive growth zone to defects in multiple segments. In this way, the long germ type has been confirmed for *Drosophila* (e.g., Lohs-Schardin et al. 1979), some lepidopterans including *Bombyx* (Lüscher 1944; Myohara 1994), and the intermediate germ type for *Acheta* (Kanellis 1952) and *Platycnemis* (Seidel 1935). There is no unequivocal experimental proof for the existence of the extreme short and even the short germ mode of development. Most assumptions are based on the observation of normal development or complex experimental manipulations (centrifugations, ligations, fragmentations, cauterizations, etc.), the results of which are open to various interpretations (Sander 1976a). On the basis of such evidence, extreme short germ development was proposed for some polyneopterans, like walking sticks (Phasmatodea) and termites (Isoptera), and less extreme short germ development for some orthopterans (e.g., *Schistocerca*) and some beetles (e.g., *Atrachya*; for review, see Anderson 1972; Sander 1976a). However, a look at *Schistocerca* shows that this classification may be questioned. Prior to ventral furrow formation, the germ anlage of the *Schistocerca* embryo shows only expression of one *wingless* stripe, indicating that the specification of the first gnathal segment has been initiated. However, shortly afterward, and still before the germ anlage starts to elongate substantially, the three thoracic and then the two remaining gnathal *wingless* stripes appear (Dearden and Akam 2001). Subsequent to this, the posterior growth zone grows and abdominal segments are formed. Thus, in *Schistocerca* it is likely that the germ anlage contains proportionate representations of the head and thorax segments (Fig. 3B) and that *Schistocerca* should be classified as intermediate germ type following the above definition.

In contrast to this fate-map-based approach, the most common criterion to assign a germ type is the timing of segment specification relative to ventral furrow formation (Davis and Patel 2002). Segment specification is defined as the onset of *engrailed* expression, which approximately marks the stage of irreversible segment determination in *Drosophila*. If one follows this strategy, *Schistocerca* would have to be classified as extreme short germ type, since no *engrailed* stripes are visible prior to ventral furrow formation (Patel et al. 1989, 1994). However, the timing of *engrailed* expression in relation to the gastrulation movements has proven to be quite variable. The situation in dipterans representing the long germ type of development is instructive. In *Drosophila*, 14 *engrailed* stripes emerge almost simultaneously at the onset of gastrulation (DiNardo et al. 1985; Karr et al. 1989). However, in the psychodid midge *Clogmia*, and even more extremely in the sciarid gnat *Rynchosciara*, the *engrailed* stripes arise in an anterior-to-posterior sequence during late germ-band extension (Rohr et al. 1999; Vanario-Alonso et al. 1996; Carvalho et al. 1999). It is very likely that in these dipterans, all segment anlagen are present at the blastoderm stage, but

that segment determination is delayed and occurs nonsimultaneously. Also in lepidopterans and hymenopterans, both classified as long germ type, pronounced anterior–posterior gradients and wide variations in the timing of *engrailed* expression relative to ventral furrow formation have been observed (for review, see Davis and Patel 2002). In *Tribolium* embryos, at the beginning of gastrulation only one *engrailed* stripe is present. At the beginning of differential elongation, at most three *engrailed* stripes are present (Brown et al. 1994). Therefore, *Tribolium* was classified as a short germ beetle (Patel et al. 1994). However, in *Tribolium*, the early germ anlage covers a substantial part of the blastoderm. Early expression of the gap gene *Krüppel* at the posterior pole and the appearance of three *hairy* stripes indicate a representation of gnathal and thoracic regions at the blastoderm stage (Sommer and Tautz 1993). Therefore, with reference to the fate map, and not the time of segment specification, *Tribolium* should be classified as an intermediate germ type. However, a definite answer has to await fate-mapping experiments using transplantation, irradiation, or in vivo labeling methods.

Given the problems in defining the different germ types and in providing experimental proof for the existence of forms with very few segment primordia, for the rest of this chapter I distinguish only between two types, the short germ and long germ types. The term short germ will be generically used for all embryos that form at least their abdominal segments from a growth zone, thus including those formerly called intermediate germ type. The latter germ type, characterized by the presence of gnathal and thoracic segment primordia at the germ anlage stage before onset of elongation, is most common among winged insects (Fig. 2). Judging from its occurrence in Ephemeroptera (Bohle 1969) and Odonata (Seidel 1935), it might represent the ancestral condition (Anderson 1972; Tautz et al. 1994). Long germ type embryos, on the other hand, are only found among the holometabolous insects.

The variations in the relative size of the germ anlagen are linked to a cell biological phenomenon: the initial thickness of the blastoderm layer, which depends on the amount of cytoplasm (as opposed to yolk granules and lipid droplets) contained in the egg (Fig. 3A) (Krause 1939). The relative richness in cytoplasm apparently correlates with the oogenesis type (Bier 1969; Büning 1994). Only eggs derived from ovaries with nurse cells are rich in cytoplasm. Yolk precursors can be taken up by the growing oocyte directly from the hemolymph. However, the increased delivery of cytoplasmic components (mRNAs, membrane components, ribosomes, entire organelles) requires nurse cells, which are connected to the oocyte by cytoplasmic bridges (nutritive cords in telotrophic, ring canals in polytrophic, ovaries). Nurse cells are a common feature of holometabolous insect orders. But two exceptions exist that support the link between the amount of cortical cytoplasm and the type of oogenesis. The primitive Ephemeroptera and the Dermaptera have independently acquired nurse cells, and in accordance, their embryos possess germ anlagen with a high columnar epithelium (Fig. 3A) (Krause 1939; Bohle 1969; Gottanka and Büning 1992; Büning 1994). High amounts of cortical cytoplasm allow both an expansion of the germ anlage relative to the serosa and, under suitable environmental conditions, a faster progression of embryonic development. This tendency culminates in the embryos of higher dipterans, like *Drosophila*, in which the germ anlage (embryo plus amnioserosa) occupies the entire blastoderm and the embryo develops within one day. The speed of development is clearly one parameter that influences the mode of gastrulation, as shown below for the process of mesoderm internalization.

FORMATION OF THE EXTRAEMBRYONIC COVERINGS

Whereas the distinction between presumptive serosa and germ anlage is the first differentiation process within the uniform blastoderm cell layer, the formation of the serosa as the outer extraembryonic membrane enveloping yolk and embryo is the result of complex morphogenetic movements usually linked to the formation of the amnion. This is a unique feature of pterygote insects not found in other arthropods or even apterygota. Yolk-rich eggs and superficial cleavage are common to arachnids and many crustaceans, and this character is linked to the appearance of extraembryonic ectoderm covering the yolk. However, the formation of two complete extraembryonic membranes evolved only in the lineage leading to the winged insects (Zeh et al. 1989; Machida and Ando 1998). The silverfish (Zygentoma) represent a transitional stage since they have two extraembryonic coverings (proserosa and proamnion), which, however, do not execute the full repertoire of morphogenetic changes found in winged insects (Jura 1972).

The presumptive amnion cells of the winged insects map to the margin of the germ anlage (Fig. 3B) (Anderson 1972). In contrast to the serosa cells, they do not become polyploid at an early stage, and in *Tribolium*, remain mitotically active while undergoing morphogenetic movements (S. Roth, unpubl.). Usually, at the time of mesoderm invagination and at the beginning of elongation of the germ anlage, the presumptive amnion cells fold ventrally over the embryonic ectoderm carrying the margin of the serosa cells with them (Figs. 1, 4I, and 5E). This process of amniotic fold formation is more pronounced at the posterior pole. In *Tribolium*, it starts with formation of a small cup-shaped invagination, called the primitive pit (Fig. 4C,D) (Handel et al. 2000). The dorsal margin of the primitive pit gives rise to the posterior amniotic fold, which moves anteriorly over the ventral face of the germ anlage. Paired amniotic folds then arise at the margins of the head lobes, and the folding spreads along the lateral edges of the germ anlage (Fig. 4F–H). This leads to the formation of a serosal "window" that contracts in purse-string fashion until it is closed (Fig. 4H). At that point, the epithelial continuity between serosa and amnion is severed so that the two epithelia represent separate extraembryonic membranes. The serosa surrounds the entire egg and closely abuts the eggshell. In most insect species, it secretes a serosal cuticle during later embryogenesis. The amnion covers the embryo at the ventral side. The fluid-filled cavity, which is delimited ventrally by the amnion and dorsally by the germ band, is called the "amniotic cavity" (Figs. 1 and 4I). A comparison between *Tribolium* and *Schistocerca* shows the remarkable similarity in the formation of extraembryonic covers in insects that considerably differ in egg structure and mode of blastoderm formation (Fig. 5E) (Dearden et al. 2000).

In *Drosophila*, amnioserosa formation depends on a Decapentaplegic (Dpp) activity gradient, which has peak levels along the dorsal midline (Arora and Nüsslein-Volhard 1992; Rushlow and Roth 1996; Dorfman and Shilo 2001). High levels of Dpp are required to maintain the expression of *zerknüllt* (*zen*) in a narrow stripe of dorsal cells that comprises the presumptive amnioserosa cells. In the absence of *zerknüllt*, the amnioserosa is transformed into dorsal ectoderm (Arora and Nüsslein-Volhard 1992). *dpp* and *zerknüllt* homologs have been found in *Tribolium* and *Schistocerca* (Falciani et al. 1996; Sanchez-Salazar et al. 1996; Dearden and Akam 2001). In *Tribolium*, both genes are expressed in the presumptive serosa, and later, *dpp* becomes restricted to a narrow stripe of cells of the presumptive amnion region bordering the serosa (Falciani et al. 1996; Sanchez-Salazar et al. 1996; Chen et al. 2000). In *Schistocerca*, *zen* is first expressed in all energids that reach the cell surface (Dearden et al. 2000). When the germ anlage has formed, expression starts in a ring of "necklace cells" that surround the forming embryo and demarcate the boundary between the amnion and the serosa (Fig. 5B). From there, expression spreads to all serosa cells. When serosa and amnion have separated, *zen* expression is also initiated in the amnion (Falciani et al. 1996; Dearden et al. 2000).

Phylogenetic reconstruction shows that the *zerknüllt* gene, located in the Antennapedia complex of *Drosophila*, and at corresponding positions in the Hox cluster in *Tribolium*, is a derived class 3 Hox gene (Falciani et al. 1996). Whereas Hox genes are expressed after germ-band formation in a fashion that is colinear with their neighbors, the *zen* genes have escaped this regulatory constraint. They are expressed before germ anlage formation and are spatially independent from their neighbors in the complex. The only reminiscence to Hox gene regulation is that in most insects their expression levels vary along the anterior–posterior body axis. The decision between serosa and germ anlage is predominantly, but not exclusively, the result of anterior–posterior patterning. Judging from the high conservation of their expression pattern in species evolutionarily as far apart as *Tribolium* and *Schistocerca*, it is conceivable that the *zen* genes play an ancestral role in the first anterior–posterior patterning event in the insect embryo, the decision between serosa and germ anlage. Only in the lineage leading to the higher dipterans do *zen* regulation and extraembryonic membrane formation become the predominant target of dorsoventral patterning. Although in the primitive nematoceran *Clogmia zen* expression is still slightly tilted to the anterior pole, its expression is entirely shifted to the dorsal side in higher flies (Fig. 2) (Stauber et al. 2002).

The lineage to the cyclorraphan flies, to which *Drosophila* belongs, also shows the reduction of serosa and amnion to one single extraembryonic membrane, the amnioserosa (Schmidt-Ott 2000). Surprisingly, this evolutionary change in the structure of the extraembryonic membranes is linked to a major innovation in axis formation in higher flies: the appearance of maternal Bicoid as a morphogen organizing head and thorax (Stauber et al. 1999, 2002). Phylogenetic reconstruction reveals that *bicoid* is derived from a Hox3/zen-like gene (Stauber et al. 1999; Brown et al. 2001). The *zen* genes in non-cyclorrhaphan flies show both maternal and zygotic expression. Therefore, it was proposed that in the stem lineage of the cyclorrhaphan flies, a duplication of the *zen* gene occurred (Stauber et al. 2002). One of the gene copies lost zygotic expression and evolved as *bicoid,* whereas the other lost maternal expression and evolved to the *zen* gene we know from *Drosophila*.

Have the evolutionary comparisons presented so far contributed to the question of whether the amnioserosa of flies is more closely related to the serosa or to the amnion of other insects? Homology with the serosa could be based on the observations that the amnioserosa cells stop dividing very early, flatten, and do not cover the embryo, but rather the yolk (Campos-Ortega and Hartenstein 1997). However, like the amnion, the amnioserosa is derived from the margin of the germ anlage and differentiates only during germ-band extension. At that stage, the amnion cells of *Tribolium* also flatten and assume cell shapes similar to those of serosa cells (Fig. 4I). This is the stage in which the amnion cells express *zen* in *Schistocerca* (Falciani et al. 1996). With regard to topology, the comparison to *Apis* is instructive (Fleig and Sander 1988). In *Apis*, the formation of amnion and serosa are not coupled. The presumptive serosa separates early from the germ anlage and wraps the entire yolk and embryo. The amnion forms after mesoderm invagination as an outgrowth of the dorsal rim of the ectoderm and, as in

Drosophila, it covers the yolk on the dorsal side, rather than the embryo on the ventral side. On the basis of these comparisons and the timing of differentiation, the amnioserosa appears to be more closely related to the amnion (this assumption underlies the schematic drawings of the *Calliphora* and *Drosophila* eggs in Figs. 2 and 3 in which the germ anlagen fill out the entire egg space). However, it is also possible that the amnioserosa still carries characters of both epithelia, as was recently suggested on the basis of molecular markers (Patel et al. 2001).

MESODERM FORMATION

The germ anlage is a single-layered epithelium. During early stages of amnion formation, and frequently prior to anterior–posterior elongation, the germ anlage is transformed into the multilayered germ band through the internalization of mesodermal cells (Fig. 1). In all cases studied, the presumptive mesoderm represents a straight stripe of cells that bisects the germ anlage (Figs. 3B and 5B,D) (Anderson 1972). One could imagine that the spatial information for mesoderm formation is derived from the earliest developmental boundary that forms in short germ embryos, the boundary between serosa and germ anlage. Indeed, an important dorsoventral patterning gene, *dpp*, is expressed at this boundary in *Schistocerca* and *Tribolium* (Sanchez-Salazar et al. 1996; Chen et al. 2000; Dearden and Akam 2001). Spreading from this expression domain, Dpp might act as a long-range morphogen, which not only patterns the ectoderm of the germ anlage, but also delimits the region of the presumptive mesoderm. However, the shapes of different germ anlagen can vary considerably (Figs. 2 and 3). They are often heart-shaped, or at least much broader in the presumptive head regions than at their posterior extremities. Therefore, there is no simple way to retrieve the spatial information for the straight mesodermal stripe from the boundary between serosa and germ anlage. Separate processes, which most likely, as in *Drosophila*, are linked to the overall bilateral egg structure, have to be postulated to provide the spatial cues for mesoderm formation.

Very little is known about such processes outside the higher dipterans. However, comparative molecular work with *Tribolium* indicates that the maternal gene network involved in dorsoventral patterning in *Drosophila* is conserved between coleopterans and dipterans and should therefore be common to all holometabolous insects (see phylogeny, Fig. 2) (Maxton-Küchenmeister et al. 1999; Chen et al. 2000). In *Drosophila*, the nuclear gradient of the rel/NF-κB transcription factor Dorsal initiates dorsoventral patterning of the embryo by its differential activating or repressing effects on a number of zygotic target genes (Stathopoulos and Levine 2002). Among those are *twist* and *snail*, which are required for specifying the mesodermal anlagen (see Chapters 7 and 46). It is crucial for the long-germ type of development exhibited by *Drosophila* that the gradient forms at all positions of the anterior–posterior axis such that peak levels of nuclear Dorsal are found in a ventral stripe spanning the entire length of the blastoderm. Surprisingly, a nuclear Dorsal gradient of similar extension along the entire egg length exists in *Tribolium* embryos at the uniform blastoderm stage (Fig. 6) (Chen et al. 2000). The gradient forms in a more dynamic way and disappears when the blastoderm differentiates into serosa and germ anlage. Thus, in contrast to *Drosophila*, it is not maintained until gastrulation takes place. The likely Dorsal target genes *twist* and *snail* are faintly expressed along the entire anterior–posterior axis at the uniform blastoderm stage, but during blastoderm differentiation their expression disappears in the presumptive serosa (Sommer and Tautz 1994; Chen et al. 2000; K. Handel and S. Roth, unpubl.). They become restricted to and highly expressed in the ventral stripe of the germ anlage, which will give rise to the mesoderm (Fig. 6).

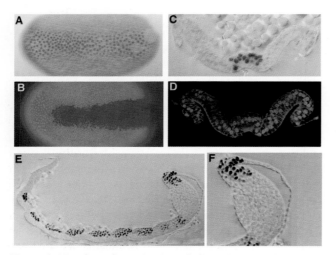

Figure 6. Mesoderm formation in *Tribolium*. (*A*) A nuclear Dorsal protein gradient forms at the ventral side of the uniform blastoderm embryo. The gradient has peak levels within a broad ventral stripe extending along the entire length of the egg. (*B*) At the differentiated blastoderm stage, Twist protein accumulates in a ventral stripe of the germ anlage, but is excluded from presumptive serosa cells at the anterior pole. (*C*) Twist protein accumulates in the cells of the ventral furrow. (*D*) Ventral furrow formation. Staining for actin (*red*) and DNA (*green*) shows the cell outlines and the position of the nuclei. (*E*) Longitudinal section through a germ-band stage embryo which is about to form the fourth abdominal segment. Mesodermal cells expressing Twist protein form clusters in each segment. (*F*) Twist expression is absent in the growth zone except in some cells close to the posterior amniotic fold. The growth zone is multilayered. The layer facing the amniotic cavity presumably corresponds to the ectoderm. The cell layers facing the yolk might comprise mesodermal precursor cells. (*A*, Reprinted, with permission, from Chen et al. 2000 [© Company of Biologists].)

These data illustrate two points. First, a short germ embryo, like *Tribolium*, may have transient features of long germ development that, as development proceeds, are adapted to the short germ mode. Second, the temporal overlap between the maternal transcription factor gradient and its zygotic target genes is reduced compared to *Drosophila*. This might allow a greater autonomy at the level of zygotic pattern formation (Chen et al. 2000).

Dorsal gradient formation in *Drosophila* depends on an extraembryonic signal that forms at the ventral side of the egg and is transmitted to the embryo via the transmembrane receptor Toll (Morisato and Anderson 1995). The spatial cues restricting this signal in turn go back to asymmetries arising during oogenesis and ultimately depend on the asymmetric movement of the oocyte nucleus (Roth 2003). In *Tribolium*, we know very little about the link between embryonic axis and polarity of the egg chamber except that the oocyte nucleus is asymmetrically localized (Roth 2003). However, a *Toll* homolog exists in *Tribolium* that, interestingly, shows zygotic rather than the strong maternal expression observed in *Drosophila* (Maxton-Küchenmeister et al. 1999; Chen et al. 2000). Furthermore, its expression is regulated through positive feedback mediated by Dorsal, such that *Toll* itself becomes expressed in a ventral-to-dorsal gradient. Positive feedback at the maternal level of axis initiation and a greater independence of the zygotic patterning network from maternal inputs might explain the larger regulative capacity along the dorsoventral axis found in many insect orders (Sander 1971, 1976a). For example, the chrysomelide beetle *Atrachya menetriesi* may form up to four complete germ bands within one egg after low-temperature treatment or ligation of pre-blastoderm embryos (Miya and Kobayashi 1974). In hemimetabolous insects, regulation along the dorsoventral axis is well documented by a variety of experimental approaches (Sander 1976a).

The actual morphogenetic movements that lead to mesoderm internalization have not been studied in detail in many insect orders. This is mainly due to the fact that they take place in a short time interval compared to other morphogenetic events in the early embryo. In some insect groups (Plecoptera, Isoptera), mesoderm internalization is a very early event, completed before the germ anlage shows any overt sign of elongation (Anderson 1972). This underscores the fact that ventral furrow formation is not linked to the process of segment specification except that it never occurs after the initiation of segmentation.

Three basic modes of internalization can be distinguished (Fig. 7) (Johannsen and Butt 1941; Schwalm 1988): Type I resembles *Drosophila*. The presumptive mesoderm forms a well-defined ventral furrow, which is then converted into a tube-like structure when the edges of the lateral ectoderm approach each other at the ventral midline and fuse. This type of mesoderm internalization is found in

Type I: Chrysopa (Neuroptera)

Type II: Apis (Hymenoptera)

Type III: Locusta (Orthoptera)

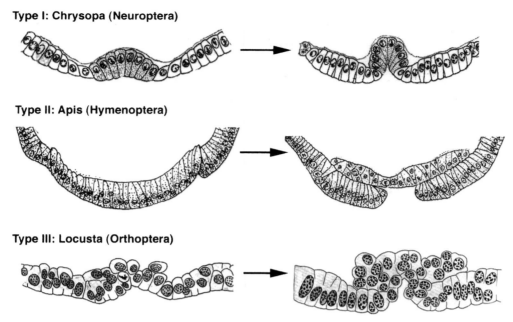

Figure 7. The three types of mesoderm internalization found in winged insects. Type I: Ventral furrow formation similar to *Drosophila*. The presumptive mesodermal cells appear to play the active role. The coordinated constriction of their apical, and expansion of their basal, surfaces results in the transient formation of a tube-like structure. (Reprinted from Bock 1939.) Type II: The presumptive mesodermal cells appear to play a more passive role and show few cell-shape changes during internalization. The ventral ectodermal plates crawl on top of the mesoderm toward the ventral midline, where they fuse with each other. (Reprinted from Sauer 1954 [© Springer-Verlag].) Type III: A weakly pronounced and irregular invagination furrow forms. The presumptive mesodermal cells are mitotically active during internalization. (Reprinted from Roonwal 1936.)

many dipterans, coleopterans (Fig. 6), and neuropterans (Fig. 7) (Bock 1939). Type II occurs in hymenopterans and some lepidopterans. Here, the epithelial continuity at the border between presumptive mesoderm and ectoderm is disrupted. The mesoderm remains as a stiff plate while the ectoderm moves over its surface toward the ventral midline (Fig. 7) (Sauer 1954; Fleig and Sander 1988). Type III has been described for many hemimetabolous insects (Anderson 1972). Here, either no ventral furrow, or only a weakly pronounced ventral furrow, is visible. Cell divisions begin early within the mesodermal region and produce an irregular mass of cells, which is pushed inside when the ectodermal plates meet at the ventral midline. Since this type of mesoderm internalization is found in Odonata (Ando 1962) and in Ephemeroptera (Ando and Kawana 1956), and since within apterygote insects the mesoderm forms by cell proliferation and immigration (Jura 1972), it is likely that the third type represents the ancestral condition for winged insects. The high degree of coordination and accuracy of cell movements required for type I and type II appear to be part of the extremely fast modes of embryogenesis found in some holometabolous insects. However, it should be kept in mind that considerable variations in mesoderm formation can be found within single insect orders. For example, it has been reported that some lower

dipterans form their mesoderm without a ventral furrow, by proliferation or immigration of single cells (Moretti and Larsen 1973). Furthermore, a careful investigation of gastrulation in *Tenebrio* and *Tribolium* shows that within a single embryo, the modes of mesoderm internalization can vary along the anterior–posterior axis with aspects of type II anteriorly, type I in the middle, and type III at posterior positions (Ullmann 1964; K. Handel and S. Roth, unpubl.).

For short-germ embryos, the question arises of how the mesoderm forms in the posterior growth zone. Is there a continuous process of gastrulation by which new mesodermal cells are generated when new segments arise, or does the growth zone possess mesodermal precursor cells that stem from early gastrulation? On the basis of morphological observations, the second alternative appears to be more likely. First, the ventral furrow runs along the entire germ anlage extending up to the posterior amniotic fold in both *Locusta* and *Tribolium* (Roonwal 1936; K. Handel and S. Roth, unpubl.). Thus, mesoderm internalization also takes place in the presumptive growth zone. Second, the growth zone in *Tribolium* remains clearly multilayered throughout the process of segment formation and thus might harbor mesodermal precursor cells that continuously give rise to differentiated mesoderm (Fig. 6E,F) (K. Handel and S. Roth, unpubl.). Fate-mapping experiments and a more careful analysis of the expression of mesodermal genes in the growth zone are required to provide a definite answer to this interesting question.

FORMATION OF THE ALIMENTARY CANAL AND THE ENIGMA OF THE INSECT ENDODERM

In insects, the alimentary canal is composed of four major components: the foregut, the midgut, the Malphigian tubules, and the hindgut (Fig. 8A) (Johannsen and Butt 1941; Schwalm 1988; Skaer 1993). The foregut and hindgut, together with the Malphigian tubules, originate from tube-like invaginations of ectodermal origin, called stomodeum and proctodeum, respectively (Fig. 8A). Only the midgut is of endodermal origin (see below). As pointed out at the beginning of this chapter, endoderm formation, and hence gastrulation, was a matter of much debate in the early days of insect embryology. Since the old arguments (Johannsen and Butt 1941) are still illuminating, far from resolved, and continue to influence the writing on comparative insect embryology, they are briefly summarized here and put into the context of current genetic and molecular findings. The controversies revolved around the unusual spatial (1), temporal (2), and cellular (3) aspects that characterize endoderm formation in insects.

1. In all winged insects, the entire endoderm, or at least substantial parts of it, are derived from two disjoined primordia, one at the anterior and one at the posterior pole of the germ anlage, giving rise to the anterior and posterior midgut, respectively. To harmonize this finding with classical views on gastrulation, the insect embryo at the time of ventral furrow formation was compared to a stretched-out gastrula (Johannsen and Butt 1941). It was proposed that stretching resulted in an elongated, slit-like blastopore and pulled the endoderm primordium into two halves. The cells invaginating through the slit-like blastopore are endodermal at the two ends and mesodermal in between. Probably influenced by this view, up to this date the fate maps presented for the germ anlagen of hemimetabolous insects show the anterior and posterior midgut as midventral groups of cells at the ends of the presumptive mesoderm (Anderson 1972). The stomodeum and proctodeum are each placed anteri-

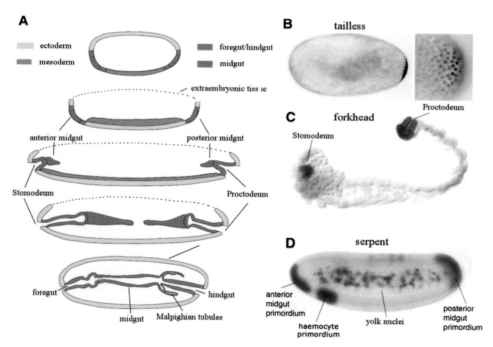

Figure 8. Formation of the alimentary canal. (*A*) Schematic representation of gut anlagen and morphogenetic movements leading to gut formation. Increase and decrease in length of the embryo reflect germ-band elongation and retraction, respectively. (*B*) Tailless protein expression in *Tribolium* at the uniform blastoderm stage. The magnified view of the posterior pole shows a nuclear gradient of Tailless. (*B*, Modified, with permission, from Schröder et al. 2000 [© National Academy of Sciences].) (*C*) *forkhead* expression in a fully segmented *Tribolium* embryo marks the stomodeal and proctodeal invaginations and the presumptive endoderm which apparently has not gone through the epithelial–mesenchymal transition. (Courtesy of Berns 2001.) (*D*) *serpent* expression in a *Drosophila* embryo at the cellular blastoderm stage. *serpent* is expressed in the primordia of the anterior and posterior midgut, of the hemocytes and in the yolk nuclei. (Courtesy of Rolf Reuter, see Rehorn et al. 1996.)

or or posterior to the respective midgut primordia. However, this spatial arrangement has not been analyzed experimentally except in *Drosophila,* where it has been shown that the anlagen of the endoderm are not contiguous with those of the mesoderm, but rather occupy the extremities of the fate map (Rehorn et al. 1996; Campos-Ortega and Hartenstein 1997). The stomodeum primordium separates the anterior midgut, that of the proctodeum the posterior midgut, from the mesoderm (Fig. 8A,D). These findings seem to refute the theory of the split-like blastopore. However, *Drosophila* might present a derived situation. Fate-mapping experiments and comparative molecular work are required to reconstruct the ancestral condition for insects.

2. Not only the spatial, but also the temporal, aspects of midgut formation pose an obvious problem for the classical view of gastrulation. The morphogenesis of the ectodermal parts of the gut, the stomodeum and proctodeum, precedes that of the midgut (Fig. 8A,C). The anterior and posterior midgut primordia are frequently pushed inward together with the stomodeum and proctodeum. They lose their epithelial organization and form a mesenchymal cell mass, which undergoes morphogenesis only after completion of segmentation at the beginning of germ-band retraction. This delay is particularly obvious in the posterior region of short-germ embryos in which both hindgut and midgut formation appear to be postponed until the activity of the growth zone ceases (Johannsen and Butt 1941; Anderson 1972; Schwalm 1988; Kispert et al. 1994; Berns 2001). At this stage, the cell groups of the future anterior and posterior midgut begin to migrate toward each other as paired lateral strands that abut the surface of the yolk. When they meet on the sides of the yolk, they adopt an epithelial organization and spread dorsally and ventrally to cover the yolk completely. Consequently, the formation of a continuous endodermal cell layer is one of the latest morphogenetic processes occurring during insect development. At this stage, many other aspects of organogenesis, like appendage formation, are well under way, and both muscle and nervous system have gained considerable cellular complexity (Fig. 8C). As the unusual spatial aspects of endoderm formation were rationalized with the concept of the slit-like blastopore, the temporal aspects were explained with the concept of the latent or dormant endoderm (Johannsen and Butt 1941). The assumption was made that the endoderm cells are specified early and stay dormant until late stages when gut morphogenesis ensues.

This view has been clearly supported by molecular work on terminal patterning in *Tribolium* (Schröder et al. 2000). The specification of the endodermal

anlagen in *Drosophila* depends on maternal signals at the termini of the early embryo which lead to the activation of the receptor tyrosine kinase Torso (see Chapter 7). Torso activation in turn allows the expression of the terminal gap genes *huckebein* and *tailless.* These are two transcription factors that specify the anlagen of the ectodermal and endodermal parts of the gut by regulating the expression of genes like *forkhead, serpent,* and *brachyenteron* (Weigel et al. 1989; Kispert et al. 1994; Rehorn et al. 1996). The MAP kinase activation pattern suggests that in *Tribolium* the terminal system is also active at both poles of the blastoderm embryo (Schröder et al. 2000). However, early *tailless* expression is only observed at the posterior pole in accordance with the fate map of *Tribolium,* in which the anteriormost region gives rise to the serosa and not to terminal embryonic structures as in *Drosophila* (Fig. 8B). This, however, is not the most striking change of *tailless* expression compared to *Drosophila. tailless* in *Tribolium* is only transiently expressed at the posterior pole, and Tailless protein has disappeared when *forkhead* and *brachyenteron* transcription is initiated (Schröder et al. 2000). Thus, Tailless cannot function as a direct transcriptional regulator of *forkhead* and *brachyenteron* as in *Drosophila,* but instead appears to be involved in the early specification of a group of terminal cells that remain quiescent and start to express gut-specific genes only after much of the abdominal segmentation has been completed. This example shows that heterochronic shifts linked to changes in the gene regulatory network constitute important differences in gut formation between *Drosophila* and *Tribolium.* It also provides a molecular illustration for the delay between early specification and late differentiation of endoderm and hindgut in a short-germ insect and thus revives the old concept of the "dormant endoderm" (Johannsen and Butt 1941).

3. The early authors who denied the endodermal character of the anterior and posterior midgut anlage believed that the vitellophages represent the endoderm (Johannsen and Butt 1941). The vitellophages populate the yolk in all orders of winged insects and are believed to participate in yolk consumption. In most cases, they are derived from cleavage energids, which remain in the yolk mass when the majority of energids have migrated to the periphery to form the blastoderm cell layer. They are then called primary vitellophages, but since they are not real cells, they should be more correctly called primary yolk nuclei. If all energids migrate to the surface, and only after several rounds of nuclear division, some nuclei move back to repopulate the yolk, they are called secondary yolk nuclei (or vitellophages). The yolk nuclei often

move to the surface of the yolk during later stages of development. In *Drosophila*, this occurs during early germ-band extension and results in the formation of a syncytial layer with nuclei at the surface of the yolk (Rickoll and Counce 1980). In many other insect orders, this process is more prominent and leads to the formation of a yolk-sac epithelium (Johannsen and Butt 1941; Anderson 1972; Schwalm 1988). Sometimes a transient type of midgut tissue is formed that functions during early stages of larval life and is replaced by cells originating from the proctodeal and stomodeal invagination only after molting of the larva. This course of events is particularly well documented for the phylogenetically basal Odonata (Ando 1962) but has also been observed in phasmids (Leuzinger et al. 1926) and even in holometabolous orders (lepidopterans; Kobayashi and Ando 1983). Furthermore, the yolk nuclei seem to contribute to the midgut in apterygote insects (Jura 1972). However, since there is no proof for winged insects that cells can emerge from yolk nuclei, which permanently contribute to the midgut, the designation of the yolk nuclei as endoderm has been widely dismissed.

Compared to the early discussions, which were based exclusively on morphological grounds, today we may add molecular and functional genetic data to assign a particular structure to a germ layer. Thus, the finding that a transcription factor (*serpent*) exists in *Drosophila* which is required to specify both the anterior and posterior midgut by suppressing the development of ectodermal cell fates provides functional support to the argument that the midgut belongs to a different germ layer than the fore- and hindgut (Reuter 1994). However, even more important is the evolutionary consideration that *serpent* belongs to the GATA family of transcription factors which are involved in the formation of endodermal organs in nematodes and vertebrates (for review, see Stainier 2002). This validates the germ layer concept, suggesting that the endoderm has a common origin in bilaterian organisms reflected in a common gene network required for its specification. In this context, it is interesting that the yolk nuclei express gut-specific genes in *Drosophila*: at early stages *serpent*, and later *forkhead*, which although responsible for hind/foregut and midgut development in *Drosophila*, belongs to the conserved group of transcription factors required for endoderm specification in other organisms (Weigel et al. 1989; Rehorn et al. 1996). Furthermore, a recent functional study has shown that the yolk nuclei are required for endoderm morphogenesis in *Drosophila* even though they might only provide a substrate for the spreading of the endodermal cells on the yolk surface (Walker et al. 2000). Since the yolk nuclei are linked to the midgut

with respect to both gene expression and morphogenesis, they indeed might be regarded as belonging to the endoderm and might be the vestige of an ancient mode of endoderm formation found in the ancestors of the insects. These observations provide another example of a recurring theme in this book: how results from classical embryology regain interest in the light of recent molecular findings.

EVOLUTIONARY STABILITY AND PLASTICITY OF EARLY INSECT DEVELOPMENT

With the arrival of winged insects, a remarkably stable mode of early embryonic development emerged that is characterized by a syncytial pre-blastoderm stage, the formation of two extraembryonic coverings (amnion and serosa), the internalization of a ventral stripe of cells as mesoderm, and the bipartite origin of the endoderm from separate terminal regions of the embryo. These features must have evolved within the apterygote insects, where they are only partially realized or display considerable variability. Thus, not all apterygote insects have syncytial pre-blastoderm stages. In springtails (Collembola), a series of holoblastic cleavages is followed by the formation of a syncytium and subsequent secondary cellularization (Jura 1972; Schwalm 1988 and references therein). As mentioned before, apterygote insects lack the elaborate extraembryonic membranes found in pterygota, and large variations occur with regard to mesoderm and endoderm formation.

The evolutionary stability of the modes of early embryonic development in pterygota is particularly striking if one considers the phylogenetic age (350 million years) and the morphological and ecological diversity of winged insects. On this background, it is even more astonishing that certain changes in reproductive strategy and lifestyle found in a limited number of pterygote insect families have brought about dramatic alterations in early embryogenesis. They are linked to a particular type of viviparity and parasitism.

Viviparity is a common phenomenon among insects (Hagan 1951). About half of the pterygote insect orders contain viviparous species. The most frequent form of viviparity is easily derived from the oviparous condition. It results from a delay in egg deposition such that the larva hatches within the female reproductive tract. It either leaves the body shortly after hatching (ovoviviparity) or is nourished through specific gland-like structures within the uterus before birth (adenotrophic viviparity). Not surprisingly, these forms of viviparity have little influence on early embryogenesis. However, this also applies to the more complex forms of pseudoplacental viviparity. In these cases, embryogenesis is initiated already in ovarian egg chambers as best characterized for some species of earwigs (Dermaptera), cockroaches (Blattoptera), and the aphids (for a recent description of aphid development, see Miura et al. 2003). The structure of the ovary has undergone dramatic changes. The embryo and/or the female develop a particular organ, called the "pseudoplacenta," which allows the uptake of maternal nutrients during embryogenesis. For example, the pseudoplacenta of viviparous earwigs is a functional unit derived from the amnion and the serosa of the embryo and the follicular epithelium of the mother (Hagan 1951). Frequently, no eggshells are secreted, and the yolk content of the egg and the number of early syncytial cleavages are greatly reduced. Despite that, blastoderm formation and gastrulation, so far described, resemble that of oviparous species, underscoring the resistance of early insect embryogenesis toward morphological change.

The only form of viviparity that affects the mode of blastoderm formation has been classified as hemocoelous (Hagan 1951). Here, development also starts within the ovary. However, the ovary disintegrates, and individual egg chambers containing developing embryos are released from the ovary and move freely in the hemolymph of the mother. In the order Strepsiptera, the egg chambers floating in the hemolymph can be very small (43–56 μm) with the oocytes almost free of yolk. In *Stylops*, only the first two nuclear divisions lack cytokinesis. Most of the subsequent pre-blastoderm cleavages are holoblastic. A ball of blastomeres forms that resembles the morula of mammalian embryos. The blastomeres arrange themselves in a hollow sphere and form a monolayered epithelium that represents the blastoderm. The germ anlage emerges at one side of the sphere, while the opposite side gives rise to the extraembryonic covering. Not much is known about mesoderm and endoderm formation, but this short and very incomplete description already indicates that early development has changed dramatically. Loss of the egg shells and holoblastic cleavage seem to prepare the stage for an even more surprising phenomenon that has been described for one strepsipteran species: the development of up to 50 embryos from a single egg (Ivanova-Kasas 1972). This process has been termed polyembryony and results in strepsipterans from the subdivision of the blastoderm into several germ anlagen, each giving rise to one embryo.

Interestingly, the only other case in which polyembryony has been observed among insects is linked to similar environmental conditions as represented by hemocoelous viviparity. Endoparasitic wasps deposit their eggs into the body of the host, where they develop floating in the hemolymph (Ivanova-Kasas 1972; Strand and Grbic 1997; Grbic 2000). Whereas closely related ectoparasitic wasps retain the large yolky eggs and the typical mode of embryogenesis found in free-living hymenopterans, the endoparasitic wasps show a trend toward small eggs and reduction of

yolk. Most importantly, several independent endoparasitic lineages have lost syncytial pre-blastoderm stages and acquired holoblastic cleavage. Furthermore, within four families with monoembryonic endoparasitic wasps, polyembryonic species have been described (Ivanova-Kasas 1972; Strand and Grbic 1997). One of the most extreme cases is *Copidosoma floridanum*, which produces about 2000 embryos from a single egg. The egg injected into the hemocoel of the host forms a primary morula, a ball of blastomeres produced by total cleavage. The primary morula splits to create many proliferative morulae that give rise to individual embryos. This mode of development almost certainly excludes a predetermination of the embryonic body axes during oogenesis that in one or the other way is characteristic for most other insect species. A fascinating question is how in these species the signals for mesoderm and endoderm formation arise, as these are maternally provided in the extraembryonic space in dipterans, coleopterans, and most likely also in the remaining holometabolous insect orders including the hymenopterans (Fig. 2).

Taken together, on the entire scale of insect diversity, hemocoelous viviparity and endoparasitism certainly represent rare exceptions. However, they clearly demonstrate that under particular environmental constraints, early insect development can display remarkable evolutionary plasticity.

ACKNOWLEDGMENTS

I am indebted to Klaus Sander, Maria Leptin, Patrik Kalscheuer, and Abidin Basal for critical reading of and comments on the manuscript. I am particularly thankful to Klaus Sander for continued discussions about insect embryology and invaluable advice for reading and understanding the older literature.

REFERENCES

Anderson D.T. 1972. The development of hemimetabolous insects. In *Developmental systems: Insects* (ed. S.J. Counce and C.H. Waddington), vol. 1, pp. 96–163. Academic Press, London, United Kingdom.

Ando H. 1962. The comparative embryology of Odonata with special reference to a relic dragonfly *Epiophlebia superstes* Seyles. Japanese Society for the Promotion of Science, Tokyo, Japan.

Ando H. and Kawana T. 1956. Embryology of the mayfly (Epemera strigata Eaton) as studied by external observation. *Kontyu* **24**: 224–232.

Arora K. and Nüsslein-Volhard C. 1992. Altered mitotic domains reveal fate map changes in *Drosophila* embryos mutant for zygotic dorsoventral patterning genes. *Development* **114**: 1003–1024.

Berns N. 2001. "Untersuchungen zur Struktur und Funktion des *Tribolium Brachyury*-Homologs Tcbyn." Ph.D. thesis. Universität

Tübingen, Germany.

Bier K. 1969. Oogenesetypen bei Insekten und Vertebraten, ihre Bedeutung für die Embryogenese und Phylogenese. *Verh. Dtsch. Zool. Ges. Zool. Anz. Suppl.* **33:** 7–29.

Bock E. 1939. Bildung und Differenzierung der Keimblätter bei *Chrysopa perla* (L). *Z. Morph. Ökol. Tiere* **35:** 615–702.

Bohle H.W. 1969. Untersuchungen über die Embryonalentwicklung und die embryonale Diapause bei *Baetis venus* Curtis und *Baetis rhodani* (Pictet) (Baetidae, Ephemeroptera). *Zool. Jahrb. Anat. Ont.* **86:** 493–575.

Brown S.J., Patel N.H., and Denell R.E. 1994. Embryonic expression of the single *Tribolium* engrailed homolog. *Dev. Genet.* **15:** 7–18.

Brown S., Fellers J., Shippy T., Denell R., Stauber M., and Schmidt-Ott U. 2001. A strategy for mapping bicoid on the phylogenetic tree. *Curr. Biol.* **11:** R43–44.

Büning J. 1994. *The insect ovary*. Chapman & Hall, London, United Kingdom.

Campos-Ortega J.A. and Hartenstein V. 1997. *The embryonic development of* Drosophila melanogaster. Springer, Berlin.

Carvalho J.C., Vanario-Alonso C.E., Silva T.M. and Abdelhay E. 1999. Specialized features of *Rhynchosciara americana* embryogenesis. *Int. J. Insect. Morphol. Embryol.* **28:** 309–319.

Chen G., Handel K., and Roth S. 2000. The maternal NF-κB/Dorsal gradient of *Tribolium castaneum*: Dynamics of early dorsoventral patterning in a short-germ beetle. *Development* **127:** 5145–5156.

Davis G.K. and Patel N.H. 2002. Short, long, and beyond: molecular and embryological approaches to insect segmentation. *Annu. Rev. Entomol.* **47:** 669–699.

Dearden P.K. and Akam M. 2001. Early embryo patterning in the grasshopper, *Schistocerca gregaria*: Wingless, decapentaplegic and caudal expression. *Developmeam* **128:** 3435–3444.

Dearden P., Grbic M., Falciani F., and Akam M. 2000. Maternal expression and early zygotic regulation of the Hox3/zen gene in the grasshopper *Schistocerca gregaria*. *Evol. Dev.* **2:** 261–270.

DiNardo S., Kuner J.M., Theis J., and O'Farrell P.H. 1985. Development of embryonic pattern in *D. melanogaster* as revealed by accumulation of the nuclear engrailed protein. *Cell* **43:** 59–69.

Dorfman R. and Shilo B.Z. 2001. Biphasic activation of the BMP pathway patterns the *Drosophila* embryonic dorsal region. *Development* **128:** 965–972.

Falciani F., Hausdorf B., Schröder R., Akam M., Tautz D., Denell R., and Brown S. 1996. Class 3 Hox genes in insects and the origin of zen. *Proc. Natl. Acad. Sci.* **93:** 8479–8484.

Fleig R. and Sander K. 1988. Honeybee morphogenesis: Embryonic cell movements that shape the larval body. *Development* **103:** 525–534.

Grbic M. 2000. "Alien" wasps and evolution of development. *BioEssays* **22:** 920–932.

Gutzeit H.O. and Sander K. 1985. Establishment of polarity in the insect egg. In *Biology of fertilization* (ed. C.H. Metz and A. Montroy), pp 347–377. Academic Press, San Diego, California.

Hagan H.R. 1951. *Embryology of the viviparous insects*. Ronald Press, New York.

Handel K., Grunfelder C.G., Roth S., and Sander K. 2000. *Tribolium* embryogenesis: A SEM study of cell shapes and movements from blastoderm to serosal closure. *Dev. Genes Evol.* **210:** 167–179.

Hinton H.E. 1981. *Biology of insect eggs*. Pergamon Press, Oxford, United Kingdom.

Ho K., Dunin-Borkowski O.M., and Akam M. 1997. Cellularization in locust embryos occurs before blastoderm formation. *Development* **124:** 2761–2768.

Ivanova-Kasas O.M. 1972. Polyembryony in insects. In *Developmental systems: Insects* (ed. S.J. Counce and C.H. Waddington), vol. 1, pp. 243–271. Academic Press, London, United Kingdom.

Johannsen O.A. and Butt F.H. 1941. *Embryology of insects and myriapods*. McGraw-Hill, New York.

Jura C. 1972. Development of apterygote insects. In *Developmental systems: Insects* (ed. S.J. Counce and C.H. Waddington), vol. 1, pp. 49–94. Academic Press, London, United Kingdom.

Kanellis A. 1952. Anlageplan und Regulationserscheinungen in der Keimanlage des Eies von *Gryllus domesticus*. *Wilhelm Roux's Arch. Entwicklungsmech. Org.* **145:** 417–461.

Karr T.L., Weir M.P., Ali Z., and Kornberg T. 1989. Patterns of engrailed protein in early *Drosophila* embryos. *Development* **105:** 605–612.

Kispert A., Herrmann B.G., Leptin M., and Reuter R. 1994. Homologs of the mouse Brachyury gene are involved in the specification of posterior terminal structures in *Drosophila*, *Tribolium*, and *Locusta*. *Genes Dev.* **8:** 2137–2150.

Kobayashi Y. and Ando H. 1983. Embryonic development of the alimentary canal and ectodermal derivatives in the primitive moth, *Neomicropteryx nipponensis*Issiki (Lepidoptera, Micropterygidae). *J. Morphol.* **176:** 289–314.

Krause G. 1939. Die Eitypen der Insekten. *Biol. Zbl.* **59:** 495–536.

Leuzinger H., Wiesmann R., and Lehmann F.E. 1926. *Zur Kenntnis der Anatomie und Entwicklungsgeschichte der Stabheuschrecke Carausius morosus Br.* Gustav Fischer, Jena, Germany.

Lohs-Schardin M., Cremer C., and Nüsslein-Volhard C. 1979. A fate map for the larval epidermis of *Drosophila melanogaster*: Localized cuticle defects following irradiation of the blastoderm with an ultraviolet laser microbeam. *Dev. Biol.* **73:** 239–255.

Lüscher M. 1944. Experimentelle Untersuchungen über die larvale und die imaginale Determination im Ei der Kleidermotte. *Rev. Suisse Zool.* **51:** 531–627.

Machida R. and Ando H. 1998. Evolutionary changes in developmental potentials of the embryo proper and embryonic membranes along with the derivative structures in atelocerata, with special reference to hexapoda (arthropoda). *Proc. Arthropod. Embryol. Soc. Jpn.* **33:** 1–13.

Maxton-Küchenmeister J., Handel K., Schmidt-Ott U., Roth S., and Jäckle H. 1999. Toll homolog expression in the beetle *Tribolium* suggests a different mode of dorsoventral patterning than in *Drosophila* embryos. *Mech. Dev.* **83:** 107–114.

Miura T., Braendle C., Shingleton A., Sisk G., Kambhampati S., and Stern D.L. 2003. A comparison of parthenogenetic and sexual embryogenesis of the pea aphid *Acyrthosiphon pisum* (Hemiptera: Aphidoidea). *J. Exp. Zoolog. Part B Mol. Dev. Evol.* **295:** 59–81.

Miya K. and Kobayashi Y. 1974. The embryonic development of Atrachya menetriesi Faldermann (Coleoptera. Chrysomelidae). *J. Fac. Agric. Iwate Univ.* **12:** 39–55.

Moretti L.J. and Larsen J.R. 1973. Embryology. In *Bionomics and embryology of the inland floodwater mosquito Aedes vexans* (ed. W.R. Horsfall et al.), pp. 135–206. University of Illinois Press, Chicago.

Morisato D. and Anderson K.V. 1995. Signaling pathways that establish the dorsal-ventral pattern of the *Drosophila* embryo. *Annu. Rev. Genet.* **29:** 371–399.

Myohara M. 1994. Fate mapping of the silkworm, *Bombyx mori*, using localized UV irradiation of the egg at fertilisation. *Development* **120:** 2869–2877.

Patel N.H., Condron B.G., and Zinn K. 1994. Pair-rule expression patterns of even-skipped are found in both short- and long-germ beetles. *Nature* **367:** 429–434.

Patel N.H., Kornberg T.B., and Goodman C.S. 1989. Expression of engrailed during segmentation in grasshopper and crayfish. *Development* **107:** 201–212.

Patel N.H., Hayward D.C., Lall S., Pirkl N.R., DiPietro D., and Ball E.E. 2001. Grasshopper hunchback expression reveals conserved and novel aspects of axis formation and segmentation. *Development* **128:** 3459–3472.

Rehorn K.P., Thelen H., Michelson A.M., and Reuter R. 1996. A molecular aspect of hematopoiesis and endoderm development common to vertebrates and *Drosophila*. *Development* **122:** 4023–4031.

Reuter R. 1994. The gene serpent has homeotic properties and specifies endoderm versus ectoderm within the *Drosophila* gut. *Development* **120:** 1123–1135.

Rickoll W.L. and Counce S.J. 1980. Morphogenesis in the embryo of *Drosophila melanogaster*—germ band extension. *Wilhelm Roux's Arch. Entwicklungsmech. Org.* **188:** 163–177.

Rivera-Pomar R. and Jäckle H. 1996. From gradients to stripes in *Drosophila* embryogenesis: filling in the gaps. *Trends Genet.* **12:** 478–483.

Rohr K.B., Tautz D., and Sander K. 1999. Segmentation gene expression in the mothmidge *Clogmia albipunctata* (Diptera, psychodidae) and other primitive dipterans. *Dev. Genes Evol.* **209:** 145–154.

Roonwal M.L. 1936. Studies on the embryology of the African migratory locus, *Locusta migratoria migratoroides*. I. The early development, with a new theory of multiphase gastrulation among insects. *Philos. Trans. R. Soc. Ser. B Biol. Sci.* **226:** 391–421.

Roth S. 2003. The origin of dorsoventral polarity in *Drosophila*. *Philos. Trans. R. Soc. Ser. B Biol. Sci.* **358:** 1317–1329.

Rushlow C. and Roth S. 1996. The role of the dpp-group genes in dorsoventral patterning of the *Drosophila* embryo. *Adv. Dev. Biol.* **4:** 27–82.

Sanchez-Salazar J., Plether M.T., Bennett R.L., Brown S.J., Dandamudi T.J., Denell R.E., and Doctor J.S. 1996. The *Tribolium* decapentaplegic gene is similar in sequence, strcture, and expression to the *Drosophila* dpp gene. *Dev. Genes Evol.* **206:** 237–246.

Sander K. 1971. Pattern formation in longitudinal halves of leaf hopper eggs (Homoptera) and some remarks on the definition of embryonic regulation. *Wilhelm Roux's Arch. Entwicklungsmech. Org.* **167:** 336–352.

———. 1976a. Morphogenetic movements in insect embryogenesis. In *Insect Development*, (ed. P.A. Lawrence), pp. 35–52. Blackwell Scientific, Oxford, United Kingdom.

———. 1976b. Specification of the basic body pattern in insect embryogenesis. *Adv. Insect Physiol.* **12:** 125–238.

Sauer E. 1954. Keimblätterbildung und Differenzierungsleistungen in isolierten Eiteilen der Honigbiene. *Wilhelm Roux's Arch. Entwicklungsmech. Org.* **147:** 302–354.

Schmidt-Ott U. 2000. The amnioserosa is an apomorphic character of cyclorrhaphan flies. *Dev. Genes Evol.* **210:** 373–376.

Schröder R., Eckert C., Wolff C., and Tautz D. 2000. Conserved and divergent aspects of terminal patterning in the beetle *Tribolium castaneum*. *Proc. Natl. Acad. Sci.* **97:** 6591–6596.

Schwalm F.E. 1988. *Insect morphogenesis*. Karger, Basel, Switzerland.

Seidel F. 1935. Der Anlageplan im Libellenei, zugleich eine Untersuchung über die allgemeinen Bedingungen für defekte Entwicklung und Regulation bei dotterreichen Eiern. *Wilhelm Roux's Arch. Entwicklungsmech. Org.* **132:** 671–751.

Skaer H. 1993. The alimentary canal. In *The development of Drosophila melanogaster* (ed. M. Bate and A. Martinez Arias), vol. 2, pp. 941–1012. Cold Spring Harbor Laboratory Press, Cold Spring Harbor, New York.

Sommer R.J. and Tautz D. 1993. Involvement of an orthologue of the *Drosophila* pair-rule gene hairy in segment formation of the short germ-band embryo of *Tribolium* (Coleoptera). *Nature* **361:** 448–450.

———. 1994. Expression patterns of twist and snail in *Tribolium* (Coleoptera) suggest a homologous formation of mesoderm in long and short germ band insects. *Dev. Genet.* **15:** 32–37.

Stainier D.Y. 2002. A glimpse into the molecular entrails of endoderm formation. *Genes Dev.* **16:** 893–907.

Stanley H.P. and Grundman A.W. 1970. The embryonic development of *Tribolium confusum*. *Ann. Entomol. Soc. Am.* **63:** 1248–1256.

Stathopoulos A. and Levine M. 2002. Dorsal gradient networks in the *Drosophila* embryo. *Dev. Biol.* **246:** 57–67.

Stauber M., Jackle H., and Schmidt-Ott U. 1999. The anterior determinant bicoid of *Drosophila* is a derived Hox class 3 gene. *Proc. Natl. Acad. Sci.* **96:** 3786–3789.

Stauber M., Prell A., and Schmidt-Ott U. 2002. A single Hox3 gene with composite bicoid and zerknullt expression characteristics in non-Cyclorrhaphan flies. *Proc. Natl. Acad. Sci.* **99:** 274–279.

Strand M.R. and Grbic M. 1997. The development and evolution of polyembryonic insects. *Curr. Top. Dev. Biol.* **35:** 121–159.

Takesue S., Keino H., and Onitake K. 1980. Blastoderm formation in the silkworm egg (*Bombyx mori* L). *J. Embryol. Exp. Morphol.* **60:** 117–124.

Tautz D., Friedrich M., and Schroder R. 1994. Insect embryogenesis— What is ancestral and what is derived? *Dev. Suppl.* **1994:** 193–199.

Ullmann S.L. 1964. The origin and structure of the mesoderm and the formation of the coelomic sacs in *Tenebrio molitor* L. (Insecta, Coleoptera). *Philos. Trans. R. Soc. Ser. B Biol. Sci.* **252:** 1–25.

Vanario-Alonso C.E., Sutton R., Carvalho J. C., Yussa M., Silva T.M., and Abdelhay E. 1996. Embryonic expression of the *engrailed* homologue of *Rhynchosciara*. *Wilhelm Roux's Arch. Entwicklungsmech. Org.* **205:** 432–436.

Walker J.J., Lee K.K., Desai R.N., and Erickson J.W. 2000. The *Drosophila melanogaster* sex determination gene sisA is required in yolk nuclei for midgut formation. *Genetics* **155:** 191–202.

Weigel D., Jurgens G., Kuttner F., Seifert E., and Jackle H. 1989. The homeotic gene fork head encodes a nuclear protein and is expressed in the terminal regions of the *Drosophila* embryo. *Cell* **57:** 645–658.

Willmann R. 2003. Phylogenese und System der Insekten. In *Wirbellose Tiere: Insecta* (ed. H.H. Dathe), vol. 1, part 5. Spektrum Akademischer Verlag, Heidelberg, Germany.

Zeh D.W., Zeh J.A., and Smith R.L. 1989. Ovipositor, amnions and eggshell architecture in the diversification of terrestrial arthropods. *Q. Rev. Biol.* **64:** 147–168.

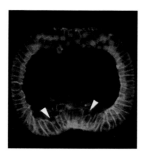

SEA URCHIN GASTRULATION

D.R. McClay, J.M. Gross, R. Range, R.E. Peterson,
and C. Bradham

Department of Biology, Duke University, Durham, North Carolina 27708

INTRODUCTION

Gastrulation in the sea urchin is the archetypal model for deuterostome morphogenesis (Fig. 1). The gut invaginates from the vegetal pole, the tip of the invaginated gut comes in contact with the oral ectoderm, and the mouth or stomodeum forms secondarily. This prototypical developmental sequence is modified by chordates, usually as an adaptive mechanism to accommodate huge amounts of yolk, and/or production of extraembryonic membranes (see Chapter 51). This chapter focuses on the molecular sequence of events that launch gastrulation and drive the simple cell rearrangements during sea urchin morphogenesis. Other recent reviews have described in detail cellular changes that occur during sea urchin gastrulation (Hardin 1996; Ettensohn and Sweet 2000) and have speculated on the mechanisms of those movements (Hardin 1996). On the molecular side, recent reviews have explored the specification of the endomesoderm (Davidson et al. 2002a) and organization of the embryonic axes (Angerer and Angerer 1999, 2000, 2003). In this review, we attempt to tie new molecular information to the mechanisms and phenomena of gastrulation movements.

The transparency of the sea urchin embryo greatly facilitates observation of gastrulation movements. Since many of the morphogenetic movements are separated in time, e.g., events of skeletal mesenchyme ingression occur before gut invagination, those events can be studied independently of other morphogenetic events (Fig. 1). The skeletogenic mesenchyme (primary mesenchyme cells, PMCs) undergoes a precocious epithelial–mesenchymal transition to provide the first visible display of gastrulation movement. About 10 hours after fertilization in *Lytechinus variegatus* (or 20 hours after sperm entry in *Strongylocentrotus purpuratus*), PMCs leave the monolayer of cells that constitute the blastula, ingress through the basal lamina, and migrate

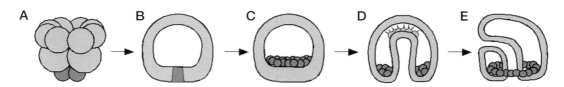

Figure 1. Early sea urchin development. An unequal 4th cleavage (*A*) leads to the formation of micromeres at the vegetal pole (*gray*) with a tier of four macromeres above them. As cleavage progresses, the micromere progeny remain at the vegetal pole (*B*) and then ingress into the blastocoel at the mesenchyme blastula stage (*C*). Shortly thereafter, the archenteron invaginates from the vegetal pole. The first two-thirds of invagination is driven by convergent extension, and the last third of that invagination utilizes secondary mesenchyme cells at the tip of the archenteron to pull this structure to its final length (*D*). Finally, the stomodeum opens and the PMCs form a syncytial ring around the archenteron prior to production of the larval skeleton (*E*).

into the blastocoel (Fig. 1B,C) where they begin a new series of behaviors as mesenchyme cells. The PMCs eventually produce the larval skeleton. Several hours after ingression, endomesoderm progeny at the vegetal plate begin the process of invagination (Fig. 1C,D). First, in the center of the vegetal plate a group of mesenchyme cells called secondary mesenchyme cells (SMCs) initiate an inward fold. The inward fold quickly extends laterally on the vegetal plate to involve endoderm cells of the future foregut and midgut. The gut then elongates by three different kinds of movements. Convergent-extension movements change the packing of epithelial cells so that the diameter of the archenteron narrows as it elongates, involution movements provide additional endoderm cells to the gut from regions lateral to the vegetal plate, and SMCs stretch the length of the archenteron through deployment of contractile filopodia late in the invagination process.

In the hours leading up to these morphogenetic events, there are no cell movements other than the cell divisions that characterize the cleavage stages and the expansion of the blastocoel cavity. Yet within this seemingly simple hollow blastula monolayer, each cell is involved in an intricate specification process to establish cell fates and set the stage for morphogenesis. By the time morphogenesis begins, each cell is programmed for its role in gastrulation movements and has deployed the gene regulatory apparatus necessary for differentiation of larval tissues. To investigate cell movements, a key first step is to understand how the cells are specified during cleavage. The cell movements themselves can be examined at many levels. The first views of morphogenetic movements in the sea urchin came from careful observation and analysis of time-lapse films (Gustafson and Wolpert 1962, 1963, 1967). Following those descriptive studies, experimental analyses explored the mechanistic basis of each cell movement (Ettensohn 1985, 1992; Hardin 1996). Adhesive changes, cytoskeletal changes, intracellular motility, coordination mechanisms, and cell polarity each are necessary for cells to move in a directed fashion. Furthermore, cells must have an appropriate receptor and signal transduction apparatus to respond to directional cues. Specification mechanisms prepare the cells for these morphogenetic events, and a number of dynamic molecular changes accompany the morphogenesis.

SPECIFICATION AND INGRESSION OF THE PRIMARY MESENCHYME CELLS

Specification of the Skeletogenic Lineage

Micromeres appear at the vegetal pole of embryos at the 16-cell stage, the result of an unequal 4th cleavage. At 5th cleavage, a second unequal division leaves the vegetal pole with four "large" micromeres and four "small" micromeres. The large micromeres become the skeletogenic PMCs. Between their birth at 4th cleavage, and ingression 15 hours later (in *S. purpuratus*), micromeres divide three more times and become specified as skeletogenic cells. From their birth, micromeres are committed to the skeletogenic fate. This has been shown in several ways. Okazaki (1975) isolated micromeres after 4th cleavage, then grew them in culture where they became skeletogenic cells and produced spicule rods, thus demonstrating the developmental independence of the micromere lineage. Later, micromeres were transplanted to other positions in the embryo. It was found that they ingressed at the correct time and became PMCs (Ettensohn and McClay 1986; Peterson and McClay 2003).

Micromeres supply at least two induction signals that have an impact on other lineages of the embryo. An early induction signal (ES) is released from micromeres beginning at the 16-cell stage (Fig. 2A) (Ransick and Davidson 1995). To demonstrate that signal, micromeres were removed from embryos after the 4th, 5th, or 6th cleavages, and the resulting embryos were later examined for expression of an endoderm marker, *endo16*. Ransick and Davidson (1995) found that micromeres are necessary for macromeres to be specified correctly as endomesoderm, beginning with 4th-cleavage micromeres. A few years later, it was learned that a second signal presented by micromeres induces SMCs (Sherwood and McClay 1999). SMCs are diverted from general endomesoderm specification by a micromere-presented Delta ligand, which activates the Notch signaling pathway in a subset of endomesoderm cells (Fig. 2B). This signal subdivides progeny of the macromeres into SMCs and endoderm. Endomesoderm cells that fail to receive Delta-Notch signaling between 7th and 9th cleavage become endoderm (Sherwood and McClay 1999; Sweet et al. 2002). With these inductions established experimentally, a number of labs then explored in some detail how specification proceeds in micromeres and endomesoderm.

The search for a mechanism of specification began with β-catenin. Three labs learned that the Wnt pathway is essential for micromere specification (Emily-Fenouil et al. 1998; Wikramanayake et al. 1998; Logan et al. 1999). β-Catenin enters micromere nuclei just after 4th cleavage (Logan et al. 1999). If that nuclear entry is prevented experimentally by expression of the cadherin cytoplasmic tail Δ-cadherin, which binds the available pool of cellular β-catenin, micromeres are not specified, fail to induce either through the ES or by production of Delta, and do not become PMCs (Wikramanayake et al. 1998; Logan et al. 1999). Specification is rescued by injection of RNA encoding a stable form of β-catenin along with the inhibiting concentration of Δ-*cadherin* RNA. In excess, β-catenin vastly expands endomesoderm specification, and experiments show that all endomesodermal lineages require β-catenin for specification, although the timing and consequences of

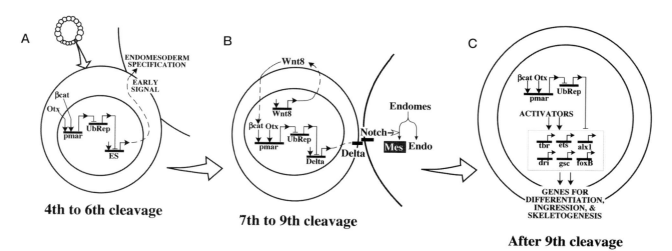

Figure 2. Micromere gene regulatory network. (*A*) At 4th cleavage, micromeres begin specification by activation of *pmar1* by Otx and β-catenin. Pmar1 represses a ubiquitous repressor and activates the synthesis of the early induction signal (ES). The ES, in turn, induces endomesoderm specification in the sister macromere. (*B*) Between 7th and 9th cleavage, the early specification network in micromeres synthesizes two induction factors. Wnt8 acts in an autocrine manner apparently to boost β-catenin, and Delta activates Notch in the adjacent macromere endomesoderm progeny. The subset of endomesoderm cells receiving an activated Notch signal is diverted to a mesoderm fate. (*C*) Later micromere specification activates a group of transcription factors in a network, and these are responsible for activation of the group of differentiation genes of PMCs, including those involved in ingression.

that transcriptional activation are distinct in different lineages. β-Catenin was chosen as a starting point for further studies because it enters micromere nuclei at the 16-cell stage and has profound effects on micromere and endomesoderm specification of macromeres.

Additional components of the micromere specification network were discovered in many laboratories (Chuang et al. 1996; Kurokawa et al. 1999; Fuchikami et al. 2002; Oliveri et al. 2002; Amore et al. 2003). Perturbation studies and quantitative PCR analysis then asked how these molecules work in the specification process (Fig. 2C) (Oliveri et al. 2002). The micromere *gene regulatory network* (GRN; Oliveri et al. 2002) assembles data from many laboratories into a logic model that describes predicted gene interactions based on a large series of biological and molecular experiments (see Chapter 48). Although this is not yet completed, it is of value for this discussion for several reasons. Most important, the micromere GRN provides a working model for events upstream of ingression, and aspects of that model are summarized in Figure 2. β-Catenin activates a paired homeodomain transcriptional repressor called *pmar1* (Fig. 2A) (Oliveri et al. 2003). Pmar1 protein activates micromere specification by repressing a ubiquitous repressor (as yet unidentified). Once the micromere repression is lifted, a series of transcription factors (*tbr, ets, alx1 dri, fox B*, and *gsc*; Fig. 2C) is activated, and these in turn activate a number of downstream genes. The downstream genes include proteins necessary for production of the spicule matrix, cytoskeletal

genes, adhesion molecules, and a number of other molecules that have been identified in recent screens (Zhu et al. 2001; Illies et al. 2002). The predictions of the micromere GRN model were tested experimentally in several ways. First, specific molecules in the network were perturbed with morpholino antisense oligonucleotides or by misexpression. Then, molecules modeled to be downstream of the perturbed molecule were examined by quantitative PCR (Davidson et al. 2002a). Second, biological assays tested those placements. For example, *pmar1* is placed as the target for β-catenin and as the upstream transcription factor necessary for initiating micromere specification (Oliveri et al. 2002, 2003). Both sets of experiments supported the GRN.

To appreciate the relationship between initial specification and ingression, consider a timeline between 4th cleavage and the beginning of ingression. *pmar1* is activated at 4th cleavage and *pmar1* RNA is present for about four or five cleavages, or from about 5 to 12 hours of development in *S. purpuratus*. Beginning at about 6–8 hours of development, the earliest transcription factors downstream of *pmar1* begin to be activated, although not all of them at the same time, and these begin to target activation of other downstream genes. By about 12 hours, *goosecoid* (*gsc*) and *deadringer* (*dri*) are activated in micromeres (Angerer et al. 2001; Amore et al. 2003). These transcription factors are then extinguished in micromeres a few hours later, but subsequently appear in other lineages of the embryo. Another important micromere transcription factor, *ets*, is expressed ubiquitously, then after hatching (about 15 hours), *ets* expression becomes restricted to the micromere lineage (Kurokawa et al. 1999). Still another, *tbr*, is activated transiently at early blastula stage then becomes highly activated in PMCs at the time of ingression (Fuchikami et al. 2002). Although they are expressed in different patterns, the six transcription factors appear responsible for expression of many of the differentiation genes of the PMC lineage. Each, however, is directly or indirectly activated by *pmar1*. One observation from the many experiments that established the GRN is that there are relatively few steps of gene activation between the initial events activating specification, and differentiation of the skeletogenic cells. The GRN model as depicted in Figure 2 is incomplete, however, since additional genes and additional network connections will almost certainly be added. Even so, the general conclusion that specification involves relatively few steps of gene activation will probably hold.

The micromere GRN sets the stage for PMC differentiation by activating genes necessary for PMC function. PMC-specific proteins that functionally participate in ingression movements begin to accumulate in micromeres by 12–16 hours, or 4–8 hours in advance of the ingression movements. Some of these proteins are immediately deployed by the hatched blastula, and other proteins are synthesized, stored in intracellular compartments, only to be activated or inserted into their target compartment coincident with the beginning of ingression. For example, LvαSU2, an epithelial integrin, is inserted into the basal membrane of micromeres when they are still epithelial cells at about hatched blastula stage (Hertzler and McClay 1999). LvαSU2 is used by micromeres as a cell–basal lamina adhesion molecule as long as micromeres have an epithelial phenotype, but the integrin is removed from the PMCs at the time of ingression. Another protein, MSP 130, is also synthesized by epithelial micromeres beginning around hatched blastula stage, but it remains in intracellular vesicles until ingression movements begin, at which time it is glycosylated and then exocytosed (Wessel and McClay 1985; Leaf et al. 1987). In both cases, the upstream micromere GRN governs the synthesis of the molecule, then contributes to cellular mechanisms that govern its storage, recruitment to the cell surface and then its involvement in ingression.

Many other downstream genes are also expressed in a variety of patterns and with distinct timing profiles. *SM50*, a spicule matrix protein, for example, is expressed well before ingression, and its regulation is governed by the transcription factors of the GRN (Kirchhamer et al. 1996; Kitajima et al. 1996; Kurokawa et al. 1999; Oliveri et al. 2002). In contrast, *SM30*, encoding another spicule matrix protein, is not transcribed until PMCs begin assembling the skeleton. *SM30* activation is controlled, in part by the GRN, and in part by signal transduction inputs from ectodermal-

ly supplied positional information (Guss and Ettensohn 1997). Thus, ingression and patterning should not be viewed as precisely timed accumulations of proteins that are suddenly called into action at the time of epithelial–mesenchymal transition. Rather, micromeres begin to accumulate some of the necessary proteins very early in micromere specification, and others accumulate in a variety of patterns along the way. From a cell biological point of view, however, ingression is temporally controlled. All the proteins and functional properties of the micromere-to-PMC transition must be in place prior to that time, and then the timing of ingression is controlled autonomously by micromeres. After ingression, PMCs begin to receive nonautonomous input from other cells of the embryo so that migration pathways and synthesis of the skeleton occur correctly.

Ingression Is Precisely Timed and Occurs as a Specific Consequence of Specification

Close examination of a population of embryos reveals that ingression is precisely controlled as all embryos cultured together begin ingression at about the same time and all PMCs begin their epithelial–mesenchymal transition at about the same time. The trigger to initiate ingression is autonomous and appears to be a consequence of a characteristic specification sequence. First, if micromeres are placed in ectopic positions in the embryo, ingression still occurs at the correct time (Peterson and McClay 2003). These data augment earlier findings indicating that micromere development is quite autonomous (Okazaki 1975). Second, if specification steps are blocked, ingression, as might be expected, is affected, but the blocks are not uniform. For example, expression of a dominant negative *ets1* blocks ingression (Kurokawa et al. 1999), as does a morpholino to an *aristalless* transcription factor, *alx1* (Ettensohn et al. 2003). In contrast, a morpholino to a *dri* transcription factor only slightly retards ingression (Amore et al. 2003), and a morpholino to the transcription factor *tbr* fails to block ingression at all (Fuchikami et al. 2002). All of these transcription factors are necessary for skeletogenesis, however. This suggests that components of the trigger for ingression are downstream of *ets1* and *alx1,* but not downstream of *dri* or *tbr.*

Although the timing mechanism is not understood, a PMC expressed sequence tag (EST) project has uncovered a large number of candidates that may be involved in this mechanism (Zhu et al. 2001). The EST project, and differential screens designed to select molecules expressed by PMCs, provide a large list of candidates, but few of these have been tested for any function, let alone the trigger function. Unfortunately, molecules expressed uniquely by PMCs may not necessarily reveal the trigger mechanism, given the possibility that the mechanism may employ ubiquitous molecules such as kinases. Thus, differential or subtractive screens may not be a panacea for the molecular discovery process.

Ingression Movements—An Epithelial to Mesenchymal Transition

Time-lapse sequences reveal that presumptive PMCs begin to move in the plane of the blastula cell monolayer prior to overt ingression movements (see supplementary movie 9-1 at www.gastrulation.org). Then, a stereotypic series of cell shape changes and ingression movements begins. These have been described in some detail previously (Katow and Solursh 1980) and are illustrated in Figure 3. In addition to becoming motile and changing cell shape, the nascent PMCs change cell adhesions. Once in motion, the PMCs pass through the basal lamina, and as they emerge into the blastocoel, they begin to behave as mesenchymal cells. PMCs engage in a characteristic sequence of cell migrations that culminates in the formation of a syncytial ring of PMCs surrounding the archenteron (Fig. 1D,E). The larval skeleton is produced by this structure. It should be noted that although skeletogenesis occurs in a syncytium, the nuclei of the PMCs remain spaced in a characteristic way, suggesting a cytoskeletal architecture for nuclear distribution, as has been shown in other syncytia (Starr and Han 2002). Based on inhibitor studies, the early motile behavior of PMCs prior to syncytial fusion is attributed to the actin cytoskeleton and not to microtubule changes (Anstrom 1989, 1992). Since that work was published, little has been added experimentally to characterize stereotypic cell shape changes or details of the motility necessary for the ingression movements.

A number of cell adhesion changes accompany ingression (Fink and McClay 1985). The cells lose affinity for hyalin and echinonectin, extracellular matrix proteins covering the outside of the embryo (Fink and McClay 1985; Burdsal et al. 1991). At the same time, the PMCs lose cell–cell adhesiveness (Fink and McClay 1985), a property that is largely attributable to a rapid endocytic removal of cell surface cadherin from the micromeres (Miller and McClay 1997b). Finally, the PMCs shift their affinities for the basal lamina from an epithelial-basal lamina adhesion via a laminin-binding integrin (Hertzler and McClay 1999), to a fibronectin-binding activity (Fink and McClay 1985). Several candidate fibronectin-binding integrins are expressed by PMCs at about the correct time (Marsden and Burke 1997, 1998; Rise and Burke 2002), although as yet, none of these has been linked experimentally to the mesenchymal migratory behavior. A striking feature of the cell adhesion changes is that they take place quite rapidly (~1 hour) and seemingly in a coordinated fashion. Since all of the cell adhesion changes occur during that time (Fink and McClay 1985), ingression movements are accompanied by a complete change in all known adhesion mechanisms in these cells. This wholesale cell-surface switch in adhesive mechanisms appears to involve a drastic change in the entire cell surface. Cadherin, for example, is completely endocytosed and removed from the cell surface as ingression proceeds (Miller and McClay 1997b). The laminin-binding integrin is removed from PMCs at the same time (Hertzler and McClay 1999). Coincidentally, mesenchymal cell-surface molecules that include MSP130 are recruited to the cell surface from prior storage in the trans-Golgi compartment (Wessel and McClay 1985; Anstrom et al. 1987; Miller

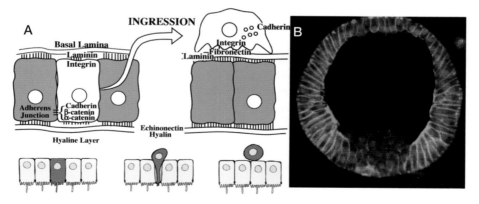

Figure 3. Primary mesenchyme cell ingression. (*A*) The lower diagram shows the integration sequence as PMCs transit from an epithelium to become a mesenchymal cell. The adhesion changes documented are shown in the upper diagram. At ingression, cadherin, β-catenin, and α-catenin disassemble the adherens junction. The laminin-binding integrin is released, and both adhesion systems are endocytosed. At the same time, a new integrin that binds to fibronectin is shuttled to the surface of the PMC along with other PMC membrane components. (*B*) An embryo at mesenchyme blastula. The egg was injected with GFP attached to a transmembrane sequence. At gastrulation, the PMCs ingress from the vegetal pole, and as they ingress, the GFP is endocytosed along with other epithelial components.

and McClay 1997b;). It appears therefore that the PMCs replace their entire epithelial adhesion apparatus with a mesenchymal adhesion apparatus at the same time by a process that involves simultaneous exocytosis and endocytosis.

Later Movements of Primary Mesenchyme Cells

An extensive literature describes the behavior of PMCs as they move within the blastocoel, organize into a syncytial ring, produce the spicule matrix, and biomineralize to form the calcareous skeleton (Ettensohn 1992; McClay et al. 1992; Ettensohn and Sweet 2000; Wilt 2000; also see supplementary movies 9-2 and 9-3 at www.gastrulation.org). Once the PMCs arrive in the blastocoel, they begin to move toward the animal pole on the basal laminar surface. The movements are nonrandom, as all PMCs follow trajectories along longitudinal lines between the vegetal and animal pole. When the PMCs reach the equator, or sometimes beyond, they reverse their movement and move to a position near the future ectoderm–endoderm boundary of the surrounding cells, and form a ring of cells surrounding the invaginating archenteron (Ettensohn and McClay 1986; Malinda and Ettensohn 1994; Peterson and McClay 2003). The movements of PMCs are fairly typical of fibroblasts during this migratory phase of behavior; they extend filopodia and lamellapodia, then the cell center moves toward the leading edge (Malinda et al. 1995). When the PMCs move in the vegetal direction and approach the ectoderm–endoderm boundary, they stop migrating. If the ectoderm–endoderm boundary is displaced in animalized or vegetalized embryos, the position of the PMC ring corresponds to the displaced boundary (Adelson et al. 1992; Sherwood and McClay 1999), suggesting that the boundary vicinity provides the spatial cues for PMC ring localization. The PMCs in the ring fuse to form a single syncytium, and their behavior changes dramatically. Each PMC begins to extend multiple thin filopodia (200–300 nm in diameter) to contact the substrate (Malinda et al. 1995; Miller et al. 1995). These thin filopodia are similar in size and behavior to other thin filopodia, later termed cytonemes in *Drosophila* (Ramirez-Weber and Kornberg 1999), and to thin filopodia extended from the tips of neuronal growth cones (Kuwada 1986; Tosney 1991; Hotary and Tosney 1996). The behavior and kinetics of thin filopodia correlate with a sensory rather than a motile function. When thin filopodia are extended, the PMCs no longer translocate on the substrate, and in the absence of thin filopodia, PMCs fail to make a skeleton (McClay 1999).

PMCs depend on cues provided by the substrate to pattern the skeleton (Armstrong et al. 1993). During early migration, the PMCs use cues along the animal–vegetal axis

only, since each PMC tends to restrict its migration to a longitudinal path. The first evidence of lateral migration occurs late in gastrulation when a subpopulation of aboral PMC nuclei migrate to the oral side of the syncytium and participate in ventrolateral cluster formation (Peterson and McClay 2003). As the skeleton grows, PMC nuclei require the ectoderm to be properly organized and specified. If the ectodermal cell–substrate interaction is disturbed, the ectoderm fails to provide proper positional information (Zito et al. 1998, 2000). If the PMCs are properly specified but the ectoderm is not, spiculogenesis fails to occur normally (Hardin et al. 1992).

For one of many examples of how the PMCs use the positional information from ectoderm to pattern, consider changes in embryo size (Horstadius 1973). If cells are separated at the 4-cell stage, each of the four blastomeres makes a 1/4-scale pluteus larva, complete with a 1/4-size skeleton (this famous observation was made first by Driesch in 1893, then confirmed by Boveri in 1901 and later by Horstadius in 1928; for review, see Horstadius 1973). That 1/4 size embryo has 1/4 the number of PMCs. If one experimentally augments the number of PMCs in the embryo, the skeleton produced is still 1/4 size (Ettensohn 1990; Armstrong et al. 1993). This, along with many other experiments, shows that positional information is provided by, and scaled by, the ectoderm. Positional cues restrict skeleton synthesis to the scale of the ectoderm surrounding them, and patterning cues within that scaled source are also provided by ectoderm. In molecular terms, this property has several consequences. The cues must be specified and distributed by oral–aboral ectodermal and animal–vegetal tiers of cells. This distribution requires constant cell–cell communication because the embryo is capable of adjusting the spatial cues to a smaller size, even when experimentally halved or quartered later in development. If halved prior to the beginning of gastrulation, the embryo regulates patterning to form two normal pluteus larvae of reduced scale (Horstadius and Wolsky 1936; D.R. McClay, unpubl.). Consider what happens when the embryo is halved. Prior to the cut, patterning information is distributed in a 360° circumference around the embryo, and the cut separates any information present into two 180° arcs. Each of the arcs rounds up and fills in the missing information for the embryo to be patterned normally. But, since the embryo is now half-size, the 180° arc has to scale down preexisting cues, redistribute them, and add cues for the other 180°, and this could only occur if the cells are in constant communication around each arc.

A second consequence of patterning is that information received by PMCs from the ectoderm must go through an integration process. Again, in molecular terms this process must be fairly complicated. The thin filopodia of each PMC are capable of sampling a large area of ectoderm, and the positional information obtained must involve signal transduction and integration of multiple patterning signals. Each PMC nucleus in the syncytium responds to locally arriving information obtained from the ectoderm (Armstrong and McClay 1994), so the incoming information must be insulated somehow from distant PMC nuclei in the syncytium. PMCs respond by producing the skeleton, which requires them to secrete the spicule matrix and supply biomineralization components. In addition, there is evidence that a subset of patterning genes is activated in PMCs depending on where a PMC is located in the syncytium. The *Tbx 2/3* gene is activated, for example, and confined to aboral PMCs (Gross et al. 2003). To generate the three-dimensional shape of the skeleton, the integrated positional inputs must be translated, somehow, into directed biomineralization of calcium carbonate (Wilt 2002). There must be a code of molecular information supplied to the biomineralization process to direct the patterning of the skeleton. As an example of the complications in this mechanism, our naive idea of how this might work was based, at first, on the simple hypothesis that the ectoderm lays down a directive substrate in a pattern that is simply followed by the growing skeletogenic cells as they deposit calcium carbonate (Armstrong and McClay 1994). This hypothesis proved to be wrong. Instead, in studies involving PMCs from an embryo of one species and ectoderm produced by another species, it was learned that positional cues only are provided by the ectoderm. The PMCs then produce the correct skeleton as dictated by their own genotype, even in regions where the host species produces no skeleton, and therefore could not be expected to provide a directive extracellular matrix. This means that positional cues are integrated by PMCs, which in turn use that information to direct the correct pattern of biomineralization.

INVAGINATION OF THE ARCHENTERON

Specification of the Endomesoderm

During archenteron invagination, the vegetal plate thickens, with cells switching from a cuboidal to a columnar shape, the PMCs ingress, then the vegetal plate flattens and bends inward. A group of 20–30 cells, called bottle cells, appears to initiate the inward bending process. Once invagination has started, the invaginating epithelial sheet changes shape. Cells within the epithelium shift position relative to one another to narrow the lumen of the archenteron, which in turn extends the archenteron. Finally, in a final phase of archenteron formation, SMCs at its tip make contact with the substrate at or near the animal pole, and pull the archenteron to its final length. These processes are examined in mechanistic detail below. Current questions focus on the molecular basis of steps in the process. The archenteron is built from

endomesoderm cells that are specified during cleavage, and many details of that specification have been established in the last two years (Howard et al. 2001; Kenny et al. 2001; Davidson et al. 2002b; Oliveri et al. 2002). These and additional data are summarized below.

Macromeres, the large daughter cells of the unequal 4th cleavage, are specified initially as endomesoderm. Maternal components begin endomesoderm specification at the 16- to 32-cell stage, and, immediately thereafter, an early inductive input from micromeres (ES) provides additional input necessary for initiation of endomesoderm specification. Several cleavages later, a second induction signal from micromeres (the Delta-Notch signal, see above) diverts some partially specified endomesoderm into a secondary mesoderm (SMCs) specification network, thus separating the paths of mesoderm specification from subsequent endoderm specification (Fig. 4).

The ES is released by micromeres at 4th cleavage, immediately after the unequal cleavage. Macromeres and their progeny receive the ES at least until the 6th cleavage (Ransick and Davidson 1995). Although the molecular identity of the ES is not yet known, the ES is necessary for macromere progeny to invaginate on time, for activation of some and inactivation of other early transcription factors (Kenny et al. 1999; Oliveri et al. 2003), and for activation of downstream endoderm genes (Ransick and Davidson 1995). Several maternal factors in macromeres also participate in early specification events starting at 4th cleavage (Davidson et al. 2002b). At 7th cleavage, micromeres provide the Delta ligand (Sweet et al. 2002). Delta activates Notch receptors on a subset of macromere progeny to induce SMCs (Sherwood and McClay 1997, 1999). Macromere progeny that do not receive the Delta signal become specified as endoderm (with the exception of a small number of cells that become ectoderm, depending on the location of the ectoderm/endoderm boundary). The embryo at mesenchyme blastula stage consists of micromeres at the vegetal pole surrounded by a ring of SMCs, and concentrically around the SMCs are the cells of the endoderm, an arrangement that is confirmed by fate mapping studies (Ruffins and Ettensohn 1996; Logan and McClay 1997).

Downstream Specification That Prepares SMCs and Endoderm for Invagination

The current SMC GRN (Davidson et al. 2002b) (see Chapter 48) is an oversimplification, since the SMC lineage later subdivides into four different cell types: pigment cells, coelomic cells, muscle, and blastocoel cells. For the purpose of gastrulation behavior, however, the SMC sublineages behave in different ways. SMCs that become pigment cells

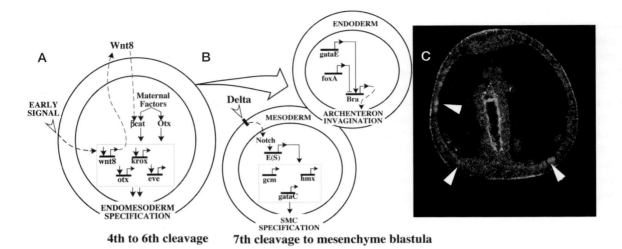

Figure 4. Endomesoderm specification. (*A*) An early induction from the micromeres induces the macromeres to begin endomesoderm specification. A series of early transcription factors are activated to begin this process, and the specification is reinforced by expression of *Wnt8*, which augments maternal β-catenin in the early specification. (*B*) Several cleavages later, the endomesoderm is divided into secondary mesenchyme and endoderm. Endomesoderm cells that activate Notch via expression of Delta from micromeres become SMCs while macromere progeny that do not receive the Delta signal become endoderm. Among the many endodermal genes activated, gata E participates in activation of *brachyury* (*Bra*). Shortly after that, *brachyury* is inactivated by FoxA. This sequence appears to be important for the morphogenetic events of gastrulation that follow. (*C*) Brachyury is shown in nuclei to either side of the archenteron (*lower arrowheads*), and also surrounding the area that will be the site of stomodeal invagination (*upper arrowhead*).

are specified soon after Notch signal activation, and the future pigment cells delaminate from the cluster of SMCs throughout gastrulation. The migratory pigment cells cross the blastocoel on collagen-like cables, invade the ectoderm by digging through the basal lamina, and emerge on the basal side of the ectoderm. They continue to migrate within the basal side of the ectoderm to space themselves throughout the aboral ectoderm using an avoidance behavior that prevents clustering of pigment cells. Disruption of any part of this sequence can leave pigment cells piled up either at the vegetal pole or at the animal pole, depending on the insult. The remaining SMC sublineages reside at the tip of the future archenteron until gastrulation is completed, and these cells participate in the last third of archenteron elongation (see below). Between 9th cleavage (12 hours), when the Notch signal ends, and the beginning of invagination (about 30 hours in *S. purpuratus*), SMCs accumulate a number of molecules that presumably are necessary for invagination, although experiments have yet to show that components of SMCs, distinct from the endomesoderm, are necessary for invagination.

The population of endomesoderm cells that did not activate the Notch signal continues to be specified, but as endoderm. If the Delta signal fails to activate Notch in the vegetal plate, the entire vegetal plate is specified as endoderm. Since endomesoderm specification is induced by the ES and maternal factors, SMC specification also requires

these initial steps prior to Delta–Notch induction. After the Notch induction, cells that continue toward endoderm fates retain the ability to regulate and replace all SMCs except pigment cells (McClay and Logan 1996). If Notch signaling is absent between 7th and 9th cleavage, the early specification of the endomesoderm lineage is entirely in the endoderm direction, and only much later, after gastrulation (48–72 hours), do regulative properties divert some of those cells to an SMC fate.

The activation of *brachyury* is important for the morphogenesis that follows endomesoderm specification (Gross and McClay 2001). *brachyury* is activated on the endoderm side of the endoderm/SMC boundary in echinoderms. Before the foregut cells or any subsequent cells begin to invaginate, *brachyury* expression is extinguished from those cells, and activated in the cells that will invaginate later in time (Fig. 4). The sequential activation and repression of *brachyury* requires *gatac* for activation of its sweeping pattern of expression, and *foxA* to turn off *brachyury* in its wave of expression. Brachyury, in turn, is involved in activation of a number of downstream genes that are necessary for invagination. Experiments showed that conversion of Brachyury from an activator to a repressor by addition of an engrailed repressor sequence (*brachy-EnR*) leads to a failure of archenteron invagination (Conlon et al. 1996; Gross and McClay 2001). Based on these data, a differential cDNA screen using *brachy-EnR*-subtracted cDNA identified a number of *brachyury* downstream genes potentially involved in motility and cytoskeleton dynamics of morphogenesis (Rast et al. 2002). *brachyury* is also expressed in a torus of cells around the stomodeum later in gastrulation, and in some cases, embryos expressing *brachy-EnR* fail to form the stomodeum (Gross and McClay 2001). Thus, in the endoderm there is a sweep of gene activation and repression necessary for the morphogenetic apparatus of invagination. *brachyury* is one gene that is important for the sweeping preparation of cells for the movements of gastrulation, although much of the detailed network involved in preparing endoderm cells for invagination remains to be discovered.

The Trigger for Invagination

Invagination begins 10 hours after ingression in *S. purpuratus* (Fig. 5). The proximal trigger for invagination is not known and is a controversial issue because many disruptions block gastrulation. Some of those disruptions are due to known failures in upstream specification of endomesoderm (Wikramanayake et al. 1998; Logan et al. 1999; Gross and McClay 2001; Howard et al. 2001), but others occur for no known reason. This has made it difficult to identify the actuator of the invagination apparatus. Although normally SMCs begin the invagination sequence, expression of a dominant-negative form of Notch that eliminates much of SMC specification fails to significantly delay the onset of invagination (Sherwood and McClay 1999). This suggests that initiation of invagination depends on the endomesoderm specification sequence shared by SMCs and endoderm.

Steps of Archenteron Invagination

The observed behaviors of SMCs and endoderm cells during archenteron invagination are well documented. The embryo is transparent and easy to image at a relatively high resolution, so a number of time-lapse recordings, theoretical treatments, experiments, and even two-photon analyses have been produced over the past 40 years (Gustafson and Wolpert 1967; Odell et al. 1981; Ettensohn 1984a,b; Nakajima and Burke 1996; Kimberly and Hardin 1998; Piston et al. 1998; Davidson et al. 1999). A brief summary of these and other observations follows, emphasizing aspects of invagination with known molecular explanations; that is, those components that are causal for the morphogenesis.

SMCs at the center of the vegetal plate begin involuting as the first sign of archenteron formation. The cause of that involution has been a matter of debate. One hypothesis tested suggests that secretion of a hygroscopic extracellular matrix between the apical surface of the cells and the hya-

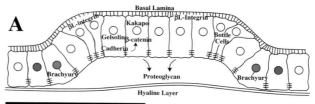

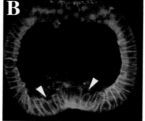

Figure 5. Invagination. (*A*) Following *brachyury* expression in the surrounding endoderm, bottle cells around the perimeter of the initial invagination provide the initial inward bending (also *arrowheads* in *B*). During gastrulation, both the adherens junction and the basal integrin adhesion remain intact. Following *brachyury* expression, the endoderm up-regulates a number of cytoskeletal proteins, including gelsolin and kakapo, that appear to participate in the invagination processes. At the same time, there is evidence for secretion of a proteoglycan between apical cell surface and the hyaline layer. (*B*) An embryo expressing GFP coupled to a transmembrane sequence delineates cell boundaries at the beginning of gastrulation at a time equivalent to the diagram in *A*. The arrowheads point to bottle cells.

line layer provides the force for the initial involution of the cells at the vegetal plate (Lane et al. 1993). Further support of this hypothesis comes from mechanical tests of stiffness. For the hygroscopic model to be supported, a stiff extracellular matrix is necessary, and Davidson et al. (1999) find that the hyaline layer provides that stiffness. Earlier models suggested that either actin or microtubule networks provide the means for inward bending using cell shape changes to accomplish this (Gustafson and Wolpert 1967; Odell et al. 1981). Actin purse-string models have been suggested in other systems and modeled in the sea urchin (Burnside 1971; Edwards et al. 1997), and cortical actin structures have been observed in the first cells to initiate gastrulation movements of the sea urchin (Nakajima and Burke 1996; Marsden and Burke 1998), but published reports of inhibitors of actin microfilaments and microtubule formation suggest that these inhibitors fail to block the initial inward bending of the SMCs at the vegetal plate (Lane et al. 1993). Thus, although it is likely that microfilaments participate in altered cell shape maintenance, the degree to which microfilament contraction is causal for driving the involution is questionable, as the inhibitor experiments provide an incomplete understanding of how actin or microtubule processes might function.

Bottle cell formation at the vegetal plate provides another early mechanism for the involution movements (Nakajima and Burke 1996; Kimberly and Hardin 1998). Twenty to thirty bottle cells appear at the beginning of invagination and surround the center of the vegetal plate. If some of the bottle cells are laser-ablated around a portion of the vegetal plate, invagination of that area is severely compromised (Kimberly and Hardin 1998). Actin microfilaments are associated with the apical constrictions of the bottle cells (Nakajima and Burke 1996), prompting again the speculation that cortical actin contractions contribute to the initial inbending, but the inhibitor studies mentioned above are in conflict with this hypothesis. Bottle cell formation is inhibited by an antibody to hyalin (Kimberly and Hardin 1998), a treatment that had earlier been shown to interfere with cell–substrate adhesion and would prevent gastrulation (Adelson and Humphreys 1988). Interference of bottle cell formation by preventing cell–substrate interactions suggests that cell–extracellular traction is involved in compressing the central vegetal plate, perhaps in much the same way that epiboly provides a mechanical force for amphibian and zebrafish gastrulation. This does not mean that cells translocate on the rigid hyaline layer as they move toward the blastopore, however. This is because the hyaline layer itself is incorporated into the archenteron (Citkowicz 1971, 1972; McClay et al. 1989). Furthermore, isolated vegetal halves of the embryos, which cannot generate global forces directed toward the vegetal plate, like epiboly, still initiate invagination (Moore and Burt 1939; Ettensohn 1984b). Although invagination begins in the absence of global forces, traction by vegetal cells on the remaining substrate of half-embryos could provide circumferential forces for inward bending to occur. Thus, although the precise mechanism that initiates invagination is not settled, currently available data support the notion that traction on the extraembryonic matrix is necessary to initiate the process. In addition to traction, the cells in the vegetal plate secrete extracellular matrix onto their apical surface that occupies space and appears to force the cells beneath to bend inward. The bottle cells change shape and move inward to contribute to the invagination. It is unclear what causes the bottle cells to change their shape and elongate into the blastocoel. Whatever the mechanism, an apical integrin, attached to the actin cytoskeleton beneath, is most likely a stabilizing component, if not a driving component of that movement (Marsden and Burke 1998).

Beyond the initial involution, it is safe to say that cell shape changes, movement of cells within cell sheets, contractile forces, filopodial contraction, and other motile mechanisms all contribute to archenteron formation. To place these movements into a sequence misrepresents the actual process, as many of the movements occur with temporal overlaps. With that in mind, however, the several different mechanisms that have been examined in some detail will be considered sequentially, simply to keep the story manageable. Following the initial involution of the center of the vegetal plate, more cells become included in the nascent archenteron. The blastopore narrows as the broad arc of cells bends into an inward arching dome and then a cylinder. As gastrulation proceeds, the archenteron narrows in diameter and elongates. The narrowing of the diameter is primarily a consequence of convergent extension. Endoderm epithelial cells shift in position relative to one another, and in the process, extend the length of the gut. Later, involution of more endoderm cells contributes to archenteron elongation. Finally, SMCs extend filopodia from the tip of the archenteron to the basal lamina inside the animal pole, the filopodia contract and stretch the length of the archenteron to its final length.

How does all of this work? A small number of experiments have offered a few clues, and a growing number of molecules have been identified that participate in this gastrulation process. As described above, the archenteron tube forms from a broad group of cells initially found in the vegetal plate (Ettensohn 1984a,b), and later incorporates additional endoderm cells that become part of the hindgut (Logan and McClay 1997; Martins et al. 1998). If the endomesoderm is appropriately specified prior to invagination, invagination proceeds. Expression of a dominant negative form of Notch blocks SMC specification, but invagination proceeds in the absence of the SMCs, indicating that SMCs do not have a unique property that allows initiation of invagination (Sherwood and McClay 1999). However, the endomesoderm must be specified and endoderm produced. If, for example, one cell at the 2-cell stage is injected with Δ-cadherin to eliminate β-catenin activation of endomesoderm specification (Logan et al. 1999), the unspecified half fails to contribute to the invagination at all, but the control half of the embryo proceeds to invaginate normally, with the mechanical limitation that only 180° of the vegetal plate invaginates. This suggests that the mechanical forces of invagination are local and do not require a global circumferential set of forces for the process to succeed. A different consequence is seen if genes farther downstream of β-catenin are blocked. Blockade of Brachyury function, for example, reveals both an autonomous and a nonautonomous influence on invagination. Injection of *brachy-EnR* RNA into one blastomere at the 2-cell stage is later followed by an apparently normal gastrulation of the entire gut (Gross and McClay 2001). The inhibited half invaginates with the control half-archenteron. Since whole embryos with *brachy-EnR* RNA fail to gastrulate, these data suggest that a molecule downstream of Brachyury signals nonautonomously, and even in the absence of Brachyury, the inhibited side must be able to receive that signal and join into the invagination movements. Support for a signal downstream of Brachyury comes from another approach; a differential screen following altered *brachyury* expression in the endoderm identified a number of genes expressed in the SMCs, even though there is no evidence that *brachyury* is expressed in SMCs (Rast et al. 2002).

Given the experiment above, the cytoskeletal elements downstream of *brachyury* expression are apparently not essential for gastrulation, unless Brachyury function normally operates through the nonautonomous signal for synthesis of the cytoskeletal molecules involved in invagination. The list of cytoskeletal elements included in the *brachyury* pathway is growing, but none of these components has yet been perturbed to determine a potential role in invagination. Blockade of other upstream endoderm specifiers also inhibits this phase of gastrulation, by operating either through *brachyury,* or through blockage of other downstream contributors to morphogenesis (Li et al. 1999; Howard et al. 2001).

At the end of the early involution phase, the archenteron has a diameter that includes about 30 cells in a typical circumference measurement (Ettensohn 1985; Hardin and Cheng 1986). Over the next few hours, the archenteron elongates and the lumen narrows until about 8 cells circumscribe the lumen (Ettensohn 1985). This observation suggested that cells actively rearrange to narrow and elongate the archenteron, a process that was observed first in *Drosophila* (Fristrom and Fristrom 1975; Fristrom 1976), and later called convergent extension (Keller et al. 1985). In the sea urchin, individual cells within labeled patches were directly observed to rearrange by the convergent extension process (Hardin 1989). For the first two-thirds of gastrulation, endoderm cells undergo convergent extension movements to become a column of cells about 2/3 to 3/4 the final length of the archenteron. At a molecular level, adherens junctions are altered during these convergent extension movements through loss of much of the β-catenin in the junctions (Miller and McClay 1997a). Cadherin levels remain unchanged throughout gastrulation to maintain cell–cell adhesions (Miller and McClay 1997b), but the altered structure of the adherens junctions may be permissive for the endoderm epithelial cells to slide past one another, although this hypothesis has not been tested directly. Importantly, for convergent extension to elongate the gut properly, there must be a directional component in the cell movements. Cells in the endodermal sheet must converge laterally rather than along the animal–vegetal axis. That convergence intercalates cells between other cells and elongates the gut. If the movement in the cell sheet lacks directionality, the sliding mechanism would occur randomly with no net change in length of the archenteron. The cues that provide that directional information are not known. Furthermore, the several species of sea urchin that have been examined closely gastrulate in slightly different patterns. *S. purpuratus*, for example, gastrulates on the oral side of the vegetal plate, and *L. variagatus* invaginates in the middle. The extension of the *S. purpuratus* gut goes toward upper reaches of the oral side whereas the gut of *L. variagatus* heads toward the animal pole. To account for these genotypic differences, cells of the vegetal plate are recruited differentially in the two species. Obviously, populations of cells are cued differently, although the nature of those cues is not known.

The final phase of gastrulation involves extension of thick filopodia by SMCs followed by contraction to pull the gut to its final length. Filopodia extend from the tip of the archenteron throughout the invagination process, but laser ablation experiments show that until the final 1/3 to 1/4 of gut length elongation, the filopodial extensions have negligible influence on gut elongation (Hardin 1988). During that final elongation, however, if filopodia are destroyed by laser ablation, the gut fails to reach its normal length (Hardin 1988). Filopodia attach strongly to a target site near the animal pole (Hardin and McClay 1990). If the embryo is inserted into a capillary tube so that the target is out of reach of the filopodia, SMCs at the tip of the gut will continue to extend for hours in an apparently futile effort to reach the target. If, on the other hand, the animal pole is pushed to the top of the archenteron when gastrulation is half completed, the SMC filopodia attach to the target precociously and exhibit an end-of-gastrulation behavior (Hardin and McClay 1990). These data suggest that there is a special patch of cells at the animal pole where the SMC filopodia anchor to provide the anatomically correct endpoint of gastrulation. Recently, a transcription factor, nkx1.2, has been found to be expressed in the target territory and to be involved in the specification of that region of the oral ectoderm (A. Ransick and E.H. Davidson, pers. comm.). This factor could be part of the molecular apparatus that places the target recognition signal at the animal pole, although this hypothesis remains untested. *brachyury* is expressed just orally to the target region, and experiments with expression of *brachy-EnR* indicate that Brachyury may be necessary for specification of the stomodeal fusion region of ectoderm (Gross and McClay 2001).

The opening of the stomodeum is the last event of archenteron formation. This occurs many hours after the gut reaches its final length and involves a poorly understood series of events. What is known is that the stomodeal and gut epithelia prepare for fusion and opening independently of one another. Exogastrulae develop a stomodeal tissue in both the ectoderm and endoderm at the correct time, and ectodermal stomodeum invagination begins without contact with the foregut. Fusion and opening of the foregut and ectoderm tissues, however, requires adhesive contact between the two layers. The stomodeum forms as a result of oral specification (Gross and McClay 2001), and the field of cells competent to form the stomodeum initially is large, but that field of cells decreases in size as development proceeds (Hardin and Armstrong 1997). The fusion apparatus in the foregut develops independently as well, as is seen by the shape assumed by the foregut in exogastrulae. The SMCs move out of the way prior to fusion of foregut with the stomodeum (Gustafson and Wolpert 1963). When the two tissues meet, an unknown fusion mechanism joins the tissues and opens the mouth.

This brief review failed to mention a number of molecules that are involved in gastrulation. Because most of these are part of the GRN for endomesoderm, the mechanism of function of these is at least tentatively understood. Other transcription factors disrupt gastrulation in ways that still are not understood. *gsc*, for example (Angerer et al. 2001), is expressed in micromeres as part of the micromere specification sequence and is expressed in oral ectoderm, presumably as part of that specification, yet, misexpression of *gsc* affects endodermal movements at gastrulation. Others, such as *msx* (Dobias et al. 1997), are present in the endoderm by the blastula stage, but currently have no known role in specification of endomesoderm. There are also a number of molecules that may be involved in motility, which could be important participants in gastrulation, but only anecdotal information is available about these at present. Several examples include a procollagen protease (Huggins and Lennarz 2001), the inhibition of which has a failed-gastrulation phenotype. Earlier, an inhibitor of collagen crosslinking was shown to block gastrulation as well (Wessel and McClay 1987). The procollagen protease is necessary for assembly of collagen filaments. This implies that collagen is an important component of the extracellular matrix during gastrulation, but how the cells doing the morphogenetic movements use this substrate is unknown. A more complete treatment of cell–substrate interactions is a subject dealt with in greater detail elsewhere in this volume (see Chapter 21). The cells of the archenteron also accumulate proteins involved in motility. For example, relatively large quantities of unconventional myosins (Sirotkin et al. 2000), and tubulin (Suprenant et al. 2000), are expressed in the archenteron. Both of these observations suggest explicit motile properties of invaginating cells. Exactly how these proteins function in sea urchin morphogenesis has not been explored in detail, unfortunately. Thus, there are many loose ends and orphan molecules that have yet to be tied into a complete picture of the mechanisms of specification and invagination of the archenteron. The picture should not be perceived as gloomy, however. With the sequencing of the sea urchin genome, the development of new optical and new micro-methods of analysis, a much higher level of understanding of this fascinating embryonic process will soon be realized.

REFERENCES

Adelson D.L. and Humphreys T. 1988. Sea urchin morphogenesis and cell-hyalin adhesion are perturbed by a monoclonal antibody specific for hyalin. *Development* 104: 391–402.

Adelson D.L., Alliegro M.C., and McClay D.R. 1992. On the ultrastructure of hyalin, a cell adhesion protein of the sea urchin embryo extracellular matrix. *J. Cell Biol.* 116: 1283–1289.

Amore G., Yavrouian R.G., Peterson K.J., Ransick A., McClay D.R., and Davidson E. H. 2003. *Spdeadringer*, a sea urchin embryo gene required separately in skeletogenic and oral ectoderm gene regulatory networks. *Dev. Biol.* 261: 55–81.

Angerer L.M. and Angerer R.C. 1999. Regulative development of the sea urchin embryo: Signalling cascades and morphogen gradients. *Semin. Cell Dev. Biol.* 10: 327–334.

———. 2000. Animal-vegetal axis patterning mechanisms in the early sea urchin embryo. *Dev. Biol.* 218: 1–12.

———. 2003. Patterning the sea urchin embryo: Gene regulatory networks, signaling pathways, and cellular interactions. *Curr. Top. Dev. Biol.* 53: 159–198.

Angerer L.M., Oleksyn D.W., Levine A.M., Li X., Klein W.H., and Angerer R.C. 2001. Sea urchin goosecoid function links fate specification along the animal-vegetal and oral-aboral embryonic axes. *Development* 128: 4393–4404.

Anstrom J.A. 1989. Sea urchin primary mesenchyme cells: Ingression occurs independent of microtubules. *Dev. Biol.* 131: 269–275.

———. 1992. Microfilaments, cell shape changes, and the formation of primary mesenchyme in sea urchin embryos. *J. Exp. Zool.* 264: 312–322.

Anstrom J.A., Chin J.E., Leaf D.S., Parks A.L., and Raff R.A. 1987. Localization and expression of msp130, a primary mesenchyme lineage-specific cell surface protein in the sea urchin embryo. *Development* 101: 255–265.

Armstrong N. and McClay D.R. 1994. Skeletal pattern is specified autonomously by the primary mesenchyme cells in sea urchin embryos. *Dev. Biol.* 162: 329–338.

Armstrong N., Hardin J., and McClay D.R. 1993. Cell-cell interactions regulate skeleton formation in the sea urchin embryo. *Development* 119: 833–840.

Burdsal C., Alliegro M.C., and McClay D.R. 1991. Echinonectin as a substrate for adhesion during development of the sea urchin embryo. *Dev. Biol.* 144: 327–334.

Burnside B. 1971. Microtubules and microfilaments in newt neurulation. *Dev. Biol.* 26: 416–441.

Chuang C.K., Wikramanayake A.H., Mao C.A., Li X., and Klein W.H. 1996. Transient appearance of *Strongylocentrotus purpuratus* Otx in micromere nuclei: Cytoplasmic retention of SpOtx possibly mediated through an alpha-actinin interaction. *Dev. Genet.* 19: 231–237.

Citkowicz E. 1971. The hyaline layer: Its isolation and role in echinoderm development. *Dev. Biol.* 24: 348–362.

———. 1972. Analysis of the isolated hyaline layer of sea urchin embryos. *Dev. Biol.* 27: 494–503.

Conlon F.L., Sedgwick S.G., Weston K.M., and Smith J.C. 1996. Inhibition of Xbra transcription activation causes defects in mesodermal patterning and reveals autoregulation of Xbra in dorsal mesoderm. *Development* 122: 2427–2435.

Davidson E.H., Rast J.P., Oliveri P., Ransick A., Calestani C., Yuh C.H., Minokawa T., Amore G., Hinman V., Arenas-Mena C., Otim O.,

Brown C.T., Livi C.B., Lee P.Y., Revilla R., Rust A.G., Pan Z., Schilstra M.J., Clarke P.J., Arnone M.I., Rowen L., Cameron R.A., McClay D.R., Hood L., and Bolouri H. 2002a. A genomic regulatory network for development. *Science* **295:** 1669–1678.

Davidson E.H., Rast J.P., Oliveri P., Ransick A., Calestani C., Yuh C.H., Minokawa T., Amore G., Hinman V., Arenas-Mena C., Otim O., Brown C.T., Livi C.B., Lee P.Y., Revilla R., Schilstra M.J., Clarke P.J., Rust A.G., Pan Z., Arnone M.I., Rowen L., Cameron R.A., McClay D.R., Hood L., and Bolouri H. 2002b. A provisional regulatory gene network for specification of endomesoderm in the sea urchin embryo. *Dev. Biol.* **246:** 162–190.

Davidson L.A., Oster G.F., Keller R.E., and Koehl M.A. 1999. Measurements of mechanical properties of the blastula wall reveal which hypothesized mechanisms of primary invagination are physically plausible in the sea urchin *Strongylocentrotus purpuratus*. *Dev. Biol.* **209:** 221–238.

Dobias S.L., Ma L., Wu H., Bell J.R., and Maxson R. 1997. The evolution of Msx gene function: expression and regulation of a sea urchin Msx class homeobox gene. *Mech. Dev.* **61:** 37–48.

Edwards K.A., Demsky M., Montague R.A., Weymouth N., and Kiehart D.P. 1997. GFP-moesin illuminates actin cytoskeleton dynamics in living tissue and demonstrates cell shape changes during morphogenesis in *Drosophila*. *Dev. Biol.* **191:** 103–117.

Emily-Fenouil F., Ghiglione C., Lhomond G., Lepage T., and Gache C. 1998. GSK3beta/shaggy mediates patterning along the animal-vegetal axis of the sea urchin embryo. *Development* **125:** 2489–2498.

Ettensohn C.A. 1984a. Mechanisms of epithelial invagination. *Q. Rev. Biol.* **60:** 289–307.

———. 1984b. Primary invagination of the vegetal plate during sea urchin gastrulation. *Am. Zool.* **24:** 571–588.

———. 1985. Gastrulation in the sea urchin embryo is accompanied by the rearrangement of invaginating epithelial cells. *Dev. Biol.* **112:** 383–390.

———. 1990. The regulation of primary mesenchyme cell patterning. *Dev. Biol.* **140:** 261–271.

———. 1992. Cell interactions and mesodermal cell fates in the sea urchin embryo. *Dev. Suppl.* **1992:** 43–51.

Ettensohn C.A. and McClay D.R. 1986. The regulation of primary mesenchyme cell migration in the sea urchin embryo: Transplantations of cells and latex beads. *Dev. Biol.* **117:** 380–391.

Ettensohn C.A. and Sweet H.C. 2000. Patterning the early sea urchin embryo. *Curr Top Dev. Biol.* **50:** 1–44.

Ettensohn, C. A., Illies, M. R., Oliveri, P. and De Jong, D. L. 2003. Alx1, a member of the Cart1/Alx3/Alx4 subfamily of Paired-class homeodomain proteins, is an essential component of the gene network controlling skeletogenic fate specification in the sea urchin embryo. *Development* **130:** 2917–2928.

Fink R.D. and McClay D.R. 1985. Three cell recognition changes accompany the ingression of sea urchin primary mesenchyme cells. *Dev. Biol.* **107:** 66–74.

Fristrom D. 1976. The mechanism of evagination of imaginal discs of *Drosophila melanogaster*. III. Evidence for cell rearrangement. *Dev. Biol.* **54:** 163–171.

Fristrom D. and Fristrom J.W. 1975. The mechanism of evagination of imaginal discs of *Drosophila melanogaster*. I. General considerations. *Dev. Biol.* **43:** 1–23.

Fuchikami T., Mitsunaga-Nakatsubo K., Amemiya S., Hosomi T., Watanabe T., Kurokawa D., Kataoka M., Harada Y., Satoh N.,

Kusunoki S., Takata K., Shimotori T., Yamamoto T., Sakamoto N., Shimada H., and Akasaka K. 2002. T-brain homologue (HpTb) is involved in the archenteron induction signals of micromere descendant cells in the sea urchin embryo. *Development* **129:** 5205–5216.

Gross J.M., and McClay D.R. 2001. The role of Brachyury (T) during gastrulation movements in the sea urchin *Lytechinus variegatus*. *Dev. Biol.* **239:** 132–147.

Gross J.M., Peterson R.E., Wu S.-Y., and McClay D.R. 2003. LvTbx2/3, a T-box family transcription factor involved in formation of the oral/aboral axis of the sea urchin embryo. *Development* **130:** 1989–1999.

Guss K.A. and Ettensohn C.A. 1997. Skeletal morphogenesis in the sea urchin embryo: Regulation of primary mesenchyme gene expression and skeletal rod growth by ectoderm-derived cues. Development **124:** 1899–1908.

Gustafson T. and Wolpert L. 1962. Cellular mechanisms in the formation of the sea urchin larva. Change in shape of cell sheets. *Exp. Cell Res.* **27:** 260–279.

———. 1963. Studies on the cellular basis of morphogenesis in the sea urchin embryo. Formation of the coelom, the mouth, and the primary pore-canal. *Exp. Cell Res.* **29:** 561–582.

———. 1967. Cellular movement and contact in sea urchin morphogenesis. *Biol. Rev.* **42:** 442–498.

Hardin J. 1988. The role of secondary mesenchyme cells during sea urchin gastrulation studied by laser ablation. *Development* **103:** 317–324.

———. 1989. Local shifts in position and polarized motility drive cell rearrangement during sea urchin gastrulation. *Dev. Biol.* **136:** 430–445.

———. 1996. The cellular basis of sea urchin gastrulation (review). *Curr. Top. Dev. Biol.* **33:** 159–262.

Hardin J. and Armstrong N. 1997. Short-range cell-cell signals control ectodermal patterning in the oral region of the sea urchin embryo. *Dev. Biol.* **182:** 134–149.

Hardin J.D. and Cheng L.Y. 1986. The mechanisms and mechanics of archenteron elongation during sea urchin gastrulation. *Dev. Biol.* **115:** 490–501.

Hardin J. and McClay D.R. 1990. Target recognition by the archenteron during sea urchin gastrulation. *Dev. Biol.* **142:** 86–102.

Hardin J., Coffman J.A., Black S.D., and McClay D.R. 1992. Commitment along the dorsoventral axis of the sea urchin embryo is altered in response to NiCl2. *Development* **116:** 671–685.

Hertzler P.L. and McClay D.R. 1999. αSU2, an epithelial integrin that binds laminin in the sea urchin embryo. *Dev. Biol.* **207:** 1–13.

Horstadius S. 1973. *Experimental embryology of echinoderms*. Clarendon Press, Oxford, United Kingdom.

Horstadius S. and Wolsky A. 1936. Studien uber die determination der bilateralsymmetrie des jungen seeigelkeimes. *Wilhelm Roux' Arch. Entwicklungsmech. Org.* **135:** 69–113.

Hotary K.B. and Tosney K.W. 1996. Cellular interactions that guide sensory and motor neurites identified in an embryo slice preparation. *Dev. Biol.* **176:** 22–35.

Howard E.W., Newman L.A., Oleksyn D.W., Angerer R.C., and Angerer L.M. 2001. SpKrl: A direct target of β-catenin regulation required for endoderm differentiation in sea urchin embryos. *Development* **128:** 365–375.

Huggins L.G. and Lennarz W.J. 2001. Inhibitors of procollagen C-terminal proteinase block gastrulation and spicule elongation in the sea urchin embryo. *Dev. Growth Differ.* **43:** 415–424.

Illies M.R., Peeler M.T., Dechtiaruk A.M., and Ettensohn C.A. 2002. Identification and developmental expression of new biomineralization proteins in the sea urchin *Strongylocentrotus purpuratus*. *Dev. Genes Evol.* **212:** 419–431.

Katow H. and Solursh M. 1980. Ultrastructure of primary mesenchyme cell ingression in the sea urchin *Lytechinus pictus*. *J. Exp. Zool.* **213:** 231–246.

Keller R.E., Danilchik M., Gimlich R., and Shih J. 1985. The function and mechanism of convergent extension during gastrulation of *Xenopus laevis*. *J. Embryol. Exp. Morph.* (suppl.) **89:** 185–209.

Kenny A.P., Angerer L.M., and Angerer R.C. 2001. SpSoxB1 serves an essential architectural function in the promoter *SpAN*, a *tolloid/BMP1*-related gene. *Gene Expr.* **9:** 283–290.

Kenny A.P., Kozlowski D., Oleksyn D.W., Angerer L.M., and Angerer R.C. 1999. SpSoxB1, a maternally encoded transcription factor asymmetrically distributed among early sea urchin blastomeres. *Development* **126:** 5473–5483.

Kimberly E.L. and Hardin J. 1998. Bottle cells are required for the initiation of primary invagination in the sea urchin embryo. *Dev. Biol.* **204:** 235–250.

Kirchhamer C.V., Bogarad L.D., and Davidson E.H. 1996. Developmental expression of synthetic *cis*-regulatory systems composed of spatial control elements from two different genes. *Proc. Natl. Acad. Sci.* **93:** 13849–13854.

Kitajima T., Tomita M., Killian C.E., Akasaka K., and Wilt F.H. 1996. Expression of spicule matrix protein gene SM30 in embryonic and adult mineralized tissues of sea urchin *Hemicentrotus pulcherrimus*. *Dev. Growth Differ.* **38:** 687–695.

Kurokawa D., Kitajima T., Mitsunaga-Nakatsubo K., Amemiya S., Shimada H., and Akasaka K. 1999. HpEts, an ets-related transcription factor implicated in primary mesenchyme cell differentiation in the sea urchin embryo. *Mech. Dev.* **80:** 41–52.

Kuwada J. 1986. Cell recognition by neuronal growth cones in a simple verterbrate embryo. *Science* **233:** 740–746.

Lane M.C., Koehl M.A., Wilt F., and Keller R. 1993. A role for regulated secretion of apical extracellular matrix during epithelial invagination in the sea urchin. *Development* **117:** 1049–1060.

Leaf D.S., Anstrom J.A., Chin J.E., Harkey M.A., and Raff R.A. 1987. A sea urchin primary mesenchyme cell surface protein, msp130, defined by cDNA probes and antibody to fusion protein. *Dev. Biol.* **121:** 29–40.

Li X., Wikramanayake A.H., and Klein W.H. 1999. Requirement of SpOtx in cell fate decisions in the sea urchin embryo and possible role as a mediator of beta-catenin signaling. *Dev. Biol.* **212:** 425–439.

Logan C.Y. and McClay D.R. 1997. The allocation of early blastomeres to the ectoderm and endoderm is variable in the sea urchin embryo. *Development* **124:** 2213–2223.

Logan C.Y., Miller J.R., Ferkowicz M.J., and McClay D.R. 1999. Nuclear β-catenin is required to specify vegetal cell fates in the sea urchin embryo. *Development* **126:** 345–357.

Malinda K.M. and Ettensohn C.A. 1994. Primary mesenchyme cell migration in the sea urchin embryo: Distribution of directional cues. *Dev. Biol.* **164:** 562–578.

Malinda K.M., Fisher G.W., and Ettensohn C.A. 1995. Four-dimensional microscopic analysis of the filopodial behavior of primary mesenchyme cells during gastrulation in the sea urchin embryo. *Dev. Biol.* **172:** 552–566.

Marsden M. and Burke R.D. 1997. Cloning and characterization of novel beta integrin subunits from a sea urchin. *Dev. Biol.* **181:** 234–245.

———. 1998. The βL integrin subunit is necessary for gastrulation in sea urchin embryos. *Dev. Biol.* **203:** 134–148.

Martins G.G., Summers R.G., and Morrill J.B. 1998. Cells are added to the archenteron during and following secondary invagination in the sea urchin *Lytechinus variegatus*. *Dev. Biol.* **198:** 330–342.

McClay D.R. 1999. The role of thin filopodia in motility and morphogenesis. *Exp. Cell Res.* **253:** 296–301.

McClay D.R. and Logan C.Y. 1996. Regulative capacity of the archenteron during gastrulation in the sea urchin. *Development* **122:** 607–616.

McClay D.R., Alliegro M.C., and Black S.D. 1989. The hyaline layer in early development of the sea urchin: Cell-extracellular matrix interactions. *In Proceedings of the Electron Microscopy Society of America* (ed. G.W. Bailey), pp. 836–837. San Francisco Press, San Francisco, California.

McClay D.R., Armstrong N.A., and Hardin J. 1992. Pattern formation during gastrulation in the sea urchin embryo. *Dev Suppl.* **1992:** 33–41.

Miller J. and McClay D.R. 1997a. Changes in the pattern of adherens junction-associated β-catenin accompany morphogenesis in the sea urchin embryo. *Dev. Biol.* **192:** 310–322.

———. 1997b. Characterization of the role of cadherin in regulating cell adhesion during sea urchin development. *Dev. Biol.* **192:** 323–339.

Miller J., Fraser S.E., and McClay D.R. 1995. Dynamics of thin filopodia during sea urchin gastrulation. *Development* **121:** 2501–2511.

Moore A.R. and Burt A.S. 1939. On the locus and nature of the forces causing gastrulation in the embryos of *Dendraster excentricus*. *J. Exp. Zool.* **82:** 159–171.

Nakajima Y. and Burke R.D. 1996. The initial phase of gastrulation in sea urchins is accompanied by the formation of bottle cells. *Dev. Biol.* **179:** 436–446.

Odell G.M., Oster G., Alberch P., and Burnside B. 1981. The mechanical basis of morphogenesis. I. Epithelial folding and invagination. *Dev. Biol.* **85:** 446–462

Okazaki K. 1975. Spicule formation by isolated micromeres of the sea urchin embryo. *Am. Zool.* **15:** 567–581.

Oliveri P., Carrick D.M., and Davidson E.H. 2002. A regulatory gene network that directs micromere specification in the sea urchin embryo. *Dev. Biol.* **246:** 209–228.

Oliveri P., Davidson E., and McClay D.R. 2003. Activation of *pmar1* controls specification of micromeres in the sea urchin embryo. *Dev Biol.* **258:** 32–43.

Peterson R.E. and McClay D.R. 2003. Primary mesenchyme cell patterning during the early stages following ingression. *Dev. Biol.* **254:** 68–78.

Piston D.W., Summers R.G., Knobel S.M., and Morrill J.B. 1998. Characterization of involution during sea urchin gastrulation using two-photon excited photorelease and confocal microscopy. *Microsc. Microanal.* **4:** 404–414.

Ramirez-Weber F.A. and Kornberg T.B. 1999. Cytonemes: Cellular processes that project to the principal signaling center in *Drosophila* imaginal discs. *Cell* **97:** 599–607.

Ransick A. and Davidson E.H. 1995. Micromeres are required for normal vegetal plate specification in sea urchin embryos. *Development* **121:** 3215–3222.

Rast J.P., Cameron R.A., Poustka A.J., and Davidson E.H. 2002. *brachyury* Target genes in the early sea urchin embryo isolated by differential macroarray screening. *Dev. Biol.* **246:** 191–208.

Rise M. and Burke R.D. 2002. SpADAM, a sea urchin ADAM, has conserved structure and expression. *Mech. Dev.* **117:** 275–281.

Ruffins S.W., and Ettensohn C.A. 1996. A fate map of the vegetal plate of the sea urchin (*Lytechinus variegatus*) mesenchyme blastula. *Development* **122:** 253–263.

Sherwood D.R. and McClay D.R. 1997. Identification and localization of a sea urchin Notch homologue: Insights into vegetal plate regionalization and Notch receptor regulation. *Development* **124:** 3363–3374.

———. 1999. LvNotch signaling mediates secondary mesenchyme specification in the sea urchin embryo. *Development* **126:** 1703–1713.

Sirotkin V., Seipel S., Krendel M., and Bonder E.M. 2000. Characterization of sea urchin unconventional myosins and analysis of their patterns of expression during early embryogenesis. *Mol. Reprod. Dev.* **57:** 111–126.

Starr D.A. and Han M. 2002. Role of ANC-1 in tethering nuclei to the actin cytoskeleton. *Science* **298:** 406–409.

Suprenant K.A., Tuxhorn J.A., Daggett M.A., Ahrens D.P., Hostetler A., Palange J.M., VanWinkle C.E., and Livingston B.T. 2000. Conservation of the WD-repeat, microtubule-binding protein, EMAP, in sea urchins, humans, and the nematode *C. elegans. Dev. Genes Evol.* **210:** 2–10.

Sweet H. C., Gehring M., and Ettensohn C.A. 2002. LvDelta is a mesoderm-inducing signal in the sea urchin embryo and can endow blastomeres with organizer-like properties. *Development* **129:** 1945–1955.

Tosney K.W. 1991. Cells and cell-interactions that guide motor axons in the developing chick embryo. *BioEssays* **13:** 17–23.

Wessel G.M. and McClay D.R. 1985. Sequential expression of germ-layer specific molecules in the sea urchin embryo. *Dev. Biol.* **111:** 451–463.

———. 1987. Gastrulation in the sea urchin embryo requires the deposition of crosslinked collagen within the extracellular matrix. *Dev. Biol.* **121:** 149–165.

Wikramanayake A.H., Huang L., and Klein W.H. 1998. β-Catenin is essential for patterning the maternally specified animal-vegetal axis in the sea urchin embryo. *Proc. Natl. Acad. Sci.* **95:** 9343–9348.

Wilt F.H. 2002. Biomineralization of the spicules of sea urchin embryos. *Zool. Sci.* **19:** 253–261.

Zhu X., Mahairas G., Illies M., Cameron R.A., Davidson E.H., and Ettensohn C.A. 2001. A large-scale analysis of mRNAs expressed by primary mesenchyme cells of the sea urchin embryo. *Development* **128:** 2615–2627.

Zito F., Nakano E., Sciarrino S., and Matranga V. 2000. Regulative specification of ectoderm in skeleton disrupted sea urchin embryos treated with monoclonal antibody to Pl-nectin. *Dev. Growth Differ.* **42:** 499–506.

Zito F., Tesoro V., McClay D.R., Nakano E., and Matranga V. 1998. Ectoderm cell-ECM interaction is essential for sea urchin embryo skeletogenesis. *Dev. Biol.* **196:** 184–192.

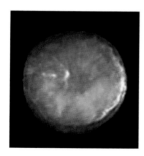

Protochordate Gastrulation
Lancelets and Ascidians

B.J. Swalla

Biology Department and Center for Developmental Biology,
University of Washington, Seattle, Washington 98195

INTRODUCTION

Ascidians and lancelets (*Amphioxus*) are invertebrate chordates, and their embryos are useful for examining questions of how chordate gastrulation evolved (Cameron et al. 2000). Ascidians and lancelets are frequently referred to as protochordates, suggesting that they may resemble ancestral chordates, even though the extant species have evolved along their separate ways since at least the Cambrian Period. Several of the early cell-fate determination mechanisms in ascidian and *Amphioxus* embryos are similar to well-characterized echinoderm embryos, and many of the signaling pathways necessary for vertebrate development are also critical to ascidian and *Amphioxus* gastrulation. However, several features of ascidian embryos make them unique. First, they have a number of maternal factors that specify cell-fate determination directly, and therefore have been known classically as a "mosaic" embryo. In modern terms, this means that they have a number of maternal genes that are necessary for determination already expressed in the cytoplasm as mRNA or protein, likely an evolutionarily successful strategy to facilitate rapid development. The second unique feature of solitary ascidian embryos is not frequently mentioned in papers describing developmental studies; that is, they are non-feeding larvae that do not have a heart, blood, immunity, gut, pharyngeal slits, a mouth, or anus. They are merely a dispersal phase of the ascidian life history. This has important consequences for whether larval cells should be considered "differentiated" or not. The third unique feature of ascidians is that both the dorsal–ventral and anterior–posterior axes are determined before first cleavage by movements of the egg cytoplasm. Although the molecular nature of many of the ascidian gastrulation events is understood, the molecular events underlying axis determination remain unsolved. In this chapter, recent gastrulation studies in ascidian and lancelet embryos are critically reviewed and areas where further research needs to be done are suggested. Since gastrulation is similar in these embryos, I focus specifically on ascidian development and mention comparisons with lancelet, or cephalochordate, embryos when appropriate.

LANCELETS

Lancelets, or Cephalochordates, are small invertebrates that burrow in the sand and resemble in appearance one of the first fossil chordates found in the Cambrian Period, *Pikia* (Briggs et al. 1994). Hence, the term protochordate has been used to describe these animals. Lancelets, also sometimes called *Amphioxus*, have a short and capricious breeding season, hampering attempts at functional experiments until recently when morpholino injection has allowed knockouts of zygotic gene expression (Yu et al. 2002). Lancelet development is elegantly described by Whittaker (1997) and is only briefly reviewed here, in comparison with ascidian gastrulation.

ASCIDIANS

The solitary ascidian egg develops into a non-feeding tad-pole larva that is a relatively simple organism with only a

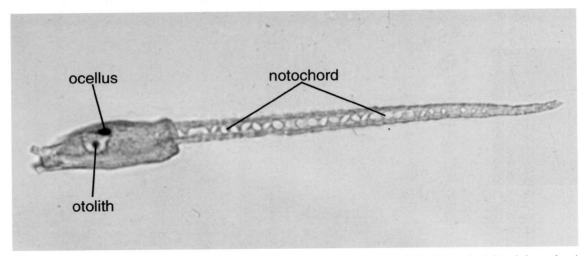

Figure 1. An ascidian tadpole larvae, *Ascidia mentula*, showing the chordate features. The head is to the left and shows the pigmented otolith, a balancing organ, in the center of a cerebral cavity. The ocellus, a light-sensing organ, is located dorsal to the otolith. The palps protruding at the anterior end of the larva aid in attachment of the larva to a substrate at metamorphosis. Forty notochord cells are seen stacked in a single row running down the center of the tail to the right. The CNS is just dorsal to the notochord, and a small endodermal strand is found ventrally.

few cell types (Satoh 1994; Jeffery and Swalla 1997). Colonial ascidians are much larger and complex tadpoles, usually brooded in the colonies and spawned ready to metamorphose (Davidson et al. 2004). In this review, I focus on gastrulation in solitary ascidian embryos, because they have been used extensively for developmental studies, and the genome has been recently sequenced in *Ciona intestinalis* (Dehal et al. 2002) and *Ciona savignyi* (http://www-genome.wi.mit.edu/annotation/ciona/background.html). The head of the solitary larva contains the endoderm, brain, mesenchymal cells, and pigmented sensory cells, and the tail contains a central notochord flanked on each side by three rows of muscle cells, while the epidermis covers the entire larva (Fig. 1) (Katz 1983; Satoh 2003). Ascidians begin gastrulation during the seventh cleavage by invagination at the vegetal pole of the yolky endodermal cells (Conklin 1905; Mancuso 1973; Satoh 1978; Swalla 1993; Nishida 1997). Ascidian gastrulae have many fewer cells than vertebrate embryos, making it easier to identify cell lineages and to follow individual cells (Jeffery 2001). Cleavage is deterministic and invariant in solitary species of ascidians, and the cell lineage is known, allowing direct comparison of developmental studies across species (Conklin 1905; Nishida and Satoh 1983, 1985; Nicol and Meinertzhagen 1988a,b, 1991; Hirano and Nishida 1997, 2000). The following sections first describe ooplasmic segregation, cytoplasmic movements that redistribute maternal factors after fertilization, then review the current cellular and molecular understanding of some of the determinants found in the ascidian

myoplasm, axial determinants and muscle determinants. Next, the process of gastrulation and molecular mechanisms of determining the three germ layers are discussed. The cellular and molecular mechanism(s) of ascidian notochord formation are reviewed in detail, as it has been studied extensively. The development of the central nervous system is reviewed briefly, because it has been reviewed recently elsewhere (Meinertzhagen and Okamura 2001; Lemaire et al. 2002). Finally, a few suggestions for future studies are put forth.

Ascidian Maternal Determinants

Many species of ascidians have colored egg cytoplasms that contain yellow or orange pigments, allowing early embryonic events to be viewed live under a microscope (Fig. 2). Edwin G. Conklin (1905) used *Styela partita*, a species with five distinctly colored cytoplasms, to trace the cell lineage of ascidians at the Marine Biological Laboratory in Woods Hole, MA. Since that time, ascidian embryos are frequently used as an example of a "mosaic" embryo, an embryo where each cell has a particular cell fate that cannot be changed easily after fertilization (Jeffery 2001). In modern terms, we know that this means there are a number of maternal factors in ascidian eggs which are directing cell fate during and following gastrulation. The myoplasm, the unique cortical egg cytoplasm that eventually makes up the muscles of the tail of the ascidian larvae (Fig. 2K,L), contains axial, cleavage, and muscle determinants (Fig. 2A–J) (Bates and Jeffery

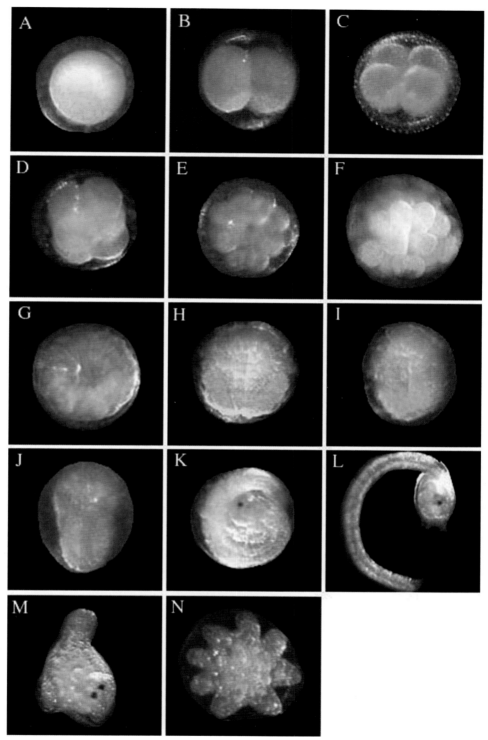

Figure 2. Development in *Boltenia villosa*, an ascidian with orange myoplasm. (*A*) First ooplasmic segregation. The orange myoplasm has moved to the vegetal pole at the bottom of the zygote. (*B*) 2-Cell embryo. First cleavage bisects the myoplasm, which marks the future posterior. (*C*) 4-Cell embryo. (*D*) 8-Cell embryo, with the two vegetal cells containing the myoplasm at the posterior (to the left). (*E*) 16-Cell embryo viewed from the vegetal pole; 4 myoplasm-containing cells mark the posterior pole, at the bottom of the embryo. (*F*) 32-Cell embryo, animal view, 6 blastomeres contain myoplasm. (*G*) Late gastrula embryo, showing the blastopore formed at the vegetal pole. (*H*) Vegetal or dorsal view of an early neurula embryo, focused on the neural plate forming anterior (*top*) and the orange muscle cells developing posterior (*bottom*). (*I*) Later neurula embryo, viewed from the vegetal (dorsal) side, as the blastopore is closing. Anterior is up. (*J*) Early tailbud embryo, with the head at the anterior (*top*), the notochord converging and extending in the center of the tail. The orange muscle cells surround the notochord. (*K*) Late tailbud embryo. The tail has converged and extended and the sensory pigment spots are beginning to form. (*L*) Hatched tadpole larvae, shows the white head and the orange muscle cells in the tail. (*M*) After swimming, the tadpole settles on a substrate, anterior palps down, and resorbs the tail. (*N*) A flattened *Boltenia villosa* larva in the process of metamorphosis viewed from the top of the animal.

1987, 1988; Jeffery and Swalla 1990; Swalla et al. 1991; Jeffery 1995; Nishikata et al. 1999; Roegiers et al. 1999; Nishida and Sawada 2001). The first event of gastrulation in solitary ascidians is the invagination of endodermal cells at (Conklin 1905) or near (Speksnijder et. al. 1990; Roegiers et al. 1999) the vegetal pole of the embryo (Fig. 2E,H). The site of gastrulation is both maternally determined by the animal–vegetal axis and zygotically determined by the point of sperm entry (Speksnijder et al. 1990; Roegiers et al. 1999). In the following section, the process of ooplasmic segregation is reviewed and the nature of the separate myoplasmic determinants is discussed.

Ascidian Ooplasmic Segregation

During the first ooplasmic movement, the cortex of the egg moves to the vegetal pole through a microfilament-driven contraction wave (Fig. 2A) (Zalokar 1974; Sawada and Osanai 1981, 1985; Swalla et al. 1991; Jeffery 1995; Roegiers et al. 1999). The first ooplasmic movement has been called the contraction phase (Roegiers et al. 1999) and is activated by a calcium (Ca^{++}) flux propagated through the egg following fertilization (Sardet et al. 1989; Speksnijder et al. 1989b, 1990; Brownlee and Dale 1990; Roegiers et al. 1999). Recently, it has been shown that the vegetal hemisphere contains more endoplasmic reticulum (ER) than the animal hemisphere, and the ER is the site of Ca^{++} sequestration and subsequent release after fertilization (Sardet et al. 1992; Roegiers et al. 1999). Following fertilization, there is an increase in free Ca^{++} at the point of sperm entry (Speksnijder et al. 1989b, 1990; Brownlee and Dale 1990), followed by a calcium wave that is propagated by the release of calcium stores in the egg (Roegiers et al. 1999). This increase in free Ca^{++} facilitates the contraction of microfilaments in the cortical cytoplasm with the contraction center near the vegetal pole (Sawada and Osanai 1981, 1985; Brownlee and Dale 1990; Jeffery 1995), but offset slightly by moving opposite the site of sperm entry (Roegiers et al. 1999).

The myoplasm contains a unique, complex cytoskeletal domain (Swalla et al. 1991; Jeffery 1995) and is made up of at least three layers (Sardet et al. 1992; Roegiers et al. 1999). The outer layer, directly next to the egg cell membrane, is rich in ER (Sardet et al. 1992; Roegiers et al. 1999). The next layer contains the actin lamina that is the driving force behind the first movements of ooplasmic segregation (Sawada and Osanai 1981, 1985; Jeffery 1995). The third inner, cortical region contains various filaments in which pigment granules, mitochondria, and more ER are embedded (Sardet et al. 1992; Jeffery 1995; Roegiers et al. 1999). The ER is found intercalated throughout the various layers of the myoplasm (Sardet et al. 1992) and the ER that is next to the membrane is critical for the calcium wave oscillations seen during the first cell cycle (Roegiers et al. 1999).

Following the first ooplasmic movement, calcium oscillations are seen as the egg completes its meiotic division and the polar bodies are extruded at the animal pole (Speksnijder et. al. 1989b, 1990; Roegiers et al. 1999). The calcium waves and oscillations cease at the end of the meiotic cycle (Speksnijder et al. 1989b, 1990; Brownlee and Dale 1990; Roegiers et al. 1999). In *Phallusia mammillata*, a vegetal button then forms at the vegetal pole that contains a distinct granule (Roegiers et al. 1999), described as a vegetal body, which contains the germ plasm of the ascidian egg (C. Djediat et al., in prep.).

During the second phase of ooplasmic segregation, the anterior–posterior axis becomes fixed as the myoplasm moves to the future posterior pole of the embryo (Conklin 1905; Jeffery 1995; Nishikata et al. 1999; Roegiers et al. 1999). The second ooplasmic movement consists of the myoplasm being pulled upward in the egg, toward the future posterior end of the embryo by the formation and movement of the sperm aster toward the animal pole (Zalokar 1974; Sawada and Schatten 1988, 1989; Nishikata et al. 1999; Roegiers et al. 1999). It has been shown by chalk-marking experiments that the myoplasm is no longer linked to the egg membrane during second ooplasmic segregation (Bates and Jeffery 1987). Instead, the sperm aster pulls the myoplasmic cytoskeleton to the equator of the zygote and then the aster begins to move toward the animal pole to fuse with the egg pronucleus (Sawada and Schatten 1989; Nishikata et al. 1999; Roegiers et al. 1999). A cortical cytoplasmic structure appears in the vegetal posterior cytoplasm, the centrosomal attracting body (CAB), which defines the posterior of the ascidian embryo by directing asymmetric cell divisions to the posterior pole (Hibino et al. 1998; Nishikata et al. 1999).

Ascidian Axis and Cleavage Determinants

Embryonic deletion experiments (Ortolani 1958; Bates and Jeffery 1987; Nishida 1997) have shown that the dorsal–ventral axis of the ascidian embryo becomes fixed after fertilization, at the time of first ooplasmic segregation (for review, see Bates 1993; Jeffery 1995). Removal of the vegetal pole region at the end of the first ooplasmic movement results in the formation of a radialized embryo that never gastrulates (Bates and Jeffery 1987; Nishida 1997; Hibino et al. 1998; Nishikata et al. 1999). The axial determinants are UV sensitive (Jeffery 1990), but have not yet been definitively identified.

Recent experiments have highlighted the role of the CAB in determining asymmetric divisions in the posterior part of the ascidian embryo (Hibino et al. 1998; Nishikata et al. 1999). The CAB appears shortly after the sperm aster forms during first cleavage and is a vegetal, cortical part of the egg that defines the posterior end (Hibino et al. 1998;

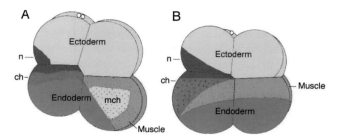

Figure 3. Comparison of the fate maps of ascidian (*A*) and lancelet (*B*) embryos at the 8-cell stage. Ascidians begin gastrulation between the 64- and 128-cell stage and lancelets do not gastrulate until about the 400-cell stage. Both embryos gastrulate by the invagination of the endoderm (*green*) at the vegetal pole. Ectoderm (*yellow*) is located at the animal pole and will cover the embryo by epiboly after the invagination of endoderm, followed by the mesoderm (*red*). In ascidians, the mesenchyme (mch; *pink*) of the larvae will remain quiescent until it forms the mesoderm of the adult after metamorphosis. (n; *blue*) presumptive neural tissue. (ch; *dark red*) presumptive notochord. (Redrawn from Whittaker 1997 and colored to show presumptive cell fates.)

Nishikata et al. 1999). The CAB attracts the centrosome during successive cleavages, causing smaller cells to form on the exact posterior pole of the embryo (Fig. 3) (Hibino et al. 1998). The CAB also sequesters a number of maternal mRNAs during successive cleavages (Yoshida et al. 1996, 1997; Satou and Satoh 1997; Sasakura et al. 1998a,b; Makabe et al. 2001). It is not yet clear when the localized RNAs are translated, or whether the proteins then become localized, but it is intriguing that *Wnt-5* (Sasakura et al. 1998a) and other mRNAs (Yoshida et al. 1996, 1997; Satou and Satoh 1997; Sasakura et al. 1998b) are tightly localized posteriorly, to a region that will eventually end up as the endodermal strand in the tail of the larva. This is also the location of *vasa*, suggesting that the germ cells are situated posteriorly in the embryo (Fujimura and Takamura 2000; Takamura and Fujimura 2002). Although much progress has been made in identifying the nature of maternal mRNAs in ascidian eggs (Makabe et al. 2001), it is still not clear whether one of these maternal messages is responsible for the formation of the dorsal–ventral and anterior–posterior axes in ascidians. In contrast, more progress has been made in identifying muscle determinants.

Ascidian Muscle Determinants

The role of the myoplasm in directing muscle development was first suggested by Conklin in his cell lineage analyses (Conklin 1905). Since then, both surgical and cleavage arrest experiments have suggested that only those cells that contain myoplasm will make muscle cells (Bates and Jeffery 1987, 1988; Jeffery and Swalla 1990; Satoh et al. 1990; Nishida 1997; Nishida and Sawada 2001). Recently, two

separate transcription factors have been identified as maternal messages in myoplasm, and both have been hailed as the long-sought "myoplasmic determinant" that is necessary for muscle determination (Erives and Levine 2000; Nishida and Sawada 2001). The first gene reported is a T-box gene, *CiVegTR* (Erives and Levine 2000). The *CiVegTR* mRNA is localized to the vegetal cytoplasm of fertilized eggs and activates the zygotic minimal *snail* promoter, suggesting it activates early transcription factors necessary for muscle development (Erives and Levine 2000). The other gene is a zinc finger transcription factor, *Macho-1*, that was isolated from two different ascidian species, *Ciona intestinalis* and *Halocynthia roretzi* (Nishida and Sawada 2001). Depletion of maternal *Macho-1* mRNA from the myoplasm with morpholinos inhibits muscle development in eggs, suggesting that the maternal message is normally necessary for muscle determination (Nishida and Sawada 2001). In addition, ectopic expression of *Macho-1* causes expression of muscle myosin, suggesting that this single transcript can induce muscle in blastomeres that would not normally make it (Nishida and Sawada 2001). It is possible that the myoplasm contains multiple muscle determinants and several transcription factors that normally act in concert to induce muscle. Furthermore, it will not be definitively proved that these candidate genes are the determinants until their protein localization and distribution are examined in ascidian eggs and embryos.

Early Gastrulation

Gastrulation begins in ascidian embryos between the 64- and 128-cell stage, whereas lancelet embryos gastrulate later, when the embryo has about 400 cells (Yasui et al. 2002). Gastrulation in both embryos involves three stages. The first stage is the invagination of the endoderm, followed by the involution of the mesoderm, and, finally, the epiboly of the ectoderm (Fig. 3) (Satoh 1978; Swalla 1993; Jeffery and Swalla 1997; Whittaker 1997; DiGregorio and Levine 1998). The site of invagination, called the "gastrulation center" in ascidians (Jeffery 1992), is determined during ooplasmic segregation as the spot where the myoplasm caps at the vegetal pole of the egg, opposite the site of sperm entry where the vegetal button forms (Roegiers et al. 1999). At the 8-cell stage, the fate map of ascidian embryos (Fig. 3A) and *Amphioxus* embryos (Fig. 3B) is quite similar. The main difference in cleavage is that ascidians undergo bilateral cleavage (Fig. 2A–F), whereas lancelet embryos undergo radial cleavage (Whittaker 1997), much like echinoderm embryos. Gastrulation in both ascidian and lancelet embryos begins by invagination of the endodermal cells that contain the vegetal cytoplasm. In ascidians, β-catenin moves into the nucleus of all of the vegetal cells, but especially the endodermal cells, which then change their shape and begin gastrulation (Imai et al. 2000). Sea urchin embryos also begin gastrulation by the movement of β-catenin protein into the nuclei of endodermal cells (Chapter 9), suggesting that this is a developmental character of deuterostome embryos. β-Catenin is necessary to specify endodermal tissue in the ascidian tadpole larvae and to allow the invagination of the endodermal cells toward the animal pole (Imai et al. 2000). Overexpression of β-catenin in ascidian embryos can convert notochord tissue to endoderm, but does not affect muscle development (Imai et al. 2000). These results strengthen those outlined in the previous section, that muscle determination in ascidians is cell autonomous. Conversely, when *Ciona* β-catenin levels are reduced by overexpression of a cadherin, both endoderm and notochord differentiation are repressed (Imai et al. 2000), confirming results that suggest the notochord is induced by FGF signaling from the endoderm (Shimauchi et al. 2001).

It is interesting that, in lancelet embryos, β-catenin moves into the nucleus at about the same stage as ascidians, the 64-cell stage (Yasui et al. 2002). However, the stabilization of β-catenin quickly is seen primarily on one side of the embryo, the dorsal side. When treated with LiCl to perturb β-catenin localization on the dorsal side, the embryos still were able to establish embryonic polarity, suggesting that β-catenin is not the primary dorsalizing factor in lancelet

embryos (Yasui et al. 2002). Invagination in both ascidians and lancelets continues until the endodermal cells contact the animal hemisphere ectodermal cells, and an archenteron is formed at the vegetal pole (Fig. 2G,H). As invagination of the endoderm cells proceeds in ascidians, shape changes in the surrounding cells cause the embryo to become cup-shaped (Fig. 4) (Satoh 1978; Jeffery and Swalla 1997; DiGregorio and Levine 1998). At the completion of invagination in ascidian embryos, the blastopore contains presumptive notochord anteriorly, presumptive muscle posteriorly, and presumptive mesenchyme on its lateral margins (Fig. 4). At this initial stage of gastrulation, all solitary ascidians examined have the same cell lineages located in the same places in the embryo, which simplifies comparisons between species (Conklin 1905; Nishida and Satoh 1983, 1985; Nicol and Meinertzhagen 1988a,b, 1991; Jeffery and Swalla 1997; Hirano and Nishida 1997, 2000; Satoh 2003).

Notochord

Cephalochordates, or lancelets, retain their notochord throughout their adult life, which is spent with the animal partially buried, filter-feeding in sand. When disturbed, the adult is a fairly strong swimmer. Notochord development in lancelets depends on expression of *Brachyury* (*T*), similar to

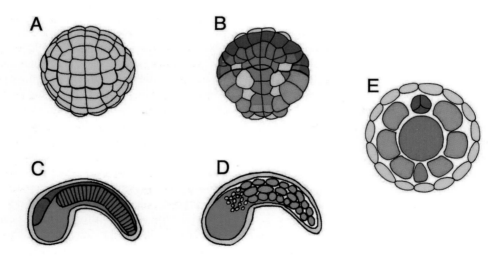

Figure 4. (*Top row*) An animal (*left*) and vegetal (*right*) view of an early gastrulating ascidian embryo, showing the different cell lineages. (*A*) Ectoderm is covering the animal pole (*yellow*). (*B*) Viewed from the vegetal pole, neural tissue is blue, muscle cells are red, notochord is dark red, and endoderm is green. (*Bottom row*) Tailbud embryos showing the various fates of internal cells in the larva. (*C*) A longitudinal section of the larva shows the notochord cells (*red*) stacked down the center of the tail, and the CNS (*blue*) is dorsal to the notochord. (*D*) On either side of the notochord, the muscle cells (*red*) of the tail surround the notochord and the mesenchyme (*pink*) will make up the mesoderm of the adult. (*E*) To the lower right is a cross-section of an ascidian larva tail, showing the central notochord (*dark red*), muscle cells on either side of the notochord (*red*), the dorsal neural tube (*blue*), and the endodermal strand (*green*). (Redrawn from Nishida 1992 and colored to show presumptive cell fates.)

ascidians and vertebrates (Holland et al. 1995). In contrast, lancelet notochords have also been called "myochords" due to the myofilaments that were seen by microscopy, and the fact that myosin could be extracted from the notochord (Flood et al. 1969). Recent EST analyses suggest that about 11% of the transcripts made during lancelet notochord development are muscle-specific genes (Suzuki and Satoh 2000). Even more recently, it has been shown that lancelets express two different MyoD genes and other muscle-specific genes in the notochord during development (Urano et al. 2003). The significance of these results is not yet well understood, but it is clear that the lancelet notochord has evolved a musculature that is not seen in other chordates, either in ascidians or in vertebrates.

All solitary ascidian embryos studied have a notochord composed of 40 cells, stacked in a row down the center of the tail (Figs. 1 and 4). These 40 cells come from two sources in ascidian embryos. The primary notochord cells, 8 cells in the anterior part of the embryo just posterior to the neural cell precursors (Fig. 4), are induced to become notochord just before gastrulation. Then, two secondary cells are added during the next division (Fig. 4). These 10 cells are easily seen in the 64-cell embryo by examining the expression of *Brachyury*, the earliest notochord-specific gene that is expressed in ascidian (Yasuo and Satoh 1998). The 10 precursor cells undergo one division after gastrulation begins and then undergo a second round of division to make 40 cells right before convergence and extension. This precise pattern of notochord precursor development and later cell movements has allowed a detailed understanding of the cellular and molecular events underlying notochord formation in solitary ascidians.

Before gastrulation, at the 32-cell stage, the presumptive notochord cells are induced to become notochord by FGF signals from the underlying endoderm (Kim et al. 2000; Minokawa et al. 2001). Mesoderm is also induced by FGF signals from the endoderm in vertebrate embryos (Chapters 32 and 25); however, in vertebrates it seems that Nodal is an equally important player. Without FGF, the notochord cells will adopt a neural fate (Minokawa et al. 2001; see also next section). During this period, cell division and intercalation of cells result in a strip of presumptive notochord underlying the neural tube (Figs. 1 and 2). During tail formation, precursor cells located on either side of the embryo intercalate to form a single row of cells (Fig. 4) (Cloney 1964; Miyamoto and Crowther 1985; Munro and Odell 2002a,b). A typical solitary tailed ascidian tadpole larva has 40 notochord cells located in the center of the tail (Figs. 1 and 4) (Conklin 1905; Nishida and Satoh 1983, 1985; Jeffery and Swalla 1997). Most of the notochord cells (32) come from a pair of anterior vegetal cells, which also give rise to endoderm, spinal cord, brain, and a few muscle cells in the posterior tail. Eight notochord cells at the tip of the tail arise

from a posterior vegetal cell pair, which also contribute to larval muscle cells. Time-lapse studies suggest that both radial and mediolateral intercalation of prospective notochord cells occur simultaneously (Miyamoto and Crowther 1985), causing the sheet of cells to get thinner and longer, via convergent extension.

In a fascinating series of studies, Ed Munro and Gary Odell have characterized the cellular forces generated during notochord formation by using time-lapse, confocal analysis, and experimental embryology (Munro and Odell 2002a,b). They show that notochord formation begins by the cells interacting in the center of the embryo following the last notochord division to make 40 cells (Munro and Odell 2002a). The notochord cells appear to be differentially adhesive to each other, and they begin to form processes and to move toward the center of the notochord cell mass (Munro and Odell 2002a). When notochord cells are isolated from the intact embryo, they do not converge and extend in isolation unless they are wrapped in a tissue of a different kind (Munro and Odell 2002b). Surprisingly, the source of the surrounding tissue was not important; instead, it appears that the cells need to be surrounded by a different tissue to be able to detect the center of the notochordal mass of cells (Munro and Odell 2002a). Once surrounded by other tissue, the notochord cells took a variety of different ways of converging and extending into a single sheet of cells, showing that this cell behavior is surprisingly plastic, yet robust in the final outcome (Munro and Odell 2002a,b).

Brachyury is found exclusively in the notochord cells in ascidian larvae (Yasuo and Satoh 1998), whereas in lancelets, it is expressed early in all mesoderm, then restricted to the notochord (Holland et al. 1995). Surprisingly, *Brachyury* is expressed in notochord lineage cells in the tail-less ascidians, *Molgula occulta* and *Molgula tectiformis*, which never make a notochord (Takada et al. 2002). Several potential targets of *Brachyury* in ascidian notochords have been identified by differential screening after overexpression of *Brachyury* and subsequent in situ hybridization in *Ciona intestinalis* (Hotta et al. 2000). Most of the *Brachyury* downstream genes are expressed later in notochord development, just before and during convergence and extension (Hotta et al. 2000). It will be interesting to compare expression of downstream genes in experimental conditions where notochord is induced to develop by other tissue types in culture (Miyamoto and Crowther 1985; Munro and Odell 2002a,b). Alternatively, knockdown experiments with antisense RNAs (Swalla and Jeffery 1996) or with morpholinos (Satou et al. 2001) of individual genes may allow insight into one of the most interesting questions about morphology: How do single genes act within a cell to create complex morphology? In this case, how do notochord cells discriminate other notochord cells from surrounding tissue?

Neurulation and CNS Development in Ascidians

Neurulation, which includes the formation, meeting, and subsequent closure of the neural folds, begins during the ninth cleavage in ascidian embryos (Nishida 1986; Nicol and Meinertzhagen 1988a,b). The ascidian embryo begins elongation during neural plate formation (Fig. 2). The mesoderm on the surrounding margins of the blastopore involutes into the blastocoel (Fig. 2G–I), as the ectoderm undergoes epiboly, surrounding the entire embryo and covering the blastopore. While ectoderm epiboly is occurring externally, the neural plate on the dorsal side of the embryo first thickens, forming the neural plate, then the neural folds appear and roll toward each other, meet, and form the neural tube (Fig. 2I). After completion of neurulation, the early tailbud embryo is formed, with a distinct head and tail (Fig. 2J).

Neural cell fate is partly cell-autonomous and partly induced by neighboring cells in ascidian embryos (Meinertzhagen and Okamura 2001; Lemaire et al. 2002). The molecular mechanisms that specify neural cell fate have been reviewed elsewhere (Lemaire et al. 2002) and are summarized briefly here as part of the morphological events that are occurring during neurulation. The larval CNS is derived from 10 cells of the 64-cell embryo, 5 on each side of the plane of bilateral symmetry (Fig. 3) (Meinertzhagen and Okamura 2001; Lemaire et al. 2002). Because ascidians have invariant cleavage, it has been possible to document the entire cell lineage of the larval central nervous system (CNS) in an elegantly detailed series of papers by Nicol and Meinertzhagen (1988a,b, 1991). The CNS in ascidian larvae is quite simple, consisting of only about 355 cells that mostly make up the larval brain (Meinertzhagen and Okamura 2001). There are about 100 neurons and 250 glial cells (Meinertzhagen and Okamura 2001). About 65 neurons are found in the nerve chord of the tail and 40–50 in the neurohypophysis, located in the larval head, which will elaborate the adult nervous system after metamorphosis (Nicol and Meinertzhagen 1991).

Two distinct changes in the neural plate occur during neurulation: the rolling up of the neural tube from the neural folds, which occurs dorsally, and the lengthening of the neural tube, which occurs along the anterior–posterior axis. The closing of the neural tube occurs in a posterior-to-anterior sequence and is likely to result from cell-shape changes (Mancuso 1973; Nicol and Meinertzhagen 1988a,b), whereas the anterior–posterior lengthening of the neural tube appears to be a result of cell rearrangements and oriented cell division (Nicol and Meinertzhagen 1988a,b). From the formation of the neural fold to closure of the neural tube, individual cells undergo a shape change from columnar to wedge-shaped. Microtubules form parallel to the long axis of the cells (Mancuso 1973), but microfilament bundles have not been observed at the apical surface of neural plate cells (Mancuso 1973), as has been reported in vertebrate embryos (Colas and Schoenwolf 2001). Lengthening of the neural tube along the anterior–posterior axis is due to intercalation of cells into neighboring cells at the midline of the embryo (Nicol and Meinertzhagen 1988b).

In ascidians, the anterior part of the neural plate will form the dorsoanterior epidermis, which includes the anterior palps, head sensory neurons, and the pharynx (Lemaire et al. 2002), all structures that will be required during and after metamorphosis (Jeffery and Swalla 1997). Regionalization in the neural tube is seen both along the anterior–posterior and dorsoventral axes, using ascidian homologs of genes that have been used as markers of CNS patterning in vertebrates (Meinertzhagen and Okamura 2001; Imai et al. 2002; Lemaire et al. 2002). Basically, ascidian larvae appear to have a single-celled floor plate and also show anterior-to-posterior expression of genes that are expressed in a similar manner in vertebrates (Imai et al. 2002). However, there are also obvious differences, as the ascidian CNS is greatly reduced in cell number and complexity (Meinertzhagen and Okamura 2001; Lemaire et al. 2002; Satoh 2003). There are some neurons of the CNS that appear to be autonomously specified (Okada et al. 1997; Nagahora et al. 2000), but most of the rest of the neurons are induced by FGF or vegetal blastomeres (Wada et al. 1999; Ohtsuka et al. 2001). Since FGF is also necessary to specify notochord (see preceding section), it is not clear whether FGF is acting directly or indirectly on the neural tissue from the 16-cell to the 64-cell stage (Darras and Nishida 2001; see Chapter 29).

The final ascidian larval CNS consists of about 355 cells, most of which are in the head of the larvae, and the tail has only a small CNS composed of three cells in cross section (Fig. 4). The CNS is important for larval swimming and undergoes programmed changes in the expression of ion channels and in electrical activity during development (Meinertzhagen and Okamura 2001). The CNS may also be important in transferring the signals for competence and settlement of the larvae to begin metamorphosis (Kimura et al. 2003). Most of the larval CNS undergoes apoptosis before metamorphosis, and the adult CNS comes from a small part of the larval brain (Meinertzhagen and Okamura 2001).

Future Prospects

Ascidian embryos undergo gastrulation and neurulation early during embryonic development, allowing examination of how early cleavage events affect later morphogenetic processes. Ascidians and lancelets have very similar fate maps at the 8-cell stage, and early gastrulation is quite similar. Both have larvae that develop quickly, but the ascidian embryo undergoes invariant cleavage, the cell lineage is known, and the ascidian larva contains only a few cell types, facilitating experimental manipulations and analysis. The ascidian genome has been sequenced, and much is known about the molecules that are expressed during different developmental stages. However, there has been little done to understand the basis of morphogenetic movements at gastrulation, and we know little about protein expression patterns of maternal transcription factors. The notochord is currently the best-understood larval tissue mechanistically, but much more needs to be done to understand the morphogenetic mechanisms of neural tube development and elaboration of the gut after metamorphosis. Finally, ascidians provide the opportunity to explore different developmental modes and the evolution of species with different life histories (Swalla and Jeffrey 1990, 1996; Huber et al. 2000; Davidson et al. 2004). Ascidian species that have anural, or tailless, larval development, offer the unique opportunity to investigate how developmental changes can profoundly affect morphology of the individual, thereby affecting the evolution of the species (Swalla and Jeffery 1990, 1996; Takada et al. 2002). Colonial ascidians may undergo gastrulation by different strategies because of their varying size and number of cells (Davidson et al. 2004). This is an untapped resource of variability in ascidian embryos and colonial ascidians may be the best embryos to examine for vestiges of neural crest and somites, two embryonic structures that are missing in solitary ascidian tadpole larvae when compared to other chordate embryos, such as lancelets and vertebrates.

REFERENCES

Bates W.R. 1993. Evolutionary modifications of morphogenetic mechanisms and alternate life history strategies in ascidians. *Microsc. Res. Tech.* **26:** 285–300.

Bates W.R. and Jeffery W.R. 1987. Localization of axial determinants in the vegetal pole region of ascidian eggs. *Dev. Biol.* **124:** 65–76.

———. 1988. Polarization of ooplasmic segregation and dorsal-ventral axis determination in ascidian embryos. *Dev. Biol.* **130:** 98–107.

Briggs D.E.G., Erwin D.H., and Collier F.J. 1994. In *The fossils of the Burgess Shale.* Smithsonian Institution Press, Washington, D.C.

Brownlee C. and Dale B. 1990. Temporal and spatial correlation of fertilization current, calcium waves, and cytoplasmic contraction in eggs of *Ciona intestinalis. Proc. R. Soc. Lond. Biol. Sci.* **239:** 321–328.

Cameron C.B., Garey J.R., and Swalla B.J. 2000. Evolution of the chordate body plan: New insights from phylogenetic analyses of deuterostome phyla. *Proc. Natl. Acad. Sci.* **97:** 4469–4474.

Cloney R.A. 1964. Development of the ascidian notochord. *Acta Embryol. Morphol. Exp.* **7:** 111–130.

Colas J.F. and Schoenwolf G.C. 2001. Towards a cellular and molecular understanding of neurulation. *Dev. Dyn.* **221:** 117–145.

Conklin E.G. 1905. The organization and cell lineage of the ascidian egg. *J. Acad. Nat. Sci. Phil.* **13:** 1–119.

Darras S. and Nishida H. 2001. The BMP/Chordin antagonism controls sensory pigment cell specification and differentiation in the ascidian embryo. *Dev. Biol.* **236:** 271–288.

Davidson B., Jacobs M.J., and Swalla B. J. 2004. The individual as a module: Metazoan evolution and coloniality. In *Modularity in development and evolution* (ed. G. Schlosser and G. Wagner). University of Chicago Press, Illinois.

Dehal P., Satou Y., Campbell R.K., Chapman J., Degnan B., De Tomaso A., Davidson B., Di Gregorio A., Gelpke M., Goodstein D.M., Harafuji N., Hastings K.E., Ho I., Hotta K., Huang W., Kawashima T., Lemaire P., Martinez D., Meinertzhagen I.A., Necula S., Nonaka M., Putnam N., Rash S., Saiga H., Satake M., Terry A., Yamada L., Wang H.G., Awazu S., Azumi K., Boore J., Branno M., Chin-Bow S., DeSantis R., Doyle S., Francino P., Keys D.N., Haga S., Hayashi H, Hino K., Imai K.S., Inaba K., Kano S., Kobayashi K., Kobayashi M., Lee B.I., Makabe K.W., Manohar C., Matassi G., Medina M., Mochizuki Y., Mount S., Morishita T., Miura S., Nakayama A., Nishizaka S., Nomoto H., Ohta F., Oishi K., Rigoutsos I., Sano M., Sasaki A., Sasakura Y., Shoguchi E., Shin-i T., Spagnuolo A., Stainier D., Suzuki M.M., Tassy O., Takatori N., Tokuoka M., Yagi K., Yoshizaki F., Wada S., Zhang C., Hyatt P.D., Larimer F., Detter C., Doggett N., Glavina T., Hawkins T., Richardson P., Lucas S., Kohara Y., Levine M., Satoh N., and Rokhsar D.S. 2002. The draft genome of *Ciona intestinalis*: Insights into chordate and vertebrate origins. *Science* **298:** 2157–2167.

DiGregorio A. and Levine M. 1998. Ascidian embryogenesis and the origins of the chordate body plan. *Curr. Opin. Gen. Dev.* **8:** 457–463.

Erives A. and Levine M. 2000. Characterization of a maternal *T-Box* gene in *Ciona intestinalis. Dev Biol.* **225:**169–178.

Flood P.R., Guthrie D.M., and Banks J.R. 1969. Paramyosin muscle in the notochord of *Amphioxus. Nature* **222:** 87–88.

Fujimura M. and Takamura K. 2000. Characterization of an ascidian DEAD-box gene, Ci-DEAD1: Specific expression in the germ cells and its mRNA localization in the posterior-most blastomeres in early embryos. *Dev. Genes Evol.* **210:** 64–72.

Hibino T., Nishikata T., and Nishida H. 1998. Centrosome-attracting body: A novel structure closely related to unequal cleavages in the ascidian embryo. *Dev. Growth Differ.* **40:** 85–95.

Hirano T. and Nishida H. 1997. Developmental fates of larval tissues after metamorphosis in ascidian *Halocynthia roretzi*. I. Origin of mesodermal tissues of the juvenile. *Dev. Biol.* **192:** 199–210.

———. 2000. Developmental fates of larval tissues after metamorphosis in the ascidian, *Halocynthia roretzi*. II. Origin of endodermal tissues of the juvenile. *Dev. Genes Evol.* **210:** 55–63.

Holland P.W.H., Koschorz B., Holland L.Z., and Herrmann B.G. 1995. Conservation of *Brachyury (T)* genes in *Amphioxus* and vertebrates: Developmental and evolutionary implications. *Development* **121:** 4283–4291.

Hotta K., Takahashi H., Asakura T., Saitoh, B., Takatori N., Satou Y., and Satoh N. 2000. Characterization of Brachyury-downstream notochord genes in the *Ciona intestinalis* embryo. *Dev. Biol.* **224:** 69–80.

Huber J.L., Burke da Silva K., Bates W.R., and Swalla B.J. 2000. The evolution of anural larvae in molgulid ascidians. *Semin. Dev. Biol.* **11:** 419–426.

Imai K.S., Satoh N., and Satou Y. 2002. Region specific gene expressions in the central nervous system of the ascidian embryo. *Gene Expr. Patterns* **2:**319–321.

Imai K.S., Takada N., Satoh N., and Satou Y. 2000. β-catenin mediates the specification of endoderm cells in ascidian embryos. *Development* 127: 3009–3020.

Jeffery W.R. 1990. Ultraviolet irradiation during ooplasmic segregation prevents gastrulation, sensory cell induction, and axis formation in the ascidian embryo. *Dev. Biol.* 140: 388–400.

———. 1992. A gastrulation center in the ascidian egg. *Development Suppl.* 53–63.

———. 1995. Development and evolution of an egg cytoskeletal domain in ascidians. *Curr. Topics Dev. Biol.* 31: 243–276.

———. 2001. Determinants of cell and positional fate in ascidian embryos. *Int. Rev. Cytol.* 203: 3–63.

Jeffery W.R. and Swalla B.J. 1990. The myoplasm of ascidian eggs: A localized cytoskeletal domain with multiple roles in embryonic development. *Semin. Cell Biol.* 1: 373–381.

———. 1997. Embryology of the tunicates. In *Embryology: Constructing the organism* (ed. S. Gilbert and A.M. Raunio), pp. 331–364. Sinauer, Sunderland, Massachusetts.

Katz M.J. 1983. Comparative anatomy of the tunicate tadpole, *Ciona intestinalis*. *Biol. Bull.* 164: 1–27.

Kim G.J., Yamada A., and Nishida H. 2000. An FGF signal from endoderm and localized factors in the posterior-vegetal egg cytoplasm pattern the mesodermal tissues in the ascidian embryo. *Development* 127: 2853–2862.

Kimura Y., Yoshida M., and Morisawa M. 2003. Interaction between noradrenaline or adrenaline and the beta 1-adrenergic receptor in the nervous system triggers early metamorphosis of larvae in the ascidian, *Ciona savignyi*. *Dev. Biol.* 258: 129–140.

Lemaire P., Bertrand V., and Hudson C. 2002. Early steps in the formation of neural tissue in ascidian embryos. *Dev. Biol.* 252: 151–169.

Makabe K.W., Kawashima T., Kawashima S., Minokawa T., Adachi A., Kawamura H., Ishikawa H., Yasuda R., Yamamoto H., Kondoh K., Arioka S., Sasakura Y., Kobayashi A., Yagi K., Shojima K., Kondoh Y., Kido S., Tsujinami M., Nishimura N., Takahashi M., Nakamura T., Kanehisa M., Ogasawara M., Nishikata T., and Nishida H. 2001. Large-scale cDNA analysis of the maternal genetic information in the egg of *Halocynthia roretzi* for a gene expression catalog of ascidian development. *Development* 128: 2555–2567.

Mancuso V. 1973. Ultrastructural changes in the *Ciona intestinalis* egg during the stage of gastrula and neurula. *Arch. Biol.* 84: 181–204.

Meinertzhagen I.A. and Okamura Y. 2001. The larval ascidian nervous system: The chordate brain from its small beginnings. *Trends Neurosci.* 24: 401–410.

Minokawa, T., Yagi K., Makabe K.W., and Nishida H. 2001. Binary specification of nerve cord and notochord cell fates in ascidian embryos. *Development* 129: 1–12.

Miyamoto D.M. and Crowther R.J. 1985. Formation of the notochord in living ascidian embryos. *J. Embryol. Exp. Morphol.* 86: 1–17.

Munro E.M. and Odell G. 2002a. Morphogenetic pattern formation during ascidian notochord formation is regulative and highly robust. *Development* 129: 1–12.

———. 2002b. Polarized basolateral cell motility underlies invagination and convergent extension of the ascidian notochord. *Development* 129: 13–24.

Nagahora H., Okada T., Yahagi N., Chong J.A., Mandel G., and Okamura Y. 2000. Diversity of voltage-gated sodium channels in the ascidian larval nervous system. *Biochem. Biophys. Res. Commun.* 275: 558–564.

Nicol D. and Meinertzhagen I.A. 1988a. Development of the central nervous system of the larva of the ascidian, *Ciona intestinalis* L. I. The early lineages of the neural plate. *Dev. Biol.* 130: 721–736.

———. 1988b. Development of the central nervous system of the larva of the ascidian, *Ciona intestinalis* L. II. Neural plate morphogenesis and cell lineages during neurulation. *Dev. Biol.* 130: 737–766.

———. 1991. Cell counts and maps in the larval central nervous system of the ascidian *Ciona intestinalis* L. *J. Comp. Neurol.* 309: 415–429.

Nishida H. 1986. Cell division pattern during gastrulation of the ascidian, *Halocynthia roretzi*. *Dev. Growth Differ.* 28: 191–201.

———. 1992. Determination of developmental fates of blastomeres in ascidian embryos. *Dev. Growth Differ.* 34: 253–262.

———. 1997. Cell fate specification by localized cytoplasmic determinants and cell interactions in ascidian embryos. *Int. Rev. Cytol.* 176: 245–306.

Nishida H. and Satoh N. 1983. Cell lineage analysis in ascidian embryos by intracellular injection of a tracer enzyme. I. Up to the eight-cell stage. *Dev. Biol.* 99: 382–394.

———. 1985. Cell lineage analysis in ascidian embryos by intracellular injection of a tracer enzyme. II. The 16- and 32-cell stages. *Dev. Biol.* 110: 440–454.

Nishida H. and Sawada K. 2001. *Macho-1* encodes a localized mRNA in ascidian eggs that specifies muscle fate during embryogenesis. *Nature* 409: 724–729.

Nishikata T., Hibino T., and Nishida H. 1999. The centrosome-attracting body, micrtubule system, and posterior egg cytoplasm are involved in positioning of cleavage planes in the ascidian embryo. *Dev. Biol.* 209: 75–85.

Ohtsuka Y., Obinata T., and Okamura Y. 2001. Induction of ascidian peripheral neuron by vegetal blastomeres. *Dev. Biol.* 239: 107–117.

Okada T., Hirano H., Takahashi K., and Okamuras Y. 1997. Distinct neuronal lineages of the ascidian embryo revealed by expression of a sodium channel gene. *Dev. Biol.* 190: 257–272.

Ortolani G. 1958. Cleavage and development of egg fragments in ascidians. *Acta Embryol. Morphol. Exp.* 1: 247–272.

Roegiers F., Djediat C., Dumollard R., Rouviére C., and Sardet C. 1999. Phases of cytoplasmic and cortical reorganizations of the ascidian zygote between fertilization and first division. *Development* 126: 3101–3117.

Sardet C., Speksnijder J.E., Inoué S., and Jaffe L. 1989. Fertilization and ooplasmic movements in the ascidian egg. *Development* 105: 237–249.

Sardet C., Speksnijder J.E., Terasaki M., and Chang P. 1992. Polarity of the ascidian egg cortex before fertilization. *Development* 115: 221–237.

Sasakura Y., Ogasawara M., and Makabe K.W. 1998a. *Hr-Wnt-5*: A maternally expressed ascidian *Wnt* gene with posterior localization in early embryos. *Int. J. Dev. Biol.* 42: 573–580.

———. 1998b. Maternally localized RNA encoding a serine/threonine protein kinase in the ascidian *Halocynthia roretzi*. *Mech. Dev.* 76: 161–163.

Satoh N. 1978. Cellular morphology and architecture during early morphogenesis of the ascidian egg: An SEM study. *Biol. Bull.* 155: 608–614.

———. 1994. *Developmental biology of ascidians*. Cambridge University Press, New York.

———. 2003. The ascidian tadpole larva: Comparative molecular development and genomics. *Nat. Rev. Genet.* 4: 285–295.

Satoh N., Deno T., Nishida H., Nishikata T., and Makabe K.W. 1990.

Cellular and molecular mechanisms of muscle cell differentiation in ascidian embryos. *Int. Rev. Cytol.* **122:** 221–258.

Satou Y. and Satoh N. 1997. *Posterior end mark 2* (*pem-2*), *pem-4*, *pem-5* and *pem-6*: Maternal genes with a localized mRNA in the ascidian embryo. *Dev. Biol.* **192:** 467–481.

Satou Y., Imai K.S., and Satoh N. 2001. Action of morpholinos in *Ciona* embryos. *Genesis* **30:** 103–106.

Sawada T. and Osanai K. 1981. The cortical contraction related to ooplasmic segregation in *intestinalis* eggs. *Wilhelm Roux's Arch. Dev. Biol.* **190:** 208–214.

———. 1985. Distribution of actin filaments in fertilized eggs of the ascidian *intestinalis*. *Dev. Biol.* **111:** 260–265.

Sawada T. and Schatten G. 1988. Microtubules in ascidian eggs during meiosis, fertilization, and mitosis. *Cell Motil. Cytoskel.* **9:** 219–230.

———. 1989. Effects of cytoskeletal inhibitors on ooplasmic segregation and microtubule organization during fertilization and early development of the ascidian *Molgula occidentalis*. *Dev. Biol.* **132:** 331–342.

Shimauchi Y., Murakami S. D. and Satoh N. 2001. FGF signals are involved in the differentiation of notochord cells and mesenchyme cells of the ascidian *Halocynthia roretzi*. *Development* **128:** 2711–2721.

Speksnijder J.E., Jaffe L.F., and Sardet C. 1989a. Polarity of sperm entry in the ascidian egg. *Dev. Biol.* **133:** 180–184.

Speksnijder J.E., Corson D.W., Sardet C., and Jaffe L.F. 1989b. Free calcium pulses following fertilization in the ascidian egg. *Dev. Biol.* **135:** 182–190.

Speksnijder J.E., Sardet C., and Jaffe L.F. 1990. The activation wave of calcium in the ascidian egg and its role in ooplasmic segregation. *J. Cell Biol.* **110:** 1589–1598.

Suzuki M.M. and Satoh N. 2000. Genes expressed in the *Amphioxus* notochord revealed by EST analysis. *Dev. Biol.* **224:** 168–177.

Swalla B.J. 1993. Mechanisms of gastrulation and tail formation in ascidians. *Microsc. Res. Tech.* **26:** 274–284.

Swalla B.J. and Jeffery W.R. 1990. Interspecific hybridization between an anural and urodele ascidian: Differential expression of urodele features suggests multiple mechanisms control anural development. *Dev. Biol.* **142:** 319–334.

———. 1996. Requirement of the *manx* gene for expression of chordate features in a tailless ascidian larva. *Science* **274:** 1205–1209.

Swalla B.J., Badgett M.R., and Jeffery W.R. 1991. Identification of a cytoskeletal protein localized in the myoplasm of ascidian eggs: Localization is modified during anural development. *Development* **111:** 425–436.

Takada N., York J., Davis J.M., Schumpert B., Yasuo H., Satoh N., and Swalla B.J. 2002. *Brachyury* expression in tailless molgulid ascidian embryos. *Dev. Evol.* **4:** 205–211.

Takamura K. and Fujimura M. 2002. Primordial germ cells originate from the endodermal strand cells in the ascidian *Ciona intestinalis*. *Dev. Genes Evol.* **212:** 11–18.

Urano A., Suzuki M.M., Zhang P., Satoh N., and Satoh G. 2003. Expression of muscle-related genes and two MyoD genes during *Amphioxus* notochord development. *Evol. Dev.* **5:** 447–458.

Wada S. Katsuyama Y., and Saiga H. 1999. Anteroposterior patterning of the epidermis by inductive influences from the vegetal hemisphere cells in the ascidian embryo. *Development* **126:** 4955–4963.

Whittaker J.R. 1997. Cephalochordates, the lancelets. In *Embryology: Constructing the organism* (ed. S. Gilbert and A.M. Raunio), pp. 331–364. Sinauer, Sunderland, Massachusetts.

Yasui K., Guorong L., Wang Y., Saiga H., Zhang P., and Aizawa S. 2002. β-catenin in early development of the lancelet embryo indicates specific determination of embryonic polarity. *Dev. Growth Differ.* **44:** 467–475.

Yasuo H., and Satoh N. 1998. Conservation of the developmental role of *Brachyury* in notochord formation in a Urochordata, the ascidian *Halocynthia roretzi*. *Dev. Biol.* **200:** 158–170.

Yoshida S., Marikawa Y., and Satoh N. 1996. *Posterior end mark*, a novel maternal gene encoding a localized factor in the ascidian embryo. *Development* **122:** 2005–2012.

Yoshida S., Satou Y., and Satoh N. 1997. Maternal genes with localized mRNA and pattern formation in the ascidian embryo. *Cold Spring Harbor Symp. Quant. Biol.* **62:** 89–96.

Yu J.K., Holland L.Z., and Holland N.D. 2002. An *Amphioxus* nodal gene (AmphiNodal) with early symmetrical expression in the organizer and mesoderm and later asymmetrical expression associated with left-right axis formation. *Evol. Dev.* **4:** 418–425.

Zalokar M. 1974. Effect of colchicine and cytochalasin B on ooplasmic segregation in ascidian eggs. *Wilhelm Roux's Arch. Dev. Biol.* **175:** 243–248.

Gastrulation in a Chondrichthyan, the Dogfish *Scyliorhinus canicula*

T. Sauka-Spengler, J.-L. Plouhinec, and S. Mazan

UMR 8080, Université Paris-Sud, Orsay, France

INTRODUCTION

Even though the early development of Chondrichthyans (or cartilaginous fishes) had been the subject of a number of reports by the end of the 19th and the beginning of the 20th century, this taxon has been almost completely neglected by embryologists over the past 50 years. However, the study of Chondrichthyans, which, as a sister group of Osteichthyans, occupy a key phylogenetic position, could be important from an evolutionary standpoint. In particular, together with the study of lampreys, it will be important to identify the genetic and cellular interactions that are conserved among all vertebrates, and thus define a "gastrulation Bauplan" characterizing this phylum. Such an analysis is a prerequisite to understanding how gastrulation mechanisms evolved at the transition between cephalochordates and vertebrates, and possibly diversified among the latter.

THE DOGFISH *SCYLIORHINUS CANICULA* AS A CHONDRICHTHYAN MODEL ORGANISM

Very few studies of gastrulation have been performed in Chondrichthyans, and almost all of them have focused on the spotted dogfish *S. canicula*, a species that is abundantly present in both the Mediterranean Sea and the Atlantic Ocean. This species is well-suited to comparative analyses of gastrulation for several reasons (Ballard et al. 1993). First, fertilization is internal in the dogfish, but eggs are laid

at early stages of development, before the formation of the blastocoel. Once laid, they can go on developing normally in the laboratory, simply in oxygenated seawater. A second advantage is the size and accessibility of the embryo. Just as in the chick, the egg undergoes a meroblastic pattern of cleavage, and by the onset of gastrulation, the blastoderm, lying on a large yolk mass, reaches 4–5 mm. Finally, a detailed developmental table (Ballard et al. 1993), which we refer to here, is available for this species, and five relatively well-characterized stages can be distinguished between the onset of gastrulation (stage 11) and the beginning of neurulation (stage 15).

MORPHOLOGICAL ASPECTS OF GASTRULATION IN THE DOGFISH

In the dogfish, the clearest indication of the anterior–posterior axis becomes visible at stage 10, with the formation of a thickening at the presumptive posterior margin of the blastoderm. At this stage, two cell populations can be distinguished at the posterior part of the blastoderm: an external cell layer, which is about 2–3 cell diameters thick, and which we refer to as the epiblast, and an inner mesenchymal cell population, whose mode of formation and later fates are unclear and which we call hypoblast (see Fig. 2A,A′) (Vandebroek 1936). Most authors consider that gastrulation begins at stage 11, when an overhang consisting of two epithelial layers becomes visible over a 60° sector of the posterior margin of the blastoderm (Figs. 1A and 2B,B′). At

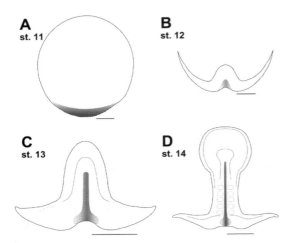

Figure 1. Dorsal views of dogfish embryos during gastrulation. *A, B, C,* and *D* correspond to stages 11, 12, 13, and 14, respectively. The overhang in *A* (stage 11) is shown shaded. Stage 12 (*B*) is characterized by the morphological appearance of the embryonic axis. The axis extends caudally at later stages. At stage 13, the notochord becomes morphologically distinct (with its location shaded in *C*), and the first somites appear at stage 14, also characterized by the head enlargement, visible at the rostral end of the embryo (*D*). Anterior is to the top in all views. Bar, 1 mm.

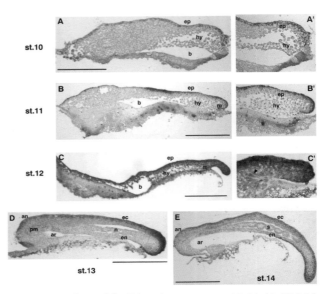

Figure 2. Histology of dogfish embryos at stages 10–14. (*A–D*) Midsagittal sections; (*E*) para-sagittal section. *A′, B′,* and *C′* show magnifications at the level of the posterior margin at stages 10, 11, and 12, respectively. Anterior is to the left. A black arrowhead points to cells showing a bottle-like morphology. (an) Anterior neuroectoderm; (a) archenteron; (b) blastocoel; (ec) ectoderm; (en) endoderm; (ep) epiblast; (hy) hypoblast; (m) mesendoderm; (n) notochord; (pm) prechordal mesoderm; (s) somites. Bar, 1 mm.

this stage, a new, more compact mesenchymal cell population arises at the posterior margin. According to Vandebroek (1936), this layer consists both of presumptive extraembryonic tissues and of embryonic mesendoderm. In the absence of more accurate data, we refer to it as mesendoderm (Fig. 2B,C). Stage 12 is characterized by the caudal extension of the overhang, which results in the formation of the embryonic axis (Figs. 1B and 2C,C′). This extension, which takes place in a strict rostral to caudal progression, continues during subsequent stages. The embryonic axis and well-defined germ layers become visible at stage 13 (Figs. 1C and 2D), and stage 14 is characterized by the formation of the first somites and the appearance of an enlargement characteristic for the head region (Figs. 1D and 2E).

FATE MAPS AND MORPHOGENETIC MOVEMENTS

Only two lineage analyses of the dogfish early gastrula have been carried out, both more than 50 years ago (Vandebroek 1936; Kopsch 1950), and only one of these has led to fate maps of the early stages (Vandebroek 1936). Despite the relative inaccuracy inherent to the technique used (vital dye labeling of broad territories, with relatively short culture times), this analysis provides the first clues as to the morphogenetic movements that take place during gastrulation in the dogfish. Starting from stage 11, two distinct phases are

shown by this study. During early gastrulation (stages 11–12), the blastoderm expands over the yolk by epiboly, with an area of minimal cell displacements located roughly at its center (Fig. 3A,A′). These cell movements do not display radial symmetry, and epiblast cells located in the posterior half of the blastoderm tend to converge to the posterior margin. Epiblast cells later contributing to mesendoderm also start internalizing, over a 60° sector of the posterior margin (Fig. 3A). This internalization, which may involve both involution and ingression movements, results in the formation of the overhang with its bilayered structure. After internalization, the direction of cell movements radically changes, cells in the lower layer being displaced both anteriorly and toward the midline (Fig. 3A′). Another phase of gastrulation can be observed starting from stage 12 (Fig. 3B,B′). During this period, cells located laterally converge posteriorly to the midline, which results in an apparent rostral extension of the axis (Fig. 3B). In parallel, cell internalizations continue at the posterior margin, now excluding the midline (Fig. 3B′). These movements of convergence and extension are maintained up to the beginning of neurulation, by stage 15. These labeling analyses also led Vandebroek to propose fate maps of gastrulating dogfish embryos. According to these data, in late blastula embryos (stage 10, Fig. 3C), the presumptive neuroectoderm is contained in a broad half-circle territory in a central area of the

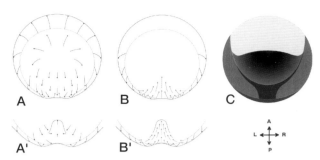

Figure 3. Cell movements during the dogfish gastrulation and fate map at stage 10. *A-B* and *A′-B′* depict cell movements in the upper epiblast layer (*black arrows*) and in the lower hypoblast and mesendoderm layer (*red arrows*), respectively. *A* and *A′* show cell displacements between stages 11 and 12; *B* and *B′* show cell displacements between stages 12 and 13. Dots show the shape of the embryo at the initial stage (11 in *A-A′*, 12 in *B-B′*). In *C*, the presumptive neuroectoderm is in blue (darker in anteriormost regions), the presumptive paraxial, intermediate, and lateral mesoderm is in red, the axial mesoderm is in dark red and the endoderm is in green. (Redrawn from Vandebroek 1936.)

blastoderm. The presumptive endoderm appears as a thin crescent located over a 60° sector at the posterior margin of the blastoderm. The presumptive mesoderm is located between the neuroectoderm and endoderm presumptive territories, with the prechordal plate and notochord located on either side of the midline and extending laterally along the presumptive neuroectoderm.

This pioneering study should be confirmed and refined using modern single cell labeling techniques, as well as longer embryo culture times. As in other vertebrates, the use of such techniques could reveal much more extensive cell mixing than is suggested by the sharp borders shown in this first study. It nevertheless provides an initial idea of the distribution of fates and probable cell movements in Chondrichthyans, which can be compared with the results obtained in Osteichthyans (see Chapters 12 and 20). This comparison leads to three main observations. First, even though the site of mesoderm internalization differs morphologically between the dogfish (posterior marginal zone) and Osteichthyan model organisms, the cell movements described by Vandebroek are clearly reminiscent of those described in the latter. Second, the fate map reported by this author shows striking similarities with those for the zebrafish (Kimmel et al. 1990) and *Xenopus* (Dale and Slack 1987), with very similar relative positions of the presumptive neuroectoderm, axial mesoderm, paraxial and lateral mesoderm, and gut. Finally, several chronological aspects may also be conserved in the dogfish. As in other vertebrates (Warga and Kimmel 1990; Lawson et al. 1991; Psychoyos and Stern 1996), presumptive endoderm cells, which are located closer to the margin, tend to internalize earlier than cells later contributing to mesoderm. As reported in the chick (Psychoyos and Stern 1996), there is also a

relationship between the timing of mesendoderm cell internalization and the position of their progeny along the anterior–posterior axis. For instance, the progeny of mesendoderm cells internalized by stages 12+ to 13 are distributed along the whole embryonic axis caudal to the head enlargement, and the rostral limit of this domain is displaced posteriorly at progressively later stages.

MOLECULAR CHARACTERIZATION OF THE DOGFISH GASTRULA

In the absence of recent cell labeling or experimental embryology studies, another way to characterize dogfish embryos is to use molecular markers that display conserved expression characteristics among Osteichthyans or even chordates. We have used five such markers: *Brachyury*, a marker of nascent mesendoderm (Wilkinson et al. 1990; Smith et al. 1991; Schulte-Merker et al. 1992; Holland et al. 1995; Kispert et al. 1995; see Chapter 41); *Lim1* (see Chapter 43) and *HNF-3β* (see Chapter 40), which are expressed in the organizer (Taira et al. 1992; Ruiz i Altaba et al. 1993; Strähle et al. 1993; Shawlot and Behringer 1995; Toyama et al. 1995; Chapman et al. 2002); and the two paralogous *Otx* forms, *Otx2* and *Otx5*, an Otx expression being present in the early organizer and presumptive anterior neuroectoderm in all vertebrates studied thus far (Simeone et al. 1993; Bally-Cuif et al. 1995; Blitz and Cho 1995; Pannese et al. 1995; see Chapter 39). Three main conclusions can be drawn from this analysis (Sauka-Spengler et al. 2003; Sauka-Spengler and Mazan, in prep.). First, the dogfish *Brachyury* and *Otx2* expression patterns support several aspects of Vandebroek fate maps and description of cell movements. As in zebrafish (Schulte-Merker et al. 1992), *Brachyury* in the dogfish is initially transcribed along the whole circumference of the blastoderm (stage 12, Fig. 4A). However, although in the zebrafish this marginal signal persists throughout gastrulation, the transcripts regress to a posterior sector of the margin of the dogfish gastrula at stage 13 and remain restricted throughout gastrulation to this location, where the tailbud, still expressing *Brachyury,* later forms (Fig. 4B,C). This expression pattern suggests that, as proposed by Vandebroek, presumptive endoderm and mesoderm cells are initially displaced laterally toward the posterior margin, where they are internalized. Similarly, as in Osteichthyans, the dogfish *Otx2* gene is transcribed in anterior-most parts of the neuroectoderm at stages 13–14 (Fig. 4O,P), and this fully confirms the rostral to caudal pattern of axis extension proposed by Vandebroek. At earlier stages (11–12), it is also expressed over broad epiblast territories (Fig. 4K,L), which as expected, largely overlap those depicted as presumptive neuroectoderm in Vandebroek fate maps, thus supporting the location of this prospective territory.

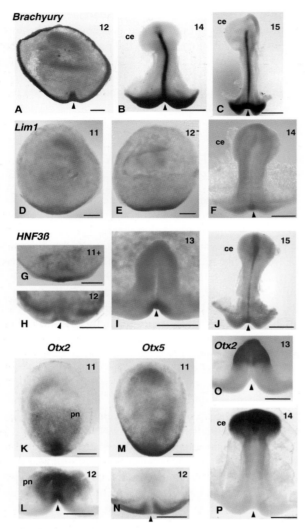

Figure 4. Molecular characterization of the dogfish gastrula. Whole-mount hybridizations of dogfish embryos, using *Brachyury* (A–C), *Lim1* (D–F), HNF-3β (G–J), Otx2 (K,L,O,P), and Otx5 (M,N) probes, Dorsal views. Only the posterior part of the blastoderm is shown in G, H, L, and N. An arrowhead points to a group of cells, which exhibit organizer genetic properties. The stages are indicated, anterior is to the top. (pn) Presumptive neuroectoderm; (ce) cephalic enlargement.

A second outcome of the molecular characterization is the identification of putative organizer regions in the dogfish gastrula, using *Lim1* and HNF-3β probes. As in the mouse, *Xenopus,* or zebrafish, HNF-3β in the dogfish is transcribed in the floor plate by the end of gastrulation (Fig. 4J). In addition, at earlier stages, it shares with *Lim1* a highly specific expression territory located in the midline, at the posterior margin of the blastoderm, where the left and right arms of the extending embryonic axis fuse (Fig. 4F,H,I). This suggests that this labeled cell population may correspond to the organizer in the dogfish gastrula. This hypothesis is also sup-

ported by the expression of *Otx2,* another marker of the early organizer in Osteichthyans, at a very similar location at stage 12, but not later (Fig. 4L,O). This further suggests that in Chondrichthyans, as in Osteichthyans and other vertebrates (see Chapter 29), the genetic and inducing properties of the organizer change during gastrulation, with only early organizers being able to induce, or stabilize, anterior neuroectoderm cell fates. However, one difference between *Lim1* and HNF-3β expression patterns in Osteichthyans and in the dogfish is the persistence in the former of the signal in newly internalized mesendoderm, anterior to the organizer (Fig. 4F,I). Whether this reflects a functional change in these genes or differences in the chronology of mesoderm and endoderm differentiation, which in the dogfish may be restricted to a very narrow time window, remains to be assessed.

The last point, which arises from this molecular characterization, concerns the early induction of the organizer. It is noticeable that the two organizer markers studied in the dogfish (HNF-3β, *Lim1*) as well as an *Otx* paralog, *Otx5,* share a narrow crescent-like expression domain lining the posterior margin of the blastoderm at very early stages (10–11), preceding the onset of *Brachyury* expression (first observed at stage 12). This marginal expression is reminiscent of the early expression observed for another node marker in the chick (*goosecoid:* Izpisùa-Belmonte et al. 1993; see Chapter 42). It suggests that some genetic mechanisms which play a key role in organizer induction and maintenance during gastrulation in other vertebrates (Harland and Gerhart 1997; Joubin and Stern 2001) may also be active in the dogfish as early as stage 10, at the level of the posterior margin. In this hypothesis, this territory could correspond to the *Xenopus* Nieuwkoop center or its putative equivalent in the chick or zebrafish (posterior marginal zone and yolk syncytial layer, respectively; Bachvarova et al. 1998; Koos and Ho 1998). It should also be noted that, at least in the dogfish, the marginal expression of *Lim1* and HNF-3β markedly fades in intensity during a transient time window (stage 12+). Whether this could be related to the rise of different cell populations at the margin, such as the chick hypoblast and endoblast, remains an open question that will have to be addressed by more accurate lineage analyses and the use of additional genetic markers.

CONCLUSION

Taken as a whole, these data suggest that despite extensively divergent morphologies, important genetic and cellular aspects of gastrulation may be conserved between Chondrichthyans and Osteichthyans. In line with this conclusion, the gastrulation pattern that emerges in the dogfish is strikingly similar to the one inferred by Arendt and Nübler-Jung (1999) for a hypothetical amniote ancestor, despite the evolutionary distance between amniotes and cartilaginous fishes. Important issues include determining how far the conservation applies among vertebrates, and addressing the species- or taxa-specific features, which account for adaptive modalities, like the presence of yolk or extraembryonic tissues.

REFERENCES

Arendt D. and Nübler-Jung K. 1999. Rearranging gastrulation in yolk-rich amniotes eggs. *Mech. Dev.* **81:** 3–22.

Bachvarova R.F., Skromme I., and Stern C.D. 1998. Induction of primitive streak and Hensen's node by the posterior marginal zone in the early chick embryo. *Development* **125:** 3521–3534.

Ballard W.W., Mellinger J., and Léchenault H.A. 1993. A series of normal stages for development of *Scyliorhinus canicula*, the lesser spotted dogfish (Chondrichthyes: Scyliorhinidae*). J. Exp. Zool.* **267:** 318–336.

Bally-Cuif L., Gulisano M., Broccoli V., and Boncinelli E. 1995. c-otx2 is expressed in two different phases of gastrulation and is sensitive to retinoic acid treatment in chick embryo. *Mech. Dev.* **49:** 49–63.

Blitz I.L. and Cho K.W. 1995. Anterior neurectoderm is progressively induced during gastrulation: The role of the *Xenopus* homeobox gene orthodenticle. *Development* **121:** 993–1004.

Chapman S. C., Schubert F.R., Schoenwolf G.C., and Lumsden A. 2002. Analysis of spatial and temporal gene expression patterns in blastula and gastrula stage chick embryos. *Dev. Biol.* **245:** 187–199.

Dale L. and Slack J.M.W. 1987. Fate map for the 32-cell stage of *Xenopus laevis. Development* **99:** 527–551.

Harland R. and Gerhart J. 1997. Formation and function of Spemann's organizer. *Annu. Rev. Cell Dev. Biol.* **13:** 611–667.

Holland P.W.H., Koschorz B., Holland L.Z., and Herrmann B.G. 1995. Conservation of *Brachyury* (*T*) genes in amphioxus and vertebrates: Developmental and evolutionary implications. *Development* **121:** 4283–4291.

Izpisua-Belmonte J.C., De Robertis E.M., Storey K.G., and Stern C.D. 1993. The homeobox gene goosecoid and the origin of organizer cells in the early chick blastoderm. *Cell* **74:** 645–659.

Joubin K. and Stern C.D. 2001. Formation and maintenance of the organizer among the vertebrates. *Int. J. Dev. Biol.* **45:** 165–175.

Kimmel C.B., Warga R.M., and Schilling T.F. 1990. Origin and organization of the zebrafish fate map. *Development* **108:** 581–594.

Kispert A., Ortner H., Cooke J., and Herrmann B.G. 1995. The chick *Brachyury* gene: Developmental expression pattern and response to axial induction by localized activin. *Dev. Biol.* **168:** 406–415.

Koos D.S. and Ho R.K. 1998. The *nieuwkoid* gene characterizes and mediates a nieuwkoop-center-like activity in the zebrafish. *Curr. Biol.* **8:** 1199–1206.

Kopsch F. 1950. Bildung und Längenswachstum des Embryons, Gastrulation und Konkreszenz bei *Scyllium canicula* und *Scyllium*

catulus. Z. Mikrosk.-Anat. Forsch. **56:** 1–101.

Lawson K.A., Meneses J.J., and Pedersen R.A. 1991. Clonal analysis of epiblast fate during germ layer formation in the mouse embryo. *Development* **113:** 891–911.

Pannese M., Polo C., Andreazzoli M., Vignali R., Kablar B., Barsacchi G., and Boncinelli E. 1995. The *Xenopus* homologue of *Otx2* is a maternal homeobox gene that demarcates and specifies anterior body regions. *Development* **121:** 707–720.

Psychoyos D. and Stern C.D. 1996. Restoration of the organizer after radical ablation of Hensen's node and the anterior primitive streak in the chick embryo. *Development* **122:** 3263–3273.

Ruiz i Altaba A., Prezioso V.R., Darnell J.E., and Jessel T.M. 1993. Sequential expression of *HNF-3α* and *HNF-3β* by embryonic organizing centers: the dorsal lip/node, notochord, and floor plate. *Mech. Dev.* **44:** 91–108.

Sauka-Spengler T., Baratte B., Lepage M., and Mazan S. 2003. Characterization of *Brachyury* genes in the dogfish *S. canicula* and the lamprey *L. fluviatilis*. Insights into gastrulation in a chondrichthyan. *Dev. Biol.* **263:** 296–307.

Schulte-Merker S., Ho R.K., Herrmann B.G., and Nüsslein-Volhard C. 1992. The protein product of the zebrafish homologue of the mouse T gene is expressed in nuclei of the germ ring and the notochord of the early embryo. *Development* **116:** 1021–1032.

Shawlot W. and Behringer R.R. 1995. Requirement for *Lim1* in head-organizer function. *Nature* **374:** 425–430.

Simeone A., Acampora D., Mallamaci A., Stornaiuolo A., D'Apice M.R., Nigro V., and Boncinelli E. 1993. A vertebrate gene related to orthodenticle contains a homeodomain of the bicoid class and demarcates anterior neuroectoderm in the gastrulating mouse embryo. *EMBO J.* **12:** 2735–2747.

Smith J.C., Proce B.M.J., Green J.B.A., Weigel D., and Herrmann B.G. 1991. Expression of a *Xenopus* homolog of *Brachyury* (*T*) is an immediate-early response to mesoderm induction. *Cell* **67:** 79–87.

Strähle U., Blader P., Henrique D., and Ingham P.W. 1993. *Axial*, a zebrafish gene expressed along the developing body axis, shows altered expression in *cyclops* mutant embryos. *Genes Dev.* **7:** 1436–1446.

Taira M., Jamrich M., Good P.J., and Dawid I.B. 1992. The LIM domain-containing homeo box gene *Xlim-1* is expressed specifically in the organizer region of *Xenopus* gastrula embryos. *Gene Dev.* **6:** 356–366.

Toyama R., O'Connell M.L., Wright C.V., Kuehn M.R., and Dawid I.B. 1995. Nodal induces ectopic goosecoid and lim1 expression and axis duplication in zebrafish. *Development* **121:** 383–391.

Vandebroek G. 1936. Les mouvements morphogénétiques au cours de la gastrulation chez *Scyllium canicula*. *Arch. Biol. (Bruxelles)* **47:** 499–582.

Warga R.M. and Kimmel C.B. 1990. Cell movements during epiboly and gastrulation in zebrafish. *Development* **108:** 569–580.

Wilkinson D.G., Bhatt S., and Herrmann B.G. 1990. Expression pattern of the mouse *T* gene and its role in mesoderm formation. *Nature* **343:** 657–659.

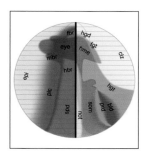

CHAPTER 12

TELEOST GASTRULATION

D.A. Kane and R.M. Warga

Department of Biology, University of Rochester, Rochester, New York 14627

INTRODUCTION

The teleost system is especially well suited for the study of gastrulation, because genetic control and cell movement can be studied together in the same system. Although still incomplete, a good description of gastrulation in the zebrafish is emerging; the aim of this chapter is to chronicle these recent advances in the context of the early normal development of the zebrafish.

In the following sections we describe teleost development from the early cleavage stages to the mid-segmentation stages, dividing this time into periods that are based on the predominant morphological movement. For each period, we first explain morphology and morphogenesis. Then we summarize what is known of the underlying developmental biology. Last, we summarize the mutants that are known to have phenotypes which are expressed during the period. To aid in this last aim, we include Table 1, which summarizes the mutants and the time of appearance of their phenotypes.

CLEAVAGE: CELL DIVISION

At fertilization the egg is a 500- to 600-μm sphere, with the chorion still closely apposed to the cell membrane. Unlike the crystalline zebrafish embryos seen just 30 minutes later, the zygote is a translucent mixture of yolk and cytoplasm, and is not visually clear. At the presumptive animal pole, there is a small cavity, the vestige of the germinal vesicle, and in it, the maternal nucleus can sometimes be seen with careful use of Nomarski DIC optics.

The biology of the first cell cycle was exploited by Streisinger and coworkers to produce parthenogenetic diploids from the maternal pronucleus. A common genetic manipulation in zebrafish is to fertilize eggs stripped from an adult female with inactivated sperm, which would normally yield haploid embryos (Streisinger et al. 1981). If high hydrostatic pressure is applied to the eggs within 6 minutes of fertilization, the nucleus of the second polar body fails to escape from the egg and, instead, is incorporated into the zygote, producing diploid progeny. Since the female pronucleus and the second polar body share one set of sister chromatids, that is, a half tetrad, recombination events are uncovered whenever loci are heterozygous. Thus, the distance from any given locus to the centromere can be estimated by the frequency of heterozygous individuals in a clutch of embryos (Streisinger et al. 1986).

Commencing at the first mitosis, the embryo begins synchronous and rapid divisions into blastomeres. The separation of yolk platelets from the cytoplasm, the so-called "bipolar segregation" (Roosen-Runge 1938), clears the embryo, giving the zebrafish embryo its crystalline appearance. Throughout this period, the embryo is staged by the number of cells in the embryo, from the 2-cell stage at approximately 30 minutes of development until the 64-cell stage at 2 hours of development.

The cell cycle is relatively simple during cleavage, being evenly divided between mitosis and interphase. This serves as an excellent demonstration for students, allowing them to watch the nuclei disappear after a 7-minute interphase, the cells round for mitosis and divide, and the nucleus reform 7 or 8 minutes later. The early cleavages are incomplete, or meroblastic. Each furrow begins at the animal pole of the dividing cell and stops at the condensed body of yolk platelets. Thus, until the 8-cell stage, all of the blastomeres are cytoplasmically continuous with the yolk; there is neither a cell membrane nor any other obvious cellular feature separating the yolk from the forming blastomeres.

157

Table 1. *Mutants that affect gastrulation movements in zebrafish*

Locus	Allele	Gene	Onset	Process
squint (*sqt*)	cz35	Zebrafish Nodal-Related2	blastula/early gastrula	specfication of mesoderm and endoderm
cyclops (*cyc*)	b16, b213, b229, tf219, te262e, m101, m294	Zebrafish Nodal-Related1	blastula/early gastrula	specfication of dorsal mesoderm, endoderm, and neuroectoderm
one-eyed pinhead (*oep*)	tz257, m134	EGF-CFC	blastula/early gastrula	cell motility, specification of mesoderm and endoderm
spadetail (*spt*)	b104, tm41, tq5, m423,	Tbx16	blastula/early gastrula	dorsal compaction, specification of lateral mesoderm and endoderm
no tail (*ntl*)	b160, b195, tb244e, tc41, ts260, m147, m550	Brachyury	blastula/early gastrula	specification of notochord specification of mesoderm?
floating head (flh)	n1, tk241, tm299, m614	Znot		specification of notochord
bozozok (*boz*)	m168	Dharma/Neiuwkoid	blastula/early gastrula	specification of dorsal
dino (*din*)	tm84, tt350	Chordin	blastula/early gastrula	specification of ventral
ogon (ogo), mercedes, short tail	tz209, tm305, m60	Sizzled	blastula/early gastrula	specification of ventral
swirl (*swl*)	ta72, tc300	Bmp2b	blastula/early gastrula	specification of ventral
snailhouse (*snh*)	ty68a	Bmp7	blastula/early gastrula	specification of ventral
somitabun (*sbn*)	dtc24	Smad5	blastula/early gastrula	specification of ventral
lost-a-fin (*lof*)	tm110b m100	Activin receptor-like kinase8	blastula/early gastrula	specification of ventral
half-baked (hab), weg, lawine, avalanche	dtv43, tx230, ts18, tm94	E-cadherin	mid-gastrula	epiboly, convergence of ectoderm
knypek (*kny*)	m119	Heparin sulfate proteoglycan	late gastrula	convergence of ectoderm and possibly mesoderm
trilobite (*tri*)	m144, m209, m747, m778	Strabismus	late gastrula	convergence of ectoderm and mesoderm
silberblick (*slb*)	tx226, tz216	Wnt11	late gastrula	convergence of mesoderm
pipetail (*pip*)	te1c,ta98, th278, ti265,	Wnt5	late gastrula	convergence of posterior mesoderm

Although the mechanism for the segregation of the yolk from the cytoplasm is unclear, it is a very useful feature, for the yolk is the route of choice for early injections into the entire blastoderm.

The cleavage patterns of these early meroblastic divisions are extremely stereotypic. The cleavage furrow of the second cleavage division aligns at 90° to the first to form a 2 × 2 array of cells at the animal pole. The furrows of the third cleavage align parallel to the plane of the first, to form a 2 × 4 array of blastomeres at the 8-cell stage. The furrows of the fourth and fifth cleavages repeat this pattern, to form a 4 × 4 array at the 16-cell stage and then a 4 × 8 array of blastomeres at the 32-cell stage. A scattering of equatorial divisions occurs at the 32- to 64-cell stage; however, which divisions are equatorial cannot be predicted.

No fate map exists for teleosts during the cleavage stages, such as has been accomplished for *Xenopus* at the 32-cell stage (Dale and Slack 1987; Moody 1987). In zebrafish, the cleavage and early blastula fate maps have produced only broad probabilistic maps (Kimmel and Law 1985b; Strehlow and Gilbert 1993). For example, although an animal pole cell marked at the 64-cell stage contributes primarily to the forebrain and eyes, it contributes to trunk muscle as well (Kimmel and Law 1985b; Kimmel and Warga 1987). This anterior–posterior spread of cell *fates* results from the mixing of cell *positions* during stages subsequent to cleavage, mainly during the radial intercalations of early epiboly, a process very much more dramatic in the fish than in the frog (Wetts and Fraser 1989; Warga and Kimmel 1990; Wilson et al. 1995). Hence, there is no information in terms of cell restrictions or cell fate that would lead one to suspect that cell determination has yet begun. In this respect, the zebrafish embryo resembles those of amniotes (see Chapters 14–18) more closely than amphibians (Chapter 13).

Nevertheless, broad decisions are being made at these early stages to determine the location of the dorsal side of the embryo. Oppenheimer, working with *Fundulus*, found that when the blastoderm was removed from the embryo before the 32-cell stage, no visible dorsal structures formed in the resulting embryoids (Oppenheimer 1936a,b). If removed afterward, the isolated blastoderms formed embryoids that contained notochord fragments. In more recent experiments in zebrafish, treatments that remove the vegetal region of the yolk cell during early cleavage stages cause dorsal patterning deficiencies that are not seen when the same treatments are done later (Mizuno et al. 1999; Ober and Schulte-Merker 1999). Because the yolk cell is cytoplasmically continuous with the 20 or so marginal cells of the 32-cell-stage blastoderm, it is quite possible that many of the types of specification events that occur in frog embryos during the first cell cycle are still occurring during these "late" stages. Consistent with this idea, an oriented microtubule array is visible at the vegetal pole as early as the 2-cell stage (Jesuthasan and Strähle 1997), and treatments that interfere with microtubules and actin often cause radialized embryos which contain no dorsal side. Hence, molecules may be in the process of being moved toward the future dorsal side of the yolk cell to allow for its role as the Nieuwkoop center in the blastula stage.

BLASTULA STAGE: REGIONAL SPECIFICATIONS

This period encompasses the midblastula transition to a longer cell cycle and the early stages of epiboly, covering a time from 2.3 to 5.3 hours of development. During this period, blastomeres become motile and morphogenesis begins.

The first motile cells appear at division 10 as the cell cycle slows, signaling the zebrafish mid-blastula transition (MBT) (Kane and Kimmel 1993). During the next two divisions, the average cell cycle length doubles and, because some cells have a longer cycle than others, synchrony is rapidly lost. This fish MBT is similar in many respects to that in the frog, which begins in cycle 12, and that in the fly, which begins in cycle 9 or 10. For example, the timing of the fish MBT is controlled by the ratio of nucleus to cytoplasm, as in frogs and flies, and more notably, in terms of zygotic gene expression, the embryo first becomes transcriptionally active at this time.

Although zygotic transcripts are present in the embryo after the MBT, they are not necessary until the beginning of epiboly. Embryos injected with α-amanitin, a potent inhibitor of Pol II transcription, do not initiate doming (Kane et al. 1996). However, other processes continue in the α-amanitin-injected embryos: The cell cycle continues until division 14, which normally occurs in the gastrula.

Maternal supplies are necessary for developmental decisions during the first day of development, indicating that the maternal to zygotic transition is quite drawn out. For example, dominant alleles of the *piggy tail* and *half baked* mutants have a dominant zygotic maternal phenotype that is only expressed if the mother is heterozygous and the embryo is heterozygous (Kane et al. 1996; Mullins et al. 1996). This phenotype indicates that the gene products are provided maternally as well as zygotically, and that, in both these cases, the embryos respond to the effects of the loss of this maternal gene expression long after MBT. It is interesting that many processes of early morphogenesis have been relatively resistant to mutagenesis in the zebrafish; if these processes were controlled by maternal zygotic genes that have no zygotic phenotype, they would be very difficult to identify in simple mutagenesis screens.

Despite the apparent uniformity of the blastomeres of the mid-blastula stage, careful inspection reveals differences among the cells in terms of cell division, cell adhesion, and cell motility that define three mitotic domains (Kane et al. 1992). The first of these domains is the yolk cell. As mentioned above, divisions are incomplete in the early cleavage stages, leaving the yolk mass undivided and the cells cytoplasmically continuous with the yolk (Fig. 1A). Subsequent horizontal divisions separate the majority of the blastomeres from the yolk; however, there always remains a ring of blastomeres, sometimes termed Wilson cells, at the yolk–blastoderm margin (Kimmel and Law 1985a), and these are cytoplasmically continuous with the yolk. At divi-

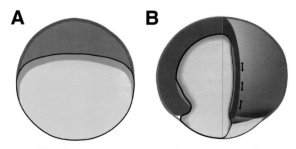

Figure 1. The teleost blastula and gastrula. (*A*) Blastula, showing the epithelial enveloping layer (*brown*), the yolk cell (mainly *yellow*), and its syncytial layer (*green*). The deep cells of the blastula, which comprise the embryo proper, are shown in blue. The yolk cell membrane is black. (*B*) Gastrula. The deep cells have thinned into the epiblast, and cells at the edge of the epiblast involute into the hypoblast (*red*). The non-involuting forerunner cells are shown in purple.

sion 10 or 11, as the cell cycle begins to slow, the Wilson cells collapse into the large yolk cell to form a syncytial layer, the yolk cell domain. These nuclei continue to divide rapidly for several more rounds and then arrest at division 13 or 14 (Kane et al. 1992). A similar mitotic arrest also occurs in *Fundulus* (Trinkaus 1993).

The second domain is the enveloping layer, the cells on the outside surface of the blastoderm. At divisions 10 and 11 these "outside" cells begin to form a thin epithelium, and shortly thereafter, the cells acquire a long asynchronous cell cycle. When cell divisions occur, daughters remain within the plane of the epithelium (Kimmel et al. 1990). The cells of this lineage-restricted epithelium adhere to each other and prevent intrusion of the extraembryonic medium into the blastoderm. This barrier is highly protective of the zebrafish embryo, as in other aquatic species (Keller and Trinkaus 1987). When added to the bathing solution, even small molecules such as α-amanitin or ethanol will not readily pass into the embryo. The enveloping layer ultimately becomes the embryo's first skin, the periderm, to be later replaced by a true ectodermal–mesodermal derived skin.

Inside the blastoderm, covered by the enveloping layer and underlain by the yolk cell, is the third domain, the deep cells of the blastoderm. After the MBT, the deep cells acquire a rhythm of divisions intermediate between that of the enveloping layer and the yolk cell in terms of both synchrony and length. Cycle 12 is reached during the high blastula stage and cycle 13 is reached at about the beginning of epiboly. Although the cells become very asynchronous as divisions progress in the zebrafish, cells, on the average, tend to enter cycle 14 when the embryo is in early gastrulation, a situation remarkably parallel to the relationship of cycle 14 coinciding with gastrulation in flies (Foe 1989). Additionally, as in flies, the cycle 14 length is the first cycle that seems to not be sensitive to the ratio of nucleus to cytoplasm (Kane et al. 1992).

At cycle 10, the deep cells of blastoderm become highly motile, extruding large blebs and moving in seemingly random directions, producing a fine mixing effect at the cellular level. Small clones produced from cells labeled at cycles 9 or 10 mix readily with non-clonal neighbors, and this is before the extreme mixing due to the radial intercalations of early epiboly. On close examination of the deep cells in the blastula stage, there is an impression of a loose aggregation of cells with ample intracellular space, and thus it is no surprise that the stages between the mid-blastula and gastrulation are a favorite time to perform cell transplantations in the zebrafish.

At this stage, it is not possible to distinguish the dorsal side of the embryo by morphological criteria. However, during the mid-blastula, β-catenin enters the nuclei of the dorsal region of the yolk cell, consistent with a role in the formation of the Nieuwkoop center of the embryo (Schneider et al. 1996; Chen and Kimelman 2000; Kelly et al. 2000; see Chapters 13 and 36). Interestingly, there is also staining in the nuclei of cells of the rim of the blastoderm. These cells may be induced by the yolk cell, or they may be induced by the same process that induced the yolk cell, because the dorsal cells of the blastoderm and the dorsal nuclei of the yolk syncytial layer are clonally related. A subset of these cells gives rise to the forerunner cells, mentioned below.

Later in the blastula stage, the process of establishing the organizer is well under way. In *swirl*, *snailhouse*, *minifin*, and *somitabun* mutants, ventral structures disappear to various degrees. *bozozok* and *dino* mutants show an opposite phenotype, with the dorsal and anterior regions reduced in a manner similar to the reduction in head structures seen after UV-induced ventralization in the frog (Hammerschmidt and Mullins 2002). The first group of mutants are either BMPs or BMP protagonists, which act to specify ventral and lateral structures, whereas the second group are BMP antagonists that act to protect the organizer from the effects of BMPs (Table 1). In situ hybridization with probes that mark the organizer or the ventral region of the embryo shows that all of these mutants are acting during the blastula stage to set up the large regional domains such as the organizer and the lateral/ventral side of the embryo. Interestingly, although individual cells are becoming specified toward general fates, they have not yet committed to specific fates, as we discuss below.

Epiboly, the spreading of the blastoderm to enclose the yolk cell completely, begins in the late blastula at about 5 hours after fertilization and continues until 10 hours, at which time the yolk cell is covered (Fig. 2). At dome stage, the formal beginning of epiboly, the cells begin to spread over the yolk and the yolk cell pokes, or "domes" upward into the blastoderm (Fig 2A'). The blastoderm then begins a slow vegetalward advance, about 100 μm per hour, hesitating only for a short time at the equator of the yolk cell, coinciding with the formation of the germ ring.

The mechanism of early epiboly seems to require an interplay between the enveloping layer of the blastoderm, the deep cells of the blastoderm, and the yolk cell, although the role for each is not completely known. The function of the zebrafish yolk cell during epiboly seems similar to that of the more thoroughly examined *Fundulus* yolk cell. Before epiboly begins, a portion of the yolk nuclei move out from under the edge of the blastoderm and, as epiboly begins, "lead" the blastoderm to the vegetal pole. In a classic experiment, John Trinkaus removed the blastoderm in the mid-blastula stage and the vegetal movement of the yolk syncytial nuclei continued, demonstrating that the movement of the yolk syncytial layer is independent of the blastoderm and autonomous to the yolk cell (Trinkaus 1951). Elaborate con-

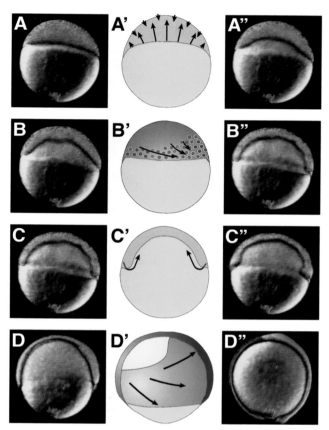

Figure 2. The major cell movements of early zebrafish development. (A′–D′) Schematics of the respective cell movement indicating the movement of cells within the blastoderm (*arrows*) and the resulting change in morphology, flanked by live embryos delimiting the stages when these movements peak (*A–D* and *A″–D″*). (*A*) Radial intercalation, from oblong stage through 40% epiboly. (*B*) Dorsal compaction, from 40% epiboly to 50% epiboly. (*C*) Involution, from 50% epiboly stage to shield stage. (*D*) Convergence and extension, from 60% epiboly to tail-bud stage.

nections develop between the yolk cell and the enveloping layer of the blastoderm, suggesting that the yolk cell is towing the enveloping layer (Betchaku and Trinkaus 1978). Consistent with this finding, the zebrafish yolk cell has microtubules and microfilaments in the proper orientation to effect such movement, and treatments that interfere with these structures impair epiboly (Strähle and Jesuthasan 1993; Solnica-Krezel and Driever 1994). Such a model predicts that large amounts of cell membrane must be recovered from the vegetal side of the yolk cell as the margin of the blastoderm sweeps vegetalward, and indeed, endocytosis occurs vegetal to the blastoderm margin in both *Fundulus* and zebrafish (Betchaku and Trinkaus 1978; Solnica-Krezel and Driever 1994). This endocytosis may be part of the motor itself, for removal of cell membrane vegetal to the blastoderm would draw the cortex of the yolk cell vegetally.

The enveloping layer of the blastoderm seems to be a passive component of epiboly, given its tight attachment to the yolk cell and its stretched-out appearance late in epiboly (Fig. 1A). This epithelium becomes lineage-restricted by division 13, a timing that coincides with the rounding of the embryo from high blastula to sphere stage. The lineage restriction may reflect the increasing tension on the enveloping layer cells, as we imagine that the mitotic apparatus can be aligned by forces external to the cell, forcing tangential divisions that give rise to daughters within the enveloping layer. Although it may be a passive partner in epiboly, the enveloping layer maintains an active process of cell rearrangement within the epithelium. As the margin narrows over the *Fundulus* yolk cell, many enveloping layer cells leave the margin and intercalate among more animally located cells (Keller and Trinkaus 1987; Fink and Cooper 1996), while constantly maintaining a barrier against the outside environment.

No experiment has demonstrated the role of the deep layers of the blastoderm in epiboly. Before epiboly begins, the cells of the deep layer are motile, but this motility seems random and non-directed. The deep cells seem merely to fill the space between the yolk cell and the enveloping layer. As epiboly proceeds, the yolk cell domes toward the animal pole, thinning the blastoderm, and a dramatic radial intercalation of cells occurs, as deeply placed deep cells mix with those more shallow (Kimmel and Warga 1987; Wilson et al. 1995). One interesting aspect of this process is that the blastoderm thins quite uniformly, from about 5 cells thick at 30% epiboly to about 2–3 cells thick at 50% epiboly. It is unclear what mechanism could explain this remarkable uniformity.

A precise zebrafish fate map becomes possible as radial intercalations of early epiboly become less extreme (Kimmel et al. 1990). Such a map (Fig. 3A), made at the 50% epiboly stage, overtly resembles the gastrula fate map of *Xenopus* (Keller 1975, 1976). The cells at the margin of the blastoderm make mesodermal and endodermal structures, and cells away from the margin make ectodermal structures. Cells on the dorsal side of the blastoderm make axial structures and tend to contribute more to anterior structures; cells on the ventral side make lateral and paraxial structures and tend to contribute more to posterior structures.

Descendants of cells in the late blastula tend to become restricted to particular tissues, suggesting the possibility that some cells are becoming committed to their fates at this time. However, cell transplantations within the blastoderm at 50% epiboly indicate that these cells are all pluripotent with respect to fates of the deep cells. Marginal cells transplanted to the animal pole become ectoderm, and animal pole cells transplanted to the margin become mesoderm (Ho and Kimmel 1993). Perhaps, when these uncom-

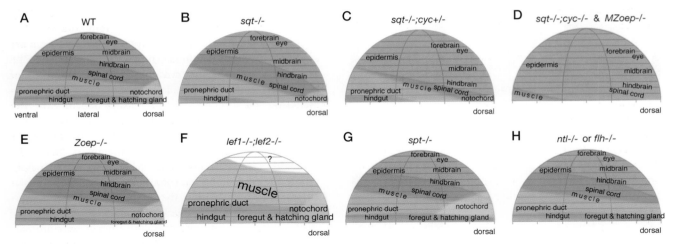

Figure 3. Shield stage fate maps of mutants that affect cell specification. These maps are derived from classic cell-tracing experiments, with the exception of *lefty1;lefty2* (*F*), which is based on gene expression data. (*A*) Wild-type fate map. (*B–F*) Mutations that alter the allocation of cell fate along the animal–vegetal axis. (*B–D*) The endodermal and mesodermal fields disappear and the ectodermal field drops closer to the margin with increasing loss of Nodal function. (*B*) *squint* homozygote, (*C*) *squint* homozygote;*cyclops* heterozygote, and (*D*) *squint*;*cyclops* double homozygote or maternal and zygotic *one-eyed pinhead* mutant. (*E*) The endodermal field disappears with loss of zygotic *one-eyed pinhead* function. (*F*) The ectodermal field disappears and the mesendodermal field expands with complete loss of Lefty function. (*G, H*) Mutations that alter the allocation of cell fate along the dorsoventral axis. (*G*) In the *spadetail* mutant, the fate map field of paraxial tissue contracts and the fate map field of axial tissue expands within the presumptive endoderm and mesoderm. (*H*) In either the *no tail* or *floating head* mutant, the fate map field of notochord is missing. In the case of *no tail*, cells in this field retain characteristics of dorsal mesoderm, but never differentiate, and in the case of *floating head,* cells in this field acquire more lateral mesodermal characteristics.

mitted blastula cells divide, their progeny remain adjacent to one another and reside in the same developmental field of the gastrula. Hence, in contrast to the random mixing of early epiboly, a degree of order has been imposed after the blastula stage that tends to keep closely related cells together.

A more subtle movement that occurs during late blastula stage is the compaction of deep cells on the dorsal side of the embryo, termed dorsal compaction (Fig. 2B). During the mid-blastula, deep cells begin to compact together at the dorsal margin, extending in an arc of about 30° from the dorsal midline. As cells are compacting, a slow dorsalward drift of cells occurs along the lateral edge of the blastula margin.

At least two mutants perturb the process of dorsal compaction, *one-eyed-pinhead (oep)* and *spadetail*. In *oep* mutants, there is compaction throughout the blastula (Warga and Kane 2003). In this mutant, cells do not drift dorsalward in the blastula stage, and endodermal fields narrow slightly toward the blastoderm margin (Fig. 3E). On the other hand, in *spadetail* mutants, compaction is weak or does not occur at all (Warga and Nüsslein-Volhard 1998). When tested by fate mapping (Fig. 3G), fates of dorsal regions of *spadetail* mutants tend to spread both toward the animal pole and toward the lateral side of the embryo. For example, mutant notochord cells will originate occasionally from positions 90° from the dorsal side, an occurrence

that never happens in normal embryos. Interestingly, in *spadetail* mutants the animal pole spread of the notochord fate region extends to that seen for the lateral and ventral sides of the mesodermal fate regions, suggesting that compaction is one of the factors that controls the animal poleward spread of hypoblast fates, especially in the dorsal region.

EARLY GASTRULA: INVOLUTION

Gastrulation begins at about 6 hours of development, when deep cells at the margin of the blastoderm turn under the edge of the blastoderm to form a new tissue layer, the hypoblast (Figs. 1B and 2C). This process begins simultaneously around the margin of the embryo, and the newly formed hypoblast causes a thickening termed the germ ring. As soon as involution begins, the outer layer of the deep cells of the blastoderm is typically termed the epiblast.

Involution in zebrafish seems similar to the "active" involution of *Xenopus* (Ibrahim and Winklbauer 2001), a movement that is the result of individual cell movements (Carmany-Rampey and Schier 2001) but appears as a coordinated movement of a sheet of cells (see Chapters 13, 19, and 20). Around the lateral and ventral edges of the blastoderm, the marginal-most layer of cells takes a mesenchymal morphology and rolls into the hypoblast (Warga and

Kimmel 1990). After moving into the hypoblast, it forms a loose single-cell-thick layer of cells and begins quickly to move away from the blastoderm margin toward the animal pole. At this time there is no hint of convergence toward the dorsal side of the embryo.

Appearing with the germ ring, or sometimes slightly afterward, is the shield, a thickening of the germ ring that is homologous to the dorsal lip of frogs and Hensen's node in the chick. The dorsal rim of the gastrula differs in several respects from the lateral and ventral rim of the blastoderm. Already stiffened by cell compaction, the lip seems to fold or flip into the hypoblast, transiently pushing almost 100 μm into the yolk cell before flattening against the inside of the dorsal epiblast. This has always been a confusion for students of zebrafish: The shield is so easily seen at shield stage, and then suddenly, only 15 minutes later, identifying the dorsal side is quite difficult, although fortunately, not impossible.

Also on the dorsal side, there is a group of cells that do not involute, termed the "forerunner cells" (Fig. 1B). Comprising a group of some 20–30 cells, these forerunners move with the margin of the enveloping layer cells (Cooper and D'Amico 1996; Melby et al. 1996). Although the function of these forerunner cells is not completely understood, it seems that they may have a role in left–right asymmetry (Essner et al. 2002).

The hypoblast is fated to become endoderm and mesoderm (Kimmel et al. 1990). On the ventral and lateral edges of the blastoderm, the first cells that involute tend to become endoderm, and later involuting cells become mesoderm, although at this early time there are no obvious morphological differences between these cells (Warga and Nüsslein-Volhard 1999). On the dorsal edge of the blastoderm, where cells are better organized, the first cells to involute are prechordal plate, a mixture of endoderm and mesoderm, and these are followed by the notochord anlage, a mesodermal fate.

Mutants that interfere with cell movement at the beginning of gastrulation tend to change cell specification as their primary consequence, and cell movements are affected secondarily. This is in line with the observation that during early gastrulation, cells in the hypoblast are becoming committed (Ho and Kimmel 1993). For example, of the mutants that affect involution, most have been shown to affect the specification of the hypoblast by blocking or interfering with the Nodal pathway. The mutations in *cyclops* and *squint*, the fish homologs of *Nodal*, cause defects in the specification of the mesendoderm (Rebagliati et al. 1998; Feldman et al. 2000). Neither of these mutants by itself completely blocks mesendoderm formation (Fig. 2B) or involution. However, the dosage of these genes can be reduced, in various double mutant combinations, until almost no mesendoderm is specified (Fig. 3C,D), and invo-

lution is almost completely abolished (Dougan et al. 2003). *one-eyed-pinhead*, mentioned above, is the zebrafish *EGF-CFC* gene, known as *Cryptic* in mice. This gene is necessary for the reception of the Nodal ligand (Gritsman et al. 1999), a function that may or may not be related to its earlier role in cell motility. In zebrafish, when the maternal and zygotic function of this gene is lost, mesendoderm is almost completely reduced (Fig. 3D) and involution is arrested. Last, morphants that knock down the Lefty1/2 ligand, which normally represses or competes with the Nodal signal, make more mesoderm than normal (Fig. 3F) and, consequently, have a higher rate of involution (Chen and Schier 2002; Feldman et al. 2002).

There are also mutants with phenotypes that do not affect involution, and these too affect cell specification. Mutations in *casanova*, homologous to Sox transcription factors, ablate the forerunner cells and most of the endodermal germ layer (Alexander et al. 1999). Similar but weaker phenotypes are seen in mutations in *bonnie and clyde* and *faust*. Mutations in *no tail*, the fish homolog of mouse *Brachyury*, and mutations in *floating head*, the fish homolog of *Xnot*, abolish the differentiation of the notochord anlage (Fig. 3H) (Talbot et al. 1995; Melby et al. 1997; Glickman et al. 2003). In all these cases, misspecification events exhibit subtle defects in morphogenesis, although their effects on subsequent inductions are substantial.

MID-GASTRULA: CONVERGENCE AND CONGRUENCE

The mid-gastrula is time of cell convergence toward the dorsal midline (Fig. 2C). Although involution continues during the mid-gastrula stage, some aspect of convergence is found in every cell movement in the embryo. Note that when involution had begun, hypoblast cells tracked directly toward the animal pole, such that the cells of the epiblast and the hypoblast are moving in opposition. At 60% epiboly, both of these trajectories begin to swing toward the dorsal side of the embryo, and by 70%, the cells of the epiblast and the hypoblast begin to align themselves with each other. At 80%, the trajectories of both layers match one another congruently throughout much of the embryo.

In the mid-gastrula, the shield reaches and extends past the animal pole. This dorsal hypoblast migrates as a large group of cells, 1–2 cells thick. These cells were compacted before they involuted, and they continue to remain compacted as they move. At the leading edge of the shield, cells are extending filopodial-like projections, although the function of this interesting activity is unclear. The late blastula fate map territory of the shield is much wider than it is high; this group of cells narrows and lengthens to form a narrow strip of cells that extends the entire length of the embryo (Melby et al. 1996). As in frogs, medial lateral cell intercalation is occurring, and it seems that this process might be driving movement (Glickman et al. 2003). As in the leading edge of the dorsal lip of the frog (see Chapter 13), the leading edge of the shield becomes prechordal plate, and this is followed by the notochord anlage.

Cells of the epiblast thin into a single-cell-thick epithelium as they begin to converge toward the dorsal side. The dorsal epiblast is fated to become central nervous system (Woo and Fraser 1995), and this region of the epithelium undergoes extensive convergence and extension as it forms the neural keel (Concha and Adams 1998). Moreover, changes in cell shape may help to move epiblast cells dorsally: Cells of the neural plate lengthen perpendicular to the surface of the epiblast, forming a columnar epithelium by the bud stage (Papan and Campos-Ortega 1994). This thickening of the dorsal epiblast further draws lateral cells closer to the axis. The ventral side of the epiblast is fated to become epidermis, and cells in this region flatten into a thin epithelium of cells as they cover the forming embryo.

In the mid-gastrula, the germ layers of the hypoblast become morphologically distinguishable and begin to move dorsally. Endodermal cells sort into a lower layer of thin flat cells against the yolk cell, and mesodermal cells form a loose layer of round cells that are located between the epiblast and the endoderm (Warga and Nüsslein-Volhard 1999). Initially, shortly after involution, the cells were migratory and moved as individuals, and it is not clear whether there is an exclusive or even preferred substrate for cell movement. At first, the cells move on the yolk cell membrane and on the inner layer of the epiblast, and later, as an endodermal layer becomes more distinct, on that too. Later, as cells move dorsally, the cells begin to compact together and seem to move in a manner similar to the epiblast cells above them, and may in fact have begun to adhere to the overlying ectoderm.

Within the epiblast, there is a sweep of different trajectories from anterior-dorsal to posterior-ventral. Near the animal pole, lateral and ventral cells move toward the animal pole, whereas on the ventral vegetal side, cells migrate directly toward the vegetal pole. Perhaps it is confusing to use "dorsal and ventral" as global coordinates as in this example; both groups of cells are actually moving directly toward a dorsal midline that now curves more than 180° around the embryo. Also, from embryo to embryo, the exact trajectory that cells use to move dorsal is variable. Some of this variability may be experimental, due to the difficulty of orienting a sphere with few landmarks. However, perhaps like driving to work in traffic, the arrival at the axis of the embryo is more important than the path traveled.

Consistent with this interpretation—and in contrast to mutants that act during early gastrulation—mutants that affect convergence tend to affect convergence primarily and cell fate secondarily. The mutants *trilobite* and *knypek* affect the convergence of the lateral and ventral mesoderm, and both later have a characteristic extreme dorsoventral lengthening of the somites during the segmentation period (Hammerschmidt et al. 1996; Marlow et al. 1998; Henry et al. 2000; Sepich et al. 2000). *pipetail* affects the number of cells that move to the ventral region of the embryo and later affects the organization of the tail (Rauch et al. 1997). *silberblick* affects the extension of the dorsal mesoderm and later affects head development, often resulting in cyclopia (Heisenberg et al. 2000). Despite the differences in phenotypes, all of these mutants partially block the Wnt planar signaling pathway (Myers et al. 2002), and all the mutants cause a slowing of convergence. Also, the mutants do not tend to affect cell fate, as shown in the fate maps in Figure 4, F and G.

Cadherins, known for their role in cell–cell adhesion, also play an essential role in convergence. The oldest of the zebrafish morphogenesis mutants, *spadetail*, blocks the convergence of the trunk paraxial mesoderm, sending all these cells to form a "spade" in the tail (Kimmel et al. 1989). Although *spadetail* is *tbx16*, a transcription factor (Griffin et al. 1998), the spade phenotype of the mutant may be largely the result of the down-regulation of paraxial *protocadherinC*, a downstream target of the *spadetail* gene (Yamamoto et al. 1998). Another mutant that has a great effect on convergence during this stage is *half-baked* (Fig. 4A), which is mainly known for its epiboly arrest phenotype, and is now known to be a mutation in *E-cadherin*. In contrast with the other convergence mutants, *half-baked* primarily slows the convergence of the epiblast (Fig. 4C) rather than the hypoblast (Figure 4D), and does so dramatically, slowing convergence to 30% of wild-type levels. Nevertheless, despite the remarkable slowing of convergence in *spadetail* and *half-baked*, cells that manage to arrive at the correct domain of the embryo differentiate appropriately, characteristic of the convergence mutants. In *half-baked* this is very striking, for development occurs somewhat normally around the stalled germ ring of the embryo (Fig. 4B,H).

Epiboly continues in the mid-gastrula, after its short hiatus in the early gastrula. The functions of the yolk cell and the enveloping layer continue as before; however, the dependence of the deep cells of the blastoderm on a yolk cell motor becomes less clear, especially given the transient and changing connections between the deep cells and the yolk cell. For instance, at any one time, individual deep cells are moving relative to the enveloping layer, relative to the margin of the blastoderm, relative to the yolk cell, and relative to other deep cells. It remains quite possible that other types of forces could aid or assist epiboly. In *Xenopus*, which has no enveloping layer and no yolk syncytial layer, epiboly occurs, albeit to a lesser extent; this movement is thought to be mediated initially by radial intercalation (Keller 1980), and later by a little-understood purse-string-like movement

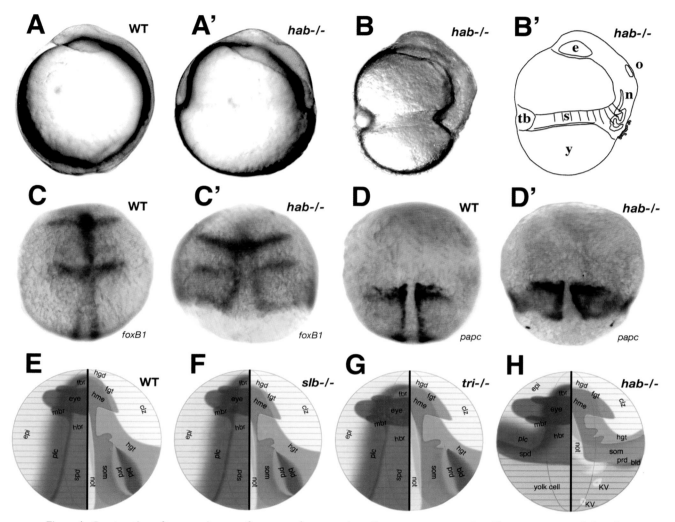

Figure 4. Construction of two-somite-stage fate maps of mutants that affect convergent extension. These maps are partly based on published gene expression data. (*A, B*) phenotype of the *half-baked* (*hab*) mutation. Wild-type (*A*) and *half-baked* homozygote mutant sibling (*A′*) at the two-somite stage and a rare surviving homozygote (*B*) at 24 hours. (*B′*) Drawing of embryo in *B*. (*e*) eye; (*o*) otic vesicle; (*n*) notochord; (*s*) somite; (*tb*) tail bud; (*y*) yolk. (*C, D*) Alterations in the expression patterns of representative genes in the *half-baked* mutation. (*C, C′*) *foxB1* expression in the neuroectoderm at the 2-somite stage in a wild-type and *half-baked* homozygote embryo. (*D, D′*) *papc* expression in the paraxial mesoderm at the two-somite stage in a wild-type and *half-baked* homozygote embryo. (*E–H*) 2-somite-stage fate maps. The left side of each map shows the position of ectodermal fates and the right side shows the position of endodermal and mesodermal fates. (*E*) Wild-type, (*F*) *silberblick*, (*G*) *trilobite*, and (*H*) *half-baked*. In all of these mutants, specification of cell fates appears normal, but the domain of many fates on the map is altered as a result of perturbations in convergence movement. In the *sliberblick* mutant (*F*), the domain of axial mesendoderm is less extended and slightly broader than wild type. In the *trilobite* mutant (*G*), the domains of most fates are less extended along the AP axis and are correspondingly broader along the DV axis than wild-type. In the *half-baked* (*H*) mutation, the domain of fates along the DV extends around the margin of the blastoderm because of the defects in epiboly. (fbr) forebrain; (mbr) midbrain; (hbr) hindbrain; (spd) spinal cord; (epi) epidermis; (plc) placode; (hgd) hatching gland; (hme) head mesoderm; (not) notochord; (som) somite; (prd) pronephric duct; (bld) blood; (fgt) foregut; (hgt) hindgut; (clz) clearing zone; (KV) Kupfer's vesicle.

of the ventral and lateral edges of the lip. Indeed, in zebrafish, cellular morphology and microtubule staining show that enveloping cells at the vegetal pole exhibit characteristic shape changes typical of a purse-string mechanism at the end of epiboly (Zalik et al. 1999).

Mutations in *half-baked*, when homozygous, throw an interesting light on the capacities of the deep cell domain during epiboly, because the mutants arrest epiboly of the deep cell domain but not that of the yolk cell (Kane et al. 1996). The yolk syncytial layer and the enveloping layer undergo epiboly normally, and complete it at 10 hours, as in wild-type embryos. What arrests is the deep cell portion of the blastoderm, stopping at 70–80% of epiboly. This arrest occurs slightly after the convergence defect. This is the first genetic evidence showing on one hand that epiboly of the deep cells is separable from that of the yolk cell, and on the other, that convergence and epiboly of the deep cells may use common genes. Further studies on the epiboly mutants will help clarify the mechanical connections, if any, between epiboly and convergence.

SEGMENTATION: THE TAIL BUD

In the absence of a blastoderm margin for involuting cells, the bud stage in the zebrafish embryo has been considered to be the end of gastrulation. However, there is evidence that gastrulation movements continue after this traditional ending point (see Chapter 15). In the older literature, the tail bud is sometimes compared to an appendage that "grows" out of the end of the embryo. However, even in those days, alternative views argued that the morphogenesis of the tail bud was a continuation of the earlier morphogenesis of gastrulation (Pasteels 1943). Supporting this latter view are studies in zebrafish which argue that the tail bud arises from a large accumulation of cells on the posterior side of the tail bud that interacts with axial tissues originating anterior to the yolk plug (Kanki and Ho 1997), a process that continues well into the segmentation stages.

The formation of the tail begins with the extension of neural ectoderm and the axial mesoderm across the closed yolk plug (Fig. 5A). The extension of the axial tissues is probably driven by the mediolateral intercalation occurring along the entirety of the dorsal axis. This tissue both pushes and slides over cells that are accumulated on the posterior side of the yolk plug. This movement tends to roll the cells out so that deeper cells contribute more to anterior tail and superficial cells contribute to the posterior tail (Fig. 5 B, C). Not obvious in Figure 5 is the fact that the axial tissues are dividing the cells as well. Interestingly, lineages of cells from the ventral region are often split by this process, so that clonal siblings are distributed on both sides of the tail (as in the chick and mouse; Chapters 15 and 16.

Cells just posterior to the yolk plug appear not to contribute to ectodermal fates. This is perplexing because, presumably, there are both involuted and non-involuted cells in that position, and it would seem that non-involuted cells should take ectodermal fates. Also, the expression of the ectodermal markers, such as *dlx3* (Akimenko et al. 1994), which approximates the border between epidermal fates and neural plate fates, indicates that such a border does indeed exist on the posterior side of the yolk plug, suggesting that the superficial tissue posterior of the yolk plug is prospective neural plate. However, when these superficial cells are labeled, they later take up positions in the paraxial mesoderm. Hence, as the neural plate and axial mesoderm extend across the tail bud, the superficial cells of the non-involuted cells are internalized. Thus, some form of gastrulation continues in the tail bud.

Many mutants have effects on the tail—we do not list them all—but without exception, all of the mutants listed in Table 1 affect the tail. The major reason for this is that any disparity in the number of cells that move to the tail—either

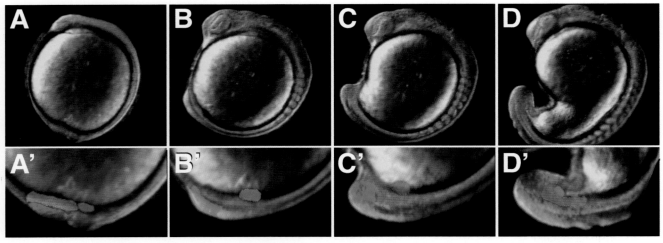

Figure 5. Cell movement in the tail bud. (*A*) Tail bud formation, about 1-somite stage. (*B*) Tail bud extension, about 5-somite stage. (*C*) Tail bud protrusion, about 10-somite stage. (*D*) Tail eversion, about 15-somite stage-somite stage. Colors (*A'* to *D'*) indicate Kupfer's vesicle (*purple*), axial tissues that extend across the yolk plug (*blue*), non-involuted cells (*green*), and involuted cells (*yellow*) on the ventral edge of the yolk plug.

because of fate mis-specification or because of failure to move normally—affects the balance of tail germ layers. In some cases, this results in missing fins (as in the case of missing epidermal fates in *minifin* and *snailhouse* mutants) or in large bulges of dying cells (as in the case of *spadetail*) or in a shortened tail (as in the case of *no tail* mutants), or a curly tail (as in *snailhouse* mutants). Interestingly, in mutants that seem to lack axial tissues, a tail forms anyway, often with many of the fates that are normally present. For example, in Nodal-type mutants, such as the *squint/cyclops* double mutant, there is no axial mesoderm, yet a small axial-free tail forms containing mesoderm that is clearly segmented.

Such findings have led to the hypothesis that the ventral margin of the blastopore has "tail organizer" properties (Agathon et al. 2003). Remarkably, when this region is transplanted to the region just ventral of the animal pole, axial-free tail structures form on the belly of the embryo, free of the influence of the midline of the embryo. Here it is interesting to imagine what types of morphogenesis are occurring in these little axial-free tails and how this compares to tail formation in normal embryos, for at this time we do not know. What would gastrulation have looked like in an ancient ancestor of chordates, before axial tissues existed as we know them and before the embryo became so yolk laden? These axial-free fish tails may give some insight into these ancient processes.

REFERENCES

Agathon A., Thisse C., and Thisse B. 2003. The molecular nature of the zebrafish tail organizer (comment). *Nature.* **424:** 448–452.

Akimenko M.A., Ekker M., Wegner J., Lin W., and Westerfield M. 1994. Combinatorial expression of three zebrafish genes related to distal-less: Part of a homeobox gene code for the head. *J. Neurosci.* **14:** 3475–3486.

Alexander J., Rothenberg M., Henry G.L., and Stainier D.Y. 1999. *casanova* plays an early and essential role in endoderm formation in zebrafish. *Dev. Biol.* **215:** 343–357.

Betchaku T. and Trinkaus J.P. 1978. Contact relations, surface activity, and cortical microfilaments of marginal cells of the enveloping layer and of the yolk syncytial and yolk cytoplasmic layers of *Fundulus* before and during epiboly. *J. Exp. Zool.* **206:** 381–426.

Carmany-Rampey A. and Schier A.F. 2001. Single-cell internalization during zebrafish gastrulation. *Curr. Biol.* **11:** 1261–5.

Chen S. and Kimelman D. 2000. The role of the yolk syncytial layer in germ layer patterning in zebrafish. *Development* **127:** 4681–4689.

Chen Y. and Schier A.F. 2002. Lefty proteins are long-range inhibitors of squint-mediated nodal signaling. *Curr. Biol.* **12:** 2124–2128.

Concha M.L. and Adams R.J. 1998. Oriented cell divisions and cellular morphogenesis in the zebrafish gastrula and neurula: A time-lapse analysis. *Development* **125:** 983–994.

Cooper M.S. and D'Amico L.A. 1996. A cluster of noninvoluting endocytic cells at the margin of the zebrafish blastoderm marks the site of embryonic shield formation. *Dev. Biol.* **180:** 184–198.

Dale L. and Slack J.M.W. 1987. Fate map for the 32-cell stage of *Xenopus laevis*. *Development* **99:** 527–551.

Dougan S.T., Warga R.M., Kane D.A., Schier A.F., and Talbot W.S. 2003. The role of the zebrafish nodal-related genes squint and cyclops in patterning of mesendoderm. *Development* **130:** 1837–1851.

Essner J.J., Vogan K.J., Wagner M.K., Tabin C.J., Yost H.J., and Brueckner M. 2002. Conserved function for embryonic nodal cilia (comment). *Nature.* **418:** 37–38.

Feldman B., Dougan S.T., Schier A.F., and Talbot W.S. 2000. Nodal-related signals establish mesendodermal fate and trunk neural identity in zebrafish. *Curr. Biol.* **10:** 531–534.

Feldman B., Concha M.L., Saude L., Parsons M.J., Adams R.J., Wilson S.W., and Stemple D.L. 2002. Lefty antagonism of Squint is essential for normal gastrulation. *Curr. Biol.* **12:** 2129–2135.

Fink R.D. and Cooper M.S. 1996. Apical membrane turnover is accelerated near cell-cell contacts in an embryonic epithelium. *Dev. Biol.* **174:** 180–189.

Foe V.E. 1989. Mitotic domains reveal early commitment of cells in *Drosophila* embryos. *Development* **107:** 1–22.

Glickman N.S., Kimmel C.B., Jones M.A., and Adams R.J. 2003. Shaping the zebrafish notochord. *Development* **130:** 873–887.

Griffin K.J., Amacher S.L., Kimmel C.B., and Kimelman D. 1998. Molecular identification of *spadetail*: Regulation of zebrafish trunk and tail mesoderm formation by T-box genes. *Development* **125:** 3379–3388.

Gritsman K., Zhang J., Cheng S., Heckscher E., Talbot W.S., and Schier A.F. 1999. The EGF-CFC protein one-eyed pinhead is essential for nodal signaling. *Cell* **97:** 121–132.

Hammerschmidt M. and Mullins M.C. 2002. Dorsoventral patterning in the zebrafish: bone morphogenetic proteins and beyond. *Results Probl. Cell Differ.* **40:** 72–95.

Hammerschmidt M., Pelegri F., Mullins M.C., Kane D.A., Brand M., van Eeden F.J., Furutani-Seiki M., Granato M., Haffter P., Heisenberg C.P. et al. 1996. Mutations affecting morphogenesis during gastrulation and tail formation in the zebrafish, *Danio rerio. Development* **123:** 143–151.

Heisenberg C.P., Tada M., Rauch G.J., Saude L., Concha M.L., Geisler R., Stemple D.L., Smith J.C., and Wilson S.W. 2000. Silberblick/Wnt11 mediates convergent extension movements during zebrafish gastrulation. *Nature* **405:** 76–81.

Henry C.A., Hall L.A., Burr Hille M., Solnica-Krezel L., and Cooper M.S. 2000. Somites in zebrafish doubly mutant for *knypek* and *trilobite* form without internal mesenchymal cells or compaction. *Curr. Biol.* **10:** 1063–1066.

Ho R.K. and Kimmel C.B. 1993. Commitment of cell fate in the early zebrafish embryo. *Science* **261:** 109–111.

Ibrahim H. and Winklbauer R. 2001. Mechanisms of mesendoderm internalization in the *Xenopus* gastrula: Lessons from the ventral side. *Dev. Biol.* **240:** 108–122.

Jesuthasan S. and Strähle U. 1997. Dynamic microtubules and specification of the zebrafish embryonic axis. *Curr. Biol.* **7:** 31–42.

Kane D.A. and Kimmel C.B. 1993. The zebrafish midblastula transition. *Development* **119:** 447–456.

Kane D.A., Warga R.M., and Kimmel C.B. 1992. Mitotic domains in the early embryo of the zebrafish. *Nature* **360:** 735–737.

Kane D.A., Hammerschmidt M., Mullins M.C., Maischein H.M., Brand M., van Eeden F.J., Furutani-Seiki M., Granato M., Haffter P., Heisenberg C.P. et al. 1996. The zebrafish epiboly mutants. *Development* **123:** 47–55.

Kanki J.P. and Ho R.K. 1997. The development of the posterior body in zebrafish. *Development* **124:** 881–893.

Keller R.E. 1975. Vital dye mapping of the gastrula and neurula of *Xenopus laevis*. I. Prospective areas and morphogenetic movements of the superficial layer. *Dev. Biol.* **42**: 222–241.

———. 1976. Vital dye mapping of the gastrula and neurula of *Xenopus laevis*. II. Prospective areas and morphogenetic movements in the deep region.

———. 1980. The cellular basis of epiboly: an SEM study of deep cell rearrangement during gastrulation in *Xenopus laevis*. *J. Embryol. Exp. Morphol.* **60**: 201–234.

Keller R.E. and Trinkaus J.P. 1987. Rearrangement of enveloping layer cells without disruption of the epithilial permeability barrier as a factor in *Fundulus* epiboly. *Dev. Biol.* **120**: 12–24.

Kelly C., Chin A.J., Leatherman J.L., Kozlowski D.J., and Weinberg E.S. 2000. Maternally controlled (-catenin-mediated signaling is required for organizer formation in the zebrafish. *Development* **127**: 3899–3911.

Kimmel C.B. and Law R.D. 1985a. Cell lineage of zebrafish blastomeres. II. Formation of the yolk syncytial layer. *Dev. Biol.* **108**: 86–93.

———. 1985b. Cell lineage of zebrafish blastomeres. III. Clonal analyses of the blastula and gastrula stages. *Dev. Biol.* **108**: 94–101.

Kimmel C.B. and Warga R.M. 1987. Indeterminate cell lineage of the zebrafish embryo. *Dev. Biol.* **124**: 269–280.

Kimmel C.B., Warga R.M., and Schilling T.F. 1990. Origin and organization of the zebrafish fate map. *Development* **108**: 581–594.

Kimmel C.B., Kane D.A., Walker C., Warga R.M., and Rothman M.B. 1989. A mutation that changes cell movement and cell fate in the zebrafish embryo. *Nature* **337**: 358–362.

Marlow F., Zwartkruis F., Malicki J., Neuhauss S.C., Abbas L., Weaver M., Driever W., and Solnica-Krezel L. 1998. Functional interactions of genes mediating convergent extension, *knypek* and *trilobite*, during the partitioning of the eye primordium in zebrafish. *Dev. Biol.* **203**: 382–399.

Melby A.E., Kimelman D., and Kimmel C.B. 1997. Spatial regulation of *floating head* expression in the developing notochord. *Dev. Dyn.* **209**: 156–165.

Melby A.E., Warga R.M., and Kimmel C.B. 1996. Specification of cell fates at the dorsal margin of the zebrafish gastrula. *Development* **122**: 2225–2237.

Mizuno T., Yamaha E., Kuroiwa A., and Takeda H. 1999. Removal of vegetal yolk causes dorsal deficencies and impairs dorsal-inducing ability of the yolk cell in zebrafish. *Mech. Dev.* **81**: 51–63.

Moody S.A. 1987. Fates of the blastomeres of the 32-cell-stage *Xenopus* embryo. *Dev. Biol.* **122**: 300–319.

Mullins M.C., Hammerschmidt M., Kane D.A., Odenthal J., Brand M., van Eeden F.J., Furutani-Seiki M., Granato M., Haffter P., Heisenberg C.P. et al. 1996. Genes establishing dorsoventral pattern formation in the zebrafish embryo: the ventral specifying genes. *Development* **123**: 81–93.

Myers D.C., Sepich D.S., and Solnica-Krezel L. 2002. Convergence and extension in vertebrate gastrulae: Cell movements according to or in search of identity? *Trends Genet.* **18**: 447–455.

Ober E.A. and Schulte-Merker S. 1999. Signals from the yolk cell induce mesoderm, neuroectoderm, the trunk organizer, and the notochord in zebrafish. *Dev. Biol.* **215**: 167–81.

Oppenheimer J.M. 1936a. The development of isolated blastoderms of *Fundulus heteroclitus*. *J. Exp. Zool.* **72**: 247–279.

———. 1936b. Processes of localiziation in developing *Fundulus*. *J. Exp. Zool.* **73**: 405–444.

Papan C. and Campos-Ortega J.A. 1994. On the formation of the neural keel and neural tube in the zebrafish *Danio* (*Brachydanio*) *rerio*.

Roux's Arch. Dev. Biol. **203**: 178–186.

Pasteels J. 1943. Proliferations et croissance dans la gastrulation et la formation de la queue des Vertebres. *Arch. Biol.* **54**: 1–51.

Rauch G.J., Hammerschmidt M., Blader P., Schauerte H.E., Strahle U., Ingham P.W., McMahon A.P., and Haffter P. 1997. Wnt5 is required for tail formation in the zebrafish embryo. *Cold Spring Harbor Symp. Quant. Biol.* **62**: 227–234.

Rebagliati M.R., Toyama R., Haffter P., and Dawid I.B. 1998. *cyclops* encodes a nodal-related factor involved in midline signaling. *Proc. Natl. Acad. Sci.* **95**: 9932–9937.

Roosen-Runge E. 1938. On the early development—bipolar differentiation and cleavage-of the zebra fish, *Brachydanio rerio*. *Biol. Bull.* **75**: 119–133.

Schneider S., Steinbeisser H., Warga R.M., and Hausen P. 1996. (-catenin translocation into nuclei demarcates the dorsalizing centers in frog and fish embryos. *Mech. Dev.* **57**: 191–198.

Sepich D.S., Myers D.C., Short R., Topczewski J., Marlow F., and Solnica-Krezel L. 2000. Role of the zebrafish *trilobite* locus in gastrulation movements of convergence and extension. *Genesis* **27**: 159–173.

Solnica-Krezel L. and Driever W. 1994. Microtubule arrays of the zebrafish yolk cell: Organization and function during epiboly. *Development* **120**: 2443–2455.

Strähle U. and Jesuthasan S. 1993. Ultraviolet irradiation impairs epiboly in zebrafish embryos: Evidence for a microtubule-dependent mechanism of epiboly. *Development* **119**: 909–919.

Strehlow D. and Gilbert W. 1993. A fate map for the first cleavages of the zebrafish. *Nature* **361**: 451–453.

Streisinger G., Singer F., Walker C., and Dower N. 1986. Segregation analysis and gene-centromere distances in zebrafish. *Genetics.* **112**: 311–319.

Streisinger G., Walker C., Dower N., Knauber D., and Singer F. 1981. Production of clones of homozygous diploid zebrafish, *Brachydanio rerio*. *Nature* **291**: 293–296.

Talbot W.S., Trevarrow B., Halpern M.E., Melby A.E., Farr G., Postlethwait J.H., Jowett T., Kimmel C.B., and Kimelman D. 1995. A homeobox gene essential for zebrafish notochord development. *Nature* **378**: 150–157.

Trinkaus J.P. 1951. A study of the mechanism of epiboly in the egg of *Fundulus heteroclitus*. *J. Experimental Zoology* **118**: 269–320.

———. 1993. The yolk syncytial layer of *Fundulus*: Its origin and history and its significance for early embryogenesis. *J. Exp. Zool.* **265**: 258–284.

Warga R.M. and Kane D.A. 2003. One-eyed pinhead regulates cell motility independent of Squint/Cyclops signaling. *Dev. Biol.* **261**: 391–411.

Warga R.M. and Kimmel C.B. 1990. Cell movements during epiboly and gastrulation in zebrafish. *Development* **108**: 569–580.

Warga R.M. and Nüsslein-Volhard C. 1998. *spadetail*-dependent cell compaction of the dorsal zebrafish blastula. *Dev. Biol.* **203**: 116–121.

———. 1999. Origin and development of the zebrafish endoderm. *Development* **126**: 827–838.

Wetts R. and Fraser S.E. 1989. Slow intermixing of cells during *Xenopus* embryogenesis contibutes to the consistency of the blastomere fate map. *Development* **105**: 9–16.

Wilson E.T., Cretekos C.J., and Helde K.A. 1995. Cell mixing during early epiboly in the zebrafish embryo. *Developmental Genetics* **17**: 6–15.

Woo K. and Fraser S.E. 1995. Order and coherence in the fate map of the zebrafish nervous system. *Development* **121:** 2595–609.

Yamamoto A., Amacher S.L., Kim S.H., Geissert D., Kimmel C.B., and De Robertis E.M. 1998. Zebrafish *paraxial protocadherin* is a downstream target of *spadetail* involved in morphogenesis of gas-

trula mesoderm. *Development* **125:** 3389–3397.

Zalik S.E., Lewandowski E., Kam Z., and Geiger B. 1999. Cell adhesion and the actin cytoskeleton of the enveloping layer in the zebrafish embryo during epiboly. *Biochem. Cell Biol.* **77:** 527–542.

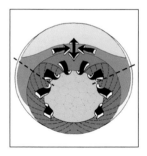

GASTRULATION IN AMPHIBIANS

R. Keller and D. Shook

Department of Biology, University of Virginia, Charlottesville, Virginia 22903

INTRODUCTION

Among vertebrates, gastrulation has been most intensely studied in the amphibians, particularly in regard to the cell and tissue movements. Refinement of the fate maps and deeper analyses of the various cell motilities underlying these movements, and their regulation, suggest that there is still much to be learned. We review what is known of the presumptive fates and the major movements of gastrulation in selected species of Anurans (tailless amphibians), Urodeles (Caudata, tailed amphibians), and Gymnophionans (Caecilians, or legless amphibians), as well as selected species in all groups that have large yolky eggs. Gastrulation evolves and varies with the evolution of reproductive strategy and the attendant differences in egg size, content, and distribution of yolk, and rate of development (see Chapter 51). Gastrulation in large, slow-developing eggs differs substantially from the common model systems, and analysis of these variations broadens our knowledge of how morphogenetic machines function and adapt to the different biomechanical loads placed on them by variations in yolk. We discuss the cell motilities driving gastrulation here, and we review these behaviors and their relation to several of the major tissue movements in more detail elsewhere (see Chapter 19).

Understanding amphibian gastrulation requires understanding the differences among amphibians in the representation of the presumptive tissues in the surface, or superficial, epithelial layer versus the deep, mesenchymal region. The common form of amphibian gastrulation involves involution of the presumptive mesodermal and endodermal tissues. However, involution rolls the superficially located presumptive mesodermal cells over the blastoporal lip and onto the roof of the future gut cavity but does not bring them into the middle layer of the body plan. For this to occur, these cells must undergo an epithelial-to-mesenchymal transition (EMT) and leave the epithelium to join the deep region. Some species, such as *Xenopus*, have few superficially derived mesodermal cells, and the morphogenetic, force-generating consequences of their EMT are negligible. Others have many superficial presumptive mesodermal cells, and when and where they undergo EMT serves an important biomechanical function in gastrulation, in addition to placing them in the middle layer.

ANURAN GASTRULATION

Gastrulation in tailless amphibians (Anura) is best understood in the common model system, *Xenopus laevis*, which rose to prominence during the late 1960s because it is easy to raise, in plentiful supply, and can be stimulated to mate or to shed eggs for in vitro fertilization with readily available hormones (Sive et al. 2000). Its mechanism of gastrulation is representative of those anurans with moderate egg sizes and amounts of yolk, a type of egg that has been the predominant model in embryological work, but it lies at the extreme within this category in having the least superficially derived presumptive mesoderm.

Cell Fates and Movements in *Xenopus*

Xenopus gastrulation centers on the movements of the involuting marginal zone (IMZ, Fig. 1, top row). The IMZ is an annulus of tissue consisting of an outer, superficial epithelial sheet of cells (Fig. 1M) and a thicker, underlying deep region of mesenchymal cells (Fig. 1I) lying at the margin of the floor of the blastocoel at the end of the blastula period, before the so-called pre-gastrulation movements begin (Nieuwkoop and Florshutz 1950; Keller 1975, 1976, 1991;

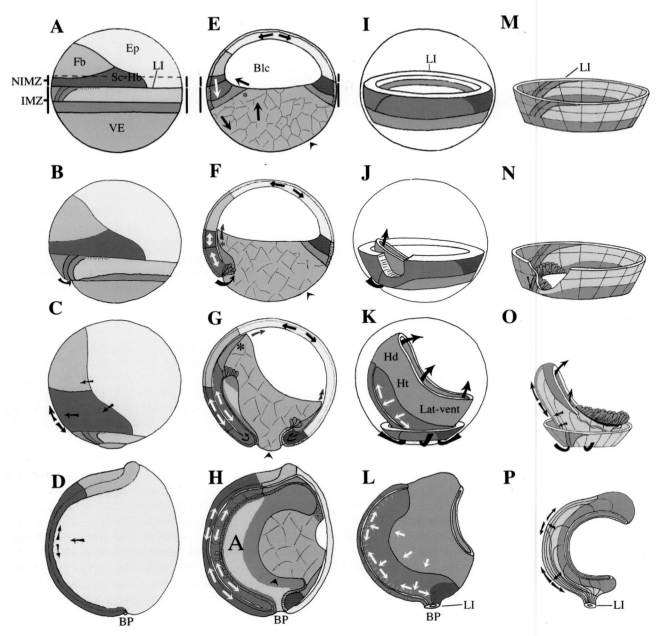

Figure 1. Presumptive fates and movements of *Xenopus* gastrulation are shown in right lateral view (*A–D*), in midsagittal plane (*E–H*), in a 3-dimensional diagram of the deep involuting marginal zone (IMZ; *I–L*), and in a 3-dimensional view of the superficial epithelial layer of the IMZ (*M–P*). The superficial IMZ in the far right row fits over the deep IMZ in the second to last row. Prospective fates are shown as follows: epidermis (*light yellow*, Ep), forebrain (*blue*, Fb), spinal cord–hindbrain (*dark blue*, Sc-Hb), superficial endoderm (*yellow*), superficial endoderm-bottle cells (*green*), vegetal endoderm (*light green*, VE), superficially derived notochord (*light magenta*), deep-derived notochord (*magenta*), superficially derived somitic mesoderm (*light red*), deep-derived somitic mesoderm (*red*), hypochord (*gray*), leading-edge mesendoderm (*orange*) of the head (Hd), heart (Ht), and lateral-ventral body (Lat-vent). Note that the light-colored, superficially derived components join the darker-colored deep components in the course of gastrulation and neurulation. This color scheme is continued in all figures. Dorsal is to the left and vegetal at the bottom. The arrows indicate the major movements as described in the text (Keller 1975, 1976, 1991; Winklbauer and Schürfeld 1999).

Minsuk and Keller 1997). The superficial IMZ consists of presumptive notochord (light magenta), a bit of presumptive somitic mesoderm (light red), presumptive hypochord (gray) (Shook et al. 2004), and the presumptive suprablastoporal endoderm. The suprablastoporal endoderm consists of a subclass of the presumptive endoderm, the presumptive bottle cells (Fig. 1, green), which are characterized by a stereotyped cell shape change during gastrulation, and the remaining non-bottle-cell suprablastoporal endoderm (Fig. 1, yellow). The suprablastoporal endoderm of the IMZ is distinguished from the subblastoporal or vegetal endoderm, which is composed of larger cells and lies below the site of blastopore formation (Fig. 1, light green). Beneath the superficial layer of the IMZ lies the multilayered deep region of the IMZ, which consists of presumptive notochord (magenta), presumptive somitic mesoderm (red), and presumptive head, heart, and lateroventral mesoderm (orange) (Fig. 1) (Keller 1976). Above the IMZ lies the animal region or "animal cap," which also consists of an outer epithelial layer and several tiers of deep mesenchymal cells. It is composed of presumptive ectoderm, including the presumptive epidermis (light yellow), the presumptive forebrain (blue), and presumptive spinal cord and hindbrain (dark blue) (Fig. 1) (Keller 1975, 1976).

Gastrulation begins with what are often called "pregastrula movements" but are actually the first gastrulation movements. These movements initiate the involution of the double-layered IMZ, as it turns inward and back on its inner surface and moves across the animal cap (blastocoel roof) (Fig. 1, three columns on the right). The events surrounding the involution of the IMZ are first represented on the surface with the formation of a local depression, the blastoporal groove on the dorsal side, which is formed by invagination, a bending inward of an epithelial sheet (Baker 1965; Perry and Waddington 1966; Keller 1978, 1981; Hardin and Keller 1988). This bending occurs as the cuboidal presumptive bottle cells undergo apical constriction, elongate in the apical–basal axis, expand basally, and thereby become wedge-shaped (Figs. 1E, F; and 2A, B; see supplemental Movie 13_1 at www.gastrulation.org). The shallow blastoporal groove marks the site of the initiation of gastrulation, but its contribution, and thus the contribution of the formation of the bottle cells to the total depth of the archenteron, is small. As the bottle cells form, the IMZ rotates and begins involution (lower curved, black arrows, Figs. 1F and 2A,B). The initiation of this massive movement of the entire IMZ is driven by the vegetal rotation movements, which occur within the large mass of vegetal endoderm (Fig. 1E, F, straight, black arrows; Fig. 2A, dashed arrows) (Winklbauer and Schürfeld 1999) and begin before and extend through the period of bottle cell formation. The central vegetal endodermal cells move animally toward the blastocoel floor while the floor moves dorsally and comes to lie against the dorsal blastocoel roof (Fig. 2A, dashed arrows), where it begins migrating toward the animal cap (gray arrows, Figs. 1F and 2C); the outer vegetal endoderm moves vegetally, pulling the attached IMZ vegetally and initiating its involution (lower black arrows, Figs. 1F and 2A,B). Like bottle cell formation, vegetal rotation begins dorsally, proceeds laterally, and reaches the ventral sector by mid-gastrulation (Ibrahim and Winklbauer 2001). The region of the blastocoel floor and inner IMZ that vegetal rotation brings into

contact with the overlying blastocoel roof is presumptive mesoderm and endoderm. It forms the leading edge of the "mesodermal" mantle (asterisks, second column, Figs. 1 and 2C), and after attaching to the overlying blastocoel roof, it migrates directionally on this substrate toward the animal pole (gray arrows, Figs. 1F,G and Fig. 2C,D) (Holtfreter 1944; Nakatsuji 1974, 1975a, 1976; Winklbauer 1990; Winklbauer and Nagel 1991; Winklbauer and Selchow 1992; Winklbauer et al. 1992, 1996; Winklbauer and Keller 1996; Davidson et al. 2002). The most advanced point of initial contact of the mesendoderm with the overlying roof of the blastocoel, before it begins migration, is beneath the anterior hindbrain/posterior forebrain, and thus is actually quite far anterior in the future body plan (Poznanski and Keller 1997). The cleft that forms rapidly with this initial involution is called the Cleft of Brachet (CB, Fig. 2B).

As involution begins, the dorsal IMZ and the presumptive posterior neural tissue animal to it undergo a rapid increase in area, which occurs anisotropically and is directed vegetally (white arrows, middle columns, Fig. 1; small arrows, Fig. 2A,B) (Keller 1975, 1978, 1980; Wilson and Keller 1991). This is due to radial intercalation of multiple layers of deep mesenchymal cells in these regions, which produces a thinner but longer array (thinning and extension) (small arrows, Fig. 2A,B) (Keller 1980; Wilson et al. 1989; Wilson and Keller 1991; see supplemental Movie 13_1 at www.gastrulation.org). This rapid thinning and vegetal extension is an active, force-producing process, which may aid the initial involution of the IMZ but is not necessary for it (see below). Radial intercalation thins the dorsal IMZ, and the posterior neural tissue thins from about 5–6 deep cell layers to about 2 layers from late blastula to the mid-gastrula stage (Keller 1980).

In mid-gastrula, the dorsal IMZ, consisting of the presumptive notochord and somitic mesoderm, and posterior neural tissue just animal to it, begins to narrow in the mediolateral axis (convergence) and elongate in the anterior–posterior axis (extension) (white arrows, Fig. 1G,H, K,L; black arrows, Fig. 2C,D) (Vogt 1929; Keller 1975, 1976, 1984, 1986; Jacobson and Gordon 1976; Keller et al. 1985). These paired movements of convergence and extension simultaneously narrow the mediolateral or the circumblastoporal aspect of the IMZ and elongate it in the anterior–posterior direction. Combined, these movements also close the blastopore and aid the continued involution of the IMZ as well (Fig. 1, two right columns; see supplemental Movie 13_1 at www.gastrulation.org). These functions involve complex three-dimensional aspects of gastrulation. The dorsal IMZ and posterior neural tissue extend posteriorly across the yolk plug while converging over it primarily from the dorsal side (third column, Fig. 1), such that the point of blastopore closure is on the ventral side of the vegetal endoderm (arrowheads, second column, Fig. 1). This anisotropic convergence and extension of the notochordal and somitic

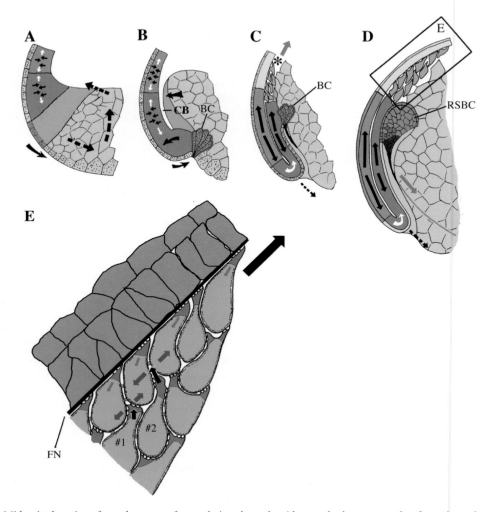

Figure 2. Midsagittal sections from the onset of gastrulation through mid-gastrula show events that form the archenteron and elongate the dorsal body axis (*A–D*; *E* is enlargement of box in *D*). The arrows indicate movements of tissues and cells as described in the text. Presumptive tissues indicated are axial mesoderm (*magenta*), bottle cells (BC, *green*), bottle cell attachment to mesoderm (*gray in B, C*), (respreading) bottle cells (RSBC, *D*), fibronectin-containing matrix layer (Fn, *black line, E*), IMZ (*orange* and *magenta*), IMZ superficial layer (*light magenta*), IMZ deep layer (*magenta*), mesendoderm (*orange*), posterior neural tissue (*blue*), vegetal endodermal cells (*light green*). (Based on Keller 1978, 1981; Hardin and Keller 1988; Winklbauer and Schürfeld 1999; Ibrahim and Winklbauer 2001; Davidson et al. 2002.)

mesoderm carries the suprablastoporal endoderm with it to form the elongated archenteron roof (cf. the two right columns, Fig. 1). The dorsal bottle cells are carried inside and anteriorly because of their attachment to the anteriorly migrating mesendoderm (Fig. 2B,C), and the remaining superficial IMZ extends behind them because of its attachment to the underlying, actively extending dorsal mesoderm of the IMZ (Fig. 2C,D). These dorsal bottle cells then spread once again, beginning at the late mid-gastrula stage, to form a large area of the archenteron lining (respreading bottle cells, RSBC, Fig. 2ED). This process of "respreading" progresses laterally and ventrally in the same order that the bottle cells formed (follow green areas, second and fourth columns, Fig. 1). The deformation of the suprablastoporal

region corresponds to the convergence and extension movements of the underlying dorsal (presumptive notochordal and somitic) mesoderm (cf. two right columns, Fig. 1), whereas the presumptive head, heart, and lateroventral mesoderm first fans out across the animal cap as a sheet, and then converges on the animal pole (orange, Fig. 1, third column). The bottle cells form in the superficial IMZ above the boundary between the converging and extending mesoderm and the spreading head, heart, and lateroventral mesoderm. Later they respread across the inner surface of these latter tissues (cf. third and fourth columns, Fig. 1).

The posterior neural tissue (dark blue), located immediately animal to the dorsal IMZ, also undergoes the initial thinning and extension in the first half of gastrulation (white arrows, Fig. 1E, F; small arrows, Fig. 2A,B) and convergence and extension in the second half of gastrulation (white arrows, Fig. 1G,H; black arrows, Fig. 2C,D). Its convergence and extension parallel that of the underlying IMZ after the IMZ involutes (black arrows, Fig. 2C,D). These movements in both tissues increase to a maximum at the limit of involution (LI, Fig. 1), which begins as a very large circle separating the IMZ and presumptive neural/epidermal tissue, and decreases tremendously by the end of gastrulation (first column, Fig. 1). The corresponding extension movements are likewise very large, with the neural tissue being only 5–7 cells in length at the dorsal midline at the outset of gastrulation, and hundreds of cells in length at the end of neurulation (see supplemental Movie 13_1 at www.gastrulation.org). We believe that the neural tissue extends somewhat faster than the underlying axial and paraxial mesoderm, resulting in a net force for involution of the remaining IMZ (white curved arrows, Fig. 2C,D).

The annular region just above the IMZ has special properties that distinguish it from the rest of the animal cap and is called the non-involuting marginal zone (NIMZ, Fig. 1A), meaning that it comes to lie at the margin of the closed blastopore but does not involute (Keller and Danilchik 1988). The dorsal sector of the NIMZ undergoes an active convergence and extension independent of other tissues (Keller and Danilchik 1988). These movements are induced in the NIMZ by planar contact with the dorsal IMZ (the Spemann organizer) (Keller et al. 1992c) and also contribute to make neural tissue (Doniach et al. 1992; Keller et al. 1992c; Sater et al. 1993). The ventral sector of the NIMZ, consisting of the presumptive ventral, posterior epidermal tissue, may show additional special behaviors corresponding to the convergence and thickening movements of the corresponding, ventral IMZ.

The Mechanism and Function of Convergence and Extension

We separate these movements into a convergence component and an extension component, because convergence is not necessarily linked to extension but is a mechanical force generator that can drive cells into either extension or thickening, or more commonly, both. Conversely, extension is not always driven by convergence but can also be driven by thinning, as described above. In addition, the geometric convergence of cells toward a point or a midline does not necessarily mean that extension will result. "Convergence and extension" or "convergent extension" refers to biomechanical linkage of the convergence of cells with extension in the active, force-producing mechanism, and likewise "convergence and thickening" or "convergent thickening"

refers to the linkage of convergence with thickening of the tissue. Many convergences of cells toward lines or points in morphogenesis have little or nothing in common with these biomechanically specialized, force-producing movements (see Chapter 19).

Convergent Extension Is an Active, Force-producing Process That Forms a Dynamic "Embryonic Skeleton" That Both Shapes and Supports the Embryo

In the whole embryo, the convergence component constricts the blastoporal lips and closes the blastopore while the extension component lengthens the dorsal axial and paraxial structures and defines the anterior–posterior axis of the body plan (Fig. 3A,B; see supplemental Movie 13_1 at www.gastrulation.org). Are these tissue deformations active force-producing processes, or are they generated by forces produced elsewhere in the embryo? In explants of the dorsal sector of the gastrula, both the neural and mesodermal tissues actively converge and extend without an external substratum or exogenous source of deforming force (Fig. 3C–E) (Keller 1984; Keller et al. 1985; Keller and Danilchik 1988; see supplemental Movie 13_2 at www.gastrulation.org). The axial and paraxial mesoderm stiffen by a factor of 3–4 during convergence and extension (Moore et al. 1995), and explants of these tissues can exert a pushing force during extension of about a half micronewton (Moore 1994). These tissues form a stiff but dynamic structure that can extend the dorsal region of the embryo without the overlying blastocoel roof serving as a substratum, and they are not easily diverted in their direction of extension by external forces, but instead shape themselves independent of the substrate (Keller and Jansa 1992). When offered this substrate off-axis, on one side or the other, or both, they do not diverge to correspond to the substrates, but instead the "substrates" move, or are pulled into congruence with the extending somitic and notochordal tissue (Keller and Jansa 1992). Thus, the axial and paraxial mesoderm form a dynamic, constantly changing "skeleton" that is able to support and also shape the embryonic body plan.

Extension of both the mesodermal and neural tissues involves two processes: *thinning and extension* and *convergence and extension* (Keller and Danilchik 1988). In the first half of gastrulation, multiple deep layers of the dorsal mesoderm and neural tissue intercalate along the radius of the embryo (radial intercalation) to form a tissue array that is primarily thinner and longer (Fig. 3F,G) (Keller and Danilchik 1988; Wilson et al. 1989; Wilson 1990; Wilson and Keller 1991). In the dorsal IMZ and NIMZ, radial intercalation is biased such that spreading occurs anisotropically, and the resulting extension occurs in the anterior– posterior direction, rather than in all directions (Wilson 1990; Wilson and Keller 1991). In contrast, radial intercalation in the ani-

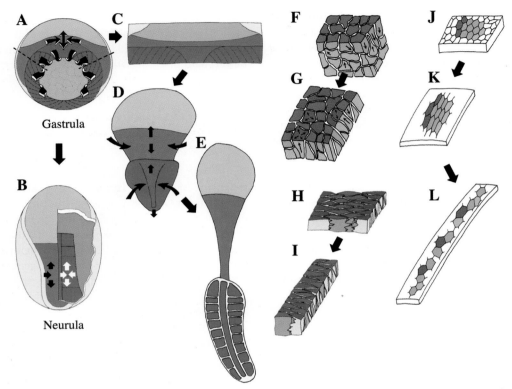

Figure 3. Diagrams show the role of the convergence and extension movements in the closure of the blastopore, involution, and body axis elongation in the embryo (*A,B*), the active, force-producing nature of these movements in explants (*C–E*), active radial intercalation (*F,G*) and mediolateral intercalation (*H,I*) of deep cells, and passive intercalation of epithelial cells (*J–L*). The arrows indicate the major movements as described in the text. Presumptive fates in *A–E* are axial mesoderm (*magenta*), somitic mesoderm (*red*), lateroventral mesoderm (*orange*), posterior neural tissue (*dark blue*), and forebrain (*light blue*) (Keller 1978, 1980; Keller and Danilchik 1988; Keller and Tibbetts 1989; Keller et al. 1989; Wilson and Keller 1991; Shih and Keller 1992a).

mal cap results in spreading in all directions (Keller 1978, 1980), although perhaps not completely isotropically (Bauer et al. 1994). Little is known of the cellular and molecular basis of radial intercalation. The early radial intercalation, occurring before the deposit of extracellular matrix on the roof of the blastocoel, may be independent of this matrix, but the later radial intercalations driving the late epibolic movements and the extension of the NIMZ and IMZ depend on integrin-mediated interactions of the deep cells with the fibronectin-containing extracellular matrix at their basal surfaces (Marsden and DeSimone 2001).

The second process, convergence and extension, occurs as a result of intercalation of cells between one another along the mediolateral axis to form a narrower, longer array (Fig. 3H,I) (Keller 1984, 1986; Keller and Hardin 1987; Keller and Tibbetts 1989; Keller et al. 1989; see supplemental Movie 13_3 at www.gastrulation.org). Because the convergence movements are organized around the circumference of the blastopore, it is squeezed shut, and at the same time, the extension movements push the blastopore and the head apart and elongate the axis (Fig. 3A,B). In the dorsal sector,

the bulk of radial intercalation occurs in the first half of gastrulation, and mediolateral intercalation begins dorsally at the mid-gastrula stage and progresses laterally. However, mediolateral intercalation tends to thicken the tissue as well as extend it, and radial intercalation may counter this tendency by continuing cryptically during medioalteral intercalation (Chapter 19). Ventrally, the presumptive posterior somitic mesoderm (dark red, center columns, Fig. 1) undergoes a "convergent thickening" during involution to form a thick collar of mesoderm just inside the ventral lip (dark red, Fig. 1H, L). Then in the neurula stages, when the notochord extends posteriorly and pushes the blastopore through this ring of tissue, the mesodermal cells in this ring enter a sequence of radial intercalation, followed by mediolateral intercalation as they join the posterior segmental plate as posterior somitic mesoderm (see Fig. 5) (Keller et al. 1989; Wilson et al. 1989; Wilson 1990; Wilson and Keller 1991).

During convergence and extension, the superficial epithelial cells of the dorsal IMZ and NIMZ (presumptive posterior neural tissue) elongate temporarily in the axis of extension to an aspect ratio of 2–2.5 (Fig. 3J,K), then intercalate to form a longer, narrower array as they return to an isodiametric apical shape (Fig. 3K,L). They also divide and spread to form a thinner epithelium (Keller 1980) of larger area (Keller 1978). This behavior suggests that they are passively stretched by active convergence and extension of the underlying deep region, as these deep cells actively intercalate to form a longer, narrower array. Likewise, their involution is passive and due to the active movement of the underlying deep cells. Replacing the native epithelium of the dorsal IMZ/posterior neural region with animal cap epithelium has a minimal effect on involution and convergence and extension movements, whereas replacing the deep cells with those from the animal cap blocks both involution and convergence and extension (Keller 1981). Both neural and mesodermal regions can converge and extend after the epithelium of these regions has been replaced at the mid-gastrula stage with animal cap epithelium (Keller and Danilchik 1988).

The epithelial cells of the gastrula are linked circumferentially at their apices by a junctional complex, including tight junctions (Sanders and Dicaprio 1976; Fesenko et al. 2000), but those in the dorsal IMZ/NIMZ may be specialized in a way that allows them to rearrange without breaking the continuity of the epithelium. Animal cap epithelial cells grafted to the dorsal IMZ/NIMZ do not rearrange during convergent extension but elongate to an aspect ratio of about 5 or 6, and then often tear apart from one another (Keller 1981). The capacity of the native cells to accommodate convergence and extension by rearranging may be an important, developmentally regulated property.

The epithelial layer appears to have a role in patterning the motility of the underlying deep cells. Removal of the epithelium of the dorsal IMZ prior to the onset of mediolateral intercalation at the mid-gastrula stage blocks its convergence and extension, whereas removal after this stage does not (Wilson 1990). Grafts of dorsal IMZ epithelium to the ventral IMZ induce both dorsal tissue types and convergence and extension in the ventral IMZ, showing that the epithelial layer has Spemann organizer activity in regard to induction of both tissue type and cell movements (Shih and Keller 1992b).

Polarized Cell Motility Is Thought to Drive Mediolateral Intercalation and Convergent Extension

The mediolateral cell intercalation during convergence and extension appears to occur as a result of polarized cell motility called mediolateral intercalation behavior (MIB). Mesodermal cells of the early gastrula are multipolar, isodiametric, and pleiomorphic (Fig. 4A), but at the mid-gastrula stage, they become polarized such that they have large lamelliform protrusions at their medial and lateral ends and fine filiform protrusions or fine contact points at their anterior and posterior sides (Fig. 4B) (Keller et al. 1989, 2000; Shih and Keller 1992a,c; see supplemental Movies 13_3,

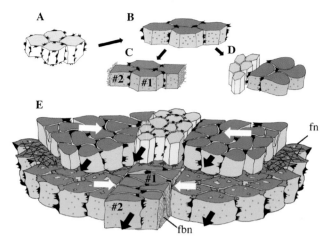

Figure 4. The differentiation of the cell motility underlying mesodermal and neural cell intercalation during convergence and extension is illustrated (*A–D*). Early gastrula, deep IMZ cells show multipolar, protrusive activity (*A*), bipolar mediolaterally polarized protrusive activity during intercalation of somitic and internal notochordal cells (*B*, #1 in *C*), monopolar protrusive activity and boundary capture of an external notochord cell (#2 in *C*), and monopolar, medially polarized protrusive activity of a deep neural cell (*blue, D*). Intercalating cells are adherent at their elongated anterior and posterior sides by small contacts (*gray dots, B–D*). A diagram (*E*) shows monopolar deep neural cells (*blue*), the internal (#1) and boundary-contacting (#2) axial (notochordal) mesodermal cells (*magenta*), the paraxial (somitic) mesodermal cells (*red*), the notoplate cells (*yellow*), the fibronectin (fn)-containing matrix separating neural and mesodermal cells, and the fibrillin fibrils (fbn) that separate notochordal from somitic mesoderm. (From Keller and Tibbetts 1989; Wilson et al. 1989; Wilson 1990; Keller et al. 1991, 1992a,b; Wilson and Keller 1991; Shih and Keller 1992a,c; Elul et al. 1997a, 1998; Davidson and Keller 1999; Elul and Keller 2000.)

13_4, and 13_5 at www.gastrulation.org). The large lamellipodia appear to exert traction on adjacent cells in the mediolateral axis, and generate mediolaterally oriented tensile forces. As a result, the cells become elongated and aligned along the mediolateral axis (Fig. 4B) and pull themselves between one another, generating a narrower, longer array (Fig. 3H,I). In the course of this process, a notochordal–somitic boundary develops and the notochordal cells become "captured" at this boundary (see supplemental Movie 13_6 at www.gastrulation.org). If an internal, bipolar notochord cell (#1, Fig. 4C) contacts this boundary, its invasive lamelliform protrusive activity ceases on the boundary-contacting end; that end of the cell spreads on the boundary but does not cross it, and it rarely leaves the boundary (#2, Fig. 4C) (Shih and Keller 1992c; see supplemental Movie 13_7 at www.gastrulation.org). This "boundary capture" contributes to the extension of the boundary (Keller et al. 1989, 1992a, 2000; Shih and Keller 1992c). Together the bipolar internal behavior and the monopolar, boundary capture mechanism produce mediolateral intercalation within the notochord (magenta, Fig. 4E). The bipolar mode, indistinguishable from that seen in the notochordal mesoderm, appears to drive intercalation of the somitic mesoderm (red, Fig. 4E). Boundary capture has not been described in the somitic mesoderm adjacent to the notochordal–somitic mesoderm boundary.

Mediolateral intercalation in the neural region may be driven by a bipolar mechanism, similar to that seen in the mesoderm, acting very early and perhaps for a very short time, but most of the convergence and extension of the posterior neural tissue is driven by a second type of monopolar mechanism. Deep neural explants made at the late midgastrula stage are deprived of further contact with the underlying notochord, and midline tissues (notoplate/floor plate) do not develop. In their absence, the deep neural cells express a bipolar mode of cell intercalation similar to that seen in the axial and paraxial mesoderm (Fig. 4B) (Elul et al. 1997, 1998). However, when explanted with the underlying mesoderm at the same stage, midline notoplate/floor plate develops, and the protrusive activity of the cells becomes regionally differentiated. The notoplate is distinguished by rounded cells that are unbiased in their protrusive activity, although they may have a slight bias toward the posterior (yellow, Fig. 4D,E) (Elul et al. 1997; Elul and Keller 2000). These cells appear to be carried posteriorly by riding passively on the dorsal surface of the underlying notochord. The protrusive activity of deep cells lateral to the midline, in the neural plate, is biased to one side of their perimeter and directed toward the midline notoplate/notochordal region (blue, Fig. 4D,E) (Elul and Keller 2000; Ezin et al. 2002). This medially biased "monopolar mode" of cell intercalation is dependent on the midline tissues of notochord and notoplate, and it can be redirected toward ectopic grafts of these tissues (Ezin et al. 2002). The midline-dependent monopolar behavior begins in late gastrulation (Elul and Keller 2000), suggesting that the midline-independent, bipolar mode of neural cell intercalation probably only exists for a short time, if at all, between the mid and late gastrula stages, before the midline develops and monopolarizes the deep neural cells. The neural bipolar mode may be an atavistic mechanism that no longer functions alone in the normal embryo but is necessary for the function of the monopolar mode that has replaced it (Elul and Keller 2000). The neural bipolar mode is unbalanced in that the cells move first one way and then another, producing a promiscuous pattern of excessive intercalation that is a poor driver of convergent extension (Elul et al. 1997).

Convergence and Thickening (Convergent Thickening) of the Ventral IMZ

The ventral IMZ also converges at the blastoporal lip and thereby contributes to its closure, but instead of extending during convergence, this region thickens to form the collar of

presumptive somitic mesoderm lying at the ventral lip of the blastopore at the end of gastrulation (Fig. 1) (Keller 1975, 1976). The leading edge, early involuting mesoderm in the ventral sector, which consists of presumptive ventrolateral mesoderm, involutes and then migrates and spreads across the blastocoel roof (orange, Fig. 1, third column) (Davidson et al. 2002). In this regard, it resembles the early involuting mesoderm in the dorsal and lateral sectors, which form the head and heart mesoderm (orange, Fig. 1, third column). However, the late involuting mesoderm, located more animally in the ventral IMZ, consisting of presumptive posterior somitic tissue, converges and thickens to form the collar of mesoderm just over the ventral lip (Fig. 1H,L). This convergent thickening appears to be an active, force-producing process, since it occurs in explants of the ventral IMZ (Keller and Danilchik 1988). Convergent thickening was thought to occur transiently and to precede convergent extension, at least in explants, but it persists in the ventral sector and contributes to closing this sector of the blastopore (Keller and Danilchik 1988). Then this tissue undergoes radial intercalation, followed by mediolateral intercalation and convergent extension in the neurula stage as it joins the posterior segmental plate (Fig. 5D,E) (Wilson et al. 1989; Wilson 1990). Neither the cell movements nor the force-generating mechanisms of convergent thickening are known. Recruitment of the mesoderm out of this collar and into participation in radial and mediolateral intercalation, convergent extension, and somite differentiation requires contact with more dorsal, or anterior, structures. Ventral IMZ explants mimic the gastrula behavior of the ventral IMZ in their convergence and thickening, but they do not undergo the later, convergent extension behavior that occurs when this tissue is swept alongside the posteriorly extending notochord (see next section) (Keller and Danilchik 1988).

Xenopus, and amphibians having similar types of eggs and gastrulation mechanisms, appear to use this convergent thickening only on the ventral sector of the gastrula, the majority of the convergence being part of convergent extension. Other amphibians, such as *Gastrotheca*, appear to close the blastopore using only convergence, perhaps convergent thickening, with convergent extension and body axis elongation to post-gastrular stages. Convergent thickening may also close the blastopore of UV-ventralized *Xenopus* embryos, which close the blastopore symmetrically and do not show convergent extension (Scharf and Gerhart 1980).

Recruitment of the Ventrally Derived Somitic Mesoderm into the Dorsal Body Axis

There is much misunderstanding of how the major features and axes are mapped onto the early gastrula. Part of the confusion arises from the fact that referring to presumptive areas on the early gastrula fate map as dorsal or ventral has

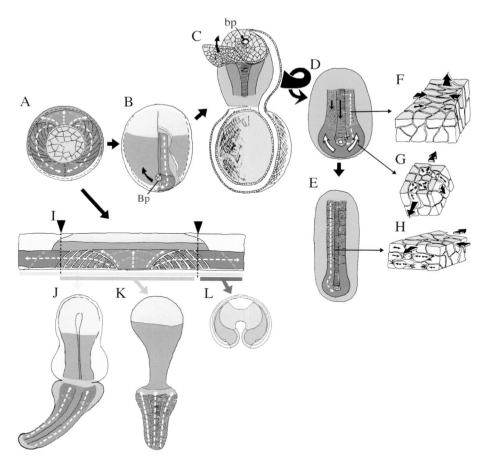

Figure 5. Diagrams show the regional expression of convergence and extension and correlated tissue patterning and orientation of the major body axes. Shown are a vegetal view of the deep tissues of the early gastrula (*A*), an early neurula with a cutaway showing involuted deep tissues (*B*), the process of making a Wilson explant of the dorsal mesoderm/neural tissue (*C,D*), the development of this explant (*D,E*), regional radial (*F*) and mediolateral (*G*) intercalation of the somitic mesoderm, and medio-lateral intercalation of notochordal mesoderm (*H*). Also shown is the morphogenesis of a giant explant of the IMZ/NIMZ (*I*), a dorsal explant (*K*), and a ventral explant (*L*). Presumptive fates are notochord (*magenta*), anterior somitic mesoderm (*red*), posterior somitic mesoderm (*dark red*), ventrolateral mesoderm (*orange*), posterior neural (*dark blue*), and forebrain (*light blue*). The presumptive anterior–posterior axes (*white, dashed arrows*, pointing posteriorly) and presumptive mediolateral axes (*green dashed arrows*, pointing medially) and the blastopore (*Bp*) are shown (Keller and Danilchik 1988; Keller et al. 1989; Wilson et al. 1989; Wilson 1990; Keller 1991; Lane and Smith 1999; Lane and Sheets 2002).

also been taken to mean that dorsally located areas in the gastrula form only dorsal tissues of the body plan and that areas located initially on the ventral aspect of the gastrula form only ventral parts of the body plan. This is not the case, at least in *Xenopus* (Keller 1991).

The early-involuting, leading-edge cell population of the IMZ (orange, middle and right columns, Fig. 1) forms the future ventral tissues, and all these cells migrate and spread across the blastocoel roof (middle and right columns, Fig. 1). The dorsal leading-edge mesendoderm or "head" mesoderm (Fig. 1F,J) migrates beneath the ventral part of the head and into the future liver region (Fig. 1G,H,K,L); the lateral leading edge migrates laterally across the roof and then ventrally where it forms the heart (Fig. 1K,L); finally the leading-edge

ventral mesendoderm migrates over the ventral lip and forms the ventral body wall, blood, and mesenchyme (Fig. 1G,H,K,L) (Keller 1991). All of this tissue shows the characteristic mesendodermal migration behavior (Winklbauer 1990; Winklbauer and Nagel 1991; Davidson et al. 2002; see supplemental Movie 13_8 at www.gastrulation.org). It borders the paraxial (somitic) mesoderm in a region that is more or less congruent with the area in which bottle cells form and later respread (Fig. 1), and this same region comes to underlie the boundary of the neural plate at the mid-gastrula stage (Fig. 1) (see Keller 1975, 1976, 1981; Poznanski and Keller 1997). Blood is formed from all dorsoventral sectors of this leading-edge mesendoderm (Lane and Smith 1999; Lane and Sheets 2002).

In contrast, the presumptive mesoderm in the animal half of the IMZ involves late, follows the leading-edge migratory cells inside, forms dorsal (notochordal and somitic) mesoderm, and undergoes convergent extension rather than migration (magenta and red, Fig. 1F–H, J–L). In the IMZ, the presumptive notochord is located dorsally and the anterior somites laterally, but the posterior ones are found ventrally (third column, Fig. 1; Fig. 5A). How do tissues derived from dorsal, lateral, and ventral sectors of the IMZ all come to lie next to one another in the dorsal aspect of the body plan?

The presumptive notochord involves over the dorsal lip and converges and extends; the presumptive anterior somitic mesoderm involves over the dorsolateral and lateral lips and also converges and extends alongside the notochord to form anterior somites (red, Fig. 5A,B). The presumptive posterior somitic mesoderm involves over the ventral and ventrolateral lips, undergoes convergent thickening, and comes to lie in the ventrolateral collar of mesoderm (brown, Fig. 5A,B). Fate mapping (Keller 1976) and time-lapse recordings (Keller et al. 1989; Wilson et al. 1989; Wilson 1990) show that this collar of mesoderm streams along both sides of the notochord, as the notochord extends posteriorly, and joins the posterior segmental plate. Both the notochordal and somitic mesoderm extend posteriorly, but the notochord extends faster (cf. black arrows, Fig. 5D) and shears posteriorly with respect to the somitic mesoderm (cf. blue marks in Fig. 5D,E). As a result, the notochord pushes the blastopore posteriorly through the ventral collar of presumptive somitic mesoderm (dark red, Fig. 5D), which streams around both sides of the blastopore and posterior notochord (white arrows, Fig. 5D) where it joins the posterior segmental plate (Fig. 5E). This posterior shearing brings the notochord into apposition with the medial aspect of the posterior somitic mesoderm (dark red, Fig. 5A), which it does not initially contact; in the early gastrula, the notochord bounds only the medial aspect of the presumptive anterior somites (Fig. 5A).

Time-lapse recording of Wilson explants (Fig. 5C,E) shows that these ventrally derived somitic mesodermal cells undergo radial and mediolateral intercalation in joining the posterior segmental plate (Fig. 5F,G) (Wilson et al. 1989; Wilson 1990). In the circumblastoporal zone, the cells undergo radial intercalation and thinning and extension as they move out of this region into the posterior segmental plate (Fig. 5G). Then, in a second zone in the posterior segmental plate, they follow with mediolateral intercalation and convergence, a convergence that leads to both extension (convergent extension) and thickening (convergent thickening) (Fig. 5F). Meanwhile, the notochord continues its convergence and extension with less of the convergence contributing to thickening, which is probably one of several reasons that it extends faster than the somitic mesoderm (Fig. 5H). The proportion of convergence causing extension or thickening may be developmentally regulated by the persistence of radial intercalation, which counteracts the mechanical tendency of convergence to produce increases in both dimensions transverse to the axis of convergence (see Keller et al. 2003; Chapter 19).

Mapping the Presumptive Body Axes onto the Early Gastrula

Fate mapping (Keller 1975, 1976) and time-lapse recording of progressions of cell behaviors and displacements (Keller et al. 1989; Wilson et al. 1989; Wilson 1990; Wilson and Keller 1991; Shih and Keller 1992c) allow mapping of the presumptive anterior–posterior and mediolateral axes of the late neurula tissues (Fig. 5B,D,E) or of later stages (Lane and Smith 1999; Lane and Sheets 2002) back onto the IMZ at the onset of gastrulation (Fig. 5A). The anterior–posterior axis at the midline of the notochord is oriented parallel to the animal–vegetal axis, with posterior being animal-most. Near the notochordal–somitic boundaries, the anterior–posterior axis passes laterally on both sides as diverging arcs, parallel to the boundary of the notochord (white dashed arrows, pointing to the presumptive posterior, Fig. 5A). In the adjacent presumptive somitic mesoderm, the presumptive anterior–posterior axis arcs laterally around the blastopore and ventrally with respect to the gastrula geometry (white dashed arrows, pointing to the presumptive posterior, Fig. 5A). The posterior ends (arrowheads) of the presumptive anterior–posterior axes are brought together at the blastopore by convergence, partly due to convergent extension in the dorsal sectors occupied by presumptive notochord and anterior somitic mesoderm (magenta and red territory, Fig. 5A) and partly due to the convergent thickening in ventral sectors occupied by presumptive posterior mesoderm (dark red, Fig. 5A). The mediolateral axes run roughly perpendicular to the anterior–posterior axes (green dashed arrows, pointing medially, Fig. 5A,B,D,E). The lateral aspect of the somitic mesoderm lies next to the vegetal endoderm, and the medial aspect lies next to the notochord in the case of the anterior somitic mesoderm, and next to epidermal tissue in the case of the posterior somitic mesoderm (green dashed arrows, pointing medially, Fig. 5A). The somitic mesoderm shears anteriorly as the notochord pushes posteriorly in the neurula stage (Fig. 5D,E), explaining why the notochord does not bound posterior somitic mesoderm early (Fig. 5A) but does so later (Fig. 5E). Together, the presumptive mediolateral axes of the notochordal and somitic tissue lie perpendicular to the presumptive anterior–posterior axes and describe arcs that span across the dorsal lip of the blastopore (Fig. 5A). It is along these arcs that convergence occurs (Figs. 6 and 7). Lane and colleagues (Lane and Smith 1999; Lane and Sheets 2002) argue that the presumptive posterior end of the embryo lies in the ventral sector of the gastrula, which is largely correct in that the posterior somitic mesoderm lies ventrally. Only the posterior end of the notochord lies mid-dorsal, and it is

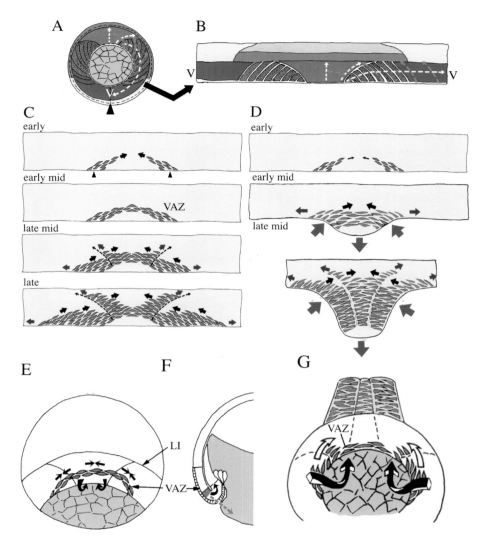

Figure 6. The pattern of expression of mediolateral cell intercalation behavior (MIB) is progressive, beginning in the presumptive anterior regions and proceeding posteriorly, and beginning in the presumptive lateral regions and proceeding medially in both the notochord and somitic domains. A vegetal view of the early gastrula (*A*) shows the tissues explanted to assay these progressions (*B*). Progression of mediolateral intercalation behavior (MIB) is shown by red bipolar cells in explants that were not allowed to extend (*C*) and those allowed to extend (*D*). The vegetal alignment zone (VAZ) is shown in a dorsal (*E*) and midsagittal section (*F*) of the early gastrula. Posterior progression of MIB to the lip of the blastopore of the gastrula is shown (*G*). The presumptive anterior–posterior (*white, dashed arrows, A, B*) and mediolateral (*green, dashed arrows*) are shown in the embryo (*A*) and in giant explants (*B*). Lateral to medial progression (*black arrows*) and anterior-to-posterior progression (*blue arrows*) of MIB are shown. Other arrows indicate movements as described in the text. Presumptive fates are notochord (*magenta*), anterior somitic mesoderm (*red*), posterior somitic mesoderm (*dark red*), posterior neural (*dark blue*), and forebrain (*blue*) (Keller 1984, 1991; Keller et al. 1991, 1992a; Shih and Keller 1992c; Lane and Keller 1997).

united with the posterior somitic mesoderm by the convergence and shearing motions described above.

Patterning of Convergent Extension and Convergent Thickening

In giant explants that include all dorsoventral sectors (Fig. 5A,I), nearly all the IMZ and the neural tissue undergo convergent extension (Fig. 5I,J). Explants of the dorsal sector, which consist of presumptive notochord and anterior somitic mesoderm, undergo convergent extension (Fig. 5I–K), whereas explants of the ventral sector, which include the presumptive posterior somitic mesoderm, do not (Fig. 5I,L) (Keller and Danilchik 1988). Moreover, the presumptive somitic mesoderm in these explants does not develop as expected but instead forms ventral mesoderm (orange, Fig. 5I). This suggests that the ventral sector depends on contact with dorsal sector tissues to undergo convergent extension and differentiation as somitic tissue. In the absence of contact with the dorsal sector, the ventral sector undergoes convergent thickening instead of convergent extension. Ventral explants involute a small amount at their vegetal edges, and form a thick ring of mesoderm, with more mesoderm spread up the inner surface of the animal region, and often form a short "archenteron" there (Fig. 5L). The effect of dorsal sectors on ventral tissues requires more than just continuity of dorsal and ventral sectors. MIB and differentiation of posterior somitic mesoderm do not spread ventrally into the ventral, presumptive posterior somitic regions of giant explants that have been mechanically prevented from deforming (Shih and Keller 1992c). These explants offer continuity but not proximity of dorsal and ventral tissues. Autonomous convergent extension of the dorsally located regions brings them closer to ventral tissues. Where and when this signal is delivered, and how it relates to the putative, third, dorsalization signal postulated to exist on the basis of differentiation of large pieces of the gastrula alone and in combination (Slack and Tannahill 1992; Slack et al. 1992) are not clear.

Patterning of Mediolateral Intercalation Behavior

MIB occurs progressively (Shih and Keller 1992c) in a pattern that allows specific morphogenetic functions, without which gastrulation fails (Keller et al. 1992a, 2000). Giant explants of the IMZ (Fig. 6A,B) cultured under mechanical restraints such that they cannot extend show that MIB occurs progressively from anterior to posterior and from lateral to medial in the IMZ (Fig. 6C) (Shih and Keller 1992c). MIB begins in the presumptive anterior somitic mesoderm and proceeds toward the dorsal midline, just above the blastopore at the early mid-gastrula, to form an arc, the vegetal alignment zone (VAZ) (Fig. 6C, early, early mid). Then the notochordal-somitic mesodermal boundary forms within the VAZ and proceeds posteriorly, separating the future somitic and notochordal fields (dashed, black arrows, Fig. 6C, late mid). As the notochord boundaries form, MIB progresses along their medial aspect, beginning from its origin in anterior VAZ and proceeding posteriorly (blue arrows, Fig. 6C, late mid, late). Meanwhile, from its origin in the VAZ in the anterior somitic mesoderm, MIB progresses posteriorly along the lateral boundaries of the somitic mesoderm (blue arrows, Fig. 6C, late mid, late) (see supplemental Movie 13_6 at www.gastrulation.org). From this origin in the lateral pre-somitic mesoderm, MIB spreads medially toward the notochord, and from its origin in the lateral notochord, it spreads medially toward the midline (black arrows, Fig. 6C, late mid, late). If the explant is unconstrained, the progressive arcs of tension developing as a result of the MIB pull the boundaries of the explant inward and extend it anterior–posteriorly (gray arrows, Fig. 6D).

Mapped back onto the gastrula, the VAZ develops across the dorsal blastoporal lip of the early gastrula, and as it shortens by mediolateral intercalation, it is thought to exert a hoop stress that would tend to pull the IMZ over the lip, thus aiding its involution (Fig. 6E, F) (Lane and Keller 1997). This occurs in early mid-gastrula stage and causes a rapid movement of the dorsal lip vegetally, a "snap down" movement that converts the initial dorsoventrally elongated aspect of the blastoporal lips to a circular one (see two forms of stage 10.5, Nieuwkoop and Faber 1967). Thereafter, the posterior progression of MIB forms a posteriorly moving hoop stress at the blastoporal lip, which would continue driving involution (Fig. 6G). These mechanisms are described below and diagramed in Figures 7 and 8.

Progression of MIB occurs cell by cell, with each successive tier in both the mediolateral and anterior–posterior axes adopting MIB in order (Shih and Keller 1992c), suggesting that the process may be organized during gastrulation, rather than pre-programed by events in the blastula stage. When labeled notochord cells from one embryo are scattered at random through the notochordal territory of a second, unlabeled explant at the early gastrula stage, the labeled cells undergo MIB, proceeding from anterior to posterior and lateral to medial (Domingo and Keller 1995). Thus, a patterning process capable of organizing MIB during gastrulation exists within the notochordal domain. Moreover, this patterning process is capable of inducing MIB and mesodermal fate in prospective epidermal cells, which would not have normally expressed MIB or mesodermal fates at all. Thus, both mesodermal tissue fate and MIB can be induced by organizing signals during gastrulation (Domingo and Keller 1995, 2000). These inducing signals and competence to respond to them are quite potent when assayed on small groups of cells in the relevant tissue environment, more so than would be expected from competence tests in artificial circumstances (Domingo and Keller 1995, 2000). This suggests that competence tests are relevant to what happens in the embryo only when the right signals are presented in a relevant context.

To visualize how MIB might drive involution and close the blastopore asymmetrically, we plot the pattern of progression of MIB onto the vegetal view of the gastrula, as it would appear if the IMZ did not move (indicated by highlighted cells, Fig. 7A–F). MIB expression can be viewed as forming consecutive arcs, anchored at both ends near the boundary of the IMZ with the vegetal endoderm (Fig. 7A–F). MIB is expressed first in the form of the VAZ, prior to formation of the notochordal-somitic mesodermal boundary (Fig. 7A,B). This boundary forms, beginning anteriorly and progressing posteriorly, and thereafter MIB is expressed in both the notochordal (magenta, Fig. 7) and somitic (red, Fig. 7) tissues as a series of continuous, progressively more posterior arcs. The progress of MIB is actu-

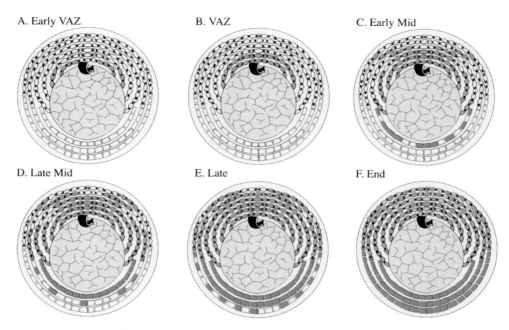

A. Early VAZ   B. VAZ   C. Early Mid

D. Late Mid   E. Late   F. End

Figure 7. The progressive pattern of expression of MIB, indicated by colored, bipolar cells, is mapped back onto the gastrula IMZ (dorsal side at the top) as it would appear if the IMZ did not move. Expression of MIB is shown in the VAZ (vegetal alignment zone) and somitic mesoderm (*red*) and the notochord (*magenta*). Expression of convergent thickening behavior, characteristic of the posterior somitic mesoderm, is indicated by dark red rectangles.

ally continuous but represented by discrete arcs here for clarity. In the ventral sector, where the presumptive posterior somitic mesoderm resides, convergent thickening takes place (see above and Keller and Danilchik 1988) (dark red rectangles, Fig. 7). The more posterior arcs of MIB are thought to link up with arcs of convergent thickening behavior somewhere in the lateroventral marginal zone to form continuous arcs finally encircling the blastopore (Fig. 7C–F). The convergent thickening of this ventral sector of presumptive posterior somitic mesoderm is followed by radial and mediolateral intercalation alongside the posterior notochord as described above (Fig. 6).

Biomechanics of Progressive MIB: Hoop Stress Drives Involution and Blastopore Closure

We believe that the progressive pattern of MIB described above has the following biomechanical consequences: As they shorten by MIB, these arc-like patterns of MIB generate a "hoop stress" across the dorsal lip, which tends to pull the IMZ over the lip to form a narrower tissue that bridges across the vegetal endoderm as the gastrocoel roof (Fig. 8A,B); at the same time the tissue becomes longer by extension, with extension increasing toward the posterior (Fig. 8A,B). The arcs of convergence are generated by MIB dorsally (elongated, bipolar cell symbols, Fig. 8A) and finished out as a continuous circle ventrally by merging

mechanically with the convergent thickening behavior (rectangles, Fig. 8A). The process begins with the tightening of the VAZ across the dorsal lip at the early mid-gastrula stage (Fig. 6E,F). Again, MIB and whatever behavior underlies convergent thickening, and the resulting convergence and hoop stress, are all expressed as continuous phenomena, cell by cell, but are depicted as discrete arcs for simplicity.

During MIB, the bipolar, mediolaterally elongated cells move between one another to form a shorter, narrower array, and thus shorten a chain of cells while making it longer (Fig. 8C,D). There is no evidence that this involves a bias of protrusive activity toward the dorsal midline. Nor is there any evidence that the cells move "toward" the dorsal midline, or that they need to know where the midline is in order to converge and extend. If the dorsal midline (notochord) is removed, the laterally situated somitic mesoderm continues to extend, whereas the isolated notochord does not; it seems dependent on contact with the somitic mesoderm, or something associated with the somitic mesoderm in order to continue convergent extension (Wilson et al. 1989; Wilson 1990). The same appears to be true in ventralized embryos that have no notochord but do have somitic mesoderm (Youn and Malacinski 1981). Cell intercalation by the bipolar mode of MIB shortens a chain that is anchored on both ends, and more cells are brought closer to the midline, but it is because of arc-shortening by intercalation, not midline-directed migration. This is quite different

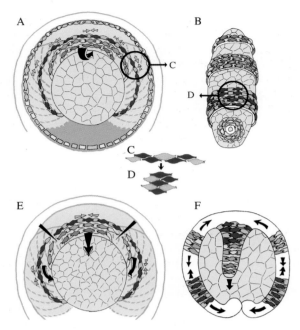

Figure 8. The biomechanical aspects of how progressive hoop stress produces involution and blastopore closure in *Xenopus* can be visualized as shortening or arcs of intercalating cells from the early gastrula (*A*, in vegetal view, dorsal at the top), where they arc across the dorsal blastopore in the IMZ, to the neurula stage (*B*, posterior-dorsal view) in which they have involuted and span the roof of the archenteron. Presumptive anterior (*green*), anterior-mid (*light blue*), posterior-mid (*dark blue*), and posterior (*yellow*) arcs are illustrated. The arcs of MIB are represented by bipolar cell symbols, and convergent thickening by boxes. Arc shortening by cell intercalation requires that bipolar cells intercalate locally to form a narrower, longer array (*C,D*); no dorsally directed activity has been observed in the mesoderm. If the arcs are broken dorsolaterally by microsurgical means (*E*), the isolated dorsal sector will extend straight out without involuting, while the lateral sectors, consisting mostly of somitic mesoderm, will converge against the vegetal endoderm, extend parallel to the margin of the blastopore, and thus enlarge it to form the classic "ring embryo" (*E,F*). (Based on Schechtman 1942; Keller 1981, 1984.)

cally integrated into the context of the whole. The information necessary to accomplish this type of morphogenesis is distributed, and input from the genome is multilevel and contextual, and it includes specifying the pattern of MIB, specifying the details of MIB that result in traction and cell intercalation, specifying the cell adhesions to link the cells into chains, and specifying the biomechanical properties of the tissues (viscoelastic properties, stiffness) that can transmit the local traction forces into a global tension (hoop stress).

The timing of MIB, notably its progressive expression, is biomechanically important for involution of the IMZ. The axial and paraxial tissues become stiffer by a factor of 3–4 with convergent extension (Moore et al. 1995), and thus, with MIB. If MIB took place simultaneously throughout the IMZ in the early gastrula stage, the IMZ would stiffen outside the blastopore and possibly would not turn the corner during involution. This may be one factor that causes the IMZ to extend outward as a proboscis of notochord, rather than involute, in embryos dorso–anteriorized by lithium treatment, which results in simultaneous extension of all sectors of the IMZ (Kao and Elinson 1988). If post-involution tissue already undergoing convergent extension is microsurgically removed and placed outside the blastopore of another embryo, it cannot involute a second time but sits on the blastoporal lip like a canoe on the edge of a waterfall, seemingly unable to bend around the lip (R. Keller, unpubl.). Genetic regulation of tissue mechanical properties is an important but largely ignored aspect of research in morphogenesis.

Role of the Bottle Cells: Anchorage and Respreading

To complete the story of the behavior of the IMZ in gastrulation, we must return to the bottle cells for a moment. The bottle cells form progressively, first dorsally (Fig. 1F,N), then laterally, and finally, ventrally at the mid-gastrula stage (Fig. 1G,O). At this stage, the dorsal bottle cells begin to spread out again (Fig. 1G,O). This process of respreading is also progressive and occurs in the same dorsal-to-ventral order as bottle cell formation. The respread cells form a large area of the lining of the archenteron (Fig. 1H; see RSBC, Fig. 2D) (Keller 1975, 1981; Hardin and Keller 1988). When bottle cells or their progenitors are removed, the archenteron is truncated to the degree that these cells contribute to the fate map (Keller 1981; Hardin and Keller 1988). Although bottle cell respreading appears to generate substantial forces in expanding the periphery of the archenteron, nothing is known about its mechanism. The formation of the bottle cells seems to be a way to pack a lot of surface area into a small space and transport it inside. There, during their respreading, the bottle cells of *Xenopus* may have their greatest effect on morphogenesis.

from the cell motility behavior in the neural plate, which is directed toward the dorsal midline (Elul et al. 1998; Elul and Keller 2000; Ezin et al. 2002). These differences are discussed further in Chapter 19.

Breaking the arcs of convergence is the most certain way to stop gastrulation. Schechtman (1942), in one of the best but most ignored papers about gastrulation, and Keller (1984), in *Hyla regilla* and in *Xenopus laevis*, respectively, broke the continuity of these arcs by insertions of other tissues in the dorsolateral IMZ. The isolated dorsal sector of the IMZ extends straight out, while the lateral and ventral sectors converge against the vegetal endoderm, extend parallel to it, and enlarge the blastopore rather than close it (Fig. 8E,F). Thus, it is not enough for the local cell behavior to be properly specified locally; the behavior must be mechani-

The bottle cells form an adhesive anchor-point of the epithelial layer and suprablastoporal endodermal layer to the underlying mesendodermal cells. When their progenitors, or the bottle cells themselves, are removed microsurgically, there is little effect on gastrulation except that the epithelium slips posteriorly as the underlying mesendoderm moves forward and the posterior axial and paraxial mesoderm extends posteriorly (Keller 1981). The epithelium of the IMZ/roof of the gastrocoel peels off the underlying mesoderm easily except in the region of the bottle cells (R. Keller, unpubl.). The molecular basis of this strong adhesion (gray outline, BC, Fig. 2B,C) is not known, and no increased concentration or activity of an adhesion molecule or a specialized adhesion molecule has been associated with these cells. Modulation of this adhesion to deeper tissue, perhaps increasing it even more, could be part of the mechanism of respreading.

The formation of the bottle cells may contribute to rotation and initiation of involution of the IMZ (Hardin and Keller 1988), but the vegetal rotation movements appear to play a larger role (Winklbauer and Schürfeld 1999). The apical constriction of the bottle cells is anisotropic in that it is oriented in the animal–vegetal direction, and it results in movement of the IMZ vegetally rather than movement of the vegetal endoderm animally, thereby aiding initiation of involution. This anisotropic behavior is not intrinsic but is a result of the mechanical properties of the surrounding tissues. Again, the genetic input into the function of these cells is a product of specification of their intrinsic behavior, specification of their geometric configuration, and specification of the deformability of surrounding tissues.

Mesendoderm Migration

The leading edge of the mesendodermal mantle becomes apposed to the underside of the blastocoel roof by the vegetal rotation movements in all sectors of the embryo, dorsally first, in the early gastrula (Fig. 1F,J), then laterally (Fig. 1G–K), and finally in the ventral sector by the mid-gastrula. It does not converge and extend like the later-involuting, posterior tissues, but instead migrates across the blastocoel roof in a directional manner, toward the animal cap as a cell stream (Fig. 1F,G,J,K). These cells act as a stream in the sense that the cells in contact with the overlying roof actively migrate directionally to the animal pole, while the deeper cells ride along on their backs (Fig. 2E) (Nakatsuji 1974, 1975a,b, 1976; Winklbauer 1990; Winklbauer and Nagel 1991; Winklbauer and Selchow 1992; Winklbauer et al. 1992; Davidson et al. 2002; see supplemental Movie 13_8 at www.gastrulation.org). The migrating cells become "shingled," with each cell underlapping the posterior portion of its anterior neighbor (Winklbauer and Nagel 1991; Winklbauer et al. 1992; Davidson et al. 2002).

This cell stream not only migrates, but it also thins and extends as the deeper cells intercalate radially between the migrating cells (Fig. 2E) (Winklbauer and Schürfeld 1999; Ibrahim and Winklbauer 2001; Davidson et al. 2002); this radial intercalation or "mesodermal surface insertion" occurs on contact with the overlying blastocoel roof (Winklbauer and Schürfeld 1999; Ibrahim and Winklbauer 2001). In addition, the advancing edge of the migrating mesendoderm toward the animal pole requires lateral (circumferential) integrity in order to move at the normal rate. Explants consisting of separated sectors of the advancing mesendoderm mantle migrate abnormally slowly in culture, but when their circumferential continuity is restored, the normal rate of advance is also restored (Davidson et al. 2002).

These cells migrate on a fibronectin-containing matrix that is deposited on the roof of the blastocoel progressively from vegetal to animal, beginning in the early gastrula, and it is essential for migration of the involuted mesendodermal cells in *Xenopus* as well as other anurans (Nakatsuji and Johnson 1982, 1983a,b; Nakatsuji 1984, 1986; Darribere et al. 1988; Shi et al. 1989; Johnson et al. 1990, 1993; Winklbauer 1990; Winklbauer and Nagel 1991; Wang et al. 1995; Ramos and Desimone 1996; Ramos et al. 1996; Darribere and Schwarzbauer 2000). The pattern and conditions for deposition of the fibronectin fibrils (Winklbauer and Stolz 1995; Winklbauer 1998), and the integrin-mediated cell interactions with these fibrils (Smith et al. 1990; Ransom et al. 1993; Ramos and Desimone 1996; Ramos et al. 1996), have been described previously. The fibrillar matrix appears to be necessary but not sufficient for guidance of the mesodermal cell population toward the animal pole (Winklbauer and Nagel 1991; Nagel and Winklbauer 1999). In culture, the matrix deposited by blastocoel roof explants on the culture substrate biases the migration explants of mesoderm, but not individual cells, toward the animal cap end of the conditioned substrate. The fibronectin must be deposited as fibrils to support cell guidance; if the explants are treated with cytochalasin D, or peptides blocking integrin-mediated interactions with fibronectin during deposition, fibrils are not organized, and cell guidance fails (Winklbauer and Nagel 1991). This matrix may be aligned parallel to the axis of migration in urodeles, but it is not in *Xenopus* (Nakatsuji and Johnson 1983a, 1984; Nakatsuji 1984; Johnson et al. 1992).

The interface between the involuting, migrating mesoderm on the inside and the overlying animal cap (blastocoel roof) begins as the Cleft of Brachet (CB, Fig. 2B) and elongates as the mesendoderm migrates animally and the posterior neural and mesodermal tissue extend posteriorly. The integrity of this interface is maintained by development of two behaviors, a separation behavior in the post-involution tissue that develops coincident with involution and prevents reintegration of the cells into the overlying blastocoel roof, and a repulsive behavior in the roof that prevents integration (Wacker et al. 2000). Mix.1, goosecoid, FGF activity, and modulation of cadherin function are involved in regulation of this behavior (Wacker et al. 2000). A Frizzled-7-dependent PKC signaling cascade may also be involved in establishing tissue separation behavior (Winklbauer et al. 2001; see Chapter 49).

Epiboly of the Animal Cap

The animal cap undergoes an epibolic, or spreading, movement throughout the blastula stage and into the gastrula stage (Fig. 1A–D) (Keller 1978). This epiboly also occurs by radial intercalation of the deep cells (Fig. 9A), which reduces the number of cell layers from three in the late-blastula stage to two (one epithelial, one deep) in the mid-gastrula stage, compared to reduction from five or six to two in the IMZ/dorsal NIMZ over same period (Keller 1980). Radial intercalation in the animal cap produces more uniform spreading in all directions than that in the IMZ/dorsal NIMZ (Keller 1978), although it may be somewhat anisotropic (Bauer et al. 1994). On the dorsal side, an area of specialized radial intercalation produces a partial interdigitation of two layers of radially elongated deep cells to form a thickened, crescent-shaped area that maps to the anterior neural ectoderm and immediately overlies the anterior dorsal mesendoderm (Fig. 9C) (Keller 1980). This

behavior occurs very early, by stage 11 (mid-gastrula), just after the anterior migrating mesendoderm touches the inner surface of this region, and it contrasts strongly with the surrounding regions, which are reduced to a single layer of deep cells by this time (Keller 1980). This may be the first morphological response of the presumptive anterior neural plate to contact with the underlying mesoderm. The cellular and molecular basis of this modified, partial radial interdigitation is not understood, and its probable role in subsequent brain morphogenesis has not been studied.

There appear to be two phases of radial intercalation that are regulated differently. Some of the epiboly by radial intercalation occurs before the deposition of extracellular matrix on the roof of the blastocoel in the late blastula/early gastrula, and this phase must therefore be matrix-independent. The later stages of epiboly, shortly before and during early gastrulation, are dependent on the formation of the matrix on the inner surface of the blastocoel roof (Marsden and DeSimone 2001).

Ingression of Superficial Mesoderm in *Xenopus*

The superficial epithelium of the IMZ involutes as a sheet and forms the roof of the gastrocoel (Fig. 1G,H) (Nieuwkoop and Florshutz 1950; Keller 1975). It consists of a small amount of presumptive notochordal (light magenta, Fig. 10A,B) (Minsuk and Keller 1997), hypochordal (gray, Fig. 10A,B), and somitic mesoderm (light red, Fig. 10A,B) (D. Shook, unpubl.), and a substantial amount of presumptive endoderm of the archenteron roof (yellow, Fig. 10A,B), including the respreading bottle cells (green, Figs. 1 and 10A,B) (Keller 1975). The definitive archenteron, defined as an endodermally lined gut tube, is formed from the gastrocoel by movement of this presumptive mesoderm from the epithelial layer to the underlying deep region by an EMT and ingression (Fig. 10B,C) (Minsuk and Keller 1997; Shook et al. 2004). Vital dye marking experiments detected little or no surface presumptive mesoderm in *Xenopus*, and because this technique often causes cells to constrict apically and ingress, the little dye that was found deep was interpreted as an artifact (Keller 1975). Better marking methods revealed a small amount of mesoderm in the surface layer in some, but not all, embryos (Minsuk and Keller 1997), including up to 30 or 40 presumptive notochordal cells in the superficial layer and about 1–3 presumptive somitic cells per future somite (Shook et al. 2004). The presumptive hypochord cells lie lateral to the superficial presumptive notochord cells, and the presumptive somitic cells lie lateral to the hypochord (Figs. 1A and 10A) (Shook et al. 2004). It is not known whether strain differences or loss of label in the surface mesoderm in some experiments accounts for the fact that not all *Xenopus* can be shown to have surface mesoderm (Minsuk and Keller 1997).

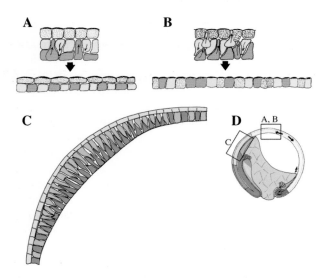

Figure 9. In *Xenopus*, epiboly of the animal cap involves radial intercalation of deep cells and spreading, flattening, and division of superficial epithelial cells (*A*). In at least some urodeles, epiboly of the animal cap involves radial intercalation of the deep and the superficial cells (*B*). In *Xenopus*, a partial interdigitation of deep cells occurs in the future anterior neural ectoderm at the mid-gastrula stage (*C*). The locations of these events in the embryo are indicated (*D*). (From Keller 1980.)

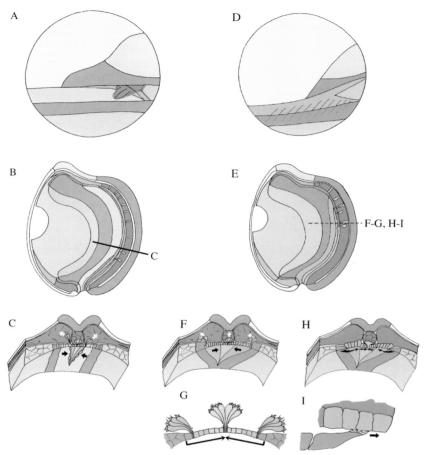

Figure 10. The fate map of the *Xenopus* early gastrula (*A*) has presumptive notochord (*light magenta*) and somitic mesoderm (*light red*), as well as hypochord (*gray*) in the superficial layer of the IMZ, in addition to presumptive endoderm of the archenteron roof (*yellow, green*). After involution, these tissues are found on the roof of the gastrocoel, shown in midsagittal (*B*) and transverse (*C*) views, but move out of the surface epithelial layer to join their deep counterparts in the neurula stages in an anterior-to-posterior progression by undergoing an EMT and ingression (*C*, shown in posterior-ventral views of a cross-section). The corresponding fate map of most other anurans mapped thus far (*Bufo, Bombina, Ceratophrys, Rana*) have larger amounts of notochordal, and particularly somitic, mesoderm on the surface of the early gastrula (*D*), and broader zones of these tissues are found on the roof of the gastrocoel (*E*, midsagittal section). Removal of these cells by EMT and ingression in *Ceratophrys* (*F,G*), and by relamination in *Hymenochirus,* a relative of *Xenopus* (*H,I*), is illustrated in ventral-posterior views of cross sections (Vogt 1929; Pasteels 1942; Purcell and Keller 1993; Minsuk 1995; Minsuk and Keller 1996, 1997).

These presumptive mesodermal cells begin ingression in an anterior-to-posterior progression from the mid-neurula stage onward through the end of neurulation when most of these cells have joined the deep region, thereby forming the definitive, endodermally lined archenteron (Fig. 10C) (Minsuk and Keller 1997; Shook et al. 2004). The superficial presumptive somitic cells layer alongside the superficial prospective hypochord/notochord and move posteriorly with respect to the deep layer-derived somitic cells, as the notochord shears posteriorly with respect to the deep somitic mesoderm in the neurula stage (Fig. 5D,E), and they contribute to somites posterior to their level of origin. These cells move posteriorly with the notochord as it shears posteriorly in the movements described above (Fig. 5D,E) (Shook et al. 2004). The surface-derived presumptive somitic cells resemble the adaxial cells or slow muscle pioneer cells of the teleost fish in morphology and position, but it is not known whether they are the equivalent cells, or play a role in patterning the somitic tissues in the frog, as they do in other vertebrates (Sporle 2001). These cells do not express *slow muscle myosin* or *engrailed*, which are molecular markers for adaxial cells, and the other marker used in

fish, early expression of *myoD* (Devoto et al. 1996; Sporle 2001), is not relevant in *Xenopus*, which expresses *myoD* everywhere in the somitic mesoderm much earlier (Hopwood et al. 1989).

Removal of the notochordal, hypochordal, and somitic mesoderm from the surface layer in *Xenopus* occurs by apical constriction, EMT, and ingression. As this occurs, the edges of the endoderm meet in the midline (Fig. 10C). Whether the endoderm migrates dorsally as the mesodermal tissues ingress, or whether it is pulled by apical constriction of the ingressing mesodermal cells, or both, is not known. In any case, the amount of mesoderm in *Xenopus* is small, and its ingression probably has little biomechanical significance other than removing the little mesoderm that is present in the superficial layer. Other anurans have larger amounts of presumptive somitic mesoderm in the superficial layer, but its ingression is also likely not to have a function in gastrulation, since the EMT and ingression occur during the neurula stage. In some urodeles, a massive amount of somitic mesoderm undergoes EMT and ingression during gastrulation, which appears to function in convergence and blastopore closure.

Gastrulation of Anurans with Eggs Similar to *Xenopus*

A number of anurans with eggs similar to *Xenopus* in yolk content and distribution have been studied. *Bombina* (Vogt 1929) and *Discoglossus* have been fate-mapped in detail (Vogt 1929; Pasteels 1942). Some fate-mapping, developmental anatomical studies, and experimental studies, have been done on *Bufo* (King 1902), *Rana* (Delarue et al. 1994, 1996), *Ceratophrys* (Purcell and Keller 1993), and *Hymenochirus* (Minsuk and Keller 1997). Most appear to have substantially more presumptive mesoderm in the superficial epithelial layer of the IMZ, and it is located in the animal half of the IMZ (Fig. 10D) (Vogt 1929). In these species, the IMZ involutes, as in *Xenopus*, resulting in a gastrocoel roof composed of a central notochordal plate (light magenta, Fig. 10E), bounded on both sides by broad areas of presumptive somitic mesoderm (light red, Fig. 10E), separating the presumptive notochord from the presumptive endoderm of the archenteron roof (green, Fig. 10E); the hypochord has not been mapped in these species but presumably lies lateral to the notochord, as in *Xenopus*. These zones begin to narrow from anterior to posterior during the neurula stage, as the cells in all three regions undergo an EMT and leave the superficial layer, bringing the medial edges of the presumptive endoderm in apposition, and thus forming the definitive archenteron (Fig. 10F–I).

The mechanism of moving the superficial mesoderm into the deep region varies with the species. *Ceratophrys ornata*, the Argentine horned frog, resembles *Xenopus*; the cells undergo apical constriction and form bottle cells, which undergo EMT, detach from the epithelium, and ingress. However, there are broader zones of mesoderm on the roof of the gastrocoel, a central zone of notochordal mesoderm and two lateral zones of somitic which extend around the blastopore (Fig. 10F,G) (Purcell and Keller 1993). Instead of undergoing apical constriction and formation of bottle cells, *Hymenochirus boetegei*, an anuran closely related to *Xenopus*, internalizes its superficial mesoderm by "relamination": the superficial mesodermal cells remain in place, and their deep, basolateral parts become integrated into the underlying deep-derived population of mesenchymal somitic cells (Fig. 10H) (Minsuk and Keller 1996). As this occurs, the endodermal epithelia on both sides seem to move across the somitic mesodermal cells, as if they were a substratum, and finally cover them over, thereby regulating them to a deep position (Fig. 10I). The mechanism of this relamination is not understood. Perhaps the apical surfaces of the superficial presumptive somitic cells become adhesive to the endodermal cells and actually serve as a substrate for their migration. Alternatively, the somitic cells may actively pull the endodermal epithelium across their apices by some undescribed mechanism. The degree to which the physiological integrity of the epithelium is maintained during relamination is also not known.

Based on the species studied thus far, it appears that anurans vary in the amount of superficially derived somitic and notochordal mesoderm, with *Xenopus* being at the minimum and other species having greater amounts. EMT and ingression, or relamination of mesoderm, occurs in the neurula stages and thus does not function in gastrulation. These processes may function to narrow the archenteron in anurans having more superficial mesoderm. In contrast, EMT and ingression have a strong morphogenetic role in urodele gastrulation, where the amount of superficial mesoderm is much larger and the EMT/ingression occurs in the gastrula as well as in the neurula stage.

URODELE GASTRULATION

General Features of Urodele Gastrulation Movements

Late blastula/early gastrula fate maps have been made for *Triturus* (Vogt 1929), *Ambystoma* (Pasteels 1942), and *Pleurodeles* (Delarue et al. 1992, 1994, 1995), and the developmental anatomy of gastrulation has been studied in a number of species, notably *Triturus (Taricha) torosus* (Daniel and Yarwood 1939), *Splerpes bilineatus* (Goodale 1911), and *Cynops pyrrhogaster* (Imoh 1988) (for others, see Keller 1986). The fate maps and morphogenetic movements of the commonly studied urodeles appear similar to those of the commonly studied anurans (Fig. 11). However, there are significant differences between the anurans and urodeles studied thus far, chiefly because there is much more superficially derived mesoderm, especially presumptive somitic mesoderm (Fig. 11A), and most of it ingresses during gastrulation.

The fate maps generally differ from those of anurans in that the area of the IMZ is much larger and that of the presumptive epidermis and neural plate is much smaller. Therefore, greater area, although not necessarily greater tissue mass, involutes during gastrulation. The IMZ of urodeles is relatively thinner than the IMZ of most anurans at the outset of gastrulation. This applies to the commonly studied anurans and urodeles, which are the ones readily available that have a relatively small amount of yolk and small to medium-sized eggs. Among the urodeles, *Ambystoma*, *Triturus*, and *Taricha* have relatively large eggs with relatively less yolk, and *Pleurodeles* is smaller but has a larger yolk mass relative to its size.

The urodele blastula cleaves in a similar manner to the anuran, although a much larger blastocoel is formed in the typical case. The circumapical junctional complex connecting the apices of the epithelial cells forms in urodeles and binds the surface cells into an epithelial pavement indistinguishable from that in anurans. Desmosomes form early in anurans, at stage 7 (early mid-blastula) in *Xenopus* (Sanders and Dicaprio 1976), for example, but only late, during neurulation in urodeles (Perry and Waddington 1966). As in

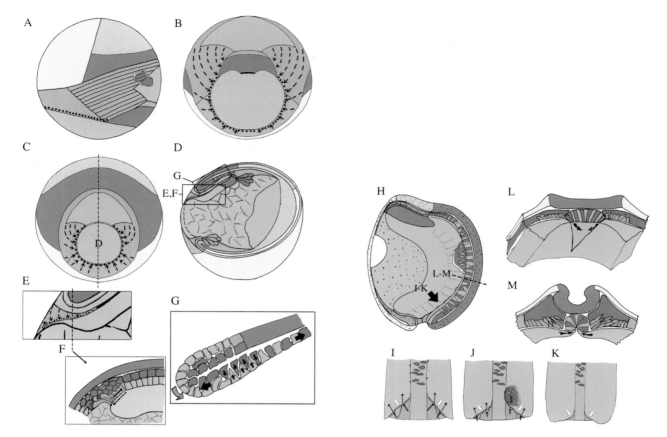

Figure 11. The fate map of the lateral aspect (*A*) and vegetal aspect (*B*) of the early gastrula of the urodele embryo shows presumptive somitic (*light red*) and ventrolateral mesoderm (*orange*) in the superficial IMZ. During gastrulation, superficial mesoderm moves toward the boundary with the vegetal endoderm (*dashed arrows, B, C*) where it undergoes EMT and ingresses beneath the endoderm (*green*), in a "subduction zone" (*dotted lines, A, B, C*), next to the vegetal endoderm. Subduction occurs at the bottom of the blastoporal groove (*arrows in D*, a ventral, quartering view, and enlargement, box *E*). A sectional view of *E* shows how subduction brings the edge of the vegetal endoderm in apposition to the lateral edge of the presumptive notochord (*F*). Two layers of presumptive notochordal mesoderm involute over the dorsal lip (*magenta, D*; enlarged in box *G*), and in *Pleurodeles*, its deep and superficial cells intercalate radially to form one layer (*small arrows, G*). Involution of the IMZ then yields a gastrocoel roof with nearly all the somitic mesoderm removed but containing a midline notochord (*magenta, H*). Posteriorly, the remaining posterior superficial somitic mesodermal (*light red*) and lateroventral mesodermal (*orange*) cells move toward the subduction zone (*black arrows*), where they undergo EMT and ingress beneath the endoderm (*green*) and move off beneath (*dashed arrows*) it to join the deep somitic mesoderm (*red*, in cutaway) (*I–K*). The notochord is removed during neurulation by apical constriction and ingression, bringing the endodermal regions together (*L,M*).

anurans, the animal cap spreads continuously throughout the blastula stage as the blastocoel inflates, and the vegetal region shrinks in area in the same period, similar to that seen in anurans (*Xenopus*, described above) (Schechtman 1934, 1935; Ballard 1955). Whether this vegetal contraction represents the surface component of vegetal rotation movements (Winklbauer and Schürfeld 1999) in these species is not known.

The blastopore forms as the bottle cells undergo apical constriction and produces a shallow invagination, the blastoporal groove (Holtfreter 1943a,b; Perry and Waddington 1966). The blastoporal groove spreads laterally and ventrally as the bottle cells form there, but rather than a continuous

and rapid spread to the ventral side, the groove may stop for a time in the lateral regions of the gastrula, and considerable involution occurs over the dorsal and dorsolateral lips of this horseshoe-shaped blastoporal groove (see supplemental Movie 13_9 at www.gastrulation.org). In most species, the blastopore completes itself ventrally, but in some, bottle cells and a blastoporal lip never form midventrally or only at very late stages near the time of blastopore closure. The bottle cells of the dorsal sector also appear to respread and form the anterior archenteron lining (Vogt 1929), much as in *Xenopus*. However, in the lateral and ventral regions, bottle cell formation represents the first step in a massive EMT and ingression of superficially derived mesodermal cells

(Holtfreter 1943b, 1944; Lewis 1952; Lundmark 1986; Shook et al. 2002). The suprablastoporal epithelium just above the dorsal lip is presumptive endoderm of the pharyngeal/liver region, and it covers the anterior tip of the presumptive notochord and the prechordal plate mesoderm; it is the only region that respreads as definitive epithelial endoderm (Vogt 1929). All amphibians appear to have this presumptive anterior, pharyngeal/liver endoderm in the dorsal, suprablastoporal position overlying the anterior mesoderm of the Spemann organizer, whereas they vary in the amount of superficial endoderm in the remaining IMZ, with *Xenopus* having the most and urodeles the least (dark green, Fig. 11A,B; cf. Fig. 10A; Fig. 1). The commonly studied urodeles appear to have no suprablastoporal endoderm in the lateral and ventral sectors, and the bottle cells forming in these regions are all presumptive mesoderm undergoing EMT and ingression (see below).

It is likely that these dorsal bottle cells serve as an anchor for the epithelial layer, like those in *Xenopus*. Holtfreter proposed that they actively crawl inside and pull everything else with them. He observed that protrusive activity of the inner, basal ends of bottle cells is stimulated by alkali when they are cultured in vitro, and he postulated that the bottle cells migrate inward, toward the alkaline blastocoel environment (Holtfreter 1943a; Gillespie 1983), and drag the rest of the outer cells of the marginal zone inside (Holtfreter 1943a,b). However, the bottle cells in the dorsal sector do not move very much with respect to adjacent mesodermal cells, and thus, this mechanism seems unlikely (Keller 1981; Hardin and Keller 1988). It is during respreading that bottle cells dramatically advance with regard to the surrounding tissues and make by far their greatest contribution to deepening the archenteron. However, Holtfreter's idea of mechanical integration of a local motility to produce a global effect was a pioneering one, appreciated somewhat at the time, long ignored, and deserving of recognition now.

It is not known whether the vegetal rotation movements occur in urodeles. It seems likely that they do, based on the fact that, as in anurans, the surface of the peripheral blastocoel floor seems to develop an increasingly acute angle with the adjacent roof of the blastocoel, and then seamlessly forms the cleft of Brachet (Vogt 1929; Imoh 1988). The IMZ involutes as it does in *Xenopus* in that it moves vegetally, turns around the blastoporal lip, and then moves in the opposite direction across the blastocoel roof by mesendoderm migration (Fig. 11C,D) (Vogt 1929; Imoh 1988). The dorsal IMZ also undergoes convergence and extension, beginning at various stages from the early gastrula to the early neurula, depending on the species (Vogt 1929; Jacobson and Gordon 1976; Imoh 1988; Delarue et al. 1996, 1998; Shook et al. 2002). These movements occur in explants and are active, force-producing movements in the IMZ of *Taricha* and *Ambystoma* (Shook

et al. 2002). The combination of neural plate and dorsal mesoderm of *Taricha* actively converges and extends and shapes the neural plate during neurulation (Jacobson and Gordon 1976).

EMT and Ingression of Somitic Mesoderm in the Commonly Studied Urodeles

A massive EMT and ingression occur in the IMZ of urodeles as it involutes (Fig. 11A–D). Vogt's vital dye mapping studies show that the presumptive superficial somitic mesoderm disappears beneath the endoderm, specifically beneath what were termed the "lateral endodermal crests," after the IMZ involutes over the blastoporal lip (Vogt 1929). Time-lapse recordings of whole embryos and explants show that the cells in the presumptive somitic and ventrolateral mesodermal regions move rapidly toward the blastoporal lip and disappear, those closest to the blastopore disappearing first (dashed arrows, Fig. 11B,C) (Lewis 1948, 1952; Shook et al. 2002). In the whole embryo, this behavior appears superficially similar to what occurs in *Xenopus* (see supplemental Movie 13_9 at www.gastrulation.org). However, the advance of the cells toward the lateral lips of the blastopore is relatively faster (cf. supplemental Movies 13_1 and 13_9 at www.gastrulation.org), and time-lapse recordings of explants reveal remarkably different events at the bottom of the shallow blastoporal groove (see supplemental Movies 3_10, 3_11, and 3_12 at www.gastrulation.org), which is normally hidden from view. The presumptive somitic cells undergo apical constriction and an EMT, and then ingress at the edge of the vegetal endoderm right at the blastoporal lip (Fig. 11D, boxes E and F). The site of EMT and ingression, called the "subduction zone," is at the bottom of the blastoporal groove in the lateral and ventral sectors of the gastrula (dotted lines, Fig. 11). Apical constriction, EMT, and ingression occur progressively, away from the endoderm (solid arrow, Fig. 11F). As a result of the apical constriction, and the eventual departure of the constricted cells into the deep layer, the lateral edge of the notochord and the lateral edge of the vegetal endoderm are pulled together (dashed lines, Fig. 11E,F) (Shook et al. 2002).

In contrast to the removal of the somitic and lateroventral mesoderm from the superficial layers at the lateral and ventral aspect of the blastoporal groove, both the deep and superficial layers of the presumptive notochord approach the lip and involute together and intact over the dorsal blastoporal lip (Fig. 11D and gray arrow, box G). Immediately after involution in at least one urodele, *Pleurodeles waltl*, the two layers of the notochord intercalate radially to form one layer (Fig. 11D, box G) (Delarue et al. 1992, 1996). This process of involution in the dorsal sector, and involution and subduction in the lateral and ventral sectors, continues to bring the edges of endoderm in apposition to the notochord though gastrulation and early neurulation; by mid-neurula, the roof of the gastrocoel is lined largely by endoderm and notochord with only a few presumptive somitic cells remaining between them (Fig. 11H). A dorsal explant of the late gastrula, imaged from the ventral side (view I–K in H), shows that the cells of superficial presumptive somitic and lateroventral mesoderm approaching the subduction zone (solid arrows, Fig. 11I,J) disappear beneath the edge of the endoderm, and emerge in the deep layer as mesenchymal cells (dashed arrows, Fig. 11I,J; cut out, Fig. 11J) (see supplemental Movies 13_12 and 13_13 at www.gastrula tion.org). This process continues until these tissues are removed from the roof of the gastrocoel (Fig. 11K) and have joined the deep region (Fig. 11L,M). In the late neurula, the notochord anlage, which in *Pleurodeles* has become a single layer by the radial intercalation mentioned above, undergoes apical constriction such that its lateral boundaries are brought medially. As this occurs, or after it occurs, the cells undergo EMT and form the deep notochord (Fig. 11M) (Vogt 1929; Lofberg 1974; Brun and Garson 1984). This EMT appears to occur by a process of cell wedging and generating a fan-shaped array of notochordal cells, which brings the lateral boundaries of notochord near one another. Whether they undergo EMT first, or the edges fuse with one another first, thereby making a nearly lumenless "invagination," is not always clear in a given species (Vogt 1929; Lofberg 1974; Brun and Garson 1984; Imoh 1988). In either case, the wedging assures that the edges to be united are close to one another and removal of all mesoderm from the superficial layer is completed with minimal disruption of the epithelium (Fig. 11L,M). This type of notochord formation, or variants of it, also appear in the large, yolky-egged urodeles and in the Gymnophionans. Elements of it also appears in formation of the anterior notochord of *Xenopus* (Novoselov 1995).

Radial Intercalation of the Presumptive Notochordal Region

In *Pleurodeles*, double labeling of the deep and superficial regions shows that after involution of the dorsal sector of the IMZ, the deep and superficial layers merge and form a single layer by radial intercalation of the deep cells into the superficial layer (Fig. 11G) (Delarue et al. 1996). How this process occurs is not known. It involves insertion of deep mesenchymal cells into an apparently intact epithelial sheet, a process that also appears to occur in the animal cap of urodeles and in gastrulation of other organisms such as *Petromyzon*, the lamprey (Hatta 1891). The result is a single layer of notochord at the dorsal midline (Fig. 11G, H). Whether this process of forming a transiently, single-layered superficial notochordal anlagen occurs in other urodeles is not known. In the axolotl, *Ambystoma mexicanum*, no large

population of deep mesodermal cells are seen to insert themselves into the notochordal plate in time-lapse recordings of dorsal explants (Shook et al. 2002). Potentially, the intercalation of deep and superficial cells into one layer would have morphogenetic consequences if it were an active process; for example, it could be expected to contribute to extension (large black arrows, Fig. 11G). Alternatively, other forces could stretch the axial tissues, and the intercalation could be a passive accommodation to these forces. These mechanisms are long overdue for further study with high-resolution cell-marking and video-imaging methods.

Morphogenesis and Patterning of the Urodele IMZ

Time-lapse recording of "giant" explants of the IMZ, which include the presumptive neural/epidermal tissue, the presumptive mesoderm, and part of the presumptive endoderm of the early urodele (*Ambystoma mexicanum, A. maculatum, Taricha granulosus*), characterize the morphogenesis of the urodele IMZ (Fig. 12A–F) (Shook et al. 2002). Giant explants show convergent extension of the notochordal region from late mid-gastrulation through neurulation (white arrows, Fig. 12C–F) (see supplemental Movie 13_11 at www.gastrula tion.org). As the notochord extends, the presumptive somitic (light red) and lateroventral mesoderm (orange) move toward and disappear beneath the edge of the vegetal endoderm (black arrows, Fig. 12D,E) until the boundaries of the notochord are apposed to the vegetal endodermal region everywhere except in the far posterior region (Fig. 12F) by the end of gastrulation (see supplemental Movie 13_11 at www.gastrulation.org). Recording of fluorescently labeled cells shows that the superficial presumptive mesodermal cells move toward the subduction zone (black arrows, Fig. 12D, inset) and enter the subduction zone where they undergo apical constriction (progressively darker shading, Fig. 12D, inset), and are covered over by the edge of the endoderm (green, green arrows, Fig. 12D, inset), as they undergo EMT, ingress (white arrows, Fig. 12D, inset), and reappear in the deep region as mesenchymal cells (Shook et al. 2002; see supplemental Movies 13_12 and 13_13 at www.gastrula tion.org). As the cells undergo apical constriction, and then leave the epithelium, they shorten the distance between the edge of the endoderm and the notochord, as described above (dashed arrows, Fig. 11E, F). The progression of the apical constriction and EMT in the tissue (solid arrow, Fig. 11F) is opposite the movement of the tissue (dashed arrow, Fig. 11F). The force-generating potential of this type of behavior during an EMT is discussed in Chapter 19.

Patterning of EMT in the IMZ of the Urodele

EMT, as assayed from the progress of apical constriction seen in recordings of explants (Shook et al. 2002), spreads

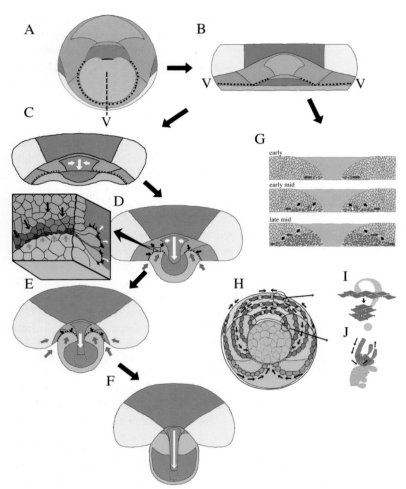

Figure 12. EMT and subduction of the superficial presumptive somitic mesoderm generates forces of convergence in the urodele gastrula. Time-lapse recordings of giant explants of *Ambystoma* or *Taricha* (*A,B*) show extension of the notochord (*white arrows, C–F*) and movement of presumptive somitic cells toward the subduction zone (*black arrows, D–F*), where they undergo apical constriction, EMT, and ingression (*inset, E*). As they do so, the lateral boundaries of the explant move toward the notochord (*gray arrows, E*). EMT and ingression (subduction), represented as red highlighted cells, are expressed progressively, beginning anteriorly in the somitic mesoderm and proceeding posteriorly (*blue arrows, C*), and from this lateral origin, progressing medially (*black arrows, G*), similar to the pattern of MIB expression in *Xenopus* (cf. Fig. 6C). Arcs of hoop stress, generated by mediolateral cell intercalation and convergent extension of the notochord (*I*) and subduction of the lateroventral and somitic mesoderm (*J*), also function in blastopore closure and involution in the urodele (*H*).

rapidly from the dorsal to the lateroventral aspect of the IMZ at its vegetal edge (mapped onto the explant as blue arrows, Fig. 12G). Apical constriction then progresses in an animal direction from this vegetal origin, similar to the pattern in *Xenopus* (black arrows, Fig. 12G) (Shook et al. 2002). Progress of EMT was also visualized by pressing explants of the IMZ between two coverslips, leaving any cells that undergo EMT without a tissue into which they can ingress. EMT, which is revealed as rapid protrusive activity of freshly liberated, individual mesenchymal cells in this preparation, is progressive from the vegetal origin toward the animal end of the explant (black arrows, Fig. 12G) (see supplemental Movie 13_14 at www.gastrulation.org) (Shook et al. 2002).

Overall, this pattern of progression of apical constriction, EMT, and subduction is similar to the progression of MIB in the *Xenopus* IMZ (cf. Figs. 12G and 6C). EMT in the urodele may share with MIB in *Xenopus* the property of progressing from anterior to posterior along the lateral boundaries of the somitic fields, and medially from this lateral origin. However, the configuration of the presumptive axes in the urodele IMZ is unclear. Vogt's fate map of the urodele shows the presumptive anterior–posterior axis of the somitic

mesoderm aligned most closely with the dorsoventral aspect of the gastrula IMZ (Vogt 1929), much as in *Xenopus*, but Pasteel's map shows the presumptive anterior–posterior axis oriented more closely with the animal–vegetal aspect of the early gastrula (Pasteels 1942). Also, lateroventral mesoderm appears to occupy the entire ventral sector of the IMZ, including its animal-most, late-involuting part, the part that forms posterior somitic mesoderm in *Xenopus* (cf. Fig. 11A–C and Fig. 1). The degree of similarity of the progressions of MIB and EMT in these groups should be resolved, as it may bear on the question of whether a common patterning pathway underlies the expression of both these behaviors. The underlying regulatory machinery may constitute a master pattern generator with different downstream effectors of cell behavior. Differences in patterning MIB and EMT could reflect fundamentally different patterning devices, or the diversification of a common one.

After a certain time, EMT becomes independent of contact with the endoderm and is regionally autonomous along the dorsoventral aspect of the IMZ. In IMZs explanted at the early gastrula stage, EMT is somewhat delayed, suggesting that normally the cells respond to cues from adjacent regions, perhaps the endoderm, to regulate its onset. When EMT does begin, it occurs in both dorsal and lateral sectors of the IMZ, showing that continuity along this axis is no longer necessary for propagation of EMT along the dorsal to ventral (possibly the presumptive anterior–posterior) axis (blue arrows, Fig. 12G) (Shook et al. 2002). In both dorsal and lateral explants, EMT progresses in the normal pattern from vegetal to animal, which demonstrates that this aspect of its patterning is independent of the endoderm at the early gastrula stage. This appears to rule out regulation of the progressive aspect of the EMT by proximity to the tissue under which subduction occurs, the endoderm. However, the delay in onset of EMT in its absence suggests that the endoderm may tune the exact time of EMT and ingression in the embryo, and in the absence of this control in isolation from the endoderm, the system defaults to a locally autonomous progression. These issues are important for regulation, biomechanics, and physiological aspects of progressive EMTs (Shook and Keller 2003).

Biomechanics of Blastopore Closure and Involution: Same Geometry and Biomechanics, Different Cell Behaviors

Removal of large amounts of superficial mesoderm from the surface layer has biomechanical consequences for gastrulation. As EMT begins at the vegetal edge of the IMZ in the subduction zone, the first step in these processes, the apical constriction (increasingly dark cells, inset, Fig. 12D) appears to pull the remaining, as yet uninvolved, superficial cells toward the subduction zone (black arrows, inset, Fig. 12D) (Shook et al. 2002). The endodermal cells may also reach across the constricted apices and pull the apically constricted cells toward themselves (green arrows, inset, Fig. 12D) (see Chapter 19). The apically constricted cells then undergo ingression (white arrows, inset, Fig. 12D). As a result, the endodermal and notochordal boundaries are pulled together as the intervening somitic tissue is "reeled in" toward the endoderm by subduction (black arrows, Fig. 12D–F). In explants, the edges are unconstrained, and thus are pulled medially (gray arrows, Fig. 12D–F). In the spherical geometry of the embryo, the integrity of the vegetal endoderm is maintained, and the reverse happens: As the somitic tissue is removed by subduction, the notochord boundaries are pulled vegetally to the edge of the vegetal endoderm (dashed arrows, Fig. 12H, J; Fig. 11B, E, F). Meanwhile, the notochord is converging and extending (see next section), probably by cell intercalation (Fig. 12H,I). Therefore, in the urodele, two cell behaviors, the convergence of the notochord (Fig. 12I) and the removal of the presumptive somitic mesoderm from the epithelial surface layer by EMT and ingression (Fig. 12J), conspire to form arcs of hoop stress, similar to those in Xenopus. These movements drive involution and blastopore closure (Fig. 12H) (Lewis 1948, 1952; Shook et al. 2002). Overall, this mechanism is very similar in form and function to that in Xenopus where only MIB

and cell intercalation are involved (Fig. 8). The biomechanical principle of using hoop stress on the surface of a sphere to close the blastopore and aid involution is preserved, but two cell behaviors are used in the urodele, whereas one is used in Xenopus.

Convergence and Extension in the Urodeles

Convergence and extension movements were first described in detail in the urodele amphibian (Vogt 1929). Are these movements active, force-producing processes in urodeles? These movements were shown to constitute a force-producing process that is important in shaping the neural plate of urodeles (Jacobson and Gordon 1976). Explants of the Pleurodeles IMZ converge and extend in the late gastrula stage and through neurulation (Shi et al. 1987). Explants of Ambystoma and Taricha show convergent extension of the notochord beginning in early gastrulation and continuing through neurulation (white arrows, Fig. 12C–F) (Shook et al. 2002; see supplemental Movie 13_11 at www.gastrulation.org). It is likely that the convergent extension of the urodele notochord is due to mediolateral cell intercalation, as in Xenopus. Cell labeling studies in Pleurodeles show that the elongation of the notochord and body axis is accompanied by mediolateral mixing (intercalation) of cells (Delarue et al. 1998). In addition, there is the potential for the radial intercalation of deep and superficial cells in the notochord to contribute to extension in Pleurodeles (Delarue et al. 1996) and perhaps in other urodeles as well. As the notochordal region extends, the deep somitic region, including those cells added from the superficial layer at the blastopore margins, follows the notochord by converging and extending somewhat (Vogt 1929; Imoh 1988), but whether it converges by MIB is not known. It is also not known whether the deep region of the presumptive somitic mesoderm actively converges in concert with the convergence in the superficial layer mediated by EMT. Much more work should be done to resolve the regional cell behaviors in urodeles.

Mesendodermal/Mesodermal Cell Migration

As the IMZ involutes, the leading-edge mesendoderm attaches to and migrates across the inner surface of the blastocoel roof toward the animal cap (Vogt 1929; Holtfreter 1944; Nakatsuji 1975b; Kubota and Durston 1978; Nakatsuji et al. 1982; Shi et al. 1989). This cell stream in Ambystoma appears to be much less cohesive than the corresponding stream in Xenopus and seems to involve migration of individual cells to a greater extent than in Xenopus (R. Keller, unpubl.). However, it clearly is a cell stream in that the cell movements are coordinated in a common direction and flow pattern, and the cells form a 3-dimensional array. Those next to the substratum are migrating on an external substratum, the blastocoel roof, and those deeper in the tissue are presumably riding along on the backs of those with access to the substratum. As in *Xenopus*, these cells are moving across a fibronectin-containing, fibrillar extracellular matrix deposited from the early gastrula stage onward, and as in *Xenopus*, this matrix is essential for migration (Nakatsuji et al. 1982; Boucaut and Darribere 1983a,b; Boucaut et al. 1984a,b, 1985; Darribere et al. 1984, 1986, 1988; Johnson et al. 1990; Darribere and Schwarzbauer 2000). The matrix also imparts an animal directionality to the migration (Nakatsuji et al. 1982; Nakatsuji and Johnson 1983b, 1984; Nakatsuji 1984; Johnson et al. 1992). The native fibrils may be aligned animal–vegetally, and experimental manipulations of this orientation in culture can orient mesendodermal cell movements (Nakatsuji and Johnson 1983a,b, 1984; Johnson et al. 1992). A blastocoel roof will deposit this matrix in an oriented fashion if it is allowed to slide down the coverslip during the deposition, and migration of cells cultured on this substrate is oriented along the same axis (Nakatsuji and Johnson 1984). However, not all amphibians have an oriented matrix (Nakatsuji and Johnson 1983a), and in *Xenopus*, where directed migration has been studied extensively, the fibrils are necessary but not sufficient for polarized migration (see Chapter 49). Much remains to be learned about the mechanism of mesendodermal guidance in both urodeles and anurans.

Epiboly in Urodeles

Epiboly of the animal cap of urodeles appears to occur by radial intercalation. The animal cap gets thinner and greater in area in the blastula and gastrula stages, and the overall amount of epiboly appears to be substantially greater than in the typical anuran (see Vogt 1929). The cellular mechanism differs in that the deep cells are thought to intercalate radially into the superficial epithelial layer, at least in many species, to form a single layer of much greater area (Fig. 9B). Holtfreter (1943b) observed that the uniformly dark pigmentation of the animal cap gave way to a speckled array of light and dark cells, an event that he attributed to movement of the less pigmented deep cells out into the epithelial layer. Labeling the superficial cells with Bolton-Hunter reagent also suggested invasion of unlabeled putative deep cells into the outer epithelial layer (Smith and Malacinski 1983). However, in *Xenopus*, time-lapse recordings show that superficial cells often divide their apices unevenly, and subsequently the apices of the smaller cells expand an exceptional amount and dilute any natural label, such as pigment, or artificial labeling, making it appear as though unlabeled cells from the deep region have appeared (Keller 1978). This phenomenon could account

for the appearance of unlabeled cells in the urodele. However, histology of amphibian embryos of several species, and also embryos of the lamprey, *Petromyzon* (Hatta 1891), shows a decrease from multiple layers to only one by the mid-gastrula. Notable among these are the California newts, *Taricha torosus* and *Taricha granulosus*, which have been exploited for studying neurulation because of the simplicity of interpreting experiments done on a single-layered epithelium (Burnside and Jacobson 1968; Jacobson and Gordon 1976).

GASTRULATION OF EMBRYOS FROM LARGE, YOLKY EGGS, INCLUDING THOSE OF THE GYMNOPHIONA

Little is known of gastrulation in the third, and much ignored, order of Amphibia, the Gymnophiona (the caecilians, or legless amphibians). The eggs are difficult to get, and little effort has been made to study their gastrulation since the late 1800s. All have internal fertilization, most bear live young, and of those that lay eggs, most lay terrestrial eggs near water; some develop into aquatic larvae and others develop directly without a larval stage (Duellman and Trueb 1994). However, the gastrulation of the species that have been studied shares features with gastrulation of anurans and urodeles with large, yolky eggs, making it useful to discuss them all together.

Why Study Gastrulation of Large, Yolky Eggs?

The urodeles and anurans discussed above gastrulate around the yolky vegetal endoderm and enclose it, or most of it, during gastrulation. These eggs are generally relatively small (0.6–2.3 mm in diameter), have moderate yolk, and are available in relatively large numbers, all features that have made several species, particularly *Xenopus*, "model" systems. However, many amphibians have large eggs with relatively more yolk. Amphibians vary greatly, more so than any other chordate, in their reproductive strategies, and many of them lay relatively small numbers of very large eggs (up to about 1 cm) with large stores of yolk, and the adults hide, brood, and guard them in a myriad of exotic ways (Duellman and Trueb 1994). Some develop directly to the adult stage without a larval stage (direct development) (Elinson 1994, 2001). Eggs with large amounts of yolk present a challenge for the machinery of gastrulation. In eggs with more yolk, the large-celled, vegetal endodermal region expands, and because of the nature of the patterning mechanisms in the early embryo, the IMZ tends to develop on the animal edge of the vegetal endoderm, placing it farther toward one pole of the egg with much more yolk to enclose

(Elinson and Beckham 2002). This presents a biomechanical challenge to the standard gastrulation machinery, as we know it from the small-egged model systems. As large eggs evolve as part of the evolution of reproductive strategy, the strategy of gastrulation also evolves. Thus, gastrulation is a diversified process rather than a conservative one. Analysis of large eggs promises to show how gastrulation mechanisms have diversified under these circumstances. Variants of the standard morphogenetic machines of gastrulation will be characterized, new combinations and timings of mechanisms uncovered, and possibly new mechanisms discovered. The full performance range of the morphogenetic machines seen in common model systems will be evaluated and their limitations discovered. Currently, however, not enough is known about gastrulation in large eggs to draw general conclusions. We discuss selected examples that may represent several different ways of gastrulating in the presence of a relatively large yolk, and speculate on possible mechanisms.

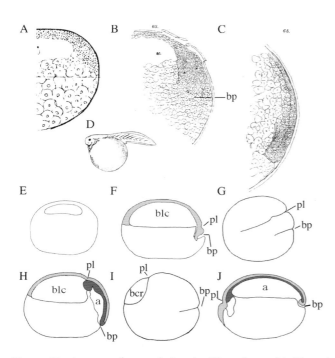

Figure 13. Aspects of gastrulation in *Rhacophorus* (*A–D*) and *Megalobatrachus maximus*, the gigantic salamander (*E–J*) are illustrated. Shown are sagittal sections of the late blastula (*A*), early gastrula (*B*), and mid-gastrula (*C*) of *Rhacophorus*. A small embryo is produced on one side of a large mass of yolky vegetal cells (*D*). Midsagittal sections (*E*, *F*, *H*, *J*) and surface views (*G*, *I*) show the major features of *Megalobatrachus* gastrulation. The blastocoel roof (*light gray*), involuting marginal zone (*dark gray*), archenteron (a), "partition line" (pl), blastopore (bp), and blastocoel (blc) are shown. *A–D* redrawn from Ikeda (1902), and *E–J* redrawn from Ishikawa (1905).

Radical, Radial Intercalation and Epibolic Extension: Gastrulation in *Rhacophorus* and *Megalobatrachus maximus*, the Gigantic Salamander

In *Rhacophorus*, an arboreal frog, a radical radial intercalation and very strong epiboly spreads the animal cap and IMZ over most of the yolk by the early mid-gastrula, at which time a conventional, *Xenopus*-style gastrulation appears to occur. The egg of *Rhacophorus* is about 3 mm in diameter but is relatively yolky and forms a small embryo on the side of a large yolk (Fig. 13A–D) (Ikeda 1902). The late mid-blastula consists of an animal cap about 3 or 4 layers thick surrounding a small blastocoel and a very thick marginal zone of 10–13 layers of cells lying well above the equator (Fig. 13A) (Ikeda 1902). When the blastopore forms at or near the equator, the animal cap has thinned to about 2 cells in thickness, but the marginal zone is still 10–12 cells thick (Fig. 13B). By the mid-gastrula stage, the marginal zone has thinned to half or less the number of cell layers seen at the onset of gastrulation and moves to within 30 degrees of the vegetal pole without a large amount of involution (Fig. 13C). At this point, convergence, extension, involution, and expansion of the archenteron occur in a manner similar to *Xenopus*.

The decrease in number of cell layers, probably by radial intercalation, suggests that extension of the animal cap and marginal zone vegetally is probably twice that seen in *Xenopus*. This extensive radial intercalation probably moves, or allows the movement of, the thick supra-equatorial IMZ well below the equator before convergent extension and involution complete gastrulation. Vegetal rotation may also play a role in this dramatic epiboly, either by actively pulling the IMZ vegetally or by moving the vegetal region out of the way of epiboly. In many aspects, gastrulation in this amphibian resembles gastrulation in the sturgeon, *Acipenser transmontanus* (Bolker 1993a,b). In both these organisms, the strategy of enclosing the large yolk seems to be to move the IMZ substantially below the equator by radial intercalation, thinning, and epiboly before initiating convergent extension, which would result in exogastrulation if exercised circumblastoporally at the equatorial regions where the IMZ originates (Bolker 1993a,b).

In the gigantic salamander, *Megalobatrachus maximus*, the animal cap also undergoes a large amount of radial intercalation and thinning and spreading (Ishikawa 1905). The roof of the blastocoel is originally 3 or 4 layers thick and the marginal zone about 10 layers thick. By the onset of gastrulation, both these regions have thinned dramatically to about 2 cells in thickness. But instead of moving the IMZ vegetally, this spreading expands the animal cap and inflates a large blastocoel (Fig. 13E,F) (Ishikawa 1905). The first external sign of gastrulation is the formation of the "partition line," a crease on the dorsal surface of the gastrula that represents the contact and migration of the anterior leading-edge mesendoderm on the inner surface of the thin blastocoel roof (pl, Fig. 13F,G). The mesendoderm migrates all the way across the large blastocoel roof and eventually closes the "blastocoel window" at the opposite side (Fig. 13H,I) (Ishikawa 1905). The blastopore forms sometime after the partition line and slightly vegetal to it (Fig. 13F,G). The marginal zone involutes over the dorsal lip, which initially curves vegetally a bit, but later reverses its curvature such that its medial part leads its advance; this probably reflects a strong midline extension. There is no ventral lip and the mesoderm and archenteron are involuted dorsally, into the space formerly occupied by the blastocoel.

One explanation of this gastrulation is as follows. Very strong epiboly and thinning and extension occur by multiple rounds of radial intercalation. But the vegetal rotation movements do not occur, or they are weak, and instead of the animal cap and IMZ moving vegetally, they expand upward, creating a large blastocoel. Then the IMZ involutes over the dorsal lip into this large vault and occupies it, using mesendoderm migration and convergent extension. The ventral lip never forms, at least not before the gastrocoel has reached its full length and expanded, and most of the mesoderm appears to involute dorsally, beneath the blastocoel roof. The vegetal region is uninvolved and unenclosed at the end of gastrulation, to be encircled later by mechanisms similar to those in *Elutherodactylus* (Elinson and Fang 1998). *Megalobatrachus* gastrulation is similar to that of *Petromyzon*, a lamprey (Hatta 1891), with respect to the formation of a large blastocoel and then gastrulating into it, while largely ignoring the endoderm. The main difference between this gastrulation and that of *Rhacophorus* appears to be the lack of vegetal rotation and gastrula-stage enclosure of the yolk.

Involution on One Side of the Yolk: Gastrulation of the Salamander *Desmognathus fusca* and the Gymnophionans, *Hypogeophis rostratus* and *Hypogeophis alternans*

These organisms share with *Megalobatrachus* the property of involuting the blastoderm on one side of a large vegetal yolk mass without enclosing the yolk during gastrulation, but differ in that they do so without using a substantial blastocoel. *Desmognathus*, a small salamander, lays large, unpigmented eggs, about 3.5 mm in diameter, in small bunches (Hilton 1909). Cleavage is holoblastic (complete) but uneven, such that it produces a central vegetal region of very large yolky cells and a roof of small cells over a small blastocoel consisting of interconnected, intercellular spaces (Fig. 14A). By the early gastrula stage, the blastocoel has disappeared and the embryo consists of large internal cells nearly covered by a thin epithelial layer of small blastomeres (Fig. 14B) (Hilton 1909). Some thinning of the small-celled region occurs, and it seems likely that radial intercalation and thinning play some role in this epiboly. Whether vegetal rotation occurs is not known. The blastopore forms within the small-celled region, and the outer small-celled sheet is said to invaginate directly into the larger yolky cell mass to form a crescentic blastoporal groove (Fig. 14B–D) and then involutes and extends on the dorsal side into the interface between the outer small-celled layer and the large yolk cells (Fig. 14D–F). The initial crescent blastopore completes itself to form a ventral lip, but not much involution occurs there (Fig. 14E). Most of the mesoderm involutes over the dorsal lip and occupies a narrow area on both sides of the axis beneath the neural plate (Fig. 14F) and around the blastopore (Fig. 14E). The endoderm is not enclosed by the mesoderm during gastrulation, and this must occur later. The mesoderm becomes multilayered, and a gut cavity forms below the dorsal mesoderm by unknown means (Fig. 14F).

This type of gastrulation features a mechanism of involution that rolls the mesoderm back onto the inner surface of the small-celled blastoderm, into the interface between this layer and the vegetal cells, without a blastocoel. This movement resembles the movement of the hypoblast cells back on the inner surface of the epiblast, between the epiblast and the yolk syncytial layer, in the teleost embryo (see Chapter 20). It is not known how the forces for these movements are produced. Such a movement could be driven by traction of the involuted cells on the basal surface of the outer layer, thereby generating a shearing that would roll cells over the lip. Again, the yolk appears to be ignored during gastrulation and enclosed later.

Brauer (1897, 1899) studied the development of the Gymnophionans, or legless amphibians, *Hypogeophis rostratus*, which has an egg 7–8 mm in diameter, and *H. alternans*, which has an egg 4–5 mm in diameter; most of the work on gastrulation is on the latter species (Brauer 1897). Cleavage produces a blastula much like *Desmognathus* with a thin shell of small cells, consisting of an outer epithelium of "animal" cells and one or more deep layers of somewhat larger, yolky "vegetative" cells around a large central yolk mass. The blastocoel is lacking except for small spaces below the animal cells, among the vegetative cells, and above the large yolk. The animal cells involute over a dorsal lip and move toward the presumptive anterior, largely displacing the vegetative cells, which are thought to form endoderm (Fig. 14G). These involuted cells form the roof of the gastrocoel, which appears to consist of presumptive mesoderm. The edges of the vegetative, presumptive endodermal cell layer move beneath the mesoderm from both sides (Fig. 14H,I), and eventually meet the boundaries of the presumptive notochord (Fig. 14J,K), which then rolls up, bringing the endodermal anlage in from both sides to form the

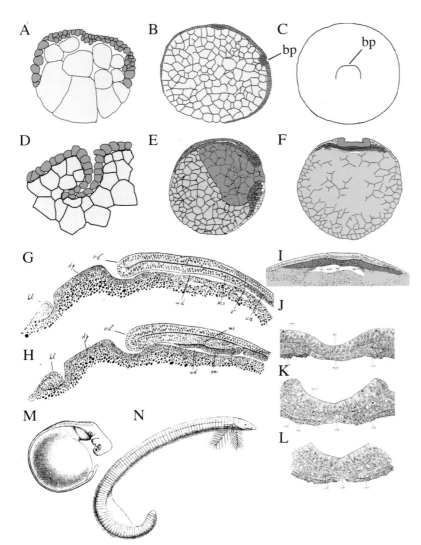

Figure 14. Features of gastrulation in *Desmognathus*, a salamander (*A–F*) and the gymnophionan, *Hypogeophis alternans* (*G–N*) are illustrated. Sectional views of *Desmognathus* (*A, B*) show a cap of small cells (*dark shading*) around one end of a large mass of vegetal endoderm (*light shading*). Sectional views (*B*, enlargement *D*) and a surface view (*C*) show invagination of the small cells at a blastopore (*bp*) in the midst of the small cells. The small-celled epithelium appears to move directly into the deep region and spreads between itself and the underlying vegetal cells to form the mesoderm (*red*, sagittal section, *E* and cross-section *F*). Probable tissue fates are neural epithelium (*blue*), mesoderm (*red*), and endoderm (*green*). Sagittal sections (*G,H*), cross-sections (*I–L*), and surface views (*M,N*) show the gastrulation and development of *Hypogeophis*, a gymnophionan, as described in the text. Shown are the Dotterpfropf (dp), yolk plug; entoderm (en), endoderm; mesoderm (ms), mesoderm; Beruhrungsstelle vom Entoderm und Mesoderm (x), place of contact between endoderm and mesoderm; hintere Lippe des Blastoporus (hl), ventral lip of the blastopore; vordere Lippe (vl), dorsal lip; Urdarmhohle (ud), archenteron. (From Brauer 1897.)

definitive archenteron (Fig. 14K,L). The gross features of closing the roof of the archenteron and removing the mesoderm from the surface resemble those in the commonly studied anuran and urodele embryos. The blastopore is completed ventrally and the yolk plug closes as in the more commonly studied amphibians, although there appears to be little involution on the ventral side (Fig. 14G,H). As in *Desmognathus*, the involuted mesoderm is found in the dorsal sector and does not appear to enclose the yolk (Fig. 14I). A slender body axis forms on one side of the yolk (Fig. 14M) and elongates dramatically as the yolk is metabolized (Fig. 14N). Overall, *Hypogeophis* shares many features of gastrulation with *Desmognathus*, including the involution of a blastoderm into an interface rather than a blastocoel, confinement of the involuted mesoderm to a narrow, dorsal strip, and failure of this mesoderm to enclose the yolk during gastrulation.

Gastrulation in Gastrotheca: A Symmetric Blastodisc with Delayed Extension

Cleavage, gastrulation, and other aspects of development of *Gastrotheca rhiobambae* have been described previously (Fig. 15A–C) (del Pino and Escobar 1981; del Pino 1982, 1996; del Pino and Elinson 1983; Elinson and del Pino 1985; del Pino and Loor 1990; del Pino and Medina 1998). *Gastrotheca*, a marsupial frog, broods its large, pale yellow eggs (2.1 mm diameter at fertilization, 3.6 mm diameter after blastocoel formation) in a pouch on the mother's back (Elinson and del Pino 1985). Cleavage is holoblastic but unequal; the yolk-poor cytoplasm at the animal cap cleaves rapidly and forms small cells, whereas the yolk-rich remainder cleaves slowly to form large cells over three-quarters of the embryonic surface (Elinson and del Pino 1985). The animal region expands as the blastocoel forms there, and

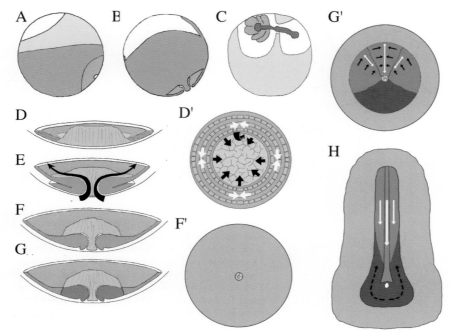

Figure 15. Features of gastrulation in *Gastrotheca rhiobambae* are illustrated in surface (*A*) and sectional (*B*) views of the gastrula, and in a surface view of the postneurula (*C*). Diagrams of sectional (*D–G*) and vegetal surface views (*D′– F′*) show symmetrical involution of the symmetrical embryonic disc (*orange*) and its role in convergence and blastopore closure. Surface diagrams of the involuted embryonic disc show one possible pattern of post-gastrula convergent extension. Probable tissue identities of notochord (*magenta*), somitic mesoderm (*red*), lateroventral mesoderm (*orange*), and regional cell behavior (convergence, *black arrows*; extension, *white arrows*) are shown. (*A–C*, Redrawn from del Pino and Elinson 1983.)

eventually it consists of a single layer of cells (Fig. 15A). This blastocoel roof region, a marginal region just vegetal to it, and a vegetal region can be distinguished in the blastula stage by differential staining with silver. Dye marking shows that prior to gastrulation the marginal zone moves vegetally in an epibolic movement, probably due to a vegetal contraction (Elinson and del Pino 1985). Whether this is associated with internal vegetal rotation movements is not known. Gastrulation begins with migration of cells across the thin roof of the blastocoel, and at the other end of the large egg, a blastopore forms at the boundary of the vegetal and marginal region near the vegetal pole, and the latter involutes to form a symmetrical "embryonic disc" and a short symmetrical gastrocoel (Fig. 15A,B). After blastopore closure, the embryo elongates and develops on one side of a large egg (Fig. 15C) (del Pino and Elinson 1983; Elinson and del Pino 1985).

The hallmark of this type of gastrulation is the symmetric involution of a symmetric IMZ (Fig. 15D–F). Prior to blastopore closure, the body axis does not elongate (del Pino and Elinson 1983), no notochord or axial extension is detected, and *Brachyury* is expressed symmetrically and circumblastoporally (del Pino 1996). The involution and blastopore closure appear similar to movements in the ventral sector of normal *Xenopus* gastrulae and in all sectors of UV-ventralized *Xenopus* embryos (Scharf and Gerhart 1980; Gerhart et al. 1983), including the formation of bottle cells, the formation of a shallow invagination, and then involution of a collar of mesoderm. The "embryonic disc" of *Gastrotheca* is initially quite thick,

about 15 cells (del Pino 1996), suggesting that the convergent thickening seen in the ventral IMZ of *Xenopus* may occur throughout the IMZ of *Gastrotheca*. A convergent thickening machinery spanning the IMZ would set up a progressive symmetrical hoop stress and drive involution and blastopore closure (Fig. 15D′,F′).

After blastopore closure, *Brachyury* expression becomes asymmetric with the development of an elongated area on one side of the blastopore, the dorsal side, which extends and forms the notochord and the body axis (del Pino and Elinson 1983; del Pino 1996). The embryonic disc also thins during this period from 15 cells to about 4 by the time the elongated archenteron expands (Elinson and del Pino 1985). These facts are consistent with a delayed convergent extension driving the elongation of the post-involution mesoderm (Fig. 15G, G′,H). Similarly, *Brachyury* expression is delayed until after blastopore closure in *Clolesthus manchalilla*, a dendrobatid frog (Benitez and del Pino 2002), and in the large egg (3.5 mm) of *Eluthrodactylus coqui*, a direct developing frog (Ninomiya et al. 2001), and probably represents delayed convergent extension in these organisms.

Gastrotheca appears to move the IMZ to the vegetal pole of the egg by an enhanced epiboly, but the mechanism has not been described. The symmetric gastrulation and blastopore closure, and convergent extension and body axis elongation delayed until after blastopore closure may be features allowed by the slow gastrulation of this species, which takes about a week. In contrast, the fast-developing *Xenopus* embryo, which gastrulates in 4 hours, telescopes the beginnings of body elongation into gastrulation and uses the process for anisotropic involution and blastopore closure. Isotropic convergent thickening movements may also have advantages over anisotropic convergent extension movements in the geometry and mechanics of the vegetal region of *Gastrotheca*. The IMZ of *Gastrotheca* involutes into a region of the embryo that essentially has no blastocoel, the blastocoel being a long way away at the other end of the large embryo, whereas the IMZ of *Xenopus* rolls directly into the blastocoel.

Comparison of different types of gastrulation provides mechanistic insights in a number of ways, and one of the most important is establishing the identity and distinctness, the separateness of mechanisms (del Pino 1996). A mechanism displayed alone, in pure, robust form in one organism may be easier to analyze than the same movement in modified, weak form, and perhaps partially masked by others, in another organism. These relatively large, relatively yolky, and slow-developing eggs may display some of the mechanisms in the standard model systems in stronger form, and perhaps offer previously unknown mechanisms, and new uses for previously described mechanisms as well.

Several strategies of gastrulation emerge from studies of large, yolky eggs. One major strategy of gastrulating with a large yolk is to move the IMZ far vegetally, below the equator, before expressing convergent extension or convergent thickening. This potentially involves enhancement of radial intercalation and epiboly and vegetal rotation. Use of vegetal rotation is a particularly effective strategy, since the yolky vegetal endoderm would then become part of the solution to the problem its increasing size had generated. Vegetal rotation and radial intercalation should be studied in these species, since little is known about them. Another strategy is to abandon the encircling type of annular IMZ and gastrulate from a patch of IMZ on one side of the yolk, without enclosing the yolk in mesoderm during gastrulation. It appears that this type of asymmetric gastrulation can occur into a blastocoel, into an interfacial zone between layers, or into interconnected, small cavities. The mechanisms involved in the latter are uncharacterized and may involve specialization of the mechanisms seen in the common model systems, such as mesendodermal migration.

These are speculations from the data available, and to understand gastrulation in these specialized eggs, a number of things are needed: better fate mapping, high-resolution live imaging of cell behaviors, use of explants to probe the mechanical autonomy and force-generating contribution of various components, and molecular markers to describe tissue differentiation. This requires adaptation of methods worked out on more convenient species. Also needed are corroborative studies of reproductive ecology, endocrinology, and behavior, and reproductive strategies of these organisms. Hypotheses about the performance parameters of specific morphogenetic machines in different contexts should be tested by making interspecific combinations and by engineering eggs with more or less yolk (see Sinervo and McEdward 1988; Sinervo 1990).

ACKNOWLEDGMENTS

We thank Toshiyasu Goto for translation of articles in Japanese and members of the Keller laboratory, past and present, for their contributions to this work. We also thank Claudio Stern, David Crotty, and Inez Sialiano for their help in bringing this work into final form.

REFERENCES

Baker P.C. 1965. Fine structure and morphogenic movements in the gastrula of the treefrog, *Hyla regilla*. *J. Cell Biol.* **24:** 95–116.

Ballard W.W. 1955. Cortical ingression during cleavage of amphibian eggs, studied by means of vital dyes. *J. Exp. Zool.* **129:** 77–97.

Bauer D.V., Huang S., and Moody S.A. 1994. The cleavage stage origin of Spemann's Organizer: Analysis of the movements of blastomere clones before and during gastrulation in *Xenopus*. *Development* **120:** 1179–1189.

Benitez M.-S. and del Pino E.M. 2002. Expression of Brachyury during development of the dendrobatid frog *Colostethus machalilla*. *Dev. Dyn.* **225:** 592–596.

Bolker J.A. 1993a. Gastrulation and mesoderm morphogenesis in the white sturgeon. *J. Exp. Zool.* **266:** 116–131.

———. 1993b. The mechanism of gastrulation in the white sturgeon. *J. Exp. Zool.* **266:** 132–145.

Boucaut J.C. and Darribere T. 1983a. Fibronectin in early amphibian embryos. Migrating mesodermal cells contact fibronectin established prior to gastrulation. *Cell Tissue Res.* **234:** 135–145.

———. 1983b. Presence of fibronectin during early embryogenesis in amphibian *Pleurodeles waltlii*. *Cell Differ.* **12:** 77–83.

Boucaut J.C., Darribere T., Boulekbache H., and Thiery J.P. 1984a. Prevention of gastrulation but not neurulation by antibodies to fibronectin in amphibian embryos. *Nature* **307:** 364–367.

Boucaut J.C., Darribere T., Li S.D., Boulekbache H., Yamada K.M., and Thiery J.P. 1985. Evidence for the role of fibronectin in amphibian gastrulation. *J. Embryol. Exp. Morphol.* **89:** 211–227.

Boucaut J.C., Darribere T., Poole T.J., Aoyama H., Yamada K.M., and Thiery J.P. 1984b. Biologically active synthetic peptides as probes of embryonic development: A competitive peptide inhibitor of fibronectin function inhibits gastrulation in amphibian embryos and neural crest cell migration in avian embryos. *J. Cell Biol.* **99:** 1822–1830.

Brauer A. 1897. I. Beiträge zur Kenntniss der Entwicklungsgeschichte und der Anatomie der Gymnophionen. *Zool. Jahrb.* **10:** 389–470.

———. 1899. Beitrage zur Kenntniss der Entwicklung und Anatomie der Gymnophionen II. Die Entwicklung der aussern Form. *Zool. Jarb. Anat.* **12:** 477–508.

Broders F. and Thiery J.P. 1995. Contribution of cadherins to directional cell migration and histogenesis in *Xenopus* embryos. *Cell Adhes. Commun.* **3:** 419–440.

Brun R.B. and Garson J.A. 1984. Notochord formation in the Mexican salamander (*Ambystoma mexicanum*) is different from notochord formation in *Xenopus laevis*. *J. Exp. Zool.* **229:** 235–240.

Burnside M.B. and Jacobson A.G. 1968. Analysis of morphogenetic movements in the neural plate of the newt *Taricha torosa*. *Dev. Biol.* **18:** 537–552.

Daniel J.F. and Yarwood E.A. 1939. The early embryology of *Triturus torosus*. *Univ. Calif. Publ. Zool.* **43:** 321–356.

Darribere T. and Schwarzbauer J.E. 2000. Fibronectin matrix composition and organization can regulate cell migration during amphibian development. *Mech. Dev.* **92:** 239–250.

Darribere T., Boucher D., Lacroix J.C., and Boucaut J.C. 1984. Fibronectin synthesis during oogenesis and early development of the amphibian *Pleurodeles waltlii*. *Cell Differ.* **14:** 171–177.

Darribere T., Yamada K.M., Johnson K.E., and Boucaut J.C. 1988. The 140-kDa fibronectin receptor complex is required for mesodermal cell adhesion during gastrulation in the amphibian *Pleurodeles waltlii*. *Dev. Biol.* **126:** 182–194.

Darribere T., Riou J.F., Shi D.L., Delarue M., and Boucaut J.C. 1986. Synthesis and distribution of laminin-related polypeptides in early amphibian embryos. *Cell Tissue Res.* **246:** 45–51.

Davidson L.A. and Keller R.E. 1999. Neural tube closure in *Xenopus laevis* involves medial migration, directed protrusive activity, cell intercalation and convergent extension. *Development* **126:** 4547–4556.

Davidson L.A., Hoffstrom B.G., Keller R., and DeSimone D.W. 2002. Mesendoderm extension and mantle closure in *Xenopus laevis* gastrulation: Combined roles for integrin alpha(5)beta(1), fibronectin, and tissue geometry. *Dev. Biol.* **242:** 109–129.

del Pino E.M. 1982. Multinucleate oogenesis, embryonic development and other adaptations for reproduction on land in egg-brooding hylid frogs. *Prog. Clin. Biol. Res.* **85:** 397–405.

———. 1996. The expression of brachyury (T) during gastrulation in the marsupial frog *Gastrotheca riobambae*. *Dev. Biol.* **177,** 64–72.

del Pino E.M. and Elinson R.P. 1983. A novel developmental Pattern for frogs: Gastrulation produces an embryonic disk. *Nature* **306:** 589–561.

del Pino E.M. and Escobar B. 1981. Embryonic stages of *Gastrotheca riobambae* (Fowler) during maternal incubation and comparison of development with that of other egg-brooding hylid frogs. *J. Morphol.* **167:** 277–295.

del Pino E.M. and Loor V.-S. 1990. The pattern of early cleavage of the marsupial frog *Gastrotheca riobambae*. *Development* **110:** 781–790.

del Pino E.M. and Medina A. 1998. Neural development in the marsupial frog *Gastrotheca riobambae*. *Int. J. Dev. Biol.* **42:** 723–731.

Delarue M., Johnson K.E., and Boucaut J.C. 1994. Superficial cells in the early gastrula of *Rana pipiens* contribute to mesodermal derivatives. *Dev. Biol.* **165:** 702–715.

———. 1996. Anteroposterior segregation of superficial and deep cells during gastrulation in *Pleurodeles waltl* and *Rana pipiens* embryos. *J. Exp. Zool.* **276:** 345–360.

Delarue M., Saez F.-J., Johnson K.-E., and Boucaut J.-C. 1995. Restriction of cell fate of superficial cells in the marginal zone of the amphibian embryo *Pleurodeles waltlii*. *J. Exp. Zool.* **273:** 303–316.

Delarue M., Saez F.-J., Boucaut J.-C., Thiery J.-P., and Broders F. 1998. Medial cell mixing during axial morphogenesis of the amphibian embryo requires cadherin function. *Dev. Dyn.* **213:** 248–260.

Delarue M., Sanchez S., Johnson K.E., Darribere T., and Boucaut J. C. 1992. A fate map of superficial and deep circumblastoporal cells in the early gastrula of *Pleurodeles waltlii*. *Development* **114:** 135–146.

Devoto S.-H., Melancon E., Eisen J.-S., and Westerfield M. 1996. Identification of separate slow and fast muscle precursor cells in vivo, prior to somite formation. *Development* **122:** 3371–3380.

Domingo C. and Keller R. 1995. Induction of notochord cell intercalation behavior and differentiation by progressive signals in the gastrula of *Xenopus laevis*. *Development* **121:** 3311–3321.

———. 2000. Cells remain competent to respond to mesoderm-inducing signals present during gastrulation in *Xenopus laevis*. *Dev. Biol.* **225:** 226–240.

Doniach T., Phillips C.R., and Gerhart J.C. 1992. Planar induction of anteroposterior pattern in the developing central nervous system of *Xenopus laevis*. *Science* **257:** 542–545.

Duellman W.E. and Trueb L. 1994. *Biology of amphibians*. The Johns Hopkins University Press, Baltimore, Maryland.

Elinson R.P. 1994. Leg development in a frog without a tadpole (*Eleutherodactylus coqui*). *J. Exp. Zool.* **270:** 202–210.

———. 2001. Direct development: An alternative way to make a frog. *Genesis* **29:** 91–95.

Elinson R.P. and Beckham Y. 2002. Development in frogs with large eggs and the origin of amniotes. *Zoology* **105:** 1–13.

Elinson R.P. and del Pino E.M. 1985. Cleavage and gastrulation in the egg-brooding, marsupial frog, *Gastrotheca riobambae*. *J. Embryol. Exp. Morphol.* **90:** 223–232.

Elinson R.P. and Fang H. 1998. Secondary coverage of the yolk by the body wall in the direct developing frog, *Eleutherodacytlus coqui*: An unusual process for amphibian embryos. *Dev. Genes Evol.* **20:** 457–466.

Elul T. and Keller R. 2000. Monopolar protrusive activity: A new morphogenic cell behavior in the neural plate dependent on vertical interactions with the mesoderm in *Xenopus*. *Dev. Biol.* **224:** 3–19.

Elul T., Koehl M.A., and Keller R. 1997. Cellular mechanism underlying neural convergent extension in *Xenopus laevis* embryos. *Dev. Biol.* **191,** 243–58.

———. 1998. Patterning of morphogenetic cell behaviors in neural ectoderm of *Xenopus laevis*. *Ann. N.Y. Acad. Sci.* **857:** 248–251.

Ezin M., Skoglund P., and Keller R. 2002. The midline (notochord and notoplate) patterns cell motility underlying convergence and extension of the *Xenopus* neural plate. *Dev. Biol.* **256:** 100–113.

Fesenko I., Kurth T., Sheth B., Fleming T.P., Citi S., and Hausen P. 2000. Tight junction biogenesis in the early *Xenopus* embryo. *Mech. Dev.* **96:** 51–65.

Gerhart J., Black S., and Scharf S. 1983. Cellular and pancellular organization of the ambhibian embryo. In *Modern cell biology*, vol. 2, pp. 483–507. Alan R. Liss, New York.

Gillespie J.I. 1983. The distribution of small ions during the early development of *Xenopus laevis* and *Ambystoma mexicanum* embryos. *J. Physiol.* **344:** 359–377.

Goodale H.D. 1911. The early development of *Spelerpes bilineatus* (Green). *Am. J. Anat.* **12:** 173–247.

Hardin J. and Keller R. 1988. The behaviour and function of bottle cells during gastrulation of *Xenopus laevis*. *Development* **103:** 211–230.

Hatta S. 1891. On the gastrulation in *Petromyzon*. *J. Coll. Sci.* **21:** 1–44.

Hilton W.A. 1909. General features of the early development of *Desmognathus fusca*. *J. Morphol.* **20:** 533–547.

Holtfreter J. 1943a. Properties and function of the surface coat in amphibian embryos. *J. Exp. Zool.* **93:** 251–323.

———. 1943b. A study of the mechanics of gastrulation. Part I. *J. Exp. Zool.* **94:** 261–318.

———. 1944. A study of the mechanics of gastrulation. Part II. *J. Exp. Zool.* **95:** 171–212.

Hopwood N.D., Pluck A., and Gurdon J.B. 1989. MyoD expression in the forming somites is an early response to mesoderm induction in *Xenopus* embryos. *EMBO J.* **8:** 3409–3418.

Ibrahim H. and Winklbauer R. 2001. Mechanisms of mesendoderm internalization in the *Xenopus* gastrula: Lessons from the ventral side. *Dev. Biol.* **240:** 108–122.

Ikeda S. 1902. Contribution to the embryology of amphibia: The mode of blastopore closure and the position of the embryonic body. *J. Coll. Sci.* **17:** 81–90.

Imoh H. 1988. Formation of germ layers and roles of the dorsal lip of the blastopore in normally developing embryos of the newt *Cynops pyrrhogaster. J. Exp. Zool.* **246:** 258–270.

Ishikawa C. 1905. The gastrulation of the gigantic salamander, *Megalobatrachus maximus. Zool. Mag.* **17:** 26–28.

Jacobson A.G. and Gordon R. 1976. Changes in the shape of the developing vertebrate nervous system analyzed experimentally, mathematically and by computer simulation. *J. Exp. Zool.* **197:** 191–246.

Johnson K.E., Darribere T., and Boucaut J.C. 1992. *Ambystoma maculatum* gastrulae have an oriented, fibronectin-containing extracellular matrix. *J. Exp. Zool.* **261:** 458–471.

———. 1993. Mesodermal cell adhesion to fibronectin-rich fibrillar extracellular matrix is required for normal *Rana pipiens* gastrulation. *J. Exp. Zool.* **265:** 40–53.

Johnson K.E., Boucaut J.-C., Darribere T., Shi D.-L., Riou J.-F., Boulek Bache H., and Delarue M. 1990. Fibrohectin-rich fibrillar extracellular matrix controls cell migration during amphibian gastrulation. *Int. J. Dev. Biol.* **34:** 139–147.

Kao K.R. and Elinson R.P. 1988. The entire mesodermal mantle behaves as Spemann organizer in dorsoanterior enhanced *Xenopus laevis* embryos. *Dev. Biol.* **127:** 64–77.

Kay B.K. and Peng H.B., eds. 1991. Xenopus laevis: *Practical uses in cell and molecular biology. Methods Cell Biol.*, vol. 36. Academic Press, New York.

Keller R. 1975. Vital dye mapping of the gastrula and neurula of *Xenopus laevis.* I. Prospective areas and morphogenetic movements of the superficial layer. *Dev. Biol.* **42:** 222–241.

———. 1976. Vital dye mapping of the gastrula and neurula of *Xenopus laevis.* II. Prospective areas and morphogenetic movements of the deep layer. *Dev. Biol.* **51:** 118–137.

———. 1978. Time lapse cinematographic of superficial cell behaviour during and prior to gastrulation in *Xenopus laevis. J. Morphol.* **157:** 223–247.

———. 1980. The cellular basis of epiboly: an SEM study of deep-cell rearrangement during gastrulation in *Xenopus laevis. J. Embryol. Exp. Morphol.* **60:** 201–34.

———. 1981. An experimental analysis of the role of bottle cells and the deep marginal zone in gastrulation of *Xenopus laevis. J. Exp. Zool.* **216:** 81–101.

———. 1984. The cellular basis of gastrulation in *Xenopus laevis*: Active, postinvolution convergence and extension by mediolateral interdigitation. *Am. Zool.* **24:** 589–603.

———. 1986. The cellular basis of amphibian gastrulation. In *Developmental biology: A comprehensive synthesis. 2. The cellular basis of morphogenesis* (ed. L. Browder), pp. 241–327. Plenum, New York.

———. 1991. Early embryonic development of *Xenopus laevis. Methods Cell Biol.* **36:** 61–113.

Keller R. and Danilchik M. 1988. Regional expression, pattern and timing of convergence and extension during gastrulation of *Xenopus laevis. Development* **103:** 193–209.

Keller R. and Hardin J. 1987. Cell behaviour during active cell rearrangement: evidence and speculations. *J. Cell Sci.* (suppl.) **8:** 369–393.

Keller R. and Jansa S. 1992. *Xenopus* gastrulation without a blastocoel roof. *Dev. Dyn.* **195:** 162–176.

Keller R. and Tibbetts P. 1989. Mediolateral cell intercalation in the dorsal, axial mesoderm of *Xenopus laevis. Dev. Biol.* **131:** 539–549.

Keller R., Davidson L.A., and Shook D.R. 2003. How we are shaped: The biomechanics of gastrulation. *Differentiation* **71**, 171–205.

Keller R., Shih J., and Domingo C. 1992a. The patterning and functioning of protrusive activity during convergence and extension of the *Xenopus* organiser. *Development Suppl.* 81–91.

Keller R., Shih J., and Sater A. 1992b. The cellular basis of the convergence and extension of the *Xenopus* neural plate. *Dev. Dyn.* **193:** 199–217.

Keller R.E., Danilchik M., Gimlich R., and Shih J. 1985. The function and mechanism of convergent extension during gastrulation of *Xenopus laevis. J. Embryol. Exp. Morphol.* **89:** 185–209.

Keller R., Shih J., Sater A.K., and Moreno C. 1992c. Planar induction of convergence and extension of the neural plate by the organizer of *Xenopus. Dev. Dyn.* **193:** 218–234.

Keller R., Shih J., Wilson P.A., and Sater A.K. 1991. Patterns of cell motility, cell interactions, and mechanisms during convergent extension in *Xenopus.* In *Cell-cell interactions in early development* (49th Symposium of the Society for Developmental Biology) (ed. G.C. Gerhart), pp. 31–62. Wiley-Liss, New York.

Keller R., Cooper M.S., Danilchik M., Tibbetts P., and Wilson P.A. 1989. Cell intercalation during notochord development in *Xenopus laevis. J. Exp. Zool.* **251:** 134–154.

Keller R., Davidson L., Edlund A., Elul T., Ezin M., Shook D., and Skoglund P. 2000. Mechanisms of convergence and extension by cell intercalation. *Philos. Trans. R. Soc. Lond. B Biol. Sci.* **355:** 897–922.

King H. D. 1902. The gastrulation of the egg of *Bufo lentiginosus. Am. Nat.* **36**, 527–548.

Kubota H.Y. and Durston A.J. 1978. Cinematographical study of cell migration in the opened gastrula of *Ambystoma mexicanum. J. Embryol. Exp. Morphol.* **44:** 71–80.

Lane M.C. and Keller R. 1997. Microtubule disruption reveals that Spemann's organizer is subdivided into two domains by the vegetal alignment zone. *Development* **124:** 895–906.

Lane M.C. and Sheets M.D. 2002. Rethinking axial patterning in amphibians. *Dev. Dyn.* **225:** 434–447.

Lane M.C. and Smith W.C. 1999. The origins of primitive blood in *Xenopus*: Implications for axial patterning. *Development* **126:** 423–34.

Lewis W.H. 1948. Mechanics of *Amblystoma gastrulation. Anat. Rec.* **101:** 700.

———. 1952. Gastrulation of *Amblystoma punctatum. Anat. Rec.* **112:** 473.

Lofberg J. 1974. Apical surface topography of invaginating and noninvaginating cells. A scanning-transmission study of amphibian neurulae. *Dev. Biol.* **36:** 311–329.

Lundmark C. 1986. Role of bilateral zones of ingressing superficial cells during gastrulation of *Ambystoma mexicanum. J. Embryol. Exp. Morphol.* **97:** 47–62.

Marsden M. and DeSimone D.W. 2001. Regulation of cell polarity, radial intercalation and epiboly in *Xenopus*: Novel roles for integrin and fibronectin. *Development* **128:** 3635–3647.

Minsuk S.B. 1995. "A comparative study of gastrulation and mesoderm formation in pipid frogs." Ph.D. thesis, University of California, Berkeley.

Minsuk S.B. and Keller R.E. 1996. Dorsal mesoderm has a dual origin and forms by a novel mechanism in *Hymenochirus*, a relative of

Xenopus. Dev. Biol. **174:** 92–103.

———. 1997. Surface mesoderm in *Xenopus*: A revision of the stage 10 fate map. *Dev. Genes Evol.* **207:** 389–401.

Moore S.W. 1994. A fiber optic system for measuring dynamic mechanical properties of embryonic tissues. *IEEE Trans. Biomed. Eng.* **41:** 45–50.

Moore S.W., Keller R.E., and Koehl M.A.R. 1995. The dorsal involuting marginal zone stiffens anisotropically during its convergent extension in the gastrula of *Xenopus laevis*. *Development* **121:** 3131–3140.

Nagel M. and Winklbauer R. 1999. Establishment of substratum polarity in the blastocoel roof of the *Xenopus* embryo. *Development* **128:** 1975–1984.

Nakatsuji N. 1974. Studies on the gastrulation of amphibian embryos: Pseudopodia in the gastrula of *Bufo bufo japonicus* and their significance to gastrulation. *J. Embryol. Exp. Morphol.* **32:** 795–804.

———. 1975a. Studies on the gastrulation of amphibian embryos: Cell movement during gastrulation in *Xenopus laevis* embryos. *Wilhelm Roux' Arch.* **178:** 1–14.

———. 1975b. Studies on the gastrulation of amphibian embryos: Light and electron microscopic observation of a urodele *Cynops pyrrhogaster*. *J. Embryol. Exp. Morphol.* **34:** 669–685.

———. 1976. Studies on the gastrulation of amphibian embryos: Ultrastructure of the migrating cell of anurans. *Wilhelm Roux' Arch.* **180:** 229–240.

———. 1984. Cell locomotion and contact guidance in amphibian gastrulation. *Am. Zool.* **24:** 615–627.

———. 1986. Presumptive mesoderm cells from *Xenopus laevis* gastrulae attach to and migrate on substrata coated with fibronectin or laminin. *J. Cell Sci.* **86:** 109–118.

Nakatsuji N. and Johnson K.E. 1982. Cell locomotion in vitro by *Xenopus laevis* gastrula mesodermal cells. *Cell Motil.* **2:** 149–161.

———. 1983a. Comparative study of extracellular fibrils on the ectodermal layer in gastrulae of five amphibian species. *J. Cell Sci.* **59:** 61–70.

———. 1983b. Conditioning of a culture substratum by the ectodermal layer promotes attachment and oriented locomotion by amphibian gastrula mesodermal cells. *J. Cell Sci.* **59:** 43–60.

———. 1984. Experimental manipulation of a contact guidance system in amphibian gastrulation by mechanical tension. *Nature* **307:** 453–455.

Nakatsuji N., Gould A.C., and Johnson K.E. 1982. Movement and guidance of migrating mesodermal cells in *Ambystoma maculatum* gastrulae. *J. Cell Sci.* **56:** 207–222.

Nieuwkoop P.D., and Faber J. 1967. *Normal table of* Xenopus laevis *(Daudin)*. North Holland Publishing, Amsterdam, The Netherlands.

Nieuwkoop P. and Florshutz P. 1950. Quelques caracteres specieux de la gastrulation et de la neurulation de l'ouef de *Xenopus laevis*, Daud. et de quelques autres Anoures. l'ere partie. Etude descriptive. *Arch. Biol.* **61:** 113–150.

Ninomiya H., Zhang Q., and Elinson R.P. 2001. Mesoderm formation in *Eleutherodactylus coqui*: Body patterning in a frog with a large egg. *Dev. Biol.* **236:** 109–123.

Novoselov V.V. 1995. Notochord formation in amphibians: Two directions and two ways. *J. Exp. Zool.* **271:** 296–306.

Pasteels J. 1942. New observations concerning the maps of presumptive areas of the young amphibian gastrula. (*Amblystoma* and *Discoglossus*). *J. Exp. Zool.* **89:** 255–281.

Perry M.M. and Waddington C.H. 1966. Ultrastructure of the blastopore cells in the newt. *J. Embryol. Exp. Morphol.* **15:** 317–330.

Poznanski A. and Keller R. 1997. The role of planar and early vertical signaling in patterning the expression of Hoxb-1 in *Xenopus*. *Dev. Biol.* **184:** 351–356.

Purcell S.M. and Keller R. 1993. A different type of amphibian mesoderm morphogenesis in *Ceratophrys ornata*. *Development* **117:** 307–317.

Ramos J.W. and Desimone D.W. 1996. *Xenopus* embryonic cell adhesion to fibronectin: Position-specific activation of RGD/synergy site-dependent migratory behavior at gastrulation. *J. Cell Biol.* **134:** 227–240.

Ramos J.W., Whittaker C.A., and DeSimone D.W. 1996. Integrin-dependent adhesive activity is spatially controlled by inductive signals at gastrulation. *Development* **122:** 2873–2883.

Ransom D.G., Hens M.D., and DeSimone D.W. 1993. Integrin expression in early amphibian embryos: cDNA cloning and characterization of *Xenopus* beta 1, beta 2, beta 3, and beta 6 subunits. *Dev. Biol.* **160:** 265–75.

Sanders E.J., and Dicaprio R.A. 1976. Intercellular junctions in the *Xenopus* embryo prior to gastrulation. *J. Exp. Zool.* **197:** 415–421.

Sater A.-K., Steinhardt R.-A., and Keller R. 1993. Induction of neuronal differentiation by planar signals in *Xenopus* embryos. *Dev. Dyn.* **197:** 268–280.

Scharf S.R. and Gerhart J.C. 1980. Determination of the dorsal-ventral axis in eggs of *Xenopus laevis*: Complete rescue of uv-impaired eggs by oblique orientation before first cleavage. *Dev Biol.* **79:** 181–98.

Schechtman A.M. 1934. Unipolar ingression in *Triturus torosus*: Hitherto undescribed movement in the pregastrular stages of a urodele. *Univ. Calif. Publ. Zool.* **36:** 277–292.

———. 1935. Mechanism of ingression in the egg of *Triturus torosus*. *Proc. Soc. Exptl. Biol. Med.* **32:** 1072–1073.

———. 1942. The mechanism of amphibian gastrulation. I. Gastrulation-promoting interactions between various region of an anuran egg (*Hyla regilla*). *Univ. Calif. Publ. Zool.* **51:** 1–39.

Shi D.L., Darribere T., Johnson K.E., and Boucaut J.C. 1989. Initiation of mesodermal cell migration and spreading relative to gastrulation in the urodele amphibian *Pleurodeles waltlii*. *Development* **105:** 351–364.

Shi D.L., Delarue M., Darribere T., Riou J.F., and Boucaut J.C. 1987. Experimental analysis of the extension of the dorsal marginal zone in *Pleurodeles waltlii* gastrulae. *Development* **100:** 147–162.

Shih J. and Keller R. 1992a. Cell motility driving mediolateral intercalation in explants of *Xenopus laevis*. *Development* **116:** 901–914.

———. 1992b. The epithelium of the dorsal marginal zone of *Xenopus* has organizer properties. *Development* **116:** 887–99.

———. 1992c. Patterns of cell motility in the organizer and dorsal mesoderm of *Xenopus laevis*. *Development* **116:** 915–930.

Shook D. and Keller R. 2003. Mechanisms, mechanics, and function of epithelial-mesenchymal transitions in early development. *Mech. Dev.* **120:** 1351–1383.

Shook D.R., Majer C., and Keller R. 2002. Urodeles remove mesoderm from the superficial layer by subduction through a bilateral primitive streak. *Dev. Biol.* **248:** 220–239.

Shook D.R., Majer C., and Keller R. 2004. Pattern and morphogenesis of presumptive superficial mesoderm in two closely related species, *Xenopus laevis* and *Xenopus tropicalis*. *Dev. Biol.* **270:** 163–185.

Sinervo B. 1990. The evolution of maternal investment in lizards: An experimental and comparative analysis of egg size and its effect on offspring performance. *Evolution* **44:** 279–294.

Sinervo B. and McEdward L.R. 1988. Developmental consequences of an evolutionary change in egg size: An experimental test. *Evolution* **42:** 885–899.

Slack J.M. and Tannahill D. 1992. Mechanism of anteroposterior axis specification in vertebrates. Lessons from the amphibians. *Development* **114:** 285–302.

Sive H.L., Grainger R.M., and Harland R.M. 2000. *Early development of* Xenopus Laevis: *A laboratory manual.* Cold Spring Harbor Laboratory Press, Cold Spring Harbor, New York.

Slack J.M., Isaacs H.V., Johnson G.E., Lettice L.A., Tannahill D., and Thompson J. 1992. Specification of the body plan during *Xenopus* gastrulation: Dorsoventral and anteroposterior patterning of the mesoderm. *Development Suppl.* **1992:** 143–149.

Smith J.C. and Malacinski G.M. 1983. The origin of the mesoderm in an anuran, *Xenopus laevis* and a urodele, *Ambystoma mexicanum*. *Dev. Biol.* **98:** 250–254.

Smith J.C., Symes K., Hynes R.O., and DeSimone D. 1990. Mesoderm induction and the control of gastrulation in *Xenopus laevis*: The roles of fibronectin and integrins. *Development* **108:** 229–238.

Sporle R. 2001. Epaxial-adaxial-hypaxial regionalisation of the vertebrate somite: Evidence for a somitic organiser and a mirror-image duplication. *Dev. Genes Evol.* **211:** 198–217.

Vogt W. 1929. Gestaltungsanalyse am Amphibienkiem Mit Ortlicher Vitalfarbung. II. Teil. Gastrulation und Mesoderbildung Bei Urodelen und Anuren. *Wilhelm Roux' Arch. Entwicklungsmech. Org.* **120:** 384–601.

Wacker S., Grimm K., Joos T., and Winklbauer R. 2000. Development and control of tissue separation at gastrulation in *Xenopus*. *Dev. Biol.* **224:** 428–439.

Wang X., Lessman C.A., Taylor D.B., and Gartner T.K. 1995. Fibronectin peptide DRVPHSRNSIT and fibronectin receptor peptide DLYYLMDL arrest gastrulation of *Rana pipiens*. *Experientia* **51:** 1097–102.

Wilson P. 1990. "The development of the axial mesoderm in *Xenopus laevis*." Ph.D. thesis, University of California, Berkeley.

Wilson P. and Keller R. 1991. Cell rearrangement during gastrulation of *Xenopus*: Direct observation of cultured explants. *Development* **112:** 289–300.

Wilson P.A., Oster G., and Keller R. 1989. Cell rearrangement and segmentation in *Xenopus*: Direct observation of cultured explants. *Development* **105:** 155–166.

Winklbauer R. 1990. Mesodermal cell migration during *Xenopus* gastrulation. *Dev. Biol.* **142:** 155–168.

———. 1998. Conditions for fibronectin fibril formation in the early *Xenopus* embryo. *Dev. Dyn.* **212:** 335–345.

Winklbauer R. and Keller R.E. 1996. Fibronectin, mesoderm migration, and gastrulation in *Xenopus*. *Dev. Biol.* **177:** 413–426.

Winklbauer R. and Nagel M. 1991. Directional mesoderm cell migration in the *Xenopus* gastrula. *Dev. Biol.* **148:** 573–589.

Winklbauer R. and Schürfeld M. 1999. Vegetal rotation, a new gastrulation movement involved in the internalization of the mesoderm and endoderm in *Xenopus*. *Development* **126:** 3703–3713.

Winklbauer R. and Selchow A. 1992. Motile behavior and protrusive activity of migratory mesoderm cells from the *Xenopus* gastrula. *Dev. Biol.* **150:** 335–351.

Winklbauer R. and Stolz C. 1995. Fibronectin fibril growth in the extracellular matrix of the *Xenopus* embryo. *J. Cell Sci.* **108:** 1575–1586.

Winklbauer R., Medina A., Swain R.K., and Steinbeisser H. 2001. Frizzled-7 signalling controls tissue separation during *Xenopus* gastrulation. *Nature* **413:** 856–60.

Winklbauer R., Nagel M., Selchow A., and Wacker S. 1996. Mesoderm migration in the *Xenopus* gastrula. *Int. J. Dev. Biol.* **40:** 305–311.

Winklbauer R., Selchow A., Nagel M., and Angres B. 1992. Cell interaction and its role in mesoderm cell migration during *Xenopus* gastrulation. *Dev. Dyn.* **195:** 290–302.

Youn B.W. and Malacinski G. 1981. Axial structure development in ultraviolet-irradiated (notochord-defective) amphibian embryos. *Dev. Biol.* **83:** 339–352.

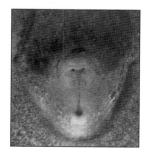

GASTRULATION IN REPTILES

E.H. Gilland[1] and A.C. Burke[2]

[1]*Department of Physiology and Neuroscience, NYU School of Medicine, New York, New York 10016;* [2]*Biology Department, Wesleyan University, Middletown, Connecticut 06480*

INTRODUCTION

Reptiles, as traditionally defined, represent a paraphyletic group including non-avian archosaurs (crocodilians), lepidosaurs (tuatara, lizards, and snakes), and turtles. This assemblage includes all the amniotes except mammals and birds. Since both mammalian and avian ontogenies are derived from reptilian patterns somewhere within the amniote tree, studies of gastrulation in different reptilian lineages are prerequisite to a broader understanding of the evolution of this important developmental process in higher vertebrates.

A broad definition of gastrulation includes the suite of morphogenetic movements and signaling interactions that result in definitive germ-layer formation and establishment of the primary body axes. In amniotes it must also include mechanisms for establishing the precursors of extraembryonic membranes (see Chapter 31). In recent years, our understanding of gastrulation has reached new levels through the identification of molecular signals associated with morphogenetic processes. In particular, gene and protein expression patterns provide valuable new characters in comparative studies where morphological landmarks may be obscure. Unfortunately, gastrulation in reptiles is known almost solely from descriptive studies. Although isolated molecular data for later stages are available (see, e.g., Locascio et al. 2002), no reptilian taxa have ever been used as model species for early developmental studies. This is largely due to the many difficulties associated with using wild populations of seasonally reproducing animals as sources of material. Moreover, the classic descriptive data from whole mounts and serial sec-

tions have not been synthesized in a modern comparative summary.

This chapter is not intended to be such a synthesis, but instead outlines a phylogenetic framework for species comparisons and attempts to identify some central questions and possible approaches for answering them. After brief phylogenetic and historical sections, we present a description of gastrulation in turtles. The majority of descriptive work on reptiles has focused on turtles, and this group represents a good comparative baseline for the scarcer studies of other reptiles. We then discuss data from other reptilian groups and raise certain issues in light of data from mammals and birds. We also present a discussion on the evolution of the extraembryonic membranes and suggest how they may have affected amniote gastrulation. Finally, we suggest some avenues of future research to expand comparisons with model systems and increase overall understanding of the evolution of gastrulation in amniotes.

PHYLOGENETIC FRAMEWORK

The most commonly accepted vertebrate phylogenies are based on fossil and recent morphological data (Laurin and Reisz 1995; Ruta et al. 2003), supplemented by molecular sequence studies (Cotton and Page 2002). Some of the traditionally accepted amniote subgroups, Anapsida and Theria (Gauthier et al. 1988; Laurin and Reisz 1995), have recently been challenged by alternative trees generated with molecular sequence data. Turtles have been relocated to various locations at the base of or within the archosaurs based on mitochondrial, ribosomal, and other nuclear gene sequences

(Zardoya and Meyer 1998; Hedges and Poling 1999; Cao et al. 2000; Janke et al. 2001). Similar analyses have also challenged the traditional view that marsupial and placental mammals share a common history as "therians," excluding monotremes (Janke et al. 1997, 2001). In the latter case, a more detailed analysis of complete mitochondrial genome data has supported the traditional mammal subgroups (Phillips and Penny 2003), and new analyses may likewise refute the proposed archosaur affinities of turtles. Morphological characters have also been used to place turtles amid the diapsids (Rieppel and deBraga 1996), within the lepidosaurs instead of the archosaurs. Given these uncertainties, we use the more widely accepted pattern of relationships depicted in Figure 1 (from Gauthier et al. 1988).

Within this classification, questions of reptile gastrulation ultimately reduce to a comparison of cellular, molecular, and architectural features seen in extant representatives of Diapsida (birds, crocodilians, lizards, snakes, and tuatara), Anapsida (turtles), and Synapsida (mammals). Our group analysis depends on gastrulation conditions in Amphibia (frogs, salamanders, caecilians), other sarcopterygians (lungfish, coelacanth), and other gnathostomes (actinopterygians and elasmobranchs). Although the common ancestor of all gnathostomes is believed to have had a holoblastic egg, all of the major gnathostome branches include members that have independently evolved megalecithal eggs and various forms of meroblastic cleavage.

Ideally, morphological, molecular, and fate-mapping data should be analyzed within a phylogenetic framework to produce a series of corroborated primitive or stem gastrulation types surrounding each branch point of the amniote tree (Fig. 1); e.g., proto-diapsid, proto-eureptilian, etc. The chick serves as a useful comparison with crocodilians to reconstruct a proto-archosaur pattern. This pattern can be compared with lepidosaurs to reconstruct proto-diapsid characters, which can be compared to turtles, if they are accepted as

extant anapsids. A hypothesized common gastrulation pattern for anapsids and diapsids would lie very close to the base of the other great branch of amniotes, the Synapsida (mammals), and differences between therian mammals and monotremes can provide a potential intermediate state.

HISTORY OF STUDIES ON REPTILIAN GASTRULATION

Despite the current dearth of work on early reptilian development, this was not always the case. Early monographs on snake and turtle development by Rathke (1839, 1848) established high standards for subsequent workers. The beginnings of modern microscopical analysis of reptile gastrulation were papers by Balfour (1879) and Kuppfer (1882), both of which were intended to elucidate the origins of the mammalian and avian primitive streaks by examining the potentially more primitive condition in reptiles. These two authors came to opposite conclusions, with Balfour proposing that the primitive plate (see below) of lizards was in fact a primitive streak, and Kuppfer denying that identity. In many ways, that issue has not yet been fully resolved. A torrent of detailed studies aimed at bridging the gaps between amphibians, mammals, and birds soon followed. Notable among these were studies on gastrulation in turtles by Will (1892), Mitsukuri (1886, 1891b, 1893, 1896), and Mehnert (1892, 1895); in geckos and common lizards by Will (1896); and in snakes by Will (1898), Ballowitz (1901), and Gerhardt (1901). Descriptions of crocodilian gastrulation are particularly scarce, because the eggs are not deposited by the females until well beyond this stage (Ferguson 1985). Clarke (1891), Voeltzkow (1899, 1901), and Reese (1908, 1915) illustrated a few early stages harvested from gravid females, including so-called "primitive streak" stages. Unfortunately, the quality of the early-stage embryos in

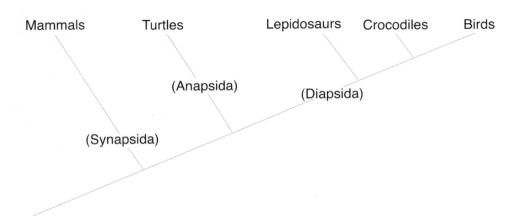

Figure 1. Cladogram representing the relationships of the major groups of extant amniotes, after Gauthier et al. (1988). Alternative views place the turtles at various positions within the Diapsida (Rieppel and deBraga 1996; Hedges and Poling 1999).

these studies was not comparable to the studies cited above, and the results are difficult to interpret. Finally, a unique series of embryos of *Sphenodon*, the tuatara, were described by Dendy (1899; for review, see Moffat 1985) and later by Tribe and Brambell (1932) and Fisk and Tribe (1949).

Studies by only four workers encompass most of the later literature on reptile gastrulation. Peter published detailed descriptive studies of early development in a chameleon (Peter 1934, 1935) and a comparative study of endoderm formation in lizards, turtles, and snakes (Peter 1938). Pasteels (1937) carried out vital dye-mapping experiments and constructed a preliminary fate map for turtles. He later published studies on the origin of germ cells in lizards (Pasteels 1953), on amnion formation in chameleons (Pasteels 1957c), and on hypoblast and endoderm formation in turtles and lepidosaurs (Pasteels 1957a). He also published a staging series for turtle and lizard gastrulation comprising ten stages beginning just before the appearance of the blastopore invagination and ending when the neural plate becomes visible (Pasteels 1957b). A pair of papers by Chandrasekharan Nayar (1959, 1966) replicated Pasteels's early dye-mapping study and reached similar conclusions (see below). Hubert (1962) examined gastrulation in a live-bearing lizard and later compared endoderm and germ-cell formation between snakes and lizards (Hubert 1970). Except for series of normal stages in *Chelydra serpentina* by Yntema (1968) and in marine turtles by Miller (1985), and work on mesodermal patterning in the segmental plate of *C. serpentina* (Packard 1980; Packard and Meier 1984), the works cited for these four authors comprise essentially the entire primary literature on reptile gastrulation for the past seven decades. A number of excellent reviews provide panoramic overviews of development in different reptile groups (Ewert 1985; Fergusson 1985; Miller 1985; Moffat 1985), whereas those by Pasteels (1970), Hubert (1985), and Arndt and Nübler-Jung (1999) include special emphasis on reptile gastrulation.

MORPHOLOGY OF GASTRULATION IN TURTLES

Turtles exhibit much less diversity than lepidosaurs both in terms of reproductive styles and in the morphology of gastrulation. They thus afford a convenient starting point for assessing early development in reptiles. The following brief outline of turtle gastrulation is based mainly on studies by Mitsukuri (1886, 1891b, 1893, 1896), Pasteels (1937, 1957a), and Chandrasekharan Nayar (1959, 1966). Mitsukuri's studies of early development in a number of species of freshwater (*Pelodiscus sinensis* and *Mauremys japonica*) and marine (*Caretta caretta*) turtles are the most extensive and well-illustrated works available from a single author. Pasteels (1937) produced a fate map of the early gastrula of the Mediterranean pond turtle, *Mauremys leprosa*, based on vital dye-mapping experiments and analysis of serial sections. Chandrasekharan Nayar (1959, 1966) essentially replicated Pasteels's study in the Indian black turtle (*Melanochelys trijuga*) and the Indian flapshell turtle (*Lissemys punctata*). Although the results of these two dye-mapping experiments must be considered incomplete, they do provide rough baseline descriptions of involution and migration of cells through and away from the blastopore.

All turtles are oviparous, and their large, yolky eggs are very similar to those of birds, except that turtle eggs generally have less albumin, lack a chalaza, and have thinner, less mineralized outer shells (Ewert 1985). Cleavage and initial blastoderm formation occur before oviposition and, unfortunately, there are only a few studies documenting these stages. Agassiz (1857) observed cleavage stages in *Clemmys insculpta* (Fig. 2), and Miller (1985) described surface views

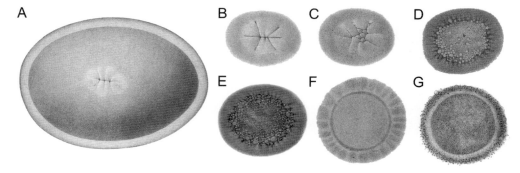

Figure 2. Cleavage in a turtle embryo (*Clemmys insculpta*), from Plate X, Agassiz (1857). (*A*) Egg without shell, removed from oviduct and showing cleavage furrows at the 8-cell stage. (*B–G*) A series of blastoderms spanning cleavage stages from 8-cell to just before the onset of gastrulation. Large marginal blastomeres are visible in *C* and *D*, whereas in *E* and *F* the entire blastoderm surface comprises a uniform sheet of small cells. The pattern of evenly spaced lines radiating away from the blastoderm in *E* have been noted for many species and may be indications of a cytoskeletal system associated with yolk syncytial nuclei. Some of the eggs used for Agassiz's 1857 studies were collected by Henry David Thoreau and are the subject of a wonderful article in the *Atlantic Monthly* (1916) by D. L. Sharpe.

of cleavage in the green turtle, *Chelonia mydas*. Cleavage in these species follows a pattern roughly similar to that of birds, with a cytoplasmic disc being subdivided by vertical cleavage furrows into a central region of smaller blastomeres and a marginal region of delayed division with larger blastomeres (Fig. 2D–F). Although not well described for any turtle, horizontal cleavages probably separate smaller superficial cells from deeper, yolky cells, as in birds and lepidosaurs. Incompletely partitioned deep and peripheral blastomeres contribute nuclei to a yolk syncytial layer (Mehnert 1892; Pasteels 1970), although, again, no detailed description is available for turtles. Further subdivision of the marginal blastomeres produces a blastoderm whose upper surface comprises small, uniformly sized cells that form a well-defined epiblast epithelium (Fig. 2G). The epiblast is thicker centrally, where columnar cells form an oval embryonic shield, and thinner peripherally, where it extends across an area pellucida and area opaca as in birds (Fig. 2H). A subgerminal cavity and germ wall underlie these respective areas. The epiblast in turtles is underlain by only a very slight hypoblast, described by Mitsukuri (1893) as a discontinuous layer of ameboid cells.

The appearance of an opaque thickening at the caudal, midline edge of the embryonic shield marks the visible onset of gastrulation (Figs. 3A and 4A). This stratified mass of cells, known as the primitive or blastoporal plate, forms in close association with a thickened portion of the embryonic shield and extends caudally from there to meet the germ wall at the edge of the area opaca. A transverse invagination forms in the embryonic shield overlying the anterior part of the blastoporal plate. This is the beginning stage of blastopore formation. The edges of the blastopore subsequently extend laterally, caudally, and then medially, forming a horseshoe-shaped invagination whose caudal ends eventually oppose each other near the midline in the central or posterior part of the blastoporal plate (Fig. 3A–C). Mitsukuri (1893) illustrated bilateral thickenings in the floor of the blastoporal opening during later stages in the sea turtle.

In longitudinal section, the anterior blastopore lip appears as an involution of the pseudostratified epiblast layer (Fig. 4A–D). The posterior rim and floor of the blastopore do not present such a clear picture of an inwardly folded sheet (Fig. 4B,C) and have been sometimes proposed to correspond not to a ventral blastopore lip, but instead to a vestigial yolk plug (Mitsukuri and Ishikawa 1886). When viewed from the ventral side, the blastopore involution appears as an elongating sac protruding between the thick epiblast and the thin hypoblast. As the involuted sac extends rostrally, it is accompanied by a spreading sheet of endoblast

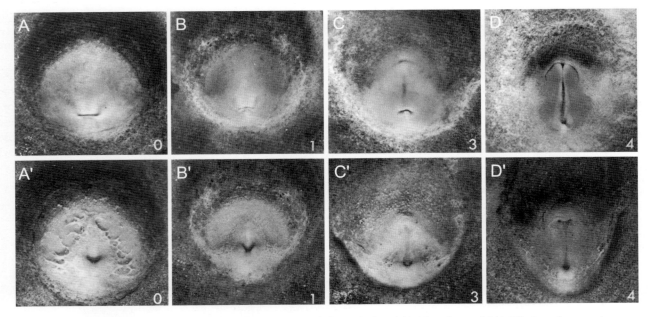

Figure 3. Blastoderms of snapping turtle (*Chelydra serpentina*) shown in dorsal (*A–D*) and ventral (*A'–D'*) views, from staging series of Yntema (1968). Stage numbers, 0–4, are in lower right of each frame. (*A, A'*) Blastoderm at time of laying, stage 0. The blastopore is visible as a horizontal slit in the posterior region of the embryonic shield. In ventral view, the blastoporal plate appears as a crescentic mound and the spreading involuted cells appear as a rostrally pointing arrow-shaped layer. (*B, B'*) At stage 1, the blastopore opening has become narrower and the blastoporal plate seen in ventral view shows prominent bilateral lobes. (*C, C'*) By stage 3, the head fold is visible anteriorly and the blastopore opening appears as a rounded, inverted "U" surrounding a central bulge in the blastopore floor. (*D, D'*) The head fold is now well formed and the amniotic fold is already enveloping the rostral region. In ventral view, the blastoporal plate forms a more compact mass than at earlier stages.

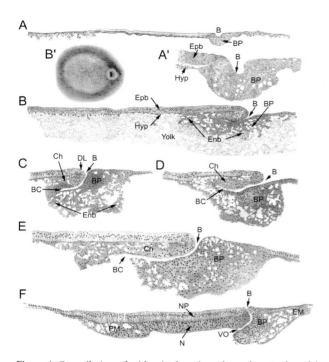

Figure 4. Compilation of midsagittal sections through gastrulae of the pond turtle *Emys orbicularis* (*A, A′*) from Will (1892) and the logger-head turtle *Carreta carreta* (*B–F*) from Mitsukuri (1893). The anterior direction of the forming embryonic axis is to the left. The underlying yolk is only shown in *B*, although it would have been present in all of these sections. (*A, A′*) The blastopore (B) appears as an invagination toward the anterior edge of the thickened blastoporal plate (BP). The hypoblast (Hyp) is a thin monolayer of cells underlying a columnar epiblast (Epb). (*B*) Endoblast cells (Enb), arising from the blastoporal plate and from the epiblast by migration through the blastoporal fur-row, form a loosely organized cell layer that displaces the hypoblast toward the periphery of the embryonic shield. (*C, D*) The epiblast rolls in at the dorsal blastoporal lip (DL), giving rise to the chordame-soderm (Ch) and forming the blastoporal canal (BC). (*E*) The blasto-poral canal reaches its greatest length at this stage. The opening of the floor of the blastoporal canal in this section is an artifact due to loss of part of the ventral endoblast layer. (*F*) This specimen was at a stage slightly younger than Yntema stage 3 (Fig. 3C). The floor of the blastoporal canal has disappeared, presumably due to lateral migra-tion of the ventral endoblast, leaving only a short distance between the external and internal (VO) openings of the blastopore. After neurula-tion, this short passage will become the neurenteric canal. The neural plate (NP) now overlies the notochord (N), and more rostrally, the prechordal mesendoderm (PM).

that appears to migrate out from deeper portions of the blastoporal plate (Fig. 4B–D). After elongating under half or more of the thickened region of the overlying embryonic shield, the floor of the involuted sac opens, bringing the blastopore cavity into direct communication with the subgerminal cavity (Figs. 4F and 5E,F). At this point, the invo-luted cell sheet comprises a vault-like layer of cranial chor-damesoderm, with its side walls confluent with the adjoin-ing endoblastic sheet. Mitsukuri (1891b) described in some

detail how cells emerged from the involuted sheet on either side of the midline to form two paramedian bands of meso-derm, with the remaining midline cells forming notochord and the lateral portions of the sheet forming the future pha-ryngeal endoderm. The mesodermal bands that ingressed from the vault roof were described as being continuous cau-dally with mesodermal cells emerging from the sides of the blastoporal plate. A similar description of mesoderm for-mation in turtles was later provided by Brachet (1914).

The epiblast overlying the involuted region rapidly thickens to form the cephalic neural plate. By the time the first somites form, the anterior end of the involuted vault and the overlying neural epithelium have begun to sink downward toward the yolk, forming the initial head fold (Fig. 3C). Just in front of the head fold, the proamnion, or ectamnion, begins to grow out from the epiblast as a horse-shoe-shaped ridge (Fig. 3D, and see Mitsukuri 1891a). As the spinal neural folds become prominent, the blastopore lips form a very narrow opening surrounding a small pro-jection of the blastoporal plate. At the 3-somite stage, the neural folds surround the remaining anterior and lateral portions of the blastopore to form a neurenteric canal (Fig. 3D). A tail bud (more properly, trunk-tail bud) forms just behind the neurenteric canal and just rostral to the small remaining caudal end of the blastoporal plate. Transverse sections through this caudal zone resemble those through a mammalian or avian primitive streak, with epi-blast, mesoderm, and endoderm merged together in an apparent midline ingression zone (Mitsukuri 1896).

In the studies of Pasteels (1937) and Chandrasekharan Nayar (1966), dye marks applied to the blastoderm sur-face near the blastopore lips indicated that considerable areas of the embryonic shield and blastoporal plate con-verge toward and involute through the blastoporal open-ing (Fig. 5A–C). Since movements of cells toward the blastopore occurred from all directions, this suggested that the portion of the blastoporal plate that formed the caudal floor of the blastopore constituted a genuine ven-tral blastopore lip as in amphibians. Pasteels (1937) believed that the earliest involuted cells, as well as cells ingressing from the deeper parts of the blastoporal plate, were all fated to form endoderm. Cells that involuted through the lateral and ventral blastopore lips at later stages were proposed to comprise the intra- and extraem-bryonic mesoderm, whereas those involuting at the dorsal lip formed notochord. Although Pasteels summarized his work in the form of a fate map (Fig. 5G), he clearly stated (1937, p. 177) that this scheme was not the direct result of marks applied at gastrulation and followed through to definitive organ stages, a feat neither he nor Chandrasekharan Nayar (1966) was able to accomplish. Rather, the map was inferred indirectly from short dura-tion marking experiments carried out at successive stages.

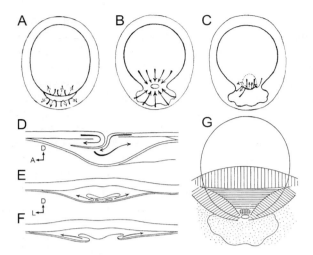

Figure 5. Morphogenetic movements and fate map for turtles as proposed by Pasteels (1937, 1970). (*A–C*) Schematic dorsal views at three successive stages showing proposed movements of epiblast and blastoporal plate cells into deeper layers. In *A*, the arrows indicate an early phase of hypoblast/endoderm production that occurs before formation of the blastoporal invagination. Presumably, these cells arise from the early blastoporal plate and form the leading edge of the primary endoblast, which displaces a more yolky endophylle layer formed during cleavage. Arrows in *B* show regions of epiblast and the surface layer of the blastoporal plate converging toward the early blastopore from all directions. In *C*, epiblast and blastoporal plate cells are involuting through the blastopore to form the roof, side walls, and floor of the blastoporal canal. (*D*) A schematic midsagittal section through an embryo at a stage slightly younger than the one in *C* indicates how the forward movement of the surface layer of the blastoporal plate contributes both to the floor of the blastoporal canal and to continued spreading of the endoblast toward the periphery of the embryonic shield. (*E, F*) Schematic transverse sections at later stages showing Pasteels's conception of how the floor of the blastoporal canal spreads laterally to contribute to the primary endoblast. Formation of the definitive gut tube occurs later, when thin bilateral sheets of endoderm separate from the mesendodermal layer and converge toward the midline. (*G*) A map of presumptive territories on the surface of the epiblast and blastoporal plate at a stage similar to *B*. Presumptive areas are coded as follows: (*vertical lines*) neural ectoderm; (*horizontal lines*) notochord; (*diagonal lines*) somitic mesoderm; (*cross-hatched area* above blastopore) prechordal plate; (*dotted area*) ventrolateral and extraembryonic mesoderm; (*unshaded portion* of embryonic shield) nonneural ectoderm.

The most that can be safely concluded from the studies of Pasteels and Chandrasekharan Nayar is that endodermal, mesodermal, and notochordal precursors likely arise from the epiblast and adjacent blastoporal plate and enter the deeper layers by some combination of ingression and involution at the blastopore lips and possibly from the deeper portions of the blastoporal plate. Although the prospective areas on the surface of the chelonian early gastrula may well correspond closely to those of salamanders, as proposed by Pasteels, this possibility remains to be experimentally confirmed.

MORPHOLOGY OF GASTRULATION IN OTHER REPTILES

Comparison of gastrulation between turtles and other reptiles raises many difficulties that stem only partly from genuine taxonomic differences. In many cases, written descriptions of gastrulation in reptiles appear widely divergent. In part, this is because most of the studies were performed before Vogt (1929), Graper (1929), and Wetzel (1929) showed unequivocally that gastrulation is dominated by morphogenetic movements. Thus, the blastopore rim, instead of being described as a dynamic zone of involution and ingression, was often viewed as a cell proliferation center, much like a regenerating limb blastema. Alternatively, the emergence of the chordamesoderm beneath the epiblast could be seen as a process of delamination (for critiques of earlier studies, see Wetzel 1931; Pasteels 1937). In addition, many of the studies were explicitly undertaken to refute or support specific theories of chordate gastrulation (see, e.g., Gehrardt 1901; Brachet 1914), resulting in otherwise very similar observations being obscured by differing emphasis or terminology. It is thus somewhat surprising that when the surface views and serial sections of well-illustrated studies are compared, most aspects of gastrulation appear to be quite uniform throughout the reptiles. For example, all reptiles studied, including crocodilians and rhyncocephalians, seem to form a genuine blastopore through which an archenteric or chordamesodermal canal arises by involution of epiblast. Likewise, despite numerous statements that imply such structures, no reptile has been found to possess an elongated primitive streak or Hensen's node, as are found in birds and mammals. The major differences between reptiles seem to lie in the origin of hypoblast and endoderm, and in the timing of epiboly and amniogenesis relative to blastopore formation and axiation.

CLEAVAGE AND BLASTODERM FORMATION

Other egg-laying reptiles share with turtles the delay in oviposition until after completion of cleavage and blastoderm formation. In many groups, freshly laid eggs contain embryos at limb bud stages (Pasteels 1970). Cleavage stages have not been described at all for crocodilians, tuatara, or amphisbaenians. In addition to the few reports on turtles mentioned above, there are accounts of early cleavage for only a few species of snakes and lizards (Nicholas 1904; Pasteels 1970; Hubert 1985). Cleavage in these lepidosaurs does not appear to differ much from cleavage in turtles. The general avian-like pattern of early vertical cleavage planes subdividing a discoidal cytoplasmic region, followed by horizontal cleavages separating a population of yolky deep blastomeres, has been reported for both snakes and lizards (Hubert 1985). Likewise, incomplete deep and peripheral blastomeres were described as sources of syncytial nuclei (Pasteels 1970).

Unlike the general similarity in cleavage stages, late-blastula stages differ considerably among reptiles with regard to the histological appearance of the epiblast and hypoblast. Although the epiblast is likely to be a single cohesive epithelial sheet in all cases, it can vary greatly in thickness. In turtles, the epiblast is generally quite thin, sometimes with only a single stratum of low cuboidal cells present (Pasteels 1937, 1970). The epiblast in lizards and snakes appears much thicker, especially in chameleons, where four or five layers of nuclei are visible in histological sections (Peter 1934, 1938; Hubert 1985).

In all species described, the central region of the blastoderm forms an embryonic shield of columnar epiblast cells surrounded by an area pellucida and area opaca as in turtles. The relative timing of subgerminal cavity formation seems to vary in different groups and roughly correlates with the pattern of hypoblast formation (see below). The blastoderm peripheral to the area opaca forms a spreading sheet that envelops the yolk mass. This enveloping sheet appears to be only one cell thick in reptiles (Virchow 1892) and thus is probably composed solely of extraembryonic ectoderm and an underlying syncytium as in avians (Blount 1909; Bellairs 1967). The rate at which the enveloping sheet surrounds the yolk is variable, being completed before gastrulation in some chameleons (Peter 1934; Pasteels 1957c; Blanc 1974) and tuatara (Dendy 1899), but only much later in most turtles. Precocious formation of the amnion seems to correlate with this, since both chameleons and tuatara initiate amniogenesis earlier than do turtles, even before the onset of gastrulation in chameleons (Peter 1934; Pasteels 1957c).

BLASTOPORAL PLATE AND BLASTOPORE

A structure similar to the blastoporal plate of turtles forms at the posterior edge of the embryonic shield in all reptiles. Earlier workers often referred to this structure as the primitive plate, or even primitive streak, implying direct homology with conditions in birds and mammals. In some species, the blastoporal plate is described as forming solely from ingression of epiblast cells (Hubert 1985), whereas in others it is described as originating from a continuous cell mass joining the epiblast and posterior germ wall (Mitsukuri 1893). Will (1892, 1896) described an initial sickle-shaped thickening as being the first indication of blastoporal plate formation. Such early stages were not available in most studies, thus it is not clear whether this structure, which resembles the avian Koller's sickle, is a general feature in reptiles. An initial midline invagination, which extends caudo-laterally to form the dorsal and lateral lips of a well-defined blastopore, has been described in numerous species of snakes and lizards (Pasteels 1970; Hubert 1985). The early stages of blastopore formation have not been observed in tuatara or

crocodilians, but the form of the blastoporal canal and the remnants of the caudal blastoporal plate described in slightly later stages in these groups (Dendy 1899; Voeltzkow 1899; Reese 1908, 1915) suggest the presence of a typical reptilian blastopore. A well-developed, sac-like blastoporal canal that gives rise to at least notochord and mesoderm is also present in all reptilian groups. Kuppfer (1882) described paired medial thickenings in the floor of the closing blastopore opening in *Lacerta*, similar to ones found by Mitsukuri in turtle.

For some workers, especially Mitsukuri (1891b, 1893) and Will (1896), the blastoporal invagination was seen as a genuine archenteron (Urmund), homologous to that of *Amphioxus*. Mitsukuri (1891b) described the walls and roof of the turtle "archenteron" as giving rise not only to notochordal material, but also to cranial mesoderm and endoderm. Others believed that the involution produced largely or solely notochord (hence, "notochordal canal") with cranial mesoderm and endoderm arising separately from the blastoporal plate and hypoblast, respectively (Hubert 1985). Because Pasteels (1937) believed that the endoderm arose by a separate, precocious ingression from the blastoporal plate rather than from the involuted blastopore lips, he denied any complete homology between the invaginated archenteron of *Amphioxus* and the involuted sac of reptiles. He insisted that the latter should be called a chordamesodermal or blastoporal canal rather than an "archenteric canal." In his much later review, however, Pasteels (1980) conceded that the archenteric nature of the reptilian blastoporal canal would have to be reconsidered in light of evidence for an epiblast/primitive streak origin of endoderm in birds (Vakaet 1962; Lawson and Schoenwolf 2003; see also Lawson et al. 1991 for mice).

THE HYPOBLAST AND ORIGIN OF THE ENDODERM

While the compact epiblast is forming, another, more loosely organized layer of cells forms beneath it. This "hypoblast" layer appears quite different in turtles, lizards, and snakes. As with studies of avian and mammalian gastrulation, the origin of embryonic endoderm and the existence of earlier, transitory hypoblast layers in reptiles have been hotly contested (Peter 1934, 1938; Pasteels 1937, 1957a).

All authors seem to agree that the early hypoblast is generally quite thin in turtles (Fig. 4A,B), somewhat thicker, but still diffuse, in snakes (Fig. 6E–H), and very thick in some species of lizards (Fig. 6A–D). Pasteels (1937) described the hypoblast layer in turtles as originating by forward spreading of cells from the posterior edge of the blastoderm (e.g., Enb in Fig. 4B,C). Peter (1938) objected strongly to that view and claimed that the hypoblast layer formed by delamination of

Lacerta vivipera   *Vipera aspis*

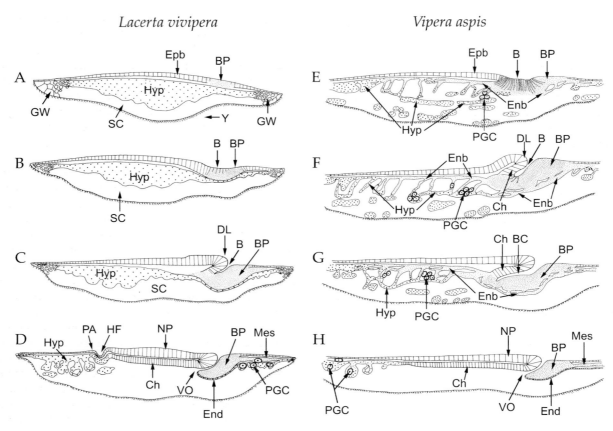

Figure 6. Schematic drawings of midsagittal sections of the embryonic shield at different gastrulation stages of a lizard, *Lacerta vivipara* (*A–D*), and a snake, *Vipera aspis* (*E–H*). The anterior direction of the forming embryonic axis is to the left. (*A*) The blastoporal plate (BP) forms by thickening of the epiblast (Epb) in the posterior region of the embryonic shield. A compact mass of hypoblast (Hyp) lies deep to the epiblast. (GW) Germ wall; (Y) yolk; (SC) subgerminal cavity. (*B*). The blastopore (B) appears as a small invagination in the anterior part of the blastoporal plate (BP). (*C*) The blastopore is well developed and the epiblast begins to roll in at the dorsal blastoporal lip (DL), giving rise to the chordamesoderm (Ch). (*D*) The midline hypoblast has been pushed into the anterior extraembryonic area by the chordamesoderm (Ch). Endoderm (End) is seen under the blastoporal plate, and the primordial germ cells (PGC) lie just posterior to the blastoporal plate. The blastoporal canal opens ventrally (VO) into the subgerminal cavity (SC). Posterior extraembryonic mesoderm (Mes) ingresses from the blastoporal plate. Rostrally, the headfold (HF) and proamnion (PA) are forming. (*E*) At the beginning of gastrulation in *Vipera*, the hypoblast (Hyp, equivalent to endophylle) is much thinner than in *Lacerta* and appears as islands of cells. Epiblastic cells migrate through the blastoporal furrow toward the anterior region of the embryonic shield, forming cellular cords of endoblast (Enb) that hang into the subgerminal cavity above remnants of the hypoblast/endophylle. The gonocytes (PGC) lie close to these hypoblastic cells, just anterior to the blastopore (B). (*F, G*) The epiblast begins to roll in at the dorsal blastoporal lip (DL), forming the blastoporal canal (BC) and giving rise to the chordamesoderm (Ch). The migration of endoblast cells from the blastopore toward the anterior region carries the hypoblast/endophylle (Hyp) and the gonocytes (PGC) in the same direction. (*H*) In the axial part of the shield, the blastoporal canal opens ventrally (VO), and the hypoblast/endophylle (Hyp) and gonocytes (PGC) have been pushed into the anterior extraembryonic area by the chordamesoderm (Ch). The only definitive endoderm (End) visible at the midsagittal plane is a thin cellular sheet on the ventral surface of the blastoporal plate. Figure and legend modified from Hubert (1970, 1985).

the early blastoderm into superficial and deep layers, not only in lepidosaurs, but also in turtles. Pasteels (1957a) later reexamined the origin of the hypoblast in a few turtle and lizard species. He concluded that in turtles an early layer of yolky cells, l'endophylle (hyp in Fig. 4B), did arise by delamination of deep blastomeres, but that this layer was displaced into extraembryonic regions by a second layer of less yolky cells, l'endoblaste (enb in Fig. 4B), that arose by "gastrular immigration" from the epiblast and blastoporal plate.

He could not distinguish these two phases of hypoblast formation in lizards and tentatively agreed with earlier accounts that the thick lepidosaurian hypoblast arose in this group by delamination of deep blastomeres (Fig. 6A–D). He attributed the differences between turtles and lizards to a supposed polyphyletic origin of reptiles. Hubert (1962, 1970, 1985) compared hypoblast formation in snakes and lizards and concluded that Pasteels's description for turtles also was true in general for the snake *Vipera aspis*. In this case, a thin hypoblast (endophylle) comprising scattered groups of cells produced by delamination of deep blastomeres was later displaced by a new layer (endoblast) of migratory epiblast cells ingressing near the blastopore and from the blastoporal plate (hyp and enb in Fig. 6E–H). Hubert (1970) confirmed previous descriptions of hypoblast formation in lizards and concluded that definitive endoderm (and primordial germ cells) arose in a unique manner in that group.

It is tempting to interpret these descriptions in terms of the much more sophisticated analyses of germ-layer formation in the chick embryo (see Chapter 15). The initial hypoblast formed in chick by diffuse polyingression from the overlying blastoderm nicely fits the descriptions of endophylle formation in turtles and snakes (Pasteels 1957a; Hubert 1985). The displacement of the chick hypoblast by a spreading sheet of endoblast cells derived from the posterior marginal wall of the blastoderm seems very similar to the earliest phase of endoblast migration described by Pasteels (1957a) and Mitsukuri (1893); namely, a layer of ameboid cells extending rostrally from the blastoporal plate. Finally, the entry of definitive embryonic endoderm into the deep layer by ingression from the primitive streak in chick would correspond to the origin of pharyngeal and gut endoderm by involution at the blastopore lips in turtle (Pasteels 1937, 1957a). Given the difficulties of carrying out in vivo manipulations in pregastrulation embryos, comparative analysis of molecular markers appears to be the best choice for resolving the origins of hypoblast and endoderm in reptiles (see below).

REPTILIAN GASTRULATION MORPHOLOGY COMPARED WITH THAT OF OTHER VERTEBRATES

A few general conclusions about reptilian gastrulation seem warranted. First, aside from the mode of endoderm formation, the morphology of gastrulation in non-avian/non-mammalian amniotes is very conservative. There does not appear to be any significant departure from the blastoporal plate/blastopore system described for any reptile, although examination of the earliest gastrulation stages in crocodilians and tuatara are needed to verify that invagination/involution at a proper "dorsal" blastopore lip initiates axis formation in all major reptilian taxa. The blastoporal plate, which forms a seemingly undifferentiated reservoir of hypoblast, endoderm, and mesoderm at the posterior end of the embryonic shield, seems a likely phylogenetic precursor of both the avian Koller's sickle and the posterior portions of the mammalian and avian primitive streaks. Claims by Will (1892, 1896) that the blastoporal plate arises from a crescentic thickening at the caudal edge of the germ wall and that distinct subdivisions can be identified at the interface of blastoporal plate and embryonic shield need to be reexamined.

Although further studies are also needed to test claims that a proper primitive streak structure forms behind the neurenteric canal at later stages in turtles (Mitsukuri 1896) and at somewhat earlier stages in crocodilians (Reese 1908), it seems clear that for the cranial region of the embryo, neither avian nor mammalian forms of primitive streak and Hensen's node are present in reptiles. This does not, however, mean that comparisons between blastoporal plate and primitive streak, and dorsal blastopore lip and Hensen's node, are invalid. Rather, it suggests that intermediate stages between these structures should be sought in terms of quantitative differences in size, shape, and the timing of cell movements and fate, rather than in absolute qualitative differences in overall form. Thus, although the reptilian blastopore complex seems quite distant structurally from the chicken or mouse pattern, it remains to be seen whether a fundamental modification of gastrulation mechanisms separates these groups, or whether they instead simply reside at opposite ends of a spectrum of variation in lesser features such as timing of migration and closeness of cell contacts. Molecular and ultrastructural data on gastrulae of turtles, crocodilians, monotremes, and marsupials should provide much better definitions of the basic similarities and differences between amniote "streaks," "plates," "nodes," and "blastopores." The potential value of such a search is emphasized by the seemingly inescapable phylogenetic conclusion that the avian and mammalian primitive streaks represent independent, parallel modifications of a blastoporal plate system.

YOLK, EXTRAEMBRYONIC MEMBRANES, AND THE ORIGIN OF AMNIOTES

Textbook descriptions of variation in vertebrate gastrulation patterns generally emphasize the response to selective pressures for increased quantities of yolk (see, e.g., Gilbert 2003). Arndt and Nübler-Jung (1999) provide a detailed summary and excellent discussion of the changes required to evolve from a primitive (amphibian-like) pattern of holoblastic cleavage and gastrulation, to that of gastrulation in discoidal blastulae characteristic of amniotes. An increase in yolk size requires a different disposition of embryonic tissues and rearrangement of the blastopore relative to the vegetal pole of the egg. Arendt and Nübler-Jung (1999) describe a key event wherein an amphibian-like blastopore becomes physically separated into embryonic and extraembryonic regions as a consequence of increased yolk size. They propose an intermediate form of gastrulation that resembles the process in modern turtles (as described by Pasteels 1937). This intermediate also resembles certain aspects of gastrulation in elasmobranchs (see Chapter 11).

It appears likely that increased yolk size was not the only factor driving changes in the dynamics of gastrulation. Many different taxa have succeeded in increasing yolk volume dramatically, and this has clearly evolved independently many times, as seen in elasmobranchs, teleosts, and amphibians. The shared derived character uniting mammals, reptiles, and birds is the invention of the cleidoic egg and its extraembryonic membranes: the amnion, chorion, and allantois. Discussions of the evolution of gastrulation and discussions of the evolution of extraembryonic membranes are generally nonoverlapping (cf. Fisk and Tribe 1949; Luckett 1977; Stewert, 1997; Chapter 31). We propose a causal link between the evolution of the extraembryonic membranes in amniotes and the specifics of gastrulation in this group. We briefly outline the nature of these membranes and summarize how their history may have affected the process of gastrulation.

The investing membranes of amniotes arise as specialized outgrowths of the blastoderm immediately surrounding the embryo (for review, see Luckett 1977). The unique adaptations of these membranes in mammals are derived from the pattern seen in reptiles (Mossman 1987; Zeller 1999) and are not discussed here. Fisk and Tribe (1949) review the work of numerous authors who identify a pro- or ectamnion (see, e.g., Mitsukuri 1891a; Peters 1935), which they described as ". . . a circular solid ridge at the periphery of the embryonal shield, formed anteriorly and laterally of ectoderm, *posteriorly of material originating from the blastoporal plate*." (Fisk and Tribe 1949, p.97, italics added). Ultimately, this mesendodermal contribution from the blastoporal plate plays a substantial role in all the membranes. The mature amnion and chorion are made from somatopleure (ectoderm and somatic mesoderm). The allantois and so-called "definitive" yolk sac are made from splanchnopleure (endoderm and splanchnic mesoderm). Together these membranes play numerous crucial physiological roles (Chapter 31).

Although the membranes listed above are unique to amniotes, a "primitive" trilaminar yolk sac is found in all vertebrates with megalecithal eggs to provide hematopoiesis and nutrient flow. In elasmobranchs, some teleosts, and some amphibians, it forms by epiboly around the edges of the blastodisc (the extraembryonic blastopore of Arendt and Nübler-Jung 1999). The leading edge of the blastodisc leaves in its wake cells or nuclei that will constitute the mesodermal and endodermal layers of the primitive yolk sac. As noted above, the definitive yolk sac of reptiles is built of endoderm and mesoderm that arise from the blastoporal plate. Furthermore, epiboly at the edges of the blastodisc in amniotes is described as only producing ectoderm (Blount 1909; Bellairs et al. 1967; Gilbert 2003). Thus, there is a strong correlation between the presence of membranes and the source of extraembryonic mesendoderm from the primitive plate/streak in amniotes. Supporting this correlation are relevant differences in the expression of certain genes in the blastodisc of amniotes and anamniotes. The expression of *Brachyury*, a gene consistently linked to mesoderm formation, extends around the full circumference of the blastodisc in the shark *Scyliorhinus* (see Fig. 3A in Chapter 11), consistent with the generation of mesendodermal cells during epiboly. The expression of the *no tail* gene in zebrafish is very similar, extending around the epibolizing blastoderm margin and indicating the generation of mesendodermal cells (Obrect-Pflumio et al. 2002). *Brachyury* expression in amphibian gastrulae is also similar, being expressed in the marginal zone, all the way around the blastopore (Smith et al. 1991). In contrast, *Brachyury* is expressed only in the primitive streak in chickens (Kispert et al. 1995) and rabbits (and "posterior gastrular extension" of Viebahn et al. 2002; Chapter 17). It is conspicuously absent from the sites of ectodermal epiboly, consistent with a lack of mesoderm in the spreading epiblast margins in avians and mammals.

The earliest fossil amniotes are identified by osteological characters. These characters are maintained in some living amniotes and are assumed to correlate with the key embryological characters that define the group (Carroll 1988). A reconstruction of the ancestral stages of extraembryonic membrane evolution is unlikely to find any definitive proof in the fossil record. What can be conjectured, however, are perhaps several plausible series of selective regimes that may have led to the suite of characters that define the cleidoic egg.

The first amniotes appeared in the Carboniferous and were extremely small compared to many of their amphibian contemporaries. Romer (1957) proposed that these early "reptiles" were at least partially aquatic as adults, but laid their eggs on land, driven by selection for decreased predation in the terrestrial environment. The deletion of an aquatic larval stage places specific demands on the makeup of the egg. Yolk size is limited by requirements of gas exchange and water retention on land. Direct developing salamanders of the family plethodontidae represent the maximum egg size relative to adult body size in Anamnia (Salthe 1969). These salamanders push the limit of surface area and yolk volume in order to prolong development and hatch as miniature adults. The earliest amniotes were comparable in size to these modern plethodontid salamanders, and it is possible that their eggs were similar—equipped with protective layers of jelly and relatively large yolks to support direct development (Carroll 1988). Once released from predation pressure, fewer eggs with bigger yolks would be selected. The extracellular segregation of the increased yolk allowed meroblastic cleavage. In turn, this would cause the changes in gastrulation associated with the separation of a "typical" amphibian blastopore into a "pouch-like invagination" (blastoporal canal) and an extraembryonic blastopore, as described by Arendt and Nübler-Jung (1999).

The intermediate of Arendt and Nübler-Jung (1999) with the separation of extra- and intraembryonic blastopore is a common response to a large yolk. The important step for amniotes is the switch from generating a trilaminar yolk sac by epiboly, to recruiting cells from the "embryonic" blastopore to provide mesendodermal cells for extraembryonic membranes. A simple mechanism for this change would entail an alteration in regulation of *Brachyury* and related genes that would restrict expression to the embryonic blastopore and initiate ingression at the primitive, or blastoporal, plate.

Fisk and Tribe (1949) hypothesize that the amnion was the first of the specialized membranes to arise, primarily as a means of protecting the blastoderm from the dangers of adhesion to the outer shell membrane. Whereas the expanding extraembryonic blastoderm margin provides a fairly simple evolutionary source for the distal portions of the chorion, the emergence of the proamniotic fold from a ridge of epiblast just distal to the embryonic epidermal ectoderm does not have any obvious parallels or ancestral precedents among anamniotes. This structure appears to be a truly unique morphogenetic innovation of early amniotes. Szarski (1968) argues that the allantois, an elaboration of the hindgut, arose first, providing a means of storing nitrogen waste that could no longer be simply excreted by diffusion. Regardless of the sequence of these events, it seems clear that a large amount of all three germ layers is required in order to generate these membranes instead of a simple, primitive yolk sac. Selection pressure for the early production of specialized extraembryonic tissues would drive changes in the dynamics of gastrulation that increased the overall availability of extraembryonic cell populations. The inception of a primitive plate in addition to the blastopore as a source of ingressing cells would increase the rate and amount of mesendoderm formation enabling precocious generation of the extraembryonic membranes.

CONCLUSIONS AND SUGGESTIONS FOR FUTURE STUDIES

It seems possible that evolution of the cleidoic egg was largely focused on problems of desiccation and waste removal rather than with any special problems associated with large yolk mass. The latter problem has been solved numerous times independently within elasmobranch, osteichthyan, and amphibian lineages, and does not seem to have any strong impact on gastrulation mechanisms other than on the means of localizing some form of a Nieuwkoop center. The morphology and scale of head–tail axis formation is highly conserved within vertebrates, despite the widest extremes of egg size. This implies that the conserved mechanisms not only of craniogenesis, but also of trunk and tail development, are rather neutral with respect to yolk content, extraembryonic membranes, or oviparity versus viviparity. The roles played by vegetal axial determinants demonstrated in model species are similar whether a yolk syncytial layer (teleost), caudal subblastodermal mesenchyme (avian), or extraembryonic endodermal source (mammal) is involved, since all act at similar scales to trigger a largely autonomous primary organizer cascade. As egg size exceeds one millimeter in diameter, the ultimate vegetal end of large holoblastic or meroblastic eggs generally plays less of a role in axial induction and patterning. Instead, signals leading to formation of a Nieuwkoop center capable of inducing an organizer become restricted to a marginal zone near the vegetal–animal interface. As yolk sizes reach greater than five millimeters, epiboly of future embryonic cells becomes largely uncoupled from marginal cells that spread over the yolk mass to form the yolk sac. The yolk-sac walls of most meroblastic species of elasmobranchs and amniotes are eventually resorbed or discarded without being incorporated into any part of the gut proper. In these cases, the true gut wall is produced during involution or ingression at a blastopore or primitive streak.

The similarities seen between turtle and elasmobranch gastrulation noted above are convergent solutions to the same problem (the yolk). The main difference between them is the lack of mesoderm production at the blastodisc margins in turtles (and other reptiles) and the correlated existence of a primitive plate as a source of ingressing cells. We suggest these differences may have been driven by selection for increased and rapid production of the extraembryonic tissue that generates the extraembryonic membranes and the essential adaptation they provide.

Given the wealth of detailed molecular data from chicks and mice, it will be straightforward to test some of the predictions made here by looking at gene expression patterns in reptilian embryos. Because turtles have been the most extensively studied and are still the most easily acquired embryos of all reptiles, they are the most appropriate "model" species to represent the reptiles.

The equivalence of the blastoporal plate and primitive streak can be given molecular credence by mapping the expression domains of specific genes in the blastoporal plate of turtles. These expression patterns can be directly compared to the early primitive streak in the chick (reviewed in Chapter 15). Lawson et al. (2001) have begun a system of classifying genes expressed in pre-streak stages in chick. Group 1 genes, including *Nodal*, *Brachyury*, and *Fgf8*, are found in the early primitive streak. One can predict that the surface cells of the reptilian primitive plate will be positive for *Nodal*, for instance. Cells expressing *Chordin* and FGFs should be found ingressing between the endoblast and the epiblast (Chapter 15, Fig. 5). The relationship between the hypoblast and definitive endoderm in turtles versus lizards can be approached by looking at expression of *goosecoid* and *Cerberus*. Cells expressing these hypoblast markers should be displaced by non-expression cells from the primitive streak as in avians.

The exciting promise of a comprehensive study of reptilian gastrulation is for a better understanding of the primitive pattern of gastrulation for all amniotes. A survey of reptilian ontogenies will expose the full diversity of features adapted to large yolk masses and accommodating extraembryonic membranes. Both these factors influence the blastopore and should illuminate the origins of the "primitive streak." Clarifying the starting point from which the major branches of amniotes have diverged will provide a fuller understanding of the history and diversification of early developmental events.

REFERENCES

Agassiz L. 1857. *Contributions to the natural history of the United States*, vol. II, part III. *Embryology of the turtle*. Little, Brown, Boston. Massachusetts.

Arendt D. and Nübler-Jung K. 1999. Rearranging gastrulation in the name of yolk: Evolution of gastrulation in yolk-rich amniote eggs. *Mech. Dev.* **81:** 3–22.

Balfour F.M. 1879. On the early development of the Lacertilia, together with some observations on the nature and relations of the primitive streak. *Q. J. Microsc. Sci.* **19:** 421–430.

Ballowitz E. 1901. Die Gastrulation bei der Ringelnatter (*Tropidonotus natrix* Boie) bis zum Auftreten der Falterform der Embryonalanlage. *Z. Wiss Zool.* **70:** 675–732.

Bellairs R., Bromham D.R., and Wylie C.C. 1967. The influence of the area opaca on the development of the young chick embryo. *J. Embryol. Exp. Morphol.* **17:** 195–212.

Blanc F. 1974. Table de développement de *Chameleo lateralis* Gray 1831. *Ann. Embryol. Morphog.* **7:** 99–115.

Blount M. 1909. The early development of the pigeon's egg, with especial reference to polyspermy and the origin of the periblast nuclei. *J. Morphol.* **20:** 1–64.

Brachet A. 1914. Recherches sur l'embryologie des reptiles. Acrogénèse, Céphalogénèse et Cormogénèse chez *Chrysemys marginata*. *Arch. Biol.* **29:** 501–577.

Cao Y., Sorenson M.D., Kumazawa Y., Mindell D.P., and Hasegawa M. 2000. Phylogenetic position of turtles among amniotes: Evidence from mitochondrial and nuclear genes. *Gene.* **259:**139–148.

Carroll R.L. 1988. *Vertebrate paleontology and evolution*. W.H. Freeman, New York.

Chandrasekharan Nayar M. 1959. On gastrulation in Chelonia. *Prod. Ind. Acad. Sci.* **49B:** 402–407.

———. 1966. *In vitro* vital staining of chelonian blastoderms. *Indian J. Exp. Biol.* **4:** 131–134.

Clarke S.F. 1891. The habits and embryology of the American alligator. *J. Morphol.* **5:** 181–205.

Cotton J.A. and Page R.D. 2002. Going nuclear: Gene family evolution and vertebrate phylogeny reconciled. *Proc. R. Soc. Lond. B Biol. Sci.* **269:** 1555–1561.

Dendy A. 1899. Outlines of the development of the tuatara, *Spenodon (Hatteria) punctatus*. *Q. J. Microsc. Sci.* **42:** 1–87.

Ewert M.A. 1985. Embryology of turtles. In *Biology of the reptilia* (ed. C. Gans), vol. 14, pp. 75–268. Wiley, New York.

Ferguson M.W. J. 1985. Reproductive biology and embryology of the crocodilians. In *Biology of the reptilia* (ed. C. Gans), vol. 14, pp. 329–492. Wiley, New York.

Fisk A. and Tribe M. 1949. The development of the amnion and chorion of reptiles. *Proc. Zool. Soc. Lond.* **119:** 83–114.

Gauthier J.A., Kluge A.G., and Rowe T. 1988. The early evolution of the amniota. In *The phylogeny and classification of tetrapods* (ed. M.J. Benton), pp. 103–155. Clarendon, Oxford, United Kingdom.

Gerhardt U. 1901. Die Keimblättbildung bei *Tropinonotus natrix*. *Anat. Anz.* **20:** 241–261.

Gilbert S.F. 2003. *Developmental biology*, 7th edition. Sinauer, Sunderland, Massachusetts.

Gräper L. 1929. Die Primitiventwicklung des Hühnchens nach stereokinematographischen Untersuchungen, kontrolliert durch vitalen Farbmarkierung und verglichen mit der Entwicklung anderer Wirbeltiere. *Wilhelm Roux' Arch. Entwicklungsmech. Org.* **116:** 382–429.

Hedges S.B., and Poling L.L. 1999. A molecular phylogeny of reptiles. *Science* **283:** 998–1001.

Hubert J. 1962. Etude histologique des jeunes stades du développement embryonnaire du lézard vivipare (*Lacerta vivipara* Jacquin). *Arch. Anat. Microsc. Morphol. Exp.* **51:** 11–26.

———. 1970. Développement precoce de l'embryon et localisation extra-embryonnaire des gonocytes chez les reptiles. *Arch. Anat. Microsc. Morphol. Exp.* **59:** 253–270.

———. 1985. Embryology of the Squamata. In *Biology of the reptilia* (ed. C. Gans and F. Billett), vol. 15, pp. 1–34. Wiley, New York.

Janke A., Xu X., and Arnason U. 1997. The complete mitochondrial genome of the wallaroo (*Macropus robustus*) and the phylogenetic relationship among Monotremata, Marsupialia, and Eutheria. *Proc. Natl. Acad. Sci.* **94:** 1276–1281.

Janke A., Erpenbeck D., Nilsson M., and Arnason U. 2001. The mitochondrial genomes of the iguana (*Iguana iguana*) and the caiman (*Caiman crocodylus*): Implications for amniote phylogeny. *Proc. R. Soc. Lond. B Biol. Sci.* **268:** 623–631.

Kispert A., Ortner H., Cooke J., and Herrmann B.G. 1995. The chick *Brachyury* gene: Developmental expression pattern and response to axial induction by localized activin. *Dev. Biol.* **168:** 406–415.

Kupffer C. 1882. Die Gastrulation an den meroblastischen Eiern der Wirbeltiere und die Bedeutung des Primitivstreifs. *Arch. Anat. Phys.* **1882:** 1–30.

Laurin M. and Reisz R.R. 1995. A reevaluation of early amniote phylogeny. *Zool. J. Linn. Soc.* **113:** 165–223.

Lawson A. and Schoenwolf G. 2003. Epiblast and primitive-streak origins of the endoderm in the gastrulating chick embryo. *Development* **130:** 3491–3501.

Lawson A., Colas J.F., and Schoenwolf G.C. 2001. Classification scheme for genes expressed during formation and progression of the avian primitive streak. *Anat. Rec.* **262:** 221–226.

Lawson K., Meneses J., and Pederson R. 1991. Clonal analysis of epiblast fate during germ layer formation in the mouse embryo. *Development* **113:** 891–911.

Locascio A., Manzanares M., Blanco M.-J., and Nieto M. 2002. Modularity and reshuffling of Snail and Slug expression during vertebrate evolution. *Proc. Natl. Acad. Sci.* **99:** 16841–16846.

Luckett W.P. 1977. Ontogeny of amniote foetal membranes and their application to phylogeny. In *Major patterns in vertebrate evolution* (ed. M.K. Hecht et al.), pp. 439–516. Plenum Press, New York.

Mehnert E. 1892. Gastrulation und Keimblätterbildung der *Emys lutaria taurica*. *Morph. Arb.* **1:** 365–495.

———. 1895. Über Entwicklung, Bau und Function des Amnion und Amnionganges nach Untersuchungen an *Emys lutaria taurica*. *Morph. Arb.* **4:** 207–274.

Miller J.D. 1985. Embryology of marine turtles. In *Biology of the reptilia* (ed. C. Gans), vol. 14, pp. 269–328. Wiley, New York.

Mitsukuri K. 1891a. On the foetal membranes of Chelonia. *J. Coll. Sci. Imp. Univ. Tokyo* **4:** 1–53.

———. 1891b. Further studies on the formation of germinal layers in Chelonia. *J. Coll. Sci. Imp. Univ. Tokyo* **5:** 35–52.

————. 1893. On the process of gastrulation in Chelonia. *J. Coll. Sci. Imp. Univ. Tokyo* **6**: 227–277.

————. 1896. On the fate of the blastopore, the relations of the primitive streak, and the formation of the posterior end of the embryo in Chelonia, together with remarks on the nature of meroblastic ova in vertebrates. *J. Coll. Sci. Imp. Univ. Tokyo* **10**: 1–119.

Mitsukuri K. and Ishikawa C. 1886. On the formation of the germinal layers in Chelonia. *Q. J. Microsc. Sci.* **27**: 17–48.

Moffat L.A. 1985. Embryonic development and aspects of reproductive biology in the tuatara, *Sphenodon punctatus*. In *Biology of the reptilia* (ed. C. Gans), pp. 493–521. Wiley, New York.

Mossman H.W. 1987. *Vertebrate fetal membranes*. Rutgers University Press, New Brunswick, New Jersey.

Nicholas A. 1904. Recherche sur l'embryologie des reptiles. IV. La segmentation chez l'orvet. *Arch. Biol.* **20**: 611–658.

Obrect-Pflumio S., Thisse B., and Thisse C. 2002. No Tail (ntl) expression in zebrafish. ZFinID ZDB-XPAT-020802-10 (http://ZFIN.org).

Packard D.S. Jr. 1980. Somite formation in cultured embryos of the snapping turtle, *Chelydra serpentina*. *J. Embryol. Exp. Morphol.* **59**: 113–130.

Packard D.S. Jr. and Meier S. 1984. Morphological and experimental studies of the somitomeric organization of the segmental plate in snapping turtle embryos. *J. Embryol. Exp. Morphol.* **84**: 35–48.

Pasteels J. 1937. Etude sur la gastrulation des vértébres méroblastiques. II. Reptiles. *Arch. Biol.* **48**: 105–184.

————. 1953. Contribution à l'étude du développement des Reptiles. I. Origine et migration des gonocytes chez deux lacertiliens (*Mabuia megalura* et *Chamaeleo bitaeniatis*). *Arch. Biol.* **78**: 637–668.

————. 1957a. La formation de l'endophylle et de l'endoblaste vitellin chez les reptiles, chéloniens et lacertiliens. *Acta Anat.* **30**: 601–612.

————. 1957b. Une table analytique du développement des reptiles. 1. Stades de gastrulation chez les Chéloniens et les Lacertiliens. *Ann. Soc. Roy. Zool. Belg.* **87**: 217–241.

————. 1957c. La formation de l'amnios chez les cameleons. *Ann. Soc. R. Zool. Belg.* **87**: 243–246.

————. 1970. Développement embryonnaire. In *Traité de Zoologie* (ed. P.P. Grasse), vol. 14, pp. 893–971. Masson, Paris, France.

————. 1980. La gastrulation des vertébrés revue et corrigée depuis cinquante ans. *Arch. Biol.* **91**: 193–221.

Peter K. 1934. Die erste Entwicklung des Chamäleon (*Chamaeleo vulgaris*) verglichen mit der Eidechse (Ei, Keimbildung, Furchung, Entodermbildung). *Z. Anat. Entwicklungsgesch.* **103**: 147–188.

————. 1935. Die innere Entwicklung des Chamäleonkeimes nach der Furchung bis zum Durchbruch des Urdarms. *Z. Anat. Entwicklungsgesch.* **104**: 1–60.

————. 1938. Untersuchungen über die Entwicklung des Dotterentoderms. 3. Die Entwicklung des Entoderms bei Reptilien. *Z. Mikrosk.-Anat. Forsch.* **44**: 498–531.

Phillips M.J. and Penny D. 2003. The root of the mammalian tree inferred from whole mitochondrial genomes. *Mol. Phylogenet. Evol.* **28**: 171–185.

Rathke H. 1839. *Entwickelungsgeschichte der Natter* (Coluber natrix). Bornträger, Koenigsberg.

————. 1848. *Über die Entwicklung der Schildkröten*. Friedrich Vieweg, Braunschweig.

Reese A. 1908. The development of the American alligator (*A. mississipiensis*). *Smithsonian Misc. Collect.* **51**: 1–66.

————. 1915. *The alligator and its allies*. G.P. Putnam's Sons, New York.

Rieppel O. and deBraga M. 1996. Turtles as diapsid reptiles. *Nature* **384**: 453–455.

Romer A.S. 1957. Origin of the amniote egg. *Sci. Monthly* **85**: 57–63.

Ruta M., Jeffrey J.E., and Coates M.I. 2003. A supertree of early tetrapods. *Proc. R. Soc. Lond. B Biol. Sci.* **270**: 2507–2516.

Salthe S.N. 1969. Reproductive modes and the number and size of ova in the Urodeles. *Am. Midl. Nat.* **81**: 467–490.

Sharpe D.L. 1916. Turtle eggs for Agassiz. In *Atlantic classics*, pp. 23–44. The Atlantic Monthly Company, Boston, Massachusetts.

Smith J.C., Price B.M., Green J.B., Weigel D., and Herrmann B.G. 1991. Expression of a *Xenopus* homolog of *Brachyury* (*T*) is an immediate-early response to mesoderm induction. *Cell* **67**: 79–87.

Stewart J.R. 1997. Morphology and evolution of the egg of oviparous amniotes. In *Amniote origins* (ed. S.S. Sumida and K.L.M. Martin), pp. 291–326. Academic Press, San Diego.

Szarski H. 1968. The origin of vertebrate foetal membranes. *Evolution* **22**: 211–214.

Tribe M. and Bramwell F. 1932. The origin and migration of the primordial germ cells of *Sphenodon punctatus*. *Q. J. Microsc. Sci.* **75**: 251–282.

Vakaet L. 1962. Some new data concerning the formation of the definitive endoblast in the chick embryo. *J. Embryol. Exp. Morphol.* **10**: 38–55.

Viebahn C., Stortz C., Mitchell S.A., and Blum B. 2002. Low proliferative and high migratory activity in the area of *Brachyury* expressing mesoderm progenitor cells in the gastrulating rabbit embryo. *Development* **129**: 2355–2365.

Virchow H. 1892. Das Dotterorgan der Wirbelthiere. *Z. Wiss. Zool.* (suppl.) **53**: 161–206.

Voeltzkow A. 1899. Beiträge zur Entwicklungsgeschichte der Reptilien. I Biologie und Entwicklung der aüsseren Körperform von *Crocodilus madagascariensis*. *Senckenberg Naturforsch. Gesellsch. Abhandl.* **26**: 1–150.

————. 1901. Beiträge zur Entwicklungsgeschichte der Reptilien. IV. Keimblatter, Dottersack und erste Anlage des Blutes und der Gefasse bei *Crocodilus madagascariensis* Grand. *Senckenberg Naturforsch. Gesellsch. Abhandl.* **26**: 337–418.

Vogt W. 1929. Gestaltungsanalyse am Amphibien keim mit örtlicher Vitalfarung. II. Gastrulation und Mesoderm bildung bei Urodelen und Anuren. *Arch. Entwicklungsmech.* **120**: 385–707.

Wetzel R. 1929. Untersuchungen am Hühnchen. Die Entwicklung des Keims während der ersten beiden Bruttage. *Arch. Entwicklungsmech.* **119**: 188–320.

————. 1931. Urmund und Primitivstreigen. *Ergeb. Anat. Entwicklungsgesch.* **29**: 1–24.

Will L. 1892. Beiträge zur Entwicklungsgeschichte der Reptilien. 2. Die Anlage der Keimblätter bei der menorquinischen Sumpfsschildkröte (*Cistudo lutaria* Gesn.). *Zool. Jahrb. Abt. Anat.* **6**: 529–615.

————. 1896. Beiträge zur Entwicklungsgeschichte der Reptilien. 3. Die Anlage der Keimblätter der Eidechse (*Lacerta*). *Zool. Jahrb. Abt. Anat.* **9**: 1–91.

————. 1898. Über die Verhältnisse des Urdarms und des Canalis neurentericus bei der Ringelnatter (*Tropidonotus natrix*). Kön preuss Akad Wissensch Sitzungsb *1898*, 609–618.

Yntema C.L. 1968. A series of stages in the embryonic development of Chelydra serpentina. *J. Morphol.* **125**: 219–251.

Zardoya R. and Meyer A. 1998. Complete mitochondrial genome suggests diapsid affinities of turtles. *Proc. Natl. Acad. Sci.* **95**: 14226–14231.

Zeller U. 1999. Mammalian reproduction: Origin and evolutionary transformations. *Zool. Anz.* **238**: 117–130.

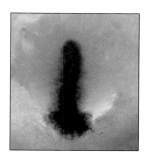

GASTRULATION IN THE CHICK

C.D. Stern

Department of Anatomy & Developmental Biology, University College London,
Gower Street, London WC1E 6BT, United Kingdom

EARLY STAGES: BLASTODERM FORMATION AND EARLY POLARITY

Avian eggs and early embryos differ in several respects from the majority of vertebrates, yet many of the principles governing the early steps of development are very similar. Unlike amphibians and some other embryos, the point of sperm entry is not a crucial determinant of polarity of the early embryo, because avian embryos are highly polyspermic—as many as 5–26 sperm may enter the egg in domestic fowl (chicken), and turkey eggs may be entered by several hundred sperm heads (Waddington et al. 1998; Stepinska and Olszanska 2003). Cleavage, as in most species containing a large amount of yolk (Figs. 1 and 2) (Arendt and Nübler-Jung 1999), is meroblastic—that is, cleavage occurs within a planar disc, and new cell membranes open into the yolk generating small cells in the center and large, yolk-laden, open cells at the periphery (Fig. 2B,C) (Bellairs et al. 1978; see also Chapter 51). Unfortunately, these stages are difficult to study because cleavage occurs while the egg is still within the maternal oviduct and before the shell is deposited (Fig. 1). In the chicken, laying occurs some 20 hours postfertilization, when the embryo is a flat disc (blastodisc or blastoderm) containing at least 20,000 cells.

A rule of thumb (von Baer's rule) can help to predict the orientation of the head–tail axis of the embryo from the outside of the egg: With the egg lying along its long axis, and with its blunt end to the operator's left, the axis of the embryo will run at right angles to the long egg axis with the head pointing away from the operator. However, this rule only applies in about 60–70% of cases.

As in teleosts, which are also meroblastic, maternal determinants are likely to exist also in avian embryos. One such determinant is called δ-ooplasm (or subgerminal ooplasm). This is contained in the thin, "white" yolk that makes up the latebra and the nucleus of Pander (Fig. 2A,C). The former is a funnel-like structure extending from just under the blastoderm to the center of the yolk (Callebaut et al. 1998a, 1999a, 2000a). We do not know the molecular nature or the functions of this ooplasm or whether it plays any role in polarity, although it has been suggested that it determines the position from which the endoblast and Koller's sickle (see below) will form (Callebaut 1993; Callebaut et al. 1998a, 2001).

Avian embryos appear to generate bilateral symmetry under the influence of gravity (Kochav and Eyal-Giladi 1971; Callebaut 1978, 1993; Eyal-Giladi and Fabian 1980; Eyal-Giladi et al. 1994; Callebaut et al. 2001). As the egg descends along the oviduct, it rotates with the blastoderm remaining at an angle of about 45° to the vertical—the lower edge of the blastoderm will become the future head end. However, we are still completely ignorant about the mechanism by which gravity breaks radial symmetry. It was suggested that opposite (upper and lower) poles of the disc are exposed to gravitational forces of different magnitude and that this causes differential amounts of cell shedding (Kochav and Eyal-Giladi 1971; Eyal-Giladi and Kochav 1976; Eyal-Giladi and Fabian 1980; Eyal-Giladi et al. 1994). However, this has never been demonstrated, and the alternative hypothesis that rotation exposes the poles of the blastoderm to subgerminal ooplasm to different extents (Callebaut 1993) seems much more likely.

Importantly, neither gravity nor maternal determinants irreversibly fix bilateral symmetry until gastrulation starts, because avian embryos are highly regulative; right up to the time of appearance of the primitive streak, blastoderms can

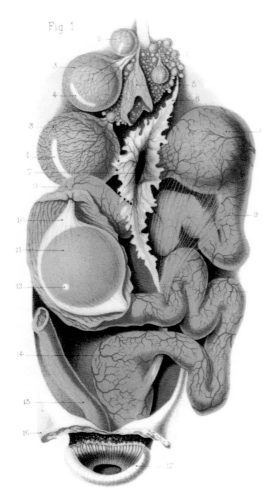

Figure 1. Formation of the hen's egg and its descent along the maternal oviduct. (Reproduced from Duval 1889.)

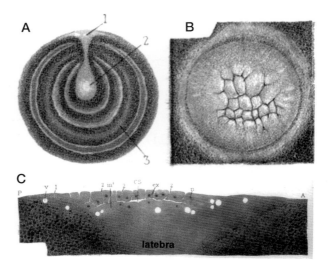

Figure 2. Cleavage in the chick embryo. (*A*) A section through the yolk reveals concentric rings (*3*) of dense (*darker*) and white (*lighter*) yolk. Under the blastoderm (*1*), a sub-blastodermic space filled with white yolk forms a funnel (the latebra) that extends deep into the center of the yolk mass, where it forms a small cavity, the Nucleus of Pander (*2*). (*B*) Cleavage in the chick embryo is meroblastic: The cleavage planes open into the surrounding yolk mass. (*C*) A section through the blastoderm and the surrounding yolk reveals the subgerminal cavity and latebra. (Adapted from Duval 1889.)

be split into several pieces (pie slices), each of which can spontaneously generate a complete embryonic axis (Lutz 1949; Spratt and Haas 1960a; Callebaut and Van Nueten 1995). Therefore, gravity and localized maternal components can, at best, have only bias polarity but do not act as definitive determinants (see also Chapter 24).

THE BLASTODERM STAGE

By the time the egg is laid, the embryo is an almost-flat disc in which an inner area pellucida can be distinguished from a more peripheral ring, the area opaca. Closest to the acellular vitelline membrane that envelops the yolk ("dorsal" side), a simple, one-cell-thick epithelium is continuous over both areas (Fig. 3) (Bancroft and Bellairs 1974; Bellairs et al. 1975). This is the epiblast. At this stage, the cells of the epiblast of the two concentric areas are almost indistinguishable mor-

phologically, except that at the very edge of the area opaca the cells are flattened and contact the vitelline membrane, against which they will later spread and help expand the blastoderm. The center of the disc is not attached to the membrane. At later stages, however, cells of the area opaca epiblast become less columnar than those of the area pellucida.

Deep (facing the yolk) to the epiblast, the cellular composition is more complex. The area opaca contains several layers of large (up to 150–200 μm) yolky cells; those closest to the epiblast are firmly attached to it. This is the germ wall. In contrast, the center of the disc (area pellucida) does not yet contain a continuous cell layer, but is peppered with small islands of about 5–10 cells each. These are also yolky, but not as large as the deep part of the area opaca (about up to 100 μm). The islands may arise by a process of polyingression (or shedding) that occurs throughout the area pellucida shortly before laying (Peter 1938; Kochav et al. 1980; Fabian and Eyal-Giladi 1981; Eyal-Giladi 1984), but the fate of the shed cells has never been studied experimentally. The islands later fuse with each other to generate the primitive endodermal layer, or hypoblast ("entophyll" or "primary hypoblast" in the earlier literature).

Between the area opaca and the area pellucida is a narrow region, known as the marginal zone. The epiblast of this region (to which the term refers) is not distinguishable from other regions of epiblast, except for the expression of *Vg1* at its posterior end (posterior marginal zone; see below)

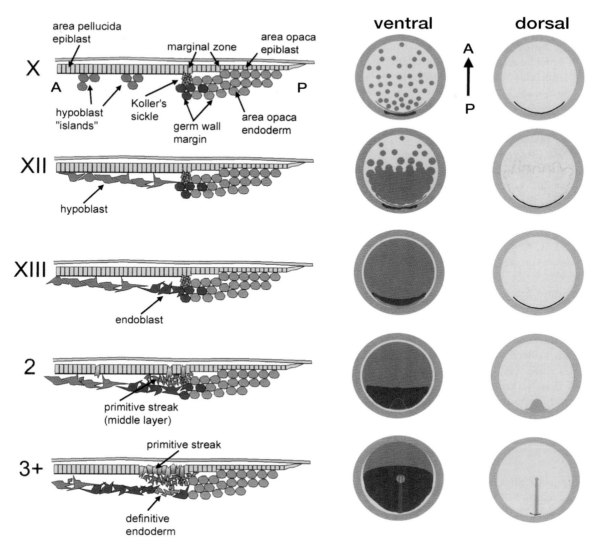

Figure 3. Stages of gastrulation in the chick embryo. Stages are indicated on the left: Roman numerals denote pre-primitive streak stages (X–XIV) (Eyal-Giladi and Kochav 1976) and Arabic numbers the stages from the appearance of the primitive streak (stage 2 onward) (Hamburger and Hamilton 1951). The left column shows diagrammatic midsagittal sections through part of the blastoderm, posterior to the right. The middle column depicts whole embryos viewed from the ventral (endodermal) side, and the rightmost column embryos viewed from the dorsal (epiblast) aspect. (*Gray*) Vitelline membrane; (*yellow*) epiblast; (*dark green*) hypoblast; (*light brown*) area opaca endoderm (germ wall and its margin); (*dark brown*) endoblast; (*red*) primitive streak mesendoderm; (*purple*) Koller's sickle; (*bright green*) definitive (gut) endoderm.

(Seleiro et al. 1996; Shah et al. 1997) and a slight gradient of *cWnt8C* decreasing from posterior to anterior (Skromne and Stern 2001). The only morphological landmark is that the deep part (germ wall) is not strongly attached to the epiblast, unlike the area opaca. This region is known as the germ wall margin. In carefully dissected blastoderms, it forms a lip that protrudes under the area pellucida for a few cell diameters (Stern and Ireland 1981; Stern 1990).

The boundary between area pellucida and marginal zone is marked, at the future posterior edge, by a crescent-shaped ridge of small cells, tightly adherent to the epiblast—Koller's sickle (also known as Rauber's sickle; for a profile of Rauber, see Chapter 1) (Koller 1882; Callebaut and Van Nueten 1994), which expresses *goosecoid* (Izpisua-Belmonte et al. 1993). Together, these components define a blastoderm of stage X (Roman numerals from I to XIV are used to classify stages before formation of the primitive streak according to Eyal-Giladi and Kochav [1976]; Arabic numerals from 2 onward are used for post-streak embryos following Hamburger and Hamilton [1951]).

In the few hours following incubation, the islands of hypoblast gradually fuse together, probably by a process of flattening of the cells, which proceeds from posterior to anterior to generate a continuous but relatively loose layer, the hypoblast proper (Vakaet 1970; Stern 1990) (see supplementary Movies 15_1 and 15_2 at www.gastrulation.org). This layer covers half of the area pellucida at stage XII and almost all of it at stage XIII (Fig. 3). Shortly after, two changes take place: first, the posterior germ wall margin cells and their progeny start to move centripetally (Stern 1990) and displace the hypoblast anteriorly; this new layer is the endoblast (or "sickle endoblast" or "secondary hypoblast" in the earlier literature). Hypoblast and endoblast can be distinguished by several markers, including *goosecoid* (in White Leghorns and some other strains), *Hex, Hesx1/Rpx, Cerberus/Caronte, Otx2,* and *Crescent,* all of which are expressed in the hypoblast but not in the endoblast (Bachvarova et al. 1998; Foley et al. 2000; Bertocchini and Stern 2002). The hypoblast is therefore similar to the anterior visceral endoderm (AVE) of the mouse embryo (see Chapters 16 and 17). At the same time, a posterior thickening (the posterior bridge), apparently derived from Koller's sickle, appears. This transient structure defines stage XIV, and the primitive streak starts to form immediately thereafter. None of the components of the deep layer (hypoblast, endoblast, germ wall or its margin) contributes to any embryonic tissues—they only generate extraembryonic membranes such as the yolk sac stalk, and later disappear.

Fate Maps and Cell Movements at the Blastoderm Stage

Many authors have constructed fate and specification maps of the epiblast of the chick at the blastoderm stage (Rudnick 1935, 1938; Hatada and Stern 1994; Callebaut et al. 1996; Bachvarova et al. 1998). The most detailed ones (Hatada and Stern 1994) reveal an orderly arrangement of prospective embryonic tissues, which gradually changes with time (Fig. 4), due to extensive morphogenetic movements of the epiblast that begin well before primitive streak formation (Gräper 1929; Vakaet 1970; Izpisua-Belmonte et al. 1993; Callebaut et al. 1999b; Foley et al. 2000). At stage X, future "dorsal" tissues (prospective organizer and its derivatives: endoderm, prechordal mesoderm, notochord) are found just central to and adjacent to Koller's sickle (Izpisua-Belmonte et al. 1993; Hatada and Stern 1994; Bachvarova et al. 1998; Streit et al. 2000). These territories quickly move toward the center of the blastoderm and gradually become replaced in their original position by more lateral regions of epiblast (progressively more "ventral" fates like somite, intermediate mesoderm, etc.) (see supplementary Movies 15_1 and 15_3 at www.gastrulation.org). Therefore, at stage X the posterior margin of the area pellucida contains a

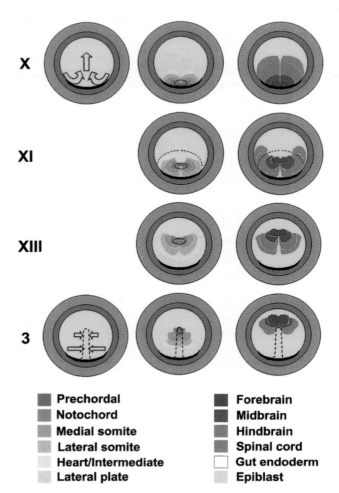

Prechordal
Notochord
Medial somite
Lateral somite
Heart/Intermediate
Lateral plate
Forebrain
Midbrain
Hindbrain
Spinal cord
Gut endoderm
Epiblast

Figure 4. Summary fate maps of the epiblast and cell movement patterns at different stages. The two diagrams in the left-hand column show the major movements in the epiblast: Polonaise movements before primitive streak formation, and convergence of the epiblast to the streak (which is strongest posteriorly) during gastrulation. The middle column of diagrams summarizes the locations of territories of cells that give rise to different mesodermal tissues and the gut endoderm. The rightmost column summarizes the locations of subdivisions of the neural plate. The dashed line at stage XI indicates the most anterior extent of spread of the hypoblast layer at this stage, and the dashed outline at stage 3 is the profile of the primitive streak.

bilateral gradation of dorsal-to-ventral fates, dorsal at the posterior midpoint. Subsequent movements "fold" this arrangement into a posterior midline presaging the future primitive streak, so that the most dorsal fates become located most anteriorly along this line (see below). These movements comprise convergence of epiblast toward the posterior midpoint and extension along the midline, but do not seem to occur by the process normally called "convergent extension" (see Chapter 19) in that it is not accompanied by significant cell-shape changes. The combination of posterior midpoint convergence and midline extension resembles a Polish dance (Polonaise) (see supplementary Movies15_1 and 15_2 at www.gastrulation.org), the name given by Gräper (1929) to these epiblast movements after his remarkable stereo-pair time-lapse films of labeled embryos, made as early as 1926 (see also Chapter 24).

It is truly remarkable that cells can move horizontally within a relatively tight epithelium, the epiblast. The mechanics of such migration, including the degree to which it is truly "active" (rather than a consequence of mechanical propagation of a remote event like cell loss through ingression), is not yet understood. However, recent observations of living, Bodipy-ceramide-labeled chick embryos using two-photon microscopy have started to reveal that individual cells within the epithelium translocate by "bobbing" up and down, as if each individual cell is a foot in a giant millipede (O. Voiculescu et al., unpubl.).

We have already described briefly above the movements in the lower layer (Waddington 1932; Spratt and Haas 1960b; Vakaet 1970; Rosenquist 1972; Stern and Ireland 1981; Stern 1990; Bakst et al. 1997; Callebaut et al. 1997a; Bachvarova et al. 1998; Foley et al. 2000; Bertocchini and Stern 2002). Essentially, the hypoblast expands as the islands fuse from posterior to anterior, and the newly formed hypoblast sheet is then displaced further anteriorly by the incoming endoblast) (see supplementary Movies 15_1 and 15_2 at www.gastrulation.org). The speed at which the hypoblast/endoblast layer spreads is similar to the midline extension in the epiblast (Hatada and Stern 1994), and recent experiments have shown that there is a causal link: Rotation of the deep layer generates a new set of Polonaise movements in the adjacent epiblast (Foley et al. 2000), although the mechanisms by which the two layers communicate are unknown.

These movements also deform Koller's sickle, which appears to be subjected to a large amount of shear. Its anterior (centrally facing) midpoint will later migrate anteriorly as the primitive streak forms, its lateral extremes converge to the midline, and the posterior aspect (facing the marginal zone) remains posterior and eventually becomes extraembryonic (Izpisua-Belmonte et al. 1993; Bachvarova et al. 1998; Streit et al. 2000).

CELL INTERACTIONS LEADING TO PRIMITIVE STREAK FORMATION

The fact that isolated fragments of blastodiscs can spontaneously initiate axis formation (Lutz 1949; Spratt and Haas 1960a) indicates that cell interactions, rather than definitive determinants, must be involved. What are the signals, and where do they come from? Three main sources have been proposed: the hypoblast and/or endoblast, Koller's sickle, and the posterior marginal zone (PMZ).

The idea that the hypoblast/endoblast layer regulates the site of primitive streak formation comes from important experiments by Waddington (1932, 1933) in which he demonstrated that rotation of the deep layer (which he called endoderm) influences the orientation of the primitive streak. When it was rotated by 90°, the streak arose from its original site but developed a gradual bend. After 180° rotation of the hypoblast, a few embryos had formed an ectopic streak arising from the opposite (anterior) side. This led Waddington to suggest that the hypoblast layer induces the primitive streak (Waddington 1933), but he was cautious to avoid ruling out a contribution from cell movements. Subsequent studies were more forceful in proposing induction by the hypoblast (Azar and Eyal-Giladi 1979, 1981, 1983; Mitrani and Eyal-Giladi 1981; Mitrani et al. 1983) but did not provide direct evidence with molecular markers or markers for different cell populations. Later, two studies repeated Waddington's original observations and highlighted the fact that since 90° hypoblast rotations do not induce primitive streak formation from a new site, this is unlikely to be a true inductive event (Khaner 1995; Foley et al. 2000). Moreover, it was shown (Foley et al. 2000) that the hypoblast layer influences the movements of the overlying epiblast—when rotated, it initiates a new set of Polonaise movements at 90° to the original. These compete with the original movements, causing the streak to bend, but cells destined for different tissue types do not change their fates.

Recently, a new emphasis has been placed on the second component of the deep layer, the endoblast (Callebaut and Van Nueten 1995; Callebaut et al. 1998b, 1999b, 2000b; Bertocchini and Stern 2002). Specifically (Bertocchini and Stern 2002), it was shown that complete removal of the hypoblast leads to the formation of multiple streaks at random positions, suggesting that the hypoblast emits an antagonist of axis formation. Analysis of expression patterns and misexpression experiments then suggested that Cerberus, a Nodal antagonist, is responsible. *Cerberus* is expressed in the hypoblast but not in the endoblast, which is consistent with the fact that the primitive streak starts to form precisely at the time when the hypoblast is displaced away from the posterior edge of the area pellucida by the incoming endoblast (Bertocchini and Stern 2002). Finally, it should be mentioned that the hypoblast does have some inducing activity, which can be revealed by assessing the expression of several epiblast genes after grafting a hypoblast ectopically. The homeobox gene *Not1/GNOT* (Knezevic and Mackem 2001) and the early "pre-neural" markers *ERNI*, *Sox3*, and *Otx2* are induced transiently by grafts of the hypoblast to ectopic sites (Foley et al. 1997; Streit et al. 2000; see Chapter 29). The induction of *Not1/GNOT* may be mediated by retinoids, whereas induction of *ERNI* and *Sox3* is mediated by FGF. We do not yet know the factors responsible for inducing *Otx2*.

The second component suggested as playing a role in primitive streak initiation is Koller's sickle (Izpisua-

Belmonte et al. 1993; Callebaut and Van Nueten 1994; Callebaut et al. 1997a, 1998b, 2003). This structure has been said to give rise to the endoblast (hence the alternative name of sickle endoblast) and has even been proposed to act as a passage for posterior marginal zone cells from the epiblast to the lower layer (Azar and Eyal-Giladi 1979; Eyal-Giladi 1997). However, higher-resolution fate mapping using different techniques has suggested instead that the sickle contributes cells to the primitive streak itself but not significantly to the endoblast (Izpisua-Belmonte et al. 1993; Bachvarova et al. 1998). Furthermore, although grafts of the sickle can indeed generate a second primitive streak upon transplantation, the extensive cellular contribution to the ectopic streak and particularly to definitive (gut) endoderm cannot be dissociated from the inductive effect (Izpisua-Belmonte et al. 1993; Bachvarova et al. 1998).

The third and final tissue involved in induction of primitive streak formation is the posterior marginal zone (PMZ) (Spratt and Haas 1960a; Azar and Eyal-Giladi 1979; Eyal-Giladi and Khaner 1989; Khaner and Eyal-Giladi 1989; Callebaut et al. 1997b; Bachvarova et al. 1998; Bachvarova 1999; Skromne and Stern 2001, 2002). Even though its activity as a streak inducer has been challenged (Callebaut et al. 1997b, 1998b), there is no question that when grafted into an ectopic position, the PMZ is able to induce the formation of a second axis without making a cellular contribution to it, as long as the host is younger than stage XI (Eyal-Giladi and Khaner 1989; Khaner and Eyal-Giladi 1989; Bachvarova et al. 1998). These properties of the PMZ have likened it to the amphibian Nieuwkoop center (see Chapters 13 and 45). Like the Nieuwkoop center, whose activity appears to depend on the overlap of TGF-β and Wnt pathways, the inducing ability of the PMZ can be mimicked by misexpression of *cVg1* in regions where *Wnt8C* is expressed (Fig. 5) (Seleiro et al. 1996; Shah et al. 1997; Skromne and Stern 2001, 2002). Surprisingly, however, unlike PMZ grafts, misexpression of *cVg1* in the anterior marginal zone will generate a full axis as late as stage XIII.

In conclusion, all three tissues (hypoblast/endoblast, Koller's sickle, and PMZ) proposed to have axis-inducing activity do indeed have the ability to influence primitive streak formation. However, their mechanisms of action and relative importance differ. The sickle has inducing ability, but this is probably only because it contains some of the cells fated to form Hensen's node (the avian organizer; see below). The earliest influences appear to come from the PMZ, where Vg1 and Wnt activities overlap. Vg1+Wnt induce expression of Nodal in the neighboring area pellucida epiblast, but Nodal can only act (presumably to induce mesendoderm) when the hypoblast has been displaced by the incoming endoblast.

In addition to Vg1+Wnt and Nodal, it is likely that FGFs (emanating from the hypoblast and/or Koller's sickle)

(Chapman et al. 2002; Karabagli et al. 2002) also play a role in primitive streak initiation because inhibitors of FGF block this process (Mitrani et al. 1990; Streit et al. 2000) and because misexpression of FGF can generate an ectopic streak (F. Bertocchini et al., unpubl.). It is likely that FGF acts in concert with Nodal, as in amphibians (Kimelman and Kirschner 1987; Cornell and Kimmelman 1994; LaBonne and Whitman 1994; Latinkic et al. 1997). Finally, BMP activity also regulates primitive streak formation, since ectopic expression of the antagonist Chordin (but not Noggin) is sufficient to induce a streak, even as late as stage 3, and misexpression of BMP4 near the streak causes the streak to disappear (Streit and Stern 1999). Chordin is normally expressed in Koller's sickle at stages XI–XIV.

Several important questions still remain unanswered. They include: What positions *Vg1* expression in the PMZ? What molecular mechanisms underlie regulation when the posterior half of the blastoderm is removed?

PRIMITIVE STREAK FORMATION AND ELONGATION

We know surprisingly little about the cellular details of how the primitive streak forms and elongates. Time-lapse films

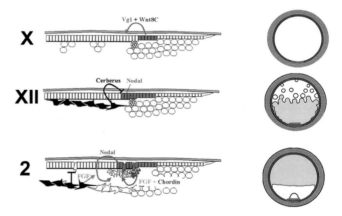

Figure 5. Molecular interactions implicated in the initiation of primitive streak formation, shown at three successive stages (indicated on the left), in sections (*left* column) and in whole mounts (*right*). At stage X, Vg1 (*red*; expressed in the posterior marginal zone) cooperates with Wnt8C (*blue*; expressed throughout the marginal zone) to induce Nodal (*bright green*) in the neighboring epiblast of the area pellucida. However, Nodal cannot act further because it is inhibited by Cerberus (*black*) produced by the underlying hypoblast (stage XII). Shortly before primitive streak formation (stages XIV–2), the displacement of the hypoblast by the non-Cerberus-expressing endoblast allows Nodal signaling to act. Nodal, in cooperation with FGF (*light brown*; emanating from the hypoblast and from Koller's sickle) and Chordin (*dark green*; produced by Koller's sickle) then induce ingression of cells from the epiblast to form the primitive streak. (Based on data from several sources, mainly Skromne and Stern 2001; Bertocchini and Stern 2002.)

(see supplementary Movies 15_1 and 15_3 at www.gastrulation.org) show that the initial appearance of the streak is extremely rapid—the embryo goes from having no visible axial structures (stage XIV) to developing a triangular, dense streak (stage 2; Fig. 3) in about 30 minutes, suggesting that primitive streak initiation is accompanied by massive ingression. However, although the basement membrane under the epiblast does partially dissolve during streak formation, this early stage does not involve the loss of epithelial continuity of the epiblast in the region of the forming streak, which happens much later (stage 3[+]) (Vakaet 1982; Andries et al. 1983; Sanders and Prasad 1989; Harrisson et al. 1991). This suggests that the formation of the early, triangular-shape streak is the result of rapid poly-ingression of individual cells through the basal lamina of the epiblast. This process may therefore be analogous to the formation of primary mesenchyme in echinoderms (see Chapter 9).

Ingression of early "pioneer" cells from the epiblast to the interior of the embryo can be seen starting as early as stage XII by staining either with the HNK-1 antibody (Canning and Stern 1988; Stern and Canning 1990; Canning et al. 2000; Mogi et al. 2000) or for activity of the enzyme acetylcholinesterase (AChE) (Drews 1975; Valinsky and Loomis 1984; Laasberg et al. 1986; Parodi and Falugi 1989). Indeed, the HNK-1 epitope is carried on a subunit of AChE (Bon et al. 1987), and AChE activity correlates very well with cell ingression and invasiveness in a variety of species (Drews 1975). It is puzzling that HNK-1/AChE expression differs in different strains of fowl; the salt-and-pepper expression in the epiblast is seen in the Rhode Island Red/Light Sussex crossbreed but not in the more inbred strain, White Leghorn. It also seems clear that HNK-1/AChE expression does not correlate completely with cells destined to form the primitive streak (Cooke 1993). A few of these cells do ingress but do not contribute to the streak (their fate remains unknown), other HNK-1-postive cells remain in the germ wall margin from where they contribute to the lower layer (Canning and Stern 1988; Stern and Canning 1990; Cooke 1993), and yet others do ingress to the primitive streak but later seem to disappear. Careful fate maps examining the origin of cells that will form the primitive streak have revealed that the definitive primitive streak is largely derived from a relatively small population of epiblast cells local to the site of streak formation and from Koller's sickle (Bachvarova et al. 1998; Wei and Mikawa 2000).

The early triangular streak (stage 2) is made up of a dense accumulation of middle layer cells between epiblast and endoblast (Fig. 3); however, it rapidly straightens, to become a mesenchymal rod of parallel sides (stage 3). At this stage there is still no groove in the overlying epiblast, and the basement membrane is largely intact. Soon afterward, however, two processes take place more or less simultaneously (Vakaet 1970): the appearance of a longitudinal

groove in the epiblast overlying the streak and the start of lateral migration of the mesenchyme of the streak, at right angles to the axis of the streak, to establish the lateral plate. These processes define stage 3[+]. Since grafts of early (stage 3) streak to a new area can generate an ectopic streak containing an epiblast groove (Vakaet 1973), and by analogy to the interactions between primary and secondary mesenchyme in echinoderms, it seems likely that the early streak cells induce the formation of a groove (and subsequent invagination) in the overlying epiblast. The signals that mediate this interaction are unknown, but FGF and/or Chordin are likely candidates since both can induce a streak at stage 3 (see above).

Stage 4 is marked by the appearance of a distinct bulge at the tip of the streak, encompassing all three layers—this is Hensen's node (Hensen 1876) (see below). We consider this to be the last phase of the "gastrula stage" in avian embryos. Shortly afterward (stage 4[+]), a small triangular mass of cells starts to protrude anteriorly from the node (the emerging tip of the head process, which contains precursors for the prechordal mesendoderm; see below). At this time, the future neural plate starts to become morphologically and molecularly (Sox2-positive) distinct, and stage 4[+] can therefore be considered the beginning of the "neurula stage."

We also know virtually nothing about the mechanics of primitive streak elongation. Time-lapse films reveal that less than 2 hours elapse between the early short streak at stage 2 and the almost fully elongated (1.5 mm long) stage 3 streak, making it very unlikely (L. Bodenstein, unpublished computer simulations) that cell division alone is the main force driving this elongation (Wei and Mikawa 2000). Streak elongation most likely involves a process of cell reorganization and changes in cell shape similar to those seen in amphibian convergent extension (see Chapter 19).

HENSEN'S NODE

The function of Hensen's node (the avian organizer) is described elsewhere in this volume (Chapters 26 and 29). Here we concentrate on the origin, maintenance, and subdivision of the node into different cellular territories.

Various fate-mapping techniques have established that Hensen's node arises from two distinct populations of cells (Izpisua-Belmonte et al. 1993; Hatada and Stern 1994; Bachvarova et al. 1998; Streit et al. 2000; Lawson and Schoenwolf 2001). One cell population (called posterior cells by Streit et al. 2000) resides deep to the epiblast, at the midpoint of Koller's sickle from stage X. It remains in this position until the primitive streak starts to form (stage 2) and then moves anteriorly with the tip of the advancing streak. The second population (called central cells by Streit et al. 2000) resides in the epiblast during these early stages—at stage X it is found immediately adjacent to the first population (in the *Nodal*-expressing territory; see Fig. 5 at stage X). However, the Polonaise movements almost immediately move these cells to the middle of the blastoderm, when these movements stop (stage XIII). As the primitive streak elongates, the posterior cells soon regain contact with the central cells (stages 3–3⁺). At this time, a morphological node forms (stage 4), and this is accompanied by the acquisition of expression of *Sonic hedgehog* (*Shh*) (Levin et al. 1995).

Neither the posterior nor the central cells possess full neural inducing ability by themselves (although posterior cells can induce transient expression of the pre-neural genes *Sox3* and *ERNI*), but acquire this ability when combined (Streit et al. 2000). When transplanted ectopically into the area pellucida of a host embryo, posterior cells can induce neighboring epiblast cells to acquire expression of *goosecoid*, a marker of the organizer (Izpisua-Belmonte et al. 1993). These findings suggest that during their migration anteriorly from stages 2–3, the posterior cells recruit adjacent epiblast cells to form part of the organizer.

The cellular composition of Hensen's node remains dynamic throughout the early stages of development. Even after stage 3⁺, neighboring epiblast cells migrate to the node, acquire the expression of organizer markers, and later migrate out again to emerge in the underlying layers as endoderm, notochord, prechordal mesendoderm, or medial somites (Joubin and Stern 1999). Inducing signals from within the streak (again, Vg1+Wnt, perhaps Nodal) and inhibitory signals from the node itself (ADMP) and from surrounding regions of the blastoderm (BMPs) form a complex network regulating the spatial and temporal expression of node markers including *Chordin*, *Goosecoid*, *Shh*, *Not1*, *HNF3β*, and others (Joubin and Stern 1999). These results account for the fact that primitive-streak-stage embryos from which the node has been extirpated can generate a new node (Grabowski 1956; Psychoyos and Stern 1996b; Joubin and Stern 1999; Yuan and Schoenwolf 1999).

Despite the dynamic composition of the node at these stages, single-cell lineage analysis has suggested that the node also contains a small population of resident cells with stem-cell characteristics (Selleck and Stern 1991, 1992b). It was proposed that when these cells divide, one daughter remains in the node while the other leaves to contribute to notochord and/or medial somite (Selleck and Stern 1992b; Stern et al. 1992).

The node is not a uniform structure, either molecularly or by tissue fate. At a molecular level, it displays left–right asymmetry of expression of a number of genes. The earliest of these are *Activin receptor IIA* (more likely to be a receptor for Nodal), which is expressed on the right, and the transcription factor *HNF3β*, which is expressed on the left (Levin et al. 1995; Stern et al. 1995), from stage 3⁺. By stage 4–4⁺, while the node starts to develop slight morphological

asymmetry, *Shh* appears on the left, *FGF8* on the right, and *Nodal* just to the left of the node (Levin et al. 1995; Dathe et al. 2002). Different regions of the node also give rise preferentially to different structures (Fig. 6), although the boundaries between these fates are not sharp. Specifically, the tip of the node contains mainly prospective notochord, prechordal mesendoderm, and floor plate cells; the sides and posterior aspect have mainly prospective medial somite and endodermal precursors (Selleck and Stern 1991). Transplantation experiments have revealed that the prospective notochord region contains cells that are already committed to this fate, whereas the lateral regions are more plastic (Selleck and Stern 1992a), but that all regions of the node are indistinguishable in their ability to induce and pattern neural tissue (Storey et al. 1995).

ESTABLISHMENT AND SUBDIVISION OF EMBRYONIC ENDODERM AND MESODERM

The study of endoderm formation in avian embryos, as in many other species, has been hindered considerably by the lack of any exclusive, permanent markers for the endoderm

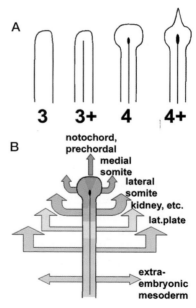

Figure 6. Fate maps of the primitive streak and Hensen's node. (*A*) Morphology of the anterior tip of the primitive streak at different stages: stage 3 (no groove, parallel sides), 3⁺ (groove, parallel sides), 4 (distinct node), 4⁺ (incipient head process, elongated pit). (*B*) Fates and movement patterns of mesoderm emerging from different portions of the streak at stage 4. Note that the anterior–posterior axis of the streak corresponds not to the head–tail axis of the embryo but rather to the mediolateral (axial–lateral, or dorsoventral) axis of the mesodermal organs. Based on data from several sources, mainly Selleck and Stern (1991) and Psychoyos and Stern (1996a).

lineage. This is even more inconvenient because the endoblast (see above) also lacks specific molecular markers, which makes it very difficult to distinguish these two neighboring tissues except that the endoblast contains typical intracellular inclusions which can be seen under phase contrast in explanted tissues (Stern and Ireland 1981). It was not until 1953 when Bellairs first recognized that the definitive endoderm is derived from the epiblast via the primitive streak (Bellairs 1953a,b, 1955, 1957), rather than from the hypoblast layer as was previously thought (see Chapter 30). The endoderm probably starts to insinuate itself into the lower layer at the early primitive streak stage (stage 2), and this insertion process ends by stage 4 (Vakaet 1962; Nicolet 1965, 1970; Modak 1966; Gallera and Nicolet 1969; Selleck and Stern 1991). We are still ignorant about the signals that induce and pattern the endoderm, from where they arise and at what stage, but based on studies in other species, it seems likely that Nodal will turn out to play a major role (reviewed in Chapter 30).

By the time the endoderm inserts into the deep layer (stage 3$^+$–4), the original hypoblast cells have become confined to the most anterior part of the blastoderm, a region called the "germinal crescent" because it also contains the primordial germ cells (Ginsburg and Eyal-Giladi 1986, 1987, 1989; Ginsburg et al. 1989; Tsunekawa et al. 2000). Since the surface of the hypoblast is greater than that of the germinal crescent, the tissue often develops blister-like projections extending ventrally from the surface of the epiblast. The fate of the hypoblast cells after this stage has not been examined thoroughly, but it is generally assumed that they contribute to the stalk of the yolk sac. It is equally likely, however, that a large proportion of hypoblast cells undergo apoptosis, since TUNEL staining at this stage shows heavy labeling in the hypoblast of the germinal crescent (A. Gibson and C.D. Stern, unpubl.).

The bulk of the middle layer of the stage 3$^+$ primitive streak will give rise to mesoderm: the notochord in the midline with prechordal mesoderm at its tip, the somites, intermediate mesoderm (prospective mesonephric kidney and its duct), heart, and lateral plate mesoderm (which includes both embryonic and extraembryonic components). It is important to recognize that the long axis of the primitive streak does not correspond to the future head–tail axis of the embryo but rather to the future dorsoventral axis of the mesoderm: Anterior streak (node) gives rise to the most dorsal/axial structures, with more ventral (lateral) structures arising from progressively more posterior streak positions (Schoenwolf et al. 1992; Psychoyos and Stern 1996a; Sawada and Aoyama 1999; Freitas et al. 2001; Lopez-Sanchez et al. 2001). This can be understood most easily by looking at the patterns of cell migration from the streak (Figs. 4 and 6), and the same relationship is seen in the mouse primitive streak (see Chapter 16).

It is likely that the major player in imparting specific dorsoventral identity to prospective mesoderm is BMP signaling. The node, which emits BMP antagonists, can transform lateral plate mesoderm into somitic mesoderm (Nicolet 1968), and this has been shown to be mimicked by Noggin (but not Chordin) (Streit and Stern 1999). Since Noggin is not expressed in the chick until about stage 4$^+$, it is likely that somite identity is not fixed until after this stage. Indeed, competence for lateral-to-medial transformation and vice versa remains until at least the early-somite stage (Tonegawa et al. 1997; Streit and Stern 1999; James and Schultheiss 2003).

Ingression from the epiblast to the deeper layers to form endoderm and the most medial (axial) mesoderm ends around the end of stage 4 (Vakaet 1962; Nicolet 1965, 1970; Modak 1966; Gallera and Nicolet 1969; Selleck and Stern 1991; Joubin and Stern 1999). A recent study has identified a zinc finger transcriptional activator, *Churchill*, which regulates the cessation of ingression at the primitive streak by activating *Sip1*, an antagonist of *Brachyury* (Sheng et al. 2003). This is described in more detail in Chapter 29; here we only point out that the expression and activities of *Churchill* and *Sip1* regulate the transition from the end of gastrular ingression to the start of neurulation opposite the anterior levels of the primitive streak.

Time-lapse films show that formation of the head process (the name given to the cranial portion of the notochord, rostral to the future level of the otic vesicle) begins at stage 4$^+$ by forward migration of cells from the node (Spratt 1947; Bellairs 1953b). After a short delay (to the end of stage 5), these movements stop and the primitive streak starts to regress (see below), which continues to extend the notochord caudally. Elongation of the notochord appears to include both a process of convergent extension (as in amphibians and fish, see Chapters 19 and 20) and the gradual deposition of progeny from resident stem cells (see above), but the major ingression movements from the epiblast opposite the anterior primitive streak have ceased by this stage. At more posterior levels, however, ingression to form lateral mesoderm continues for some time.

The emigration of prospective somite and lateral plate mesoderm from the primitive streak is controlled by chemorepulsion by FGF (perhaps FGF8) expressed in the streak (Yang et al. 2002; see Chapter 22). Yang et al. (2002) also proposed that after emerging from the streak, prospective somite tissue is then attracted back to the midline, specifically to FGF4 expressed in the notochord. However, since the entire embryo elongates and narrows at this stage, it is difficult to determine whether the migration of somitic mesoderm toward the midline is as active a process as was proposed by Yang et al. (2002; see also Easton et al. 1990). Furthermore, embryos lacking a notochord make a midline row of somites underlying the neural tube, raising the question of what would attract cells to the midline if this model is indeed correct (Stern and Bellairs 1984).

ENDING GASTRULATION AND REGRESSION OF THE PRIMITIVE STREAK

The main period of gastrulation is characterized by massive movement of epiblast into the primitive streak to generate mesoderm and endoderm. These movements gradually stop from stages 4–4$^+$ at the most anterior levels of the streak (prospective notochord and medial somite) and progressively more caudally. As mentioned above, the end of ingression through the anterior streak is regulated by *Churchill* and *Sip1* (Sheng et al. 2003). Soon after this (between stages 5 and 6), the primitive streak starts to regress) (see supplementary Movie 15_2 at www.gastrulation.org).

Several studies have attempted to establish the main cellular forces driving regression of the primitive streak. The earliest (Spratt 1947) made the important discovery that shortening of the streak is predominantly a morphological change, rather than a migration of node cells. However, convergent extension also plays a major role in the process as mentioned earlier (Spratt 1947; Bellairs 1963; Lepori 1966; Stern and Bellairs 1984; Schoenwolf et al. 1992; Catala et al. 1996; Colas and Schoenwolf 2001). We still know nothing, however, about the signals that regulate the timing, the speed, or the specific changes in cell behavior that control regression.

THE TAIL BUD: A CONTINUATION OF GASTRULATION?

While regression continues, the deposition of axial and paraxial mesoderm continues as the whole embryo narrows and elongates caudally to generate the tail bud (Sanders et al. 1986; Catala et al. 1996; Knezevic et al. 1998; Charrier et al. 2002). It was therefore proposed that the tail bud is a continuation of the process of gastrulation (Knezevic et al. 1998). Although it is true that several processes characteristic of gastrulation do continue in the regressing streak and later in the forming tail bud, other critical processes do not. Specifically, massive ingression of epiblast to form axial tissues (notochord and somites) has ceased (except perhaps at the most caudal end), the formation of new endoderm from the streak has also ended, and regression of the streak is accompanied by cell depletion from this structure. Furthermore, the node starts to lose its neural inducing ability just after stage 4 (Dias and Schoenwolf 1990; Storey et al. 1995; see Chapter 29). Together with the fact that the neural plate starts to elevate at about stage 4$^+$ (Bancroft and

Bellairs 1975), we consider that the end of gastrulation (as a stage) occurs between stages 4 and 4$^+$.

REFERENCES

Andries L., Vakaet L., and Vanroelen C. 1983. The dorsal surface of the animal pole of the just laid quail egg, studied with SEM. *Anat. Embryol.* **166:** 135–147.

Arendt D. and Nübler-Jung K. 1999. Rearranging gastrulation in the name of yolk: Evolution of gastrulation in yolk-rich amniote eggs. *Mech. Dev.* **81:** 3–22.

Azar Y. and Eyal-Giladi H. 1979. Marginal zone cells—the primitive streak-inducing component of the primary hypoblast in the chick. *J. Embryol. Exp. Morphol.* **52:** 79–88.

———. 1981. Interaction of epiblast and hypoblast in the formation of the primitive streak and the embryonic axis in chick, as revealed by hypoblast-rotation experiments. *J. Embryol. Exp. Morphol.* **61:** 133–144.

———. 1983. The retention of primary hypoblastic cells underneath the developing primitive streak allows for their prolonged inductive influence. *J. Embryol. Exp. Morphol.* **77:** 143–151.

Bachvarova R.F. 1999. Establishment of anterior-posterior polarity in avian embryos. *Curr. Opin. Genet. Dev.* **9:** 411–416.

Bachvarova R.F., Skromne I., and Stern C.D. 1998. Induction of primitive streak and Hensen's node by the posterior marginal zone in the early chick embryo. *Development* **125:** 3521–3534.

Bakst M.R., Gupta S.K., and Akuffo V. 1997. Comparative development of the turkey and chicken embryo from cleavage through hypoblast formation. *Poult. Sci.* **76:** 83–90.

Bancroft M. and Bellairs R. 1974. The onset of differentiation in the epiblast of the chick blastoderm (SEM and TEM). *Cell Tissue Res.* **155:** 399–418.

———. 1975. Differentiation of the neural plate and neural tube in the young chick embryo. A study by scanning and transmission electron microscopy. *Anat. Embryol.* **147:** 309–335.

Bellairs R. 1953a. Studies on the development of the foregut in the chick blastoderm. 1. The presumptive foregut area. *J. Embryol. Exp. Morphol.* **1:** 115–124.

———. 1953b. Studies on the development of the foregut in the chick blastoderm. 2. The morphogenetic movements. *J. Embryol. Exp. Morphol.* **1:** 369–385.

———. 1955. Studies on the development of the foregut in the chick embryo. 3. The role of mitosis. *J. Embryol. Exp. Morphol.* **3:** 242–250.

———. 1957. Studies on the development of the foregut in the chick embryo. 4. Mesodermal induction and mitosis. *J. Embryol. Exp. Morphol.* **5:** 340–350.

———. 1963. The development of somites in the chick embryo. *J. Embryol. Exp. Morphol.* **11:** 697–714.

Bellairs R., Breathnach A.S., and Gross M. 1975. Freeze-fracture replication of junctional complexes in unincubated and incubated chick embryos. *Cell Tissue Res.* **162:** 235–252.

Bellairs R., Lorenz F.W., and Dunlap T. 1978. Cleavage in the chick embryo. *J. Embryol. Exp. Morphol.* **43:** 55–69.

Bertocchini F. and Stern C.D. 2002. The hypoblast of the chick embryo positions the primitive streak by antagonizing nodal signalling. *Dev. Cell* **3:** 735–744.

Bon S., Meflah K., Musset F., Grassi J., and Massoulie J. 1987. An immunoglobulin M monoclonal antibody, recognizing a subset of acetylcholinesterase molecules from electric organs of

Electrophorus and Torpedo, belongs to the HNK-1 anti-carbohydrate family. *J. Neurochem.* **49:** 1720–1731.

Callebaut M. 1978. Effects of centrifugation on living oocytes from adult Japanese quails. *Anat. Embryol.* **153:** 105–113.

———. 1993. Early eccentricity in gravitationally oriented quail germs. *Eur. J. Morphol.* **31:** 5–8.

Callebaut M. and Van Nueten E. 1994. Rauber's (Koller's) sickle: The early gastrulation organizer of the avian blastoderm. *Eur. J. Morphol.* **32:** 35–48.

———. 1995. Gastrulation inducing potencies of endophyll and Rauber's sickle in isolated caudocranially oriented prestreak avian blastoderm quadrants (or fragments) in vitro. *Eur. J. Morphol.* **33:** 221–235.

Callebaut M., Harrisson F., and Bortier H. 2001. Effect of gravity on the interaction between the avian germ and neighbouring ooplasm in inverted egg yolk balls. *Eur. J. Morphol.* **39:** 27–38.

Callebaut M., Van Nueten E., Bortier H., and Harrisson F. 1999a. Interaction of central subgerminal ooplasm with the elementary tissues (endophyll, Rauber's sickle and upper layer) of unincubated avian blastoderms in culture. *Reprod. Nutr. Dev.* **39:** 589–605.

———. 2003. Positional information by rauber's sickle and a new look at the mechanisms of primitive streak initiation in avian blastoderms. *J. Morphol.* **255:** 315–327.

Callebaut M., Van Nueten E., Harrisson F., and Bortier H. 2000a. Activation of avian embryo formation by unfertilized quail germ discs: Comparison with early amphibian development. *Reprod. Nutr. Dev.* **40:** 597–606.

Callebaut M., Van Nueten E., Bortier H., Harrisson F., and Van Nassauw L. 1996. Map of the Anlage fields in the avian unincubated blastoderm. *Eur. J. Morphol.* **34:** 347–361.

Callebaut M., Van Nueten E., Harrisson F., Van Nassauw L., and Bortier H. 1999b. Endophyll orients and organizes the early head region of the avian embryo. *Eur. J. Morphol.* **37:** 37–52.

Callebaut M., Van Nueten E., Harrisson F., Van Nassauw L., and Bortier H. 2000b. Avian junctional endoblast has strong embryo-inducing and -dominating potencies. *Eur. J. Morphol.* **38:** 3–16.

Callebaut M., Van Nueten E., Harrisson F., Van Nassauw L., and Schrevens A. 1998a. Induction of (pre) gastrulation and/or (pre) neurulation by subgerminal ooplasm and Rauber's sickle in cultured anti-sickle regions of avian unincubated blastoderms. *Eur. J. Morphol.* **36:** 1–10.

Callebaut M., Van Nueten E., Van Nassauw L., Bortier H., and Harrisson F. 1998b. Only the endophyll-Rauber's sickle complex and not cells derived from the caudal marginal zone induce a primitive streak in the upper layer of avian blastoderms. *Reprod. Nutr. Dev.* **38:** 449–463.

Callebaut M., Van Nueten E., Bortier H., Harrisson F., Van Nassauw L., and Schrevens A. 1997a. Spatial relationship between endophyll, primordial germ cells, sickle endoblast and upper layer in cultured avian blastoderms. *Reprod. Nutr. Dev.* **37:** 293–304.

Callebaut M., van Nueten E., Harrisson F., van Nassauw L., Schrevens A. and Bortier H. 1997b. Avian gastrulation and neurulation are not impaired by the removal of the marginal zone at the unincubated blastoderm stage. *Eur. J. Morphol.* **35:** 69–77.

Canning D.R. and Stern C.D. 1988. Changes in the expression of the carbohydrate epitope HNK-1 associated with mesoderm induction in the chick embryo. *Development* **104:** 643–655.

Canning D.R., Amin T., and Richard E. 2000. Regulation of epiblast cell movements by chondroitin sulfate during gastrulation in the chick. *Dev. Dyn.* **219:** 545–559.

Catala M., Teillet M.A., De Robertis E.M., and Le Douarin M.L. 1996. A

spinal cord fate map in the avian embryo: while regressing, Hensen's node lays down the notochord and floor plate thus joining the spinal cord lateral walls. *Development* **122:** 2599–2610.

Chapman S.C., Schubert F.R., Schoenwolf G.C., and Lumsden A. 2002. Analysis of spatial and temporal gene expression patterns in blastula and gastrula stage chick embryos. *Dev. Biol.* **245:** 187–199.

Charrier J.B., Lapointe F., Le Douarin N.M., and Teillet M.A. 2002. Dual origin of the floor plate in the avian embryo. *Development* **129:** 4785–4796.

Colas J.F. and Schoenwolf G.C. 2001. Towards a cellular and molecular understanding of neurulation. *Dev. Dyn.* **221:** 117–145.

Cooke J. 1993. Expression of the HNK-1 epitope is unaltered among early chick epiblast cells despite behavioral transformation by inducing factors in vitro. *Int. J. Dev. Biol.* **37:** 479–486.

Cornell R.A. and Kimmelman D. 1994. Activin-mediated mesoderm induction requires FGF. *Development* **120:** 453–462.

Dathe V., Gamel A., Manner J., Brand-Saberi B., and Christ B. 2002. Morphological left-right asymmetry of Hensen's node precedes the asymmetric expression of Shh and Fgf8 in the chick embryo. *Anat. Embryol.* **205:** 343–354.

Dias M.S. and Schoenwolf G.C. 1990. Formation of ectopic neurepithelium in chick blastoderms: age-related capacities for induction and self-differentiation following transplantation of quail Hensen's nodes. *Anat. Rec.* **228:** 437–448.

Drews U. 1975. Cholinesterase in embryonic development. *Prog. Histochem. Cytochem.* **7:** 1–52.

Duval M. 1889. *Atlas d'embryologie.* G. Masson, Paris, France.

Easton H.S., Bellairs R., and Lash J.W. 1990. Is chemotaxis a factor in the migration of precardiac mesoderm in the chick? *Anat. Embryol.* **181:** 461–468.

Eyal-Giladi H. 1984. The gradual establishment of cell commitments during the early stages of chick development. *Cell Differ.* **14:** 245–255.

———. 1997. Establishment of the axis in chordates: Facts and speculations. *Development* **124:** 2285–2296.

Eyal-Giladi H. and Fabian B.C. 1980. Axis determination in uterine chick blastodiscs under changing spatial positions during the sensitive period for polarity. *Dev. Biol.* **77:** 228–232.

Eyal-Giladi H. and Khaner O. 1989. The chick's marginal zone and primitive streak formation. II. Quantification of the marginal zone's potencies—temporal and spatial aspects. *Dev. Biol.* **134:** 215–221.

Eyal-Giladi H. and Kochav S. 1976. From cleavage to primitive streak formation: A complementary normal table and a new look at the first stages of the development of the chick. I. General morphology. *Dev. Biol.* **49:** 321–337.

Eyal-Giladi H., Goldberg M., Refael H., and Avner O. 1994. A direct approach to the study of the effect of gravity on axis formation in birds. *Adv. Space Res.* **14:** 271–279.

Fabian B. and Eyal-Giladi H. 1981. A SEM study of cell shedding during the formation of the area pellucida in the chick embryo. *J. Embryol. Exp. Morphol.* **64:** 11–22.

Foley A.C., Skromne I.S., and Stern C.D. 2000. Reconciling different models of forebrain induction and patterning: A dual role for the hypoblast. *Development* **127:** 3839–3854.

Foley A.C., Storey K.G., and Stern C.D. 1997. The prechordal region lacks neural inducing ability, but can confer anterior character to more posterior neuroepithelium. *Development* **124:** 2983–2996.

Freitas C., Rodrigues S., Charrier J.B., Teillet M.A., and Palmeirim I. 2001. Evidence for medial/lateral specification and positional information within the presomitic mesoderm. *Development* **128:** 5139–5147.

Gallera J. and Nicolet G. 1969. Le pouvoir inducteur de l'endoblaste présomptif contenu dans la ligne primitive jeune de l'embryon de poulet. *J. Embryol. Exp. Morphol.* **21:** 105–118.

Ginsburg M. and Eyal-Giladi H. 1986. Temporal and spatial aspects of the gradual migration of primordial germ cells from the epiblast into the germinal crescent in the avian embryo. *J. Embryol. Exp. Morphol.* **95:** 53–71.

———. 1987. Primordial germ cells of the young chick blastoderm originate from the central zone of the area pellucida irrespective of the embryo-forming process. *Development* **101:** 209–219.

———. 1989. Primordial germ cell development in cultures of dispersed central disks of stage X chick blastoderms. *Gamete Res.* **23:** 421–427.

Ginsburg M., Hochman J., and Eyal-Giladi H. 1989. Immunohistochemical analysis of the segregation process of the quail germ cell lineage. *Int. J. Dev. Biol.* **33:** 389–395.

Grabowski C.T. 1956. The effects of the excision of Hensen's node on the early development of the chick embryo. *J. Exp. Zool.* **133:** 301–344.

Gräper L. 1929. Die Primitiventwicklung des Hünchens nach stereokinematographischen Untersuchungen, kontrolliert durch vitale Farbmarkierung und verglichen mit der Entwicklung anderer Wirbeltiere. *Arch. Entwicklungsmech. Org.* **116:** 382–429.

Hamburger V. and Hamilton H.L. 1951. A series of normal stages in the development of the chick embryo. *J. Morphol.* **88:** 49–92.

Harrisson F., Callebaut M., and Vakaet L. 1991. Features of polyingression and primitive streak ingression through the basal lamina in the chicken blastoderm. *Anat. Rec.* **229:** 369–383.

Hatada Y. and Stern C.D. 1994. A fate map of the epiblast of the early chick embryo. *Development* **120:** 2879–2889.

Hensen V. 1876. Beobachtungen über die Befruchtung und Entwicklung des Kaninchens und Meerschweinchens. *Z. Anat. Entwicklungsesch.* **1:** 353–423.

Izpisua-Belmonte J.C., De Robertis E.M., Storey K.G., and Stern C.D. 1993. The homeobox gene goosecoid and the origin of organizer cells in the early chick blastoderm. *Cell* **74:** 645–659.

James R.G. and Schultheiss T.M. 2003. Patterning of the avian intermediate mesoderm by lateral plate and axial tissues. *Dev. Biol.* **253:** 109–124.

Joubin K. and Stern C.D. 1999. Molecular interactions continuously define the organizer during the cell movements of gastrulation. *Cell* **98:** 559–571.

Karabagli H., Karabagli P., Ladher R.K., and Schoenwolf G.C. 2002. Comparison of the expression patterns of several fibroblast growth factors during chick gastrulation and neurulation. *Anat. Embryol.* **205:** 365–370.

Khaner O. 1995. The rotated hypoblast of the chicken embryo does not initiate an ectopic axis in the epiblast. *Proc. Natl. Acad. Sci.* **92:** 10733–10737.

Khaner O. and Eyal-Giladi H. 1989. The chick's marginal zone and primitive streak formation. I. Coordinative effect of induction and inhibition. *Dev. Biol.* **134:** 206–214.

Kimelman D. and Kirschner M. 1987. Synergistic induction of mesoderm by FGF and TGF-beta and the identification of an mRNA coding for FGF in the early *Xenopus* embryo. *Cell* **51:** 869–877.

Knezevic V. and Mackem S. 2001. Activation of epiblast gene expression by the hypoblast layer in the prestreak chick embryo. *Genesis* **30:** 264–273.

Knezevic V., De Santo R., and Mackem S. 1998. Continuing organizer function during chick tail development. *Development* **125:** 1791–1801.

Kochav S. and Eyal-Giladi H. 1971. Bilateral symmetry in chick embryo determination by gravity. *Science* **171:** 1027–1029.

Kochav S., Ginsburg M., and Eyal-Giladi H. 1980. From cleavage to primitive streak formation: a complementary normal table and a new look at the first stages of the development of the chick. II. Microscopic anatomy and cell population dynamics. *Dev. Biol.* **79:** 296–308.

Koller C. 1882. Untersuchungen über die Blätterbildung im Hühnerkeim. *Arch. Mikrosk. Anat.* **20:** 174–211.

Laasberg T., Neuman T., and Langel U. 1986. Acetylcholine receptors in the gastrulating chick embryo. *Experientia* **42:** 439–440.

LaBonne C. and Whitman M. 1994. Mesoderm induction by activin requires FGF-mediated intracellular signals. *Development* **120:** 463–472.

Latinkic B.V., Umbhauer M., Neal K.A., Lerchner W., Smith J.C., and Cunliffe V. 1997. The *Xenopus Brachyury* promoter is activated by FGF and low concentrations of activin and suppressed by high concentrations of activin and by paired-type homeodomain proteins. *Genes Dev.* **11:** 3265–3276.

Lawson A. and Schoenwolf G.C. 2001. Cell populations and morphogenetic movements underlying formation of the avian primitive streak and organizer. *Genesis* **29:** 188–195.

Lepori N.G. 1966. An analysis of the shortening process of the primitive streak in the blastodisc of the chicken and duck. *Acta Embryol. Morphol. Exp.* **9:** 61–68.

Levin M., Johnson R.L., Stern C.D., Kuehn M., and Tabin C. 1995. A molecular pathway determining left-right asymmetry in chick embryogenesis. *Cell* **82:** 803–814.

Lopez-Sanchez C., Garcia-Martinez V., and Schoenwolf G.C. 2001. Localization of cells of the prospective neural plate, heart and somites within the primitive streak and epiblast of avian embryos at intermediate primitive-streak stages. *Cells Tissues Organs* **169:** 334–346.

Lutz H. 1949. Sur la production experimentale de la polyembryonie et de la monstruosite double chez les oiseaux. *Arch. Anat. Microsc. Morphol. Exp.* **39:** 79–144.

Mitrani E. and Eyal-Giladi H. 1981. Hypoblastic cells can form a disk inducing an embryonic axis in chick epiblast. *Nature* **289:** 800–802.

Mitrani E., Shimoni Y., and Eyal-Giladi H. 1983. Nature of the hypoblastic influence on the chick embryo epiblast. *J. Embryol. Exp. Morphol.* **75:** 21–30.

Mitrani E., Gruenbaum Y., Shohat H., and Ziv T. 1990. Fibroblast growth factor during mesoderm induction in the early chick embryo. *Development* **109:** 387–393.

Modak S.P. 1966. Analyse experimental de l'origine de l'endoblaste embryonnaire chez les oiseaux. *Rev. Suisse Zool.* **73:** 877–908.

Mogi K., Toyoizumi R., and Takeuchi S. 2000. Correlation between the expression of the HNK-1 epitope and cellular invasiveness in prestreak epiblast cells of chick embryos. *Int. J. Dev. Biol.* **44:** 811–814.

Nicolet G. 1965. Autoradiographic study of the fate of cells invaginating through Hensen's node in the chick embryo at the definitive streak stage. *Acta Embryol. Morphol. Exp.* **8:** 213–220.

———. 1968. Role of Hensen's node in the differentiation of somites in birds. *Experientia* **24:** 263–264.

———. 1970. An autoradiographic study of the presumptive fate of the primitive streak in chick embryos. *J. Embryol. Exp. Morphol.* **23:** 70–108.

Parodi M. and Falugi C. 1989. Effects of acetylcholinesterase specific inhibitors on the development of chick embryos. *Boll. Soc. Ital.*

Biol. Sper. **65:** 839–845.

Peter K. 1938. Untersuchungen über die Entwicklung des Dotterentoderms. 1. Die Entwicklung des Entoderms beim Hühnchen. *Z. Mikrosk.-Anat. Forsch.* **43:** 362–415.

Psychoyos D. and Stern C.D. 1996a. Fates and migratory routes of primitive streak cells in the chick embryo. *Development* **122:** 1523–1534.

———. 1996b. Restoration of the organizer after radical ablation of Hensen's node and the anterior primitive streak in the chick embryo. *Development* **122:** 3263–3273.

Rosenquist G.C. 1972. Endoderm movements in the chick embryo between the early short streak and head process stages. *J. Exp. Zool.* **180:** 95–103.

Rudnick D. 1935. Regional restriction of potencies in the chick during embryogenesis. *J. Exp. Zool.* **71:** 83–99.

———. 1938. Differentiation in culture of pieces of the early chick blastoderm. *Anat. Rec.* **70:** 351–368.

Sanders E.J. and Prasad S. 1989. Invasion of a basement membrane matrix by chick embryo primitive streak cells in vitro. *J. Cell Sci.* **92:** 497–504.

Sanders E.J., Khare M.K., Ooi V.C., and Bellairs R. 1986. An experimental and morphological analysis of the tail bud mesenchyme of the chick embryo. *Anat. Embryol.* **174:** 179–185.

Sawada K. and Aoyama H. 1999. Fate maps of the primitive streak in chick and quail embryo: ingression timing of progenitor cells of each rostro-caudal axial level of somites. *Int. J. Dev. Biol.* **43:** 809–815.

Schoenwolf G.C., Garcia-Martinez V., and Dias M.S. 1992. Mesoderm movement and fate during avian gastrulation and neurulation. *Dev. Dyn.* **193:** 235–248.

Seleiro E.A., Connolly D.J., and Cooke J. 1996. Early developmental expression and experimental axis determination by the chicken Vg1 gene. *Curr. Biol.* **6:** 1476–1486.

Selleck M.A.J. and Stern C.D. 1991. Fate mapping and cell lineage analysis of Hensen's node in the chick embryo. *Development* **112:** 615–626.

———. 1992a. Commitment of mesoderm cells in Hensen's node of the chick embryo to notochord and somites. *Development* **114:** 403–415.

———. 1992b. Evidence for stem cells in the mesoderm of Hensen's node and their role in embryonic pattern formation. In *Formation and differentiation of early embryonic mesoderm* (ed. R. Bellairs et al.), pp. 23–31. Plenum Press, New York.

Shah S.B., Skromne I., Hume C.R., Kessler D.S., Lee K.J., Stern C.D., and Dodd J. 1997. Misexpression of chick Vg1 in the marginal zone induces primitive streak formation. *Development* **124:** 5127–5138.

Sheng G., Dos Reis M., and Stern C.D. 2003. Churchill, a zinc finger transcriptional activator, regulates the transition from gastrulation to neurulation. *Cell* **115:** 603–613.

Skromne I. and Stern C.D. 2001. Interactions between Wnt and Vg1 signalling pathways initiate primitive streak formation in the chick embryo. *Development* **128:** 2915–2927.

———. 2002. A hierarchy of gene expression accompanying induction of the primitive streak by Vg1 in the chick embryo. *Mech. Dev.* **114:** 115–118.

Spratt N.T. 1947. Regression and shortening of the primitve streak in the explanted chick blastoderm. *J. Exp. Zool.* **104:** 69–100.

Spratt N.T. and Haas H. 1960a. Integrative mechanisms in development. I. Regulative potentiality of separated parts. *J. Exp. Zool.* **145:** 97–137.

———. 1960b. Morphogenetic movements in the lower surface of the unincubated and early chick blastoderm. *J. Exp. Zool.* **144:** 139–157.

Stepinska U. and Olszanska B. 2003. DNase I and II present in avian oocytes: A possible involvement in sperm degradation at polyspermic fertilisation. *Zygote* **11:** 35–42.

Stern C.D. 1990. The marginal zone and its contribution to the hypoblast and primitive streak of the chick embryo. *Development* **109:** 667–682.

Stern C.D. and Bellairs R. 1984. The roles of node regression and elongation of the area pellucida in the formation of somites in avian embryos. *J. Embryol. Exp. Morphol.* **81:** 75–92.

Stern C.D. and Canning D.R. 1990. Origin of cells giving rise to mesoderm and endoderm in chick embryo. *Nature* **343:** 273–275.

Stern C.D. and Ireland G.W. 1981. An integrated experimental study of endoderm formation in avian embryos. *Anat. Embryol.* **163:** 245–263.

Stern C.D., Hatada Y., Selleck M.A., and Storey K.G. 1992. Relationships between mesoderm induction and the embryonic axes in chick and frog embryos. *Dev. Suppl.* **1992:** 151–156.

Stern C.D., Yu R.T., Kakizuka A., Kintner C.R., Mathews L.S., Vale W.W., Evans R.M., and Umesono K. 1995. Activin and its receptors during gastrulation and the later phases of mesoderm development in the chick embryo. *Dev. Biol.* **172:** 192–205.

Storey K.G., Selleck M.A., and Stern C.D. 1995. Neural induction and regionalisation by different subpopulations of cells in Hensen's node. *Development* **121:** 417–428.

Streit A. and Stern C.D. 1999. Mesoderm patterning and somite formation during node regression: Differential effects of chordin and noggin. *Mech. Dev.* **85:** 85–96.

Streit A., Berliner A., Papanayotou C., Sirulnik A., and Stern C.D. 2000. Initiation of neural induction by FGF signalling before gastrulation. *Nature* **406:** 74–78.

Tonegawa A., Funayama N., Ueno N., and Takahashi Y. 1997. Mesodermal subdivision along the mediolateral axis in chicken controlled by different concentrations of BMP-4. *Development* **124:** 1975–1984.

Tsunekawa N., Naito M., Sakai Y., Nishida T., and Noce T. 2000. Isolation of chicken vasa homolog gene and tracing the origin of primordial germ cells. *Development* **127:** 2741–2750.

Vakaet L. 1962. Some data concerning the formation of the definitive endoblast in the chick embryo. *J. Embryol. Exp. Morphol.* **10:** 38–57.

———. 1970. Cinephotomicrographic investigations of gastrulation in the chick blastoderm. *Arch. Biol.* **81:** 387–426.

———. 1973. Inductions by the nodus posterior of the primitive streak of birds. *C.R. Seances Soc. Biol. Fil.* **167:** 1053–1055.

———. 1982. Experimental study of the ingression during gastrulation in the chick blastoderm. *Verh. K. Acad. Geneeskd. Belg.* **44:** 419–437.

Valinsky J.E. and Loomis C. 1984. The cholinergic system of the primitive streak chick embryo. *Cell Differ.* **14:** 287–294.

Waddington C.H. 1932. Experiments on the development of chick and duck embryos cultivated in vitro. *Philos. Trans. R. Soc. Lond. B Biol. Sci.* **221:** 179–230.

———. 1933. Induction by the endoderm in birds. *Wilhelm Roux' Arch. Entwicklungsmech. Org.* **128:** 502–521.

Waddington D., Gribbin C., Sterling R.J., Sang H.M., and Perry M.M. 1998. Chronology of events in the first cell cycle of the polyspermic egg of the domestic fowl (*Gallus domesticus*). *Int. J. Dev. Biol.* **42:** 625–628.

Wei Y. and Mikawa T. 2000. Formation of the avian primitive streak from spatially restricted blastoderm: Evidence for polarized cell division in the elongating streak. *Development* **127:** 87–96.

Yang X., Dormann D., Munsterberg A.E., and Weijer C.J. 2002. Cell movement patterns during gastrulation in the chick are con-

trolled by positive and negative chemotaxis mediated by FGF4 and FGF8. *Dev. Cell* **3:** 425–437.

Yuan S. and Schoenwolf G.C. 1999. Reconstitution of the organizer is both sufficient and required to re-establish a fully patterned body plan in avian embryos. *Development* **126:** 2461–2473.

GASTRULATION IN THE MOUSE EMBRYO

P.P.L. Tam and J.M. Gad

Embryology Unit, Children's Medical Research Institute, University of Sydney, Wentworthville, New South Wales 2145, Australia

INTRODUCTION: A CLASS OF THEIR OWN

The Unique Anatomy of Rodent Gastrulae

Mammalian embryos arrive at gastrulation in different sizes and shapes and at different times in relation to implantation. Most of the eutherian and metatherian (marsupial) embryos form a round to oblong discoid gastrula, similar to that of the avian counterpart. The size varies from 4 to 10 mm in diameter for the marsupials (Selwood 1992) to 200–600 μm for the eutherians (Chapters 17 and 18). Some mammalian embryos (e.g., the marsupials) implant relatively late, after the formation of the blastocyst, and gastrulation may start during or even before implantation (Selwood 1992). However, in the higher primates and the rodents, implantation occurs well ahead of the onset of gastrulation.

Although ~40% of all living mammalian species are in the order Rodentia, the early embryology of only a few laboratory rodents (e.g., mouse, rat, hamster, guinea pig, and gerbil) has been studied in some detail. The blastocyst of the laboratory rodents adopts a vesicular morphology similar to that of most other eutherian embryos, with the trophectoderm forming the vesicular wall and the inner cell mass (ICM) located asymmetrically on the polar side of the blastocyst. The trophectoderm can be distinguished into two subpopulations: the polar trophectoderm associated with the ICM in the embryonic compartment and the mural trophectoderm in the abembryonic (cavity) compartment of the blastocyst. The polar trophectoderm of the rodent blastocyst proliferates to form a column of extraembryonic ectoderm that grows into the blastocyst cavity and carries the ICM at its distal pole (Long and Burlingame 1938; Copp 1979; ten Donkelaar et al. 1979; Ilgren 1981; Jollie 1990). It is not entirely clear what morphogenetic forces drive the formation of the cylindrical embryo. It has been postulated that the spatial constraint imposed by the uterine tissues around the implanting blastocyst restricts the lateral growth of the cap of polar trophectoderm. To accommodate the increase in tissue mass, the polar trophectoderm would have to grow along the path of least resistance by protruding into the blastocyst cavity (Copp 1981; Ilgren 1981), resulting in the formation of a proximal ectoplacental cone and a distally extending cylindrical column of extraembryonic ectoderm.

Prior to gastrulation, the cluster of ICM cells is organized into an epithelial tissue, the epiblast, which takes the shape of a cup with its rim juxtaposed to the distal pole of the extraembryonic ectoderm (Fig. 1A). The adoption of a cup configuration might be a morphogenetic adaptation to accommodate the growth of the epithelium when the rim of the epiblast could not increase adequately due to the limited growth in the girth of the cylindrical extraembryonic ectoderm. Further growth of the epiblast is therefore achieved more by increasing the depth, rather than the diameter, of the cup. The rodent embryo retains the cup shape during gastrulation as it develops from a bilaminar cup (epiblast and visceral endoderm) to a trilaminar cup (Fig. 1A) consisting of an inner layer of ectoderm, a middle layer of mesoderm, and an outer layer of endoderm. This results in the so-called "inversion" of germ layers at the conclusion of gastrulation. This unique configuration of the

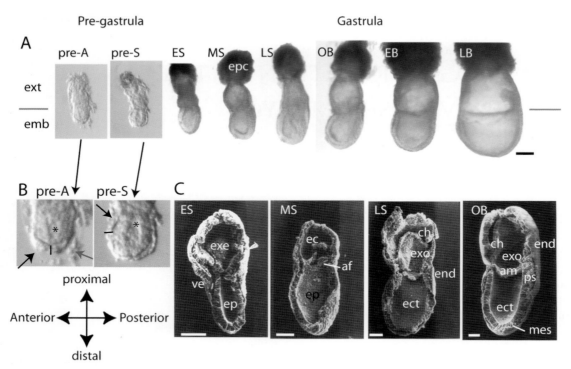

Figure 1. (*A, B*) Growth and morphogenesis of postimplantation mouse embryos prior to and at gastrulation (5.5 dpc–7.75 dpc). Pre-gastrula embryos are designated as pre-axis-aligned (pre-A, 5.5 dpc) stage, characterized by distal thickened visceral endoderm (*B*: *black arrow*, thickness indicated by the vertical bar) and proamniotic cavity only in the epiblast (*), and pre-streak (*B*: pre-S, 6.0 dpc) stage showing thickened visceral endoderm on one side and proamniotic cavity extending into the extraembryonic ectoderm) stages. 6.5- to 7.5-dpc embryos are staged as early-streak (ES), mid-streak (MS), late-streak (LS), no-bud (OB), early-bud (EB), and late-bud (LB) embryos (Downs and Davies 1993). (*C*) Scanning electron micrographs of ES, MS, LS, and OB embryos showing the formation of the extraembryonic and embryonic structures during gastrulation. Abbreviations: (am) amnion; (af) amniotic fold; (ch) chorion; (ec) ectoplacental cavity; (end) endoderm; (epc) ectoplacental cone; (ect) ectoderm; (ep) epiblast; (exe) extraembryonic ectoderm; (exo) exocoelom; (mes) mesoderm; (ps) primitive streak; (ve) visceral endoderm. (*A–C*) Anterior side of the embryo to the left and proximal side to the top. Red lines in *A* mark the boundary between the extraembryonic (ext) and embryonic (emb) regions. Red arrow in *B*, Reichert's membrane. Bar, 50 μm. (C, Reprinted, with permission, from Tam and Meier 1982; Tam et al. 1993 [© Wiley].)

gastrula of the laboratory rodents clearly diverges from the more conventional discoid morphology of the gastrula of other eutherian and metatherian mammals and the avian gastrula. The cup-shaped rodent gastrula is ~100–170 μm wide and 100–280 μm deep. However, if the cup could be flattened, the rodent gastrula would form a disc about 300–400 μm in diameter, which is remarkably similar to the dimension of the discoid gastrula of other eutherian mammals. Subsequent embryonic development follows this fundamental germ layer organization with the allocation of ectoderm to the dorsal and exterior tissues (skin and neural tube) of the embryo, and the endoderm to the ventral and later interior (gut) derivatives. The mesoderm gives rise to tissues (skeletal, muscular, and connective tissues) between the ectoderm and endoderm.

As the rodent embryo undergoes neurulation, the ectodermal neural plate is formed inside the cup and the gut endoderm forms on the outside. This necessitates a rotation

of the embryo to switch from the "lordotic" to the fetal shape and at the same time brings the ectoderm to the outside and the endoderm to the inside of the body (Kaufman 1992). The rotation of the body axis is a highly stereotypic morphogenetic process and appears to be part of the global mechanism for the acquisition of laterality (left/right asymmetry) of the body that is regulated by activity of signaling pathways and downstream transcription factors (Hamada et al. 2002; Chapter 28). However, other than the different proliferative activity found between tissues on contralateral sides of the embryo during turning (Miller and Runner 1978; Poelmann et al. 1987), little is known of the cellular and signaling mechanism of this rodent-specific morphogenetic process.

There are, however, variations from this general theme among the rodent gastrulae. Whereas the epiblast derived from the ICM remains in direct contact with the extraembryonic ectoderm in the mouse and rat embryo until gastrulation commences, the extraembryonic ectoderm of the guinea pig embryo only extends partially along the cylindrical embryo and is separated from the ICM before the onset of gastrulation (Ilgren 1981). Some variations in the organization of layers of the extraembryonic ectoderm, extraembryonic mesoderm, and visceral endoderm in the visceral yolk sac are found in the gastrulae of the laboratory rodents (ten Donkelaar et al. 1979; Jollie 1990). Last, it is not known whether gastrulae of every rodent species, particularly those outside the subfamily Muridae, will adopt the cylindrical murine configuration.

The extensive knowledge of the developmental biology and the genetics of the mouse results in its adoption as the model for studying rodent embryology and, more broadly, mammalian development. This chapter therefore focuses on the morphogenesis of the gastrula of the laboratory mouse. Specifically, the establishment of the body plan will be examined in the context of our current understanding of the regionalization of cell fate, the coordination of tissue movement, and the repertoire of tissue interactions mediated by signaling activity and lineage-specific transcriptional function.

THE ANATOMY OF GERM LAYER MORPHOGENESIS

Mouse blastocysts implant in the uterus at about 4.5 days after fertilization (4.5 dpc, days postcoitum). By 5.0 dpc, the embryo comprises a cylindrical core consisting of a proximal solid segment of extraembryonic ectoderm and a distal segment of a compact cellular cluster descended from the ICM, and an epithelial layer of visceral endoderm enveloping the entire cylindrical core. At 5.5 dpc, the epiblast adopts an epithelial morphology, and a proamniotic cavity is formed in the core of the epiblast but does not extend into the extraembryonic ectoderm (Downs and Davies 1993).

Gastrulation commences at about 6.5 dpc following the formation of a primitive streak in the posterior region of the embryo. In the primitive streak, epiblast cells lose their polarized epithelial morphology to become a loose mesenchyme (Figs. 2A,C and 3A,B) (Hashimoto et al. 1987; Tam et al. 1993). Epithelial-mesenchyme transformation of the epiblast can be triggered by FGF activity (Burdsal et al. 1998) and is accompanied by the local disappearance of the basement membrane (Tam et al. 1993), changes in the expression profiles of adhesion molecules (Takeichi 1988), and glycosyltransferase (Shur 1982). The mesenchymal cells also acquire the morphology typical of motile cells (Solursh and Revel 1978). Prior to gastrulation, molecular markers (e.g., *Wnt3* and *T*) that are associated with the primitive streak are initially expressed in the epiblast in the proximity of the extraembryonic ectoderm (Fig. 2A, pre-S). Expression of these markers becomes lopsided shortly

before the appearance of the primitive streak and subsequently is localized in the primitive streak (Fig. 2A, ES to E-som). This raises the possibility that first, the potential for the formation of the primitive streak is initially established in the proximal epiblast and second, cells with primitive streak potency may be recruited directly for the formation of the primitive streak. It is not known whether primitive streak formation is initiated by a few "pioneer" epiblast cells whose ingression triggers a localized disruption of the basement membrane which is then amplified into the ingression en masse of the epiblast cells (Chapters 4, 9, and 15).

As gastrulation continues, the primitive streak extends distally along the posterior side of the embryo until it spans the full length to the distal tip of the cylindrical embryo (Fig. 2A, B). Mesenchymal cells that are formed in the primitive streak spread distally as a new layer of cells (the mesoderm) between the epiblast and the visceral endoderm (Figs. 2C and 3B,C) (Tam and Meier 1982). The mesodermal cells also spread proximally across the proximal border of the epiblast and displace the extraembryonic ectoderm, which retracts proximally and anteriorly. Finally, by 7.5 dpc, a complete layer of mesoderm is formed between the ectoderm and the endoderm (Figs. 3D, E). In the extraembryonic region, the mesoderm expands to occupy the space between the receding extraembryonic ectoderm and the visceral endoderm (Figs. 1C and 3B, E). A cavity is formed within the mass of extraembryonic mesoderm to produce the exocoelom (this later becomes the cavity of the visceral yolk sac; Fig. 1C, MS to OB). The mesoderm lining the exocoelom differentiates into the endothelial and blood-forming cells of the yolk sac (Chapter 31). The extraembryonic mesoderm, together with ectoderm that extends inwardly from the margin of the proximal epiblast, contributes to the formation of the amniotic folds. The folds stretch across the proamniotic cavity to form the amnion that separates the proamniotic cavity into the amniotic cavity and exocoelom, which is formed within the extraembryonic mesoderm (Fig. 1C). The mouse embryo therefore adopts a strategy for the formation of the amnion which is a compromise between those that develop amniotic folds reaching over the flat embryonic disc to produce the amnion (e.g., rabbit, ungulates, carnivores, lemur, and marsupials) and those that form an amniotic cavity by splitting the ICM into the embryonic disc and the amnion (e.g., bat, primates, and insectivores) (Wimsatt 1975; Viebahn 1999).

During gastrulation, the epiblast layer remains as a pseudostratified epithelium that merges posteriorly into the primitive streak. Within 24 hours of gastrulation, the embryo undergoes rapid cell proliferation such that cell number increases from about 660 to 15,000, with the cell population doubling every 7–8 hours (Snow 1977; Power and Tam 1993). No distinctive changes in tissue morphology are discerned in the transition from epiblast to ectoderm

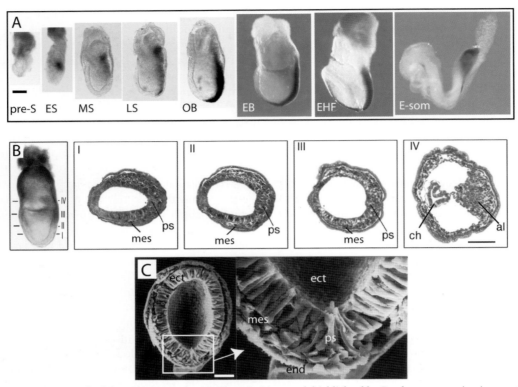

Figure 2. Primitive streak of the mouse gastrula. (*A*) The primitive streak highlighted by *Brachyury* expression in pre-streak to early-somite (E-som) embryos. (*B*) Histological sections of a LB embryo at different transverse planes along the (I–III) length of the primitive streak and (IV) the extraembryonic region showing the germ layer organization. (Pictures courtesy of Dr. Lorraine Robb, Walter Eliza Hall Institute of Medical Research.) (*C*) Scanning electron micrograph of the distal part of a LS embryo viewed from the proximal side. The embryo was fractured at the transverse plane (B,II) to show the primitive streak (*box*, shown in a magnified view) where epiblast cells undergo ingression (*red arrows*) and epithelial-mesenchymal transformation. Embryonic stages: (pre-S) pre-streak; (ES) early-streak; (MS) mid-streak; (LS) late-streak; (OB) no-bud; (EB) early-bud; (LB) late-bud; (EHF) early-head-fold; (E-som) early-somite embryos. Abbreviations: (al) allantois; (ch) chorion; (end) endoderm; (ect) ectoderm; (mes) mesoderm; (ps) primitive streak. Anterior side of the embryo to the left (*A, B*) and to the top (*C*). Bar, 50 μm. (*C*, Reprinted, with permission, from Tam et al. 1993 [© Wiley].)

(Tam et al. 1993). In contrast, the endoderm undergoes a significant change in cellular morphology during gastrulation. Endodermal cells change from a low columnar epithelium with a dense microvillous apical surface into a squamous epithelium with sparse microvilli (Hogan and Tilly 1981). This morphological change begins with the endoderm in the distal and posterior region of the embryo near the distal end of the primitive streak and gradually extends to other regions of the embryo (Poelmann 1981; Tam et al. 1993). The changes in the epithelial height and cell volume in conjunction with the proliferation of cells may enable the endodermal cells to cover a wider area on the outside of the growing cylindrical embryo.

MODELING THE MOUSE GASTRULA INTO A DISCOID CONFIGURATION

The cylindrical configuration of the pre-gastrula and gastrula-stage mouse embryo poses a unique anatomical rela-

tionship of germ layers and their derivatives that is different from the conventional discoid morphology of gastrulae of other mammalian species. The elucidation of the nature of tissue interactions that are instrumental to embryonic patterning requires the knowledge of (1) the molecules involved in intercellular signaling and its transduction to the transcriptional machinery and (2) the physical conduit through which tissue interaction is achieved. It is particularly crucial that the spatial relationship of interacting tissues is clearly understood and can be unambiguously related to and compared with other model systems.

To highlight the sites of potential tissue interaction, a "flattened" model of the mouse embryo with the various tissue compartments laid out in a two-dimensional configuration (Behringer et al. 2000) would be useful for the appreciation of the directionality and the localization of the inductive interactions between tissue compartments. We propose to extend this concept by presenting the cup-shaped mouse embryo in a discoid configuration that is

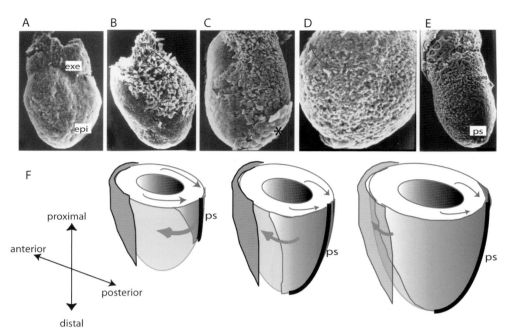

Figure 3. Morphogenesis of the mesodermal layer in the mouse gastrula. (*A*) Basal aspect of the epiblast of the pre-streak embryo (the overlying visceral endoderm has been removed) before the formation of the primitive streak. (*B–E*) Recruitment of epiblast cells to the primitive streak and expansion (*red curved arrow*) of the mesodermal layer from the posterior to the anterior region of (*B*) early-streak, (*C*) mid-streak, (*D*) late-streak to no-bud, (*E*) late-bud embryo. Asterisk in *C* indicates a layer of endoderm (presumptive definitive endoderm) that adheres tightly to the underlying tissues in the anterior region of the primitive streak. Dashed oval in *E* shows the vascular anastomosis formed in the extraembryonic mesoderm underneath the visceral (yolk sac) endoderm, which has been partially removed. (*F*) The direction of tissue movement in the epiblast (*blue arrows*) toward the primitive streak and expansion (*red arrows*) of the mesoderm toward the distal and anterior regions of the embryo between the epiblast and the endoderm (only the anterior part is shown). Abbreviations: (exe) extraembryonic ectoderm; (epi) epiblast; (ps) primitive streak.

more in line with the shape of the gastrulae of other mammals and birds (Fig. 4). Each of the three definitive germ layers is depicted as a *disc* (a full disc for the ectoderm and endoderm, and either a partial or a whole disc for the mesoderm) that is encircled by a *ring* of extraembryonic tissues, and the disc is subdivided into *sectors* of specialized tissue. This *disc–ring–sector* model captures the key features of the mouse embryo at the beginning (early-streak) and final (late-streak) stage of gastrulation:

1. The early-streak embryo (Fig. 4A,I) is converted from a cup-shaped configuration (Fig. 4A,II) to a stack of two complete discs, the upper ectoderm disc and the lower endoderm disc (Fig. 4A,III). The ectodermal disc comprises a circumferential zone of extraembryonic ectoderm, the central epiblast, and the posterior primitive streak sector. In the endoderm disc, the anterior and posterior sectors represent the anterior and posterior visceral endoderm, respectively. The nascent mesoderm forms a crescent that is sandwiched between the posterior primitive streak sector of the epiblast and the posterior visceral endoderm sector of the endoderm.

2. Upon almost complete formation of the germ layers, the late-streak gastrula (Fig. 4B, I, II) could be modeled as a stack of three discs corresponding to, from top to bottom, the ectoderm, mesoderm, and endoderm (Fig. 4B, III). The periphery of the ectoderm disc is now surrounded by the amnion ectoderm. In the ectoderm disc, the primitive streak sector has extended to the center of the disc where Hensen's node (or the node, for brevity) is located. The mesoderm layer consists of the peripheral extraembryonic mesoderm ring and the central embryonic mesoderm disc, in which the anterior axial mesoderm occupies the anterior sector. The endoderm disc comprises the peripheral extraembryonic endoderm and the central definitive endoderm.

The proposed model enables a direct comparison of the fate maps and morphogenetic movement of the cup-shaped mouse gastrula with other discoid gastrulae. It also provides a spatial framework for evaluating the effects of tissue interaction on germ layer patterning. In the ring–disc–sector model, interaction between the extraembryonic and embryonic tissues within a disc (the epiblast/ectoderm,

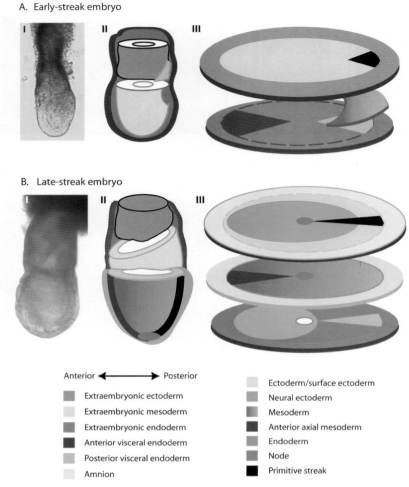

A. Early-streak embryo

I   II   III

B. Late-streak embryo

I   II   III

Anterior ◄————► Posterior

- Extraembryonic ectoderm
- Extraembryonic mesoderm
- Extraembryonic endoderm
- Anterior visceral endoderm
- Posterior visceral endoderm
- Amnion

- Ectoderm/surface ectoderm
- Neural ectoderm
- Mesoderm
- Anterior axial mesoderm
- Endoderm
- Node
- Primitive streak

Figure 4. Schematic representation of the germ layer tissues of (*A*,I) the early-streak embryo and (*B*,I) late-streak embryo in (*A*,II and *B*,II) the natural cup-shaped configuration and (*A*,III and *B*,III) the flattened ring–disc–sector model.

mesoderm, or visceral/definitive endoderm) will take place either centrifugally or centripetally. For the early-streak embryo, the proximal epiblast now occupies the peripheral position of the ectodermal disc, and the distal epiblast corresponds to the cell population in the central region of the disc. Germ layer tissues closer to the circumference of the disc will therefore be expected to experience a more powerful interaction with the extraembryonic tissues than those in the central part of the disc, on the assumption that the distance from the interface between the two interacting tissues grades the level of interactivity. Interaction between germ layers may also involve neighboring discs in the stack, and the effect of interaction between the upper and lower disc of the stack could be relayed, reinforced, or counteracted by the middle layer. Regionalized interaction between tissues of the adjacent discs may be mediated by the activity of cells in specific sectors of the disc, and this may be enhanced or relayed by the extraembryonic tissue in the vicinity of the

sector. Within each disc, planar inductive interaction, which may change significantly as the embryo develops, can occur locally between specific sectors or globally in either graded or uniform levels in the cell population.

DEVELOPMENT IN ANTICIPATION OF GASTRULATION

Maintaining the Growth and Differentiation of Extraembryonic Ectoderm

Implantation of the mouse blastocyst is achieved by the attachment of the mural trophectoderm, followed shortly by the polar trophectoderm, to the uterine tissue, and the invasion of trophoblasts derived from both types of trophectoderm (Carson et al. 2000). The success of implantation depends critically on the ability of the trophectoderm to differentiate into functional trophoblasts. The establishment of implantation is an absolute prerequisite for the embryo to develop beyond the blastocyst stage (Rossant and Cross 2001). Embryos that lack *Cdx2* and *Eomes* activity are deficient of trophoblasts and fail to implant (Russ et al. 2000; van den Akker et al. 2002).

Following implantation, the trophectoderm proliferates actively to form the extraembryonic ectoderm. The proliferation and differentiation of the trophectodermal derivatives are influenced by the inductive interaction with the ICM and the epiblast (Rossant and Ofer 1977; Rossant and Cross 2001). Blastocysts that lack *Fgf4* activity in the epiblast (Feldman et al. 1995) or *Fgfr2* function (Arman et al. 1998; Haffner-Krausz et al. 1999) in the trophectodermal tissues are depleted of trophoblasts and fail to implant properly. The mutant phenotype implicates a critical role for FGF signaling in sustaining proliferation of the trophoblast. It is therefore postulated that extraembryonic ectodermal cells located in the proximity of the epiblast are subject to strong FGF signaling activity. These cells will remain proliferative to provide a constant source of cells for the growth of extraembryonic ectoderm. As the extraembryonic ectoderm grows, cells that are displaced away from the epiblast experience a diminishing level of FGF signaling and thereby lose proliferative potency and undergo terminal differentiation (Tanaka et al. 1998).

Early arrest of embryonic development and failure of trophoblast differentiation at peri-implantation are also associated with the loss of *Err2* (Luo et al. 1997), *Nodal* (Brennan et al. 2001), and both *Nodal* downstream *Otx2* and cofactor *Cripto* activity (Kimura et al. 2001). Specifically, the combined loss of *Otx2* and *Cripto* activity in the epiblast leads to losses of *Bmp4* and *Eomes* activity in the extraembryonic ectoderm, which is essential for the differentiation of this trophectodermal tissue. These findings suggest that the interaction with the epiblast is critical for trophoblast proliferation and differentiation. Furthermore, in addition to FGF, other signaling activity involving EGF and Nodal also plays a role in the maintenance and differentiation of the extraembryonic ectoderm.

Proliferation of the Epiblast

During the development of the mouse embryo from implantation to gastrulation, cell number increases from about 15 in the 3.5-dpc ICM, 40 in 4.5-dpc ICM, 100 in 5.5-dpc epiblast, 200 in 6.0-dpc epiblast, to 660 in 6.5-dpc epiblast (Snow 1976, 1977). The increase in cell number indicates that the cell population doubles every 24 hours initially, then every 12 hours closer to the onset of gastrulation. The building up of a critical population of cells in the epiblast is intransigent for the initiation of gastrulation. There is also detectable cell death in the ICM and the epiblast of the pre-gastrula and gastrula (Copp 1979; Manova

et al. 1998; Poelmann 1980), which may be instrumental to the shaping of the epiblast during embryogenesis.

For mutant embryos that have lost the function of genes such as *Brca1*, *Evx1*, *Fgf4*, *fug1*, *rad51*, *Sox2*, *Alk2*, *Alk4*, *Smad1*, *–2*, *–4*, and *Hnf4*, development is often arrested during the immediate postimplantation period and does not proceed to gastrulation (DeGregori et al. 1994; Spyropoulos and Capecchi 1994; Hakem et al. 1996; Lim and Hasty 1996; Tam and Behringer 1997; Arman et al. 1998; Gu et al. 1998; Sirard et al. 1998; Waldrip et al. 1998; Beppu et al. 2000; Goumans and Mummery 2000; Tremblay et al. 2001). The epiblast of these mutant embryos shows inadequate cell proliferation, extensive cell necrosis, and lack of epithelial organization. Disorganization of the epiblast is also found in embryos that are heterozygous for null mutations of the *Foxa2* and *Lhx1* genes (Perea-Gomez et al. 1999). In addition, the epiblast cells acquire a mesenchymal phenotype revealed by the inappropriate expression of mesodermal markers (e.g., *T, Fgf8, Mesp2,* and *Lefty2*). No ectodermal markers are expressed, indicating that the epiblast has lost the ability to undergo ectodermal differentiation.

Of specific significance is that many of these genes (e.g., *Hnf4, Lhx1, Foxa2, Smad2,* and *Smad4*) are also expressed in the visceral endoderm (Morrisey et al. 1998; Sirard et al. 1998; Perea-Gomez et al. 1999, 2001a; Shawlot et al. 1999; Tremblay et al. 2000; Hallonet et al. 2002). The defect in the epiblast is often accompanied by morphological abnormality of the visceral endoderm (e.g., *Smad2* and *Smad4* mutants). It has been shown that the epiblast defects of the compound *Foxa2;Lhx1* mutant embryo and those of *Hnf4⁻/⁻*, *Smad2⁻/⁻*, and *Smad4⁻/⁻* embryos can be ameliorated by the restoration of normal gene function in the visceral endoderm (Chen et al. 1994; Sirard et al. 1998; Perea-Gomez et al. 1999). These findings therefore strongly argue for a critical supportive role of the visceral endoderm in the proliferation and organization of the epiblast.

Epithelialization of the Epiblast

A major milestone in the development of the ICM is the formation of a pseudostratified epithelium of epiblast. After epithelialization, the progenitors of definitive germ layers are placed in a two-dimensional/planar configuration where cells are potentially able to perceive polarity and positional information by virtue of their location in the epithelium. This may be the most crucial prerequisite for the regionalization of cell fates that heralds the establishment of the body plan.

Concomitant with the epithelialization of the epiblast, isolated intercellular lacunae are formed that coalesce to a proamniotic cavity (Coucouvanis and Martin 1995). Cavitation can be induced by BMP that triggers cell death in the core of aggregates of embryonic stem cells (the embryoid body) (Coucouvanis and Martin 1999). It is not known, however, whether BMP activity is essential for cavitation in the intact embryo. Loss of *Bmp2* and *Bmp4* activity from the visceral endoderm and the extraembryonic ectoderm, respectively, and that of *Alk3* in the epiblast do not prevent cavitation of the epiblast (Winnier et al. 1995; Zhang and Bradley 1996; Ying and Zhao 2001).

Cavitation of the ectodermal core is dependent on the formation of visceral endoderm on the outer surface of the embryoid body (Coucouvanis and Martin 1995). This raises the possibility that the visceral endoderm plays a critical role in the morphogenesis of the epiblast in the intact embryo by regulating cell death and epithelialization of the epiblast. Disruption of epithelialization and cavitation, however, occurs in embryos that lack the signal transduction adapter encoded by *Dab2* (D.H. Yang et al. 2002). *Dab2* is expressed in the visceral endoderm of the mouse embryo, and its activity is associated with the activation of Ras/MAPK during the differentiation of F9 embryonal carcinoma cells into visceral endoderm (Smith et al. 2001). In the *Dab2*-deficient embryo, cells that express markers for visceral endoderm mingle with the epiblast cells and do not organize into an epithelium (Norris et al. 2002; D.H. Yang et al. 2002). In the absence of the endodermal layer, the epiblast fails to acquire any polarized characteristics, does not form an epithelium or cavitate, and does not proliferate properly (D.H. Yang et al. 2002). Physical interaction and signaling activities of the visceral endoderm are therefore crucial for the epithelialization of the epiblast.

Patterning of the Epiblast by Interaction with Extraembryonic Tissues

A Potential Repository of Patterning Cues

Essential elements that define the body plan of the embryo include the orientation and the polarity of the primary body axes: anterior (head, A); posterior (tail, P); dorsal (back, D); ventral (front, V); and transverse (left–right and medial–lateral) axes, and the spatial arrangement of the progenitors of tissues and organs relative to these primary body axes. Critical to the delineation of the primary axes is the generation of asymmetry in the early embryo (Chapter 24). By tracking the distribution of the descendants of ICM cells in the visceral endoderm, it was found that cells in this tissue layer are not randomly distributed. Cells derived from the ICM close to and away from the polar body occupy different positions in the visceral endoderm (Weber et al. 1999). A necessary condition for the generation of tissue patterning is the presence of a stable and orderly source of patterning information that directs the assembly of embryonic tissues. It is likely that the repository of patterning information resides in a tissue that provides a constant spatial distribution of signaling activity to other tissues in the embryo during the period when the body plan is established. The visceral endoderm in which the distribution of clonal populations is regionalized during early embryogenesis is the most likely source of such patterning activity.

Formation of the Anterior Visceral Endoderm

The primitive endoderm that initially covers the luminal surface of the ICM of the blastocyst expands to become the visceral endoderm, which completely envelops the epiblast and the extraembryonic ectoderm of the pre-gastrula. The visceral endoderm forms a tall columnar epithelium of vacuolated cells displaying apical microvilli, characteristic of actively endocytotic cells, and may serve a trophic function for the embryo (Bielinska et al. 1999). In the distal region of the 5.5-dpc pre-gastrula, the visceral endoderm cells acquire a tall pseudostratified columnar or stratified cuboidal epithelial morphology, which is distinct from the visceral endoderm in other regions of the pre-gastrula by its exceptional thickness (Fig. 1A, B). As the embryo develops, the distal visceral endoderm cells lose their distinct morphological features (Kimura et al. 2000), but the visceral endoderm on one side of the pre-streak becomes evidently thickened (Fig. 1A, B). Lineage analysis reveals that the visceral endoderm of the pre-gastrula embryo is fated for the endoderm of the visceral yolk sac (Gardner and Rossant 1979; Lawson et al. 1986; Lawson and Pedersen 1987) and only makes a minor contribution to the embryonic foregut (Tam and Beddington 1992). The visceral endoderm, although derived from the ICM, is principally an extraembryonic tissue.

A genealogical relationship of the apical and anterior population of thickened visceral endoderm is strongly hinted at by the continuity of the expression of the *Hex* gene. *Hex* is expressed in the visceral endoderm first at the distal tip of the 5.0-dpc embryo (Thomas et al. 1998). By the pre-streak stage, *Hex*-expressing cells are found in the visceral endoderm on one side of the embryo. These cells also express *Hesx1*, which later marks the visceral endoderm in the anterior side of the early-streak embryo (Thomas and Beddington 1996; Thomas et al. 1998). The shift in the domain of *Hex* expression and the overlap with *Hesx1* activity suggest that the visceral endoderm in the distal region of the pre-gastrula is the precursor of visceral endoderm that is found in the anterior region of the early streak (ES) embryo, the so-called anterior visceral endoderm (AVE). In addition to *Hex* and *Hesx1*, other genes, such as *Otx2* (Ang et al. 1994; Acampora et al. 1995), *Cer1, Gsc, Lhx1* (Belo et al. 1997; Biben et al. 1998), *Foxa2* (Perea-Gomez et al. 1999), *Smad2* (Waldrip et al. 1998), *Lefty1* (Perea-Gomez et al. 2001b) and *Dkk1* (Monaghan et al. 1999) are expressed in the AVE at various stages of development. An open question regarding the activity of different genes is whether they are expressed uniformly in all AVE cells or restricted to subsets of cells.

Formation of the AVE is dependent on signaling activity (Fig. 5A, B). AVE is absent when the signaling adapter molecule encoded by *Dab2* is missing in the visceral endoderm precursors (D.H. Yang et al. 2002). In the *Nodal*, *Smad2*, and *Cripto* mutants, molecular markers of the AVE (e.g., *Hex*, *Cerl*, *Lhx1*, and *Hesx1*) are not expressed in the visceral endoderm (Varlet et al. 1997; Ding et al. 1998; Waldrip et al. 1998; Brennan et al. 2001). The visceral endoderm of these mutant embryos also displays structural abnormalities such as excessive folding and loss of epithelial integrity. In contrast, formation of the AVE is apparently independent of two other signaling activities, BMP and WNT, which are expressed concurrently with Nodal in the pre-gastrula embryo. Molecularly defined AVE and proper anterior patterning of the epiblast are found in mutant embryos in which components of the WNT pathway are disrupted (e.g., *Wnt3*⁻/⁻, *Mesd*⁻/⁻, and *Dkk1*⁻/⁻) (Liu et al. 1999; Mukhopadhyay et al. 2001; Hsieh et al. 2003). Mutant embryos with defective BMP signaling (e.g., *Bmp2, -4*, *Smad1, -4, Alk2, -4*) generally display deficiency in mesoderm formation and primitive streak function that may be secondary to the functional defect in visceral endoderm and not specifically due to the absence of AVE (Zhang and Bradley 1996; Gu et al. 1998, 1999; Sirard et al. 1998; Takaku et al. 1998; Yang et al. 1998; Mishina et al. 1999; Ying et al. 2000; Lechleider et al. 2001; Tremblay et al. 2001; Ying and Zhao 2001).

Anterior Displacement of the Visceral Endoderm: Implications for Axis Orientation

By tracking the movement of carbocyanine dye-labeled visceral endoderm (VE) and *Hex-GFP*-expressing cells, it has been shown that cells in the distal region of the visceral endoderm of the pre-axis aligned (pre-A) embryo are displaced asymmetrically to the prospective anterior side of the pre-streak (pre-S) embryo (Thomas and Beddington 1996; Thomas et al. 1998; Beddington and Robertson 1999; Rivera-Perez et al. 2003; S. Srinivas, pers. comm.). The anterior displacement of the distal visceral endoderm may be a part of the general movement of clonal populations of ICM-derived visceral endoderm from the proximal posterior to the anterior distal region of the pre-gastrula (Weber et al. 1999). However, it is not yet known what drives AVE movement and whether the AVE cells move individually or en masse as a coherent community (see supplemental Movie 16_1 at www.gastrulation.org) (Srinivas et al. 2004).

The anterior movement of distal endodermal cells and the concurrent shift in gene expression domains offer a compelling argument that the anterior–posterior polarity of the visceral endoderm is initially aligned with the proximal–distal axis of the pre-gastrula and is subsequently reoriented with the transverse plane of the embryo (Beddington and Robertson 1999; Lu et al. 2001). It is postulated that a corresponding realignment of the anterior–posterior axis also takes place in the epiblast in response to interaction with the visceral endoderm. The expression of *Nodal* and *Cripto*, which initially is widespread in the

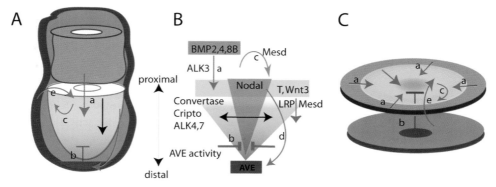

Figure 5. Inductive interaction and modulation of signaling activity leading to patterning of the epiblast of the pre-gastrula. (*A*, *B*) The epiblast is patterning proximo-distally by (*a*) BMP activity in the extraembryonic ectoderm, and the graded Nodal activity (shown as an inverted triangle whose width indicates the level of activity) in the epiblast, which is enhanced (*double arrow* in *B*) by the activity of proprotein convertase (encoded by *Spc*), coreceptor (encoded by *Cripto*), and receptors (ALK) and (*b*) the putative antagonistic AVE (anterior visceral endoderm, which is specified by [*d*] Nodal activity) activity and WNT signaling (*black arrow*) mediated by MESD and LRPs. Nodal function in the epiblast is activated by (*e*) Nodal activity in the visceral endoderm and is maintained by (*c*) an autoregulatory pathway that requires *Mesd* and *Cripto* function. Specification of the AVE is dependent on (*d*) the Nodal activity presumably from the epiblast, which is also mediated by MESD. The proximo-distal pattern in (*A*, *B*) the cup-shaped epiblast is equivalent to the radially graded mesodermal (*pink*) and ectodermal (*blue*) potency in (*C*) the epiblast disc.

epiblast, is progressively restricted to the proximal epiblast where *Brachyury* and *Wnt3* expression is localized (Fig. 2) (Ding et al. 1998). Concomitant with the movement of the *Hex*-expressing visceral endoderm from the distal to one side of the embryo, expression of genes in the epiblast shifts from the proximal to the side that presumably is opposite to the *Hex*-expressing AVE (Thomas and Beddington 1996; Varlet et al. 1997; Ding et al. 1998; Liu et al. 1999; Brennan et al. 2001). These changes in gene expression domains are consistent with the hypothesis that the proximal side of the epiblast marks the prospective posterior pole of the body axis.

Furthermore, by tracing the distribution of clonal descendants of single epiblast cells marked in the anterior side of the proximal epiblast at the pre-S stage, it has been shown that the labeled cell population does spread to the primitive streak and therefore is consistent with the movement predicted by the gene expression pattern (Lawson et al. 1991). This putative proximal to posterior–distal movement of epiblast and the overlying visceral endoderm (Weber et al. 1999), and the distal to anterior–proximal displacement of the distal visceral endoderm (Thomas and Beddington 1996), are the foundation of the concept of "orthogonal rotation" of the axis during the development of the pre-gastrula embryo (Beddington and Robertson 1999). However, when the embryo is viewed in the disc–ring–sector model (Fig. 4), the so-called proximal–distal orientation of the axis is in fact aligned radially from the periphery to the center of the disc. The "realignment" of the anterior–posterior axis could be consequential to the breaking of the radial symmetry by the polarized displacement of the epiblast and the visceral endoderm, resulting in the emergence of anterior–posterior patterning of tissues along the prospective longitudinal embryonic axis (Tam et al. 2001). It has yet to be verified that there is concerted movement of cells in the epiblast and the visceral endoderm (as found in the chick; Foley et al. 2000) that coordinates the alignment of the prospective anterior–posterior axis in these two tissue layers of the mouse pre-gastrula.

Specification of Cell Fate in the Proximal Epiblast

There is mounting evidence supporting the notion that specification of extraembryonic and posterior mesodermal fate in the proximal epiblast comes from the adjacent extraembryonic tissues and that this is mediated by TGF-β signaling (Fig. 5A, B). In the pre-streak embryo, *Bmp2* is initially expressed throughout the visceral endoderm but becomes localized in the area over the boundary between the epiblast and the extraembryonic ectoderm. *Bmp4* and *-8b* are expressed in the distal extraembryonic ectoderm next to the proximal epiblast. In the ring–disc–sector configuration, cells in the periphery of the epiblast disc are therefore exposed to a local source of BMP activity (Fig.

5C). Embryos that lose one or more of these *Bmp* genes either lack, or have a reduced population of, primordial germ cells that are derived from the proximal epiblast. Some of the mutant embryos also lack extraembryonic tissues such as allantois and amnion (Winnier et al. 1995; Zhang and Bradley 1996; Dunn et al. 1997; Lawson et al. 1999; Ying et al. 2000; Ying and Zhao 2001).

Reduction of germ cell population and poor extraembryonic mesoderm differentiation are also found in *Smad1* and *-5* mutant embryos in which TGF-β signal transduction activity may be affected (Chang et al. 2000; Tremblay et al. 2001). BMP4 activity alone is not sufficient to fully specify or maintain this unique cell population. *Lhx1*, *Ldb1*, *Foxa2*, and *Brachyury* (*T*) mutant embryos lack germ cells, even though *BMP4* is still expressed in the extraembryonic ectoderm and the extraembryonic mesoderm (Tsang et al. 2001; Mukhopadhyay et al. 2003). *Lhx1* mutants also show defective differentiation of derivatives of the proximal epiblast such as the allantois and lateral mesoderm (Tsang et al. 2000). TGF-β/BMP/Nodal signaling and the transcriptional activity of *Lhx1 Foxa2* and *T* may act independently, but loss of either function leads to similar defects in the proximal epiblast.

Integration of Patterning Activities and Cell Fates in the Epiblast

Accumulating evidence suggests that patterning activity of the AVE and its precursor may impart anterior (neural) characteristics and suppress posterior fates in the adjacent epiblast, and thereby establish the anterior–posterior polarity in the embryo. The current focus on the AVE function has not considered the possibility that there may be counteracting patterning activities from extraembryonic and embryonic tissues in other regions of the embryo. On the basis of the expression pattern of the *Nodal-lacZ* transgene, it has been postulated that there is a graded Nodal activity in the epiblast that both specifies the distal visceral endoderm to be the AVE precursor and patterns the epiblast (Brennan et al. 2001). However, regional variation in the level of ligand activity may not be the sole factor for generating graded signals. The activation of the Nodal protein requires the post-secretory modification by proprotein convertases (Furin, encoded by *Spc*). Potentially, protease activity that activates Nodal factor locally may lead to elevated levels of Nodal signaling in the epiblast, and a graded pattern of signaling may be generated by the distribution of protease activity (Fig. 5B) (Beck et al. 2002). Individual convertases are functionally redundant, and only when both *Spc1* and *Spc4* are lost is a deficiency in Nodal signaling revealed. *Spc1;Spc4* mutants fail to activate the Nodal protein, and the embryo displays a *Nodal*-null phenocopy. In the *Nodal* mutant, the extraembryonic ectoderm loses *Eomes* and *Bmp4* expression as the embryo develops

(Brennan et al. 2001). The widespread expression of *Otx2* and *Foxa2* in the *Nodal*-deficient epiblast might reflect the consequence of diminished TGF-β signaling. It has been shown that the activation of Nodal in the visceral endoderm is dependent on normal Smad4 function (Brennan et al. 2001). Embryos lacking Smad4 function (transducer for TGF-β signaling) show abnormal visceral endoderm differentiation and no mesoderm formation (Sirard et al. 1998). This suggests that signals from the visceral endoderm may synergize the graded patterning activity in the epiblast.

In addition to TGF-β/BMP signaling, cells in the epiblast may also be subject to different levels of WNT activity. In the pre-streak embryo, *Wnt3* is expressed in the proximal epiblast (Liu et al. 1999). The proximal region of the embryo may be the source of TGF-β and WNT signaling activity creating a descending gradient in the proximal–distal direction in the epiblast (Fig. 5A,B), or in a centripetal direction from the ring of extraembryonic tissues to the center of the disc (Fig. 5C). Mesd function, which is required for the trafficking of lipoprotein receptors from the endoplasmic reticulum to the cell surface, is essential for normal WNT signaling and the maintenance of Nodal activity in the proximal epiblast (Hsieh et al. 2003). Loss of *Mesd* activity leads to the absence of primitive streak and mesoderm formation, and a concomitant expansion of the AVE and the acquisition of anterior neural fate by the epiblast. Loss of *Wnt3* activity from the proximal epiblast and the posterior visceral endoderm produces a *Mesd* phenocopy of mesodermal deficiency (Liu et al. 1999). In contrast, loss of *Ldb1* function results in the loss of anterior structures, which is comparable to the loss of *Lhx1* function. The phenotypic effect is accompanied by the downregulation of several known factors (*Frzb, Sfrp1, Sfrp2, Cerl*) that inhibit WNT signaling (Mukhopadhyay et al. 2003). This suggests that WNT signaling constitutes an important posterior patterning activity that is modulated by the antagonistic activity of the AVE (Fig. 5B,C).

The counteracting activities of the extraembryonic ectoderm and proximal epiblast versus the AVE may generate a gradient of TGF-β/WNT signaling that leads to the regionalization of cell fates in the epiblast of pre-S embryo. Cells in the proximal (peripheral in the disc model) epiblast that are subject to a high level of signaling activity are fated for extraembryonic tissues (amnion ectoderm, yolk sac, and allantoic mesoderm) and germ cells. Intermediate signaling activity is associated with the specification of embryonic mesoderm and definitive endoderm, and the weakest signaling is found in the distal (central in the disc) epiblast destined to be ectodermal tissues (Fig. 5C). Lineage analysis reveals that individual epiblast cells can contribute to multiple tissue types. This finding suggests that there is no distinctive restriction of the lineage potency among the epiblast cells, which leads to a fate map showing extensive overlaps of the domain of germ layer precursors for the epiblast of the pre-streak embryo (Lawson et al. 1991; Lawson and Hage 1994).

The First Hint of Patterning: Regionalization of Gene Activity and Cell Fates

The formation of the body plan requires the production of cell populations displaying different fates and the correct assembly of these progenitor cells in relation to the primary axes of the embryo. Analyses of genetic activity in the embryo reveal regionalization of gene expression occurring well before the morphogenesis of germ layer derivatives (Fig. 6). The presence of discrete expression domains in both the extraembryonic and embryonic tissues strongly suggests that different cell populations in the embryo have acquired lineage-specific characteristics and/or functions during gastrulation (Lawson et al. 1991). Whereas the activity of some genes may herald the specification of specific tissue lineages by marking the progenitor population and its derivatives, other genes are expressed in tissues of totally different germ layer origin and are unlikely to be related by lineage (Fig. 6). Genetic activity often occurs only transiently and does not persist in all lineage descendants. Such activity tends to reflect the concurrent status of tissue differentiation and not the history of the lineage. Extensive fate-mapping studies performed by either clonal analysis or testing for lineage contribution have generated fate maps of the germ layers of the mouse embryo at successive stages of the gastrula. These fate maps provide a graphical display of the body plan and the prospective fates of cell populations in defined regions of the embryo. They also enable a reconstruction of the pattern of morphogenetic tissue movements associated with germ layer formation.

Regionalization of the Epiblast/Ectoderm

Changes in Cell Fates Reflect the Expansion of Ectodermal Lineages

The earliest indication of a blueprint of the body plan is that cells at different positions in the germ layer display regular and predictable differences in their developmental fate (Fig. 7A, epiblast). In the early-streak (ES) embryo, the proximal one-third of the epiblast (~220 of the total population of 660 cells) contains the precursors of the extraembryonic (yolk sac and allantoic) mesoderm and the primordial germ cells (Lawson and Hage 1994; Parameswaran and Tam 1995; Kinder et al. 1999). Cells in the anterior sector of the proximal epiblast are allocated principally to the ectoderm of the amnion (Kinder et al. 1999). The distal two-thirds of the epiblast are broadly divided into an

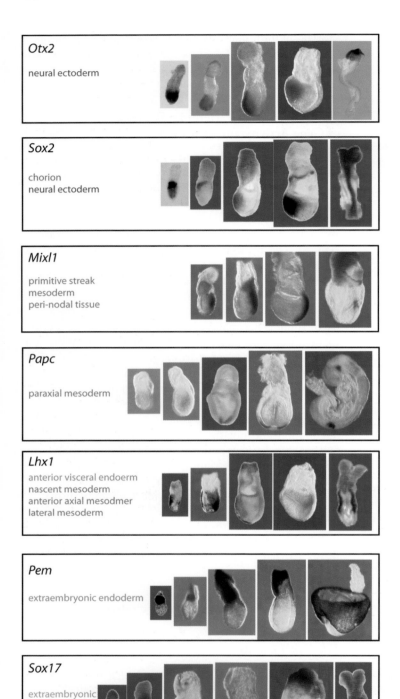

Otx2

neural ectoderm

Sox2

chorion
neural ectoderm

Mixl1

primitive streak
mesoderm
peri-nodal tissue

Papc

paraxial mesoderm

Lhx1

anterior visceral endoerm
nascent mesoderm
anterior axial mesodmer
lateral mesoderm

Pem

extraembryonic endoderm

Sox17

extraembryonic
endoderm
gut endoderm

Cerl

anterior visceral
endoderm
foregut endoderm
paraxial mesoderm

Figure 6. Examples of molecular heterogeneity of tissue compartments in the gastrulating embryo revealed by regionalized genetic activity. *Otx2* and *Sox2* expression mark the emergence of the neural tube tissues. Formation of mesoderm is marked by *Mixl1* activity in the primitive streak. Development of the paraxial mesoderm and the lateral mesoderm is associated with *Papc* and *Lhx1* activity. *Pem* activity marks the extraembryonic (yolk sac) endoderm throughout gastrulation. *Sox17* activity is initially found in the extraembryonic visceral endoderm prior to gastrulation (Kanai-Azuma et al. 2002). Both *Sox17* and *Cerl* are associated with the definitive (gut) endoderm during gastrulation and early gut development. *Cerl* and *Lhx1* are both expressed in the anterior visceral endoderm at early gastrulation.

A. Early-streak embryo

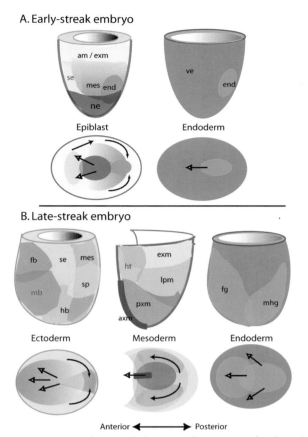

B. Late-streak embryo

Figure 7. Fate maps showing the location of precursors of embryonic and extraembryonic tissue lineages in the epiblast/ectoderm, mesoderm, and endoderm of the (*A*) early-streak and (*B*) late-streak embryo in the natural anatomical and the flattened disc configurations. The arrows in the disc models indicate the direction of morphogenetic movement of tissues in the germ layers. The hole in the center of the endoderm disc represents the hiatus through which the node pit opens to the extraembryonic parietal yolk sac. Abbreviations: (am/exm) amnion/extraembryonic ectoderm; (axm) axial mesoderm; (end) endoderm; (fb) forebrain; (fg) foregut; (hb) hindbrain; (ht) heart; (lpm) lateral plate mesoderm; (mb) midbrain; (mes) mesoderm; (mhg) mid- and hindgut; (ne) neural ectoderm; (pxm) paraxial mesoderm; (se) surface ectoderm; (sp) spinal cord; (ve) visceral endoderm.

anterior–distal ectodermal domain and a posterior–lateral mesodermal domain. The precursors of the entire neural tube are in the ectodermal domain containing about 125 cells (the distal region of the cup and the center of the epiblast disc; Fig. 7A, epiblast). Within the domain of neural tube precursor, progenitors for the cranial and trunk segments of neural tube are aligned with the anterior–posterior axis of the gastrula (Quinlan et al. 1995). Anterior to the neuroectodermal domain are cells that will contribute to the surface ectoderm of the embryo.

During gastrulation, cells depart from the epiblast through the primitive streak to form the mesoderm and the endoderm. Consequently, the remaining cell population is fated to become the ectoderm of the neural primordium and the surface ectoderm. By late gastrulation, over 70% of cells in the anterior half of ectoderm of the late-streak/early-bud (LS/EB) embryo are allocated to form the neural tube (Tam 1989). The majority of this population contributes to the prospective fore-, mid-, and hindbrain (Fig. 7B, ectoderm). Precursors of the spinal cord are found more posteriorly in the epiblast adjacent to the primitive streak. Cells in the proximal region of the ectoderm (~15% of the total population) are allocated to the surface ectoderm and the neural crest cells.

Tissue Movement in the Epiblast/Ectoderm

The domain of neuroectoderm precursors progressively expands from the distal (i.e., center of the disc) to the anterior region of the embryo. Coupled to this is the displacement of the precursor of surface ectoderm to the position previously occupied by the amnion ectoderm and extraembryonic mesoderm. There is therefore a general movement of the distal (central) ectodermal cells toward the anterior–proximal region of the embryo (toward the anterior half of the disc). Cells in the proximal epiblast (anterior and lateral epiblast of the disc) move posteriorly (along the periphery of the disc) to be recruited into the primitive streak or form the amnion (Fig. 7A, epiblast; 7B, ectoderm). Direct visualization of cell movement has not been performed in the gastrulating embryo. However, reconstruction of the pattern of tissue movement based on the series of fate maps is compatible with the observation of the distribution of trails of clonal descendants of cells stretching from the anterior region of the epiblast posteriorly to the primitive streak (Lawson et al. 1991; Lawson and Pedersen 1992a,b; Tam et al. 2001).

Mesoderm Formation

Polarity of Precursors in the Epiblast

In the epiblast of the early-streak embryo (Fig. 7A, epiblast), precursors of the paraxial and lateral mesoderm span the lateral to posterior region of the epiblast, and those of the cranial mesoderm and the heart are localized mainly in the posterior region (Parameswaran and Tam 1995). In the mesodermal domain, precursors for cranial and heart mesoderm are localized more posteriorly than the paraxial and lateral mesoderm of the trunk (Parameswaran and Tam 1995; Tam et al. 1997a). The anterior–posterior polarity of the mesodermal precursor is therefore opposite to the orientation of their descendant in the mesodermal tissues of the embryo after gastrulation (Fig. 7B, mesoderm).

Cell Fates in the Mesodermal Layer

Concomitant with the formation of primitive streak, mesoderm is formed as a nascent population of mesenchyme between the epiblast and the visceral endoderm in the posterior region of the embryo. Cells in the newly formed mesoderm of the early-streak embryo are mostly fated to be the extraembryonic mesoderm and give rise to the hematopoietic cells and the endothelium of the vitelline (yolk sac) blood vessels (Kinder et al. 1999). The mesoderm expands toward the ectoplacental cone in the extraembryonic region and toward the anterior and distal sides to envelope the lateral aspects of the epiblast (Fig. 3A–C). Cells in the more distal (hence more medial) portion of the mesoderm are precursors of the cranial and heart mesoderm. By the late-streak/no-bud (LS/OB) stage, a complete layer of mesoderm is formed between the ectoderm and endoderm (Fig. 3D,E). The embryonic layer of mesoderm now primarily contains the precursors of the heart mesoderm, cranial and upper trunk paraxial and lateral mesoderm (Fig. 7B, mesoderm) (Parameswaran and Tam 1995). By this stage of development, very few precursors of the extraembryonic mesoderm remain in the mesodermal layer associated with the ectoderm except in the proximal (lateral) and posterior regions.

Concerted Morphogenetic Movement of Cells in the Mesoderm

The location of precursors of the extraembryonic tissues in the expanding mesoderm strongly suggests that these cells are being displaced from the embryonic to the extraembryonic region of the embryo during gastrulation (Parameswaran and Tam 1995). The anterior displacement of the heart mesoderm and cranial paraxial mesoderm (Fig. 7B, mesoderm) has been studied by tracking the movement of labeled cells in live embryos in vitro (Tam et al. 1997a). The heart mesoderm initially moves laterally away from the primitive streak to reach the anterior region of the embryo (Kinder et al. 2001a). The paraxial mesoderm takes a more paramedial course and stays close to the midline, moving anteriorly (Fig. 7B) (Kinder et al. 2001c). Time-lapse tracking of cell positions over a brief 5-hour period in mouse embryos kept under in vitro conditions reveals that the mesoderm cells generally move as a cohesive sheet, except for those in the leading edge of the expanding cell sheet, which may move as individual cells (Nakatsuji et al. 1986).

The morphogenetic forces driving the displacement of the mesoderm have not been identified. The inherent tendency of the cells to disperse due to repulsive interaction with other cells, the propulsion generated by the incoming cells recruited at the primitive streak, or the traction force exerted by the neighboring germ layers are all possible mechanisms for moving the mesoderm. It has been shown that in the avian embryo, movement of mesodermal cells may depend on FGF signaling activity (X. Yang et al. 2002). In the Fgf8-deficient embryo, mesodermal cells are accumulated in the vicinity of the primitive streak as if they cannot move away after ingression (Sun et al. 1999). A similar failure of cell migration is encountered by Fgfr1-deficient cells in the primitive streak of the $Fgfr1^{-/-} \leftrightarrow$wild-type chimera (Green et al. 1992; Ciruna and Rossant 2001). Whether the loss of FGF function is the primary cause of the defective migratory activity is not known.

In embryos lacking Mixl1 (Brachyury induced Mix-like homeobox gene) activity, there is an accumulation of cells in the primitive streak accompanied by a deficiency in mesoderm and endoderm (Hart et al. 2002). The apparent defect of cell migration is, however, not associated with any significant reduction in Fgf8 activity. Similar to the Mixl1 mutant, embryos lacking Lhx1 function are deficient in cranial mesoderm (Shawlot and Behringer 1995). Lhx1-deficient mesodermal cells show poor migratory ability after they are transplanted to the primitive streak of both the wild-type and Lhx1 mutant embryos, suggesting that Lhx1 has a cell-autonomous function on mesoderm migration. Interestingly, loss of Lhx1 is associated with the down-regulation of paraxial protocadherin (Papc) in both Xenopus and mouse embryos (Hukriede et al. 2003), indicating that Papc is a common potential downstream target. However, although Papc activity may overcome the lack of Lhx1 function in Xenopus and ameliorate the migration defects of the axial and paraxial mesoderm, it is not known whether a gain of PAPC function will correct the Lhx1 defects in the mouse. Loss of Papc function in the mouse has no effect on cell migration or embryogenesis, as would be predicted from Lhx1 deficiency (Yamamoto et al. 2000). Some of the more than 80 types of protocadherins (Frank and Kemler 2002) may compensate for the loss of PAPC in the $Lim1^{-/-}$ and $Papc^{-/-}$ mouse embryos.

Definitive Endoderm

Replacement of Visceral Endoderm by Definitive Endoderm

Precursors of the definitive endoderm that contribute to the formation of the gut are localized in the posterior epiblast among the cells of the gastrula organizer (Lawson et al. 1991; Tam et al. 1997b; Kinder et al. 2001c). Cells of the definitive endoderm first appear at the mid-streak (MS) stage following the recruitment of epiblast cells (Tam and Beddington 1992). Cells destined for definitive endoderm are presumed to ingress together with the mesoderm through the anterior region of the primitive streak. Little is known of the mechanism or the process that sorts the definitive endodermal precursors from other cell types, although the differential expression of cadherins in the fully formed mesoderm and endoderm suggests homotypic cellular interactions may underpin the separation of germ-layer progenitors (Takeichi 1988). By tracking the movement of cells during germ-layer morphogenesis, it has been shown that cells ingressed through the primitive streak are incorporated directly into the epithelium of visceral endoderm. The newly recruited cells intercalate between the visceral endoderm cells and establish proper intercellular junctions with preexisting cells in the epithelium (Tam et al. 1993). Incorporation of the definitive endoderm occurs in the endoderm immediately subadjacent to the anterior segment of the primitive streak, which sometimes contains apoptotic cells (Poelmann 1980; Tam et al. 1993). The incoming population progressively displaces the preexisting visceral endoderm away from the primitive streak.

Definitive endoderm cells recruited early in gastrulation are allocated to the anterior region of the late-streak (LS) embryo and populate the foregut, whereas those recruited later populate the more posterior region of the gut (Lawson et al. 1986; Lawson and Pedersen 1987; Tam and Beddington 1992). Recruitment of definitive endoderm continues throughout gastrulation so that the visceral endoderm is almost completely replaced by the incoming cell population. Analysis of cell fate reveals that some descendants of the visceral endoderm are still present in the prospective foregut and hindgut of the LS embryo (Tam and Beddington 1992). Fate-mapping studies so far have focused on cell fate in the mid-line of the gastrula and on the sequential recruitment of epiblast cells to the definitive endoderm. A more comprehensive study is required to reveal the finer regionalization of cell fates in the entire endodermal layer and the allocation of endoderm to specific segments of the embryonic gut and the associated endodermal structures. Preliminary results on the developmental fate of the endoderm of the LS embryo indicate that the precursors for the entire embryonic gut are already present at the completion of gastrulation, and further recruitment is likely to be minimal during subsequent organogenesis (J. Gad and P. Tam, unpubl.). Furthermore, the majority of the endoderm of the LS embryo is allocated to the foregut (Fig. 7B) (Wells and Melton 1999), raising the possibility that the rest of the embryonic gut has to be derived by the expansion of a disproportionately small progenitor population through cell proliferation and tissue expansion.

Movement of the Definitive Endoderm

Fate-mapping results show that the domain occupied by the definitive endoderm expands anteriorly and laterally from the area immediately subadjacent to the anterior segment of the primitive streak to cover an increasingly larger area of the growing embryo (Fig. 7). The visceral endoderm is displaced by the expanding definitive endoderm anteriorly and proximally toward the extraembryonic region of the gastrula. The definitive endoderm appears about 8–10 hours after the onset of gastrulation (during which the extraembryonic mesoderm is formed) and coincides with the exit of the embryonic (cranial and heart) mesoderm from the primitive streak and its incorporation in the mesodermal layer (Lawson et al. 1986; Lawson and Pedersen 1987; Tam and Beddington 1992; Parameswaran and Tam 1995; Tam et al. 1997a; Kinder et al. 2001a). A correlation of the fate maps suggests that mesoderm and definitive endoderm may comigrate during gastrulation. Although this notion is yet to be tested by tracking the movement of the mesoderm and endoderm simultaneously, it raises the possibility that the primary function of gastrulation is to assemble primordial tissues which can then participate in the morphogenesis of specific structures by coordinating their allocation and migration. The comigration of the definitive endoderm and mesoderm further suggests that a common morphogenetic force may be directing the movement of both layers. During gastrulation, the cavity enclosed by the extraembryonic mesoderm (the exocoelom) expands rapidly, which may generate a traction force that pulls the visceral endoderm away from the embryo over to the surface of the exocoelom. This traction, coupled with continuous addition of cells to the posterior sides of the germ layer at the primitive streak, may provide the force driving the migration of the mesoderm and the endoderm.

Lineage Potency Is Not Irreversibly Restricted following Germ-Layer Formation

Although regionalization of tissue fate can be demonstrated in the epiblast of the ES embryo, single-cell and heterotopic cell-lineage analyses reveal a remarkable plasticity in the lineage potency of epiblast cells. Individual cells in the epiblast of pre-streak and early-streak embryos show no restriction in their ability to contribute to multiple tissue lineages, which often include derivatives of more than one germ layer (Lawson et al. 1991). Furthermore, cells localized in different regions of the epiblast, the mesoderm, and the primitive streak also display no restriction in lineage potency (Beddington 1982; Tam and Beddington 1987; Parameswaran and Tam 1995). Cells that are transplanted heterotopically to other regions in the germ layers adopt different fates and contribute to tissues that are typically derived from the cells of the ectopic sites. For example, epiblast cells that are predestined to be neural tissues can differentiate to form germ cells and extraembryonic mesoderm when transplanted to the region of epiblast that gives rise to these tissues and vice versa (Tam and Zhou 1996).

Furthermore, it has been shown that the extraembryonic ectoderm is able to induce the formation of primordial germ cells (PGCs) in distal epiblast when in culture (Ying et al. 2000). Anterior epiblast cells that normally do not form mesoderm can participate in ingression and colonize the mesodermal tissues of the embryo if they are transplanted to the posterior epiblast. Conversely, posterior epiblast cells colonize the neural plate of the embryo after they are transplanted to the anterior epiblast (Parameswaran and Tam 1995).

With the establishment of fine regionalization of cell fate during late gastrulation, cells of the primitive streak and the germ layers still display a substantial degree of developmental plasticity. Ectoderm cells from the prospective forebrain (anterior ectoderm) can undergo ingression to form axial mesoderm and gut endoderm when transplanted into the anterior region of the primitive streak (Beddington 1982). The same cells can also contribute to the allantoic and posterior mesoderm when introduced to the posterior region of the primitive streak. Reciprocal transplantation of anterior or posterior primitive streak cells to anterior ectoderm results in the colonization of neural plate by the graft-derived cells (Beddington 1982). In essence, despite the difference in cell fates, the epiblast/ectoderm cells are far from irreversibly restricted in lineage options.

The predictability of the prospective fate of cells in the epiblast might be the result of a stereotypic pattern of morphogenetic tissue movement that leads to the expected allocation of cells to specific tissue types by the position of the precursors in the germ layers. Mesoderm cells that have passed through the primitive streak can repeat the gastrulation movement and contribute to mesoderm tissues (except for the lateral mesoderm) (Parameswaran and Tam 1995). Ectodermal cells can be transplanted directly to the cardiogenic mesoderm, and they will differentiate into cardiomyocytes without any previous experience of gastrulation (Tam et al. 1997a). These findings indicate that gastrulation tissue movement does not impose any discernible restriction on mesodermal potency. Whether the plasticity in cell fate extends to all mesodermal and endodermal lineages has not been tested.

THE PRIMITIVE STREAK: AN INSTRUMENT OF GASTRULATION

Tissue Mass and Chronological Age Influence but Do Not Control the Formation of the Primitive Streak

The formation of the primitive streak at 6.5 dpc marks the onset of gastrulation. Studies on the development of postimplantation embryos with altered cell number or tissue mass reveal that formation of the primitive streak does not follow a strict developmental schedule (Rands 1986a; Power and Tam 1993). When the number of cells in the embryo is reduced by 25–50% at the preimplantation stage, the embryo is reduced to about half the normal size at the pre-gastrula stage. These embryos do not initiate gastrulation at 6.5 dpc with their normal-size littermates. Instead, the embryo postpones the formation of primitive streak by about 12 hours (to 7.0 dpc) until the epiblast has built up a population of about 600–1000 cells. Initiation of gastrulation therefore seems to be determined by the cell number and not by the chronological age of the embryo. This concept of a prerequisite cell number (or tissue mass) is supported by the failure of many mutant embryos to initiate gastrulation when there is significant reduction in the epiblast population due to poor proliferative activity or excessive cell death. An immediate prediction of this hypothesis is that gastrulation should commence whenever the threshold cell number or tissue mass is present in the epiblast.

A pre-gastrula embryo with an increased size (greater cell number and/or tissue mass) can be produced by aggregating two or three preimplantation embryos together. In this case, one might predict that the oversized embryo will commence gastrulation precociously about a day earlier (5.5 dpc). However, this is not observed. The embryo compensates for the excessive size by cessation of epiblast growth, and gastrulation is initiated at the normal age of 6.5 dpc at a normal embryonic size (Rands 1986b). The obligatory delay in the onset of gastrulation of the oversize chimeras argues strongly that cell number (or tissue mass) alone is not a critical determinant for initiating germ layer morphogenesis. The different impact of cell number and the chronological age on the development of under- and oversized pre-gastrula embryos implies that the initiation of gastrulation (or the formation of the primitive streak) requires the coexistence of a threshold mass of the epiblast and morphogenetic instructions that are independent of epiblast growth but are triggered on a correct chronological cue.

TGF-β and WNT Signaling Are Required for the Formation of the Primitive Streak

The formation of the primitive streak is dependent on the molecular activity that restricts TGF-β signaling first to the proximal and later to the posterior region of the embryo. In the pre-gastrula embryo, expression of Wnt3 activity and its downstream target, T (Arnold et al. 2000), occurs uniformly in the proximal epiblast, the equivalent of the peripheral region of the epiblast disc. As the embryo progresses to gastrulation, T and $Wnt3$ expression is down-regulated in the prospective anterior sector (where the AVE is localized) and up-regulated and expanded in the prospective posterior sector (Fig. 8A–C). Finally, the expression of

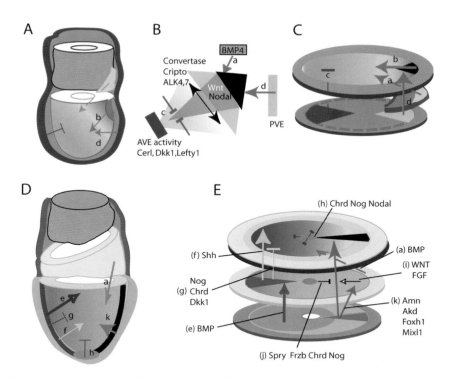

Figure 8. Inductive interaction and modulation of signaling activity leading to patterning in (*A–C*) the early-streak and (*D, E*) the late-streak embryo. (*A, B*) In the early-streak embryo, the epiblast is patterning in the anterior–posterior dimension by (*a*) BMP, (*b*) WNT and Nodal signaling activity from the primitive streak, the posterior germ layers and extraembryonic tissues, and by (*c*) the AVE activity that antagonizes BMP and WNT signaling. The anterior and distal epiblast (*A*), or the anterior and central region of (*C*) the epiblast disc, acquires an ectodermal fate (*blue*) while the posterior tissues are allocated to the mesoderm and endoderm (*pink/red*). The combined actions of (*c*) the AVE activity and (*d*) the posterior visceral endoderm (PVE) determine the position and maintain the function of the primitive streak (*black*). (*D, E*) In the late-streak embryo, the regionalization of tissue fates in the ectoderm and the mesoderm is further refined by the inductive interaction with (*a*) the extraembryonic mesoderm, (*e*) the anterior definitive endoderm, the combination of enhancing (*f*) and suppressive (*g*) activities of the axial mesoderm (*green* sector) and the antagonistic activity of (*h*) the node. The primitive streak and the associated mesoderm are the source of (*i*) FGF and WNT signaling activity which is modulated by (*j*) *Spry* and *Frzb* activity in the same tissues. The primitive streak is patterned by (*k*) the activity of the underlying endoderm.

both genes, as well as that of *Nodal*, is consolidated in the primitive streak. In the *Wnt3*-null mutant, the epiblast remains viable and proliferative, but no primitive streak is ever formed (Liu et al. 1999). It is likely that initially the potency of primitive streak formation is distributed along the circumference of the epiblast similar to that in the chick (Skromne and Stern 2001; Bertocchini and Stern 2002).

In *Lhx1, Foxa2, Cripto,* and *Otx2* mutants, the restriction of the primitive streak gene activity in the proximal region of the epiblast is accompanied by the lack of anterior displacement of AVE or its precursor. Genes such as *Dkk1, Cerl,* and *Lefty1*, which are expressed in the visceral endoderm, encode factors that reputedly antagonize or modulate TGF-β (BMP, Nodal) and WNT signaling. Loss of *Dkk1, Cerl,* or *Lefty1* function does not seem to affect gastrulation or primitive streak formation. In contrast, simultaneous loss of *Cerl* and *Lefty1* leads to deregulation of

Nodal signaling and results in the formation of extra primitive streaks as shown by ectopic expression of *Gsc, Foxa2,* and *T* (Perea-Gomez et al. 2002). *Foxa2;Lhx1* compound mutant embryos form an enlarged primitive streak and show ectopic localization of *Mesp1*- and *Lefty2*-expressing mesoderm (Perea-Gomez et al. 2002). Similarly, *Otx2;Cripto* and *Smad2* mutants also fail to express AVE markers and show an expanded domain of *T* and *Fgf8* expression in the epiblast as if there is an uncontrolled regulation of primitive streak potency (Brennan et al. 2001; Kimura et al. 2001). Embryos lacking Nodal or Cripto activity fail to establish a primitive streak (Conlon et al. 1994; Ding et al. 1998; Brennan et al. 2001). Loss of individual components of TGF-β signaling activity such as the ligand (*Nodal, BMP2*), the receptor (*Alk,* Mishina et al. 1995; *Alk4,* Gu et al. 1998; *BmprII,* Beppu et al. 2000), the transducer (*Smad4*), and various combinations of them (e.g.,

ActRIIA and *ActRIIB*, *ActRIIA;Nodal*, and *Nodal;Smad2*, Liu et al. 1995; Attisano et al. 1992; Song et al. 1999) leads to disruption of, or delay in the formation of, the primitive streak. The phenotypic consequences of disruptions of TGF-β signaling are generally consistent with an essential role of BMP4 and Nodal activity in the formation of the primitive streak.

Chimera analysis reveals that primitive streak formation depends critically on TGF-β signaling from the extraembryonic tissues and the activation of WNT, T, and Nodal pathways in the adjacent peripheral epiblast (Mishina et al. 1995; Gu et al. 1998; Sirard et al. 1998; Waldrip et al. 1998; Yang et al. 1998). A potential source of signaling activity for establishing primitive streak potency in the proximal epiblast is the extraembryonic ectoderm and visceral endoderm. The position of the primitive streak is determined by a restriction of the morphogenetic potency to the posterior sector of the epiblast as a result of the suppressive activities of the AVE (Fig. 8A–C). Development of the primitive streak may be further enhanced by the posteriorizing FGF activity in the posterior epiblast and posterior visceral endoderm. Evidence in support of the posteriorizing activity comes from the induction of primitive streak formation in explanted rabbit embryos and the upregulation of *sprouty* and *T* expression in the anterior epiblast of the mouse embryo by elevated FGF activity (Hrabe and Kirchner 1993; Davidson et al. 2000; Wells and Melton 2000).

Conduit for Lineage Allocation and Tissue Patterning

At different stages of gastrulation, the primitive streak is shown to contain precursor cells of different mesodermal and endodermal lineages that are destined for different parts of the body (Kinder et al. 1999). Interestingly, the types of tissue precursors found in the primitive streak at any instance match closely with those that have recently been recruited to the mesodermal layer (Parameswaran and Tam 1995). These findings point to several important features in the allocation and patterning of the mesoderm.

First, the order of recruitment of the mesoderm tissues is dependent on the association of the primitive streak with the specific tissue precursor in the epiblast. The order is therefore determined by the position of the precursor in the anterior–posterior axis of the gastrula and the progression of streak elongation. This may account for the earlier recruitment of the extraembryonic mesoderm than that seen for the embryonic structures.

Second, the mesoderm destined for more anterior regions, such as the heart and head of the embryo, is recruited earlier during gastrulation than that fated for the trunk mesoderm. Allocation of mesoderm to the embryo

therefore takes place in an anterior–posterior manner determined by the timing and order of recruitment through the primitive streak. The order of recruitment may be related to the apparent reverse anterior–posterior arrangement of the mesodermal precursors in the epiblast relative to the primitive streak. That is, the more distant posterior precursors will be recruited later into the primitive streak. However, a correlation of the order of recruitment of cells to the primitive streak with the distance of the cells from the primitive streak has not been shown (Lawson and Pedersen 1992b), suggesting that the position of the precursor cells in the epiblast is not the sole determinant of the recruitment process.

Third, it is noted that cells which are allocated from the primitive streak to the paraxial mesoderm tend to be distributed over several consecutive somites (Tam 1986; Tam and Beddington 1987; Tam and Tan 1992; Mathis and Nicolas 2002). Lineage analysis shows that cells distributed over different somite segments are clonally related. Furthermore, the allocation of progenitors to specific clones is likely to take place in the primitive streak in order to account for the remarkable concordance of the spatial distribution of clones that are found bilaterally in the paraxial mesoderm (Eloy-Trinquet and Nicolas 2002).

Finally, the site of ingression in the primitive streak determines the position of the mesodermal derivative in the dorsoventral plane of the embryo. As a result of the association of different anterior–posterior segments of the primitive streak to different types of mesodermal precursors in the epiblast, precursors of dorsal (the axial mesoderm) and the paraxial mesoderm ingress through the anterior segment of the primitive streak while the ventral (lateral plate) mesoderm ingresses through the more posterior segment of the primitive streak (Tam and Beddington 1987; Kinder et al. 1999). The primitive streak provides the conduit not simply for epithelio-mesenchymal transformation and the recruitment of cells for germ layer formation, but, more importantly, for the translation of the anterior–posterior distribution of the mesodermal precursors into the dorsoventral pattern.

TGF-β Signaling Activity Is Essential for Primitive Streak Function

Tracking the recruitment of definitive endoderm into the preexisting visceral endoderm layer revealed that the incoming cells replace nearly all the visceral endoderm during gastrulation except for those cells that are subadjacent to the primitive streak (Tam and Beddington 1992). The persistent association of the posterior visceral endoderm with the primitive streak raises the possibility that this population of visceral endoderm may play a role in the maintenance of the primitive streak (Fig. 8D,E). Concomitant to the initial expression of *Wnt3* in the proximal epiblast, activity is also found in the posterior visceral endoderm of the pre-streak embryo, and following the formation of the primitive streak, *Wnt3* expression becomes restricted to the primitive streak and the mesoderm (Liu et al. 1999). Additional sources of interactive signals may also be derived from the newly recruited definitive endoderm. Like the AVE, endodermal cells subadjacent to the anterior segment of the primitive streak of the MS embryo also express *Cerl* and *Lefty1* (Perea-Gomez et al. 2001a, 2002) and therefore are potential sources of activity that modulate TGF-β signaling in the primitive streak (Fig. 8D,E).

Significant insight into the role of TGF-β signaling in the maintenance of the primitive streak (Fujiwara et al. 2001) is provided by the phenotypic outcome of loss-of-function mutations of different molecules acting downstream to modulate TGF-β activity. *Amnionless* (*Amn*) which is expressed in the visceral endoderm encodes a cysteine-rich domain transmembrane protein (Kalantry et al. 2001). The protein has cysteine-rich (CR) domains that may modulate BMP signaling like other CR-containing factors such as Chordin. *Amn*-null mutants form a morphologically normal primitive streak but lack the paraxial and lateral mesoderm that is recruited through the middle segment of the primitive streak (Kalantry et al. 2001). *Amn* activity therefore maintains the production of the embryonic mesoderm from the primitive streak. Loss of *Cerl* and *Lefty1* activity that antagonizes TGF-β signaling leads to the absence of foregut endoderm and the paraxial and lateral mesoderm cell populations (Perea-Gomez et al. 2002), but an increase of mesendodermal derivatives. More primitive streak cells seem to be allocated to the axial mesendoderm (an anterior primitive streak derivative) at the expense of the definitive endoderm and other mesodermal derivatives. *Arkadia* (*Akd*), which is widely expressed in the gastrula, encodes an intracellular Nodal signaling regulator (Episkopou et al. 2001; Niederlander et al. 2001). *Akd* function is required in the extraembryonic tissue for maintaining primitive streak function. The *Akd*-null embryo lacks axial mesendoderm and paraxial mesoderm that are normally derived from the anterior primitive streak (Episkopou et al. 2001). Loss of the *Lefty2* (Meno et al. 2001) or *Drap1* (Iratni et al. 2002) activity that antagonizes Nodal signaling, or of both *Foxa2* and *Lhx1* which act downstream of Nodal activity (Perea-Gomez et al. 1999), leads to formation of excessive mesoderm by an expanded primitive streak. Proper TGF-β signaling activity is therefore essential for the function of the primitive streak.

FGF Signaling Influences Tissue Movement and Mesoderm Allocation

FGF signaling is essential for normal function of the primitive streak (Fig. 8D,E). *Fgf8*-deficient embryos form a prim-

itive streak. However, cells ingressing through the anterior and middle segments of the primitive streak fail to migrate properly, leading to a deficiency in embryonic mesoderm but not extraembryonic mesoderm (Sun et al. 1999). An intriguing outcome of loss of *Fgf8* is the concomitant loss of *Fgf4* expression in the streak. Similar loss of Fgf4 activity but not Fgf8 is found in embryos homozygous for a null allele of *eed*, an ortholog of the *Drosophila extra sex combs* gene (Faust et al. 1995; Schumacher et al. 1996). In the *eed* mutant, epiblast cells ingress through the anterior streak, but newly formed mesoderm is unable to migrate anteriorly and laterally and becomes mislocalized to the extraembryonic compartment (Faust et al. 1998). Whether the altered FGF activity has resulted in a bias toward differentiation of extraembryonic mesoderm is not known.

A less severe mesodermal phenotype associated with FGF signaling is observed with loss of FGF receptor1 (Fgfr1) function in the developing mouse (Deng et al. 1994; Yamaguchi and Rossant 1995). *Fgfr1* mutant embryos also have an accumulation of cells in the streak. However, these embryos form an excess of axial mesoderm and a shortage of paraxial mesoderm (Yamaguchi and Rossant 1995; Ciruna et al. 1997). Deficiency in the paraxial and lateral mesoderm but normal formation of axial mesoderm is also found in the *Ednrb*[-/-] mutant embryo (Welsh and O'Brien 2000). It is postulated that the loss of EDNRB function has an impact on FGF signaling, which affects the normal function of the primitive streak. FGF function is therefore critical for the specification of paraxial mesoderm, and it has been shown that it acts by positively regulating *T* and *Tbx6* expression and acts on WNT3A signaling through the cadherin/β-catenin pathway (Ciruna and Rossant 2001).

T-Box Gene Activity Is Required for Cell Ingression and Mobility in the Primitive Streak

Brachyury (*T*) mutant embryos form a primitive streak but display poor ability to form the axial mesendoderm, the mesoderm of the posterior part of the trunk and the extraembryonic structures (Rashbass et al. 1991; Beddington et al. 1992; Kispert and Herrmann 1994). As the embryonic axis develops, there is progressive accumulation of mesoderm cells at the primitive streak, suggesting that T function is required for deployment of cells emerging from the primitive streak (Wilson et al. 1993, 1995). A similar requirement of another T-box gene, *Eomes*, is shown in the development of chimeras containing wild-type extraembryonic and mutant embryonic tissues (Russ et al. 2000). The *Eomes*-deficient cells are unable to ingress through the primitive streak and consequently do not form the node and are unable to specify the neuroectoderm. Embryos that lack the activity of the *Mixl1* gene, which is potentially downstream of T-box gene activity, also show the failure of cells to leave the primitive streak, leading to the accumulation of cells in the axial tissues (Hart et al. 2002).

ROLE OF THE GASTRULA ORGANIZER IN AXIS PATTERNING

Identifying the Mouse Gastrula Organizer

In the gastrulae of zebrafish, *Xenopus,* and the bird, a specific population of cells is found to be able to initiate the formation of embryonic axes after transplantation to another embryo (Shih and Fraser 1995, 1996; Harland and Gerhart 1997; Stern 2001). These cells express a unique repertoire of transcription factors and molecules that antagonize TGF-β and WNT signaling activity (Harland and Gerhart 1997; De Robertis et al. 2000; Saude et al. 2000; Stern 2001). The axial mesoderm of the late-streak and early-organogenesis-stage mouse embryos expresses a multitude of genes encoding transcription factors, signaling molecules, and factors that modulate or antagonize BMP and WNT signaling activity (Fig. 8D,E) (Camus and Tam 1999; Davidson and Tam 2000). By tracing the expression pattern back to earlier developmental stages, three of these genes (*Foxa2, Lim1,* and *Gsc*) are found to be expressed first in the posterior epiblast anterior to the newly formed primitive streak of the early-streak embryo (the EGO), then in the anterior end of the elongating primitive streak of the mid-streak embryo (the mid-gastrula organizer, MGO), and finally at the anterior end of the fully extended primitive streak (Hensen's node) of the late-streak to early-bud embryo (Fig. 9A) (Beddington 1994; Tam and Quinlan 1996; Tam et al. 1997b; Camus and Tam 1999; Davidson and Tam 2000).

Specification of the Gastrula Organizer

In the amphibian embryo, the dorsal vegetal cells act as a source of signaling activity (the Nieuwkoop center) that specifies the gastrula organizer (Harland and Gerhart 1997; De Robertis et al. 2000). A potential source of organizer-inducing signals may be present in the posterior marginal zone of the avian embryo (Bachvarova et al. 1998). In contrast, the equivalent of a Nieuwkoop center has yet to be found in the mouse embryo. However, there is no a priori reason that an anatomically defined or localized source of signals should be present. Specification of the gastrula organizer can be achieved by interaction of a combination of signals in a specific location that culminate in the formation of the organizer. The absence of a node in the *Akd*-null mutant suggests that Nodal signaling in the extraembryonic tissues may be essential for specification of the mouse gastrula organizer (Episkopou et al. 2001).

Most mutants that fail to form a primitive streak also lack the node, suggesting that the formation of both structures may be coupled. For example, embryos without Foxa2 function fail to elongate the primitive streak and lack a definitive node and notochord. The gene *Foxh1* encodes the forkhead DNA-binding protein (FAST) that partners with SMAD and is essential for mounting a transcriptional response to activin and Nodal signaling (Norris et al. 2002). Ablation of this gene results in losing the mesodermal and endodermal precursors of the anterior primitive streak including the node (Hoodless et al. 2001; Yamamoto et al. 2001). However, studies of other genetic mutants have shown that the formation of these two structures can be uncoupled. In the *Akd* mutant, a primitive streak is formed, but the node is absent (Episkopou et al. 2001). In the chimeras containing wild-type extraembryonic tissues and *Foxa2*-deficient embryonic cells, a primitive streak is formed, but not the node (Dufort et al. 1998). Despite the absence of a morphologically distinct node in *eed* mutant mice, some axial mesoderm cells that express *T, Shh,* and *Foxa2* are present, and the definitive endoderm, which is another derivative of the anterior primitive streak, is formed (Faust et al. 1995). Furthermore, in *Lhx1, Otx2,* and *Foxa2*-null mutant embryos, cells that display the molecular characteristics of the organizer are localized ectopically and can be separated from the aberrant primitive streak (Klingensmith et al. 1999; Kinder et al. 2001b). These findings strongly suggest that specification of the gastrula organizer and of the primitive streak are independently controlled.

A Dynamic Population of Precursors for the Axial Mesoderm

Cells in the organizer display different tissue fates during gastrulation (Kinder et al. 2001c). Cells in the EGO produce descendants in the anterior axial mesoderm (the prechordal mesoderm, the notochord of the midbrain and upper hindbrain), the cranial mesenchyme, and the foregut endoderm in close proximity to the prechordal mesoderm. There is very little contribution of the EGO cells to the notochord of the lower hindbrain and the trunk. Finer mapping of the epiblast population within and around the EGO has revealed that (1) progenitors of the anterior axial mesoderm are sequestered to the anterior part of the EGO, whereas the rest is occupied by cells that will contribute to the extraembryonic mesoderm, (2) epiblast located anterior to the EGO contains a substantial population of progenitors of the anterior axial mesoderm in addition to the gut endoderm and cranial mesoderm, and (3) progenitors of the axial mesoderm are clustered in the midline, and more laterally located epiblast does not contribute to any axial mesoderm.

During the early stages of gastrulation of the avian embryo, cells that finally constitute Hensen's node are initially found as two separate populations, one in the proximity of the posterior marginal zone and the other in the central area of the blastoderm (Izpisua-Belmonte et al. 1993; Lemaire et al. 1997). As the primitive streak extends anteriorly from the posterior margin, it incorporates the posterior population into the tip of the advancing primitive streak and eventually brings the posterior population to merge with the central population when Hensen's node is formed. A similar morphogenetic process might also take place in the mouse embryo. As the primitive streak extends anteriorly, cells of the EGO are apparently incorporated into the advancing tip of the primitive streak. Subsequently, the EGO cells appear to congregate with epiblast cells that are found previously anterior to the EGO to form the MGO (Kinder et al. 2001c). The merger of the EGO with the epiblast cells anterior to the EGO has yet to be demonstrated directly by tracking the movement of the two populations independently.

In the mid-streak embryo, cells originating from the EGO are found both in the anterior end of the primitive streak (the MGO) and in the nascent mesoderm adjacent to the anterior end of the primitive streak (Kinder et al. 2001c). EGO descendants in the mesoderm are subsequently found in the cranial paraxial mesoderm and the heart. In contrast, EGO descendants that remain in the MGO contribute to the anterior axial mesoderm. By the late-streak stage, EGO-derived cells that emerge from the MGO form the axial mesendoderm in the prospective forebrain, midbrain, and upper hindbrain of the late-streak embryo, but few of them remain in the node. By the early-somite stage, descendants of the EGO cells are found predominantly in the cranial portion of the axial mesoderm, such as the prechordal mesoderm, endoderm in the roof of the foregut, and the notochord of the diencephalon, midbrain, and hindbrain, and only a minor contribution is found in the notochord of the trunk (Fig. 9C).

Analysis of cell fates of the MGO population reveals that they can contribute to the entire axial mesoderm underlying the midline of the brain including the prechordal mesoderm and to the notochord of the upper trunk (Fig. 9C). However, in contrast to the EGO, the MGO contains all the progenitors of the anterior axial mesoderm. Precursors of the notochord of the lower trunk are found outside the MGO in the anterior segment of the primitive streak among the precursors of the trunk paraxial mesoderm. This raises the possibility that the majority of the cells found in the EGO and the MGO are progenitors of the anterior axial mesoderm and that only a minority constitutes that of the trunk notochord (Kinder et al. 2001c).

The changing composition of the precursors in the EGO and MGO strongly suggests that not all precursors of every part of the axial mesoderm are initially present in the gastrula organizer, and as the embryo develops, additional precursors are recruited from outside the organizer. The origin of these precursors is currently unknown, but studies on the avian embryo reveal that they are localized in the epiblast near Hensen's node and its predecessor (Psychoyos

and Stern 1996; Yuan and Schoenwolf 1998). By late gastrulation, precursors of the anterior axial mesoderm have essentially departed from the organizer, and cells in the node contribute almost exclusively to the notochord of the trunk (Lemaire et al. 1997; Camus and Tam 1999; Kinder et al. 2001c). However, there may still be further recruitment of axial mesoderm, since some cells in the anterior segment of the primitive streak remain capable of colonizing the notochord of the lower trunk. Cell fate analysis of the gastrula organizer therefore reveals that the organizer comprises a dynamic precursor population where cells are constantly incorporated by recruitment from outside the organizer and are progressively allocated to different segments of the axial mesoderm (Fig. 9C).

The presence of anterior axial mesoderm and trunk notochord progenitors outside the EGO and MGO is reminiscent of the situation in the avian embryo, where there is continuous recruitment of epiblast cells into the gastrula organizer for the formation of axial mesoderm during gastrulation (Joubin and Stern 1999). In the zebrafish embryo, progenitors of the axial mesendoderm are found both within the dorsal embryonic shield (Shih and Fraser 1995, 1996) and in the adjacent blastoderm (Saude et al. 2000). Ablation of the shield per se does not deplete all the precursors of the axial mesoderm, and the remaining progenitor cells can sustain normal axis development. Cells from outside the embryonic shield are later incorporated into the shield during gastrulation and axis development (Shih and Fraser 1995,1996; Saude et al. 2000).

Morphogenetic Movement of the Axial Mesoderm

When the movement of the MGO cells is tracked over time in embryos grown in vitro, cells are found to exit the anterior aspect of the MGO and move anteriorly in a tight column until they reach the most rostral end of the anterior–posterior axis. The MGO descendants do not stay as a cohesive population but are interspersed between nontransgenic cells also derived from the MGO. This behavior of cells in the axial mesoderm and the neuroectoderm is consistent with the outcome of convergence and extension of cell populations that is characteristic of morphogenetic tissue movement in the ectoderm, axial and paraxial mesoderm in other vertebrate gastrulae (Sausedo and Schoenwolf 1994; X. Yang et al. 2002)(Chapters 15, 19, and 20). Cell tracking experiments reveal that cells in the axial mesoderm derived from the MGO and those in the paraxial mesoderm move in unison as one cohesive tissue but remain segregated from one another (Tam et al. 2000; Kinder et al. 2001c). The expansion of the mesoderm is therefore a wholesale anterior movement of different types of tissues emerging from the node and the primitive streak (Fig. 9B). Molecular activity leading to the establishment of

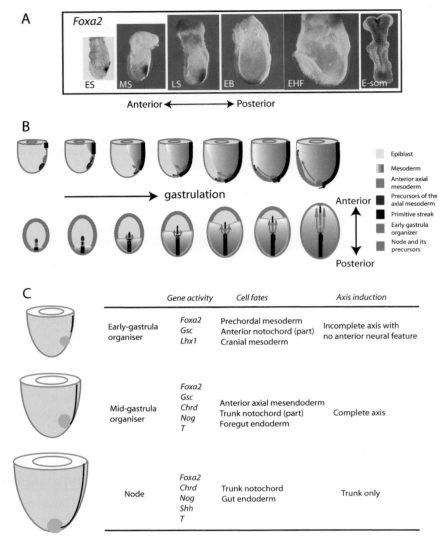

Figure 9. The gastrula organizer of the mouse embryo, identified by (*A*) the regionalized expression of the *Foxa2* gene in the epiblast and cells in the anterior end of the primitive streak. Red line shows the length of the primitive streak. Embryonic stages: (ES) early-streak, (MS) mid-streak, (LS) late-streak; (EB) early-bud; (EHF) early-head-fold; (E-som) early-somite. (*B*) Concerted movement of the axial mesoderm derived from the gastrula organizer and the paraxial mesoderm during gastrulation (for details, see Kinder et al. 2000). (*C*) Summary of changes in the genetic activity, cell fates, and the axis-inducing activity of cells of the organizer during gastrulation.

planar cell polarity has been shown to play a critical role in regulating the directionality of morphogenetic cell movement in the mesoderm of the zebrafish and frog (Mlodzik 2002; Myers et al. 2002). A similar role of planar cell polarity has not yet been shown in the mouse gastrula.

Incomplete Axis-inducing Activity of the Early Gastrula Organizer and the Node

The EGO can induce the formation of a secondary axis in a host embryo, but the new ectopic axis lacks the morpholo-

gy and molecular properties of anterior neural tissues (Tam et al. 1997b; Tam and Steiner 1999). Although this may be due to the lack of a full complement of precursors of the anterior axial mesoderm (Kinder et al. 2001c), it may also be noted that the EGO lacks the activity of genes such as *Chrd* and *Nogn* that are required for the morphogenesis of anterior (head) structures (Camus and Tam 1999; Bachiller et al. 2000). Induction of forebrain-like tissues in the secondary axis can be achieved when the EGO is combined with the AVE and the anterior epiblast of the early-streak embryo (Tam and Steiner 1999). The precise mechanism of the synergistic interaction of these three tissues that leads to axis induction is not fully understood. In the intact embryo, a possible scenario is that the interaction of AVE with the adjacent epiblast results in the anteriorization of epiblast. Subsequently, the inductive interaction of the anterior axial mesoderm derived from the EGO with the primed epiblast may lead to the formation of anterior structures (Fig. 8A–C) (Beddington and Robertson 1999; Tam and Steiner 1999; de Souza and Niehrs 2000). Like the EGO, the node also induces a partial axis without anterior characteristics (Beddington 1994; Tam et al. 1997b). This is consistent with the presence in the node of only the trunk notochord precursors and not those of anterior axial mesoderm (Tam et al. 1997b; Kinder et al. 2001c). Whether the anterior inducing activity may be reconstituted by interaction with other tissues such as AVE has not been tested.

Patterning Function of the AVE May Not Be Absolutely Essential for Anterior Development

The AVE has been reputed to be an independent source of signaling activity for anterior specification. This notion is strongly supported by similar repertoires of genetic activity expressed by the AVE and the gastrula organizer (Davidson and Tam 2000; Perea-Gomez et al. 2001a,b). Current concepts on the morphogenetic role of the AVE focus on its activity (1) to promote anterior differentiation (revealed by the expression of anterior neuroectoderm markers, e.g., *Otx2, Hesx1*) and (2) the suppression of posterior differentiation in the epiblast (indicated by the expression of mesodermal markers, e.g., *T, Fgf8, Cripto*). It is hypothesized that the AVE and its precursors act as the "head organizer" by specifying anterior (neuroectodermal) characteristics in the epiblast (Beddington and Robertson 1999; de Souza and Niehrs 2000). However, AVE alone does not display any axis-inducing activity typical of the gastrula organizer (Tam and Steiner 1999).

Evidence supporting the role of the AVE in anterior development comes from tissue ablation and genetic mutant studies. Surgical removal of the *Hesx1*-expressing AVE from the ES embryo leads to the loss of *Hesx1* activity in the prospective forebrain tissues and a smaller cephalic neural plate (Thomas and Beddington 1996). Loss of function of genes that are expressed by the AVE has, however, led to different phenotypic outcomes. Although null mutations of several AVE genes (*Cerl, Gsc,* and *Lefty1*) have no phenotypic consequences for anterior development (Rivera-Perez et al. 1995; Belo et al. 1997, 1998; Meno et al. 1998; Shawlot et al. 1999), others, such as *Dkk1, Hesx1,* and *Hex,* cause a mild deficiency with minor loss of the rostral-most part of the forebrain (Martinez Barbera et al. 2000a,b; Martinez-Barbera and Beddington 2001; Mukhopadhyay et al. 2001).

Loss of *Lhx1* or *Otx2* leads to deletion of the anterior structures (Matsuo et al. 1995; Shawlot and Behringer 1995; Ang et al. 1996; Acampora et al. 1998; Rhinn et al. 1998; Suda et al. 1999; Kimura et al. 2000). Loss of *Foxa2* activity results in a rudimentary neural tube that is still, however, correctly patterned in the anterior–posterior axis (Dufort et al. 1998; Klingensmith et al. 1999). Partial loss of *Foxa2* and *Otx2* function results in mild disruption of head development (Jin et al. 2001). Results of chimera studies reveal that the presence of normal *Dkk1* (Mukhopadhyay et al. 2001), *Hesx1, Hex* (Martinez Barbera et al. 2000a,b; Martinez-Barbera and Beddington 2001), *Foxa2* (Filosa et al. 1997; Dufort et al. 1998), *Lhx1* (Shawlot et al. 1999), and *Otx2* (Rhinn et al. 1998) function in the visceral endoderm cannot ameliorate the anterior defects, suggesting that the activity of these genes in the AVE are not essential for anterior patterning.

A key issue regarding the role of the AVE is whether its position in the embryo is critical for patterning the epiblast. The localization of the AVE is influenced by gene activity that acts in parallel (e.g., Cripto EGF-CFC cofactor) or downstream of Nodal signaling (e.g., *Otx2, Lhx1*) (Schier and Shen 2000) (Chapter 34). In the *Otx2* and *Lhx1* mutants, defects in anterior development are accompanied by the failure of AVE precursors to move from the distal to the anterior position (Shawlot et al. 1999; Kimura et al. 2000, 2001; Perea-Gomez et al. 2001b). Embryos lacking *Cripto* also cannot mobilize the AVE to the anterior region. However, unlike the *Otx2* mutants, the epiblast of *Cripto*-deficient embryos acquires the molecular properties of anterior neuroectoderm (Ding et al. 1998), indicating that the ectopic AVE is still functional. In contrast, despite the absence of any recognizable AVE, molecular markers for anterior hindbrain (*Sox1, Gbx2*) are expressed in the epiblast of *Otx2;Cripto* mutants (Kimura et al. 2001), suggesting that AVE function may not be absolutely essential for patterning the epiblast. Last, the correct positioning of the AVE does not guarantee a competent patterning activity. In the *Wnt3* and *Mesd* mutants, the presence of AVE-like endoderm is not accompanied by any anterior neural differentiation of the epiblast (Liu et al. 1999; Hsieh et al. 2003).

The AVE may promote the differentiation of anterior tissues by counteracting the activity that specifies posterior tissues. The AVE can repress the expression of posterior tissue markers (*T* and *Cripto*) in the epiblast in vitro (Kimura et al. 2000; Perea-Gomez et al. 2001b). This repressive effect is unique to the AVE and cannot be replaced by definitive endoderm of the anterior gut (ADE). The defective anterior development and overexpression of mesodermal markers (*T* and *Fgf8*) in the epiblast of the *Otx2⁻/⁻, Smad2, Otx2;Cripto,* and *Foxa2;Lhx1* mutants may be caused by lack of molecular activity to antagonize TGF-β/Nodal and Wnt signaling (Waldrip et al. 1998; Perea-Gomez et al.

1999, 2001b; Kimura et al. 2000, 2001; Kinder et al. 2001b). It is, however, not possible to pinpoint whether the loss of function of any of these genes has a specific impact on AVE function, or whether a fully operational AVE is formed in these mutant embryos.

Mid-gastrula Organizer Displays a Full Axis-inducing Activity

In contrast to the EGO and the node, the MGO alone can induce an axis with both anterior and posterior molecular characteristics (Tam and Steiner 1999; Kinder et al. 2001c). The morphogenetic function of the gastrula organizer may be determined by the composition and the inductive activity of the precursor tissues it contains. The MGO is shown by cell-fate analysis to contain precursors of the whole anterior axial mesoderm and much of the trunk notochord (Kinder et al. 2001c). The anterior axial mesoderm has been shown to be essential for the induction and morphogenesis of the anterior neural structures (Fig. 8D,E) (Camus et al. 2000). The complete assembly of progenitors is evident by the concurrent expression of an abundant number of genes, including *Gsc* (prechordal mesoderm), *Nog and Chrd* (head process giving rise to the notochord of the cranial region), *T* (notochord), and *Foxa2* (entire axial mesoderm). This may primarily underpin the ability of the MGO to induce a complete secondary axis.

Analysis of the fate of cells in the MGO reveals that they contribute not only to the axial mesoderm, but also to the cranial paraxial mesoderm and the ADE for the foregut (Kinder et al. 2001c). The foregut endoderm has been shown to play a role in patterning the brain in avian embryos (Withington et al. 2001). In the mouse, it is possible that the endoderm and mesoderm together may direct the specification of the cranial neural primordium in the ectoderm (Fig. 8D,E). Results of explant coculture experiments have shown that the combined mesoderm and endoderm in the anterior region of the gastrula can induce and maintain the expression of anterior neuroectoderm markers (*Otx2, Six3,* and *En2*) in the epiblast, whereas the posterior mesoderm and endoderm suppress *Otx2* expression (Ang and Rossant 1993; Ang et al. 1994). Whether the ADE or its precursors have independent patterning activity is not known.

The ability to induce a more complete axis by the MGO implies that synergistic interactions between organizer derivatives and the AVE/anterior epiblast may not always be necessary. It also demonstrates that the tissues in the late-streak host embryo are adequately primed and are not restrained in their response to full axis-organizing activity. Furthermore, it supports the notion that the failure to induce anterior tissues by the EGO or the node alone is a genuine reflection of the extent of axis-inducing activity of the organizer of embryos at the different stages of gastrulation.

THE MORPHOGENETIC OUTCOME OF GASTRULATION: ESTABLISHING THE BODY PLAN

The embryo at the beginning of gastrulation is already richly endowed with positional information derived from the inductive and suppressive interactions with the visceral endoderm and the extraembryonic ectoderm. This results in the establishment of anterior–posterior polarity, manifested by the regionalization of gene activity and the localization of the primitive streak. The generation of definitive germ layers during gastrulation provides the embryo with the basic building blocks for the establishment of the fetal body plan. Concomitant with germ layer formation, there is progressively finer delineation of the precursors for specific ectodermal and mesodermal derivatives in the epiblast, such as those for surface ectoderm, neural ectoderm, heart, and lateral, paraxial, and axial mesoderm. The regionalization of tissue precursors in the germ layers is the result of lineage-specific molecular activity and position-dependent inductive signaling. The temporal and spatial pattern of recruitment of these precursors from the epiblast through the primitive streak to the mesoderm and endoderm determines the developmental fates and the destination of the cells in the fetal body. A primary function of gastrulation is therefore the correct allocation and distribution of primordial tissues that will participate subsequently in the formation of specific body parts or organs. Finally, the translation of the body plan into embryonic structures requires the coordination of morphogenetic processes through the activity of the gastrula organizer and inductive interactions between germ-layer derivatives.

ACKNOWLEDGMENTS

We thank S. L. Ang, R. R. Behringer, E. M. De Robertis, B. Herrmann, Y. Kanai, R. Lovell-Badge, L. Robb, and M. Wilkinson for the gift of riboprobes; Lorraine Robb for providing Figure 2B; Kirsten Steiner, Nicole Wong, and Poh Lynn Khoo for performing in situ hybridization; and Peter Rowe, Claudio Stern, and Christoph Viebahn for comments on the manuscript. Our work is supported by the National Health and Medical Research Council (NMHRC) of Australia and by Mr. James Fairfax. P.P.L.T. is a NMHRC Senior Principal Research Fellow.

REFERENCES

Acampora D., Avantaggiato V., Tuorto F., Briata P., Corte G., and Simeone A. 1998. Visceral endoderm-restricted translation of Otx1 mediates recovery of Otx2 requirements for specification of anterior neural plate and normal gastrulation. *Development* **125:** 5091–5104.

Acampora D., Mazan S., Lallemand Y., Avantaggiato V., Maury M., Simeone A., and Brulet P. 1995. Forebrain and midbrain regions are deleted in Otx2$^{-/-}$ mutants due to a defective anterior neuroectoderm specification during gastrulation. *Development* **121:** 3279–3290.

Ang S.L. and Rossant J. 1993. Anterior mesendoderm induces mouse Engrailed genes in explant cultures. *Development* **118:** 139–149.

Ang S.L., Conlon R.A., Jin O., and Rossant J. 1994. Positive and negative signals from mesoderm regulate the expression of mouse Otx2 in ectoderm explants. *Development* **120:** 2979–2989.

Ang S.L., Jin O., Rhinn M., Daigle N., Stevenson L., and Rossant J. 1996. A targeted mouse Otx2 mutation leads to severe defects in gastrulation and formation of axial mesoderm and to deletion of rostral brain. *Development* **122:** 243–252.

Arman E., Haffner-Krausz R., Chen Y., Heath J.K., and Lonai P. 1998. Targeted disruption of fibroblast growth factor (FGF) receptor 2 suggests a role for FGF signaling in pregastrulation mammalian development. *Proc. Natl. Acad. Sci.* **95:** 5082–5087.

Arnold S.J., Stappert J., Bauer A., Kispert A., Herrmann B.G., and Kemler R. 2000. Brachyury is a target gene of the Wnt/beta-catenin signaling pathway. *Mech. Dev.* **91:** 249–258.

Attisano L., Wrana J.L., Cheifetz S., and Massague J. 1992. Novel activin receptors: Distinct genes and alternative mRNA splicing generate a repertoire of serine/threonine kinase receptors. *Cell* **68:** 97–108.

Bachiller D., Klingensmith J., Kemp C., Belo J.A., Anderson R.M., May S.R., McMahon J.A., McMahon A.P., Harland R.M., Rossant J., and De Robertis E.M. 2000. The organizer factors Chordin and Noggin are required for mouse forebrain development. *Nature* **403:** 658–661.

Bachvarova R.F., Skromne I., and Stern C.D. 1998. Induction of primitive streak and Hensen's node by the posterior marginal zone in the early chick embryo. *Development* **125:** 3521–3534.

Beck S., Le Good J.A., Guzman M., Ben Haim N., Roy K., Beermann F., and Constam D.B. 2002. Extraembryonic proteases regulate Nodal signaling during gastrulation. *Nat. Cell Biol.* **4:** 981–985.

Beddington R.S. 1982. An autoradiographic analysis of tissue potency in different regions of the embryonic ectoderm during gastrulation in the mouse. *J. Embryol. Exp. Morphol.* **69:** 265–285.

———. 1994. Induction of a second neural axis by the mouse node. *Development* **120:** 613–620.

Beddington R.S. and Robertson E.J. 1999. Axis development and early asymmetry in mammals. *Cell* **96:** 195–209.

Beddington R.S., Rashbass P., and Wilson V. 1992. Brachyury—a gene affecting mouse gastrulation and early organogenesis. *Dev. Suppl.* **1992:** 157–165.

Behringer R.R., Wakamiya M., Tsang T.E., and Tam P.P.L. 2000. A flattened mouse embryo: Leveling the playing field. *Genesis* **28:** 23–30.

Belo J.A., Leyns L., Yamada G., and De Robertis E.M. 1998. The prechordal midline of the chondrocranium is defective in Goosecoid-1 mouse mutants. *Mech. Dev.* **72:** 15–25.

Belo J.A., Bouwmeester T., Leyns L., Kertesz N., Gallo M., Follettie M., and De Robertis E.M. 1997. Cerberus-like is a secreted factor with neutralizing activity expressed in the anterior primitive endoderm of the mouse gastrula. *Mech. Dev.* **68:** 45–57.

Beppu H., Kawabata M., Hamamoto T., Chytil A., Minowa O., Noda T., and Miyazono K. 2000. BMP type II receptor is required for gastrulation and early development of mouse embryos. *Dev. Biol.* **221:** 249–258.

Bertocchini F. and Stern C.D. 2002. The hypoblast of the chick embryo positions the primitive streak by antagonizing Nodal signaling. *Dev. Cell* **3:** 735–744.

Biben C., Stanley E., Fabri L., Kotecha S., Rhinn M., Drinkwater C., Lah M., Wang C.C., Nash A., Hilton D., Ang S.L., Mohun T., and Harvey R.P. 1998. Murine cerberus homologue mCer-1: A candidate anterior patterning molecule. *Dev. Biol.* **194:** 135–151.

Bielinska M., Narita N., and Wilson D.B. 1999. Distinct roles for visceral endoderm during embryonic mouse development. *Int. J. Dev. Biol.* **43:** 183–205.

Brennan J., Lu C.C., Norris D.P., Rodriguez T.A., Beddington R.S., and Robertson E.J. 2001. Nodal signaling in the epiblast patterns the early mouse embryo. *Nature* **411:** 965–969.

Burdsal C.A., Flannery M.L., and Pedersen R.A. 1998. FGF-2 alters the fate of mouse epiblast from ectoderm to mesoderm in vitro. *Dev. Biol.* **198:** 231–244.

Camus A. and Tam P.P.L. 1999. The organizer of the gastrulating mouse embryo. *Curr. Top. Dev. Biol.* **45:** 117–153.

Camus A., Davidson B.P., Billiards S., Khoo P., Rivera-Perez J.A., Wakamiya M., Behringer R.R., and Tam P.P.L. 2000. The morphogenetic role of midline mesendoderm and ectoderm in the development of the forebrain and the midbrain of the mouse embryo. *Development* **127:** 1799–1813.

Carson D.D., Bagchi I., Dey S.K., Enders A.C., Fazleabas A.T., Lessey B.A., and Yoshinaga K. 2000. Embryo implantation. *Dev. Biol.* **223:** 217–237.

Chang H., Zwijsen A., Vogel H., Huylebroeck D., and Matzuk M.M. 2000. Smad5 is essential for left-right asymmetry in mice. *Dev. Biol.* **219:** 71–78.

Chen W.S., Manova K., Weinstein D.C., Duncan S.A., Plump A.S., Prezioso V.R., Bachvarova R.F., and Darnell J.E., Jr. 1994. Disruption of the HNF-4 gene, expressed in visceral endoderm, leads to cell death in embryonic ectoderm and impaired gastrulation of mouse embryos. *Genes Dev.* **8:** 2466–2477.

Ciruna B. and Rossant J. 2001. FGF signaling regulates mesoderm cell fate specification and morphogenetic movement at the primitive streak. *Dev. Cell* **1:** 37–49.

Ciruna B.G., Schwartz L., Harpal K., Yamaguchi T.P., and Rossant J. 1997. Chimeric analysis of fibroblast growth factor receptor-1 (Fgfr1) function: A role for FGFR1 in morphogenetic movement through the primitive streak. *Development* **124:** 2829–2841.

Conlon F.L., Lyons K.M., Takaesu N., Barth K.S., Kispert A., Herrmann B., and Robertson E.J. 1994. A primary requirement for Nodal in the formation and maintenance of the primitive streak in the mouse. *Development* **120:** 1919–1928.

Copp A.J. 1979. Interaction between inner cell mass and trophectoderm of the mouse blastocyst. II. The fate of the polar trophectoderm. *J. Embryol. Exp. Morphol.* **51:** 109–120.

———. 1981. The mechanism of mouse egg-cylinder morphogenesis in vitro. *J. Embryol. Exp. Morphol.* **61:** 277–287.

Coucouvanis E. and Martin G.R. 1995. Signals for death and survival: A two-step mechanism for cavitation in the vertebrate embryo. *Cell* **83:** 279–287.

———. 1999. BMP signaling plays a role in visceral endoderm differentiation and cavitation in the early mouse embryo. *Development* **126:** 535–546.

Davidson B.P. and Tam P.P.L. 2000. The node of the mouse embryo. *Curr. Biol.* **10:** R617–R619.

Davidson B.P., Cheng L., Kinder S.J., and Tam P.P.L. 2000. Exogenous FGF-4 can suppress anterior development in the mouse embryo during neurulation and early organogenesis. *Dev. Biol.* **221:** 41–52.

DeGregori J., Russ A., von Melchner H., Rayburn H., Priyaranjan P., Jenkins N.A., Copeland N.G., and Ruley H.E. 1994. A murine homolog of the yeast RNA1 gene is required for postimplantation development. *Genes Dev.* **8:** 265–276.

De Robertis E.M., Larrain J., Oelgeschlager M., and Wessely O. 2000. The establishment of Spemann's organizer and patterning of the vertebrate embryo. *Nat. Rev. Genet.* **1:** 171–181.

de Souza F.S. and Niehrs C. 2000. Anterior endoderm and head induction in early vertebrate embryos. *Cell Tissue Res.* **300:** 207–217.

Deng C.X., Wynshaw-Boris A., Shen M.M., Daugherty C., Ornitz D.M., and Leder P. 1994. Murine FGFR-1 is required for early postimplantation growth and axial organization. *Genes Dev.* **8:** 3045–3057.

Ding J., Yang L., Yan Y.T., Chen A., Desai N., Wynshaw-Boris A., and Shen M.M. 1998. Cripto is required for correct orientation of the anterior-posterior axis in the mouse embryo. *Nature* **395:** 702–707.

Downs K.M. and Davies T. 1993. Staging of gastrulating mouse embryos by morphological landmarks in the dissecting microscope. *Development* **118:** 1255–1266.

Dufort D., Schwartz L., Harpal K., and Rossant J. 1998. The transcription factor HNF3beta is required in visceral endoderm for normal primitive streak morphogenesis. *Development* **125:** 3015–3025.

Dunn N.R., Winnier G.E., Hargett L.K., Schrick J.J., Fogo A.B., and Hogan B.L. 1997. Haploinsufficient phenotypes in Bmp4 heterozygous null mice and modification by mutations in Gli3 and Alx4. *Dev. Biol.* **188:** 235–247.

Eloy-Trinquet S. and Nicolas J.F. 2002. Cell coherence during production of the presomitic mesoderm and somitogenesis in the mouse embryo. *Development* **129:** 3609–3619.

Episkopou V., Arkell R., Timmons P.M., Walsh J.J., Andrew R.L., and Swan D. 2001. Induction of the mammalian node requires Arkadia function in the extraembryonic lineages. *Nature* **410:** 825–830.

Faust C., Schumacher A., Holdener B., and Magnuson T. 1995. The eed mutation disrupts anterior mesoderm production in mice. *Development* **121:** 273–285.

Faust C., Lawson K.A., Schork N.J., Thiel B., and Magnuson T. 1998. The Polycomb-group gene eed is required for normal morphogenetic movements during gastrulation in the mouse embryo. *Development* **125:** 4495–4506.

Feldman B., Poueymirou W., Papaioannou V.E., DeChiara T.M., and Goldfarb M. 1995. Requirement of FGF-4 for postimplantation mouse development. *Science* **267:** 246–249.

Filosa S., Rivera-Perez J.A., Gomez A.P., Gansmuller A., Sasaki H., Behringer R.R., and Ang S.L. 1997. Goosecoid and HNF-3beta genetically interact to regulate neural tube patterning during mouse embryogenesis. *Development* **124:** 2843–2854.

Foley A.C., Skromne I., and Stern C.D. 2000. Reconciling different models of forebrain induction and patterning: A dual role for the hypoblast. *Development* **127:** 3839–3854.

Frank M. and Kemler R. 2002. Protocadherins. *Curr. Opin. Cell Biol.* **14:** 557–562.

Fujiwara T., Dunn N.R., and Hogan B.L. 2001. Bone morphogenetic protein 4 in the extraembryonic mesoderm is required for allantois development and the localization and survival of primordial germ cells in the mouse. *Proc. Natl. Acad. Sci.* **98:** 13739–13744.

Gardner R.L. and Rossant J. 1979. Investigation of the fate of 4-5 day post-coitum mouse inner cell mass cells by blastocyst injection. *J. Embryol. Exp. Morphol.* **52:** 141–152.

Goumans M.J. and Mummery C. 2000. Functional analysis of the TGFbeta receptor/Smad pathway through gene ablation in mice. *Int. J. Dev. Biol.* **44:** 253–265.

Green A.R., Lints T., Visvader J., Harvey R., and Begley C.G. 1992. SCL is coexpressed with GATA-1 in hemopoietic cells but is also expressed in developing brain. *Oncogene* **7:** 653–660.

Gu Z., Nomura M., Simpson B.B., Lei H., Feijen A., van den Eijnden-van Raaij, Donahoe P.K., and Li E. 1998. The type I activin receptor ActRIB is required for egg cylinder organization and gastrulation in the mouse. *Genes Dev.* **12:** 844–857.

Gu Z., Reynolds E.M., Song J., Lei H., Feijen A., Yu L., He W., MacLaughlin D.T., van den Eijnden-van Raaij, Donahoe P.K., and Li E. 1999. The type I serine/threonine kinase receptor ActRIA (ALK2) is required for gastrulation of the mouse embryo. *Development* **126:** 2551–2561.

Haffner-Krausz R., Gorivodsky M., Chen Y., and Lonai P. 1999. Expression of Fgfr2 in the early mouse embryo indicates its involvement in preimplantation development. *Mech. Dev.* **85:** 167–172.

Hakem R., de la Pompa J.L., Sirard C., Mo R., Woo M., Hakem A., Wakeham A., Potter J., Reitmair A., Billia F., Firpo E., Hui C.C., Roberts J., Rossant J., and Mak T.W. 1996. The tumor suppressor gene Brca1 is required for embryonic cellular proliferation in the mouse. *Cell* **85:** 1009–1023.

Hallonet M., Kaestner K.H., Martin-Parras L., Sasaki H., Betz U.A., and Ang S.L. 2002. Maintenance of the specification of the anterior definitive endoderm and forebrain depends on the axial mesendoderm: A study using HNF3beta/Foxa2 conditional mutants. *Dev. Biol.* **243:** 20–33.

Hamada H., Meno C., Watanabe D., and Saijoh Y. 2002. Establishment of vertebrate left-right asymmetry. *Nat. Rev. Genet.* **3:** 103–113.

Harland R. and Gerhart J. 1997. Formation and function of Spemann's organizer. *Annu. Rev. Cell Dev. Biol.* **13:** 611–667.

Hart A.H., Hartley L., Sourris K., Stadler E.S., Li R., Stanley E.G., Tam P.P.L., Elefanty A.G., and Robb L. 2002. Mixl1 is required for axial mesendoderm morphogenesis and patterning in the murine embryo. *Development* **129:** 3597–3608.

Hashimoto K., Fujimoto H., and Nakatsuji N. 1987. An ECM substratum allows mouse mesodermal cells isolated from the primitive streak to exhibit motility similar to that inside the embryo and reveals a deficiency in the T/T mutant cells. *Development* **100:** 587–598.

Hogan B.L. and Tilly R. 1981. Cell interactions and endoderm differentiation in cultured mouse embryos. *J. Embryol. Exp. Morphol.* **62:** 379–394.

Hoodless P.A., Pye M., Chazaud C., Labbe E., Attisano L., Rossant J., and Wrana J.L. 2001. FoxH1 (Fast) functions to specify the anterior primitive streak in the mouse. *Genes Dev.* **15:** 1257–1271.

Hrabe d.A. and Kirchner C. 1993. Fibroblast growth factor induces primitive streak formation in rabbit pre-implantation embryos in vitro. *Anat. Embryol.* **187:** 269–273.

Hsieh J.C., Lee L., Zhang L., Wefer S., Brown K., Rosenquist T.A., and Holdener B.C. 2003. Mesd encodes a novel LRP5/6 chaperone essential for specification of mouse embryonic polarity. *Dev. Cell* **112:** 355–367.

Hukriede N.A., Tsang T.E., Habas R., Khoo P.-L., Steiner K.A., Weeks D.L., Tam P.P.L., and Dawid I.B. 2003. Conserved requirement of Lim1 function for cell movements during gastrulation. *Dev. Cell* **4:** 83–94

Ilgren E.B. 1981. The in vitro morphogenesis of the guinea pig egg cylinder. *Anat. Embryol.* **163:** 351–365.

Iratni R., Yan Y.T., Chen C., Ding J., Zhang Y., Price S.M., Reinberg D.,

and Shen M.M. 2002. Inhibition of excess Nodal signaling during mouse gastrulation by the transcriptional corepressor DRAP1. *Science* **298:** 1996–1999.

Izpisua-Belmonte J.C., De Robertis E.M., Storey K.G., and Stern C.D. 1993. The homeobox gene goosecoid and the origin of organizer cells in the early chick blastoderm. *Cell* **74:** 645–659.

Jin O., Harpal K., Ang S.L., and Rossant J. 2001. Otx2 and HNF3beta genetically interact in anterior patterning. *Int. J. Dev. Biol.* **45:** 357–365.

Jollie W.P. 1990. Development, morphology, and function of the yolk-sac placenta of laboratory rodents. *Teratology* **41:** 361–381.

Joubin K. and Stern C.D. 1999. Molecular interactions continuously define the organizer during the cell movements of gastrulation. *Cell* **98:** 559–571.

Kalantry S., Manning S., Haub O., Tomihara-Newberger C., Lee H.G., Fangman J., Disteche C.M., Manova K., and Lacy E. 2001. The amnionless gene, essential for mouse gastrulation, encodes a visceral-endoderm-specific protein with an extracellular cysteine-rich domain. *Nat. Genet.* **27:** 412–416.

Kanai-Azuma M., Kanai Y., Gad J.M., Tajima Y., Taya C., Kurohmaru M., Sanai Y., Yonekawa H., Yazaki K., Tam P.P.L., and Hayashi Y. 2002. Depletion of definitive gut endoderm in Sox17-null mutant mice. *Development* **129:** 2367–2379.

Kaufman M.H. 1992. *The atlas of mouse development*. Academic Press, London.

Kimura C., Shen M.M., Takeda N., Aizawa S., and Matsuo I. 2001. Complementary functions of Otx2 and Cripto in initial patterning of mouse epiblast. *Dev. Biol.* **235:** 12–32.

Kimura C., Yoshinaga K., Tian E., Suzuki M., Aizawa S., and Matsuo I. 2000. Visceral endoderm mediates forebrain development by suppressing posteriorizing signals. *Dev. Biol.* **225:** 304–321.

Kinder S.J., Loebel D.A., and Tam P.P.L. 2001a. Allocation and early differentiation of cardiovascular progenitors in the mouse embryo. *Trends Cardiovasc. Med.* **11:** 177–184.

Kinder S.J., Tsang T.E., Ang S.L., Behringer R.R., and Tam P.P.L. 2001b. Defects of the body plan of mutant embryos lacking Lim1, Otx2 or Hnf3beta activity. *Int. J. Dev. Biol.* **45:** 347–355.

Kinder S.J., Tsang T.E., Quinlan G.A., Hadjantonakis A.K., Nagy A., and Tam P.P.L. 1999. The orderly allocation of mesodermal cells to the extraembryonic structures and the anteroposterior axis during gastrulation of the mouse embryo. *Development* **126:** 4691–4701.

Kinder S.J., Tsang T.E., Wakamiya M., Sasaki H., Behringer R.R., Nagy A., and Tam P.P.L. 2001c. The organizer of the mouse gastrula is composed of a dynamic population of progenitor cells for the axial mesoderm. *Development* **128:** 3623–3634.

Kispert A. and Herrmann B.G. 1994. Immunohistochemical analysis of the Brachyury protein in wild-type and mutant mouse embryos. *Dev. Biol.* **161:** 179–193.

Klingensmith J., Ang S.L., Bachiller D., and Rossant J. 1999. Neural induction and patterning in the mouse in the absence of the node and its derivatives. *Dev. Biol.* **216:** 535–549.

Lawson K.A. and Hage W.J. 1994. Clonal analysis of the origin of primordial germ cells in the mouse. *Ciba Found. Symp.* **182:** 68–84.

Lawson K.A. and Pedersen R.A. 1987. Cell fate, morphogenetic movement and population kinetics of embryonic endoderm at the time of germ layer formation in the mouse. *Development* **101:** 627–652.

———. 1992a. Clonal analysis of cell fate during gastrulation and early neurulation in the mouse. *Ciba Found. Symp.* **165:** 3–21.

———. 1992b. Early mesoderm formation in the mouse embryo. In *Formation and differentiation of early embryonic mesoderm. NATO ASI Ser.* **231:** 33–46.

Lawson K.A., Meneses J.J., and Pedersen R.A. 1986. Cell fate and cell lineage in the endoderm of the presomite mouse embryo, studied with an intracellular tracer. *Dev. Biol.* **115:** 325–339.

———. 1991. Clonal analysis of epiblast fate during germ layer formation in the mouse embryo. *Development* **113:** 891–911.

Lawson K.A., Dunn N.R., Roelen B.A., Zeinstra L.M., Davis A.M., Wright C.V., Korving J.P., and Hogan B.L. 1999. Bmp4 is required for the generation of primordial germ cells in the mouse embryo. *Genes Dev.* **13:** 424–436.

Lechleider R.J., Ryan J.L., Garrett L., Eng C., Deng C., Wynshaw-Boris A., and Roberts A.B. 2001. Targeted mutagenesis of Smad1 reveals an essential role in chorioallantoic fusion. *Dev. Biol.* **240:** 157–167.

Lemaire L., Roeser T., Izpisua-Belmonte J.C., and Kessel M. 1997. Segregating expression domains of two goosecoid genes during the transition from gastrulation to neurulation in chick embryos. *Development* **124:** 1443–1452.

Lim D.S. and Hasty P. 1996. A mutation in mouse rad51 results in an early embryonic lethal that is suppressed by a mutation in p53. *Mol. Cell. Biol.* **16:** 7133–7143.

Liu F., Ventura F., Doody J., and Massague J. 1995. Human type II receptor for bone morphogenic proteins (BMPs): Extension of the two-kinase receptor model to the BMPs. *Mol. Cell. Biol.* **15:** 3479–3486.

Liu P., Wakamiya M., Shea M.J., Albrecht U., Behringer R.R., and Bradley A. 1999. Requirement for Wnt3 in vertebrate axis formation. *Nat. Genet.* **22:** 361–365.

Long J.A. and Burlingame P.L. 1938. The development of the external form of the rat with observations on the origin of the extraembryonic coelom and foetal membranes. *U. Calif. Publ. Zool.* **43:** 143–184.

Lu C.C., Brennan J., and Robertson E.J. 2001. From fertilization to gastrulation: Axis formation in the mouse embryo. *Curr. Opin. Genet. Dev.* **11:** 384–392.

Luo J., Sladek R., Bader J.A., Matthyssen A., Rossant J., and Giguere V. 1997. Placental abnormalities in mouse embryos lacking the orphan nuclear receptor ERR-beta. *Nature* **388:** 778–782.

Manova K., Tomihara-Newberger C., Wang S., Godelman A., Kalantry S., Witty-Blease K., De L, V, Chen W.S., Lacy E., and Bachvarova R.F. 1998. Apoptosis in mouse embryos: Elevated levels in pregastrulae and in the distal anterior region of gastrulae of normal and mutant mice. *Dev. Dyn.* **213:** 293–308.

Martinez-Barbera J.P. and Beddington R.S. 2001. Getting your head around Hex and Hesx1: Forebrain formation in mouse. *Int. J. Dev. Biol.* **45:** 327–336.

Martinez-Barbera J.P., Rodriguez T.A., and Beddington R.S. 2000a. The homeobox gene Hesx1 is required in the anterior neural ectoderm for normal forebrain formation. *Dev. Biol.* **223:** 422–430.

Martinez Barbera J.P., Clements M., Thomas P., Rodriguez T., Meloy D., Kioussis D., and Beddington R.S. 2000b. The homeobox gene Hex is required in definitive endodermal tissues for normal forebrain, liver and thyroid formation. *Development* **127:** 2433–2445.

Mathis L. and Nicolas J.F. 2002. Cellular patterning of the vertebrate embryo. *Trends Genet.* **18:** 627–635.

Matsuo I., Kuratani S., Kimura C., Takeda N., and Aizawa S. 1995. Mouse Otx2 functions in the formation and patterning of rostral head. *Genes Dev.* **9:** 2646–2658.

Meno C., Shimono A., Saijoh Y., Yashiro K., Mochida K., Ohishi S.,

Noji S., Kondoh H., and Hamada H. 1998. lefty-1 is required for left-right determination as a regulator of lefty-2 and Nodal. *Cell* **94:** 287–297.

Meno C., Takeuchi J., Sakuma R., Koshiba-Takeuchi K., Ohishi S., Saijoh Y., Miyazaki J., ten Dijke P., Ogura T., and Hamada H. 2001. Diffusion of Nodal signaling activity in the absence of the feedback inhibitor Lefty2. *Dev. Cell* **1:** 127–138.

Miller S.A. and Runner M.N. 1978. Tissue specificity for incorporation of [3H]thymidine by the 10- to 12-somite mouse embryo: Alteration by acute exposure to hydroxyurea. *J. Embryol. Exp. Morphol.* **44:** 181–189.

Mishina Y., Crombie R., Bradley A., and Behringer R.R. 1999. Multiple roles for activin-like kinase-2 signaling during mouse embryogenesis. *Dev. Biol.* **213:** 314–326.

Mishina Y., Suzuki A., Ueno N., and Behringer R.R. 1995. Bmpr encodes a type I bone morphogenetic protein receptor that is essential for gastrulation during mouse embryogenesis. *Genes Dev.* **9:** 3027–3037.

Mlodzik M. 2002. Planar cell polarization: Do the same mechanisms regulate Drosophila tissue polarity and vertebrate gastrulation? *Trends Genet.* **18:** 564–571.

Monaghan A.P., Kioschis P., Wu W., Zuniga A., Bock D., Poustka A., Delius H., and Niehrs C. 1999. Dickkopf genes are co-ordinately expressed in mesodermal lineages. *Mech. Dev.* **87:** 45–56.

Morrisey E.E., Tang Z., Sigrist K., Lu M.M., Jiang F., Ip H.S., and Parmacek M.S. 1998. GATA6 regulates HNF4 and is required for differentiation of visceral endoderm in the mouse embryo. *Genes Dev.* **12:** 3579–3590.

Mukhopadhyay M., Teufel A., Yamashita T., Agulnick A.D., Chen L., Downs K.M., Schindler A., Grinberg A., Huang S.P., Dorward D., and Westphal H. 2003. Functional ablation of the mouse Ldb1 gene results in severe patterning defects during gastrulation. *Development* **130:** 495–505.

Mukhopadhyay M., Shtrom S., Rodriguez-Esteban C., Chen L., Tsukui T., Gomer L., Dorward D.W., Glinka A., Grinberg A., Huang S.P., Niehrs C., Izpisua Belmonte J.C., and Westphal H. 2001. Dickkopf1 is required for embryonic head induction and limb morphogenesis in the mouse. *Dev. Cell* **1:** 423–434.

Myers D.C., Sepich D.S., and Solnica-Krezel L. 2002. Convergence and extension in vertebrate gastrulae: Cell movements according to or in search of identity? *Trends Genet.* **18:** 447–455.

Nakatsuji N., Snow M.H., and Wylie C.C. 1986. Cinemicrographic study of the cell movement in the primitive-streak-stage mouse embryo. *J. Embryol. Exp. Morphol.* **96:** 99–109.

Niederlander C., Walsh J.J., Episkopou V., and Jones C.M. 2001. Arkadia enhances Nodal-related signaling to induce mesendoderm. *Nature* **410:** 830–834.

Norris D.P., Brennan J., Bikoff E.K., and Robertson E.J. 2002. The Foxh1-dependent autoregulatory enhancer controls the level of Nodal signals in the mouse embryo. *Development* **129:** 3455–3468.

Parameswaran M. and Tam P.P.L. 1995. Regionalisation of cell fate and morphogenetic movement of the mesoderm during mouse gastrulation. *Dev. Genet.* **17:** 16–28.

Perea-Gomez A., Rhinn M., and Ang S.L. 2001a. Role of the anterior visceral endoderm in restricting posterior signals in the mouse embryo. *Int. J. Dev. Biol.* **45:** 311–320.

Perea-Gomez A., Shawlot W., Sasaki H., Behringer R.R., and Ang S. 1999. HNF3β and Lim1 interact in the visceral endoderm to regulate primitive streak formation and anterior-posterior polarity in the mouse embryo. *Development* **126:** 4499–4511.

Perea-Gomez A., Lawson K.A., Rhinn M., Zakin L., Brulet P., Mazan S., and Ang S.L. 2001b. Otx2 is required for visceral endoderm movement and for the restriction of posterior signals in the epiblast of the mouse embryo. *Development* **128:** 753–765.

Perea-Gomez A., Vella F.D., Shawlot W., Oulad-Abdelghani M., Chazaud C., Meno C., Pfister V., Chen L., Robertson E., Hamada H., Behringer R.R., and Ang S.L. 2002. Nodal antagonists in the anterior visceral endoderm prevent the formation of multiple primitive streaks. *Dev. Cell* **3:** 745–756.

Poelmann R.E. 1980. Differential mitosis and degeneration patterns in relation to the alterations in the shape of the embryonic ectoderm of early post-implantation mouse embryos. *J. Embryol. Exp. Morphol.* **55:** 33–51.

———. 1981. The head-process and the formation of the definitive endoderm in the mouse embryo. *Anat. Embryol.* **162:** 41–49.

Poelmann R.E., Mentink M.M., and van Leeuwen J.L. 1987. Axial rotation of murine embryos, a study of asymmetric mitotic activity in the neural tube of somite stages. *Anat. Embryol.* **176:** 99–103.

Power M.A. and Tam P.P.L. 1993. Onset of gastrulation, morphogenesis and somitogenesis in mouse embryos displaying compensatory growth. *Anat. Embryol.* **187:** 493–504.

Psychoyos D. and Stern C.D. 1996. Restoration of the organizer after radical ablation of Hensen's node and the anterior primitive streak in the chick embryo. *Development* **122:** 3263–3273.

Quinlan G.A., Williams E.A., Tan S.S., and Tam P.P.L. 1995. Neuroectodermal fate of epiblast cells in the distal region of the mouse egg cylinder: Implication for body plan organization during early embryogenesis. *Development* **121:** 87–98.

Rands G.F. 1986a. Size regulation in the mouse embryo. I. The development of quadruple aggregates. *J. Embryol. Exp. Morphol.* **94:** 139–148.

———. 1986b. Size regulation in the mouse embryo. II. The development of half embryos. *J. Embryol. Exp. Morphol.* **98:** 209–217.

Rashbass P., Cooke L.A., Herrmann B.G., and Beddington R.S. 1991. A cell autonomous function of Brachyury in T/T embryonic stem cell chimaeras. *Nature* **353:** 348–351.

Rhinn M., Dierich A., Shawlot W., Behringer R.R., Le Meur M., and Ang S.L. 1998. Sequential roles for Otx2 in visceral endoderm and neuroectoderm for forebrain and midbrain induction and specification. *Development* **125:** 845–856.

Rivera-Perez J.A., Mager J., and Magnuson T. 2003. Dynamic morphogenetic events characterize the mouse visceral endoderm. *Dev. Biol.* **261:** 470–487.

Rivera-Perez J.A., Mallo M., Gendron-Maguire M., Gridley T., and Behringer R.R. 1995. Goosecoid is not an essential component of the mouse gastrula organizer but is required for craniofacial and rib development. *Development* **121:** 3005–3012.

Rossant J. and Cross J.C. 2001. Placental development: Lessons from mouse mutants. *Nat. Rev. Genet.* **2:** 538–548.

Rossant J. and Ofer L. 1977. Properties of extra-embryonic ectoderm isolated from postimplantation mouse embryos. *J. Embryol. Exp. Morphol.* **39:** 183–194.

Russ A.P., Wattler S., Colledge W.H., Aparicio S.A., Carlton M.B., Pearce J.J., Barton S.C., Surani M.A., Ryan K., Nehls M.C., Wilson V., and Evans M.J. 2000. Eomesodermin is required for mouse trophoblast development and mesoderm formation. *Nature* **404:** 95–99.

Saude L., Woolley K., Martin P., Driever W., and Stemple D.L. 2000. Axis-inducing activities and cell fates of the zebrafish organizer. *Development* **127:** 3407–3417.

Sausedo R.A. and Schoenwolf G.C. 1994. Quantitative analyses of cell behaviors underlying notochord formation and extension in mouse embryos. *Anat. Rec.* **239:** 103–112.

Schier A.F. and Shen M.M. 2000. Nodal signaling in vertebrate development. *Nature* **403:** 385–389.

Schumacher A., Faust C., and Magnuson T. 1996. Positional cloning of a global regulator of anterior-posterior patterning in mice. *Nature* **383:** 250–253.

Selwood L. 1992. Mechanisms underlying the development of pattern in marsupial embryos. *Curr. Top. Dev. Biol.* **27:** 175–233.

Shawlot W. and Behringer R.R. 1995. Requirement for Lim1 in head-organizer function. *Nature* **374:** 425–430.

Shawlot W., Wakamiya M., Kwan K.M., Kania A., Jessell T.M., and Behringer R.R. 1999. Lim1 is required in both primitive streak-derived tissues and visceral endoderm for head formation in the mouse. *Development* **126:** 4925–4932.

Shih J. and Fraser S.E. 1995. Distribution of tissue progenitors within the shield region of the zebrafish gastrula. *Development* **121:** 2755–2765.

———. 1996. Characterizing the zebrafish organizer: Microsurgical analysis at the early-shield stage. *Development* **122:** 1313–1322.

Shur B.D. 1982. Cell surface glycosyltransferase activities during normal and mutant (T/T) mesenchyme migration. *Dev. Biol.* **91:** 149–162.

Sirard C., de la Pompa J.L., Elia A., Itie A., Mirtsos C., Cheung A., Hahn S., Wakeham A., Schwartz L., Kern S.E., Rossant J., and Mak T.W. 1998. The tumor suppressor gene Smad4/Dpc4 is required for gastrulation and later for anterior development of the mouse embryo. *Genes Dev.* **12:** 107–119.

Skromne I. and Stern C.D. 2001. Interactions between Wnt and Vg1 signaling pathways initiate primitive streak formation in the chick embryo. *Development* **128:** 2915–2927.

Smith E.R., Smedberg J.L., Rula M.E., Hamilton T.C., and Xu X.X. 2001. Disassociation of MAPK activation and c-Fos expression in F9 embryonic carcinoma cells following retinoic acid-induced endoderm differentiation. *J. Biol. Chem.* **276:** 32094–32100.

Snow M.H.L. 1976. Embryo growth during the immediate postimplantation period. *Ciba Found. Symp.* **40:** 66–70.

———. 1977. Gastrulation in the mouse: Growth and regionalisation of the epiblast. *J. Embryol. Exp. Morphol.* **42:** 293–303.

Solursh M. and Revel J.P. 1978. A scanning electron microscope study of cell shape and cell appendages in the primitive streak region of the rat and chick embryo. *Differentiation* **11:** 185–190.

Song J., Oh S.P., Schrewe H., Nomura M., Lei H., Okano M., Gridley T., and Li E. 1999. The type II activin receptors are essential for egg cylinder growth, gastrulation, and rostral head development in mice. *Dev. Biol.* **213:** 157–169.

Spyropoulos D.D. and Capecchi M.R. 1994. Targeted disruption of the even-skipped gene, evx1, causes early postimplantation lethality of the mouse conceptus. *Genes Dev.* **8:** 1949–1961.

Srinivas S., Rodriguez T., Clements M., Smith J.C., and Beddington R.S.P. 2004. Active cell migration drives the unilateral movements of the anterior visceral endoderm. *Development* **131:** 1157–1164.

Stern C.D. 2001. Initial patterning of the central nervous system: How many organizers? *Nat. Rev. Neurosci.* **2:** 92–98.

Suda Y., Nakabayashi J., Matsuo I., and Aizawa S. 1999. Functional equivalency between Otx2 and Otx1 in development of the rostral head. *Development* **126:** 743–757.

Sun X., Meyers E.N., Lewandoski M., and Martin G.R. 1999. Targeted disruption of Fgf8 causes failure of cell migration in the gastrulating mouse embryo. *Genes Dev.* **13:** 1834–1846.

Takaku K., Oshima M., Miyoshi H., Matsui M., Seldin M.F., and

Taketo M.M. 1998. Intestinal tumorigenesis in compound mutant mice of both Dpc4 (Smad4) and Apc genes. *Cell* **92:** 645–656.

Takeichi M. 1988. The cadherins: Cell-cell adhesion molecules controlling animal morphogenesis. *Development* **102:** 639–655.

Tam P.P.L. 1986. A study of the pattern of prospective somites in the presomitic mesoderm of mouse embryos. *J. Embryol. Exp. Morphol.* **92:** 269–285.

———. 1989. Regionalisation of the mouse embryonic ectoderm: Allocation of prospective ectodermal tissues during gastrulation. *Development* **107:** 55–67.

Tam P.P.L. and Beddington R.S. 1987. The formation of mesodermal tissues in the mouse embryo during gastrulation and early organogenesis. *Development* **99:** 109–126.

———. 1992. Establishment and organization of germ layers in the gastrulating mouse embryo. *Ciba Found. Symp.* **165:** 27–41.

Tam P.P.L. and Behringer R.R. 1997. Mouse gastrulation: The formation of a mammalian body plan. *Mech. Dev.* **68:** 3–25.

Tam P.P.L. and Meier S. 1982. The establishment of a somitomeric pattern in the mesoderm of the gastrulating mouse embryo. *Am. J. Anat.* **164:** 209–225.

Tam P.P.L. and Quinlan G.A. 1996. Mapping vertebrate embryos. *Curr. Biol.* **6:** 104–106.

Tam P.P.L. and Steiner K.A. 1999. Anterior patterning by synergistic activity of the early gastrula organizer and the anterior germ layer tissues of the mouse embryo. *Development* **126:** 5171–5179.

Tam P.P.L. and Tan S.-S. 1992. The somitogenetic potential of cells in the primitive streak and the tail bud of the organogenesis-stage mouse embryo. *Development* **115:** 703–715.

Tam P.P.L. and Zhou S.X. 1996. The allocation of epiblast cells to ectodermal and germ-line lineages is influenced by the position of the cells in the gastrulating mouse embryo. *Dev. Biol.* **178:** 124–132.

Tam P.P.L., Williams E.A., and Chan W.Y. 1993. Gastrulation in the mouse embryo: Ultrastructural and molecular aspects of germ layer morphogenesis. *Microsc. Res. Tech.* **26:** 301–328.

Tam P.P.L., Goldman D., Camus A., and Schoenwolf G.C. 2000. Early events of somitogenesis in higher vertebrates: Allocation of precursor cells during gastrulation and the organization of a meristic pattern in the paraxial mesoderm. *Curr. Top. Dev. Biol.* **47:** 1–32.

Tam P.P.L., Parameswaran M., Kinder S.J., and Weinberger R.P. 1997a. The allocation of epiblast cells to the embryonic heart and other mesodermal lineages: The role of ingression and tissue movement during gastrulation. *Development* **124:** 1631–1642.

Tam P.P.L., Steiner K.A., Zhou S.X. and Quinlan G.A. 1997b. Lineage and functional analyses of the mouse organizer. *Cold Spring Harbor Symp. Quant. Biol.* **62:** 135–44.

Tam P.P.L., Gad J.M., Kinder S.J., Tsang T.E., and Behringer R.R. 2001. Morphogenetic tissue movement and the establishment of body plan during development from blastocyst to gastrula in the mouse. *BioEssays* **23:** 508–517.

Tanaka S., Kunath T., Hadjantonakis A.K., Nagy A., and Rossant J. 1998. Promotion of trophoblast stem cell proliferation by FGF4. *Science* **282:** 2072–2075.

ten Donkelaar H.J., Geysberts L.G., and Dederen P.J. 1979. Stages in the prenatal development of the Chinese hamster (*Cricetulus griseus*). *Anat. Embryol.* **156:** 1–28.

Thomas P. and Beddington R. 1996. Anterior primitive endoderm may be responsible for patterning the anterior neural plate in the mouse embryo. *Curr. Biol.* **6:** 1487–1496.

Thomas P.Q., Brown A., and Beddington R.S. 1998. Hex: A homeobox gene revealing peri-implantation asymmetry in the mouse embryo and an early transient marker of endothelial cell precursors. *Development* **125**: 85–94.

Tremblay K.D., Dunn N.R., and Robertson E.J. 2001. Mouse embryos lacking Smad1 signals display defects in extra-embryonic tissues and germ cell formation. *Development* **128**: 3609–3621.

Tremblay K.D., Hoodless P.A., Bikoff E.K., and Robertson E.J. 2000. Formation of the definitive endoderm in mouse is a Smad2-dependent process. *Development* **127**: 3079–3090.

Tsang T.E., Khoo P.L., Jamieson R.V., Zhou S.X., Ang S.L., Behringer R., and Tam P.P.L. 2001. The allocation and differentiation of mouse primordial germ cells. *Int. J. Dev. Biol.* **45**: 549–555.

Tsang T.E., Shawlot W., Kinder S.J., Kobayashi A., Kwan K.M., Schughart K., Kania A., Jessell T.M., Behringer R.R., and Tam P.P.L. 2000. Lim1 activity is required for intermediate mesoderm differentiation in the mouse embryo. *Dev. Biol.* **223**: 77–90.

van den Akker E., Forlani S., Chawengsaksophak K., de Graaff W., Beck F., Meyer B.I., and Deschamps J. 2002. *Cdx1* and *Cdx2* have overlapping functions in anteroposterior patterning and posterior axis elongation. *Development* **129**: 2181–2193.

Varlet I., Collignon J., and Robertson E.J. 1997. Nodal expression in the primitive endoderm is required for specification of the anterior axis during mouse gastrulation. *Development* **124**: 1033–1044.

Viebahn C. 1999. The anterior margin of the mammalian gastrula: Comparative and phylogenetic aspects of its role in axis formation and head induction. *Curr. Top. Dev. Biol.* **46**: 63–103.

Waldrip W.R., Bikoff E.K., Hoodless P.A., Wrana J.L., and Robertson E.J. 1998. Smad2 signaling in extraembryonic tissues determines anterior-posterior polarity of the early mouse embryo. *Cell* **92**: 797–808.

Weber R.J., Pedersen R.A., Wianny F., Evans M.J., and Zernicka-Goetz M. 1999. Polarity of the mouse embryo is anticipated before implantation. *Development* **126**: 5591–5598.

Wells J.M. and Melton D.A. 1999. Vertebrate endoderm development. *Annu. Rev. Cell Dev. Biol.* **15**: 393–410

———. 2000. Early mouse endoderm is patterned by soluble factors from adjacent germ layers. *Development* **127**: 1563–1572.

Welsh I.C. and O'Brien T.P. 2000. Loss of late primitive streak mesoderm and interruption of left-right morphogenesis in the Ednrb(s-1Acrg) mutant mouse. *Dev. Biol.* **225**: 151–168.

Wilson V., Rashbass P., and Beddington R.S. 1993. Chimeric analysis of T (Brachyury) gene function. *Development* **117**: 1321–1331.

Wilson V., Manson L., Skarnes W.C., and Beddington R.S. 1995. The T gene is necessary for normal mesodermal morphogenetic cell movements during gastrulation. *Development* **121**: 877–886.

Wimsatt W.A. 1975. Some comparative aspects of implantation. *Biol. Reprod.* **12**: 1–40.

Winnier G., Blessing M., Labosky P.A., and Hogan B.L. 1995. Bone morphogenetic protein-4 is required for mesoderm formation and patterning in the mouse. *Genes Dev.* **9**: 2105–2116.

Withington S., Beddington R., and Cooke J. 2001. Foregut endoderm is required at head process stages for anteriormost neural patterning in chick. *Development* **128**: 309–320.

Yamaguchi T.P. and Rossant J. 1995. Fibroblast growth factors in mammalian development. *Curr. Opin. Genet. Dev.* **5**: 485–491.

Yamamoto A., Kemp C., Bachiller D., Geissert D., and De Robertis E.M. 2000. Mouse paraxial protocadherin is expressed in trunk mesoderm and is not essential for mouse development. *Genesis* **27**: 49–57.

Yamamoto M., Meno C., Sakai Y., Shiratori H., Mochida K., Ikawa Y., Saijoh Y., and Hamada H. 2001. The transcription factor FoxH1 (FAST) mediates Nodal signaling during anterior-posterior patterning and node formation in the mouse. *Genes Dev.* **15**: 1242–1256.

Yang D.H., Smith E.R., Roland I.H., Sheng Z., He J., Martin W.D., Hamilton T.C., Lambeth J.D., and Xu X.X. 2002. Disabled-2 is essential for endodermal cell positioning and structure formation during mouse embryogenesis. *Dev. Biol.* **251**: 27–44.

Yang X., Li C., Xu X., and Deng C. 1998. The tumor suppressor SMAD4/DPC4 is essential for epiblast proliferation and mesoderm induction in mice. *Proc. Natl. Acad. Sci.* **95**: 3667–3672.

Yang X., Dormann D., Munsterberg A.E., and Weijer C.J. 2002. Cell movement patterns during gastrulation in the chick are controlled by positive and negative chemotaxis mediated by FGF4 and FGF8. *Dev. Cell* **3**: 425–437.

Ying Y. and Zhao G.Q. 2001. Cooperation of endoderm-derived BMP2 and extraembryonic ectoderm-derived BMP4 in primordial germ cell generation in the mouse. *Dev. Biol.* **232**: 484–492.

Ying Y., Liu X.M., Marble A., Lawson K.A., and Zhao G.Q. 2000. Requirement of Bmp8b for the generation of primordial germ cells in the mouse. *Mol. Endocrinol.* **14**: 1053–1063.

Yuan S. and Schoenwolf G.C. 1998. De novo induction of the organizer and formation of the primitive streak in an experimental model of notochord reconstitution in avian embryos. *Development* **125**: 201–213.

Zhang H. and Bradley A. 1996. Mice deficient for BMP2 are nonviable and have defects in amnion/chorion and cardiac development. *Development* **122**: 2977–2986.

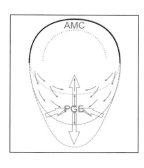

GASTRULATION IN THE RABBIT

C. Viebahn

Zentrum Anatomie, Georg-August-Universität, Kreuzbergring 36, 37075 Göttingen, Germany

INTRODUCTION: GASTRULATING AS A NON-RODENT AND NON-PRIMATE

One of the most common shapes in which mammalian embryos arrive at gastrulation is the "expanded blastocyst." Having entered the typical blastocyst stage, many embryos do not prepare to implant immediately as rodents (cf. Chapter 16) and higher primates (cf. Chapter 18) do. Instead, the blastocyst cavity enlarges considerably prior to implantation through the secretory activity of the uterine glands in concert with transcytotic activity of the trophoblast (Gray et al. 2002) until trophoblast attachment and the developing placenta take over nutrition of the conceptus (at the start of neurulation as in the rabbit, or at advanced somite stages in the case of the cow) (cf. Hue et al. 2001). As a consequence, the expanded blastocyst measures anything from several millimeters to decimeters in diameter (rabbit) or length (sheep, cow, pig), whereas the embryo proper remains comparatively small (~500 μm diameter) in all these species (Evans and Sack 1973). For reasons of mechanical stability (and probably to prevent premature attachment and implantation) the zona pellucida does not disintegrate as in the early-implanting species. Additional "soft" extracellular matrix (Leiser and Denker 1988; Herrler and Beier 2000) is laid down transiently on the outer surface. Blastocyst expansion is typically rapid also due to trophoblast proliferation, while proliferation in the inner cell mass, the area of the embryo proper, cannot "keep up." Considerable attenuation of the inner cell mass (now consisting of hypoblast and epiblast layers) turns the embryo proper into a flat bilaminar disc (also called blastodisc). Exceptions are species with a prolonged arrest of pregastrulation development (diapause), such as the deer or lactating rodents in the wild (Renfree and Shaw 2000).

In all mammals, early trophoblast falls into two categories with fates contrasting between species: In most late-implanting species, and also in rabbit, the *polar* trophoblast, which covers the inner cell mass, degenerates (through apoptosis and phagocytosis by epiblast cells; Williams and Biggers 1990) and the *mural* trophoblast, which forms the wall of the blastocyst cavity, is responsible for attachment and placenta formation. The dorsal surface of the embryonic disc, i.e., the epiblast, is thus exposed to the zona pellucida and, if implantation starts shortly after gastrulation has commenced (rabbit, pig), gastrulating tissues (neural plate ectoderm, epidermal ectoderm) may come into direct contact with maternal tissues. Gastrulation and neurulation signals seem to be "inert" against the direct contact to uterine environment, as the folding amnion creates a separation between the two during the course of neurulation only, in many species. In higher primates and rodents, however, the polar trophoblast does not disintegrate, but, on the contrary, forms the site of attachment and placenta formation (Chapter 18) while the mural trophoblast is not needed for this purpose and degenerates. Depending on the speed of polar trophoblast enlargement relative to the speed of inner cell mass enlargement, the embryo stays relatively flat throughout neurulation (e.g., in primates). Alternatively, as in rodents, the inner cell mass enlarges "prematurely" and bulges into the blastocyst cavity with the shape of Selenka's egg cylinder (Selenka 1884). This "unbalanced" proliferation may also be the reason for the appearance of the so-called "embryocyst" occurring at earlier stages of blastocyst enlargement in some mammals such as *Tupaia* (Kuhn and Schwaier 1973), bat (van Beneden 1911), deer (Keibel 1902), and rhesus monkey (Hill 1932; Heuser and Streeter 1941). However, this premature egg cylinder is flattened out again before the onset of gastrulation. Rodents are thus the only species that have to manage the unusual topographical relationships of the egg cylinder throughout gastrulation (Chapter 16).

The superficial arrangement of the two-layered embryo proper within the wall of the blastocyst cavity (which is not to be confused with the amphibian blastocoel lying between primary germ layers) can be regarded as archetypical for mammals as it can be found in "lower" forms such as scandentia (*Tupaia*: Kuhn and Schwaier 1973) through to ungulates (Evans and Sack 1973) (cf. Assheton 1899, for excellent comparison of many mammals). As a result, gastrulation may be observed directly after the blastocyst is flushed from the uterine cavity, and this may be the reason that rabbit and sheep, as well as bat and shrew, were once the prototype model organisms for mammalian embryology (Kölliker 1879). The discovery of the node, a structure that later turned out to be no less than the mammalian equivalent of the amphibian organizer, was made in the rabbit (Hensen 1876; Viebahn 2001). Since then, the rabbit embryo in particular was "revived" several times for the study of gastrulation and neurulation, partly due to the ease of morphological analysis (van Beneden 1912; Rabl 1915; Aasar 1931; Waterman 1943; Viebahn et al. 1995a,b) and partly due to its amenability to embryological experimentation (Waddington 1936; Daniel and Olson 1966; Knoetgen et al. 1999; Viebahn et al. 2002). Since it does not belong to the rodent class, takes an intermediate position among early- and late-implanting mammals, and gastrulates with a flat embryonic disc as most

amniotes do, the rabbit is a valuable and typical model organism for mammalian gastrulation.

TWO EARLY AXIAL STRUCTURES HELP STAGING MAMMALIAN GASTRULATION

Expanded blastocysts of the rabbit about to engage in gastrulation (at 6 days post conception, dpc) are just starting to bulge the uterine horns as they find their site of attachment at equal intervals between the distal oviduct opening and the insertion of the partes cervicales into the vaginal fornix (Böving 1956). In freshly isolated unfixed blastocysts with intact protein coverings, the embryonic disc stands out, if viewed *en face* under darkfield illumination, in the center of the blastocyst as an area of high cellular density (Fig. 1A).

The edge of the embryonic disc area in 6.0 to 6.5 dpc blastocysts (Fig. 1B) has different characteristics in different quadrants, which become more pronounced with further development. Initially, a stretch with a relatively sharp contour contrasts with another stretch on the opposite side of the disc displaying a more irregular, ragged, and therefore more indistinct contour (Fig. 1B). In later stages that show obvious primitive streak formation (Fig. 1D), the sharp and distinct contour marks the anterior pole of the embryo. This represents a clear morphological sign of the anterior–posterior axis preceding primitive streak formation, and was named the anterior marginal crescent (AMC) (Viebahn et al. 1995a).

At an intermediate stage toward primitive streak formation (at 6.3 dpc, Fig. 1C), the posterior contour becomes more definite and a sickle-shaped area within this more definite posterior contour is enlarged, but retains the reduced density of the preceding stage, while most of the anterior part acquires an increased density. Due to the posterior cellular movement (see below), the enlargement

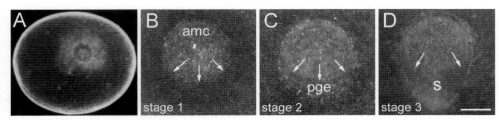

Figure 1. *En face* views of living rabbit blastocysts between 6.0 and 6.5 dpc under darkfield illumination at low (*A*) and higher (*B*) magnification. Embryonic discs at initial gastrulation stages show differentiation of the anterior marginal crescent (amc in *B*), the posterior gastrula extension (pge in *C*), and the primitive streak (s in *D*). Arrows in *B–D* point to the posterior border at stage 1 (*B*) and the remnant of this border at later stages (*C, D*). Bar, 700 μm in *A*, 200 μm in *B–D*. (*A*, Reprinted, with permission, from Viebahn et al. 1995a [©Springer-Verlag]; *B–D* reprinted, with permission, from Viebahn et al. 2002.)

effected by it and due to the involvement of these movements in the forthcoming gastrulation events, this posterior area was coined posterior gastrula extension (PGE) (Viebahn et al. 2002). Its anterior transverse limiting line (arrows in Fig. 1C) is retained even after the primitive streak has formed (arrows in Fig. 1D) and had been used before to delineate the anterior border of the *area triangularis* on both sides of the primitive streak (Rabl 1915).

The unifying aim of the gastrulation process during animal development is considered to be the establishment of the germ layers; however, creating a unifying staging system, if only for vertebrates, is as futile now as it was 50 years ago (Witschi 1956) as long as the main functional cascades of germ layer determination and formation have not been defined. However, in an attempt to be compatible with the late gastrulation events in the chick (the gastrulating amniote system best described so far: Chapter 15), the numbering of primitive streak and node stages as stage 3 and 4, respectively (Hamburger and Hamilton 1993), is adopted here for the rabbit as well (cf. Figs. 1 and 6). Discriminating features between earlier stages are the first appearance of the AMC (stage 1, Fig. 1B) and the PGE (stage 2, Figs. 1C, 2A). These two early axial structures are used as landmarks for staging the beginning of gastrulation as they are found in many mammalian (AMC: cf. Viebahn 1999; PGE: cf. Hill and Florian 1963; Hue et al. 2001) but not avian species. The stage without overt axial differentiation is designated "stage 0" and is expected to be replaced eventually, possibly based on gene expression studies, by a new row of stages, in a manner similar to the stages I to XIII of Eyal-Giladi and Kochav (1976) that replaced stage 1 HH.

ANTERIOR MARGINAL CRESCENT

The two cell layers of the late pre-streak embryonic disc at 6.0 dpc may be distinguished in *en face* views using Nomarski optics even at low magnification (10X lens). The squamous epithelium of the hypoblast on the ventral side (facing the blastocyst cavity) has large polygonal cellular contours and large round nuclei, whereas the high columnar pseudostratified epithelium of the epiblast on the dorsal side has small nuclei, and it is virtually impossible to focus on epiblast cellular contours. Young stages still have remnants of an additional layer, the polar trophoblast consisting of isolated dense cells intercalated between epiblast cells (Williams and Biggers 1990). The morphological characteristics of epiblast vs. hypoblast become apparent most clearly in histological preparations; e.g., after fixation with glutaraldehyde and osmium tetroxide followed by embedding in epoxy resins and sectioning at 1 μm (cf. Fig. 2A–F) (Viebahn et al. 1995a). Apart from obvious cell shape differences, epiblast cells have adherens-type intercellular

junctions (arrowhead in Fig. 2G, Fig. 2I–K) whereas hypoblast cells are connected by desmosome-like differentiations with intracellular connections to ample intermediate filaments (Viebahn et al. 1995b). This provides the hypoblast with the rigidity needed when acting as the reference plane for axial differentiation in the epiblast. This rigidity makes isolation and transplantation of hypoblast segments technically feasible (Knoetgen et al. 1999), whereas similar procedures with the loosely coherent epiblast cells seem, at present, virtually impossible. On the other hand, the loose connection of epiblast cells is a prerequisite for their extensive intercellular movements during gastrulation (see Fig. 4). Careful dehydration procedures preceding embedding of specimens to be analyzed in the electron microscope help to retain the physiologically close apposition between epiblast and hypoblast (Fig. 2B,C). However, artificial separation of the two layers during dehydration and embedding (cf. Fig. 2D, F, H) reveals a thin continuous basement membrane that is closely associated with the basal surface of epiblast cells but not with hypoblast cells.

Biochemically, epiblast and hypoblast differ with regard to the cell-surface molecule E-cadherin, which is expressed in the epiblast only (C. Viebahn, unpubl.) as in the mouse (Vestweber and Kemler 1984; Damjanov et al. 1986). Apart from differential cytokeratin expression (cytokeratins 8 and 18 are expressed much more strongly in hypoblast than in epiblast) (Viebahn et al. 1992), there is a striking mutually exclusive expression of gap junction proteins (Liptau and Viebahn 1999): Connexin 43 (cx43) is specifically expressed in the epiblast, whereas connexin 32 (cx32) is specific for the hypoblast and the yolk sac epithelium adjacent to it. At the border between the two tissues, i.e., straddling the border of the embryonic disc in the epiblast layer, there is an increase of cx32 expression in a belt-like area, 3–4 cells wide, of hitherto unknown significance.

The sharp anterior contour of the 6.0 dpc embryonic disc (Fig. 1B) is the correlate of distinct morphological differences between (extraembryonic) trophoblast and (embryonic) epiblast (Fig. 2B,C). Immediately posterior to the anterior contour, the embryonic disc shows a higher density in both its layers (Fig. 2C). On the opposite side of the disc contrasting characteristics are found: decreased cellular density and no difference in cellular height between epiblast and trophoblast (Fig. 2D). The hallmark of the AMC is thus increased cellular density combined with increased cellular height, both in the epiblast and in the hypoblast.

The posterior limit of the AMC is morphologically inconspicuous and may be found at different transverse levels in the epiblast and the hypoblast. However, a fundamental difference exists functionally between anterior and posterior hypoblast, as transplantation of anterior but not posterior hypoblast thirds from stage-2 rabbit embryos into stage-3 chick embryos initiated the formation of an anteri-

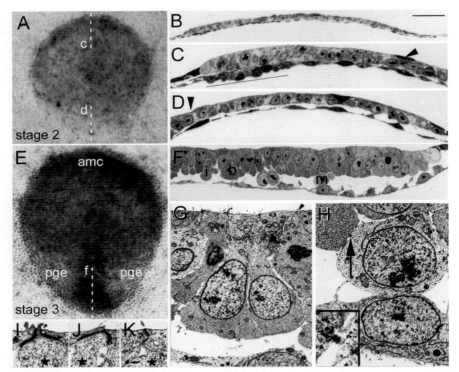

Figure 2. Correlation of light microscope (*B–D*) or ultrastructural (*F–K*) details with the "gross morphology" (*A, E*) in aldehyde- and osmium-fixed embryonic discs embedded in epoxy resin immediately prior to (6.25 dpc, *A–D*) and after (6.5 dpc, *E–K*) the onset of mesoderm formation in dorsal views (*A, E*) and sagittal sections (*B–D, F–K*); the position of the sagittal sections is indicated by vertical hatched lines in *A* and *E*. *C* and *D* show the anterior and the posterior margin, respectively, of the embryo in *A* at high light microscopical magnification; the line in *C* indicates the width of the AMC, the arrowhead points to a cell of Rauber's layer (polar trophoblast); the overview in *B* is taken from a different embryo of the same stage. The two prospective mesoderm cells ("bottle cells") about to ingress and to leave the epiblast compartment shown at the electron microscopic level in *G* are marked (b) in the semithin section shown in *F* (i.e., the thin section shown in *G* was prepared from a semithin section next to that shown in *F*); the definite mesoderm cell shown in *H* is marked (m) in *F*. The arrow in *H* points to adherens-type junction (inset in *H*) fixing mesoderm cells to neighboring epiblast cells while complex apical cell contacts (*arrowhead* in *G*) are being rearranged for ingression to become complete (*I–K*). Asterisks in *I–K* point to the same cell in neighboring thin sections. Ingressing mesoderm cell (i), anterior marginal crescent (amc), posterior gastrula extension (pge). Bar, 100 μm in *A* and *E*, 50 μm in *B*, 25 μm in *C, D,* and *F*, 6 μm in *G* and *H*, 1.5 μm in *I–K*, and inset in *H*. (*A, C, D,* and *E* reprinted, with permission, from Viebahn et al. 1995a [©Springer-Verlag]; *F–K* reprinted, with permission, from Viebahn et al. 1995b [©Karger].)

or neural plate structure and appropriate gene expression (*Sox3* and *ANF*) even in extraembryonic tissues (Fig. 3A–C) (Knoetgen et al. 1999). Obviously, full head formation needs further tissue and molecular interaction (cf. Stern 2001), but this experiment proved that the cytological differentiation in AMC hypoblast is a sign of a specific functional capacity. For the anterior visceral endoderm (AVE)—the mouse equivalent of the hypoblast part of the AMC—a similar role in head and neural plate induction was also shown experimentally, but only by loss-of-function experiments, i.e., physical (Thomas and Beddington 1996) or genetic ablation (Rhinn et al. 1998; Martinez-Barbera and Beddington 2001; Perea-Gomez et al. 2002). Here, too, the size of the tissue endowed with this property is not entirely clear, but may eventually be defined by the expression domains of secreted inhibitory molecules (see below). Remarkably, neural induction prior to mesoderm forma-

tion has been shown in the chick (Streit et al. 2000), and the appearance of the AMC as a head-inducing tissue in pre-streak stages in the mammalian embryo fits this unexpected sequence of events at the start of gastrulation.

In the search for factors responsible for the head-inducing capacity in the anterior hypoblast, a number of different gene classes have been analyzed with regard to their in situ expression at the mRNA or protein level. The list of genes and proteins expressed anteriorly in the lower layer in the mouse is impressive (cf. Beddington and Robertson 1999; Perea-Gomez et al. 2001a). However, the size and shape of the hypoblast area with a given expression activity is still less well defined in the mouse compared to the delineation of the AMC in the rabbit. Expression studies in the rabbit begin to bring some clarity into this field. The mRNA expression of the secreted inhibitory molecules Dkk1 (Glinka et al. 1998) and cerberus-like (Cer-l) (Piccolo

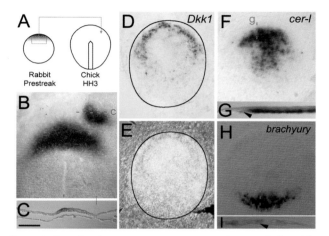

Figure 3. Functional compartments at pre-streak stages in hypoblast (*A, D–G*) and epiblast (*H, I*): Anterior subdomains in the hypoblast have signs of head-inducing capacity as shown by heterospecific transplantation (*A–C*) or with expression of the secreted signaling molecules Dkk1 (*D, E*) or Cer-l (*F*): Anterior hypoblast from a rabbit stage-2 embryo transplanted ectopically into a stage-3 chick embryo (*A*) elicits expression of the neural plate marker *ANF* (*B, C*). *C* shows section of the ectopic expression domains in *B*. Comparison of bright (*D*) and dark (*E*) field views of the same rabbit embryonic disc shows that the arch-like expression domain of *Dkk1* mRNA in the hypoblast of the AMC does not reach the margin (*black line*) of the disc. *Cer-l* mRNA (*F*) is expressed in hypoblast cells (*G*) overlapping with, and central to, the *Dkk1* domain. In contrast and conceptually complementing the *Dkk1* and *Cer-l* hypoblast domains, a crescent-shaped salt-and-pepper pattern of *Brachyury* expression in the epiblast characterizes the PGE (*H, I*). Dashed red lines in *B, F,* and *H* indicate positions of sections shown in *C, G,* and *I*, respectively. Bar, 230 μm in *B,E*; 140 μm in *C, F*; 100 μm in *G–J*. (*A–C* reprinted, with permission, from Knoetgen et al. 1999.)

et al. 1999) are examples of how specific expression patterns help to define critical functional compartments in the hypoblast (Fig. 3D–F). *Dkk1* mRNA is expressed in an arch-like area within the anterior circumference of the disc sparing the outermost row of 3–4 cells (Fig. 3D,E). *Dkk1*-expressing cells lie in the anterior third of the hypoblast that was used for the cross-species transplants demonstrating head-inducing capacity (Fig. 3A) (Knoetgen et al. 1999). The posterior border of this *Dkk1*-expressing arch may be taken as the ill-defined posterior limit of the AMC. However, there is a mushroom-shaped area, also sparing the outermost row of hypoblast cells, expressing *Cer-l* (arrowhead in Fig. 3G), but *Cer-l*-expressing cells extend farther caudally than the *Dkk1* expression domain (cf. *Cer-l* expression in the mouse) (Stanley et al. 2000). Because both molecules are inhibiting factors in the Wnt signaling cascade, a likely proposition is that these two areas cooperate and complement each other in axial differentiation, e.g., neural induction (Knoetgen et al. 1999). Moreover, recent experimental data suggest that Cer-l acts to inhibit primi-

tive streak formation by repressing the action of nodal in mouse (cf. Perea-Gomez et al. 2002) and chick (Bertocchini and Stern 2002). The definition of the posterior AMC limit is thus a complex task. However, these expression patterns suggest that neural induction or priming prior to mesoderm formation, shown so far only for the chick and FGF8 signaling (Streit et al. 2000), may also occur in the mammalian embryo and may involve, in addition, the Wnt and nodal signaling pathways.

An inhibitory role of the hypoblast, which may be executed through inhibitors other than Dkk1 and Cer-l, has been suggested by elegant studies involving knockout mice for the transcription factor *Otx2* (Kimura et al. 2000; Perea-Gomez et al. 2001b) and double knockout experiments involving *Otx2* and the EGF-signaling molecule Cripto (Kimura et al. 2001). In these mutants, inhibitory molecules are not expressed anteriorly. Instead, mesoderm differentiation, which is a posterior characteristic of the early embryo, is widespread throughout the embryo. One could argue that this is much the same thing because head (anterior) induction is frequently seen to be the result of suppressing the default pathway leading to trunk (posterior) differentiation (Glinka et al. 1998; Piccolo et al. 1999; Pera et al. 2001). However, neural differentiation was also shown to be the result of active induction (Knoetgen et al. 1999; Streit et al. 2000). Therefore, anterior hypoblast may have dual and sequential functions (maybe mirrored by the shape of expression domains) initiating neural differentiation prior to mesoderm formation and, later, protecting epiblast from posteriorizing signals.

POSTERIOR GASTRULA EXTENSION

Only hours prior to overt primitive streak formation, a mosaic appears in the posterior quadrant of the embryonic disc and covers a sickle-shaped area coincident with the posterior gastrula extension (PGE) and seemingly complementing the Dkk1 and Cer-l hypoblast domains (Fig. 3H). About 30 epiblast cells express the transcription factor *Brachyury*, which is generally considered the "master switch" for mesoderm formation (Willison 1990). These cells stand out singly or in small groups among epiblast cells which do not show, or show only weak, expression of the gene. *Brachyury* expression thus marks the presumed area of primitive streak formation (Viebahn et al. 2002) and together with other functional characteristics (see below) contributed to the qualification of this area as PGE. The *Brachyury* mosaic is reminiscent of the expression pattern and mesodermal fate of HNK1-expressing chick epiblast cells (Stern and Canning 1990), although it is clear now that not all HNK1-positive cells, which are distributed over most of the blastodisc (Canning and Stern 1988), have a mesodermal fate (Chapter 15).

The PGE shows up morphologically in sections, where one sees that the epiblast is higher in the center part of the disc than in the posterior quadrant. The position of the "step" between these two zones (arrowhead in Fig. 2D) coincides with the border between the denser center of the disc and the lesser density near the posterior border seen in darkfield views of living embryos (arrows in Fig. 1C). Difference in cellular height and density in the hypoblast are not of the same quality as in the epiblast. If anything, hypoblast cells are more numerous posteriorly than they are in the center of the disc (Fig. 2D). Most remarkable in the morphology of the PGE is the relative scarcity of cells just prior to mesoderm formation, which starts in that very area and brings with it absolutely contrasting cytological characteristics—high density of (mesoderm-forming) epiblast cells with a high columnar shape (Fig. 2F).

This sudden change in histological characteristics seems all the more surprising at first sight, as the PGE shows reduced proliferative activity prior to primitive streak formation (Fig. 4B,C) (Viebahn et al. 2002). Single short-pulse (10-minute) BrdU incorporation during suspension culture of whole blastocysts leads to labeling of about 80% of epiblast cells (which is equivalent to a cell cycle time of about 3 hours) (cf. Mac Auley et al. 1993) in the AMC, in a belt-like area within the anterior dense part of the stage-2 disc and in the center of the disc (Fig. 4C). In the PGE, "only" 60% of epiblast cells were labeled. Interestingly, this differential proliferation pattern reveals, again, subcompartments in the embryonic disc hitherto unknown, but, more importantly, it develops within hours from the rather uniform distribution found at stage 1 (Fig. 4A). This invites the (testable) speculation that the start of *Brachyury* mRNA

transcription in the PGE at stage 2 is the result of a slight and transient reduction of proliferative activity similar to the mechanism described for mesoderm formation in *Drosophila* (cf. Seher and Leptin 2000).

Local proliferation in the presumed area of primitive streak formation is apparently not an option for accumulating the high cell numbers needed for mesoderm ingression. Instead, epiblast cell migration from anterior regions of the embryonic disc has been shown to accomplish this task. Using the AMC as landmark under darkfield illumination, prospective mesoderm cells in the PGE were labeled with the live cell marker DiI at the beginning of a 12-hour suspension culture in stage-2 embryos (Fig. 5) (Viebahn et al. 2002). Integration of migratory paths deduced from a total of 59 DiI injections shows classical convergent extension movements (Keller 1992) that result in an elongation and narrowing of the posterior quadrant of the stage-2 embryonic disc and thus in the transformation of the round disc shape into the elongated pear shape at stage 3 (Fig. 5F). This convergent extension is likely also a result of pre-gastrulation cell movements shown in the mouse to be under the control of the *Otx2* gene (Zakin et al. 2000; Perea-Gomez et al. 2001b).

PRIMITIVE STREAK

The posterior quadrant of the embryonic disc changes in a matter of a few hours from the attenuated PGE to a distinct localized cellular density along the posterior stretch of the midline (Fig. 2E). The epiblast changes from a flat or cuboidal epithelium into a high columnar pseudostratified one with single cells deformed to the bottle shape (Fig. 2G), which is typical for ingressing mesoderm cells in all animals from sea urchin (Nakajima and Burke 1996) to *Xenopus* (Hardin and Keller 1988) and mouse (Poelmann 1981). Electron microscopy of nascent mesoderm cells reveals an intracellular polarized organelle distribution (Fig. 2G), as all mitochondria accumulate near the apex of the bottle cells until definitive ingression—and thus epithelial–mesenchymal transition—is complete (Fig. 2H). In addition, the basement membrane is discontinuous in the area of the primitive streak (Viebahn et al. 1995b), and this seems to be a prerequisite for transformation of epiblast cells into mesoderm as in mouse (Poelmann 1981) and chick (Harrisson et al. 1991). Concomitantly with developing basal bulging, the apex of ingressing cells shows complex transformation of cellular adherens-type contacts, thus allowing neighboring cells to come into contact and seal the gap left behind by the ingressing mesoderm cell (Fig. 2I–K). Isolated focal contacts seem to hold these early mesoderm cells in place for a time (Fig. 2H, inset). A small peripheral region immediately posterior to the primitive streak never contains bottle-shaped ingressing cells. Here, the basement

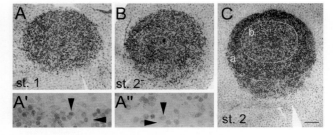

Figure 4. Concentric epiblast compartments developing at stage 2 as defined by differential proliferative activity: BrdU incorporation in vitro at stages 1 (*A*), 2⁻ (*B*), and 2 (*C*) labels between 80% and 60% of nuclei near the anterior (*A′*) and the posterior (*A′′*) margins of the embryonic disc shown in *A* (*vertical arrowheads*: labeled nuclei, *horizontal arrowheads*: unlabeled nuclei). An almost uniform proliferation pattern at stage 1 transforms into a differential distribution of graded proliferation at stage 2, which consists of a ring-like structure of strong proliferation (between lines *a* and *b* in *C*) surrounding a central area of weaker proliferation (inside line *b* in *C*) and delineating the crescent-shaped area of the PGE (posterior to line *a*) with still weaker proliferation. Bar, 60 μm in *A–C*; 12 μm in *A′*, *A′′*. (Reprinted, with permission, from Viebahn et al. 2002.)

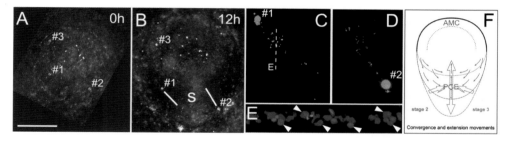

Figure 5. Migrating epiblast cells in the posterior half of late pre-streak embryonic discs as demonstrated by labeling intact blastocysts with deposits of DiI suspended in corn oil (#1, #2, and #3) and culturing for 12 hours. The same embryonic disc before (*A*) and after (*B*) suspension culture, under darkfield optics (*A* and *B*) and using fluorescence microscopy (*C–E*). Hatched line in *C* indicates orientation and length of frozen section segment shown in *E*. (*F*) Composite drawing indicating the convergence and extension movements (*blue arrows*) at stage 3 resulting from cell movements in the epiblast (*red arrows*) at stage 2. Bar, 300 μm in *A* and *B*, 75 μm in *C* and *D*, 20 μm in *E*. (*A–E* reprinted, with permission, from Viebahn et al. 2002.)

membrane remains intact (Viebahn et al. 1995b), and hypoblast cells are not in direct contact with epiblast cells; contrary to common suggestions for mammalian embryos (Florian 1933; Hendrickx 1971), signs of a cloacal membrane are not found at these early stages. Lateral to the emerging density of the primitive streak, the posterior quadrant of the embryonic disc (the *area triangularis* of Rabl 1915) remains as flat as the PGE (Fig. 2E) but shows strong *Brachyury* expression (Fig. 6B) indicating that in this area epiblast cells are primed and recruited for mesoderm formation in the primitive streak.

Preceding the morphological changes typical for mesoderm ingression, epiblast cells within the streak lose E-cadherin expression (C. Viebahn, unpubl.) as in the mouse (Vestweber and Kemler 1984; Damjanov et al. 1986). In contrast, the biochemical constitution of the cytoskeleton does not change immediately as was initially deduced from data obtained in the mouse (Franke et al. 1982). Ingressing mesoderm cells retain the cytokeratin expression until they have migrated near the lateral edges of the embryonic disc (Viebahn et al. 1992) and only then switch from cytokeratin to vimentin expression (Fig. 6C,D). Epiblast cells seem to lose cx43 expression specifically, first in the emerging node and later, to a lesser extent, in the primitive streak (J. Idkowiak, pers. comm.).

Rather than "growing" entirely at its anterior extremity, as described in most textbooks, the primitive streak also elongates by appositional growth at the posterior end, as suggested by the convergent extension movements (Fig. 5). Anterior and posterior modes of streak elongation are matched by differential gene expression in anterior and posterior halves of the streak. In particular, *Brachyury* expression is confined transiently (during stages 3+ and 4-) to the posterior half and to the tip of the streak (Fig. 6B), creating a gap of *Brachyury* expression in the anterior half of the streak, a situation which is similarly found in the chick (Knezevic et al. 1997). This differential *Brachyury* expression

is remarkable for two reasons. First, epithelial–mesenchymal transition in the anterior part of the primitive streak occurs morphologically in the same way as posteriorly (Viebahn 1995), but anteriorly this is apparently independent of the *Brachyury* expression program. Second, morphological formation of Hensen's node, i.e., the cellular density of the node, is preceded and perhaps heralded molecularly by "precocious" *Brachyury* expression similar to the mosaic of *Brachyury* expression in mesoderm precursors among PGE epiblast. The differential *Brachyury* expression also supports the functional separation of the streak into at least two compartments (anterior and posterior half) which have different fates (mouse: Lawson et al. 1991; Kinder et al. 1999), (chick: Schoenwolf et al. 1992, cf. Chapter 15) and function (Forlani et al. 2003).

In parallel with primitive streak elongation, mesoderm cells spread between epiblast and hypoblast in a concentric manner, with the posterior extremity of the streak representing the center of the mesoderm circle (Fig. 6B). Subsequently, mesoderm cells within the embryonic area proper move anteriorly in a broad front (Fig. 6A) to reach the anterior margin together with the prechordal mesoderm.

HENSEN'S NODE

Primitive streak morphology changes by elongation and differentiation. Its anterior and posterior extremities remain dense, forming Hensen's node and the "end node," respectively, at about 6.75 dpc. Presence of Hensen's node is assumed (and taken as the beginning of stage 4) when mesoderm cells appear anterior to the streak. These mesoderm cells are the prechordal mesoderm soon followed by the "chordal mesoderm" proper, i.e., the notochordal process. In contrast to the chick (Levin et al. 1995), rabbit node structure and gene expression in and around the node are symmetrical at all stages (Fischer et al. 2002) of gastru-

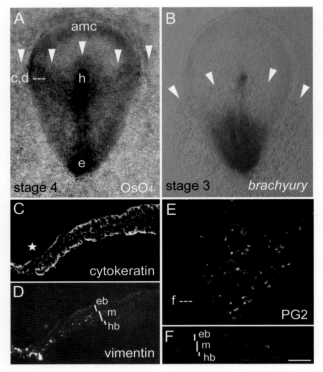

Figure 6. Rabbit embryos between 6.5 and 6.75 dpc with a full-length primitive streak in dorsal views (*A, B, E*) and transverse sections (*C, D, F*) after aldehyde and osmium tetroxide fixation and resin embedding (*A*), in situ hybridization for *Brachyury* expression (*B*), or immunofluorescent staining on a whole mount (*E*) or on transverse frozen sections (*C, D, F*). Mesodermal cells in the embryo shown in *A*, but not in that shown in *B*, have advanced (*arrowheads*) beyond the anterior tip of the primitive streak, thus defining the presence of Hensen's node in *A. Brachyury* expression in the anterior tip of the streak (*B*) indicates the molecular differentiation of node precursor cells. Nascent mesoderm cells continue to express cytokeratin (*C*) and acquire vimentin expression (*D*, double immunofluorescence on same frozen section as shown in *C*) following migration toward the lateral margins (*asterisks*) of the embryonic disc only. Primordial germ cells (labeled with the germ cell marker PG2) emerge in the area of the end node and in the posterior half of the primitive streak at stage 4 (*E*); labeled PGCs are found near the epiblast (eb), within the mesodermal (m) compartment but not yet near the lower layer (hypoblast, hb). Anterior marginal crescent (amc), Hensen's node (h), end node (e); hatched lines in *A* and *E* indicate the plane of the frozen sections shown in *C, D*, and *F*, respectively. Bar, 125 μm in *A* and *B*, 40 μm in *C* and *D*, 20 μm in *E*, 16 μm in *F*. (*C* and *D* reprinted, with permission, from Viebahn et al. 1992 [©Springer-Verlag]; *E* and *F* reprinted, with permission, from Schäfer-Haas and Viebahn 2000.)

lation (and neurulation), suggesting that different programs for molecular left–right differentiation are involved in birds and mammals.

The existence of a specific thickening at the posterior extremity of the primitive streak had been noted before in several mammalian species (rabbit: Rabl 1915; Viebahn et al. 1992) (dog: Bonnet 1918), including the human embryo (Stieve 1926; Florian 1933) and was termed sickle node

(Rabl 1915; Stieve 1926), end node (Florian 1933) or—in view of the analogy of the primitive streak with the amphibian blastoporus and its dorsal and ventral blastopore lips (van Beneden 1880; Kollmann 1886; see also De Robertis et al. 1994)—posterior lip ("Hinterlippe": Bonnet 1918). With the description of the morphology and functional characteristics of PGE at the pre-primitive streak stage (Viebahn et al. 2002), it is now clear that Koller's sickle of the avian embryo, which led to the term "sickle node," is not present in the mammalian embryo, nor is there any known homologous mammalian structure. The term "end node" seems more appropriate (Viebahn 1999), as the late primitive streak is generally accepted to be the origin of the tail bud (cf. Griffith et al. 1992). Even though structure and function of the end node are as yet ill defined, allocating a special term to the underlying structure may be warranted also in view of the origin of primordial germ cells from this region (Fig. 6E,F).

During differentiation within the primitive streak, there is further morphological differentiation in the rest of the embryonic disc. A semi-circular area anterior to the level of Hensen's node now shows reduced density while the margin containing the AMC remains dense. On both sides of the streak, the PGE (now *area triangularis*, see above) appears less translucent due to the spread of mesoderm cells. Originating at all levels of the primitive streak, these mesoderm cells spread centripetally with the end node providing the geometric center. Anteriorly, there is an almost transverse "front line" at the level of the node, which indicates the level to which the mesoderm cells have spread (Fig. 6A) (Viebahn et al. 1992). In the midline, anterior to the node, this front line bulges anteriorly and contains mesoderm cells that can be addressed as prechordal mesoderm (Adelmann 1926; Sulik et al. 1994). Anterior to this front line, the embryo is still devoid of mesoderm, which leaves molecular or morphological interactions between epiblast and hypoblast "undisturbed" in this area.

As initiation of mesoderm formation in the posterior and anterior half of the streak is apparently under the control of the different molecular master switches (indicated by differential *Brachyury* expression; Fig. 6B), different fates in the primitive streak are apparently also mirrored by different molecular programs accomplishing mesoderm formation and subsequent migration. One cell type emerging specifically from the posterior half of the primitive streak is concerned with the germ line. Primordial germ cells arise here under the control of BMP4 secreted by the extraembryonic mesoderm (Lawson et al. 1999; Fujiwara et al. 2001). A monoclonal antibody raised against primordial germ cells in the rabbit (Viebahn et al. 1998) helped to identify this cell line within the mesoderm of the posterior primitive streak half (Fig. 6E,F) (Schäfer-Haas and Viebahn 2000) comparatively earlier than the expression of tissue nonspecific alkaline phosphatase (Ginsburg et al. 1990) or differential *Oct4* expression (Nichols et al. 1998) does in the mouse. In contrast to the mouse, *BMP4* is specifically expressed in the posterior half of the streak and not in strictly extraembryonic tissues underlying the trophoblast (C. Viebahn and A. Fischer, unpubl.). This suggests that molecular control of primordial germ cell formation may be regulated differently in the rabbit, or that the posterior half of the streak (and possibly the PGE, also) should be considered extraembryonic structures at this stage.

CONCLUSIONS

Germ Layer Formation

Among the definite germ layers, the mesoderm is the first to be morphologically distinguishable. This occurs both on the "macroscopic" level through the appearance of the primitive streak (Figs. 1D, 2E) and on the "microscopic" level through the fundamental change in the cellular characteristics from the polarized epithelial type to a seemingly unpolarized mesenchymal type (Fig. 2H). In contrast, establishment of the endoderm and ectoderm is less conspicuous. The former is inserted (at the level of the node) into an existing layer, the hypoblast, and shares with this layer most morphological characteristics (cf. Sulik et al. 1994; Lawson and Schoenwolf 2003). The latter seems to consist simply of the cells left behind in the epiblast plane during and after the ingression of mesoderm and endoderm cells. However, differentiation of epiblast cells toward an ectodermal fate may occur earlier than the decision toward the mesodermal fate, as was indeed shown in the chick (Streit et al. 2000), even if the "precocious" *Brachyury* expression in epiblast cells (Fig. 3H) is taken into account. In addition, the formation of the AMC (and the AVE in mouse) shows that axis formation—another hallmark of gastrulation—also starts well ahead of mesoderm appearance.

Functional Compartments during Gastrulation

Prior to primitive streak formation, a subdivision of the embryonic disc into novel, morphologically inconspicuous compartments can be drawn up on the basis of in situ analysis of protein and gene expression in mouse (cf. Tsang et al. 1999) and chick (Chapman et al. 2002). Mostly, these subcompartments are confined to a single layer of the two-layered embryonic disc. In a few cases, however, such as with *nodal* (Zhou et al. 1993; Conlon et al. 1994; Varlet et al. 1997) or *Otx2* (Simeone 1998), these novel "cross-layer" compartments do not fit the classical view of the epiblast and hypoblast as morphologically separated and functionally different entities, the hypoblast being the inducer and the epiblast the induced tissue in the context of primitive streak and axis formation in the chick (Waddington 1933; Azar and Eyal-Giladi 1981; Foley et al. 2000). The expression of proteins responsible for direct intercellular communication such as connexins does indeed show these layers to be apparently "uncoupled" (Liptau and Viebahn 1999), and this may be a prerequisite for reciprocal inducers. "Cross-layer" expression may be unexpected, but it mirrors the need for molecular networks to be present "on both sides of the fence." In the majority of expression patterns, however, new subdivisions become apparent within these layers, and the rabbit, with its flat embryonic disc and its larger size, offers the chance to distinguish finer details of expression, and hence subcompartments, within the hypoblast or the epiblast.

The Rabbit as a Model Organism for Mammalian Gastrulation?

In the face of the plethora of molecular and experimental data obtained recently in the mouse, it may seem unnecessary to reestablish an old model organism. However, the enlarging number of cascades of molecules interacting extra- and intercellularly put morphology and topographical relationships between signaling centers back into the focus of attention. Demonstrating gene expression in specific areas of the flat mammalian embryonic disc at high resolution may answer the question whether scenarios devised for the unique anatomy of the mouse egg cylinder would still be plausible in other species. The "simple" anatomy of the rabbit leaves little doubt on the number and distribution of cells expressing specific molecules or showing differential proliferation and directed cellular movement. This helps in most cases to understand the complex interactions needed for the orchestration of gastrulation as one of the critical phases in embryonic life. Looking at mammals with a flat embryonic disc will also enable direct interspecific comparison between mammals and birds.

ACKNOWLEDGMENTS

Special thanks go to Gunnar Weisheit and Jan Idkowiak for providing unpublished gene expression data (Fig. 3D–F) and to Hans-Georg Sydow for excellent help with graphic work. This work was supported by grants of the Deutsche Forschungsgemeinschaft (Vi 151/3-4, Vi 151/6-3) and the Deutsche Akademische Austauschdienst (Procope program D/9910412).

REFERENCES

Aasar Y.H. 1931. The history of the protochordal plate in the rabbit. *J. Anat.* **66:** 14–45.

Adelmann H.B. 1926. The development of the premandibular head cavities and the relations of the anterior end to the notochord in the chick and robin. *J. Morphol. Physiol.* **42:** 371–439.

Assheton R. 1899. The segmentation on the ovum of the sheep, with observations on the hypotheses of a hypoblastic origin for the trophoblast. *Q. J. Microsc. Sci.* **41:** 205–262.

Azar Y. and Eyal-Giladi H. 1981. Interaction of epiblast and hypoblast in the formation of the primitive streak and the embryonic axis in chick, as revealed by hypoblast rotation experiments. *J. Embryol. Exp. Morphol.* **61:** 133–141.

Beddington R.S. and Robertson E.J. 1999. Axis development and early asymmetry in mammals. *Cell* **96:** 195–209.

Bertocchini F. and Stern C.D. 2002. The hypoblast of the chick embryo positions the primitive streak by antagonizing nodal signaling. *Dev. Cell* **3:** 735–744.

Bonnet R. 1918. *Lehrbuch der Entwicklungsgeschichte,* 3rd edition. Paul Parey, Berlin.

Böving B.G. 1956. Rabbit blastocyst distribution. *Am. J. Anat.* **98:** 403–434.

Canning D.R. and Stern C.D. 1988. Changes in the expression of the carbohydrate epitope HNK-1 associated with mesoderm induction in the chick embryo. *Development* **104:** 643–655.

Chapman S.C., Schubert F.R., Schoenwolf G.C., and Lumsden A. 2002. Analysis of spatial and temporal gene expression patterns in blastula and gastrula stage chick embryos. *Dev. Biol.* **245:** 187–199.

Conlon F.L., Lyons K.M., Takaesu N., Barth K.S., Kispert A., Herrmann B., and Robertson E.J. 1994. A primary requirement for nodal in the formation and maintenance of the primitive streak in the mouse. *Development* **120:** 1919–1928.

Damjanov I., Damjanov A., and Damsky C.H. 1986. Developmentally regulated expression of the cell-cell adhesion glycoprotein Cell-CAM 120/80 in peri-implantation mouse embryos and extraembryonic membranes. *Dev. Biol.* **116:** 194–202.

Daniel J.C. and Olson J.D. 1966. Cell movement, proliferation and death in the formation of the embryonic axis of the rabbit. *Anat. Rec.* **156:** 123–128.

De Robertis E.M., Fainsod A., Gont L.K., and Steinbeisser H. 1994. The evolution of vertebrate gastrulation. *Dev. Suppl.* **1994:** 117–124.

Evans H.E. and Sack W.O. 1973. Prenatal development of domestic and laboratory mammals: Growth curves, external features and selected references. *Anat. Histol. Embryol.* **2:** 11–45.

Eyal-Giladi H. and Kochav S. 1976. From cleavage to primitive streak formation: A complementary normal table and a new look at the first stages of the development of the chick. *Dev. Biol.* **49:** 321–337.

Fischer A., Viebahn C., and Blum M. 2002. FGF8 acts as a right determinant during establishment of the left-right axis in the rabbit. *Curr. Biol.* **12:** 1807–1816.

Florian J. 1933. The early development of man, with special reference to the development of the mesoderm and the cloacal membrane. *J. Anat.* **67:** 263–276.

Foley A.C., Skromne I., and Stern C.D. 2000. Reconciling different models of forebrain induction and patterning: A dual role for the hypoblast. *Development* **127:** 3839–3854.

Forlani S., Lawson K.A., and Deschamps J. 2003. Acquisition of Hox codes during gastrulation and axial elongation in the mouse embryo. *Development* **130:** 3807–3819.

Franke W.W., Grund C., Kuhn C., Jackson B.W., and Illmensee K. 1982. Formation of cytoskeletal elements during mouse embryogenesis. III. Primary mesenchymal cells and the first appearance of vimentin filaments. *Differentiation* **23:** 43–59.

Fujiwara T., Dunn N.R., and Hogan B.L. 2001. Bone morphogenetic protein 4 in the extraembryonic mesoderm is required for allantois development and the localization and survival of primordial

germ cells in the mouse. *Proc. Natl. Acad. Sci.* **98:** 13739–13744.

Ginsburg M., Snow M.H.L., and McLaren A. 1990. Primordial germ cells in the mouse embryo during gastrulation. *Development* **110:** 521–528.

Glinka A., Wu W., Delius H., Monaghan A.P., Blumenstock C., and Niehrs C. 1998. Dickkopf-1 is a member of a new family of secreted proteins and functions in head induction. *Nature* **391:** 357–362.

Gray C.A., Burghardt R.C., Johnson G.A., Bazer F.W., and Spencer T.E. 2002. Evidence that absence of endometrial gland secretions in uterine gland knockout ewes compromises conceptus survival and elongation. *Reproduction* **124:** 289–300.

Griffith C.M., Wiley M.J., and Sanders E.J. 1992. The vertebrate tail bud: Three germ layers from one tissue. *Anat. Embryol.* **185:** 101–113.

Hamburger V. and Hamilton H.L. 1993. A series of normal stages in the development of the chick embryo. *Dev. Dyn.* **195:** 231–272.

Hardin J. and Keller R. 1988. The behaviour and function of bottle cells during gastrulation of *Xenopus laevis. Development* **103:** 211–230.

Harrisson F., Callebaut M., and Vakaet L. 1991. Features of polyingression and primitive streak ingression through the basal lamina in the chicken blastoderm. *Anat. Rec.* **229:** 369–383.

Hendrickx A.G. 1971. *Embryology of the baboon.* University of Chicago Press, Chicago, Illinois.

Hensen V. 1876. Beobachtungen über die Befruchtung und Entwicklung des Kaninchens und Meerschweinchens. *Z. Anat. Entwickl. Gesch.* **1:** 213–273, 353–423.

Herrler A. and Beier H.M. 2000. Early embryonic coats: Morphology, function, practical applications. An overview. *Cells Tissues Organs* **166:** 233–246.

Heuser C.H. and Streeter G.L. 1941. Development of the macaque embryo. *Contrib. Embryol. Carnegie Instn.* **29:** 15–55.

Hill J. 1932. The developmental history of the primates. *Philos. Trans. R. Soc. Lond. B* **221:** 45–165.

Hill J.P. and Florian J. 1963. The development of the primitive streak, head-process and annular zone in *Tarsius,* with comparative notes on *Loris.* In *Bibliotheca primatologica* (ed. H. Hofer et al.), vol. 2, p. 1–90. Karger, Basel, Switzerland.

Hue I., Renard J.P., and Viebahn C. 2001. Brachyury is expressed in gastrulating bovine embryos well ahead of implantation. *Dev. Genes Evol.* **211:** 157–159.

Keibel F. 1902. Die Entwickelung des Rehes bis zur Anlage des Mesoblast. *Arch. Anat. Physiol. Anat. Abt.* **1902:** 292–314.

Keller R., Shih J., and Sater A. 1992. The cellular basis of the convergence and extension of the *Xenopus* neural plate. *Dev. Dyn.* **193:** 199–217.

Kimura C., Shen M.M., Takeda N., Aizawa S., and Matsuo I. 2001. Complementary functions of Otx2 and Cripto in initial patterning of mouse epiblast. *Dev. Biol.* **235:** 12–32.

Kimura C., Yoshinaga K., Tian E., Suzuki M., Aizawa S., and Matsuo I. 2000. Visceral endoderm mediates forebrain development by suppressing posteriorizing signals. *Dev. Biol.* **225:** 304–321.

Kinder S.J., Tsang T.E., Quinlan G.A., Hadjantonakis A.K., Nagy A., and Tam P.P. 1999. The orderly allocation of mesodermal cells to the extraembryonic structures and the anteroposterior axis during gastrulation of the mouse embryo. *Development* **126:** 4691–4701.

Knezevic V., De Santo R., and Mackem S. 1997. Two novel chick T-box genes related to mouse *Brachyury* are expressed in different, nonoverlapping mesodermal domains during gastrulation.

Development **124:** 411–419.

Knoetgen H., Viebahn C., and Kessel M. 1999. Head induction in the chick by primitive endoderm of mammalian, but not avian origin. *Development* **126:** 815–825.

Kölliker A. 1879. *Entwicklungsgeschichte des Menschen und der höheren Thiere.* Wilhelm Engelmann, Leipzig.

Kollmann J.E. 1886. "Gastrulasitzung" der 59. Versammlung deutscher Naturforscher und Ärzte zu Berlin. *Anat. Anz.* **1:** 281–294.

Kuhn H.J. and Schwaier A. 1973. Implantation, early placentation, and the chronology of embryogenesis in *Tupaia belangeri. Z. Anat. Entwickl. Gesch.* **142:** 315–340.

Lawson A. and Schoenwolf G.C. 2003. Epiblast and primitive-streak origins of the endoderm in the gastrulating chick embryo. *Development* **130:** 3491–3501.

Lawson K.A., Meneses J.J., and Pedersen R.A. 1991. Clonal analysis of epiblast fate during germ layer formation in the mouse embryo. *Development* **113:** 891–911.

Lawson K.A., Dunn N.R., Roelen B.A., Zeinstra L.M., Davis A.M., Wright C.V., Korving J.P., and Hogan B.L. 1999. BMP4 is required for the generation of primordial germ cells in the mouse embryo. *Genes Dev.* **13:** 424–436.

Leiser R. and Denker H.W. 1988. The dynamic structure of rabbit blastocyst coverings. II. Ultrastructural evidence for a role of the trophoblast in neozona formation. *Anat. Embryol.* **179:** 129–134.

Levin M., Johnson R.L., Stern C.D., Kuehn M., and Tabin C. 1995. A molecular pathway determining left-right asymmetry in chick embryogenesis. *Cell* **82:** 803–814.

Liptau H. and Viebahn C. 1999. Expression patterns of gap junctional proteins connexin 32 and 43 suggest new communication compartments in the gastrulating rabbit embryo. *Differentiation* **65:** 209–219.

Mac Auley A., Werb Z., and Mirkes P.E. 1993. Characterization of the unusually rapid cell cycles during rat gastrulation. *Development* **117:** 873–883.

Martinez-Barbera J.P. and Beddington R.S. 2001. Getting your head around Hex and Hesx1: Forebrain formation in mouse. *Int. J. Dev. Biol.* **45:** 327–336.

Nakajima Y. and Burke R.D. 1996. The initial phase of gastrulation in sea urchins is accompanied by the formation of bottle cells. *Dev. Biol.* **179:** 436–446.

Nichols J., Zevnik B., Anastassiadis K., Niwa H., Klewe-Nebenius D., Chambers I., Schöler H., and Smith A. 1998. Formation of pluripotent stem cells in the mammalian embryo depends on the POU transcription factor Oct4. *Cell* **95:** 379–391.

Pera E.M., Wessely O., Li S.Y., and De Robertis E.M. 2001. Neural and head induction by insulin-like growth factor signals. *Dev. Cell* **5:** 655–665.

Perea-Gomez A., Rhinn M., and Ang S.L. 2001a. Role of the anterior visceral endoderm in restricting posterior signals in the mouse embryo. *Int. J. Dev. Biol.* **45:** 311–320.

Perea-Gomez A., Lawson K.A., Rhinn M., Zakin L., Brulet P., Mazan S., and Ang S.L. 2001b. Otx2 is required for visceral endoderm movement and for the restriction of posterior signals in the epiblast of the mouse embryo. *Development* **128:** 753–765.

Perea-Gomez A., Vella F.D., Shawlot W., Oulad-Abdelghani M., Chazaud C., Meno C., Pfister V., Chen L., Robertson E., Hamada H., Behringer R.R., and Ang S.L. 2002. Nodal antagonists in the anterior visceral endoderm prevent the formation of multiple primitive streaks. *Dev. Cell* **3:** 745–756.

Piccolo S., Agius E., Leyns L., Bhattacharyya S., Grunz H., Bouwmeester T., and De Robertis E.M. 1999. The head inducer Cerberus is a multifunctional antagonist of Nodal, BMP and Wnt signals. *Nature* **397:** 707–710.

Poelmann R.E. 1981. The formation of the embryonic mesoderm in the early post-implantation mouse embryo. *Anat. Embryol.* **162:** 29–40.

Rabl C. 1915. Edouard van Beneden und der gegenwärtige Stand der wichtigsten von ihm behandelten Probleme. *Arch. Mikr. Anat.* **88:** 3–470.

Renfree M.B. and Shaw G. 2000. Diapause. *Annu. Rev. Physiol.* **62:** 353–375.

Rhinn M., Dierich A., Shawlot W., Behringer R.R., Le Meur M., and Ang S.L. 1998. Sequential roles for Otx2 in visceral endoderm and neuroectoderm for forebrain and midbrain induction and specification. *Development* **125:** 845–856.

Schäfer-Haas A. and Viebahn C. 2000. The germ cell epitope PG-2 is expressed in primordial germ cells and in hypoblast cells of the gastrulating rabbit embryo. *Anat. Embryol.* **202:** 13–23.

Schoenwolf G.C., Garcia-Martinez V., and Dias M.S. 1992. Mesoderm movement and fate during avian gastrulation and neurulation. *Dev. Dyn.* **193:** 235–248.

Seher T.C. and Leptin M. 2000. Tribbles, a cell-cycle brake that coordinates proliferation and morphogenesis during *Drosophila* gastrulation. *Curr. Biol.* **10:** 623–629.

Selenka E. 1884. *Studien über Entwickelungsgeschichte der Thiere. 3. Die Blätterumkehr im Ei der Nagethiere.* Kreidel, Wiesbaden, Germany.

Simeone A. 1998. Otx1 and Otx2 in the development and evolution of the mammalian brain. *EMBO J.* **17:** 6790–6798.

Stanley E.G., Biben C., Allison J., Hartley L., Wicks I.P., Campbell I.K., McKinley M., Barnett L., Koentgen F., Robb L., and Harvey R.P. 2000. Targeted insertion of a lacZ reporter gene into the mouse Cer1 locus reveals complex and dynamic expression during embryogenesis. *Genesis* **26:** 259–264.

Stern C.D. 2001. Initial patterning of the central nervous system: How many organizers? *Nat. Rev. Neurosci.* **2:** 92–98.

Stern C.D. and Canning D. 1990. Origin of cells giving rise to mesoderm and endoderm in chick embryo. *Nature* **343:** 273–275.

Stieve H. 1926. Ein 13 1/2-Tage altes, in der Gebärmutter erhaltenes und durch Eingriff gewonnenes menschliches Ei. *Z. Mikr. Anat. Forsch.* **7:** 295–402.

Streit A., Berliner A.J., Papanayotou C., Sirulnik A., and Stern C.D. 2000. Initiation of neural induction by FGF signaling before gastrulation. *Nature* **406:** 74–78.

Sulik K., Dehart D.B., Inagaki T., Carson J.L., Vrablic T., Gesteland K., and Schoenwolf G.C. 1994. Morphogenesis of the murine node and notochordal plate. *Dev. Dyn.* **201:** 260–278.

Thomas P.Q. and Beddington R.S.P. 1996. Anterior primitive endoderm may be responsible for patterning the anterior neural plate in the mouse embryo. *Curr. Biol.* **6:** 1487–1496.

Tsang T.E., Kinder S.J., and Tam P.P. 1999. Experimental analysis of the emergence of left-right asymmetry of the body axis in early postimplantation mouse embryos. *Cell. Mol. Biol.* **45:** 493–503.

van Beneden E. 1880. Recherches sur l'embryologie des mammifères. La formation des feuillets chez le lapin. *Arch. Biol.* **1:** 137–224.

———. 1911. Recherches sur l'embryologie des Mammifères. De la segmentation, de la formation de la cavité blastodermique et de l'embryon didermique chez le murin. *Arch. Biol.* **26:** 1–63.

———. 1912. Recherches sur l'embryologie des mammifères. II. De la ligne primitive, du prolongement céphalique, de la notocorde et du mésoblaste chez le lapin et chez le murin. *Arch. Biol.* **27:** 191–401.

Varlet I., Collignon J., and Robertson E.J. 1997. Nodal expression in the primitive endoderm is required for specification of the anterior axis during mouse gastrulation. *Development* **124:** 1033–1044.

Vestweber D. and Kemler R. 1984. Rabbit antiserum against a purified surface glycoprotein decompacts mouse preimplantation embryos and reacts with specific adult tissues. *Exp. Cell Res.* **152:** 169–178.

Viebahn C. 1995. Epithelio-mesenchymal transformation during formation of the mesoderm in the mammalian embryo. *Acta Anat.* **154:** 79–97.

———. 1999. The anterior margin of the mammalian gastrula: Comparative and phylogenetic aspects of its role in axis formation and head induction. *Curr. Top. Dev. Biol.* **46:** 63–103.

———. 2001. Hensen's node. *Genesis* **29:** 96–103.

Viebahn C., Lane E.B., and Ramaekers F.C.S. 1992. Intermediate filament protein expression and mesoderm formation in the rabbit. A double-labeling immunofluorescence study. *Roux's Arch. Dev. Biol.* **201:** 45–60.

Viebahn C., Mayer B., and Hrabé de Angelis M. 1995a. Signs of the principal body axes prior to primitive streak formation in the rabbit embryo. *Anat. Embryol.* **192:** 159–169.

Viebahn C., Mayer B., and Miething A. 1995b. Morphology of incipient mesoderm formation in the rabbit embryo: A light- and retrospective electron-microscopic study. *Acta Anat.* **154:** 99–110.

Viebahn C., Miething A., and Wartenberg H. 1998. Primordial germ cells of the rabbit are specifically recognized by a monoclonal antibody labeling the perimitochondrial cytoplasm. *Histochem. Cell Biol.* **109:** 49–58.

Viebahn C., Stortz C., Mitchell S.M., and Blum M. 2002. Low proliferative and high migratory activity in the area of *Brachyury* expressing mesoderm progenitor cells in the gastrulating rabbit embryo. *Development* **129:** 2355–2365.

Waddington C.H. 1933. Induction by the endoderm in birds. *Roux's Arch. Dev. Biol.* **128:** 502–521.

Waddington C.H. 1936. Experiments on determination in the rabbit embryo. *Arch. Biol.* **48:** 273–290.

Waterman A.J. 1943. Studies of normal development of the New Zealand white strain of rabbit. *Am. J. Anat.* **72:** 473–515.

Williams B.S. and Biggers J.D. 1990. Polar trophoblast (Rauber's layer) of the rabbit blastocyst. *Anat. Rec.* **227:** 211–222.

Willison K. 1990. The mouse Brachyury gene and mesoderm formation. *Trends Genet.* **6:** 104–105.

Witschi E. 1956. Proposals for an international agreement on normal stages in vertebrate embryology. In *Proceedings of the XIV International Congress of Zoology, 1953,* pp. 260–262. Copenhagen, Denmark.

Zakin L., Reversade B., Virlon B., Rusniok C., Glaser P., Elalouf J.M., and Brulet P. 2000. Gene expression profiles in normal and Otx2-/- early gastrulating mouse embryos. *Proc. Natl. Acad. Sci.* **97:** 14388–14393.

Zhou X., Sasaki H., Lowe L., Hogan B.L.M., and Kuehn M.R. 1993. *Nodal* is a novel TGF-beta-like gene expressed in the mouse node during gastrulation. *Nature* **361:** 543–547.

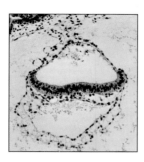

GASTRULATION IN OTHER MAMMALS AND HUMANS

G.S. Eakin and R.R. Behringer

Program in Developmental Biology, Baylor College of Medicine and Department of Molecular Genetics, University of Texas M.D. Anderson Cancer Center, Houston, Texas 77030

INTRODUCTION

The class Mammalia is composed of morphologically diverse animals that live on the land, swim in the sea, or fly in the air. Adults are characterized by a number of traits, including the presence of body hair as well as the mammary glands from which their name is derived. There are three subclasses of mammals, including monotremes or prototherian mammals, marsupial or metatherian mammals, and eutherian mammals, representing over 4,800 living species (Nowak 1997). The three species of Monotremata, the platypus and two echidnas, are remarkable among mammals because they are oviparous; i.e, their young are born from eggs that are laid and incubated to term outside the mother's body. Marsupials are viviparous, giving birth to live progeny. However, marsupials have gestations that are short relative to lactation, resulting in the birth of very immature young. Much of monotreme and marsupial organogenesis occurs after hatching or birth, respectively, as the immature young acquire nutrition and grow during a protracted period of lactation. In contrast to monotremes and marsupials, eutherian mammals, like the mouse (see Chapter 16), rabbit (see Chapter 17), and human, give birth to more mature progeny.

As mammals exhibit differing adult morphologies and modes of reproduction, so do their appearances vary during early embryonic development (Bininda-Emonds et al. 2003). Most of our knowledge about gastrulation in mammals is derived from the mouse (see Chapter 16), which forms a gastrula with a cup-shaped structure called the egg cylinder (Fig. 1A,C). In contrast, the human gastrula assumes a planar morphology like most non-rodent and marsupial mammals (Fig. 1B,D). Thus, one merely has to compare the disparate morphologies of mouse and human gastrula-stage embryos to realize mammalian gastrulation cannot be understood by the study of only one species. In this chapter, we provide an overview of the early embryonic development of mammals that are not routinely studied in modern laboratories. We highlight the similarities and differences between the patterns of germ-layer formation and gastrulation between mammalian species (Fig. 2) and suggest that important mechanistic insights can be gained by studying gastrulation using "nontraditional" mammalian model systems. Ultimately, this knowledge may be useful for understanding human embryonic development and birth defects caused by alterations in germ-layer formation and gastrulation.

GERM-LAYER FORMATION AND GASTRULATION IN MONOTREMES

Due to captive breeding difficulties and protected status, most monotreme embryos and tissues available for study are limited to wild-caught specimens collected during the latter half of the 19th century (Griffiths 1968, 1978; Hughes and Hall 1998). Like birds and reptiles, monotreme oocytes contain proportionately greater amounts of yolk than other mammals and, upon fertilization, undergo meroblastic

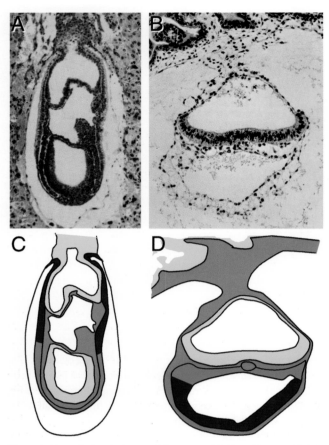

Figure 1. Gastrulating embryos of the mouse (*A,C*) and human (*B,D*). A midsaggital section of a mouse embryo at 7.5 days of development (*A,C*). A frontal section of a human embryo with visible notochordal process presumed to be at 16 days of gestation. Carnegie No. 7802 section (*B,D*). Ectoderm (*yellow*), mesoderm (*red*), hypoblast (*green*), extraembryonic endoderm (*brown*).

cleavages at one pole of the developing embryo. These cleavages ultimately result in a single cell sheath of ectoderm encapsulating the yolk. Prior to complete encapsulation, the prospective endoderm cells differentiate within the ectoderm. These prospective endodermal cells eventually migrate inwardly. Here they proliferate, forming a complete endodermal layer subjacent to the overlying epiblast (Griffiths 1968; Hughes and Hall 1998). Aside from the appearance of an epiblast produced by meroblastic cleavages, monotremes follow a similar pattern of development as some marsupials, albeit without implantation into the uterus (see below). Although monotremes do not implant, the yolk sac must absorb uterine secretions prior to being laid. These secretions support the nutritional needs of the embryo during gestation (Caldwell 1887; Flynn and Hill 1942, 1947; Luckett 1977; Griffiths 1978; Hughes 1993; Hughes and Hall 1998). Molecular knowledge of gastrulation in monotremes has not been reported.

GERM-LAYER FORMATION AND GASTRULATION IN MARSUPIALS

Gastrulation in marsupials comprises roughly two-thirds of gestation and, unlike the mouse, occurs prior to the attachment of the embryo to the uterus. Indeed, implantation does not occur until after a complete body axis and many somites have formed. This suggests that the marsupial embryo, like the monotreme embryo, has the intrinsic information required to initiate gastrulation and axial patterning. Thus, gastrulae can be isolated from the uterus and cultured in vitro, providing experimental access to the developing embryo (New and Mizell 1972; New et al. 1977; Yousef and Selwood 1993; Renfree and Lewis 1996). This has been used in conjunction with techniques such as vital cell labeling with the lipophilic dye DiI (Cruz et al. 1996). Marsupial embryos are surrounded by a zona pellucida. Like some eutherians (e.g., rabbits), marsupials also acquire additional acellular coats. During oviductal transport, marsupial embryos are enveloped by a glycoprotein-rich mucoid layer and a fibrous shell coat. Adherence of blastomeres to the zona is required for blastocyst development. However, cleavage-stage embryos and blastocysts can be

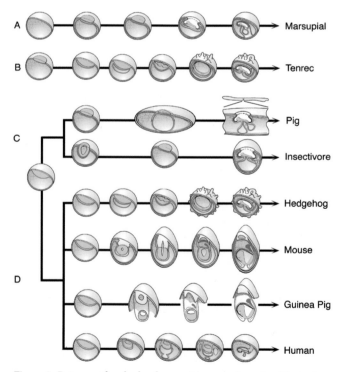

Figure 2. Patterns of early development in several species, illustrating development between blastulation and somitogenesis. (*A*) Metatherian development from a unilaminar blastocyst. (*B*) Eutherian development from a unilaminar blastocyst. (*C*) Examples of mammals that develop from an exteriorized ICM. (*D*) Representative mammals that develop from blastocysts with interiorized ICMs. Ectoderm (*yellow*), endoderm (*green*), mesoderm (*red*).

successfully cultured for limited times without their shell coats (dunnart, *Sminthopsis macroura*; Hickford and Selwood 2003, and tammar wallaby *Macropus eugenii*; Renfree and Lewis 1996). Marsupial embryo transfer into surrogate mothers has been achieved, although pregnancies were not brought to term for experimental reasons (Tyndale-Biscoe 1963, 1970; Renfree 1972; Renfree and Tyndale-Biscoe 1973, 1978). Molecular data on marsupials are accumulating. Recently, expression patterns of molecular markers including *Pou5f1* (*Oct3/4*) (Frankenberg et al. 2001), alkaline phosphatase (Ullmann et al. 1996), and members of the TGF-β signaling pathway (Hickford and Selwood 2002) were reported for marsupials. Thus, marsupials provide a very tractable platform for experimental and molecular embryology. Furthermore, given the 165–185 million years of estimated divergence between marsupial and eutherian mammals (Thomas and Touchman 2002), comparative studies between the two subclasses are likely to illuminate shared mechanisms representing the most indispensable aspects of mammalian gastrulation.

The blastocyst of many marsupials is immediately distinguishable from those of most eutherian mammals in that it forms by obligate adherence of the blastomeres to the zona pellucida, forming a hollow sphere of blastomeres (Fig. 2A) (Selwood 1989) without going through a morula stage or ever possessing an inner cell mass (ICM) (Hill 1910; Hartman 1916; McCrady 1938; Selwood and Young 1983; Selwood 1986a,b, 1992; for review, see Tyndale-Biscoe and Renfree 1987; Ward and Renfree 1987; Cruz and Pedersen 1991; Renfree 1994; Renfree and Lewis 1996; Cruz 1997). This unilaminar blastocyst is, so far as is known, present in all marsupials (Smith 1981; Tyndale-Biscoe and Renfree 1987) and may be apparent as early as the 8-cell stage (Tyndale-Biscoe 1979; Selwood 1992). The lack of a morula stage or ICM presents a significant caveat to the "inside-out" model of determination of the embryonic and trophoblast cell lineages in cleavage-stage mouse embryos. In this model, the interiorly and exteriorly located cells of the morula are predicted to contribute to embryonic and trophoblastic tissues, respectively (Tarkowski and Wroblewska 1967). It is important to note that whereas the presence of a unilaminar blastocyst can be generalized to all described marsupials, macropodids (kangaroos and wallabies) and acrobatid marsupials (such as the feathertail glider) may not display obligate adherence of blastomeres to the zona as seen in dasyurid marsupials (such as the native cat and Tasmanian devil). Thus, the macropodid and acrobatid marsupials may possess a morula-like stage before formation of the unilaminar blastocyst (Ward and Renfree 1987; Renfree 1994; Renfree and Lewis 1996). The unilaminar blastocyst is not unique to the marsupials, as a lack of a distinguishable ICM is also found in some eutherian mammals, including insectivores such as the Indian musk shrew (*Crocidura caerulea*) (Sansom 1937),

elephant shrews (Macroscelididae) (Van der Horst 1942; Tripp 1971), and tenrecs (Tenrecidae) (Fig. 2B) (Bluntschi 1938; Goetz 1938; for review, see Renfree 1982). A similar condition is also seen in a prosimian, the dwarf galago (*Galagoides demidoff*) (Gerard 1932).

The parietal endoderm and hypoblast of the marsupial are formed by the inward migration of ectoderm to line the interior of the embryo, forming the bilaminar blastocyst. This is accomplished in one of three ways (Selwood 1992). The first scheme is characteristic of the Virginia opossum (*Didelphis virginiana*) (Selenka 1887; Hartman 1919) and Bennett's wallaby (*Macropus rufogriseus*) (Walker and Rose 1981; Selwood 1986a). In these species, a subset of the blastocyst cells, termed endoderm mother cells, swell to several times their normal size. The enlarged cells then migrate into the blastocyst cavity, whereupon they undergo several asymmetric mitotic divisions. The daughters of the endoderm mother cells proliferate, forming a clump of hypoblast cells that rests at one pole of the blastocyst. The epithelium that overlies the new endoderm becomes cuboidal and becomes recognizable as the epiblast. The second scheme for the formation of a bilaminar blastocyst is found in the bandicoot (*Perameles*) (Hill 1910; Lyne and Hollis 1977) and native cat (*Dasyurus viverrinus*) (Hill 1910). Cells originating from the entire surface of the epithelium ingress to form a small proliferative mass at one pole of the blastocyst. Again, the epithelium overlying the endodermal mass develops as the epiblast. In both cases, those cells not in contact with the hypoblast develop as trophoblast. Although it is likely that communication exists between the three tissue types, there is evidence to suggest that epiblast determination occurs prior to the formation of the hypoblast. Indeed, the endoderm mother cells of the Virginia opossum appear from the margins of the presumptive epiblast (Hartman 1919; McCrady 1938), suggesting a prior acquisition of epiblast fate. Finally, the third scheme for endoderm formation is observed in the Tasmanian bettong (*Bettongia gaimardi*) (Kerr 1935, 1936). The endoderm is produced from the already-present epiblast and its margins. This lends support to the idea that the epiblast is determined prior to the appearance of endoderm (Hickford and Selwood 2003).

The size of the marsupial bilaminar blastocyst ranges from approximately 0.75 mm (Hartman 1919; McCrady 1938; Walker and Rose 1981; Selwood 1986a,b, 1992) to as large as 8.5 mm in the native cat (Hill 1910; for review, see Renfree and Tyndale-Biscoe 1973; Tyndale-Biscoe and Renfree 1987). There appears to be a correlation between the size of the blastocyst and the type of endoderm formation. Smaller blastocysts produce endoderm via endoderm mother cells, whereas direct division of ingressing endoderm is characteristic of larger blastocysts (Selwood 1992; Cruz 1997).

As endoderm formation finishes, one end of the epiblast thickens due to proliferation of the epiblast. This signifies the future posterior of the embryo. During this period, the embryonic disc begins to form a pear-shaped or pyriform morphology. Primitive streak and mesoderm formation proceeds from the narrower posterior portion of the embryo and is coincident with a rostral expansion of the epiblast. Amniogenesis in marsupials occurs by folding of the yolk sac, reproducing a topology similar to an equivalently staged chicken embryo.

GERM-LAYER FORMATION AND GASTRULATION IN EUTHERIAN MAMMALS

Unlike the marsupial, most eutherian mammals have blastocysts that develop as a single-celled layer of trophoblast enveloping an ICM anchored to the trophoblast at one side of a fluid-filled cavity. Once the ICM begins to differentiate, it either remains internalized or breaks the surface of the trophoblast and becomes exposed to the exterior of the conceptus (Fig. 2C,D). This positioning of the ICM is generally correlated with one of two types of amniogenesis: folding or cavitation.

The Internalized Inner Cell Mass

An ICM that remains entirely within the trophoblast layer of the blastocyst is representative of the Old World mice and rats, porcupines, hedgehogs, as well as most bats and hominoid primates (great apes and man) (Fig. 2D). In these cases, amniogenesis occurs by cavitation. This process can take a number of forms.

In humans and the great apes, the embryonic disc undergoes a process of delamination to form the amniotic cavity within the former ICM. This results in the embryonic ectoderm being displaced toward the blastocyst cavity. The blastocyst cavity then begins to fill with a fine lattice of mesodermal cells originating from the hypoblast (Enders and King 1988) and coating the extraembryonic ectoderm.

Myomorph and hystricomorph rodent (mice, rats, guinea pigs, and hedgehogs) embryos produce the amnion by slightly different manners of cavitation. Shortly after implantation in the guinea pig, the ICM loses connection with its overlying trophectoderm (the ectoplacental cone, or träger) and forms a gap between the epiblast and overlying trophectoderm (Wilson 1928; Sansom and Hill 1931; Roberts and Perry 1974). A midsaggital section would appear to the viewer as if the epiblast were dangling in a disintegrating yolk cavity, supported by a "pouch" of hypoblast. This is quite different from the morphologies of similarly staged mouse and rat embryos, whose epiblast

does not lose contact with the extraembryonic ectoderm. In the guinea pig, a proamniotic cavity appears in the center of the epiblast. The proximal cells at the roof of this cavity flatten to form the ectodermal component of the amnion. The formation of the proamniotic cavity is coincident with the appearance of mesoderm, which lines the extraembryonic coelom and displaces the hypoblast.

In the mouse and rat, cavitation of the epiblast also occurs, but in these rodents a separate epamniotic, or false, cavity arises in the extraembryonic ectoderm. This cavity soon becomes contiguous with the developing proamniotic cavity in the epiblast. Epamniotic cavities also arise in the extraembryonic ectoderm of other myo- and hystricomorphs, as well as armadillos, but are never contiguous with the amniotic or exocoelomic cavities of these species.

In the mouse pre-gastrula, a band of extraembryonic cells at the junction of the extraembryonic and embryonic ectoderm transiently expresses *Bmp4*. BMP4 function in this region is considered to be a likely means of inducing the expression of genes in the adjacent epiblast (Beddington and Robertson 1998, 1999; Behringer et al. 2000). As described previously, the guinea pig extraembryonic and embryonic ectoderm are not in physical contact at similar stages (Fig. 2D). This separation of tissues may thereby prevent the function of BMP4 as theorized in the mouse, and may require alternative strategies for the specification of other BMP4-dependent cell types, such as primordial germ cells and the allantois (Fujiwara et al. 2001). Experimental consideration of the differences between these early events in both guinea pig and mouse may therefore provide insight into the evolutionary plasticity of the molecular pathways governing gastrulation between rodents.

Some species combine aspects of both folding and cavitation. During the first step of hedgehog amniogenesis, the inner aspect of the ICM loses contact with the trophoblast, forming a cavity between the two tissues. The edges of the ICM remain closely opposed to the overlying trophoblast. However, once cavitation has occurred, the ICM edges undergo an inward folding to complete the amnion. Similarly to other folding species, the hedgehog's amnion does not close until after the appearance of the somites as in marsupials. For a short time beyond this point, the amniotic ectoderm remains juxtaposed against the trophectoderm until the mesodermal lining of the yolk sac coats and separates the two tissues (Hubrecht 1912; Morris 1953).

Amniogenesis by either of these mechanisms tends to correlate with the size of the blastocyst and depth of implantation. Those embryos with smaller blastocysts and deep (interstitial) implantation generally undergo cavitation (Mossman 1987). Compare the interstitially implanting mouse and superficially implanting rabbit blastocysts. At implantation, their diameters are 85 and 4000 μM, respectively (for review, see Davies and Hesseldahl 1971). It is thought that the uterine tissue places physical and bio-chemical constraints on interstitially implanting blastocysts which necessitate smaller embryos at the time of implanta-tion. The process of cavitation generally results in embryos with proportionately less yolk sac. Although correlation does not necessarily imply causation, those embryos that implant deeply may be benefited by more compact sizes associated with cavitation (Mossman 1987).

The complete "inversion of the germ layers," observed in many rodents, including mice and rats, is an additional example of reduction of space required for deeply implant-ing embryos. The thirteen-striped ground squirrel (*Spermophilus tridecemlineatus*) is a superficially implant-ing rodent with incomplete inversion of the germ layers. A comparison of *S. tridecemlineatus* and similarly staged mouse and rat embryos reveals an impressive 1:50 reduc-tion of total volume of the yolk sac, with very little differ-ence in the sizes of the embryos proper (Mossman and Weisfeldt 1939; Snell and Stevens 1966).

The Exteriorized Inner Cell Mass

An exteriorized ICM is representative of ungulates, most carnivores, insectivores, and the sciurid rodents (including squirrels and chipmunks). The disappearance of Rauber's layer, a layer of trophoblast covering the ICM, reveals the ICM to the exterior of the blastocyst at the beginning of gastrulation (Fig. 2C). Folding of the periphery of the embryonic ectoderm envelops the embryonic disc, thus establishing the amnion and causing the embryonic por-tions to be localized deeper within the blastocyst cavity. This mode of development generally occurs concurrently or subsequently to gastrulation, such that closure of the amnion is often completed at the somite stages. In pigs (Heuser and Streeter 1929), sheep (Boyd and Hamilton 1952), and rabbits (Kölliker 1880) the ICM simply appears to flatten, and while doing so, penetrates the overlying tro-phoblast, forming an embryonic knob. Although arriving at the same final topology, the ICM of the majority of insecti-vores first forms a deep cleft, the periphery of which forms attachments to the overlying trophoblast. Upon relaxation of the fold, the ICM comes to rest in a position similar to the rabbit and pig (Hubrecht 1889). For purposes of exper-imental embryology, those animals that undergo folding (e.g., squirrel, rabbit, carnivores, artiodactyls) present a period of time during which the embryo proper is exposed to the exterior of the conceptus (Fig. 3C). In some cases, gastrulation has even begun prior to implantation (Table 1). The gastrula of these species may be easily accessible and amenable to embryological techniques such as time-lapse confocal microscopy or gene transfer. Molecular data on gastrulation-stage artiodactyl (hoofed mammal) embryos is beginning to accumulate (Simpson et al. 1999; Meijer et

al. 2000; Hue et al. 2001; Fléchon et al. 2004) and is, in some cases, revealing differential gene expression patterns between these species and the mouse (Hue et al. 2001).

UNIQUE PROCESSES IN THE EARLY DEVELOPMENT OF SELECT MAMMALIAN SPECIES

Blastocyst Growth in Artiodactyls

Artiodactyls represent many agriculturally important species, including pigs, cattle, sheep, and goats. Gastrulation and even somitogenesis in these animals occurs before implantation. Although the early blastocyst of these animals is reminiscent of other eutherians in terms of size and morphology, later blastocysts display considerable disparity. In some artiodactyls, this late blastocyst, now termed the chorionic vesicle, may grow to a meter in length (for review, see Guillomot 1995).

In the pig, hatching of the blastocyst from the zona pel-lucida occurs between days 6 and 7 (for review, see Stroband and Van der Lende 1990). As in other ungulates, the blastocyst remains free in the uterine lumen for a pro-tracted period of time, until day 13 in pigs. By day 10, the cells of Rauber's layer begin to shed, revealing the embryon-ic disc to the uterine environment (Fig. 3B). This is fol-lowed by a 48-hour period of extreme elongation of the blastocyst, in which the conceptus may extend at 30–45 mm/hour (Fig. 3A). This initial growth is mainly achieved by cytoskeletal remodeling of individual cells, rather than by cell division. At later stages of growth, mitotic contribu-tion to the thread-like elongating blastocyst is evident. Because elongation is limited to the trophoblast cells, the embryonic disc maintains a relatively constant size until maximal elongation is achieved around day 12.

During implantation near day 13, the pig blastocyst attaches within the folds of the uterus, such that the entire meter length of the pig blastocyst may occupy only 10–15 centimeters of the uterus (Patten 1959). In sheep, transient trophoblast papillae may play a role in the initial anchoring of the conceptus to the uterine epithelium. These projec-tions disappear within 4 days postimplantation (Wooding et al. 1982). Unlike mice and humans, the epitheliochorial placenta of artiodactyls is not highly invasive and must, therefore, rely on the elongated blastocyst to establish max-imal contact area between mother and fetus.

Timing of Implantation and Gastrulation

In the mouse, gastrulation begins only after interstitial implantation of the conceptus deep within the uterine lin-ing. As in marsupials, some eutherian mammals such as

Table 1. *Reproductive statistics of several mammals*

	Common name	Genus	Number of young	Length of gestation (days)	Day of blastulation	Number of cells of blastocyst	Size of blastocyst at implantation	Implantation	Day of gastrulation	Stage at implantation	Depth of implantation
Prototherian subclass											
Monotremata	echidna	*Tachyglossus*	1–2	18 at laying; 28–29 at hatching						19–20 somite neurula at laying	
Metatherian subclass											
Didelphimorphia	opossum	*Didelphis*	3–13	13-Dec		32	@ expansion 0.11 mm	7–8	7	28 somites, organogenesis	very superficial
Dasyuromorphia	brown antechinus	*Antechinus*	6–10	27		20	1.0 mm	23	21	8–12 somites; neurula	very superficial
	stripe-faced dunnart	*Smithopsis*	8–10	12.5–16		32			17	organogenesis	very superficial
	native cat	*Dasyurus*	5	19			5.0 mm			late primitive streak	very superficial
Diprotodontia	brush-tailed possum	*Trichosurus*	1	16–17			0.75 mm		11	blastocyst	very superficial
	tammar wallaby	*Macropus*	1	29					14	25–30 somites	very superficial
Eutherian subclass											
Edentata	armadillo	*Dasypus*	4	150*			0.5 mm			bilaminar blastocyst	superfical

Order		Genus									
Primate	macaque	*Macaca*	1	165	7–8		0.4 mm	9–11	11–13	blastocyst	superficial
	human	*Homo*	1	280	4	64–107	0.1–2 mm	8–13	13–15	blastocyst	interstitial
	baboon	*Papio*	1	175	5–8		0.04 mm	9–10	16–18	blastocyst	superficial
Perissodactyla	horse	*Equus*	1	350	6–8	64	75 mm @ day 28	28		limb bud	superficial
Artiodactyla	pig	*Sus*	9–11	113	3.5–5	16	up to 100 cm	11–20	7	early somite	superficial
	cow	*Bos*	1.3	282	7–8	32	8 mm @ day 14	30–35	13	early somite	superficial
	goat	*Capra*	1.75	148	6	32	18–20		elongated chorionic vesicle with allantois	superficial
	sheep	*Ovis aries*	1.5	148	4.5–6	64	68 mm @ day 15	15–17	9	early somite	superficial
Rodentia	mouse	*Mus*	6–12	18–21	3.5	32		4 dpf	6	blastocyst, some endoderm	interstitial
	rat	*Rattus*	7–11	22	4.5		0.06–0.08 mm	5 dpf	6–7	blastocyst, some endoderm	interstitial
	guinea pig	*Cavia*	2–4	68	4.75			6 dpf	7–8+	unilaminar blastocyst	interstitial
	golden hamster	*Cricetus*	5–10	16				4.3 dpf	5.5	blastocyst, some endoderm	interstitial
	woodchuck	*Marmota*	2–3	31–32	5	128	0.2 mm	6–7 dpf		blastocyst	superficial
Lagomorpha	rabbit	*Oryctolagus*	4–10	28	3–4	128	4 mm	7 dpf	8–9	bilaminar blastocyst	superficial

Adapted from Wimsatt (1975); Tyndale-Biscoe and Renfree (1987); Cruz and Pedersen (1991). * Delayed implantation.

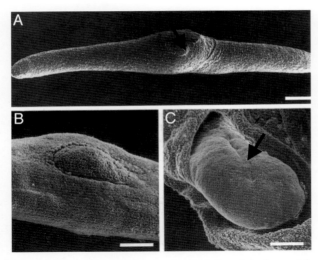

Figure 3. Electron micrographs of preimplantation pig embryos at day 14. (*A*) Elongated blastocyst with exposed embryonic disc (*arrow*). (*B*) Magnified view of *A*, detailing the embryonic knob. (*C*) Folding of the amnion to enclose the somite-stage embryo. Arrow denotes the location of the primitive streak. Bars: *A*, 500 μM; *B*, *C*, 100 μM. (Reprinted, with permission, from Guillmot 1995).

horses (*Perissodactyla*) and artiodactyls complete gastrulation and begin somitogenesis prior to implantation (Table 1; Fig. 3C). In the case of the horse, implantation begins after the embryo has formed both sets of limb buds (Ewart 1915). This suggests that, in these species, all information required for initiation of gastrulation and establishment of polarity are present prior to implantation.

Polyembryony in the Armadillo

The incidence of monozygotic twinning in humans is increased after assisted reproduction (Wenstrom et al. 1993) and is associated with obstetrical pathology. Additionally, some types of twinning may have a genetic basis (Bamforth et al. 2003). However, unassisted monozygotic twinning in the mouse is rare (McLaren et al. 1995). Monoamniotic twinning has occurred as a result of manipulations. The occurrence of monoamniotic twinning after in vitro culture (Hsu and Gonda 1980) suggests that the rarity of twinning in mice may be, in part, due to physical constraints on the conceptus imposed by the uterine environment (Enders 2002).

The nine-banded (*Dasypus novemcinctus*) and long-nosed (*D. hybridis*) armadillos are the only mammals known to exhibit obligate monozygotic twinning. Both species produce litters of four genetically identical monozygotic quadruplets (Fig. 4A). In armadillos, the four fetuses share a common villous hemochorial placenta, convergently similar to the human. The common placenta becomes

partially subdivided at later gestational ages (Mossman 1987). Mechanisms leading to polyembryony in armadillos have been elaborated upon in several reports (Fernandez 1909; Newman and Patterson 1909, 1910; Patterson 1913; Enders et al. 1958; Enders 1960, 1962, 1963, 2002; Benirschke et al. 1964; Galbreath 1985; Enders and Welsh 1993; Loughry et al. 1998).

After implantation, the abembryonic trophoblast disintegrates, revealing the hypoblast and parietal endoderm to the uterine lumen, thus leaving the conceptus with inverted germ layers. Following the disintegration of the trophoblast, the embryonic disc proliferates, bulging into the uterine lumen. During this proliferation, the cells of the epiblast begin to segregate into two bulging embryonic shields covered by a common amnion. Following the first partitioning, a second segregation divides the two embryonic discs again into four regions of epiblast overlying a continuous sheet of hypoblast (Fig. 4C). At the same time, the single large amnion is gradually restricted to four smaller amnions covering each of the four epiblast bulges, except for four small canals of amniotic membrane that communicate between the developing embryos (Fig. 4B). Four primitive streaks begin at the outer rim of the epiblastic disc and extend rostrally toward the disc center. Thus, anterior–posterior pattern is established such that each embryo

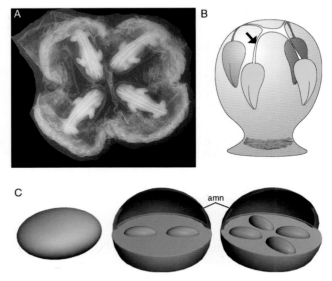

Figure 4. (*A*) Armadillo monozygotic quadruplets sharing a common hemochorial placenta. Anterior/posterior pattern is established such that all embryos develop in a "head to head" configuration (Reprinted, with permission, from Benirschke 2003). (*B*) Thin canals (*arrow*) connect the amniotic sacs of each embryo preventing conjoining (after Patterson 1913). (*C*) Monozygotic twinning occurs progressively, such that the ectoderm of the implanted blastocyst (*yellow*) proceeds through amniogenesis by cavitation, developing two embryonic bulges within a shared amnion (amn). These two bulges subsequently divide again, producing four genetically identical quadruplets.

develops with the future head regions pointed centrally (Fig. 4A). The molecular and cellular mechanisms controlling anterior–posterior patterning are still poorly understood. If a functional homolog of the mouse anterior visceral endoderm (AVE) exists, it would be interesting to know whether each armadillo embryo possesses a unique "AVE," or whether a common AVE provides instructive signals to each of the four embryos. Such a distinction might suggest whether the AVE is intrinsic to the hypoblast, or whether its position might be determined by a previously patterned overlying epiblast.

Human Gastrulation

To those familiar with only the mouse gastrula, the human embryo at this stage of development may seem like an alien landscape (Figs. 1B, 2D). Even in some modern embryology textbooks, gastrulation is described only by reference to other species, often the chick. What is known about early human embryology comes largely from descriptions of serially sectioned specimens fortuitously discovered during hysterectomies, autopsies, or abortions. These rare samples lack many transitional states, especially in stages just prior to the appearance of the primitive streak. Thus, much of human gastrulation relies on studies of cellular morphology from which lineage must be hypothesized. Molecular data are unreported for gastrulation stages, although some expression studies have been described on later gestational stages. Notably, species-specific differences in the spatial and temporal expression pattern of *WNT7A* and other genes have been documented between mouse and human embryos (Fougerousse et al. 2000). The logistics and expense of obtaining closely related nonhuman primate embryos have also limited this field primarily to descriptions of serially sectioned embryos (Hendrickx 1971; Heuser and Streeter 1941). The isolation of several nonhuman primate embryonic stem (ES) cell lines may provide the reagents necessary to develop new technologies for modeling early primate embryology (Thomson et al. 1995, 1996; Suemori et al. 2001). Indeed, under rare conditions, embryoid bodies (EB) produced from marmoset (*Callithrix jacchus*) ES cells appear capable of producing an in vitro-derived primitive streak–stage embryo. In this case, the EBs possessed a pyriform-shaped "embryo and hypoblast," as well as extraembryonic structures including the amnion, primary yolk sac, and a body stalk connecting the EB to the culture dish (Thomson et al. 1996).

The development of the human embryo produces an initial blastocyst similar to most eutherians, containing an internal ICM surrounded by a layer of trophoblast cells. Around implantation, between 7 and 12 days of gestation, a set of cells from the ICM differentiates into a layer of hypoblast, or primary endoderm, which forms on the blas-

tocyst cavity side of the ICM, while the remaining epiblast cells become columnar. At this point, amniogenesis occurs by cavitation as described above. Beginning around day 9, the primary endoderm cells begin to proliferate to line the entire blastocyst cavity, forming the primary yolk sac or umbilical vesicle. Coincident with the development of the primary yolk sac, some endodermal cells differentiate into a loose mesodermal network of cells. In its most mature form, this network is sometimes referred to as the *magma recticulare*. These cells serve to separate the embryonic disc and primary yolk sac from the trophoblast. A thin layer of these mesodermal cells coats the endodermal component of the yolk sac surface completing Heuser's membrane, a term reserved for the portion of the yolk sac that does not include the hypoblast. The endodermal origin of the primary mesoderm has been experimentally determined by examination of naturally occurring chromosomal mosaics during chorionic villus sampling (Enders and King 1988; Bianchi et al. 1993). This unusual mesodermal lattice is also found in the chimpanzee but not in other primates, such as the baboon or Rhesus macaque.

The development of human extraembryonic mesoderm occurs in sharp contrast to the origin of the extraembryonic mesoderm of mice, which is derived exclusively from epiblast cells that have migrated through the primitive streak (Tam et al. 1993). It is believed that primitive streak–derived mesoderm in the human may eventually replace much of the original hypoblast-derived mesoderm (Robinson et al. 2002), but to what degree remains a matter of uncertainty.

Similarly to the rabbit, in some specimens (e.g., No. 8330) of the Carnegie Collection of Human Embryos housed at the National Museum of Health and Medicine of the Armed Forces Institute, a thickening of the presumptive anterior endoderm, the prechordal plate, may be noted at this stage. Although it cannot, at this time, be known whether this structure is homologous to the AVE of the mouse, a structure involved in the establishment of embryonic polarity (see Chapter 16), the prechordal plate does represent the first indication of established anterior–posterior identity in the human.

The spaces in the loose mesodermal reticulum begin to enlarge, ultimately merging, creating a single exocoelomic cavity. By day 12, the embryonic disc and primary yolk sac are suspended in the exocoelom by condensed mesodermal tissue, the body stalk, that attaches the cells of the amnion to the overlying cytotrophoblast. A continued proliferation of the hypoblast causes the primary yolk sac to expand. By day 13, the majority of the primary yolk sac disintegrates into small vesicles. A smaller, secondary, or definitive, yolk sac is left in its place. Vesicular remnants of the primary yolk sac may persist in the extraembryonic coelom.

The primitive streak becomes apparent at the narrow posterior of the embryonic disc between days 13 and 15.

Epiblast cells migrate through the primitive streak and initially replace the hypoblast with definitive endoderm. The epiblast, which ingresses between the definitive endoderm and ectoderm, forming the embryonic mesoderm, moves both rostrally and laterally, displacing the hypoblast at the limits of the epiblast, thereby becoming contiguous with the primary mesoderm of the definitive yolk sac. A subset of the secondary mesodermal cells migrates caudally, infiltrating the body stalk. A small endodermal canal extends from the caudal aspect of the embryonic disc into the body stalk, forming the allantoic diverticulum, a structure not found in mice but weakly evident in the sciurid rodents (Mossman 1987). By day 16, the notochordal process ingresses through the node or primitive pit. A hollow notochordal process temporarily embeds into the definitive endoderm, forming the notochordal plate. Unlike mice, the act of intercalation of a hollow notochordal process creates a neuroenteric canal, communicating the yolk sac to the amniotic cavity. By day 22–24, this process is reversed, and the notochordal plate detaches from the endoderm, emerging as a solid notochord. All of these pivotal events in human embryonic development occur within the first month of gestation, when many women are not even aware that they are pregnant.

CONCLUSIONS

We have attempted to highlight some of the major differences between the morphologies of gastrula-stage mammalian embryos and, furthermore, to illustrate that many fundamental and exciting questions concerning comparative mammalian gastrulation remain unanswered. Indeed, simple descriptions of early embryonic development are unreported in many common species.

The early development of mammals is richly varied. However, mammalian germ-layer formation and gastrulation have been studied primarily in a small number of experimentally favorable "model systems." These traditional model systems are preferred largely due to a highly adapted reproductive biology that allows smaller body sizes, rapid reproduction, and large numbers of offspring. These are, of course, derived traits that neither humans nor many major agricultural species share with models such as the mouse. One must therefore consider the possibility that when a traditional model system is used to study problems in patterning and reproduction, it may be in exactly these processes that the model is most different from the species for which it is believed to be a model. Due to the amount of research that has been performed on traditional models, they, of course, bring tremendously powerful techniques to the study of their own embryology. Consequently, a great temptation exists to extrapolate these findings to other species. As evidence of

this, one must only consider the number of citations of mouse research that begin, "In mammals, it has been shown that ...." However, these models provide only hypotheses about the development of other species that, unfortunately, are rarely tested. Furthermore, the focus of developmental biology on molecular pathways elucidated from essentially one mammalian model has further removed awareness of these differences from the field. Indeed, the low frequency of reviews on the subject of comparative mammalian gastrulation within the last 20 years suggests that this important subject is passing from either memory or interest of contemporary developmental biologists.

Recent advances in transgenesis (e.g., Asano et al. 2002; Lois et al. 2002) and ES cell technologies may help to level the playing field such that tests of hypotheses produced from years of research on traditional model systems may now be applied to understand mechanisms employed during the early stages of development in other less studied mammals, including ourselves.

A myriad of diseases, of which many are surely uncharacterized, affect the early development of humans. Understanding human obstetrical pathology requires knowledge of both normal and abnormal development. As ethical, biological, and increasingly political restrictions limit the types of permissible human embryonic research, this necessitates the use of animal models of human development. Nonetheless, the use of any mammalian model must be assessed from the viewpoint that it represents only one of over 4,800 unique mammalian species. Our ability to understand human development and to remedy human disease is therefore constrained by the degree to which we understand the limitations and potentials of our experimental models.

ACKNOWLEDGMENTS

We thank Marilyn Renfree, Lynne Selwood, and Patrick Tam for helpful comments. In the interest of space and due to the fact that some aspects of mammalian development have been described in several sources, a great many references that contributed to this field were not included in the bibliography. This work was supported by National Institutes of Health grant HD30284 to R.R.B.

REFERENCES

Asano T., Hanazono Y., Ueda Y., Muramatsu S., Kume A., Suemori H., Suzuki Y., Kondo Y., Harii K., Hasegawa M., Nakatsuji N., and Ozawa K. 2002. Highly efficient gene transfer into primate embryonic stem cells with a simian lentivirus vector. *Mol. Ther.* **6:** 162-168.

Bamforth F., Brown L., Senz J., and Huntsman D. 2003. Mechanisms of monozygotic (MZ) twinning: A possible role for the cell adhesion molecule, E-cadherin. *Am. J. Med. Genet.* **120A:** 59–62.

Beddington R.S. and Robertson E.J. 1998. Anterior patterning in mouse. *Trends Genet.* **14:** 277–284.

———. 1999. Axis development and early asymmetry in mammals. *Cell.* **96:** 195–209.

Behringer R.R., Wakamiya M., Tsang T.E., and Tam P.P. 2000. A flattened mouse embryo: leveling the playing field. *Genesis* **28:** 23–30.

Benirschke K. 2003. Comparative placentation. http://www.medicine.ucsd.edu/cpa/homefs.html

Benirschke K., Sullivan M.M., and Marin-Padilla M. 1964. Size and number of umbilical vessels: A study of multiple pregnancy in man and the armadillo. *Obstet. Gynecol.* **24:** 819–834.

Bianchi D.W., Wilkins-Haug L.E., Enders A.C., and Hay E.D. 1993. The origin of extraembryonic mesoderm in experimental animals: Relevance to chorionic mosaicism in humans. *Am. J. Med. Genet.* **46:** 542–550.

Bininda-Emonds O.R., Jeffery J.E., and Richardson M.K. 2003. Inverting the hourglass: quantitative evidence against the phylotypic stage in vertebrate development. *Proc. R. Soc. Lond. B Biol. Sci.* **270:** 341–346.

Bluntschi H. 1938. Le developpement primaire et l'implantation chez un centetine (Hemicentetes). *C.R. Assoc. Anat.* **44:** 39–46

Boyd J.D. and Hamilton W.J. 1952. Cleavage, early development and implantation of the egg. In *Marshall's physiology of reproduction*, 3rd edition (ed. A.S. Parks), vol. 2, pp. 1–126. Longmans, Green and Co., Ltd., London, England.

Caldwell W.H. 1887. The embryology of Monotremata and Marsupialia: Part 1. *Philos. Trans. R. Soc. Lond. B* **178:** 463–468.

Cruz Y.P. 1997. Mammals. In *Embryology: Constructing the organism* (ed. S.F. Gilbert and A.M. Raunio), pp. 459–489. Sinauer, Sunderland, Massachusetts.

Cruz Y.P. and Pedersen R.A. 1991. Origin of embryonic and extraembryonic cell lineages in mammalian embryos. *Curr. Commun. Cell Mol. Biol.* **4:** 147–204.

Cruz Y.P., Yousef A., and Selwood L. 1996. Fate-map analysis of the epiblast of the dasyurid marsupial *Sminthopsis macroura* (Gould). *Reprod. Fertil. Dev.* **8:** 779–788.

Davies J. and Hesseldahl H. 1971. Comparative embryology of mammalian blastocysts. In *The biology of the blastocyst* (ed. R.J. Blandau), pp. 27–48. University of Chicago Press, Illinois.

Enders A.C. 1960. Development and structure of the villous haemochorial placenta of the nine-banded armadillo (*Dasypus novemcinctus*). *J. Anat.* **94:** 34–45.

———. 1962. The structure of the armadillo blastocyst. *J. Anat.* **96:** 39–48.

———. 1963. Fine structural studies of implantation in the armadillo. In *Delayed implantation* (ed. A.C. Enders), pp. 281–290. University of Chicago Press, Illinois.

———. 2002. Implantation in the nine-banded armadillo: How does a single blastocyst form four embryos? *Placenta* **23:** 71–85.

Enders A.C. and King B.F. 1988. Formation and differentiation of extraembryonic mesoderm in the rhesus monkey. *Am. J. Anat.* **181:** 327–340.

Enders A.C. and Welsh A.O. 1993. Structural interactions of trophoblast and uterus during haemochorial placenta formation. *J. Exp. Zool.* **266:** 578–587.

Enders A.C., Buchanan G.D., and Talmage R.V. 1958. Histological and histochemical observations on the armadillo uterus during the delayed and post-implantation periods. *Anat. Rec.* **130:** 639–657.

Ewart J.C. 1915. Studies on the development of the horse. I. The development during the third week. *Trans. Roy. Soc. Edin.* **51:** 287–329.

Fernandez M. 1909. Beitrage zur Embryologie der Gurteltiere. I. Zur Keimblatter-inversion und spezifischen Polyembryonie der Mulita (Tatusia hybrida Desm.). *Morph. Jahrb.* **39:** 302–333.

Fléchon J.E., Degrouard J., and Fléchon B. 2004. Gastrulation events in the pre-streak pig embryo. Ultrastructure and cell markers. *Genesis* **38:** 13–25.

Flynn T.T. and Hill J.P. 1942. The later stages of cleavage and the formation of the primary germ-layers in the Monotremata. *Proc. Zool. Soc. Lond. A* **3:** 233–253.

———. 1947. The development of the Monotremata. 6. The later stages of cleavage and the formation of the primary germ-layers. *Trans. Zool. Soc. Lond.* **26:** 1–151.

Fougerousse F., Bullen P., Herasse M., Lindsay S., Richard I., Wilson D., Suel L., Durand M., Robson S., Abitbol M., Beckmann J.S., and Strachan T. 2000. Human-mouse differences in the embryonic expression patterns of developmental control genes and disease genes. *Hum. Mol. Genet.* **9:** 165–173.

Frankenberg S., Tisdall D., and Selwood L. 2001. Identification of a homologue of POU5F1 (OCT3/4) in a marsupial, the brushtail possum. *Mol. Reprod. Dev.* **58:** 255–261.

Fujiwara T., Dunn N.R., and Hogan B.L. 2001. Bone morphogenetic protein 4 in the extraembryonic mesoderm is required for allantois development and the localization and survival of primordial germ cells in the mouse. *Proc. Natl. Acad. Sci.* **98:** 13739–13744.

Galbreath G.J. 1985. The evolution of monozygotic polyembryony in *Dasypus*. In *The evolution and ecology of armadillos, sloths and vermilinguas* (ed. G.G. Montgomery), pp. 243–246. Smithsonian Institution Press, Washington, D.C.

Gerard P. 1932. Etudes sur l'ovogenese et l'ontogenese chea les lemuriens de genre *Galago*. *Arch. Biol.* **43:** 93–151.

Goetz R.H. 1938. On the early development of Tenrecoidea (Hemicentetes semispinous). *Biomorphosis* **1:** 67–79.

Griffiths M. 1968. *Echidnas*. Pergamon Press, Oxford, United Kingdom.

———. 1978. *The biology of the monotremes*. Academic Press, New York.

Guillomot M. 1995. Cellular interactions during implantation in domestic ruminants. *J. Reprod. Fertil. Suppl.* **49:** 39–51.

Hartman C.G. 1916. Studies in the development of the opossum *Didelphys virginiana*. I. History of the early cleavage. II. Formation of the Blastocyst. *J. Morphol.* **27:** 1–83.

———. 1919. On the development of the opossum *Didelphys virginiana*. III. Description of new material on maturation, cleavage, and endoderm formation. IV. The bilaminar blastocyst. *J. Morphol.* **32:** 1–139.

Hendrickx A. 1971. *Embryology of the baboon*. University of Chicago Press, Illinois.

Heuser C.H. and Streeter G.L. 1929. Early stages in the development of pig embryos from the period of initial cleavage to the time of the appearance of limb-buds. *Contrib. Embryol. Carnegie Instn.* **20:** 1–30.

———. 1941. Development of the macaque embryo. *Contrib. Embryol. Carnegie Instn.* **29:** 17–55.

Hickford D. and Selwood L. 2002. Localization of transforming growth factor β2 (TGFβ2) and its receptors, TβRI and TβRII, in uteri and blastocysts of the stripe-faced dunnart (*Sminthopsis macroura*) during gastrulation. *Reproduction* **124:** 417–426.

———. 2003. Peri-gastrulation of the dasyurid marsupial *Sminthopsis macroura* (stripe faced dunnart) in vitro and evidence for patterning of the epiblast prior to gastrulation. *Mol. Reprod. Dev.* **4:** 402–419.

Hill J.P. 1910. The early development of the marsupialia, with special reference to the native cat (*Dasyurus viverrinus*). Contributions to the embryology of the marsupialia. IV. The bilaminar blastocyst. *Q. J. Microsc. Sci.* **56:** 1–134.

Hsu Y.C. and Gonda M.A. 1980. Monozygotic twin formation in mouse embryos in vitro. *Science.* **209:** 605–606.

Hubrecht A.A.W. 1889. Ueber die Entwicklung der Placenta von Tarsius und Tupaja nebst Bemerkungen über deren Bedeutung als haematopoietische Organe, pp. 343–411. *Proc. 5th Int. Congr. Zool.* Cambridge, United Kingdom.

———. 1912. Frühe Entwicklungs-stadien des Igels und ihre Bedeutung fur die Vorgeschichte (Phylogenese) des Amnions. *Zool. Jahrb.* (suppl. 15). **2:** 739–774.

Hue I., Renard J.P., and Viebahn C. 2001. Brachyury is expressed in gastrulating bovine embryos well ahead of implantation. *Dev. Genes Evol.* **211:** 157–159.

Hughes R.L. 1993. Monotreme development with particular reference to the extraembryonic membranes. *J. Exp. Zool.* **266:**480–494.

Hughes R.L. and Hall L.S. 1998. Early development and embryology of the platypus. *Philos. Trans. R. Soc. Lond. B Biol. Sci.* **353:** 1101–1114.

Kerr T. 1935. Notes on the development of the germ layer in diprodont marsupials. *Q. J. Microsc. Sci.* **77:** 305–315.

———. 1936. On the primitive streak and associated structures in the marsupial *Bettongia cuniculus. Q. J. Microsc. Sci.* **78:** 687–715.

Kölliker A. 1880. Die Entwick. d. Keimblätter des Kaninchens. *Zool. Anz.* nos. 61, 62, vol. III.

Lois C., Hong E.J., Pease S., Brown E.J., and Baltimore D. 2002. Germline transmission and tissue-specific expression of transgenes delivered by lentiviral vectors. *Science.* **295:** 868–872.

Loughry W.J., Prodohl P.A., McDonough C.M., and Avise J.C. 1998. Polyembryony in armadillos. *Am. Sci.* **86:** 274–279.

Luckett W.P. 1977. Ontogeny of amniote fetal membranes and their application to phylogeny. In *Major patterns in vertebrate evolution* (ed. M.K. Hecht et al.), pp. 439–516. Plenum Press, New York.

Lyne A.G. and Hollis D.E. 1977. The early development of marsupials with special reference to bandicoots. In *Reproduction and evolution* (ed. J.H. Calaby and C.H. Tyndale-Biscoe), pp. 293–302. Australian Academy of Science, Canberra.

McCrady E., Jr. 1938. The embryology of the opossum. *Am. Anat. Mem.* **16:** 1–233.

McLaren A., Molland P., and Signer E. 1995. Does monozygotic twinning occur in mice? *Genet Res.* **66:** 195–202.

Meijer H.A., Van de Pavert S.A., Stroband H.W.J., and Boerjan M.L. 2000. Expression of the organizer specific homeobox gene *goosecoid (gsc)* in porcine embryos. *Mol. Reprod. Dev.* **55:** 1–7.

Morris B. 1953. The yolk sacs of *Erinaceus europea* and *Putorius furo. J. Embryol. Exp. Morphol.* **5:** 184–200.

Mossman H.W. 1987. *Vertebrate fetal membranes.* Rutgers University Press, New Brunswick, New Jersey.

Mossman H.W. and Weisfeldt L.A. 1939. The fetal membranes of a primitive rodent, the thirteen striped ground squirrel. *Am. J. Anat.* **64:** 59–109.

New D.A.T. and Mizell M. 1972. Opossum fetuses grown in culture. *Science* **175:** 533–536.

New D.A.T., Mizell M., and Cockroft D.L. 1977. Growth of opossum embryos in vitro during organogenesis *J. Exp. Morphol.* **41:** 111–123.

Newman H.H. and Patterson J.T. 1909. A case of normal identical

quadruplets in the nine-banded armadillo, and its bearing on the problems of identical twins and of sex determination. *Biol. Bull.* **17:** 181–187.

———. 1910. The development of the nine-banded armadillo from primitive streak till birth, with special reference to the question of specific polyembryony. *J. Morphol.* **21:** 359–423.

Nowak R.N. 1997. *Walker's mammals of the world* (online version 5.1). Johns Hopkins University Press, Baltimore, Maryland.

Patten B.M. 1959. *Embryology of the pig*, 3rd edition. McGraw-Hill, New York.

Patterson J.T. 1913. Polyembryonic development in *Tatusia novemcincta. J. Morphol.* **24:** 559–684.

Renfree M.B. 1972. Influence of the embryo on the marsupial uterus. *Nature* **240:** 475–477.

———. 1982. Implantation and placentation. In *Reproduction in mammals,* 2nd edition (Embryonic and fetal development) (ed. C.R. Austin and R.V. Short), pp 26–69. Cambridge University Press, New York.

———. 1994. Monotreme and marsupial reproduction. *Reprod. Fertil. Dev.* **7:** 1003–1020.

Renfree M.B. and Lewis A.M. 1996. Cleavage in vivo and in vitro in the marsupial *Macropus eugenii. Reprod. Fertil. Dev.* **8:** 725–742.

Renfree M.B. and Tyndale-Biscoe C.H. 1973. Intrauterine development after diapause in the marsupial, *Macropus eugenii. Dev. Biol.* **32:** 28–40.

———. 1978. Manipulation of marsupial embryos and pouch young. In *Methods of mammalian reproduction* (ed. J.C. Daniel), pp. 307–331. Academic Press, New York.

Roberts C.M. and Perry J.S. 1974. Histricomorph embryology. *Symp. Zool. Soc. Lond.* **34:** 333–360.

Robinson W.P., McFadden D.E., Barrett I.J., Kuchinka B., Penaherrera M.S., Bruyere H., Best R.G., Pedreira D.A., Langlois S., and Kalousek D.K. 2002. Origin of amnion and implications for evaluation of the fetal genotype in cases of mosaicism. *Prenatal Diagn.* **22:** 1076–1085.

Sansom G.S. 1937. The placentation of the Indian musk-shrew (*Crocidura caerula*). *Trans. Zool. Soc. Lond.* **23:** 267–314.

Sansom G.S. and Hill J.P. 1931. Observations on the structure and mode of implantation of the blastocyst of Cavia. *Trans. Zool. Soc. Lond.* **21:** 295–354.

Selenka E. 1887. Studien uber Entwickelungs-geschichte der Thiere: Part 4. Das Opossum (*Didelphys virginiana*). C.W. Kreidels, Wiesbaden.

Selwood L. 1986a. The marsupial blastocyst—A study of the blastocysts in the Hill collection. *Aust. J. Zool.* **34:** 177–187.

———. 1986b. Cleavage in vitro following destruction of some blastomeres in the marsupial *Antechinus stuarti* (Macleay). *J. Embryol. Exp. Morphol.* **92:** 71–84.

———. 1989. Development in vitro of investment-free marsupial embryos during cleavage and early blastocyst formation. *Gamete Res.* **23:** 399–413.

———. 1992. Mechanisms underlying the development of pattern in marsupial embryos. *Curr. Top. Dev. Biol.* **27:** 175–233.

Selwood L. and Young G.J. 1983. Cleavage in vivo and in culture in the dasurid marsupial *Antechinus stuarti* (Macleay). *J. Morphol.* **176:** 43–60.

Simpson K.S., Adams M.H., Behrendt-Adam C.Y., Baker C.B., and McDowell K.J. 1999. Identification and initial characterization of calcyclin and phospholipase a2 in equine conceptuses. *Mol. Reprod. Dev.* **53:** 179-187.

Smith M.J. 1981. Morphological observations on the diapausing blastocyst of some macropodid marsupials. *J. Reprod. Fertil.* **61:** 141–150.

Snell G.D. and Stevens L.C. 1966. Early embryology. In *Biology of the laboratory mouse*, 2nd edition (ed. E.L. Green), pp. 205–245. Dover Publications, New York.

Stroband H.W.J. and Van der Lende T. 1990. Embryonic and uterine development during early pregnancy in pigs. *J. Reprod. Fertil. Suppl.* **40:** 261–277.

Suemori H., Tada T., Torii R., Hosoi Y., Kobayashi K., Imahie H., Kondo Y., Iritani A., and Nakatsuji N. 2001. Establishment of embryonic stem cell lines from cynomolgus monkey blastocysts produced by IVF or ICSI. *Dev. Dyn.* **222:** 273–289.

Tam P.P.L., Williams E.S., and Chan W.Y. 1993. Gastrulation in the mouse embryo: Ultrastructural and molecular aspects of germ layer morphogenesis. *Microsc. Res. Tech.* **26:** 301–328.

Tarkowski A.K. and Wroblewska J. 1967. Development of the blastomeres of mouse eggs isolated at the 4- and 8-cell stage. *J. Embryol. Exp. Morphol.* **18:** 155–180.

Thomas J.W. and Touchman J.W. 2002. Vertebrate genome sequencing: building a backbone for comparative genomics. *Trends Genet.* **18:** 104–108.

Thomson J.A., Kalishman J., Golos T.G., Durning M., Harris C.P., and Hearn J.P. 1996. Pluripotent cell lines derived from common marmoset (*Callithrix jacchus*) blastocysts. *Biol. Reprod.* **55:** 254–259.

Thomson J.A., Kalishman J., Golos T.G., Durning M., Harris C.P., Becker R.A., and Hearn J.P. 1995. Isolation of a primate embryonic stem cell line. *Proc. Natl. Acad. Sci.* **92:** 7844–7848.

Tripp H.R.H. 1971. Reproduction in the elephant shrew (Macroscelidae) with special reference to ovulation and implantation. *J. Reprod. Fertil.* **26:** 149–159.

Tyndale-Biscoe C.H. 1963. Blastocyst transfer in the marsupial *Setonix brachyurus*. *J. Reprod. Fertil.* **6:** 41–48.

———. 1970. Resumption of development by quiescent blastocysts transferred to primed, ovariectomized recipients in the marsupial *Macropus eugenii*. *J. Reprod. Fertil.* **23:** 25–32.

———. 1979. Hormonal control of embryonic diapause and reactivation in the tammar wallaby. *Ciba Found. Symp.* **64:** 173–190.

Tyndale-Biscoe H. and Renfree M. 1987. *Reproductive physiology of marsupials*. Cambridge University Press, United Kingdom.

Ullmann S.L., Shaw G., Alcorn G., and Renfree M.B. 1996. Migration of primordial germ cells to the developing gonadal ridges in the tammar wallaby *Macropus eugenii*. *J. Reprod. Fertil.* **110:** 135–143.

Van der Horst C.J. 1942. Early stages in the embryonic development of *Elephantulus*. *S. Afr. J. Med. Sci. Biol.* (suppl.) **7:** 55–65.

Walker M.T. and Rose R. 1981. Prenatal development after diapause in the marsupial *Macropus rufogriseus*. *Aust. J. Zool.* **29:** 167–187.

Ward S.J. and Renfree M.B. 1987. Reproduction in females of the feathertail gliders, *Acrobates pygmaeus* (Marsupialia). *J. Zool.* **216:** 225–240.

Wenstrom K.D., Syrop C.H., Hammitt D.G., and Van Voorhis B.J. 1993. Increased risk of monochorionic twinning associated with assisted reproduction. *Fertil. Steril.* **60:** 510–514.

Wilson J.T. 1928. On the question of the interpretation of the structural features of the early blastocyst of the guinea-pig. *J. Anat.* **62:** 346.

Wimsatt W.A. 1975. Some comparative aspects of implantation. *Biol. Reprod.* **12:** 1–40.

Wooding F.B., Staples L.D., and Peacock M.A. 1982. Structure of trophoblast papillae on the sheep conceptus at implantation. *J. Anat.* **134:** 507–516.

Yousef A. and Selwood L. 1993. Embryonic development in culture of the marsupials *Antechinus stuartii* (Macleay) and *Sminthopsis macroura* (Spencer) during preimplantation stages. *Reprod. Fertil. Dev.* **5:** 445–458.

SUGGESTED READINGS

In preparing this manuscript, we have relied heavily on several textbook sources. Much of the information was described in several sources, and is treated as "common knowledge" with regard to citation as the text themselves often did not cite individual research articles. In stating this, these texts were essential to the authors' own understanding of the processes described in this chapter. These texts are listed here.

Balfour F.M. 1895. *A treatise on comparative embryology*, vols. I-II. MacMillan, London.

Huettner A.F. 1942. *Fundamentals of comparative embryology of the vertebrates*. MacMillan, New York.

McEwan R.S. 1949. *Vertebrate embryology*, 3rd edition. Henry Holt and Company, New York.

Nelsen O.E. 1953. *Comparative embryology of the vertebrates*. McGraw-Hill, New York.

Patten B.M. 1958. *Foundations of embryology*. McGraw-Hill, New York.

Richards A. 1931. *Outline of comparative embryology*. John Wiley and Sons, Inc., New York.

Witschi E. 1956. *Development of vertebrates*. W.B. Saunders, Philadelphia, Pennsylvania.

Part II

Cellular Events of Gastrulation

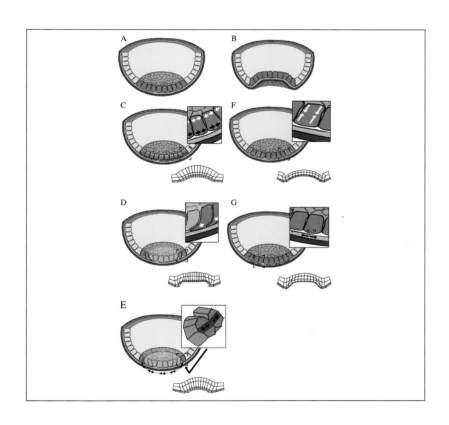

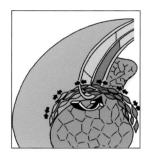

CELL MOVEMENTS
OF GASTRULATION

R. Keller[1] and L. Davidson[2]

[1]Department of Biology, University of Virginia, Charlottesville, Virginia 22903;
[2]Departments of Biology and of Cell Biology, Health Science Center, University of Virginia, Charlottesville, Virginia 22903

INTRODUCTION: CELLS AND MORPHOGENESIS

Here we discuss the general properties of some of the major cell movements of gastrulation. The developmental mechanics (Entwicklungsmechanik) school of embryology in the 1920s and 1930s sought to understand the cellular mechanisms underlying the inherited reiteration of form in each generation, but many embryologists were intimidated by the fact that gastrulation involved the coordinated behavior of hundreds or thousands of cells. In the words of Walter Vogt (1923):

> It does not appear at all as if cells were walking, in the sense that single part movements were combining to form the movements of the masses; for even the most natural and plausible explanation by means of amoeboid moving of single cells fails utterly. We evidently have not the wandering of cells before us, but rather a passive obedience to a superior force.
>
> (translated by Holtfreter, 1939)

Nevertheless, one of the bolder embryologists of that era, Johannes Holtfreter, founded several paradigms for emergence of tissue morphogenesis from local cell behavior. In tissue and cell recombination experiments, he showed that cells recognize one another as alike or different and that this recognition is expressed as "affinities" between cells, which are positive or negative, graded in intensity, and developmentally regulated. These affinities guide the sorting of cells into groups of like and unlike cells, and also function in assembly of cell populations into their proper relationships during gastrulation and early development (Holtfreter 1939; Townes and Holtfreter 1955; Steinberg 1964, 1970). Holtfreter also developed a paradigm for generating global tissue movements by mechanical integration of local, region-specific cell behaviors. He postulated that the polarized motility of the basal ends of the flask or bottle cells of the amphibian blastopore groove in response to the alkaline blastocoel environment results in the directed migration of these cells inward, and that they drag the marginal zone along as a result of their common apical connection (Holtfreter 1943a,b). Much more than this is going on at the blastopore lip (see Chapter 13), but Holtfreter's notion of global mechanical integration of a local cell behavior was a pioneering one. Time-lapse studies of cell behavior, notably in the echinoderm (Gustafson and Wolpert 1963) and teleost fish embryos (Trinkaus 1973a,b), revealed regional cell shape changes, motilities, and contact behavior, and generated plausible hypotheses of how these behaviors are integrated to produce tissue movement. Microsurgical manipulations of teleost epiboly (Trinkaus 1951) and amphibian gastrulation (Schechtman 1942) revealed mechanical relationships between regional tissue movements within the gastrula, and mechanical models (Lewis 1947), and eventually computer models (Jacobson and Gordon 1976; Odell et al. 1981; Davidson et al. 1995), related cell behavior to tissue movements. In a pioneering paper, Jacobson and Gordon (1976) combined analysis of cell movements, mechanics, and computer modeling. They showed experimentally that regional cell shape changes in amphibian neurulation are autonomous, and that these changes, along with an independent midline elongation force, account for the shaping of the amphibian neural plate. Using high-resolution time-lapse recordings of cultured cells, Abercrombie and his associates (Abercrombie and Heaysman 1953, 1954; Abercrombie et al. 1970a,b,c; Abercrombie 1980) pioneered the idea that cells "behave" and have a "sociobiology" comprising rules of contact and interaction, and that these parameters are not generic or nonspecific, but are specialized in a myriad of ways that imply underlying specificity of control and specific consequences at the tissue level. In his book, *Cells into Organs, The Forces that Shape the Embryo*, Trinkaus (1976, 1984) brought the rigor of the Abercrombie school to the study of embryonic cellular morphogenesis by careful definition of cell and tissue behavior, quantitative analysis, and formulation of plausible, testable hypotheses as to how these activities develop the forces that shape the embryo. Abercrombie and Trinkaus established the standards of analysis and the definitions of motility and contact behavior and laid the framework for modern analysis of cell behavior in embryos (Trinkaus 1984). However, it was with the development of molecular genetics and modern biochemistry that informative manipulations of molecules regulating morphogenesis could be made. Even then, the phenotypes were analyzed in terms of macroscopic defects, such as "arrested" or "inhibited" gastrulation, which perhaps identifies participating molecules but reveals little about mechanism. With the recent development of fluorescent dyes and green fluoroscent protein (GFP) as contrast-enhancing agents (Gimlich and Braun 1985; Honig and Hume 1986; Chalfie et al. 1994), powerful imaging and image processing methods (Inoué and Spring 1997; Periasamy 2001) and biomechanical methods of analyzing forces at the molecular, cellular, and tissue level, and cell and tissue material properties (Adams et al. 1990; Choquet et al. 1997; Galbraith and Sheetz 1998; Sheetz et al. 1998; Davidson et al. 1999; Hutson et al. 2003), the cell and its behavior have returned to center stage of morphogenesis, and of gastrulation in particular. However, to make the best use of these advances, we must have clear concepts of the various morphogenetic movements. Here we define some of the major morphogenetic processes of gastrulation, following the lead of Trinkaus (1984). We explore the limits of these definitions, discuss characteristics that might be important in how we think about these movements, design experiments to probe their mechanisms, and interpret the results of these experiments.

INVAGINATION

Invagination, the formation of tubes and grooves by bending cell sheets, is a common morphogenetic mechanism (Fig. 1A,B) (Gustafson and Wolpert 1963; Burnside and

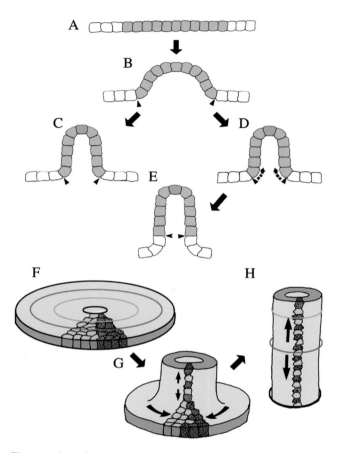

Figure 1. Invagination is depicted in sectional view (*A–E*) and in 3-dimensional view (*F–H*). Invagination can occur without involution (*A–C*) or with, or followed by, involution (*B,D,E*). Movement of the colored hoops shows convergence and extension during formation of a cylinder from a disc (*F–H*), due to intercalation of cells (*yellow* and *green*).

Jacobson 1968; 1976; Ettensohn 1985b) (see supplementary Movie 19_1 at www.gastrulation.org). The initial bending at the periphery of the invaginating tissue defines the limit of the process (arrowheads, Fig. 1A–C). However, in some cases, such as in amphibian gastrulation (Chapter 13) and in some echinoderms (Logan and McClay 1997), the initial invagination is followed by an involution, as cells initially outside the lip of the invagination roll around it and move inside (dashed arrows, arrowheads, Fig. 1D,E). Invaginations generally occur in epithelial or epitheloid sheets of cells (Ettensohn 1984, 1985b) and play substantial roles in gastrulation of amphibians (Holtfreter 1943a,b; Baker 1965; Perry and Waddington 1966), echinoderms (Gustafson and Wolpert 1963; Ettensohn 1984; Davidson et al. 1995), *Drosophila* (see Leptin and Grunewald 1990; Sweeton et al. 1991; Leptin et al. 1992; Leptin 1999; Oda and Tsukita 2000), ascidians (Conklin 1905, 1928; Satoh 1978; Munro and Odell 2002), and crustaceans (Hertzler and

Clark 1992). Often invagination precedes involution, as mentioned above, or ingression. For example, in amniote gastrulation, the apical constriction of epiblast cells forms a shallow trough, the primitive groove, where these cells subsequently undergo an epithelial–mesenchymal transition (EMT) and ingress (see Chapter 15). The same occurs in the bilateral subduction zones of the urodele amphibian (Shook et al. 2002).

The cells, or a subset of cells in the invaginating tissue, are thought to generate the forces that cause the movement, but this has been demonstrated only in a few cases. Moore and Burt (1939) and Ettensohn (1984) showed that isolated vegetal plates of echinoderms undergo the initial (primary) invagination, indicating that this is an active movement on the part of the participating cells. Likewise, in amphibians, bottle cell formation, and the resulting blastopore groove formation, occur in explants of presumptive bottle cells, showing that this is an active process (Holtfreter 1943a,b; Hardin and Keller 1988). Invagination of the neural plate of urodele amphibians is driven by regionally autonomous changes in cell shape (Burnside and Jacobson 1968), but external forces contribute to shaping the neural plate (Jacobson and Gordon 1976) and may contribute to its invagination in some species (Schroeder 1970; Smith and Schoenwolf 1991; Alvarez and Schoenwolf 1992; Hackett et al. 1997). It is not known which of the many invaginations occurring during gastrulation of various organisms are autonomous and which are driven, or assisted, by external forces.

Invaginations are often depicted in sectional view where changes in cell shape figure heavily in concepts of mechanism. Transition from a cuboidal to a wedge shape at the bend points could bring about bending of the cell sheet (Fig. 1A,B). Two-dimensional mechanical models using rubber and brass (Lewis 1948), and computer models (Odell et al. 1981), verify that cell wedging will cause bending of the sheet. Cell wedging is observed in some invaginations, including blastopore groove formation (Holtfreter 1943a; Baker 1965; Perry and Waddington 1966; Hardin and Keller 1988) and neurulation (Burnside and Jacobson 1968) in amphibians, and in formation of the primitive groove of birds (see Chapter 15). In other cases, such as gastrulation in some echinoderms, this relationship is less obvious (Ettensohn 1984), probably because mechanisms other than cell wedging are involved.

Considering the three-dimensional aspects of invagination suggests what these additional mechanisms might be. Forming a tube from a flat disc involves changes in tissue shape not apparent in sectional view (Fig. 1F–H). The circumference of the disc must undergo progressively greater decreases toward the periphery to form a cylindrical tube (follow the colored bands, Fig. 1F–H). What is the mechanism of such a progressive decreasing in circumference? Circumferential cell intercalation throughout the disc, or only along its perimeter, could

produce compressive forces that might drive the inward buckling of the central part of the disc, and perhaps subsequent involution over the lip of the invagination (Fig. 1D,E). Circumferential intercalation occurs during echinoderm gastrulation but only after an initial (primary) invagination forms a short archenteron; it drives the initial stages of secondary invagination, the elongation and reduction in diameter (convergent extension) of the short primary invagination (Ettensohn 1985a; Hardin and Cheng 1986; Hardin 1988, 1989). Initiating circumferential cell intercalation at an earlier stage might be sufficient to drive primary invagination as well, but there is no evidence for this mechanism thus far. However, there is evidence that other cell activities act in the spherical geometry of the embryo to drive invagination.

MULTIPLE MECHANISMS OF INVAGINATION ARE PLAUSIBLE

Studies on echinoderm primary invagination show that several mechanisms could drive invagination. Echinoderm primary invagination involves a single-layered epithelium and an extracellular matrix (ECM) consisting of two layers, an inner apical lamina (light gray, Fig. 2A) to which the cells are attached, and an outer hyaline layer (dark gray, Fig. 2A). Preceding primary invagination, the vegetal epithelium first flattens and thickens to form the vegetal plate (not shown), and then it bends inward to form a depression between 5 and 10 microns deep (Fig. 2A,B) (Ettensohn 1984, 1985a) (see supplementary Movie 19_1 at www.gastrulation.org). Microsurgically isolated vegetal plates undergo primary invagination (Moore and Burt 1939), so the forces are generated within the isolated plates. Five plausible mechanisms have been proposed to account for the observed behaviors. According to the *apical constriction* model (Lewis 1947), cells of the vegetal plate contract apically, which drives their cytoplasm basally and expands the basal ends of the cells (Fig. 2C), thereby resulting in wedge-shaped cells. Summed over the vegetal plate, this wedging is thought to produce bending of the epithelium (Nakajima and Burke 1996; Kimberly and Hardin 1998). In the *cell tractor* model (Burke et al. 1991), the apical surfaces of epithelial cells arrayed in an annular ring are thought to exert traction on the overlying ECM, and thus generate a force directed toward the center of the vegetal plate (Fig. 2D); this centripetal force is thought to buckle the plate inward. In the *apical contractile ring* model, Davidson and others (1995) propose that a similar, centripetal force is generated by a contractile cytoskeletal ring that runs through the periphery of the vegetal plate. As the ring contracts, it reduces the circumference of the tissue at that level and causes buckling (Fig. 2E) (Davidson et al. 1995). In the *apical–basal shortening* model (Gustafson and Wolpert 1963), the cells are thought to contract their api-

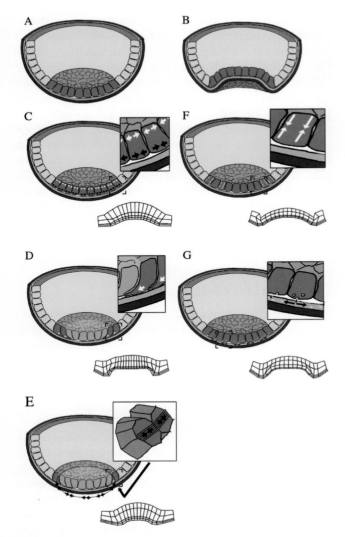

Figure 2. Mechanisms of invagination (*A*,*B*) are illustrated in diagrams, including the apical constriction (*C*), the cell tractor (*D*), the apical ring contraction (*E*), apical–basal contraction (*F*), and the gel-swelling mechanisms (*G*). Cell behavior is magnified in the upper insets, and changes in cell shapes predicted by finite element modeling are shown in the lower insets. (Modified from Davidson et al. 1995.)

structure of interest into subunits (finite elements) (Zienkiewicz and Taylor 1991). Given a certain stiffness of the cell layer (20 Pa), and estimates of the forces generated by each mechanism, each of these models will work for some set of values for the stiffness of the inner and outer matrix layers but not for others (Fig. 3A) (Davidson et al. 1995). Invagination by the cell tractor, apical–basal contraction, and gel swelling overlap substantially in a two-dimensional parameter space (see position I, Fig. 3B). The apical constriction and cell tractor modes can occur at position II, and the apical constriction and apical ring contraction are successful at position III (Fig. 3B). FEM also predicts the changes in cell shape during invagination by each of these methods (bottom insets, Fig. 2C–G). Actual measurements of the stiffness of the cells and the matrix layers in *Strongylocentrotus purpuratus* suggest that for this species, where the matrix was found to be nearly six times stiffer than the cell layer, all the models except the cell tractor and gel swelling are ruled out (Davidson et al. 1999).

These studies show that there are several plausible mechanisms for invagination and identify the important parameters, including cell behaviors, the magnitude and orientation of the forces generated, the mechanical properties of the tissue and ECM, and the global geometry of expression of these activities and properties. These properties are genetically encoded in the sense that they are the result of specific cytoskeletal functions, cell polarities, and the type of matrix that is organized by the cells. Most models involve a contractile force generated by the cytoskeleton, but in some cases it is oriented tangentially (parallel) to the plane of the tissue and in others it is oriented radially (perpendicularly) with respect to the tissue surface. The cell tractor and apical contractile ring both involve a centripetal force, but this force is generated by polarized traction on ECM in the first and by coordinated circumferential contraction of the cytoskeleton in the second. Contraction is tangential to the tissue plane in the apical constriction model and normal to it in the apical–basal contraction model. The patterning of the shape of the area of cell activity is also important. The contractile ring and cell tractor models invoke cell activity in a ring of cells, whereas the apical constriction model involves all the invaginating tissue, as does the gel-swelling model. The gel-swelling model does not invoke cytoskeletal or cellular forces directly, but depends on swelling of an ECM deposited by polarized secretion, rather than on polarized motility or contraction.

It is clear that the functional morphogenetic "pathway" of invagination involves genetically encoded information in a number of locations and formats, controlling a number of cellular parameters, and that interpretation of genetic and molecular instructions will be complex. Each mechanism depends on specification of a local force-generating event, the local polarization of this event, the global geometry of the area of expression of this event, and specification of

cal–basal aspect and generate compressive forces due to the expansion of the cell bodies in the planar dimension, which results in buckling of the sheet (Fig. 2G). Finally, in the *gel-swelling* model, Lane and others (1993) provide evidence that a proteoglycan is secreted into the inner, apical lamina layer of the double-layered hyaline layer, which swells with a force typical of such hydration events, and bends the epithelium (Fig. 2G). Each of these models is consistent with or is supported by observations and experimental data in one or another species of echinoderm (Davidson et al. 1995).

Finite element modeling (FEM) shows that each of these models will produce an invagination under certain conditions. FEM is a numerical approach to solving complex problems in the mechanics of structures that involves dividing the

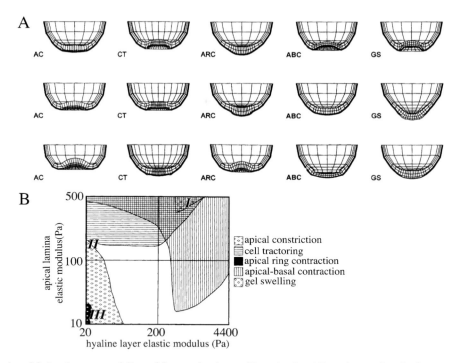

Figure 3. Results of finite element modeling of five mechanisms of invagination (*A*) at three points in the parameter space of elastic modulus for apical lamina and the hyaline layer of echinoderms (*B*). (AC) Apical constriction; (CT) cell tractor; (ARC) apical ring contraction; (ABC) apical–basal contraction; (GS) gel swelling. (Reprinted from Davidson et al. 1995.)

mechanical properties of tissues or matrix that harness local forces to produce tissue movements. A few cellular, force-generating processes can be used in a number of geometric and mechanical contexts to generate multiple mechanisms of invagination, and the evolvability of these mechanisms is defined, in part, by the constraints revealed in modeling the cell and tissue mechanical parameter spaces.

INVOLUTION: TISSUE ROLLING OVER AN INFLECTION POINT

Involution consists of the rolling of tissue over an inflection point, or around a lip, and often, but not always, back on itself (Fig. 4A). Involution resembles the movement of a bulldozer tread viewed from the front of the bulldozer; the track or tissue moves toward the front of the machine, goes around the wheel (the blastoporal lip) and disappears under the bulldozer (inside the embryo) (see supplementary Movie 19_2 at www.gastrulation.org). In some cases, the tissue rolls over an inflection point but does not become apposed to the inner surface of the pre-involution or non-involuting tissue (Fig. 4B). An example of the first type are the movements of the involuting tissue in amphibians (Chapter 13), and an example of the second type occurs in some echinoderms where involution occurs late, over a blastoporal lip initially formed by invagination (Fig. 1C,D)

(Logan and McClay 1997; Ransick and Davidson 1998). Trinkaus (1984, 1996) defines involution as the movement of a coherent sheet of cells over the edge of an inpocketing or invagination, and distinguishes it from an ingression of cells, in which cells move inward without necessarily retaining their coherence, their order within the sheet. In the amphibian, both an epithelial cell sheet and one or more deep mesenchymal cell layers involute together (Fig. 4D). The epithelial layer maintains its coherence, but the details of the cell movements in the deep region are unknown. In the teleost, a single layer, the epiblast, which is apparently epitheloid but perhaps not a tight-junctioned epithelial sheet, undergoes involution-like movements that also have elements of ingression behavior (Shih and Fraser 1995; Trinkaus 1996; Glickman et al. 2003). Involution-like movements also may result from other morphogenetic movements. For example, in the bird, epiblast cells describe an overall involution movement as they move toward the primitive streak, undergo EMT and ingression to become deep mesenchymal cells, and then most of them reverse their direction and migrate laterally beneath the epiblast (red arrows, Fig. 4C; see also Chapter 15). In this case, as with most involution-like movements, the amount of coherence during involution is difficult to determine.

As with invagination, there are a number of plausible ways to drive involution. Traction of the post-involution cells (red cells, Fig. 4D) on the undersurface of the pre-involution

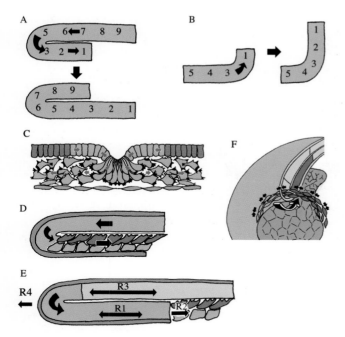

Figure 4. Diagrams show tissue movements during an involution in which the tissue turns back on its inner surface (*A*) or not (*B*). The involution component of ingression at the primitive streak of amniotes is illustrated (*red arrows*, *C*). Some involutions, such as the one in *Xenopus* gastrulation, involve both an epithelial layer (*gray*) and a deep mesenchymal layer (*D*). Differential rates of extension (R3>R1–R2) can drive involution (*curved arrow*) and extend tissue in the opposite direction (R4) at the same time (*E*). Transverse hoop-stress generated by mediolateral intercalation (*red cells*) produces convergence (*black arrows*), which can drive involution (*curved black arrow*) in amphibians (*F*).

Differential rates of extension could also drive involution. In the amphibian, the posterior neural plate extends posteriorly at a faster rate (R3, Fig. 4E) than the involuted axial and paraxial mesoderm (R1, Fig. 4E), and part of this extension is absorbed by anterior migration of the mesendoderm (R2, Fig. 4E) in the opposite direction, giving these tissues a lower overall rate of posterior extension (R4, Fig. 4E). As a result, the outer tissue may tend to push the IMZ over the lip (curved arrow, Fig. 4E). Extending tissues can push. The IMZ pushes with a force of at least 0.5–0.6 micronewtons (Moore 1994) and the NIMZ (posterior neural) probably does likewise (Keller and Danilchik 1988; Moore et al. 1995), although the force has not been measured. However, neural extension is not necessary for involution of the IMZ, since it will still occur after removal of the animal cap and neural tissue (Keller and Jansa 1992). Forces in the transverse dimension allow the IMZ to bootstrap its way over the lip without tissue behind it and without a substrate for its post-involution migration. The posterior progression of mediolateral intercalation behavior (MIB) generates convergence, or shortening of arcs of tension spanning the dorsal blastoporal lip, and this hoop stress serves to drive involution (Fig. 4F). As MIB progresses posteriorly, the easiest route to accommodate the resulting arc shortening is for the IMZ to roll over the lip. This mechanism appears to drive involution of the IMZ in absence of all other, outer tissues (Keller and Jansa 1992). Progressive constriction and hoop stress, perhaps driven by circumferential cell intercalation, could also reduce the circumferences of a disc of cells in a way that would drive involution of the disc to form a tube (Fig. 1F–H), and it would work without re-apposition of the pre- and post-involution layers (Fig. 1D,E).

Involution could also be a passive response to forces generated elsewhere. Vegetal rotation rolls the IMZ inward and initiates its involution in early gastrulation (Winklbauer and Schürfeld 1999). Vegetal rotation seems like a better initiator than a sustainer of these movements, however, because after the initial rolling, the archenteron lies between the endoderm and the overlying involuting tissue, making it unlikely that forces for involution are transmitted around this obstacle.

Little is known about what cells actually do during involution, how their behavior changes, and how these changes contribute to involution. In the amphibian, the deep cells have a pre-involution behavior in which they re-integrate into a stable, pre-involution tissue when placed on its inner surface; in contrast, after transition into the post-involution region, the cells spread and migrate on the surface of the pre-involution region instead of re-integrating into it (Winklbauer and Keller 1996). This behavioral transition, called separation behavior, may coincide with the posterior progression of MIB in *Xenopus* (Shih and Keller 1992b), which appears to occur at or near the point of involution. A similar transition to a migratory phenotype occurs during involution in teleost fish (Ulrich et al. 2003).

tissue could generate a shearing force that would contribute to or perhaps drive involution (Fig. 4D). The mesendoderm of the amphibian, which migrates directionally as a cell stream toward the animal pole (Fig. 4D) (Winklbauer and Nagel 1991; Davidson et al. 2002), could play this role. However, removal of this overlying substrate does not block involution of the involuting marginal zone (IMZ) (Keller and Jansa 1992), and thus this mechanism is not absolutely necessary, but it may make a contribution in the normal embryo. Similar shearing forces could contribute to involution in fish gastrulation and could also have cryptic functions in the EMT and ingression in amniote gastrulation. Traction of the ingressed, laterally migrating mesoderm might exert a medially directed force on the overlying epiblast and thereby generate a shearing of the epiblast medially relative to the lateral movement of the mesoderm (Fig. 4C). However, removal of a patch of ingressed mesoderm does not affect movement of the overlying epiblast cells toward the primitive streak, suggesting that in the bird, such traction plays no role (Stern and Ireland 1981).

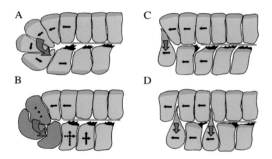

Figure 5. Diagrams illustrate possible changes in cell polarity during involution. Apical–basal polarity is indicated by gray apices and red basal ends; planar polarity is indicated with black arrows within cells. Movements are shown with gray arrows. Cells could cartwheel over the lip (A), lose their polarity and reestablish a new, unrelated polarity after involution (B), move radially in serial order (C), or move radially without a strict serial order (D).

Cells could roll over a lip in a number of ways in terms of behavior and polarity. Involution in its purest form is defined as the movement of a coherent cell sheet over a lip (Trinkaus 1996), and thus the elements of tissue coherence, such as cell polarity and the order in which cells undergo their movements, are key issues in understanding involution, ingression, and the differences between them. Cell polarity could change in a number of ways between pre- and post-involution cells. Assume that pre-involution cells have an apical–basal polarity and also an anterior–posterior polarity (Fig. 5). The involuting cells could "roll" or "cartwheel" over the blastopore lip and maintain their initial polarity such that the former outer ends are now innermost and the inner ends are outermost and apposed to the inner pre-involution cells (Fig. 5A). In this case, their animal–vegetal polarity would be reversed with respect to the pre-involution tissue but maintained within the sheet. Alternatively, any pre-involution polarization might be disassembled and a new, post-involution polarity established (Fig. 5B). Cells could also drop directly out of the pre-involution layer and join the deep layer in a movement that locally would be described as an ingression (Fig. 5C,D). In this case, the animal–vegetal polarity in the pre-and post-involution tissues would be the same, and the outer end would be apposed to the inner pre-involution region (Fig. 5C,D). The "ingression" might occur in a strict progression, and the serial order of cells would be maintained within the sheet but reversed between the pre- and post-involution regions (Fig. 5C). Alternatively, they might not ingress in serial order, and mixing would occur in the animal–vegetal axis (Fig. 5D).

It is not clear which of these occur in nonepithelial systems. A rolling type of motion has been recorded in the teleost fish (Thorogood and Wood 1987), and in *Xenopus* the appearance of cells in scanning electron microscopy is often consistent with the rolling model (Fig. 5A,B) (Keller and

Schoenwolf 1977; Keller 1984), but there is no direct evidence on the order of movement over the lip or the maintenance of polarity. Other preparations reveal a broad zone of involution up to 11 or 12 cells in animal–vegetal extent, suggesting that the order of moving over the lip may not be tightly regulated (L. Davidson, unpubl.). In the zebrafish (Shih and Fraser 1995) and *Fundulus* (Trinkaus 1996), the passage of the cells over the lip has been described as an ingression because the cells do not always drop out of the epiblast in a tightly organized, sequential fashion, but as individuals here and there. These movements have also been described as an involution (Glickman et al. 2003) because their overall path is to move vegetally on the outside, drop to a deeper level in some fashion, and then move animally relative to the epiblast. More work should be done to determine the details of cell behavior during involution in various species.

CONVERGENCE AND EXTENSION

The cell motility, polarity, biomechanics, and the molecular aspects of convergence and extension have been reviewed previously (Keller et al. 1991, 1992a,b, 2000; Keller 2002; Myers et al. 2002b; Wallingford et al. 2002). Convergence and extension were described as the narrowing and lengthening, respectively, of the dorsal tissues of the amphibian gastrula and neurula (see supplementary Movies 19_1 and 19_3 at www.gastrulation.org), revealed by distortion of vital dye marks (Vogt 1929). It was not clear whether these movements are active or passive, and the mechanistic relationship between convergence and extension was not understood. Explantation experiments showed that these movements are active, force-producing movements (Schechtman 1942; Keller 1984; Keller and Danilchik 1988; Moore 1994; Moore et al. 1995) (see supplementary Movie 19_3 at www.gastrulation.org), and tracing of fluorescently labeled cells and imaging their behavior showed that convergence was biomechanically coupled to extension by cell intercalation (Keller 1984; Keller and Danilchik 1988; Keller et al. 1985, 1989). Convergence (narrowing) occurs as cells actively wedge between one another (cell intercalation) along one axis (the mediolateral) (see supplementary Movie 19_4 at www.gastrulation.org), and thereby produce a pushing force in the transverse axis (the anterior–posterior), which results in extension (Wilson et al. 1989; Wilson 1990; Wilson and Keller 1991; Shih and Keller 1992a,b). This biomechanical coupling of convergence to extension, and also to thickening of the tissue, distinguishes this active form of convergence and extension from other examples during morphogenesis in which cells converge toward a point or a line but do not produce extension. Here we explore some of the properties of active, force-producing convergence and extension by cell intercalation and compare these movements to other, superficially similar processes.

Convergence and extension can be generated by the activities of the cells within the tissue or as a passive response to external forces. The superficial epithelial sheet of the dorsal mesodermal and posterior neural tissue of the *Xenopus* gastrula appears to be passively distorted by the active convergence and extension of the underlying deep, mesenchymal region (Moore 1994; Moore et al. 1995; Shih and Keller 1992a; Keller et al. 2000). It is not clear which of the many examples of convergence and extension in embryonic development are active, internally driven processes and which are passive responses (Keller et al. 2003).

The mechanical coupling of convergence and extension differs depending on whether the process is active or passive. In the active form, cells forcibly wedge between one another, which generates a force in the transverse direction, and the summation of these local exchanges of neighbors drives the extension of the cell population as a whole (Fig. 6A). In the passive form, the tissue is stretched (extended) by external forces, and the cells accommodate this distortion first by elongating parallel to the axis of extension and narrowing perpendicularly (Fig. 6C,D), and then by intercalating passively in this axis while returning to an isodiametric shape (Fig. 6D,E) (Keller 1978). In this case, convergence does not drive extension, but passive stretching (extension) occurs in response to external forces and drives convergence.

During the active process of convergence by cell intercalation, convergence is also mechanically coupled to tissue thickening. As the cells wedge between one another along one axis, and the tissue comes under compression in the transverse axis, it tends to extend, but it also tends to thicken, as cells either become taller, or pile up and become multilayered (Fig. 6B). This multilayering constitutes "de-intercalation," the reverse of the radial intercalation that causes thinning and extension at an earlier stage (Wilson and Keller 1991; Keller et al. 2003). Tissues differ in their efficiency of transducing convergence into extension as opposed to thickening, the notochord being efficient and the somitic mesoderm less so, for example (Keller et al. 1989; Wilson et al. 1989). Thus, convergence is biomechanically linked to extension (convergent extension) and to thickening (convergent thickening). It should be understood that the term "convergent extension" does not necessarily preclude some component of convergent thickening.

Active extension is not always driven by active convergence. The initial phase of extension of both mesodermal and neural tissues in the early amphibian gastrula occurs by radial intercalation and thinning, a thinning of the tissue that is channeled largely into extension of the anterior–posterior axis (Keller 1980; Keller and Danilchik 1988; Wilson et al. 1989; Wilson 1990; Wilson and Keller 1991).

The linkage of convergence to extension by cell intercalation has been described reasonably well only in the dorsal

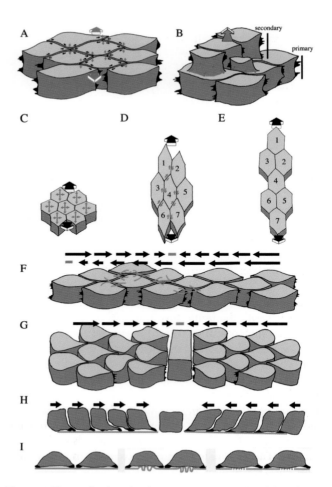

Figure 6. The mechanics of active convergent extension (*A*) and convergent thickening (*B*) by cell intercalation are illustrated. Local shearing of bipolar cells between one another (*black arrows, A*) results in compressive forces (*small red arrows, A*), which are summed to produce extension (*large red arrows, A*). Compression in the anterior–posterior axis results in a tendency to thicken (convergent thickening), either due to the cells becoming taller or by multilayering (*red arrows, B*). In passive convergent extension, the cell layer is stretched and the cells elongate (*C,D*) and then passively shear past one another (*red arrows, D*) and return to isodiametric shapes (*E*). Bipolar cell-on-cell traction (*F*) results in an increase in rate of movement (*black arrows*) with distance from any fixed point (*red bar*). Monopolar cell-on-cell traction (*G*) results in increased rates of approach (*black arrows*) to the fixed point (the midline, *red bar*). Cell-on-external substrate traction (*H*) results in uniform rates of approach to the midline (*arrows*), unless other factors come in to play. Potential results of balanced, bipolar traction on a rigid substrate (*I, left*), a flexible substrate (*I, middle*), and a digestible one (*I, right*) are illustrated. A flexible substrate could also be used to pull cells between one another, as tethers between cells (*green, F*).

tissues of the amphibian (Keller et al. 1992a, 2000) and in teleost fish (Glickman et al. 2003); in both these organisms, growth does not occur during pre-larval stages, and cell division and changes in cell volume do not play significant roles. In amniotes, where embryonic growth occurs, both cell intercalation and cell divisions oriented along the anterior–posterior axis have been described in the bird and the mouse, and both may play a role in extension (Schoenwolf and Alvarez 1989, 1992; Sausedo and Schoenwolf 1993, 1994; Sausedo et al. 1997; Wei and Mikawa 2000).

In active convergence and extension by cell intercalation, it is the relative movement of the cells with respect to one another that is important. The most direct way to bring this about is for the cells to exert traction directly on one another, rather than on an external substrate. Therefore, we proposed the cell-traction/cell-substrate model (Keller et al. 1992a, 2000). In this model, cells form filo-lamelliform protrusions in the medio-lateral axis, which exert traction on the cortices of cells lateral and medial to themselves (see supplementary Movie 19_5 at www.gastrulation.org). This mutual traction is thought to form mediolaterally continuous chains of cells that are under mediolateral tension and are shortened by mediolateral inter-calation. The local paths of tension are self-defining and remodeled with the progress of intercalation. In this model, the shortening produced by the local cell intercalation is summed over the chain, and cells move faster toward any point chosen as a reference within the chain, the farther they are from it (Fig. 6F). The bipolar, mediolaterally oriented pro-trusive activity observed during mediolateral intercalation is most consistent with the cells pulling themselves between one another by local traction on their neighbors. Bipolar traction on a deformable, extracellular substrate might also drive inter-calation. The deformable substrate could form tethers that transmit traction to neighboring cells, and thus pull cells between one another. However, it is difficult to see how bipo-lar traction on a rigid external substrate, or on a deformable one that did not somehow serve as an intercellular tether, could drive intercalation.

Cell traction on one another and traction on an external substrate would produce different rates of cell displacement within a tissue, all other things being the same. In the dorsal regions of both amphibians (Shih and Keller 1992a,b) and teleost fish (Glickman et al. 2003), cells approach the mid-line faster the farther they are from the midline. This is con-sistent with the cells crawling on one another, and their relative rates becoming additive with distance from the mid-line. However, in the case of the bipolar cells, cell intercala-tion simply shortens an array of cells, and the cells move toward any point chosen as a reference at increasing rates the farther they are from it. In the frog embryo, mediolater-al intercalation is organized in an arc pattern, spanning the blastoporal lip and anchored at each end near the vegetal endoderm. As a result, cell intercalation actually brings the dorsal tissues toward the vegetal endoderm, with the points farther away from the endoderm (and nearer the dorsal midline) moving fastest toward the endoderm (Chapter 13). In giant sandwich explants, made with free mid-ventral

edges, however, the dorsal midline becomes the fixed point, and cells farther from it move toward it more rapidly than those closer to it. In lateral, giant sandwich explants, cut with free middorsal and midventral edges, the ventral becomes the fixed point and dorsal is pulled toward the ven-tral with cells farther dorsally moving fastest toward the ventral (Keller and Danilchik 1988; Shih and Keller 1992a,b; Poznanski et al. 1997).

In contrast, in teleost fish, there is no lateral anchorage to a vegetal endoderm, and the cells in the embryonic shield converge on the dorsal midline (Glickman et al. 2003), rather than pulling the dorsal tissue toward endoderm as in amphibians. Farther laterally and ventrally, the cells of the germ ring migrate directionally toward the shield, with those farther away moving more slowly (Trinkaus 1998; Trinkaus et al. 1992). The biomechanics of frog and teleost fish gastrulation differ in that in *Xenopus* vegetal rotation and convergent extension are used to enclose the yolk, whereas in the zebrafish the epibolic towing by the yolk cytoplasmic layer (Betchaku and Trinkaus 1978) seems to play a larger role (Keller et al. 2003).

The difference between a directed cell migration on an external substratum and a directed cell intercalation is illus-trated by the behavior of the deep neural cells during conver-gence and extension of the neural plate of amphibians. Instead of a bipolar mediolaterally directed protrusive activity, the deep neural cells have a midline directed protrusive activity (Elul et al. 1997, 1998; Elul and Keller 2000; Ezin et al. 2003). If the medially directed protrusive activity reflects traction on more medial cells, and a constant rate of translocation on their surfaces, these local rates of translocation become additive; the cells farthest laterally benefit from the translocation of all those medial to themselves, and the rate of convergence increases with distance from the midline (Fig. 6G). Lateral cells move relatively faster toward the midline, intercalate between medial neighbors, and generate mediolateral inter-calation and extension over a broad area. In contrast, if this activity reflects a directed migration on an external substra-tum, then the local rate of translocation would bring all cells toward the midline at the same rate, no local differential rate of movement would occur and, therefore, no mediolateral intercalation, because each cell's medial progress would be matched, on average, to that of the cells more medial (Fig. 6H). Mediolateral intercalation would occur only at the boundary of the notoplate, where each succeeding tier of neural cells could proceed no farther (Elul and Keller 2000). The rate of convergence would be uniform overall with a rapid drop just lateral to the midline notoplate where the rate of intercalation and extension would be high.

Cell intercalation could occur by traction on a rigid, external substrate as well, but each tier farther laterally would have to move medially faster than its medial neighbors in order to intercalate between them and produce extension.

This could occur by progressive up-regulation of migration speed with distance from the midline, or progressive down-regulation as cells approach the midline. In addition, cells could encounter progressively greater mechanical resistance, which would slow their medial movement, as they near the midline.

It is important not to lump all convergence movements into the convergent extension category. Directed migration on an external substratum, without increasing rates of translocation, can produce convergence toward a midline but not patterned cell intercalation and extension. The cells of the germ ring of the teleost converge toward the embryonic shield, and although they occasionally intercalate, there is no organized intercalation that produces extension (Trinkaus et al. 1992; Trinkaus 1998; Myers et al. 2002a). These cells appear to migrate as individuals in a cell stream, and they move faster toward the midline as they approach the midline (Trinkaus et al. 1992; Trinkaus 1998), a behavior most easily explained as a migration on an external substrate, in this case, possibly the overlying enveloping layer or the underlying yolk syncytial layer. Near the midline, in the embryonic shield, the cells of teleost fish do undergo convergent extension, and they express the velocity gradient of approach to the midline, increasing with distance from the midline (Glickman et al. 2003), suggesting that at that point they may switch to a cell-on-cell traction, similar to that postulated for amphibians. Likewise, the oriented, bipolar protrusive activity, if it results in balanced traction on the external substratum, could either remain motionless (Fig. 6I, left), pile up a deformable substrate beneath the cell (Fig. 6I, middle), or break down the substrate there (Fig. 6I, right). In the latter two cases, reeling in the substrate toward the cell could be used as a tether to exert forces on other cells, and thus perhaps function in intercalation.

EMT AND INGRESSION

Gastrulation often involves an EMT (Hay 1995; Hay and Zuk 1995), and EMT is usually followed by ingression of the mesenchymal cells, which is the process of the cell crawling out of the epithelium, or being excluded from it, and entering the deep mesenchymal region (Fig. 7A–J) (Shook and Keller 2003) (see supplementary Movie 19_6 at www.gastrulation.org). The ingression of the primary mesenchyme cells in the echinoderm (Chapter 9), the ingression of a large area of superficial presumptive endoderm and mesoderm from the epiblast through the primitive streak of the amniote (Chapters 15 and 16), and ingression of mesoderm through the bilateral subduction zones of the urodele amphibian (Shook et al. 2002) are examples. Ingression does not necessarily follow an EMT since the cells could stay

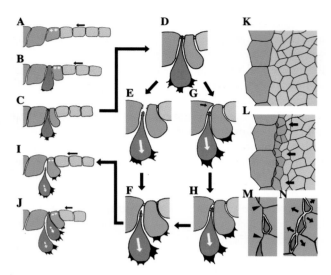

Figure 7. Apical constriction, EMT, and ingression are illustrated in sectional view (*A–J*), including ingression without prior sealing (*D–F*) and with sealing over the ingressing cell (*D,G,H*), and in surface view (*K–N*). (Modified from Shook and Keller 2003.)

in place, and EMT is sometimes followed by other processes such as relamination, a process in which cells join the deep mesenchymal region at their basal ends, appear to take on mesenchymal characteristics, and become covered over by the epithelium (Minsuk and Keller 1996). Ingression is usually preceded by constriction of the apical ends of the epithelial cells, which is usually accompanied by a cell shape change, most commonly an elongation in the apical–basal axis and a bulging of the opposite ends of the cells to form the classic "bottle" or "flask" shape. Such cell shape changes result in an invagination, such as the trough or groove at the primitive streak, for example. However, grooves of this type are incidental invaginations in the course of what is primarily a process of ingression, in contrast to invaginations, such as neural tube formation, in which the primary process is change in cell shape and rolling of a sheet into a trough.

The apical constriction component of ingression may have several functions. It reduces the perimeter of the apical junctional contact with adjacent epithelial cells and thereby may facilitate loss of adhesion and ingression. Apical constriction may also pull cells farther away toward the zone of ingression as the apical area decreases (black arrows, Fig. 7A–J). Such a towing function could pull the epiblast toward the primitive streak of amniotes, and the superficial mesoderm of the amphibian toward the subduction zone (see supplementary Movies 19_7 and 19_8 at www.gastrulation.org). There, EMT and ingression of these cells would make room for more cells to enter the primitive streak, or subduction zone, where they repeat the process. To work, such a mechanism would require a tension-bearing, mechanical continuity of the epithelium during EMT. That could be done by temporarily breaking continuity locally, as individual cells ingress (Fig. 7D–F). This mechanism requires that as apical constriction occurs and tension in the sheet increases (Fig. 7K,L), not all cells would ingress at once, but would do so as individuals, and thus continuity is maintained (Fig. 7M). Simultaneous EMTs in adjacent cells would leave large lesions in the cell sheet (Fig. 7N). Alternatively, the adjacent cell or cells could bridge across the ingressing epithelial cells before the latter lose their adhesions, and thus maintain continuity (Fig. 7D,F–H).

OTHER MOVEMENTS

Other movements of gastrulation have been less problematic in terms of establishing a common terminology and understanding. Delamination is the mass separation of an entire layer from a preexisting layer, such as the formation of splanchnic and somatic mesodermal sheets from the lateral mesodermal layer (Nelsen 1953). Ingression is sometimes referred to as "delamination," which is accurate in that the cells leave a lamina or sheet, but the term is best not used this way, because delamination was and still is used to refer to the mass separation described above. It is also in this sense of a mass reassociation of one layer with another that "relamination" was invoked (Minsuk and Keller 1996). Epiboly refers to the spreading of a cell population, usually a sheet of cells. It occurs among epithelial cells of amphibians by cell spreading and flattening, along with cell division, and among mesenchymal, deep cells by radial intercalation, a radial intercalation that sometimes involves the epithelial layer as well (Chapter 13).

In those organisms that show embryonic growth, one might expect oriented cell division coupled with cell growth to be a major component of this movement as well. The reverse of epiboly, the shrinking of surface area without bending, does not have a formal, useful name. "Emboly" refers to a general, inward movement (Nelsen 1953), but it includes so many more specific processes (ingression, invagination, involution, vegetal rotation) that the term is not very useful. Decrease in surface area usually involves thickening of a layer of squamous or cuboidal cells by columnarization, or palisading, to form a flattened plate, often just prior to cell wedging and invagination. Examples include formation of the vegetal plate preceding primary invagination in echinoderm gastrulation, the neural plate preceding neural tube formation in vertebrates, the thickened ventral region prior to mesoderm invagination in *Drosophila*, and the sensory placodes of the vertebrate head (Keller et al. 2003). "Concrescence" has been used to refer to the fusion of the blastoporal lips along the midline, which does not actually occur (Keller et al. 1991).

ACKNOWLEDGMENTS

We thank the members of the Keller laboratory, past and present, for their contributions to this work. We also thank Claudio Stern, David Crotty, and Inez Sialiano for their help in bringing this work into final form.

REFERENCES

Abercrombie M. 1980. The crawling of metazoan cells. *Proc. R. Soc. Lond. B* **207:** 129–147.

Abercrombie M. and Heaysman J.E.M. 1953. Observations on the social behaviour of cells in tissue culture I. Speed of movement of chick heart fibroblasts in relation to their mutual contacts. *Exp. Cell Res.* **5:** 111–131.

———. 1954. Observations on the social behavior of cells in tissue culture II. 'Monolayering' of fibroblasts. *Exp. Cell Res.* **6:** 293–306.

Abercrombie M., Heaysman J.E.M., and Pegrum S.M. 1970a. The locomotion of fibroblasts in culture I. Movements of the leading edge. *Exp. Cell Res.* **59:** 393–398.

———. 1970b. The locomotion of fibroblasts in culture II. 'Ruffling'. *Exp. Cell Res.* **60:** 437–444.

———. 1970c. The locomotion of fibroblasts in culture III. Movements of particles on the dorsal surface of the leading lamella. *Exp. Cell Res.* **62:** 389–398.

Adams D., Keller R.E., and Koehl M.A.R. 1990. The mechanics of notochord elongation, straightening, and stiffening in the embryo of *Xenopus laevis*. *Development* **110:** 115–130.

Alvarez I.S. and Schoenwolf G.C. 1992. Expansion of the surface epithelium provides the major extrinsic force for bending of the neural plate. *J. Exp. Zool.* **261:** 340–348.

Baker P.C. 1965. Fine structure and morphogenic movements in the gastrula of the treefrog, *Hyla regilla*. *J. Cell Biol.* **24:** 95–116.

Betchaku T. and Trinkaus J.P. 1978. Contact relations, surface activity, and cortical microfilaments of marginal cells of the enveloping layer and of the yolk syncytial and yolk cytoplasmic layers of fundulus before and during epiboly. *J. Exp. Zool.* **206:** 381–426.

Burke R.D., Myers R.L., Sexton T.L., and Jackson C. 1991. Cell movements during the initial phase of gastrulation in the sea urchin. *Dev. Biol.* **26:** 542–557.

Burnside M.B. and Jacobson A.G. 1968. Analysis of morphogenetic movements in the neural plate of the newt *Taricha torosa*. *Dev. Biol.* **18:** 537–552.

Chalfie M., Tu Y., Euskirchen G., Ward W., and Prasher D. 1994. Green fluorescent protein as a marker for gene expression. *Science* **263:** 802–804.

Choquet D., Felsenfeld D.P., and Sheetz M.P. 1997. Extracellular matrix rigidity causes strengthening of integrin-cytoskeleton linkages. *Cell* **88:** 39–48.

Conklin E.G. 1905. The organization and cell lineage of the ascidian egg. *J. Acad Nat. Sci.* **13:** 1–119.

———. 1928. The embryology of amphioxus. *J. Morphol.* **54:** 69–151.

Davidson L.A., Hoffstrom B.G., Keller R., and DeSimone D.W. 2002. Mesendoderm extension and mantle closure in *Xenopus laevis* gastrulation: Combined roles for integrin alpha(5)beta(1), fibronectin, and tissue geometry. *Dev. Biol.* **242:** 109–129.

Davidson L.A., Koehl M.A., Keller R., and Oster G.F. 1995. How do sea urchins invaginate? Using biomechanics to distinguish between mechanisms of primary invagination. *Development* **121:** 2005–2018.

Davidson L.A., Oster G.F., Keller R.E., and Koehl M.A. 1999. Measurements of mechanical properties of the blastula wall reveal which hypothesized mechanisms of primary invagination are physically plausible in the sea urchin *Strongylocentrotus purpuratus*. *Dev. Biol.* **209:** 221–238.

Elul T. and Keller R. 2000. Monopolar protrusive activity: A new morphogenic cell behavior in the neural plate dependent on vertical interactions with the mesoderm in *Xenopus*. *Dev. Biol.* **224:** 3–19.

Elul T., Koehl M.A., and Keller R. 1997. Cellular mechanism underlying neural convergent extension in *Xenopus laevis* embryos. *Dev. Biol.* **191:** 243–258.

———. 1998. Patterning of morphogenetic cell behaviors in neural ectoderm of *Xenopus laevis*. *Ann. N.Y. Acad. Sci.* **857:** 248–251.

Ettensohn C. 1984. Primary invagination of the vegetal plate during sea urchin gastrulation. *Am. Zool.* **24:** 571–588.

———. 1985a. Gastrulation in the sea urchin embryo is accompanied by the rearrangement of invaginating epithelial cells. *Dev. Biol.* **112:** 383–390.

———. 1985b. Mechanisms of epithelial invagination. *Q. Rev. Biol.* **60:** 289–307.

Ezin M., Skoglund P., and Keller R. 2003. The midline (notochord and notoplate) patterns cell motility underlying convergence and extension of the *Xenopus* neural plate. *Dev. Biol.* **256:** 100–114.

Galbraith C.G. and Sheetz M.P. 1998. Forces on adhesive contacts affect cell function. *Curr. Opin. Cell Biol.* **10:** 566–571.

Gimlich R. and Braun J. 1985. Improved fluorescent compounds for tracing cell lineage. *Dev. Biol.* **109:** 509–514.

Glickman N., Kimmel C., Jones M., and Adams R. 2003. Shaping the zebrafish notochord. *Development* **130:** 873–887.

Gustafson T. and Wolpert L. 1963. The cellular basis of morphogenesis and sea urchin development. *Int. Rev. Cytol.* **15:** 139–214.

Hackett D.A., Smith J.L., and Schoenwolf G.C. 1997. Epidermal ectoderm is required for full elevation and for convergence during bending of the avian neural plate. *Dev. Dyn.* **210:** 397–406.

Hardin J.D. 1988. The role of secondary mesenchyme cells during sea urchin gastrulation studied by laser ablation. *Development* **103:** 317–324.

———. 1989. Local shifts in position and polarized motility drive cell rearrangement during sea urchin gastrulation. *Dev. Biol.* **136:** 430–445.

Hardin J.D. and Cheng L.Y. 1986. The mechanisms and mechanics of archenteron elongation during sea urchin gastrulation. *Dev. Biol.* **115:** 490–501.

Hardin J. and Keller R. 1988. The behaviour and function of bottle cells during gastrulation of *Xenopus laevis*. *Development* **103:** 211–230.

Hay E.D. 1995. An overview of epithelio-mesenchymal transformation. *Acta Anatomica* **154:** 8–20.

Hay E.D. and Zuk A. 1995. Transformations between epithelium and mesenchyme: Normal, pathological, and experimentally induced. *Am. J. Kidney Dis.* **26:** 678–690.

Hertzler P. and Clark W. 1992. Cleavage and gastrulation in the the shrimp *Sicyonia ingentis*: Invagination is accompanied by oriented cell division. *Development* **116:** 127–140.

Holtfreter J. 1939. Gewebeaffinität, ein Mittel der embryonalen Formbildung. *Arch. Exp. Zellforschung* **23:** 169–209.

———. 1943a. Properties and function of the surface coat in amphibian embryos. *J. Exp. Zool.* **93:** 251–323.

———. 1943b. A study of the mechanics of gastrulation. Part I. *J. Exp. Zool.* **94:** 261–318.

Honig M.G. and Hume R.I. 1986. Fluorescent carbocyanine dyes allow living neurons of identified origin to be studied in long-term cultures. *J. Cell Biol.* **103:** 171–187.

Hutson M.S., Tokutake Y., Chang M.-S., Bloor J.W., Venakides S., Kiehart D.P., and Edwards G.S. 2003. Forces for morphogenesis investigated with laser microsurgery and quantitative modeling. *Science* **300:** 145–149.

Inoué S. and Spring K.R. 1997. *Video microscopy*. Plenum Press, New York.

Jacobson A.G. and Gordon R. 1976. Changes in the shape of the developing vertebrate nervous system analyzed experimentally, mathematically and by computer simulation. *J. Exp. Zool.* **197:** 191–246.

Keller R.E. 1978. Time lapse cinematographic of superficial cell behaviour during and prior to gastrulation in *Xenopus laevis*. *J. Morphol.* **157:** 223–247.

———. 1980. The cellular basis of epiboly: An SEM study of deep-cell rearrangement during gastrulation in *Xenopus laevis*. *J. Embryol. Exp. Morphol.* **60:** 201–234.

———. 1984. The cellular basis of gastrulation in *Xenopus laevis*: Active, postinvolution convergence and extension by mediolateral interdigitation. *Am. Zool.* **24:** 589–603.

———. 2002. Shaping the vertebrate body plan by polarized embryonic cell movements. *Science* **298:** 1950–1954.

Keller R. and Danilchik M. 1988. Regional expression, pattern and timing of convergence and extension during gastrulation of *Xenopus laevis*. *Development* **103:** 193–209.

Keller R. and Jansa S. 1992. *Xenopus* gastrulation without a blastocoel roof. *Dev. Dyn.* **195:** 162–176.

Keller R.E. and Schoenwolf G.C. 1977. An SEM study of cellular morphology, contact, and arrangement, as related to gastrulation in *Xenopus laevis*. *Wilhelm Roux's Arch.* **182:** 165–186.

Keller R., Davidson L.A., and Shook D.R. 2003. How we are shaped: The biomechanics of gastrulation. *Differentiation* **71:** 171–205.

Keller R., Shih J., and Domingo C. 1992a. The patterning and functioning of protrusive activity during convergence and extension of the *Xenopus* organiser. *Dev. Suppl.* **1992:** 81–91.

Keller R., Shih J., and Sater A. 1992b. The cellular basis of the convergence and extension of the *Xenopus* neural plate. *Dev. Dyn.* **193:** 199–217.

Keller R.E., Danilchik M., Gimlich R., and Shih J. 1985. The function and mechanism of convergent extension during gastrulation of *Xenopus laevis*. *J. Embryol. Exp. Morphol.* **89:** 185–209.

Keller R., Shih J., Wilson P.A., and Sater A.K. 1991. Patterns of cell motility, cell interactions, and mechanisms during convergent extension in *Xenopus*. In *Cell-cell interactions in early development* (49th Symposium of the Society for Developmental Biology) (ed. G.C. Gerhart), pp. 31–62. Society for Developmental Biology, 49th Symposium.

Keller R., Cooper M.S., Danilchik M., Tibbetts P., and Wilson P.A. 1989. Cell intercalation during notochord development in *Xenopus laevis*. *J. Exp. Zool.* **251:** 134–154.

Keller R., Davidson L., Edlund A., Elul T., Ezin M., Shook D., and Skoglund P. 2000. Mechanisms of convergence and extension by cell intercalation. *Philos Trans. R. Soc. Lond. B. Biol. Sci.* **355:** 897–922.

Kimberly E.L. and Hardin J. 1998. Bottle cells are required for the initiation of primary mesenchyme invagination in the sea urchin embryo. *Dev. Biol.* **204:** 235–250.

Lane M.C., Koehl M.A., Wilt F., and Keller R. 1993. A role for regulated secretion of apical extracellular matrix during epithelial invagination in the sea urchin. *Development* **117:** 1049–1060.

Leptin M. 1999. Gastrulation in *Drosophila*: The logic and the cellular mechanisms. *EMBO J.* **18:** 3187–3192.

Leptin M. and Grunewald B. 1990. Cell shape changes during gastrulation in *Drosophila*. *Development* **110:** 73–84.

Leptin M., Casal J., Grunewald B., and Reuter R. 1992. Mechanisms of early *Drosophila* mesoderm formation. *Dev. Suppl.* **1992:** 23–31.

Lewis W.H. 1947. The mechanics of invagination. *Anat. Rec.* **97:** 139–156.

———. 1948. Mechanics of Amblystoma gastrulation. *Anat. Rec.* **101:** 700.

Logan C.Y. and McClay D.R. 1997. The allocation of early blastomeres to the ectoderm and endoderm is variable in the sea urchin embryo. *Development* **124:** 2213–2223.

Minsuk S.B. and Keller R.E. 1996. Dorsal mesoderm has a dual origin and forms by a novel mechanism in *Hymenochirus*, a relative of *Xenopus*. *Dev. Biol.* **174:** 92–103.

Moore A.R. and Burt A.S. 1939. On the locus and nature of the forces causing gastrulation in the embryos of *Dendraster excentricus*. *J. Exp. Zool.* **82:** 159–171.

Moore S.W. 1994. A fiber optic system for measuring dynamic mechanical properties of embryonic tissues. *IEEE Trans. Biomed. Eng.* **41:** 45–50.

Moore S.W., Keller R.E., and Koehl M.A.R. 1995. The dorsal involuting marginal zone stiffens anisotropically during its convergent extension in the gastrula of *Xenopus laevis*. *Development* **121:** 3131–3140.

Munro E.M. and Odell G.M. 2002. Polarized basolateral cell motility underlies invagination and convergent extension of the ascidian notochord. *Development* **129:** 13–24.

Myers D.C., Sepich D.S., and Solnica-Krezel L. 2002a. BMP activity gradient regulates convergent extension during zebrafish gastrulation. *Dev. Biol.* **243:** 81–98.

———. 2002b. Convergence and extension in vertebrate gastrulae: Cell movements according to or in search of identity. *Trends Genet.* **18:** 447–455.

Nakajima Y. and Burke R.D. 1996. The initial phase of gastrulation in sea urchins is accompanied by the formation of bottle cells. *Dev. Biol.* **179:** 436–446.

Nelsen O.E. 1953. *Comparative embryology of the vertebrates*. The Blakiston Company, New York.

Oda H. and Tsukita S. 2000. Real-time imaging of cell-cell adherens junctions reveals that *Drosophila* mesoderm invagination begins with two phases of apical constriction of cells. *J. Cell Sci.* **114:** 493–501.

Odell G.M., Oster G., Alberch P., and Burnside B. 1981. The mechanical basis of morphogenesis. *Dev. Biol.* **85:** 446–462.

Periasamy A. 2001. *Methods in cellular imaging*. Oxford University Press, Oxford, United Kingdom.

Perry M.M. and Waddington C.H. 1966. Ultrastructure of the blastopore cells in the newt. *J. Embryol. Exp. Morphol.* **15:** 317–330.

Poznanski A., Minsuk S., Stathopoulos D., and Keller R. 1997. Epithelial cell wedging and neural trough formation are induced planarly in *Xenopus*, without persistent vertical interactions with mesoderm. *Dev. Biol.* **189:** 256–269.

Ransick A. and Davidson E. 1998. Late specification of veg1 lineages to endodermal fate in the sea urchin embryo. *Dev. Biol.* **195:** 38–48.

Satoh N. 1978. Cellular morphology and architecture during early morphogenesis of the ascidian egg: An SEM study. *Biol. Bull.* **155:** 608–614.

Sauseodo R.A. and Schoenwolf G.C. 1993. Cell behaviors underlying notochord formation and extension in avian embryos: Quantitative and immunocytochemical studies. *Anat. Rec.* **237:** 58–70.

———. 1994. Quantitative analysis of cell behaviors underlying notochord formation and extension in mouse embryos. *Anat. Rec.* **239:** 103–112.

Sauseodo R.A., Smith J.L., and Schoenwolf G.C. 1997. Role of nonrandomly oriented cell division in shaping and bending of the neural plate. *J. Comp. Neurol.* **381:** 473–488.

Schechtman A.M. 1942. The mechanism of amphibian gastrulation. I. Gastrulation-promoting interactions between various regions of an anuran egg (*Hyla regilla*). *Univ. Calif. Publ. Zool.* **51:** 1–39.

Schoenwolf G.C. and Alvarez I.S. 1989. Roles of neuroepithelial cell rearrangement and division in shaping of the avian neural plate. *Development* **106:** 427–439.

———. 1992. Role of cell rearrangement in axial morphogenesis. *Curr. Top. Dev. Biol.* **27:** 129–173.

Schroeder T.E. 1970. Neurulation in *Xenopus laevis*. An analysis and model based upon light and electron microscopy. *J. Embryol. Exp. Morphol.* **23:** 427–462.

Sheetz M.P., Felsenfeld D.P., and Galbraith C.G. 1998. Cell migration: Regulation of force on extracellular-matrix-integrin complexes. *Trends Cell Biol.* **8:** 51–54.

Shih J. and Fraser S.E. 1995. Distribution of tissue progenitors within the shield region of the zebrafish gastrula. *Development* **121:** 2755–2765.

Shih J. and Keller R. 1992a. Cell motility driving mediolateral intercalation in explants of *Xenopus laevis*. *Development* **116:** 901–914.

———. 1992b. Patterns of cell motility in the organizer and dorsal mesoderm of *Xenopus laevis*. *Development* **116:** 915–930.

Shook D. and Keller R. 2003. Mechanisms, mechanics, and function of epithelial-mesenchymal transitions in early development. *Mech. Dev.* **120:** 1351–1383.

Shook D.R., Majer C., and Keller R. 2002. Urodeles remove mesoderm from the superficial layer by subduction through a bilateral primitive streak. *Dev. Biol.* **248:** 220–239.

Smith J.L. and Schoenwolf G.C. 1991. Further evidence of extrinsic forces in bending of the neural plate. *J. Comp. Neurol.* **307:** 225–236.

Steinberg M. 1964. The problem of adhesive selectivity in cellular interactions. In *Cellular membranes in development* (ed. M. Locke), pp. 321–366. Academic Press, New York.

———. 1970. Does differential adhesion govern self-assembly processes in histogenesis? Equilibrium configurations and the emergence of a hierarchy among populations of embryonic cells. *J. Exp. Zool.* **173:** 395–434.

Stern C.D. and Ireland G.W. 1981. An integrated experimental study of endoderm formation in avian embryos. *Anat. Embryol.* **163:** 245–263.

Sweeton D., Parks S., Costa M., and Wieschaus E. 1991. Gastrulation in *Drosophila*: The formation of the ventral furrow and posterior midgut invaginations. *Development* **112:** 775–89.

Thorogood P. and Wood A. 1987. Analysis of in vivo cell movement using transparent tissue systems. *J. Cell Sci. Suppl.* **8:** 395–413.

Townes P. and Holtfreter J. 1955. Directed movements and selective adhesion of embryonic amphibian cells. *J. Exp. Zool.* **128:** 53–120.

Trinkaus J.P. 1951. A study of the mechanism of epiboly in the egg of *Fundulus heteroclitus*. *J. Exp. Zool.* **118:** 269–320.

———. 1973a. Modes of cell locomotion in vivo. *Ciba Found. Symp.* **14:** 233–249.

————. 1973b. Surface activity and locomotion of *Fundulus* deep cells during blastula and gastrula stages. *Dev. Biol.* **30:** 69–103.

————. 1976. On the mechanism of metazoan cell movements. In *The cell surface in animal embryogenesis and development* (ed. G. Poste and G.I. Nicolson), pp. 225–329. North-Holland Publishing Company, Amsterdam, The Netherlands.

————. 1984. *Cells into organs: The forces that shape the embryo.* Prentice-Hall, Englewood Cliffs, New Jersey.

————. 1996. Ingression during early gastrulation of *Fundulus*. *Dev. Biol.* **177:** 356–370.

————. 1998. Gradient in convergent cell movement during *Fundulus* gastrulation. *J. Exp. Zool.* **281:** 328–335.

Trinkaus J.P., Trinkaus M., and Fink R.D. 1992. On the convergent cell movements of gastrulation in *Fundulus*. *J. Exp. Zool.* **261:** 40–61.

Ulrich F., Concha M., Heid P., Voss E., Witzel S., Roehl H., Tada M., Wilson S., Adams R., Soll D., and Heisenberg C.-P. 2003. Slb/Wnt11 controls hypoblast cell migration and morphogenesis at the onset of zebrafish gastrulation. *Development* **130:** 5375–5384.

Vogt W. 1929. Gestaltungsanalyse am Amphibienkiem Mit Ortlicher Vitalfarbung. II. Teil Gastrulation und Mesoderbildung Bei Urodelen und Anuren. *Wilhelm Roux' Arch. Entwicklungsmech. Org.* **120:** 384–601.

Wallingford J.B., Fraser S.E., and Harland R.M. 2002. Convergent extension: The molecular control of polarized cell movement during embryonic development. *Dev. Cell* **2:** 695–706.

Wei Y. and Mikawa T. 2000. Formation of the avian primitive streak from spatially restricted blastoderm: Evidence for polarized cell division in the elongating streak. *Development* **127:** 87–96.

Wilson P. 1990. The development of the axial mesoderm in *Xenopus laevis*. Ph.D. thesis. University of California, Berkeley.

Wilson P. and Keller R. 1991. Cell rearrangement during gastrulation of *Xenopus*: Direct observation of cultured explants. *Development* **112:** 289–300.

Wilson P.A., Oster G., and Keller R. 1989. Cell rearrangement and segmentation in *Xenopus*: Direct observation of cultured explants. *Development* **105:** 155–166.

Winklbauer R. and Keller R.-E. 1996. Fibronectin, mesoderm migration, and gastrulation in *Xenopus*. *Dev. Biol.* **177:** 413–426.

Winklbauer R. and Nagel M. 1991. Directional mesoderm cell migration in the *Xenopus* gastrula. *Dev. Biol.* **148:** 573–589.

Winklbauer R. and Schürfeld M. 1999. Vegetal rotation, a new gastrulation movement involved in the internalization of the mesoderm and endoderm in *Xenopus*. *Development* **126:** 3703–3713.

Zienkiewicz O.C. and Taylor R.L. 1991. *The finite element method: Solid and fluid mechanics dynamics and non-linearity.* McGraw-Hill, New York.

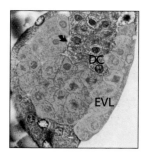

Morphogenetic Cellular Flows during Zebrafish Gastrulation

R.J. Adams[1] and C.B. Kimmel[2]

[1]*Department of Anatomy, University of Cambridge, Cambridge CB2 3DY, United Kingdom;* [2]*Institute of Neuroscience, University of Oregon, Eugene, Oregon 97403-1254*

INTRODUCTION

We focus this review on the cellular dynamics that drive two largely separable tissue rearrangements occurring during gastrulation in zebrafish. The first is the way that the blastoderm undergoes internalization to form an inner layer, the hypoblast. The second is the way that a subset of the hypoblast cells, which is at first much wider than it is long, reorganizes by convergence and extension to form an embryonic axis that is much longer than it is wide.

Internalization initiates germ-layer formation, for the zebrafish hypoblast (or "mesendoderm") is the rudiment of the endoderm and mesoderm (Kimmel et al. 1990). Other terms (invagination, involution, ingression; see below) used to describe internalization emphasize different aspects of the cell dynamics involved, as we consider in detail. We use the term internalization only to describe the tissue rearrangement, without implication as to the underlying cellular mechanism.

Gastrulation includes not only germ-layer formation, but also the ensemble of processes that lay out and begin to shape the primary organ rudiments. The coordinated narrowing (convergence) and elongation (extension) of the embryonic or anterior–posterior axis are crucial among these processes. The dorsal side of the embryo becomes prominent (Fig. 1), and, as we review, convergence and extension reshape the dorsal organ fields. Cells might also

relocate dorsally during gastrulation by migration (as apparently in *Fundulus*; Trinkaus et al. 1992; Trinkaus 1998), a different process that we do not address. Neither do we discuss mechanisms of blastoderm epiboly, reviewed recently by Kane and Adams (2002).

INTERNALIZATION

Internalization in the Neighborhood of the Blastoderm Margin

At the beginning of gastrulation in zebrafish, the blastoderm consists of several thousand cells positioned immediately adjacent to the yolk syncytial layer (YSL) and covering half of the uncleaved yolk cell (Fig. 1A). The blastoderm consists of two parts, a thin outer epithelial monolayer, the enveloping layer (EVL), which tightly covers a region of motile deep cells that are arranged in a multilayer approximately uniform in thickness. In *Fundulus*, adhesive junctions attach the marginal EVL to the YSL (Betchaku and Trinkaus 1978, 1986; Trinkaus 1984b). However, marginal deep cells are not firmly anchored in this way, and can move relatively freely with respect to both the EVL above and the YSL below. Zebrafish internalization begins on the dorsal side (Schmitz and Campos-Ortega 1994), just as in *Xenopus*. The inward-moving cells turn animalward to form the hypoblast, and later internalizing cells join into the hypoblast. A boundary appears between the hypoblast and outer epiblast, and the marginal region thickens into a distinctive annulus, the germ

We dedicate this chapter to the memory of J. P. Trinkaus, a pioneer in the study of teleost gastrulation.

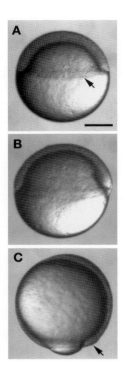

Figure 1. Representative views of the zebrafish embryo during gastrulation. (*A*) 50% epiboly stage, the onset of gastrulation, side view, with the animal pole to the top and the yolk to the bottom at 5 1/4 h (h = hours post fertilization). The arrow indicates the blastoderm margin. (*B*) Shield stage, left side view during early gastrulation at 6 h. The shield is visible as local dorsal thickening of germ ring (to the right in the figure). (*C*) 90% epiboly stage, left side view of the mid-late gastrula left side at 9 h. The arrow indicates the tail bud. Bar (in *A*), 200 μm. (Reprinted, with permission, from Kimmel et al. 1995.)

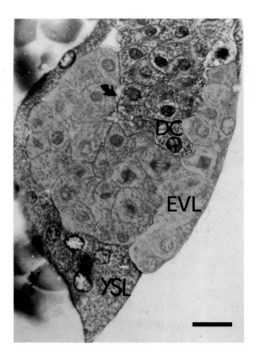

Figure 2. Histological section through the blastoderm margin at shield stage (in the rosy barb, a cyprinid relative of the zebrafish, and with a very similar embryo). The orientation is as for the dorsal side of the shield stage embryo in Fig. 1B. The arrow indicates Brachet's cleft separating epiblast and hypoblast. Colors as in Fig. 3. (EVL) Enveloping layer, (YSL) yolk syncytial layer, (DC) deep cells, (Y) yolk. Bar, 15 μm. (Reprinted, with permission, from Wood and Timmermans 1988.)

ring (Fig. 2). The dorsal marginal region, the embryonic shield (Fig. 1B), is thicker than elsewhere, and its cells are more densely packed. As gastrulation continues, the dorsal side remains distinctively thickened (Fig. 1C).

Internalization was first directly observed in zebrafish by following dye-labeled clones of muscle progenitors (Kimmel and Warga 1987). From earlier studies in trout, neither internalization nor a special role of the margin was expected in teleost gastrulation; germ layers were supposed to arise by delamination—a simple splitting of the multilayered blastoderm into sheets (Ballard 1966a,b, 1973). Yet zebrafish analyses revealed that prospective muscle cells relocate to deeper levels within the nascent germ ring with the clonally related cells inverting their order relative to the margin as they internalized. Cells destined to form mesendodermal fates (muscle, notochord, gut) invariably internalized, and cells in clones forming ectodermal fates (e.g., neural) did not (Warga and Kimmel 1990). An *involution* model explained the behavior, including the observed inversion of cellular order of the inter-

nalizing cells. Involution is the "*flowing of a sheet over the edge of an inpocketing.... Cells that are some distance from the point of invagination move to the margin of the site of invagination, flow over it, and move inside*" (Trinkaus 1984a; p. 11). In zebrafish, blastoderm cells first flow toward the blastoderm margin (which serves as the invagination site in this model), internalize, and, in the newly forming hypoblast layer, they move away from the margin (Warga and Kimmel 1990). However, observing the behavior of labeled clones of cells, scattered within a large unlabeled field, does not address whether the population flows and internalizes coherently, as a cellular sheet. Hence the study does not tell us whether the flow is an involution.

A time-lapse DIC study in another small teleost gastrula, the rosy barb, also supported the involution model (Wood and Timmermans 1988). Here, at the margin of the germ ring, whole ranks of cells were shown to disappear from the superficial edge, to be replaced there by new ranks of cells, which disappeared in turn. Furthermore, superb imaging allowed time-lapse analyses in sectional view of the germ ring (the view in Fig. 2). Individual cells traverse pathways that D'Amico and Cooper (2001) describe as "curvilinear" as they relocate from epiblast to hypoblast. Such motion is predicted by an involution model, but again, as in the zebrafish study, coherence of cellular movement was not addressed.

Shih and Fraser (1995) proposed that cells internalizing along the dorsal side of the zebrafish embryo might not all do so just at the margin. Rather, some cells located a distance away might *ingress*, entering the hypoblast "*as individuals, changing their relations to each other in the process*" (Trinkaus 1984a; p. 11). Internalization at a location well away from the margin would be at odds with the view from the studies just discussed, and indeed, Shih and Frasier provided no direct evidence for their proposal. However, in support of it, Trinkaus (1996) described what might be nonmarginal ingression in *Fundulus*. He followed cells in time-lapse recordings and showed that cells located up to 10–13 cell diameters away from the margin disappeared, as individuals, from the superficial layer. The work also emphasizes the marginal *region* in *Fundulus* as being crucial, because most internalization occurred within the nearest four-cell ranks of the margin. A caveat (Solnica-Krezel et al. 1995) is that these recordings were made at only a single plane of focus, such that it is impossible to decide whether the disappearing cells actually entered the hypoblast or only moved to a deeper location within the epiblast. As discussed below, single cells transplanted into the epiblast far from the zebrafish margin can indeed move as individuals into the hypoblast; however, there is no evidence that this occurs normally.

Recent studies lead to the strong inference that internalization in zebrafish is limited to the marginal region. D'Amico and Cooper (2001) followed internalizing cells using confocal microscopy. Their dye-labeling methods and the multilevel recordings permitted unambiguous recordings of many cells from the epiblast into the hypoblast, and revealed that essentially all (98%) of the internalizing cells are within three cell diameters of the margin. The majority internalized just at the margin, but significantly, some cells internalized at a position one or two cell diameters away from the margin (25% and 12%, respectively). Hence, as in *Fundulus*, a position exactly at the margin is not required for cells to internalize, but being in its immediate vicinity seems critical.

R.J. Adams et al. (in prep.) tracked internalization of large, locally contiguous populations of cells present in the superficial-most deep cells of the early gastrula dorsal marginal zone. Internalization was limited to cells located within about two cells (25 μm) from the margin. Again the marginal region appears essential, and the major part of the behavior occurs in just the marginal cell row. Tracking essentially all of the cells in a local field has the advantage that one can learn whether marginal cells uniformly and coherently internalize. Adams et al. found no case of marginal cells that remain superficial for an extended period in the early gastrula germ ring. They also report a previously undescribed behavior: Anywhere within the germ ring epiblast (including cells up to

about ten diameters away from the margin), cells transiently move out of the superficial epiblast and relocate to a deeper position within the epiblast. This behavior does not constitute internalization; except at the margin the deep-moving cells do not enter the hypoblast. The significance of this bobbing behavior is not clear, but it could well have a bearing on the interpretation that ingression occurs well away from the margin in *Fundulus* (Trinkaus 1996).

A Set of Noninternalizing Marginal Cells

In zebrafish, a subset of the most dorsal and superficially located deep cells does not internalize, but separates entirely from the blastoderm to form a "forerunner" cell cluster immediately vegetal to the margin (see Fig. 4). These cells are intensely endocytic even before separation, and in the late blastula they are present exactly at the dorsal

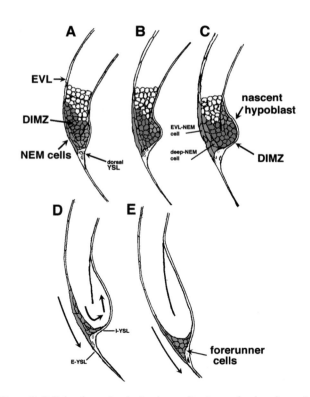

Figure 3. Cellular dynamics during internalization at the dorsal margin, occurring over about 1.5 hours at the beginning of gastrulation (from 50–60% epiboly). NEM-forerunner cells, shown in green, do not internalize (see text). The deep cells shown in orange are at first at the margin in contact with the YSL (*A*), and are the first to internalize (*B*), maintaining association with the YSL as they move toward the animal pole (toward the top of the figure; *C*). Cells initially underlying the NEM-forerunner cells (*red*) first move toward the margin (*B*) and then internalize (*C*), contributing to the nascent hypoblast that continues to expand toward the animal pole (*D, E*). (DIMZ) Dorsal internalizing marginal zone. (Reprinted, with permission, from D'Amico and Cooper 1997.)

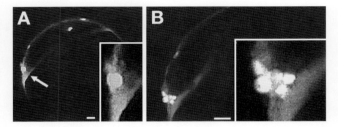

Figure 4. NEM-forerunner cells (labeled with SYTO-11) are first detected superficially in the dorsal marginal zone in the late blastula (*arrow* in *A*, 30% epiboly stage at 4.7 h), and during early gastrulation they migrate to occupy a position just vegetal to the margin (*B*, 65% epiboly stage, 2.2 hr later). Bars, 15 μm. (Reprinted, with permission, from Cooper and D'Amico 1996.)

margin (Fig. 4A). The noninvoluting endocytic marginal (NEM) cells begin to undergo rearrangement at a stage (55% epiboly) that could coincide with the beginning of internalization. The cells relocate as a loosely associated band to a position just past the blastoderm margin, and then undergo a compaction at that location, to form a cluster that looks wedge-shaped in side view, capping the leading face of the germ ring (Figs. 3E and 4B). After separation, these NEM-forerunner cells migrate toward the vegetal pole (Fig. 4B), perhaps adherent to the YSL. Fate mapping shows that they transiently form the lining of Kuppfer's vesicle (D'Amico and Cooper 1997) and then tail mesodermal derivatives (Melby et al. 1996).

The special behavior of the NEM-forerunner cells shows clearly that cellular flow in the zebrafish dorsal marginal region at the outset of gastrulation cannot be entirely coherent in the fashion we might imagine of an involuting cell sheet. At least two morphogenetic domains of cells are present, one that will internalize and one that will migrate (D'Amico and Cooper 1997, 2001).

Synchronized Ingression along the YSL

Even after the forerunners separate and gastrulation continues, the geometry of the germ ring dictates that flow from epiblast to hypoblast cannot be coherent. This is because the thickness of the epiblast and hypoblast differ (about three cells deep versus one cell deep). Hence, the population must locally reorganize. We suggest that even if internalization is restricted to the margin, cells internalize as individuals, rather than involute (see also Chapters 9 and 15). The study of R.J. Adams et al. (in prep.) supports this view. Superficial deep cells approaching the margin become compactly arranged and move very coherently. Just at the margin, the cells abruptly and individually begin to exhibit protrusions along their leading edges that are adjacent to the YSL. The change is in the fashion of an epithelial-to-mesenchymal transition. Critically, coherence is lost. As neighboring cells

internalize and enter the hypoblast, they begin to move at different rates and in different directions. Adams et al. conclude that internalization is better characterized as a "synchronized ingression" at the margin rather than as an involution.

The distinctive behaviors of internalizing cells might be influenced by the YSL, which contacts blastoderm cells at the edge of the germ ring (Figs. 2 and 3). D'Amico and Cooper (2001) discovered by tracking the movements of YSL nuclei that the YSL does not behave as a single unit; rather it consists of a number of separate morphogenetic domains. There are two dorsal domains, initially external to the margin. One band of nuclei remains external, moving to the vegetal pole during epiboly. Other nuclei move in the opposite direction, beneath the blastoderm and toward the animal pole. Hence, if the nuclear movement reflects activity of the YSL surface, the YSL could provide a moving substrate for internalization. R.J. Adams et al. (in prep.) independently observed the same inward and animalward YSL nuclear movement. Additionally, they found that just when internalization is beginning, YSL epiboly pauses, perhaps indicating a reorganization of its motor machinery. Coincident with the beginning of the pause, the velocity of the superficial deep cells moving toward the margin increases. Such an increase could be due to a pull coming from within the marginal region, perhaps from the YSL (R.J. Adams et al., in prep.).

Both R.J. Adams et al. (in prep.) and D'Amico and Cooper (2001) speculate that the YSL in zebrafish might be playing a role in early gastrulation similar to a well-known population of endodermal cells in *Xenopus*, the bottle cells (see Chapter 13). The bottle cells are the first cells to internalize, and appear to exert a pull on cells just behind them, thus facilitating the inward movement of the follower cells through the blastopore. A difference between the two species could be that the participation of the YSL in internalization persists during zebrafish gastrulation, whereas amphibian bottle cells seem to play an early role only. As YSL epiboly resumes in zebrafish, following the early pause (R.J. Adams et al., in prep.), YSL nuclei continue to move underneath the blastoderm.

Autonomous and Nonautonomous Roles of Nodal Signaling

Mutational analyses suggest that the velocity increase is related to internalization and requires Nodal signaling. If Nodal signaling is abolished, as in either *squint(ndr1): cyclops (ndr2)* double mutants or mutants in which the Nodal cofactor *one-eyed pinhead* is removed both maternally and zygotically (MZ*oep* mutants), the hypoblast fails to develop (Feldman et al. 2000; for review, see Schier 2001; see also Chapter 34). R.J. Adams et al. (in prep.) confirmed, by time-lapse analyses of M*zoep* mutants, that involution is completely blocked, accounting for the absence of hypoblast. In the MZ*oep* mutant, the YSL pause occurs as in the wild type, but the velocities of the superficial blastoderm cells, rather than increasing as in the wild type, decrease nearly to zero. Hence, the high velocities seem to depend on internalization. Supporting evidence comes from *lefty* mutants, in which Nodal signaling is up-regulated, and both the period of rapid internalization at the dorsal margin and the period of high velocity of movement of superficial cells toward the margin are prolonged (Feldman et al. 2002). Hence the mutational analyses show the correlation between internalization and cellular velocities of movement in the neighborhood of the margin—whether internalization is turned off or turned up.

Mosaic analyses with Nodal pathway mutants have provided further insight into the mechanism of internalization. Carmany-Rampey and Schier (2001) asked the important question of whether marginal cells receiving a Nodal signal can internalize directly and cell-autonomously, or alternatively, whether internalization of Nodal-responsive marginal cells requires other cell–cell interactions. The question gets at the very heart of the cellular basis of the internalization mechanism in zebrafish.

To examine cell autonomy, Carmany-Rampey and Schier (2001) constructed mosaic embryos by transplanting single MZ*oep* mutant cells into the marginal region of wild-type late-blastula embryos and vice versa. Function of the *oep* gene clearly is required in cells receiving and responding to Nodal, rather than in cells sending the signal (Gritsman et al. 1999). The use of single-cell transplants in these experiments, technically demanding, is a crucial part of this elegant study—clearly essential to understand how individual cells behave while completely surrounded by neighbors that differ in genotype.

A key finding from this study is that an MZ*oep* mutant cell at the margin, and in the company of wild-type neighbors, moves with these neighbors to take up a deep marginal location, apparently just as wild-type cells themselves do (studied as control transplantations). Hence, the critical inward movement of gastrulation does not require cell-autonomous responsiveness to Nodal signaling.

Carmany-Rampey and Schier also identified the exact step at which such responsiveness (or, more precisely, cell-autonomous *oep* function) is essential. Normally, as discussed above, internalizing cells, after moving inward, immediately move away from the margin and are replaced at their deep marginal position by newly internalized cells arriving there. The displaced cells maintain a deep position in the blastoderm as newly forming hypoblast, moving along the surface of the YSL (toward the animal pole during the early part of the gastrula period) or very close to the YSL. In the mosaics, on the other hand, MZ*oep* mutant cells

in the company of wild-type neighbors relocate completely abnormally. They "egress" to a superficial location just at the margin and remain there. It is a remarkable observation that the relocated cells remain in this superficial location stably, considering that normally, cells are only present transiently at the margin as they internalize. The isolated mutant cells must be continuously "bucking" the cellular flow of internalizing wild-type neighbors moving past them.

That the MZ*oep* mutant cells do not join into the hypoblast in any stable way fits another finding of the same study—that the mutant cells do not begin to express markers, *axial*(*foxA2*) and *sox17*, which are normally up-regulated in newly internalized hypoblast cells. The unavoidable interpretation is that cells autonomously require function of *oep* for at least some key features of mesendodermal identity. On the other hand, internalization itself seems independent.

The converse experiment also produced a very interesting result. Wild-type cells transplanted singly near the margins of MZ*oep* mutant hosts move internally, without first moving to the margin, to take up stable positions adjacent to the YSL, underneath the mutant blastoderm. One can easily interpret the position to which they arrive to be appropriate, for this is where the hypoblast normally would be located. In accord with this interpretation, the relocated cell up-regulates expression of the hypoblast marker *axial*(*foxA2*). However, the direct pathway the cells take to get to this position is abnormal. The authors interpret the finding to mean that "*in wild-type embryos cell-cell interactions prevent* [nonmarginal] *cells from internalizing directly*" (Carmany-Rampey and Schier 2001; p. 1265).

What is the nature of such interactions? We propose that it is cellular recognition, as might be based on cell–cell adhesion. According to this scenario, marginal cells responding to Nodal signaling in the wild-type embryo autonomously up-regulate signaling or adhesion molecules that cause them to maintain contact with one another (and/or with the YSL) during internalization and hypoblast formation. In the mosaic situation (Carmany-Rampey and Schier 2001), the transplanted wild-type cell carries out this response because Nodal signals are present as usual in the host MZ*oep* marginal region, but its mutant neighbors, lacking Nodal responsiveness, cannot. The differences between the wild-type cell and its mutant neighbors cause the wild-type cell to sort away from the mutant cells to a location it finds compatible, the surface of the YSL.

This interpretation is supported by a related study showing that cells do not need to be located anywhere near the margin in order to undergo an ingression-like inward movement if the Nodal-response pathway is up-regulated (by overexpression of Tar*, an activated form of the Nodal-receptor Taram-A; David and Rosa 2001). *Tar**-expressing cells transplanted to the animal pole "ingress" at that location, far from the margin, taking up a position adjacent to the YSL. David and Rosa (2001) also favor an interpretation based on cell sorting to explain the result. The behavior would be like that in salt-and-pepper mixtures of cells taken from different germ layers that sort out to reestablish the germ layers (Townes and Holtfreter 1955). Ho and Kimmel (1993) showed that sorting between germ layers is an autonomous feature of committed cells in zebrafish. It may be due to adhesive differences (Steinberg 1963, 1996; Steinberg and Takeichi 1994; Kim et al. 1998). However, other explanations are possible; e.g., cell sorting can also be driven by signaling (Eph-ephrin) differences rather than adhesive differences between cells (for review, see Cooke and Moens 2002). Identifying the downstream targets of the Nodal signaling pathway that produce the sorting between wild type and MZoep or Tar* cells would be of considerable interest. Nothing is known of this in zebrafish, but there is a clue from work in Xenopus: Overexpression of the transcription factor Xsox-17 can cause cells to acquire an endodermal fate and to sort out of the ectoderm independently of Nodal signaling (Clements and Woodland 2000). This homeodomain protein is a genetic target of Nodal.

That a single MZoep mutant cell in the company of wildtype neighbors internalizes (at least transiently) with these neighbors (Carmany-Rampey and Schier 2001), while a community of MZoep mutant cells itself does not show internalization (R.J. Adams et al., in prep.), is one of the marvelous mysteries of gastrulation that future work needs to address. The "synchronized ingression" model of R.J. Adams et al. (in prep.) implies that internalizing cells are not acting completely as individuals, but that something—margin-dependent—is coordinating their behaviors. Community-based interaction dependent on Nodal signaling is a strong candidate.

CONVERGENCE AND EXTENSION OF AXIAL MESODERM

Soon after cells have internalized at the germ ring, movements toward the dorsal face of the embryo begin. This quickly produces a recognizable accumulation of cells, the embryonic shield. Within this domain are the precursors of anterior and axial mesoderm, the prechordal plate, the notochord, floor plate, and endoderm. This domain, as has been described in other vertebrates (Keller et al. 2000), is originally short in its anterior–posterior extent but will dramatically reduce in width and extend in length during the course of gastrulation. The cellular rearrangements that we have encountered so far in this chapter have been largely concerned with the migrations of cells relative either to a substrate or to adjacent cell layers. In this domain, we believe that another class of movements, one similar in many respects to those described in amphibia (Keller et al. 2000; Keller 2002; Glickman et al. 2003; see Chapter 19), becomes the predomi-

nant form of tissue shape change. This is the process of convergence—narrowing—and extension of the notochord domain during gastrulation that gives rise to the characteristic vertebrate embryo shape. We and others have postulated (Keller et al. 2000; Keller 2002; Solnica-Krezel and Cooper 2002; Glickman et al. 2003) that in the normal animal, tissue changes in shape by a distributed and conservative local rearrangement of neighboring cells, most likely using a mechanism reminiscent of mediolateral intercalation behavior (MIB) previously described in amphibians (Keller et al. 2000). The result of these behaviors can rapidly produce dramatic changes in shape of the entire tissue by the accumulative contributions of all the local changes.

How Do the Cells of the Axial Mesoderm Reorganize during Convergence and Extension?

The approach we have taken to ask how tissues change shape is to follow in detail the individual movements of a large fraction—ideally all—of the cells in a region of the embryo throughout the time of its morphogenesis. These data, collected from fine-resolution 3D confocal volumes (see supplemental Movie 20_1 at www.gastrulation.org), reveal a wealth of information about the patterns of cell movement (Fig. 5 and supplemental Movie 20_2 at www.gastrulation.org) (Glickman et al. 2003). The resulting parameters can be used in a quantitative way to assay morphogenesis. Thus, we have been able to begin experimentally to dissect morphogenetic development in the intact zebrafish embryo. A simple but elegant prediction from the work of Keller and colleagues (1992) on the convergence and extension of axial mesoderm in Xenopus is that the process is driven at the cellular level by a tension produced by the medial and lateral extension of processes from cells that adhere to and generate traction upon the cells immediately adjacent to them, a behavior called MIB (see Chapter 13). This polarized behavior is performed by all cells in the field, and collectively they contribute to the generation of a global and oriented tension within the tissue. At a cellular level, progressive and repeated intrusions of cellular processes pull neighboring cells closer such that over time they gradually become intercalated between these neighboring cells on either side, and in so doing push those cells apart along the anterior–posterior axis. This cycle has the dual effect of narrowing the lateral span of each local neighborhood of cells while simultaneously extending it along the anterior–posterior axis. This pattern is repeated across the entire tissue to cause a global convergence and extension (Keller et al. 2000; Glickman et al. 2003). It is important to note from this description that convergence and extension can be accomplished entirely by interactions and forces generated from within the tissue itself. Indeed, in Xenopus (Keller et al. 1985), convergence and extension can be achieved by explants of tissue isolated from neighboring tissue, substrates and support. In this way, the process

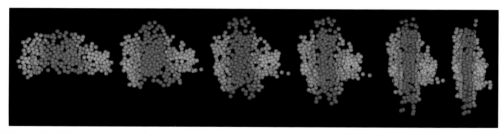

Figure 5. Tracked cells from the zebrafish hypoblast undergoing convergence and extension movements. Approximately 200 cells in each frame are colored according to their fates: notochord (*red*), adaxial cells (*dark pink*), cells forming somite 2 (*yellow*), and cell of other somites (*pink*). Images depict the locations of cells at times 7.3, 8.3, 8.8, 9.3, 10.3, and 11.3 hours after fertilization. (Reprinted, with permission, from Glickman et al. 2003.)

differs from simple "migration" and we therefore suggest that the term migration is misleading. That is not to say that, in situ, other forces do not also contribute to the net change in shape of the embryo. Indeed, as we show below, it is quite likely that they do (Glickman et al. 2003).

Surprisingly, although an apparently quite complex pattern of movements takes place across the domain of the notochord, the movements of cells in space over time is very simple (Glickman et al. 2003). On average, any cell is moving with mediolateral and anterior–posterior velocities that are predictable in a simple linear way just from knowing its current location in the embryo. Thus, as a cell moves, its location changes and consequently its velocity changes (because location determines velocity), but this pattern for the cell and the tissue as a whole remains predictable. Furthermore, we can predict that this is just the kind of relationship one might expect from the description of the MIB model given above. We propose that over any short period of time, each cell is actively pulling on its immediate neighbors, left and right, with a given tension. Of course, its neighbors are doing the same, and indeed their neighbors likewise, and so on. This tension gradually draws those cells adjacent to our reference cell a little closer, and those neighbors have pulled their neighbors a little closer, too. The result is that the neighbors of our neighbors have moved toward us by twice that increment. Furthermore, the next rank of neighbors across the field has moved closer by three whole increments. Of course, their movement relative to their own neighbors is just the same as those around our cell. The velocity of any cell relative to our cell of interest—or any other cell—is proportional to the mediolateral distance between those two cells, proportional to how much tissue under traction is able to narrow during our period of measurement. Thus, if we plot a graph of velocity in the mediolateral direction versus the location in that direction, we find a simple linear relationship, and the line that we can fit to these data provides a very valuable estimate of the rate of shape change over that period of time. We call this parameter the *rate of convergence* k_c (Glickman et al. 2003). This is powerful, since it is generated as an accumulated measurement from a population of

cells and can therefore overcome the noise inherent in any single velocity measurement over short distances and times. This mode of tissue shape change has some consequences that are not apparent from looking at the tracks of cell movement. When viewed locally, any group of cells are behaving in the same way and this movement is subtle; we estimate that relative displacements are approximately 0.6% of a cell's displacement per minute. That means that two cells separated by one cell diameter (say 15 μm) will move closer to each other by about 100 nm in one minute, too little to perceive easily. Consequently, convergence is barely detectable at the midline where movements are focused, but the velocities of cells more lateral in the field are much greater. A cell, ten cell diameters from the midline, will be initially moving toward the midline at an easily measurable 1.0 μm/minute. This gives the impression that cell movements are much greater in the lateral than the medial field, but it misses the essence of the process. All of these movements are the consequence of local reorganizations, and these are the transformations of most interest; local neighbor–neighbor intercalations that are uniform across the field. Indeed, we can show that shape changes of a *simulated* field of cells—moving just according to the average linear gradient of velocity measured from a wild-type embryo—have almost identical dynamics to the changes in the *actual* width of the notochord domain as tracked in the movie of the developing embryo. To give an indication of the rate of shape change that is taking place in the normal animal, the notochord field halves in width about every two hours. Interestingly, this measure is independent of the actual width at any time; if we had simply measured, say, the movement of the boundary over time, we would see that the velocity of the boundary was forever changing and thus comparison of boundary velocities would be very sensitive to the timing of the experiment. Not so the measurement of the gradients of velocities; over each small time interval, the convergence rate constant gives a very sensitive measure of tissue shape change in that period; thus, we can follow the dynamics of morphogenesis very precisely (Glickman et al. 2003).

A similar relationship must hold for the extension of the field along the anterior–posterior axis. If the tissue were behaving simply as a two-dimensional sheet of cells, it must extend to the same degree as it narrows in order for its area to remain constant. We know in practice that this is not the case, but the pattern of reorganization during extension is again linear in the relationship between the rate of divergence of any two cells and their displacement along the anterior–posterior axis. The reason for the gradient is again easy to see: If the model of MIB is correct, the narrowing of the field by active convergence over time drives cell processes between previously adjacent neighbors. These intrusive protrusions and cell movements actively push apart cells along the anterior–posterior axis. A cell and its anterior neighbor will move apart by some small distance d over a small time interval. The neighbor and the next cell down the line will also be displaced, on average, by the same amount so that the cell will have moved away by $2d$. Again, velocity plotted against location varies linearly, but this time with a positive gradient; the farther apart two cells are, the faster they recede from each other. The slope of this relationship gives a measure of the shape change of the tissue along the anterior–posterior axis, k_e, *the rate of extension* (Glickman et al. 2003). Together, k_c and k_e capture most of the variance in cell velocity versus location across the entire notochord domain for most of the period of gastrulation, starting long before any morphologically recognizable structure can be seen.

Although in a perfect two-dimensional sheet of cells it might be possible for cells to reorganize while retaining a constant planar tissue area, this is not seen in normal axial development. There are temporal variations in the measured rates of convergence and extension (k_c and k_e), and these variations are not coordinated in an obvious way (N.S. Glickman et al., unpubl.). Convergence precedes extension in the notochord field during gastrulation. There is also a net excess in the amount of convergence compared with the amount of extension, giving an ~ 20% reduction in planar area. Some of this lost "space" may be accounted for by a reduction in intercellular space, although the cells already appear to be densely packed throughout this time. Cells may also be reorganizing in the dorsoventral direction: Axial tissue may thicken over time. This effect is most pronounced in the somite field where convergence is comparable to that seen in the notochord field, but the extension rate is only about 35% that seen in the notochord. We speculate in this case that most of the repacking associated with convergence contributes to an increasing thickness of the presomitic mesoderm rather than a lengthening (Wood and Thorogood 1994; Glickman et al. 2003). This difference in the rate of change in shape across the somite–notochord boundary produces a shear between the cells of each tissue. Interestingly, although the somite–notochord boundary is distinct and seemingly devoid of cell processes, it does not produce any detectable dislocation in a velocity–location plot of cell movements spanning it, suggesting that it is a mechanically continuous structure with presumably matrix material providing a mechanical continuity across the gap. The temporal dissociation in the rates of convergence and extension are the subject of ongoing study and may reflect the overlapping influence of multiple force-generating mechanisms intrinsic and extrinsic to the mesoderm, as well as potentially more complicated influences of cell shape changes and packing arrangements.

What Do Quantitative Descriptions of Convergence and Extension Tell Us?

Metrics are useful for a number of reasons and powerful because they allow the distillation of a large body of information into a few parameters that can be used as shorthand to describe and compare experimental conditions. For example, we can interpret the narrowing of the field of cells in terms of MIB in the notochord domain. We have some evidence that cells do have some of the shape changes consistent with this (Glickman 2000), but the changes are not large and not significant for much of the time over which narrowing is seen. There are some indications (Shih and Keller 1992; Topczewski and Solnica-Krezel 1999; Wallingford et al. 2000) that directed cell protrusions may be present, but we cannot exclude that at least some narrowing is due simply to a reduction in the size of cells in the mediolateral direction, maybe by a deepening in the dorsoventral direction. Without more checks and tallies on the parameters of morphogenesis, we are unable to distinguish this possibility by the data presented so far, which are based solely on dynamics.

MIB produces a predictable pattern of cell rearrangement within the notochord field. If our contention is correct that cells are, just as in *Xenopus*, intercalating mediolaterally and consequently pushing their neighbors apart anterior–posteriorly, we would expect a local motif of cell rearrangement to be detectable in the relative movements of cells in the field over time. Since we have a record of all of these motions, we can ask whether there really is an underlying pattern of rearrangement by constructing a scheme of the average local reorganization of cells. Furthermore, we can keep track of which cells are neighbors within the sheet and ask whether changes in this neighbor connectivity matrix are consistent with our underlying model for the morphogenetic algorithm. We find in both cases that the results obtained are entirely consistent with this conservative model of local cell remodeling with cells immediately lateral and medial converging toward the central cell and cells above and below diverging away from it (Glickman et al. 2003). This causes a change in local connectivity that, on average, leads new neighbors to be made mediolaterally and to break apart anterior–posteriorly, as cells intercalate between them. In summary, from quantitative measures of both tissue shape change and local cellular rearrangement, we find no evidence to cause us to doubt that an incremental cellular intercalation mechanism of convergence and extension could entirely account for the morphogenesis we have seen in the normal embryo.

There are now a number of zebrafish mutants that are deficient in axial mesoderm morphogenesis, most of which are collectively known as convergence mutants (for review, see Solnica-Krezel and Cooper 2002). We will look more carefully at what is known of their lesions in a later section. We have looked in detail at another mutant, *no tail* (*ntl*), which is mutated in the zebrafish homolog of the mouse T-box gene *Brachyury* (Schulte-Merker et al. 1994). In this mutant, axial tissue is initially specified but subsequently fails to differentiate into notochord. During gastrulation, the axial domain of cells fails to express notochord makers and the field remains broader than the normal developing tissue (Melby et al. 1997). We performed the same analyses we have outlined in the previous section to the cells in this axial stripe and asked how the movements of these cells are affected in this mutant. To our surprise, we found that for much of the period of gastrulation, there was little evidence for coordinated convergent movements between these cells. There is no relationship between the velocity of movement and the location of the cells and there is little or no net narrowing of the field of cells (Fig. 6) (Glickman et al. 2003). For

much of the period of gastrulation, there remains only the residual mediolateral movement of cells that is seen as "noise" in the patterned reorganizations of cells in the normally converging wild-type notochord. There is little evidence for organized convergent cell movements, nor is there any suggestion that there is augmentation of autonomous migrations in the mediolateral direction instead. However, when we look at the pattern of axial tissue shape change in the anterior–posterior direction, we find that there is no obvious lesion and, as far as we can tell, the tissue extends just as efficiently as in the normal notochord field (Glickman et al. 2003). However, in this case, the field is getting longer but not narrower. This raises a number of interesting questions and indicates the importance of studying morphogenetic mechanisms using complementary techniques. The planar area of this field of cells increases threefold over time, in contrast to the normal notochord, which decreases in area by about 20% over the same period. We would expect the thickness of the tissue to decrease as a consequence. This is more difficult to measure precisely.

Do Convergence Movements Actively Direct Extension in the Notochord Field of the Zebrafish?

We can conclude that convergence movements are not essential for the extension of this field of cells, but that is not the whole story. We do not know the mechanical consequences of changing the fates of cells in the axial mesoderm in the *ntl* mutant. The reorganization of cells in the normal notochord field is geometrically and dynamically entirely consistent with the convergence and extension mode of morphogenesis by the evidence we explored above. We currently lack direct evidence in the zebrafish for the generation of tension by this cellular behavior; we know only the pattern of rearrangement, not the distribution of forces that produces it. As an extreme example, we cannot exclude the possibility that convergence is a response to extension; perhaps as gaps between cells open up due to an anterior–posterior stretching of the tissue, cells respond by moving into those spaces, thus leading to a net convergence. We do not believe this to be true. It is more likely that there are several mechanical consequences of the *ntl* mutation. For instance, if tension within the notochord domain is reduced as a consequence of the removal of normal convergence movements, the net stiffness of this tissue may be lower than in the wild-type notochord. This tissue would then be much easier to stretch under the influence of other embryonic forces that may not ordinarily be sufficient to generate anything like the same shape change in the wild-type embryo. Convergence might then be an essential component of normal axial development, given its normal mechanical

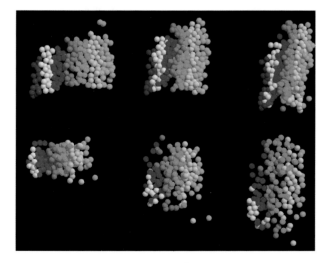

Figure 6. Axial mesodermal cells of wild type (*upper row*) and no-tail (*lower row*) embryos differ in their behaviors during gastrulation. Vertical stripes have been "painted" onto the cells to show how convergence and extension narrows and lengthens the field in a uniform way in the wild-type embryo while cells disperse mediolaterally but stretch in the anterior–posterior direction in the mutant. (Reprinted, with permission, from Glickman et al. 2003.)

properties. We also know that there are other mesoderm and ectodermal tissues, as well as the yolk cell, that are also undergoing extension in the normal animal (Kimmel et al. 1994; Concha and Adams 1998; D'Amico and Cooper 2001) and the *ntl* mutant (N.S. Glickman et al., unpubl.). This finding highlights the likely explanation that in the developing embryo, local morphogenetic mechanisms are rarely performing tasks in mechanical isolation. It does seem likely that parallel and overlapping mechanisms should ideally produce redundancy, since the converse situation of conflicting actions—crudely, one mechanism pulling while another is pushing—is not a stable developmental or evolutionary situation. A small variation in the efficacy of one of those mechanisms could significantly affect the resultant form. Parallel and redundant mechanisms would be far more tolerant of variation in efficacy of their parts.

How Does the Model So Far Fit with Other Evidence of Mesodermal Development?

The basic model of notochord development described above might consist of two modes of tissue shape change. The first is the normally stereotypical convergence of cells in the mediolateral direction, and the second is the extension of the tissue along the axis. Together, we would refer to these as *convergence* and *extension* but do not yet have evidence that it is *convergent-extension*—the latter being an extension actively driven by convergence. At its simplest, the model of convergence requires that cells have some knowledge of their orientation within the embryo—mediolateral from anterior–posterior, but not necessarily direction, medial from lateral. From this orientation, cell processes can become polarized and tension generation directed to give rise to dorsal-ward convergence. This has followed directly from the descriptions of cell behaviors in *Xenopus* (Shih and Keller 1992; Keller 2002; Solnica-Krezel and Cooper 2002; Glickman et al. 2003) but is still an active field of analysis. It does not need to be the case that any individual cell has a very precise measure of orientation, since even a loosely oriented field may rapidly self-organize into a well-aligned population average. We do know that other aspects of cell behavior in the early embryo are also consistent with such a model, notably the patterns of oriented cell divisions (Kimmel et al. 1994; Concha and Adams 1998). In recent years, the analyses of other convergence mutants (Heisenberg et al. 2000; Solnica-Krezel and Cooper 2002; Carreira-Barbosa et al. 2003) have led to the suggestion of a planar cell polarity (PCP) mechanism utilizing a noncanonical Wnt signaling mechanism that might be responsible for at least some orientation of cell behavior. Consistently, these mutants have reduced efficacy of both convergence and extension

and reduced polarization of cell behavior (Wallingford et al. 2000; Topczewski et al. 2001; Keller 2002). A growing number of components along the Wnt signaling pathway are now identified and, when modulated by either mutation or misexpression, show a consistent role in the establishment of efficient convergence and extension (for review, see Keller 2002; Solnica-Krezel and Cooper 2002). The effects of most of these are quite mild when applied singly, but in combination (e.g., as in double mutants of PCP genes) can result in a significant reduction in development (Henry et al. 2000). We are not quite yet at an understanding of how directional or orientational information is transmitted within the cell fields, how it is transduced, nor how cell behavior is directed as a consequence, but progress in this field is adding new components at a steady pace.

It may be noted that the phenotypes of convergence mutants differ in an important respect from the description of the *ntl* mutant we gave above. Convergence mutants and disruption of other known components of the noncanonical Wnt PCP pathway reduce both the convergence and extension of dorsal mesoderm, whereas the *ntl* mutant undergoes apparently normal extension. There is one significant difference in how our data on extension and the data of other investigators have been obtained: We have directly measured the rate of extension of the axial tissue rather than the final shapes and locations of expression domains. These two measures may not be equivalent, because intrinsic shape changes of the tissue may have superimposed upon them a global movement of that region of tissue. This would result in different relative locations of markers, but due to factors "extrinsic" to the tissue itself. We have postulated that extension is still seen in the axial domain of *ntl* because a parallel mechanism is exerting a redundant stretching tension upon the tissue. Why is this not also the case in the convergence mutants? One possible explanation may be that the axial domain in *ntl* mutants is disrupted far more significantly than is the same region in convergence mutants. Many aspects of cell behavior, including cell–cell adhesion, protrusive activity, and force generation, might very well be aberrant in the axial cells of the *ntl* mutant. However, when the PCP pathway is disrupted in the convergence mutants, this lesion may be more selective such that only the "compass" is affected, while the force-generating mechanisms and adhesive properties of the cells remain intact. Cells might then be fully able to generate tension between themselves, but tension is no longer directed in such a way that an effective narrowing of the tissue occurs. Furthermore, if tension is generated but is no longer polarized in the direction of tissue convergence, some component of that force will be directed anterior–posteriorly, effectively shortening the axis. This "misplaced" tension would oppose the influence of extrinsic extending forces and thus diminish both the convergence and extension of the tissue. We would predict that the stiffness of axial tissue is higher in convergence mutants than in *ntl* mutant fish. Careful quantitative analyses in these mutant animals will be required to test the validity of this model.

Prospects

Many details of this important but complex phase of gastrulation are not yet understood. These include how and when cells in the early gastrula are first incorporated into the dorsal convergent domain. We have a growing indication that Wnt ligands mediating a noncanonical signaling pathway(s) are central to the organization of the domain (for recent reviews, see Keller 2002; Solnica-Krezel and Cooper 2002). Many of the details of where and when the relevant signals are propagated and how the cells respond are still to be discovered. There must be a transition from the early autonomous cell migrations that form the hypoblast, as we describe in the first section of this chapter, into the collective convergence movements of this second section. At early times, certainly as the cells enter the germ ring and then move toward the embryonic shield, autonomous cell movements relative to a substrate are likely to be most significant, if only to accomplish the accumulation of cells into a tissue capable of supporting intrinsic convergence. Little is known of this transition, but some understanding of the resulting cell domains has come from the study of the influence of BMP signals on convergence and extension during gastrulation (Myers et al. 2002). On the dorsal side of the embryo, where our analyses were also carried out, there is evidence for both convergence and extension. More lateral to this, there is evidence for changes in the global patterns of movement such that the ventral-most cells show little sign of either convergence or extension (Myers et al. 2002). Manipulation of the levels of BMP signaling alters the relative sizes of these domains and perhaps the magnitude of the behaviors within them (Myers et al. 2002). The relationship between the fate-determining function of BMP and its role in influencing cell behavior is not yet clear, but it has been shown that the expression of members of the Wnt family involved in the vertebrate planar cell polarity pathway, *Wnt11* (Heisenberg et al. 2000) and *Wnt 5a* (Yamaguchi et al. 1999), are both modulated by BMP signaling (Myers et al. 2002).

Rapid progress is being made on several fronts and in the domain of cellular morphogenesis and tissue shaping. We believe that it is time to agree on a uniform and precise terminology, and to use consistent and comparable methods of analysis from which the greatest insight into the underlying mechanics can be achieved.

REFERENCES

Ballard W.W. 1966a. Origin of the hypoblast in *Salmo*. I. Does the blastodisc edge turn inward? *J. Exp. Zool.* **161:** 201–210.

———. 1966b. Origin of the hypoblast in *Salmo*. II. Outward movement of deep central cells. *J. Exp. Zool.* **161:** 211–220.

———. 1973. Morphogenetic movements in *Salmo gairdneri* Richardson. *J. Exp. Zool.* **184:** 381–426.

Betchaku T. and Trinkaus J.P. 1978. Contact relations, surface activity, and cortical microfilaments of marginal cells of the enveloping layer and of the yolk syncytial and yolk cytoplasmic layers of *Fundulus* before and during epiboly. *J. Exp. Zool.* **206:** 27–48.

———. 1986. Programmed endocytosis during epiboly of *Fundulus heteroclitus*. *Am. Zool.* **26:** 193–199.

Carmany-Rampey A. and Schier A.F. 2001. Single-cell internalization during zebrafish gastrulation. *Curr. Biol.* **11:** 1261–1265.

Carreira-Barbosa F., Concha M.L., Takeuchi M., Ueno N., Wilson S.W., and Tada M. 2003. Prickle 1 regulates cell movements during gastrulation and neuronal migration in zebrafish. *Development* **130:** 4037–4046.

Clements D. and Woodland H.R. 2000. Changes in embryonic cell fate produced by expression of an endodermal transcription factor, Xsox17. *Mech. Dev.* **99:** 65–70.

Concha M.L. and Adams R.J. 1998. Oriented cell divisions and cellular morphogenesis in the zebrafish gastrula and neurula: A time-lapse analysis. *Development* **125:** 983–994.

Cooke J.E. and Moens C.B. 2002. Boundary formation in the hindbrain: Eph only it were simple . . . *Trends Neurosci.* **25:** 260–267.

Cooper M.S. and D'Amico L.A. 1996. A cluster of noninvoluting endocytic cells at the margin of the zebrafish blastoderm marks the site of embryonic shield formation. *Dev. Biol.* **180:** 184–198.

D'Amico L.A. and Cooper M.S. 1997. Spatially distinct domains of cell behavior in the zebrafish organizer region. *Biochem. Cell Biol.* **75:** 563–577.

———. 2001. Morphogenetic domains in the yolk syncytial layer of axiating zebrafish embryos. *Dev. Dyn.* **222:** 611–624.

David N. and Rosa F. 2001. Cell autonomous commitment to an endodermal fate and behavior by activation of Nodal signaling. *Development* **128:** 3937–3947.

Feldman B., Dougan S.T., Schier A.F., and Talbot W.S. 2000. Nodal-related signals establish mesendodermal fate and trunk neural identity in zebrafish. *Curr. Biol.* **10:** 531–534.

Feldman B., Concha M.L., Saude L., Parsons M.J., Adams R.J., Wilson S.W., and Stemple D.L. 2002. Lefty antagonism of Squint is essential for normal gastrulation. *Curr. Biol.* **12:** 2129–2135.

Glickman N. 2000. "Cell behaviors driving convergence and extension of the dorsal mesoderm of zebrafish." Ph.D. thesis, Dissertation. University of Oregon.

Glickman N.S., Kimmel C.B., Jones M.A., and Adams R.J. 2003. Shaping the zebrafish notochord. *Development* **130:** 873–887.

Gritsman K., Zhang J., Cheng S., Heckscher E., Talbot W.S., and Schier A.F. 1999. The EGF-CFC protein one-eyed pinhead is essential for nodal signaling. *Cell* **97:** 121–132.

Heisenberg C.P., Tada M., Rauch G.J., Saude L., Concha M.L., Geisler R., Stemple D.L., Smith J.C., and Wilson S.W. 2000. Silberblick/Wnt11 mediates convergent extension movements during zebrafish gastrulation. *Nature* **405:** 76–81.

Henry C.A., Hall L.A., Burr Hille M., Solnica-Krezel L., and Cooper M.S. 2000. Somites in zebrafish doubly mutant for knypek and trilobite form without internal mesenchymal cells or compaction. *Curr. Biol.* **10:** 1063–1066.

Ho R.K. and Kimmel C.B. 1993. Commitment of cell fate in the early zebrafish embryo. *Science* **261:** 109–111.

Kane D. and Adams R. 2002. Life at the edge: Epiboly and involution in the zebrafish. *Results Probl. Cell Differ.* **40:** 117–35.

Keller R. 2002. Shaping the vertebrate body plan by polarized embryonic cell movements. *Science* **298:** 1950–1954.

Keller R., Shih J., and Domingo C. 1992. The patterning and functioning of protrusive activity during convergence and extension of the *Xenopus* organiser. *Dev. Suppl.* **1992:** 81–91.

Keller R.E., Danilchik M., Gimlich R., and Shih J. 1985. The function and mechanism of convergent extension during gastrulation of *Xenopus laevis. J. Embryol. Exp. Morphol.* (suppl.) **89:** 185–209.

Keller R., Davidson L., Edlund A., Elul T., Ezin M., Shook D., and Skoglund P. 2000. Mechanisms of convergence and extension by cell intercalation. *Philos. Trans. R. Soc. Lond. B Biol. Sci.* **355:** 897–922.

Kim S.-H., Yamamoto A., Bouwmeester T., Agius E., and De Robertis E.M. 1998. The role of Paraxial Protocadherin in selective adhesion and cell movements of the mesoderm during *Xenopus* gastrulation. *Development* **125:** 4681–4691.

Kimmel C.B. and Warga R.M. 1987. Cell lineages generating axial muscle in the zebrafish embryo. *Nature* **234:** 234–237.

Kimmel C.B., Warga R.M., and Kane D.A. 1994. Cell cycles and clonal strings during formation of the zebrafish central nervous system. *Development* **120:** 265–276.

Kimmel C.B., Warga R.M., and Schilling T.F. 1990. Origin and organization of the zebrafish fate map. *Development* **108:** 581–594.

Kimmel C.B., Ballard W.W., Kimmel S.R., Ullmann B., and Schilling T.F. 1995. Stages of embryonic development of the zebrafish. *Dev. Dyn.* **203:** 253–310.

Melby A.E., Kimelman D., and Kimmel C.B. 1997. Spatial regulation of floating head expression in the developing notochord. *Dev. Dyn.* **209:** 156–165.

Melby A.E., Warga R.M., and Kimmel C.B. 1996. Specification of cell fates at the dorsal margin of the zebrafish gastrula. *Development* **122:** 2225–2237.

Myers D.C., Sepich D.S., and Solnica-Krezel L. 2002. Bmp activity gradient regulates convergent extension during zebrafish gastrulation. *Dev. Biol.* **243:** 81–98.

Schier A.F. 2001. Axis formation and patterning in zebrafish. *Curr. Opin. Genet. Dev.* **11:** 393–404.

Schmitz B. and Campos-Ortega J.A. 1994. Dorso-ventral polarity of the zebrafish embryo is distinguishable prior to the onset of gastrulation. *Roux's Arch. Dev. Biol.* **203:** 374–380.

Schulte-Merker S., van Eeden F.J., Halpern M.E., Kimmel C.B., and Nusslein-Volhard C. 1994. no tail (ntl) is the zebrafish homologue of the mouse T (Brachyury) gene. *Development* **120:** 1009–1015.

Shih J. and Fraser S.E. 1995. Distribution of tissue progenitors within the shield region of the zebrafish gastrula. *Development* **121:** 2755–2765.

Shih J. and Keller R. 1992. Cell motility driving mediolateral intercalation in explants of *Xenopus laevis. Development* **116:** 901–914.

Solnica-Krezel L. and Cooper M.S. 2002. Cellular and genetic mechanisms of convergence and extension. *Results Probl. Cell Differ.* **40:** 136–165.

Solnica-Krezel L., Stemple D.L., and Driever W. 1995. Transparent things: Cell fates and cell movements during early embryogenesis of zebrafish. *BioEssays* **17:** 931–939.

Steinberg M.S. 1963. Reconstruction of tissues by dissociated cells. *Science* **141:** 401–408.

———. 1996. Adhesion in development: An historical overview. *Dev. Biol.* **180:** 377–388.

Steinberg M.S. and Takeichi M. 1994. Experimental specification of cell sorting, tissue spreading, and specific spatial patterning by quantitative differences in cadherin expression. *Proc. Natl. Acad. Sci.* **91:** 206–209.

Topczewski J. and Solnica-Krezel L. 1999. Cytoskeletal dynamics of the zebrafish embryo. *Methods Cell Biol.* **59:** 205–226.

Topczewski J., Sepich D.S., Myers D.C., Walker C., Amores A., Lele Z., Hammerschmidt M., Postlethwait J., and Solnica-Krezel L. 2001. The zebrafish glypican knypek controls cell polarity during gastrulation movements of convergent extension. *Dev. Cell* **1:** 251–264.

Townes P.L. and Holtfreter J. 1955. Directed movements and selective adhesion of embryonic amphibian cells. *J. Exp. Zool.* **128:** 53–120.

Trinkaus J.P. 1984a. *Cells into organs: The forces that shape the embryo.* Prentice-Hall, Englewood Cliffs, New Jersey.

———. 1984b. Mechanisms of *Fundulus* epiboly—A current view. *Am. Zool.* **24:** 673–688.

———. 1992. The midblastula transition, the YSL transition and the onset of gastrulation in *Fundulus. Dev. Suppl.* **1992:** 75–80.

———. 1996. Ingression during early gastrulation of *Fundulus. Dev. Biol.* **177:** 356–370.

———. 1998. Gradient in convergent cell movement during *Fundulus* gastrulation. *J. Exp. Zool.* **281:** 328–335.

Trinkaus J.P., Trinkaus M., and Fink R.D. 1992. On the convergent cell movements of gastrulation in *Fundulus. J. Exp. Zool.* **261:** 40–61.

Wallingford J.B., Rowning B.A., Vogeli K.M., Rothbacher U., Fraser S.E., and Harland R.M. 2000. Dishevelled controls cell polarity during *Xenopus* gastrulation. *Nature* **405:** 81–85.

Warga R.M. and Kimmel C.B. 1990. Cell movements during epiboly and gastrulation in zebrafish. *Development* **108:** 569–580.

Wood A. and Thorogood P. 1994. Patterns of cell behavior underlying somitogenesis and notochord formation in intact vertebrate embryos. *Dev. Dyn.* **201:** 151–167.

Wood A. and Timmermans L.P.M. 1988. Teleost epiboly: A reassessment of deep cell movement in the germ ring. *Development* **102:** 575–585.

Yamaguchi T.P., Bradley A., McMahon A.P., and Jones S. 1999. A Wnt5a pathway underlies outgrowth of multiple structures in the vertebrate embryo. *Development* **126:** 1211–1223.

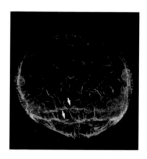

Cell–Substrate Interactions during Deuterostome Gastrulation

C.A. Ettensohn[1] and R. Winklbauer[2]

[1]Carnegie Mellon University, Department of Biological Sciences, Pittsburgh, Pennsylvania 15213; [2]University of Toronto, Department of Zoology, Toronto, Ontario, Canada M55 3G5

INTRODUCTION

The directional migration of mesoderm cells is an essential feature of gastrulation in triploblastic animals. In this chapter, we review our current understanding of mesoderm migration in selected deuterostomes in which this process has been particularly well studied. Our principal focus is on cell–substrate interactions and their role in supporting and guiding mesoderm migration. The composition and organization of the extracellular matrix (ECM) in early deuterostome embryos is reviewed separately in Chapter 49.

AMPHIBIANS

General Features of Mesoderm Migration

Mesoderm cell migration has been analyzed extensively in both anurans (*Xenopus*) and urodeles (*Pleurodeles* and *Ambystoma*) (for review, see Boucaut et al. 1991; Johnson et al. 1992; Winklbauer et al. 1996). Mesoderm and endoderm are internalized at the blastopore, and a mantle of mesendoderm spreads from all sides of the embryo toward the animal pole (AP). Anterior mesoderm and adjacent endoderm cells migrate directionally along the blastocoel roof (BCR). These cells, which move as a partially coherent, multilayered mass, are generally unipolar and extend actin-rich filopodia and lamellipodia preferentially toward the AP (Fig. 1) (Kubota and Durston 1978; Nakatsuji et al. 1982;

Boucaut et al. 1991; Winklbauer and Nagel 1991, Davidson et al. 2002). In *Xenopus*, the cells show an overlapping arrangement, with the lamellipodium of a given cell underlying the posterior part of the cell lying anterior to it (Fig. 1) (Winklbauer and Nagel 1991).

Although directional migration of mesoderm cells takes place in both anurans and urodeles, its relative role in gastrulation differs in these two groups of amphibians. Inhibition of mesoderm migration completely arrests gastrulation in *Pleurodeles*, but has less dramatic effects in *Xenopus* and *Rana* (Boucaut et al. 1984a,b; Saint-Jeannet and Dawid 1994; Ramos and De Simone 1996; Ramos et al. 1996; Winklbauer and Keller 1996). It appears that in *Xenopus*, other cell behaviors, including convergent extension movements of the involuting marginal zone, contribute to the overall expansion of the mesendodermal mantle toward the animal pole (Shi et al. 1987; Winklbauer and Keller 1996).

The BCR serves as the substrate for anterior mesoderm and endoderm cells. In anurans such as *Xenopus*, it is composed of ectoderm cells arranged in a layer 2–3 cells thick, whereas urodeles possess a single-layered BCR. Extirpation and transplantation experiments show that the apical surface of the BCR is nonadhesive and a poor substrate for mesoderm cells (Boucaut et al. 1984a; Johnson 1986; Winklbauer and Keller 1996). Only the inner, basal surface of the BCR is effective as a substrate for migration.

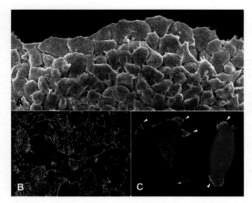

Figure 1. Aspects of mesoderm migration in the *Xenopus* embryo. (*A*) The leading edge of the dorsal mesodermal mantle, as viewed from the substrate side (with the BCR substratum removed) in the scanning electron microscope. Migration is toward the top. (*B*) Fibronectin fibril network on the basal surface of the BCR, as seen after labeling with antibody to fibronectin and fluorescent secondary antibody. (*C*) Actin cytoskeleton of migratory mesoderm cell spread on FN substratum in vitro, phalloidin-rhodamin staining. (*Arrowheads*) Lamellipodia; (*arrow*) trailing edge.

Fibronectin and Integrins in Mesoderm Migration

In both anuran and urodele embryos, the basal surface of the BCR is partially covered by a sparse network of ECM fibrils. These fibrils first form at the blastula stage and increase in abundance during gastrulation (Nakatsuji et al. 1982). The ECM fibrils contain fibronectin (FN) and other ECM molecules (Fig. 1) (see Chapter 49).

The role of FN has been extensively studied, and considerable evidence points to this ECM molecule as an essential substrate. Migration of mesoderm cells in vivo, on BCR explants, or on BCR-conditioned substrates is inhibited by Arg-Gly-Asp (RGD)-containing peptides and anti-FN antibodies (Boucaut et al. 1984a,b; Darribere et al. 1988; Shi et al. 1989; Winklbauer 1990; Saint-Jeannet and Dawid 1994; Ramos and DeSimone 1996; Ramos et al. 1996; Winklbauer and Keller 1996). These agents interfere with FN–integrin binding but may have other effects; for example, under some conditions they block the assembly of FN-containing fibrils (Darribere et al. 1992; Winklbauer and Stoltz 1995). Mesoderm cells are still able to attach to the BCR in the presence of anti-FN antibodies and RGD-containing peptides. Apparently, FN-independent adhesion, possibly involving proteins like Cyr61 in the BCR matrix (Latinkic et al. 2003), contributes to mesoderm–BCR attachment. In the absence of interactions with FN, however, the cells remain spherical and do not extend lamellipodia. This has led to the suggestion that the essential function of FN is not to promote cell adhesion to the BCR but rather to control protrusion formation and hence the motility of mesoderm cells (Winklbauer 1990; Winklbauer and Keller 1996).

Several regions of the FN molecule have been implicated in the binding, spreading, and migration of mesoderm cells, including the RGD, Hep II, and "synergy" sites within the central cell-binding domain (Alfandari et al. 1996; Ramos and DeSimone 1996; Darribere and Schwartzbauer 2000).

There are region-specific differences in the ability of gastrula cells to interact with FN. Although cells from many regions of the embryo are able to adhere to FN substrates composed of either the complete FN molecule or the central cell-binding region, only anterior mesoderm and endoderm cells are able to spread and migrate on such substrates. Ectoderm cells acquire mesoderm-specific spreading and migratory behavior when treated with the inducing factor, activin, a member of the TGF-β superfamily (Winklbauer 1988, 1990; Smith et al. 1990; Ramos et al. 1996; Wacker et al. 1998). Rac, a small GTPase, may be involved in mediating this effect (Hens et al. 2002). Presumably, the pattern of responsiveness to FN substrates shown by mesoderm cells is normally regulated by endogenous inducing factors in the early embryo. Mesoderm cells require the RGD-binding motif and synergy site for attachment, and mesoderm migration is associated with a developmentally regulated capacity to interact with the synergy site (Ramos et al. 1996; Davidson et al. 2002).

Several integrin subunits are expressed during gastrulation (Darribere et al. 1988; DeSimone and Hynes 1988; Gawantka et al. 1992; Ransom et al. 1993; Whittaker and DeSimone 1993; Alfandari et al. 1995; Joos et al. 1995, 1998; Meng et al. 1997). The α5 and β1 subunits are most abundant, and integrin α5β1 appears to be the major isoform in all gastrula cells. The αv subunit is less abundant, and α3 subunit expression is restricted to the dorsal, notochord-forming mesoderm. The α5β1 integrin has been postulated to play a role both in the assembly of FN fibrils on the BCR and in the attachment of mesoderm cells to these fibrils (Darribere et al. 1988). Function-blocking studies with antibodies against α5β1 inhibit both processes. In *Pleurodeles*, integrin αv protein is expressed selectively on the surfaces of mesoderm cells (Alfandari et al. 1995) and is required for cell spreading (and possibly migration) on the FN matrix (Skalski et al. 1998). Changes in integrin expression that would parallel changes in adhesion to FN during gastrulation have not been detected in mesoderm cells or in animal cells treated with mesoderm-inducing factors. This suggests that posttranslational regulation of integrin function may modulate adhesiveness (Smith et al. 1990; Ramos et al. 1996).

Tissue Separation during Mesoderm Migration

The BCR is only partially covered by ECM fibrils, and migrating mesoderm cells also contact the basal surfaces of BCR cells (Nakatsuji 1976). In *Xenopus*, both tissues express many of the same cell–cell adhesion molecules, EP/C- and XB/U-cadherin (Choi et al. 1990; Angres et al. 1991; Ginsberg et al. 1991; Herzberg et al. 1991; Heasman et al. 1994; Muller et al. 1994; Munchberg et al. 1997). Thus, mechanisms must be in place that prevent the fusion and intermixing of the migratory mesodermal cell mass and the multilayered BCR substratum.

Two complementary cell behaviors underlie this gastrula-stage tissue separation mechanism. First, the mesoderm develops "separation behavior" as it involutes. This can be demonstrated by placing a small explant of mesoderm on the BCR from a gastrula-stage embryo. Such an explant migrates on the BCR surface and remains separate from it. In contrast, a small BCR cell aggregate rapidly sinks into the BCR substrate layer and merges with it. Second, the capacity to exclude mesoderm is developmentally regulated within the BCR cells. When placed on BCR explants from pre-gastrula-stage embryos, both mesoderm and BCR explants sink inward (Winklbauer and Keller 1996; Wacker et al. 2000). Separation behavior in the anterior mesoderm is regulated by the homeodomain transcription factors Mix.1 and goosecoid, and by noncanonical Wnt signaling through the frizzled-7 receptor (Wacker et al. 2000; Winklbauer et al. 2001).

BCR Guidance Cues

The BCR matrix contains guidance information which is polarized along the animal–vegetal axis. The ECM can be transferred to a glass or plastic surface by culturing an explant of the BCR basal-side-down on the substrate (Nakatsuji and Johnson 1983). Remarkably, mesoderm cells move directionally toward the original AP on such a conditioned substrate, demonstrating that the ECM fibrils contain information that guides the cells in a vegetal-to-animal direction (Nakatsuji and Johnson 1983; Shi et al. 1989; Riou et al. 1990; Winklbauer 1990; Winklbauer and Nagel 1991). In *Xenopus*, this directional movement is shown by explants consisting of many cells, but single cells generally show little orientation of locomotion, either on conditioned ECM substrates or on intact BCRs (Nakatsuji and Johnson 1983; Winklbauer 1990).

In vivo, the mass of mesoderm cells is initially concentrated near the blastopore and has a free margin only at its anterior, leading edge. This geometry might be sufficient to cause a net displacement of the mesoderm mantle away from the blastopore and toward the animal pole. Indeed, the autonomous "closure" of circular mesendodermal explants on artificial substrates supports the notion that guidance information from the blastocoel roof is dispensable, at least for the later stages of mantle closure (Davidson et al. 2002). The polarized guidance information in the

BCR matrix, however, probably reinforces other mechanisms that direct the expansion of the mesoderm mantle.

In *Ambystoma*, but not in *Xenopus,* the ECM fibrils that cover the BCR become preferentially aligned along the animal–vegetal axis early in gastrulation (Nakatsuji et al. 1982; Winklbauer and Nagel 1991). The mechanism of the alignment is not known; it may result from traction exerted by the migrating mesoderm cells or from expansion of the ectoderm. Experimental realignment of the fibrils orthogonal to the animal–vegetal axis by mechanical tension redirects mesoderm migration along the new fibril axis (Nakatsuji and Johnson 1984). The orientation of cell movement by fiber alignment in *Ambystoma* may result from both the topography and chemistry of the ECM fibers (Wojciak-Stothard et al. 1997; Ahmed and Brown 1999; Curtis and Wilkinson 1999).

In *Xenopus*, the BCR becomes polarized at the late-blastula stage under the influence of planar signals from the vegetal region of the embryo (Nagel and Winklbauer 1999). Disruption of FN fibrillogenesis with RGD-containing peptide or cytochalasin B randomizes mesoderm movement on conditioned substrates, suggesting that FN-containing fibrils are required for guidance (Winklbauer and Nagel 1991). The fibrils are apparently not sufficient, however. If BCR is cultured without adjacent polarizing tissue, morphologically normal, FN-containing fibers are deposited but the substrate lacks guidance information (Nagel and Winklbauer 1999).

There is no evidence of a gradient in ECM fibril density or in substrate adhesiveness along the animal–vegetal axis that might account for the polarity of mesoderm migration (Winklbauer and Nagel 1991). It was originally proposed that individual FN-containing fibrils might have an intrinsic, structural polarity (Nakatsuji and Johnson 1983; Winklbauer and Nagel 1991). More recent studies, however, suggest that, at least in *Xenopus*, platelet-derived growth factor (PDGF) associated with the ECM provides directional information. Tissue culture studies have shown that PDGF acts as a chemoattractant for various mesenchymal cell types (Heldin and Westermark 1999). In the amphibian embryo, *PDGF-AA* is expressed by cells of the BCR (Mercola et al. 1988) and *PDGF receptor* α (*PDGFRα*) by anterior mesoderm cells (Jones et al. 1993). PDGF-AA greatly enhances the spreading of activin-treated cells on FN substrates (Symes and Mercola 1996). Moreover, overexpression of a dominant-negative form of PDGFRα perturbs mesoderm cell migration and gastrulation in *Xenopus* (Ataliotis et al. 1995). Recent studies have shown that the directed movement of mesoderm cells toward the AP on conditioned ECM, and mesoderm cell orientation and migration in vivo, are perturbed when PDGF function is inhibited or when *PDGF-AA* is overexpressed (M. Nagel et al., unpubl.).

BIRDS

General Features of Mesoderm Migration

In the chick embryo, mesoderm cells are derived from the epiblast and ingress through the primitive streak (see Chapter 15) (for review, see Sanders 1986; Harrisson et al. 1988). After emerging from the streak, they migrate laterally and anteriorly between the epiblast and a deep (ventral) layer of endoderm. Examination of fixed embryos indicates that the mesoderm migrates as a loose association of stellate, mesenchymal cells that extend numerous lamellipodia and filopodia. A major substrate used by these cells is the basal lamina that underlies the epiblast. Both epiblast and deep cells contribute to the formation of the basal lamina, which appears prior to gastrulation (Harrisson et al. 1985). Migrating mesoderm cells also contact the dorsal surface of the endoderm, which is covered by a sparse ECM, and move within the matrix between the two cell layers.

The medial movement of epiblast cells toward the primitive streak raises the question of whether there is also an overall movement of the underlying basal lamina in that direction. Such a movement of the basal lamina would be in a direction opposite that of the migrating mesoderm cells. Labeling studies using concanavalin A–ferritin initially suggested that at least some components of the basal lamina move with the epiblast toward the primitive streak (Sanders 1984). Recent grafting experiments using ^{3}H-glucosamine-labeled epiblast fragments indicate, however, that epiblast cells slide over the basal lamina as they move toward the primitive streak (Bortier et al. 2001).

ECM Interactions and FGFs in Mesoderm Migration

The basal lamina underlying the chick epiblast, like the ECM that lines the amphibian BCR, contains FN, laminin, collagens, and proteoglycans (see Chapter 49). Immunolabeling studies indicate that the basal lamina near the primitive streak is relatively FN-deficient (Sanders 1982), and some workers have described an increasing gradient of FN from the primitive streak to the lateral regions of the blastoderm (Harrisson et al. 1993). Microinjection of anti-FN antibodies (intact IgG or Fab fragments) or Arg-Gly-Asp-Ser (RGDS) peptide adjacent to the primitive streak causes mesoderm cells to round up partially and interferes with their movement away from the primitive streak, although ingression is not affected (Brown and Sanders 1991; Harrisson et al. 1993). The same agents inhibit spreading of mesoderm cells on FN substrates in vitro (Sanders 1980; Brown and Sanders 1991). Together, these studies support the view that FN plays a role in migration in vivo.

FN and other ECM components are concentrated in the "fibrous band," a region of densely packed, parallel fibers at the border between the area pellucida and area opaca in the anterior region of the gastrula-stage embryo (Critchley et al. 1979). Somewhat surprisingly, this region of the basal matrix does not support mesoderm cell spreading or migration (Andries et al. 1985; Harrisson et al. 1992). Instead, it may function in limiting the expansion of the mesoderm layer in the anterior region of the embryo.

A major component of the interstitial matrix between the migrating mesoderm cells is hyaluronic acid (HA) (see Chapter 49). Microinjection of hyaluronidase causes a collapse of this space and compaction and rounding of mesoderm cells (Fisher and Solursh 1977; Van Hoof et al. 1986). These findings have generally been interpreted in the context of the space-filling properties of HA (see Toole 2002), although hyaluronidases would also disrupt interactions between HA and its cell-surface receptors (Isacke and Yarwood 2002). Although hyaluronidases affect mesoderm cell morphology and packing, it has not been determined whether lateral migration from the primitive streak is affected.

FGFs may act as chemoattractants and chemorepellents that regulate mesoderm cell movements in the chick (Yang et al. 2002). Primitive streak cells are attracted by heparin beads soaked in FGF4 and repelled by FGF8-coated beads (see Chapters 15 and 22). These findings, coupled with the normal patterns of expression of these growth factors during gastrulation, suggest that FGF8 might normally promote dispersal of cells away from the primitive streak whereas FGF4 produced by the notochord might mediate the medial migration of anterior mesoderm cells at a later stage. In the mouse, mutations in FGF8 and FGF-receptor-1 result in defects in mesoderm cell movements during gastrulation that are consistent with this model (Ciruna et al. 1997; Sun et al. 1999; Ciruna and Rossant 2001). FGF signaling has also been implicated in mesoderm cell movements in *Xenopus* (Kroll and Amaya 1996; Nutt et al. 2001). A major unanswered question is whether FGFs affect cell movement in a direct fashion (e.g., by acting as chemoattractants/chemorepellents) or indirectly by altering gene expression and cell fate.

MAMMALS

General Features of Mesoderm Migration

General features of mesoderm migration are similar in mammals and other amniotes, and it seems very likely that the mechanisms are conserved. In the primitive-streak-stage mouse embryo, mesoderm cells migrate away from the primitive streak through an extracellular space between the embryonic ectoderm and visceral endoderm. As in the chick, the ectodermal layer is supported by a basal lamina which is continuous everywhere except beneath the streak itself. The endoderm is initially covered by a sparser, less organized ECM, and later by a basal lamina. Immunochemical studies indicate that FN, laminin, and other ECM molecules are present in the migratory environment (see Chapter 49). Time-lapse studies of mesoderm cell migration in whole mouse embryo cultures show that, as in the chick, mesoderm cells are typically spindle-shaped or stellate rather than flattened, and move as a loose collection of individual cells (Nakatsuji et al. 1986). In vitro, these cells show contact inhibition of movement, a behavior that may contribute to their dispersal away from the primitive streak in vivo (Hashimoto et al. 1987).

ECM Interactions and PDGF in Mesoderm Migration

In vitro, mouse mesoderm cells are capable of spreading and migrating on substrates coated with FN, laminin, vitronectin, or collagen IV in an integrin-dependent fashion. The cells express $\beta 1$ integrin subunits and an $\alpha 6/\beta 1$ laminin receptor (Burdsal et al. 1993; Klinowska et al. 1994). These observations are consistent with the view that ECM–integrin interactions play a role in mesoderm migration in mammals.

Targeted disruptions of the genes encoding several ECM proteins and integrin subunits, however, have not yet provided definitive evidence that these proteins are required for mesoderm migration in vivo. Disruption of the fibronectin gene results in embryonic lethality (George et al. 1993; Georges-Labouesse et al. 1996a). Surprisingly, in early embryos homozygous for such gene disruptions, mesoderm cells ingress through the primitive streak and spread extensively between the ectoderm and endoderm. Defects in the mesoderm appear only later in development. Whether these late effects are a consequence of perturbing mesoderm cell migration, rather than cell proliferation, survival, or differentiation, is not known. Disruption of the FN gene might lead to compensatory changes in the expression of other ECM components, and the possible role of other FN-related genes in the mouse has not yet been tested. Nevertheless, these gene targeting studies indicate that FN is not essential for extensive mesoderm migration during gastrulation in the mouse.

Disruption of the *laminin γ1 chain* gene blocks development prior to the onset of gastrulation (Smyth et al. 1999), leaving it unclear whether laminin plays a role in mesoderm migration. Likewise, *integrin β1*-deficient embryos do not survive to gastrulation (Stephens et al. 1995). On the other hand, targeted disruption of the integrin α6 subunit, which together with the β1 subunit forms a laminin receptor, yields no embryonic phenotype (Georges-Labouesse et al. 1996b). Mutations in the α5 and α*v integrin* genes, and the respective double mutation, have only late effects on mesoderm development, comparable to FN gene disruption (Yang et al. 1999). Given the demonstrated ability of mouse mesoderm cells to interact with multiple matrix proteins (Burdsal et al. 1993), a possible explanation for the lack of specific migration defects in these knockout experiments is a redundancy in the function of ECM components.

As discussed above, recent data point to a role for PDGF in the guidance of anterior mesoderm cells in *Xenopus*. The early patterns of expression of *PDGF-AA* and *PDGFRα* in mouse are similar to those observed in *Xenopus* (Orr-Utreger and Lonai 1992; Palmieri et al. 1992). Embryos homozygous for the *Patch* mutation (a deletion of the *PDGFRα* gene) show a variety of defects, including disorganization and aberrant movement of mesoderm cells (Schatteman et al. 1992). These findings suggest that PDGF may play a similar role in gastrulation in mouse and *Xenopus* embryos.

SEA URCHINS

General Features of Mesoderm Migration

Mesoderm migration can be observed directly in living echinoderm embryos because of their optical transparency. In most commonly studied sea urchins that exhibit indirect development (i.e., those that develop via a transient, feeding larva stage), a subpopulation of mesoderm cells ingresses shortly before the vegetal plate begins to invaginate. These are the primary mesenchyme cells (PMCs), which give rise to the embryonic skeleton. Later in gastrulation, secondary mesenchyme cells (SMCs) ingress from the tip of the archenteron and give rise to two major cell types, blastocoelar cells and pigment cells. The three major migratory mesodermal cell types show different behaviors and patterns of cell movement during gastrulation.

PMC Migration

PMCs execute a sequence of behaviors that includes ingression (epithelial-mesenchymal transition), migration, cell–cell fusion, and skeletogenesis (Ettensohn et al. 1997) (Chapter 9). A quantitative cell adhesion assay has been used to show that the onset of PMC migration is accompanied by loss of adhesion to two hyaline layer components, hyalin and echinonectin, and an increased affinity for purified blastocoelar ECM and (vertebrate) FN (Fink and McClay 1985; Burdsal et al. 1991). These changes in adhesive properties occur in parallel with an extensive remodeling of the PMC surface at ingression by endocytosis (Miller and McClay 1997; Kanoh et al. 2001).

PMCs migrate on the blastocoel wall during gastrulation and gradually become arranged in a characteristic ring-like pattern between the vegetal pole and the equator (Fig. 2). The cells move by means of filopodia, the dynamic activity of which has been analyzed in vivo (Gustafson and Wolpert 1967; Malinda et al. 1995; Miller et al. 1995). Studies in the sea urchin and in other systems show that filopodia exert tension (Gustafson and Wolpert 1961; Wood and Martin 2002). The cytoskeletal mechanism of force-generation is poorly understood but presumably involves the actin filaments that form the core of the filopodium. As there is no apparent cytoplasmic flow into PMC filopodial protrusions, these cells must translocate through the active pulling (or pushing) of filopodia. PMCs can be stimulated to move directionally when a filopodium contacts a latex bead coated with blastocoelar ECM, even if the bead is not anchored to the substrate (Solursh and Lane 1988). This suggests that interaction between a filopodium and the ECM can polarize cell motility by altering the adhesive or contractile behavior of other filopodia extended by the same cell.

PMCs are guided by nondiffusible (substrate-bound) cues associated with the ectoderm (Ettensohn et al. 2000; Katow et al. 2000). Directional information is widely distributed in the blastocoel; i.e., it is present even in the animal hemisphere, where PMCs rarely migrate normally (Malinda and Ettensohn 1994). Ectodermal cues arise progressively during development. For example, cues that establish the position of the subequatorial PMC ring along the anterior–ventral axis arise prior to those that polarize the ring along the dorsoventral axis (Ettensohn and McClay 1986; Peterson and McClay 2003). PMCs are specifically competent to respond to this guidance information; SMCs transplanted into early-gastrula-stage embryos do not migrate to the position of the subequatorial ring (Ettensohn and McClay 1986).

Although in amphibians it is clear that the ECM provides directional information to migrating mesoderm cells, in echinoderms it remains uncertain whether cells receive guidance cues from the ECM, the ectoderm cell layer, or both. PMC filopodia appear to interact primarily with the basal lamina that lines the blastocoel wall but can also penetrate the basal lamina and interact directly with the lateral surfaces of ectoderm cells (Ettensohn et al. 2000). PMCs can be isolated in intact basal lamina "bags" that have no adherent ectoderm cells (Harkey and Whiteley 1980). PMCs exhibit filopodial activity in such bags but do not form a ring pattern unless the bags are reassociated with epithelial cells. Although the ECM bags may be altered in some way during isolation, these experiments suggest that essential information is provided by the ectoderm cells. It has long been recognized that the position of the PMC ring is closely correlated with a distinctive, fan-like arrangement of epithelial cells in the subequatorial region, although the origin and significance of the fan pattern are unknown (Okazaki et al. 1962; Galileo and Morrill 1985).

Because PMCs are intimately associated with the basal lamina during their migration, many investigators have sought to identify ECM molecules that support or direct PMC movements. Collagens do not appear to be required, as a well-formed subequatorial PMC ring forms even when

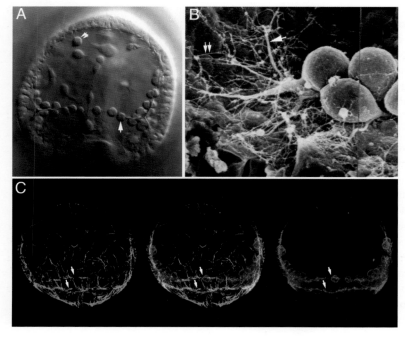

Figure 2. Aspects of mesoderm migration in the sea urchin embryo. (*A*) A living, late-gastrula-stage embryo viewed with differential interference contrast optics. PMCs (*arrow*) are arranged in a characteristic subequatorial ring pattern, and migrating SMCs (*double arrow*) are scattered throughout the blastocoel. (*B*) Scanning electron micrograph showing PMCs migrating on the wall of the blastocoel. The bodies of the PMCs are nearly spherical. Numerous filopodia (*arrow*) extend from each cell and interact with ECM fibers (*double arrow*) that line the blastocoel wall and fill the cavity. Photograph courtesy of John Morrill. (*C*) Interaction of PMCs with ECM3-containing fibers in a late-gastrula-stage embryo. Images are projections of z-stacks collected by confocal microscopy. (*Left panel, green fluorescence*) Immunostaining with a monoclonal antibody against ECM3. (*Right panel, red fluorescence*) Immunostaining with a monoclonal antibody against the PMC-specific cell-surface glycoprotein, MSP130. (*Center panel,* overlay of green and red images.) ECM3-containing fibers line the wall of the blastocoel and are more abundant in the vegetal region. During gastrulation, PMCs arrange vegetal, ECM3-containing fibers into a circumferential belt that colocalizes with the subequatorial PMC ring (*arrows*) (Hodor et al. 2000).

the posttranslational processing of collagen is inhibited (Wessel and McClay 1987). Treatment of embryos with various inhibitors of glycoprotein and proteoglycan biosynthesis inhibits PMC migration, as well as other cell movements during gastrulation (for review, see Ettensohn and Ingersoll 1992). Because of possible indirect effects, however, these findings must be interpreted with caution. Studies by Lane and Solursh (1988, 1991) indicate that at least some of these inhibitors act by interfering with the synthesis of urea-extractable, cell-surface proteoglycans which are produced by PMCs and required for migration.

Early studies devoted considerable attention to FN as a possible PMC substrate. These studies relied largely on antibodies against vertebrate FN, synthetic peptides based on the sequence of vertebrate FN, or cell migration/adhesion assays using vertebrate FN (for review, see Ettensohn and Ingersoll 1992). Studies examining the distribution of FN in early embryos yielded inconsistent results, however, and despite biochemical evidence for a FN-like molecule(s) (Iwata and Nakano 1983; DeSimone et al. 1985), initial efforts to clone FN from sea urchins were unsuccessful. The reported effects of RGDS peptide on PMC migration, originally thought to point to a role for FN, have more recently been attributed to a different molecule, pamlin (Katow 1995). The recent identification of sea urchin FN (Zhu et al. 2001) will allow a more direct analysis of the role of this ECM molecule in echinoderm gastrulation.

One promising candidate for a PMC substrate molecule is ECM3, a secreted protein with multiple putative calcium-binding domains and a large amino-terminal domain similar to the mammalian chondroitin sulfate proteoglycan core protein NG2 (Hodor et al. 2000). ECM3 has a carboxy-terminal transmembrane domain, although the protein undergoes posttranslational processing that results in its release into the extracellular space (Wessel and Berg 1995; Hodor et al. 2000). *ECM3* mRNA is expressed by ectodermal cells in the vegetal–lateral walls of the early embryo, and fibrils containing ECM3 first appear on the basal surface of the ectoderm at the blastula stage. As PMCs migrate, they redistribute ECM3-containing fibrils into a striking, circumferential belt that colocalizes with the PMC ring (Fig. 2). Foci of fibril bundling are often observed in close association with the thickened, proximal trunks of filopodia. The structure of ECM3 is similar in several respects to that of MAFp4, a protein implicated in calcium-dependent cell sorting in sponges. Recently, a protein highly similar to ECM3, but with additional furin and von Willebrand factor type C domains at the amino terminus, has been identified in mice and humans. Mutations in this protein, FRAS1, have been shown to underlie Fraser syndrome, a human genetic disorder, and the epithelial blistering phenotype of the *blebbed* (*bl*) mutant in the mouse (McGregor et al. 2003).

The role of PDGF signaling in mesoderm migration in sea urchins is not yet clear. Perturbation of PDGF signaling throughout the embryo has complex, widespread consequences, including effects on mesoderm (Govindarajan et al. 1995; Ramachandran et al. 1997), but some of these effects may be indirect.

SMC Movements

The movements of SMCs are regulated by specific cell–substrate interactions, but these have not yet been analyzed at the molecular level. Late in gastrulation, cells at the tip of the archenteron (probably presumptive SMCs) extend filopodia that make attachments near the animal pole. Force exerted by these filopodia is responsible for pulling the tip of the archenteron to the site of the future mouth opening (Dan and Okazaki 1956). Two lines of evidence show that the region of the ectoderm with which SMC filopodia interact represents a specific target site. First, filopodia that contact this local site are highly stable, whereas contacts with other regions of the blastocoel wall are short-lived. Second, when the animal–vegetal axis of the gastrula is elongated by constraining the embryo in a capillary, cells leave the archenteron tip and migrate over a considerable distance before eventually concentrating at the target site (Hardin and McClay 1990). Recognition of the target site is a specific property of SMCs; PMCs fail to recognize it even when placed in close proximity at the late gastrula stage (Hodor and Ettensohn 1998).

During gastrulation, pigment cells are released from the vegetal epithelium. The timing of ingression and migratory pathways of the pigment cells varies among species (Kominami et al. 2001), but in all cases the pigment cells subsequently become localized in the aboral ectoderm. This pattern of localization probably involves multiple substrate cues. For example, there must be differences between PMCs and pigment cells that account for the fact that the former use the basal lamina as a substrate, whereas the latter penetrate this ECM and migrate within the epithelial layer. Little is known about the molecular control of pigment cell migration, although these cells express relatively high levels of a specific β-integrin subunit, βL (Marsden and Burke 1998), and a 75-kD membrane protein that binds the YTGIR motif found in the laminin β chain (B. Crawford and B. Brandhorst, pers. comm.)

Blastocoelar cells form a network of fibroblast-like cells within the blastocoel (Tamboline and Burke 1992). They become concentrated around the gut, along the skeletal rods, and within the larval arms. Like PMCs, these cells undergo cell–cell fusion to form a syncytium (Kaneko et al. 1990: Hodor and Ettensohn 1998). Blastocoelar cells may require interaction with a fibrillar component of the blastocoel ECM recognized by the monoclonal antibody Sp14 for their efficient release or dispersal from the archenteron tip (Burke and Tamboline 1990).

THEMES AND VARIATIONS IN MESODERM MIGRATION

Many important features of mesoderm migration are shared among deuterostomes. Mesoderm cells are internalized at the onset of gastrulation, when they initiate a developmental program of cell motility that includes protrusive activity and a transformation from an epithelial to mesenchymal phenotype. Inside the embryo, migrating mesoderm cells interact with a fibrillar ECM that covers the basal surface of the overlying ectoderm and fills the blastocoel. The evidence is strong in both sea urchins and amphibians that nondiffusible (i.e., substrate-bound) cues provide directional information to the migrating cells. In amphibians, this directional information is provided at least in part by the ECM, whereas in echinoderms, it is not clear whether it originates from the ECM, overlying ectoderm cells, or both.

There are differences, however, in the scale of the movements and the morphology of the migrating cells among different deuterostomes. In vertebrates, large populations of mesodermal cells migrate over distances of hundreds of micrometers. In contrast, in the small embryos of sea urchins, a relatively small number of mesoderm cells (<200) migrate over much shorter distances to find their various target sites. In echinoderms and birds, mesoderm cells migrate as scattered individuals, whereas in *Xenopus* they move as a dense, multilayered mass. Sea urchin mesoderm cells move almost exclusively by means of filopodia, and these are extended apparently randomly. In contrast, amphibian mesoderm cells extend flattened lamellipodia which are preferentially oriented toward the AP.

At present, by far the strongest evidence concerning a molecular mechanism of mesoderm cell–substrate interactions points to a critical role of FN and integrin receptors in amphibians and other vertebrates. Even here, however, significant questions remain (see the discussion of FN-deficient mouse embryos above), and it is likely that other molecules and pathways are also involved. Tantalizing early work implicates growth factors (PDGF and FGFs) in mesoderm cell guidance in vertebrates. There are promising, but different, candidates in sea urchins (ECM3 and pamlin). An important and exciting goal for the future will be to uncover conserved substrate molecules and molecular pathways of mesoderm guidance that are shared among deuterostomes.

ACKNOWLEDGMENTS

The authors thank Gary Wessel for his critical reading of an early version of this manuscript, and Martina Nagel for preparing Figure 1.

REFERENCES

Ahmed Z. and Brown R.A. 1999. Adhesion, alignment, and migration of cultured Schwann cells on ultrathin fibronectin fibres. *Cell Motil. Cytoskelet.* **42:** 331–343.

Alfandari D., Whittaker C.A., DeSimone D.W., and Darribere T. 1995. Integrin αv subunit is expressed on mesoderm cell surfaces during amphibian gastrulation. *Dev. Biol.* **170:** 249–261.

Alfandari D., Ramos J., Clavilier L., DeSimone D.W., and Darribere T. 1996. The RGD-dependent and the Hep II binding domains of fibronectin govern the adhesive behaviors of amphibian embryonic cells. *Mech. Dev.* **56:** 83–92.

Andries L., Vanroelen C., Van Hoof J., and Vakaet L. 1985. Inhibition of cell spreading on the band of extracellular fibres in early chick and quail embryos. *J. Cell Sci.* **74:** 37–50.

Angres B., Muller A.H.J., Kellerman J., and Hausen P. 1991. Differential expression of two cadherins in *Xenopus laevis*. *Development* **111:** 8229–8244.

Ataliotis P., Symes K., Chou M.M., Ho L., and Mercola M. 1995. PDGF signalling is required for gastrulation of *Xenopus laevis*. *Development* **121:** 3099–3110.

Bortier H., Callebaut M., van Nueten E., and Vakaet L. 2001. Autoradiographic evidence for the sliding of the upper layer over the basement membrane in chicken blastoderms during gastrulation. *Eur. J. Morphol.* **39:** 91–98.

Boucaut J.C., Darribere T., Boulekbache H., and Thiery J.P. 1984a. Prevention of gastrulation but not neurulation by antibodies to fibronectin in amphibian embryos. *Nature* **307:** 364–367.

Boucaut J.C., Darribere T., Poole T.J., Aoyama H., Yamada K.M., and Thiery J.P. 1984b. Biologically active synthetic peptides as probes of embryonic development: A competitive peptide inhibitor of fibronectin function inhibits gastrulation in amphibian embryos and neural crest cell migration in avian embryos. *J. Cell Biol.* **99:** 1822–1830.

Boucaut J.C., Darribere T., Shi D.L., Riou J.F., Johnson K.E., and Delarue M. 1991. Amphibian gastrulation: The molecular basis of mesoderm cell migration in urodele embryos. In *Gastrulation: Movements, patterns, and molecules* (ed. R. Keller et al.), pp. 169–184. Plenum Press, New York.

Brown A.J. and Sanders E.J. 1991. Interactions between mesoderm cells and the extracellular matrix following gastrulation in the chick embryo. *J. Cell Sci.* **99:** 431–441.

Burdsal C.A., Alliegro M.C., and McClay D.R. 1991. Tissue-specific, temporal changes in cell adhesion to echinonectin in the sea urchin embryo. *Dev. Biol.* **144:** 327–334.

Burdsal C.A., Damsky C.H., and Pedersen R.A. 1993. The role of E-cadherin and integrins in mesoderm differentiation and migration at the mammalian primitive streak. *Development* **118:** 829–844.

Burke R.D. and Tamboline C.R. 1990. Ontogeny of an extracellular matrix component of sea urchins and its role in morphogenesis. *Dev. Growth Differ.* **32:** 461–471.

Choi Y.S., Seghal R., McCrea P. and Gumbiner B. 1990. A cadherin-like protein in eggs and cleaving embryos of *Xenopus laevis* is expressed in oocytes in response to progesterone. *J. Cell Biol.* **110:** 1575–1582.

Ciruna B. and Rossant J. 2001. FGF signaling regulates mesoderm cell fate specification and morphogenetic movement at the primitive streak. *Dev. Cell* **1:** 37–49.

Ciruna B.G., Schwartz L., Harpal K., Yamaguchi T.P., and Rossant J. 1997. Chimeric analysis of fibroblast growth factor receptor-1 (Fgfr1) function: A role for FGFR1 in morphogenetic movement through the primitive streak. *Development* **124:** 2829–2841.

Critchley D.R., England M.A., Wakely J., and Hynes R.O. 1979. Distribution of fibronectin in the ectoderm of gastrulating chick embryos. *Nature* **280:** 498–500.

Curtis A. and Wilkinson C. 1999. New depths in cell behaviour: Reactions of cells to nanotopography. *Biochem. Soc. Symp.* **65:** 15–26.

Dan K. and Okazaki K. 1956. Cyto-embryological studies of sea urchins. III. Role of the secondary mesenchyme cells in the formation of the primitive gut in sea urchin larvae. *Biol. Bull.* **110:** 29–42.

Darribere T. and Schwarzbauer J.E. 2000. Fibronectin matrix composition and organization can regulate cell migration during amphibian development. *Mech. Dev.* **92:** 239–250.

Darribere T., Yamada K.M., Johnson K.E., and Boucaut J.C. 1988. The 140-kDa fibronectin receptor complex is required for mesoderm cell adhesion during gastrulation in the amphibian *Pleurodeles waltlii*. *Dev. Biol.* **126:** 182–194.

Darribere T., Koteliansky V.E., Chernousov M.A., Akiyama S.K., Yamada K.M., Thiery, J.P., and Boucaut J.C. 1992. Distinct regions of human fibronectin are essential for fibril assembly in an in vivo developing system. *Dev. Dyn.* **194:** 63–70.

Davidson L.A., Hoffstrom B.G., Keller R., and DeSimone D.W. 2002. Mesendoderm extension and mantle closure in *Xenopus laevis* gastrulation: Combined roles for integrin α5β1, fibronectin, and tissue geometry. *Dev. Biol.* **242:** 109–129.

DeSimone D.W. and Hynes R.O. 1988. *Xenopus laevis* integrins. *J. Biol. Chem.* **263:** 5333–5340.

DeSimone D.W., Spiegel E., and Spiegel M. 1985. The biochemical identification of fibronectin in the sea urchin embryo. *Biochem. Biophys. Res. Commun.* **27:** 183–188.

Ettensohn C.A. and Ingersoll E.P. 1992. Morphogenesis of the sea urchin embryo. In *Morphogenesis: An analysis of the development of biological form* (ed. E.F. Rossomando and S. Alexander), pp. 189–262. Marcel Dekker, New York.

Ettensohn C.A. and McClay D.R. 1986. The regulation of primary mesenchyme cell migration in the sea urchin embryo: Transplantations of cells and latex beads. *Dev. Biol.* **117:** 380–391.

Ettensohn, C.A., Guss K.A., Hodor P.G., and Malinda K.M. 1997. The morphogenesis of the skeletal system of the sea urchin embryo. In *Reproductive biology of invertebrates: Progress in developmental biology* (ed. J.R. Collier), vol. 7, pp. 225–265. Wiley, New York.

Ettensohn C.A., Malinda K.M., Sweet H.C., and Zhu X. 2000. The ontogeny of cell guidance information in an embryonic epithelium. *Wenner-Gren Int. Ser.* **75:** 31–48.

Fink R.D. and McClay D.R. 1985. Three cell recognition changes accompany the ingression of sea urchin primary mesenchyme cells. *Dev. Biol.* **107:** 66–74.

Fisher M. and Solursh M. 1977. Glycosaminoglycan localization and role in maintenance of tissue spaces in the early chick embryo. *J. Embryol. Exp. Morphol.* **42:** 195–207.

Galileo D.S and Morrill J.B. 1985. Patterns of cells and extracellular material of the sea urchin *Lytechinus variegatus* (Echinodermata; Echinoidea) embryo, from hatched blastula to late gastrula. *J. Morphol.* **185:** 387–402.

Gawantka V., Ellinger-Ziegelbauer H., and Hausen P. 1992. β1-integrin is a maternal protein that is inserted into all newly formed membranes during early *Xenopus* embryogenesis. *Development* **115:** 595–605.

George E.L., Georges-Labouesse E.N., Patel-King R.S., Rayburn H., and Hynes R.O. 1993. Defects in mesoderm, neural tube and vascular development in mouse embryos lacking fibronectin.

Development **119:** 1079–1091.

Georges-Labouesse E.N., George E.L., Rayburn H., and Hynes R.O. 1996a. Mesoderm development in mouse embryos mutant for fibronectin. *Dev. Dyn.* **207:** 145–156.

Georges-Labouesse E.N., Messaddeq N., Yehia G., Cadalbert L., Dierich A., and Le Meur M. 1996b. Absence of integrin alpha 6 leads to epidermolysis bullosa and neonatal death in mice. *Nat. Genet.* **13:** 370–373.

Ginsberg D., DeSimone D. and Geiger B. 1991. Expression of a novel cadherin (EP-cadherin) in unfertilized eggs and early *Xenopus* embryos. *Development* **111:** 315–325.

Govindarajan V., Ramachandran R.K., George J.M., Shakes D.C., and Tomlinson C.R. 1995. An ECM-bound, PDGF-like growth factor and a TGF-alpha-like growth factor are required for gastrulation and spiculogenesis in the *Lytechinus* embryo. *Dev. Biol.* **172:** 541–551.

Gustafson T. and Wolpert L. 1961. Studies on the cellular basis of morphogenesis in the sea urchin embryo. Directed movements of primary mesenchyme cells in normal and vegetalized larvae. *Exp. Cell Res.* **24:** 64–79.

———. 1967. Cellular movement and contact in sea urchin morphogenesis. *Biol. Rev. Camb. Philos. Soc.* **42:** 442–498.

Hardin J. and McClay D.R. 1990. Target recognition by the archenteron during sea urchin gastrulation. *Dev. Biol.* **142:** 86–102.

Harkey M.A, and Whiteley A.H. 1980. Isolation, culture, and differentiation of echinoid primary mesenchyme cells. *Wilheim Roux' Arch. Dev. Biol.* **189:** 111–122.

Harrisson F., Andries L., and Vakaet L. 1988. The chicken blastoderm: Current views on cell biological events guiding intercellular communication. *Cell Differ.* **22:** 83–105.

———. 1992. The arrest of cell migration in the chicken blastoderm: Experimental evidence for the involvement of a band of extracellular fibrils associated with the basal lamina. *Int. J. Dev. Biol.* **36:** 123–137.

Harrisson F., Van Hoof J., Vanroelen C., and Vakaet L. 1985. Transfer of extracellular matrix components between germ layers in chimaeric chicken-quail blastoderms. *Cell Tissue Res.* **239:** 643–649.

Harrisson F., Van Nassauw L., Van Hoof J., and Foidart J.M. 1993. Microinjection of antifibronectin antibodies in the chicken blastoderm: Inhibition of mesoblast cell migration but not of cell ingression at the primitive streak. *Anat. Rec.* **236:** 685–696.

Hashimoto K., Fujimoto H., and Nakatsuji N. 1987. An ECM substratum allows mouse mesoderm cells isolated from the primitive streak to exhibit motility similar to that inside the embryo and reveals a deficiency in the T/T mutant cells. *Development* **100:** 587–598.

Heasman J., Ginsberg D., Geiger B., Goldstone K., Pratt T., Yoshida-Noro C. and Wylie C. 1994. A functional test for a maternally inherited cadherin in *Xenopus* shows its importance in cell adhesion at the blastula stage. *Development* **120:** 49–57.

Heldin C.H. and Westermark B. 1999. Mechanism of action and in vivo role of platelet-derived growth factor. *Physiol. Rev.* **79:** 1283–1316.

Hens M.D., Nikolic I., and Woolcock C.M. 2002. Regulation of *Xenopus* embryonic cell adhesion by the small GTPase, rac. *Biophys. Res. Commun.* **298:** 364–370.

Herzberg F., Wildermuth V. and Wedlich D. 1991. Expression of Xbcad, a novel cadherin, during oogenesis and early development of *Xenopus*. *Mech. Dev.* **35:** 33–42.

Hodor P.G. and Ettensohn C.A. 1998. The dynamics and regulation of mesenchymal cell fusion in the sea urchin embryo. *Dev. Biol.* **199:** 111–124.

Hodor P.G., Illies M.R., Broadley S., and Ettensohn C.A. 2000. Cell-substrate interactions during sea urchin gastrulation: Migrating primary mesenchyme cells interact with and align extracellular matrix fibers that contain ECM3, a molecule with NG2-like and multiple calcium-binding domains. *Dev. Biol.* **222:** 181–194.

Isacke C.M. and Yarwood H. 2002. The hyaluronan receptor, CD44. *Int. J. Biochem. Cell Biol.* **34:** 718–721.

Iwata M. and Nakano E. 1983. Characterization of sea-urchin fibronectin. *Biochem. J.* **215:** 205–208.

Johnson K.E. 1986. Transplantation studies to investigate mesoderm-ectoderm adhesive cell interactions during gastrulation. *J. Cell Sci.* **82:** 99–117.

Johnson K.E., Boucaut J.C., and DeSimone D.W. 1992. Role of the extracellular matrix in amphibian gastrulation. *Curr. Top. Dev. Biol.* **27:** 91–127.

Jones S.D., Ho L., Smith J.C., Yordan C., Stiles C.D., and Mercola M. 1993. The *Xenopus* platelet-derived growth factor alpha receptor: cDNA cloning and demonstration that mesoderm induction establishes the lineage-specific pattern of ligand and receptor gene expression. *Dev. Genet.* **14:** 185–193.

Joos T.O., Reintsch W.E., Brinker A., Klein C., and Hausen P. 1998. Cloning of the *Xenopus* integrin alpha(v) subunit and analysis of its distribution during early development. *Int. J. Dev. Biol.* **42:** 171–179.

Joos T.O., Whittaker C.A., Meng F., DeSimone D.W., Gnau V., and Hausen P. 1995. Integrin α5 during early development of *Xenopus laevis. Mech. Dev.* **50:** 187–199.

Kaneko H., Takaichi S., Yamamoto M., and Dan-Sohkawa M. 1990. Acellularity of starfish embryonic mesenchyme cells as shown in vitro. *Development* **109:** 129–138.

Kanoh K., Aizu G., and Katow H. 2001. Disappearance of an epithelial cell surface-specific glycoprotein (Epith-1) associated with epithelial-mesenchymal conversion in sea urchin embryogenesis. *Dev. Growth Differ.* **43:** 83–95.

Katow H. 1995. Pamlin, a primary mesenchyme cell adhesion protein, in the basal lamina of the sea urchin embryo. *Exp. Cell Res.* **218:** 469–478.

Katow H., Nakajima Y., and Uemura I. 2000. Primary mesenchyme cell-ring pattern formation in 2D-embryos of the sea urchin. *Dev. Growth Differ.* **42:** 9–17.

Klinowska T.C., Ireland G.W., and Kimber S.J. 1994. A new in vitro model of murine mesoderm migration: The role of fibronectin and laminin. *Differentiation* **57:** 7–19.

Kominami T., Takata H., and Takaichi M. 2001. Behavior of pigment cells in gastrula-stage embryos of *Hemicentrotus pulcherrimus* and *Scaphechinus mirabilis. Dev. Growth Differ.* **43:** 699–707.

Kroll K.L. and Amaya E. 1996. Transgenic *Xenopus* embryos from sperm nuclear transplantations reveal FGF signaling requirements during gastrulation. *Development* **122:** 3173–3183.

Kubota H.Y. and Durston A.J. 1978. Cinematographical study of cell migration in the opened gastrula of *Ambystoma mexicanum. J. Embryol. Exp. Morphol.* **44:** 71–80.

Lane M.C. and Solursh M. 1988. Dependence of sea urchin primary mesenchyme cell migration on xyloside- and sulfate-sensitive cell surface-associated components. *Dev. Biol.* **127:** 78–87.

———. 1991. Primary mesenchyme cell migration requires a chondroitin sulfate/dermatan sulfate proteoglycan. *Dev. Biol.* **143:** 389–397.

Latinkic B.V., Mercurio S., Bennett B., Hirst E.M.A., Xu Q., Lau L.F., Mohun T.J., and Smith J.C. 2003. *Xenopus* Cyr61 regulates gastrulation movements and modulates Wnt signaling. *Development* **130:** 2429–2441.

Malinda K.M. and Ettensohn C.A. 1994. Primary mesenchyme cell migration in the sea urchin embryo: Distribution of directional cues. *Dev. Biol.* **164:** 562–578.

Malinda K.M.. Fisher G.W., and Ettensohn C.A. 1995. Four-dimensional microscopic analysis of the filopodial behavior of primary mesenchyme cells during gastrulation in the sea urchin embryo. *Dev. Biol.* **172:** 552–566.

Marsden M. and Burke R.D. 1998. The βL integrin subunit is necessary for gastrulation in sea urchin embryos. *Dev. Biol.* **203:** 134–148.

McGregor L., Makela V., Darling S.M., Vrontou S., Chalepakis G., Roberts C., Smart N., Rutland P., Prescott N., Hopkins J., Bentley E., Shaw A., Roberts E., Mueller R., Jadeja, S., Philip N., Nelson J., Francannet C., Perez-Aytes A., Megarbane A., Kerr B., Wainwright B., Woolf A.S., Winter R.M., and Scambler P.J. 2003. Fraser syndrome and mouse blebbed phenotype caused by mutations in FRAS1/Fras1 encoding a putative extracellular matrix protein. *Nat. Genet.* **34:** 203–208.

Meng F., Whittaker C.A., Ransom D.G., and DeSimone D.W. 1997. Cloning and characterization of cDNAs encoding the integrin alpha2 and alpha3 subunits from *Xenopus laevis. Mech. Dev.* **67:** 141–155.

Mercola M., Melton D.A., and Stiles C.D. 1988. Platelet-derived growth factor A chain is maternally encoded in *Xenopus* embryos. *Science* **241:** 1223–1225.

Miller J.R. and McClay D.R. 1997. Characterization of the role of cadherin in regulating cell adhesion during sea urchin development. *Dev. Biol.* **192:** 323–339.

Miller J., Fraser S.E., and McClay D. 1995. Dynamics of thin filopodia during sea urchin gastrulation. *Development* **121:** 2501–2511.

Muller A.H.J., Kuhl M., Finnemann S., van der Poel S., Schneider S., Hausen P., and Wedlich D. 1994. *Xenopus* cadherins: The maternal pool comprises distinguishable members of the family. *Mech. Dev.* **64:** 213–223.

Munchberg F.E., Spieker T.P., Joos T.O., and Hausen P. 1997. A paired oocyte adhesion assay reveals the homophilic binding properties of the *Xenopus* maternal cadherins, XB/U- and EP/C-cadherin. *Mech. Dev.* **64:** 87–94.

Nagel M. and Winklbauer R. 1999. Establishment of substratum polarity in the blastocoel roof of the *Xenopus* embryo. *Development* **126:** 1975–1984.

Nakatsuji N. 1976. Studies on the gastrulation of amphibian embryos: Ultrastructure of the migrating cells of anurans. *Wilhelm Roux's Arch. Dev. Biol.* **180:** 229–240.

Nakatsuji N. and Johnson K.E. 1983. Conditioning of a culture substratum by the ectodermal layer promotes attachment and oriented locomotion by amphibian gastrula mesoderm cells. *J. Cell Sci.* **59:** 43–60.

———. 1984. Experimental manipulation of a contact guidance system in amphibian gastrulation by mechanical tension. *Nature* **307:** 453–455.

Nakatsuji N., Gould A.C., and Johnson K.E. 1982. Movement and guidance of migrating mesoderm cells in *Ambystoma maculatum* gastrulae. *J. Cell Sci.* **56:** 207–222.

Nakatsuji N., Snow M.H., and Wylie C.C. 1986. Cinemicrographic study of the cell movement in the primitive-streak-stage mouse embryo. *J. Embryol. Exp. Morphol.* **96:** 99–109.

Nutt S.L., Dingwell K.S., Holt C.E., and Amaya E. 2001. *Xenopus*

Sprouty2 inhibits FGF-mediated gastrulation movements but does not affect mesoderm induction and patterning. *Genes Dev.* **15:** 1152–1166.

Okazaki K., Fukushi T., and Dan K. 1962. Cyto-embryological studies of sea urchins. IV. Correlation between the shape of the ectodermal cells and the arrangement of the primary mesenchyme cells in sea urchin larvae. *Acta Embryol. Morphol. Exp.* **5:** 17–31.

Orr-Urtreger A. and Lonai P. 1992. Platelet-derived growth factor-A and its receptor are expressed in separate, but adjacent cell layers of the mouse embryo. *Development* **115:** 1045–1058.

Palmieri S.L., Payne J., Stiles C.D., Biggers J.D., and Mercola M. 1992. Expression of mouse PDGF-A and PDGF α-receptor genes during pre- and post-implantation development: Evidence for a developmental shift from an autocrine to a paracrine mode of action. *Mech. Dev.* **39:** 181–191.

Peterson R.E. and McClay D.R. 2003. Primary mesenchyme cell patterning during the early stages following ingression. *Dev. Biol.* **254:** 68–78.

Ramachandran R.K., Wikramanayake A.H., Uzman J.A., Govindarajan V., and Tomlinson C.R. 1997. Disruption of gastrulation and oral-aboral ectoderm differentiation in the *Lytechinus pictus* embryo by a dominant/negative PDGF receptor. *Development* **124:** 2355–2364.

Ramos J.W. and DeSimone D.W. 1996. *Xenopus* embryonic cell adhesion to fibronectin: Position-specific activation of RGD/synergy site-dependent migratory behavior at gastrulation. *J. Cell Biol.* **134:** 227–240.

Ramos J.W., Whittaker C.A., and DeSimone D.W. 1996. Integrin-dependent adhesive activity is spatially controlled by inductive signals at gastrulation. *Development* **122:** 2873–2883.

Ransom D.G., Hens M.D., and DeSimone D.W. 1993. Integrin expression in early amphibian embryos: cDNA cloning and characterization of *Xenopus* β1, β2, β3, and β6 subunits. *Dev. Biol.* **160:** 265–275.

Riou J.-F., Shi D.-L., Chiquet M., and Boucaut J.-C. 1990. Exogenous tenascin inhibits mesodermal cell migration during amphibian gastrulation. *Dev. Biol.* **137:** 305–317.

Saint-Jeannet J.-P. and Dawid I.B. 1994. Vertical versus planar neural induction in *Rana pipiens* embryos. *Proc. Natl. Acad. Sci.* **91:** 3049–3053.

Sanders E.J. 1980. The effect of fibronectin and substratum-attached material on the spreading of chick embryo mesoderm cells in vitro. *J. Cell Sci.* **44:** 225–242.

———. 1982. Ultrastructural immunocytochemical localization of fibronectin in the early chick embryo. *J. Embryol. Exp. Morphol.* **71:** 155–170.

———. 1984. Labelling of basement membrane constituents in the living chick embryo during gastrulation. *J. Embryol. Exp. Morphol.* **79:** 113–123.

———. 1986. Mesoderm migration in the early chick embryo. In *Developmental biology: A comprehensive synthesis* (ed. L.W. Browder), vol. 2, pp. 449– 480. Plenum Press, New York.

Schatteman G.C., Morrison-Graham K., van Koppen A., Weston J.A., and Bowen-Pope D.F. 1992. Regulation and role of PDGF receptor α-subunit expression during embryogenesis. *Development* **115:** 123–131.

Shi D.-L., Darribere T., Johnson K.E., and Boucaut J.C. 1989. Initiation of mesoderm cell migration and spreading relative to gastrulation in the urodele amphibian *Pleurodeles waltl*. *Development* **105:** 351–363.

Shi D.-L., Delarue M., Darribere T., Riou J.F., and Boucaut J.C. 1987. Experimental analysis of the extension of the dorsal marginal zone in *Pleurodeles waltl* gastrulae. *Development* **100:** 147–161.

Skalski M., Alfandari D., and Darribere T. 1998. A key function for αv containing integrins in mesoderm cell migration during *Pleurodeles waltl* gastrulation. *Dev. Biol.* **195:** 158–173.

Smith J.C., Symes K., Hynes R.O., and DeSimone D. 1990. Mesoderm induction and the control of gastrulation in *Xenopus laevis*: The roles of fibronectin and integrins. *Development* **108:** 229–238.

Smyth N., Vatansever H.S., Murray P., Meyer M., Frie C., Paulsson M., and Edgar D. 1999. Absence of basement membranes after targeting the LAMC1 gene results in embryonic lethality due to failure of endoderm differentiation. *J. Cell Biol.* **144:** 151–160.

Solursh M. and Lane M.C. 1988. Extracellular matrix triggers a directed cell migratory response in sea urchin primary mesenchyme cells. *Dev. Biol.* **130:** 397–401.

Stephens L.E., Sutherland A.E., Klimanskaya I.V., Andrieux A., Meneses J., Pedersen R.A., and Damsky C.H. 1995. Deletion of beta 1 integrins in mice results in inner cell mass failure and peri-implantation lethality. *Genes Dev.* **9:** 1883–1895.

Sun X., Meyers E.N., Lewandoski M., and Martin G.R. 1999. Targeted disruption of Fgf8 causes failure of cell migration in the gastrulating mouse embryo. *Genes Dev.* **13:** 1834–1846.

Symes K. and Mercola M. 1996. Embryonic mesoderm cells spread in response to platelet-derived growth factor and signaling by phosphatidylinositol 3-kinase. *Proc. Natl. Acad. Sci.* **93:** 9641–9644.

Tamboline C.R. and Burke R.D. 1992. Secondary mesenchyme of the sea urchin embryo: Ontogeny of blastocoelar cells. *J. Exp. Zool.* **262:** 51–60.

Toole B.P. 2002. Hyaluronan promotes the malignant phenotype. *Glycobiology* **12:** 37R–42R.

Van Hoof J., Harrisson F., Andries L., and Vakaet L. 1986. Microinjection of glycosaminoglycan-degrading enzymes in the chicken blastoderm. An ultrastructural study. *Differentiation* **31:** 14–19.

Wacker S., Brodbeck A., Lemaire P., Niehrs C., and Winklbauer R. 1998. Patterns and control of cell motility in the *Xenopus* gastrula. *Development* **125:** 1931–1942.

Wacker S., Grimm K., Joos T., and Winklbauer R. 2000. Development and control of tissue separation at gastrulation in *Xenopus. Dev. Biol.* **224:** 428–439.

Wessel G.M. and Berg L. 1995. A spatially restricted molecule of the extracellular matrix is contributed both maternally and zygotically in the sea urchin embryo. *Dev. Growth Differ.* **37:** 517–527.

Wessel G.M. and McClay D.R. 1987. Gastrulation in the sea urchin embryo requires the deposition of crosslinked collagen within the extracellular matrix. *Dev. Biol.* **121:** 149–165.

Whittaker C.A. and DeSimone D.W. 1993. Integrin α subunit mRNAs are differentially expressed in early *Xenopus* embryos. *Development* **117:** 1239–1249.

Winklbauer R. 1988. Differential interaction of *Xenopus* embryonic cells with fibronectin in vitro. *Dev. Biol.* **130:** 175–183.

———. 1990. Mesoderm cell migration during *Xenopus* gastrulation. *Dev. Biol.* **142:** 155–168.

Winklbauer R. and Keller R. 1996. Fibronectin, mesoderm migration, and gastrulation in *Xenopus. Dev. Biol.* **177:** 413–426.

Winklbauer R. and Nagel M. 1991. Directional mesoderm cell migration in the *Xenopus* gastrula. *Dev. Biol.* **148:** 573–589.

Winklbauer R. and Stoltz C. 1995. Fibronectin fibril growth in the extracellular matrix of the *Xenopus* embryo. *J. Cell Sci.* **108:** 1575–1586.

Winklbauer R., Medina A., Swain R.K., and Steinbeisse H. 2001. Frizzled-7 signalling controls tissue separation during *Xenopus* gastrulation. *Nature* **413:** 856–860.

Winklbauer R., Nagel M., Selchow A., and Wacker S. 1996. Mesoderm migration in the *Xenopus* gastrula. *Int. J. Dev. Biol.* **40:** 305–311.

Wojciak-Stothard B., Denyer M., Mishra M., and Brown R.A. 1997. Adhesion, orientation, and movement of cells cultured on ultrathin fibronectin fibers. *In Vitro Cell Dev. Biol. Anim.* **33:** 110–117.

Wood W. and Martin P. 2002. Structures in focus–filopodia. *Int. J. Biochem. Cell Biol.* **34:** 726–730.

Yang J.T., Bader B.L., Kreidberg J.A., Ullman-Cullere M., Trevithick J.E., and Hynes R.O. 1999. Overlapping and independent functions of fibronectin receptor integrins in early mesodermal development. *Dev. Biol.* **215:** 264–277.

Yang X., Dormann D., Munsterberg A.E., and Weijer C.J. 2002. Cell movement patterns during gastrulation in the chick are controlled by positive and negative chemotaxis mediated by FGF4 and FGF8. *Dev. Cell.* **3:** 425–437.

Zhu X., Mahairas G., Illies M., Cameron R.A., Davidson E.H., and Ettensohn C.A. 2001. A large-scale analysis of mRNAs expressed by primary mesenchyme cells of the sea urchin embryo. *Development* **128:** 2615–2627.

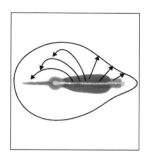

CHEMOTAXIS IN COORDINATING CELL MOVEMENTS DURING GASTRULATION

C.J. Weijer

School of Life Sciences, Wellcome Trust Biocentre, University of Dundee, Dundee, DD1 5EH, United Kingdom

GASTRULATION INVOLVES EXTENSIVE CELL MOVEMENT

Gastrulation is characterized by extensive cell movements that are precisely regulated in time and in space. Detailed knowledge of gastrulation movements was derived from cell lineage studies, in which groups of cells were vitally labeled before gastrulation and their fate was investigated after gastrulation had been completed using direct observation and sectioning techniques. The movement paths taken by the cells were inferred from changes in their positions recorded at a few selected time points (Stern and Fraser 2001). Time-lapse studies also have contributed greatly to our knowledge of cell movement patterns (Graeper 1929; Vakaet 1970), but due to much improved abilities to label cells in vivo, most recently by transfecting them with spectral variants of the green fluorescent protein (GFP), it has now become possible to track their movement during gastrulation using high-resolution imaging techniques (Keller 2002). This has resulted in a tremendous increase in our knowledge of the migration behavior of individual cells and groups of cells during gastrulation in various organisms. This, in combination with the ability to manipulate genes in a number of model organisms using either forward and/or reverse genetics, has led to the identification of numerous signaling pathways that directly or indirectly affect cell movement during gastrulation (Keller 2002; Wallingford et al. 2002). However, the mechanisms that control these movements are not yet well understood; i.e., it

is not clear which signals control cell polarization and movement, how they are detected, and how they are translated in directed movement.

CHEMOTAXIS IS INVOLVED IN THE CONTROL OF CELL MOVEMENT IN MANY DIFFERENT DEVELOPMENTAL PROCESSES

Chemotaxis has been shown to play an important role in the control of directed cell movement in a wide variety of developmental processes. One of the first chemoattractants to be identified in a eukaryote was cAMP, which controls the starvation-induced aggregation of thousands of individual social *Dictyostelium discoideum* amoebae (Konijn et al. 1967). Starving *Dictyostelium* cells aggregate guided by cAMP signals emitted periodically by cells in the aggregation center and relayed by the rest of the population, resulting in cAMP waves propagating from the center outward, thus directing the movement of the cells toward the aggregation center (Kessin 2001). The interaction between these cAMP waves, which are made by the cells, and the resulting cell movement can explain much of the morphogenesis observed during *Dictyostelium* development (Weijer 1999; Dormann and Weijer 2001). Since these initial studies, *Dictyostelium* has become a prime model system to investigate the molecular basis of signal detection and the transduction of these signals to polarization of cells and directed chemotactic cell movement (Iijima et al. 2002).

Chemotaxis also controls the branching morphogenesis occurring during the development of the tracheal system in *Drosophila* and during lung formation in vertebrates (Metzger and Krasnow 1999; Weaver et al. 2000). Both processes involve the attraction of epithelial cells to localized sources of FGFs located in the surrounding tissues. The formation of the trachea in *Drosophila* involves chemotaxis of cells at the tips of the trachea, directed toward an FGF signal (Branchless) secreted by surrounding tissue. During the branching of the primary lung buds, FGF10 is a potent chemoattractant. In vertebrates, chemotaxis toward FGF is also involved in branching morphogenesis during lung development. FGF10 secreted by mesenchymal cells will attract endodermal cells, and furthermore, FGF10 results in the induction of BMP4 in the endodermal cells, which is then involved in turning off the chemotactic response to FGF10 (Weaver et al. 2000). An FGF signal (Egl17) secreted by the gonads guides the migration of the sex-myoblast toward the gonads during *Caenorhabditis elegans* development (Branda and Stern 1998, 2000; Montell 1999). Furthermore, FGFs have been implicated in the movement of mesenchyme cells toward the apical ectodermal ridge during limb formation (Li and Muneoka 1999). In *Drosophila,* the migration of border cells toward the oocytes is controlled by chemotactic signals, EGF and VEGF secreted by the oocytes (Duchek and Rorth 2001; Duchek et al. 2001). Finally, it is well established that during the outgrowth of axons from the forming central nervous system, the growth cone is directed by a combination of attractive and repulsive signals originating in the midline (Brose et al. 1999; Guthrie 2001). The attractive signals are often mediated by netrins, whereas the repulsive signals are generally mediated by members of the slit family of diffusible extracellular ligands (Yu and Bargmann 2001; Wong et al. 2002). It has been proposed that the balance between attraction and repulsion is achieved by a modulation of signaling of the netrin receptor DCC (deleted in colorectal cancer) to the cytoskeleton by activated slit receptors of the Robo (roundabout) family (Emerson and Van Vactor 2002; Rhee et al. 2002).

In view of all these examples, where chemotaxis controls the migration of cells over long distances in the embryo, it is surprising that there have only been a few experiments performed to directly test whether chemotaxis plays a role in the guidance of cell movement during gastrulation, a process that in many organisms involves long-range directed cell movement of individual mesenchymal cells. In this chapter, I investigate the evidence that chemotaxis, i.e., the directed movement up or down of a chemical gradient of a signaling molecule (chemoattractant or repellent) may play a role in the control of cell movement during gastrulation. I review recent results implicating chemotaxis as an important mechanism controlling movement of mesodermal cells during gastrulation in the chick embryo (Yang et al. 2002), as well as evidence for a role of chemotaxis in gastrulation in other vertebrate and invertebrate systems, and I briefly review what is known about the signaling pathways involved.

SIGNALS CONTROLLING CELL MOVEMENT DURING GASTRULATION IN *DROSOPHILA*

During *Drosophila* gastrulation, the mesodermal cells in the ventral side of the embryo undergo an epithelial to mesenchymal transition (EMT), after which the mesodermal cells migrate dorsally and laterally along the inside of the body wall (Leptin 1999). The mesenchymal transition involves a down-regulation of *E-cadherin* by snail and an up-regulation of *N-cadherin* by *Twist*. The dorsolateral spreading of the mesodermal cells is dependent on the expression of the FGF receptor *heartless* (Beiman et al. 1996; Gisselbrecht et al. 1996). The *heartless* receptor is expressed in the mesodermal primordium and in the migrating mesoderm cells, and its expression is also dependent on the expression of both *twist* and *snail*, and necessary for the formation of heart, visceral, and somatic muscles (Shishido et al. 1993, 1997). The ligand for this receptor is not known, and consequently, there is no insight into its localization; however, it has been possible to infer the localization of the FGF signaling pathway by monitoring the Heartless-dependent activating phosphorylation of *Drosophila* ERK (dpERK), using phospho-specific antibodies (Gabay et al. 1997). dpERK is a downstream target of several receptor tyrosine kinases; however, during gastrulation, dpERK activation is restricted to the most dorsal 3–4 rows of migrating mesoderm cells in wild-type embryos and is completely absent in a *heartless* null mutant. This dynamic pattern of dpERK activation suggests an instructive role of FGF signaling in the process of mesoderm migration and, although not shown directly, is compatible with FGF functioning as a chemoattractant.

There have been some interesting observations of how an FGF signal may be detected by chemotactically responding cells. In the case of the formation of the tracheal system, as well as during the formation of the air sacs, cells associated with the flight muscles, migrating tip cells respond to an FGF signal by extending long filopodia in the direction of the signal, suggesting that filopodial extension is a part of the chemotactic detection mechanism (Petit et al. 2002; Ribeiro et al. 2002; Sato and Kornberg 2002). It has been shown that FGFs can induce cytonemes, long, thin cytoplasmic projections in epithelial cells of the imaginal wing disc, and in vitro *Drosophila* imaginal disc cells form cytonemes in the direction of heparin beads coated with human FGF4 (Kornberg 1999; Ramirez-Weber and Kornberg 1999). It seems likely that cells use these processes to explore their environment and that, in case of a positive signal, they translocate in the direction of the signal. The formation of long cellular processes during signal detection may not be limited to FGF signaling, since it has also been shown recently that the PDGF/VEGF guided migration of border cells seems to involve the extension of very long cellular protrusions in the direction of the signal (Fulga and Rorth 2002).

The pathways linking FGF signaling to cell polarization and directed motion are not well understood. The cells that express the Heartless receptor also express an essential component of the signal transduction pathway known as Dof (downstream of FGF), Stumps, or Heartbroken (Michelson et al. 1998; Vincent et al. 1998; Imam et al. 1999). Mutants in *dof* resemble mutants in *heartless* in that the mesodermal cells undergo an EMT but do not move away from the ventral midline, and as a result, cardiac visceral and dorsal muscle fates are not induced by Decapentaplegic (Dpp), a TGF-β family member that is produced and secreted by dorsal ectoderm. Dof also has been shown to be an essential component of FGF signaling through the *Breathless* receptor during the chemotactic migration of tip cells during trachea formation. By the use of chimeric receptors under the control of the *twist* enhancer, it was shown that chimeric receptors with an extracellular Heartless and intracellular EGF or torso (tor) receptor kinase domain could rescue the migration of cardiac cells in a *dof*-independent manner, showing that the signaling pathways regulating movement use generic components (Dossenbach et al. 2001).

FROGS AND FISH

The cellular basis underlying gastrulation has been intensively investigated in frogs. It is now mostly seen to be driven by a combination of cell behaviors such as radial intercalation and vegetal rotation that drive the early phases of gastrulation, and involution of mesoderm cells followed by medio-lateral intercalation and convergent extension of cells in the involuting mesoderm and overlying neural plate (Keller et al. 2000). The signals controlling vegetal rotation and radial intercalation are unknown, but convergent extension has been shown to involve signaling through the Wnt planar polarity pathway (for review, see Keller et al. 2000; Keller 2002; Wallingford et al. 2002; Chapter 13). Intercalating mesodermal cells polarize their protrusive behavior, become bipolar, and align themselves. Cells in the overlying neural plate show monopolar protrusive activity that is directed toward the midline during their intercalation (for review, see Keller et al. 2000). This mode of behavior is strictly dependent on an activity residing in the axial mesoderm, and in many ways resembles the polarization of cells in gradients of a chemoattractant, where cells first

polarize in response to an external signal and then move in the direction of polarization. *eFGF*, a *Xenopus FGF4* homolog, is expressed in the dorsal midline (Isaacs et al. 1992, 1995), and it is possible that this acts as an attractant. Another candidate is Sonic hedgehog, which has recently been shown to be a potent midline chemoattractant during axon guidance (Charron et al. 2003)

Embryos expressing a dominant negative FGF receptor do not form notochord or ordered somites, but the formation of more anterior head structures is more normal (see Chapter 32). The lateral and ventral mesoderm cells do not involute properly, and the blastopore does not close properly (Amaya et al. 1991; Kroll and Amaya 1996). These findings could point to an instructive role for FGF signaling in the control of cell movement. Recently, it has been shown that Sprouty2 acts as a negative downstream regulator of the FGF signaling pathway leading to movement (Nutt et al. 2001). It has been reported to selectively block convergent extension in the frog embryo without affecting the expression of mesoderm-specific downstream genes such as *brachyury*. The movement phenotype is very similar to that caused by expression of the dominant negative FGF receptor. However, to interpret this effect, it will be necessary to investigate the changes in cell behavior in these mutants in more detail. *brachyury* expression may control the movement behavior of the mesodermal cells. Cells expressing *brachyury* undergo convergent extension, whereas nonexpressing cells, such as prechordal mesoderm, migrate as individual mesenchymal cells, which require an interaction with a fibronectin-rich extracellular matrix. *brachyury* expression appears to inhibit this interaction through an unknown pathway (Kwan and Kirschner 2003).

In *Fundulus*, the cells in the germ band move largely as individuals or as small groups of mesenchymal-like cells (Trinkaus et al. 1992). The cells move by filopodial extensions and the cells accumulate dorsally by relatively directed movement, resulting in a thickening of the shield. It has been suggested that an obvious explanation for this type of behavior would be an attraction of cells to the dorsal edge by a signal originating in the shield; however, the molecular basis of this signal has not been elucidated (Trinkaus et al. 1992; Concha and Adams 1998; Trinkaus 1998). In zebrafish it has recently been shown that the movement of these cells may partly be guided by PDGF signals emitted by the overlying ectoderm acting to polarize the cells, and perhaps acting as chemoattractants (Montero et al. 2003.)

FGF SIGNALING CONTROLS GASTRULATION MOVEMENTS IN MICE

Mutants in components of the FGF signaling pathway in mice have shown that FGF signaling plays a major role in the control of movement of cells going through the primitive streak in mouse embryos. During early mouse development, at least six FGFs have been shown to be expressed at the gastrulation stages: *FGF2, FGF3, FGF4, FGF5, FGF8,* and *FGF17* (Sun et al. 1999). Only deletions of *FGF4* and *FGF8* give rise to recognizable phenotypes before or during the gastrulation stages of development. Deletion of *FGF4* results in the death of the embryos immediately after implantation and it has been suggested that FGF4 is required for the growth of the trophoblast stem cells (Feldman et al. 1995; Tanaka et al. 1998). The *FGF8* mutant displays a strong phenotype during gastrulation (Sun et al. 1999). The mesoderm cells undergo EMT, but then do not move away from the streak, showing that FGF signaling is required for the cells to be able to migrate away from the primitive streak. Instead, they form extensive extraembryonic mesodermal structures. The expression of *FGF3* and *FGF5* is normal in these mutants, but there is no expression of *FGF4* in the primitive streak. Therefore, it is difficult to conclude whether the defect to migrate away is due to the absence of FGF8 or FGF4 signaling, but the chick findings described above make it more likely that the defect is due to the absence of FGF8 signaling.

FGF Receptor Mutants

The initial observations that FGF signaling might be involved in controlling mesoderm differentiation and movement during gastrulation in mice came from the study of knockout mutants of the FGF receptors. The *FGFR1* receptor is widely expressed during gastrulation in the mouse (Yamaguchi et al. 1992). Knockout of the *FGFR1* receptor results in early death of the embryo shortly after implantation, and these embryos lack all axial and paraxial mesodermal structures (Deng et al. 1994; Yamaguchi et al. 1994). However, a subsequent chimeric analysis using embryonic stem cells from *FGFR1* knockout mice has shown that most cells lacking the *FGFR1* receptor do not traverse the primitive streak and therefore are deficient in formation of the heart somites and notochord (Ciruna et al. 1997). The cells accumulate at the primitive streak and tend to form ectopic neural tubes. On the basis of these observations, it was suggested that the failure of the *FGFR1* receptor null cells to traverse the primitive streak was due to a defect in the down-regulation of *E-cadherin*, since down-regulation of *E-cadherin* had been shown to be an important event in the EMT (Burdsal et al. 1993). Subsequently, it was shown that in chimeric embryos, cells lacking *FGFR1* accumulated in the streak and failed to down-regulate *E-cadherin* and up-regulate *N-cadherin* (Ciruna and Rossant 2001). Using explants of pieces of chimeric primitive streaks, it was shown that *FGFR1⁻/⁻* cells segregated from wild-type cells and did not move very well. However, upon

treatment of these streak explants with E-cadherin function blocking antibodies, it was found that the *FGFR1⁻/⁻* cells could move as well as wild-type cells, but their chemotactic ability was not tested (Ciruna and Rossant 2001).

FGF receptors exist in different splice variants due to alternative splicing in the first extracellular IgG domain generating FGFR1α and FGF1β splice variants. In normal development, only FGFR1α variant receptors are detectable, and deletion of *FGFR1α* results in severe defects of posterior mesodermal structures. The somites after somite 10 are absent and the notochord is mostly absent or severely disorganized, suggesting that FGFR1α is responsible for node regression and essential for posterior mesoderm development (Xu et al. 1999). Deletion of the *FGFR1α* receptor resulted in an up-regulation of *FGFR1β* at the gastrulation stage, and this up-regulation of *FGFR1β* apparently rescues the development of the more anterior mesodermal structures, which are absent in constitutive *FGFR1* receptor knockout strains.

Alternative splicing in the third extracellular IgG domain results in FGFR1b and FGFR1c isomers, respectively. These receptors have also been shown to possess different FGF receptor-binding specificities (Ornitz et al. 1996). An analysis of the expression patterns of these subtypes has shown that the FGFR1c is the main receptor expressed during early development, and FGF1b is expressed much less often. Deletion of *FGFR1b* results in only minor defects in gastrulation movements, suggesting that most signaling involves the FGFR1c variant (Partanen et al. 1998).

Deletion of the *FGFR2* receptor has revealed that its function is needed early in development, immediately after implantation, and it seems likely that FGF4 is the ligand for this receptor at this stage of development (Arman et al. 1998). The role of FGFR2 in later development using conditional inactivation of the receptor has not yet been analyzed. It has been shown, however, that deletion of *FGFR2b* does not result in major defects during gastrulation, making it likely that the function of FGFR2c is more important in early development (Revest et al. 2001). Deletion of *FGFR3* has been shown to have no major effect in early gastrulation stages of development. It has, however, been shown to be an important regulator of long bone development (Deng et al. 1996). Deletion of *FGFR4* does result in the development and birth of apparently normal fertile mice (Weinstein et al. 1998), and even *FGFR3/FGFR4* double null mutants give rise to viable offspring, showing that signaling through the FGFR3 and FGFR4 receptors is not necessary for the control of cell movements during gastrulation.

Downstream of FGF Receptors, *Shp2* Mutants

Specific tyrosine phosphatases are often found downstream of many tyrosine kinase receptors. One prominent phosphatase is Shp2. It is very interesting to see that the phenotype of the mouse *Shp2* knockout mutant resembles that of the *FGFR1* knockout rather closely (Saxton et al. 1997; Saxton and Pawson 1999). A chimeric analysis of *Shp2* knockout cells in a wild-type background has shown that these cells accumulate at the posterior primitive streak and contribute excessively to the neural tube (Saxton et al. 2000). The cells that manage to progress through the streak are strongly underrepresented in the extending mesodermal wings and the forming somites. The fact that these cells can migrate through the streak, but then do not migrate out as well as wild-type cells, suggests that *Shps2$^{-/-}$* cells are defective in responding to the correct guidance signals, or alternatively, are defective in migration. It has been shown that *Shp2$^{-/-}$* embryo fibroblasts are defective in a chemotactic response to acidic FGF, a generic FGF that binds to all receptors (Ornitz et al. 1996), whereas their chemotactic response to another well-known chemoattractant, PDGF, is unaffected (Saxton and Pawson 1999). This makes it likely that Shp2 is needed for an FGF-mediated chemotactic response. It is not known whether *Shp2$^{-/-}$* cells are defective in chemotaxis toward FGF4 and whether Shp2 is involved in chemorepulsion by FGF8 as well as in chemoattraction.

In conclusion, the mouse mutants show that FGF signaling through the FGFR1 receptor is essential for the control of EMT and that signaling by FGF4 and/or FGF8 through FGFR1 and FGFR2 is very important for the control of migration of mesoderm cells. Therefore, these findings are compatible with the notion that FGF8 may act as a chemorepellent and FGF4 as a chemoattractant.

GASTRULATION IN THE CHICK EMBRYO

The chick embryo is a widely used experimental system to investigate the mechanisms of primitive streak formation and cell movement during gastrulation. The primitive streak is initiated in the posterior region of the area opaca by the transformation of a sickle-shaped group of cells into an elongated thickening structure, the streak, which extends in an anterior direction (Bachvarova et al. 1998; Chapman et al. 2002; see Chapter 15). This process involves extensive cell rearrangements, known as polonaise movements. However, to date there is little insight into the mechanisms underlying these movements (Graeper 1929; Lawson and Schoenwolf 2001). When the streak is halfway extended, cells in the streak undergo an EMT, ingress, and move out anteriorly and laterally in the space between the epiblast and the hypoblast. Extensive fate map studies have been performed at these stages, and the fate of the different cell populations moving through the streak is known in considerable detail (Selleck and Stern 1991; Psychoyos and Stern 1996). Recently, the migration pathways of cells emerging at

various anterior to posterior positions from the fully extended streak (HH stage 4 embryo) have been analyzed by tracing GFP-labeled cells for extended periods of time (Yang et al. 2002). The node lays down the notochord during its regression, while cells emerging from the streak, just posterior to the node, move laterally away from the streak. However, as soon as the node regresses past them, they appear to move back inward toward the midline to form medial somites (Fig. 1). Cells emerging from the middle streak also first move away from the streak, followed by an inward movement after the node has regressed past them. These cells will form lateral plate mesoderm. Cells emerging from the posterior streak move outward toward the boundary of the area pellucida and area opaca. The fates of these populations of cells are in good agreement with published fate maps (Psychoyos and Stern 1996). The movements of the cells leaving the anterior and posterior streak are controlled by distinct signals in the anterior and posterior embryo. Anterior streak cells, which normally move around the node to form medial somites, move outward to the boundary between the area pellucida and area opaca when transplanted in the posterior streak (Garcia-Martinez and Schoenwolf 1992; Schoenwolf et al. 1992; Yang et al. 2002). The inward movements observed appear to be active; however, it is possible that they are passive and result from the narrowing of the embryo observed at this stage. On the other hand, chemotaxis would be an attractive mechanism to explain these long-range directed cell movements in response to signals from surrounding tissues.

FGFs in the Control of Cell Movement

The phenotypes of knockout mutants in mice show that deletions in *FGF8* and/or *FGF4* result in defects of the migration of axial and paraxial mesoderm (Sun et al. 1999). In the chick embryo, both *FGF4* and *FGF8* are expressed in the primitive streak during the early stages of gastrulation. *FGF4* is expressed in the early embryo in the anterior primitive streak, the forming head process, and the precursor of the notochord (Shamim and Mason 1999; Karabagli et al. 2002). *FGF8* is expressed in the forming streak, but expression in the node is lost just before it starts to regress (Lawson et al. 2001; Chapman et al. 2002). Middle primitive streak cells are very strongly attracted toward beads soaked in FGF4. This attraction is completely blocked by the expression of a dominant negative FGF receptor, FGFR1c, in the responding cells, while they are strongly repelled by FGF8 (Yang et al. 2002). A hypothesis to explain these observations was put forward: FGF8 acts as a chemorepellent and instructs the cells to move away from the streak. Cells leaving the anterior streak are attracted back in toward the midline by FGF4 produced in the head process and forming notochord, while cells leaving the posterior

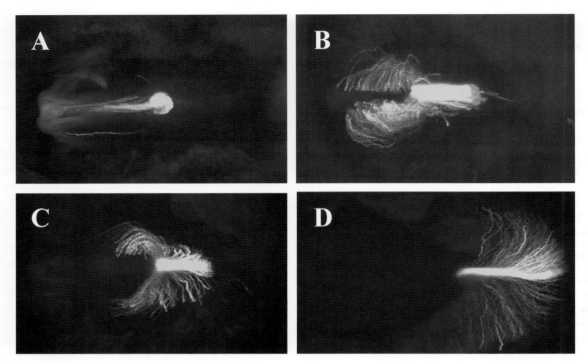

Figure 1. Cell movement patterns of cells leaving the primitive streak at a stage-HH4 chick embryo. The epiblast of a stage-HH3 donor embryo was electroporated with GFP, and after 3–4 hours, the cells had started to express GFP and move toward the primitive streak. An appropriate piece of GFP-expressing streak was cut out and transplanted in an unlabeled host embryo, of which the equivalent piece of streak had been removed. The movement of the cells was followed over the next 12–18 hours until the embryos reached the 4- to 6-somite stage. The lines in the photographs represent the movement tracks of individual cells and they have been processed to show the direction of movement. The initial part of the track is yellow, and the movement during the last 20 images is shown in green. The anterior of the embryo is left in all panels. (*A*) Movement of GFP-labeled cells in node during 15 hours of development. (*B*) Movement of cells emerging from the anterior streak during 15 hours of the experiment. It is clearly shown that the cells move out laterally from the streak, then back in toward the midline and forward after the node has regressed past them. (*C*) Movement tracks of cells emerging from the middle primitive streak recorded for the 15 hours from the time of transplantation. It can be seen that the cells move out from the streak laterally and then most move toward the midline to form lateral plate mesoderm. (*D*) Movement trajectories of cells transplanted to the posterior streak during 15 hours of the experiment. The cells move out in a fan-like pattern and form mostly hematopoietic cells and extraembryonic tissues.

streak are attracted by an unknown signal possibly generated by the boundary between the area pellucida and area opaca (Fig. 2). In support of this idea, it was shown that small pieces of head process, transplanted in the epiblast of a stage-4 embryo, could attract mesoderm cells in a manner similar to an FGF4-coated bead transplanted in the epiblast, whereas the boundary region of the area opaca and area pellucida also produces an attractant for middle streak cells (Yang et al. 2002).

This hypothesis has raised a number of questions. In the posterior part of the streak, both *FGF4* and *FGF8* are expressed; however, the cells move away. If the above stated chemotactic hypothesis is correct, then the repulsive effect of FGF8 must be dominant over the attractive effect of FGF4. The molecular basis for this effect is not yet known; however, it indicates that other signaling pathways are in operation that modulate the FGF-mediated response in a context-sensitive manner, as has been shown for other sys-

tems where migration is controlled by chemotactic signaling (Lu et al. 2001; Schmucker and Zipursky 2001). The node expresses *FGF4* but not *FGF8*; however, the node does not attract mesoderm cells during its regression. The mesoderm cells move around it, suggesting that the attractive effect by FGF4 could be counteracted by another repellent. A possible candidate may be a member of the slit family of repellents. Some of these are expressed well before axon guidance becomes important. For instance, *slit2* is known to be expressed in the node during these very early stages of development and would be a good candidate for a node-repellent molecule (Li et al. 1999).

OTHER CHEMOATTRACTANTS?

It seems likely that when chemotaxis plays a role in directing the movement of cells toward different goals, their action will not be limited to responding to FGFs. Other

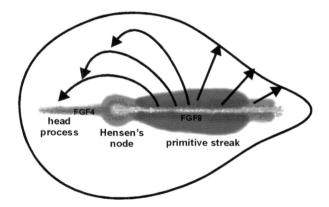

Figure 2. Model for the FGF-mediated control of chemotactic cell movement of mesoderm cells in the early chick embryo. The *FGF4* expression pattern is shown schematically in red and the *FGF8* expression pattern in blue. The area in the middle primitive streak, which expresses both *FGF4* and *FGF8*, appears dark blue in the overlay. The black arrows show the trajectories of cell movement observed in relation to the *FGF4* and *FGF8* expression patterns. The yellow arrow indicates the direction of node regression. The movement away from the streak is proposed to result from negative chemotaxis in response to FGF8 produced by cells in the middle primitive streak, whereas the inward movement of the cells emerging from the anterior streak is the result of attraction by FGF4 secreted by cells in the head process and notochord. The cells in the caudal part of the embryo are attracted by a signal of unknown identity, secreted in the border region between the area opaca and area pellucida.

candidate molecules controlling chemotactic movements of cells in the nervous systems are the family of slit molecules, which can act as repellents or attractants for nerve growth cones and mesoderm cells (Li et al. 1999; Brose and Tessier-Lavigne 2000; Kramer et al. 2001; Rao et al. 2002; Wong et al. 2002). It has been shown that slits are secreted diffusible molecules that play a major role in preventing the premature crossing of axons across the midline (Brose and Tessier-Lavigne 2000; Piper and Little 2002). They are detected by a family of receptors first identified in *Drosophila*: The Roundabout (or Robo) receptors and their signaling pathways seem to interact with netrins and Sonic Hedgehog, attractive cues for axons (Guthrie 2001; Charron et al. 2003).

Although the functions of Slits and Robos have been characterized best in later development and maturation of the nervous system, it has been noted recently that members of the Slit family start to be expressed during gastrulation and during the initial formation of the nervous system. In zebrafish, *slit2* and *slit3* start to be expressed halfway through gastrulation. *slit2* is expressed in the anterior part of the neural plate, and *slit3* is expressed in the anterior end of the prechordal mesoderm (Yeo et al. 2001). *slit2* overexpression in the early zebrafish embryo from the 1- to 2-cell stage onward results in strong defects in convergent exten-

sion movements and impairs the rostral migration of cells in the diencephalon and, thus, cyclopia. In the chick embryo, *slit2* is expressed during gastrulation, and in stage-4 embryos, it has been detected in Hensen's node. From stage 5 to stage 8 it is expressed in Hensen's node, the prechordal plate, notochord, and paraxial mesoderm (Li et al. 1999). The expression of the other slits has not yet been described, but their role in these early developmental stages will need further investigation.

Other putative chemoattractants expressed during gastrulation are VEGF and PDGF. Although both of these molecules have been shown to be potent chemoattractants in a variety of developmental contexts, little is known about their role in gastrulation.

PDGFa is known to be expressed in the ectoderm in *Xenopus* and mouse, whereas the PDGFR1α receptor is expressed in migrating mesoderm cells. Changes in expression of PDGFR result in severe gastrulation defects. It has been suggested that this is mostly the result of a defective adhesion and spreading of the mesoderm cells on the blastocoel roof, but the possibility that they function as chemoattractants has not been investigated (Ataliotis et al. 1995; Symes and Mercola 1996).

VEGF has been shown to have an important function in directing the chemotactic migration of hemocytes in *Drosophila*, and in the migration of border cells during oocyte maturation (Duchek et al. 2001; Cho et al. 2002; Rorth 2002). However, again, their role in controlling gastrulation movement has not been investigated. It has been shown that mutants in *vegfA* and the vegf receptor *flk1* are defective in extraembryonic mesoderm that normally gives rise to blood cells, suggesting a potential role for the components in the migration of the posterior mesoderm cells to their targets (Shalaby et al. 1997; Damert et al. 2002). Another potentially interesting factor is hepatocyte growth factor/scatter factor, which is expressed in the node during gastrulation (Thery et al. 1995), and which has been shown to act as a chemoattractant for muscle cells and for axonal growth cones at later stages (Niemann et al. 2000; Kurz et al. 2002).

SIGNAL TRANSDUCTION PATHWAYS CONTROLLING CELL POLARIZATION AND CHEMOTACTIC MOVEMENT

Signal detection and cell polarization have been studied in detail for the chemotactic movement of *Dictyostelium* cells and neutrophils (Iijima et al. 2002). In these cases, the signal pathway involves signal detection through G-protein-coupled serpentine receptors, resulting in localized phosphatidylinositol-triphosphate ($PI_{[3,4,5]}P_3$) production involving localized activation of PI3 kinase and inactivation of the PIP3-degrading enzyme PTEN. This allows the localization and activation of racGEFs, which contain specific $PI_{[3,4,5]}P_3$-binding domains, to the leading edge of the cell. These localized racGEFs then activate rac, and through rac, target molecules of the Scar and Wasp families, activate the ARP2/3 complex and localize actin polymerization. Actin polymerization is then the driving force for pseudopod and lamellopod extension, essential mechanisms for cell movement (Etienne-Manneville and Hall 2002). It has been possible to measure the spatiotemporal dynamics of PIP3 accumulation during cell polarization in response to a localized chemoattractant, both in single chemotactically moving cells, as well as in individual cells in the multicellular structures of *Dictyostelium* (Dormann et al. 2002; Iijima et al. 2002). It will be interesting to see whether these techniques to monitor chemotactic signal transduction in vivo can be used to study chemotactic signaling during gastrulation.

Little is known about how the FGF signal transduction pathway results in chemotactic movement. It is not yet known how local gradients of FGFs are detected and how they are translated in cell polarization, localized actin polymerization, and movement. Work on cell lines has shown that PI3 kinase signaling can act downstream of the FGF receptors (Schlessinger 2000; Ong et al. 2001), and it seems worthwhile to investigate these signaling pathways in more detail during gastrulation using some of the techniques outlined above.

Another chemotactic signaling pathway that is beginning to be elucidated plays a role in the chemotactic movement of border cells during *Drosophila* oocyte maturation. This movement requires the activation of the tyrosine kinase PDGF/VEGF receptor and involves the extension of long cellular processes in the direction of the chemotactic gradient, thus showing some similarity to FGF-mediated chemotaxis during tracheal and air sac development. Through the analysis of mutants, it has been shown that this signaling pathway does not require activation of PI3-kinase signaling, but is absolutely dependent on the function of the racGEF *myoblast city*, a homolog of the mammalian *Dock180* also resulting in localized actin polymerization (Duchek et al. 2001; Fulga and Rorth 2002). It will therefore be interesting to investigate whether this signaling pathway is involved in FGF-mediated chemotactic cell movement during gastrulation.

CONCLUSION

Chemotaxis has been shown to be an important mechanism in various developmental contexts that involve cell rearrangements such as lung morphogenesis, the wiring of the nervous system, sex cell migration, limb development, and the functioning of the immune system. Evidence is accumulating that chemotaxis will turn out to be an impor-

tant cell guidance mechanism during gastrulation. The top candidate signaling molecules appear to be members of the FGF family, but it seems likely that members of other families of molecules will be involved as well. Once the molecules have been identified, it will be necessary to understand the signal transduction pathways involved. It is also important to understand the interactions between signaling systems that control the localized production of chemoattractants, and their receptors in responding cells.

REFERENCES

Amaya E., Musci T.J., and Kirschner M.W. 1991. Expression of a dominant negative mutant of the FGF receptor disrupts mesoderm formation in *Xenopus* embryos. *Cell* **66:** 257–270.

Arman E., Haffner-Krausz R., Chen Y., Heath J.K., and Lonai P. 1998. Targeted disruption of fibroblast growth factor (FGF) receptor 2 suggests a role for FGF signaling in pregastrulation mammalian development. *Proc. Natl. Acad. Sci.* **95:** 5082–5087.

Ataliotis P., Symes K., Chou M.M., Ho L., and Mercola M. 1995. PDGF signalling is required for gastrulation of *Xenopus laevis*. *Development* **121:** 3099–3110.

Bachvarova R.F., Skromne I., and Stern C.D. 1998. Induction of primitive streak and Hensen's node by the posterior marginal zone in the early chick embryo. *Development* **125:** 3521–3534.

Beiman M., Shilo B.Z., and Volk T. 1996. Heartless, a *Drosophila* FGF receptor homolog, is essential for cell migration and establishment of several mesodermal lineages. *Genes Dev.* **10:** 2993–3002.

Branda C.S. and Stern M.J. 1998. Molecular mechanisms of sex myoblast migration in *C. elegans*. *Dev. Biol.* **198:** 393.

———. 2000. Mechanisms controlling sex myoblast migration in *Caenorhabditis elegans* hermaphrodites. *Dev. Biol.* **226:** 137–151.

Brose K. and Tessier-Lavigne M. 2000. Slit proteins: Key regulators of axon guidance, axonal branching, and cell migration. *Curr. Opin. Neurobiol.* **10:** 95–102.

Brose K., Bland K.S., Wang K.H., Arnott D., Henzel W., Goodman C.S., Tessier-Lavigne M., and Kidd T. 1999. Slit proteins bind Robo receptors and have an evolutionarily conserved role in repulsive axon guidance. *Cell* **96:** 795–806.

Burdsal C.A., Damsky C.H., and Pedersen R.A. 1993. The role of E-cadherin and integrins in mesoderm differentiation and migration at the mammalian primitive streak. *Development* **118:** 829–844.

Chapman S.C., Schubert F.R., Schoenwolf G.C., and Lumsden A. 2002. Analysis of spatial and temporal gene expression patterns in blastula and gastrula stage chick embryos. *Dev. Biol.* **245:** 187–199.

Charron F., Stein E., Jeong J., McMahon A.P., and Tessier-Lavigne M. 2003. The morphogen sonic hedgehog is an axonal chemoattractant that collaborates with netrin-1 in midline axon guidance. *Cell* **113:** 11–23.

Cho N.K., Keyes L., Johnson E., Heller J., Ryner L., Karim F., and Krasnow M.A. 2002. Developmental control of blood cell migration by the *Drosophila* VEGF pathway. *Cell* **108:** 865–876.

Ciruna B. and Rossant J. 2001. FGF signaling regulates mesoderm cell fate specification and morphogenetic movement at the primitive streak. *Dev. Cell* **1:** 37–49.

Ciruna B.G., Schwartz L., Harpal K., Yamaguchi T.P., and Rossant J. 1997. Chimeric analysis of fibroblast growth factor receptor-1

(Fgfr1) function: A role for FGFR1 in morphogenetic movement through the primitive streak. *Development* 124: 2829–2841.

Concha M. and Adams R. 1998. Oriented cell divisions and cellular morphogenesis in the zebrafish gastrula and neurula: A time-lapse analysis. *Development* 125: 983–994.

Damert A., Miquerol L., Gertsenstein M., Risau W., and Nagy A. 2002. Insufficient VEGFA activity in yolk sac endoderm compromises haematopoietic and endothelial differentiation. *Development* 129: 1881–1892.

Deng C., Wynshaw-Boris A., Zhou F., Kuo A., and Leder P. 1996. Fibroblast growth factor receptor 3 is a negative regulator of bone growth. *Cell* 84: 911–921.

Deng C.X., Wynshaw-Boris A., Shen M.M., Daugherty C., Ornitz D.M., and Leder P. 1994. Murine FGFR-1 is required for early postimplantation growth and axial organization. *Genes Dev.* 8: 3045–3057.

Dormann D. and Weijer C.J. 2001. Propagating chemoattractant waves coordinate periodic cell movement in *Dictyostelium* slugs. *Development* 128: 4535–4543.

Dormann D., Weijer G., Parent C.A., Devreotes P.N., and Weijer C.J. 2002. Visualizing PI3 kinase-mediated cell-cell signaling during *Dictyostelium* development. *Curr. Biol.* 12: 1178–1188.

Dossenbach C., Rock S., and Affolter M. 2001. Specificity of FGF signaling in cell migration in *Drosophila*. *Development* 128: 4563–4572.

Duchek P. and Rorth P. 2001. Guidance of cell migration by EGF receptor signaling during *Drosophila* oogenesis. *Science* 291: 131–133.

Duchek P., Somogyi K., Jekely G., Beccari S., and Rorth P. 2001. Guidance of cell migration by the *Drosophila* PDGF/VEGF receptor. *Cell* 107: 17–26.

Emerson M.M. and Van Vactor D. 2002. Robo is Abl to block N-Cadherin function. *Nat. Cell Biol.* 4: E227–230.

Etienne-Manneville S. and Hall A. 2002. Rho GTPases in cell biology. *Nature* 420: 629–635.

Feldman B., Poueymirou W., Papaioannou V.E., Dechiara T.M., and Goldfarb M. 1995. Requirement of FGF-4 for postimplantation mouse development. *Science* 267: 246–249.

Fulga T.A. and Rorth P. 2002. Invasive cell migration is initiated by guided growth of long cellular extensions. *Nat. Cell Biol.* 4: 715–719.

Gabay L., Seger R., and Shilo B.Z. 1997. MAP kinase in situ activation atlas during *Drosophila* embryogenesis. *Development* 124: 3535–3541.

Garcia-Martinez V. and Schoenwolf G.C. 1992. Positional control of mesoderm movement and fate during avian gastrulation and neurulation. *Dev. Dyn.* 193: 249–256.

Gisselbrecht S., Skeath J.B., Doe C.Q., and Michelson A.M. 1996. heartless encodes a fibroblast growth factor receptor (DFR1/DFGF-R2) involved in the directional migration of early mesodermal cells in the *Drosophila* embryo. *Genes Dev.* 10: 3003–3017.

Graeper L. 1929. Die Primitiventwicklung des Hunchens nach stereokinematischen Untersuchungen, kontoliert durch vitale Farbmarkierung und verglichen mitt der Entwicklung andere Wirbeltiere. *Wilhelm Roux' Arch.* 116: 382–429.

Guthrie S. 2001. Axon guidance: Robos make the rules. *Curr. Biol.* 11: R300–303.

Iijima M., Huang Y.E., and Devreotes P. 2002. Temporal and spatial regulation of chemotaxis. *Dev. Cell* 3: 469–478.

Imam F., Sutherland D., Huang W., and Krasnow M.A. 1999. *stumps*, a Drosophila gene required for fibroblast growth factor (FGF)-directed migrations of tracheal and mesodermal cells. *Genetics* 152: 307–318.

Isaacs H.V., Pownall M.E., and Slack J.M. 1995. eFGF is expressed in the dorsal midline of *Xenopus laevis*. *Int. J. Dev. Biol.* 39: 575–579.

Isaacs H.V., Tannahill D., and Slack J.M. 1992. Expression of a novel FGF in the *Xenopus* embryo. A new candidate inducing factor for mesoderm formation and anteroposterior specification. *Development* 114: 711–720.

Karabagli H., Karabagli P., Ladher R.K., and Schoenwolf G.C. 2002. Comparison of the expression patterns of several fibroblast growth factors during chick gastrulation and neurulation. *Anat. Embryol.* 205: 365–370.

Keller R. 2002. Shaping the vertebrate body plan by polarized embryonic cell movements. *Science* 298: 1950–1954.

Keller R., Davidson L., Edlund A., Elul T., Ezin M., Shook D., and Skoglund P. 2000. Mechanisms of convergence and extension by cell intercalation. *Philos. Trans. Roy. Soc. Lond. B Biol. Sci.* 355: 897–922.

Kessin R. 2001. Dictyostelium: *Evolution, cell biology, and the development of multicellularity*. Cambridge University Press, Cambridge, United Kingdom.

Konijn T.M., van de Meene J.G.C., Bonner J.T., and Barkley D.S. 1967. The acrasin activity of adenosine-3′,5′-cyclic phosphate. *Proc. Natl. Acad. Sci.* 58: 1152–1154.

Kornberg T. 1999. Pictures in cell biology. Cytonemes. *Trends Cell Biol.* 9: 434.

Kramer S.G., Kidd T., Simpson J.H., and Goodman C.S. 2001. Switching repulsion to attraction: changing responses to slit during transition in mesoderm migration. *Science* 292: 737–740.

Kroll K.L. and Amaya E. 1996. Transgenic *Xenopus* embryos from sperm nuclear transplantations reveal FGF signaling requirements during gastrulation. *Development* 122: 3173–3183.

Kurz S.M., Diebold S.S., Hieronymus T., Gust T.C., Bartunek P., Sachs M., Birchmeier W., and Zenke M. 2002. The impact of c-met/scatter factor receptor on dendritic cell migration. *Eur. J. Immunol.* 32: 1832–1838.

Kwan K.M. and Kirschner M.W. 2003. Xbra functions as a switch between cell migration and convergent extension in the *Xenopus* gastrula. *Development* 130: 1961–1972.

Lawson A. and Schoenwolf G.C. 2001. Cell populations and morphogenetic movements underlying formation of the avian primitive streak and organizer. *Genesis* 29: 188–195.

Lawson A., Colas J.F., and Schoenwolf G.C. 2001. Classification scheme for genes expressed during formation and progression of the avian primitive streak. *Anat. Rec.* 262: 221–226.

Leptin M. 1999. Gastrulation in *Drosophila*: The logic and the cellular mechanisms. *EMBO J.* 18: 3187–3192.

Li H.S., Chen J.H., Wu W., Fagaly T., Zhou L., Yuan W., Dupuis S., Jiang Z.H., Nash W., Gick C., Ornitz D.M., Wu J.Y., and Rao Y. 1999. Vertebrate slit, a secreted ligand for the transmembrane protein roundabout, is a repellent for olfactory bulb axons. *Cell* 96: 807–818.

Li S.G. and Muneoka K. 1999. Cell migration and chick limb development: Chemotactic action of FGF-4 and the AER. *Dev. Biol.* 211: 335–347.

Lu Q., Sun E.E., Klein R.S., and Flanagan J.G. 2001. Ephrin-B reverse signaling is mediated by a novel PDZ-RGS protein and selectively inhibits G protein-coupled chemoattraction. *Cell* 105: 69–79.

Metzger R.J. and Krasnow M.A. 1999. Genetic control of branching morphogenesis. *Science* **284:** 1635–1639.

Michelson A.M., Gisselbrecht S., Buff E., and Skeath J.B. 1998. Heartbroken is a specific downstream mediator of FGF receptor signalling in *Drosophila*. *Development* **125:** 4379–4389.

Montell D.J. 1999. The genetics of cell migration in *Drosophila melanogaster* and *Caenorhabditis elegans* development. *Development* **126:** 3035–3046.

Montero J.A., Kilian B., Chan J., Bayliss P.E., and Heisenberg C.P. 2003. Phosphoinositide 3-kinase is required for process outgrowth and cell polarization of gastrulating mesendodermal cells. *Curr. Biol.* **13:** 1279–1289.

Niemann C., Brinkmann V., and Birchmeier W. 2000. Hepatocyte growth factor and neuregulin in mammary gland cell morphogenesis. *Adv. Exp. Med. Biol.* **480:** 9–18.

Nutt S.L., Dingwell K.S., Holt C.E., and Amaya E. 2001. *Xenopus* Sprouty2 inhibits FGF-mediated gastrulation movements but does not affect mesoderm induction and patterning. *Genes Dev.* **15:** 1152–1166.

Ong S.H., Hadari Y.R., Gotoh N., Guy G.R., Schlessinger J., and Lax I. 2001. Stimulation of phosphatidylinositol 3-kinase by fibroblast growth factor receptors is mediated by coordinated recruitment of multiple docking proteins. *Proc. Natl. Acad. Sci.* **98:** 6074–6079.

Ornitz D.M., Xu J.S., Colvin J.S., McEwen D.G., MacArthur C.A., Coulier F., Gao G.X., and Goldfarb M. 1996. Receptor specificity of the fibroblast growth factor family. *J. Biol. Chem.* **271:** 15292–15297.

Partanen J., Schwartz L., and Rossant J. 1998. Opposite phenotypes of hypomorphic and Y766 phosphorylation site mutations reveal a function for Fgfr1 in anteroposterior patterning of mouse embryos. *Genes Dev.* **12:** 2332–2344.

Petit V., Ribeiro C., Ebner A., and Affolter M. 2002. Regulation of cell migration during tracheal development in *Drosophila melanogaster*. *Int. J. Dev. Biol.* **46:** 125–132.

Piper M. and Little M. 2002. Movement through Slits: Cellular migration via the Slit family. *BioEssays* **25:** 32–38.

Psychoyos D. and Stern C.D. 1996. Fates and migratory routes of primitive streak cells in the chick embryo. *Development* **122:** 1523–1534.

Ramirez-Weber F.A. and Kornberg T.B. 1999. Cytonemes: Cellular processes that project to the principal signaling center in *Drosophila* imaginal discs. *Cell* **97:** 599–607.

Rao Y., Wong K., Ward M., Jurgensen C., and Wu J.Y. 2002. Neuronal migration and molecular conservation with leukocyte chemotaxis. *Genes Dev.* **16:** 2973–2984.

Revest J.M., Spencer-Dene B., Kerr K., De Moerlooze L., Rosewell I., and Dickson C. 2001. Fibroblast growth factor receptor 2-IIIb acts upstream of Shh and Fgf4 and is required for limb bud maintenance but not for the induction of Fgf8, Fgf10, Msx1, or Bmp4. *Dev. Biol.* **231:** 47–62.

Rhee J., Mahfooz N.S., Arregui C., Lilien J., Balsamo J., and VanBerkum M.F. 2002. Activation of the repulsive receptor Roundabout inhibits N-cadherin-mediated cell adhesion. *Nat. Cell Biol.* **4:** 798–805.

Ribeiro C., Ebner A., and Affolter M. 2002. In vivo imaging reveals different cellular functions for FGF and Dpp signaling in tracheal branching morphogenesis. *Dev. Cell* **2:** 677–683.

Rorth P. 2002. Initiating and guiding migration: Lessons from border cells. *Trends Cell Biol.* **12:** 325–331.

Sato M. and Kornberg T.B. 2002. FGF is an essential mitogen and chemoattractant for the air sacs of the *Drosophila* tracheal system. *Dev. Cell* **3:** 195–207.

Saxton T.M., Ciruna B.G., Holmyard D., Kulkarni S., Harpal K., Rossant J., and Pawson T. 2000. The SH2 tyrosine phosphatase shp2 is required for mammalian limb development. *Nat. Genet.* **24:** 420–423.

Saxton T.M., Henkemeyer M., Gasca S., Shen R., Rossi D.J., Shalaby F., Feng G.S., and Pawson T. 1997. Abnormal mesoderm patterning in mouse embryos mutant for the SH2 tyrosine phosphatase Shp-2. *EMBO J.* **16:** 2352–2364.

Saxton T.M. and Pawson T. 1999. Morphogenetic movements at gastrulation require the SH2 tyrosine phosphatase Shp2. *Proc. Natl. Acad. Sci.* **96:** 3790–3795.

Schlessinger J. 2000. Cell signaling by receptor tyrosine kinases. *Cell* **103:** 211–225.

Schmucker D. and Zipursky S.L. 2001. Signaling downstream of Eph receptors and ephrin ligands. *Cell* **105:** 701–704.

Schoenwolf G.C., Garcia-Martinez V., and Dias M.S. 1992. Mesoderm movement and fate during avian gastrulation and neurulation. *Dev. Dyn.* **193:** 235–248.

Selleck M.A.J. and Stern C.D. 1991. Fate mapping and cell lineage aanalysis of Hensen node in the chick embryo. *Development* **112:** 615–626.

Shalaby F., Ho J., Stanford W.L., Fischer K.D., Schuh A.C., Schwartz L., Bernstein A., and Rossant J. 1997. A requirement for Flk1 in primitive and definitive hematopoiesis and vasculogenesis. *Cell* **89:** 981–990.

Shamim H. and Mason I. 1999. Expression of Fgf4 during early development of the chick embryo. *Mech. Dev.* **85:** 189–192.

Shishido E., Higashijima S., Emori Y., and Saigo K. 1993. Two FGF-receptor homologues of *Drosophila*: One is expressed in mesodermal primordium in early embryos. *Development* **117:** 751–761.

Shishido E., Ono N., Kojima T., and Saigo K. 1997. Requirements of DFR1/Heartless, a mesoderm-specific *Drosophila* FGF-receptor, for the formation of heart, visceral and somatic muscles, and ensheathing of longitudinal axon tracts in CNS. *Development* **124:** 2119–2128.

Stern C.D. and Fraser S.E. 2001. Tracing the lineage of tracing cell lineages. *Nat. Cell Biol.* **3:** E216–218.

Sun X., Meyers E.N., Lewandoski M., and Martin G.R. 1999. Targeted disruption of Fgf8 causes failure of cell migration in the gastrulating mouse embryo. *Genes Dev.* **13:** 1834–1846.

Symes K. and Mercola M. 1996. Embryonic mesoderm cells spread in response to platelet-derived growth factor and signaling by phosphatidylinositol 3-kinase. *Proc. Natl. Acad. Sci.* **93:** 9641–9644.

Tanaka S., Kunath T., Hadjantonakis A.K., Nagy A., and Rossant J. 1998. Promotion of trophoblast stem cell proliferation by FGF4. *Science* **282:** 2072–2075.

Thery C., Sharpe M.J., Batley S.J., Stern C.D., and Gherardi E. 1995. Expression of HGF/SF, HGF1/MSP, and c-met suggests new functions during early chick development. *Dev. Genet.* **17:** 90–101.

Trinkaus J.P. 1998. Gradient in convergent cell movement during *Fundulus* gastrulation. *J. Exp. Zool.* **281:** 328–335.

Trinkaus J.P., Trinkaus M., and Fink R.D. 1992. On the convergent cell movements of gastrulation in *Fundulus*. *J. Exp. Zool.* **261:** 40–61.

Vakaet L. 1970. Cinephotomicrographic investigations of gastrulation in the chick blastoderm. *Arch. Biol.* **81:** 387–426.

Vincent S., Wilson R., Coelho C., Affolter M., and Leptin M. 1998. The *Drosophila* protein Dof is specifically required for FGF signaling. *Mol. Cell* **2:** 515–525.

Wallingford J.B., Fraser S.E., and Harland R.M. 2002. Convergent extension: The molecular control of polarized cell movement during embryonic development. *Dev. Cell* **2:** 695–706.

Weaver M., Dunn N.R., and Hogan B.L.M. 2000. Bmp4 and Fgf10 play opposing roles during lung bud morphogenesis. *Development* **127:** 2695–2704.

Weijer C.J. 1999. Morphogenetic cell movement in *Dictyostelium*. *Semin. Cell Dev. Biol.* **10:** 609–619.

Weinstein M., Xu X., Ohyama K., and Deng C.X. 1998. FGFR-3 and FGFR-4 function cooperatively to direct alveogenesis in the murine lung. *Development* **125:** 3615–3623.

Wong K., Park H.T., Wu J.Y., and Rao Y. 2002. Slit proteins: Molecular guidance cues for cells ranging from neurons to leukocytes. *Curr. Opin. Genet. Dev.* **12:** 583–591.

Xu X., Li C., Takahashi K., Slavkin H.C., Shum L., and Deng C.X. 1999. Murine fibroblast growth factor receptor 1alpha isoforms mediate node regression and are essential for posterior mesoderm development. *Dev. Biol.* **208:** 293–306.

Yamaguchi T.P., Conlon R.A., and Rossant J. 1992. Expression of the fibroblast growth factor receptor FGFR-1/flg during gastrulation and segmentation in the mouse embryo. *Dev. Biol.* **152:** 75–88.

Yamaguchi T.P., Harpal K., Henkemeyer M., and Rossant J. 1994. fgfr-1 is required for embryonic growth and mesodermal patterning during mouse gastrulation. *Genes Dev.* **8:** 3032–3044.

Yang X., Dormann D., Munsterberg A., and Weijer C. 2002. Cell movement patterns during gastrulation in the chick are controlled by positive and negative chemotaxis mediated by FGF4 and FGF8. *Dev. Cell* **3:** 425–437.

Yeo S.Y., Little M.H., Yamada T., Miyashita T., Halloran M.C., Kuwada J.Y., Huh T.L., and Okamoto H. 2001. Overexpression of a slit homologue impairs convergent extension of the mesoderm and causes cyclopia in embryonic zebrafish. *Dev. Biol.* **230:** 1–17.

Yu T.W. and Bargmann C.I. 2001. Dynamic regulation of axon guidance. *Nat. Neurosci.* (suppl.) **4:** 1169–1176.

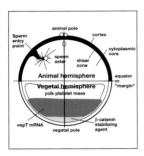

SYMMETRY BREAKING IN THE EGG OF *XENOPUS LAEVIS*

J. Gerhart

Department of Molecular and Cell Biology, University of California, Berkeley, California 94720-3200

INTRODUCTION

The unfertilized frog egg has cylindrical symmetry around an animal–vegetal axis, visible in terms of the polar body spot at the animal pole, the uniformly pigmented animal hemisphere, and the nonpigmented vegetal hemisphere. The tadpole, of course, has a bilateral symmetry. Thus, at some time during embryonic development the cylindrical symmetry of the egg gives way to bilateral symmetry. Symmetry is broken. From the time of the observations of Newport (1854) and Roux (1887), the site of sperm entry into the egg has been known to predict the orientation of the future bilateral body axis of the tadpole. Namely, it predicts the locus of ventral and posterior development in the egg, whereas opposite positions are destined for anterior and dorsal development. The sperm can be experimentally applied from a needle tip at any point randomly chosen in the animal hemisphere, and the prediction still holds (Elinson 1975). Thus, the egg has the capacity to undertake ventral–posterior or dorsal–anterior development at any place around its circumference, and normally the entering sperm is the sufficient input for the choice. In what sense, however, does the sperm determine bilateral symmetry? There are many ways to disrupt the sperm–body axis relationship, such as by tipping the egg 90° on its side and holding it there for 10 minutes in the middle of the first cell cycle after fertilization. Then the body axis eventually forms in relation to whatever equatorial point was uppermost during the tipping period, irrespective of the sperm. Thus, the sperm seems unlikely on its own to break the egg's symmetry, although it must at least have a biasing effect, as discussed below.

In some amphibia, the egg changes its coloration late in the first cell cycle, in one quadrant at the border of the pigmented and unpigmented hemispheres. This area of displacement of pigment, which occurs in the cortex and in the adjacent subcortical cytoplasm of the egg, has been called the "grey crescent" (for historical review, see Gerhart 1980). The cortex is defined as the egg's plasma membrane plus associated actin microfilament meshwork, and various materials attached to the meshwork. In frog eggs, the cortex is 5–10 μm thick. Eggs of some species show a crescent, but others do not, presumably depending on the abundance and depth of their pigment granules. The crescent is located at the equatorial sector, farthest from the sperm entry point. The blastopore will first indent at the vegetal border of the crescent. At this site, Spemann's organizer will later develop, and with the organizer's mediation, dorso–anterior development will eventually occur there. By the time of first cleavage, the egg has a bilateral symmetry defined by a plane running through the animal and vegetal poles, the sperm entry point, the grey crescent, and the first cleavage furrow (if the egg is undisturbed).

The grey crescent has long been ascribed more importance than the sperm entry point in axis specification because it coincides with the future organizer site and more reliably foretells the orientation of the body axis (Morgan and Boring 1903; Schechtman 1936). The sequence of events, as classically understood, is (1) the sperm entry point is randomly located on one side of the animal hemisphere, (2) the grey crescent forms at an equatorial site farthest from the sperm entry point, (3) the organizer forms at or near the grey crescent site in the late blastula, and (4) the organizer patterns the body axis. In this way, the sperm entry point is related to the orientation of the body axis. Dalcq and Pasteels (1937) assigned great significance to the grey crescent in their double gradient model of egg patterning. It was the center of a gradient of influence in the cortex of the egg (the cortical gradient). The second gradient, orthogonal to the first, was a cytoplasmic gradient (the vitelline gradient), running from the vegetal to the animal pole. At the high point of input from the two gradients (CxV), a substance "organisine" was produced, causing organizer formation at that site. This was their way to connect the grey crescent to organizer formation. As discussed later, their interpretation has formal similarities to the modern understanding involving β-catenin and VegT proteins.

Curtis (1960) obtained evidence that the grey crescent cortex contains determinants for axis specification. He succeeded in microdissecting squares of cortex from the grey crescent region of 8-cell donor embryos and grafting these squares into the cortex of the opposite side of 2- or 4-cell recipient embryos. Each recipient developed a complete secondary body axis at the site of the graft. More recently, Kageura (1997) has repeated these experiments in a modified form, by inserting the square of cortex into the subcortical cytoplasm in the recipient rather than grafting it into the cortex. He, too, found that recipients developed a secondary axis at the site where the cortical fragment had been introduced. Furthermore, Kageura ran various controls missing from the Curtis experiments, such as showing that non-grey crescent cortex was not active. Related results have been obtained by others who withdrew cytoplasm by micropipette from just under the grey crescent cortex and injected this into host eggs (Yuge et al. 1990; Holowacz and Elinson 1995; Marikawa and Elinson 1997).Thus, the grey crescent cortex did indeed have a determinative role in axis specification, and credence was given to the Dalcq-Pasteels hypothesis.

At the time, there was little information about the process of crescent formation and no hint about the cortex-associated agents important for forming the organizer and for axis specification. In the most extensive early study of the grey crescent, Ancel and Vintemberger (summarized in 1948) observed the egg of *Rana fusca* to establish that (1) the crescent can form even without the sperm; that is, in electrically activated eggs, (2) the crescent forms in the second half of the period between fertilization and first cleavage, (3) its location of formation can be controlled by tipping the egg obliquely at the time of activation, and (4) its formation involves a displacement of the cortex relative to the core cytoplasm, and they called the reorganization process a "rotation of symmetrization."

The recent era of study began in 1980, when R. Elinson and our own group sought to elucidate the process of grey crescent formation in the eggs of *Rana pipiens* and *Xenopus laevis*. These results are reviewed in the next sections. We call the process "cortical rotation," since it does indeed involve a rotational displacement of the cortex relative to the underlying cytoplasmic core, although it is not sufficient by itself to establish lasting bilateral symmetry. Coupled processes are needed as well. It occurs in the first cell cycle of the egg after fertilization. Since gene expression does not begin until the 4000-cell stage, the process must rely entirely on maternal RNAs and proteins.

In the 1990s, insights were finally gained about the axis-specifying agents, namely, the β-catenin protein that accumulates in the cytoplasm close to the gray crescent cortex and contributes to organizer formation. In *Xenopus* eggs, the crucial experiments were (1) the discovery by McMahon and Moon (1989) that an injection of mRNA for a Wnt ligand could provoke the development of a second body axis. This led to the analysis of Wnt pathway intermediates in the egg, and in turn to (2) the observation by Wylie et al. (1996) that the depletion of the maternal β-*catenin* mRNA from oocytes prevents later development of the organizer, and (3) the recognition by Zhang et al. (1998) that a depletion of maternal *vegT* mRNA in the oocyte precludes later endo-mesoderm induction by nodal-related secreted proteins and hence precludes organizer formation. As discussed elsewhere in this volume (see Chapters 26, 34, 36, and 45), it is now understood that the organizer forms in the late blastula at an equatorial location where two conditions are met: β-catenin protein is present at high levels in cells of the region and various nodal-related protein signals are produced nearby, which diffuse into the region. The organizer forms where β-catenin modifies endo-mesoderm induction. If either condition is not met, the organizer does not form. Symmetry breaking, as a lasting effect, is now understood as the accumulation of β-catenin protein on one side of the fertilized *Xenopus* egg.

COMPONENTS LOCALIZED IN OOGENESIS

During oogenesis, the amphibian oocyte develops animal–vegetal polarity. The enormous germinal vesicle (nucleus) occupies the animal hemisphere. Yolk platelets are deposited densely in the vegetal half. The pattern of deposition of mRNAs and proteins in the vegetal hemisphere is of importance for later cortical rotation and organizer formation. At the vegetal pole is a material, of still unknown identity, involved eventually in the stabilization of β-catenin protein and thereby in the formation of Spemann's organizer, as explained below. Also in this hemisphere, *vegT* mRNA is deposited, eventually to be involved in the induction of endoderm and mesoderm in the mid and late blastula stages, and thereby involved in the formation of the organizer.

The β-catenin stabilizing component is probably localized to a fairly tight spot at the vegetal pole of the unfertilized egg. This was shown by the cortex transplantation experiments of Kageura (1997). As noted above, he succeeded in microdissecting squares of cortex from various locations on fertilized eggs and implanting the fragments in the ventral equatorial zone of 2- or 4-cell host embryos, to test for their capacity to provoke the development of a partial or complete secondary body axis. He also assessed unfertilized eggs. Cortex from the vegetal pole had the greatest activity when taken from unfertilized eggs, or even from eggs until the middle of the first cell cycle. Thereafter, activity declined at the vegetal pole and increased at the equatorial level on one side of the egg. Similar results were obtained by others who withdrew subcortical cytoplasm by micropipette and injected this into host eggs (Yuge et al. 1990; Holowacz and Elinson 1995; Marikawa and Elinson 1997, 1999). Thus, the component seems initially located at the vegetal pole.

If the β-catenin stabilizing agent is really deposited in a small spot at the oocyte's vegetal pole, it may have been translocated there by the METRO pathway (message transport organizer; Kloc and Etkin 1995), which operates early in oogenesis. Certain mRNAs such as *vasa*, *Xlsirts*, *cat2*, *wnt11*, *boule*, and *otx1* depend on this pathway for their translocation and deposition. They initially associate with the mitochondrial cloud, which is a large cluster of mitochondria and endoplasmic reticulum located to one side of the nucleus. By stage 3, before yolk deposition begins, the cloud migrates to a place on the cortex, the future vegetal pole. It is not known whether this is a predesignated site or a random spot. The cloud spreads out on the cortex, and various organelles, mRNAs, and proteins are deposited there. In the full-grown oocyte, they occupy a spot spreading approximately 30° of arc from the pole. Most of the mRNAs are associated with germ plasm granules which are involved in the eventual formation of the germ line. The β-catenin stabilizing agent shares this locale in unfertilized eggs, although no one has yet located it in stage-3 oocytes. From Kageura's results, it seems associated with the cortex, strongly enough to withstand transplantation.

The *vegT* mRNA may be translocated by a different mechanism acting later in oogenesis (Kloc and Etkin 1995). It becomes associated with the cortex of the vegetal hemisphere, but in a larger area, extending perhaps 60° of arc from the pole. Deposition continues until oocyte stage 5 and seems to be independent of the mitochondrial cloud. Other mRNAs, such as *Vg1*, are also deposited in this manner.

Many other materials important for the symmetry-breaking process are distributed uniformly in the oocyte. β-Catenin mRNA and protein are everywhere, as are all components of the Wnt pathway. This ubiquity is known because, as first shown by McMahon and Moon (1989), an injection of mRNA for a Wnt ligand into any equatorial site of the fertilized egg leads to the development of a secondary axis at that site. All components of the pathway must be present for β-catenin stabilization to occur at the arbitrarily chosen site. The components would include the Wnt receptor, Dsh, Axin, β-catenin itself, GSK3, APC, casein kinase I, the protein degradation machinery, and the repressive transcription factor, Tcf (and its mRNA). All microtubule components such as α- and β-tubulin, various microtubule stabilizing and destabilizing factors, and kinesin and dynein motor molecules are also ubiquitous. In fact, the only nonuniformly distributed factors are the β-catenin stabilizing agent and *vegT* mRNA. Both occupy cylindrically symmetric locations close to the vegetal pole, as shown in Figure 1A.

CORTICAL ROTATION IN THE FIRST CELL CYCLE

There are at least two steps in the cortical rotation process. The first is the formation of a thin mat of microtubules at the interface of the cortex and the subcortical cytoplasm of the egg. All the microtubules point in the same direction, in an animal–vegetal orientation. Their plus ends point toward the grey crescent in the vegetal hemisphere and away from it in the animal hemisphere. As shown in Figure 2A, this parallel polarized array has bilateral symmetry. The second step is the translocation of materials along the parallel microtubules, toward their plus ends. Maternal materials from the vegetal pole travel to the equatorial level on one side of the egg, thereby acquiring a bilateral disposition. As they persist, symmetry remains broken. Other materials may move from the opposite equatorial sector toward the vegetal pole.

Step 1: Formation of the Bilateral Microtubule Array

Cortical rotation can be followed in living *X. laevis* eggs by marking the surface with a pattern of spots of a fluorescent lectin, or by marking the internal yolk platelets with a pattern of spots of nile blue dye or nile red dye (fluorescent), and then recording the shifting of patterns relative to one another. The overall patterns on the surface and core remain stable during rotation, indicating the rigidity of the cortical layer and of the cytoplasmic core. In the first cell cycle, which lasts ~100 minutes, rotation begins slowly at 45 minutes, accelerates to full speed by 50 minutes, and continues at this speed until about 90 minutes when it stops abruptly (Vincent et al. 1986; Vincent and Gerhart 1987; Gerhart et al. 1989). The cortex behaves as a rigid unit; it moves over the solid core by 30° of arc, or 400 μm, in 40

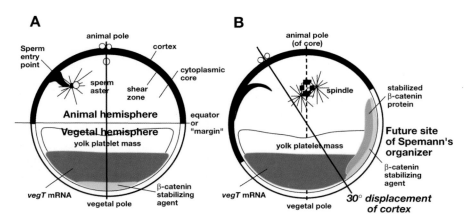

Figure 1. Symmetry breaking in the egg of *Xenopus laevis*. (*A*) Before cortical rotation (20 minutes after fertilization). The egg has cylindrical symmetry except for the sperm aster on the left, near the point of sperm entry. The egg consists of two layers, a rigid cortex surrounding a solidified cytoplasmic core. Note the shear zone at which the two layers will move relative to each other. The β-catenin stabilizing agent (*yellow*) and *vegT* mRNA (*blue*), both associated with the cortex, are located at the vegetal pole. (*B*) After cortical rotation (90 minutes after fertilization, 10 minutes before first cleavage). The egg now has bilateral symmetry. The cortex has moved 30° relative to the core, and the β-catenin stabilizing agent has moved 60–120° (*yellow*). β-Catenin protein (*red*) has accumulated in the area due to the agent's inhibition of breakdown of β-catenin protein. The cortex and the agent have both translocated along the microtubule tracks. Kinesin, anchored to the cortex or to the stabilizing agent, probably enforces translocation. After translocation, the stabilizing agent again attaches to the cortex, now at the equatorial level of the egg. Spemann's organizer will later form at this location. The *vegT* mRNA does not move (*blue*).

minutes (10 μm/min). In a population of eggs, some move as much as 40° and some as little as 20°, yet all develop normally. Movement is always in the animal–vegetal direction (except after certain experimental manipulations). The direction of rotation is strongly predictive of the future site of the organizer and of the orientation of the embryonic axis, even in eggs in which the sperm entry relationship is completely reversed. Considered grossly, the rotation movement involves two rigid units, a spherical shell (the cortex) rotating over a solid ball (the cytoplasmic core). Since the shell and ball initially share cylindrical symmetry along the same axis, this is a very simple geometric operation for transforming cylindrical to bilateral symmetry. As explained below, there is more to the event than mere rotation of the cortex.

When eggs are fixed during rotation and later stained to reveal microtubules, the parallel array can be seen as a thin layer 4–8 μm thick, at the interface of the cortex and core, about 4 μm beneath the plasma membrane (Elinson and Rowning 1988; Rowning et al. 1997). The plus ends of the microtubules point mostly or entirely in one direction (Houliston and Elinson 1991). The parallel array is wrapped on a sphere, with opposed hubs at the equator, as shown in Figure 2A. Since the microtubules point in one direction, the array has bilateral symmetry. It can form in any of 360° of orientations. Rotation occurs in the direction of the microtubules. Rotation begins as the array forms and stops as the array disappears.

Cortical rotation occurs at a special time in the cell cycle. Meiosis II and polar body formation are over, and the

DNA of the pronuclei has been replicated. The cell cycle then stalls in a G_2-like state at 40 minutes postfertilization, for a period of 40–50 minutes, and cortical rotation occurs in this period. It resumes at about 80 minutes, leading to mitosis and first cleavage at 100 minutes. This 40- to 50-minute period is absent from the subsequent 12 cell cycles, which are shorter. At 45 minutes into the first cycle, a microtubule stabilizing agent, XMAP230, is phosphorylated; at 80 minutes it is dephosphorylated (Cha and Gard 1999). It may stabilize microtubules in this special interval.

Microtubules are nucleated throughout the cytoplasmic core at 40 minutes postfertilization, and they extend toward the cortex, bending over and running along the interface of the cortex and core (Fig. 2A). In timed samples, later fixed and stained, the microtubules appear at first short and poorly oriented at the interface. The cortex at this time begins to move slowly in one direction. Then microtubules are seen to become longer and better oriented in the direction of movement. Movement accelerates as more microtubules are seen well oriented in the direction of movement, and reaches top speed within 5 minutes of starting. (These observations of array formation could be improved by the use of new reagents such as the GFP-Clip170 protein.) The parallel array moves with the core, consistent with the fact that microtubules originate from within it. The cortex moves along the microtubules toward their plus ends. The plus-end-directed motor of choice would be kinesin, anchored directly or indirectly to the cortex. The speed of translocation of the cortex is about 10 μm/min, which is well within the speeds known for kinesin-based

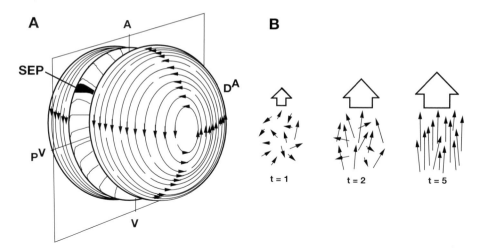

Figure 2. The bilateral parallel array of microtubules during cortical rotation. (*A*) Diagram of an egg at 50 minutes postfertilization, with a fully formed parallel array of microtubules on the surface of its solidified core, and interior to the cortex. The cortex has been omitted in order to reveal the surface of the core. The egg is cut on the bilateral plane to show microtubules originating in the core and extending to the cortex (spoke-like arrangement) where they bend over. Also, the sperm trail is shown, a track of pigment granules from the point of sperm entry (SEP). The animal pole (A) is at the top, and the vegetal pole (V) at the bottom. Note that the microtubules all point in the same direction, with plus ends in the direction indicated by the arrowheads. Note the hub formed at the equatorial level as parallel microtubules traverse ever smaller circumferences. The cortex, to which kinesin is attached, will move toward the plus ends of microtubules. The side of the egg on which dorsal anterior structures of the tadpole develop is indicated by "DA"; and of posterior ventral structures by "PV." (*B*) Model of the formation of the parallel array by a self-organizing process based on reciprocal feedbacks. When microtubules begin to form at the shear zone (*t* = 1), they are short and nearly random in their orientation (*small arrows*). However, they are not perfectly cylindrically symmetric in their directionalities. One direction is slightly in excess. The cortex moves in this direction (*large open arrowhead*). By *t* = 2, the slowly moving cortex has oriented materials in the shear zone, and these feed back on microtubules, stabilizing those growing in the direction of movement of the cortex. With the additional aligned microtubules, the cortex moves faster and orients more materials in the shear zone. By *t* = 5 (5 minutes after the start of cortical rotation), the microtubules are predominantly aligned in one direction, and the speed of cortical movement is maximal. Movement will continue for 40 minutes, at which time the array disintegrates and it stops. In this model, cortical rotation is needed for the alignment of microtubules in a parallel array, not for the translocation of vegetal materials, although translocation of attached materials could be accomplished to the extent of 30° of arc. Once the array is formed, vegetal materials translocate much farther on the microtubule tracks.

movements of materials along microtubules (>40 μm/min). Anti-kinesin antibodies disrupt movement, a result indicating a role for kinesin; anti-dynein antibodies also disrupt movement (Marrari et al. 2003). They suggest that dynein, which is a minus-end-directed motor, also contributes to movement by being anchored to the core and pushing elongating microtubules out of the core and into the interface, where they carry the cortex along. The relative contributions of the two motors remain to be evaluated.

The question for symmetry-breaking is, How do microtubules become oriented into a single parallel array? We hypothesize that cortical movement and microtubule elongation are coupled in a reciprocal positive feedback loop (Gerhart et al. 1989). This loop creates a self-organizing symmetry-breaking process that amplifies any small random departure from perfect cylindrical symmetry into a large and singular bilateral outcome. First we discuss the case of the artificially activated egg, which is simpler than the fertilized egg, since it is free of the biasing influences of

the sperm aster. As noted above, Ancel and Vintemberger (1948) studied grey crescent formation in artificially activated *R. fusca* eggs. A *X. laevis* egg can be artificially activated by puncturing it with a needle, or by electrical activation. The cortex rotates by 30°, and the array of microtubules forms on time, aligned with movement. However, the direction of movement is unrelated to the site of puncture. Delayed nuclear transplant experiments show that these eggs can develop normally with the embryonic body axis in the plane of rotation (Rowning and Gerhart 1990). Even if the egg is given ambiguous inputs, it still breaks symmetry. For example, eggs artificially activated at the animal pole, to avoid a lateral bias, succeed in generating bilaterality, and polyspermic *X. laevis* eggs (3–4 sperm per egg), which may receive multiple biases, undergo a single direction of rotation. It does not seem possible for the activated or fertilized egg to remain in a cylindrically symmetric state.

As summarized in Figure 2B, we propose that in artificially activated eggs, the plus ends of the microtubules, as they reach the cortex and bend over at 45 minutes, are short and poorly aligned, but as a population, they randomly depart from perfect cylindrical symmetry. A few more microtubules are pointed one way than other ways. The rigid cortex, with kinesin motors anchored to it, is the perfect agency to sum the vectors of the directions of all the microtubules under it. Whatever direction is slightly in excess, the cortex moves in that direction, toward the positive ends. Movement is at first slow because the microtubules are short, and only a few are in excess in one direction.

Cortical movement, we hypothesize (see Fig. 2B), then enhances microtubule formation and/or stabilization in the direction of rotation. Those microtubules aligned with movement would grow longer while others die back and regrow in a new direction, each with a chance of extending in the direction of movement. As movement occurs and more microtubules form in the direction of that movement, rotation accelerates. A positive feedback loop is established. A small amount of movement in one direction leads to more movement in that direction. According to this hypothesis, movement of the cortex by 30° serves the function of aligning microtubules in one global parallel array rather than of transporting vegetal materials. The formation of only a single global array is important because the egg has the capacity to form multiple organizers or even a large continuous organizer around the entire equator, leading to exaggerated anterior–dorsal development. Indeed, double centrifugations of the egg, which move the core first in one direction and then in the opposite, relative to the cortex, lead to twins at high frequency (Black and Gerhart 1986). Thus, the cortex moving as a rigid unit over a rigid core may assure a single bilateral outcome and a single body axis.

Within this hypothesis, symmetry-breaking is readily biased by various inputs, now discussed. When the sperm enters the animal hemisphere, it delivers a centrosome (and centriole pair) on one side. Microtubules of the sperm aster are nucleated from this centrosome. Aster microtubules reach the cortex well before the parallel array begins to form, and in some way exert a bias. The aster might seem to have inherent radial symmetry, but its effect has to be asymmetric in order to exert a bias. Astral microtubules presumably reach the vegetal cortex near the sperm entry point and run along the cortex, with their plus ends away from the centrosome, perhaps giving a head start for rotation in that one direction.

In artificially activated eggs, which lack a sperm aster, the direction of rotation can be biased by gravity. This curious effect fits well with the positive feedback model presented above. Ancel and Vintemberger (1948) in their classical studies of grey crescent formation examined this directly: They suspended an unfertilized *R. fusca* egg with its vegetal pole rotated 120° from the gravitationally stable downward position. Then the egg was activated by electroshock. As the fertilization envelope lifted from the egg surface over a period of minutes, the egg slowly became free to rotate 120° into gravitational equilibrium. This gravity-driven rotation, which should not be confused with the microtubule-dependent cortical rotation, is completed within 15 minutes after activation, and at least 30 minutes before the grey crescent forms. When the crescent eventually forms, it is invariably located at the equatorial site that had been uppermost in the gravitational field at the time of activation. These studies have been repeated with artificially activated *X. laevis* eggs and extended by analyzing the direction of the subsequent cortical rotation and of the microtubule array. If the egg is tipped 90° on its side briefly before 40 minutes postactivation, and then returned to the vertical even after only 5 minutes, rotation will follow a direction predicted by which site of the equator was uppermost in the gravitational field. Fertilized eggs can also be biased by gravity, the effects of which readily override the sperm aster bias. What is the remarkable biasing effect of gravity, laid down as the entire egg rotates into gravitational equilibrium 30 minutes before cortical rotation occurs?

When eggs are marked with dyes, the yolky core can be observed to slip relative to the cortex in the period of oblique orientation. The vegetal half of the core is weighted by abundant large yolk platelets, and it tends to move toward gravitational equilibrium, vegetal pole down. If the cortex is restrained from moving, the core slips relative to it. This was the case for the tipped activated eggs of Ancel and Vintemberger in the brief period before the fertilization envelope had fully lifted. The cortex can be artificially restrained for an hour or more, by embedding tipped eggs in gelatin or agarose, and then the slippage of the core can be complete, a full 90° or 120°. In a period of a few minutes, the core may slip only a few degrees, but this is sufficient for a biasing effect. We assume that the slipping of the vegetal yolk mass along the cortex by a few degrees, driven by gravity, resembles in some way the event of cortical rotation in which the cortex slips along the core driven by kinesin walking on microtubules, in both cases leaving oriented traces in the shear zone that can influence the direction of microtubule formation. Hence, when the array begins to form at 45 minutes postfertilization in a previously tipped egg, a bias is already present in the shear zone, favoring microtubule elongation and stabilization in the direction of the prior slippage. We assume that this bias is the same kind that builds up in cortical rotation and serves in the feedback loop discussed above. The nature of the lasting bias is not known, but it may involve aligned endoplasmic reticulum and vesicles, as has been found to affect microtubule directionality in other cell types (Schulze and Kirschner 1987).

Another way to control the direction of rotation is to compress the egg laterally between two glass slides before rotation begins. It then occurs in the long axis of the egg (Black and Vincent 1988). This compression, we assume, favors movement of the cortex in one direction, which then feeds back on microtubule polymerization in that direction. After compression is released shortly before first cleavage, the egg develops to a tadpole with its body axis aligned with the previous compression plane, regardless of the sperm entry site.

Step 2: Movement of Materials along the Microtubule Tracks

Although the polarized array of microtubules is itself bilateral, it is transient, lasting only until the end of the first cell cycle. What is the lasting symmetry-breaking consequence? Indirect evidence supports the explanation that a β-catenin stabilizing agent moves on the microtubule tracks, toward their plus ends, from a vegetal to an equatorial location. As shown by immunofluorescence, β-catenin protein does indeed accumulate on one side of the egg as cortical rotation proceeds, the side toward which the plus ends of microtubules point (Larabell et al. 1997; Rowning et al. 1997), indicating the presence of a stabilizing agent (diagrammed in Fig. 1B). Stabilization rather than de novo synthesis is likely because β-catenin mRNA is present uniformly in the egg and is translated at a steady rate. The protein degradation machinery is also uniformly distributed, so that in most places in the egg, degradation exceeds synthesis, leaving only a low level of β-catenin protein. In one location, however, β-catenin accumulates, indicating that its breakdown is retarded there. The elevated level of β-catenin protein persists at least until the 4000-cell stage when zygotic gene expression begins (Schneider et al. 1996). Presumably, the persistence of the protein over those many hours reflects continued inhibition of breakdown due to the continued presence of the stabilizing agent that remains stuck to the equatorial cortex after its transport along microtubules. β-Catenin protein presumably accumulates in the cytoplasm immediately adjacent to the equatorial cortical patch of stabilizing agent. The agent seems to have the capacity to remain immobile, attached to the cortex at the vegetal pole or equator, or to be translocated on microtubules, attached to plus-end-directed motor molecules.

What is the stabilizing agent? It is not known, but various attempts have been made to observe materials moving on microtubules of the parallel array and to identify components capable of stabilizing β-catenin. From the distribution of β-catenin and from the cortical transplantations assays of Kageura (1997), the stabilizing agent appears to move 90–120° of arc from the vegetal pole, whereas the cortex itself moves but 30° of arc. Thus, transport of the agent on the cortex would not be sufficient. The microtubule array itself might support a longer distance movement. In an artificial test of movement, fluorescent microspheres were injected in the vegetal region of *X. laevis* eggs. These microspheres carried negative charges that facilitate adsorption of cytoplasmic components including motor

proteins. These indeed moved along the microtubule array, some as much as 90–120° of arc, enough to reach the animal hemisphere (Rowning et al. 1997). When endogenous vesicles were stained with $DiOC_6(3)$ in living eggs, they too were seen to translocate at the cortex–core interface along microtubules of the array. Their speeds reach 30–40 μm/min, sufficient for 90°–120° of arc displacement from the vegetal pole within the 40-minute lifetime of the array, a speed substantially greater than that of the large rigid cortex (10 μm/min). The speed is still within the range known for kinesin-based translocations. These results indicate that vegetal materials, including the stabilizing agent, could move on the array, as diagrammed in Figure 1B. Not all materials move, however; there is selectivity. *vegT* mRNA, for example, stays in its cylindrically symmetric distribution in the vegetal hemisphere, as do the germ plasm granules.

In a more directed assay of possible stabilizing agents, Miller et al. (1999) injected oocytes with an mRNA encoding the disheveled protein (Dsh) coupled to green fluorescent protein (GFP). This mRNA was translated by the oocyte, which was later matured into an egg and fertilized. During cortical rotation, GFP-Dsh-labeled fluorescent vesicles moved toward the plus ends of microtubules at 30–40 μm/min. Of course, Dsh was chosen because it is the intermediate of the Wnt signaling pathway known to inhibit β-catenin breakdown when a Wnt ligand binds to the transmembrane receptor. β-Catenin accumulation could then be explained by a directional transport of Dsh protein on vesicles. However, since *Dsh* mRNA seems to be uniformly distributed in the egg, one would have to posit that the Dsh protein is localized at the vegetal pole if it is to be the agent.

Other candidates have been considered. The GSK3-binding protein (GBP/Frat) binds to the axin scaffold and antagonizes the GSK3 enzyme that phosphorylates β-catenin prior to its ubiquitination and breakdown. Dominguez and Green (2000) have noted that the concentration of GSK3 protein is low in the area of β-catenin accumulation, and they attribute the effect to GBP. Furthermore, GBP/Frat binds to kinesin and is translocated directionally on microtubules (Weaver et al. 2003). Only recently realized as a further possibility, axin itself undergoes degradation, and it might be the target of the translocated agent. Locally reduced axin would lead to locally accumulated β-catenin (Lee et al. 2003). As another possibility, an endogenous Wnt ligand may be the agent, localized in vesicles at the vegetal pole. Such vesicles would then serve as signaling sources translocated along the microtubule array (perhaps also carrying active Dsh). This possibility seems contradicted by the result that dominant negative Wnt ligands, introduced by way of injected mRNAs, do not interfere with axis formation. However, if the endogenous Wnt ligand were localized early in oogenesis (as the *wnt11* mRNA is, at stages 2 and 3 by the METRO pathway), the late-injected dominant negative form might not affect it. In favor of an endogenous ligand, when Wnt transmembrane receptor proteins are depleted by morpholino and antisense methods, axis formation is indeed reduced (Sumanas et al. 2000). To conclude the candidate studies, several possibilities exist for the endogenous β-catenin stabilizing agent, but its identity is still unknown.

β-Catenin accumulation is not the only early effect of cortical rotation. As noted above, pigment granules may be displaced at the cortical-core interface (depending on the amphibian species), leaving a patch of lessened color, the grey crescent, which correlates with the site of β-catenin accumulation and later with organizer formation. There are two other early effects. Slack and Palmer (1969) observed at the 32-cell stage of *X. laevis* eggs that injected dyes pass much more rapidly from cell to cell on the grey crescent side than on the other side. Gap junction connections between cells are apparently more abundant on the crescent side. Although this effect has not been adequately explained, it seems plausible that the buildup of β-catenin on one side stimulates adherens junction formation locally between cells, and that these adherent cells form more gap junctions. (This could be tested further with localized injections of β-catenin mRNA.) Finally, Levin et al. (2002) find that the maternal mRNA encoding the H^+/K^+-ATPase is localized to the one of four blastomeres (non-grey-crescent containing), after being present uniformly at the 1-cell stage. They consider this an early step in the pathway establishing left–right asymmetry of the embryo (see Chapter 28). The dependence of this localization on cortical rotation remains uncertain.

THE NORMAL REQUIREMENT FOR CORTICAL ROTATION

The need for cortical rotation has been tested by eliminating, reducing, or exaggerating it to see what is altered in development. These tests uniformly support the conclusion that cortical rotation is crucial for the subsequent formation of Spemann's organizer (Gerhart et al. 1989). If rotation does not occur, there is no organizer, and hence no embryonic body axis. On the other hand, endo-mesoderm induction occurs even when rotation fails.

No Rotation, No Body Axis

If microtubule polymerization is blocked early in the first cell cycle, rotation fails to take place, and embryos later fail to form an organizer. The treatments for *X. laevis* eggs include brief exposure of eggs to nocodazole, brief UV irradiation on the vegetal hemisphere surface of the egg, brief exposure of eggs to low temperatures (5°C, 10 minutes), or high hydrostatic pressure (Scharf and Gerhart 1980, 1983). Without a parallel array of microtubules, the eggs do not accumulate β-catenin in the normal manner. At the vegetal

pole of these eggs, some β-catenin stabilization does occur, but it does not reach equatorial levels where the organizer would normally form (Medina et al. 1997). Treated eggs develop as posteriorized-ventralized embryoids containing a posterior gut, coelomic mesoderm, red blood cells (perhaps 15–20 times more), and ciliated epidermis. The embryoid retains cylindrical symmetry. It lacks a notochord, gill slits, a dorsal hollow nerve cord, and a post-anal tail, the four key traits of chordates. All of these depend on the organizer for induction or self-differentiation. The embryoid resembles the "belly piece" that Spemann produced by isolating and rearing the non-grey-crescent pair of blastomeres at the 4-cell stage (Spemann 1902).

These treated eggs can be rescued in various ways:

1. If the egg is tipped obliquely 90° in the first cell cycle or early second to force the movement of the core relative to the cortex by gravity, it develops normally (Scharf and Gerhart 1980), without the mediation of the microtubule array. The stabilizing agents perhaps stay attached to the vegetal cortex and, as the core slides past, driven by gravity, they are brought into contact with equatorial cytoplasm where the organizer can form.

2. If an artificial β-catenin stabilizing agent is injected at the equatorial level on one side, such as the mRNA for a dominant negative GSK3 or for GPB or for a nondegradable form of β-catenin (Farr et al. 2000), an asymmetric accumulation of β-catenin protein is thereby achieved. The embryo develops normally, without ever having had a parallel array of microtubules.

3. If lithium ion (as LiCl) is injected on one side at the equatorial level, an embryonic axis develops. β-Catenin accumulates locally because lithium inhibits the GSK3 enzyme that phosphorylates β-catenin, marking it for breakdown (Kao and Elinson 1988). If any of these treatments, 1, 2, or 3, is done ubiquitously in the egg, rather than on one side, the embryo develops as an anterior-dorsal exaggerated form that is cylindrically symmetric, as discussed below.

4. If a patch of vegetal cortex is removed from the vegetal pole and implanted at the equator, to deliver the natural stabilizing agent without cortical rotation (Kageura 1997), a partial or complete body axis is formed.

Partial Rotation, Partial Body Axis

The above treatments can be applied to *X. laevis* eggs at different times during cortical rotation to disrupt the microtubule array, thereby controlling how long vegetal materials are translocated. The later cortical rotation is stopped, the less truncated is the eventual body axis from its anterior end (Vincent and Gerhart 1987). With 20° of rotation, instead of the normal 30°, the axis is missing only the most anterior parts, such as forebrain, eyes, and nasal pits. With 15° of rotation, the axis is missing the entire head. With only 5–10°, the trunk is also missing, and only a tail forms. As noted above, with no rotation, no body axis is formed. Similar results are obtained if late-blastula-stage embryos are dissected to remove different amounts of the organizer (Stewart and Gerhart 1990). With only a quarter organizer, for example, the eventual body axis lacks the head but still forms the trunk and tail. Thus, we think that partial rotation, which is related to a partial accumulation of β-catenin protein, leads to the formation of a partial organizer, which leads to a partial body axis truncated from the anterior end.

The existence of this anterior–posterior series of phenotypes in *X. laevis* is one of the arguments that cortical rotation and organizer formation in the equatorial zone really concern the prospective anterior–posterior axis of the mesendoderm of the animal, not the dorsoventral axis (Gerhart 2002; Kumano and Smith 2002; Lane and Sheets 2002). As Lane and Sheets (2002) note, the prospective dorsoventral axis of mesendoderm is oriented along the animal–vegetal axis; that is, vertically across the equatorial zone (see Chapter 13). The older literature, in stating that the grey crescent or sperm entry point set up the dorsoventral axis, was referring more to the ectoderm rather than mesendoderm; but the recent discoveries about *vegT* mRNA and β-catenin indicate that early axis specification concerns the mesendoderm, not the ectoderm. Thus, it is more informative to say that cortical rotation sets up the anterior–posterior axis (of the mesendoderm) rather than the dorsoventral axis (of the ectoderm).

Multiple Translocations and Excessive Anterior–Dorsal Development

If the *X. laevis* egg is treated with D_2O or other agents that force microtubules to polymerize precociously and rapidly in the first cell cycle, a random array is formed at the cortex–core interface (Scharf et al. 1989; Rowning et al. 1997). The array never becomes aligned and cortical rotation does not occur, perhaps because the precocious array cannot be remodeled. Nonetheless, vegetal materials translocate away from the vegetal pole toward the positive ends of microtubules, but in all directions. Such translocations have been detected with $DiOC_6(3)$-labeled vesicles. The egg eventually develops to an exaggerated anterior-dorsal form, an embryoid with a large knot of notochord, a large heart, circumferential bands of eye pigment and adhesive gland cells, and an anterior gut. Cylindrical symmetry is never broken. The early gastrula forms an organizer around its entire equator. In these interesting cases, microtubules are present and translocations of vegetal materials do occur, but the first symmetry-breaking event has failed, that of forming the bilateral parallel array of microtubules. These anterior-dorsal embryoids also show that the egg has sufficient

maternal components to form an organizer much wider than the normal organizer, which occupies only a 60° sector of the equator.

FUTILE SYMMETRY BREAKING

Even though cortical rotation may occur, and even though a parallel array of microtubules may be present, the embryo may nonetheless fail to form Spemann's organizer at a later time. Heasman, Wylie, and colleagues have performed a series of experiments to deplete maternal mRNAs in *X. laevis* oocytes, such as those encoding proteins involved in β-catenin stabilization. Of course, the depletion of β-*catenin* mRNA itself is the most telling (Wylie et al. 1996). When such oocytes are matured and fertilized, the egg eventually develops into a cylindrically symmetric posteriorized-ventralized embryoid. The organizer does not form even though cortical rotation has occurred. After the midblastula transition, zygotic expression of β-*catenin* genes provides protein for essential Wnt signaling and for intercellular junction formation. By that time, however, the requirement of maternal β-catenin for axis specification has passed. Normal development of the axis in such depleted eggs can be rescued by the injection of β-*catenin* mRNA. In a reciprocal experiment, excess levels of mRNA for *GSK3,* axin, or adenopolyposis coli (APC) can be introduced by injection. All of these proteins promote β-catenin degradation, overwhelming whatever stabilization occurs on one side due to the translocated stabilizing agent. In these cases, the egg also develops as a posteriorized-ventralized form despite cortical rotation.

If the vegetal pole of the full-grown stage-VI oocyte is briefly exposed to UV irradiation (at low levels compared to those used on fertilized eggs to block microtubule polymerization), the eventual fertilized egg from this oocyte will undergo cortical rotation, but a posteriorized-ventralized embryo nonetheless develops from it (Elinson and Pasceri 1989). The current interpretation is that the endogenous stabilizing agent, or a precursor needed for its activation or positioning at the vegetal pole, has been damaged. Therefore, even though a parallel array is formed, the vegetal agent is never translocated in an active form capable of stabilizing β-catenin.

BYPASSING THE NEED FOR CORTICAL ROTATION

Houston et al. (2002) have used morpholino and antisense methods to deplete *tcf* mRNA from the full-grown *X. laevis* oocyte. After the oocyte is matured and fertilized, it develops into a cylindrically symmetric exaggerated anterior-dorsal form, as if it had formed an organizer around the entire equator. The phenotype is not as extreme as with lithium treatment or with the knockout of *axin* mRNA, but is nonetheless clear. The current interpretation of this effect is that *tcf* mRNA, which is uniformly distributed in the normal egg, is translated to the Tcf protein, a transcription factor that represses many or all aspects of organizer formation at the midblastula stage when new gene expression begins.

After β-catenin protein accumulates on one side in the normal egg, it complexes with the Tcf protein and prevents it from entering the nuclei of blastomeres, thereby precluding its repressive effects in that one locale. Thereafter, in the β-catenin-rich region, various genes are expressed which encode proteins contributing to organizer formation and function. These include *siamois* and *xnr3* in *X. laevis*. Genes related to BMP and Wnt signaling are not expressed there. Thus, the organizer forms at a location where β-catenin has been present and where endo-mesoderm induction occurs, a double requirement.

The egg, therefore, has an intrinsic capacity to form the organizer anywhere and everywhere around its circumference, due to ubiquitous transcription factors of yet unknown identity, but this capacity is repressed everywhere by the ubiquitous Tcf repressor protein. Then the repression is locally relieved via β-catenin, whose accumulation depends on cortical rotation, the symmetry-breaking event.

In summary, cortical rotation is a symmetry-breaking process operating in the first cell cycle of the *Xenopus* egg. It is normally essential for the transformation of the egg's initial cylindrical symmetry into a bilateral symmetry determining the location of Spemann's organizer in the gastrula and hence the orientation of the eventual body axis of the tadpole. During cortical rotation, microtubules become aligned in a bilateral parallel array in a thin sheet at the interface of the egg's cortex and cytoplasmic core, with their plus ends all pointed in the same direction. It is hypothesized that movement of the cortex is needed to align the microtubules and that the array is largely self-organizing due to reciprocal positive feedback between cortical movement and microtubule formation. On this transient bilateral array, a β-catenin stabilizing agent of yet unknown identity is translocated from the vegetal pole region to the equator on one side, giving the egg a long-lasting bilateral symmetry and determining the site at which Spemann's organizer will be formed.

REFERENCES

Ancel P. and Vintemberger P. 1948. Recherches sur le déterminisme de la symmetrie bilaterale dans l'oeuf des Amphibiens. *Bull. Biol. Fr. Belg.* (suppl.) **31:** 1–182.

Black S.D. and Gerhart J.C. 1986. High-frequency twinning of *Xenopus laevis* eggs centrifuged before first cleavage. *Dev. Biol.* **116:** 228–240.

Black S.D. and Vincent J.-P. 1988. The first cleavage plane and the embryonic axis are determined by separate mechanisms in *Xenopus laevis*. II. Experimental dissociation by lateral compression of the egg. *Dev. Biol.* **128:** 65–71.

Cha B.J. and Gard D.L.1999. XMAP230 is required for the organization of cortical microtubules and patterning of the dorsoventral axis in fertilized *Xenopus* eggs. *Dev. Biol.* **205:** 275–286.

Curtis A.S.G. 1960. Cortical grafting in *Xenopus laevis*. *J. Embryol. Exp. Morphol.* **8:** 163–173.

Dalcq A. and Pasteels J. 1937. Une conception nouvelle des bases physiologiques de la morphogenèse. *Arch. Biol.* **48:** 699–710.

Dominguez I. and Green J.B.A. 2000. Dorsal downregulation of GSK3 by a non-Wnt-like mechanism is an early molecular consequence of cortical rotation in early *Xenopus* embryos. *Development* **127:** 861–888.

Elinson R.P. 1975. Site of sperm entry and a cortical contraction associated with egg activation in the frog *Rana pipiens*. *Dev. Biol.* **47:** 257–268.

Elinson R.P. and Pasceri P. 1989. Two UV-sensitive targets in dorsoanterior specification of frog embryos. *Development* **106:** 511–518.

Elinson R.P. and Rowning B. 1988. A transient array of parallel microtubules in frog eggs: Potential tracks for a cytoplasmic rotation that specifies the dorso-ventral axis. *Dev. Biol.* **128:** 185–197.

Farr G.H., III, Ferkey D.M., Yost C., Pierce S.B., Weaver C., and Kimelman D. 2000. Interaction among GSK-3, GBP, axin, and APC in *Xenopus* axis specification. *J. Cell Biol.* **148:** 691–702.

Gerhart J. 1980. Mechanisms regulating pattern formation in the amphibian egg and early embryo. In *Biological regulation and development* (ed. R.F. Goldberger), pp. 131–316. Plenum Press, New York.

———. 2002. Changing the axis changes the perspective. *Dev. Dynamics* **225:** 380–383.

Gerhart J.C., Danilchik M., Doniach T., Roberts S., Rowning B., and Stewart R.M. 1989. Cortical rotation of the *Xenopus* egg: Consequences for the anteroposterior pattern of embryonic dorsal development. *Development* (suppl.) **107:** 37–51.

Holowacz T. and Elinson R.P. 1995. Properties of the dorsal activity found in the vegetal cortical cytoplasm of *Xenopus* eggs. *Development* **121:** 2789–2798.

Houliston E. and Elinson R.P. 1991. Patterns of microtubule polymerization relating to cortical rotation in *Xenopus laevis* eggs. *Development* **112:** 107–118.

Houston D.W., Kofron M., Resnik E., Langland R., Destree O., Wylie C., and Heasman J. 2002. Repression of organizer genes in dorsal and ventral *Xenopus* cells mediated by maternal XTcf3. *Development* **129:** 4015–4025.

Kageura H. 1997. Activation of dorsal development by contact between the cortical dorsal determinant and the equatorial core cytoplasm in eggs of *Xenopus laevis*. *Development* **124:** 1543–1551.

Kao K.R. and Elinson R.P. 1988. The entire mesodermal mantle behaves as Spemann's organizer in dorsoanterior enhanced *Xenopus laevis* embryos. *Dev. Biol.* **127:** 64–77.

Kloc M. and Etkin L.D. 1995. Two distinct pathways of the localization of RNAs at the vegetal cortex in *Xenopus* oocytes. *Development* **121:** 287–297.

Kumano G., and Smith W.C. 2002. Revisions to the *Xenopus* gastrula fate map: Implications for mesoderm induction and patterning. *Dev. Dynamics* **225:** 409–421.

Lane M.C. and Sheet M.D. 2002. Rethinking axial patterning in amphibians. *Dev. Dynamics* **225:** 434–447.

Larabell C.A., Torre M., Rowning B.A., Yost C., Miller J.R., Wu M., Kimelman D., and Moon R.T. 1997. Establishment of the dorso–ventral axis in *Xenopus* embryos is prestaged by early asymmetries in β-catenin that are modulated by the Wnt signaling pathway. *J. Cell Biol.* **136:** 1123–1136.

Lee E., Salic A., Kruger R., Heinrich R., and Kirschner M.W. 2003. The roles of APC and axin derived from experimental and theoretical analysis of the Wnt pathway. *PLoS Biol.* **1:** 116–132.

Levin M., Thorlin T., Robinson K.R., Nogi T., and Mercola M. 2002. Asymmetries in H+/K+-ATPase and cell membrane potentials comprise a very early step in left-right patterning. *Cell* **111:** 77–89.

Marikawa Y. and Elinson R.P. 1999. Relationship of vegetal cortical dorsal factors in the *Xenopus* egg with the Wnt/β-catenin signaling pathway. *Mech. Dev.* **89:** 93–102.

Marikawa Y., Li Y., and Elinson R.P. 1997. Dorsal determinants in the *Xenopus* egg are firmly associated with the vegetal cortex and behave like activators of the Wnt pathway. *Dev. Biol.* **191:** 69–79.

Marrari Y., Clarke E.J., Rouviere C., and Houliston E. 2003. Analysis of microtubule movement on isolated *Xenopus* egg cortices provides evidence that the cortical rotation involves dynein as well as kinesin related proteins and is regulated by local microtubule polymerisation. *Dev. Biol.* **257:** 55–70.

McMahon A.P. and Moon R.T. 1989. Ectopic expression of the proto-oncogene Int-1 in *Xenopus* embryos leads to duplication of the embryonic axis. *Cell* **58:** 1075–1084.

Medina A., Wendler S.R., and Steinbeisser H. 1997. Cortical rotation is required for the correct spatial expression of *nr3, sia* and *gsc* in *Xenopus* embryos. *Int. J. Dev. Biol.* **41:** 741–745.

Miller J.R., Rowning B.A., Larabell C.A., Yang-Snyder J.A., Bates R.L., and Moon R.T. 1999. Establishment of the dorsal-ventral axis in *Xenopus* embryos coincides with the dorsal enrichment of Dishevelled that is dependent on cortical rotation. *J. Cell Biol.* **146:** 427–437.

Morgan T.H. and Boring A.M. 1903. The relation of the first plane of cleavsage and the grey crescent to the median plane of the embryo of the frog. *Roux Arch.* **16:** 680–690.

Newport G. 1854. Researches on the impregnation of the ovum in Amphibia, and on the early stages of development of the embryo. *Philos. Trans. R. Soc. Lond. Ser. B Biol. Sci.* **144:** 229–244.

Roux W. 1887. IV. Die Richtungsbestimmung der Medianebene des Froschembryo durch die Copulationsrichtung des Eikernes und des Spermakernes. *Arch. Mikroskop. Anat.* **29:** 157–211.

Rowning B.A. and Gerhart J.C. 1990. Microtubule mediated cortical rotation and axis specification in the *Xenopus* egg. *UCLA Symp. Mol. Cell. Biol.* **125:** 61–78.

Rowning B.A, Wells J., Wu M., Gerhart J.C., Moon R.T., and Larabell C.A. 1997. Microtubule-mediated transport of organelles and localization of β-catenin to the future dorsal side of *Xenopus* eggs. *Proc. Nat. Acad. Sci.* **94:** 1224–1229.

Scharf S.R. and Gerhart J.C. 1980. Determination of the dorsal-ventral axis in eggs of *Xenopus laevis*: Complete rescue of UV-impaired eggs by oblique orientation before first cleavage. *Dev. Biol.* **79:** 181–198.

———. 1983. Axis determination in eggs of *Xenopus laevis*: A critical period before first cleavage, identified by the common effects of cold, pressure and UV-irradiation. *Dev. Biol.* **99:** 75–87.

Scharf S.R., Rowning B., Wu M., and Gerhart J.C. 1989. Hyper-dorsoanterior embryos from *Xenopus* eggs treated with D$_2$0. *Dev. Biol.* **134:** 175–188.

Schechtman A.M. 1936. Relation between the grey crescent and the organizer center in the urodele egg (*Triturus torosus*) *Roux Arch.* **134:** 207–208.

Schneider S., Steinbeisser H., Warga R.M., and Hausen P. 1996. β-Catenin translocation into nuclei demarcates the dorsalizing centers in frog and fish embryos. *Mech. Dev.* **57:** 191–198.

Schulze E. and Kirschner M. 1987. New features of microtubule behavior observed in vivo. *Nature* **334:** 356–359.

Slack C. and Palmer J.F. 1969. The permeability of intercellular junctions in the early embryo of *Xenopus laevis* studied with a fluorescent tracer. *Exp. Cell Res.* **55:** 416–419.

Spemann H. 1902. Entwicklungsphysiologische Studien am Triton-Ei. *Roux Arch.* **15:** 448–534.

Stewart R.M. and Gerhart J.C. 1990. The anterior extent of dorsal development of the *Xenopus* embryonic axis depends on the quantity of organizer in the late blastula. *Development* **109:** 363–372.

Sumanas S., Strege P., Heasman J., and Ekker S.C. 2000. The putative Wnt receptor *Xenopus* frizzled-7 functions upstream of β-catenin in vertebrate dorsoventral mesoderm patterning. *Development* **127:** 1981–1990.

Vincent J.-P. and Gerhart J.C. 1987. Subcortical rotation in *Xenopus* eggs: An early step in embryonic axis specification. *Dev. Biol.* **123:** 526–539.

Vincent J.-P., Oster G.F., and Gerhart J.C. 1986. Kinematics of gray crescent formation in the amphibian egg: Displacement of subcortical cytoplasm relative to the egg surface. *Dev. Biol.* **113:** 484–500.

Weaver C., Farr III, G.H., Pan W., Rowning B.A., Wang J., Mao J., Wu D., Li L., Larabell C.A., and Kimelman D. 2003. GBP binds kinesin light chain and translocates during cortical rotation in *Xenopus* eggs. *Development* **130:** 5425–5436.

Wylie C., Kofron M., Payne C., Anderson R., Hosobuchi M., Joseph E., and Heasman J. 1996. Maternal β-catenin establishes a 'dorsal signal' in early *Xenopus* embryos. *Development* **122:** 2987–2996.

Yuge M., Kobayakawa Y., Fujisue M., and Yamana K. 1990. A cytoplasmic determinant for dorsal axis formation in an early embryo of *Xenopus laevis*. *Development* **110:** 1051–1056.

Zhang J., Houston D.W., King M.L., Payne C., Wylie C., and Heasman J. 1998. The role of maternal VegT in establishing the primary germ layers in *Xenopus* embryos. *Cell* **94:** 515–524.

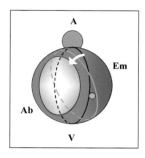

Breaking Radial Symmetry
Amniote Type

S. Frankenberg and M. Zernicka-Goetz

Wellcome Trust/Cancer Research Gurdon Institute, University of Cambridge,
Cambridge CB2 1QR, United Kingdom

REGULATION AND PATTERNING: HARMONY OF COEXISTENCE

Why do we have separate chapters about the ways in which symmetry is broken in embryos of amphibians or fish versus those seen in amniotes? One reason is that in the former there is virtually no flexibility through which the embryo can respond to any disturbance of its early organization, and in the latter such flexibility has considerable power. Indeed, one striking conclusion that has emerged from studies on early amniote embryos is that they are able to develop normally following a variety of experimental manipulations, often involving drastic modifications in cell arrangement or cell number (for review, see McLaren 1976; Bachvarova 1999). The impressive ability of amniote embryos to retain this flexibility to highly multicellular stages of development, known as regulative development, deeply influenced ideas about how embryonic patterning might be orchestrated. Specifically, it encouraged a view that patterning in such regulative embryos could not follow any spatial information from the beginning of embryonic development. Accordingly, such spatially provided cues were thought not to exist.

The regulative ability of an embryo allows both the fate of individual early embryonic cells and the embryo as a whole to remain adaptable when development is perturbed. However, it does not explain how the embryo develops its pattern under normal circumstances, when it is not perturbed. So, does the unperturbed embryo have any early spatial cues that can guide developmental progress of the zygote, and, if so, what might such early cues be? An answer

to both of these questions requires a means of observing the embryo as it develops with minimal, if any, intervention. Such an observation-driven approach is possible from the very beginning of development of the model mammalian embryo, the mouse, as the first few days of its life can be followed in culture. However, to recognize pattern, should it exist, and to position in space any potential cues for its development, the experimental approach requires markers to enable the fate of individual cells and groups of cells to be followed as development proceeds. As both the egg and embryonic cells of the mouse are deprived of natural markers, such as the pigment found in *Xenopus* eggs, we have had to develop non- or minimally invasive ways to introduce exogenous markers (for review, see Zernicka-Goetz 1999).

Through this route of observing the behavior of marked cells or specific regions of the egg surface, it is now evident that normal mouse development does not take place by cells acquiring identity randomly once a particular size or stage has been reached, but rather it follows a defined pattern. A series of cytoplasmic ablation and transplantation experiments has demonstrated the possibility of the existence of localized cytoplasmic determinants in the egg, determinants not for developmental fate but for establishing the cleavage pattern (Plusa et al. 2002b). Such preferred development of pattern does not deny the embryo its impressive regulative ability at the early embryonic stages. Thus, the two concepts—regulative development and early patterning—once thought to be in conflict, can now be viewed in harmony. These studies make us realize that although the major body axes emerge relatively late morphologically, at the time the embryo initiates gastrulation, asymmetries that may provide the foundation for them are laid down very much earlier.

What can provide the initial radial symmetry-breaking information and spatial cues in such an embryo? A symmetry-breaking event does not have to be drastic but can be a subtle trigger culminating in the greatly amplified and focused effect of different developmental pathways across the embryo. Perhaps the minuteness of the trigger allows sufficient lability for contrasting mechanisms to evolve among different groups of animals. It can also make identification of the trigger difficult. From the several options that can be considered as a source of an asymmetric signal, we could include maternal effects (derived directly from the surrounding maternal tissues), sperm-derived signals, environmental cues (e.g., light, gravity, magnetism), internal mechanisms (molecular chirality), or random events. Which of these might be operating in an amniote embryo?

Although there is evidence for an early specification of polarity in avian embryos (see Chapter 15), development of the bird egg does not lend itself to studies of the very earliest stages of the embryo. Hence, although gravity has been postulated as the basis for axial organization in the chick embryo, it remains unclear how this initiates asymmetry (Eyal-Giladi 1984). Most studies of the events that precede emergence of the axes in amniotes have been performed in mammalian systems, especially in the mouse, where evidence points toward a number of potential contributors to steps that introduce asymmetry to the embryo. These include the intrinsic asymmetric properties of the oocyte; the position at which a sperm enters the egg; asymmetries in the early cleavage pattern; and, possibly, the interactions between the embryo and the mother that occur upon implantation (for review, see Zernicka-Goetz 2002). So what are the relative contributions of each of these and what is the mechanism by which they define early embryonic polarity and the final body plan?

The emphasis in this chapter is on early processes that are uniquely accessible for study in mammalian species. We outline the evidence for early events leading to asymmetries that establish the embryonic–abembryonic axis, the proximal–distal axis, and asymmetries in cell movements that define the site at which the primitive streak forms. Thus, by the time of gastrulation, the embryo has been exposed to several polarizing influences. Although this highlights the importance of symmetry-breaking events at the outset of development, it also emphasizes that there is still very little understanding at the molecular level of how these early cues form and convey their effects to later stages.

EMBRYONIC–ABEMBRYONIC AXIS

Axial differentiation depends on spatial cues that arise at earlier stages of development. These cues, in turn, depend on symmetry-breaking events that occur during either oogenesis or early embryogenesis. In the oviparous amniotes—birds, reptiles, and monotreme mammals—a large yolk ensures that the cue for at least one axis is specified from the beginning of development (see Chapter 51). This is because yolk accumulates asymmetrically in the oocyte as it develops in the ovary and is segregated during cleavage. Similarly, in marsupials (see Chapter 18) an analogous material, deutoplasm (often called yolk but more gelatinous in nature), displaces the nuclear material asymmetrically either during oogenesis or shortly after fertilization and is also expelled asymmetrically during early cleavage (for review, see Selwood 1992). Consequently, in all of the above groups of animals, the subsequent asymmetric position of the embryonic area with respect to yolk or yolk-like material occurs before differentiation of any tissues and defines the embryonic–abembryonic axis. The epiblast (which will build the future body) forms superficially at the embryonic pole, with the hypoblast (an extraembryonic tissue) forming underneath. The embryonic–abembryonic axis can also be considered to specify the relative positions of all the future germ layers, such as the hypoblast underlying the epiblast.

In eutherian mammals, there is an absence of asymmetric yolk, and a presumptive embryonic–abembryonic axis is not recognizable in the fertilized egg, at least at the level of resolution we have thus far. It was previously believed that this axis can first be recognized only from the time of blastocyst formation, by the relative positions of the inner cell mass (ICM), trophoblast, and blastocyst cavity. However, in contrast to a previous suggestion, it now appears that the embryonic–abembryonic axis is predicted by much earlier events (Gardner 1997, 2001; Piotrowska et al. 2001; Piotrowska and Zernicka-Goetz 2001, 2002). Although the molecular mechanisms that orient the embryonic–abembryonic axis are still not known, some features of the early mouse embryo allow its orientation to be predicted. One of these is the first cleavage plane, which has been found to tend to align approximately, although with some variability in angle, with the plane of the equator of the blastocyst (Gardner 2001; Piotrowska and Zernicka-Goetz 2001; Fujimori et al. 2003). Consistent with this, the second polar body, which is normally aligned with the first cleavage plane (Plusa et al. 2002b), also comes to lie near the equator of the blastocyst (Gardner 1997). Moreover, as shown by a variety of surface marking studies and by time-lapse imaging of intact eggs, the position on the mouse egg surface at which the sperm enters the egg has a tendency to align with the first cleavage furrow (Piotrowska and Zernicka-Goetz 2001; Plusa et al. 2002a). These two separate sites on the egg surface are usually sufficient to predict the orientation of the first cleavage and hence the orientation of the embryonic–abembryonic axis. However, whereas the first cleavage plane might define approximately the orientation of the embryonic–abembryonic axis, it does not define its polarity per se.

From the foregoing, one would expect that differences between blastomeres at the 2-cell stage might predict the polarity of the embryonic–abembryonic axis. At the 2-cell stage there is asynchrony between the blastomere divisions, so that one blastomere divides earlier than its sister. It has long been known that the earlier dividing cell gives rise to more ICM cells than the later dividing cell, thus giving rise to more cells of the embryonic hemisphere of the blastocyst (Kelly et al. 1978; Spindle 1982; Surani and Barton 1984; Garbutt et al. 1987). However, this in itself does not necessarily predict the polarity of the embryonic–abembryonic axis because cavitation could occur on either side of the inner cells. The results of lineage-tracing experiments in which embryos were preselected at the 3-cell stage and their pattern was examined at the blastocyst stage (Piotrowska et al. 2001) suggest that polarity at the 2-cell stage tends to predict the polarity of the embryonic–abembryonic axis, implying that cavitation normally occurs on the side of the later-dividing 2-cell blastomere (Fig. 1). However, this relationship is clear in those embryos in which the asynchrony of divisions from the 2- to 4-cell stage is significant, whereas in embryos in which there is little asynchrony, the relationship between the earlier dividing blastomere and its embryonic fate might not be so obvious (Piotrowska et al. 2001; Fujimori et al. 2003). The source of this polarity is still unknown. It might in part be linked to the sperm entry position, because the blastomere that divides earlier has a greater tendency to inherit the cortex of the egg that was penetrated by the sperm (Bennett 1982; Piotrowska and Zernicka-Goetz 2001). This potential involvement of fertilization in establishing embryonic polarity is emphasized by the findings that each 2-cell blastomere of a parthenogenetic egg, which has never been approached by sperm, no longer

shows a bias to follow an embryonic or abembryonic fate (Piotrowska and Zernicka-Goetz 2002). Furthermore, a similar dissociation of the embryonic–abembryonic axis from the first cleavage is observed in embryos from which the cortical cytoplasm has been removed from the site of the egg that is approached by sperm. Thus, the orientation of the first cleavage division of the zygote and the order of the second cleavage division tend to predict the orientation and polarity of the embryonic–abembryonic axis—the axis that corresponds to the proximo-distal axis of the embryo after implantation.

The embryonic–abembryonic axis provides positional information for patterning in two ways: directly through the conceptus and meridianally (Fig. 2). The former is essential for specification of the germ layers because the endoderm forms as a deep layer underlying the epiblast (which also underlies the polar trophoblast in eutherian mammals). The latter provides a two-dimensional radial symmetry that provides essential positional information for patterning of the

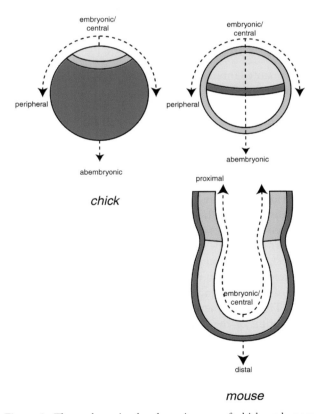

Figure 2. The embryonic–abembryonic axes of chick and mouse (showing late blastocyst and egg cylinder) provide positional information both through the conceptus and radially (central versus peripheral). In the mouse, the embryonic–abembryonic axis of the blastocyst is thus equivalent to the proximal–distal axis of the egg cylinder. (*Brown*) Yolk or endoderm. (*Yellow*) Area pellucida or epiblast. (*Orange*) Area opaca or extraembryonic ectoderm.

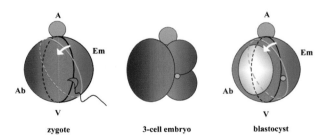

Figure 1. Model for relationship between the first cleavage plane, marked by the second polar body (*green*), the sperm entry position (*red*), and orientation and polarity of the embryonic–abembryonic axis. The equator of the blastocyst tends to be approximately aligned with the plane of first cleavage, but often with some deviation (*arrow*). (A) Animal pole. (V) Vegetal pole. (Em) Embryonic pole. (Ab) Abembryonic pole.

embryo during gastrulation. The cylindrical shape ("egg cylinder") of rodent conceptuses represents a morphological distortion of the essentially two-dimensional disc typical of most amniotes. The "proximo-distal axis" of the egg cylinder thus directly translates to the peripheral-central radial symmetry of the embryonic disc. In birds, reptiles, and non-eutherian mammals, two-dimensional radial symmetry is also essential for the differentiation of extraembryonic ectoderm peripherally from a central pluripotent population. In eutherian mammals, a modification of this central-peripheral mechanism occurs in a three-dimensional framework, formulated by Tarkowski and Wroblewska (1967) as the "inside–outside" hypothesis. In this scheme, the trophoblast emerges as an epithelium that entirely envelops the asymmetrically positioned inner cell mass. The generation of inner and outer cells is usually initiated before cavitation, but in some species can occur afterward. Eutherian mammals are thus unique in that the earliest differentiation event is intimately linked with establishment of the embryonic–abembryonic axis.

PROXIMO–DISTAL POLARITY AND ASYMMETRIC CELL MOVEMENTS

The implanted mouse embryo appears radially symmetrical at day 5.5 of its development (E5.5). At this stage it is known as the egg cylinder. Embryonic patterning is now evident in the proximo-distal axis, with ectoderm located proximally, epiblast distally, and visceral endoderm enveloping both tissues. Within these tissues, gene expression patterns further define subdomains of asymmetry in the proximo-distal axis (for review, see Beddington and Robertson 1999) (see also Chapter 16). *Bmp4* is restricted to the distal part of the ectoderm (Lawson et al. 1999), *Otx2* is concentrated more distally in the epiblast and visceral endoderm (Simeone et al. 1993; Ang et al. 1994; Lawson et al. 1999), and *Hex* and *Cer-like* are specifically expressed in a discrete population of visceral endoderm cells at the distal tip of the egg cylinder (Thomas et al. 1998; Stanley et al. 2000). Thus, the emergence of molecular markers of axial organization of the postimplantation mouse embryo can first be detected along the proximo-distal axis. This asymmetry in gene expression pattern relates to the future anterior–posterior axis, and asymmetric cell movements play a key role in this relationship. How does it come about?

At about E5.5, the distal visceral endoderm cells initiate asymmetric migration toward the site that will become the future anterior (Thomas et al. 1998; Weber et al. 1999; Kimura et al. 2000; Perea-Gomez et al. 2001). As this change occurs, genes previously expressed radially in the proximal epiblast of the egg cylinder become restricted in expression toward the opposite side of the egg cylinder where the

primitive streak forms. Asymmetric cell movements thus give rise to the earliest stage of anterior–posterior asymmetry in terms of known gene expression patterns. They also appear critical for the future anterior parts of the embryo and the streak to emerge in the right place and in the right orientation (Ding et al. 1998).

Pre-gastrulation asymmetric cell movements relating to streak formation were first described in the embryo of another amniote, the chick. They were described as occurring in the epiblast, and their characteristic pattern of movements was named "Polonaise" because they resemble a traditional Polish dance (Gräper 1929). In the chick embryo, such movements begin, as in the mouse, with anterior migration of the endoderm, or hypoblast and endoblast (Vakaet 1970; Stern and Ireland 1981; Bachvarova et al. 1998) (see Chapter 15). These movements are essential for correct orientation of Polonaise movements in the epiblast and therefore for axial polarity of the embryo (Foley et al. 2000). We refer to the similar cell movements in the mouse embryo as "Polonaise-like," but thus far they have only been described in the visceral endoderm, and not in the epiblast. Yet it cannot be precluded that epiblast cells also follow a similar pattern of behavior. The mechanism of these morphogenetic cell movements is unknown. In the mouse they might reflect not only cell migration, but also differential growth or change in shape of the embryo. Moreover, the mechanisms underlying these movements may differ in different parts of the embryo.

Asymmetric endoderm migration can also be seen as one of the earliest signs of morphological asymmetry in *Xenopus*. The vegetal floor of the blastocoel forms a structure known as Brachet's cleft, which is at first radially uniform but soon after skews toward the side of the future blastopore and then toward the anterior or animal pole (Winklbauer and Schurfeld 1999). This asymmetric migration of the blastocoel floor toward the future anterior, described as "vegetal rotation" (see Chapter 23), can be viewed as analogous to distal visceral endoderm migration in the mouse and hypoblast migration in the chick. All such movements occur before any asymmetry in mesoderm induction is evident. This strongly suggests that the former is essential for the latter, possibly by transporting factors to a location that allows the organizer to form.

ESTABLISHING THE SITE OF PRIMITIVE STREAK FORMATION

Localization of the streak appears to be achieved as a consequence of these asymmetric cell movements. Growing evidence shows that in the mouse embryo an important consequence of the anterior movement of the distal visceral endoderm is to maintain the overlying epiblast in a state receptive to later head induction by protecting it from posterior caudalizing signals (Foley et al. 2000; Kimura et al. 2000; Stern 2001; Perea-Gomez et al. 2002; for review, see Zernicka-Goetz 2002). This is achieved by the production and delivery of Cer-like and Lefty1, antagonists of Nodal, and perhaps other signaling cues to the epiblast, so as to restrict streak formation to the posterior edge (Bertocchini and Stern 2002; Perea-Gomez et al. 2002). In the chick, streak formation is inhibited by the hypoblast, which expresses the Wnt-, Nodal-, and BMP-antagonist *Cerberus*. Displacement of the hypoblast anteriorly by the emerging endoblast (which does not express *Cerberus*) is suggested to allow streak formation by removal of its inhibition (Bertocchini and Stern 2002). Thus, endoderm migration may be essential for streak formation because it both removes the inhibition of streak formation in the posterior and prevents its formation anteriorly.

Asymmetric cell migration must depend on an earlier symmetry-breaking cue. In mammals, no molecular asymmetry has been definitively identified that predicts the direction of endoderm migration. In amphibians, however, the direction of this migration (vegetal rotation) is predicted by the position of the Nieuwkoop center, which induces the organizer without contributing cells to it, within the vegetally located presumptive endoderm (see Chapters 14, 24, and 46). Its equivalent in the chick is not so clearly defined but lies within the posterior marginal zone (Bachvarova 1999). One candidate tissue to consider in the mouse is visceral endoderm because, in contrast to the epiblast, it remains coherent as it proliferates (Gardner and Cockroft 1998; Weber et al. 1999) and would thus be more likely to retain positional information acquired from earlier stages (Weber et al. 1999). In the mouse, one can therefore hypothesize that such a center resides in the visceral endoderm overlying the extraembryonic ectoderm, close to the epiblast. However, at present, one cannot exclude positional information being provided by the extraembryonic ectoderm adjacent to the site of streak formation or a part of the epiblast itself.

EARLY CUES: CAN THEY PREDICT DIRECTIONALITY OF THE ASYMMETRIC CELL MOVEMENTS?

The transport of dorsalizing determinants away from the site of sperm entry by cytoskeletal mechanisms is known to determine the dorsoventral axis in the *Xenopus* embryo (Gerhart et al. 1989; see Chapter 24). Thus far, this mechanism has no known equivalent in the chick or mouse embryo, which are more flexible in their development and in which such a strong "determinative" mechanism might not necessarily be expected. However, as indicated earlier,

there are known asymmetries in the early embryos of these species whose collective role might be to induce an axis organizing center. So what are they?

In birds and reptiles, which in general have the largest and yolkiest eggs of all animals, the initial stages of anterior–posterior axis specification usually occur in utero. Classic experiments in the chick have shown that gravity is essential for the orientation and polarity of this axis (Eyal-Giladi 1984). In Galliform birds, the egg rotates slowly on its long axis within the uterus, with its central yolk remaining stationary. This is believed to cause the developing blastoderm to be held in an oblique position, presumably exposing it asymmetrically to some unknown determinant(s), such that the uppermost edge of the blastoderm becomes the future posterior (see Chapter 15). It is possible that this could involve the polarization of the cytoskeleton to define the asymmetric localization of intrinsic axis determinants, as demonstrated in other developing embryos. Kochav and Eyal-Giladi (1971) showed that the anterior–posterior axis is determined gradually during a critical 2-hour period a few hours before laying. It is not yet clear whether there are additional mechanisms that would help to account for this asymmetry.

In the mouse, increasing evidence suggests that the future anterior–posterior axis might also stem from asymmetries accumulating from the beginning of development. In addition to the embryonic–abembryonic axis, morphology of the early blastocyst reveals another axis—in this case an axis of bilateral symmetry—such that when the blastocyst is viewed from the embryonic pole, one axis is significantly shorter in diameter than the other (Fig. 3). It has also been noted that the axis of blastocyst bilateral symmetry is marked by the position of the second polar body, which remains tethered to the embryo from the zygote stage throughout cleavage (Gardner 1997). Subsequent lineage-tracing studies of the progeny of blastomeres that descend from either the animal or the vegetal pole of the egg have shown that their relative axial positions are preserved by coherent growth during early development (Ciemerych et al. 2000). Therefore, the distribution of any axial information along the animal–vegetal axis of the zygote could be preserved in the blastocyst.

It is not clear what defines the molecular nature of the animal–vegetal axis of the mouse egg; but this axis is aligned with the migratory path of the meiotic spindle in the maturing oocyte, which morphologically defines the animal (or meiotic) pole. Whether the animal–vegetal axis of the egg is established randomly through the process of spindle migration or whether it originates from an even earlier polarity arising during oocyte growth remains unknown. If the latter were true, this would suggest the possibility of asymmetric interactions between the oocyte and surrounding follicle cells. Such asymmetric interac-

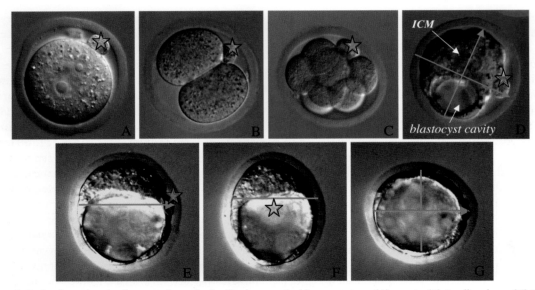

Figure 3. Preimplantation development from the fertilized egg to the blastocyst stage. (*A*) zygote, (*B*) 2-cell embryo, (*C*) 8-cell embryo, and (*D*) blastocyst. The blastocyst (*D*) has two axes of symmetry: long axis of ICM (*red*) that corresponds to the animal–vegetal axis and embryonic–abembryonic axis (*blue*). *E–G* show the same blastocyst at 3 different orientations. The blastocyst in *E* is viewed as it would normally lie in a culture dish, with its embryonic–abembryonic axis and animal–vegetal axis (*red*) lying horizontally. A shorter axis (*green*) is revealed by rotation about the embryonic–abembryonic axis (*F*) and about the animal–vegetal axis (*G*). The position of the second polar body is marked by a yellow star. (Reprinted, with permission, from Zernicka-Goetz 2002 [© The Company of Biologists Ltd.].)

tions have not yet been identified. A developmental role for the meiotic pole of the egg is indicated by a series of transplantation and ablation experiments (Plusa et al. 2002b), which indicate the existence of localized cytoplasmic/cortical determinants at the meiotic pole of the egg that direct the plane of the first cleavage. How intrinsic oocyte polarity is influenced by the act of fertilization is also unclear, but it may be achieved either by altering the cortical region of the egg through which the sperm enters or by defining the path of male pronuclear migration. Whatever the mechanism, both oocyte asymmetry and fertilization appear to provide positional information to the mouse embryo, as is the case in developing embryos of many other species. In the mouse, the oocyte's meiotic pole, possibly together with sperm entry position, appears to contribute to setting the cleavage pattern and this, in turn, establishes the form and polarity of the blastocyst (Piotrowska and Zernicka-Goetz 2001; Piotrowska et al. 2001).

A series of lineage-tracing experiments approached the question of whether the animal–vegetal axis provides some positional information for postimplantation cell fate (for review, see Zernicka-Goetz 2002). In this study, surface ICM cells from both ends of the animal–vegetal axis of the blastocyst were labeled by injection of mRNA for a modified form of GFP (Zernicka-Goetz et al. 1997; Weber et al. 1999). When the distribution of visceral endoderm progeny of labeled cells was analyzed at the early postimplanta-

tion stage, either before or after the onset of gastrulation, it was discovered that the fate of the cells depended on their origin. Thus, cells located at one end of the blastocyst axis (descendants of the meiotic pole of the egg) tended to give visceral endoderm progeny that localized to the more distal portion of the egg cylinder. In contrast, the cells descending from the opposite pole tended to occupy the more proximal portion of the egg cylinder. This result suggests that, in addition to the embryonic–abembryonic axis, there is intrinsic polarity within the blastocyst (already before implantation) that relates the morphological axis of asymmetry described above to the subsequent positions of cells postimplantation. In addition, the characteristic patterns of labeled cells indicated that there is either asymmetric cell movement or growth of the embryo at the prospective anterior and posterior sides: the coherent visceral endoderm clones extended diagonally from the anterior/proximal to the posterior/distal orientation (for review, see Zernicka-Goetz 2002). This study therefore demonstrates that there are spatial cues present even before implantation that might influence postimplantation cell behavior. The extent to which early preimplantation polarity is carried through implantation and the extent to which it might be modified by additional symmetry-breaking events are not fully understood.

The relationships of blastocyst polarity with implantation and later body axes were explored in the histological studies of Smith (1980, 1985) (Fig. 4), who showed that in utero, the blastocyst lies and initially attaches to the uterine wall with its embryonic–abembryonic axis horizontal. This is relevant considering that subsequent growth of the blastocyst after implantation results in a reorientation of the embryonic–abembryonic axis vertically, such that the egg cylinder forms with the polar trophoblast-derived ectoplacental cone toward the mesometrial (dorsal) side of the uterus and the yolk sac toward the antimesometrial (ven-

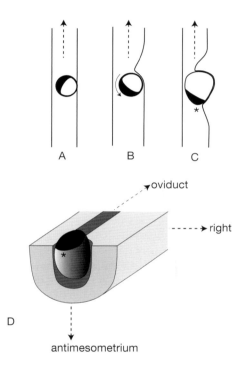

Figure 4. Implantation according to Smith (1980, 1985). *A*, *B*, and *C* represent frontal views of a uterine horn viewed from the mesometrial (dorsal) side with the oviductal end toward the top of the page. All images show type R implantation as described by Smith. Type L implantation would be represented if the same images represented a view from the antimesometrial (ventral) side. (*A*) The bilaterally symmetrical blastocyst is initially horizontal as it moves down the uterine horn. Initial contact with the right wall (or the left wall for type L) is made via the mural trophectoderm so that the embryonic–abembryonic axis is approximately parallel to the left–right axis of the horn. Asymmetric bulging of the uterine wall reorients (*curved arrow*) the embryonic–abembryonic axis to approximately parallel with the long axis of the horn so that the embryonic pole is directed toward the cervical end (or oviductal end for type L). (*C*) With blastocyst growth, the blastocyst cavity develops an asymmetry such that the side in contact with the uterine wall becomes more rounded and the opposite side more pointed. This asymmetry is also reflected by a tilting of the ICM. (*D*) A cut-away representation of a uterine horn shows the expansion of both the blastocyst cavity and the implantation chamber toward the antimesometrial wall of the uterine horn. The tilt thus becomes (partially) reoriented toward both the cervical end and the left wall of the horn. The same side of the blastocyst in *C* and *D* is marked with an asterisk.

tral) side. Whereas gravity serves to orient the blastocyst horizontally with the long axis of the uterine horn, growth of the egg cylinder toward the mesometrial side, at least in the rat, appears to be due to differences in properties of the mesometrial and antimesometrial walls (Kirby et al. 1967). A consequence of egg cylinder growth as described by Smith is that the ectoplacental cone becomes tilted (Fig. 4). Tracing of this tilt to later stages of streak formation led Smith to conclude that the lower side of the tilt corresponds to the side of the streak and thus the future posterior. From these asymmetries, Smith concluded that the anterior–posterior axis relates to an earlier asymmetry observed in the bilaterally symmetrical implanting blastocyst. In an attempt to validate Smith's findings, Gardner and colleagues (1992) recorded the side toward which the ectoplacental cone tilted in dissected streak-stage embryos, but found no correlation with the side of the streak. It is thus unclear whether the asymmetries observed during implantation truly correspond to asymmetry in the ectoplacental cone, or whether later randomly asymmetric growth of the cone masks this. Because Smith's observations were made on static stages of development, it was not possible to prove continuity in the features described. What seems clear is that different parts of the ICM–polar trophectoderm complex of the implanting blastocyst are readily distinguishable in their morphology and their proximity to regions of uterine contact. Thus, a role for the uterus in enhancing asymmetries that relate to the future anterior–posterior axis cannot be discounted. If indeed uterine interactions do contribute, what might the role be for asymmetries observed before implantation? They may influence the polarity of implantation and eventually the asymmetric growth of the egg cylinder. Thus, the mouse embryo does appear to implant with respect to its axis of bilateral symmetry, and the early asymmetries that relate to the animal–vegetal polarity could take a role in asymmetric early postimplantation growth.

A new insight into the relationship between the morphological axes of the embryo and uterus and their link to the development of the anterior–posterior axis came from studies of Mesnard and colleagues (Mesnard et al. 2004). These authors showed that although initially (at E5.0) the egg cylinder is bilaterally symmetrical in shape, with long and short axes perpendicular to the proximal–distal axis, by E5.5 this bilateral symmetry is lost. It appears that when bilateral symmetry in embryo shape reemerges after E5.5, it shows no tendency to be aligned with the uterine axes until shortly before gastrulation. This study (Mesnard et al. 2004) thus was able to demonstrate that the emerging anterior–posterior axis, as defined by molecular markers showing distal-to-anterior cell migration, tends to respect embryo morphology, rather than the uterine axes. Unexpectedly, the emerging anterior–posterior axis is initially aligned more with the short, rather than long, axis of the embryo (Mesnard et al. 2004; Perea-Gomez et al. 2004). Then, whether the embryo is allowed to develop in vitro or in the uterus, the anterior–posterior axis becomes aligned with the long morphological axis of the embryo just prior to gastrulation. Of three mechanisms that could account for this apparent shift in orientation of the anterior–posterior axis—cell migration, spatial change of gene expression, or a change in embryo shape—lineage-tracing studies on E6.0 embryos favor a change in shape accompanied by a restriction of the expression domain of anterior markers.

The fact that embryos cultured in vitro can undertake distal-to-anterior cell movement (Thomas et al. 1998; Weber et al. 1999) and also align their axes with the long morphological axis of the embryo offers some additional insight into development of the anterior–posterior axis. It suggests that the initial orientation and then apparent "reorientation" of the anterior–posterior axis can occur independently of the uterus. However, although the embryo might have the intrinsic potential to position the anterior–posterior axis, a role for the uterus cannot be fully excluded. Indeed, a mechanism must exist for the embryo to become aligned with respect to the uterine axes from the time of gastrulation. The study of Mesnard and colleagues (Mesnard et al. 2004) suggests a fine interdependence between the expression of anterior–posterior markers, the shape of the embryo, and the axes of the uterus.

It is unclear whether bilateral symmetries observed at various stages of development relate to each other and to formation of the anterior–posterior axis. The embryo flattening at E5.0 may correspond to the bilateral symmetry ascribed by Smith (1985) to the implanting blastocyst. Smith claimed that the asymmetries she saw were directly related to the final orientation of the anterior–posterior axis. However, the complexity of cell movement and growth from the blastocyst to the egg cylinder stage revealed by the studies of Weber and colleagues (1999) indicates that this relationship might not be so straightforward. Although these authors showed that progeny of ICM cells from opposite ends of the blastocyst axis of bilateral symmetry tended to differ in their spatial distribution along the proximal–distal axis, they did not occupy exclusively anterior or posterior positions. Thus, it is difficult to relate these findings to each other, since we do not yet know how bilateral symmetry before implantation relates to that after implantation and, in turn, to that of the E5.0 egg cylinder as described by Mesnard and colleagues (Mesnard et al. 2004). The latter study adds a further complication that should be taken into account in understanding these relationships: namely, the embryo passes through an intermediate stage of almost radial symmetry and we have no molecular markers at these stages to which the shape changes can be referred.

From all that we have learned thus far (notwithstanding regulative abilities), in normal, unperturbed development, the mouse embryo does not establish its pattern randomly. Rather, its "patterning" program appears to follow early spatial cues that we are only beginning to understand. It has often been argued that the full developmental capacity of an individual 2-cell blastomere (Tarkowski 1959), and occasionally even of both 2-cell-stage blastomeres (Tsunoda and McLaren 1983; Papaioannou et al. 1989), precludes the possibility of asymmetrically localized determinants influencing later specification of axes. But why should it do so? If, for example, one considers that the specification depends on a gradient of such a determinant, dividing an embryo in two could still preserve such a gradient in both resultant embryos. In seeking an explanation for the distinction between the very early appearance of the anterior–posterior axis in amphibians, and its relatively late appearance in amniotes, it can be argued that the longer period from zygote to gastrulation in amniotes, involving extensive growth, favors a more labile and gradual mechanism for axis determination. Thus, the initial presence of an asymmetric determinant can be considered as a "penciling in" of the future body plan, with definitive specification occurring much later in development.

REFERENCES

Ang S.L., Conlon R.A., Jin O., and Rossant J. 1994. Positive and negative signals from mesoderm regulate the expression of mouse Otx2 in ectoderm explants. *Development* **120:** 2979–2989.

Bachvarova R.F. 1999. Establishment of anterior-posterior polarity in avian embryos. *Curr. Opin. Genet. Dev.* **9:** 411–416.

Bachvarova R.F., Skromne I., and Stern C.D. 1998. Induction of primitive streak and Hensen's node by the posterior marginal zone in the early chick embryo. *Development* **125:** 3521–3534.

Beddington R.S. and Robertson E.J. 1999. Axis development and early asymmetry in mammals. *Cell* **96:** 195–209.

Bennett J. 1982. Sperm entry point is related to early division of mouse blastomeres. *J. Cell Biol.* **95:** 163a.

Bertocchini F. and Stern C.D. 2002. The hypoblast of the chick embryo positions the primitive streak by antagonizing nodal signaling. *Dev. Cell* **3:** 735–744.

Ciemerych M.A., Mesnard D., and Zernicka-Goetz M. 2000. Animal and vegetal poles of the mouse egg predict the polarity of the embryonic axis, yet are nonessential for development. *Development* **127:** 3467–3474.

Ding J., Yang L., Yan Y.T., Chen A., Desai N., Wynshaw-Boris A., and Shen M.M. 1998. Cripto is required for correct orientation of the anterior-posterior axis in the mouse embryo. *Nature* **395:** 702–707.

Eyal-Giladi H. 1984. The gradual establishment of cell commitments during the early stages of chick development. *Cell Differ.* **14:** 245–255.

Foley A.C., Skromne I., and Stern C.D. 2000. Reconciling different models of forebrain induction and patterning: a dual role for the hypoblast. *Development* **127:** 3839–3854.

Fujimori T., Kurotaki Y., Miyazaki J.I., and Nabeshima Y.I. 2003.

Analysis of cell lineage in two- and four-cell mouse embryos. *Development.* **130:** 5113–5122.

Garbutt C.L., Chisholm J.C., and Johnson M.H. 1987. The establishment of the embryonic-abembryonic axis in the mouse embryo. *Development* **100:** 125–134.

Gardner R.L. 1997. The early blastocyst is bilaterally symmetrical and its axis of symmetry is aligned with the animal-vegetal axis of the zygote in the mouse. *Development* **124:** 289–301.

———. 2001. Specification of embryonic axes begins before cleavage in normal mouse development. *Development* **128:** 839–847.

Gardner R.L. and Cockroft D.L. 1998. Complete dissipation of coherent clonal growth occurs before gastrulation in mouse epiblast. *Development* **125:** 2397–2402.

Gardner R.L., Meredith M.R., and Altman D.G. 1992. Is the anterior-posterior axis of the fetus specified before implantation in the mouse? *J. Exp. Zool.* **264:** 437–443.

Gerhart J., Danilchik M., Doniach T., Roberts S., Rowning B., and Stewart R. 1989. Cortical rotation of the *Xenopus* egg: Consequences for the anteroposterior pattern of embryonic dorsal development. *Development* **107:** 37–51.

Gräper L. 1929. Die Primitiventwicklung des Hühnchens nach stereokipnematographischen Untersuchungen kontrolliert durch vitale Farbmarkierung und verglichennit der Entwicklung anderer Wirbeltiere. *Roux's Arch. Entw. Mech. Organ.* **116:** 382–429.

Kelly S.J., Mulnard J.G., and Graham C.F. 1978. Cell division and cell allocation in early mouse development. *J. Embryol. Exp. Morphol.* **48:** 37–51.

Kimura C., Yoshinaga K., Tian E., Suzuki M., Aizawa S., and Matsuo I. 2000. Visceral endoderm mediates forebrain development by suppressing posteriorizing signals. *Dev. Biol.* **225:** 304–321.

Kirby D.R., Potts D.M., and Wilson I.B. 1967. On the orientation of the implanting blastocyst. *J. Embryol. Exp. Morphol.* **17:** 527–532.

Kochav S. and Eyal-Giladi H. 1971. Bilateral symmetry in chick embryo determination by gravity. *Science* **171:** 1027–1029.

Lawson K.A., Dunn N.R., Roelen B.A., Zeinstra L.M., Davis A.M., Wright C. V., Korving J.P., and Hogan B.L. 1999. Bmp4 is required for the generation of primordial germ cells in the mouse embryo. *Genes Dev.* **13:** 424–436.

McLaren A. 1976. *Mammalian chimaeras.* Cambridge University Press, Cambridge, United Kingdom

Mesnard D., Filipe M., Belo J.A., and Zernicka-Goetz M. 2004. The anterior-posterior axis emerges respecting the morphology of the mouse embryo that changes and aligns with the uterus before gastrulation. *Curr. Biol.* **14:** 184–196.

Papaioannou V.E., Mkandawire J., and Biggers J.D. 1989. Development and phenotypic variability of genetically identical half mouse embryos. *Development* **106:** 817–827.

Perea-Gomez A., Lawson K.A., Rhinn M., Zakin L., Brulet P., Mazan S., and Ang S.L. 2001. Otx2 is required for visceral endoderm movement and for the restriction of posterior signals in the epiblast of the mouse embryo. *Development* **128:** 753–765.

Perea-Gomez A., Camus A., Moreau A., Grieve K., Moneron G., Dubois A., Cibert C., and Collignon J. 2004. Initiation of gastrulation in the mouse embryo is preceded by a shift in the orientation of the anterior-posterior axis. *Curr. Biol.* **14:** 197–207.

Perea-Gomez A., Vella F.D., Shawlot W., Oulad-Abdelghani M., Chazaud C., Meno C., Pfister V., Chen L., Robertson E., Hamada H. et al. 2002. Nodal antagonists in the anterior visceral endoderm prevent the formation of multiple primitive streaks. *Dev. Cell* **3:** 745–756.

Piotrowska K. and Zernicka-Goetz M. 2001. Role for sperm in spatial patterning of the early mouse embryo. *Nature* **409:** 517–521.

———. 2002. Early patterning of the mouse embryo - contributions of sperm and egg. *Development* **129:** 5803–5813.

Piotrowska K., Wianny F., Pedersen R.A., and Zernicka-Goetz M. 2001. Blastomeres arising from the first cleavage division have distinguishable fates in normal mouse development. *Development* **128:** 3739–3748.

Plusa B., Piotrowska K., and Zernicka-Goetz M. 2002a. Sperm entry position provides a surface marker for the first cleavage plane of the mouse zygote. *Genesis* **32:** 193–198.

Plusa B., Grabarek J.B., Piotrowska K., Glover D.M., and Zernicka-Goetz M. 2002b. Site of the previous meiotic division defines cleavage orientation in the mouse embryo. *Nat. Cell Biol.* **4:** 811–815.

Selwood L. 1992. Mechanisms underlying the development of pattern in marsupial embryos. *Curr. Top. Dev. Biol.* **27:** 175–233.

Simeone A., Acampora D., Mallamaci A., Stornaiuolo A., D'Apice M.R., Nigro V., and Boncinelli E. 1993. A vertebrate gene related to orthodenticle contains a homeodomain of the bicoid class and demarcates anterior neuroectoderm in the gastrulating mouse embryo. *EMBO J.* **12:** 2735–2747.

Smith L.J. 1980. Embryonic axis orientation in the mouse and its correlation with blastocyst relationships to the uterus. Part 1. Relationships between 82 hours and 4 1/4 days. *J. Embryol. Exp. Morphol.* **55:** 257–277.

———. 1985. Embryonic axis orientation in the mouse and its correlation with blastocyst relationships to the uterus. II. Relationships from 4 1/4 to 9 1/2 days. *J. Embryol. Exp. Morphol.* **89:** 15–35.

Spindle A. 1982. Cell allocation in preimplantation mouse chimeras. *J. Exp. Zool.* **219:** 361–367.

Stanley E.G., Biben C., Allison J., Hartley L., Wicks I.P., Campbell I.K., McKinley M., Barnett L., Koentgen F., Robb L., et al. 2000. Targeted insertion of a lacZ reporter gene into the mouse Cer1 locus reveals complex and dynamic expression during embryogenesis. *Genesis* **26:** 259–264.

Stern C.D. 2001. Initial patterning of the central nervous system: How many organizers? *Nat. Rev. Neurosci.* **2:** 92–98.

Stern C.D. and Ireland G.W. 1981. An integrated experimental study of endoderm formation in avian embryos. *Anat. Embryol.* **163:** 245–263.

Surani M.A. and Barton S.C. 1984. Spatial distribution of blastomeres is dependent on cell division order and interactions in mouse morulae. *Dev. Biol.* **102:** 335–343.

Tarkowski A.K. 1959. Experiments on the development of isolated blastomeres of mouse eggs. *Nature* **184:** 1286–1287.

Tarkowski A.K. and Wroblewska J. 1967. Development of blastomeres of mouse eggs isolated at the 4- and 8-cell stage. *J. Embryol. Exp. Morphol.* **18:** 155–180.

Thomas P.Q., Brown A., and Beddington R.S. 1998. Hex: A homeobox gene revealing peri-implantation asymmetry in the mouse embryo and an early transient marker of endothelial cell precursors. *Development* **125:** 85–94.

Tian E., Kimura C., Takeda N., Aizawa S., and Matsuo I. 2002. Otx2 is required to respond to signals from anterior neural ridge for forebrain specification. *Dev. Biol.* **242:** 204–223.

Tsunoda Y. and McLaren A. 1983. Effect of various procedures on the viability of mouse embryos containing half the normal number of blastomeres. *J. Reprod. Fertil.* **69:** 315–322.

Vakaet L. 1970. Cinephotomicrographic investigations of gastrulation in the chick blastoderm. *Arch. Biol.* **81:** 387–426.

Weber R.J., Pedersen R.A., Wianny F., Evans M.J., and Zernicka-Goetz M. 1999. Polarity of the mouse embryo is anticipated before implantation. *Development* **126:** 5591–5598.

Winklbauer R. and Schurfeld M. 1999. Vegetal rotation, a new gastrulation movement involved in the internalization of the mesoderm and endoderm in *Xenopus. Development* **126:** 3703–3713.

Zernicka-Goetz M. 1999. Green fluorescent protein: A new approach to understanding spatial patterning and cell fate in early mammalian development. In *Cell lineage and fate determination* (ed. S.A. Moody), pp. 521–527. Academic Press, San Diego, California.

———. 2002. Patterning of the embryo: The first spatial decisions in the life of a mouse. *Development* **129:** 815–829.

Zernicka-Goetz M., Pines J., McLean Hunter S., Dixon J.P., Siemering K.R., Haseloff J., and Evans M.J. 1997. Following cell fate in the living mouse embryo. *Development* **124:** 1133–1137.

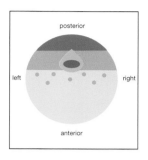

VERTEBRATE MESODERM INDUCTION
FROM FROGS TO MICE

D. Kimelman and C. Bjornson

Department of Biochemistry, University of Washington, Seattle, Washington 98195

INTRODUCTION

The mesoderm comprises the middle of the three primary germ layers and contributes to a wide variety of different tissue types, including the muscle, the blood, the heart, and the dermis. In the vertebrate embryo, there are generally two separate, but connected, components to mesoderm formation. The mesoderm is first induced, or differentiated, from the rest of the embryo by intercellular signaling, and then the mesoderm is polarized through the generation of an organizing center. Although the concepts of mesoderm induction and organizer formation have been around for many decades, the molecular nature of these processes has only been determined in the last 15 years. This review focuses on the initial steps in mesoderm formation from the perspectives of four currently studied model organisms—*Xenopus*, zebrafish, chick, and mouse.

A BRIEF HISTORY OF MESODERM INDUCTION

Because many (although by no means all) of the key initial discoveries in the field of mesoderm induction were made in the frog, this review begins with studies in *Xenopus*. In the frog embryo, the mesoderm forms at the equator of the embryo with Spemann's organizer located on what is classically called the dorsal side (Fig. 1A). The organizer plays a critical role from the gastrula stage onward as a source of intercellular signals that inhibit the bone morphogenetic protein (BMP) and Wnt signaling pathways (Chapters 26, 29, and 36), and it will eventually form head mesoderm and notochord as the embryo develops.

In 1969, Nieuwkoop performed seminal experiments that introduced the concept of mesoderm induction in frog. Whereas Spemann had considered the organizer as the first signaling center in the frog embryo, Nieuwkoop demonstrated that the organizer, as well as the rest of the mesoderm, is induced by signals (Nieuwkoop 1969a). In Nieuwkoop's experiment, the upper part of the embryo (the "animal cap") was separated from the rest of the embryo, as was the vegetal cap (Fig. 1B). In isolation, the two explants developed as ectoderm and endoderm, respectively, exactly as predicted from the fate map (Fig. 1A). When combined, however, the vegetal cap induced mesoderm to form in the adjacent region of the animal cap. Nieuwkoop also demonstrated that the dorsal vegetal region induced dorsal mesoderm, whereas the ventral vegetal region induced ventral mesoderm (Nieuwkoop 1969b). These and later studies were elegantly brought together in the 3-signal model (Smith and Slack 1983; Dale and Slack 1987). In this view, a signal emanating from the most dorsal part of the vegetal pole induces the formation of the organizer, whereas the rest of the vegetal pole induces homogeneous non-organizer mesoderm (Fig. 1C). Beginning at the late-blastula stages, the organizer emits a gradient of signals that pattern the non-organizer mesoderm, leading to a variety of different mesodermal fates.

Although Nieuwkoop's results and the 3-signal model provided an appealing framework for mesoderm induc-

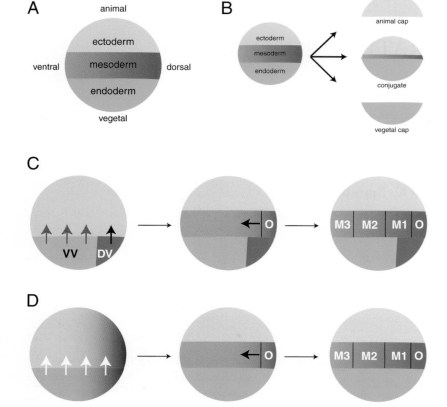

Figure 1. History of mesoderm induction. (*A*) Fate map of a pregastrula *Xenopus* embryo. (*B*) Nieuwkoop's experiment. Only conjugates of the animal cap and vegetal cap form mesoderm. (*C*) In the 3-signal model, the vegetal pole emits a dorsal vegetal (DV) signal dorsally that induces the organizer (O) and a ventral vegetal (VV) signal elsewhere that induces ventral mesoderm. The organizer then emits signals that pattern the nearby mesoderm. (*D*) In the synergistic model, all of the vegetal pole emits a mesoderm-inducing signal. A dorsally located Wnt-type signal (*blue*) modifies the response to the signal to produce the organizer. The tissues are colored: ectoderm (*yellow*), mesoderm (*red → pink* from dorsal → ventral), and endoderm (*green*).

tion, they raised a question that had arisen in Spemann's era: What are the signals? Although Spemann seemed very uninterested in the "chemicals" involved in inductive processes, an enormous amount of time went into the search for these molecules over the course of many decades (Grunz 1996). The big breakthrough came in the mid-1980s. Instead of trying to biochemically purify the signal from embryos, as others had done unsuccessfully, Smith gambled that *Xenopus* embryonic cell lines might produce the same signal found in the embryo. His demonstration that the XTC cell line secreted a mesoderm-inducing factor was a major advance, and it led to speculation that soluble, secreted "growth" factors might be mesoderm-inducing signals (Smith 1987). This led two groups to discover that fibroblast growth factor (FGF; Chapter 32) could act as a mesoderm-inducing factor (Kimelman and Kirschner 1987; Slack et al. 1987), and others found that members of the transforming growth factor-β (TGF-β; Chapters 34 and 35) superfamily were also very potent mesoderm inducers (Rosa et al. 1988; Thomsen et al. 1990). Eventual purification of the XTC factor demonstrated that it was activin, a member of the TGF-β superfamily (Smith et al. 1990).

The discovery of two families of molecules that can induce mesoderm raised the question, Are TGF-β and FGF equivalent? Using dominant-negative receptors, it was shown that whereas the TGF-βs are needed for all mesoderm formation, FGF is needed in frog only for the formation of the posterior mesoderm (Amaya et al. 1991; Hemmati-Brivanlou and Melton 1992). Furthermore, these results indicated that there is a major difference between the anterior (head) and posterior (trunk and tail) mesoderm: TGF-βs can induce anterior mesoderm without FGF signaling, whereas induction of posterior mesoderm by the TGF-β pathway needs an active FGF signaling pathway (Cornell and Kimelman 1994; LaBonne and Whitman 1994).

A final pathway involved in mesoderm induction is the Wnt pathway (Chapter 36), and more specifically, the branch of the Wnt pathway that operates through β-catenin stabilization (called the canonical Wnt pathway). As first shown by McMahon and Moon (1989), Wnts can induce the formation of the organizer, suggesting that they play a key role in regulating where the organizer forms. Many studies supported this idea, including those demonstrating that β-catenin is required to form the endogenous dorsoventral axis (Heasman et al. 1994, 2000) and that β-catenin is stabilized early in embryogenesis in the region where the organizer will form (Schneider et al. 1996; Larabell et al. 1997). The Wnt/β-catenin pathway differs from the FGF and TGF-β pathways in that it does not directly induce mesoderm, but instead acts synergistically with other factors to regulate organizer formation (Christian et al. 1992).

The demonstration that Wnts act synergistically with mesoderm-inducing factors led to a modification of the 3-signal model. In this view, the mesoderm was proposed to be induced throughout the equator by vegetally expressed mesoderm-inducing signals of the FGF and TGF-β families (Fig. 1D: Kimelman et al. 1992). On the dorsal side, the Wnt pathway was proposed to act together with the vegetal signals to change the type of mesoderm induced to that of organizer tissue, which subsequently patterns the rest of the mesoderm.

MESODERM INDUCTION IN FROG
THE CURRENT VIEW

The restriction of the mesoderm-inducing signal to the vegetal hemisphere suggested that this signal was spatially restricted during oogenesis. Although initial attention focused on vegetally localized *vg1* (Weeks and Melton 1987), a TGF-β member now implicated in establishing the left–right axis (Hyatt et al. 1996) (Chapter 35), the more likely mesoderm-inducing factors are TGF-β family members belonging to the *Nodal* subfamily: *Xnr1, Xnr2, Xnr4, Xnr5,* and *Xnr6* (Chapter 34). The Xnrs are transcribed in the vegetal region in response to the T-box transcription factor VegT (Fig. 2) (Clements et al. 1999; Kofron et al. 1999; Agius et al. 2000; Hyde and Old 2000; Takahashi et al. 2000)

(Chapter 41). Maternal *vegT* transcripts are localized to the vegetal region of the *Xenopus* embryo (Fig. 2) (Lustig et al. 1996; Stennard et al. 1996; Zhang and King 1996; Horb and Thomsen 1997), and elimination of *vegT* transcripts eliminates the mesoderm (Zhang et al. 1998). An additional important target of VegT is *derriere*, a vegetally expressed member of the TGF-β family that is important for the induction of posterior mesoderm (Kofron et al. 1999; Sun et al. 1999; White et al. 2002). The Xnrs and Derriere, together with the ubiquitously expressed Activin (Marchant et al. 1998), are responsible for mesoderm formation.

The Xnrs are expressed at higher levels dorsally than ventrally. The increased dorsal levels of Xnr expression are due to dorsally stabilized β-catenin, which most likely acts as a direct transcriptional activator of Xnr gene expression (Fig. 2) (Agius et al. 2000; Hyde and Old 2000; Takahashi et al. 2000; Xanthos et al. 2002; Yang et al. 2002). Stabilized β-catenin has additional targets, including the homeobox gene *siamois*, a transcription factor that activates organizer genes (Fig. 2) (Lemaire et al. 1995; Brannon et al. 1997) (Chapter 45). The combination of high-level Xnr signaling and the expression of genes like *siamois* leads to the formation of the organizer in the dorsal equatorial region (Fig. 2).

FGF signaling is currently understood to act subsequent to the Xnr signals within the posterior mesoderm in frog, most likely to keep a mesodermal progenitor population alive as the trunk and tail grow out (Chapter 32). Although FGF was identified as a factor that could induce mesoderm, this may instead reflect its normal ability to maintain the mesodermal precursor population through the regulation of key mesodermal genes such as *Brachyury* (Amaya et al. 1993; Northrop and Kimelman 1994) (Chapter 41). FGFs are also important for the terminal differentiation of mesodermal cells in specific tissues such as the muscle, a phenomenon known as the "community effect" (Standley et al.

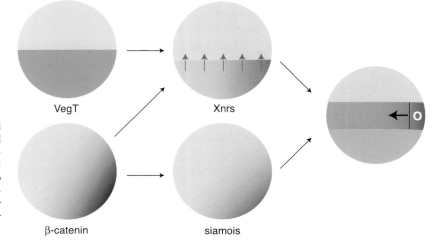

Figure 2. Mesoderm induction in frog. Maternal *vegT* (*teal*) transcripts are localized to the vegetal pole, whereas β-catenin (*dark blue*) is dorsally stabilized. VegT and β-catenin activate *Xnr* (*orange*) expression in a gradient as shown. β-Catenin also activates specific organizer genes like *siamois* (*purple*). The Xnrs induce the mesoderm and, together with factors such as Siamois, activate the organizer (O). The tissues are colored as in Fig. 1.

2001). Because the focus of this review is on the initial steps in mesoderm formation, the role of FGF in mesodermal patterning will not be dealt with further here.

MESODERM INDUCTION IN FISH

The zebrafish and frog pre-gastrula fate maps resemble one another because mesoderm is derived from blastomeres at the equator of each embryo at this stage; however, there are a few important architectural differences. First, the zebrafish embryo develops on top of an uncleaved yolk cell. Second, an extraembryonic structure called the yolk syncytial layer (YSL), which is a syncytium of cytoplasm and nuclei that forms during the 10th cleavage (Kimmel and Law 1985), forms between the yolk cell and the embryonic blastomeres. Last, whereas the endoderm is separate from the mesoderm in the frog fate map (Fig. 1A), in fish, the prospective endoderm is intermixed with the prospective mesoderm in the region closest to the yolk cell (Fig. 3A) (Kimmel et al. 1990).

Explant experiments have demonstrated that the YSL is a source of mesoderm- and endoderm-inducing signals (Mizuno et al. 1996; Rodaway et al. 1999), and elimination of transcripts from the YSL eliminates the formation of all endoderm and all but the most dorsal mesoderm (Chen and Kimelman 2000). The YSL signal acts upstream of the zebrafish Nodals (Chapter 34), *squint* and *cyclops* (Fig. 3B) (Chen and Kimelman 2000), and possibly the pan-mesodermal gene *no tail*, but the identity of this signal is not yet known. Squint and Cyclops both act as mesoderm inducers (Erter et al. 1998; Feldman et al. 1998; Rebagliati et al. 1998). Consistent with their overlapping early expression patterns, their early function is mostly redundant, since a mutation in either gene has only little or no effect on mesoderm formation (Feldman et al. 1998; Gritsman et al. 1999). When both factors are nonfunctional, however, the mesoderm is formed only on the ventral side of the early embryo, which will become tail mesoderm. Therefore, Nodal signaling is critical for the proper formation of mesoderm in both fish and frog.

Similar to the situation found in frog, the β-catenin pathway plays an essential role in establishing the fish organizer. The stabilization of β-catenin in the dorsal marginal blastomeres (Schneider et al. 1996) activates genes such as *bozozok* (Fig. 3B) (Fekany et al. 1999; Koos and Ho 1999; Kelly et al. 2000; Shimizu et al. 2000; Ryu et al. 2001), and is likely to also regulate dorsal *squint* expression (Chen and Kimelman 2000; Kelly et al. 2000). Both these functions of β-catenin act in parallel to promote the formation of the zebrafish organizer (Fig. 3B) (Shimizu et al. 2000; Sirotkin et al. 2000). As in *Xenopus*, the organizer acts during the gastrula stages to pattern the mesoderm along the dorsoventral axis.

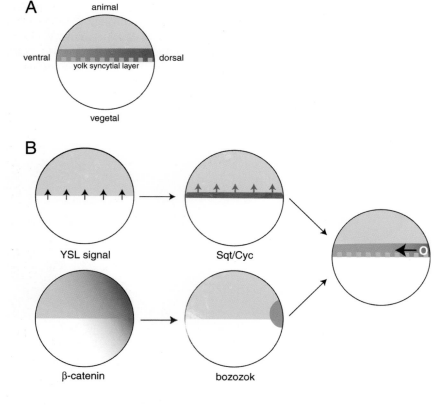

Figure 3. Mesoderm induction in fish. (*A*) Fate map of the pre-gastrula embryo. Cells closest to the yolk syncytial layer (YSL) are a mixture of mesoderm and endoderm. (*B*) An unknown signal (*black arrows*) from the YSL activates the expression of *squint* and *cyclops* (Sqt/Cyc; *orange*), which induce the mesoderm (and endoderm) at the equator, and act with the β-catenin (*blue*)-induced gene *bozozok* (*teal*) to establish the organizer (O) on the dorsal side. The tissues are colored as in Fig. 1. The yolk and YSL are white.

Although Nodals and β-catenin play important roles in mesoderm induction in both fish and frog, the initial step in this process appears to be completely different. Whereas maternally localized VegT is critically required for mesoderm formation in frog as an upstream transcriptional activator of the Xnrs, the zebrafish ortholog of *vegT* is not maternally expressed and instead has a later role in the mesoderm (Griffin et al. 1998). Since no maternal T-box gene has been identified as a likely activator of the YSL signal, a unique mechanism appears to control the YSL signal, which then activates *squint* and *cyclops* expression (Fig. 3) (Chen and Kimelman 2000). How a signal is even restricted to the YSL is far from clear, since the YSL does not separate from the embryonic blastomeres until the 10th cleavage, and the allocation of nuclei and cytoplasm between the YSL and the overlying blastomeres is random (Kimmel and Law 1985). It is possible that a transcription factor or mRNA for the YSL signal is tethered to the yolk until the 10th cleavage, but there is no evidence to support such a mechanism.

MESODERM INDUCTION IN CHICK

The mesoderm in the chick embryo forms from the epiblast, which is an epithelial sheet overlying a small part of the large yolk mass (Lawson and Schoenwolf 2001) (Chapter 15). A second layer, the hypoblast, is extraembryonic and forms underneath the epiblast from cells delaminating from the epiblast and migrating in from the edges, forming a bilaminar structure. The beginning of gastrulation is marked by a thickening of the posterior epiblast to produce the primitive streak, a region where cells undergo an epithelial to mesenchymal transition (EMT) and dive between the epiblast and hypoblast, then subsequently move away from the site of ingression. Cells that undergo the EMT through the primitive streak give rise to the mesoderm and endoderm, with the anterior end becoming prechordal mesoderm and axial tissue, whereas the more posterior end gives rise to paraxial and lateral fates (Fig. 4A) (Eyal-Giladi and Kochav 1976).

As in fish and frog, Nodal signaling is likely important for mesoderm induction since the chick *Nodal* (*cNodal*) gene is expressed in the early mesodermal territory (Fig. 4B) (Lawson et al. 2001), and TGF-β factors can induce mesoderm in the chick epiblast (Mitrani and Shimoni 1990), similar to the studies in the frog explants. The expression of *cNodal* appears to be under the control of the TGF-β factor Vg1 (Chapter 35), because *vg1* is expressed in the posterior region before *cNodal* (Seleiro et al. 1996; Shah et al. 1997; C. Stern, pers. comm.) and because *vg1*-expressing cells can induce the expression of ectopic *cNodal* in the epiblast (Skromne and Stern 2001). As in frog and fish, the formation of the chick organizer (the primitive streak) requires a Wnt/β-catenin signal acting together with a TGF-β signal (Skromne and Stern 2001).

In frog and fish, β-catenin is stabilized on the dorsal side of the embryo by maternal factors to break radial symmetry and establish the organizer (Chapters 23 and 36). In chick, *wnt8C* is zygotically expressed in the posterior region (Hume and Dodd 1993; Skromne and Stern 2001), indicating that β-catenin in the chick embryo is zygotically regulated. In addition, since *wnt8C* expression is restricted to the margins of the epiblast (Fig. 4B), and Vg1 requires Wnt signaling to induce *cNodal* expression, the marginal expression of *wnt8C* limits the zone in which the mesoderm can form (Hume and Dodd 1993; Skromne and Stern 2001).

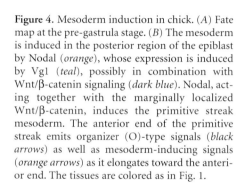

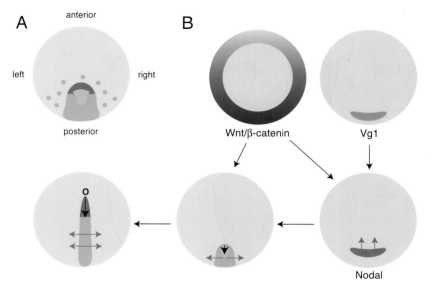

Figure 4. Mesoderm induction in chick. (*A*) Fate map at the pre-gastrula stage. (*B*) The mesoderm is induced in the posterior region of the epiblast by Nodal (*orange*), whose expression is induced by Vg1 (*teal*), possibly in combination with Wnt/β-catenin signaling (*dark blue*). Nodal, acting together with the marginally localized Wnt/β-catenin, induces the primitive streak mesoderm. The anterior end of the primitive streak emits organizer (O)-type signals (*black arrows*) as well as mesoderm-inducing signals (*orange arrows*) as it elongates toward the anterior end. The tissues are colored as in Fig. 1.

The combination of Wnt and TGF-β-type signaling results in the formation of the primitive streak, which expresses at the anterior end the same factors that are found in the fish and frog organizer, resulting in the gastrula stage patterning of the mesoderm (Chapter 26). During the gastrula stages, the streak expands toward the anterior region and then contracts, emitting organizer-type signals during these stages (Chapter 15). Unlike the frog, where the ability of mesoderm-inducing factors to function drops off during the gastrula stages (Green et al. 1990), the chick primitive streak continues to express *cNodal* and *vg1* (Lawson et al. 2001), and epiblast cells remain responsive to the inducing factors (Stern et al. 1995). These results suggest that the period of mesoderm formation in chick continues throughout the gastrula stages (Fig. 4B).

During the period the egg is in the oviduct, it rotates along its long axis at the rate of 10–12 turns per hour, and this establishes the anterior–posterior axis (Eyal-Giladi 1997) (Chapter 24). How this leads to the restricted expression of *vg1* and *wnt8C* and the consequent formation of mesoderm in the posterior regions is not known. Nor is it known what restricts the expression of *vg1* and *wnt8C* to the margins of the epiblast, which is responsible for limiting the prospective mesodermal territory to the marginal zone.

MESODERM INDUCTION IN MOUSE

Perhaps not surprisingly, the factors discussed above are also involved in mesoderm induction in the mouse, although there are a number of intriguing differences. Unlike other species, a large portion of the mouse blastula becomes extraembryonic tissue, and only a subset of the blastula cells become the definitive embryo. At preimplantation stages, the mouse blastocyst resembles a sphere that contains two cell populations, the trophectoderm and the inner cell mass. After implantation, the embryo begins to resemble a cup, with the top of the cup, which is derived from the trophectoderm lineage, giving rise to the so-termed extraembryonic ectoderm that will ultimately form the placenta. The bottom of the cup is ensheathed by extraembryonic visceral endoderm derived from the inner cell mass (Fig. 5A). The inner layer of the cup, which also forms from the inner cell mass, will become the epiblast and will form the ectoderm, endoderm, and mesoderm of the embryo (Fig. 5A,B). The mesoderm forms on one side of the cup, which will become the posterior end of the embryo, with the opposite side of the cup producing anterior fates (Fig. 5A).

As in the other species, Nodal plays a key role in forming the mesoderm (Chapter 34). Embryos lacking *Nodal*,

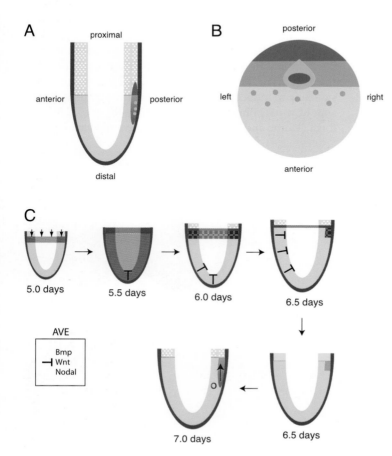

Figure 5. Mesoderm induction in mouse. (*A*) Fate map at the pre-gastrula (6.5 days postcoitum [dpc]) stage shown as a section through the middle of the embryo. (*B*) The pre-gastrula fate map shown with the embryo unfolded and flattened into a 2-dimensional disc. (*C*) At 5.0 dpc, *Nodal* may be induced in the proximal epiblast by an unknown signal. At 5.5 dpc, *Nodal* expression expands throughout the epiblast. The anterior visceral endoderm (AVE, indicated by T) begins to secrete Nodal inhibitors restricting *Nodal* to the proximal epiblast at 6.0 dpc. The movement of the AVE cells further restricts *Nodal* to the future posterior side by 6.5 dpc. Nodal, together with BMP4 (*purple*) and Wnt3 (*dark blue*), induces mesoderm. The AVE also secretes inhibitors of Wnts and BMPs, further restricting the mesoderm-inducing function of these factors to the posterior side of the embryo. The organizer (O) is located at the distal end of the primitive streak and emits signals (*black arrows*). The tissues are colored: definitive ectoderm (*yellow*), extraembryonic ectoderm (*speckled bright yellow*), dorsal mesoderm (*red*), ventral lateral mesoderm (*pink*), extraembryonic mesoderm (*purple*), definitive endoderm (*green*), and visceral (extraembryonic) endoderm (*brown*). (*B,* Adapted from Stern 1992.)

which was first discovered in the mouse, fail to form most, but not all, mesoderm (Zhou et al. 1993; Conlon et al. 1994). The regulation of *Nodal* expression is highly dynamic in the mouse, and it is only restricted to the mesodermal territory at the start of gastrulation. In a 5.0-day post-coitum (dpc) mouse, *Nodal* is first expressed in the proximal epiblast (Brennan et al. 2001), possibly in response to a signal from the overlying extraembryonic ectoderm (proximal and distal are defined relative to the extraembryonic ectoderm; Fig. 5C); however, the nature of the signal is unknown. Although Vg1 has been proposed to be the initiator of *Nodal* expression in chick embryos (Skromne and Stern 2001), deletion of the closest relative of *vg1* in mouse embryos disrupts left/right axis formation but has no effect on mesoderm induction (Wall et al. 2000). By 5.5 dpc, *Nodal* is found through the epiblast and visceral endoderm (Varlet et al. 1997), potentially due to Nodal activating its own expression from neighboring cells in an autocatalytic loop.

Once *nodal* expression fills the epiblast and visceral endoderm, the anterior visceral endoderm (AVE), which is required for the orientation of the anterior–posterior axis, is established at the distal tip of the visceral endoderm (Fig. 5C). The AVE begins to secrete the Nodal inhibitors Cerberus-like (Cerl) and Lefty1, which are proposed to block the expression of *Nodal* by inhibiting the autocatalytic loop (Perea-Gomez et al. 2002). The AVE begins to migrate proximally to establish the future anterior side of the embryo at 5.5–6.0 dpc (Chapter 16), and the zone of Nodal-inhibitory activity moves with the AVE, thus progressively restricting *Nodal* expression and function to the proximal posterior region (Fig. 5C). Interestingly, Nodal inhibitors secreted by the extraembryonic hypoblast also limit the region in which Nodal functions in chick (Bertocchini and Stern 2002).

Despite the fact that *Nodal* is at one point expressed throughout much of the embryo, mesoderm only forms in the posterior region, suggesting that other factors regulate mesoderm formation in mouse. One of these factors is BMP (Chapter 33), which has an early unique role in mouse. In frog and fish, where BMP has been extensively studied, it plays a critical role in dorsoventral and anterior–posterior patterning (Chapters 26 and 27). In mouse, BMP signaling is critically required for all mesoderm induction (Mishina et al. 1995; Winnier et al. 1995), which is clearly not the case in fish and frog. *Bmp4* is expressed in the extraembryonic ectoderm as early as 5.5 dpc (Fig. 5C) (Lawson et al. 1999), and signals to the underlying *Nodal*-expressing cells, setting up a signaling cycle that maintains the expression of *Bmp4* (Brennan et al. 2001). Unfortunately, the targets of BMP signaling are not yet known. BMP4 could be required for the expression of key mesodermal genes, for the expression of *Nodal*, or for the

expression of the other key mesodermal gene, *Wnt3* (Fig. 5C).

Wnt3 is expressed in the proximal epiblast at 6.25 dpc and then in the proximal posterior-lateral region at 6.5 dpc (Fig. 5C) (Liu et al. 1999). Overexpression of a Wnt in mouse embryos leads to ectopic primitive streaks (Popperl et al. 1997), as was also found in a mouse mutant that lacks a key inhibitor of the canonical Wnt pathway (Zeng et al. 1997), paralleling the observations in fish, frog, and chick that the Wnt/β-catenin pathway regulates the early embryonic axis (Chapter 36). However, Wnts also have a fundamental role in forming mesoderm in the mouse, since the *Wnt3* knockout does not express *Nodal* at 7.5 dpc, and fails to form mesoderm (Liu et al. 1999). The importance of Wnt3 in mesoderm induction is likely related to *Nodal* maintenance during streak formation and not the early establishment of the AVE by Nodal signaling, since markers of the AVE are still expressed in *Wnt3*⁻/⁻ animals (Liu et al. 1999). This suggests that Wnts are not required for the early *Nodal*-dependent patterning of the AVE, but are required to maintain *Nodal* expression in the primitive streak.

In addition to limiting the domain of Nodal signaling, the AVE also secretes inhibitors of BMPs and Wnts (for review, see Beddington and Robertson 1999). Since both BMPs and Wnts are involved in mesoderm induction, this aids in limiting the formation of mesoderm to the future posterior side of the embryo (Fig. 5C).

CONCLUSIONS

The field of mesoderm induction has advanced tremendously since the original studies of Nieuwkoop over 40 years ago. It is clear that members of four families of signaling factors, Vg1/Nodal, BMP, Wnt, and FGF, are involved in inducing and maintaining the mesoderm in all vertebrates. Some aspects of mesoderm induction, such as the primary role of Nodal signaling, are conserved among all vertebrates, whereas other functions, such as the role of BMPs, may differ between species. The major future challenge for the field is to determine how each embryo integrates these various signals to activate the genes that are required for the formation of the mesoderm.

ACKNOWLEDGMENTS

The analysis of mesoderm induction in vertebrates is built upon studies from a large number of excellent scientists. We apologize to all those whose work was omitted due to space limitations or oversight on the part of the authors. We thank Kevin Griffin, Liz Robertson, and Henk Roelink for comments on the manuscript. This work was supported by grant IBN-0078303 from the National Science Foundation to D.K.

REFERENCES

Agius E., Oelgeschlager M., Wessely O., Kemp C., and De Robertis E. M. 2000. Endodermal nodal-related signals and mesoderm induction in *Xenopus*. *Development* **127:** 1173–1183.

Amaya E., Musci T.J., and Kirschner M. W. 1991. Expression of a dominant negative mutant of the FGF receptor disrupts mesoderm formation in *Xenopus* embryos. *Cell* **66:** 257–270.

Amaya E., Stein P.A., Musci T.J., and Kirschner M. W. 1993. FGF signaling in the early specification of mesoderm in *Xenopus*. *Development* **118:** 477–487.

Beddington R.S. and Robertson E.J. 1999. Axis development and early asymmetry in mammals. *Cell* **96:** 195–209.

Bertocchini F. and Stern C. D. 2002. The hypoblast of the chick embryo positions the primitive streak by antagonizing nodal signaling. *Dev. Cell* **3:** 735–44.

Brannon M., Gomperts M., Sumoy L., Moon R.T., and Kimelman D. 1997. A β-catenin/XTcf-3 complex binds to the *siamois* promoter to regulate specification of the dorsal axis in *Xenopus*. *Genes Dev.* **11:** 2359–2370.

Brennan J., Lu C.C., Norris D.P., Rodriguez T.A., Beddington R.S., and Robertson E.J. 2001. Nodal signalling in the epiblast patterns the early mouse embryo. *Nature* **411:** 965–969.

Chen S. and Kimelman D. 2000. The role of the yolk syncytial layer in germ layer patterning in zebrafish. *Development* **127:** 4681–4689.

Christian J. L., Olson D. J., and Moon R.T. 1992. Xwnt-8 modifies the character of mesoderm induced by bFGF in isolated *Xenopus* ectoderm. *EMBO J.* **11:** 33–41.

Clements D., Friday R.V., and Woodland H.R. 1999. Mode of action of VegT in mesoderm and endoderm formation. *Development* **126:** 4903–4911.

Conlon F. L., Lyons K.M., Takaesu N., Barth K.S., Kispert A., Herrmann B., and Robertson E.J. 1994. A primary requirement for nodal in the formation and maintenance of the primitive streak in the mouse. *Development* **120:** 1919–1928.

Cornell R.A. and Kimelman D. 1994. Activin-mediated mesoderm induction requires FGF. *Development* **120:** 453–462.

Dale L. and Slack J.M.W. 1987. Regional specification within the mesoderm of early embryos of *Xenopus laevis*. *Development* **100:** 279–295.

Erter C.E., Solnica-Krezel L., and Wright C.V. 1998. Zebrafish *nodal-related 2* encodes an early mesendodermal inducer signaling from the extraembryonic yolk syncytial layer. *Dev. Biol.* **204:** 361–372.

Eyal-Giladi H. 1997. Establishment of the axis in chordates: Facts and speculations. *Development* **124:** 2285–2296.

Eyal-Giladi H. and Kochav S. 1976. From cleavage to primitive streak formation: A complementary normal table and a new look at the first stages of the development of the chick. I. General morphology. *Dev. Biol.* **49:** 321–337.

Fekany K., Yamanaka Y., Leung T., Sirotkin H.I., Topczewski J., Gates M.A., Hibi M., Renucci A., Stemple D., Radbill A. et al. 1999. The zebrafish *bozozok* locus encodes Dharma, a homeodomain protein essential for induction of gastrula organizer and dorsoanterior embryonic structures. *Development* **126:** 1427–1438.

Feldman B., Gates M.A., Egan E.S., Dougan S.T., Rennebeck G., Sirotkin H.I., Schier A.F., and Talbot W.S. 1998. Zebrafish organizer development and germ-layer formation require nodal-related signals. *Nature* **395:** 181–185.

Green J. B., Howes G., Symes K., Cooke J., and Smith J.C. 1990. The biological effects of XTC-MIF: Quantitative comparison with *Xenopus* bFGF. *Development* **108:** 173–183.

Griffin K.J.P., Amacher S.L., Kimmel C.B., and Kimelman D. 1998. Molecular identification of *spadetail*: Regulation of zebrafish trunk and tail mesoderm formation by T-box genes. *Development* **125:** 3379–3388.

Gritsman K., Zhang J., Cheng S., Heckscher E., Talbot W.S., and Schier A.F. 1999. The EGF-CFC protein one-eyed pinhead is essential for nodal signaling. *Cell* **97:** 121–132.

Grunz H. 1996. The long road to chemical and molecular embryology. What the amphibians can teach us on differentiation (an intereview with Professor Heinz Tiedemann). *Int. J. Dev. Biol.* **40:** 113–122.

Heasman J., Kofron M., and Wylie C. 2000. β-catenin signaling activity dissected in the early *Xenopus* embryo: A novel antisense approach. *Dev. Biol.* **222:** 124–134.

Heasman J., Crawford A., Goldstone K., Garner-Hamrick P., Gumbiner B., McCrea P., Kintner C., Noro C.Y., and Wylie C. 1994. Overexpression of cadherins and underexpression of β-catenin inhibit dorsal mesoderm induction in early *Xenopus* embryos. *Cell* **79:** 791–803.

Hemmati-Brivanlou A. and Melton D.A. 1992. A truncated activin receptor inhibits mesoderm induction and formation of axial structures in *Xenopus* embryos. *Nature* **359:** 609–614.

Horb M.E. and Thomsen G.H. 1997. A vegetally localized T-box transcription factor in *Xenopus* eggs specifies mesoderm and endoderm and is essential for embryonic mesoderm formation. *Development* **124:** 1689–1698.

Hume C.R. and Dodd J. 1993. *Cwnt-8C*: A novel *Wnt* gene with a potential role in primitive streak formation and hindbrain organization. *Development* **119:** 1147–1160.

Hyatt B.A., Lohr J.L., and Yost H.J. 1996. Initiation of vertebrate left-right axis formation by maternal Vg1. *Nature* **384:** 62–65.

Hyde C.E. and Old R.W. 2000. Regulation of the early expression of the *Xenopus nodal-related 1* gene, *Xnr1*. *Development* **127:** 1221–1229.

Kelly C., Chin A.J., Leatherman J.L., Kozlowski D.J., and Weinberg E.S. 2000. Maternally controlled β-catenin-mediated signaling is required for organizer formation in the zebrafish. *Development* **127:** 3899–3911.

Kimelman D. and Kirschner M. 1987. Synergistic induction of mesoderm by FGF and TGF-β and the identification of an mRNA coding for FGF in the early *Xenopus* embryo. *Cell* **51:** 869–877.

Kimelman D., Christian J.L., and Moon R.T. 1992. Synergistic principles of development: Overlapping patterning systems in *Xenopus* mesoderm induction. *Development* **116:** 1–9.

Kimmel C.B. and Law R.D. 1985. Cell lineage of zebrafish blastomeres. II. Formation of the yolk syncytial layer. *Dev. Biol.* **108:** 86–93.

Kimmel C.B., Warga R.M., and Schilling T.F. 1990. Origin and organization of the zebrafish fate map. *Development* **108:** 581–594.

Kofron M., Demel T., Xanthos J., Lohr J., Sun B., Sive H., Osada S., Wright C., Wylie C., and Heasman J. 1999. Mesoderm induction in *Xenopus* is a zygotic event regulated by maternal VegT via TGFβ growth factors. *Development* **126:** 5759–5770.

Koos D.S. and Ho R.K. 1999. The *nieuwkoid/dharma* homeobox gene is essential for *bmp2b* repression in the zebrafish pregastrula. *Dev. Biol.* **215:** 190–207.

LaBonne C. and Whitman M. 1994. Mesoderm induction by activin requires FGF mediated intracellular signals. *Development* **120:** 463–472.

Larabell C.A., Torres M., Rowning B.A., Yost C., Miller J.R., Wu M., Kimelman D., and Moon R.T. 1997. Establishment of the dorso-ventral axis in *Xenopus* embryos is presaged by early asymmetries in β-catenin that are modulated by the Wnt signaling pathway. *J. Cell Biol.* **136:** 1123–1136.

Lawson A. and Schoenwolf G.C. 2001. Cell populations and morphogenetic movements underlying formation of the avian primitive streak and organizer. *Genesis* **29:** 188–195.

Lawson A., Colas J.F., and Schoenwolf G.C. 2001. Classification scheme for genes expressed during formation and progression of the avian primitive streak. *Anat. Rec.* **262:** 221–226.

Lawson K.A., Dunn N.R., Roelen B.A., Zeinstra L.M., Davis A.M., Wright C.V., Korving J.P., and Hogan B.L. 1999. Bmp4 is required for the generation of primordial germ cells in the mouse embryo. *Genes Dev.* **13:** 424–436.

Lemaire P., Garrett N., and Gurdon J.B. 1995. Expression cloning of *Siamois*, a *Xenopus* homeobox gene expressed in dorsal-vegetal cells of blastulae and able to induce a complete secondary axis. *Cell* **81:** 85–94.

Liu P., Wakamiya M., Shea M.J., Albrecht U., Behringer R.R., and Bradley A. 1999. Requirement for Wnt3 in vertebrate axis formation. *Nat. Genet.* **22:** 361–365.

Lustig K.D., Kroll K.L., Sun E.E., and Kirschner M.W. 1996. Expression cloning of a *Xenopus* T-related gene (*Xombi*) involved in mesodermal patterning and blastopore lip formation. *Development* **122:** 4001–4012.

Marchant L., Linker C., and Mayor R. 1998. Inhibition of mesoderm formation by follistatin. *Dev. Genes Evol.* **208:** 157–160.

McMahon A.P. and Moon R.T. 1989. Ectopic expression of the proto-oncogene *int-1* in *Xenopus* embryos leads to duplication of the embryonic axis. *Cell* **58:** 1075–1084.

Mishina Y., Suzuki A., Ueno N., and Behringer R.R. 1995. Bmpr encodes a type I bone morphogenetic protein receptor that is essential for gastrulation during mouse embryogenesis. *Genes Dev.* **9:** 3027–3037.

Mitrani E. and Shimoni Y. 1990. Induction by soluble factors of organized axial structures in chick epiblasts. *Science* **247:** 1092–1094.

Mizuno T., Yamaha E., Wakahara M., Kuroiwa A., and Takeda H. 1996. Mesoderm induction in zebrafish. *Nature* **383:** 131–132.

Nieuwkoop P.D. 1969a. The formation of the mesoderm in urodelean amphibians I. The induction by the endoderm. *Wilhelm Roux's Arch. Entwicklungsmech. Org.* **162:** 341–373.

———. 1969b. The formation of the mesoderm in urodelean amphibians II. The origin of the dorso-ventral polarity of the mesoderm. *Wilhelm Roux's Arch. Entwicklungsmech. Org.* **163:** 298–315.

Northrop J. and Kimelman D. 1994. Dorsal-ventral differences in response to FGF mediated induction in *Xenopus. Dev. Biol.* **161:** 490–503.

Perea-Gomez A., Vella F.D., Shawlot W., Oulad-Abdelghani M., Chazaud C., Meno C., Pfister V., Chen L., Robertson E., Hamada H. et al. 2002. Nodal antagonists in the anterior visceral endoderm prevent the formation of multiple primitive streaks. *Dev. Cell* **3:** 745–756.

Popperl H., Schmidt C., Wilson V., Hume C.R., Dodd J., Krumlauf R., and Beddington R.S. 1997. Misexpression of Cwnt8C in the mouse induces an ectopic embryonic axis and causes a truncation of the anterior neuroectoderm. *Development* **124:** 2997–3005.

Rebagliati M.R., Toyama R., Fricke C., Haffter P., and Dawid I.B. 1998. Zebrafish nodal-related genes are implicated in axial patterning and establishing left-right asymmetry. *Dev. Biol.* **199:** 261–272.

Rodaway A., Takeda H., Koshida S., Broadbent J., Price B., Smith J.C., Patient R., and Holder N. 1999. Induction of the mesendoderm in the zebrafish germ ring by yolk cell-derived TGF-β family signals and discrimination of mesoderm and endoderm by FGF. *Development* **126:** 3067–3078.

Rosa F., Roberts A.B., Danielpour D., Dart L.L., Sporn M.B., and Dawid I.B. 1988. Mesoderm induction in Amphibians: The role of TGF-β2-like factors. *Science* **239:** 783–785.

Ryu S. L., Fujii R., Yamanaka Y., Shimizu T., Yabe T., Hirata T., Hibi M., and Hirano T. 2001. Regulation of *dharma/bozozok* by the Wnt pathway. *Dev. Biol.* **231:** 397–409.

Schneider S., Steinbeisser H., Warga R.M., and Hausen P. 1996. β-catenin translocation into nuclei demarcates the dorsalizing centers in frog and fish embryos. *Mech. Dev.* **57:** 191–198.

Seleiro E.A., Connolly D.J., and Cooke J. 1996. Early developmental expression and experimental axis determination by the chicken Vg1 gene. *Curr. Biol.* **6:** 1476–1486.

Shah S.B., Skromne I., Hume C.R., Kessler D.S., Lee K.J., Stern C.D., and Dodd J. 1997. Misexpression of chick Vg1 in the marginal zone induces primitive streak formation. *Development* **124:** 5127–5138.

Shimizu T., Yamanaka Y., Ryu S.L., Hashimoto H., Yabe T., Hirata T., Bae Y.K., Hibi M., and Hirano T. 2000. Cooperative roles of Bozozok/Dharma and Nodal-related proteins in the formation of the dorsal organizer in zebrafish. *Mech. Dev.* **91:** 293–303.

Sirotkin H.I., Dougan S.T., Schier A.F., and Talbot W.S. 2000. *bozozok* and *squint* act in parallel to specify dorsal mesoderm and anterior neuroectoderm in zebrafish. *Development* **127:** 2583–2592.

Skromne I. and Stern C.D. 2001. Interactions between Wnt and Vg1 signalling pathways initiate primitive streak formation in the chick embryo. *Development* **128:** 2915–2927.

Slack J.M.W., Darlington B.G., Heath J.K., and Godsave S.F. 1987. Mesoderm induction in early *Xenopus* embryos by heparin-binding growth factors. *Nature* **326:** 197–200.

Smith J.C. 1987. A mesoderm-inducing factor is produced by a *Xenopus* cell line. *Development* **99:** 3–14.

Smith J.C. and Slack J.M.W. 1983. Dorsalization and neural induction: Properties of the organizer in *Xenopus laevis. J. Embryol. Exp. Morphol.* **78:** 299–317.

Smith J.C., Price B.M.J., Van Nimmen K., and Huylebroeck D. 1990. Identification of a potent *Xenopus* mesoderm-inducing factor as a homologue of activin A. *Nature* **345:** 729–731.

Standley H.J., Zorn A.M., and Gurdon J.B. 2001. eFGF and its mode of action in the community effect during *Xenopus* myogenesis. *Development* **128:** 1347–1357.

Stennard F., Carnac G., and Gurdon J.B. 1996. The *Xenopus* T-box gene, *Antipodean*, encodes a vegetally localised maternal mRNA and can trigger mesoderm formation. *Development* **122:** 4179–4188.

Stern C.D. 1992. Mesoderm induction and development of the embryonic axis in amniotes. *Trends Genet.* **8:** 158–163.

Stern C.D., Yu R.T., Kakizuka A., Kintner C.R., Mathews L.S., Vale W.W., Evans R.M., and Umesono K. 1995. Activin and its receptors during gastrulation and the later phases of mesoderm development in the chick embryo. *Dev. Biol.* **172:** 192–205.

Sun B.I., Bush S.M., Collins-Racie L.A., LaVallie E.R., DiBlasio-Smith E.A., Wolfman N.M., McCoy J.M., and Sive H.L. 1999. *derrière*: A TGF-β family member required for posterior development in *Xenopus. Development* **126:** 1467–1482.

Takahashi S., Yokota C., Takano K., Tanegashima K., Onuma Y., Goto J., and Asashima M. 2000. Two novel *nodal*-related genes initiate early inductive events in *Xenopus* Nieuwkoop center. *Development* **127:** 5319–5329.

Thomsen G., Woolf T., Whitman M., Sokol S., Vaughan J., Vale W., and Melton D.A. 1990. Activins are expressed in *Xenopus* embryogenesis and can induce axial mesoderm and anterior structures. *Cell* **63:** 485–493.

Varlet I., Collignon J., and Robertson E.J. 1997. Nodal expression in the primitive endoderm is required for specification of the anterior axis during mouse gastrulation. *Development* **124:** 1033–1044.

Wall N.A., Craig E.J., Labosky P.A., and Kessler D.S. 2000. Mesendoderm induction and reversal of left-right pattern by mouse Gdf1, a Vg1-related gene. *Dev. Biol.* **227:** 495–509.

Weeks D.L. and Melton D.A. 1987. A maternal mRNA localized to the vegetal hemisphere in *Xenopus* eggs codes for a growth factor related to *TGF-β*. *Cell* **51:** 861–867.

White R.J., Sun B.I., Sive H.L., and Smith J.C. 2002. Direct and indirect regulation of *derrière*, a *Xenopus* mesoderm-inducing factor, by VegT. *Development* **129:** 4867–4876.

Winnier G., Blessing M., Labosky P.A., and Hogan B.L. 1995. Bone morphogenetic protein-4 is required for mesoderm formation and patterning in the mouse. *Genes Dev.* **9:** 2105–2116.

Xanthos J.B., Kofron M., Tao Q., Schaible K., Wylie C., and Heasman J. 2002. The roles of three signaling pathways in the formation and function of the Spemann Organizer. *Development* **129:** 4027–4043.

Yang J., Tan C., Darken R.S., Wilson P.A., and Klein P.S. 2002. β-Catenin/Tcf-regulated transcription prior to the midblastula transition. *Development* **129:** 5743–5752.

Zeng L., Fagotto F., Zhang T., Hsu W., Vasicek T.J., Perry III, W.L., Lee J.J., Tilghman S.M., Gumbiner B.M., and Costantini F. 1997. The mouse *Fused* locus encodes Axin, an inhibitor of the Wnt signaling pathway that regulates embryonic axis formation. *Cell* **90:** 181–192.

Zhang J. and King M.L. 1996. *Xenopus VegT* RNA is localized to the vegetal cortex during oogenesis and encodes a novel T-box transcription factor involved in mesodermal patterning. *Development* **122:** 4119–4129.

Zhang J., King M.L., Houston D., Payne C., Wylie C., and Heasman J. 1998. The role of maternal VegT in establishing the primary germ layers in *Xenopus* embryos. *Cell* **94:** 515–524.

Zhou X., Sasaki H., Lowe L., Hogan B.L., and Kuehn M.R. 1993. Nodal is a novel TGF-β-like gene expressed in the mouse node during gastrulation. *Nature* **361:** 543–547.

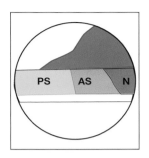

DORSOVENTRAL PATTERNING
OF THE MESODERM

R. Harland

Department of Molecular and Cell Biology,
University of California, Berkeley, California 94720-3204

INTRODUCTION

Immediately following gastrulation, the embryonic regions that produce definitive tissues of the vertebrate adult can be identified with some precision. The germ layers have formed and segregated, and as organogenesis begins, the relatively crude fate map of the early embryo is replaced with a clear picture of prospective tissues of the organism. The fate map describes developmental outcomes for cells left undisturbed in their normal locations, and prior to gastrulation, cells are not committed to follow these fates. The potentials of such cells can be revealed by the challenge of transplantation, and during gastrulation, cells become progressively restricted in their potential fates. Here I review the signaling processes that occur during gastrulation and produce the broad outlines of the adult tissues.

At the onset of neurulation, the organization of different vertebrate embryos is topologically equivalent (Fig. 1). At the trunk level of mesoderm, the notochord occupies the medial or most dorsal position and is flanked by the somites. The organization of the rest of the mesoderm can appear to be radically different, since it is constrained by the amount of yolk and by the way in which the animals evolved. Thus, in the chick, the embryo is arranged on top of an enormous yolk, such that the mesoderm is organized in a mediolateral distribution from notochord, through somites, to lateral plate, and with blood forming in islands far distant from the embryo midline. In frogs and fish, the yolk is quickly enclosed and so the mediolateral organization becomes a dorsoventral one, with the blood islands

forming opposite the notochord in the ventral midline. However, the blood is still distant from the notochord, and the tissues occupy the same relative positions. Many mammals resemble the chick in having a flat gastrula; however, the mouse gastrula is cup-shaped, and the shape of the neurulating embryo is so odd that dorsoventral organization is not often mentioned; nevertheless, a section through the trunk shows an organization of tissues similar to those of the chick and frog, although the distant blood islands form in a posterior and extraembryonic region. It is not obvious that signaling events during gastrulation are homologous between vertebrates; therefore, before comparing amniote embryos with those of fish and frog, I review some of the signaling events that impose pattern on the mesoderm of *Xenopus* during gastrulation.

The previous chapter discussed mesoderm induction and left the late blastula with an equatorial ring of mesoderm that is split into two zones, an organizer and a non-organizer mesoderm. The organizer is the precursor of the axial mesoderm, which occupies the dorsal domain in the neurula and is a dominant source of signals that act on neighboring cells. During gastrulation, the mesoderm spreads, and a large proportion of the mesodermal mantle comes into proximity with and under the influence of the organizer. The properties of the organizer are justifiably famous, starting with Spemann and Mangold's grafting experiments in the 1920s. When an organizer was grafted to the ventral side of a host embryo, it was shown to recruit host tissues into an organized secondary axis. This conclusion relied on using newts of different pigmentation, so that the contributions of donor and host

A

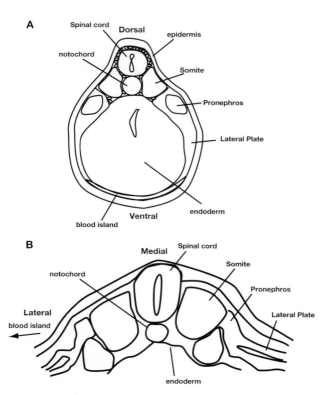

Figure 1. Comparison of (*A*) the dorsal to ventral organization of the tail bud tadpole of *Xenopus* and (*B*) the mediolateral organization of the mesoderm in the chick embryo.

could be distinguished. The secondary axis contained organizer-derived midline tissues, but host tissues were recruited to form the somites and other mesodermal derivatives (Spemann 1938; Hamburger 1988). The experiment has been repeated in *Xenopus*, with enzyme-based and fluorescent lineage tracing techniques, to solidify the idea that the organizer has dominant properties (Gimlich and Cooke 1983; Smith and Slack 1983). The converse experiment, of transplanting pieces of non-organizer mesoderm into the organizer region, showed that these "ventral" tissues had little effect on dorsal fates (Smith and Slack 1983). However, this is not to say that the ventrolateral mesoderm is a passive tissue that can only develop with simple pattern by itself. As discussed below, the rest of the marginal zone differentiates considerable pattern without any organizer signals and is a potent source of secondary signals. These signals caudalize the neural tube and influence the development of the mesoderm. One surprise of recent years is that in order to suppress the caudalizing effects of the ventrolateral marginal zone, the organizer secretes a cocktail of antagonists, which block BMP and Wnt signals from the marginal zone and allow the more dorsal regions to realize their fates (Harland and Gerhart 1997; De Robertis et al. 2001).

DOES DORSOVENTRAL PATTERNING IN THE MESODERM NEED ORGANIZER SIGNALS?

Early experiments from Nieuwkoop suggested that the mesoderm-inducing signals are graded, such that a range of fates would be induced in the blastula prior to organizer signaling (Boterenbrood and Nieuwkoop 1973). The idea that the early mesoderm inducers provide extensive pattern to the marginal zone is attractive, but in *Xenopus* is not supported by experimental embryology. Instead, the non-organizer mesoderm, often called ventrolateral mesoderm, must receive signals from the organizer in order to assume its definitive fates. For simplicity, I refer here to the organizer as being dorsal mesoderm, since it populates the most dorsal mesodermal tissue of the neurula, namely the notochord. Although I discuss controversies over how to assign the dorsoventral and anterior–posterior axes later in the chapter, the important aspect of the non-organizer region (often referred to as the ventrolateral marginal zone) is that it has not yet received the information that specifies paraxial or intermediate mesoderm. This is clear from two kinds of experiments.

The first experiments are those that explant regions of the blastula into simple buffered salts and ask what differentiation ensues. In the late blastula, Dale and Slack (1987b) concluded that the ventrolateral explants are specified as blood, mesenchyme, and mesothelium. Although some explants from all regions do occasionally develop paraxial (muscle) fates, the striking finding was that many explants made no muscle, and hence there cannot be robust paraxial specification early on. The explants that do differentiate muscle may result from errors in dissection or the release of FGF-like inducing factors by wounding of the embryo during dissection (LaBonne and Whitman 1997).

The second kind of experiment used donor embryos bisected along vertical meridians (through the animal and vegetal poles), with the resultant half embryos then grafted onto "naïve" half embryos (Stewart and Gerhart 1990). The naïve hosts were derived from embryos where the organizer had been ablated by early irradiation with ultraviolet light. The manipulation tested how the grafted halves developed, and thus measured their state of specification, as well as how they signaled to naïve halves. In these experiments, half embryos cut just 30° away from the dorsal midline (and hence lacking the 60° wide organizer) developed in the same way as embryos cut 90° away from the midline (considered to be ventral halves). Although a few of the recombinants developed muscle, the striking result was how often embryos lacking the organizer had no paraxial or intermediate mesoderm (Stewart and Gerhart 1990). Thus, the entire ventrolateral mesoderm (or non-organizer mesoderm) in the late blastula is specified as ventral types of tissue. Since the fate maps show that much of this mesoderm would normally give rise to muscle, the inevitable conclusion is that ventrally specified mesoderm must receive further signals that enable it to develop other fates. Indeed, by the onset of gastrulation, the prospective paraxial mesoderm, just next to the organizer, has received these signals and is specified as muscle (Dale and Slack 1987b).

All of these signals that pattern the mesoderm act from the blastula stage onward. They include BMP and Wnt antagonists (Chapter 33 and Chapter 36). In addition, some of these signals may be the continuation of mesoderm induction (Chapter 25), since the Nodal-related (Xnr) genes are expressed at higher levels in the organizer than in the non-organizer mesoderm (Jones et al. 1995; Agius et al. 2000). The Xnr proteins have multiple functions, acting directly as mesoderm inducers, but also as dorsalizing agents that are mechanistically different from mesoderm inducers (Jones et al. 1995); the dorsalizing properties come about by dimerization with BMPs, thereby blocking BMP activity (Yeo and Whitman 2001; Eimon and Harland 2002). The latter function may be the most important during gastrulation, since embryonic cells lose their ability to respond to mesoderm inducers (Green et al. 1990; Grimm and Gurdon 2002), and new signals from the organizer become the dominant theme (Smith and Slack 1983; Gimlich 1986). Thus, patterning of the mesoderm continues during gastrulation, and from several sources. Organizer signaling is the best characterized, but is reinforced by dorsalizing signals from the undifferentiated neural plate, which acts during gastrulation. In the absence of neural plate, the amount of paraxial mesoderm is reduced (Mariani et al. 2001).

ORGANIZER SIGNALING IN A "THREE-SIGNAL MODEL" FOR MESODERM INDUCTION AND PATTERNING

Based on the behavior of explants and whole embryos manipulated by grafting, Slack and colleagues proposed a three-signal model for patterning the early embryo (Smith and Slack 1983; Dale and Slack 1987b). This was a very successful working model for molecular biologists to build upon, but not surprisingly, over the last 20 years the model has had to adapt to the results of both experimental embryology and molecular embryology. The original scheme involved signaling by a vegetal dorsalizing center that induced the organizer, and a more generic ventral mesoderm-inducing signal from the rest of the vegetal region (Smith and Slack 1983; Gimlich and Gerhart 1984; Dale and Slack 1987b). These first two signals led to the binary choice of mesodermal cell fates inferred from explant experiments. Further patterning relied on the dorsalizing action of the organizer (Gimlich 1986; Lettice and Slack 1993). These experiments were valuable in establishing the timing and

the kinds of signals that are generated in the embryo. However, the necessity of signaling from the vegetal cells is now usually considered an experimental phenomenon only demonstrable in recombinants, and the same signals have now been shown to be made in the marginal zone and largely act cell-autonomously within this context (Chapter 25). However, the net output is the same, and the binary decision of specification in the late blastula marginal zone is now well-supported by molecular experiments; these show that the late blastula has essentially two zones of gene activity around the ring of mesoderm. Interestingly, there is additional complexity in the radial direction, with different gene expression in deep and superficial zones (Harland and Gerhart 1997). Because the deep tissues are the first to migrate across the blastocoel roof, this radial difference in gene expression translates into an anterior–posterior difference, and the anterior expression of antagonists has a particular role in formation of the head (Chapter 27).

Although most transcripts that are differently expressed around the radius of the embryo show two distinct zones of expression, their boundaries of expression are somewhat graded, and this may contribute to a small extent to patterning the marginal zone. The principle that different doses of a transcription factor might contribute to the initial patterning of the marginal zone was shown for Goosecoid (Niehrs et al. 1994). However, in normal development, expression of these transcription factors does not extend far around the marginal zone; therefore, graded expression cannot account for most of the pattern of the marginal zone. Furthermore, the experimental embryology provides no good evidence for graded fates very far around the initial late blastula marginal zone. All these observations highlight the importance of the third signal, a dorsalizing signal from the organizer. As originally envisioned (Smith and Slack 1983; Dale and Slack 1987b), this signal would diffuse from the organizer and act over a considerable range to induce different fates of mesoderm, such as heart, somites, pronephros, and blood (with the mesenchyme and mesothelium being tissues that are poorly defined molecularly and analyzed by few workers histologically).

The signals that induce the initial mesodermal territories of the late blastula are qualitatively different from those that can pattern mesoderm during gastrulation. Mesoderm inducers can be pieces of the embryo, or soluble TGF-β or FGF, and are operationally defined by their ability to induce mesoderm in ectodermal (animal cap) tissue taken from the blastula (see Chapters 25, 32, 34, and 35). Such mesoderm inducers are unable to induce mesoderm when applied at gastrula stages. Furthermore, in contrast to their ability to induce muscle when applied at higher doses to blastula explants, mesoderm inducers are unable to induce paraxial mesoderm in ventrolateral explants from the gastrula. In contrast, signals from the organizer are able to repattern ventrolateral explants in a process of dorsalization and are able to act on the mesoderm throughout gastrulation (Lettice and Slack 1993). Soluble BMP antagonists such as Noggin can mimic, and are likely to mediate, this effect of the organizer (Smith et al. 1993; Chapter 33). Experimentally, the difference is clearly demonstrated by using recombinant protein that can be applied at specific times to explants, rather than injection of mRNAs whose protein products are expressed from early times.

IS THERE GRADED SIGNALING FROM THE ORGANIZER?

The marginal zone forms the notochord and prechordal plate, as well as the heart, somatic muscle, pronephros, other mesenchyme and mesothelium, and the blood. The idea of a single diffusible morphogen that may induce these different tissues at different doses is inherently attractive (Gurdon and Bourillot 2001). Many experiments have provided considerable support for the idea that morphogens might in principle organize the mesoderm. For example, BMP antagonists that are ectopically expressed at high levels can act over several cell diameters (Jones and Smith 1998; Blitz et al. 2000). In addition, graded blocking of BMP signals in whole embryos results in formation of different fates (Dosch et al. 1997). However, whether these elevated doses of proteins are recapitulating the normal range of protein diffusion is doubtful, and the overexpressed protein may be saturating the cell surface and extracellular matrix that normally restricts the range of diffusion (Blitz et al. 2000). Despite the general attractiveness of the ideas and principles, evidence that any particular endogenous molecule acts physiologically as a morphogen in the mesoderm is still poor.

To establish a molecule as a physiologically active morphogen, it must be shown not only to be distributed in a graded concentration, but also to elicit multiple different fates at different concentrations and to be required at different concentrations to elicit these actions (Gurdon and Bourillot 2001; see also Chapter 35). These are difficult standards to satisfy. The soluble signaling molecules in early embryos have not yet been convincingly detected, so inferences about physiological concentration and diffusion are indirect. Comparisons of mutation of the BMP antagonists *Noggin* and *Gremlin* in the mouse limb showed essentially nonoverlapping phenotypes and therefore suggest that each antagonist acts in its own territory of expression and does not diffuse far (Khokha et al. 2003). In contrast, in the zebrafish, *dino* (*chordin*) mutant embryos are substantially rescued by the transplantation of a few wild-type cells into the shield region, suggesting at the least a non-autonomous effect, and possibly a longer-range effect of Chordin (Hammerschmidt et al. 1996). Thus, some of the signaling

molecules may diffuse over several cell diameters to elicit patterning.

There are no general rules for the diffusion range of ligands, and diffusion is regulated not only by the extracellular matrix, but also by intrinsic properties of the molecule. Perhaps the greatest variation has been found among the TGF-β family members, where a surprising variation in diffusion has been reported (Jones et al. 1996a; Cui et al. 2001; Eimon and Harland 2002; Ohkawara et al. 2002; Hashimoto-Partyka et al. 2003). The differences depend on motifs that may specifically bind the extracellular matrix, as is the case for the amino terminus of BMP2 (Ohkawara et al. 2002). Another interesting observation is that different pro-regions that are cleaved from TGF-βs lead to different ranges of action, suggesting that different pro regions may act as chaperones to facilitate the secretion, cleavage, diffusion, and release of the ligand to receptors (Jones et al. 1996a; Cui et al. 2001; Eimon and Harland 2002).

Finally, it has even been difficult to determine whether different concentrations of the signals emitted during gastrulation induce multiple fates in a direct way; instead, it is becoming clearer that patterning of the mesoderm during and following gastrulation is mediated by sequential and combinatorial signaling from a variety of signaling centers. These are set up by the radial (deep to superficial) organization of the blastula and by the animal to vegetal organization of the embryo. The signals then act over a considerable time period, exploiting the increasing complexity of the embryo to produce and juxtapose new signaling centers. Arguably, only the muscle and notochord are set up and maintained by early signals, whereas the heart, kidney, and blood are set aside at a later time and by different signaling centers. Even the early induction of muscle in amphibians and fish may be a special case; in amniotes the muscles are induced from the somite during neurulation, and therefore muscle induction, like the formation of other organ rudiments, is a later process of organogenesis.

A discussion of the different fates of the mesoderm and how they are affected by different manipulations will illustrate the molecular signals that lead to different fates in the mesoderm.

Notochord Maintenance

The notochord is the most dorsal (medial) mesoderm in the early neurula and is directly descended from the organizer. It is specified during mesoderm induction (Chapter 25), and its identity is maintained by expression of transcription factors and secreted antagonists. Notochord can be induced by manipulations that suppress only zygotic signaling, for example, in the marginal zone, by a combination of Wnt and BMP antagonists. By themselves, single BMP antagonists do not induce notochord efficiently when introduced opposite the organizer, although in many cases they induce some notochord and head structures (Hsu et al. 1998). However, combinations of antagonists, or combinations of truncated receptors, are more effective, at least in inducing head-containing secondary axes, and therefore presumably must induce notochord efficiently (Yamamoto et al. 2001). In embryos deprived of organizer by UV irradiation, or by inhibition of β-catenin translation, BMP antagonists induce complete secondary axes efficiently, and in that context induce notochord (Smith and Harland 1992; Sasai et al. 1994; Xanthos et al. 2002). This difference in inducing behavior of BMP antagonists on the ventrolateral marginal zone versus the marginal zone of UV-ventralized embryos is not understood.

BMP expression ventralizes the embryo (Dale et al. 1992; Jones et al. 1992) and suppresses notochord formation. BMPs have multiple effects on different target tissues, but in the context of the mesoderm, the action is one that occurs during gastrulation. Thus, the early organizer-specific transcripts are still expressed in the context of excess BMP signaling, but their transcripts fade during gastrulation (Jones et al. 1996b; Laurent and Cho 1999). Similarly, the dorsalizing properties of BMP antagonists are only evident after the onset of gastrulation, with a delayed expression of organizer genes (Eimon and Harland 1999).

Wnts are effective inhibitors of notochord formation when expressed in the gastrula from injected plasmids (Christian and Moon 1993). Results of this late activation of Wnt signaling contrast markedly with the early effects of activation of the Wnt signal transduction system, which stabilizes β-catenin. As reviewed by Kimelman and Bjornson, (Chapter 25), this program activates dorsal identities during mesoderm induction, by synergy with both T-box transcription factors and Smad2 complexes. When expressed zygotically in prospective notochord, Wnts cause those cells to become paraxial mesoderm (Christian and Moon 1993). Thus, to maintain the notochord program, both Wnt and BMP activity must be inhibited locally. This is achieved at the transcriptional level in the early marginal zone by the repressive activity of Goosecoid, whose principal role appears to be to suppress *Wnt8* expression (Yao and Kessler 2001). BMP expression is also cleared from the prospective dorsal region of the embryo by the activity of the early-acting β-catenin pathway (Baker et al. 1999; see Chapter 36), which may act through induction of a transcriptional repressor similar to Bozozok in the zebrafish (Koos and Ho 1999; Fekany-Lee et al. 2000; Leung et al. 2003).

In addition to the transcriptional control of BMPs and Wnts to promote notochord development, the organizer protects itself from these signals by producing a cocktail of Wnt and BMP antagonists. The BMP antagonists include Noggin, Chordin, Xnr3, Follistatin, and the multifunctional antagonist Cerberus (Smith and Harland 1992; Hemmati-Brivanlou et al. 1994; Sasai et al. 1994; Smith et al. 1995; Bouwmeester et al. 1996; Piccolo et al. 1996, 1999; Zimmerman et al. 1996; Hansen et al. 1997). The Wnt antagonists include Frzb, Dkk1, Cerberus, and Crescent (Bouwmeester et al. 1996; Leyns et al. 1997; Wang et al. 1997; Glinka et al. 1998; Piccolo et al. 1999; Pera and De Robertis 2000; Shibata et al. 2000). The presence of all these overlapping pathways to suppress activity of zygotic Wnt and BMP activity is a puzzle, and the direct requirements of each for notochord formation have not yet been resolved.

Prechordal Plate

The prechordal plate is a special example of dorsoanterior mesendoderm that derives from the deep region of the gastrula. In most species, this is considered to include the midline tissues of the mesoderm and endoderm rostral to the notochord (Nieuwkoop and Faber 1967), although in other organisms it has been defined more narrowly as endoderm-only (Seifert et al. 1993). It is assumed that its induction is similar to that of the notochord, although expression of transcription factors such as *goosecoid* and *hex* is maintained there, and in addition, it locally expresses several of the Wnt antagonists (Schneider and Mercola 1999; Zorn et al. 1999). These are thought to have a function in maintaining a Wnt-free region of the overlying neural plate. The induction of the prechordal mesendoderm is usually considered to be an extreme output of the signals that also induce notochord, although this has not been examined in detail in *Xenopus*. The sharp boundary between prechordal mesoderm and notochord may represent a case where cells interpret their position on a morphogen gradient of Smad2 activating TGF-βs, or it may be a reflection of the deep origins of this tissue, and their history of expressing different transcription factors from the superficial layer. The subsequent development of the prechordal plate has not been studied in *Xenopus*, although experiments in the chick suggest that the transient expression of BMPs around, and then in, the prechordal mesoderm is sufficient and necessary for the separate differentiation of prechordal plate from the notochord (Dale et al. 1997; Vesque et al. 2000). Finally, as discussed below, FGF signaling plays into the decision, since FGF activity suppresses anterior development (Pownall et al. 1996), and blockade of FGF signaling prevents notochord, but not prechordal plate, formation (Amaya et al. 1993; Cornell and Kimelman 1994; LaBonne and Whitman 1994).

Somite Development

There are substantial differences between the early development of the somites in frogs and fish, compared to the amniotes. In frogs and fish, which form swimming larvae, the bulk of the somite develops as muscle immediately and almost exclusively, and muscle induction is an event of gastrulation. The sclerotome and dermomyotome develop later, and initially are difficult to recognize without molecular markers (Morin-Kensicki and Eisen 1997; Goto et al. 2000). It is not known in *Xenopus* when the sclerotome and dermatome are induced, although it is likely that, as in the chick, it follows gastrulation. They are not considered further here.

Muscle Induction

Induction of the muscle during gastrulation was the first dorsalizing event recapitulated by a single secreted signal, Noggin, which could be applied as protein to ventral marginal zones, and shown to induce muscle during gastrulation (Smith et al. 1993). The distinct effects of Noggin, an antagonist of BMPs, from mesoderm-inducing activities such as Activin or FGF, highlight the changes in competence of embryonic tissues and the difference between early mesodermal induction and patterning of the mesoderm during gastrulation. Not only is the competence of cells to respond to mesoderm inducers lost at the onset of gastrulation, but cells increase their sensitivity to signals such as Noggin, at least in the ectoderm (Knecht and Harland 1997); as measured by induction of neural markers, there is a dramatic increase in sensitivity to Noggin at the onset of gastrulation. Such changes in responsiveness can contribute to the apparent properties of a substance as a morphogen. A classic morphogen will induce different gene responses at different threshold concentrations, but if the responsive tissue increases its responsiveness over time, the persistence of the ligand will also affect the response. In cases like Noggin, where the ligand binds avidly to the extracellular matrix (Mann 1995; Paine-Saunders et al. 2002), higher doses of Noggin will persist for a long time and be able to act over an extended period. Thus, the dose-dependent effects of Noggin may in part be due to activity as a morphogen, but could just as well be explained by higher doses persisting and acting at a more effective time, while lower doses are neutralized more quickly and act only transiently (Knecht and Harland 1997). The difficulty of pigeonholing Noggin as the morphogen in such assays illustrates the difficulty of understanding how such proteins act over space and time.

BMP antagonists do not convert the entire marginal zone into muscle, and from experiments discussed below, it is likely that only the more animal part is competent to become muscle (Kumano and Smith 2002). The marginal zone expresses a thin stripe of *MyoD*, and although this is not enough to cause cells to develop as muscle in the absence of dorsalizing signals (Frank and Harland 1991; Rupp and Weintraub 1991; Hopwood et al. 1992), it may

mark the cells that are competent to differentiate into muscle (Kumano and Smith 2000).

The additional signals in the marginal zone that interact with BMP antagonists are the caudalizing signals, which include FGFs and Wnts. If Wnt signaling is strongly suppressed, muscle fails to differentiate (Hoppler et al. 1996; Leyns et al. 1997; Wang et al. 1997). This part of the signaling system is therefore analogous to the later induction of muscle in the avian somite, where a combination of BMP antagonists and Wnts induces the epaxial muscle from the undifferentiated somite (Hirsinger et al. 1997; Marcelle et al. 1997; Reshef et al. 1998).

FGF signaling through the Ras and PI3 kinase pathways is crucial for muscle formation (see Chapter 32). Development of muscle and notochord is prevented by either dominant negative FGF receptors or their downstream signaling components. Several FGFs are expressed in the marginal zone, and their transcription localizes to the region of the blastopore as it closes, thus restricting signaling to the prospective posterior end of the embryo (Isaacs et al. 1995; Christen and Slack 1997; Lombardo et al. 1998). FGF signaling maintains a feedback loop of the T-box transcription factor Brachyury in the mesoderm, and this has been proposed as a mechanism to integrate signaling locally such that individual "rogue" cells do not accidentally set off and stabilize cell-autonomous transcriptional programs (Isaacs et al. 1994; Schulte-Merker and Smith 1995). The signaling by FGF is therefore likely to be the basis of the "community effect" (Gurdon 1988), where a cohort of prospective muscle cells can maintain a program of differentiation (see Chapter 32). Indeed, in culture, FGF can provide the necessary permissive environment for muscle cells to differentiate (Standley et al. 2001).

Whereas the expression of FGF mRNAs has been well described, the location of active FGF signaling has been more difficult to localize and quantify in embryos. FGF signaling spreading from the animal hemisphere has been proposed to mediate the restriction of mesoderm to the marginal zone and may overlap there with the vegetal mesoderm-inducing signals (Cornell et al. 1995). The proposal was based on observations that increasing FGF signals in prospective endoderm would induce mesoderm. Initial biochemical assays of dissected embryos indicated that no obvious MAP kinase activity differences could be found along the animal vegetal axis (LaBonne and Whitman 1997). Despite this, antibodies to biphosphorylated and active MAP kinase show that the highest levels of activated MAP kinase are found in the animal part of the marginal zone, where they maintain a zone of competence to form muscle (Christen and Slack 1999; Curran and Grainger 2000; Kumano et al. 2001). In the absence of these signals, the marginal zone produces blood in preference to muscle (Kumano and Smith 2000). These observations have led to the proposal that much of the patterning of the mesoderm to provide muscle (more dorsal) and blood (more ventral) depends on a signal that is asymmetric in the animal–vegetal axis, and that this animal–vegetal axis therefore defines the dorsoventral organization of the non-axial mesoderm. As described above, the axial and most dorsal mesoderm derives from the organizer.

Heart Induction

The heart originates in two fields adjacent to the organizer, and after gastrulation these fields migrate ventrally to fuse into the heart tube. The heart is induced by organizer signals, acting in concert with the deep endomesoderm, on the marginal zone adjacent to the organizer (Sater and Jacobson 1990a,b; Nascone and Mercola 1995). This induction can be mimicked by the secreted Wnt antagonists Dkk1 and Crescent, although not by others such as Frzb or Sizzled (Marvin et al. 2001; Schneider and Mercola 2001). In addition, Wnt-11, which activates a noncanonical pathway, is also implicated in heart induction (Pandur et al. 2002). These results suggest that a certain ratio of Wnt-like activities may be important in induction of the heart field.

The Pronephros

The pronephros is the primitive kidney formed in the tadpole. Pronephric induction has also been suggested to result from graded signals directly from the organizer, perhaps under the influence of intermediate levels of BMPs (Dosch et al. 1997). However, attenuation of BMP signals in marginal zone explants does not induce kidney, just as it does not induce differentiated heart tissue, and instead the kidney forms as a result of combinatorial signals (Carroll and Vize 1999; Seufert et al. 1999). In this case, anterior somites combine with more lateral mesoderm to induce the kidney (Seufert et al. 1999; Mitchell and Sheets 2001). The nature of this anterior somite signal remains obscure. Ultimately, the kidney forms in a field defined by the overlap of two transcription factors, namely *Xlim1* and *Pax8*, which begin their expression shortly after gastrulation (Carroll and Vize 1999).

Blood

Because blood is the most ventral mesodermal tissue formed in the tail-bud tadpole, understanding its formation has influenced, and been influenced by, various fate maps of the embryo. Circulating blood cells derive from blastomeres distributed broadly through the embryo, and all of these have the common property that some of their progeny cells finish gastrulation residing in the ventral part of the tail-bud-stage tadpole (Lane and Smith 1999; Lane and Sheets 2002). Formation of the blood is a relatively late process, and although regions of the blastula and gastrula can be said to be specified to form blood (Dale and Slack 1987b), the

definitive markers of blood differentiation are not activated until the late neurula to tail-bud stages (Kelley et al. 1994). The signaling center that is most important in blood formation is the ventral ectoderm (Maeno et al. 1994; Kikkawa et al. 2001), which continues to be a powerful source of BMP4. Blood formation is an output of BMP4 signaling, but the sensitive period for this induction extends past gastrulation, such that BMP signals are required to set up the ventral mesoderm, then again in the neurula to maintain blood fates. The distinct nature of these signals over time is evident from use of a hormone-inducible version of the cell-autonomous BMP antagonist Smad6 (Schmerer and Evans 2003). It is likely that isolated cells can respond to the second set of signals, and cells populating the blood islands can be derived from any tissues that move during morphogenesis to be far away from the organizer. Thus, the blood cells that trace from the dorsal region of the blastula near the organizer become separated from the organizer during the extensive movements of gastrulation (Lane and Sheets 2002). In doing so, they are separated from the BMP antagonists that are expressed in the axial mesoderm and are instead influenced by the BMPs that are expressed in the ventral ectoderm. Some blood cells even derive from animal blastomeres of the cleavage-stage embryo, and although most of the volume of these blastomeres contributes to the epidermis, some of the progeny cells form blood after the movements of epiboly bring them to the ventral side, where they may be displaced from the ectoderm into the blood islands. Even there, the amount of blood is regulated by feedback mechanisms, with the amount of ventral mesoderm being limited by the function of Sizzled (Collavin and Kirschner 2003).

In general, cells that lose their original neighbors and become surrounded by another tissue remain labile and differentiate according to their new position. This principle has been established particularly clearly for non-notochordal cells from different gastrula stages, which are transplanted into the notochord territory (Domingo and Keller 2000). Cells from such regions as the epidermis and ventral mesoderm gradually lose their ability to accommodate to the new notochord fate, as they age through gastrulation. However, they retain the ability to be reprogrammed for much longer than a larger tissue graft or explant. The converse, the failure of cells to develop according to their prospective fate unless they are in a community of like cells, has been particularly well documented for muscle cells (Gurdon 1988; Zorn et al. 1999).

These discussions of cell fate choices during gastrulation highlight the deficits of a naïve model of dorsoventral patterning driven by a graded activity of BMPs. Instead, it has become clear that BMP dose during gastrulation is just one influence on fate choices, and additional influences are the effects of other signaling pathways, particularly the FGF and

Wnt signaling pathways. Furthermore, there is now an improved understanding of the effects of signaling at different times, where the competence of cells to respond to any one signal changes over time.

All Fate Maps Are Wrong

Recently, there has been a renewal of a debate about what to call the early axes of the amphibian embryo (Fig. 2) (Tracey et al. 1998; Lane and Smith 1999; Gerhart 2002; Kumano and Smith 2002; Lane and Sheets 2002). It would certainly be useful to have a rigorous way to label the early embryo, especially if it would lead to a coherent view of how the gastrulae of different vertebrates relate to one another. Such comparisons have been made in both classical and modern times and have become more rigorous with widespread use of new lineage-tracing tools in the 1990s (Kimmel et al. 1990; Lawson et al. 1991; Stern et al. 1992; Beddington and Smith 1993). Although the similarities in gene expression are striking, it is not easy to arrive at a simple designation of the dorsoventral or anterior–posterior axes. For amphibians, the convention has been that a dorsoventral axis can be drawn through the organizer on one side and a ventral territory on the other. Given the enormous amount of rearrangement of cells in gas-

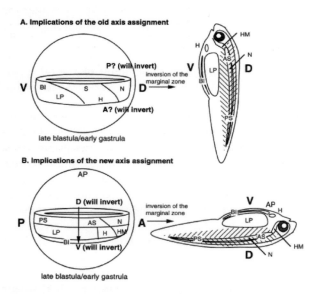

Figure 2. Reassignment of the axes in amphibian embryos. (*A*) The prospective mesoderm is the ring of tissue, and is shown with a dorsal to ventral organization from right to left. In *B*, the new proposal is illustrated, with an emphasis on the animal to vegetal correspondence of prospective muscle to lateral plate and blood islands, and an anterior to posterior organization from right to left in the non-organizer mesoderm. (A) Anterior; (AP) animal pole; (AS) anterior somites; (BI) blood islands; (D) dorsal; (H) heart; (HM) head mesoderm; (LP) lateral plate; (N) notochord; (P) posterior; (PS) posterior somites; (V) ventral. (Reprinted, with permission, from Gerhart 2002 [©Wiley-Liss].)

trulation, there is no simple formula for nomenclature, and the difficulties are compounded by the choice of whether the end of gastrulation, or a later tadpole stage, is chosen for anatomical reference. Furthermore, descriptions of regions of the embryo are not necessarily explicit about whether they refer to the fate map or the state of specification of tissues. For example, the ventral state of specification of the ventrolateral marginal zone has contributed to the view that the side of the embryo opposite the organizer should be viewed as ventral (Dale and Slack 1987b). However, during normal development, this tissue populates the tail bud (Smith and Slack 1983), and much of it is recruited by the caudal organizer into paraxial tissues (Dale and Slack 1987a; Moody 1987; Lane and Sheets 2000). A further historical complication derives from the use of molecular markers; muscle markers were among the first obtained and have been considered dorsal markers, although an argument could be made that the marker should be more rigorously used to define paraxial fates. If muscle is considered dorsal, and muscle also derives from the ventral marginal zone, the logical inconsistency is all too apparent.

Furthermore, the origins of the blood were not defined in early fate maps and were inferred based on states of specification of small explants. Recently, the blood, which is undeniably ventral in the blood islands of the tadpole, has been shown to have very mixed origins, but with contributions from around the entire marginal zone, with a tendency to come from the vegetal part of the marginal zone (Tracey et al. 1998; Lane and Smith 1999; Lane and Sheets 2002). Thus, some of the blood comes from blastomeres in the prospective organizer territory. Again, it is important to realize that gastrulation movements spread the original organizer territory over a considerable region by the end of gastrulation; whereas most cells remain in the axial derivatives, some of the leading edge of the original organizer region will migrate into the future ventral region of the tadpole, where it populates not only anterior endodermal derivatives such as the liver (Bouwmeester et al. 1996), but also some of the mesoderm in the anterior blood islands (Lane and Smith 1999). The cells that adopt this extreme movement and become separated from organizer influences can become blood, a ventral tissue, even though most of their original neighbors in the late blastula remain in the dorsal and axial region. This extreme difference in behavior illustrates the danger of affixing dorsal and ventral labels literally to the blastula stages.

In the most extreme revision of the fate map, an anterior–posterior axis is drawn through the marginal zone, and a dorsoventral axis along the animal–vegetal axis (Gerhart 2002; Lane and Sheets 2002). Although the posterior designation of the marginal zone works well (Smith and Slack 1983), the fates of the cells nearer the organizer are more mixed and spread over a considerable distance from anterior to posterior in the trunk and tail of the tadpole. Once again, it becomes difficult to be too literal about an anterior–posterior distinction mapped back onto the blastula, and whereas an anterior designation is somewhat defensible for the paraxial mesoderm next to the organizer, it breaks down entirely for the organizer, whose axial derivatives span the entire rostrocaudal extent of the tadpole. At the onset of gastrulation, the organizer also possesses intrinsic anterior–posterior patterning and inductive properties (Zoltewicz and Gerhart 1997). Furthermore, the revised anterior–posterior description of the embryo does not accommodate the ectoderm well, since the organizer side of the embryo forms most of the neural plate, which is entirely dorsal at the end of gastrulation.

The new fate maps are all made at the early cleavage stages, and lines on the fate maps have been drawn by inference from these data. There is a strong case to be made for repeating the classic fate maps of the *Xenopus* blastula and gastrula stages (Keller 1975, 1976), so that the gastrula-stage origins of the various tissues could be drawn with greater certainty. For the cleavage stages, cell mixing and the sequential set of movements over time conspire to make any simple set of labels imperfect. It could also be argued that the state of specification, as well as the ultimate fate, can be used to describe the states of the marginal zone. Finally, there is no unambiguously obvious stage that should be chosen for reference. Morphogenesis continues, and even within one region, such as the somite, the mediolateral organization of the early neurula changes, so that the most medial somite remains near the notochord, while the more lateral somite unfolds to lie both dorsal and ventral to the notochord (Fig. 3). Any designation of fates will be wrong in detail, so the main questions to be addressed are the usefulness of terms for convenience and for understanding developmental mechanisms. In this respect, the recent papers have made a good case that the naïve dorsal/ventral labels have obscured important mechanisms. In particular, the animal–vegetal organization of the marginal zone that is driven by MAP kinase signaling specifies muscle and blood and is a clear example where rethinking the axes has led to new mechanistic insights (Kumano and Smith 2002).

While embryologists argue about what should be considered anterior or posterior mesoderm, dorsal and ventral, the tadpole mocks us by ignoring the debate and undergoing such morphogenesis that we should take the labels somewhat lightly. The debate has been useful, and will continue to be, but there also needs to be some labeling of the gastrula for convenience. My own view derives from the primacy of organizer signaling, and the organizer derivation of the notochord, which seems to be a patently dorsal tissue. At the late blastula and gastrula stage, small explants of the ventral marginal zone produce more blood and mesenchyme than small explants of the dorsal side and only differentiate blood when they are able to get away from the direct influence of the organizer by morphogenesis. It seems reasonable that states of specification, and not just ultimate

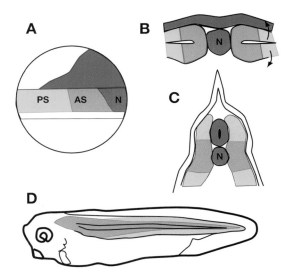

Figure 3. This series of fate maps is compiled from the variety of fate maps referenced in the text, and illustrates the difficulty of assigning either an anterior–posterior or dorsoventral axis to the somites of the marginal zone. *A* marks the somitic region close to the organizer (*pink*) or distant from it (*pale pink*). (*B*) Schematic section of the dorsal region of a neurula; following gastrulation, the pink region remains close to the notochord—medially. (*C*) Schematic section of a tadpole; as the somites unfold, the pink region remains medial, but the pale pink region lies both dorsal and ventral to the medial somite. (*D*) Lateral view of a tadpole. Although the region distant from the organizer occupies posterior regions, there is considerable spread along the anterior–posterior axis. For the somites that derive from the organizer region, there is spread along the entire anterior–posterior axis.

fates in normal development, should be permitted in a simplistic description. Thus, I will continue to refer to the dorsal marginal zone, and since an axis is defined by two ends, I will refer to a dorsoventral axis of the gastrula (although dorsal-posterior would be equally defensible). However, when certain of my colleagues are within earshot, I shall try to refer to organizer and non-organizer tissues.

ZEBRAFISH

If the fate map of *Xenopus* is a mess, then that of the zebrafish is more so, since there is extensive cell mixing during cleavage and blastula stages (Warga and Kimmel 1990). Nonetheless, the broad fates of regions of these embryos are similar, and to date the mechanisms employed to partition mesodermal cell fates during gastrulation have been the same. The dominant lessons from the zebrafish have come from loss-of-function genetics and have given rigorous insights into the signaling pathways and the transcriptional networks used to specify different tissues.

Mutations in Signaling Components

A large number of mutants affect the amount of tissue allocated to different dorsoventral fates, and the molecular identification of the mutations showed that these lie in the BMP and Wnt pathways and therefore prove that the mechanisms proposed to be important in *Xenopus* are indeed conserved and important. The BMP pathway is affected by mutations in the BMPs (*swirl* and *snailhouse*), a BMP receptor, *alk8* (*lost-a-fin*), and the intracellular transducer *smad5* (*somitabun*). Finally, two transcription factor-encoding targets of BMP signaling, *vox* and *vent*, when removed by a deletion, have profound effects on pattern (Imai et al. 2001). All of these mutations lead to animals with deficits in the amount of tail and blood, with varying effects on the amount of somite, notochord, and head. In the opposite direction, mutations in *chordin* (*dino* or *chordino*) identify this antagonist as an essential modulator of the strength of Bmp signaling. In addition to mutations in *chordin*, a mutation in *sizzled* (*ogon* or *mercedes*) also blocks BMP signaling. This is an interesting example, because the exact molecular mechanism and target of Sizzled action have been difficult to determine. By molecular analogy with its relatives, which encode Wnt antagonists, the target should be a Wnt ligand, but Sizzled does not inhibit known Wnts (Bradley et al. 2000; Collavin and Kirschner 2003; Yabe et al. 2003). Indeed, for Ogon to function, Chordin is required, suggesting the possibility that Ogon may interact more directly with the BMP modulators (Yabe et al. 2003). Whatever the molecular mechanism, the loss of *ogon* function shows that Ogon's normal role is to promote dorsal development.

For the Wnt pathway, mutations that act during gastrulation include a deficiency of the *wnt8* locus, which is transcribed into a bicistronic transcript encoding two functional Wnt8 proteins (Lekven et al. 2001). In the absence of these, the organizer expands, and subsequent development includes an expansion of the notochord.

Mutations in Transcription Factors

Mutations have identified crucial roles for a variety of transcription factors and particularly those with T-box domains or homeodomains. The mutations particularly affect the segregation of mesodermal tissues into notochord and somite. These transcription factors are homologous to factors that had first been isolated from mouse or frog, but their roles had not been elucidated there. In the zebrafish, mutations showed the involvement of transcription factors in tissue fates in a way that overexpression experiments in *Xenopus* or fish could not. When transcription factors are expressed at elevated levels, they lose their specificity for their physiological binding targets, since the difference between specific and nonspecific binding is not great. This is in contrast to overexpression of signaling components, which generally retain considerable specificity. Thus, the genetic approach has been crucial in rigorously showing the physiological roles of transcription factors, which can only be suggested from experiments with ectopic expression of wild-type or antimorphic factors (Conlon et al. 1996; for review, see Sive et al. 1999).

The properties of transcription factors that regulate mesodermal segregation can be appreciated by considering the properties of *no tail*, *spadetail*, and *floating head*. These encode T-box domain proteins and homeobox proteins. Floating head is a homeobox protein that is expressed in the notochord and is required for notochord development (Talbot et al. 1995). In its absence, the notochord is lost, and somites meet at the midline. In turn, somite identity requires Spadetail, and at least part of the mechanism by which Floating head specifies notochord is to inhibit *spadetail* function (Amacher and Kimmel 1998). In the double mutant, trunk muscle is not formed, but some anterior notochord does form. Therefore, Floating head does not so much promote notochord fate as repress paraxial fates.

no tail encodes a T-box transcription factor and is the homolog of *brachyury* (Schulte-Merker et al. 1994), which was first identified through positional cloning of the *T* locus in the mouse (Herrmann et al. 1990). In *no tail* mutants, the anterior trunk notochord and somites are still present, but posterior morphogenesis and differentiation are aberrant. Interestingly, these factors also affect the specification of floor plate, with *no tail* mutants having an excess of floor plate. Indeed, zebrafish mutants have presented some of the clearest evidence that the notochord and floor plate must share a common lineage, with proteins such as No Tail repressing floor plate fates. In this picture, the floor plate can be considered a part of the mesoderm, also induced early in gastrulation, and requiring Nodal/Smad2 signaling either for its induction or for the fate of an early inducing population of cells that subsequently develops into prechordal plate (Rebagliati et al. 1998; Sampath et al. 1998; Amacher et al. 2002; Tian et al. 2003). Whether this mechanism is universally true in the vertebrates is still debated (Le Douarin and Halpern 2000; Placzek et al. 2000).

Combinatorial Functions of T-Box Genes

Transcription factors interact with one another on DNA, and this endows them with new properties. Such properties are beginning to be worked out for T-box function in the trunk of the zebrafish. The *no tail*, *spadetail*, and *tbx6* expression domains overlap with one another, and where they overlap, there is evidence that they do not act alone, but act additively, or can interact to provide new specificity, or can mutually antagonize one another's activity (Goering et al. 2003). Thus, where *no tail* and *spadetail* overlap, *mesogenin* turns on and requires both *no tail* and *spadetail* function to do so; it does not turn on in the domain occupied by just *no tail* or just *spadetail*.

The interesting effects of these mutations demonstrate a large gap in understanding between the signaling pathways and the final effectors of differentiation. This area is ripe for genomic approaches.

CHICK

Manipulations That Alter Specification of Regions of the Gastrula

It has been long established that the lateral mesoderm of the chick embryo can be respecified to somitic tissues during and after gastrulation, either by grafts of notochord or Hensen's node (Nicolet 1970; Hornbruch et al. 1979). In this respect, the ability of the organizer to change the fate of mesoderm is similar to the observations in *Xenopus*, although the lability may extend even further into neurulation stages. As in the other vertebrates discussed previously, BMP signaling is important in subdividing the tissues of the mediolateral axis. BMP doses have been manipulated by implantation of beads soaked in BMPs and their antagonists, or cells expressing BMPs and their antagonists (Tonegawa et al. 1997; Tonegawa and Takahashi 1998; Streit and Stern 1999). These experiments show that increased amounts of somites can be induced by blocking BMP activity, with the converse also true. Somites can be induced outside their normal position by blocking BMPs, although notochord is not (Tonegawa and Takahashi 1998; Streit and Stern 1999). Interestingly, experiments in the chick document different properties for the BMP antagonists Noggin and Chordin; whereas Noggin is able to induce somite formation in the lateral plate, Chordin is not. In contrast, at an earlier stage, Chordin is able to induce some streak formation whereas Noggin is not (Streit and Stern 1999). Such experiments suggest that different BMP family members, which are differentially inhibited by the antagonists, may have specific roles during early patterning of the embryo. Experiments in *Xenopus* have not revealed such differences; instead, the antagonists appear to have similar properties. Thus, experiments in the chick may help to address why there are so many different antagonists, suggesting that different antagonists have qualitatively different activities, rather than additive effects.

Clearly, BMPs are not the only important players in signaling during chick embryogenesis, but for the signals involved in mediolateral patterning of the mesoderm during gastrulation, they have been the prominent activities to be identified. Slightly later in development, a greater collection of signaling pathways participates in partitioning the somites, intermediate mesoderm, and lateral plate into definitive tissues. However, these signals act during neurulation and are outside the scope of this chapter.

MOUSE

Just as the frog and fish provide extreme versions of gastrulation, where cell division is modest and even not required for gastrulation movements (Harris and Hartenstein 1991), the mouse provides the opposite extreme, where cell division is a paramount mechanism to provide new material for gastrulation movements (Hogan et al. 1994). This increase in mass occurs initially by proliferation and cell migration from the primitive streak and node, and further mass increase occurs throughout the embryo so that the streak rapidly appears to be a fairly insignificant part of the whole embryo. Even though gastrulation continues over days, the precise temporal sequence of tissue specifications is not well described, and this is due to the difficulty of manipulating the embryo in vitro. Fate maps show that the relative organization of tissues is topologically equivalent to that in the other vertebrates, and as in the chick (Psychoyos and Stern 1996), the node produces midline tissues, whereas the rest of the streak produces paraxial and other mesodermal tissues (Lawson et al. 1991; Hogan et al. 1994; Kinder et al. 1999). Much of the mediolateral organization derives from the position in the streak where the cells emerge, but the dominant appearance of the embryo is as an anterior-to-posterior series of tissues produced by the node and streak. Thus, in considering mouse embryos, more consideration is given to anterior–posterior patterning and the consequences of changes in the levels of mesoderm inducers (see, e.g., Vincent et al. 2003). However, important insights have come from the analysis of targeted mutations, which often affect the mediolateral patterning of the mesoderm.

Mutations in Transcription Factors

The simplest changes in mesoderm patterning arise as a result of loss of particular transcription factors. For example, if *Foxa2/HNF-3β* is removed, the node is not maintained and the notochord does not form (Ang and Rossant 1994; Weinstein et al. 1994). The somites fuse across the midline, and there are defects that result from the lack of the patterning activity of the node derivatives. Perhaps the most surprising effect of the mutation is the lack of effects on anterior–posterior patterning (Klingensmith et al. 1999). Had the loss of the node been equivalent to an early loss of the organizer in a frog or fish, rostral defects would be expected. Perhaps this shows that there is a very early resolution of anterior identities in the very early streak stage, but this discussion belongs in a different chapter.

One of the most spectacular consequences of removing a transcription factor is seen when *Tbx6* is removed. In this mutant, the somites transfate to neural tube (Chapman and Papaioannou 1998). This illustrates the crucial role of the T-box family of transcription factors in maintaining mesodermal identities, at least in the mouse. No equivalent mutant has been found in the zebrafish, and indeed, knockdowns of *tbx6* function have relatively mild effects (Goering et al. 2003).

Mutations in Signaling Pathways

By analogy with manipulations in the frog and the chick, it might be expected that changes in BMP signaling would affect the allocation of cells to the somite compartment, but this allocation does not appear to be grossly affected in mutants for the BMP antagonists *Chordin* and *Noggin* (Bachiller et al. 2000). Instead, and in contrast to the effect of removing *FoxA2*, there is a failure to maintain full organizer function, as seen through a failure to maintain anterior identities, and ultimately loss of head structures.

Mutation of *BMP4* illustrates the complexity of mouse embryonic patterning, where a series of interactions take place between embryonic and extraembryonic tissues during mesoderm formation (Fujiwara et al. 2002; for review, see Lu et al. 2001). Such complexity makes it difficult to assess the degree to which mediolateral patterning of the mesoderm depends on the same signals as in the frog and fish. In addition, small changes in cell fate allocation may be rapidly compensated by cell division.

FGFR1

Mutations in *FGFR1* lead to an increase in the size of the notochord at the expense of the somites (Yamaguchi et al. 1994). These observations make clear that changes in allocation of cells to somite and notochord can, in principle, be observed in the mouse. Mechanistically, this observation does not fit easily with paradigms from other vertebrates and is therefore telling us something quite different about patterning of the mesoderm. One way to look at the effect is by considering the movements of cells out of the streak, and it has been proposed that the main phenotypic consequences of this mutation may result from changes in morphogenesis, particularly as a result of changes in adhesion (Ciruna et al. 1997; Ciruna and Rossant 2001).

CONCLUSION

In surveying the vertebrates, the dorsoventral or mediolateral organization of the mesoderm is influenced by several signaling pathways. Where it has been easy to analyze this experimentally, there is a period of lability in mesodermal fates during gastrulation, and BMP signaling emerges as a dominant pathway in this aspect of patterning. It has been attractive to consider that many fates are specified by a morphogen gradient of BMPs or other ligands, but evidence for this proposition is still elusive, and instead it is likely that combinations of pathways conspire to allocate the mesoderm to different tissues.

Furthermore, it is crucial to consider the activities of pathways over time, since the responses of tissues become restricted or change as the cells mature. Different experimental systems have provided their own insights into the mechanisms of action of signaling pathways and transcriptional networks in specifying mesodermal fates. There is an assumption that the principal mechanisms should be conserved, but as specific cases are analyzed, some important differences are evident. Finally, the pattern of the vertebrate embryo is not specified by tidy dorsoventral and anterior–posterior Cartesian coordinates, but rather emerges by more complex interactions, and it is difficult to impose a simplistic description of early fate maps.

REFERENCES

Agius E., Oelgeschlager M., Wessely O., Kemp C., and De Robertis E.M. 2000. Endodermal Nodal–related signals and mesoderm induction in *Xenopus. Development* **127:** 1173–1183.

Amacher S.L. and Kimmel C.B. 1998. Promoting notochord fate and repressing muscle development in zebrafish axial mesoderm. *Development* **125:** 1397–1406.

Amacher S.L., Draper B.W., Summers B.R., and Kimmel C.B. 2002. The zebrafish T-box genes *no tail* and *spadetail* are required for development of trunk and tail mesoderm and medial floor plate. *Development* **129:** 3311–3323.

Amaya E., Stein P.A., Musci T.J., and Kirschner M.W. 1993. FGF signalling in the early specification of mesoderm in *Xenopus. Development* **118:** 477–487.

Ang S.L. and Rossant J. 1994. HNF-3 beta is essential for node and notochord formation in mouse development. Cell **78:** 561–574.

Bachiller D., Klingensmith J., Kemp C., Belo J.A., Anderson R.M., May S.R., McMahon J.A., McMahon A.P., Harland R.M., Rossant J., and De Robertis E.M. 2000. The organizer factors Chordin and Noggin are required for mouse forebrain development. *Nature* **403:** 658–661.

Baker J.C., Beddington R.S., and Harland R.M. 1999. Wnt signaling in *Xenopus* embryos inhibits bmp4 expression and activates neural development. *Genes Dev.* **13:** 3149–3159.

Beddington R.S. and Smith J.C. 1993. Control of vertebrate gastrulation: Inducing signals and responding genes. *Curr. Opin. Genet. Dev.* **3:** 655–661.

Blitz I.L., Shimmi O., Wunnenberg-Stapleton K., O'Connor M.B., and Cho K.W. 2000. Is chordin a long-range- or short-range-acting factor? Roles for BMP1-related metalloproteases in chordin and BMP4 autofeedback loop regulation. *Dev. Biol.* **223:** 120–138.

Boterenbrood E.C. and Nieuwkoop P.D. 1973. The formation of the mesoderm in Urodelan amphibians: V. Its regional induction by the endoderm. *Wilhelm Roux' Arch.Entwicklungsmech. Org.* **173:** 319–332.

Bouwmeester T., Kim S., Sasai Y., Lu B., and De Robertis E.M. 1996. Cerberus is a head-inducing secreted factor expressed in the anterior endoderm of Spemann's organizer. *Nature* **382:** 595–601.

Bradley L., Sun B., Collins-Racie L., LaVallie E., McCoy J., and Sive H. 2000. Different activities of the frizzled-related proteins *frzb2* and *sizzled2* during *Xenopus* anteroposterior patterning. *Dev. Biol.* **227:** 118–132.

Carroll T.J. and Vize P.D. 1999. Synergism between Pax-8 and lim-1 in embryonic kidney development. *Dev. Biol.* **214:** 46–59.

Chapman D.L. and Papaioannou V.E. 1998. Three neural tubes in mouse embryos with mutations in the T-box gene Tbx6. *Nature* **391:** 695–697.

Christen B. and Slack J.M. 1997. FGF-8 is associated with anteroposterior patterning and limb regeneration in *Xenopus*. *Dev. Biol.* **192:** 455–466.

———. 1999. Spatial response to fibroblast growth factor signalling in *Xenopus* embryos. *Development* **126:** 119–125.

Christian J.L. and Moon R.T. 1993. Interactions between Xwnt-8 and Spemann organizer signaling pathways generate dorsoventral pattern in the embryonic mesoderm of *Xenopus*. *Genes Dev.* **7:** 13–28.

Ciruna B. and Rossant J. 2001. FGF signaling regulates mesoderm cell fate specification and morphogenetic movement at the primitive streak. *Dev. Cell* **1:** 37–49.

Ciruna B.G., Schwartz L., Harpal K., Yamaguchi T.P., and Rossant J. 1997. Chimeric analysis of *fibroblast growth factor receptor-1* (*Fgfr1*) function: A role for FGFR1 in morphogenetic movement through the primitive streak. *Development* **124:** 2829–2841.

Collavin L. and Kirschner M.W. 2003. The secreted Frizzled-related protein Sizzled functions as a negative feedback regulator of extreme ventral mesoderm. *Development* **130:** 805–816.

Conlon F.L., Sedgwick S.G., Weston K.M., and Smith J.C. 1996. Inhibition of Xbra transcription activation causes defects in mesodermal patterning and reveals autoregulation of Xbra in dorsal mesoderm. *Development* **122:** 2427–2435.

Cornell R.A. and Kimelman D. 1994. Activin-mediated mesoderm induction requires FGF. *Development* **120:** 453–462.

Cornell R.A., Musci T.J., and Kimelman D. 1995. FGF is a prospective competence factor for early activin-type signals in *Xenopus* mesoderm induction. *Development* **121:** 2429–2437.

Cui Y., Hackenmiller R., Berg L., Jean F., Nakayama T., Thomas G., and Christian J.L. 2001. The activity and signaling range of mature BMP-4 is regulated by sequential cleavage at two sites within the prodomain of the precursor. *Genes Dev.* **15:** 2797–2802.

Curran K.L. and Grainger R.M. 2000. Expression of activated MAP kinase in *Xenopus laevis* embryos: Evaluating the roles of FGF and other signaling pathways in early induction and patterning. *Dev. Biol.* **228:** 41–56.

Dale J.K., Vesque C., Lints T.J., Sampath T.K., Furley A., Dodd J., and Placzek M. 1997. Cooperation of BMP7 and SHH in the induction of forebrain ventral midline cells by prechordal mesoderm. *Cell* **90:** 257–269.

Dale L. and Slack J.M. 1987a. Fate map for the 32-cell stage of *Xenopus laevis*. *Development* **99:** 527–551.

———. 1987b. Regional specification within the mesoderm of early embryos of *Xenopus laevis*. *Development* **100:** 279–295.

Dale L., Howes G., Price B.M., and Smith J.C. 1992. Bone morphogenetic protein 4: A ventralizing factor in early *Xenopus* development. *Development* **115:** 573–585.

De Robertis E.M., Wessely O., Oelgeschlager M., Brizuela B., Pera E., Larrain J., Abreu J., and Bachiller D. 2001. Molecular mechanisms of cell-cell signaling by the Spemann-Mangold organizer. *Int. J. Dev. Biol.* **45:** 189–197.

Domingo C. and Keller R. 2000. Cells remain competent to respond to mesoderm-inducing signals present during gastrulation in *Xenopus laevis*. *Dev. Biol.* **225:** 226–240.

Dosch R., Gawantka V., Delius H., Blumenstock C., and Niehrs C. 1997. Bmp-4 acts as a morphogen in dorsoventral mesoderm patterning in *Xenopus*. *Development* **124:** 2325–2334.

Eimon P.M. and Harland R.M. 1999. In *Xenopus* embryos, BMP heterodimers are not required for mesoderm induction, but BMP activity is necessary for dorsal/ventral patterning. *Dev. Biol.* **216:** 29–40.

———. 2002. Effects of heterodimerization and proteolytic processing on Derriere and Nodal activity: Implications for mesoderm induction in *Xenopus*. *Development* **129:** 3089–3103.

Fekany-Lee K., Gonzalez E., Miller-Bertoglio V., and Solnica-Krezel L. 2000. The homeobox gene *bozozok* promotes anterior neuroectoderm formation in zebrafish through negative regulation of BMP2/4 and Wnt pathways. *Development* **127:** 2333–2345.

Frank D. and Harland R.M. 1991. Transient expression of XMyoD in non-somitic mesoderm of *Xenopus* gastrulae. *Development* **113:** 1387–1393.

Fujiwara T., Dehart D.B., Sulik K.K., and Hogan B.L. 2002. Distinct requirements for extra-embryonic and embryonic bone morphogenetic protein 4 in the formation of the node and primitive streak and coordination of left-right asymmetry in the mouse. *Development* **129:** 4685–4696.

Gerhart J. 2002. Changing the axis changes the perspective. *Dev. Dyn.* **225:** 380–383.

Gimlich R.L. 1986. Acquisition of developmental autonomy in the equatorial region of the *Xenopus* embryo. *Dev. Biol.* **115:** 340–352.

Gimlich R.L. and Cooke J. 1983. Cell lineage and the induction of second nervous systems in amphibian development. *Nature* **306:** 471–473.

Gimlich R.L. and Gerhart J.C. 1984. Early cellular interactions promote embryonic axis formation in *Xenopus laevis*. *Dev. Biol.* **104:** 117–130.

Glinka A., Wu W., Delius H., Monaghan A.P., Blumenstock C., and Niehrs C. 1998. Dickkopf-1 is a member of a new family of secreted proteins and functions in head induction. *Nature* **391:** 357–362.

Goering L.M., Hoshijima K., Hug B., Bisgrove B., Kispert A., and Grunwald D.J. 2003. An interacting network of T-box genes directs gene expression and fate in the zebrafish mesoderm. *Proc. Natl. Acad. Sci.* **100:** 9410–9415.

Goto T., Katada T., Kinoshita T., and Kubota H.Y. 2000. Expression and characterization of *Xenopus type I collagen alpha 1* (*COL1A1*) during embryonic development. *Dev. Growth Differ.* **42:** 249–256.

Green J.B., Howes G., Symes K., Cooke J., and Smith J.C. 1990. The biological effects of XTC-MIF: Quantitative comparison with *Xenopus* bFGF. *Development* **108:** 173–183.

Grimm O.H. and Gurdon J.B. 2002. Nuclear exclusion of Smad2 is a mechanism leading to loss of competence. *Nat. Cell Biol.* **4:** 519–522.

Gurdon J.B. 1988. A community effect in animal development. *Nature* **336:** 772–774.

Gurdon J.B. and Bourillot P.Y. 2001. Morphogen gradient interpretation. *Nature* **413:** 797–803.

Hamburger V. 1988. *The heritage of experimental embryology: Hans Spemann and the organizer.* Oxford University Press, New York.

Hammerschmidt M., Serbedzija G.N., and McMahon A.P. 1996. Genetic analysis of dorsoventral pattern formation in the zebrafish: Requirement of a BMP-like ventralizing activity and its dorsal repressor. *Genes Dev.* **10:** 2452–2461.

Hansen C.S., Marion C.D., Steele K., George S., and Smith W.C. 1997. Direct neural induction and selective inhibition of mesoderm and epidermis inducers by Xnr3. *Development* **124:** 483–492.

Harland R. and Gerhart J. 1997. Formation and function of Spemann's organizer. *Annu. Rev. Cell Dev. Biol.* **13:** 611–667.

Harris W.A. and Hartenstein V. 1991. Neuronal determination without cell division in *Xenopus* embryos. *Neuron* **6:** 499–515.

Hashimoto–Partyka M.K., Yuge M., and Cho K.W. 2003. Nodal signaling in *Xenopus* gastrulae is cell-autonomous and patterned by beta-catenin. *Dev. Biol.* **253:** 125–138.

Hemmati-Brivanlou A., Kelly O.G., and Melton D.A. 1994. Follistatin, an antagonist of activin, is expressed in the Spemann organizer and displays direct neuralizing activity. *Cell* **77:** 283–295.

Herrmann B.G., Labeit S., Poustka A., King T.R., and Lehrach H. 1990. Cloning of the T gene required in mesoderm formation in the mouse. *Nature* **343:** 617–622.

Hirsinger E., Duprez D., Jouve C., Malapert P., Cooke J., and Pourquie O. 1997. Noggin acts downstream of Wnt and Sonic Hedgehog to antagonize BMP4 in avian somite patterning. *Development* **124:** 4605–4614.

Hogan B.L., Beddington R., Costantini F., and Lacey E. 1994. *Manipulating the mouse embryo: A laboratory manual*, 2nd edition. Cold Spring Harbor Laboratory Press, Cold Spring Harbor, New York.

Hoppler S., Brown J.D., and Moon R.T. 1996. Expression of a dominant-negative Wnt blocks induction of MyoD in *Xenopus* embryos. *Genes Dev.* **10:** 2805–2817.

Hopwood N.D., Pluck A., Gurdon J.B., and Dilworth S.M. 1992. Expression of XMyoD protein in early *Xenopus laevis* embryos. *Development* **114:** 31–38.

Hornbruch A., Summerbell D., and Wolpert L. 1979. Somite formation in the early chick embryo following grafts of Hensen's node. *J. Embryol. Exp. Morphol.* **51:** 51–62.

Hsu D.R., Economides A.N., Wang X., Eimon P.M., and Harland R.M. 1998. The *Xenopus* dorsalizing factor Gremlin identifies a novel family of secreted proteins that antagonize BMP activities. *Mol. Cell* **1:** 673–683.

Imai Y., Gates M.A., Melby A.E., Kimelman D., Schier A.F., and Talbot W.S. 2001. The homeobox genes *vox* and *vent* are redundant repressors of dorsal fates in zebrafish. *Development* **128:** 2407–2420.

Isaacs H.V., Pownall M.E., and Slack J.M. 1994. eFGF regulates Xbra expression during *Xenopus* gastrulation. *EMBO J.* **13:** 4469–4481.

———. 1995. eFGF is expressed in the dorsal midline of *Xenopus laevis. Int. J. Dev. Biol.* **39:** 575–579.

Jones C.M. and Smith J.C. 1998. Establishment of a BMP-4 morphogen gradient by long-range inhibition. *Dev. Biol.* **194:** 12–17.

Jones C.M., Armes N., and Smith J.C. 1996a. Signalling by TGF-β family members: Short-range effects of Xnr-2 and BMP-4 contrast with the long-range effects of activin. *Curr. Biol.* **6:** 1468–1475.

Jones C.M., Dale L., Hogan B.L., Wright C.V., and Smith J.C. 1996b. Bone morphogenetic protein-4 (BMP-4) acts during gastrula stages to cause ventralization of *Xenopus* embryos. *Development* **122:** 1545–1554.

Jones C.M., Kuehn M.R., Hogan B.L., Smith J.C., and Wright C.V. 1995. Nodal-related signals induce axial mesoderm and dorsalize mesoderm during gastrulation. *Development* **121:** 3651–3662.

Jones C.M., Lyons K.M., Lapan P.M., Wright C.V., and Hogan B.L. 1992. DVR-4 (bone morphogenetic protein-4) as a posterior-ventralizing factor in *Xenopus* mesoderm induction. *Development*

115: 639–647.

Keller R.E. 1975. Vital dye mapping of the gastrula and neurula of *Xenopus laevis*. I. Prospective areas and morphogenetic movements of the superficial layer. *Dev. Biol.* **42:** 222–241.

———. 1976. Vital dye mapping of the gastrula and neurula of *Xenopus laevis*. II. Prospective areas and morphogenetic movements of the deep layer. *Dev. Biol.* **51:** 118–137.

Kelley C., Yee K., Harland R., and Zon L.I. 1994. Ventral expression of GATA-1 and GATA-2 in the *Xenopus* embryo defines induction of hematopoietic mesoderm. *Dev. Biol.* **165:** 193–205.

Khokha M.K., Hsu D., Brunet L.J., Dionne M.S., and Harland R.M. 2003. Gremlin is the BMP antagonist required for maintenance of Shh and Fgf signals during limb patterning. *Nat. Genet.* **34:** 303–307.

Kikkawa M., Yamazaki M., Izutsu Y., and Maeno M. 2001. Two-step induction of primitive erythrocytes in *Xenopus laevis* embryos: Signals from the vegetal endoderm and the overlying ectoderm. *Int. J. Dev. Biol.* **45:** 387–396.

Kimmel C.B., Warga R.M., and Schilling T.F. 1990. Origin and organization of the zebrafish fate map. *Development* **108:** 581–594.

Kinder S.J., Tsang T.E., Quinlan G.A., Hadjantonakis A.K., Nagy A., and Tam P.P. 1999. The orderly allocation of mesodermal cells to the extraembryonic structures and the anteroposterior axis during gastrulation of the mouse embryo. *Development* **126:** 4691–4701.

Klingensmith J., Ang S.L., Bachiller D., and Rossant J. 1999. Neural induction and patterning in the mouse in the absence of the node and its derivatives. *Dev. Biol.* **216:** 535–549.

Knecht A.K. and Harland R.M. 1997. Mechanisms of dorsal-ventral patterning in noggin-induced neural tissue. *Development* **124:** 2477–2488.

Koos D.S. and Ho R.K. 1999. The *nieuwkoid/dharma* homeobox gene is essential for *bmp2b* repression in the zebrafish pregastrula. *Dev. Biol.* **215:** 190–207.

Kumano G. and Smith W.C. 2000. FGF signaling restricts the primary blood islands to ventral mesoderm. *Dev. Biol.* **228:** 304–314.

———. 2002. Revisions to the *Xenopus* gastrula fate map: implications for mesoderm induction and patterning. *Dev. Dyn.* **225:** 409–421.

Kumano G., Ezal C., and Smith W.C. 2001. Boundaries and functional domains in the animal/vegetal axis of *Xenopus* gastrula mesoderm. *Dev. Biol.* **236:** 465–477.

LaBonne C. and Whitman M. 1994. Mesoderm induction by activin requires FGF-mediated intracellular signals. *Development* **120:** 463–472.

———. 1997. Localization of MAP kinase activity in early *Xenopus* embryos: implications for endogenous FGF signaling. *Dev. Biol.* **183:** 9–20.

Lane M. C. and Sheets M. D. 2000. Designation of the anterior/posterior axis in pregastrula *Xenopus laevis. Dev. Biol.* **225:** 37–58.

———. 2002. Primitive and definitive blood share a common origin in *Xenopus*: A comparison of lineage techniques used to construct fate maps. *Dev. Biol.* **248:** 52–67.

Lane M.C. and Smith W.C. 1999. The origins of primitive blood in *Xenopus*: Implications for axial patterning. *Development* **126:** 423–434.

Laurent M.N. and Cho K.W. 1999. Bone morphogenetic protein antagonism of Spemann's organizer is independent of Wnt signaling. *Dev. Biol.* **206:** 157–162.

Lawson K.A., Meneses J.J., and Pedersen R.A. 1991. Clonal analysis of epiblast fate during germ layer formation in the mouse embryo. *Development* **113:** 891–911.

Le Douarin N.M. and Halpern M.E. 2000. Discussion point. Origin and specification of the neural tube floor plate: Insights from the chick and zebrafish. *Curr. Opin. Neurobiol.* **10**: 23–30.

Lekven A.C., Thorpe C.J., Waxman J.S., and Moon R.T. 2001. Zebrafish *wnt8* encodes two wnt8 proteins on a bicistronic transcript and is required for mesoderm and neurectoderm patterning. *Dev. Cell* **1**: 103–114.

Lettice L.A. and Slack J.M.W. 1993. Properties of the dorsalizing signal in gastrulae of *Xenopus laevis. Development* **117**: 263–271.

Leung T., Bischof J., Soll I., Niessing D., Zhang D., Ma J., Jackle H., and Driever W. 2003. *bozozok* directly represses *bmp2b* transcription and mediates the earliest dorsoventral asymmetry of *bmp2b* expression in zebrafish. *Development* **130**: 3639–3649.

Leyns L., Bouwmeester T., Kim S.-H., Piccolo S., and De Robertis E.M. 1997. Frzb-1 is a secreted antagonist of Wnt signaling expressed in the Spemann organizer. *Cell* **88**: 747–756.

Lombardo A., Isaacs H.V., and Slack J.M. 1998. Expression and functions of FGF-3 in *Xenopus* development. *Int. J. Dev. Biol.* **42**: 1101–1107.

Lu C.C., Brennan J., and Robertson E.J. 2001. From fertilization to gastrulation: Axis formation in the mouse embryo. *Curr. Opin. Genet. Dev.* **11**: 384–392.

Maeno M., Ong R.C., Xue Y., Nishimatsu S., Ueno N., and Kung H.F. 1994. Regulation of primary erythropoiesis in the ventral mesoderm of *Xenopus* gastrula embryo: Evidence for the expression of a stimulatory factor(s) in animal pole tissue. *Dev. Biol.* **161**: 522–529.

Mann R.W. 1995. "Characterization of neural-specific genes and mutagenesis of the growth factor noggin in embryogenesis of *Xenopus laevis.*" Ph.D. thesis. University of California, Berkeley.

Marcelle C., Stark M.R., and Bronner-Fraser M. 1997. Coordinate actions of BMPs, Wnts, Shh and Noggin mediate patterning of the dorsal somite. *Development* **124**: 3955–3963.

Mariani F.V., Choi G.B., and Harland R.M. 2001. The neural plate specifies somite size in the *Xenopus laevis* gastrula. *Dev. Cell* **1**: 115–126.

Marvin M.J., Di Rocco G., Gardiner A., Bush S.M., and Lassar A.B. 2001. Inhibition of Wnt activity induces heart formation from posterior mesoderm. *Genes Dev.* **15**: 316–327.

Mitchell T.S. and Sheets M.D. 2001. The FGFR pathway is required for the trunk-inducing functions of Spemann's organizer. *Dev. Biol.* **237**: 295–305.

Moody S. A. 1987. Fates of the blastomeres of the 32-cell-stage *Xenopus* embryo. *Dev. Biol.* **122**: 300–319.

Morin-Kensicki E.M. and Eisen J.S. 1997. Sclerotome development and peripheral nervous system segmentation in embryonic zebrafish. *Development* **124**: 159–167.

Nascone N. and Mercola M. 1995. An inductive role for the endoderm in *Xenopus* cardiogenesis. *Development* **121**: 515–523.

Nicolet G. 1970. Is the presumptive notochord responsible for somite genesis in the chick? *J. Embryol. Exp. Morphol.* **24**: 467–478.

Niehrs C., Steinbeisser H., and De Robertis E.M. 1994. Mesodermal patterning by a gradient of the vertebrate homeobox gene goosecoid. *Science* **263**: 817–820.

Nieuwkoop P.D. and Faber J. 1967. *Normal table of* Xenopus laevis *(Daudin). A systematical and chronological survey of the development from the fertilized egg till the end of metamorphosis.* North-Holland Publishing, Amsterdam, The Netherlands (reprinted Garland Publishing, New York [1994]).

Ohkawara B., Iemura S., ten Dijke P., and Ueno N. 2002. Action range of BMP is defined by its N-terminal basic amino acid core. *Curr.*

Biol. **12**: 205–209.

Paine-Saunders S., Viviano B.L., Economides A.N., and Saunders S. 2002. Heparan sulfate proteoglycans retain Noggin at the cell surface: A potential mechanism for shaping bone morphogenetic protein gradients. *J. Biol. Chem.* **277**: 2089–2096.

Pandur P., Lasche M., Eisenberg L.M., and Kuhl M. 2002. Wnt-11 activation of a non-canonical Wnt signalling pathway is required for cardiogenesis. *Nature* **418**: 636–641.

Pera E.M. and De Robertis E.M. 2000. A direct screen for secreted proteins in *Xenopus* embryos identifies distinct activities for the Wnt antagonists Crescent and Frzb-1. *Mech. Dev.* **96**: 183–195.

Piccolo S., Sasai Y., Lu B., and De Robertis E.M. 1996. Dorsoventral patterning in *Xenopus*: Inhibition of ventral signals by direct binding of chordin to BMP-4. *Cell* **86**: 589–598.

Piccolo S., Agius E., Leyns L., Bhattacharya S., Grunz H., Bouwmeister T., and DeRobertis E. 1999. The head inducer Cerberus is a multifunctional antagonist of Nodal, BMP and Wnt signals. *Nature* **397**: 707–710.

Placzek M., Dodd J., and Jessell T.M. 2000. Discussion point. The case for floor plate induction by the notochord. *Curr. Opin. Neurobiol.* **10**: 15–22.

Pownall M.E., Tucker A.S., Slack J.M., and Isaacs H.V. 1996. *eFGF, Xcad3* and Hox genes form a molecular pathway that establishes the anteroposterior axis in *Xenopus. Development* **122**: 3881–3892.

Psychoyos D. and Stern C.D. 1996. Fates and migratory routes of primitive streak cells in the chick embryo. *Development* **122**: 1523–1534.

Rebagliati M.R., Toyama R., Fricke C., Haffter P., and Dawid I.B. 1998. Zebrafish nodal-related genes are implicated in axial patterning and establishing left-right asymmetry. *Dev. Biol.* **199**: 261–272.

Reshef R., Maroto M., and Lassar A.B. 1998. Regulation of dorsal somitic cell fates: BMPs and Noggin control the timing and pattern of myogenic regulator expression. *Genes Dev.* **12**: 290–303.

Rupp R.A. and Weintraub H. 1991. Ubiquitous MyoD transcription at the midblastula transition precedes induction-dependent MyoD expression in presumptive mesoderm of *X. laevis. Cell* **65**: 927–937.

Sampath K., Rubinstein A.L., Cheng A.M., Liang J.O., Fekany K., Solnica-Krezel L., Korzh V., Halpern M.E., and Wright C.V. 1998. Induction of the zebrafish ventral brain and floorplate requires cyclops/nodal signalling. *Nature* **395**: 185–189.

Sasai Y., Lu B., Steinbeisser H., Geissert D., Gont L.K., and De Robertis E.M. 1994. *Xenopus* chordin: A novel dorsalizing factor activated by organizer-specific homeobox genes. *Cell* **79**: 779–790.

Sater A.K. and Jacobson A.G. 1990a. The restriction of the heart morphogenetic field in *Xenopus laevis. Dev. Biol.* **140**: 328–336.

———. 1990b. The role of the dorsal lip in the induction of heart mesoderm in *Xenopus laevis. Development* **108**: 461–470.

Schmerer M. and Evans T. 2003. Primitive erythropoiesis is regulated by Smad-dependent signaling in postgastrulation mesoderm. *Blood* **102**: 3196–3205.

Schneider V.A. and Mercola M. 1999. Spatially distinct head and heart inducers within the *Xenopus* organizer region. *Curr. Biol.* **9**: 800–809.

———. 2001. Wnt antagonism initiates cardiogenesis in *Xenopus laevis. Genes Dev.* **15**: 304–315.

Schulte-Merker S. and Smith J.C. 1995. Mesoderm formation in response to *Brachyury* requires FGF signalling. *Curr. Biol.* **5**: 62–67.

Schulte-Merker S., van Eeden F.J.M., Halpern M.E., Kimmel C.B., and Nüsslein-Volhard C. 1994. *no tail (ntl)* is the zebrafish homologue of the mouse *T* (*Brachyury*) gene. *Development* **120**: 1009–1015.

Seifert R., Jacob M., and Jacob H.J. 1993. The avian prechordal head region: A morphological study. *J. Anat.* **183**: 75–89.

Seufert D.W., Brennan H.C., DeGuire J., Jones E.A., and Vize P.D. 1999. Developmental basis of pronephric defects in *Xenopus* body plan phenotypes. *Dev. Biol.* **215**: 233–242.

Shibata M., Ono H., Hikasa H., Shinga J., and Taira M. 2000. *Xenopus* crescent encoding a Frizzled-like domain is expressed in the Spemann organizer and pronephros. *Mech. Dev.* **96**: 243–246.

Sive H.L., Grainger R.M., and Harland R.M. 1999. *Early development of Xenopus laevis: A laboratory manual.* Cold Spring Harbor Laboratory Press, Cold Spring Harbor, New York.

Smith J.C. and Slack J. M. 1983. Dorsalization and neural induction: Properties of the organizer in *Xenopus laevis. J. Embryol. Exp. Morphol.* **78**: 299–317.

Smith W.C. and Harland R.M. 1992. Expression cloning of noggin, a new dorsalizing factor localized to the Spemann organizer in *Xenopus* embryos. *Cell* **70**: 829–840.

Smith W.C., Knecht A.K., Wu M., and Harland R.M. 1993. Secreted noggin protein mimics the Spemann organizer in dorsalizing *Xenopus* mesoderm. *Nature* **361**: 547–549.

Smith W.C., McKendry R., Ribisi S., Jr., and Harland R.M. 1995. A nodal-related gene defines a physical and functional domain within the Spemann organizer. *Cell* **82**: 37–46.

Spemann H. 1938. *Embryonic development and induction.* Yale University Press, New Haven, Connecticut.

Standley H.J., Zorn A.M., Gurdon J.B., Standley H., Dyson S., Butler K., Langon T., Ryan K., Stennard F., Shimizu K., and Zorn A. 2001. eFGF and its mode of action in the community effect during *Xenopus* myogenesis. *Development* **128**: 1347–1357.

Stern C.D., Hatada Y., Selleck M.A., and Storey K.G. 1992. Relationships between mesoderm induction and the embryonic axes in chick and frog embryos. *Dev. Suppl.* **1992**: 151–156.

Stewart R.M. and Gerhart J.C. 1990. The anterior extent of dorsal development of the *Xenopus* embryonic axis depends on the quantity of organizer in the late blastula. *Development* **109**: 363–372.

Streit A. and Stern C.D. 1999. Mesoderm patterning and somite formation during node regression: Differential effects of chordin and noggin. *Mech. Dev.* **85**: 85–96.

Talbot W.S., Trevarrow B., Halpern M.E., Melby A.E., Farr G., Postlethwait J.H., Jowett T., Kimmel C.B., and Kimelman D. 1995. A homeobox gene essential for zebrafish notochord development. *Nature* **378**: 150–157.

Tian J., Yam C., Balasundaram G., Wang H., Gore A., and Sampath K. 2003. A temperature-sensitive mutation in the nodal-related gene cyclops reveals that the floor plate is induced during gastrulation in zebrafish. *Development* **130**: 3331–3342.

Tonegawa A., Funayama N., Ueno N., and Takahashi Y. 1997. Mesodermal subdivision along the mediolateral axis in chicken controlled by different concentrations of BMP–4. *Development* **124**: 1975–1984.

Tonegawa A. and Takahashi Y. 1998. Somitogenesis controlled by Noggin. *Dev. Biol.* **202**: 172–182.

Tracey W.D., Jr., Pepling M.E., Horb M.E., Thomsen G.H., and Gergen J.P. 1998. A *Xenopus* homologue of *aml-1* reveals unexpected patterning mechanisms leading to the formation of embryonic blood. *Development* **125**: 1371–1380.

Vesque C., Ellis S., Lee A., Szabo M., Thomas P., Beddington R., and Placzek M. 2000. Development of chick axial mesoderm: Specification of prechordal mesoderm by anterior endoderm-derived TGFβ family signalling. *Development* **127**: 2795–2809.

Vincent S.D., Dunn N.R., Hayashi S., Norris D.P., and Robertson E.J. 2003. Cell fate decisions within the mouse organizer are governed by graded Nodal signals. *Genes Dev.* **17**: 1646–1662.

Wang S., Krinks M., Lin K., Luyten F.P., and Moos M., Jr. 1997. Frzb, a secreted protein expressed in the Spemann organizer, binds and inhibits Wnt-8. *Cell* **88**: 757–766.

Warga R.M. and Kimmel C.B. 1990. Cell movements during epiboly and gastrulation in zebrafish. *Development* **108**: 569–580.

Weinstein D.C., Ruiz i Altaba A., Chen W.S., Hoodless P., Prezioso V.R., Jessell T.M., and Darnell J.E., Jr. 1994. The winged-helix transcription factor HNF-3 beta is required for notochord development in the mouse embryo. *Cell* **78**: 575–588.

Xanthos J.B., Kofron M., Tao Q., Schaible K., Wylie C., and Heasman J. 2002. The roles of three signaling pathways in the formation and function of the Spemann organizer. *Development* **129**: 4027–4043.

Yabe T., Shimizu T., Muraoka O., Bae Y.K., Hirata T., Nojima H., Kawakami A., Hirano T., and Hibi M. 2003. Ogon/Secreted Frizzled functions as a negative feedback regulator of Bmp signaling. *Development* **130**: 2705–2716.

Yamaguchi T.P., Harpal K., Henkemeyer M., and Rossant J. 1994. fgfr-1 is required for embryonic growth and mesodermal patterning during mouse gastrulation. *Genes Dev.* **8**: 3032–3044.

Yamamoto T.S., Takagi C., Hyodo A.C., and Ueno N. 2001. Suppression of head formation by Xmsx-1 through the inhibition of intracellular nodal signaling. *Development* **128**: 2769–2779.

Yao J. and Kessler D.S. 2001. Goosecoid promotes head organizer activity by direct repression of Xwnt8 in Spemann's organizer. *Development* **128**: 2975–2987.

Yeo C. and Whitman M. 2001. Nodal signals to Smads through Cripto-dependent and Cripto-independent mechanisms. *Mol. Cell* **7**: 949–957.

Zimmerman L.B., De Jesus-Escobar J.M., and Harland R.M. 1996. The Spemann organizer signal noggin binds and inactivates bone morphogenetic protein 4. *Cell* **86**: 599–606.

Zoltewicz J.S. and Gerhart J.C. 1997. The Spemann organizer of *Xenopus* is patterned along its anteroposterior axis at the earliest gastrula stage. *Dev. Biol.* **192**: 482–491.

Zorn A.M., Butler K., and Gurdon J.B. 1999. Anterior endomesoderm specification in *Xenopus* by Wnt/β catenin and TGF-β signalling pathways. *Dev. Biol.* **209**: 282–297.

EARLY ROSTROCAUDAL PATTERNING OF THE MESODERM AND NEURAL PLATE

S.E. Fraser[1] and C.D. Stern[2]

[1]*Beckman Institute (139-74), California Institute of Technology, Pasadena, California 91125;* [2]*Department of Anatomy and Developmental Biology, University College London, London WC1E 6BT, United Kingdom*

INTRODUCTION

This chapter and those on either side of it (Chapters 26 and 28) survey the events that contribute to the establishment of the major embryonic axes during vertebrate gastrulation: dorsoventral (axial/nonaxial), rostrocaudal (head–tail) and left–right. Of these three axes, the establishment of the rostrocaudal axis is the most protracted, with the events taking place during gastrulation reflecting only the start of the process. By "rostrocaudal axis" we mean the orderly arrangement of cells that will eventually contribute to the long axis of the body, extending from the head (rostral) to the tail (caudal). Gastrulation proper sets up an embryonic plan where the size of the head territory is greatly exaggerated with respect to the rest of the body, and it is later events, including complex cell movements and differential growth, which set up and expand more caudal regions in a progressive way. These processes affect both the mesoderm and the ectoderm (primarily nervous system). We know much less about how the rostrocaudal axis and local regional identities are established in endodermal derivatives (see Chapter 30).

At the outset, it is worth mentioning that in the current literature the word "anteroposterior" is used variously to refer to the head–tail (rostrocaudal) axis and to the transient axis which relates the primitive streak ("posterior") to the diametrically opposite pole of the embryo ("anterior"), particularly in mouse embryos (see Chapters 16 and 24). However, although these positions will eventually correspond to tail and head regions, respectively, the cells that occupy them correspond more closely to ventral and dorsal fates, respectively. Thus, the "posterior" primitive streak contains ventral mesoderm, and the "anterior" streak contains axial/dorsal mesoderm (see also Chapter 15). Another source of frequent confusion arises from findings in amphibians, and to some extent in zebrafish, that certain treatments (for example, lithium) which affect dorsoventral patterning also affect rostrocaudal character in a predictable way, and the literature often talks about "dorsoanterior" and "ventral-posterior" (see Stern et al. 1992) (Chapter 26).

As outlined in the individual chapters on different vertebrate systems (Chapters 11–18), the axial mesoderm is laid down progressively, in head–tail sequence. Much of this occurs (or is refined) after the end of gastrulation proper. Something similar seems to be true for the central nervous system (CNS). By the end of gastrulation, the territory designated "prospective neural plate" in most vertebrate classes comprises mainly the head (from forebrain to mid- or caudal hindbrain); the spinal cord is laid down progressively at later, post-gastrulation stages. For these reasons, this chapter deals mainly with the very early events, inasmuch as they relate to gastrulation.

MODELS FOR ROSTROCAUDAL PATTERNING OF THE NEURAL PLATE

The "Head/Trunk/Tail Organizer" Model

When Hilde Mangold and Hans Spemann performed their famous "organizer experiment" in 1924 (Spemann and Mangold 1924), they were struck by the observation that the second axis generated by transplanting the dorsal lip of a newt gastrula into the ventral side of a host generated a complete axis extending from the tip of the nose to the tip of the tail. How can a single cell population generate such complexity? One of the earliest ideas was the proposal that the dorsal lip region is not homogeneous, but instead contains different cell populations, each capable of inducing one part of the axis. Furthermore, as descendants of each of these populations exit the dorsal lip after gastrulation, each set of descendants retains its ability to induce specific regions of the axis. Thus (Fig. 1), explants of rostral mesoderm from the neurula-stage embryo can induce mainly head structures when placed inside the blastocoel of a younger host, and progressively more caudal mesoderm will induce more and more caudal trunk and tail regions (Mangold 1933). Equally striking were the findings that certain heterologous inducers (including killed tissues, or pieces of adult organs from the same or different species, or particular chemical substances; Holtfreter 1933, 1934; Chuang 1938, 1939, 1940) can induce specific regions of

the axis. The question immediately arises: How many distinct inducing tissues, or substances, are required to generate the full complexity of the axis? Holtfreter, Otto Mangold, and their contemporaries went as far as suggesting that even smaller parts of the organizer territory could induce individual sensory placodes and other small head structures (Mangold 1933; Spemann 1938; Saxén and Toivonen 1962).

Nieuwkoop's "Activation–Transformation" Model

One of the most widely considered alternatives to the "multiple organizers" model is that proposed by Nieuwkoop (Nieuwkoop et al. 1952; Eyal-Giladi 1954; Nieuwkoop and Nigtevecht 1954; Nieuwkoop 1963), which postulates that induction and patterning of the nervous system occurs in two steps: A first step ("activation") both induces a neural fate and specifies forebrain, and later some of these cells receive other signals from the organizer ("transformation", also called posteriorization or caudalization) which gradually cause them to acquire more caudal character (Fig. 2). This model has received some support, but certain data are difficult to explain within these proposals. In favor of the model is the finding that when prospective forebrain territories are combined with the tail regions containing the remnants of the organizer at the neurula stage, the intermediate regions are pro-

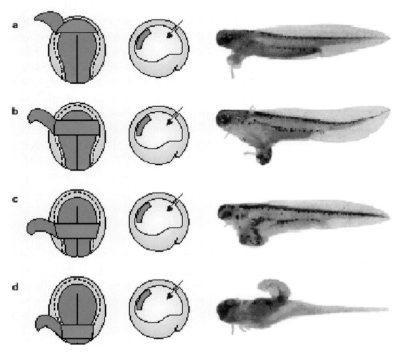

Figure 1. Otto Mangold's (1933) experiment. Grafts of different levels of the mesoderm underlying the neural plate from a neurula-stage newt embryo are grafted into the blastocoel of a host embryo at the gastrula stage (*Einsteck* experiment). The structures induced by such grafts are consistent with the level of origin of the transplant: Rostral mesoderm induces head, more caudal mesoderm induces progressively more caudal structures (trunk, tail). (Adapted, with permission, from Stern 2001 [© Nature Publishing Group].)

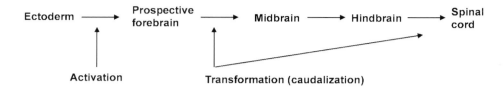

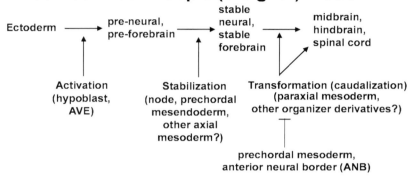

Figure 2. Nieuwkoop's "activation/transformation" model and a modified version (3-signal model) of it, which introduces an intermediate (stabilization) step.

duced from the rostral piece (Cox and Hemmati-Brivanlou 1995). Three candidate molecules have been proposed as mediating the transformation process: FGFs (Cox and Hemmati-Brivanlou 1995; Kengaku and Okamoto 1995; Kroll and Amaya 1996; McGrew et al. 1997; Xu et al. 1997; Storey et al. 1998; Holowacz and Sokol 1999; Ribisi et al. 2000; Kudoh et al. 2002; Villanueva et al. 2002), Retinoic acid (Durston et al. 1989; Boncinelli et al. 1991; Papalopulu et al. 1991; Ruiz i Altaba and Jessell 1991a,b; Conlon and Rossant 1992; Dekker et al. 1992; Marshall et al. 1992, 1994; Yamada 1994; Bally-Cuif et al. 1995; Hill et al. 1995; Simeone et al. 1995; Avantaggiato et al. 1996; Bang et al. 1997; Blumberg et al. 1997; Muhr et al. 1997, 1999; Gavalas et al. 1998; Maden 2000; Escriva et al. 2002; Grandel et al. 2002; Diez del Corral et al. 2003; Novitch et al. 2003; Oosterveen et al. 2003; Sockanathan et al. 2003) and Wnts, particularly Wnt3A (McGrew et al. 1995, 1997; Domingos et al. 2001; Erter et al. 2001; Kiecker and Niehrs 2001; Niehrs et al. 2001; Yamaguchi 2001; Diez del Corral et al. 2002; Kudoh et al. 2002; Nordstrom et al. 2002; Villanueva et al. 2002; Momoi et al. 2003; Nambiar and Henion 2004), all of which can promote, to a greater or lesser extent, the formation of posterior structures from anterior ones. However, this model appears particularly difficult to reconcile with the finding that grafts of older organizers in amphibians, chick, or mouse generate a truncated nervous system that lacks a forebrain (Mangold 1933; Gallera and Nicolet 1969; Gallera 1970, 1971a; Dias and Schoenwolf 1990; Storey et al. 1992; Beddington 1994). Even though the concept of multiple (head and trunk–tail) organizers is formally incompatible with the activation–transformation model, several studies, particularly in amphibians, have been interpreted with a mixture of the two; for example, by implying that the head organizer emits activating signals and the trunk–tail organizer emits both neuralizing and transforming signals.

Other Models

Other models have also been proposed, among them the double-gradient model (Saxén and Toivonen 1962; Toivonen 1978) and the evocation-individuation model (Waddington and Needham 1936). Neither model ever received overwhelming support, but it is also true that neither has been tested enough to dismiss it completely.

MOUSE AVE AND CHICK HYPOBLAST—HOW MANY ORGANIZERS?

Eyal-Giladi and Wolk (1970) pioneered the idea that forebrain-inducing signals may come not from the organizer, but from tissue that only has extraembryonic fate, the hypoblast of the chick (see Chapter 15). This was not pursued, but findings apparently supporting this idea started to sprout in quick succession some 25 years later, following the findings that the mouse organizer (Hensen's node, often called "the node") generates an axis lacking the forebrain when transplanted to a host embryo (Beddington 1994); instead, formation of a forebrain requires the presence of the anterior visceral endoderm (AVE), a layer of cells with extraembryonic fate that is never part of the node (Thomas and Beddington 1996). Soon, a number of groups discovered that wild-type function of a number of genes, including *Nodal*, *Otx2*, *Lim1*, *Hesx1/Rpx*, and *Hex* is required within the AVE for normal forebrain formation (Matsuo et al. 1995; Shawlot and Behringer 1995; Varlet et al. 1997; Acampora et al. 1998; Dufort et al. 1998; Rhinn et al. 1998, 1999; Perea-Gomez et al. 1999; Shawlot et al. 1999; Martinez-Barbera et al. 2000; Martinez-Barbera and Beddington 2001). At about the same time, amphibian experiments had uncovered genes whose early misexpression causes the formation of either an ectopic head (Bouwmeester et al. 1996; Glinka et al. 1998; Hashimoto et al. 2000) or more caudal parts of the axis.

The connection between these findings and the concept of a "head organizer" was only circumstantial, but very tempting. Although most of the laboratories using mouse embryos were careful to avoid this logical leap, by the end of the last decade, others interpreted the data as suggesting the existence of at least two separate organizers, one responsible for inducing the forebrain and another (or more than one) for the rest of the axis (Knoetgen et al. 1999a,b; Niehrs 1999; de Souza and Niehrs 2000). Indeed, many were starting to assume that the head organizer is a physically separate entity and that, at the early gastrula stage, it resides outside the classical territory defined as the organizer by Spemann and his followers.

Amphibian embryos do not have obvious extraembryonic tissue, and some workers sought to identify an equivalent cell population, separate from the organizer at the early gastrula stage, that could induce an isolated head. No such population could be found (Schneider and Mercola 1999). Furthermore, in the mouse, transplantation of the AVE adjacent to (presumably nonneural) ectoderm does not generate an ectopic forebrain—this can only be obtained when the AVE is combined with both prospective forebrain epiblast and the tip of the early primitive streak ("early gastrula organizer") (Tam and Steiner 1999). Another report suggested that the rabbit hypoblast is able to induce forebrain markers when grafted into a chick host, whereas the

chick counterpart cannot (Knoetgen et al. 1999b), which was coupled to the proposal that mammals had evolved a new way of specifying the forebrain (Knoetgen et al. 1999a,b; Niehrs 1999; de Souza and Niehrs 2000). Furthermore, a series of findings in several laboratories suggested that the prechordal mesendoderm (a tissue derived from the node) can specify forebrain fates at least in cells that had already received neural inducing signals (Ang and Rossant 1993; Shawlot and Behringer 1995; Dale et al. 1997; Foley et al. 1997; Muhr et al. 1997; Pera and Kessel 1997; Schier et al. 1997; Shimamura and Rubenstein 1997; Camus et al. 2000). By now, the only possible conclusion from these findings is that the AVE (and perhaps its equivalent, the hypoblast, in the chick) (Eyal-Giladi and Wolk 1970; Foley et al. 2000) may indeed be required for forebrain development, but that its role is unlikely to be as simple as a direct induction of forebrain or head character.

Two more recent studies further strengthened this conclusion. Matsuo, Aizawa, and their colleagues (Kimura et al. 2000, 2001), and subsequently Ang and colleagues (Perea-Gomez et al. 2001), suggested that the mouse AVE may serve a protective function against caudalization of the nervous system by the organizer, and that its anterior movements regulate the movement of the overlying epiblast (including the prospective forebrain) in the same direction. In another study, it was shown that the chick hypoblast also directs the movements of the overlying forebrain territory in the epiblast; although in naïve cells, it is capable of transient induction of markers that define forebrain at a later stage (*Otx2*, *Sox3*), this expression is unstable and the interaction not sufficient to generate a recognizable forebrain or expression of definitive prosencephalic markers (Foley et al. 2000). Indeed, as in amphibians, the chick node is able to induce a complete axis including the forebrain even when grafted into a region that has never been in contact with hypoblast, provided that the node is obtained from an embryo at the full primitive-streak stage but before the emergence of the head process (Gallera and Nicolet 1969; Gallera 1970, 1971b; Dias and Schoenwolf 1990; Storey et al. 1992). All this argues rather strongly against the existence of a true head organizer residing outside the classic organizer region (the node).

A THREE-STEP MODEL

If, as Nieuwkoop proposed (Nieuwkoop et al. 1952; Nieuwkoop and Nigtevecht 1954), there are two steps to neural induction, when do these steps take place? It has been widely assumed that neural induction begins during gastrulation, when the organizer can first be identified morphologically, and mesoderm induction has probably ended (Gurdon 1987; Smith 1995; Harland and Gerhart 1997). However, more recent work suggested that an early step may occur much earlier, before gastrulation begins (Streit et al. 2000; Wilson et al. 2000; see also Chapter 29). In both chick and mouse embryos, expression of early neural markers (e.g., *Sox3* and *ERNI* in the chick, *Sox1* in mouse) begins before the appearance of the primitive streak, in a broad domain of the epiblast. In chick, the expression of these markers coincides with the spread of the hypoblast layer, suggesting that the hypoblast could be responsible for inducing their expression. Indeed, this conclusion is supported by the finding that the hypoblast transiently induces expression of *Sox3* as well as *Otx2* (Foley et al. 2000). At later stages of development, *Otx2* is a marker for the forebrain and anterior midbrain (Bally-Cuif et al. 1995). Before primitive-streak formation, however, it is expressed very broadly in the epiblast, throughout the prospective neural plate in both chick and mouse embryos. The broad, early coexpression of markers that at later stages of development identify all neural tissue (*Sox3*) and fore- and midbrain (*Otx2*) is suggestive that this state may have been "activated," as proposed in Nieuwkoop's model. However, this activation is not sufficient for the cells to acquire a forebrain, or even a neural state, because when the inducing tissue (hypoblast) is allowed to remain in contact for a longer time, expression of both markers is lost and no neural or forebrain structures develop (Foley et al. 2000).

What are the early "activating" signals? Like grafts of the hypoblast or of organizer precursor cells, FGF-8 can induce *Sox3*, *ERNI*, and *Otx2*, and loss-of-function approaches suggest that FGF signaling is required for this induction (Streit et al. 2000; Wilson et al. 2000; Bertrand et al. 2003). In addition, other studies mentioned above have also implicated FGF signaling as a transforming step (Cox and Hemmati-Brivanlou 1995; Kengaku and Okamoto 1995; Kroll and Amaya 1996; McGrew et al. 1997; Xu et al. 1997; Storey et al. 1998; Holowacz and Sokol 1999; Ribisi et al. 2000; Kudoh et al.

2002; Villanueva et al. 2002). It is conceivable, therefore, that FGF signaling represents both the activation signal and a transforming step. However, other "maintenance" signals are also required, because when either FGF, precursors of the organizer, or hypoblast grafts are allowed to stay in place, they do not elicit expression of later markers, even when combined with BMP antagonists (see Chapter 29). Moreover, FGF is probably not the only transforming signal because injection of a dominant-negative FGF receptor in *Xenopus* does not cause the complete absence of posterior nervous system (Amaya et al. 1991, 1993). Therefore, the activation step of Nieuwkoop's model may be characterized by the transient establishment of a "*pre*-forebrain" state, upon which maintenance signals must act to allow cells to acquire neural fate, and some of which will also receive caudalizing signals to generate more posterior parts of the CNS (Figs. 2 and 3).

In addition to providing an early activating step similar to that proposed by Nieuwkoop, the chick hypoblast and mouse AVE appear to have a separate role: directing the movement of at least some of the pre-forebrain cells anteriorly, away from the node and its transforming/caudalizing influence (Foley et al. 2000; Kimura et al. 2000, 2001). This view explains how the forebrain is induced, since in both chick and mouse embryos it never lies close to the node (see Chapters 15 and 16). What, then, provides the maintenance signals? One possibility is that the anterior head process (Foley et al. 1997; Rowan et al. 1999) and/or prechordal mesendoderm (Ang and Rossant 1993; Dale et al. 1997; Foley et al. 1997; Muhr et al. 1997; Pera and Kessel 1997; Schier et al. 1997; Shimamura and Rubenstein 1997; Camus et al. 2000) provide this maintenance function, as well as some degree of further "protection" against caudalizing signals, because grafts of prechordal mesendoderm placed adjacent to prospective hindbrain can elicit expression of forebrain markers.

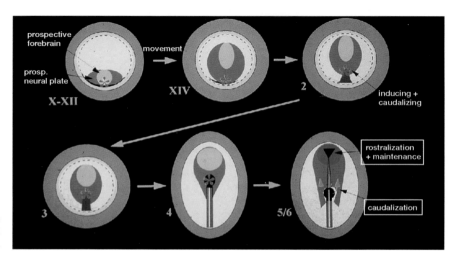

Figure 3. Diagrammatic representation of the three steps of the modified (3-signal) model shown in Fig. 2, as applied to the chick embryo.

ESTABLISHMENT OF THE ROSTROCAUDAL AXIS OF THE CNS IN FISH EMBRYOS

Much of the preceding discussion was based on data obtained in amniote embryos. In these embryos, the elongation/caudalization of the nervous system occurs over a very protracted period. Are the processes that set up rostrocaudal regional identity similar in the anamnia?

In the zebrafish, detailed fate maps (Woo and Fraser 1995) of the region around the embryonic shield show considerable order as early as the time when the embryonic shield appears (6-hour [h] fate map; Fig. 4). In almost all cases, the descendants of single labeled cells contributed to only one brain region, and the cells contributing to a given brain area were contiguous. However, at this time point the fate map was not fully elaborated, as cells contributing to different brain regions were somewhat intermixed. A similarly constructed neural fate map from cells labeled at 10 h shows increased order: The number of injections yielding descendants in a single brain area increased and the overlap between the domains decreased. This homogeneity in fate and the reduced overlap between domains suggest that regional patterning has become significantly more pronounced in the 10-h fate map (Fig. 4). The spatial relationships of the domains change significantly between 6 h and 10 h. The entire neuroectoderm has narrowed, and tissue movements have brought the domains fated to make given

CNS regions into roughly the same order found later in the neural tube (Woo and Fraser 1995; Varga et al. 1999).

The localized tissue domains observed in the 6-h and 10-h fate maps (Woo and Fraser 1995) might be taken to argue that cell fate is assigned to the neuroectoderm around these stages. However, fate maps cannot be used to pinpoint the time of cell fate decisions, just as analyses of gene expression patterns are not sufficient to understand the state of commitment of cell groups. Testing for cell fate assignments requires confronting cells with a foreign set of cell interactions; if the cells remain true to their original fates, they are said to be committed (see Glossary). The first tests of regional commitment (Woo and Fraser 1998) showed clearly that the regularity of the 6-h fate map cannot be taken as evidence of specification or commitment to either neural or regional fates (Fig. 5). Hindbrain progenitors were selected because their fate map location was clearly defined and it was possible to isolate these cells free of axial-mesoderm contamination. The recipient site was the ventral-most ectoderm of a shield-stage host, selected because it seemed to represent a "neutral" environment. When scored at 24–36 h, the grafted cells integrated seamlessly into the host and failed to develop into neural tissue. Instead, the labeled descendants of the grafts differentiated as expected for the site in the host: ectodermal cells of the trunk and tail, somites in the tail, and cells in the fin. In contrast, similar transplantations performed with donor cells from embryos

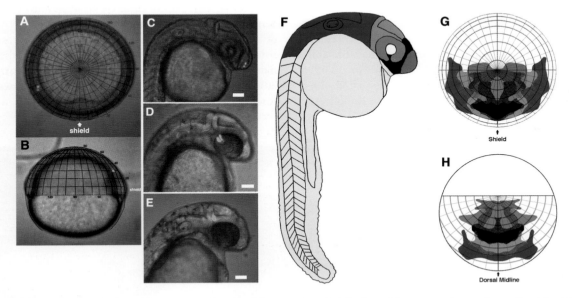

Figure 4. Fate maps of the nervous system in the zebrafish. Iontophoretic injection of fluorescent dextran into cells of the 6-h zebrafish embryo resulted in one or two cells labeled (injected cells pseudocolored red). The positions of the cells were imaged from the animal pole (*A*) and in side view (*B*) to define the positions of the cells unambiguously. After allowing time for development, the positions of the descendants of the labeled cells were determined with fluorescence microscopy. The examples here show pseudocolored cells in the telencephalon (*C*), the retina (*D*), the midbrain (*E*). The results were summarized by shading in the regions that give rise to the different brain regions (*F*) to create fate maps of the 6-h (*G*) and 10-h (*H*) embryos. (Adapted, with permission, from Woo and Fraser 1995.)

allowed to develop for an additional 2 h showed clear evidence for commitment to both neural and hindbrain fates (Woo and Fraser 1998). The grafted hindbrain precursor cells integrated into the host site, and only later gave rise to neural tissue that expressed *Krox20* (Fig. 5), a marker for rhombomeres 3 and 5 of the hindbrain (Oxtoby and Jowett 1993). Further studies are needed to define the interactions that take place during the 2-h period (shield stage to 80% epiboly) during which hindbrain progenitors become committed to their neuronal and regional fates. Because this commitment takes place before the hindbrain progenitors contact the axial mesoderm, alternative sources of patterning information must be considered. The source of this patterning signal resides, at least in part, outside the shield region, as microsurgical deletion of the shield does not prevent hindbrain formation (Shih and Fraser 1996; Saude et al. 2000).

The above results show that the fates of hindbrain progenitors become fixed at 80% epiboly, while the progenitors are still lateral to the forming embryonic axis and have yet to establish vertical contact with the axial mesoderm. Thus, the ectoderm is becoming neuralized and regionalized before contact is established with axial tissues, suggesting that positional signals can be transmitted by a planar mechanism. Based on the 6-h neural fate map (Woo and Fraser 1995), cells in the lateral germ ring are ideally positioned for conveying rostrocaudal patterning information. Forebrain progenitors are at the dorsal midline near the animal pole and far from the germ ring, while precursors of the midbrain are in contact with the germ ring briefly, until the normal motions of early epiboly move them rostrally. Hindbrain and spinal progenitors remain close to the germ ring for a much longer period during epiboly. This direct correlation between the length of time that the cells remain near the germ ring and their caudal identity suggests a role for signaling from the non-axial mesoderm. Grafting experiments confirm this. First, when labeled prospective forebrain cells are grafted to the presumptive hindbrain region, they adopt a hindbrain fate (Woo and Fraser 1997). Second, grafts of non-axial germ ring tissue to the animal pole of shield-stage embryos (prospective forebrain; Figs. 4 and 5) is sufficient to caudalize the host cells into hindbrain-like structures expressing *Krox20* as well as ectopic otic vesicles (Woo and Fraser 1997). Lateral and ventral germ ring (normally fated to become somitic mesoderm, posterior mesoderm, and endoderm) possessed transforming activities; in contrast, the embryonic shield appeared to have little if any. These results are consistent with those described above for amniotes. An "activating" influence from the embryonic shield (or its precursors) would neuralize the presumptive neuroectoderm at or before the onset of shield formation (Fig. 6). The "transforming" signal would emanate radially from the germ ring of the gastrula to caudalize the activated tissue, and little, if any, of the transforming signal is present in the zebrafish embryonic shield.

In zebrafish, as in amniotes, the transforming signal is very strong and can override the pre-forebrain character of initially induced tissue, arguing for the importance of providing active protection against caudalizing signals in regions destined to form the forebrain. Apart from the prechordal mesendoderm, a different group of cells with a protecting function has been identified in the zebrafish (called "row 1" at early stages, whose descendants come to occupy a very rostral position, known as the anterior neural border; ANB) (Fig. 2) (Houart et al. 1998, 2002; Heisenberg et al. 2001; Wilson and Houart 2004). The ANB cells act by counteracting Wnt signaling through the specific Wnt-antagonist Tlc (Houart et al. 2002). It is not yet clear

Commitment test:

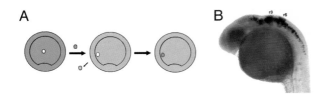

Transformation test:

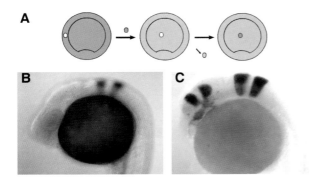

Figure 5. Tests for commitment and transforming signals in the 6-h embryo. ***Commitment test:*** (*A*) Cells were grafted from the region of an embryo fated to become forebrain to the regions fated to be hindbrain. The donor was labeled with a mix of fluorescent dextran and biotin dextran permitting the descendants of the grafted cells (*brown*) to be recognized. (*B*) After neural differentiation, the descendants of the grafted cell contributed to the hindbrain of the host, as shown by *Krox20* expression in the grafted cells. ***Transformation test:*** (*A*) Cells were grafted from an embryo labeled with fluorescent dextran and biotin dextran into an unlabeled host to confront cells fated to become forebrain with transforming signals. (*B*) In situ hybridization for *Krox20* after a control graft shows the expected two domains of gene expression in the hindbrain. (*C*) In contrast, there are supernumerary *Krox20* domains in the embryos receiving a germ ring graft (descendants of the graft are stained brown), indicating that the forebrain has been transformed to a hindbrain fate. (Based on data from Woo and Fraser 1997, 1998.)

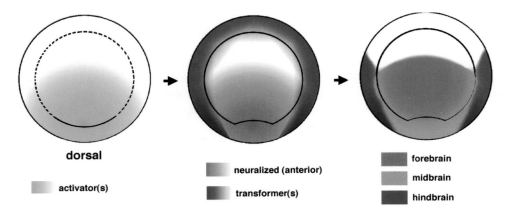

Figure 6. An activation-transformation model for patterning the zebrafish neuraxis. An activator signal (*yellow*) originating from the dorsal midline acts to induce neural tissue with a broad anterior character (*red*). A transforming signal (*blue*) from the non-axial germ ring promoted more caudal axial character of the nearby tissue that has previously been neuralized. Thus, the forebrain remains untransformed (*deep red*); by proximity to the signal and/or the length of time exposed to the signal, the midbrain (*orange*), and hindbrain (*blue*) territories are promoted to more caudal character.

whether other vertebrates possess an equivalent region within the rostral neural plate or its border, but the principle that antagonists of the signaling pathways (Wnt, Retinoid, FGF) with caudalizing activity might account for the "rostralization" phenotype observed by other workers (Ang and Rossant 1993; Foley et al. 1997) allows such observations to remain compatible with Nieuwkoop's model.

ELONGATION OF THE CNS AND REFINEMENT OF ROSTROCAUDAL IDENTITY

In summary, therefore, the evidence is starting to favor the concepts of Nieuwkoop's model rather strongly, and to implicate FGF signaling both in the activation and in the transformation step, although it is also clear that other factors are required for both steps. It appears that activation is a very early event, occurring even before gastrulation begins and before either a gastrulation center (shield/marginal zone/germ wall ring, blastopore, primitive streak) or a classic Spemann organizer (shield, dorsal lip, Hensen's node) can be identified morphologically. By the end of gastrulation, the embryo has set up a large territory whose fate is to contribute to the head portion of the CNS, and a very small region (at least in amniotes, adjacent to the anterior primitive streak) destined to give rise to the spinal cord (Henrique et al. 1997; Pituello 1997; Goriely et al. 1999; Pituello et al. 1999; Bertrand et al. 2000; Brown and Storey 2000; Diez del Corral et al. 2002, 2003; Eblaghie et al. 2003). This small region now proceeds to elongate and gradually to establish a differentiating spinal cord in a rostral to caudal direction. In birds and mammals this process takes several days.

Another peculiarity of the early fate maps is that the most ventral regions of the CNS arise in quite a different way from more lateral and dorsal regions (Charrier et al. 1999, 2002; Gunhaga et al. 2000; Lopez-Sanchez et al. 2001; Jeong and Epstein 2003; Patten et al. 2003). Cells in the ventral midline along most of the length of the neural tube appear from a small region of epiblast within or adjacent to Hensen's node and rapidly extend along the axis, separating the original neural territory into left and right sides (Patten et al. 2003). This probably explains why the floor plate region seems to lack expression of many regional markers that are expressed in more dorsal parts of the CNS, including many Hox genes.

Eventually, the major part of the CNS is marked by distinct patterns of expression of genes, many of which are transcription factors that contribute to specify rostrocaudal identity. Caudal to rhombomere 1-2 in the hindbrain, the major effectors of this are the Hox genes (McGinnis and Krumlauf 1992). Many of these begin their expression caudally and rather early in development, from a common territory from which they expand rostrally. For example, *Hoxb1* starts to be expressed around the primitive streak at the gastrula stage in chick and mouse embryos and quickly expands rostrally. The characteristic regional expression in rhombomere 4 and the more caudal domain do not arise until long after gastrulation (Sundin et al. 1990; Sundin and Eichele 1992). Rostral to the hindbrain, other transcription factors delineate different rostrocaudal territories (see Shimamura et al. 1995; Rubenstein et al. 1998). We are only just starting to understand how these different territories are set up; in any case, much of this process occurs after gastrulation has ended and is therefore outside the scope of this review. It seems likely that the caudalizing factors retinoic acid, Wnts, and FGFs are clearly implicated in the process, and that they interact in complex ways (see, e.g., Bertrand et al. 2000; Zakany et al. 2001; Diez del Corral et al. 2002, 2003; Eblaghie et al. 2003; Itasaki et al. 2003; Sockanathan et al. 2003). Furthermore, the three caudalizing signals appear to have different importance at different levels of the CNS (see, e.g., Gould et al. 1998; Bel-Vialar et al. 2002). At these stages at least, the main regionalizing signals appear to be produced by the paraxial, rather than the axial, mesoderm (Gould et al. 1998; Itasaki et al. 2003).

ROSTROCAUDAL PATTERNING OF THE MESODERM AND EMBRYONIC TIME CLOCKS

Within the mesoderm, it seems to be the paraxial (future somites and paraxial head mesoderm) that is most clearly regionalized along the rostrocaudal axis. The intermediate mesoderm and lateral plate are regionalized (for example, in the former, regions destined for pronephros and mesonephros; in the latter, limb-forming regions), but this seems to be acquired much later in development. In contrast, there is little or no evidence for regionalization of the axial mesoderm (notochord) apart from its obvious subdivision into prechordal and chordal mesoderm. This is not to say that the axial tissues are not regionalized at all, but may be a measure of our current level of ignorance—indeed, a recent study (Fleming et al. 2004) suggested that the notochord may impart segmental information to the neighboring somites.

In the paraxial mesoderm, especially in the trunk (somites), timing plays a central role in the acquisition of rostrocaudal positional identity. A molecular clock regulates the rate of segmentation (Pourquie 2003), and FGF plays a critical role in regulating both the clock and positional identity of the somites (Dubrulle et al. 2001; Vasiliauskas and Stern 2001; Zakany et al. 2001; Dubrulle and Pourquie 2004): *FGF8* is strongly expressed in the most caudal presomitic mesoderm and in the primitive streak region from which this arises, and is down-regulated as the presomitic mesoderm matures. The boundary of *FGF8* expression correlates not only with the time at which prospective somitic cells become "committed to segment," but also with the expression of *Pax6* in the neighboring elongating spinal cord, which provides a link between the process of segmentation and neurogenesis (Pituello 1997; Pituello et al. 1999; Diez del Corral et al. 2002, 2003).

Taken together, these findings are now starting to suggest that the "transforming" (caudalizing) signal suggested by Nieuwkoop's model is not so much a series of different signals present at different levels of the axis, each of which specifies a different level (which would be more consistent with the "head/trunk/tail organizers" model), but rather a consequence of translating time into space. Cells become more caudal the longer they have been exposed to the region with caudalizing activity, with FGF signaling being

the major transducer between time and space. A major challenge for the future will be to understand how a single signaling pathway, FGF, can provide both the activation and the transformation instructions, and what are the necessary cooperating signals that might account for the stabilization step(s).

REFERENCES

Acampora D., Avantaggiato V., Tuorto F., Briata, P., Corte G., and Simeone A. 1998. Visceral endoderm-restricted translation of Otx1 mediates recovery of Otx2 requirements for specification of anterior neural plate and normal gastrulation. *Development* **125:** 5091–5104.

Amaya E., Musci T.J., and Kirschner M.W. 1991. Expression of a dominant negative mutant of the FGF receptor disrupts mesoderm formation in *Xenopus* embryos. *Cell* **66:** 257–270.

Amaya E., Stein P.A., Musci T.J., and Kirschner M.W. 1993. FGF signaling in the early specification of mesoderm in *Xenopus*. *Development* **118:** 477–487.

Ang S.L. and Rossant J. 1993. Anterior mesendoderm induces mouse *Engrailed* genes in explant cultures. *Development* **118:** 139–149.

Avantaggiato V., Acampora D., Tuorto F., and Simeone A. 1996. Retinoic acid induces stage-specific repatterning of the rostral central nervous system. *Dev. Biol.* **175:** 347–357.

Bally-Cuif L., Gulisano M., Broccoli V., and Boncinelli E. 1995. c-otx2 is expressed in two different phases of gastrulation and is sensitive to retinoic acid treatment in chick embryo. *Mech. Dev.* **49:** 49–63.

Bang A.G., Papalopulu N., Kintner C., and Goulding M.D. 1997. Expression of *Pax-3* is initiated in the early neural plate by posteriorizing signals produced by the organizer and by posterior non-axial mesoderm. *Development* **124:** 2075–2085.

Beddington R.S. 1994. Induction of a second neural axis by the mouse node. *Development* **120:** 613–620.

Bel-Vialar S., Itasaki N., and Krumlauf R. 2002. Initiating Hox gene expression: In the early chick neural tube differential sensitivity to FGF and RA signaling subdivides the *HoxB* genes in two distinct groups. *Development* **129,** 5103–5115.

Bertrand N., Medevielle F., and Pituello F. 2000. FGF signalling controls the timing of *Pax6* activation in the neural tube. *Development* **127:** 4837–4843.

Bertrand V., Hudson C., Caillol D., Popovici C., and Lemaire P. 2003. Neural tissue in ascidian embryos is induced by FGF9/16/20, acting via a combination of maternal GATA and Ets transcription factors. *Cell* **115:** 615–627.

Blumberg B., Bolado J., Jr., Moreno T.A., Kintner C., Evans R.M., and Papalopulu N. 1997. An essential role for retinoid signaling in anteroposterior neural patterning. *Development* **124:** 373–379.

Boncinelli E., Simeone A., Acampora D., and Mavilio F. 1991. HOX gene activation by retinoic acid. *Trends Genet.* **7:** 329–334.

Bouwmeester T., Kim S.-H., Sasai Y., Lu B., and De Robertis E. M. 1996. Cerberus is a head-inducing secreted factor expressed in the anterior endoderm of Spemann's organizer. *Nature* **382:** 595–601.

Brown J.M. and Storey K.G. 2000. A region of the vertebrate neural plate in which neighbouring cells can adopt neural or epidermal fates. *Curr. Biol.* **10:** 869–872.

Camus A., Davidson B.P., Billiards S., Khoo P., Rivera-Perez J.A., Wakamiya M., Behringer R.R., and Tam P.P. 2000. The morphogenetic role of midline mesendoderm and ectoderm in the development of the forebrain and the midbrain of the mouse embryo. *Development* **127**: 1799–1813.

Charrier J.B., Lapointe F., Le Douarin N.M., and Teillet M.A. 2002. Dual origin of the floor plate in the avian embryo. *Development* **129**: 4785–4796.

Charrier J.B., Teillet M.A., Lapointe F., and Le Douarin N.M. 1999. Defining subregions of Hensen's node essential for caudalward movement, midline development and cell survival. *Development* **126**: 4771–4783.

Chuang H.-H. 1938. Spezifische Induktionsleistungen von Leber und Niere im Explantationsversuch. *Biol. Zentbl.* **58**: 472–480.

———. 1939. Induktionsleistungen von frischen und gekochten Organteilen (Niere, Leber) nach ihrer verpflanzung in Explantate und verschiedene Wirtsregionen von Tritonkeimen. *Arch. Entwicklungsmech. Org.* **139**: 556–638.

———. 1940. Weitere Versuch über die Veränderung der Induktionsleistungen von gekochten Organteilen. *Arch. Entwicklungsmech. Org.* **140**: 25–38.

Conlon R.A. and Rossant J. 1992. Exogenous retinoic acid rapidly induces anterior ectopic expression of murine Hox-2 genes in vivo. *Development* **116**: 357–368.

Cox W.G. and Hemmati-Brivanlou A. 1995. Caudalization of neural fate by tissue recombination and bFGF. *Development* **121**: 4349–4358.

Dale J.K., Vesque C., Lints T.J., Sampath T.K., Furley A., Dodd J., and Placzek M. 1997. Cooperation of BMP7 and SHH in the induction of forebrain ventral midline cells by prechordal mesoderm. *Cell* **90**: 257–269.

de Souza F.S. and Niehrs C. 2000. Anterior endoderm and head induction in early vertebrate embryos. *Cell Tissue Res.* **300**: 207–217.

Dekker E.J., Pannese M., Houtzager E., Timmermans A., Boncinelli E., and Durston A. 1992. *Xenopus* Hox-2 genes are expressed sequentially after the onset of gastrulation and are differentially inducible by retinoic acid. *Dev Suppl.* **1992**: 195–202.

Dias M.S. and Schoenwolf G.C. 1990. Formation of ectopic neurepithelium in chick blastoderms: Age-related capacities for induction and self-differentiation following transplantation of quail Hensen's nodes. *Anat. Rec.* **228**: 437–448.

Diez del Corral R., Breitkreuz D.N., and Storey K.G. 2002. Onset of neuronal differentiation is regulated by paraxial mesoderm and requires attenuation of FGF signalling. *Development* **129**: 1681–1691.

Diez del Corral R., Olivera-Martinez I., Goriely A., Gale E., Maden M., and Storey K. 2003. Opposing FGF and retinoid pathways control ventral neural pattern, neuronal differentiation, and segmentation during body axis extension. *Neuron* **40**: 65–79.

Domingos P.M., Itasaki N., Jones C.M., Mercurio S., Sargent M.G., Smith J.C., and Krumlauf R. 2001. The Wnt/β-catenin pathway posteriorizes neural tissue in *Xenopus* by an indirect mechanism requiring FGF signalling. *Dev. Biol.* **239**: 148–160.

Dubrulle J. and Pourquie O. 2004. *fgf8* mRNA decay establishes a gradient that couples axial elongation to patterning in the vertebrate embryo. *Nature* **427**: 419–422.

Dubrulle J., McGrew M.J., and Pourquie O. 2001. FGF signaling controls somite boundary position and regulates segmentation clock control of spatiotemporal *Hox* gene activation. *Cell* **106**: 219–232.

Dufort D., Schwartz L., Harpal K., and Rossant J. 1998. The transcription factor HNF3β is required in visceral endoderm for normal primitive streak morphogenesis. *Development* **125**, 3015–3025.

Durston A.J., Timmermans J.P., Hage W.J., Hendriks H.F., de Vries N.J.,

Heideveld M., and Nieuwkoop P.D. 1989. Retinoic acid causes an anteroposterior transformation in the developing central nervous system. *Nature* **340**: 140–144.

Eblaghie M.C., Lunn J.S., Dickinson R.J., Munsterberg A.E., Sanz-Ezquerro J.J., Farrell E.R., Mathers J., Keyse S.M., Storey K., and Tickle C. 2003. Negative feedback regulation of FGF signaling levels by Pyst1/MKP3 in chick embryos. *Curr. Biol.* **13**: 1009–1018.

Erter C.E., Wilm T.P., Basler N., Wright C.V., and Solnica-Krezel L. 2001. Wnt8 is required in lateral mesendodermal precursors for neural posteriorization in vivo. *Development* **128**, 3571–3583.

Escriva H., Holland N.D., Gronemeyer H., Laudet V., and Holland L.Z. 2002. The retinoic acid signaling pathway regulates anterior/posterior patterning in the nerve cord and pharynx of amphioxus, a chordate lacking neural crest. *Development* **129**: 2905–2916.

Eyal-Giladi H. 1954. Dynamic aspects of neural induction in amphibia. *Arch. Biol.* **65**: 179–259.

Eyal-Giladi H. and Wolk M. 1970. The inducing capacities of the primary hypoblast as revealed by transfilter induction studies. *Wilhelm Roux' Arch.* **165**: 226–241.

Fleming A., Keynes R., and Tannahill D. 2004. A central role for the notochord in vertebral patterning. *Development* **131**: 873–880.

Foley A.C., Skromne I., and Stern C.D. 2000. Reconciling different models of forebrain induction and patterning: A dual role for the hypoblast. *Development* **127**: 3839–3854.

Foley A.C., Storey K.G., and Stern C.D. 1997. The prechordal region lacks neural inducing ability, but can confer anterior character to more posterior neuroepithelium. *Development* **124**: 2983–2996.

Gallera J. 1970. Inductions cérébrales et médullaires chez les Oiseaux. *Experientia* **26**: 886–887.

———. 1971a. Différence de la reactivité à l'inducteur neurogène entre l'ectoblaste de l'aire opaque et celui de l'aire pellucide chez le poulet. *Experientia* **26**: 1953–1954.

———. 1971b. Le pouvoir inducteur de la jeune ligne primitive et les différences régionales dans les compétences de l'ectoblaste chez les oiseaux. *Arch. Biol.* **82**: 85–102.

Gallera J. and Nicolet G. 1969. Le pouvoir inducteur de l'endoblaste presomptif contenu dans la ligne primitive jeune de l'embryon de poulet. *J. Embryol. Exp. Morphol.* **21**: 105–118.

Gavalas A., Studer M., Lumsden A., Rijli F.M., Krumlauf R., and Chambon P. 1998. *Hoxa1* and *Hoxb1* synergize in patterning the hindbrain, cranial nerves and second pharyngeal arch. *Development* **125**, 1123–1136.

Glinka A., Wu W., Delius H., Monaghan A. P., Blumenstock C., and Niehrs C. 1998. Dickkopf-1 is a member of a new family of secreted proteins and functions in head induction. *Nature* **391**: 357–362.

Goriely A., Diez del Corral R., and Storey K. G. 1999. c-Irx2 expression reveals an early subdivision of the neural plate in the chick embryo. *Mech. Dev.* **87**: 203–206.

Gould A., Itasaki N., and Krumlauf R. 1998. Initiation of rhombomeric *Hoxb4* expression requires induction by somites and a retinoid pathway. *Neuron* **21**: 39–51.

Grandel H., Lun K., Rauch G.J., Rhinn M., Piotrowski T., Houart C., Sordino P., Kuchler A.M., Schulte-Merker S., Geisler R., Holder N., Wilson S.W., and Brand M. 2002. Retinoic acid signalling in the zebrafish embryo is necessary during pre-segmentation stages to pattern the anterior-posterior axis of the CNS and to induce a pectoral fin bud. *Development* **129**: 2851–2865.

Gunhaga L., Jessell T.M., and Edlund T. 2000. Sonic hedgehog signaling at gastrula stages specifies ventral telencephalic cells in the chick embryo. *Development* **127**: 3283–3293.

Gurdon J.B. 1987. Embryonic induction–molecular prospects. *Development* **99**: 285–306.

Harland R. and Gerhart J. 1997. Formation and function of Spemann's organizer. *Annu. Rev. Cell Dev. Biol.* **13**: 611–667.

Hashimoto H., Itoh M., Yamanaka Y., Yamashita S., Shimizu T., Solnica-Krezel L., Hibi M., and Hirano T. 2000. Zebrafish dkk1 functions in forebrain specification and axial mesendoderm formation. *Dev. Biol.* **217**, 138–152.

Heisenberg C.P., Houart C., Take-Uchi M., Rauch G.J., Young N., Coutinho P., Masai I., Caneparo L., Concha M.L., Geisler R., Dale T.C., Wilson S.W., and Stemple D.L. 2001. A mutation in the Gsk3-binding domain of zebrafish Masterblind/Axin1 leads to a fate transformation of telencephalon and eyes to diencephalon. *Genes Dev.* **15**: 1427–1434.

Henrique D., Tyler D., Kintner C., Heath J.K., Lewis J.H., Ish-Horowicz D., and Storey K.G. 1997. Cash4, a novel achaete-scute homolog induced by Hensen's node during generation of the posterior nervous system. *Genes Dev.* **11**: 603–615.

Hill J., Clarke J.D., Vargesson N., Jowett T., and Holder N. 1995. Exogenous retinoic acid causes specific alterations in the development of the midbrain and hindbrain of the zebrafish embryo including positional respecification of the Mauthner neuron. *Mech. Dev.* **50**: 3–16.

Holowacz T. and Sokol S. 1999. FGF is required for posterior neural patterning but not for neural induction. *Dev. Biol.* **205**: 296–308.

Holtfreter J. 1933. Eigenschaften und Verbreitung induzierender Stoffe. *Naturwissenschaften* **21**: 766–770.

———. 1934. Der Einfluss thermischer, mechanischer und chemischer Eingreffe auf die Induzierfähigkeit von Triton-Keimteilen. *Wilhelm Roux' Arch. Entwicklungsmech. Org.* **132**: 225–306.

Houart C., Westerfield M., and Wilson S.W. 1998. A small population of anterior cells patterns the forebrain during zebrafish gastrulation. *Nature* **391**: 788–792.

Houart C., Caneparo L., Heisenberg C., Barth K., Take-Uchi M., and Wilson S. 2002. Establishment of the telencephalon during gastrulation by local antagonism of Wnt signaling. *Neuron* **35**: 255–265.

Itasaki N., Jones C.M., Mercurio S., Rowe A., Domingos P.M., Smith J.C., and Krumlauf R. 2003. Wise, a context-dependent activator and inhibitor of Wnt signalling. *Development* **130**, 4295–4305.

Jeong Y. and Epstein D.J. 2003. Distinct regulators of *Shh* transcription in the floor plate and notochord indicate separate origins for these tissues in the mouse node. *Development* **130**: 3891–3902.

Kengaku M. and Okamoto H. 1995. bFGF as a possible morphogen for the anteroposterior axis of the central nervous system in *Xenopus*. *Development* **121**: 3121–3130.

Kiecker C. and Niehrs C. 2001. A morphogen gradient of Wnt/β-catenin signalling regulates anteroposterior neural patterning in *Xenopus*. *Development* **128**: 4189–4201.

Kimura C., Shen M.M., Takeda N., Aizawa S., and Matsuo I. 2001. Complementary functions of *Otx2* and *Cripto* in initial patterning of mouse epiblast. *Dev. Biol.* **235**: 12–32.

Kimura C., Yoshinaga K., Tian E., Suzuki M., Aizawa S., and Matsuo I. 2000. Visceral endoderm mediates forebrain development by suppressing posteriorizing signals. *Dev. Biol.* **225**: 304–321.

Knoetgen H., Teichmann U., and Kessel M. 1999a. Head-organizing activities of endodermal tissues in vertebrates. *Cell Mol. Biol. (Noisy-Le-Grand)* **45**: 481–492.

Knoetgen H., Viebahn C., and Kessel M. 1999b. Head induction in the chick by primitive endoderm of mammalian, but not avian origin. *Development* **126**: 815–825.

Kroll K.L. and Amaya E. 1996. Transgenic *Xenopus* embryos from sperm nuclear transplantations reveal FGF signaling requirements during gastrulation. *Development* **122**: 3173–3183.

Kudoh T., Wilson S.W., and Dawid I.B. 2002. Distinct roles for Fgf, Wnt and retinoic acid in posteriorizing the neural ectoderm. *Development* **129**: 4335–4346.

Lopez-Sanchez C., Garcia-Martinez V., and Schoenwolf G. C. 2001. Localization of cells of the prospective neural plate, heart and somites within the primitive streak and epiblast of avian embryos at intermediate primitive-streak stages. *Cells Tissues Organs* **169**: 334–346.

Maden M. 2000. The role of retinoic acid in embryonic and post-embryonic development. *Proc. Nutr. Soc.* **59**: 65–73.

Mangold O. 1933. Über die Induktionsfähigkeit der verschiedenen Bezirke der Neurula von Urodelen. *Naturwissenhaften* **21**: 761–766.

Marshall H., Nonchev S., Sham M.H., Muchamore I., Lumsden A., and Krumlauf R. 1992. Retinoic acid alters hindbrain *Hox* code and induces transformation of rhombomeres 2/3 into a 4/5 identity. *Nature* **360**: 737–741.

Marshall H., Studer M., Popperl H., Aparicio S., Kuroiwa A., Brenner S., and Krumlauf R. 1994. A conserved retinoic acid response element required for early expression of the homeobox gene *Hoxb-1*. *Nature* **370**: 567–571.

Martinez-Barbera J.P. and Beddington R.S. 2001. Getting your head around *Hex* and *Hesx1*: Forebrain formation in mouse. *Int. J. Dev. Biol.* **45**: 327–336.

Martinez-Barbera J.P., Rodriguez T.A., and Beddington R.S. 2000. The homeobox gene *Hesx1* is required in the anterior neural ectoderm for normal forebrain formation. *Dev. Biol.* **223**, 422–430.

Matsuo I., Kuratani S., Kimura C., Takeda N., and Aizawa S. 1995. Mouse Otx2 functions in the formation and patterning of rostral head. *Genes Dev.* **9**: 2646–2658.

McGinnis W. and Krumlauf R. 1992. Homeobox genes and axial patterning. *Cell* **68**: 283–302.

McGrew L.L., Hoppler S., and Moon R.T. 1997. Wnt and FGF pathways cooperatively pattern anteroposterior neural ectoderm in *Xenopus*. *Mech. Dev.* **69**: 105–114.

McGrew L.L., Lai C.J., and Moon R.T. 1995. Specification of the anteroposterior neural axis through synergistic interaction of the Wnt signaling cascade with *noggin* and *follistatin*. *Dev. Biol.* **172**: 337–342.

Momoi A., Yoda H., Steinbeisser H., Fagotto F., Kondoh H., Kudo A., Driever W., and Furutani-Seiki M. 2003. Analysis of Wnt8 for neural posteriorizing factor by identifying Frizzled 8c and Frizzled 9 as functional receptors for Wnt8. *Mech. Dev.* **120**: 477–489.

Muhr J., Jessell T.M., and Edlund T. 1997. Assignment of early caudal identity to neural plate cells by a signal from caudal paraxial mesoderm. *Neuron* **19**: 487–502.

Muhr J., Graziano E., Wilson S., Jessell T.M., and Edlund T. 1999. Convergent inductive signals specify midbrain, hindbrain, and spinal cord identity in gastrula stage chick embryos. *Neuron* **23**: 689–702.

Nambiar R.M. and Henion P.D. 2004. Sequential antagonism of early and late Wnt-signaling by zebrafish *colgate* promotes dorsal and anterior fates. *Dev. Biol.* **267**: 165–180.

Niehrs C. 1999. Head in the WNT: The molecular nature of Spemann's head organizer. *Trends Genet.* **15**: 314–319.

Niehrs C., Kazanskaya O., Wu W., and Glinka A. 2001. Dickkopf1 and the Spemann-Mangold head organizer. *Int. J. Dev. Biol.* **45**: 237–240.

Nieuwkoop P.D. 1963. Pattern formation in artificially activated ectoderm (*Rana pipiens* and *Ambystoma punctatum*). *Dev. Biol.* **7**: 255–279.

Nieuwkoop P.D. and Nigtevecht G.V. 1954. Neural activation and transformation in explants of competent ectoderm under the influence of fragments of anterior notochord in urodeles. *J. Embryol. Exp. Morphol.* **2**: 175–193.

Nieuwkoop P.D., Botternenbrood E.C., Kremer A., Bloesma F.F.S.N., Hoessels E.L.M.J., Meyer G., and Verheyen F.J. 1952. Activation and organization of the central nervous system in amphibians. *J. Exp. Zool.* **120**: 1–108.

Nordstrom U., Jessell T.M., and Edlund T. 2002. Progressive induction of caudal neural character by graded Wnt signaling. *Nat. Neurosci.* **5**: 525–532.

Novitch B.G., Wichterle H., Jessell T.M., and Sockanathan S. 2003. A requirement for retinoic acid-mediated transcriptional activation in ventral neural patterning and motor neuron specification. *Neuron* **40**: 81–95.

Oosterveen T., Niederreither K., Dolle P., Chambon P., Meijlink F., and Deschamps J. 2003. Retinoids regulate the anterior expression boundaries of 5′ *Hoxb* genes in posterior hindbrain. *EMBO J.* **22**: 262–269.

Oxtoby E. and Jowett T. 1993. Cloning of the zebrafish *krox-20* gene (*krx-20*) and its expression during hindbrain development. *Nucleic Acids Res.* **21**: 1087–1095.

Papalopulu N., Clarke J.D., Bradley L., Wilkinson D., Krumlauf R., and Holder N. 1991. Retinoic acid causes abnormal development and segmental patterning of the anterior hindbrain in *Xenopus* embryos. *Development* **113**: 1145–1158.

Patten I., Kulesa P., Shen M.M., Fraser S., and Placzek M. 2003. Distinct modes of floor plate induction in the chick embryo. *Development* **130**: 4809–4821.

Pera E.M. and Kessel M. 1997. Patterning of the chick forebrain anlage by the prechordal plate. *Development* **124**: 4153–4162.

Perea-Gomez A., Rhinn M., and Ang S.L. 2001. Role of the anterior visceral endoderm in restricting posterior signals in the mouse embryo. *Int. J. Dev. Biol.* **45**: 311–320.

Perea-Gomez A., Shawlot W., Sasaki H., Behringer R.R., and Ang S. 1999. HNF3β and *Lim1* interact in the visceral endoderm to regulate primitive streak formation and anterior-posterior polarity in the mouse embryo. *Development* **126**: 4499–4511.

Pituello F. 1997. Neuronal specification: Generating diversity in the spinal cord. *Curr. Biol.* **7**: R701–704.

Pituello F., Medevielle F., Foulquier F., and Duprat A.M. 1999. Activation of *Pax6* depends on somitogenesis in the chick embryo cervical spinal cord. *Development* **126**, 587–596.

Pourquie O. 2003. The segmentation clock: Converting embryonic time into spatial pattern. *Science* **301**: 328–330.

Rhinn M., Dierich A., Le Meur M., and Ang S. 1999. Cell autonomous and non-cell autonomous functions of *Otx2* in patterning the rostral brain. *Development* **126**, 4295–4304.

Rhinn M., Dierich A., Shawlot W., Behringer R.R., Le Meur M., and Ang S.L. 1998. Sequential roles for *Otx2* in visceral endoderm and neuroectoderm for forebrain and midbrain induction and specification. *Development* **125**: 845–856.

Ribisi S., Jr., Mariani F.V., Aamar E., Lamb T.M., Frank D., and Harland R.M. 2000. Ras-mediated FGF signaling is required for the formation of posterior but not anterior neural tissue in *Xenopus laevis*. *Dev. Biol.* **227**: 183–196.

Rowan A.M., Stern C.D., and Storey K.G. 1999. Axial mesendoderm refines rostrocaudal pattern in the chick nervous system. *Development* **126**: 2921–2934.

Rubenstein J.L., Shimamura K., Martinez S., and Puelles L. 1998. Regionalization of the prosencephalic neural plate. *Annu. Rev. Neurosci.* **21**: 445–477.

Ruiz i Altaba A. and Jessell T. 1991a. Retinoic acid modifies mesodermal patterning in early *Xenopus* embryos. *Genes Dev.* **5**: 175–187.

———. 1991b. Retinoic acid modifies the pattern of cell differentiation in the central nervous system of neurula stage *Xenopus* embryos. *Development* **112**: 945–958.

Saude L., Woolley K., Martin P., Driever W., and Stemple D.L. 2000. Axis-inducing activities and cell fates of the zebrafish organizer. *Development* **127**: 3407–3417.

Saxén L. and Toivonen S. 1962. *Primary embryonic induction*. Logos Press, Englewood Cliffs, New Jersey, Academic Press, London, United Kingdom

Schier A.F., Neuhauss S.C., Helde K.A., Talbot W.S., and Driever W. 1997. The *one-eyed pinhead* gene functions in mesoderm and endoderm formation in zebrafish and interacts with *no tail*. *Development* **124**: 327–342.

Schneider V.A. and Mercola M. 1999. Spatially distinct head and heart inducers within the *Xenopus* organizer region. *Curr. Biol.* **12**: 800–809.

Shawlot W. and Behringer R.R. 1995. Requirement for *Lim1* in head-organizer function. *Nature* **374**: 425–430.

Shawlot W., Wakamiya M., Kwan K.M., Kania A., Jessell T.M., and Behringer R.R. 1999. *Lim1* is required in both primitive streak-derived tissues and visceral endoderm for head formation in the mouse. *Development* **126**: 4925–4932.

Shih J. and Fraser S.E. 1996. Characterizing the zebrafish organizer: Microsurgical analysis at the early-shield stage. *Development* **122**: 1313–1322.

Shimamura K. and Rubenstein J.L. 1997. Inductive interactions direct early regionalization of the mouse forebrain. *Development* **124**: 2709–2718.

Shimamura K., Hartigan D.J., Martinez S., Puelles L., and Rubenstein J.L. 1995. Longitudinal organization of the anterior neural plate and neural tube. *Development* **121**: 3923–3933.

Simeone A., Avantaggiato V., Moroni M.C., Mavilio F., Arra C., Cotelli F., Nigro V., and Acampora D. 1995. Retinoic acid induces stage-specific antero-posterior transformation of rostral central nervous system. *Mech. Dev.* **51**: 83–98.

Smith J.C. 1995. Mesoderm-inducing factors and mesodermal patterning. *Curr. Opin. Cell Biol.* **7**: 856–861.

Sockanathan S., Perlmann T., and Jessell T.M. 2003. Retinoid receptor signaling in postmitotic motor neurons regulates rostrocaudal positional identity and axonal projection pattern. *Neuron* **40**: 97–111.

Spemann H. 1938. *Embryonic development and induction*. Yale University Press.

Spemann H. and Mangold H. 1924. Über Induktion von Embryonalanlagen durch Implantation artfremder Organisatoren. *Wilhelm Roux' Arch. Entwicklungsmech. Org.* **100**: 599–638.

Stern C.D. 2001. Initial patterning of the central nervous system. How many organizers? *Nat. Rev. Neurosci.* **2**: 92–98.

Stern C.D., Hatada Y., Selleck M.A., and Storey K.G. 1992. Relationships between mesoderm induction and the embryonic axes in chick and frog embryos. *Dev. Suppl.* **1992**: 151–156.

Storey K.G., Crossley J. M., De Robertis E.M., Norris W.E., and Stern C.D. 1992. Neural induction and regionalisation in the chick embryo. *Development* **114**: 729–741.

Storey K.G., Goriely A., Sargent C.M., Brown J.M., Burns H.D., Abud

H.M., and Heath J.K. 1998. Early posterior neural tissue is induced by FGF in the chick embryo. *Development* **125**: 473–484.

Streit A., Berliner A.J., Papanayotou C., Sirulnik A., and Stern C.D. 2000. Initiation of neural induction by FGF signalling before gastrulation. *Nature* **406**: 74–78.

Sundin O. and Eichele G. 1992. An early marker of axial pattern in the chick embryo and its respecification by retinoic acid. *Development* **114**: 841–852.

Sundin O.H., Busse H.G., Rogers M.B., Gudas L.J., and Eichele G. 1990. Region-specific expression in early chick and mouse embryos of Ghox-lab and Hox 1.6, vertebrate homeobox-containing genes related to *Drosophila* labial. *Development* **108**: 47–58.

Tam P.P. and Steiner K.A. 1999. Anterior patterning by synergistic activity of the early gastrula organizer and the anterior germ layer tissues of the mouse embryo. *Development* **126**: 5171–5179.

Thomas P. and Beddington R. 1996. Anterior primitive endoderm may be responsible for patterning the anterior neural plate in the mouse embryo. *Curr. Biol.* **6**: 1487–1496.

Toivonen S. 1978. Regionalization of the embryo. In *Organizer— A milestone of a half-century since Spemann* (ed. O. Nakamura and S. Toivonen), pp. 119–156. Elsevier/North Holland, Amsterdam, The Netherlands.

Varga Z.M., Wegner J., and Westerfield M. 1999. Anterior movement of ventral diencephalic precursors separates the primordial eye field in the neural plate and requires *cyclops*. *Development* **126**: 5533–5546.

Varlet I., Collignon J., and Robertson E.J. 1997. *nodal* expression in the primitive endoderm is required for specification of the anterior axis during mouse gastrulation. *Development* **124**: 1033–1044.

Vasiliauskas D. and Stern C.D. 2001. Patterning the embryonic axis: FGF signaling and how vertebrate embryos measure time. *Cell* **106**: 133–136.

Villanueva S., Glavic A., Ruiz P., and Mayor R. 2002. Posteriorization by FGF, Wnt, and retinoic acid is required for neural crest induction. *Dev. Biol.* **241**: 289–301.

Waddington C.H. and Needham J. 1936. Evocation and individuation and competence in amphibian organizer action. *Proc. K. Ned. Akad Wet.* **39**: 887–891.

Wilson S.I., Graziano E., Harland R., Jessell T.M., and Edlund T. 2000. An early requirement for FGF signalling in the acquisition of neural cell fate in the chick embryo. *Curr. Biol.* **10**: 421–429.

Wilson S.W. and Houart C. 2004. Early steps in the development of the forebrain. *Dev. Cell* **6**: 167–181.

Woo K. and Fraser S.E. 1995. Order and coherence in the fate map of the zebrafish nervous system. *Development* **121**: 2595–2609.

———. 1997. Specification of the zebrafish nervous system by nonaxial signals. *Science* **277**: 254–257.

———. 1998. Specification of the hindbrain fate in the zebrafish. *Dev. Biol.* **197**: 283–296.

Xu R.H., Kim J., Taira M., Sredni D., and Kung H. 1997. Studies on the role of fibroblast growth factor signaling in neurogenesis using conjugated/aged animal caps and dorsal ectoderm-grafted embryos. *J. Neurosci.* **17**: 6892–6898.

Yamada T. 1994. Caudalization by the amphibian organizer: *brachyury*, convergent extension and retinoic acid. *Development* **120**, 3051–3062.

Yamaguchi T.P. 2001. Heads or tails: Wnts and anterior–posterior patterning. *Curr. Biol.* **11**: R713–724.

Zakany J., Kmita M., Alarcon P., de la Pompa J.L., and Duboule D. 2001. Localized and transient transcription of *Hox* genes suggests a link between patterning and the segmentation clock. *Cell* **106**: 207–217.

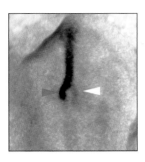

EARLY PATTERNING OF THE LEFT/RIGHT AXIS

D.S. Adams and M. Levin

Cytokine Biology Department, The Forsyth Institute, and Department of Oral and Developmental Biology, Harvard School of Dental Medicine, Boston, Massachusetts 02115

INTRODUCTION

Vertebrates have a bilaterally symmetrical body plan, but this symmetry is broken by the consistently asymmetric placement of various internal organs such as the heart, liver, spleen, and gut, or the asymmetric development of paired organs (such as brain hemispheres and lungs). The establishment of left/right (LR) asymmetry raises a number of fascinating biological questions. Why does asymmetry exist at all? What are the implications of asymmetry for the normal structure and physiology of the heart, gut, and brain? Why are all normal individuals not only asymmetric, but asymmetric in the same direction (i.e., why a consistent bias and not a 50%/50% racemic population)? When, during evolution, did handed asymmetry appear, and were there true bilaterally symmetrical organisms prior to the invention of oriented asymmetry? Is it connected to chirality in lower forms (such as snail shell coiling and chirality in some plants)? At what developmental stages is asymmetry initiated in vertebrate embryos? How conserved are the molecular mechanisms establishing correct asymmetry in animals with drastically different modes of gastrulation? How can the left/right axis be consistently oriented with respect to the anterior-posterior and dorsoventral axes in the absence of any macroscopic feature of chemistry or physics that distinguishes left from right? None of these questions can be properly addressed until we have a detailed understanding, at the molecular, genetic, and biochemical levels, of the formation of biased asymmetry in embryos.

Whereas in most species all normal individuals are asymmetrical in the same direction, animals with complete mirror reversal of internal organs can arise (situs inversus totalis) and are otherwise phenotypically unimpaired. Thus, although it is possible to come up with plausible evolutionary reasons that organisms might be asymmetric in the first place (optimal packing, fluid dynamics, maximizing surface area of tubes, etc.), there is no obvious reason they should all be asymmetric to the same direction. It is, after all, much easier to imagine a developmental mechanism for generating asymmetry (such as positive feedback and amplification of stochastic biochemical differences) than for biasing it to a given direction.

Although mechanisms underlying anterior–posterior and dorsoventral asymmetry have been studied in detail with the advent of molecular genetics, the mechanistic basis for LR asymmetry was, until recently, completely unknown. However, within the last 10–15 years, significant advances in embryonic asymmetry have been made by a number of groups (Levin and Mercola 1998a; Burdine and Schier 2000; Mercola and Levin 2001; Yost 2001; Mercola 2003). Tables 1–3 summarize the molecular players in the LR pathway and show which ones are conserved among various model systems. Gene products in the LR pathway have been identified in forward and reverse genetics approaches (exemplified by the zebrafish mutants and Sonic Hedgehog, respectively), and almost all have roles in embryonic processes other than LR asymmetry. Although a few of the components have no known homology or function

Table 1. *Asymmetrically expressed genes in embryos that have been the focus of a paper on LR asymmetry*

Gene	Species	Product/Role	Side	Reference
lefty	mouse, chick, frog	TGF-β-family signaling molecule	left	Meno et al. (1996, 1998); Branford et al. (2000); Cheng et al. (2000); Essner et al. (2000)
Activin βB	chick	TGF-β-family signaling molecule	right	Levin et al. (1997)
cAct-RIIa	chick	Activin receptor	right	Levin et al. (1995)
Shh	chick	signaling molecule	left	Levin et al. (1995)
cSnR	chick	zinc finger protein	right	Isaac et al. (1997)
Nodal	chick, mouse, frog	TGF-β-family signaling molecule	left	Levin et al. (1995); Collignon et al. (1996); Lowe et al. (1996); Lohr et al. (1997); Morokuma et al. (2002)
cPTC	chick	Shh receptor	left	Levin (1998b); Pagan-Westphal and Tabin (1998)
Cerberus/Caronte	chick	signaling molecule	left	Yokouchi et al. (1999); Zhu et al. (1999)
BMP-4	zebrafish, chick	BMP family signaling molecule	left	Chen et al. (1997); Monsoro-Burq and LeDouarin (2000)
Pitx-2	chick, frog, mouse	transcription factor	left	Logan et al. (1998); Ryan et al. (1998); Morokuma et al. (2002)
NKX3.2	chick, mouse	transcription factor	left in chick, right in mice	Schneider et al. (1999)
Follistatin	chick	signaling molecule	right	Levin (1998a)
FGF-8	chick	growth factor	right	Boettger et al. (1999)
flectin	chick	extracellular matrix molecule	left	Tsuda et al. (1996)
dHAND	chick, mouse, frog	bHLH transcription factor	right	Srivastava (1995); Angelo et al. (2000)
eHAND	chick, mouse, frog	bHLH transcription factor	left	Cserjesi et al. (1995); Srivastava (1995); Biben and Harvey (1997); Sparrow et al. (1998); Angelo et al. (2000)
N-Cadherin	chick	adhesion molecule	right node, left groove	Garcia-Castro et al. (2000)
Cx43	chick	gap junction protein	right	Levin and Mercola (1999)
Islet-1	chick	LIM homeobox gene	left	Yuan and Schoenwolf (2000)
H⁺/K⁺-ATPase	frog, chick	H⁺ and K⁺ ion pump	left	Levin et al. (2002)
PKI-a	chick	PKA inhibitor	right	Kawakami and Nakanishi (2001); Rodriguez-Esteban et al. (2001)
NCX-1	chick, mouse	sodium–calcium exchanger	right	Linask et al. (2001)
HoxC-8	frog	transcription factor	left	Thickett and Morgan (2002)
Xin	mouse	?	right	Wang et al. (1999)
Southpaw	zebrafish	TGF-β family	left	Long et al. (2003)
cMid-1	chick	microtubule-associated protein	right	Granata and Quaderi (2003)
lsy-6	C. elegans	micro RNA repressor	left	Johnston and Hobert (2003)
Dll1	chick	delta-like signaling molecule	left	Raya et al. (2004)
14-3-3E	frog	14-3-3 family	right	Bunney et al. (2003)
Kif5C	chick	kinesin motor	right	Dathe et al. (2004)

Bold entries indicate genes expressed during gastrulation.

(such as INV), the remainder form a fairly diverse group of molecules including secreted signaling factors, regulators of ion flux, motor proteins, and transcription factors. Some are asymmetrically expressed at the level of mRNA or protein, whereas others appear to have no asymmetry with respect to their localization.

Conceptually, LR patterning is divided into three phases. In the final phase, individual organs utilize cell migration, differential proliferation, and other mechanisms to achieve asymmetries in their location or morphogenesis. Upstream of these processes lies a pathway of asymmetric genes: genes which are expressed in cell fields

only on one side of the embryonic midline, and which propagate signals that dictate sidedness for the organs undergoing asymmetric morphogenesis. These cascades of asymmetric gene expression form the middle phase of LR patterning. However, for whichever asymmetric gene is at the top of the pathway, it is necessary to ask what determines its asymmetry. Thus, in the first phase of LR patterning, an as-yet-unknown mechanism must orient the LR axis with respect to the other two axes (Brown and Wolpert 1990).

The developmental timing of each phase differs among species, although asymmetric gene expression almost always

Table 2. *Asymmetrically expressed genes that have not been the focus of a LR paper*

Gene	Species	Product/Role	Side	Reference
HNF-3β	chick, mouse	winged-helix transcription factor	left	Levin et al. (1995); Collignon et al. (1996)
cWnt-8C	chick	wnt-family signaling molecule	right	Levin (1998b); Pagan-Westphal and Tabin (1998)
hLAMP-1[a]	chick	extracellular matrix molecule	left	Smith et al. (1997)
JB3[a]	chick	extracellular matrix molecule	right	Wunsch et al. (1994); Smith et al. (1997)
HGF	chick	kringle signaling molecule	left	Streit et al. (1995)
Hrlim	ascidian	LIM-family signaling molecule	right	Wada et al. (1995)
Rtk2	zebrafish	Eph receptor	right	Schilling et al. (1999)
Fli-1	zebrafish	transcription factor	left	Schilling et al. (1999)
DM-GRASP	zebrafish	adhesion protein		Schilling et al. (1999)
Xbap	frog	transcription factor	left	Newman et al. (1997)
Hest1	zebrafish	ASIC ion channel	left	Concha et al. (2003)

Bold entries indicate genes expressed during gastrulation.
[a]Antibody epitopes.

Table 3. *Genes involved in LR patterning that are not asymmetrically expressed*

Gene	Species	Product/Role	Reference
Iv	mouse	dynein (cytoplasmic transport or ciliary motor)	Lowe et al. (1996); Supp et al. (1997, 1999, 2000)
Inv	mouse	?	Mochizuki et al. (1998); Morgan et al. (1998, 2002)
Vg-1	frog	TGF-β-family signaling molecule	Hyatt et al. (1996); Hyatt and Yost (1998)
Connexins	frog, chick, human	system of gap-junctional cell–cell signaling	Britz-Cunningham et al. (1995); Levin and Mercola (1998b); Levin and Mercola (1999)
No turning	mouse	midline patterning	Melloy et al. (1998)
SIL	mouse	midline patterning	Izraeli et al. (1999)
KIF-3	mouse	component of ciliary motor	Nonaka et al. (1998); Takeda et al. (1999)
Polaris	mouse	?	Murcia et al. (2000)
HFH-4	mouse	transcription factor	Chen et al. (1998); Brody et al. (2000)
Lin-12	C. elegans	Notch signaling molecule	Hermann et al. (2000)
Delta-1	mouse	Notch signaling molecule	Przemeck et al. (2003)
Notch	mouse, zebrafish	Notch signaling molecule	Krebs et al. (2003); Raya et al. (2003)
Smo	mouse	membrane protein involved in hedgehog signaling	Zhang et al. (2001)
Ihh	mouse	member of hedgehog signaling proteins	Zhang et al. (2001)
GDF-1	mouse	TGF-β-family signaling molecule	Rankin et al. (2000)
Lrd	mouse	Dynein	Supp et al. (1997, 1999)
DNAH5	human	Dynein	Ibanez-Tallon et al. (2002); Olbrich et al. (2002)
PCKD-2	mouse	Polycystin-2 ion channel	Pennekamp et al. (2002)
ZIC3	human, mouse, frog	zinc-finger protein	Gebbia et al. (1997); Kitaguchi et al. (2000); Purandare et al. (2002)
EGF-CFC	mouse, fish	extracellular receptor	Yan et al. (1999)
Furin	mouse	pro-protein convertase	Roebroek et al. (1998); Constam and Robertson (2000)
Brachyury	mouse	transcription factor	King et al. (1998)
Ednrb	mouse	piebald deletion complex	Welsh and O'Brien (2000)
Rotatin	mouse	transmembrane protein	Faisst et al. (2002)
PDI-P5	zebrafish	protein disulfide isomerase	Hoshijima et al. (2002)
Pol-1	mouse	DNA polymerase	Kobayashi et al. (2002)
PA26	human	sestrin-family	Peeters et al. (2003)
Cryptic	mouse, human, zebrafish	EGF-CFC gene	Gaio et al. (1999); Yan et al. (1999); Bamford et al. (2000)

begins at or shortly after gastrulation. The LR axis is probably specified after the anterior–posterior (AP) and dorsoventral (DV) axes, and is determined with respect to them (McCain and McClay 1994; Danos and Yost 1995).

The timing of the initiation of LR asymmetry is particularly controversial. In the following text, we review the most important data on mechanisms of asymmetry elucidated in a number of model systems.

FISH

Flatfishes acquire a profound asymmetry in eye location (and scale/skin pigmentation) during metamorphosis from bilaterally symmetric larvae (Matsumoto and Seikai 1992; Okada et al. 2001; Hashimoto et al. 2002). Analysis of mutants in the zebrafish embryo has identified a number of loci which, when altered, cause aberrant LR patterning (Yost 1998), although some of these are likely to represent secondary LR effects of disrupted notochord or AP/DV patterning. In zebrafish, asymmetric markers such as *lefty*, *nodal*, and *pitx2* exhibit well-conserved asymmetric expression during neurulation and somitogenesis (Cheng et al. 2000; Essner et al. 2000; Liang et al. 2000). Unfortunately, almost nothing is known about early, upstream mechanisms in this model system.

FROGS

Embryos of the frog *Xenopus laevis* are analogous to the fish and chick with respect to a number of asymmetrically expressed left-sided genes (e.g., *Nodal*, *Lefty*, and *Pitx-2*) that function after neurulation (Levin 2004b). Although the mechanisms that process LR information during gastrulation in amphibia are largely unknown, the *Xenopus* embryo has allowed discovery of a number of mechanisms that underlie asymmetry at the earliest stages known in any species. Experiments in *Xenopus* were the first to suggest that the LR axis might be established extremely early, and to be intimately linked with DV axis formation (Yost 1991). The DV axis is initiated by sperm entry during fertilization, followed by a cytoplasmic rotation during the first cell cycle, driven by a microtubule array at the vegetal cortex (Gerhart et al. 1989). Work from the Yost lab showed that embryos in which the microtubule array was blocked, but which were tilted manually to rescue the DV axis, exhibited laterality defects, suggesting that the LR axis may be dependent on the transient microtubule array during the first cell cycle. The microtubule-associated motor proteins kinesin and dynein have been linked with LR asymmetry in mammals (see below). The appearance of LR asymmetry between fertilization and the first cell division is also consistent with the recent work on ion fluxes and the appearance of asymmetric mRNA and 14-3-3 protein localization during early cleavage (Levin et al. 2002; Bunney et al. 2003).

Syndecans

Localized perturbation of a small patch of extracellular matrix (ECM) by microsurgery, as well as global perturbation of the ECM by microinjection of Arg-Gly-Asp peptides or heparinase into the blastocoel, resulted in randomization of LR asymmetry. This work provided the first molecular entry point into LR asymmetry and suggested that the ECM participated in transfer of LR information in development. Inhibition of proteoglycan synthesis with the drug p-nitrophenyl-β-D-xylopyranoside prevents heart looping in *Xenopus* (Yost 1990). The sensitivity window was between stages 12 and 15—just after gastrulation.

On the basis of the proposal that heparan sulfate proteoglycans (HSPGs) or the ECM on the basal surface of the ectoderm transmits LR information to mesodermal primordia during gastrulation (Yost 1992), Teel and Yost examined the roles of the syndecan family; syndecan-1 and -2 are maternally expressed HSPGs specifically located in the animal cap ectoderm (Teel and Yost 1996). Using dominant-negative and loss-of-function approaches, it was shown that syndecan-2 is involved in LR asymmetry (Kramer and Yost 2002) in *Xenopus*. A cytoplasmic domain of syndecan-2 is phosphorylated in cells on the right but not the left half of the frog embryo during gastrulation. Moreover, they showed that attachment of multiple heparan sulfate glycosaminoglycans on syndecan-2 and functional interaction of these sites with the cytoplasmic domain are an obligate part of LR patterning during gastrulation, immediately prior to the migration of mesoderm across ectoderm. Kramer and Yost also presented biochemical data on the direct interaction of syndecan-2 with Vg1, suggesting that these two molecules function together during LR patterning at gastrulation.

Vg1 and the "Coordinator"

Another key finding in *Xenopus* was the discovery of an experimental perturbation that can produce almost full situs inversus; this is especially interesting, since almost every other reported manipulation results in heterotaxia—an independent randomization of situs and not full reversal (or loss of asymmetry). The active form of Vg1, a TGF-β family member, can almost completely invert the LR axis when misexpressed on the right side (R3 blastomere) of a *Xenopus* embryo (Hyatt et al. 1996; Hyatt and Yost 1998). This can be interpreted as signifying that Vg1 normally acts in descendants of the L3 blastomere, which contribute to the left lateral plate mesoderm, and the model suggests signaling through ALK2 and mutual antagonism with BMP on the right side of the embryo (Ramsdell and Yost 1999). Axial inversion is specific to the activated Vg1, as it cannot be mimicked by Activin. Although these data are consistent with an early LR pattern in the pre-gastrula-stage *Xenopus* embryo, the precise timing remains uncertain, since the persistence of the injected mRNA to later stages raises the possibility that the injected Vg1 persists in the embryo and mimics a later signal. Confirmation of the role of endogenous Vg1 in this process remains uncertain pending characterization of processed, endogenous Vg1 in early *Xenopus* embryos (and especially, asymmetries therein) (see Chapter 35).

Gap Junctional Communication

Gap junctions are channels connecting adjacent cells which allow the direct transfer of small molecule signals. The cell biology of gap junctions has been described in several excellent recent reviews (Falk 2000), and gap junctional flow is involved in a number of important patterning events in embryonic development and tumor progression (Lo 1996; Levin 2001). Briefly, the most frequently studied gap junction channel is formed by the assembly and docking of hexamers of proteins from the connexin family (one hexamer in each of two adjacent cell membranes). Functional gap junctional communication (GJC) is dependent on the existence of compatible hemichannels on the cells' surfaces, the permeability of the hemichannels to the substance, and the open status of the gap junction.

On the basis of a report that several unrelated patients with viscero-atrial heterotaxia contain potential mutations within *Connexin43* (Britz-Cunningham et al. 1995), and data from frog embryos that indicated asymmetric patterns of GJC in early blastomeres (Guthrie 1984; Guthrie et al. 1988), Levin and Mercola (1998b) tested the hypothesis that gap junctional paths are a mechanism by which LR information is communicated across large cell fields. *Xenopus* embryos at early cleavage stages were shown to contain a junctional path across the dorsal blastomeres, and a zone of junctional isolation on the ventral midline (confirming with a double-dye system previous observation using a small-molecule probe [Guthrie 1984; Olson et al. 1991; Brizuela et al. 2001], but see Landesman et al. [2000]). Injection of mRNA encoding a dominant negative connexin protein into dorsal blastomeres or wild-type connexins into ventral blastomeres both resulted in heterotaxia and randomization of *Nodal* expression in the absence of other developmental defects (Levin and Mercola 1998b).

These results indicated that an endogenous path of GJC between dorsal and lateral blastomeres, as well as the isolation across the ventral midline, is necessary for normal LR asymmetry in *Xenopus*. Pharmacological blocker experiments suggested that the gap junctional system begins to function in LR asymmetry during cleavage stages and is upstream of asymmetric XNR-1 and heart tube looping. These data have led to the hypothesis that a circumferential path of GJC, around a zone of isolation, could be the mechanism that bridges asymmetry at the level of a cell (step 1) to the embryo-wide cascades of asymmetric gene expression (step 2). It was proposed (Levin and Nascone 1997; Levin and Mercola 1998b) that small-molecule determinants are initially randomly distributed but traverse the circumferential

GJC path unidirectionally, accumulating on one side of the midline, and then induce asymmetric gene expression in conventional ways. Similar data were later obtained in the chick embryo (see below). The identities of the putative low-molecular-weight determinants remain unknown.

Ion Flux

One key aspect of the GJC model is that the net junctional flow must be unidirectional in order to derive a LR asymmetry from the existing DV difference in GJC. Hypothesizing that a voltage difference might provide an electromotive force which can be used to electrophorese charged molecules in preferred directions through GJC paths, Levin et al. (2002) tested the model that ion fluxes (needed to generate the standing voltage gradients) might be an obligatory aspect of early LR patterning in *Xenopus*.

A pharmacological screen of hundreds of various types of ion channels, pumps, and co-transporters (Levin et al. 2002) specifically implicated four target genes involved in H^+ and K^+ flux. One of these, the H^+/K^+-ATPase, functions during early cleavage stages. Moreover, maternal H^+/K^+-ATPase mRNA is asymmetrically localized during the first two cell divisions, demonstrating that asymmetry is generated by 2 hours postfertilization. Examination of the situs of asymmetric genes (*xNR-1*, *xLefty*, and *xPitx-2*) following early exposure to blockers of the H^+/K^+-ATPase revealed that, consistently with the early asymmetrical expression, the ion flux mechanism is upstream of the asymmetric expression of those genes. Gain-of-function experiments using H^+/K^+-ATPase and K^+ channel overexpression constructs also demonstrated that equalizing H^+ and K^+ flux on either side of the midline randomizes the LR axis.

Taken together, these data demonstrate that the *Xenopus* embryo assigns L and R identities to cells during the first few cleavages. This conclusion is also confirmed by the finding of asymmetric 14-3-3E protein localization, which is crucial for normal LR asymmetry (Bunney et al. 2003). However, a key series of experiments demonstrated that under some circumstances, ectopic organizers induced much later are still able to impose correct LR identity on nearby tissue (Nascone and Mercola 1997). Thus, the *Xenopus* embryo is likely to contain an endogenous very early mechanism for aligning the LR axis, but also the capacity for regulatory patterning of the LR axis at later stages.

CHICK

The first morphological asymmetry in the chick embryo is a subtle tilt of Hensen's node toward the end of gastrulation (Kölliker 1879; Hertwig 1902; Dathe et al. 2002). The first obvious sign of asymmetry is the looping of the heart tube, which has been shown to be determined during gastrulation in transplantation experiments (Hoyle et al. 1992). The chick was the first system in which asymmetric gene expression was demonstrated, and this organism provides the most detailed picture of left/right mechanisms functioning during gastrulation.

Asymmetric Genes

Characterizing the expression of a number of known genes during early chick embryogenesis, Levin et al. found that several had consistently asymmetric expression patterns during gastrulation and at the beginning of neurulation (Levin et al. 1995, 1997; Levin 1998a). *Sonic Hedgehog* (*Shh*) encodes a signaling molecule that is also involved in patterning of the limb and the neural tube (Capdevila and Johnson 2000) and is expressed symmetrically within the ectoderm of Hensen's node (the chick organizer; see Chapters 15 and 29) before stage 4, at which time it becomes restricted to the left side of the node (Fig. 1A). This is followed at stage 7 by the left-sided expression of *Nodal* (a TGF-β family member, originally called *cNR-1* in the chick). *Nodal* is first expressed in a small domain of endoderm cells directly adjacent to the ectoderm cells expressing *Shh*, and then in a large domain in the lateral plate mesoderm.

The juxtaposition of the proximal domain of *Nodal* to the cells expressing *Shh* suggested an inductive interaction, and indeed, implanting cells expressing *Shh* on the right side of Hensen's node is sufficient to induce an ectopic domain of *Nodal* expression on the right side. The Activin-inducible

gene *Activin Receptor IIa* (*cAct-RIIa*) becomes expressed on the right side of Hensen's node at the same time that *Shh* becomes restricted to the left (Fig. 1B). This suggested the right-sided presence of an Activin-like repressor upstream of *Shh*; it was then shown that a local source of Activin protein implanted on the left side is able to induce *cAct-RIIa* there, and to repress the expression of left-sided *Shh* (Levin et al. 1995). Although right-sided asymmetric expression of *Activin βB* has been reported in the early chick streak (Levin et al. 1997; Levin 1998a), Act-RIIa is now thought more likely to be a receptor for Nodal-related ligands than Activin (see Chapter 35); thus, the details of these interactions remain to be elucidated, and it is still unknown whether cAct-RIIa itself plays a causal role in LR patterning.

Many more asymmetric genes have been identified in chick embryos (Levin 1998b); these factors participate in cascades of induction and repression of asymmetric gene pathways taking place on the left and right of the midline (see Tables 1 and 2). The signaling molecules functioning during gastrulation dictate heart and gut situs as well as embryonic turning through control of the expression of the highly conserved left-sided *Nodal*.

Gap Junctional Communication

The fairly dense pathway of LR cascade members in chick embryos suggests an immediate question: What mechanism is upstream of the very first asymmetrically expressed gene? Interestingly, contrary to the paradigm of genetically separate L and R compartments which begins during mid-gastrulation, it was observed that events occurring on the far R side were required for establishment of L identity on the left side at the beginning of streak initiation (Levin and Mercola 1999). Thus, GJC was examined in the chick embryo as a candidate for a mechanism that would enable cells to communicate across large distances along the LR axis and assign LR identities to cell fields.

Similar to the results in *Xenopus*, it was discovered that differential GJC is required upstream of asymmetric *Shh* expression in the node, and one connexin, Cx43, was implicated by treatment with specific antisense oligonucleotides or blocking antibodies (Levin and Mercola 1999). Interestingly, *Cx43* mRNA is broadly expressed in the epiblast of streak-stage embryos, but not in the streak itself. Thus, GJC required for LR asymmetry may propagate signals throughout the epiblast but not across an insulating zone at the streak. In support of this model, surgical incisions made along various radii emanating from the developing node abolish node asymmetry. Although a topological transformation is required to map the GJC system onto the different embryonic architectures of the chick and *Xenopus*, the basic schematic of this system is the same in both systems: Correct laterality determination upstream of

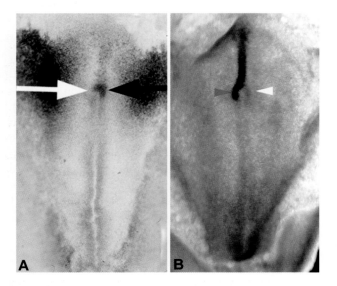

Figure 1. Asymmetric gene expression in chick embryos. During gastrulation, a number of genes are asymmetrically expressed. Two of the best characterized are *Activin Receptor 2a* (*cAct-RIIa*) on the right side of Hensen's node (*A*), and *Sonic Hedgehog* (*Shh*) on the left (*B*). (Reprinted, with permission, from Levin et al. 1995 [© Elsevier].)

asymmetric gene expression appears to depend on an uninterrupted contiguous region of GJC around a small zone of junctional isolation.

An essential feature of the GJC model in both *Xenopus* and chick is circumferential GJC around a zone of junctional insulation (the streak in chick and the ventral midline in *Xenopus*). Although consistent with the idea that the epiblast influences node asymmetry, this set of findings also indicates that the information does not originate from a single source, but that contiguity of the blastodisc on both sides of the midline is necessary (Levin and Mercola 1999). The GJC model predicts that the midline cells receive LR information from lateral tissue during gastrulation. In the chick, current data strongly indicate that, indeed, Hensen's node is instructed with respect to the LR axis by adjacent lateral cell groups (Psychoyos and Stern 1996; Pagan-Westphal and Tabin 1998; Yuan and Schoenwolf 1998; Levin and Mercola 1999). Important open areas of research include identification of upstream signals that orient GJC in embryos, characterization of the determinants that traverse gap junctions and downstream target genes that they regulate, and the targets that are immediately downstream of GJC flow.

Ion Flux

Because the GJC system has been shown to be conserved to both chick and *Xenopus*, Levin et al. tested whether embryonic laterality was dependent on ion flux in the chick as well (Levin et al. 2002). Analysis of the chick embryo using an in vivo membrane voltage reporter dye (Fig. 2A) indicated the existence of a consistently biased depolarization of cells on one side of the early primitive streak (*prior* to the formation of Hensen's node). This indicates that the chick embryo has assigned L and R identities by stage 3^-—prior to the earliest

known asymmetric gene. Similar to the data in *Xenopus*, specific inhibition of the H^+/K^+-ATPase prior to gastrulation equalized the depolarization of cells across the midline and randomized the asymmetric expression of *Shh*, *cWnt-8C*, and other markers (including *Cerberus*—a marker of head asymmetry). Interestingly, whereas the H^+/K^+-ATPase is expressed, as predicted by the GJC model (which requires the motive force battery to be located in the zone of isolation), in the primitive streak during early gastrulation, no asymmetry in pump localization has been observed in the chick at the level of mRNA. This echoes a theme that highlights an important difference between species. Although both chick and *Xenopus* appear to use GJC and ion flux to pattern the LR axis, there are differences in how this mechanism is regulated. The dorsoventral difference in GJC in frog embryos takes place posttranslationally (by gating control of existing gap junctions). In contrast, the chick embryo seems to establish the zone of isolation at the level of mRNA (by not transcribing *Cx43* mRNA in the primitive streak). Similarly, whereas asymmetric ion flux is provided by asymmetric localization of mRNA in early frog embryos, it appears to be established in the chick embryo by a posttranslational mechanism (such as gating of electrogenic activity of mature pump complexes).

The most interesting future data are likely to come from pursuing the asymmetric gene cascade upstream and determining how it interfaces with the GJC and ion flux systems. What are the *first* asymmetrically expressed genes on the left and right sides in the chick embryo? Some of the details of this process have recently been provided by a study which showed that an H^+/K^+-ATPase-dependent extracellular calcium accumulation on the left side of Hensen's node is sensed by a Notch pathway mechanism (Raya et al. 2004). Does asymmetric gene expression begin prior to gastrulation? It has previously been suggested (Levin and Mercola 1998a) that the computation which aligns the LR axis with the DV and AP axes takes place at the initiation of gastrulation, at the base of the primitive streak (which reliably progresses from the periphery to the center of the blastoderm). However, the molecular mechanism of this process cannot be elucidated until we have a good understanding of how (and whether) individual cells in the chick blastoderm have an anterior–posterior polarity.

MAMMALS

Errors of LR patterning during embryogenesis are relevant to the clinical considerations of several fairly common human birth defects: syndromes including Kartagener's and Ivemark's (Winer-Muram 1995), dextrocardia, situs inversus (a complete mirror-image reversal of the sidedness of asymmetrically positioned organs and asymmetric

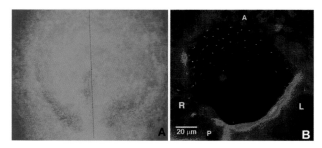

Figure 2. Asymmetric ion flux during gastrulation. A voltage-sensitive fluorescent dye allowed in vivo detection of an endogenous asymmetry in the steady-state membrane voltage levels of cells on the left and right sides of the early primitive streak (*A, red line* indicates midline of streak). The left side of the streak is depolarized with respect to the right. (*B*) In the mature node of the mouse embryo, an asymmetric Ca^{++} signal is detected. (*A*, Reprinted, with permission, from Levin et al. 2002; *B*, reprinted, with permission, from McGrath and Brueckner 2003 [© Elsevier].)

paired organs), heterotaxia (a loss of concordance where each organ makes an independent decision as to its situs), and right or left isomerism (where the organism is completely symmetrical; for example, polysplenia or asplenia); these alterations of normal asymmetry are recapitulated in a number of animal models (Bisgrove and Yost 2001). Of these, only the complete (and rare) situs inversus totalis is not associated with physiological difficulties. The rest, especially heterotaxia, often result in serious health problems for the patient (Burn 1991). Laterality defects can arise in a single individual (Winer-Muram 1995; Kosaki and Casey 1998), but are especially associated with monozygotic twinning (see below).

One crucial question in mammalian embryos concerns when LR information is first generated. Mouse embryos have been shown to be able to reconstitute normal morphology after significant experimental manipulation—early blastomeres can be removed or added without affecting normal development. This has been suggested to signify that the patterning of axes in mammalian embryos takes place later than in other species such as *Xenopus*. However, a number of recent studies have suggested that the polar body may indicate the future axis of bilateral symmetry in fertilized mouse eggs (Gardner 2001; Johnson 2001). Although the extent of LR patterning (if any) during early cell divisions in mammals remains unknown, recent findings in mammalian embryos have shed light on processes that may generate or transmit LR information.

Cilia

The observation that human Kartagener's syndrome patients exhibited randomization of visceral situs (heterotaxia) and had ultrastructural defects in the dynein component of cilia (Afzelius 1976, 1985) was of great interest because it suggested that asymmetry could be bootstrapped from molecular chirality of some ciliary component. This idea was supported by the finding that the murine *iv* mutation, which unbiases laterality (Singh et al. 1991; Schreiner et al. 1993; Lowe et al. 1996), encodes a dynein called Left-Right Dynein (LRD) that is expressed in cells of the mouse node (Supp et al. 1997). Axonemal dynein is a component of the motor driving ciliary motion; the chirality of this motion is intrinsic to the protein components. Genetic deletions of KIF3-A or KIF3-B, two microtubule-dependent kinesin motor proteins, resulted in randomization of the situs of the viscera, and this finding is also often interpreted as evidence for a primary role for cilia in LR determination (Vogan and Tabin 1999). Most importantly, following the first observation of cilia in the murine node (Sulik et al. 1994), elegant experiments have revealed a clockwise rotation of monocilia extending ventral to the node that produces a localized net right-to-left fluid flow of fluorescent beads placed in the extraembryonic space

(Nonaka et al. 1998; Marszalek et al. 1999; Okada et al. 1999; Takeda et al. 1999). Thus, it was proposed that vortical action of cilia (coupled with the wedge shape of the node) may initiate asymmetry by moving an extracellular signaling molecule to one side, where it can induce asymmetric gene expression (Nonaka et al. 1998; Vogan and Tabin 1999). A more sophisticated version of this model, invoking two kinds of cilia (motile and sensory), was later proposed, to account for discrepancies between data from observations of ciliary beating in cultured mouse embryos and the molecular and morphological phenotype observed in certain LR mutants (Tabin and Vogan 2003). In addition to kinesin and dynein, a number of other proteins have also been linked to asymmetry that has been interpreted to result from impaired ciliary function. These include Inversin (Morgan et al. 1998, 2002; Otto et al. 2003; Watanabe et al. 2003), Polaris (Murcia et al. 2000; Taulman et al. 2001), and Polycystin (Pennekamp et al. 2002).

The strongest version of this model (McGrath and Brueckner 2003) hypothesizes that LR asymmetry is initiated by the motion of the cilia in the mature node (toward the end of gastrulation). Consistent with this idea, no upstream LR mechanisms have yet been described in rodents although the rodent embryo is unusual in its architecture, compared to more typical mammalian embryos (such as rabbit and human). Despite the existence of cilia in many organisms (Essner et al. 2002), no functional data implicate cilia in establishment of asymmetry in any organism other than rodents. Because embryos in which molecular motors have been mutated are also likely to have impaired cytoplasmic function of motor transport, it has not yet been possible to separate the ciliary functions of the LR-relevant motors from cytoplasmic roles. Thus, whereas a function for motor proteins in LR patterning is fairly certain, the mechanisms by which they control laterality and the role of cilia in asymmetry remain controversial (Levin 2003, 2004a).

The earliest known endogenous LR mechanisms (Syndecans, GJC, H^+/K^+ flux, Vg1 coordinator) have not been found in mammals. No mouse mutants in gap junction genes have as yet reported a true LR phenotype. Since many different Connexin genes exist in mouse embryos, there is the potential for compensation during single-gene-deletion experiments, so knockins of dominant-negative constructs will be required to determine whether GJC plays a role in LR asymmetry of rodents. Significant insight into the evolutionary conservation of GJC mechanisms is expected from analysis of GJC in rabbits; the rabbit embryo exhibits circumferential patterns of Connexin expression (Liptau and Viebahn 1999), and functional analysis of GJC in a mammal with a more prototypical flat gastrulation architecture is likely to shed significant light on the evolutionary conservation and origin of the GJC system as it participates in LR patterning.

However, ion flux has been implicated in mouse asymmetry. A genetic deletion experiment suggested that the ion channel Polycystin is required for normal asymmetry in the mouse (Pennekamp et al. 2002). More directly, it has recently been shown that asymmetric calcium signaling (Fig. 2B) appears at the left margin of the node at the time of nodal flow (McGrath et al. 2003); this cytoplasmic Ca^{++} gradient may be related to the extracellular Ca^{++} flux recently demonstrated in the chick at gastrulation (Raya et al. 2004). Although it is still unknown whether flows of ions other than calcium play a role in rodents and other mammals, and whether Ca^{++} flow is important for LR patterning prior to mature node stages, future studies of the conservation of ion flow mechanisms among embryos with very different gastrulation modes (frog, chick, rabbit, rodents) are likely to teach us much about asymmetry and basic development.

Conjoined Twins

It is a long-known but puzzling fact that conjoined twins of armadillo (Newman 1916), fish (Morrill 1919), frog (Spemann and Falkenberg 1919), and man (Aird 1959; Burn 1991; Winer-Muram 1995) often exhibit alterations of situs in one of the twins. It has been proposed that an explanation for the laterality defects might be found in consideration of interactions between signaling molecules in two adjacent primitive streaks. Analysis of spontaneous twins of chick embryos (Levin et al. 1996) by in situ hybridization with probes to asymmetric signaling factors such as *Shh* and *Nodal* have given rise to two models that are predictive with respect to which classes of conjoined twins should exhibit laterality defects, and which twin should be affected. For example, parallel streaks during early gastrulation could result in the right-sided Activin of the left embryo inhibiting the expression of *Shh* in the left side of the right embryo. This would result in a normal left embryo, but the right embryo would have no expression of *Shh* in the node, leading to lack of *Nodal* expression and, ultimately, randomized morphological situs. These models have yet to be tested directly in mammalian embryos.

OPEN QUESTIONS AND EVOLUTIONARY PARADIGMS

Because no macroscopic force distinguishes right from left, a powerful paradigm has been proposed to leverage large-scale asymmetry from the chirality of subcellular components (Brown and Wolpert 1990; Brown et al. 1991). In this class of models, some molecule or organelle with a fixed chirality is oriented with respect to the anterior–posterior and dorsoventral axes, and its chiral nature is thus able to nucleate

asymmetric processes such as transport (Levin and Mercola 1998a). Thus, the first developmental event that distinguishes left from right would take place on a subcellular scale. However, a mechanism must then exist to transduce subcellular signals to cell fields (Levin and Nascone 1997; Levin and Mercola 1998a). Asymmetric gene expression in embryos requires that fairly large fields of cells know on which side of the midline they are located (e.g., the cells on the right side of the chick node express *Activin*, but those on the left side do not). In contrast, proposed mechanisms of step 1 of asymmetry (such as the F-molecule model) rely on subcellular mechanisms for determining which direction is left and which is right. Thus, one key question concerns how orientation can be turned into information on a cell's location, relative to the midline, within the context of the whole embryo. This information flow must take place between cells, and cell–cell communication via gap junctions is a natural candidate for such a signal exchange (Levin and Nascone 1997). The extracellular matrix, membrane voltage, and Ca^{++} signaling are also likely to play a role in this process.

One crucial open question in the field concerns the conservation of the early members of the asymmetric gene cascade. The earliest asymmetric gene known in *Xenopus* is *Nodal*, which is detected at somite stages. None of the early genes known to be asymmetric during chick gastrulation (*Shh, cAct-RIIa, cHNF-3β, Follistatin, cWnt-8C*, etc.) has been reported to be asymmetric in *Xenopus* despite searches by a number of labs (Ekker et al. 1995; Stolow and Shi 1995). Interestingly, misexpression of Hedgehog proteins in frog embryos is known to randomize asymmetry (Sampath et al. 1997), raising the possibility that the asymmetric Hedgehog signal exists in amphibia but perhaps utilizes an as-yet-uncharacterized family member. It is possible that the asymmetry in expression exists but has not been detected; it may also be that in *Xenopus* the asymmetries in Hedgehog signaling exist at the level of protein, and not mRNA. The situation with respect to the early asymmetric genes is the same in mouse, where genetic deletions have suggested roles for some of the same molecules (Oh and Li 1997; Tsukui et al. 1999), but no consistent asymmetric gene expression has been reported upstream of *Nodal* (although the Notch pathway is known to direct *Nodal* laterality in mice [Krebs et al. 2003; Raya et al. 2003]).

A difference in mechanisms upstream of *Nodal* may exist between chicks and *Xenopus*. Although in chick embryos, the default state is lack of *Nodal* expression (Shh signaling is required to induce *Nodal* transcription on the left side [Levin et al. 1995]), it was reported that explants of right lateral mesoderm from *Xenopus* embryos turn on *XNR-1* expression (Lohr et al. 1997), arguing for an endogenous repressive influence from the midline. However, it was later demonstrated that explanted lateral tissue induces ectopic notochord-like structures containing Shh (in both frog and chick embryos), suggesting that an inductive pathway upstream of *Nodal* may actually be conserved in both species. Regardless of the details of this possible difference between chick and frog embryos, other asymmetric factors definitely exhibit reversed laterality among species. Asymmetry of *FGF-8* is opposite in chicks versus mice, as are some downstream events such as asymmetry of *Nkx3.2* expression (Meyers and Martin 1999; Schneider et al. 1999).

Vertebrates thus initiate left/right asymmetry by various mechanisms that all, nonetheless, converge on the apparently invariant mechanism of left-sided Nodal signaling. In other words, Nodal may be a "stable point" in the establishment of pattern along the vertebrate left/right axis, whereas the pathways leading to that expression pattern have been free to diverge. This may seem an unlikely result; however, there are other examples of apparent stable points reached by different pathways. One example is the three-layered embryo created by the wide variety of gastrulation movements: The generation of three germ layers is a given, but the suites of movements that result in that organization vary significantly. It is as if stabilizing selection acted on a midpoint, but the means to that midpoint were free to evolve, as were the subsequent events. Of course, each pattern of gastrulation has become important; that is, the characteristic magnitude, direction, and rate of the movements are necessary for normal development of any given species. It is, nonetheless, interesting to speculate as to why having three distinct layers is the sine qua non of this stage of development.

Another well-known example of stable points is the pharyngula stage of vertebrate development (Collazo 2000). Despite very different patterns of cleavage and gastrulation, all vertebrates pass through what Gilbert has called a "bottleneck" in the period following neurulation during which diverse species have a very similar appearance regardless of the mechanisms by which they achieved that appearance (Gilbert 2000). This similarity is at the root of von Baer's principles. Raff has suggested that the pharyngula stage is less able to evolve (i.e., is more stable) because only at that stage in development are there whole-embryo-scale inductive events, and thus a need for a whole-embryo-scale geometry that puts inducing and induced tissues in the correct relative orientation. Prior to gastrulation, there are few inductive events; after early organogenesis, induction occurs, but on a localized scale (Raff 1994). The pharyngula stage of vertebrates is particularly interesting in the context of left/right asymmetry, because universal left-sided *Nodal* expression overlaps with the pharyngula stage. It is tempting to ask whether the stability of *Nodal* laterality is linked to the morphological stability of this stage in development. It does seem to be an example of a molecular bottleneck.

ACKNOWLEDGMENTS

This review is dedicated to Benjamin Levin. The authors gratefully acknowledge the grant support of the American Cancer Society (Research Scholar Grant RSG-02-046-01), March of Dimes (Basil O'Connor grant #5-FY01-509), and the National Institutes of Health (1-R01-GM-06227) to M.L., and of the National Institutes of Health grant T32-DE-07327 to D.S.A.

REFERENCES

Afzelius B. 1976. A human syndrome caused by immotile cilia. *Science* **193:** 317–319.

———. 1985. The immotile cilia syndrome: A microtubule-associated defect. *CRC Crit. Rev. Biochem.* **19:** 63–87.

Aird I. 1959. Conjoined twins. **1:** 1313–1315.

Angelo S., Lohr J., Lee K. H., Ticho B. S., Breitbart R. E., Hill S., Yost H. J., and Srivastava D. 2000. Conservation of sequence and expression of *Xenopus* and zebrafish dHAND during cardiac, branchial arch and lateral mesoderm development. *Mech. Dev.* **95:** 231–237.

Bamford R.N., Roessler E., Burdine R.D., Saplakoglu U., dela Cruz J., Splitt M., Goodship J.A., Towbin J., Bowers P., Ferrero G.B., Marino B., Schier A.F., Shen M.M., Muenke M., and Casey B. 2000. Loss-of-function mutations in the EGF-CFC gene CFC1 are associated with human left-right laterality defects. *Nat. Genet.* **26:** 365–369.

Biben C. and Harvey R. 1997. Homeodomain factor Nkx2-5 controls left/right asymmetric expression of bHLH gene eHand during murine heart development. *Genes Dev.* **11:** 1357–1369.

Bisgrove B.W. and Yost H.J. 2001. Classification of left-right patterning defects in zebrafish, mice, and humans. *Am. J. Med. Genet.* **101:** 315–323.

Boettger T., Wittler L., and Kessel M. 1999. FGF8 functions in the specification of the right body side of the chick. *Curr. Biol.* **9:** 277–280.

Branford W.W., Essner J.J., and Yost H.J. 2000. Regulation of gut and heart left-right asymmetry by context-dependent interactions between *Xenopus* lefty and BMP4 signaling. **223:** 291–306.

Britz-Cunningham S., Shah M., Zuppan C., and Fletcher W. 1995. Mutations of the connexin-43 gap-junction gene in patients with heart malformations and defects of laterality. *New Engl. J. Med.* **332:** 1323–1329.

Brizuela B.J., Wessely O., and De Robertis E.M. 2001. Overexpression of the *Xenopus* tight-junction protein claudin causes randomization of the left-right body axis. *Dev. Biol.* **230:** 217–229.

Brody S.L., Yan X.H., Wuerffel M.K., Song S.K., and Shapiro S.D. 2000. Ciliogenesis and left-right axis defects in forkhead factor HFH-4-null mice. *Am. J. Respir. Cell Mol. Biol.* **23:** 45–51.

Brown N. and Wolpert L. 1990. The development of handedness in left/right asymmetry. *Development* **109:** 1–9.

Brown N., McCarthy A., and Wolpert L. 1991. Development of handed body asymmetry in mammals. *CIBA Found. Symp.* **162:** 182–196.

Bunney T.D., De Boer A.H., and Levin M. 2003. Fusicoccin signaling reveals 14-3-3 protein function as a novel step in left-right patterning during amphibian embryogenesis. *Development* **130:** 4847–4858.

Burdine R. and Schier A. 2000. Conserved and divergent mechanisms in left-right axis formation. *Genes Dev.* **14:** 763–776.

Burn J. 1991. Disturbance of morphological laterality in humans. *CIBA Found. Symp.* **162:** 282–296.

Capdevila J. and Johnson R.L. 2000. Hedgehog signaling in vertebrate and invertebrate limb patterning. *Cell Mol. Life Sci.* **57:** 1682–1694.

Chen J., Knowles H.J., Hebert J.L., and Hackett B.P. 1998. Mutation of the mouse hepatocyte nuclear factor/forkhead homologue 4 gene results in an absence of cilia and random left-right asymmetry. *J. Clin. Investig.* **102:** 1077–1082.

Chen J., Eeden F.V., Warren K., Chin A., Nusslein-Volhard C., Haffter P., and Fishman M. 1997. Left-right pattern of cardiac BMP4 may drive asymmetry of the heart in zebrafish. **124:** 4373–4382.

Cheng A.M., Thisse B., Thisse C., and Wright C.V. 2000. The lefty-related factor Xatv acts as a feedback inhibitor of nodal signaling in mesoderm induction and L-R axis development in *Xenopus*. *Development* **127:** 1049–1061.

Collazo A. 2000. Developmental variation, homology, and the pharyngula stage. *Syst. Biol.* **49:** 3–18.

Collignon J., Varlet I., and Robertson E. 1996. Relationship between asymmetric nodal expression and the direction of embryonic turning. *Nature* **381:** 155–158.

Concha M.L., Russell C., Regan J.C., Tawk M., Sidi S., Gilmour D.T., Kapsimali M., Sumoy L., Goldstone K., Amaya E., Kimelman D., Nicolson T., Grunder S., Gomperts M., Clarke J.D., and Wilson S.W. 2003. Local tissue interactions across the dorsal midline of the forebrain establish CNS laterality. *Neuron* **39:** 423–438.

Constam D. and Robertson E. 2000. Tissue-specific requirements for the proprotein convertase Furin/SPC1 during embryonic turning and heart looping. *Development* **127:** 245–254.

Cserjesi P., Brown D., Lyons G., and Olson E. 1995. Expression of novel basic helix-loop-helix gene eHAND in neural crest derivatives and extraembryonic membranes during mouse development. *Dev. Biol.* **170:** 664–678.

Danos M.C. and Yost H.J. 1995. Linkage of cardiac left-right asymmetry and dorsal-anterior development in *Xenopus*. *Development* **121:** 1467–1474.

Dathe V., Prols F., and Brand-Saberi B. 2004. Expression of kinesin kif5c during chick development. *Anat. Embryol.* **207:** 475–480.

Dathe V., Gamel A., Manner J., Brand-Saberi B., and Christ B. 2002. Morphological left-right asymmetry of Hensen's node precedes the asymmetric expression of Shh and Fgf8 in the chick embryo. *Anat. Embryol.* **205:** 343–354.

Ekker S.C., McGrew L.L., Lai C.J., Lee J.J., von Kessler D.P., Moon R.T., and Beachy P.A. 1995. Distinct expression and shared activities of members of the hedgehog gene family of *Xenopus laevis*. *Development* **121:** 2337–2347.

Essner J.J., Branford W.W., Zhang J., and Yost H.J. 2000. Mesendoderm and left-right brain, heart and gut development are differentially regulated by pitx2 isoforms. *Development* **127:** 1081–1093.

Essner J.J., Vogan K., Wagner M., Tabin C., Yost H., and Brueckner M. 2002. Conserved function for embryonic nodal cilia. *Nature* **418:** 37–38.

Faisst A.M., Alvarez-Bolado G., Treichel D., and Gruss P. 2002. Rotatin is a novel gene required for axial rotation and left-right specification in mouse embryos. *Mech. Dev.* **113:** 15–28.

Falk M.M. 2000. Biosynthesis and structural composition of gap junction intercellular membrane channels. *Eur. J. Cell Biol.* **79:** 564–574.

Gaio U., Schweickert A., Fischer A., Garratt A.N., Muller T., Ozelik C., Lankes W., Strehle M., Britsch S., Blum M., and Birchmeier C. 1999. A role of the cryptic gene in the correct establishment of the left-right axis. *Curr. Biol.* **9:** 1339–1342.

Garcia-Castro M., Vielmetter E., and Bronner-Fraser E. 2000. N-cadherin, a cell adhesion molecule involved in establishment of embryonic left-right asymmetry. *Science* **288:** 1047–1051.

Gardner R.L. 2001. Specification of embryonic axes begins before cleavage in normal mouse development. *Development* **128:** 839–847.

Gebbia M., Ferrero G.B., Pilia G., Bassi M.T., Aylsworth A., Penman-Splitt M., Bird L.M., Bamforth J.S., Burn J., Schlessinger D., Nelson D.L., and Casey B. 1997. X-linked situs abnormalities result from mutations in ZIC3. *Nat. Genet.* **17:** 305–308.

Gerhart J., Danilchik M., Doniach T., Roberts S., Rowning B., and Stewart R. 1989. Cortical rotation of the *Xenopus* egg: Consequences for the anteroposterior pattern of embryonic dorsal development. *Development* (suppl.) **107:** 37–51.

Gilbert S.F. 2000. *Developmental biology*, 6th edition. Sinauer, Sunderland, Massachusetts.

Granata A. and Quaderi N.A. 2003. The Opitz syndrome gene *MID1* is essential for establishing asymmetric gene expression in Hensen's node. *Dev. Biol.* **258:** 397–405.

Guthrie S. 1984. Patterns of junctional communication in the early amphibian embryo. *Nature* **311:** 149–151.

Guthrie S., Turin L., and Warner A. 1988. Patterns of junctional communication during development of the early amphibian embryo. *Development* **103:** 769–783.

Hashimoto H., Mizuta A., Okada N., Suzuki T., Tagawa M., Tabata K., Yokoyama Y., Sakaguchi M., Tanaka M., and Toyohara H. 2002. Isolation and characterization of a Japanese flounder clonal line, reversed, which exhibits reversal of metamorphic left-right asymmetry. *Mech. Dev.* **111:** 17–24.

Hermann G.J., Leung B., and Priess J.R. 2000. Left-right asymmetry in *C. elegans* intestine organogenesis involves a LIN-12/Notch signaling pathway. *Development* **127:** 3429–3440.

Hertwig O. 1902. *Lehrbuch der Entwicklungsgeschichte des Menschen und der Wirbelthiere*, vol. 7 of Aufl. Fischer, Jena, Germany.

Hoshijima K., Metherall J.E., and Grunwald D.J. 2002. A protein disulfide isomerase expressed in the embryonic midline is required for left/right asymmetries. *Genes Dev.* **16:** 2518–2529.

Hoyle C., Brown N., and Wolpert L. 1992. Development of left/right handedness in the chick heart. *Development* **115:** 1071–1078.

Hyatt B. and Yost H. 1998. The left-right coordinator: the role of Vg1 in organizing left-right axis. *Cell* **93:** 37–46.

Hyatt B., Lohr J., and Yost H. 1996. Initiation of left-right axis formation by maternal Vg1. *Nature* **384:** 62–65.

Ibanez-Tallon I., Gorokhova S., and Heintz N. 2002. Loss of function of axonemal dynein *Mdnah5* causes primary ciliary dyskinesia and hydrocephalus. *Hum. Mol. Genet.* **11:** 715–721.

Isaac A., Sargent M.S., and Cooke J. 1997. Control of vertebrate left-right asymmetry by a *snail*-related zinc finger gene. *Science* **275:** 1301–1304.

Izraeli S., Lowe L., Bertness V., Good D., Dorward D., Kirsch I., and Kuehn M. 1999. The SIL gene is required for mouse embryonic axial development and left-right specification. *Nature* **399:** 691–694.

Johnson M.H. 2001. Mammalian development: Axes in the egg? *Curr. Biol.* **11:** R281–284.

Johnston R.J. and Hobert O. 2003. A microRNA controlling left/right neuronal asymmetry in *Caenorhabditis elegans*. *Nature* **426:** 845–849.

Kawakami M. and Nakanishi N. 2001. The role of an endogenous PKA inhibitor, PKIα, in organizing left-right axis formation. *Development* **128**: 2509–2515.

King T., Beddington R. S., and Brown N. A. 1998. The role of the brachyury gene in heart development and left-right specification in the mouse. *Mech. Dev.* **79**: 29–37.

Kitaguchi T., Nagai T., Nakata K., Aruga J., and Mikoshiba K. 2000. Zic3 is involved in the left-right specification of the *Xenopus* embryo. *Development* **127**: 4787–4795.

Kobayashi Y., Watanabe M., Okada Y., Sawa H., Takai H., Nakanishi M., Kawase Y., Suzuki H., Nagashima K., Ikeda K., and Motoyama N. 2002. Hydrocephalus, situs inversus, chronic sinusitis, and male infertility in DNA polymerase λ-deficient mice: Possible implication for the pathogenesis of immotile cilia syndrome. *Mol. Cell. Biol.* **22**: 2769–2776.

Kölliker A. 1879. *Entwicklungsgeschichte des Menschen und höheren Thiere*. Wilhelm Engelmann, Leipzig.

Kosaki K. and Casey B. 1998. Genetics of human left-right axis malformations. *Semin. Cell Dev. Biol.* **9**: 89–99.

Kramer K.L. and Yost H.J. 2002. Ectodermal syndecan-2 mediates left-right axis formation in migrating mesoderm as a cell-nonautonomous Vg1 cofactor. *Dev. Cell* **2**: 115–124.

Krebs L.T., Iwai N., Nonaka S., Welsh I.C., Lan Y., Jiang R., Saijoh Y., O'Brien T.P., Hamada H., and Gridley T. 2003. Notch signaling regulates left-right asymmetry determination by inducing *Nodal* expression. *Genes Dev.* **17**: 1207–1212.

Landesman Y., Goodenough D. A., and Paul D. L. 2000. Gap junctional communication in the early *Xenopus* embryo. *J. Cell Biol.* **150**: 929–936.

Levin M. 1998a. Follistatin mimics the endogenous streak inhibitory activity in early chick embryos. *Int. J. Dev. Biol.* **42**: 553–559.

———. 1998b. Left-right asymmetry and the chick embryo. *Semin. Cell Dev. Biol.* **9**: 67–76.

———. 2001. Isolation and community: The role of gap junctional communication in embryonic patterning. *J. Membr. Biol.* **185**: 177–192.

———. 2003. Hypothesis: Motor proteins and ion pumps, not ciliary motion, initiate LR asymmetry. *BioEssays* **25**: 1002–1010.

———. 2004a. Embryonic origins of left-right asymmetry. *Crit. Rev. Oral Biol. Med.* (in press).

———. 2004b. Left-right asymmetry in amphibian embryogenesis. In *Biology of the Amphibia* (ed. H. Heatwole and B. Brizuela). Surrey Beatty & Sons, Australia. (In press.)

Levin M. and Mercola M. 1998a. The compulsion of chirality: Toward an understanding of left-right asymmetry. *Genes Dev.* **12**: 763–769.

———. 1998b. Gap junctions are involved in the early generation of left right asymmetry. *Dev. Biol.* **203**: 90–105.

———. 1999. Gap junction-mediated transfer of left-right patterning signals in the early chick blastoderm is upstream of Shh asymmetry in the node. *Development* **126**: 4703–4714.

Levin M. and Nascone N. 1997. Two models of initial LR determination. *Med. Hypotheses* **49**: 429–435.

Levin M., Roberts D., Holmes L., and Tabin C. 1996. Laterality defects in conjoined twins. *Nature* **384**: 321.

Levin M., Johnson R., Stern C., Kuehn M., and Tabin C. 1995. A molecular pathway determining left-right asymmetry in chick embryogenesis. *Cell* **82**: 803–814.

Levin M., Pagan S., Roberts D., Cooke J., Kuehn M., and Tabin C. 1997. Left/right patterning signals and the independent regulation of different aspects of *situs* in the chick embryo. *Dev. Biol.* **189**: 57–67.

Levin M., Thorlin T., Robinson K., Nogi T., and Mercola M. 2002. Asymmetries in H⁺/K⁺-ATPase and cell membrane potentials comprise a very early step in left-right patterning. *Cell* **111**: 77–89.

Liang J.O., Etheridge A., Hantsoo L., Rubinstein A.L., Nowak S.J., Izpisua Belmonte J.C., and Halpern M.E. 2000. Asymmetric nodal signaling in the zebrafish diencephalon positions the pineal organ. *Development* **127**: 5101–5112.

Linask K.K., Han M.D., Artman M., and Ludwig C.A. 2001. Sodium-calcium exchanger (NCX-1) and calcium modulation: NCX protein expression patterns and regulation of early heart development. *Dev. Dyn.* **221**: 249–264.

Liptau H. and Viebahn C. 1999. Expression patterns of gap junctional proteins connexin 32 and 43 suggest new communication compartments in the gastrulating rabbit embryo. *Differentiation* **65**: 209–219.

Lo C.W. 1996. The role of gap junction membrane channels in development. *J. Bioenerg. Biomemb.* **28**: 379–85.

Logan M., Pagan-Westphal S., Smith D., Paganessi L., and Tabin C. 1998. The transcription factor Pitx2 mediates situs-specific morphogenesis in response to left-right asymmetric signals. *Cell* **94**: 307–317.

Lohr J., Danos M., and Yost H. 1997. Left-right asymmetry of a *nodal*-related gene is regulated by dorsoanterior midline structures during *Xenopus* development. *Development* **124**: 1465–1472.

Long S., Ahmad N., and Rebagliati M. 2003. The zebrafish *nodal*-related gene *southpaw* is required for visceral and diencephalic left-right asymmetry. **130**: 2303–2316.

Lowe L., Supp D., Sampath K., Yokoyama T., Wright C., Potter S., Overbeek P., and Kuehn M. 1996. Conserved left-right asymmetry of nodal expression and alterations in murine situs inversus. *Nature* **381**: 158–161.

Marszalek J., Ruiz-Lozano P., Roberts E., Chien K., and Goldstein L. 1999. Situs inversus and embryonic ciliary morphogenesis defects in mouse mutants lacking the KIF3A subunit of kinesin-II. *Proc. Natl. Acad. Sci.* **96**: 5043–5048.

Matsumoto J. and Seikai T. 1992. Asymmetric pigmentation and pigment disorders in pleuronectiformes (flounders). *Pigm. Cell Res.* (suppl.) **2**: 275–282.

McCain E. and McClay D. 1994. The establishment of bilateral asymmetry in sea urchin embryos. *Development* **120**: 395–404.

McGrath J. and Brueckner M. 2003. Cilia are at the heart of vertebrate left-right asymmetry. *Curr. Opin. Genet. Dev.* **13**: 385–392.

McGrath J., Somlo S., Makova S., Tian X., and Brueckner M. 2003. Two populations of node monocilia initiate left-right asymmetry in the mouse. *Cell* **114**: 61–73.

Melloy P., Ewart J., Cohen M., Desmond M., Kuehn M., and Lo C. 1998. No turning, a mouse mutation causing left-right and axial patterning defects. *Dev. Biol.* **193**: 77–89.

Meno C., Saijoh Y., Fujii H., Ikeda M., Yokoyama T., Yokoyam M., Toyoda Y., and Hamada H. 1996. Left-right asymmetric expression of the TGFβ-family member *lefty* in mouse embryos. *Nature* **381**: 151–155.

Meno C., Shimono A., Saijoh Y., Yashiro K., Mochida K., Ohishi S., Noji S., Kondoh H., and Hamada H. 1998. *lefty-1* is required for left-right determination as a regulator of *lefty-2* and *nodal*. *Cell* **94**: 287–297.

Mercola M. 2003. Left-right asymmetry: Nodal points. *J. Cell Sci.* **116:** 3251–3257.

Mercola M. and Levin M. 2001. Left-right asymmetry determination in vertebrates. *Annu. Rev. Cell Dev. Biol.* **17:** 779–805.

Meyers E.N. and Martin G.R. 1999. Differences in left-right axis pathways in mouse and chick: Functions of FGF8 and SHH. **285:** 403–406.

Mochizuki T., Saijoh Y., Tsuchiya K., Shirayoshi Y., Takai S., Taya C., Yonekawa H., Yamada K., Nihei H., Nakatsuji N., Overbeek P.A., Hamada H., and Yokoyama T. 1998. Cloning of *inv*, a gene that controls left/right asymmetry and kidney development. *Nature* **395:** 177–181.

Monsoro-Burq A. and LeDouarin N. 2000. Left-right asymmetry in BMP4 signalling pathway during chick gastrulation. *Mech. Dev.* **97:** 105–108.

Morgan D., Goodship J., Essner J. J., Vogan K. J., Turnpenny L., Yost H. J., Tabin C. J., and Strachan T. 2002. The left-right determinant inversin has highly conserved ankyrin repeat and IQ domains and interacts with calmodulin. *Hum. Genet.* **110:** 377–384.

Morgan D., Turnpenny L., Goodship J., Dai W., Majumder K., Matthews L., Gardner A., Schuster G., Vien L., Harrison W., Elder F.F., Penman-Splitt M., Overbeek P., and Strachan T. 1998. Inversin, a novel gene in the vertebrate left-right axis pathway, is partially deleted in the *inv* mouse. *Nat. Genet.* **20:** 149–56.

Morokuma J., Ueno M., Kawanishi H., Saiga H., and Nishida H. 2002. *HrNodal*, the ascidian *nodal*-related gene, is expressed in the left side of the epidermis, and lies upstream of *HrPitx*. *Dev. Genes Evol.* **212:** 439–446.

Morrill C. 1919. Symmetry reversal and mirror imaging in monstrous trout and a comparison with similar conditions in human double monsters. *Anat. Rec.* **16:** 265–292.

Murcia N.S., Richards W.G., Yoder B.K., Mucenski M.L., Dunlap J.R., and Woychik R.P. 2000. The *Oak Ridge Polycystic Kidney* (*orpk*) disease gene is required for left-right axis determination. *Development* **127:** 2347–2355.

Nascone N. and Mercola M. 1997. Organizer induction determines left-right asymmetry in *Xenopus*. *Dev. Biol.* **189:** 68–78.

Newman C., Grow M., Cleaver O., Chia F., and Krieg P. 1997. *Xbap*, a vertebrate gene related to *bagpipe*, is expressed in developing craniofacial structures and in anterior gut muscle. *Dev. Biol.* **181:** 223–233.

Newman H. 1916. Heredity and organic symmetry in armadillo quadruplets. *Biol. Bull.* **XXX:** 173–203.

Nonaka S., Tanaka Y., Okada Y., Takeda S., Harada A., Kanai Y., Kido M., and Hirokawa N. 1998. Randomization of left-right asymmetry due to loss of nodal cilia generating leftward flow of extraembryonic fluid in mice lacking KIF3B motor protein. *Cell* **95:** 829–837.

Oh S. and Li E. 1997. The signaling pathway mediated by the type IIB activin receptor controls axial patterning and lateral asymmetry in the mouse. *Genes Dev.* **11:** 1812–1826.

Okada N., Takagi Y., Seikai T., Tanaka M., and Tagawa M. 2001. Asymmetrical development of bones and soft tissues during eye migration of metamorphosing Japanese flounder, *Paralichthys olivaceus*. *Cell Tissue Res.* **304:** 59–66.

Okada Y., Nonaka S., Tanaka Y., Saijoh Y., Hamada H., and Hirokawa N. 1999. Abnormal nodal flow precedes situs inversus in *iv* and *inv* mice. *Mol. Cell* **4:** 459–468.

Olbrich H., Haffner K., Kispert A., Volkel A., Volz A., Sasmaz G., Reinhardt R., Hennig S., Lehrach H., Konietzko N., et al. 2002.

Mutations in DNAH5 cause primary ciliary dyskinesia and randomization of left-right asymmetry. *Nat. Genet.* **30:** 143–144.

Olson D.J., Christian J.L., and Moon R.T. 1991. Effect of wnt-1 and related proteins on gap junctional communication in *Xenopus* embryos. *Science* **252:** 1173–1176.

Otto E.A., Schermer B., Obara T., O'Toole J.F., Hiller K.S., Mueller A.M., Ruf R.G., Hoefele J., Beekmann F., Landau D., et al. 2003. Mutations in *INVS* encoding inversin cause nephronophthisis type 2, linking renal cystic disease to the function of primary cilia and left-right axis determination (comment). *Nat. Genet.* **34:** 413–420.

Pagan-Westphal S. and Tabin C. 1998. The transfer of left-right positional information during chick embryogenesis. *Cell* **93:** 25–35.

Peeters H., Debeer P., Bairoch A., Wilquet V., Huysmans C., Parthoens E., Fryns J.P., Gewillig M., Nakamura Y., Niikawa N., Van de Ven W., and Devriendt K. 2003. PA26 is a candidate gene for heterotaxia in humans: Identification of a novel PA26-related gene family in human and mouse. *Hum. Genet.* **112:** 573–80.

Pennekamp P., Karcher C., Fischer A., Schweickert A., Skryabin B., Horst J., Blum M., and Dworniczak B. 2002. The ion channel polycystin-2 is required for left-right axis determination in mice. *Curr. Biol.* **12:** 938–943.

Przemeck G.K., Heinzmann U., Beckers J., and Hrabe de Angelis M. 2003. Node and midline defects are associated with left-right development in *Delta1* mutant embryos. *Development* **130:** 3–13.

Psychoyos D. and Stern C. 1996. Restoration of the organizer after radical ablation of Hensen's node and the anterior primitive streak in the chick embryo. *Development* **122:** 3263–3273.

Purandare S.M., Ware S.M., Kwan K.M., Gebbia M., Bassi M.T., Deng J.M., Vogel H., Behringer R.R., Belmont J.W., and Casey B. 2002. A complex syndrome of left-right axis, central nervous system and axial skeleton defects in *Zic3* mutant mice. *Development* **129:** 2293–2302.

Raff R.A. 1994. Developmental mechanisms in the evolution of animal form: Origins and evolvability of body plans. In *Early life on earth* (ed. S. Bengston). Columbia University Press, New York.

Ramsdell A.F. and Yost H.J. 1999. Cardiac looping and the vertebrate left-right axis: Antagonism of left-sided Vg1 activity by a right-sided ALK2-dependent BMP pathway. *Development* **126:** 5195–205.

Rankin C.T., Bunton T., Lawler A.M., and Lee S.J. 2000. Regulation of left-right patterning in mice by growth/differentiation factor-1. *Nat. Genet.* **24:** 262–265.

Raya A., Kawakami Y., Rodriguez-Esteban C., Buscher D., Koth C.M., Itoh T., Morita M., Raya R.M., Dubova I., Bessa J.G., de la Pompa J.L., and Belmonte J.C. 2003. Notch activity induces *Nodal* expression and mediates the establishment of left-right asymmetry in vertebrate embryos. *Genes Dev.* **17:** 1213–1218.

Raya A., Kawakami Y., Rodriguez-Esteban C., Ibanes M., Rasskin-Gutman D., Rodriguez-Leon J., Buscher D., Feijo J. A., and Izpisua Belmonte J. C. 2004. Notch activity acts as a sensor for extracellular calcium during vertebrate left-right determination. *Nature* **427:** 121–128.

Rodriguez-Esteban C., Capdevila J., Kawakami Y., and Izpisua Belmonte J. C. 2001. Wnt signaling and PKA control *Nodal* expression and left-right determination in the chick embryo. *Development* **128:** 3189–3195.

Roebroek A.J., Umans L., Pauli I.G., Robertson E.J., Van Leuven F., Van De Ven W.J., and Constam D.B. 1998. Failure of ventral closure and axial rotation in embryos lacking the proprotein convertase Furin. *Development* **125:** 4863–4876.

Ryan A., Blumberg B., Rodriguez-Esteban C., Yonei-Tamura S., Tamura K., Tsukui T., de la Pena J., Sabbagh W., Greenwald J., Choe S., et al. 1998. Pitx2 determines left-right asymmetry of internal organs in vertebrates. *Nature* **394:** 545–51.

Sampath K., Cheng A. M., Frisch A., and Wright C. V. 1997. Functional differences among *Xenopus nodal-related* genes in left-right axis determination. *Development* **124:** 3293–3302.

Schilling T., Concordet J., and Ingham P. 1999. Regulation of left-right asymmetries in the zebrafish by Shh and BMP4. *Dev. Biol.* **210:** 277–287.

Schneider A., Mijalski T., Schlange T., Dai W., Overbeek P., Arnold H., and Brand T. 1999. The homeobox gene NKX3.2 is a target of left-right signalling and is expressed on opposite sides in chick and mouse embryos. *Curr. Biol.* **9:** 911–914.

Schreiner C.M., Scott W.J., Jr., Supp D.M., and Potter S.S. 1993. Correlation of forelimb malformation asymmetries with visceral organ situs in the transgenic mouse insertional mutation, *legless. Dev. Biol.* **158:** 560–562.

Singh G., Supp D., Schreiner C., McNeish J., Merker H., Copeland N., Jenkins N., Potter S., and Scott W. 1991. *legless* insertional mutation: Morphological, molecular, and genetic characterization. *Genes Dev.* **5:** 2245–2255.

Smith S., Dickman E., Thompson R., Sinning A., Wunsch A., and Markwald R. 1997. Retinoic acid directs cardiac laterality and the expression of early markers of precardiac asymmetry. *Dev. Biol.* **182:** 162–171.

Sparrow D.B., Kotecha S., Towers N., and Mohun T.J. 1998. *Xenopus eHAND*: A marker for the developing cardiovascular system of the embryo that is regulated by bone morphogenetic proteins. *Mech. Dev.* **71:** 151–163.

Spemann H. and Falkenberg H. 1919. Über Asymmetrische Entwicklung und Situs inversus viscerum bei Zwillingen und Doppelbildungen. *Arch. Entwicklungsmech. Org.* **45:** 371–422.

Srivastava D. 1995. A subclass of bHLH proteins required for cardiac morphogenesis. *Science* **270:** 1995–1999.

Stolow M.A. and Shi Y.B. 1995. *Xenopus* sonic hedgehog as a potential morphogen during embryogenesis and thyroid hormone-dependent metamorphosis. *Nucleic Acids Res.* **23:** 2555–2562.

Streit A., Stern C.D., Thery C., Ireland G.W., Aparicio S., Sharpe M.J., and Gherardi E. 1995. A role for HGF/SF in neural induction and its expression in Hensen's node during gastrulation. *Development* **121:** 813–824.

Sulik K., Dehart D.B., Iangaki T., Carson J.L., Vrablic T., Gesteland K., and Schoenwolf G.C. 1994. Morphogenesis of the murine node and notochordal plate. *Dev. Dyn.* **201:** 260–278.

Supp D., Potter S., and Brueckner M. 2000. Molecular motors: The driving force behind mammalian left-right development. *Trends Cell Biol.* **10:** 41–45.

Supp D.M., Witte D.P., Potter S.S., and Brueckner M. 1997. Mutation of an axonemal dynein affects left-right asymmetry in inversus viscerum mice. *Nature* **389:** 963–966.

Supp D.M., Brueckner M., Kuehn M.R., Witte D.P., Lowe L.A., McGrath J., Corrales J., and Potter S.S. 1999. Targeted deletion of the ATP binding domain of left-right dynein confirms its role in specifying development of left-right asymmetries. *Development* **126:** 5495–5504.

Tabin C.J. and Vogan K.J. 2003. A two-cilia model for vertebrate left-right axis specification. *Genes Dev.* **17:** 1–6.

Takeda S., Yonekawa Y., Tanaka Y., Nonaka Y.O.S., and Hirokawa N. 1999. Left-right asymmetry and kinesin superfamily protein KIF3A: New insights in determination of laterality and mesoderm induction by kif3A$^{-/-}$ mice analysis. *J. Cell Biol.* **145:** 825–836.

Taulman P.D., Haycraft C. J., Balkovetz D. F., and Yoder B. K. 2001. Polaris, a protein involved in left-right axis patterning, localizes to basal bodies and cilia. *Mol. Biol. Cell* **12:** 589–599.

Teel A.L. and Yost H.J. 1996. Embryonic expression patterns of *Xenopus* syndecans. *Mech. Dev.* **59:** 115–127.

Thickett C. and Morgan R. 2002. Hoxc-8 expression shows left-right asymmetry in the posterior lateral plate mesoderm. *Gene Expr. Patterns* **2:** 5–6.

Tsuda T., Philp N., Zile M.H., and Linask K.K. 1996. Left-right asymmetric localization of flectin in the extracellular matrix during heart looping. *Dev. Biol.* **173:** 39–50.

Tsukui T., Capdevila J., Tamura K., Ruiz-Lozano P., Rodriguez-Esteban C., Yonei-Tamura S, Magallon J., Chandraratna R.A., Chien K., Blumberg B., Evans R.M., and Belmonte J.C. 1999. Multiple left-right asymmetry defects in *Shh$^{-/-}$* mutant mice unveil a convergence of the Shh and retinoic acid pathways in the control of *Lefty-1. Proc. Natl. Acad. Sci.* **96:** 11376–11381.

Vogan K.J. and Tabin C.J. 1999. A new spin on handed asymmetry. *Nature* **397:** 295, 297–298.

Wada S., Katsuyama Y., Yasugi S., and Saiga H. 1995. Spatially and temporally regulated expression of the LIM class homeobox gene Hrlim suggests multiple distinct functions in development of the ascidian, *Halocynthia roretzi. Mech. Dev.* **51:** 115–126.

Wang D.Z., Reiter R.S., Lin J.L., Wang Q., Williams H.S., Krob S.L., Schultheiss T.M., Evans S., and Lin J.J. 1999. Requirement of a novel gene, *Xin*, in cardiac morphogenesis. *Development* **126:** 1281–1294.

Watanabe D., Saijoh Y., Nonaka S., Sasaki G., Ikawa Y., Yokoyama T., and Hamada H. 2003. The left-right determinant Inversin is a component of node monocilia and other 9+0 cilia. **130:** 1725–1734.

Welsh I.C. and O'Brien T.P. 2000. Loss of late primitive streak mesoderm and interruption of left-right morphogenesis in the *Ednrb$^{s-1Acrg}$* mutant mouse. *Development* **225:** 151–168.

Winer-Muram H. 1995. Adult presentation of heterotaxic syndromes and related complexes. *J. Thoracic Imag.* **10:** 43–57.

Wunsch A., Little C.D., and Markwald R.R. 1994. Cardiac endothelial heterogeneity defines valvular development as demonstrated by the diverse expression of JB3, an antigen of the endocardial cushion tissue. *Dev. Biol.* **165:** 585–601.

Yan Y.T., Gritsman K., Ding J., Burdine R.D., Corrales J.D., Price S.M., Talbot W.S., Schier A.F., and Shen M.M. 1999. Conserved requirement for *EGF-CFC* genes in vertebrate left-right axis formation. *Genes Dev.* **13:** 2527–2537.

Yokouchi Y., Vogan K., Pearse R., and Tabin C. 1999. Antagonistic signaling by *Caronte*, a novel *Cerberus*-related gene, establishes Left-Right asymmetric gene expression. *Cell* **98:** 573–583.

Yost H.J. 1990. Inhibition of proteoglycan synthesis eliminates left-right asymmetry in *Xenopus laevis* cardiac looping. *Development* **110:** 865–874.

———. 1991. *Development of the left-right axis in amphibians.* Vol. 162 of *Ciba Foundation Symposium.*

———. 1992. Regulation of vertebrate left-right asymmetries by extracellular matrix. *Nature* **357:** 158–161.

———. 1998. Left-right development in *Xenopus* and zebrafish. **9:** 61–66.

———. 2001. Establishment of left-right asymmetry. *Int. Rev. Cytol.* **203:** 357–381.

Yuan S. and Schoenwolf G. 1998. De novo induction of the organizer and formation of the primitive streak in an experimental model of notochord reconstitution in avian embryos. *Development* **125:** 201–213.

———. 2000. Islet-1 marks the early heart rudiments and is asymmetrically expressed during early rotation of the foregut in the chick embryo. *Anat. Rec.* **260:** 204–207.

Zhang X.M., Ramalho-Santos M., and McMahon A.P. 2001. Smoothened mutants reveal redundant roles for Shh and Ihh signaling including regulation of L/R symmetry by the mouse node (republished from *Cell* [2001] **105:** 781–92). *Cell* **106:** 781–792.

Zhu L., Marvin M., Gardiner A., Lassar A., Mercola M., Stern C., and Levin M. 1999. Cerberus regulates left-right asymmetry of the embryonic head and heart. *Curr. Biol.* **9:** 931–938.

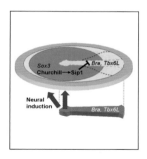

NEURAL INDUCTION

C.D. Stern

Department of Anatomy & Developmental Biology, University College London,
Gower Street, London WC1E 6BT, United Kingdom

INTRODUCTION

What Is Neural Induction?

Embryonic induction has been defined by John Gurdon as "... an interaction between one (inducing) tissue and another (responding) tissue, as a result of which the responding tissue undergoes a change in its direction of differentiation" (Gurdon 1987). Neural induction is therefore the process by which cells acquire a neural fate in response to appropriate signals during development or after embryonic manipulations that bring two dissimilar cell types together. During normal vertebrate development, neural induction is generally believed to occur around the time of gastrulation, directed at least in part by signals emanating from a special region of the embryo, the organizer. The organizer resides in the embryonic shield of teleosts, the dorsal lip of the blastopore in amphibians, and the tip of the primitive streak (Hensen's node) in amniotes. This chapter takes an historical approach to trace the development of our understanding of this process at the cellular and molecular levels.

EARLY HISTORY

The concept of induction originated in Karl Ernst von Baer's work (von Baer 1828) and was further developed at the turn of the 20th century, notably by Curt Herbst (Oppenheimer 1991). But the concept of neural induction really evolved from the pioneering experiments of Warren Lewis (Lewis 1907) and the better-known work of Hans Spemann and Hilde Mangold (Figs. 1 and 2) (Spemann and Mangold 1924; Hamburger 1988; De Robertis and Aréchaga 2001). As part of an effort to resolve an ongoing controversy about whether embryos are "regulative" or

"mosaic," Lewis found that transplantation of the dorsal lip of the blastopore of *Rana* to an ectopic position caused a second axis to form. However, he interpreted this as self-differentiation of the graft, and it was not until Spemann's use of interspecies grafts between three differently pigmented species of newts (*Triturus taeniatus, T. cristatus,* and *T. alpestris*) (Spemann 1921; Spemann and Mangold 1924), allowing the cells of the donor and host to be distinguished, that this could be concluded to be an example of an inductive interaction. Spemann and Mangold termed the dorsal lip of the blastopore "the organizer," because it could direct the formation of a coherently organized, ectopic axis from cells whose fate was other than to form axial structures.

It took only a few years for these findings to be extended to other vertebrates, including amniotes: first to avian species (chick and duck) (Hunt 1929; Waddington 1930, 1932, 1933b) and shortly afterward to mammalian embryos (rabbit), by interspecies grafts in all combinations (Waddington 1932, 1934, 1936, 1937). In all these cases, the

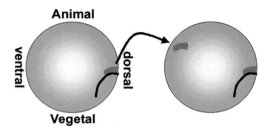

Figure 1. Diagram of the "organizer graft" experiment of Spemann and Mangold (1924). The dorsal lip (*red*) of the blastopore (*thick black line*) of a donor newt at mid-gastrula stage is transplanted to the opposite (ventral) side of a host.

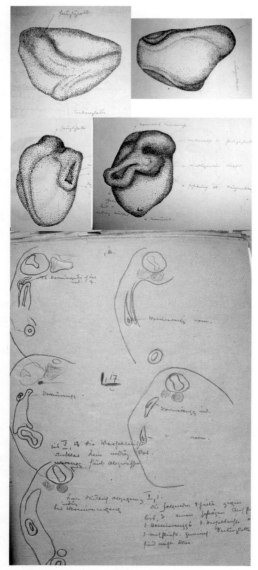

Figure 2. Diagrams by Hilde Mangold, illustrating the results of her organizer graft experiments (Fig. 1). The upper four figures are India ink drawings prepared as for publication, showing her embryos (Um25b, Um27a, and two views of Um16) in whole mount. The lower part of the figure shows sketches from the sections of her most famous grafted embryo (Um132) where the donor tissue is colored red. Note that the mesoderm including somite and notochord is derived from the donor, whereas the adjacent nervous system is not. Access to the notebook and permission to reproduce them by courtesy of Jenny Narraway and the Embryological Collection of the Hubrecht Laboratory, Utrecht.

primitive streak, and specifically Hensen's node at its anterior end, was found to contain the "organizer activity."

The ability of the organizer to induce a nervous system is coupled with its ability to pattern the induced structures, the property that led to its name. Anterior–posterior patterning is discussed elsewhere in this book (Chapter 27);

suffice it to say here that several models have been proposed to account for this activity of the organizer, the main ones being the head/trunk/tail model most clearly formulated by Otto Mangold (1933), which proposes the existence of separate inducing activities for the head, trunk, and tail portions of the axis, and the activation/transformation model of Nieuwkoop (Nieuwkoop et al. 1952; Nieuwkoop and Nigtevecht 1954), which proposes that the nervous system initially induced is of anterior (forebrain) character and that later signals transform parts of it to more caudal fates. Currently, there is evidence both for and against both opposing models, and the issue has not yet been fully resolved (see Stern 2001). The rest of this chapter will concentrate on neural induction proper—the cellular and molecular mechanisms leading to the specification of neural fate regardless of its rostrocaudal character.

SEVEN FRUITLESS DECADES

Following the identification of the organizer and of neural induction, the hunt began for the organizing principles. Spemann himself favored a vitalistic explanation, whereas several laboratories (most notably those of Holtfreter and O. Mangold, and later, Tiedemann and Grunz in Germany, Toivonen and Saxén in Finland, Dorothy and Joseph Needham and Waddington in England, Nakamura and Yamada in Japan, and Brachet in Belgium) embarked on trying to identify a chemical inducer (Holtfreter 1933, 1934, 1945; Waddington 1933a; Needham et al. 1934; Spemann 1938; Toivonen 1938, 1940; Chuang 1939, 1940; Waddington 1940; Saxén and Toivonen 1962; Toivonen et al. 1975; Rollhauser-ter Horst 1977a,b; Saxén 1980; Chen and Solursh 1992; for review, see Nakamura and Toivonen 1978). Early indications for a steroid, then for various protein or RNA extracts, led to transient flurries of excitement, which quickly waned as a result of the discovery that numerous "heterologous," or nonspecific inducers (including killed organizers, high or low pH, alcohol, histological dyes) were just as effective as an organizer graft in inducing a second axis in amphibians. Essentially no progress was made until well into the 1990s.

A TURNING POINT: BMP ANTAGONISM AND THE DEFAULT MODEL

Several seemingly unrelated observations gradually led to a new concept, commonly known as the "default model" for neural induction (Fig. 3) (Hemmati-Brivanlou and Melton 1997). First, several groups had observed that in amphibians, dissociation of gastrula-stage animal caps into single cells for a short time before reaggregating them leads to the formation of neural tissue (Born et al. 1989; Godsave and Slack 1989; Grunz and Tacke 1989; Sato and Sargent 1989;

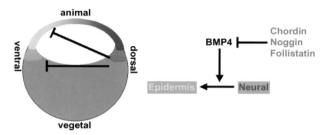

Figure 3. The "default model" in *Xenopus*. On the left is a rough fate map of a blastula-stage embryo (organizer in *red*, ventral mesoderm in *pink*, neural in *blue*, epidermis in *yellow*, yolky endoderm in *green*). The inhibitory arrows represent BMP antagonist activity emanating from the organizer. On the right is a "genetic" diagram of the inductive interactions proposed by the model: Ectoderm cells (represented by the gray boxes) have an autonomous tendency to differentiate into neural tissue, but are prevented from doing this and directed instead to epidermis by BMP4, which is expressed ubiquitously. Near the organizer, BMP antagonists block BMP4 signaling, allowing neighboring ectoderm cells to develop according to their "default" neural fate.

Saint-Jeannet et al. 1990). A few years later, it was found that misexpression of a dominant-negative "activin" receptor (it was later discovered that this construct inhibits several TGF-β-related factors) into *Xenopus* embryos not only blocks mesoderm formation, but also generates ectopic neural tissue (Hemmati-Brivanlou and Melton 1992, 1994). At about the same time, it was discovered that BMP4 is a ventralizing factor in *Xenopus* (Dale et al. 1992; Jones et al. 1992). Several of these authors speculated that neural induction might be induced mediated by removal of some inhibitory substance (Hemmati-Brivanlou and Melton 1994), but direct evidence was still lacking.

Soon, three genes encoding proteins with neuralizing activity were isolated and found to be expressed in the organizer: *Noggin* (Smith and Harland 1992; Lamb et al. 1993; Smith et al. 1993), *Follistatin* (Hemmati-Brivanlou et al. 1994), and *Chordin* (Sasai et al. 1994, 1995). However, several other findings were required before the connections were established firmly: The turning points were the finding that BMP4 is an effective inhibitor of neural fate while promoting epidermal differentiation (even in dissociated cells) (Hawley et al. 1995; Wilson and Hemmati-Brivanlou 1995) and the observations that all three neuralizing/dorsalizing proteins, Noggin, Chordin, and Follistatin, are binding partners and antagonists of BMP signaling (De Robertis and Sasai 1996; Piccolo et al. 1996; Zimmerman et al. 1996; Fainsod et al. 1997). The *Drosophila* homolog of Chordin (Short gastrulation, or Sog) also binds and inactivates the BMP4 homolog Decapentaplegic (Dpp), and vertebrate Chordin can even rescue *sog* mutants (Francois and Bier 1995; Holley et al. 1995; Schmidt et al. 1995; Biehs et al. 1996; De Robertis and Sasai 1996; Ferguson 1996).

Together, these findings led to the default model (Hemmati-Brivanlou and Melton 1997), which proposes

that cells within the ectoderm layer of the frog gastrula have an autonomous tendency to differentiate into neural tissue. This tendency is inhibited by bone morphogenetic proteins—in particular, BMP4, which acts as an epidermal inducer (Fig. 3).

Consistent with this model (Fig. 4), neuralization does not occur after dissociation of animal caps obtained from embryos previously injected with RNA encoding effectors of BMP4 (*Msx1, Smad1,* or *Smad5*; Suzuki et al. 1997a,b; Wilson et al. 1997), consistent with the view that the neural pathway is inhibited by an endogenous BMP-like activity. Moreover, the expression pattern of *BMP4* in *Xenopus* conforms to its proposed anti-neural function: In the early gastrula, *BMP4* transcripts are widely expressed in the entire ectoderm and then clear from the future neural plate at the time when the organizer appears (Fainsod et al. 1994).

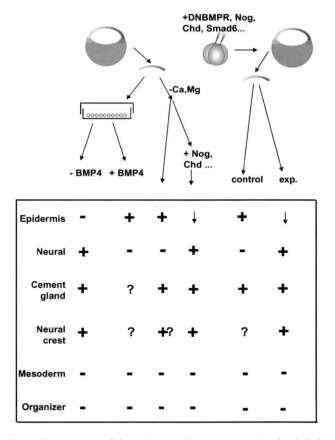

Figure 4. Summary of the main experiments supporting the default model (Fig. 3) as done in *Xenopus*. The leftmost two columns illustrate the results of cell dissociation experiments, the next two show the effects of incubating animal caps with BMP antagonists, and the last two columns summarize the most common type of "animal cap" experiment. The lower box shows the usual results of these experiments, where + implies expression of markers for the tissue shown, – means no expression, and up- or down-arrows represent up- or down-regulation, respectively.

Transcription of *BMP* RNA is maintained by the activity of BMP protein (Biehs et al. 1996), which accounts for the disappearance of *BMP4* and *-7* expression from the vicinity of the organizer (which secretes BMP inhibitors) at the gastrula stage (Fainsod et al. 1994; Hawley et al. 1995).

The model is further supported by the effects of treatments that inhibit the BMP signaling pathway. Animal caps cut from embryos injected with either RNA encoding dominant-negative receptors that bind BMPs (Hemmati-Brivanlou and Melton 1994; Xu et al. 1995), or noncleavable forms of BMP4 or -7 (Hawley et al. 1995), or antisense *BMP4* RNA (Sasai et al. 1995) adopt a neural fate instead of epidermis (Fig. 4). Finally, Chordin and Noggin protein can neuralize isolated animal caps (provided that these have been exposed briefly to low Ca^{++}/Mg^{++}-medium—effectively a partial dissociation, although the rationale given for this is that it helps the protein penetrate between the cells).

In addition to its role in neural induction, the organizer can pattern the mesoderm at the gastrula stage (dorsalization; see Chapter 26). This activity can also be attributed to BMP inhibition. BMPs can modify dorsal mesoderm to give ventral cell types (Dale et al. 1992; Fainsod et al. 1994; Jones et al. 1996), and their inhibitors can generate notochord and muscle from ventral mesoderm (Smith et al. 1993; Sasai et al. 1994; Tonegawa et al. 1997; Tonegawa and Takahashi 1998; Streit and Stern 1999b). BMP inhibitors can also regulate the dorsoventral polarity of the whole embryo before gastrulation. For example, UV-irradiated embryos lack dorsoventral polarity and fail to gastrulate but can be rescued fully by injection of RNA encoding any of the BMP inhibitors: The blastopore (dorsal) will form close to the site of injection (Smith and Harland 1992; Sasai et al. 1994).

In addition to Chordin, Noggin, and Follistatin, other secreted molecules that antagonize BMP signaling have been found and several of these are expressed in, or close to, the organizer. These include Cerberus (Bouwmeester et al. 1996; Belo et al. 1997), Gremlin, Dan, and Drm (Hsu et al. 1998; Pearce et al. 1999; Dionne et al. 2001; Eimon and Harland 2001; Khokha et al. 2003), Ogon/Sizzled (Wagner and Mullins 2002; Yabe et al. 2003b), and Twisted gastrulation (Chang et al. 2001).

Finally, a recent study by the De Robertis lab demonstrated that organizer activity, or at least dorsalization, requires functional Chordin (Oelgeschlager et al. 2003). Together, these findings provide compelling evidence that BMPs and their modulation by endogenous inhibitors are involved in the activities of the organizer, including the establishment of neural and nonneural domains in *Xenopus*. This model is very attractive both because of its simplicity and because it provides the first truly coherent model to explain neural induction since the discovery of the organizer by Spemann and Mangold.

MORE COMPLEXITY?

What followed was perhaps a little reminiscent of the events in the 1940s that eventually led to a temporary loss of interest in identifying neural inducers (see above)—a flurry of papers reporting neural inducing activity of a variety of other molecules (Otte et al. 1988, 1990, 1991; Bolce et al. 1992; Kengaku and Okamoto 1995; Lamb and Harland 1995; Sokol et al. 1995; Witta et al. 1995; Hansen et al. 1997; Rodriguez-Gallardo et al. 1997; Xu et al. 1997; Alvarez et al. 1998; Barnett et al. 1998; Mariani and Harland 1998; Storey et al. 1998; Baker et al. 1999; Hongo et al. 1999; Kato et al. 1999; Leclerc et al. 1999; Matsuo-Takasaki et al. 1999; Beanan et al. 2000; Fekany-Lee et al. 2000; Hardcastle et al. 2000; Ishimura et al. 2000; Strong et al. 2000; Kim and Nishida 2001; Pera et al. 2001, 2003; Sullivan et al. 2001; Wessely et al. 2001; Borchers et al. 2002; Peng et al. 2002; Tsuda et al. 2002; Osada et al. 2003; Yabe et al. 2003a). Some of these clearly act by inhibition of the BMP pathway at some level, whereas others do not obviously act that way. Many of those that do not, however, seem to act by regulating the pattern of the whole embryo at earlier stages of development. Misexpression experiments conducted by injection of RNA at early cleavage stages give results that are difficult to interpret in terms of whether the encoded protein acts directly (as a true neural inducer) or through prior induction of a cell fate that can emit an inducer.

To some extent, this explains some contradictory findings. For example, the Wnt pathway has been reported to act as a neural inducer by some investigators (Sokol et al. 1995; Baker et al. 1999; Wessely et al. 2001) but to inhibit neural induction by others (Wilson et al. 2001). The difference could be accounted for if at early stages Wnt dorsalizes the embryo (inducing tissues with organizer properties), while at later stages Wnt activity somehow antagonizes other signals from the organizer (Bainter et al. 2001; Wilson and Edlund 2001). It becomes important to use reagents that work in a cell-autonomous manner and to express them in a stage- and position-controlled way (as appropriate to the specific inductive event being studied).

A more complex literature surrounds the role of FGF signaling in neural induction. A first study implicated this pathway indirectly when Suramin (which inhibits FGF, among other related proteins) was found to block neural induction in *Xenopus* (Grunz 1992). Later, several labs found that FGF can induce neural tissue under certain circumstances (Lamb and Harland 1995; Rodriguez-Gallardo et al. 1997; Alvarez et al. 1998; Barnett et al. 1998; Storey et al. 1998; Hongo et al. 1999; Hardcastle et al. 2000; Ishimura et al. 2000; Wilson et al. 2000; Kim and Nishida 2001; Hudson et al. 2003), and other studies suggested that FGF is not a sufficient signal for neural induction (Amaya et al. 1991; Cox and Hemmati-Brivanlou 1995; Kroll and Amaya

1996; Holowacz and Sokol 1999; Ribisi et al. 2000; Pownall et al. 2003). One explanation for this discrepancy is the observation that different FGF receptors are required to mediate different activities of FGF: FGFR1 is required for the mesoderm-inducing function of FGF, whereas FGFR4 appears to mediate its role in neuralization (Hardcastle et al. 2000; Umbhauer et al. 2000). The involvement of FGF signals in neural induction are discussed further below.

FGF, WNT, AND BMP IN NEURAL INDUCTION

Despite the obvious attraction of the default model, several observations in different organisms do not fit its proposals so neatly. In *Xenopus*, inhibition of FGF signaling by a dominant-negative version of the FGF-receptor-1 (XFD) blocks the neuralizing activity of both Noggin and Chordin (Launay et al. 1996; Sasai et al. 1996). Furthermore, mere cutting of the animal cap activates MAP kinase by phosphorylation, at least transiently (LaBonne and Whitman 1997), which could explain the finding made by several labs that "control" animal caps express markers for the cement gland (Fig. 4). These observations suggest that FGF signaling is required for neural induction in addition to BMP inhibition.

In chick (Figs. 5 and 6), the patterns of expression of components of the BMP pathway do not agree with the model: *Chordin* continues to be expressed in organizer at stages when this has lost its neural inducing activity, and *Noggin* and *Follistatin* are not expressed in the organizer at the appropriate stages at all, whereas *BMP4* and *BMP7* are expressed only weakly (if at all) in the ectoderm before neural induction begins, and their expression increases at the border of the neural plate starting from the end of gastrulation (stage 4) (Streit et al. 1998). Moreover, misexpression of *Chordin* or *Noggin* in competent epiblast does not neuralize the epiblast (Fig. 6) (Streit et al. 1998; Streit and Stern 1999b; C. Linker et al., in prep.), and dissociation of the epiblast leads to differentiation of muscle rather than neurons (George-Weinstein et al. 1996). In zebrafish, neither *Noggin* nor *Follistatin* is expressed in the organizer (Bauer et al. 1998), and *Chordin* (*chordino*) mutants, although ventralized, still have a neural plate (Hammerschmidt et al. 1996a,b; Kishimoto et al. 1997; Schulte-Merker et al. 1997; Bauer et al. 1998). In mouse, *BMP4* mutants are uninformative (the embryos die too early with mesoderm and other generalized defects (Winnier et al. 1995), but *BMP2* and *BMP7* mutants lack an early neural phenotype (Dudley et al. 1995; Zhang and Bradley 1996) and *Chordin*, *Noggin*, and even *Chordin-Noggin* double mutants have a respectable neural plate (Brunet et al. 1998; McMahon et al. 1998; Bachiller et al. 2000). In the urochordate *Ciona*, FGF

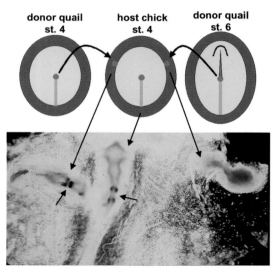

Figure 5. Organizer graft experiment in the chick, also demonstrating the changes in inducing ability of the organizer with increasing age of the donor. The middle diagram shows a host chick embryo, at stage 4. This embryo simultaneously receives a graft of a quail stage-4 node on its left and a quail stage-6 node on its right. The lower panel shows the result of this experiment, after in situ hybridization (*purple*) for the hindbrain marker *Krox-20* (expressed in rhombomeres 3 and 5, *arrows*) and staining with an anti-quail antibody (*brown*). The young graft has induced a complete axis including the entire head, while the older graft on the right has generated a short axis, mostly derived from the graft itself and lacking rostral structures including the *Krox-20*-expressing region. Experiment performed by Kate Storey (Storey et al. 1992).

signaling through the MEK pathway, but not BMP inhibition, appears to be responsible for neural induction (Darras and Nishida 2001; Hudson and Lemaire 2001; Kim and Nishida 2001; Bertrand et al. 2003; Hudson et al. 2003).

FGF signaling now appears to be a prerequisite for neural induction, but this step occurs (or at least begins) very early, before gastrulation (Streit et al. 2000; Wilson et al. 2000). However, FGF does not appear to be a sufficient or direct neural inducer in vertebrates (Streit et al. 2000). Wilson and colleagues (Wilson et al. 2001) suggested that FGF only shows neural inducing activity when Wnt signaling is also blocked and proposed that there are two divergent pathways both involving FGF: for "medial epiblast cells" (prospective neural plate), FGF signaling alone is sufficient to repress the BMP pathway and thus cause neuralization. For "lateral epiblast cells" (prospective epidermis), both FGF signaling and Wnt inhibition are required to block the BMP pathway and neuralize (Wilson and Edlund 2001; Wilson et al. 2001). It has been proposed (Bainter et al. 2001; Wilson and Edlund 2001) that the critical event involves regulation of *BMP4* transcription. Most of these experiments have been conducted using explants, and in

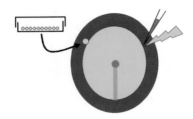

Factor/tissue	Sox3	Sox2	Otx2	ERNI	ChCh	Bra
node	+ 2h	+ 9h	+ 2h	+ 1h	+ 4h	-
hypoblast	+ 2h	-	+ 2h	+ 1h	-	-
FGF8	+ 2h	-	?	+ 1h	+ 4h	-
Chordin	-	-	-	-	-	-†
Noggin	-	-	-	-	-	-
Smad6	-	-	-	-	-	-
BMP4 elec.*	n.e.	-	?	n.e.	?	?
Smad6+Chordin	-	-	-	-	-	-
Smad6+Noggin	-	-	-	-	-	-
Smad6+FGF8	+	-	+	+	+	-
Smad6+αWnt	-	-	-	-	-	-
Smad6+FGF8+αWnt	+	-	+	+	+	-
FGF8+αWnt	+	-	+	+	+	-

Figure 6. Results of chick misexpression experiments. The upper diagram shows the two main types of misexpression experiments usually done in whole chick embryos: a graft of cultured COS cells that had been transfected with an expression plasmid encoding a secreted factor (*left*), and in vivo electroporation of an expression plasmid encoding the desired protein (which may be a transcription factor, secreted protein, or any other construct) directly into a test region of the epiblast. The lower table summarizes the results of the main experiments done in whole chick embryos. + indicates induction, – no induction, n.e., no effect. In the first three examples (node, hypoblast, FGF8) the time (hours) of exposure required to obtain induction of the marker is shown. *Brachyury* (*Bra*) is a marker for mesoderm. *Sox2* is a definitive neural plate marker, and the remaining markers (*ERNI*, *Otx2*, *Sox3*, *Churchill* [*ChCh*]) are expressed in the early epiblast including the prospective neural territory but do not indicate commitment of the cells to a neural fate. "αWnt" is a mixture of three different Wnt antagonists (Crescent, NFz8, and Dkk1) and a multifunctional antagonist of Wnt, BMP, and Nodal (Cerberus). Note that no combination of factors can mimic the induction of *Sox2* by the node. (†) Chordin induces an ectopic primitive streak when misexpressed inside the embryo (Streit et al. 1998) but does not induce neural markers in competent epiblast. (*) BMP4 misexpressed by electroporation within the neural plate inhibits *Sox2* but not *Sox3*. When using COS cells there is no effect. Based on experiments by Streit et al. (1998, 2000), Streit and Stern (1999a,b), and C. Linker et al. (in prep.).

our own experiments in intact embryos, we are unable to neuralize competent epiblast by any combination of FGF, Wnt antagonists, and/or BMP antagonists at any stage of development (Streit et al. 2000; C. Linker et al., in prep.). Furthermore, we find that misexpression of the intracellular BMP antagonist *Smad6* in chick embryos is not sufficient to cause neuralization of competent epiblast even when combined with secreted BMP antagonists (Chordin and Noggin), FGF, secreted Wnt antagonists (NFz8, Dkk1, and Crescent) and a multifunctional antagonist (Cerberus) (C. Linker et al., in prep.). We therefore believe that not all the required signals have been identified, and that although down-regulation of BMP is almost certainly involved in the specification of the neural plate, this is not a sufficient signal, even in combination with FGF and Wnt inhibition.

LOOKING UPSTREAM FROM A CRITICAL PROMOTER

To date, therefore, we have not yet arrived at a full understanding of the molecular signals that trigger the acquisition of neural fate by the ectoderm. One reason for this may be the diversity of approaches used in different "model" systems, and the fact that many of the approaches used have not taken full account of the issue of developmental timing, which is particularly important when studying molecules with multiple, and sometimes opposing, functions at different times. One might, however, gain further insight by changing the viewpoint to the promoter of a target gene. This has recently been attempted for the first time by analysis of the *Sox2* promoter, which is a good marker for committed, developing neural plate in the chick (unlike the mouse where the early functions of *Sox3* and *Sox2* appear to have been exchanged). Kondoh and colleagues have identified many enhancers both upstream and downstream of the *cSox2* gene, which are conserved in mouse and human (Uchikawa et al. 2003). Two of these enhancers, N1 and N2, appear to be responsible for the onset of expression of *Sox2* in the early neural plate—they contain binding sites for several identified transcription factors, including a Sox-related protein (perhaps encoded by *Sox3*, which in the chick is expressed earlier in a similar domain to *Sox2*), TCF/LEF (Wnt pathway), homeodomain-containing proteins, and an E-box sequence shown to be a target of Sip1/δEF1 (Verschueren et al. 1999). *Sip1* was recently identified as a target of the zinc finger protein Churchill, and morpholino-mediated down-regulation of Churchill function leads to loss of the neural plate, whereas misexpression of *Churchill* can confer or maintain the competence of epiblast to neural inducing signals from the node (Sheng et al. 2003; see also Chapter 15).

A VIEW FROM THE STREAK/BLASTOPORE

Churchill was first isolated from a molecular screen designed to identify genes that are regulated by 5 hours of signaling from a graft of the organizer, Hensen's node, in the chick embryo (Sheng et al. 2003). This was done because previous studies had revealed that 5 hours' exposure to a node is required for epiblast cells to become sensitive to *Chordin* misexpression (by maintaining *Sox3* expression, which is otherwise only transiently induced by a node graft) (Streit et al. 1998). *Churchill* is expressed in the prospective neural plate from the late gastrula stage and thereafter persists in the forming neural plate in a pattern similar to that of *Sox2*. Both a node graft and misexpression of *FGF* induce *Churchill* expression in about 4 hours, as expected from the screen (Fig. 7) (Sheng et al. 2003). *Churchill* misexpression close to the streak of the chick embryo or the blastopore of frog embryos causes downregulation of the mesodermal marker *Brachyury*; however, Churchill is a transcriptional activator, which suggested that one of its targets may be a repressor of *Brachyury*. A selection strategy and gel mobility shift assays identified the sequence CGGGRR as a binding target of Churchill, and analysis of the putative regulatory regions of *Sip1* identified numerous occurrences of this sequence. Indeed, *Sip1* is expressed identically to *Churchill*, and morpholino-knockdown of the latter causes loss of *Sip1* expression (Sheng et al. 2003). Sip1 is a good candidate to mediate the down-regulation of *Brachyury* by Churchill, since this is one of its known functions (Verschueren et al. 1999; Lerchner et al. 2000; Papin et al. 2002).

Misexpression of *Churchill* near the primitive streak causes not only loss of *Brachyury*, but also a failure of cells to continue to ingress through the streak to form mesendoderm. Since *Churchill* begins to be expressed at about the time this ingression stops through the anterior primitive streak (see Chapter 15), this raised the possibility that one of its functions may be to end the process of gastrulation to keep some epiblast cells on the surface. This hypothesis is supported by the finding that morpholino down-regulation of *Churchill* at the late gastrula/early neurula stage causes cells to continue to ingress through the streak, and leads to ectopic contribution to the mesoderm rather than neural plate. This effect can be rescued by co-electroporation of either *Churchill* or its target *Sip1* (Sheng et al. 2003). Thus, Churchill regulates the end of ingression through the streak, as well as the competence of the cells that express it to respond to neural inducing signals from the node. This raises the possibility that this is a critical protein in the neural induction process, and particularly in regulating the transition from gastrulation to neurulation. This finding drew attention to the rather overlooked fact that at the gastrula stage, the prospective mesoderm territory lies adjacent to the prospective neural plate in all vertebrate classes, and to the possibility that during normal development, the decisions leading to the acquisition of neural fate involve a switch between these two identities in addition to the choice between epidermis and neural plate, as suggested by the Spemann/Mangold transplantation experiment and its equivalent in other species. One possibility, therefore, is that two separate decisions lead to the establishment of the neural plate: one (a decision between mesendoderm and neural fates) at the medial edge of the neural plate, involving Churchill and Sip1 medially, which prevent further gastrular ingression, and the other (neural versus epidermis) at its lateral/anterior edges, which could involve inhibition of BMP signaling.

A VIEW FROM THE BORDER

The same screen that led to the identification of *Churchill* also identified another early response to an organizer graft: *ERNI* (*Early Response to Neural Induction*) (Streit et al. 2000). Like *Churchill*, *ERNI* is expressed in the prospective neural plate but its expression begins much earlier, before gastrulation. At the end of gastrulation, *ERNI* is down-regulated starting from the center of the neural domain until it is expressed only at the neural–epidermis border, and then it disappears from this domain after stage 7. Also like *Churchill*, *ERNI* is induced by a graft of the node and by FGF, but this induction is much more rapid (just 1 hour). Analysis of the expression of different FGFs and comparison with these early "pre-neural" markers suggested that

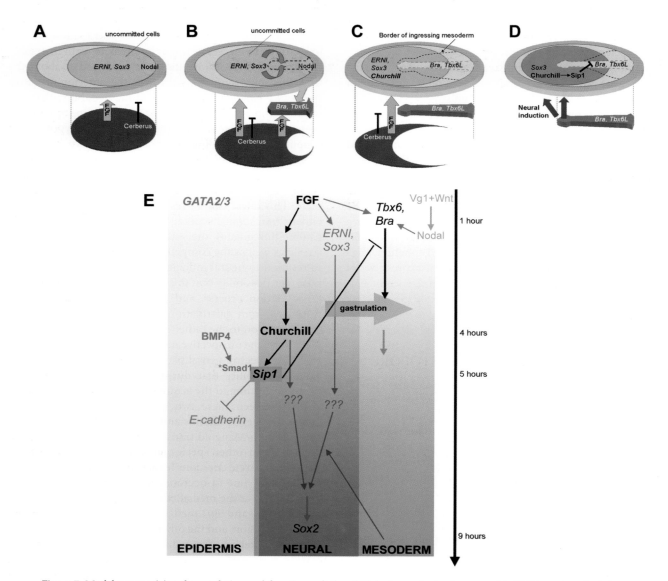

Figure 7. Model summarizing the regulation and functions of Churchill during early development. (*A–D*) The embryologist's view; (*E*) the geneticist's view. In *A–D*, embryos are shown at four stages, with their germ layers exploded. (*A*) At stages XI–XII, the hypoblast (*brown*) emits FGF8, which induces the early pre-neural genes *ERNI* and *Sox3* (*orange*) in the overlying epiblast (*yellow*), but the cells in this domain are still uncommitted. At this stage *Nodal* is expressed in the posterior (*right*) epiblast but is inhibited by Cerberus secreted by the hypoblast. (*B*) At stages XIII–2, the hypoblast is displaced from the posterior part of the embryo by the endoblast (*white*), which allows Nodal signaling, in synergy with FGF, to induce *Brachyury* and *Tbx6L* and ingression (*red arrows*) to form the primitive streak (*red*). (*C*) At stages 3⁺–4, continued FGF signaling induces *Churchill* in a domain of the epiblast (*turquoise*). The border of the epiblast territory destined to ingress to form mesoderm is shown with a dashed black line. (*D*) At the end of stage 4, Churchill induces *Sip1*, which blocks *Brachyury*, *Tbx6L,* and further ingression of epiblast into the streak. The epiblast remaining outside the streak (*blue*) is now sensitized to neural inducing signals emanating from the node (*blue arrows*). (*E*) The same model shown as a genetic cascade. Interactions described in this chapter are shown as black lines; those from the literature are faint. The time axis runs vertically, wherein the color gradients indicate progressive commitment to epidermis (*yellow*), neural (*blue*), and mesoderm (*red*). BMP/Smad/Sip1 interactions regulate the epidermis–neural plate border, whereas ChCh/Sip1/FGF/Bra/Tbx6 regulate the mesoderm–neural decision. (Reproduced, with permission, from Sheng et al. 2003 [©Elsevier].)

ERNI and *Sox3* are first induced before gastrulation by FGF emanating either from the hypoblast or from prospective organizer cells at the posterior end of the blastodisc, or both

(Streit et al. 2000). Indeed, transplantation of either of these tissues can induce *ERNI* and *Sox3* just like a node or FGF, and blocking the FGF pathway abolishes their induction by any of these tissues. These and other findings (Streit and Stern 1999a) led to the view that an early response to neural induction (and FGF signaling) is the specification of a region with "border-like" character, which is responsive to BMP and its antagonists (Streit et al. 1998; Streit and Stern 1999a,b). Subsequent events confine these properties exclusively to the future neural/epidermal border, as the neural plate proper becomes insensitive to BMP during gastrulation. By stage 3^+–4, the only region sensitive to BMP signaling is the border itself: Up-regulation of BMP here moves the border toward the midline (but only by a modest amount), narrowing the neural plate, while down-regulation of BMP at the border widens the neural plate (but again only slightly). Sip1 was first identified by its interaction with phosphorylated Smad1, a target of BMP (Verschueren et al. 1999), raising the possibility that Churchill, through Sip1, contributes to sensitizing cells to BMP antagonists after 5 hours' exposure to FGF.

The situation in amphibians may not be different from that in birds. Based on at least some fate maps from early (32–64 cells) stages in both *Xenopus* (Jacobson and Hirose 1981; Moody 1987; Moody and Kline 1990) and other amphibians (Moury and Jacobson 1989, 1990; Saint-Jeannet and Dawid 1994; Delarue et al. 1997), most animal blastomeres (A2 and A3) will contribute progeny to the neural crest (i.e., the border of the neural plate). Since most injections designed to test the ability of BMP antagonists and other factors to induce a neural plate are placed in the animal pole, it is likely that they target, at least in part, what may be the most sensitive region: the neural/epidermal border. It is also possible that the observations that cutting an animal cap activates MAPK as well as inducing expression of cement gland markers (the cement gland is part of the anteriormost border of the neural plate; see above) are causally connected. In agreement with this view, injection of the downstream inhibitory component of the BMP pathway, *Smad6* in *Xenopus* at the 32-cell stage causes axis duplications and/or expansion of the neural plate when placed into the A1–A3 animal blastomeres, but no ectopic neural plate when placed into the most ventral, A4 animal blastomere (E. Delaune et al.; C. Linker et al., both in prep.). Together, these observations raise the possibility that inhibition of the BMP pathway may only be effective in generating ectopic neural plate (expansion of the endogenous neural plate) within or close to the border between neural and epidermal territories, but not within a region wholly destined to give rise to epidermis. The results also point to the border of the neural plate as a special region, distinct from both neural plate and epidermis.

THE TIMING OF NEURAL INDUCTION

Organizer grafts are technically easiest after the start of gastrulation, when the blastopore, primitive streak, or shield can be identified morphologically. Experiments in which the stage of the host and donor embryos were varied in different vertebrate classes established that neural induction is likely to end by the end of the gastrula stage. For example, in the chick, a Hensen's node taken from an embryo up to stage 4 can induce a complete nervous system, but older donors gradually lose their inducing ability, whereas hosts rapidly lose their competence between stages 4 and 4^+ (Damas 1947; Gallera and Ivanov 1964; Gallera 1971; Dias and Schoenwolf 1990; Storey et al. 1992; Streit et al. 1997). Experiments such as these have suggested that neural induction by the organizer is likely to end by the end of gastrulation, but they do not give insight into when the process starts. As mentioned above, at least some of the early signals may be present and active before the start of gastrulation (Streit et al. 2000; Wilson et al. 2000). At this time, a number of "pre-neural" genes are expressed in the epiblast (including *ERNI*, *Sox3*, and *Otx2*), in a fairly broad domain that includes but is not restricted to the future neural plate. Some cells expressing all three markers are destined to contribute to mesendoderm as well as neural/epidermis border and some epidermis. Further signals and other refining mechanisms are required downstream of this initial "pre-neural" specification to commit cells to a neural fate.

In the chick, grafts of the hypoblast (which expresses both *FGF8* and the Wnt antagonists *Dkk-1*, *Crescent*, and *Cerberus*) can induce all three "pre-neural" genes, but only transiently (Foley et al. 2000; Streit et al. 2000). It has been suggested (Stern 2001) that signals from the node or from its derivatives like the head process/notochord and/or the prechordal mesendoderm may be required to stabilize this early expression and to drive cells to *Sox2* expression and commitment to a neural plate fate. The hypoblast is equivalent (in terms of fate as well as expression of various markers) to the anterior visceral endoderm (AVE) of the mouse (see Chapters 15 and 16), which has been shown to be required for normal development of the mouse forebrain (Thomas and Beddington 1996; Beddington and Robertson 1998, 1999). However, it is important to point out that neither the chick hypoblast nor the mouse AVE is equivalent to Spemann's organizer in that neither can induce the formation of a neural plate/axis when grafted to an ectopic site. In the mouse, it was shown that forebrain induction requires a combination of the AVE, the "early gastrula organizer" (EGO, which may contain some of the precursors of the later node) as well as the appropriate responding part of the epiblast (prospective forebrain) (Tam and Steiner 1999).

The findings that grafts of a mouse node to a lateral site induce a nervous system lacking the most rostral structures (Beddington 1994) and that homozygous HNF3β mutants, which lack a node (Klingensmith et al. 1999), still have a fairly acceptable neural tube have been used by some to argue that the node is not essential for neural induction. Indeed, it is clear that the most rostral portions of the neural plate (prospective forebrain) are never adjacent to the mature node in either mouse or chick embryos. However, the former finding can be explained because mouse embryos are very small and it is virtually impossible to find a site to graft the node allowing a complete axis to form without fusing with the host, and the latter can be explained by the possibility that although no morphological node or its derivatives form, some of its properties are also expressed by other tissues. Furthermore, by the time a "node" can be defined morphologically in the mouse, the embryos are starting to produce a head process; chick nodes at the equivalent stage (4[+]/5) have already lost their ability to induce the most rostral parts of the CNS (Dias and Schoenwolf 1990; Storey et al. 1992). Taking this evidence together, the most parsimonious view is therefore that during normal development, signals from the AVE and/or other areas initiate some of the earliest events of neural induction, but are not sufficient. As development proceeds to gastrulation, "maintenance" signals as well as regional identity are imparted by the node and/or its derivatives (prechordal and chordal tissues) (Stern 2001). However, the node itself (provided that it is taken from an embryo at the full primitive streak stage but before any prechordal/head process cells emerge) contains sufficient signals to induce a complete axis when grafted far enough from the host neural plate. In the chick this can be done in the inner third of the area opaca, which only has extraembryonic fate (Gallera 1971; Dias and Schoenwolf 1990; Storey et al. 1992; Streit et al. 1998); in the mouse (and perhaps also in *Xenopus*), it is impossible to find a site far enough from the host neural plate to avoid recruitment of host neural plate cells.

CONCLUSIONS

Although very substantial progress has been made in understanding neural induction since the pioneering experiments of the 1920s, there are still substantial gaps in our knowledge of both the cellular and molecular aspects of this important process. It is now becoming more likely that rather than a single inducing factor, a whole cascade of molecular events and converging pathways are required to specify the neural plate, and that different mechanisms may contribute to define this territory in different embryonic locations at different times. Our view is that a full understanding will only come when the embryological/cellular processes can be fully correlated with their molecular basis—much the same as what Wolpert may have had in mind when criticizing excessive emphasis on signaling events (inductions) at the expense of understanding the resulting patterns: "... induction and its related concepts, which have so dominated embryological thinking, have completely obscured the problems of pattern formation by emphasizing the information coming from some other tis-

sue rather than the response in the tissue which gives rise to the pattern ... [a] failure of inductive theory to consider the problem of spatial organization" (Wolpert 1970; quoted from Horder 2001).

REFERENCES

Alvarez I.S., Araujo M., and Nieto M.A. 1998. Neural induction in whole chick embryo cultures by FGF. *Dev. Biol.* **199:** 42–54.

Amaya E., Musci T.J., and Kirschner M.W. 1991. Expression of a dominant negative mutant of the FGF receptor disrupts mesoderm formation in *Xenopus* embryos. *Cell* **66:** 257–270.

Bachiller D., Klingensmith J., Kemp C., Belo J.A., Anderson R.M., May S.R., McMahon J.A., McMahon A.P., Harland R.M., Rossant J., and De Robertis E.M. 2000. The organizer factors Chordin and Noggin are required for mouse forebrain development. *Nature* **403:** 658–661.

Bainter J.J., Boos A., and Kroll K.L. 2001. Neural induction takes a transcriptional twist. *Dev. Dyn.* **222:** 315–327.

Baker J.C., Beddington R.S., and Harland R.M. 1999. Wnt signaling in *Xenopus* embryos inhibits bmp4 expression and activates neural development. *Genes Dev.* **13:** 3149–3159.

Barnett M.W., Old R.W., and Jones E.A. 1998. Neural induction and patterning by fibroblast growth factor, notochord and somite tissue in *Xenopus. Dev. Growth Differ.* **40:** 47–57.

Bauer H., Meier A., Hild M., Stachel S., Economides A., Hazelett D., Harland R.M., and Hammerschmidt M. 1998. Follistatin and noggin are excluded from the zebrafish organizer. *Dev. Biol.* **204:** 488–507.

Beanan M.J., Feledy J.A., and Sargent T.D. 2000. Regulation of early expression of Dlx3, a *Xenopus* anti-neural factor, by beta-catenin signaling. *Mech. Dev.* **91:** 227–235.

Beddington R.S. 1994. Induction of a second neural axis by the mouse node. *Development* **120:** 613–620.

Beddington R.S. and Robertson E.J. 1998. Anterior patterning in mouse. *Trends Genet.* **14:** 277–284.

———. 1999. Axis development and early asymmetry in mammals. *Cell* **96:** 195–209.

Belo J.A., Bouwmeester T., Leyns L., Kertesz N., Gallo M., Follettie M., and De Robertis E.M. 1997. Cerberus-like is a secreted factor with neutralizing activity expressed in the anterior primitive endoderm of the mouse gastrula. *Mech. Dev.* **68:** 45–57.

Bertrand V., Hudson C., Caillol D., Popovici C., and Lemaire P. 2003. Neural tissue in ascidian embryos is induced by FGF9/16/20, acting via a combination of maternal GATA and Ets transcription factors. *Cell* **115:** 615–627.

Biehs B., Francois V., and Bier E. 1996. The *Drosophila* short gastrulation gene prevents Dpp from autoactivating and suppressing neurogenesis in the neuroectoderm. *Genes Dev.* **10:** 2922–2934.

Bolce M.E., Hemmati-Brivanlou A., Kushner P.D., and Harland R.M. 1992. Ventral ectoderm of *Xenopus* forms neural tissue, including hindbrain, in response to activin. *Development* **115:** 681–688.

Borchers A.G., Hufton A.L., Eldridge A.G., Jackson P.K., Harland R.M., and Baker J.C. 2002. The E3 ubiquitin ligase GREUL1 anteriorizes ectoderm during *Xenopus* development. *Dev. Biol.* **251:** 395–408.

Born J., Janeczek J., Schwarz W., and Tiedemann H. 1989. Activation of masked neural determinants in amphibian eggs and embryos and their release from the inducing tissue. *Cell Differ. Dev.* **27:** 1–7.

Bouwmeester T., Kim S.-H., Sasai Y., Lu B., and De Robertis E.M. 1996. Cerberus is a head-inducing secreted factor expressed in the anterior endoderm of Spemann's Organizer. *Nature* **382:** 595–601.

Brunet L.J., McMahon J.A., McMahon A.P., and Harland R.M. 1998. Noggin, cartilage morphogenesis, and joint formation in the mammalian skeleton. *Science* **280:** 1455–1457.

Chang C., Holtzman D.A., Chau S., Chickering T., Woolf E.A., Holmgren L.M., Bodorova J., Gearing D.P., Holmes W.E., and Brivanlou A.H. 2001. Twisted gastrulation can function as a BMP antagonist. *Nature* **410:** 483–487.

Chen Y. and Solursh M. 1992. Comparison of Hensen's node and retinoic acid in secondary axis induction in the early chick embryo. *Dev. Dyn.* **195:** 142–151.

Chuang H.-H. 1939. Induktionsleistungen vond frischen und gekochten Organteilen (Niere, Leber) nach ihrer Verpflanzung in Explantate und verschiedene Wirtsregionen von Tritonkeimen. *Arch. Entwicklungsmech. Org.* **139:** 556–638.

———. 1940. Weitere Versuch über die Veränderung der Induktionsleistungen von gekochten Organteilen. *Arch. Entwicklungsmech. Org.* **140:** 25–38.

Cox W.G. and Hemmati-Brivanlou A. 1995. Caudalization of neural fate by tissue recombination and bFGF. *Development* **121:** 4349–4358.

Dale L., Howes G., Price B.M.J., and Smith J.C. 1992. Bone morphogenetic protein 4: A ventralizing factor in early *Xenopus* development. *Development* **115:** 573–585.

Damas H. 1947. Effet de la suspension précoce du flux inducteur sur la détermination du neurectoblaste médullaire. *Arch. Biol.* **58:** 15–57.

Darras S. and Nishida H. 2001. The BMP/CHORDIN antagonism controls sensory pigment cell specification and differentiation in the ascidian embryo. *Dev. Biol.* **236:** 271–288.

De Robertis E.M. and Aréchaga J. 2001. The Spemann–Mangold organizer. *Int. J. Dev. Biol.* **45:** 1–378.

De Robertis E.M. and Sasai Y. 1996. A common plan for dorso-ventral patterning in Bilateria. *Nature* **380:** 37–40.

Delarue M., Saez F.J., Johnson K.E., and Boucaut J.C. 1997. Fates of the blastomeres of the 32-cell stage Pleurodeles waltl embryo. *Dev. Dyn.* **210:** 236–248.

Dias M.S. and Schoenwolf G.C. 1990. Formation of ectopic neurepithelium in chick blastoderms: Age-related capacities for induction and self-differentiation following transplantation of quail Hensen's nodes. *Anat. Rec.* **228:** 437–448.

Dionne M.S., Skarnes W.C., and Harland R.M. 2001. Mutation and analysis of Dan, the founding member of the Dan family of transforming growth factor beta antagonists. *Mol. Cell. Biol.* **21:** 636–643.

Dudley A.T., Lyons K.M., and Robertson E.J. 1995. A requirement for bone morphogenetic protein-7 during development of the mammalian kidney and eye. *Genes Dev.* **9:** 2795–2807.

Eimon P.M. and Harland R.M. 2001. *Xenopus* Dan, a member of the Dan gene family of BMP antagonists, is expressed in derivatives of the cranial and trunk neural crest. *Mech. Dev.* **107:** 187–189.

Fainsod A. Steinbeisser H., and De Robertis E. M. 1994. On the function of BMP-4 in patterning the marginal zone of the *Xenopus* embryo. *EMBO J.* **13:** 5015–5025.

Fainsod A., Deissler K., Yelin R., Marom K., Epstein M., Pillemer G., Steinbeisser H., and Blum M. 1997. The dorsalizing and neural inducing gene follistatin is an antagonist of BMP-4. *Mech. Dev.* **63:** 39–50.

Fekany-Lee K., Gonzalez E., Miller-Bertoglio V., and Solnica-Krezel L. 2000. The homeobox gene bozozok promotes anterior neuroectoderm formation in zebrafish through negative regulation of BMP2/4 and Wnt pathways. *Development* **127:** 2333–2345.

Ferguson E.L. 1996. Conservation of dorsal-ventral patterning in arthropods and chordates. *Curr. Opin. Genet. Dev.* **6:** 424–431.

Foley A.C., Skromne I.S., and Stern C.D. 2000. Reconciling different models of forebrain induction and patterning: A dual role for the hypoblast. *Development* **127:** 3839–3854.

Francois V. and Bier E. 1995. *Xenopus* chordin and *Drosophila* short gastrulation genes encode homologous proteins functioning in dorsal–ventral axis formation. *Cell* **80:** 19–20.

Gallera J. 1971. Différence de la reactivité à l'inducteur neurogène entre l'ectoblaste de l'aire opaque et celui de l'aire pellucide chez le poulet. *Experientia* **26:** 1953–1954.

Gallera J. and Ivanov I. 1964. La compétence neurogène du feuillet externe du blastoderme de Poulet en fonction du facteur 'temps'. *J. Embryol. Exp. Morphol.* **12:** 693.

George-Weinstein M., Gerhart J., Reed R., Flynn J., Callihan B., Mattiacci M., Miehle C., Foti G., Lash J.W., and Weintraub H. 1996. Skeletal myogenesis: the preferred pathway of chick embryo epiblast cells in vitro. *Dev. Biol.* **173:** 279–291.

Godsave S.F. and Slack J.M. 1989. Clonal analysis of mesoderm induction in *Xenopus laevis*. *Dev. Biol.* **134:** 486–490.

Grunz H. 1992. Suramin changes the fate of Spemann's organizer and prevents neural induction in *Xenopus laevis*. *Mech. Dev.* **38:** 133–141.

Grunz H. and Tacke L. 1989. Neural differentiation of *Xenopus laevis* ectoderm takes place after disaggregation and delayed reaggregation without inducer. *Cell Differ. Dev.* **28:** 211–217.

Gurdon J.B. 1987. Embryonic induction—molecular prospects. *Development* **99:** 285–306.

Hamburger V. 1988. *The heritage of experimental embryology: Hans Spemann and the Organizer*. Oxford University Press, Oxford, United Kingdom.

Hammerschmidt M., Serbedzija G.N., and McMahon A.P. 1996a. Genetic analysis of dorsoventral pattern formation in the zebrafish: Requirement of a BMP-like ventralizing activity and its dorsal repressor. *Genes Dev.* **10:** 2452–2461.

Hammerschmidt M., Pelegri F., Mullins M.C., Kane D.A., van Eeden F.J., Granato M., Brand M., Furutani-Seiki M., Haffter P., Heisenberg C.P., Jiang Y.J., Kelsh R.N., Odenthal J., Warga R.M., and Nusslein-Volhard C. 1996b. *dino* and *mercedes*, two genes regulating dorsal development in the zebrafish embryo. *Development* **123:** 95–102.

Hansen C.S., Marion C.D., Steele K., George S., and Smith W.C. 1997. Direct neural induction and selective inhibition of mesoderm and epidermis inducers by Xnr3. *Development* **124:** 483–492.

Hardcastle Z., Chalmers A.D., and Papalopulu N. 2000. FGF-8 stimulates neuronal differentiation through FGFR-4a and interferes with mesoderm induction in *Xenopus* embryos. *Curr. Biol.* **10:** 1511–1514.

Hawley S.H., Wunnenberg-Stapleton K., Hashimoto C., Laurent M.N., Watabe T., Blumberg B.W., and Cho K. W. 1995. Disruption of BMP signals in embryonic *Xenopus* ectoderm leads to direct neural induction. *Genes Dev.* **9:** 2923–2935.

Hemmati-Brivanlou A. and Melton D. 1992. A truncated activin receptor inhibits mesoderm induction and formation of axial structures in *Xenopus* embryos. *Nature* **359:** 609–614.

———. 1994. Inhibition of activin receptor signaling promotes neuralization in *Xenopus*. *Cell* **77:** 273–281.

———. 1997. Vertebrate embryonic cells will become nerve cells unless told otherwise. *Cell* **88:** 13–17.

Hemmati-Brivanlou A., Kelly O.G., and Melton D.A. 1994. Follistatin, an antagonist of activin, is expressed in the Spemann organizer and displays direct neuralizing activity. *Cell* **77:** 283–295.

Holley S.A., Jackson P.D., Sasai Y., Lu B., De Robertis E.M., Hoffmann F.M., and Ferguson E.L. 1995. A conserved system for dorsal-ventral patterning in insects and vertebrates involving sog and chordin. *Nature* **376:** 249–253.

Holowacz T. and Sokol S. 1999. FGF is required for posterior neural patterning but not for neural induction. *Dev. Biol.* **205:** 296–308.

Holtfreter J. 1933. Eigenschaften und Verbreitung induzierender Stoffe. *Naturwissenschaften* **21:** 766–770.

———. 1934. Der Einfluss thermischer, mechanischer und chemischer Eingreffe auf die Induzierfaehigkeit von Triton-Keimteilen. *Wilhem Roux' Archiv. Entwicklingsmech. Org.* **132:** 225–306.

———. 1945. Neuralization and epidermalization of gastrula ectoderm. *J. Exp. Zool.* **98:** 161–210.

Hongo I., Kengaku M., and Okamoto H. 1999. FGF signaling and the anterior neural induction in *Xenopus*. *Dev. Biol.* **216:** 561–581.

Horder T.J. 2001. The organizer concept and modern embryology: Anglo-American perspectives. *Int. J. Dev. Biol.* **45:** 97–132.

Hsu D.R., Economides A.N., Wang X., Eimon P.M., and Harland R.M. 1998. The *Xenopus* dorsalizing factor Gremlin identifies a novel family of secreted proteins that antagonize BMP activities. *Mol. Cell* **1:** 673–683.

Hudson C. and Lemaire P. 2001. Induction of anterior neural fates in the ascidian *Ciona intestinalis*. *Mech. Dev.* **100:** 189–203.

Hudson C., Darras S., Caillol D., Yasuo H., and Lemaire P. 2003. A conserved role for the MEK signalling pathway in neural tissue specification and posteriorisation in the invertebrate chordate, the ascidian *Ciona intestinalis*. *Development* **130:** 147–159.

Hunt T.E. 1929. Hensen's node as an organizer in the formation of the chick embryo. *Anat. Rec.* **42:** 22.

Ishimura A., Maeda R., Takeda M., Kikkawa M., Daar I.O., and Maeno M. 2000. Involvement of BMP-4/msx-1 and FGF pathways in neural induction in the *Xenopus* embryo. *Dev. Growth Differ.* **42:** 307–316.

Jacobson M. and Hirose G. 1981. Clonal organization of the central nervous system of the frog. II. Clones stemming from individual blastomeres of the 32- and 64-cell stages. *J. Neurosci.* **1:** 271–284.

Jones C.M., Dale L., Hogan B.M., Wright C.V.E., and Smith J.C. 1996. Bone Morphogenetic Protein-4 (BMP-4) acts during gastrulation stages to cause ventralization of *Xenopus* embryos. *Development* **122:** 1545–1554.

Jones C.M., Lyons K.M., Lapan P.M., Wright C.V.E., and Hogan B.M. 1992. DVR-4 (bone morphogentic protein-4) as a posterior-ventralizing factor in *Xenopus* mesoderm induction. *Development* **115:** 639–647.

Kato Y., Shi Y., and He X. 1999. Neuralization of the *Xenopus* embryo by inhibition of p300/ CREB-binding protein function. *J. Neurosci.* **19:** 9364–9373.

Kengaku M. and Okamoto H. 1995. bFGF as a possible morphogen for the anteroposterior axis of the central nervous system in *Xenopus*. *Development* **121:** 3121–3130.

Khokha M.K., Hsu D., Brunet L.J., Dionne M.S., and Harland R.M. 2003. Gremlin is the BMP antagonist required for maintenance of Shh and Fgf signals during limb patterning. *Nat. Genet.* **34:** 303–307.

Kim G.J. and Nishida H. 2001. Role of the FGF and MEK signaling pathway in the ascidian embryo. *Dev. Growth Differ.* **43:** 521–533.

Kishimoto Y., Lee K.H., Zon L., Hammerschmidt M., and Schulte-Merker S. 1997. The molecular nature of zebrafish swirl: BMP2 function is essential during early dorsoventral patterning. *Development* **124:** 4457–4466.

Klingensmith J., Ang S.L., Bachiller D., and Rossant J. 1999. Neural induction and patterning in the mouse in the absence of the node and its derivatives. *Dev. Biol.* **216:** 535–549.

Kroll K.L. and Amaya E. 1996. Transgenic *Xenopus* embryos from sperm nuclear transplantations reveal FGF signaling requirements during gastrulation. *Development* **122:** 3173–3183.

LaBonne C. and Whitman M. 1997. Localization of MAP kinase activity in early *Xenopus* embryos: Implications for endogenous FGF signaling. *Dev. Biol.* **183:** 9–20.

Lamb T.M. and Harland R.M. 1995. Fibroblast growth factor is a direct neural inducer, which combined with noggin generates anterior–posterior neural pattern. *Development* **121:** 3627–3636.

Lamb T.M., Knecht A.K., Smith W.C., Stachel S.E., Economides A.N., Stahl N., Yancopolous G.D., and Harland R.M. 1993. Neural induction by the secreted polypeptide noggin. *Science* **262:** 713–718.

Launay C., Fromentoux V., Shi D.L., and Boucaut J.C. 1996. A truncated FGF receptor blocks neural induction by endogenous *Xenopus* inducers. *Development* **122:** 869–880.

Leclerc C., Duprat A.M., and Moreau M. 1999. Noggin upregulates Fos expression by a calcium-mediated pathway in amphibian embryos. *Dev. Growth Differ.* **41:** 227–238.

Lerchner W., Latinkic B.V., Remacle J.E., Huylebroeck D., and Smith J.C. 2000. Region-specific activation of the *Xenopus Brachyury* promoter involves active repression in ectoderm and endoderm: A study using transgenic frog embryos. *Development* **127:** 2729–2739.

Lewis W.H. 1907. Transplantation of the lips of the blastopore in Rana pipiens. *Am. J. Anat.* **7:** 137–141.

Mangold O. 1933. Über die Induktionsfähigkeit der verschiedenen Bezirke der Neurula von Urodelen. *Naturwissenshaften* **21:** 761–766.

Mariani F.V. and Harland R.M. 1998. XBF-2 is a transcriptional repressor that converts ectoderm into neural tissue. *Development* **125:** 5019–5031.

Matsuo-Takasaki M., Lim J.H., and Sato S.M. 1999. The POU domain gene, XlPOU 2 is an essential downstream determinant of neural induction. *Mech. Dev.* **89:** 75–85.

McMahon J.A., Takada S., Zimmerman L.B., Fan C.M., Harland R.M., and McMahon A.P. 1998. Noggin-mediated antagonism of BMP signaling is required for growth and patterning of the neural tube and somite. *Genes Dev.* **12:** 1438–1452.

Moody S.A. 1987. Fates of the blastomeres of the 32-cell-stage *Xenopus* embryo. *Dev. Biol.* **122:** 300–319.

Moody S.A. and Kline M.J. 1990. Segregation of fate during cleavage of frog (*Xenopus laevis*) blastomeres. Anat. Embryol. **182:** 347–362.

Moury J.D. and Jacobson A.G. 1989. Neural fold formation at newly created boundaries between neural plate and epidermis in the axolotl. *Dev. Biol.* **133:** 44–57.

———. 1990. The origins of neural crest cells in the axolotl. *Dev. Biol.* **141:** 243–253.

Nakamura O. and Toivonen S., eds. 1978. *Organizer: A milestone of a half-century from Spemann*. Elsevier, Amsterdam, The Netherlands.

Needham J., Waddington C.H., and Needham D.M. 1934. Physico-chemical experiments on the amphibian organizer. *Proc. Roy. Soc. Lond. B Biol. Sci.* **114:** 393–422.

Nieuwkoop P.D. and Nigtevecht G.V. 1954. Neural activation and transformation in explants of competent ectoderm under the influence of fragments of anterior notochord in urodeles. *J. Embryol. Exp. Morphol.* **2:** 175–193.

Nieuwkoop P.D., Botternenbrood E.C., Kremer A., Bloesma F.F.S.N., Hoessels E.L.M.J., Meyer G., and Verheyen F.J. 1952. Activation and organization of the central nervous system in amphibians. *J. Exp. Zool.* **120:** 1–108.

Oelgeschlager M., Kuroda H., Reversade B., and De Robertis E.M. 2003. Chordin is required for the Spemann organizer transplantation phenomenon in *Xenopus* embryos. *Dev. Cell* **4:** 219–230.

Oppenheimer J.M. 1991. Curt Herbst's contributions to the concept of embryonic induction. In *A conceptual history of modern embryology* (ed. S.F. Gilbert), pp. 83–90. Plenum Press, New York.

Osada S., Ohmori S.Y., and Taira M. 2003. XMAN1, an inner nuclear membrane protein, antagonizes BMP signaling by interacting with Smad1 in *Xenopus* embryos. *Development* **130:** 1783–1794.

Otte A.P., Kramer I.M., and Durston A.J. 1991. Protein kinase C and regulation of the local competence of *Xenopus* ectoderm. *Science* **251:** 570–573.

Otte A.P., Koster C.H., Snoek G.T., and Durston A.J. 1988. Protein kinase C mediates neural induction in *Xenopus laevis*. *Nature* **334:** 618–620.

Otte A.P., Kramer I.M., Mannesse M., Lambrechts C., and Durston A.J. 1990. Characterization of protein kinase C in early *Xenopus* embryogenesis. *Development* **110:** 461–470.

Papin C., van Grunsven L.A., Verschueren K., Huylebroeck D., and Smith J.C. 2002. Dynamic regulation of *Brachyury* expression in the amphibian embryo by XSIP1. *Mech. Dev.* **111:** 37–46.

Pearce J.J., Penny G., and Rossant J. 1999. A mouse cerberus/Dan-related gene family. *Dev. Biol.* **209:** 98–110.

Peng Y., Xu R.H., Mei J.M., Li X.P., Yan D., Kung H.F., and Phang J.M. 2002. Neural inhibition by c-Jun as a synergizing factor in bone morphogenetic protein 4 signaling. *Neuroscience* **109:** 657–664.

Pera E., Ikeda A., Eivers E., and De Robertis E.M. 2003. Integration of IGF, FGF and anti-BMP signals via Smad1 phosphorylation in neural induction. *Genes Dev.* **17:** 3023–3028.

Pera E.M., Wessely O., Li S.Y., and De Robertis E.M. 2001. Neural and head induction by insulin-like growth factor signals. *Dev. Cell* **1:** 655–665.

Piccolo S., Sasai Y., Lu B., and De Robertis E.M. 1996. Dorsoventral patterning in *Xenopus*: Inhibition of ventral signals by direct binding of chordin to BMP-4. *Cell* **86:** 589–598.

Pownall M.E., Welm B.E., Freeman K.W., Spencer D.M., Rosen J.M., and Isaacs H.V. 2003. An inducible system for the study of FGF signalling in early amphibian development. *Dev. Biol.* **256:** 89–99.

Ribisi Jr., S., Mariani F.V., Aamar E., Lamb T.M., Frank D., and Harland R.M. 2000. Ras-mediated FGF signaling is required for the formation of posterior but not anterior neural tissue in *Xenopus laevis*. *Dev. Biol.* **227:** 183–196.

Rodriguez-Gallardo L., Climent V., Garcia-Martinez V., Schoenwolf G.C., and Alvarez I.S. 1997. Targeted over-expression of FGF in chick embryos induces formation of ectopic neural cells. *Int. J. Dev. Biol.* **41:** 715–723.

Rollhauser-ter Horst J. 1977a. Artificial neural induction in amphibia. I. Sandwich explants. *Anat. Embryol.* **151:** 309–316.

———. 1977b. Artificial neural induction in amphibia. II. Host embryos. *Anat. Embryol.* **151:** 317–324.

Saint-Jeannet J.P. and Dawid I.B. 1994. A fate map for the 32-cell stage of *Rana pipiens*. *Dev. Biol.* **166:** 755–762.

Saint-Jeannet J.P., Huang S., and Duprat A.M. 1990. Modulation of neural commitment by changes in target cell contacts in *Pleurodeles waltl*. *Dev. Biol.* **141:** 93–103.

Sasai Y., Lu B., Piccolo S., and De Robertis E.M. 1996. Endoderm

induction by the organizer-secreted factors chordin and noggin in *Xenopus* animal caps. *EMBO J.* **15:** 4547–4555.

Sasai Y., Lu B., Steinbeisser H., and De Roberti E.M. 1995. Regulation of neural induction by the Chd and Bmp-4 antagonistic patterning signals in *Xenopus*. *Nature* **376:** 333–336.

Sasai Y., Lu B., Steinbeisser H., Geissert D., Gont L.K., and De Robertis E.M. 1994. *Xenopus* chordin: A novel dorsalizing factor activated by organizer-specific homeobox genes. *Cell* **79:** 779–790.

Sato S.M. and Sargent T.D. 1989. Development of neural inducing capacity in dissociated *Xenopus* embryos. *Dev. Biol.* **134:** 263–266.

Saxén L. 1980. Neural induction: Past, present, and future. *Curr. Top. Dev. Biol.* **15:** 409–418.

Saxén L. and Toivonen S. 1962. *Primary embryonic induction*. Logos Press, London, United Kingdom.

Schmidt J., Francois V., Bier E., and Kimelman D. 1995. *Drosophila* short gastrulation induces an ectopic axis in *Xenopus*: Evidence for conserved mechanisms of dorsal–ventral patterning. *Development* **121:** 4319–4328.

Schulte-Merker S., Lee K.J., McMahon A.P., and Hammerschmidt M. 1997. The zebrafish organizer requires chordino. *Nature* **387:** 862–863.

Sheng G., dos Reis M., and Stern C.D. 2003. Churchill, a zinc finger transcriptional activator, regulates the transition from gastrulation to neurulation. *Cell* **115:** 603–613.

Smith W.C. and Harland R.M. 1992. Expression cloning of noggin, a new dorsalizing factor localized to the Spemann organizer in *Xenopus* embryos. *Cell* **70:** 829–840.

Smith W.C., Knecht A.K., Wu M., and Harland R.M. 1993. Secreted noggin protein mimics the Spemann organizer in dorsalizing *Xenopus* mesoderm. *Nature* **361:** 547–549.

Sokol S.Y., Klingensmith J., Perrimon N., and Itoh K. 1995. Dorsalizing and neuralizing properties of Xdsh, a maternally expressed *Xenopus* homolog of dishevelled. *Development* **121:** 1637–1647.

Spemann H. 1921. Die Erzeugung tierischer Chimären durch heteroplastische Transplantation zwischen *Triton cristatus* und *taeniatus*. *Wilhelm Roux' Arch. Entwicklungsmech. Org.* **48:** 533–570.

———. 1938. *Embryonic development and induction*. Yale University Press, New Haven, Connecticut.

Spemann H. and Mangold H. 1924. Über induktion von Embryonalanlagen durch implantation artfremder Organisatoren. *Wilhelm Roux' Arch. Entwicklungsmech. Org.* **100:** 599–638.

Stern C.D. 2001. Initial patterning of the central nervous system: How many organizers? *Nat. Rev. Neurosci.* **2:** 92–98.

Storey K.G., Crossley J.M., De Robertis E.M., Norris W.E., and Stern C.D. 1992. Neural induction and regionalisation in the chick embryo. *Development* **114:** 729–741.

Storey K.G., Goriely A., Sargent C.M., Brown J.M., Burns H.D., Abud H.M., and Heath J.K. 1998. Early posterior neural tissue is induced by FGF in the chick embryo. *Development* **125:** 473–484.

Streit A. and Stern C.D. 1999a. Establishment and maintenance of the border of the neural plate in the chick: Involvement of FGF and BMP activity. *Mech. Dev.* **82:** 51–66.

———. 1999b. Mesoderm patterning and somite formation during node regression: Differential effects of chordin and noggin. *Mech. Dev.* **85:** 85–96.

Streit A., Berliner A.J., Papanayotou C., Sirulnik A., and Stern C.D. 2000. Initiation of neural induction by FGF signalling before gastrulation. *Nature* **406:** 74–78.

Streit A., Lee K.J., Woo I., Roberts C., Jessell T.M., and Stern C.D. 1998. Chordin regulates primitive streak development and the stability of induced neural cells, but is not sufficient for neural induction in the chick embryo. *Development* 125: 507–519.

Streit A., Sockanathan S., Perez L., Rex M., Scotting P.J., Sharpe P.T., Lovell-Badge R., and Stern C.D. 1997. Preventing the loss of competence for neural induction: HGF/SF, L5 and Sox-2. *Development* 124: 1191–1202.

Strong C.F., Barnett M.W., Hartman D., Jones E.A., and Stott D. 2000. Xbra3 induces mesoderm and neural tissue in *Xenopus laevis*. *Dev. Biol.* 222: 405–419.

Sullivan S.A., Akers L., and Moody S.A. 2001. foxD5a, a *Xenopus* winged helix gene, maintains an immature neural ectoderm via transcriptional repression that is dependent on the C-terminal domain. *Dev. Biol.* 232: 439–457.

Suzuki A., Ueno N., and Hemmati-Brivanlou A. 1997a. *Xenopus* msx1 mediates epidermal induction and neural inhibition by BMP4. *Development* 124: 3037–3044.

Suzuki A., Chang C., Yingling J.M., Wang X.F., and Hemmati-Brivanlou A. 1997b. Smad5 induces ventral fates in *Xenopus* embryo. *Dev. Biol.* 184: 402–405.

Tam P.P. and Steiner K.A. 1999. Anterior patterning by synergistic activity of the early gastrula organizer and the anterior germ layer tissues of the mouse embryo. *Development* 126: 5171–5179.

Thomas P. and Beddington R. 1996. Anterior primitive endoderm may be responsible for patterning the anterior neural plate in the mouse embryo. *Curr. Biol.* 6: 1487–1496.

Toivonen S. 1938. Spezifische Induktionsleistungen von abnormen Induktoren im Implantatversuch. *Ann. Soc. Zool.-Bot. Fenn. Vanamo* 6: 1–12.

———. 1940. Über die Leistungsspezifität der abnormen Induktoren im Implantatversuch bei Triton. *Ann. Acad. Sci. Fenn. Ser. A IV Biol.* 55: 1–150.

Toivonen S., Tarin D., Saxen L., Tarin P.J., and Wartiovaara J. 1975. Transfilter studies on neural induction in the newt. *Differentiation* 4: 1–7.

Tonegawa A. and Takahashi Y. 1998. Somitogenesis controlled by Noggin. *Dev. Biol.* 202: 172–182.

Tonegawa A., Funayama N., Ueno N., and Takahashi Y. 1997. Mesodermal subdivision along the mediolateral axis in chicken controlled by different concentrations of BMP-4. *Development* 124: 1975–1984.

Tsuda H., Sasai N., Matsuo-Takasaki M., Sakuragi M., Murakami Y., and Sasai Y. 2002. Dorsalization of the neural tube by *Xenopus* tiarin, a novel patterning factor secreted by the flanking nonneural head ectoderm. *Neuron* 33: 515–528.

Uchikawa M., Ishida Y., Takemoto T., Kamachi Y., and Kondoh H. 2003. Functional analysis of chicken Sox2 enhancers highlights an array of diverse regulatory elements that are conserved in mammals. *Dev. Cell* 4: 509–519.

Umbhauer M., Penzo-Mendez A., Clavilier L., Boucaut J., and Riou J. 2000. Signaling specificities of fibroblast growth factor receptors in early *Xenopus* embryo. *J. Cell Sci.* 113: 2865–2875.

Verschueren K., Remacle J.E., Collart C., Kraft H., Baker B.S., Tylzanowski P., Nelles L., Wuytens G., Su M. T., Bodmer R., Smith J.C., and Huylebroeck D. 1999. SIP1, a novel zinc finger/homeodomain repressor, interacts with Smad proteins and binds to 5′-CACCT sequences in candidate target genes. *J. Biol. Chem.* 274: 20489–20498.

von Baer K.E. 1828. *Über Entwickelungsgeschichte der Thiere. Beobachtung und Reflexion.* Bornträger, Königsberg.

Waddington C.H. 1930. Developmental mechanics of chick and duck embryos. *Nature* 125: 924–925.

———. 1932. Experiments on the development of chick and duck embryos cultivated in vitro. *Philos. Trans. R. Soc. Lond. B* 221: 179–230.

———. 1933a. Induction by coagulated organisers in the chick embryo. *Nature* 131: 275.

———. 1933b. Induction by the primitive streak and its derivatives in the chick. *J. Exp. Biol.* 10: 38–48.

———. 1934. Experiments on embryonic induction. *J. Exp. Biol.* 11: 211–227.

———. 1936. Organizers in mammalian development. *Nature* 138: 125.

———. 1937. Experiments on determination in the rabbit embryo. *Arch. Biol.* 48: 273–290.

———. 1940. *Organizers and genes.* Cambridge University Press, London, United Kingdom.

Wagner D.S. and Mullins M.C. 2002. Modulation of BMP activity in dorsal–ventral pattern formation by the chordin and ogon antagonists. *Dev. Biol.* 245: 109–123.

Wessely O., Agius E., Oelgeschlager M., Pera E.M., and De Robertis E.M. 2001. Neural induction in the absence of mesoderm: beta-catenin-dependent expression of secreted BMP antagonists at the blastula stage in *Xenopus*. *Dev. Biol.* 234: 161–173.

Wilson P.A. and Hemmati-Brivanlou A. 1995. Induction of epidermis and inhibition of neural fate by Bmp-4. *Nature* 376: 331–333.

Wilson P.A., Lagna G., Suzuki A., and Hemmati-Brivanlou A. 1997. Concentration-dependent patterning of the *Xenopus* ectoderm by BMP4 and its signal transducer Smad1. *Development* 124: 3177–3184.

Wilson S.I. and Edlund T. 2001. Neural induction: Toward a unifying mechanism. *Nat. Neurosci.* 4: 1161–1168.

Wilson S.I., Graziano E., Harland R., Jessell T.M., and Edlund T. 2000. An early requirement for FGF signalling in the acquisition of neural cell fate in the chick embryo. *Curr. Biol.* 10: 421–429.

Wilson S.I., Rydstrom A., Trimborn T., Willert K., Nusse R., Jessell T.M., and Edlund T. 2001. The status of Wnt signalling regulates neural and epidermal fates in the chick embryo. *Nature* 411: 325–330.

Winnier G., Blessing M., Labosky P.A., and Hogan B.M. 1995. Bone morphogenetic protein-4 is required for mesoderm formation and patterning in the mouse. *Genes Dev.* 9: 2105–2116.

Witta S.E., Agarwal V.R., and Sato S.M. 1995. XIPOU 2, a noggin-inducible gene, has direct neuralizing activity. *Development* 121: 721–730.

Xu R.H., Kim J., Taira M., Zhan S., Sredni D., and Kung H.F. 1995. A dominant negative bone morphogenetic protein 4 receptor causes neuralization in *Xenopus* ectoderm. *Biochem. Biophys. Res. Commun.* 212: 212–219.

Xu R.H., Kim J., Taira M., Lin J.J., Zhang C.H., Sredni D., Evans T., and Kung H.F. 1997. Differential regulation of neurogenesis by the two *Xenopus* GATA-1 genes. *Mol. Cell. Biol.* 17: 436–443.

Yabe S., Tanegashima K., Haramoto Y., Takahashi S., Fujii T., Kozuma S., Taketani Y., and Asashima M. 2003a. FRL-1, a member of the EGF-CFC family, is essential for neural differentiation in *Xenopus* early development. *Development* 130: 2071–2081.

Yabe T., Shimizu T., Muraoka O., Bae Y.K., Hirata T., Nojima H., Kawakami A., Hirano T., and Hibi M. 2003b. Ogon/Secreted Frizzled functions as a negative feedback regulator of Bmp signaling. *Development* 130: 2705–2716.

Zhang H. and Bradley A. 1996. Mice deficient for BMP2 are nonviable and have defects in amnion/chorion and cardiac development. *Development* 122: 2977–2986.

Zimmerman L.B., De Jesus-Escobar J.M., and Harland R.M. 1996. The Spemann organizer signal noggin binds and inactivates bone morphogenetic protein 4. *Cell* 86: 599–606.

ENDODERM DEVELOPMENT

A. Grapin-Botton and D. Constam

ISREC, CH-1066 Epalinges s/Lausanne, Switzerland

INTRODUCTION

Endoderm differentiation and movements are of fundamental importance, not only for subsequent morphogenesis of the digestive tract and other internal organs involved in nutrient, gas, and waste exchange, but also to enable normal patterning of the adjacent germ layers. The molecular signaling pathways that specify and pattern the endoderm during gastrulation have not been investigated until relatively recently. The limited available data have been summarized in several excellent reviews (Alexander and Stainier 1999; Shivdasani 2002; Stainier 2002) and are also covered to some extent elsewhere in this book. In this chapter, we ask how signaling pathways are integrated and lead to differentiation of endoderm lineages and their characteristic gastrulation movements in different species.

The endoderm is classically defined as the inner germ layer of diplo- and triploblastic animals. Although it is internalized only during gastrulation, its differentiation starts much earlier in many species. Its main derivative is the epithelial lining of the gut, although in all species a portion of the gut epithelium is of ectodermal origin. The definition of endoderm stands quite clearly as far as invertebrates are concerned. In yolk-rich eggs of vertebrates, the vegetal hemisphere also contains voluminous nutritive granules that are internalized during gastrulation in amphibians but remain extraembryonic and uncleaved in birds and fish. The so-called frog endoderm includes these yolk-rich cells. In amniotes, several endoderm-specific genes are expressed before gastrulation in extraembryonic lineages, which transiently assume the position and nutritive function of the prospective inner layer. Even though they eventually do not contribute to embryonic structures, these extraembryonic tissues have also been termed "endoderm," which leads to some confusion (Fig. 1). Their

ontogeny, function, and evolution have been extremely well described by Arendt and Nübler-Jung (1999; see also Chapters 15, 16, 31, and 51).

In most species, with the notable exception of sea urchin, ectoderm precursors become segregated early from progenitors that give rise to endoderm and mesoderm. Nieuwkoop's experiments associating vegetal and animal hemispheres indicated that in frogs mesoderm is induced in the animal hemisphere by vegetal cells (Nieuwkoop 1997). More recent observations in *Caenorhabditis elegans*, sea urchin, and zebrafish showed that mesoderm and endoderm derive from bipotential progenitors. There are several arguments for such a population in other vertebrates: lineage analysis (Nieuwkoop 1997), coexpression of endoderm and mesoderm markers (Rodaway et al. 1999), and the observation that certain signaling cascades induce both types of cells (Lemaire et al. 1998; Reiter et al. 1999; Rodaway and Patient 2001).

ENDODERM ORIGIN

Origin of Endoderm Prior to Gastrulation

Fate-mapping experiments that located endoderm precursors prior to or during gastrulation are summarized in Figure 2. In sea urchin, the endoderm derives from 16 cells at the 6th cleavage stage. The most ventral aspect of the embryo consists of micromeres that give rise to adult mesoderm and skeletogenic cells (Chapter 9). A ring of 8 cells, the veg2 cells, is located above the micromeres and contributes to endoderm and mesoderm. The 8 veg1 cells of the tier above contribute to endoderm and ectoderm (Khaner and Wilt 1991; Logan and McClay 1997). The segregation of veg1 daughters into ectoderm or endoderm occurs relatively late during development and is unpre-

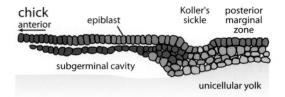

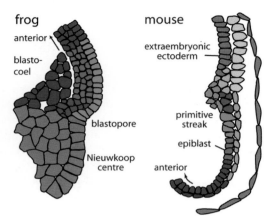

● definitive endoderm (c, f; m?)
● prechordal plate (c, f; m?)
● chordal mesoderm (c, f; m)
● hypoblast (c), deep endoderm (f),
  anterior visceral endoderm (m)
● endoblast (c), posterior visceral endoderm (m)
● extraembryonic visceral endoderm (m)
● yolk (c, f), parietal endoderm (m)
● extraembryonic mesoderm (c?, m)
● neurectoderm

Figure 1. Comparison of the multilayered structures surrounding the organizer region at the onset of gastrulation in chick, frog, and mouse embryos. Extraembryonic endoderm lineages and the fate of epiblast cells are color-coded. For the sake of simplification, the territories of prospective endoderm, mesoderm, and neurectoderm are separated by sharp boundaries in this diagram. In birds, one of the extraembryonic endoderm lineages is derived from islands of polyingressing epiblast cells that assemble to form the pregastrulation-stage hypoblast layer. The other, termed endoblast (previously known as secondary hypoblast), arises from deep cells in the posterior marginal zone, which at the onset of gastrulation start moving in an antero-lateral direction behind the hypoblast (Eyal-Giladi et al. 1992; Bachvarova et al. 1998). In the mouse, posterior visceral endoderm (VE) is speculated to be the equivalent of the chick endoblast due to its analogous position. Likewise in the proximal epiblast, prechordal plate progenitor cells are predicted to reside between definitive endoderm precursors and posterior VE, in analogy with chick and frog, even though the limited resolution of current fate maps cannot distinguish two separate populations in the mouse epiblast. In mammals, primitive endoderm (PrE) segregates from the inner cell mass (ICM) at the blastocyst stage as a squamous epithelium. Whereas some PrE cells remain attached to the basement membrane of the ICM and differentiate into cuboidal VE, others undergo an epithelial–mesenchymal transition to become parietal endoderm (PE). PE cells migrate along the basement membrane of trophectoderm (TE) cells, which gives rise to Reichert's membrane of the parietal yolk sac. Until placentation, PE and VE lineages together are responsible for nutrient and waste exchange between maternal tissue and the fetus, a function which in lower organisms is partly fulfilled by the yolk. Note that only parts of the embryos are drawn, with the anterior pole facing to the left, and that the situation depicted in mouse corresponds to a slightly more advanced stage (early primitive streak). (Adapted from Arendt and Nübler-Jung 1999.)

dictable, indicating that cell position is more important than the early cleavage pattern in determining ectodermal and archenteron cell fates (Logan and McClay 1997).

In *C. elegans*, the 20 cells that constitute the gut come from the E-blastomere at the 8-cell stage (Sulston 1983; Leung et al. 1999) (Chapter 4). This blastomere generates endoderm exclusively and is itself derived from EMS, a 4-cell-stage blastomere that contributes to both endoderm and mesoderm. EMS and E initially reside on the surface of the embryo, but eventually become internalized prior to mesoderm at the 28-cell stage.

In *Drosophila*, the endoderm forms the so-called midgut epithelium and develops from anterior and posterior tips of the embryo (Technau and Campos Ortega 1985; Reuter and Leptin 1994) (Chapter 7). Prospective anterior midgut territories are located ventral and anterior to mesoderm precursors prior to invagination in the ventral furrow. The stomodeum, which is of ectodermal origin, is located more anteriorly and invaginates separately. The proctodeum (also of ectodermal origin) invaginates together with the posterior endoderm that contributes to the posterior midgut.

In zebrafish, endoderm and mesoderm cells originate near the blastoderm margin (Kimmel et al. 1990) (Chapter 12). Endoderm progenitors are mostly less than 2, but maximally 4, cell diameters away from the yolk. Although most cells give rise to clones in both layers at the 1000- to 4000-cell-stage, clones that give rise to endoderm only are more abundant in dorsal and lateral margins (Warga and Nusslein-Volhard 1999). Early-involuting cells predominantly form endoderm, and later-involuting cells form more mesoderm (Warga and Kimmel 1990).

In amphibians, the endoderm mainly derives from the vegetal blastomeres (Vogt 1929; Keller 1975, 1976) (Chapter 19). At the equivalent stage in the ascidian *Ciona* (Urochordate), the four vegetal blastomeres also give rise exclusively to endoderm (Nishida and Satoh 1983; Nishida 1987). In the 32-cell blastula of *Xenopus*, most endoderm cells come from the four vegetal blastomeres, but there is also some contribution from the row above (Dale and Slack 1987; Moody 1987). Cells that give rise to the archenteron roof later (dorsal endoderm) originate from superficial cells that invaginate at the suprablastoporal lip, whereas the archenteron floor (ventral endoderm) is derived from more vegetal superficial cells and from the cone of deep cells extending into the blastocoel. These cells invaginate at the sub-blastoporal lip (Keller 1975, 1976).

Fate-mapping experiments in chick and mouse led to similar conclusions (Chapters 15 and 16). The definitive endoderm progenitors are located in the epiblast prior to gastrulation (Rudnick 1932; Sandstrom 1934; Rawles 1936; Hunt 1937a,b; Rosenquist 1971). At pre-streak and early-streak stages, endoderm precursors are located in the posterior half of the epiblast in a wing-shaped territory hinged at the posterior midline (Eyal-Giladi et al. 1992; Hatada and Stern 1994). The cells that form the early primitive streak originate from a small epiblast region extending anteriorly 200–300 μm from Koller's sickle, which is located immediately rostral to the posterior marginal zone (Fig. 2) (Callebaut et al. 1996; Bachvarova et al. 1998; Wei and Mikawa 2000). The cells in Koller's sickle are organized: Superficial prospective endoderm overlies deeper prospective prechordal plate (Fig. 1) (Bachvarova et al. 1998). By mid-late-streak stage, endoderm precursors become restricted to the anterior streak and node. By the time the streak has reached its maximal extension, there appear to be no more endodermal progenitors either in the epiblast, streak, or node (Fig.2) (Psychoyos and Stern 1996).

Specification of Endodermal Fate before or during Gastrulation

In *C. elegans*, endoderm cells are specified prior to gastrulation. EMS and E are induced through cell–cell interactions, but the E blastomere when separated from other blastomeres makes endoderm in an autonomous fashion (Fig. 2) (Laufer et al. 1980; Priess and Thomson 1987; Schroeder and McGhee 1998; Leung et al. 1999) (Chapter 4). Laser ablations that expose the E blastomere to contact with other cells further suggest that E is already determined at the 4-cell stage. However, in the related nematode *Acrobeloides nanus*, all four blastomeres make endoderm when isolated. Thus, cell interactions repress endodermal fate (Wiegner and Schierenberg 1998). Similarly, when chick epiblast is isolated at stage XII and cultured, it becomes gut irrespective of its original location (Butler 1935). In normal development, only the posterior third normally becomes gut (Hatada and Stern 1994). Thus, at least in these two species, cell interactions must exist that prevent endoderm differentiation.

As in *C. elegans*, endodermal fate in sea urchin is also specified prior to gastrulation. veg2-cell specification depends on maternal factors and an unknown signal from the micromeres at fourth to sixth cleavage (Chapters 9 and 48). Similarly, in *Xenopus*, ectopic transplantation of individual blastomeres revealed that vegetal cells are already determined to become endoderm by the beginning of gas-

trulation (Heasman et al. 1985) (Chapter 19). Endodermal cells in zebrafish are determined by early (50% epiboly) to mid-gastrulation (Ho and Kimmel 1993; David and Rosa 2001) (Chapter 12). Since the cells are not internalized at the same time, it is possible that endoderm commitment occurs prior to internalization of individual cells. Timing in birds and mammals has not yet been determined.

MOLECULAR CONTROL OF ENDODERM DIFFERENTIATION

A comprehensive analysis of the regulatory gene network responsible for endoderm differentiation has recently been carried out in sea urchin (Davidson et al. 2002a,b) (Chapter 48). On the basis of these data, the inputs of intrinsic maternal determinants and secreted proteins from other blastomeres appear to be integrated into a complex cell-autonomous transcriptional network. Although this work is the most comprehensive so far about endoderm specification, data from other model organisms suggest that some components in this network have not yet been found, or function in a species-specific manner. Figure 3 highlights the most conserved aspects of the network.

Maternal Signals

There are three families of maternal signals contributing to the initiation of endoderm specification: β-catenin (Catnb), VegT, and Otx.

In sea urchin, initial activation of mesendoderm is triggered by an unknown signal from the micromeres to the parents and grandparents of veg2 cells as well as autonomous nuclearization of maternal β-catenin in veg2 cells (Davidson et al. 2002a,b). Although β-catenin is clearly involved in endoderm development in many other species also, its maternal contribution to cell-fate determination has only been documented in *Xenopus* (Zorn et al. 1999; Schohl and Fagotto 2002; Hashimoto-Partyka et al. 2003) and *Ciona* (Imai et al. 2000). Before mid-blastula transition, however, nuclear uptake of β-catenin is undetectable both in *Xenopus* and in the yolk syncytial layer of zebrafish blastulae, and thus may depend on zygotic signals (Schneider et al. 1996; Wylie et al. 1996; Jesuthasan and Strähle 1997) (see also Chapter 36).

In sea urchin, Otx is provided both as a maternal and a zygotic protein (Li et al. 1997; Davidson et al. 2002b). Zygotic Otx is induced by β-catenin and is responsible for shutting down further Wnt signaling. In mouse *Otx2* mutants, the Wnt3 domain is expanded, suggesting that the feedback may be conserved (Perea-Gomez et al. 2001). An *Otx* homolog in *Ciona* is also downstream of β-catenin, but its function does not appear to affect endoderm develop-

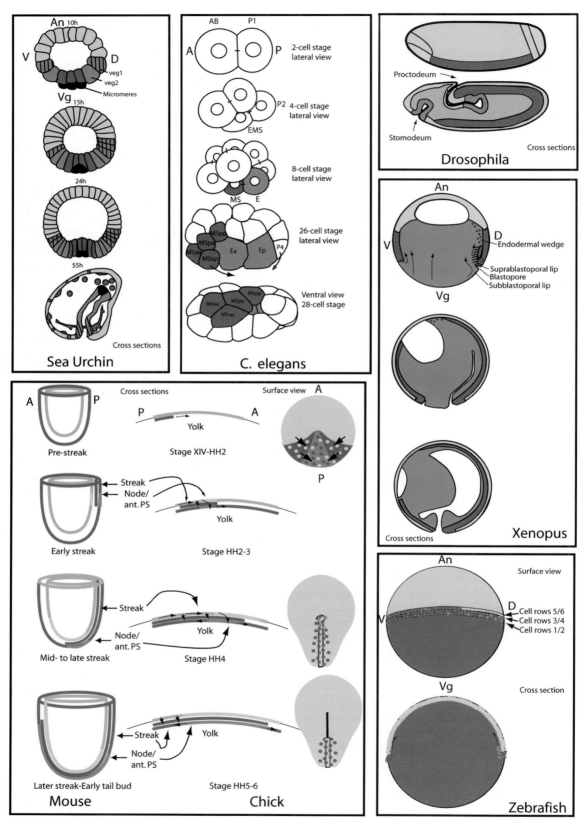

Figure 2. (See *facing page for legend.*)

ment (Satou et al. 2001). In vertebrates, *Otx2* is expressed in the anterior endoderm, but a maternal function has not been demonstrated (Chapter 39).

In *Xenopus*, the localization of the maternal RNA encoding the T-box transcription factor VegT to the vegetal pole is essential for the correct spatial organization and identity of endoderm and mesoderm (Horb and Thomsen 1997; Zhang et al. 1998). VegT synergizes with β-catenin to activate the Nodal pathway (Yasuo and Lemaire 1999). Several candidate *VegT* homologs have been identified in zebrafish, chick, and mouse, but they do not seem to be involved in endoderm development. *Eomesodermin* (*Eomes*), although involved in mesoderm development in *Xenopus*, may be the T-box gene instrumental in endoderm induction in mouse, zebrafish, and *Amphioxus* (Horton and Gibson-Brown 2002). It is maternal in the latter two. In mouse, genetic inactivation of *Eomes* results in embryonic failure prior to gastrulation (Russ et al. 2000), and expression of a mouse homolog of the VegT target gene *Mix* is abolished specifically in *Eomes* mutants (see also Chapter 41).

Signaling between Blastomeres

Wnts

In sea urchin, veg2 cells secrete the wingless-related Wnt8 protein, which not only amplifies the Wnt pathway already activated maternally, but also induces veg1 cells to become endoderm. Target genes of Wnt8 in veg1 cells include *Eve* and a *LIM* family gene (Ransick and Davidson 1998). A *LIM* family gene required for endoderm induction is also a Wnt target in *Ciona* (Satou et al. 2001). The Wnt loop progressively increases β-catenin levels above the thresholds needed to activate late endodermal Tcf target genes (Davidson et al. 2002b). The Wnt pathway is not specific for the endoderm lineage but instead induces bipotential mesendoderm progenitors as seen in sea urchin (Davidson et al. 2002b).

In *C. elegans*, Wnt signaling is best characterized in the segregation of EMS daughter cells into endoderm (E) and mesoderm (MS) (see Chapter 4). However, from their function in sea urchin and mouse, one would expect Wnts to be involved in mesendoderm progenitor induction (EMS). SKN-1, a maternal transcription factor responsible for EMS induction, is prevented from inducing mesendoderm in other blastomeres expressing by GSK3 kinase, a critical component in the Wnt pathway, indicating that a Wnt sig-

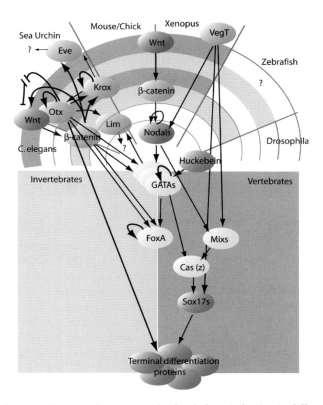

Figure 3. Gene-regulatory network of endoderm induction in different species. The upper half shows divergence in the initiation of endoderm differentiation whereas in the bottom, more downstream elements are more conserved. Final differentiation targets are numerous and variable and are thus not specified. Concentric circles of similar shading show the conservation of a protein function in the species represented. Arrows represent induction and bar-ended lines represent inhibition. Color codes: (*red*) secreted, (*orange*) transcription factors, (*yellow*) highly conserved downstream transcription factors, (*green*) endoderm-specific.

nal may be involved in mesendoderm specification (Maduro et al. 2001).

Nodal, a Vertebrate Signal

Insights into the inductive mechanisms underlying endoderm formation in vertebrates initially came from studies in *Xenopus* using a dominant-negative activin receptor that blocks secreted TGF-β-related activities, including those of activin, Vg1, and *Xenopus* Nodal-related proteins (Xnrs). Vegetal pole explants of embryos injected with this con-

Figure 2. Position of endoderm precursors before and during gastrulation. Endoderm precursors are green, mesoderm is red, and ectoderm is yellow. In mouse and chick diagrams, brown marks visceral endoderm and gray marks epiblast. Circles with a green center and red periphery mark bipotential mesendodermal progenitors in zebrafish. Arrows refer to cell movements. (PS) Primitive streak. (Adapted from Keller 1975, 1976; Schoenwolf and Alvarez 1991; Garcia-Martinez et al. 1993; Reuter and Leptin 1994; Warga and Nusslein-Volhard 1999; Lopez-Sanchez et al. 2001; Davidson et al. 2002b; Nance and Priess 2002.)

struct express mesodermal and ectodermal marker genes at the expense of the pancreatic endoderm marker *Xlhbox8* (Henry et al. 1996). On the basis of this observation, it was proposed that the choice between endoderm and mesodermal cell fates is controlled by a graded TGF-β signal mediated by Vg1 and/or Xnrs. Several lines of evidence suggest that the same is likely to be true in other vertebrates also. In light of these findings, it is critical to understand how peak levels of Nodal signaling are generated and how they can be distinguished by cells fated to form endoderm (see Chapter 34).

Nodal Expression Is Induced by the Canonical Wnt Pathway and Positive Feedback Signaling

Among the signals that activate *Nodal* expression is the Wnt pathway. Mouse embryos lacking Nodal or β-catenin fail to form a primitive streak (Fig. 4) (Conlon et al. 1994; Huelsken et al. 2000), suggesting that the canonical Wnt pathway and Nodal are epistatic, or act in parallel to specify definitive endoderm. Furthermore, embryos lacking Wnt3 fail to maintain expression of *Nodal* (Liu et al. 1999). Thus, the canonical Wnt pathway may promote endoderm formation in mammals primarily via its effect on a graded Nodal signal by locally stimulating Nodal feedback signaling. Residual Nodal signaling in *Wnt3* and β-*catenin* mutants indicates, however, that additional signals exist which induce *Nodal* expression. In addition to Wnts, T-box transcription factors as well as FoxH1 and Cripto are required to generate peak levels of Nodal signaling (see Chapter 34).

Similar to other TGF-β-related proteins, Nodal is activated posttranslationally upon in vitro cleavage by soluble forms of several subtilisin-like proprotein convertases. Of these, Spc1 and Spc4 function redundantly to activate Nodal signaling in the mouse embryo. They are both expressed in extraembryonic ectoderm adjacent to cells secreting Nodal. Localized Nodal cleavage near the source of Spc activities has thus been proposed to establish a graded signal that patterns the embryo even prior to primitive streak formation (Beck et al. 2002).

Network of Transcription Factors

A Conserved Core Involving GATA and Forkhead Transcription Factors

GATA factors. Forkhead transcription factors of the FoxA family and GATA factors are key components of the network in all triploblasts studied so far. Several family members are expressed in endomesoderm, endoderm, or mesoderm. Their inactivation prevents endoderm formation in several species, either completely as in *Drosophila* (Reuter 1994; Rehorn et al. 1996), or partially when several GATAs

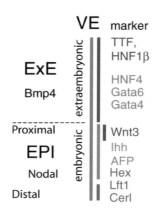

Figure 4. Patterning of the murine VE along the proximal-distal axis anticipates A-P polarity within the embryo. Signals from the adjacent embryonic and extraembryonic ectoderm including Nodal and possibly BMP4 specify distinct VE populations according to their relative positions. In return, specific gene products in the VE such as Lefty1 (Lft1) and Cerberus-like (Cerl) signal back to the embryo to position the primitive streak (see text for details).

of redundant function are present as in zebrafish or mouse (Soudais et al. 1995; Morrisey et al. 1998; Koutsourakis et al. 1999; Reiter et al. 1999; Jacobsen et al. 2002). *Gata* genes often function in chains in which they autoactivate and activate each other. In *Drosophila*, a second GATA, dGATAc, acts downstream of Serpent (Lin et al. 1995). *Gata6* knockout mice have reduced expression of *Gata4* and endoderm differentiation markers (Morrisey et al. 1998). The largest number of GATA genes is found in *C. elegans*, which contains seven GATAs that are sequentially activated (Patient and McGhee 2002). In *C. elegans* and *Drosophila*, some GATAs have specialized in regulating endoderm differentiation, whereas in vertebrates they control either mesendoderm or mesoderm development. They are not always sufficient to induce endoderm. Although in *Xenopus*, *Gata4* and *Gata5* are expressed in endoderm and convert ectomesoderm into endoderm (Jiang and Evans 1996; Gao et al. 1998; Weber et al. 2000), zebrafish Fau/Gata5 induces later endoderm markers only in the presence of the Sox family member *cas* (see below) (Reiter et al. 2001). Their inducers are quite variable. In sea urchin, they are induced by the Wnt pathway, Otx, and the Notch pathway via an unknown *Hes* gene (Nocente-McGrath et al. 1989; Ransick and Davidson 1993; Harada et al. 1996; Yuh et al. 1998). In *Drosophila*, Huckebein, a Zn finger protein, acts upstream of Serpent (Bronner et al. 1994). In *Xenopus* and in zebrafish, Gata5/fau acts downstream of Nodal (Alexander and Stainier 1999; Rodaway et al. 1999; Weber et al. 2000; Reiter et al. 2001). They control a wide panel of downstream targets, including genes carrying adult endodermal functions (Maeda et al. 1996; Bossard and Zaret 1998; Gao et al. 1998; Morrisey et al. 1998). Thus, in sea urchin, GataE activates six endomesoderm regulatory genes and feeds back into Otx2, which locks the system (Yuh et al. 1998; Davidson et al. 2002b). In many systems, members of the GATA family interact with members of the NKX (NK2 transcription factor related) family of transcription factors. In *Ciona*, *Ci-ttf1* is an *NK2* gene expressed in endoderm downstream of β-catenin, which upon misexpression, induces endoderm (Ristoratore et al. 1999; Satou et al. 2001; Spagnuolo and Di Lauro 2002). In addition to their role in differentiation, Gata4 and Serpent play a role in gut folding (Narita et al. 1997).

Forkhead factors. *Forkhead* genes of the *FoxA* class are also expressed in sea urchin endoderm (Harada et al. 1996), *C. elegans* (Horner et al. 1998; Kalb et al. 1998), *Drosophila* (Weigel et al. 1989), zebrafish (Strähle et al. 1993; Schier et al. 1997), *Xenopus*, mouse, and chick (Ang et al. 1993; Monaghan et al. 1993; Sasaki and Hogan 1993; Odenthal and Nusslein-Volhard 1998; Kaestner et al. 2000) (see also Chapter 40). Their inactivation perturbs, but does not abolish, endoderm development. In *Drosophila*, posterior endoderm invagination begins normally in their absence, but the mid- and hindgut decay soon and the anterior gut fails to invaginate altogether. Interestingly, *forkhead* genes are often expressed in a subpopulation of endodermal cells, and their inactivation often inhibits the development of parts of the gut. In sea urchin and zebrafish, *FoxA* expression marks the fore- and midgut (Alexander and Stainier 1999). In mouse, *FoxA2*, which is expressed at the onset of gastrulation, is required for fore- and midgut formation (Ang et al. 1993; Sasaki and Hogan 1993; Weinstein et al. 1994; Dufort et al. 1998). In *C. elegans*, *PHA-4/FoxA2* is required for pharynx and rectum formation but not in the endodermal midgut where its expression is lower (Mango et al. 1994; Azzaria et al. 1996; Horner et al. 1998).

FoxA targets have been studied comprehensively in *C. elegans* (Gaudet and Mango 2002). Beyond compiling a list of targets, Mango and co-authors show that the late targets have lower-affinity binding sites. Several studies have proposed that GATAs and HNFs together form a preinitiation complex which is required but not sufficient for endoderm gene transcription (Bossard and Zaret 1998; Cirillo et al. 2002).

In zebrafish, *gata5/fau* is expressed before *foxA2*, suggesting that it is upstream, as described in sea urchin. The promoter of *PHA-4/foxA2* is bound and *trans*-activated by two GATAs (Mango et al. 1994; Azzaria et al. 1996; Horner et al. 1998; Kalb et al. 1998). In *Drosophila*, Serpent regulates Forkhead in a dual fashion: It down-regulates it in the midgut and activates it in the yolk (Weigel et al. 1989; Casanova 1990). FoxA is also regulated by Otx and Tcf in sea urchin (Davidson et al. 2002b). In *Ciona*, *FoxA2* expression in endoderm depends on a 44-bp sequence that contains overlapping T-box and Snail-binding sites. Putative snail-binding sites prevent expression in lateral mesoderm regions. Mutations in the T-box site prevent expression in endoderm but not notochord (Di Gregorio et al. 2001). Ci-VegT is a likely candidate to bind this element. Furthermore, the presence of Smad2-binding elements in the *Xenopus Foxa2* promoter raises the possibility that it is a direct Nodal target (Howell and Hill 1997). Autoregulatory loops have been demonstrated in different species. In sea urchin, FoxA represses its own expression, which results in an oscillation due to counteracting positive inputs (Davidson et al. 2002b). *Ciona* features a positive autoregulation that appears to be direct (Di Gregorio et al. 2001).

Sox and Mix, a Vertebrate Invention?

In vertebrates, other key components of the network include *Sox17*, *Mix*, and several related genes, the homologs of which cannot be unambiguously distinguished in other organisms. Sox17 is a high mobility group transcription factor of SoxF class that was first implicated in endoderm development in *Xenopus* (Hudson et al. 1997). Overexpression, together with *Mixer*, induces endoderm in animal caps, and Sox17-ENR prevents endoderm formation in embryo or vegetal pole explants. In zebrafish, *sox17* is also expressed in endodermal progenitors before gastrulation. Another protein of the same family encoded by *casanova* (*cas*) is required upstream of Sox17 (Alexander and Stainier 1999; Alexander et al. 1999; Dickmeis et al. 2001; Kikuchi et al. 2001). Cas is required for the Mix family protein Bonnie-and-Clyde (Bon/Mixer) or Fau/Gata5 to induce Sox17 (Alexander and Stainier 1999; Reiter et al. 1999, 2001). Cas may control the switch between endoderm and mesoderm. *cas* is initially expressed in the yolk syncytial layer (YSL) in a Nodal-independent manner, but later is induced in endoderm in response to Nodal (Sakaguchi et al. 2001; Aoki et al. 2002a,b). When overexpressed, it converts mesoderm into endoderm, whereas in its absence, endoderm forms at the expense of mesoderm (Dickmeis et al. 2001). In ectoderm, it only induces *sox17* without any further endoderm differentiation (Kikuchi et al. 2001). In mouse, *Sox17* is first expressed in visceral endoderm nearest to the ectoplacental cone at 6.0 days post coitum (dpc) and progressively spreads to the entire extraembryonic VE. It is also expressed in definitive endoderm until 8.5 dpc (Kanai-Azuma et al. 2002). In *sox17* knockout mice, definitive endoderm is replaced by visceral endoderm-like tissue in the most posterior and lateral regions (Kanai-Azuma et al. 2002). In chimeras, *Sox17* mutant cells can contribute to some extent to the foregut but not mid- and hindgut (like *Foxh1* and *smad2*). On the contrary, *Foxa2* knockout cells can form hindgut but not fore- and midgut (Dufort et al. 1998). Elevated levels of apoptosis in the foregut later lead to foregut reduction, suggesting that Sox17 is also a maintenance factor for endoderm (Kanai-Azuma et al. 2002). Sox7 and -17 may be redundant in extraembryonic visceral endoderm. There is no correlation between the expression patterns of *Drosophila* and vertebrate *SoxF* (Cremazy et al. 2001), and there is no gene of the *SoxF* family in *C. elegans* (Bowles et al. 2000).

The *Mix* family encodes homeodomain proteins initially described in *Xenopus*. *Mixer* is predominantly expressed at the endoderm/mesoderm boundary, an expression pattern similar to that of *Mix1* and *Bix2/Milk* (Ecochard et al. 1998; Henry and Melton 1998; Lemaire et al. 1998). *Bix1–4* are expressed in endoderm and mesoderm throughout gastrulation (Tada et al. 1998; Casey et al. 1999). Members of this gene family can induce either endoderm (*Bix2/Milk, Mix3/Mixer*, and *Bix4*) or mesoderm. Mixer appears to induce endoderm specifically (Henry and Melton 1998). At high levels, Mix1 and Bix1/Mix4 induce endoderm and can repress mesodermal genes like *Xbra*, whereas at low levels, they induce mesoderm (Henry and Melton 1998; Lemaire et al. 1998; Tada et al. 1998; Latinkic and Smith 1999). Only one *Mix* gene has been found in chick, mouse, and human (Peale et al. 1998; Stein et al. 1998; Pearce and Evans 1999; Robb et al. 2000). *C-mix* is likely to be expressed in mesendoderm precursors, since it disappears from primitive streak and node when they lose the ability to generate endoderm (our unpublished observations). Mouse *Mixl1* is first detected in the visceral endoderm and later in nascent primitive streak but not node or definitive endoderm. *Mixl1* mutants have reduced endoderm (Hart et al. 2002). The anterior intestinal portal is rudimentary and no caudal intestinal portal ever forms. The mesendoderm marker *FoxA2* is maintained and *Sox17* and *Cer-l* remain expressed, showing that definitive endoderm forms, but the amount of definitive endoderm is reduced and mid-hindgut lies at the level of the foregut. However, visceral endoderm is displaced normally to the periphery. *Mixl1* mutant cells in chimeras contribute to all organs but the hindgut. In zebrafish, the gene whose mutation causes the bonnie and clyde (*bon*) phenotype is most similar to *Xenopus Mixer* (Kikuchi et al. 2000). *Mezzo* is another zebrafish gene of the family whose homeobox is most similar to *Mix1*, and its mutation is very similar to Bon. They are expressed in mesendoderm, are induced by Nodal, and are able to induce *cas* and *Sox17* (Alexander and Stainier 1999; Alexander et al. 1999). *Sox17* expression is abolished when both genes are rendered inactive (Poulain and Lepage 2002). *Nodal* expression is expanded in mouse *Mixl1* mutants, suggesting a feedback loop. Mezzo also induces mesodermal markers like *ntl* (Poulain and Lepage 2002). SMAD2/4 dimers bind the activin-responsive element of the *mix2* promoter (Howell et al. 1999). Mixer, Bix2/Milk, and Bix3 recruit SMAD2/4 to activin-responsive elements of mesendodermal genes such as *Gsc* (Germain et al. 2000).

CELL MOVEMENTS

In many embryos, such as those of sea urchins and amphibians, the early embryonic cleavages generate a cluster of cells called the blastula, and a central cavity called the blastocoel. Surface cells are internalized and enter the blastocoel either individually (ingression) or as a layer (invagination and involution). In other species, gastrulation occurs without formation of a blastocoel. In ctenophores, for example, ectodermal cells spread over and thus internalize endodermal cells (epiboly) (Martindale and Henry 1999).

In sea urchin, isolated vegetal plates invaginate autonomously (Ettensohn 1999; Wessel and Wikramanayake 1999) (Chapter 9). A few mesodermal cells surrounding the micromeres located at the center of the vegetal plate become bottle-shaped (Nakajima and Burke 1996; Kimberly and Hardin 1998). They constitute the moving force at the tip of the archenteron and are followed by endoderm cells with which they invaginate as a sheet of cells. There is no cell mixing, and adherens and septate junctions are maintained throughout the process (Burke et al. 1991). Cell rearrangements initiated after partial internalization of the archenteron lead to further dorsal extension of the gut tube (Ettensohn 1985). This requires the extension of filopodia that provide traction forces on ectoderm.

Gastrulation in embryos like *Drosophila* initially involves sheets of cells that invaginate to form furrows and pockets (Leptin 1999) (Chapter 7). Adherens junctions link the surface cells of embryos throughout gastrulation (Tepass and Hartenstein 1994; Oda et al. 1998). A few minutes after the ventral furrow that gives rise to mesoderm has begun to invaginate, cells of the posterior endoderm primordium also constrict at their apical ends and eventually invaginate, while at the same time the posterior end of the embryo is pushed dorsally by independent ectodermal cell movements (Fig. 2). The endoderm later disperses into individual cells. These cells then use the mesodermal cell layer as substratum for migration toward the middle of the embryo, where they meet up with the cells of the anterior endoderm to form the continuous endodermal cell layer that becomes the midgut epithelium (Hartenstein et al. 1992; Reuter et al. 1993; Tepass and Hartenstein 1994, 1995). Posterior endoderm gastrulates from *snail* negative areas, and the terminal gene *hkb* is responsible for *snail* repression (Reuter and Leptin 1994). Anteriorly, endoderm gastrulation requires *snail*, like the rest of the ventral furrow. Anterior Hkb is prevented from repressing *snail* by the presence of Bicoid (Reuter and Leptin 1994). Initial invagination of posterior endoderm and mesoderm requires the maternal heterotrimeric G subunit Concertina (Parks and Wieschaus 1991) and Rho-GTPase (Barrett et al. 1997; Hacker and Perrimon 1998).

In *C. elegans* embryos, there are no adherens junctions before the end of gastrulation (Costa et al. 1998) (Chapter

4). This may facilitate ingression by small groups of cells. Lee and Goldstein (2003) recently found that the process of endoderm cell gastrulation in *C. elegans* depends on microfilament and myosin but not microtubules. It does not require anchorage on the eggshell. Ingressing cells in *C. elegans* embryos show an apical flattening and an apical accumulation of non-muscle myosin (Fig. 2) (Nance and Priess 2002). The two E cells invaginate together. A likely possibility is that the midbody linkage between sister cells prevents either sister from ingressing alone (Nance and Priess 2002).

In zebrafish, although cells are internalized as a group, introduction of normal cells in an *MZoep* background that fails to form ectoderm shows that single cells can move autonomously. The converse experiment shows that *MZoep*[−/−] cells can be internalized following the movement of wild-type neighbors (Carmany-Rampey and Schier 2001).

In *Xenopus*, endoderm invagination at the time of gastrulation has been described thoroughly by Nieuwkoop (1997) and Keller (1975, 1976) (Chapter 13). The first author showed that the endodermal lining of the gut is invaginated as a continuous layer superficial to the mesoderm throughout gastrulation. Suprablastoporal and subblastoporal cells that contribute to dorsal and ventral aspects of the gut, respectively, invaginate at the same time. The rate of convergence and extension is higher in the suprablastoporal area that extends anteriorly in the blastocoel cavity forming the archenteron roof. This phenomenon requires Bra and one of its target genes *Wnt11* in zebrafish and *Xenopus* (Heisenberg et al. 2000; Tada and Smith 2000). Evidence from *Xenopus* and *Drosophila* suggests that Wnt11 may be directly linked to Rho and subsequently E-cadherin and the cytoskeleton (Burdsal et al. 1993; Boutros et al. 1998; Djiane et al. 2000; Takeichi et al. 2000). Recent experiments showed that movement of subblastoporal lip is not passive but actually initiates gastrulation by crawling on the blastocoel roof (Winklbauer and Schurfeld 1999; Ibrahim and Winklbauer 2001). Internalized cells are prevented from fusing to the adjacent layer (Wacker et al. 2000).

Avian and mammalian mesoderm and endoderm progenitor cells first move within the primitive ectoderm to form a streak-shaped anlage, the primitive streak, where they ingress to disperse between endoderm and epiblast or to incorporate into the preexisting primitive endodermal layer (Poelmann 1981; Lawson et al. 1986; Tam et al. 1993) (Chapters 15 and 16). Their movement is accompanied by epithelial–mesenchymal transition and requires *Snail* or the related gene *Slug* that repress E-cadherin (Chapter 47). Definitive endoderm cells in chick seeded on primitive endoderm/hypoblast reproduce their in vivo behavior and insert within this layer (Sanders et al. 1978). Definitive endoderm cells insert massively in the recipient layer, but there is a certain extent of intermingling in both mouse and chick (Sanders et al. 1978; Stern 1990). The cells then form

junctional complexes with primitive endoderm cells and reestablish epithelial continuity, eventually displacing the visceral endoderm to the extraembryonic region. The relative contribution of active cell migration, versus passive movement due to cell proliferation, is unknown (Poelmann 1980; Lawson et al. 1986; Tam and Beddington 1992).

In the mouse, it is thought that endoderm and mesoderm cells exiting early from the streak migrate together laterally and anteriorly away from the streak and, therefore, simultaneously ingressing cells remain juxtaposed later on (Stern 1979; Tam et al. 1993, 2001; Tam and Behringer 1997). Mesodermal cells juxtaposed to the endoderm are more packed and closely associated with basal lamina that they may use to migrate with the endodermal sheet (Spiegelman and Bennett 1974; Poelmann 1981; Tam and Meier 1982).

In zebrafish, cells fated to become endoderm during their movement are exposed to signals from the YSL. Likewise in chick, endoderm progenitors in Koller's sickle undergoing their characteristic "Polonaise movement" are thought to become specified as they receive signals from the posterior marginal zone that activate the Nodal signaling pathway, including Wnt ligands (Skromne and Stern 2001). At a later stage, mesoderm and endoderm in passing may receive instructive patterning signals from the node (Pagan-Westphal and Tabin 1998; Brennan et al. 2002). By sensing their position relative to a more or less stationary signaling center, moving cells thus might coordinate their differentiation. This does by no means imply that signaling centers are necessarily static in terms of their cellular composition (Joubin and Stern 1999).

COUPLING BETWEEN DIFFERENTIATION AND MOVEMENT

In principle, gastrulation could position a subset of unspecified cells inside the embryo, where subsequent events would specify the endodermal and mesodermal fates. Instead, *C. elegans* and many other animals appear to specify the fates of endodermal and mesodermal progenitors when these progenitors are still on the surface. Cell-fate specification may thus induce ingression movements. Indeed, in *mex-1* mutants, ectopic MS cells differentiate and ingress at the same time as endogenous MS cells (Mello et al. 1992). In a mosaic environment, subpopulations of *Xenopus* animal pole cells forced to express *Sox17* either migrate toward the endodermal layer or die, suggesting that if the cell does not move appropriately there is a feedback check on its differentiation status (Clements and Woodland 2000). In fish, Ho and Kimmel (1993) showed that endodermal cells are determined by mid-gastrulation. Cells transplanted ectopically may use atypical migration routes to get inserted in the proper layer. Likewise, single wild-type cells transplanted to the margin of *MZoep* mutant embryos autonomously internalize and can express the mesendodermal markers *axial/foxa2* and *sox17* (Carmany-Rampey and Schier 2001). In agreement, David and Rosa (2001) showed that single cells do not need to pass through the germ ring to become endoderm once they have activated the Nodal pathway. However, proper differentiation is not mandatory for ingression. *MZoep* mutant single-cell transplants in the context of a wild-type host internalize, but eventually fail to become mesendoderm (Carmany-Rampey and Schier 2001). The movement of surrounding cells is thus sufficient for ingression, but not absolutely required.

PATTERNING

Before Gastrulation: Prospective Endodermal Cells from Different Areas Are Assigned to Different Parts of the Gut

In *Drosophila*, anterior and posterior parts of the midgut originate from different primordia (Fig. 2). In sea urchin, veg2 cells contribute to the archenteron tip and veg1 cells to its base (Logan and McClay 1997) and accordingly gastrulate only at the mid to late gastrula stage (Fig. 2). However, veg1 and veg2 descendants give rise to overlapping subsets of endodermal cell types (Logan and McClay 1997).

In *C. elegans*, the E cell divides along the anterior–posterior (A-P) axis to give rise to Ea and Ep (Fig. 2) (Chapter 4). Ea eventually gives rise to anterior gut derivatives and Ep to the posterior gut. There is thus an A-P pattern prior to gastrulation. This pattern is intrinsic to the cells, since ablation of Ea or Ep ablates anterior or posterior gut, respectively (Schroeder and McGhee 1998). All but one division in the E lineage occur along the A-P axis. These divisions are asymmetric such that only the anterior cell exposed to lower levels of Wnt ligand inherits the Tcf-related protein POP-1 (Lin et al. 1998).

In a 1000- to 4000-cell zebrafish embryo, pharynx and esophagus cells are closer to the dorsal side of the embryo and involute first, whereas stomach and intestinal cells are more lateral and involute later, but there is generally a great overlap between domains (Warga and Nusslein-Volhard 1999) (Chapter 12).

Lineage-tracing experiments in *Xenopus* revealed that certain blastomeres (B3, B4, C4, and D4) contribute specifically to the hindgut as early as the 32-blastomere stage (Moody 1987) (Chapter 13). Unfortunately, the dyes do not persist long enough to permit analysis of progeny after all organs have formed. Just before gastrulation, cells fated to form the dorsal and ventral aspects of the gut tube are in distinct locations: Suprablastopore lip cells form the archenteron roof, and subblastopore cells contribute to the floor (Keller 1975, 1976). Cells located more laterally contribute to more posterior tissues and gastrulate later.

In mouse and chick, many genes that are expressed anteriorly in definitive endoderm are also expressed in anterior visceral endoderm (AVE) prior to gastrulation. These genes include *Hex*, asymmetrically expressed as soon as 5.5 dpc, *Hesx1* (*Homeobox expressed in ES cells*), and *Otx2* (Thomas and Beddington 1996; Thomas et al. 1998). Interestingly, a single promoter element controls anterior expression of *Hex* in AVE and anterior definitive endoderm (Rodriguez et al. 2001). Expression of these genes confers to AVE specific inductive properties of the anterior nervous system that prevent the posteriorizing influence of Nodal (Thomas and Beddington 1996; Martinez-Barbera and Beddington 2001; Perea-Gomez et al. 2001).

Patterning Just after Gastrulation: Crude Regionalization and No Determination

In zebrafish, *foxa2* is expressed in anterior but not posterior cells at gastrulation (Alexander and Stainier 1999). *her5* is an anterior endodermal marker that controls A-P patterning in endoderm, suggesting an involvement of the Notch pathway as in other species (Bally-Cuif et al. 2000). Its expression domain is controlled by antagonizing functions of BMPs and Chordin, two proteins that usually control dorsoventral patterning in mesoderm and ectoderm in vertebrates (Tiso et al. 2002) (see Chapter 33). Retinoic acid, a well-known regulator of A-P patterning in ectoderm, was recently shown to control A-P patterning in endoderm at the time of gastrulation (Stafford and Prince 2002) (Chapter 38). It also controls A-P patterning of endoderm in *Amphioxus* (Escriva et al. 2002) and mouse (Huang et al. 1998, 2002).

In the mouse, as soon as the nascent definitive endoderm displaces the primitive endoderm to extraembryonic regions (6.5–7 dpc), certain genes are already asymmetrically expressed. For instance, transcripts for the secreted protein Cer1, and the homeobox transcription factors Otx2, Hesx1, and Hex are restricted to anterior regions in primitive and definitive endoderm. In addition, functional A-P asymmetry of the endoderm at the same stages is demonstrated by the specific ability of the anterior endoderm of the chick to induce heart differentiation in the mesoderm. This early pattern is, however, not determined since association of the anterior endoderm half with posterior mesoderm induces posterior genes in endoderm and vice versa (Wells and Melton 2000).

The rough regional patterning is progressively refined by signaling between endoderm and mesoderm after gastrulation in *Drosophila* and vertebrates (Bienz 1997; Grapin-Botton and Melton 2000).

OPEN QUESTIONS

It is difficult at this point to decipher whether the apparent species differences in endoderm differentiation are due to true divergence or to gaps in our knowledge. Moreover, knowledge of the gene network that activates endoderm formation is far from complete. One question of particular interest concerns the segregation of endoderm and mesoderm. The Notch/Delta pathway is important in this respect in sea urchin (Chapter 37) and also in *Ciona* (Sherwood and McClay 1997; Corbo et al. 1998; Davidson et al. 2002b; Imai et al. 2002). Expression of *Notch* in the *Amphioxus* blastopore and comparison with β-*catenin* expression suggest that its role in mesoderm and endoderm segregation as well as A-P patterning may be conserved in chordates (Holland et al. 2001). In zebrafish, DeltaA regulates the ratio between notochord, floor plate, and midline endoderm (hypochord) cells at the midline (Appel et al. 1999). In *Drosophila*, it has an apparently different role. Notch is required in endoderm to maintain an epithelial sheet at the beginning of gastrulation and to re-epithelialize after ingression (Hartenstein et al. 1992). Such a function has not yet been characterized in vertebrates.

Diseases that affect endodermal organs may benefit from the development of cell-replacement therapies based on stem cells. Unless adult endodermal stem cells are targeted, a major challenge consists in directing ES cells toward the endodermal lineage. ES cells spontaneously form endoderm, including definitive endoderm, but in a small proportion (Itskovitz-Eldor et al. 2000). Although several growth factors applied to embryonic stem cells favor the development of ectoderm and mesoderm, none was shown to favor endoderm development (Schuldiner et al. 2000). It was recently shown that forced expression of *GATA4* and *GATA6* induces cells to form extraembryonic endoderm (Fujikura et al. 2002). Since these genes are also pivotal in definitive endoderm induction, one may wonder what is missing. Forced expression of *Foxa2* induces endoderm, but it does not fully differentiate (Levinson-Dushnik and Benvenisty 1997). Forced expression of other key players of definitive endoderm differentiation, in particular *Sox17*, is awaited.

REFERENCES

Alexander J. and Stainier D.Y. 1999. A molecular pathway leading to endoderm formation in zebrafish. *Curr. Biol.* **9:** 1147–1157.

Alexander J., Rothenberg M., Henry G.L., and Stainier D.Y. 1999. casanova plays an early and essential role in endoderm formation in zebrafish. *Dev. Biol.* **215:** 343–357.

Ang S.L., Wierda A., Wong D., Stevens K.A., Cascio S., Rossant J., and Zaret K.S. 1993. The formation and maintenance of the definitive endoderm lineage in the mouse: Involvement of HNF3/forkhead proteins. *Development* **119:** 1301–1315.

Aoki T.O., Mathieu J., Saint-Etienne L., Rebagliati M.R., Peyrieras N., and Rosa F.M. 2002a. Regulation of nodal signalling and mesendoderm formation by TARAM-A, a TGFbeta-related type I receptor. *Dev. Biol.* **241:** 273–288.

Aoki T.O., David N.B., Minchiotti G., Saint-Etienne L., Dickmeis T., Persico G.M., Strahle U., Mourrain P., and Rosa F.M. 2002b. Molecular integration of casanova in the Nodal signalling path-

way controlling endoderm formation. *Development* **129:** 275–286.

Appel B., Fritz A., Westerfield M., Grunwald D.J., Eisen J.S., and Riley B.B. 1999. Delta-mediated specification of midline cell fates in zebrafish embryos. *Curr. Biol.* **9:** 247–256.

Arendt D. and Nubler-Jung K. 1999. Rearranging gastrulation in the name of yolk: Evolution of gastrulation in yolk-rich amniote eggs. *Mech Dev.* **81:** 3–22.

Azzaria M., Goszczynski B., Chung M.A., Kalb J.M., and McGhee J.D. 1996. A fork head/HNF-3 homolog expressed in the pharynx and intestine of the *Caenorhabditis elegans* embryo. *Dev. Biol.* **178:** 289–303.

Bachvarova R.F., Skromne I., and Stern C.D. 1998. Induction of primitive streak and Hensen's node by the posterior marginal zone in the early chick embryo. *Development* **125:** 3521–3534.

Bally-Cuif L., Goutel C., Wassef M., Wurst W., and Rosa F. 2000. Coregulation of anterior and posterior mesendodermal development by a hairy-related transcriptional repressor. *Genes Dev.* **14:** 1664–1677.

Barrett K., Leptin M., and Settleman J. 1997. The Rho GTPase and a putative RhoGEF mediate a signaling pathway for the cell shape changes in *Drosophila* gastrulation. *Cell* **91:** 905–915.

Beck S., Le Good J.A., Guzman M., Haim N.B., Roy K., Beermann F., and Constam D.B. 2002. Extraembryonic proteases regulate Nodal signalling during gastrulation. *Nat. Cell Biol.* **4:** 981–985.

Bienz M. 1997. Endoderm induction in *Drosophila*: The nuclear targets of the inducing signals. *Curr. Opin. Genet. Dev.* **7:** 683–688.

Bossard P. and Zaret K.S. 1998. GATA transcription factors as potentiators of gut endoderm differentiation. *Development* **125:** 4909–4917.

Boutros M., Paricio N., Strutt D.I., and Mlodzik M. 1998. Dishevelled activates JNK and discriminates between JNK pathways in planar polarity and wingless signaling. *Cell* **94:** 109–118.

Bowles J., Schepers G., and Koopman P. 2000. Phylogeny of the SOX family of developmental transcription factors based on sequence and structural indicators. *Dev. Biol.* **227:** 239–255.

Brennan J., Norris D.P., and Robertson E.J. 2002. Nodal activity in the node governs left-right asymmetry. *Genes Dev.* **16:** 2339–2344.

Bronner G., Chu-LaGraff Q., Doe C.Q., Cohen B., Weigel D., Taubert H., and Jackle H. 1994. Sp1/egr-like zinc-finger protein required for endoderm specification and germ-layer formation in *Drosophila*. *Nature* **369:** 664–668.

Burdsal C.A., Damsky C.H., and Pedersen R.A. 1993. The role of E-cadherin and integrins in mesoderm differentiation and migration at the mammalian primitive streak. *Development* **118:** 829–844.

Burke R.D., Myers R.L., Sexton T.L., and Jackson C. 1991. Cell movements during the initial phase of gastrulation in the sea urchin embryo. *Dev. Biol.* **146:** 542–557.

Butler E. 1935. The developmental capacity of regions of the unincubated chick blastoderm as tested in chorio-allantoic grafts. *J. Exp. Zool.* **70:** 387–388.

Callebaut M., van Nueten E., Bortier H., Harrisson F., and van Nassauw L. 1996. Map of the Anlage fields in the avian unincubated blastoderm. *Eur. J. Morphol.* **34:** 347–361.

Carmany-Rampey A. and Schier A.F. 2001. Single-cell internalization during zebrafish gastrulation. *Curr. Biol.* **11:** 1261–1265.

Casanova J. 1990. Pattern formation under the control of the terminal system in the *Drosophila* embryo. *Development* **110:** 621–628.

Casey E.S., Tada M., Fairclough L., Wylie C.C., Heasman J., and Smith J.C. 1999. Bix4 is activated directly by VegT and mediates endoderm formation in *Xenopus* development. *Development* **126:**

4193–4200.

Cirillo L.A., Lin F.R., Cuesta I., Friedman D., Jarnik M., and Zaret K.S. 2002. Opening of compacted chromatin by early developmental transcription factors HNF3 (FoxA) and GATA-4. *Mol. Cell* **9:** 279–289.

Clements D. and Woodland H.R. 2000. Changes in embryonic cell fate produced by expression of an endodermal transcription factor, Xsox17. *Mech. Dev.* **99:** 65–70.

Conlon F.L., Lyons K.M., Takaesu N., Barth K.S., Kispert A., Herrmann B., and Robertson E.J. 1994. A primary requirement for nodal in the formation and maintenance of the primitive streak in the mouse. *Development* **120:** 1919–1928.

Corbo J.C., Fujiwara S., Levine M., and Di Gregorio A. 1998. Suppressor of hairless activates brachyury expression in the *Ciona* embryo. *Dev. Biol.* **203:** 358–368.

Costa M., Raich W., Agbunag C., Leung B., Hardin J., and Priess J.R. 1998. A putative catenin-cadherin system mediates morphogenesis of the *Caenorhabditis elegans* embryo. *J. Cell Biol.* **141:** 297–308.

Cremazy F., Berta P., and Girard F. 2001. Genome-wide analysis of Sox genes in *Drosophila melanogaster*. *Mech. Dev.* **109:** 371–375.

Dale L. and Slack J.M. 1987. Fate map for the 32-cell stage of *Xenopus laevis*. *Development* **99:** 527–551.

David N.B. and Rosa F.M. 2001. Cell autonomous commitment to an endodermal fate and behaviour by activation of Nodal signalling. *Development* **128:** 3937–3947.

Davidson E.H., Rast J.P., Oliveri P., Ransick A., Calestani C., Yuh C.H., Minokawa T., Amore G., Hinman V., Arenas-Mena C. et al. 2002a. A genomic regulatory network for development. *Science* **295:** 1669–1678.

———. 2002b. A provisional regulatory gene network for specification of endomesoderm in the sea urchin embryo. *Dev. Biol.* **246:** 162–190.

Di Gregorio A., Corbo J.C., and Levine M. 2001. The regulation of fork-head/HNF-3beta expression in the *Ciona* embryo. *Dev. Biol.* **229:** 31–43.

Dickmeis T., Mourrain P., Saint-Etienne L., Fischer N., Aanstad P., Clark M., Strahle U., and Rosa F. 2001. A crucial component of the endoderm formation pathway, CASANOVA, is encoded by a novel sox-related gene. *Genes Dev.* **15:** 1487–1492.

Djiane A., Riou J., Umbhauer M., Boucaut J., and Shi D. 2000. Role of frizzled 7 in the regulation of convergent extension movements during gastrulation in *Xenopus laevis*. *Development* **127:** 3091–3100.

Dufort D., Schwartz L., Harpal K., and Rossant J. 1998. The transcription factor HNF3beta is required in visceral endoderm for normal primitive streak morphogenesis. *Development* **125:** 3015–3025.

Ecochard V., Cayrol C., Rey S., Foulquier F., Caillol D., Lemaire P., and Duprat A.M. 1998. A novel *Xenopus* mix-like gene milk involved in the control of the endomesodermal fates. *Development* **125:** 2577–2585.

Escriva H., Holland N.D., Gronemeyer H., Laudet V., and Holland L.Z. 2002. The retinoic acid signaling pathway regulates anterior/posterior patterning in the nerve cord and pharynx of amphioxus, a chordate lacking neural crest. *Development* **129:** 2905–2916.

Ettensohn C.A. 1985. Gastrulation in the sea urchin embryo is accompanied by the rearrangement of invaginating epithelial cells. *Dev. Biol.* **112:** 383–390.

———. Cell movements in the sea urchin embryo. *Curr. Opin. Genet. Dev.* **9:** 461–465.

Eyal-Giladi H., Debby A., and Harel N. 1992. The posterior section of the chick's area pellucida and its involvement in hypoblast and primitive streak formation. *Development* **116:** 819–830.

Fujikura J., Yamato E., Yonemura S., Hosoda K., Masui S., Nakao K., Miyazaki Ji J., and Niwa H. 2002. Differentiation of embryonic stem cells is induced by GATA factors. *Genes Dev.* **16:** 784–789.

Gao X., Sedgwick T., Shi Y.B., and Evans T. 1998. Distinct functions are implicated for the GATA-4, -5, and -6 transcription factors in the regulation of intestine epithelial cell differentiation. *Mol. Cell. Biol.* **18:** 2901–2911.

Garcia-Martinez V., Alvarez I.S., and Schoenwolf G.C. 1993. Locations of the ectodermal and nonectodermal subdivisions of the epiblast at stages 3 and 4 of avian gastrulation and neurulation. *J. Exp. Zool.* **267:** 431–446.

Gaudet J. and Mango S.E. 2002. Regulation of organogenesis by the *Caenorhabditis elegans* FoxA protein PHA-4. *Science* **295:** 821–825.

Germain S., Howell M., Esslemont G.M., and Hill C.S. 2000. Homeodomain and winged-helix transcription factors recruit activated Smads to distinct promoter elements via a common Smad interaction motif. *Genes Dev.* **14:** 435–451.

Grapin-Botton A. and Melton D.A. 2000. Endoderm development: From patterning to organogenesis. *Trends Genet.* **16:** 124–130.

Hacker U. and Perrimon N. 1998. DRhoGEF2 encodes a member of the Dbl family of oncogenes and controls cell shape changes during gastrulation in *Drosophila*. *Genes Dev.* **12:** 274–284.

Harada Y., Akasaka K., Shimada H., Peterson K.J., Davidson E.H., and Satoh N. 1996. Spatial expression of a forkhead homologue in the sea urchin embryo. *Mech. Dev.* **60:** 163–173.

Hart A.H., Hartley L., Sourris K., Stadler E.S., Li R., Stanley E.G., Tam P.P., Elefanty A.G., and Robb L. 2002. Mixl1 is required for axial mesendoderm morphogenesis and patterning in the murine embryo. *Development* **129:** 3597–5608.

Hartenstein A.Y., Rugendorff A., Tepass U., and Hartenstein V. 1992. The function of the neurogenic genes during epithelial development in the *Drosophila* embryo. *Development* **116:** 1203–1220.

Hashimoto-Partyka M.K., Yuge M., and Cho K.W. 2003. Nodal signaling in *Xenopus* gastrulae is cell-autonomous and patterned by beta-catenin. *Dev. Biol.* **253:** 125–138.

Hatada Y. and Stern C.D. 1994. A fate map of the epiblast of the early chick embryo. *Development* **120:** 2879–2889.

Heasman J., Snape A., Smith J., and Wylie C.C. 1985. Single cell analysis of commitment in early embryogenesis. *J. Embryol. Exp. Morphol.* (suppl.) **89:** 297–316.

Heisenberg C.P., Tada M., Rauch G.J., Saude L., Concha M.L., Geisler R., Stemple D.L., Smith J.C., and Wilson S.W. 2000. Silberblick/Wnt11 mediates convergent extension movements during zebrafish gastrulation. *Nature* **405:** 76–81.

Henry G.L. and Melton D.A. 1998. Mixer, a homeobox gene required for endoderm development. *Science* **281:** 91–96.

Henry G.L., Brivanlou I.H., Kessler D.S., Hemmati-Brivanlou A., and Melton D.A. 1996. TGF-β signals and a pattern in *Xenopus laevis* endodermal development. *Development* **122:** 1007–1015.

Ho R.K. and Kimmel C.B. 1993. Commitment of cell fate in the early zebrafish embryo. *Science* **261:** 109–111.

Holland L.Z., Rached L.A., Tamme R., Holland N.D., Kortschak D., Inoko H., Shiina T., Burgtorf C., and Lardelli M. 2001. Characterization and developmental expression of the amphioxus homolog of Notch (AmphiNotch): Evolutionary conservation of multiple expression domains in amphioxus and vertebrates. *Dev. Biol.* **232:** 493–507.

Horb M.E. and Thomsen G.H. 1997. A vegetally localized T-box transcription factor in *Xenopus* eggs specifies mesoderm and endoderm and is essential for embryonic mesoderm formation. *Development* **124:** 1689–1698.

Horner M.A., Quintin S., Domeier M.E., Kimble J., Labouesse M., and Mango S.E. 1998. pha-4, an HNF-3 homolog, specifies pharyngeal organ identity in *Caenorhabditis elegans*. *Genes Dev.* **12:** 1947–1952.

Horton A.C. and Gibson-Brown J.J. 2002. Evolution of developmental functions by the Eomesodermin, T-brain-1, Tbx21 subfamily of T-box genes: Insights from amphioxus. *J. Exp. Zool.* **294:** 112–121.

Howell M. and Hill C.S. 1997. XSmad2 directly activates the activin-inducible, dorsal mesoderm gene XFKH1 in *Xenopus* embryos. *EMBO J.* **16:** 7411–7421.

Howell M., Itoh F., Pierreux C.E., Valgeirsdottir S., Itoh S., ten Dijke P., and Hill C.S. 1999. *Xenopus* Smad4β is the co-Smad component of developmentally regulated transcription factor complexes responsible for induction of early mesodermal genes. *Dev. Biol.* **214:** 354–469.

Huang D., Chen S.W., and Gudas L.J. 2002. Analysis of two distinct retinoic acid response elements in the homeobox gene Hoxb1 in transgenic mice. *Dev. Dyn.* **223:** 353–370.

Huang D., Chen S.W., Langston A.W., and Gudas L.J. 1998. A conserved retinoic acid responsive element in the murine Hoxb-1 gene is required for expression in the developing gut. *Development* **125:** 3235–3246.

Hudson C., Clements D., Friday R.V., Stott D., and Woodland H.R. 1997. Xsox17alpha and -beta mediate endoderm formation in *Xenopus. Cell* **91:** 397–405.

Huelsken J., Vogel R., Brinkmann V., Erdmann B., Birchmeier C., and Birchmeier W. 2000. Requirement for beta-catenin in anterior-posterior axis formation in mice. *J. Cell Biol.* **148:** 567–578.

Hunt T.E. 1937a. The development of gut and its derivatives from the mesectoderm and mesentoderm of early chick blastoderms. *Anat. Rec.* **68:** 349–363.

———. 1937b. The origin of entodermal cells from the primitive streak of the chick embryo. *Anat. Rec.* **68:** 449–460.

Ibrahim H. and Winklbauer R. 2001. Mechanisms of mesendoderm internalization in the *Xenopus* gastrula: Lessons from the ventral side. *Dev. Biol.* **240:** 108–122.

Imai K.S., Satoh N., and Satou Y. 2002. An essential role of a FoxD gene in notochord induction in *Ciona* embryos. *Development* **129:** 3441–3453.

Imai K., Takada N., Satoh N., and Satou Y. 2000. β-catenin mediates the specification of endoderm cells in ascidian embryos. *Development* **127:** 3009–3020.

Itskovitz-Eldor J., Schuldiner M., Karsenti D., Eden A., Yanuka O., Amit M., Soreq H., and Benvenisty N. 2000. Differentiation of human embryonic stem cells into embryoid bodies compromising the three embryonic germ layers. *Mol. Med.* **6:** 88–95.

Jacobsen C.M., Narita N., Bielinska M., Syder A.J., Gordon J.I., and Wilson D.B. 2002. Genetic mosaic analysis reveals that GATA-4 is required for proper differentiation of mouse gastric epithelium. *Dev. Biol.* **241:** 34–46.

Jesuthasan S. and Strähle U. 1997. Dynamic microtubules and specification of the zebrafish embryonic axis. *Curr. Biol.* **7:** 31–42.

Jiang Y. and Evans T. 1996. The *Xenopus* GATA-4/5/6 genes are associated with cardiac specification and can regulate cardiac-specific transcription during embryogenesis. *Dev. Biol.* **174:** 258–270.

Joubin K. and Stern C.D. 1999. Molecular interactions continuously define the organizer during the cell movements of gastrulation. *Cell* **98:** 559–571.

Kaestner K.H., Knochel W., and Martinez D.E. 2000. Unified nomen-

clature for the winged helix/forkhead transcription factors. *Genes Dev.* **14:** 142–146.

Kalb J.M., Lau K.K., Goszczynski B., Fukushige T., Moons D., Okkema P.G., and McGhee J.D. 1998. pha-4 is *Ce-fkh-1*, a *fork head*/HNF-3α,β,γ homolog that functions in organogenesis of the *C. elegans* pharynx. *Development* **125:** 2171–2180.

Kanai-Azuma M., Kanai Y., Gad J.M., Tajima Y., Taya C., Kurohmaru M., Sanai Y., Yonekawa H., Yazaki K., Tam P.P. et al. 2002. Depletion of definitive gut endoderm in Sox17-null mutant mice. *Development* **129:** 2367–2379.

Keller R.E. 1975. Vital dye mapping of the gastrula and neurula of *Xenopus laevis*. I. Prospective areas and morphogenetic movements of the superficial layer. *Dev. Biol.* **42:** 222–241.

———. 1976. Vital dye mapping of the gastrula and neurula of *Xenopus laevis*. II. Prospective areas and morphogenetic movements of the deep layer. *Dev. Biol.* **51:** 118–137.

Khaner O. and Wilt F. 1991. Interactions of different vegetal cells with mesomeres during early stages of sea urchin development. *Development* **112:** 881–890.

Kikuchi Y., Trinh L.A., Reiter J.F., Alexander J., Yelon D., and Stainier D.Y. 2000. The zebrafish *bonnie* and *clyde* gene encodes a Mix family homeodomain protein that regulates the generation of endodermal precursors. *Genes Dev.* **14:** 1279–1289.

Kikuchi Y., Agathon A., Alexander J., Thisse C., Waldron S., Yelon D., Thisse B. and Stainier D.Y. 2001. *casanova* encodes a novel Sox-related protein necessary and sufficient for early endoderm formation in zebrafish. *Genes Dev.* **15:** 1493–1505.

Kimberly E.L. and Hardin J. 1998. Bottle cells are required for the initiation of primary invagination in the sea urchin embryo. *Dev. Biol.* **204:** 235–250.

Kimmel C.B., Warga R.M., and Schilling T.F. 1990. Origin and organization of the zebrafish fate map. *Development* **108:** 581–594.

Koutsourakis M., Langeveld A., Patient R., Beddington R., and Grosveld F. 1999. The transcription factor GATA6 is essential for early extraembryonic development. *Development* **126:** 723–732.

Latinkic B.V. and Smith J.C. 1999. Goosecoid and mix.1 repress Brachyury expression and are required for head formation in *Xenopus. Development* **126:** 1769–1779.

Laufer J.S., Bazzicalupo P., and Wood W.B. 1980. Segregation of developmental potential in early embryos of *Caenorhabditis elegans. Cell* **19:** 569–577.

Lawson K.A., Meneses J.J., and Pedersen R.A. 1986. Cell fate and cell lineage in the endoderm of the presomite mouse embryo, studied with an intracellular tracer. *Dev. Biol.* **115:** 325–339.

Lee J.Y. and Goldstein B. 2003. Mechanisms of cell positioning during *C. elegans* gastrulation. *Development* **130:** 307–320.

Lemaire P., Darras S., Caillol D., and Kodjabachian L. 1998. A role for the vegetally expressed *Xenopus* gene Mix.1 in endoderm formation and in the restriction of mesoderm to the marginal zone. *Development* **125:** 2371–2380.

Leptin M. 1999. Gastrulation in *Drosophila*: The logic and the cellular mechanisms. *EMBO J.* **18:** 3187–3192.

Leung B., Hermann G.J., and Priess J.R. 1999. Organogenesis of the *Caenorhabditis elegans* intestine. *Dev. Biol.* **216:** 114–134.

Levinson-Dushnik M. and Benvenisty N. 1997. Involvement of hepatocyte nuclear factor 3 in endoderm differentiation of embryonic stem cells. *Mol. Cell. Biol.* **17:** 3817–3822.

Li X., Chuang C.K., Mao C.A., Angerer L.M., and Klein W.H. 1997. Two Otx proteins generated from multiple transcripts of a single gene in *Strongylocentrotus purpuratus. Dev. Biol.* **187:** 253–266.

Lin R., Hill R.J., and Priess J.R. 1998. POP-1 and anterior-posterior fate decisions in *C. elegans* embryos. *Cell* **92:** 229–239.

Lin W.H., Huang L.H., Yeh J.Y., Hoheisel J., Lehrach H., Sun Y.H., and Tsai S.F. 1995. Expression of a *Drosophila* GATA transcription factor in multiple tissues in the developing embryos. Identification of homozygous lethal mutants with P-element insertion at the promoter region. *J. Biol. Chem.* **270:** 25150–25158.

Liu P., Wakamiya M., Shea M.J., Albrecht U., Behringer R.R., and Bradley A. 1999. Requirement for Wnt3 in vertebrate axis formation. *Nat. Genet.* **22:** 361–365.

Logan C.Y. and McClay D.R. 1997. The allocation of early blastomeres to the ectoderm and endoderm is variable in the sea urchin embryo. *Development* **124:** 2213–2223.

Lopez-Sanchez C., Garcia-Martinez V., and Schoenwolf G.C. 2001. Localization of cells of the prospective neural plate, heart and somites within the primitive streak and epiblast of avian embryos at intermediate primitive–streak stages. *Cells Tissues Organs* **169:** 334–346.

Maduro M.F., Meneghini M.D., Bowerman B., Broitman-Maduro G., and Rothman J.H. 2001. Restriction of mesendoderm to a single blastomere by the combined action of SKN-1 and a GSK-3β homolog is mediated by MED-1 and -2 in *C. elegans*. *Mol. Cell* **7:** 475–485.

Maeda M., Kubo K., Nishi T., and Futai M. 1996. Roles of gastric GATA DNA-binding proteins. *J. Exp. Biol.* **199:** 513–520.

Mango S.E., Lambie E.J., and Kimble J. 1994. The pha-4 gene is required to generate the pharyngeal primordium of *Caenorhabditis elegans*. *Development* **120:** 3019–3031.

Martindale M.Q. and Henry J.Q. 1999. Intracellular fate mapping in a basal metazoan, the ctenophore *Mnemiopsis leidyi*, reveals the origins of mesoderm and the existence of indeterminate cell lineages. *Dev. Biol.* **214:** 243–257.

Martinez-Barbera J.P. and Beddington R.S. 2001. Getting your head around Hex and Hesx1: Forebrain formation in mouse. *Int. J. Dev. Biol.* **45:** 327–336.

Mello C.C., Draper B.W., Krause M., Weintraub H., and Priess J.R. 1992. The pie-1 and mex-1 genes and maternal control of blastomere identity in early *C. elegans* embryos. *Cell* **70:** 163–176.

Monaghan A.P., Kaestner K.H., Grau E., and Schutz G. 1993. Postimplantation expression patterns indicate a role for the mouse *forkhead*/HNF-3 α, β and γ genes in determination of the definitive endoderm, chordamesoderm and neuroectoderm. *Development* **119:** 567–578.

Moody S.A. 1987. Fates of the blastomeres of the 32-cell-stage *Xenopus* embryo. *Dev. Biol.* **122:** 300–319.

Morrisey E.E., Tang Z., Sigrist K., Lu M.M., Jiang F., Ip H.S., and Parmacek M.S. 1998. GATA6 regulates HNF4 and is required for differentiation of visceral endoderm in the mouse embryo. *Genes Dev.* **12:** 3579–3590.

Nakajima Y. and Burke R.D. 1996. The initial phase of gastrulation in sea urchins is accompanied by the formation of bottle cells. *Dev. Biol.* **179:** 436–446.

Nance J. and Priess J.R. 2002. Cell polarity and gastrulation in *C. elegans*. *Development* **129:** 387–397.

Narita N., Bielinska M., and Wilson D.B. 1997. Wild-type endoderm abrogates the ventral developmental defects associated with GATA-4 deficiency in the mouse. *Dev. Biol.* **189:** 270–274.

Nieuwkoop P.D. 1997. Short historical survey of pattern formation in the endo-mesoderm and the neural anlage in the vertebrates: The role of vertical and planar inductive actions. *Cell. Mol. Life Sci.* **53:** 305–318.

Nishida H. 1987. Cell lineage analysis in ascidian embryos by intracel-

lular injection of a tracer enzyme. III. Up to the tissue restricted stage. *Dev. Biol.* **121:** 526–541.

Nishida H. and Satoh N. 1983. Cell lineage analysis in ascidian embryos by intracellular injection of a tracer enzyme. I. Up to the eight-cell stage. *Dev. Biol.* **99:** 382-394.

Nocente-McGrath C., Brenner C.A., and Ernst S.G. 1989. Endo16, a lineage-specific protein of the sea urchin embryo, is first expressed just prior to gastrulation. *Dev. Biol.* **136:** 264–272.

Oda H., Tsukita S., and Takeichi M. 1998. Dynamic behavior of the cadherin-based cell-cell adhesion system during *Drosophila* gastrulation. *Dev. Biol.* **203:** 435–450.

Odenthal J. and Nusslein-Volhard C. 1998. fork head domain genes in zebrafish. *Dev. Genes Evol.* **208:** 245–258.

Pagan-Westphal S.M. and Tabin C.J. 1998. The transfer of left-right positional information during chick embryogenesis. *Cell* **93:** 25–35.

Parks S. and Wieschaus E. 1991. The *Drosophila* gastrulation gene concertina encodes a Gα-like protein. *Cell* **64:** 447–458.

Patient R.K. and McGhee J.D. 2002. The GATA family (vertebrates and invertebrates). *Curr. Opin. Genet. Dev.* **12:** 416–422.

Peale Jr., F. V., Sugden L., and Bothwell M. 1998. Characterization of CMIX, a chicken homeobox gene related to the *Xenopus* gene mix.1. *Mech. Dev.* **75:** 167–170.

Pearce J.J. and Evans M.J. 1999. Mml, a mouse Mix-like gene expressed in the primitive streak. *Mech. Dev.* **87:** 189–192.

Perea-Gomez A., Lawson K.A., Rhinn M., Zakin L., Brulet P., Mazan S., and Ang S.L. 2001. Otx2 is required for visceral endoderm movement and for the restriction of posterior signals in the epiblast of the mouse embryo. *Development* **128:** 753–765.

Poelmann R.E. 1980. Differential mitosis and degeneration patterns in relation to the alterations in the shape of the embryonic ectoderm of early post-implantation mouse embryos. *J. Embryol. Exp. Morphol.* **55:** 33–51.

———. 1981. The head-process and the formation of the definitive endoderm in the mouse embryo. *Anat. Embryol.* **162:** 41–49.

Poulain M. and Lepage T. 2002. Mezzo, a *paired-like* homeobox protein is an immediate target of Nodal signalling and regulates endoderm specification in zebrafish. *Development* **129:** 4901–4914.

Priess J.R. and Thomson J.N. 1987. Cellular interactions in early *C. elegans* embryos. *Cell* **48:** 241–250.

Psychoyos D. and Stern C.D. 1996. Restoration of the organizer after radical ablation of Hensen's node and the anterior primitive streak in the chick embryo. *Development* **122:** 3263–3273.

Ransick A. and Davidson E.H. 1993. A complete second gut induced by transplanted micromeres in the sea urchin embryo. *Science* **259:** 1134–1138.

———. 1998. Late specification of Veg1 lineages to endodermal fate in the sea urchin embryo. *Dev. Biol.* **195:** 38–48.

Rawles M.E. 1936. A study in the localization of organ-forming areas in the chick blastoderm of the head-process stage. *J. Exp. Zool.* **72:** 271–315.

Rehorn K.P., Thelen H., Michelson A.M., and Reuter R. 1996. A molecular aspect of hematopoiesis and endoderm development common to vertebrates and *Drosophila*. *Development* **122:** 4023–4031.

Reiter J.F., Kikuchi Y., and Stainier D.Y. 2001. Multiple roles for Gata5 in zebrafish endoderm formation. *Development* **128:** 125–135.

Reiter J.F., Alexander J., Rodaway A., Yelon D., Patient R., Holder N., and Stainier D.Y. 1999. Gata5 is required for the development of the heart and endoderm in zebrafish. *Genes Dev.* **13:** 2983–2995.

Reuter R. 1994. The gene *serpent* has homeotic properties and specifies endoderm versus ectoderm within the *Drosophila* gut. *Development* **120**: 1123–1135.

Reuter R. and Leptin M. 1994. Interacting functions of *snail, twist* and *huckebein* during the early development of germ layers in *Drosophila*. *Development* **120**: 1137–1150.

Reuter R., Grunewald B., and Leptin M. 1993. A role for the mesoderm in endodermal migration and morphogenesis in *Drosophila*. *Development* **119**: 1135–1145.

Ristoratore F., Spagnuolo A., Aniello F., Branno M., Fabbrini F., and Di Lauro R. 1999. Expression and functional analysis of *Cititf1*, an ascidian *NK-2* class gene, suggest its role in endoderm development. *Development* **126**: 5149–5159.

Robb L., Hartley L., Begley C.G., Brodnicki T.C., Copeland N.G., Gilbert D.J., Jenkins N.A., and Elefanty A.G. 2000. Cloning, expression analysis, and chromosomal localization of murine and human homologues of a *Xenopus* mix gene. *Dev. Dyn.* **219**: 497–504.

Rodaway A. and Patient R. 2001. Mesendoderm: An ancient germ layer? *Cell* **105**: 169–172.

Rodaway A., Takeda H., Koshida S., Broadbent J., Price B., Smith J. C., Patient R., and Holder N. 1999. Induction of the mesendoderm in the zebrafish germ ring by yolk cell-derived TGF-β family signals and discrimination of mesoderm and endoderm by FGF. *Development* **126**: 3067–3078.

Rodriguez T.A., Casey E.S., Harland R.M., Smith J.C., and Beddington R.S. 2001. Distinct enhancer elements control Hex expression during gastrulation and early organogenesis. *Dev. Biol.* **234**: 304–316.

Rosenquist G.C. 1971. The origin and movements of the hepatogenic cells in the chick embryo as determined by radioautographic mapping. *J. Embryol. Exp. Morphol.* **25**: 97–113.

Rudnick D. 1932. Regional restrictions of potencies in the chick during embryogenesis. *J. Exp. Zool.* **71**: 83–99.

Russ A.P., Wattler S., Colledge W.H., Aparicio S.A., Carlton M.B., Pearce J.J., Barton S.C., Surani M.A., Ryan K., Nehls M.C. et al. 2000. *Eomesodermin* is required for mouse trophoblast development and mesoderm formation. *Nature* **404**: 95–99.

Sakaguchi T., Kuroiwa A., and Takeda H. 2001. A novel sox gene, *226D7*, acts downstream of Nodal signaling to specify endoderm precursors in zebrafish. *Mech. Dev.* **107**: 25–38.

Sanders E.J., Bellairs R., and Portch P.A. 1978. In vivo and in vitro studies on the hypoblast and definitive endoblast of avian embryos. *J. Embryol. Exp. Morphol.* **46**: 187–205.

Sandstrom R.H. 1934. The differentiation of the hepatic and pancreatic tissues of the chick embryo in chorioallantoic grafts. *Physiol. Zool.* **7**: 226–246.

Sasaki H. and Hogan B.L. 1993. Differential expression of multiple fork head related genes during gastrulation and axial pattern formation in the mouse embryo. *Development* **118**: 47–59.

Satou Y., Imai K.S., and Satoh N. 2001. Early embryonic expression of a LIM-homeobox gene *Cs-lhx3* is downstream of β-catenin and responsible for the endoderm differentiation in *Ciona savignyi* embryos. *Development* **128**: 3559–3570.

Schier A.F., Neuhauss S.C., Helde K.A., Talbot W.S., and Driever W. 1997. The *one-eyed pinhead* gene functions in mesoderm and endoderm formation in zebrafish and interacts with *no tail*. *Development* **124**: 327–342.

Schneider S., Steinbeisser H., Warga R.M., and Hausen P. 1996. β-catenin translocation into nuclei demarcates the dorsalizing centers in frog and fish embryos. *Mech. Dev.* **57**: 191–198.

Schoenwolf G.C. and Alvarez I.S. 1991. Specification of neurepithelium and surface epithelium in avian transplantation chimeras. *Development* **112**: 713–722.

Schohl A. and Fagotto F. 2002. β-catenin, MAPK and Smad signaling during early *Xenopus* development. *Development* **129**: 37–52.

Schroeder D.F. and McGhee J.D. 1998. Anterior-posterior patterning within the *Caenorhabditis elegans* endoderm. *Development* **125**: 4877–4887.

Schuldiner M., Yanuka O., Itskovitz-Eldor J., Melton D.A., and Benvenisty N. 2000. Effects of eight growth factors on the differentiation of cells derived from human embryonic stem cells. *Proc. Natl. Acad. Sci.* **97**: 11307–11312.

Sherwood D.R. and McClay D.R. 1997. Identification and localization of a sea urchin Notch homologue: Insights into vegetal plate regionalization and Notch receptor regulation. *Development* **124**: 3363–3374.

Shivdasani R.A. 2002. Molecular regulation of vertebrate early endoderm development. *Dev. Biol.* **249**: 191–203.

Skromne I. and Stern C.D. 2001. Interactions between Wnt and Vg1 signalling pathways initiate primitive streak formation in the chick embryo. *Development* **128**: 2915–2927.

Soudais C., Bielinska M., Heikinheimo M., MacArthur C.A., Narita N., Saffitz J.E., Simon M.C., Leiden J.M., and Wilson D.B. 1995. Targeted mutagenesis of the transcription factor GATA-4 gene in mouse embryonic stem cells disrupts visceral endoderm differentiation in vitro. *Development* **121**: 3877–3888.

Spagnuolo A. and Di Lauro R. 2002. *Cititf1* and endoderm differentiation in *Ciona intestinalis*. *Gene* **287**: 115–119.

Spiegelman M. and Bennett D. 1974. Fine structural study of cell migration in the early mesoderm of normal and mutant mouse embryos (T-locus: t-9/t-9). *J. Embryol. Exp. Morphol.* **32**: 723–728.

Stafford D. and Prince V.E. 2002. Retinoic acid signaling is required for a critical early step in zebrafish pancreatic development. *Curr. Biol.* **12**: 1215–1220.

Stainier D.Y. 2002. A glimpse into the molecular entrails of endoderm formation. *Genes Dev.* **16**: 893–907.

Stein S., Roeser T., and Kessel M. 1998. CMIX, a paired-type homeobox gene expressed before and during formation of the avian primitive streak. *Mech. Dev.* **75**: 163–165.

Stern C.D. 1979. A re-examination of mitotic activity in the early chick embryo. *Anat. Embryol.* **156**: 319–329.

———. 1990. The marginal zone and its contribution to the hypoblast and primitive streak of the chick embryo. *Development* **109**: 667–682.

Strähle U., Blader P., Henrique D., and Ingham P.W. 1993. Axial, a zebrafish gene expressed along the developing body axis, shows altered expression in cyclops mutant embryos. *Genes Dev.* **7**: 1436–1446.

Sulston J.E. 1983. Neuronal cell lineages in the nematode *Caenorhabditis elegans*. *Cold Spring Harbor Symp. Quant. Biol.* **48**: 443–452.

Tada M. and Smith J.C. 2000. *Xwnt11* is a target of *Xenopus* Brachyury: Regulation of gastrulation movements via Dishevelled, but not through the canonical Wnt pathway. *Development* **127**: 2227–2238.

Tada M., Casey E.S., Fairclough L., and Smith J.C. 1998. *Bix1*, a direct target of *Xenopus* T-box genes, causes formation of ventral mesoderm and endoderm. *Development* **125**: 3997–4006.

Takeichi M., Nakagawa S., Aono S., Usui T., and Uemura T. 2000. Patterning of cell assemblies regulated by adhesion receptors of the cadherin superfamily. *Philos. Trans. R. Soc. Lond. B Biol. Sci.* **355**: 885–890.

Tam P.P. and Beddington R.S. 1992. Establishment and organization of germ layers in the gastrulating mouse embryo. *CIBA Found. Symp.* **165**: 27–49.

Tam P.P. and Behringer R.R. 1997. Mouse gastrulation: The formation of a mammalian body plan. *Mech. Dev.* **68**: 3–25.

Tam P.P. and Meier S. 1982. The establishment of a somitomeric pattern in the mesoderm of the gastrulating mouse embryo. *Am. J. Anat.* **164**: 209–225.

Tam P.P., Williams E.A., and Chan W.Y. 1993. Gastrulation in the mouse embryo: Ultrastructural and molecular aspects of germ layer morphogenesis. *Microsc. Res. Tech.* **26**: 301–328.

Tam P.P., Gad J.M., Kinder S.J., Tsang T.E., and Behringer R.R. 2001. Morphogenetic tissue movement and the establishment of body plan during development from blastocyst to gastrula in the mouse. *BioEssays* **23**: 508–517.

Technau G.M. and Campos Ortega J.A. 1985. Fate mapping in wild type *Drosophila melanogaster*: II: Injections of horseradish peroxidase in cells of the early gastrula stage. *Wilhelm Roux's Arch. Dev. Biol.* **194**: 196–212.

Tepass U. and Hartenstein V. 1994. Epithelium formation in the *Drosophila* midgut depends on the interaction of endoderm and mesoderm. *Development* **120**: 579–590.

———. 1995. Neurogenic and proneural genes control cell fate specification in the *Drosophila* endoderm. *Development* **121**: 393–405.

Thomas P. and Beddington R. 1996. Anterior primitive endoderm may be responsible for patterning the anterior neural plate in the mouse embryo. *Curr. Biol.* **6**: 1487–1496.

Thomas P.Q., Brown A., and Beddington R.S. 1998. *Hex*: A homeobox gene revealing peri-implantation asymmetry in the mouse embryo and an early transient marker of endothelial cell precursors. *Development* **125**: 85–94.

Tiso N., Filippi A., Pauls S., Bortolussi M., and Argenton F. 2002. BMP signalling regulates anteroposterior endoderm patterning in zebrafish. *Mech. Dev.* **118**: 29–37.

Vogt W. 1929. Gestaltunganalyse am Amphibienkeim mit örtlicher Vitalfärbung. II. Teil. Gastrulation und Mesodermbildung bei Urodelen und Anuren. *Wihlelm Roux' Arch. Entwicklungsmech. Org.* **166**: 189–204.

Wacker S., Grimm K., Joos T., and Winklbauer R. 2000. Development and control of tissue separation at gastrulation in *Xenopus. Dev. Biol.* **224**: 428–439.

Warga R.M. and Kimmel C.B. 1990. Cell movements during epiboly and gastrulation in zebrafish. *Development* **108**: 569–580.

Warga R.M. and Nusslein-Volhard C. 1999. Origin and development of the zebrafish endoderm. *Development* **126**: 827–838.

Weber H., Symes C.E., Walmsley M.E., Rodaway A.R., and Patient R.K. 2000. A role for GATA5 in *Xenopus* endoderm specification. *Development* **127**: 4345–4360.

Wei Y. and Mikawa T. 2000. Formation of the avian primitive streak from spatially restricted blastoderm: Evidence for polarized cell division in the elongating streak. *Development* **127**: 87–96.

Weigel D., Jurgens G., Kuttner F., Seifert E., and Jackle H. 1989. The homeotic gene fork head encodes a nuclear protein and is expressed in the terminal regions of the *Drosophila* embryo. *Cell* **57**: 645–658.

Weinstein D.C., Ruiz i Altaba A., Chen W.S., Hoodless P., Prezioso V.R., Jessell T.M., and Darnell Jr., J.E. 1994. The winged-helix transcription factor HNF-3 beta is required for notochord development in the mouse embryo. *Cell* **78**: 575–588.

Wells J.M. and Melton D.A. 2000. Early mouse endoderm is patterned by soluble factors from adjacent germ layers. *Development* **127**: 1563–1572.

Wessel G.M. and Wikramanayake A. 1999. How to grow a gut: Ontogeny of the endoderm in the sea urchin embryo. *BioEssays* **21**: 459–471.

Wiegner O. and Schierenberg E. 1998. Specification of gut cell fate differs significantly between the nematodes *Acrobeloides nanus* and *Caenorhabditis elegans. Dev. Biol.* **204**: 3–14.

Winklbauer R. and Schurfeld M. 1999. Vegetal rotation, a new gastrulation movement involved in the internalization of the mesoderm and endoderm in *Xenopus. Development* **126**: 3703–3713.

Wylie C., Kofron M., Payne C., Anderson R., Hosobuchi M., Joseph E., and Heasman J. 1996. Maternal β-catenin establishes a 'dorsal signal' in early *Xenopus* embryos. *Development* **122**: 2987–2996.

Yasuo H. and Lemaire P. 1999. A two-step model for the fate determination of presumptive endodermal blastomeres in *Xenopus* embryos. *Curr. Biol.* **9**: 869–879.

Yuh C.H., Bolouri H., and Davidson E.H. 1998. Genomic cis-regulatory logic: Experimental and computational analysis of a sea urchin gene. *Science* **279**: 1896–1902.

Zhang J., Houston D.W., King M.L., Payne C., Wylie C., and Heasman J. 1998. The role of maternal VegT in establishing the primary germ layers in *Xenopus* embryos. *Cell* **94**: 515–524.

Zorn A.M., Butler K., and Gurdon J.B. 1999. Anterior endomesoderm specification in *Xenopus* by Wnt/β-catenin and TGF-β signalling pathways. *Dev. Biol.* **209**: 282–297.

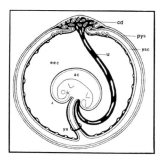

EXTRAEMBRYONIC TISSUES

K.M. Downs

Department of Anatomy, University of Wisconsin-Madison Medical School, Madison, Wisconsin 53706

INTRODUCTION

The months spent in our mothers' wombs must surely be the most eventful in our lives. Whether we emerge successfully is largely dependent on correct formation of our fetal membranes; i.e., those extraembryonic tissues required to sustain and nourish the fetus throughout gestation but which are shed at birth. The fertilized eggs of amniotes, including reptiles, birds, and mammals, give rise to the definitive adult organism, as well as to fetal membranes: the chorion, the allantois, the yolk sac, and the amnion. All of these extraembryonic tissues play roles of varying importance in fetal survival and development. Although the steps leading to their formation are highly variable and dependent on the taxa and the time in gestation, ultimately each fetal membrane carries out similar functions between species. Fetal membranes exhibit an astonishing variety of arrangement and form, especially in mammals (Mossman 1937). Unfortunately, anything other than presentation of basic principles, typically in the mouse, is beyond the scope of this chapter.

Origin of Extraembryonic Tissues: Anatomy of the Blastocyst

Formation of the extraembryonic tissues in all eutherian (placental) mammals begins with two distinct populations of founder cells in the blastocyst (Fig. 1): an outer epithelial layer, the trophectoderm (TE) (Hubrecht 1888), and a group of nonpolarized inner cells, the inner cell mass (ICM). TE, while initially regulating the blastocyst's internal environment, will embark upon a protracted series of poorly understood differentiation events, becoming tro-

phoblast cells (Hubrecht 1888). Trophoblast, meaning "nourishing layer" (Perry 1981), mediates implantation of the blastocyst into the maternal uterus, and contributes to the chorionic component of the embryo's transient and definitive placentae. In contrast, the ICM produces the fetus, the mesodermal component of the chorionic plate, and the entire yolk sac, amnion, and allantois/umbilicus. In contrast to mammals, the chick forms a blastodisc, or blastoderm, rather than a blastocyst, and thus, has no trophectoderm. Instead, the chorion of the chick shares its origin with the amnion in the extraembryonic somatopleure, which is external to the embryo (Fig. 2).

Organ of Exchange: The Placenta

The general term given to organs of exchange, either in the eggshell or within the mother's uterine environment, is "placenta," which is established by the fertilized egg, or conceptus (Hubrecht 1889; Gardner 1972; Gardner et al. 1973, 1985; Gardner and Papioannou 1975; Rossant et al. 1978; Gardner and Rossant 1979; Rossant and Lis 1979). The placenta of mammals anatomically separates the maternal and fetal circulations, and thereby forces the exchange of nutrients, wastes, and gases across its interface. The chorioallantoic placenta, or definitive placenta of eutherian mammals, serves collectively as the alimentary, pulmonary, renal, hepatic, endocrine, and excretory organs for the fetus (Fig. 3) (Beaconsfield et al. 1980). It also regulates many maternal functions by synthesizing a large number of hormones (Allen 1975). In addition, the choriovitelline placenta, a composite of chorion and yolk sac, is recognized, but is often transitory or limited in its capacity for exchange (Fig. 4).

449

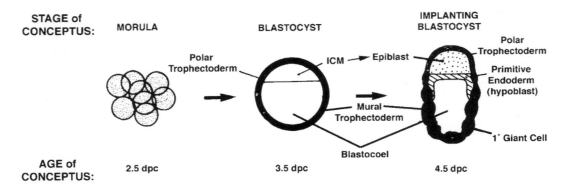

STAGE of CONCEPTUS: MORULA   BLASTOCYST   IMPLANTING BLASTOCYST

Polar Trophectoderm

ICM → Epiblast

Polar Trophectoderm

Primitive Endoderm (hypoblast)

Mural Trophectoderm

1° Giant Cell

Blastocoel

AGE of CONCEPTUS: 2.5 dpc   3.5 dpc   4.5 dpc

Figure 1. Formation of the murine blastocyst. The fertilized egg undergoes a series of cleavage divisions, with outer cells of the 8- to 16-cell morula forming trophectoderm, and the inner cells forming the inner cell mass (ICM) of the blastocyst by about 3.5 days postcoitum (dpc; of an ~19- to 20-day gestation period). (Modified, with permission, from Gardner and Rossant 1976.)

Trophectoderm and Formation of the Chorionic Disc of the Definitive Placenta

The ultimate function of trophectoderm and its derivative trophoblast cells is to serve as a selective barrier between the fetus and its mother. Normal trophoblast cells are highly proliferative (Avery and Hunt 1969; Barlow and Sherman 1972) and, although not metastatic, they are migratory (Fawcett et al. 1947) and invasive to varying degrees (Salamonsen 1999). During their development, trophoblast cells acquire distinct endocrinological and immunological properties (Ilgren 1983).

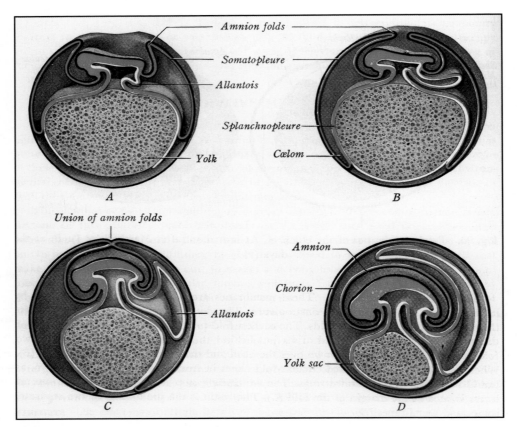

Figure 2. Stages in the development of the fetal membranes of the chick. Sagittal hemisections. Ectoderm (*black*); mesoderm (*red*); endoderm (*white*). (Reprinted, with permission, from Arey 1966.)

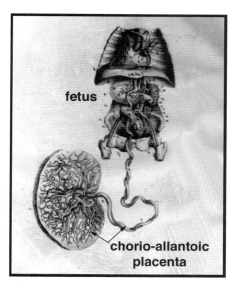

Figure 3. Eighteenth-century Dutch engraving of unknown origin. This engraving, found in an antiques print shop in Utrecht, The Netherlands, by the author, depicts the exposed circulatory system of a human fetus connected to its chorioallantoic placenta, composed of the flattened chorionic disc and umbilical vasculature (also common to rodents and nonhuman primates).

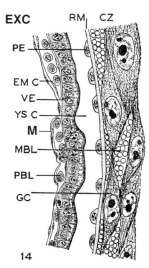

Figure 4. The choriovitelline placenta of the mouse conceptus, day 10. The relationship between the parietal yolk sac and the visceral yolk sacs is shown on day 10 in this camera lucida drawing. The parietal endoderm (PE) and its associated Reichert's membrane (RM) are complexed with trophoblast giant cells (GC) and maternal blood (MBL), and together constitute the "central zone" (CZ). The visceral yolk sac is separated from the parietal yolk sac by the yolk sac cavity (YS C), and contains extraembryonic visceral endoderm (VE), and its associated yolk sac vasculature (EM C, embryonic capillary) surrounding peripheral blood cells (PBL), overlaid with a layer of simple squamous mesothelium (M) that lines the exocoelom (EXC). The entire conceptus is surrounded by trophoblast giant cells (not shown), which are complexed with the parietal yolk sac. (Modified, with permission, from Everett 1935 [©Wiley-Liss].)

Trophoblast giant cells (TGCs) are the first differentiated derivatives of trophectoderm in all mammalian species and mediate implantation of the blastocyst into the maternal uterus. They were first observed in grafts of mouse ova to the anterior chamber of the eye (Fawcett et al. 1947). TGCs may be multinucleate and/or mononucleate (Ilgren 1983), they may form syncytia with maternal tissue (Enders and Schlafke 1971), and they may cease cell proliferation (Brodskii and Uryvaeva 1977) and endoreduplicate their genomes up to 500-fold (Zybina 1970; Zybina and Grischenko 1970; Barlow and Sherman 1972), with sister chromatids becoming polytene (Varmuza et al. 1988). Recent findings have suggested that suppression of maturation promoting factor (MPF), through inhibition of translation of *cyclin B1* transcripts, is a key mechanism in the trophoblast endocycle (Palazon et al. 1998).

In the mouse, mural primary TGCs begin the process of implantation in advance of secondary TGCs (Figs. 1 and 5A). Then, postimplantation, polar trophectoderm proliferates and forms multiple layers of diploid trophoblast, called extraembryonic ectoderm, which overlie the ICM-derived epiblast. At a distance of about six diploid extraembryonic ectodermal cell layers from the epiblast, secondary TGCs invade the maternal uterus, and both remain centrally located atop the extraembryonic ectoderm and migrate to lie laterally, forming a "central zone" (Everett 1935) filled with maternal blood (Everett 1935). There, with the parietal and visceral yolk sacs, they form a temporary choriovitelline pla-

centa for fetal nutrition (Fig. 4). The genetic control of TGC formation is poorly understood but may involve *I-mfa* (Kraut et al. 1998), *Mash2* (Guillemot et al. 1994; Tanaka et al. 1997), and *Hand1* (Firulli et al. 1998; Riley et al. 1998). Thus, extraembryonic ectoderm (also called the "trager") overlies and is in direct contact with the epiblast (Fig. 5B). It contains bone morphogenetic protein-4 (BMP-4), which may play an important role just before and at the onset of gastrulation in differentiation of the embryo (Beddington and Robertson 1999; Lawson et al. 1999).

The mesometrial aspect of extraembryonic ectoderm differentiates into the ectoplacental cone (epc), which overlies the extraembryonic ectoderm and consists of tightly packed small polyhedral cells grouped into islets (Fig. 5C) (Jollie 1964b), between which maternal blood circulates (Hernandez-Verdun 1974). Then, as gastrulation commences, epiblast begins to differentiate into the three definitive germ layers, including extraembryonic mesoderm (Lawson et al. 1991), which complexes with extraembryonic visceral endoderm, forming the visceral yolk sac (Fig. 6A). Extraembryonic ectoderm, now displaced toward the epc, forms a new structure, the chorionic ectoderm, sepa-

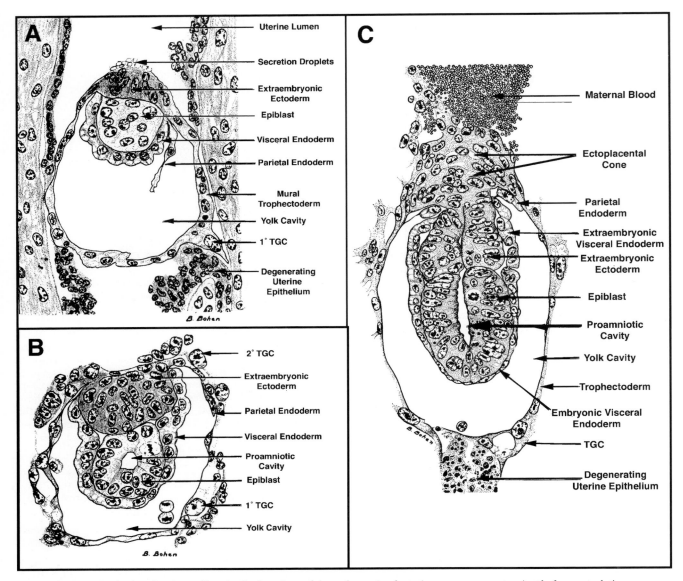

Figure 5. Projection drawings of longitudinal sections of the early postimplantation mouse conceptus, just before gastrulation. (*A*) 4.5 dpc. Major components of the implanting blastocyst are trophectoderm (mural and polar), primary giant cells (1° TGC), ICM-derived epiblast, and visceral (in contact with the egg cylinder) and parietal endoderms. (*B*) 5–6 dpc. Polar trophectoderm becomes extraembryonic ectoderm, over which secondary giant cells (2° TGC) are forming. The egg cylinder consists of epiblast and extraembryonic ectoderm surrounded by visceral endoderm. (*C*) 5.5 dpc. A portion of the extraembryonic ectoderm has formed the ectoplacental cone, and visceral endoderm becomes extraembryonic visceral endoderm and embryonic visceral endoderm, distinguishable by their varying height. (Modified, with permission, from Snell and Stevens 1966.)

rated from the ectoplacental cone by an ectoplacental cavity (Fig. 6B).

The highly proliferative chorionic ectoderm (Hernandez-Verdun and Legrand 1975; Ellington 1987; Uy et al. 2002) ultimately complexes with extraembryonic mesoderm (Fig. 6B), forming the chorionic plate, or chorion (Fig. 6C). Shortly thereafter, the central region of the chorion domes (Snell and Stevens 1966; Uy et al. 2002), and union is initiated between the chorionic ectoderm and the

ectoplacental cone. Ultimately, union spreads radially, becoming complete by about the 9- to 10-somite-pair stage (Uy et al. 2002). The chorionic plate is initially flat, but, due to the gene product of *Gcm1*, it folds into a highly branched network of chorionic villi, the depths of which will be penetrated by the allantoic vasculature (Anson-Cartwright et al. 2000; Schreiber et al. 2000). Tissue union between the epc, chorion, and allantois gives rise to the nascent chorionic disc, whose mesometrial surface contains giant cells, analogous to polyploid extravillous cytotrophoblast in the human (Rossant and Cross 2001), which overlie the junctional zone, which, in turn, overlies the chorionic labyrinth (Fig. 7).

Cells of the junctional zone (spongiotrophoblast, basal zone, or trophospongium; Fig. 7) evolve by differentiation of the epc into small basophilic cells, called spongiotrophoblasts, or spongioblasts (Jollie 1964b), rich in ATPase and acid phosphatases (Padykula 1958), and glycogen cells. Spongioblasts were originally thought to be rich in glycogen, and therefore, possibly serve as a storage site prior to liver formation (Wislocki and Padykula 1961). That conclusion has been called into question (Jollie 1965), making their function unclear. The junctional zone, which contains only maternal blood and is without fetal blood vessels, may be analogous in humans to the basal plate (Georgiades et al. 2002) or column cytotrophoblast (Rossant and Cross 2001). Transport and exchange, steroid production, and serving as a reservoir of proliferating trophoblast for the labyrinth have all been suggested as functions for the junctional zone (Jollie 1964a,b, 1965; Davies and Glasser 1968).

Union between the allantois and chorion results in formation of the chorionic labyrinth, containing maternal vascular channels and fetal umbilical vessels (Fig. 7); it constitutes the definitive chorioallantoic placental barrier, where all fetal/maternal exchange takes place. The labyrinth is functionally analogous to the fetal placenta of humans, which contains chorionic villi surrounding the fetal capillaries, as well as intervillus spaces that contain the maternal blood. In the mouse labyrinth, maternal blood cells are separated from the fetal (umbilical) endothelium by three layers of trophoblast cells ("couche plasmode": Duval 1891), possibly derived from both the chorion and the ectoplacental trophoblast (Hernandez-Verdun 1974). Those two layers closest to the allantoic endothelium are syncytial (syncytiotrophoblasts), whereas the third layer, in contact with maternal blood, is diploid (cytotrophoblast) and represents the only mitotic cell type in this zone (Fig. 7) (Davies and Glasser 1968). Thus, the rodent placenta is often referred to as hemo-trichorial (Enders 1965). In contrast, the human fetal placenta is hemo-monochorial, as the fetal capillaries are surrounded by a single layer of syncytiotrophoblast, with an intervening discontinuous layer of diploid cytotrophoblast (Langhan's cells), enabling direct contact between syncytiotrophoblast and fetal endothelium (Steven 1975).

Trophoblast stem cells (TSCs), dependent on FGF-4 for immortalization, have recently been isolated from blastocysts, extraembryonic ectoderm, and the chorion (Tanaka et al. 1998; Uy et al. 2002). Maintenance of TSCs in the implanting blastocyst is thought to involve proximity to the epiblast (Copp 1978, 1979, 1982; Gardner 1983) and paracrine signaling by Oct-4 and FGF-4 (Nichols et al. 1998), whereas in the chorionic ectoderm, it may involve factors contained within the ectoplacental cavity (Uy et al. 2002).

Formation of the Umbilical Component of the Definitive Placenta

The allantois is common to reptiles, birds, and mammals. In mammals, the allantois/body stalk is the precursor of the mature umbilicus, which consists of connective tissue into which is embedded a cord of blood vessels whose function is to shuttle fetal blood to and from the chorionic disc. Along with the vitelline (yolk sac) and cardiovascular systems, the umbilicus forms one of three major circulatory systems in the conceptus.

In all species, including birds and reptiles, the allantois is initially well separated from the chorion. In most, but not all, species (Morriss 1975), the allantois grows toward the chorion and ultimately unites with it. The murine allantois is derived from ICM-derived epiblast, the same tissue that gives rise to the fetus (Gardner et al. 1985), and is unique in that it consists wholly of mesoderm, whereas in other species, such as the chick, allantoic mesoderm surrounds a core of endoderm. In humans, the term "allantois" is applied to an endodermal evagination of the hindgut, which becomes entrapped in the mesodermal body stalk that forms the umbilical vasculature; the body stalk may therefore be analogous to the rodent allantois (for discussion, see Airhart et al. 1996).

Once it appears during early gastrulation, the murine allantoic bud expands in the exocoelomic cavity by cell proliferation (Ellington 1985; Downs and Bertler 2000), sustained mesodermal contribution from the primitive streak (Tam and Beddington 1987; Downs and Bertler 2000), and cavitation (Ellington 1985; Brown and Papaioannou 1993; Downs 2002). The genetic control of formation of the allantois is largely unknown, although a deficiency in BMP-4 may result either in complete absence of the allantois or in a foreshortened allantois, whose vasculature may be affected (Winnier et al. 1995; Fujiwara et al. 2001).

Prior to union with the chorion, the allantois acquires two morphologically distinguishable cell populations: an outer layer of simple squamous epithelium, the mesothelium, and an inner core of vascularizing mesoderm. A third cell type becomes visible which overlaps that of distally situated mesothelium: Chorio-adhesive cells contain vascular cell adhesion molecule (VCAM-1), which is required for enduring chorioallantoic union (Gurtner et al. 1995; Kwee et al. 1995; Downs 2002). Because the allantois will form a major artery and vein in the mouse, it must eventually acquire vascular smooth muscle and connective tissue through which all blood vessels flow. However, when these cell types appear, or whether they are even allantois-derived rather than chorion-derived, is not known.

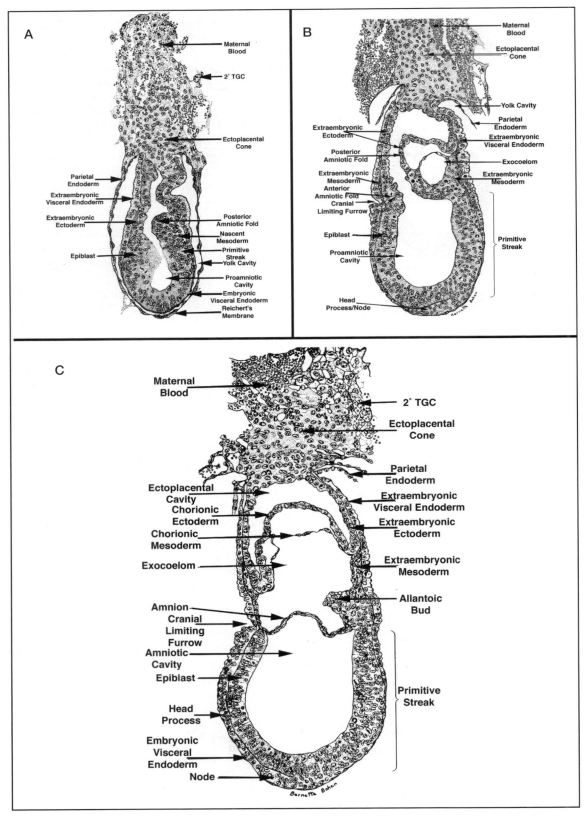

Figure 6. (*See facing page for legend.*)

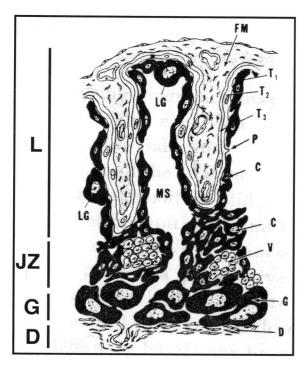

Figure 7. Schematic diagram of the structural organization of the mature rat placenta at about day 17. The labyrinth is made up of trophoblastic septa containing a core of fetal (allantoic/umbilical) blood vessels (FM, fetal mesenchyme). Layer 1, in solid black, is composed of a discontinuous layer (T₁) of diploid cytotrophoblasts (C) and forms the wall of the maternal sinuses (MS). Discontinuity permits contact between the underlying syncytiotrophoblasts (T₂) of layer 2 and maternal blood. T₃ is the third layer of the septum and also consists of syncytiotrophoblasts. The junctional zone (JZ) is composed of spongiotrophoblasts (also C), glycogen cells (V, vesicular glycogen cells), above which are giant cells (G) toward the deciduum (D). Occasionally these authors noted giant cells within T₁, as well as in the junctional zone. (Modified, with permission, from Davies and Glasser 1968 [©S. Karger AG, Basel].)

Differentiation of allantoic mesoderm into endothelium, mesoderm, and chorio-adhesive cells is not dependent on contact with the chorion (Downs et al. 1998, 2001, 2004). Moreover, despite its preeminence as the allantois's birthplace and its continuity with allantoic mesoderm, the primitive streak is not involved in specification of allantoic

endothelium (Downs et al. 2004) but may play a signaling role in establishment of chorio-adhesive cells (Downs et al. 2004). The allantois vascularizes by vasculogenesis that is not accompanied by erythropoiesis (Downs et al. 1998).

Differentiation into endothelium begins in the distal allantoic region (Downs and Harmann 1997; Downs et al. 1998, 2001), where mesoderm is oldest (Lawson et al. 1991; Lawson and Pedersen 1992; Downs and Harmann 1997; Kinder et al. 1999). From there, formation of the allantois's nascent vascular plexus spreads proximally and amalgamates with the yolk sac and fetal vasculatures, creating a vascular continuum within the conceptus.

Chorioallantoic union is mediated by allantoic mesothelium and is dependent on this tissue's developmental maturity (Downs and Gardner 1995; Downs 1998, 2002). Outer cells exhibit junctional contacts as soon as the allantoic bud is visible (Downs et al. 2004), and contain Ahnak (Downs et al. 2002) and cytokeratins (Downs et al. 2004), all of which may be involved in epithelialization of allantoic mesothelium.

Union requires a receptor/counterreceptor complex, VCAM-1, in the allantois (Gurtner et al. 1995; Kwee et al. 1995) and α4-integrin in the chorion (Yang et al. 1995). Although other investigators (see, e.g., Rossant and Cross 2001) have suggested a role for a large number of gene products in chorioallantoic union (e.g., *Brachyury (T)*: Glueksohn-Shoenheimer 1944; *Lim1*: Shawlot and Behringer 1995; *Mrj*: Hunter et al. 1999), a defect in chorioallantoic union was not distinguished from a defect in allantoic enlargement in these studies.

Chorioallantoic union is, in many ways, similar to the process of implantation (Enders and Schlafke 1969), taking place in three major and similar steps (Downs 2002). In the first, called contact, low levels of VCAM-1 are found in distal allantoic mesothelium, whereas levels of α4-integrin, VCAM-1's counterreceptor, are high throughout chorionic mesoderm (Downs 2002). In the second step, chorioallantoic union is more enduring, with the fusing surfaces becoming intimately juxtaposed; levels of VCAM-1 are robust, while levels of α4-integrin remain relatively unchanged in chorionic mesoderm. In the last step, the fusing surfaces appear to break down, engendering intimate juxtaposition of the nascent allantoic vasculature and chorionic ectoderm, possibly mediated by VCAM-1, which

Figure 6. Projection drawings of sagittal sections of a mouse conceptus during gastrulation. (*A*) 6.5 dpc. Gastrulation has begun, as evidenced by the appearance of the primitive streak and mesoderm. (*B*) 7 dpc. Extraembryonic mesoderm displaces extraembryonic ectoderm from its association with extraembryonic visceral endoderm, resulting in the formation of the posterior amniotic fold whose mesometrial aspect will contribute to the chorion, and whose anti-mesometrial aspect will contribute to the amnion. (*C*) 7.25 dpc. Formation of the exocoelomic cavity is now complete. Its roof is the chorion (chorionic ectoderm plus chorionic mesoderm), its walls the yolk sac (extraembryonic visceral endoderm plus extraembryonic mesoderm), and its floor the amnion (extraembryonic ectoderm plus extraembryonic mesoderm). The allantoic bud has appeared and is continuous with the primitive streak. (Modified, with permission, from Snell and Stevens 1966.)

is associated with core endothelium (Downs 2002). *junB*'s gene product may affect penetration of the allantoic vasculature into the nascent chorionic labyrinth, as *junB* homozygous null conceptuses exhibit a flat chorionic plate and direct apposition of the allantois onto the plate but no penetration (Schorpp et al. 1999).

The Yolk Sac

In vertebrate evolution, the primary function of the yolk sac has been to nourish the embryo until it is sufficiently developed to feed itself. In primitive species of the vertebrate phylum, as well as in oviparous species, including mammals, nourishment is in the form of yolk, which is originally deposited in the maternal egg, then stored in the yolk sac, digested there, and delivered to the embryo via the yolk sac circulation (Fig. 2). Thus, not surprisingly, the vitelline (yolk sac) vasculature is not only the first vascular system to develop in amniotes, but it is also the first site of hematopoiesis (Sabin 1920; Haar and Ackerman 1971). Even in eutherian mammals where it no longer functions to digest yolk, the yolk sac is hematopoietic and retains both digestive and absorptive functions as well.

The yolk sac of the rodent conceptus consists of visceral and parietal components (Snell and Stevens 1966) which, together with trophoblast giant cells, make up a single choriovitelline placenta (Fig. 4) (Brunschwig 1927; Jollie 1990). The rodent yolk sac is an important nutritive organ during gestation ("yolk sac placenta") (Brunschwig 1927; Everett 1933, 1935; Boe 1951), as the chorioallantoic placenta is not operative in rodents until about the limb-bud stage (Jollie 1990). It is due to the yolk sac that postimplantation rodent conceptuses are amenable to whole-embryo culture during gastrulation (New 1978).

Common to both parietal and visceral yolk sacs is extraembryonic endoderm, called, respectively, parietal and visceral endoderm, whose immediate precursor tissue, primitive endoderm (thought to be the equivalent of the chick's hypoblast), came from the ICM at implantation (Figs. 1 and 5A). Parietal endoderm ultimately forms a complex with mural trophectoderm, onto which it crawls (Fig. 5B), but does not form a continuous layer (Jollie 1968). Rather, it consists of a layer of small, isolated cells capable of amoeboid movement and phagocytosis (Gérard 1925). In addition, parietal endoderm is involved in embryo nutrition (Payne and Deuchar 1972), and it secretes a thick extracellular membrane, called Reichert's membrane (Duval 1891), which comes to lie between it and trophoblast giant cells. This membrane, rich in collagen and noncollagenous glycoproteins, is peculiar to rodents (Snell and Stevens 1966) and may act as a filter, allowing free access of nutrients to the embryo while excluding maternal cells (Gardner 1983).

Visceral endoderm forms a continuous layer of simple epithelium, which remains complexed with the epiblast once it is formed. Then, as postimplantation development commences, it becomes associated with both extraembryonic ectoderm and the epiblast and is called, respectively, extraembryonic visceral endoderm and embryonic visceral endoderm (Fig. 5C), which are morphologically different: Extraembryonic endoderm is columnar and gradually becomes squamous in the embryonic region (Fig. 6A). These morphological differences likely reflect differences in biochemical activities, as evidenced by distinct patterns of gene expression (see, e.g., Beddington and Robertson 1999).

Extraembryonic visceral endoderm forms a complex with extraembryonic mesoderm, becoming the visceral yolk sac (Fig. 6B,C); yolk sac mesoderm will differentiate into the vitelline vasculature and primitive erythroid cells, possibly under the influence of the overlying endoderm (Wilt 1965; Miura and Wilt 1970; Belaoussoff et al. 1998; Vokes and Krieg 2002). Embryonic visceral endoderm will either be displaced, via the genetic activity of *angiomotin* (*amot*) (Shimono and Behringer 2003), to the yolk sac, contribute to the foregut, or be replaced by definitive endoderm (Lawson et al. 1986; Lawson and Pedersen 1987). The fine structure of the visceral yolk sac endoderm is remarkably similar in all species (Jollie 1990). The most notable morphological features are its brush border (Dempsey 1953), associated glycocalyx (Calarco and Moyer 1966), cytoplasmic granules (Branca 1923; Gérard 1925), and extensive cytoplasmic lipid droplets (Dempsey 1953). It is extensively involved in absorption (Morriss 1975), manufacture (Shi and Heath 1984), and secretion (Jollie 1990).

Extraembryonic visceral and parietal endoderms are separated from each other by the blastocoel-derived yolk sac cavity (Figs. 4 and 8) which, over time, increases in size so that it is continuous with the labyrinth (Everett 1935). Thus, in that way, materials that go through the parietal yolk sac gain access not only to the visceral yolk sac for transport into the conceptus, but also to the labyrinth of the chorioallantoic placenta (Everett 1935), which may also utilize these compounds.

The Amnion

Of all the fetal membranes, the amnion is most morphologically conserved among amniotes (Arey 1966). The amnion of all species is composed of ectoderm and mesoderm, both of which are derived from the epiblast. In reptiles, birds, and rodents, the amnion forms folds, which ultimately fuse (Figs. 2 and 6B) and surround the embryo. The amniotic cavity produces fluid, which, through the action of smooth muscle derived from amniotic mesoderm, is gently agitated and possibly prevents adhesions during embryonic devel-

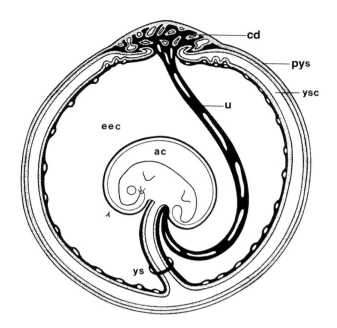

Figure 8. Schematic diagram of the relationship between the definitive fetal membranes in the rat. (cd) Chorionic disk; (u) umbilicus; (ys) visceral yolk sac; (pys) parietal yolk sac; (ysc) yolk sac cavity; (eec) exocoelom; (ac) amniotic cavity. The amnion separates the eec and ac. Some authors (e.g., Arey 1966; Airhart et al. 1996) report that the mature umbilicus is covered in the amniotic membrane. (Modified, with permission, from Jollie 1990 [©Wiley-Liss].)

opment. Thus, by about 12.5 days postcoitum, formation of murine fetal membranes is complete (Fig. 8).

ACKNOWLEDGMENT

K.M.D. was supported by grants (HD-036847 and HD-042706) from the National Institutes of Child Health and Human Development.

REFERENCES

Airhart M.J., Robbins C.M., Knudsen T.B., Church J.K., and Skalko R.G. 1996. Developing allantois is a primary site of 2′-deoxycoformycin toxicity. *Teratology* **53:** 361–373.

Allen W.R. 1975. Endocrine functions of the placenta. In *Comparative placentation: Essays in structure and function* (ed. D. H. Steven), pp. 214–267. Academic Press, London, United Kingdom.

Anson-Cartwright L., Dawson K., Holmyard D., Fisher S.J., Lazzarini R.A., and Cross J.C. 2000. The glial cells missing-1 protein is essential for branching morphogenesis in the chorioallantoic placenta. *Nat. Genet.* **25:** 311–314.

Arey L.B. 1966. *Developmental anatomy.* W.B. Saunders, Philadelphia, Pennsylvania.

Avery G.B. and Hunt C.V. 1969. The differentiation of the trophoblast giant cells in the mouse studied in kidney capsule grafts. *Transplant. Proc.* **1:** 61–66.

Barlow P.W. and Sherman M.I. 1972. The biochemistry of differentiation of mouse trophoblast: Studies on polyploidy. *J. Embryol. Exp. Morphol.* **27:** 447–465.

Beaconsfield P., Birdwood G., and Beaconsfield R. 1980. The placenta. *Sci. Amer.* **243:** 94–102.

Beddington R.S.P. and Robertson E.J. 1999. Axis development and early asymmetry in mammals. *Cell* **96:** 195–209.

Belaoussoff M., Farrington S.M., and Baron M.H. 1998. Hematopoietic induction and respecification of A-P identity by visceral endoderm signaling in the mouse embryo. *Development* **125:** 5009–5018.

Boe F. 1951. Studies on placental circulation in rats. III. Vascularization of the yolk sac. *Acta Endocrinol.* **7:** 42–53.

Branca A. 1923. Recherches sur la vésicule ombilicale. III. La vésicule ombilicale des rongeurs. *Arch. Biol.* **33:** 605–670.

Brodskii V.Y. and Uryvaeva I.V. 1977. Cell polyploidy: Its relation to growth and function. *Int. Rev. Cytol.* **50:** 275–349.

Brown J.J. and Papaioannou V.E. 1993. Ontogeny of hyaluronan secretion during early mouse development. *Development* **117:** 483–492.

Brunschwig A.E. 1927. Notes on experiments in placental permeability. *Anat. Rec.* **34:** 237–244.

Calarco P.G. and Moyer F.H. 1966. Structural changes in the murine yolk sac during gestation: Cytochemical and electron microscope observations. *J. Morphol.* **119:** 341–357.

Copp A.J. 1978. Interaction between inner cell mass and trophectoderm of the mouse blastocyst. I. A study of cellular proliferation. *J. Embryol. Exp. Morphol.* **48:** 109–125.

———. 1979. Interaction between inner cell mass and trophectoderm of the mouse blastocyst. II. The fate of the polar trophectoderm. *J. Embryol. Exp. Morphol.* **51:** 109–120.

———. 1982. Effect of implantation delay on cellular proliferation in the mouse blastocyst. *J. Reprod. Fertil.* **66:** 681–685.

Davies J. and Glasser S.R. 1968. Histological and fine structural observations on the placenta of the rat. Part I. Organization of the normal placenta. *Acta Anat.* **69:** 542–608.

Dempsey E.W. 1953. Electron microscopy of the visceral yolk-sac epithelium of the guinea pig. *Am. J. Anat.* **93:** 331–363.

Downs K.M. 1998. The murine allantois. *Curr. Top. Dev. Biol.* **39:** 1–33.

———. 2002. Early placentation in the mouse. *Placenta* **23:** 116–131.

Downs K.M. and Bertler C. 2000. Growth in the pre-fusion murine allantois. *Anat. Embryol.* **202:** 323–331.

Downs K.M. and Gardner R.L. 1995. An investigation into early placental ontogeny: Allantoic attachment to the chorion is selective and developmentally regulated. *Development* **121:** 407–416.

Downs K.M. and Harmann C. 1997. Developmental potency of the murine allantois. *Development* **124:** 2769–2780.

Downs K.M., Gifford S., Blahnik M., and Gardner R.L. 1998. The murine allantois undergoes vasculogenesis that is not accompanied by erythropoiesis. *Development* **125:** 4507–4521.

Downs K.M., McHugh J., Copp A.J., and Shtivelman E. 2002. Multiple developmental roles of Ahnak are suggested by localization to sites of placentation and neural plate fusion in the mouse conceptus. *Mech. Dev.* (suppl.) **119:** S31–S38.

Downs K.M., Temkin R., Gifford S., and McHugh J. 2001. Study of the murine allantois by allantoic explants. *Dev. Biol.* **233:** 347–364.

Downs K.M., Hellman E.R., McHugh J., Barrickman K., and Inman K. 2004. Investigation into a role for the primitive streak in development of the murine allantois. *Development* **131:** 37–55.

Duval M. 1891. Le placenta des rongeurs. Troisième partie. Le placenta de la souris et du rat. *J. Anat. Physiol. Norm. Pathol. Homme Animaux* **27:** 24–73, 344–395, 515–612.

Ellington S.K.L. 1985. A morphological study of the development of the allantois of rat embryos in vivo. *J. Anat.* **142:** 1–11.

———. 1987. A morphological study of the development of the chorion of rat embryos. *J. Anat.* **150:** 247–263.

Enders A.C. 1965. A comparative study of the fine structure of the trophoblast in several hemochorial placentas. *Am. J. Anat.* **116:** 29–67.

Enders A.C. and Schlafke S. 1969. Cytological aspects of trophoblast-uterine interaction in early implantation. *Am. J. Anat.* **125:** 1–30.

———. 1971. Penetration of the uterine epithelium during implantation in the rabbit. *Am. J. Anat.* **132:** 219–239.

Everett J.W. 1933. Structure and function of the the yolk sac placenta in Mus norvegicus albinus. *Proc. Soc. Exp. Biol. Med.* **31:** 77–79.

———. 1935. Morphological and physiological studies of the placenta in the albino rat. *J. Exp. Zool.* **70:** 243–287.

Fawcett D.W., Wislocki G.B., and Waldo C.M. 1947. The develoment of mouse ova in the anterior chamber of the eye and in the abdominal cavity. *Am. J. Anat.* **81:** 413–443.

Firulli A.B., McFadden D.G., Lin Q., Srivastava D., and Olson E.N. 1998. Heart and extra-embryonic mesodermal defects in mouse embryos lacking the bHLH transcription factor Hand1. *Nat. Genet.* **18:** 266–270.

Fujiwara T., Dunn N.R., and Hogan B.L. 2001. Bone morphogenetic protein 4 in the extraembryonic mesoderm is required for allantois development and the localization and survival of primordial germ cells in the mouse. *Proc. Natl. Acad. Sci.* **98:** 13739–13744.

Gardner R.L. 1972. An investigation of inner cell mass and trophoblast tissue following their isolation from the mouse blastocyst. *J. Embryol. Exp. Morph.* **28:** 279–312.

———. 1983. Origin and differentiation of extraembryonic tissues in the mouse. *Int. Rev. Exp. Pathol.* **24:** 63–143.

Gardner R.L. and Papioannou V.E. 1975. Differentiation in the trophectoderm and inner cell mass. In *The early development of mammals: Second symposium of the British Society for Developmental Biology* (ed. M. Balls and A. E. Wild), pp. 107–132. Cambridge University Press, Cambridge, United Kingdom.

Gardner R.L. and Rossant J. 1976. Determination during embryogenesis. In *Embryogenesis in mammals. Ciba Found. Symp.* **40:** 5–25. Elsevier, Amsterdam, The Netherlands.

———. 1979. Investigation of the fate of 4.5 day post coitum mouse inner cell mass cells by blastocyst injection. *J. Embryol. Exp. Morphol.* **52:** 141–152.

Gardner R.L., Papaioannou V.E., and Barton S.C. 1973. Origin of the ectoplacental cone and secondary giant cells in mouse blastocysts reconstituted from isolated trophoblast and inner cell mass. *J. Embryol. Exp. Morphol.* **30:** 561–572.

Gardner R.L., Lyon M.F., Evans E.P., and Burtenshaw M.D. 1985. Clonal analysis of X-chromosome inactivation and the origin of the germ line in the mouse embryo. *J. Embryol. Exp. Morphol.* **52:** 141–152.

Georgiades P., Ferguson-Smith A.C., and Burton G.J. 2002. Comparative developmental anatomy of the murine human definitive placentae. *Placenta* **23:** 3–19.

Gérard P. 1925. Recherches morphologiques et expérimentales sur la vésicule des rongeurs à feuillets inversés. *Arch. Biol.* **35:** 269–293.

Glueksohn-Shoenheimer S. 1944. The development of normal and homozygous brachy (T/T) mouse embryos in the extraembryonic coelom of the chick. *Proc. Natl. Acad. Sci.* **30:** 134–140.

Guillemot F., Nagy A., Auerbach A., Rossant J., and Joyner A.L. 1994. Essential role of Mash-2 in extraembryonic development. *Nature* **371:** 333–336.

Gurtner G.C., Davis V., Li H., McCoy M.J., Sharpe A., and Cybulsky M.I. 1995. Targeted disruption of the murine VCAM1 gene: Essential role of VCAM-1 in chorioallantoic fusion and placentation. *Genes Dev.* **9:** 1–14.

Haar J.L. and Ackerman G.A. 1971. A phase and electron microscopic study of vasculogenesis and erythropoiesis in the yolk sac of the mouse. *Anat. Rec.* **170:** 199–223.

Hernandez-Verdun D. 1974. Morphogenesis of the syncytium in the mouse placenta. *Cell Tissue Res.* **148:** 381–396.

Hernandez-Verdun D. and Legrand C. 1975. In vitro study of chorionic and ectoplacental trophoblast differentiation in the mouse. *J. Embryol. Exp. Morphol.* **34:** 633–644.

Hubrecht A.A.W. 1888. Keimblätterbildung und Placentation des Igels. *Anat. Anz.* **3:** 510–515.

———. 1889. Studies in mammalian embryology. I. The placentation of the hedgehog (Erinaceus europaeus). *Q. J. Microsc. Sci.* **XXXI.**

Hunter P.J., Swanson B.J., Haendel M.A., Lyons G.E., and Cross J.C. 1999. Mrj encodes a DNAJ-related co-chaperone that is essential for murine placental development. *Development* **126:** 1247–1258.

Ilgren E.B. 1983. Review article: Control of trophoblastic growth. *Placenta* **4:** 307–328.

Jollie W.P. 1964a. Fine structural changes in placental labyrinth of the rat with increasing gestational age. *J. Ultrastruct. Res.* **10:** 27–47.

———. 1964b. Radiographic observations on variations in desoxyribonucleic acid synthesis in rat placenta with increasing gestational age. *Am. J. Anat.* **114:** 161–171.

———. 1965. Fine structural changes in the junctional zone of the rat placenta with increasing gestational age. *J. Ultrastruct. Res.* **12:** 420–438.

———. 1968. Changes in the fine structure of the parietal yolk sac of the rat placenta with increasing gestational age. *Am. J. Anat.* **122:** 513–531.

———. 1990. Development, morphology, and function of the yolk-sac placenta of laboratory rodents. *Teratology* **41:** 361–381.

Kinder S.J., Tsang T.E., Quinlan G.A., Hadjantonakis A.-K., Nagy A., and Tam P.P.L. 1999. The orderly allocation of mesodermal cells to the extraembryonic structures and the anteroposterior axis during gastrulation of the mouse embryo. *Development* **126:** 4691–4701.

Kraut N., Snider L., Chen C.-M.A., Tapscot S.J., and Groudine M. 1998. Requirement of the mouse I-mfa gene for placental development and skeletal patterning. *EMBO J.* **17:** 6276–6288.

Kwee L., Baldwin H.S., Shen H.M., Steward C.L., Buck C., Buck C.A., and Labow M.A. 1995. Defective development of the embryonic and extraembryonic circulatory systems in vascular cell adhesion molecule (VCAM-1) deficient mice. *Development* **121:** 489–503.

Lawson K.A. and Pedersen R.A. 1987. Cell fate, morphogenetic movement and population kinetics of embryonic endoderm at the time of germ layer formation in the mouse. *Development* **101:** 627–652.

———. 1992. Early mesoderm formation in the mouse embryo. *NATO ASI Ser. Ser. A Life Sci.* **231:** 33–46.

Lawson K.A., Meneses J.J., and Pedersen R.A. 1986. Cell fate and cell lineage in the endoderm of the presomite mouse embryo, studied with an intracellular tracer. *Dev. Biol.* **115:** 325–339.

Lawson K.A., Meneses J., and Pedersen R.A. 1991. Clonal analysis of epiblast fate during germ layer formation in the mouse embryo.

Development **113:** 891–911.

Lawson K.A., Dunn N.R., Roelen B.A., Zeinstra L.M., Davis A.M., Wright C.V., Korving J.P., and Hogan B.L. 1999. Bmp4 is required for the generation of primordial germ cells in the mouse embryo. *Genes Dev.* **13:** 424–436.

Miura Y. and Wilt F.H. 1970. The formations of blood islands in dissociated-reaggregated chick embryo yolk sac cells. *Exp. Cell Res.* **59:** 217–226.

Morriss G.M. 1975. Placental evolution and embryonic nutrition. In *Comparative placentation: Essays in structure and function* (ed. D.H. Steven), pp. 87–107. Academic Press, London, United Kingdom.

Mossman H.W. 1937. Comparative morphogenesis of the fetal membranes and accessory uterine structures. *Contr. Embryol.* **26:** 133–247.

New D.A.T. 1978. Whole-embryo culture and the study of mammalian embryos during organogenesis. *Biol. Rev.* **53:** 81–122.

Nichols J., Zevnik B., Anastassiadis K., Niwa H., Klewe–Nebenius D., Chambers I., Schoeler H., and Smith A. 1998. Formation of pluripotent stem cells in the mammalian embryo depends on the POU transcription factor Oct 4. *Cell* **95:** 379–391.

Padykula H. 1958. A histochemical and quantitative study of enzymes of the rat placenta. *J. Anat.* **92:** 118–128.

Palazon L.S., Davies T.J., and Gardner R.L. 1998. Translation inhibition of cyclin B1 and appearance of cyclin D1 very early in the differentiation of mouse trophoblast giant cells. *Mol. Hum. Reprod.* **4:** 1013–1020.

Payne G.S. and Deuchar E.M. 1972. An in vitro study of functions of embryonic membranes in the rat. *J. Embryol. Exp. Morphol.* **27:** 533–542.

Perry J.S. 1981. The mammalian fetal membranes. *J. Reprod. Fertil.* **62:** 321–335.

Riley P., Anson-Cartwright L., and Cross J.C. 1998. The Hand1 bHLH transcription factor is essential for placentation and cardiac morphogenesis. *Nat. Genet.* **18:** 271–275.

Rossant J. and Cross J.C. 2001. Placental development: Lessons from mouse mutants. *Nat. Genet. Rev.* **2:** 538–548.

Rossant J. and Lis W.T. 1979. Potential of isolated mouse inner cell masses to form trophectoderm derivatives in vivo. *Dev. Biol.* **70:** 255–261.

Rossant J., Gardner R.L., and Alexandre H.L. 1978. Investigation of the potency of cells from the postimplantation mouse embryo by blastocyst injection: A preliminary report. *J. Embryol. Exp. Morphol.* **48:** 239–247.

Sabin F.R. 1920. Studies on the origin of blood-vessels and of red blood-corpuscles as seen in the living blastoderm of chicks during the second day of incubation. *Contr. Embryol.* **9:** 215–262.

Salamonsen L.A. 1999. Role of proteases in implantation. *Rev. Reprod.* **4:** 11–22.

Schorpp M., Wang A.-Q., Angel P., and Wagner E.F. 1999. JunB is essential for mammalian placentation. *EMBO J.* **18:** 934–948.

Schreiber J., Riethmacher-Sonnenberg E., Riethmacher D., Tuerk E.E., Enderich J., Bosl M.R., and Wegner M. 2000. Placental failure in mice lacking the mammalian homolog of glial cells missing, GCMa. *Mol. Cell. Biol.* **20:** 2466–2474.

Shawlot W. and Behringer R.R. 1995. Requirement for Lim1 in head-organizer function. *Nature* **374:** 425–430.

Shi W.-K. and Heath J.K. 1984. Apolipoprotein expression by murine visceral yolk sac endoderm. *J. Embryol. Exp. Morphol.* **81:** 143–152.

Shimono A. and Behringer R.R. 2003. Angiomotin regulates visceral endoderm movements during mouse embryogenesis. *Curr. Biol.* **13:** 613–617.

Snell G.B. and Stevens L.C. 1966. Early embryology. In *Biology of the laboratory mouse, by the staff of the Jackson laboratory* (ed. E.L. Green), pp. 205–245. McGraw-Hill, New York.

Steven D.H. 1975. Anatomy of the placental barrier. In *Comparative placentation: Essays in structure and function* (ed. D. Steven), pp. 25–57. Academic Press, London, United Kingdom.

Tam P.P.L. and Beddington R.S.P. 1987. The formation of mesodermal tissues in the mouse embryo during gastrulation and early organogenesis. *Development* **99:** 109–126.

Tanaka M., Gertsenstein M., Rossant J., and Nagy A. 1997. Mash2 acts cell autonomously in mouse spongiotrophoblast development. *Dev. Biol.* **190:** 55–65.

Tanaka S., Kunath T., Hadjantonakis A.-K., Nagy A., and Rossant J. 1998. Promotion of trophoblast stem cell proliferation by FGF4. *Science* **282:** 2072–2075.

Uy G.D., Downs K.M., and Gardner R.L. 2002. Inhibition of trophoblast stem cell potential in chorionic ectoderm coincides with occlusion of the ectoplacental cavity in the mouse. *Development* **129:** 3913–3924.

Varmuza S., Prideaux V., Kothary R., and Rossant J. 1988. Polytene chromosomes in mouse trophoblast giant cells. *Development* **102:** 127–134.

Vokes S.A. and Krieg P.A. 2002. Endoderm is required for vascular endothelial tube formation, but not for angioblast specification. *Development* **129:** 775–785.

Wilt F.H. 1965. Erythropoiesis in the chick embryo: The role of endoderm. *Science* **147:** 1588–1590.

Winnier G., Blessing M., Labosky P.A., and Hogan B.L.M. 1995. Bone morphogenetic protein-4 is required for mesoderm formation and patterning in the mouse. *Genes Dev.* **9:** 2105–2116.

Wislocki G.B. and Padykula H. 1961. Histochemistry and electron microscopy of the placenta. In *Sex and internal secretions* (ed. W.C. Young), pp. 883–957. Williams and Wilkins, Baltimore, Maryland.

Yang J.T., Rayburn H., and Hynes R.O. 1995. Cell adhesion events mediated by a4integrins are essential in placental and cardiac development. *Development* **121:** 549–560.

Zybina E. 1970. Anomalies of polyploidization of the cells of the trophoblast. *Tsitologiya* **12:** 1081–1093.

Zybina E. and Grischenko T. 1970. Polyploid cells of the trophoblast in various sections of the white rat placenta. *Tsitologiya* **12:** 585–595.

PART III

THE MOLECULAR BIOLOGY OF GASTRULATION

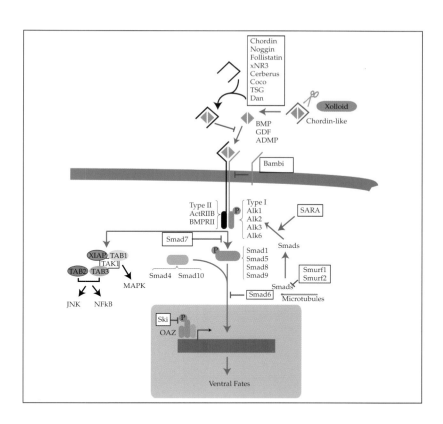

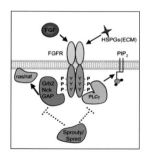

FGF Signaling during Gastrulation

J. Sivak and E. Amaya

Wellcome Trust/Cancer Research UK Institute, University of Cambridge, Cambridge CB2 1QR, United Kingdom

INTRODUCTION

The induction, maintenance, and patterning of the mesoderm and neural ectoderm occur around the time of gastrulation, in response to the positive and negative influences of a handful of growth factor signals. Among these signals are members of the fibroblast growth factor (FGF) family. A role for this family of growth factors in early vertebrate development was first suggested in the late 1980s when it was demonstrated that purified mammalian FGF was able to induce mesoderm in competent animal cap tissue, normally fated to become epidermis in *Xenopus* (Kimelman and Kirschner 1987; Slack et al. 1987). Prior to these experiments, the molecular nature of inducing molecules in vertebrate development had been elusive, despite efforts to purify these factors over several decades (see Chapters 25 and 35). For this reason, the description of the inducing capacity of purified FGF in animal caps received much attention and can be credited for much of the great resurgence in the study of inducing factors in embryogenesis over the past 15 years.

Although initially described as mesoderm-inducing factors, members of the FGF family have subsequently been shown to take part in many other events during early development. In this chapter, we present an overview of the modes of FGF action and review work suggesting a role for this family of growth factors in mesoderm formation, the coordination of cell movements, induction of the nervous system, and anterior–posterior patterning of the embryo. Finally, we discuss the emerging issues and unanswered questions that need to be addressed in order to achieve a

more complete understanding of how FGF signaling accomplishes its multiple roles in the embryo.

MECHANISMS OF FGF SIGNALING

FGFs and Their Receptors

FGFs are a family of peptide factors that are known to elicit a variety of mitogenic and morphogenetic cellular responses upon binding to their receptors. The FGF family now contains over 20 members, which share between 15% and 70% identity (Ornitz and Itoh 2001). Receptor binding is dependent on cooperative interactions with the extracellular matrix component, heparin sulfate proteoglycans (HSPGs) (Yayon et al. 1991). FGF receptors (FGFRs) belong to the family of receptor tyrosine kinases (RTKs). There are four known FGFRs in vertebrates, each containing a single transmembrane domain, an extracellular ligand-binding domain, with two or three immunoglobulin-like domains and an intracellular tyrosine kinase domain (Dickson et al. 2000). Upon ligand binding, the receptor dimerizes, resulting in the initiation of signaling through the trans-phosphorylation of intracellular tyrosine residues within the receptor (Mohammadi et al. 1996; Schlessinger 2000). These phosphorylated tyrosines then act both to activate the catalytic activity of the receptor and to recruit a variety of intracellular signaling molecules (Fig. 1).

Post-Receptor Signaling Pathways

Once the receptor is activated, a number of downstream signaling events are initiated (Fambrough et al. 1999;

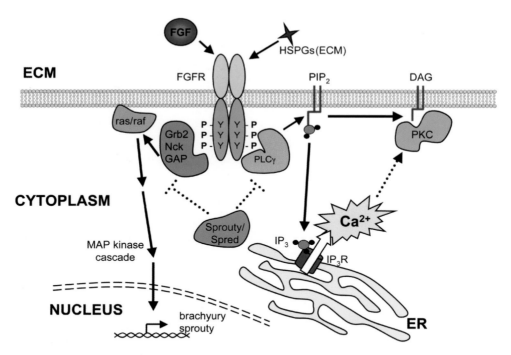

Figure 1. Diagram of two prominent signaling pathways activated in response to FGF in cells. FGF binds to FGF receptors (FGFRs) in cooperation with heparin sulfate proteoglycans (HSPGs) in the extracellular matrix (ECM). Ligand binding induces dimerization and activation of the receptor, which leads to the trans-phosphorylation of tyrosines in the intracellular domain of the receptor. The phosphorylated tyrosines serve as docking sites for a number of SH2-containing proteins, such as Grb2, RasGAP, Nck, and PLCγ. These in turn activate different signaling pathways, such as the ras/MAPK pathway, and Ca++ signaling and activation of PKC. In addition, FGF leads to the transcriptional activation of a number of targets, including genes that feed back on the downstream pathways activated by the receptor. One example of these types of genes is the Sprouty family.

Simon 2000). Src homology 2 (SH2) domain–containing molecules (Songyang and Cantley 1995), such as Nck and Grb2 adapter proteins (Gupta and Mayer 1998), bind to specific phosphorylated residues in the cytoplasmic domain of the receptor. Phosphatidylinositol 3′ kinase (PI3K), shp2, Ras-GTPase activating protein (GAP) (Ryan et al. 1998), and phospholipase Cγ (PLCγ) also interact with the intracellular region of the FGFR. PLCγ hydrolyzes phosphatidylinositol into inositol triphosphate (IP3) and diacylglycerol (DAG) (Mohammadi et al. 1992; Peters et al. 1992). IP3 binds to its receptor on the endoplasmic reticulum, resulting in the release of Ca++ into the cytoplasm, and DAG activates protein kinase C (PKC).

A much-studied signal transduction pathway downstream of the FGFR in gastrulation is the ras/MAPK cascade. This cascade is initiated by the activation of the small GTPase Ras. Ras activates Raf, also known as Map kinase kinase kinase (MAPKKK or MEKK), which then phosphorylates and activates MAP kinase kinase (MAPKK or MEK), which phosphorylates and activates mitogen-activated protein kinase (MAPK, also known as ERK). Activated MAPK (ERK) translocates into the nucleus, where it phosphorylates a subset of transcription factors, ultimately leading to

the transcriptional activation of downstream target genes (Schlessinger 2000).

Regulation of FGFR Signaling

FGFR signaling is regulated at multiple levels, including ligand availability, receptor availability, kinase/phosphatase activity, and activity of Ras inhibitory proteins (Schlessinger 2000). Recently, a number of signaling modulators have been identified in embryos, which are themselves targets of FGFR signaling. These include sef (Furthauer et al. 2002; Tsang et al. 2002; Kovalenko et al. 2003) and Pyst1/MKP3, a dual specificity protein phosphatase, which specifically binds and deactivates MAPK (Eblaghie et al. 2003). Of particular interest during early development is a new family of FGF signaling inhibitors, of which the prototypic member is Sprouty (Spry). Sprouty was first identified in *Drosophila* through its role in development of apical branching of *Drosophila* airways (Hacohen et al. 1998), a process known to require FGF signaling. The *sprouty* mutation leads to growth of multiple fine branches from the stalks of primary ones. This effect mirrors that of overactive FGF signaling and suggests that

Sprouty is an inhibitor of the FGF signaling pathway (Hacohen et al. 1998). In addition, Hacohen et al. (1998) showed that FGF signaling induces *sprouty* expression, which then feeds back and inhibits FGF signaling. Further studies in *Drosophila* have revealed that Sprouty has broader functions, as it can inhibit signaling from a variety of RTKs, including the EGF receptor (Casci et al. 1999). Sprouty family members have been identified in vertebrates, with at least four known mammalian homologs (Christofori 2003; Guy et al. 2003). Vertebrates also contain a family of Sprouty-related molecules called Spreds (Wakioka et al. 2001; Hashimoto et al. 2002; Kato et al. 2003), which similarly function as antagonists of receptor tyrosine kinase signaling.

Sprouty family members are now known to act intracellularly near the plasma membrane to affect MAPK signaling. Several mechanisms of Sprouty activity have been proposed, including interactions with Ras, Grb2, Raf, or molecules farther downstream to affect MAPK signaling (Casci et al. 1999; Reich et al. 1999; Sasaki et al. 2001; Hanafusa et al. 2002). However, evidence in the context of gastrulation in the frog has suggested that Sprouty proteins may inhibit other downstream signaling pathways in addition to, or instead of, the MAPK pathway (Nutt et al. 2001). Whereas Sprouty proteins modulate FGF signaling primarily through inhibition, other proteins, such as c-Cbl, interact with and down-regulate the Sprouty molecules, thus relieving the inhibition on the FGF signaling pathway (Egan et al. 2002; Hall et al. 2003; also see Christofori 2003). Such interactions highlight the intricate and complex regulatory processes within the cells of the embryo.

FGF SIGNALING IN MESODERMAL FORMATION

Early Roles for FGF Signaling in Development

Although FGF1 and FGF2 were first purified in the mid 1970s (Gospodarowicz 1975; Gospodarowicz et al. 1978), an embryological role for these growth factors was not discovered for another 12 years. In 1987, two groups independently reported that purified mammalian FGF1 and FGF2 were capable of changing the fate of naive animal cap tissue isolated from *Xenopus laevis* embryos from ectoderm into mesoderm, a process referred to as mesoderm induction (Kimelman and Kirschner 1987; Slack et al. 1987). Subsequently, several FGF family members have been shown to induce mesoderm in animal cap tissue (Kimelman and Griffin 2000). Inhibition of FGFR signaling through the overexpression of a dominant negative FGFR in both *Xenopus* and zebrafish results in a severe disruption in mesoderm formation (Fig. 2) (Amaya et al. 1991, 1993; Griffin et al. 1995). A similar requirement for FGF signaling during mesoderm formation has also been shown in

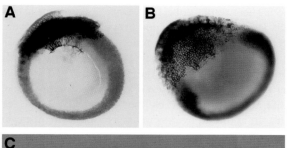
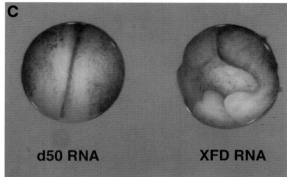
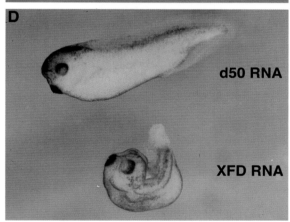

Figure 2. Phenotypes in *Xenopus* embryos overexpressing a dominant negative *FGFR1* construct (XFD) and a control *FGFR1* construct (HAVØ or d50). One blastomere at the 1-cell stage was injected with either HAVØ (*A*) or XFD (*B*) and allowed to develop to the mid-gastrula stage. The embryos were subjected to whole-mount in situ hybridization for *Xenopus brachyury* (*Xbra*), an early mesoderm marker (*blue* stain). The embryos were simultaneously stained for the overexpressed FGF receptor constructs using an FGFR1 antibody (*red* stain). Note in *A* that overexpression of the control construct, HAVØ, does not affect the expression of *Xbra*. However, in *B* the cells expressing the dominant negative construct, XFD, fail to express *Xbra*. (*C*) Phenotype of mid-neurula-stage embryos expressing the control construct, d50 (*left*) and the dominant negative *FGFR1* construct, XFD (*right*). The embryo on the right has failed to complete gastrulation. (*D*) Phenotype of tail-bud-stage embryos expressing the control construct, d50 (*upper*), and the dominant negative *FGFR1* construct, XFD (*lower*). The lower embryo failed to complete gastrulation, but it retains anterior neural structures, such as brain and eyes.

mouse, where targeted deletions of *FGFR1* and *FGF8* result in severe defects in cell migration and mesodermal patterning during gastrulation (Yamaguchi et al. 1994; Sun et al.

1999). These results collectively suggest that some aspect of FGF signaling is essential for proper formation of mesoderm in vertebrates. In mammals, however, a pre-gastrula requirement for FGF signaling has also been described. Mouse embryos lacking either *FGFR2* or *FGF4* die soon after implantation, with severe defects in the proliferation of the inner cell mass (Feldman et al. 1995; Arman et al. 1998). In addition, overexpression of a dominant negative *FGFR1* construct in early mouse embryos results in a cell-autonomous failure to divide beyond the fifth division (Chai et al. 1998). *FGF4* has also been implicated in the maintenance of trophectoderm and primitive endoderm at embryonic day 4.5 (Goldin and Papaioannou 2003). Therefore, although a developmental role for FGF signaling was first described during mesoderm induction, at least in mammals, FGF signaling is also important earlier in development, where it plays an essential role in the control of cell proliferation and in the maintenance of trophectoderm and primitive endoderm.

FGF Signaling in Mesoderm Maintenance

Although FGFs were first shown to induce mesoderm in animal cap tissues from *Xenopus* blastulae, there is still some debate as to whether FGFs play this role in vivo. However, it is clear that FGF signaling is essential for the maintenance of mesodermal fate during the gastrula stages. If mesodermal explants from gastrula embryos are cultured, the explants will continue to form mesodermal derivatives. However, when cells from these explants are dissociated, the cells are no longer capable of differentiating into mesoderm. This suggests that a secreted factor is needed for the maintenance of mesodermal fate (Isaacs et al. 1994). This factor can be substituted by the addition of purified FGF (Isaacs et al. 1994). The process of mesoderm maintenance is similar to, if not the same as, the process described by Gurdon as the "community effect" (Gurdon 1988). The community effect refers to the phenomenon whereby similarly specified cells must be present in a group for them to differentiate (Gurdon 1988). Interestingly, it has been shown recently that the community-effect signal during mesoderm formation in the frog is also FGF (Standley et al. 2001), further suggesting that the processes of mesoderm maintenance and community effect may be one and the same. A definitive role for FGF signaling in mesoderm maintenance was revealed by experiments where a dominant negative FGFR was overexpressed in *Xenopus* embryos during gastrulation, after mesoderm induction had occurred. Although the embryos express early mesodermal markers, the expression of these markers is only transient, and most mesodermal derivatives fail to form (Kroll and Amaya 1996). This maintenance role of FGFs appears to be propagated through induction of an autocrine loop in

which FGFR signaling leads to the expression of the T-box gene *Brachyury* (*Xbra*), which in turn induces eFGF to feed back and reinforce mesodermal fate (Isaacs et al. 1994; Schulte-Merker and Smith 1995). This maintenance role for FGF signaling also explains why TGF-β-related molecules, such as nodal and activin, are unable to "induce" mesoderm in the absence of FGF signaling (Cornell and Kimelman 1994; LaBonne and Whitman 1994). In these cases, mesoderm induction is unperturbed, but maintenance is severely disrupted.

Intracellular Signaling Pathways during Mesoderm Formation

In *Xenopus*, FGF-mediated mesoderm formation requires activity of the MAPK pathway. Addition of dominant negative constructs to any of the signaling molecules along this pathway leads to a blockage of downstream *Xbra* expression and mesoderm formation. For example, dominant negative forms of Ras (Whitman and Melton 1992), Raf (MacNicol et al. 1993), the src-like kinase Laloo (Weinstein et al. 1998), the p85 subunit of PI3K (Carballada et al. 2001), the adapter molecules Nck and Grb2 (Gupta and Mayer 1998), and the tyrosine phosphatase Shp-2 (SH-PTP2) (Tang et al. 1995) all block expression of downstream early mesodermal markers. Conversely, activation of the MAPK pathway alone is sufficient to lead to mesoderm formation (Gotoh et al. 1995; Umbhauer et al. 1995). As noted above, FGFR activation can lead to signaling through a variety of downstream pathways. So far, data have only supported a role for MAPK signaling as a direct regulator of mesoderm formation. In particular, binding of PLCγ to its phosphorylated FGFR site is not necessary for *Xbra* expression and mesoderm formation (Muslin et al. 1994).

FGF SIGNALING AND THE COORDINATION OF GASTRULATION MOVEMENTS

FGFs and Gastrulation Movements in the Frog

During gastrulation, a series of coordinated morphogenetic cell movements are initiated, which ultimately result in the correct positioning of the different germ layers: ectoderm on the outside, mesoderm in the middle, and endoderm on the inside. A potential role for FGF signaling in gastrulation movements was first suggested by analyzing the phenotype of embryos overexpressing a dominant negative FGFR in *Xenopus* (Fig. 2) (Amaya et al. 1991). Although bottle cell and blastopore lip formation is undisturbed in these embryos, gastrulation movements fail soon after, leaving the embryos with an open blastopore (Amaya et al. 1991). The point when gastrulation begins to fail correlates very well with the point when convergent extension movements begin. In *Xenopus*, convergent extension movements, a process by which mesodermal cells intercalate mediolaterally, leading to elongation of the embryo along an anterior–posterior axis, are one of the major driving forces of gastrulation (Gerhart and Keller 1986; Keller 1991; Keller et al. 2000; see Chapter 19). However, since embryos lacking FGFR signaling also fail to form mesoderm, it was not possible to distinguish whether the failure in gastrulation was a direct consequence of disrupted FGF signaling, or whether it was indirectly caused by the failure of the embryos to form mesoderm, the germ layer primarily responsible for the movements of gastrulation. Recent work on the RTK inhibitor, Xsprouty2, has resolved this issue (Nutt et al. 2001). Xsprouty2 is a target of FGF signaling, but has the interesting property that it also feeds back and modulates FGF signaling. However, rather than inhibiting all signaling downstream of the receptor, it only inhibits one or a subset of pathways downstream. For example, Xsprouty2 leaves MAPK signaling largely unaffected, while severely inhibiting signaling leading to Ca++ mobilization (Nutt et al. 2001). As a consequence, Xsprouty2 does not affect mesoderm maintenance, which is dependent on the MAPK pathway, but it does affect the cell movements of gastrulation; in particular, convergent extension movements. This finding is of particular interest because it suggests that bifurcation of signaling pathways downstream of the FGFR may be important in allowing the receptor to take distinct roles in mesoderm formation and morphogenesis. In addition, this work suggests that signaling other than through the ras/MAPK pathway is important for coordinating the movements of gastrulation.

FGFs and Gastrulation in Other Organisms

Work on flies, worms, mouse, and chick, as well as on *Xenopus,* is now beginning to support the hypothesis that FGF signaling plays a direct role in the coordination of cell movements during development.

Drosophila melanogaster *and* Caenorhabditis elegans

D. melanogaster has two FGFRs, and mutations in either lead to defects in morphogenesis (Skaer 1997; Wilson and Leptin 2000). Mutations in the *Drosophila FGFR2* homolog *heartless* (*htl*) result in the correct early delineation of mesodermal fate at the ventral midline. However, although these cells invaginate normally at the beginning of gastrulation, they subsequently fail to migrate properly away from the ventral midline (Beiman et al. 1996; Gisselbrecht et al. 1996). Mutations in the *Drosophila FGFR1* gene, *breathless* (*blt*), result in a disrupted tracheal system, due to a failure of tracheoblasts to migrate in response to the FGF ligand

encoded by the *Drosophila branchless* (*bnl*) gene (Klambt et al. 1992). In this case, FGF acts as a chemoattractant, to which the tracheoblasts respond by extending filopodia and moving toward the FGF signal source (Ribeiro et al. 2002; Sato and Kornberg 2002). A chemoattractant role for FGF signaling has also been shown in the worm, *C. elegans*, during the directed migration of sex myoblasts (DeVore et al. 1995; Burdine et al. 1997, 1998).

Mouse

In mouse, *FGFR1*−/− and *FGF8*−/− embryos display severe gastrulation defects (Deng et al. 1994; Yamaguchi et al. 1994; Sun et al. 1999). Analysis of chimeric mice containing *FGFR1*−/− cells indicates that the primary defect in these cells is their inability to traverse and migrate away from the primitive streak (Ciruna et al. 1997), a defect similar to the one seen in *Drosophila* embryos with a mutation in the *FGFR2* homolog *heartless* (*htl*) gene, as described above (Beiman et al. 1996; Gisselbrecht et al. 1996). The *FGFR1*−/− cells in mouse ultimately fail to contribute to mesodermal lineages and often form ectopic secondary neural tubes (Ciruna et al. 1997). A follow-up study revealed that *FGFR1*−/− cells failed to down-regulate *E-cadherin* expression, possibly as a consequence of the loss of FGF-mediated expression of the transcription factor *mSnail*, and thus, the cells cannot undergo the necessary epithelial to mesenchymal transition (EMT) required for the proper progression of the cells through the primitive streak (Ciruna and Rossant 2001). In addition, *lazy mesoderm* (*lzme*) mutant embryos, which have a mutation in the mouse gene encoding UDP-glucose dehydrogenase required for glycosaminoglycan synthesis, arrest during gastrulation with a similar defect in cell migration (Garcia-Garcia and Anderson 2003). Because these embryos also have disrupted FGF signaling, this result further implicates FGF signaling in the proper migration of mesodermal cells away from the primitive streak in the mouse (Garcia-Garcia and Anderson 2003). Although it is well established that FGFR signaling is essential for the migration of cells through the primitive streak in the mouse, this region does not stain with antibodies specific for activated MAPK (Corson et al. 2003). Therefore, it is unlikely that the movement of cells through the primitive streak in this system depends on MAPK signaling, and thus must be mediated by another pathway downstream of the receptor.

Chick

In the chick, cells follow a characteristic pattern of movements during gastrulation (Yang et al. 2002). Cells in the anterior and middle primitive streak move out laterally, away from the streak, and as the node regresses posteriorly, the cells begin to move back toward the midline. Therefore, anterior and middle streak cells essentially move in a big arc around the node (see Chapter 15). Recent work has suggested that this movement may be coordinated by different FGFs (Yang et al. 2002). Yang et al. (2002) showed that streak cells move toward a source of FGF4, but away from a source of FGF8b. FGF4 acts as a chemoattractant, and FGF8b acts as a chemorepellent. Interestingly, the cells failed to respond significantly to either FGF2 or FGF8c, suggesting that the type of FGF, and even the variant, are very important (Yang et al. 2002). Since FGF4 is expressed in the primitive streak, node, and head process, whereas FGF8 is expressed in the streak only, Yang et al. (2002) postulate that the movement of cells out of the anterior and middle primitive streak is controlled first by a chemorepellent response to FGF8 produced by the streak, followed, once the node has regressed, by a chemoattractant response to FGF4 produced by the head process. Consistent with these results, expression of a dominant negative *FGFR1* construct or treatment with the FGFR1 inhibitor, SU5402, disrupts the movement of streak cells in the chick (Yang et al. 2002).

In summary, studies on a variety of model organisms strongly support a role for FGFs in the regulation of the cell movements of gastrulation, and the details of the postreceptor processes are now coming into focus. Of particular importance are the bifurcation of the pathway and the association of MAPK-independent mechanisms with the regulation of gastrulation movements.

FGF SIGNALING IN NEURAL INDUCTION AND ANTERIOR–POSTERIOR PATTERNING

FGFs and Neural Induction—The Frog

In addition to their role in mesoderm formation and morphogenesis, FGFs have also been implicated in neural induction. The first experiments suggesting a direct role for FGFs in neural induction were performed in *Xenopus*. It was discovered that, whereas FGF induced mesoderm in blastula-stage animal caps, it induced neural tissue in gastrula-stage animal cap cells (Kengaku and Okamoto 1993, 1995; Lamb and Harland 1995). In addition, it appeared that high concentrations of FGF could induce posterior neural tissue, whereas lower concentrations induced anterior neural tissue (Kengaku and Okamoto 1995). Therefore, it was postulated that FGF acts as a morphogen during neural induction (Kengaku and Okamoto 1995). However, these experiments were carried out under conditions of partial or complete dissociation of the gastrula-stage ectoderm cells. Given that simple dissociation of gastrula-stage ectodermal cells is sufficient to induce them to become neural (Godsave and Slack 1989; Grunz and Tacke 1989;

Sato and Sargent 1989), due to attenuation of BMP signaling (Wilson and Hemmati-Brivanlou 1995), it is unclear whether FGF signaling is sufficient to induce neural fate, or indeed, whether it is necessary.

To help answer this question, a number of groups tested the effect of expressing the dominant negative *FGFR1* construct in embryos and animal caps in the context of neural induction. Unfortunately, the results from these experiments were in many cases contradictory. Launay et al. (1996) reported that neural induction in animal caps by either Noggin (a BMP antagonist) or organizer tissue required FGF signaling. Similarly, Sasai et al. (1996) showed that Chordin (another BMP antagonist) induced animal caps to form endoderm when FGF signaling was blocked, instead of neural tissue, suggesting that neural induction by Chordin required FGF signaling. However, other groups have shown that neural induction per se is not dependent on FGFR1 signaling. For example, anterior neural structures form in embryos expressing the dominant negative *FGFR1* construct (Fig. 2) (Amaya et al. 1991, 1993). In addition, several groups have reported that the ability of Noggin to induce anterior neural markers is not dependent on FGFR1 signaling (McGrew et al. 1997; Barnett et al. 1998; Ribisi et al. 2000). Finally, Keller explants (explants isolated from gastrula-stage embryos containing prospective mesoderm and neural tissue) from embryos expressing the dominant negative *FGFR1* express anterior neural markers (Holowacz and Sokol 1999). Therefore, the weight of evidence supports the conclusion that FGFR1 signaling is not required for neural induction in the frog. This conclusion is consistent with evidence in the mouse, where cells mutant for the *FGFR1* gene form predominantly secondary neural tubes, suggesting that not only is signaling through this receptor not required for neural fate, but lack of signaling through this receptor correlates positively with neural fate (Ciruna et al. 1997). However, questions have been raised as to whether neural induction by FGFs is mediated through FGFR1 at all. Hongo et al. (1999) challenged embryos with a dominant negative *FGFR4a* construct in *Xenopus* and found that it was considerably more effective in inhibiting anterior neural structures than the dominant negative *FGFR1* construct, suggesting that FGF signaling is essential in the induction of anterior neural structure, but that the signaling is dependent on FGFR4, rather than FGFR1. Differential roles for FGFR1 and FGFR4 have also been described in the context of neurogenesis and neural patterning in the frog (Hardcastle et al. 2000; Umbhauer et al. 2000).

FGFs and Neural Induction—The Chick

The strongest evidence for a role for FGF signaling during neural induction comes from work on the chick. Recent evidence in the chick suggests that neural induction begins before gastrulation, at a time equivalent to the early blastula in frogs (Streit et al. 2000; Wilson et al. 2000). Explants taken from the medial region of the area pellucida at stage XII are specified to become neural (Wilson et al. 2000). Furthermore, this early specification is dependent on FGF signaling, as incubating these explants in the FGFR inhibitor, SU5402, will inhibit their ability to become neural (Wilson et al. 2000). The role for FGF signaling during this early neural specification phase in the chick appears to be, in part, due to its role in down-regulating the expression of BMPs within the area fated to become neural (Wilson et al. 2000). Further evidence for an early role for FGF signaling in the induction of neural fate in the chick comes from experiments on whole embryos. The region of the epiblast, which is specified early to become neural, expresses the early neural marker *cERNI* (*e*arly *r*esponse to *n*eural *i*nduction) (Streit et al. 2000). *cERNI* is induced by FGFs, but not by the BMP antagonists, Chordin, Noggin, or Cerberus (Streit et al. 2000). In addition, the induction of *cERNI* by the node is dependent on FGFR signaling (Streit et al. 2000). However, FGF signaling is not sufficient for the induction of late neural markers. Therefore, it appears as though FGF signaling plays a very early role during neural induction, perhaps to nudge the medial epiblast toward a neural fate, but maintenance of neural fate requires additional signals (Streit et al. 2000; Wilson et al. 2000). To this end, it is interesting to note that a later role for FGF signaling in the maintenance of neural stem cell precursors within the node has also been described (Mathis et al. 2001).

FGFs and Anterior–Posterior Patterning

Although a role for FGF signaling in "preparing" the ectoderm for neural induction has only been shown in the chick, there is clear evidence from several organisms that FGF signaling is important during posterior neural development. In *Xenopus*, Keller explants expressing the dominant negative *FGFR1* construct lack posterior neural markers (Holowacz and Sokol 1999), and FGF signaling, through Ras, is essential for posterior neural fate in whole embryos (Ribisi et al. 2000). In zebrafish, FGF signaling is also required for posterior neural development (Koshida et al. 2002; Kudoh et al. 2002). In chick, FGF-soaked beads are capable of inducing ectopic posterior neural tissue (Alvarez et al. 1998; Storey et al. 1998). Whether FGFs have a direct role in inducing posterior neural tissue, or an indirect role, is not clear. However, based on its ability to "posteriorize" neural tissue (Cox and Hemmati-Brivanlou 1995; Doniach 1995; Lamb and Harland 1995), it has been suggested that FGF may be part of the transforming signal, postulated by Pieter Nieuwkoop (Nieuwkoop et al. 1952), that caudalizes anterior neurectoderm, thus establishing anterior–posterior patterning of the nervous system.

A potential role for FGF signaling is not limited to anterior–posterior patterning of the nervous system. In fact, it has been suggested that FGFs play an important part in establishing anterior–posterior patterning of the whole embryo, in part through its regulation of the caudal and Hox genes (Ruiz i Altaba and Melton 1989; Cho and De Robertis 1990; Isaacs et al. 1992, 1998; Kolm and Sive 1995; Pownall et al. 1996, 1998; Slack et al. 1996; Lohnes 2003). A role for FGF signaling in anterior–posterior patterning through regulation of Hox genes may be conserved, as hypermorphic and hypomorphic mutations in *FGFR1* in the mouse cause shifts in anterior–posterior patterning (Partanen et al. 1998).

EMERGING ISSUES

As described in this chapter, FGFs have been implicated in a number of events during early development. One overriding question is, How can FGFs perform so many different functions essentially at the same time, and often within the same cells? One possible explanation may be that there are several ligands and receptors expressed during early development. Therefore, one only needs to consider that the cells respond in different ways, simply because they discriminate between the different ligands. Therefore, although FGFs, in general, provide many functions, different ligands are used more specifically for one function or another. One good example of this is the effect on the motility of streak cells in the chick when exposed to FGF4 and FGF8b (Yang et al. 2002). Although streak cells move toward an FGF4 source, the same cells move away from an FGF8b source, and completely ignore FGF2. Clearly, these cells can easily discriminate between the different FGFs. It is not yet known how they discriminate between them, but it is probably due to differential affinities of the ligands to the receptors. This, however, does not explain why the cells behave differently. If the difference is simply due to one ligand stimulating a receptor more effectively than another, why is it that stimulating one receptor results in chemoattraction, whereas stimulating another results in chemorepulsion? Might this observation be a result of a qualitative difference in the activation of different intracellular signaling pathways, or a quantitative difference? If so, how would such a mechanism ultimately result in different (and in this case, opposite) responses in the behaviors of cells? A related question is, How much input does the extracellular matrix (ECM) have in the interpretation of the signal? FGFs require the cooperation of heparin-binding proteoglycans to efficiently bind their receptors, so perhaps the ECM might affect ligand-binding specificities and, ultimately, the responses of cells to different FGFs.

Differential responses of cells to the same signal can also be dictated by differences in competence of the cells to the signal, either in space or time. In these cases, the cells respond differently not because they discriminate between different FGFs, but because the cells are initially different. Recently, work in several systems has begun to reveal some of the molecules responsible for these differences in competence of cells. A good example of this is seen in the ascidian. Here, FGF can induce either mesenchyme or notochord (Kobayashi et al. 2003). Whether the cells respond to the signal by becoming mesenchyme or notochord depends on the cells inheriting a maternally localized zinc finger transcription factor, Macho-1. Cells that inherit Macho-1 respond to FGF by becoming mesenchyme, and cells not inheriting Macho-1 respond to FGF by becoming notochord (Kobayashi et al. 2003).

Interestingly, in the chick, another zinc finger transcription factor, Churchill, has been implicated in the differential response of cells to FGF (Sheng et al. 2003), except that in this case, the difference is in time, rather than in space. Churchill functions in the switch between the responses of cells to FGF during mesoderm formation versus neural formation, which occur at slightly different times in development. Interestingly, Churchill is itself induced slowly by FGF. Once it accumulates, this transcription factor changes the responsiveness to FGF, so that, rather than the cells responding in mesoderm formation, the cells now become sensitized to neural induction. It has therefore been postulated that Churchill plays an important role in the transition between gastrulation and neurulation (Sheng et al. 2003). Here, an important aspect of the cellular response to FGF is the induction of genes that ultimately feed back and change the responsiveness of the cells to FGF. In this way, FGF can have one role at one point, and another role later. A similar mechanism has also been described for the Sproutys. Sprouty genes are induced by FGF, and once translated, Sprouty proteins feed back and modulate FGF signaling (Christofori 2003; Guy et al. 2003). It has been proposed that by having these two properties, Sprouty2 in the frog helps cells distinguish between the separate roles for FGF in mesoderm formation and in the coordination of cell movements during gastrulation (Nutt et al. 2001).

FGF signaling in the embryo does not occur in isolation from other signaling events. One important issue that needs to be addressed is how the different pathways influence each other, and how much interplay between them is essential for the correct coordination of developmental processes. Interactions between the Wnt and FGF pathways have been proposed. Ciruna and Rossant (2001) have suggested that changes in *E-Cadherin* expression downstream of FGFR1 during gastrulation in the mouse result in indirect regulation of Wnt3a signaling via alterations in the levels of cytoplasmic β-catenin (Ciruna and Rossant 2001). Another link is through *wnt11* (zebrafish *silberblick* (*slb*), which is involved in zebrafish and *Xenopus* gastrulation

through a noncanonical signaling pathway (Tada and Smith 2000). Mutations in *wnt11* result in convergence and extension defects in both systems, and Wnt11 signals are necessary for the formation of cell polarity during *Xenopus* gastrulation (Heisenberg et al. 2000; Tada and Smith 2000; Wallingford et al. 2002; Ulrich et al. 2003). *Xwnt11* is itself a target of Xbra, a prominent downstream effector of FGF signaling (Tada and Smith 2000), therefore linking FGF signaling to noncanonical Wnt pathways. This link, however, is transcriptional. How much interplay between these signaling pathways is direct? For example, both the noncanonical Wnt pathway and the FGF pathway can stimulate Ca^{++} signaling (Schlessinger 2000; Wang and Malbon 2003). Both pathways can stimulate or be affected by protein kinase C activity (Schlessinger 2000; Kinoshita et al. 2003; Sheldahl et al. 2003). What is the molecular basis of this interplay? Does stimulation of the pathway by one signal affect the response of the cell to the other signal? If so, how, and what role does this play in the coordination of the two signaling pathways? Similar questions can be raised in the interplay between FGF signaling and TGF-β signaling, including nodal, activin, and BMP signaling. This cross-talk in signaling pathways adds yet another layer of complexity that must be taken into account before we fully understand how the embryo develops.

REFERENCES

Alvarez I.S., Araujo M., and Nieto M.A. 1998. Neural induction in whole chick embryo cultures by FGF. *Dev. Biol.* **199:** 42–54.

Amaya E., Musci T.J., and Kirschner M.W. 1991. Expression of a dominant negative mutant of the FGF receptor disrupts mesoderm formation in *Xenopus* embryos. *Cell* **66:** 257–270.

Amaya E., Stein P.A., Musci T.J., and Kirschner M.W. 1993. FGF signalling in the early specification of mesoderm in *Xenopus*. *Development* **118:** 477–487.

Arman E., Haffner-Krausz R., Chen Y., Heath J.K., and Lonai P. 1998. Targeted disruption of fibroblast growth factor (FGF) receptor 2 suggests a role for FGF signaling in pregastrulation mammalian development. *Proc. Natl. Acad. Sci.* **95:** 5082–5087.

Barnett M.W., Old R.W., and Jones E.A. 1998. Neural induction and patterning by fibroblast growth factor, notochord and somite tissue in *Xenopus*. *Dev. Growth. Differ.* **40:** 47–57.

Beiman M., Shilo B.Z., and Volk T. 1996. Heartless, a *Drosophila* FGF receptor homolog, is essential for cell migration and establishment of several mesodermal lineages. *Genes Dev.* **10:** 2993–3002.

Burdine R.D., Branda C.S., and Stern M.J. 1998. EGL-17(FGF) expression coordinates the attraction of the migrating sex myoblasts with vulval induction in *C. elegans*. *Development* **125:** 1083–1093.

Burdine R.D., Chen E.B., Kwok S.F., and Stern M.J. 1997. egl-17 encodes an invertebrate fibroblast growth factor family member required specifically for sex myoblast migration in *Caenorhabditis elegans*. *Proc. Natl. Acad. Sci.* **94:** 2433–2437.

Carballada R., Yasuo H., and Lemaire P. 2001. Phosphatidylinositol-3 kinase acts in parallel to the ERK MAP kinase in the FGF pathway during *Xenopus* mesoderm induction. *Development* **128:** 35–44.

Casci T., Vinós J., and Freeman M. 1999. Sprouty, an intracellular inhibitor of Ras signaling. *Cell* **96:** 655–665.

Chai N., Patel Y., Jacobson K., McMahon J., McMahon A., and Rappolee D.A. 1998. FGF is an essential regulator of the fifth cell division in preimplantation mouse embryos. *Dev. Biol.* **198:** 105–115.

Cho K.W. and De Robertis E.M. 1990. Differential activation of *Xenopus* homeo box genes by mesoderm-inducing growth factors and retinoic acid. *Genes Dev.* **4:** 1910–1916.

Christofori G. 2003. Split personalities: The agonistic antagonist Sprouty. *Nat. Cell Biol.* **5:** 377–379.

Ciruna B. and Rossant J. 2001. FGF signaling regulates mesoderm cell fate specification and morphogenetic movement at the primitive streak. *Dev. Cell* **1:** 37–49.

Ciruna B.G., Schwartz L., Harpal K., Yamaguchi T.P., and Rossant J. 1997. Chimeric analysis of fibroblast growth factor receptor-1 (Fgfr1) function: A role for FGFR1 in morphogenetic movement through the primitive streak. *Development* **124:** 2829–2841.

Cornell R.A. and Kimelman D. 1994. Activin-mediated mesoderm induction requires FGF. *Development* **120:** 453–462.

Corson L.B., Yamanaka Y., Lai K.M., and Rossant J. 2003. Spatial and temporal patterns of ERK signaling during mouse embryogenesis. *Development* **130:** 4527–4537.

Cox W.G. and Hemmati-Brivanlou A. 1995. Caudalization of neural fate by tissue recombination and bFGF. *Development* **121:** 4349–4358.

Deng C.X., Wynshaw-Boris A., Shen M.M., Daugherty C., Ornitz D.M., and Leder P. 1994. Murine FGFR-1 is required for early postimplantation growth and axial organization. *Genes Dev.* **8:** 3045–3057.

DeVore D.L., Horvitz H.R., and Stern M.J. 1995. An FGF receptor signaling pathway is required for the normal cell migrations of the sex myoblasts in *C. elegans* hermaphrodites. *Cell* **83:** 611–620.

Dickson C., Spencer-Dene B., Dillon C., and Fantl V. 2000. Tyrosine kinase signalling in breast cancer: Fibroblast growth factors and their receptors. *Breast Cancer Res.* **2:** 191–196.

Doniach T. 1995. Basic FGF as an inducer of anteroposterior neural pattern. *Cell* **83:** 1067–1070.

Eblaghie M.C., Lunn J.S., Dickinson R.J., Munsterberg A.E., Sanz-Ezquerro J.J., Farrell E.R., Mathers J., Keyse S.M., Storey K., and Tickle C. 2003. Negative feedback regulation of FGF signaling levels by Pyst1/MKP3 in chick embryos. *Curr. Biol.* **13:** 1009–1018.

Egan J.E., Hall A.B., Yatsula B.A., and Bar-Sagi D. 2002. The bimodal regulation of epidermal growth factor signaling by human Sprouty proteins. *Proc. Natl. Acad. Sci.* **99:** 6041–6046.

Fambrough D., McClure K., Kazlauskas A., and Lander E.S. 1999. Diverse signaling pathways activated by growth factor receptors induce broadly overlapping, rather than independent, sets of genes. *Cell* **97:** 727–741.

Feldman B., Poueymirou W., Papaioannou V.E., DeChiara T.M., and Goldfarb M. 1995. Requirement of FGF-4 for postimplantation mouse development. *Science* **267:** 246–249.

Furthauer M., Lin W., Ang S.L., Thisse B., and Thisse C. 2002. Sef is a feedback-induced antagonist of Ras/MAPK-mediated FGF signalling. *Nat. Cell Biol.* **4:** 170–174.

Garcia-Garcia M.J. and Anderson K.V. 2003. Essential role of glycosaminoglycans in FGF signalling during mouse gastrulation. *Cell* **114:** 727–737.

Gerhart J. and Keller R. 1986. Region-specific cell activities in amphibian gastrulation. *Annu. Rev. Cell Biol.* **2:** 201–229.

Gisselbrecht S., Skeath J.B., Doe C.Q., and Michelson A.M. 1996.

heartless encodes a fibroblast growth factor receptor (DFR1/DFGF-R2) involved in the directional migration of early mesodermal cells in the *Drosophila* embryo. *Genes Dev.* **10:** 3003–3017.

Godsave S.F. and Slack J.M. 1989. Clonal analysis of mesoderm induction in *Xenopus laevis*. *Dev. Biol.* **134:** 486–490.

Goldin S.N. and Papaioannou V.E. 2003. Paracrine action of FGF4 during periimplantation development maintains trophectoderm and primitive endoderm. *Genesis* **36:** 40–47.

Gospodarowicz D. 1975. Purification of a fibroblast growth factor from bovine pituitary. *J. Biol. Chem.* **250:** 2515–2520.

Gospodarowicz D., Bialecki H., and Greenburg G. 1978. Purification of the fibroblast growth factor activity from bovine brain. *J. Biol. Chem.* **253:** 3736–3743.

Gotoh Y., Masuyama N., Suzuki A., Ueno N., and Nishida E. 1995. Involvement of the MAP kinase cascade in *Xenopus* mesoderm induction. *EMBO J.* **14:** 2491–1498.

Griffin K., Patient R., and Holder N. 1995. Analysis of FGF function in normal and no tail zebrafish embryos reveals separate mechanisms for formation of the trunk and the tail. *Development* **121:** 2983–2994.

Grunz H. and Tacke L. 1989. Neural differentiation of *Xenopus laevis* ectoderm takes place after disaggregation and delayed reaggregation without inducer. *Cell Differ. Dev.* **28:** 211–217.

Gupta R.W. and Mayer B.J. 1998. Dominant-negative mutants of the SH2/SH3 adapters Nck and Grb2 inhibit MAP kinase activation and mesoderm-specific gene induction by eFGF in *Xenopus*. *Oncogene* **17:** 2155–2165.

Gurdon J.B. 1988. A community effect in animal development. *Nature* **336:** 772–774.

Guy G.R., Wong E.S., Yusoff P., Chandramouli S., Lo T.L., Lim J., and Fong C.W. 2003. Sprouty: How does the branch manager work? *J. Cell Sci.* **116:** 3061–3068.

Hacohen N., Kramer S., Sutherland D., Hiromi Y., and Krasnow M.A. 1998. *sprouty* encodes a novel antagonist of FGF signaling that patterns apical branching of the *Drosophila* airways. *Cell* **92:** 253–263.

Hall A.B., Jura N., DaSilva J., Jang Y.J., Gong D., and Bar-Sagi D. 2003. hSpry2 is targeted to the ubiquitin-dependent proteasome pathway by c-Cbl. *Curr. Biol.* **13:** 308–314.

Hanafusa H., Torii S., Yasunaga T., and Nishida E. 2002. Sprouty1 and Sprouty2 provide a control mechanism for the Ras/MAPK signalling pathway. *Nat. Cell Biol.* **4:** 850–858.

Hardcastle Z., Chalmers A.D., and Papalopulu N. 2000. FGF-8 stimulates neuronal differentiation through FGFR-4a and interferes with mesoderm induction in *Xenopus* embryos. *Curr. Biol.* **10:** 1511–1514.

Hashimoto S., Nakano H., Singh G., and Katyal S. 2002. Expression of Spred and Sprouty in developing rat lung. *Gene Expr. Patterns* **2:** 347–353.

Heisenberg C.P., Tada M., Rauch G.J., Saude L., Concha M.L., Geisler R., Stemple D.L., Smith J.C., and Wilson S.W. 2000. Silberblick/Wnt11 mediates convergent extension movements during zebrafish gastrulation. *Nature* **405:** 76–81.

Holowacz T. and Sokol S. 1999. FGF is required for posterior neural patterning but not for neural induction. *Dev. Biol.* **205:** 296–308.

Hongo I., Kengaku M., and Okamoto H. 1999. FGF signaling and the anterior neural induction in *Xenopus*. *Dev. Biol.* **216:** 561–581.

Isaacs H.V., Pownall M.E., and Slack J.M. 1994. eFGF regulates Xbra expression during *Xenopus* gastrulation. *EMBO J.* **13:** 4469–4481.

———. 1998. Regulation of Hox gene expression and posterior development by the *Xenopus* caudal homologue Xcad3. *EMBO J.* 17: 3413–3427.

Isaacs H.V., Tannahill D., and Slack J.M. 1992. Expression of a novel FGF in the *Xenopus* embryo. A new candidate inducing factor for mesoderm formation and anteroposterior specification. *Development* 114: 711–720.

Kato R., Nonami A., Taketomi T., Wakioka T., Kuroiwa A., Matsuda Y., and Yoshimura A. 2003. Molecular cloning of mammalian Spred-3 which suppresses tyrosine kinase-mediated Erk activation. *Biochem. Biophys. Res. Commun.* 302: 767–772.

Keller R. 1991. Early embryonic development of *Xenopus laevis*. *Methods Cell Biol.* 36: 61–113.

Keller R., Davidson L., Edlund A., Elul T., Ezin M., Shook D., and Skoglund P. 2000. Mechanisms of convergence and extension by cell intercalation. *Philos. Trans. R. Soc. Lond. B Biol. Sci.* 355: 897–922.

Kengaku M. and Okamoto H. 1993. Basic fibroblast growth factor induces differentiation of neural tube and neural crest lineages of cultured ectoderm cells from *Xenopus* gastrula. *Development* 119: 1067–1078.

———. 1995. bFGF as a possible morphogen for the anteroposterior axis of the central nervous system in *Xenopus*. *Development* 121: 3121–3130.

Kimelman D. and Griffin K.J. 2000. Vertebrate mesendoderm induction and patterning. *Curr. Opin. Genet. Dev.* 10: 350–356.

Kimelman D. and Kirschner M. 1987. Synergistic induction of mesoderm by FGF and TGF-beta and the identification of an mRNA coding for FGF in the early *Xenopus* embryo. *Cell* 51: 869–877.

Kinoshita N., Iioka H., Miyakoshi A., and Ueno N. 2003. PKCδ is essential for Dishevelled function in a noncanonical Wnt pathway that regulates *Xenopus* convergent extension movements. *Genes Dev.* 17: 1663–1676.

Klambt C., Glazer L., and Shilo B.Z. 1992. breathless, a *Drosophila* FGF receptor homolog, is essential for migration of tracheal and specific midline glial cells. *Genes Dev.* 6: 1668–1678.

Kobayashi K., Sawada K., Yamamoto H., Wada S., Saiga H., and Nishida H. 2003. Maternal macho-1 is an intrinsic factor that makes cell response to the same FGF signal differ between mesenchyme and notochord induction in ascidian embryos. *Development* 130: 5179–5190.

Kolm P.J. and Sive H.L. 1995. Regulation of the *Xenopus* labial homeodomain genes, HoxA1 and HoxD1: Activation by retinoids and peptide growth factors. *Dev. Biol.* 167: 34–49.

Koshida S., Shinya M., Nikaido M., Ueno N., Schulte-Merker S., Kuroiwa A., and Takeda H. 2002. Inhibition of BMP activity by the FGF signal promotes posterior neural development in zebrafish. *Dev. Biol.* 244: 9–20.

Kovalenko D., Yang X., Nadeau R.J., Harkins L.K., and Friesel R. 2003. Sef inhibits fibroblast growth factor signaling by inhibiting FGFR1 tyrosine phosphorylation and subsequent ERK activation. *J. Biol. Chem.* 278: 14087–14091.

Kroll K.L. and Amaya E. 1996. Transgenic *Xenopus* embryos from sperm nuclear transplantations reveal FGF signaling requirements during gastrulation. *Development* 122: 3173–3183.

Kudoh T., Wilson S.W., and Dawid I.B. 2002. Distinct roles for Fgf, Wnt and retinoic acid in posteriorizing the neural ectoderm. *Development* 129: 4335–4346.

LaBonne C. and Whitman M. 1994. Mesoderm induction by activin requires FGF-mediated intracellular signals. *Development* 120: 463–472.

Lamb T.M. and Harland R.M. 1995. Fibroblast growth factor is a direct neural inducer, which combined with noggin generates anterior-posterior neural pattern. *Development* 121: 3627–3636.

Launay C., Fromentoux V., Shi D.L., and Boucaut J.C. 1996. A truncated FGF receptor blocks neural induction by endogenous *Xenopus* inducers. *Development* 122: 869–880.

Lohnes D. 2003. The Cdx1 homeodomain protein: An integrator of posterior signaling in the mouse. *BioEssays* 25: 971–980.

MacNicol A.M., Muslin A.J., and Williams L.T. 1993. Raf-1 kinase is essential for early *Xenopus* development and mediates the induction of mesoderm by FGF. *Cell* 73: 571–583.

Mathis L., Kulesa P.M., and Fraser S.E. 2001. FGF receptor signalling is required to maintain neural progenitors during Hensen's node progression. *Nat. Cell Biol.* 3: 559–566.

McGrew L.L., Hoppler S., and Moon R.T. 1997. Wnt and FGF pathways cooperatively pattern anteroposterior neural ectoderm in *Xenopus*. *Mech. Dev.* 69: 105–114.

Mohammadi M., Dikic I., Sorokin A., Burgess W.H., Jaye M., and Schlessinger J. 1996. Identification of six novel autophosphorylation sites on fibroblast growth factor receptor 1 and elucidation of their importance in receptor activation and signal transduction. *Mol Cell Biol* 16: 977–989.

Mohammadi M., Dionne C.A., Li W., Li N., Spivak T., Honegger A.M., Jaye M., and Schlessinger J. 1992. Point mutation in FGF receptor eliminates phosphatidylinositol hydrolysis without affecting mitogenesis. *Nature* 358: 681–684.

Muslin A.J., Peters K.G., and Williams L.T. 1994. Direct activation of phospholipase C-gamma by fibroblast growth factor receptor is not required for mesoderm induction in *Xenopus* animal caps. *Mol. Cell. Biol.* 14: 3006–3012.

Nieuwkoop P.D., Boterenbrood E.C., Kremer A., Bloesma F.F.S.N., Hoessels E.L.M.J., Meyer G., and Verheyen F.J. 1952. Activation and organization of the central nervous system in amphibians. *J. Exp. Zool.* 120: 1–108.

Nutt S.L., Dingwell K.S., Holt C.E., and Amaya E. 2001. *Xenopus* Sprouty2 inhibits FGF-mediated gastrulation movements but does not affect mesoderm induction and patterning. *Genes Dev.* 15: 1152–1166.

Ornitz D.M. and Itoh N. 2001. Fibroblast growth factors. *Genome Biol.* 2: REVIEWS3005.

Partanen J., Schwartz L., and Rossant J. 1998. Opposite phenotypes of hypomorphic and Y766 phosphorylation site mutations reveal a function for Fgfr1 in anteroposterior patterning of mouse embryos. *Genes Dev.* 12: 2332–44.

Peters K.G., Marie J., Wilson E., Ives H.E., Escobedo J., Del Rosario M., Mirda D., and Williams L.T. 1992. Point mutation of an FGF receptor abolishes phosphatidylinositol turnover and Ca2+ flux but not mitogenesis. *Nature* 358: 678–681.

Pownall M.E., Isaacs H.V., and Slack J.M. 1998. Two phases of Hox gene regulation during early *Xenopus* development. *Curr. Biol.* 8: 673–676.

Pownall M.E., Tucker A.S., Slack J.M., and Isaacs H.V. 1996. eFGF, Xcad3 and Hox genes form a molecular pathway that establishes the anteroposterior axis in *Xenopus*. *Development* 122: 3881–3892.

Reich A., Sapir A., and Shilo B. 1999. Sprouty is a general inhibitor of receptor tyrosine kinase signaling. *Development* 126: 4139–4147.

Ribeiro C., Ebner A., and Affolter M. 2002. In vivo imaging reveals different cellular functions for FGF and Dpp signaling in tracheal branching morphogenesis. *Dev. Cell* 2: 677–683.

Ribisi S. Jr., Mariani F.V., Aamar E., Lamb T.M., Frank D., and Harland R.M. 2000. Ras-mediated FGF signaling is required for the formation of posterior but not anterior neural tissue in *Xenopus laevis*. *Dev. Biol.* 227: 183–196.

Ruiz i Altaba A. and Melton D.A. 1989. Interaction between peptide growth factors and homoeobox genes in the establishment of antero-posterior polarity in frog embryos. *Nature* 341: 33–38.

Ryan P.J., Paterno G.D., and Gillespie L.L. 1998. Identification of phosphorylated proteins associated with the fibroblast growth factor receptor type I during early *Xenopus* development. *Biochem. Biophys. Res. Commun.* 244: 763–767.

Sasai Y., Lu B., Piccolo S., and De Robertis E.M. 1996. Endoderm induction by the organizer-secreted factors chordin and noggin in *Xenopus* animal caps. *EMBO J.* 15: 4547–4555.

Sasaki A., Taketomi T., Wakioka T., Kato R., and Yoshimura A. 2001. Identification of a dominant negative mutant of Sprouty that potentiates fibroblast growth factor- but not epidermal growth factor-induced ERK activation. *J. Biol. Chem.* 276: 36804–36808.

Sato M. and Kornberg T.B. 2002. FGF is an essential mitogen and chemoattractant for the air sacs of the *Drosophila* tracheal system. *Dev. Cell* 3: 195–207.

Sato S.M. and Sargent T.D. 1989. Development of neural inducing capacity in dissociated *Xenopus* embryos. *Dev. Biol.* 134: 263–266.

Schlessinger J. 2000. Cell signaling by receptor tyrosine kinases. *Cell* 103: 211–225.

Schulte-Merker S. and Smith J.C. 1995. Mesoderm formation in response to *Brachyury* requires FGF signalling. *Curr. Biol.* 5: 62–67.

Sheldahl L.C., Slusarski D.C., Pandur P., Miller J.R., Kuhl M., and Moon R.T. 2003. Dishevelled activates Ca2+ flux, PKC, and CamKII in vertebrate embryos. *J. Cell Biol.* 161: 769–777.

Sheng G., dos Reis M., and Stern C.D. 2003. Churchill, a zinc finger transcriptional activator, regulates the transition between gastrulation and neurulation. *Cell* 115: 603–613.

Simon M.A. 2000. Receptor tyrosine kinases: Specific outcomes from general signals. *Cell* 103: 13–15.

Skaer H. 1997. Morphogenesis: FGF branches out. *Curr. Biol.* 7: R238–241.

Slack J.M., Darlington B.G., Heath J.K., and Godsave S.F. 1987. Mesoderm induction in early *Xenopus* embryos by heparin-binding growth factors. *Nature* 326: 197–200.

Slack J.M., Isaacs H.V., Song J., Durbin L., and Pownall M.E. 1996. The role of fibroblast growth factors in early *Xenopus* development. *Biochem. Soc. Symp.* 62: 1–12.

Songyang Z. and Cantley L.C. 1995. Recognition and specificity in protein tyrosine kinase-mediated signalling. *Trends Biochem. Sci.* 20: 470–475.

Standley H.J., Zorn A.M., and Gurdon J.B. 2001. eFGF and its mode of action in the community effect during *Xenopus* myogenesis. *Development* 128: 1347–1357.

Storey K.G., Goriely A., Sargent C.M., Brown J.M., Burns H.D., Abud H.M., and Heath J.K. 1998. Early posterior neural tissue is induced by FGF in the chick embryo. *Development* 125: 473–484.

Streit A., Berliner A.J., Papanayotou C., Sirulnik A., and Stern C.D. 2000. Initiation of neural induction by FGF signalling before gastrulation. *Nature* 406: 74–78.

Sun X., Meyers E.N., Lewandoski M., and Martin G.R. 1999. Targeted disruption of Fgf8 causes failure of cell migration in the gastrulating mouse embryo. *Genes Dev.* 13: 1834–1846.

Tada M. and Smith J.C. 2000. *Xwnt11* is a target of *Xenopus* Brachyury: Regulation of gastrulation movements via Dishevelled, but not through the canonical Wnt pathway. *Development* 127: 2227–2238.

Tang T.L., Freeman R.M. Jr., O'Reilly A.M., Neel B.G., and Sokol S.Y. 1995. The SH2-containing protein-tyrosine phosphatase SH-PTP2 is required upstream of MAP kinase for early *Xenopus* development. *Cell* 80: 473–483.

Tsang M., Friesel R., Kudoh T., and Dawid I.B. 2002. Identification of Sef, a novel modulator of FGF signalling. *Nat. Cell Biol.* 4: 165–169.

Ulrich F., Concha M.L., Heid P.J., Voss E., Witzel S., Roehl H., Tada M., Wilson S.W., Adams R.J., Soll D.R., and Heisenberg C.-P. 2003. Slb/Wnt11 controls hypoblast cell migration and morphogenesis at the onset of zebrafish gastrulation. *Development* 130: 5375–5384.

Umbhauer M., Marshall C.J., Mason C.S., Old R.W., and Smith J.C. 1995. Mesoderm induction in *Xenopus* caused by activation of MAP kinase. *Nature* 376: 58–62.

Umbhauer M., Penzo-Mendez A., Clavilier L., Boucaut J., and Riou J. 2000. Signaling specificities of fibroblast growth factor receptors in early *Xenopus* embryo. *J. Cell Sci.* 113: 2865–2875.

Wakioka T., Sasaki A., Kato R., Shouda T., Matsumoto A., Miyoshi K., Tsuneoka M., Komiya S., Baron R., and Yoshimura A. 2001. Spred is a Sprouty-related suppressor of Ras signalling. *Nature* 412: 647–651.

Wallingford J.B., Fraser S.E., and Harland R.M. 2002. Convergent extension: The molecular control of polarized cell movement during embryonic development. *Dev. Cell* 2: 695–706.

Wang H.Y. and Malbon C.C. 2003. Wnt signaling, Ca2+, and cyclic GMP: Visualizing Frizzled functions. *Science* 300: 1529–1530.

Weinstein D.C., Marden J., Carnevali F., and Hemmati-Brivanlou A. 1998. FGF-mediated mesoderm induction involves the Src-family kinase Laloo. *Nature* 394: 904–908.

Whitman M. and Melton D.A. 1992. Involvement of p21ras in *Xenopus* mesoderm induction. *Nature* 357: 252–254.

Wilson P.A. and Hemmati-Brivanlou A. 1995. Induction of epidermis and inhibition of neural fate by Bmp-4. *Nature* 376: 331–333.

Wilson R. and Leptin M. 2000. Fibroblast growth factor receptor-dependent morphogenesis of the *Drosophila* mesoderm. *Philos. Trans. R. Soc. Lond. B Biol. Sci.* 355: 891–895.

Wilson S.I., Graziano E., Harland R., Jessell T.M., and Edlund T. 2000. An early requirement for FGF signalling in the acquisition of neural cell fate in the chick embryo. *Curr. Biol.* 10: 421–429.

Yamaguchi T.P., Harpal K., Henkemeyer M., and Rossant J. 1994. fgfr-1 is required for embryonic growth and mesodermal patterning during mouse gastrulation. *Genes Dev.* 8: 3032–3044.

Yang X., Dormann D., Munsterberg A.E., and Weijer C.J. 2002. Cell movement patterns during gastrulation in the chick are controlled by positive and negative chemotaxis mediated by FGF4 and FGF8. *Dev. Cell* 3: 425–437.

Yayon A., Klagsbrun M., Esko J.D., Leder P., and Ornitz D.M. 1991. Cell surface, heparin-like molecules are required for binding of basic fibroblast growth factor to its high affinity receptor. *Cell* 64: 841–848.

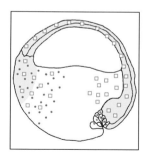

Modulation of BMP Signaling during Vertebrate Gastrulation

I. Muñoz-Sanjuán and A.H. Brivanlou

Laboratory of Molecular Vertebrate Embryology,
The Rockefeller University, New York, New York 10021

INTRODUCTION

Recent years have seen a great advancement in the understanding of the molecular signals that specify cell fate decisions during gastrulation. Among the signal transduction pathways, transforming growth factor β (TGF-β) signaling plays a central role in all aspects of gastrulation, from regulating cell fate to morphogenetic movements. In the embryo, patterning of the germ layers during gastrulation occurs largely from the interplay of two opposing morphogen gradients, one dorso-anterior to ventro-posterior and mediated by Nodal/Activin-type signals, and the second, in the opposite direction, mediated by bone morphogenetic protein (BMP) signals (Harland and Gerhart 1997; Muñoz-Sanjuán and Brivanlou 2001; Myers et al. 2002a). Recently, it has become apparent that both processes, cell migration and cell fate, are intrinsically connected (Ip and Gridley 2002; Myers et al. 2002a,b). Indeed, the elucidation of the molecular basis of both processes is essential to understanding how the complex cellular behaviors that take place in gastrulation are translated into the specification of diverse cellular fates. It is still unclear, however, to what extent the regulation of cellular behavior and cell fate decisions are connected mechanistically. In short, are there independent but parallel pathways that control both processes, does cell fate specification precede migratory behavior, or is migration required for cell fate acquisition? In this review, we focus on the role of the BMP subfamily of TGF-β signals during gastrulation. Because the role of BMPs in dorsoventral patterning and neural induction is covered elsewhere in this volume, we concentrate on the molecular aspects of the regulation of BMP signaling, how that regulation affects the generation of BMP morphogen gradients in the gastrulating embryo, and how these gradients are interpreted to generate a diverse array of cell fates in all three germ layers.

The BMPs and growth and differentiation factors (GDFs) comprise a separate branch of the TGF-β superfamily of ligands. BMPs and GDFs are grouped together on the basis of amino acid homology, and typically they activate similar signal transducers (Massague and Chen 2000). BMP signals during gastrulation have been generally described as ventralizing signals in *Xenopus* embryos, in both the ectoderm (specifying epidermal fates versus neural fates) and the mesoderm (Dale et al. 1992; Jones et al. 1992; Wilson and Hemmati-Brivanlou 1995; Suzuki et al. 1997; Muñoz-Sanjuán and Brivanlou 2002). Because BMPs appear to act as morphogens in a variety of assays in *Xenopus* and zebrafish (Dosch et al. 1997; Wilson et al. 1997; Jones and Smith 1998; Dale and Wardle 1999), the integration of BMP responses becomes important for understanding fate specification. Recently, a myriad of secreted and intracellular regulators of BMP signaling have been described (Massague and Chen 2000; Muñoz-Sanjuán and Brivanlou 2002) (see Table 1 and Fig. 1) and their activities act to modulate the extent and quality of BMP signals.

Although we have a good understanding of what BMPs and their inhibitors can do in the embryo, based on gain-of-function experiments, we still lack a clear picture of how BMP signals act as morphogens in the specification of cell

Table 1. *Secreted inhibitors of the BMP pathway*

Gene	Inhibits	Species	Gastrula expression?[a]	Features, comments	References
Chordin	BMP-2,4,7	mouse *Xenopus* zebrafish chicken	node (m) organizer (x,z) node and rostral mesendoderm (c)		Sasai et al. (1994) Schulte-Merker et al. (1997) Piccolo et al. (1996) Streit et al. (1998)
CHL/chordin-like	BMP-4,5,6	mouse	no	3 CR-domains	Nakayama et al. (2001)
Noggin	BMP-2,4,7 GDF-5	mouse *Xenopus* zebrafish	node (m) organizer (x,z) axial mesendoderm (c)	3-noggin-like genes found in zebrafish	Smith and Harland (1992) Zimmerman et al. (1996) Furthauer et al. (1999) Connolly et al. (1997) McMahon et al. (1998) Chapman et al. (2002)
Follistatin	BMP-2,4,7,11 GDF-8,11 activin	mouse *Xenopus* chick	node (m) organizer (x) node, mesendoderm, caudal neural plate (c)		Hemmati-Brivanlou et al. (1994) Nakamura et al. (1990) Yamashita et al. (1995) Iemura et al. (1998) Chapman et al. (2002)
FSRP proteins: FLRG, Flik	BMP-2,6,7 activin	mouse chicken	FLRG: e7.0 by northern (m) Flik-1: node (c)	follistatin related	Shibanuma et al. (1993) Patel et al. (1996) Hayette et al. (1998) Schneyer et al. (2001) Tortoriello et al. (2001)
Cerberus	BMP-4 xNr-1,2 Wnt-8	*Xenopus* mouse (Cer1) chicken	anterior endoderm (x) anterior visceral endoderm (m) hypoblast, ant. endoderm, prechordal plate (c)		Bouwmeester et al. (1996) Piccolo et al. (1999) Biben et al. (1998) Belo et al. (1997) Hsu et al. (1998)
Coco	BMP-4 activin xNr-1 Wnt-8	*Xenopus*	gradient from animal to vegetal strongest expression in ectoderm	cerberus/dan related	Bell et al. (2003)
Dan	BMP-2,4,7 GDF-5,6,7	mouse *Xenopus*	no no		Ozaki et al. (1996) Hsu et al. (1998) Pearce et al. (1999) Eimon and Harland (2001)
Caronte	BMP-4,7	chicken	mesoderm flanking the node		Rodriguez Esteban et al. (1999) Yokouchi et al. (1999)
Lefty1 Lefty2	BMP2	mouse chicken	notochord/midline (Lefty1; m,c) mesoderm (Lefty2; m,c)		Meno et al. (1996, 1997)
Dante	N.D.	mouse	node	no full-length cDNA reported	Pearce et al. (1999)
PRDC	N.D.	mouse	N.D.	cerberus/Dan-like	Minabe-Saegusa et al. (1998)
Drm/Gremlin	BMP-2,4	mouse *Xenopus*	no no		Hsu et al. (1998) Topol et al. (1997) Pearce et al. (1999)
Neuralin-1	BMP-4,5 TGF-β1,2	mouse	emerging neural plate	3 CR-domains	Coffinier et al. (2001) Nakayama et al. (2001)
CTGF-	BMP-4 TGF-β1	*Xenopus*	weak expression	1 CR-domain	Abreu et al. (2002)
Kielin	N.D	*Xenopus*	axial mesoderm	27 CR-domains	Matsui et al. (2000)
TSG	BMP-4	*Xenopus* mouse	ventral region (x) N.D.	reported to act as both an antagonist and an agonist of BMP signaling	Oelgeschlager et al. (2000) Chang et al. (2001) Ross et al. (2001) Scott et al. (2001) Larrain et al. (2001)

Table 1. (Continued.)

Gene	Inhibits	Species	Gastrula expression?[a]	Features, comments	References
Amnionless	N.D.	mouse	visceral endoderm	1 CR-domain 1 TM-domain	Kalantry et al. (2001)
CRIM-1	N.D.	mouse	no	6 CR-domains 1 IGFBP motif 1 TM-domain	Kolle et al. (2000)
Nell family	N.D.	mouse (NELL1,2) chicken	no N.D.	multiple CR-domains. multiple EGF-domains; some contain TM-domains	Matsuhashi et al. (1995) Watanabe et al. (1996) Kuroda and Tanizawa (1999)
Xnr3	BMP-4	*Xenopus*	organizer	nodal-related gene	Hansen et al. (1997)
Sclerostin/SOST	BMP-5,6	mouse	no		Balemans et al. (2001); Balemans and Van Hul (2002) Brunkow et al. (2001)
Sclerostin-like	N.D.	mouse	N.D.		Balemans and Van Hul (2002)

[a]Expression as measured by RNA localization. Species expression domains are described as follows:
(m) mouse; (c) chicken; (x) *Xenopus laevis*; (f) zebrafish.
(N.D.) Not determined.

fates, and how a certain threshold of BMP signals is linked to a permanent fate acquisition. Loss-of-function studies in the mouse and in zebrafish (Kodjabachian et al. 1999; Massague and Chen 2000; Zhao 2003) have revealed the requirement of some components of the BMP pathway in gastrulation (see Table 2). Ultimately, if we are to understand the role of BMPs during gastrulation, we must try to make sense of how single cells within a given position and migratory trajectory in the embryo acquire a certain fate as a *consequence* of integrating the level of BMP signals, in the context of other pathways that play important roles in modulating a cell's response to BMPs.

THE BMP PATHWAY: REGULATION AND INTEGRATION OF SIGNALING

BMP and GDF ligands bind and activate a subset of type I and type II TGF-β receptors (Fig. 1) (Massague and Chen 2000), which leads to the phosphorylation of the carboxy-terminal domain of the signal transducers Smads-1, -5, -8, and -9. These form a complex with Smad-4 or Smad-4b and translocate to the nucleus where they associate with a variety of developmentally regulated transcription factors, such as OAZ, to direct gene expression (Fig. 1). Although rarely included in the BMP pathway, the TGF-β-activated kinase (TAK1) has been shown to be activated following BMP signaling, and it is likely to play an important role in the regulation of signaling cross-talk between Smad activation and other pathways (see below) (Ishitani et al. 1999; Meneghini et al. 1999; Kimura et al. 2000; Takatsu et al. 2000; Goswami et al. 2001; Wang et al. 2001). There are 42 TGF-β ligands identified in the human genome, although the actual num-

ber of effective ligands is unknown, given the promiscuity of heterodimer formation among BMP, GDF, or TGF-β proteins (Nishimatsu and Thomsen 1998; Massague and Chen 2000).

Extracellular Modulation of BMP Signals

Although usually shown as a relatively simple linear pathway, BMP signaling is regulated at multiple levels, both extra- and intracellularly (Figs. 1 and 2). In fact, the extracellular modulation of BMP signaling is extremely complex (Pearce et al. 1999; Abreu et al. 2002; Balemans and Van Hul 2002; Muñoz-Sanjuán and Brivanlou 2002), highlighting the notion that the robustness of morphogen gradients arises from a tight regulation in the extracellular space. This regulation likely involves controlling the rates of synthesis, diffusion, degradation, and differential affinity for distinct receptor complexes. A number of secreted BMP inhibitors act to bind BMPs and prevent receptor activation. These inhibitors were initially identified as molecules produced by the vertebrate organizer which eliminated activity of BMPs in the dorsoanterior region of the embryo in order to promote dorsal fates in the mesendoderm and neural fates in the ectoderm (Harland and Gerhart 1997; Muñoz-Sanjuán and Brivanlou 2002). The majority of the secreted inhibitors, which include Noggin, Chordin, Follistatin, Cerberus, and XNR3, have restricted expression domains in dorsal tissues, and they are thought to act locally to eliminate BMP signals. Although much progress has been made in identifying inhibitors that display neural inducing properties (owing to their ability to block BMP signals in the anterior ectoderm; Wilson et al. 1997), it is still unclear whether these inhibitors play qualitatively different roles in

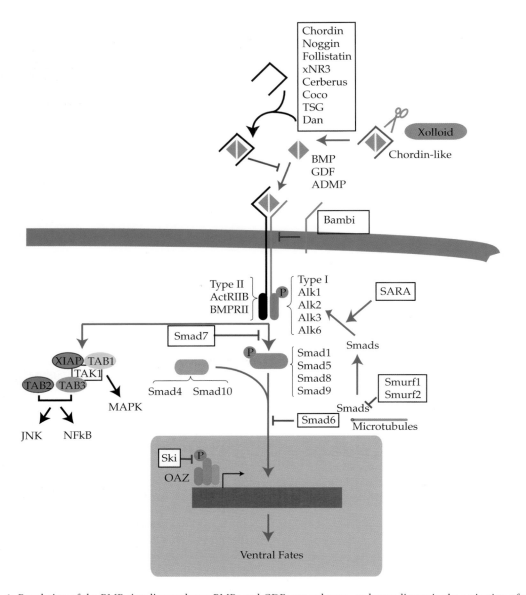

Figure 1. Regulation of the BMP signaling pathway. BMPs and GDFs act as homo- or heterodimers in the activation of type I and type II TGF-β receptors. A variety of secreted extracellular binding proteins act to inhibit or modulate the extent of receptor activation by BMPs. Following receptor dimerization, the Smad signal transducers become activated by direct phosphorylation by the receptor complex. Once activated, the Smad1/5/8/9 molecules bind to Smad4 and translocate into the nucleus where they can activate transcription in association with a variety of tissue-specific factors that largely dictate the specificity of the transcriptional targets. Intracellularly, there is negative regulation of the pathway by Smad6 and Smad7, and by the activity of the Smurf-ubiquitin ligases. In addition to Smad activation, there is cumulative evidence for a role of the TGF-β-activated kinase (TAK1) complex to signal downstream of BMP receptors during gastrulation. How Smad and TAK-dependent signaling events are integrated remains largely unexplored during early development. (*Red arrows*) Inhibitory interactions; (*green arrows*) stimulatory interactions.

the embryo during gastrulation. On the basis of homology searches, a large number of secreted inhibitors have been identified (Table 1), many of which are also expressed during gastrulation. In particular, there are many molecules with Chordin-like repeats (CR-repeats), which supposedly also bind BMPs and are likely to play important roles in the

generation of BMP graded signals (Abreu et al. 2002). CR-domains are also found in many extracellular matrix (ECM)-proteins, such as pro-collagens and the Nell-family members (Table 1) (Abreu et al. 2002).

Interestingly, a novel Cerberus-like molecule has been identified that shows a ubiquitous pattern of expression in

Table 2. *Genetics of BMP components in the mouse and zebrafish during gastrulation*

Mutant	Species	Gastrulation phenotype	References
Bmp2	m	delayed primitive streak	Zhang and Bradley (1996); Ying and Zhao (2001)
Bmp4	m	lack of mesoderm and posterior truncation	Winnier et al. (1995)
Bmp5	m	lack of mesoderm; lethality at e8.5–e9.5	Pfendler et al. (2000)
Lefty2	m	increased mesodermal derivatives and enlarged primitive streak	Meno et al. (1999)
Alk2/ ActRIA	m	defects in mesoderm and visceral endoderm formation; gastrulation defect linked to *Alk2* expression in extraembryonic visceral endoderm	Gu et al. (1999); Mishina et al. (1999)
Alk4	m	lack of mesoderm and epiblast differentiation	Gu et al. (1998)
Alk3/BmprIA	m	lack of mesoderm; lethality at e8.5–e9.5	Mishina et al. (1995)
BmprII	m	lack of mesoderm and epiblast differentiation; lethality at e9.5	Beppu et al. (2000)
Smad1	m	defects in visceral endoderm and extraembryonic mesoderm	Tremblay et al. (2001); Lechleider et al. (2001)
Smad2	m	defects in visceral endoderm; abnormal mesoderm patterning	Waldrip et al. (1998); Weinstein et al. (1998)
Smad4	m	lack of mesoderm; lethality at e8.5–e9.5	Takaku et al. (1998); Sirard et al. (1998); Yang et al. (1998)
Smad5	m	reduced mesoderm; lethality at e9.5–e11.5	Chang et al. (1999, 2000, 2001)
swirl/bmp2b	z	dorsalization phenotype	Kishimoto et al. (1997); Nguyen et al. (1998)
snailhouse/bmp7	z	dorsalization phenotype	Mullins et al. (1996)
chordino/chordin	z	ventralization phenotype	Schulte-Merker et al. (1997); Fisher et al. (1997); Miller-Bertoglio et al. (1999)
Bozozok	z	lack of shield/organizer formation	Solnica-Krezel et al. (1996); Fekany et al. (1999)
mini fin/tolloid	z	dorsalization phenotype	Connors et al. (1999)
ogon	z	ventralization phenotype; mutated gene not cloned	Miller-Bertoglio et al. (1997)
somitabun/Smad5	z	dorsalization phenotype	Hild et al. (1999)
swirl/chordino	z	phenotype similar to *swirl* fish	Hammerschmidt et al. (1996)
lost-a-fin/Alk8	z	dorsalization phenotype	Mintzer et al. (2001); Bauer et al. (2001)

the animal region of the *Xenopus* gastrula (Fig. 2) (Bell et al. 2003). This gene, termed *Coco*, is maternally encoded and can act as a potent mesoderm inhibitor in the embryo, promoting neural fates (Bell et al. 2003). This widespread expression challenges the notion that secreted BMP inhibitors are expressed in discrete domains, and poses the question of what the interplay is among these different inhibitors in the generation of patterning information. Recently, data from *Drosophila* suggest that an additional role for the BMP inhibitor Chordin/Sog might be to regulate the diffusion rate of BMP/Dpp (Eldar et al. 2002). The implications of this finding for morphogen generation are discussed below, but it appears that some secreted BMP-binding molecules, such as Chordin (Eldar et al. 2002; Myers et al. 2002b) and Twisted Gastrulation (TSG) (Oelgeschlager et al. 2000; Chang et al. 2001; Larrain et al. 2001; Ross et al. 2001; Scott et al. 2001), might not always be inhibitory but can act as both agonists and antagonists of BMP signals. TSG is a secreted factor initially identified in *Drosophila* (Yu et al. 2000). Although the biology of TSG is

complex, in *Drosophila* it is thought to act to modulate BMP signaling in an inhibitory fashion (Yu et al. 2000). In *Xenopus*, TSG activities have been shown to block BMP signals (Chang et al. 2001; Ross et al. 2001; Scott et al. 2001), although it has also been shown that, in some contexts, it can promote BMP signaling through the recruitment of Xolloid to the complexes. Xolloid is a zinc metalloprotease known to cleave Chordin as part of a larger Chordin/BMP/TSG complex, which can therefore act to free BMP molecules (Larrain et al. 2001). Whether this level of complexity extends to BMP-binding molecules other than Chordin is unknown.

Intracellular Modulation of the BMP Pathway

After receptor activation, the activation of Smad signal transducers is regulated by SARA, a molecule that facilitates the interaction between the Smads and the receptor complex (Fig. 1) (Tsukazaki et al. 1998). However, an involvement of SARA during gastrulation has not been

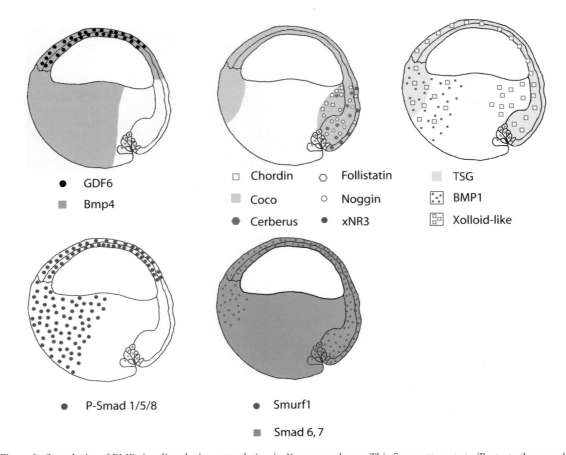

Figure 2. Complexity of BMP signaling during gastrulation in *Xenopus* embryos. This figure attempts to illustrate the complex regulation of the BMP pathway during early gastrulation in *Xenopus*, where the expression and activity of various BMP components have been better illustrated. The various panels depict approximate RNA expression patterns of the BMP/GDF ligands, extracellular inhibitors, intracellular inhibitors, and extracellular modulators of the pathway in a stage 10.5 (early gastrula) embryo. We also depict recent data on the level of phosphorylated Smad1/5/8 at this time in *Xenopus*. Ultimately, a cell's response to a BMP signal depends on the expression of the various components and also on the integration of BMP receptor activation with inputs from other signaling pathways.

addressed. In the cytosol, a number of inhibitory molecules regulate the pathway (Massague and Chen 2000). Smad6 competes with Smad4 for binding to activated Smad1 (Nakayama et al. 1998), whereas Smad7 is thought to exert part of its inhibitory activities by preventing the activation of Smad1 and 2 (Casellas and Brivanlou 1998). In addition, the ubiquitin ligases Smurf1 and 2 target Smad1 and 2 and the TGF-β receptors for degradation by the proteosome (Zhu et al. 1999; Kavsak et al. 2000). Interestingly, Smad6 and 7 are expressed ubiquitously in the gastrula embryo (Casellas and Brivanlou 1998; Nakayama et al. 1998), although their role in modulating the BMP pathway, and their interplay with the secreted neural inducers, have not been addressed. It would be interesting to evaluate whether cells lacking these inhibitors shift the threshold of activation to BMP signals

during mesoderm induction, which could be studied in the context of cell fate decisions from embryonic stem cells isolated from mice lacking Smad6 or 7, which do not show obvious gastrulation phenotypes (Zhao 2003).

Finally, within the nucleus, BMP signaling can be inhibited by the activity of a variety of transcriptional repressor complexes such as those recruited by the Ski oncoprotein, among others (Fig. 1) (Massague and Chen 2000; Wotton and Massague 2001). The specificity of BMP signal transduction at the level of gene expression resides in the association of the activated Smad complexes with a variety of cell-type-specific transcription factors (Massague and Chen 2000). In the gastrulating embryo, OAZ appears to be a critical determinant of ventral mesodermal fates, through the regulation of the ventral mesodermal genes *Vent-1* and *-2* (Hata et al. 2000). However, there is a need to identify further transcriptional regulators that regulate BMP responses in a cell-specific manner. The responsiveness of a cell to a given amount of initial BMP signal, or to an initial level of Smad1/5/8 phosphorylation, will be largely dependent on the intracellular milieu of factors that interact with the activated Smads. Therefore, the translation of the initial BMP signal depends on the status of the cell at the time and position in which it receives a signal.

Integration of Signals Downstream of BMP Receptor Activation

BMP receptor engagement activates the TAK1 kinase (Shibuya et al. 1998; Yamaguchi et al. 1999). TAK1 can interact with a variety of TAK1-binding proteins (TABs), which are required to modulate the activity of the kinase and direct its substrate specificity. TAK1 is an important enzyme during development, as it regulates the Wnt pathway in *Caenorhabditis elegans* and mammalian cells (Ishitani et al. 1999; Meneghini et al. 1999) and MAPK and JNK pathways in vertebrate cells (Kimura et al. 2000; Takatsu et al. 2000; Goswami et al. 2001; Wang et al. 2001). In *Xenopus*, BMP signaling regulates MAPK signaling through TAK1/TAB1 complexes, and inhibition of TAK1 activity pharmacologically results in neural fates as a consequence of a blockade of BMP signals (Goswami et al. 2001). Furthermore, a dominant negative TAB3 molecule can promote epidermal fates in animal cap explants even in the presence of neural inducers (Muñoz-Sanjuán et al. 2002). These results suggest that TAK1 plays a critical role in the modulation of BMP signaling in the *Xenopus* gastrula, and that it acts in the intracellular signaling cross-talk between Smads, MAPK, JNK, and Wnt signals. Additionally, Smad1 activity can be regulated directly by the MAPK (Kretzschmar et al. 1997) and STAT pathways (Nakashima et al. 1999). Although these findings need to be evaluated in vivo, the relevance of such signal transduction cross-talk for the gastrulating embryo is unknown.

In Vivo Patterns of Smad1/5/8- Phosphorylation during Gastrulation

With carboxy-terminal phospho-Smad-specific antibodies, it is now possible to monitor activated Smads in the embryo. Although these observations only communicate a static picture, they do allow us to follow Smad1/5/8 activated molecules. Two recent papers, on *Xenopus* and chick embryos, have shed light on the level of active BMP signaling during gastrulation (Faure et al. 2000, 2002). However, several important caveats must be taken into consideration when analyzing these results. First, the status of Smad1/5/8-P does not convey any information on whether BMP receptor activation leads to an activation or repression of certain developmental programs, as this is mostly dictated by transcrip-

tional cofactors present in the cell at the time of signaling (Massague and Chen 2000). Second, it is unclear whether signaling initiated by BMP receptor activation results in the phosphorylation of Smad1 versus Smad5 or 8, and whether differentially activating these Smads plays a role in the responses following the initial BMP signal. Third, Smad6, a negative regulator of BMP signaling, has been shown to bind to Smad1/5/8-P and to act as a transcriptional repressor (Hata et al. 1998; Bai et al. 2000), independently of the association of Smad1/5/8-P with Smad4. Therefore, it is important to keep in mind that the presence of Smad1/5/8-P does not automatically imply a transcriptionally active complex.

In *Xenopus*, the levels of Smad1/5/8-P do not allow one to distinguish the germ layers (Faure et al. 2000), although Smad1/5/8-P levels appear higher by a few fold in the ventral than in the dorsal marginal zones (Faure et al. 2000). The pattern of Smad1/5/8-P changes as gastrulation commences, from a symmetric expression pattern (Faure et al. 2000) to an asymmetric pattern with higher levels in the ventral side. This initial lack of asymmetry has been used to postulate that the major role of BMP signaling in the early embryo is to oppose dorsalizing/neuralizing signals (Faure et al. 2000), although this did not take into account the possibility that the responsiveness of a cell to a given amount of Smad1/5/8-P might depend on additional cofactors expressed asymmetrically prior to and during gastrulation. As shown for BMPs in embryonic frog explants (Suzuki et al. 1997; Wilson et al. 1997), a small increase in Smad1/5/8-P might be sufficient to account for fate choices in response to BMPs, although this has not been fully demonstrated in vivo.

In the chick, Smad1/5/8-P immunoreactivity is detected in the entire epiblast, similar to the expression of Bmp-4, prior to primitive streak formation. This expression becomes asymmetric as this immunoreactivity decreases at the caudal end of the area pellucida during primitive streak initiation (Faure et al. 2002), and there is no detectable Smad1/5/8-P in the entire streak subsequently. Later, this immunoreactivity reappears in the area pellucida at HH stage 3, with the exception of the node and primitive streak, where, presumably, the secreted BMP inhibitors prevent Smad1/5/8 phosphorylation (Faure et al. 2002). Altogether, these results are not entirely consistent with the expression of the various BMP ligands and secreted inhibitors in the chick at this time. As is the case with *Xenopus*, additional regulators must be present to account for the levels of Smad1/5/8-P. These results suggest that mechanisms other than receptor activation regulate the readout of BMP signaling in fate specification. In the context of neural induction in the chick, in the pre-streak stages there is ample Smad1/5/8-P, which has been used to support the arguments that BMP inhibition is not required in the prospective neural territory prior to gastrulation (Streit et al. 1998;

Wilson et al. 2000; Faure et al. 2002). However, the complete lack of phospho-Smad1 immunoreactivity in the region overlaying the node and in the neural plate, and the inverse correlation between the distribution of secreted BMP antagonists and Smad1/5/8-P also supports models which postulate that BMP cessation mediated by the secreted inhibitors is critical for neural tissue specification in the chick as well (see Chapter 29).

THE ESTABLISHMENT OF THE BMP MORPHOGEN GRADIENT

In all species studied, there is a decreasing gradient of Bmp ligand expression from the ventral side of the embryo, and an opposing (presumed) gradient of secreted antagonists in the dorsal side (Harland and Gerhart 1997; Harland 2000), a view that is also supported by existing data on Smad1/5/8 activation during gastrulation in *Xenopus* (Faure et al. 2000). Traditionally, it has been postulated that the graded expression of these soluble molecules could explain the morphogenic effects of BMPs in dorsoventral specification in all germ layers (Muñoz-Sanjuán and Brivanlou 2001). However, recent evidence from flies and zebrafish suggests that additional components are required to refine the morphogen model of action of BMPs (Piccolo et al. 1997; Goodman et al. 1998; Blitz et al. 2000; Eldar et al. 2002; Wagner and Mullins 2002). Recently, Eldar and colleagues (Eldar et al. 2002) presented a model that could account for the robustness of the BMP morphogen gradient in *Drosophila*. In this model, soluble BMP/Dpp or Chordin/Sog does not diffuse, although BMP/Chordin complexes can diffuse greatly. Therefore, if this model holds true, a second role for Chordin would be to act as a carrier of inactive BMP molecules to a distant site of action, following cleavage by Tolloid-like proteases. Similarly, in this model, Tolloid can only cleave Chordin/Sog as part of a complex with BMP/Dpp and TSG. The degradation of Chordin by the Tolloid protease would allow a peak of BMP signaling away from BMP-producing cells. Consistently, diffusion of the *Drosophila* BMP ligand Dpp was dependent on the expression of *sog/chordin* or *tsg*, as there was little diffusion of Dpp (as measured by a P-mad antibody [Eldar et al. 2002]) in flies mutant for either one of these genes. This work, although it needs to incorporate the greater complexity of vertebrate development, has important implications for morphogen generation in vertebrates. An aspect seldom addressed is the rate of diffusion of the various molecular complexes containing BMPs, and how diffusion and the rates of synthesis, degradation, and liberation of BMPs by *tolloid*-like proteases act to refine the morphogen gradient of BMPs during gastrulation. However, in the zebrafish, *tolloid/minifin* function is required only post-gastrulation and

therefore cannot play a role in refining the BMP gradient during gastrulation (Connors et al. 1999). The apparent enhancement of BMP signaling within the tail fin in zebrafish heterozygotes for *bmp2/swirl* and homozygotes for *chordin/dino* (Wagner and Mullins 2002) could arise from an increased local concentration of BMPs in the absence of Chordin, akin to what has recently been reported in *Drosophila* (Eldar et al. 2002). However, a potential agonist role of Chordin in BMP signaling needs to be demonstrated in vertebrates. The *Drosophila* model could be tested in the *chordino* mutant zebrafish, which should show a restricted diffusion of exogenous BMPs.

Twisted Gastrulation is a good candidate to refine the activity of BMP/Chordin complexes. Although the literature on the biochemical activities of TSG is controversial regarding whether it acts to inhibit or promote BMP signaling (Oelgeschlager et al. 2000; Chang et al. 2001; Larrain et al. 2001; Ross et al. 2001; Scott et al. 2001), the formation of BMP/Chordin/TSG complexes and the degradation of Chordin and Chordin CR-fragments by the Tolloid-like proteases in these complexes appears as a critical regulator in the refinement of BMP signals. Furthermore, the ability of Chordin to diffuse over long distances has been postulated in *Xenopus* (Blitz et al. 2000), although the dependence of diffusion of Chordin/BMP complexes has not been evaluated. Similarly, given the large array of BMP ligands and their inhibitors in vertebrate genomes (see Table 1), this model may only hold true for a subset of BMP/inhibitor complexes. Indeed, in the case of Noggin, which is structurally unrelated to the cysteine-knot family of BMP antagonists (Chordin/Cerberus/Dan), BMP binding to the receptor complex is inhibited by Noggin due to strong conformational changes following Noggin binding to BMPs (Groppe et al. 2002). Therefore, the model presented for Chordin (Eldar et al. 2002) might not be applicable to Noggin/BMP complexes. Whether there are additional proteases that can cleave complexes other than those containing Chordin-like molecules, or whether only Chordin complexes act as the major determinants of the morphogen gradient during gastrulation, is unknown.

BMPS IN ECTODERMAL FATE DECISIONS

The role of BMPs as ventral/epidermal inducers in the ectoderm has been well established (Wilson and Hemmati-Brivanlou 1995; Muñoz-Sanjuán and Brivanlou 2002). Although the role of BMP inhibition in neural fate determination is still contested (Streit et al. 2000; Wilson et al. 2001; Muñoz-Sanjuán and Brivanlou 2002; see Chapter 29), it is clear that a loss of BMP signaling correlates with the acquisition of neural fates in all species studied. This might occur through clearance of BMP ligands by the secreted inhibitors, or through a transcriptional down-regulation of BMP messages from the prospective neuroectoderm (Harland 2000; Wilson et al. 2000, 2001). So far, there has been no report of neural fate acquisition in the absence of BMP signaling inhibition, although the interactions between BMP-activated pathways and other pathways known to influence neuronal generation, such as Wnts and FGFs, remain to be fully addressed (Muñoz-Sanjuán and Brivanlou 2002).

BMP SIGNALING IN MESODERM SPECIFICATION AND MIGRATION DURING GASTRULATION

BMPs in Mesendodermal Fate Specification during Gastrulation

BMPs have been identified as ventrolateral mesoderm inducers in both *Xenopus* and zebrafish embryos (Muñoz-Sanjuán and Brivanlou 2001), and there are abundant genetic data to support the notion that BMP signaling is the major determinant of ventral mesendodermal fates during gastrulation (Table 2). Interestingly, although mutant screens in the zebrafish have underscored the prevalent role of BMP signaling during dorsoventral patterning, loss-of-function analyses in mice have revealed that BMP signaling also acts during mesoderm specification (Table 2). Many of the defects revealed in the mouse arise in the visceral endoderm and extraembryonic mesoderm, reflecting a role for these tissues during mesoderm specification in the mouse (Table 2 and references within; see Chapter 16). Therefore, BMP signals likely act during both specification and patterning of the mesendoderm. Although the role of BMPs in endodermal patterning is less strong, there is recent evidence to suggest that BMP signaling also promotes ventroposterior fates in the endoderm as well (Tiso et al. 2002). Therefore, BMP activities appear to act to pattern all germ layers.

BMPs have been shown to suppress the maintenance of the organizer and therefore act to ventralize the *Xenopus* embryo (Jones et al. 1996). This inhibition of organizer formation by BMPs is due to the loss of expression of organizer genes (Laurent and Cho 1999). Several components of the BMP pathway have been shown to play a critical role in posterior and ventral patterning of the mesoderm, including the ligands, BMP-2, BMP-4, and BMP-7. The expression patterns of *BMP-4* genes, together with results in zebrafish and *Xenopus*, suggest that BMP-4 is a strong candidate to be the major morphogenic determinant of ventral and posterior fates (Dale et al. 1992; Jones et al. 1992, 1996; Nikaido et al. 1999; Muñoz-Sanjuán and Brivanlou 2001). Loss of

BMP-4 function in the mouse yields severe defects in the hematopoietic system and posterior region of the embryos (Table 2) (Winnier et al. 1995), consistent with the results obtained in zebrafish and *Xenopus*, where overexpression of *BMP-4* causes an expansion of ventroposterior structures (Dale et al. 1992; Neave et al. 1997). Loss of *BMP-2* and *BMP-5* affects mesoderm formation in the mouse (Table 2), and in the zebrafish, loss of *BMP-2* function in the *swirl* mutation leads to a phenotype similar to that of the mouse *BMP-4* loss of function (Mullins et al. 1996; Kishimoto et al. 1997). Zebrafish *BMP-2* appears to be the functional counterpart of the *Xenopus* and mouse *BMP-4* genes (Kishimoto et al. 1997), and it appears that BMP2/BMP7 heterodimers act as the major determinants in ventroposterior fates in the zebrafish (Schmid et al. 2000). Genetic analysis has shown that mutations in the zebrafish type-I BMP-receptor *Alk-8-/lost-a-fin* lead to embryos dorsalized in a fashion similar to loss of function of *BMP-2/swirl*, *BMP-7/snailhouse* embryos (Bauer et al. 2001; Mintzer et al. 2001). Whether additional receptors are implicated as well is unknown. In the mouse, genetic loss of several type I and type II receptors, *Alk-2,3,4* and *BMPRII*, leads to a loss of mesodermal derivatives, strongly enforcing the role of BMP signaling during mesoderm specification across vertebrates.

Similarly, loss of the intracellular transducers leads to defects in gastrulation and mesoderm formation (Table 2). The zebrafish mutant *somitabun* results from a dominant-negative missense mutation in the *Smad5* gene (Hild et al. 1999). Although this mutation implicates Smad5 as an endogenous mediator of BMP signaling in ventral fates, many results suggest that this is unlikely to be the case prior to gastrulation in zebrafish. The expression patterns of *Smad1* and *Smad5* vary among vertebrates, and it is likely that they carry out distinct functions during embryogenesis (Dick et al. 1999). In *Xenopus*, there is maternal *Smad1* contribution (Graff et al. 1996), whereas in the zebrafish, *Smad1* is not expressed until mid-gastrulation (Hild et al. 1999), and *Smad5* is expressed at all stages. Still lacking is genetic evidence for the identification of the Smad transcriptional partners, such as OAZ (Hata et al. 2000), that might act during mesoderm specification and patterning during gastrulation.

BMPs and Convergent Extension Movements during Gastrulation

The role of BMPs in fate determination has been well established. However, only recently has evidence been obtained that BMPs can also specify cell migratory behaviors during convergent extension (Ip and Gridley 2002; Myers et al. 2002a,b; see Chapter 19). During gastrulation, the movements of convergence (along the mediolateral axis) and extension (along the anterior–posterior axis) lead to the establishment and patterning of the germ layers through the involution of the mesendoderm (Keller 2000; Myers et al. 2002b). In terms of migratory behavior, cells located in the dorsoanterior region of the embryo show the highest level of extension, whereas ventroposterior cells show minimal convergent extension (Myers et al. 2002a). Although the patterns of migratory behavior might differ among vertebrates, the position of cells along the dorsoventral axis acts as a good predictor of how they will migrate during gastrulation. The nature of the signals that drive the direction of migration in the dorsoanterior side of the embryo is unknown. However, in the zebrafish, these appear to be downstream of Stat3 activation in the organizer region (Yamashita et al. 2002).

In addition, it has long been established that signaling by the TGF-β superfamily can affect both cell fate and the ability of cells to undergo convergent extension in explants and in vivo (Graff et al. 1994; Keller 2000). The fundamental question that remains is whether and how cell fate determination and migratory movements are linked at a molecular level. Because of the antagonism between the organizer, which secretes BMP inhibitors, and the ventroposterior regions of the embryo, with highest BMP activity, BMPs are likely candidates to also regulate cell movements. In *Drosophila*, a role for Dpp in dorsal closure during embryogenesis has been clearly established (Martin-Blanco 1998; Jacinto and Martin 2001; Reed et al. 2001). During dorsal closure, the activities of the Dpp and JNK pathways function to regulate the actin cytoskeleton and migratory movements. In light of the finding that the JNK pathway can be connected to BMP signaling through TAK1 activity (Takatsu et al. 2000), it is likely that BMP signaling might affect movement behavior through a similar mechanism during gastrulation.

In the zebrafish mutants *chordino*, *bozozok/chordino*, and *smad5/somitabun*, convergent extension movements are affected as predicted if BMP activity acts to specify the different domains of convergent extension behaviors (Myers et al. 2002a). This work has shown that high BMP activity downregulates expression of zebrafish *wnt-5a* and *wnt-11*, genes required for convergent extension, although not for cell fate (Myers et al. 2002a,b). Similarly, work in *Xenopus* had previously demonstrated that one can dissociate fate specification in ectodermal explants from convergent extension movements (Sokol 1996; Wallingford et al. 2000), and that Noggin can affect not only cell fate, but also morphogenetic movements in animal caps (Knecht and Harland 1997). The finding that Noggin can lead to an increase in the motility of ectodermal cells supports current findings in the zebrafish for a role of the gradient of BMP signaling in morphogenesis. Although cell fate acquisition is thought to occur during late gastrulation, cells along the dorsoventral axis show distinct

migratory behaviors from the late blastula/early gastrula. These arguments have been used to postulate that the mechanisms controlling cell behavior and cell fate determination act in parallel, rather than linearly, even if specified by similar secreted factors (Myers et al. 2002b). However, because we cannot usually assess fate determination until we detect the expression of cell-specific markers, it is impossible to determine whether cells showing a certain migratory pattern in early gastrulation are already committed to a particular fate, since we still lack understanding of how a cell measures the total BMP input (in terms of concentration and length of exposure) to specify a certain fate. In the case of ectodermal cells, it has been shown that epidermal gene expression depends on the total amount of exposure to BMP protein in explants and in vivo (Wilson and Hemmati-Brivanlou 1995; Barth et al. 1999). Therefore, the more time a cell spends in a high-BMP activity region, the more likely it is to adopt a ventrolateral fate. Additional cell transplantation experiments should address how movements can affect fate, particularly with ectodermal cells, whose period of competence to respond to mesodermal signals is clearly established (Green et al. 1990; Domingo and Keller 2000).

Whatever the mechanism of action, BMPs do act as major determinants of both fate and migration. The challenge remains to integrate BMP inputs with other signals regulating these processes, and to identify components that affect migration but not cell fate. Evolutionarily, BMP signals have been used as a major determinant of cellular diversity during gastrulation. Despite many advances in this field, there is still plenty of work for the future. The next frontier resides in understanding how a cell, given a certain position and migratory trajectory, integrates BMP inputs intracellularly to activate a specified gene expression program.

ACKNOWLEDGMENTS

The authors acknowledge Dr. Alin Vonica for comments on the manuscript, and Blaine Cooper for help with the figures. I.M.-S. is a fellow of the Helen Hay Whitney Foundation.

REFERENCES

Abreu J.G., Ketpura N.I., Reversade B., and De Robertis E.M. 2002. Connective-tissue growth factor (CTGF) modulates cell signalling by BMP and TGF-β. *Nat. Cell Biol.* **4:** 599–604.

Bai S., Shi X., Yang X., and Cao X. 2000. Smad6 as a transcriptional corepressor. *J. Biol. Chem.* **275:** 8267–8270.

Balemans W. and Van Hul W. 2002. Extracellular regulation of BMP signaling in vertebrates: A cocktail of modulators. *Dev. Biol.* **250:** 231–250.

Balemans W., Ebeling M., Patel N., Van Hul E., Olson P., Dioszegi M., Lacza C., Wuyts W., Van Den Ende J., Willems P., Paes-Alves A.F., Hill S., Bueno M., Ramos F.J., Tacconi P., Dikkers F.G., Stratakis C.,

Lindpaintner K., Vickery B., Foernzler D., and Van Hul W. 2001. Increased bone density in sclerosteosis is due to the deficiency of a novel secreted protein (SOST). *Hum. Mol. Genet.* **10:** 537–543.

Barth K.A., Kishimoto Y., Rohr K.B., Seydler C., Schulte-Merker S., and Wilson S.W. 1999. Bmp activity establishes a gradient of positional information throughout the entire neural plate. *Development* **126:** 4977–4987.

Bauer H., Lele Z., Rauch G.J., Geisler R., and Hammerschmidt M. 2001. The type I serine/threonine kinase receptor Alk8/Lost-a-fin is required for Bmp2b/7 signal transduction during dorsoventral patterning of the zebrafish embryo. *Development* **128:** 849–858.

Bell E., Muñoz-Sanjuán I., Altmann C., and Brivanlou A.H. 2003. Cell fate specification and competence by Coco, a maternal BMP, TGFβ and WNT inhibitor. *Development* **130:** 1381–1389.

Belo J.A., Bouwmeester T., Leyns L., Kertesz N., Gallo M., Follettie M., and De Robertis E.M. 1997. *Cerberus-like* is a secreted factor with neutralizing activity expressed in the anterior primitive endoderm of the mouse gastrula. *Mech. Dev.* **68:** 45–57.

Beppu H., Kawabata M., Hamamoto T., Chytil A., Minowa O., Noda T., and Miyazono K. 2000. BMP type II receptor is required for gastrulation and early development of mouse embryos. *Dev. Biol.* **221:** 249–258.

Biben C., Stanley E., Fabri L., Kotecha S., Rhinn M., Drinkwater C., Lah M., Wang C.C., Nash A., Hilton D., Ang S.L., Mohun T., and Harvey R.P. 1998. Murine cerberus homologue mCer-1: A candidate anterior patterning molecule. *Dev. Biol.* **194:** 135–151.

Blitz I.L., Shimmi O., Wunnenberg-Stapleton K., O'Connor M.B., and Cho K.W. 2000. Is chordin a long-range or short-range-acting factor? Roles for BMP1-related metalloproteases in chordin and BMP4 autofeedback loop regulation. *Dev. Biol.* **223:** 120–138.

Bouwmeester T., Kim S., Sasai Y., Lu B., and De Robertis E.M. 1996. Cerberus is a head-inducing secreted factor expressed in the anterior endoderm of Spemann's organizer. *Nature* **382:** 595–601.

Brunkow M.E., Gardner J.C., Van Ness J., Paeper B.W., Kovacevich B.R., Proll S., Skonier J.E., Zhao L., Sabo P.J., Fu Y., Alisch R.S., Gillett L., Colbert T., Tacconi P., Galas D., Hamersma H., Beighton P., and Mulligan J. 2001. Bone dysplasia sclerosteosis results from loss of the SOST gene product, a novel cystine knot-containing protein. *Am. J. Hum. Genet.* **68:** 577–589.

Casellas R. and Brivanlou A.H. 1998. *Xenopus* Smad7 inhibits both the activin and BMP pathways and acts as a neural inducer. *Dev. Biol.* **198:** 1–12.

Chang C., Holtzman D. A., Chau S., Chickering T., Woolf E.A., Holmgren L.M., Bodorova J., Gearing D.P., Holmes W.E. and Brivanlou A.H. 2001. Twisted gastrulation can function as a BMP antagonist. *Nature* **410:** 483–487.

Chang H., Zwijsen A., Vogel H., Huylebroeck D., and Matzuk M.M. 2000. Smad5 is essential for left-right asymmetry in mice. *Dev. Biol.* **219:** 71–78.

Chang H., Huylebroeck D., Verschueren K., Guo Q., Matzuk M.M., and Zwijsen A. 1999. Smad5 knockout mice die at mid-gestation due to multiple embryonic and extraembryonic defects. *Development* **126:** 1631–1642.

Chapman S.C., Schubert F.R., Schoenwolf G.C., and Lumsden A. 2002. Analysis of spatial and temporal gene expression patterns in blastula and gastrula stage chick embryos. *Dev. Biol.* **245:** 187–199.

Coffinier C., Tran U., Larrain J., and De Robertis E.M. 2001. Neuralin-1 is a novel Chordin-related molecule expressed in the mouse neural plate. *Mech. Dev.* **100:** 119–122.

Connolly D.J., Patel K., and Cooke J. 1997. Chick noggin is expressed in the organizer and neural plate during axial development, but offers no evidence of involvement in primary axis formation. *Int. J. Dev. Biol.* **41:** 389–396.

Connors S.A., Trout J., Ekker M., and Mullins M.C. 1999. The role of tolloid/mini fin in dorsoventral pattern formation of the zebrafish embryo. *Development* **126:** 3119–3130.

Dale L. and Wardle F.C. 1999. A gradient of BMP activity specifies dorsal-ventral fates in early *Xenopus* embryos. *Semin. Cell Dev. Biol.* **10:** 319–326.

Dale L., Howes G., Price B.M., and Smith J.C. 1992. Bone morphogenetic protein 4: A ventralizing factor in early *Xenopus* development. *Development* **115:** 573–585.

Dick A., Meier A., and Hammerschmidt M. 1999. Smad1 and Smad5 have distinct roles during dorsoventral patterning of the zebrafish embryo. *Dev. Dyn.* **216:** 285–298.

Domingo C. and Keller R. 2000. Cells remain competent to respond to mesoderm-inducing signals present during gastrulation in *Xenopus laevis*. *Dev. Biol.* **225:** 226–240.

Dosch R., Gawantka V., Delius H., Blumenstock C., and Niehrs C. 1997. Bmp-4 acts as a morphogen in dorsoventral mesoderm patterning in *Xenopus*. *Development* **124:** 2325–2334.

Eimon P.M. and Harland R.M. 2001. *Xenopus Dan*, a member of the Dan gene family of BMP antagonists, is expressed in derivatives of the cranial and trunk neural crest. *Mech. Dev.* **107:** 187–189.

Eldar A., Dorfman R., Weiss D., Ashe H., Shilo B.Z., and Barkai N. 2002. Robustness of the BMP morphogen gradient in *Drosophila* embryonic patterning. *Nature* **419:** 304–308.

Faure S., de Santa Barbara P., Roberts D.J., and Whitman M. 2002. Endogenous patterns of BMP signaling during early chick development. *Dev. Biol.* **244:** 44–65.

Faure S., Lee M.A., Keller T., ten Dijke P., and Whitman M. 2000. Endogenous patterns of TGF-β superfamily signaling during early *Xenopus* development. *Development* **127:** 2917–2931.

Fekany K., Yamanaka Y., Leung T., Sirotkin H.I., Topczewski J., Gates M.A., Hibi M., Renucci A., Stemple D., Radbill A., Schier A.F., Driever W., Hirano T., Talbot W.S., and Solnica-Krezel L. 1999. The zebrafish *bozozok* locus encodes Dharma, a homeodomain protein essential for induction of gastrula organizer and dorsoanterior embryonic structures. *Development* **126:** 1427–1438.

Fisher S., Amacher S.L., and Halpern M.E. 1997. Loss of cerebum function ventralizes the zebrafish embryo. *Development* **124:** 1301–1311.

Furthauer M., Thisse B., and Thisse C. 1999. Three different noggin genes antagonize the activity of bone morphogenetic proteins in the zebrafish embryo. *Dev. Biol.* **214:** 181–196.

Goodman S.A., Albano R., Wardle F. C., Matthews G., Tannahill D., and Dale L. 1998. BMP1-related metalloproteinases promote the development of ventral mesoderm in early *Xenopus* embryos. *Dev. Biol.* **195:** 144–157.

Goswami M., Uzgare A.R., and Sater A.K. 2001. Regulation of MAP kinase by the BMP-4/TAK1 pathway in *Xenopus* ectoderm. *Dev. Biol.* **236:** 259–270.

Graff J.M., Bansal A., and Melton D.A. 1996. *Xenopus* Mad proteins transduce distinct subsets of signals for the TGF-β superfamily. *Cell* **85:** 479–487.

Graff J.M., Thies R.S., Song J.J., Celeste A.J., and Melton D.A. 1994. Studies with a *Xenopus* BMP receptor suggest that ventral mesoderm-inducing signals override dorsal signals in vivo. *Cell*

79: 169–179.

Green J.B., Howes G., Symes K., Cooke J., and Smith J.C. 1990. The biological effects of XTC-MIF: Quantitative comparison with *Xenopus* bFGF. *Development* **108:** 173–83.

Groppe J., Greenwald J., Wiater E., Rodriguez-Leon J., Economides A.N., Kwiatkowski W., Affolter M., Vale W.W., Belmonte J.C., and Choe S. 2002. Structural basis of BMP signalling inhibition by the cystine knot protein Noggin. *Nature* **420:** 636–642.

Gu Z., Nomura M., Simpson B.B., Lei H., Feijen A., van den Eijnden-van Raaij J., Donahoe P.K., and Li E. 1998. The type I activin receptor ActRIB is required for egg cylinder organization and gastrulation in the mouse. *Genes Dev.* **12:** 844–857.

Gu Z., Reynolds E.M., Song J., Lei H., Feijen A., Yu L., He W., MacLaughlin D.T., van den Eijnden-van Raaij J., Donahoe P.K., and Li E. 1999. The type I serine/threonine kinase receptor ActRIA (ALK2) is required for gastrulation of the mouse embryo. *Development* **126:** 2551–2561.

Hammerschmidt M., Serbedzija G.N., and McMahon A.P. 1996. Genetic analysis of dorsoventral pattern formation in the zebrafish: Requirement of a BMP-like ventralizing activity and its dorsal repressor. *Genes Dev.* **10:** 2452–2461.

Hansen C.S., Marion C.D., Steele K., George S., and Smith W.C. 1997. Direct neural induction and selective inhibition of mesoderm and epidermis inducers by Xnr3. *Development* **124:** 483–492.

Harland R. 2000. Neural induction. *Curr. Opin. Genet. Dev.* **10:** 357–362.

Harland R. and Gerhart J. 1997. Formation and function of Spemann's organizer. *Annu. Rev. Cell Dev. Biol.* **13:** 611–667.

Hata A., Lagna G., Massague J., and Hemmati-Brivanlou A. 1998. Smad6 inhibits BMP/Smad1 signaling by specifically competing with the Smad4 tumor suppressor. *Genes Dev.* **12:** 186–197.

Hata A., Seoane J., Lagna G., Montalvo E., Hemmati-Brivanlou A., and Massague J. 2000. OAZ uses distinct DNA- and protein-binding zinc fingers in separate BMP-Smad and Olf signaling pathways. *Cell* **100:** 229–240.

Hayette S., Gadoux M., Martel S., Bertrand S., Tigaud I., Magaud J.P., and Rimokh R. 1998. FLRG (follistatin-related gene), a new target of chromosomal rearrangement in malignant blood disorders. *Oncogene* **16:** 2949–2954.

Hemmati-Brivanlou A., Kelly O.G., and Melton D.A. 1994. Follistatin, an antagonist of activin, is expressed in the Spemann organizer and displays direct neuralizing activity. *Cell* **77:** 283–295.

Hild M., Dick A., Rauch G.J., Meier A., Bouwmeester T., Hafffter P., and Hammerschmidt M. 1999. The smad5 mutation somitabun blocks Bmp2b signaling during early dorsoventral patterning of the zebrafish embryo. *Development* **126:** 2149–2159.

Hsu D.R., Economides A.N., Wang X., Eimon P.M., and Harland R.M. 1998. The *Xenopus* dorsalizing factor Gremlin identifies a novel family of secreted proteins that antagonize BMP activities. *Mol. Cell* **1:** 673–683.

Iemura S., Yamamoto T.S., Takagi C., Uchiyama H., Natsume T., Shimasaki S., Sugino H., and Ueno N. 1998. Direct binding of follistatin to a complex of bone-morphogenetic protein and its receptor inhibits ventral and epidermal cell fates in early *Xenopus* embryo. *Proc. Natl. Acad. Sci.* **95:** 9337–9342.

Ip Y.T. and Gridley T. 2002. Cell movements during gastrulation: Snail dependent and independent pathways. *Curr. Opin. Genet. Dev.* **12:** 423–429.

Ishitani T., Ninomiya-Tsuji J., Nagai S., Nishita M., Meneghini M., Barker N., Waterman M., Bowerman B., Clevers H., Shibuya H., and Matsumoto K. 1999. The TAK1-NLK-MAPK-related pathway antagonizes signalling between β-catenin and transcription factor TCF. *Nature* **399:** 798–802.

Jacinto A. and Martin P. 2001. Morphogenesis: Unravelling the cell biology of hole closure. *Curr. Biol.* **11:** R705–707.

Jones C.M. and Smith J.C. 1998. Establishment of a BMP-4 morphogen gradient by long-range inhibition. *Dev. Biol.* **194:** 12–17.

Jones C.M., Dale L., Hogan B.L., Wright C.V., and Smith J.C. 1996. Bone morphogenetic protein-4 (BMP-4) acts during gastrula stages to cause ventralization of *Xenopus* embryos. *Development* **122:** 1545–1554.

Jones C.M., Lyons K.M., Lapan P.M., Wright C.V., and Hogan B.L. 1992. DVR-4 (bone morphogenetic protein-4) as a posterior-ventralizing factor in *Xenopus* mesoderm induction. *Development* **115:** 639–647.

Kalantry S., Manning S., Haub O., Tomihara-Newberger C., Lee H.G., Fangman J., Disteche C.M., Manova K., and Lacy E. 2001. The amnionless gene, essential for mouse gastrulation, encodes a visceral-endoderm-specific protein with an extracellular cysteine-rich domain. *Nat. Genet.* **27:** 412–416.

Kavsak P., Rasmussen R.K., Causing C.G., Bonni S., Zhu H., Thomsen G.H., and Wrana J.L. 2000. Smad7 binds to Smurf2 to form an E3 ubiquitin ligase that targets the TGF-β receptor for degradation. *Mol. Cell* **6:** 1365–1375.

Keller R. 2000. The origin and morphogenesis of amphibian somites. *Curr. Top. Dev. Biol.* **47:** 183–246.

Kimura N., Matsuo R., Shibuya H., Nakashima K., and Taga T. 2000. BMP2-induced apoptosis is mediated by activation of the TAK1-p38 kinase pathway that is negatively regulated by Smad6. *J. Biol. Chem.* **275:** 17647–17652.

Kishimoto Y., Lee K.H., Zon L., Hammerschmidt M., and Schulte-Merker S. 1997. The molecular nature of zebrafish swirl: BMP2 function is essential during early dorsoventral patterning. *Development* **124:** 4457–66.

Knecht A.K. and Harland R.M. 1997. Mechanisms of dorsal-ventral patterning in noggin-induced neural tissue. *Development* **124:** 2477–2488.

Kodjabachian L., Dawid I.B., and Toyama R. 1999. Gastrulation in zebrafish: What mutants teach us. *Dev. Biol.* **213:** 231–245.

Kolle G., Georgas K., Holmes G.P., Little M.H., and Yamada T. 2000. CRIM1, a novel gene encoding a cysteine-rich repeat protein, is developmentally regulated and implicated in vertebrate CNS development and organogenesis. *Mech. Dev.* **90:** 181–193.

Kretzschmar M., Doody J., and Massague J. 1997. Opposing BMP and EGF signalling pathways converge on the TGF-β family mediator Smad1. *Nature* **389:** 618–622.

Kuroda S. and Tanizawa K. 1999. Involvement of epidermal growth factor-like domain of NELL proteins in the novel protein-protein interaction with protein kinase C. *Biochem. Biophys. Res. Commun.* **265:** 752–757.

Larrain J., Oelgeschlager M., Ketpura N. I., Reversade B., Zakin L., and De Robertis E.M. 2001. Proteolytic cleavage of Chordin as a switch for the dual activities of Twisted gastrulation in BMP signaling. *Development* **128:** 4439–4447.

Laurent M.N. and Cho K.W. 1999. Bone morphogenetic protein antagonism of Spemann's organizer is independent of Wnt signaling. *Dev. Biol.* **206:** 157–162.

Lechleider R.J., Ryan J.L., Garrett L., Eng C., Deng C., Wynshaw-Boris A., and Roberts A.B. 2001. Targeted mutagenesis of Smad1 reveals an essential role in chorioallantoic fusion. *Dev. Biol.* **240:** 157–167.

Martin-Blanco E. 1998. Regulatory control of signal transduction during morphogenesis in *Drosophila*. *Int. J. Dev. Biol.* **42:** 363–368.

Massague J. and Chen Y.G. 2000. Controlling TGF-β signaling. *Genes Dev.* **14:** 627–644.

Matsuhashi S., Noji S., Koyama E., Myokai F., Ohuchi H., Taniguchi S., and Hori K. 1995. New gene, nel, encoding a M(r) 93 K protein with EGF-like repeats is strongly expressed in neural tissues of early stage chick embryos. *Dev. Dyn.* **203:** 212–222.

Matsui M., Mizuseki K., Nakatani J., Nakanishi S., and Sasai Y. 2000. *Xenopus* kielin: A dorsalizing factor containing multiple chordin-type repeats secreted from the embryonic midline. *Proc. Natl. Acad. Sci.* **97:** 5291–5296.

McMahon J.A., Takada S., Zimmerman L.B., Fan C.M., Harland R.M., and McMahon A.P. 1998. Noggin-mediated antagonism of BMP signaling is required for growth and patterning of the neural tube and somite. *Genes Dev.* **12:** 1438–1452.

Meneghini M.D., Ishitani T., Carter J.C., Hisamoto N., Ninomiya-Tsuji J., Thorpe C.J., Hamill D.R., Matsumoto K., and Bowerman B. 1999. MAP kinase and Wnt pathways converge to downregulate an HMG-domain repressor in *Caenorhabditis elegans*. *Nature* **399:** 793–797.

Meno C., Ito Y., Saijoh Y., Matsuda Y., Tashiro K., Kuhara S., and Hamada H. 1997. Two closely-related left-right asymmetrically expressed genes, lefty-1 and lefty-2: Their distinct expression domains, chromosomal linkage and direct neuralizing activity in *Xenopus* embryos. *Genes Cells* **2:** 513–24.

Meno C., Saijoh Y., Fujii H., Ikeda M., Yokoyama T., Yokoyama M., Toyoda Y., and Hamada H. 1996. Left-right asymmetric expression of the TGF-β-family member lefty in mouse embryos. *Nature* **381:** 151–155.

Meno C., Gritsman K., Ohishi S., Ohfuji Y., Heckscher E., Mochida K., Shimono A., Kondoh H., Talbot W.S., Robertson E.J. Schier A.F., and Hamada H. 1999. Mouse Lefty2 and zebrafish antivin are feedback inhibitors of nodal signaling during vertebrate gastrulation. *Mol. Cell* **4:** 287–298.

Miller-Bertoglio V.E., Fisher S., Sanchez A., Mullins M.C., and Halpern M.E. 1997. Differential regulation of chordin expression domains in mutant zebrafish. *Dev. Biol.* **192:** 537–550.

Miller-Bertoglio V., Carmany-Rampey A., Furthauer M., Gonzalez E.M., Thisse C., Thisse B., Halpern M.E., and Solnica-Krezel L. 1999. Maternal and zygotic activity of the zebrafish ogon locus antagonizes BMP signaling. *Dev. Biol.* **214:** 72–86.

Minabe-Saegusa C., Saegusa H., Tsukahara M., and Noguchi S. 1998. Sequence and expression of a novel mouse gene PRDC (protein related to DAN and cerberus) identified by a gene trap approach. *Dev. Growth Differ.* **40:** 343–353.

Mintzer K.A., Lee M.A., Runke G., Trout J., Whitman M., and Mullins M.C. 2001. Lost-a-fin encodes a type I BMP receptor, Alk8, acting maternally and zygotically in dorsoventral pattern formation. *Development* **128:** 859–869.

Mishina Y., Crombie R., Bradley A., and Behringer R.R. 1999. Multiple roles for activin-like kinase-2 signaling during mouse embryogenesis. *Dev. Biol.* **213:** 314–326.

Mishina Y., Suzuki A., Ueno N., and Behringer R.R. 1995. Bmpr encodes a type I bone morphogenetic protein receptor that is essential for gastrulation during mouse embryogenesis. *Genes Dev.* **9:** 3027–3037.

Mullins M.C., Hammerschmidt M., Kane D.A., Odenthal J., Brand M., van Eeden F.J., Furutani-Seiki M., Granato M., Haffter P., Heisenberg C.P., Jiang Y.J., Kelsh R.N., and Nusslein-Volhard C. 1996. Genes establishing dorsoventral pattern formation in the zebrafish embryo: The ventral specifying genes. *Development* **123:** 81–93.

Muñoz-Sanjuán I. and Brivanlou A.H. 2001. Early posterior/ventral fate specification in the vertebrate embryo. *Dev. Biol.* **237:** 1–17.

———. 2002. Neural induction, the default model and embryonic stem cells. *Nat. Rev. Neurosci.* **3:** 271–280.

Muñoz-Sanjuán I., Bell E., Altmann C.R., Vonica A., and Brivanlou A.H. 2002. Gene profiling during neural induction in *Xenopus laevis*: Regulation of BMP signaling by post-transcriptional mechanisms and TAB3, a novel TAK1-binding protein. *Development* **129:** 5529–5540.

Myers D.C., Sepich D.S., and Solnica-Krezel L. 2002a. Bmp activity gradient regulates convergent extension during zebrafish gastrulation. *Dev. Biol.* **243:** 81–98.

———. 2002b. Convergence and extension in vertebrate gastrulae: Cell movements according to or in search of identity? *Trends Genet.* **18:** 447–455.

Nakamura T., Takio K., Eto Y., Shibai H., Titani K., and Sugino H. 1990. Activin-binding protein from rat ovary is follistatin. *Science* **247:** 836–838.

Nakashima Y., Sun D.H., Trindade M.C., Maloney W.J., Goodman S.B., Schurman D.J., and Smith R.L. 1999. Signaling pathways for tumor necrosis factor-alpha and interleukin-6 expression in human macrophages exposed to titanium-alloy particulate debris in vitro. *J. Bone Jt. Surg. Am. Vol.* **81:** 603–615.

Nakayama N., Han C.E., Scully S., Nishinakamura R., He C., Zeni L., Yamane H., Chang D., Yu D., Yokota T., and Wen D. 2001. A novel chordin-like protein inhibitor for bone morphogenetic proteins expressed preferentially in mesenchymal cell lineages. *Dev. Biol.* **232:** 372–387.

Nakayama T., Gardner H., Berg L.K., and Christian J.L. 1998. Smad6 functions as an intracellular antagonist of some TGF-β family members during *Xenopus* embryogenesis. *Genes Cells* **3:** 387–394.

Neave B., Holder N., and Patient R. 1997. A graded response to BMP-4 spatially coordinates patterning of the mesoderm and ectoderm in the zebrafish. *Mech. Dev.* **62:** 183–195.

Nguyen M., Park S., Marques G., and Arora K. 1998. Interpretation of a BMP activity gradient in *Drosophila* embryos depends on synergistic signaling by two type I receptors, SAX and TKV. *Cell* **95:** 495–506.

Nikaido M., Tada M., Takeda H., Kuroiwa A., and Ueno N. 1999. In vivo analysis using variants of zebrafish BMPR-IA: Range of action and involvement of BMP in ectoderm patterning. *Development* **126:** 181–190.

Nishimatsu S. and Thomsen G.H. 1998. Ventral mesoderm induction and patterning by bone morphogenetic protein heterodimers in *Xenopus* embryos. *Mech. Dev.* **74:** 75–88.

Oelgeschlager M., Larrain J., Geissert D., and De Robertis E.M. 2000. The evolutionarily conserved BMP-binding protein Twisted gastrulation promotes BMP signalling. *Nature* **405:** 757–763.

Ozaki T., Ma J., Takenaga K., and Sakiyama S. 1996. Cloning of mouse DAN cDNA and its down-regulation in transformed cells. *Jpn. J. Cancer Res.* **87:** 58–61.

Patel K., Connolly D.J., Amthor H., Nose K., and Cooke J. 1996. Cloning and early dorsal axial expression of Flik, a chick follistatin-related gene: Evidence for involvement in dorsalization/neural induction. *Dev. Biol.* **178:** 327–342.

Pearce J.J., Penny G., and Rossant J. 1999. A mouse cerberus/Dan-related gene family. *Dev. Biol.* **209:** 98–110.

Pfendler K.C., Yoon J., Taborn G.U., Kuehn M.R., and Iannaccone P.M. 2000. Nodal and bone morphogenetic protein 5 interact in murine mesoderm formation and implantation. *Genesis* **28:** 1–14.

Piccolo S., Sasai Y., Lu B., and De Robertis E.M. 1996. Dorsoventral patterning in *Xenopus*: Inhibition of ventral signals by direct binding of chordin to BMP-4. *Cell* **86:** 589–598.

Piccolo S., Agius E., Lu B., Goodman S., Dale L., and De Robertis E.M. 1997. Cleavage of Chordin by Xolloid metalloprotease suggests a role for proteolytic processing in the regulation of Spemann organizer activity. *Cell* **91**: 407–416.

Piccolo S., Agius E., Leyns L., Bhattacharyya S., Grunz H., Bouwmeester T., and De Robertis E.M. 1999. The head inducer Cerberus is a multifunctional antagonist of Nodal, BMP and Wnt signals. *Nature* **397**: 707–710.

Reed B.H., Wilk R., and Lipshitz H.D. 2001. Downregulation of Jun kinase signaling in the amnioserosa is essential for dorsal closure of the *Drosophila* embryo. *Curr. Biol.* **11**: 1098–1108.

Rodriguez Esteban C., Capdevila J., Economides A.N., Pascual J., Ortiz A., and Izpisua Belmonte J.C. 1999. The novel Cer-like protein Caronte mediates the establishment of embryonic left-right asymmetry. *Nature* **401**: 243–251.

Ross J.J., Shimmi O., Vilmos P., Petryk A., Kim H., Gaudenz K., Hermanson S., Ekker S.C., O'Connor M.B., and Marsh J.L. 2001. Twisted gastrulation is a conserved extracellular BMP antagonist. *Nature* **410**: 479–483.

Sasai Y., Lu B., Steinbeisser H., Geissert D., Gont L.K., and De Robertis E.M. 1994. *Xenopus* chordin: A novel dorsalizing factor activated by organizer-specific homeobox genes. *Cell* **79**: 779–90.

Schmid B., Furthauer M., Connors S.A., Trout J., Thisse B., Thisse C., and Mullins M.C. 2000. Equivalent genetic roles for *bmp7/snailhouse* and *bmp2b/swirl* in dorsoventral pattern formation. *Development* **127**: 957–967.

Schneyer A., Tortoriello D., Sidis Y., Keutmann H., Matsuzaki T., and Holmes W. 2001. Follistatin-related protein (FSRP): A new member of the follistatin gene family. *Mol. Cell. Endocrinol.* **180**: 33–38.

Schulte-Merker S., Lee K.J., McMahon A.P., and Hammerschmidt M. 1997. The zebrafish organizer requires chordino. *Nature* **387**: 862–3.

Scott I.C., Blitz I.L., Pappano W.N., Maas S.A., Cho K.W., and Greenspan D.S. 2001. Homologues of Twisted gastrulation are extracellular cofactors in antagonism of BMP signalling. *Nature* **410**: 475–478.

Shibanuma M., Mashimo J., Mita A., Kuroki T., and Nose K. 1993. Cloning from a mouse osteoblastic cell line of a set of transforming-growth-factor-beta 1-regulated genes, one of which seems to encode a follistatin-related polypeptide. *Eur. J. Biochem.* **217**: 13–19.

Shibuya H., Iwata H., Masuyama N., Gotoh Y., Yamaguchi K., Irie K., Matsumoto K., Nishida E., and Ueno N. 1998. Role of TAK1 and TAB1 in BMP signaling in early *Xenopus* development. *EMBO J.* **17**: 1019–1028.

Sirard C., de la Pompa J.L., Elia A., Itie A., Mirtsos C., Cheung A., Hahn S., Wakeham A., Schwartz L., Kern S.E., Rossant J., and Mak T.W. 1998. The tumor suppressor gene *Smad4/Dpc4* is required for gastrulation and later for anterior development of the mouse embryo. *Genes Dev.* **12**: 107–119.

Smith W.C. and Harland R.M. 1992. Expression cloning of noggin, a new dorsalizing factor localized to the Spemann organizer in *Xenopus* embryos. *Cell* **70**: 829–840.

Sokol S.Y. 1996. Analysis of Dishevelled signalling pathways during *Xenopus* development. *Curr. Biol.* **6**: 1456–1467.

Solnica-Krezel L., Stemple D.L., Mountcastle-Shah E., Rangini Z., Neuhauss S.C., Malicki J., Schier A.F., Stainier D.Y., Zwartkruis F., Abdelilah S., and Driever W. 1996. Mutations affecting cell fates and cellular rearrangements during gastrulation in zebrafish.

Development **123**: 67–80.

Streit A., Berliner A.J., Papanayotou C., Sirulnik A., and Stern C.D. 2000. Initiation of neural induction by FGF signalling before gastrulation. *Nature* **406**: 74–78.

Streit A., Lee K.J., Woo I., Roberts C., Jessell T.M., and Stern C.D. 1998. Chordin regulates primitive streak development and the stability of induced neural cells, but is not sufficient for neural induction in the chick embryo. *Development* **125**: 507–519.

Suzuki A., Ueno N., and Hemmati-Brivanlou A. 1997. *Xenopus* msx1 mediates epidermal induction and neural inhibition by BMP4. *Development* **124**: 3037–3044.

Takaku K., Oshima M., Miyoshi H., Matsui M., Seldin M.F., and Taketo M.M. 1998. Intestinal tumorigenesis in compound mutant mice of both Dpc4 (Smad4) and Apc genes. *Cell* **92**: 645–656.

Takatsu Y., Nakamura M., Stapleton M., Danos M.C., Matsumoto K., O'Connor M.B., Shibuya H., and Ueno N. 2000. TAK1 participates in c-Jun N-terminal kinase signaling during *Drosophila* development. *Mol. Cell Biol.* **20**: 3015–3026.

Tiso N., Filippi A., Pauls S., Bortolussi M., and Argenton F. 2002. BMP signalling regulates anteroposterior endoderm patterning in zebrafish. *Mech. Dev.* **118**: 29–37.

Topol L.Z., Marx M., Laugier D., Bogdanova N.N., Boubnov N.V., Clausen P.A., Calothy G., and Blair D.G. 1997. Identification of drm, a novel gene whose expression is suppressed in transformed cells and which can inhibit growth of normal but not transformed cells in culture. *Mol. Cell Biol.* **17**: 4801–4810.

Tortoriello D.V., Sidis Y., Holtzman D.A., Holmes W.E., and Schneyer A.L. 2001. Human follistatin-related protein: A structural homologue of follistatin with nuclear localization. *Endocrinology* **142**: 3426–3434.

Tremblay K.D., Dunn N.R., and Robertson E.J. 2001. Mouse embryos lacking Smad1 signals display defects in extra-embryonic tissues and germ cell formation. *Development* **128**: 3609–3621.

Tsukazaki T., Chiang T.A., Davison A.F., Attisano L., and Wrana J.L. 1998. SARA, a FYVE domain protein that recruits Smad2 to the TGF-β receptor. *Cell* **95**: 779–791.

Wagner D.S. and Mullins M.C. 2002. Modulation of BMP activity in dorsal-ventral pattern formation by the chordin and ogon antagonists. *Dev. Biol.* **245**: 109–123.

Waldrip W.R., Bikoff E.K., Hoodless P.A., Wrana J.L., and Robertson E.J. 1998. Smad2 signaling in extraembryonic tissues determines anterior-posterior polarity of the early mouse embryo. *Cell* **92**: 797–808.

Wallingford J.B., Rowning B.A., Vogeli K.M., Rothbacher U., Fraser S.E., and Harland R.M. 2000. Dishevelled controls cell polarity during *Xenopus* gastrulation. *Nature* **405**: 81–85.

Wang C., Deng L., Hong M., Akkaraju G.R., Inoue J., and Chen Z.J. 2001. TAK1 is a ubiquitin-dependent kinase of MKK and IKK. *Nature* **412**: 346–351.

Watanabe T.K., Katagiri T., Suzuki M., Shimizu F., Fujiwara T., Kanemoto N., Nakamura Y., Hirai Y., Maekawa H., and Takahashi E. 1996. Cloning and characterization of two novel human cDNAs (NELL1 and NELL2) encoding proteins with six EGF-like repeats. *Genomics* **38**: 273–276.

Weinstein M., Yang X., Li C., Xu X., Gotay J., and Deng C.X. 1998. Failure of egg cylinder elongation and mesoderm induction in mouse embryos lacking the tumor suppressor smad2. *Proc. Natl. Acad. Sci.* **95**: 9378–9383.

Wilson P.A. and Hemmati-Brivanlou A. 1995. Induction of epidermis and inhibition of neural fate by Bmp-4. *Nature* **376**: 331–333.

Wilson P.A., Lagna G., Suzuki A., and Hemmati-Brivanlou A. 1997. Concentration-dependent patterning of the *Xenopus* ectoderm by BMP4 and its signal transducer Smad1. *Development* **124:** 3177–3184.

Wilson S.I., Graziano E., Harland R., Jessell T.M., and Edlund T. 2000. An early requirement for FGF signalling in the acquisition of neural cell fate in the chick embryo. *Curr. Biol.* **10:** 421–429.

Wilson S.I., Rydstrom A., Trimborn T., Willert K., Nusse R., Jessell T.M., and Edlund T. 2001. The status of Wnt signalling regulates neural and epidermal fates in the chick embryo. *Nature* **411:** 325–330.

Winnier G., Blessing M., Labosky P.A., and Hogan B.L. 1995. Bone morphogenetic protein-4 is required for mesoderm formation and patterning in the mouse. *Genes Dev.* **9:** 2105–2116.

Wotton D. and Massague J. 2001. Smad transcriptional corepressors in TGF-β family signaling. *Curr. Top. Microbiol. Immunol.* **254:** 145–164.

Yamaguchi K., Nagai S., Ninomiya-Tsuji J., Nishita M., Tamai K., Irie K., Ueno N., Nishida E., Shibuya H., and Matsumoto K. 1999. XIAP, a cellular member of the inhibitor of apoptosis protein family, links the receptors to TAB1-TAK1 in the BMP signaling pathway. *EMBO J.* **18:** 179–187.

Yamashita H., ten Dijke P., Huylebroeck D., Sampath T.K., Andries M., Smith J. C., Heldin C.H., and Miyazono K. 1995. Osteogenic protein-1 binds to activin type II receptors and induces certain activin-like effects. *J. Cell Biol.* **130:** 217–226.

Yamashita S., Miyagi C., Carmany-Rampey A., Shimizu T., Fujii R., Schier A.F., and Hirano T. 2002. Stat3 controls cell movements during zebrafish gastrulation. *Dev. Cell* **2:** 363–375.

Yang X., Li C., Xu X., and Deng C. 1998. The tumor suppressor SMAD4/DPC4 is essential for epiblast proliferation and mesoderm induction in mice. *Proc. Natl. Acad. Sci.* **95:** 3667–3672.

Ying Y. and Zhao G.Q. 2001. Cooperation of endoderm-derived BMP2 and extraembryonic ectoderm-derived BMP4 in primordial germ cell generation in the mouse. *Dev. Biol.* **232:** 484–492.

Yokouchi Y., Vogan K.J., Pearse II R.V., and Tabin C.J. 1999. Antagonistic signaling by *Caronte*, a novel *Cerberus*-related gene, establishes left-right asymmetric gene expression. *Cell* **98:** 573–583.

Yu K., Srinivasan S., Shimmi O., Biehs B., Rashka K.E., Kimelman D., O'Connor M.B., and Bier E. 2000. Processing of the *Drosophila* Sog protein creates a novel BMP inhibitory activity. *Development* **127:** 2143–2154.

Zhang H. and Bradley A. 1996. Mice deficient for BMP2 are nonviable and have defects in amnion/chorion and cardiac development. *Development* **122:** 2977–2986.

Zhao G.Q. 2003. Consequences of knocking out BMP signaling in the mouse. *Genesis* **35:** 43–56.

Zhu H., Kavsak P., Abdollah S., Wrana J.L., and Thomsen G.H. 1999. A SMAD ubiquitin ligase targets the BMP pathway and affects embryonic pattern formation. *Nature* **400:** 687–693.

Zimmerman L.B., De Jesus-Escobar J.M., and Harland R.M. 1996. The Spemann organizer signal noggin binds and inactivates bone morphogenetic protein 4. *Cell* **86:** 599–606.

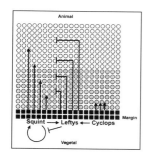

NODAL SIGNALING DURING GASTRULATION

A.F. Schier

Developmental Genetics Program, and Skirball Institute of Biomolecular Medicine, Department of Cell Biology; New York University School of Medicine, New York, New York 10016

INTRODUCTION

The TGF-β signal Nodal was discovered more than 10 years ago through genetic and molecular studies in mouse (Conlon et al. 1991, 1994; Iannaccone et al. 1992; Zhou et al. 1993). Since then, it has become clear that Nodals are the major group of signals that induce mesendoderm in vertebrates (Jones et al. 1995; Toyama et al. 1995; Feldman et al. 1998; Sampath et al. 1998; Rebagliati et al. 1998a,b; Agius et al. 2000). In addition, components of the Nodal signaling pathway have been recognized as essential determinants of left–right asymmetry and neural patterning (Hatta et al. 1991; Levin et al. 1995; Collignon et al. 1996; Lowe et al. 1996; Sampath et al. 1997; Schier et al. 1997; Strahle et al. 1997; Meno et al. 1998, 2001; Gaio et al. 1999; Yan et al. 1999; Bamford et al. 2000; Mathieu et al. 2002). The study of the Nodal signaling pathway has also highlighted the importance of coreceptors and feedback antagonists in TGF-β signaling (Bisgrove et al. 1999; Gritsman et al. 1999; Meno et al. 1999; Thisse and Thisse 1999; Reissmann et al. 2001; Yeo and Whitman 2001; Cheng et al. 2003). Here, I summarize our current understanding of the Nodal signaling pathway and its role in development. For a more detailed discussion, the reader is referred to a more extensive review (Schier 2003).

THE NODAL SIGNALING PATHWAY

Nodals constitute a group within the TGF-β superfamily that is apparently restricted to chordates. They include mouse, human, and chick Nodal; zebrafish Cyclops, Squint, and Southpaw; *Xenopus* Xnr1, 2, 4, 5, and 6; *Amphioxus*

AmphiNodal; and a Nodal in ascidians (Zhou et al. 1993; Conlon et al. 1994; Jones et al. 1995; Levin et al. 1995; Lustig et al. 1996; Joseph and Melton 1997; Feldman et al. 1998; Sampath et al. 1998; Rebagliati et al. 1998a,b; Takahashi et al. 2000; Dehal et al. 2002; Morokuma et al. 2002; Yu et al. 2002; Long et al. 2003). An additional *Xenopus* Nodal-like protein, Xnr3, shares some of the hallmarks of Nodal proteins but does not seem to have the same mesendoderm-inducing activity as other Nodals (Smith et al. 1995; Yokota et al. 2003). Instead, this signal has been implicated in convergent-extension movements and neural induction (see Chapter 29).

The role of Nodals has been dissected both genetically and biochemically. The main conclusion of these studies is that Nodals act as canonical TGF-β family members but that additional pathway-specific factors regulate Nodal signaling (Fig. 1). Like most TGF-β signals, Nodals are processed by convertases and form dimers that bind type I and II TGF-β Ser/Thr kinase receptors. This results in the phosphorylation of Smad proteins, their association with additional transcription factors, and the regulation of specific downstream genes. Whereas this general outline is shared by most TGF-β signals, several proteins such as EGF-CFC co-receptors and Lefty antagonists have been found that appear to have more specific roles in Nodal signaling.

Processing

Like most TGF-βs, Nodals are synthesized as pro-proteins that are proteolytically processed by subtilisin-like pro-protein convertases (Constam and Robertson 1999). Genetic analysis indicates that Spc1 and Spc4 (also called Furin and Pace4, respectively) are two convertases that are essential for

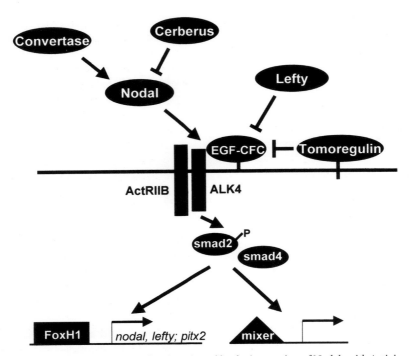

Figure 1. The Nodal signaling pathway. Nodal signaling is activated by the interaction of Nodals with Activin receptors (ActRIIB, ALK4) and EGF-CFC coreceptors and inhibited by Leftys, Tomoregulin, and Cerberus. Convertases process Nodal pro-proteins. Nodal signaling is transmitted intracellularly by phosphorylation of Smad2 and its association with Smad4 and transcription factors (FoxH1; Mixer) that determine which genes are regulated by P-Smad2. (Reprinted, with permission, from Schier 2003 [© Annual Reviews].)

the maturation of Nodal (Beck et al. 2002). *Spc1* and *4* double mutant embryos resemble mutants with strong *Nodal* loss-of-function phenotypes, and double mutant ES cells cannot process Nodal pro-proteins. Intriguingly, these convertases appear to cleave Nodal not only intracellularly, but also extracellularly (Beck et al. 2002). These results suggest that active and mature Nodal can be generated both at the site of Nodal synthesis and at a distance.

Dimerization

Nodals are not only processed, but they also dimerize. It has been suggested that heterodimerization of Nodals with other TGF-βs might lead either to synergistic activities or to mutual inhibition. For example, Xnrs and the mesoderm-inducing TGF-β Derriere can heterodimerize (Eimon and Harland 2002), potentially changing signaling activity or specificity. In addition, Nodals can heterodimerize with BMPs (BMP4, BMP7), leading to mutual inhibition (Yeo and Whitman 2001). However, the in vivo contribution of this mechanism to Nodal activities remains unresolved. For example, overexpression of Nodals in zebrafish mutant for the coreceptor *one-eyed pinhead* does not block BMP signaling (Gritsman et al. 1999).

Range

Misexpression studies have suggested that Nodals have similar activities (Jones et al. 1995; Toyama et al. 1995). For instance, heterologous Nodals can act as mesoderm inducers in *Xenopus* and zebrafish, similar to the endogenous signals. Hence, the functional specificity of Nodals is predominantly controlled by regulating the temporal and spatial expression of *Nodal* genes. However, Nodals also appear to differ in their range of activity (Jones et al. 1996; Chen and Schier 2001; Niederlander et al. 2001). For example, *cyclops*-expressing clones only induce downstream genes at a short range (Chen and Schier 2001). In contrast, clones of *squint*-expressing cells can induce downstream genes in distant cells. It has been demonstrated that this effect of Squint is direct and not mediated via a secondary (relay) signal (Chen and Schier 2001). In particular, Squint can be made in nonresponsive cells and travel through a field of nonresponsive cells to activate Nodal signaling in distant responsive cells (Chen and Schier 2001). It is unclear whether the short-range activity of Cyclops or the long-range activity of Squint applies to other Nodals. One study suggests that *Xenopus* Nodals might act purely cell-autonomously (Hashimoto-Partyka et al. 2003), but others indicate that Xnrs can have nonautonomous effects (Jones et al. 1996; Niederlander et al. 2001). Chimera studies show that mouse *Nodal* mutant cells can contribute to all mesendodermal tissues, indicating that mouse Nodal can at least act at a short range (Conlon et al. 1991; Varlet et al. 1997).

Reception

TGF-β ligands activate intracellular events by assembling a complex consisting of type I and type II receptor Ser/Thr kinases (Attisano and Wrana 2002; Shi and Massague 2003). Assembly results in the phosphorylation and activation of the type I by the type II receptor. This is followed by phosphorylation of downstream Smads by the type I receptor. In the case of Nodal, signaling is mediated by the Activin receptors ALK4 (type I) and ActRII and ActRIIB (type II). This conclusion is supported by genetic and biochemical studies. Mutant analyses in the mouse have revealed phenotypic similarities of *Nodal* and *Activin receptor* mutants (Oh and Li 1997; Gu et al. 1999; Song et al. 1999; Schier and Shen 2000), and in vitro studies demonstrate that Nodals can be part of a complex with Activin Receptors (Reissmann et al. 2001; Yeo and Whitman 2001; Bianco et al. 2002; Sakuma et al. 2002; Yan et al. 2002).

Additional type I receptors have been implicated as Nodal receptors. ALK7 can confer Nodal responsiveness to nonresponsive cells, bind to Xnr1, and dominant negative forms can block the effects of Nodal and Xnr1 (Reissmann et al. 2001). Intriguingly, Xnr2 and 4 appear not to interact with ALK7 in *Xenopus* assays. However, no genetic data are available to support a requirement for ALK7 in Nodal signaling. The zebrafish type I receptor TARAM-A might also mediate aspects of Nodal signaling (Renucci et al. 1996). Dominant negative receptors block Nodal-mediated induction of endoderm, and TARAM-A can potentiate the effects of Nodals (Aoki et al. 2002a). However, no biochemical or genetic studies have been performed which would unequivocally demonstrate that TARAM-A is a Nodal receptor.

Coreception

EGF-CFC proteins are important components of the Nodal signaling pathway by acting as coreceptors for Nodal ligands. This gene family includes zebrafish *one-eyed pinhead*, frog *FRL-1*, chick *CFC*, and mouse and human *Cripto* and *Cryptic* (Ciccodicola et al. 1989; Kinoshita et al. 1995; Shen et al. 1997; Zhang et al. 1998; Colas and Schoenwolf 2000; Shen and Schier 2000; Schlange et al. 2001). EGF-CFC proteins are extracellular and membrane-attached via GPI linkage and contain two cysteine-rich domains, one with homology to EGF, the other called CFC (cripto-FRL-1-Cryptic) domain (Zhang et al. 1998; Minchiotti et al. 2000, 2001; Shen and Schier 2000). In addition, an O-linked fucose modification has been identified in the EGF domain

of Cripto (Schiffer et al. 2001; Yan et al. 2002). Mutating this site appears to decrease signaling activity. Both genetic and biochemical studies are consistent with the model that EGF-CFC proteins serve as Nodal coreceptors. For instance, zebrafish embryos lacking all *one-eyed pinhead* activity have a phenotype identical to *cyclops;squint* double mutants, which lack all Nodal signals (Gritsman et al. 1999). These maternal-zygotic *one-eyed pinhead* mutants are unresponsive to Nodals but can be partially rescued by activated ALK4. Similarly, *Cripto* and *Nodal* mutants share phenotypic characteristics (Ding et al. 1998). Biochemical studies indicate that Nodals can assemble Activin receptor complexes only in the presence of EGF-CFC proteins (Reissmann et al. 2001; Yeo and Whitman 2001). Cripto binds to the type I ALK4 receptor, mediated by the CFC domain, and Nodal can bind to Cripto, mediated by the EGF domain. This suggests that Nodal activates intracellular signaling by assembling a complex of EGF-CFC proteins, ALK4 and ActRIIB (Reissmann et al. 2001; Yeo and Whitman 2001; Bianco et al. 2002; Sakuma et al. 2002; Yan et al. 2002). The requirement for EGF-CFC proteins is particularly intriguing when one considers that Activin can activate the very same receptors without the need for EGF-CFC proteins (Gritsman et al. 1999). EGF-CFC proteins thus appear to be pathway-specific coreceptors.

Although clearly involved in Nodal signaling, EGF-CFC proteins are likely to have additional roles. Genetic and biochemical studies in zebrafish and *Xenopus* indicate that the TGF-β signals Vg1 and GDF1 also signal via EGF-CFC and Activin receptors (Cheng et al. 2003). These findings and the observation that mutations in *GDF1*, *Nodal*, and *Cryptic* lead to very similar left–right defects (Gaio et al. 1999; Yan et al. 1999; Rankin et al. 2000; Lowe et al. 2001; Brennan et al. 2002; Norris et al. 2002; Saijoh et al. 2003; Watanabe et al. 2003; Yamamoto et al. 2003) have led to the suggestion that Cryptic might integrate signaling by both Nodal and GDF1 (Cheng et al. 2003). Antisense and misexpression approaches suggest that FRL-1 might be involved in neural induction in *Xenopus* (Kinoshita et al. 1995; Yabe et al. 2003), and cell culture studies have indicated that EGF-CFC proteins might also activate non-TGF-β signaling pathways (Adamson et al. 2002); however, there is no genetic evidence to support these suggestions.

Antagonism

Leftys

Lefty molecules are divergent members of the TGF-β family. They have been identified only in chordates, ranging from ascidians with an apparently single *lefty* gene to zebrafish and mouse, which contain two *lefty* genes, *lefty1* and *lefty2* (Meno et al. 1996, 1997; Bisgrove et al. 1999;

Thisse and Thisse 1999; Cheng et al. 2000; Ishimaru et al. 2000; Tanegashima et al. 2000). Unlike other TGF-β-related factors, Lefty proteins appear to be monomeric (Sakuma et al. 2002). Overexpression of *lefty* in zebrafish induces phenotypes strongly resembling *cyclops;squint* double mutants or maternal-zygotic *one-eyed pinhead* mutants (Bisgrove et al. 1999; Meno et al. 1999; Thisse and Thisse 1999; Thisse et al. 2000). Conversely, loss of *Lefty* function leads to enhanced Nodal signaling during mesendoderm induction and left–right development in mouse, frog, and zebrafish (Meno et al. 1998, 1999, 2001; Agathon et al. 2001; Branford and Yost 2002; Chen and Schier 2002; Feldman et al. 2002). These studies have established that Lefty proteins act as antagonists of the Nodal signaling pathway. *lefty* expression generally follows *Nodal* expression both temporally and spatially, and in most tissues *lefty* expression is dependent on Nodal signaling (Bisgrove et al. 1999; Meno et al. 1999; Thisse and Thisse 1999; Cheng et al. 2000). Hence, Lefty proteins act as classical feedback inhibitors (Schier and Talbot 2001; Hamada et al. 2002).

Some of the molecular mechanisms of Lefty activity have recently been elucidated (Cheng et al. 2004). Genetic and biochemical studies in zebrafish and frog indicate that Lefty proteins bind to EGF-CFC coreceptors and block their interaction with Activin receptors and Nodal signals. First, overexpression of EGF-CFC proteins can suppress the effects of Lefty misexpression in zebrafish. Second, Lefty proteins only block TGF-β signals that require EGF-CFC coreceptors for signaling. Third, Lefty binds to the EGF-CFC protein Cripto and blocks its interaction with Nodal and Activin receptors. These results reveal that coreceptors are the targets not only of agonists, but also of antagonists and can thus contribute to the diversification of signaling pathways.

Leftys appear to restrict the duration and range of Nodal signaling by two mechanisms. First, Leftys inhibit *Nodal* autoregulation (Bisgrove et al. 1999; Meno et al. 1999, 2001; Chen and Schier 2002; Feldman et al. 2002). In the absence of zebrafish Lefty, the *squint* expression domain extends and perdures for a longer time than in wild type (Chen and Schier 2002; Feldman et al. 2002). Similarly, *Nodal* expression perdures and is expanded in mouse *lefty2* mutants (Meno et al. 1999, 2001). However, blocking *Nodal* autoregulation and transcription cannot be the only mechanism of Lefty-mediated inhibition. Specifically, depletion of Lefty can extend the range of Squint activity even in the absence of *squint* autoregulation and transcription (Chen and Schier 2002). A second role of Lefty appears to be the long-range inhibition of Nodal signaling in distant cells (Branford and Yost 2002; Chen and Schier 2002). For example, ectopic expression of *lefty* at the animal pole can block Nodal signaling at the margin of the zebrafish embryo (Chen and Schier 2002). Overexpression of Nodal-GFP and Lefty-GFP fusion proteins in chick suggests that Leftys can move farther away from the source of synthesis than Nodals (Sakuma et al. 2002). Hence, Lefty proteins negatively modulate Nodal signaling activity by dampening the generation of Nodal signals and blocking the activation of the Nodal signaling pathway.

Cerberus/DAN

Xenopus Cerberus and Coco, and mouse Cerberus-like, are members of the DAN family. These three proteins directly bind and block Nodals. Cerberus and Coco can also block Wnts and BMPs, whereas Cerberus-like also blocks BMP signaling (Piccolo et al. 1999; Belo et al. 2000; Bell et al. 2003). A truncated form of Cerberus, Cer-S, specifically blocks Nodals by direct binding to the ligand (Piccolo et al. 1999) and, in *Xenopus*, induces phenotypes that resemble the loss of Nodal signaling in zebrafish and mouse (Agius et al. 2000). Direct genetic evidence for an important role of these inhibitors comes from genetic studies in mouse. Mouse Cerberus-like acts redundantly with Lefty1 to restrict Nodal signaling in the epiblast, thus limiting mesoderm (primitive streak) formation to the future posterior side of the embryo (Perea-Gomez et al. 2002).

Tomoregulin

Tomoregulin (TMEFF1; Eib and Martens 1996) is a transmembrane protein with two follistatin domains and an EGF motif. Experiments in *Xenopus* and tissue culture have shown that Tomoregulin binds to the EGF-CFC protein Cripto and blocks Nodal signaling (Chang et al. 2003; Harms and Chang 2003), but its in vivo functions are not yet known.

Transduction

Nodal signaling leads to the phosphorylation of the transcription factors Smad2 and Smad3 (Kumar et al. 2001; Lee et al. 2001; Yeo and Whitman 2001). P-Smads can then bind to Smad4, translocate into the nucleus, and activate or inhibit downstream genes by interaction with specific transcription factors and DNA elements. It is not yet clear whether Smads mediate all aspects of Nodal signaling. Mouse *Smad4*[-/-] mutants develop mesoderm, indicating that there is a Smad4-independent way to transduce Nodal signaling (Sirard et al. 1998; Yang et al. 1998). Mouse *Smad2*[-/-] mutants affect only a subset of Nodal targets, and *Smad3*[-/-] mutants do not have Nodal signaling-related phenotypes (Nomura and Li 1998; Waldrip et al. 1998; Weinstein et al. 1998; Zhu et al. 1998; Datto et al. 1999; Heyer et al. 1999; Whitman 2001; Vincent et al. 2003). Overlapping roles of Smad2 and Smad3 are suggested by the phenotypic enhancement caused by removing one copy of *Smad3* in a homozygous and epiblast-specific mutant of *Smad2* (Vincent et al. 2003). The phenotype of *Smad2; Smad3* double mutants has not been reported yet.

FoxH1 (initially called Fast1) is the best examined transcription factor binding to P-Smad2 to regulate target genes (Chen et al. 1996; Watanabe and Whitman 1999; Osada et al. 2000; Whitman 2001). However, FoxH1 cannot account for all Nodal signaling, because *FoxH1* mutants in zebrafish and mouse affect only a subset of Nodal targets (Brand et al. 1996; Schier et al. 1996; Solnica-Krezel et al. 1996; Pogoda et al. 2000; Sirotkin et al. 2000b; Hoodless et al. 2001; Yamamoto et al. 2001). FoxH1 activity is regulated not only by interaction with P-Smad2, but also by the transcriptional repressor DRAP1 (Iratni et al. 2002). Mutants in *DRAP1* display ectopic mesendoderm formation, reflecting increased Nodal signaling. Drap1 is thought to dampen Nodal signaling by binding to FoxH1 and blocking DNA binding. ARC105, a component of the Mediator complex, is an additional factor involved in the regulation of Nodal target genes (Kato et al. 2002). Overexpression of *ARC105* stimulates Nodal signaling in *Xenopus* embryos, whereas depletion of *ARC105* inhibits Nodal signaling. Although ARC105 protein binds to Smad2/Smad4 complexes on Nodal-responsive elements, it seems to have a very general role in linking all Smad2-mediated TGF-β signaling to transcriptional activation.

The limited role of FoxH1 suggests that one or several other transcription factors that mediate Nodal signaling and interact with P-Smad2, P-Smad3, and ARC105 remain to be identified. Biochemical studies in *Xenopus* indicate that the homeodomain protein Mixer might be such a factor. Mixer/P-Smad2 complexes can regulate the expression of *goosecoid* (Germain et al. 2000). Although mutations in the zebrafish *mixer*-related gene *bonnie & clyde* (*bon*) (Kikuchi et al. 2000) affect *goosecoid* expression only weakly (Trinh et al. 2003), double mutants for both *FoxH1/sur* and *bon* lack *goosecoid* expression (Kunwar et al. 2003). Since *bon* mutants lack most endoderm, it is conceivable that Bon/P-Smad2 complexes also mediate the endoderm-inducing activity of Nodals. Analysis of Nodal-regulated proteins reveals that Bon and Sur have both distinct and overlapping regulatory roles. Some genes are regulated by both Bon and Sur, and others by either Bon or Sur (Kunwar et al. 2003). Complete loss of Nodal signaling results in a more severe phenotype than loss of both Bon and Sur (Kunwar et al. 2003; Trinh et al. 2003), indicating that additional Smad-associated transcription factors remain to be identified that act as components of the Nodal signaling pathway.

The tumor suppressor p53 is another candidate transcription factor that might contribute to Nodal signaling (Cordenonsi et al. 2003; Takebayashi-Suzuki et al. 2003).

Overexpression of *p53* can phenocopy aspects of Nodal signaling, and depleting p53 activity affects mesoderm formation in *Xenopus*. P-Smad2 and p53 bind to the control regions of Nodal-regulated genes such as *mix2* and can physically interact. In light of loss-of-function studies in mouse and zebrafish, the proposed role for p53 in Nodal signaling is surprising. *p53* mouse mutants, or zebrafish depleted of p53, do not show defects in mesendoderm induction (Donehower et al. 1992; Langheinrich et al. 2002). It is not known whether this reflects redundancy among p53 family members or different roles in mesoderm induction in different vertebrates. Hence, it remains unclear which and how many different transcription factors in addition to FoxH1 and Mixer are involved in the regulation of Nodal targets.

The RING domain-containing protein Arkadia, a potential ubiquitin ligase, is an additional intracellular factor involved in modulating Nodal signaling. *Arkadia*$^{-/-}$ mutant mice display loss of anterior structures and the node (Episkopou et al. 2001). Chimera experiments indicate that Arkadia is required extraembryonically to enhance or allow Nodal signaling in the epiblast. Similarly, Arkadia can enhance Nodal signaling nonautonomously, but also autonomously, in *Xenopus* (Niederlander et al. 2001). The molecular mechanisms underlying this activity are unknown.

NODAL SIGNALING IN DEVELOPMENT

Having reviewed the components of the Nodal signaling pathway, we can now discuss how Nodals regulate development. As described below, the roles of Nodals in zebrafish, frog, and mouse patterning exemplify three developmental properties of Nodals. First, Nodals are generated locally and act as long- and short-range signals to pattern a field of cells. Second, differential Nodal signaling activity can induce different cell fates. Third, positive and negative autoregulatory loops modulate the activity of Nodal signaling in a given cell.

Mesendoderm Induction in Zebrafish

Loss-of-function experiments using mutations or inhibitors have established that Nodals are essential for the induction of most mesodermal and endodermal cell types in zebrafish, frog, chick, and mouse. The zebrafish Nodal signals *cyclops* and *squint* are expressed in blastomeres closest to the margin (Fig. 2) (Erter et al. 1998; Feldman et al. 1998; Sampath et al. 1998; Rebagliati et al. 1998a,b; Gritsman et al. 2000; Chen and Schier 2002; Dougan et al. 2003). These cells and their neighbors give rise to mesendodermal fates (Kimmel et al. 1990; Warga and Nüsslein-Volhard 1999; Gritsman et al. 2000; Schier 2001; Dougan et

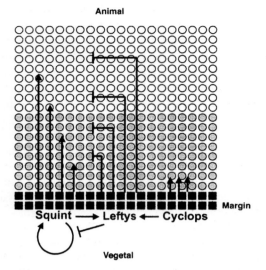

Figure 2. Model of mesendoderm induction in zebrafish. The interplay of the Nodal signals Squint and Cyclops with antagonists belonging to the Lefty family determines the extent of mesendoderm progenitors (*gray*) at the onset of zebrafish gastrulation. These secreted factors are expressed close to the margin (*black squares*). Squint is a long-range activator, Leftys are long-range inhibitors, and Cyclops is a short-range activator. Cross- and autoregulatory interactions control the expression of *squint, cyclops,* and *leftys*. As described in the text, prechordal plate cells are found in marginal-most tiers on the dorsal side; notochord progenitors are present farther away from the dorsal margin. Endoderm progenitors are located in marginal-most tiers all around the margin, overlapping with the expression domains of *squint, cyclops,* and *leftys* (*black squares*). (Reprinted, with permission, from Schier 2003 [© Annual Reviews].)

al. 2003). Absence of Nodal signaling in *cyclops;squint* double mutants or maternal-zygotic *one-eyed pinhead* mutants results in lack of all endoderm and head and trunk mesoderm, including notochord, heart, kidney, blood, liver, pancreas, and gut (Feldman et al. 1998; Gritsman et al. 1999). Abnormal development in Nodal signaling mutants is already revealed at blastula stages by the abnormal expression of genes that mark the mesendodermal progenitor territory located at the margin of the embryo (Feldman et al. 1998; Gritsman et al. 1999). For example, *bhikhari*, a marker expressed in all marginal blastomeres (Vogel and Gerster 1999) in wild type, is not expressed in *cyclops;squint* double mutants or maternal-zygotic *one-eyed pinhead* mutants (Chen and Schier 2001). Fate mapping has shown that in the absence of Nodal signaling, progenitors acquire fates inappropriate for their position (Feldman et al. 2000; Gritsman et al. 2000; Carmany-Rampey and Schier 2001). This is illustrated on the dorsal blastula margin, where cells give rise to prechordal plate anteriorly and notochord more posteriorly (Gritsman et al. 2000). The prechordal plate progenitors reside closer to the margin, whereas notochord progenitors are located at a distance from the margin. In the

absence of Nodal signaling, dorsal blastomeres give rise to hindbrain and midbrain fates (Schier and Talbot 2001). This defect is also reflected in the loss of expression of *floating head*, a marker for notochord progenitors, and *goosecoid*, a gene expressed in prechordal plate progenitors (Feldman et al. 1998; Gritsman et al. 1999, 2000). The abnormal specification of cells is accompanied by abnormal gastrulation movements. Instead of internalizing to give rise to mesoderm and endoderm, marginal cells stay on the outside and contribute to tail fates and neuroectoderm (Feldman et al. 2000; Carmany-Rampey and Schier 2001). This feature has been used to test whether single cells can internalize in the background of a noninternalizing mutant or whether a community effect is required for internalization. Single-cell transplantation experiments revealed that wild-type cells transplanted into a maternal-zygotic *one-eyed pinhead* mutant can internalize and express endodermal markers (Carmany-Rampey and Schier 2001). Similarly, activation of the Nodal signaling pathway using an activated form of the TARAM-A receptor can induce the internalization of cells located in the animal region (David and Rosa 2001). It is not known how Nodal signaling is translated into changes in morphogenesis.

The extent and pattern of mesendoderm are dependent on the strength of Nodal signaling. For example, increasing Nodal signaling by blocking Lefty1 and Lefty2 function results in the formation of an expanded domain of mesendoderm progenitors and an enlarged germ ring (Agathon et al. 2001; Chen and Schier 2002; Feldman et al. 2002). Nodal signaling can also be partially blocked by overexpressing low levels of Lefty or reducing the levels of One-eyed pinhead (Schier et al. 1997; Strahle et al. 1997; Bisgrove et al. 1999; Meno et al. 1999; Thisse and Thisse 1999; Gritsman et al. 2000; Thisse et al. 2000). In these cases, dorsal blastomeres that normally contribute to the prechordal plate acquire notochord fates instead (Gritsman et al. 2000). This change of fate is reflected in an expansion of *floating head* expression and loss of *goosecoid* expression. Hence, it has been postulated that high levels of Nodal signaling are required to induce *goosecoid* and prechordal plate fates and lower levels induce *floating head* and notochord progenitors (Gritsman et al. 2000). These results are consistent with misexpression assays which show that high levels of Nodal signaling induce *goosecoid*, whereas lower levels induce *floating head* (Gritsman et al. 2000) and that *floating head* can be activated farther from a source of Squint than *goosecoid* (Chen and Schier 2001). Similarly, high levels of Nodal signaling are also required for the induction of endoderm progenitors, which are located closest to the margin (Schier et al. 1997; Feldman et al. 1998; Alexander and Stainier 1999; Gritsman et al. 1999; Thisse and Thisse 1999; Thisse et al. 2000; Aoki et al. 2002b). In addition, reduction of Nodal signaling leads to a vegetal shift in the location of mesodermal progenitors at the ventral and lateral margin (Dougan et al. 2003). These results suggest that differential Nodal signaling at the zebrafish margin patterns cell fates along the animal–vegetal axis.

It is not fully understood how differential Nodal signaling is achieved. One possibility is that there is a vegetal-to-animal concentration gradient of Nodal signaling activity. This possibility is raised by the expression patterns of *squint*, *cyclops*, *lefty1*, and *lefty2*, and the range of activity of their gene products. In particular, mesendodermal progenitors are found in up to 10 tiers from the margin (Kimmel et al. 1990; Warga and Nüsslein-Volhard 1999; Gritsman et al. 2000; Dougan et al. 2003), whereas *cyclops*, *squint*, *lefty1*, and *lefty2* mRNA are detectable in the 1–3 cell tiers closest to the margin (Erter et al. 1998; Feldman et al. 1998, 2002; Sampath et al. 1998; Rebagliati et al. 1998a,b; Gritsman et al. 2000; Chen and Schier 2002; Dougan et al. 2003). Moreover, Squint and Leftys can act at a long range, whereas Cyclops is thought to be short range (Chen and Schier 2001, 2002). Hence, it is likely that there is a higher concentration of Nodal signals and signaling activity closer to the margin than more animally. This scenario is supported by experiments where an ectopic source of Squint can regenerate the pattern observed at the margin (Chen and Schier 2001). However, the situation is complicated by the fact that *cyclops* mutants display only minor defects in prechordal plate formation (Hatta et al. 1991; Thisse et al. 1994; Dougan et al. 2003) and *squint* mutants have rather mild defects in axial mesoderm and endoderm formation (Heisenberg and Nüsslein-Volhard 1997; Feldman et al. 1998; Sirotkin et al. 2000a; Dougan et al. 2003). Hence, both short- and long-range Nodals can apparently orchestrate most aspects of mesendoderm patterning. Although it is obvious how a long-range signal can set up an activity gradient, it is unclear how a short-range signal could generate a gradient of Nodal signaling activity. It is possible that secondary cell–cell interactions or cell movements allow patterning in *squint* mutants. In addition, the temporal regulation of Nodal signaling might contribute to graded effects. Studies in zebrafish have shown that blocking Nodal signaling for a limited period of time results in the loss of fates that require high levels of Nodal signaling (Gritsman et al. 2000; Aoki et al. 2002b). Moreover, loss of Lefty leads to persistent Nodal signaling and the expansion of fates that require higher levels of Nodal signaling (Agathon et al. 2001; Chen and Schier 2002; Feldman et al. 2002). These effects are likely to be due to abnormal duration of Nodal signaling, not because of a strict requirement to receive Nodal signals at a certain time (Dougan et al. 2003). Specifically, in *squint* mutants, *cyclops* expression is delayed, but dorsal mesodermal fates can still be induced, albeit at a later stage than in wild type. Conversely, in *cyclops* mutants, *squint* expression before gastrulation is sufficient for induc-

tion of dorsal mesoderm. This suggests that the exact timing of the response to Nodal signals is not a crucial factor in the specification of dorsal mesendoderm in zebrafish. Taken together, these observations suggest, but do not prove, that the graded distribution of Nodal signals and the differential duration of Nodal signaling are major contributors to differential fate specification.

Mesendoderm Induction in *Xenopus*

Inhibitors of Nodal signaling such as Cer-S and Lefty block aspects of mesendoderm formation in *Xenopus*, indicating a similar role as in zebrafish and mouse (Piccolo et al. 1999; Agius et al. 2000; Cheng et al. 2000; Tanegashima et al. 2000). Conversely, misexpression of Nodals in presumptive ectoderm can induce cells to become mesoderm or endoderm (Jones et al. 1995, 1996; Lustig et al. 1996; Joseph and Melton 1997). Although the study of *Nodal* genes in *Xenopus* is complicated by the lack of genetics and the large number of *Xnr* genes, the regulation of *Nodal* genes and their effect on Smad2 phosphorylation are best understood in *Xenopus*. *Xnr* gene expression is initiated by maternal factors. The vegetally localized T-box transcription factor VegT induces *Xnr* expression in vegetal cells of the blastula (Clements et al. 1999; Kofron et al. 1999; Yasuo and Lemaire 1999; Takahashi et al. 2000; Xanthos et al. 2001). In addition, the local stabilization of β-catenin contributes to *Xnr* expression on the dorsal side (Agius et al. 2000; Hyde and Old 2000), similar to its role in zebrafish (Shimizu et al. 2000; Dougan et al. 2003). The interplay of VegT, β-catenin and Nodal autoregulation is thus thought to control *Xnr* expression and P-Smad2 distribution. Supporting this view, the analysis of P-Smad2 expression has suggested that Xnrs and related signals activate the pathway first on the dorsal side, and then activation spreads ventrally and disappears dorsally (Lee et al. 2001). In addition, P-Smad2 is present at higher levels vegetally than equatorially (Faure et al. 2000; Lee et al. 2001). These results suggest that Nodal signaling in *Xenopus* might pattern mesendoderm along the animal–vegetal and/or dorsoventral axis (Agius et al. 2000; Hashimoto-Partyka et al. 2003). Indeed, higher levels of Nodal signaling appear to be required to induce dorsal mesoderm than ventral mesoderm (Agius et al. 2000). It is unclear, however, whether this reflects a patterning role of Nodal along the dorsoventral axis. It is equally possible that higher levels of Nodal signaling are required dorsally to block the development of ectoderm. This scenario would be more compatible with results in zebrafish which have shown that Nodal signaling is required both ventrally and dorsally to promote mesendoderm formation and block ectoderm development (Feldman et al. 2000; Gritsman et al. 2000; Carmany-Rampey and Schier 2001; Dougan et al. 2003).

Axis Formation and Mesendoderm Induction in Mouse

As in zebrafish and frog, loss of Nodal signaling in mouse blocks mesendoderm induction, as evidenced by lack of primitive streak formation and mesendodermal marker expression (Conlon et al. 1991, 1994; Iannaccone et al. 1992; Zhou et al. 1993). In addition, and in contrast to zebrafish and frog, mouse Nodal signaling has an even earlier role and initiates a cascade of reciprocal interactions that patterns the early embryo (Fig. 3) (Beddington and Robertson 1999). *Nodal* expression is initiated in the epiblast long before gastrulation (5 days post coitum [dpc]) and maintained and amplified by autoregulation (Brennan et al. 2001; Norris et al. 2002). Nodal maturation in the epiblast also depends on the convertases Spc1 and Spc4, which are expressed and required in the extraembryonic ectoderm (Beck et al. 2002). Nodal from the epiblast then induces *Nodal* and other downstream genes such as *Cerberus-like* and *Lefty1* in the visceral endoderm (Waldrip et al. 1998; Brennan et al. 2001; Hoodless et al. 2001; Yamamoto et al. 2001; Norris et al. 2002; Perea-Gomez et al. 2002). In addition, Nodal in the epiblast is also required for the anterior movement of the visceral endoderm, whereby the distal visceral endoderm gives rise to the anterior visceral endoderm at the prospective anterior side of the embryo (Ding et al. 1998; Brennan et al. 2001; Lu et al. 2001; Norris et al. 2002; see Chapter 16). The induction of *Lefty1* and *Cerberus-like* and the rotation of the visceral endoderm thus creates an anterior source of Nodal inhibitors that is thought to restrict Nodal signaling activity to the proximal and posterior epiblast, where the primitive streak will form (Varlet et al. 1997; Perea-Gomez et al. 2002). A similar mechanism also seems to restrict primitive streak formation in chick (Bertocchini and Stern 2002) and appears to allow forebrain formation in *Xenopus* (Bouwmeester et al. 1996). The initial restriction of Nodal signaling and primitive streak formation to the posterior-proximal epiblast is then followed by *Nodal* autoregulation (Norris and Robertson 1999; Brennan et al. 2001; Norris et al. 2002), resulting in the anterior expansion of the streak and the induction of *Lefty2* in the streak, which in turn restricts the spread of Nodal signaling (Meno et al. 1999). Hence, as in zebrafish, *Nodal* autoregulation and Lefty-mediated inhibition control the extent of mesendoderm formation in mouse.

In the mouse, evidence for concentration-dependent effects of Nodals comes from partial loss-of-function and gain-of-function studies. Nodal signaling is dampened in various alleles of *Nodal* or *Smad2* and increased in mutants for the antagonists *lefty2* and *DRAP1* (Meno et al. 1999; Lowe et al. 2001; Iratni et al. 2002; Norris et al. 2002; Vincent et al. 2003). The phenotypes of these mutants indicate that the induction of anterior definitive endoderm and prechordal plate requires higher levels of Nodal signaling than formation of the node and its midline derivatives. Lower levels yet are sufficient for the induction of posterior mesoderm. Again, this scenario is reminiscent of the situation in zebrafish, where differential Nodal signaling patterns mesendoderm.

Left/Right Axis Formation

Nodal and *Lefty* genes are expressed in the left lateral plate of chordates, including ascidians, *Amphioxus*, zebrafish (*cyclops* and *southpaw*), *Xenopus* (*Xnr1*), chick, and mouse

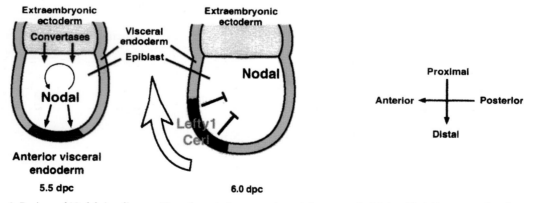

Figure 3. Reciprocal Nodal signaling positions the anterior–posterior axis in mouse. At 5.5 dpc, *Nodal* is expressed in the mouse epiblast and is maintained and enhanced by autoregulation. Convertases are expressed in the extraembryonic ectoderm and thought to process Nodal predominantly in the proximal epiblast. Nodal signaling induces *Lefty1* and *Cerberus-like* in the distal visceral endoderm, which will become the anterior visceral endoderm (AVE). Nodal signaling is also required for the displacement of the AVE, determining the position of the anterior–posterior axis (6.0 dpc). Nodal signaling becomes restricted by Lefty1 and Cerberus-like (Cerl) to the posterior region, where the primitive streak will form. (Reprinted, with permission, from Schier 2003 [©2003 by Annual Reviews].)

(Levin et al. 1995; Collignon et al. 1996; Lowe et al. 1996; Lustig et al. 1996; Meno et al. 1997; Sampath et al. 1998; Rebagliati et al. 1998a,b; Bisgrove et al. 1999; Rodriguez Esteban et al. 1999; Thisse and Thisse 1999; Tsukui et al. 1999; Yokouchi et al. 1999; Branford et al. 2000; Cheng et al. 2000; Ishimaru et al. 2000; Tanegashima et al. 2000; Morokuma et al. 2002; Yu et al. 2002; Long et al. 2003). This asymmetric expression is in part controlled by autoregulatory interactions similar to the ones described above during axis formation and mesoderm induction. In particular, Nodal in the node is required to induce *Nodal* in the left lateral plate mesoderm (Fig. 4) (Brennan et al. 2002; Saijoh et al. 2003). In addition, Nodal also induces *lefty2* and more *Nodal* in the left lateral plate and *lefty1* in the left midline (Adachi et al. 1999; Norris and Robertson 1999; Saijoh et al. 1999, 2000; Yamamoto et al. 2003). Despite the conserved left-sided expression pattern, direct evidence for a role of Nodal expression on the left is still elusive, because no left-side-specific knockout of a *Nodal* gene has been described yet. Four pieces of evidence are consistent with a role for asymmetric *Nodal* expression in left/right axis formation (Burdine and Schier 2000; Hamada et al. 2002). First, ectopic expression of *Nodal* on the right can induce the expression of left-side-specific genes (Levin et al. 1997;

Sampath et al. 1997; Yamamoto et al. 2003). Second, elimination of *lefty2* from the left lateral plate or *lefty1* from the midline results in perdurance of *Nodal* expression and the inappropriate activation of Nodal downstream genes on the right (Meno et al. 1998, 2001; Feldman et al. 2002). Although these results suggest that Nodal signaling has to be limited to the left, they do not directly prove a requirement for Nodal signaling on the left. Moreover, Lefty proteins also antagonize signaling by the TGF-β signal GDF1 (Cheng et al. 2004), suggesting that *lefty* mutant phenotypes could be caused by increased Nodal signaling. Third, left/right defects are caused by depletion of the Nodal protein Southpaw in zebrafish (Long et al. 2003) and by hypomorphic *Nodal* mutations in mouse (Lowe et al. 2001; Norris et al. 2002). However, both *Nodal* and *southpaw* are expressed in additional regions, suggesting that they might (also) act indirectly in left/right development. This is clearly the case for mouse *Nodal*, whose elimination from the node induces left/right laterality defects (Brennan et al. 2002; Saijoh et al. 2003). Fourth, mutations in Nodal signaling components (mouse *ActRIIB*, *Cryptic*, and *FoxH1*; and zebrafish *one-eyed pinhead* and *schmalspur*) result in left/right defects (Chen et al. 1997; Oh and Li 1997; Gaio et al. 1999; Yan et al. 1999; Bamford et al. 2000; Concha et al. 2000; Liang et al. 2000; Pogoda et al. 2000; Sirotkin et al. 2000b; Hoodless et al. 2001; Yamamoto et al. 2001, 2003). However, as described above, these factors also mediate GDF1 signaling (Cheng et al. 2003), and *GDF1* mutants have very similar left/right defects as *FoxH1* or *cryptic* mutants (Rankin et al. 2000).

PROSPECTS

Although several important features of the Nodal signaling pathway and its role during development have now been identified, three main areas of research need further exploration. First, many components and molecular interactions within the Nodal signaling pathway are likely to remain to be identified. Second, it is still unclear how different levels of Nodal signaling are generated and interpreted in target cells. Third, the molecular link between Nodal signaling and gastrulation movements is unknown. Only when we understand how Nodal signaling induces the internalization of cells will we have gained a deeper understanding of gastrulation.

ACKNOWLEDGMENTS

I thank present and former members of the lab for invaluable contributions, and the McKnight Endowment Fund for Neuroscience, Irma T. Hirschl Trust, American Heart Association, and National Institutes of Health for support.

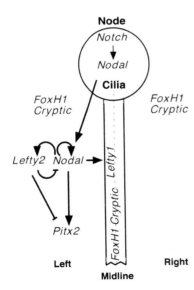

Figure 4. A regulatory cascade establishes left/right asymmetry. Notch signaling induces *Nodal* expression in the mouse node. Nodal in the node is required for *Nodal* expression in the left lateral plate. Cilia in the node are required for the left-side-specific activation of *Nodal*. Nodal in the left lateral plate autoregulates, activates *Pitx2* and *Lefty2*, and induces *Lefty1* in the left midline. The coreceptor Cryptic and the phospho-Smad2 associated transcription factor FoxH1 are expressed symmetrically and in the midline to allow Nodal signaling in these regions. (Reprinted, with permission, from Schier 2003 [© Annual Reviews].)

REFERENCES

Adachi H., Saijoh Y., Mochida K., Ohishi S., Hashiguchi H., Hirao A., and Hamada H. 1999. Determination of left/right asymmetric expression of nodal by a left side-specific enhancer with sequence similarity to a lefty-2 enhancer. *Genes Dev.* **13:** 1589–1600.

Adamson E.D., Minchiotti G., and Salomon D.S. 2002. Cripto: A tumor growth factor and more. *J. Cell Physiol.* **190:** 267–278.

Agathon A., Thisse B., and Thisse C. 2001. Morpholino knock-down of antivin1 and antivin2 upregulates nodal signaling. *Genesis* **30:** 178–182.

Agius E., Oelgeschlager M., Wessely O., Kemp C., and DeRobertis E.M. 2000. Endodermal Nodal-related signals and mesoderm induction in *Xenopus*. *Development* **127:** 1173–1183.

Alexander J. and Stainier D.Y. 1999. A molecular pathway leading to endoderm formation in zebrafish. *Curr. Biol.* **9:** 1147–1157.

Aoki T.O., Mathieu J., Saint-Etienne L., Rebagliati M.R., Peyrieras N., and Rosa F.M. 2002a. Regulation of nodal signalling and mesendoderm formation by TARAM-A, a TGFbeta-related type I receptor. *Dev. Biol.* **241:** 273–288.

Aoki T.O., David N.B., Minchiotti G., Saint-Etienne L., Dickmeis T., Persico G.M., Strahle U., Mourrain P., and Rosa F.M. 2002b. Molecular integration of casanova in the Nodal signalling pathway controlling endoderm formation. *Development* **129:** 275–286.

Attisano L. and Wrana J.L. 2002. Signal transduction by the TGF-β superfamily. *Science* **296:** 1646–1647.

Bamford R.N., Roessler E., Burdine R.D., Saplakoglu U., dela Cruz J., Splitt M., Goodship J.A., Towbin J., Bowers P., Ferrero G.B., Marino B., Schier A.F., Shen M.M., Muenke M., and Casey B. 2000. Loss-of-function mutations in the EGF-CFC gene CFC1 are associated with human left-right laterality defects. *Nat. Genet.* **26:** 365–369.

Beck S., Le Good J.A., Guzman M., Ben Haim N., Roy K., Beermann F., and Constam D.B. 2002. Extraembryonic proteases regulate Nodal signalling during gastrulation. *Nat. Cell Biol.* **4:** 981–985.

Beddington R.S. and Robertson E.J. 1999. Axis development and early asymmetry in mammals. *Cell* **96:** 195–209.

Bell E., Munoz-Sanjuan I., Altmann C.R., Vonica A., and Brivanlou A.H. 2003. Cell fate specification and competence by Coco, a maternal BMP, TGFβ and Wnt inhibitor. *Development* **130:** 1381–1389.

Belo J.A., Bachiller D., Agius E., Kemp C., Borges A.C., Marques S., Piccolo S., and De Robertis E.M. 2000. *Cerberus-like* is a secreted BMP and nodal antagonist not essential for mouse development. *Genesis* **26:** 265–270.

Bertocchini F. and Stern C.D. 2002. The hypoblast of the chick embryo positions the primitive streak by antagonizing nodal signaling. *Dev. Cell* **3:** 735–744.

Bianco C., Adkins H.B., Wechselberger C., Seno M., Normanno N., De Luca A., Sun Y., Khan N., Kenney N., Ebert A., Williams K.P., Sanicola M., and Salomon D.S. 2002. Cripto-1 activates nodal- and ALK4-dependent and -independent signaling pathways in mammary epithelial cells. *Mol. Cell. Biol.* **22:** 2586–2597.

Bisgrove B.W., Essner J.J., and Yost H.J. 1999. Regulation of midline development by antagonism of lefty and nodal signaling. *Development* **126:** 3253–3262.

Bouwmeester T., Kim S., Sasai Y., Lu B., and De Robertis E.M. 1996. Cerberus is a head-inducing secreted factor expressed in the anterior endoderm of Spemann's organizer. *Nature* **382:** 595–601.

Brand M., Heisenberg C.P., Warga R.M., Pelegri F., Karlstrom R.O., Beuchle D., Picker A., Jiang Y.J., Furutani-Seiki M., van Eeden F.J. Granato M., Haffter P., Hammerschmidt M., Kane D.A., Kelsh R.N., Mullins M.C., Odenthal J., and Nüsslein-Volhard C. 1996. Mutations affecting development of the midline and general body shape during zebrafish embryogenesis. *Development* **123:** 129–142.

Branford W.W. and Yost H.J. 2002. Lefty-dependent inhibition of nodal- and wnt-responsive organizer gene expression is essential for normal gastrulation. *Curr. Biol.* **12:** 2136–2141.

Branford W.W., Essner J.J., and Yost H.J. 2000. Regulation of gut and heart left-right asymmetry by context-dependent interactions between *Xenopus* lefty and BMP4 signaling. *Dev. Biol.* **223:** 291–306.

Brennan J., Norris D.P., and Robertson E.J. 2002. Nodal activity in the node governs left–right asymmetry. *Genes Dev.* **16:** 2339–2344.

Brennan J., Lu C.C., Norris D.P., Rodriguez T.A., Beddington R.S., and Robertson E.J. 2001. Nodal signalling in the epiblast patterns the early mouse embryo. *Nature* **411:** 965–969.

Burdine R.D. and Schier A.F. 2000. Conserved and divergent mechanisms in left–right axis formation. *Genes Dev.* **14:** 763–776.

Carmany-Rampey A. and Schier A.F. 2001. Single-cell internalization during zebrafish gastrulation. *Curr. Biol.* **11:** 1261–1265.

Chang C., Eggen B.J., Weinstein D.C., and Brivanlou A.H. 2003. Regulation of nodal and BMP signaling by tomoregulin-1 (X7365) through novel mechanisms. *Dev. Biol.* **255:** 1–11.

Chen J.N., van Eeden F.J., Warren K.S., Chin A., Nüsslein-Volhard C., Haffter P., and Fishman M.C. 1997. Left-right pattern of cardiac BMP4 may drive asymmetry of the heart in zebrafish. *Development* **124:** 4373–4382.

Chen J.N., Haffter P., Odenthal J., Vogelsang E., Brand M., van Eeden F.J., Furutani-Seiki M., Granato M., Hammerschmidt M., Heisenberg C.P., Jiang Y.J., Kane D.A., Kelsh R.N., Mullins M.C., and Nüsslein-Volhard C. 1996. Mutations affecting the cardiovascular system and other internal organs in zebrafish. *Development* **123:** 293–302.

Chen Y. and Schier A.F. 2001. The zebrafish Nodal signal Squint functions as a morphogen. *Nature* **411:** 607–610.

———. 2002. Lefty proteins are long-range inhibitors of squint-mediated nodal signaling. *Curr. Biol.* **12:** 2124–2128.

Cheng A.M., Thisse B., Thisse C., and Wright C.V. 2000. The lefty-related factor Xatv acts as a feedback inhibitor of nodal signaling in mesoderm induction and L-R axis development in *Xenopus*. *Development* **127:** 1049–1061.

Cheng S.K., Olale F., Brivanlou A.H., and Schier A.F. 2004. Lefty blocks a subset of TGF-β signals by antagonizing EGF-CFC coreceptors. *Public Library of Science Biology* **2:** 215–226.

Cheng S.K., Olale F., Bennett J.T., Brivanlou A.H., and Schier A.F. 2003. EGF-CFC proteins are essential coreceptors for the TGF-beta signals Vg1 and GDF1. *Genes Dev.* **17:** 31–36.

Ciccodicola A., Dono R., Obici S., Simeone A., Zollo M., and Persico M.G. 1989. Molecular characterization of a gene of the 'EGF family' expressed in undifferentiated human NTERA2 teratocarcinoma cells. *EMBO J.* **8:** 1987–1991.

Clements D., Friday R.V., and Woodland H.R. 1999. Mode of action of VegT in mesoderm and endoderm formation. *Development* **126:** 4903–4911.

Colas J.F. and Schoenwolf G.C. 2000. Subtractive hybridization identifies chick-cripto, a novel EGF-CFC ortholog expressed during gastrulation, neurulation and early cardiogenesis. *Gene* **255:** 205–217.

Collignon J., Varlet I., and Robertson E.J. 1996. Relationship between asymmetric nodal expression and the direction of embryonic turning. *Nature* **381:** 155–158.

Concha M.L., Burdine R.D., Russell C., Schier A.F., and Wilson S.W. 2000. A nodal signaling pathway regulates the laterality of neuroanatomical asymmetries in the zebrafish forebrain. *Neuron* **28:** 399–409.

Conlon F.L., Barth K.S., and Robertson E.J. 1991. A novel retrovirally induced embryonic lethal mutation in the mouse: assessment of the developmental fate of embryonic stem cells hemozygous for the 413.d proviral integration. *Development* **111:** 969–981.

Conlon F.L., Lyons K.M., Takaesu N., Barth K.S., Kispert A., Herrmann B., and Robertson E.J. 1994. A primary requirement for nodal in the formation and maintenance of the primitive streak in the mouse. *Development* **120:** 1919–1928.

Constam D.B. and Robertson E.J. 1999. Regulation of bone morphogenetic protein activity by pro domains and proprotein convertases. *J. Cell Biol.* **144:** 139–149.

Cordenonsi M., Dupont S., Maretto S., Insinga A., Imbriano C., and Piccolo S. 2003. Links between tumor suppressors: p53 is required for TGF-beta gene responses by cooperating with Smads. *Cell* **113:** 301–314.

Datto M.B., Frederick J.P., Pan L., Borton A.J., Zhuang Y., and Wang X.F. 1999. Targeted disruption of Smad3 reveals an essential role in transforming growth factor beta-mediated signal transduction. *Mol. Cell. Biol.* **19:** 2495–2504.

David N.B. and Rosa F.M. 2001. Cell autonomous commitment to an endodermal fate and behaviour by activation of Nodal signalling. *Development* **128:** 3937–3947.

Dehal P., Satou Y., Campbell R.K., Chapman J., Degnan B., De Tomaso A., Davidson B., Di Gregorio A., Gelpke M., Goodstein D.M. et al. 2002. The draft genome of *Ciona intestinalis*: Insights into chordate and vertebrate origins. *Science* **298:** 2157–2167.

Ding J., Yang L., Yan Y. T., Chen A., Desai N., Wynshaw-Boris A., and Shen M.M. 1998. Cripto is required for correct orientation of the anterior–posterior axis in the mouse embryo. *Nature* **395:** 702–707.

Donehower L.A., Harvey M., Slagle B.L., McArthur M.J., Montgomery C.A. Jr., Butel J.S., and Bradley A. 1992. Mice deficient for p53 are developmentally normal but susceptible to spontaneous tumours. *Nature* **356:** 215–221.

Dougan S.T., Warga R.M., Kane D.A., Schier A.F., and Talbot W.S. 2003. The role of the zebrafish nodal-related genes squint and cyclops in patterning of mesendoderm. *Development* **130:** 1837–1851.

Eib D.W. and Martens G.J. 1996. A novel transmembrane protein with epidermal growth factor and follistatin domains expressed in the hypothalamo-hypophysial axis of *Xenopus laevis*. *J. Neurochem.* **67:** 1047–1055.

Eimon P.M. and Harland R.M. 2002. Effects of heterodimerization and proteolytic processing on Derrière and Nodal activity: Implications for mesoderm induction in *Xenopus*. *Development* **129:** 3089–3103.

Episkopou V., Arkell R., Timmons P.M., Walsh J.J., Andrew R.L., and Swan D. 2001. Induction of the mammalian node requires Arkadia function in the extraembryonic lineages. *Nature* **410:** 825–830.

Erter C.E., Solnica-Krezel L., and Wright C.V. 1998. *Zebrafish nodal-related 2* encodes an early mesendodermal inducer signaling from the extraembryonic yolk syncytial layer. *Dev. Biol.* **204:** 361–372.

Faure S., Lee M.A., Keller T., ten Dijke P., and Whitman M. 2000. Endogenous patterns of TGFβ superfamily signaling during early

Xenopus development. *Development* **127:** 2917–2931.

Feldman B., Dougan S.T., Schier A.F., and Talbot W.S. 2000. Nodal-related signals establish mesendodermal fate and trunk neural identity in zebrafish. *Curr. Biol.* **10:** 531–534.

Feldman B., Concha M.L., Saude L., Parsons M.J., Adams R.J., Wilson S.W., and Stemple D.L. 2002. Lefty antagonism of squint is essential for normal gastrulation. *Curr. Biol.* **12:** 2129–2135.

Feldman B., Gates M.A., Egan E.S., Dougan S.T., Rennebeck G., Sirotkin H.I., Schier A.F., and Talbot W.S. 1998. Zebrafish organizer development and germ-layer formation require nodal-related signals. *Nature* **395:** 181–185.

Gaio U., Schweickert A., Fischer A., Garratt A.N., Muller T., Ozcelik C., Lankes W., Strehle M., Britsch S., Blum M., and Birchmeier C. 1999. A role of the cryptic gene in the correct establishment of the left- right axis. *Curr. Biol.* **9:** 1339–1342.

Germain S., Howell M., Esslemont G.M., and Hill C.S. 2000. Homeodomain and winged-helix transcription factors recruit activated Smads to distinct promoter elements via a common Smad interaction motif. *Genes Dev.* **14:** 435–451.

Gritsman K., Talbot W.S., and Schier A.F. 2000. Nodal signaling patterns the organizer. *Development* **127:** 921–932.

Gritsman K., Zhang J., Cheng S., Heckscher E., Talbot W.S., and Schier A.F. 1999. The EGF-CFC protein one-eyed pinhead is essential for nodal signaling. *Cell* **97:** 121–132.

Gu Z., Reynolds E.M., Song J., Lei H., Feijen A., Yu L., He W., MacLaughlin D.T., van den Eijnden-van Raaij J., Donahoe P.K., and Li E. 1999. The type I serine/threonine kinase receptor ActRIA (ALK2) is required for gastrulation of the mouse embryo. *Development* **126:** 2551–2561.

Hamada H., Meno C., Watanabe D., and Saijoh Y. 2002. Establishment of vertebrate left-right asymmetry. *Nat. Rev. Genet.* **3:** 103–113.

Harms P.W. and Chang C. 2003. Tomoregulin-1 (TMEFF1) inhibits nodal signaling through direct binding to the nodal coreceptor Cripto. *Genes Dev.* **17:** 2624–2629.

Hashimoto-Partyka M.K., Yuge M., and Cho K.W. 2003. Nodal signaling in *Xenopus* gastrulae is cell-autonomous and patterned by beta-catenin. *Dev. Biol.* **253:** 125–138.

Hatta K., Kimmel C.B., Ho R.K., and Walker C. 1991. The *cyclops* mutation blocks specification of the floor plate of the zebrafish central nervous system. *Nature* **350:** 339–341.

Heisenberg C.P. and Nüsslein-Volhard C. 1997. The function of silberblick in the positioning of the eye anlage in the zebrafish embryo. *Dev. Biol.* **184:** 85–94.

Heyer J., Escalante-Alcalde D., Lia M., Boettinger E., Edelmann W., Stewart C.L., and Kucherlapati R. 1999. Postgastrulation Smad2-deficient embryos show defects in embryo turning and anterior morphogenesis. *Proc. Natl. Acad. Sci.* **96:** 12595–12600.

Hoodless P.A., Pye M., Chazaud C., Labbe E., Attisano L., Rossant J., and Wrana J.L. 2001. FoxH1 (Fast) functions to specify the anterior primitive streak in the mouse. *Genes Dev.* **15:** 1257–1271.

Hyde C.E. and Old R.W. 2000. Regulation of the early expression of the *Xenopus* nodal-related 1 gene, Xnr1. *Development* **127:** 1221–1229.

Iannaccone P.M., Zhou X., Khokha M., Boucher D., and Kuehn M.R. 1992. Insertional mutation of a gene involved in growth regulation of the early mouse embryo. *Dev. Dyn.* **194:** 198–208.

Iratni R., Yan Y.T., Chen C., Ding J., Zhang Y., Price S.M., Reinberg D., and Shen M.M. 2002. Inhibition of excess nodal signaling during mouse gastrulation by the transcriptional corepressor DRAP1. *Science* **298:** 1996–1999.

Ishimaru Y., Yoshioka H., Tao H., Thisse B., Thisse C., Wright C.V.E.,

Hamada H., Ohuchi H., and Noji S. 2000. Asymmetric expression of *antivin/lefty1* in the early chick embryo. *Mech. Dev.* **90:** 115–118.

Jones C.M., Armes N., and Smith J.C. 1996. Signalling by TGF-β family members: Short-range effects of Xnr-2 and BMP-4 contrast with the long-range effects of activin. *Curr. Biol.* **6:** 1468–1475.

Jones C.M., Kuehn M.R., Hogan B.L., Smith J.C., and Wright C.V. 1995. Nodal-related signals induce axial mesoderm and dorsalize mesoderm during gastrulation. *Development* **121:** 3651–3662.

Joseph E.M. and Melton D.A. 1997. Xnr4: A *Xenopus* nodal-related gene expressed in the Spemann organizer. *Dev. Biol.* **184:** 367–372.

Kato Y., Habas R., Katsuyama Y., Naar A.M., and He X. 2002. A component of the ARC/Mediator complex required for TGF beta/Nodal signalling. *Nature* **418:** 641–646.

Kikuchi Y., Trinh L.A., Reiter J.F., Alexander J., Yelon D., and Stainier D.Y. 2000. The zebrafish *bonnie and clyde* gene encodes a Mix family homeodomain protein that regulates the generation of endodermal precursors. *Genes Dev.* **14:** 1279–1289.

Kimmel C.B., Warga R.M., and Schilling T.F. 1990. Origin and organization of the zebrafish fate map. *Development* **108:** 581–594.

Kinoshita N., Minshull J., and Kirschner M.W. 1995. The identification of two novel ligands of the FGF receptor by a yeast screening method and their activity in *Xenopus* development. *Cell* **83:** 621–630.

Kofron M., Demel T., Xanthos J., Lohr J., Sun B., Sive H., Osada S., Wright C., Wylie C., and Heasman J. 1999. Mesoderm induction in *Xenopus* is a zygotic event regulated by maternal VegT via TGFβ growth factors. *Development* **126:** 5759–5770.

Kumar A., Novoselov V., Celeste A.J., Wolfman N.M., ten Dijke P., and Kuehn M.R. 2001. Nodal signaling uses activin and transforming growth factor-β receptor-regulated Smads. *J. Biol. Chem.* **276:** 656–661.

Kunwar P.S., Zimmerman S., Bennett J.T., Chen Y., Whitman M., and Schier A.F. 2003. Mixer/Bon and FoxH1/Sur have overlapping and divergent roles in Nodal signaling and mesendoderm induction. *Development* **130:** 5589–5599.

Langheinrich U., Hennen E., Stott G., and Vacun G. 2002. Zebrafish as a model organism for the identification and characterization of drugs and genes affecting p53 signaling. *Curr. Biol.* **12:** 2023–2028.

Lee M.A., Heasman J., and Whitman M. 2001. Timing of endogenous activin-like signals and regional specification of the *Xenopus* embryo. *Development* **128:** 2939–2952.

Levin M., Johnson R.L., Stern C.D., Kuehn M., and Tabin C. 1995. A molecular pathway determining left–right asymmetry in chick embryogenesis. *Cell* **82:** 803–814.

Levin M., Pagan S., Roberts D.J., Cooke J., Kuehn M.R., and Tabin C.J. 1997. Left/right patterning signals and the independent regulation of different aspects of situs in the chick embryo. *Dev. Biol.* **189:** 57–67.

Liang J.O., Etheridge A., Hantsoo L., Rubinstein A.L., Nowak S.J., Izpisua-Belmonte J.C., and Halpern M.E. 2000. Asymmetric nodal signaling in the zebrafish diencephalon positions the pineal organ. *Development* **127:** 5101–5112.

Long S., Ahmad N., and Rebagliati M. 2003. The zebrafish nodal-related gene southpaw is required for visceral and diencephalic left–right asymmetry. *Development* **130:** 2303–2316.

Lowe L.A., Yamada S., and Kuehn M.R. 2001. Genetic dissection of nodal function in patterning the mouse embryo. *Development* **128:** 1831–1843.

Lowe L.A., Supp D.M., Sampath K., Yokoyama T., Wright C.V., Potter S.S., Overbeek P., and Kuehn M.R. 1996. Conserved left–right asymmetry of nodal expression and alterations in murine situs inversus. *Nature* **381:** 158–161.

Lu C.C., Brennan J., and Robertson E.J. 2001. From fertilization to gastrulation: Axis formation in the mouse embryo. *Curr. Opin. Genet. Dev.* **11:** 384–392.

Lustig K.D., Kroll K., Sun E., Ramos R., Elmendorf H., and Kirschner M.W. 1996. A *Xenopus* nodal-related gene that acts in synergy with noggin to induce complete secondary axis and notochord formation. *Development* **122:** 3275–3282.

Mathieu J., Barth A., Rosa F.M., Wilson S.W., and Peyrieras N. 2002. Distinct and cooperative roles for Nodal and Hedgehog signals during hypothalamic development. *Development* **129:** 3055–3065.

Meno C., Ito Y., Saijoh Y., Matsuda Y., Tashiro K., Kuhara S., and Hamada H. 1997. Two closely-related left-right asymmetrically expressed genes, *lefty-1* and *lefty-2*: Their distinct expression domains, chromosomal linkage and direct neuralizing activity in *Xenopus* embryos. *Genes Cells* **2:** 513–524.

Meno C., Saijoh Y., Fujii H., Ikeda M., Yokoyama T., Yokoyama M., Toyoda Y., and Hamada H. 1996. Left-right asymmetric expression of the TGFβ-family member *lefty* in mouse embryos. *Nature* **381:** 151–155.

Meno C., Shimono A., Saijoh Y., Yashiro K., Mochida K., Ohishi S., Noji S., Kondoh H., and Hamada H. 1998. *lefty-1* is required for left-right determination as a regulator of *lefty-2* and *nodal*. *Cell* **94:** 287–297.

Meno C., Takeuchi J., Sakuma R., Koshiba-Takeuchi K., Ohishi S., Saijoh Y., Miyazaki J., ten Dijke P., Ogura T., and Hamada H. 2001. Diffusion of nodal signaling activity in the absence of the feedback inhibitor Lefty2. *Dev. Cell* **1:** 127–138.

Meno C., Gritsman K., Ohishi S., Ohfuji Y., Heckscher E., Mochida K., Shimono A., Kondoh H., Talbot W.S., Robertson E.J., Schier A.F., and Hamada H. 1999. Mouse Lefty2 and zebrafish antivin are feedback inhibitors of nodal signaling during vertebrate gastrulation. *Mol. Cell* **4:** 287–298.

Minchiotti G., Manco G., Parisi S., Lago C.T., Rosa F., and Persico M.G. 2001. Structure-function analysis of the EGF-CFC family member Cripto identifies residues essential for nodal signalling. *Development* **128:** 4501–4510.

Minchiotti G., Parisi S., Liguori G., Signore M., Lania G., Adamson E.D., Lago C.T., and Persico M.G. 2000. Membrane-anchorage of Cripto protein by glycosylphosphatidylinositol and its distribution during early mouse development. *Mech. Dev.* **90:** 133–142.

Morokuma J., Ueno M., Kawanishi H., Saiga H., and Nishida H. 2002. HrNodal, the ascidian nodal-related gene, is expressed in the left side of the epidermis, and lies upstream of HrPitx. *Dev. Genes Evol.* **212:** 439–446.

Niederlander C., Walsh J.J., Episkopou V., and Jones C.M. 2001. Arkadia enhances nodal-related signalling to induce mesendoderm. *Nature* **410:** 830–834.

Nomura M. and Li E. 1998. Smad2 role in mesoderm formation, left–right patterning and craniofacial development. *Nature* **393:** 786–790.

Norris D.P. and Robertson E.J. 1999. Asymmetric and node-specific *nodal* expression patterns are controlled by two distinct *cis*-acting regulatory elements. *Genes Dev.* **13:** 1575–1588.

Norris D.P., Brennan J., Bikoff E.K., and Robertson E.J. 2002. The Foxh1-dependent autoregulatory enhancer controls the level of Nodal signals in the mouse embryo. *Development* **129:** 3455–3468.

Oh S.P. and Li E. 1997. The signaling pathway mediated by the type IIB activin receptor controls axial patterning and lateral asymmetry in the mouse. *Genes Dev.* **11:** 1812–1826.

Osada S.I., Saijoh Y., Frisch A., Yeo C.Y., Adachi H., Watanabe M., Whitman M., Hamada H., and Wright C.V. 2000. Activin/nodal responsiveness and asymmetric expression of a *Xenopus nodal*-related gene converge on a FAST-regulated module in intron 1. *Development* **127:** 2503–2514.

Perea-Gomez A., Vella F.D., Shawlot W., Oulad-Abdelghani M., Chazaud C., Meno C., Pfister V., Chen L., Robertson E., Hamada H., Behringer R.R., and Ang S.L. 2002. Nodal antagonists in the anterior visceral endoderm prevent the formation of multiple primitive streaks. *Dev. Cell* **3:** 745–756.

Piccolo S., Agius E., Leyns L., Bhattacharyya S., Grunz H., Bouwmeester T., and De Robertis E.M. 1999. The head inducer Cerberus is a multifunctional antagonist of Nodal, BMP and Wnt signals. *Nature* **397:** 707–710.

Pogoda H.M., Solnica-Krezel L., Driever W., and Meyer D. 2000. The zebrafish forkhead transcription factor FoxH1/Fast1 is a modulator of nodal signaling required for organizer formation. *Curr. Biol.* **10:** 1041–1049.

Rankin C.T., Bunton T., Lawler A.M., and Lee S.J. 2000. Regulation of left–right patterning in mice by growth/differentiation factor-1. *Nat. Genet.* **24:** 262–265.

Rebagliati M.R., Toyama R., Haffter P., and Dawid I.B. 1998a. *cyclops* encodes a nodal-related factor involved in midline signaling. *Proc. Natl. Acad. Sci.* **95:** 9932–9937.

Rebagliati M.R., Toyama R., Fricke C., Haffter P., and Dawid I.B. 1998b. Zebrafish nodal-related genes are implicated in axial patterning and establishing left–right asymmetry. *Dev. Biol.* **199:** 261–272.

Reissmann E., Jörnvall H., Blokzijl A., Andersson O., Chang C., Minchiotti G., Persico M.G., Ibáñez C.F., and Brivanlou A.H. 2001. The orphan receptor ALK7 and the Activin receptor ALK4 mediate signaling by Nodal proteins during vertebrate development. *Genes Dev.* **15:** 2010–2022.

Renucci A., Lemarchandel V., and Rosa F. 1996. An activated form of type I serine/threonine kinase receptor TARAM-A reveals a specific signalling pathway involved in fish head organiser formation. *Development* **122:** 3735–3743.

Rodriguez Esteban C., Capdevila J., Economides A.N., Pascual J., Ortiz A., and Izpisua Belmonte J.C. 1999. The novel Cer-like protein Caronte mediates the establishment of embryonic left–right asymmetry. *Nature* **401:** 243–251.

Saijoh Y., Oki S., Ohishi S., and Hamada H. 2003. Left–right patterning of the mouse lateral plate requires nodal produced in the node. *Dev. Biol.* **256:** 161–73.

Saijoh Y., Adachi H., Mochida K., Ohishi S., Hirao A., and Hamada H. 1999. Distinct transcriptional regulatory mechanisms underlie left–right asymmetric expression of *lefty-1* and *lefty-2*. *Genes Dev.* **13:** 259–269.

Saijoh Y., Adachi H., Sakuma R., Yeo C. Y., Yashiro K., Watanabe M., Hashiguchi H., Mochida K., Ohishi S., Kawabata M., Miyazono K., Whitman M., and Hamada H. 2000. Left–right asymmetric expression of *lefty2* and *nodal* is induced by a signaling pathway that includes the transcription factor FAST2. *Mol. Cell* **5:** 35–47.

Sakuma R., Ohnishi Yi Y., Meno C., Fujii H., Juan H., Takeuchi J., Ogura T., Li E., Miyazono K., and Hamada H. 2002. Inhibition of Nodal signalling by Lefty mediated through interaction with common receptors and efficient diffusion. *Genes Cells* **7:** 401–412.

Sampath K., Cheng A.M., Frisch A., and Wright C.V. 1997. Functional differences among *Xenopus* nodal-related genes in left-right axis determination. *Development* **124:** 3293–3302.

Sampath K., Rubinstein A.L., Cheng A.M., Liang J.O., Fekany K., Solnica-Krezel L., Korzh V., Halpern M.E., and Wright C.V. 1998. Induction of the zebrafish ventral brain and floorplate requires cyclops/nodal signalling. *Nature* **395:** 185–189.

Schier A.F. 2001. Axis formation and patterning in zebrafish. *Curr. Opin. Genet. Dev.* **11:** 393–404.

———. 2003. Nodal signaling in vertebrate development. *Annu. Rev. Cell Dev. Biol.* **19:** 589–621.

Schier A.F. and Shen M.M. 2000. Nodal signalling in vertebrate development. *Nature* **403:** 385–389.

Schier A.F. and Talbot W.S. 2001. Nodal signaling and the zebrafish organizer. *Int. J. Dev. Biol.* **45:** 289–297.

Schier A.F., Neuhauss S.C., Helde K.A., Talbot W.S., and Driever W. 1997. The *one-eyed pinhead* gene functions in mesoderm and endoderm formation in zebrafish and interacts with *no tail*. *Development* **124:** 327–342.

Schier A.F., Neuhauss S.C., Harvey M., Malicki J., Solnica-Krezel L., Stainier D.Y., Zwartkruis F., Abdelilah S., Stemple D.L., Rangini Z., Yang H., and Driever W. 1996. Mutations affecting the development of the embryonic zebrafish brain. *Development* **123:** 165–178.

Schiffer S.G., Foley S., Kaffashan A., Hronowski X., Zichittella A.E., Yeo C.Y., Miatkowski K., Adkins H.B., Damon B., Whitman M., Salomon D., Sanicola M., and Williams K.P. 2001. Fucosylation of Cripto is required for its ability to facilitate nodal signaling. *J. Biol. Chem.* **276:** 37769–37778.

Schlange T., Schnipkoweit I., Andree B., Ebert A., Zile M.H., Arnold H.H., and Brand T. 2001. Chick CFC controls Lefty1 expression in the embryonic midline and nodal expression in the lateral plate. *Dev. Biol.* **234:** 376–389.

Shen M.M. and Schier A.F. 2000. The EGF-CFC gene family in vertebrate development. *Trend Genet.* **16:** 303–309.

Shen M.M., Wang H., and Leder P. 1997. A differential display strategy identifies Cryptic, a novel EGF-related gene expressed in the axial and lateral mesoderm during mouse gastrulation. *Development* **124:** 429–442.

Shi Y. and Massague J. 2003. Mechanisms of TGF-β signaling from cell membrane to the nucleus. *Cell* **113:** 685–700.

Shimizu T., Yamanaka Y., Ryu S.L., Hashimoto H., Yabe T., Hirata T., Bae Y.K., Hibi M., and Hirano T. 2000. Cooperative roles of Bozozok/Dharma and Nodal-related proteins in the formation of the dorsal organizer in zebrafish. *Mech. Dev.* **91:** 293–303.

Sirard C., de la Pompa J.L., Elia A., Itie A., Mirtsos C., Cheung A., Hahn S., Wakeham A., Schwartz L., Kern S.E., Rossant J., and Mak T.W. 1998. The tumor suppressor gene *Smad4/Dpc4* is required for gastrulation and later for anterior development of the mouse embryo. *Genes Dev.* **12:** 107–119.

Sirotkin H.I., Dougan S.T., Schier A.F., and Talbot W.S. 2000a. *bozozok* and *squint* act in parallel to specify dorsal mesoderm and anterior neuroectoderm in zebrafish. *Development* **127:** 2583–2592.

Sirotkin H.I., Gates M.A., Kelly P.D., Schier A.F., and Talbot W.S. 2000b. *fast1* is required for the development of dorsal axial structures in zebrafish. *Curr. Biol.* **10:** 1051–1054.

Smith W.C., McKendry R., Ribisi S. Jr., and Harland R.M. 1995. A nodal-related gene defines a physical and functional domain within the Spemann organizer. *Cell* **82:** 37–46.

Solnica-Krezel L., Stemple D.L., Mountcastle-Shah E., Rangini Z., Neuhauss S.C., Malicki J., Schier A.F., Stainier D.Y., Zwartkruis F., Abdelilah S., and Driever W. 1996. Mutations affecting cell fates and cellular rearrangements during gastrulation in zebrafish. *Development* **123:** 67–80.

Song J., Oh S. P., Schrewe H., Nomura M., Lei H., Okano M., Gridley T., and Li E. 1999. The type II activin receptors are essential for egg cylinder growth, gastrulation, and rostral head development in mice. *Dev. Biol.* **213:** 157–169.

Strahle U., Jesuthasan S., Blader P., Garcia-Villalba P., Hatta K., and Ingham P.W. 1997. *one-eyed pinhead* is required for development of the ventral midline of the zebrafish (*Danio rerio*) neural tube. *Genes Funct.* **1:** 131–148.

Takahashi S., Yokota C., Takano K., Tanegashima K., Onuma Y., Goto J. and Asashima M. 2000. Two novel *nodal*-related genes initiate early inductive events in *Xenopus* Nieuwkoop center. *Development* **127:** 5319–5329.

Takebayashi-Suzuki K., Funami J., Tokumori D., Saito A., Watabe T., Miyazono K., Kanda A., and Suzuki A. 2003. Interplay between the tumor suppressor p53 and TGFβ signaling shapes embryonic body axes in *Xenopus*. *Development* **130:** 3929–3939.

Tanegashima K., Yokota C., Takahashi S., and Asashima M. 2000. Expression cloning of Xantivin, a *Xenopus* lefty/antivin-related gene, involved in the regulation of activin signaling during mesoderm induction. *Mech. Dev.* **99:** 3–14.

Thisse B., Wright C.V., and Thisse C. 2000. Activin- and Nodal-related factors control antero-posterior patterning of the zebrafish embryo. *Nature* **403:** 425–428.

Thisse C. and Thisse B. 1999. Antivin, a novel and divergent member of the TGFβ superfamily, negatively regulates mesoderm induction. *Development* **126:** 229–240.

Thisse C., Thisse B., Halpern M.E., and Postlethwait J.H. 1994. Goosecoid expression in neurectoderm and mesendoderm is disrupted in zebrafish cyclops gastrulas. *Dev. Biol.* **164:** 420–429.

Toyama R., O'Connell M.L., Wright C.V., Kuehn M.R., and Dawid I.B. 1995. Nodal induces ectopic goosecoid and lim1 expression and axis duplication in zebrafish. *Development* **121:** 383–391.

Trinh L.A., Meyer D., and Stainier D.Y. 2003. The Mix family homeodomain gene *bonnie and clyde* functions with other components of the Nodal signaling pathway to regulate neural patterning in zebrafish. *Development* **130:** 4989–4998.

Tsukui T., Capdevila J., Tamura K., Ruiz-Lozano P., Rodriguez-Esteban C., Yonei-Tamura S., Magallon J., Chandraratna R.A., Chien K., Blumberg B., Evans R.M., and Belmonte J.C. 1999. Multiple left–right asymmetry defects in *Shh*⁻/⁻ mutant mice unveil a convergence of the shh and retinoic acid pathways in the control of *Lefty-1*. *Proc. Natl. Acad. Sci.* **96:** 11376–11381.

Varlet I., Collignon J., and Robertson E.J. 1997. Nodal expression in the primitive eododerm is required for the specification of the anterior axis during mouse gastrulation. *Development* **124:** 1033–1044.

Vincent S.D., Dunn N.R., Hayashi S., Norris D.P., and Robertson E.J. 2003. Cell fate decisions within the mouse organizer are governed by graded Nodal signals. *Genes Dev.* **17:** 1646–1662.

Vogel A.M. and Gerster T. 1999. Promoter activity of the zebrafish bhikhari retroelement requires an intact activin signaling pathway. *Mech. Dev.* **85:** 133–146.

Waldrip W.R., Bikoff E.K., Hoodless P.A., Wrana J.L., and Robertson E.J. 1998. Smad2 signaling in extraembryonic tissues determines anterior–posterior polarity of the early mouse embryo. *Cell* **92:** 797–808.

Warga R.M. and Nüsslein-Volhard C. 1999. Origin and development of the zebrafish endoderm. *Development* **126:** 827–838.

Watanabe D., Saijoh Y., Nonaka S., Sasaki G., Ikawa Y., Yokoyama T., and Hamada H. 2003. The left–right determinant Inversin is a component of node monocilia and other 9+0 cilia. *Development* **130:** 1725–1734.

Watanabe M. and Whitman M. 1999. FAST-1 is a key maternal effector of mesoderm inducers in the early *Xenopus* embryo. *Development* **126:** 5621–5634.

Weinstein M., Yang X., Li C., Xu X., Gotay J., and Deng C.X. 1998. Failure of egg cylinder elongation and mesoderm induction in mouse embryos lacking the tumor suppressor *smad2*. *Proc. Natl. Acad. Sci.* **95:** 9378–9383.

Whitman M. 2001. Nodal signaling in early vertebrate embryos: Themes and variations. *Dev. Cell* **1:** 605–617.

Xanthos J.B., Kofron M., Wylie C., and Heasman J. 2001. Maternal VegT is the initiator of a molecular network specifying endoderm in *Xenopus laevis*. *Development* **128:** 167–180.

Yabe S.I., Tanegashima K., Haramoto Y., Takahashi S., Fujii T., Kozuma S., Taketani Y., and Asashima M. 2003. FRL-1, a member of the EGF-CFC family, is essential for neural differentiation in *Xenopus* early development. *Development* **130:** 2071–2081.

Yamamoto M., Meno C., Sakai Y., Shiratori H., Mochida K., Ikawa Y., Saijoh Y., and Hamada H. 2001. The transcription factor FoxH1 (FAST) mediates Nodal signaling during anterior–posterior patterning and node formation in the mouse. *Genes Dev.* **15:** 1242–1256.

Yamamoto M., Mine N., Mochida K., Sakai Y., Saijoh Y., Meno C., and Hamada H. 2003. Nodal signaling induces the midline barrier by activating Nodal expression in the lateral plate. *Development* **130:** 1795–1804.

Yan Y.T., Gritsman K., Ding J., Burdine R.D., Corrales J.D., Price S.M., Talbot W.S., Schier A.F., and Shen M.M. 1999. Conserved requirement for EGF-CFC genes in vertebrate left–right axis formation. *Genes Dev.* **13:** 2527–2537.

Yan Y.T., Liu J.J., Luo Y., E C., Haltiwanger R.S., Abate-Shen C., and Shen M.M. 2002. Dual roles of Cripto as a ligand and coreceptor in the nodal signaling pathway. *Mol. Cell. Biol.* **22:** 4439–4449.

Yang X., Li C., Xu X., and Deng C. 1998. The tumor suppressor SMAD4/DPC4 is essential for epiblast proliferation and mesoderm induction in mice. *Proc. Natl. Acad. Sci.* **95:** 3667–3672.

Yasuo H. and Lemaire P. 1999. A two-step model for the fate determination of presumptive endodermal blastomeres in *Xenopus* embryos. *Curr. Biol.* **9:** 869–879.

Yeo C. and Whitman M. 2001. Nodal signals to Smads through Cripto-dependent and Cripto-independent mechanisms. *Mol. Cell* **7:** 949–957.

Yokota C., Kofron M., Zuck M., Houston D.W., Isaacs H., Asashima M., Wylie C.C., and Heasman J. 2003. A novel role for a nodal-related protein; Xnr3 regulates convergent extension movements via the FGF receptor. *Development* **130:** 2199–2212.

Yokouchi Y., Vogan K.J., Pearse R.V., and Tabin C.J. 1999. Antagonistic signaling by *Caronte*, a novel *Cerberus*-related gene, establishes left–right asymmetric gene expression. *Cell* **98:** 573–583.

Yu J.K., Holland L.Z., and Holland N.D. 2002. An amphioxus nodal gene (AmphiNodal) with early symmetrical expression in the organizer and mesoderm and later asymmetrical expression associated with left–right axis formation. *Evol. Dev.* **4:** 418–425.

Zhang J., Talbot W.S., and Schier A.F. 1998. Positional cloning identifies zebrafish *one-eyed pinhead* as a permissive EGF-related ligand required during gastrulation. *Cell* **92:** 241–251.

Zhou X., Sasaki H., Lowe L., Hogan B.L., and Kuehn M.R. 1993. Nodal is a novel TGF-beta-like gene expressed in the mouse node during gastrulation. *Nature* **361:** 543–547.

Zhu Y., Richardson J.A., Parada L.F., and Graff J.M. 1998. Smad3 mutant mice develop metastatic colorectal cancer. *Cell* **94:** 703–714.

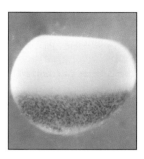

ACTIVIN AND VG1 AND THE SEARCH FOR EMBRYONIC INDUCERS

D.S. Kessler

Department of Cell & Developmental Biology, University of Pennsylvania School of Medicine, Philadelphia, Pennsylvania 19104-6058

INTRODUCTION

Defining the nature of embryonic induction has been a major goal of embryologists for nearly a century. In the amphibian embryo, the pioneering studies of Pieter Nieuwkoop in the late 1960s provided a conceptual framework for the pursuit of embryonic inducers of the mesodermal germ layer (Nieuwkoop 1969a,b; Chapter 25). Following Nieuwkoop's studies and continuing today, much energy is focused on the identification and characterization of mesoderm-inducing factors and pathways. The study of mesoderm induction addresses questions fundamental to all aspects of embryonic development: What are the endogenous inducing molecules? How is the action of an inducer spatially and temporally restricted? How do multiple inducers cooperate to generate properly patterned tissues?

In the amphibian embryo, an endogenous mesoderm inducer is predicted to fulfill several expectations: localization or expression in the vegetal hemisphere, an ability to induce mesodermal genes and tissues, and an essential role in endogenous mesoderm formation. Of the early successes in identifying candidate mesoderm inducers, studies of Activin and Vg1, members of the TGF-β family, were especially prominent. Whereas the initial investigation of these proteins suggested roles for Activin and Vg1 in endogenous mesoderm formation, evidence arguing against a requirement for these factors has also been obtained. Despite some uncertainty and occasional controversy, the study of Activin and Vg1 has revealed much about mesoderm induction, including the elucidation of an essential TGF-β signaling pathway, a role for threshold responses in determining mesodermal cell fate,

and the posttranslational regulation of TGF-β ligand processing and activity. Beyond these important advances, the study of Activin and Vg1 highlights the considerable challenges of defining the embryonic function of an individual inducer when multiple factors can each elicit an identical signaling response. Significantly, recent work in several model systems provides evidence that Activin and Vg1 may, in fact, play essential roles in the establishment and patterning of the germ layers. Here, I provide a brief overview of what is known about the molecular and embryological functions of Activin and Vg1 in the gastrula and identify some of the important questions that remain to be resolved.

MESODERM INDUCTION AND THE *XENOPUS* "ANIMAL CAP"

The recombination studies of Nieuwkoop provided two fundamental insights into the process of mesoderm induction: that the vegetal pole is a source of mesoderm-inducing signals, and that animal pole cells can respond to vegetal inducing signals and form differentiated mesoderm (Chapter 25). When cultured as isolated explants of the blastula, cells of the animal and vegetal poles form ectoderm and endoderm, respectively. However, animal pole cells will form differentiated mesoderm when cultured in recombination with vegetal pole cells (Nieuwkoop 1969a,b, 1973; Sudarwati and Nieuwkoop 1971). The mesodermal competence of the animal pole explant offers a simple yet powerful assay for the identification of mesoderm-inducing substances. In this assay, a source of inducing factors, such as crude extract, cell supernatant, or purified protein, substitutes for the vegetal signal, and mesoderm induction is assessed by morphology, histology, and tissue-specific gene expression. Early studies demonstrated the mesoderm-inducing activity of heterologous tissues, such as fish swim bladder, guinea pig bone marrow, and chicken embryos (Nieuwkoop 1985; Gurdon 1987), and more recent efforts with purified proteins reported the ability of basic FGF and TGF-β2 to induce mesoderm formation (Slack et al. 1987; Rosa et al. 1988). Although these factors induce muscle differentiation, they do not induce notochord, the dorsalmost mesoderm, as do the signals produced by vegetal pole cells.

MIF, PIF, AND THE IDENTIFICATION OF ACTIVIN AS A MESODERM INDUCER

In an attempt to identify the endogenous inducers of mesoderm in the amphibian embryo, investigators turned to extracts of *Xenopus laevis* embryos, as well as supernatants of cultured *Xenopus* cells. Although mesoderm-inducing activity can be detected in embryonic extracts (Faulhaber 1970), this material is too limited for purification of the factor. In a series of breakthrough experiments in the late 1980s, Jim Smith identified a potent mesoderm-inducing factor in supernatant of the XTC cell line. Animal explants treated with this factor, referred to as XTC-MIF, form differentiated muscle and notochord, and undergo convergent extension, the morphogenetic movements of dorsal mesoderm (Smith 1987; Symes and Smith 1987; Chapter 19). Upon purification, XTC-MIF was found to be a secreted protein composed of two ~15 kD subunits covalently linked to form an ~23.5 kD biologically active dimer (Smith et al. 1988). With these properties, XTC-MIF is strikingly similar in its size and subunit structure to the active ligand for TGF-β family members, including TGF-β2, which has weak mesoderm-inducing activity (Rosa et al. 1988). In a parallel effort from Doug Melton and colleagues, a potent mesoderm-inducing activity was identified in the supernatant of P388D1 cells, a mouse macrophage cell line. The P388D1-derived inducing factor (PIF) and XTC-MIF are very similar in their biochemical properties, with active PIF consisting of a dimer of ~15 kD subunits (Sokol et al. 1990).

The identification of the active component of MIF and PIF resulted from the convergence of several approaches, including microsequencing of purified protein, functional comparison with mammalian TGF-β proteins, and cloning of *Xenopus* TGF-β genes. The initial clue came from the finding that mammalian Activin A, a TGF-β-related protein, was a potent inducer of mesoderm in the animal cap assay (Asashima et al. 1990a; Thomsen et al. 1990; van den Eijnden-Van Raaij et al. 1990). Activin A was previously shown in mammalian systems to regulate follicle-stimulating hormone (FSH) release from pituitary cells and ery-

throleukemia cell differentiation (EDF) (Petraglia 1997). When tested in the FSH and EDF assays, MIF had strong Activin-like activity (Albano et al. 1990; Smith et al. 1990; van den Eijnden-Van Raaij et al. 1990). Consistent with these results, purification and microsequencing identified the active component of XTC-MIF as the *Xenopus* ortholog of Activin A (Smith et al. 1990). In the case of purified PIF, an Activin-specific antibody inhibited mesoderm induction and recognized the major protein band of ~27 kD (Thomsen et al. 1990). Finally, the genes encoding *Xenopus* Activin A and Activin B were cloned and shown to have potent mesoderm-inducing activity as recombinant protein or microinjected mRNA (Fig. 1) (Thomsen et al. 1990). Animal explants treated with *Xenopus* Activin B in the picomolar range form muscle, notochord, neural tube, and eyes, with these tissues organized into an "embryoid" with rudimentary axial and AP pattern. The ability of Activin protein to induce axial mesoderm and the morphogenetic movements of dorsal mesoderm fulfills a subset of the characteristics predicted for an endogenous mesoderm inducer. In a satisfying bit of closure, Tiedemann's "vegetalizing" factor, one of the original

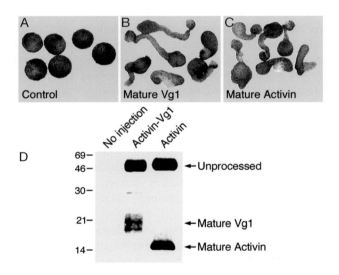

Figure 1. Mesoderm induction by mature Activin and Vg1 proteins. (*A–C*) *Xenopus* animal pole explants were prepared at the blastula stage and cultured in oocyte-conditioned supernatant until the tailbud stage. Explants cultured in control supernatant remained spherical (*A*), while supernatants containing mature Vg1 (*B*) or mature Activin (*C*) induced a dramatic elongation of the explants, a morphogenetic response indicative of dorsal mesoderm formation. (*D*) SDS-PAGE analysis of conditioned supernatants of *Xenopus* oocytes injected with mRNAs encoding Activin-Vg1 or Activin B. Secreted proteins were metabolically labeled with ^{35}S-methionine and detected by autoradiography. Mature Vg1 was secreted as a pair of glycosylated proteins of ~18 kD and mature Activin as a ~14 kD unglycosylated protein. For both Vg1 and Activin, high level expression also resulted in secretion of unprocessed precursor (~46 kD). No major labeled proteins were detected in the supernatant of uninjected oocytes (No injection). (Modified, with permission, from Yao and Kessler 1999.)

mesoderm-inducing substances derived from chick embryo extracts, was also identified as Activin A protein (Tiedemann and Tiedemann 1959; Born et al. 1972; Geithe et al. 1981; Asashima et al. 1990b; Tiedemann et al. 1992).

The *Xenopus Activin A* and *Activin B* genes are expressed zygotically, and maternal transcripts are undetectable (Thomsen et al. 1990; Asashima et al. 1991). *Activin A* mRNA expression is first detected at the end of gastrulation, with very low levels present until the swimming tadpole stage when expression levels increase. *Activin B* mRNA is expressed earlier, just after the mid-blastula transition, but is also present at low levels until the swimming tadpole stage. During the blastula and gastrula stages, *Activin B* mRNA is uniformly distributed throughout the embryo (Dohrmann et al. 1993). Despite the absence of maternal *Activin A* or *B* mRNA, an Activin-like activity is present in extracts of unfertilized *Xenopus* eggs. This Activin-like factor is functional in EDF and mesoderm-induction assays, and was inhibited by Follistatin, an Activin antagonist (Asashima et al. 1991). Western blotting of partially purified egg extracts confirmed that maternal Activin protein was present as Activin A and Activin B homodimers, as well as Activin AB heterodimers (Fukui et al. 1994). This discrepancy—maternal protein without maternal mRNA—may be explained by the finding that the follicle cells surrounding the oocyte express *Activin A* and *B* transcripts and protein, and during oogenesis Activin protein is transported from the follicle cells into the oocyte (Dohrmann et al. 1993; Rebagliati and Dawid 1993; Fukui et al. 1999). An additional, and more likely, source of maternal protein is *Activin D*, a third *Xenopus* Activin gene. Maternal *Activin D* transcripts are present in the oocyte (Oda et al. 1995), but it is uncertain whether maternal Activin D protein accumulates, or if the antibody used to detect Activin A and B also detects Activin D. Therefore, Activins are present both before and during the period of mesoderm induction in the *Xenopus* embryo. However, Activin mRNAs and proteins are uniformly distributed until the end of gastrulation, and the presence of functional Activin protein throughout the blastula, particularly in the animal pole region, is not compatible with mesoderm formation in the marginal zone. Additional mechanisms would be required to spatially restrict Activin activity or the cellular response to Activin.

Activin and Mesodermal Patterning: Threshold Responses and Morphogen Action

Beyond the potent mesoderm-inducing activity of Activin, the early experiments with MIF and PIF revealed that the type of mesodermal tissues induced reflects the dosage of Activin protein. Mesenchyme and mesothelium are induced at low doses, muscle and pronephros at intermediate doses, and muscle and notochord at high doses (Cooke et al. 1987;

Smith et al. 1988; Green et al. 1990; Sokol et al. 1990). The order of mesodermal cell types induced with increasing Activin dose corresponds to the ventral to dorsal pattern of the gastrula marginal zone: Ventrolateral cells form mesenchyme and mesothelium, dorsolateral cells form pronephros and muscle, and dorsalmost cells form notochord (see Chapter 26). This dose response suggested that a gradient of Activin protein or activity may be responsible for both the induction and initial patterning of mesodermal lineages.

The response of animal explants to Activin is not homogeneous, and multiple mesodermal cell types are induced at a single dose. This complex response appears to be due to the presence of several layers of responsive cells in the animal explant. The superficial layer of cells (blastocoel roof) receives a full Activin dose while internal cells receive a reduced dose that diffuses into the tissue, resulting in distinct induced cell populations within a single explant. To overcome this problem, Jeremy Green and Jim Smith examined the mesodermal response of individual animal pole cells. When cultured in calcium and magnesium free medium, animal pole cells become nonadherent and the explant dissociates into individual cells (Sargent et al. 1986). Treatment of disaggregated cells with Activin, each cell receiving an identical dosage, results in a homogeneous, but highly dose-sensitive, response. Increasing Activin concentration by as little as 1.5-fold results in a dramatic alteration in mesodermal gene expression and tissue differentiation (Green and Smith 1990; Green et al. 1992). Animal pole cells can discriminate between these small changes in Activin concentration, resulting in the homogeneous differentiation of distinct mesodermal cell types, including posterior mesoderm, muscle, notochord, or organizer. The observation that threshold responses to Activin can establish distinct mesodermal identities raised the possibility that Activin functions as an endogenous morphogen to establish mesodermal pattern directly.

While the threshold response to Activin offers a simple, appealing mechanism for patterning of the mesodermal lineages, subsequent studies revealed that precise thresholds are not established as an immediate response to Activin dosage. In the studies described above, mesodermal gene expression and differentiation are examined many hours after Activin treatment and reaggregation of cells, which is required for differentiation of mesodermal tissues (Gurdon 1988; Symes et al. 1988). When examined within 3 hours of Activin treatment, threshold responses are not observed. Instead, mesodermal genes representing the high and low threshold responses show overlapping expression at multiple Activin dosages (Green et al. 1994; Symes et al. 1994; Wilson and Melton 1994; Gurdon et al. 1999; Papin and Smith 2000). The emergence of precise, nonoverlapping thresholds following the initial exposure to Activin requires secondary interactions involving new protein synthesis and communication between reaggregated cells (Green et al. 1994; Symes et al. 1994; Wilson and Melton 1994; Papin and Smith 2000). Therefore, the development of a multithreshold pattern does not merely reflect the ability of an individual cell to measure the extracellular concentration of Activin, but requires the refinement of initial responses to establish a precise pattern.

The potential role of Activin as a mesodermal morphogen is supported by the elegant studies of John Gurdon. To assess the morphogen character of Activin, a local source of Activin was introduced into explants by mRNA injection or implanting an Activin-soaked bead, and mesodermal gene expression was examined in cells adjacent to or distant from the Activin source. Consistent with the formation of an Activin protein gradient, genes induced by high Activin were expressed close to the source, while genes induced by low Activin were expressed as far as 10 cell diameters away (Gurdon et al. 1994). Exogenous Activin can induce distant cells even when a barrier of nonresponsive tissue is placed between the source and the responsive cells, suggesting that diffusing Activin protein acts directly on distant cells, rather than through a relay mechanism involving secondary signals (Gurdon et al. 1994; McDowell et al. 1997). The distribution of exogenous radiolabeled Activin in explants confirmed the formation of an Activin gradient by protein diffusion, with protein detected at least 7 cell diameters (120 μm) from the source (McDowell et al. 1997). In addition to describing the morphogen-like behavior of Activin, an important aspect of the cellular response to a dynamic protein gradient was revealed in these studies. When Activin beads are removed from explants and replaced with either higher or lower dosage beads, the gene expression response reflects the highest dose of Activin received by the cells (Gurdon et al. 1995). This "ratchet" mechanism offers stability in the response of a field of cells to a protein gradient as it emerges and decays. Biochemical studies demonstrated that Activin binds to its receptor with high affinity and stability (80% binding maintained after 2 hours) and that mesodermal gene response reflects the absolute number of occupied Activin receptors, providing a molecular mechanism for the "ratchet" effect (Dyson and Gurdon 1998). Therefore, exogenous Activin diffuses within embryonic tissues, acts directly over long distances, and is stably received and quantified by responsive cells, each a predicted property of a morphogen (Fig. 2).

Is endogenous Activin a mesodermal morphogen? The studies with exogenous protein provide compelling evidence that Activin can function as a morphogen. However, a number of observations argue against a morphogen function for endogenous Activin. Prior to the

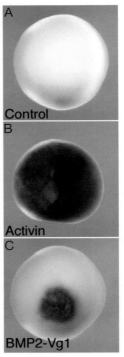

Figure 2. Differential diffusion of mature Activin and Vg1 in the *Xenopus* embryo. A single animal pole blastomere was injected with *Activin B* or *BMP2-Vg1* mRNA at the 32-cell stage, and at the early gastrula stage ectopic expression of *Brachyury*, a pan-mesodermal marker, was analyzed by whole mount in situ hybridization. Mature Activin induced high levels of *Brachyury* expression throughout the animal pole (*B*), while mature Vg1 induced expression in a domain limited to the site of injection (*C*). Functionally equivalent doses of *Activin* and *BMP2-Vg1* were injected, suggesting that the differing extent of mesoderm induction reflects a difference in the range of action for mature Activin and mature Vg1. No *Brachyury* expression was detected in the animal pole of uninjected embryos (*A*). Animal pole views of albino embryos are shown.

neurula stage, mRNA or protein for Activin A and Activin B are not spatially restricted by localization or expression, and a restricted source of Activin protein would be required to establish a protein gradient. Although it is possible that a uniform distribution of Activin protein could be transformed into an activity gradient by positive or negative regulators of signaling, there is no evidence of such an activity gradient for Activin. It should also be noted that, unlike Activin, other *Xenopus* TGF-β-related proteins (Nodal, BMP, and Vg1) have a limited range of action and only induce cells adjacent to the source of protein (Jones et al. 1996; Reilly and Melton 1996), although one of the zebrafish Nodal proteins, Squint, has been shown to act directly on distant cells (Chen and Schier 2001). Even if no endogenous Activin gradient exists, or if other TGF-β ligands do not display morphogen behavior, the work on Activin has revealed much about the activity and behavior of embryonic inducers, as well as the response of embryonic cells.

Is Activin an Essential Inducer of *Xenopus* Mesoderm?

The presence of Activin protein in the oocyte and the expression of Activin mRNAs soon after the mid-blastula transition suggest that Activin may be an endogenous regulator of mesoderm induction. Efforts to determine the requirement for Activin function in *Xenopus* mesodermal development have relied on strategies to interfere with ligand function, receptor binding, or signal transduction. Activin enters the secretory pathway as a pro-protein, forms covalently linked homodimers or heterodimers, and is proteolytically cleaved to release a biologically active dimer of the carboxy-terminal mature domain (Vale et al. 1990). Mature Activin binds to a heterodimeric receptor complex consisting of type II and type I receptors that are single-pass serine–threonine kinases (Wrana et al. 1994). The Activin receptor subunits expressed in the early *Xenopus* embryo are the type II receptor ActRIIB/XAR1 and the type I receptor ActRIB/Alk4, which form a functional Activin receptor complex (Kondo et al. 1991; Hemmati-Brivanlou et al. 1992; Mathews et al. 1992; Chang et al. 1997). Activin binding to ActRIIB leads to the recruitment and phosphorylation of ActRIB and subsequent phosphorylation of the receptor-associated signal transducer, Smad2 (Massague and Chen 2000). Activated Smad2 binds to Smad4, translocates to the nucleus, and associates with Fast1, or other DNA-binding factors, to activate target gene transcription (Whitman 1998, 2001). Each of these components of the Activin signaling pathway are expressed maternally and zygotically and are ubiquitous in their distribution before the neurula stage. It should be noted that the EGF-CFC proteins, Cripto and One-eyed pinhead, are essential coreceptors for Nodal, Vg1, and Gdf1, but are not required for Activin signal transduction (Gritsman et al. 1999; Shen and Schier 2000; Reissmann et al. 2001; Yeo and Whitman 2001; Yan et al. 2002; Cheng et al. 2003; Chapter 34). In fact, Cripto can inhibit Activin function by forming a receptor complex incapable of signaling in response to Activin binding (Gray et al. 2003). This result suggests that cofactors of the Activin receptor complex may confer response specificity by facilitating signaling by certain ligands and inhibiting signaling by others.

The requirement for Activin in endogenous mesoderm induction has been examined using mutant, dominant negative components of the Activin signaling pathway, as well as natural inhibitors of Activin function. Truncated type I and type II Activin receptors lacking the intracellular serine–threonine kinase domain function as dominant negatives to inhibit signal transduction. In animal explants, a truncated form of ActRIIB blocks mesoderm induction by Activin. When targeted to the marginal zone of the intact embryo, dominant negative ActRIIB inhibits mesodermal gene expression at the gastrula and tailbud stages, and blocks axis formation and the differentiation of axial mesodermal tissues (Hemmati-Brivanlou and Melton 1992). Similarly, a truncated from of ActRIB inhibits axial and mesodermal development in embryos, as well as Activin induction of mesoderm in animal explants (Chang et al. 1997). Mutated forms of Smad2 that can bind to ActRIB, but cannot be phosphorylated, reduce mesoderm induction by Activin and inhibit dorsoanterior axial development, consistent with a partial block to mesoderm formation (Hoodless et al. 1999). A fusion protein consisting of the Engrailed repressor domain and the Fast1 DNA-binding domain acts as a dominant negative factor that blocks mesoderm induction by Activin in explants, as well as mesoderm and axis formation in the embryo (Watanabe and Whitman 1999). Therefore, inhibition of the Activin signaling pathway in several independent ways results in a block to endogenous mesoderm induction.

The dominant negative inhibitor studies argue that Activin is required for mesoderm formation in *Xenopus*. However, the mesoderm inducers Nodal, Vg1, and Derriere, also signal via the Activin signaling pathway, raising the possibility that the dominant negative inhibitors block the function of multiple TGF-β ligands to inhibit mesoderm formation. Indeed, truncated ActRIIB can inhibit mesoderm induction by Nodal and Vg1, as well as BMP4, an activator of Smad1 signaling (Schulte-Merker et al. 1994; Kessler and Melton 1995; Sasai et al. 1995; Wilson and Hemmati-Brivanlou 1995; Dyson and Gurdon 1997; Wall et al. 2000). Truncated ActRIB inhibits signaling by Nodal and Vg1, but not by BMP4 (Chang et al. 1997). In addition, dominant negative Smad2 and Fast1 inhibit Nodal and Vg1 signaling (Hoodless et al. 1999; Watanabe and Whitman 1999; Whitman 2001). Although these studies demonstrate a requirement for the signaling pathway downstream of Activin in mesoderm formation, the lack of ligand specificity for the dominant negative inhibitors precludes a definitive conclusion regarding the role of Activin ligand in mesoderm formation.

In subsequent studies, Follistatin, a secreted protein that binds to Activin ligand and prevents receptor binding and signaling, was used in an attempt to inhibit Activin ligand function more specifically (Fukui et al. 1993; Hemmati-Brivanlou et al. 1994). In the animal explant assay, Follistatin inhibits mesoderm induction in response to Activin, but not to Vg1 or Nodal (Hemmati-Brivanlou et al. 1994; Schulte-Merker et al. 1994; Kessler and Melton 1995). However, when expressed throughout the marginal zone, Follistatin did not inhibit mesoderm formation and no reduction of early mesodermal gene expression was observed (Kessler and Melton 1995; Fainsod et al. 1997). High doses of Follistatin resulted in a shortened body axis, but axial mesoderm and head structures were present (Schulte-Merker et al. 1994; Kessler and Melton 1995; Fainsod et al. 1997). The body axis phenotype is a consequence of the ability of Follistatin to inhibit BMP signaling as well, which is required in *Xenopus* for ventral patterning of the mesoderm, but not for mesoderm induction (Fainsod et al. 1997; Chapter 33).

In an additional attempt to block endogenous Activin function, a mutant ActRIIB receptor lacking the intracellular kinase and the transmembrane domains was generated. For this truncated receptor, it is predicted that the extracellular ligand-binding domain will be secreted from cells and the released protein will bind to and inhibit ligands in the extracellular space. In contrast to the ActRIIB mutant discussed above, this freely secreted receptor mutant blocked mesoderm induction by Activin, but not by Vg1 or Nodal, in animal explants (Dyson and Gurdon 1997). The Activin specificity of this dominant negative ActRIIB is likely to be due to the ability of Activin to bind receptor in the absence of Cripto/One-eyed pinhead, while this coreceptor is required for Nodal and Vg1 binding (Reissmann et al. 2001; Yeo and Whitman 2001; Cheng et al. 2003). In the embryo, the ActRIIB ligand-binding domain caused a delay in mesodermal gene expression and a disorganized axis (Dyson and Gurdon 1997), and these results led the authors to conclude that Activin plays an essential role in endogenous mesoderm formation. However, unlike the previous dominant negative studies, no obvious mesodermal deficit (absence of mesodermal markers or tissues) is observed. At most, the results suggest that endogenous Activin is a minor contributor to mesoderm induction. The defects observed could also result from a weak inhibition of Nodal, or other inducers, by the ActRIIB extracellular domain. The failure of Follistatin and the ActRIIB extracellular domain to inhibit endogenous mesoderm induction argues against a role for Activin. Taken together, the complete set of inhibitor studies suggests that inducers other than Activin play the predominant role in stimulating the Activin signaling pathway to induce mesoderm. In fact, an abundance of evidence accumulated in multiple model systems suggests that Nodal ligands are the key activators of the Activin signaling pathway during early vertebrate embryogenesis (Whitman 2001; Chapter 34).

The study of Activin in the *Xenopus* embryo provides an exceptionally careful and revealing analysis of inducer behavior and function, and the precision of cellular responses. While the weight of the evidence argues against an essential role for Activin ligands in mesoderm induction and other aspects of germ layer formation, the published work does not definitively exclude such a role for Activin in the early embryo. As new technologies arise for gain of function and loss of function, the Activin question will undoubtedly continue to receive attention. Whether a requirement for Activin is ultimately demonstrated, valuable insights will be revealed by those studying Activin and its signaling pathway.

Activin Function in the Chick, Fish, and Mouse

Activin orthologs have been isolated and studied in each of the vertebrate model systems. In the chick, the epiblast receives signals that induce mesoderm, primitive streak formation, and axial development (Chapter 15). Epiblast explants cultured in isolation from hypoblast tissue do not form axial mesoderm, but treatment of epiblast tissue with PIF or XTC-MIF induces axis formation and the differentiation of somitic muscle and notochord (Mitrani and Shimoni 1990; Mitrani et al. 1990). Activin beads or Activin-expressing cell pellets implanted into the marginal zone or epiblast induce an ectopic primitive streak and differentiation of axial mesoderm (Ziv et al. 1992; Cooke et al. 1994). In addition, Activin treatment of explants of the area opaca and the posterior primitive streak induces a range of mesodermal tissues in a dose-dependent manner (Stern et al. 1995). *Activin B* mRNA is detected in the hypoblast (albeit very weakly) before and during primitive streak formation and in the marginal zone adjacent to the site of primitive streak initiation (Mitrani et al. 1990; Connolly et al. 1995; Levin 1998). Treatment of the chick embryo with exogenous Follistatin did not prevent formation of the primitive streak or mesoderm induction, but did perturb the morphology of the streak once formed (Levin 1998). Although the chick studies are limited in scope compared to the *Xenopus* studies, the basic conclusions are similar. In the chick, Activin is present at the appropriate time and place to regulate mesoderm formation and exogenous Activin can induce axial mesodermal tissues. However, Activin function does not appear to be essential for the induction of endogenous mesoderm.

In Medaka and the zebrafish, *activin B* mRNA is maternally expressed and zygotic mRNA expression is first detected at the blastula stage. Activin protein is detected in the oocyte and in the embryo from the one-cell stage through the blastula stage (Wittbrodt and Rosa 1994). Ectopic expression of Activin in the early embryo results in an expansion of mesodermal markers and the formation of ectopic axial mesoderm (Gritsman et al. 1999). Mutant Activin proteins were generated that prevent proteolytic cleavage of pro-protein or that form an inactive mature dimer that binds receptor without activating signaling. Interestingly, the cleavage mutant had no detectable effect on development, while the inactive dimer mutant disrupts axis formation, inhibits gastrulation, and blocks mesodermal gene expression (Wittbrodt and Rosa 1994). While the cleavage mutant is predicted to affect only newly synthesized ligand, the inactive dimer mutant is expected to inhibit maternal Activin protein as well as newly synthesized zygotic protein. Given the predicted mechanisms of inhibition, it was concluded that maternal Activin protein, but not zygotic protein, is required for mesoderm and axis formation. Although the inactive dimer mutant did not inhibit the activity of BMP4, the interaction with ligands that signal via the Activin signaling pathway (Nodal or Vg1) was not tested. So while these results suggest that maternal Activin may be required for mesoderm formation, the specificity of inhibition is uncertain, as is the main conclusion of the study.

Activin receptors (ActRIIB and ActRIB) and ligands (Activin A and Activin B) are expressed in the mouse embryo before implantation (Paulusma et al. 1994; Manova et al. 1995). Loss-of-function analysis for *ActRIIB* or *ActRIB* resulted in embryos with profound defects in gastrulation and mesoderm formation (Oh and Li 1997; Gu et al. 1998; Song et al. 1999). In contrast, single mutants in *Activin A* or *Activin B*, or the double mutant, have no significant early phenotype, survive to term, and die soon after birth with craniofacial defects, but have no mesodermal defects (Matzuk et al. 1995). While it is possible that other Activin genes expressed in the early embryo may compensate for the loss of *Activin A* and *Activin B* function, embryos mutant for both *Activin C* and *Activin E* are viable and survive to adulthood (Lau et al. 2000; Chang et al. 2001). As for the *Xenopus* studies, this contrast between the receptor and ligand phenotypes argues that the Activin signaling pathway is essential for early mesodermal development, but that ligands other than Activin are the predominant stimulators of this pathway.

Vg1 Localization to the *Xenopus* Vegetal Pole

In the animal–vegetal recombination experiments of Nieuwkoop, animal pole cells were induced to form differentiated mesoderm when cultured in contact with vegetal pole cells (Nieuwkoop 1969a,b; Sudarwati and Nieuwkoop 1971; Nieuwkoop 1973). This identified the vegetal pole as a source of mesoderm-inducing signals, and offered a strategy for the identification of endogenous mesoderm inducers (Chapter 25). Given that vegetal cells secrete a mesoderm-inducing factor (Slack 1991), these cells must contain and/or express the inducer, while the inducer is predicted to be absent from marginal and animal cells. It should be noted, however, that Nieuwkoop's results do not require that "the factor" be restricted in its localization or expression to vegetal pole cells. For example, the activity of a ubiquitous factor could be restricted if only vegetal cells were capable of secreting an active form or if only a subset of cells in the embryo were competent to respond.

In an attempt to identify an endogenous inducer of mesoderm, Doug Melton and colleagues reasoned that such an inducer would be found in a population of mRNAs enriched in vegetal pole cells. A differential screening strategy was pursued and mRNA isolated from the animal or vegetal pole of a fertilized egg was used to generate labeled cDNA probes for screening of a maternal cDNA library. Four localized clones were isolated, with three enriched in the animal pole (An1-3), and one enriched in the vegetal pole (Vg1) (Rebagliati et al. 1985). *Vg1*, a member of the TGF-β family, is a maternal mRNA encoding a pre-pro-protein

consisting of a signal sequence, pro-domain, and carboxy-terminal mature domain (Weeks and Melton 1987). Maternal *Vg1* mRNA is synthesized early during oogenesis and persists until the gastrula stage, but no zygotic expression of *Vg1* is detected (Rebagliati et al. 1985). *Vg1* mRNA is uniformly distributed in the cytoplasm of early oocytes, translocated to the vegetal pole during mid-oogenesis, and tightly localized to the vegetal cortex by the end of oogenesis (Fig. 3) (Melton 1987; Weeks and Melton 1987). With oocyte maturation, *Vg1* mRNA is released from the cortex, but remains restricted to the vegetal pole region and is "captured" by vegetal cells during cleavage so that *Vg1* mRNA is present in all vegetal cells of the early blastula (Weeks and Melton 1987). The expression, localization, and predicted structure of *Vg1* match the predicted characteris-

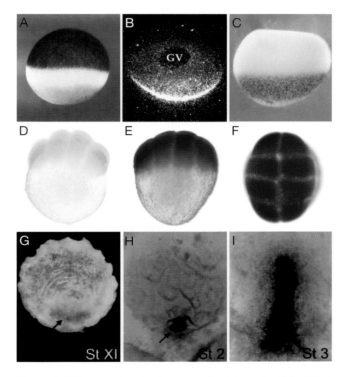

Figure 3. *Vg1* expression in *Xenopus*, zebrafish, and chick. (*A*) A mature *Xenopus* oocyte. (*B*) Radioactive in situ hybridization analysis of a sagittal section of a *Xenopus* oocyte reveals the restricted localization of *Vg1* mRNA to the cortex of the vegetal pole. (*C*) Whole mount in situ hybridization of an albino oocyte showing localization of *Vg1* transcripts to the vegetal domain. Lateral views with the animal pole up are shown. (*D–F*) Whole mount immunohistochemistry of an eight-cell stage zebrafish embryo using a Vg1 monoclonal antibody reveals expression of the zebrafish Vg1 precursor in all blastomeres. (*D*) Staining with secondary antibody only results in minimal background staining. Lateral (*D–E*) and dorsal (*F*) views are shown. (*G–I*) Whole mount in situ hybridization analysis of *Vg1* expression in the early chick embryo. (*G*) In the prestreak embryo, *Vg1* expression is localized to the posterior marginal zone (*arrow*). (*H,I*) As the primitive streak forms (*arrow*) and extends, *Vg1* is localized to the cells of the primitive streak. Ventral views are shown. (Modified, with permission, from Dohrmann et al. 1996 [©Elsevier]; Shah et al. 1997.)

tics of an endogenous mesoderm inducer. As an mRNA localized to vegetal blastomeres before and during the blastula stage, and encoding a member of the TGF-β family of secreted inducing proteins, Vg1 was proposed as a strong candidate for an endogenous mesoderm inducer. Furthermore, soon after *Vg1* was identified it was shown that TGF-β2 protein could induce mesoderm (Rosa et al. 1988), suggesting that an endogenous TGF-β-related factor, such as Vg1, may regulate *Xenopus* mesoderm induction.

Posttranslation Regulation of Vg1 Protein

TGF-β family proteins are synthesized as pre-pro-proteins that are cotranslationally inserted into the secretory pathway, form disulfide-linked dimers, and are proteolytically cleaved to release a bioactive dimer of the carboxy-terminal mature domain (Sha et al. 1989; Massague et al. 1994; Dubois et al. 1995). In agreement with the expression and localization of *Vg1* mRNA, Vg1 protein is present in the vegetal pole of oocytes and embryos through the blastula stage, and Vg1 protein is not detected outside of the vegetal pole region (Dale et al. 1989; Tannahill and Melton 1989). Vg1 protein persists until the tailbud stage, nearly 1 day after *Vg1* mRNA is no longer present, indicating that endogenous Vg1 protein is very stable. Vg1 is a glycoprotein and associates with membranes in vivo, consistent with entry into the secretory pathway (Dale et al. 1989; Tannahill and Melton 1989). Endogenous Vg1 protein is abundantly expressed and accumulates as an ~46 kD unprocessed pro-protein, but mature Vg1 protein (predicted size ~18 kD) is not consistently detected (Dale et al. 1989, 1993; Tannahill and Melton 1989; Thomsen and Melton 1993). Overexpression of Vg1 protein by mRNA injection resulted in a similar production of unprocessed protein without accumulation of mature protein (Tannahill and Melton 1989; Dale et al. 1993; Thomsen and Melton 1993). The biological activity of TGF-β-related proteins requires cleavage and release of mature protein and therefore, Vg1 processing may be tightly regulated to limit formation of active ligand or, alternatively, mature Vg1 produced may be rapidly turned over to limit the accumulation of active protein. Either way, if mature Vg1 protein is generated, it is at levels below the limit of detection by western blot or immunoprecipitation.

The mesoderm-inducing activity of Vg1 was examined in the animal cap assay. The animal cap assay provides a very sensitive bioassay, and for some inducers, a biological response can be detected at protein levels below the limit of immunodetection. Although *Vg1* mRNA injection produces high levels of Vg1 precursor protein, no evidence of mesoderm induction—either by gene expression, tissue differentiation, or morphogenesis—was observed (Dale et al. 1993; Thomsen and Melton 1993). Therefore, if mature Vg1 is generated, it accumulates at levels too low for biochemical detection or biological response. This suggests that the proteolytic processing of Vg1 precursor is tightly regulated to minimize or completely prevent the production of mature protein. In contrast, overexpression of *Activin B* results in the efficient production of mature protein and the induction of mesoderm in animal cap explants (Thomsen et al. 1990; Kessler and Melton 1995). The efficiency of proteolytic processing of TGF-β precursors is influenced by the pro-domain, the sequence of the tetrabasic cleavage site, and Subtilisin-like proprotein convertases (SPCs) (Gentry et al. 1988; Sha et al. 1989; Gray and Mason 1990). Specificity in the proteolytic processing of TGF-β precursors has been demonstrated, and two SPCs, Furin and Pace 4, are required for cleavage and activation of Nodal and BMP4 (Cui et al. 1998; Constam and Robertson 1999, 2000; Beck et al. 2002). It is not yet clear why the processing of Vg1 is inefficient or prevented altogether. It may be that Vg1 processing requires a specific proprotein convertase or a specific cofactor for a common proprotein convertase. Alternatively, an inhibitor of Vg1 processing may be widely distributed in the *Xenopus* embryo.

Mesoderm Induction by Mature Vg1

To bypass the posttranslational regulation of Vg1 and determine the biological activity of mature Vg1, the pro-domain and cleavage site of Vg1 were replaced with the corresponding domains of efficiently processed TGF-β family members. Since the pro-domain and cleavage site of BMP2 direct efficient cleavage and release of mature BMP2, it was predicted that replacing the mature domain of BMP2 with that of Vg1 would facilitate production of mature Vg1. A chimeric molecule containing the signal sequence, pro-domain, and cleavage site of BMP2 fused to the carboxy-terminal mature domain of Vg1 (*BVg1*) was expressed in *Xenopus* embryos and explants to assess precursor processing and biological activity (Thomsen and Melton 1993). In embryos injected with *BVg1* mRNA, two overexpressed proteins are detected by Western blotting, and these match the predicted sizes of unprocessed chimeric precursor (~46 kD) and mature Vg1 (~18 kD). In contrast to native Vg1, which has no mesoderm- or axis-inducing activity, BVg1 is a potent inducer of mesoderm and axis formation. *BVg1*-injected embryos develop ectopic axes with trunk structures containing differentiated muscle and notochord, and head structures with eyes (Thomsen and Melton 1993; Kessler and Melton 1995). *BVg1*-expressing animal cap explants express mesodermal genes, undergo the convergent extension movements of dorsal mesoderm, and form notochord and muscle (Thomsen and Melton 1993). Therefore, the BMP2 pro-domain and cleavage site facilitate the processing of chimeric precursor to release mature Vg1, and mature Vg1 is a strong inducer of mesoderm and axis formation (Figs. 1 and 2). A second chimera containing the BMP4 signal sequence and pro-domain fused to the Vg1 cleavage site and mature domain was analyzed (Dale et al. 1993). BMP4-Vg1 is also processed to release mature Vg1, but it is a weaker inducer of mesoderm than BMP2-Vg1. Animal explants induced with BMP4-Vg1 occasionally contain muscle and never contain notochord. Similarly, BMP4-Vg1 induces only partial axial duplication, generating trunk structures with muscle, but no notochord. This difference in the mesoderm-inducing activity of the two chimeras suggests that both the pro-domain and cleavage site contribute to the processing of the chimeric proteins, and that the presence of the Vg1 cleavage site in BMP4-Vg1 reduces processing efficiency.

The chimera studies suggest that if native Vg1 is processed, secreted mature Vg1 protein will induce mesoderm. However, despite the potent inducing activity of the chimeras, very low levels of mature Vg1 protein are secreted into the medium of *BVg1*-expressing *Xenopus* oocytes or tissue culture cells (Dale et al. 1993; Kessler and Melton 1995). This is likely due to the poor secretion of native BMP2 from oocytes and certain cell lines. In an attempt to produce soluble mature Vg1 protein, an Activin-Vg1 chimera (*AVg1*), containing the Activin B signal sequence, pro-domain, and cleavage site, was generated (Kessler and Melton 1995). Activin B, unlike BMP2, is efficiently processed and secreted from *Xenopus* oocytes. Similar to *BVg1*, injection of *AVg1* mRNA results in mesoderm and axis induction. When expressed in oocytes, AVg1 precursor is processed and mature Vg1 protein accumulates in the culture medium at substantial levels (~0.5 μg/ml). Treatment of animal explants with soluble mature Vg1 strongly induces mesoderm, resulting in the expression of mesodermal genes, convergent extension movements, and differentiation of muscle and notochord (Fig. 1) (Kessler and Melton 1995). Mesoderm is induced by picomolar concentrations of mature Vg1 (~40 pM), and at higher doses "embryoids" form that contain muscle, notochord, neural tube, and eyes organized in a rudimentary axial and AP pattern. Therefore, the *AVg1* chimera bypasses the tight posttranslation regulation of Vg1 processing, and the activity of soluble mature Vg1 protein confirms that mature Vg1 is a potent inducer of mesoderm.

The mesoderm-inducing activity of mature Vg1 is indistinguishable from Activin in most respects. The profile of mesodermal genes induced by increasing doses of mature Vg1 is identical to that observed with Activin, and the types of differentiated mesodermal tissues formed and their organization is also the same. These similarities suggest that mature Vg1 activates the same signaling pathway as Activin, and this was confirmed by the identification of downstream signaling components that mediate the response to mature Vg1. Like Activin, mature Vg1 can stimulate the phosphorylation of Smad2, but has no effect on Smad1 activation (Faure et al. 2000; Lee et al. 2001). In addition, mesoderm

induction by mature Vg1 is inhibited by dominant negative forms of ActRIIB, ActRIB, Smad2, and Fast1 (Schulte-Merker et al. 1994; Kessler and Melton 1995; Dyson and Gurdon 1997; Hoodless et al. 1999; Wall et al. 2000; D.S. Kessler, unpubl.). Therefore, if endogenous mature Vg1 is present in the embryo it is predicted to have the same signaling and mesoderm induction activities as Activin.

There are, however, several significant differences between mature Activin and Vg1. Unlike Activin, which can act at a distance from its source via protein diffusion, mature Vg1 protein and the cells it induces are found only in close proximity to the source of mature Vg1 (Fig. 2) (Reilly and Melton 1996; D.S. Kessler, unpubl.). This apparent inability of mature Vg1 to diffuse in the embryo is similar to the behavior of other *Xenopus* TGF-β ligands, including Xnr2 and BMP4 (Jones et al. 1996), and may reflect a difference in the glycosylation state of mature Activin and Vg1. While the mature domain of Activin lacks consensus sites for N-linked glycosylation and appears to be unglycosylated (Thomsen et al. 1990), mature Vg1 is glycosylated (Dale et al. 1989, 1993; Tannahill and Melton 1989), raising the possibility that mature Vg1 binding to cell surface or extracellular matrix components limits movement within the embryo. Another distinction between Activin and Vg1 is the dependence of mature Vg1 signaling, but not Activin, on the EGF-CFC coreceptor proteins Cripto and One-eyed pinhead (Gritsman et al. 1999; Reissmann et al. 2001; Yeo and Whitman 2001; Yan et al. 2002; Cheng et al. 2003; Chapter 34). Consistent with the inability of mature Vg1 to induce mesoderm in a zebrafish, *one-eyed pinhead* mutant, a protein complex of Cripto, ActRIIB, and ActRIB, is required for receptor binding and signaling by mature Vg1 (Cheng et al. 2003). There is also a difference in the timing of Smad2 activation in response to mature Activin and Vg1. Activin stimulates Smad2 phosphorylation before and after the mid-blastula transition, but mature Vg1 induces phospho-Smad2 only following the mid-blastula transition (Faure et al. 2000). Although the mid-blastula transition marks the onset of most zygotic transcription, inhibition of transcription with α-amanitin did not prevent Smad2 activation by Vg1, suggesting that mature Vg1 requires a signaling component that is not active until the mid-blastula transition because of posttranscriptional regulation. Therefore, while the gene expression and tissue differentiation responses to mature Activin and Vg1 are the same, there are important differences in ligand behavior and signal transduction for these mesoderm inducers.

The diffusion and signaling properties of mature Vg1 are strikingly similar to the Nodal-related members of the TGF-β family. Most Nodal-related genes do not appear to diffuse from their site of expression, are dependent on Cripto/One-eyed pinhead for receptor binding and signaling, and activate Smad2 only after the mid-blastula transition in *Xenopus* assays (Whitman 2001; Chapter 34). Given these similarities, it is predicted that processing of endogenous Vg1 would result in a signaling output indistinguishable from that of the endogenous Nodal-related proteins. This raises the intriguing possibility that endogenous mature Vg1, produced at low levels and/or for a limited developmental period, could act together with mature Nodal proteins to activate the Smad2 signaling pathway that is essential for germ layer formation in vertebrates.

Vg1 and *Vg1*-related Genes in the Fish, Chick, and Mouse

Vg1 orthologs and *Vg1*-related genes have been isolated and studied in the zebrafish, chick, and mouse. In the zebrafish, *Vg1* (also called *zDVR-1*) is a maternal mRNA that is uniformly distributed in developing oocytes and cleaving embryos (Helde and Grunwald 1993). Similar to *Xenopus Vg1*, zebrafish *Vg1* mRNA is detected through the gastrula stage, but not afterward, and Vg1 protein persists in the zebrafish embryo long after the mRNA is turned over (Helde and Grunwald 1993; Dohrmann et al. 1996). The uniform distribution of *Vg1* mRNA and protein in the zebrafish contrasts with the vegetal localization in *Xenopus* and likely reflects differences in the regional organization of the early zebrafish and *Xenopus* embryos (Fig. 3). Unlike the fixed cell positions during *Xenopus* cleavage, the early zebrafish embryo has no predictable fate map because of extensive cell mixing and, therefore, localization of maternal mRNAs or proteins would not result in stable regional differences. This suggests that mechanisms other than mRNA or protein localization would be required to spatially restrict the activity of Vg1 in the zebrafish embryo.

Zebrafish Vg1 is similar to *Xenopus* Vg1 in posttranslational regulation and the inducing activity of mature protein. Endogenous zebrafish Vg1 is an abundant protein with the predicted size of unprocessed precursor and no mature protein is detected at any stage (Dohrmann et al. 1996). Overexpression of native Vg1 in the zebrafish embryo results in the accumulation of high levels of precursor protein, but no mature protein is produced and ectopic mesoderm is not induced. A chimeric construct containing the *Xenopus* BMP2 pro-domain and cleavage site and the zebrafish Vg1 mature domain was processed to release mature Vg1 and induced ectopic mesodermal gene expression in the zebrafish embryo (Dohrmann et al. 1996). These studies indicate that the posttranslational regulation of Vg1 is conserved in zebrafish, and when bypassed to allow proteolytic processing, mature zebrafish Vg1 is a potent mesoderm inducer. Unexpectedly, when expressed in *Xenopus* embryos, native zebrafish Vg1 is processed to generate mature protein, albeit at very low levels, and this results in a weak, but reproducible, induction of mesoderm.

In *Xenopus* oocytes, native zebrafish Vg1 is processed and secreted mature protein accumulates in the culture medium as a heat-stable disulfide-linked dimer with potent mesoderm-inducing activity (Dohrmann et al. 1996). Therefore, posttranslational control of precursor processing is a conserved aspect of Vg1 regulation, but there may be species-specificity to this regulatory mechanism. *Xenopus* and zebrafish Vg1 are not processed in their natural hosts even when overexpressed, but the zebrafish protein can be processed in a heterologous host. What accounts for this regulatory difference is not clear, but may involve the ability of regulatory proteins to recognize sequences in the vicinity of the cleavage site that differ between *Xenopus* and zebrafish Vg1 (Dohrmann et al. 1996).

Chick *Vg1*, like the *Xenopus* and zebrafish orthologs, is expressed prior to the onset of gastrulation, is posttranslationally regulated, and has potent inducing activity as mature protein (Seleiro et al. 1996; Shah et al. 1997). In the chick pregastrula embryo, *Vg1* mRNA is first detected strongly in the epiblast layer of the posterior marginal zone (Chapter 15). With the initiation of gastrulation, *Vg1* mRNA is concentrated in cells of the primitive streak and is maintained in cells of the streak and Hensen's node during extension of the primitive streak (Fig. 3). As the streak regresses, *Vg1* expression is lost from the anterior streak and node, but continues in the posterior streak and in the prospective axial and paraxial mesoderm emerging from the streak. The prestreak and primitive streak embryo contains high levels of Vg1 protein and lower levels are present for an extended period following the end of gastrulation. At all stages, endogenous Vg1 protein has the size of unprocessed precursor protein, and no mature protein is detected (Seleiro et al. 1996). Unlike the other *Vg1* orthologs, chick *Vg1* is expressed in mesodermal and neural structures at later stages. *Vg1* is expressed in the unsegmented paraxial mesoderm and is maintained in the caudal half of the epithelial somite. Additional sites of mesodermal expression include the myotome, branchial arch mesenchyme, myocardium, and notochord. In the nervous system, *Vg1* is first expressed in the hindbrain, is then restricted to rhombomeres 3 and 5, and at later stages is expressed transiently in the hindbrain, spinal cord, diencephalon, and sympathetic ganglia (Shah et al. 1997).

The posterior marginal zone, the initial site of *Vg1* expression, is an important signaling center that induces adjacent epiblast cells to form the primitive streak (Khaner and Eyal-Giladi 1986; Bachvarova et al. 1998; Chapter 15). When transplanted to lateral or anterior positions of the prestreak embryo, the posterior marginal zone induces the formation of an ectopic primitive streak and node, with the posterior end of the ectopic streak positioned adjacent to the transplanted tissue. Lineage-labeling and chick-quail recombinant studies indicate that the ectopic primitive streak is not derived from cells of the posterior marginal zone transplant, but consists of epiblast cells that change their fate, gene expression, and morphogenetic movements in response to inducing signals from the posterior marginal zone (Bachvarova et al. 1998). The expression of *Vg1* in the posterior marginal zone before the formation of the primitive streak raises the possibility that Vg1 may mediate the inducing function of this early organizing center.

To assess the role of chick Vg1 in primitive streak induction, pellets of mammalian cells transfected with native or chimeric forms of *Vg1* were transplanted into the anterior marginal zone of prestreak embryos (Seleiro et al. 1996; Shah et al. 1997). Like in the chick embryo, native Vg1 was not processed in transfected cells, and the implanted cells had no effect on primitive streak formation. Chimeric constructs containing the pro-domain and cleavage site of BMP4 or Dorsalin fused to the mature domain of chick Vg1 were processed and mature protein was secreted by transfected cells. Pellets of cells expressing mature Vg1 induced formation of an ectopic primitive streak in the epiblast adjacent to the implant. The ectopic primitive streak induced by mature Vg1 is similar to the endogenous primitive streak in gene expression patterns, morphogenetic movements, and the ability to form an organized axis with notochord, somites, and neural tube (Seleiro et al. 1996; Shah et al. 1997; Skromne and Stern 2002). Treatment of epiblast explants with purified mature Vg1 protein induced axial mesodermal gene expression, consistent with the axis-inducing activity of mature Vg1 (Shah et al. 1997). In *Xenopus* explants and embryos, chimeric forms of chick Vg1 were processed and induced mesodermal gene expression, muscle differentiation, and axis formation (Seleiro et al. 1996; Shah et al. 1997). In addition, as is the case for zebrafish Vg1, native chick Vg1 is processed in *Xenopus*, resulting in the formation and secretion of mature chick Vg1 protein with strong mesoderm-inducing activity (Shah et al. 1997). Therefore, mature chick Vg1 shares the potent mesoderm- and axis-inducing activities of the other Vg1 orthologs. Given the presence of endogenous Vg1 in a primitive streak-organizing center of the prestreak embryo, the results suggest that Vg1 may play a role in mesoderm induction and axis formation in the early chick embryo.

The search for a mammalian ortholog of *Vg1*, by low stringency hybridization and degenerate PCR, has been unsuccessful. Two *Vg1*-related genes, *Vgr1* and *Vgr2*, were identified in the mouse, but despite the relation the nomenclature implies, low sequence relatedness indicates that these genes are not *Vg1* orthologs (Lyons et al. 1989; Jones et al. 1992). The cross-reaction of zebrafish and chick Vg1 with a *Xenopus* Vg1 mature domain antiserum offered an alternative approach for identifying mouse Vg1 by recogni-

tion of a conserved epitope. A single major protein was detected by the Vg1 antiserum in mouse embryo extracts, and this protein had the approximate size of an unprocessed TGF-β precursor (~44 kD). Screening of an embryonic mouse cDNA expression library with the Vg1 antiserum identified a single cross-reacting protein, Gdf1, a previously identified member of the TGF-β superfamily (Lee 1990, 1991; Wall et al. 2000). Mature domain sequence comparisons indicated that *Gdf1* is highly related to *Xenopus*, zebrafish, and chick *Vg1* (54–57% identity), and that *Gdf1* is more closely related to *Vg1* than other known TGF-β-related genes in the mouse. However, relatedness between *Vg1* orthologs was significantly higher (69–75% identity) and, therefore, *Gdf1* is not a mouse ortholog of *Vg1*.

Gdf1 mRNA is expressed ubiquitously in the embryonic region of the pregastrula and gastrula embryo at 5.5–6.5 dpc, but is excluded from extraembryonic tissues. Ubiquitous expression continues at 7.5–8.0 dpc with more robust expression in the developing head folds. The ubiquitous expression of *Gdf1* prior to and during gastrulation suggests a role for Gdf1 in the early mouse embryo. At later stages (8.5–10.5 dpc), *Gdf1* is expressed in a variety of mesodermal and neural tissues, including the node, notochord, paraxial and lateral plate mesoderm, tailbud, midbrain, spinal cord, and limb bud (Rankin et al. 2000; Wall et al. 2000). These sites of *Gdf1* expression are similar to the sites of chick *Vg1* expression in the later stage chick embryo.

In addition to the antigenic relation with Vg1, Gdf1 processing, signal transduction, and inducing activity is strikingly similar to Vg1 (Wall et al. 2000). In *Xenopus* embryos, native Gdf1 is not processed and has no inducing activity. A chimeric BMP2-Gdf1 protein, like the chimeric forms of Vg1, is processed and mature Gdf1 has potent mesoderm- and axis-inducing activity. The signaling pathway that mediates response to mature Gdf1 is identical to the Vg1 pathway, and includes ActRIIB, ActRIB, Cripto, Smad2, and Fast1 (Wall et al. 2000; Cheng et al. 2003; D.S. Kessler, unpubl.). Despite the potent mesoderm-inducing activity of mature Gdf1, *Gdf1* null mice had no detectable defects in germ layer formation or gastrulation (Rankin et al. 2000). This suggests that either *Gdf1* is not essential for early mouse development, or that redundant genes compensate for *Gdf1* loss of function. However, *Gdf1* null mice do display left/right patterning defects, including situs inversus of the visceral organs, pulmonary isomerism, and cardiac defects, as well as misregulation of asymmetric gene expression in the lateral plate mesoderm (Rankin et al. 2000; Chapter 28). Consistent with a left/right patterning role of Gdf1, right-sided overexpression of mature Gdf1 in *Xenopus* results in misregulation of asymmetric gene expression and reversal of intestinal and heart looping (Wall et al. 2000). In light of this left/right patterning activity of Gdf1, it should be noted that a role for Vg1 in *Xenopus* left/right patterning has been proposed (Hyatt et al. 1996; Hyatt and Yost 1998; Chapter 28), suggesting yet another functional similarity of Gdf1 and Vg1. Therefore, Gdf1 and Vg1 are equivalent in biochemical and functional assays, suggesting that *Gdf1* may be a functional homolog of *Vg1* in the mouse embryo. Further efforts will be required to isolate the elusive mouse ortholog of *Vg1*.

Is Vg1 a Functional Embryonic Inducer?

The requirement for Vg1 in early *Xenopus* development has been tested by dominant negative and knockdown approaches. Dominant negative components of the Activin signaling pathway were tested for the ability to inhibit mature Vg1 activity. In animal explants, mesoderm induction by mature Vg1 was inhibited by truncated forms of ActRIIB and ActRIB, by Smad2 mutants, by the Smad2-interaction domain of Fast1, and by an Engrailed-Fast1 fusion protein, confirming that mature Vg1 signals via components of the Activin pathway (Schulte-Merker et al. 1994; Kessler and Melton 1995; Dyson and Gurdon 1997; Hoodless et al. 1999; Watanabe and Whitman 1999; Wall et al. 2000; D.S. Kessler, unpubl.). Although these signaling components are essential for endogenous mesoderm formation, the Activin pathway mediates signaling by several other TGF-β ligands, including Nodal, and therefore, it cannot be concluded from these studies that endogenous Vg1 in required for mesoderm formation.

In an attempt to inhibit more specifically the activity of endogenous Vg1, inactivating point mutations were introduced into the mature domain of Vg1 (Joseph and Melton 1998). Expression of mutated protein is predicted to inhibit endogenous Vg1 either by forming a mutated homodimer that competes with wild-type homodimer for receptor binding, or by forming an inactive heterodimer, containing wild-type and mutated subunits, preventing formation of an active wild-type homodimer. In animal explants, mutated Vg1 had no inducing activity and, when expressed at a 100-fold excess, inhibited mesoderm induction by BVg1. Mutated Vg1 did not inhibit mesoderm induction by overexpression of Activin or Nodal-related proteins, suggesting that the mature domain mutations resulted in a Vg1-specific dominant negative protein. Expression of mutated Vg1 in intact embryos resulted in a disruption of gastrulation, severe axial defects, and a partial inhibition of mesodermal gene expression, suggesting a requirement for endogenous Vg1 in mesoderm induction and axis formation (Joseph and Melton 1998).

Despite the apparent strength of the conclusions offered by the studies with mutated Vg1, there are a number of issues to consider in interpreting the results. The mature domain mutations have inhibitory activity only when introduced into BVg1, which contains the pro-domain and cleav-age site of BMP2. Given the role of the pro-domain in dimer formation, and the potential for heterodimerization in the TGF-β family, it is difficult to predict which endogenous TGF-β ligands will interact with mutated BVg1. Heterodimers can form for Nodal-BMP, Nodal-Derriere, and Derriere-BMP, as well as other subunit combinations (Eimon and Harland 1999; Yeo and Whitman 2001). Mutated BVg1 may therefore inhibit the function of TGF-β ligands other than Vg1 by forming inactive heterodimers. Although overexpressed Nodal is not inhibited by mutated Vg1 in animal explants, it is possible that endogenous Nodal, present at far lower levels, may form inactive heterodimers with mutated Vg1, a scenario that could account for the embryonic defects resulting from mutated Vg1. Furthermore, mature Vg1 signaling requires ActRIIB, ActRIB, and Cripto, and this receptor complex is also required for Nodal signaling (Gritsman et al. 1999; Reissmann et al. 2001; Yeo and Whitman 2001; Yan et al. 2002; Cheng et al. 2003; Chapter 34). Therefore, if mutated Vg1 acts as a homodimer to compete for receptor binding and activation, it is predicted that both Vg1 and Nodal signaling would be inhibited. Given that the mechanism of action and the targets of inhibition of mutated Vg1 are undefined, it is uncertain whether these studies reveal an essential role for endogenous Vg1, or perhaps for another TGF-β ligand.

In *Xenopus*, loss-of-function analysis for maternal mRNAs can be accomplished by antisense oligonucleotide-mediated depletion. In this approach, oocytes are injected with a modified oligonucleotide that targets a specific maternal mRNA for degradation. Depleted oocytes are then matured in vitro, surgically implanted into a host mother, and the developmental consequences of target depletion are evaluated following fertilization (Zuck et al. 1999). This approach, pioneered by Janet Heasman and colleagues, was used to assess the developmental function of Vg1 (J. Heasman, pers. comm.). Injection of oocytes with a *Vg1*-specific antisense oligonucleotide resulted in a near complete depletion of *Vg1* mRNA (~95% reduction). Despite the dramatic reduction of *Vg1* mRNA, depleted embryos displayed no significant developmental defects, a result that argues against an essential role for endogenous Vg1 in the early embryo. However, Vg1 protein levels were reduced only ~50% in depleted embryos. This discrepancy in the depleted levels of *Vg1* mRNA and protein is likely to be due to the translation of Vg1 protein during early oogenesis (Dale et al. 1989; Tannahill and Melton 1989), allowing the accumulation of substantial levels of Vg1 protein prior to oligonucleotide injection into late oocytes. Given that mature Vg1 is active in the picomolar range (Kessler and Melton 1995), only a small fraction of Vg1 precursor need be processed to generate bioactive levels of endogenous mature Vg1, suggesting that a 50% reduction in Vg1 precursor protein is unlikely to result in a significant loss of function for Vg1. Therefore, as a result of the early accumulation of Vg1 protein, this maternal depletion approach does not provide a definitive analysis of Vg1 function.

Taken together, the studies evaluating the requirement for endogenous Vg1 in early *Xenopus* development are inconclusive. For the dominant negative approaches, uncertain specificity of inhibition makes it difficult to conclude that the developmental defects observed are due to interference only with endogenous Vg1. Conversely, the maternal depletion approach is highly specific, but does not result in the substantial reduction of maternal Vg1 protein required for true loss-of-function analysis. No doubt further attempts will be forthcoming as more effective strategies become available. It should be noted that Joe Yost has obtained evidence that Vg1 or Vg1-like factors may play a role is establishing left/right asymmetry in *Xenopus*, and the pursuit of this idea may ultimately reveal an essential developmental function for Vg1 (Hyatt et al. 1996; Hyatt and Yost 1998; Chapter 28).

CONCLUSIONS

A predominant view among developmental biologists is that a gene is developmentally "important" if it is shown by loss-of-function analysis to be essential for normal embryogenesis (i.e., loss of function results in an obvious abnormal phenotype). However, the term "essential" has an operational definition that depends on the details of experimental design, including culture conditions, genetic background, or other factors. Moreover, redundancy and compensation within gene families can further confound the evaluation of gene function by loss-of-function approaches. The story of Activin and Vg1 in the vertebrate embryo provides an alternative definition of developmental importance. Despite the lack of evidence demonstrating that Activin and Vg1 serve essential functions in the early embryo, the study of these factors has yielded profound insight into mechanisms of embryonic induction that are essential and conserved in vertebrate embryos. Furthermore, the book is not yet closed on the embryonic roles of Activin and Vg1, and many are still drawn to the challenge of defining the embryonic functions of these intriguing factors. Whether future studies reveal essential roles for Activin and Vg1 in the early embryo, it is certain that these inducers will continue to reveal new and unexpected facets of embryonic induction.

ACKNOWLEDGMENTS

I am grateful to Peter Klein, Jerry Thomsen, and Malcolm Whitman for valuable discussion, and to Janet Heasman for sharing unpublished results. This work was supported by a grant from the National Institutes of Health (HD35159).

REFERENCES

Albano R.M., Godsave S.F., Huylebroeck D., Van Nimmen K., Isaacs H.V., Slack J.M.W., and Smith J.C. 1990. A mesoderm inducing factor produced by WEHI-3 murine myelomonocytic leukemia cells is activin A. *Development* **110:** 435–443.

Asashima M., Nakano H., Shimada K., Kinoshita K., Ishii K., Shibai H., and Ueno N. 1990a. Mesodermal induction in early amphibian embryos by activin A (erythroid differentiation factor). *Roux's Arch. Dev. Biol.* **198:** 330–335.

Asashima M., Nakano H., Uchiyama H., Davids M., Plessow S., Loppnow-Blinde B., Hoppe P., Dau H., and Tiedemann H. 1990b. The vegetalizing factor belongs to a family of mesoderm inducing proteins related to erythroid differentiation factor. *Naturwissenschaften* **77:** 389–391.

Asashima M., Nakano H., Uchiyama H., Sugino H., Nakamura T., Eto Y., Ejima D., Nishimatsu S.I., Ueno N., and Kinoshita K. 1991. Presence of activin (erythroid differentiation factor) in unfertilized eggs and blastulae of *Xenopus laevis*. *Proc. Natl. Acad. Sci.* **88:** 6511–6514.

Bachvarova R.F., Skromne I., and Stern C.D. 1998. Induction of primitive streak and Hensen's node by the posterior marginal zone in the early chick embryo. *Development* **125:** 3521–3534.

Beck S., Le Good J.A., Guzman M., Ben Haim N., Roy K., Beermann F., and Constam D.B. 2002. Extraembryonic proteases regulate Nodal signalling during gastrulation. *Nat. Cell Biol.* **4:** 981–985.

Born J., Geithe H.P., and Tiedemann H. 1972. Isolation of a vegetalizing factor. *Hoppe-Seylers Z. Physiol. Chem.* **353:** 1075–1084.

Chang C., Wilson P.A., Mathews L.S., and Hemmati-Brivanlou A. 1997. A *Xenopus* type I activin receptor mediates mesodermal but not neural specification during embryogenesis. *Development* **124:** 827–837.

Chang H., Lau A.L., and Matzuk M.M. 2001. Studying TGF-β superfamily signaling by knockouts and knockins. *Mol. Cell. Endocrinol.* **180:** 39–46.

Chen Y. and Schier A.F. 2001. The zebrafish Nodal signal Squint functions as a morphogen. *Nature* **411:** 607–610.

Cheng S.K., Olale F., Bennett J.T., Brivanlou A.H., and Schier A.F. 2003. EGF-CFC proteins are essential coreceptors for the TGF-β signals Vg1 and GDF1. *Genes Dev.* **17:** 31–36.

Connolly D.J., Patel K., Seleiro E.A., Wilkinson D.G., and Cooke J. 1995. Cloning, sequencing, and expressional analysis of the chick homologue of follistatin. *Dev. Genet.* **17:** 65–77.

Constam D.B. and Robertson E.J. 1999. Regulation of bone morphogenetic protein activity by pro domains and proprotein convertases. *J. Cell Biol.* **144:** 139–149.

———. 2000. SPC4/PACE4 regulates a TGFβ signaling network during axis formation. *Genes Dev.* **14:** 1146–1155.

Cooke J., Takada S., and McMahon A. 1994. Experimental control of axial pattern in the chick blastoderm by local expression of Wnt and activin: the role of HNK-1 positive cells. *Dev. Biol.* **164:** 513–527.

Cooke J., Smith J.C., Smith E.J., and Yaqoob M. 1987. The organization of mesodermal pattern in *Xenopus laevis*: Experiments using a *Xenopus* mesoderm-inducing factor. *Development* **101:** 893–908.

Cui Y., Jean F., Thomas G., and Christian J.L. 1998. BMP-4 is proteolytically activated by furin and/or PC6 during vertebrate embryonic development. *EMBO J.* **17:** 4735–4743.

Dale L., Matthews G., and Colman A. 1993. Secretion and mesoderm-inducing activity of the TGF-β-related domain of *Xenopus* Vg1. *EMBO J.* **12:** 4471–4480.

Dale L., Matthews G., Tabe L., and Colman A. 1989. Developmental expression of the protein product of Vg1, a localized maternal mRNA in the frog *Xenopus laevis. EMBO J.* **8:** 1057–1065.

Dohrmann C.E., Kessler D.S., and Melton D.A. 1996. Induction of axial mesoderm by zDVR-1, the zebrafish orthologue of *Xenopus* Vg1. *Dev. Biol.* **175:** 108–117.

Dohrmann C.E., Hemmati B.A., Thomsen G.H., Fields A., Woolf T.M., and Melton D.A. 1993. Expression of activin mRNA during early development in *Xenopus laevis. Dev. Biol.* **157:** 474–483.

Dubois C.M., Laprise M.H., Blanchette F., Gentry L.E., and Leduc R. 1995. Processing of transforming growth factor β1 precursor by human furin convertase. *J. Biol. Chem.* **270:** 10618–10624.

Dyson S. and Gurdon J.B. 1997. Activin signalling has a necessary function in *Xenopus* early development. *Curr. Biol.* **7:** 81–84.

———. 1998. The interpretation of position in a morphogen gradient as revealed by occupancy of activin receptors. *Cell* **93:** 557–568.

Eimon P.M. and Harland R.M. 1999. In *Xenopus* embryos, BMP heterodimers are not required for mesoderm induction, but BMP activity is necessary for dorsal/ventral patterning. *Dev. Biol.* **216:** 29–40.

Fainsod A., Deissler K., Yelin R., Marom K., Epstein M., Pillemer G., Steinbeisser H., and Blum M. 1997. The dorsalizing and neural inducing gene follistatin is an antagonist of BMP-4. *Mech. Dev.* **63:** 39–50.

Faulhaber I. 1970. Enrichment of the vegetalizing induction factor from the clawed toad (*Xenopus laevis*) gastrula and determination of the molecular weight range by gradient centrifugation. *Hoppe-Seylers Z. Physiol. Chem.* **351:** 588–594.

Faure S., Lee M.A., Keller T., ten Dijke P., and Whitman M. 2000. Endogenous patterns of TGFβ superfamily signaling during early *Xenopus* development. *Development* **127:** 2917–2931.

Fukui A., Shiurba R., and Asashima M. 1999. Activin incorporation into vitellogenic oocytes of *Xenopus laevis. Cell Mol. Biol.* **45:** 545–554.

Fukui A., Nakamura T., Uchiyama H., Sugino K., Sugino H., and Asashima M. 1994. Identification of activins A, AB, and B and follistatin proteins in *Xenopus* embryos. *Dev. Biol.* **163:** 279–281.

Fukui A., Nakamura T., Sugino K., Takio K., Uchiyama H., Asashima M., and Sugino H. 1993. Isolation and characterization of *Xenopus* follistatin and activins. *Dev. Biol.* **159:** 131–139.

Geithe H.P., Asashima M., Asahi K.I., Born J., Tiedemann H., and Tiedemann H. 1981. A vegetalizing inducing factor. Isolation and chemical properties. *Biochim. Biophys. Acta* **676:** 350–356.

Gentry L.E., Lioubin M.N., Purchio A.F., and Marquardt H. 1988. Molecular events in the processing of recombinant type 1 pre-pro-transforming growth factor beta to the mature polypeptide. *Mol. Cell. Biol.* **8:** 4162–4168.

Gray A.M. and Mason A.J. 1990. Requirement for activin A and transforming growth factor-β1 pro-regions in homodimer assembly. *Science* **247:** 1328–1330.

Gray P.C., Harrison C.A., and Vale W. 2003. Cripto forms a complex with activin and type II activin receptors and can block activin signaling. *Proc. Natl. Acad. Sci.* **100:** 5193–5198.

Green J.B.A. and Smith J.C. 1990. Graded changes in dose of a *Xenopus* activin A homologue elicit stepwise transitions in embryonic cell fate. *Nature* **347:** 391–394.

Green J.B.A., New H.V., and Smith J.C. 1992. Responses of embryonic

Xenopus cells to activin and FGF are separated by multiple dose thresholds and correspond to distinct axes of the mesoderm. *Cell* **71:** 731–739.

Green J.B.A., Smith J.C., and Gerhart J.C. 1994. Slow emergence of a multithreshold response to activin requires cell-contact-dependent sharpening but not prepattern. *Development* **120:** 2271–2278.

Green J.B., Howes G., Symes K., Cooke J., and Smith J.C. 1990. The biological effects of XTC-MIF: Quantitative comparison with *Xenopus* bFGF. *Development* **108:** 173–183.

Gritsman K., Zhang J., Cheng S., Heckscher E., Talbot W.S., and Schier A.F. 1999. The EGF-CFC protein one-eyed pinhead is essential for nodal signaling. *Cell* **97:** 121–132.

Gu Z., Nomura M., Simpson B.B., Lei H., Feijen A., van den Eijnden-van Raaij J., Donahoe P.K., and Li E. 1998. The type I activin receptor ActRIB is required for egg cylinder organization and gastrulation in the mouse. *Genes Dev.* **12:** 844–857.

Gurdon J. 1987. Embryonic induction-molecular prospects. *Development* **99:** 285–306.

———. 1988. A community effect in animal development. *Nature* **336:** 772–774.

Gurdon J.B., Mitchell A., and Mahony D. 1995. Direct and continuous assessment by cells of their position in a morphogen gradient. *Nature* **376:** 520–521.

Gurdon J.B., Harger P., Mitchell A., and Lemaire P. 1994. Activin signalling and response to a morphogen gradient. *Nature* **371:** 487–492.

Gurdon J.B., Standley H., Dyson S., Butler K., Langon T., Ryan K., Stennard F., Shimizu K., and Zorn A. 1999. Single cells can sense their position in a morphogen gradient. *Development* **126:** 5309–5317.

Helde K.A. and Grunwald D.J. 1993. The DVR-1 (Vg1) transcript of zebrafish is maternally supplied and distributed throughout the embryo. *Dev. Biol.* **159:** 418–426.

Hemmati-Brivanlou A. and Melton D.A. 1992. A truncated activin receptor inhibits mesoderm induction and formation of axial structures in *Xenopus* embryos. *Nature* **359:** 609–614.

Hemmati-Brivanlou A., Kelly O.G., and Melton D.A. 1994. Follistatin, an antagonist of activin, is expressed in the Spemann organizer and displays direct neuralizing activity. *Cell* **77:** 283–295.

Hemmati-Brivanlou A., Wright D.A., and Melton D.A. 1992. Embryonic expression and functional analysis of a *Xenopus* activin receptor. *Dev. Dyn.* **194:** 1–11.

Hoodless P.A., Tsukazaki T., Nishimatsu S., Attisano L., Wrana J.L., and Thomsen G.H. 1999. Dominant-negative Smad2 mutants inhibit activin/Vg1 signaling and disrupt axis formation in *Xenopus. Dev. Biol.* **207:** 364–379.

Hyatt B.A. and Yost H.J. 1998. The left-right coordinator: The role of Vg1 in organizing left-right axis formation. *Cell* **93:** 37–46.

Hyatt B.A., Lohr J.L., and Yost H.J. 1996. Initiation of vertebrate left-right axis formation by maternal Vg1. *Nature* **384:** 62–65.

Jones C.M., Armes N., and Smith J.C. 1996. Signalling by TGF-β family members: Short-range effects of Xnr-2 and BMP-4 contrast with the long-range effects of activin. *Curr. Biol.* **6:** 1468–1475.

Jones C.M., Simon-Chazottes D., Guenet J.L., and Hogan B.L. 1992. Isolation of Vgr-2, a novel member of the transforming growth factor-β-related gene family. *Mol. Endocrinol.* **6:** 1961–1968.

Joseph E.M. and Melton D.A. 1998. Mutant Vg1 ligands disrupt endoderm and mesoderm formation in *Xenopus* embryos. *Development* **125:** 2677–2685.

Kessler D.S. and Melton D.A. 1995. Induction of dorsal mesoderm by soluble, mature Vg1 protein. *Development* **121:** 2155–2164.

Khaner O. and Eyal-Giladi H. 1986. The embryo-forming potency of the posterior marginal zone in stages X through XII of the chick. *Dev. Biol.* **115:** 275–281.

Kondo M., Tashiro K., Fujii G., Asano M., Miyoshi R., Yamada R., Muramatsu M., and Shiokawa K. 1991. Activin receptor mRNA is expressed early in *Xenopus* embryogenesis and the level of the expression affects the body axis formation. *Biochem. Biophys. Res. Commun.* **181:** 684–690.

Lau A.L., Kumar T.R., Nishimori K., Bonadio J., and Matzuk M.M. 2000. Activin βC and βE genes are not essential for mouse liver growth, differentiation, and regeneration. *Mol. Cell. Biol.* **20:** 6127–6137.

Lee M.A., Heasman J., and Whitman M. 2001. Timing of endogenous activin-like signals and regional specification of the *Xenopus* embryo. *Development* **128:** 2939–2952.

Lee S.J. 1990. Identification of a novel member (GDF-1) of the transforming growth factor-beta superfamily. *Mol. Endocrinol.* **4:** 1034–1040.

———. 1991. Expression of growth/differentiation factor 1 in the nervous system: Conservation of a bicistronic structure. *Proc. Natl. Acad. Sci.* **88:** 4250–4254.

Levin M. 1998. The roles of activin and follistatin signaling in chick gastrulation. *Int. J. Dev. Biol.* **42:** 553–559.

Lyons K., Graycar J.L., Lee A., Hashmi S., Lindquist P.B., Chen E.Y., Hogan B.L.M., and Derynck R. 1989. Vgr-1, a mammalian gene related to *Xenopus* Vg-1, is a member of the transforming growth factor-β superfamily. *Proc. Natl. Acad. Sci.* **86:** 4554–4558.

Manova K., De Leon V., Angeles M., Kalantry S., Giarre M., Attisano L., Wrana J., and Bachvarova R.F. 1995. mRNAs for activin receptors II and IIB are expressed in mouse oocytes and in the epiblast of pregastrula and gastrula stage mouse embryos. *Mech. Dev.* **49:** 3–11.

Massague J. and Chen Y.G. 2000. Controlling TGF-β signaling. *Genes Dev.* **14:** 627–644.

Massague J., Attisano L., and Wrana J.L. 1994. The TGF-β family and its composite receptors. *Trends Cell Biol.* **4:** 172–178.

Mathews L.S., Vale W.W., and Kintner C.R. 1992. Cloning of a second type of activin receptor and functional characterization in *Xenopus* embryos. *Science* **255:** 1702–1705.

Matzuk M.M., Kumar T.R., Vassalli A., Bickenbach J.R., Roop D.R., Jaenisch R., and Bradley A. 1995. Functional analysis of activins during mammalian development. *Nature* **374:** 354–356.

McDowell N., Zorn A.M., Crease D.J., and Gurdon J.B. 1997. Activin has direct long-range signalling activity and can form a concentration gradient by diffusion. *Curr. Biol.* **7:** 671–681.

Melton D.A. 1987. Translocation of a localized maternal mRNA to the vegetal pole of *Xenopus* oocytes. *Nature* **328:** 80–82.

Mitrani E. and Shimoni Y. 1990. Induction by soluble factors of organized axial structures in chick epiblasts. *Science* **247:** 1092–1094.

Mitrani E., Ziv T., Thomsen G., Shimoni Y., Melton D.A., and Bril A. 1990. Activin can induce the formation of axial structures and is expressed in the hypoblast of the chick. *Cell* **63:** 495–501.

Nieuwkoop P.D. 1969a. The formation of mesoderm in urodelean amphibians. I. Induction by the endoderm. *Wilhelm Roux' Arch. Entwicklungsmech. Org.* **162:** 341–373.

———. 1969b. The formation of the mesoderm in urodelean Amphibians II. The origin of the dorso-ventral polarity of the mesoderm. *Wilhelm Roux' Arch. Entwicklungsmech. Org.* **163:** 298–315.

———. 1973. The "organisation center" of the amphibian embryo: Its origin, spatial organisation and morphogenetic action. *Adv. Morphogen.* **10:** 1–39.

———. 1985. Inductive interactions in early amphibian development and their general nature. *J. Embryol. Exp. Morphol.* (suppl.) **89:** 333–347.

Oda S., Nishimatsu S., Murakami K., and Ueno N. 1995. Molecular cloning and functional analysis of a new activin β subunit: A dorsal mesoderm-inducing activity in *Xenopus. Biochem. Biophys. Res. Commun.* **210:** 581–588.

Oh S.P. and Li E. 1997. The signaling pathway mediated by the type IIB activin receptor controls axial patterning and lateral asymmetry in the mouse. *Genes Dev.* **11:** 1812–1826.

Papin C. and Smith J.C. 2000. Gradual refinement of activin-induced thresholds requires protein synthesis. *Dev. Biol.* **217:** 166–172.

Paulusma C.C., Van der Kruijssen C.M., and Van den Eijnden-Van Raaij A.J. 1994. Localization of activin subunits in early murine development determined by subunit-specific antibodies. *J. Immunol. Methods* **169:** 143–152.

Petraglia F. 1997. Inhibin, activin and follistatin in the human placenta—A new family of regulatory proteins. *Placenta* **18:** 3–8.

Rankin C.T., Bunton T., Lawler A.M., and Lee S.J. 2000. Regulation of left-right patterning in mice by growth/differentiation factor-1. *Nat. Genet.* **24:** 262–265.

Rebagliati M.R. and Dawid I.B. 1993. Expression of activin transcripts in follicle cells and oocytes of *Xenopus laevis. Dev. Biol.* **159:** 574–580.

Rebagliati M.R., Weeks D.L., Harvey R.P., and Melton D.A. 1985. Identification and cloning of localized maternal mRNAs from *Xenopus* eggs. *Cell* **42:** 769–777.

Reilly K.M. and Melton D.A. 1996. Candidate morphogens of the TGFβ family are restricted to their source: Implications for the mechanism of patterning the embryo. *Cell* **86:** 743–754.

Reissmann E., Jornvall H., Blokzijl A., Andersson O., Chang C., Minchiotti G., Persico M.G., Ibanez C.F., and Brivanlou A.H. 2001. The orphan receptor ALK7 and the Activin receptor ALK4 mediate signaling by Nodal proteins during vertebrate development. *Genes Dev.* **15:** 2010–2022.

Rosa F., Roberts A.B., Danielpour D., Dart L.L., Sporn M.B., and Dawid I.B. 1988. Mesoderm induction in amphibians: The role of TGF-β2 like factors. *Science* **239:** 783–785.

Sargent T.D., Jamrich M., and Dawid I.B. 1986. Cell interactions and the control of gene activity during early development of *Xenopus laevis. Dev. Biol.* **114:** 238–246.

Sasai Y., Lu B., Steinbeisser H., and De Robertis E.M. 1995. Regulation of neural induction by the Chd and Bmp-4 antagonistic patterning signals in *Xenopus. Nature* **376:** 333–336.

Schulte-Merker S., Smith J.C., and Dale L. 1994. Effects of truncated activin and FGF receptors and of follistatin on the inducing activities of BVg1 and activin: Does activin play a role in mesoderm induction? *EMBO J.* **13:** 3533–3541.

Seleiro E.A., Connolly D.J., and Cooke J. 1996. Early developmental expression and experimental axis determination by the chicken Vg1 gene. *Curr. Biol.* **6:** 1476–1486.

Sha X., Brunner A.M., Purchio A.F., and Gentry L.E. 1989. Transforming growth factor beta 1: Importance of glycosylation and acidic proteases for processing and secretion. *Mol. Endocrinol.* **3:** 1090–1098.

Shah S.B., Skromne I., Hume C.R., Kessler D.S., Lee K. J., Stern C.D., and Dodd J. 1997. Misexpression of chick Vg1 in the marginal zone induces primitive streak formation. *Development* **124:** 5127–5138.

Shen M.M. and Schier A.F. 2000. The EGF-CFC gene family in verte-brate development. *Trends Genet.* **16:** 303–309.

Skromne I. and Stern C.D. 2002. A hierarchy of gene expression accompanying induction of the primitive streak by Vg1 in the chick embryo. *Mech. Dev.* **114:** 115–118.

Slack J.M.W. 1991. The nature of the mesoderm-inducing signal in *Xenopus*: A transfilter induction study. *Development* **113:** 661–669.

Slack J.M.W., Darlington B.G., Heath J.K., and Godsave S.F. 1987. Mesoderm induction in early *Xenopus* embryos by heparin-bind-ing growth factors. *Nature* **326:** 197–200.

Smith J.C. 1987. A mesoderm-inducing factor is produced by a *Xenopus* cell line. *Development* **99:** 3–14.

Smith J.C., Yaqoob M., and Symes K. 1988. Purification, partial char-acterization and biological effects of the XTC mesoderm-inducing factor. *Development* **103:** 591–600.

Smith J.C., Price B.M.J., Van Nimmen K., and Huylebroeck D. 1990. Identification of a potent *Xenopus* mesoderm-inducing factor as a homolog of activin A. *Nature* **345:** 729–731.

Sokol S., Wong G.G., and Melton D.A. 1990. A mouse macrophage fac-tor induces head structures and organizes a body axis in *Xenopus*. *Science* **249:** 561–564.

Song J., Oh S.P., Schrewe H., Nomura M., Lei H., Okano M., Gridley T., and Li E. 1999. The type II Activin receptors are essential for egg cylinder growth, gastrulation, and rostral head development in mice. *Dev. Biol.* **213:** 157–169.

Stern C.D., Yu R.T., Kakizuka A., Kintner C.R., Mathews L.S., Vale W.W., Evans R.M., and Umesono K. 1995. Activin and its receptors during gastrulation and the later phases of mesoderm develop-ment in the chick embryo. *Dev. Biol.* **172:** 192–205.

Sudarwati S. and Nieuwkoop P.D. 1971. Mesoderm formation in the anuran *Xenopus laevis* (Daudin). *Wilhem Roux' Arch. Entwicklungsmech. Org.* **166:** 189–204.

Symes K. and Smith J.C. 1987. Gastrulation movements provide an early marker of mesoderm induction in *Xenopus laevis*. *Develop-ment* **101:** 339–349.

Symes K., Yaqoob M., and Smith J.C. 1988. Mesoderm induction in *Xenopus laevis*: Responding cells must be in contact for mesoderm formation but suppression of epidermal differentiation can occur in single cells. *Development* **104:** 609–618.

Symes K., Yordan C., and Mercola M. 1994. Morphological differences in *Xenopus* embryonic mesodermal cells are specified as an early response to distinct threshold concentrations of activin. *Development* **120:** 2339–2346.

Tannahill D. and Melton D.A. 1989. Localized synthesis of the Vg1 protein during early *Xenopus* development. *Development* **106:** 775-785.

Thomsen G.H. and Melton D.A. 1993. Processed Vg1 protein is an axial mesoderm inducer in *Xenopus*. *Cell* **74:** 433–441.

Thomsen G., Woolf T., Whitman M., Sokol S., Vaughan J., Vale W., and Melton D.A. 1990. Activins are expressed early in *Xenopus* embryogenesis and can induce axial mesoderm and anterior structures. *Cell* **63:** 485–493.

Tiedemann H. and Tiedemann H. 1959. Versuche zur gewinnung eines mesodermalen induktionsstoffes aus huhnerembryonen. *Hoppe-Seylers Z. Physiol. Chem.* **314:** 156–176.

Tiedemann H., Lottspeich F., Davids M., Knochel S., and Hoppe P. 1992. The vegetalizing factor. A member of the evolutionarily highly conserved activin family. *FEBS Lett.* **300:** 123–126.

Vale W., Hsueh A., Rivier C., and Yu J. 1990. The inhibin/activin family of hormones and growth factors. In *Peptide growth factors and their receptors* (ed. M.B. Sporn and A.B. Roberts), vol. 1, pp. 211–248. Springer Verlag, Berlin.

van den Eijnden-Van Raaij A.J.M., van Zoelent E.J.J., van Nimmen K., Koster C.H., Snoek G.T., Durston A.J., and Huylebroeck D. 1990. Activin-like factor from a *Xenopus* cell line responsible for meso-derm induction. *Nature* **345:** 732–734.

Wall N.A., Craig E.J., Labosky P.A., and Kessler D.S. 2000. Mesendoderm induction and reversal of left-right pattern by mouse Gdf1, a Vg1-related gene. *Dev. Biol.* **227:** 495–509.

Watanabe M. and Whitman M. 1999. FAST-1 is a key maternal effector of mesoderm inducers in the early *Xenopus* embryo. *Development* **126:** 5621–5634.

Weeks D.L. and Melton D.A. 1987. A maternal mRNA localized to the vegetal hemisphere in *Xenopus* eggs codes for a growth factor related to TGF-β. *Cell* **51:** 861–867.

Whitman M. 1998. Smads and early developmental signaling by the TGFβ superfamily. *Genes Dev.* **12:** 2445–2462.

———. 2001. Nodal signaling in early vertebrate embryos: Themes and variations. *Dev. Cell* **1:** 605–617.

Wilson P.A. and Hemmati-Brivanlou A. 1995. Induction of epidermis and inhibition of neural fate by Bmp-4. *Nature* **376:** 331–333.

Wilson P.A. and Melton D.A. 1994. Mesodermal patterning by an inducer gradient depends on secondary cell-cell communication. *Curr. Biol.* **4:** 676–686.

Wittbrodt W. and Rosa F.R. 1994. Disruption of mesoderm and axis formation in fish by ectopic expression of activin variants: The role of maternal activin. *Genes Dev.* **8:** 1448–1462.

Wrana J.L., Attisano L., Wieser R., Ventura F., and Massague J. 1994. Mechanism of activation of the TGF-β receptor. *Nature* **370:** 341–347.

Yan Y.T., Liu J.J., Luo Y., Chaosu E., Haltiwanger R.S., Abate-Shen C., and Shen M.M. 2002. Dual roles of Cripto as a ligand and core-ceptor in the nodal signaling pathway. *Mol. Cell. Biol.* **22:** 4439–4449.

Yao J. and Kessler D.S. 1999. Mesoderm induction in *Xenopus*. Oocyte expression system and animal cap assay. *Methods Mol. Biol.* **137:** 169–178.

Yeo C. and Whitman M. 2001. Nodal signals to Smads through Cripto-dependent and Cripto-independent mechanisms. *Mol. Cell* **7:** 949–957.

Ziv T., Shimoni Y., and Mitrani E. 1992. Activin can generate ectopic axial structures in chick blastoderm explants. *Development* **115:** 689–694.

Zuck M., Wylie C.C., and Heasman J. 1999. Studying the function of maternal mRNAs in *Xenopus* embryos: An antisense approach. In *A comparative methods approach to the study of oocytes and embryos* (ed. J.D. Richter), pp. 341–354. Oxford University Press, Oxford, United Kingdom.

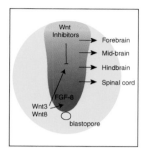

THE ROLE OF WNTS IN GASTRULATION

D.W. Houston[1] and C. Wylie

Cincinnati Children's Hospital Medical Center, Division of Developmental Biology, MLC 7007, Cincinnati, Ohio 45229-3039

INTRODUCTION

The Wnts are a large family of secreted proteins with diverse roles in the regulation of cell growth and differentiation. The prototype family member, *Wnt1*, previously called *int-1*, was identified as a genomic site of retroviral integration (Nusse and Varmus 1982). Although *Wnt1* was originally identified as an oncogene, the idea that oncogenes might regulate normal cell growth and differentiation led to the investigation of its role in development. *Wnt1* expression was found to be restricted to the neural plate and neural tube of mouse embryos (Shackleford and Varmus 1987; Wilkinson et al. 1987). Subsequently, *Wnt1* was discovered to be homologous to the *Drosophila* segment polarity gene *wingless* (*wg*), a gene required for segmentation and wing development (Cabrera et al. 1987; Rijsewijk et al. 1987). Functional analysis showed first that injection of *Wnt1* RNA into *Xenopus* embryos caused primary axis duplication (McMahon and Moon 1989), and later that brain defects were caused by targeted mutations of *Wnt1* in mice (McMahon and Bradley 1990; Thomas and Capecchi 1990). Similar brain defects were known to occur in the classic mouse mutant *swaying*, and it was subsequently found that these mutants harbored a single base-pair deletion in the *Wnt1* gene (Thomas et al. 1991). Since then, numerous Wnt genes have been identified in many organisms, and a vast array of experimental evidence has established the Wnts as important regulators of development in different organs and organisms.

[1]Present address: Department of Biological Sciences, University of Iowa, Iowa City, IA 52242.

With the availability of whole-genome sequence data, it has been possible to determine the level of diversity of Wnts present in a given organism. The human genome contains 19 WNT genes. Phylogenetic analysis has grouped Wnts into 12 subfamilies, 9 of which were probably present in the common ancestor of bilaterian animals (Sidow 1992; Schubert et al. 2000; Prud'homme et al. 2002). Interestingly, the genes of one subfamily, the *wg*, *DWnt6*, *DWnt10* group of paralogs, are colinear on the chromosome in *Drosophila*, and this gene order is conserved in humans (Nusse 2001). This observation led Nusse (2001) to speculate that this cluster of Wnt genes resembles Hox genes in their evolutionary conservation and potential co-regulation. Wnts and components of the Wnt signal transduction pathway are present in *Hydra* (Cnidaria) (Hobmayer et al. 2000), suggesting that cell–cell communication via Wnts was likely to occur in the earliest of multicellular organisms.

Wnt proteins are secreted, N-glycosylated (Brown et al. 1987; Papkoff et al. 1987), and require palmitate modification on a conserved cysteine residue for efficient signaling (Willert et al. 2003). Palmitoylation of Wnts may be mediated by the Porcupine protein, a putative acyl transferase (Hofmann 2000) that binds to Wnts (Tanaka et al. 2000) and is required for their efficient secretion (van den Heuvel et al. 1993; Kadowaki et al. 1996). However, neither acyl transferase activity nor direct modification of Wnt proteins by Porcupine has yet been demonstrated. Once secreted, Wnts act as both long- and short-range signaling molecules and can induce different responses in cells as a function of concentration, thus fulfilling the classic definition of a morphogen (Zecca et al. 1996). How Wnt signals are received and transduced by cells and the biological responses elicited by Wnt signals have been the subject of intense investigation.

In this chapter, the role of Wnts in gastrulation is reviewed. First the different mechanisms of Wnt signaling are briefly summarized. Then the roles of different Wnts in cell specification, patterning, and morphogenesis during gastrulation are examined.

WNT SIGNALING PATHWAYS

The Wnt–β-Catenin Pathway

The Wnt signal transduction pathway leading to the stabilization of β-catenin (Wnt–β-catenin or "canonical" pathway) has been extensively studied. The basic components were found through *Drosophila* genetics (Siegfried et al. 1992; Klingensmith et al. 1994; Noordermeer et al. 1994; Siegfried et al. 1994), in vitro studies (Papkoff et al. 1996), and embryology (Zeng et al. 1997), and have since been shown to be almost absolutely conserved through evolution. Essentially, in the absence of Wnt signals, a multiprotein complex containing APC (adenomatosis polyopsis coli), GSK-3β, and Axin constitutively phosphorylates β-catenin (for review, see Wodarz and Nusse 1998; Polakis 2000). Loss of these components results in stabilized β-catenin and mimics Wnt overexpression. Phosphorylated β-catenin is subsequently targeted for ubiquitination by members of the βTrcp family (Jiang and Struhl 1998; C. Liu et al. 1999) and degraded by proteasomes (Aberle et al. 1997). Wnt signals inhibit this "β-catenin destruction machinery" through multiple mechanisms involving the Dishevelled proteins (Dvl), a family of cytoskeletal adapter molecules (for review, see Boutros and Mlodzik 1999). Inhibition of β-catenin phosphorylation and degradation permits the accumulation of β-catenin protein and its subsequent nuclear localization. Once stabilized, β-catenin binds proteins of the *lymphoid enhancer factor/T-cell factor* (LEF/TCF) family of high mobility group (HMG) domain transcription factors (Molenaar et al. 1996; Brunner et al. 1997) and either activates or derepresses target genes through a conserved nuclear complex containing Pygopus and Legless/BCL9 proteins (Fig. 1A) (Belenkaya et al. 2002; Kramps et al. 2002; Parker et al. 2002).

Wnts that stimulate the β-catenin pathway bind to a cell-surface receptor complex consisting of a seven-transmembrane domain protein of the Frizzled (Fz) family and a lipoprotein receptor-related protein Lrp5 or Lrp6 (Lrp5/6; Arrow in *Drosophila* [for review, see Pandur and Kuhl 2001]). Lrp5/6 proteins bind Wnts and are required for Wnt signaling in a variety of organisms, and mutants generally resemble Wnt loss-of-function mutants (Pinson et al. 2000; Tamai et al. 2000; Wehrli et al. 2000; Gong et al. 2001). Frizzled proteins are nearly as diverse as Wnts (11 human genes, including *SMOOTHENED*) and can also

activate both β-catenin-dependent and -independent signaling pathways (Strapps and Tomlinson 2001). Wnts bind directly to Fz receptors via a cysteine-rich domain (CRD) at the amino terminus (Bhanot et al. 1996); however, Lrp5/6 can associate with Fz receptors only in the presence of Wnts and appear to be coreceptors specifically for β-catenin signaling (Tamai et al. 2000). Members of the heparan sulfate proteoglycan family also bind Wnts and may function as low-affinity receptors and modulators of Wnt gradients across tissues (for review, see Lin and Perrimon 2000). In this coreceptor model, association of the cytoplasmic domain of Lrp5/6 with the carboxy-terminal tail of Frizzled proteins appears important for Wnt signal transduction, as fusion of the Arrow endodomain to DFz2 creates a receptor that is active in the absence of ligand (Tolwinski et al. 2003). In vivo formation of a trimeric Wnt–Fz–Lrp5/6 complex has yet to be observed and is critical to establishing the validity of this model.

Initiation of Wnt signaling following binding of Wnts to the cell surface requires the activation of Dvl to inhibit β-catenin degradation (for review, see Boutros and Mlodzik 1999). Although the nature of this activation is not known, the Dvl PDZ domain interacts directly, and possibly transiently, with a conserved motif in the carboxy-terminal region of Fz (Wong et al. 2003). Additionally, residues in the DIX domain of Dvl that mediate phospholipid binding are critical for Dvl to activate Wnt signaling (Capelluto et al. 2002). Dvl can also bind to members of the arrestin family of G-protein-coupled receptor (GPCR) regulatory proteins (Chen et al. 2003). Frizzleds have been shown to activate heterotrimeric G-protein signaling in response to Wnts (Slusarski et al. 1997), suggesting that Dvl might regulate or be activated by GPCR signals. Recently, it has been shown that binding of Wnts to the Lrp5/6 coreceptor also stimulates the relocalization of Axin to the plasma membrane and fosters its subsequent degradation in a Dvl-dependent manner (J. Mao et al. 2001; Cliffe et al. 2003; Tolwinski et al. 2003). Axin is critically involved in the recruitment of GSK-3β and its priming kinase, CKIα, into the β-catenin degradation complex (Amit et al. 2002; Liu et al. 2002) and has an additional role as a cytoplasmic anchor for β-catenin (Tolwinski and Wieschaus 2001). Axin is likely to be a focal point for Wnt-β-catenin signaling, and more needs to be learned about the regulation of Axin subcellular localization and degradation, and its relationship to Dvl and other signaling components.

The Wnt-PCP Pathway

Genetic studies in *Drosophila* have identified a unique role for the *fz* gene in mediating planar cell polarity (PCP) signaling during imaginal disc development (Vinson and Adler 1987; Krasnow and Adler 1994). In flies, Fz and DFz2 play redundant roles in mediating β-catenin/Arm signal-

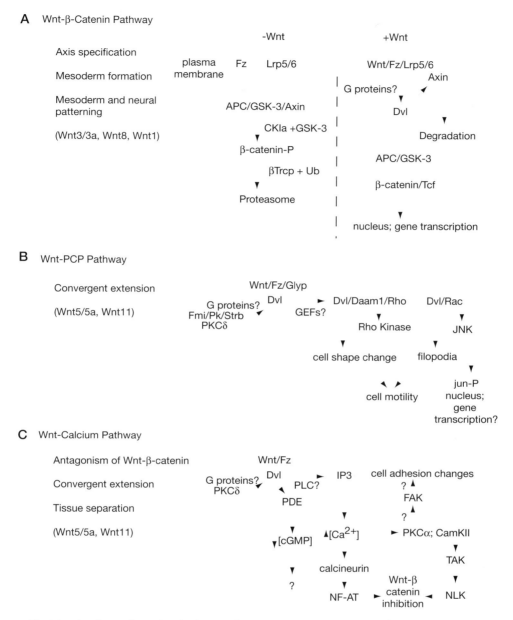

Figure 1. The Wnt signaling pathways involved in vertebrate gastrulation. (*A*) The Wnt–β-catenin pathway. In the absence of a Wnt signal, β-catenin is phosphorylated first by a priming kinase, casein Iα (CKIα), and then by glycogen synthase kinase-3β (GSK-3). Phosphorylated β-catenin is then recognized by β-transducin repeat-containing protein (βTrcp) and targeted for ubiquitination (Ub) and degradation by the proteasome. In the presence of a Wnt ligand, Frizzled (Fz) and lipoprotein receptor-related proteins 5/6 (Lrp5/6) associate and activate Dishevelled (Dvl) proteins, possibly via heterotrimeric G proteins. Dvl helps to recruit Axin to the plasma membrane, causing its degradation. Dvl also inhibits the action of the β-catenin phosphorylation complex resulting in stabilized β-catenin, which is free to associate with T-cell factor (Tcf) proteins and translocate to the nucleus to mediate gene transcription. (*B*) The Wnt–planar cell polarity (PCP) pathway. Activation of Wnt–PCP causes recruitment of Dvl to the plasma membrane, which may activate as yet unknown guanine nucleotide exchange factors (GEFs) to induce distinct complexes containing Dvl and Rac or Dvl and Rho with associated Daam1. These complexes then mediate JNK activation and filopodial extensions, and Rho activation and bipolar cell morphology, respectively. (*C*) The Wnt–calcium pathway. Binding of Wnts to Frizzled receptors activates Dvl, leading to phospholipase C activation, cleavage of PIP_2 to IP_3 and DAG, followed by release of calcium and protein kinase C (PKC) activation, respectively. Increased calcium levels may activate calcineurin, leading to nuclear localization of nuclear factor of activated cells (NF-AT), transcription factors that may inhibit Wnt–β-catenin signaling. Activation of PKCα can lead to focal adhesion kinase (FAK) activation and remodeling of focal adhesion complexes. Stimulation of calmodulin-dependent protein kinase II may activate TGF-β-activated kinase and NEMO-like kinase to inhibit Wnt–β-catenin signaling. Alternatively, phosphodiesterase (PDE) can be activated causing decreased levels of cGMP.

ing, but only Fz is required for generating polarity signals (Kennerdell and Carthew 1998; Bhanot et al. 1999; Chen and Struhl 1999; Muller et al. 1999). Regulation of polarity is thought not to involve a Wnt ligand (Lawrence et al. 2002) but is rather controlled by asymmetrical localization of Fz protein (Strutt 2001). A core set of additional proteins, including Dishevelled, Flamingo (a seven transmembrane pass cadherin), Prickle (a LIM and PET domain protein), strabismus (a four-transmembrane protein with a PDZ motif), and Diego (an ankyrin repeat protein), are all critical in establishing and maintaining asymmetric Fz localization (for review, see Strutt 2003). Importantly, these proteins are conserved in vertebrates and are involved in a range of processes including the regulation of gastrulation movements, neural tube closure, and sensory hair-cell polarity in the inner ear (for review, see Veeman et al. 2003a). In contrast to flies, the Wnt–PCP pathway in vertebrates appears to require activation by different Wnt ligands including Wnt4, Wnt5a, Wnt7a, and Wnt11 (for review, see Veeman et al. 2003a). Asymmetrical localization of Fz proteins has not yet been observed in vertebrates, suggesting that directional information may be provided by Wnts, rather than localized Fz proteins.

Activation of the Wnt–PCP pathway has several outcomes, all of which require regulation of Dvl and its accumulation at the plasma membrane (for review, see Boutros and Mlodzik 1999). In contrast to Wnt–β-catenin signaling, where the Dvl DIX domain is critical, it is the carboxy-terminal DEP domain that is most important for the PCP pathway (Axelrod et al. 1998; Boutros et al. 1998). Dvl has been shown to activate the small GTPases Rho, Rac and Cdc42, as well as jun kinase (JNK) and numerous other proteins in overexpression assays (for review, see Wharton 2003). Whereas the Wnt–β-catenin pathway is thought to primarily affect gene transcription and cell fate, the Wnt–PCP pathway predominantly regulates cytoskeletal rearrangements, cell polarity, and morphogenesis (Fig. 1B). Whether the PCP pathway also affects gene regulation directly has not been well studied, but evidence suggests that nuclear events may also be required for PCP. First, activation of Rho kinase and JNK are associated with transcription factor regulation (Davis 1995; Miralles et al. 2003). Furthermore, in flies, *jun* is required to mediate polarity defects induced by *dsh* overexpression in the eye, suggesting that transcriptional responses are involved in establishing polarity (Boutros et al. 1998). This situation is complicated as effectors such as JNK might regulate both nuclear and cytoskeletal events, or feedback mechanisms from the cytoskeleton might subsequently affect gene regulation.

The Wnt–Calcium Pathway

Overexpression of a subset of Wnt and Frizzled proteins has been shown to elicit calcium release in several experimental models (for review, see Kuhl et al. 2000b). Activation of this pathway requires heterotrimeric G proteins and initiates responses typical of GPCRs, including phosphoinositide turnover (Slusarski et al. 1997), cGMP-phosphodiesterase activation (Ahumada et al. 2002), and stimulation of the calcium-regulated kinases, calmodulin-dependent protein kinase II (CamKII) and protein kinase C (PKCα) (Fig. 1C) (Sheldahl et al. 1999; Kuhl et al. 2000a). Dvl also appears to be involved in the Wnt–calcium pathway, as overexpression of Dvl can stimulate calcium release and activate CamKII and PKCα (Sheldahl et al. 2003). In this situation, Dvl is likely to act downstream of heterotrimeric G proteins, as its effects are insensitive to pertussis toxin (Sheldahl et al. 2003).

The consequences of Wnt–calcium activation are likely to be diverse given the importance of calcium signaling in cells. Overexpression of *Xwnt5a* is known to antagonize Wnt–β-catenin signaling and to alter cell adhesion in frog embryos (Torres et al. 1996). Recently, an in vivo role for *Wnt5* in β-catenin antagonism has been demonstrated in zebrafish (Westfall et al. 2003). How this inhibition occurs is not known, but Wnt5 ligands could potentially regulate a Nemo-like kinase (NLK) pathway (Ishitani et al. 1999, 2003; Meneghini et al. 1999) or NF-AT activation (Saneyoshi et al. 2002) to block β-catenin signaling. It is not clear whether the Wnt–calcium pathway is wholly distinct from the Wnt–PCP pathway due to their shared components, such as Dvl. Additionally, the observation that a calcium-regulated PKC isoform, PKCδ, is required for Dvl membrane accumulation (Kinoshita et al. 2003) suggests significant overlap and cross-talk between the Wnt–PCP and Wnt–calcium pathways.

Alternative Pathways (Dwnt4, Derailed)

Recent work in *Drosophila* has uncovered additional signal transduction mechanisms for Wnt signaling. DWnt4 (homologous to vertebrate Wnt9/9a) is required for proper cell movements during ovarian morphogenesis, and requires Fz2, Dsh, and PKC in the regulation of focal adhesion kinase (FAK) and the formation of focal adhesions (Cohen et al. 2002). The DWnt4 pathway uses components common to other Wnt pathways but results in distinct outcomes. Another Wnt signaling pathway that requires none of the components of other Wnt pathways was identified in studies of axon guidance in flies. *DWnt5* was found in a suppressor screen to rescue overexpression of *derailed*, a molecule important in controlling axon midline crossing in the ventral nerve cord (Yoshikawa et al. 2003). Derailed, a potential receptor tyrosine kinase of the RYK family (for review, see Halford and Stacker 2001), contains an extracellular Wnt inhibitory factor (WIF; Hsieh et al. 1999) domain, and may serve as a receptor for DWnt5. Signaling via Derailed does not require any of the well-known Wnt components, suggesting that it operates by a different mechanism. Whether Wnts can signal through Derailed-like molecules in vertebrates remains to be determined. It should be kept in mind that the factors controlling cell responses to different Wnt signals are not well understood. It is likely that temporal and/or spatial constraints will determine which signaling mechanisms are utilized.

WNT SIGNALING DURING GASTRULATION

Axis Induction

In vertebrates, gastrulation is initiated asymmetrically, along the future dorsal axis of the embryo. Wnt–β-catenin signaling has a well-demonstrated role in specifying the dorsal axis in many model organisms. Overexpression of Wnt components in *Xenopus* embryos causes the formation of a secondary dorsal blastopore lip during gastrulation and induces a second axis (for review, see Heasman 1997). β-catenin was also found enriched in dorsal nuclei of frog and fish blastulae prior to dorsal-specific gene expression (Schneider et al. 1996). Loss-of-function experiments in early *Xenopus* embryos demonstrated that Wnt signaling via the β-catenin pathway is required for dorsal axis development (Heasman et al. 1994). Depletion of the maternal store of β-*catenin* RNA from frog oocytes using antisense oligonucleotides yields embryos lacking dorsal structures, including the notochord, somites, and nervous system. Although gastrulation does occur in β-*catenin*-depleted embryos, it is delayed, and not initiated on the dorsal side, but uniformly around the embryo, and the expression of many dorsal-specific genes is lost or severely delayed (Heasman et al. 1994). Additional maternal depletion studies in frog oocytes have shown that other components of the Wnt–β-catenin signaling cascade are also required for proper axis formation and dorsal gene expression (Yost et al. 1998; Sumanas et al. 2000; Kofron et al. 2001; Belenkaya et al. 2002; Houston et al. 2002).

Mouse embryos also use Wnt–β-catenin signaling to specify the embryonic axes. Gene targeting of β-*catenin* in the mouse results in an early, pre-gastrulation loss of anterior–posterior polarity followed by a failure of mesoderm formation and embryonic lethality (Haegel et al. 1995; Huelsken et al. 2000). Defects in anterior–posterior polarity are evident at the blastula stage (E6.0–E7.0) by a reduction or loss of anterior visceral endoderm (AVE) markers such as *Hex* and *Otx2*, and mis-localization distally of *Cer-l* and *Lim1* (Huelsken et al. 2000). Chimera analysis showed that β-catenin is primarily required in the epiblast, suggesting a non-cell-autonomous role in AVE specification (Huelsken et al. 2000). Due to technical considerations, expression of nuclear β-catenin, asymmetric or otherwise, has not yet been observed in the mouse embryo. It is inter-

esting to note that depletion of β-catenin in mice and frogs does not compromise cell adhesion (Heasman et al. 1994; Huelsken et al. 2000), due to the apparent redundant function of plakoglobin in adherens junctions.

Zebrafish are similar to frogs in that the dorsal axis is specified as a result of maternally regulated Wnt–β-catenin signaling (Pelegri and Maischein 1998; Sumoy et al. 1999). Recently, it has been shown that zebrafish use *Wnt5* to antagonize the dorsal β-catenin pathway. Fish carrying mutations in both maternal and zygotic *Wnt5* are hyper-dorsalized, and *Wnt5* was found to inhibit Wnt–β-catenin signaling via a calcium-dependent pathway (Westfall et al. 2003). How this occurs is still unknown, but it may result from activation of calcium-sensitive kinases such as PKC or CaMKII (Kuhl et al. 2000b) or via the NLK pathway (Ishitani et al. 2003). Also, overexpression of *Xwnt5a* in frog embryos is sufficient to inhibit Wnt–β-catenin signaling (Torres et al. 1996), possibly through the NF-AT family of transcription factors via calcineurin activation (Saneyoshi et al. 2002), suggesting that this mechanism of Wnt–β-catenin antagonism may be conserved during axis specification in frogs and fish. Axis formation appears normal in *Wnt5a*-deficient mouse embryos (Yamaguchi et al. 1999a), and *Wnt5a/5b* double mutants have not been described, so it remains to be seen whether similar mechanisms exist in different vertebrate species.

How these early Wnt–β-catenin pathways are initially established and what processes they mediate are not well understood. In frog eggs, sperm entry determines the future ventral side of the embryo (Chapter 23) and the cytoskeletal rearrangements of cortical rotation are thought to localize vesicles containing Dvl (Miller et al. 1999) and GSK3-binding protein (GBP) dorsally (Weaver et al. 2003). These dorsal determinants could either mediate β-catenin stabilization directly, or sensitize the cells that receive them to respond to Wnt signals (Fig. 2A). In support of the latter possibility, several Wnts, including *Wnt5a*, *Wnt7b*, *Wnt8b*, and *Wnt11*; numerous Fzs (for review, see Gradl et al. 1999); and the Wnt coreceptors *Lrp5/6* (Tamai et al. 2000; Houston and Wylie 2002) are expressed maternally. Additionally, *Wnt11* appears to be preferentially polyadenylated on the dorsal side of the early zygote (Schroeder et al. 1999). Detailed loss-of-function studies are necessary to distinguish between these hypotheses. In mice and chickens, where maternal cytoplasmic localization does not occur, the events of axis induction upstream of β-catenin are completely unknown, although several Wnts are expressed in the early mouse and chick blastula (Skromne and Stern 2001; Wilson et al. 2001; Lloyd et al. 2003).

The enrichment of β-catenin in the dorsal nuclei of the blastula has several major consequences for subsequent gene expression. First, β-catenin can interact with TCF proteins either to activate or to derepress target gene expression (for review, see Brantjes et al. 2002). Maternal mRNA

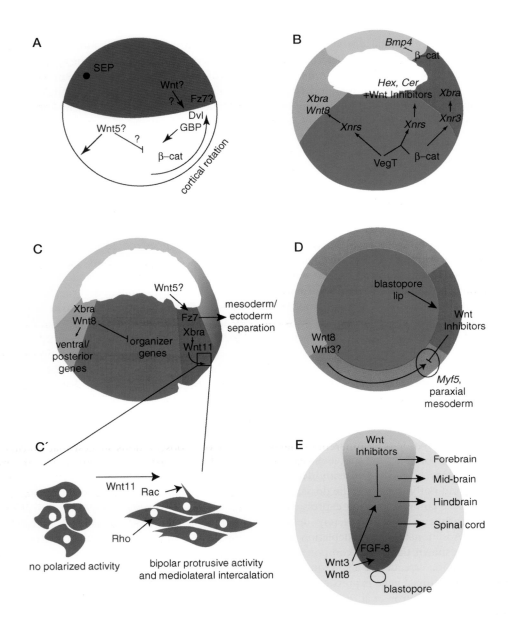

Figure 2. Roles of Wnt signaling pathways during *Xenopus* gastrulation. (*A*) Egg to early cleavage stages; initiation of dorsal–ventral polarity. Cortical rotation is initiated opposite to the sperm entry point (SEP) and localizes Dvl and GBP proteins dorsally, leading to dorsal stabilization of β-catenin. The roles of Wnts in initiating this pathway dorsally or inhibiting it ventrally have not been clarified. (*B*) Mid-blastula stage; mesoderm induction and dorsal patterning. Stabilized β-catenin dorsally leads directly to activation of dorsal genes (*Siamois* and *Xnr3*) and synergizes with VegT to induce *Nodal*-related genes (Xnrs) and to induce mesoderm. (*C*) Early gastrula stage; ventral mesoderm patterning and regulation of gastrulation movements. Ventrally expressed Wnt8 induces ventral/posterior genes and inhibits organizer gene expression. Dorsally, Fz7, potentially activated by Wnt5a, induces separation of the involuting mesoderm from the blastocoel roof ectoderm. Wnt11 regulates cell shape changes and gastrulation movements. (*C´*) In the dorsal mesoderm (*inset* from *C*) Wnt11 activates small GTPases Rho and Rac to generate bipolar cell morphology and filopodia, respectively. (*D,E*) Mid-late gastrulation; mesoderm and neural plate patterning. Dorsal-anteriorly expressed Wnt inhibitors (Cer, Dkk1, sFRPs) control activity of Wnts regionally by antagonizing ventral-posteriorly expressed Wnt ligands. In the mesoderm (*D*, vegetal view of a mid-gastrula), Wnt signaling is excluded from the dorsal lip leading to expression of Wnt target genes (e.g., *Myf5*) in the paraxial tissue. In the neural plate (*E*, late gastrula, dorsal view) posterior Wnts induce spinal cord fates (in the presence of FGF-8). Wnt activity is inhibited in a graded fashion anteriorly leading to the development of anterior neural fates. (*Brown*) Animal pole pigment; (*yellow*) epidermal ectoderm; (*blue*) neural ectoderm (darker is more posterior); (*dark red*) dorsal mesoderm; (*pink*) ventral mesoderm; (*green*) endoderm. In *A–D*, dorsal is to the right, ventral to the left, animal pole to the top, and vegetal pole to the bottom. In *E*, anterior is toward the top and posterior at the bottom.

depletion studies in *Xenopus* identified derepression of XTcf3-mediated inhibition as a major factor in the activation of Wnt-responsive genes (Houston et al. 2002). Two of these genes, *Siamois* and *Xnr3*, have distinct roles in further regulating the development of the organizer (Kodjabachian and Lemaire 2001) and in controlling convergent extension (Yokota et al. 2003), respectively. Other TCF proteins, including XTcf4 and XTcf1, are also expressed maternally in *Xenopus* (Houston et al. 2002; Roel et al. 2003), although their roles in development remain to be determined. Second, the Wnt–β-catenin pathway can act in concert with other cell-specification pathways, namely those involving the vegetal T-domain protein VegT (Agius et al. 2000; Xanthos et al. 2002) and *Xenopus* Nodal-related proteins (Xnrs) (Zorn et al. 1999; Labbe et al. 2000; Nishita et al. 2000), to direct mesoderm induction and formation of the organizer (Fig. 2B).

The interaction of β-catenin and Nodal signals is likely to be a conserved mechanism for axis specification. Similar to frogs, a role for *Tcf3* in repressing the dorsal axis has been found in the mouse. Deletion of *Tcf3* causes duplications of node and notochord, as well as up-regulation of organizer markers (Merrill et al. 2004). These data suggest that repression of Wnt target genes during axis specification is a conserved function of Tcf3. In the mouse, Nodal, in addition to β-catenin, is required for the specification of the anterior–posterior axis (Chapter 34). Additionally, graded doses of Nodal signaling are required for different cell fates in the embryo proper (Vincent et al. 2003), as is the case in *Xenopus* and zebrafish. Whether *Nodal* expression in mice is spatiotemporally controlled by Wnt signaling in a similar manner to *Xenopus* is a subject that should be investigated. In addition, Nodal signaling is autoregulatory (Brennan et al. 2001), suggesting that Wnt signaling might also be involved in the response to Nodals. In support of this idea, it has recently been discovered that mouse *Cripto*, a required coreceptor for Nodal, is regulated by β-catenin (Morkel et al. 2003). In pre-gastrula chick embryos, *Wnt8C* and the nodal-related *cVg1* are expressed in the axis-inducing region of the primitive streak (Joubin and Stern 1999). Studies with Wnt inhibitors suggested that Wnt signaling is required for the axis-inducing properties of cVg1 (Skromne and Stern 2001). A similar synergism of Wnts with *Vg1* occurs in the frog egg (Cui et al. 1996), but a role for endogenous Vg1 in axis induction has not been demonstrated.

Mesoderm Specification

As mentioned previously, mouse embryos deficient in β-catenin fail to form an anterior–posterior axis, primitive streak, and mesoderm. Genetic studies have provided evidence that the latter two aspects of the β-catenin phenotype are most likely due to signaling via Wnt3 (P. Liu et al. 1999).

Wnt3 is expressed at the proper place and time to function in mesoderm specification (i.e., the proximal epiblast at E6.25–7.0), and its ablation by gene targeting leads to a failure of primitive streak formation and a loss of mesodermal molecular markers, including *T* (*Brachyury*), *Goosecoid*, and *Nodal* (P. Liu et al. 1999). Anterior–posterior pattern was normally generated in these embryos, suggesting that Wnt3 is unlikely to mediate the early role of β-catenin in axis formation. Interestingly, mutations in the paralogous gene *Wnt3a*, typified by the classic mouse mutant *vestigial tail* (Greco et al. 1996), allow formation of the primitive streak and have excess neural tissue arising from loss of paraxial mesoderm (Takada et al. 1994; Yoshikawa et al. 1997). Further evidence for Wnt signaling in mouse mesoderm formation came from studies on the mouse mutant *mesoderm deficient* (*Mesd*), which lacks a primitive streak and mesoderm similar to *Wnt3* mutants (Holdener et al. 1994; Hsieh et al. 2003). The gene disrupted in this mutation encodes a protein-folding chaperone whose substrates include the Wnt coreceptors Lrp5 and Lrp6 (Hsieh et al. 2003). These results are only suggestive, however, as mesoderm and anterior–posterior patterning are normal in *Lrp5* and *Lrp6* single mutant embryos (Pinson et al. 2000; Gong et al. 2001); double mutants have not yet been described. Mesd may also be required for the proper folding of several extracellular molecules and thus may affect a wide variety of signaling pathways.

Several observations suggest that *Brachyury* is likely to be a direct target of Wnt signaling in the mesoderm. Its expression is reduced in both *Wnt3* (P. Liu et al. 1999) and *Wnt3a* (Yamaguchi et al. 1999b) mutants, and the *T* promoter contains a pair of consensus TCF sites (Yamaguchi et al. 1999b). These sites are necessary and sufficient to direct *T* expression in the primitive streak in transgenic *LacZ*-expressing embryos (Yamaguchi et al. 1999b). Interestingly, earlier (E7.5) *Brachyury* expression is maintained in *Lef1/Tcf1* double mutant embryos, whereas later expression is lost, suggesting that these TCFs mediate the later activity of *Wnt3a* signaling in the paraxial mesoderm (Galceran et al. 1999, 2001). *Tcf3* is a likely candidate to mediate the early action of *Wnt3* in mesoderm formation, as it is expressed prior to mesoderm formation and is subsequently lost from prospective paraxial mesoderm (Korinek et al. 1998). The role of Brachyury in development of the mesoderm is well characterized and is covered in Chapter 41.

Wnt signaling also has a role in mesoderm formation in other organisms. Expression of the frog *Brachyury* homolog (*Xbra*) is reduced in experiments using dominant-negative *XTcf3* constructs (Vonica and Gumbiner 2002) or antisense reagents against β-*catenin* (Xanthos et al. 2002; Schohl and Fagotto 2003). The effect on *Xbra* expression in frogs is subtle, and mesoderm does form in embryos depleted of β-catenin, suggesting that the requirement for Wnt signals is less stringent in frogs. Alternatively, because β-catenin-depleted embryos are not genetically null and retain residual β-catenin, *Xbra* expression may be merely delayed and reduced. It is possible that complete loss of β-catenin in frogs might more closely resemble the situation in the mouse. These data concerning β-catenin must also be reconciled with strong evidence for a primary role for maternal VegT in mesoderm formation in frogs (Zhang et al. 1998; Kofron et al. 1999). Although less well studied, Wnt signaling also acts during paraxial mesoderm formation in frogs. Zygotic Wnt signals directly induce *Myf5* expression in paraxial mesoderm (Shi et al. 2002), suggesting conservation of the genetic pathways involved in somite formation.

Ventral Mesoderm and Posterior Neural Patterning

Early Wnt signals prior to gastrulation are critical in the specification of dorsal cells to form the organizer. Subsequent to gastrulation, Wnts are expressed in ventral/posterior regions of embryos and act to limit the extent of the organizer and to pattern neural tissue induced by the organizer. Overexpression of Wnt–β-catenin pathway ligands zygotically in frogs and fish produce anteriorly truncated embryos (Christian and Moon 1993; Kelly et al. 1995; Hoppler et al. 1996). Conversely, injection of dominant-negative constructs or endogenous Wnt inhibitors causes anteriorization (Hoppler et al. 1996; Leyns et al. 1997; Wang et al. 1997; Glinka et al. 1998; Piccolo et al. 1999). In these studies, it was generally difficult to distinguish whether the effects on neural patterning were direct. Recent studies have clarified this issue, and evidence has accumulated to suggest that ventral/posterior mesoderm patterning and posterior neural patterning are distinct activities mediated by Wnt–β-catenin signaling (Fig. 2D,E).

Mesodermal Patterning

Mice deficient in *Wnt3a* exhibit posterior paraxial mesoderm defects and display excess neural tissue, which was correctly patterned (Yoshikawa et al. 1997). Zebrafish *tcf3* (*hdl*) mutants lack anterior structures and show expanded ventrolateral mesoderm (Kim et al. 2000). This phenotype arises from derepression of Wnt target genes and is consistent with Tcf3 acting as a repressor. In *tcf3* mutants, Wnt targets are ectopically active and appear to specify ventral fates. It is unclear whether the organizer is expanded in *tcf3* mutants, as expression of *chordin* and *goosecoid* is normal. However, expression of *dickkopf*, a prechordal plate marker, is expanded, suggesting that Wnts may antagonize prechordal plate regions of the organizer. Fish mutant for *wnt8* have expanded organizer gene expression and lack expression of the T-box gene *tbx6*, a ventrolateral mesoderm marker (Lekven et al. 2001). This apparent discrepancy might be resolved if

wnt8 regulates multiple Tcf proteins, whereas Tcf3 is only critical in specific regions of the organizer. Importantly, anterior neural markers are reduced, and posterior markers expanded, in *tcf3* mutant embryos, whereas the converse occurs in *wnt8* mutants (Kim et al. 2000; Lekven et al. 2001). Although these studies inferred that Wnt signaling is directly responsible for anterior–posterior patterning across the three germ layers, there is no evidence that Wnts act directly on neural tissue during gastrulation.

Posterior Neural Patterning

Much attention has been focused on the role of signaling molecules during neural induction following the classic studies of Nieuwkoop, Saxen, Toivonen, and others, demonstrating the activation–transformation nature of neural induction. In this model, neural induction generates cells with anterior (or rostral) neural identity. These cells then retain anterior fates unless exposed to a caudalizing signal from the prospective posterior regions of the embryo during gastrulation (Chapter 27).

The role of Wnts in mediating anterior–posterior patterning in general was strongly suggested by the discovery of endogenous Wnt inhibitors expressed in the organizer. These include secreted Frizzled-like molecules (sFRPS; for review, see Jones and Jomary 2002), Cerberus (Cer; Bouwmeester et al. 1996), and Dickkopf1 (Dkk1; Glinka et al. 1998). sFRPs and Cer inhibit Wnt signaling by binding to and possibly sequestering Wnt ligands (Leyns et al. 1997; Wang et al. 1997; Piccolo et al. 1999), whereas Dkk1 acts by antagonizing Lrp5/6 Wnt coreceptors (Bafico et al. 2001; B. Mao et al. 2001; Semenov et al. 2001). Ectopic expression of *Cer* in frogs induces secondary heads (Bouwmeester et al. 1996). *Cer* encodes a multifunctional antagonist of Wnt, BMP, and Nodal signaling (Piccolo et al. 1999), and analysis of *Cer* overexpression in *Xenopus* led to the hypothesis that anterior specification occurs by inhibition of multiple signaling pathways. Similar results were obtained using inhibitors of BMP and Wnt together (Glinka et al. 1997), demonstrating that blocking these two pathways is sufficient for anterior development. Gene targeting has shown that Wnt inhibition is required for head formation, providing in vivo loss-of-function evidence for the role of Wnt antagonism in anterior–posterior patterning. Genetic deletion of mouse *Dkk1* causes head truncations and loss of anterior neural markers (Mukhopadhyay et al. 2001), suggesting that Wnt inhibitors have nonoverlapping roles in neural patterning and head induction.

Although these studies demonstrate the requirement for Wnts in promoting posterior fates across germ layers, direct evidence for caudalization of neural fates was only suggestive (McGrew et al. 1997). In *Xenopus*, nuclear β-catenin, a marker for active Wnt signaling, is present in an anterior–posterior gradient in the prospective neural plate, with the highest levels posteriorly (Kiecker and Niehrs 2001). Importantly, if activated β-catenin is expressed during the late-gastrula stage, using an inducible construct, posterior fates are induced in neural ectoderm explants in the absence of dorsal mesoderm (Domingos et al. 2001). Activation of this construct early in gastrulation still induces posterior neural markers, but dorsal mesoderm and anterior neural markers were also expressed. Posterior neural markers are induced in cells that do not receive activated β-catenin, suggesting cell nonautonomous regulation. FGFs, particularly FGF8, are induced by late β-catenin activation, and FGF signaling is required for the generation of caudal neural fates (Domingos et al. 2001). In related experiments, Wnts in combination with FGFs caudalize chick neural plate explants in a graded fashion (Nordstrom et al. 2002). Critically, Wnts were shown to be the caudalizing agents, whereas a permissive role was demonstrated for FGF signaling. The levels of added Wnt, but not FGF, ligands determined the rostro-caudal character of neural tissue expressed in the explants (Nordstrom et al. 2002). The role of Wnts in neural caudalization is generally studied at the level of the major subdivisions of the CNS into forebrain, midbrain, hindbrain, and spinal cord. However, the mechanism of Wnt inhibition in rostral specification appears to be used reiteratively in patterning subdivisions of the brain. For instance, Wnt signals during gastrulation have a role in specifying the diencephalon versus the more anterior forebrain, the telencephalon (Heisenberg et al. 2001; Houart et al. 2002). A similar role for Wnts in caudalizing different subregions of the brain and spinal cord should be investigated, although sophisticated genetic manipulations will be needed to isolate early from late patterning events.

Wnts may also have roles in the early events of neural induction. Overexpression of Wnts or β-catenin in Xenopus animal caps induces neural fate by repressing Bmp4, an epidermal inducer (Baker et al. 1999). In chick blastulae, Wnts (of the Wnt–β-catenin class) inhibit FGF signal transduction to induce epidermal fates and allow Bmp expression (Wilson et al. 2001; Chapter 29). Clearly more data, especially loss-of-function analyses, are needed to reconcile these observations and elucidate the role of Wnt signals in neural induction.

Control of Morphogenetic Movements during Gastrulation

During vertebrate gastrulation, cells of the dorsal mesoderm undergo morphogenetic movements that result in axial elongation of the embryo (Chapter 19). The role of Wnts in controlling gastrulation movements was first recognized by the distinct activities of different Wnts when overexpressed. Wnt4, Wnt5a, and Wnt11 alter axial morphogenesis in fish and frog embryos (Du et al. 1995; Ungar et al. 1995), inhibit activin-induced elongation of frog animal caps without affecting cell fate (Moon et al. 1993), and antagonize signaling by Wnt1 family members (Torres et al. 1996). Additionally, expression of a dominant-negative Dvl (Xdd1) in Xenopus embryos inhibits dorsal mesoderm morphogenesis and activin-induced animal cap elongation (Sokol 1996). Subsequently, studies in Drosophila found that Dvl/Dsh is a mediator of planar polarity, independent of Wg signaling (Axelrod et al. 1998; Boutros et al. 1998), hinting at a role for signaling in vertebrate gastrulation movements.

Convergent Extension

Direct evidence for the role of Wnt–PCP signaling in convergent extension was found through analyses of fish mutants and frog dominant-negative constructs. Mutations in the wnt11 locus were found to underlie the convergent extension defects in silberblick (slb) mutant zebrafish (Heisenberg et al. 2000). Furthermore, a glypican heparan-sulfate proteoglycan protein (glypican-4), encoded by the knypek gene, is required to mediate Wnt11 function during convergent extension in zebrafish (Topczewski et al. 2001). Interference with Xenopus Xwnt11 (Tada and Smith 2000) or Dvl (Wallingford et al. 2000) function using dominant-negative constructs inhibited convergent extension in embryos as well as in animal caps treated with activin (Tada and Smith 2000) and dorsal mesoderm explants (Wallingford et al. 2000). Work in the fly and frog identified domains in Dvl critical for participation in either Wnt–β-catenin or Wnt–PCP signaling (Axelrod et al. 1998; Boutros et al. 1998; Rothbacher et al. 2000). Dvl mutants defective for Wnt–β-catenin signaling but competent for Wnt–PCP signaling can rescue Wnt11 deficiency but do not block convergent extension, whereas Wnt–PCP-compromised Dvl constructs do inhibit convergent extension and fail to rescue Wnt11 loss of function (Heisenberg et al. 2000; Tada and Smith 2000; Wallingford et al. 2000).

Convergent extension involves the lengthening of a tissue in one axis with concomitant narrowing in the perpendicular axis. How cells in a tissue accomplish convergent extension can vary greatly. In Xenopus, cells in the dorsal mesoderm extend stable mediolateral protrusions to generate traction on neighboring cells, resulting in the mediolateral cell intercalation that drives convergent extension (Keller et al. 1985; Keller and Tibbetts 1989; Shih and Keller 1992; see Chapters 19 and 20). Wallingford et al. (2000) showed in Xenopus that Dvl, and by extension Wnt–PCP signaling, is required in dorsal marginal zone cells for the proper stable extension of mediolateral lamellipodia, as well as in the neural plate for convergent extension of neural tissue (Wallingford and Harland 2001). It is likely that Dvl acts by controlling the stability of lamellipodia in this case as well, although monopolar protrusive activity, a hallmark of neural convergent extension (Elul and Keller 2000), was not directly examined. In fish, migrating prechordal plate cells in *wnt11* mutant embryos (Ulrich et al. 2003) and paraxial mesendoderm cells in *wnt5* mutants (Kilian et al. 2003) also lack stable, directed protrusions, suggesting that regulation of lamellipodial activity might be a key outcome of Wnt–PCP signaling during gastrulation movements.

Wnt–PCP signaling uses a set of proteins conserved from flies to vertebrates to regulate gastrulation movements. Recent evidence suggests that Dvl acts via Rac and Rho to activate JNK and potentially Rho kinase (ROCK or Rok), respectively, and ultimately to regulate different cell behaviors. Other Wnt–PCP proteins such as Strabismus (Darken et al. 2002; Goto and Keller 2002; Jessen et al. 2002), Prickle (Carreira-Barbosa et al. 2003; Takeuchi et al. 2003; Veeman et al. 2003b), and PKCδ (Sheldahl et al. 2003) are required for normal gastrulation and appear to modulate Dvl membrane localization and activation of effector molecules. Dvl binds to both Rac and Rho in a Wnt-dependent manner and can form independent complexes in Wnt-stimulated cells (Habas et al. 2001, 2003). Dvl associates with Rac via the DEP domain and mediates JNK activation (Habas et al. 2003), whereas in Rho activation, Dvl binds a formin homology protein Daam1 through the PDZ and DEP domains and recruits Rho (Habas et al. 2001). Both Rac and Rho are required for convergent extension in *Xenopus* (Habas et al. 2003) and appear to mediate filopodium formation and mediolateral cell elongation, respectively (Fig. 2C′) (Tahinci and Symes 2003). Zebrafish Rok2, an effector of active Rho, is also required for cell elongation (Marlow et al. 2002), suggesting that Rho activates Rok2 during convergent extension, although this has not been directly established. It is not known whether JNK activation by Rac is required for formation of filopodia; however, JNK is also required for convergent extension (Yamanaka et al. 2002).

Frizzled proteins with potential roles in mediating Wnt–PCP signals during convergent extension include Fz2, Fz7, and Fz8. Inhibition of *Fz2* translation causes defects in axial morphogenesis in fish (Sumanas et al. 2001). The *Fz2* phenotype resembles the *pipetail/wnt5* mutation and functions downstream of *wnt5* (Kilian et al. 2003) in fish, suggesting that Wnt5 and Fz2 might act as a ligand–receptor pair. A dominant-negative Fz8 blocks convergent extension in *Xenopus* Keller (dorsal blastopore lip) explants (Deardorff et al. 1998; Itoh et al. 1998; Wallingford et al. 2001a). Some reports suggested that Fz7, a putative Wnt11 receptor, is required for convergent extension (Djiane et al. 2000; Sumanas and Ekker 2001). However, Winklbauer et

al. (2001) showed that Fz7 is required for tissue separation, whereas convergent extension (as assayed by Keller explant elongation) is unaffected. Other reports relied on inhibition of activin-induced animal cap elongation by *Fz7* dominant-negative constructs or antisense reagents (Djiane et al. 2000; Sumanas and Ekker 2001). Activin-treated animal caps exhibit many types of morphogenesis and should not be used as the sole indicator of convergent extension. Keller explants are probably a more reliable assay for this purpose, particularly if cellular morphology and cell behavior are also visualized.

Taken together, the above observations suggest that the function of Wnt–PCP signaling during gastrulation is to activate cells into a motile state by regulating cell process formation and cell shape changes. Whether a Wnt–calcium pathway is also active is not clear. The involvement of Wnt5 and Fz2 suggest a role for Wnt–calcium signaling in convergent extension, as Wnt5 is a major activator of this pathway. However, although calcium flux is required for convergent extension in frogs, overexpression of a dominant-negative Fz8 capable of blocking convergent extension does not fully inhibit calcium waves in dorsal mesoderm explants (Wallingford et al. 2001b). What causes cells to extend bipolar protrusions perpendicular to the anterior–posterior axis, or monopolar extensions in neural convergent extension and fish dorsal migration, is unknown. Experiments to identify the molecules responsible could be designed to look for factors that induce cells to orient and elongate parallel to the anterior–posterior axis while exhibiting stable protrusions in the same direction.

Tissue Separation

The Wnt–calcium pathway, acting via Fz7, heterotrimeric G proteins, and PKCα, is thought to regulate the ability of involuted anterior mesoderm to maintain differential adhesion from gastrula blastocoel roof ectoderm during gastrulation, a process known as tissue separation (Wacker et al. 2000). Inhibition of *Fz7* translation in *Xenopus* inhibits this process and was further shown to be required specifically in the mesoderm, as tissue separation occurs normally when Fz7 is depleted in the ectoderm (Fig. 2C) (Winklbauer et al. 2001). Components of the Wnt–β-catenin and Wnt–PCP pathways are not sufficient to rescue separation behavior of Fz7-depleted mesoderm, whereas expression of *PKCα*, which mimics activation of the Wnt–calcium pathway, efficiently rescues Fz7 depletion. Treatment with pertussis toxin abrogates endogenous separation behavior in explants as well as in animal caps induced with FGF and Fz7, suggesting that heterotrimeric G proteins are involved. In frog embryos, it is possible that Wnt5a mediates this process, as interference with Wnt11 function, another potential calcium-stimulating Wnt, does not inhibit separation behavior (Winklbauer et al. 2001). How Wnt–calcium signaling causes the differential adhesion necessary for tissue separation is currently unknown. However, several possibilities exist, including differential expression of adhesion molecules such as eph/ephrins or protocadherins, or differential changes in focal adhesion complexes. With respect to the latter possibility, signaling via Dwnt4 was shown to active FAK, although the involvement of calcium signaling cannot be ruled out (Cohen et al. 2002). Interestingly, the hypothesis that Dwnt4 induces a unique pathway partially stemmed from the observation that Dvl cannot rescue Fz7 depletion in separation behavior (Winklbauer et al. 2001), a conclusion called into question by subsequent Dvl loss-of-function data showing the involvement of Dvl in Wnt–calcium signaling (Sheldahl et al. 2003). Future work will help resolve whether there is additional similarity between these two pathways, as well other potential mechanisms involved in Wnt regulation of tissue separation.

CONCLUSIONS

Wnt signals play diverse roles, both in the initiation and in the process of gastrulation. Wnt signaling is required prior to gastrulation to generate the dorsal axis and induce the dorsal mesoderm to initiate a dorsal-specific type of movement. Subsequently, Wnt signaling in the mesoderm induces posterior fates in both the mesoderm and overlying neural plate. Wnt signaling during gastrulation also enables cells to undergo the proper morphogenetic movements to complete gastrulation. These different Wnt-mediated signals use different downstream signaling pathways and different downstream targets. How they are coordinated in time and space, both with each other and with other signaling pathways required for gastrulation, is still a subject of active research. A key example of this continuity of Wnt signaling is the induction of *Xbra* in *Xenopus* mesoderm by maternal β-catenin. Xbra then induces *Wnt11*, first in the dorsal mesoderm and then throughout the entire mesoderm, to mediate convergent extension. Wnt signaling potentially regulates the expression or activity of other signaling molecules, particularly Nodal/TGF-β signals. There are also interactions between different kinds of Wnt signals; namely, the inhibitory action of Wnt–calcium-type signals on the Wnt–β-catenin pathway.

There are also outstanding questions concerning the control of Wnt signaling in gastrulation. First, how are the major asymmetries in Wnt activity established to regulate axis formation during early embryogenesis? Particularly in frog and fish eggs, how does fertilization result in β-catenin accumulation on the opposite side? Some Wnt pathway components are carried dorsally by cortical rotation, but it

is not known whether this is causative or permissive for Wnt signaling. In addition, a calcium wave is propagated after fertilization; does this wave require the Wnt–calcium pathway? In the mouse, as in *Xenopus*, the identities of Wnts that might act upstream of β-catenin in the blastula are unknown. Second, Wnt–β-catenin signals are critically involved in posteriorization of both mesoderm and neuroectoderm, but it is not known whether this occurs by similar or divergent mechanisms in different germ layers. Again, dissecting the interaction of Wnt signals with other factors, particularly FGFs, is important in this regard. Last, it is important to understand how Wnt signaling by both the PCP and calcium pathways regulates the complicated morphogenetic events of convergent extension and tissue separation. How might a uniformly distributed Wnt ligand, in the case of Wnt11, provide polarity information? Because many different stimuli can activate small GTPases such as Rac and Rho, how does Wnt-mediated activation of these molecules differ from other stimuli? This last aspect is likely to be more experimentally accessible with the emergence of technologies such as live cell imaging of fluorescent molecules and fluorescent resonance energy transfer (FRET) to analyze molecular interactions in vivo.

REFERENCES

Aberle H., Bauer A., Stappert J., Kispert A., and Kemler R. 1997. β-catenin is a target for the ubiquitin-proteasome pathway. *EMBO J.* **16**: 3797–3804.

Agius E., Oelgeschlager M., Wessely O., Kemp C., and De Robertis E.M. 2000. Endodermal Nodal-related signals and mesoderm induction in *Xenopus*. *Development* **127**: 1173–1183.

Ahumada A., Slusarski D.C., Liu X., Moon R.T., Malbon C.C., and Wang H.Y. 2002. Signaling of rat Frizzled-2 through phosphodiesterase and cyclic GMP. *Science* **298**: 2006–2010.

Amit S., Hatzubai A., Birman Y., Andersen J. S., Ben-Shushan E., Mann M., Ben-Neriah Y., and Alkalay I. 2002. Axin-mediated CKI phosphorylation of β-catenin at Ser 45: A molecular switch for the Wnt pathway. *Genes Dev.* **16**: 1066–1076.

Axelrod J.D., Miller J.R., Shulman J.M., Moon R.T., and Perrimon N. 1998. Differential recruitment of Dishevelled provides signaling specificity in the planar cell polarity and Wingless signaling pathways. *Genes Dev.* **12**: 2610–2622.

Bafico A., Liu G., Yaniv A., Gazit A., and Aaronson S.A. 2001. Novel mechanism of Wnt signalling inhibition mediated by Dickkopf-1 interaction with LRP6/Arrow. *Nat. Cell Biol.* **3**: 683–686.

Baker J.C., Beddington R.S., and Harland R.M. 1999. Wnt signaling in *Xenopus* embryos inhibits bmp4 expression and activates neural development. *Genes Dev.* **13**: 3149–3159.

Belenkaya T.Y., Han C., Standley H.J., Lin X., Houston D.W., and Heasman J. 2002. *pygopus* encodes a nuclear protein essential for wingless/Wnt signaling. *Development* **129**: 4089–4101.

Bhanot P., Fish M., Jemison J.A., Nusse R., Nathans J., and Cadigan K.M. 1999. Frizzled and Dfrizzled-2 function as redundant receptors for Wingless during *Drosophila* embryonic development. *Development* **126**: 4175–4186.

Bhanot P., Brink M., Samos C.H., Hsieh J.C., Wang Y., Macke J.P., Andrew D., Nathans J., and Nusse R. 1996. A new member of the frizzled family from *Drosophila* functions as a Wingless receptor. *Nature* **382:** 225–230.

Boutros M. and Mlodzik M. 1999. Dishevelled: at the crossroads of divergent intracellular signaling pathways. *Mech. Dev.* **83:** 27–37.

Boutros M., Paricio N., Strutt D.I., and Mlodzik M. 1998. Dishevelled activates JNK and discriminates between JNK pathways in planar polarity and wingless signaling. *Cell* **94:** 109–118.

Bouwmeester T., Kim S., Sasai Y., Lu B., and De Robertis E.M. 1996. Cerberus is a head-inducing secreted factor expressed in the anterior endoderm of Spemann's organizer. *Nature* **382:** 595–601.

Brantjes H., Barker N., van Es J., and Clevers H. 2002. TCF: Lady Justice casting the final verdict on the outcome of Wnt signalling. *Biol. Chem.* **383:** 255–261.

Brennan J., Lu C.C., Norris D.P., Rodriguez T.A., Beddington R.S., and Robertson E.J. 2001. Nodal signalling in the epiblast patterns the early mouse embryo. *Nature* **411:** 965–969.

Brown A.M., Papkoff J., Fung Y.K., Shackleford G.M., and Varmus H.E. 1987. Identification of protein products encoded by the proto-oncogene int-1. *Mol. Cell. Biol.* **7:** 3971–3977.

Brunner E., Peter O., Schweizer L., and Basler K. 1997. pangolin encodes a Lef-1 homologue that acts downstream of Armadillo to transduce the Wingless signal in *Drosophila*. *Nature* **385:** 829–833.

Cabrera C.V., Alonso M.C., Johnston P., Phillips R.G., and Lawrence P.A. 1987. Phenocopies induced with antisense RNA identify the wingless gene. *Cell* **50:** 659–663.

Capelluto D.G., Kutateladze T.G., Habas R., Finkielstein C.V., He X., and Overduin M. 2002. The DIX domain targets dishevelled to actin stress fibres and vesicular membranes. *Nature* **419:** 726–729.

Carreira-Barbosa F., Concha M.L., Takeuchi M., Ueno N., Wilson S.W., and Tada M. 2003. Prickle 1 regulates cell movements during gastrulation and neuronal migration in zebrafish. *Development* **130:** 4037–4046.

Chen C.M. and Struhl G. 1999. Wingless transduction by the Frizzled and Frizzled2 proteins of *Drosophila*. *Development* **126:** 5441–5452.

Chen W., ten Berge D., Brown J., Ahn S., Hu L.A., Miller W.E., Caron M.G., Barak L.S., Nusse R., and Lefkowitz R.J. 2003. Dishevelled 2 recruits beta-arrestin 2 to mediate Wnt5A-stimulated endocytosis of Frizzled 4. *Science* **301:** 1391–1394.

Christian J.L. and Moon R.T. 1993. Interactions between Xwnt-8 and Spemann organizer signaling pathways generate dorsoventral pattern in the embryonic mesoderm of *Xenopus*. *Genes Dev.* **7:** 13–28.

Cliffe A., Hamada F., and Bienz M. 2003. A role of Dishevelled in relocating Axin to the plasma membrane during wingless signaling. *Curr. Biol.* **13:** 960–966.

Cohen E.D., Mariol M.C., Wallace R.M., Weyers J., Kamberov Y.G., Pradel J., and Wilder E.L. 2002. DWnt4 regulates cell movement and focal adhesion kinase during *Drosophila* ovarian morphogenesis. *Dev. Cell* **2:** 437–448.

Cui Y., Tian Q., and Christian J.L. 1996. Synergistic effects of Vg1 and Wnt signals in the specification of dorsal mesoderm and endoderm. *Dev. Biol.* **180:** 22–34.

Darken R.S., Scola A.M., Rakeman A.S., Das G., Mlodzik M., and Wilson P.A. 2002. The planar polarity gene strabismus regulates convergent extension movements in *Xenopus*. *EMBO J.* **21:** 976–985.

Davis R.J. 1995. Transcriptional regulation by MAP kinases. *Mol Reprod. Dev.* **42:** 459–467.

Deardorff M.A., Tan C., Conrad L.J., and Klein P.S. 1998. Frizzled-8 is expressed in the Spemann organizer and plays a role in early morphogenesis. *Development* **125:** 2687–2700.

Djiane A., Riou J., Umbhauer M., Boucaut J., and Shi D. 2000. Role of frizzled 7 in the regulation of convergent extension movements during gastrulation in *Xenopus laevis*. *Development* **127:** 3091–3100.

Domingos P.M., Itasaki N., Jones C.M., Mercurio S., Sargent M.G., Smith J.C., and Krumlauf R. 2001. The Wnt/beta-catenin pathway posteriorizes neural tissue in *Xenopus* by an indirect mechanism requiring FGF signalling. *Dev. Biol.* **239:** 148–160.

Du S.J., Purcell S.M., Christian J.L., McGrew L.L., and Moon R.T. 1995. Identification of distinct classes and functional domains of Wnts through expression of wild-type and chimeric proteins in *Xenopus* embryos. *Mol. Cell. Biol.* **15:** 2625–2634.

Elul T. and Keller R. 2000. Monopolar protrusive activity: A new morphogenic cell behavior in the neural plate dependent on vertical interactions with the mesoderm in *Xenopus*. *Dev. Biol.* **224:** 3–19.

Galceran J., Hsu S.C., and Grosschedl R. 2001. Rescue of a Wnt mutation by an activated form of LEF-1: Regulation of maintenance but not initiation of Brachyury expression. *Proc. Natl. Acad. Sci.* **98:** 8668–8673.

Galceran J., Farinas I., Depew M.J., Clevers H., and Grosschedl R. 1999. *Wnt3a$^{-/-}$*-like phenotype and limb deficiency in *Lef1$^{-/-}$ Tcf1$^{-/-}$* mice. *Genes Dev.* **13:** 709–717.

Glinka A., Wu W., Onichtchouk D., Blumenstock C., and Niehrs C. 1997. Head induction by simultaneous repression of Bmp and Wnt signalling in *Xenopus*. *Nature* **389:** 517–519.

Glinka A., Wu W., Delius H., Monaghan A.P., Blumenstock C., and Niehrs C. 1998. Dickkopf-1 is a member of a new family of secreted proteins and functions in head induction. *Nature* **391:** 357–362.

Gong Y., Slee R.B., Fukai N., Rawadi G., Roman-Roman S., Reginato A.M., Wang H., Cundy T., Glorieux F.H., Lev D. et al. 2001. LDL receptor-related protein 5 (LRP5) affects bone accrual and eye development. *Cell* **107:** 513–523.

Goto T. and Keller R. 2002. The planar cell polarity gene strabismus regulates convergence and extension and neural fold closure in *Xenopus*. *Dev. Biol.* **247:** 165–181.

Gradl D., Kuhl M., and Wedlich D. 1999. Keeping a close eye on Wnt-1/wg signaling in *Xenopus*. *Mech. Dev.* **86:** 3–15.

Greco T.L., Takada S., Newhouse M.M., McMahon J.A., McMahon A.P., and Camper S.A. 1996. Analysis of the vestigial tail mutation demonstrates that Wnt-3a gene dosage regulates mouse axial development. *Genes Dev.* **10:** 313–324.

Habas R., Dawid I.B., and He X. 2003. Coactivation of Rac and Rho by Wnt/Frizzled signaling is required for vertebrate gastrulation. *Genes Dev.* **17:** 295–309.

Habas R., Kato Y., and He X. 2001. Wnt/Frizzled activation of Rho regulates vertebrate gastrulation and requires a novel Formin homology protein Daam1. *Cell* **107:** 843–854.

Haegel H., Larue L., Ohsugi M., Fedorov L., Herrenknecht K., and Kemler R. 1995. Lack of beta-catenin affects mouse development at gastrulation. *Development* **121:** 3529–3537.

Halford M.M. and Stacker S.A. 2001. Revelations of the RYK receptor. *BioEssays* **23:** 34–45.

Heasman J. 1997. Patterning the *Xenopus* blastula. *Development* **124:** 4179–4191.

Heasman J., Crawford A., Goldstone K., Garner-Hamrick P., Gumbiner B., McCrea P., Kintner C., Noro C.Y., and Wylie C. 1994. Overexpression of cadherins and underexpression of β-catenin inhibit dorsal mesoderm induction in early *Xenopus* embryos. *Cell* **79:** 791–803.

Heisenberg C.P., Tada M., Rauch G.J., Saude L., Concha M.L., Geisler R., Stemple D.L., Smith J.C., and Wilson S.W. 2000. Silberblick/Wnt11 mediates convergent extension movements during zebrafish gastrulation. *Nature* **405:** 76–81.

Heisenberg C.P., Houart C., Take-Uchi M., Rauch G.J., Young N., Coutinho P., Masai I., Caneparo L., Concha M.L., Geisler R., Dale T.C., Wilson S.W., and Stemple D.L. 2001. A mutation in the Gsk3-binding domain of zebrafish Masterblind/Axin1 leads to a fate transformation of telencephalon and eyes to diencephalon. *Genes Dev.* **15:** 1427–1434.

Hobmayer B., Rentzsch F., Kuhn K., Happel C.M., von Laue C.C., Snyder P., Rothbacher U., and Holstein T.W. 2000. WNT signalling molecules act in axis formation in the diploblastic metazoan Hydra. *Nature* **407:** 186–189.

Hofmann K. 2000. A superfamily of membrane-bound O-acyltransferases with implications for wnt signaling. *Trends Biochem. Sci.* **25:** 111–112.

Holdener B.C., Faust C., Rosenthal N.S., and Magnuson T. 1994. msd is required for mesoderm induction in mice. *Development* **120:** 1335–1346.

Hoppler S., Brown J.D., and Moon R.T. 1996. Expression of a dominant-negative Wnt blocks induction of MyoD in *Xenopus* embryos. *Genes Dev.* **10:** 2805–2817.

Houart C., Caneparo L., Heisenberg C., Barth K., Take-Uchi M., and Wilson S. 2002. Establishment of the telencephalon during gastrulation by local antagonism of Wnt signaling. *Neuron* **35:** 255–265.

Houston D.W. and Wylie C. 2002. Cloning and expression of *Xenopus* Lrp5 and Lrp6 genes. *Mech. Dev.* **117:** 337–342.

Houston D.W., Kofron M., Resnik E., Langland R., Destree O., Wylie C., and Heasman J. 2002. Repression of organizer genes in dorsal and ventral *Xenopus* cells mediated by maternal XTcf3. *Development* **129:** 4015–4025.

Hsieh J.C., Kodjabachian L., Rebbert M.L., Rattner A., Smallwood P.M., Samos C.H., Nusse R., Dawid I.B., and Nathans J. 1999. A new secreted protein that binds to Wnt proteins and inhibits their activities. *Nature* **398:** 431–436.

Hsieh J.C., Lee L., Zhang L., Wefer S., Brown K., DeRossi C., Wines M.E., Rosenquist T., and Holdener B.C. 2003. Mesd encodes an LRP5/6 chaperone essential for specification of mouse embryonic polarity. *Cell* **112:** 355–367.

Huelsken J., Vogel R., Brinkmann V., Erdmann B., Birchmeier C., and Birchmeier W. 2000. Requirement for beta-catenin in anterior–posterior axis formation in mice. *J. Cell Biol.* **148:** 567–578.

Ishitani T., Kishida S., Hyodo-Miura J., Ueno N., Yasuda J., Waterman M., Shibuya H., Moon R.T., Ninomiya-Tsuji J., and Matsumoto K. 2003. The TAK1-NLK mitogen-activated protein kinase cascade functions in the Wnt-5a/Ca(2+) pathway to antagonize Wnt/beta-catenin signaling. *Mol. Cell. Biol.* **23:** 131–139.

Ishitani T., Ninomiya-Tsuji J., Nagai S., Nishita M., Meneghini M., Barker N., Waterman M., Bowerman B., Clevers H., Shibuya H. et al. 1999. The TAK1-NLK-MAPK-related pathway antagonizes signalling between beta-catenin and transcription factor TCF. *Nature* **399:** 798–802.

Itoh K., Jacob J., and Sokol S.Y. 1998. A role for *Xenopus* Frizzled 8 in dorsal development. *Mech. Dev.* **74:** 145–157.

Jessen J.R., Topczewski J., Bingham S., Sepich D.S., Marlow F., Chandrasekhar A., and Solnica-Krezel L. 2002. Zebrafish trilobite

identifies new roles for Strabismus in gastrulation and neuronal movements. *Nat. Cell Biol.* **4:** 610–615.

Jiang J. and Struhl G. 1998. Regulation of the Hedgehog and Wingless signalling pathways by the F-box/WD40-repeat protein Slimb. *Nature* **391:** 493–496.

Jones S.E. and Jomary C. 2002. Secreted Frizzled-related proteins: Searching for relationships and patterns. *Bioessays* **24:** 811–820.

Joubin K. and Stern C.D. 1999. Molecular interactions continuously define the organizer during the cell movements of gastrulation. *Cell* **98:** 559–571.

Kadowaki T., Wilder E., Klingensmith J., Zachary K., and Perrimon N. 1996. The segment polarity gene porcupine encodes a putative multitransmembrane protein involved in Wingless processing. *Genes Dev.* **10:** 3116–3128.

Keller R. and Tibbetts P. 1989. Mediolateral cell intercalation in the dorsal, axial mesoderm of *Xenopus laevis. Dev. Biol.* **131:** 539–549.

Keller R.E., Danilchik M., Gimlich R., and Shih J. 1985. The function and mechanism of convergent extension during gastrulation of *Xenopus laevis. J. Embryol. Exp. Morphol.* (suppl.) **89:** 185–209.

Kelly G.M., Greenstein P., Erezyilmaz D.F., and Moon R.T. 1995. Zebrafish wnt8 and wnt8b share a common activity but are involved in distinct developmental pathways. *Development* **121:** 1787–1799.

Kennerdell J.R. and Carthew R.W. 1998. Use of dsRNA-mediated genetic interference to demonstrate that frizzled and frizzled 2 act in the wingless pathway. *Cell* **95:** 1017–1026.

Kiecker C. and Niehrs C. 2001. A morphogen gradient of Wnt/beta-catenin signalling regulates anteroposterior neural patterning in *Xenopus. Development* **128:** 4189–4201.

Kilian B., Mansukoski H., Barbosa F.C., Ulrich F., Tada M., and Heisenberg C.P. 2003. The role of Ppt/Wnt5 in regulating cell shape and movement during zebrafish gastrulation. *Mech. Dev.* **120:** 467–476.

Kim C.H., Oda T., Itoh M., Jiang D., Artinger K.B., Chandrasekharappa S.C., Driever W., and Chitnis A.B. 2000. Repressor activity of Headless/Tcf3 is essential for vertebrate head formation. *Nature* **407:** 913–916.

Kinoshita N., Iioka H., Miyakoshi A., and Ueno N. 2003. PKC delta is essential for Dishevelled function in a noncanonical Wnt pathway that regulates *Xenopus* convergent extension movements. *Genes Dev.* **17:** 1663–1676.

Klingensmith J., Nusse R., and Perrimon N. 1994. The *Drosophila* segment polarity gene dishevelled encodes a novel protein required for response to the wingless signal. *Genes Dev.* **8:** 118–130.

Kodjabachian L. and Lemaire P. 2001. Siamois functions in the early blastula to induce Spemann's organiser. *Mech. Dev.* **108:** 71–79.

Kofron M., Klein P., Zhang F., Houston D.W., Schaible K., Wylie C., and Heasman J. 2001. The role of maternal axin in patterning the *Xenopus* embryo. *Dev. Biol.* **237:** 183–201.

Kofron M., Demel T., Xanthos J., Lohr J., Sun B., Sive H., Osada S., Wright C., Wylie C., and Heasman J. 1999. Mesoderm induction in *Xenopus* is a zygotic event regulated by maternal VegT via TGFβ growth factors. *Development* **126:** 5759–5770.

Korinek V., Barker N., Willert K., Molenaar M., Roose J., Wagenaar G., Markman M., Lamers W., Destree O., and Clevers H. 1998. Two members of the Tcf family implicated in Wnt/β-catenin signaling during embryogenesis in the mouse. *Mol. Cell. Biol.* **18:** 1248–1256.

Kramps T., Peter O., Brunner E., Nellen D., Froesch B., Chatterjee S., Murone M., Zullig S., and Basler K. 2002. Wnt/wingless signaling requires BCL9/legless-mediated recruitment of pygopus to the nuclear β-catenin-TCF complex. *Cell* **109:** 47–60.

Krasnow R.E. and Adler P.N. 1994. A single frizzled protein has a dual function in tissue polarity. *Development* **120:** 1883–1893.

Kuhl M., Sheldahl L.C., Malbon C.C., and Moon R.T. 2000a. Ca2+/calmodulin-dependent protein kinase II is stimulated by Wnt and Frizzled homologs and promotes ventral cell fates in *Xenopus. J. Biol. Chem.* **275:** 12701–12711.

Kuhl M., Sheldahl L.C., Park M., Miller J.R., and Moon R.T. 2000b. The Wnt/Ca2+ pathway: A new vertebrate Wnt signaling pathway takes shape. *Trends Genet.* **16:** 279–283.

Labbe E., Letamendia A., and Attisano L. 2000. Association of Smads with lymphoid enhancer binding factor 1/T cell-specific factor mediates cooperative signaling by the transforming growth factor-β and wnt pathways. *Proc. Natl. Acad. Sci.* **97:** 8358–8363.

Lawrence P.A., Casal J., and Struhl G. 2002. Towards a model of the organisation of planar polarity and pattern in the *Drosophila* abdomen. *Development* **129:** 2749–2760.

Lekven A.C., Thorpe C.J., Waxman J.S., and Moon R.T. 2001. Zebrafish wnt8 encodes two wnt8 proteins on a bicistronic transcript and is required for mesoderm and neurectoderm patterning. *Dev. Cell* **1:** 103–114.

Leyns L., Bouwmeester T., Kim S.H., Piccolo S., and De Robertis E.M. 1997. Frzb-1 is a secreted antagonist of Wnt signaling expressed in the Spemann organizer. *Cell* **88:** 747–756.

Lin X. and Perrimon N. 2000. Role of heparan sulfate proteoglycans in cell-cell signaling in *Drosophila. Matrix Biol.* **19:** 303–307.

Liu C., Kato Y., Zhang Z., Do V.M., Yankner B.A., and He X. 1999. β-Trcp couples β-catenin phosphorylation-degradation and regulates *Xenopus* axis formation. *Proc. Natl. Acad. Sci.* **96:** 6273–6278.

Liu C., Li Y., Semenov M., Han C., Baeg G.H., Tan Y., Zhang Z., Lin X., and He X. 2002. Control of β-catenin phosphorylation/degradation by a dual-kinase mechanism. *Cell* **108:** 837–847.

Liu P., Wakamiya M., Shea M.J., Albrecht U., Behringer R.R., and Bradley A. 1999. Requirement for Wnt3 in vertebrate axis formation. *Nat. Genet.* **22:** 361–365.

Lloyd S., Fleming T.P., and Collins J.E. 2003. Expression of Wnt genes during mouse preimplantation development. *Gene Expr. Patterns* **3:** 309–312.

Mao B., Wu W., Li Y., Hoppe D., Stannek P., Glinka A., and Niehrs C. 2001. LDL-receptor-related protein 6 is a receptor for Dickkopf proteins. *Nature* **411:** 321–325.

Mao J., Wang J., Liu B., Pan W., Farr G.H. III, Flynn C., Yuan H., Takada S., Kimelman D., Li L. et al. 2001. Low-density lipoprotein receptor-related protein-5 binds to Axin and regulates the canonical Wnt signaling pathway. *Mol. Cell* **7:** 801–809.

Marlow F., Topczewski J., Sepich D., and Solnica-Krezel L. 2002. Zebrafish Rho kinase 2 acts downstream of Wnt11 to mediate cell polarity and effective convergence and extension movements. *Curr. Biol.* **12:** 876–884.

McGrew L.L., Hoppler S., and Moon R.T. 1997. Wnt and FGF pathways cooperatively pattern anteroposterior neural ectoderm in *Xenopus. Mech. Dev.* **69:** 105–114.

McMahon A.P. and Bradley A. 1990. The Wnt-1 (int-1) proto-oncogene is required for development of a large region of the mouse brain. *Cell* **62:** 1073–1085.

McMahon A.P. and Moon R.T. 1989. Ectopic expression of the proto-oncogene int-1 in *Xenopus* embryos leads to duplication of the embryonic axis. *Cell* **58:** 1075–1084.

Meneghini M.D., Ishitani T., Carter J.C., Hisamoto N., Ninomiya-Tsuji J., Thorpe C.J., Hamill D.R., Matsumoto K., and Bowerman B. 1999. MAP kinase and Wnt pathways converge to downregulate an HMG-domain repressor in *Caenorhabditis elegans. Nature* **399:** 793–797.

Merrill B.J., Pasolli H.A., Polak L., Rendl M., Garcia-Garcia M.J., Anderson K.V., and Fuchs E. 2004. Tcf3: A transcriptional regulator of axis induction in the early embryo. *Development* **131:** 263–274.

Miller J.R., Rowning B.A., Larabell C.A., Yang-Snyder J.A., Bates R.L., and Moon R.T. 1999. Establishment of the dorsal-ventral axis in *Xenopus* embryos coincides with the dorsal enrichment of dishevelled that is dependent on cortical rotation. *J. Cell Biol.* **146:** 427–437.

Miralles F., Posern G., Zaromytidou A.I., and Treisman R. 2003. Actin dynamics control SRF activity by regulation of its coactivator MAL. *Cell* **113:** 329–342.

Molenaar M., van de Wetering M., Oosterwegel M., Peterson-Maduro J., Godsave S., Korinek V., Roose J., Destree O., and Clevers H. 1996. XTcf-3 transcription factor mediates β-catenin-induced axis formation in *Xenopus* embryos. *Cell* **86:** 391–399.

Moon R.T., Campbell R.M., Christian J.L., McGrew L.L., Shih J., and Fraser S. 1993. Xwnt-5A: A maternal Wnt that affects morphogenetic movements after overexpression in embryos of *Xenopus laevis. Development* **119:** 97–111.

Morkel M., Huelsken J., Wakamiya M., Ding J., Van De Wetering M., Clevers H., Taketo M.M., Behringer R.R., Shen M.M., and Birchmeier W. 2003. β-Catenin regulates Cripto- and Wnt3-dependent gene expression programs in mouse axis and mesoderm formation. *Development* **130:** 6283–6294.

Mukhopadhyay M., Shtrom S., Rodriguez-Esteban C., Chen L., Tsukui T., Gomer L., Dorward D.W., Glinka A., Grinberg A., Huang S.P., Niehrs C., Belmonte J.C., and Westphal H. 2001. *Dickkopf1* is required for embryonic head induction and limb morphogenesis in the mouse. *Dev. Cell* **1:** 423–434.

Muller H., Samanta R., and Wieschaus E. 1999. Wingless signaling in the *Drosophila* embryo: Zygotic requirements and the role of the frizzled genes. *Development* **126:** 577–586.

Nishita M., Hashimoto M.K., Ogata S., Laurent M.N., Ueno N., Shibuya H., and Cho K.W. 2000. Interaction between Wnt and TGF-β signalling pathways during formation of Spemann's organizer. *Nature* **403:** 781–785.

Noordermeer J., Klingensmith J., Perrimon N., and Nusse R. 1994. *dishevelled* and *armadillo* act in the wingless signalling pathway in *Drosophila. Nature* **367:** 80–83.

Nordstrom U., Jessell T.M., and Edlund T. 2002. Progressive induction of caudal neural character by graded Wnt signaling. *Nat. Neurosci.* **5:** 525–532.

Nusse R. 2001. An ancient cluster of Wnt paralogues. *Trends Genet.* **17:** 443.

Nusse R. and Varmus H.E. 1982. Many tumors induced by the mouse mammary tumor virus contain a provirus integrated in the same region of the host genome. *Cell* **31:** 99–109.

Pandur P. and Kuhl M. 2001. An arrow for wingless to take-off. *Bioessays* **23:** 207–210.

Papkoff J., Brown A.M., and Varmus H.E. 1987. The int-1 proto-oncogene products are glycoproteins that appear to enter the secretory pathway. *Mol. Cell. Biol.* **7:** 3978–3984.

Papkoff J., Rubinfeld B., Schryver B., and Polakis P. 1996. Wnt-1 regulates free pools of catenins and stabilizes APC-catenin complexes. *Mol. Cell. Biol.* **16:** 2128–2134.

Parker D.S., Jemison J., and Cadigan K.M. 2002. Pygopus, a nuclear

PHD-finger protein required for Wingless signaling in *Drosophila. Development* **129:** 2565–2576.

Pelegri F. and Maischein H.M. 1998. Function of zebrafish β-catenin and TCF-3 in dorsoventral patterning. *Mech. Dev.* **77:** 63–74.

Piccolo S., Agius E., Leyns L., Bhattacharyya S., Grunz H., Bouwmeester T., and De Robertis E.M. 1999. The head inducer Cerberus is a multifunctional antagonist of Nodal, BMP and Wnt signals. *Nature* **397:** 707–710.

Pinson K.I., Brennan J., Monkley S., Avery B.J., and Skarnes W.C. 2000. An LDL-receptor-related protein mediates Wnt signalling in mice. *Nature* **407:** 535–538.

Polakis P. 2000. Wnt signaling and cancer. *Genes Dev.* **14:** 1837–1851.

Prud'homme B., Lartillot N., Balavoine G., Adoutte A., and Vervoort M. 2002. Phylogenetic analysis of the *Wnt* gene family. Insights from lophotrochozoan members. *Curr. Biol.* **12:** 1395–1400.

Rijsewijk F., Schuermann M., Wagenaar E., Parren P., Weigel D., and Nusse R. 1987. The *Drosophila* homolog of the mouse mammary oncogene int-1 is identical to the segment polarity gene wingless. *Cell* **50:** 649–657.

Roel G., van den Broek O., Spieker N., Peterson-Maduro J., and Destree O. 2003. Tcf-1 expression during *Xenopus* development. *Gene Expr. Patterns* **3:** 123–126.

Rothbacher U., Laurent M.N., Deardorff M.A., Klein P.S., Cho K.W., and Fraser S.E. 2000. Dishevelled phosphorylation, subcellular localization and multimerization regulate its role in early embryogenesis. *EMBO J.* **19:** 1010–1022.

Saneyoshi T., Kume S., Amasaki Y., and Mikoshiba K. 2002. The Wnt/calcium pathway activates NF-AT and promotes ventral cell fate in *Xenopus* embryos. *Nature* **417:** 295–299.

Schneider S., Steinbeisser H., Warga R.M., and Hausen P. 1996. β-catenin translocation into nuclei demarcates the dorsalizing centers in frog and fish embryos. *Mech. Dev.* **57:** 191–198.

Schohl A. and Fagotto F. 2003. A role for maternal beta-catenin in early mesoderm induction in *Xenopus. EMBO J.* **22:** 3303–3313.

Schroeder K.E., Condic M.L., Eisenberg L.M., and Yost H.J. 1999. Spatially regulated translation in embryos: asymmetric expression of maternal Wnt-11 along the dorsal-ventral axis in *Xenopus. Dev. Biol.* **214:** 288–297.

Schubert M., Holland L.Z., Holland N.D., and Jacobs D.K. 2000. A phylogenetic tree of the Wnt genes based on all available full-length sequences, including five from the cephalochordate amphioxus. *Mol. Biol. Evol.* **17:** 1896–1903.

Semenov M.V., Tamai K., Brott B.K., Kuhl M., Sokol S., and He X. 2001. Head inducer Dickkopf-1 is a ligand for Wnt coreceptor LRP6. *Curr. Biol.* **11:** 951–961.

Shackleford G.M. and Varmus H.E. 1987. Expression of the proto-oncogene int-1 is restricted to postmeiotic male germ cells and the neural tube of mid-gestational embryos. *Cell* **50:** 89–95.

Sheldahl L.C., Park M., Malbon C.C., and Moon R.T. 1999. Protein kinase C is differentially stimulated by Wnt and Frizzled homologs in a G-protein-dependent manner. *Curr. Biol.* **9:** 695–698.

Sheldahl L.C., Slusarski D.C., Pandur P., Miller J.R., Kuhl M., and Moon R.T. 2003. Dishevelled activates Ca2+ flux, PKC, and CamKII in vertebrate embryos. *J. Cell Biol.* **161:** 769–777.

Shi D.L., Bourdelas A., Umbhauer M., and Boucaut J.C. 2002. Zygotic Wnt/β-catenin signaling preferentially regulates the expression of *Myf5* gene in the mesoderm of *Xenopus. Dev. Biol.* **245:** 124–135.

Shih J. and Keller R. 1992. Cell motility driving mediolateral intercalation in explants of *Xenopus laevis. Development* **116:** 901–914.

Sidow A. 1992. Diversification of the Wnt gene family on the ancestral lineage of vertebrates. *Proc. Natl. Acad. Sci.* **89:** 5098–5102.

Siegfried E., Chou T.B., and Perrimon N. 1992. wingless signaling acts through zeste-white 3, the *Drosophila* homolog of glycogen synthase kinase-3, to regulate engrailed and establish cell fate. *Cell* **71:** 1167–1179.

Siegfried E., Wilder E.L., and Perrimon N. 1994. Components of wingless signalling in *Drosophila. Nature* **367:** 76–80.

Skromne I. and Stern C.D. 2001. Interactions between Wnt and Vg1 signalling pathways initiate primitive streak formation in the chick embryo. *Development* **128:** 2915–2927.

Slusarski D.C., Corces V.G., and Moon R.T. 1997. Interaction of Wnt and a Frizzled homologue triggers G-protein-linked phosphatidylinositol signalling. *Nature* **390:** 410–413.

Sokol S.Y. 1996. Analysis of Dishevelled signalling pathways during *Xenopus* development. *Curr. Biol.* **6:** 1456–1467.

Strapps W.R. and Tomlinson A. 2001. Transducing properties of *Drosophila* Frizzled proteins. *Development* **128:** 4829–4835.

Strutt D. 2001. Asymmetric localization of frizzled and the establishment of cell polarity in the *Drosophila* wing. *Mol. Cell* **7:** 367–375.

———. 2003. Frizzled signalling and cell polarisation in *Drosophila* and vertebrates. *Development* **130:** 4501–4513.

Sumanas S. and Ekker S.C. 2001. *Xenopus* frizzled-7 morphant displays defects in dorsoventral patterning and convergent extension movements during gastrulation. *Genesis* **30:** 119–122.

Sumanas S., Kim H.J., Hermanson S., and Ekker S.C. 2001. Zebrafish frizzled-2 morphant displays defects in body axis elongation. *Genesis* **30:** 114–118.

Sumanas S., Strege P., Heasman J., and Ekker S.C. 2000. The putative wnt receptor *Xenopus* frizzled-7 functions upstream of β-catenin in vertebrate dorsoventral mesoderm patterning. *Development* **127:** 1981–1990.

Sumoy L., Kiefer J., and Kimelman D. 1999. Conservation of intracellular Wnt signaling components in dorsal-ventral axis formation in zebrafish. *Dev. Genes Evol.* **209:** 48–58.

Tada M. and Smith J.C. 2000. *Xwnt11* is a target of *Xenopus* Brachyury: Regulation of gastrulation movements via Dishevelled, but not through the canonical Wnt pathway. *Development* **127:** 2227–2238.

Tahinci E. and Symes K. 2003. Distinct functions of Rho and Rac are required for convergent extension during *Xenopus* gastrulation. *Dev. Biol.* **259:** 318–335.

Takada S., Stark K.L., Shea M.J., Vassileva G., McMahon J.A., and McMahon A.P. 1994. Wnt-3a regulates somite and tailbud formation in the mouse embryo. *Genes Dev.* **8:** 174–189.

Takeuchi M., Nakabayashi J., Sakaguchi T., Yamamoto T.S., Takahashi H., Takeda H., and Ueno N. 2003. The prickle-related gene in vertebrates is essential for gastrulation cell movements. *Curr. Biol.* **13:** 674–679.

Tamai K., Semenov M., Kato Y., Spokony R., Liu C., Katsuyama Y., Hess F., Saint-Jeannet J.P., and He X. 2000. LDL-receptor-related proteins in Wnt signal transduction. *Nature* **407:** 530–535.

Tanaka K., Okabayashi K., Asashima M., Perrimon N., and Kadowaki T. 2000. The evolutionarily conserved porcupine gene family is involved in the processing of the Wnt family. *Eur. J. Biochem.* **267:** 4300–4311.

Thomas K.R. and Capecchi M.R. 1990. Targeted disruption of the murine int-1 proto-oncogene resulting in severe abnormalities in midbrain and cerebellar development. *Nature* **346:** 847–850.

Thomas K.R., Musci T.S., Neumann P.E., and Capecchi M.R. 1991.

Swaying is a mutant allele of the proto-oncogene Wnt-1. *Cell* **67:** 969–976.

Tolwinski N.S. and Wieschaus E. 2001. Armadillo nuclear import is regulated by cytoplasmic anchor Axin and nuclear anchor dTCF/Pan. *Development* **128:** 2107–2117.

Tolwinski N.S., Wehrli M., Rives A., Erdeniz N., DiNardo S., and Wieschaus E. 2003. Wg/Wnt signal can be transmitted through arrow/LRP5,6 and Axin independently of Zw3/Gsk3β activity. *Dev. Cell* **4:** 407–418.

Topczewski J., Sepich D.S., Myers D.C., Walker C., Amores A., Lele Z., Hammerschmidt M., Postlethwait J., and Solnica-Krezel L. 2001. The zebrafish glypican knypek controls cell polarity during gastrulation movements of convergent extension. *Dev. Cell* **1:** 251–264.

Torres M.A., Yang-Snyder J.A., Purcell S.M., DeMarais A.A., McGrew L.L., and Moon R.T. 1996. Activities of the Wnt-1 class of secreted signaling factors are antagonized by the Wnt-5A class and by a dominant negative cadherin in early *Xenopus* development. *J. Cell Biol.* **133:** 1123–1137.

Ulrich F., Concha M.L., Heid P.J., Voss E., Witzel S., Roehl H., Tada M., Wilson S.W., Adams R.J., Soll D.R., and Heisen C.-P. 2003. Slb/Wnt11 controls hypoblast cell migration and morphogenesis at the onset of zebrafish gastrulation. *Development* **130:** 5375–5384.

Ungar A.R., Kelly G.M., and Moon R.T. 1995. Wnt4 affects morphogenesis when misexpressed in the zebrafish embryo. *Mech. Dev.* **52:** 153–164.

van den Heuvel M., Harryman-Samos C., Klingensmith J., Perrimon N., and Nusse R. 1993. Mutations in the segment polarity genes wingless and porcupine impair secretion of the wingless protein. *EMBO J.* **12:** 5293–5302.

Veeman M.T., Axelrod J.D., and Moon R.T. 2003a. A second canon. Functions and mechanisms of β-catenin-independent Wnt signaling. *Dev. Cell* **5:** 367–377.

Veeman M.T., Slusarski D.C., Kaykas A., Louie S.H., and Moon R.T. 2003b. Zebrafish prickle, a modulator of noncanonical Wnt/Fz signaling, regulates gastrulation movements. *Curr. Biol.* **13:** 680–685.

Vincent S.D., Dunn N.R., Hayashi S., Norris D.P., and Robertson E.J. 2003. Cell fate decisions within the mouse organizer are governed by graded Nodal signals. *Genes Dev.* **17:** 1646–1662.

Vinson C.R. and Adler P.N. 1987. Directional non-cell autonomy and the transmission of polarity information by the frizzled gene of *Drosophila*. *Nature* **329:** 549–551.

Vonica A. and Gumbiner B.M. 2002. Zygotic Wnt activity is required for Brachyury expression in the early *Xenopus laevis* embryo. *Dev. Biol.* **250:** 112–127.

Wacker S., Grimm K., Joos T., and Winklbauer R. 2000. Development and control of tissue separation at gastrulation in *Xenopus*. *Dev. Biol.* **224:** 428–439.

Wallingford J.B. and Harland R.M. 2001. *Xenopus* Dishevelled signaling regulates both neural and mesodermal convergent extension: parallel forces elongating the body axis. *Development* **128:** 2581–2592.

Wallingford J.B., Vogeli K.M., and Harland R.M. 2001a. Regulation of convergent extension in *Xenopus* by Wnt5a and Frizzled-8 is independent of the canonical Wnt pathway. *Int. J. Dev. Biol.* **45:** 225–227.

Wallingford J.B., Ewald A.J., Harland R.M., and Fraser S.E. 2001b. Calcium signaling during convergent extension in *Xenopus*. *Curr. Biol.* **11:** 652–661.

Wallingford J.B., Rowning B.A., Vogeli K.M., Rothbacher U., Fraser S.E., and Harland R.M. 2000. Dishevelled controls cell polarity during *Xenopus* gastrulation. *Nature* **405:** 81–85.

Wang S., Krinks M., Lin K., Luyten F.P., and Moos M., Jr. 1997. Frzb, a secreted protein expressed in the Spemann organizer, binds and inhibits Wnt-8. *Cell* **88:** 757–766.

Weaver C., Farr G.H., III, Pan W., Rowning B.A., Wang J., Mao J., Wu D., Li L., Larabell C.A., and Kimelman D. 2003. GBP binds kinesin light chain and translocates during cortical rotation in *Xenopus* eggs. *Development* **130:** 5425–5436.

Wehrli M., Dougan S.T., Caldwell K., O'Keefe L., Schwartz S., Vaizel-Ohayon D., Schejter E., Tomlinson A., and DiNardo S. 2000. arrow encodes an LDL-receptor-related protein essential for Wingless signalling. *Nature* **407:** 527–530.

Westfall T.A., Brimeyer R., Twedt J., Gladon J., Olberding A., Furutani-Seiki M., and Slusarski D.C. 2003. Wnt-5/*pipetail* functions in vertebrate axis formation as a negative regulator of Wnt/β-catenin activity. *J. Cell Biol.* **162:** 889–898.

Wharton K.A., Jr. 2003. Runnin' with the Dvl: Proteins that associate with Dsh/Dvl and their significance to Wnt signal transduction. *Dev. Biol.* **253:** 1–17.

Wilkinson D.G., Bailes J.A., and McMahon A.P. 1987. Expression of the proto-oncogene int-1 is restricted to specific neural cells in the developing mouse embryo. *Cell* **50:** 79–88.

Willert K., Brown J.D., Danenberg E., Duncan A.W., Weissman I.L., Reya T., Yates J.R. III, and Nusse R. 2003. Wnt proteins are lipid-modified and can act as stem cell growth factors. *Nature* **423:** 448–452.

Wilson S.I., Rydstrom A., Trimborn T., Willert K., Nusse R., Jessell T.M., and Edlund T. 2001. The status of Wnt signalling regulates neural and epidermal fates in the chick embryo. *Nature* **411:** 325–330.

Winklbauer R., Medina A., Swain R.K., and Steinbeisser H. 2001. Frizzled-7 signalling controls tissue separation during *Xenopus* gastrulation. *Nature* **413:** 856–860.

Wodarz A. and Nusse R. 1998. Mechanisms of Wnt signaling in development. *Annu. Rev. Cell Dev. Biol.* **14:** 59–88.

Wong H.C., Bourdelas A., Krauss A., Lee H.J., Shao Y., Wu D., Mlodzik M., Shi D.L., and Zheng J. 2003. Direct binding of the PDZ domain of Dishevelled to a conserved internal sequence in the C-terminal region of Frizzled. *Mol. Cell* **12:** 1251–1260.

Xanthos J.B., Kofron M., Tao Q., Schaible K., Wylie C., and Heasman J. 2002. The roles of three signaling pathways in the formation and function of the Spemann Organizer. *Development* **129:** 4027–4043.

Yamaguchi T.P., Bradley A., McMahon A.P., and Jones S. 1999a. A *Wnt5a* pathway underlies outgrowth of multiple structures in the vertebrate embryo. *Development* **126:** 1211–1223.

Yamaguchi T.P., Takada S., Yoshikawa Y., Wu N., and McMahon A.P. 1999b. T (Brachyury) is a direct target of Wnt3a during paraxial mesoderm specification. *Genes Dev.* **13:** 3185–3190.

Yamanaka H., Moriguchi T., Masuyama N., Kusakabe M., Hanafusa H., Takada R., Takada S., and Nishida E. 2002. JNK functions in the non-canonical Wnt pathway to regulate convergent extension movements in vertebrates. *EMBO Rep.* **3:** 69–75.

Yokota C., Kofron M., Zuck M., Houston D.W., Isaacs H., Asashima M., Wylie C.C., and Heasman J. 2003. A novel role for a nodal-related protein; Xnr3 regulates convergent extension movements via the FGF receptor. *Development* **130:** 2199–2212.

Yoshikawa S., McKinnon R.D., Kokel M., and Thomas J.B. 2003. Wnt-mediated axon guidance via the *Drosophila* Derailed receptor. *Nature* **422:** 583–588.

Yoshikawa Y., Fujimori T., McMahon A.P., and Takada S. 1997. Evidence that absence of Wnt-3a signaling promotes neuralization instead of paraxial mesoderm development in the mouse. *Dev. Biol.* **183:** 234–242.

Yost C., Farr G.H. III, Pierce S.B., Ferkey D.M., Chen M.M., and Kimelman D. 1998. GBP, an inhibitor of GSK-3, is implicated in *Xenopus* development and oncogenesis. *Cell* **93:** 1031–1041.

Zecca M., Basler K., and Struhl G. 1996. Direct and long-range action of a wingless morphogen gradient. *Cell* **87:** 833–844.

Zeng L., Fagotto F., Zhang T., Hsu W., Vasicek T.J., Perry W.L. III, Lee J.J., Tilghman S.M., Gumbiner B.M., and Costantini F. 1997. The mouse Fused locus encodes Axin, an inhibitor of the Wnt signaling pathway that regulates embryonic axis formation. *Cell* **90:** 181–192.

Zhang J., Houston D.W., King M.L., Payne C., Wylie C., and Heasman J. 1998. The role of maternal VegT in establishing the primary germ layers in *Xenopus* embryos. *Cell* **94:** 515–524.

Zorn A.M., Butler K., and Gurdon J.B. 1999. Anterior endomesoderm specification in *Xenopus* by Wnt/beta-catenin and TGF-beta signalling pathways. *Dev. Biol.* **209:** 282–297.

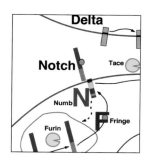

ROLES OF NOTCH DURING GASTRULATION

D.R. McClay, R. Peterson, and R. Range

Department of Biology, Duke University, Durham, North Carolina 27708

INTRODUCTION

Pioneering studies on *Drosophila* and *Caenorhabditis elegans* established the role of Notch in an important signaling pathway in development. Notch operates in every organism studied, and at many stages of development. A number of reviews address these activities (Artavanis-Tsakonas et al. 1991, 1999; Kimble and Simpson 1997; Kopan 2002; Pourquie 2003). Although Notch/Glp-1/LIN-12 and their ligands Delta, Serrate, LAG-2 (DSL) (and others) have been known for many years, recent information has clarified how the ligands are produced and processed during the signal presentation, how Notch is processed prior to signal reception, and how Notch responds to the signal. Furthermore, recent information on the entire Notch pathway provides comparative insights into involvement of the pathway in an evolutionary context. The purpose of this review is to examine Notch in gastrulation, wherever it might occur in the animal kingdom. Most studies on the function of Notch have concentrated on stages other than gastrulation, but evidence is now accumulating to suggest important roles in this process.

Generally, embryos use Notch for cell fate decisions and for defining and reinforcing boundaries. Cell fate decisions attributable to Notch were first noted in the proneural cluster of cells in the *Drosophila* epithelium (Heitzler and Simpson 1991). In *Drosophila,* epithelial cells initially form an equivalence group with roughly equal amounts of Delta and Notch in each cell. Notch–Delta signaling somehow switches a subset of cells to a neural fate. Some cells downregulate Notch and become neural while the surrounding cells express high levels of Notch and remain epithelial.

Mutants in *notch* or the Notch pathway that reduce Notch function lead to an excess of neural cells. In *C. elegans*, a different sort of cell fate decision uses Notch signaling to keep germ-line precursors mitotic (Austin and Kimble 1987). The distal tip cell of the somatic gonad expresses LAG-2 (Henderson et al. 1994), a protein containing the same DSL motif as Delta and Serrate, both Notch ligands. On the receiving end of that signal, germ-line precursor cells of the gonad express GLP-1, a receptor protein containing the signature motifs of Notch (Austin and Kimble 1989; Yochem and Greenwald 1989). GLP-1 activation leads to expression of target genes, through LAG-1, a protein in the Suppressor of Hairless (Su(H)) family of proteins (Christensen et al. 1996). In this signal transduction, the function of the pathway is to keep the germ-line precursor cells mitotic. As the number of germ-line precursor cells increases, some cells lose contact with the distal tip cell. Those cells no longer receive the LAG-2 signal, and as a result they become meiotic. Loss-of-function mutants in the GLP-1 pathway are sterile because of a reduction in the number of germ-line precursor cells.

The classic example of Notch–Delta signaling at a boundary is in the dorsoventral boundary of the *Drosophila* wing. In the wing imaginal disc, the future margin of the adult wing is a line of cells across the middle of the disc. That line of cells establishes the dorsoventral boundary well before pupation and expansion of the disc to form the wing at the end of pupation. At the dorsoventral boundary, Delta, present throughout the ventral compartment, signals to Notch in dorsal boundary cells. Serrate, another DSL protein located throughout the dorsal compartment, signals to Notch at the ventral compartment boundary to set up that side of the boundary (Diaz-Benjumea and Cohen 1995; Fleming et al. 1997). Notch is found throughout the wing but signals only at the dorsoventral boundary due to the presence and function of Fringe, an enzyme that glycosylates Notch. Delta is recognized as a ligand by Notch if Fringe glycosylates Notch. Serrate is recognized as a Notch ligand if Fringe is absent from the Notch-expressing cell. Fringe is present in the entire dorsal compartment along with Serrate and Notch (Blair 1997; Fleming et al. 1997). Glycosylation of Notch by Fringe in the dorsal compartment prevents Notch from using Serrate in that compartment, so the only dorsal cells able to use the Notch pathway are at the boundary where dorsal Fringe-modified Notch-bearing cells receive Delta from ventral cells. In the ventral compartment, Fringe is absent so cells with Notch cannot receive the Delta signal there. Only at the ventral boundary can the Fringe-minus-Notch-bearing cells receive the Serrate ligand from the dorsal compartment (Fleming et al. 1997). The assumption is that in both cases downstream target genes are activated which reinforce the dorsal or ventral boundary. Thus, activation of Notch along a defined border can produce a developmental boundary.

Genetic screens in *Drosophila, C. elegans,* fish, and mouse added a number of genes to the Notch signaling system and provided many details that illuminate the function and mechanisms of action of the pathway. These data show that the pathway is highly conserved with all Notch ligands bearing the DSL domain, and all Notch receptors bearing EGF repeats, enzyme-processing sites, and DNA-binding function. In addition, the pathway collaborates with Su(H) homologs to activate a family of hairy-related bHLH transcription factors in most, if not all, systems examined. The following summarizes the components of the Notch pathway and how they function.

THE NOTCH SIGNAL TRANSDUCTION PATHWAY: HOW DOES IT WORK?

The LNG family of receptors (LIN-12/Notch/GLP-1) interact with transmembrane DSL ligands to receive a short-range cell–cell signal and initiate the transduction of that signal. The Notch receptor is the central molecule in the pathway, and its sequential processing is important for the sequence of signaling events. Notch is a 300-kD protein that contains 34–36 extracellular EGF repeats (a smaller number in *C. elegans glp1* and *lin12* genes [Austin and Kimble 1989; Yochem and Greenwald 1989]), and 3 lin12 repeats near the transmembrane region (Artavanis-Tsakonas et al. 1983; Wharton et al. 1985). The intracellular domain of Notch contains 6 tandem ankyrin repeats and a pest domain near the carboxyl terminus. The protein is processed into two fragments before the mature protein

appears at the membrane (Fig. 1). In the trans-Golgi, Notch is cleaved by the protease Furin about 69 amino acids outside the membrane at a site called the S1 site (Logeat et al. 1998). Although apparently not essential for Notch function (Bush et al. 2001; Kidd and Lieber 2002), this cleavage is followed by a heterodimerization that keeps the extracellular peptide in contact with the intracellular peptide as the mature receptor is inserted into the plasma membrane.

The pathway appears quite simple: Notch is a receptor that becomes a transcription factor when the pathway is activated. Closer examination, however, reveals a number of proteins that contribute to regulation of the Notch pathway, and many of them, if differentially regulated, influence the functionality of the entire pathway. To begin the signaling process, a Delta, Serrate, Jagged, or another ligand binds

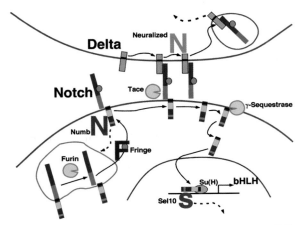

Figure 1. Model of the Notch pathway. In the Notch-bearing cell (*lower cell*), Notch is processed in several ways prior to insertion of the molecule onto the cell surface. The Furin protease cuts Notch, and the two fragments associate with one another in intracellular compartments. At a late stage of processing, prior to insertion into the membrane, the Fringe glycosyltransferase, if present, glycosylates Notch, which allows Notch to recognize and bind the Delta ligand. The product of the *numb* gene, if present, may increase the rate of endocytosis of Notch by binding to an α-adaptin, and thereby negatively regulating Notch availability. When Notch binds the Delta ligand presented by another cell, a TACE protease (or Kuzbanian) cuts Notch 69 amino acids from the transmembrane sequence. This somehow potentiates a third Notch cleavage by γ-sequestrase, which releases the Notch intracellular domain (NICD). The free NICD goes to the nucleus, where it forms a complex with Suppressor of Hairless (Su(H)) to convert the complex from a repressor to an activator. In most Notch signaling systems, the target of this gene activation is the synthesis of a bHLH transcription factor. To terminate the Notch-directed gene activation, the product of *sel10* helps regulate polyubiquitination of the NICD, which is then destroyed by nuclear proteasomes. In the Delta-presenting cell, Delta presentation is modified by the product of the *neuralized* gene, a ubiquitin 3 ligase, that is thought to potentiate the TACE proteolytic activity in Notch activation, and the product of this activity leads first to Delta endocytosis, which in at least some cases, includes the trans-endocytosis of the Notch extracellular domain.

to the extracellular region of Notch (see below). Ligand recognition by Notch depends on the presence or absence of a glycosylation step that adds an N-acetyl glucosamine to an O-linked fucose on the Notch extracellular domain, using the glycosyl transferase protein Fringe (Bruckner et al. 2000; Moloney et al. 2000). This posttranslational modification greatly augments the ability of Notch to receive the Delta signal, and it tends to inhibit the ability of Notch to be activated by Serrate (Fleming et al. 1997; Klein and Arias 1998).

Ligand binding somehow exposes a second proteolytic cleavage site, the S2 site, in the exposed 69 amino acids of Notch near the plasma membrane. The S2 site is cleaved either by a mammalian TACE protease (Brou et al. 2000), or by Kuzbanian, an ADAMs family metalloprotease (Pan and Rubin 1997; Lieber et al. 2002). The cleavage of the S2 site enables a third protease, operating within the hydrophobic region of the membrane, to free the Notch intracellular domain (NICD) by cutting the Notch protein at the S3 site located in the transmembrane region of the protein (De Strooper et al. 1999; Mumm and Kopan 2000). The intramembranous S3 Notch site is cut by the same γ-secretase activity that is associated with the Presenilin protein and Alzheimer's disease (De Strooper et al. 1999; Mumm and Kopan 2000). Once freed of the membrane, the NICD moves to the nucleus where it interacts with the Su(H) protein in a complex and converts Su(H) from a repressor to an activator. Among the conserved known targets of Notch function is the activation of the *hairy/enhancer of split* genes of the bHLH family (Jarriault et al. 1995). Following its transduction activity, the NICD is eliminated from the nucleus by a phosphorylation step that triggers an association with the SEL10 protein, which, in turn, increases the rate of polyubiquitination and subsequent degradation of the NICD by nuclear proteasomes (Gupta-Rossi et al. 2001; Oberg et al. 2001).

Ligands that activate Notch share the DSL domain. All DSL proteins are transmembrane proteins that *trans*-activate Notch on the adjacent cell. The Delta ligand binds to Notch (Fehon et al. 1990), at a specific EGF repeat (usually 11–12) (Rebay et al. 1991), to activate the Notch pathway. Binding of Delta triggers endocytosis and degradation of Delta through the protein Neuralized, an E3 ubiquitin ligase (Boulianne et al. 1991; Deblandre et al. 2001; Lai et al. 2001). In the *Drosophila* proneural signal system, Neuralized is segregated to one of two cells involved in an asymmetric cell division. As a consequence, the Neuralized-containing cell ubiquitinates Delta, and in the process enhances the endocytosis of Delta, and perhaps surprisingly, potentiates the activation of Notch (Le Borgne and Schweisguth 2003). It is suggested that the released Notch fragment, still bound to Delta, is trans-endocytosed by the Delta-bearing cell.

Binding of Numb to the intracellular domain of Notch further regulates the pathway (Uemura et al. 1989 ; Rhyu et al. 1994). Recent data suggest that Numb, which binds to the Notch intracellular domain and to an α-adaptin, allows Numb to increase the endocytosis of Notch (Berdnik et al. 2002). Numb binding to Notch is then modeled to speed Notch endocytosis, even in the absence of ligand reception, thereby negatively regulating Notch.

Additional mechanisms and signal transduction systems also affect Notch function. Most notably, Dishevelled and the Wnt pathway intersect with Notch in many systems to modulate the Notch signal (Couso and Martinez Arias 1994; De Strooper and Annaert 2001; Aulehla et al. 2003). The intersection with other signaling systems adds to the complexity of Notch pathway regulation and often makes it difficult to know how, precisely, the Notch pathway fits into the gene regulatory network operating in a cell at a given time. Nevertheless, much has been learned about Notch, since it usually is near the top of the regulatory sequence in the progression toward differentiation. Other patterning mechanisms tend to appear downstream of the Notch pathway. As an example, Hox gene expression, important for anterior–posterior patterning, is shown to be downstream of Notch in segmentation (Beck and Slack 1999; Dubrulle and Pourquie 2002).

NOTCH FUNCTION IN GASTRULATION

For gastrulation to occur, the mesoderm, endoderm, and ectoderm must be specified and the appropriate morphogenetic mechanisms put into place. In *Drosophila*, where Notch was first found, it was originally proposed that Notch signaling occurs in many locations where cell fate decisions take place (Kooh et al. 1993). Although there is a tendency for Notch signaling to keep cells in an undifferentiated state if they express high levels of Notch, as is the case in the proneural signaling system, Notch is also used in other ways, such as in segmentation of vertebrate somites and for cell fate decisions. The following section examines the Notch pathway in gastrulation where, especially in the deuterostome line, it participates in early specification and morphogenesis. The degree to which this pathway is conserved during gastrulation among the deuterostomes is still uncertain, but data so far appear to support the concept that all deuterostomes make use of this pathway in gastrulation.

NOTCH AND SEA URCHIN GASTRULATION

Notch signaling is essential for specification and for morphogenesis in sea urchin development (Fig. 2) (Sherwood and McClay 1999). The pathway exhibits a complicated pattern of expression and regulation during cleavage stages

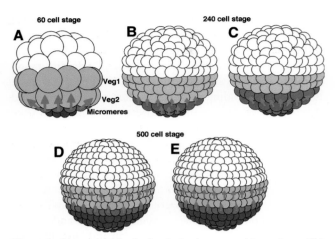

Figure 2. Sequential Notch signals in the sea urchin embryo. This model depicts the sequential Notch signaling that occurs during endomesoderm specification of the sea urchin embryo. (*A*) Beginning just after the 6th cleavage, the Delta ligand is presented by the micromeres (*red*) to Veg2 cells just above. The Veg1 and Veg2 cells are the progeny of the macromeres that begin to be specified as endomesoderm beginning about two cleavages earlier. (*B*) As a result of the first Delta–Notch signal, by about two cleavages later, the next tier of cells is switched to a mesoderm fate (*dark red*), and these cells become pigment cells. (*C*) As a consequence of their initial specification, and reception of the first Delta signal, the lower Veg2 tier (*dark red*) begins to synthesize Delta, which then signals to the next tier during this and subsequent cleavages. As a consequence, another tier of cells is diverted from endomesoderm to become muscle and blastocoelar cells. (*D*) By one or two cleavages later, evidence indicates that Delta then signals from the nascent mesoderm compartment into the remaining endomesodermal compartment as it becomes specified further as endoderm. Although several kinds of evidence indicate sequential Notch signaling in the endoderm (*yellow* in *E*), this later phase of Notch signaling is not well understood (based on Sherwood and McClay 1999, 2001; McClay et al. 2000; Davidson et al. 2002; Sweet et al. 2002; Oliveri et al. 2003; R.E. Peterson and D.R. McClay, unpubl.).

and during gastrulation (Sherwood and McClay 1997, 1999, 2001). In the sea urchin, a vegetal-to-animal signaling system activates a gene regulatory network that specifies the endomesoderm (Davidson et al. 2002). Several important specification events occur prior to activation of the Notch pathway in this system that are necessary for its functionality. The Notch receptor and Fringe are present maternally on all blastomeres early in cleavage, but the pathway does not function until 7th cleavage, coincident with the appearance of Delta on micromeres (Sherwood and McClay 1997; McClay et al. 2000; Sweet et al. 2002). Micromere specification begins at 4th cleavage through activation of the micromere gene regulatory network (Oliveri et al. 2002, 2003). At this cleavage, an unequal cell division separates the micromeres at the vegetal pole from the macromeres just above them (see Chapter 9). The micromeres begin their specification assisted by several inputs. Maternal β-catenin enters the nucleus of the micromeres at the 16-cell

stage to activate the micromere gene regulatory network along with another maternal transcription factor Otx (Chuang et al. 1996; Logan et al. 1999). These activate *pmar1* expression, which indirectly is necessary for activation of *delta* (Oliveri et al. 2003). Pmar1 activates *delta* by a double repression in which Pmar1 represses a repressor of *delta*, allowing ubiquitous activators to transcribe the *delta* gene. Between the end of the 4th cleavage and the beginning of the 7th cleavage, several hours later, enough Delta is present on the micromere cell surface to initiate a signal to Notch in the macromere progeny just above the micromeres.

Between 4th and 7th cleavage, while the micromeres begin their specification trajectory, the macromeres also begin to be specified, again initiated, at least in part, by activation of β-catenin-directed specification of the endomesoderm (Davidson et al. 2002). By 7th cleavage, the macromere progeny are prepared to receive the Delta signal. Even though these cells bear maternal Notch, zygotic endomesodermal specification must be initiated for the Notch pathway to successfully transduce the Delta signal.

As with all Notch signaling, direct contact between the Delta-bearing cell and the Notch-bearing cell is required for signal transduction (McClay et al. 2000). Any one of the macromere progeny is competent to receive the Delta ligand, but normally only those cells immediately surrounding the micromeres at the vegetal pole receive the ligand. In the absence of micromeres, but with addition of an activated *notch* RNA construct (Rebay et al. 1993; Sherwood and McClay 1999), the embryo activates the Notch pathway in the endomesoderm with a bias at the vegetal pole, indicating that although ectopic endomesoderm cells can be activated by placement of micromeres in ectopic locations, the endomesoderm cells normally in contact with the micromeres have the greatest sensitivity to activation by the Delta ligand. This sensitivity also requires prior specification of the endomesoderm network, which also began at 4th cleavage (Davidson et al. 2002).

This first Delta–Notch signal is the cue that subdivides the endomesoderm into mesoderm and endoderm in the sea urchin. Until the 7th to 9th cleavage, the entire macromere lineage is specified as endomesoderm. Those endomesoderm cells that activate the Notch receptor become secondary mesenchyme cells (SMCs), while those that fail to activate Notch during this time become endoderm. The pathway is temporally controlled rather precisely because later Notch signaling of a different kind occurs in the SMCs (Sweet et al. 2002) and then the endoderm (Sherwood and McClay 2001).

After Notch is activated in this first signal transduction, all remaining Notch protein is eliminated from the SMCs. Elimination of Notch occurs regardless of whether a particular Notch receptor has bound to Delta, because addition of activated Notch to the presumptive SMCs leads to Notch disappearance from those cells, even when no Delta is present in the system (Sherwood and McClay 1999). Thus, at 10th cleavage, Notch is present on all cells except the SMCs; yet several cleavages earlier, it was Notch on the SMCs that was necessary for their fate specification. This is reminiscent of the disappearance of Notch from sensory organ precursors in *Drosophila,* where the neural cells lose Notch as they become committed to their neural fate.

Macromere endomesoderm progeny that do not activate Notch between 7th and 9th cleavage become endoderm, and shortly thereafter up-regulate zygotic expression of Notch. Experimentally, if micromeres are removed at the 16-cell stage, the Delta signal fails to be available between the 7th and 9th cleavage, so as a consequence, the entire endomesoderm up-regulates Notch at the time endoderm normally up-regulates Notch (McClay et al. 2000). This later Notch expression is necessary for gastrulation by mechanisms that are still not well understood. The timing and function of the later Notch signal are unknown; however, it is known that following their induction by Delta, SMCs synthesize Delta (Sweet et al. 2002). Since the nearest Notch at this time is at the SMC–endoderm boundary, it is likely that there is a Notch signal, as yet not well characterized, which functions at the mesoderm/endoderm boundary. Thus, although many molecular details remain unknown, gastrulation in the sea urchin requires an early Delta–Notch induction, first to subdivide the endomesoderm into mesoderm and endoderm, and later to activate genes in the endoderm that are necessary for the morphogenetic movements of gastrulation. The Notch pathway is crucial for cell fate decisions between endoderm and mesoderm.

NOTCH AND GASTRULATION IN ECHINODERMS

Sea urchins are in the deuterostome clade, and textbooks describe sea urchin gastrulation as the prototype for all deuterostomes, including vertebrates (see Chapters 5 and 9). A recent analysis of Notch and the Wnt pathway supports the notion that the Wnt–Notch–Brachyury expression sequence is highly conserved among deuterostomes, and suggests that the blastopore represents the ancestral posterior pole of the anterior–posterior axis, based on studies of *Amphioxus* (Holland 2002). This statement is based on in situ analysis of *wnt, notch,* and *brachyury* during gastrulation in *Amphioxus.* The pattern and timing of expression matches that of the sea urchin for each of the three genes, and that pattern is conserved (Fig. 3). Hemichordates were recently shown to share the same ordered spatial expression patterns of many anterior–posterior genes with

an alignment parallel to the archenteron as in chordates, further supporting a conserved spatial organization of deuterostome gene expression (Lowe et al. 2003). Recent phylogenies, based on rDNA analysis, suggest that hemichordate and echinoderm ancestors are more closely related than originally thought; echinoderms and hemichordates diverged from each other more recently than their progenitors diverged from the chordate clade (for review, see Holland 2002). If these notions are correct, it means that the location of the blastopore in the sea urchin is likely aligned along the same axis as in chordates, since hemi-

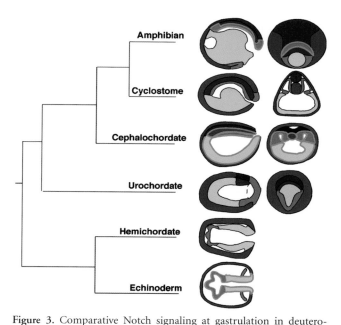

Figure 3. Comparative Notch signaling at gastrulation in deuterostomes. Six representative deuterostomes are displayed according to their evolutionary relationships based on rDNA analysis. Colors depict ectoderm (*blue*), neural ectoderm (*dark blue*), mesoderm (*red*), chordamesoderm (*dark red*), and endoderm (*yellow*). Notch location and/or signaling has been demonstrated in three of these embryos (amphibian, cephalochordate, and echinoderm; *green outlines*). The embryos are shown at late gastrula stage in each case. They are shown in the same relative position based on location of the blastopore, which is modeled to be posterior (Holland 2002). For embryos where there are two diagrams, the left embryo is a longitudinal view and the right is either a surface view of the blastoporal area (amphibian, *Xenopus,* and urochordate, *Ciona*), or a cross-section as seen about midway along the gut (cyclostome, lamprey, and cephalochordate, *Amphioxus*). In all embryos shown except the amphibians, the coelomic pouches (later they become somites in the cephalochordates and vertebrates) initially form as pockets off the archenteron. Notch signaling is observed sequentially throughout the endomesoderm of the sea urchin. In cephalochordates, Notch is localized in the endomesodermal tissues, although no functional studies have been conducted. In amphibians, Notch is located in the presomitic and somite tissue as well as in other tissues later. It functions in the segmentation of the somites. (Data based on Lehman 1977; Jen et al. 1999; Sherwood and McClay 1999; Holland et al. 2001; Holland 2002; Lowe et al. 2003.)

chordates and echinoderms share a highly conserved process of gastrulation. For that reason, Figure 3 models sea urchin gastrulation 90° from the usual display of vegetal pole down and displays amphibian gastrulation in a similar orientation. Superimposed on amphibians, cephalochordates, and echinoderm gastrulae is the expression of *notch* (green outlines), where its expression pattern is known at gastrulation. The other embryos shown in Figure 3 have *notch*, but its location during gastrulation has not been determined as yet.

Does Notch play a role in gastrulation in other deuterostomes that might be similar to that known in sea urchins? A small body of literature is beginning to suggest Notch pathway utilization in chordate gastrulation. Notch has not yet been explored in hemichordates, but it would not be surprising for Notch to play a similar role there, as well. Starfish, separated from the sea urchin lineage for about 500 million years, appear to use Notch in a very similar way as sea urchins (Hinman et al. 2003). At the base of the chordates, *Amphioxus* expresses *notch* in a pattern that at least partially overlaps that of the sea urchin, and the *notch* expression pattern has a relationship both with *wnts* and with *brachyury*, again in a way that appears similar to sea urchins, although no functional analyses have been conducted in *Amphioxus* (Holland et al. 2001; Holland 2002). In *Amphioxus*, the earliest reported expression of *notch* is in a ring surrounding the blastopore at the time of gastrulation, at the site of *brachyury* expression (Holland et al. 1995, 2001). In the sea urchin, the expression of *notch* and of *brachyury* occurs in the same region around the blastopore and at about the same time as in *Amphioxus* (Gross and McClay 2001). Furthermore, there is evidence in sea urchins that *brachyury* expression is downstream of Wnt/Notch signaling (Davidson et al. 2002). Later in gastrulation, expression of *notch* in *Amphioxus* is associated with the notochord and with somitogenesis, a feature that links gastrular expression of *notch* in *Amphioxus* to known vertebrate expression patterns. In *Ciona*, a urochordate, the tunicate genome project has uncovered many genes in the Notch pathway (Satou et al. 2003), but localized expression of few of those genes has been characterized as yet. However, the *brachyury* gene, expressed in the notochord, contains a Su(H)-binding site in its enhancer region (Corbo et al. 1997, 1998), and this was shown experimentally to bind a Su(H) protein; if the Su(H) protein is modified to express a hairy repression domain, notochord differentiation is compromised in the tunicate tadpole. Furthermore, ectopic expression of *Xenopus notch* in *Ciona* causes abnormal tail development. All of this suggests that the Notch pathway participates in notochord development, and that *brachyury* is responsive to Notch signals. Whether urochordates also express *notch* earlier in the endoderm remains to be seen, but if the pattern in *Amphioxus* holds

for urochordates as it does for cephalochordates, a prediction is that *notch* will be present in that location at the blastopore. Later in gastrulation of *Amphioxus*, Notch is associated first with the coelomic pouches that evaginate from the archenteron in the primitive mode of chordates (Holland et al. 2001), and then Notch becomes associated with the notochord and the somites. In *Ciona*, Notch is apparently associated only with the notochord (Corbo et al. 1997, 1998; Imai et al. 2002). In the sea urchin, Notch signaling is required for specification of SMCs (see above), and a subset of those cells form the coelomic pouches as an outpocketing of the anterior archenteron. Coelomic pouches, the somitic precursors in cephalochordates and cyclostomes, also form by an outpocketing from the archenteron, and in cephalochordates there is an associated expression of *notch* at the time of coelom formation, and also later during segmentation of the somites, structures derived from the coelomic pouches (Holland 2001). Thus, an evolutionary connection exists between echinoderms, hemichordates, and cephalochordates in archenteron and somite formation. As functional data accumulate, and as more genes associated with Notch are discovered in these species, it will be interesting to learn the degree to which Notch function in the sea urchin endomesoderm is conserved.

Xenopus and zebrafish appear to use the Notch pathway in a very similar way during early development. In both, Notch is present very early during gastrulation in the organizer. In *Xenopus*, manipulation of expression of either the *notch* or *delta* gene affects the distribution of cells into notochord or floor plate (Lopez et al. 2003). An increase in Notch activity expands the floor plate at the expense of the notochord, and the opposite occurs if Notch activity is reduced experimentally. Lopez et al. showed that Notch expands the region of *sonic hedgehog* expression in the floor plate and represses the notochord markers *brachyury* and *chordin*. In zebrafish, *notch* is expressed in the shield, and a similar reciprocal relationship exists between notochord and floor plate: A mutation in *deltaA* causes a deficit in floor plate and hypochord cells with an increase in notochord, whereas overexpression of *deltaA* causes the opposite phenotype (Appel et al. 1999). Later, prospective hypochord cells express *Notch5* and *her4* and receive a Delta C and/or Delta D signal from lateral mesoderm to specify the hypochord from a subset of *notail* (*ntl*)-expressing cells at the dorsal midline (Latimer et al. 2002). All of this indicates a conserved role for the Notch pathway in notochord formation.

How early is Notch present in vertebrate endomesoderm, and when is the earliest function? A recent in situ hybridization study reveals that both *Notch 1* and *Delta* are present prior to gastrulation in the chick embryo epiblast (Caprioli et al. 2002). Whether these genes function in specification as they do in the sea urchin, or during gastrulation, is not yet known, as functional studies at this stage have not been done. An additional complication in vertebrates is that most species contain duplicated genes encoding Notch and Delta to generate families with multiple members, so manipulation of one of the genes may have little or no early phenotype due to compensation by another gene family member expressed at the same time. In mouse this is especially true where there are four *notch* genes, and single knockouts have no reported consequences early in development.

SOMITOGENESIS

Vertebrate somites form by serial segmentation of lateral plate mesoderm during late gastrulation. Segmentation occurs sequentially from anterior to posterior as lateral plate mesenchyme is subdivided and transformed into a mesodermal epithelium that hollows out as the coelomic cavities of the vertebrate body. As mentioned above, the likely mechanistic antecedent of segments and vertebrate coelom formation is represented by *Amphioxus* and in cyclostome fish gastrulation by an outpocketing of the dorsal archenteron that pinches off to becomes the coelomic pouches, and these are then segmented to form the somites (Fig. 3). Thus, coelom formation and somitogenesis appear to be part of an evolutionary continuum, and related to processes that occur during gastrulation of all deuterostomes. Since Notch is an important component of somitogenesis, how do its appearance and regulation relate to other deuterostomes?

In vertebrates, just after chordamesoderm separates from the ingressed mesenchyme to become the notochord, the paraxial mesoderm, now called presomitic mesoderm, begins to generate segments in a wave that moves posteriorly along the anterior–posterior axis in the wake of gastrulation (Cooke and Zeeman 1976; Stern and Vasiliauskas 2000). The anterior-most somites develop slightly differently from the more posterior somites, but in the posterior group an oscillator mechanism is somehow superimposed upon this anterior–posterior wave to cause segmentation of a new somite at a regular frequency (Stern and Vasiliauskas 2000), or about every 90 minutes in the chick, for example. Prior to the physical segmentation of somites, several undivided "somitomeres" show gene expression patterns that predict subsequent overt segmentation. The Notch pathway is centrally involved in this segmentation process both upstream *and* downstream of the oscillatory mechanism. In the presomitic mesoderm, Notch and Delta are present in the same cells, and in a mouse knockout of *Dll1*, somitogenesis is eliminated (Hrabe de Angelis et al. 1997; Jouve et al. 2000), indicating a required Notch signal through Delta at

this time. Notch perturbations also indicate that Notch is active in the presomitic mesoderm of zebrafish, perhaps to set up the oscillator complex (Takke and Campos-Ortega 1999).

The oscillator follows the anterior–posterior wave and, as a consequence, *hairy* and *lunatic fringe* (plus other genes) are expressed with a relatively constant periodicity. Experimentally, the expression of both of these genes is eliminated when the early Notch–Delta interaction is lost (Palmeirim et al. 1997; McGrew et al. 1998). In fish, chick, and mouse, Notch is upstream of a *hairy* gene homolog that is associated with the oscillator (for review, see Pourquie 2003; Weinmaster and Kintner 2003).

Lunatic Fringe appears as part of the oscillation mechanism in response to the early Notch signal and is associated with the downstream Notch function, as well as the dynamic change in Notch distribution between the anterior and posterior somitomere (Prince et al. 2001; Dale et al. 2003). The precise mechanisms by which hairy gene products and Lunatic Fringe provide positive and negative feedback regulation for Notch in the two halves of the somitomere are still unclear, although accomplishment of that regulation is essential for the later segmentation of somites. As a consequence of oscillator function, *notch* continues to be expressed in the anterior half of the first somitomere but is repressed in the posterior half. In *Xenopus*, two *Enhancer of split*-related genes apparently control the presence and absence of Notch in each somite by positive and negative feedback activation and repression (Jen et al. 1999). At this point in somitogenesis, the Notch pathway becomes functional downstream of the oscillator, and dependent on it for spatial resolution. In the presence of Lunatic Fringe from the oscillatory mechanism, Notch in the anterior half-segment receives the Delta signal and activates a series of downstream genes through Su(H) that include *thylacine*, *Esr4*, *Esr5*, *Mesp* genes, and other bHLH factors (Saga et al. 1997; Sparrow et al. 1998; Jen et al. 1999; Sawada et al. 2000). These in turn, presumably, activate the morphogenetic mechanism that subdivides the somites. Repressors downstream of the oscillator block *notch* expression in the posterior half-somitomere, allowing a different network of genes to operate. Positive feedback loops keep the Notch pathway active in the anterior half-somitomere. This positive and negative feedback reinforces the compartments, using the Notch pathway as the reinforcing device. This reinforcement lasts through several oscillatory cycles and leads to overt somite formation when the oldest somitomere splits off from the next oldest. As part of the somite formation process in *Xenopus*, genes downstream of the Notch pathway in the anterior somitomere activate a protocadherin that participates in the adhesive changes associated with this morphogenetic movement (Kim et al. 2000). Although much remains to be learned about the gene regulatory networks governing somitogenesis, it is clear that the Notch pathway is adapted to subdivide each somitomere into anterior and posterior compartments. Upstream, in the presomitic mesoderm, the Notch pathway drives initiation of a wave-like oscillator activity from anterior to posterior. The oscillator in turn activates the zones of *notch* up- and down-regulation, and thus the Notch pathway controls the formation of each somite at several points in time. An unusual aspect of this *notch* regulation is that in the anterior somitomere where *notch* is expressed and reinforced, it is in a large patch of cells. Generally when this occurs, as in *Drosophila*, either Notch becomes involved in a lateral repression signal as occurs in the proneural system, or Notch function occurs only at the border cells with another compartment. Since in the anterior somite Notch does not appear to subdivide the cells of the half-somite into different fates, it is likely that Notch functions at the boundary, perhaps to reinforce that boundary and promote somite formation.

It may be more than coincidental that Notch functions serially during gastrulation in the sea urchin. Although initially modeled as a single signal subdividing endomesoderm, close inspection suggests that a second Notch signal follows in the SMCs (Sweet et al. 2002), and a short time later a third, then a fourth, and possibly more follow in later development (Sherwood and McClay 1999, 2001; R.E. Peterson and D.R. McClay, unpubl.). In the sea urchin, these signals do not participate in the same kind of serial segmentation as occurs with somites, but a closely spaced temporal and spatial sequence in use of the Notch pathway during gastrulation may have arisen from a similar oscillatory mechanism that now drives both processes.

SUMMARY

The Notch pathway is clearly an important component of gastrulation in many organisms. In the sea urchin, the pathway is used to subdivide endomesoderm into mesoderm and endoderm. The patterns of expression of *notch* and of many Notch pathway components in a variety of deuterostomes suggest that the role of Notch in gastrulation is conserved.

REFERENCES

Appel B., Fritz A., Westerfield M., Grunwald D.J., Eisen J.S., and Riley B.B. 1999. Delta-mediated specification of midline cell fates in zebrafish embryos. *Curr. Biol.* **9:** 247–256.

Artavanis-Tsakonas S., Delidakis C., and Fehon R.G. 1991. The Notch locus and the cell biology of neuroblast segregation. *Annu. Rev. Cell Biol.* **7:** 427–452.

Artavanis-Tsakonas S., Muskavitch M.A., and Yedvobnick B. 1983. Molecular cloning of Notch, a locus affecting neurogenesis in *Drosophila melanogaster. Proc. Natl. Acad. Sci.* **80:** 1977–1981.

Artavanis-Tsakonas S., Rand M.D., and Lake R.J. 1999. Notch signaling: Cell fate control and signal integration in development. *Science* **284:** 770–776.

Aulehla A., Wehrle C., Brand-Saberi B., Kemler R., Gossler A., Kanzler B., and Herrmann B.G. 2003. Wnt3a plays a major role in the segmentation clock controlling somitogenesis. *Dev. Cell* **4:** 395–406.

Austin J. and Kimble J. 1987. glp-1 is required in the germ line for regulation of the decision between mitosis and meiosis in *C. elegans*. *Cell* **51:** 589–599.

———. 1989. Transcript analysis of glp-1 and lin-12, homologous genes required for cell interactions during development of *C. elegans*. *Cell* **58:** 565–571.

Beck C.W. and Slack J.M. 1999. A developmental pathway controlling outgrowth of the *Xenopus* tail bud. *Development* **126:** 1611–1620.

Berdnik D., Torok T., Gonzalez-Gaitan M., and Knoblich J.A. 2002. The endocytic protein α-Adaptin is required for numb-mediated asymmetric cell division in *Drosophila*. *Dev. Cell* **3:** 221–231.

Blair S.S. 1997. Limb development: Marginal fringe benefits. *Curr. Biol.* **7:** R686–690.

Boulianne G.L., de la Concha A., Campos-Ortega J.A., Jan L.Y., and Jan Y.N. 1991. The *Drosophila* neurogenic gene neuralized encodes a novel protein and is expressed in precursors of larval and adult neurons. *EMBO J.* **10:** 2975–2983.

Brou C., Logeat F., Gupta N., Bessia C., LeBail O., Doedens J.R., Cumano A., Roux P., Black R.A., and Israel A. 2000. A novel proteolytic cleavage involved in Notch signaling: The role of the disintegrin-metalloprotease TACE. *Mol Cell* **5:** 207–216.

Bruckner K., Perez L., Clausen H., and Cohen S. 2000. Glycosyltransferase activity of Fringe modulates Notch–Delta interactions. *Nature* **406:** 411–415.

Bush G., diSibio G., Miyamoto A., Denault J.B., Leduc R., and Weinmaster G. 2001. Ligand-induced signaling in the absence of furin processing of Notch1. *Dev. Biol.* **229:** 494–502.

Caprioli A., Goitsuka R., Pouget C., Dunon D., and Jaffredo T. 2002. Expression of Notch genes and their ligands during gastrulation in the chicken embryo. *Mech. Dev.* **116:** 161–164.

Christensen S., Kodoyianni V., Bosenberg M., Friedman L., and Kimble J. 1996. *lag-1*, a gene required for *lin-12* and *glp-1* signaling in *Caenorhabditis elegans*, is homologous to human CBF1 and *Drosophila* Su(H). *Development* **122:** 1373–1383.

Chuang C.K., Wikramanayake A.H., Mao C.A., Li X., and Klein W.H. 1996. Transient appearance of *Strongylocentrotus purpuratus* Otx in micromere nuclei: Cytoplasmic retention of SpOtx possibly mediated through an alpha-actinin interaction. *Dev. Genet.* **19:** 231–237.

Cooke J. and Zeeman E.C. 1976. A clock and wavefront model for control of the number of repeated structures during animal morphogenesis. *J. Theor. Biol.* **58:** 455–476.

Corbo J.C., Levine M., and Zeller R.W. 1997. Characterization of a notochord-specific enhancer from the *Brachyury* promoter region of the ascidian, *Ciona intestinalis. Development* **124:** 589–602.

Corbo J.C., Fujiwara S., Levine M., and Di Gregorio A. 1998. Suppressor of hairless activates brachyury expression in the *Ciona* embryo. *Dev. Biol.* **203:** 358–368.

Couso J.P. and Martinez Arias A. 1994. Notch is required for wingless signaling in the epidermis of *Drosophila*. *Cell* **79:** 259–272.

Dale J.K., Maroto M., Dequeant M.L., Malapert P., McGrew M., and Pourquie O. 2003. Periodic notch inhibition by lunatic fringe underlies the chick segmentation clock. *Nature* **421:** 275–278.

Davidson E.H., Rast J.P., Oliveri P., Ransick A., Calestani C., Yuh C.H., Minokawa T., Amore G., Hinman V., Arenas-Mena C., et al. 2002.

A provisional regulatory gene network for specification of endomesoderm in the sea urchin embryo. *Dev. Biol.* **246:** 162–190.

Deblandre G.A., Lai E.C., and Kintner C. 2001. *Xenopus* neuralized is a ubiquitin ligase that interacts with XDelta1 and regulates Notch signaling. *Dev. Cell* **1:** 795–806.

De Strooper B. and Annaert W. 2001. Where Notch and Wnt signaling meet. The presenilin hub. *J. Cell Biol.* **152:** F17–20.

De Strooper B., Annaert W., Cupers P., Saftig P., Craessaerts K., Mumm J.S., Schroeter E.H., Schrijvers V., Wolfe M.S., Ray W.J., Goate A., and Kopan R. 1999. A presenilin-1-dependent γ-secretase-like protease mediates release of Notch intracellular domain. *Nature* **398:** 518–522.

Diaz-Benjumea F.J. and Cohen S.M. 1995. Serrate signals through Notch to establish a Wingless-dependent organizer at the dorsal/ventral compartment boundary of the *Drosophila* wing. *Development* **121:** 4215–4225.

Dubrulle J. and Pourquie O. 2002. From head to tail: Links between the segmentation clock and antero-posterior patterning of the embryo. *Curr. Opin. Genet. Dev.* **12:** 519–523.

Fehon R.G., Kooh P.J., Rebay I., Regan C.L., Xu T., Muskavitch M.A., and Artavanis-Tsakonas S. 1990. Molecular interactions between the protein products of the neurogenic loci Notch and Delta, two EGF-homologous genes in *Drosophila*. *Cell* **61:** 523–534.

Fleming R.J., Gu Y., and Hukriede N.A. 1997. *Serrate*-mediated activation of *Notch* is specifically blocked by the product of the gene *fringe* in the dorsal compartment of the *Drosophila* wing imaginal disc. *Development* **124:** 2973–2981.

Gross J.M. and McClay D.R. 2001. The role of Brachyury (T) during gastrulation movements in the sea urchin *Lytechinus variegatus*. *Dev. Biol.* **239:** 132–147.

Gupta-Rossi N., Le Bail O., Gonen H., Brou C., Logeat F., Six E., Ciechanover A., and Israel A. 2001. Functional interaction between SEL-10, an F-box protein, and the nuclear form of activated Notch1 receptor. *J. Biol. Chem.* **276:** 34371–34378.

Heitzler P. and Simpson P. 1991. The choice of cell fate in the epidermis of *Drosophila*. *Cell* **64:** 1083–1092.

Henderson S.T., Gao D., Lambie E.J., and Kimble J. 1994. *lag-2* may encode a signaling ligand for the GLP-1 and LIN-12 receptors of *C. elegans*. *Development* **120:** 2913–2924.

Hinman V.F., Nguyen A.T., Cameron R.A., and Davidson E.H. 2003. Developmental gene regulatory network architecture across 500 million years of echinoderm evolution. *Proc. Natl. Acad. Sci.* **100:** 13356–13361.

Holland L.Z. 2002. Heads or tails? Amphioxus and the evolution of anterior-posterior patterning in deuterostomes. *Dev. Biol.* **241:** 209–228.

Holland L.Z., Rached L.A., Tamme R., Holland N.D., Kortschak D., Inoko H., Shiina T., Burgtorf C., and Lardelli M. 2001. Characterization and developmental expression of the amphioxus homolog of Notch (AmphiNotch): Evolutionary conservation of multiple expression domains in amphioxus and vertebrates. *Dev. Biol.* **232:** 493–507.

Holland P.W., Koschorz B., Holland L.Z., and Herrmann B.G. 1995. Conservation of *Brachyury* (T) genes in amphioxus and vertebrates: developmental and evolutionary implications. *Development* **121:** 4283–4291.

Hrabe de Angelis M., McIntyre J., II, and Gossler A. 1997. Maintenance of somite borders in mice requires the *Delta* homologue *DII1*. *Nature* **386:** 717–721.

Imai K.S., Satoh N., and Satou Y. 2002. An essential role of a FoxD gene in notochord induction in *Ciona* embryos. *Development* **129:** 3441–3453.

Jarriault S., Brou C., Logeat F., Schroeter E.H., Kopan R., and Israel A. 1995. Signalling downstream of activated mammalian Notch. *Nature* **377:** 355–358.

Jen W.C., Gawantka V., Pollet N., Niehrs C., and Kintner C. 1999. Periodic repression of Notch pathway genes governs the segmentation of *Xenopus* embryos. *Genes Dev.* **13:** 1486–1499.

Jouve C., Palmeirim I., Henrique D., Beckers J., Gossler A., Ish-Horowicz D., and Pourquie O. 2000. Notch signalling is required for cyclic expression of the hairy-like gene *HES1* in the presomitic mesoderm. *Development* **127:** 1421–1429.

Kidd S. and Lieber T. 2002. Furin cleavage is not a requirement for *Drosophila* Notch function. *Mech. Dev.* **115:** 41–51.

Kim S.H., Jen W.C., De Robertis E.M., and Kintner C. 2000. The protocadherin PAPC establishes segmental boundaries during somitogenesis in *Xenopus* embryos. *Curr. Biol.* **10:** 821–830.

Kimble J. and Simpson P. 1997. The LIN-12/Notch signaling pathway and its regulation. *Annu. Rev. Cell Dev. Biol.* **13:** 333–361.

Klein T. and Arias A.M. 1998. Interactions among Delta, Serrate and Fringe modulate Notch activity during *Drosophila* wing development. *Development* **125:** 2951–62.

Kooh P.J., Fehon R.G., and Muskavitch M.A. 1993. Implications of dynamic patterns of Delta and Notch expression for cellular interactions during *Drosophila* development. *Development* **117:** 493–507.

Kopan R. 2002. Notch: A membrane-bound transcription factor. *J. Cell Sci.* **115:** 1095–1097.

Lai E.C., Deblandre G.A., Kintner C., and Rubin G.M. 2001. *Drosophila* neuralized is a ubiquitin ligase that promotes the internalization and degradation of delta. *Dev. Cell* **1:** 783–794.

Latimer A.J., Dong X., Markov Y., and Appel B. 2002. Delta-Notch signaling induces hypochord development in zebrafish. *Development* **129:** 2555–2563.

Le Borgne R. and Schweisguth F. 2003. Unequal segregation of Neuralized biases Notch activation during asymmetric cell division. *Dev. Cell* **5:** 139–148.

Lehman H.E. 1977. *Chordate development*. Hunter Press, Winston-Salem, North Carolina.

Lieber T., Kidd S., and Young M.W. 2002. *kuzbanian*-mediated cleavage of *Drosophila* Notch. *Genes Dev.* **16:** 209–221.

Logan C.Y., Miller J.R., Ferkowicz M.J., and McClay D.R. 1999. Nuclear β-catenin is required to specify vegetal cell fates in the sea urchin embryo. *Development* **126:** 345–357.

Logeat F., Bessia C., Brou C., LeBail O., Jarriault S., Seidah N.G., and Israel A. 1998. The Notch1 receptor is cleaved constitutively by a furin-like convertase. *Proc. Natl. Acad. Sci.* **95:** 8108–8112.

Lopez S.L., Paganelli A.R., Siri M.V., Ocana O.H., Franco P.G., and Carrasco A.E. 2003. Notch activates sonic hedgehog and both are involved in the specification of dorsal midline cell-fates in *Xenopus*. *Development* **130:** 2225–2238.

Lowe C.J., Wu M. Salic A., Evans L., Lander E., Stange-Thomann N., Gruber C.E., Gerhart J., and Kirschner M. 2003. Anteroposterior patterning in hemichordates and the origins of the chordate nervous system. *Cell* **113:** 853–865.

McClay D.R., Peterson R.E., Range R.C., Winter-Vann A.M., and Ferkowicz M.J. 2000. A micromere induction signal is activated by β-catenin and acts through Notch to initiate specification of secondary mesenchyme cells in the sea urchin embryo. *Development* **127:** 5113–5122.

McGrew M.J., Dale J.K., Fraboulet S., and Pourquie O. 1998. The *lunatic Fringe* gene is a target of the molecular clock linked to somite segmentation in avian embryos. *Curr. Biol.* **8:** 979–982.

Moloney D.J., Panin V.M., Johnston S.H., Chen J., Shao L., Wilson R., Wang Y., Stanley P., Irvine K.D., Haltiwanger R.S., and Vogt T.F. 2000. Fringe is a glycosyltransferase that modifies Notch. *Nature* **406:** 369–375.

Mumm J.S. and Kopan R. 2000. Notch signaling: From the outside in. *Dev. Biol.* **228:** 151–165.

Oberg C., Li J., Pauley A., Wolf E., Gurney M., and Lendahl U. 2001. The Notch intracellular domain is ubiquitinated and negatively regulated by the mammalian Sel-10 homolog. *J. Biol. Chem.* **276:** 35847–35853.

Oliveri P., Carrick D. M., and Davidson E.H. 2002. A regulatory gene network that directs micromere specification in the sea urchin embryo. *Dev. Biol.* **246:** 209–228.

Oliveri P., Davidson E., and McClay D.R. 2003. Activation of *pmar1* controls specification of micromeres in the sea urchin embryo. *Dev. Biol.* **258:** 32–43.

Palmeirim I., Henrique D., Ish-Horowicz D., and Pourquie O. 1997. Avian *hairy* gene expression identifies a molecular clock linked to vertebrate segmentation and somitogenesis. *Cell* **91:** 639–648.

Pan D. and Rubin G.M. 1997. Kuzbanian controls proteolytic processing of Notch and mediates lateral inhibition during *Drosophila* and vertebrate neurogenesis. *Cell* **90:** 271–280.

Pourquie O. 2003. The segmentation clock: Converting embryonic time into spatial pattern. *Science* **301:** 328–330.

Prince V.E., Holley S.A., Bally-Cuif L., Prabhakaran B., Oates A.C., Ho R.K., and Vogt T.F. 2001. Zebrafish *lunatic fringe* demarcates segmental boundaries. *Mech. Dev.* **105:** 175–180.

Rebay I., Fehon R.G., and Artavanis-Tsakonas S. 1993. Specific truncations of *Drosophila* Notch define dominant activated and dominant negative forms of the receptor. *Cell* **74:** 319–329.

Rebay I., Fleming R.J., Fehon R.G., Cherbas L., Cherbas P., and Artavanis-Tsakonas S. 1991. Specific EGF repeats of Notch mediate interactions with Delta and Serrate: Implications for Notch as a multifunctional receptor. *Cell* **67:** 687–699.

Rhyu M.S., Jan L.Y., and Jan Y.N. 1994. Asymmetric distribution of numb protein during division of the sensory organ precursor cell confers distinct fates to daughter cells. *Cell* **76:** 477–491.

Saga Y., Hata N., Koseki H., and Taketo M.M. 1997. Mesp2: A novel mouse gene expressed in the presegmented mesoderm and essential for segmentation initiation. *Genes Dev.* **11:** 1827–1839.

Satou Y., Sasakura Y., Yamada L., Imai K.S., Satoh N., and Degnan B.

2003. A genomewide survey of developmentally relevant genes in *Ciona intestinalis*. V. Genes for receptor tyrosine kinase pathway and Notch signaling pathway. *Dev. Genes Evol.* **213:** 254–263.

Sawada A., Fritz A., Jiang Y., Yamamoto A., Yamasu K., Kuroiwa A., Saga Y., and Takeda H. 2000. Zebrafish Mesp family genes, *mesp-a* and *mesp-b* are segmentally expressed in the presomitic mesoderm, and Mesp-b confers the anterior identity to the developing somites. *Development* **127:** 1691–1702.

Sherwood D.R. and McClay D.R. 1997. Identification and localization of a sea urchin Notch homologue: Insights into vegetal plate regionalization and Notch receptor regulation. *Development* **124:** 3363–3374.

———. 1999. LvNotch signaling mediates secondary mesenchyme specification in the sea urchin embryo. *Development* **126:** 1703–1713.

———. 2001 Notch signaling regulates the position of the ectoderm-endoderm boundary in the sea urchin embryo. *Development* **128:** 2221–2232.

Sparrow D.B., Jen W.C., Kotecha S., Towers N., Kintner C., and Mohun T.J. 1998. *Thylacine 1* is expressed segmentally within the paraxial mesoderm of the *Xenopus* embryo and interacts with the Notch pathway. *Development* **125:** 2041–2051.

Stern C.D. and Vasiliauskas D. 2000. Segmentation: A view from the border. *Curr. Top. Dev. Biol.* **47:** 107–129.

Sweet H.C., Gehring M., and Ettensohn C.A. 2002. LvDelta is a mesoderm-inducing signal in the sea urchin embryo and can endow blastomeres with organizer-like properties. *Development* **129:** 1945–1955.

Takke C. and Campos-Ortega J.A. 1999. *her1*, a zebrafish pair-rule like gene, acts downstream of notch signalling to control somite development. *Development* **126:** 3005–3014.

Uemura T., Shepherd S., Ackerman L., Jan L.Y., and Jan Y.N. 1989. *numb*, a gene required in determination of cell fate during sensory organ formation in *Drosophila* embryos. *Cell* **58:** 349–360.

Weinmaster G. and Kintner C. 2003. Modulation of Notch signaling during somitogenesis. *Ann. Rev. Cell Dev. Biol.* **19:** 367–395.

Wharton K.A., Johansen K.M., Xu T., and Artavanis-Tsakonas S. 1985. Nucleotide sequence from the neurogenic locus notch implies a gene product that shares homology with proteins containing EGF-like repeats. *Cell* **43:** 567–81.

Yochem J. and Greenwald I. 1989. *glp-1* and *lin-12*, genes implicated in distinct cell-cell interactions in *Caenorhabditis elegans*, encode similar transmembrane proteins. *Cell* **58:** 553–563.

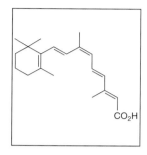

RETINOID SIGNALING DURING GASTRULATION

M. Maden

MRC Centre for Developmental Neurobiology, King's College London, Guy's Campus, London Bridge, London SE1 1UL, United Kingdom

INTRODUCTION

Retinoids comprise a family of molecules derived from vitamin A (retinol). The cells of the developing embryo receive retinol from the maternal circulation (in mammals) or from the stored yolk (in birds and lower vertebrates). Inside the cells, retinol undergoes two enzymatic reactions, first by the retinol or alcohol dehydrogenases to retinaldehyde, and then by the retinaldhyde dehydrogenases to retinoic acid (RA) (Duester 2000). RA is the active retinoid in developing embryos and exists in two forms, all-*trans*-retinoic acid and its isomer, 9-*cis*-retinoic acid. RA is active because it enters the nucleus and activates or represses genes by binding to ligand-activated transcription factors. There are two types of these transcription factors, the retinoic acid receptors (RARs), which bind both all-*trans*-retinoic acid, and 9-*cis*-retinoic acid and the retinoid X receptors (RXRs), which bind only 9-*cis*-retinoic acid. Each receptor exists as three subtypes, α, β, and γ, each coded for by a different gene. The RARs and RXRs heterodimerize and together bind to specific DNA sequences known as retinoic acid response elements (RAREs) in the regulatory sequences of RA-responsive genes (Chambon 1996).

RA is a low-molecular-weight substance that is soluble in organic solvents. Thus, if one wishes to investigate its role in any developing system, one can simply add it ectopically to look for an effect on gene expression or anatomy. Further investigative experiments invariably involve determining whether it is present endogenously, whether it can be synthesized by the system in question, whether there are any of the RA-synthesizing enzymes (the retinaldehyde dehydrogenases [RALDHs]) present, and whether any of the RARs or RXRs are expressed appropriately. Most of these types of

experiments have been performed with regard to gastrulation and have led to the concept that the node is a source of RA which participates in subsequent patterning events, as described below.

RA AND THE NODE

The action of RA in patterning developmental systems was first described in the developing chick limb bud and the regenerating amphibian limb (Maden 1982; Tickle et al. 1982; Summerbell 1983). In the former, RA duplicated the anterior–posterior axis of the limb bud, producing a double-posterior limb with six digits instead of the normal three digits. When endogenous measurements were made, RA was found to be concentrated on the posterior side of the limb bud (Thaller and Eichele 1987). This behavior mimicked precisely the behavior of the organizer of the limb bud, a region on the posterior margin known as the zone of polarizing activity (ZPA) (Summerbell et al. 1973).

Two observations then turned the attention of experimenters to Hensen's node. The first was the concept that embryonic organizing regions of both primary and secondary fields might have been conserved during evolution, both in their properties and in their molecular mechanism of action. The second was the observation that Hensen's node or the mouse node, when grafted to the anterior margin of the chick limb bud, would induce extra digits in the same fashion as a ZPA graft or the application of RA (Hornbruch and Wolpert 1986; Wagner et al. 1990; Hogan et al. 1992).

With this background, Chen and Solursh (1992) showed that a bead which had been soaked in RA could substitute for a grafted Hensen's node and induce a secondary axis. This phenomenon was reported as only occurring when the bead was placed anterior to, or on the left side of, the host node and not when placed posterior to, or on the right side of, the node, but this has not proved repeatable and may have been due to differential damage to the streak. Nevertheless, since RA seemed to substitute completely for Hensen's node, this stimulated Chen et al. (1992) to ask whether RA is normally present in the node. They isolated 750 stage-4 nodes and 700 stage-6 nodes from chick embryos, determined their volumes and protein content, and extracted the retinoids. An F9 reporter cell system that can detect RA by the induction of the luciferase gene was used for the assay of RA, and by comparison with a standard RA curve, it was found that each Hensen's node contained 0.24 pg of RA, and the remaining part of the embryo minus node (area pellucida) contained 0.27 pg of RA. Thus, half of all the RA in the early embryo is concentrated in the node. Since the protein content of the area pellucida was 22 times higher than in the node and correcting for tissue volume, it was concluded that the concentration of RA in the node is 33 nM and in the area pellucida, 1.6 nM. A similar result was obtained for the stage-6 node, and by comparison with the limb bud, it was revealed that the node contains as much RA as one limb bud.

In the mouse node, RA synthesis was measured directly. Ten to 17 pieces of node, anterior neural tissue, or tissue posterior to the node were cultured for 5 hours in [³H]retinol and the rate of synthesis of [³H] RA was determined by HPLC (Hogan et al. 1992). When corrections were made for the amount of DNA in the samples, it was found that anterior neural tissue produced only background levels, the posterior tissue made significant levels of RA, and the node synthesized fourfold more RA than the posterior tissue. On the basis of these results, Hogan et al. proposed that RA generated within the node is responsible for anterior–posterior (AP) patterning such that cells which leave the node early have been exposed to less RA than those which leave later. Cells could remember the amount of RA to which they have been exposed by, for example, having their Hox code set.

Most recently, these concepts have been given a considerable boost by the demonstration that there are several RA-synthesizing and -catabolizing enzymes in and around Hensen's node (Blentic et al. 2003). *Raldh3*, which encodes a RA-synthesizing enzyme, is expressed in the node itself (Fig. 1A) and sections reveal that it is present in the epiblast of the node (Fig. 1E). In the mesoderm below the stage-4 node, transcripts for the RA-catabolizing enzyme *Cyp26A1* are present, and expression also extends anteriorly in the epiblast layers (Fig. 1B,C). Posterior to the node in the mesoderm, another gene encoding a synthesizing enzyme, *Raldh2,* is active (Fig. 1D), although this is not present in

the midline of the primitive streak itself. These distributions around the node are diagrammed in Figure 1E and show that the machinery is present to generate substantial RA activity in the node, which could explain the measurements of endogenous RA and RA synthesis described above.

Assuming that RA is synthesized by the node, it is possible to test its function by examining the anatomy of embryos that develop in the absence of RA. The RA-deficient quail embryo system was specifically designed for this, as these embryos do not contain detectable levels of RA or any other retinoid (Dong and Zile 1995). Such embryos display multiple abnormalities but are not grossly defective in

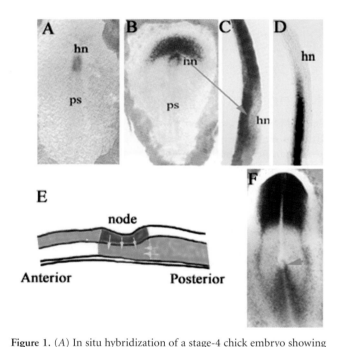

Figure 1. (*A*) In situ hybridization of a stage-4 chick embryo showing expression of *Raldh3* only in Hensen's node. (*B*) In situ hybridization of a stage-4⁺ chick embryo showing expression of *Cyp26A1* in the anterior epiblast (presumptive anterior neural plate) and in an arc around Hensen's node. The red arrow points to where Hensen's node is on the longitudinal section in C. (*C*) Longitudinal section of the embryo in *B* showing *Cyp26A1* expression in the anterior epiblast; then in Hensen's node it is expressed in the mesoderm and then fades out posteriorly. (*D*) Longitudinal section through a stage-4 chick embryo showing *Raldh2* expression in the mesoderm posterior to Hensen's node. (*E*) Summary of the expression patterns of the three enzymes that have been described in and around the node. (*Red*) *Cyp26A1* in the epiblast anterior to the node and in the mesoderm below the node. (*Blue*) *Raldh3* in the epiblast of the node. (*Green*) *Raldh2* in the mesoderm posterior to the node. (*Yellow arrows*) Directions of possible RA signaling in the node. (*F*) In situ hybridization of a stage-6⁺ quail embryo showing expression of *Cyp26A1* in the anterior neural plate, spreading posteriorly in a W-shape and on the right side of the regressing node (*red arrowhead*). (Hn) Hensen's node, (ps) primitive streak.

the AP axis, as might be expected if RA in the node is responsible for AP organization using a Hox code. There are defects in the AP axis, but these are at the level of the first few somites/posterior hindbrain where a stripe of CNS and mesoderm tissue fails to develop correctly (Maden et al. 1996) due to a localized zone of apoptosis (Maden et al. 1997). However, the lateral halves of all of the somites undergo apoptosis beginning at stage 11 (Maden et al. 2000). Since the medial and lateral halves of the somites originate from different regions around the node, medial halves from the node itself and lateral halves from posterior to the node (Selleck and Stern 1991), this could be the function of nodal RA synthesis—to specify the medial halves of the somites at high RA concentrations with the lateral halves specified at lower RA concentrations. The somite population most sensitive to decreases in RA levels would thus be the lateral halves of the somites.

RA AND LEFT/RIGHT ASYMMETRY

The node and its surrounding domains of gene expression regulate left/right asymmetry of the early embryo (see Chapter 28). Although rarely mentioned in conventional descriptions of the gene cascades, which are thought to give rise to left/right asymmetry (Capdevila et al. 2000), there is considerable evidence that RA plays a role in this process. For example, a bead soaked in RA placed on the right of Hensen's node results in randomization of heart looping, ectopic expression of *Nodal* and *Pitx2*, and ectopic expression of two extracellular matrix proteins, hLAMP1 and a fibrillin-related protein (Smith et al. 1997; Tsukui et al. 1999). Conversely, application of an RA antagonist to the left side of Hensen's node abolishes endogenous *Nodal* and *Pitx2* expression and results in randomization of heart looping (Tsukui et al. 1999). In the mouse, RA treatment at head-fold stages perturbs cardiac looping and induces bilateral expression of the normally left-sided genes *Nodal, Lefty-1, Lefty-2, Nkx3.2,* and *Pitx-2* (Chazaud et al. 1999; Wasiak and Lohnes 1999) and, conversely, treatment of mice with an RAR antagonist to inhibit RA signaling results in the randomization of heart looping and the down-regulation of these laterality genes (Chazaud et al. 1999; Wasiak and Lohnes 1999). A final piece of evidence is that the RA-deficient quail embryos have abnormal heart looping (Heine et al. 1985; Dersch and Zile 1993), and a survey of 529 embryos revealed that 72% of these had a reversal of cardiac asymmetry (Zile et al. 2000). However, this heart asymmetry could be due to other factors, such as an abnormality in the notochord or floor plate, which act as a midline barrier for asymmetry signals (Danos and Yost 1996; Lohr et al. 1997).

RA is considered to cooperate with *Shh*, and both act to establish *Lefty-1* and *Pitx2* expression on the left side of the embryo. One might therefore predict an asymmetric distribution of an RA-synthesizing enzyme on the left side of the embryo or an RA-catabolizing enzyme on the right side of the embryo, and very recently this is precisely what we have observed (E. Gale et al., unpubl.). For a brief period of time around stage 6/7, one can see an asymmetric expression of the gene for the RA-catabolizing enzyme, *Cyp26A1*, on the right side of the node (Fig. 1F). This therefore could allow RA to act on the left side of the node, exactly where the experiments described above would predict.

SUMMARY

RA and RA synthesis have been detected in the early Hensen's node. These observations can be explained by the local presence of two RA-synthesizing enzymes, Raldh3 in the epiblast of the node and Raldh2 in the mesoderm posterior to the node, and an RA-catabolizing enzyme, Cyp26A1, in the mesoderm below the node and in the epiblast anterior to the node. The function of this RA may be in the specification of the lateral halves of the somites all along the AP axis of the embryo. RA may also be involved in the cascade of genes that establishes left/right asymmetry as gene alterations and heart situs alterations are observed after experimental interventions involving increasing or decreasing the supply of RA. Indeed, an asymmetric, right-sided distribution of transcripts for an RA-catabolizing enzyme, *Cyp26A1*, has been observed which could explain this phenomenon.

REFERENCES

Blentic A., Gale E., and Maden M. 2003. Retinoic acid signaling centers in the avian embryo identified by sites of expression of synthesizing and catabolizing enzymes. *Dev. Dyn.* **227:** 114–127.

Capdevila J., Vogan K.J., Tabin C.J., and Izpisua-Belmonte J.-C. 2000. Mechanisms of left-right determination in vertebrates. *Cell* **101:** 9–21.

Chambon P. 1996. A decade of molecular biology of retinoic acid receptors. *FASEB J.* **10:** 940–954.

Chazaud C., Chambon P., and Dolle P. 1999. Retinoic acid is required in the mouse embryo for left-right determination and heart morphogenesis. *Development* **126:** 2589–2596.

Chen Y.-P. and Solursh M. 1992. Comparison of Hensen's node and retinoic acid in secondary axis induction in the early chick embryo. *Dev. Dyn.* **195:** 142–151.

Chen Y.-P., Huang L., Russo A.F., and Solursh M. 1992. Retinoic acid is enriched in Hensen's node and is developmentally regulated in the early chick embryo. *Proc. Natl. Acad. Sci.* **89:** 10056–10059.

Danos M.C. and Yost H.J. 1996. Role of notochord in specification of cardiac left-right orientation in zebrafish and *Xenopus. Dev. Biol.* **177:** 96–103.

Dersch H. and Zile M.H. 1993. Induction of normal cardiovascular development in the vitamin A-deprived quail embryo by natural retinoids. *Dev. Biol.* **160:** 424–433.

Dong D. and Zile M.H. 1995. Endogenous retinoids in the early avian embryo. *Biochem. Biophys. Res. Commun.* **217:** 1026–1031.

Duester G. 2000. Families of retinoid dehydrogenases regulating vitamin A function. Production of visual pigment and retinoic acid. *Eur. J. Biochem.* **267:** 4315–4324.

Heine U.I., Roberts A.B., Munoz E.F., Roche N.S., and Sporn M.B. 1985. Effects of retinoid deficiency on the development of the heart and vascular system of the quail embryo. *Virchow's Arch. B Cell Pathol.* **50:** 135–152.

Hogan B.L.M., Thaller C., and Eichele G. 1992. Evidence that Hensen's node is a site of retinoic acid synthesis. *Nature* **359:** 237–241.

Hornbruch A. and Wolpert L. 1986. Positional signaling by Hensen's node when grafted to the chick limb bud. *J. Embryol. Exp. Morphol.* **94:** 257–265.

Lohr J.L., Danos M.C., and Yost H.J. 1997. Left-right asymmetry of a *nodal*-related gene is regulated dy dorsoanterior midline structures during *Xenopus* development. *Development* **124:** 1465–1472.

Maden M. 1982. Vitamin A and pattern formation in the regenerating limb. *Nature* **295:** 672–675.

Maden M., Gale E., Kostetskii I., and Zile M. 1996. Vitamin A-deficient quail embryos have half a hindbrain and other neural defects. *Curr. Biol.* **6:** 417–426.

Maden M., Graham A., Zile M., and Gale E. 2000. Abnormalities of somite development in the absence of retinoic acid. *Int. J. Dev. Biol.* **44:** 151–159.

Maden M., Graham A., Gale E., Rollinson C., and Zile M. 1997. Positional apoptosis during vertebrate CNS development in the absence of endogenous retinoids. *Development* **124:** 2799–2805.

Niazi I.A. and Saxena S. 1978. Abnormal hindlimb regeneration in tadpoles of the toad, *Bufo andersoni*, exposed to excess Vitamin A. *Folia Biol. (Krakow)* **26:** 3–8.

Selleck M.A.J. and Stern C.D. 1991. Fate mapping and cell lineage analysis of Hensen's node in the chick embryo. *Devlopment* **112:** 615–626.

Smith S.M., Dickman E.D., Thompson R.P., Sinning A.R., Wunsch A.M., and Markwald R.R. 1997. Retinoic acid directs cardiac laterality and the expression of early markers of precardiac asymmetry. *Dev. Biol.* **182:** 162–171.

Summerbell D. 1983. The effects of local application of retinoic acid to the anterior margin of the developing chick limb. *J. Embryol. Exp. Morphol.* **78:** 269–289.

Summerbell D., Lewis J.H., and Wolpert L. 1973. Positional information in chick limb morphogenesis. *Nature* **244:** 492–496.

Thaller C. and Eichele G. 1987. Identification and spatial distribution of retinoids in the developing chick limb bud. *Nature* **327:** 625–628.

Tickle C., Alberts B., Wolpert L., and Lee J. 1982. Local application of retinoic acid to the limb bond mimics the action of the polarizing region. *Nature* **296:** 564–566.

Tsukui T., Capdevila J., Tamura K., Ruiz-Lozano P., Rodriguez-Esteban C., Yonei-Tamura S., Magallon J., Chandraratna R.A., Chien K.R., Blumberg B., Evans R.M., and Izpisua-Belmonte J.-C. 1999. Multiple left-right asymmetry defects in Shh$^{-/-}$ mutant mice unveil a convergence of the Shh and retinoic acid pathways in the control of *Lefty-1*. *Proc. Natl. Acad. Sci.* **96:** 11376–11381.

Wagner M., Thaller C., Jessell T.M., and Eichele G. 1990. Polarizing activity and retinoid synthesis in the floor plate of the neural tube. *Nature* **345:** 819–822.

Wasiak S. and Lohnes D. 1999. Retinoic acid affects left-right patterning. *Dev. Biol.* **215:** 332–342.

Zile M.H., Kostetskii I., Yuan S., Kostetskaia E., St. Amand T.R., Chen Y.-P., and Jiang W. 2000. Retinoid signaling is required to complete the vertebrate cardiac left/right asymmetry pathway. *Dev. Biol.* **223:** 323–338.

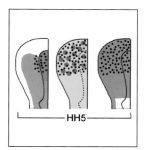

ROLE OF *OTX2* DURING GASTRULATION

A. Simeone

MRC Centre for Developmental Neurobiology, King's College, London SE1 1UL, United Kingdom; Institute of Genetics and Biophysics "A. Buzzati-Traverso", CNR, 80125 Naples, Italy

INTRODUCTION

Fate and patterning of tissues depend on the activity of organizer cells emanating signals to a responding tissue, which undergoes morphogenetic changes resulting in a specific differentiated fate (Spemann and Mangold 1924; Waddington 1933; Gurdon 1987).

Based on the cell fate and inductive properties of the grafted tissue, homologous organizers have been described in all vertebrates: the embryonic shield in zebrafish, Hensen's node in chick, and the node in mouse (Waddington 1933; Storey et al. 1992; Zoltewicz and Gerhart 1997; Saude et al. 2000).

In mouse, transplantation of early and full-length streak nodes is able to induce secondary neural axes, but the forebrain and midbrain regions are absent (Beddington 1994; Tam and Steiner 1999). In contrast, when donor tissue is obtained from mid-streak embryos, expression of forebrain and midbrain markers is induced in the ectopic neural axis (Kinder et al. 2001). These studies indicate that the mammalian node shares neural inducing properties with other vertebrate organizers, but they also suggest that tissues other than the node should be involved in anterior neural patterning in the mouse (Davidson et al. 1999; Klingensmith et al. 1999; Episkopou et al. 2001).

Much evidence has now accumulated that in mammalian development, a separate signaling center, distinct from the classic organizer, is directly or indirectly required prior to and during early gastrulation for early anterior neural patterning (Beddington and Robertson 1999). This signaling center, the anterior visceral endoderm (AVE) is composed of a group of cells destined to populate the visceral yolk sac. By early head-fold stage, node-derived tissues such as the anterior definitive endoderm (ADE) and the axial mesendoderm (AME), which gives rise to the prechordal plate, replace most of the AVE, which moves toward the extraembryonic region (Lawson and Pedersen 1987; Tam and Beddington 1992; Thomas and Beddington 1996).

Pioneering evidence for a role of the AVE in anterior neural patterning has been provided from elegant ablation experiments in mouse (Thomas and Beddington 1996; Beddington and Robertson 1999). The idea that the synergistic action of the AVE and early node is required for induction of anterior neural plate has been further supported by the analysis of mouse mutants and tissue recombination experiments (Ding et al. 1998; Beddington and Robertson 1999; Sun et al. 1999; Stern 2001). However, there is clear evidence that the murine node and its derivatives are able to emanate neuralizing signals as well as to induce and stabilize the expression of forebrain, midbrain, and hindbrain markers (Ang and Rossant 1993; Ruiz i Altaba 1993, 1994; Ang et al. 1994; Beddington and Robertson 1999).

Embryological and genetic studies indicate an essential role of the AME and/or ADE in brain formation (Shawlot et al. 1999; Bachiller et al. 2000; Camus et al. 2000; Hoodless et al. 2001; Withington et al. 2001; Hallonet et al. 2002). Patterning of anterior neuroectoderm is also controlled by additional mechanisms involving planar signals acting through the neuroectodermal plane (Doniach 1993; Ruiz i Altaba 1993, 1994) and emitted by specific groups of cells having organizer properties. In zebrafish, a small group of ectodermal cells located in the anteriormost head region (Houart et al. 1998), and in mouse the anterior neural ridge (ANR) at the anterior ectodermal–neuroectodermal border as well as the isthmic organizer (IsO) at the early midbrain–hindbrain boundary, play a remarkable role in early CNS patterning events (Crossley et al. 1996; Wurst and Bally-Cuif 2001). Among the genes required in the specification of the rostral neural plate during gastrulation, the homeobox-containing gene *Otx2* is an important factor in both activation and maintenance of the anterior neural patterning.

OTX2 EXPRESSION IN VERTEBRATE AND PROTOCHORDATE GASTRULATING EMBRYOS

Otx2 is expressed in the mouse embryo throughout the entire epiblast and visceral endoderm (VE) prior to the onset of gastrulation. At the onset of gastrulation, *Otx2* expression is down-regulated from the posterior pole of the embryo, where the primitive streak forms and is maintained in the epiblast, VE (including the AVE), and the leading cells of the primitive streak (Fig.1). As gastrulation proceeds, *Otx2* transcripts are detected in the node derivatives, the ADE and rostral portion of the AME, as well as in the anterior neuroectoderm (ANE) (Simeone et al. 1993; Ang et al. 1994). In particular in the ANE, the *Otx2* expression domain includes territory fated to give rise to the forebrain and midbrain and defines a sharp boundary at the future midbrain–hindbrain border where the IsO will subsequently form (Fig. 1). Therefore, in mouse, *Otx2* is expressed in tissues that are relevant for anterior patterning, where it plays a crucial role in specification and maintenance of anterior character.

The expression of *Otx2* cognate genes has been studied in detail in many species. In vertebrates other than mammals, *Otx2* follows a very similar expression profile (Li et al. 1994; Mori et al. 1994; Bally-Cuif et al. 1995; Mercier et al. 1995; Pannese et al. 1995). Indeed, in chick pregastrulating embryos, *c-Otx2* is transcribed in the hypoblast and epiblast. As gastrulation starts and a primitive streak is visible, *c-Otx2* transcripts disappear from the more posterior region of epiblast and hypoblast and are detected in the anteriormost region of primitive streak (Fig.1). Noteworthy at Hamburger and Hamilton stage 4⁺, *c-Otx2* transcripts disappear from the epiblast and are transiently concentrat-

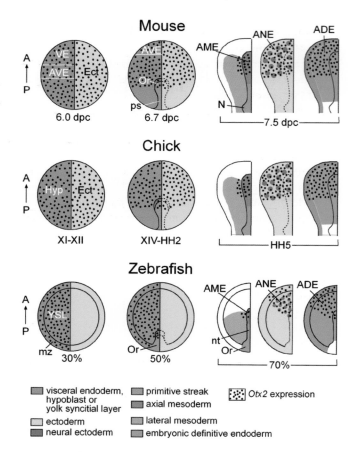

Figure 1. Schematic representation of *Otx2* expression in mouse, chick, and zebrafish embryos at pre-, early, and late gastrula stages. The *Otx2* expression is indicated by dotted areas. Stages and expression patterns for chick and zebrafish are according to Bally-Cuif et al. (1995) and Li et al. (1994), respectively. For the pregastrula and early gastrula stages, expression is indicated in a lower layer (*left*) and an upper layer (*right*). At the pregastrula stage, the lower layer represents the visceral endoderm (VE) or the hypoblast (Hyp) or the yolk syncytial layer (YSL) and, at the early gastrula stage, indicates the VE or the Hyp or the YSL as well as the early mesodermal components (axial mesoderm and primitive streak). At the same stages, the upper layer corresponds to the ectoderm. For late gastrulating embryos, the mesodermal components (*left*), ectoderm with differentiating neuroectoderm (*center*), and endoderm (*right*) are shown separately. For mouse and chick embryos at 7.5 dpc and HH5 stage, the region posterior to the node is not represented. Abbreviations: (ANE) Differentiating anterior neuroectoderm; (AME) axial mesoderm; (ADE) anterior definitive endoderm; (AVE) anterior visceral endoderm; (mz) marginal zone; (ps) primitive streak; (N) node; (Or) organizer; (nt) notochord. The stages in zebrafish are percentage of epiboly.

ed in the Hensen's node (Bally-Cuif et al. 1995). Afterward, *c-Otx2* is transcribed in the anterior neuroectoderm and prechordal mesendoderm (Fig. 1).

In zebrafish pregastrulating embryos, *zOtx* genes are expressed in the yolk syncytial layer (YSL), at early-mid gastrula (~50% of epiboly) in the axial mesendoderm cells also expressing the *goosecoid* gene, and in the definitive endoderm. *zOtx1* and *zOtx2* are also detected in a triangularly shaped area corresponding to the future forebrain and midbrain neuroectoderm (Fig. 1) (Li et al. 1994).

In protochordates and, in particular, in pregastrulating ascidian embryos, *Hroth* is transcribed in the blastomeres of the endoderm cell lineage and sensory vesicle as well as in the precursors of the mesenchyme and trunk lateral cells. During gastrulation, *Hroth* expression is maintained in the involuting endoderm and anterior ectoderm. From the neurula stage, expression is maintained only in the neuroectoderm (Wada et al. 1996). Interestingly, in the endoderm precursors, *Hroth* expression dynamically coincides with the leading edge of involuting cells, thus suggesting a role for *Hroth* in the initiation and progression of involution movement (Wada et al. 1996). A similar role has been previously suggested for the *Xenopus* and chick *Otx2* genes (Bally-Cuif et al. 1995; Pannese et al. 1995), and this is supported by the phenotypic analysis of *Otx2* null mice (see below and Acampora et al. 1995; Matsuo et al. 1995; Ang et al. 1996). In cephalochordates, the *amphi Otx* gene is also expressed in the AME and ANE equivalent tissues (Williams and Holland 1996).

Together these studies have suggested that *Otx2* may play a common role in protochordates and vertebrates, controlling the specification of anterior neuroectoderm during gastrulation in different cell types (AVE, AME, ADE, ANE). Moreover, these expression data are consistent with the idea that the ground territory of forebrain and midbrain differentiation evolved before the separation of urochordates and vertebrates, and it may be controlled by a similar molecular mechanism.

OTX2 IN EARLY ANTERIOR NEURAL PATTERNING

The first indication that *Otx2* is responsive to inductive interactions comes from explant-recombination experiments in gastrulating mouse embryos, showing that a positive signal from the AME of head-fold stage embryos is able to maintain *Otx2* expression in the anterior ectoderm of early-streak embryos. A negative signal from the posterior mesendoderm, mimicked by exogenous retinoic acid (RA), represses *Otx2* expression in the anterior ectoderm of late streak embryos (Ang et al. 1994). Similar interactions have been also demonstrated in *Xenopus* (Blitz and Cho 1995).

The possibility that RA might contribute to the early distinction between fore-midbrain and hindbrain by controlling *Otx2* expression is supported by the finding that administration of exogenous RA at mid-late streak stage represses *Otx2* expression in both the AME and the posterior region of the anterior neural plate (Simeone et al. 1995; Avantaggiato et al. 1996). This repression correlates with the appearance of microcephalic embryos showing loss of forebrain molecular and morphological landmarks, and gain of midbrain molecular markers in the rostralmost neuroectoderm (Simeone et al. 1995; Avantaggiato et al. 1996). Moreover, *Otx2* responsiveness to RA application is a common feature in different species including *Xenopus* and chick (Bally-Cuif et al. 1995; Pannese et al. 1995). Nevertheless, the question of whether endogenous RA plays a physiological role in rostral CNS demarcation by contributing to the establishment of the posterior border of *Otx2* expression still remains open.

The evidence that *Otx2* may play an important role in rostral CNS specification derives from in vivo genetic manipulation experiments performed in mouse and *Xenopus*, which, to some extent, complement each other. In *Xenopus*, microinjection of synthetic *Otx2* RNA results in an abnormal reduction in size of tail and trunk structures, and in the appearance of a second cement gland (Blitz and Cho 1995; Pannese et al. 1995). These phenotypes have been interpreted either as a possible *Otx2*-mediated interference with movements of extension and convergence during gastrulation (for trunk and tail reduction) and/or as an *Otx2* requirement in the specification of anteriormost head structures (for the ectopic cement gland) (Gammill and Sive 1997).

In mouse, *Otx2* null embryos die early in embryogenesis, lack the rostral neuroectoderm fated to become forebrain, midbrain, and rostral hindbrain, and show heavy abnormalities in their body plan (Fig. 2) (Acampora et al. 1995; Matsuo et al. 1995; Ang et al. 1996).

OTX2 IS REQUIRED FOR SPECIFICATION OF ANTERIOR NEUROECTODERM AND GASTRULATION

Otx2 null mouse embryos show pre-gastrulation and gastrulation defects as well as severe abnormalities in anterior neural patterning (Acampora et al. 1995; Matsuo et al. 1995; Ang et al. 1996). Pre-streak and early-streak *Otx2*$^{-/-}$ mutant embryos fail to anteriorize the AVE, as evidenced by the distal expression of AVE markers such as *Cerrl*, *Hex*, *Hesx1*, and *Lim1* (Acampora et al. 1998; Kimura et al. 2000; Perea-Gomez et al. 2001). Primitive streak markers, such as *T*, *Fgf8*, and *Cripto* are not restricted to the posterior proximal epiblast as in wild-type embryos, and they are ectopically expressed in a ring around the entire proximal epiblast at 6.5 days postcoitum (dpc) (Fig. 2). During gastrulation, formation of the primitive streak, node, and node derivatives such as ADE and AME is severely impaired in *Otx2*$^{-/-}$ mutants. At 8.5 dpc, mutant embryos are delayed and lack neural tissue rostral to rhombomere 2 (Fig. 2). Analysis of early neural markers during gastrulation indicates that the anterior defects have an early conception owing to a failure of induction of forebrain and midbrain at late-streak stages (Acampora et al. 1998 and unpubl.).

Chimeric studies have been conducted to better understand the etiology of the anterior defects and to address when and where *Otx2* is required during mouse development. Indeed, *Otx2* might be required primarily either in the VE or in the epiblast and its derivatives (ADE, AME,

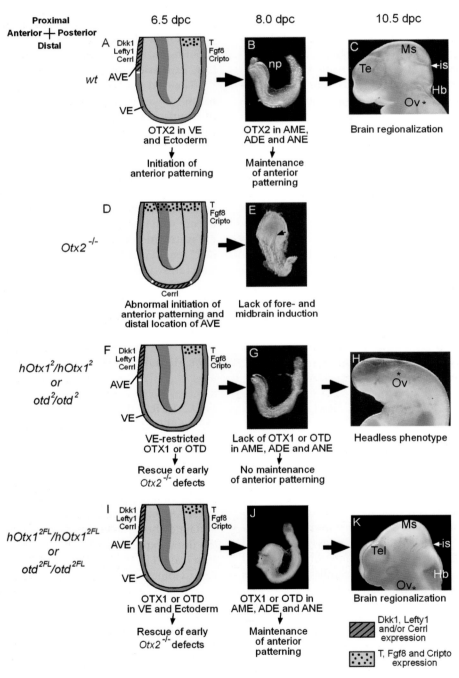

Figure 2. (*See facing page for legend.*)

and ANE), or in all of these tissues. Chimeras in which only the epiblast is wild-type for *Otx2* function, while the VE is composed of *Otx2*⁻/⁻ cells, and vice versa, have been analyzed. Chimeric embryos composed predominantly of *Otx2*⁺/⁺ epiblast cells and *Otx2*⁻/⁻ VE exhibit the same neural defects observed in the *Otx2*⁻/⁻ mutants. In contrast, chimeras obtained from injection of *Otx2*⁻/⁻ ES cells into wild-type blastocysts exhibit a normal induction of the fore- and midbrain markers at 7.5 dpc (Rhinn et al. 1998). Similar conclusions have been drawn from the analysis of another mouse model where the *Otx2* gene is replaced by the human *Otx1* cDNA (*hOtx1²*) (Acampora et al. 1998). In homozygous mutant embryos for the *hOtx1* allele (*hOtx1²/hOtx1²*), hOTX1 protein is detectable only in the VE, but not in the epiblast or its derivatives (ADE, AME, and ANE). Interestingly, the VE-restricted hOTX1 protein is sufficient to rescue the early neural and gastrulation defects observed in *Otx2*⁻/⁻ mutants at 7.5 dpc, but *hOtx1²/hOtx1²* mutants invariably showed a headless phenotype at 9.5 dpc, as hOTX1 protein was absent in the epiblast derivatives (Fig. 2). Together, chimera experiments and mouse models indicate that at the pre- and early-streak stages, *Otx2* is required in the VE for the specification of the early anterior neural plate and proper gastrulation, and later in epiblast derivatives for the maintenance of forebrain and midbrain territories.

Recently, the function of *Otx2* in the AVE has been investigated in more detail. Using cell-labeling techniques, it has been confirmed, as previously suggested (Acampora et al. 1995), that in *Otx2*⁻/⁻ mutants the AVE antecedents fail to move anteriorly by the onset of gastrulation (Perea-Gomez et al. 2001). This is concomitant with the ectopic expression of the primitive streak and mesodermal markers *Cripto*, *Fgf8*, *Lefty2*, *Mesp1*, and *T* in a ring around the proximal epiblast (Kimura et al. 2000; Perea-Gomez et al. 2001). Interestingly, the distally located AVE fails to express the *Wnt* and *nodal* antagonists *Dkk1* and *Lefty1*, although other AVE markers such as *Lim1*, *Hesx1*, and *Cerrl* are expressed in *Otx2*⁻/⁻ mutants. As Wnt and Nodal signaling are involved in primitive streak formation (Lu et al. 2001), the absence of *Dkk1* and *Lefty1* is likely the reason for the persistent expression of primitive streak and mesodermal markers in the antero-proximal epiblast of *Otx2*⁻/⁻ mutants at 6.5 dpc. Fate-map studies have shown that at the onset of gastrulation, the antero-proximal epiblast is destined to populate the anterior neural and nonneural ectoderm in wild-type embryos, but in *Otx2*⁻/⁻ embryos, there is a tendency of these cells to give rise to mesoderm (Perea-Gomez et al. 2001). This supports the idea that one of the functions of *Otx2* in the VE, and specifically in the AVE, might be to protect the antero-proximal epiblast from posteriorizing signals involved in the formation of the primitive streak. The failure of AVE antecedents to move anteriorly, and the lack of *Dkk1* and *Lefty1* in the AVE, might leave the anterior epiblast exposed to the action of posteriorizing factors. In fact, tissue recombination experiments have shown that the AVE is able to repress the posterior markers *T* and *Cripto*, and that *Otx2* is specifically required in the AVE for this repression (Kimura et al. 2000). In this context, it has been shown that in ascidian embryos, the ectodermal expression of *Hroth* requires an inductive influence from cells of the vegetal hemisphere and that in the vegetal hemisphere, *Hroth* is expressed in the endoderm precursors where it may be required to antagonize notochord and muscle differentiation (Wada and Saiga 1999).

Together these findings support the model that the AVE, and specifically an *Otx2*-positive AVE, is required to restrict expression of posterior genes involved in mesoderm

---

Figure 2. Molecular and morphological defects in *Otx2*⁻/⁻, *hOtx1²/hOtx1²*, *otd²/otd²*, *otd²ᶠᴸ/otd²ᶠᴸ*, and *hOtx1²ᶠᴸ/hOtx1²ᶠᴸ* mutant embryos at different stages of development. (*A–C*) In wild-type embryos at the onset of gastrulation (6.5 dpc), genes such as *Dkk1*, *Lefty1*, and *Cerrl* are expressed in the AVE whereas *T*, *Fgf8*, and *Cripto* are expressed in the early primitive streak cells (*A*). One function of the AVE is to protect the overlying epiblast from posteriorizing signals by secreting BMP, Wnt, and Nodal antagonists. This is required for normal neural induction of forebrain and midbrain territories. This anterior neural character is maintained subsequently by the activities of the ADE and AME (*B*). Finally, the brain is divided into forebrain, midbrain, and hindbrain regions (*C*). (*D,E*) In *Otx2*⁻/⁻ pregastrulating mutants, cell movements are impaired and the AVE is located at the distal tip of the embryo (*D*). Furthermore, the activity of the AVE is also severely affected as no *Dkk1* or *Lefty1* expression is detected. As a consequence, the primitive streak markers *T*, *Fgf8*, and *Cripto* are expressed ectopically in a ring around the proximal epiblast (*D*). This leads to a failure in the initial induction of the forebrain and midbrain territories in *Otx2*⁻/⁻ mutants (*E*). (*F–H*) In contrast, in *hOtx1²/hOtx1²* and *otd²/otd²* mutants, the early requirement of *Otx2* is compensated for, and a normal induction of the anterior neural plate takes place (*F*). However, this anterior neural character is maintained until the presomitic stage (*G*). Subsequently, lack of maintenance of the anterior character results in a headless phenotype at 10.5 dpc (*H*). (*I–K*) Maintenance of forebrain and midbrain territories (*K*) occurs when the *hOtx1²ᶠᴸ* or *otd²ᶠᴸ* mRNA is translated in the epiblast and its derivatives such as AME, ADE, and ANE (*I,J*). This indicates that OTD and hOTX1 proteins exhibit a relevant functional equivalence with OTX2 and that posttranscriptional control of *hOtx1²ᶠᴸ* and *otd²ᶠᴸ* mRNAs depend on the 5′UTR and 3′UTR of *Otx2*. The arrow in *E* points to the rostral limit of the neuroectoderm in an *Otx2*⁻/⁻ embryo. (Hb) Hindbrain; (is) isthmus; (Te) telencephalon; (Ms) mesencephalon; (np) neural plate; (Ov) otic vesicle. Anterior is to the left for all the embryos. The coordinates (*top left*) refer to 6.5-dpc embryos.

induction, thereby allowing the anterior epiblast to remain receptive to later anterior neural induction and patterning. The change of fate of the anterior epiblast might also explain why in *Otx2* null embryos replacing the wild-type *Otx2* locus with the *lacZ* gene, *lacZ* transcription occurs in the AVE, but not in the overlying anterior epiblast (Acampora et al. 1995). Indeed, this is the expected result if the anterior epiblast acquires a posterior fate in *Otx2−/−* mutants.

ROLE OF *OTX2* IN MAINTENANCE OF FORE- AND MIDBRAIN IDENTITIES

As mentioned above, *Otx2* is required during mid-late gastrulation for maintenance of forebrain and midbrain identity (Acampora et al. 1998; Rhinn et al. 1998). Indeed, *Otx2* is expressed in the ADE and AME, which are both sources of signals involved in neural patterning, as well as in the ANE, where it could be required for providing competence to respond to these signals.

At the head-fold stage, in *hOtx1²/hOtx1²* mutants, the expression of several ADE and AME markers such as *Cerrl, Lim1, Gsc, T, Hesx1, Chd,* and *Noggin* is indistinguishable from wild-type embryos and, similarly, the anterior patterning of the rostral neural plate is correctly defined by the expression of forebrain, midbrain, and hindbrain markers such as *Six3, Gbx2,* and *Hoxb1* (Acampora et al. 1998; Martinez-Barbera et al. 2001). Nevertheless, in these mutant embryos, at the early-somite stage, the forebrain markers *BF1* and the *hOtx1²* mRNA are not detected in the rostral neuroectoderm where mid-hindbrain markers such as *Wnt1, En1, Fgf8, Gbx2,* and *Pax2* appear to be coexpressed.

In this respect, the phenotype observed at the early-somite stage appears to be the consequence of an anterior to posterior repatterning involving the entire anterior neural plate (fore- and midbrain) which, in the absence of any OTX gene product, adopts a more posterior fate (hindbrain) (Fig. 2) (Acampora et al. 1998; Acampora and Simeone 1999). As seen in the epiblast of *Otx2−/−* embryos, this repatterning may be mediated by the failure in antagonizing posteriorizing determinants such as *Gbx2* (Martinez-Barbera et al. 2001). Therefore, the *hOtx1²/hOtx1²* mouse model allows us to uncouple two distinct phases, both requiring *Otx2*: early specification of anterior neural patterning and proper gastrulation when it is required in the VE, and subsequent maintenance of the anterior patterning when it is required in epiblast-derived cells (AME and ANE).

However, because hOTX1 protein is undetectable in the ADE, AME, and ANE, this analysis does not clarify in which of these tissues *Otx2* is required for maintenance properties. Similarly, the *Otx2* chimeric analyis could not address

this issue, because it is not possible to generate chimeras where only one of these tissues is of *Otx2−/−* genotype (Rhinn et al. 1998, 1999).

Tissue recombination experiments using wild-type AME and anterior ectoderm isolated from *Otx2* mutants carrying a hypomorphic allele have suggested a specific function of OTX2 in the anterior ectoderm for providing competence to respond to signals emanating from the AME and ANR (Tian et al. 2002).

In this context, it has been shown that *Lim1, Chordin,* and *Noggin* are required in the AME for specification and/or maintenance of the anterior patterning (Shawlot et al. 1999; Bachiller et al. 2000). In *hOtx1²/hOtx1²* mutants and chimera experiments, gastrulation and normal expression of *Lim1, Chordin,* and *Noggin* are recovered even though the anterior patterning is established but not maintained (Acampora et al. 1998). On this basis, it can be speculated that OTX2 might confer to the ANE the competence to respond to *Lim1-, Chordin-,* and/or *Noggin*-dependent signals emitted from the AME and required for maintenance and/or specification of anterior neural character. In the absence of OTX2 these signals might be ineffective.

In conclusion, although genetic and embryological evidence suggests that *Otx2* is required in the ANE for the maintenance of forebrain and midbrain regions, its function in the ADE and AME remains unclear. A definitive answer for the role of *Otx2* in these tissues awaits further experiments, such as the conditional inactivation of *Otx2* in these specific tissues.

FUNCTIONAL EQUIVALENCE OF OTD/OTX PROTEINS AND REGULATORY CONTROL OF OTX2

To assess whether *Otx2* encodes unique functional properties that are important during gastrulation, mouse models replacing *Otx2* with the human *Otx1* gene have been generated (Acampora et al. 1998; Suda et al. 1999). As previously discussed, *hOtx1²/hOtx1²* embryos recover early specification of anterior neural plate and proper gastrulation but fail to maintain forebrain and midbrain identities. This is due to VE-restricted translation of *hOtx1* mRNA. As a consequence, this mutant fails to address the question of whether *Otx1* is functionally equivalent to *Otx2* in the AME and ANE. A second mouse model replacing *Otx2* with *Otx1* provides evidence that *Otx1* may partially rescue *Otx2* functions in AME and ANE (Suda et al. 1999).

However, it is unclear whether the level of OTX1 is similar to that of OTX2 in this mutant. This is a crucial aspect, since a reduction below a critical threshold of Otx gene product in the ANE is invariably reflected in an abnormal development of forebrain and midbrain. A third mouse model replacing *Otx2* with *Otx1* has been recently generated (Acampora et al. 2003). In this latter mutant (*hOtx1²FL/Otx1²FL*), the *Otx1* mRNA is translated in epiblast, AME, and ANE, and the level of OTX1 protein is comparable to that of OTX2 in these tissues (Fig. 2). Homozygous mutants are healthy and fertile. This mutant differs from the *hOtx1²* mutant due to the presence of the complete 5′ and 3′ untranslated regions (UTRs). This strongly supports the possibility that the regulatory control of *Otx2* expression in the epiblast, AME, and ANE requires 5′ and/or 3′ UTRs.

Similar mouse models replacing *Otx2* with the *Drosophila otd* cDNA(*otd²/otd²* and *otd²FL/otd²FL*) have been generated in order to assess whether, despite the huge divergence in coding sequence, OTD and OTX2 share common genetic functions (Acampora et al. 2001a,b). As seen in mouse models replacing *Otx2* with *hOtx1*, the translation of the *otd* mRNA and maintenance of forebrain and midbrain identities is strictly dependent on the presence of 5′ and 3′ UTRs of *Otx2*. Together these findings indicate that OTD and OTX1 are functionally equivalent to OTX2 in the VE, AME, and ANE, and that this equivalence is subordinate to *Otx2* regulatory control (Fig. 2). The molecular nature of this extended OTD/OTX equivalence is still unclear and remains to be investigated in detail.

This aspect, however, raises the question whether functional equivalence means that downstream targets or morphogenetic pathways controlled by otd/Otx genes are the same, or are different but functionally redundant. It is necessary to define whether the functional equivalence is only an operative definition, or whether it underlines the presence of different pathways, each one specific for each member of the otd/Otx gene family, potentially resulting in the accomplishment of the same final morphogenetic program. In other words, it should be assessed whether otd/Otx genes act through the same target(s) or operate through different but equivalent pathways converging on the same final result. The answer to this question will clarify whether there is a real functional equivalence, or whether the equivalence among otd/Otx members does not exist at the molecular level.

Our experimental data support the notion that otd/Otx functions have been established in a common ancestor of fly and mouse and retained throughout evolution, while the regulatory control of their expression has been modified and re-adapted by evolutionary events that might have contributed to the specification of the vertebrate brain (Sharman and Brand 1998; Reichert and Simeone 1999; Acampora et al. 2001a,b).

This morphogenetic event might have coincided with the duplication, recruitment, and stabilization of conserved genetic functions into new cell types that, in turn, have modified or created new versions of preexisting developmental pathways (Holland 1999; Acampora et al. 2001a). Indeed, it is possible that conserved functions such as those encoded by OTD/OTX proteins acquired new roles because they became expressed in new cell types over the course of evolution. On the basis of this hypothesis, it is expected that drastic evolutionary events should act on the regulatory control (transcription, processing, and translation) of *Otx*-related genes, and the functional studies previously mentioned support this possibility. On this basis, it is tempting to speculate that, once established, the regulatory control of *Otx* genes has been recruited and maintained throughout vertebrate head evolution, thus contributing to the establishment of the morphogenetic pathway(s) required during gastrulation to specify and maintain the anterior neural plate.

ACKNOWLEDGMENTS

The author thanks members of his group for stimulating discussions, and S. Boscolo for manuscript preparation. The author also apologizes for those whose work has not been cited because of space constraints.

REFERENCES

Acampora D. and Simeone A. 1999. Understanding the roles of *Otx1* and *Otx2* in controlling brain morphogenesis. *Trends Neurosci.* **22:** 116–122.

Acampora D., Gulisano M., Broccoli V., and Simeone A. 2001a. *Otx* genes in brain morphogenesis. *Prog. Neurobiol.* **64:** 69–95.

Acampora D., Avantaggiato V., Tuorto F., Briata P., Corte G., and Simeone A. 1998. Visceral endoderm-restricted translation of *Otx1* mediates recovering of *Otx2* requirements for specification of anterior neural plate and proper gastrulation. *Development* **125:** 5091–5104.

Acampora D., Annino A., Puelles E., Alfano I., Tuorto F., and Simeone A. 2003. OTX1 compensates for OTX2 requirement in regionalisation of anterior neuroectoderm. *Gene Expr. Patterns* **3:** 497–501.

Acampora D., Mazan S., Lallemand Y., Avantaggiato V., Maury M., Simeone A., and Brûlet, P. 1995. Forebrain and midbrain regions are deleted in *Otx2⁻/⁻* mutants due to a defective anterior neuroectoderm specification during gastrulation. *Development* **121:** 3279–3290.

Acampora D., Pilo Boyl P., Signore M., Martinez Barbera J.P., Ilengo C., Puelles E., Annino A., Reichert H., Corte G., and Simeone A. 2001b. OTD/OTX2 functional equivalence depends on 5′ and 3′ UTR-mediated control of *Otx2* mRNA for nucleo-cytoplasmic export and epiblast-restricted translation. *Development* **128:** 4801–4813.

Ang S.-L. and Rossant J. 1993. Anterior mesendoderm induces mouse *Engrailed* genes in explant cultures. *Development* **118:** 139–149.

Ang S.-L., Conlon R.A., Jin O., and Rossant J. 1994. Positive and negative signals from mesoderm regulate the expression of mouse *Otx2* in ectoderm explants. *Development* **120:** 2979–2989.

Ang S.L., Jin O., Rhinn M., Daigle N., Stevenson L., and Rossant J.

1996. A targeted mouse *Otx2* mutation leads to severe defects in gastrulation and formation of axial mesoderm and to deletion of rostral brain. *Development* **122:** 243–252.

Avantaggiato V., Acampora D., Tuorto F., and Simeone A. 1996. Retinoic acid induces stage-specific repatterning of the rostral central nervous system. *Dev. Biol.* **175:** 347–357.

Bachiller D., Klingensmith J., Kemp C., Belo J.A., Anderson R.M., May S.R., McMahon J.A., McMahon A.P., Harland R.M., Rossant J., and De Robertis E.M. 2000. The organizer factors *Chordin* and *Noggin* are required for mouse forebrain development. *Nature* **403:** 658–661.

Bally-Cuif L., Gulisano M., Broccoli V., and Boncinelli E. 1995. c-*otx2* is expressed in two different phases of gastrulation and is sensitive to retinoic acid treatment in chick embryo. *Mech. Dev.* **49:** 49–63.

Beddington R.S.P. 1994. Induction of a second neural axis by the mouse node. *Development* **120:** 613–620.

Beddington R.S.P. and Robertson E.J. 1999. Axis development and early asymmetry in mammals. *Cell* **96:** 195–209.

Blitz I.L. and Cho K.W.Y. 1995. Anterior neuroectoderm is progressively induced during gastrulation: The role of the *Xenopus* homeobox gene *orthodenticle.* *Development* **121:** 993–1004.

Camus A., Davidson B.P., Billiards S., Khoo P.-L., Rivera-Pérez J.A., Wakamiya M., Behringer R.R., and Tam P.P.L. 2000. The morphogenetic role of midline mesendoderm and ectoderm in the development of the forebrain and the midbrain of the mouse embryo. *Development* **127:** 1799–1813.

Crossley P.H., Martinez S., and Martin G.R. 1996. Midbrain development induced by FGF8 in the chick embryo. *Nature* **380:** 66–68.

Davidson B.P., Kinder S.J., Steiner K., Schoenwolf G.C., and Tam P.P. 1999. Impact of node ablation on the morphogenesis of the body axis and the lateral asymmetry of the mouse embryo during early organogenesis. *Dev. Biol.* **1:** 11–26.

Ding J., Yang L., Yan Y.T., Chen A., Desai N., Wynshaw-Boris A., and Shen M.M. 1998. *Cripto* is required for correct orientation of the anterior-posterior axis in the mouse embryo. *Nature* **395:** 125–133.

Doniach T. 1993. Planar and vertical induction of antero-posterior pattern during the development of the amphibian central nervous system. *J. Neurobiol.* **24:** 1256–1276.

Episkopou V., Arkell R., Timmons P.M., Walsh J.J., Andrew R.L., and Swan D. 2001. Induction of the mammalian node requires *Arkadia* function in the extraembryonic lineages. *Nature* **12:** 825–830.

Gammill L.S. and Sive H. 1997. Identification of *otx2* target genes and restrictions ectodermal competence during *Xenopus* cement gland formation. *Development* **124:** 471–481.

Gurdon J.B. 1987. Embryonic induction-molecular prospects. *Development* **99:** 285–306.

Hallonet M., Kaestner K.H., Martin-Parras L., Sasak, H., Betz U.A., and Ang S.L. 2002. Maintenance of the specification of the anterior definitive endoderm and forebrain depends on the axial mesendoderm: A study using HNF3beta/Foxa2 conditional mutants. *Dev. Biol.* **1:** 20–33.

Holland P.W.H. 1999. The future of evolutionary developmental biology. *Nature* (suppl.) **402:** C41–C44.

Hoodless P.A., Pye M., Chazaud C., Labbe E., Attisano L., Rossant J., and Wrana J.L. 2001. *FoxH1* (Fast) functions to specify the anterior primitive streak in the mouse. *Genes Dev.* **15:** 1257–1271.

Houart C., Westerfield M., and Wilson S.W. 1998. A small population of anterior cells patterns the forebrain during zebrafish gastrulation. *Nature* **391:** 788–792.

Kimura C., Yoshinaga K., Tian E., Suzuki M., Aizawa S., and Matsuo I. 2000. Visceral endoderm mediates forebrain development by suppressing posteriorizing signals. *Dev. Biol.* **225:** 304–321.

Kinder S.J., Tsang T.E., Wakamiya M., Sasaki H., Behringer R.R., Nagy A., and Tam P.P. 2001. The organizer of the mouse gastrula is composed of a dynamic population of progenitor cells for the axial mesoderm. *Development* **128:** 3623–3634.

Klingensmith J., Ang S.L., Bachiller D., and Rossant J. 1999. Neural induction and patterning in the mouse in the absence of the node and its derivatives. *Dev. Biol.* **216:** 535–549.

Lawson K.A. and Pedersen R.A. 1987. Cell fate, morphogenetic movement and population kinetics of embryonic endoderm at the time of germ layer formation in the mouse. *Development* **101:** 627–652.

Li Y., Allende M.L., Finkelstein R., and Weinberg E.S. 1994. Expression of two zebrafish *orthodenticle*-related genes in the embryonic forebrain. *Mech. Dev.* **48:** 229–244.

Lu C.C., Brennan J., and Robertson E.R. 2001. From fertilization to gastrulation: Axis formation in the mouse embryo. *Curr. Opin. Genes. Dev.* **11:** 384–392.

Martinez Barbera J.P., Signore M., Pilo Boyl P., Puelles E., Acampora D., Gogoi R., Schubert F., Lumsden A., and Simeone A. 2001. Regionalisation of anterior neuroectoderm and its competence in responding to forebrain and midbrain inducing activities depend on mutual antagonism between OTX2 and GBX2. *Development* **128:** 4789–4800.

Matsuo I., Kuratani S., Kimura C., Takeda N., and Aizawa S. 1995. Mouse *Otx2* functions in the formation and patterning of rostral head. *Genes Dev.* **9:** 2646–2658.

Mercier P., Simeone A., Cotelli F., and Boncinelli E. 1995. Expression patterns of two *Otx* genes suggest a role in specifying anterior body structures in zebrafish. *Int. J. Dev. Biol.* **3:** 559–573.

Mori H., Miyazaki S., Morita I., Nitta H., and Mishina M. 1994. Different spatio-temporal expressions of three *otx* homeoprotein transcripts during zebrafish embryogenesis. *Mol. Brain. Res.* **27:** 221–231.

Pannese M., Polo C., Andreazzoli M., Vignali R., Kablar B., Barsacchi G., and Boncinelli E. 1995. The *Xenopus* homologue of *Otx2* is a maternal homeobox gene that demarcates and specifies anterior body regions. *Development* **121:** 707–720.

Perea-Gomez A., Lawson K.A., Rhinn M., Zakin L., Brulet P., Mazan S., and Ang S.L. 2001. *Otx2* is required for visceral endoderm movement and for the restriction of posterior signals in the epiblast of the mouse embryo. *Development* **128:** 753–765.

Reichert H. and Simeone A. 1999. Conserved usage of gap and homeotic genes in patterning the CNS. *Curr. Opin. Neurobiol.* **9:** 589–595.

Rhinn M., Dierich A., Le Meur M., and Ang S.-L. 1999. Cell autonomous and non-cell autonomous functions of *Otx2* in patterning the rostral brain. *Development* **126:** 4295–4304.

Rhinn M., Dierich A., Shawlot W., Behringer R.R., Le Meur M., and Ang S.-L. 1998. Sequential roles for *Otx2* in visceral endoderm and neuroectoderm for forebrain and midbrain induction and specification. *Development* **125:** 845–856.

Ruiz i Altaba A. 1993. Induction and axial patterning of the neural plate: Planar and vertical signals. *J. Neurobiol.* **24:** 1276–1304.

———. 1994. Pattern formation in the vertebrate neural plate. *Trends Neurosci.* **17:** 233–243.

Saude L., Woolley K., Martin P., Driever W., and Stemple D.L. 2000. Axis-inducing activities and cell fates of the zebrafish organizer. *Development* **127:** 3407–3417.

Sharman A.C. and Brand M. 1998. Evolution and homology of the nervous system: Cross-phylum rescues of *otd/Otx* genes. *Trends Genet.* **14:** 211–214.

Shawlot W., Wakamiya M., Kwan M.K., Kania A., Jessell T.M., and Behringer R.R. 1999. *Lim1* is required in both primitive streak-derived tissues and visceral endoderm for head formation in the mouse. *Development* **126:** 4925–4932.

Simeone A., Acampora D., Mallamaci A., Stornaiuolo A., D'Apice M.R., Nigro V., and Boncinelli E. 1993. A vertebrate gene related to *orthodenticle* contains a homeodomain of the *bicoid* class and demarcates anterior neuroectoderm in the gastrulating mouse embryo. *EMBO J.* **12:** 2735–2747.

Simeone A., Avantaggiato V., Moroni M.C., Mavilio F., Arra C., Cotelli F., Nigro V., and Acampora F. 1995. Retinoic acid induces stage-specific antero-posterior transformation of rostral central nervous system. *Mech. Dev.* **51:** 83–98.

Spemann H. and Mangold H. 1924. Über induktion von Embryonanlagen durch implantation artfremder Organisatoren. *Wilhem Roux's Arch. Entwicklungsmech. Org.* **100:** 599–638.

Stern C. 2001. Initial patterning of the central nervous system: How many organizers? *Nat. Rev. Neurosc.* **2:** 92–98.

Storey K.G., Crossley J.M., De Robertis E.M., Norris W.E., and Stern C.D. 1992. Neural induction and regionalisation in the chick embryo. *Development* **114:** 729–741.

Suda Y., Nakababayashi J., Matsuo I., and Aizawa S. 1999. Functional equivalency between *Otx2* and *Otx1* in development of the rostral head. *Development* **126:** 743–757.

Sun X., Meyers E.N., Lewandoski M., and Martin G.R. 1999. Targeted disruption of *Fgf8* causes failure of cell migration in the gastrulating mouse embryo. *Genes Dev.* **13:** 1834–1846.

Tam P.P. and Beddington R.S.P. 1992. Establishment and organization of germ layers in the gastrulating mouse embryo. *Ciba Found. Symp.* **165:** 27–41.

Tam P.P.L. and Steiner K.A. 1999. Anterior patterning by synergistic activity of the early gastrula organizer and the anterior germ layer tissues of the mouse embryo. *Development* **126:** 5171–5179.

Thomas P. and Beddington R.S.P. 1996. Anterior primitive endoderm may be responsible for patterning the anterior neural plate in the mouse embryo. *Curr. Biol.* **6:** 1487–1496.

Tian E., Kimura C., Takeda N., Aizawa S., and Matsuo I. 2002. *Otx2* is required to respond to signals from anterior neural ridge for forebrain specification. *Dev. Biol.* **242:** 204–223.

Wada S. and Saiga H. 1999. Vegetal cell fate specification and anterior neuroectoderm formation by *Hroth*, the ascidian homologue of *orthodenticle/otx*. *Mech. Dev.* **82:** 67–77.

Wada S., Katsuyama Y., Sato Y., Itoh C., and Saiga H. 1996. *Hroth* an *orthodenticle*-related homeobox gene of the ascidian, *Halocynthia roretzi*: Its expression and putative roles in the axis formation during embryogenesis. *Mech. Dev.* **60:** 59–71.

Waddington C.H. 1933. Induction by the primitive streak and its derivatives in the chick. *J. Exp. Biol.* **10:** 38–46.

Williams N.A. and Holland P.W.H. 1996. Old head on young shoulders. *Nature* **383:** 490.

Withington S., Beddington R., and Cooke J. 2001. Foregut endoderm is required at head process stages for anteriormost neural patterning in chick. *Development* **128:** 309–320.

Wurst W. and Bally-Cuif L. 2001. Neural plate patterning: Upstream and downstream of the isthmic organizer. *Nat. Rev. Neurosc.* **2:** 99–108.

Zoltewicz J.S. and Gerhart J.C. 1997. The Spemann organizer of *Xenopus* is patterned along its anteroposterior axis at the earliest gastrula stage. *Dev. Biol.* **192:** 482–491.

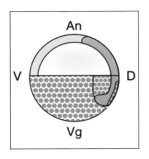

FUNCTION OF THE WINGED HELIX TRANSCRIPTION FACTOR *HNF3β/FoxA2* DURING GASTRULATION

D.C. Weinstein

Department of Pharmacology and Biological Chemistry, Mount Sinai School of Medicine, New York, New York 10029

INTRODUCTION

The hepatocyte nuclear factor 3 (HNF3) family consists of three related transcription factors, HNF3α, HNF3β, and HNF3γ, first identified by their ability to regulate liver-specific gene expression (Lai et al. 1990, 1991). HNF3 proteins contain a highly conserved, roughly 110-amino acid DNA-binding domain, first identified in the *Drosophila fork head* gene, which itself plays a critical role in the development of the anterior and posterior gut of the fly (Weigel et al. 1989; Weigel and Jackle 1990). This now-sizeable class of proteins is referred to variously as the Forkhead, Fox (for Forkhead box), or winged helix family of transcription factors. The term "winged helix" derives from the crystal structure of the HNF3γ DNA-binding domain bound, as a monomer, to DNA—this domain folds into a helix-turn-helix variant, characterized by three α-helices and two wing-like loops (Clark et al. 1993). HNF3β, the focus of this chapter, is also called FOXA2 (human)/Foxa2 (mouse)/FoxA2 (all other chordates), based on the phylogenetic grouping of chordate Fox genes according to the amino acid sequence of their DNA-binding domains (Kaestner et al. 2000).

Several years after its initial characterization as a "liver-enriched" transcription factor, expression of *HNF3β* was detected in embryonic organizing centers of the gastrula-stage frog, fish, chicken, and mouse. Much of our under-standing of *HNF3β* function during early development comes from genetic studies in the mouse—mammalian *HNF3β* is required cell-autonomously for the formation of axial mesodermal structures, the subsequent dorsoventral patterning of the neural tube, and the formation of the embryonic gut, and is involved in the non-cell-autonomous positioning and morphogenesis of the primitive streak. We focus here on the role of *HNF3β* during vertebrate gastrulation, the signals that govern *HNF3β* expression, and the mechanisms by which *HNF3β* regulates the formation and patterning of multiple cell types in the early embryo.

EXPRESSION STUDIES

Vertebrates

Embryonic Expression of HNF3β

Localization of embryonic *HNF3β* transcripts was first described in the mouse. Although *HNF3β* transcripts are present in both embryonic and extraembryonic lineages during early murine development, initial reports focused on *HNF3β* expression in the embryo proper (Ang et al. 1993; Monaghan et al. 1993; Ruiz i Altaba et al. 1993b; Sasaki and Hogan 1993). At the onset of gastrulation (E6.5), *HNF3β*-positive cells are detected at the anterior tip of the primitive streak (Fig. 1). During early-streak stages, this expression domain expands to include a broad region at the anterior third of the streak, where *HNF3β* transcripts are found both in epiblast cells and in delaminating mesoderm. By E7.5, *HNF3β* is found in both layers of the definitive node, as well as in midline mesodermal and endodermal cells that have migrated anteriorly. These expression patterns are consistent with those observed in mice containing a heterozygous *lacZ* knock-in at the *HNF3β* locus (Weinstein et al. 1994).

Similar expression domains have been described for both chick *HNF3β* and *axial*, the zebrafish *HNF3β* gene (Fig. 1). At stage XII, chick *HNF3β* transcripts are detected in Koller's sickle and, subsequently, in the anterior third of the primitive streak. By stage 4, chick *HNF3β* is expressed in Hensen's node, both in the superficial ectoderm and in the deeper layers (Ruiz i Altaba et al. 1995). *axial* transcripts are

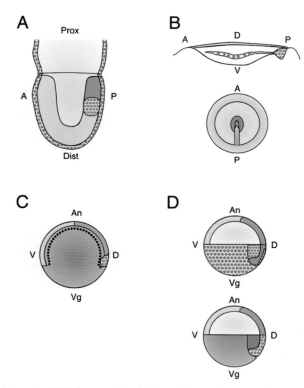

Figure 1. *HNF3β* expression during early gastrula stages of representative vertebrate organisms. (A) *Mus musculus*, E6.5; (B) *Gallus gallus*, stage XII, lateral view (*top*), stage 3, dorsal view (*bottom*); (C) *Danio rerio*, 50% epiboly; (D) *Xenopus laevis*, stage 10+. Bottom panel in *D* shows gastrula-stage expression of the related winged helix factor *Pintallavis*. (*Yellow*) Ectoderm/epidermis; (*blue*) neuroectoderm; (*red*) mesoderm; (*green*) embryonic endoderm; (*brown*) extraembryonic endoderm. *HNF3β* transcripts are shown as dots coded to match the tissues in which they are expressed; mesendodermal *HNF3β* expression in organizer tissue is shown in red; see text for details. (Prox) proximal; (Dist) distal; (A) anterior; (P) Posterior; (D) dorsal; (V) ventral; (An) animal; (Vg) vegetal.

first detected at 40% epiboly in a thin band on the dorsal side of the embryo. By 60% epiboly, *axial* expression is seen in the cells of the zebrafish hypoblast, which will give rise to embryonic mesendodermal derivatives (Strahle et al. 1993). At the end of gastrulation, mouse, chick, and zebrafish *HNF3β* are all expressed in the notochord, the embryonic gut, and the floor plate of the neural tube.

Surprisingly, *Xenopus HNF3β* (*XFD3*) (Knochel et al. 1992; Ruiz i Altaba et al. 1993b) is not highly expressed prior to or during gastrulation. However, dorsal marginal zone expression of *Xenopus HNF3β* appears to be limited to the cells of the suprablastoporal endoderm (Fig. 1) (Ruiz i Altaba et al. 1993b); these cells are fated to give rise to the pharyngeal endoderm (Keller 1975). Consistently, *Xenopus HNF3β* is never observed in the notochord (Ruiz i Altaba et al. 1993b). Thus, in an apparent departure from patterns observed in fish and amniotes, *Xenopus HNF3β* expression appears to be excluded from the mesoderm, both during and after gastrulation. Based on this divergent expression pattern and on functional studies (described below), it has been suggested that the related winged helix factor Pintallavis/XFKH1XFD1/FoxA4, expressed in the dorsal lip at stage 10 and subsequently in axial mesoderm and the ventral midline of the neural tube, functions as the *Xenopus* equivalent of the mesodermal and neurectodermal component of HNF3β during early development (Fig. 1) (Dirksen and Jamrich 1992; Knochel et al. 1992; Ruiz i Altaba and Jessell 1992; Ruiz i Altaba et al. 1993b).

Extraembryonic Expression of HNF3β

In addition to its expression in early embryonic lineages, mouse *HNF3β* transcripts are also found in the visceral endoderm (VE), an extraembryonic structure, prior to and during gastrulation (Ang et al. 1993; Monaghan et al. 1993; Ruiz i Altaba et al. 1993b; Sasaki and Hogan 1993; Filosa et al. 1997; Perea-Gomez et al. 1999) (see Chapter 16). The VE is thought to serve as a source both of nutrients and of signals that regulate early embryonic polarity (Chapter 31; see below). *HNF3β* is expressed in the VE lining both embryonic and extraembryonic tissues at E5.5, prior to gastrulation. This expression is maintained through E6.5, when embryonic *HNF3β* expression is first detected (Fig. 1). In the chick, *HNF3β* is first observed throughout the forming hypoblast, the avian equivalent of the mouse anterior visceral endoderm (AVE) (Foley et al. 2000) (see Chapter 15). Using a reverse transcription-polymerase chain reaction (RT-PCR) assay, we have found that *Xenopus HNF3β* is also expressed at low levels throughout the vegetal pole during early gastrulation. This region shares functional similarities with the mouse VE and includes the deep cells of the anterior endoderm (AE), the potential AVE equivalent (Zorn et al. 1999; T. Haremaki and D. Weinstein, unpubl.) (Chapter 31). *Axial* expression, however, has not been reported in the cells of the yolk syncytial layer, which may represent the zebrafish equivalent of the mouse VE or chick hypoblast and endoblast (Fekany et al. 1999; Ho et al. 1999).

Protochordates, Flies, Worms, and Diploblasts

HNF3β-related genes are expressed in the notochord, floor plate, and endoderm of at least two ascidian species, *Ciona intestinalis* and *Molgula oculata*, suggesting a conserved role for HNF3β in lower chordates and vertebrates (Corbo et al. 1997; Olsen and Jeffery 1997). However, although these ascidian proteins (Cifkh/FoxA5 and MocuFH1/FoxA5, respectively) clearly belong to the FoxA class of winged helix factors, they cannot be definitively classified as HNF3β homologs.

Although none of the following can be said to represent true *HNF3β* homologs, *Drosophila fork head* (the founding winged helix/forkhead transcription factor), the *Caenorhabditis elegans* gene *pha-4* (Ce-fkh-1), and *budhead*, a winged helix gene from the cnidarian *Hydra vulgaris*, share high homology with the FoxA chordate genes in both their DNA-binding domains and in one (or two, for *fork head* and *budhead*) carboxy-terminal region implicated in *trans*-activation (Weigel et al. 1989; Azzaria et al. 1996; Martinez et al. 1997). Early *fork head* expression patterns (high levels of expression in foregut and hindgut) do not closely resemble the range of *HNF3β* expression in chordates. *fork head* mutations result in homeotic transformation of ectodermal portions of the gut into head structures, demonstrating the importance of the FoxA proteins in early development of both flies and chordates (Weigel et al. 1989). The early expression of *pha-4*, and its requirement for pharynx and rectum formation, suggest a marked functional similarity between the role of HNF3-related factors in flies and worms (Azzaria et al. 1996; Horner et al. 1998; Kalb et al. 1998). Finally, *budhead* is expressed during axis formation in hydra bud development, suggesting conserved roles for FoxA proteins throughout the metazoan subkingdom (Martinez et al. 1997).

HNF3β FUNCTION DURING EARLY DEVELOPMENT

Mouse Knockouts

Much of our appreciation of *HNF3β* function during gastrulation comes from genetic studies in the mouse. Targeted deletion of *HNF3β* leads to embryonic lethality and a number of pronounced developmental defects (Fig. 2) (Ang and Rossant 1994; Weinstein et al. 1994). Abnormalities include a failure of primitive streak elongation, the absence of a definitive node and notochord, defects in gut development,

the morphological and functional lack of a floor plate and subsequent disruption of dorsoventral polarity in the neural tube, and the loss of anterior neural structures in a subset of mutant embryos. Anterior truncations have also been observed in embryos that are heterozygous for *HNF3β* and either lack *goosecoid*, or are additionally heterozygous for *Otx2* or *nodal* (Filosa et al. 1997; Varlet et al. 1997; Jin et al. 2001). Finally, the normally asymmetric expression of several markers along the left–right axis is either absent or symmetric in *HNF3β* mutant embryos (Dufort et al. 1998), although this effect may be largely secondary to the loss of midline structures. A role for *HNF3β* in the asymmetric expression of *Nodal* has also been suggested in analyses of *HNF3β*$^{+/-}$; *nodal*$^{+/-}$ mice (Collignon et al. 1996).

Although examination of HNF3β mutant mice clearly demonstrated a central role for this gene in early mam-

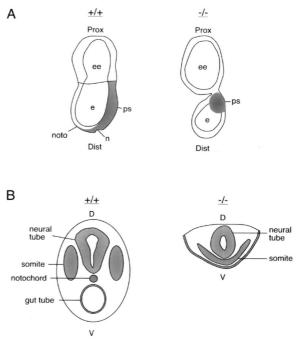

Figure 2. Phenotype of *HNF3β*$^{-/-}$ mutant mouse embryos. (*A*) Schematic lateral views of E7.5 wild-type (*left*) and *HNF3β*$^{-/-}$ (*right*) embryos. *T*, expressed in the primitive streak, node, and notochord of wild-type embryos, is shown in purple to highlight defects in the mutant mice, including the lack of a definitive node and notochord, and a shortened primitive streak. Mutant embryos also display a constriction between embryonic and extraembryonic regions. (*B*) Schematic transverse sections from E9.5 wild-type (*left*) and *HNF3β*$^{-/-}$ (*right*) embryos. Mutant embryos lack a notochord (*dark red* in wild-type embryo); fusion of somites (*red*) across the midline is sometimes observed. Additional defects include disruption of dorsoventral polarity in the neural tube (*blue*) and the failure to form fore- or midgut (*green*). A loss of anterior neural structures is also observed in some *HNF3β*$^{-/-}$ embryos (not shown); see text for details. (Prox) proximal; (Dist) distal; (D) dorsal; (V) ventral; (e) embryonic region; (ee) extraembryonic region; (ps) primitive streak; (n) node; (noto) notochord.

malian development, these studies did not distinguish between the relative contribution of embryonic and extraembryonic *HNF3β* expression. Subsequent experiments, in which *HNF3β* was selectively eliminated from either embryonic or extraembryonic tissues, resolved this issue. Chimeric embryos were generated in which wild-type or homozygous mutant *HNF3β* ES cells were fused with tetraploid mutant or wild-type embryos—in these mice, tetraploid cells contribute solely to extraembryonic lineages (Dufort et al. 1998). Chimeric embryos with wild-type VE still lack a node and notochord, fail to form foregut and midgut endoderm, and show disruption of DV patterning in the neural tube; these embryos do, however, show a rescue both of anterior neural plate specification and of normal primitive streak morphogenesis. The analysis of *HNF3β⁻/⁻; Lim1⁻/⁻* mice, described below, also demonstrated a clear role for extraembryonic HNF3β in patterning of the epiblast (Perea-Gomez et al. 1999). These studies with chimeric *HNF3β* mutant mice thus revealed a functional division between the embryonic expression of *HNF3β*, required for node and notochord formation, as well as foregut and midgut endoderm, and extraembryonic *HNF3β*, required for normal development of the primitive streak.

Xenopus Pintallavis as a Functional Equivalent of Embryonic Mouse *HNF3β*

As described above, it has been proposed that the winged helix *Xenopus* gene *Pintallavis* functions as an *HNF3β* equivalent during early development (Ruiz i Altaba et al. 1993b). Gain-of-function studies have demonstrated that both *Pintallavis* and mouse *HNF3β* can stimulate the expression of floor plate markers at ectopic locations in the neural tube, suggesting a conservation of function at later stages of development (Ruiz i Altaba et al. 1993a; Sasaki and Hogan 1994). Furthermore, co-injection of *Pintallavis* and *Brachyury* mRNA induces notochord formation in animal cap explants, pointing to a similar role for *Pintallavis* and the embryonic component of murine *HNF3β* during dorsal mesoderm formation (O'Reilly et al. 1995).

Loss-of-function studies have not shown a precise correlation between the requirement of *Pintallavis* and mammalian *HNF3β* during early development. Expression of a Pintallavis-Engrailed repressor fusion protein (Pintallavis-EnR) disrupts convergent extension movements during *Xenopus* gastrulation, similar to effects seen following antisense oligonucleotide-mediated disruption of the ascidian *HNF3β*-related gene *MocuFH1* (Olsen and Jeffery 1997; Saka et al. 2000). On the other hand, the notochord, although truncated, is clearly present in embryos expressing Pintallavis-EnR, suggesting that *Pintallavis* and mouse *HNF3β* may not play entirely equivalent roles during notochord formation (Saka et al. 2000). It is possible, however, that *Pintallavis* is involved with, but alone is not required

for, axial mesoderm formation in *Xenopus*. Loss-of-function experiments have not been described for *Xenopus HNF3β*. Thus, it has not been possible to make direct functional comparisons between *HNF3β* and *Pintallavis* in the frog. As suggested by the early expression of *Xenopus HNF3β*, however, this gene is more likely to function in a manner analogous to that of the extraembryonic (VE) component of mouse *HNF3β*.

MECHANISMS OF HNF3β ACTIVITY

Biochemical and cell culture studies suggest that HNF3β functions as a transcriptional activator (Cereghini 1996). It should be kept in mind, however, that HNF3β might also function as a de facto repressor in some contexts during early development, via competition with other *trans*-activators for overlapping binding sites. For example, in cell culture assays, HNF3β has been shown to compete with, and thus antagonize, *trans*-activation of the liver-specific aldolase B gene by the unrelated and, in this case, more potent *trans*-activator HNF1; this effect is due to competition for binding to mutually exclusive, partly overlapping sites (Gregori et al. 1994). Furthermore, as suggested by the similarity between the winged helix structure of the HNF3β DNA-binding domain and that of linker histone (Clark et al. 1993; Ramakrishnan et al. 1993), studies by the Zaret group have demonstrated that HNF3 proteins may function in some contexts primarily as "pioneer" factors, binding to and opening compacted chromatin (Cirillo et al. 2002). This activity does not, however, necessarily preclude a subsequent role for HNF3β in transcriptional activation.

HNF3β TARGET GENES

Mediators of HNF3β Activity in the Mouse Visceral Endoderm

The nature of the early targets of HNF3β in the visceral endoderm can be inferred from compound mutants of *HNF3β* and the LIM homeodomain gene *Lim1/Lhx1*, the latter required for anterior neural development (Shawlot and Behringer 1995; Perea-Gomez et al. 1999). *HNF3β⁻/⁻; Lim1⁻/⁻* mice have a greatly enlarged primitive streak and excess production of ventral mesoderm, coupled with an absence of both dorsal mesoderm and embryonic ectoderm. The primary defect in these compound mutants is due to the loss of both HNF3β and Lim1 activity in the visceral endoderm (Perea-Gomez et al. 1999). Several reports have demonstrated a fundamental role for the AVE in positioning of the primitive streak, with the head formed in the anterior epiblast, adjacent to the AVE, and primitive streak formation occurring at the posterior end (Beddington and Robertson 1999) (see Chapters 16, 27, and 31). Recent studies in the mouse and chick point to Nodal antagonists, expressed in the AVE and hypoblast, respectively, as key mediators of early embryonic polarity, restricting initially widespread, mesoderm-inducing, Nodal signaling to the eventual posterior site of primitive streak formation (Bertocchini and Stern 2002; Perea-Gomez et al. 2002). The enlarged streak observed in the $HNF3\beta^{-/-};Lim1^{-/-}$ compound mutant mice may thus reflect a role for these genes in the restriction of Nodal activity, perhaps in part via stimulation of Nodal antagonists in the AVE.

A potential mediator of antagonist activation by HNF3β is the homeobox gene *Hex*, one of the earliest genes expressed in the mouse AVE (Thomas et al. 1998). The *Xenopus Hex* homolog, *Xhex*, induces the Nodal, Wnt, and bone morphogenetic protein (BMP) antagonist *Cerberus* expression in endodermal explants when overexpressed (Zorn et al. 1999). *Hex* expression is lost in $HNF3\beta^{-/-}$; $Lim1^{-/-}$ mice (Perea-Gomez et al. 1999). Furthermore, the mouse *Hex* promoter contains sites that bind HNF3β and confer HNF3β responsiveness in both the *Hex* promoter and heterologous promoters in cell culture assays (Denson et al. 2000). *Hex* alone is not required for extraembryonic *Cerberus-like* (*Cerl*) expression in the mouse, however, since *Cerl* expression in the AVE is maintained (as is AVE function) in *Hex* $^{-/-}$ mice (Martinez Barbera et al. 2000).

HNF3β may also regulate embryonic polarity through protein–protein, as well as protein–DNA, interactions. HNF3β has been shown to bind the transcription factor Otx2, the latter required in the AVE for the initiation of anterior neural patterning (Ang et al. 1996; Rhinn et al. 1998; Nakano et al. 2000) (see Chapter 39). This interaction may contribute to the genetic interaction observed between *HNF3β* and *Otx2* during anterior development, although in vitro studies suggest that binding of HNF3β inhibits, rather than stimulates, *trans*-activation by Otx2 (Nakano et al. 2000; Jin et al. 2001). Physical interaction has also been demonstrated between HNF3β and the transducin-like Enhancer of Split (TLE) proteins, mammalian homologs of the *Drosophila* transcriptional corepressor Groucho (Wang et al. 2000). Overexpression of TLE proteins inhibits *trans*-activation by HNF3β in cell culture, suggesting that TLE proteins may function as negative regulators of HNF3β activity in vivo. The developmental relevance of this interaction, however, remains to be explored (Wang et al. 2000).

Mediators of HNF3β in the Node and Its Derivatives

Several candidate regulators of embryonic *HNF3β* have been proposed. *Hex*, described above, is also expressed in the mouse anterior definitive endoderm (ADE) (Thomas et al. 1998); the relatively subtle early effects of *Hex* deletion, however, suggest that *Hex* is not a central effector of *HNF3β* function in axial mesendoderm (Martinez Barbera et al. 2000). A more likely primary mediator of *HNF3β* function in the mouse node is *Nodal*. A node-specific enhancer in an

upstream region of the *Nodal* promoter includes a predicted HNF3β-binding site (Overdier et al. 1994; Brennan et al. 2002). Furthermore, this site is contained within a 110-bp stretch that shares high identity with the corresponding region of the human *Nodal* locus (Brennan et al. 2002). These data, coupled with the absence of *nodal* expression in chimeric mice lacking embryonic *HNF3β* (Dufort et al. 1998), suggest that *HNF3β* is a direct activator of *nodal* expression in the mouse node.

A third potential mediator of *HNF3β* in the node and its derivatives is *Sonic hedgehog* (*Shh*). *Shh* is co-expressed with *HNF3β*, in lower and higher vertebrates, in a number of embryonic tissues including the late node or its equivalents, notochord and floor plate (Echelard et al. 1993; Krauss et al. 1993; Riddle et al. 1993; Chang et al. 1994; Roelink et al. 1994; Ekker et al. 1995a,b). The zebrafish *shh* promoter contains binding sites for Axial (HNF3β), which are required for HNF3β-dependent reporter gene activation in cell culture (Chang et al. 1997). Transgenic studies in the mouse demonstrate that deletions encompassing HNF3β-binding sites in the *Shh* promoter correlate with a loss of reporter gene expression in the notochord (Epstein et al. 1999). However, notochord defects in *Shh*$^{-/-}$ mice are less severe than those seen in *HNF3β* mutants, indicating that *Shh* is not the sole mediator of *HNF3β* activity in this tissue. Furthermore, reporter studies in zebrafish embryos suggest that essential enhancer regions for *shh* expression in the notochord do not contain HNF3β-binding sites, providing evidence that direct early regulation of *shh* by HNF3β is not a conserved feature among vertebrates (Muller et al. 1999).

HNF3β also contains an HNF3-binding site within its own promoter (Pani et al. 1992). Consistent with a model of *HNF3β* autoactivation, ectopic *HNF3β* induces *HNF3β* expression in the midbrain/hindbrain of transgenic mouse embryos (Sasaki and Hogan 1994). DNA-binding and mutagenesis studies with FoxA5/Cifkh in *Ciona* also suggest that FoxA autoregulation plays a role in maintaining expression in the *Ciona* notochord, endoderm, and CNS (Di Gregorio et al. 2001).

Upstream Activators of *HNF3β*

Genetic studies point to a number of genes that function upstream of *HNF3β* in both embryonic and extraembryonic tissue. The *Nodal* effector Smad2 probably stimulates *HNF3β* expression in the visceral endoderm, as mice that lack extraembryonic Smad2 activity phenocopy $HNF3\beta^{-/-}$; $Lim1^{-/-}$ compound mutants (Waldrip et al. 1998; Perea-Gomez et al. 1999). Modulators of embryonic *HNF3β* expression include the nuclear protein Arkadia, whose extraembryonic activity is required for embryonic expression of *HNF3β* (Episkopou et al. 2001; Niederlander et al. 2001), and the winged helix protein FoxH1/Fast, whose activity is required within the cells of the mouse epiblast for embryonic expression of *HNF3β* (Hoodless et al. 2001; Yamamoto et al. 2001). Finally, widespread ectopic expression of *HNF3β* throughout the epiblast of *Nodal* mutant embryos suggests that early Nodal signaling is also involved in the inhibition of embryonic *HNF3β* expression (Brennan et al. 2001).

We lack, at present, a precise understanding of the molecular mechanisms that link Nodal/Smad signaling to regulation of *HNF3β* expression. Several groups have demonstrated, however, that induction of *Pintallavis* in response to stimulation by the TGF-β ligand Activin does not require protein synthesis (Dirksen and Jamrich 1992; Knochel et al. 1992; Ruiz i Altaba and Jessell 1992). This direct response to Activin, or to a constitutively active *Smad2* construct, can be conferred by a 107-bp region in the first intron of *Pintallavis* (Howell and Hill 1997). These results suggest that *HNF3β* may be a direct target of Nodal/Smad signaling in some embryonic contexts.

Two distinct enhancers have been identified that mediate early embryonic *HNF3β* expression in the mouse—an upstream sequence, which governs node/notochord expression, and a downstream region, which regulates gene expression in the floor plate, posterior notochord, and posterior gut epithelium (Sasaki and Hogan 1996). Regulatory sequences that govern expression of *HNF3β* in the visceral endoderm have not been described. Although the factors governing activation of the floor plate enhancer appear to include members of the Gli family of zinc finger proteins (Sasaki et al. 1997, 1999), factors that regulate the node/notochord enhancer have not been identified. Sasaki and colleagues have, however, characterized regions within the node/notochord enhancer that are conserved between the mouse, chicken, and the dwarf gourami (*Colisa lalia*) fish that are required for node/notochord expression in transgenic mouse embryos (Nishizaki et al. 2001). Sequence analysis of these regions does not reveal consensus binding sites for any known regulators of node/notochord development, suggesting that the gastrula-stage expression of *HNF3β* is regulated by one or more novel DNA-binding proteins. Promoter studies performed on upstream elements of the *Ciona* gene *Ci-fkh/FoxA5* identified a region responsible for notochord/endoderm expression of *Ci-fkh* within this domain. A T-box recognition sequence was identified that, when mutated, disrupts *Ci-fkh* expression in the endoderm (Di Gregorio et al. 2001). Similar sequences have not, however, been reported in vertebrate *HNF3β* promoters.

CONCLUSIONS

HNF3β plays several distinct and fundamental roles during early mouse embryogenesis. Expression of *HNF3β* in the visceral endoderm during pre-gastrula and gastrula devel-

opment contributes to the positioning and morphogenesis of the primitive streak. Embryonic expression of *HNF3β* in the node and its derivatives is required for the formation of the definitive node and notochord, the development of the gut, and the dorsoventral patterning of the neural tube. Expression studies suggest that HNF3β, and/or a closely related protein, may function in one or both of these capacities in all chordates examined. Despite a long-held appreciation of the requirement for HNF3β function during gastrulation, there is still relatively little known about the factors that directly regulate, or are regulated by, this transcription factor. Functional screening and subtractive hybridization approaches in amenable species may be useful for the identification, respectively, of developmentally relevant *HNF3β* activators and target genes.

ACKNOWLEDGMENTS

I thank T. Haremaki for assistance with the figures, and P. Hoodless for critical reading of the manuscript.

REFERENCES

Ang S.L. and Rossant J. 1994. HNF-3 beta is essential for node and notochord formation in mouse development. *Cell* **78:** 561–574.

Ang S.L. Jin O., Rhinn M., Daigle N., Stevenson L., and Rossant J. 1996. A targeted mouse Otx2 mutation leads to severe defects in gastrulation and formation of axial mesoderm and to deletion of rostral brain. *Development* **122:** 243–252.

Ang S.L., Wierda A., Wong D., Stevens K.A., Cascio S., Rossant J., and Zaret K.S. 1993. The formation and maintenance of the definitive endoderm lineage in the mouse: Involvement of HNF3/forkhead proteins. *Development* **119:** 1301–1315.

Azzaria M., Goszczynski B., Chung M. A., Kalb J.M., and McGhee J.D. 1996. A fork head/HNF-3 homolog expressed in the pharynx and intestine of the *Caenorhabditis elegans* embryo. *Dev. Biol.* **178:** 289–303.

Beddington R.S. and Robertson E.J. 1999. Axis development and early asymmetry in mammals. *Cell* **96:** 195–209.

Bertocchini F. and Stern C.D. 2002. The hypoblast of the chick embryo positions the primitive streak by antagonizing nodal signaling. *Dev. Cell* **3:** 735–744.

Brennan J., Norris D.P., and Robertson E.J. 2002. Nodal activity in the node governs left-right asymmetry. *Genes Dev.* **16:** 2339–2344.

Brennan J., Lu C.C., Norris D.P., Rodriguez T.A., Beddington R.S.P., and Robertson E.J. 2001. Nodal signalling in the epiblast patterns of the early mouse embryo. *Nature* **411:** 965–969.

Cereghini S. 1996. Liver-enriched transcription factors and hepatocyte differentiation. *FASEB J.* **10:** 267–282.

Chang B.E., Blader P., Fischer N., Ingham P.W., and Strahle U. 1997. Axial (HNF3beta) and retinoic acid receptors are regulators of the zebrafish sonic hedgehog promoter. *EMBO J.* **16:** 3955–3964.

Chang D.T., Lopez A., von Kessler D.P., Chiang C., Simandl B.K., Zhao R., Seldin M.F., Fallon J.F., and Beachy P.A. 1994. Products, genetic linkage and limb patterning activity of a murine hedgehog gene. *Development* **120:** 3339–3353.

Cirillo L.A., Lin F.R., Cuesta I., Friedman D., Jarnik M., and Zaret K.S. 2002. Opening of compacted chromatin by early developmental transcription factors HNF3 (FoxA) and GATA-4. *Mol. Cell* **9:** 279–289.

Clark K.L., Halay E.D., Lai E., and Burley S.K. 1993. Co-crystal structure of the HNF-3/fork head DNA-recognition motif resembles histone H5. *Nature* **364:** 412–420.

Collignon J., Varlet I., and Robertson E.J. 1996. Relationship between asymmetric nodal expression and the direction of embryonic turning. *Nature* **381:** 155–158.

Corbo J.C., Erives A., Di Gregorio A., Chang A., and Levine M. 1997. Dorsoventral patterning of the vertebrate neural tube is conserved in a protochordate. *Development* **124:** 2335–2344.

Denson L.A., McClure M.H., Bogue C.W., Karpen S.J., and Jacobs H.C. 2000. HNF3beta and GATA-4 transactivate the liver-enriched homeobox gene, Hex. *Gene* **246:** 311–320.

Di Gregorio A., Corbo J.C., and Levine M. 2001. The regulation of forkhead/HNF-3beta expression in the Ciona embryo. *Dev. Biol.* **229:** 31–43.

Dirksen M.L. and Jamrich M. 1992. A novel, activin-inducible, blastopore lip-specific gene of *Xenopus laevis* contains a fork head DNA-binding domain. *Genes Dev.* **6:** 599–608.

Dufort D., Schwartz L., Harpal K., and Rossant J. 1998. The transcription factor HNF3beta is required in visceral endoderm for normal primitive streak morphogenesis. *Development* **125:** 3015–3025.

Echelard Y., Epstein D.J., St-Jacques B., Shen L., Mohler J., McMahon J.A., and McMahon A.P. 1993. Sonic hedgehog, a member of a family of putative signaling molecules, is implicated in the regulation of CNS polarity. *Cell* **75:** 1417–1430.

Ekker S.C., McGrew L.L., Lai C.J., Lee J.J., von Kessler D.P., Moon R.T., and Beachy P.A. 1995a. Distinct expression and shared activities of members of the hedgehog gene family of *Xenopus laevis*. *Development* **121:** 2337–2347.

Ekker S.C., Ungar A.R., Greenstein P., von Kessler D.P., Porter J.A., Moon R.T., and Beachy P.A. 1995b. Patterning activities of vertebrate hedgehog proteins in the developing eye and brain. *Curr. Biol.* **5:** 944–955.

Episkopou V., Arkell R., Timmons P.M., Walsh J.J., Andrew R.L., and Swan D. 2001. Induction of the mammalian node requires Arkadia function in the extraembryonic lineages. *Nature* **410:** 825–830.

Epstein D.J., McMahon A.P., and Joyner A.L. 1999. Regionalization of Sonic hedgehog transcription along the anteroposterior axis of the mouse central nervous system is regulated by Hnf3-dependent and -independent mechanisms. *Development* **126:** 281–292.

Fekany K., Yamanaka Y., Leung T., Sirotkin H.I., Topczewski J., Gates M.A., Hibi M., Renucci A., Stemple D., Radbill A. et al. 1999. The zebrafish bozozok locus encodes Dharma, a homeodomain protein essential for induction of gastrula organizer and dorsoanterior embryonic structures. *Development* **126:** 1427–1438.

Filosa S., Rivera-Perez J.A., Gomez A.P., Gansmuller A., Sasaki H., Behringer R.R., and Ang S.L. 1997. Goosecoid and HNF-3beta genetically interact to regulate neural tube patterning during mouse embryogenesis. *Development* **124:** 2843–2854.

Foley A.C., Skromne I., and Stern C.D. 2000. Reconciling different models of forebrain induction and patterning: A dual role for the hypoblast. *Development* **127:** 3839–3854.

Gregori C., Kahn A., and Pichard A.L. 1994. Activity of the rat liver-specific aldolase B promoter is restrained by HNF3. *Nucleic Acids Res.* **22:** 1242–1246.

Ho C.Y., Houart C., Wilson S.W., and Stainier D.Y. 1999. A role for the extraembryonic yolk syncytial layer in patterning the zebrafish embryo suggested by properties of the hex gene. *Curr. Biol.* **9:**

1131–1134.

Hoodless P.A., Pye M., Chazaud C., Labbe E., Attisano L., Rossant J., and Wrana J.L. 2001. FoxH1 (Fast) functions to specify the anterior primitive streak in the mouse. *Genes Dev.* **15:** 1257–1271.

Horner M.A., Quintin S., Domeier M.E., Kimble J., Labouesse M., and Mango S.E. 1998. pha-4, an HNF-3 homolog, specifies pharyngeal organ identity in *Caenorhabditis elegans*. *Genes Dev.* **12:** 1947–1952.

Howell M. and Hill C.S. 1997. XSmad2 directly activates the activin-inducible, dorsal mesoderm gene XFKH1 in *Xenopus* embryos. *EMBO J.* **16:** 7411–7421.

Jin O., Harpal K., Ang S.L., and Rossant J. 2001. Otx2 and HNF3beta genetically interact in anterior patterning. *Int. J. Dev. Biol.* **45:** 357–365.

Kaestner K.H., Knochel W., and Martinez D.E. 2000. Unified nomenclature for the winged helix/forkhead transcription factors. *Genes Dev.* **14:** 142–146.

Kalb J.M., Lau K.K., Goszczynski B., Fukushige T., Moons D., Okkema P.G., and McGhee J.D. 1998. pha-4 is Ce-fkh-1, a fork head/HNF-3alpha,beta,gamma homolog that functions in organogenesis of the *C. elegans* pharynx. *Development* **125:** 2171–2180.

Keller R.E. 1975. Vital dye mapping of the gastrula and neurula of *Xenopus laevis*. I. Prospective areas and morphogenetic movements of the superficial layer. *Dev. Biol.* **42:** 222–241.

Knochel S., Lef J., Clement J., Klocke B., Hille S., Koster M., and Knochel W. 1992. Activin A induced expression of a fork head related gene in posterior chordamesoderm (notochord) of *Xenopus laevis* embryos. *Mech. Dev.* **38:** 157–165.

Krauss S., Concordet J.P., and Ingham P.W. 1993. A functionally conserved homolog of the *Drosophila* segment polarity gene hh is expressed in tissues with polarizing activity in zebrafish embryos. *Cell* **75:** 1431–1444.

Lai E., Prezioso V.R., Tao W.F., Chen W.S., and Darnell J.E. Jr. 1991. Hepatocyte nuclear factor 3 alpha belongs to a gene family in mammals that is homologous to the *Drosophila* homeotic gene fork head. *Genes Dev.* **5:** 416–427.

Lai E., Prezioso V.R., Smith E., Litvin O., Costa R.H., and Darnell J.E. Jr. 1990. HNF-3A, a hepatocyte-enriched transcription factor of novel structure is regulated transcriptionally. *Genes Dev.* **4:** 1427–1436.

Martinez D.E., Dirksen M.L., Bode P.M., Jamrich M., Steele R.E., and Bode H.R. 1997. Budhead, a fork head/HNF-3 homologue, is expressed during axis formation and head specification in hydra. *Dev. Biol.* **192:** 523–536.

Martinez Barbera J.P., Clements M., Thomas P., Rodriguez T., Meloy D., Kioussis D., and Beddington R.S. 2000. The homeobox gene *Hex* is required in definitive endodermal tissues for normal forebrain, liver and thyroid formation. *Development* **127:** 2433–2445.

Monaghan A.P., Kaestner K.H., Grau E., and Schutz G. 1993. Postimplantation expression patterns indicate a role for the mouse forkhead/HNF-3 alpha, beta and gamma genes in determination of the definitive endoderm, chordamesoderm and neuroectoderm. *Development* **119:** 567–578.

Muller F., Chang B., Albert S., Fischer N., Tora L., and Strahle U. 1999. Intronic enhancers control expression of zebrafish sonic hedgehog in floor plate and notochord. *Development* **126:** 2103–2116.

Nakano T., Murata T., Matsuo I., and Aizawa S. 2000. Otx2 directly interacts with Lim1 and HNF3β. *Biochem. Biophys. Res. Commun.* **267:** 64–70.

Niederlander C., Walsh J.J., Episkopou V., and Jones C.M. 2001. Arkadia enhances nodal-related signalling to induce mesendoderm. *Nature* **410**: 830–834.

Nishizaki Y., Shimazu K., Kondoh H., and Sasaki H. 2001. Identification of essential sequence motifs in the node/notochord enhancer of Foxa2 (Hnf3beta) gene that are conserved across vertebrate species. *Mech. Dev.* **102**: 57–66.

O'Reilly M.A., Smith J.C., and Cunliffe V. 1995. Patterning of the mesoderm in *Xenopus*: Dose-dependent and synergistic effects of Brachyury and Pintallavis. *Development* **121**: 1351–1359.

Olsen C.L. and Jeffery W.R. 1997. A forkhead gene related to HNF-3beta is required for gastrulation and axis formation in the ascidian embryo. *Development* **124**: 3609–3619.

Overdier D.G., Porcella A., and Costa R.H. 1994. The DNA-binding specificity of the hepatocyte nuclear factor 3/forkhead domain is influenced by amino-acid residues adjacent to the recognition helix. *Mol. Cell. Biol.* **14**: 2755–2766.

Pani L., Quian X.B., Clevidence D., and Costa R.H. 1992. The restricted promoter activity of the liver transcription factor hepatocyte nuclear factor 3 beta involves a cell-specific factor and positive autoactivation. *Mol. Cell. Biol.* **12**: 552–562.

Perea-Gomez A., Shawlot W., Sasaki H., Behringer R.R., and Ang S. 1999. HNF3beta and Lim1 interact in the visceral endoderm to regulate primitive streak formation and anterior-posterior polarity in the mouse embryo. *Development* **126**: 4499–4511.

Perea-Gomez A., Vella F.D., Shawlot W., Oulad-Abdelghani M., Chazaud C., Meno C., Pfister V., Chen L., Robertson E., Hamada H. et al. 2002. Nodal antagonists in the anterior visceral endoderm prevent the formation of multiple primitive streaks. *Dev. Cell* **3**: 745–756.

Ramakrishnan V., Finch J.T., Graziano V., Lee P.L., and Sweet R.M. 1993. Crystal structure of globular domain of histone H5 and its implications for nucleosome binding. *Nature* **362**: 219–223.

Rhinn M., Dierich A., Shawlot W., Behringer R.R., Le Meur M., and Ang S.L. 1998. Sequential roles for Otx2 in visceral endoderm and neuroectoderm for forebrain and midbrain induction and specification. *Development* **125**: 845–856.

Riddle R.D., Johnson R.L., Laufer E., and Tabin C. 1993. Sonic hedgehog mediates the polarizing activity of the ZPA. *Cell* **75**: 1401–1416.

Roelink H., Augsburger A., Heemskerk J., Korzh V., Norlin S., Ruiz i Altaba A., Tanabe Y., Placzek M., Edlund T., Jessell T. M. et al. 1994. Floor plate and motor neuron induction by vhh-1, a vertebrate homolog of hedgehog expressed by the notochord. *Cell* **76**: 761–775.

Ruiz i Altaba A. and Jessell T.M. 1992. Pintallavis, a gene expressed in the organizer and midline cells of frog embryos: Involvement in the development of the neural axis. *Development* **116**: 81–93.

Ruiz i Altaba A., Cox C., Jessell T.M., and Klar A. 1993a. Ectopic neural expression of a floor plate marker in frog embryos injected with the midline transcription factor Pintallavis. *Proc. Natl. Acad. Sci.* **90**: 8268–8272.

Ruiz i Altaba A., Prezioso V.R., Darnell J.E., and Jessell T.M. 1993b. Sequential expression of HNF-3 beta and HNF-3 alpha by embryonic organizing centers: The dorsal lip/node, notochord and floor plate. *Mech. Dev.* **44**: 91–108.

Ruiz i Altaba A., Placzek M., Baldassare M., Dodd J., and Jessell T.M. 1995. Early stages of notochord and floor plate development in the chick embryo defined by normal and induced expression of HNF-3 beta. *Dev. Biol.* **170**: 299–313.

Saka Y., Tada M., and Smith J.C. 2000. A screen for targets of the *Xenopus* T-box gene Xbra. *Mech. Dev.* **93**: 27–39.

Sasaki H. and Hogan B.L. 1993. Differential expression of multiple fork head related genes during gastrulation and axial pattern formation in the mouse embryo. *Development* **118**: 47–59.

———. 1994. HNF-3 beta as a regulator of floor plate development. *Cell* **76**: 103–115.

———. 1996. Enhancer analysis of the mouse HNF-3 beta gene: Regulatory elements for node/notochord and floor plate are independent and consist of multiple sub-elements. *Genes Cells* **1**: 59–72.

Sasaki H., Hui C., Nakafuku M., and Kondoh H. 1997. A binding site for Gli proteins is essential for HNF-3beta floor plate enhancer activity in transgenics and can respond to Shh in vitro. *Development* **124**: 1313–1322.

Sasaki H., Nishizaki Y., Hui C., Nakafuku M., and Kondoh H. 1999. Regulation of Gli2 and Gli3 activities by an amino-terminal repression domain: Implication of Gli2 and Gli3 as primary mediators of Shh signaling. *Development* **126**: 3915–3924.

Shawlot W. and Behringer R.R. 1995. Requirement for Lim1 in head-organizer function. *Nature* **374**: 425–430.

Strahle U., Blader P., Henrique D., and Ingham P.W. 1993. Axial, a zebrafish gene expressed along the developing body axis, shows altered expression in cyclops mutant embryos. *Genes Dev.* **7**: 1436–1446.

Thomas P.Q., Brown A., and Beddington R.S. 1998. Hex: A homeobox gene revealing peri-implantation asymmetry in the mouse embryo and an early transient marker of endothelial cell precursors. *Development* **125**: 85–94.

Varlet I., Collignon J., Norris D.P., and Robertson E.J. 1997. Nodal signaling and axis formation in the mouse. *Cold Spring Harbor Symp. Quant. Biol.* **62**: 105–113.

Waldrip W.R., Bikoff E. K., Hoodless P.A., Wrana J.L., and Robertson E.J. 1998. Smad2 signaling in extraembryonic tissues determines anterior-posterior polarity of the early mouse embryo. *Cell* **92**: 797–808.

Wang J.C., Waltner-Law M., Yamada K., Osawa H., Stifani S., and Granner D.K. 2000. Transducin-like enhancer of split proteins, the human homologs of *Drosophila* groucho, interact with hepatic nuclear factor 3beta. *J. Biol. Chem.* **275**: 18418–18423.

Weigel D. and Jackle H. 1990. The fork head domain: A novel DNA binding motif of eukaryotic transcription factors? *Cell* **63**: 455–456.

Weigel D., Jurgens G., Kuttner F., Seifert E., and Jackle H. 1989. The homeotic gene fork head encodes a nuclear protein and is expressed in the terminal regions of the *Drosophila* embryo. *Cell* **57**: 645–658.

Weinstein D.C., Ruiz i Altaba A., Chen W.S., Hoodless P., Prezioso V.R., Jessell T.M., and Darnell J.E. Jr. 1994. The winged-helix transcription factor HNF-3 beta is required for notochord development in the mouse embryo. *Cell* **78**: 575–588.

Yamamoto M., Meno C., Sakai Y., Shiratori H., Mochida K., Ikawa Y., Saijoh Y., and Hamada H. 2001. The transcription factor FoxH1 (FAST) mediates Nodal signaling during anterior-posterior patterning and node formation in the mouse. *Genes Dev.* **15**: 1242–1256.

Zorn A.M., Butler K., and Gurdon J.B. 1999. Anterior endomesoderm specification in *Xenopus* by Wnt/beta-catenin and TGF-beta signalling pathways. *Dev. Biol.* **209**: 282–297.

ROLE OF T-BOX GENES DURING GASTRULATION

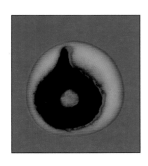

J.C. Smith

Wellcome Trust/Cancer Research UK Gurdon Institute and Department of Zoology,
University of Cambridge, Cambridge CB2 1QR, United Kingdom

INTRODUCTION

The *T*, or *Brachyury*, mutation was first described over 75 years ago as a mutation that affected tail length in heterozygous animals and was lethal in homozygotes, causing defects in notochord differentiation and the absence of structures posterior to the forelimb bud (Dobrovolskaïa-Zavadskaïa 1927; Chesley 1935; Gluecksohn-Schoenheimer 1944). The affected gene product proved to encode a transcription factor, and if any transcription factor could lay claim to be intimately involved in the regulation of gastrulation, a process often regarded as being controlled by the cytoskeleton, cell adhesion molecules, and the extracellular matrix, that transcription factor would be Brachyury. In this chapter, I describe how Brachyury, and other members of what came to be known as the T-box gene family (Bollag et al. 1994; Papaioannou 1997; Papaioannou and Silver 1998), regulate the cell movements of gastrulation.

THE T-BOX FAMILY

The mouse *Brachyury* gene was cloned by Bernhard Herrmann and colleagues (Herrmann et al. 1990). The gene proved to encode a protein of 436 amino acids, and it soon became clear that an amino-terminal region of ~229 residues was homologous to a domain in the *Drosophila* gene product optomotor-blind, and that this domain is involved in binding DNA (Pflugfelder et al. 1992). Further work demonstrated that Brachyury and optomotor-blind were the founder members of a family of proteins, all of which contained a conserved DNA-binding domain, or T domain (Bollag et al. 1994; Agulnik et al. 1995). The family, as families do, has grown extensively, and members have been isolated throughout the metazoans, with 18 T-box genes in mammals, 10 in *Ciona intestinalis,* and 20 in *Caenorhabditis elegans.*

Binding-site-selection experiments revealed that the Brachyury T domain binds the near-palindromic sequence T(C/G)ACACCTAGGTGTGAAATT (Kispert and Herrmann 1993). Although more recent experiments suggest that a half-site comprising the sequence TCACACCT is sufficient (Casey et al. 1998, 1999; Tada et al. 1998), more efficient binding does occur with more than one half-site, with different T-box proteins preferring different orientations and spacings (Conlon et al. 2001). The significance of this observation will remain unclear, however, unless T-box target genes are isolated whose promoters contain the sites in question. Such promoters have not yet been identified.

The structure of the *Xenopus* Brachyury T domain bound to a 24-nucleotide palindromic sequence has been solved, and it reveals that the domain binds as a dimer to the palindromic sequence, interacting with both the major and the minor grooves of DNA. A carboxy-terminal helix of the T domain is deeply embedded into an enlarged minor groove of the DNA without causing it to bend, and there is a small dimer interface between the two T domains (Muller and Herrmann 1997).

In mouse, *Xenopus,* and zebrafish, the carboxy-terminal half of Brachyury functions as an activator of transcription (Kispert et al. 1995a; Conlon et al. 1996). This region of the protein is only weakly conserved between different species, but the ability of *Xenopus* Brachyury (Xbra), at least, to function normally depends on its activation activity; if the activation domain is replaced by the transcriptional repressor domain from *Drosophila* engrailed, the resulting protein acts in a dominant-negative fashion (Conlon et al. 1996 and see below). Other T-box proteins are even less well conserved outside of the T domain, and although most behave predominantly as transcriptional activators, one should note that Tbx2, at least in some contexts, functions as a repressor (Carreira et al. 1998).

BRACHYURY EXPRESSION PATTERNS

The tissues affected in *Brachyury* homozygous mutant mouse embryos are those that express the gene at the highest levels and for the longest time. Thus, in situ hybridization demonstrated that *Brachyury* is expressed in the primitive streak at the onset of gastrulation and persists in the streak and then in the tail bud. Expression also occurs in the node and notochord (Wilkinson et al. 1990; Kispert and Herrmann 1994). This expression pattern is conserved in other vertebrates such as *Xenopus* (see Fig. 1) (Smith et al. 1991), zebrafish (Schulte-Merker et al. 1992), and chick (Kispert et al. 1995b), but as I discuss below, other metazoans differ quite dramatically, with *Brachyury* homologs being expressed in all three germ layers of the embryo (Marcellini et al. 2003). For example, it is expressed in the

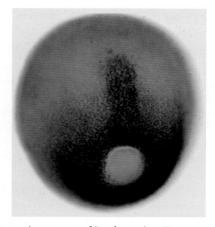

Figure 1. Expression pattern of Brachyury in a *Xenopus* embryo at the mid-gastrula stage. Brachyury protein is visualized using a specific antibody. Note that expression occurs around the blastopore and in the notochord and that reduced levels of protein are detectable anterior to the circumblastoporal region where transcription has declined. Brachyury is present in the nuclei of embryonic cells.

notochord of ascidians (Corbo et al. 1997), the presumptive endoderm of sea urchins (Gross and McClay 2001), and the ectodermal hindgut of *Drosophila* (Kispert et al. 1994).

BRACHYURY AND GASTRULATION

I describe the influence of *Brachyury* on gastrulation in a variety of species before attempting to elucidate the molecular basis of its mode of action.

Mouse *Brachyury*

At first sight, the mouse *T* phenotype did not encourage the view that the gene might be involved in the regulation of gastrulation; homozygous mutant embryos develop reasonably normally anterior to the forelimb bud, suggesting that the early stages of gastrulation occur relatively normally, even if the much later phases, which give rise to posterior tissues, are disrupted. This impression was revised, however, by careful chimeric analysis in which *T/T* mutant embryonic stem (ES) cells, marked with *lacZ*, were introduced into wild-type embryos. During most of gastrulation, the mutant cells, lacking *T* function, accumulated in the primitive streak, leading to their eventual incorporation into the tail bud (Wilson et al. 1995). The behavior of the cells suggested that one role of the T-gene product is to affect the adhesive properties of cells as they pass through the primitive streak (Wilson et al. 1995), and consistent with this idea, the phenotype of homozygous *Brachyury* embryos resembles that of embryos lacking *integrin* α5 (Yang et al. 1993). Only the most anterior mesodermal cells in the mouse embryo appeared not to require T function (Wilson et al. 1995), and as I discuss below, these may be equivalent to those cells in *Xenopus* and zebrafish embryos in which Goosecoid causes the down-regulation of *Brachyury* and which move by migration rather than convergent extension.

To confirm that the aberrant behavior of mutant cells in the primitive streak is a consequence only of the loss of Brachyury function, Wilson and Beddington introduced a construct expressing *T* under the control of its own promoter into *T/T* mutant cells and found that the cells no longer accumulated in the tail bud and were able to contribute to all types of mesoderm (Wilson and Beddington 1997). They also obtained evidence for dose-dependent effects of *Brachyury*, because two lines of ES cells that expressed particularly high levels of *Brachyury* tended to leave the streak prematurely and showed a reduced ability to colonize posterior mesoderm. These dose-dependent effects of *Brachyury* are reminiscent of, but differ from, the concentration-dependent effects of *Brachyury* in isolated *Xenopus* ectodermal tissue, where low concentrations induce ventral mesoderm and high concentrations induce somitic muscle (Cunliffe and Smith 1992; O'Reilly et al. 1995).

Xenopus Brachyury

An equally extensive analysis of the role of *Brachyury* in gastrulation has been carried out in *Xenopus laevis*, where gastrulation has been described in enormous and careful detail, particularly by Ray Keller and his colleagues (Keller et al. 2000; Keller et al. 2003). Investigations of the role of *Xbra* have made use of overexpression strategies and the use of dominant-negative constructs. Overexpression can be carried out quite simply by injecting RNA encoding Xbra into *Xenopus* embryos (Cunliffe and Smith 1992). If desired, one can make use of a hormone-inducible construct in which the ligand-binding domain of the human glucocorticoid receptor is fused to Xbra (Tada et al. 1997). Such constructs are inactive unless dexamethasone is added to the culture medium, thus allowing temporal control of Xbra activation. The design of dominant-negative constructs has taken advantage of the fact that Xbra functions as a transcription activator, and has involved replacing the activation domain of the protein with the repressor domain of *Drosophila* engrailed, to create the fusion protein Xbra-En[R]. This protein very effectively inhibits the ability of wild-type Brachyury to activate transcription (Conlon et al. 1996).

Expression of Xbra-En[R] in *Xenopus* embryos causes a severe disruption of gastrulation; embryos fail to close their blastopores and go on to display posterior truncations and, in some cases, impairment of notochord differentiation (Conlon et al. 1996). In these respects, they resemble *Brachyury* homozygous mutant mouse embryos, suggesting that in the regulation of gastrulation, the function of *Brachyury* is conserved.

More detailed investigations reveal that *Brachyury* acts as a "switch" to decide what sort of gastrulation behavior cells should undergo. Dorsal mesodermal cells in *Xenopus* exhibit two distinct types of movements. Those that involute first undergo active cell migration and go on to form prechordal mesoderm, whereas those that involute later undergo convergent extension and form notochord and somite (Chapter 13). One can study these two types of cell behaviors by dissecting tissue from the embryo and allowing it to develop in isolation, or by treating isolated ectodermal tissue with the mesoderm-inducing factor Activin (Smith et al. 1990a), which causes that tissue to undergo convergent extension (Symes and Smith 1987) and its constituent cells to acquire the ability to spread and migrate on fibronectin (Smith et al. 1990b).

Experiments of this sort reveal that Xbra activity is required for convergent extension but not for migration on a fibronectin substrate. Thus, animal caps derived from embryos injected with RNA encoding Xbra-En[R] do not undergo convergent extension movements in response to Activin (Conlon and Smith 1999; Kwan and Kirschner

2003), and neither does isolated dorsal marginal zone tissue (Kwan and Kirschner 2003). In contrast, Activin-treated animal pole cells derived from such embryos adhere to and migrate on a fibronectin-coated substrate (Conlon and Smith 1999; Kwan and Kirschner 2003), as do cells from the dorsal marginal zone (Kwan and Kirschner 2003).

Far from being required for adhesion to fibronectin and cell migration, *Xbra* actually inhibits this behavior, both in Activin-treated cells derived from prospective ectodermal tissue and in cells from the dorsal marginal zone (Kwan and Kirschner 2003). This observation suggests, therefore, that the default behavior of mesodermal cells is to migrate; expression of Xbra suppresses cell migration and permits them to undergo convergent extension. In this respect, Xbra functions as a switch to ensure that cell migration and convergent extension are mutually exclusive types of cell movement during *Xenopus* gastrulation. During this strategic role, it is important that *Xbra* expression is tightly regulated. This is discussed below.

Zebrafish

The zebrafish homolog of *Brachyury* (Schulte-Merker et al. 1992) is encoded by the gene *no tail* (*ntl*) (Schulte-Merker et al. 1994), the homozygous phenotype of which is remarkably similar to that of mouse *T* mutants and of *Xenopus* embryos expressing the *Xbra-En[R]* construct (Halpern et al. 1993). The early movements of involution and convergence therefore appear rather normal in *ntl* mutants, but a properly defined notochord primordium does not form, and the morphogenetic movements required for the formation of the tail do not occur (Halpern et al. 1993). More detailed analysis has now revealed that convergence is significantly blocked during gastrula stages, but that significant extension can occur through epiboly, which may function redundantly with mediolateral intercalation to cause extension (Glickman et al. 2003).

Other Species

The most detailed analyses of *Brachyury* function during gastrulation have been carried out in the vertebrates. Studies in other species are consistent with the idea that *Brachyury* plays a role in the regulation of gastrulation, but it is difficult to draw general conclusions because, as indicated above, expression patterns of *Brachyury* differ in different species, and because few functional studies have been carried out. Nevertheless, evolutionary analysis does suggest that the ancestral function of *Brachyury* was indeed to specify a circumblastoporal region that plays a role in axial elongation and specification of hindgut and posterior mesoderm (Peterson et al. 1999a,b; Technau 2001; Scholz and Technau 2003). This circumblastoporal expression has been lost independently in insects and tunicates (Yasuo and Satoh 1993; Kispert et al. 1994; Corbo et al. 1997; Bassham and Postlethwait 2000), and it may be significant that whereas Brachyury orthologs from most species, including vertebrates, induce isolated *Xenopus* animal pole regions to form only mesodermal tissues, those from *Drosophila* and ascidians are able to induce both mesoderm and endoderm (Marcellini et al. 2003). This derived behavior is associated in part with the loss of an amino-terminal peptide that restricts the inducing activity of Brachyury; a chimeric version of Xbra in which the amino-terminal amino acids are replaced by those of the ascidian *Halocynthia roretzi* is able, like *Drosophila* and ascidian Brachyury, to induce endoderm as well as mesoderm (Marcellini et al. 2003).

One example in which it has been possible to interfere with the function of Brachyury in a nonvertebrate species concerns the sea urchin *Lytechinus variegatus*, where *Brachyury* (*LvBrac*) is expressed both circumferentially and in a second domain in the oral opening (Gross and McClay 2001). Interference with *LvBrac* function by means of an Engrailed repressor construct analogous to that of Conlon and colleagues (Conlon et al. 1996) resulted in a disruption of morphogenetic movements in both expression domains without disrupting the expression of mesodermal and endodermal marker genes. This suggests that the major role of *LvBrac* is to regulate gastrulation movements rather than to specify cell type (Gross and McClay 2001). Perhaps surprisingly, the requirement for *LvBrac* is non-cell-autonomous, in the sense that if just one blastomere is injected with LvBrac-En^R at the two-cell stage, the injected side gastrulates normally, while invagination is blocked in the other. This suggests that the targets of LvBrac include some secreted proteins, such as growth factors or extracellular matrix molecules, and this is discussed in more detail below, in the search for Brachyury target genes.

REGULATION OF *BRACHYURY* EXPRESSION

The experiments described above indicate that *Brachyury* plays an essential role in the successful prosecution of gastrulation. How is expression of *Brachyury* regulated? This question has been addressed in the mouse, in *Xenopus*, and in *Ciona intestinalis* (Corbo et al. 1997). I concentrate here on the two vertebrate species, where *Brachyury* plays an important role in gastrulation.

Regulation of Mouse *Brachyury*

Analysis of the mouse *Brachyury* regulatory region has revealed that constructs consisting of 430 nucleotides 5′ of the transcription start site are sufficient to recapitulate expression of *T* in the primitive streak, but expression was

not detectable in axial mesoderm (Clements et al. 1996). The same was true of constructs containing 8.3 kb of 5′ sequence, even when such constructs included intron and 3′ sequence, indicating that the regulatory sequences required for expression in the two *Brachyury* domains are distinct. Multiple elements are required within the 430 bp of sequence necessary for primitive streak expression of *Brachyury*, because three internal deletions all resulted in loss of stable expression. One construct, however, appeared to be defective in maintenance rather than activation, because expression was observed in the first cells to move through the streak, but not in later cells.

This initial work did not identify specific elements that are required for *Brachyury* expression, but two groups have implicated Wnt signaling in the regulation of *T*. Yamaguchi and colleagues (Yamaguchi et al. 1999) noted that *Wnt3a* mutant embryos, like embryos lacking *Brachyury*, lack paraxial mesoderm. They observed that *Brachyury* expression is not maintained in *Wnt3a* mutant embryos and that two canonical Tcf-binding sites are present in the mouse *T* promoter. These binding sites are conserved in the *Xbra* promoter (see below) and are required for expression of *T* in the primitive streak (Yamaguchi et al. 1999). Together, these experiments suggest that expression of *T* in the primitive streak is regulated, at least in part, by the Wnt3a pathway and that this interaction is required for development of paraxial mesoderm.

The same conclusion was reached by Arnold and colleagues, who noted that ES cells cultured on Wnt-expressing feeder cells were induced to express *Brachyury* (Arnold et al. 2000). Significantly, such activation could be induced by Wnt1, 3a, and 4, but not by Wnt5a, 7a, 7b, or 11, and consistent with the idea that activation occurs as a direct response to Wnt signaling, *Brachyury* could also be induced by a stable form of β-catenin.

Regulation of *Xenopus Brachyury*

The regulation of *Brachyury* has been extensively studied in *Xenopus*. One approach has been to take advantage of the fact that expression of *Xenopus Brachyury* can be activated in isolated ectodermal tissue by the mesoderm-inducing factors FGF and Activin (Smith et al. 1991). FGF activates *Xbra* expression through the MAP kinase pathway (Gotoh et al. 1995; LaBonne et al. 1995; Umbhauer et al. 1995), although the main role of FGF signaling appears to be in the maintenance of *Xbra* transcription (Kroll and Amaya 1996). This occurs through an autocatalytic loop in which Xbra activates the expression of *eFGF* and eFGF maintains expression of *Xbra* (Isaacs et al. 1994; Schulte-Merker and Smith 1995). Xbra acts directly on the *eFGF* promoter, which contains one "perfect" T-box half-site TCACACCT positioned 936 nucleotides upstream of the transcription start site, as well as a related sequence CCACACCT positioned 123 nucleotides downstream of the transcription start site (Casey et al. 1998). The requirement for eFGF signaling in the maintenance of expression of *Xbra* may be related to its later requirement for muscle differentiation (Standley et al. 2001), in a phenomenon known as the community effect (Gurdon 1988; Gurdon et al. 1993). It may also be significant that the Xbra/eFGF autocatalytic loop may maintain *Xbra* expression in closely apposed cells that are undergoing convergent extension, but will not allow maintenance in actively migrating cells that go on to form prechordal mesoderm (see above).

Interestingly, there is no evidence for such an autocatalytic loop in the mouse embryo, where expression of *Fgf-4*, the putative homolog of *eFGF*, does not require *T* function during the first 48 hours of gastrulation, and where the *T* promoter is active in the primitive streak and tail bud in the absence of functional T protein (Schmidt et al. 1997). It remains possible, however, that an FGF/Brachyury autoregulatory loop does exist in the node, head process, and primitive streak (Schmidt et al. 1997). Indeed, in *Xenopus*, Xbra function is required for maintenance of *Xbra* expression in dorsal mesoderm but not in ventral or lateral tissue (Conlon et al. 1996).

Also of importance in *Xenopus* is the observation that expression of *Xbra* in response to Activin depends critically on the concentration of inducing factor that cells experience. Thus, low levels of Activin do not induce expression, intermediate concentrations activate it strongly, and high concentrations do not induce the gene (Green et al. 1992; Gurdon et al. 1994, 1995). A similar Activin concentration range proves to induces posterior and then anterior mesoderm (Green et al. 1992), such that *Xbra* is induced in cells that do undergo convergent extension and is not expressed in cells that do not. This observation is consistent with the proposed role of *Xbra* as a switch that determines whether cells undergo convergent extension or migration (Kwan and Kirschner 2003).

How is *Xbra* down-regulated at high concentrations of Activin, thus allowing cells to undergo active migration? One clue comes from the observation that the high concentrations of Activin that do not induce expression of *Xbra* do activate the homeobox-containing gene *goosecoid* (Green et al. 1992; Gurdon et al. 1994, 1995). This raises the possibility that goosecoid, which functions as a transcriptional repressor (Ferreiro et al. 1998; Smith and Jaynes 1996; Jimenez et al. 1999; Latinkic and Smith 1999), causes downregulation of *Xbra*. Consistent with this idea, careful timecourse studies show that high concentrations of Activin do cause transient activation of *Xbra* expression, and that this transient activation becomes stable in cells that are treated with cycloheximide, which presumably inhibits the translation of *Xbra* repressors (Papin and Smith 2000). It is also

known that goosecoid can directly repress expression of *Xbra* (Artinger et al. 1997; Latinkic and Smith 1999), but goosecoid is unlikely to be the only gene product responsible for the down-regulation of *Xbra* at high Activin concentrations, because specific inhibition of goosecoid function, in contrast to widespread inhibition of protein synthesis, does not allow stable activation of *Xbra* at high Activin doses (Papin and Smith 2000). Nevertheless, it is noteworthy that ectopic expression of *goosecoid* in *Xenopus* embryos leads to cell migration toward the anterior of the embryo, consistent with the idea that goosecoid represses *Xbra* and thereby leads to the promotion of migration (Niehrs et al. 1993).

Studies of this sort are complemented by analyses of the *Xenopus Brachyury* promoter. In a first analysis, 381 nucleotides 5′ of the *Xbra* transcription start site proved to be sufficient to recapitulate the Activin dose-response profile. Thus, following injection of reporter constructs into *Xenopus* embryos and the dissection of isolated ectodermal tissue, low concentrations of Activin were unable to activate expression, intermediate concentrations induced high reporter gene activity, and high concentrations suppressed expression (Latinkic et al. 1997). This promising start was not, however, followed by the definition of FGF or Activin response elements, and further work has, in general, succeeded only in emphasizing the complexity of *Xbra* regulation (see Fig. 2). Work in transgenic embryos, for example, has identified two repressor modules that exert their effects at different stages during gastrulation (Lerchner et al. 2000). One module is defined by a bipartite binding site for a Smad-interacting protein (SIP1) of the δEF1 repressor family (Remacle et al. 1999; Verschueren et al. 1999; van Grunsven et al. 2000). This element acts to confine *Xbra* expression to the marginal zone early in gastrulation, and mutation of either site causes widespread expression of reporter constructs (Fig. 2) (Lerchner et al. 2000). The other module is defined by two homeodomain-binding sites and is responsible for repression in dorsal mesoderm and ectoderm at mid-gastrula stages. It is possible that it is through these sites that goosecoid, and probably other transcriptional repressors, cause the repression of *Xbra* at high concentrations of Activin (Latinkic et al. 1997; Lerchner et al. 2000).

One notable similarity between the mouse *T* and *Xbra* promoters is that the *Xenopus* regulatory regions isolated to date are only capable of driving stable expression in lateral and ventral mesoderm; activation is not observed in dorsal tissues such as the prospective notochord (Lerchner et al. 2000).

These promoter analyses emphasize the importance of region-specific repression in establishing the correct spatial expression pattern of *Xbra*, but recent work (Vonica and Gumbiner 2002) has implicated the canonical Wnt pathway

A

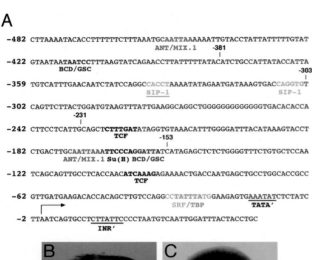

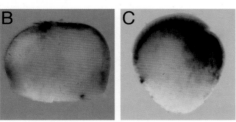

Figure 2. Regulation of *Xenopus Brachyury*. (*A*) Sequence of the proximal *Xbra* promoter region. Transcription factor-binding sites previously studied by Lerchner and colleagues (2000) are in color, and additional binding sites with putative regulatory functions are in **bold**. (SIP1) δEF1 and SIP1 half-site; (ANT/MIX.1) Antennapedia-type homeodomain-binding site; (BCD/GSC) Bicoid-type homeodomain-binding site; (SRF) serum response factor-binding site; (TBP) putative TATA box; (Su(H)) putative Suppressor of Hairless-binding site; (TCF) TCF-binding site. The transcription start site is marked by an arrow; potential alternative TATA box and initiator region are indicated TATA′ and INR′. (*B*) Bisected transgenic *Xenopus* embryo showing appropriate activation of a reporter gene under the control of 2.1 kb of proximal sequence. Note that expression of the reporter gene is restricted to the marginal zone of the embryo. (*C*) Bisected transgenic *Xenopus* embryo showing ectopic activation of a reporter gene under the control of 2.1 kb of proximal sequence in which the underlined SIP1 half-site in *A* is mutated from CACCT to CATGG. Note expression of the reporter gene in the animal hemisphere of the embryo.

as a region-specific activator, thus creating parallels with the regulation of *Brachyury* in the mouse embryo (Yamaguchi et al. 1999; Arnold et al. 2000). Thus, as observed in the mouse *Brachyury* gene, the *Xbra* promoter contains TCF-binding sites (Latinkic et al. 1997; Lerchner et al. 2000; Vonica and Gumbiner 2002), and loss of Wnt function, by means of a dominant-negative *Xenopus Tcf-3* construct, causes a down-regulation of *Xbra* expression in the marginal zone (Vonica and Gumbiner 2002). As one has come to expect in analyses of *Xbra* regulation, however, the role of Wnt signaling in *Xbra* expression is complicated and involves additional pathways. Thus, Wnt signaling (in the

form of a *VP16 Xtcf-3* construct) is able to induce ectopic *Xbra* expression (as it does in mouse ES cells), but this activation requires an intact Activin/nodal signaling pathway (Vonica and Gumbiner 2002). On the other hand, activation of *Xbra* by Activin or nodal signaling does not require Wnt activity. The regulation of *Xbra* expression therefore involves the input and coordination of multiple signaling pathways, belying its apparently simple expression pattern.

T-TARGETS—*Wnt11*

How does Brachyury, a transcription factor, play such an important role in the regulation of gastrulation? In all reported assays, Brachyury behaves as an activator of transcription (Kispert et al. 1995a; Conlon et al. 1996), and it is a reasonable assumption that among the genes that are regulated by Brachyury are some that are involved in the control of gastrulation. Extensive searches for targets of Brachyury have been carried out in *Xenopus* (Tada et al. 1998; Saka et al. 2000; Tada and Smith 2000), and among the genes that have been isolated is *Xenopus Wnt11* (*Xwnt11*) (Tada and Smith 2000). Like *Xbra*, *Xwnt11* is expressed in the marginal zone of the *Xenopus* embryo (Fig. 3), and inhibition of Xbra function causes dramatic downregulation of *Xwnt11*. The regulation of *Xwnt11* has not been studied in detail, but use of a hormone-inducible *Xbra* construct (Tada et al. 1997) shows that activation of *Xwnt11* in isolated *Xenopus* animal caps (prospective ectodermal tissue) can occur in the absence of protein synthesis (Tada and Smith 2000), suggesting that Xbra acts directly on *Xwnt11* regulatory regions.

That Xwnt11 plays a role in the regulation of gastrulation is suggested by the use of a dominant-negative *Xwnt11* construct, itself based on a dominant-negative version of Xwnt8 (Hoppler et al. 1996). This dominant-negative

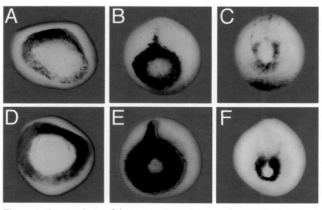

Figure 3. Comparison of the expression patterns of *Xwnt11* (*A–C*) and *Xbra* (*D–F*) at early gastrula (*A,D*), late gastrula (*B,E*), and early neurula (*C,F*) stages.

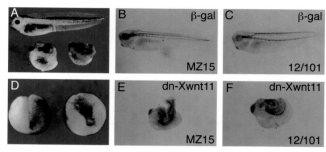

Figure 4. Inhibition of Xwnt11 signaling disrupts gastrulation without affecting mesodermal differentiation. (*A*) Embryos injected with β-*gal* RNA (*above*) or with *dn-Xwnt11* RNA (*below*) and cultured to the tadpole stage. Note that *dn-Xwnt11* inhibits formation of posterior structures. (*D*) Control β-gal-expressing embryo (*left*) or dn-Xwnt11-expressing embryo (*right*) cultured to the neurula stage. Note that *dn-Xwnt11* inhibits gastrulation movements. (*B,C*) Control embryos injected with β-*gal* RNA were stained at the tadpole stage with the notochord-specific monoclonal antibody MZ15 (*B*) or the muscle-specific monoclonal antibody 12/101 (*C*). (*E,F*) Embryos injected with *dn-Xwnt11* RNA stained with MZ15 (*E*), or 12/101 (*F*). *Dn-Xwnt11* does not affect notochord differentiation and, although it reduces the size of the somites, it does not inhibit muscle differentiation.

Xwnt11 is specific in the sense that it inhibits the action of Xwnt11 but not Xwnt8. Inhibition of Xwnt11 function, like inhibition of Xbra, causes a severe disruption of gastrulation without interfering with mesodermal differentiation (Fig. 4) (Tada and Smith 2000).

This conclusion was confirmed in the zebrafish embryo, where the *silberblick* (*slb*) locus encodes Wnt11, and Slb/Wnt11 activity proves to be necessary for cells to undergo convergent extension movements (Heisenberg et al. 2000). The requirement for Slb/Wnt11 is non-cell-autonomous, and analysis of cell behavior in *slb*[−/−] embryos suggests that the extension of axial tissue depends in part on medio-lateral cell intercalation in paraxial tissue.

GASTRULATION AND THE PLANAR CELL POLARITY PATHWAY

The discovery that a member of the Wnt family plays a role in the regulation of gastrulation raised the question as to which Wnt signal transduction pathway might be involved. Two lines of evidence in *Xenopus* suggested that the pathway is not the canonical one, involving GSK-3 and β-catenin, and is more likely to involve the planar cell polarity pathway (Tada and Smith 2000). First, a dominant-negative *Tcf-3* construct failed to inhibit Activin-induced elongation of *Xenopus* animal caps. Second, the Dishevelled construct *Dsh-DEP+* is able to inhibit Activin-induced elongation of animal caps without inhibiting the canonical Wnt pathway. This construct is also able to restore elongation to Activin-treated animal caps expressing a dominant-negative *Xwnt11* construct (Tada and Smith 2000).

Similar experiments in zebrafish *slb*[−/−] embryos also indicate that Wnt11 signals through a noncanonical pathway, and subsequent work in both *Xenopus* and zebrafish has indicated that components of the planar cell polarity pathway are involved in the regulation of gastrulation (Jessen et al. 2002; Marlow et al. 2002; Park and Moon 2002; Tada et al. 2002; Wallingford et al. 2002; Yamanaka et al. 2002; Takeuchi et al. 2003; Veeman et al. 2003). This work is discussed in Chapter 36.

OTHER T-BOX GENES

Finally, what of other T-box genes in the regulation of gastrulation? Most work addressing the role of T-box genes in gastrulation has concerned itself with *Brachyury*, but other T-box genes are expressed in the mesoderm of vertebrate embryos during gastrulation, and there is evidence that these too play a role in gastrulation. In the mouse these genes include *eomesodermin* (Ciruna and Rossant 1999; Hancock et al. 1999) and *Tbx6* (Chapman et al. 1996); in the chick *TbxT* and *Tbx6L* (Knezevic et al. 1997); in *Xenopus Antipodean* and *eomesodermin;* and in zebrafish *spadetail* (Griffin et al. 1998; Warga and Nusslein-Volhard 1998; Amacher et al. 2002) and *Tbx6* (Hug et al. 1997).

Of these genes, the best-characterized with respect to its role in gastrulation is zebrafish *spadetail*. Spadetail function is first required for the flattening and maximization of intercellular contacts of future mesendodermal cells. This behavior may suppress cell intermingling and thereby control the way that cells coalesce to form the organizer (Warga and Nusslein-Volhard 1998). At later stages, *spadetail* mutant embryos show defects in convergence movements in prospective trunk somitic mesoderm. Transplantation experiments show that the mutation acts in a cell-autonomous fashion in this respect (Ho and Kane 1990). Little is known about the mechanism of action of Spadetail, but one prospective target gene is *paraxial protocadherin* (*PAPC*) (Yamamoto et al. 1998), which encodes a transmembrane cell adhesion molecule expressed during gastrulation in prospective trunk mesoderm. Use of a dominant-negative *PAPC* construct suggests that PAPC is required for dorsal convergence movements during gastrulation, and genetic analysis indicates that *PAPC* is a downstream target of Spadetail (Yamamoto et al. 1998). It is not yet clear, however, whether *PAPC* is a direct target of Spadetail, and indeed, *Xenopus PAPC* appears not to be a direct target of the related T-box protein VegT (White et al. 2002).

CONCLUSIONS

The work described in this chapter indicates that the T-box genes play an essential role in the regulation of gastrulation, but there is clearly an enormous amount of work needed to understand how the T-box genes actually do it. First, and most obviously, one needs to know how the spatial and temporal expression patterns of the T-box genes are regulated. Next, and also rather obviously, one has to identify the spectra of genes that are activated by the different T-box genes, and understand the mechanism by which different T-box genes activate different targets (Conlon et al. 2001; Marcellini et al. 2003). Then the hard bit is to work out how all these genes act in concert to regulate the different types of cell behavior that occur in different regions of the embryo at different times in development: A daunting task, but as this volume makes clear, an enormous amount of progress has been made in the study of gastrulation in the last five years, and there's more to come. We shouldn't be too pessimistic.

REFERENCES

Agulnik S.I., Bollag R.J., and Silver L.M. 1995. Conservation of the T-box gene family from Mus musculus to *Caenorhabditis elegans*. *Genomics* **25**: 214–219.

Amacher S.L., Draper B.W., Summers B.R., and Kimmel C.B. 2002. The zebrafish T-box genes *no tail* and *spadetail* are required for development of trunk and tail mesoderm and medial floor plate. *Development* **129**: 3311–3323.

Arnold S.J., Stappert J., Bauer A., Kispert A., Herrmann B.G., and Kemler R. 2000. *Brachyury* is a target gene of the Wnt/b-catenin signaling pathway. *Mech. Dev.* **91**: 249–258.

Artinger M., Blitz I., Inoue K., Tran U., and Cho K.W. 1997. Interaction of *goosecoid* and *brachyury* in *Xenopus* mesoderm patterning. *Mech. Dev.* **65**: 187–196.

Bassham S. and Postlethwait J. 2000. *Brachyury* (*T*) expression in embryos of a larvacean urochordate, *Oikopleura dioica*, and the ancestral role of *T*. *Dev. Biol.* **220**: 322–332.

Bollag R.J., Siegfried Z., Cebra-Thomas J.A., Garvey N., Davison E.M., and Silver L.M. 1994. An ancient family of embryonically expressed mouse genes sharing a conserved protein motif with the T locus. *Nat. Genet.* **7**: 383–389.

Carreira S., Dexter T.J., Yavuzer U., Easty D.J., and Goding C.R. 1998. Brachyury-related transcription factor Tbx2 and repression of the melanocyte-specific TRP-1 promoter. *Mol. Cell. Biol.* **18**: 5099–5108.

Casey E.S., O'Reilly M.A., Conlon F.L., and Smith J.C. 1998. The T-box transcription factor Brachyury regulates expression of *eFGF* through binding to a non-palindromic response element. *Development* **125**: 3887–3894.

Casey E.S., Tada M., Fairclough L., Wylie C.C., Heasman J., and Smith J.C. 1999. *Bix4* is activated directly by VegT and mediates endoderm formation in *Xenopus* development. *Development* **126**: 4193–4200.

Chapman D.L., Agulnik I., Hancock S., Silver L.M., and Papaioannou V.E. 1996. *Tbx6*, a mouse T-Box gene implicated in paraxial mesoderm formation at gastrulation. *Dev. Biol.* **180**: 534–542.

Chesley P. 1935. Development of the short-tailed mutant in the house mouse. *J. Exp. Zool.* **70**: 429–459.

Ciruna B.G. and Rossant J. 1999. Expression of the T-box gene *Eomesodermin* during early mouse development. *Mech. Dev.* **81**: 199–203.

Clements D., Taylor H.C., Herrmann B.G., and Stott D. 1996. Distinct regulatory control of the *Brachyury* gene in axial and non-axial mesoderm suggests separation of mesoderm lineages early in mouse gastrulation. *Mech. Dev.* **56**: 139–149.

Conlon F.L. and Smith J.C. 1999. Interference with Brachyury function inhibits convergent extension, causes apoptosis, and reveals separate requirements in the FGF and activin signaling pathways. *Dev. Biol.* **213**: 85–100.

Conlon F.L., Sedgwick S.G., Weston K.M., and Smith J.C. 1996. Inhibition of Xbra transcription activation causes defects in mesodermal patterning and reveals autoregulation of Xbra in dorsal mesoderm. *Development* **122**: 2427–2435.

Conlon F.L., Fairclough L., Price B.M., Casey E.S., and Smith J.C. 2001. Determinants of T box protein specificity. *Development* **128**: 3749–3758.

Corbo J.C., Levine M., and Zeller R.W. 1997. Characterization of a notochord-specific enhancer from the *Brachyury* promoter region of the ascidian *Ciona intestinalis*. *Development* **124**: 589–602.

Cunliffe V. and Smith J.C. 1992. Ectopic mesoderm formation in *Xenopus* embryos caused by widespread expression of a Brachyury homologue. *Nature* **358**: 427–430.

Dobrovolskaïa-Zavadskaïa N. 1927. Sur la mortification spontanée de la queue chez la souris nouveau-née et sur l'existence d'un caractère heriditaire "non-viable". *C. R. Seances Soc. Biol. Fil.* **97**: 114–116.

Ferreiro B., Artinger M., Cho K., and Niehrs C. 1998. Antimorphic *goosecoids*. *Development* **125**: 1347–1359.

Glickman N.S., Kimmel C.B., Jones M.A., and Adams R.J. 2003. Shaping the zebrafish notochord. *Development* **130**: 873–887.

Gluecksohn-Schoenheimer S. 1944. The development of normal and homozygous *brachy* (*T/T*) mouse embryos in the extraembryonic coelem of the chick. *Proc. Natl. Acad. Sci.* **30**: 134–140.

Gotoh Y., Masuyama N., Suzuki A., Ueno N., and Nishida E. 1995. Involvement of the MAP kinase cascade in *Xenopus* mesoderm induction. *EMBO J.* **14**: 2491–2498.

Green J.B.A., New H.V., and Smith J.C. 1992. Responses of embryonic *Xenopus* cells to activin and FGF are separated by multiple dose thresholds and correspond to distinct axes of the mesoderm. *Cell* **71**: 731–739.

Griffin K.J., Amacher S.L., Kimmel C.B., and Kimelman D. 1998. Molecular identification of *spadetail*: Regulation of zebrafish trunk and tail mesoderm formation by T-box genes. *Development* **125**: 3379–3388.

Gross J.M. and McClay D.R. 2001. The role of Brachyury (T) during gastrulation movements in the sea urchin *Lytechinus variegatus*. *Dev. Biol.* **239**: 132–147.

Gurdon J.B. 1988. A community effect in animal development. *Nature* **336**: 772–774.

Gurdon J.B., Lemaire P., and Kato K. 1993. Community effects and related phenomena in development. *Cell* **75**: 831–834.

Gurdon J.B., Mitchell A., and Mahony D. 1995. Direct and continuous assessment by cells of their position in a morphogen gradient. *Nature* **376**: 520–521.

Gurdon J.B., Harger P., Mitchell A., and Lemaire P. 1994. Activin signaling and response to a morphogen gradient. *Nature* **371**: 487–492.

Halpern M.E., Ho R.K., Walker C., and Kimmel C.B. 1993. Induction of muscle pioneers and floor plate is distinguished by the zebrafish *no tail* mutation. *Cell* **75**: 99–111.

Hancock S.N., Agulnik S.I., Silver L.M., and Papaioannou V.E. 1999. Mapping and expression analysis of the mouse ortholog of *Xenopus Eomesodermin*. *Mech. Dev.* **81:** 205–208.

Heisenberg C.P., Tada M., Rauch G.J., Saude L., Concha M.L., Geisler R., Stemple D.L., Smith J.C., and Wilson S.W. 2000. Silberblick/Wnt11 mediates convergent extension movements during zebrafish gastrulation. *Nature* **405:** 76–81.

Herrmann B.G., Labeit S., Poutska A., King T.R., and Lehrach H. 1990. Cloning of the T gene required in mesoderm formation in the mouse. *Nature* **343:** 617–622.

Ho R.K. and Kane D.A. 1990. Cell-autonomous action of zebrafish spt-1 mutation in specific mesodermal precursors. *Nature* **348:** 728–730.

Hoppler S., Brown J.D., and Moon R.T. 1996. Expression of a dominant-negative Wnt blocks induction of MyoD in *Xenopus* embryos. *Genes Dev.* **10:** 2805–2817.

Hug B., Walter V., and Grunwald D.J. 1997. tbx6, a *Brachyury*-related gene expressed by ventral mesendodermal precursors in the zebrafish embryo. *Dev. Biol.* **183:** 61–73.

Isaacs H.V., Pownall M.E., and Slack J.M.W. 1994. eFGF regulates *Xbra* expression during *Xenopus* gastrulation. *EMBO J.* **13:** 4469–4481.

Jessen J.R., Topczewski J., Bingham S., Sepich D.S., Marlow F., Chandrasekhar A., and Solnica-Krezel L. 2002. Zebrafish trilobite identifies new roles for Strabismus in gastrulation and neuronal movements. *Nat. Cell Biol.* **4:** 610–615.

Jimenez G., Verrijzer C.P., and Ish-Horowicz D. 1999. A conserved motif in goosecoid mediates groucho-dependent repression in *Drosophila* embryos. *Mol. Cell. Biol.* **19:** 2080–2087.

Keller R., Davidson L.A., and Shook D.R. 2003. How we are shaped: The biomechanics of gastrulation. *Differentiation* **71:** 171–205.

Keller R., Davidson L.A., Edlund A., Elul T., Ezin M., Shook D., and Skoglund P. 2000. Mechanisms of convergence and extension by cell intercalation. *Philos. Trans. R.. Soc. Lond. B Biol. Sci.* **355:** 897–922.

Kispert A. and Herrmann B.G. 1993. The Brachyury gene encodes a novel DNA binding protein. *EMBO J.* **12:** 3211–3220.

———. 1994. Immunohistochemical analysis of the *Brachyury* protein in wild-type and mutant mouse embryos. *Dev. Biol.* **161:** 179–193.

Kispert A., Korschorz B., and Herrmann B.G. 1995a. The T protein encoded by *Brachyury* is a tissue-specific transcription factor. *EMBO J.* **14:** 4763–4772.

Kispert A., Herrmann B.G., Leptin M., and Reuter R. 1994. Homologs of the mouse *Brachyury* gene are involved in the specification of posterior terminal structures in *Drosophila*, *Tribolium* and *Locusta*. *Genes Dev.* **8:** 2137–2150.

Kispert A., Ortner H., Cooke J., and Herrmann B.G. 1995b. The chick *Brachyury* gene: Developmental expression pattern and response to axial induction by localized activin. *Dev. Biol.* **168:** 406–415.

Knezevic V., De Santo R., and Mackem S. 1997. Two novel chick T-box genes related to mouse *Brachyury* are expressed in different, non-overlapping mesodermal domains during gastrulation. *Development* **124:** 411–419.

Kroll K.L. and Amaya E. 1996. Transgenic *Xenopus* embryos from sperm nuclear transplantations reveal FGF signaling requirements during gastrulation. *Development* **122:** 3173–3183.

Kwan K.M. and Kirschner M.W. 2003. Xbra functions as a switch between cell migration and convergent extension in the *Xenopus* gastrula. *Development* **130:** 1961–1972.

LaBonne C., Burke B., and Whitman M. 1995. Role of MAP kinase in

mesoderm induction and axial patterning during *Xenopus* development. *Development* **121:** 1475–1486.

Latinkic B.V. and Smith J.C. 1999. *Goosecoid* and *Mix.1* repress *Brachyury* expression and are required for head formation in *Xenopus*. *Development* **126:** 1769–1779.

Latinkic B.V., Umbhauer M., Neal K., Lerchner W., Smith J.C., and Cunliffe V. 1997. The *Xenopus Brachyury* promoter is activated by FGF and low concentrations of activin and suppressed by high concentrations of activin and by paired-type homeodomain proteins. *Genes Dev.* **11:** 3265–3276.

Lerchner W., Latinkic B.V., Remacle J.E., Huylebroeck D., and Smith J.C. 2000. Region-specific activation of the *Xenopus Brachyury* promoter involves active repression in ectoderm and endoderm: A study using transgenic frog embryos. *Development* **127:** 2729–2739.

Marcellini S., Technau U., Smith J.C., and Lemaire P. 2003. Evolution of Brachyury proteins: Identification of a novel regulatory domain conserved within Bilateria. *Dev. Biol.* **260:** 352–361.

Marlow F., Topczewski J., Sepich D., and Solnica-Krezel L. 2002. Zebrafish Rho kinase 2 acts downstream of Wnt11 to mediate cell polarity and effective convergence and extension movements. *Curr. Biol.* **12:** 876–884.

Muller C.W. and Herrmann B.G. 1997. Crystallographic structure of the T domain-DNA complex of the *Brachyury* transcription factor. *Nature* **389:** 884–888.

Niehrs C., Keller R., Cho K.W., and De Robertis E.M. 1993. The homeobox gene *goosecoid* controls cell migration in *Xenopus* embryos. *Cell* **72:** 491–503.

O'Reilly M.-A. J., Smith J.C., and Cunliffe V. 1995. Patterning of the mesoderm in *Xenopus*: Dose-dependent and synergistic effects of *Brachyury* and *Pintallavis*. *Development* **121:** 1351–1359.

Papaioannou V.E. 1997. T-box family reunion. *Trends Genet.* **13:** 212–213.

Papaioannou V.E. and Silver L.M. 1998. The T-box gene family. *BioEssays* **20:** 9–19.

Papin C. and Smith J.C. 2000. Gradual refinement of activin-induced thresholds requires protein synthesis. *Dev. Biol.* **217:** 166–172.

Park M. and Moon R.T. 2002. The planar cell-polarity gene stbm regulates cell behavior and cell fate in vertebrate embryos. *Nat. Cell Biol.* **4:** 20–25.

Peterson K.J., Harada Y., Cameron R.A., and Davidson E.H. 1999a. Expression pattern of *Brachyury* and *Not* in the sea urchin: Comparative implications for the origins of mesoderm in the basal deuterostomes. *Dev. Biol.* **207:** 419–431.

Peterson K.J., Cameron R.A., Tagawa K., Satoh N., and Davidson E.H. 1999b. A comparative molecular approach to mesodermal patterning in basal deuterostomes: The expression pattern of *Brachyury* in the enteropneust hemichordate *Ptychodera flava*. *Development* **126:** 85–95.

Pflugfelder G.O., Roth H., and Poeck B. 1992. A homology domain shared between *Drosophila* optomotor-blind and mouse Brachyury is involved in DNA binding. *Biochem. Biophys. Res. Commun.* **186:** 918–925.

Remacle J.E., Kraft H., Lerchner W., Wuytens G., Collart C., Verschueren K., Smith J.C., and Huylebroeck D. 1999. New mode of DNA binding of multi-zinc finger transcription factors: δEF1 family members bind with two hands to two target sites. *EMBO J.* **18:** 5073–5084.

Saka Y., Tada M., and Smith J.C. 2000. A screen for targets of the *Xenopus* T-box gene Xbra. *Mech. Dev.* **93:** 27–39.

Schmidt C., Wilson V., Stott D., and Beddington R.S. 1997. *T* promoter activity in the absence of functional T protein during axis formation and elongation in the mouse. *Dev. Biol.* **189:** 161–173.

Scholz C.B. and Technau U. 2003. The ancestral role of Brachyury: Expression of *NemBra1* in the basal cnidarian *Nematostella vectensis* (Anthozoa). *Dev. Genes Evol.* **212:** 563–570.

Schulte-Merker S. and Smith J.C. 1995. Mesoderm formation in response to *Brachyury* requires FGF signaling. *Curr. Biol.* **5:** 62–67.

Schulte-Merker S., Ho R.K., Herrmann B.G., and Nüsslein-Volhard C. 1992. The protein product of the zebrafish homologue of the mouse *T* gene is expressed in nuclei of the germ ring and the notochord of the early embryo. *Development* **116:** 1021–1032.

Schulte-Merker S., van Eeden F.J., Halpern M.E., Kimmel C.B., and Nusslein-Volhard C. 1994. *no tail (ntl)* is the zebrafish homologue of the mouse *T* (*Brachyury*) gene. *Development* **120:** 1009–1015.

Smith J.C., Price B.M., Van Nimmen K., and Huylebroeck D. 1990a. Identification of a potent *Xenopus* mesoderm-inducing factor as a homologue of activin A. *Nature* **345:** 729–731.

Smith J.C., Symes K., Hynes R.O., and DeSimone D. 1990b. Mesoderm induction and the control of gastrulation in *Xenopus laevis*: The roles of fibronectin and integrins. *Development* **108:** 229–238.

Smith J.C., Price B.M.J., Green J.B.A., Weigel D., and Herrmann B.G. 1991. Expression of a *Xenopus* homolog of *Brachyury* (*T*) is an immediate-early response to mesoderm induction. *Cell* **67:** 79–87.

Smith S.T. and Jaynes J.B. 1996. A conserved region of engrailed, shared among all en-, gsc-, Nk1-, Nk2- and msh-class homeoproteins, mediates active transcriptional repression in vivo. *Development* **122:** 3141–3150.

Standley H.J., Zorn A.M., and Gurdon J.B. 2001. eFGF and its mode of action in the community effect during *Xenopus* myogenesis. *Development* **128:** 1347–1357.

Symes K. and Smith J.C. 1987. Gastrulation movements provide an early marker of mesoderm induction in *Xenopus laevis*. *Development* **101:** 339–349.

Tada M. and Smith J.C. 2000. Xwnt11 is a target of *Xenopus* Brachyury: Regulation of gastrulation movements via Disheveled, but not through the canonical Wnt pathway. *Development* **127:** 2227–2238.

Tada M., Concha M.L., and Heisenberg C.P. 2002. Non-canonical Wnt signaling and regulation of gastrulation movements. *Semin. Cell Dev. Biol.* **13:** 251–260.

Tada M., O'Reilly M.-A.J. and Smith J.C. 1997. Analysis of competence and of *Brachyury* autoinduction by use of hormone-inducible *Xbra*. *Development* **124:** 2225–2234.

Tada M., Casey E.S., Fairclough L., and Smith J.C. 1998. *Bix1*, a direct target of *Xenopus* T-box genes, causes formation of ventral mesoderm and endoderm. *Development* **125:** 3997–4006.

Takeuchi M., Nakabayashi J., Sakaguchi T., Yamamoto T.S., Takahashi H., Takeda H., and Ueno N. 2003. The prickle-related gene in vertebrates is essential for gastrulation cell movements. *Curr. Biol.* **13:** 674–679.

Technau U. 2001. *Brachyury*, the blastopore and the evolution of the mesoderm. *BioEssays* **23:** 788–794.

Umbhauer M., Marshall C.J., Mason C.S., Old R.W., and Smith J.C. 1995. Mesoderm induction in *Xenopus* caused by activation of MAP kinase. *Nature* **376:** 58–62.

van Grunsven L.A., Papin C., Avalosse B., Opdecamp K., Huylebroeck D., Smith J.C., and Bellefroid E.J. 2000. *XSIP1*, a *Xenopus* zinc finger/homeodomain encoding gene highly expressed during early neural development. *Mech. Dev.* **94:** 189–193.

Veeman M.T., Slusarski D.C., Kaykas A., Louie S.H., and Moon R.T. 2003. Zebrafish prickle, a modulator of noncanonical wnt/fz signaling, regulates gastrulation movements. *Curr. Biol.* **13:** 680–685.

Verschueren K., Remacle J.E., Collart C., Kraft H., Baker B.S., Tylzanowski P., Nelles L., Wuytens G., Su M.T., Bodmer R., Smith J.C., and Huylebroeck D. 1999. SIP1, a novel zinc finger/homeodomain repressor, interacts with Smad proteins and binds to 5′-CACCT sequences in candidate target genes. *J. Biol. Chem.* **274:** 20489–20498.

Vonica A. and Gumbiner B.M. 2002. Zygotic Wnt activity Is required for *Brachyury* expression in the early *Xenopus laevis* embryo. *Dev. Biol.* **250:** 112–127.

Wallingford J.B., Goto T., Keller R., and Harland R.M. 2002. Cloning and expression of *Xenopus Prickle*, an orthologue of a *Drosophila* planar cell polarity gene. *Mech. Dev.* **116:** 183–186.

Warga R.M. and Nusslein-Volhard C. 1998. *spadetail*-dependent cell compaction of the dorsal zebrafish blastula. *Dev. Biol.* **203:** 116–121.

White R.J., Sun B.I., Sive H.L., and Smith J.C. 2002. Direct and indirect regulation of *derrière*, a *Xenopus* mesoderm-inducing factor, by VegT. *Development* **129:** 4867–4876.

Wilkinson D.G., Bhatt S., and Herrmann B.G. 1990. Expression pattern of the mouse T gene and its role in mesoderm formation. *Nature* **343:** 657–659.

Wilson V. and Beddington R. 1997. Expression of T protein in the primitive streak is necessary and sufficient for posterior mesoderm movement and somite differentiation. *Dev. Biol.* **192:** 45–58.

Wilson V., Manson L., Skarnes W.C., and Beddington R.S. 1995. The T gene is necessary for normal mesodermal morphogenetic cell movements during gastrulation. *Development* **121:** 877–886.

Yamaguchi T.P., Takada S., Yoshikawa Y., Wu N., and McMahon A.P. 1999. *T* (*Brachyury*) is a direct target of Wnt3a during paraxial mesoderm specification. *Genes Dev.* **13:** 3185–3190.

Yamamoto A., Amacher S.L., Kim S.H., Geissert D., Kimmel C.B., and De Robertis E.M. 1998. Zebrafish *paraxial protocadherin* is a downstream target of *spadetail* involved in morphogenesis of gastrula mesoderm. *Development* **125:** 3389–3397.

Yamanaka H., Moriguchi T., Masuyama N., Kusakabe M., Hanafusa H., Takada R., Takada S., and Nishida E. 2002. JNK functions in the non-canonical Wnt pathway to regulate convergent extension movements in vertebrates. *EMBO Rep.* **3:** 69–75.

Yang J.T., Rayburn H., and Hynes R.O. 1993. Embryonic mesodermal defects in α5 integrin-deficient mice. *Development* **119:** 1093–1105.

Yasuo H. and Satoh N. 1993. Function of vertebrate T gene. *Nature* **364:** 582.

GOOSECOID AND GASTRULATION

E.M. De Robertis

Howard Hughes Medical Institute, University of California, Los Angeles, California 90095-1662

INTRODUCTION

For many years, the transplantation of Spemann's organizer into a host embryo was the only way of revealing the remarkable biological activity of the dorsal lip of the blastopore. The discovery of *goosecoid* (*gsc*) initiated the molecular era of research on Spemann's organizer (Cho et al. 1991). It became possible to visualize, by following *gsc* expression, the region of the embryo that corresponds to the inductive activity. The Spemann organizer became a concrete group of cells, rather than an embryological concept. Furthermore, since microinjection of *gsc* synthetic mRNA is able to execute some of the properties of Spemann's organizer, these initial studies implicated homeobox genes as key elements in the function of the organizer. Subsequently, many other homeobox genes were found to be expressed in Spemann's organizer. These transcription factors control the expression of secreted growth factors and growth factor antagonists that mediate the inductive activities of the organizer on neighboring cells. In this chapter, I review studies on *goosecoid* and what they have taught us about the organizer, how embryonic signaling pathways are integrated at the level of the promoter, and the comparative anatomy of gastrulation.

XENOPUS GOOSECOID EXPRESSION

goosecoid was isolated from a cDNA library constructed from dissected dorsal lips. The library was screened with degenerate oligonucleotides complementary to a region conserved in most homeobox genes, including Hox genes and *bicoid* (Blumberg et al. 1991). The gene was named *goosecoid* to reflect the similarity of its homeodomain region to both *Drosophila gooseberry* and *bicoid*, two members of the larger paired-homeodomain family. The *gsc* homeobox contains a lysine in position 50 of the homeobox (instead of glutamine as in Hox genes). This change is seen in other anterior homeobox genes such as *Otx*, *Siamois*, and *Xtwn*, and correlates with binding to a *bicoid*-type DNA sequence, rather than an *Antennapedia*-Hox target sequence (Blumberg et al. 1991).

At early gastrula, *goosecoid* is expressed in a 60° arc of the dorsal marginal zone, providing an excellent marker for Spemann's organizer (Fig. 1A). Expression is maximal at stage 10, when the dorsal lip first appears. In hybridizations on histological sections, *gsc* expression is seen in involuting cells, extending almost to the leading edge (Fig. 1B). The *gsc*-positive tissue corresponds to the future dorsal mesendoderm, including foregut and pharyngeal mesoderm, prechordal plate, and notochord. As involution proceeds, by mid-gastrula, the *gsc*-expressing region narrows and leaves the blastopore lip (Fig. 2A,B). By the end of gastrulation, *gsc* expression is seen in the prechordal plate and anterior endoderm, ahead of the Hox gene border provided by *Xlabial* that separates the head and trunk regions (Fig. 2C). By the late-neurula stage, *goosecoid* expression is seen in the prechordal plate and in the anterior endomesoderm that will give rise to the pharynx and foregut (Fig. 2D; Fig. 3A). Thus, although *gsc* initially is expressed in the territory that gives rise to the notochord, it is then switched off in this structure.

As shown in Figure 2, the *Xlabial* gene, a member of the Hox family, provides a good marker for the formation of the sharp border that demarcates the head and trunk regions of the embryo. This anterior–posterior border originates from the circumblastoporal involution of *Xlab*-positive cells

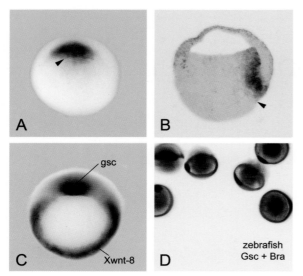

Figure 1. Expression of *goosecoid* in Spemann's organizer in *Xenopus* and zebrafish. (*A*) *goosecoid* in the dorsal lip of an early *Xenopus* gastrula. (*B*) *Gsc* in situ hybridization to a histological section of a *Xenopus* stage 10 1/2 gastrula. Expression is seen in the involuted endomesoderm, extending from the dorsal lip (*arrowhead*) to the leading edge of the endoderm. (*C*) Expression of *Gsc* and *Xwnt-8* at stage 10 3/4. Gsc is a transcriptional repressor of *Xwnt-8,* and these two genes acting together pattern the marginal zone. (*D*) Zebrafish embryos at 50% epiboly hybridized for *Gsc* (in *black*) and stained with a Brachyury antibody (in *brown*). This is the stage of maximal *Gsc* expression in zebrafish; Brachyury demarcates the trunk mesoderm. We thank Uyen Tran for the embryos shown in *A–C.* (*D,* Reprinted, with permission, from Schulte-Merker et al. 1994.)

(marked by a small circle in Fig. 2 A–C), whereas the border of *Xlab* expression on the sides of the notochord derives from the convergence toward the midline of *Xlab*-expressing cells flanking the organizer (marked by a small asterisk). Recently, a revision of the dorsal–ventral and anterior–posterior designation of the axes in the *Xenopus* fate map was proposed (Kumano and Smith 2002; Lane and Sheets 2002) (see Chapter 19). The expression domains of *goosecoid* and *Xlab* during *Xenopus* gastrulation do not support this new proposal, since the anterior–posterior border of *Xlab* originates from the blastopore circumference (Fig. 2A–C).

EFFECTS OF *GOOSECOID* OVEREXPRESSION

Microinjection of *gsc* mRNA into two ventral blastomeres at the 4-cell stage causes the formation of secondary axes lacking head structures (Cho et al. 1991; Yao and Kessler 2001). When *gsc* is overexpressed in ventral marginal zone explants, twofold increases in *gsc* mRNA concentration are sufficient to cause dorsalization of mesoderm, and at least three thresholds of dorsal histotypic differentiation can be triggered (Niehrs et al. 1994). Microinjected *gsc* mRNA has

non-cell-autonomous effects, recruiting neighboring uninjected cells into the twinned dorsal axis (Niehrs et al. 1993). These non-cell-autonomous effects of *gsc* mRNA are mediated by the induction of secreted proteins such as Chordin and Frzb-1 (Sasai et al. 1994; Leyns et al. 1997).

When overexpressed in dorsal blastomeres, *goosecoid* promotes cell migration in the dorso-anterior direction (Niehrs et al. 1993). In normal embryos, C1 blastomeres from the 32-cell stage contribute progeny to the leading edge of the head endomesoderm (including pharyngeal endoderm, foregut, and liver), prechordal plate, and the entire length of the notochord (Fig. 3B). The notochord contribution of the C1 blastomere includes its posterior-most end, called the chordoneural hinge, which is homologous to the regressing Hensen's node in amniotes (Gont et al. 1993). In *gsc*-injected embryos, the C1 progeny undergo a change in cell fate, contributing predominantly to the leading edge involuting cells, while the notochord becomes devoid of labeled cells (Fig. 3C). Since these embryos are phenotypically normal, other cells must have taken the place of C1 progeny in the trunk dorsal axis. Thus, when a dorsal blastomere is dorsalized further, it populates the anterior-most endomesoderm, which occupies the anterior foregut region (Fig. 3). Some of these cells of dorsal origin have recently been found also to contribute to the anterior blood island (Lane and Smith 1999). This is in agreement with earlier observations that dorsal-most endomesodermal cells end up in the liver and foregut region (Fig. 3A) (Niehrs et al. 1993; Bouwmeester et al. 1995).

GOOSECOID AND THE PATTERNING OF THE MARGINAL ZONE

In a loss-of-function situation, *Xenopus goosecoid* is required for the formation of the head region. Anterior head defects have been obtained by a variety of methods, including antisense *gsc* mRNA (Steinbeisser et al. 1995), an antimorphic Gsc resulting from the addition of epitope tags (Ferreiro et al. 1998), and fusions of Gsc with the activation domain of VP16 (Latinkic and Smith 1999; Yao and Kessler 2001). Goosecoid functions as a transcriptional repressor, and contains in its amino-terminal domain a conserved heptapeptide (sequence FSIDNIL) that is also present in *Drosophila engrailed.* This sequence provides a binding site for transcriptional corepressors (Mailhos et al. 1998). Morpholino knockdown experiments of *gsc* in *Xenopus* or zebrafish embryos have not been reported yet.

Gsc overexpression has been shown to repress the expression of genes that control the differentiation of the ventral marginal zone, in particular *Xwnt-8* and *BMP4,* which are antagonistic to organizer function (Christian and Moon 1993; Fainsod et al. 1994; Steinbeisser et al. 1995; Yao and Kessler 2001). Xwnt-8 is a crucial regulator of ventro-

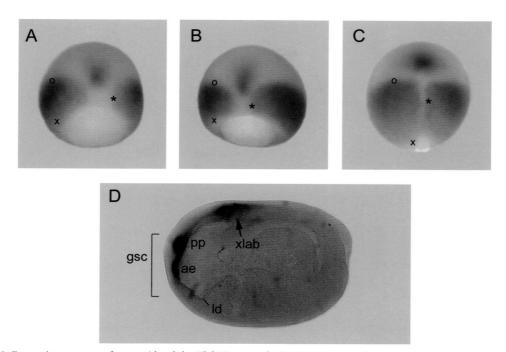

Figure 2. Expression patterns of *goosecoid* and the *Xlab* Hox gene during *Xenopus* gastrulation. (*A–C*) *Gsc*-expressing cells leave the dorsal lip and move into the prechordal plate and anterior endoderm region, while Xlab is observed in involuted mesoderm surrounding the circular blastopore lip. Markings have been placed in the blastopore (*small x*), anterior border of *Xlab* expression (*small o*), and midline border (*small asterisk*) of the notochord to illustrate the relative movements of the two expression domains. Embryos are at stage 11, 11 1/4, and 13. (*D*) Side view of a late-neurula (stage 15) *Xenopus* embryo showing expression of *gsc* in the prechordal plate (pp) just behind the eye anlage and in anterior endoderm (ae). The liver diverticulum (ld) is indicated; expression of *Xlab* in mesoderm, including the first somites, is also indicated. This embryo was cleared after whole-mount in situ hybridization. Photographs courtesy of Sung Kim and Peter Pfeffer.

lateral mesoderm, which is expressed in mesoderm in a complementary pattern to that of *goosecoid* (Fig. 1C). Dorsalization of *Xenopus* embryos with lithium chloride results in the repression of *Xwnt-8* transcription, and this requires the *gsc* gene product (Steinbeisser et al. 1995). The importance of the repression of *Xwnt-8* by Gsc is illustrated by the work of Yao and Kessler (2001), which showed that the anterior truncation caused by VP16-Gsc could be fully rescued by *Frzb-1* mRNA, a secreted Wnt inhibitor that can block Xwnt-8 activity. Wild-type Gsc represses *Xwnt-8* transcription directly, through four Gsc-binding sites located in the *Xwnt-8* promoter (Yao and Kessler 2001). In addition, the *gsc* promoter has binding sites for Vent-2, which mediate the transcriptional repression of *gsc* by BMP (Trinidade et al. 1999). These studies suggest that a main function of Gsc is to exclude *Xwnt-8* and *BMP* transcription from Spemann's organizer, and that this repression is required for normal *Xenopus* development.

The mutual repressive interactions between *gsc* in the organizer and *BMP* and *Xwnt-8* in the ventrolateral marginal zone are supported by studies on notochord development. Overexpression of *gsc* together with a dominant-negative BMP receptor (tBR) causes the formation of noto-

chords at higher frequency than either component alone (Yasuo and Lemaire 2001). This effect of *gsc* can be mimicked by microinjection of *Frzb-1* mRNA and *tBR* (Yasuo and Lemaire 2001). Since the expression of *Frzb-1* and *chordin* is activated by Gsc (Sasai et al. 1994; Leyns et al. 1997), these growth factor antagonists provide a molecular mechanism for the formation of dorsal mesoderm, such as notochord, in ventral mesoderm explants microinjected with *gsc* mRNA (Niehrs et al. 1994). Marginal zone mesoderm is regulated by long-range mutually repressive interactions by which cells in the dorsal or ventral poles of the embryo communicate to ensure that a perfectly proportioned embryo develops each time.

GOOSECOID TRANSCRIPTIONAL CONTROL

The *gsc* promoter has provided a paradigm for understanding how embryonic patterning signals are integrated at the level of individual genes, resulting in spatial patterns of gene expression. Watabe et al. (1995) showed that the *goosecoid* promoter has two growth-factor-responsive elements. A proximal element mediates activation by β-Catenin/Wnt and a distal element by a TGF-β signal of the

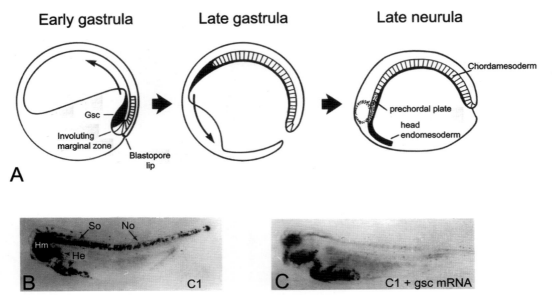

Figure 3. Overexpression of *goosecoid* mRNA promotes dorso-anterior migratory movements. (*A*) Diagram illustrating the movements of the *goosecoid*-expressing region (in *black*) during gastrulation. The notochord region is hatched, and the position of the eye anlage is indicated. Note that the *goosecoid* region contributes to head endomesoderm, including the entire foregut. (*B*) Normal progeny of a C1 blastomere using colloidal gold as lineage tracer. Note staining of the entire length of the notochord (No) including the posterior tip, medial somites (So), prechordal plate (Hm, head mesoderm), and head endomesoderm (He) up to the level of the liver and beyond. (*C*) When *gsc* mRNA is injected into C1 blastomeres, cell progeny is lost from notochord and accumulates in the head endomesoderm. Dorsal blastomeres contribute to the entire head region, including cells that populate the ventral head and trunk. (Modified, with permission, from Niehrs et al. 1993.)

Activin/Vg1/Nodal-related family. As shown in Figure 4, the β-Catenin signal is mediated by binding of the related homeodomain proteins Siamois or Xtwn, which are downstream transcriptional targets of the maternal β-Catenin signal (Laurent et al. 1997). The Activin/Nodal-related signal is transduced by the distal element, which requires a combination of Mixer and Smad2/4 (Germain et al. 2000). Mixer is a paired-homeodomain family transcription factor that is induced in endomesoderm by Activin. Smad2 is phosphorylated by the TGF-β receptors of the Activin family, allowing binding of the cofactor Smad4 and translocation into the nucleus. Mixer has a conserved carboxy-terminal Smad2-binding motif that mediates their mutual interaction on the distal promoter element (Germain et al. 2000). Therefore, Activin/Nodal-related must induce both *Mixer* transcription and Smad2 phosphorylation to turn on the distal promoter element (Germain et al. 2000). Since both the distal and proximal promoter elements are essential for high-level transcription of the *gsc* gene, the expression of *gsc* in the dorsal lip of the gastrula results from the synergistic input of the Activin/Vg1/Nodal-related and β-catenin pathways (Fig. 4).

Many additional transcription factors further modulate *gsc* transcription, and their binding sites on the *Xenopus gsc* promoter have been mapped. These include sites for the transcriptional activators Xlim-1 and Otx-2 (Mochizuki et al. 2000) and negative autoregulatory sites for goosecoid itself (Danilov et al. 1998) and for the ventral repressor Vent-2 (Trinidade et al. 1999). The negative regulation of *gsc* by *Vent* genes plays a fundamental role in the zebrafish embryo. The *Vent/vega2* and *vox/vega1* genes are expressed in complementary patterns to that of *goosecoid* (Kawahara et al. 2000). *Vent* genes are transcriptionally activated by BMP signals. They ventralize the embryo and repress *goosecoid* when overexpressed. When both zebrafish *vent* genes (*vega1* and *vega2*) are knocked down with morpholinos, zebrafish embryos become dorsalized, and *goosecoid*

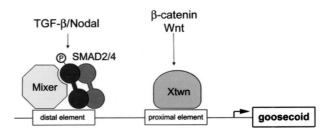

Figure 4. Diagram of the *goosecoid* promoter region, showing how signals from the β-catenin and the TGF-β pathway are integrated at the level of the DNA. Diagram courtesy of Oliver Wessely and Uyen Tran.

expression is greatly expanded (Imai et al. 2001). In summary, the *goosecoid* promoter provides one of the best-studied systems for understanding how multiple signaling pathways are integrated at the level of the DNA during early vertebrate development.

GOOSECOID IN THE ZEBRAFISH

The identification of *goosecoid* homologs was instrumental in the visualization of tissues with organizer properties in a number of organisms. Particularly when used in combination with the marginal zone/primitive streak marker *Brachyury* (*Bra*), which marks all trunk mesoderm, comparative studies have facilitated a unified view of vertebrate gastrulation. In zebrafish, *gsc* expression starts at the blastula stage, forming a dorsoventral gradient (Stachel et al. 1993; Schulte-Merker et al. 1994). The moment of maximal *gsc* expression corresponds to 50% epiboly, a stage that may be considered homologous to early gastrula (stage 10) of *Xenopus*. Figure 1D shows zebrafish embryos with intense staining of *gsc* mRNA in the organizer (or embryonic shield), double-stained with a Brachyury antibody. Brachyury protein is seen throughout the ring comprising both dorsal and ventral marginal zone. Initially, individual organizer cells express both *gsc* and *Brachyury*, but once the goosecoid-positive cells migrate anteriorly, *Brachyury* expression is turned off in the prechordal plate (Schulte Merker et al. 1994). This process, by which *Brachyury* expression is restricted to the notochord, is mediated by repressive Gsc-binding sites in the *Xbra* promoter (Artinger et al. 1997; Latinik et al. 1997).

In zebrafish, *gsc* transcription requires Nodal-related signals provided by two genes, *cyclops* and *squint* (Feldman et al. 1998; Gritsman et al. 2000; Shimizu et al. 2000). In *Xenopus*, Agius et al. (2000) used a specific inhibitor of Nodal-related signals, a secreted protein called Cerberus-short, to show that the *gsc*-inducing signals released by the endodermal "Nieuwkoop center" require Nodal. Thus, both in zebrafish and in *Xenopus*, Nodal-related signals are required for the induction of the gastrula organizer.

GOOSECOID IN THE MOUSE

The mouse gastrula develops from a cup-shaped epiblast. The cloning of mouse *Gsc* (Blum et al. 1992) helped to identify the location of the mouse organizer. As shown in Figure 5A, at the time of maximal *Gsc* expression (mid-streak day 6 1/2), *Gsc* transcripts are found in the anterior primitive streak and, at lower levels, in the anterior visceral endoderm (AVE) (Blum et al. 1992; Belo et al. 1997). The AVE corresponds to the chick hypoblast and also expresses

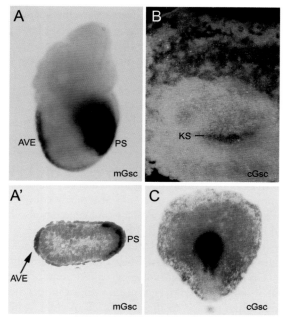

Figure 5. Expression of *GSC* in mouse and chick embryos. (*A*) Mouse mid-streak gastrula at the time of maximal *Gsc* expression. Note expression in anterior primitive streak (PS) and in anterior visceral endoderm (AVE). (*A´*) Transverse section through the same egg cylinder at the level of the anterior primitive streak. (*B*) Stage XII chick embryo in which the hypoblast has covered the posterior half of the area pellucida. Note *Goosecoid* expression in Koller's sickle (KS) and more weakly in the hypoblast anterior to it. (*C*) Expression of chick *Goosecoid* in the anterior third of the primitive streak, at stage 3⁺, when *Gsc* expression reaches its maximum. (*A*, *A´*, courtesy of J.A. Belo; *B*,*C*, reproduced, with permission, from Izpisúa-Belmonte et al. 1993.)

other organizer-specific genes such as *Lim-1* and *HNF3β* (Belo et al. 1997). The discovery that the anterior primitive streak marked the mouse organizer met with some initial skepticism, for mouse embryologists at the time believed that the organizer resided in Hensen's node, which is formed later. Once the mouse Hensen's node starts its regression, *Gsc* expression is down-regulated in the node but remains in the prechordal plate and foregut (Belo et al. 1998). At later stages (day 10 1/2 and later), mouse *Gsc* has a second phase of expression in the neural crest of pharyngeal arches 1 and 2, floor of the diencephalon, limb buds, and other sites (Gaunt et al. 1993).

Mouse knockouts for *Gsc* are born alive but die shortly after birth (Rivera-Pérez et al. 1995; Yamada et al. 1995). Homozygous mutants present numerous craniofacial malformations resulting from the late phase of *Gsc* expression in neural crest. In addition, bone reductions and fusions in the base of the cranium are observed in the midline region anterior to the pituitary, which develops in close association with the *Gsc*-expressing cells of the prechordal plate (Belo et al. 1998). The lack of a gastrulation phenotype was surprising, and contrasted with the results obtained with antimorphic or antisense *Gsc* constructs in *Xenopus* (Steinbeisser et al. 1995; Ferreiro et al. 1998; Yao and Kessler 2001). Perhaps the slower developmental pace of the mouse allows for compensation by other redundant repressive mechanisms. One gene that has been demonstrated to compensate in part is *HNF3β*. Filosa et al. (1997) showed that in *Gsc*−/−; *HNF3β*+/− compound mutants, dorsoventral patterning of the CNS is severely disrupted at an early stage. A second mouse homolog, *Gsc-like*, may also partly compensate (Funke et al. 1997). A second *Goosecoid* gene has also been found in chick (Lemaire et al. 1997).

In contrast to the mild phenotype in intact mice, a stronger requirement is seen when the inducing ability of *Gsc*−/− cells is tested in embryological experiments. Zhu et al. (1999) showed that wild-type mouse nodes induce neural markers when grafted to the area opaca of chick, whereas *Gsc*−/− nodes are severely impaired in their neural inducing activity. Even *Gsc*+/− nodes showed decreased inducing activity. Similarly, the expression of *Nkx2.1* in ventral forebrain in response to anterior mesoderm ablation is altered in *Gsc*−/− (Camus et al. 2000). Thus, a requirement for *Gsc* in the mouse organizer can be uncovered by embryological manipulations even when it is not apparent in the intact mouse embryo.

GOOSECOID IN THE CHICK

Because the chick blastoderm is flat and translucent, transcript detection can be achieved with a much better resolution than in *Xenopus* or mouse. Chick *Goosecoid* expression has been studied in considerable detail (Izpisúa-Belmonte et al. 1993; Chapman et al. 2002; Skromme and Stern 2002) and is first detectable in the unincubated egg, which already contains several thousand cells. Expression starts in a thickening of the posterior edge called Koller's sickle, where it is confined to a group of cells located in a middle layer between the epiblast and the forming hypoblast. The existence of this cell population had been overlooked by embryologists, even though Koller's sickle had been described almost 100 years earlier. Its importance for the gastrulation process was revealed by the *Goosecoid* marker. Figure 5B shows *Goosecoid* expression in Koller's sickle in an embryo incubated for a few hours (stage XII, when the forming sheet of hypoblast covers 50% of the area pellucida). At slightly later stages (and only in the White Leghorn strain), *Gsc* becomes stronger in the hypoblast (Hume and Dodd 1993; Foley et al. 2000; Chapman et al. 2002), a tissue that is homologous to the mouse AVE (see Chapter 15).

As development continues, the primitive streak forms, and as it progresses in the anterior direction, *Goosecoid* is expressed in its anterior region. Transcripts become more abundant as the streak elongates, and by stage 3+, before the streak reaches its maximal extension, *Goosecoid* attains its maximal expression in the anterior third of the streak (Fig. 5C). This is also the stage at which the anterior end of the primitive streak, the young Hensen's node, presents its maximal neural inducing activity. Once the primitive streak reaches its full extension at stage 4, *Gsc* expression becomes confined to the morphological node, and immediately afterward (stage 4+) is down-regulated except in cells that move anteriorly out of the node, ingressing to form the endomesoderm of the head process and the prechordal plate that underlies the future forebrain.

The second chick *Goosecoid* gene, *GSX*, is initially expressed together with *Goosecoid* in Koller's sickle and early primitive streak. Later in development, the two expression domains separate. *GSX* is expressed in the primitive streak, excluding Hensen's node, and in the neural plate, in which it remains (Lemaire et al. 1997).

The inducing potency of the chick organizer correlates with cells having expressed *Goosecoid* during their development. Grafts of Koller's sickle cells are able to induce the formation of partial ectopic axes and can induce neighboring cells to express *Gsc* (Izpisúa-Belmonte et al. 1993; Streit et al. 2000). At stage 3+, grafts from anterior or posterior primitive streak induce mesodermal structures, but only the anterior third of the streak has the ability to induce neural tissues and *Gsc*-positive cells. At stage 4+, when *Gsc*-positive cells leave the node, the neural inducing activity of the node itself drastically decreases. The neural inducing activity is found in the head process, both in cells that continue to express *Gsc* and in cells that expressed *Gsc* at earlier stages (Storey et al. 1992; Izpisúa-Belmonte et al. 1993).

The chick embryo, due to its flat anatomy, has provided some of the best information on the sequence of events that lead to organizer formation. Studies on chick *Gsc* expression allowed a precise analysis of the inductive powers of tissues with organizer activity. Unexpectedly, these studies led to the conclusion that the organizer starts much earlier than previously thought, in Koller's sickle of the unincubated egg. Perhaps the most important contribution of these studies was to our understanding of the general architecture of gastrulation in vertebrates, as discussed below.

GOOSECOID AND THE MOLECULAR ANATOMY OF VERTEBRATE GASTRULATION

By analyzing the expression of *Goosecoid* and of the primitive-streak marker *Brachyury*, it is possible to discern the homologous elements of vertebrate gastrulation. This is best illustrated by analyzing the chick embryo. At stage 3⁺, expression of *Gsc* is maximal and located in the anterior streak (Fig. 5C). *Brachyury* is expressed throughout the length of the streak, including its posterior end (Kispert et al. 1995). Shortly afterward, *Gsc*-positive cells move out of the streak and into the head process endomesoderm, while *Brachyury* remains in the streak. The equivalent stage of maximal expression of *Gsc*, before it exits into the head process, corresponds to stage 10 in *Xenopus*, 50% epiboly in zebrafish, and mid-streak in mouse (Figs. 1A,D and 5A). In the mouse, *Brachyury* is also expressed throughout the length of the primitive streak mesoderm. In zebrafish and *Xenopus*, however, *Brachyury* is expressed as a ring around yolk-laden cells, in what is called the marginal zone (Fig. 1D) (Smith et al. 1991; Schulte-Merker et al. 1994). The main difference between the ring-shaped marginal zone of *Xenopus* and the primitive streak of the chick or mouse is the interposition of yolky cells in the midline. Both sides of the marginal zone are joined in the midline of the primitive streak of amniotes (De Robertis et al. 1994).

GOOSECOID IS A DORSAL MARKER

As mentioned above, a revision in the nomenclature of the *Xenopus* marginal zone has been proposed recently, by which the vertical short axis across the mesoderm ring would be renamed dorsal-ventral (see Fig. 1B in Gerhart 2002). If one accepts the view of the primitive streak of amniotes being homologous to the *Brachyury*-expressing marginal zone of *Xenopus*, then the revised *Xenopus* proposal (Kumano and Smith 2002; Lane and Sheets 2002) could not apply to chick or mouse gastrulation. This is because extensive lineage analyses have shown that the primitive streak has a very different organization. Mesodermal cells in the chick primitive streak are arranged in a rostro-caudal sequence so that the more rostral parts of the streak contribute to dorsal elements such as somites, intermediate ones to kidney, and the more caudal primitive streak to ventral elements such as lateral plate mesoderm (Rosenquist 1966; Nicolet 1970; Schoenwolf et al. 1992; García-Martínez et al. 1993; Catala et al. 1996: Psychoyos and Stern 1996). In the mouse, a similar movement of ventral divergence is seen in the mesoderm emerging from the primitive streak (Tam and Beddington 1987; Lawson et al. 1991; Smith et al. 1994; Wilson and Beddington 1996). In *Xenopus*, at the end of gastrulation (stage 13), the fate map of the mesoderm of the slit blastopore has the same arrangement as in chick or mouse (Gont et al. 1993). At earlier stages of gastrulation in *Xenopus*, this issue will have to be resolved by a detailed lineage analysis of the marginal zone at mid gastrula, after *gsc*-positive cells move into the head process (Fig. 2A) (the use of 32-cell blastomere lineag-

es may confuse the issue for the lineages of the trunk, because the entire head region arises from dorsal blastomeres in *Xenopus*; Fig. 3). In the meantime, the common elements of vertebrate gastrulation indicate that *gsc* is a dorsal marker.

CONCLUSIONS AND PERSPECTIVES

Twelve years after the isolation of *Goosecoid*, much has been learned about the molecular nature of the vertebrate organizer. *Gsc* is part of a gene hierarchy that executes organizer activity, and is downstream of β-Catenin and Nodal-related signals. We have learned that many other genes participate in organizer formation, resulting in convergent redundant mechanisms. It is noteworthy that *goosecoid* is found in all metazoans examined, from Hydra up (e.g., Broun et al. 1999). In general, in the invertebrates, *gsc* is also expressed during gastrulation and marks the foregut (see, e.g., Arendt et al. 2001). In the vertebrates, *gsc* provides an excellent marker first for Spemann's organizer, and later for prechordal plate and foregut. Comparative analyses of *gsc* expression patterns during gastrulation in *Xenopus*, zebrafish, mouse, and chick have helped uncover the homologous components and mechanisms of vertebrate gastrulation. Despite progress in the molecular exploration of gastrulation, much remains to be understood, as explained in this volume.

REFERENCES

Agius E., Oelgeschläger M., Wessely O., Kemp C., and De Robertis E.M. 2000. Endodermal nodal-related signals and mesoderm induction in *Xenopus*. *Development* **127:** 1173-1183.

Arendt D., Technau U., and Wittbrodt J. 2001. Evolution of the bilaterian larval foregut. *Nature* **409:** 81–85.

Artinger M., Blitz I., Inoue K., Tran U., and Cho K.W.Y. 1997. Interaction of *goosecoid* and *brachyury* in *Xenopus* mesoderm patterning. *Mech. Dev.* **65:** 187–196.

Belo J.A., Leyns L., Yamada G., and De Robertis E.M. 1998. The prechordal midline of the chondrocranium is defective in *Goosecoid*-1 mouse mutants. *Mech. Dev.* **72:** 15–25.

Belo J.A., Bouwmeester T., Leyns L., Kertesz N., Gallo M., Follettie M., and De Robertis E.M. 1997. *Cerberus-like* is a secreted factor with neuralizing activity expressed in the anterior primitive endoderm of the mouse gastrula. *Mech. Dev.* **68:** 45–57.

Blum M., Gaunt S.J., Cho K.W.Y., Steinbeisser H., Blumberg B., Bittner D., and De Robertis E.M. 1992. Gastrulation in the mouse: The role of the homeobox gene *goosecoid*. *Cell* **69:** 1097–1106.

Blumberg B., Wright C.V.E., De Robertis E.M., and Cho K.W.Y. 1991. Organizer-specific homeobox genes in *Xenopus laevis* embryos. *Science* **253:** 194–196.

Bouwmeester T., Kim S.H., Sasai Y., Lu B., and De Robertis E.M. 1996. Cerberus is a head-inducing secreted factor expressed in the anterior endoderm of Spemann's organizer. *Nature* **382:** 595–601.

Broun M., Sokol S., and Bode H.R. 1999. Cngsc, a homologue of goosecoid, participates in the patterning of the head, and is expressed in the organizer region of Hydra. *Development* **126:** 5245–5254.

Camus A., Davidson B.P., Billiards S., Khoo P., Rivera-Pérez J.A., Wakamiya M., Behringer R.R., and Tam P.P. 2000. The morphogenetic role of midline mesendoderm and ectoderm in the development of the forebrain and the midbrain of the mouse embryo. *Development* **127:** 1799–1813.

Catala M., Teillet M.A., De Robertis E.M., and Le Douarin N.M. 1996. A spinal cord fate map in the avian embryo: While regressing, the Hensen's node lays down the notochord and floor plate thus joining the spinal cord lateral walls. *Development* **122:** 2599–2610.

Chapman S.C., Schubert F.R., Schoenwolf G.C., and Lumsden A. 2002. Analysis of spatial and temporal gene expression patterns in blastula and gastrula stage chick embryos. *Dev. Biol.* **245:** 187–199.

Cho K.W.Y., Blumberg B., Steinbeisser H., and De Robertis E.M. 1991. Molecular nature of Spemann's organizer: The role of the *Xenopus* homeobox gene *goosecoid*. *Cell* **67:** 1111–1120.

Christian J.L. and Moon R.T. 1993. Interactions between *Xwnt-8* and Spemann organizer signaling pathways generate dorsoventral pattern in the embryonic mesoderm of *Xenopus*. *Genes Dev.* **7:** 13–28.

Danilov V., Blum M., Schweickert A., Campione M., and Steinbeisser H. 1998. Negative autoregulation of the organizer-specific homeobox gene goosecoid. *J. Biol. Chem.* **273:** 627–635.

De Robertis E.M., Fainsod A., Gont L.K., and Steinbeisser H. 1994. The evolution of vertebrate gastrulation. *Dev. Suppl.* **1994:** 117–124.

Fainsod A., Steinbeisser H., and De Robertis E.M. 1994. On the function of BMP-4 in patterning the marginal zone of the *Xenopus* embryo. *EMBO J.* **13:** 5015–5025.

Feldman B., Gates M.A., Egan E.S., Dougan S.T., Rennebeck G., Sirotkin H.I., Schier A.F., and Talbot W.S. 1998. Zebrafish organizer development and germ-layer formation require nodal-related signals. *Nature* **395:** 181–185.

Ferreiro B., Artinger M., Cho K., and Niehrs C. 1998. Antimorphic goosecoids. *Development* **125:** 1347–1359.

Filosa S., Rivera-Pérez J.A., Gómez A.P., Gansmuller A., Sasaki H., Behringer R.R., and Ang S.-L. 1997. *goosecoid* and *HNF-3β* genetically interact to regulate neural tube patterning during mouse embryogenesis. *Development* **124:** 2843–2854.

Foley A.C., Skromme I., and Stern C.D. 2000. Reconciling different models for forebrain induction and patterning: A dual role for the hypoblast. *Development* **127:** 3839–3854.

Funke B., Saint-Jore B., Puech A., Sirotkin H., Edelmann L., Carlson C., Raft. S., Pandita R.K., Kucherlapati R., Skoultchi A., and Morrow B.E. 1997. Characterization and mutation analysis of goosecoid-like (GSCL), a homeodomain-containing gene that maps to the critical region for VCFS/DGS on 22q11. *Genomics* **46:** 364–372.

García-Martínez V., Alvarez I.S., and Schoenwolf G.C. 1993. Locations of the ectodermal and non-ectodermal subdivisions of the epiblast at stages 3 and 4 of avian gastrulation and neurulation. *J. Exp. Zool.* **267:** 431–446.

Gaunt S.J., Blum M., and De Robertis E.M. 1993. Expression of the mouse *goosecoid* gene during mid-embryogenesis may mark mesenchymal cell lineages in the developing head, limbs and body wall. *Development* **117:** 769–778.

Gerhart J. 2002. Changing the axis changes the perspective. *Dev. Dyn.* **225:** 380–384.

Germain S., Howell M., Esslemont G.M., and Hill C.S. 2000. Homeodomain and winged-helix transcription factors recruit activated Smads to distinct promoter elements via a common Smad interaction motif. *Genes Dev.* **14:** 435–451.

Gont L.K., Steinbeisser H., Blumberg B., and De Robertis E.M. 1993. Tail formation as a continuation of gastrulation: The multiple cell populations of the *Xenopus* tailbud derive from the late blastopore lip. *Development* **119:** 991–1004.

Gritsman K., Talbot W.S., and Schier A.F. 2000. Nodal signaling patterns the organizer. *Development* **127:** 921–932.

Hume C.R. and Dodd J. 1993. Cwnt-8C: A novel Wnt gene with a potential role in primitive streak formation and hindbrain organization. *Development* **119:** 1147–1160.

Imai Y., Gates M.A., Melby A.E., Kimelman D., Schier A.F., and Talbot W.S. 2001. The homeobox genes vox and vent are redundant repressors of dorsal fates in zebrafish. *Development* **128:** 2407–2420.

Izpisúa-Belmonte J.C., De Robertis E.M., Storey K.G., and Stern C.D. 1993. The homeobox gene *goosecoid* and the origin of organizer cells in the early chick blastoderm. *Cell* **74:** 645–659.

Kawahara A., Wilm T., Solnica-Krezel L., and Dawid I.B. 2000. Functional interaction of vega2 and goosecoid homeobox genes in zebrafish. *Genesis* **28:** 58–67.

Kispert A., Ortner H., Cooke J., and Herrmann B.G. 1995. The chick Brachyury gene: Developmental expression pattern and response to axial induction by localized activin. *Dev. Biol.* **168:** 406–415.

Kumano G. and Smith W.C. 2002. Revisions to the *Xenopus* gastrula fate map implications for mesoderm induction and patterning. *Dev. Dyn.* **225:** 409–421.

Lane M.C. and Sheets M.D. 2002. Rethinking axial patterning in amphibians. *Dev. Dyn.* **225:** 434–447.

Lane M.C. and Smith W.C. 1999. The origins of primitive blood in *Xenopus*: Implications for axial patterning. *Development* **126:** 423–434.

Latinik B.V. and Smith J.C. 1999. Goosecoid and mix.1 represses Brachyury expression and are required for head formation in *Xenopus*. *Development* **125:** 1769–1779.

Latinik B.V., Umbhauer M., Neal K.A., Lerchner W., Smith J.C., and Cunliffe V. 1997. The *Xenopus* Brachyury promoter is activated by FGF and low concentrations of activin and suppressed by high concentrations of activin and by paired-type homeodomain proteins. *Genes Dev.* **11:** 3265–3276.

Laurent M.N., Blitz I.L., Hashimoto C., Rothbächer U., and Cho K.W.Y. 1997. The *Xenopus* homeobox gene *Twin* mediates Wnt induction of *Goosecoid* in establishment of Spemann's organizer. *Development* **124:** 4905–4916.

Lawson K.A., Meneses J.J., and Pedersen R.A. 1991. Clonal analysis of epiblast fate during germ layer formation in the mouse embryo. *Development* **113:** 891–911.

Lemaire L., Roeser T., Izpisúa-Belmonte J.C., and Kessel M. 1997. Segregating expression domains of two *goosecoid* genes during the transition from gastrulation to neurulation in chick embryos. *Development* **124:** 1443–1452.

Leyns L., Bouwmeester T., Kim S.-H., Piccolo S., and De Robertis E.M. 1997. Frzb-1 is a secreted antagonist of Wnt signaling expressed in the Spemann Organizer. *Cell* **88:** 747–756.

Mailhos C., Andre S., Mollereau B., Goriely A., Hemmati-Brivanlou A., and Desplan C. 1998. *Drosophila* Goosecoid requires a conserved heptapeptide for repression of paired-class homeoprotein activators. *Development* **125:** 937–947.

Mochizuki T., Karanov A.A., Curtiss P.E., Ault K.T., Sugimoto N., Watabe T., Shiokawa K., Jamrich M., Cho K.W., Dawid I.B., and Taira M. 2000. Xlim-1 and LIM domain binding protein 1 cooperate with various transcription factors in the regulation of the goosecoid promoter. *Dev. Biol.* **224:** 470–485.

Nicolet G. 1970. Analyse autoradiographique de la localization des différentes ébauches présomptives dans la ligne primitive de l'embryon de poulet. *J. Embryol. Exp. Morphol.* **23:** 79–108.

Niehrs C., Steinbeisser H., and De Robertis E.M. 1994. Mesodermal patterning by a gradient of the vertebrate homeobox gene *goosecoid*. *Science* **263:** 817–820.

Niehrs C., Keller R., Cho K.W.Y., and De Robertis E. 1993. The homeobox gene *goosecoid* controls cell migration in *Xenopus* embryos. *Cell* **72:** 491–503.

Psychoyos D. and Stern C.D. 1996. Fates and migratory routes of primitive streak cells in the chick embryo. *Development* **122:** 1523–1534.

Rivera-Pérez J.A., Mallo M., Gendron-Maguire M., Gridley T., and Behringer R.R. 1995. *goosecoid* is not an essential component of the mouse gastrula organizer but is required for craniofacial and rib development. *Development* **121:** 3005–3012.

Rosenquist G.C. 1966. A radioautographic study of labeled grafts in the chick blastoderm. Development from primitive-streak to stage 12. *Contrib. Embryol. Carnegie Instn.* **38:** 73–110.

Sasai Y., Lu B., Steinbeisser H., Geissert D., Gont L.K., and De Robertis E.M. 1994. *Xenopus chordin*: A novel dorsalizing factor activated by organizer-specific homeobox genes. *Cell* **79:** 779–790.

Schoenwolf G.C., García-Martínez V., and Dias M.S. 1992. Mesoderm movement and fate during avian gastrulation and neurulation. *Dev. Dyn.* **193:** 235–248.

Schulte-Merker S., Hammerschmidt M., Beuchle D., Cho K.W., De Robertis E.M., and Nüsslein-Volhard C. 1994. Expression of zebrafish *goosecoid* and *no tail* gene products in wild-type and mutant *no tail* embryos. *Development* **120:** 843–852.

Shimizu T., Yamanaka Y., Ryu S.L., Hashimoto H., Yabe T., Hirata T., Bae Y.K., Hibi M., and Hirano T. 2000. Cooperative roles of Bozozok/Dharma and Nodal-related proteins in the formation of the dorsal organizer in zebrafish. *Mech. Dev.* **91:** 293–303.

Skromne I. and Stern C.D. 2002. A hierarchy of gene expression accompanying induction of the primitive streak by Vg1 in the chick embryo. *Mech. Dev.* **114:** 115–118.

Smith J.L., Gesteland K.M., and Schoenwolf G.C. 1994. Prospective fate map of the mouse primitive streak at 7.5 days of gestation. *Dev. Dyn.* **201:** 279–289.

Smith J.C., Price B.M.J., Green J.B.A., Weigel D., and Herrmann B.G.

1991. Expression of a *Xenopus* homolog of *Brachyury* (*T*) is an immediate-early response to mesoderm induction. *Cell* **67:** 79–87.

Stachel S.E., Grunwald D.J., and Myers P.Z. 1993. Lithium perturbation and *goosecoid* expression identify a dorsal specification pathway in the pregastrula zebrafish. *Development* **117:** 1261–1274.

Steinbeisser H., Fainsod A., Neihrs C., Sasai Y., and De Robertis E.M. 1995. The role of *Gsc* and BMP-4 in dorsal-ventral patterning of the marginal zone in *Xenopus*: A loss of function study using antisense RNA. *EMBO J.* **14:** 5230–5243.

Storey K.G., Crossley J.M., De Robertis E.M., Norris W.E., and Stern C.D. 1992. Neural induction and regionalisation in the chick embryo. *Development* **114:** 729–741.

Streit A., Berliner A.J., Papanayotou C., Sirulnik A., and Stern C.D. 2000. Initiation of neural induction by FGF signalling before gastrulation. *Nature* **406:** 74–78.

Tam P.P.L. and Beddington R.S.P. 1987. The formation of mesodermal tissues in the mouse embryo during gastrulation and early organogenesis. *Development* **99:** 109–126.

Trinidade M., Tada M., and Smith J.C. 1999. DNA-binding specificity and embryological function of Xom (Xvent-2). *Dev. Biol.* **216:** 442–456.

Watabe T., Kim S., Candia A., Rothbächer U., Hashimoto C., Inoue K., and Cho K.W.Y. 1995. Molecular mechanisms of Spemann's organizer formation: Conserved growth factor synergy between *Xenopus* and mouse. *Genes Dev.* **9:** 3038–3050.

Wilson V. and Beddington R.S.P. 1996. Cell fate and morphogenetic movement in the late mouse primitive streak. *Mech. Dev.* **55:** 79–89.

Yamada G., Mansouri A., Torres M., Stuart E.T., Blum M., Schultz M., De Robertis E.M., and Gruss P. 1995. Targeted mutation of the murine *goosecoid* gene results in craniofacial defects and neonatal death. *Development* **121:** 2917–2922.

Yao J. and Kessler D.S. 2001. Goosecoid promotes head organizer activity by direct repression of Xwnt8 in Spemann's organizer. *Development* **128:** 2975–2987.

Yasuo H. and Lemaire P. 2001. Role of Goosecoid, Xnot and Wnt antagonists in the maintenance of the notochord genetic programme in *Xenopus* gastrulae. *Development* **128:** 3783–3793.

Zhu L., Belo J.A., De Robertis E.M., and Stern C.D. 1999. *Goosecoid* regulates the neural inducing strength of the mouse node. *Dev. Biol.* **216:** 276–281.

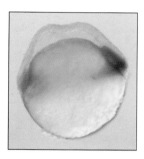

ROLES OF LIM1 DURING GASTRULATION

W. Shawlot[1] and R.R. Behringer[2]

[1]*Department of Genetics, Cell Biology, and Development, University of Minnesota, Minneapolis, Minnesota 55455;* [2]*Department of Molecular Genetics, University of Texas M. D. Anderson Cancer Center, Houston, Texas 77030*

INTRODUCTION

For almost 70 years, the search for molecules responsible for the unique properties of the organizer has been a primary goal for embryologists (Waddington et al. 1935). In the past 10+ years a number of genes have been identified that are specifically expressed in the gastrula organizer of *Xenopus*, chick, mouse, and fish embryos. One of these is the homeobox gene *Lim1*, which was identified in a screen for LIM class homeobox genes expressed during early embryogenesis in *Xenopus* (Taira et al. 1992). Molecular and genetic studies demonstrate that *Lim1* plays an essential role in regulating gastrulation and vertebrate head development.

LIM1 STRUCTURE

The *Lim1* gene was initially isolated from *Xenopus* embryonic cDNA libraries using degenerate PCR primers designed on the basis of conserved amino acid sequences present in the homeodomain of several LIM class homeodomain proteins (Taira et al. 1992). *Lim1* orthologs have been isolated in mouse (*Lim1/Lhx1*), chick (*lim1*), and zebrafish (*lim1*) (Barnes et al. 1994; Fujii et al. 1994; Tsuchida et al. 1994; Toyama et al. 1995). In addition, *Fugu* (Ensembl SINFRUG00000142325) and human (*LIM1/LHX1*) orthologs are present in the database. The Lim1 protein sequence is highly conserved among vertebrate species (>90% identity throughout the entire protein). *Lim1* encodes a protein that contains two tandemly repeated cysteine–histidine motifs (termed LIM domains) at its amino terminus, a LIM class homeodomain, and a proline-rich transcriptional activation domain at its carboxyl terminus (Fig. 1) (Taira et al.

1992; Breen et al. 1998). LIM domains were first recognized as conserved motifs in the proteins encoded by the transcription factor genes *lin-11* in *Caenorhabditis elegans*, *Isl-1* in rat, and *mec-3* in *C. elegans* (Way and Chalfie 1988; Freyd et al. 1990; Karlsson et al. 1990). Although LIM domains were first recognized in homeodomain proteins, they are not unique to them but are found in a variety of proteins that are located in both the nuclear and cytoplasmic compartments of the cell and are involved in mediating protein–protein interactions (Dawid et al. 1998).

EXPRESSION IN MODEL ORGANISMS

The expression pattern of *Lim1* during gastrulation is generally evolutionarily conserved in *Xenopus*, zebrafish, mouse, and chick embryos.

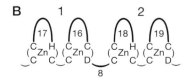

Figure 1. Lim1 structure. (*A*) *Lim1* encodes a LIM class homeodomain protein. There are two LIM domains that are located at the amino terminus. The homeodomain is located approximately in the middle of the protein. (*B*) Each LIM domain contains two zinc fingers. Numbers indicate the number of amino acids.

Xenopus

In stage-11.5 *Xenopus* embryos, *Xlim-1* is expressed in the dorsal lip and in the head mesoderm that migrates anteriorly (Fig. 2A) (Taira et al. 1992). By stage 12.5, *Xlim-1* is expressed in the notochord, but expression subsequently declines in this tissue by the mid-neurula stage (Taira et al. 1994a). In explant experiments, both Activin A and retinoic acid can induce *Xlim-1* expression (Taira et al. 1992). Induction of *Xlim-1* by retinoic acid is somewhat paradoxical because retinoic acid is thought to induce posterior gene expression, but this finding may be explained by the observation that *Xlim-1* is also expressed in the nephric duct and kidney later in development (Taira et al. 1994a).

Zebrafish

The initial expression of the zebrafish *lim1* gene occurs at the margin of 30% epiboly embryos with a region of more intense expression on one side of the embryo corresponding to the forming embryonic shield (Fig. 2B). Expression is restricted to cells of the deep layer. This appears to be slightly different from the expression pattern of *Xlim-1* in *Xenopus* in that *Xlim-1* mRNA is strongly concentrated on the dorsal side of the embryo (Fig. 2A). It is not clear whether this represents a higher sensitivity of detection of *lim1* mRNA in zebrafish embryos or a biological difference

in how the organizer develops in zebrafish versus *Xenopus* embryos. During subsequent development, *lim1* expression becomes restricted to the dorsal shield and the axial mesoderm (Toyama et al. 1995). Injection of *nodal* mRNA into a single random blastomere of an 8-cell-stage zebrafish embryo induces ectopic expression of *lim1*, suggesting that *nodal* acts upstream of *lim1* (Toyama et al. 1995).

Mouse

Lim1 is expressed throughout the visceral endoderm that contacts the epiblast at embryonic day (E) 5.5 (Perea-Gomez et al. 1999). The visceral endoderm is an extraembryonic tissue that does not contribute to the embryo proper (see Chapter 16). At E6.5, *Lim1* is expressed in the anterior visceral endoderm and the newly forming primitive streak (Barnes et al. 1994; Shawlot and Behringer 1995; Belo et al. 1997). Between E7.0 and E7.5, *Lim1* is expressed in the anterior primitive streak, the mesodermal wings and the anterior definitive mesendoderm (Fig. 2D) (Shawlot and Behringer 1995; Shimono and Behringer 1999). At E7.5, expression of *Lim1* in the anterior mesendoderm is downregulated and becomes restricted to the node, where it is expressed in a horseshoe-shaped pattern (Barnes et al. 1994).

Chick

Lim1 is strongly expressed in the hypoblast at stages XI–XIV (Chapman et al. 2002). Morphological and molecular marker studies in chick suggest that the hypoblast is likely to be the equivalent of the mouse anterior visceral endoderm (Foley et al. 2000). During primitive-streak stages, *Lim1* is expressed in the anterior primitive streak prior to the formation of Hensen's node. At stage 4+, *Lim1* is expressed in Hensen's node and the axial mesoderm, including the prechordal mesendoderm (Fig. 2C) (Chapman et al. 2002).

GAIN-OF-FUNCTION EXPERIMENTS IN *XENOPUS*

Injection of mRNA for mutant forms of *Xlim-1* into *Xenopus* embryos can mimic properties of the Spemann organizer (Taira et al. 1994b). Injection of *Xlim-1* mRNAs with point mutations in both LIM domains (*Xlim-1/3m*), or in which the LIM domains have been deleted into 2-cell embryos, induces the expression of the neural markers *NCAM* and *en-2* in animal caps. In explant recombination experiments, animal caps from embryos injected with *Xlim-1/3* mRNA can induce *NCAM* expression in apposed uninjected animal caps. This indicates that the neural-

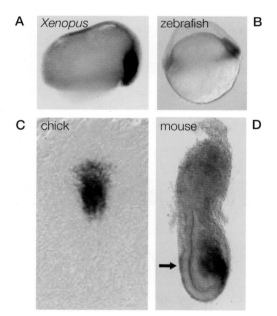

Figure 2. *Lim1* expression in amniotes. All embryos are shown as lateral views with anterior to the left except for chick in which anterior is located at the top of the panel. (*A*) *Xenopus* stage 11, (*B*) zebrafish 50% epiboly, (*C*) chick stage 4+, and (*D*) mouse late-streak stage. (*Arrow*) *Lim1* expression in anterior visceral endoderm. (*C*, Reprinted from Chapman et al. 2002 [©Elsevier].)

inducing effect involves intercellular signaling. Injection of wild-type *Xlim-1* mRNA does not induce neural gene expression, suggesting that the LIM domains exert a negative regulatory role on Xlim-1 activity and may require a cofactor(s) for activity (see below). Although the *Xlim-1/3m* mRNA induces neural gene expression, *Xlim-1/3m* mRNA by itself is unable to induce muscle α-*actin*-gene expression. However, co-injection of *Xlim-1/3m* with *Xbra* substantially increases muscle α-actin expression over injection of *Xbra* alone. This implies that *Xlim-1/3m* enhances the formation of dorsal mesoderm. Also, in whole embryos, injection of *Xlim-1/3m* mRNA into the ventral equatorial region of 4-cell-stage embryos can induce the formation of secondary axes with neural tissue and somites (Taira et al. 1994b).

LOSS-OF-FUNCTION EXPERIMENTS IN MICE

Gene targeting experiments in the mouse demonstrate that *Lim1* is required for the formation of anterior head structures (Shawlot and Behringer 1995; Cheah et al. 2000; Kania et al. 2000). E9.5 *Lim1⁻/⁻* embryos lack forebrain, midbrain, and anterior hindbrain tissue, but form posterior hindbrain tissue (Fig. 3A). Most *Lim1*-deficient embryos die at approximately E10.0, presumably due to defects in allantois formation, but several stillborn headless pups have been recovered (Fig. 3B). In addition to head defects, these mice lack kidneys and reproductive tract organs, indicating that *Lim1* also has an essential role in urogenital development (Shawlot and Behringer 1995; Kobayashi et al. 2004). In addition to the head phenotype, approximately 10–20% of *Lim1* mutant embryos develop secondary body axes (Fig. 3C).

Lim1⁻/⁻ embryos display a number of morphological defects during early gastrulation. E7.5 *Lim1* mutant embryos have a prominent constriction between the embryonic–extraembryonic region, lack a morphologically distinct node, and have a clump of visceral endoderm at the distal tip of the embryo (Shawlot and Behringer 1995). Consistent with the morphological defects, molecular marker studies indicate that the position of the primitive streak and presumptive node is abnormal. In E6.5 *Lim1⁻/⁻* embryos, *goosecoid* (a marker for the anterior primitive streak; see Chapters 16 and 42) is expressed throughout the proximal region of the embryo adjacent to the embryonic–extraembryonic junction instead of at the anterior portion of the streak that normally extends down the lateral aspect of wild-type embryos (Fig. 4A). Both *HNF3β* and *nodal*, markers of the node, are expressed in *Lim1* mutants on the lateral aspect of the embryos instead of at the distal tip region (Fig. 4B and not shown). In addition,

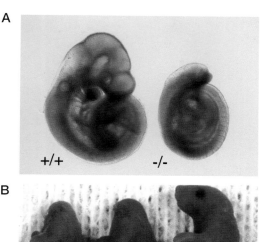

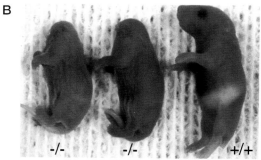

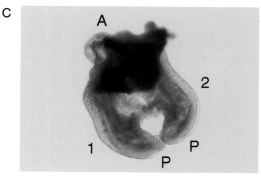

Figure 3. *Lim1* mutant mice. (*A*) E9.5 Wild-type (+/+) and *Lim1* mutant (–/–) embryos; (*B*) *Lim1* mutant (–/–) and wild-type (+/+) neonates; (*C*) axis duplication in a *Lim1⁻/⁻* embryo. (1) Axis #1; (2) axis #2; (A) anterior; (P) posterior ends of the duplicated axes.

expression of the *Secreted frizzled-related protein 5* gene, a marker for anterior visceral endoderm cells, remains at the distal tip region of E6.5 *Lim1⁻/⁻* embryos (Finley et al. 2003). These findings suggest that conversion of the initial proximal–distal axis to an anterior–posterior axis is impaired in *Lim1⁻/⁻* embryos.

Mouse chimera experiments using *lacZ*-tagged embryonic stem (ES) cells indicate that *Lim1* is required in both the primitive-streak-derived cells and extraembryonic visceral endoderm for head development (Shawlot et al. 1999). ES cells have a developmental bias: they contribute predominantly to embryonic tissues but poorly to extraembryonic tissues (Beddington and Robertson 1989). *Lim1⁻/⁻* ES cells injected into wild-type blastocysts are able to con-

A

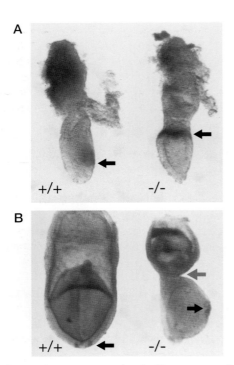

B

Figure 4. Organizer gene expression in *Lim1* mutant mice. (*A*) E6.5 *Gsc* expression (*black arrows*); (*B*) E7.5 *nodal* expression (*black arrows*), constriction between the embryonic and extraembryonic regions (*red arrow*). (+/+) wild type; (–/–) *Lim1* homozygous mutant.

this fate is labile and must be stabilized by a second signal(s) from the definitive anterior mesendoderm that replaces the visceral endoderm. Consistent with this hypothesis is the observation that the chick hypoblast, which is likely to be the equivalent of mouse anterior visceral endoderm, can induce anterior neural gene expression when analyzed 4–6 hours after being placed in contact with the inner third region of the area opaca. However, no anterior neural gene expression was detected when analyzed 18–24 hours after grafting (Foley et al. 2000). This suggests that the hypoblast transiently induces neural development in the overlying epiblast. In contrast, transplantation experiments in the mouse indicate that the anterior visceral endoderm cells alone do not induce anterior neural gene expression (Tam and Steiner 1999). However, in these experiments, gene expression was analyzed after 24 hours of embryo culture, thus a transient induction might not have been detected. Anterior neural gene expression was only observed in the mouse transplantation experiments when the anterior visceral endoderm, the early gastrula organizer region (EGO), and epiblast tissue underlying the anterior visceral endoderm were transplanted together. These results are consistent with the two-step model in that the visceral endoderm provides an initial signal to the overlying epiblast tissue and the EGO-derived cells provide a second signal to reinforce or stabilize the initial induction event.

INTERACTING PROTEINS

LIM domains mediate protein–protein interactions (Dawid et al. 1998). A protein named Ldb1 that can interact with the LIM domains of Lim1 and Xlim-1 was isolated by expression cDNA library screening and yeast two-hybrid screening (Agulnick et al. 1996). The same protein was isolated by its ability to interact with the LIM domains of other proteins and named NLI and CLIM2 (Jurata et al. 1996; Bach et al. 1997). Ldb1 can also interact with Otx, GATA, and bHLH proteins (Bach et al. 1997; Wadman et al. 1997). Ldb1 proteins are highly conserved. The mouse, fish, and frog Ldb1 proteins are 95% identical over their length of 375 amino acids, and related proteins are found in human, *C. elegans,* and *Drosophila* (Agulnick et al. 1996; Dawid et al. 1998; Toyama et al. 1998). Although Ldb1 has no similarity to other proteins, deletion analysis studies indicate that Ldb1 contains two functional regions. Ldb1 binds LIM domains through it carboxy-terminal region and forms homodimers via its amino-terminal region (Breen et al. 1998). In the mouse, *Ldb1* is widely expressed in the embryo and in adult tissues. In *Xenopus* embryos, *Xldb1* is ubiquitously expressed in the early gastrula embryo, including the dorsal lip region where *Xlim1* is expressed (Agulnick et al. 1996).

tribute extensively to the embryonic tissues of E7.5 embryos including the distal tip region and the anterior mesendoderm. E7.5 chimeras with a high percentage of *Lim1⁻/⁻* cells do not develop a constriction between the embryonic and extraembryonic region, indicating that the constriction phenotype is caused by a defect in the visceral endoderm. E9.5 chimeric embryos that contain greater than 50% *Lim1⁻/⁻* cells have head defects. These defects range from holoprosencephaly to complete loss of anterior head structures similar to *Lim1⁻/⁻* embryos. Mouse explant recombination experiments indicate that *Lim1* mutant anterior mesendoderm is unable to maintain the expression of the anterior neural marker gene *Otx2* in wild-type ectoderm. *Lim1* function is also required in the extraembryonic visceral endoderm for head development. Chimeric embryos that develop from *Lim1⁻/⁻* blastocysts in which the extraembryonic visceral endoderm is composed of predominantly *Lim1⁻/⁻* cells have head defects that are identical to *Lim1⁻/⁻* embryos even when the embryo proper is composed of nearly all wild-type cells.

A two-step induction model has been proposed to explain the requirement for *Lim1* in the visceral endoderm and primitive-streak-derived cells (Shawlot et al. 1999). In this model, *Lim1* function in the visceral endoderm is necessary for the epiblast to adopt an anterior neural fate, but

The role of Ldb1 in the regulation of Lim1 transcriptional activity has been examined in gain-of-function experiments in *Xenopus*. Co-injection of wild-type *Xlim-1* and *Ldb1* mRNA into *Xenopus* embryos induces the expression of *goosecoid, chordin, NCAM,* and the cement gland marker *XCG7* in explants, and the formation of secondary axes in whole embryos (Agulnick et al. 1996). Injection of either *Ldb1* or *Xlim1* mRNA alone does not produce this effect. These results are similar to the results found by injecting *Xlim1/3m*, which contains point mutations in the LIM domains. Interestingly, co-injection of mRNA for the LIM-binding domain portion of *Ldb1* with *Xlim-1* did not activate organizer gene expression (Breen et al. 1998). Thus, Ldb1 likely functions as a linker to bring together additional Xlim-1 molecules or other cofactors such as Otx2 in a complex to activate expression of target genes in the organizer.

Mice with mutations in the LIM domains of *Lim1* have been generated by gene targeting in mouse ES cells (Cheah et al. 2000). Amino acid substitutions were engineered in the conserved cysteine residues of both LIM domains to disrupt the zinc finger motifs and model the changes that disrupt the interaction of Ldb1 with Lim1 (Taira et al. 1994b; Agulnick et al. 1996). The phenotype of mice homozygous for a *Lim1* allele that disrupts both LIM domains is identical to the headless phenotype that occurs in the original *Lim1* deletion allele. Thus, the integrity of the LIM domains is essential for the activity of Lim1. Interestingly, no dominant phenotype was seen in mice heterozygous for the mutant LIM domain allele, suggesting that inappropriate activation of Lim1 by mutation of the LIM domains does not cause alterations in gastrulation when expressed at physiological levels.

Gene targeting in the mouse indicates that *Ldb1* is required for early embryonic development (Mukhopadhyay et al. 2003). *Ldb1*-deficient mice have head defects and develop posterior axis duplications similar to *Lim1*-deficient mice. This implies that Ldb1 is a necessary cofactor for Lim1 function. Other phenotypes present in *Ldb1*-deficient mice include altered heart development and defects in mesodermally derived extraembryonic structures, including the allantois, blood islands of the yolk sac, primordial germ cells, and the amnion. These phenotypes may result from the loss of activity of other transcription factors that interact with Ldb1.

REGULATION OF *LIM1* EXPRESSION

Activin can induce *Xlim-1* expression in animal cap explants (Taira et al. 1992). Studies using explants from embryos injected with various reporter constructs have shown that *Xlim-1* contains an Activin response element (ARE) in the first intron, that cooperates with an upstream promoter to regulate transcription (Rebbert and Dawid 1997). Transcriptional regulation by Activin and Nodal signals are mediated by transcriptional effectors FAST-1/FoxH1 and Smad4 (Whitman 2001). The *Xlim-1* ARE contains four FAST-1/FoxH1 and three Smad4 recognition sites. Mutation of the two FAST-1 sites that most closely match the consensus binding sequence is able to eliminate Activin responsiveness (Watanabe et al. 2002). A similar ARE is also present in the first intron of the zebrafish *lim1* gene (Watanabe et al. 2002). The *one-eyed pinhead* (*oep*) and the *schmalspur* (*sur*) mutants interfere with Nodal signaling in zebrafish embryos (Whitman 2001). The *oep* mutation affects a gene encoding an obligatory cofactor required for Nodal signaling and the *schmalspur* (*sur*) mutation disrupts the gene encoding *FoxHI*, the zebrafish homolog of *Xenopus FAST-1*. In embryos that lack both maternal and zygotic *one-eyed pinhead* (*oep*) function, *lim1* is not expressed, and in *sur* mutant embryos, *lim1* expression is severely reduced.

The regulation of the *Lim1* gene in mouse is less clear. Mouse embryos that lack *nodal* specifically in the visceral endoderm have a headless phenotype (Varlet et al. 1997). This phenotype is similar to *Lim1*, suggesting that *nodal* and *Lim1* are in the same genetic pathway. Consistent with Nodal signaling activating *Lim1* is the observation that *Lim1* is not expressed in the visceral endoderm of *nodal* and *Smad2* mutant embryos (Brennan et al. 2001).

DOWNSTREAM TARGETS

In gain-of-function experiments in *Xenopus*, a mutant form of *Xlim-1*, *Xlim-1/3m*, is able to activate expression of *goosecoid* in animal caps (see above and Taira et al. 1994b). A region between –493 to –413 upstream of the start of transcription start site of *goosecoid* has been identified that is responsive to Xlim-1/3m (Rebbert and Dawid 1997). Electrophoretic mobility shift assays (EMSA) and footprinting experiments indicate that the Xlim-1/Ldb1 complex recognizes several TAATXY core elements in the responsive element. EMSA data indicate that Xlim-1 preferentially binds to the sequences TAATTA, TAATTG, TAAT-CA, and TAATGG. Although Xlim-1 can bind the *goosecoid* promoter, it appears not to be required for *goosecoid* expression. *Xenopus* embryos depleted of *Xlim-1* using N,N-diethylethylenediamine (DEED) antisense oligonucleotides still express *goosecoid* (Hukriede et al. 2003). Consistent with this finding is that *goosecoid* is expressed in *Lim1*$^{-/-}$ embryos (Shawlot and Behringer 1995; Shawlot et al. 1999).

The mouse *Cer1* is a candidate downstream target of *Lim1*. *Cer1* is related to the *Xenopus cerberus* gene, and encodes a secreted factor expressed in the anterior endoderm of *Xenopus* embryos (Bouwmeester et al. 1996). Injection of *cerberus* mRNA into *Xenopus* embryos can induce the formation of ectopic head structures. *cerberus* appears to induce head development in *Xenopus* by blocking BMP, nodal, and Wnt signaling (Piccolo et al. 1999). In the mouse, *Cerl* is expressed in both the anterior visceral endoderm and definitive anterior mesendoderm and is down-regulated in most *Lim1⁻/⁻* embryos (Belo et al. 1997; Biben et al. 1998; Shawlot et al. 1998). However, mice lacking *Cer1* do not have any obvious mutant phenotype (Simpson et al. 1999; Belo et al. 2000; Shawlot et al. 2000; Stanley et al. 2000). Recently, mice with compound mutations in *Cerl* and *Lefty1* have been generated (Perea-Gomez et al. 2002). *Lefty1* encodes a secreted factor expressed in the anterior visceral endoderm that can antagonize Nodal signaling (Perea-Gomez et al. 1999; Sakuma et al. 2002). *Cer1/Lefty1* compound mutant mice form ectopic primitive streaks leading to the formation of secondary body axes (Perea-Gomez et al. 2002). Thus, the secondary axes that form in some *Lim1* mutant embryos may be the result of the loss of *Cer1* and *Lefty1* expression in the anterior visceral endoderm that normally serves to block Nodal signaling on the anterior side of the embryo. Because the *Cerl/Lefty1* compound mutant mice still develop anterior neural tissue, *Lim1* must regulate an additional factor(s) involved in head induction.

Xenopus embryos depleted of *Xlim-1* using DEED anti-sense oligonucleotides lack anterior head structures and have a shortened embryonic axis due to defects in gastrulation movements (Hukriede et al. 2003). Although the expression of most organizer genes is fairly normal in the *Xlim-1*-depleted embryos, *Paraxial Protocadherin* (*PAPC*) expression is strongly inhibited in early-gastrula-stage embryos. *PAPC* encodes a cell adhesion molecule initially expressed in the presumptive organizer at the late blastula stage and has a role in cell movements (Kim et al. 1998). Interestingly, head formation and axis formation can be rescued to a large extent by injection of *PAPC* mRNA into *Xlim-1*-depleted embryos, suggesting that *PAPC* is a key downstream target of Xlim-1. In the mouse, orthotopic transplantation experiments indicate that *Lim1⁻/⁻* cells from the primitive streak contribute to more posterior mesodermal structures than corresponding wild-type cells, suggesting that morphogenetic movements may be impaired in *Lim1⁻/⁻* embryos (Hukriede et al. 2003). *PAPC* is not expressed in the newly formed mesoderm in E7.5 *Lim1⁻/⁻* embryos (Hukriede et al. 2003). However, *PAPC*-deficient mice have no obvious phenotype (Yamamoto et al. 2000).

The mouse *Angiomotin* (*Amot*) gene was isolated by subtractive hybridization between wild-type anterior mesendoderm and *Lim1⁻/⁻* anterior mesendoderm cells (Shimono and Behringer 1999). *Amot* encodes an angiostatin-binding protein that localizes to the leading edge of migrating endothelial cells (Troyanovsky et al. 2001).

Originally named *SII6*, *Amot* is expressed at E6.5 in the anterior visceral endoderm in a manner similar to *Lim1* and is absent or significantly reduced in *Lim1⁻/⁻* embryos. At E7.5, *Amot* is expressed in the visceral endoderm associated with the extraembryonic ectoderm (Shimono and Behringer 2003). Most *Amot*-deficient mice generated by gene targeting die shortly after gastrulation and display distinct furrows of visceral endoderm at the embryonic–extraembryonic junction, suggesting defects in the migration of visceral endoderm (Shimono and Behringer 2003). Cell labeling studies indicate that the distal to anterior movement of visceral endoderm cells still occurs in *Amot* mutant embryos, but lateral migration of visceral endoderm cells is greatly reduced. Consistent with this observation, *Lim1* and *Cerl*, markers of the anterior visceral endoderm, are expressed at the anterior end of the embryo. In addition, the anterior neural markers *Otx2* and *Six3* are also expressed in *Amot* mutant embryos. These results indicate that *Amot* is essential for regulating the morphogenetic movements of subregions of the visceral endoderm.

FUTURE DIRECTIONS

The transcription factor Lim1 has an essential role in extraembryonic and organizer tissues for the formation of anterior head structures. One of the challenges for the future will be to determine how this protein fits into a regulatory network that directs axial patterning of the vertebrate embryo. One goal will be to define the *cis*-regulatory sequences in the mouse that can direct *Lim1* transcription in the extraembryonic visceral endoderm and the anterior primitive streak to identify upstream regulators. Comparison of DNA sequences from the *Lim1* locus of different mammals may help in the identification of evolutionary conserved noncoding sequences that may serve as regulatory sequences. A second challenge will be to identify *Lim1* downstream genes encoding secreted proteins involved in anterior neural induction and development.

ACKNOWLEDGMENTS

We thank Igor Dawid for providing frog and zebrafish images and helpful comments on the manuscript. This work was supported by National Institutes of Health grant HD30284 to R.R.B.

REFERENCES

Agulnick A.D., Taira M., Breen J.J., Tanaka T., Dawid I.B., and Westphal H. 1996. Interactions of the LIM-domain-binding factor Ldb1 with LIM homeodomain proteins. *Nature* **384:** 270–272.

Barnes J.D., Crosby J.L., Jones C.M., Wright C.V.E., and Hogan B.L.M. 1994. Embryonic expression of *Lim-1*, the mouse homolog of

Xlim-1, suggests a role in lateral mesoderm differentiation and neurogenesis. *Dev. Biol.* **161:** 168–178.

Bach I., Carriere C., Ostendorff H.P., Andersen B., and Rosenfeld M.G. 1997. A family of LIM domain-associated cofactors confer transcriptional synergism between LIM and Otx homeodomain proteins. *Genes Dev.* **11:** 1370–1380.

Beddington R.S. and Robertson E.J. 1989. An assessment of the developmental potential of embryonic stem cells in the midgestation mouse embryo. *Development* **105:** 733–737.

Belo J.A., Bouwmeester T., Leyns L., Kertesz N., Gallo M., Follettie M., and De Robertis E.M. 1997. *Cerberus-like* is a secreted factor with neutralizing activity expressed in the anterior primitive endoderm of the mouse gastrula. *Mech. Dev.* **68:** 45–57.

Belo J.A., Bachiller D., Agius E., Kemp C., Borges A.C., Marques S., Piccolo S., De Robertis E.M. 2000. *Cerberus-like* is a secreted BMP and nodal antagonist not essential for mouse development. *Genesis* **26:** 265–270.

Biben C., Stanley E., Fabri L., Kotecha S., Rhinn M., Drinkwater C., Lah M., Wang C.C., Nash A., Hilton D., Ang S.L., Mohun T., and Harvey R.P. 1998. Murine cerberus homologue mCer-1: A candidate anterior patterning molecule. *Dev. Biol.* **194:** 135–151.

Breen J.J., Agulnick A.D., Westphal H., and Dawid I.B. 1998. Interactions between LIM domains and the LIM domain-binding protein Ldb1. *J. Biol. Chem.* **273:** 4712–4717.

Brennan J., Lu C.C., Norris D.P., Rodriguez T.A., Beddington R.S., and Robertson E.J. 2001. Nodal signalling in the epiblast patterns the early mouse embryo. *Nature* **411:** 965–969.

Bouwmeester T., Kim S., Sasai Y., Lu B., and De Robertis E.M. 1996. Cerberus is a head-inducing secreted factor expressed in the anterior endoderm of Spemann's organizer. *Nature* **382:** 595–601.

Chapman S.C., Schubert F.R., Schoenwolf G.C., and Lumsden A. 2002. Analysis of spatial and temporal gene expression patterns in blastula and gastrula stage chick embryos. *Dev. Biol.* **245:** 187–199.

Cheah S.S., Kwan K.M., and Behringer R.R. 2000. Requirement of LIM domains for LIM1 function in mouse head development. *Genesis* **27:** 12–21.

Dawid I.B., Breen J.J., and Toyama R. 1998. LIM domains: Multiple roles as adapters and functional modifiers in protein interactions. *Trends Genet.* **14:** 156–162.

Finley K. R., Tennessen J., and Shawlot W. 2003. The mouse *Secreted frizzled-related protein 5* gene is expressed in the anterior visceral endoderm and foregut endoderm during early post-implantation development. *Gene Exp. Patterns* **3:** 681–684.

Foley A.C., Skromne I., and Stern C.D. 2000. Reconciling different models of forebrain induction and patterning: A dual role for the hypoblast. *Development* **127:** 3839–3854.

Freyd G., Kim S.K., and Horvitz H.R. 1990. Novel cysteine-rich motif and homeodomain in the product of the *Caenorhabditis elegans* cell lineage gene *lin-11*. *Nature* **344:** 876–879.

Fujii T., Pichel J.G., Taira M., Toyama R., Dawid I.B., and Westphal H. 1994. Expression patterns of the murine LIM class homeobox gene *lim1* in the developing brain and excretory system. *Dev. Dyn.* **199:** 73–83.

Hukriede N.A., Tsang T.E., Habas R., Khoo P.L., Steiner K., Weeks D.L., Tam P.P., and Dawid I.B. 2003. Conserved requirement of Lim1 function for cell movements during gastrulation. *Dev. Cell* **4:** 83–94.

Jurata L.W., Kenny D.A., and Gill G.N. 1996. Nuclear LIM interactor, a rhombotin and LIM homeodomain interacting protein, is expressed early in neuronal development. *Proc. Natl. Acad. Sci.* **93:** 11693–11698.

Kania A., Johnson R.L., and Jessell T.M. 2000. Coordinate roles for LIM homeobox genes in directing the dorsoventral trajectory of motor axons in the vertebrate limb. *Cell* **102:** 161–173.

Karlsson O., Thor S., Norberg T., Ohlsson H., and Edlund T. 1990. Insulin gene enhancer binding protein Isl-1 is a member of a novel class of proteins containing both a homeo- and a Cys-His domain. *Nature* **344:** 879–882.

Kim S.H., Yamamoto A., Bouwmeester T., Agius E., and Robertis E.M. 1998. The role of paraxial protocadherin in selective adhesion and cell movements of the mesoderm during *Xenopus* gastrulation. *Development* **125:** 4681–4690.

Kobayashi A., Shawlot W., Kania A., and Behringer R.R. 2004. Requirement of Lim1 for female reproductive tract development. *Development* **131:** 539–549.

Mukhopadhyay M., Teufel A., Yamashita T., Agulnick A.D., Chen L., Downs K.M., Schindler A., Grinberg A., Huang S.P., Dorward D., and Westphal H. 2003. Functional ablation of the mouse *Ldb1* gene results in severe patterning defects during gastrulation. *Development* **130:** 495–505.

Perea-Gomez A., Shawlot W., Sasaki H., Behringer R.R., and Ang S. 1999. *HNF3* β and *Lim1* interact in the visceral endoderm to regulate primitive streak formation and anterior-posterior polarity in the mouse embryo. *Development* **126:** 4499–4511.

Perea-Gomez A., Vella F.D., Shawlot W., Oulad-Abdelghani M., Chazaud C., Meno C., Pfister V., Chen L. Robertson E., Hamada H., Behringer R.R., and Ang S.L. 2002. Nodal antagonists in the anterior visceral endoderm prevent the formation of multiple primitive streaks. *Dev. Cell* **3:** 745–756.

Piccolo S., Agius E., Leyns L., Bhattacharyya S., Grunz H., Bouwmeester T., and De Robertis E.M. 1999. The head inducer Cerberus is a multifunctional antagonist of Nodal, BMP and Wnt signals. *Nature* **397:** 707–710.

Rebbert M.L. and Dawid I.B. 1997. Transcriptional regulation of the *Xlim-1* gene by activin is mediated by an element in intron I. *Proc. Natl. Acad. Sci.* **94:** 9717–9722.

Sakuma R., Ohnishi, Yi, Y., Meno C., Fujii H., Juan H., Takeuchi J., Ogura T., Li E., Miyazono K., and Hamada H. 2002. Inhibition of Nodal signalling by Lefty mediated through interaction with common receptors and efficient diffusion. *Genes Cells* **7:** 401–412.

Shawlot W. and Behringer R.R. 1995. Requirement for *Lim1* in head-organizer function. *Nature* **374:** 425–430.

Shawlot W., Deng J.M., and Behringer R.R. 1998. Expression of the mouse *cerberus-related* gene, *Cerr1*, suggests a role in anterior neural induction and somitogenesis. *Proc. Natl. Acad. Sci.* **95:** 6198–6203.

Shawlot W., Deng J., Wakamiya M., and Behringer R.R. 2000. The *cerberus-related* gene, *Cerr1*, is not essential for mouse head formation. *Genesis* **26:** 253–258.

Shawlot W., Wakamiya M., Kwan K.M., Kania A., Jessell T.M., and Behringer RR. 1999. *Lim1* is required in both primitive streak-derived tissues and visceral endoderm for head formation in the mouse. *Development* **126:** 4925–4932.

Shimono A. and Behringer R.R. 1999. Isolation of novel cDNAs by subtractions between the anterior mesendoderm of single mouse gastrula stage embryos. *Dev Biol.* **209:** 369–380.

———. 2003. Angiomotin regulates visceral endoderm movements during mouse embryogenesis. *Curr. Biol.* **13:** 613–617.

Simpson E.H., Johnson D.K., Hunsicker P., Suffolk R., Jordan S.A., and Jackson I.J. 1999. The mouse *Cer1* (*Cerberus related* or homologue) gene is not required for anterior pattern formation. *Dev. Biol.* **213:** 202–206.

Stanley E.G., Biben C., Allison J., Hartley L., Wicks I.P., Campbell I.K., McKinley M., Barnett L., Koentgen F., Robb L., and Harvey R.P. 2000. Targeted insertion of a *lacZ* reporter gene into the mouse *Cer1* locus reveals complex and dynamic expression during embryogenesis. *Genesis* **26:** 259–264.

Tam P.P. and Steiner K.A. 1999. Anterior patterning by synergistic activity of the early gastrula organizer and the anterior germ layer tissues of the mouse embryo. *Development* **126:** 5171–5179.

Taira M, Jamrich M., Good P.J., and Dawid I.B. 1992. The LIM domain containing homeobox gene *Xlim-1* is expressed specifically in the organizer region of *Xenopus* gastrula embryos. *Genes Dev.* **6:** 356–366.

Taira M., Otani H., Jamrich M., and Dawid I.B. 1994a. Expression of the LIM class homeobox gene *Xlim-1* in pronephros and CNS cell lineages of *Xenopus* embryos is affected by retinoic acid and exogastrulation. *Development* **20:** 1525–1536.

Taira M., Otani H., Saint-Jeannet J.-P., and Dawid I.B. 1994b. Role of the homeodomain protein Xlim-1 in neural and muscle induction by the Spemann organizer in *Xenopus*. *Nature* **372:** 677–679.

Toyama R., Kobayashi M., Tomita T., and Dawid I.B. 1998. Expression of LIM-domain binding protein (*ldb*) genes during zebrafish embryogenesis. *Mech. Dev.* **71:** 197–200.

Toyama R., O'Connell M.L., Wright C.V., Kuehn M.R., and Dawid I.B. 1995. Nodal induces ectopic *goosecoid* and *lim1* expression and axis duplication in zebrafish. *Development* **121:** 383–391.

Troyanovsky B., Levchenko T., Mansson G., Matvijenko O., and Holmgren L. 2001. Angiomotin: An angiostatin binding protein that regulates endothelial cell migration and tube formation. *J.*

Cell Biol. **152:** 1247–1254.

Tsuchida T., Ensini M., Morton S.B., Baldassare M., Edlund T., Jessell T.M., and Pfaff S.L. 1994. Topographic organization of embryonic motor neurons defined by expression of LIM homeobox genes. *Cell* **79:** 957–970.

Varlet I., Collignon J., and Robertson E.J. 1997. *nodal* expression in the primitive endoderm is required for specification of the anterior axis during mouse gastrulation. *Development* **124:** 1033–1044.

Waddington C.H. Needham J., Nowinski W.W. and Lemberg R. 1935. Studies on the amphibian organization centre I. Chemical properties of the evocator. *Proc. Roy. Soc. Lond. B Biol. Sci.* **117:** 289–310.

Wadman I.A., Osada H., Grutz G.G., Agulnick A.D., Westphal H., Forster A., and Rabbitts T.H. 1997. The LIM-only protein Lmo2 is a bridging molecule assembling an erythroid, DNA-binding complex which includes the TAL1, E47, GATA-1 and Ldb1/NLI proteins. *EMBO J.* **16:** 3145–3157.

Watanabe M., Rebbert M.L., Andreazzoli M., Takahashi N., Toyama R., Zimmerman S., Whitman M., and Dawid I.B. 2002. Regulation of the *Lim-1* gene is mediated through conserved FAST-1/FoxH1 sites in the first intron. *Dev. Dyn.* **225:** 448–456.

Way J.C. and Chalfie M. 1988. *mec-3*, a homeobox-containing gene that specifies differentiation of the touch receptor neurons in *C. elegans*. *Cell* **54:** 5–16.

Whitman M. 2001. Nodal signaling in early vertebrate embryos: themes and variations. *Dev. Cell* **1:** 605–617.

Yamamoto A., Kemp C., Bachiller D., Geissert D., and De Robertis E.M. 2000. *Mouse paraxial protocadherin* is expressed in trunk mesoderm and is not essential for mouse development. *Genesis* **27:** 49–57.

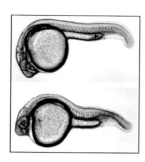

NOT HOMEOBOX GENE FUNCTION IN AXIAL MESODERM DEVELOPMENT

S. Mackem

Laboratory of Pathology, Center for Cancer Research, National Cancer Institute, National Institutes of Health, Bethesda, Maryland 20892

INTRODUCTION

The vertebrate gastrula organizer has several key properties regulated by "organizer" genes. It functions as a signaling center for induction and organization of surrounding tissues, and it differentiates to give rise to both midline (axial) mesendoderm and the floor plate of the neural tube (for review, see Harland and Gerhart 1997; Le Douarin and Halpern 2000). The axial mesoderm lineage, including prechordal plate in the head region, and notochord in the trunk–tail region, subsequently plays an important role as a signaling center that patterns the overlying neural tube, adjacent paraxial mesoderm, and underlying gut endoderm (for review, see Ruiz i Altaba 1993; Cleaver and Krieg 2001; Fleming et al. 2001). Among the organizer genes, the small class of Not homeobox genes (named for *not*ochord-specific; von Dassow et al. 1993) has been characterized most extensively in vertebrates (including teleosts, amphibians, and birds), where they function primarily in regulating the fate of the organizer lineage to promote notochord formation. However, definitive homologs have yet to be found in mammals, raising the question of why a gene necessary for notochord formation might become dispensable in higher vertebrates. The vertebrate Not genes are also broadly expressed considerably before gastrulation and are highly regulated, but the significance of this conserved early expression domain is unclear.

STRUCTURAL FEATURES AND CONSERVATION OF NOT HOMEOBOX GENES

The vertebrate Not genes form a distinct homeobox class homologous with a little-studied *Drosophila* gene designat-

ed *90Bre* (Dessain and McGinnis 1993; see Flybase CG18599 and also E103, at www.flybase.org). The next most closely related class is the Ems/Emx homeobox family. Whereas a single Not gene has been found in *Xenopus* (*Xnot* or *Xnot2*, which are likely to represent duplicated alleles in this pseudo-tetraploid organism; von Dassow et al. 1993; Gont et al. 1993) and in zebrafish (*floating head*, *flh*; Talbot et al. 1995), chicks have two different genes that are linked physically and are only about 5 kb apart (Ranson et al. 1995; Stein et al. 1996). Despite the close linkage, the dual avian genes probably arose through an ancient rather than a recent duplication event, with subsequent gene loss in fish and amphibian ancestors, because the avian *Cnot1* (or *Gnot1*) is most closely related to the *Xenopus Xnot/Xnot2* gene, whereas *Cnot2* (or *Gnot2*) is most closely related to the zebrafish *flh* gene (Fig. 1A). Not proteins are far more divergent between species than many other homeoproteins, having only isolated foci of conservation outside the homeodomain; in comparison, *Xenopus* and mouse Goosecoid (Gsc) are 98% identical in the homeodomain and 78% identical overall (see Blum et al. 1992). Vertebrate Not proteins all share a short amino-terminal domain with similarity to the Engrailed (Eng) homology-1 domain (Eh1, Fig. 1B) implicated in transcriptional repression (Smith and Jaynes 1996), suggesting that they are transcriptional repressors. Direct evidence supporting this comes from misexpression experiments in *Xenopus*. Fusion of the Xnot homeodomain to the potent VP16 activator domain produced an antimorphic *Xnot* chimera, whereas fusion to a heterologous Eng repressor domain produced a hypermorphic chimera (Yasuo and Lemaire 2001). Aside from the Eh1-like motif, the Not members share an "extended"

A

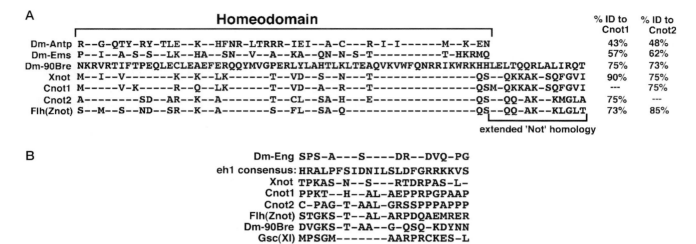

B

Figure 1. Conserved "extended" homeodomain and amino-terminal repressor motif in Not-class homeobox proteins. (*A*) Extended homeodomain homology in Not genes. Dashes indicate residues that are identical to *Drosophila* 90Bre as reference. Homeodomains of the next most closely related class, Empty Spiracles (Ems), and of the prototypical Antennapedia (Antp) are also shown for comparison. Percent identities with Cnot1 and Cnot2 homeodomains illustrate the closer evolutionary relationship of chick Cnot1 to *Xenopus* Xnot, and Cnot2 to zebrafish floating head (Flh, Znot). (*B*) Amino-terminal eh1 repressor motif in Not proteins. Dashes indicate residues identical to the engrailed homology-1 (eh1) consensus sequence (see Smith and Jaynes 1996) that was derived from the prototypical engrailed repressor (Eng) and related motifs present in several other homeobox gene classes that also function as transcriptional repressors (such as Gsc, also shown for comparison). The 90Bre sequences (homeodomain extension and amino-terminal repressor motif) are taken from gene CG18599 in Flybase (www.flybase.org), which is located near the cytogenetic map position of 90Bre and is 100% identical in the homeodomain sequence.

homeodomain homology of about 14 residues (Fig. 1A) and are rich in acidic residues at the carboxyl terminus.

EXPRESSION OF VERTEBRATE NOT GENES

In contrast to their protein sequence, the expression patterns of the different vertebrate Not genes show a remarkable degree of conservation, validating their classification as likely homologs or orthologs (see Fig. 2). *Xnot* is present as a maternal transcript and is expressed uniformly throughout blastula-stage embryos, becoming restricted to notochordal progenitors located above the blastopore dorsal lip (Chapter 13) by early gastrula stages, but is absent from prechordal plate (Gont et al. 1993; von Dassow et al. 1993). Levels of zygotic *Xnot* RNA rise and peak at the onset of gastrulation (Fig. 2G), and by late gastrula, expression localizes to the notochord and the overlying dorsal midline ectoderm that becomes the floor plate of the neural tube (Fig. 2H,I). By tail-bud stages (Fig. 2J,K), expression is limited to the most posterior (newly formed) notochord, the ventral neural tube and floor plate, and also continues in the tail chordoneural hinge, the derivative of the dorsal lip "organizer" and site of ongoing notochord formation during tail elongation (Gont et al. 1993). *Xnot* is also expressed in mature epiphysis (pineal) and its progenitors in the anterior neural plate (Fig. 2I–K) (von Dassow et al. 1993).

Zebrafish *flh* expression is very similar to *Xnot* in most respects, except that the earliest reported expression occurs at the blastula stage and is not uniform, but includes the entire marginal zone (Fig. 2A) (Talbot et al. 1995). Expression becomes restricted to dorsal marginal cells of the embryonic shield (teleost organizer-equivalent; see Chapter 12) by the early gastrula stage (Fig. 2B), and at this point correlates well with cell fate contributions to the axial mesoderm (Melby et al. 1996). During gastrulation, expression continues in notochord and floor plate precursors (axial hypoblast and epiblast layers in fish) but is not sustained in mature floor plate, and later expression is restricted to the most posterior notochord and the tail-bud chordoneural hinge (Fig. 2C–F) (Talbot et al. 1995; Melby et al. 1997). *Flh* is also the earliest marker for epiphyseal progenitors (Masai et al. 1997) in anterior neural plate (Fig. 2D–F).

The expression patterns of the chick Not genes (*Cnot1*, *Cnot2*) are largely similar to each other (as well as to fish and amphibian Not genes), except that *Cnot1* is expressed more widely than *Cnot2*, in early neural plate epiphyseal precursors (Fig. 2O,P), and anterior-distal limb bud (Knezevic et al. 1995; Ranson et al. 1995; Stein and Kessel 1995; Stein et al. 1996). Both *Cnot1* and *Cnot2* are expressed throughout the future embryo proper (epiblast) before gastrulation (Fig. 2L), but this uniform expression arises through progressive activation of Not genes in the epiblast

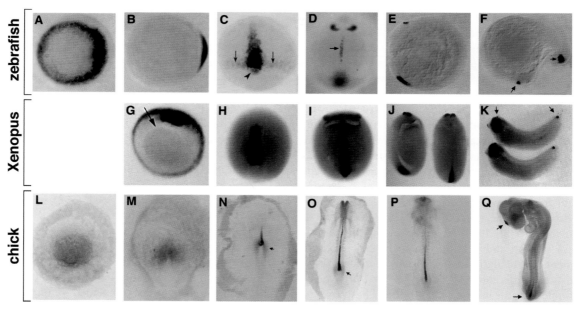

Figure 2. Expression of vertebrate Not genes from blastula/gastrula through tail-bud stages. Dynamic expression is broad in blastula–early gastrula stages (*A,G,L*), and then becomes limited to notochordal precursors in organizer (*B,C,H,M,N*), and a new anterior neural plate domain appears (*D,E*), or later anterior neural fold domain in chick (*O,P*). Later expression during segmentation continues in posterior notochord, tail bud, and epiphysis (*E,F,J,K,P,Q*). All views are dorsal with anterior at the top, except panels *A,B*: animal pole view with dorsal side right; panels *E,F,K*: side views with anterior to left; and panel *G*: vegetal view with dorsal at top. Zebrafish: (*A*) blastula, dome stage (4.3 h); (*B*) early gastrula, shield stage (6 h); (*C*) late gastrula, 90% epiboly (9 h), arrowhead shows forerunner cells (tail bud progenitor), arrows show residual lateral, marginal bands; (*D*) 1–2 somites (10.7 h), arrow shows weak residual in anterior notochord; (*E*) 9–10 somites (14 h); (*F*) 17 somites, arrows show epiphysis and tail. Xenopus: (*G*) early gastrula, stage 10.5, arrow shows dorsal blastopore lip; (*H*) late gastrula, stage 12; (*I*) neural plate stage, stage 16; (*J*) tailbud stage 23; (*K*) stage 30, arrows show epiphysis and tail. Chick: *Cnot2* (*L,M*) and *Cnot1* (*N–Q*) are expressed very similarly except that *Cnot2* is absent from early anterior neural folds. (*L*) late blastula, stage XIII; (*M*) early gastrula, stage 3; (*N*) late gastrula, stage 5 (head process); (*O*) 5 somites, stage 8+, arrow shows sr; (*P*) 10 somites, stage 10; (*Q*) 27 somites, stage 16+, arrows show epiphysis and tail. (*A,B,E,* Modified, with permission, from Talbot et al. 1995 [©Nature Publishing Group]; *C,D,F,* modified, with permission, from Melby et al. 1997 [©Wiley-Liss]; *G–K,* modified, with permission, from von Dassow et al. 1993; *L–Q,* courtesy of V. Knezevic and S. Mackem.)

during hypoblast layer formation (Knezevic and Mackem 2001; discussed under Regulation of Not genes, below). During gastrulation, *Cnot1* and *Cnot2* expression becomes restricted to Hensen's node (chick organizer-equivalent; Chapter 15) and the emerging notochord (Fig. 2M,N). *Cnot1* and *Cnot2* transcripts are transiently expressed in ectodermal floor plate progenitors just anterior to Hensen's node but are absent from neural tube floor plate. As in fish and amphibians, later expression during segmentation is limited to the posterior notochord and the tail bud chordoneural hinge (Fig. 2P,Q). *Cnot1* and *Cnot2* are also expressed in the sinus rhomboidalis ectoderm (Fig. 2N,O), immediately posterolateral to Hensen's node, which contributes to the tail neural plate during nodal regression in chick (Catala et al. 1996). Late Xenopus (Gont et al. 1993) and zebrafish (Fig. 2E) (Talbot et al. 1995) expression in tail neural tube may be related, raising the possibility that this gene may also play a role in posterior neural progenitors.

FUNCTION OF NOT GENES IN SPECIFYING AXIAL MESODERM

Much of our understanding of Not gene function comes from analysis of the zebrafish *flh* mutant, a null allele of the fish Not gene (Talbot et al. 1995), and is corroborated by misexpression analyses in Xenopus (Gont et al. 1996; Yasuo and Lemaire 2001). The *flh* mutant is mainly defective in trunk and tail development, and most of the phenotypic changes are due to loss of notochord formation and its signaling functions (Halpern et al. 1995; Talbot et al. 1995). The prechordal plate, which does not express *flh* (but see Melby et al. 1997), is unaffected, and the head region is spared, aside from independent defects in epiphyseal neurogenesis (Masai et al. 1997). In the trunk/tail, notochord precursors become diverted to form somites instead and the notochord is absent, with fused somites crossing the midline (Fig. 3). Consistent with a key role for Not genes in antagonizing muscle fate, expansion of notochord in *Xenopus* embryos overexpressing *Xnot* occurs at the expense of muscle progenitors (Gont et al. 1996). Notochord progenitors expressing appropriate axial markers are present initially in *flh* embryos (perhaps accounting for the preserved anterior floor plate formation), but they later lose their notochord-specific gene expression and become re-specified toward muscle after gastrulation, indicating a greater role in maintaining than in initiating the notochord genetic program. Floor-plate formation is disrupted in the posterior trunk and tail; this phenotype behaves nonautonomously in genetic mosaics and presumably results from lost notochord signals.

The fish T-box gene, *spadetail (spt)/Tbx16* (Griffin et al. 1998; Ruvinsky et al. 1998), is expressed in pre-somitic paraxial mesoderm and is critical for paraxial mesoderm development (particularly in the trunk), regulating both movement and differentiation of progenitors. Analysis of *flh/spt* double mutants indicates that the basis for trans-fating to muscle is due to unopposed *spadetail (spt)* activity in *flh* mutants. Notochord formation is largely restored in *flh/spt* double mutants, particularly the anterior notochord, suggesting that antagonism of *spt/Tbx16* function may be the sole or major function of Flh anteriorly (Fig. 4A–H) (Amacher and Kimmel 1998). However, in the tail region other T-box genes, in addition to *spt/Tbx16*, play essential roles in paraxial mesoderm formation and the targets of flh in tail development are less clear. Flh function is required to repress *spt/Tbx16* expression in axial mesoderm, which may occur directly, since Flh probably acts as a transcriptional repressor like Xnot (Yamamoto et al. 1998). Of interest in

this regard, chick *Tbx16* (*Tbx6L*) and Not gene expression are likewise complementary and mutually exclusive in nascent axial and paraxial mesoderm, and the loss of *Cnot* gene expression from extending notochord coincides spatially with down-regulation of *Tbx16*/Tbx6L expression in the adjacent segmenting paraxial mesoderm (Knezevic et al. 1995, 1997, 1998; Charrier et al. 1999).

Surprisingly, Flh function in notochord is not strictly cell-autonomous in genetic mosaics. When large tissue blocks of *flh* mutant cells are transplanted orthotopically to wild-type embryos, notochord progenitors form muscle (Halpern et al. 1995), whereas small cell clusters still form notochord (Amacher and Kimmel 1998), suggesting that Flh may activate an as-yet-unknown secreted signal important for maintaining notochord fate. However, the role of Flh in repressing the muscle genetic program raises an alternative possibility. Muscle development requires a nonautonomous "community effect" (Gurdon 1988) and may fail if too few *flh* cells trans-fated to muscle are transplanted into notochord territory. In this situation, Flh function may be unnecessary to maintain notochord cell fate.

What is the relationship of Not genes to other key transcriptional regulators of notochord fate? Two factors with direct roles in notochord development have been characterized in vertebrates: the Forkhead family gene *Foxa2/ HNF3β*, and the prototype T-box gene *Brachyury (T, Xbra)*. *Foxa2/HNF3β* homologs are expressed in the organizer and its derivatives (Dirksen and Jamrich 1992; Ruiz i Altaba and Jessell 1992), and in *Xenopus*, *Xnot* misexpression induces *Foxa2/HNF3β* (Gont et al. 1996). In mouse, no node or notochord forms in the absence of *Foxa2/ HNF3β* function, placing this factor early in the regulatory cascade determining axial mesoderm identity (Ang and Rossant 1994; Weinstein et al. 1994). However, the likely zebrafish *Foxa2/HNF3β* homolog (*axial*) is essential mainly for floor-plate formation, suggesting the involvement of other factors in fish (Rastegar et al. 2002). *Brachyury* is expressed in the organizer, notochord, and transiently in all nascent mesoderm during gastrulation, and is critical for notochord differentiation as well as posterior/tail mesoderm formation and morphogenesis (Schulte-Merker et al. 1994; Wilson et al. 1995; for review, see Herrmann and Kispert 1994; see also Chapter 41). In *Xenopus*, *Foxa2/HNF3β* (*FKH-1*, *pintallavis*) and *Xbra* cooperate to form axial mesoderm when coexpressed, whereas neither can promote notochord formation by itself (O'Reilly et al. 1995; see also Chapter 40). *Xbra* can also synergize with BMP antagonists to induce notochord (Cunliffe and Smith 1994). *Xnot* has been variously reported to induce notochord by itself (Gont et al. 1996), or to require co-inhibition of BMP signaling, similar to *Xbra* (Yasuo and Lemaire 2001). The reason for these different results is uncertain, but in the former study, secondary notochord induction was very patchy and

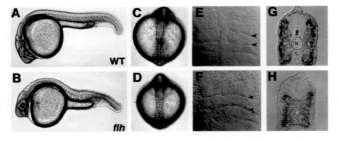

Figure 3. Phenotype of the *floating head* (*flh*) null mutation in zebrafish Not gene: conversion of notochord to somite resulting in fused somites across midline. (Panels *A,C,E,G*) Wild-type (WT) embryos; (panels *B,D,F,H*) *flh* mutant embryos. (*A,B*) Side views, pharyngula stage (24 h); (*C,D*) dorsal views, segmentation stage (14 h); (*E,F*) dorsal close-up, 3 somites (11 h), arrows show second somite, fused across midline in *flh*; (*G,H*) transverse trunk sections (24 h) with keratin sulfate staining (*brown*) labeling notochord (N) rim, and muscle marker (*blue*) labeling somite (S). (*A–D*, Modified, with permission, from Amacher and Kimmel 1998 [©Company of Biologists Ltd.]; *E–H*, modified, with permission, from Talbot et al. 1995 [©Nature Publishing Group].)

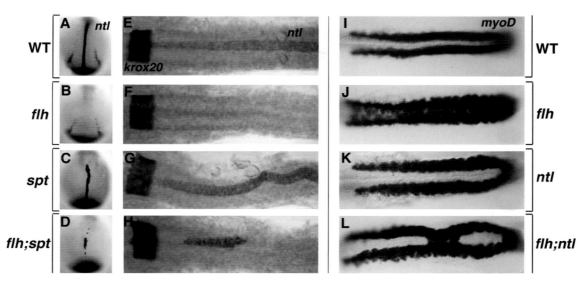

Figure 4. Genetic interactions between zebrafish Not (*flh*) and *Tbx16* (*spt*), and between Not (*flh*) and *T* (*ntl*) genes. Loss of notochord in *flh* mutant embryos is partially restored in the anterior trunk in *flh;spt* double mutants (demonstrated by midline *ntl* expression). (*A–D*) dorsal views with anterior at top, 4–6 somites (12 h); (*E–H*) dorsal views, 22 somites (20 h), *krox20* at left marks hindbrain and trunk region extends to right. Variance in epistasis is seen in *flh;ntl* double mutants, which resemble *ntl* in some regions and *flh* in other focal regions of trunk (fused midline somites, demonstrated by *myoD* expression). (*I–L*) Dorsal views with anterior at left, 20 somites (19 h). (*A–H*, Modified, with permission, from Amacher and Kimmel 1998 [©Company of Biologists Ltd.]; *I–L*, modified, with permission, from Halpern et al. 1997 [©Elsevier].)

perhaps actually dependent on endogenous BMP antagonists.

In zebrafish, the epistatic relationship between *flh* and *no tail* (*ntl*), the fish *Brachyury* homolog (Schulte-Merker et al. 1994), has been examined. Neither *flh* nor *ntl* is functionally required to initiate the other's expression during gastrulation. Subsequently, each is necessary to maintain expression of the other in prospective notochord (Talbot et al. 1995), and functional flh protein is required to maintain its own expression as well (Melby et al. 1997). In chick, misexpression of *Cnot1* in epiblast culture strongly induces *Brachyury* expression (V. Knezevic and S. Mackem, unpubl.). The cross-regulation suggests a complex positive feedback circuit between *flh* and *ntl* in notochord progenitors, rather than a simple hierarchical relationship. This complexity is further evident in *ntl/flh* double mutants, which show variability in epistasis (Fig. 4I–L) (Halpern et al. 1997), although the *ntl/flh* double mutants tend to resemble *ntl* more frequently than *flh* in trunk phenotype, suggesting that *ntl* functions largely upstream of *flh* in maintaining notochord identity. flh/Not may up-regulate *ntl* indirectly by repressing *spt/Tbx16* expression, since genetic evidence implicates *spt/Tbx16* in antagonizing *ntl* function (Amacher and Kimmel 1998).

Altogether, several genes (including *flh*/Not) have been identified that are essential, but none that are alone sufficient for notochord formation; the results are most consistent with a program of multiple genes acting in concert to specify notochord fate, rather than a single "master switch" and simple linear hierarchy. Multiple, reinforcing feedback loops operating between these factors may further ensure maintenance of notochord identity (see Fig. 5).

REGULATION OF NOT GENE EXPRESSION IN EARLY EMBRYOS

Not gene expression is highly dynamic at both early (pregastrula stage) and later times, apparently governed by a fine interplay of positive and negative inputs. Genetic approaches, in vitro tissue recombination and coculture studies, and factor additions to evaluate inductive interactions have implicated several stimulatory and inhibitory signals in dynamic Not gene regulation (see Fig. 5).

The Nodal signaling pathway, transduced through the Forkhead factor FoxH1, is required for formation of the gastrula organizer (for review, see Whitman 2001; see Chapter 34). The concentration and duration of nodal signals also patterns mesoderm (Jones et al. 1995; Agius et al. 2000), regulating both prechordal plate and notochord formation (Gritsman et al. 2000; for review, see Schier and Shen 2000). Genetic manipulations of nodal signaling levels in zebrafish reveal differential activation of *flh* or of *Gsc* in a concentration- and time-dependent fashion, such that transient/lower levels induce mainly *flh*/Not expression (notochord precursors), whereas persistent/higher levels are needed to induce *Gsc* expression (prechordal plate). It is likely

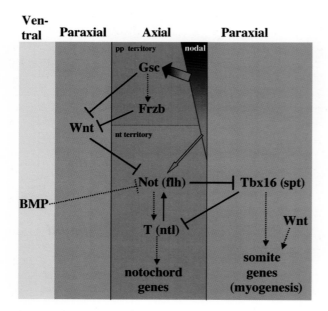

Figure 5. Not gene regulation during gastrulation. Summary of proposed genetic interactions regulating vertebrate Not gene expression during gastrula mesoderm formation, as discussed in text (based on conclusions from Halpern et al. 1997; Amacher and Kimmel 1998; Gritsman et al. 2000; Lekven et al. 2001; Yasuo and Lemaire 2001). Arrows indicate genetic relationships and do not necessarily represent direct effects. Where it is very likely that the effects are indirect, dashed arrows or bars are used. For simplicity, T (ntl) function in promoting paraxial mesoderm formation is not shown. (pp) Prechordal plate territory, (nt) notochordal territory in gastrula.

that Not genes are direct targets of nodal-type signals. In both *Xenopus* and chick, the related TGF-β family members, activins, strongly activate Not gene expression directly as an immediate-early effect, with no need for new protein synthesis (von Dassow et al. 1993; Knezevic et al. 1995; Knezevic and Mackem 2001). Although the normal role of Activins in gastrulation remains uncertain (for review, see Smith 1995; Harland and Gerhart 1997), they do signal through the same intracellular pathways as nodal (for review, see Schier and Shen 2000) and most likely mimic the effects of nodal signals on Not gene expression in these studies.

Although Activin/nodal type TGF-β signals are strong activators of all vertebrate Not genes, positive regulation by other signals, which may be grouped together as "caudalizing" factors (see Chapters 27, 32, and 38), appears to be less conserved. In *Xenopus*, FGF stimulates *Xnot* expression mildly, whereas retinoic acid (RA) has no effect (von Dassow et al. 1993). Conversely in chick, RA activates strongly, but FGF only activates in conjunction with RA and even then has only a modest effect (Knezevic et al. 1995; Knezevic and Mackem 2001). The effects of FGF may be mediated by *Brachyury*, which is regulated by FGF (see

Smith et al. 1997) and in turn regulates Not gene expression. In chick, Activin and RA synergize very strongly, suggesting that they may cooperate to maintain high Not expression more posteriorly.

Maintenance of the genetic program for "trunk organizer" mesoderm (prospective notochord), like head induction, requires co-inhibition of both the BMP and the Wnt canonical β-catenin pathway (Yasuo and Lemaire 2001; see also Chapters 26, 33, and 36). Consequently, regulation of Not expression by BMP and Wnt signals is anticipated. Previous work in *Xenopus* (Hoppler and Moon 1998) has shown that whereas BMP signals repress dorsal genes in general, Xwnt8 appears to repress *Xnot* expression very specifically, which may serve to distinguish dorsal and dorsolateral (somitic) mesoderm fates. Conversely, loss of zebrafish *Wnt8* function results in expanded *flh*/Not expression and organizer region (Lekven et al. 2001). In *Xenopus*, BMP antagonism alone induces several organizer genes transiently, but is insufficient to maintain notochord formation stably without concurrent Wnt antagonism. However, *Xnot* misexpression can bypass the need for co-repression of Wnt signals and induce ectopic notochord in a cell-autonomous manner, circumventing the repression of endogenous *Xnot* by Xwnt8 (Yasuo and Lemaire 2001). Gsc functions upstream of Xnot in this pathway, acting nonautonomously in the neighboring prospective prechordal plate territory, both by up-regulating expression of Wnt antagonists (particularly *Frzb*) and by repressing *Xwnt8* expression directly.

Broad expression at blastula stages and subsequent restriction to the gastrula organizer appear to be common features of early Not gene expression in vertebrates. The blastula expression does not strictly correlate with notochordal fate, and hence down-regulation at gastrulation does not simply reflect movement of notochordal precursors. Rising levels of BMP and Wnt signals in ventrolateral regions are likely to play a role in restricting Not gene expression by gastrula stages. Negative regulation in vivo may also have a relation to the striking effects of cycloheximide on Not gene expression seen both in vivo and in vitro (von Dassow et al. 1993; Knezevic et al. 1995; Knezevic and Mackem 2001). In *Xenopus*, cycloheximide prevents the normal loss of *Xnot* RNA outside of the dorsal blastopore lip organizer region and maintains broad *Xnot* expression during gastrulation. In *Xenopus* animal cap assays or in chick blastoderm cultures, Not RNA is likewise highly elevated by cycloheximide treatment (~50-fold increase), indicating that a labile regulator either represses Not gene transcription or enhances Not RNA degradation, perhaps acting downstream of inhibitory signals such as BMPs or Wnts . In chick, *Cnot1* and *Cnot2* transcripts each occur in two sizes, whose relative proportion is also regulated by cycloheximide (by splicing within the 3′UTR in the case of

Cnot1), suggesting that this regulation may occur post-transcriptionally.

Uniform blastula-stage Not gene expression (in the epiblast layer) in chick is achieved by progressive activation from posterior to anterior (Fig. 6A) (Knezevic et al. 1995), concomitant with formation of the hypoblast layer (homologous to murine anterior visceral endoderm layer, see Foley et al. 2000; Bertocchini and Stern 2002; Chapter 15). Expression is later extinguished from posterior to anterior (Fig. 6A), as the hypoblast becomes displaced by a distinct endoblast layer (Bachvarova et al. 1998). Epiblast–hypoblast separation and recombination cultures show that the early induction of Not gene expression in epiblast requires the hypoblast layer, and hypoblast coculture is sufficient to restore Not activation (Fig. 6B). The nature of the hypoblast signal(s) is unknown. Chick hypoblast expresses a number of candidate factors, including Activins, Fgfs, and several signal antagonists (Mitrani et al. 1990a,b; Foley et al. 2000; Bertocchini and Stern 2002). Although Wnt antagonists are expressed in the hypoblast, canonical β-catenin-mediated Wnt activity (implicated in *Xnot* regulation) is restricted to the margins around the epiblast edge at this stage (Skromne and Stern 2001; Wilson et al. 2001). BMP antagonists are also expressed in the hypoblast and could be involved in Not gene activation, since BMP signaling is already active in early epiblast (Streit et al. 1998; Faure et al.

2002). The gradual loss of Not gene expression from the epiblast, which occurs as the hypoblast becomes replaced by the endoblast layer, may be related to unopposed BMP or Wnt signaling, since the endoblast is relatively devoid of signal antagonists (Bertocchini and Stern 2002). At gastrulation, local up-regulation of Not expression in forming Hensen's node may also involve release of nodal signaling from hypoblast antagonists that are lost when the endoblast replaces the hypoblast during streak formation.

The functional significance of such dynamic pre-gastrula regulation remains unknown. In zebrafish, the *flh* mutant has no obvious pre-gastrula phenotype, although Not expression apparently begins at blastula stage (Talbot et al. 1995). However, low-level maternal transcripts (if present) could rescue a potential early phenotype in the zygotic *flh* mutant. In chick, the hypoblast layer, intimately involved in Not gene regulation, plays a key role in preventing formation of multiple embryonic axes and in fixing polarity at the onset of gastrulation (Bertocchini and Stern 2002). It is tempting to speculate that a "temporal" gradient of Not gene activation, generated as the hypoblast forms, may reflect or even contribute to the progressive restriction of inherent axis-forming ability that is present radially in the early blastoderm (Spratt and Haas 1960).

EVOLUTIONARY ENIGMA OF NOT GENES

Not gene homologs are present in fruit fly and have also been isolated from cnidarians and echinoderms; their expression in these organisms is diverse, and function has not been evaluated (Dessain and McGinnis 1993; Peterson et al. 1999; Gauchat et al. 2000). In vertebrates, Not genes have been co-opted for a new function in notochord development, but their role in primitive chordates has not been explored despite an intense focus on genes regulating notochord formation in these organisms (Takahashi et al. 1999; Suzuki and Satoh 2000; Dehal et al. 2002). Surprisingly, Not genes may have disappeared from mammals, based on genome sequences that are available to date, despite a clear role in notochord development in non-mammalian vertebrates. Recent database entries of purported "Gnot1 homologs" in human and mouse (see gi30147787 and gi38084604) are in fact less closely related to other vertebrate Not genes than is the *Drosophila* homolog. It may be noteworthy that the T-box gene *Tbx16* also has not been identified in mammalian genomes thus far, since a major function of Not/*flh* in zebrafish notochord development is antagonism of *Tbx16*/*spt* function. If indeed *Tbx16* has been lost in mammals through functional redundancy with other factors, this raises the possibility that an entire regulatory circuit has been lost during the evolution of mammals. However, the loss of potential Not functions in other

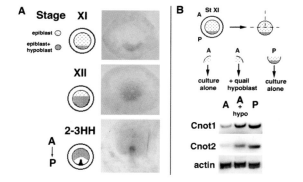

Figure 6. Regulation of chick Not expression by hypoblast signals. (*A*) *Cnot1* and *Cnot2* expression is progressively up-regulated in epiblast as underlying hypoblast forms from posterior to anterior and then declines from the epiblast during streak induction, as a new endoblast layer replaces the hypoblast from posterior to anterior. Schematic diagrams of early chick embryo stages show the extent of the extending hypoblast layer (*dark stippling*) relative to the epiblast (*light stippling*). During streak extension, the anteriorly moving hypoblast is replaced by a different type of endoderm layer (*white areas:* from Bachvarova et al. 1998). Whole-mount embryos in situ hybridized with a *Cnot1* probe are shown dorsal side up and are oriented with their anterior border at the top; stages are as indicated in the schematic. (*B*) Coculture of anterior epiblast fragments with hypoblast (as in schematic) up-regulates *Cnot* expression. (Modified, with permission, from Knezevic and Mackem 2001 [©Wiley-Liss].)

tissues (such as epiphysis) remains unaccounted for. As the roster of mammalian genes becomes complete, these questions will be resolved.

REFERENCES

Agius E., Oelgeschlager M., Wessely O., Kemp C., and De Robertis E.M. 2000. Endodermal nodal-related signals and mesoderm induction in *Xenopus*. *Development* **127:** 1173–1183.

Amacher S.L. and Kimmel C.B. 1998. Promoting notochord fate and repressing muscle development in zebrafish axial mesoderm. *Development* **125:** 1397–1406.

Ang S.L. and Rossant J. 1994. HNF-3-beta is essential for node and notochord formation in mouse development. *Cell* **78:** 561–574.

Bachvarova R.F., Skromne I., and Stern C.D. 1998. Induction of primitive streak and Hensen's node by the posterior marginal zone in the early chick embryo. *Development* **125:** 3521–3534.

Bertocchini F. and Stern C.D. 2002. The hypoblast of the chick embryo positions the primitive streak by antagonizing nodal signaling. *Dev. Cell* **3:** 735–744.

Blum M., Gaunt S.J., Cho K.W.Y., Steinbeisser H., Blumberg B., Bittner D., and DeRobertis E.M. 1992. Gastrulation in the mouse: The role of the homeobox gene goosecoid. *Cell* **69:** 1097–1106.

Catala M., Teillet M.A., De Robertis E.M., and Le Douarin N.M. 1996. A spinal cord fate map in the avian embryo: while regressing, Hensen's node lays down the notochord and floor plate thus joining the spinal cord lateral walls. *Development* **122:** 2599–2610.

Charrier J.B., Teillet M.A., Lapointe F., and Le Douarin N.M. 1999. Defining subregions of Hensen's node essential for caudalward movement, midline development and cell survival. *Development* **126:** 4771–4783.

Cleaver O. and Krieg P.A. 2001. Notochord patterning of the endoderm. *Dev. Biol.* **234:** 1–12.

Cunliffe V. and Smith J.C. 1994. Specification of mesodermal pattern in *Xenopus laevis* by interactions between Brachyury, noggin and XWnt-8. *EMBO J.* **13:** 349–359.

Dehal P., Satou Y., Campbell R.K., Chapman J., Degnan B., De Tomaso A., Davidson B., Di Gregorio A., Gelpke M., Goodstein D.M., et al. 2002. The draft genome of Ciona intestinalis: Insights into chordate and vertebrate origins. *Science* **298:** 2157–2167.

Dessain S. and McGinnis W. 1993. *Drosophila* homeobox genes. *Adv. Dev. Biochem.* **2:** 1–55.

Dirksen M.L. and Jamrich M. 1992. A novel, activin-inducible, blastopore lip-specific gene of *Xenopus laevis* contains a fork head DNA-binding domain. *Genes Dev.* **6:** 599–608.

Faure S., de Santa Barbara P., Roberts D.J., and Whitman M. 2002. Endogenous patterns of BMP signaling during early chick development. *Dev. Biol.* **244:** 44–65.

Fleming A., Keynes R.J., and Tannahill D. 2001. The role of the notochord in vertebral column formation. *J. Anat.* **199:** 177–180.

Foley A.C., Skromne I., and Stern C.D. 2000. Reconciling different models of forebrain induction and patterning: a dual role for the hypoblast. *Development* **127:** 3839–3854.

Gauchat D., Mazet F., Berney C., Schummer M., Kreger S., Pawlowski J., and Galliot B. 2000. Evolution of Antp-class genes and differential expression of Hydra Hox/paraHox genes in anterior patterning. *Proc. Natl. Acad. Sci.* **97:** 4493–4498.

Gont L.K., Fainsod A., Kim S.H., and DeRobertis E.M. 1996. Overexpression of the homeobox gene Xnot2 leads to notochord formation in *Xenopus*. *Devel. Biol.* **174:** 174–178.

Gont L.K., Steinbeisser H., Blumberg B., and De Robertis E.M. 1993. Tail formation as a continuation of gastrulation: The multiple cell-populations of the *Xenopus* tailbud derive from the late blastopore lip. *Development* **119:** 991–1004.

Griffin K.J.P., Amacher S.L., Kimmel C.B., and Kimelman D. 1998. Molecular identification of spadetail: regulation of zebrafish trunk and tail mesoderm formation by T-box genes. *Development* **125:** 3379–3388.

Gritsman K., Talbot W.S., and Schier A.F. 2000. Nodal signaling patterns the organizer. *Development* **127:** 921–932.

Gurdon J.B. 1988. A community effect in animal development. *Nature* **336:** 772–774.

Halpern M.E., Hatta K., Amacher S.L., Talbot W.S., Yan Y.L., Thisse B., Thisse C., Postlethwait J.H., and Kimmel C.B. 1997. Genetic interactions in zebrafish midline development. *Dev. Biol.* **187:** 154–170.

Halpern M.E., Thisse C., Ho R.K., Thisse B., Riggleman B., Trevarrow B., Weinberg E.S., Postlethwait J.H., and Kimmel C.B. 1995. Cell-autonomous shift from axial to paraxial mesodermal development in zebrafish floating head mutants. *Development* **121:** 4257–4264.

Harland R. and Gerhart J. 1997. Formation and function of Spemann's organizer. *Annu. Rev. Cell Dev. Biol.* **13:** 611–667.

Herrmann B.G. and Kispert A. 1994. The T-genes in embryogenesis. *Trends Genet.* **10:** 280–286.

Hoppler S. and Moon R.T. 1998. BMP-2/-4 and Wnt-8 cooperatively pattern the *Xenopus* mesoderm. *Mech. Dev.* **71:** 119–129.

Jones C.M., Kuehn M.R., Hogan B.L.M., Smith J.C., and Wright C.V.E. 1995. Nodal-related signals induce axial mesoderm and dorsalize mesoderm during gastrulation. *Development* **121:** 3651–3662.

Knezevic V. and Mackem S. 2001. Activation of epiblast gene expression by the hypoblast layer in the prestreak chick embryo. *Genesis* **30:** 264–273.

Knezevic V., De Santo R., and Mackem S. 1997. Two novel chick *T*-box genes related to mouse *Brachyury* are expressed in different, non-overlapping mesodermal domains during gastrulation. *Development* **124:** 411–419.

———. 1998. Continuing organizer function during chick tail development. *Development* **125:** 1791–1801.

Knezevic V., Ranson M., and Mackem S. 1995. The organizer-associated chick homeobox gene, Gnot1, is expressed before gastrulation and regulated synergistically by activin and retinoic acid. *Dev. Biol.* **171:** 458–470.

Le Douarin N.M. and Halpern M.E. 2000. Origin and specification of the neural tube floor plate: insights from the chick and zebrafish. *Curr. Opin. Neurobiol.* **10:** 23–30.

Lekven A.C., Thorpe C.J., Waxman J.S., and Moon R.T. 2001. Zebrafish wnt8 encodes two wnt8 proteins on a bicistronic transcript and is required for mesoderm and neurectoderm patterning. *Dev. Cell* **1:** 103–114.

Masai I,. Heisenberg C.P., Barth K.A., Macdonald R., Adamek S., and Wilson S.W. 1997. Floating head and masterblind regulate neuronal patterning in the roof of the forebrain. *Neuron* **18:** 43–57.

Melby A.E., Kimelman D., and Kimmel C.B. 1997. Spatial regulation of floating head expression in the developing notochord. *Dev. Dyn.* **209:** 156–165.

Melby A.E., Warga R.M., and Kimmel C.B. 1996. Specification of cell fates at the dorsal margin of the zebrafish gastrula. *Development* **122:** 2225–2237.

Mitrani E., Gruenbaum Y., Shohat H., and Ziv T. 1990a. Fibroblast growth factor during mesoderm induction in the early chick embryo. *Development* **109**: 387–393.

Mitrani E., Ziv T., Thomsen G., Shimoni Y, Melton D.A., and Bril A. 1990b. Activin can induce the formation of axial structures and is expressed in the hypoblast of the chick. *Cell* **63**: 495–501.

O'Reilly M.A.J., Smith J.C., and Cunliffe V. 1995. Patterning of the mesoderm in *Xenopus*: Dose-dependent and synergistic effects of *Brachyury* and *Pintallavis*. *Development* **121**: 1351–1359.

Peterson K.J., Harada Y., Cameron R.A., and Davidson E.H. 1999. Expression pattern of Brachyury and not in the sea urchin: Comparative implications for the origins of mesoderm in the basal deuterostomes. *Dev. Biol.* **207**: 419–431.

Ranson M., Tickle C., Mahon K.A., and Mackem S. 1995. Gnot1, a member of a new homeobox gene subfamily, is expressed in a dynamic, region-specific domain along the proximodistal axis of the developing limb. *Mech. Dev.* **51**: 17–30.

Rastegar S., Albert S., Le Roux I., Fischer N., Blader P., Muller F., and Strahle U. 2002. A floor plate enhancer of the zebrafish netrin1 gene requires Cyclops (Nodal) signalling and the winged helix transcription factor FoxA2. *Dev. Biol.* **252**: 1–14.

Ruiz i Altaba A. 1993. Induction and axial patterning of the neural plate: Planar and vertical signals. *J. Neurobiol.* **24**: 1276–1304.

Ruiz i Altaba A. and Jessell T.M. 1992. *Pintallavis*, a gene expressed in the organizer and midline cells of frog embryos: Involvement in the development of the neural axis. *Development* **116**: 81–93.

Ruvinsky I., Silver L.M., and Ho R.K. 1998. Characterization of the zebrafish Tbx16 gene and evolution of the vertebrate T-box family. *Dev. Genes Evol.* **208**: 94–99.

Schier A.F. and Shen M.M. 2000. Nodal signalling in vertebrate development. *Nature* **403**: 385–389.

Schulte-Merker S., Vaneeden F.J.M., Halpern M.E., Kimmel C.B., and Nusslein-Volhard C. 1994. *no tail* (*ntl*) is the zebrafish homolog of the mouse-T (*Brachyury*) gene. *Development* **120**: 1009–1015.

Skromne I. and Stern C.D. 2001. Interactions between Wnt and Vg1 signalling pathways initiate primitive streak formation in the chick embryo. *Development* **128**: 2915–2927.

Smith J.C. 1995. Developmental biology: Angles on activin's absence. *Nature* **374**: 311–312.

Smith J.C., Armes N.A., Conlon F.L., Tada M., Umbhauer M., and Weston K.M. 1997. Upstream and downstream from Brachyury, a gene required for vertebrate mesoderm formation. *Cold Spring Harbor Symp. Quant. Biol.* **62**: 337–346.

Smith S.T. and Jaynes J.B. 1996. A conserved region of engrailed, shared among all en-, gsc-, Nk1-, Nk2- and msh-class homeoproteins, mediates active transcriptional repression in vivo. *Development* **122**: 3141–3150.

Spratt Jr. N.T. and Haas H. 1960. Integrative mechanisms in development of the early chick blastoderm. I. Regulative potentiality of separated parts. *J. Exp. Zool.* **145**: 97–137.

Stein S. and Kessel M. 1995. A homeobox gene involved in node, notochord and neural plate formation of chick-embryos. *Mech. Dev.* **49**: 37–48.

Stein S., Niss K., and Kessel M. 1996. Differential activation of the clustered homeobox genes CNOT2 and CNOT1 during notogenesis in the chick. *Dev. Biol.* **180**: 519–533.

Streit A., Lee K.J., Wool I., Roberts C., Jessell T.M., and Stern C.D. 1998. Chordin regulates primitive streak development and the stability of induced neural cells, but is not sufficient for neural induction in the chick embryo. *Development* **125**: 507–519.

Suzuki M.M. and Satoh N. 2000. Genes expressed in the amphioxus notochord revealed by EST analysis. *Dev. Biol.* **224**: 168–177.

Takahashi H., Hotta K., Erives A., Di Gregorio A., Zeller R.W., Levine M., and Satoh N. 1999. *Brachyury* downstream notochord differentiation in the ascidian embryo. *Gene Dev.* **13**: 1519–1523.

Talbot W.S., Trevarrow B., Halpern M.E., Melby A.E., Farr G., Postlethwait J.H., Jowett T., Kimmel C.B., and Kimelman D. 1995. A homeobox gene essential for zebrafish notochord development. *Nature* **378**: 150–157.

von Dassow G., Schmidt J.E., and Kimelman D. 1993. Induction of the *Xenopus* organizer: Expression and regulation of Xnot, a novel FGF and activin-regulated homeo box gene. *Genes Dev.* **7**: 355–366.

Weinstein D.C., Ruiz i Altaba A., Chen W.S., Chen W.S., Hoodless P., Prezioso V.R., Jessell T.M., and Darnell J.E. 1994. The winged-helix transcription factor HNF-3-beta is required for notochord development in the mouse embryo. *Cell* **78**: 575–588.

Whitman M. 2001. Nodal signaling in early vertebrate embryos: Themes and variations. *Dev. Cell* **1**: 605–617.

Wilson S.I., Rydstrom A., Trimborn T., Willert K., Nusse R., Jessell T.M., and Edlund T. 2001. The status of Wnt signalling regulates neural and epidermal fates in the chick embryo. *Nature* **411**: 325–330.

Wilson V., Manson L., Skarnes W.C., and Beddington R.S.P. 1995. The T-gene is necessary for normal mesodermal morphogenetic cell movements during gastrulation. *Development* **121**: 877–886.

Yamamoto A., Amacher S.L., Kim S.H., Geissert D., Kimmel C.B., and De Robertis E.M. 1998. Zebrafish *paraxial protocadherin* is a downstream target of *spadetail* involved in morphogenesis of gastrula mesoderm. *Development* **125**: 3389–3397.

Yasuo H. and Lemaire P. 2001. Role of Goosecoid, Xnot and Wnt antagonists in the maintenance of the notochord genetic programme in *Xenopus* gastrulae. *Development* **128**: 3783–3793.

ROLE OF *SIAMOIS* BEFORE AND DURING GASTRULATION

L. Kodjabachian and P. Lemaire

Institut de Biologie du Développement de Marseille, Laboratoire de Génétique et Physiologie du Développement, CNRS-Université de la Méditerranée, 13288 Marseille Cedex 9, France

INTRODUCTION

Axis formation in amphibians requires the coordinate action in the blastula of several signaling systems, which contribute to position prospective germ layers and to generate an inductive center known as Spemann's organizer (for review, see Harland and Gerhart 1997). Transplantation of Spemann's organizer into ventral posterior tissues generates a novel embryonic axis, and it is expected that molecules responsible for organizer establishment or function will provoke a similar effect when overexpressed. The paired-like homeobox gene *Siamois* was identified in an expression screen aimed at isolating such new axis-generating molecules in *Xenopus laevis* (Lemaire et al. 1995). Subsequently, a close relative, *twin*, was independently found by the same approach (Laurent et al. 1997).

Organizer formation implicates transcriptional regulation by maternal β-catenin, which directly turns on *Siamois* expression in dorsal nuclei of the blastula (Heasman et al. 1994; Brannon et al. 1997; Laurent et al. 1997; Fan et al. 1998). Remarkably, ventral overexpression of *Siamois*, like that of β-catenin, generates complete secondary axes at a high frequency, whereas all other organizer genes yield incomplete axes in the same assay (Sasai et al. 1994; Lemaire et al. 1995; Bouwmeester et al. 1996). This suggests that Siamois functions immediately downstream of β-catenin to initiate the genetic cascade responsible for body axis formation and patterning. Consistent with this idea, *Siamois* is one of the earliest zygotic genes expressed within and around the prospective Spemann organizer, and ventral

injection of *Siamois* RNA activates the expression of all tested organizer genes (Lemaire et al. 1995; Carnac et al. 1996; Darras et al. 1997; Engleka and Kessler 2001). Hence, *Siamois* provides us with a good model to gain insight into how the organizer is established and how it functions.

ONTOGENY OF SPEMANN'S ORGANIZER

Currently, Spemann's organizer is thought to be induced in dorsal equatorial (marginal) cells by the Nieuwkoop center, which forms earlier in the blastula, in response to the combinatorial activity of two sets of maternal determinants (for review, see Gerhart et al. 1991; Kodjabachian and Lemaire 1998; Moon and Kimelman 1998; Gerhart 2001). The dorsal determinants constitute one of these two sets. Prior to fertilization, these determinants are located in the vegetal cortical cytoplasm of the egg (Marikawa et al. 1997). Following fertilization, they are translocated, probably through the rapid movement of vesicles along a microtubular network, toward the prospective dorsal region, opposite the sperm entry site (Marikawa et al. 1997; Rowning et al. 1997). The main consequence of the mobilization of the dorsal determinants is the enrichment of the maternal β-catenin protein in dorsal nuclei at cleavage and blastula stages (Larabell et al. 1997; Rowning et al. 1997). In the absence of β-catenin, Spemann's organizer does not form and embryos completely lack axial tissues (Heasman et al. 1994).

The second set of determinants shapes the animal–vegetal axis. One of its major components is encoded by maternal VegT, a T-box transcription factor, localized in the vegetal hemisphere of the blastula (Lustig et al. 1996a; Stennard et al. 1996, 1999; Zhang and King 1996; Horb and Thomsen 1997). Depletion of VegT showed that it is required for the formation of endoderm and mesoderm in the vegetal and marginal regions of the embryo, respectively (Zhang et al. 1998; Kofron et al. 1999). Consequently, the equatorially located Spemann organizer does not form in VegT-depleted embryos.

It is thought that in vegetal and marginal cells, where β-catenin and VegT activities functionally overlap (Nieuwkoop center), organizer-inducing properties are being produced, which help recruit more animal cells into the future Spemann organizer. It must be noted, however, that some of the cells in the Nieuwkoop center autonomously become organizer cells.

REGULATION OF *SIAMOIS* EXPRESSION

Spatial Regulation of *Siamois* Expression

This section describes how information provided by β-catenin and VegT is integrated at the molecular level, using *Siamois* as a paradigm. At the early gastrula stage, in situ hybridization on sagittal sections showed that *Siamois* is expressed primarily in deep dorsal vegetal cells (Lemaire et al. 1995). However, *Siamois* is a very rare message, difficult to detect with non-biochemical methods, particularly before gastrulation, which suggests that the distribution observed in situ might reflect only the sites of high-level expression. In agreement with this, RT-PCR and in situ hybridization on dorsal explants suggested that *Siamois* is also expressed in more animal cells (Ding et al. 1998; Nagano et al. 2000). The early and broad dorsal expression of *Siamois* suggests that this gene is a good marker during the blastula stages for the prospective organizer territory.

Regulation of Siamois *by* β-Catenin

In all experiments aimed at changing the position of the dorsal determinants in the embryo, it was found that the expression of *Siamois* was changed accordingly (Darras et al. 1997; Laurent et al. 1997; Marikawa et al. 1997). Moreover, overexpression or activation of β-catenin is sufficient to turn on *Siamois* expression in animal cells, which lack VegT protein (Carnac et al. 1996; Crease et al. 1998), and β-catenin-depleted embryos no longer express *Siamois* (Heasman et al. 1994, 2000). Thus, the maternal dorsal determinants help stabilize β-catenin in dorsal cells, which leads to the activation of *Siamois* at the blastula stage (see Chapter 36).

It is known that β-catenin can interact directly with factors of the TCF/Lef family, which function as sequence-specific DNA-binding transcription factors (for review, see van Noort and Clevers 2002). In *Xenopus*, TCF3 is expressed maternally and is likely to be the cognate partner of β-catenin in dorsal cells of the blastula (Molenaar et al. 1996; Houston et al. 2002). In agreement with this idea, several TCF-binding sites clustered in the vicinity of the transcription start site are essential for the activation of *Siamois* reporters by β-catenin (Brannon et al. 1997; Laurent et al. 1997; Fan et al. 1998). However, it was found that mutations of these TCF binding sites lead to a general derepression of the reporters in all cells of the embryo, suggesting that, in the absence of β-catenin, TCF3 functions as a transcriptional repressor (Brannon et al. 1997). The consequence of the depletion of TCF3 has been recently examined and confirms that it acts as a repressor of *Siamois* expression (Houston et al. 2002). These data suggest that *Siamois* expression and organizer formation depend primarily on the ability of β-catenin to alleviate the repression exerted by TCF3, as was initially speculated by Klymkowsky and colleagues (Merriam et al. 1997).

Integration of VegT and β-Catenin Activities on the Siamois Promoter

On the basis of observations made with TCF3-deficient embryos, the requirement for VegT in organizer establishment appears unclear. However, it was recently shown that depletion of VegT also dramatically reduces *Siamois* expression, indicating that β-catenin activity is not sufficient to generate a functional organizer (Xanthos et al. 2002). Furthermore, concomitant depletion of VegT and TCF3 showed that *Siamois,* as well as several other organizer genes that are overexpressed in the absence of TCF3, is down-regulated when VegT is also missing (Houston et al. 2002). This suggests that VegT could function synergistically with β-catenin to enhance *Siamois* expression in vegetal and equatorial cells. Unlike β-catenin, VegT is not able to activate *Siamois* expression in ectodermal explants (Crease et al. 1998; Agius et al. 2000). Moreover, there is no T-box-binding site in the cloned regulatory regions of *Siamois* and *twin*, which were shown to be sufficient to drive high-level expression in organizer territories (Brannon et al. 1997; Laurent et al. 1997; Fan et al. 1998). These results suggest that VegT is unlikely to directly participate in the control of *Siamois* expression and that, perhaps, this function is carried out by VegT transcriptional targets. These targets include mesoderm inducers such as Activin βB, derrière, and nodal-related, which all belong to the TGF-β superfamily, as well as FGF (Zhang et al. 1998; Clements et al. 1999; Kofron et al. 1999; Yasuo and Lemaire 1999; Hyde and Old 2000; Rex et al. 2002; White et al. 2002). It is important to note, however, that some TGF-β ligands, such as Activins and Vg1, are expressed maternally and could regulate *Siamois* independently of VegT (Fig. 1) (Brannon and Kimelman 1996; Cui et al. 1996). Interestingly, TGF-βs are not sufficient to activate *Siamois* expression in animal cells (Carnac et al. 1996; Crease et al. 1998; Agius et al. 2000), but they can synergize with β-catenin to enhance it (Brannon and Kimelman 1996; Cui et al. 1996). The possibility that such secreted factors participate in *Siamois* regulation is consistent with the observation that *Siamois* expression is reduced if cell communication is prevented (Lemaire et al. 1995). The precise molecular nature of the TGF-β factor(s) responsible for the synergistic activation of *Siamois* in vivo remains to be determined.

Nodal and Activin signaling is mediated by the transcription factors Smad2, Smad3, and Smad4 (for review, see Attisano and Wrana 2002). Analysis of the *twin* promoter revealed the presence of Smad-binding sites, distal to the TCF-binding sites, which are necessary for maximum expression in dorsal vegetal cells (Nishita et al. 2000). Additional experiments in cell culture revealed that the promoter responds to TGF-β/Smad2 signaling but not to BMP/Smad1, provided that a TCF/Lef factor is coexpressed (Labbe et al. 2000). This apparent synergistic activation of *twin* seems to depend on the formation of a multimeric complex containing β-catenin, TCF/Lef, Smad2 or 3, and Smad4 (Labbe et al. 2000; Nishita et al. 2000). These data are also consistent with the synergistic action of Smad2 and β-catenin on *Siamois* expression (Crease et al. 1998).

In summary, largely owing to the analysis of *Siamois* regulation, in the fertilized egg, maternal β-catenin and VegT collaborate to position the prospective organizer in dorsal vegetal and marginal cells (Fig. 1). β-catenin relieves the repression exerted by TCF3 upon *Siamois* expression, which allows basal-level expression in dorsal cells. VegT activates the expression of TGF-β factors, which in turn trigger the Smad-dependent synergistic activation of *Siamois* dorsally in vegetal and marginal cells. Finally, it must be noted that *Siamois* expression, although strongly diminished, is not abolished in TCF3/VegT-deficient embryos, suggesting the existence of yet additional regulators required for organizer establishment.

Temporal Control of *Siamois* Expression

In the previous section, we have seen that pre-mid-blastula transition (MBT) β-catenin signaling leads to *Siamois* activation. It is known that post-MBT Wnt signaling, acting via β-catenin, is functional in the ventral-lateral region of the late blastula/early gastrula embryo, and yet does not lead to *Siamois* activation (Hamilton et al. 2001). One possible reason for this difference is that, depending on the developmental stage, β-catenin interacts with different TCF/Lef nuclear partners, thus generating distinct responses such as activating *MyoD* instead of *Siamois* (Hamilton et al. 2001; Roel et al. 2002). *TCF3* expression is, however, maintained at this stage, and it is difficult to understand why β-catenin could not interact with TCF3 and allow *Siamois* expression

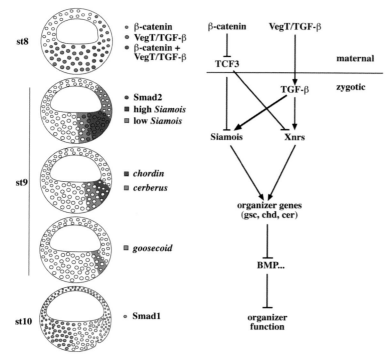

Figure 1. Schematic representation of a simplified molecular pathway leading to Spemann's organizer formation. Embryos are represented at the mid-blastula (st 8), late blastula (st 9), and early gastrula (st 10) stages, as sagittal sections with dorsal to the right. Circles represent nuclei. Colored circles represent nuclei expressing the corresponding proteins. Dark blue sectors represent high-level, and light blue sectors represent low-level, expression of the corresponding genes. Organizer formation is initiated by the localized transcriptional activities of maternal β-catenin and VegT, which leads to the activation of *Siamois* and *Xnrs*, via the relief of TCF3-mediated repression and the activation of zygotic TGF-β ligands. Note that maternally produced TGF-βs may work along VegT to shape the animal–vegetal axis. *Siamois* and *Xnrs* then collaborate to activate the expression of several critical organizer factors, such as *goosecoid*, *chordin*, and *cerberus*. Note that these last two genes are expressed in distinct domains due to the action of localized additional regulators, such as *Mix.1* (see text). This genetic cascade leads to the repression of BMP signaling, as well as other ventralizing signals, such as Wnt, allowing dorso-anterior structures to form.

in ventral blastula/gastrula cells. Experiments using a hormone-inducible form of TCF3, which activates transcription constitutively in the absence of β-catenin, have shed light on this problem. Injection of this chimeric form of TCF3 can stimulate *Siamois* expression when hormonal induction takes place prior to MBT, but not by the onset of gastrulation (Darken and Wilson 2001). Moreover, it was shown that a reporter construct containing the β-catenin-responsive region of the *Siamois* promoter does not respond to late TCF3 activity either. This loss of response might be brought about by the presence of a repressor accumulating after MBT and binding to the *Siamois* promoter. Support for this idea comes from the observation that Xcad2, a transcriptional repressor expressed ventrally in the late blastula, antagonizes organizer gene expression and is required to restrict the temporal window of competence to axis formation by *Siamois* (Levy et al. 2002). Whether Xcad2 is itself responsible for the repression of *Siamois* in late-blastula ventral cells remains to be determined, but it is interesting to note that several putative binding sites for Xcad2 are clustered in the distal region of the reported *Siamois* regulatory region (L. Kodjabachian, unpubl.).

ROLE(S) OF *SIAMOIS*

Injection of *Siamois* RNA leads to the development of a complete secondary body axis, accompanied by the activation of numerous organizer genes (Lemaire et al. 1995; Carnac et al. 1996; Darras et al. 1997; Engleka and Kessler 2001). Conversely, overexpression of a dominant-negative chimera made of a transcriptional repressor domain fused to the Siamois homeodomain provokes the loss of all axial structures, and the extinction of all organizer genes tested (Darras et al. 1997; Fan and Sokol 1997; Kessler 1997). Thus, *Siamois* appears to be necessary and sufficient for organizer establishment and axis formation. We note, however, that more specific knock-down experiments, perhaps using morpholino antisense injection, are required to confirm these conclusions based solely on the use of dominant-negative constructs.

With this restriction in mind, we examine the developmental stages and the cells in which *Siamois* functions, to gain knowledge on how Spemann's organizer patterns the embryo.

Developmental Stages of *Siamois* Function

Siamois transcripts first accumulate soon after MBT, reach peak levels in the late blastula, and remain detectable for an additional few hours during gastrulation, after which expression ceases (Lemaire et al. 1995). A hormone-inducible form of Siamois (Sia-GR) has been used to determine the stage that this factor triggers organizer develop-

ment and axis formation (Kodjabachian and Lemaire 2001). When *Sia-GR* is injected ventrally and activated before or at MBT, the axis generated is complete and contains a head as well as a notochord. This is similar to the axis produced by injection of wild-type *Siamois* RNA. When Sia-GR is activated past MBT but prior to the onset of gastrulation, the induced axis contains trunk somites and a neural tube, but lacks a head and a notochord. Finally, when Sia-GR is activated after the onset of gastrulation, only tail tissue is formed (Fig. 2). These results indicate that the response of ventral cells to the presence of Siamois changes with time. This is further exemplified by the fact that all organizer genes tested were turned on in injected cells upon early activation of Sia-GR (prior or at MBT), but not upon late activation (late blastula) (Kodjabachian and Lemaire 2001). This raises the question of what is responsible for the production of trunk dorsal structures upon late activation of Siamois, as the organizer program is not turned on under these conditions. Unexpectedly, BMP target genes were repressed by Sia-GR regardless of the moment of activation, suggesting that the ventral program was still antagonized by late Siamois activity, in a way sufficient to allow dorsal fates to be generated (Kodjabachian and Lemaire 2001). Because Siamois functions as a transcriptional activator, we must envisage the existence of an unknown repressor activated by

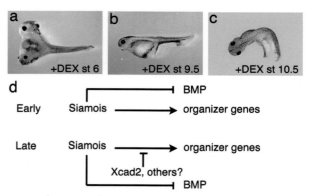

Figure 2. The temporal response to *Siamois*. Embryos were injected at the 4-cell stage in the two ventral blastomeres, with RNA encoding the hormone-inducible fusion Sia-GR. The hormone substitute dexamethasone (DEX) was then added to the culture medium at the 32-cell (st 6, *a*), late blastula (st 9.5, *b*), and early gastrula (st 10.5, *c*) stages. Pre-MBT activation of Sia-GR leads to the complete duplication of the body axis. Activation of Sia-GR at the late-blastula stage leads to the formation of an incomplete secondary axis lacking a head and a notochord (not shown). Activation of Sia-GR after the onset of gastrulation leads to tail duplication. (*d*) A model for the stage-specific activity of *Siamois*. In early blastulae, *Siamois* activates the organizer genetic program in addition to repressing the function of ventralizing factors such as BMP. In late blastulae and early gastrulae, *Siamois* is unable to activate organizer genes, as inhibitory factors such as Xcad2 repress these genes. However, *Siamois* is still able to repress BMP activity, thus leading to the formation of incomplete secondary axes.

Siamois, perhaps at all stages of development, and able to repress the ventral program. This mechanism could serve to ensure the maintenance of the organizer program after its initiation or to permit the progressive spatial expansion of organizer territories.

The experiments described above suggest that as time passes, dorsal genes become less responsive to the transcriptional activity of Siamois in ventral cells. The simplest explanation for this phenomenon would be the existence of a negative influence(s) accumulating autonomously within the ventral region of the embryo that modifies the response to Siamois on some of its target promoters. As discussed in the previous section, Xcad2 is a good candidate to participate in this process. These ectopic assays, although not informative as to the endogenous temporal requirement for *Siamois*, underscore the importance of analyzing the potential of non-organizer cells, in order to understand how the body axis is laid down.

Siamois Activates Different Targets in Different Tissues

As described above, *Siamois* is expressed in a broad domain of the blastula embryo, albeit at variable levels, in a domain that exceeds the limits of the domains encompassed by most other organizer genes (Fig. 1). A crucial question is whether *Siamois* triggers a uniform program in all the cells where it is expressed, or whether it generates distinct responses in different cells. To answer this, we need to identify the relevant transcriptional targets of *Siamois* in different tissues.

At present, only one direct target of Siamois is known: *goosecoid*, which is a paired-like homeobox gene expressed primarily in the presumptive prechordal plate (Laurent et al. 1997; see Chapter 42). The promoter of *goosecoid* contains a proximal element responsive to Wnt/β-catenin signaling, and a distal element that responds to Activin or Vg1 (Watabe et al. 1995). When expressed in ectodermal cells, Siamois and Twin are able to activate *goosecoid* reporter constructs containing the Wnt-responsive element only (Laurent et al. 1997). Mutational analysis has confirmed that this element contains a functional binding site for Twin that is required for activation in animal cells. *goosecoid* encodes a transcriptional repressor that prevents the expression of genes able to oppose the organizer program, such as certain BMP targets (Ferreiro et al. 1998; Trindade et al. 1999; Yao and Kessler 2001). Thus, Siamois is believed to initiate, via the activation of genes such as *goosecoid*, as well as unknown additional repressors, the creation of an organizer territory devoid of ventralizing influences (Fig. 1).

BMP overexpression suppresses ectopic axis development by Siamois more efficiently than endogenous axis development, however (L. Kodjabachian, unpubl.), suggesting that Siamois is not the sole determinant of organizer

formation. In this respect, as the distal element in the *goosecoid* promoter responds to TGF-β signaling, this pathway also appears essential for establishing the organizer (Watabe et al. 1995). Thus, *Siamois* could constitute one of the regulatory branches acting downstream of β-catenin, while TGF-β signaling would constitute a second critical branch. This possibility becomes particularly attractive when considering the relationship between *Siamois* and Nodal signaling. *Xenopus nodal-related 1* (*Xnr1*), which is another direct target of β-catenin and VegT, can generate a secondary axis and activate the expression of *goosecoid* but not *Siamois* (Jones et al. 1995; Lustig et al. 1996b). Moreover, overexpression of Nodal inhibitors, such as Cerberus, prevents *goosecoid* but not *Siamois* expression (Agius et al. 2000). Conversely, Siamois is not able to induce *Xnr1* expression in ectodermal cells (Carnac et al. 1996) and is not required for its normal expression (Agius et al. 2000). These results suggest that these two proteins function in parallel to control common targets, acting downstream of β-catenin (Fig. 1). Interestingly, Siamois can heterodimerize with the related homeodomain protein Mix.1 (Mead et al. 1996), and members of the Mix family have been shown to bind Smad2 and mediate the action of TGF-β on the *goosecoid* promoter (Germain et al. 2000). Thus, Siamois could help to link distinct regulatory elements within the enhancers of organizer-specific genes.

Although *goosecoid* can be activated by Siamois overexpression in all embryonic cells, this is not the case for other organizer genes, which are activated by Siamois in a tissue-specific manner. For instance, the BMP inhibitor *chordin*, which is normally expressed in prechordal and chordal mesoderm, is activated by Siamois in ectodermal cells (Carnac et al. 1996), whereas the multivalent inhibitor *cerberus*, normally expressed in deep anterior endo-mesoderm, is not (Lemaire et al. 1998). Siamois can, however, activate *cerberus* expression in endodermal cells of the vegetal hemisphere, or in animal cells when it is coexpressed with the mesendodermal homeobox gene *Mix.1* (Lemaire et al. 1998). Because *Cerberus* can induce ectopic head features (Bouwmeester et al. 1996), whereas chordin only induces trunk tissue (Sasai et al. 1994), these data were interpreted as evidence for Siamois initiating a "head organizer program" in vegetal organizer cells, and a "trunk organizer program" in more animal cells of the organizer. This model was attractive, as it was reported that head-inducing activity was found in the vegetal region of the organizer prior to invagination, whereas the animal region was capable of inducing trunk structures only (Zoltewicz and Gerhart 1997). However, it is unclear whether this mechanism is relevant, as *Siamois* RNA can induce a complete secondary axis containing a well-patterned trunk when its overexpression is restricted to ventral vegetal cells (Lemaire et al. 1995; Engleka and Kessler 2001). In particular, in these embryos, *chordin* expression is activated in a non-cell-autonomous manner in animal cells (L. Kodjabachian, unpubl.). Thus, although in ectopic conditions Siamois expression in vegetal cells is sufficient to generate a fully functional organizer, it remains to be tested whether in normal conditions its activity is also required in more animal territories. It is possible, for instance, to imagine that the presence of Siamois in these cells makes them competent to become part of the organizer, in response to a secreted vegetal signal. Testing this idea will require inhibiting the function of Siamois and Twin in a tissue-specific manner.

FUNCTIONAL CONSERVATION OF *SIAMOIS* IN DIFFERENT VERTEBRATE SPECIES

The crucial role played by *Siamois* in the formation of the *Xenopus* organizer poses the question of the evolutionary conservation of both the gene and its function. As described elsewhere in this volume, embryonic domains showing properties similar to Spemann's organizer are present in other vertebrates. This is exemplified by the axis-inducing activity displayed in homotypic transplantation experiments by mouse and chick Hensen's node, or the zebrafish shield (for review, see Lemaire and Kodjabachian 1996). Furthermore, many organizer-specific genes from *Xenopus* have orthologs expressed in zebrafish and amniote organizer territories (for review, see Lemaire and Kodjabachian 1996; Stern 2001). The precise mechanisms involved in organizer formation may, however, differ between amniotes and lower vertebrates, as for instance, organizer markers such as *cerl* and *lim1* are not lost in β-catenin-deficient mouse embryos (Huelsken et al. 2000).

Taken together, these reports suggest a very strong conservation of the genetic circuitry guiding the formation of the organizer in lower vertebrates and, to a certain extent, in amniotes. Yet, in contrast to other known organizer genes, *Siamois* cannot be detected by homology search against the sequenced *Fugu*, mouse, and human genomes (L. Kodjabachian, unpubl.). Furthermore, in zebrafish, although overexpression of other components of the β-catenin pathway has similar effects as in *Xenopus*, *Siamois* only has weak axis-inducing activity, and expression of the dominant-negative version only mildly affects *goosecoid* expression (Sumoy et al. 1999).

Interestingly, in fish, both expression cloning and mutational analysis have led to the identification of Bozozok/Dharma/Nieuwkoid (Boz), another paired-like homeoprotein, specifically expressed during blastula stages in the shield precursors and the dorsal yolk syncitial layer (Koos and Ho 1998; Yamanaka et al. 1998; Fekany et al. 1999). This gene is both required and sufficient for the activation of organizer genes such as *goosecoid*, and it mediates

the action of β-catenin (for review, see Solnica-Krezel and Driever 2001). Unlike *Siamois*, however, *boz* encodes a transcriptional repressor, which is thought to act at least partially by transcriptionally repressing *BMP2b* (Koos and Ho 1999). Consistent with a different mode of action for *Siamois* and *boz*, overexpression of the latter in *Xenopus* leads to a mild effect (analogous to the repression of BMP activity by late *Siamois* activation). There is also no clear ortholog of *boz* in the mouse and human genomes, although there is one in *Fugu* (P. Lemaire, unpubl.). It is at present unclear whether there is any *boz* ortholog in *Xenopus*, but the *Xenopus tropicalis* genome project should provide an answer in the near future. Comparison of *Siamois* and *twin* in the two *Xenopus* species will also provide an estimate of the rate of divergence of these genes within amphibians.

Although it appears that zebrafish and *Xenopus* use different transcription factors to mediate the function of β-catenin, the fact that there are no clearly identifiable *Siamois* orthologs in the mouse and human genomes does not exclude the presence of a functional amniote ortholog, albeit with a derived sequence. Although this hypothesis may appear unlikely, we do not know which crucial amino acids determine the specificity of action of Siamois. Should these key amino acids be scattered throughout the homeodomain, a functional ortholog may not be easily recognizable. A prediction from this hypothesis, however, is that overexpression before gastrulation of wild-type or dominant-negative *Xenopus* Siamois should lead to predictable axial phenotypes in mouse or chick. Should this be the case, *Siamois* could be seen as a tetrapod-specific axial determinant. If it is not, then *Siamois* and perhaps *boz* may represent exceptions like *Drosophila* bicoid, another paired-like homeoprotein. They could encode crucial determinants for small groups of animals, amphibians for *Siamois*, cyclorrhaphan flies in the case of *bicoid* (Stauber et al. 2002).

bicoid is thought to have evolved from a duplication and specialization of *Hox3* in cyclorrhaphan flies (Stauber et al. 1999). The origin of *Siamois* may lie in a group of paired-like homeobox genes collectively known as the *Mix* genes. This multigene family is present in all major vertebrate groups from fish to amniotes but cannot be detected in the sequenced invertebrate genomes (L. Kodjabachian, unpubl.). They are involved in the establishment of germ layers during early embryogenesis. Mix.1, the founding member of the family, was shown to cooperate and heterodimerize with Siamois (Mead et al. 1996; Lemaire et al. 1998). *Mix* genes generally show a radially symmetric expression pattern and display no axis-inducing activity. Thus, the evolution of a *Mix*-like gene to *Siamois* has involved changes both in the expression pattern and in the function of the protein. Interestingly, phylogenetic analysis suggests that *boz* has evolved from a *bicoid*- or *goosecoid*-

like gene and not from a *Mix*-like gene. Thus, and somewhat surprisingly, whereas the early role of β-catenin and of later organizer genes has been remarkably well conserved in the vertebrate lineage, the key components that link these two levels of regulation show poor conservation.

CONCLUSION

Siamois encodes a transcriptional factor, which appears to be necessary and sufficient for axis development in *Xenopus*. Multiple signals affect its expression, which could generate regions of high-level expression, as well as regions of low-level expression. In this scheme, β-catenin induces a basal level of expression of *Siamois* wherever dorsal determinants have been inherited along the dorsal meridian, and VegT, via TGF-β signals, helps define a dorsal vegetal region showing enhanced *Siamois* expression. The synergy between β-catenin and VegT also seems to be responsible for the graded expression of *Xnr1*, another gene essential for axis development, and which may collaborate with Siamois to control downstream targets, such as *goosecoid*. It is likely that a similar mechanism is responsible for establishing an organizer expressing *nodal* in other vertebrates. However, it is unclear whether a Siamois-like function is required to collaborate with *nodal* in these species. Despite this uncertainty, it is reasonable to expect that many of the transcriptional targets of Siamois in *Xenopus* are conserved in other animals. Thus, learning more about their identity may help not only to understand the function of Siamois in *Xenopus*, but also to point toward key elements in the evolution of the organizer.

REFERENCES

Agius E., Oelgeschlager M., Wessely O., Kemp C., and De Robertis E.M. 2000. Endodermal Nodal-related signals and mesoderm induction in *Xenopus*. *Development* **127:** 1173–1183.

Attisano L. and Wrana J.L. 2002. Signal transduction by the TGF-beta superfamily. *Science* **296:** 1646–1647.

Bouwmeester T., Kim S., Sasai Y., Lu B., and De Robertis E.M. 1996. Cerberus is a head-inducing secreted factor expressed in the anterior endoderm of Spemann's organizer. *Nature* **382:** 595–601.

Brannon M. and Kimelman D. 1996. Activation of Siamois by the Wnt pathway. *Dev. Biol.* **180:** 344–347.

Brannon M., Gomperts M., Sumoy L., Moon R.T., and Kimelman D. 1997. A beta-catenin/XTcf-3 complex binds to the siamois promoter to regulate dorsal axis specification in *Xenopus*. *Genes Dev.* **11:** 2359–2370.

Carnac G., Kodjabachian L., Gurdon J.B., and Lemaire P. 1996. The homeobox gene Siamois is a target of the Wnt dorsalisation pathway and triggers organiser activity in the absence of mesoderm. *Development* **122:** 3055–3065.

Clements D., Friday R.V., and Woodland H.R. 1999. Mode of action of VegT in mesoderm and endoderm formation. *Development* **126:** 4903–4911.

Crease D.J., Dyson S., and Gurdon J.B. 1998. Cooperation between the activin and Wnt pathways in the spatial control of organizer gene expression. *Proc. Natl. Acad. Sci.* **95:** 4398–4403.

Cui Y., Tian Q., and Christian J.L. 1996. Synergistic effects of Vg1 and Wnt signals in the specification of dorsal mesoderm and endoderm. *Dev. Biol.* **180:** 22–34.

Darken R.S. and Wilson P.A. 2001. Axis induction by wnt signaling: Target promoter responsiveness regulates competence. *Dev. Biol.* **234:** 42–54.

Darras S., Marikawa Y., Elinson R.P., and Lemaire P. 1997. Animal and vegetal pole cells of early *Xenopus* embryos respond differently to maternal dorsal determinants: Implications for the patterning of the organiser. *Development* **124:** 4275–4286.

Ding X., Hausen P., and Steinbeisser H. 1998. Pre-MBT patterning of early gene regulation in *Xenopus*: The role of the cortical rotation and mesoderm induction. *Mech. Dev.* **70:** 15–24.

Engleka M.J. and Kessler D.S. 2001. Siamois cooperates with TGFbeta signals to induce the complete function of the Spemann-Mangold organizer. *Int. J. Dev. Biol.* **45:** 241–250.

Fan M.J. and Sokol S.Y. 1997. A role for Siamois in Spemann organizer formation. *Development* **124:** 2581–2589.

Fan M.J., Gruning W., Walz G., and Sokol S.Y. 1998. Wnt signaling and transcriptional control of Siamois in *Xenopus* embryos. *Proc. Natl. Acad. Sci.* **95:** 5626–5631.

Fekany K., Yamanaka Y., Leung T., Sirotkin H.I., Topczewski J., Gates M.A., Hibi M., Renucci A., Stemple D., Radbill A., Schier A.F., Driever W., Hirano T., Talbot W.S., and Solnica-Krezel L. 1999. The zebrafish bozozok locus encodes Dharma, a homeodomain protein essential for induction of gastrula organizer and dorsoanterior embryonic structures. *Development* **126:** 1427–1438.

Ferreiro B., Artinger M., Cho K., and Niehrs C. 1998. Antimorphic goosecoids. *Development* **125:** 1347–1359.

Gerhart J. 2001. Evolution of the organizer and the chordate body plan. *Int. J. Dev. Biol.* **45:** 133–153.

Gerhart J., Doniach T., and Stewart R. 1991. Organizing the *Xenopus* organizer. In *Gastrulation: Movements, patterns, and molecules* (ed. R. Keller et al.), pp. 37–51. Plenum Press, New York.

Germain S., Howell M., Esslemont G.M., and Hill C.S. 2000. Homeodomain and winged-helix transcription factors recruit activated Smads to distinct promoter elements via a common Smad interaction motif. *Genes Dev.* **14:** 435–451.

Hamilton F.S., Wheeler G.N., and Hoppler S. 2001. Difference in XTcf-3 dependency accounts for change in response to beta-catenin-mediated Wnt signalling in *Xenopus* blastula. *Development* **128:** 2063–2073.

Harland R. and Gerhart J. 1997. Formation and function of Spemann's organizer. *Annu. Rev. Cell Dev. Biol.* **13:** 611–667.

Heasman J., Kofron M., and Wylie C. 2000. Beta-catenin signaling activity dissected in the early *Xenopus* embryo: A novel antisense approach. *Dev. Biol.* **222:** 124–134.

Heasman J., Crawford A., Goldstone K., Garner-Hamrick P., Gumbiner B., McCrea P., Kintner C., Noro C.Y., and Wylie C. 1994. Overexpression of cadherins and underexpression of beta-catenin inhibit dorsal mesoderm induction in early *Xenopus* embryos. *Cell* **79:** 791–803.

Horb M.E. and Thomsen G.H. 1997. A vegetally localized T-box transcription factor in *Xenopus* eggs specifies mesoderm and endoderm and is essential for embryonic mesoderm formation. *Development* **124:** 1689–1698.

Houston D.W., Kofron M., Resnik E., Langland R., Destree O., Wylie C., and Heasman J. 2002. Repression of organizer genes in dorsal and ventral *Xenopus* cells mediated by maternal XTcf3. *Development* **129**: 4015–4025.

Huelsken J., Vogel R., Brinkmann V., Erdmann B., Birchmeier C., and Birchmeier W. 2000. Requirement for beta-catenin in anterior-posterior axis formation in mice. *J. Cell Biol.* **148**: 567–578.

Hyde C.E. and Old R.W. 2000. Regulation of the early expression of the *Xenopus* nodal-related 1 gene, Xnr1. *Development* **127**: 1221–1229.

Jones C.M., Kuehn M.R., Hogan B.L., Smith J.C., and Wright C.V. 1995. Nodal-related signals induce axial mesoderm and dorsalize mesoderm during gastrulation. *Development* **121**: 3651–3662.

Kessler D.S. 1997. Siamois is required for formation of Spemann's organizer. *Proc. Natl. Acad. Sci.* **94**: 13017–13022.

Kodjabachian L. and Lemaire P. 1998. Embryonic induction: Is the Nieuwkoop centre a useful concept? *Curr. Biol.* **8**: R918–921.

———. 2001. Siamois functions in the early blastula to induce Spemann's organiser. *Mech. Dev.* **108**: 71–79.

Kofron M., Demel T., Xanthos J., Lohr J., Sun B., Sive H., Osada S., Wright C., Wylie C., and Heasman J. 1999. Mesoderm induction in *Xenopus* is a zygotic event regulated by maternal VegT via TGFbeta growth factors. *Development* **126**: 5759–5770.

Koos D.S. and Ho R.K. 1998. The nieuwkoid gene characterizes and mediates a Nieuwkoop-center-like activity in the zebrafish. *Curr. Biol.* **8**: 1199–1206.

———. 1999. The nieuwkoid/dharma homeobox gene is essential for bmp2b repression in the zebrafish pregastrula. *Dev. Biol.* **215**: 190–207.

Labbe E., Letamendia A., and Attisano L. 2000. Association of Smads with lymphoid enhancer binding factor 1/T cell-specific factor mediates cooperative signaling by the transforming growth factor-beta and wnt pathways. *Proc. Natl. Acad. Sci.* **97**: 8358–8363.

Larabell C.A., Torres M., Rowning B.A., Yost C., Miller J.R., Wu M., Kimelman D., and Moon R.T. 1997. Establishment of the dorsoventral axis in *Xenopus* embryos is presaged by early asymmetries in beta-catenin that are modulated by the Wnt signaling pathway. *J. Cell Biol.* **136**: 1123–1136.

Laurent M.N., Blitz I.L., Hashimoto C., Rothbacher U., and Cho K.W. 1997. The *Xenopus* homeobox gene twin mediates Wnt induction of goosecoid in establishment of Spemann's organizer. *Development* **124**: 4905–4916.

Lemaire P. and Kodjabachian L. 1996. The vertebrate organizer: Sructure and molecules. *Trends Genet.* **12**: 525–531.

Lemaire P., Garrett N., and Gurdon J.B. 1995. Expression cloning of Siamois, a *Xenopus* homeobox gene expressed in dorsal-vegetal cells of blastulae and able to induce a complete secondary axis. *Cell* **81**: 85–94.

Lemaire P., Darras S., Caillol D., and Kodjabachian L. 1998. A role for the vegetally expressed *Xenopus* gene Mix.1 in endoderm formation and in the restriction of mesoderm to the marginal zone. *Development* **125**: 2371–2380.

Levy V., Marom K., Zins S., Koutsia N., Yelin R., and Fainsod A. 2002. The competence of marginal zone cells to become Spemann's organizer is controlled by Xcad2. *Dev. Biol.* **248**: 40–51.

Lustig K.D., Kroll K.L., Sun E.E., and Kirschner M.W. 1996a. Expression cloning of a *Xenopus* T-related gene (Xombi) involved in mesodermal patterning and blastopore lip formation. *Development* **122**: 4001–4012.

Lustig K.D., Kroll K., Sun E., Ramos R., Elmendorf H., and Kirschner M.W. 1996b. A *Xenopus* nodal-related gene that acts in synergy with noggin to induce complete secondary axis and notochord formation. *Development* **122**: 3275–3282.

Marikawa Y., Li Y., and Elinson R.P. 1997. Dorsal determinants in the *Xenopus* egg are firmly associated with the vegetal cortex and behave like activators of the Wnt pathway. *Dev. Biol.* **191**: 69–79.

Mead P.E., Brivanlou I.H., Kelley C.M., and Zon L.I. 1996. BMP-4-responsive regulation of dorsal-ventral patterning by the homeobox protein Mix.1. *Nature* **382**: 357–360.

Merriam J.M., Rubenstein A.B., and Klymkowsky M.W. 1997. Cytoplasmically anchored plakoglobin induces a WNT-like phenotype in *Xenopus*. *Dev. Biol.* **185**: 67–81.

Molenaar M., van de Wetering M., Oosterwegel M., Peterson-Maduro J., Godsave S., Korinek V., Roose J., Destree O., and Clevers H. 1996. XTcf-3 transcription factor mediates beta-catenin-induced axis formation in *Xenopus* embryos. *Cell* **86**: 391–399.

Moon R.T. and Kimelman D. 1998. From cortical rotation to organizer gene expression: Toward a molecular explanation of axis specification in *Xenopus*. *BioEssays* **20**: 536–545.

Nagano T., Ito Y., Tashiro K., Kobayakawa Y., and Sakai M. 2000. Dorsal induction from dorsal vegetal cells in *Xenopus* occurs after midblastula transition. *Mech. Dev.* **93**: 3–14.

Nishita M., Hashimoto M.K., Ogata S., Laurent M.N., Ueno N., Shibuya H., and Cho K.W. 2000. Interaction between Wnt and TGF-beta signalling pathways during formation of Spemann's organizer. *Nature* **403**: 781–785.

Rex M., Hilton E., and Old R. 2002. Multiple interactions between maternally-activated signalling pathways control *Xenopus* nodal-related genes. *Int. J. Dev. Biol.* **46**: 217–226.

Roel G., Hamilton F.S., Gent Y., Bain A.A., Destree O., and Hoppler S. 2002. Lef-1 and Tcf-3 transcription factors mediate tissue-specific Wnt signaling during *Xenopus* development. *Curr. Biol.* **12**: 1941–1945.

Rowning B.A., Wells J., Wu M., Gerhart J.C., Moon R.T., and Larabell C.A. 1997. Microtubule-mediated transport of organelles and localization of beta-catenin to the future dorsal side of *Xenopus* eggs. *Proc. Natl. Acad. Sci.* **94**: 1224–1229.

Sasai Y., Lu B., Steinbeisser H., Geissert D., Gont L.K., and De Robertis E.M. 1994. *Xenopus* chordin: A novel dorsalizing factor activated by organizer-specific homeobox genes. *Cell* **79**: 779–790.

Solnica-Krezel L. and Driever W. 2001. The role of the homeodomain protein Bozozok in zebrafish axis formation. *Int. J. Dev. Biol.* **45**: 299–310.

Stauber M., Jackle H., and Schmidt-Ott U. 1999. The anterior determinant bicoid of *Drosophila* is a derived Hox class 3 gene. *Proc. Natl. Acad. Sci.* **96**: 3786–3789.

Stauber M., Prell A., and Schmidt-Ott U. 2002. A single *Hox3* gene with composite *bicoid* and *zerknullt* expression characteristics in non-Cyclorrhaphan flies. *Proc. Natl. Acad. Sci.* **99**: 274–279.

Stennard F., Carnac G., and Gurdon J.B. 1996. The *Xenopus* T-box gene, *Antipodean*, encodes a vegetally localised maternal mRNA and can trigger mesoderm formation. *Development* **122**: 4179–4188.

Stennard F., Zorn A.M., Ryan K., Garrett N., and Gurdon J.B. 1999. Differential expression of VegT and Antipodean protein isoforms in *Xenopus*. *Mech. Dev.* **86**: 87–98.

Stern C.D. 2001. Initial patterning of the central nervous system: How many organizers? *Nat. Rev. Neurosci.* **2**: 92–98.

Sumoy L., Kiefer J., and Kimelman D. 1999. Conservation of intracellular Wnt signaling components in dorsal-ventral axis formation in zebrafish. *Dev. Genes Evol.* **209**: 48–58.

Trindade M., Tada M., and Smith J.C. 1999. DNA-binding specificity and embryological function of Xom (Xvent-2). *Dev. Biol.* **216**: 442–456.

van Noort M. and Clevers H. 2002. TCF transcription factors, mediators of Wnt-signaling in development and cancer. *Dev. Biol.* **244:** 1–8.

Watabe T., Kim S., Candia A., Rothbacher U., Hashimoto C., Inoue K., and Cho K.W. 1995. Molecular mechanisms of Spemann's organizer formation: Conserved growth factor synergy between *Xenopus* and mouse. *Genes Dev.* **9:** 3038–3050.

White R.J., Sun B.I., Sive H.L., and Smith J.C. 2002. Direct and indirect regulation of derriere, a *Xenopus* mesoderm-inducing factor, by VegT. *Development* **129:** 4867–4876.

Xanthos J.B., Kofron M., Tao Q., Schaible K., Wylie C., and Heasman J. 2002. The roles of three signaling pathways in the formation and function of the Spemann Organizer. *Development* **129:** 4027–4043.

Yamanaka Y., Mizuno T., Sasai Y., Kishi M., Takeda H., Kim C.H., Hibi M., and Hirano T. 1998. A novel homeobox gene, dharma, can induce the organizer in a non-cell-autonomous manner. *Genes Dev.* **12:** 2345–2353.

Yao J. and Kessler D.S. 2001. Goosecoid promotes head organizer activity by direct repression of Xwnt8 in Spemann's organizer. *Development* **128:** 2975–2987.

Yasuo H. and Lemaire P. 1999. A two-step model for the fate determination of presumptive endodermal blastomeres in *Xenopus* embryos. *Curr. Biol.* **9:** 869–879.

Zhang J. and King M.L. 1996. *Xenopus* VegT RNA is localized to the vegetal cortex during oogenesis and encodes a novel T-box transcription factor involved in mesodermal patterning. *Development* **122:** 4119–4129.

Zhang J., Houston D.W., King M.L., Payne C., Wylie C., and Heasman J. 1998. The role of maternal VegT in establishing the primary germ layers in *Xenopus* embryos. *Cell* **94:** 515–524.

Zoltewicz J.S. and Gerhart J.C. 1997. The Spemann organizer of *Xenopus* is patterned along its anteroposterior axis at the earliest gastrula stage. *Dev. Biol.* **192:** 482–491.

CONSERVED AND DIVERGENT ROLES OF *TWIST* IN GASTRULATION

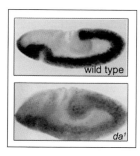

A. Tapanes-Castillo, V. Cox, and M. Baylies

Developmental Biology Program, Memorial Sloan-Kettering Cancer Center,
Weill Graduate School of Medical Sciences at Cornell University, New York,
New York 10021

INTRODUCTION

The discovery of *twist* as a key regulator of gastrulation and mesoderm differentiation in *Drosophila* sparked the search for similar genes in other species. As a result, *Twist* homologs have been found in a great number of different species, which span the evolutionary landscape from jellyfish and mollusks to mouse and man. In these species, *Twist* is expressed in mesoderm and muscle-forming cells as it is in *Drosophila*. However, in many cases, the range of *Twist* expression broadens to include mesodermally derived tissues such as sclerotome and osteoblasts as well as nonmesodermally derived cells such as myoepithelium and neural crest. Analyses of *Twist* mutations support an important role for *Twist* in tissue specification, patterning, and differentiation in organisms such as the nematode, fly, mouse, and particularly human, where it has been linked to the developmental disorder, Saethre-Chozten syndrome. Although some aspects of the Twist structure and function have been conserved (e.g., the basic-helix-loop-helix [bHLH] region and a role in mesodermal patterning), the diversity among family members in overall structure and expression suggest additional, novel aspects of Twist function across phyla. In this chapter, we discuss both conserved and unique elements of Twist structure, its regulation, and, importantly, its function, in the context of gastrulation and early mesodermal differentiation. Our reference point is *Drosophila*, in which the first *twist* mutation was uncovered in the genetic screen of Nusslein-Volhard and Wieschaus and about which the most information is available.

THE bHLH TRANSCRIPTIONAL REGULATOR *TWIST*

Drosophila Twist (DTwist) (Thisse et al. 1988) is the founding member of a subfamily of bHLH proteins based on amino acid identity and signature amino acids within the bHLH domain. The Twist bHLH domain consists of a basic amino acid–rich region, which is required for DNA binding, two conserved amphipathic helices, and a loop of 14 amino acids, which is important in maintaining the tertiary structure of the HLH moiety. The HLH region is necessary and sufficient for protein dimerization (Kadesch 1993).

Twist protein family members have been identified in species ranging from primates—human (HTwist; Wang et al. 1997), monkey, gorilla, among others (Gachot-Neveu et al. 2002); rodents—mouse (MTwist; Wolf et al. 1991) and rat (Bloch-Zupan et al. 2001); birds—chicken (CTwist; Tavares et al. 2001); amphibians—frog (XTwist; Hopwood et al. 1989); cephalochordates—lancelet (BbTwist; Yasui et al. 1998); fish—zebrafish (DrTwist; Morin-Kensicki and Eisen 1997); insects—Tribolium (TrTwist; Sommer and Tautz 1994); nematodes—*Caenorhabditis elegans* (CeTwist; Harfe et al. 1998); lophotrochozoans—leech (HroTwist; Soto et al. 1997) and mollusks (PvTwist; Nederbragt et al. 2002) to cnidarians—jellyfish (PcTwist; Spring et al. 2000). These Twist proteins share 59–85% amino acid identity in the bHLH region when compared with DTwist (Fig. 1A,C). However, the bHLH domains within vertebrate species show a higher degree of identity: human, monkey, gorilla, mouse, frog, zebrafish, and even lancelet, a cephalochordate considered a sister group of vertebrates, share 84–100% amino acid identity over the bHLH domain.

The importance of this region has been emphasized by isolation of mutations in the bHLH domain: point mutations in the human TWIST bHLH domain have been linked to Saethre-Chotzen syndrome, which is characterized by developmental abnormalities including craniosynostosis and syndactyly in hands and feet (El Ghouzzi et al. 1997; Howard et al. 1997). Two classes of mutations have been found: (1) nonsense mutations, which lead to proteins lacking the DNA-binding and dimerization domains and are thought to result in TWIST haploinsufficiency, and (2) missense mutations, which are not well understood functionally (Gripp et al. 2000). Recent work in *C. elegans*, however, has provided some insight to this second class of bHLH mutations. Both biochemical and phenotypic analyses of CeTwist proteins carrying amino acid substitutions in the basic region suggest that in humans, this syndrome may be the result of dominant-negative proteins rather than TWIST haploinsufficiency (Corsi et al. 2002).

bHLH proteins function as transcriptional regulators by binding as dimers to the consensus hexanucleotide sequence E-box, 5′-CANNTG-3′ (Ephrussi et al. 1985). Twist dimers (*Drosophila*, nematode, and mouse) bind preferentially to the E-box, 5′-CA**TA**TG-3′ as compared to 5′-CA**CC**TG-3′ for other bHLH dimers such as MyoD/E heterodimers (Ip et al. 1992a; Lee et al. 1997; Yin et al. 1997; Cripps et al. 1998; Kophengnavong et al. 2000; Oshima et al. 2002). Binding-site specificity requires amino acid residues within the basic domain, especially N6, which is conserved among all Twist family members. The highly conserved junction region between the basic domain and helix 1 is also important to site specificity. This region functions with N6 to position the basic region on the DNA (Fig. 1A) (Kophengnavong et al. 2000).

The WR motif (tryptophan-arginine motif; Spring et al. 2000), located 20–25 amino acids carboxy-terminal to the bHLH region, is the second highly conserved region in the Twist family. Among vertebrates and jellyfish, this 14-amino acid region is 100% identical but is less well conserved in lancelets and flies (79%) and almost unidentifiable in nematodes (39%) (Fig. 1B,C). No WR motif can be recognized in leeches. The function of this region is unclear. Experiments in which the DTwist WR motif was mutagenized indicate a role for this domain in protein stability (Gonzalez and Baylies 2002). Likewise, a nonsense mutation in HTwist just before the WR motif causes a mild appearance of Saethre-Chotzen syndrome, reinforcing that the carboxy-terminal domain is required either for Twist activity or for the stability of its mRNA and/or its protein (Gripp et al. 2000).

Outside these two domains, the Twist protein diverges greatly across species. For example, only 35% amino acid identity over the entire protein is shared between DTwist and MTwist, whereas 96% identity is shared between vertebrate forms of Twist. DTwist is a much larger protein (490 aa versus 202 aa [human]): It contains in its amino-terminal region an arginine/lysine-rich domain and two stretches of glutamine/histidine repeats (CAX), which are not conserved in other family members, with the exception of HroTwist. Missense mutations that specifically affect the DTwist amino-terminal region have been isolated. These mutations lead to embryos with severely disrupted mesoderm and muscles (Thisse et al. 1987; Baylies and Bate 1996), as a result of reduced transcriptional activity (Gonzalez and Baylies 2002). CAX repeats have been shown to be involved in transcriptional activation (Chung et al. 1996; Shirokawa and Courey 1997; Castanon 2002). Since the CAX transcriptional activation domains are not conserved among all Twist proteins, this suggests alternative mechanisms of transcriptional regulation employed by Twist across the evolutionary landscape.

REGULATION OF *TWIST* ACTIVITY THROUGH ITS bHLH DOMAIN

TWIST forms heterodimers with the E family of proteins in humans (El Ghouzzi et al. 2000), whereas in nematodes, flies, and mice, Twist forms both homodimers and heterodimers with E protein (Daughterless [Da] in *Drosophila*) (Spicer et al. 1996; Lee et al. 1997; Harfe et al. 1998; Castanon et al. 2001). These different dimer forms have different functions, some conserved across species. In mouse, heterodimers of vertebrate E and MTwist repress muscle reporter genes and myogenesis in transient transfection experiments. No activity has been reported for MTwist homodimers in this assay (Spicer et al. 1996). In contrast, CeTwist/CeE/Da heterodimers have been linked to activation of target genes in vivo (Harfe et al. 1998). Thus, it appears that Twist/E heterodimers can activate (nematodes) or repress (mouse) gene expression. In *Drosophila*, similar studies have been performed that support and extend these observations.

Whereas both overexpression and dose experiments showed that DTwist/E(Da) heterodimers repress myogenesis when expressed in mesoderm, these experiments also revealed that DTwist homodimers activate genes required for mesoderm specification, gastrulation, and muscle formation in vivo (Castanon et al. 2001). A direct functional test of the different dimers was performed using tethered dimers (Fig. 2). Two monomers (Twist-Twist and Twist-Da) were joined head to tail via a flexible polylinker. Overexpression of Twist-Twist–tethered dimers in mesoderm or in ectoderm led to ectopic muscle formation, indicating that Twist homodimers induced muscle develop-

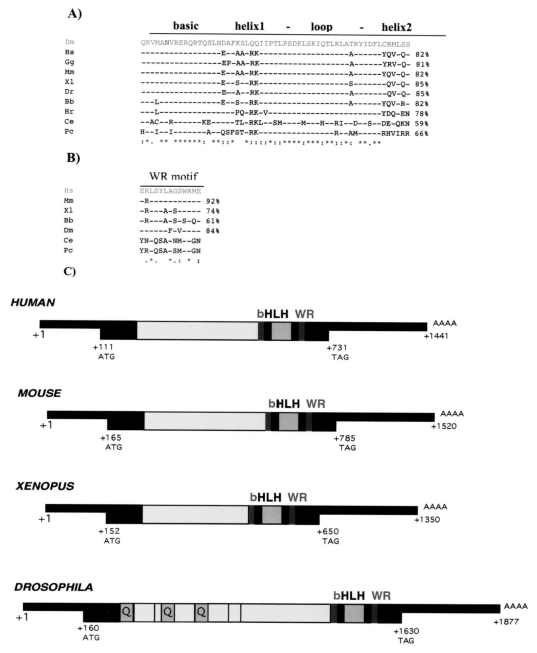

Figure 1. Alignment of selected Twist family members. (*A*) Amino acid alignment within the bHLH domain (Murre et al. 1989) and (*B*) the WR motif of Twist proteins from flies (*Drosophila melanogaster*; Dm), human (*Homo sapiens*; Hs), gorilla (*Gorilla gorilla*, Gg), mouse (*Mus musculus*; Mm), frog (*Xenopus laevis*; Xl), zebrafish (*Danio rerio*, Dr), lancelet (*B. belcheri*; Bb), leech (*Helobdella robusta*; Hr), worm (*Caenorhabditis elegans*; Ce), and cnidarians (*Podocoryne carnea*; Pc). Amino acids identical to DTwist in *A* or Htwist in *B* are indicated by dashes. Asterisks indicate conserved amino acids in all species compared. Double dots indicate identities, and single dots indicate conservative substitutions. (*A*) The bHLH domain is highly conserved among vertebrates and lancelets, which share 84–100% amino acid identity. The most distantly related Twist (CeTwist) shares only 59% identity within the bHLH domain. The conserved N6 amino acid (*green*) in the basic domain is critical for DNA-binding-site specificity. (*B*) The WR motif is highly conserved in human, mouse, *Xenopus,* and lancelets. It is less conserved in *Drosophila*, and unrecognizable in *C. elegans* and leeches. (*C*) Schematic comparing overall structure of human, mouse, frog, and fly Twist proteins. The conserved bHLH and WR motifs are located in the carboxyl terminus of the proteins. DTwist contains stretches of polyglutamine-histidine repeats (Q) that are critical for transcriptional activation.

Wildtype　Twist　Twist-Twist　Twist-Da

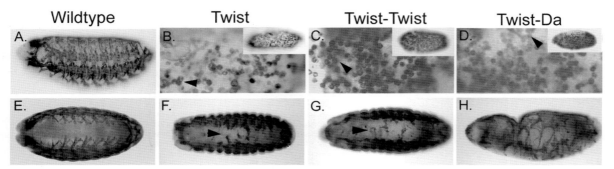

Figure 2. Activity of Twist dimers depends on context. (*A, E*) Wild-type embryos; (*B, F*) ectopic expression of *UAS-twist*; (*C, G*) ectopic expression of *UAS-twist-twist*; (*D, H*) ectopic expression of *UAS-twist-daughterless*. All embryos probed with antibodies to myosin heavy chain (Mhc). For (*A–D, H*) lateral views of stage-16 embryos, dorsal up and anterior left. (*E–G*) Ventral views of stage-16 embryos. (*A*) Wild-type multinucleated Mhc-positive muscle cells; all cells are mesodermal and internal. (*B–D*) High magnifications and whole mounts (*insets*) of embryos misexpressing *twist* (*B*), *twist-twist* (*C*), or *twist-daughterless* (*D*) using one copy of *da-GAL4* driver. Expression of Twist, Twist-Twist, or Twist-Daughterless led to the conversion of ectoderm into somatic muscle; fused di- and tri-nucleated Mhc-positive cells, which are characteristic of somatic myogenesis (*arrows*) were found externally. Since ectodermal tissues such as epidermis do not form, the muscles do not spread out to form a pattern. (*F*) Expression of *twist* or (*G*) *twist-twist* in the mesoderm using *twist-GAL4* driver led to ectopic formation of Mhc-positive muscle cells, shown here ventrally (*arrows*). Note that Mhc is found in cells where it is never usually expressed. (*H*) Embryos expressing *twist-daughterless* in mesoderm show severely reduced numbers of muscle cells. Thus, Twist homodimers activate the myogenic program, whereas Twist/Daughterless heterodimers repress or activate the myogenic program, depending on cell type.

ment. In addition, Twist-Twist homodimers rescued the *twist* null mutation, both for mesoderm induction and for allocation of cells to the myogenic pathway. Overexpression of Twist-Da heterodimers in mesoderm resulted in a loss of body muscles, as well as reduced expression of a reporter construct containing Twist E boxes. These results indicate that Twist-Da heterodimers block gene expression and thereby inhibit body muscle development (Castanon et al. 2001).

The activity of different Twist dimer forms—homodimers or heterodimers with E (Da)—varies depending on the organism and, perhaps, cell type. Underscoring the importance of context are data which indicate that Twist-Da activates mesodermal genes when expressed in non-mesodermal tissues such as ectoderm (Fig. 2) (Castanon 2002). Hence, in *Drosophila*, Twist/E (Da) dimers function both as a transcriptional activator (as in *C. elegans*) and a repressor (as in mouse), depending on the tissue in which it is expressed. Whether this switch also occurs with other species remains to be tested.

In addition, a search for other Twist dimerization partners has been undertaken in *Drosophila*. Genetic interactions were established between Twist and other HLH proteins, such as Extramacrochaete (Id homolog), L'scute (Mash homolog), and Nautilus (MyoD homolog), which are required for mesoderm development and are expressed at the right time and place to interact with Twist. However, DNA-binding data revealed that these proteins do not interact with Twist directly but with Da (J. Kass and M.K. Baylies, unpubl.). Interestingly, yeast two-hybrid experi-

ments performed with jellyfish Twist, Id, Da, and JellyD, the MyoD homolog, reveal that jellyfish Twist interacts with Id and JellyD, but not with Da or itself (Müller et al. 2003). Although future experiments need to address the functional significance of these relationships in vivo, these data suggest that species differences in Twist structure and/or context influence Twist partners.

REGULATION OF *TWIST* ACTIVITY THROUGH PROTEIN INTERACTIONS OUTSIDE THE bHLH DOMAIN

The list of Twist interacting proteins, isolated from different species, is rapidly increasing. However, because the Twist protein structure is less conserved outside the bHLH and WR domains, whether the interacting proteins discussed below cooperate with all Twist family members and regulate their different activities remains to be demonstrated.

The amino-terminal domain of MTwist directly binds two independent domains of the histone acetyltransferases (HAT), P300 and p300/CBP-associated factor (PACF), and regulates their HAT activities (Hamamori et al. 1999). Likewise, the amino-terminal region of DTwist is associated with transcription activation (Chung et al. 1996; Shirokawa and Courey 1997; Castanon 2002); particularly, this region interacts with TAF110, a subunit of the TFIID complex, mediating target gene activation (Pham et al. 1999). In addition, the DTwist amino-terminal domain cooperates with other transcriptional regulators, specifically Dorsal, a member of the Rel family of transcription factors. These two proteins synergize, activating downstream targets such as *snail* (Ip et al. 1992b; Shirokawa and Courey 1997). Although MTwist also interacts with the vertebrate Dorsal (NF-κB), both the Twist domain involved in the interaction and the activity of Twist differ: The Twist carboxyl terminus (rather than the amino terminus) mediates repression (rather than activation) of the NF-κB-dependent cytokine pathway. The authors speculate that DTwist may also be involved in a similar negative feedback loop in Dorsal regulation during dorsal–ventral patterning and the immune response of *Drosophila* (Sosic et al. 2003). Hence, the architecture of the overall relationship between Twist and Dorsal (NF-κB) appears to have been maintained during evolution, but the particular details have evolved to fit the specific context required.

TRANSCRIPTIONAL REGULATION OF *TWIST*

Twist is expressed in many different places and times during development of a variety of species. In this section, we focus on those genes that control the dynamic pattern of *twist* expression during the first two stages of *Drosophila* mesoderm development: gastrulation, when mesoderm is specified, and subdivision, when mesodermal cells are allocated to different tissues (Fig. 3). We then consider whether the regulatory relationships uncovered in *Drosophila* extend to other species.

REGULATORS OF *TWIST* EXPRESSION AT GASTRULATION

twist is first detected prior to gastrulation in an 18 to 20-cell-wide band in the ventral-most region of the embryo, where it is critical for mesoderm specification and gastrulation. Factors that determine *twist* expression are Dorsal, Snail, Twist itself, and members of the extended bHLH family—Da, Achaete, and Scute (Kosman et al. 1991; Leptin 1991). The mechanism underlying the action of these genes has been well studied through dissection of the promoters of multiple genes expressed in the ventral (mesodermal) and lateral (neuroectodermal) domains of the developing embryo (Jiang et al. 1991; Thisse et al. 1991; Ip et al. 1992a,b; Gonzalez-Crespo and Levine 1993).

The Dorsal transcription factor forms a broad gradient along the dorsoventral axis of the embryo with high levels of nuclear Dorsal at the most ventral point of the embryo and progressively lower levels of nuclear Dorsal more dorsally. This gradient is established in the egg chamber by the Toll pathway (Morisato and Anderson 1995). Mutations in the Toll pathway, such as *dorsal* itself, lead to a dorsalized embryo and loss of *twist* expression ventrally, suggesting that Dorsal is necessary for *twist* expression (Simpson 1983). Two factors, the presence of low-affinity Dorsal-binding sites in the *twist* promoter, and expression of high levels of nuclear Dorsal ventrally, determine the normal expression domain of *twist* (Jiang et al. 1991; Thisse et al. 1991).

Dosage-sensitive interactions between Dorsal and a number of bHLH factors including Da, Achaete, and Scute, are also critical to the establishment of *twist* expression (Gonzalez-Crespo and Levine 1993). Embryos carrying double heterozygous combinations of *dorsal* and *da* alleles have a reduced *twist* expression domain. Instead of 18–20 cells expressing *twist*, only 12–14 cells now express *twist*. Moreover, embryos that are triple heterozygous for *dorsal*, *twist,* and *da* or the *achaete-scute complex* essentially fail to form mesoderm. These embryos show gaps in and a narrowing of *twist* expression to 3–10 cells ventrally. Moreover, Twist regulates its own expression. Embryos homozygous for a *twist* allele that leads to the production of *twist* RNA but no protein show gaps in *twist* RNA expression reminiscent of the triple heterozygous embryos (Fig. 3A,B). Hence, the establishment of mesoderm and *twist* expression specifically requires synergistic interactions between divergent bHLH proteins.

twist expression also relies on Snail. *snail* is expressed prior to and during gastrulation in a ventral domain overlapping that of *twist* (Kosman et al. 1991; Leptin 1991). Snail represses expression of lateral, neuroectodermal genes in the ventral region of the embryo. In *snail* mutant embryos, however, *twist* expression also disappears ventrally. Whether this effect on *twist* is due to a loss of a direct, positive action by Snail or to the derepression of lateral genes ventrally, which in turn repress *twist*, is unclear (Leptin 1991).

As gastrulation begins, *twist* expression continues in the mesoderm as it involutes and migrates dorsally, forming a single-layered sheet of cells beneath the ectoderm (Bate et al. 1993). During this period, Da is required to maintain high *twist* levels (Castanon et al. 2001). In *da* maternal and zygotic loss-of-function embryos, *twist* expression is lost during mesoderm migration (Fig. 3C,D). Data suggest that Da homodimers regulate *twist* expression during this period (Tapanes-Castillo and Baylies 2004).

REGULATORS OF *TWIST* DURING MESODERM SUBDIVISION

Following gastrulation, the mesoderm is patterned so that uncommitted cells are allocated to specific tissue fates; this event requires segmentation, subdivision, and differential *twist* expression. Initially, segmentation partitions the

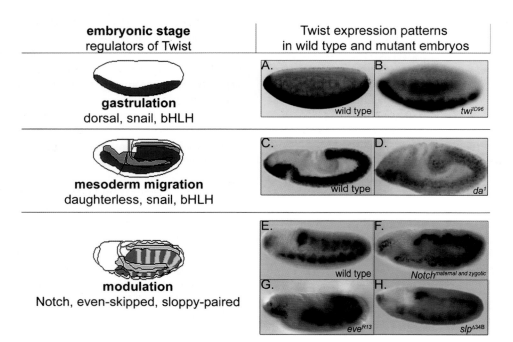

embryonic stage regulators of Twist	Twist expression patterns in wild type and mutant embryos

gastrulation
dorsal, snail, bHLH

mesoderm migration
daughterless, snail, bHLH

modulation
Notch, even-skipped, sloppy-paired

Figure 3. Regulators of Twist at gastrulation and subdivision. On left are cartoons of embryos at stages 6, 8, and 10 of *Drosophila* development. Below each cartoon, the mesodermal or *twist* expression stage is listed, as are the known regulators for each period. (*A, B*) Stage-6 embryos were hybridized with an RNA probe to *twist*. (*C–H*) Embryos at various stages of development were probed with Twist antibody (Castanon et al., 2001). (*A*) A wild-type embryo shows *twist* RNA expressed in the ventralmost portion of the embryo in the presumptive mesoderm. (*B*) Embryos homozygous for a *twist* mutation that results in RNA expression but no protein show reduced *twist* expression as compared to wild type. (*C*) A wild-type embryo (stage 8) shows high *twist* expression in the mesoderm as the germ band extends. (*D*) An embryo mutant for *daughterless* fails to maintain high *twist* expression. (*E*) A wild-type embryo (stage 10) shows a modulated *twist* expression pattern in which each segment contains a low and high domain of *twist* expression. (*F*) Embryos missing both the maternal and zygotic contributions of Notch fail to modulate *twist*, resulting in uniform high *twist* expression. The mesoderm appears undulated due to ectopic neuroblast production in this background. (*G*) *even-skipped* mutant embryos also fail to modulate *twist*, maintaining high *twist* levels throughout the mesoderm (cf. Riechmann et al. 1997). (*H*) Embryos lacking *sloppy-paired* express uniform low *twist* levels throughout the mesoderm (cf. Riechmann et al. 1997).

mesoderm along the anterior–posterior axis. Subdivision then divides each mesodermal segment across the dorsoventral axis into tissue competency domains; based on their position, cells are assigned a specific fate—visceral mesoderm, fat body, heart, or somatic muscle (body wall muscle). This process of subdivision alters *twist* expression. The characteristic uniform high *twist* expression seen earlier in development is modulated; a segmentally repeated pattern of low and high *twist*-expressing domains forms along the anterior–posterior axis of the embryo. Cells located in the low *twist* domain differentiate into visceral muscle and fat body, whereas cells in the high *twist* domain develop into somatic muscles and heart (Borkowski et al. 1995; Baylies and Bate 1996). During this period, *Twist* expression is directed by the Wingless and Hedgehog pathways, through the regulation of the pair-rule genes *sloppy-paired* and *even-skipped*, respectively, as well as by the Notch pathway, through the regulation of a bHLH transcription factor network (Fig. 3) (Azpiazu et al. 1996;

Riechmann et. al. 1997; Lee and Frasch 2000; Cox 2004; Tapanes-Castillo and Baylies 2004).

COMPARISON OF *TWIST* REGULATION ACROSS SPECIES

In *Drosophila*, *twist* expression is confined to the mesoderm. However, in other species, *Twist* is expressed both in the mesoderm and in non-mesodermal tissues. For example, *Twist* is expressed in the somites and neural crest cells of mice (Wolf et al. 1991; Hebrok et al. 1994; Gitelman 1997), and it is expressed in nonstriated muscles as well as in a small number of neuron-like cells in *C. elegans* (Harfe et al. 1998). These similarities, as well as differences, in tissue-specific *Twist* expression reflect both conserved and nonconserved mechanisms of *Twist* regulation.

Analogous to its early function in *Drosophila* mesoderm development, the NF-κB (Dorsal) family has been implicated in *Twist* activation in vertebrates, nematodes, and short-germ-band insects. In the chick limb buds, where *NF-κB* and *Twist* expression overlap, NF-κB reduction results in diminished *Twist* expression (Bushdid et al. 1998; Kanegae et al. 1998). Furthermore, studies conducted in mouse fibroblasts suggest that the NF-κB family member p65 (Rel A) is required for TNFα-mediated induction of murine *Twist* (Sosic et al. 2003). In addition, *CeTwist* promoter analysis revealed that a conserved Dorsal-binding site is required for activity within the postembryonic mesodermal blast-cell lineage (Harfe et al. 1998). Last, in the short-germ-band beetle, *Tribolium castaneum*, the transiently overlapping expression patterns of Dorsal (Tc-dl) protein and *twist* mRNA imply that *twist* expression is regulated by Tc-dl (Chen et al. 2000).

Likewise, a conserved role for Snail in positively regulating *Twist* has been described. In *Xenopus,* injection of wild-type *snail* constructs into animal caps, or inducible *snail* constructs into embryos, results in *twist* up-regulation or the expansion of neural crest *twist* expression, respectively. Correspondingly, injecting a *snail* dominant-negative construct into embryos blocks *twist* neural crest expression (Aybar et al. 2003).

Also similar to the fly, Wnt (Wingless) signaling up-regulates *Twist* expression in frogs and mice. In *Xenopus*, Xwnt7B stimulates *twist* expression (Chang and Hemmati-Brivanlou 1998). Likewise, elevated *Twist* expression is detected both in mammary tumors of *Wnt1* transgenic mice and in *Wnt1*-expressing murine mammary epithelial cells. A murine *Twist* promoter fragment was isolated that positively responds to β-*catenin* overexpression in cell culture; sequence analysis of this regulatory region revealed three putative TCF sites, suggesting that Wnt signaling directly regulates *Twist* (Howe et al. 2003).

Ras signaling has been shown to regulate *twist* during *Drosophila* muscle formation (Cox 2004) as well as in chicks, mice, and ascidians. Chick limb bud manipulations determined that FGF released from the apical ectodermal ridge—a specialized epithelial structure that forms at the apical tip of the vertebrate limb bud—is essential for proper limb outgrowth and *Twist* expression (Tavares et al. 2001). Similarly, FGF signaling positively regulates *Twist* in mice in multiple contexts. Upon treatment with FGF-2, *Twist* expression is up-regulated in murine osteoblast-like cells (Fang et al. 2001). Comparably, in murine calvarial bones, *Twist* expression overlaps with FGFR-2 (Johnson et al. 2000) and FGF ligand expression (Rice et al. 2000). Additionally, in ascidian larvae, two *twist*-like genes were identified as downstream targets of Cs-FGF9/16/20 in mesenchymal cells (Imai et al. 2003). Vertebrate tissue-culture experiments suggest a direct connection between Ras signaling and *Twist* expression: A murine *Twist* promoter contains 34 potential binding sites for transcriptional effectors of Ras signaling, many of which

are found in both human and fly *Twist* promoters (Cox and Baylies 2001; Howe et al. 2003).

In conclusion, several aspects of *Twist* regulation during mesodermal development—such as the mechanisms employed by Dorsal, Snail, Wnt, and Ras—are preserved across the phylogenetic tree. Further investigation is required to test whether other strategies for directing *Twist* expression, such as regulation by Notch signaling, are evolutionarily conserved.

ROLES OF *TWIST* DURING GASTRULATION AND MESODERMAL PATTERNING

In species from mollusk to human, *Twist* is expressed in cells contributing to mesoderm and/or its derivatives. This pattern suggests that Twist may have a common role and perhaps universal targets in mesoderm differentiation. However, non-mesodermal *Twist* expression detected in many species implies that Twist may play other roles during development. We now consider whether the roles for Twist, defined in *Drosophila* mesoderm development, are applicable to other species. Moreover, we discuss additional roles for Twist, as suggested by recent experiments.

Mesoderm Specification

twist is critical for mesoderm specification and gastrulation in *Drosophila*. Embryos that carry mutations for *twist* gastrulate abnormally, form no mesoderm, and subsequently die at the end of embryogenesis (Simpson 1983). However, analysis of *Twist* expression and function in several other species indicates that this role is not necessarily conserved across phyla. On one hand, analysis of *PcTwist* (jellyfish) reveals that it is expressed in the entocodon, among other tissues. The entocodon is a "mesoderm-like" cell layer that consists of proliferative undifferentiated cells from the dorsal ectoderm, which undergo a process comparable to gastrulation. The entocodon differentiates into the medusa-typical smooth and striated muscles (Spring et al. 2000). Perhaps in jellyfish, as in flies, Twist is targeting genes required for processes typical in gastrulation, such as cell-shape changes, cell migration, and cell proliferation, as well as muscle differentiation such as myosin. On the other hand, in mouse, for example, the Twist protein is not detectable prior to gastrulation but only after mesoderm has been specified (Wolf et al. 1991; Gitelman 1997). This observation holds true for nematodes and lancelets (Harfe et al. 1998, Yasui et al. 1998). Furthermore, loss-of-function analysis in mouse (Chen and Berhringer 1995) and in nematodes (Corsi et al. 2000) indicates that gastrulation and mesoderm specification occurs normally in the absence of Twist.

Mesodermal Patterning

In flies, Twist patterns both the mesoderm and somatic muscles. High levels (via Twist homodimers) activate somatic myogenesis, whereas low levels (via Twist/Da heterodimers) repress this fate and are permissive for gut muscle and fat body fate (Baylies and Bate 1996; Castanon et al. 2001). This patterning role has been preserved in other species. The single Ce*Twist* gene, which encodes a transcriptional activator when partnered with E (Da), is required for proper differentiation of a subset of mesodermal tissues, including the enteric muscles and the M mesoblasts and their descendants (Corsi et al. 2000). Likewise in vertebrates, a patterning role for Twist during mesoderm development has been conserved, with Twist/E heterodimers functioning in somites as myogenic repressors instead of activators (Wolf et al. 1991; Chen and Behringer 1995; Spicer et al. 1996). Analysis of *Twist* null mutant mouse embryos, however, reveals that Twist does not solely regulate somite development. A wide range of developmental defects can be detected in these mutants (Chen and Behringer 1995; O'Rourke and Tam 2002). Recent work suggests that some of these phenotypes are due to loss in Twist maintenance of the integrity of signaling centers, particularly those generated by FGF and/or Shh signaling, which are responsible for proper patterning of multiple organs, including the somite, limb bud, branchial arches, or forebrain (O'Rourke et al. 2002; Zuniga et al. 2002).

Additional Functions for *TWIST*

In addition to controlling the cell-fate specification, further roles for Twist have been suggested. For example, Twist activity has been linked to proliferation in jellyfish, fly, and mouse (Roy and VijayRaghavan 1999; Spring et al. 2000). Moreover, Twist has been identified as an antiapoptotic factor. Although increased apoptosis was noted in the somites of *Twist* null mice (Chen and Behringer 1995), a role for Twist in inhibition of the apoptotic pathway, mediated by p53 and insulin growth factor-1, has been described previously (Maestro et al. 1999; Dupont et al. 2001). However, a direct connection between the mechanisms uncovered in the tissue-culture experiments and the observed apoptosis in vivo remains to be made.

Conservation of *TWIST* Target Genes

Due to the critical involvement of Twist in the development of many different tissues, efforts have been made to identify Twist target genes (O'Rourke and Tam 2002). In this section, we discuss evolutionarily conserved target genes and comment on interesting novel target genes.

A well-characterized and conserved target of Twist-mediated transcription is the NK-class homeodomain gene *tinman/ceh-24*. In *Drosophila*, genetic and molecular analyses have shown that Twist is directly required for early *tinman* expression (Lee et al. 1997; Yin et al. 1997; Venkatesh et al. 2000). Similarly, in *C. elegans*, the expression of a *ceh-24* vulval muscle promoter also requires CeTwist directly (Harfe et al. 1998; Corsi et al. 2000).

Twist regulation of FGFRs has also been preserved throughout the phylogenetic tree. In *Drosophila*, early mesodermal activation of the FGFR *heartless* requires Twist (Shishido et al. 1993), whereas in *C. elegans*, genetic and promoter studies strongly suggest that CeTwist directly activates the FGFR-like gene *egl-15* in vivo (Harfe et al. 1998; Corsi et al. 2000). Analysis of *Twist* heterozygous and knockout mice indicate that Twist regulates the expression of FGFRs during limb, osteoblast, branchial arch, and forebrain development (Rice et al. 2000; O'Rourke et al. 2002; Soo et al. 2002; Zuniga et al. 2002). Moreover, phenotypic studies of individuals affected by Saethre-Chotzen, which is associated with mutations in *TWIST,* and Crouzon syndrome, which is associated with mutations in *FGFR-1, -2,* and *-3,* suggest that TWIST positively regulates FGFR expression in humans (Meyers et al. 1995; Neilson and Friesel 1995; Oldridge et al. 1995).

Twist regulation of Gli genes, the transcriptional effectors of the Shh signal transduction pathway, also appears to be conserved. Analysis of human *GLI1* regulatory sequences indicated that *GLI1* is a direct transcriptional TWIST target (Villavicencio et al. 2002). Furthermore, in *Twist* null mice embryos, *Gli1, Gli2,* and *Gli3* expression is either absent or reduced in limb buds (O'Rourke et al. 2002; Zuniga et al. 2002). In *Drosophila,* Twist activity affects expression of *gleeful,* a zinc finger transcription factor gene that is highly similar to vertebrate Gli proteins. *gleeful,* which is transcribed in the somatic mesoderm of wild-type embryos, is differentially expressed between wild-type, *twist*-null, and *twist*-overexpressing embryos (Furlong et al. 2001).

Gene expression profiles of *Twist* mutants in both fly and mouse have identified additional candidate Twist target genes (Furlong et al. 2001; Loebel et al. 2002). In mice, the homeobox genes *Aristaless 3* (*Alx3*) and *Aristaless 4* (*Alx4*) have been proposed to be direct targets of Twist-mediated transcription. *Alx3* and *Alx4* expression overlaps with *Twist* in wild-type embryos; both genes exhibit reduced expression in *Twist*-null mouse embryos (Loebel et al. 2002). A Bcl-2-interacting transcription factor-like gene (Kasof et al. 1999) has also been proposed to be a Twist target; this mouse gene (*TC186873*) and a weakly homologous *Drosophila* gene (*EST GH22851*) were both identified as differentially expressed between wild type and *Twist* mutants (Furlong et al. 2001; Loebel et al. 2002). Further study of Twist's regulation of this apoptotic factor may lend insight to the mechanism Twist uses to regulate apoptosis. Moreover, it will be interesting to learn whether these

CONCLUSIONS

The isolation of *twist* in *Drosophila* and the identification of its role in mesoderm differentiation have lead to the identification of *Twist* family members in other species. Like *Dtwist*, many of these family members are expressed in mesoderm and share striking structural identity in the bHLH and WR domains. In all species, Twist appears to function as a transcriptional regulator. Interestingly, Twist acts as both an activator and a repressor of downstream target genes, depending on dimer partner and tissue context. However, further investigation of the "tissue context" that enables the switch between activator and repressor is still required. Collaboration with the growing list of Twist interacting proteins may provide insight to this mechanism. Indeed, these interacting proteins may uncover new modes of regulation for the distinct differentiation programs controlled by Twist in cell types ranging from muscle to neural crest. Likewise, how Twist interacts with various signal transduction pathways, which regulate and cooperate with Twist to control downstream events, remains an open question, up and down the phylogenetic tree. Although many "cassettes" of gene activity, such as Dorsal (NF-κB)-Twist or FGFR-Twist, have been preserved across species, many new relationships await discovery. Last, identification of Twist targets remains key to understanding how Twist orchestrates its many different cell-fate decisions. The combination of genome-wide screening for candidate genes and loss- and gain-of-function analysis of these genes will uncover the circuitry that links *Twist* to the elaboration of cell fate.

ACKNOWLEDGMENTS

We thank Irene Zohn and Karen Beckett for helpful comments on the manuscript. This work has been supported by The Society of Memorial Sloan Kettering Cancer Center and National Institutes of Health (GM-56989) to M.B. V.C. was a Bruce Charles Forbes predoctoral fellow.

REFERENCES

Aybar M.J., Nieto M.A., and Mayor R. 2003. Snail precedes slug in the genetic cascade required for the specification and migration of the *Xenopus* neural crest. *Development* **130:** 483–494.

Azpiazu N., Lawrence P.A., Vincent J.P., and Frasch M. 1996. Segmentation and specification of the *Drosophila* mesoderm. *Genes Dev.* **10:** 3183–3194.

Bate M., Martinez Arias A., and Hartenstein V. 1993. *The development of* Drosophila melanogaster. Cold Spring Harbor Laboratory Press, Cold Spring Harbor, New York.

Baylies M.K. and Bate M. 1996. *twist*: A myogenic switch in *Drosophila. Science* **272:** 1481–1484.

Bloch-Zupan A., Hunter N., Manthey A., and Gibbins J. 2001. R-twist gene expression during rat palatogenesis. *Int. J. Dev. Biol.* **45:** 397–404.

Borkowski O.M., Brown N.H., and Bate M. 1995. Anterior-posterior subdivision and the diversification of the mesoderm in Drosophila. *Development* **121:** 4183–4193.

Bushdid P.B., Brantley D.M., Yull F.E., Blaeuer G.L., Hoffman L.H., Niswander L., and Kerr L.D. 1998. Inhibition of NF-κB activity results in disruption of the apical ectodermal ridge and aberrant limb morphogenesis. *Nature* **392:** 615–618.

Castanon I. 2002. "Regulation of the transcription factor Twist during muscle development in *Drosophila*." Ph.D. thesis, p. 103. Universitat de Valencia, Spain.

Castanon I., Von Stetina S., Kass J., and Baylies M.K. 2001. Dimerization partners determine the activity of the Twist bHLH protein during *Drosophila* mesoderm development. *Development* **128:** 3145–3159.

Chang C. and Hemmati-Brivanlou A. 1998. Neural crest induction by Xwnt7B in *Xenopus. Dev. Biol.* **194:** 129–134.

Chen G., Handel K., and Roth S. 2000. The maternal NF-κB/dorsal gradient of *Tribolium castaneum*: Dynamics of early dorsoventral patterning in a short-germ beetle. *Development* **127:** 5145–5156.

Chen Z.F. and Behringer R.R. 1995. *twist* is required in head mesenchyme for cranial neural tube morphogenesis. *Genes Dev.* **9:** 686–699.

Chung K.W., Lee Y.M., Park T.K., Kim, S.J., and Lee C.C. 1996. Cooperative transcriptional activation by two glutamine-rich regions of *twist* product in *Drosophila melanogaster. Mol. Cells* **6:** 197–202.

Corsi A.K., Brodigan T.M., Jorgensen E.M., and Krause M. 2002. Characterization of a dominant negative *C. elegans* Twist mutant protein with implications for human Saethre-Chotzen syndrome. *Development* **129:** 2761–2772.

Corsi A.K., Kostas S.A., Fire A., and Krause M. 2000. *Caenorhabditis elegans* twist plays an essential role in non-striated muscle development. *Development* **127:** 2041–2051.

Cox V.T. 2004. "Signal transport, levels and integration: New lessons from Wingless regulation of *Drosophila* mesoderm development." Ph.D. thesis, Cornell Medical School, New York.

Cripps R.M., Black B.L., Zhao B., Lien C.L., Schulz R.A., and Olson E.N. 1998. The myogenic regulatory gene Mef2 is a direct target for transcriptional activation by Twist during *Drosophila* myogenesis. *Genes Dev.* **12:** 422–434.

Dupont J., Fernandez A.M., Glackin C.A., Helman L., and LeRoith D. 2001. Insulin-like growth factor 1 (IGF-1)-induced twist expression is involved in the anti-apoptotic effects of the IGF-1 receptor. *J. Biol. Chem.* **276:** 26699–26707.

El Ghouzzi V., Legeai-Mallet L., Aresta S., Benoist C., Munnich A., de Gunzburg J., and Bonaventure J. 2000. Saethre-Chotzen mutations cause TWIST protein degradation or impaired nuclear location. *Hum. Mol. Genet.* **9:** 813–819.

El Ghouzzi V., Le Merrer M., Perrin-Schmitt F., Lajeunie E., Benit P., Renier D., Bourgeois P., Bolcato-Bellemin A.L., Munnich A., and Bonaventure J. 1997. Mutations of the TWIST gene in the Saethre-Chotzen syndrome. *Nat. Genet.* **15:** 42–46.

Ephrussi A., Church G.M., Tonegawa S., and Gilbert W. 1985. B lineage specific interactions of an immunoglobulin enhancer with cellular factors in vivo. *Science* **227:** 134–140.

Fang M.A., Glackin C.A., Sadhu A., and McDougall S. 2001.

Transcriptional regulation of alpha 2(I) collagen gene expression by fibroblast growth factor-2 in MC3T3-E1 osteoblast-like cells. *J. Cell. Biochem.* **80:** 550–559.

Furlong E.E., Andersen E.C., Null B., White K.P., and Scott M.P. 2001. Patterns of gene expression during *Drosophila* mesoderm development. *Science* **293:** 1629–1633.

Gachot-Neveu H., Stoetzel C., Quillet R., Dollfus H., and Perrin-Schmitt F. 2002. Natural TWIST protein variants in a panel of eleven non-human primates: Possible implications of TWIST gene-tree for primate species tree. *Dev. Genes Evol.* **212:** 496–503.

Gitelman I. 1997. Twist protein in mouse embryogenesis. *Dev. Biol.* **189:** 205–214.

Gonzalez K. and Baylies M.K. 2002. A screen to examine the transcriptional activity of Twist on multiple enhancers. In *42nd Annual Drosophila Research Conference*, San Diego. Genetics Society of America, Rockville, Maryland. (Abstr., p. 228.)

Gonzalez-Crespo S. and Levine M. 1993. Interactions between dorsal and helix-loop-helix proteins initiate the differentiation of the embryonic mesoderm and neuroectoderm in *Drosophila*. *Genes Dev.* **7:** 1703–1713.

Gripp K.W., Zackai E.H., and Stolle C.A. 2000. Mutations in the human TWIST gene. *Hum. Mutat.* **15:** 479.

Hamamori Y., Sartorelli V., Ogryzko V., Puri P.L., Wu H.Y., Wang J.Y., Nakatani Y., and Kedes L. 1999. Regulation of histone acetyltransferases p300 and PCAF by the bHLH protein twist and adenoviral oncoprotein E1A. *Cell* **96:** 405–413.

Harfe B.D., Vaz Gomes A., Kenyon C., Liu J., Krause M., and Fire A. 1998. Analysis of a *Caenorhabditis elegans* Twist homolog identifies conserved and divergent aspects of mesodermal patterning. *Genes Dev.* **12:** 2623–2635.

Hebrok M., Wertz K., and Fuchtbauer E.M. 1994. M-twist is an inhibitor of muscle differentiation. *Dev. Biol.* **165:** 537–544.

Hopwood N.D., Pluck A., and Gurdon J.B. 1989. A *Xenopus* mRNA related to *Drosophila* twist is expressed in response to induction in the mesoderm and the neural crest. *Cell* **59:** 893–903.

Howard T.D., Paznekas W.A., Green E.D., Chiang L.C., Ma N., Ortiz de Luna R.I., Garcia Delgado C., Gonzalez-Ramos M., Kline A.D., and Jabs E.W. 1997. Mutations in TWIST, a basic helix-loop-helix transcription factor, in Saethre-Chotzen syndrome. *Nat. Genet.* **15:** 36–41.

Howe L.R., Watanabe O., Leonard J., and Brown A.M. 2003. Twist is up-regulated in response to Wnt1 and inhibits mouse mammary cell differentiation. *Cancer Res.* **63:** 1906–1913.

Imai K.S., Satoh N., and Satou Y. 2003. A Twist-like bHLH gene is a downstream factor of an endogenous FGF and determines mesenchymal fate in the ascidian embryos. *Development* **130:** 4461–4472.

Ip Y.T., Park R.E., Kosman D., Bier E., and Levine M. 1992a. The dorsal gradient morphogen regulates stripes of rhomboid expression in the presumptive neuroectoderm of the *Drosophila* embryo. *Genes Dev.* **6:** 1728–1739.

Ip Y.T., Park R.E., Kosman D., Yazdanbakhsh K., and Levine M. 1992b. dorsal-twist interactions establish snail expression in the presumptive mesoderm of the *Drosophila* embryo. *Genes Dev.* **6:** 1518–1530.

Jiang J., Kosman D., Ip Y.T., and Levine M. 1991. The dorsal morphogen gradient regulates the mesoderm determinant twist in early *Drosophila* embryos. *Genes Dev.* **5:** 1881–1891.

Johnson D., Iseki S., Wilkie A.O., and Morriss-Kay G.M. 2000. Expression patterns of Twist and Fgfr1, -2 and -3 in the develop-

ing mouse coronal suture suggest a key role for twist in suture initiation and biogenesis. *Mech. Dev.* **91:** 341–345.

Kadesch T. 1993. Consequences of heteromeric interactions among helix-loop-helix proteins. *Cell Growth Differ.* **4:** 49–55.

Kanegae Y., Tavares A.T., Izpisua Belmonte J.C., and Verma I.M. 1998. Role of Rel/NF-kappaB transcription factors during the outgrowth of the vertebrate limb. *Nature* **392:** 611–614.

Kasof G.M., Goyal L., and White E. 1999. Btf, a novel death-promoting transcriptional repressor that interacts with Bcl-2-related proteins. *Mol. Cell. Biol.* **19:** 4390–4404.

Kophengnavong T., Michnowicz J.E., and Blackwell T.K. 2000. Establishment of distinct MyoD, E2A, and twist DNA binding specificities by different basic region-DNA conformations. *Mol. Cell. Biol.* **20:** 261–272.

Kosman D., Ip Y.T., Levine M., and Arora K. 1991. Establishment of the mesoderm-neuroectoderm boundary in the *Drosophila* embryo. *Science* **254:** 118–122.

Lee H.H. and Frasch M. 2000. Wingless effects mesoderm patterning and ectoderm segmentation events via induction of its downstream target sloppy paired. *Development* **127:** 5497–5508.

Lee Y.M., Park T., Schulz R.A., and Kim Y. 1997. Twist-mediated activation of the NK-4 homeobox gene in the visceral mesoderm of *Drosophila* requires two distinct clusters of E-box regulatory elements. *J. Biol. Chem.* **272:** 17531–17541.

Leptin M. 1991. twist and snail as positive and negative regulators during *Drosophila* mesoderm development. *Genes Dev.* **5:** 1568–1576.

Loebel D.A., O'Rourke M.P., Steiner K.A., Banyer J., and Tam P.P. 2002. Isolation of differentially expressed genes from wild-type and Twist mutant mouse limb buds. *Genesis* **33:** 103–113.

Maestro R., Dei Tos A.P., Hamamori Y., Krasnokutsky S., Sartorelli V., Kedes L., Doglioni C., Beach D.H., and Hannon G.J. 1999. Twist is a potential oncogene that inhibits apoptosis. *Genes Dev.* **13:** 2207–2217.

Meyers G.A., Orlow S.J., Munro I.R., Przylepa K.A., and Jabs E.W. 1995. Fibroblast growth factor receptor 3 (FGFR3) transmembrane mutation in Crouzon syndrome with acanthosis nigricans. *Nat. Genet.* **11:** 462–464.

Morin-Kensicki E.M. and Eisen J.S. 1997. Sclerotome development and peripheral nervous system segmentation in embryonic zebrafish. *Development* **124:** 159–167.

Morisato D. and Anderson K.V. 1995. Signaling pathways that establish the dorsal-ventral pattern of the *Drosophila* embryo. *Annu. Rev. Genet.* **29:** 371–499.

Müller P., Seipel K., Yanze N., Reber-Müller S., Streitwolf-Engel R., Stierwald M., Spring J., and Schmid V. 2003. Evolutionary aspects of developmentally regulated helix-loop-helix transcription factors in striated muscle of jellyfish. *Dev. Biol.* **255:** 216–229.

Murre C., McCaw P.S., and Baltimore D. 1989. A new DNA binding and dimerization motif in immunoglobulin enhancer binding, daughterless, MyoD, and myc proteins. *Cell* **56:** 777–783.

Nederbragt A.J., Lespinet O., van Wageningen S., van Loon A.E., Adoutte A., and Dictus W.J. 2002. A lophotrochozoan twist gene is expressed in the ectomesoderm of the gastropod mollusk Patella vulgata. *Evol. Dev.* **4:** 334–343.

Neilson K.M. and Friesel R.E. 1995. Constitutive activation of fibroblast growth factor receptor-2 by a point mutation associated with Crouzon syndrome. *J. Biol. Chem.* **270:** 26037–26040.

Oldridge M., Wilkie A.O., Slaney S.F., Poole M.D., Pulleyn L.J., Rutland P., Hockley A.D., Wake M.J., Goldin J.H., Winter R.M. et al. 1995. Mutations in the third immunoglobulin domain of the fibroblast growth factor receptor-2 gene in Crouzon syndrome. *Hum. Mol. Genet.* **4:** 1077–1082.

O'Rourke M.P. and Tam P.P. 2002. Twist functions in mouse development. *Int. J. Dev. Biol.* **46:** 401–413.

O'Rourke M.P., Soo K., Behringer R.R., Hui C.C.,and Tam P.P. 2002. Twist plays an essential role in FGF and SHH signal transduction during mouse limb development. *Dev. Biol.* **248:** 143–156.

Oshima A., Tanabe H., Yan T., Lowe G.N., Glackin C.A., and Kudo A. 2002. A novel mechanism for the regulation of osteoblast differentiation: Transcription of periostin, a member of the fasciclin I family, is regulated by the bHLH transcription factor, twist. *J. Cell. Biochem.* **86:** 792–804.

Pham A.D., Muller S., and Sauer F. 1999. Mesoderm-determining transcription in *Drosophila* is alleviated by mutations in TAF(II)60 and TAF(II)110. *Mech. Dev.* **84:** 3–16.

Rice D.P., Aberg T., Chan Y., Tang Z., Kettunen P.J., Pakarinen L., Maxson R.E., and Thesleff I. 2000. Integration of FGF and TWIST in calvarial bone and suture development. *Development* **127:** 1845–1855.

Riechmann V., Irion U., Wilson R., Grosskortenhaus R., and Leptin M. 1997. Control of cell fates and segmentation in the *Drosophila* mesoderm. *Development* **124:** 2915–2922.

Roy S. and VijayRaghavan K. 1999. Muscle pattern diversification in *Drosophila*: The story of imaginal myogenesis. *BioEssays* **21:** 486–498.

Shirokawa J.M. and Courey A.J. 1997. A direct contact between the dorsal rel homology domain and Twist may mediate transcriptional synergy. *Mol. Cell. Biol.* **17:** 3345–3355.

Shishido E., Higashijima S., Emori Y., and Saigo K. 1993. Two FGF-receptor homologues of *Drosophila*: One is expressed in mesodermal primordium in early embryos. *Development* **117:** 751–761.

Simpson P. 1983. Maternal-zygotic gene interactions during formation of the *Drosophila* doroventral pattern in *Drosophila* embryos. *Genetics* **105:** 615–632.

Sommer R.J. and Tautz D. 1994. Expression patterns of twist and snail in Tribolium (Coleoptera) suggest a homologous formation of mesoderm in long and short germ band insects. *Dev. Genet.* **15:** 32–37.

Soo K., O'Rourke M.P., Khoo P.L., Steiner K.A., Wong N., Behringer R.R., and Tam P.P. 2002. Twist function is required for the morphogenesis of the cephalic neural tube and the differentiation of the cranial neural crest cells in the mouse embryo. *Dev. Biol.* **247:** 251–270.

Sosic D., Richardson J.A., Yu K., Ornitz D.M., and Olson E.N. 2003. Twist regulates cytokine gene expression through a negative feedback loop that represses NF-kappaB activity. *Cell* **112:** 169–180.

Soto J.G., Nelson B.H., and Weisblat D.A. 1997. A leech homolog of twist: evidence for its inheritance as a maternal mRNA. *Gene* **199:** 31–37.

Spicer D.B., Rhee J., Cheung W.L., and Lassar A.B. 1996. Inhibition of myogenic bHLH and MEF2 transcription factors by the bHLH protein Twist. *Science* **272:** 1476–1480.

Spring J., Yanze N., Middel A.M., Stierwald M., Groger H., and Schmid V. 2000. The mesoderm specification factor twist in the life cycle of jellyfish. *Dev. Biol.* **228:** 363–375.

Tapanes-Castillo A. and Baylies M.K. 2004. Notch signaling patterns *Drosophila* mesodermal segments by regulating the bHLH transcription factor Twist. *Development* (in press).

Tavares A.T., Izpisuja-Belmonte J.C., and Rodriguez-Leon J. 2001. Developmental expression of chick twist and its regulation during limb patterning. *Int. J. Dev. Biol.* **45:** 707–713.

Thisse B., el Messal M., and Perrin-Schmitt F. 1987. The twist gene: isolation of a *Drosophila* zygotic gene necessary for the establishment of dorsoventral pattern. *Nucleic Acids Res.* **15:** 3439–3453.

Thisse B., Stoetzel C., Gorostiza-Thisse C., and Perrin-Schmitt F. 1988. Sequence of the twist gene and nuclear localization of its protein in endomesodermal cells of early *Drosophila* embryos. *EMBO J.* **7:** 2175–2183.

Thisse C., Perrin-Schmitt F., Stoetzel C., and Thisse B. 1991. Sequence-specific transactivation of the *Drosophila* twist gene by the dorsal gene product. *Cell* **65:** 1191–1201.

Venkatesh T.V., Park M., Ocorr K., Nemaceck J., Golden K., Wemple M., and Bodmer R. 2000. Cardiac enhancer activity of the homeobox gene tinman depends on CREB consensus binding sites in *Drosophila*. *Genesis* **26:** 55–66.

Villavicencio E.H., Yoon J.W., Frank D.J., Fuchtbauer E.M., Walterhouse D.O., and Iannaccone P.M. 2002. Cooperative E-box regulation of human GLI1 by TWIST and USF. *Genesis* **32:** 247–258.

Wang S.M., Coljee V.W., Pignolo R.J., Rotenberg M.O., Cristofalo V.J., and Sierra F. 1997. Cloning of the human twist gene: Its expression is retained in adult mesodermally-derived tissues. *Gene* **187:** 83–92.

Wolf C., Thisse C., Stoetzel C., Thisse B., Gerlinger P., and Perrin-Schmitt F. 1991. The M-twist gene of Mus is expressed in subsets of mesodermal cells and is closely related to the *Xenopus* X-twi and the *Drosophila* twist genes. *Dev. Biol.* **143:** 363–373.

Yasui K., Zhang S.C., Uemura M., Aizawa S., and Ueki T. 1998. Expression of a twist-related gene, Bbtwist, during the development of a lancelet species and its relation to cephalochordate anterior structures. *Dev. Biol.* **195:** 49–59.

Yin Z., Xu X.L., and Frasch M. 1997. Regulation of the twist target gene tinman by modular cis-regulatory elements during early mesoderm development. *Development* **124:** 4971–4982.

Zuniga A., Quillet R., Perrin-Schmitt F., and Zeller R. 2002. Mouse Twist is required for fibroblast growth factor-mediated epithelial-mesenchymal signalling and cell survival during limb morphogenesis. *Mech. Dev.* **114:** 51–59.

THE *SNAIL* GENE FAMILY DURING GASTRULATION

A.V. Morales and M.A. Nieto

Instituto Cajal, CSIC. Doctor Arce, 37, 28002- Madrid, Spain

INTRODUCTION

The Snail superfamily plays a central role in different morphogenetic processes during embryonic development and, in particular, during gastrulation (Nieto 2002). The first family member described, *snail*, was identified in *Drosophila* as a mesoderm determinant essential for embryonic development (Grau et al. 1984; Nüsslein-Volhard et al. 1984). Embryos functionally deficient for *snail* show defects in the invagination of the presumptive mesoderm and retraction of the germ band. Since then, a plethora of homologs have been described in different species, ranging from jellyfish to humans (Nieto 2002; Spring et al. 2002), many of which share a role in mesoderm development. Snail genes also participate in other processes that involve long-range movements, such as the development of the neural crest and the acquisition of invasive and migratory behaviors during tumor progression. In addition, different family members have been assigned roles in neural differentiation, the development of appendages, cell fate and cell survival, and the establishment of left/right asymmetry.

This chapter summarizes the roles of the different family members during mesoderm formation, a complex process that involves the coordination of cell fate determination, cell shape changes, and control of the cell cycle. All these cellular processes occur concomitantly with large-scale morphogenetic movements, the hallmark of gastrulation.

THE SNAIL SUPERFAMILY OF TRANSCRIPTIONAL REPRESSORS

The Snail gene superfamily encompasses the closely related Snail and Scratch families (Manzanares et al. 2001; Nieto 2002). They encode transcription factors of the zinc-finger type, where the number of fingers varies from four to six. The fingers are of the C2H2 type and are located near the carboxyl terminus of the protein (Fig. 1).

The zinc-finger domain is the most conserved structural feature of the Snail superfamily. Within this domain, the third and fourth fingers contain a consensus sequence that is conserved in all Snail superfamily members described to date (>40). The sequences of the second and fifth fingers serve unambiguously to assign proteins to the Snail or the Scratch families, while the first finger and the Scratch and Slug domains in the amino-terminal region are diagnostic clues to identify Scratch and vertebrate Slug subfamily members, respectively (Fig. 1) (Manzanares et al. 2001).

Like zinc fingers in other proteins, the Snail zinc fingers are involved in sequence-specific DNA binding. They recognize a six-base sequence, CAGGTG, a type of E-box that is also the binding site for basic helix-loop-helix (bHLH) transcription factors (Mauhin et al. 1993; Fuse et al. 1994; Inukai et al. 1999; Batlle et al. 2000; Cano et al. 2000). Interestingly, in vitro studies have shown that Snail proteins compete with bHLH proteins for the same binding sequences, suggesting that in vivo, competition may occur for these sites (Fuse et al. 1994; Nakayama et al. 1998; Kataoka et al. 2000; Pérez-Moreno et al. 2001).

Upon binding to the specific E-box consensus site, Snail proteins function as transcriptional repressors (for review, see Hemavathy et al. 2000a; Nieto 2002). Nevertheless, under certain circumstances, human Slug and *Drosophila* Snail seem to contain a transcriptional activation domain (Hemavathy et al. 2000b), and thus, the possibility that they act as transcriptional activators cannot be excluded.

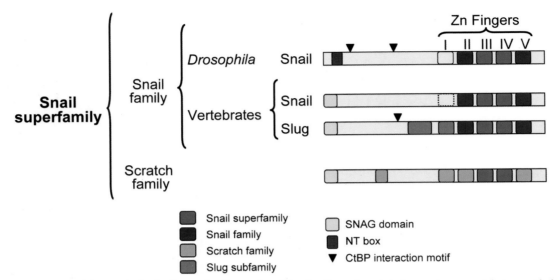

Figure 1. Conserved features in the Snail superfamily members. The zinc-finger domain is the most conserved feature, with diagnostic sequences in the third and fourth fingers that allow the identification of the whole superfamily (*purple*). The sequences of the second and fifth fingers are specific for either the Snail (*blue*) or the Scratch families (*green*). Sequences in the first finger and in a central domain of the protein define the Scratch family (*green*) or the Slug subfamily (*red*). The SNAG domain (*yellow*) is absent from *Drosophila* Snail family members. CtBP-binding sites are present in the fly Snail proteins and in the vertebrate Slug subfamily. The absence of the first finger in some vertebrate Snail subfamily members is indicated by a dotted line.

Transcriptional repression is mediated by the amino-terminal region in both vertebrates and invertebrates (Gray and Levine 1996; LaBonne and Bronner-Fraser 2000; Mayor et al. 2000). Within the amino-terminal domain, two different sequences seem to be responsible for repression, depending on the species. Snail proteins in jellyfish (Spring et al. 2002), echinoderms (Manzanares et al. 2001), cephalochordates (Langeland et al. 1998), limpet (Lespinet et al. 2002), and *Drosophila* Scratch and all vertebrate Snail and Scratch proteins (Manzanares et al. 2001) contain and likely utilize the SNAG (*Snail/Gfi-1*) domain (Fig. 1). This domain was initially described in the zinc-finger protein Gfi-1 as necessary and sufficient for repression (Grimes et al. 1996).

In contrast, *Drosophila* Snail family members (Snail, Escargot, and Worniu) lack the SNAG domain and mediate repression by interacting with the corepressor CtBP (*C-terminal Binding Protein*). This corepressor binds to a conserved sequence P-DLS-K/R that is present in two copies in the fly Snail genes (Fig. 1) (Nibu et al. 1998; Ashraf et al. 1999). Interestingly, ascidian Snail proteins also lack the SNAG motif and contain CtBP consensus sites (Corbo et al. 1997; Wada and Saiga 1999). Therefore, transcriptional repression via Snail proteins seems to be an activity that has been conserved throughout evolution, either through the use of the SNAG domain or by corepression with CtBP. Indeed, vertebrate Slug proteins may use both the SNAG domain and CtBP binding to induce repression (Fig. 1).

In vitro studies indicate that the function of Snail as a transcription factor is regulated by its subcellular distribution. A nuclear export sequence (NES) located upstream of

the zinc fingers controls its export to the cytoplasm, which occurs upon phosphorylation of a Ser-rich sequence adjacent to the NES (Domínguez et al. 2003).

EVOLUTIONARY HISTORY OF THE SNAIL GENE SUPERFAMILY

Phylogenetic analysis of the Snail gene superfamily has shed some light on the evolution of the family from ancestral genes (Manzanares et al. 2001). Members of the family have been identified from jellyfish to humans, and an early duplication of a *Snail* gene in the metazoan ancestor appears to have given rise to two highly related genes, *Snail* and *Scratch*. Since the *Scratch* genes do not seem to play roles in mesoderm formation (see Nieto 2002), we concentrate on the Snail family for the rest of the chapter.

After the early duplication, independent duplication events in different groups gave rise to a different number of Snail genes. In *Drosophila*, intrachromosomal tandem duplications may have given rise to three members—*snail, escargot,* and *worniu*. Only one Snail gene has been found in jellyfish (Spring et al. 2002), sea urchin and non-vertebrate chordates such as ascidians (Corbo et al. 1997; Wada and Saiga 1999) and *Amphioxus* (Langeland et al. 1998). However, independent duplications occurred in lophotrochozoans, since two Snail genes have been found in both the limpet and the leech (Fig. 2) (Goldstein et al. 2001; Lespinet et al. 2002).

In the vertebrate lineage, either the duplication of the whole genome (Holland et al. 1994), or a large-scale gene duplication event (Wolfe 2001), led to the generation of at

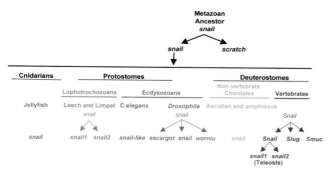

Figure 2. Evolutionary history of the Snail gene family. The early duplication of a unique *Snail* gene present in the metazoan ancestor likely gave rise to two genes: *Snail* and *Scratch*. The subsequent duplications undergone by *Snail* are shown in all groups where family members have been described to date. Names in bold indicate present genes. For a picture of the proposed duplications within the Scratch family, see Nieto (2002). (Adapted, with permission, from Nieto 2002 [©Nature Publishing Group].)

least two Snail family members: *Snail* and *Slug* (Fig. 2). A new vertebrate *Snail* homolog—*Smuc*—has been characterized in the mouse where it is not expressed during early mesodermal development (Kataoka et al. 2000). Its orthologs have been recently described in other vertebrates and renamed as *Snail 3* (Manzanares et al. 2004). The presence of two very closely related Snail genes (*snail1* and *snail2*) in zebrafish and pufferfish (Hammerschmidt and Nüsslein-Volhard 1993; Thisse et al. 1993, 1995; Smith et al. 2000) is in agreement with the proposed tetraploidization in the teleost lineage (Postlethwait et al. 1998).

ESTABLISHMENT OF THE MESODERMAL FATE

In *Drosophila*, gastrulation is first visible when a longitudinal fold appears on the ventral side of the blastoderm. The cells of the ventral furrow then invaginate and give rise to the mesoderm and part of the anterior gut. The morphogenetic movements that occur at gastrulation are discussed in detail in Chapter 7. A prerequisite for gastrulation is the establishment of the territory that will invaginate from the ventral side of the blastoderm and that represents the mesoderm primordium. The maternal gradients established by Toll and Dorsal define ventral fates. Dorsal is a transcription factor that at high concentrations binds directly to the promoters of the zygotic genes *twist* and *snail*, activating their transcription in the ventral cells (Fig. 3) (Jiang et al. 1991; Ip et al. 1992b). Twist acts as a transcriptional activator for mesodermal genes (*N-cadherin*, *PS2 α-integrin*, and *myosin*; Leptin 1991) and collaborates with Dorsal to establish high levels of *snail* expression and a sharp boundary in the presumptive mesoderm (Fig. 4) (Ip et al. 1992a). Snail functions as a repressor of nonmesodermal genes binding to the promoters of at least two

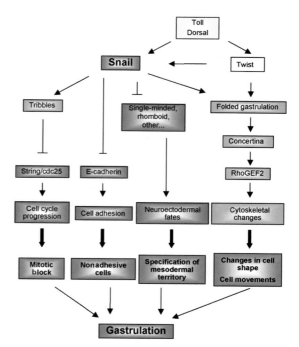

Figure 3. Genetic pathways involved in *Drosophila* gastrulation. The arrows indicate the flow of the pathway, not direct transcriptional regulation. Only *rhomboid* and *single-minded* have been shown to be directly repressed by Snail, which also directly represses *E-cadherin* transcription in vertebrates and very likely in the fly embryo. To better follow the sequence, genes and cellular processes repressed or inactive are shown in red and those active are shown in green.

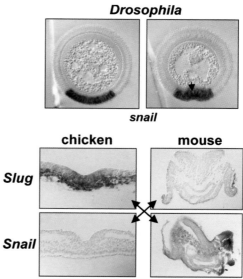

Figure 4. Expression of Snail genes in *Drosophila*, chicken, and mouse embryos during gastrulation. snail expression is observed in the mesodermal precursors prior to and during invagination in *Drosophila* embryos. In vertebrates, Snail genes are expressed in the primitive streak and the early mesodermal cells. However, the family member present in these tissues depends on the species: *Snail* in the mouse and *Slug* in the chick. (Adapted, with permission, from Nieto 2002 [©Nature Publishing Group].) Pictures of *Drosophila* embryos were kindly provided by Maria Leptin.

genes: *single-minded* (Kasai et al. 1992) and *rhomboid* (Ip et al. 1992a). Other relevant genes in this process that are de-repressed in *snail* mutants include *lethal of scute, crumbs, short gastrulation, Delta,* and genes of the *Enhancer of Split* complex (for review, see Ip and Gridley 2002).

In *snail* mutant embryos, the expression domains of genes normally confined to the ectoderm invade the mesoderm anlagen and, as in *twist* mutant embryos, the ventral invagination is largely abolished (Fig. 5). In this case, no mesodermal differentiation occurs and fully developed mutant embryos lack all mesoderm derivatives (Grau et al. 1984).

As in *Drosophila*, Snail is expressed in the mesodermal precursors of nonvertebrate chordates—ascidian and *Amphioxus* (Corbo et al. 1997; Langeland et al. 1998; Wada and Saiga 1999). In anamniote vertebrates, *Slug* expression is observed in the dorsal marginal zone in *Xenopus* (Mayor et al. 2000) and *snail-1* is expressed in the involuting cells of the germ ring in zebrafish (Hammerschmidt and Nüsslein-Volhard 1993; Thisse et al. 1993). Thus, they may also be involved in the determination of mesodermal fate in these species.

In amniotes, ectodermal cells are recruited to the primitive streak, a transient structure that forms along the midline of the embryo (see Chapters 14–17), where they undergo epithelial-to-mesenchymal transition (EMT, see below) and then ingress between the primitive ectoderm and endoderm to give rise to both mesoderm and definitive endoderm. *Snail* genes are expressed in the primitive streak, although the family member—either *Snail* or *Slug*—depends on the corresponding species (Fig. 4) (Sefton et al. 1998).

Despite their expression pattern, there is no direct evidence for the role of Snail family members in mesoderm specification in vertebrates. Mouse embryos homozygous for a null mutation in the *Snail* gene still form the three embryonic tissue layers. However, their mesodermal cells express low levels of *Brachyury,* and the expression of *Otx-2* is extended (Carver et al. 2001). The main defect observed in these mutants is that these "mesodermal" cells exhibit an epithelial character, since they fail to undergo EMT (Fig. 5).

In support of *Snail* having an ancestral function in mesoderm specification is its expression in the entocodon, a tissue that some authors regard as the equivalent of mesoderm in diploblastic animals and that gives rise to the smooth muscle tissue of the jellyfish at the medusa stage (Spring et al. 2002).

INTRACELLULAR MORPHOGENETIC CHANGES PRIOR TO MIGRATION

In the *Drosophila* blastoderm, once the mesoderm territory has been determined and prior to invagination, the ventral cells undergo coordinated changes in shape that move the mesoderm anlagen toward the interior of the embryo (discussed in Chapter 7). In *snail* mutants, the misspecified ventral epithelium becomes very thin, suggesting that a shortening of the cells occurs. However, no apical constrictions can be seen at all and, as a consequence, the furrow does not form (Fig. 5) (Leptin 1991). Furthermore, in mutant alleles with reduced *snail* function, the ventral cells express both neuroectodermal and mesodermal markers and, in this case, are able to invaginate ventrally (Hemavathy et al. 1997). This suggests that Snail, together with Twist, not only controls the specification of the mesoderm, but also regulates a separate set of molecules involved in cell shape changes in the ventral furrow (Fig. 3). One of these molecules is Folded Gastrulation (Fog), a putative secreted ligand that controls normal concerted apical constrictions during invagination and that is aberrantly expressed in *snail* mutant embryos (Morize et al. 1998). It has been postulated that Fog can exert its function by binding to a receptor associated with the Gα-like protein concertina (Costa et al. 1994) and that it interacts with the exchange factor RhoGEF2. This reveals the importance of small GTPases in the changes in cell shape that occur during invagination (Barrett et al. 1997).

In a similar manner to mesoderm invagination in *Drosophila*, the amphibian and fish prospective mesoderm, together with the endoderm, first involutes at the blastopore lip, prior to converging medially and extending longitudinally. In both these morphogenetic movements, the cells move as a single multilayered sheet (for discussion, see Locascio and Nieto 2001; Chapter 19). Convergent extension

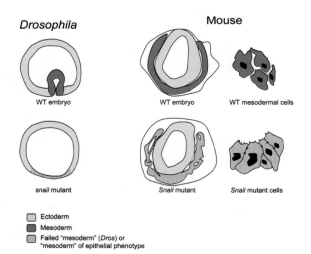

Figure 5. Schematic representation of the *Snail* mutant phenotype in *Drosophila* and mouse embryos. In *Drosophila snail* mutants, the ventral invagination is abolished and the mesoderm does not form. In the mouse, cells expressing low levels of mesodermal markers are formed. However, as in *Drosophila*, these cells fail to down-regulate *E-cadherin* expression and show epithelial characteristics, indicating that they are unable to undergo the EMT.

movements are triggered by signaling cascades involving the mesoderm-inducing factors Activin and FGF, which induce the expression of T-box transcription factors. One of these, No tail (Brachyury), has been shown to enhance *snail1* expression in zebrafish (Hammerschmidt and Nüsslein-Volhard 1993; Thisse et al. 1993).

In amniotes, convergent extension mainly influences the development of axial mesoderm and the accumulation of the epiblast cells in the primitive streak prior to the onset of gastrulation. Once formed, the cells of the primitive streak acquire a bottle-like shape and undergo EMT (Bellairs 1987). In *Snail* mutant mouse embryos, the primitive streak and mesoderm-like layer that is formed contain lacunae and the cells retain epithelial characteristics (Fig. 5) (Carver et al. 2001), changes compatible with a failure in undergoing the EMT.

Many of the cellular processes involved in gastrulation are also common to those that occur during neural crest delamination (Bellairs 1987). In chicken embryos, Slug gain of function in the neural tube causes an up-regulation of RhoB, a small GTPase involved in actin rearrangements (Del Barrio and Nieto 2002). This up-regulation is accompanied by an increase in neural-crest migration in the cranial region. In light of these results, it seems that a Rho-mediated signaling cascade is crucial for the cellular changes that occur during gastrulation in invertebrates and vertebrates (Figs. 3 and 6).

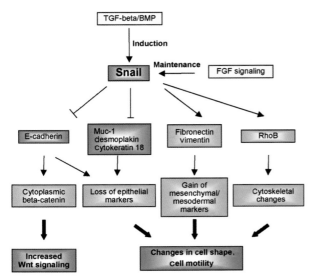

Figure 6. Genetic pathways involved in vertebrate gastrulation. Members of the TGF-β superfamily have been proposed to activate *Snail* in different vertebrate embryos, and data from *FGFR1* mutant mice indicate that FGF signaling is necessary for its maintenance (Ciruna and Rossant 2001). The repression of *E-cadherin* expression maintains cytoplasmic β-catenin available to transduce the Wnt-signaling pathway. As in Fig. 3, the arrows indicate the flow of the pathway, not direct transcriptional regulation. Again, the genes and cellular processes repressed or inactive are shown in red and those active are shown in green.

COORDINATION OF CELL DIVISION AND MIGRATION

Morphogenetic movements and cell division both involve reorganization of the cytoskeleton. This reorganization can be a cause of incompatibility between the two processes, especially during rapid developmental events such as gastrulation. As such, this might explain the mitotic arrest in the ventral cells of the gastrulating fly embryo, which only enter mitosis after mesoderm invagination is complete (Foe 1989). The cells of the ventral furrow start expressing *string*, a *cdc25* homolog essential for entry into mitosis, at the same time as the rest of the cells in the embryo. However, String function is inhibited by Tribbles and Frühstart (Grosshans and Wieschaus 2000; Mata et al. 2000; Seher and Leptin 2000) and this inhibition is dependent on Snail function (Grosshans and Wieschaus 2000). Thus, Snail seems to act as a mitotic inhibitor during gastrulation of the fly embryo.

Interestingly, mammalian epithelial cells transfected with *Snail* undergo dramatic changes in cell shape compatible with EMT and reduce their rate of proliferation (Vega et al. 2004). Similarly, the invasive areas of tumors that express *Snail* (Cano et al. 2000; Blanco et al. 2002) show a lower rate of proliferation when compared to noninvasive areas of the same tumor (Jung et al. 2001).

TRIGGERING OF EMT

Once inside the *Drosophila* embryo, the mesodermal cells must disperse into single cells that divide, attach, and migrate along the ectoderm to form a single cell layer. The mechanism by which cells become migratory is not yet fully understood, but it must involve a switch of adhesion molecules (Oda et al. 1998). In *snail* mutants the ventral cells of the embryo fail to down-regulate *E-cadherin* expression (Oda et al. 1998). Snail is also required in *Drosophila* to increase *N-cadherin* expression, indicating that it is involved in the switch from *E-cadherin* (implicated in the maintenance of stable junctions) to *N-cadherin* expression (a weak intercellular adhesion system), which is important for the movement of cells after invagination (Oda et al. 1998).

As mentioned before, in amniotes, cells that undergo EMT and delaminate give rise to the mesoderm, cells migrating as individual cells through the extracellular matrix. The EMT process implies the loss of epithelial markers and the gain or redistribution of mesenchymal markers, as well as the acquisition of a fibroblastic-like shape concomitant with that of a migratory and invasive behavior (Hay 1995; Thiery 2002).

The first indication that the Snail family was involved in EMT during gastrulation came from studies in the chick embryo. Incubation of early chicken embryos with anti-sense oligonucleotides to inhibit *Slug* function led to the formation of an abnormally cell-dense, wrinkled tissue at the primitive streak. This tissue was generated due to the incapacity of the cells to convert to mesenchyme and migrate as mesodermal precursors (Fig. 7) (Nieto et al. 1994). Since *Snail* rather than *Slug* is expressed in the primitive streak of the mouse embryo (Fig. 4), it was proposed that *Snail* might be responsible for triggering EMT in mammals (Sefton et al. 1998). Indeed, mouse *Snail* is able to induce EMT when expressed in mammalian epithelial cells (Batlle et al. 2000; Cano et al. 2000) and *Snail* mutant mice die at gastrulation due to a failure of the mesodermal cells to undergo EMT (Fig. 5) (Carver et al. 2001). As in *Drosophila*, loss of E-cadherin expression is essential for the mesodermal cells to ingress at gastrulation in mouse embryos (Burdsal et al. 1993). Indeed, as already mentioned, in *Snail* knockout mouse embryos the mesodermal layer forms, but cells maintain their epithelial character, retaining an apical–basal polarity, microvilli, and adherens junctions (Fig. 5) (Carver et al. 2001). Although they establish a low level of expression of some mesodermal markers, as in *Drosophila snail* mutants, they fail to down-regulate *E-cadherin* expression (Oda et al. 1998; Carver et al. 2001), further evidence that Snail acts as a repressor of *E-cadherin* transcription (Batlle et al. 2000; Cano et al. 2000). Thus, the repression of *E-cadherin* expression by Snail appears to have been conserved through evolution from insects to mammals.

However, the levels of E-cadherin in the mesoderm layer of *Snail* mutant mouse embryos are lower than those of the embryonic ectoderm in the same embryos, indicating that other *E-cadherin* repressors might be present. Candidates include HLH-type transcription factors such as SIP1 and E12/E47 (Comijn et al. 2001; Pérez-Moreno et al. 2001), which, although weaker repressors, are both expressed in the mesoderm.

Although convergent extension is a well-studied movement during the gastrulation process in anamniote vertebrates, examples of individual cell migration have also been reported. In *Xenopus* embryos, the first mesodermal cells to ingress during gastrulation transiently express *Snail*, they migrate as individual cells, and they give rise to the most anterior mesoderm (Fig. 8) (R. Mayor, pers. comm.). In a similar way, in zebrafish embryos, *snail2* mRNA appears in a reticular staining formed by single cells in the anterior cephalic paraxial mesoderm (Fig. 8) (Thisse et al. 1995). Interestingly, the formation of the mesendoderm layer in zebrafish has recently been shown

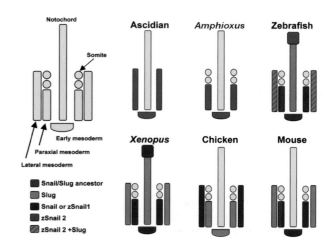

Figure 8. Schematic representation of the mesodermal expression of Snail family members in chordates. A single *Snail* gene has been identified in ascidians and *Amphioxus*, where it is expressed in the mesodermal precursors and in the paraxial mesoderm. Among vertebrates, Snail family members are also expressed in different mesodermal populations. Interestingly, the expression of *Snail* and *Slug* in these populations is largely complementary in each species and compatible with a function in the subdivision of mesodermal territories. Note that expression in the chicken is interchanged with respect to that of other vertebrates. This interchange explains why the loss of *Slug* function in the chick (Fig. 6) produces a similar mesodermal phenotype to that of mouse *Snail* or *FGFR* mutants (Figs. 5 and 6) and why the mouse *Slug* mutant does not show mesodermal defects. For a recent analysis of *Snail* and *Slug* expression in vertebrates, including reptiles, see Locascio et al. (2002). The expression of *Snail* in *Xenopus* and *Snail2* at the tip of the axial mesoderm corresponds to the first cells that ingress and will form the head mesenchyme (see text).

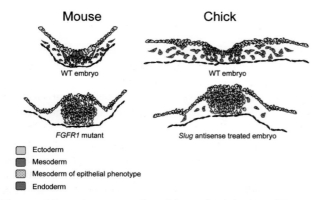

Figure 7. Schematic representation of the result of *Slug* loss of function in the chick and that of FGF signaling in the mouse. In *Slug*-antisense-treated chicken embryos, the early mesodermal cells accumulate in the region of the primitive streak and a low number of cells are able to migrate to the locations of paraxial and lateral mesoderm. Similarly, mouse embryos mutant for *FGFR1* show a similar phenotype, also reminiscent of that of the *Snail* mouse mutant. However, in the *FGFR1* null mice, *Snail* is transiently expressed in the primitive streak, allowing the first mesoderm that forms (extraembryonic) to develop. Therefore, FGF signaling seems to be essential for the maintenance of *Snail* expression and the corresponding formation of the embryonic mesoderm.

to involve a combined process of ingression of individual cells and the involution-like movement observed in *Xenopus* gastrulation (Carmany-Rampey and Schier 2001). Thus, these examples of individual cell migration are accompanied by *Snail* expression and the triggering of an epithelial mesenchymal transition. Even in a gastropod mollusk, *Patella vulgata*, *snail-2* is expressed in the involuting cells of the mantle tip, compatible with its controlling EMT and cell motility at this site. Altogether, these results have led to the suggestion that EMT and the control of cell motility are ancient functions associated with the Snail gene family (Lespinet et al. 2002; Manzanares et al. 2001). In keeping with this idea, Snail genes have been implicated in the different EMTs that occur at different sites and times during vertebrate embryonic development. This includes processes such as formation of the parietal endoderm (Velmaat et al. 2000), delamination of the neural crest (Nieto et al. 1994; Carl et al. 1999; LaBonne and Bronner-Fraser 2000; Mayor et al. 2000; Del Barrio and Nieto 2002), formation of the heart cushions (Romano and Runyan 2000), decondensation of the somites (Sefton et al. 1998), and closure of the palate (Martínez-Alvarez et al. 2004). There is also good evidence that Snail is involved in triggering EMTs in pathological situations, such as when single carcinoma cells dissociate from the site of primary tumors (Blanco et al. 2002), and that associated with the transformation of mesothelial cells after continuous peritoneal dialysis (Yáñez-Mo et al. 2003).

SIGNALING PATHWAYS IN VERTEBRATES

It is clear that different members of the TGF-β superfamily are involved in mesoderm induction and are associated with the activation of different Snail family members in vertebrates (for review, see Beddington and Robertson 1999; Nieto 2002). Additionally, other signaling pathways such as those involving Wnt and FGF have also been implicated in both mesoderm induction and *Snail* expression (Smith 1993; Kimelman and Griffin 2000; Nieto 2002). With respect to the involvement of Wnt signaling, the analysis of the promoter of *Xenopus Slug* genes has revealed a functional binding site for the transcription factor Lef1, which regulates gene expression following activation of the Wnt pathway (Vallin et al. 2001). However, this may be dependent on the cellular context, as Snail genes are not up-regulated in epithelial cells or colon carcinoma cells overexpressing *LEF* (Tan et al. 2001; Kim et al. 2002).

Interestingly, the phenotype of *FGF receptor 1* (*FGFR1*) mutant mice is reminiscent of that of *Snail* mutants (Ciruna and Rossant 2001). In an elegant study, Ciruna and Rossant

showed that FGF signaling is not involved in *Snail* induction, but rather in the maintenance of its expression. Indeed, *Snail* is transiently expressed in these mutants, permitting the formation of the extraembryonic mesoderm (the first to develop). However, subsequently, *Snail* expression disappears and the embryonic mesoderm fails to develop (see Fig. 7). In these cases, *E-cadherin* is not down-regulated and the cells remain trapped close to the primitive streak while expressing low levels of mesodermal markers, as occurs in both *Drosophila* and mouse *Snail* mutants. In relation to this, it is interesting to note that the *Drosophila FGF-R2* mutants—*heartless*—show a similar phenotype: The invaginated mesodermal cells remain aggregated along the ventral midline and fail to migrate (Beiman et al. 1996; Gisselbrecht et al. 1996).

There is an interesting connection between FGF and Wnt signaling that was also unveiled in the *FGFR1* mouse mutants. In these embryos, Wnt signaling is attenuated in the primitive streak, but this can be reverted by disrupting the abnormal high levels of E-cadherin. This result indicates that the *E-cadherin* expression attenuates Wnt signaling by sequestering cytoplasmic β-catenin, making it unavailable for binding to the Tcf/Lef family transcription factors and subsequent activation of Wnt target genes (Ciruna and Rossant 2001). The model that emerges is that the members of the TGF-β/BMP families induce Snail genes, and FGF maintains their levels allowing for Wnt signaling to be active (Fig. 6).

DISTRIBUTION OF MESODERMAL TERRITORIES

We have already mentioned that in *Drosophila snail* is expressed in the prospective mesoderm and acts as a repressor of the neuroectodermal genes *rhomboid* (*rho*) and *single-minded* (*sim*) (for review, see Ip and Gridley 2002). Thus, Snail specifies the mesodermal anlagen by preventing alternative fates from being specified. In addition, Snail controls the territory in which Notch signaling can act in the fly at early gastrulation stages and limits the expression of *single-minded* (*sim*) to the mesectoderm (Cowden and Levine 2002; Morel et al. 2003). Interestingly, Snail seems to have a conserved role in setting up boundaries between tissues even at later stages. It establishes the muscle/notochord boundary in ascidians, by repressing *Brachyury* (*T*) expression in muscles but not in the notochord (Fijiwara et al. 1998).

In vertebrates, there are indications to suggest that Snail genes may play a role in defining tissue boundaries. As discussed above, the phenotype of *Drosophila* and mouse mutants for *Snail* and FGF signaling include the accumulation of mesodermal cells in the midline. The result is an enlargement of the axial mesoderm territory at the expense of the paraxial mesoderm anlagen, something that it is also observed in zebrafish *spadetail* (*VegT*) mutants (Thisse et al. 1993, 1995). However, since there is a failure in migration from the streak, this putative role in defining territories is somehow mixed here with its role in inducing the movement of mesodermal cells after invagination or ingression. Thus, it is likely that cells acquire characteristics of the different mesodermal territories (paraxial and lateral) not only according to the time they delaminate from the primitive streak, but also upon receiving signals that do not reach them while trapped close to the primitive streak. Considering that VegT (Spadetail) lies upstream of Snail, it is tempting to speculate that the *spadetail* phenotype mentioned above could be at least partly explained by defective Snail function. Indeed, *snail1* transcripts are detected in the involuting cells of the germ ring in zebrafish and, as gastrulation begins, become restricted to the paraxial mesoderm. Undoubtedly, snail loss-of-function experiments in zebrafish will help to clarify this issue.

In *Xenopus* embryos, *Slug* is expressed in the dorsal marginal zone above the blastopore lip, a territory that includes the prospective dorsal mesoderm and endoderm. In this region, Slug controls dorsal mesoderm genes such as *Chordin* and *Cerberus* by inhibiting *BMP* transcription, a signal that normally induces ventral (lateral) mesodermal fates (Mayor et al. 2000).

In addition to zebrafish and *Xenopus*, the analysis of expression patterns of Snail genes in other vertebrates is also compatible with their function in the subdivision of mesoderm territories. Following the duplication at the base of the vertebrate lineage, the two genes—*Snail* and *Slug*—seem to have distributed their expression so that they show complementary patterns in the different mesodermal territories along the mediolateral axis (dorsoventral in *Xenopus* and fish). This is summarized in Figure 8, where the expression of Snail family members in chordates is shown in the early mesoderm and in its subsequent derivatives: axial, paraxial, and lateral mesodermal tissues.

Recent data indicate that the expression of *Snail* in the early mesoderm is also conserved in turtle and lizard embryos, as reflected by its expression in the tail bud (Locascio et al. 2002), considered to be a continuation of the blastopore or the primitive streak (Gont et al. 1993; see Chapter 14). The lack of Snail expression at the tip of the tail in ascidian embryos is consistent with this, as the ascidian tail is formed by convergent-extension movements of the notochord (Nishida 1987), and Snail genes do not seem to play a primary role in convergent extension (Locascio and Nieto 2001).

CONCLUDING REMARKS AND PERSPECTIVES

The analysis of Snail genes in invertebrates and vertebrates has unveiled their role in mesoderm specification, migration, and subsequent distribution of mesodermal territories. However, an ancestral function in the specification of the mesoderm is not supported after the recent data obtained in lophotrochozoans. It seems that its prominent role in mesoderm development is related to the establishment of tissue boundaries and the promotion of cell shape changes and cell movement.

Despite the increase in our knowledge in the inductive signals and targets and the similarities observed in different systems, further work is needed to understand the interaction between the different signaling pathways that have been implicated in these processes in both *Drosophila* and vertebrates.

Independent duplications have occurred in different groups in evolution, giving rise to a different number of genes in lophotrochozoans, arthropods, and vertebrates. The duplicated genes have adopted different sites of expression and functions in the different species and as a result, some unexpected interchanges of expression patterns have arisen, although the mechanisms that originated in these remain to be determined. Nevertheless, in general, either *Snail* or *Slug* is expressed in the relevant mesodermal populations of different species. Indeed, in the primitive streak, Slug induces EMT in the chick, whereas Snail does it in the mouse. In addition, the two can behave as functionally equivalent when ectopically expressed at the appropriate sites both intra- and interspecies (Del Barrio and Nieto 2002). These transcription factors are composed of two main domains, a DNA-binding domain that is essentially identical in all Snail and Slug vertebrate family members and a much more divergent putative protein–protein interaction domain. Thus, it will be a challenge to determine the mechanisms that these two proteins have used to fulfill the same role during evolution with respect to their target genes, their regulatory partners, and the genetic cascades in which they are involved.

REFERENCES

Ashraf S.I., Hu X., Roote J., and Ip Y.T. 1999. The mesoderm determinant snail collaborates with related zinc-finger proteins to control *Drosophila* neurogenesis. *EMBO J.* **18:** 6426–6438.

Barrett K., Leptin M., and Settleman J. 1997. The Rho GTPase and a putative RhoGEF mediate a signaling pathway for the cell shape changes in *Drosophila* gastrulation. *Cell* **91:** 905–915.

Batlle E., Sancho E., Franci C., Dominguez D., Monfar M., Baulida J., and Garcia De Herreros A. 2000. The transcription factor snail is a repressor of *E-cadherin* gene expression in epithelial tumour cells. *Nat. Cell. Biol.* **2:** 84–89.

Beddington R.S.P. and Robertson E.J. 1999. Axis development and early asymmetry in mammals. *Cell* **96:** 195–209.

Beiman M., Shilo B.Z., and Volk T. 1996. Heartless, a Drosophila FGF receptor homolog, is essential for cell migration and establishment of several mesodermal lineages. *Genes Dev.* **10**: 2993–3002.

Bellairs R. 1987. The primitive streak and the neural crest: Comparable regions of cell migration? In *Developmental and evolutionary aspects of the neural crest* (ed. P. Maderson), pp. 123–145. Wiley, New York.

Blanco M.J., Moreno-Bueno G., Sarrio D., Locascio A., Cano A., Palacios J., and Nieto M.A. 2002. Correlation of *Snail* expression with histological grade and lymph node status in breast carcinomas. *Oncogene* **21**: 3241–3246.

Burdsal C.A., Damsky C.H., and Pedersen R.A. 1993. The role of E-cadherin and integrins in mesoderm differentiation and migration at the mammalian primitive streak. *Development* **118**: 829–844.

Cano A., Pérez-Moreno M.A., Rodrigo I., Locascio A., Blanco M.J., del Barrio M.G., Portillo F., and Nieto M.A. 2000. The transcription factor snail controls epithelial-mesenchymal transitions by repressing *E-cadherin* expression. *Nat. Cell Biol.* **2**: 76–83.

Carl T.F., Dufton C., Hanken J., and Klymkowsky M.W. 1999. Inhibition of neural crest migration in *Xenopus* using antisense *slug* RNA. *Dev. Biol.* **213**: 101–115.

Carmany-Rampey A. and Schier A.F. 2001. Single-cell internalization during zebrafish gastrulation. *Curr. Biol.* **11**: 1261–1265.

Carver E.A., Jiang R., Lan Y., Oram K.F., and Gridley T. 2001. The mouse *Snail* gene encodes a key regulator of the epithelial-mesenchymal transition. *Mol. Cell. Biol.* **21**: 8184–8188.

Ciruna B. and Rossant J. 2001. FGF signalling regulates mesoderm cell fate specification and morphogenetic movement at the primitive streak. *Dev. Cell* **1**: 37–49.

Comijn J., Berx G., Vermassen P., Verschueren K., van Grunsven L., Bruyneel E., Mareel M., Huylebroeck D., and van Roy F. 2001. The two-handed E box binding zinc finger protein SIP1 downregulates E-cadherin and induces invasion. *Mol. Cell* **7**: 1267–1278.

Corbo J.C., Erives A., Di Gregorio A., Chang A., and Levine M. 1997. Dorsoventral patterning of the vertebrate neural tube is conserved in a protochordate. *Development* **124**: 2335–2344.

Costa M., Wilson E.T., and Wieschaus E. 1994. A putative cell signal encoded by the *folded gastrulation* gene coordinates cell shape changes during *Drosophila* gastrulation. *Cell* **76**: 1075–1089.

Cowden J. and Levine M. 2002. The Snail repressor positions Notch signaling in the *Drosophila* embryo. *Development* **129**: 1785–1793.

Del Barrio M.G. and Nieto M.A. 2002. Overexpression of Snail family members highlights their ability to promote chick neural crest formation. *Development* **129**: 1583–1593.

Domínguez D., Montserrat-Sentís B., Virgós-Soler A., Guaita S., Grueso S., Porta M., Puig I., Baulida J., Francí C., and García de Herreros A. 2003. Phosphorylation regulates the subcellular location and activity of Snail transcriptional repressor. *Mol. Cell. Biol.* **23**: 5078–5089.

Fijiwara S., Corbo J.C., and Levine M. 1998. The snail repressor establishes a muscle/notochord boundary in the *Ciona* embryo. *Development* **125**: 2511–2520.

Foe V.E. 1989. Mitotic domains reveal early commitment of cells in *Drosophila* embryos. *Development* **107**: 1–22.

Fuse N., Hirose S. and Hayashi S. 1994. Diploidy of *Drosophila* imaginal cells is maintained by a transcriptional repressor encoded by *escargot*. *Genes Dev.* **8**: 2270–2281.

Gisselbrecht S., Skeath J.B., Doe C.Q., and Michelson A.M. 1996. heartless encodes a fibroblast growth factor receptor (DFR1/DFGF-R2) involved in the directional migration of early mesodermal cells in the *Drosophila* embryo. *Genes Dev.* **10**: 3003–3017.

Goldstein B., Leviten M.W., and Weisblat D.A. 2001. Dorsal and Snail homologs in leech development. *Dev. Genes Evol.* **211**: 329–337.

Gont L.K., Fainsod A., Kim S.H., and De Robertis E. 1993. Tail formation as a continuation of gastrulation: The multiple populations of the *Xenopus* tailbud derive from the late blastopore lip. *Development* **119**: 991–1004.

Grau Y., Carteret C., and Simpson P. 1984. Mutations and chromosomal rearrangements affecting the expression of *snail*, a gene involved in embryonic patterning in *Drosophila melanogaster*. *Genetics* **108**: 347–360.

Gray S. and Levine M. 1996. Short-range transcriptional repressors mediate both quenching and direct repression within complex loci in *Drosophila*. *Genes Dev.* **10**: 700–710.

Grimes H.L., Chan T.O., Zweidler-McKay P.A., Tong B. and Tsichlis P.N. 1996. The Gfi-1 proto-oncoprotein contains a novel transcriptional repressor domain, SNAG, and inhibits G1 arrest induced by interleukin-2 withdrawal. *Mol. Cell. Biol.* **11**: 6263–6272.

Grosshans J. and Wieschaus E. 2000. A genetic link between morphogenesis and cell division during formation of the ventral furrow in *Drosophila*. *Cell* **101**: 523–531.

Hammerschmidt M. and Nüsslein-Volhard C. 1993. The expression of a zebrafish gene homologous to *Drosophila snail* suggests a conserved function in invertebrate and vertebrate gastrulation. *Development* **119**: 1107–1118.

Hay E.D. 1995. An overview of epithelio-mesenchymal transformation. *Acta Anat.* **154**: 8–20.

Hemavathy K., Ashraf S.I., and IP Y.T. 2000a. Snail/Slug family of repressors: Slowly going to the fast lane of development and cancer. *Gene* **257**: 1–12.

Hemavathy K., Meng X., and Ip Y.T. 1997. Differential regulation of gastrulation and neuroectodermal gene expression by Snail in the *Drosophila* embryo. *Development* **124**: 3683–3691.

Hemavathy K., Guru S.C., Harris J., Chen J.D., and Ip Y.T. 2000b. Human Slug is a repressor that localizes to sites of active transcription. *Mol. Cell. Biol.* **26**: 5087–5095.

Holland P.W.H., García-Fernández J., Williams N.A., and Sidow A. 1994. Gene duplications at the origin of vertebrate development. *Dev. Suppl.* **1994**: 125–133.

Inukai T., Inoue A., Kurosawa H., Goi K., Shinjyo T., Ozawa K., Mao M., Inaba T., and Look A.T. 1999. Slug, a ces-1-related zinc finger transcription factor gene with antiapoptotic activity is a downstream target of the E2A-HLF oncoprotein. *Mol. Cell* **4**: 343–352.

Ip Y.T. and Gridley T. 2002. Cell movements during gastrulation: Snail dependent and independent pathways. *Curr. Opin. Genet. Dev.* **12**: 423–429.

Ip Y.T., Park R., Kosman D., and Levine M. 1992a. The *dorsal* gradient morphogen regulates stripes of *rhomboid* expression in the presumptive neuroectoderm of the *Drosophila* embryo. *Genes Dev.* **6**: 1728–1739.

Ip Y.T., Park R.E., Kosman D., Yazdanbakhsh K., and Levine M. 1992b. Dorsal-twist interactions establish snail expression in the presumptive mesoderm of the *Drosophila* embryo. *Genes Dev.* **6**: 1518–1530.

Jiang J., Kosman D., Ip Y.T., and Levine M. 1991. The dorsal morphogen gradient regulates the mesoderm determinant twist in early *Drosophila* embryos. *Genes Dev.* **5**: 1881–1891.

Jung A., Schrauder M., Oswald U., Knoll C., Sellberg P., Palmqvist R., Niedobitek G., Brabletz T., and Kirchner T. 2001. The invasion front of human colorectal adenocarcinomas shows co-localization of nuclear β-catenin, cyclin D₁, and p16^INK4A and is a region of low proliferation. *Am. J. Pathol.* **159:** 1613–1617.

Kasai Y., Nambu J.R., Lieberman P.M., and Crews S.T. 1992. Dorsal-ventral patterning in *Drosophila*: DNA binding of snail protein to the *single-minded* gene. *Proc. Natl. Acad. Sci.* **89:** 3414–3418.

Kataoka H., Murayama T., Yokode M., Mori S., Sano H., Ozaki H., Yokota Y., Nishikawa S., and Kita T. 2000. A novel snail-related transcription factor Smuc regulates basic helix-loop-helix transcription factor activities via specific E-box motifs. *Nucleic Acids Res.* **28:** 626–633.

Kim K., Lu Z., and Hay E.D. 2002. Direct evidence for a role of beta-catenin/LEF-1 signaling pathway in induction of EMT. *Cell. Biol. Int.* **26:** 463–476.

Kimelman D. and Griffin K.J. 2000. Vertebrate mesendoderm induction and patterning. *Curr. Opin. Genet. Dev.* **10:** 350–356.

LaBonne C. and Bronner-Fraser M. 2000. Snail-related transcriptional repressors are required in *Xenopus* for both the induction of the neural crest and its subsequent migration. *Dev. Biol.* **221:** 195–205.

Langeland J.A., Tomsa J.M., Jackman W.R. Jr., and Kimmel C.B. 1998. An amphioxus snail gene: Expression in paraxial mesoderm and neural plate suggests a conserved role in patterning the chordate embryo. *Dev. Genes Evol.* **208:** 569–577.

Leptin M. 1991. twist and snail as positive and negative regulators during *Drosophila* mesoderm development. *Genes Dev.* **5:** 1568–1576.

Lespinet O., Nederbragt A.J., Cassan M., Dictus W.J.A.G., Van Loon A.E., and Adoutte A. 2002. Characterization of two *snail* genes in the gastropod mollusk *Patella vulgata*. Implications for understanding the ancestral function of the *snail*-related genes in *Bilateralia*. *Dev. Genes Evol.* **212:** 186–195.

Locascio A. and Nieto M.A. 2001. Cell movements during vertebrate development: Integrated tissue behaviour versus individual cell migration. *Curr. Opin. Genet. Dev.* **11:** 464–469.

Locascio A., Manzanares M., Blanco M.J., and Nieto M.A. 2002. Modularity and reshuffling of Snail and Slug expression during vertebrate evolution. *Proc. Natl. Acad. Sci.* **99:** 16841–16846.

Mayor R., Guerrero R., Young R.M., Gomez-Skarmeta J.L., and Cuellar C. 2000. A novel function for the *Xslug* gene: Control of dorsal mesendoderm development by repressing BMP-4. *Mech. Dev.* **97:** 47–56.

Martínez-Álvarez C., Blanco M.J., Pérez R., Aparicio M., Resel E., Rabadán M.A., Martínez T., and Nieto M.A. 2004. Snail family members and cell survival in physiological and pathological cleft palates. *Dev. Biol.* **265:** 207–218.

Manzanares M., Blanco M.J., and Nieto M.A. 2004. Snail3 orthologues in vertebrates: Divergent members of the Snail zinc-finger gene family. *Dev. Genes Evol.* **214:** 47–53.

Manzanares M., Locascio A., and Nieto M.A. 2001. The increasing complexity of the Snail superfamily in metazoan evolution. *Trends Genet.* **17:** 178–181.

Mata J., Curado S., Ephrussi A., and Rorth P. 2000. Tribbles coordinates mitosis and morphogenesis in *Drosophila* by regulating string/CDC25 proteolysis. *Cell* **101:** 511–522.

Mauhin V., Lutz Y., Dennefeld C., and Alberga A. 1993. Definition of the DNA-binding site repertoire for the *Drosophila* transcription factor SNAIL. *Nucleic Acids Res.* **21:** 3951–3957.

Morel V., Le Borgne R., and Schweisguth F. 2003. Snail is required for Delta endocytosis and Notch-dependent activation of *single-minded* expression. *Dev. Genes Evol.* **213:** 65–72.

Morize P., Christiansen A.E., Costa M., Parks S., and Wieschaus E.

1998. Hyperactivation of the folded gastrulation pathway induces specific cell shape changes. *Development* **125:** 589–597.

Nakayama H., Scott I.C., and Cross J.C. 1998. The transition to endoreduplication in trophoblast giant cells is regulated by the mSna zinc finger transcription factor. *Dev. Biol.* **199:** 150–163.

Nibu Y., Zhang H., Bajor E., Barolo S., Small S., and Levine M. 1998. dCtBP mediates transcriptional repression by Knirps, Kruppel and Snail in the *Drosophila* embryo. *EMBO J.* **17:** 7009–7020.

Nieto M.A. 2002. The snail superfamily of zinc-finger transcription factors. *Nat. Rev. Mol. Cell. Biol.* **3:** 155–166.

Nieto MA., Sargent M., Wilkinson D.G., and Cooke J. 1994. Control of cell behavior during vertebrate development by *Slug*, a Zinc finger gene. *Science* **264:** 835–839.

Nishida H. 1987. Cell lineage analysis in ascidian embryos by intracellular injection of a tracer enzyme: III. Up to the tissue restricted stage. *Dev. Biol.* **121:** 526–541.

Nüsslein-Volhard C., Weischaus E., and Kluding H. 1984. Mutations affecting the pattern of the larval cuticle in *Drosophila melanogaster*. I. Zygotic loci on the second chromosome. *Wilheim Roux's Arch. Dev. Biol.* **193:** 267–282.

Oda H., Tsukita S., and Takeichi M. 1998. Dynamic behavior of the cadherin-based cell-cell adhesion system during *Drosophila* gastrulation. *Dev. Biol.* **203:** 435–450.

Pérez-Moreno M., Locascio A., Rodrigo I., Dhont G., Portillo F., Nieto M.A., and Cano A. 2001. A new role for E12/E47 in the repression of E-cadherin expression and epithelial-mesenchymal transition. *J. Biol. Chem.* **276:** 27424–27431.

Postlethwait J.H., Yan Y.L., Gates M.A., Horne S., Amores A., Brownlie A., Donovan A., Egan E.S., Force A., Gong Z., Goutel C., Fritz A., Kelsh R., Knapik E., Liao E., Paw B., Ransom D., Singer A., Thomson M., Abduljabbar T.S., Yelick P., Beier D., Joly J.S., Larhammar D., Rosa F., et al. 1998. Vertebrate genome evolution and the zebrafish gene map. *Nat. Genet.* **18:** 345–349.

Romano L. and Runyan R.B. 2000. Slug is an essential target of TGF β2 signaling in the developing chicken heart. *Dev. Biol.* **223:** 91–102.

Sefton M., Sanchez S., and Nieto M.A. 1998. Conserved and divergent roles for members of the Snail family of transcription factors in the chick and mouse embryo. *Development* **125:** 3111–3121.

Seher T.C. and Leptin M. 2000. Tribbles, a cell-cycle brake that coordinates proliferation and morphogenesis during *Drosophila* gastrulation. *Curr. Biol.* **10:** 623–629.

Smith J.C. 1993. Mesoderm inducing factors in early vertebrate development. *EMBO J.* **12:** 4463–4470.

Smith S., Metcalfe J.A., and Elgar G. 2000. Identification and analysis of two *snail* genes in the pufferfish (*Fugu rubripes*) and mapping of human *SNA* to 20q. *Gene* **247:** 119–128.

Spring J., Yanze N., Josch C., Middel A.M., Winninger B., and Schmid V. 2002. Conservation of *Brachyury*, *Mef2*, and *Snail* in the myogenic lineage of jellyfish: A connection to the mesoderm of bilateria. *Dev. Biol.* **244:** 372–384.

Tan C., Costello P., Sanghera J., Domínguez D., Baulida J., García de Herreros A., and Dedhar S. 2001. Inhibition of integrin linker kinase (ILK) suppresses β-catenin-Lef/Tcf-dependent transcription and expression of the E-cadherin repressor snail, in APC–/– human colon carcinoma cells. *Oncogene* **20:** 133–140.

Thiery J.P. 2002. Epithelial-mesenchymal transitions in tumour progression. *Nat. Rev. Cancer* **2:** 442–454.

Thisse C., Thisse B., and Postlethwait J.H. 1995. Expression of *snail2*, a second member of the zebrafish snail family, in cephalic mesendoderm and presumptive neural crest of wild-type and *spadetail* mutant embryos. *Dev. Biol.* **172:** 86–99.

Thisse C., Thisse B., Schilling T.F., and Postlethwait J.H. 1993. Structure of the zebrafish *snail1* gene and its expression in wild-type, *spadetail* and *no tail* mutant embryos. *Development* **119:** 1203–1215.

Vallin J., Thuret R., Giacomello E., Faraldo M.M., Thiery J.P., and Broders F. 2001. Cloning and characterization of three *Xenopus slug* promoters reveal direct regulation by Lef/beta-catenin signaling. *J. Biol. Chem.* **276:** 30350–30358.

Vega S., Morales A.V., Ocaña O.H., Valdés F., Fabregat I., and Nieto M.A. 2004. Snail blocks the cell cycle and confers resistance to cell death. *Genes Dev.* **18:** 1131–1143.

Velmaat J. M., Orelio C.C., Ward-Van Oostwaard D., Van Rooijen M.A., Mummery C.L., and Defize L.H.K. 2000. *Snail* an immediate early target gene of parathyroid hormone related peptide signaling in parietal endoderm formation. *Int. J. Dev. Biol.* **44:** 297–307.

Wada S. and Saiga H. 1999. Cloning and embryonic expression of *Hrsna*, a snail family gene of the ascidian *Halocynthia roretzi*: Implication in the origins of mechanisms for mesoderm specification and body axis formation in chordates. *Dev. Growth Differ.* **41:** 9–18.

Wolfe K. 2001. Yesterday's polyploids and the mistery of diploidization. *Nat. Rev. Genet.* **2:** 333–341.

Yáñez-Mo M., Lara-Pezzi E., Selgas R., Ramírez-Huesca M, Domínguez-Jiménez C, Jiménez-Heffernan J.A., Aguilera A., Sánchez-Tomero J.A., Bajo M.A., Alvarez V., Castro M.A., del Peso G., Cirujeda A., Gamallo C., Sánchez-Madrid F., and López-Cabrera M. 2003. Peritoneal dialysis and epithelial-to-mesenchymal transition of mesothelial cells. *N. Engl. J. Med.* **348:** 403–413.

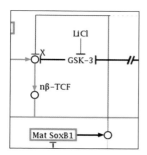

System-level Properties Revealed by a Gene Regulatory Network Analysis of Pre-gastrular Specification in Sea Urchins

V.F. Hinman and E.H. Davidson

Division of Biology, California Institute of Technology, Pasadena, California 91125

INTRODUCTION

Traditional studies of gastrulation in sea urchins have relied on cell labeling and micromanipulation procedures that in the molecular era were combined with gene expression, functional, and regulatory analyses. These now provide a wealth of information about how the relatively simple sea urchin embryo is built. Recent research demonstrates that this embryo is also amenable to the high-throughput gene regulatory analyses required for a system-level description of development (Davidson et al. 2002a,b). The aim of this approach is to provide an in-depth understanding of the properties of development, achievable only at this level, by constructing a logic model that describes the causality of the process in fullest possible detail (see, e.g., reviews by Ideker et al. 2001; Kitano 2002). In the example used here, the model is the gene regulatory network (GRN) that explains the acquisition of the transcriptional regulatory state required for normal gastrulation in sea urchins. The intention of this review is to explore what the GRN analysis of gastrulation reveals.

Gastrulation in sea urchins involves (1) ingression of a skeletogenic primary mesenchyme cell (PMC) population into the blastocoel, (2) invagination of the endomesoderm

from the vegetal epithelium, and (3) delamination of secondary mesenchyme cells (SMCs). However, before these fields of cells undertake any of the movements of gastrulation, they must be specified to their various endomesodermal fates. This specification process can be traced back to the differential mobilization of maternally loaded molecules in the early blastomeres. The GRN model considered here starts, therefore, with an examination of cytological processes in early cleavage that lead to the initial activation of lineage-restricted transcription factors, and ends with the expression of proteins responsible for the changes in cell morphology and cell migrations necessary for the movements of gastrulation.

The model and our understanding of its system properties will be most useful if as many as is possible of the regulatory molecules required are included. Even taking full advantage of the large amount of prior research on pre-gastrulation-stage sea urchin embryos, this required extensive gene discovery efforts. The development of a highly sensitive macroarray hybridization technology was instrumental in isolating many mRNAs encoding transcription factors that are often expressed at very low levels, but which turn out to be key effectors (Rast et al. 2000; Brown et al. 2002). These differential screens identified genes acting downstream of the Notch and β-catenin/Wnt8 signaling pathways (Ransick et al. 2002; Calestani et al. 2003) and of the transcription factor Brachyury (Rast et al. 2002). In addition, sea urchin orthologs to genes with known functions in gastrulation in other taxa were recovered, and expressed sequence tag (EST) screens from specific cell populations (e.g., Zhu et al. 2001) or developmental stages (e.g., Lee et al. 1999) have also contributed to the catalog of genes incorporated in the model. To date, more than 50 genes, mostly transcription factors and signaling molecules, are included in the GRN for endomesoderm specification in sea urchins. The relatively small number of differentiation gene products contained within the GRN reflects the bias toward transcription factors and signaling effectors of the gene discovery efforts: The basic objective of the GRN model is, after all, to illuminate the control circuitry. A large-scale perturbation analysis, directed particularly at regulatory gene function and the operation of signaling pathways, combined with knowledge of normal gene expression, was used to form a network of epistatic relationships among the genes. The GRN can be portrayed as a logic model of the form shown in Figure 1 (see continuously updated version at http://sugp.caltech.edu/endomes/). Here the *cis*-regulatory region of each gene is portrayed as a short line controlling the expression of the gene named beneath. Arrows extending from the genes to the *cis*-regulatory regions of other genes represent regulatory inputs into the target genes. The regulatory linkages upon which the network is built, therefore, are coded in the particular DNA sequence of the *cis*-regulatory regions of each of the genes. When each of these linkages has been decoded, we will be in a position to describe how the heritable arrangement of genomic DNA sequence prescribes the development of the sea urchin gastrula.

The GRN leading up to gastrulation in the sea urchin embryo may be considered as several interconnected subnetworks, representing initial specification, execution of the regulatory states defined by specification, and finally, the differentiation of different fields of cells that are connected by intercellular signaling (Fig. 1). The initial territories of specification may be considered as the micromere/PMC (pink block in Fig. 1) and the veg2 endomesoderm territories (green block), the latter of which in later development resolves into the SMC and endodermal (yellow block) territories. Through an examination of these sub-GRNs, several recurrent properties of the developmental process underlying gastrulation in sea urchins emerge. The number of developmental systems that have been described at the level of a GRN model is as yet too small to determine how general to development these properties are. In the final section of this review, we present a GRN of endomesodermal specification in starfish that serves the dual process of examining this issue and also of understanding how differ-

ences are incorporated into GRNs and how these relate to the evolution of morphology.

PROPERTIES OF SPECIFICATION

During early cleavage of indirectly developing sea urchin embryos, certain regulatory proteins, either maternal factors themselves or proteins under the control of maternal factors, become diffusely localized, although not necessarily confined within lineage compartments. Combinations of these proteins are sufficient to initiate the expression of a few transcription factors in a lineage-restricted manner. Thus, for example, during the specification of the micromeres to a PMC fate (Fig. 1), as-yet-unknown cell-autonomous processes result in the nuclearization of maternally loaded β-catenin (Wikramanayake et al. 1998; Logan et al. 1999) and of the *Strongylocentrotus purpuratus* orthodenticle homolog, SpOtx (Chuang et al. 1996) in the micromeres very soon after their formation. Although neither is ultimately micromere-specific (β-catenin shortly appears within the nuclei of macromeres [(Logan et al. 1999], and SpOtx is translocated to other nuclei in subsequent development [Chuang et al. 1996]), their combined presence in 4th cleavage micromeres (possibly with other as-yet-unknown factors) activates the earliest known expressed micromere-specific transcription factor, *pmar1* (Oliveri et al. 2002). Global overexpression of *pmar1* is sufficient to cause global overexpression of all other known micromere-specific lineage markers, including genes utilized as transcriptional regulators, signaling ligands, and PMC differentiation products (Oliveri et al. 2002). Expression of *pmar1* in micromeres is normally required for all of the three known functions of the micromeres; e.g., the differentiation of PMCs, signaling to veg2 to specify endomesoderm, and signaling to the inner vegetal plate to specify mesoderm (Oliveri et al. 2002, 2003). Expression of *pmar1* in the micromeres, therefore, defines a state of specification of this lineage. As shown in Figure 1, rather than activating other micromere/PMC-specific genes, *pmar1* instead acts to repress a global repressor, *r of mic*, which is yet to be isolated. The effect is to relieve the repression of a number of transcription factors and signaling molecules, which are then expressed exclusively in the micromere lineage and direct the various aspects of PMC fate. Oliveri et al. (2002) provide an intriguing evolutionary hypothesis for the reason that such an indirect means of activation may occur during specification of the PMCs. At the center of the argument lies the fact that a skeletogenic micromere lineage is a unique echinoid character. Therefore, the specification and differentiation processes of this lineage are likely to be an evolutionary "add-on." The transcription factors required for skeletogenesis, however, must be present in the skeletogenic cells of late larval and postmetamorphic stages in all echinoderms, since the development of a calcite endoskeleton is a defining feature of the phylum. In most other echinoderms, the expression of skeletogenic gene batteries is silent during embryogenesis. Thus, the establishment of just one new regulatory connection, i.e., from *pmar1* to the global *r of mic* repressor, would precociously relieve this silencing and allow differentiation of a skeletogenic state to proceed during embryogenesis. The "switchlike" mechanism achieved by turning on or off a global repressor in specific lineages may facilitate significant heterochronic evolutionary changes.

Specification of the veg2 lineage to endomesoderm also requires the suppression of a ubiquitous repressor, in this instance Soxb1 (Fig. 1). The early signal from the micromeres, which is mediated by *pmar1*, causes clearance of maternal Soxb1 from the adjacent veg2 cells (Kenny et al. 1999; Oliveri et al. 2003). When 4th-cleavage micromeres are transplanted to ectopic locations, Soxb1 is later cleared from the underlying cells, demonstrating that the early signal from the micromeres is both necessary and sufficient for this clearance (Oliveri et al. 2003). The region from which Soxb1 is normally cleared corresponds to the future endomesoderm (Kenny et al. 1999). The reason for this is that Soxb1 antagonizes the positive input provided by nuclear β-catenin/TCF (Fig. 1) (Kenny et al. 1999, 2003), and hence clearance of Soxb1 is necessary for this input to be maintained at the necessary level. The zygotic β-catenin/TCF input (Vonica et al. 2000) then triggers a series of positive autoregulatory feedback loops. The immediate target is the *wnt8* gene, which in turn drives further nuclearization of β-catenin. β-catenin/TCF, along with maternal Otx, also activates endomesodermally restricted regulatory genes such as *krox*. The *krox* gene product in turn activates the *wnt8* gene, thereby establishing a nuclear β-catenin/TCF/*wnt8*/*krox* positive feedback circuit within the endomesoderm (Fig. 1). These autoregulatory loops drive further β-catenin/TCF nuclearization, which is required for the activation of many other endomesodermally restricted regulatory genes (e.g., *bra*, *foxa*, *eve*, *gcm*, and *ui*). The establishment of *wnt8* gene expression, therefore, is one of the first indicators of a specified endomesoderm. Yet another series of autoregulatory loops is required before the endomesoderm may be considered as fully specified. The *krox* gene product activates the zygotic expression of *otx*. Otx, in turn, activates *krox*. Zygotic *otx* transcription subsequently activates the *gatae* gene in the endomesoderm, and gatae is in turn required to maintain the expression of *otx*. This establishes a three-way positive feedback loop, the end product of which is the expression of *gatae* across the prospective endomesoderm. The expression of *gatae* is required for the activation of many of the other known endomesodermally restricted transcription factors.

The stabilization of its expression defines, therefore, a state of specification of the endomesoderm.

The analysis of this part of the GRN suggests an explanation of a well-known developmental phenomenon. The GRN model shows why all cells are competent to form endomesoderm if micromeres are transplanted next to them (as shown by the experiments of Hörstadius 1939; Ransick and Davidson 1993). The signal from the micromeres is sufficient to clear the ubiquitously expressed Soxb1, and the implication is that following this, the whole stepwise endomesodermal regulatory apparatus shown in Figure 1 will be set in train. That is, the transcriptional feedback circuitry that defines endomesodermal specification will be set up in the ectopic location, and hence, all of the downstream functions required to produce the observed phenomenon of gut and mesoderm differentiation will proceed ectopically.

One more specification event has yet to occur that precedes the morphogenetic events of gastrulation. This effects the specification of an inner ring of cells in the vegetal plate as mesoderm and the outer as endoderm. The determinant event is a signal from the micromeres to the adjacent veg2 cells (Fig. 1). During the 7th to 9th cleavages, the *pmar1* double repression system causes the micromeres to express the ligand Delta (Oliveri et al. 2002, 2003; Sweet et al. 2002). Delta activates maternal Notch receptor in those veg2 cells directly adjacent to the micromeres (Sherwood and McClay 1999; Sweet et al. 1999, 2002). Downstream of Notch, the transcription factor gcm is activated exclusively in these presumptive mesodermal cells (Ransick et al. 2002). *gcm* also receives inputs from β-catenin/TCF. All of this explains at the level of the transcriptional control apparatus the phenomenon observed by Sherwood and McClay (1999), that in the absence of Notch signaling, there is an excess of endoderm and a corresponding decrease of mesoderm progenitors.

EXECUTION OF REGULATORY STATES

The GRN described so far explains how three populations of cells (the PMCs, the endoderm, and the SMCs) become specified. Specification is defined by the presence of a set of transcription factors whose expression has become stabilized within a lineage, independent of transient signals. Once specified to a particular fate, these fields of cells then execute the regulatory states defined by the specification process. That is, the expression of the specification transcription factors activates many other transcription factors that direct these fields to a particular differentiated fate. In each sub-GRN, a relatively large number of transcription factors are activated during this stage of the developmental process and the genes encoding them interact in a complex,

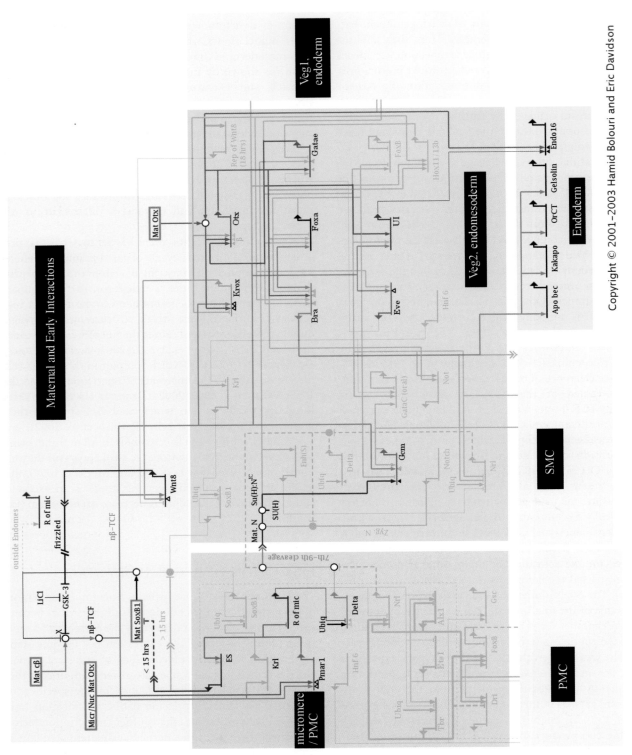

Copyright © 2001–2003 Hamid Bolouri and Eric Davidson

Figure 1. (*See facing page for legend.*)

functional network (Fig. 1). This reflects the intricate control circuitry needed to regulate downstream differentiation gene batteries that are required to build morphology, even in the sea urchin embryo, which is very simple in form compared to vertebrate and *Drosophila* embryos.

The ultimate purpose of the complex of transcription factor expressions that defines each regulatory state is to provide spatial and temporal control of downstream batteries of differentiation genes. It is the expressions of these genes that will determine morphological form. This is the least understood of the developmental processes here examined, largely because of the relatively few differentiation products that have thus far been discovered and characterized. Several recent differential (Ransick et al. 2002; Rast et al. 2002; Calestani et al. 2003) and EST (Smith et al. 1996; Lee et al. 1999; Zhu et al. 2001) screens have added to the catalog of differentiation genes, but the number included in the GRN is only a small fraction of those utilized in the embryo. This is because the focus of investigation in this study has been the specification apparatus, rather than the regulatory apparatus underlying differentiation, cell biology, or morphogenesis functions. Those differentiation genes included serve essentially as placeholders and examples of genes operating at this terminal level of the GRN.

Many of the differentiation products contained within the model were identified using a differential macroarray screening technology developed by Rast et al. (2000). In these screens, specific signaling pathways are blocked, or regulatory functions disrupted, and the pool of mRNA from the resultant embryos is subtracted against mRNA from normal embryos, or from embryos in which the same signaling pathway or regulatory function is enhanced. For example, Rast et al. (2002) perturbed the expression of *bra* (using either a morpholino antisense oligonucleotide to block its function or a ubiquitous promoter to overexpress its mRNA) and subtracted the total mRNA against that from normal embryos. Two examples of the genes isolated in this screen, *kakapo* and *gelsolin* (Fig. 1), have possible roles in controlling cell motility and shape changes. The

requirement of *bra* for the correct expression of cell-shape and motility genes explains why embryos in which the function of *bra* is blocked are unable to undergo the invagination movements of gastrulation (Gross and McClay 2001; Rast et al. 2002). No transcription factors were identified as targets of *bra*, despite the proven ability of this technique in identifying rare factors (Ransick et al. 2002). This suggests that *bra* may directly activate batteries of differentiation genes rather than acting through additional intermediary transcription factors. If this is the case, there are relatively few levels of regulatory architecture from specification through to differentiation. This can also be seen in the regulation of a very well characterized gene, *endo16*. A direct analysis of the *cis*-regulatory region of *endo16* shows that both Otx and ui are activators (Yuh et al. 1998). Otx is one of the earliest transcription factors involved in specification of endomesoderm, and *ui* is activated by gatae (Fig. 1). Otx and Gatae are two transcription factors initially involved in defining the endomesoderm specification regulatory state. This relatively shallow architecture explains why such terminal differentiation genes as *gelsolin*, *kakapo*, and *endo16* are expressed prior to gastrulation, and well before the emergence of any morphological structure (Ransick et al. 1993; Rast et al. 2002). This is the rule for many cell-type-specific genes encoding terminal differentiation products in other domains of the sea urchin embryo as well, including the aboral ectoderm and the mesodermal and skeletogenic domains (Davidson 1989; Davidson et al. 1998; Calestani et al. 2003).

CONSERVATION AND EVOLUTION OF A GRN ELEMENT

Our exploration of the GRN underlying sea urchin gastrulation illustrates how a systems biology approach can be used to reveal properties of a developmental process. Comparisons of the sub-GRNs (i.e., those of the PMCs, endomesoderm, and SMCs) have revealed several recurrent

---

Figure 1. GRN of specification processes required for gastrulation in the sea urchin. Parts of the GRN discussed in the text are highlighted; some other known elements are shown in the background. Each short horizontal line from which a bent arrow extends, to indicate transcription, represents the *cis*-regulatory element that is responsible for expression of the gene named beneath. Inputs into each *cis*-regulatory region are either positive (indicated by an arrow) or negative (indicated by a bar). Blocks represent different fields of cells; pink represents the micromere/PMC lineage, green represents veg2 endomesoderm, which in later development resolves into endoderm (*yellow*) and SMC mesoderm. Genes are placed within these colored regions according to their final loci of expression. Gene products that are present maternally are boxed. Double arrowheads represent signaling between fields of cells. Large open circles represent cytoplasmic biochemical interactions at the protein level, e.g., those responsible for nuclearization of β-catenin or the effect of Delta on Notch. This model displays the "view from the genome" (Arnone and Davidson 1997) in that it attempts to illustrate all of the linkages in all fields of cells and at all times. There is, however, a conceptual temporal component to the figure, as early interactions are indicated at the top of the diagram, the middle of the diagram represents specification processes, and the lower boxes indicate differentiation products. (Modified from Davidson et al. 2002a,b and http://sugp.caltech.edu/endomes/.)

properties of these systems; for example, its reinforcing feedback loops. Systematic, comparative examination of GRN architectures from a variety of taxa will determine which of these properties are general to development. Since evolution must proceed through architectural reorganization of developmental GRNs, comparative analyses will also reveal at a causal level the nature of evolutionary changes that affect formation of the body plan. The literature so far, however, contains no direct examples. We have generated a partial GRN for endomesodermal specification in the starfish *Asterina miniata* (Hinman et al. 2003a) in order to compare it with the orthologous sea urchin GRN.

EARLY DEVELOPMENT OF STARFISH

Starfish (Cl. Asteroidea) and sea urchins (Cl. Echinoidea) are distantly related free-living echinoderms that last shared a common ancestor around 500 million years ago (Fig. 2) (Smith 1988; Wada and Satoh 1994). As is typical for most echinoderms, but not echinoids, cleavage in starfish embryos is equal at least until formation of the blastula. Furthermore, starfish embryos do not have an early ingressing, micromere/PMC population, nor do they generate a larval skeleton such as is derived from the micromere lineage in sea urchins (Fig. 3). The presence of an early ingress-

ing micromere lineage is unique to the euechinoids. Apart from these differences, development of starfish is similar to that of sea urchins (Fig. 3). A vegetal plate forms at the vegetal pole, and the invagination movements of gastrulation proceed from the center to the periphery of this plate. Crude cell-labeling experiments demonstrate that the central cells of the vegetal plate contribute to the mesodermal coeloms that form at the top of the archenteron, whereas the more peripheral vegetal plate contributes to the later invaginating gut endoderm (Kuraishi and Osanai 1992), just as in the sea urchin embryo.

Maternal determinants contained within the vegetal ooplasm are required for invagination and differentiation

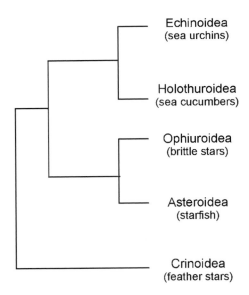

Figure 2. Evolutionary relationships of the extant classes of echinoderms. The common ancestor of the echinoderms was a stalked form existing in the Cambrian (Sprinkle and Kier 1987). The stalked Crinoidea are considered the most basal of the extant echinoderms. The remaining modern-day classes are free-living and mobile as adults and are grouped together as the Eleutherozoa. The starfish (Asteroidea) and sea urchins (Echinoidea) are two distantly related classes within the Eleutherozoa.

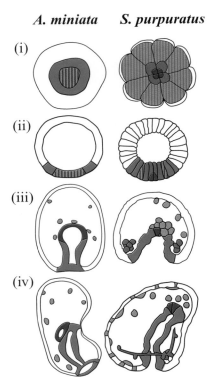

Figure 3. Fate maps of selected stages of *A. miniata* and *S. purpuratus* development. (*i*) Vegetal (posterior) pole of view of 6th-cleavage *S. purpuratus* embryo and blastula-stage *A. miniata* embryo. (*ii–iv*) Lateral optical sections; animal pole top, vegetal pole bottom. (*ii*) blastulae, (*iii*) gastrulae, and (*iv*) early larval stages. In *iv*, oral side to right and aboral to left. Color coding indicates the fate of cells through development; green with lavender marks the endomesodermal veg2 lineage of *S. purpuratus* that resolves into endodermal (*green*) and mesodermal (*lavender*) by 7th cleavage, and by blastula stage in *A. miniata*. The mesoderm lineage can be further subdivided into coelomic (*purple*) and other mesodermal cell types (*lavender*). The echinoid-specific, micromere-derived skeletogenic lineage is shown in red. (Modified, with permission, from Hinman et al. 2003a [©National Academy of Sciences].)

of the archenteron in starfish. This is supported by several experimental observations. Isolated animal cap blastomeres have essentially no capacity for archenteron formation but will invaginate if fused with the vegetal pole ooplasm from immature oocytes (Kiyomoto and Shirai 1993a,b). Kuraishi and Osanai (1994) demonstrated that removal of ~7% of the ooplasm from the vegetal pole of the starfish *Asterina pectinifera* severely suppresses archenteron formation. They further demonstrated that if one of the blastomeres of the 2-cell embryo was rotated by 180°, an archenteron would form at both poles. Recently, it has also been shown that β-catenin is localized to the nuclei of vegetal cells from the 16-cell stage through to the beginning of gastrulation (Miyawaki et al. 2003). This parallels the situation in echinoids, where maternally driven processes are responsible for the nuclearization of β-catenin in the vegetal cells of the 4th-cleavage embryo. The molecular pathway from the maternal determinants implied by these phenomenological experiments to the institution of a transcriptional regulatory apparatus for endomesoderm specification in starfish has not been explored.

GRN OF ENDOMESODERMAL SPECIFICATION IN STARFISH

The regulatory linkages between the *A. miniata* orthologs of *otx, krox, gatae, foxa,* and *bra* have been determined and organized into a GRN model (Fig. 4B). The same principles were used to construct the starfish GRN as were used to construct the orthologous region of the network shown in Figure 1 and reproduced as Figure 4A. That is, *otx, krox, gatae,* and *foxa* were perturbed in *A. miniata,* and the effect on transcript abundance of each of the genes was assessed.

A SEA URCHIN

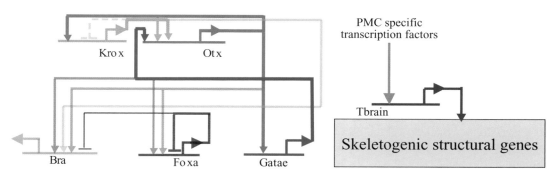

B STARFISH

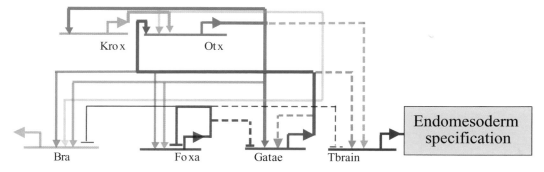

Figure 4. Architectural comparisons of the GRN for endomesodermal specification in sea urchins and starfish. The GRN for *S. purpuratus* is extracted directly from Fig. 1 but represents only the interactions between *krox, otx, gatae, foxa,* and *bra.* The orthologous GRN is shown for *A. miniata.* Comparisons of the GRN involving *tbr* are also included. In *S. purpuratus* the *tbr* gene, its downstream target genes, and the upstream regulator encoded by the *pmar1* gene are all expressed in the micromere/skeletogenic lineage and are necessary for its normal specification. The *tbr* gene in *A. miniata* is expressed within the endomesoderm and is required for its correct specification. Dashed lines indicate a regulatory connection that was observed only within the taxa for which it is drawn. The positive regulatory feedback loops between *krox, otx,* and *gatae* that are present in both taxa are highlighted in bold. (Modified, with permission, from Hinman et al. 2003a [©National Academy of Sciences].)

As we have seen, in *S. purpuratus* these genes are required for specification of the endomesoderm (Davidson et al. 2002a,b). The expression of these *A. miniata* orthologs in starfish embryos is spatially similar to that found in sea urchins; they are all expressed within parts of the prospective endomesoderm prior to and during gastrulation (Hinman and Davidson 2003a,b; Hinman et al. 2003a,b). The phenotypes resulting from disruption of *otx, krox, gatae,* and *foxa* function also indicate an essential role for these transcription factors in specification of the endomesoderm in starfish.

The most prominent feature of the network architecture in the starfish is the presence of a three-gene positive intergenic regulatory loop between *krox, otx,* and *gatae* (Fig. 4). The *krox* gene product is required for the correct expression of *otx,* and Otx is required for the correct expression of *krox.* In turn, Otx activates *gatae,* which is in turn then required for the maintained expression of *otx.* This architectural feature is precisely the same as in the sea urchin. It has been conserved in these two independently evolving lineages for half a billion years. Evidently, such a loop is an indispensable feature of the early development of the free-living echinoderms. In sea urchins, this autoregulatory loop is hypothesized to be necessary to drive specification forward into a stable state that is independent of the initial transient inputs (Davidson et al. 2002a,b). The end product of this loop is the expression of *gatae* in endomesoderm. In *S. purpuratus* and *A. miniata,* the *gatae* gene is an essential endodermal driver for many other regulatory genes; e.g., *foxa* and *bra* (Fig. 4). In fact, members of the *gata* family of transcriptional regulatory genes are required for gut development throughout the Bilateria (Patient and McGhee 2002). Ensuring endodermal *gatae* expression may thus be one of the important early functions of the endomesodermal specification GRN (Hinman et al. 2003a).

Intergenic autoregulatory loops acting immediately downstream of early and transient specification inputs also occur in other developmental systems. Examples of these are reviewed in Davidson et al. (2003). The use of intergenic autoregulatory loops may be a widespread developmental phenomenon that functions to transform transient signaling inputs into stable states of transcriptional control.

The comparison of these two GRNs also reveals differences that illustrate how evolutionary changes are incorporated into the regulatory architecture. The most striking of these changes concerns the different usage of *tbrain* (*tbr*) in the sea urchin and starfish (Fig. 4). In starfish, *tbr* is a part of the endomesodermal specification network. This gene is regulated by *gatae* and *otx,* and in later development is repressed by *foxa.* In contrast, the sea urchin orthologs of *otx, krox,* and *foxa* have no role in regulating *tbr*; it is instead under the indirect regulation of *pmar1* (Oliveri et al. 2002). Consequently, in sea urchins, *tbr* is expressed in the

micromere cell lineage (Croce et al. 2001), whereas in starfish it is expressed within the endomesoderm (Shoguchi et al. 2000; Hinman et al. 2003a). These expressions are predicted from the network architectures. Furthermore, correct *tbr* expression is required for archenteron formation in starfish (Hinman et al. 2003a), whereas if *tbr* function is perturbed, sea urchin embryos gastrulate normally but fail to develop a larval skeleton (P. Oliveri and E. Davidson, unpubl.). The different functions of *tbr* in these two taxa allude to changes that must be incorporated in the network architecture downstream of this gene. It will be extremely interesting to determine how the architecture of the GRN has evolved to accommodate the altered placement of a "high level" regulatory factor and the precise *cis*-regulatory rearrangements that lead to the altered morphologies.

Other changes are apparent when the GRNs are compared. An intergenic regulatory loop also exists between *foxa* and *gatae* in starfish, although in this instance it is not mutually positive. *gatae* activates both itself and *foxa,* whereas *foxa* represses both itself and *gatae.* In sea urchin, *gatae* does not self-activate, nor does *foxa* repress *gatae* expression, so the regulatory feedback loop seen in *A. miniata* is not established. The role of *foxa* in starfish appears to be to repress both itself and *gatae* from the mesoderm progenitors, while not affecting expression in the peripheral, endodermally fated vegetal plate (Hinman et al. 2003a). As a consequence of this changed architecture, *gatae* in starfish is not expressed within the central disk of the vegetal plate by late blastula, whereas in sea urchin, *gatae* expression is maintained in this central SMC region until the beginning of invagination (P.Y. Lee and E. Davidson, unpubl.).

The comparison of just these small orthologous regions of the GRN from starfish and sea urchin has proven to be tremendously revealing about which of the system-level properties of the development of these two taxa are conserved, and which are plastic. Prominent early-specification processes have been preserved during the 500 million years since these echinoderms last shared an ancestor. Many evolutionary changes, however, have been integrated into the GRN architecture since divergence. These changes may have profound effects on the deployment of downstream differentiation batteries, and hence, on the development of larval form.

CONCLUSIONS

The model of the GRN that describes the acquisition of the regulatory state required for gastrulation in the sea urchin *S. purpuratus* provides a new understanding of how genomes control development. Combinations of maternally loaded products initiate the expression of a few transcription factors within particular cell-lineage compartments. The expression of these factors then becomes stabilized within these lineages, independent of the initial and transient maternal cues. This stabilization often proceeds through the activation of positive genetic regulatory feedback loops. That such regulatory loops are an indispensable feature of early development is demonstrated by the fact that a three-way positive regulatory loop has been conserved in two lineages that have been evolving independently for around half a billion years. The stabilized, lineage-restricted expression of particular transcription factors specifies the cell population. Many transcription factors are activated during the execution of the regulatory state defined by this specification; however, the number of regulatory "levels" between specification and differentiation is relatively few.

The comparison of orthologous GRNs from the sea urchin *S. purpuratus* and the starfish *A. miniata* reveals the usefulness of developmental GRNs for evolutionary comparisons. Not only do conserved features suggest essential developmental properties, but novel circuitry provides precise mechanistic explanations of how evolutionary changes are incorporated into animal development.

ACKNOWLEDGMENTS

We thank Roger Revilla for thoughtful and critical comments on the manuscript. This work was supported by NASA/Ames (NAG-2-1587) and NCRR (RR-15044) to E.H.D.

REFERENCES

Arnone M.I. and Davidson E.H. 1997. The hardwiring of development: Organization and function of genomic regulatory systems. *Development* **124:** 1851–1864.

Brown C.T., Rust A.G., Clarke P.J.C., Pan Z., Schilstra M.J., De Buysscher T., Griffin G., Wold B.J., Cameron R.A., Davidson E.H., and Bolouri H. 2002. New computational approaches for analysis of cis-regulatory networks. *Dev. Biol.* **246:** 86–102.

Calestani C., Rast J.P., and Davidson E.H. 2003. Isolation of pigment cell specific genes in the sea urchin embryo by differential macroarray screening. *Development* **130:** 4587–4596.

Chuang C.K., Wikramanayake A.H., Mao C.A., Li X.T., and Klein W.H. 1996. Transient appearance of *Strongylocentrotus purpuratus* Otx in micromere nuclei: Cytoplasmic retention of SpOtx possibly mediated through an α-actinin interaction. *Dev. Genet.* **19:** 231–237.

Croce J., Lhomond G., Lozano J.C., and Gache C. 2001. ske-T, a T-box gene expressed in the skeletogenic mesenchyme lineage of the sea urchin embryo. *Mech. Dev.* **107:** 159–162.

Davidson E.H. 1989. Lineage-specific gene-expression and the regulative capacities of the sea-urchin embryo—a proposed mechanism. *Development* **105:** 421–445.

Davidson E.H., Cameron R.A., and Ransick A. 1998. Specification of cell fate in the sea urchin embryo: Summary and some proposed mechanisms. *Development* **125:** 3269–3290.

Davidson E.H., McCay D.R., and Hood L. 2003. Regulatory gene networks and the properties of the developmental process. *Proc. Natl. Acad. Sci.* **100:** 1475–1480.

Davidson E.H., Rast J.P., Oliveri P., Ransick A., Calestani C., Yuh C.H., Minokawa T., Amore G., Hinman V., Arenas-Mena C., Otim O., Brown C.T., Livi C.B., Lee P.Y., Revilla R., Rust A.G., Pan Z.J., Schilstra M.J., Clarke P.J.C., Arnone M.I., Rowen L., Cameron R.A., McClay D.R., Hood L., and Bolouri H. 2002a. A genomic regulatory network for development. *Science* **295:** 1669–1678.

Davidson E.H., Rast J.P., Oliveri P., Ransick A., Calestani C., Yuh C.H., Minokawa T., Amore G., Hinman V., Arenas-Mena C., Otim O., Brown C.T., Livi C.B., Lee P.Y., Revilla R., Schilstra M.J., Clarke P.J.C., Rust A.G., Pan Z.J., Arnone M.I., Rowen L., Cameron R.A., McClay D.R., Hood L., and Bolouri H. 2002b. A provisional regulatory gene network for specification of endomesoderm in the sea urchin embryo. *Dev. Biol.* **246:** 162–190.

Gross J.M. and McClay D.R. 2001. The role of Brachyury (T) during gastrulation movements in the sea urchin *Lytechinus variegatus*. *Dev. Biol.* **239:** 132–147.

Hinman V.F. and Davidson E.H. 2003a. Expression of a gene encoding a Gata transcription factor during embryogenesis of the starfish *Asterina miniata*. *Gene Expr. Patterns* **3:** 419–422.

———. 2003b. Expression of AmKrox, a starfish ortholog of a sea urchin transcription factor essential for endomesodermal specification. *Gene Expr. Patterns* **3:** 423–426.

Hinman V.F., Nguyen A.T., and Davidson E.H. 2003b. Expression and function of a starfish Otx ortholog, AmOtx: a conserved role for Otx proteins in endoderm development that predates divergence of the eleutherozoa. *Mech. Dev.* **120:** 1165–1176.

Hinman V.F., Nguyen A.T., Cameron R.A., and Davidson E.H. 2003a. Developmental gene regulatory network architecture across 500 million years of echinoderm evolution. *Proc. Natl. Acad. Sci.* **100:** 13356–13361.

Hörstadius S. 1939. The mechanics of sea urchin development studied by operative methods. *Biol. Rev. Camb. Philos. Soc.* **14:** 132–179.

Ideker T., Galitski T., and Hood L. 2001. A new approach to decoding life: Systems biology. *Annu. Rev. Genomics Hum. Genet.* **2:** 343–372.

Kenny A.P., Kozkowski D.J., Oleksyn D.W., Angerer L.M., and Angerer R.C. 1999. SpSoxB1, a maternally encoded transcription factor asymmetrically distributed among early sea urchin blastomeres. *Development* **126:** 5473–5483.

Kenny A.P., Oleksyn D.W., Newman L.A., Angerer R.C., and Angerer L.M. 2003. Tight regulation of SpSoxB factors is required for patterning and morphogenesis in sea urchin embryos. *Dev. Biol.* **261:** 412–425.

Kitano H. 2002. Looking beyond the details: A rise in system-oriented approaches in genetics and molecular biology. *Curr. Genet.* **41:** 1–10.

Kiyomoto M. and Shirai H. 1993a. The determinant for archenteron formation in starfish: Co-culture of an animal egg fragment-derived cluster and a selected blastomere. *Dev. Growth Differ.* **35:** 99–105.

———. 1993b. Reconstruction of starfish eggs by electric cell fusion: A new method to detect the cytoplasmic determinant for archenteron formation. *Dev. Growth Differ.* **35:** 107–114.

Kuraishi R. and Osanai K. 1992. Cell movements during gastrulation of starfish larvae. *Biol. Bull.* **183:** 258–268.

———. 1994. Contribution of maternal factors and cellular interaction to determination of archenteron in the starfish embryo.

Development **120:** 2619–2628.

Lee Y.H., Huang G.M., Cameron R.A., Graham G., Davidson E.H., Hood L., and Britten R.J. 1999. EST analysis of gene expression in early cleavage-stage sea urchin embryos. *Development* **126:** 3857–3867.

Logan C.Y., Miller J.R., Ferkowicz M.J., and McClay D.R. 1999. Nuclear β-catenin is required to specify vegetal cell fates in the sea urchin embryo. *Development* **126:** 345–357.

Miyawaki K., Yamamoto M., Saito K., Saito S., Kobayashi N., and Matsuda S. 2003. Nuclear localization of β-catenin in vegetal pole cells during early embryogenesis of the starfish *Asterina pectinifera*. *Dev. Growth Differ.* **45:** 121–128.

Oliveri P., Carrick D.M., and Davidson E.H. 2002. A regulatory gene network that directs micromere specification in the sea urchin embryo. *Dev. Biol.* **246:** 209–228.

Oliveri P., Davidson E.H., and McClay D.R. 2003. Activation of pmar1 controls specification of micromeres in the sea urchin embryo. *Dev. Biol.* **258:** 32–43.

Patient R.K. and McGhee J.D. 2002. The GATA family (vertebrates and invertebrates). *Curr. Opin. Genet. Dev.* **12:** 416–422.

Ransick A. and Davidson E.H. 1993. A complete 2nd gut induced by transplanted micromeres in the sea-urchin embryo. *Science* **259:** 1134–1138.

Ransick A., Ernst S., Britten R.J., and Davidson E.H. 1993. Whole-mount in-situ hybridization shows endo-16 to be a marker for the vegetal plate territory in sea-urchin embryos. *Mech. Dev.* **42:** 117–124.

Ransick A., Rast J.P., Minokawa T., Calestani C., and Davidson E.H. 2002. New early zygotic regulators expressed in endomesoderm of sea urchin embryos discovered by differential array hybridization. *Dev. Biol.* **246:** 132–147.

Rast J.P., Cameron R.A., Poustka A.J., and Davidson E.H. 2002. Brachyury target genes in the early sea urchin embryo isolated by differential macroarray screening. *Dev. Biol.* **246:** 191–208.

Rast J.P., Amore G., Calestani C., Livi C.B., Ransick A., and Davidson E.H. 2000. Recovery of developmentally defined gene sets from high-density cDNA macroarrays. *Dev. Biol.* **228:** 270–286.

Sherwood D.R. and McClay D.R. 1999. LvNotch signaling mediates secondary mesenchyme specification in the sea urchin embryo. *Development* **126:** 1703–1713.

Shoguchi E., Satoh N., and Maruyama Y.K. 2000. A starfish homolog of mouse T-brain-1 is expressed in the archenteron of *Asterina pectinifera* embryos: Possible involvement of two T-box genes in starfish gastrulation. *Dev. Growth Differ.* **42:** 61–68.

Smith A.B. 1988. Fossil evidence for the relationships of extant echinoderm classes and their times of divergence. In *Echinoderm phylogeny and evolutionary biology* (ed. C.R.C. Paul and A.B. Smith), pp. 85–97. Clarendon Press, Oxford, United Kingdom.

Smith L.C., Chang L., Britten R.J., and Davidson E.H. 1996. Sea urchin genes expressed in activated coelomocytes are identified by expressed sequence tags—Complement homologues and other putative immune response genes suggest immune system homology within the deuterostomes. *J. Immunol.* **156:** 593–602.

Sprinkle J. and Kier P.M. 1987. The echinodermata. In *Fossil invertebrates* (ed. R.S. Boardman et al.), pp. 550–611. Blackwell Scientific, Palo Alto, California.

Sweet H.C., Gehring M., and Ettensohn C.A. 2002. LvDelta is a mesoderm-inducing signal in the sea urchin embryo and can endow blastomeres with organizer-like properties. *Development* **129:** 1945–1955.

Sweet H.C., Hodor P.G., and Ettensohn C.A. 1999. The role of micromere signaling in Notch activation and mesoderm specification during sea urchin embryogenesis. *Development* **126:** 5255–5265.

Vonica A., Weng W., Gumbiner B.M., and Venuti J.M. 2000. TCF is the nuclear effector of the β-catenin signal that patterns the sea urchin animal-vegetal axis. *Dev. Biol.* **217:** 230–243.

Wada H. and Satoh N. 1994. Phylogenetic relationships among extant classes of echinoderms, as inferred from sequences of 18s Rdna, coincide with relationships deduced from the fossil record. *J. Mol. Evol.* **38:** 41–49.

Wikramanayake A.H., Huang L., and Klein W.H. 1998. β-Catenin is essential for patterning the maternally specified animal-vegetal axis in the sea urchin embryo. *Proc. Natl. Acad. Sci.* **95:** 9343–9348.

Yuh C.H., Bolouri H., and Davidson E.H. 1998. Genomic cis-regulatory logic: Experimental and computational analysis of a sea urchin gene. *Science* **279:** 1896–1902.

Zhu X.D., Mahairas G., Illies M., Cameron R.A., Davidson E.H., and Ettensohn C.A. 2001. A large-scale analysis of mRNAs expressed by primary mesenchyme cells of the sea urchin embryo. *Development* **128:** 2615–2627.

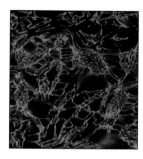

THE EXTRACELLULAR MATRIX OF THE DEUTEROSTOME GASTRULA

R. Winklbauer[1] and C.A. Ettensohn[2]

[1]University of Toronto, Department of Zoology, Toronto, Ontario M5S 3G5, Canada; [2]Carnegie Mellon University, Department of Biological Sciences, Pittsburgh, Pennsylvania 15213

INTRODUCTION

Embedding individual cells in a common ground substance is a general characteristic of multicellular organisms. In metazoans, this substance is the extracellular matrix (ECM), and it functions in parallel with direct cell–cell adhesion to mediate multicellularity. Being built from a large and diverse set of components, the composition of the ECM, and hence its structure and function, vary widely within and between organisms.

A function of the ECM is to generate and fill spaces between cells, and macromolecules particularly suited for this are the glycosaminoglycans (GAGs). Their negatively charged polysaccharide chains form hydrated gels that occupy large volumes. Most GAGs are covalently linked to core proteins to form proteoglycans, which can associate into vast complexes (Iozzo 1998). Fibrillar structures such as those formed by collagens convey tensile strength to the matrix (Prockop and Kivirikko 1995). The ECM also contains a large variety of glycoproteins that mediate cell adhesion to the matrix and control aspects of cell behavior. They typically comprise several domains, taken from a set of conserved modules, that mediate multimerization, or interactions with other matrix constituents and with cellular receptors (Hynes and Zhao 2000; Hohenester and Engel 2002).

In general, the architecture of the ECM varies greatly, depending on its function. The basal lamina, however, is a specialized structure that is conserved throughout the metazoans (Hynes and Zhao 2000; Erickson and Couchman 2000). It lines the basal side of most epithelia.

Its major constituents are the laminins, cross-shaped glycoproteins composed of three subunits, the α, β, and γ chains (Tunggal et al. 2000). During basal lamina formation, laminins bind with their long arm to cellular receptors, whereas the short arms mediate multimerization. In this way, a supramolecular sheet is formed that corresponds to the lamina densa of the basal lamina (Yurchenco and O'Rear 1994; Li et al. 2003). A second network consists of collagen IV. In contrast to fibrillar collagens, its triple-helical domains are frequently interrupted, and the collagen IV molecules interact with each other to generate multilayered sheets (Hudson et al. 1993). Direct collagen IV–laminin interaction connects the two networks. Additional cross-linking is provided by entactin/nidogen and the proteoglycan perlecan. Basal lamina assembly depends on trimeric laminin, but not on the other components (Yurchenco and O'Rear 1994; Erickson and Couchman 2000; Li et al. 2003).

Cell binding to the ECM is dominated by a single family of receptors, the integrins. Each integrin consists of an α and a β subunit, and in bilaterian metazoans, two basic sets of $\alpha\beta$ heterodimers are present. One recognizes laminins and the other binds to the Arg-Gly-Asp (RGD) sequence found in many matrix proteins. Vertebrates have additional integrins that recognize collagens or Ig-superfamily cell-surface counterreceptors, or are leukocyte specific (Brown 2000; Hynes and Zhao 2000; Hynes 2002).

Several non-integrin receptors contribute to cell–ECM adhesion, and like integrins, they also transmit signals that inform the cell about its matrix environment (e.g., Winder 2001; Isacke and Yarwood 2002). Matrix proteins in the strict sense, those contributing to ECM structure, are not the only source of signals emanating from this compartment, however. Many growth factors bind to the ECM and interact with their cellular receptors while bound to the matrix or after their controlled release (Iozzo 1998). As an intermediate between matrix proteins and growth factors, proteins like those of the CCN (connective tissue growth factor/cyr61/nov) family combine ECM glycoprotein modules with growth-factor or growth-factor-binding domains (Perbal et al. 2003). In these ways, the ECM plays an important role in intercellular signaling (see also Chapter 50).

Depending on the organism, the ECM at gastrulation varies in complexity. In principle, an *apical matrix* is present on the outer surface of the epithelial wall of the embryo, and a *basal matrix* on its inner side. The basal ECM may consist of a *basal lamina* or a less organized form of matrix. The *blastocoel matrix* fills the blastocoel or comparable cavities. Finally, an *interstitial matrix* is often conspicuous between mesendodermal cells (Fig. 1). The embryonic ECM has been studied most extensively in sea urchins, amphibians, chickens, and mice. This chapter focuses on these systems.

THE EXTRACELLULAR MATRIX OF THE SEA URCHIN EMBRYO

Sea urchin embryos hatch from their fertilization envelopes as swimming blastulae. The blastocoel wall consists of monociliated cells that are arranged into a fully differentiated epithelium. Before embryos begin to gastrulate, they already possess an elaborate ECM consisting of an apical matrix, a basal matrix in the form of a basal lamina, and a blastocoel matrix (Fig. 1A).

The Structure and Formation of the Blastula-stage ECM

The Apical Matrix

The surface of the blastula is coated by a thick (~1 μm) hyaline layer (Fig. 1A). It consists of a finely fibrillar inner zone that includes the apical lamina adjacent to the cell membrane, a separating dense layer, and an outer zone of coarse fibrils. The tips of microvilli that are embedded in this matrix are covered by clusters of granules, the microvillus-associated bodies (Wolpert and Mercer 1963; Hall and Vacquier 1982; Spiegel et al. 1989). The basic structure of the hyaline layer is conserved not only within echinoderms, but also in many marine invertebrates including cnidarians (Spiegel et al. 1989), suggesting that this apical matrix is a primitive trait of metazoans.

A major component of the inner zone is hyalin, a filamentous glycoprotein that multimerizes to form a fibrous network (Hylander and Summers 1982; Bisgrove et al. 1991; Wessel et al. 1998) (see Chapter 9). Fibropellins, gly-

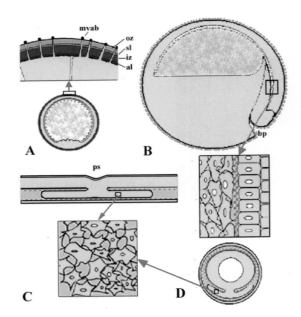

Figure 1. The extracellular matrix in the gastrula of deuterostomes. Sectional views of sea urchin (*A*), amphibian (frog) (B), avian (chick) (*C*), and mammalian (mouse) (*D*) early gastrulae are depicted. Magnified view of the sea urchin apical matrix (*rectangle*) is shown (*A, top*). Presence of fibronectin fibrils in the sea urchin gastrula (*A*) is hypothetical. Magnified views of the interstitial matrix (*rectangles*) in the frog (*B, bottom*) and chick and mouse gastrula (*C, bottom*) are shown. (*Blue*) Apical matrix (hyaline layer of sea urchin embryo); (*brown*) thick glycocalyx of amphibian embryo; (*red*) basal lamina; (*green*) fibronectin containing fibrils; (*gray*) blastocoel and interstitial matrix. (mvab) microvillus-associated body; (oz) outer zone of hyaline layer; (sl) separating layer; (iz) inner zone; (al) apical lamina; (bp) blastopore; (ps) primitive streak.

coproteins with EGF repeats and an avidin-like domain, are another constituent and make up the apical lamina (Hall and Vacquier 1982; Bisgrove et al. 1991; Bisgrove and Raff 1993; Burke et al. 1998). Additional components of the apical matrix include the Ecto-V protein of the microvillus-associated bodies (Coffman and McClay 1990), the glycoprotein apextrin (Haag et al. 1999), and pamlin (Katow and Komazaki 1996). Lectins like echinonectin (Alliegro et al. 1988) or Pl-nectin (Matranga et al. 1992) have also been regarded to be apical matrix constituents. Echinonectin has been shown to reside predominantly in vesicles below the apical cell membrane, however, rather than in the hyaline layer (Fuhrman et al. 1992).

The Basal Matrix and Blastocoel Matrix

The matrix on the basal surface of the blastula wall consists of a thin (0.1 μm) basal lamina (Fig. 1A) (Wolpert and Mercer 1963; Amemiya 1989a; Cherr et al. 1992). The blastocoel is filled with a hydrated matrix that is difficult to preserve during fixation. It shrinks when dehydrated and collapses onto the basal lamina, giving the impression that the latter is continuous with the blastocoel matrix (Amemiya 1989a; Burke and Tamboline 1990). When the blastocoel matrix is carefully preserved by freeze-substitution fixation, it appears closely attached to the basal lamina, yet not continuous with it. It consists of an interconnected, anisotropic network of fibrillar or sheet-like elements (Cherr et al. 1992).

Laminin is present in the blastula basal lamina (McCarthy and Burger 1987; McCarthy et al. 1987; Benson et al. 1999). Data on collagen IV expression are ambiguous. Antibodies against vertebrate collagen IV label the basal side of the blastula wall (Wessel et al. 1984), but genes coding for sea urchin type IV collagens, *COLL3α* and *COLL4α*, are expressed only later, and antibodies detect the respective proteins from the gastrula stage on (Angerer et al. 1988; Nemer and Harlow 1988; Wessel et al. 1991; Exposito et al. 1993, 1994; Suzuki et al. 1997). Dermatan sulfate, heparan sulfate, and chondroitin sulfate may also occur in the basal lamina (Solursh and Katow 1982), implying the presence of proteoglycans that contain these GAGs.

ECM molecules that are not exclusive basal lamina components have been proposed to reside in the basal matrix. Antibodies against vertebrate fibronectin (FN) label this structure (Spiegel et al. 1980; Wessel et al. 1984), and it would be interesting to see whether FN fibrils attach to the basal lamina surface, as in amniote gastrulae (Fig. 1A) (see below). Consistent with the presence of FN protein, a sea urchin *FN* gene has been identified (Zhu et al. 2001) and a fibronectin-like protein has been purified (DeSimone et al. 1985). Biochemical evidence suggests that fibrillar collagens occur in the blastula (Benson et al. 1990), and collagens type I and III have been localized to the basal matrix using cross-reacting antibodies (Wessel et al. 1984). Given its tendency to collapse upon fixation, fibrillar collagens could be components of the blastocoel matrix. GAGs are also present in this matrix compartment (Oguri and Yamagata 1978; Akasaka and Terayama 1980; Vilela-Silva et al. 2001).

Many components of the blastula ECM are synthesized in the oocyte, stored in vesicles, and secreted after fertilization. Different matrix molecules reside in different classes of vesicles and are released in a spatially and temporally regulated fashion (Wessel et al. 1984; Alliegro and McClay 1988; Matese et al. 1997). For example, the apical matrix constituent, hyalin, is stored in the cortical granules, which release their contents to the egg surface within seconds after fertilization (Hylander and Summers 1982; McClay and Fink 1982; Matese et al. 1997; Wessel et al. 1998). In contrast, the fibropellins are contained in "apical vesicles," whose secretion starts later (Bisgrove et al. 1991).

Components of the basal lamina and blastocoel matrix are also stored. Immediately after fertilization, "basal lamina vesicles" move to the egg surface, but at the morula stage, their contents begin to be released into the blastocoel (Wessel et al. 1984; Matese et al. 1997; Tesoro et al. 1998;

Benson et al. 1999). Based on fluorescence microscopy, it has been proposed that some of their contents are released into the apical matrix (Matese et al. 1997), but this has not yet been confirmed ultrastructurally. In any case, compartmentalization and elaborate vesicle trafficking are employed to build the basic structures of the ECM from maternal protein pools. Matrix proteins are also synthesized de novo, however. For example, *laminin α-chain* mRNA is present throughout early development and its translation leads to a continuous accumulation of the protein (Benson et al. 1999).

Remodeling and Diversification of the ECM during Gastrulation

Changes in the Apical Matrix during Gastrulation

No obvious morphological changes occur in the apical matrix during the ingression of primary mesenchyme cells (see Chapter 9). However, the apical matrix is remodeled during the invagination of the vegetal plate: The outer zone of the hyaline layer is shed in the invaginating region (Spiegel et al. 1989). Although cells partially detach from the hyaline layer (Amemiya 1989b), it is nevertheless carried into the forming archenteron (Citkowitz 1971). New hyalin, however, is added only in the ectodermal region (McClay and Fink 1982), and *hyalin* mRNA transcription becomes restricted to the aboral ectoderm (Wessel et al. 1998). A similar pattern is observed for the fibropellins (Bisgrove et al. 1991; Grimwade et al. 1991). In contrast, a putative chondroitin sulfate proteoglycan is newly secreted into the hyaline layer only at the site of invagination (Lane et al. 1993).

Changes in the Basal and Blastocoel Matrix at Gastrulation

Immediately after its formation, the basal lamina is disrupted at the site of primary mesenchyme cell ingression (Fig. 1A) (Katow and Solursh 1980; Amemiya 1989a; Benson et al. 1999). At the onset of invagination, it has disappeared from the vegetal plate (Ettensohn 1984; Amemiya 1989a), although it re-forms on the archenteron later in gastrulation (McCarthy and Burger 1987; Benson et al. 1999). These changes are matched by the expression of αSU2, a putative laminin-binding integrin subunit. Initially found on all blastula cells, it disappears from the vegetal plate at the time of ingression, to be weakly reexpressed in the archenteron (Hertzler and McClay 1999). Sea urchin integrin βG shows also lowered expression in the archenteron (Marsden and Burke 1997).

Several other matrix components are differentially expressed during gastrulation. The ECM1 antibody epitope (Ingersoll and Ettensohn 1994), the ECM3 protein which resembles amino-terminally the mammalian proteoglycan core protein NG2 (Wessel and Berg 1995; Hodor et al. 2000), and pamlin (Katow and Komazaki 1996) all become enriched in the matrix at the vegetal-most part of the ectoderm, adjacent to the site of invagination, whereas Endo16 protein is expressed on endoderm cells (Soltysik-Espanola et al. 1994; Godin et al. 1996). mRNA for the ECM18 protein is present throughout early development, but its translation starts in the mesenchyme blastula (Berg et al. 1996).

Collagen gene expression also changes at gastrulation. Sea urchin *COLL3α* and *COLL4α* encode type IV collagens (Angerer et al. 1988; Nemer and Harlow 1988; Wessel et al. 1991; Exposito et al. 1993, 1994). Both collagens are coexpressed, with mRNAs appearing in primary mesenchyme cells shortly before ingression, and later also in secondary mesenchyme (Angerer et al. 1988; Wessel et al. 1991; Exposito et al. 1994; Suzuki et al. 1997). In the gastrula, the 3α-chain protein is ubiquitously distributed in the basal matrix and on primary mesenchyme cells (Wessel et al. 1991). Of the two fibrillar collagen genes cloned from sea urchins, *COLL1α* is a type I (Saitta et al. 1989; Exposito et al. 1992b), and *COLL2α* a type II collagen (D'Alessio et al. 1990; Exposito et al. 1992a; Lethias et al. 1997). The two collagens are coexpressed. Transcripts accumulate in the late gastrula in primary and secondary mesenchyme cells (D'Alessio et al. 1990; Gambino et al. 1997; Suzuki et al. 1997). For *COLL1α*, some localized maternal mRNA is present in the egg (Gambino et al. 1997; Romancino et al. 2000). The 2α-chain protein is first detected in prism-stage embryos (Lethias et al. 1997).

In a large-scale analysis of mRNAs expressed in primary mesenchyme, additional ECM constituents were identified: the basal lamina components perlecan and entactin/nidogen, a fibrinogen-related protein, fibrillin, F-spondin, and osteonectin/SPARC (Zhu et al. 2001). Moreover, in the sea urchin, terminal cell differentiation already occurs at the early gastrula stage, when primary mesenchyme cells begin to synthesize a spicular endoskeleton. In this process, 40–50 spicule matrix proteins become expressed (Killian and Wilt 1996; Wilt 1999; Illies et al. 2002, and references therein). In conclusion, a well-developed, spatially and temporally differentiated matrix is present in the sea urchin gastrula, suggesting distinct functions in regional differentiation processes and in morphogenesis.

THE EXTRACELLULAR MATRIX OF THE AMPHIBIAN GASTRULA

The timing of many early developmental events is similar in sea urchins and amphibians. The amphibian embryo is an order of magnitude larger, however. Whereas the sea urchin embryo hatches as a free-swimming blastula, with a differentiated, ciliated epithelium surrounding the blastocoel, the amphibian embryo leaves the egg capsule much later, and differentiated epithelia develop only after gastrulation. The ECM appears to be correspondingly less developed in the amphibian blastula and gastrula (Fig. 1B). Matrix formation has just started at the onset of gastrulation. Therefore, matrix structure and development are discussed simultaneously.

The Apical Matrix

No apical matrix corresponding to the hyaline layer of invertebrates forms on the surface of the amphibian fertilized egg. In the cephalochordate *Amphioxus*, however, a trace of hyaline layer formation is observed. At fertilization, hyaline material is secreted from cortical granules. The putative hyaline layer does not attach to the egg, however, but rather to the fertilization envelope that elevates from the embryo surface (Holland and Holland 1989). This suggests that in chordates, the hyaline layer may have become inconspicuous by losing its association with the embryo surface, and hence its apical matrix function.

In *Xenopus*, an unusually thick (2 μm) "glycocalyx" covers the unfertilized egg (Fig. 1B). At fertilization, a "smooth layer" of material, probably of cortical granule origin, is added to the surface of the glycocalyx (Larabell and Chandler 1988). The glycocalyx is retained on the apical membrane of blastomeres. Its composition is not known, but in contrast to the hyaline layer, it is not sensitive to Ca^{++} removal, and it does not form a supracellular structure that is continuous across cell boundaries (Muller and Hausen 1995). The only putative matrix component localized so far to the apical surface is a collagen-VI-like molecule, which is secreted from the egg and early embryo and is required for gastrulation (Otte et al. 1990). The apical surface is nonadhesive to the basolateral membranes of embryonic cells (Byers and Armstrong 1986).

The Basal Matrix

In the late blastula, the basal surface of the blastocoel wall becomes covered by a network of ECM fibrils (Fig. 1B). It is restricted to the animal part of the wall, the blastocoel roof (Nakatsuji et al. 1982, 1985b; Nakatsuji and Johnson 1983; Darribere et al. 1985b). In all amphibians, FN is an essential constituent of these fibrils (Boucaut and Darribere 1983a,b; Darribere et al. 1985a,b; Delarue et al. 1985; Nakatsuji et al. 1985b; Johnson et al. 1992; Collazo 1994). The two disulfide-linked subunits of FN, which are identical except for splice variations, consist of domains that mediate interactions with other matrix constituents, with FN itself, and with cellular receptors. A major cell-binding site is defined by an RGD sequence, whose recognition by integrins is facilitated by an adjacent synergy site. In matrix fibrils, FN dimers are cross-linked into an insoluble multimer (Pankov and Yamada 2002).

In *Xenopus* embryos, FN protein is initially translated from maternal mRNA (Lee et al. 1984; Danker et al. 1993). During gastrulation, zygotic transcription increases the *FN* mRNA pool (DeSimone et al. 1992). A similar pattern of FN synthesis is observed in the urodele *Pleurodeles* (Darribere et al. 1985a). Although all cells of the early embryo synthesize and secrete FN, fibril assembly is restricted to the blastocoel roof. Apparently, availability of secreted FN is not limiting fibrillogenesis (Lee et al. 1984; Winklbauer 1998). Once initiated, the fibril network grows rapidly. Individual fibrils elongate in a unipolar fashion at 5 µm per minute, and many fibrils grow simultaneously on a given cell (Winklbauer and Stoltz 1995).

The mechanism of FN fibril formation in amphibian embryos is similar to that in other systems (Darribere et al. 1990, 1992; Winklbauer and Nagel 1991; Winklbauer and Stoltz 1995; Na et al. 2003). Fibrillogenesis is integrin-dependent (Darribere et al. 1990; Winklbauer and Stoltz 1995), and the major isoform in the gastrula is integrin α5β1. Maternal mRNAs coding for both subunits are stored in the egg, and during cleavage and gastrulation, this integrin is synthesized and inserted into the membranes of all cells (DeSimone and Hynes 1988; Gawantka et al. 1992; Ransom et al. 1993; Whittaker and DeSimone 1993; Joos et al. 1995). No differential expression of integrins is obvious that could explain the onset of FN fibril formation in the late blastula or its restriction to the blastocoel roof (Ramos and DeSimone 1996). Syndecans are specifically expressed in the blastocoel roof, however, and are essential for FN fibril assembly (Kramer and Yost 2002).

In urodeles, laminin is present in the gastrula, but instead of being incorporated into a basal lamina, it colocalizes with FN in its fibril network (Nakatsuji et al. 1985a; Darribere et al. 1986). Like FN, laminin is synthesized in oocytes and during cleavage, apparently from maternal mRNA. At the mid-blastula stage, synthesis increases in all blastomeres (Riou et al. 1987). In *Xenopus*, laminin β-chain is only detected at the end of gastrulation, around the notochord (Fey and Hausen 1990). However, dystroglycan, a non-integrin laminin receptor, is expressed throughout early *Xenopus* development (Lunardi and Dente 2002; Moreau et al. 2003).

A few other glycoproteins reside in the basal matrix. In *Xenopus*, a representative of the CCN family (Perbal et al. 2003), Cyr61, is expressed. Like other members of the family, it contains regions similar to IGF-binding domains, to cysteine-rich chordin domains, thrombospondin type I repeats, and slit protein. Although the respective maternal mRNA disappears before gastrulation, the protein seems to persist in the blastocoel roof matrix, where it is required for gastrulation movements (Latinkic et al. 2003). In *Bufo*

embryos, antibodies against 85- to 75-kD proteins from embryonic extracellular matrix stain dots and fine fibrils on the blastocoel roof (Genta et al. 1997).

Of the GAGs present in the amphibian gastrula, hyaluronan is enriched on the blastocoel roof surface (Müllegger and Lepperdinger 2002a). Correspondingly, in freeze-dried embryos, this surface appears covered by a smooth, amorphous material (Komazaki 1985, 1986). Syndecans are transmembrane proteoglycans, and therefore do not qualify as matrix proteins in a strict sense. Nevertheless, the syndecans Xsyn-1 and Xsyn-2 are both expressed in the inner cell layers of the blastocoel roof (Teel and Yost 1996), where they are involved in FN matrix formation (Kramer and Yost 2002).

Blastocoel Matrix and Interstitial Matrix

As in sea urchins, the hydrated blastocoel matrix of amphibians is poorly preserved by conventional fixation methods. It is visualized in freeze-dried embryos as a fibrillar mass that fills the blastocoel from cleavage stages on. As the blastocoel shrinks in the late gastrula, the matrix diminishes accordingly. At the same time, the expanding archenteron fills with similar material (Komazaki 1985, 1986). The interstitial matrix, which may be continuous with the blastocoel matrix (Komazaki 1985), can be visualized by toluidine blue or lanthanum red staining. It occupies small, occasional gaps between cells at blastula and gastrula stages (Fig. 1B) (Johnson 1977).

In *Xenopus*, hyaluronan is a main constituent of the blastocoel and interstitial matrix. It appears also in the expanding archenteron; i.e., in a space continuous with the exterior of the embryo (Müllegger and Lepperdinger 2002a). An extracellular hyaluronidase, Xhyal2 (Müllegger and Lepperdinger 2002b), and two hyaluronan synthases are expressed in the early *Xenopus* embryo. DG42/Xhas1 synthesizes intermediate length (Rosa et al. 1988), and Xhas2 high-molecular-weight hyaluronan (Koprunner et al. 2000).

Few glycoproteins of the interstitial and blastocoel matrix are known. FN is associated with the mucous material of these matrices (Danker et al. 1993; Winklbauer 1998). Furthermore, XL-43, an extracellular lectin, is localized interstitially in the blastopore region, and on the blastocoel roof in an irregular layer reminiscent of collapsed blastocoel matrix (Outenreath et al. 1988). Latent TGF-β-binding proteins (LTBPs) establish disulfide links to the propeptide of TGF-β growth factors to regulate signaling, but they also play structural roles in the matrix. In the *Xenopus* gastrula, LTBP-1 is expressed in the prospective axial mesoderm (Altmann et al. 2002). Last, fibrillin, a constitutive component of 10-nm extracellular microfibrils, is present in the interstitial matrix of the axial mesoderm during *Xenopus* gastrulation (Skoglund 1996; thesis, cited after Rongish et al. 1998).

Other Matrix Molecules in the Amphibian Gastrula

Radiolabeling suggested that in *Xenopus* (Green et al. 1968) and *Rana* (Klose and Flickinger 1971), collagen is already synthesized in the gastrula. In fact, type VI collagen-like protein is synthesized from early cleavage. In addition to the apical matrix, it accumulates between cells and probably in the blastocoel (Otte et al. 1990). A short form of collagen XVIII is also present in the gastrula (Elamaa et al. 2002), but the major fibrillar type I and II collagens are expressed only after gastrulation (Su et al. 1991; Goto et al. 2000). Radiolabeling also provided indirect evidence for the presence of GAG-containing compounds, most likely proteoglycans, in the amphibian blastula and gastrula (Kosher and Searls 1973; Hoglund and Lovtrup 1976; Brickman and Gerhart 1994; Itoh and Sokol 1994).

THE EXTRACELLULAR MATRIX OF THE CHICKEN GASTRULA

The size of the chicken blastoderm is of the same order of magnitude as an amphibian embryo. The blastoderm is initially multilayered, but the central region becomes a single-layered epiblast before gastrulation. Cells from the epiblast and from the marginal zone form a novel structure, the hypoblast layer, below the epiblast (see Chapter 15). At gastrulation, matrix formation is in an advanced state, with a basal lamina lining the epiblast, but not the hypoblast, and with a distinct interstitial matrix in the mesoderm (Fig. 1C).

The Apical Matrix

No distinct apical matrix or glycocalix is detected on the surface of the chick embryo by electron microscopy (Mestres and Hinrichsen 1974; Sanders 1979), although localization of fibrillin-1 to the epiblast apical surface has been described (Burke et al. 2000). Interestingly, the vitelline membrane is used as a substrate for migration by cells of the blastoderm edge (Downie 1976). This shows how the vitelline membrane can take on functions of an apical matrix.

The Basal Matrix

The basal matrix on the inner surface of the late-blastula epiblast consists of a basal lamina containing laminin, nidogen/entactin, and collagen IV (Fig. 1C) (Low 1967; Sanders 1979; Bortier et al. 1989; Zagris and Chung 1990;

Harrisson et al. 1991; Zagris et al. 1993). Fibrillar material is attached to the surface of the basal lamina, showing distinct local differentiation by varying the thickness and the pattern of arrangement of fibrils (Wakely and England 1979; Harrisson et al. 1992). All these fibrils contain FN (Fig. 1C) (Critchley et al. 1979; Meier and Drake 1984; Monnet-Tschudi et al. 1985; Raddatz et al. 1991; Toyoizumi and Takeuchi 1992). Another matrix glycoprotein, tenascin, appears during gastrulation in the basal matrix, between the prospective neural ectoderm and the notochord (Crossin et al. 1986). During gastrulation, the basal matrix becomes disrupted at the primitive streak (Duband and Thiery 1982; Mitrani 1982; Sanders 1982; Harrisson et al. 1991) (Chapter 15). The hypoblast does not develop a basal lamina during gastrulation, but sparse FN staining is observed, particularly on its basal surface (Fig. 1C) (Sanders 1982; Harrisson et al. 1985; Raddatz et al. 1991).

The Interstitial Matrix

For the chicken, a matrix in the subgerminal cavity or between epiblast and hypoblast has not yet been described, but the interstitial matrix of the mesoderm has been well studied. After their ingression, mesodermal cells separate from each other, although they remain in contact via cytoplasmic processes (for review, see Winklbauer 1994). The gaps between cells are filled by a GAG matrix (Fig. 1C).

From radiolabeling, it appeared that hyaluronan is the most prominent GAG in the chicken blastula and gastrula, followed by chondroitin sulfate and heparan sulfate (Solursh 1976; Vanroelen et al. 1980c; Vanroelen and Vakaet 1981). However, direct isolation of GAGs from the embryo suggested that dermatan and chondroitin sulfate are most abundant initially, but heparan sulfate and hyaluronan are increasingly synthesized in the gastrula (Skandalis et al. 2003). GAGs have been localized to the interstitial matrix (Fisher and Solursh 1977; Vanroelen et al. 1980a,b). Moreover, injection of hyaluronidases into the embryo leads to the collapse of interstitial gaps in the mesoderm, suggesting that hyaluronan is the essential space-filling substance (Fisher and Solursh 1977; Van Hoof et al. 1986). Importantly, the interstitial matrix is not stationary, with individual mesoderm cells crawling through it, but moves with the migrating cells. When mesoderm is transplanted together with its labeled ECM into unlabeled hosts, the matrix remains linked to the donor cells instead of being left behind (Vanroelen et al. 1980a; Van Hoof and Harrisson 1986). As in *Xenopus*, fibrillin is part of the chick gastrula interstitial matrix, being expressed between the ingressing cells of the primitive streak (Gallagher et al. 1993; Wunsch et al. 1994). Fibrillin-1 has also been described in the epiblast (Burke et al. 2000).

THE EXTRACELLULAR MATRIX OF THE MOUSE GASTRULA

The mouse ovum is relatively small. Cleavage proceeds extremely slowly, with gastrulation starting 7 days after fertilization, although zygotic gene expression commences at the 2-cell stage. In contrast to amphibian or sea urchin embryos, the surface of the embryo proper is not related to the egg surface; it develops de novo in the inner cell mass, and no apical matrix is present. The mesoderm layer of the mouse gastrula closely resembles that of the chick embryo (Fig. 1C,D) (see, e.g., Batten and Haar 1979), but the prominent interstitial matrix in this tissue has not been analyzed. Extraembryonic tissues form early in the mouse, but only the development of the ECM of the embryo proper is discussed below.

The Basal Matrix

Before the onset of gastrulation, epiblast and visceral endoderm of the embryo proper are single-layered epithelia. A basal lamina covers the lower surface of the epiblast. At the early primitive-streak stage, it becomes disrupted at the site of ingression. Elsewhere, a network of matrix fibrils has been deposited on the basal lamina, as in the chick embryo. On the basal surface of the visceral endoderm, a similar fibril network has developed at that time, and shortly later, a basal lamina forms there (Fig. 1D; Pierce 1966; Batten and Haar 1979; Hashimoto and Nakatsuji 1989).

Laminin (Leivo et al. 1980; Wu et al. 1983; Dziadek and Timpl 1985; Herken and Barrach 1985; Miosge et al. 1993), collagen IV (Adamson and Ayers 1979; Leivo et al. 1980; Herken and Barrach 1985), nidogen/entactin (Wu et al. 1983; Dziadek and Timpl 1985), and heparan sulfate proteoglycan (Dziadek et al. 1985), as well as FN (Wartiovaara et al. 1979), have been localized to the basal matrix of the mouse gastrula. These proteins appear early, although not synchronously, before basal lamina formation. Laminin subunits β1 and γ1 and heparan sulfate proteoglycan occur on 2-cell blastomeres (Cooper and MacQueen 1983; Dziadek and Timpl 1985; Dziadek et al. 1985). At the 8-cell stage, the laminin α1 subunit is added (Cooper and MacQueen 1983), allowing assembly of the laminin-1 trimer (Yurchenco et al. 1997). It accumulates, together with collagen IV, entactin/nidogen, and FN, in the inner cell mass of the blastocyst, and then between nascent epiblast and visceral endoderm, respectively, where the basal lamina of the gastrula forms (Adamson and Ayers 1979; Wartiovaara et al. 1979; Leivo et al. 1980; Wu et al. 1983; Dziadek and Timpl 1985; Herken and Barrach 1985; Miosge et al. 1993). At this stage, laminin is synthesized by the visceral endoderm, under the control of FGF and Akt/PKB signaling and COUP-TF transcription factors (Li et al. 2001; Murray and Edgar 2001). Targeted disruption of the gene for the γ1 laminin subunit disrupts basal lamina formation between epiblast and visceral endoderm (Smyth et al. 1999).

SUMMARY: ECM STRUCTURE AND FUNCTION IN THE DEUTEROSTOME GASTRULA

Its ubiquity among invertebrates suggests that an apical matrix in the form of a hyaline layer is a primitive metazoan feature (Spiegel et al. 1989). Within the deuterostomes, the apical matrix becomes reduced in the chordate lineage (Fig. 1). Traces of it occur in *Amphioxus*, but in amphibians it has been replaced by a thick apical glycocalyx. No apical matrix has been described in amniotes, including chicken and mouse.

In sea urchins and amniotes, the basal matrix is represented by a basal lamina (Fig. 1). It forms shortly before gastrulation, but is to be soon disrupted at the site of mesoderm and endoderm internalization. In amniotes (and perhaps in sea urchins, although this has not been explicitly examined), FN-rich fibrils line the blastocoelic surface of the basal lamina, and such fibrils are the main constituents of the basal matrix in amphibians (Fig. 1). In amphibians, as well as in insects (Yarnitzky and Volk 1995) and nematodes (Huang et al. 2003), a basal lamina develops only at the end of gastrulation. For most groups of animals, however, data are lacking, and it cannot be decided presently whether a pregastrular basal lamina is a primitive feature or whether it evolved independently several times.

In sea urchins and vertebrates, space-filling blastocoel and interstitial matrices (Fig. 1) are composed largely of GAGs, and in vertebrates, hyaluronan is the main constituent. In addition, fibrillin is expressed in sea urchin primary mesenchyme cells, and fibrillin-containing microfilaments permeate the interstitial matrix of mesodermal subregions in vertebrates.

The ECM composition of the vertebrate gastrula appears simpler than that of sea urchins. Part of this may be due to the absence of an elaborate apical matrix in vertebrate embryos. Beyond that, it is not clear whether this tendency reflects a real difference, or a bias in the information on matrix protein expression. In sea urchins, ECM constituents often have been purified biochemically and subsequently characterized. This has led to the discovery of many new matrix components in the gastrula. In contrast, studies in amphibians, chick, and mouse have employed mostly a candidate molecule approach, focusing on the expression of known vertebrate matrix molecules in the gastrula. By concentrating thus on a fraction of already characterized ECM molecules—those that are also expressed at gastrulation—the impression of a less complex matrix may be generated. On the other hand, of the 43 mutations in mouse ECM proteins listed by Gustafsson and Fassler (2000), only 2—those targeting laminin γ1 chain and fibronectin—yielded phenotypes at early embryonic stages. This would be consistent with the view that the gastrula-stage ECM of the mouse is not very complex, although alternative explanations such as redundant protein functions cannot be excluded.

The importance of the ECM in early development differs between echinoderms and amniotes on one hand, and amphibians on the other. In sea urchins, almost any disturbance of matrix function leads to gross defects in development, which typically includes abnormal cell-fate specification and differentiation, to the effect that ECM functions in morphogenetic processes are difficult to identify (Hardin 1996). In amniotes, laminin and fibronectin are the major glycoproteins of the gastrula ECM, and mutations in both lead to defects in tissue specification, and perhaps only secondarily to altered morphogenesis (George et al. 1993; Smyth et al. 1999). In amphibians, on the other hand, the function of the matrix is primarily morphogenetic, as seen, for example, when FN or collagen VI function is inhibited (Otte et al. 1990; Ramos and DeSimone 1996; Winklbauer and Keller 1996). This may partly be explained from the small set of ECM molecules studied in this respect in amphibians. However, overexpression of matrix metalloproteases in *Xenopus*, which prompts extensive matrix degradation, affects development only long after the end of gastrulation, suggesting that amphibian embryos are indeed less dependent on the ECM during early development (Damjanovski et al. 2001). This difference is correlated with the degree of tissue differentiation attained at gastrulation. Whereas sea urchin and amniote gastrulae consist of mature epithelia and other terminally differentiating cells, such as primary mesenchyme, the amphibian embryo attains a similar level of cell differentiation only after gastrulation.

ACKNOWLEDGMENT

R.W. thanks Bojan Macanovic for preparing Figure 1.

REFERENCES

Adamson E.D. and Ayers S.E. 1979. The localization and synthesis of some collagen types in developing mouse embryos. *Cell* **16:** 953–965.

Akasaka K. and Terayama H. 1980. General pattern, $^{35}SO_4$-incorporation and intracellular localization of glycans in developing sea urchin (Anthocidaris) embryos. *Dev. Growth Differ.* **22:** 749–762.

Alliegro M.C. and McClay D.R. 1988. Storage and mobilization of extracellular matrix proteins during sea urchin development. *Dev. Biol.* **125:** 208–216.

Alliegro M.C., Ettensohn C.A., Burdsal C.A., Erickson H.P., and McClay D.R. 1988. Echinonectin: A new embryonic substrate adhesion protein. *J. Cell Biol.* **107:** 2319–2327.

Altmann C.R., Chang C., Munoz-Sanjuan I., Bell E., Heke M., Rifkin D.B., and Brivanlou A.H. 2002. The latent-TGFβ-binding-protein-1 (LTBP-1) is expressed in the organizer and regulates nodal and activin signaling. *Dev. Biol.* **248:** 118–127.

Amemiya S. 1989a. Development of the basal lamina and its role in migration and pattern formation of primary mesenchyme cells in sea urchin embryos. *Dev. Growth Differ.* **31:** 131–145.

———. 1989b. Electron microscopic studies on primary mesenchyme cell ingression and gastrulation in relation to vegetal pole cell behavior in sea urchin embryos. *Exp. Cell Res.* **183:** 453–462.

Angerer L.M., Chambers S.A., Yang Q., Venkatesan M., Angerer R.C., and Simpson R.T. 1988. Expression of a collagen gene in mesenchyme lineages of the *Strongylocentrotus purpuratus* embryo. *Genes Dev.* **2:** 239–246.

Batten B.E. and Haar J.L. 1979. Fine structural differentiation of germ layers in the mouse at the time of mesoderm formation. *Anat. Rec.* **194:** 125–142.

Benson S., Smith L., Wilt F., and Shaw R. 1990. The synthesis and secretion of collagen by cultured sea urchin micromeres. *Exp. Cell Res.* **188:** 141–146.

Benson S., Page L., Ingersoll E., Rosenthal E., Dungca K., and Signor D. 1999. Developmental characterization of the gene for laminin α-chain in sea urchin embryos. *Mech. Dev.* **81:** 37–49.

Berg L., Chen S.W., and Wessel G.M. 1996. An extracellular matrix molecule that is selectively expressed during development is important for gastrulation in the sea urchin embryo. *Development* **122:** 703–713.

Bisgrove B.W. and Raff R.A. 1993. The SpEGF III gene encodes a member of the fibropellins: EGF repeat-containing proteins that form the apical lamina of the sea urchin embryo. *Dev. Biol.* **157:** 526–538.

Bisgrove B.W., Andrews M.E., and Raff R.A. 1991. Fibropellins, products of an EGF repeat-containing gene, form a unique extracellular matrix structure that surrounds the sea urchin embryo. *Dev. Biol.* **146:** 89–99.

Bortier H., De Bruyne G., Espeel M., and Vakaet L. 1989. Immunohistochemistry of laminin in early chicken and quail blastoderms. *Anat. Embryol.* **180:** 65–69.

Boucaut J.-C. and Darriere T. 1983a. Fibronectin in early amphibian embryos. *Cell Tissue Res.* **234:** 135–145.

———. 1983b. Presence of fibronectin during early embryogenesis in the amphibian *Pleurodeles waltlii. Cell Differ.* **12:** 77–83.

Brickman M.C. and Gerhart J.C. 1994. Heparitinase inhibition of mesoderm induction and gastrulation in *Xenopus laevis* embryos. *Dev. Biol.* **164:** 484–501.

Brown N.H. 2000. Cell-cell adhesion via the ECM: Integrin genetics in fly and worm. *Matrix Biol.* **19:** 191–201.

Burke R.D. and Tamboline C.R. 1990. Ontogeny of an extracellular matrix component of sea urchins and its role in morphogenesis. *Dev. Growth Differ.* **32:** 461–471.

Burke R.D., Lail M., and Nakajima Y. 1998. The apical lamina and its role in cell adhesion in sea urchin embryos. *Cell Adhes. Commun.* **5:** 97–108.

Burke R.D., Wang D., Mark S., and Martens G. 2000. Distribution of fibrillin I in extracellular matrix and epithelia during early development of avian embryos. *Anat. Embryol.* **201:** 317–326.

Byers T.J. and Armstrong P.B. 1986. Membrane protein redistribution during *Xenopus* first cleavage. *J. Cell Biol.* **102:** 2176–2184.

Cherr G.N., Summers R.G., Baldwin J.D., and Morrill J.B. 1992. Preservation and visualization of the sea urchin embryo blastocoelic extracellular matrix. *Microsc. Res. Tech.* **22:** 11–22.

Citkowitz E. 1971. The hyaline layer: Its isolation and role in echinoderm development. *Dev. Biol.* **24:** 348–362.

Coffman J.A. and McClay D.R. 1990. A hyaline layer protein that becomes localized to the oral ectoderm and foregut of sea urchin embryos. *Dev. Biol.* **140:** 93–104.

Collazo A. 1994. Molecular heterochrony in the pattern of fibronectin expression during gastrulation in amphibians. *Evolution* **48:** 2037–2045.

Cooper A.R. and MacQueen H.A. 1983. Subunits of laminin are differentially synthesized in mouse eggs and early embryos. *Dev. Biol.* **96:** 467–471.

Critchley D.R., England M.A., Wakely J., and Hynes R.O. 1979. Distribution of fibronectin in the ectoderm of gastrulating chick embryos. *Nature* **280:** 498–500.

Crossin K.L., Hoffman S., Grumet M., Thiery J.-P., and Edelman G.M. 1986. Site-restricted expression of cytotactin during development of the chicken embryo. *J. Cell Biol.* **102:** 1917–1930.

D'Alessio M., Ramirez F., Suzuki H.R., Solursh M., and Gambino R. 1990. Cloning of a fibrillar collagen gene expressed in the mesenchymal cells of the developing sea urchin embryo. *J. Biol. Chem.* **265:** 7050–7054.

Damjanovski S., Amano T., Li Q., Pei D., and Shi Y.-B. 2001. Overexpression of matrix metalloproteinases leads to lethality in transgenic *Xenopus laevis*: Implications for tissue-dependent functions of matrix metalloproteinases during late embryonic development. *Dev. Dyn.* **221:** 37–47.

Danker K., Hacke H., Ramos J., DeSimone D., and Wedlich D. 1993. V+-Fibronectin expression and localization prior to gastrulation in *Xenopus laevis* embryos. *Mech. Dev.* **44:** 155–165.

Darribere T., Boucher D., Lacroix J.-C., and Boucaut J.-C. 1985a. Fibronectin synthesis during oogenesis and early development of the amphibian *Pleurodeles waltlii*. *Cell Differ.* **14:** 171–177.

Darribere T., Boulekbache H., Shi D.L., and Boucaut J.-C. 1985b. Immuno-electron-microscopic study of fibronectin in gastrulating amphibian embryos. *Cell Tissue Res.* **239:** 75–80.

Darribere T., Riou J.-F., Shi D.L., Delarue M., and Boucaut J.-C. 1986. Synthesis and distribution of laminin-related polypeptides in early amphibian embryos. *Cell Tissue Res.* **246:** 45–51.

Darribere T., Guida K., Larjava H., Johnson K.E., Yamada K.M., Thiery J.-P., and Boucaut J.-C. 1990. In vivo analyses of integrin β1 subunit function in fibronectin matrix assembly. *J. Cell Biol.* **110:** 1813–1823.

Darribere T., Koteliansky V.E., Chernousov M.A., Akiyama S.K., Yamada K.M., Thiery J.-P., and Boucaut J.-C. 1992. Distinct regions of human fibronectin are essential for fibril assembly in an in vivo developing system. *Dev. Dyn.* **194:** 63–70.

Delarue M., Darribere T., Aimar C., and Boucaut J.-C. 1985. Bufonid nucleocytoplasmic hybrids arrested at the early gastrula stage lack a fibronectin-containing fibrillar extracellular matrix. *Roux's Arch. Dev. Biol.* **194:** 275–280.

DeSimone D.W. and Hynes R.O. 1988. *Xenopus laevis* integrins. *J. Biol. Chem.* **263:** 5333–5340.

DeSimone D.W., Norton P.A., and Hynes R.O. 1992. Identification and characterization of alternatively spliced fibronectin mRNAs expressed in early *Xenopus* embryos. *Dev. Biol.* **149:** 357–369.

DeSimone D.W., Spiegel E., and Spiegel M. 1985. The biochemical identification of fibronectin in the sea urchin embryo. *Biochem. Biophys. Res. Commun.* **133:** 183–188.

Downie J.R. 1976. The mechanism of chick blastoderm expansion. *J. Embryol. Exp. Morphol.* **35:** 559–575.

Duband J.-L. and Thiery J.-P. 1982. Appearance and distribution of fibronectin during chick embryo gastrulation and neurulation. *Dev. Biol.* **94:** 337–350.

Dziadek M. and Timpl R. 1985. Expression of nidogen and laminin in basement membranes during mouse embryogenesis and in teratocarcinoma cells. *Dev. Biol.* **111:** 372–382.

Dziadek M., Fujiwara S., Paulsson M., and Timpl R. 1985. Immunological characterization of basement membrane types of heparan sulfate proteoglycan. *EMBO J.* **4:** 905–912.

Elamaa H., Peterson J., Pihlajaniemi T., and Destree O. 2002. Cloning of three variants of type XVIII collagen and their expression patterns during *Xenopus laevis* development. *Mech. Dev.* **114:** 109–113.

Erickson A.C. and Couchman J.R. 2000. Still more complexity in mammalian basement membranes. *J. Histochem. Cytochem.* **48:** 1291–1306.

Ettensohn C.A. 1984. Primary invagination of the vegetal plate during sea urchin gastrulation. *Am. Zool.* **24:** 571–588.

Exposito J-Y., D'Alessio M., and Ramirez F. 1992a. Novel amino-terminal propeptide configuration in a fibrillar procollagen undergoing alternative splicing. *J. Biol. Chem.* **267:** 17404–17408.

Exposito J-Y., D'Alessio M., Di Liberto M., and Ramirez F. 1993. Complete primary structure of a sea urchin type IV collagen α chain and analysis of the 5′ end of its gene. *J. Biol. Chem.* **268:** 5249–5254.

Exposito J-Y., D'Alessio M., Solursh M., and Ramirez F. 1992b. Sea urchin collagen evolutionarily homologous to vertebrate pro-α2(I) collagen. *J. Biol. Chem.* **267:** 15559–15562.

Exposito J-Y., Suzuki H., Geourjon C., Garrone R., Solursh M., and Ramirez F. 1994. Identification of a cell lineage-specific gene coding for a sea urchin α2(IV)-like collagen chain. *J. Biol. Chem.* **269:** 13167–13171.

Fey J. and Hausen P. 1990. Appearance and distribution of laminin during development of *Xenopus laevis*. *Differentiation* **42:** 144–152.

Fisher M. and Solursh M. 1977. Glycosaminoglycan localization and role in maintenance of tissue spaces in the early chick embryo. *J. Embryol. Exp. Morphol.* **42:** 195–207.

Fuhrman M.H., Suhan J.P., and Ettensohn C.A. 1992. Developmental expression of echinonectin, an endogenous lectin of the sea urchin embryo. *Dev. Growth Differ.* **34:** 137–150.

Gallagher B.C., Sakai L.Y., and Little C.D. 1993. Fibrillin delineates the primary axis of the early avian embryo. *Dev. Dyn.* **196:** 70–78.

Gambino R., Romancino D.P., Cervello M., Vizzini A., Isola M.G., Virruso L., and Di Carlo M. 1997. Spatial distribution of collagen type I mRNA in *Paracentrotus lividus* eggs and embryos. *Biochem. Biophys. Res. Commun.* **238:** 334–337.

Gawantka V., Ellinger-Ziegelbauer H., and Hausen P. 1992. β1-integrin is a maternal protein that is inserted into all newly formed plasma membranes during early *Xenopus* embryogenesis. *Development* **115:** 595–605.

George E.L., Georges-Labouesse E.N., Patel-King R.S., Rayburn H., and Hynes R.O. 1993. Defects in mesoderm, neural tube and vascular development in mouse embryos lacking fibronectin. *Development* **119:** 1079–1091.

Genta S.B., Aybar M.J., Peralta M.A., and Sanchez S.S. 1997. Evidence for the presence and participation of 85-75 Kda extracellular matrix components in cell interactions of Bufo arenarum gastrulation. *J. Exp. Zool.* **277:** 181–197.

Godin R.E., Urry L.A., and Ernst S.G. 1996. Alternative splicing of the Endo16 transcript produces differentially expressed mRNAs during sea urchin gastrulation. *Dev. Biol.* **179:** 148–159.

Goto T., Katada T., Kinoshita T., and Kubota H.Y. 2000. Expression and characterization of *Xenopus* type I collagen alpha 1 (COL1A1) during embryonic development. *Dev. Growth Differ.* **42:** 249–256.

Green H., Goldberg B., Schwartz M., and Brown D.D. 1968. The synthesis of collagen during the development of *Xenopus laevis. Dev. Biol.* **18:** 391–400.

Grimwade J.E., Gagnon M.L., Yang Q., Angerer R.C., and Angerer L.M. 1991. Expression of two mRNAs encoding EGF-related proteins identifies subregions of sea urchin embryonic ectoderm. *Dev. Biol.* **143:** 44–57.

Gustafsson E. and Fassler R. 2000. Insights into extracellular matrix functions from mutant mouse models. *Exp. Cell Res.* **261:** 52–68.

Haag E.S., Sly B.J., Andrews M.E., and Raff R.A. 1999. Apextrin, a novel extracellular protein associated with larval ectoderm evolution in *Heliocidaris erythrogramma. Dev. Biol.* **211:** 77–87.

Hall H.G. and Vacquier V.D. 1982. The apical lamina of the sea urchin embryo: Major glycoproteins associated with the hyaline layer. *Dev. Biol.* **89:** 168–178.

Hardin J. 1996. The cellular basis of sea urchin gastrulation. *Curr. Top. Dev. Biol.* **33:** 159–262.

Harrisson F., Andries L., and Vakaet L. 1992. The arrest of cell migration in the chicken blastoderm: Experimental evidence for the involvement of a band of extracellular fibrils associated with the basal lamina. *Int. J. Dev. Biol.* **36:** 123–137.

Harrisson F., Callebaut M., and Vakaet L. 1991. Features of polyingression and primtive streak ingression through the basal lamina in the chicken blastoderm. *Anat. Rec.* **229:** 369–383.

Harrisson F., Vanroelen C., and Vakaet L. 1985. Fibronectin and its relation to the basal lamina and to the cell surface in the chicken blastoderm. *Cell Tissue Res.* **241:** 391–397.

Hashimoto K. and Nakatsuji N. 1989. Formation of the primitive streak and mesoderm cells in mouse embryos—Detailed scanning electron microscopical study. *Dev. Growth Differ.* **31:** 209–218.

Herken R. and Barrach H.-J. 1985. Ultrastructural localization of type IV collagen and laminin in the seven-day-old mouse embryo. *Anat. Embryol.* **171:** 365–371.

Hertzler P.L. and McClay D.R. 1999. αSU2, an epithelial integrin that binds laminin in the sea urchin embryo. *Dev. Biol.* **207:** 1–13.

Hodor P.G., Illies M.R., Broadley S., and Ettensohn C.A. 2000. Cell-substrate interactions during sea urchin gastrulation: Migrating primary mesenchyme cells interact with and align extracellular matrix fibers that contain ECM3, a molecule with NG2-like and multiple calcium-binding domains. *Dev. Biol.* **222:** 181–194.

Hoglund L.R. and Lovtrup S. 1976. Changes in acid mucopolysaccharides during the development of the frog *Rana temporaria. Acta Embryol. Exp.* **1:** 63–79.

Hohenester E. and Engel J. 2002. Domain structure and organization in extracellular matrix proteins. *Matrix Biol.* **21:** 115–128.

Holland N.D. and Holland L.Z. 1989. Fine structural study of the cortical reaction and formation of the egg coats in a lancelet (=Amphioxus), Branchiostoma floridae (Phylum Chordata: Subphylum Cephalochordata=Acrania). *Biol. Bull.* **176:** 111–122.

Huang C.C., Hall D.H., Hedgecock E.M., Kao G., Karantza V., Vogel B.E., Hutter H., Chisholm A.D., Yurchenco P.D., and Wadsworth W.G. 2003. Laminin α subunits and their role in *C. elegans* development. *Development* **130:** 3343–3358.

Hudson B.G., Reeders S.T., and Tryggvason K. 1993. Type IV collagen: Structure, gene organization, and role in human diseases. *J. Biol. Chem.* **268:** 26033–26036.

Hylander B.L. and Summers R.G. 1982. An ultrastructural immuno-cytochemical localization of hyalin in the sea urchin egg. *Dev. Biol.* **93:** 368–380.

Hynes R.O. 2002. Integrins: Bidirectional, allosteric signaling machines. *Cell* **110:** 673–687.

Hynes R.O. and Zhao Q. 2000. The evolution of cell adhesion. *J. Cell Biol.* **150:** F89–95.

Illies M.R., Peeler M.T., Dechtiaruk A.M., and Ettensohn C.A. 2002. Identification and developmental expression of new biomineralization proteins in the sea urchin *Strongylocentrotus purpuratus. Dev. Genes Evol.* **212:** 419–431.

Ingersoll E.P. and Ettensohn C.A. 1994. An N-linked carbohydrate-containing extracellular matrix determinant plays a key role in sea urchin gastrulation. *Dev. Biol.* **163:** 351–366.

Iozzo R.V. 1998. Matrix proteoglycans: From molecular design to cellular function. *Annu. Rev. Biochem.* **67:** 609–652.

Isacke C.M. and Yarwood H. 2002. The hyaluronan receptor, CD44. *Int. J. Biochem. Cell Biol.* **34:** 718–721.

Itoh K. and Sokol S.Y. 1994. Heparan sulfate proteoglycans are required for mesoderm formation in *Xenopus* embryos. *Development* **120:** 2703–2711.

Johnson K.E. 1977. Extracellular matrix synthesis in blastula and gastrula stages of normal and hybrid frog embryos. I. Toluidine blue and lanthanum staining. *J. Cell Sci.* **25:** 313–322.

Johnson K.E., Darribere T., and Boucaut J.-C. 1992. Ambystoma maculatum gastrulae have an oriented, fibronectin-containing extracellular matrix. *J. Exp. Zool.* **261:** 458–471.

Joos T.O., Whittaker C.A., Meng F., DeSimone D.W., Gnau V., and Hausen P. 1995. Integrin α5 during early development of *Xenopus laevis. Mech. Dev.* **50:** 187–199.

Katow H. and Komazaki S. 1996. Spatio-temporal expression of pamlin during early embryogenesis in sea urchin and importance of N-linked glycosylaton for the glycoprotein function. *Roux's Arch. Dev. Biol.* **205:** 371–381.

Katow H. and Solursh M. 1980. Ultrastructure of primary mesenchyme cell ingression in the sea urchin Lytechinus pictus. *J. Exp. Zool.* **213:** 231–246.

Killian C.E. and Wilt F.H. 1996. Characterization of the proteins comprising the integral matrix of *Strongylocentrotus purpuratus* embryonic spicules. *J. Biol. Chem.* **271:** 9150–9159.

Klose J. and Flickinger R.A. 1971. Collagen synthesis in frog embryo endoderm cells. *Biochim. Biophys. Acta* **232:** 207–211.

Komazaki S. 1985. Scanning electron microscopy of the extracellular matrix of amphibian gastrulae by freeze-drying. *Dev. Growth Differ.* **27:** 57–62.

———. 1986. Accumulation and distribution of extracellular matrix as revealed by scanning electron microscopy in freeze-dried newt embryos before and during gastrulation. *Dev. Growth Differ.* **28:** 285–292.

Koprunner M., Mullegger J., and Lepperdinger G. 2000. Synthesis of hyaluronan of distinctly different chain length is regulated by differential expression of Xhas1 and 2 during early development of *Xenopus laevis. Mech. Dev.* **90:** 275–278.

Kosher R.A. and Searls R.L. 1973. Sulfated mucopolysaccharide synthesis during development of *Rana pipiens. Dev. Biol.* **32:** 50–68.

Kramer K.L. and Yost H.J. 2002. Ectodermal syndecan-2 mediates left-right axis formation in migrating mesoderm as a cell-nonautonomous Vg1 cofactor. *Dev. Cell* **2:** 115–124.

Lane M.C., Koehl M.A.R., Wilt F., and Keller R. 1993. A role for regulated secretion of apical extracellular matrix during epithelial invagination in the sea urchin. *Development* **117**: 1049–1060.

Larabell C.A. and Chandler D.E. 1988. The extracellular matrix of *Xenopus laevis* eggs: A quick-freeze, deep-etch analysis of its modification at fertilization. *J. Cell Biol.* **107**: 731–741.

Latinkic B.V., Mercurio S., Bennett B., Hirst E.M.A., Xu Q., Lau L.F., Mohun T.J., and Smith J.C. 2003. *Xenopus* Cyr61 regulates gastrulation movements and modulates Wnt signalling. *Development* **130**: 2429–2441.

Lee G., Hynes R., and Kirschner M. 1984. Temporal and spatial regulation of fibronectin in early *Xenopus* development. *Cell* **36**: 729–740.

Leivo I., Vaheri A., Timpl R., and Wartiovaara J. 1980. Appearance and distribution of collagens and laminin in the early mouse embryo. *Dev. Biol.* **76**: 100–114.

Lethias C., Exposito J.-Y., and Garrone R. 1997. Collagen fibrillogenesis during sea urchin development. *Eur. J. Biochem.* **245**: 434–440.

Li S., Edgar D., Fassler R., Wadsworth W., and Yurchenco P.D. 2003. The role of laminin in embryonic cell polarization and tissue organization. *Dev. Cell* **4**: 613–624.

Li X., Talts U., Talts J.F., Arman E., Ekblom P., and Lonai P. 2001. Akt/PKB regulates laminin and collagen IV isotypes of the basement membrane. *Proc. Natl. Acad. Sci.* **98**: 14416–14421.

Low F.N. 1967. Developing boundary (basement) membranes in the chick embryo. *Anat. Rec.* **159**: 231–238.

Lunardi A. and Dente L. 2002. Molecular cloning and expression analysis of dystroglycan during *Xenopus laevis* embryogenesis. *Gene Expr. Patterns* **2**: 45–50.

Marsden M. and Burke R.D. 1997. Cloning and characterization of novel β integrin subunits from a sea urchin. *Dev. Biol.* **181**: 234–245.

Matese J.C., Black S., and McClay D.R. 1997. Regulated exocytosis and sequential construction of the extracellular matrix surrounding the sea urchin zygote. *Dev. Biol.* **186**: 16–26.

Matranga V., Di Ferro D., Zito F., Cervello M., and Nakano E. 1992. A new extracellular matrix protein of the sea urchin embryo with properties of a substrate adhesion molecule. *Roux's Arch. Dev. Biol.* **201**: 173–178.

McCarthy R.A. and Burger M.M. 1987. In vivo embryonic expression of laminin and its involvement in cell shape change in the sea urchin *Sphaerechinus granularis*. *Development* **101**: 659–671.

McCarthy R.A., Beck K., and Burger M.M. 1987. Laminin is structurally conserved in the sea urchin basal lamina. *EMBO J.* **6**: 1587–1593.

McClay D.R. and Fink R.D. 1982. Sea urchin hyalin: Appearance and function in development. *Dev. Biol.* **92**: 285–293.

Meier S. and Drake C. 1984. SEM localization of cell-surface-associated fibronectin in the cranium of chick embryos utilizing immunolatex microspheres. *J. Embryol. Exp. Morphol.* **80**: 175–195.

Mestres P. and Hinrichsen K. 1974. The cell coat in the early chick embryo. *Anat. Embryol.* **146**: 181–192.

Miosge N., Gunther E., Becker-Rabbenstein V., and Herken R. 1993. Ultrastructural localization of laminin subunits during the onset of mesoderm formation in the mouse embryo. *Anat. Embryol.* **187**: 601–605.

Mitrani E. 1982. Primitive streak-forming cells of the chick invaginate through a basement membrane. *Wilhelm Roux's Arch.* **191**: 320–324.

Monnet-Tschudi F., Favrod P., Burnand M.-B., Verdan C., and Kucera

P. 1985. Fibronectin in the area opaca of the young chick embryo. Immunofluorescence and immuno-electron-microscopic study. *Cell Tissue Res.* **241**: 85–92.

Moreau N., Alfandari D., Gaultier A., Cousin H., and Darribere T. 2003. Cloning and expression patterns of dystroglycan during the early development of *Xenopus laevis*. *Dev. Genes Evol.* **213**: 355–359.

Müllegger J. and Lepperdinger G. 2002a. Hyaluronan is an abundant constituent of the extracellular matrix of *Xenopus* embryos. *Mol. Reprod. Dev.* **61**: 312–316.

———. 2002b. Degradation of hyaluronan by a Hyal2-type hyaluronidase affects pattern formation of vitelline vessels during embryogenesis of *Xenopus laevis*. *Mech. Dev.* **111**: 25–35.

Muller H.-A.J. and Hausen P. 1995. Epithelial cell polarity in early *Xenopus* development. *Dev. Dyn.* **202**: 405–420.

Murray P. and Edgar D. 2001. Regulation of laminin and COUP-TF expression in extraembryonic endodermal cells. *Mech. Dev.* **101**: 213–215.

Na J., Marsden M., and DeSimone D.W. 2003. Differential regulation of cell adhesive functions by integrin α subunit cytoplasmic tails in vivo. *J. Cell Sci.* **116**: 2333–2343.

Nakatsuji N. and Johnson K.E. 1983. Comparative study of extracellular fibrils on the ectodermal layer in gastrulae of five amphibian species. *J. Cell Sci.* **59**: 61–70.

Nakatsuji N., Gould A.C., and Johnson K.E. 1982. Movement and guidance of migrating mesodermal cells in *Ambystoma maculatum* gastrulae. *J. Cell Sci.* **56**: 207–222.

Nakatsuji N., Hashimoto K., and Hayashi M. 1985a. Laminin fibrils in newt gastrulae visualized by the immunofluorescent staining. *Dev. Growth Differ.* **27**: 639–643.

Nakatsuji N., Smolira M.A., and Wylie C.C. 1985b. Fibronectin visualized by scanning electron microscopy immunocytochemistry on the substratum for cell migration in *Xenopus laevis* gastrulae. *Dev. Biol.* **107**: 264–268.

Nemer M., and Harlow P. 1988. Sea urchin RNAs display differences in developmental regulation and in complementarity to a collagen exon probe. *Biochim. Biophys. Acta* **950**: 445–449.

Oguri K. and Yamagata T. 1978. Appearance of a proteoglycan in developing sea urchin embryos. *Biochim. Biophys. Acta* **541**: 385–393.

Otte A.P., Roy D., Siemerink M., Koster C.H., Hochstenbach F., Timmermans A., and Durston A.J. 1990. Characterization of a maternal type VI collagen in *Xenopus* embryos suggests a role for collagen in gastrulation. *J. Cell Biol.* **111**: 271–278.

Outenreath R.L., Roberson M.M., and Barondes S.H. 1988. Endogenous lectin secretion into the extracellular matrix of early embryos of *Xenopus laevis*. *Dev. Biol.* **125**: 187–194.

Pankov R. and Yamada K.M. 2002. Fibronectin at a glance. *J. Cell Sci.* **115**: 3861–3863.

Perbal B., Brigstock D.R., and Lau L.F. 2003. Report on the second international workshop on the CCN family of genes. *J. Clin. Pathol.* **56**: 80–85.

Pierce G.B. 1966. The development of basement membranes of the mouse embryo. *Dev. Biol.* **13**: 231–249.

Prockop D.J. and Kivirikko K.I. 1995. Collagens: Molecular biology, diseases, and potentials for therapy. *Annu. Rev. Biochem.* **64**: 403–434.

Raddatz E., Monnet-Tschudi F., Verdan C., and Kucera P. 1991. Fibronectin distribution in the chick embryo during formation of the blastula. *Anat. Embryol.* **183**: 57–65.

Ramos J.W. and DeSimone D.W. 1996. *Xenopus* embryonic cell adhesion to fibronectin: Position-specific activation of RGD/synergy site-dependent migratory behavior at gastrulation. *J. Cell Biol.* **134:** 227–240.

Ransom D.G., Hens M.D., and DeSimone D.W. 1993. Integrin expression in early amphibian embryos: cDNA cloning and characterization of *Xenopus* β1, β2, β3, and β6 subunits. *Dev. Biol.* **160:** 265–275.

Riou J.-F., Darribere T., Shi D.L., Richoux V., and Boucaut J.-C. 1987. Synthesis of laminin-related polypeptides in oocytes, eggs and early embryos of the amphibian *Pleurodeles waltlii*. *Wilhelm Roux's Arch. Dev. Biol.* **196:** 328–332.

Romancino D.P., Dalmazio S., Cervello M., Montana G., Virruso L., Bonura A., Gambino R., and Di Carlo M. 2000. Localization and association to cytoskeleton of COLL1α mRNA in *Paracentrotus lividus* eggs requires *cis*- and *trans*-acting factors. *Mech. Dev.* **99:** 113–121.

Rongish B.J., Drake C.J., Argraves W.S., and Little C.D. 1998. Identification of the developmental marker, JB3-antigen, as fibrillin-2 and its de novo organization into embryonic microfibrous arrays. *Dev. Dyn.* **212:** 461–471.

Rosa F., Sargent T.D., Rebbert M.L., Michaels G.S., Jamrich M., Grunz H., Jonas E., Winkles J.A., and Dawid I.B. 1988. Accumulation and decay of DG42 gene products follow a gradient pattern during *Xenopus* embryogenesis. *Dev. Biol.* **129:** 114–123.

Saitta B., Buttice G., and Gambino R. 1989. Isolation of a putative collagen-like gene from the sea urchin *Paracentrotus lividus*. *Biochem. Biophys. Res. Commun.* **158:** 633–639.

Sanders E.J. 1979. Development of the basal lamina and extracellular materials in the early chick embryo. *Cell Tissue Res.* **198:** 527–537.

———. 1982. Ultrastructural immunocytochemical localization of fibronectin in the early chick embryo. *J. Embryol. Exp. Morphol.* **71:** 155–170.

Skandalis S.S., Theocharis A.D., Papageorgakopoulou N., and Zagris N. 2003. Glycosaminoglycans in early chick embryo. *Int. J. Dev. Biol.* **47:** 311–314.

Skoglund P.M. 1996. "The role of *Xenopus* fibrillin in the early embryo: Clues from a dominant negative approach." Ph.D. thesis. University of California, San Diego.

Smyth N., Vatansever H.S., Murray P., Meyer M., Frie C., Paulsson M., and Edgar D. 1999. Absence of basement membranes after targeting the LAMC1 gene results in embryonic lethality due to failure of endoderm differentiation. *J. Cell Biol.* **144:** 151–160.

Soltysik-Espanola M., Klinzing D.C., Pfarr K., Burke R.D., and Ernst S.G. 1994. Endo16, a large multidomain protein found on the surface and ECM of endodermal cells during sea urchin gastrulation, binds calcium. *Dev. Biol.* **165:** 73–85.

Solursh M. 1976. Glycosaminoglycan synthesis in the chick gastrula. *Dev. Biol.* **50:** 525–530.

Solursh M. and Katow H. 1982. Initial characterization of sulfated macromolecules in the blastocoels of mesenchyme blastulae of *Strongylocentrotus purpuratus* and *Lytechinus pictus*. *Dev. Biol.* **94:** 326–336.

Spiegel E., Burger M., and Spiegel M. 1980. Fibronectin in the developing sea urchin embryo. *J. Cell Biol.* **87:** 309–313.

Spiegel E., Howard L., and Spiegel M. 1989. Extracellular matrix of sea urchin and other marine invertebrate embryos. *J. Morphol.* **199:** 71–92.

Su M.W., Suzuki H.R., Bieker J.J., Solursh M., and Ramirez F. 1991. Expression of two non-allelic type II procollagen genes during *Xenopus laevis* embryogenesis is characterized by stage-specific production of alternatively spliced transcripts. *J. Cell Biol.* **115:** 565–575.

Suzuki H.R., Reiter R.S., D'Alessio M., Di Liberto M., Ramirez F., Exposito J-Y., Gambino R., and Solursh M. 1997. Comparative analysis of fibrillar and basement membrane collagen expression in embryos of the sea urchin, *Strongylocentrotus purpuratus*. *Zool. Sci.* **14:** 449–454.

Teel A.L. and Yost H.J. 1996. Embryonic expression patterns of *Xenopus* syndecans. *Mech. Dev.* **59:** 115–127.

Tesoro V., Zito F., Yokota Y., Nakano E., Sciarrino S., and Matranga V. 1998. A protein of the basal lamina of the sea urchin embryo. *Dev. Growth Differ.* **40:** 527–535.

Toyoizumi R. and Takeuchi S. 1992. Morphometry of cellular protrusions of mesodermal cells and fibrous extracellular matrix in the primitive streak stage chick embryo. *Roux's Arch. Dev. Biol.* **201:** 36–44.

Tunggal P., Smyth N., Paulsson M., and Ott M.-C. 2000. Laminins: Structure and genetic regulation. *Microsc. Res. Tech.* **51:** 214–227.

Van Hoof J. and Harrisson F. 1986. Interaction between epithelial basement membrane and migrating mesoblast cells in the avian blastoderm. *Differentiation* **32:** 120–124.

Van Hoof J., Harrisson F., Andries L., and Vakaet L. 1986. Microinjection of glycosaminoglycan-degrading enzymes in the chicken blastoderm. *Differentiation* **31:** 14–19.

Vanroelen C. and Vakaet L. 1981. Incorporation of ³⁵S-sulphate in chick blastoderms during elongation and during shortening of the primitive streak. *Wilhelm Roux's Arch.* **190:** 233–236.

Vanroelen C., Vakaet L., and Andries L. 1980a. Distribution and turnover of testicular hyaluronidase sensitive macromolecules in the primitive streak stage chick blastoderm as revealed by autoradiography. *Anat. Embryol.* **159:** 361–367.

———. 1980b. Alcian blue staining during the formation of mesoblast in the primitive streak stage chick blastoderm. *Anat. Embryol.* **160:** 361–367.

———. 1980c. Localization and characterization of acid mucopolysaccharides in the early chick blastoderm. *J. Embryol. Exp. Morphol.* **56:** 169–178.

Vilela-Silva A.E.S., Werneck C.C., Valente A.P., Vacquier V.C., and Mourao P.A.S. 2001. Embryos of the sea urchin *Strongylocentrotus purpuratus* synthesize a dermatan sulfate enriched in 4-O- and 6-O-disulfated galactosamine units. *Glycobiology* **11:** 433–440.

Wakely J. and England M.A. 1979. Scanning electron microscopical and histochemical study of the structure and function of basement membranes in the early chick embryo. *Proc. R. Soc. Lond. B Biol. Sci.* **206:** 329–352.

Wartiovaara J., Leivo I., and Vaheri A. 1979. Expression of the cell surface-associated glycoprotein, fibronectin, in the early mouse embryo. *Dev. Biol.* **69:** 247–257.

Wessel G.M. and Berg L. 1995. A spatially restricted molecule of the extracellular matrix is contributed both maternally and zygotically in the sea urchin embryo. *Dev. Growth Differ.* **37:** 517–527.

Wessel G.M., Etkin M., and Benson S. 1991. Primary mesenchyme cells of the sea urchin embryo require an autonomously produced, nonfibrillar collagen for spiculogenesis. *Dev. Biol.* **148:** 261–272.

Wessel G.M., Marchese R.B., and McClay D.R. 1984. Ontogeny of the basal lamina in the sea urchin embryo. *Dev. Biol.* **103:** 235–245.

Wessel G.M., Berg L., Adelson D.L., Cannon G., and McClay D.R. 1998. A molecular analysis of hyalin—A substrate for cell adhesion in the hyaline layer of the sea urchin embryo. *Dev. Biol.* **193:** 115–126.

Whittaker C.A. and DeSimone D.W. 1993. Integrin α subunit mRNAs are differentially expressed in early *Xenopus* embryos. *Development* **117:** 1239–1249.

Wilt F. 1999. Matrix and mineral in the sea urchin larval skeleton. *J. Struct. Biol.* **126:** 216–226.

Winder S.J. 2001. The complexities of dystroglycan. *Trends Biochem. Sci.* **26:** 118–124.

Winklbauer R. 1994. Mesoderm cell migration in the vertebrate gastrula. *Sem. Dev. Biol.* **5:** 91–99.

———. 1998. Conditions for fibronectin fibril formation in the early *Xenopus* embryo. *Dev. Dyn.* **212:** 335–345.

Winklbauer R. and Keller R. 1996. Fibronectin, mesoderm migration, and gastrulation in *Xenopus. Dev. Biol.* **177:** 413–426.

Winklbauer R. and Nagel M. 1991. Directional mesodermal cell migration in the *Xenopus* gastrula. *Dev. Biol.* **148:** 573–589.

Winklbauer R. and Stoltz C. 1995. Fibronectin fibril growth in the extracellular matrix of the *Xenopus* embryo. *J. Cell Sci.* **108:** 1575–1586.

Wolpert L. and Mercer E.H. 1963. An electron microscope study of the development of the blastula of the sea urchin embryo and its radial polarity. *Exp. Cell Res.* **30:** 280–300.

Wu T.-C., Wan Y.-J., Chung A.E., and Damjanov I. 1983. Immunohistochemical localization of entactin and laminin in mouse embryos and fetuses. *Dev. Biol.* **100:** 496–505.

Wunsch A.M., Little C.D., and Markwald R.R. 1994. Cardiac endothe-lial heterogeneity defines valvular development as demonstrated by the diverse expression of JB3, an antigen of the endocardial cushion tissue. *Dev. Biol.* **165:** 585–601.

Yarnitzky T. and Volk T. 1995. Laminin is required for heart, somatic muscles, and gut development in the *Drosophila* embryo. *Dev. Biol.* **169:** 609–618.

Yurchenco P.D. and O'Rear J.J. 1994. Basal lamina assembly. *Curr. Opin. Cell Biol.* **6:** 674–681.

Yurchenco P.D., Quan Y., Colognato H., Mathus T., Harrison D., Yamada Y., and O'Rear J.J. 1997. The α chain of laminin-1 is inde-pendently secreted and drives secretion of its β- and γ-chain part-ners. *Proc. Natl. Acad. Sci.* **94:** 10189–10194.

Zagris N. and Chung A.E. 1990. Distribution and functional role of laminin during induction of the embryonic axis in the chick embryo. *Differentiation* **43:** 81–86.

Zagris N., Stavridis V., and Chung A. 1993. Appearance and distribu-tion of entactin in the early chick embryo. *Differentiation* **54:** 67–71.

Zhu X., Mahairas G., Illies M., Cameron R.A., Davidson, E.H., and Ettensohn C.A. 2001. A large-scale analysis of mRNAs expressed by primary mesenchyme cells of the sea urchin embryo. *Development* **128:** 2615–2627.

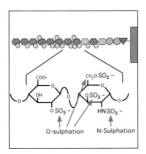

CARBOHYDRATE AND LIPID MODIFICATION OF DEVELOPMENTAL SIGNALING MOLECULES

K. Nybakken[1] and N. Perrimon[1,2]

[1]Department of Genetics and [2]Howard Hughes Medical Institute, Harvard Medical School, Boston, Massachusetts 02115

INTRODUCTION

The correct production, localization, and delivery/reception of signals are all critical for cell communication. One of the emerging themes from studies of signal transduction pathways is the importance of posttranslational modification of both the ligands and the receptors. Posttranslational modification can regulate numerous aspects of the structure and function of a protein and can affect the signal, the receptor(s), and/or the intracellular signal transduction mechanism of a particular cellular communication pathway. Whereas posttranslational modifications such as phosphorylation have long been recognized as modifiers of signaling components, only recently have lipidation and glycosylation been identified as significant modifiers of these molecules.

Many signaling pathways have been implicated in gastrulation, the focus of this book (see Chapters 32, 36, and 37). For example, members of the TGF-β family of signaling ligands affect gastrulation in both vertebrates and invertebrates, and certain members of the Wnt family of signaling molecules regulate convergent extension movements during gastrulation. Although few glycosylation and lipidation events have been shown to directly affect gastrulation, these modifications regulate signaling events during cell-type specification, and thus it seems likely that they will regulate gastrulation as well. In this chapter, we therefore introduce the glycosylation and lipidation events that have been shown to be critical in regulating specific ligand and

receptor activities. We outline how each of these modifications affects a particular developmental signaling pathway, while describing in detail two examples: (1) how glycosylation events affect Notch signaling in *Drosophila* and (2) how lipid modifications regulate Hedgehog (Hh) signaling.

GLYCOSYLATION

Glycosylation of proteins—the modification of proteins by the addition of carbohydrate or modified carbohydrate chains—is a widespread biological phenomenon and is involved in a number of different cellular processes, including protein degradation, protein trafficking, and protein localization (Alberts et al. 2002). There are two types of glycosylation: N-linked glycosylation is the result of glycans attached to asparagine residues in the core protein, and O-linked glycosylation, which is the result of glycan addition to the hydroxyl groups of serines and threonines (Varki et al. 1999). Most glycosylation events occur in the endoplasmic reticulum (ER) and the Golgi apparatus during the synthesis and processing of proteins. The carbohydrate additions, or glycans, that are attached to the core protein can be linear, as are most O-linked glycans, or can be heavily branched, like many N-linked glycans.

Linear O-linked glycans have been found to be the most important type of glycan with respect to developmental signaling pathways. The linear heparan sulfate (HS) and O-fucose glycans are especially important. HS is a linear glycan composed of repeating dimers of the modified carbohydrates N-acetylglucosamine (GlcNAc) and glucuronic acid (GlcA) linked to a protein core by a tetrasaccharide linker (Fig. 1A). HS and its close relative, chondroitin sulfate (CS), are described as glycosaminoglycans (CAGs) because of the presence of an amino group in the GlcNAc monomer, and

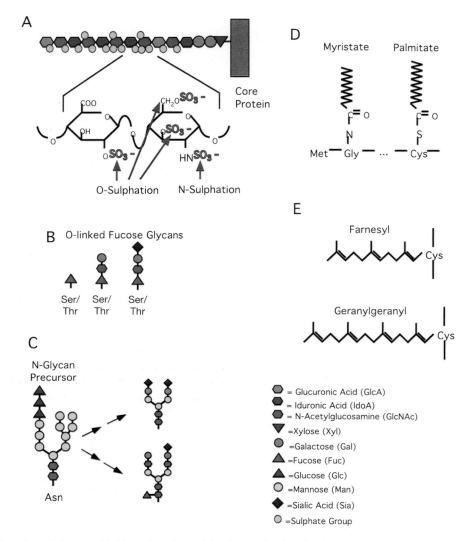

Figure 1. Overview of glycan and lipid modifications of signaling molecules. The glycan modifications most relevant to signaling pathways are diagrammed in *A* and *B*. (*A*) Structure of heparan sulfate (HS). HS consists of repeating dimers of Glucuronic acid (GlcA) and N-acetylglucosamine (GlcNAc) connected to the serine or threonine of a protein core by a tetrasaccharide linker. The GlcNAc residues are variably N-deacetylated and then N-sulfated by an NDST enzyme after polymerization. GlcA can then be converted to its epimer iduronic acid by an epimerase. After these steps, the hydroxyl groups at the 2, 3, and 6 positions in the carbohydrate rings can be modified by addition of sulfate groups. Where in a HS chain these particular modifications occur is somewhat variable, although the later modifications, such as sulfation, tend to occur more where the earlier modifications have already taken place. (*B*) O-linked fucose glycans are linear glycans connected to the core protein by the fucose, to which a variable number (or none) of other glycan subunits can be added (GlcNAc, Gal, and sialic acid are shown here). (*C*) N-linked glycans are all formed from the depicted N-glycan precursor oligosaccharide that is added to proteins in the endoplasmic reticulum. Once added, this precursor is modified by the action of numerous glycosidases and glycosyltransferases to produce a great variety of final N-linked glycan structures, illustrated here by two of many possible end products. The four main lipid modifications known to affect signaling molecules are diagrammed in *D* and *E*. (*D*) Myristic acid and palmitic acid are 14- and 16-carbon, respectively, saturated fatty acids. Myristic acid is cotranslationally added to a glycine immediately following the initiating methionine, whereas palmitic acid is added to cysteines that can occur throughout a protein. (*E*) Farnesyl and geranylgeranyl groups are 15- and 20-carbon isoprenoids, respectively, that are added to cysteines in small intracellular, signaling proteins.

proteins with HS GAG chains attached are referred to as heparan sulfate proteoglycans (HSPGs). After a HS chain is polymerized, the GlcA and GlcNAc residues can be modified further by several enzymes. An enzyme called a N-deacetylase/N-sulfotransferase (NDST) can deacetylate a portion of the GlcNAc in a HS chain and then add a sulfate group to the deacetylated amino groups. Epimerases cause a conformational change in the ring structure of GlcA, converting it to Iduronic acid (IdoA). O-sulfotransferases (OSTs) catalyze the addition of sulfate groups to hydroxyl moieties in both GlcNAc and GlcA (Fig. 1A) (Varki et al. 1999).

The O-fucose glycans are rare glycans that contain O-fucose bound to the hydroxyl group of either serine or threonine and are linearly extended by other glycan subunits. They tend to have short, linear glycan chains containing a number of modified carbohydrate residues, including GlcA, GlcNAc, galactose (Gal), and sialic acid (SiA), in addition to fucose (Fuc) (Fig 1B).

N-linked glycans are more common than O-linked glycans, as most of the proteins found in the ER are N-glycosylated as they enter the ER (Alberts et al. 2002). Proteins are N-glycosylated by the one-step addition of a 14-subunit precursor oligosaccharide from a lipid donor to an asparagine on the target protein. As the protein containing the precursor oligosaccharide makes its way through the ER and Golgi, the oligosaccharide precursor is modified by the removal of many of the modified carbohydrate subunits and the addition of new subunits in varying new patterns (Fig. 1C) (Varki et al. 1999).

MODIFICATION OF THE NOTCH SIGNALING PATHWAY BY GLYCOSYLATION

The Notch family of transmembrane proteins is a group of very large receptor molecules important in both lateral inhibitory signaling in the nervous system and in cell lineage decisions throughout developing animals, including wing development in flies, limb development in vertebrates, and hematopoetic lineage decisions in vertebrates (see Chapter 37). In the canonical pathway, Notch receptors are activated by one of two ligands—Delta and/or Serrate—which trigger two proteolytic cleavage events. The second cleavage results in the intracellular portion of Notch (*Notch intracellular domain* or NICD) being released from the membrane and translocating to the nucleus, where NICD then cooperates with other molecules to induce gene transcription (Fig. 2A) (Artavanis-Tsakonas et al. 1999). Notch is affected by glycosylation events at many different levels. The best-characterized of these modifications is the glycosylation of Notch itself.

Glycosylation of Notch

Endogenous Notch is modified by the addition of two types of short oligosaccharides—an O-fucose-linked tetrasaccharide and an O-glucose-linked trisaccharide (Moloney et al. 2000a). The O-fucose-linked tetrasaccharide is composed of a fucose residue linked to Notch, followed by a GlcNAc, a galactose, and ending in a SiA residue. For many years before the biochemical characterization of these sugar chains, it was known that the action of a gene called *fringe* is required for Notch activation in the *Drosophila* wing. Loss of *fringe* prevents activation of Notch at the edge of the *Drosophila* wing blade, resulting in the eponymous wing notching phenotype. *fringe* was found to encode a protein with similarity to a fucose-specific β 1,3 GlcNAc transferase that specifically adds GlcNAc to fucose residues (Irvine and Wieschaus 1994; Yuan et al. 1997). Subsequent in vitro experiments demonstrated that Fringe does add GlcNAc to O-linked fucoses attached to the protein core. Furthermore, it was shown that a catalytically inactive form of Fringe expressed in flies dominantly interferes with Notch signaling, but only in the wing (Bruckner et al. 2000; Moloney et al. 2000b). Modification of Notch by Fringe differentially regulates the response of Notch-expressing cells to Delta and Serrate. Cells in the wing imaginal disc that express both Fringe and Notch are less responsive to Serrate and more responsive to Delta (Panin et al. 1997).

As the Fringe GlcNAc transferase has phenotypes limited to the wing, while Notch signaling affects many other tissues, an important question arises: Do O-fucosyltransferase mutants have phenotypes similar to *fringe* phenotypes? Recently, the O-fucosyltransferase gene that modifies Notch in flies—called *Ofut1* or *neurotic* (*ntc*)—has been identified and its loss-of-function effects determined by classic mutant analysis and by using in vivo RNA interference (Okajima and Irvine 2002; Sasamura et al. 2003). Loss of *Ofut1/ntc* function leads to wing and neurogenic defects that are very similar to those of *Notch* loss-of-function mutants, and *Ofut1/ntc* loss also prevents binding of Notch to Delta (Okajima and Irvine 2002; Sasamura et al. 2003). Interestingly, elimination of *OFUT1/ntc* function affects all of the Notch functions examined so far, not just those involved in wing development, implying that O-linked fucose is required for most, if not all, of the functions of Notch. Addition of GlcNAc to O-fucose residues of Notch seems to affect only a subset of Notch activity (Okajima and Irvine 2002; Sasamura et al. 2003).

Consistent with the role of Fringe and *Ofut1/ntc* in Notch modification, mutations in *fringe connection* (*frc*), which encodes a nucleotide sugar transporter necessary to transport sugar precursors from the cytoplasm to the ER (Goto et al. 2001; Selva et al. 2001), are associated with "*fringe*-like" phenotypes. As observed in the case of *fringe*,

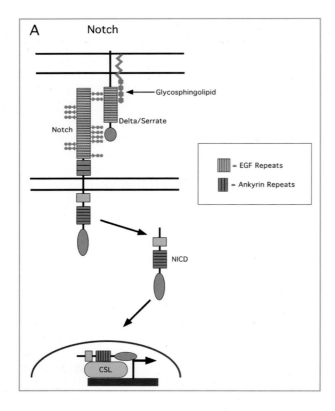

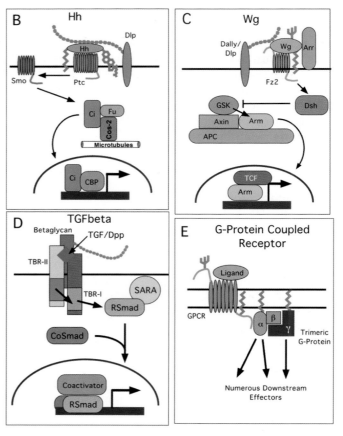

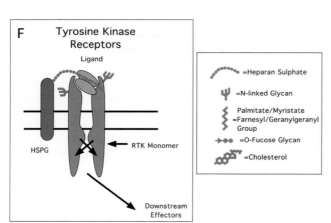

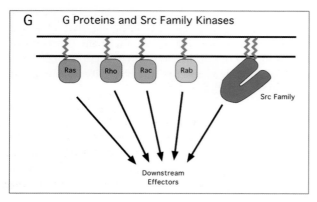

Figure 2. Schematic representation of the signaling pathways discussed in this chapter and summary of the glycosylation and lipidation events affecting the components of these pathways. Only the core components of a particular pathway are shown and therefore, for instance, the ubiquitinating enzymes important in the Hh and Wg pathways and the numerous modifiers of the Notch pathway are not shown. All glycosylation and lipid addition modifications are represented in red. (*A*) Notch signaling pathway. Activation of Notch by its ligands (Delta/Serrate) results in the cleavage of Notch into an intracellular fragment called Notch intracellular domain (NICD). The NICD moves to the nucleus where it interacts with the CBF1, Su(H), and the Lag-1 (CSL) family of transcription factors that activate Notch target genes. O-Fucose linked glycans and glycosphingolipids play important roles in Notch signaling and are indicated in red; yellow boxes indicate EGF repeats. (*B*) Hedgehog (Hh) signaling pathway. Hh binds to and inactivates the multipass transmembrane protein Patched (Ptc). Binding of Hh to Ptc releases the inhibitory effects of Ptc on a seven-pass transmembrane protein called Smoothened (Smo). Smo then stimulates a complex of the Fused (fu), Costal2 (Cos2), and Cubitus interruptus (Ci) proteins to prevent cleavage of the Ci protein, which is normally cleaved into a small repressor form in the absence of Hh signaling. Uncleaved Ci can then move to the nucleus and activate Hh target genes. Hh is modified by both palmitic acid and cholesterol moieties. The Dlp HSPG also appears to play a role in Hh binding to the cell surface. (Cos2) Costal-2, a kinesin-like molecule; (Fu) Fused, a serine/threonine kinase; (Ci) Cubitus interruptus, a Zn-finger transcription factor; (Dlp) Dally-like protein, a glycan family HSPG; (CBP) CREB binding protein, a transcriptional coactivator of Ci. (*C*) Wg signaling pathway. Binding of Wg to its

frc mutations also prevent activation of Notch at the edge of the *Drosophila* wing blade. *frc* mutants exhibit phenotypes, such as wing nicking, wing outgrowths, and leg segment fusions, similar to some *notch* and *fringe* alleles (Goto et al. 2001; Selva et al. 2001). Heterozygous *frc* alleles also enhance some wing-specific *notch* alleles (Goto et al. 2001; Selva et al. 2001). In addition to Notch signaling, *frc* mutants perturb the Wg and Hh signaling pathways, which is not surprising, given that Frc appears to transport a number of modified nucleotide sugars.

Glycosphingolipid Modulation of Notch Function

Two modifiers of Notch function in both neurogenesis and oogenesis are the *brainiac* and *egghead* genes. Loss-of-function alleles of either of these genes result in *notch*-like phenotypes in the ovary. However, whereas Notch function is required in the follicular cells that surround the oocyte, *brainiac* and *egghead* are required in the oocyte, implying that *brainiac* and *egghead* may regulate the Notch ligands Delta or Serrate and not Notch itself (Goode et al. 1996a,b; Goode and Perrimon 1997; Panin and Irvine 1998).

Cloning and characterization of both *brainiac* and *egghead* have revealed that both appear to be involved in the generation of glycosphingolipids. *brainiac* encodes a glycolipid-specific β 1,3 N-acetylglucosaminyl-transferase protein (Egan et al. 2000; Muller et al. 2002; Schwientek et al. 2002), and *egghead* encodes a glycolipid-specific β 1,4-mannosyl-transferase (Wandall et al. 2003). How glycosphingolipids might affect Delta and Serrate activity is not known, but it is intriguing to speculate that the presentation or the activity of Delta and Serrate may be regulated by glycosphingolipids in their membrane environment.

CARBOHYDRATE MODIFICATION OF OTHER SIGNALING PROTEINS

Wnt

Wnt family signaling molecules play diverse roles during development in many organisms, through the signal transduction system outlined in Figure 2C. Signaling by Wnt molecules is modulated at several levels by glycosylation. When Wnt proteins were first isolated from mammalian cells, they were shown to be glycosylated. Subsequent studies have shown that Wnts are glycosylated by the addition of two separate N-glycan chains, although the role of these modifications in Wnt function is not yet clear (Brown et al. 1987; Papkoff et al. 1987; Smolich et al. 1993; Kadowaki et al. 1996). Members of the family of Wnt receptors, the Frizzled (Fz) family of transmembrane receptors, also contain N-glycan moieties.

Studies using *Drosophila* cell lines then demonstrated that glycosaminoglycans (GAGs) appear to be required to produce active Wingless (Wg), as S2 cells expressing *wg* from a transfected expression construct are not able to produce active Wg when treated with enzymes degrading HS

receptor, Frizzled2 (Fz2), activates the Disheveled (Dsh) protein. Activation of Dsh prevents the kinase GSK3β from phosphorylating the Armadillo (Arm) protein. Phosphorylation of Arm normally leads to its degradation. In the absence of Arm degradation, Arm moves to the nucleus where it cooperates with the T-cell factor/lymphoid enhancer factor (TCF/LEF) transcription factor to activate Wg target gene transcription. Wg is glycosylated and palmitoylated and also requires the HSPGs Dally and Dlp. Arrow (Arr), a lipoprotein-like protein, and Dally and Dally-like protein (Dlp), glypican family HSPGs, act as cofactors in Wg binding. (Axin) A scaffolding molecule; (Arm) armadillo, a cadherin binding protein/transcription factor; (APC) a large protein that binds several of the intracellular transducers of Wg signaling. (D) TGF-β signaling pathway. TGF-β ligands bring together one each of a Type I TGF-β receptor (TGF-β RI) and Type II TGF-β receptor (TGF-β RII). Both TGF-β receptors are serine threonine kinases, and ligand binding causes TGF-β RII to phosphorylate and activate TGF-β RI. TGF-β RI then phosphorylates and activates a R-SMAD which then associates with a Co-SMAD and both then move to the nucleus to activate target gene transcription. (TGF-β/Dpp) Transforming growth factor β/Decapentaplegic ligands; (Betaglycan) a HSPG; (SMADs) transcriptional regulatory proteins; (R-SMAD) receptor SMAD; (Co-SMAD) Co-mediator SMAD; (SARA) a protein that anchors SMADs at the plasma membrane via a lipid-binding motif. (E) G-Protein coupled receptor signaling pathway. In the absence of ligand stimulation, the α, β, and γ subunits of the heterotrimeric G proteins are bound together and cannot stimulate downstream signaling effectors. Activation of a GPCR by its ligand causes the α and βγ subunits to dissociate, allowing either subunit to signal to downstream effectors. GPCRs are modified by glycosylation and palmitoylation, while the heterotrimeric G-proteins are also lipid-modified (see text). (GPCR) G-Protein-coupled receptor; (α, β, and γ) subunits of a heterotrimeric G-protein. There are numerous downstream effectors of the very large GPCR family, and so no specific pathways are illustrated. (F) Receptor tyrosine kinase (RTK) signaling pathway. Binding of a RTK ligand to two of the RTK receptors causes the two receptors to autophosphorylate each other on tyrosines, which then leads to activation of downstream effectors. HSPGs appear to be required for binding or optimal presentation of the ligands to the receptors. Many downstream effectors are involved in RTK signaling and are not diagrammed in this figure. RTKs are often glycosylated as discussed in the text. (HSPG) Heparan sulfate proteoglycan; (RTK) receptor tyrosine kinase. (G) Small G-proteins and Src family kinases. Most small G-proteins have only one lipid modification, whereas the majority of Src family cytoplasmic tyrosine kinases have two lipid modifications. Again, many effectors have been linked to each of these molecules and are not listed here.

and CS (Reichsman et al. 1996). As GAGs are always connected to a core protein, a proteoglycan of some sort must be involved in Wnt signaling. Two *Drosophila* proteoglycan relatives of the vertebrate glypicans, called Dally and Dally-like protein (Dlp), have been shown to have roles in Wnt signaling. *dally* mutants have a cuticular phenotype similar to that of *wg*, and *dally* overexpression can partially rescue a *wg* phenotype. No mutants exist in *dlp*, but reduction in Dlp protein by RNA interference produces a *wg*-like phenotype (Baeg et al. 2001). This *wg*-like phenotype is enhanced when *dally* is similarly eliminated, indicating some sort of synergy with respect to Wg signaling between the two *Drosophila* glypicans (Lin and Perrimon 1999; Tsuda et al. 1999).

Several pieces of evidence indicate that HS is the relevant GAG found on glypicans with respect to Wg signaling. First, loss of both the maternal and zygotic copies of the *Drosophila* NDST gene, called *sulfateless* (*sfl*), phenocopy *wg* mutants in several *Drosophila* tissues (Lin and Perrimon 1999). *sfl* mutants do not reduce the amount of Wg produced in the wing imaginal disc, but they do greatly reduce the normal extracellular accumulation of Wg (Baeg et al. 2001). Second, mutations in the *sugarless* (*sgl*) gene, which encodes a uridine-diphosphate dehydrogenase required for the production of GlcA, result in *wg* phenotypes (Binari et al. 1997; Hacker et al. 1997; Haerry et al. 1997). Third, mutation in the *frc* gene described above also gives a *wg* mutant cuticular phenotype (Goto et al. 2001; Selva et al. 2001).

In the context of this volume, it is interesting to note that a glypican termed Knypek has been shown to be involved in regulation of gastrulation in zebrafish (Topczewski et al. 2001). Embryos in which *knypek* is mutated do not undergo proper convergence and extension movements as a consequence of abnormal cell polarity (see Chapters 12 and 20). Overexpression studies and analysis of double mutants demonstrated that *knypek* mutants affect the Wnt11 signaling pathway, which had already been shown to be involved in gastrulation. Hence, this is one clear example of how glycosylation affects a signaling pathway involved in gastrulation.

Hedgehog

Hh is another signaling molecule important in many developmental decisions (Fig. 2B). Hh itself is not known to be glycosylated, but HSPGs have been shown to be important for Hh signaling. Mutations in the *tout-velu* (*ttv*) gene generate a phenotype that can be interpreted as a loss-of-function phenotype of Wg or Hh signaling. When the phenotypes of *ttv* mutants were examined in the wing imaginal disc, it was found that *ttv* only appeared to affect the Hh pathway. Hh production was not affected, but Hh secreted from the posterior compartment in the wing imaginal disc could only signal to cells immediately abutting the compartment border and not the multiple cell diameter distances that Hh can normally signal across in the wing disc. The *ttv* gene was then shown to encode a member of the HS polymerase family of EXT proteins (Bellaiche et al. 1998; The et al. 1999). This finding implies that movement of the Hh signal requires HS, presumably attached to a HSPG.

A candidate for this HSPG has recently been identified. Using an RNA interference screen to look for new components of the Hh signaling pathway, the Dlp glypican was identified as a part of the Hh reception apparatus. It appears to be required only for reception and not transduction of the Hh signal as the stimulus in these experiments was an artificial, soluble form of Hh called Hh-N. Additional experiments using full-length Hh indicated that Dlp is not required to transduce a signal by native Hh protein (Lum et al. 2003).

Receptor Tyrosine Kinases

RTKs are signaling receptors that activate their signaling pathway after dimerization induced by ligand binding (Fig. 2F). Glycosylation affects both the receptors and the ligands of many RTKs. Numerous RTK ligands are glycosylated, including nerve growth factor (NGF), platelet-derived growth factor (PDGF), and epidermal growth factor (EGF), although in most cases the role of these glycosylations is not clear. Most RTKs themselves are glycosylated, usually by N-linked glycans and by short O-linked glycans. Subcellular sorting of the NGF receptor, for instance, is regulated by glycosylation, whereas glycosylation of the EGF receptor appears to regulate its autophosphorylation and activation states (Tsuda et al. 2000).

Signaling by FGF and the FGF receptor (FGFR) is probably the best-characterized interaction of an RTK with glycans. Early studies showed that, although FGF can signal through FGFR in the absence of any glycans, signaling is maximized when HS is present (Rapraeger et al. 1991; Olwin and Rapraeger 1992). Subsequent investigations revealed that dimerization of the FGFRs is optimized when HS binds together two molecules of FGF (Pellegrini et al. 2000; Schlessinger et al. 2000). Genetic studies have since shown that the *sgl* and *sfl* genes necessary for HS synthesis are required for FGF signaling in *Drosophila*. Mutation in either of these genes disrupts the formation of the insect dorsal vessel and of the insect respiratory system, the tracheal network, both of which require FGF for their development (Lin et al. 1999).

TGF-β

Several studies in *Drosophila* have shown that signaling by the TGF-β-like molecule Decapentaplegic (Dpp) is modified by mutations in the glypican *dally* (Fig. 2D). Mutations in *dally* cause a reduction in the range of Dpp signaling, whereas overexpression of *dally* causes cells to become hypersensitive to Dpp stimulation, suggesting that Dally is a coreceptor of Dpp that traps Dpp (Fujise et al. 2003). The interactions of Dally and Dpp also seem to vary with tissue type and so may be more complicated than originally thought (Jackson et al. 1997).

LIPID MODIFICATION

Most signaling receptors are membrane bound, and so interacting signaling molecules must either reside in or be recruited to the membrane. Modification of signaling proteins by lipids is a common method used to bring interacting molecules to membranes. Proteins can be modified by several different types of lipids on one or more sites within the protein. The most common types of lipids added to signaling molecules are the saturated fatty acids myristate and palmitate, and the isoprenoid farnesyl and geranylgeranyl groups.

Myristate groups use an amide linkage to attach to a glycine immediately adjacent to the initiating methionine in the modified protein. Myristoylation takes place at the time of translation and is followed by the removal of the amino-terminal methionine by a methionine peptidase. Palmitate groups are added to cysteines that can be located throughout the modified protein, but which are often found near a myristoylated glycine. A membrane-bound palmitoyl acyl transferase, using palmitoyl-CoA as a lipid donor, posttranslationally catalyzes the addition of palmitate to a substrate protein (Fig. 1D). Palmitate groups, unlike myristate groups, can be removed from proteins by the actions of palmitoyl thioesterases.

Geranylgeranyl and farnesyl groups are added to cysteines found at or near the carboxyl terminus of a protein by the actions of dimeric farnesyl or geranylgeranyl transferases. A "CAAX" box in the carboxyl terminus of protein is usually the site of modification, with the identity of the "X" residue determining whether the protein is farnesylated or geranylgeranylated. Both groups are linked to the cysteine by a stable thioether bond (Fig. 1E). After the farnesyl or geranylgeranyl group is added to the protein, the residues carboxyl-terminal of the cysteine are proteolytically removed and the cysteine is then carboxy-methylated (for review, see Resh 1996, 1999).

LIPID MODIFICATION IN HEDGEHOG SIGNALING

Proteins of the Hh family of signaling molecules undergo perhaps the most diverse set of lipid modifications of any signaling molecule (Fig. 2B). For Hh molecules to become biologically active, they must complete a complex matura-

tion process. First, Hh molecules autoproteolytically cleave into an amino-terminal fragment, usually referred to as HhNp, and a carboxy-terminal fragment. Autoproteolysis occurs between a glycine and a cysteine. In the process of undergoing this autoproteolysis, a cholesterol molecule is covalently added to the carboxy-terminal glycine residue of HhNp via an ester linkage (Lee et al. 1994; Porter et al. 1995, 1996a,b). The amino-terminal fragment contains all the signaling functions of Hh, whereas the sole function of the carboxy-terminal fragment is to catalyze autoproteolysis and the addition of cholesterol. Mutation of the cysteine immediately carboxy-terminal to the cleavage site prevents cleavage and cysteine-mutated forms of Hh have no effect when overexpressed in flies (Porter et al. 1995).

After autoproteolysis and cholesterol addition, a palmitic acid group is added to the amino-terminal end of HhNp, such that the final, mature form of Hh has two covalently attached hydrophobic adjuncts (Pepinsky et al. 1998). Recently, the gene that regulates the attachment of the palmitic acid moieties to HhNp has been identified by four different groups and called *sightless* (*sit*), *rasp*, *central missing* (*cmn*), and *skinny hedgehog* (*ski*), respectively (Amanai and Jiang 2001; Chamoun et al. 2001; Lee and Treisman 2001; Micchelli et al. 2002). Mutations in this gene do not affect the amount of *hh* transcript or protein in imaginal discs and closely mimic *hh* loss-of-function phenotypes. When *sit* was cloned, it was found to encode a protein having strong homology with acyltransferases, and thus probably identifies the enzyme that catalyzes addition of palmitate to Hh. In the absence of *sit* function, Hh accumulates properly and is stable, but it is not active with respect to induction of target genes (Amanai and Jiang 2001; Chamoun et al. 2001; Lee and Treisman 2001; Micchelli et al. 2002).

G-Protein-coupled Receptors

A significant proportion of GPCRs are modified by palmitate addition near the carboxyl terminus of the protein (Fig. 2E) (Resh 1996, 1999). These palmitoylation events seem to have different functions depending on the particular modified GPCR. The endothelin receptor, for example, is not able to signal efficiently when its palmitoylated cysteine is mutated (Cramer et al. 2001). Loss of the palmitoylation sites of the α 2A adrenergic receptor, on the other hand, does not affect signaling ability of the receptor but rather prevents receptor down-regulation leading to overactivation of the α 2A receptor in tissue culture (Eason et al. 1994).

Src Family Cytoplasmic Tyrosine Kinases

Most Src family cytoplasmic tyrosine kinases are modified by myristoylation, and a number are also modified by palmitoylation of cysteines near the myristoylation site (Fig. 2G). Lipid modification of these Src family kinases appears to be required for their subcellular localization and function, although it does not appear to regulate their kinase activity (Buss et al. 1984, 1986; Kamps et al. 1985, 1986). As an example, mutation of the myristoylation site of Src eliminates its transforming ability, without affecting its kinase activity. Elimination of the palmitoylation site of Lck, another member of the Src tyrosine kinase family involved in lymphocyte signaling, abrogates signaling through the Fc receptor, indicating that palmitoylation is necessary for the Fc signal transduction pathway (Kabouridis et al. 1997; van't Hof and Resh 1999).

Wnt

Recently, it has been demonstrated that Wnts, like Hh, are modified by addition of palmitate groups to conserved cysteines (Fig. 2C). Surprisingly, this palmitoylation does not eliminate solubility of the modified Wnt as it can be secreted into media by tissue culture cells. Mutation of the palmitoylated cysteine in Wnts does not inhibit secretion of the Wnts, but it does completely eliminate ability of the mutated Wnt to activate signal transduction, as judged by accumulation of β-catenin (Willert et al. 2003).

There is already a good candidate for the acyltransferase that modifies Wg. The *porcupine* (*porc*) gene was identified several years ago as a gene whose loss-of-function phenotype is very similar to a *wg* mutant (Kadowaki et al. 1996). *porc* was subsequently shown to have homology with the same family of acyltransferases to which *sit/rasp* belongs and Porc also binds to Wg, strongly suggesting that it encodes the Wg acyltransferase (Hofmann 2000; Tanaka et al. 2000, 2002).

CONCLUSION

Through the examples provided in this chapter, it is apparent that posttranslational modification of either secreted signaling molecules or receptors/coreceptors by lipid or glycan addition can greatly influence the biological activity of developmental signaling molecules. Because the signaling protein modifications that we have discussed here can have profound effects on the biological events which they regulate, it is very likely that any signaling system which impacts on gastrulation will be regulated by a glycosylation or lipidation event.

ACKNOWLEDGMENTS

We apologize to those authors not directly cited due to space restrictions. K.N. is supported by National Institutes of Health postdoctoral fellowship. N.P. is an investigator of the Howard Hughes Medical Institute. This work is supported by a National Institutes of Health grant to N.P.

REFERENCES

Alberts B., Johnson A., Lewis J., Raff M., Roberts K., and Walter P. 2002. *Molecular biology of the cell*, 4th edition. Garland Science, New York.

Amanai K. and Jiang J. 2001. Distinct roles of Central missing and Dispatched in sending the Hedgehog signal. *Development* **128:** 5119–5127.

Artavanis-Tsakonas S., Rand M.D., and Lake R.J. 1999. Notch signaling: Cell fate control and signal integration in development. *Science* **284:** 770–776.

Baeg G.H., Lin X., Khare N., Baumgartner S., and Perrimon N. 2001. Heparan sulfate proteoglycans are critical for the organization of the extracellular distribution of Wingless. *Development* **128:** 87–94.

Bellaiche Y., The I., and Perrimon N. 1998. *Tout-velu* is a *Drosophila* homologue of the putative tumour suppressor *EXT-1* and is needed for Hh diffusion. *Nature* **394:** 85–88.

Binari R.C., Staveley B.E., Johnson W.A., Godavarti R., Sasisekharan R., and Manoukian A.S. 1997. Genetic evidence that heparin-like glycosaminoglycans are involved in *wingless* signaling. *Development* **124:** 2623–2632.

Brown A.M., Papkoff J., Fung Y.K., Shackleford G.M., and Varmus H.E. 1987. Identification of protein products encoded by the proto-oncogene int-1. *Mol. Cell. Biol.* **7:** 3971–3977.

Bruckner K., Perez L., Clausen H., and Cohen S. 2000. Glycosyltransferase activity of Fringe modulates Notch-Delta interactions (comments). *Nature* **406:** 411–415.

Buss J.E., Kamps M.P., and Sefton B.M. 1984. Myristic acid is attached to the transforming protein of Rous sarcoma virus during or immediately after synthesis and is present in both soluble and membrane-bound forms of the protein. *Mol. Cell. Biol.* **4:** 2697–2704.

Buss J.E., Kamps M.P., Gould K., and Sefton B.M. 1986. The absence of myristic acid decreases membrane binding of p60src but does not affect tyrosine protein kinase activity. *J. Virol.* **58:** 468–474.

Chamoun Z., Mann R.K., Nellen D., von Kessler D.P., Bellotto M., Beachy P. A., and Basler K. 2001. Skinny hedgehog, an acyltransferase required for palmitoylation and activity of the hedgehog signal. *Science* **293:** 2080–2084.

Cramer H., Schmenger K., Heinrich K., Horstmeyer A., Boning H., Breit A., Piiper A., Lundstrom K., Muller-Esterl W., and Schroeder C. 2001. Coupling of endothelin receptors to the ERK/MAP kinase pathway. Roles of palmitoylation and G(alpha)q. *Eur. J. Biochem.* **268:** 5449–5459.

Eason M.G., Jacinto M.T., Theiss C.T., and Liggett S.B. 1994. The palmitoylated cysteine of the cytoplasmic tail of alpha 2A-adrenergic receptors confers subtype-specific agonist-promoted down-regulation. *Proc. Natl. Acad. Sci.* **91:** 11178–11182.

Egan S., Cohen B., Sarkar M., Ying Y., Cohen S., Singh N., Wang W., Flock G., Goh T., and Schachter H. 2000. Molecular cloning and expression analysis of a mouse UDP-GlcNAc:Gal(beta1-4)Glc(NAc)-R beta1,3-N-acetylglucosaminyltransferase homologous to *Drosophila melanogaster* Brainiac and the beta1,3-galactosyltransferase family. *Glycoconj. J.* **17:** 867–875.

Fujise M., Takeo S., Kamimura K., Matsuo T., Aigaki T., Izumi S., and Nakato H. 2003. Dally regulates Dpp morphogen gradient formation in the *Drosophila* wing. *Development* **130:** 1515–1522.

Goode S. and Perrimon N. 1997. Brainiac and fringe are similar pioneer proteins that impart specificity to notch signaling during *Drosophila* development. *Cold Spring Harbor Symp. Quant. Biol.* **62:** 177–184.

Goode S., Melnick M., Chou T.B., and Perrimon N. 1996a. The neurogenic genes egghead and brainiac define a novel signaling pathway essential for epithelial morphogenesis during *Drosophila* oogenesis. *Development* **122:** 3863–3879.

Goode S., Morgan M., Liang Y.P., and Mahowald A.P. 1996b. *brainiac* encodes a novel, putative secreted protein that cooperates with Grk TGFα in the genesis of the follicular epithelium. *Dev. Biol.* **178:** 35–50.

Goto S., Taniguchi M., Muraoka M., Toyoda H., Sado Y., Kawakita M., and Hayashi S. 2001. UDP-sugar transporter implicated in glycosylation and processing of Notch. *Nat. Cell Biol.* **3:** 816–822.

Hacker U., Lin X., and Perrimon N. 1997. The *Drosophila sugarless* gene modulates Wingless signaling and encodes an enzyme involved in polysaccharide biosynthesis. *Development* **124:** 3565–3573.

Haerry T.E., Heslip T.R., Marsh J.L., and O'Connor M.B. 1997. Defects in glucuronate biosynthesis disrupt Wingless signaling in *Drosophila*. *Development* **124:** 3055–3064.

Hofmann K. 2000. A superfamily of membrane-bound O-acyltransferases with implications for wnt signaling. *Trends Biochem. Sci.* **25:** 111–112.

Irvine K.D. and Wieschaus E. 1994. *fringe*, a boundary-specific signaling molecule, mediates interactions between dorsal and ventral cells during Drosophila wing development. *Cell* **79:** 595–606.

Jackson S.M., Nakato H., Sugiura M., Jannuzi A., Oakes R., Kaluza V., Golden C., and Selleck S.B. 1997. *dally*, a *Drosophila* glypican, controls cellular responses to the TGF-β-related morphogen, Dpp. *Development* **124:** 4113–4120.

Kabouridis P.S., Magee A.I., and Ley S.C. 1997. S-acylation of LCK protein tyrosine kinase is essential for its signalling function in T lymphocytes. *EMBO J.* **16:** 4983–4998.

Kadowaki T., Wilder E., Klingensmith J., Zachary K., and Perrimon N. 1996. The segment polarity gene porcupine encodes a putative multitransmembrane protein involved in Wingless processing. *Genes Dev.* **10:** 3116–3128.

Kamps M.P., Buss J.E., and Sefton B.M. 1985. Mutation of NH2-terminal glycine of p60src prevents both myristoylation and morphological transformation. *Proc. Natl. Acad. Sci.* **82:** 4625–4628.

———. 1986. Rous sarcoma virus transforming protein lacking myristic acid phosphorylates known polypeptide substrates without inducing transformation. *Cell* **45:** 105–112.

Lee J.D. and Treisman J.E. 2001. Sightless has homology to transmembrane acyltransferases and is required to generate active Hedgehog protein. *Curr. Biol.* **11:** 1147–1152.

Lee J.J., Ekker S.C., von Kessler D.P., Porter J.A., Sun B.I., and Beachy P.A. 1994. Autoproteolysis in hedgehog protein biogenesis. *Science* **266:** 1528–1537.

Lin X. and Perrimon N. 1999. Dally cooperates with *Drosophila* Frizzled 2 to transduce Wingless signalling (comments). *Nature* **400:** 281–284.

Lin X., Buff E.M., Perrimon N., and Michelson A.M. 1999. Heparan sulfate proteoglycans are essential for FGF receptor signaling during *Drosophila* embryonic development. *Development* **126:** 3715–3723.

Lum L., Yao S., Mozer B., Rovescalli A., von Kessler D., Nirenberg M., and Beachy P.A. 2003. Identification of Hedgehog pathway components by RNAi in *Drosophila* cultured cells. *Science* **299:** 2039–2045.

Micchelli C.A., The I., Selva E., Mogila V., and Perrimon N. 2002. *rasp*, a putative transmembrane acyltransferase, is required for Hedgehog signaling. *Development* **129:** 843–851.

Moloney D.J., Shair L.H., Lu F.M., Xia J., Locke R., Matta K.L., and Haltiwanger R.S. 2000a. Mammalian Notch1 is modified with two unusual forms of O-linked glycosylation found on epidermal growth factor-like modules. *J. Biol. Chem.* **275:** 9604–9611.

Moloney D.J., Panin V.M., Johnston S.H., Chen J., Shao L., Wilson R., Wang Y., Stanley P., Irvine K.D., Haltiwanger R.S., and Vogt T.F. 2000b. Fringe is a glycosyltransferase that modifies Notch. *Nature* **406:** 369–375.

Muller R., Altmann F., Zhou D., and Hennet T. 2002. The *Drosophila melanogaster* brainiac protein is a glycolipid-specific β1,3N-acetylglucosaminyltransferase. *J. Biol. Chem.* **277:** 32417–32420.

Okajima T. and Irvine K.D. 2002. Regulation of notch signaling by O-linked fucose. *Cell* **111:** 893–904.

Olwin B.B. and Rapraeger A. 1992. Repression of myogenic differentiation by aFGF, bFGF, and K-FGF is dependent on cellular heparan sulfate. *J. Cell Biol.* **118:** 631–639.

Panin V.M. and Irvine K.D. 1998. Modulators of Notch signaling. *Semin. Cell Dev. Biol.* **9:** 609–617.

Panin V.M., Papayannopoulos V., Wilson R., and Irvine K.D. 1997. Fringe modulates Notch-ligand interactions. *Nature* **387:** 908–912.

Papkoff J., Brown A.M., and Varmus H.E. 1987. The int-1 proto-oncogene products are glycoproteins that appear to enter the secretory pathway. *Mol. Cell. Biol.* **7:** 3978–3984.

Pellegrini L., Burke D.F., von Delft F., Mulloy B., and Blundell T.L. 2000. Crystal structure of fibroblast growth factor receptor ectodomain bound to ligand and heparin. *Nature* **407:** 1029–1034.

Pepinsky R.B., Zeng C., Wen D., Rayhorn P., Baker D.P., Williams K.P., Bixler S.A., Ambrose C.M., Garber E.A., Miatkowski K., Taylor F.R., Wang E.A., and Galdes A. 1998. Identification of a palmitic acid-modified form of human Sonic hedgehog. *J. Biol. Chem.* **273:** 14037–14045.

Porter J.A., Young K.E., and Beachy P.A. 1996a. Cholesterol modification of hedgehog signaling proteins in animal development. *Science* **274:** 255–259.

Porter J.A., von Kessler D.P., Ekker S.C., Young K.E., Lee J.J., Moses K., and Beachy P.A. 1995. The product of *hedgehog* autoproteolytic cleavage active in local and long-range signalling. *Nature* **374:** 363–366.

Porter J.A., Ekker S.C., Park W.J., von Kessler D.P., Young K.E., Chen C.H., Ma Y., Woods A.S., Cotter R.J., Koonin E.V., and Beachy P.A. 1996b. Hedgehog patterning activity: Role of a lipophilic modification mediated by the carboxy-terminal autoprocessing domain. *Cell* **86:** 21–34.

Rapraeger A.C., Krufka A., and Olwin B.B. 1991. Requirement of heparan sulfate for bFGF-mediated fibroblast growth and myoblast differentiation. *Science* **252:** 1705–1708.

Reichsman F., Smith L., and Cumberledge S. 1996. Glycosaminoglycans can modulate extracellular localization of the wingless protein and promote signal transduction. *J. Cell Biol.* **135:** 819–827.

Resh M.D. 1996. Regulation of cellular signalling by fatty acylation and prenylation of signal transduction proteins. *Cell Signal* **8:** 403–412.

———. 1999. Fatty acylation of proteins: New insights into membrane targeting of myristoylated and palmitoylated proteins. *Biochim. Biophys. Acta* **1451:** 1–16.

Sasamura T., Sasaki N., Miyashita F., Nakao S., Ishikawa H.O., Ito M., Kitagawa M., Harigaya K., Spana E., Bilder D., Perrimon N., and Matsuno K. 2003. *neurotic,* a novel maternal neurogenic gene, encodes an *O*-fucosyltransferase that is essential for Notch-Delta interactions. *Development* **130:** 4785–4795.

Schlessinger J., Plotnikov A.N., Ibrahimi O.A., Eliseenkova A.V., Yeh B.K., Yayon A., Linhardt R.J., and Mohammadi M. 2000. Crystal structure of a ternary FGF-FGFR-heparin complex reveals a dual role for heparin in FGFR binding and dimerization. *Mol. Cell* **6:** 743–750.

Schwientek T., Keck B., Levery S.B., Jensen M.A., Pedersen J.W., Wandall H.H., Stroud M., Cohen S.M., Amado M., and Clausen H. 2002. The *Drosophila* gene braniac encodes a glycosyltransferase putatively involved in glycosphingolipid synthesis. *J. Biol. Chem.* **277:** 32421–32429.

Selva E.M., Hong K., Baeg G.H., Beverley S.M., Turco S.J., Perrimon N., and Hacker U. 2001. Dual role of the fringe connection gene in both heparan sulphate and fringe-dependent signalling events. *Nat. Cell Biol.* **3:** 809–815.

Smolich B.D., McMahon J.A., McMahon A.P., and Papkoff J. 1993. Wnt family proteins are secreted and associated with the cell surface. *Mol. Biol. Cell* **4:** 1267–1275.

Tanaka K., Kitagawa Y., and Kadowaki T. 2002. *Drosophila* segment polarity gene product porcupine stimulates the posttranslational *N*-glycosylation of wingless in the endoplasmic reticulum. *J. Biol. Chem.* **277:** 12816–12823.

Tanaka K., Okabayashi K., Asashima M., Perrimon N., and Kadowaki T. 2000. The evolutionarily conserved *porcupine* gene family is involved in the processing of the Wnt family. *Eur. J. Biochem.* **267:** 4300–4311.

The I., Bellaiche Y., and Perrimon N. 1999. Hedgehog movement is regulated through tout velu-dependent synthesis of a heparan sulfate proteoglycan. *Mol. Cell* **4:** 633–639.

Topczewski J., Sepich D.S., Myers D.C., Walker C., Amores A., Lele Z., Hammerschmidt M., Postlethwait J., and Solnica-Krezel L. 2001. The zebrafish glypican knypek controls cell polarity during gastrulation movements of convergent extension. *Dev. Cell* **1:** 251–264.

Tsuda M., Kamimura K., Nakato H., Archer M., Staatz W., Fox B., Humphrey M., Olson S., Futch T., Kaluza V., Siegfried E., Stam L., and Selleck S.B. 1999. The cell-surface proteoglycan Dally regulates Wingless signalling in *Drosophila. Nature* **400:** 276–280.

Tsuda T., Ikeda Y., and Taniguchi N. 2000. The Asn-420-linked sugar chain in human epidermal growth factor receptor suppresses ligand-independent spontaneous oligomerization. Possible role of a specific sugar chain in controllable receptor activation. *J. Biol. Chem.* **275:** 21988–21994.

van't Hof W. and Resh M.D. 1999. Dual fatty acylation of p59^{Fyn} is required for association with the T cell receptor ζ chain through phosphotyrosine-Src homology domain-2 interactions. *J. Cell Biol.* **145:** 377–389.

Varki A., Cummings R., Esko J., Freeze H., Hart G., and Marth J., eds. 1999. *Essentials of glycobiology.* Cold Spring Harbor Laboratory Press, Cold Spring Harbor, New York.

Wandall H.H., Pedersen J.W., Park C., Levery S.B., Pizette S., Cohen S.M., Schwientek T., and Clausen H. 2003. *Drosophila egghead* encodes a β1,4-mannosyltransferase predicted to form the immediate precursor glycosphingolipid substrate for brainiac. *J. Biol. Chem.* **278:** 1411–1414.

Willert K., Brown J.D., Danenberg E., Duncan A.W., Weissman I.L., Reya T., Yates J.R., III, and Nusse R. 2003. Wnt proteins are lipid-modified and can act as stem cell growth factors. *Nature* **423:** 448–452.

Yuan Y.P., Schultz J., Mlodzik M., and Bork P. 1997. Secreted fringe-like signaling molecules may be glycosyltransferases. *Cell* **88:** 9–11.

PART IV

THE EVOLUTION OF GASTRULATION

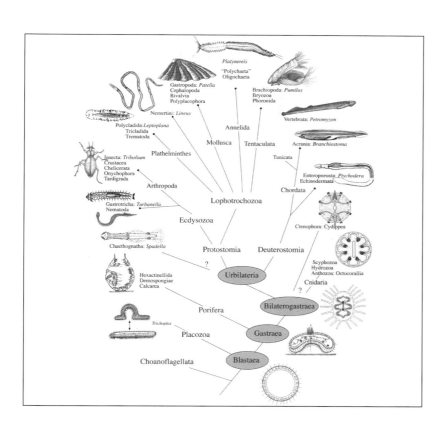

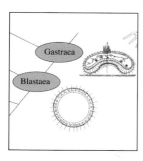

COMPARATIVE ASPECTS OF GASTRULATION

D. Arendt

Developmental Biology Programme, European Molecular Biology Laboratory, 69012 Heidelberg, Germany

INTRODUCTION

The study of gastrulation in any one given animal group can give us detailed descriptions of the events and their molecular regulation, but we cannot fully understand their true biological purpose. For example, we can describe vertebrate convergence and extension movements in cellular detail, and find the molecules involved—but we will still ask: Why do these movements exist at all? Why should tissues move from one side to the other, and not just develop in place? An evolutionary approach is the most effective way to answer these questions. In this chapter, we reconstruct the course of animal evolution to get some deeper understanding of animal gastrulation.

Ontogeny recapitulates phylogeny, as stated by Ernst Haeckel's "Biogenetic law," means that to some extent, the historical course of evolution explains the course of extant development. In his *Gastraea-Theory*, Haeckel had exemplified this on gastrulation (Haeckel 1874). This chapter follows his reasoning, with the background of another 130 years of research. It is assumed that evolution always means genetic change, which affects development, either by adding new steps to a developmental process, or by altering preexisting steps. In the main branches of evolution leading to today's animal groups, adult body plans evolved from the simple to the more complex. Initially, this involved the addition of new developmental steps to preexisting development, so that the resulting chain of developmental events then recapitulated evolution. However, since development itself is also subject to evolutionary change, "the phylogenetic readout of an ontogeny is gradually blurred, by development taking a more and more straight way from the egg to the adult, and by the struggle for life that free-living larvae have to withstand" (Haeckel 1874, p. 8, author's translation). Hence, in our comparative survey of animal gastrulation we focus on ancestral developmental stages or processes, which still resemble the initial course of development and of evolution. If, instead, development has been strongly modified and thus has departed from the course of evolution, it is derived. How can we distinguish between ancestral and derived patterns? Apart from paleontological studies (see, e.g., Chapter 53), which present problems intrinsic to the limited availability of material as well as to their interpretation, only a comparative study of embryology can help. Having detected a range of variants of a given developmental stage, we can attempt to deduce the ancestral variant as the one from which the other variants can most plausibly have arisen. Importantly, ancestral variants are not necessarily the most frequent and do not necessarily show up in the most basal groups of a tree.

Our comparative survey refers to animal groups with key positions in the tree of animal evolution, which show a gastrulation pattern prototypic for the respective group and reasonably ancestral body plans, such as *Branchiostoma* for the chordates, polychaetes for annelids, and crustaceans for arthropods (Fig. 1). Note that in this tree the Bilateria bifurcate early into Protostomia and Deuterostomia. Interestingly, this deep phylogenetic split, introduced as early as 1904 by Karl Grobben, traces back to an application of Haeckel's law on gastrulation, and was based on the observation that only in the Protostomia does the mouth form from the blastopore (Grobben 1908; see Chapter 1).

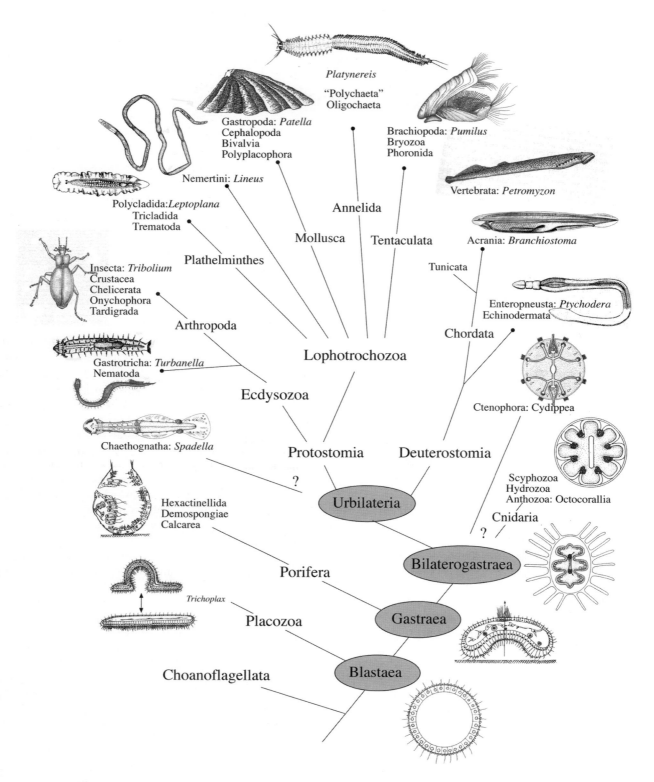

Figure 1. The phylogenetic tree of animal evolution, showing adult body plans in metazoa. Drawings adapted from various sources.

Grobben's phylogenetic grouping is still valid today (Aguinaldo et al. 1997), and its significance is reevaluated at the end of this chapter.

STARTING POINT: AN EPITHELIAL SPHERE

We begin our survey with a brief description of the state before gastrulation starts. This needs some consideration, because in metazoan animals, pre-gastrula stages can be rather divergent. However, it is most plausible that ancestrally, metazoan animals develop from a hollow sphere of cells, called a blastula. It is also assumed that this should recapitulate the metazoan ancestor which represented an epithelial sphere called a blastaea. This was the starting point for the evolution of metazoan body plans.

TOTAL CLEAVAGE PRODUCES A BLASTULA WITH RADIAL SYMMETRY

A survey of animal early development reveals (Fioroni 1992; Nielsen 2001) that in the most ancestral condition, animals enter gastrulation as a blastula, exemplified for all major bilaterian branches in Figure 2a–c. A prototype blastula is made up of one single, continuous epithelium. The cavity surrounded by the epithelium is the blastocoel. From the egg, the blastula inherits its polarity along one axis, the animal–vegetal axis. The upper pole is called "animal," and the lower pole "vegetal." Animal and vegetal egg regions differ from each other by the storage of maternal determinants and yolk, mostly in vegetal regions. Some conserved vegetal determinants are distributed to, and define, the later germ cells. Other determinants contribute to axial patterning.

The prototype blastula shows radial symmetry. Ideally, this means it can be divided at various angles into two equal halves with the animal–vegetal axis in the plane of section. Practically, there should be no fewer than two planes of symmetry (quadri-radial symmetry with only two planes results, for example, from spiral cleavage; Shankland and Seaver 2000).

Developing bilaterians tend to already deviate from radial symmetry at blastula stages, anticipating to some extent the later bilateral symmetry (with one plane of symmetry only). For example, in the polychaete *Nereis*, the blastomeres are of unequal size, with the D quadrant being largest, although the overall spiral cleavage pattern obeys quadri-radial symmetry (see Fig. 7a1). This quadrant contains determinants required for the later establishment of bilateral symmetry (Fischer and Dorresteijn 2004). In the frog, the sperm entry point sets up bilateral symmetry, via the unequal redistribution of maternal determinants (see

Fig. 7b1). These are inherited by the later organizer tissue that establishes bilaterality (Chapters 13, 23, 25, 26, 36, and 42). A very striking case of anticipation of bilateral symmetry occurs in the egg and blastula of higher insects (Fig. 2e), where the later anterior–posterior and dorsal–ventral body axes are preformed in the egg, with no trace of radial symmetry. This is a clearly derived state.

Another plausible ancestral feature of pre-gastrula stages is that the blastulae have little, or moderate, yolk and result from total cleavage. That means cleavages dissect the entire egg and are not impeded to do so by the storage of yolk. Eggs derived from that store vast amounts of yolk as a built-in nutrient supply and cleave only partially (as seen in fish, Fig. 2d) (see Chapter 12), or superficially (in pterygote insects, Fig. 2e, and also in reptiles and birds) (see Chapters 7, 14, and 15).

CONSERVED MOLECULES ACTIVE IN ANIMAL–VEGETAL PATTERNING AND GERM-LINE DEVELOPMENT

Key factors establishing the first polarity in eggs and early embryos are conserved across the animal kingdom, which makes it likely that their evolutionary birth relates to the evolution of animal–vegetal polarity and/or of the germ cells. For example, maternally supplied *nanos* RNA is stored in vegetal pole regions in the *Xenopus* egg (Forristall et al. 1995) and posteriorly in the *Drosophila* egg (Wang and Lehmann 1991). The translated Nanos protein then separates the germ-cell lineage from somatic tissue by translational repression of somatic factors active in the animal hemisphere, such as Hunchback. Other conserved translational regulators of vegetally stored maternal messengers are Pumilio (Nakahata et al. 2001) and CPEB (Chang et al. 1999). The Vasa protein is an RNA helicase that is required for efficient translation of *nanos* in *Drosophila* oogenesis and drives cells into the germ-cell fate (Saffman and Lasko 1999).

HAECKEL'S BLASTAEA

If ancestral, the hollow blastula as an initial developmental state should recapitulate an early state of animal evolution. Haeckel's blastaea concept (Fig. 3a; Haeckel 1874) is based on this and is still used today (Nielsen 2001). It views the metazoan ancestors as spherical, planktonic colonies of choanoflagellates (such as *Volvox*, although this is a case of convergent evolution). The first metazoan, blastaea, came into existence as an epithelial sphere, the simplest conceivable continuous epithelium. For the first time, the epithelium enclosed a cavity more or less completely isolated from

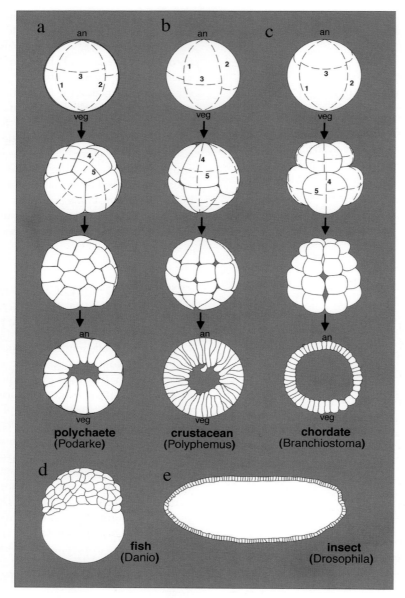

Figure 2. Ancestral and derived blastulae. (*a–c*) Total cleavage produces a hollow blastula with approximate radial symmetry, in all three major branches of the Bilateria. (*a*) *Eupomatos*, Polychaeta (spiral cleavage, quadri-radial symmetry). (*b*) *Branchiostoma*, Chordata. (*c*) *Polyphemus*, Crustacea. (*d,e*) Derived conditions: partial or superficial cleavage leads to blastula filled with yolk. Symmetry is radial in *d*, *Danio*, Teleostei, but bilateral in *e*, *Drosophila*, Insecta.

the environment, so that organic molecules could be exchanged without loss to the outside (Nielsen 2001). This allowed the first division of labor between cells. Some cells became specialized on feeding, others were kept apart as germ cells and as stem cells for asexual reproduction. This established the first body axis (the animal–vegetal axis), reflecting the initial germ-line–soma polarity (Fig. 3a). Along this axis, "vegetal" Nanos translationally repressed specific genes, thus restricting their activity to the "animal"

pole. In its flattened state, today's *Trichoplax adhaerens* (Parazoa) can be viewed as a benthic "blastaea" (Fig. 1).

To facilitate understanding of the subsequent paragraphs, a modern blastaea concept can be phrased as follows: *The invention of a continuous epithelium was the key step in early metazoan evolution.* It started as a simple sphere, flattened or not, and subsequent evolution then was a series of repeated inward folding of this epithelium, as recapitulated in various aspects of gastrulation in extant species.

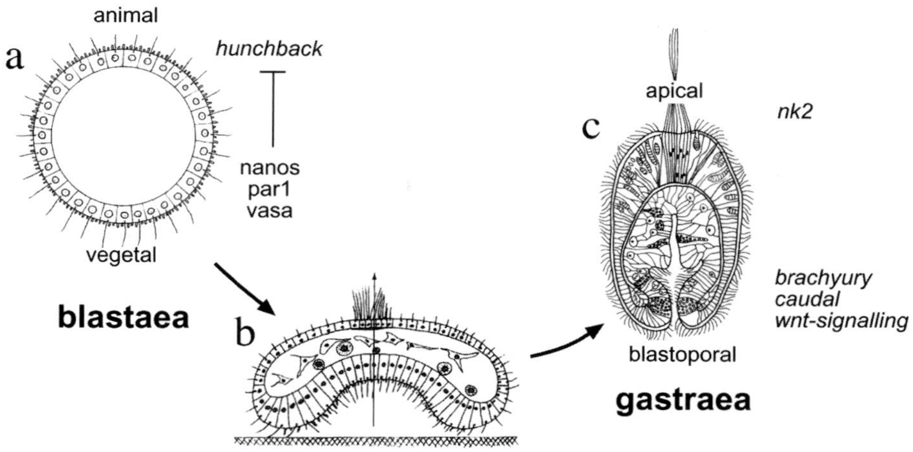

Figure 3. Haeckel's blastaea and gastraea. (Modified from Siewing 1985.)

THE FIRST FOLDING: INVAGINATION

A key characteristic of gastrulation is the internalization of the vegetal tissue, the later endoderm and mesoderm. In its ancestral form, this is achieved by the vegetal invagination of the blastula epithelium. The invaginated tissue forms the primitive gut, the archenteron. Its opening to the exterior is called the blastopore. As shown below, the initial invagination pattern has been modified to a variable extent in many groups. Moreover, it is often interwoven with other morphogenetic processes that also form part of gastrulation—such as mesoderm formation, or convergence extension movements. Paradoxically, although internalization is the key characteristic of gastrulation, the analysis of the molecular mechanisms that drive it is less advanced compared to the other aspects of gastrulation.

MODIFYING INVAGINATION IN THE NAME OF YOLK

At first sight, the tissue and cellular movements that drive internalization seem very diverse (as detailed in Part I of this book). They can be understood as recurrent modifications of a common theme: invagination (Figs. 4 and 5). It is found in species with little yolk, in all major branches of animals (Fioroni 1992, p. 212). In Lophotrochozoa, it is, for example, seen in polychaetes (*Eupomatos:* Fig. 4b, *Podarke, Polygordius*), in nemertines (*Pedicellina*), and in some mollusks (Fioroni, 1992, p. 212). In Ecdysozoa, it is found in crustaceans (Fig. 4c). Invagination is widespread in basal Deuterostomia, where it is present in *Amphioxus* (*Branchiostoma*, Fig. 4d), and in sea urchins and hemichordates. Outside the Bilateria, invagination is found, for example, in Cnidaria (Fig. 4a). Prototypic invagination is schematized in Figure 5a. All blastomeres are arranged in a

simple epithelial layer, of which the entire vegetal half folds inward. This creates a huge blastopore, which is subsequently closed. Vertebrates show a more derived variant of invagination (generalized in Fig. 5c). Here, movements have been altered by the increased storage of yolk in the most vegetal blastomeres—instead of a single epithelial layer, these blastomeres now form a massive vegetal mass of cells. Regardless of the amount of yolk, three different types of tissue movements contribute to the invagination process that are observed in three distinct regions (animal, equatorial, vegetal) of the initial blastula.

Involution ("rolling inside") is the turning inward of the blastula epithelium around the blastopore lip. This is the most obvious aspect of invagination, seen in the equatorial girdle of the blastula (red in Fig. 5). Involution requires an extension of the outer, apical surface of the epithelial cells forming the blastopore lip (red arrows in Fig. 5a). In amphibians, involution occurs at the edge of the

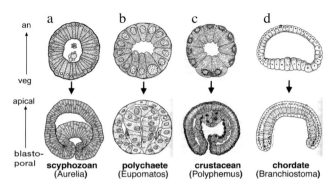

Figure 4. Ancestral invagination. (*a*) *Aurelia*, Scyphozoa. (*b*) *Eupomatos*, Polychaeta. (*c*) Crustacea. (*d*) *Branchiostoma*, Chordata. (Adapted from Nielsen 2001.)

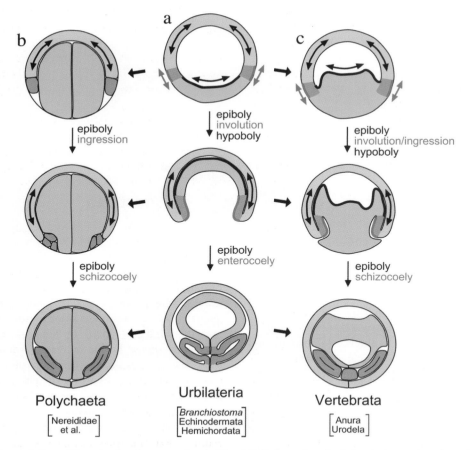

Figure 5. The evolution of invagination (cross sections). (*Gray*) Epiboly region; (*red*) involution region; (*green*) hypoboly region. Arrows visualize deformation and expansion of tissues. (*a*) Prototype invagination; (*b*) polychaete; (*c*) lower vertebrate. Groups in brackets come close to the depicted scenario.

huge mass of yolk (Fig. 5c). The inward turning movement involves the blastopore region in a wave-like fashion. As observed during sea urchin invagination (Gross and McClay 2001; see Chapter 9), and depicted by Conklin for *Branchiostoma* (Conklin 1932), cells turn inward at the blastopore lip sequentially from the more vegetal to the more animal, so that the blastopore lip comprises a changing population of cells. This is also true for the yolk-rich amphibians, where cells likewise turn inward sequentially, from vegetal to animal, around the blastopore lip. Ancestrally, involution is the movement of a simple epithelium. In anurans such as *Xenopus*, however, the blastopore lip has expanded to multiple layers, of which only the superficial cells retain epithelial integrity, while the deep layer cells migrate inward as individuals (ingression, Keller 1981; see Chapters 13 and 19).

Epiboly ("moving over") is the active expansion of ectodermal cells over the internalized tissue (gray territories with black arrows in Fig. 5). Although widely conserved in animal gastrulation, the individual contribution of epiboly to the whole process of internalization can vary greatly. In the ancestral invagination scenario, epiboly is needed for at least the closure of the blastopore. This is especially clear for the transition from the second to the third panel in Figure 5a. With increasing storage of yolk, epiboly has become essential to surround the large vegetal blastomeres (Fig. 5b,c). The cellular and molecular mechanisms of epiboly have been described in some detail only in the vertebrates (Kane et al. 1996; Hasegawa and Kinoshita 2000; Kane and Adams 2002; see Chapter 20).

Hypoboly ("moving underneath") is the active movement of endodermal cells underneath the ectoderm (Arendt and Nübler-Jung 1999) and represents the exact opposite of epiboly. This movement is exerted by the internalized vegetal hemisphere itself (green in Fig. 5) and has so far been described in more detail, for amphibians (referred to as vegetal rotation, green arrows in Fig. 5c; Winklbauer and Schurfeld 1999). Although clearly derived in its details in the amphibian variant—enrolling in a huge mass of yolk—hypoboly could well be evolutionarily ancient. If we compare the amphibian situation to prototypical invagination (green arrows in Fig. 5a), a conceptually similar vegetal movement becomes apparent. Here, prospective endodermal tissue, located "vegetal" from the initial site of appearance of the blastopore lip, has to undergo a considerable change in shape from slightly convex to strongly concave. This should imply a constriction of the outer, *apical* surface of cells leading to an extension of the inner, *basal* epithelial surface of the vegetal epithelium—as visualized in Figure 5a—which is the exact opposite of what happens in the involution region. Accordingly, the active inward movement of bottle cells, which constrict their apical surfaces at the onset of amphibian gastrulation (Chapters 13 and 19), should also be regarded as a component of the hypoboly movement. Finally, avian gastrulation shows a derived variant of hypoboly, when endodermal tissue (hypoblast) moves underneath the ectoderm (epiblast) (Arendt and Nübler-Jung 1999).

Having focused so far on the vertebrate line of evolution, modifications of invagination are also apparent in Lophotrochozoa (generalized in Fig. 5b). In almost all Lophotrochozoan groups, species showing ancestral invagination (Fig. 4b) coexist with rather derived variants. As in the vertebrates, the major trend is again the increase in yolk. Adding to this, and characteristic for spirally cleaving Lophotrochozoa, they gastrulate with fewer cells (compared to vertebrates, for example). It is conceivable that the storage of yolk in only four heavily expanded vegetal blastomeres will disrupt the integrity of the vegetal epithelium—meaning that any involution or hypoboly movement can no longer take place. Instead, cells in the blastopore region will internalize by ingression (inward migration) of individual cells. The major remaining internalization movement in yolk-rich Lophotrochozoans is then epiboly of the much smaller superficial ectodermal cells, which retain epithelial integrity, over the large, yolky, vegetal blastomeres.

THE *BRACHYURY* GENE DEFINES BLASTOPORAL AS OPPOSED TO APICAL

During invagination (Fig. 4), the former animal and vegetal poles fold onto each other. The former animal–vegetal polarity is thus replaced by a new polarity, referred to as apical–blastoporal (Nielsen 2001). The new apical pole comprises the former animal and the underlying vegetal pole tissue. The blastoporal end comprises ectodermal and endomesodermal tissue surrounding the blastopore. The overall symmetry is still radial, as inherited from the blastula.

The most prominent marker for blastoporal tissue, widely conserved in evolution, is the T-box transcription factor *brachyury* (Arendt et al. 2001; Technau 2001; see Chapter 41). During *Xenopus* gastrulation, *Xbra* is expressed in the ring of involuting (endomesodermal) tissue in the upper blastopore lip (Fig. 6a). The gene is turned off as internalized cells move away from the blastopore. Expression around the blastopore is also observed in *Branchiostoma* and in various lower deuterostomes (Fig. 6b–d) (Holland et al. 1995; Tagawa et al. 1998; Peterson et al. 1999; Shoguchi et al. 1999; Gross and McClay 2001; Takada et al. 2002), as well as in Lophotrochozoan polychaetes (Fig. 6e) and gastropods (Lartillot et al. 2002b). Conservation extends even beyond Bilateria, as best exemplified in the cnidarian *Nematostella* in Figure 6f (Scholz and Technau 2003).

Given its highly conserved blastoporal identity, it is conceivable that *brachyury* has evolved as a gene controlling involution movements in the blastopore region. Indeed, evidence for such a role has been described for the sea urchin *Lytechinus*, where *brachyury* expression shows a close, dynamic affiliation with the blastopore. Here, microinjection of a dominant negative *brachyury* repressor construct results in a total block of invagination (Gross and McClay

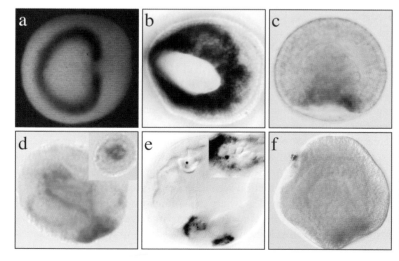

Figure 6. Expression of *brachyury* orthologs around the blastopore. (*a*) *Xenopus*; (*b*) *Branchiostoma*; (*c*) *Ptychodera*; (*d*) *Spadella*; (*e*) *Platynereis*; (*f*) *Nematostella*. (*a*, Photo by Ali Hemmati-Brivanlou in http://www.xenbase.org/xmmr/Marker_pages/earlymeso/xBra.jpg; *b*, reprinted from Holland et al. 1995 [Company of Biologists]; *c*, reprinted from Tagawa et al. 1998 [©Elsevier]; *d*, reprinted from Takada et al. 2002; *e*, D. Arendt, unpubl.; *f*, reprinted from Scholz and Technau 2003 [©Springer-Verlag].)

2001). Consistent with this finding, a screen for *brachyury* target genes in another sea urchin, *Strongylocentrotus*, has yielded cytoskeletal modulators expressed in vegetal endodermal cells where they might control involution movements (Rast et al. 2002). So far, no other studies exist on *brachyury* function for species with prototypic invagination. However, interference with *brachyury* function also affects the internalization of cells in vertebrates where the ancestral involution movement is substituted by ingression (Keller 1981, and see above). In mouse, clones of cells mutant for the *T (Brachyury)* gene fail to internalize through the primitive streak but instead accumulate externally in the ventral neural tube and axial tissue (Wilson et al. 1995; Wilson and Beddington 1997). Similarly, *Xbra* is required for proper gastrulation movements, including internalization proper, in *Xenopus* (Conlon and Smith 1999; Tada and Smith 2000).

The homeobox gene *caudal* codes for another transcription factor and shows blastoporal expression in many animal groups. It is expressed around the blastopore in vertebrates such as *Xenopus* (Northrop and Kimelman 1994; Pillemer et al. 1998), around the closed blastopore in *Branchiostoma* (Brooke et al. 1998), as well as in the polychaete *Platynereis* (B. Prudhomme, pers. comm.). A *caudal* ortholog is also coexpressed with *brachyury* in the ingression region equivalent to the blastopore in the planula larva of the cnidarian *Podocoryne* (Yanze et al. 2001; Spring et al. 2002), indicating that this pattern predates the Bilateria. In addition, comparative evidence suggests that canonical Wnt signaling plays a conserved role in activating *brachyury* (and *caudal*) around the blastopore, at least indirectly. Zygotic Wnt activity is required for blastoporal *brachyury* expression in frog (Vonica and Gumbiner 2002) and mouse (Galceran et al. 2001), and *brachyury* expression matches nuclear β-catenin, indicative of active Wnt signaling, in sea urchin (Holland 2002). Most important, the Wnt-signaling cascade is also active around the mouth in *Hydra* (Hobmayer et al. 2000). A conserved interaction of *brachyury*, *caudal*, and canonical Wnt signaling in the specification of blastoporal identity is strongly advocated by recent comparative surveys (Holland 2002; Lengyel and Iwaki 2002).

In *Xenopus*, additional factors and pathways have been identified that contribute to the initial, circumferential activation of *brachyury*, which is built upon the initial animal–vegetal polarity of the egg (for recent review, see Kumano and Smith 2002). A maternally stored and vegetally expressed transcription factor, VegT, activates Nodal-related/activin-like secreted molecules of the TGF-β superfamily (see Chapter 41). An FGF-like factor acts to define animal hemisphere tissue. At the interface of FGF and Nodal/Activin signaling, an equatorial zone of *brachyury* expression is then established. A possible evolutionary conservation of any of these players outside vertebrates, however, remains elusive.

HAECKEL'S GASTRAEA

In his *Gastraea-Theory*, Haeckel postulated a gastrula-like last common ancestor of sponges, Cnidaria, and Bilateria (Haeckel 1874). Its evolution was driven by the stepwise invention of the archenteron, as a primitive digestive chamber. It was useful to capture and to enrich for food, and locally to optimize conditions for digestion in that chamber (Fig. 3b) (Nielsen 2001). An illustrative model for the emergence of the gastraea is the feeding behavior of the blastaea-like *Trichoplax adhaerens* (Parazoa) (Fig. 1) (Grell and Ruthmann 1991; Ender and Schierwater 2003), which, by invagination, forms a temporary digestive chamber. Animals with true gastraea-like organization should have been the ancestors of the sponges (Porifera) and of the Cnidaria, although it is a yet-unresolved question as to what extent the latter already exhibit some aspects of bilateral symmetry. The evolutionary transformation of adult body plans toward the gastraea would be most closely recapitulated by invagination as observed in *Branchiostoma* (Fig. 5a). Note that the gastraea still exhibits radial symmetry.

An ancestral *brachyury* gene was found to be active around the opening of the digestive cavity. The evolution of the *brachyury* subfamily of T-box genes was then tightly coupled to the stepwise deepening of the digestive cavity on its way toward the archenteron. *brachyury* was coexpressed with *caudal*, in a region of active Wnt signaling. Genes that might have evolved to confer the opposite, apical identity, are the *nk2* homeobox genes (Fig. 3c) (Galliot 2000).

THE SECOND FOLDING: FORMATION OF MESODERMAL POUCHES

The tissue internalized during gastrulation gives rise to the gut epithelium proper (endoderm), but also to the mesoderm. In many species, the formation of the mesoderm is tightly coupled to gastrulation as, for example, in amphibians (Chapter 13). The following section tries a brief evolutionary explanation for this developmental link.

MESODERM FORMATION AS A CONTINUATION OF INVAGINATION

In the Bilateria developing via a spherical blastula, presumptive mesodermal cells locate to equatorial regions, where the blastopore forms during gastrulation (red in Fig. 5). Ancestrally, this endomesoderm anlage was probably ring-shaped (Arendt and Nübler-Jung 1997). Hence, in the prototype invagination scenario, the prospective mesoderm forms from the region where the epithelium actively turns inward.

Morphologically, mesoderm formation can vary considerably. Mesoderm and endoderm initially form a developmental unit, the endomesoderm, which separates into distinct tissues at any time point before, during, or after internalization. The presumed ancestral mode of mesoderm formation is schematized in the two lower panels in Figure 5a (seen, e.g., in *Branchiostoma*, Conklin 1932). Here, the mesoderm anlage initially forms part of the lining of the archenteron. The mesoderm then separates from the archenteron by enterocoely ("cavity formation from the gut"), meaning by the formation of mesodermal pouches that pinch off into separate vesicles. The mesodermal vesicles then give rise to somites and enlarge into the coelomic cavities. As essential in this ancestral scenario, epithelial continuity is maintained from the initial endomesoderm to the epithelial lining of the coelom—and thus, from the blastula to the coelom. Notably, as a morphogenetic process, enterocoely is equivalent to invagination. In both, pouches sink inward on the basal side of the epithelium, to finally pinch off (this separates mesoderm from endoderm in the former, endomesoderm from ectoderm in the latter). We can thus interpret mesoderm formation as a continuation of epithelial folding during invagination, at local spots close to the blastopore in the archenteron.

Derived variants of mesoderm formation are referred to as schizocoely ("cavity formation by segregation"): Instead of forming an epithelial pouch, the incipient mesoderm is a mesenchymal mass of cells that proliferates and detaches from within the blastoporal region. This mass secondarily hollows, and forms anew the epithelial linings of the coelom. In amphibians, this can occur after internalization (Fig. 5b), precisely during the inward turning of tissue, or even before gastrulation starts (as deep layer of the marginal zone in *Xenopus*). This reflects the evolutionary tendency to progressively anticipate the segregation of mesoderm and endoderm.

In Deuterostomia, enterocoely is observed not only in *Branchiostoma*, but also in echinoderms and acorn worms (Nielsen 2001). It is less widespread, however, in Protostomia. An enterocoelous mode of mesoderm formation is found in some Lophotrochozoa, namely Brachiopoda (Fig. 1) (Nielsen 2001), and in the polychaete Pogonophora (Ivanoff 1988). In Ecdysozoa, it has been described for Tardigrada (see, however, Nielsen 2001). In most Lophotrochozoa, mesoderm formation more resembles the derived vertebrate situation, with an early separation of endoderm and mesoderm, and schizocoely (as a result of similar modifications of development in convergent evolutionary lines). This is exemplified for polychaete Nereidids in Figure 5b. A common endomesoderm anlage does exist, but it separates into endodermal and mesodermal lineages as early as during spiral cleavage stages (Wilson 1892). A few mesodermal founder cells are located around the blastopore that are internalized during gastrulation and then proliferate to form mesodermal bands that split into somites, hollowing by secondary cavitation. In insects, mesoderm formation is derived, in that endoderm

and mesoderm segregate early (Chapters 7 and 8). It forms, however, by invagination, a process very much resembling enterocoelous mesoderm formation. Since this is a general insect feature (Anderson 1973), insect mesoderm invagination might very well be an evolutionary remnant of ancestral enterocoely.

The genes involved in mesoderm specification and formation in the vertebrates have been identified and characterized to a considerable extent, but are not detailed here because they are covered in other chapters in this volume. However, *brachyury* also plays a major role in specifying mesoderm in the vertebrates. This underscores the link between invagination and mesoderm specification.

THE ENTEROCOELY CONCEPT

This concept, put forward by Sedgwick (1884) and Remane (1950), views the mesoderm as an evolutionary derivative of gastric pouches as observed in today's cnidarians (Fig. 1). These pockets became progressively separated from the gut lumen proper and could finally acquire the hydrostatic function of a coelom. Gastric pouches could have evolved by the stepwise prolongation of the invagination process around the opening of the digestive cavity. This would explain the tight developmental link between mesoderm formation and gastrulation at morphological and molecular levels.

Notably, the aspects of gastrulation described so far— invagination and mesoderm formation—are built upon the initial radial symmetry of egg and blastula. Although acting in parallel, they do not themselves contribute to the establishment of the bilateral body form, which can be seen as a third and independent component of gastrulation.

THE TRANSFORMATION FROM RADIAL TO BILATERAL SYMMETRY

The worm-shaped body form is characteristic of Bilateria, with bilateral symmetry, and an anterior–posterior and a dorsal–ventral body axis. We have seen before that ancestrally, pre-gastrula stages of bilaterian embryos still exhibit (approximate) radial symmetry, so that the bilateral body form is shaped only during gastrulation. This means that the radial-to-bilateral transformation is a key feature of bilaterian gastrulation. It took place in the Urbilaterian development.

In vertebrates, the Spemann organizer initiates and drives the morphogenetic movements that establish bilateral symmetry. Do similar movements exist outside vertebrates, and what controls them? In principle, the radial-to-bilateral transformation in the lower chordate *Branchiostoma* appears to follow the same pattern as in vertebrates, in terms of morphology (Conklin 1932) and molecules (see, e.g., Yu et al. 2002). Other Deuterostomia are either more derived in that they show bilateral symmetry only transiently during larval stages and do not form a worm-shaped adult body (i.e., echinoderms), or gastrulation movements and molecules are scarcely described (i.e., hemichordates). In Protostomia, a radial-to-bilateral transformation is observed in almost all groups with bilateral, worm-shaped body forms. Unfortunately, the currently favored molecular model organisms in Protostomia either do not show radial–bilateral transformation at all (*Drosophila*, other insects) or are too derived (*Caenorhabditis*, leech) to allow reasonable comparison to the Deuterostomia with respect to the acquisition of bilateral symmetry. It is only with the molecular data emerging from Lophotrochozoan model organisms such as *Platynereis*—which do show a prototypic radial–bilateral transformation—that a sounder base emerges for comparison between Proto- and Deuterostomia, across the Bilateria.

To begin with, we compare the overall tissue movements in the radial-to-bilateral transformation between polychaetes (Protostomia) and vertebrates (Deuterostomia). This reveals striking similarities that are reflected at the molecular level by the spatially similar expression of conserved marker genes. Therefore, these similarities are likely to recapitulate the evolution of bilaterian animals.

EMERGENCE OF THE BILATERAL BODY FORM IN POLYCHAETES

Polychaete Nereidids (Fig. 7a), representative of the spirally cleaving Lophotrochozoan branch, initially exhibit quadri-radial symmetry as inherited from the spiral cleavage pattern, with apical–blastoporal polarity (Fig. 7, upper panels). Radial symmetry is soon broken by the progeny of one specific blastomere, the 2d cell. It proliferates into a mass of cells on one side of the gastrula (yellow and red in Fig. 7), so that one side of the blastula is distinct from the other, and a second polarity emerges in right angle to the apical–blastoporal axis.

The radial-to-bilateral transformation as provoked by the 2d progeny then involves two unexpected kinds of interwoven movements. First, in a strictly bilateral movement, the mass of 2d progeny expands toward the other side of the gastrula (black arrows in Fig. 7a), where its two halves meet again and fuse. Thus forms the ventral plate, a columnar epithelium of mainly neurogenic tissue. It subsequently expands to form the ventral body side (from which the CNS arises). Therefore,

1. *Cells move from one side of the gastrula toward the other to form the ventral (neural) body side.*

What about future anterior and posterior? The establishment of the anterior–posterior axis also involves unexpected rearrangements. The precursor cells of the stomodeum (mouth tissue) and of the proctodeum (anal tissue) are initially located adjacent to each other along the blastopore (light and dark gray in Fig. 7a) (Wilson 1892). They then

become separated by the ventral plate precursors moving between them. By this movement, the nascent stomodeum adjoins the future head region, while the proctodeum reaches the posterior extreme. That means that

2. *The blastoporal region comprises future anterior and posterior tissue that is driven apart by the interposition and expansion of the ventral plate.*

Since *brachyury* is a marker gene for the blastoporal tissue, its expression traces the tissue rearrangements during and after gastrulation. As outlined above, *Platynereis brachyury* starts to be expressed in stomodaeal and proctodaeal precursors along the blastopore (Fig. 6e). When these move apart from each other, driven by the formation of the ventral plate, a narrow stripe of *brachyury*-expressing cells maintains the continuity between stomodeum and proctodeum (Fig. 8a; schematized in Fig. 8c). This stripe demarcates the midline of the future ventral body side. A similar stripe of *brachyury* expression between stomodeum and proctodeum, along the ventral midline, is also observed in the gastropod mollusk *Patella* (Lartillot et al. 2002b; see Chapter 5), indicating conservation of this pattern in Lophotrochozoa. In *Patella*, the ventral midline also specifically expresses the *hedgehog* gene, encoding a secreted signaling molecule (Fig. 8e) (Nederbragt et al. 2002).

EMERGENCE OF THE BILATERAL BODY FORM IN VERTEBRATES

Morphogenetic movements that establish the bilateral body form in the vertebrates have been studied in cellular and molecular detail in amphibians, in particular *Xenopus*, and in teleost fish, and are described in Chapters 12, 13, 15, 19, 20, and 27. We refer to these movements here only for comparison. It will be seen that in essence, the vertebrate tissue movements take a similar course when compared to the polychaetes.

As in ancestral forms, the amphibian gastrula starts off with radial symmetry and apical–blastoporal axis (Fig. 7b, upper panels). Radial symmetry is broken by the formation of the Spemann organizer tissue on one side of the gastrula. A second polarity thus emerges at a right angle to the apical–blastoporal polarity. Again, formation of the two body axes involves a complicated morphogenetic transformation with two major characteristics.

Most prominent is the convergence and extension of the future neural plate (black arrows in Fig. 7b, lower panels). Directed, mediolateral intercalation of cells showing planar polarity leads to a net movement of cells toward the organizer. This means,

1. *Cells move from one side of the gastrula toward the other to form the dorsal (neural) body side.*

This is a first obvious parallel to the polychaete scenario, and it is more than superficial, given that the similar

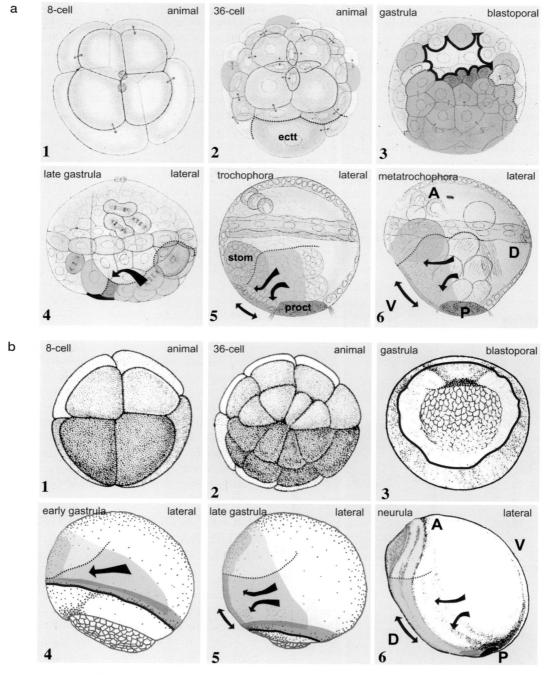

Figure 7. The radial-to-bilateral transformation. (*a*) *Platynereis.* Yellow marks 2d-cell and descendants. (Adapted from Wilson 1892.) (*b*) *Xenopus.* Ochre marks neural plate. (Adapted, in part, from Nieuwkoop and Faber 1994.) Black arrows represent convergence and extension movements. Red is the *hox*-positive neural ectoderm. (ectt) Ectodermal teloblasts; (A) anterior; (P) posterior; (D) dorsal; (V) ventral.

tissue movement in the polychaetes involves similar cellular movements (P. Steinmetz and D. Arendt, unpubl.). In addition, the fate of the converging tissue has the same molecular identity (e.g., neurogenic tissue later specified by Hox cluster gene expression), and orthologous Gbx genes demarcate the morphogenetic boundary to the nonconverging tissue (visualized by the color code in Fig. 9).

2. *The convergence and extension of the neural plate disjoins future anterior and posterior tissue of blastoporal origin.*

It has always been a puzzling observation that the vertebrate organizer tissue, blastoporal by origin, is of both anterior and posterior fate. Anterior cells give rise to ventral forebrain (Stern et al. 1992; Hatada and Stern 1994; Woo and Fraser 1995; Psychoyos and Stern 1996; Charrier et al. 2002) and prechordal plate, and posterior cells form part of the tail bud (Gont et al. 1993). These two extremes separate by the convergence and extension movements, meaning that they are forced apart by the neural plate tissue (and underlying mesoderm) moving in between them.

Again, this is an intriguing parallel to the polychaete situation, which is reflected by the similar change in shape of the blastoporal expression domain of the *brachyury* gene. From a ring around the blastopore (Fig. 6a), the mass of *brachyury*-expressing cells extends into a longitudinal stripe (the future notochord) that demarcates the midline of the

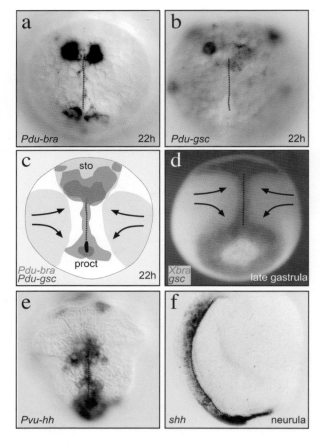

Figure 8. *brachyury*, *goosecoid*, and *hedgehog* as conserved marker for the extension of blastoporal tissue during AP axis formation. (*Black arrows*) Convergence and extension movements. (*Stippled line*) neural midline. (*a*) *Platynereis brachyury*, 22 hr, blastoporal view; (*b*) *Platynereis goosecoid*, 22 hr, blastoporal view; (*c*) *Platynereis brachyury* and *goosecoid*, 22 hr, blastoporal view; (*d*) *Xenopus brachyury* and *goosecoid*, blastoporal view; (*e*) *Patella hedgehog*, ventral view; (*f*) *Xenopus Shh*, dorsal view.

newly forming dorsal body side (Fig. 8d). Note that the population of *brachyury*-expressing midline cells behaves distinctly from the surrounding tissue, in that its elongation involves the proliferation of a resident population of cells in the node/organizer producing midline cells of increasingly posterior fate (as evidenced for amniotes; Hatada and Stern 1994; Psychoyos and Stern 1996). It is unknown whether similar midline stem cells also exist in lower vertebrates, or outside vertebrates.

Another aspect conserved between vertebrates and polychaetes is the expression of *goosecoid*, encoding a homeobox transcription factor (see Chapter 42), in blastoporal tissue that comes to lie in the future anterior, in *Platynereis* (Fig. 8b,c) (Arendt et al. 2001), *Patella* (Lartillot et al. 2002a), and vertebrates (Fig. 8d). A conserved marker for the extending midline cells is *sonic hedgehog* (expressed in future floor plate cells that overlay the notochordal precursors; Fig. 8f); compare to *hedgehog* expression in *Patella* (Fig. 8e). The shared expression of *brachyury* and *hedgehog* orthologs suggests homology of the midline cells in Protostomia and Deuterostomia.

What are the genes and pathways acting upstream, downstream, and in parallel with *brachyury*, *sonic hedgehog*, and *goosecoid*, in the radial-to-bilateral transformation? Although there is considerable knowledge about the molecular nature of the vertebrate Spemann organizer, almost nothing is known for polychaetes so far. Molecular studies focus on polychaete orthologs of vertebrate organizer antagonists such as *chordin* and *noggin* (against Dpp/*BMP* signaling), or *dickkopf* (against Wnt signaling), to identify and characterize a putative gastrula organizer in the polychaetes (P. Steinmetz and D. Arendt, unpubl.).

JAEGERSTEN'S BILATEROGASTRAEA, AND THE AMPHISTOMY CONCEPT OF BILATERIAN EVOLUTION

A common theme has emerged for the radial-to-bilateral transformation in Bilateria. If we assume that these movements recapitulate evolution, what do they tell about the historical process that created the Bilateria?

Starting from the gastraea-like organism with a single, round opening into the primitive gut (Fig. 3c), how did it evolve into a bilaterian worm? The most parsimonious explanation for the morphogenetic movements we observe is given by the amphistomy concept of evolution (amphi: both sides; stoma: opening, mouth) (Arendt and Nübler-Jung 1997; Bruce and Shankland 1998; and see below): It is assumed that the blastopore first elongated to a slit, which then closed. Two openings remained at its ends, mouth and anus. The closure of the slit thus gave rise to a tube-shaped gut. Concomitantly, the new mouth approached the apical brain to form part of the new head, while the anus moved

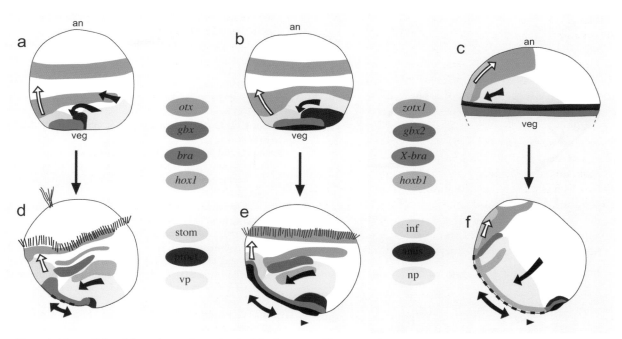

Figure 9. The radial-to-bilateral transformation in (*a*) *Platynereis*, (*b*) proposed Urbilaterian, (*c*) fish. (*Black arrows*) convergence and extension movements. (*White arrow*) Displacement of former blastoporal tissue toward the anterior.

toward the posterior of the now worm-like creature (Fig. 10). This scenario follows Jaegersten (1955) and Siewing (1985) and follows initial ideas by Sedgwick (1884). Jaegersten referred to the presumed bilaterian ancestor with slit-like blastopore as bilaterogastraea. Such open slit-like blastopore is found in extant polychaetes with little yolk (e.g., *Polygordius*), and in onychophorans (see Arendt and Nübler-Jung 1997). The advantage of forming a slit is the physical separation of the nutritive water current into an inward and an outward current.

This scenario explains the two main features of the polychaete and vertebrate radial-to-bilateral transformation, namely why (1) tissue moves toward, and along, the midline of the neural body side: this recapitulates the closure of the slit, its elongation, and the evolution of the bilaterian head conjoining blastoporal mouth and apical brain. It also explains why (2) the blastoporal tissue comprises future anterior and posterior tissue: i.e., because mouth and anus both evolved from the single blastoporal opening of the initial digestive cavity.

Finally, the amphistomy concept of bilaterian evolution sheds new light on the evolution of the different fates of the blastopore in Protostomia versus Deuterostomia (Fig. 11): In protostomy ("mouth first") the blastopore forms the mouth, and the anus breaks through secondarily. This pattern is found in Protostomia; for example, in the gastropod *Patella*. In deuterostomy ("mouth second") the blastopore

forms the anus, and the mouth breaks through secondarily. This pattern is found in all Deuterostomia, but also in a few Protostomia. The "amphistomy concept" regards both, proto- and deuterostomy, as secondary modifications of the ancestral amphistome pattern (Arendt and Nübler-Jung 1997). In line with this, the mouth regions of protostome and deuterostome larvae are homologous, regardless of

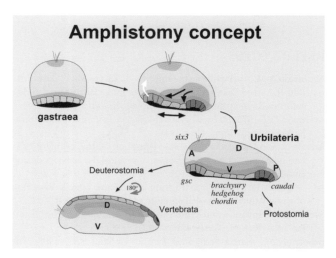

Figure 10. The amphistomy concept of bilaterian evolution.

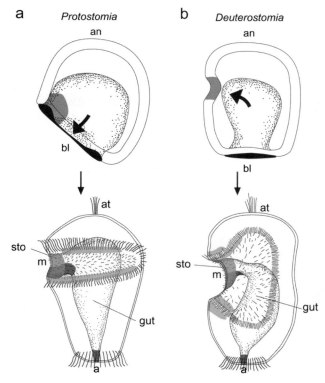

Figure 11. Homology of mouth regions. (*a*) Protostomia; (*b*) Deuterostomia. Color code visualizes expression territories of conserved genes: (*blue*) *brachyury*; (*red*) *otx*; (at) apical tuft; (an) animal; (bl) blastoporal; (m) mouth; (sto) stomodeum; (a) anus.

whether the mouth results from proto-, amphi-, or deuterostome gastrulation (Fig. 11) (Arendt et al. 2001).

Finally, it is important to note that the "amphistomy concept" implies an inversion of the dorsoventral body axis during early vertebrate evolution, and the formation of a new mouth in the vertebrates on their new ventral body side (Fig. 10).

REFERENCES

Aguinaldo A.M., Turbeville J.M., Linford L.S., Rivera M.C., Garey J.R., Raff R.A., and Lake J.A. 1997. Evidence for a clade of nematodes, arthropods and other moulting animals. *Nature* 387: 489–493.

Anderson D.T. 1973. *Embryology and phylogeny in annelids and arthropods.* Pergamon Press, New York.

Arendt D. and Nübler-Jung K. 1997. Dorsal or ventral: Similarities in fate maps and gastrulation patterns in annelids, arthropods and chordates. *Mech. Dev.* 61: 7–21.

———. 1999. Rearranging gastrulation in the name of yolk: Evolution of gastrulation in yolk-rich amniote eggs. *Mech. Dev.* 81: 3–22.

Arendt D., Technau U., and Wittbrodt J. 2001. Evolution of the bilaterian larval foregut. *Nature* 409: 81–85.

Brooke N.M., Garcia-Fernandez J., and Holland P.W. 1998. The ParaHox gene cluster is an evolutionary sister of the Hox gene cluster. *Nature* 392: 920–922.

Bruce A.E. and Shankland M. 1998. Expression of the head gene Lox22-

Otx in the leech Helobdella and the origin of the bilaterian body plan. *Dev. Biol.* 201: 101–112.

Chang J.S., Tan L., and Schedl P. 1999. The *Drosophila* CPEB homolog, orb, is required for oskar protein expression in oocytes. *Dev. Biol.* 215: 91–106.

Charrier J.B., Lapointe F., Le Douarin N.M., and Teillet M.A. 2002. Dual origin of the floor plate in the avian embryo. *Development* 129: 4785–4796.

Conklin E.G. 1932. The embryology of amphioxus. *J. Morphol.* 54: 69–118.

Conlon F.L. and Smith J.C. 1999. Interference with brachyury function inhibits convergent extension, causes apoptosis, and reveals separate requirements in the FGF and activin signaling pathways. *Dev. Biol.* 213: 85–100.

Ender A. and Schierwater B. 2003. Placozoa are not derived cnidarians: evidence from molecular morphology. *Mol. Biol. Evol.* 20: 130–134.

Fioroni P. 1992. *Allgemeine und vergleichende Embryologie.* Springer, Berlin, Germany.

Fischer A. and Dorresteijn A.W.C. 2004. The polychaete Platynereis dumerilii (Annelida): A laboratory animal with spiralian cleavage, lifelong segment proliferation and a mixed benthic/pelagic life cycle. *BioEssays* 26: 314–325.

Forristall C., Pondel M., Chen L., and King M.L. 1995. Patterns of localization and cytoskeletal association of two vegetally localized RNAs, Vg1 and Xcat-2. *Development* 121: 201–208.

Galceran J., Hsu S.C., and Grosschedl R. 2001. Rescue of a Wnt mutation by an activated form of LEF-1: regulation of maintenance but not initiation of Brachyury expression. *Proc. Natl. Acad. Sci.* 98: 8668–8673.

Galliot B. 2000. Conserved and divergent genes in apex and axis development of cnidarians. *Curr. Opin. Genet. Dev.* 10: 629–637.

Gont L.K., Steinbeisser H., Blumberg B., and Derobertis E.M. 1993. Tail formation as a continuation of gastrulation: The multiple cell populations of the *Xenopus* tailbud derive from the late blastopore lip. *Development* 119: 991–1004.

Grell K.G. and Ruthmann A. 1991. Placozoa. In *Microscopic anatomy of invertebrates* (ed. F.W. Harrison and J.A. Westfall), pp. 13–27. Wiley-Liss, New York.

Grobben K. 1908. Die systematische Einteilung des Tierreichs. *Verh. Zool. Bot. Ges. Wien* 58: 491–511.

Gross J.M. and McClay D.R. 2001. The role of Brachyury (T) during gastrulation movements in the sea urchin *Lytechinus variegatus.* *Dev. Biol.* 239: 132–147.

Haeckel E. 1874. Die Gastraea-Theorie, die phylogenetische Classification des Thierreiches und die Homologie der Keimblätter. *Jena Z. Naturwiss.* 8: 1–55.

Hasegawa K. and Kinoshita T. 2000. Xoom is required for epibolic movement of animal ectodermal cells in *Xenopus laevis* gastrulation. *Dev. Growth Differ.* 42: 337–346.

Hatada Y. and Stern C.D. 1994. A fate map of the epiblast of the early chick embryo. *Development* 120: 2879–2889.

Hobmayer B., Rentzsch F., Kuhn K., Happel C.M., von Laue C.C., Snyder P., Rothbacher U., and Holstein T.W. 2000. WNT signaling molecules act in axis formation in the diploblastic metazoan Hydra. *Nature* 407: 186–189.

Holland L.Z. 2002. Heads or tails? Amphioxus and the evolution of anterior–posterior patterning in deuterostomes. *Dev. Biol.* 241: 209–228.

Holland P.W., Koschorz B., Holland L.Z., and Herrmann B.G. 1995. Conservation of Brachyury (T) genes in amphioxus and vertebrates: Developmental and evolutionary implications. *Development* 121: 4283–4291.

Ivanoff A.V. 1988. Analysis of the embryonic development of Pogonophora in connection with the problems of phylogenetics. *Z. Zool. Syst. Evolutionforsch.* **26:** 161–185.

Jaegersten G. 1955. On the early phylogeny of the Metazoa. *Zool. Bidr. Uppsala* **30:** 321–354.

Kane D. and Adams R. 2002. Life at the edge: Epiboly and involution in the zebrafish. *Results Probl. Cell Differ.* **40:** 117–135.

Kane D.A., Hammerschmidt M., Mullins M.C., Maischein H.M., Brand M., van Eeden F.J., Furutani-Seiki M., Granato M., Haffter P., Heisenberg C.P. et al. 1996. The zebrafish epiboly mutants. *Development* **123:** 47–55.

Keller R.E. 1981. An experimental analysis of the role of bottle cells and the deep marginal zone in gastrulation of *Xenopus laevis*. *J. Exp. Zool.* **216:** 81–101.

Kumano G. and Smith W.C. 2002. Revisions to the *Xenopus* gastrula fate map: Implications for mesoderm induction and patterning. *Dev. Dyn.* **225:** 409–421.

Lartillot N., Le Gouar M., and Adoutte A. 2002a. Expression patterns of *fork head* and *goosecoid* homologues in the mollusc *Patella vulgata* supports the ancestry of the anterior mesendoderm across Bilateria. *Dev. Genes Evol.* **212:** 551–561.

Lartillot N., Lespinet O., Vervoort M., and Adoutte A. 2002b. Expression pattern of *Brachyury* in the mollusc *Patella vulgata* suggests a conserved role in the establishment of the AP axis in Bilateria. *Development* **129:** 1411–1421.

Lengyel J.A. and Iwaki D.D. 2002. It takes guts: The *Drosophila* hindgut as a model system for organogenesis. *Dev. Biol.* **243:** 1–19.

Nakahata S., Katsu Y., Mita K., Inoue K., Nagahama Y., and Yamashita M. 2001. Biochemical identification of *Xenopus* Pumilio as a sequence-specific cyclin B1 mRNA-binding protein that physically interacts with a Nanos homolog, Xcat-2, and a cytoplasmic polyadenylation element-binding protein. *J. Biol. Chem.* **276:** 20945–20953.

Nederbragt A.J., van Loon A.E., and Dictus W.J. 2002. Evolutionary biology: Hedgehog crosses the snail's midline. *Nature* **417:** 811–812.

Nielsen C. 2001. *Animal evolution. Interrelationships of the living phyla*, 2nd edition. Oxford University Press, Oxford, United Kingdom.

Nieuwkoop P.D. and Faber J. 1994. *Normal table of* Xenopus laevis (Daudin). Garland, New York.

Northrop J.L. and Kimelman D. 1994. Dorsal-ventral differences in Xcad-3 expression in response to FGF-mediated induction in *Xenopus*. *Dev. Biol.* **161:** 490–503.

Nübler-Jung K. and Arendt D. 1994. Is ventral in insects dorsal in vertebrates? A history of embryological arguments favouring axis inversion in chordate ancestors. *Roux's Arch. Dev. Biol.* **203:** 357–366.

Peterson K.J., Cameron R.A., Tagawa K., Satoh N., and Davidson E.H. 1999. A comparative molecular approach to mesodermal patterning in basal deuterostomes: The expression pattern of *Brachyury* in the enteropneust hemichordate *Ptychodera flava*. *Development* **126:** 85–95.

Pillemer G., Epstein M., Blumberg B., Yisraeli J.K., De Robertis E.M., Steinbeisser, H., and Fainsod A. 1998. Nested expression and sequential downregulation of the *Xenopus* caudal genes along the anterior–posterior axis. *Mech. Dev.* **71:** 193–196.

Psychoyos D. and Stern C.D. 1996. Fates and migratory routes of primitive streak cells in the chick embryo. *Development* **122:** 1523–1534.

Rast J.P., Cameron R.A., Poustka A.J., and Davidson E.H. 2002. *brachyury* target genes in the early sea urchin embryo isolated by differential macroarray screening. *Dev. Biol.* **246:** 191–208.

Remane A. 1950. Die Entstehung der Metamerie der Wirbellosen. *Verh.*

Dtsch. Zool. Ges. Mainz 16–23.

Saffman E.E. and Lasko P. 1999. Germline development in vertebrates and invertebrates. *Cell. Mol. Life Sci.* **55:** 1141–1163.

Scholz C.B. and Technau U. 2003. The ancestral role of *Brachyury*: Expression of *NemBra1* in the basal cnidarian *Nematostella vectensis* (Anthozoa). *Dev. Genes Evol.* **212:** 563–570.

Sedgwick A. 1884. On the origin of metameric segmentation and some other morphological questions. *Q. J. Microsc. Sci.* **24:** 43–82.

Shankland M. and Seaver E.C. 2000. Evolution of the bilaterian body plan: What have we learned from annelids? *Proc. Natl. Acad. Sci.* **97:** 4434–4437.

Shoguchi E., Satoh N., and Maruyama Y.K. 1999. Pattern of *Brachyury* gene expression in starfish embryos resembles that of hemichordate embryos but not of sea urchin embryos. *Mech. Dev.* **82:** 185–189.

Siewing R. 1985. *Lehrbuch der Zoologie. Systematik.* Gustav Fischer Verlag, Stuttgart, Germany.

Spring J., Yanze N., Josch C., Middel A.M., Winninger B., and Schmid V. 2002. Conservation of *Brachyury*, *Mef2*, and *Snail* in the myogenic lineage of jellyfish: A connection to the mesoderm of bilateria. *Dev. Biol.* **244:** 372–384.

Stern C.D., Hatada Y., Selleck M.A.J., and Storey K.G. 1992. Relationships between mesoderm induction and the embryonic axes in chick and frog embryos. *Development* (suppl.) **1992:** 151–156.

Tada M. and Smith J.C. 2000. Xwnt11 is a target of *Xenopus Brachyury*: Regulation of gastrulation movements via Dishevelled, but not through the canonical Wnt pathway. *Development* **127:** 2227–2238.

Tagawa K., Humphreys T., and Satoh N. 1998. Novel pattern of *Brachyury* gene expression in hemichordate embryos. *Mech. Dev.* **75:** 139–143.

Takada N., Goto T., and Satoh N. 2002. Expression pattern of the *Brachyury* gene in the arrow worm *paraspadella gotoi* (chaetognatha). *Genesis* **32:** 240–245.

Technau U. 2001. *Brachyury*, the blastopore and the evolution of the mesoderm. *BioEssays* **23:** 788–794.

Vonica A. and Gumbiner B.M. 2002. Zygotic Wnt activity is required for *Brachyury* expression in the early *Xenopus laevis* embryo. *Dev. Biol.* **250:** 112–127.

Wang C. and Lehmann R. 1991. Nanos is the localized posterior determinant in *Drosophila*. *Cell* **66:** 637–647.

Wilson E.B. 1892. The cell-lineage of *Nereis*. A contribution to the cytogeny of the annelid body. *J. Morphol.* **6:** 361–480.

Wilson V. and Beddington R. 1997. Expression of T protein in the primitive streak is necessary and sufficient for posterior mesoderm movement and somite differentiation. *Dev. Biol.* **192:** 45–58.

Wilson V., Manson L., Skarnes W.C., and Beddington R.S. 1995. The T gene is necessary for normal mesodermal morphogenetic cell movements during gastrulation. *Development* **121:** 877–886.

Winklbauer R. and Schurfeld M. 1999. Vegetal rotation, a new gastrulation movement involved in the internalization of the mesoderm and endoderm in *Xenopus*. *Development* **126:** 3703–3713.

Woo K. and Fraser S. 1995. Order and coherence in the fate map of the zebrafish nervous system. *Development* **121:** 2595–2609.

Yanze N., Spring J., Schmidli C., and Schmid V. 2001. Conservation of Hox/ParaHox-related genes in the early development of a cnidarian. *Dev. Biol.* **236:** 89–98.

Yu J.K., Holland L.Z., and Holland N.D. 2002. An amphioxus nodal gene (AmphiNodal) with early symmetrical expression in the organizer and mesoderm and later asymmetrical expression associated with left-right axis formation. *Evol. Dev.* **4:** 418–425.

THE EVOLUTION OF GASTRULATION
CELLULAR AND MOLECULAR ASPECTS

A.L. Price[1] and N.H. Patel[2]

[1]*Committee on Developmental Biology, University of Chicago, Chicago, Illinois 60637, and Center for Integrative Genomics, University of California, Berkeley;* [2]*Department of Integrative Biology, Department of Molecular Cell Biology, and Howard Hughes Medical Institute, University of California, Berkeley, California 94720*

INTRODUCTION

All bilateral organisms with multiple cell layers undergo gastrulation as an important step toward the development of complex body forms. Gastrulation not only includes a series of orchestrated morphogenetic movements, which culminates in the formation of germ layers, but also involves key steps in the elaboration of body axes and cell fate determination. Various steps in early development, including gastrulation, body axis formation, and cell fate determination, have been studied to some depth in model organisms, and a striking similarity in the molecules and genetic pathways used has been uncovered for at least some aspects of these events. Such results from model systems have stimulated work to characterize the expression and function of orthologous genes in more diverse animals in order to increase our knowledge of metazoan relationships and to generate more informed hypotheses as to how development may have evolved.

One of the oldest and best-known theories on the evolution of gastrulation is the Gastrea Theory put forward by Haeckel in the 1870s. According to Haeckel, the ancestor of the Bilateria was a free-swimming blastula, or ball of cells, termed the Blastaea. Eventually, the Blastaea favored digestion and uptake of nutrients on one surface, and on this digestive surface a pit evolved and transformed the Blastaea into a gastrula, or Gastraea. Thus, in Haeckel's view, the original Blastaea evolved two innovations for more efficient feeding: a distinction between two different cell types and a morphogenetic process of invagination. These two innovations laid the groundwork for the evolution of gastrulation. However, many metazoans have three germ layers, and diverse modes of gastrulation have been described, including invagination, epiboly, involution, delamination, and ingression (see Chapters 19 and 20). One could envision that divergent (non-invagination) modes were derived during the evolution of the third, mesodermal, germ layer. Still, animals which we regard as basal, or phylogenetically primitive, the Porifera (sponges) and Cnidaria (coral, hydra, etc.), typically create endoderm by delamination or ingression, as opposed to invagination, and studies of the development of a large array of invertebrates have suggested that ingression is the most ancestral mode of gastrulation (Hyman 1951; for discussion, see Willmer 1990).

Before talking further about the evolution of gastrulation, it is important to understand its phylogenetic context. The creation of metazoan phylogenies has traditionally been based on the similarity and presence or absence of morphological characters. DNA sequence comparisons are now being used to create phylogenies with the hope that this approach can unearth more ancient or problematic relationships that may not be clearly delineated by morphology. Use of 18S ribosomal DNA sequence analysis for the resolution of deep evolutionary splits has resulted in a new metazoan phylogeny (Halanych et al. 1995; Aguinaldo et al. 1997; Ruiz-Trillo et al. 1999; Adoutte et al. 2000) that has gained wide acceptance. Although many aspects of older phylogenies are supported by the new molecular data, some new relationships suggest that there may be alternate hypotheses about many evolutionary changes that were not previously recognized. Most strikingly, phylogenies created by these means, combined with developmental data, can lead to exciting new ideas about the evolution of development, and discrepancies between new and old molecular phylogenies may suggest places where developmental data from the past should be reviewed under a new light. In this chapter, we provide a survey of some of the ways molecular and developmental biology mix with evolutionary theory to create new ideas about the evolution of gastrulation.

A NEW METAZOAN PHYLOGENY

A relatively new metazoan phylogeny based initially on 18S ribosomal DNA sequences (Halanych et al. 1995; Aguinaldo et al. 1997; Ruiz-Trillo et al. 1999), and subsequently supported by additional analyses (Boyer et al. 1998; de Rosa et al. 1999; Peterson and Eernisse 2001), has some very significant changes relative to previously accepted phylogenies (Fig. 1A). This tree, like many older trees, asserts that there are two superclades within the Bilateria, the deuterostomes and the protostomes. Whereas the deuterostome clade is similar to that seen in older phylogenies, the structure of the protostome clade is quite different (compare Fig. 1A and 1B). In the new phylogeny, the protostomes are composed of two major groups, the Lophotrochozoa and the Ecdysozoa, which have subsumed the Acoelomate, Pseudocoelomate, and Lophophorate clades. The Lophotrochozoa share ciliated larval forms, and the Ecdysozoa all share the common property of molting. Accepting this phylogeny greatly changes the way we think about common ancestors that may have existed during the evolution of the Bilateria. Various aspects of this phylogeny, such as the new placement of previously "intermediate" phyla and the dissolution of the Articulata as a clade, also cause us to reconsider the relationship of diploblasts to triploblasts and even the definition of protostome versus deuterostome, all issues relevant to understanding the evolution of gastrulation.

One concept central to theories of evolution of gastrulation that is affected by the new phylogeny is the origin of mesoderm as a trait of the Bilateria. Historically, there has been a lot of opposition to the idea that the germ layers in diploblasts (Porifera, Cnidaria, and Ctenophores) are homologous to these in triploblasts (Bilateria). This is mainly due to the assertion that groups within the diploblasts do not possess a true mesoderm and derivatives usually associated with the mesoderm in triploblasts are derived from the other germ layers in diploblasts. The repositioning of previously "intermediate" phyla, the Pseudocoelomates and Acoelomates, places the position of the last common ancestor between diploblasts and triploblasts more proximal to complex bilaterians (compare Fig. 1A and 1B). This suggests that the germ layers in diploblasts may be more complex than was previously understood and that the mesenchyme in cnidarians and ctenophores may be more similar to the mesoderm in Bilateria, an idea that is further supported by molecular data (see below).

Many of the traditional views of the evolution of development have assumed steps through "lower phyla" as intermediates in the process; i.e., the traditional view of "ontogeny recapitulates phylogeny." By this method, more "simple" bilaterians were used as examples illustrating a primitive or ancestral state. The lower phyla, including acoelomates (organisms with no coelomic cavities) and the pseudocoelomates (organisms with body cavities that are not lined with mesoderm-derived tissue) were seen as stages along the path to the development of coelomates (animals with body cavities lined by mesoderm). Thus, the mesoderm of the acoelomate flatworms (such as platyhelminthes) was once thought to have been representative of an intermediate stage in the evolution of mesoderm. Obviously, the assumption that extant animals in phylogenetically "basal" positions retain primitive characteristics in development is often incorrect, but more importantly, the new phylogeny places members of these groups within the protostome clade, suggesting that the evolution of traits such as coeloms may not be as progressive as was once thought, and that the relationship of "triploblasts" with "diploblasts" is less disparate. However, there is some suggestion that the acoelomates are paraphyletic and that there are some members that are still basal to the Bilateria (Ruiz-Trillo et al. 1999), giving us hope that some "intermediate" phyla still exist.

These new interpretations also highlight the need to consider loss of complexity as important a mode of evolution as the gain of complexity, and are well demonstrated by the placement of the nematodes within the protostomes. These organisms were traditionally thought of as primitive due to their lack of coeloms and their relatively simple morphology. Now, with the advent of whole-genome sequencing, we know that *C. elegans*, a nematode, has more genes than *Drosophila melanogaster*, an arthropod, and that the loss of apparent complexity in morphological structure may be due to adaptation. In addition, analysis of Hox gene clusters among nematode species show that gene loss and rapid sequence evolution (Aboobaker and Blaxter 2003) may be a mechanism for this.

The protostome–deuterostome split has historically been interrelated with theories of the evolution of gastrulation. The terms protostome and deuterostome are based on ideas about the fate of the blastopore, which is thought to give rise to the mouth in protostomes, whereas the mouth arises secondarily in deuterostomes. Again, our current

A
"NEW"

B
"OLD"

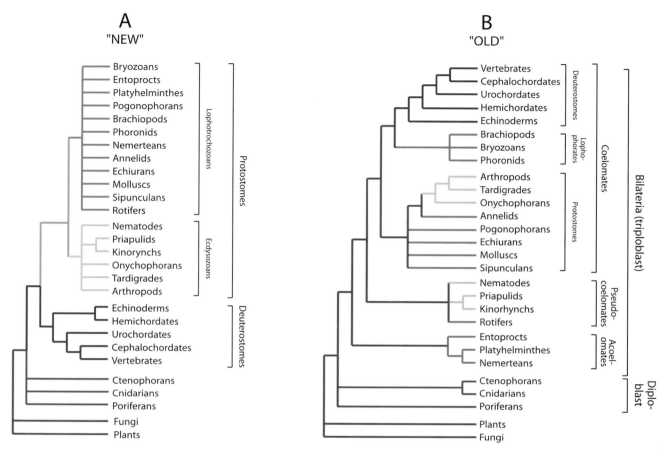

Figure 1. Comparison of "new" and "old" metazoan phylogenies. (*A*) A new metazoan phylogeny based on 18S rDNA sequences (Halanych et al. 1995; Aguinaldo et al. 1997; Ruiz-Trillo et al. 1999). (*B*) Old metazoan phylogeny. (Adapted from Adoutte et al. 2000, based originally on Hyman 1951.) Color scheme of the new phylogeny is superimposed on the old phylogeny in order to more easily see changes that have occurred. Colors: (*red*) lophotrochozoans, (*yellow*) ecdysozoans, (*orange*) protostomes, (*blue*) deuterostomes, (*brown*) bilaterians, (*purple*) diploblasts.

thinking about animal phylogeny causes us to revisit several longstanding hypotheses on the evolution of the deuterostome–protostome split. For example, the lophophorates were historically considered basal deuterostomes. This classification was due to the characters of radial cleavage, enterocoely, and derivation of the mouth secondarily, rather than from the blastopore (Brusca and Brusca 1990; Nielsen 1995). Our current thinking about animal phylogeny, however, places lophophorates squarely within the protostomes (Halanych et al. 1995; de Rosa et al. 1999), although the members of the group, aside from the phoronids, do not display the protostome mode of development. This illustrates that continued use of the terms deuterostome and protostome as descriptive features of development, especially in the context of the evolution of gastrulation, should be done with caution.

Another change that will influence the way we think about homologous structures and their development is the

dissolution of the "Articulata," which used to consist of the annelids and the arthropods, based on the common occurrence of segmentation. Not only are these phyla now not considered sisters, but they reside within separate subclades of the protostomes; the arthropods within the Ecdysozoa and the annelids within the Lophotrochozoa. A major implication of this split for studies of early development is the undermining of spiral cleavage as an ancestral trait among protostomes. However, spiral cleavage is present throughout the Lophotrochozoa and may be an innovation in this group, whereas radial cleavage is a common character of the Ecdysozoa. Early cleavage and blastula stages form the foundation for the events of axis formation, morphogenesis, and germ layer derivation. Thus, changes in the patterns of early cleavage may have an impact on the process of gastrulation. A good example of this is seen in the heterochronic shift of cell fate determination of the 4d mesentoblast, the progenitor of endoderm and mesoderm germ layers, in relation to other cells in the Mollusca (Guralnick and Lindberg 2001). It will be exciting to determine whether there exist archetype forms of blastula-to-gastrula conversions that are common within the super-clades Lophotrochozoa, Ecdysozoa, and Deuterostomia. These archetypes can then be compared with non-bilaterian groups, such as Cnidaria and Ctenophora, in order to formulate theories on the evolution of gastrulation that encompasses all these modes of development.

CELL AND MOLECULAR BIOLOGY AND GASTRULATION

Although similarities in pattern formation and gastrulation between species used for genetic and molecular studies (model systems) are not always easy to perceive on a morphological level, comparisons of gene expression patterns and functions at key points during development suggest that some general molecular pathways are used to pattern bilaterian embryos. These include the use of Wnt and TGF-β signaling pathways in the establishment of the dorsoventral axis, Hox genes in the anterior–posterior axis, and germ-layer-specific expression of genes such as T-box genes, *twist*, *snail*, cadherins, and GATA factors. There is now a push for understanding the development of an increasing number of organisms within and between multiple phyla on a molecular level, both to understand the commonalities in development and to understand the differences that have led to evolutionary change. These studies have the additional potential to extract an evolutionary signal from a process as complex as gastrulation. The common appearance of large-scale signaling pathways to generate global pattern in embryos has been discussed in depth in Chapter 51, so here we focus on two further questions that can be approached by molecular studies: the origin of the mesoderm and the evolution of the morphogenetic processes of gastrulation.

When analyzed in the appropriate phylogenetic context, the comparison of extant phyla can give us important information about how development might have worked in ancestral species that disappeared millions of years ago. One extension of such an approach is to look for the use of similar genetic pathways in phyla that are not included within the Bilateria in order to assess the role of these genes in the common bilaterian ancestor. As mentioned above, the "mesenchyme" of cnidarians is not commonly thought to be homologous to the "mesoderm" found in the Bilateria. Cnidarians are generally accepted to have bilayer tissue organization consisting of endoderm and ectoderm. Within both the endoderm and ectoderm reside specialized myoepithelial cells that contain basal contractile apparatuses at the interface of endoderm and ectoderm (Brusca and Brusca 2003), and their contractions are coordinated to

allow for the movement of the adult animal. Intriguingly, the medusae of more "complex" cnidarians contain both striated muscle for movement and smooth muscle around the gut (Boero et al. 1998). These tissues arise from the entocodon, an ectodermally derived layer of cells that forms early in medusa formation.

Although the entocodon is not considered to be a mesodermal germ layer, gene expression would suggest otherwise. The genes *twist*, *snail*, *brachyury*, and *mef2* are known to be involved in both gastrulation and mesoderm development in model bilaterian animals (Black and Olson 1998; Manzanares et al. 2001; Technau 2001; Castanon and Baylies 2002). Interestingly, these genes are all expressed as well during the process of entocodon formation in the cnidarian, *Podocoryne carnea* (Spring et al. 2000, 2002). Thus, these cnidarians have either independently evolved a third germ layer using molecules similar to those used in Bilateria, a case of convergent evolution, or the common ancestor to cnidarians and Bilateria may already have had a cell type we can call mesoderm. An in-depth study of the genes controlling myoepithelial cell specification in other Cnidaria could help resolve this issue and further our understanding of the evolution of mesoderm in Bilateria. Furthermore, the identification of downstream targets of these genes in *Podocoryne*, in comparison to similar data from Bilateria, will provide additional information as to the potential ancestral mechanism of mesoderm formation and function. Finally, the ctenophores, another "diploblastic" outgroup to the Bilateria (see Fig. 1), have a well-defined population of cells that is formed during gastrulation and gives rise to the middle layers of the adult including the musculature (Nielsen 1995). There is no molecular work to date to suggest that ctenophores may be using orthologs of bilaterian "mesodermal genes" to control differentiation of this population of cells, but this would be an important question to explore.

The second question we address concerns the evolution of morphogenetic movements. With the exception of delamination, all modes of gastrulation require that a subset of cells detach from their neighbors and acquire properties that allow them to migrate to new positions. The process by which cells do this is called epithelial-to-mesenchymal transition (EMT). Epithelia are contiguous sheets of cells that cover a surface or line a cavity and are characterized by strong cell–cell adhesion. Cells within an epithelium display apical–basal polarity, typically with the basal end in contact with an extracellular matrix (ECM) while the apical end forms a surface. Mesenchymal cells are motile and can move through the ECM as individual cells. The different characteristics of the two cell types are well displayed in cell culture (Hay 1995). In vitro, isolated epithelial cells placed on ECM are stationary, flatten their basal surfaces, become cuboidal in shape, and have a tendency to form a single cell layer. In vitro, isolated mesenchymal cells placed on the same ECM show front end–back end polarity, elongate, and invade the matrix. During EMT, a complement of structural genes gives cells the ability to release from their neighbors and take on migratory characteristics, such as membranous extensions and filopodia (Hay 1995; Batlle et al. 2000; Cano et al. 2000).

A superfamily of genes that are implicated in the process of EMT are the cadherin genes. Classic embryological studies showed that the regulation of adhesion between cells of different tissue types was important for them to sort into germ layers (Townes and Holfreter 1955). More recently, the cadherins have been shown to be a superfamily of homologous Ca^{++}-dependent cell–cell adhesion glycoproteins involved in this sorting behavior seen at gastrulation in model organisms (Yoshida and Takeichi 1982; Takeichi 1987). Integral to the formation of junctional complexes, cadherins are expressed in all cell types that form solid tissues and have been shown to be involved in cell polarization and tumor invasion. For developmental biologists, the most striking characteristic of the classic cadherins is the differential tissue distribution of the subtypes during the development of vertebrate, echinoderm, and *Drosophila* embryos (Takeichi 1987; Uemura et al. 1996; Iwai et al. 1997; Miller and McClay 1997). In both vertebrate model systems and *Drosophila*, orthologs of *E-cadherin* are expressed in both ectodermal and endodermal epithelium, whereas orthologs of *N-cadherin* are expressed in nervous system and mesoderm (Takeichi 1995; Huber et al. 1996; Batlle et al. 2000; Cano et al. 2000).

Studies of this adhesion system also reveal a link between global patterning and cell biology, in that *snail*, a zinc-finger transcription factor whose expression has been found to be associated with gastrulation, has been shown to be a direct repressor of *E-cadherin* transcription in mouse and human cells in tissue culture (Batlle et al. 2000; Cano et al. 2000; see Chapter 47). E-cadherin is one of the most important cell–cell adhesion molecules for formation of epithelia (Hay 1995), thus the down-regulation of *E-cadherin* is an important step in the progress of EMTs. The transition from blastula to gastrula in many organisms involves the transition from panembryonic *E-cadherin* expression to selective down-regulation of *E-cadherin* in subsets of gastrulating tissues (Burdsal et al. 1993; Takeichi 1995; Uemura et al. 1996; Miller and McClay 1997; Oda et al. 1998). It is tempting to consider the events occurring in gastrulating tissues as a large-scale epithelial–mesenchymal transition. Thus, the Snail family of transcription factors plays a crucial role in gastrulation by down-regulation of E-cadherins in cells undergoing gastrulation (Oda et al. 1998) and, through its regulation of the cadherin adhesion system, provides a global mechanism within the embryo for orchestrating a key step in gastrulation.

These two examples highlight how molecular biology may help us to understand evolution of both global pat-

terning and cellular and genetic mechanisms. Although this analysis is superficial at best for making evolutionary claims, it suggests directions that may be pursued as we obtain a larger taxon sampling and a more complete understanding of molecular and developmental mechanisms in multiple species.

SYNTHESIS

The phylogenetic tree of animal life presents us with a premise from which we can begin to build theories using comparative molecular developmental biology in conjunction with morphological data from the past. We should not look at extant metazoans as frozen stages in an evolutionary series, but rather as endpoints in a series of independent evolutionary experiments. The greater the number of endpoints we sample, the more confident we can become of our hypotheses about common ancestors, especially with regard to finding evolutionarily conserved developmental mechanisms. Equally important, however, are the organisms that have undergone noticeable divergences, often as a result of adaptation, such as the nematodes. Investigating the mechanisms by which these organisms have modified their development on a molecular level to allow for extreme changes in phenotype will give us exciting insights into the plasticity of developmental programs. These types of analyses are especially informative when looking at the evolution of early patterning and gastrulation in animal embryogenesis.

An important innovation in the evolution of complex organisms is the advent of multicellularity and the allocation of different roles to different cells. This separation of roles is clearly manifest during gastrulation and the formation of the germ layers. This process requires that differences exist between cells at the molecular level in stages leading up to and through gastrulation and can occur through a wide variety of mechanisms in various animals. Despite the differences we see in early patterning between different animal groups, however, there is conservation among the gene families expressed during morphogenesis. One way to reconcile these facts is to acknowledge that gene regulatory pathways may be modified; for example, different mechanisms may now exist in different species to establish *snail* expression in the cells that will play a role in gastrulation. Despite these evolutionary changes, however, the combination of transcriptional patterning and restructuring of cell morphology is a common theme in gastrulation that may rely on an evolutionarily conserved mechanism. The interaction between *snail* and *cadherin* function serves as a good example of the coupling of patterning genes to genes more directly involved in controlling cell morphology during the process of gastrulation. However, there must be many more factors involved in this process that contribute to the morphological diversity of gastrulation seen in extant animals. The analysis of the molecular mechanisms underlying gastrulation in more varied phyla will lead us to a more complete picture of both the ancestry of this process, and how it has evolved during metazoan evolution.

What about the Blastaea–Gastraea theory suggested in the 19th century by Haeckel? Although there is no evidence to suggest that invagination is not the primitive mode of gastrulation, a more straightforward suggestion is that the ancestral form of gastrulation is ingression. An ancestral mode of gastrulation by inwandering, or ingression (Fig. 2A,B), as seen in the formation of endoderm and mesoderm in more basal invertebrates, allows for the formation of mesoderm from both ectodermal and endodermal origins (Hyman 1951), which is seen throughout the Metazoa. In support of this is the observation that members of the Bilateria, as well as sponges and cnidarians, use this mechanism to generate internalized cells. In addition, ingression offers a starting point for the evolution of the different types of gastrulation. Invagination may be seen as a modification of ingression where loss of adhesion between neighboring cells is delayed. In support of this, invaginations are seen during sea urchin morphogenesis when loss of adhesion in unipolar ingression is delayed during the formation of primary mesenchyme (Wolpert 1992). Thus, it is easy to invoke an evolutionary scenario where continued adhesion between ingressing cells could lead to invagination (Fig. 2C). Likewise, involution could be achieved by a similar change in which cells at one edge of a field of ingressing cells do not lose contact, while cells along the opposing edge do (Fig. 2D). Epiboly may be a combination of migration and involution to incorporate a large yolky vegetal area.

The diversity in blastoderm forms and modes of gastrulation in extant organisms makes it difficult to adopt a simple scenario for the evolutionary changes that have occurred in the process of gastrulation. However, the powerful tools provided by molecular biology, used in conjunction with embryological data, will allow us rapidly to advance our understanding of the mechanism and evolution of gastrulation. The cloning of orthologous genes between phyla, and the subsequent comparisons of expression and function, allow us to test theories of morphological homology and to identify homologous genetic pathways. However, we must use caution in making assumptions of homology based solely on gene expression data. As we begin to increase our understanding of the molecular basis of cell and tissue morphogenesis at the molecular and genetic level, we will also begin to have a basis for further understanding the evolutionary diversification of gastrulation.

REFERENCES

Aboobaker A.A. and Blaxter M.L. 2003. Hox gene loss during dynamic evolution of the nematode cluster. *Curr. Biol.* **13:** 37–40.

Adoutte A., Balavoine G., Lartillot N., Lespinet O., Prud'homme B., and de Rosa R. 2000. The new animal phylogeny: Reliability and implications. *Proc. Natl. Acad. Sci.* **97:** 4453–4456.

Aguinaldo A.M.A., Turbeville J.M., Linford L.S., Rivera M.C., Garey J.R., Raff R.A., and Lake J.A. 1997. Evidence for a clade of nematodes, arthropods and other moulting animals. *Nature* **387:** 489–493.

Batlle E., Sancho E., Franci C., Dominguez D., Monfar M., Baulida J., and Garcia de Herreros A. 2000. The transcription factor Snail is a repressor of *E-cadherin* gene expression in epithelial tumor cells. *Nat. Cell Biol.* **2:** 84–89.

Black B.L. and Olson E.N. 1998. Transcriptional control of muscle development by myocyte enhancer factor-2 (MEF2) proteins. *Annu. Rev. Cell Dev. Biol.* **14:** 167–196.

Boero F., Gravili C., Pagliara P., Piraino S., Bouillon J., and Schmid V. 1998. The cnidarian premises of metazoan evolution: From triploblasty, to coelom formation, to metamery. *Ital. J. Zool.* **65:** 5–9.

Boyer B.C., Henry J.J., and Martindale M.Q. 1998. The cell lineage of a polyclad turbellarian embryo reveals close similarity to coelomate spiralians. *Dev. Biol.* **204:** 111–123.

Brusca R.C. and Brusca G.J. 1990. The Lophophorate Phyla: Phoronids, Ectoprocts, and Brachiopods. In *Invertebrates*. Sinauer, Sunderland, Massachusetts.

———. 2003. Phylum Cnidaria. In *Invertebrates*. Sinauer, Sunderland, Massachusetts.

Burdsal C.A., Damsky C.H., and Pedersen R.A. 1993. The role of E-cadherin and integrins in mesoderm differentiation and migration at the mammalian primitive streak. *Development* **118:** 829–844.

Cano A., Perez-Moreno M.A., Rodrigo I., Locascio A., Blanco M.J., del Barrio M.G., Portillo F., and Nieto M.A. 2000. The transcription factor snail controls epithelial-mesenchymal transitions by repressing E-cadherin expression. *Cell Biol.* **2:** 76–83.

Castanon I. and Baylies M.K. 2002. A Twist in fate: Evolutionary comparison of Twist structure and function. *Gene* **287:** 11–22.

de Rosa R., Grenier J.K., Andreeva T., Cook C.E., Adoutte A., Akam M., Carroll S.B., and Balavoine G. 1999. Hox genes in bra-

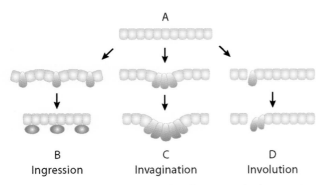

Figure 2. Modification of ingression by adhesion can lead to invagination and involution. (*A*) Undifferentiated epithelia. (*B*) Ingression of single cells. (*C*) Invagination of a group of cells without loss of adhesion. (*D*) Involution of cells by loss of adhesion at one edge of a group of cells.

chiopods and priapulids and protostome evolution. *Nature* **399:** 772–776.

Guralnick R.P. and Lindberg D.R. 2001. Reconnecting cell and animal lineages: What do cell lineages tell us about the evolution and development of Spiralia? *Evolution* **55:** 1501–1519.

Halanych K.M., Bacheller J.D., Aguinaldo A.M.A., Liva S.M., Hillis D.M., and Lake J.A. 1995. Evidence from 18s ribosomal DNA that the lophophorates are protostome animals. *Science* **267:** 1641–1643.

Hay E.D. 1995. An overview of epithelio-mesenchymal transformation. *Acta Anat.* **154:** 8–20.

Huber O., Bierkamp C., and Kemler R. 1996. Cadherins and catenins in development. *Curr. Opin. Cell Biol.* **8:** 685–691.

Hyman L.H. 1951. *The Invertebrates: Platyhelminthes and Rhynchocoela. The acoelomate Bilateria.* McGraw-Hill, New York.

Iwai Y., Usui T., Hirano S., Steward R., Takeichi M., and Uemura T. 1997. Axon patterning requires *D* N-cadherin, a novel neuronal adhesion receptor, in the *Drosophila* embryonic CNS. *Neuron* **19:** 77–89.

Manzanares M., Locascio A., and Nieto M.A. 2001. The increasing complexity of the Snail gene superfamily in metazoan evolution. *Trends Genet.* **17:** 178–181.

Miller J.R. and McClay D.R. 1997. Characterization of the role of cadherin in regulating cell adhesion during sea urchin development. *Dev. Biol.* **192:** 323–339.

Nielsen C. 1995. *Animal evolution interrelationships of the living phyla.* Oxford University Press, Oxford, United Kingdom.

Oda H., Tsukita S., and Takeichi M. 1998. Dynamic behavior of the cadherin-based cell-cell adhesion system during *Drosophila* gastrulation. *Dev. Biol.* **203:** 435–450.

Peterson K.J. and Eernisse D.J. 2001. Animal phylogeny and the ancestry of bilaterians: Inferences from morphology and 18S rDNA gene sequences. *Evol. Dev.* **3:** 170–205.

Ruiz-Trillo I., Ruitort M., Littlewood D.T.J., Herniou E.A., and Baguna J. 1999. Acoel flatworms: Earliest extant bilaterian Metazoans, not members of Platyhelminthes. *Science* **283:** 1919–1923.

Spring J., Yanze N., Josch C., Middel A. M., Winninger B., and Schmid V. 2002. Conservation of *Brachyury*, *Mef2*, and *Snail* in the myogenic lineage of jellyfish: A connection to the mesoderm of bilateria. *Dev. Biol.* **244:** 372–384.

Spring J., Yanze N., Middel A.M., Stierwald M., Groeger H., and Schmid V. 2000. The mesoderm specification factor Twist in the life cycle of the jellyfish. *Dev. Biol.* **228:** 363–375.

Takeichi M. 1987. Cadherins: A molecular family essential for selective cell-cell adhesion and animal morphogenesis. *Trends Genet.* **3:** 213–217.

———. 1995. Morphogenetic roles of classic cadherins. *Curr. Opin. Cell Biol.* **7:** 619–627.

Technau U. 2001. *Brachyury*, the blastopore and the evolution of the mesoderm. *BioEssays* **23:** 788–794.

Townes P.L. and Holtfreter J. 1955. Directed movements and selective adhesion of embryonic amphibian cells. *J. Exp. Zool.* **128:** 53–120.

Uemura T., Oda H., Kraut R., Hayashi S., Kataoka Y., and Takeichi M. 1996. Zygotic *Drosophila* E-cadherin expression is required for processes of dynamic epithelial cell rearrangement in the *Drosophila* embryo. *Genes Dev.* **10:** 659–671.

Willmer P. 1990. *Invertebrate relationships.* Cambridge University Press, Cambridge.

Wolpert L. 1992. Gastrulation and the evolution of development. *Development Suppl.* **1992:** 7–13.

Yoshida C. and Takeichi M. 1982. Teratocarcinoma cell adhesion: Identification of a cell-surface protein involved in calcium-dependent cell aggregation. *Cell* **28:** 217–224.

FOSSIL EMBRYOS

S. Conway Morris

Department of Earth Sciences, University of Cambridge, Cambridge CB2 3EQ,
United Kingdom

INTRODUCTION

With a few exceptions of asexuality, notably by partheno-genesis or fission, nearly all metazoans arrange a life cycle where sperm meets egg with a resultant embryo. Embryological information pertaining to several groups of invertebrates is available from the fossil record from skeletal material. Apart from the well-known vertebrates, e.g., dinosaurs' eggs, most obvious are those groups that secrete some sort of accretionary shell, such that the earliest ontogenetic stages allow inferences on the type of larval development. In the case of some mollusks, the type of larval or embryonic shell (known as protoconch I and prodissoconch I in snails and bivalve mollusks, respectively) allows distinction between histories that are either planktotrophic (feed in the water column) or non-planktotrophic (rely on supplied nutriments) (see, e.g., Jablonski and Lutz 1983; Jablonski 1986). Although most of this work has focused on Cretaceous (the geological interval spanning 144–165 million years [Ma]) taxa, Chaffee and Lindberg (1986) extrapolated this approach, on the basis of body size, to Cambrian (the geological interval spanning 543–490 Ma) mollusks. They concluded that planktonic development only arose by the Late Cambrian, and prior to that, taxa had non-planktonic yolk-rich larvae (lecithotrophs; see also Haszprunar 1992). The extinct hyoliths (a group with a calcareous shell consisting of elongate cone, operculum, and two curved "arms"), which are usually allied with the mollusks, also possess larval shells, and on this basis, Dzik (1978) drew specific comparison to gastropods with a trochophore stage (but see Jablonski and Lutz 1979). In the brachiopods (a lophotrochozoan phylum with bivalved shell and feeding lophophore), but only among the group known as the inarticulates, the larval shell (or protegulum) also provides early ontogenetic information that is readily preserved in fossil material. In some acrotretid brachiopods from the Cambrian and Ordovician (the geological interval spanning 490–443 Ma), a distinctive honeycomb-like structure on the larval shell has been identified as an adaptation for flotation (Biernat and Williams 1970; cf. von Bitter and Ludvigsen 1979; see also Rowell 1986). Despite the apparent evolutionary conservatism (as "living fossils"), especially in the lingulids, in this group of inarticulates at least there is evidence for significant change in the larval shells since the Devonian (415–355 Ma) (Balinski 2001).

That larval shells of gastropods, bivalve mollusks, brachiopods, and possibly hyoliths can give some insights into ancient embryologies, or at least larval forms, is not so surprising. A far more remarkable inference on larval type has been arrived at by Emlet (1985). He observed that in certain planktotrophic echinoids, the larval calcite rods found in the extended "arms" of the larvae have an optic axis that is inherited by four (of five) and two (of five) of the genital and ocular plates that in the adult are arranged around the anus. Subsequently (Emlet 1989), it was recognized that the different families of echinoids have particular arrangements of the optic axis, and that only the genital plates were reliable indicators, whereas in those groups with reduced numbers of genital plates (e.g., clypeasteroids) the method is inapplicable. Even so, this relatively simple method is an almost foolproof way of inferring a planktotrophic larva in an extinct regular echinoid. This does not apply to the non-planktotrophic larvae, which, in lacking the larval rods, have a crystallography of the apical region that typically has the optic axis perpendicular to the plates.

This distinction in larval type can of course be applied to fossils. On this basis, Jeffery (1997) presented evidence that non-planktotrophy first arose in the Cretaceous. In addition, she argued that this shift occurred independently in up to nine lineages, suggesting that environmental factors (possibly temperature and/or nutrient supply) were the driving forces. Rather surprisingly, the expectation of selective extinction (see Valentine 1986) of planktotrophs across the end-Cretaceous (K-T) extinctions is not met, despite the widely held assumption that as a consequence of massive disruption of the phytoplankton (by whatever mechanism) planktotrophic larvae would be peculiarly vulnerable. Rather, as Smith and Jeffery (1998) argue, it is the benthic adult forms pursuing particular trophic strategies that show either enhanced survival or disproportional extinction.

Despite these rather remarkable insights into larval histories in the geological past, at first sight it would seem improbable that actual embryos, as typified by a blastula or slightly later stages of the ontogeny, e.g., a ciliated larva, could be preserved by fossilization. It has, however, become clear that truly exceptional fossil preservation is possible in a wide series of circumstances and is also amenable to laboratory replication (Martin et al. 2003).

In particular, the process of phosphatization (the precipitation of phosphatic minerals either by direct replacement of the existing material or by formation of a thin coating) very early in the process of diagenesis (post-burial chemical reactions) has long been known to yield spectacular material. Among the best-known examples are the vertebrates, especially fish, from the Cretaceous Santana Formation of Brazil (see, e.g., Martill 1990, 1993), and the microarthropod assemblages, which include some larval forms (see, e.g., Müller and Walossek 1986), from the Upper Cambrian of southern Sweden and adjacent areas (see, e.g., Walossek and Müller 1998; Maas and Walossek 2001). A particular advantage of diagenetic phosphatization is that fossils contained in a calcareous sediment can be released as the carbonate matrix dissolves after immersion of the sample in weak organic acid.

Such techniques, augmented by the use of petrographic thin sections, have led to a series of remarkable discoveries of fossilized embryos, most notably from the late Neoproterozoic (ca. 570 Ma) and Cambrian. These form the topic of the rest of this chapter, but it should be noted that only one of the examples sheds any direct light on gastrulation. There are, however, sound reasons for maintaining an interest in this area. First, along with the examples given above, this fossil material may help to reveal the evolution of life histories, such as the primitiveness (or not) of direct versus indirect development. Second, evidence may emerge on the types of cell cleavage, notably the classical distinction between radial and spiral. This in turn has potentially interesting phylogenetic implications (Valentine 1997; see also van den Biggelaar 1997). Third, and related to the first two points, is the potential for these Neoproterozoic and Cambrian embryos to throw new light on the Cambrian "explosion" of early metazoans (see, e.g., Conway Morris 2000a). Existing evidence is still too sketchy, and sometimes controversial, to offer anything more than some tantalizing clues, but there is little doubt that the field of embryonic paleobiology will develop rapidly.

To date there are seven principal records of fossil animal embryos: (1) *Megasphaera* and *Parapandorina* (and possibly several other microfossil taxa) from the late Neoproterozoic Doushantuo Formation of Guizhou, South China (Xiao and Knoll 2000; Xiao 2002); (2) the probable cnidarian *Olivooides* from the Lower Cambrian Dengying Formation of Shaanxi, central China (Bengtson and Yue 1997; Yue and Bengtson 1999); (3) a possible cnidarian from the Lower Cambrian Manykay Formation of northern Siberia (Kouchinsky et al. 1999); (4) the bilaterian *Markuelia* from the Lower Cambrian Pestrotsvet Formation of Eastern Siberia (Bengtson and Yue 1997); (5) *Markuelia*-like fossils from the Middle and Upper Cambrian of China and Lower Ordovician of North America (Dong and Donoghue 2002); (6) a possible arthropod from the Middle Cambrian Gaotai Formation of Guizhou, South China (Zhang and Pratt 1994; see also Zhang 1987); and (7) purported sponge embryos from the Doushantuo Formation of Guizhou, South China (Li et al. 1998).

For convenience, I consolidate these seven discoveries under three headings: Neoproterozoic, Cambrian cnidarians, and Cambrian bilaterians.

NEOPROTEROZOIC

The Doushantuo Formation, classically exposed in the Yangtze Gorges and with extensive outcrops elsewhere in South China, e.g., Guizhou Province, is an important stratigraphic interval in the late Neoproterozoic (an interval that in total spans 1000–1545 Ma). Unfortunately, its precise age is still unresolved. It postdates a glacial interval, marked by the Nantuo tillite (present in the Yangtze Gorges, but not exposed in the embryo localities in Guizhou), but it is uncertain whether this unit is equivalent to either one of the "Snowball Earth" episodes (Hoffman et al. 1998). Given the overall stratigraphic sequence of the Neoproterozoic in South China, a correlation of the Nantuo tillite (sediments of glacial origin colloquially known as "boulder beds") with the younger Marinoan (or Varangerian) ice age is perhaps more likely. This, in turn, would suggest that the Doushantuo Formation is younger than about 600 Ma.

Isotope geochronometry, using two separate systems (Lu-Hf and Pb-Pb), however, gives dates of ca. 600 Ma (Barfod et al. 2002). If correct, then as Barfod et al. (2002) point out, not only would the Doushantuo metazoans significantly predate most of the Ediacaran age fossils (see also Knoll and Xiao 1999), but the underlying Nantuo tillite might represent a separate glaciation. The details of this discussion are beyond the scope of this review, but it is important to remember that (1) whatever the correlations (or lack thereof) between the Neoproterozoic tillite horizons, the first fossil record of animals is effectively always post-tillite (but see Hofmann et al. 1990), and (2) the notion that the Neoproterozoic episodes were so catastrophic as to impose a major bottleneck on eukaryotic diversification, and by implication, the origin of animals, seems increasingly unlikely (see, e.g., Kennedy et al. 2001; Leather et al. 2002).

The fossil embryos were first discovered in 1986 (Chen and Liu 1986; see also Xue et al. 1995), but effectively only recognized as such more than a decade later (Xiao et al. 1998). More detailed descriptions of their morphology and possible affinities (Xiao and Knoll 2000), cell division and mitotic processes (Xiao 2002), and fossil preservation (Xiao and Knoll 1999a) are now available. The area, however, is not free of controversy. This included some initial skepticism (e.g., Xue et al. 1999, but see Xiao and Knoll 1999b), and more significantly, reservations (see Xiao et al. 2000) that accompany a study (Chen et al. 2000) of one assemblage of phosphatized material that, it was claimed, included examples of gastrulation. However, although Xiao et al. (2000) concede that certain structures could be gastrulae, comparison with other phosphatized material points to the various details identified by Chen et al. (2000) as being more likely to be the products of diagenesis (i.e., chemical changes such as recrystallization and mineral growth within the sediment).

The material that can be attributed with more confidence to animal embryos consists of several form-taxa. *Megasphaera* (two species, *M. inornata* and *M. ornata*) consist of an egg envelope which, as the specific names indicate, is, respectively, smooth and strikingly ornamented in a variety of patterns that possibly represent distinct species. The envelope houses an internal spheroid (Fig. 1). The other taxon, *Parapandorina* (*P. rhaphospissa*), is more informative inasmuch as the envelope contains a variable number of cells, from 2 to probably 64 (Fig. 2). Although there is a general consensus that these embryos are animals, and thus are complementary to a remarkably preserved algal assemblage from the same unit (Zhang 1989; Zhang and Yuan 1992; Zhang et al. 1998), the material of *Parapandorina* is somewhat frustrating. This is both because of the lack of later ontogenetic stages, perhaps caused either by a shift to a pelagic mode of life or the larger embryos failing to fossilize (Xiao and Knoll 2000), and the unresolved question as to

the zoological affinities of *Parapandorina*. The cleavage pattern is somewhat unusual (Xiao 2002), but parallels can be found in some acoels (a group of flatworms) as well as certain ecdysozoans (a superphylum encompassing arthropods, nematodes, and priapulids) and lophotrochozoans (another superphylum that includes annelids, brachiopods, mollusks, nemerteans, and most flatworms). However, with the possible exception of the acoels, what we know of their phylogenetic history makes questionable a direct link to any of these groups. The affinities of *Megasphaera* are even more problematic, although Xiao and Knoll (2000) draw attention to the similarity in ornamentation of egg envelopes of this Neoproterozoic fossil and a branchiopod crustacean, but here again a direct affinity is less probable. This is because the phylogenetic and stratigraphic evidence suggests that, as a group, the branchiopods appeared substantially later in Earth history.

Several other spherical microfossils from the Doushantuo Formation are identified as animal embryos by Xiao and Knoll (2000). Of these, *Megaclonophycus* (*M. onustus*) and *Spiralicellula bulbifera* (which is only tentatively identified by Xiao and Knoll 2000) are rather problematic. So too is *Cavesphaera* (*C. costata*), although as Xiao and Knoll (2000) point out, these fossils (Fig. 3) have a resemblance to the stereoblastulas of an extant octocoral, *Dendropeththya hemiprichi* (see Dahan and Benayahu 1998), but the similarities are rather imprecise. Xiao and Knoll (2000) also remark on some similarities between *Cavesphaera* and *Praviglobus ?dentisuturalis* (see Xue et al. 1995), but the larger size of the latter may indicate a later developmental stage. Brief mention should also be made of

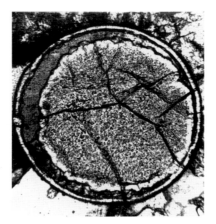

Figure 1. The metazoan embryo *Megasphaera inornata*, in transverse section, showing smooth envelope and shrunken internal body. The intervening gap is filled with the secondary mineral dolomite. From the Doushantuo Formation (latest Neoproterozoic), Shirenao section, Weng'an, Guizhou, China. Magnification, 78.5×. Photograph courtesy of Shuhai Xiao.

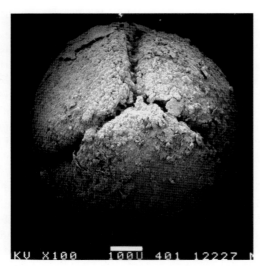

Figure 2. The metazoan embryo *Parapandorina raphospissa*, at the 4-cell stage. Locality details as in Fig. 1. Photograph courtesy of Shuhai Xiao.

putative sponge material, including what are identified as associated embryos (Li et al. 1998). Despite a variety of embryos being identified, this material is problematic. Zhang et al. (1998) note, for example, that among this material a supposed amphiblastula is algal, and a larva with flagella is actually an acritarch, whereas other features may be diagenetic (see also Xiao et al. 2000).

CAMBRIAN CNIDARIANS

Although the Neoproterozoic record of embryos gives us some tantalizing glimpses, to date they cannot rival for completeness the proposed ontogenetic series seen in *Olivooides*, from the lowermost Cambrian of Shaanxi, China (Bengtson and Yue 1997; Yue and Bengtson 1999). Effectively, three distinct stages are recognized. The earliest are embryos showing cell cleavage, with cell numbers ranging from ca. 64 to several thousand (Fig. 4), and in at least one case showing evidence for gastrulation (Fig. 5). In direct contrast to the Doushantuo material, therefore, embryos with <64 cells have not (with one possible exception) been recognized, even though at least superficially the taphonomic histories of very early phosphatization are the same. Curiously, this potentially important point was not addressed by Martin et al. (2003) in their description of experimental phosphatization of invertebrate eggs. Nevertheless, this contrast may be a clue that the nature of the life cycles, and by implication the metazoans themselves, were radically different in the Neoproterozoic versus the Cambrian. In particular, the *Olivooides* ontogeny is "normal," with the absence of earliest embryos reasonably interpreted as a reflection of their fragility. In contrast, the Doushantuo material was relatively robust at a very early stage of ontogeny and may have transformed into something very different from a "typical" metazoan, arguably even without the process of gastrulation.

In any event, this assemblage from Shaanxi may actually represent several taxa (Yue and Bengtson 1999), and the

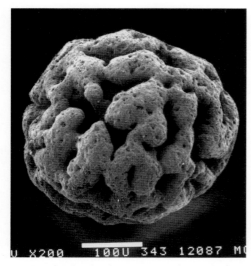

Figure 3. The possible metazoan embryo *Caveasphaera costata*, the holotype specimen. Locality details as in Fig. 1. Photograph courtesy of Shuhai Xiao.

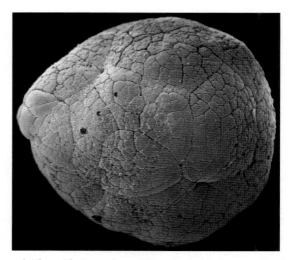

Figure 4. The cnidarian embryo *Olivooides multisulcatus* at the cleavage stage, with many thousands of cells, from Dengying Formation (earliest Cambrian), Shizhonggou section, Kuanchuanpu, Shaanxi, China. Magnification, 43.5×. Photograph courtesy of Stefan Bengtson.

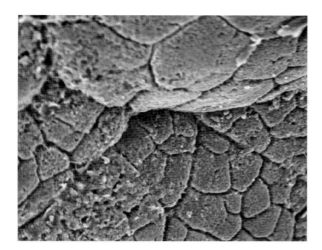

Figure 5. Detail of specimen illustrated in Fig. 4, to show possible evidence of ingression of cells as gastrulation. Magnification, 530x. Photograph courtesy of Stefan Bengtson.

tentative evidence for coeloblasty and gastrulation is in embryos that may not belong to *Olivooides*. A further point, if not difficulty, is that the next ontogenetic stage, where the embryos possess a distinctive stellate surface (Fig. 6), does not have transitional stages that link it to the cleavage embryos. There is, however, good reason to believe that these stellate embryos continued their development by unfolding into an ornamented tube that was presumably attached to the seafloor. Yue and Bengtson (1999) make a strong case for *Olivooides* being related to an extinct group of cnidarians known as the conulariids (characterized by a

phosphatic tube showing quadriradial symmetry), which in turn are probably some sort of coronate scyphozoan (a class of Cnidaria, the phylum that also includes *Hydra*). It is likely, therefore, that the tube originally housed structures known as scyphistomae, from which were budded the medusoids (ephyrae). At some later point the sexual stage produced the zygotes, but as noted, no details of this part of the ontogeny are available.

The other example of a possible cnidarian embryo concerns phosphatized spheres from the Manykay Formation (lowermost Cambrian) of northern Siberia (Kouchinsky et al. 1999). The material is relatively featureless, apart from one specimen (Fig. 7). This shows a sort of tetra-radial symmetry that consists of two sets of apical ridges, one beneath the other, that define two crosses set at 45° intervals. These are interpreted as possibly "incipient tentacles." The only other noteworthy feature, apart from a polygonal pattern that may represent cell imprints, is another specimen with radially arranged filaments. These have some resemblance to the late-stage embryos of *Olivooides*.

In addition to assigning this material to the cnidarians, Kouchinsky et al. (1999) suggest a possible link to co-occurring anabaritids. This is a distinctive group of Lower Cambrian tubicolous fossils with a characteristic tri-radial (or in some species sexa-radial) symmetry. The affinities of the anabaritids are still problematic, and although a position close to the cnidarians is most likely (see also below), shell structure is reminiscent of serpulid polychaete (phylum Annelida) tubes (Kouchinsky and Bengtson 2002). Given the distinctiveness of anabaritid tri-radiality, it is understandable why Kouchinsky et al. (1999) make some play on symmetry variants found in living cnidarians, especially the scyphozoan *Aurelia*. This was in an attempt to reconcile the characteristic tetra-radiality of scyphozoan

Figure 6. The cnidarian embryo *Olivooides multisulcatus* at the stellate stage, with folds surrounding aperture. Magnification, 90.1x. Locality details as in Fig. 4. Photograph courtesy of Stefan Bengtson.

Figure 7. A probable cnidarian (anabaritid?) embryo, from the Manykay Formation (earliest Cambrian), Bol'shaya Kuonamka River, northern Sakha (Yakutia), Russia. Magnification, 117.8x. Photograph courtesy of Stefan Bengtson.

cnidarians (and perhaps the corresponding pattern on one of the embryos; Fig. 7) with the tri-radiality of anabaritids. Although the supposition was that in the primitive state, a variety of symmetries were available, from which tetra-radial was the "chosen" (if not stable) form, a transition from tri- to tetra-radial symmetry receives some support from the fossil record. This is because of the recognition of tri-radiality in the earliest conulariids (that otherwise are tetra-radial), from the latest Neoproterozoic of northern Russia (Ivantsov and Fedonkin 2002), suggesting that a phylogenetic connection between the conulariids (and *Olivooides*) and anabaritids is a reasonable possibility.

CAMBRIAN BILATERIANS

In the same paper that first described the embryos of *Olivooides*, Bengtson and Yue (1997) provided convincing evidence that the microfossil *Markuelia* (*M. secunda*) from the Pestrotsvet Formation (Lower Cambrian) of eastern Siberia was also an embryo (Fig. 8). In this case it is seg-

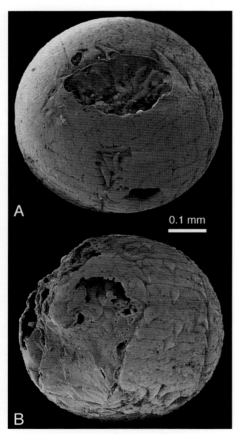

Figure 8. Two views of the bilaterian (halkieriid?) embryo *Markuelia secunda*, from the Pestrotsvet Formation (early Cambrian), Dvortsy section, eastern Siberia, Russia. Photograph courtesy of Stefan Bengtson.

mented, and looped around a sphere in a yin–yang-like arrangement. The segments bear short conical processes, whereas internal rod-like structures were tentatively proposed to be part of the nervous system. Apart from recognizing *Markuelia* as the embryo of a segmented worm and therefore a bilaterian, Bengtson and Yue (1997) made no further suggestion as to its affinities. I have suggested (Conway Morris 1998) that it might belong to the halkieriids (Conway Morris and Peel 1995), which as isolated sclerites are known to co-occur with these embryos. This view, however, may need to be revised, because Dong and Donoghue (2002) identify *Markuelia*-like embryos from the Middle and Late Cambrian of China and the Lower Ordovician of North America as a priapulid, another group with a rich Cambrian record (see, e.g., Conway Morris 1977; Hou et al. 1999).

Other bilaterian embryos (and eggs) from the Cambrian are rather more indeterminate. Zhang and Pratt (1994) described embryos from Gaotai Formation (middle Cambrian) of Guizhou, China. These show convincing blastomeres, but are otherwise relatively featureless and show no evidence for gastrulation. Zhang and Pratt (1994) suggest that the embryos derive from trilobites, possibly eodiscids.

CONCLUSIONS

Remarkable as the discoveries of fossil embryos have been, they pose a series of problems, not least their zoological affinities and hence position(s) in metazoan phylogeny. In addition, whereas mitotic cell division, and in one case probable gastrulation, give some insight into processes that apparently (and unsurprisingly) have remained unchanged for more than half a billion years, there is as yet little information on areas of classical concern such as cleavage patterns or types of gastrulation (or epiboly), let alone cell lineages. Nevertheless, the finds to date are germane to a wider debate concerning both the nature of the Cambrian "explosion" and the evolution of life cycles.

Although to varying degrees the affinities of the fossil embryos are speculative, a striking observation is that where the evidence is sufficient, development always appears to be direct (see, e.g., Bengtson and Yue 1997; Kouchinsky et al. 1999; Yue and Bengtson 1999; see also Conway Morris 1998), and therefore consistent with the view that this mode of development was primitive (see, e.g., Conway Morris 1998). This by no means rules out the coexistence of planktotrophic life cycles, the pre-metamorphic stages of which escaped fossilization. Nevertheless, as far as the early triploblasts are concerned, this would also be consistent with the proposed basal position of the acoel flatworms, which also show direct development (see, e.g., Ruiz-

Trillo et al. 1999, 2002). Such an arrangement might in turn have been derived from direct development in cnidarian ancestors. This view is, of course, at odds with the view that the biphasic life cycle is primitive to metazoans (see, e.g., Rieger 1994), and especially the extreme case of maximal indirect development where a morphologically distinct planktotrophic larva undergoes a catastrophic metamorphosis into an adult stage (e.g., Jägersten 1972). In fact, the view that planktotrophy is primitive in biphasic life cycles has already been criticized (e.g., Haszprunar et al. 1995), and this would also make the transition from direct to biphasic life cycle with lecithotrophic larvae more plausible. Still, the idea of planktotrophy as primitive dies hard, not least because of the ideas of Davidson et al. (1995; see also Peterson et al. 1997, 2000). This group has suggested that the evolution of the so-called "set-aside" cells, that is, the cells that will develop into the adult primordium in maximally indirect larvae, represents a key step in both the evolution of metazoans and thus the Cambrian explosion. Central to the hypothesis of these authors is the idea that, prior to the invention of the set-aside cells, metazoans were effectively larval. In this way, they could have had a protracted Neoproterozoic history, conveniently cryptic, that could be used to reconcile the apparent discrepancy between the divergence times as seen in the fossil record versus those based on molecular clocks (e.g., Wray et al. 1996; Bromham et al. 1998). In fact, the reliability, or otherwise, of molecular clocks is hotly debated, and it is clear there can be no simple application: Different molecules often give widely discrepant results. On the other hand, the general idea that molecular clocks uniformly point toward very ancient divergences of the metazoans is, at least in some cases, now in serious doubt (Ayala 1999; Aris-Brouson and Yang 2002). In addition, the notion of the evolution of set-aside cells being the trigger for the Cambrian "explosion," whereby the "invention" of large adults breached a taphononomic (i.e., postmortem processes involved with fossilization) threshold (i.e., became capable of fossilization, either as body remains or imprints on the sediment as traces), also suffers two cardinal difficulties. First, no convincing functional and adaptational explanation has been put forward for the original appearance of the set-aside cells (see, e.g., Conway Morris 1998, 2000b), and it seems at least as plausible to invoke their evolution as an advanced feature of the biphasic life cycle whereby the adult rudiment is transferred to the larval stage. Second, it seems likely, given the range of both planktotrophy and larval types, that set-aside cells evolved multiple times. In this sense, they evolved independently, and so join a long list of larval characters that are believed to be convergent and reflect the functional constraints of living in a very demanding environment (Strathmann 1993; Rouse 1999; Hadfield 2000; Hart 2000; Nezlin 2000). In this way, the study of larval life histories epitomizes a central problem in evolution: To what extent is it a historical process (other than trivially) versus the possibility that, although it is set in time, the constraints of life are such as to stamp strong probabilities, if not inevitabilities, on the patterns of life.

ACKNOWLEDGMENTS

Stefan Bengtson (Stockholm) and Shuhai Xiao (Tulane) kindly provided illustrations, and the former also is thanked for a critical review. Two anonymous referees provided additional comments. Sandra Last is warmly acknowledged for assistance with manuscript preparation. Cambridge Earth Sciences Publication 7704.

REFERENCES

Aris-Brouson S. and Yang Z. 2002. Effects of models of rate evolution on estimation of divergence dates with special reference to the metazoan 18S ribosomal RNA phylogeny. *Syst. Biol.* **51**: 703–714.

Ayala F.J. 1999. Molecular clock mirages. *BioEssays* **21**: 71–75.

Balinski A. 2001. Embryonic shells of Devonian linguloid brachiopods. In *Brachiopods past and present* (ed. C.H.C. Brunton et al.), pp. 91-101. Taylor & Francis, New York.

Barfod G.H., Albarède F., Knoll A.H., Xiao S.-H., Télouk P., Frei R., and Baker J. 2002. New Lu-Hf and Pb-Pb age constraints on the earliest animal fossils. *Earth Planet. Sci. Lett.* **201**: 203–212.

Bengtson S. and Yue Z. 1997. Fossilized metazoan embryos from the earliest Cambrian. *Science* **277**: 1645–1648.

Biernat G. and Williams A. 1970. Ultrastructure of the protegulum of some acrotretide brachiopods. *Palaeontology* **13**: 491–502.

Bromham L., Rambaut A., Fortey R., Cooper A., and Penny D. 1998. Testing the Cambrian explosion hypothesis by using a molecular dating technique. *Proc. Natl. Acad. Sci.* **95**: 12386–12389.

Chaffee C. and Lindberg D.R. 1986. Larval biology of early Cambrian molluscs: The implications of small body size. *Bull. Mar. Sci.* **39**: 535–549.

Chen J.-Y., Oliveri P., Li C.-W., Zhou G.-Q., Gao F., Hagadorn J.W., Peterson K.J., and Davidson E.H. 2000. Precambrian animal diversity: Putative phosphatized embryos from the Doushantuo Formation of China. *Proc. Natl. Acad. Sci.* **97**: 4457–4462.

Chen M. and Liu K. 1986. The geological significance of newly discovered microfossils from the Upper Sinian (Doushantuo age) phosphorites. *Sci. Geol. Sin.* **1986**: 46–53.

Conway Morris S. 1977. Fossil priapulid worms. *Spec. Pap. Palaeontol.* **20**: iv + 95 pp.

———. 1998. Eggs and embryos from the Cambrian. *BioEssays* **20**: 676–682.

———. 2000a. The Cambrian "explosion": Slow-fuse or megatonnage? *Proc. Natl. Acad. Sci.* **97**: 4426–4429.

———. 2000b. Evolution: Bringing molecules into the fold. *Cell* **100**: 1–11.

Conway Morris S. and Peel J.S. 1995. Articulated halkieriids from the Lower Cambrian of North Greenland and their role in early protostome evolution. *Phil. Trans. R. Soc. Lond. B Biol. Sci.* **347**: 305–358.

Dahan M. and Benayahu Y. 1998. Embryogenesis, planulae longevity, and competence in the octocoral *Dendronephthya hemiprichi*. *Invertebr. Biol.* **117:** 271–280.

Davidson E.H., Peterson K.J., and Cameron R.A. 1995. Origin of bilaterian body plans: Evolution of developmental regulatory mechanisms. *Science* **270:** 1319–1325.

Dong X.-P. and Donoghue P. 2002. Affinity of the earliest bilaterian embryos. *Palaeontol. Assoc. Newslett.* **51:** 90–91.

Dong X.-P. and Pratt B.R. 1994. Middle Cambrian embryos with blastomeres. *Science* **266:** 637–639.

Dzik J. 1978. Larval development of hyolithids. *Lethaia* **11:** 293–299.

Emlet R.B. 1985. Crystal axes in Recent and fossil adult echinoids indicate trophic mode in larval development. *Science* **230:** 937–940.

———. 1989. Apical skeletons of sea urchins (Echinodermata: Echinoidea): Two methods for inferring mode of larval development. *Paleobiology* **15:** 223–254.

Hadfield M.G. 2000. Why and how marine-invertebrate larvae metamorphose so fast. *Semin. Cell Dev. Biol.* **11:** 437–443.

Hart M. 2000. Phylogenetic analyses of mode of larval development. *Semin. Cell Dev. Biol.* **11:** 411–418.

Haszprunar G. 1992. The first molluscs—Small animals. *Boll. Zool.* **59:** 1–16.

Haszprunar G., Salvini-Plawen L.V., and Rieger R.M. 1995. Larval planktotrophy—A primitive trait in Bilateria? *Acta Zool.* **76:** 141–154.

Hofman H.J., Narbonne G.M., and Aitken J.D. 1990. Ediacaran remains from intertillite beds in northwestern Canada. *Geology* **18:** 1199–1202.

Hoffman P.F., Kaufman A.J., Halverson G.P., and Schrag D.P. 1998. A Neoproterozoic snowball Earth. *Science* **281:** 1342–1346.

Hou X.-G., Bergström J., Wang H.-F., Feng X.-H., and Chen A.-L. 1999. *The Chengjiang fauna: Exceptionally well-preserved animals from 530 million years ago.* Yunnan Science and Technology Press, Kunming, China.

Ivantsov A.Y. and Fedonkin M.A. 2002. Conulariid-like fossil from the Vendian of Russia: A metazoan clade across the Proterozoic/Palaeozoic boundary. *Palaeontology* **45:** 1219–1229.

Jablonksi D. 1986. Larval ecology and macroevolution in marine invertebrates. *Bull. Mar. Sci.* **39:** 565–587.

Jablonski D. and Lutz R.A. 1979. Larval ecology of extinct molluscs: Comment on larval development of hyolithids. *Lethaia* **12:** 306.

———. 1983. Larval ecology of marine benthic invertebrates: Paleobiological implications. *Biol. Rev.* **58:** 21–89.

Jägersten G. 1972. *Evolution of the metazoan life cycle: A comprehensive theory.* Academic Press, London, United Kingdom.

Jeffery C.H. 1997. Dawn of echinoid nonplanktotrophy: Coordinated shifts in development indicate environmental instability prior to the K-T boundary. *Geology* **25:** 991–994.

Kennedy M.J., Christie-Blick N., and Prave A.R. 2001. Carbon isotopic composition of Neoproterozoic glacial carbonates as a test of paleoceanographic models for snowball Earth phenomena. *Geology* **29:** 1135–1138.

Knoll A.H. and Xiao S.-H. 1999. On the age of the Doushantuo Formation. *Acta Micropalaeontol. Sin.* **16:** 225–236.

Kouchinsky A. and Bengtson S. 2002. The tube wall of Cambrian anabaritids. *Acta Palaeontol. Pol.* **47:** 431–444.

Kouchinsky A., Bengtson S., and Gershwin L.-A. 1999. Cnidarian-like embryos associated with first shelly fossils in Siberia. *Geology* **27:** 609–612.

Leather J., Allen P.A., Brasier M.D., and Cozzi A. 2002. Neoproterozoic

Snowball Earth under scrutiny: Evidence from the Fiq glaciation of Oman. *Geology* **30:** 891–894.

Li C.-W., Chen J.-Y., and Hua T.-E. 1998. Precambrian sponges with cellular structures. *Science* **279:** 879–882.

Maas A. and Waloszek D. 2001. Cambrian derivatives of the early arthropod stem lineage, pentastomids, tardigrades and lobopodians—An 'Orsten' perspective. *Zool. Anz.* **240:** 451–459.

Martill D.M. 1990. Macromolecular resolution of fossilized muscle tissue from an elopomorph fish. *Nature* **346:** 171–172.

———. *Fossils of the Santana and Crato Formations, Brazil.* Palaeontological Association, London, United Kingdom.

Martin D., Briggs D.E.G., and Parkes R.J. 2003. Experimental mineralization of invertebrate eggs and the preservation of Neoproterozoic embryos. *Geology* **31:** 39–42.

Müller K.J. and Walossek D. 1986. Arthropod larvae from the Upper Cambrian of Sweden. *Trans. R. Soc. Edinb. Earth Sci.* **77:** 157–179.

Nezlin L.P. 2000. Tornaria of hemichordates and other dipleurula-type larvae: A comparison. *J. Zool. Syst. Evol. Res.* **38:** 149–156.

Peterson K.J., Cameron R.A., and Davidson E.H. 1997. Set-aside cells in maximal indirect development: Evolutionary and developmental significance. *BioEssays* **19:** 623–631.

———. 2000. Bilaterian origins: Significance of new experimental observations. *Dev. Biol.* **219:** 1–17.

Rieger R.M. 1994. The biphasic life cycle—A central theme of metazoan evolution. *Am. Zool.* **34:** 484–491.

Rouse G.W. 1999. Trochophore concepts: Ciliary bands and the evolution of larvae in spiralian Metazoa. *Biol. J. Linn. Soc.* **66:** 411–464.

Rowell A.J. 1986. The distribution and inferred larval dispersion of *Rhondellina dorei*: A new Cambrian brachiopod (Acrotretida). *J. Paleontol.* **60:** 1056–1065.

Ruiz-Trillo I., Riutort M., Littlewood D.T.-J., Herniou E.A., and Baguna J. 1999. Acoel flatworms: Earliest extant bilaterian metazoans, not members of platyhelminthes. *Science* **283:** 1919–1923.

Ruiz-Trillo I., Paps J., Loukota M., Ribera C., Jondelius U., Baguña J., and Riutort M. 2002. A phylogenetic analysis of myosin heavy chain type II sequences corroborates that Acoela and Nemertodermatida are basal bilaterians. *Proc. Natl. Acad. Sci.* **99:** 11246–11251.

Smith A.B. and Jeffery C.H. 1998. Selectivity of extinction among sea urchins at the end of the Cretaceous period. *Nature* **392:** 69–71.

Strathmann R.R. 1993. Hypothesis on the origins of marine larva. *Annu. Rev. Ecol. Syst.* **24:** 89–117.

Valentine J.W. 1986. The Permian-Triassic event and invertebrate developmental modes. *Bull. Marine Sci.* **39:** 607–615.

———. 1997. Cleavage patterns and the topology of the metazoan tree of life. *Proc. Natl. Acad. Sci.* **94:** 8001–8005.

van den Biggelaar J.A.M., Dictus W.J.A.G., and van Loon A.E. 1997. Cleavage patterns, cell-lineages and cell specification are clues to phyletic lineages in Spiralia. *Sem. Cell Dev. Biol.* **8:** 367–378.

von Bitter P.H. and Ludvigsen R. 1979. Formation and function of protegular pitting in some North American acrotretid brachiopods. *Palaeontology* **22:** 705–720.

Walossek D. and Müller K.J. 1998. Early Cambrian phylogeny in light of the Cambrian "Orsten" fossils. In *Arthropod fossils and phylogeny* (ed. G.D. Edgecombe), pp. 185–231. Columbia University Press, New York.

Wray G.A., Levinton J.S., and Shapiro L.H. 1996. Molecular evidence for deep Precambrian divergences among metazoan phyla. *Science* **274:** 568–573.

Xiao S.-H. 2002. Mitotic topologies and mechanics of Neoproterozoic algae and animal embryos. *Paleobiology* **28:** 244–250.

Xiao S.-H. and Knoll A.H. 1999a. Fossil preservation in the Neoproterozoic Doushantuo phosphorite Lagerstätte, South China. *Lethaia* **32:** 219–240.

———. 1999b. Embryos and algae? A reply. *Acta Micropalaeontol. Sin.* **16:** 313–322.

———. 2000. Phosphatized animal embryos from the Neoproterozoic Doushantuo Formation at Weng'an, Guizhou, South China. *J. Paleontol.* **74:** 767–788.

Xiao S.-H., Yuan X.-L., and Knoll A.H. 2000. Eumetazoan fossils in terminal Proterozoic phosphorites? *Proc. Natl. Acad. Sci.* **97:** 13684–13689.

Xiao S.-H., Zhang Y., and Knoll A.H. 1998. Three-dimensional preservation of algae and animal embryos in a Neoproterozoic phosphorite. *Nature* **391:** 553–558.

Xue Y.-S., Zhou C.-M., and Tang T.-F. 1999. "Animal embryos", a misinterpretation of Neoproterozoic microfossils. *Acta Micropalaeontol. Sin.* **16:** 1–4.

Xue Y.-S., Tang T.-F., Yu C.-L., and Zhou C.-M. 1995. Large spheroidal Chlorophyta fossils from the Doushantuo Formation phosphoric sequence (late Sinian), central Guizhou, South China. *Acta Palaeontol. Sin.* **34:** 688–706.

Yue Z. and Bengtson S. 1999. Embryonic and post-embryonic development of the Early Cambrian cnidarian *Olivooides*. *Lethaia* **32:** 181–195.

Zhang X.-G. 1987. Moult stages and dimorphism of Early Cambrian bradoriids from Xichuan, Henan, China. *Alcheringa* **11:** 1–9.

Zhang X.-G. and Pratt B.R. 1994. Middle Cambrian embryos with blastomeres. *Science* **266:** 637–639.

Zhang Y. 1989. Multicellular thallophytes with differentiated tissues from Late Proterozoic phosphate rocks of South China. *Lethaia* **22:** 113–132.

Zhang Y. and Yuan X.-L. 1992. New data on multicellular thallophytes and fragments of cellular tissues from Late Proterozoic phosphate rocks, South China. *Lethaia* **25:** 1–18.

Zhang Y., Yin L.-M., Xiao S.-H., and Knoll A.H. 1998. Permineralized fossils from the terminal Proterozoic Doushantuo Formation, South China. *Palaeontol. Soc. Mem.* **50:** 1–52.

Part V

Open Questions

GASTRULATION OVERVIEW
OPEN QUESTIONS

L. Wolpert

Department of Anatomy & Developmental Biology, University College London, London WC1E 6BT, United Kingdom

INTRODUCTION

Looking at all these studies on gastrulation, one cannot help but be impressed, if not amazed, that this process is present in all animals and that there are both striking similarities and differences. It is a process that is fundamental to animal development and a lovely illustration of evolution of a developmental process. Just how it has evolved is a deep and little considered problem. As Leys (Chapter 2) points out, the term "gastrula" was established by Haeckel in 1872 after he had studied sponge development. He used the term to refer to a stage in embryogenesis that represented the ancestral diploblastic metazoan that had formed a primitive gut (the archenteron) and a mouth (the blastopore) by invagination. He proposed that all metazoans passed through this stage, and that the embryonic germ layers are therefore homologous among all metazoans, a concept he termed the Gastraea-Theory. Gastrulation therefore comprises two elements: the formation of a multilayered organism and the formation of a digestive system. Although there has been excellent progress in understanding gastrulation in a wide variety of organisms, many difficult problems still remain in relation to evolution, axes and patterning, and morphogenesis.

EVOLUTION

Gastrulation had its origin quite early in the evolution of multicellular animals. Multicellularity requires overall coordination of at least some metabolic or informational processes among the cells of a colony. Developmentally dif-ferentiated cell types are a sufficient condition. Little new had to be invented by evolution in the transition from single cells to multicellular organisms. The basic processes required for development were already present in the eukaryotic cell. These included a program of gene activity to regulate the cell cycle, signal transduction, cell motility, and processes similar to cell differentiation. Possible advantages of multicellularity included more efficient predation (cooperative feeding), resistance to predation, division of labor, cannibalism, nutrient storage, and more efficient dispersal during starvation. The evolution of an egg was necessary to reduce the chance of selfish cells taking over, and it could have increased evolvability by rendering development coherent. The biggest hurdle could have been to down-regulate cell division at the appropriate time and place, a problem that becomes more severe with increasing organism size and complexity (Szathmary and Wolpert 2003).

Why is gastrulation so similar in all animals? It does actually recapitulate an ancient ancestor. The first metazoan, blastaea, came into existence as an epithelial sphere, the simplest continuous epithelium we can think of. For the first time, the epithelium enclosed a cavity more or less completely isolated from the environment, so that organic molecules could be exchanged without loss to the outside. This allowed the first division of labor between cells. Some cells became specialized for feeding, others were kept apart as germ cells, where vasa was expressed. Thus, the first body axis, the animal–vegetal axis was set up, first reflecting the initial germ-line–soma polarity (Chapter 51).

The basic idea is that a very early two-layered primitive organism fed on the bottom, and it formed an infolding to

715

aid feeding. This basic idea comes from Jaegerstern (1956). In a blastaea-like organism, a hollow organism made of a single layer of cells, the feeding was encouraged by currents from cilia. Living on the bottom, it formed an invagination to sweep the food into a primitive gut where the cells would engulf the food. It takes no stretch of the imagination to see that all it had to do was to fuse this infolding with the sheet on the other side, and you then have a mouth, a gut, and an anus; or alternatively, the opening, the blastopore, elongated into a slit form, which then closed in the middle, leaving an open mouth and anus. The very simple organism *Trichoplax* is made up of just a single layer of cells and a hollow interior, rather like Haeckel's blastaea. What is remarkable in *Trichoplax* is that it undergoes a change similar to early gastrulation while feeding. Particles of food or microorganisms that it is going to eat are moved into a digestive chamber.

A very brief review of the varieties of gastrulation described illustrates both similarities and important differences. It is striking that in most cases the process is complete within a few hours. Invagination is found in species with little yolk. One of the most common features is the convergence and extension toward and along the midline and the transition of epithelial cells to mesenchymal cells.

Cleavage in the Cnidaria and Ctenophora results in the formation of a blastula that later gastrulates to produce a ciliated, cigar-shaped stage called the planula (Chapter 3). Within the phylum Cnidaria, at least nine forms of gastrulation have been described. These include two forms of ingression (unipolar and multipolar ingression), three forms of delamination (blastula, morula, and syncytial delamination), invagination, epiboly, and at least two mixed forms of gastrulation (mixed delamination and forms combining ingression with invagination).

Gastrulation in the protostomian molluscan embryo resembles gastrulation in the chordate embryo (Chapter 5). Gastrulation in mollusks necessarily leads to the conclusion that there is no essential difference in gastrulation between protostomians and deuterostomians. In both groups, the dorsal blastoporal lip has an organizer function, expresses homologous genes, and uses a conserved signaling pathway.

Depending on the species, gastrulation in crustacean embryos differs with different modes of cleavage (Chapter 6). In the case of total cleavage, cell divisions distribute all the material of the egg among the blastomeres, and then gastrulation takes place as a movement of some cells toward the center of the egg. In the case of superficial cleavage, cell divisions result in the formation of a superficial monolayer of cells overlying a yolk mass, and gastrulation creates a multilayered embryo, which still sits on the surface of the egg. For both total and superficial cleavage, gastrulation can occur through either invagination or ingression. During invagination, an inner layer of mesendoderm cells forms a pouch that is continuous with an outer layer of prospective ectoderm. When ingression occurs, single cells or groups of cells depart from the blastula layer to form the precursors of the mesoderm and endoderm. In *Sicyonia*, gastrulation takes place after five tangential cleavages have set up 32 blastomeres. Later gastrulation involves a switch of the position of their spindles during division from tangential to radial, and they divide toward the center of the egg and push the two mesendoderm cells inward. This is the only documented example of oriented cell division initiating gastrulation in crustaceans.

Gastrulation begins in ascidian embryos between the 6th and 7th cleavages and involves three stages (Chapter 10). The first stage is the invagination of the endoderm, followed by the involution of the mesoderm, and, finally, the epiboly of the ectoderm. The site of invagination, called the gastrulation center, is determined during ooplasmic segregation as the spot where the myoplasm caps at the vegetal pole of the egg, opposite the site of sperm entry.

In *Drosophila* (Chapter 7), on the ventral side, the mesoderm invaginates together with the anterior endoderm primordium, while at the posterior end, the posterior endoderm invaginates, carrying the pole cells with it. Cell intercalations in the ectoderm result in the lengthening of the trunk region, which results in the displacement of the posterior end of the embryo toward the dorsal side. These movements begin around 3 hours after fertilization and occur over a period of approximately 2 hours. Once the posterior midgut has invaginated, it disperses into single cells that migrate anteriorly along the underlying mesoderm until the cells reach their counterparts migrating from the anterior toward the center of the embryo. Unlike the creation of the mesodermal cell layer, which is complete within approximately 2 hours, endodermal cell migration takes several hours.

In insects, the alimentary canal is composed of four major components: the foregut, the midgut, the Malphighian tubules, and the hindgut. The foregut and hindgut, together with the Malphighian tubules, originate from tube-like invaginations of ectodermal origin, called stomodeum and proctodeum, respectively. Only the midgut is of endodermal origin (Chapter 8). Three basic modes of internalization can be distinguished in insects. Type I resembles *Drosophila*. The presumptive mesoderm forms a well-defined ventral furrow, which is then converted into a tube-like structure when the edges of the lateral ectoderm approach each other at the ventral midline and fuse. The mesoderm remains as a stiff plate, while the ectoderm moves over its surface toward the ventral midline. The third type has been described for many hemimetabolous insects. Here, either no, or only a weakly pronounced, ventral furrow is visible. Cell divisions begin early within the mesodermal region and produce an irregular mass of cells which is pushed inside when the ectodermal plates meet at the ventral midline. The short-germ mode of insect development poses the problem of where the mesoderm for the posterior segments originates. Is there a continuous process of gastrulation by which new mesodermal cells are generated when new segments arise, or does the growth zone possess mesodermal precursor cells that stem from early gastrulation?

Understanding the cell movements of amphibian gastrulation requires understanding the differences among amphibians in the representation of the presumptive tissues in the surface, or superficial, epithelial layer versus the deep, mesenchymal region (Chapter 13). The common method of gastrulation in amphibians involves involution of the presumptive mesodermal and endodermal tissues. However, involution of superficially located presumptive mesodermal cells only rolls them over onto the roof of the future gut cavity, and it does not bring them into the middle layer. For this to occur, these cells must undergo an epithelial-to-mesenchymal transition (EMT) and leave the epithelium to join the deep region. In some species, such as *Xenopus*, there are few superficially derived mesodermal cells, and the morphogenetic, force-generating consequences of their EMT are negligible. In other species, there are many superficial presumptive mesodermal cells, and when and where they undergo EMT serves an important biomechanical function in gastrulation, in addition to placing the presumptive mesoderm in the middle layer where it belongs. However, there are some major and significant differences between the anurans and urodeles studied thus far, chiefly in the fact that there is much more superficially derived mesoderm, especially presumptive somitic mesoderm, and most of it ingresses during gastrulation. The involuting marginal zone (IMZ) of urodeles undergoes a massive epithelial–mesenchymal transition and ingresses (leaves the superficial layer) as it involutes.

Avian eggs and early embryos differ in several respects from the majority of vertebrates, yet many of the principles governing the early steps of development are very similar. By the time the egg is laid, the embryo is an almost flat disc of some 20,000 to 50,000 cells (Chapter 15). In the following few hours of incubation, the islands of hypoblast gradually fuse together, probably by a process of flattening of the cells, which proceeds from posterior to anterior to generate a continuous but loose layer, the hypoblast proper. The initial appearance of the streak is extremely rapid—the embryo goes from having no visible axial structures (stage XIV) to developing a triangular, dense streak (stage 2) in about 30 minutes, suggesting that primitive streak initiation is accompanied by massive ingression of the future endoderm and mesoderm. Streak formation and thus gastrulation in mammals is similar and similarly involves EMTs.

Given all these patterns of gastrulation, we need to understand how all these differences evolved. This is no simple problem and has, alas, received relatively little attention. It seems reasonable to believe that they all evolved from a form like Haeckel's blastaea. But how did this happen? The real difficulty is in understanding the nature of the intermediate forms, how small or large were the evolutionary steps, and what was their adaptive significance. Worse still, it is hard to see how changes in gastrulation were linked to later stages in the development of the embryo, as they had to be.

Consider three examples. What were the intermediate stages in the evolution of the endoderm/gut in *Drosophila*, developing as it does independently from almost opposite ends of the embryo and not linked to the mesoderm? Again consider the significant differences between the anurans and urodeles where there is much more superficially derived mesoderm, and most of it ingresses during gastrulation. The IMZ of urodeles undergoes a massive EMT and ingresses as it involutes. Finally, how did extraembryonic tissues, like the endoblast of the chick, which do not contribute to the embryo, evolve and come to play an important role in gastrulation?

MORPHOGENESIS

It may be possible to see the cellular movements and cellular forces that bring about gastrulation as modifications of a few basic mechanisms. Two common themes are invagination and cell ingression, the latter involving an EMT. This is nicely illustrated by the similarity between invagination and then ingression of the mesoderm in sea urchins, fruit flies, and amniotes. There is apical constriction, and the cells leave the epithelial layer to which they belonged. Apical constriction and ingression are best understood in the ventral furrow of *Drosophila*. It is accompanied by drastic cell shape changes and a movement of the nucleus away from its original position adjacent to the apical membrane to a new position within the basal region of the cell (Chapter 7). Thus, ventral cells are converted from a columnar shape to a wedge shape. Myosin relocates from the basal part of the cells, where it is needed during cellularization, to the apical sides of constricting ventral and posterior cells. The maternally supplied components—Concertina, RhoGEF2, and Rho, cytoskeletal components, and myosin—are ubiquitously distributed throughout the egg, yet cell shape changes occur only in precisely specified regions, indicating that there must be region-specific triggers for the activation of the maternal proteins. In the case of Concertina, it is the regionally restricted expression of Fog that provides the necessary and sufficient trigger for activation, but the main trigger for RhoGEF2 is not known. Since no furrow formation occurs in the absence of Snail and Twist, it is likely that the regional activation of RhoGEF is caused by the product of one or more of their target genes. These have so far eluded identification.

A number of cell adhesion changes accompany ingression of the primary mesenchyme cells (PMCs) in the sea urchin (Chapter 9). The cells lose an affinity for hyalin and echinonectin, extracellular matrix proteins covering the outside of the embryo. At the same time, the PMCs lose cell–cell adhesiveness, a property that is largely attributable to a rapid endocytic removal of cell-surface cadherin from the micromeres. Finally, the PMCs shift their affinities toward basal lamina from an epithelial–basal lamina adhesion via a laminin-binding integrin. PMCs depend on cues provided by the substrate to pattern the skeleton.

A major morphogenetic movement is convergence of cells toward the midline and their simultaneous extension along the main body axis (Chapter 19). These movements are commonly separated into two components, the convergence component and the extension component, because convergence is not necessarily linked to extension but is a mechanical force generator that can drive cells into either extension or thickening, or more commonly, both. On the other hand, extension is not always driven by convergence but can also be driven by thinning, as described above. In addition, the geometric convergence of cells toward a point or a midline does not necessarily mean that extension will result. Convergent extension is an active, force-producing process that forms a dynamic "embryonic skeleton," which both shapes and supports the embryo. Polarized cell motility is thought to drive mediolateral intercalation and convergent extension. The mediolateral cell intercalation during convergence and extension appears to occur as a result of polarized cell motility, a behavioral transformation that is called "mediolateral intercalation behavior." Initially, in the early *Xenopus* gastrula, the mesodermal cells are multipolar, isodiametric, and pleiomorphic, but at the mid-gastrula stage, they become polarized such that they have large lamelliform protrusions at their medial and lateral ends and fine filiform protrusions or fine contact points at their anterior and posterior sides. The large lamellipodia appear to exert traction on adjacent cells in the mediolateral axis and generate mediolaterally oriented tensile forces, and thus the cells become elongated and aligned with the mediolateral axis and pull themselves between one another, generating a narrower, longer array. Again the molecular basis is poorly understood, but polarization of cells involving the Wnt pathway is a promising candidate.

The formation of the primitive streak in chick development involves both invagination and ingression in the midline, possibly similar to that seen in sea urchins (Chapter 15). These movements comprise convergence of epiblast toward the posterior midpoint and extension along the

midline, but it may not occur by a convergent extension as described in the frog, and this is as yet an unresolved problem. The initial combination of posterior midpoint convergence and midline extension resembles a Polish dance—Polonaise. To what extent these movements are due to the motion of individual cells or forces generated in the region of the streak due to apical constriction of the cells is not clear. However, although the basement membrane under the epiblast does partially dissolve during streak formation, this early stage does not involve the loss of epithelial continuity of the epiblast. We know surprisingly little about the cellular details of how the fully elongated primitive streak (stage 3$^+$) forms. Less than 2 hours elapse between the early short streak at stage 2 and the almost fully elongated (1.5 mm long) stage-3 streak, making it very unlikely that cell division is the major force driving this elongation.

Two other movements are commonly found in gastrulating embryos: involution, which is the turning inward of the epithelium around the blastopore lip, and epiboly, the expansion of external ectodermal cells over the rest of the embryo. Neither is well understood.

There is an urgent need for more information on the cellular forces responsible for all these movements. It will also be very important to develop techniques to model all these morphogenetic processes. Such techniques, possibly computer simulations, need to take into account the mechanical properties of the cells and the cell sheets, and the actual forces generated by the cells. Only in this way will it be possible to validate proposed mechanisms.

PATTERNING AND AXES

How is the pattern of cellular activities in gastrulation specified by the genes? Axes must be specified as well as cell fates and cell motility and cellular forces.

In some embryos, like *Drosophila* and the sea urchin, the axes are specified in the egg. In others, external signals play a role, like fertilization in the nematode and *Xenopus*, and gravity in the chick. In so many cases, the axes, anterior–posterior and dorsoventral, are at right angles to each other and just how this is specified is not always clear (see Chapters 13 and 26). In amniotes, the relation between the axis of the primitive streak and those future body axes is complex. There is also the key question of the specification of left/right asymmetry. Recent studies on the mouse implicate the role of motor and sensory cilia in the node, resulting in the asymmetric release of calcium (McGrath et al. 2003). There is also evidence for much earlier specification in *Xenopus* by a hydrogen ion pump (Levin et al. 2002; Chapter 28), and in other animals like the nematode, the asymmetry is present by the 3rd cleavage.

There is a wide variety of genes controlling gastrulation and how they evolved is totally unknown. The model of the gene regulatory network that describes the acquisition of the regulatory state required for gastrulation in the sea urchin *S. purpuratus* provides a new understanding of how genomes control development (Chapter 48). Combinations of maternally loaded products initiate the expression of a few transcription factors within particular cell-lineage compartments. The expression of these factors then becomes stabilized within these lineages, independent of the initial and transient maternal cues. This stabilization often proceeds through the activation of positive genetic regulatory feedback loops.

The most prominent marker for blastoporal tissue, widely conserved in evolution, is the T-box transcription factor *Brachyury*. *Brachyury* orthologs are also expressed around the blastopore in starfish, in the sea urchin *Lytechinus*, in acorn worms, in arrow worms, in Lophotrochozoa, in the polychaete *Platynereis,* and in the gastropod *Patella.* Conservation extends even beyond Bilateria, as best exemplified in the cnidarian *Nematostella.* The gene *caudal* is expressed in the blastopore region of many animal groups—vertebrates, as well as polychaetes (Chapter 51).

β-Catenin is a key player in the Wnt signaling pathway that regulates transcription. In many embryos, activation of β-catenin occurs in those early cells that will gastrulate and give rise to the endomesoderm (Wikramanayake et al. 2003). This is true not only of bilateral animals, but nuclear translocation of this factor also occurs in the sea anemone to the site of gastrulation and specifies the ectoderm. Reducing the concentration of β-catenin in both the anemone and the sea urchin blocks gastrulation.

The *snail/slug* gene family also seems to play a key role in epithelial/mesodermal transition in the mesoderm during gastrulation in many animal groups (Chapters 47 and 52).

In the chick, the earliest influences appear to come from the posterior marginal zone, where Vg1 and Wnt activities overlap (Chapter 15). Vg1+Wnt induces expression of *Nodal* in the neighboring area pellucida epiblast, but this is only free to act when the hypoblast (which produces Nodal antagonists) has been displaced by the incoming endoblast. Surprisingly, however, although *Vg1* mRNA is localized in vegetal cells in amphibians, there is no evidence that it plays a similar role as in the chick.

CONCLUSIONS

Gastrulation is a fundamental process in animal development. There has been excellent progress in understanding gastrulation in a variety of different animal embryos. Although there are striking similarities and even conserved genes controlling the process, many fundamental problems still remain to be understood. First, there is the early specification of the germ layers that will be involved in gastrulation, particularly the endoderm and the mesoderm, but also the extraembryonic tissues. Then there is the morphogenetic process that results in the tissue movements; convergent extension is a key process but it is still not clear as to just how the cells become oriented and move to bring this about. Then there is the biggest problem of all: How could all these differences in gastrulation have evolved, particularly the adaptive nature of the intermediate forms, and the coordination between early and later stages of development?

REFERENCES

Jaegerstern G. 1956. The early phylogeny of the metazoa. The bilatero-gastrea theory. *Zool. Bidr. Upps.* **30:** 321–354.

Levin M., Thorlin T., Robinson K.R., Nogi T., and Mercola M. 2002. Asymmetries in H+/K+-ATPase and cell membrane potentials comprise a very early step in left-right patterning. *Cell* **111:** 77–89.

McGrath J., Somlo S., Makova S., Tian X., and Brueckner M. 2003. Two populations of node monocilia initiate left-right asymmetry in the mouse. *Cell* **114:** 61–73.

Szathmary E. and Wolpert L. 2003. The transition from single cells to multicellulaity. In *Genetic and cultural evolution of cooperation* (ed. P. Hammerstein), pp. 271–290. MIT Press, Cambridge, Massachusetts.

INDEX